DEVELOPMENTAL MATHEMATICS

THIRD EDITION

D. FRANKLIN WRIGHT

HAWKES
LEARNING

Lead Editor:
Barbara Miller

Editors:
Allison Conger,
Ian Craig,
Jolie Even,
S. Rebecca Johnson

Creative Services Manager:
Trudy Tronco

Designers:
Lizbeth Mendoza,
Patrick Thompson,
Joel Travis

Cover Design:
Caitlin Neville

Design and Layout Assistance:
U. Nagesh,
E. Jeevan Kumar,
D. Kanthi,
K.V.S. Anil

A division of Quant Systems, Inc.

546 Long Point Road,
Mount Pleasant, SC 29464

Library of Congress Control Number: 2023936427

Printed in the United States of America

10 9 8 7 6 5 4 3 2 1

ISBN: 978-1-64277-616-4

AIE ISBN: 978-1-64277-615-7

Table of Contents

Chapter 5 Measurement

Chapter 6 Geometry

Chapter 7 Statistics, Graphs, and Probability

Chapter 8 Introduction to Algebra

Chapter 9 Solving Linear Equations and Inequalities

Chapter 10 Graphing Linear Equations and Inequalities

Chapter 11 Systems of Linear Equations

Chapter 12 Exponents and Polynomials

Chapter 13 Factoring Polynomials

Chapter 14 Rational Expressions

Chapter 15 Roots, Radicals, and Complex Numbers

Chapter 16 Quadratic Equations

Chapter 17 Exponential and Logarithmic Functions

Chapter 18 Conic Sections

Preface

Developmental Mathematics: Purpose and Style

The problem-solving and analytical skills of mathematics are essential in helping students excel during and beyond their college years. The purpose of *Developmental Mathematics* is to provide students with an all-in-one text that will prepare them for college mathematics. Its goal is to provide students with a learning tool that will help them

1. acquire learning skills necessary for succeeding in their coursework,
2. review basic arithmetic skills,
3. develop reasoning and problem-solving skills,
4. become familiar with algebraic notation,
5. understand the connections between arithmetic and algebra,
6. develop basic algebra skills,
7. provide a smooth transition from arithmetic through prealgebra to algebra, and
8. achieve satisfaction in learning so that they will be encouraged to continue their education in mathematics.

The writing style gives carefully worded, thorough explanations that are direct, easy to understand, and mathematically accurate. The use of color, boldface, subheadings, and shaded boxes helps students understand and reference important topics. Each topic is developed in a straightforward, step-by-step manner. Each section contains many detailed examples to lead students successfully through the exercises and help them develop an understanding of the related concepts.

Reading graphs and topics from geometry are integrated within the discussions and problems. In particular, Chapter 6 provides an in-depth study of geometrical concepts (perimeter, area, volume, and so on). Chapter 7 introduces statistical concepts (mean, median, mode, and range), goes deeper into graphing, and provides an interesting introduction to probability (what is the probability of getting two heads if a coin is tossed twice?). From Chapter 8 on, the text concentrates on developing useful algebraic skills and concepts.

Students are encouraged to use calculators when appropriate, and explicit directions and diagrams are provided as they relate to a simple four-function calculator, as well as to a TI-84 Plus graphing calculator.

The NCTM and AMATYC curriculum standards have been taken into consideration in the development of the topics throughout the text.

Developmental Mathematics: Content Highlights

New Features

Strategies for Academic Success

The Strategies for Academic Success lessons have been updated to provide more of a focus on time management and stress management along with success in a math course. Other topics include reading a math book and taking notes, effective study strategies, and reducing test anxiety.

vi Strategies for Academic Success 0.3 Managing Your Time Effectively

Strategies for Academic Success

0.3 Managing Your Time Effectively

Have you ever made it to the end of a day and wondered where all your time went? Sometimes it feels like there aren't enough hours in the day. Finding time to balance work, school, and home life can be difficult. Some interruptions, like unexpected traffic or family emergencies, are simply outside of your control. However, other distractions are within your control, such as watching TV or scrolling through social media. It's important to find a balance between activities you need to do (such as attending class and work) and activities you want to do (such as watching TV). Managing your time is important because you can never get time back.

Here are three strategies for managing your time more effectively.

Take Breaks

When you are working on an important project or studying for a big exam, you may feel tempted to work as long as possible without taking a break. This is especially true when you're working or studying at the last minute. While staying focused is important, working yourself for hours until you're mentally drained will lower the quality of your work and force you to take even more time recovering.

Think about the way that overworking can affect your body physically. If you're weight-training, you must take frequent breaks both between individual sets and entire workout sessions. If you don't let your muscles recover, you risk injuring yourself, which could leave you laid up for weeks.

Just like taking breaks helps your physical body recover, it will also help your brain re-energize and refocus. During study sessions, you should plan to take a short study break at least once an hour. If you usually work indoors, take this time to get a breath of fresh air outside and clear your head.

Study breaks and work breaks should usually last around five minutes. The longer the break, the harder it is to start working again. Instead of stopping for half an hour, take a five-minute break and reward yourself with some downtime when the task is complete. Similarly, if a course you are taking has a built-in break during the middle of the class period, use it to get up and move around. This little bit of physical movement can help you think more clearly.

Avoid Multitasking

Multitasking is working on more than one task at a time. When you have several assignments that need to be completed, you may be tempted to save time by working on two or three of them at once. While this strategy might seem like a time-saver, you will probably end up using more time than if you had completed each task individually. Not only will you have to switch your focus from one task to the next, but you will also make more mistakes that will need to be corrected later.

People don't multitask nearly as well as they think they do. For example, research studies have shown that multitasking while driving is similar to or even worse than driving while drunk. While multitasking on a project for school or work may not be dangerous, it can lead to wasted time and silly mistakes. Instead of trying to do two things at once, schedule yourself time to work on one task at a time.

Multitasking can also become an excuse for distractions, especially electronic ones. Have you ever tried to complete a homework assignment, watch TV, and message friends all at the same time? You probably did one of these things well and two of these things badly. That's because your brain can't give its full attention to three tasks at once. To stay focused in class or while studying, try stashing your phone in your backpack or purse and staying logged out of your computer until you need it.

Chapter Projects

New projects promote collaboration and show students the practical side of mathematics through activities using real-world applications of the concepts taught in the chapter.

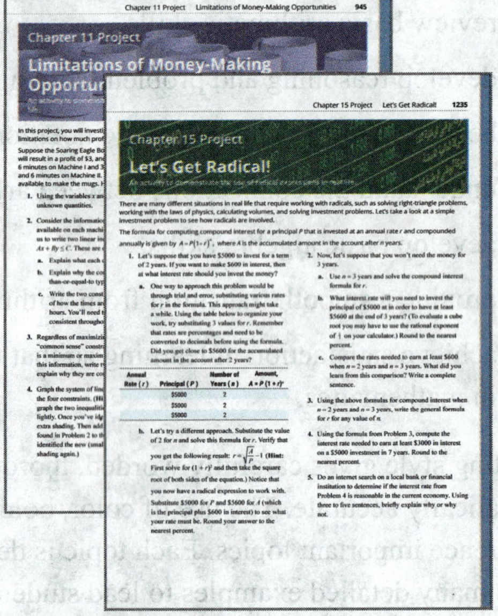
Chapter 11 Project Limitations of Money-Making Opportunities 945

Chapter 11 Project

Limitations of Money-Making Opportur

Chapter 15 Project Let's Get Radical! 1235

Chapter 15 Project

Let's Get Radical!

There are many different situations in real life that require working with radicals, such as solving right-triangle problems, working with the laws of physics, calculating volumes, and solving investment problems. Let's take a look at a simple investment problem to see how radicals are involved.

The formula for computing compound interest for a principal P that is invested at an annual rate r and compounded annually is given by $A = P(1+r)^n$, where A is the accumulated amount in the account after n years.

1. Let's suppose that you have $5000 to invest for a term of 2 years. If you want to make $600 in interest, then at what interest rate should you invest the money?
   a. One way to approach this problem would be through trial and error, substituting various rates for r in the formula. This approach might take a while. Using the table below to organize your work, try substituting 3 values for r. Remember that rates are percentages and need to be converted to decimals before using the formula. Did you get close to $5600 for the accumulated amount in the account after 2 years?

Annual Rate (r)	Principal (P)	Number of Years (n)	Amount, $A = P(1+r)^n$
	$5000	2	
	$5000	2	
	$5000	2	

   b. Let's try a different approach. Substitute the value of 2 for n and solve this formula for r. Verify that you get the following result: $r = \sqrt{\frac{A}{P}} - 1$ (**Hint:** First solve for $(1+r)^2$ and then take the square root of both sides of the equation.) Notice that you now have a radical expression to work with. Substitute $5000 for P and $5600 for A (which is the principal plus $600 in interest) to see what your rate must be. Round your answer to the nearest percent.
2. Now, let's suppose that you won't need the money for 3 years.
   a. Use $n = 3$ years and solve the compound interest formula for r.
   b. What interest rate will you need to invest the principal of $5000 at in order to have at least $5600 at the end of 3 years? (To evaluate a cube root you may have to use the rational exponent of $\frac{1}{3}$ on your calculator.) Round to the nearest percent.
   c. Compare the rates needed to earn at least $600 when $n = 2$ years and $n = 3$ years. What did you learn from this comparison? Write a complete sentence.
3. Using the above formulas for compound interest when $n = 2$ years and $n = 3$ years, write the general formula for r for any value of n.
4. Using the formula from Problem 3, compute the interest rate needed to earn at least $3000 in interest on a $5000 investment in 7 years. Round to the nearest percent.
5. Do an internet search on a local bank or financial institution to determine if the interest rate from Problem 4 is reasonable in the current economy. Using three to five sentences, briefly explain why or why not.

Additional Features

Objectives

The objectives provide students with a clear and concise list of the main concepts and methods taught in each section, enabling students to focus their time and effort on the most important topics. Objectives have corresponding labels located in the section text where the topic is introduced for ease of reference.

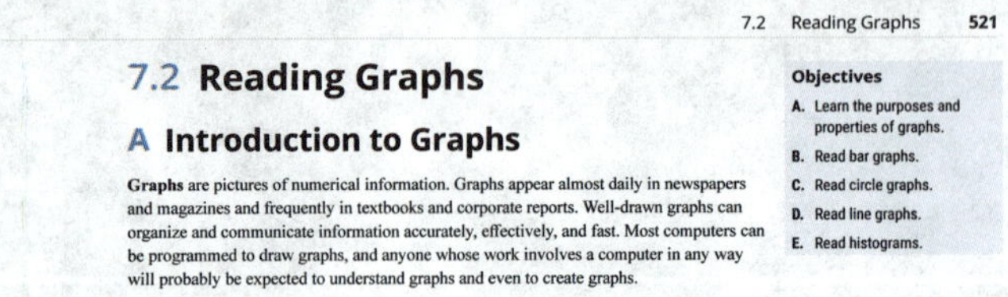
7.2 Reading Graphs 521

7.2 Reading Graphs

A Introduction to Graphs

Graphs are pictures of numerical information. Graphs appear almost daily in newspapers and magazines and frequently in textbooks and corporate reports. Well-drawn graphs can organize and communicate information accurately, effectively, and fast. Most computers can be programmed to draw graphs, and anyone whose work involves a computer in any way will probably be expected to understand graphs and even to create graphs.

Objectives

A. Learn the purposes and properties of graphs.
B. Read bar graphs.
C. Read circle graphs.
D. Read line graphs.
E. Read histograms.

Connections

Each chapter begins with a brief discussion related to a concept developed in the coming material and connects this concept to a real-life situation to improve student understanding and interest.

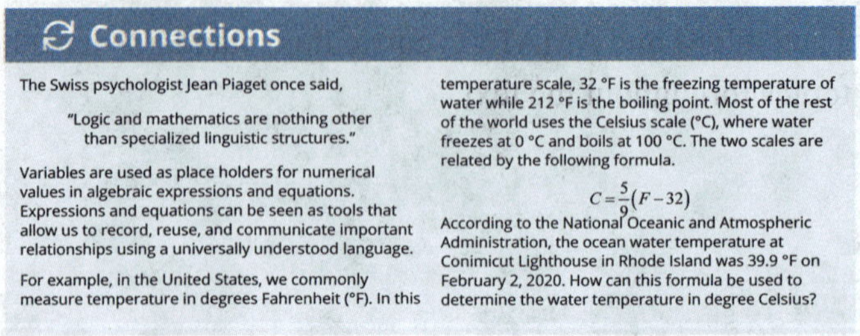
Connections

The Swiss psychologist Jean Piaget once said,

"Logic and mathematics are nothing other than specialized linguistic structures."

Variables are used as place holders for numerical values in algebraic expressions and equations. Expressions and equations can be seen as tools that allow us to record, reuse, and communicate important relationships using a universally understood language.

For example, in the United States, we commonly measure temperature in degrees Fahrenheit (°F). In this temperature scale, 32 °F is the freezing temperature of water while 212 °F is the boiling point. Most of the rest of the world uses the Celsius scale (°C), where water freezes at 0 °C and boils at 100 °C. The two scales are related by the following formula.

$$C = \frac{5}{9}(F - 32)$$

According to the National Oceanic and Atmospheric Administration, the ocean water temperature at Conimicut Lighthouse in Rhode Island was 39.9 °F on February 2, 2020. How can this formula be used to determine the water temperature in degree Celsius?

Examples

Examples are denoted with titled headers indicating the problem-solving skill being presented. Each section contains carefully explained examples with appropriate tables, diagrams, and graphs. Examples are presented in an easy-to-understand, step-by-step fashion and are annotated with notes for additional clarification.

Example 1 Multiplying Fractions

Multiply: $\frac{6}{7}\cdot\frac{8}{5}$

Solution

$$\frac{6}{7}\cdot\frac{8}{5}=\frac{6\cdot 8}{7\cdot 5}=\frac{48}{35}$$

Now work margin exercise 1.

Completion Examples

Completion examples encourage students to practice the methods taught in the lesson by providing a partial solution to an example and guiding the student to fill in the missing parts. Answers for completion examples are located at the end of the instruction for that lesson.

Example 1 Prime Factorizations

Find the prime factorization of 60.

Solution

$$60 = 6 \cdot \underline{10}$$
$$= 2 \cdot 3 \cdot \underline{2} \cdot \underline{5}$$
$$= \underline{2^2 \cdot 3 \cdot 5}$$

Now work margin exercise 1.

Completion Example Answers

1. $6 \cdot 10 = 2 \cdot 3 \cdot 2 \cdot 5 = 2^2 \cdot 3 \cdot 5$

Margin Exercises

Each example has a corresponding margin exercise to test students' understanding of what was taught in the example. Answers for the margin exercises are located at the end of the instruction for that lesson.

1. Solve: $3x + 4 = 7$

Margin Exercise Answers

1. $x = 1$

Notes

Note boxes in the margin point out important information that will help deepen students' understanding of the topics. Often these are helpful hints about subtle details in the definitions that many students do not notice upon first glance.

> **Note**
>
> Greek mathematician Euclid is often referred to as the 'Father of Geometry' for his revolutionary ideas and influential textbook called *Elements* that he wrote around the year 300 BC.

Definition Boxes

Straightforward definitions are presented in highly visible boxes for easy reference.

> **Algebra**
>
> **Algebra** is the branch of mathematics that deals with general statements of relations, utilizing letters and other symbols to represent specific sets of numbers, values, vectors, and so on, in the description of such relations.
>
> DEFINITION

Common Errors

These hard-to-miss boxes highlight common mistakes and how to avoid them.

> **Caution**
>
> Don't forget to carry the 1!

Calculators

For visual learners, key strokes and screenshots are provided when appropriate for visual reference. We also provide step-by-step instructions for using a simple four-function calculator for more basic operations, as well as a TI-84 Plus calculator for graphing skills.

CALCULATORS

Performing a task with a calculator

Press the keys 6 7 9 × 2 + 3 8 0.
Then press =.

The display will read 1738.

Exercises

Each section includes a variety of exercises to give the students much-needed practice applying and reinforcing the skills they learned in the section. The exercises progress from relatively easy problems to more difficult problems.

Concept Check

Exercises to assess students' conceptual understanding of topics and important definitions are included in every section.

1. Every square is a rectangle but not every rectangle is a ________.

Applications

Real-world application problems have been included throughout the text to challenge students to apply the concepts taught in the lesson.

66. To calculate what your maximum house payment should be, a banker multiplied your income by 0.28. Change 0.28 to a percent.

Writing and Thinking

This feature gives students an opportunity to independently explore and expand on concepts presented in the chapter. These questions foster a better understanding of the concepts learned within each section.

79. Compare and contrast the process for converting percents to decimal numbers and changing decimal numbers to percents.

Collaborative Learning

This feature encourages students to work with others to further explore and apply concepts learned in the chapter. These questions help students realize that they see many mathematical concepts in the world around them every day.

96. The class should be divided into teams of 2 or 3 students. Each team will need access to a digital camera, a printer, and a ruler.

a. Take pictures of 8 things with a defined slope. (**Suggestions:** A roof, a stair railing, a beach umbrella, a crooked tree, etc. Be creative!)

b. Print each picture.

c. Use a ruler to draw a coordinate system on

Answer Key

Located in the back of the book, the answer key provides answers to all odd numbered exercises in each section. This allows students to check their work to ensure that they are accurately applying the methods and skills they have learned.

1486 Answer Key 9.3 Exercises

Chapter 9: Solving Linear Equations and Inequalities

9.1 Exercises

Concept Check

1. equation
3. addition, equality
5. dividing
7. True
9. True

Practice

1. $x=-2$ is a solution
3. $x=4$ is not a solution
5. $x=-4$ is a solution
7. $x=-18$ is a solution
9. $x=-28$ is a solution
11. $x=7$
13. $y=-4$
15. $x=-19$

61. $y=-50.753$
63. $x=-17.214$
65. $x=246$
67. $x=-153.17$

Applications

69. 1945 *kanji* characters
71. a. Answers will vary. b. $x=12.5$ c. The garden should be 12.5 feet wide.
73. 4.24 light years
75. 11,500 words
77. 2250 students
79. a. Answers will vary. b. $x=11,550$ c. Clara will need a loan for $11,550.

19. $y=0$
21. $x=-2$
23. $y=-6$
25. $n=6$
27. $n=8$
29. $x=0$
31. $x=-7$
33. $x=-\frac{1}{8}$
35. $x=-\frac{13}{2}$
37. $x=-\frac{21}{5}$
39. $x=-\frac{8}{15}$
41. $y=\frac{28}{5}$
43. $x=2$

the 3 was distributed over the entire left-hand side of the equation; Correct answer is $x=15$.

b. 3 should be subtracted from each side, not from each term, and $5x-3$ doesn't simplify to $2x$; Correct answer is $x=\frac{8}{5}$.

9.3 Exercises

Concept Check

1. identity
3. conditional
5. real, $\mathbb{R}$

Acknowledgements

Many thanks to the following reviewers who offered their constructive and critical feedback for the current and previous editions.

Meredith Altman
Genesee Community College

Daniel Asera
University of Nevada–Las Vegas

Robert Bennett
Oklahoma State University Institute of Technology–Okmulgee

Sage Bentley

Gina Brandt
Augusta State University

Elizabeth Chu
Suffolk County Community College

Chris Copple
Northwest State Community College

Nick Counts
Culver Academies

Diane Devanney

Eric Drake
Culver Academies

Bonnie Ernst

Fenecia Foster

Andy Geary
Harper College

Richard Getso
South Texas College

Jeffrey Grell
Baltimore City Community College

Kathy Gulliver
Ancilla College

Rev. Thomas Haynes
Culver Academies

Lutchmiparsad Hazareesingh
Mesabi Range Community and Technical College

Heather Huntington
Nassau Community College

Rose Jenkins
Midlands Community College

L. John Jerome
Suffolk County Community College

Barbara Keener
North Central State College

Joanne Koratich
Muskegon Community College

Michael W. Lanstrum

Brian Leonard
Southwestern Michigan College

Laurie Lindstrom

David Lung
South Texas College

Keith Luoma
Augusta State University

Yolanda Manzano
Dallas College Richland Campus

Jo McCormick
Northwest State Community College

Nathan Mercer
Holyoke Community College

Henry Miller

Shelley Parks
Texas State Technical College–Waco

Michelle Toni Parsons
San Diego Mesa College

Leonardo Pinheiro

Pam Reising
Green River Community College

William Sargeant

Lasse Savola
Fashion Institute of Technology

Debra Shafer
University of North Carolina–Charlotte

Billie Shannon
Southwestern Oregon Community College

Jim Sheff
Spoon River College

Merrie Van Loy
South Texas College

Hao-Nhien Vu

Amy Young
Navarro College–Corsicana Campus

Kellie Zimmer

Hawkes Learning: A Clear Path to Mastery

Hawkes' software employs an adaptive, competency-based approach to knowledge mastery supported by a user-friendly interface. The student-centric platform promotes positive active learning by adapting to each student's needs through algorithmically generated questions based on an individual learner's pace, skill, and knowledge level. The real-time adaptive feedback addresses errors immediately, so that students learn from their mistakes when they make them. For each topic, the Hawkes Learning path to content mastery engages students through three simple modes: Learn, Practice, and Certify.

Competency-based learning made simple in three steps:

LEARN offers a multimedia-rich presentation of the lesson content. It includes instructional videos, interactive examples, and more.

PRACTICE engages students with algorithmically generated questions and intelligent tutoring in an ungraded, penalty-free environment.

CERTIFY requires students to demonstrate mastery of the material at a defined proficiency level without access to tutoring aids.

Explore Hawkes Learning with student access.

Sign Up:

1. Enter your info at **explore.hawkeslearning.com/signup**
2. **Check your email** for your username and password
3. Log in at **explore.hawkeslearning.com** to access the student software

Explore the Content:

1. Access available courses from the **Dashboard**
2. Select **"View Course"** to see all course content

Support

If you have questions or comments, we can be contacted as follows:

24/7 Chat: chat.hawkeslearning.com

Phone: 1-800-426-9538

Email: support@hawkeslearning.com

Web: support.hawkeslearning.com

Strategies for Academic Success

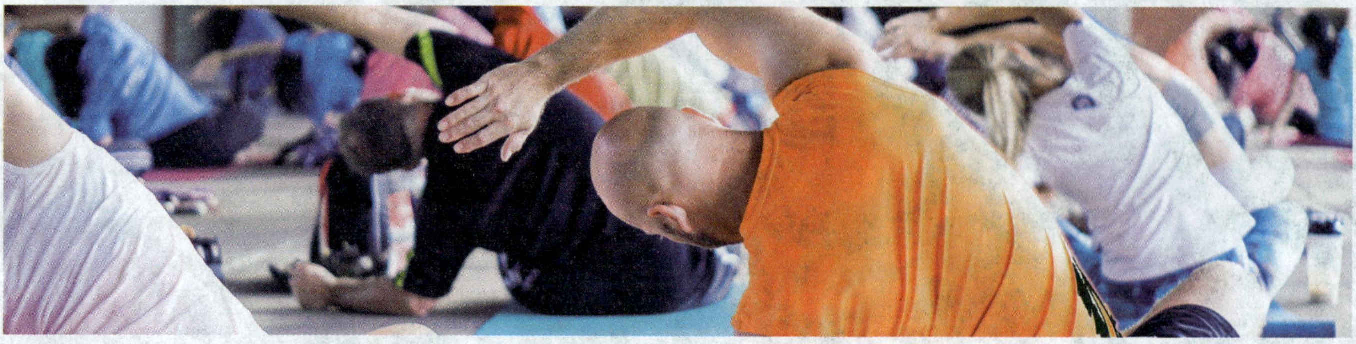

Strategies for Academic Success

0.1 Understanding and Reducing Stress

Stress is a part of daily life and comes from many different sources, including school, work, and family life. Stress in small doses can be positive, providing you with more energy and making you feel more alert. However, too much stress can start affecting you mentally and physically. You may feel overwhelmed or physically sick as the result of too much stress.

Determine the Causes of Stress

Stress can be caused by either external or internal stressors. External stressors are things outside of ourselves and are typically situations that we cannot control, such as class registration deadlines, traffic jams, or the loss of a loved one. Internal stressors happen inside of our minds and are related to our feelings, expectations, and goals, such as feeling nervous about an upcoming math exam or annoyed with a friend's decision. Internal stressors can also be the result of decisions we make. While external and internal stressors are separate, they often do affect one another. For instance, if you are feeling nervous about a presentation and run into technology issues, the technology issues can increase your nervousness, which can influence how you react to the situation.

Examples of common stressors are as follows. Consider whether each is an external stressor, internal stressor, or a combination of both.

School: disappointment about a test result, class schedule, homework

Work: hours, commute, finances, budgets, nervousness about a presentation

Family life: childcare, spending time with family, family conflict

Putting Stress into Perspective

The first step in dealing with stress is to take a step back and evaluate the situation. Try to understand what is causing your stress, and then determine which parts of the situation you can control and which parts you cannot control. Typically, we have control over internal stressors but do not have control over external stressors.

Suppose you have a math final next week that you need to study for. You discover during a meeting with your advisor that you need at least a B on the exam to maintain your scholarship. Besides finding time to study, you also have classes to attend, a job to work, and an important family event, along with all of the regular day-to-day activities in your life, such as sleeping, bathing, and eating. When viewed all together, this situation can seem stressful.

Taking a step back and carefully considering each item listed, we can assign a priority to each and put the whole situation into perspective. To assign a priority, you can use a scale of 1 to 3, in which 1 is the most important, and 3 is the least important. For instance, you might consider studying for the exam and working to be priority 1, attending classes and the family event priority 2, and some of your day-to-day activities (such as catching up on news or podcasts) to be priority 3.

Once you put your stressors into perspective, you can focus on managing your stress.

Plan to Manage Stress Effectively

Without a plan to manage the various stressors in life, it's easy to become overwhelmed. To prevent the stressors from affecting your mental and physical health, it's best to have a plan in place to effectively manage stress.

Take Action: Once the causes of stress are identified and given a priority level, you can take action to manage them. Focus on potential solutions instead of focusing on the problem. Consider creating a prioritized to-do list or creating a weekly schedule. It can also be helpful to determine if certain items are completely out of your control, and if so, letting them go.

Stay Organized: Staying organized can greatly reduce stress. This includes keeping work or school papers organized as well as keeping upcoming events and deadlines organized in a calendar.

Ask for Help: Asking for help can be a stressor itself but can also greatly reduce the stress that a situation or event is causing. The help can come in many forms; you can ask family or friends for assistance with childcare while you study for an exam. You can ask a therapist for help coping with the loss of a loved one. There are a variety of services available to help you deal with nearly any stressor in your life, regardless of whether the stressor is internal or external.

Focus on Yourself: Your physical and mental health play a role in how you respond to any stressor that arises. If you're tired or hungry, your reaction to stressors may be magnified. Getting enough rest, eating healthy meals, and exercising regularly can help. You can also find ways to manage your immediate reactions to stress by practicing breathing techniques, mediating, or doing yoga. With a clear mind and a healthy, well-rested body, you'll have an easier time managing any stressor that comes your way.

Practice Prevention: As Benjamin Franklin once said, "An ounce of prevention is worth a pound of cure." Some stressors can be prepared for in advance. For example, you can review your class notes daily and finish all homework assignments so that you don't need to cram last minute for the final exam. Similarly, you can save money each paycheck for financial emergencies instead of worrying about how to borrow money when an emergency strikes. Sometimes, prevention is simply avoiding procrastination; other times, it requires foresight of potential stressors and preventative action.

Actively Reduce Stress and Anxiety

Practicing deep breathing is an effective strategy to reduce in-the-moment stress or anxiety. During moments of high stress, several things happen in the body: blood pressure and heart rate increase, breathing becomes more shallow and rapid, and the flight-or-fight response can be triggered. By practicing deep breathing, also known as diaphragmatic breathing, you can combat these natural responses. This works because deep breathing triggers the rest and relaxation response, which reduces anxiety, lowers blood pressure, and decreases heart rate.

To practice deep breathing yourself, try the following simple technique:

1. Breathe in slowly, until your lungs feel full, while imagining drawing the air towards your belly.
2. Hold in this breath for a few seconds.
3. Slowly exhale all the air out of your lungs.
4. Repeat for several cycles.

If this method of deep breathing does not work well for you, try another. Any deep-breathing technique can trigger the same relaxation response. Try practicing deep breathing if you are experiencing high anxiety before or during a math test or during any other stressful events in your day-to-day life.

Questions

1. List two internal stressors and two external stressors in your life right now. Give each stressor a priority level.
2. Create a plan to deal with the stressors you listed in Question 1.

Strategies for Academic Success

0.2 Staying Organized

Staying organized is a key element in reducing stress and making sure you finish all your tasks. This is true whether you are a student, employee, parent, or in any other stage of life. It is especially true if you are in multiple stages at once.

While keeping a mental to-do list is an option, it's likely that you'll run into a variety of issues with this method, such as forgetting to do tasks, scheduling overlapping appointments, or missing a loan payment. Occasional issues like these might not seem like a big deal, but if they are recurring, you may eventually end up failing a class, losing your job, or paying late fees.

Keep a Planner

A planner is a place for you to organize your schedule and record any important tasks or responsibilities. By keeping all of this information in one place, you'll be less likely to forget to complete something important.

You can make a planner or you can purchase one. Planners can be made in either printed form or electronic form. Planners can be broken down into daily, weekly, or monthly pages. Daily planners allow for the most detail per day, while monthly planners only allow for a high-level overview of events. Weekly planners are a balance between the two.

When purchasing or creating a planner, consider how much detail you'll need in order to keep your life organized. If you need hourly breakdowns of each day, a daily planner will be best. If you need less precision, but still have several events or tasks each day, then a weekly planner might suit your needs. If you only need to keep track of a few events per week or month, then a monthly planner might be the option that works for you.

It is worth noting that an electronic planner can travel with you everywhere you carry your phone. There are a variety of apps available for most operating systems. At the very least, the calendar app on your phone can be utilized to remind you of the most important events in your schedule.

If you are unsure which type of planner you need, list out the regular events and tasks you have coming up, and then determine the level of precision you'll need to accomplish everything on your list.

Items on your list might include the following:

- Your work schedule
- Your class schedule
- Academic calendar dates
- Assignment deadlines
- Billing due dates
- Important holidays and birthdays
- Upcoming trips

When filling out your planner, first prioritize the events and tasks, then schedule the most important items before adding any nonessential items. Once you've filled in your planner, be sure to keep it up to date by adjusting appointments or adding in new events. And don't forget to always check your planner before scheduling any new appointments!

Use a File System

A well-organized file system is important for keeping documents sorted and easy to find at any moment. File systems can be used for a variety of documents, such as course materials, financial papers, and legal documents.

Keeping track of course materials, such as handouts, homework, exams, and notes, can help you improve your grades. If these materials are readily available, it is easier to fill free time with reviewing and studying. Other important papers to keep track of are course policies and the course syllabus. Having these documents on hand will help you know what to expect in the course along with a list of important dates.

The first step in creating a file system is determining what is important to save. If you save everything, your file system may become cluttered and unusable. If you keep too few items, your file system may not be informative enough to be helpful. A good rule of thumb is to keep recent important documents and discard older documents after a certain date. For instance, you may decide to keep all current course documents (notes, homework, tests, and so on) while the course is in progress, then trim down the files to only the notes and tests once the course is complete. For financial documents, you may decide to keep initial loan or account information, recent account statements, and account updates, then decide to discard old files after a certain number of years.

The second step in keeping an organized file system is deciding where to save files. Printed documents can be saved in binders, folders, or divided cabinets. For example, you can have a binder that holds all of your course information, with a divider tab for class information, notes, homework, and tests. Keeping the files within each section in order by date can make it easy to find what you are looking for when you need it.

Electronic files can be saved in a system of nested folders. For example, you can have a main folder titled with the course name, and within that folder, have individual subfolders for notes, tests, and course documents. It's important to clearly name files that are saved electronically.

Create a Workspace

An organized workspace can help you focus and work efficiently. If you have a dedicated workspace that is comfortable and uncluttered, it is easier to sit down and work without distractions. Having enough space is also essential. For example, if you are working on math homework, you need space for your computer, textbook, notebook, calculator, and writing utensils. If you have a clean area where you can spread out, you are less likely to misplace items when you need them.

If your workspace needs to be used for multiple purposes, having an easy-to-access storage space for your work items is essential. Keeping your items organized and together can help you set up and break down your workspace as needed.

If you live in a crowded house or dorm room, finding a workspace at home can be tricky. You may need to get a little creative and be flexible. The items you need at your workspace should ideally be easily relocated from room to room and completely stored away when not in use. This will allow the workspace to be multifunctional.

If you don't have space to work where you live, you can consider public places such as a library or a coffee shop. Consider the pros and cons of each location before making a choice. Libraries are typically free to use, but you might be limited on hours you can be in the building or how long you can use study rooms. Coffee shops typically require you to make a purchase to stay for any length of time, but the music and atmosphere can help create a focused mindset. Both options usually have free Wi-Fi, but proceed with caution when using unsecured public Wi-Fi and don't access important personal information when connected.

It is also important to figure out the level of distractions you can tolerate before picking a location to set up your workspace. Distractions can come in the form of friends and family talking to you, an open space on campus getting too crowded, or the smell of baked goods at a bakery becoming too tempting. While it's unlikely you can avoid all distractions, keeping them to a minimum can help make your study session more productive.

Questions

1. Would you benefit best from a daily, weekly, or monthly planner? Explain your reasoning.
2. Describe your current workspace and determine three changes you can make to create a more organized workspace.

Strategies for Academic Success

0.3 Managing Your Time Effectively

Have you ever made it to the end of a day and wondered where all your time went? Sometimes it feels like there aren't enough hours in the day. Finding time to balance work, school, and home life can be difficult. Some interruptions, like unexpected traffic or family emergencies, are simply outside of your control. However, other distractions are within your control, such as watching TV or scrolling through social media. It's important to find a balance between activities you need to do (such as attending class and work) and activities you want to do (such as watching TV). Managing your time is important because you can never get time back.

Here are three strategies for managing your time more effectively.

Take Breaks

When you are working on an important project or studying for a big exam, you may feel tempted to work as long as possible without taking a break. This is especially true when you're working or studying at the last minute. While staying focused is important, working yourself for hours until you're mentally drained will lower the quality of your work and force you to take even more time recovering.

Think about the way that overworking can affect your body physically. If you're weight-training, you must take frequent breaks both between individual sets and entire workout sessions. If you don't let your muscles recover, you risk injuring yourself, which could leave you laid up for weeks.

Just like taking breaks helps your physical body recover, it will also help your brain re-energize and refocus. During study sessions, you should plan to take a short study break at least once an hour. If you usually work indoors, take this time to get a breath of fresh air outside and clear your head.

Study breaks and work breaks should usually last around five minutes. The longer the break, the harder it is to start working again. Instead of stopping for half an hour, take a five-minute break and reward yourself with some downtime when the task is complete. Similarly, if a course you are taking has a built-in break during the middle of the class period, use it to get up and move around. This little bit of physical movement can help you think more clearly.

Avoid Multitasking

Multitasking is working on more than one task at a time. When you have several assignments that need to be completed, you may be tempted to save time by working on two or three of them at once. While this strategy might seem like a time-saver, you will probably end up using more time than if you had completed each task individually. Not only will you have to switch your focus from one task to the next, but you will also make more mistakes that will need to be corrected later.

People don't multitask nearly as well as they think they do. For example, research studies have shown that multitasking while driving is similar to or even worse than driving while drunk. While multitasking on a project for school or work may not be dangerous, it can lead to wasted time and silly mistakes. Instead of trying to do two things at once, schedule yourself time to work on one task at a time.

Multitasking can also become an excuse for distractions, especially electronic ones. Have you ever tried to complete a homework assignment, watch TV, and message friends all at the same time? You probably did one of these things well and two of these things badly. That's because your brain can't give its full attention to three tasks at once. To stay focused in class or while studying, try stashing your phone in your backpack or purse and staying logged out of your computer until you need it.

Use a Time Budget

Just like a financial budget shows you how you spend your money, a time budget shows you how you spend your time. A time budget can help you identify "wasted" time that could be used more productively.

To begin budgeting your time, you first need to get an idea of how you usually spend it. For one week, keep track of everything you do in fifteen-minute time blocks. Be as accurate as possible and track as you go throughout your day. Here's an example of what a partial record of activities for one day might look like.

Time	Activity
12:00 a.m. to 7:30 a.m.	Sleeping
7:30 a.m. to 8:15 a.m.	Getting ready for class
8:15 a.m. to 8:45 a.m.	Driving
8:45 a.m. to 9:00 a.m.	Walking to class
9:00 a.m. to 10:30 a.m.	Math class
10:30 a.m. to 10:45 a.m.	Walking to next class
10:45 a.m. to 11:30 a.m.	English class

Once you've recorded your entire week in fifteen-minute time blocks, you can calculate how much time you spend on different types of activities. First, review the activities you entered and assign each of them to one of the following categories: sleep, meals, work, class, study, extracurricular, exercise, personal, family, entertainment, social, and other. Then, answer the following questions.

1. Do you feel like you had enough time to fulfill all your responsibilities?
2. Were you surprised by how much time you spent on any particular activity?
3. What are some important activities you should have spent more time doing?
4. What are some activities you'd like to spend more time doing in the future?
5. What are some activities you'd like to spend less time doing in the future?

Based on your answers to these questions, create a weekly time budget. Remember that one week contains only 168 hours. If you want to spend more time on a particular activity, you'll need to find that time somewhere. Use a planner to schedule specific blocks of time for study sessions, meals, travel times, and morning/evening routines. As a general rule, you should set aside at least two hours of study time for every one hour of class time. That means that a three-credit-hour course would require at least six hours of outside study time per week.

Strategies for Academic Success

0.4 Reading a Textbook and Note-Taking

Reading a math textbook and taking notes is more intensive than casually reading a book for fun. You have to concentrate more on what you are reading and be careful how you take notes because you will likely be tested on the content. Here are some tips to help you successfully read a math textbook and take notes.

Reading a Math Textbook

Reading a math textbook requires a different approach than reading literature or history textbooks because math textbooks contain a lot of symbols and formulas in addition to words. Whether you are reading a physical math textbook or an e-book, the following tips can help you understand and retain the information presented.

Start at the Beginning: Don't start in the middle of an assigned section. Math tends to build on previously learned concepts, and you may miss an important concept or formula that is crucial to understanding the rest of the material in the section.

Don't Skim: When reading math textbooks, look at everything: titles, learning objectives, definitions, formulas, text in the margins, and any text that is highlighted, outlined, or in bold. Also, pay close attention to any tables, figures, charts, and graphs.

Work through Examples: Make sure you understand each step of an example. If you don't understand something, mark it so you can ask about it in class. Sometimes, math textbooks leave out intermediate steps to save space. Try working through the examples on your own, filling in any missing steps.

Understand the Mathematical Definitions: Many terms used in everyday English have a different meaning when used in mathematics. Some examples include *equivalent*, *similar*, *average*, *median*, and *product*. It is important to note these differences in your notebook along with other important definitions and formulas. You might find it helpful to keep a separate math glossary that contains each term with its definition along with an example of how the term is used.

Reread important information: After reading through the section, go back and reread boxed definitions, examples, or any other important information. It is also important to reread any portions of the lesson that seemed more complex or difficult compared to the rest.

Annotating a Textbook

Annotating involves marking the text and taking notes in the margin. Put your book to good use and don't be afraid to add comments and highlighting. If you don't understand something in the text, reread it a few times. If it is still not clear, note the text with a question mark or some other notation so that you can ask your instructor about it. A well-annotated text can help you find important information while completing homework assignments as well as help you review for exams.

You can use sticky notes, pens, and highlighters to annotate a physical text. Try adding special colors or symbols for different types of information. For example, questions could be in blue with a question mark, and important ideas could be in red with a star. However, if you are renting a textbook, be sure to limit annotations to sticky notes and easily removed markings. Most e-books can be annotated with mark-up tools available in the e-reader.

Some important things to annotate or highlight are definitions, key concepts, important ideas, and examples. You can also add notes about prior knowledge to the margin and summarize definitions or procedures in your own words.

General Note-Taking Tips

- Write the date and the course name at the top of each page.
- Write the notes in your own words and paraphrase.
- Use abbreviations, such as ft for foot, # for number, def for definition, and RHS for right-hand side.
- Copy all figures or examples that are presented during the lecture.
- Review your notes after class and add additional detail where needed.
- Review and rewrite your notes after class. Do this on the same day, if possible.

There are many different methods of note taking and it's always good to explore new methods. A good time to try out new note-taking methods is when you rewrite your class notes. Try each new method a few times before deciding which works best for you. Presented here are three note-taking methods you can try out. You may even find that a blend of several methods works best for you.

Note-Taking Methods

Taking notes in class is an important step toward understanding new material. Review your notes daily while you do your homework and before taking quizzes and tests. While there are several methods for taking notes, three are presented here.

Outlines

An outline consists of several topic headings, each followed by a series of indented bullet points that include subtopics, definitions, examples, and other details.

Example:

1. Ratio
    - a. Comparison of two quantities by division.
    - b. Ratio of a to b
        - i. $\frac{a}{b}$
        - ii. $a : b$
        - iii. a to b
    - c. Can be reduced
    - d. Common units can cancel

Split Page

The split page method divides the page vertically into two columns, with the left column narrower than the right column. Main topics go in the left column, and detailed comments go in the right column. The bottom of the page is reserved for a short summary of the material covered.

Example:

Keywords:	**Notes:**
Ratios	1. Comparison of two quantities by division 2. $\frac{a}{b}$, $a : b$, a to b 3. Can reduce 4. Common units can cancel
Summary: Ratios are used to compare quantities and units can cancel.	

Mapping

The mapping method is the most visual of the three methods. One common way to create a mapping is to write the main idea or topic in the center and draw lines from that main idea to smaller ideas or subtopics. Additional branches can be created from the subtopics until all of the key ideas and definitions are included. Using a different color for subtopics can help you visually organize the topics.

Example:

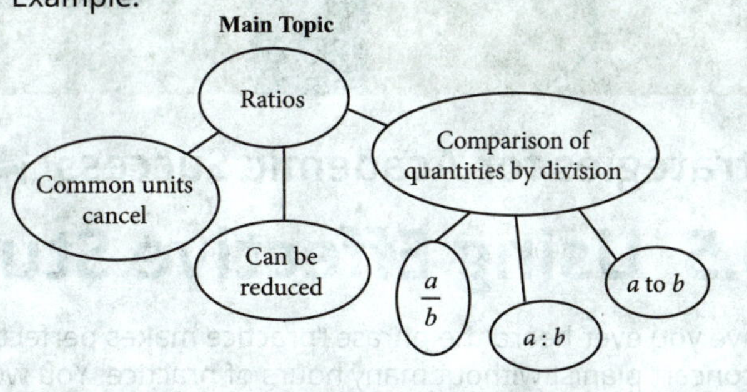

Reviewing Supplemental Content

Many textbooks have corresponding online courseware that includes lesson or example videos to supplement the text. The videos can vary in topic from reviewing the skills and concepts covered in the lesson to working through a specific example. While watching these videos, keep the following tips in mind.

- Pause the video as needed to self-assess your understanding.
- Avoid other audio or visual distractions.
- Take notes.
- Write down any additional questions asked during the video, then try to find the solution in the textbook or ask your professor for more information.
- Rewatch videos as needed for clarification.
- Work through the same or similar exercises on paper while watching the video or soon after.

In addition to videos, you might have access to other supplemental content such as PowerPoints, chapter reviews, and content created by your instructor.

Questions

1. Explain why you need to carefully read a math textbook in the order the information is presented.
2. Which note-taking method do you use? (If it isn't one listed, describe it.)

Strategies for Academic Success

0.5 Using Effective Study Strategies

Have you ever heard the phrase "practice makes perfect"? This saying applies to many things in life. You won't become a concert pianist without many hours of practice. You won't become an NBA basketball star by sitting around watching basketball on TV. You can watch all of the videos and read all of the books on how to do something, but you won't learn the skills without actually practicing. The same idea applies to math: math is not a spectator sport.

Just as you work your body through physical exercise, you have to work your brain through mental exercise. Math is an excellent subject to provide the mental exercise needed to stimulate your brain. So when doing mathematics, remember the 3 P's—Practice, Patience, and Persistence—and the positive effects they will have on your brain! A few ways to instill practice, patience, and persistence into your routine are to manage your study schedule, use creative study strategies, and diligently prepare for exams.

Manage Your Study Schedule

Creating a study routine will keep you in the habit of studying and provide personal accountability. Studying daily will also help prevent the need for emergency cram sessions before an exam. To make studying part of your daily routine, try using the following strategies.

Find a study time that works for you. If you're a morning person, wake up early for a brief study session. If you struggle to stay awake after eating lunch, don't plan to study in the early afternoon.

Schedule study times in your planner. Adding study sessions to your planner makes them harder to skip. Plan to study two to three hours outside of class for every hour spent in class. Schedule your most difficult subjects first so your mind is fresh.

Be flexible when necessary. Unexpected events are bound to happen once in a while. Don't feel guilty about rescheduling your study time around these disruptions. You should also be flexible with your study location and not be afraid to relocate if you start getting distracted by your surroundings.

Keep study times separate. When you have multiple assignments due for class, you may find yourself spending all your study time working on assignments and not reviewing. Even if you only have fifteen minutes to spare, reserve time for studying your class notes.

Review class notes as soon as possible. Reviewing class notes soon after class will help you retain the skills and concepts covered. This will also allow you to fill in any gaps in your notes for future reference.

Use Creative Study Strategies

Sometimes studying can be boring and you lose your attention after extended periods of time. Creative study strategies can help retain your attention and help you learn.

Recite information aloud. Ask yourself questions about the material to see if you can recall important facts and details. Pretend you are teaching or explaining the material to someone else.

Use mnemonics or memory techniques. For example, a mnemonic that is commonly used to remember the order of operations is "Please Excuse My Dear Aunt Sally," which uses the first letter of the words Parentheses, Exponents, Multiplication, Division, Addition, and Subtraction.

Use acronyms to help remember important concepts or procedures. An acronym is created by taking the first letter (or letters) from each word in the phrase that you want to remember and making a new word. For example, the acronym HOMES is often used to remember the five Great Lakes in North America, where each letter in the word represents the first letter of each lake: Huron, Ontario, Michigan, Erie, and Superior.

Use visual images like diagrams, charts, and pictures. You can make your own pictures and diagrams to help you recall important definitions, theorems, or concepts.

Split larger pieces of information into smaller "chunks." For example, instead of remembering a sequence of digits such as 555777213, you can break it into chunks and remember 555 777 213.

Group long lists of information into categories that make sense. For example, instead of remembering all the properties of real numbers individually, try grouping them into shorter lists by operation, such as addition and multiplication.

Associate the information with something you already know. Think about how you can make the new information personally meaningful. How does it relate to your life, your experiences, and your current knowledge? If you can link new information to existing memories, you can create "mental hooks" to help you recall the information in the future.

Tips for Preparing for an Exam

When preparing for a final math exam, it's best to take at least a week to prepare for the exam along with regular studying throughout the semester.

Determine Important Exam Information

1. What is the date, time, and location of the exam?
2. Is there a time limit to complete the exam?
3. What materials can you bring to the exam? Can you use formula sheets, calculators, or scrap paper?

One Week before the Exam

1. **Create a study schedule.** Determine where to study and whether you will study with classmates. Be sure to limit distractions when deciding. Bring snacks and take regular breaks.
2. **Organize your study materials.** Even if you have well-organized course materials, you may decide to condense them into a single study sheet, create note cards, or make a formula sheet.
3. **Follow through with your study plan.** Starting right away will give you time to ask your instructor or classmates for help with difficult topics.

Three Days before the Exam

1. Make a practice test and take it under the same constraints the final exam will have.
2. Ask your instructor or classmates about any questions you struggled with on the practice test.

Night before the Exam

1. Organize and pack all of the supplies you will need for the exam: pencils, erasers, calculator, scratch paper, and so on.
2. Review your formula sheet.
3. Avoid trying to cram in a lot of last-minute studying.
4. Go to bed early and get a good night of sleep.

Day of the Exam

1. Eat a healthy breakfast and don't drink too much caffeine, which can make you anxious.
2. Make sure you have all of your supplies with you when you leave for the exam.
3. Review your formula sheet if you are not allowed to use it on the exam.
4. Get to the exam location early so you can be organized and mentally prepared.

Questions

1. Which creative study strategy do you find most useful, and why? (If it isn't one listed, describe it.)
2. Describe your current method for preparing for an exam and determine three ways you can improve it.

Strategies for Academic Success

0.6 Reducing Test Anxiety

Approximately 93% of adults in the United States have experienced anxiety related to math at some point in their life.[1] Anxiety can come in many forms when it comes to the subject of math. You may have general anxiety about math, you may have anxiety about specific aspects of math class, or you may have test anxiety. Practicing the following strategies can help you reduce any general math anxiety and test-specific anxiety you may experience.

Reducing Math Anxiety

Depending on what causes your math anxiety, the tools to overcome it will differ. You may need to develop an effective study strategy so that you always feel prepared for your math class or any test. You may need to learn how to properly read a math textbook and take notes. You may even need to learn to manage your time, prioritizing what is most important (such as work and studying), or develop a system to keep your documents and workspace organized.

Math anxiety could also be caused by your thoughts and feelings about either your abilities or the subject itself (or both). If this is the case, try the following methods to reduce math anxiety.

1. Maintain a positive attitude and avoid negative self-talk. Remind yourself that each mistake is an opportunity to learn and improve. Set small math achievement goals to keep you moving toward bigger goals.
2. Visualize yourself doing well in math, whether it's on a quiz, test, or passing the course.
3. Think of setbacks as opportunities for growth. When you experience a setback, learn from your mistakes or the situation and keep moving forward.
4. Take deep, slow breaths, and think about the people or places that make you feel happy and peaceful.
5. Find a ritual that makes you happy. This could be listening to music, taking deep breaths, or imagining your future success. Find what works for you and practice it regularly.
6. Form a math study group. Working with others will help you feel more relaxed, and you can support each other.
7. Learn effective study skills that work with your learning style. Don't be afraid to change the way you study if it's not working for you.

Preparing for an Exam

Being prepared for an exam can reduce your anxiety leading up to and during the exam. Preparing for an exam is not just the studying that happens the few days before the exam. The preparation should take place throughout the semester. Be sure to consistently do the following to always be prepared for exams or pop quizzes.

1. Read your math textbook before class and review the lesson again after class.
2. Take notes using the method that works best for you, and review the notes regularly. Rewrite your notes if necessary to expand or condense the information covered.
3. Keep your class materials organized. Use labels or highlighting so you can easily review the content.
4. Use study aids, such as note cards, to help you remember definitions, theorems, formulas, or procedures.
5. Actively practice the skills you learn by working through the examples and doing your homework.
6. Strive to understand the material, not simply memorize it. Explain the material in your own words or look for patterns.

1 Christie Blazer, "Strategies for Reducing Math Anxiety. Information Capsule. Volume 1102," ERIC, September 2011, eric.ed.gov/?id=ED536509.

7. Plan to study two to three hours outside of class for every hour spent in class. Pick a study time and place that works well with your schedule.
8. Do not spend all of your study time working through a single exercise or figuring out a single concept. Move on to the next exercise or concept. Make notes to ask your instructor or classmates for help when needed.

» Test-Taking Strategies to Reduce Anxiety

In addition to preparing for a test by studying, you can try the following tips to reduce your test anxiety.

1. Get plenty of sleep the night before the exam and eat nutritious meals on the day of the exam. Being well-rested and avoiding hunger will help you focus.
2. Talk to your instructor about your anxiety. Your school might allow special accommodations, such as extra time on the test or taking the test in a more calming area.
3. When handed a test, immediately write your name at the top, and then perform a "brain drain" to write down all the formulas and important facts you remember on your test or scratch paper. Having this information readily available will boost your confidence and reduce your anxiety.
4. Carefully read the directions of the exam before starting any work. Then, read through the questions to make sure you understand them. You can add notation next to each question to indicate the perceived level of difficulty. It might build your confidence to start with problems you find easier and then move on to the more difficult problems.
5. If you panic or freeze during a math test, focus on a single problem you can do. Once you gain confidence, work through other problems you know how to do. Then, attempt the more difficult problems.
6. Check your solutions as you go to make sure they make sense. Then, use any extra time after you finish the exam to do a final check of your work and solutions before you turn it in.
7. If you get to a problem you don't know how to solve, skip it and come back after you finish the problems you do know how to solve. Another problem on the test may help you remember how to solve the more difficult problems.
8. If a problem is multiple choice, work the problem before looking at the answers. Alternatively, you can start by looking at the answer choices and working backwards to see if any are easily eliminated.

Reviewing Test Results

Whether you ace the test, receive a passing score, or receive a low score, you should take the time to review your work and any feedback from your instructor. This is especially true for tests and exams early in the course because math courses tend to be cumulative. That is, concepts and skills learned for one test are likely to be needed for the next test. When you receive a test back, do the following:

1. Correct any of your incorrect work so that you know the correct way to solve the problem in the future. If you are unsure what you did wrong, ask a classmate or visit your instructor during office hours.
2. Analyze the test questions to determine if most of them came from your class notes, the homework, or the textbook. This will give you an idea of how to spend your time studying for the next test.
3. Analyze the errors you made on the test. Were they careless mistakes? Did you run out of time? Did you not understand the material well enough? Were you unsure which method to use?
4. Based on your analysis of the test, adjust your study methods and schedule. It is important to adjust your study strategy or try new approaches. This may seem difficult if you are used to your current study techniques. Adjusting one or two aspects of your study methods and schedule at a time can allow you to find out what works (or does not work).

Questions

1. Describe any anxiety you may have in relation to math.
2. Create a plan to help work through your math anxiety.

CHAPTER 1

1 Whole Numbers

Connections

Numbers play a role in any job you may take in the future. If you go into sales, you will need to keep track of the number of sales you make and the profit you've earned for the company. If you go into construction, you will need to be able to accurately take measurements of the buildings and structures you help create. If you go into the medical field, numerical accuracy is important in situations such as measuring out the medicine for a sick child or recording a patient's blood pressure. If you decide to start your own business, you will need to evaluate operation costs to determine how much to charge your clients for your products or services.

During your education, numbers will also play a large role. For example, as a student striving to succeed, you will want to keep track of your progress in the courses you take. One way to do this is to find the average of your test scores throughout the semester. The average of these scores will help you determine if you are doing well in the course or if you need to seek assistance from either the instructor or a tutor. Suppose during the semester you receive the following scores on the first four tests in a course.

78, 85, 94, 83

What is your current test average for the course? Should you seek assistance to improve your grade?

Objectives

A. Understand the decimal system.

B. Write whole numbers in standard and expanded notation.

C. Read and write whole numbers in words.

D. Read tables.

1.1 Introduction to Whole Numbers

A The Decimal System

The symbols used in our number system are called **digits**, and the value of a digit depends on its position to the left of a beginning point, called a **decimal point**. Figure 1 shows the value of the first ten places in the place value system we use. Every three places constitute a **period** and periods are separated with **commas**. A number with four digits may be written with a comma or without a comma.

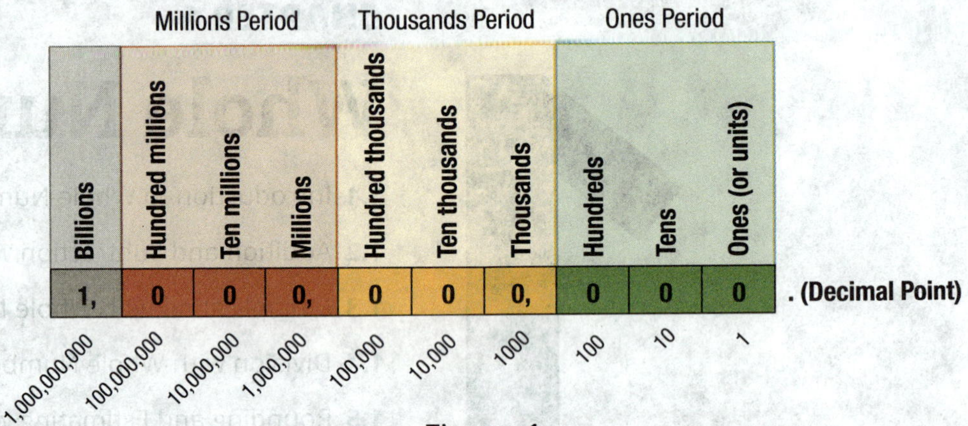

Figure 1

The **whole numbers** are the number 0 and the **natural numbers** (also called the **counting numbers**). We use ℕ to represent the set of natural numbers and 𝕎 to represent the set of whole numbers.

Whole Numbers

The **whole numbers** are the **natural numbers** (or **counting numbers**) along with the number 0.

Natural numbers = ℕ = {1, 2, 3, 4, 5, 6, 7, 8, 9, 10, 11, ...}

Whole numbers = 𝕎 = {0, 1, 2, 3, 4, 5, 6, 7, 8, 9, 10, 11, ...}

Note that 0 is a whole number but not a natural number.

DEFINITION

The three dots (called an **ellipsis**) in the definition indicate that the pattern continues without end.

To write a whole number in **standard notation** (or **standard form**) we use the **decimal system**.

The Decimal System

The **decimal system** (or base ten system) is a place value system that depends on three things.

1. the **ten digits**: 0, 1, 2, 3, 4, 5, 6, 7, 8, 9
2. the **placement** of each digit
3. the **value** of each place

DEFINITION

Example 1 Understanding Place Value

Given the number 350,472, which digit indicates the number of

a. thousands? **b.** tens? **c.** hundreds?

Solution

a. 0 **b.** 7 **c.** 4

Now work margin exercise 1.

1. Given the number 498,651, which digit indicates the number of

a. ten thousands?

b. hundreds?

c. ones?

Example 2 Understanding Place Value

For the number 37,895, state the place value of each digit.

Solution

3 is in the ten thousands place.

7 is in the thousands place.

8 is in the hundreds place.

9 is in the tens place.

5 is in the ones place.

Now work margin exercise 2.

2. State the place value of the 3 in each number.

a. 370

b. 35,499

c. 172,530

B Standard and Expanded Notation

Standard notation is the notation you are familiar with.

865 Standard notation

In **expanded notation**, the values represented by each digit are written as a sum.

865 = 800 + 60 + 5 Expanded notation

3. Write each number in expanded notation.

a. 463

b. 7300

Example 3 Writing Numbers in Expanded Notation

Write each number in expanded notation.

a. 954 **b.** 6507

Solution

a.

	9	5	4	⟵	Standard notation
	100	10	1	⟵	Place value of each digit
954 =	900	+ 50	+ 4		Expanded notation

b.

	6	5	0	7	⟵	Standard notation
	1000	100	10	1	⟵	Place value of each digit
6507 =	6000	+ 500	+ 0	+ 7		Expanded notation (**Note:** The 0 is optional and from now on will not be included.)

Now work margin exercise 3.

4. Write each number in expanded notation.

a. 29,524

b. 808,491

Completion Example 4 Writing Numbers in Expanded Notation

Write each number in expanded notation.

a. 32,081 **b.** 497,500

Solution

a. 32,081 = 30,000 + __2000__ + __80__ + 1

b. 497,500 = 400,000 + __90,000__ + __7000__ + __500__

Now work margin exercise 4.

Note: Completion Examples are answered at the end of the section where they appear.

C Reading and Writing Whole Numbers in Words

To read (or write) a whole number, consider one period at a time, moving from left to right. State the name of the number in each period (this number will be three digits or less), followed by the period name. Disregard any period containing three zeros.

Expanded notation can sometimes help in translating a whole number into words. For example,

1397 = 1000 + 300 + 90 + 7 is read as

one thousand three hundred ninety-seven.

As another example,

9,076,532 is read as

nine million, seventy-six thousand, five hundred thirty-two.

Reading and Writing Whole Numbers

You should note the following four things when reading or writing whole numbers.

1. Digits are read in periods (groups of three).
2. Commas are used to separate periods **if a number has more than four digits**. (A comma is optional if a number has exactly four digits.)
3. The word **and** does not appear in whole numbers written in words. **And** is said only when reading a decimal point. (See Chapter 3.)
4. Hyphens (-) are used to write words for the two-digit numbers from 21 to 99 except for those that end in 0. For example: twenty-one and thirty-five.

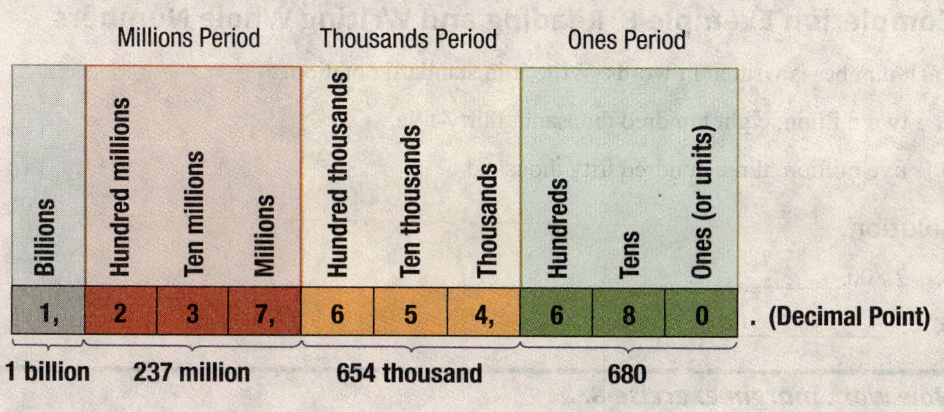

Example 5 Reading and Writing Whole Numbers

Each number is written in standard notation. Write it in words.

a. 25,380 **b.** 3,000,562

Solution

a. twenty-five thousand, three hundred eighty

b. three million, five hundred sixty-two

Now work margin exercise 5.

Note that, just as in standard notation, a comma is used to separate periods when writing a number in words.

5. Write each number in words.

a. 32,450,090

b. 5084

Completion Example 6 Reading and Writing Whole Numbers

Each number is written in standard notation. Write it in words.

a. 8407 **b.** 15,352

Solution

a. eight thousand four hundred seven

b. fifteen thousand, three hundred fifty-two

Now work margin exercise 6.

6. Write each number in words.

a. 6791

b. 23,507

7. Write each number in standard notation.

a. six thousand forty-one

b. one million, four hundred eighty-three thousand, seven

Example 7 Reading and Writing Whole Numbers

Each number is written in words. Write it in standard notation.

a. twenty-seven thousand, three hundred thirty-six

b. three hundred forty million, sixty-two thousand, forty-eight

Solution

a. 27,336

b. 340,062,048 **Note:** 0s must be used to fill out a three-digit period.

Now work margin exercise 7.

8. Write each number in standard notation.

a. six million, three hundred thousand, five hundred

b. four million, eight hundred seventy-five thousand

Completion Example 8 Reading and Writing Whole Numbers

Each number is written in words. Write it in standard notation.

a. two million, eight hundred thousand, thirty-five

b. five million, three hundred fifty thousand

Solution

a. 2,800, __035__

b. 5, __350__, __000__

Now work margin exercise 8.

D Reading Tables

Tables can be used to display numerical information. A table needs a title and headers for each column to indicate the categories listed. In Example 9, we see the title of the table is **Depths of Lakes** and the headers for the columns are **Name**, **Location**, and **Depth (in feet)**.

9. Use the "Depths of Lakes" table and your understanding of whole numbers to answer each question.

a. Where is Lake Vostok located?

b. What is the depth of the Caspian Sea?

c. Write the depth of Lake Toba in words and in expanded notation.

Example 9 Application: Reading Tables

Use the following table and your understanding of whole numbers to answer each question.

Depths of Lakes

Name	Location	Depth (in feet)
Baikal	Siberia, Russia	2487
Tanganyika	Tanzania, Africa	1870
Crater	Oregon, United States	1148
Vostok	Antarctica	1129
Tahoe	California, United States	989
Issyk Kul	Kyrgyzstan	886
Hornindalsvatnet	Sogn og Fjordane, Norway	778
Toba	Sumatra, Indonesia	707
Caspian Sea	Iran, Russia, Turkmenistan, Kazakhstan, Azerbaijan	604

Source: en.wikipedia.org/wiki/List_of_lakes_by_depth

Depths of Lakes

Name	Location	Depth (in feet)
Quesnel	British Columbia, Canada	515
Ohrid	Macedonia, Albania	508
Geneva	Switzerland, France	502
Loch Ness	Scotland, United Kingdom	436
Dead Sea	Jordan, Israel	387
Titicaca	Peru, Bolivia	351

Source: en.wikipedia.org/wiki/List_of_lakes_by_depth

a. What is the depth of Lake Tahoe in California?

b. Which lake is the deepest and what is its location?

c. Write the depth of Crater Lake in words and in expanded notation.

Solution

a. 989 feet

b. The deepest lake is Baikal Lake in Siberia, Russia.

c. One thousand one hundred forty-eight feet; 1000 + 100 + 40 + 8 feet

Now work margin exercise 9.

Completion Example Answers

4. a. 30,000 + 2000 + 80 + 1 **b.** 400,00 + 90,000 + 7000 + 500 **6. a.** eight thousand four hundred seven **b.** fifteen thousand, three hundred fifty-two **8. a.** 2,800,035 **b.** 5,350,000

Margin Exercise Answers

1. a. 9 **b.** 6 **c.** 1 **2. a.** hundreds place **b.** ten thousands place **c.** tens place **3. a.** 400 + 60 + 3 **b.** 7000 + 300 **4. a.** 20,000 + 9000 + 500 + 20 + 4 **b.** 800,000 + 8000 + 400 + 90 + 1 **5. a.** thirty-two million, four hundred fifty thousand, ninety **b.** five thousand eighty-four **6. a.** six thousand seven hundred ninety-one **b.** twenty-three thousand, five hundred seven **7. a.** 6041 **b.** 1,483,007 **8. a.** 6,300,500 **b.** 4,875,000 **9. a.** Antarctica **b.** 604 feet **c.** seven hundred seven feet; 700 + 7

1.1 Exercises

Concept Check

Fill-in-the-Blank. Complete each sentence using information found in this section.

1. Given the number 791,065, the digit ______ is in the thousands place.
2. The place value to the left of the millions place is the __________ place.
3. Another name for the natural numbers is the __________ numbers.
4. The smallest whole number is _______ while the smallest natural number is _____.
5. The number 5,012 written in expanded notation would be ________ + ________ + _________ + ________.
6. The number 40,782 written in words is ________________________.

1.1 Exercises

Concept Check

1. 1
2. Ten millions
3. Counting
4. 0, 1
5. 5000 + 0 + 10 + 2
6. Forty thousand, seven hundred eighty-two

7. True
8. False; It can be written as 50,000+6000+300+17.
9. True
10. True

Practice

1. a. 6 b. 8 c. 0
2. a. 6 b. 3 c. 1
3. a. Ten thousands
b. Hundreds
c. Ones
4. a. Thousands
b. Hundreds c. Tens
5. 2: ten thousands, 4: thousands, 6: hundreds, 8: ones
6. 3: ten thousands, 4: thousands, 7: hundreds, and 8: ones
7. 2: billions, 4: hundred millions, 3: millions, 1: hundred thousands, 8: ten thousands, 9: thousands, 5: hundreds
8. 4: ten billions, 8: billions, 5: hundred millions, 6: ten millions, 9: millions, 1: hundred thousands, 2: thousands, 3: hundreds, 4: tens
9. 30 + 7
10. 80 + 4
11. 500 + 60 + 3
12. 800 + 20 + 1
13. 1000 + 800 + 90 + 2
14. 2000 + 50 + 9
15. 20,000 + 5000 + 600 + 50 + 8
16. 30,000 + 2000 + 300 + 40 + 1
17. Eight hundred thirty-five
18. One hundred twenty-two
19. Thirty thousand, two hundred one

True/False. Determine whether each statement is true or false. If a statement is false, explain how it can be changed so the statement will be true. (**Note:** There may be more than one acceptable change.)

7. In the number 21,057, the "1" represents 1000.
8. 56,317 can be written as 56,000 + 300 + 17 in expanded notation.
9. 42,360 can be written as forty-two thousand, three hundred sixty.
10. The word "and" is not used when reading or writing whole numbers.

Practice

1. Given the number 284,065, which digit indicates the number of
   a. tens? b. ten thousands? c. hundreds?
2. Given the number 13,476,582, which digit indicates the number of
   a. thousands? b. millions? c. ten millions?
3. For the number 71,349, state the place value of
   a. the digit 7. b. the digit 3. c. the digit 9.
4. For the number 309,472, state the place value of
   a. the digit 9. b. the digit 4. c. the digit 7.
5. Name the place value of each nonzero digit in the following number: 24,608.
6. Name the place value of each nonzero digit in the following number: 34,708.
7. Name the place value of each nonzero digit in the following number: 2,403,189,500.
8. Name the place value of each nonzero digit in the following number: 48,569,102,340.

Write each number in expanded notation. See Examples 3 and 4.

9. 37
10. 84
11. 563
12. 821
13. 1892
14. 2059
15. 25,658
16. 32,341

Write each number in words. See Examples 5 and 6.

17. 835
18. 122
19. 30,201
20. 10,500
21. 683,100
22. 529,300
23. 16,302,590
24. 71,500,000

Write each number in standard notation. See Examples 7 and 8.

25. five hundred eighty
26. seven hundred fifty-seven
27. two thousand five
28. ten thousand, eleven

29. three thousand eight hundred thirty-four

30. eight thousand four hundred sixty-five

31. seventy-eight thousand, nine hundred two

32. eighty-six thousand, six hundred ten

33. sixty-three thousand, sixty-five

34. twenty-four thousand, six hundred

35. four hundred thousand, seven hundred thirty-six

36. five hundred thirty-seven thousand, eighty-two

37. eighty-two million, seven hundred thousand

38. one hundred million

39. thirty-five million, nine hundred sixty thousand, thirteen

40. two hundred eighty-one million, three hundred thousand, five hundred one

Applications

Write the numbers in each sentence in words.

41. The largest lake in the United States is Lake Superior. It takes up an area of 82,103 square kilometers.

42. The population of Los Angeles is 3,976,322.

43. The country of Chile averages about 110 miles in width and is about 2650 miles long.

44. One of the world's tallest buildings is the Taipei Financial Center in Taipei, Taiwan. The building has 101 stories and reaches 1761 feet tall.

45. The Republic of Venezuela covers an area of 352,143 square miles, or about 912,050 square kilometers.

46. The Pacific Ocean has an area of approximately 63,800,000 square miles and has an average depth of 14,040 feet.

47. In 2021, the student enrollment at College of Charleston was 10,667 and the student enrollment at the University of Minnesota was 66,496.[1]

48. The average distance from the earth to the sun is about 93,000,000 miles, or 149,730,000 kilometers.

Write the numbers in each sentence in standard form.

49. The largest collection of Joker playing cards consists of eight thousand, five hundred twenty cards amassed by Tony De Santis after inheriting a two thousand piece collection from the magician Fernando Riccardi.[2]

50. The 2022 Tour de France covered three thousand three hundred twenty-eight kilometers. The winner received five hundred thousand euros.

51. The largest playing card structure in the world was completed by Bryan Berg in Macau, China. It took him forty-four days and he used two hundred eighteen thousand, seven hundred ninety-two cards.[3]

1 Source: cofc.edu/about/ataglance and idr.umn.edu/reports-by-topic-enrollment/enrollments
2 Source: www.guinnessworldrecords.com/records-6000/largest-playing-card-joker-collection
3 Source: www.guinnessworldrecords.com/world-records/largest-playing-card-structure

20. Ten thousand, five hundred

21. Six hundred eighty-three thousand, one hundred

22. Five hundred ninety-two thousand, three hundred

23. Sixteen million, three hundred two thousand, five hundred ninety

24. Seventy-one million, five hundred thousand

25. 580

26. 757

27. 2005

28. 10,011

29. 3834

30. 8465

31. 78,902

32. 86,610

33. 63,065

34. 24,600

35. 400,736

36. 537,082

37. 82,700,000

38. 100,000,000

39. 35,960,013

40. 281,300,501

Applications

41. Eighty-two thousand, one hundred three

42. Three million, nine hundred seventy-six thousand, three hundred twenty-two

43. One hundred ten; two thousand six hundred fifty

44. One hundred one; one thousand seven hundred sixty-one

45. Three hundred fifty-two thousand, one hundred forty-three; nine hundred twelve thousand, fifty

52. Russia is the largest country in the world at seventeen million, ninety-eight thousand, two hundred forty-two square kilometers. It borders fourteen countries.[4]

53. Florida, by far, has the highest number of seasonal, recreational, or occasional use homes. Wisconsin and Arizona are in second and third place with one hundred ninety-three thousand, forty-six homes and one hundred eighty-four thousand, three hundred twenty-seven homes, respectively.[5]

54. According to the latest census, the state with the most housing units was California with one million, six hundred eighty thousand, eighty-one while the state with the fewest was Wyoming, with two hundred sixty-one thousand, eight hundred sixty-eight.[6]

55. Greece has a population of ten million, five hundred thirty-three thousand, eight hundred seventy-one, with approximately three million, one hundred fifty-four thousand of their citizens living in Athens.[7]

56. In 2020, Cedar Point in Sandusky, Ohio had one million, twenty thousand visitors. In the same year, the Magic Kingdom at Walt Disney World in Lake Buena Vista, Florida, had six million, nine hundred forty-one thousand visitors.[8]

Solve.

57. During a talk show, a guest says that his website has four hundred eighty thousand one hundred five views per day. The transcriptionist writes this number as "408,105". Is this the correct standard notation? If not, what should it be?

58. A jewelry maker made one thousand five hundred dollars during one week. She keeps track of her income and writes "$1500" in a spreadsheet. Is this the correct standard notation? If not, what should it be?

Use the table to answer the following questions.

PGA Tour Statistics 2021–2022

Rank	Player	Events	Rounds	Cuts Made	Top 10	Wins	Cup Points	Earnings
1	Scottie Scheffler	25	66	21	11	4	3556	$14,046,910
2	Cameron Smith	18	49	15	7	3	2335	$10,107,897
3	Will Zalatoris	24	57	19	9	1	1680	$9,405,082
4	Patrick Cantlay	20	52	18	12	2	2108	$9,369,605
5	Rory McIlroy	16	52	14	10	3	2104	$8,654,566
6	Xander Schauffele	21	57	19	7	3	2153	$7,427,299
7	Sam Burns	24	69	18	8	3	2275	$7,073,986
8	Matt Fitzpatrick	20	48	17	10	1	1596	$7,012,672
9	Justin Thomas	21	61	19	10	1	1783	$6,829,576
10	Cameron Young	25	70	18	7	0	1774	$6,520,598

Source: Golf Stat and Records, PGA Tour, Accessed August 29, 2022, www.pgatour.com/stats.html.

59. Which player played 69 tournament rounds during the 2021–2022 season?

4 Source: www.cia.gov/library/publications/the-world-factbook
5 Source: www.census.gov
6 Source: www.census.gov
7 Source: www.cia.gov/the-world-factbook/countries/greece
8 Source: aecom.com/theme-index

46. Sixty-three million, eight hundred thousand; fourteen thousand, forty

47. Ten thousand, six hundred sixty-seven; sixty-six thousand, four hundred ninety-six

48. Ninety-three million; one hundred forty-nine million, seven hundred thirty thousand

49. 8520; 2000

50. 3328; 500,000

51. 44; 218,792

52. 17,098,242; 14

53. 193,046; 184,327

54. 1,680,081; 261,868

55. 10,533,871; 3,154,000

56. 1,020,000; 6,941,000

What about the Women?

The 2022 leading earners on the Ladies Professional Golf Association tour, as shown on the LPGA's official website, were Minjee Lee of Australia with $3,736,526, followed by South Korea's In Gee Chun with $2,603,128 and Canada's Brooke M. Henderson with $2,218,426. The highest career earner in the LPGA is Sweden's Annika Sorenstam with $22,583,603.

Source: LPGA, www.lpga.com

60. Which player won more money than Cameron Smith during the 2021-2022 season? Write the amount that he won in words and in expanded notation.

61. Which player had 9 top ten finishes?

62. How many cup points did Rory McIlroy win? Write this number of points in words.

63. Write the earnings of Matt Fitzpatrick in words.

Use the table to answer the following questions.

Top Attended Broadway Shows for the Week Ending March 27, 2022

Broadway Show	Attendance	Gross
The Music Man	12,092	$3,451,724
Hamilton	10,559	$2,272,219
Wicked	14,640	$1,944,704
The Lion King	13,208	$1,778,641
Plaza Suite	7791	$1,657,090
Harry Potter and the Cursed Child	11,191	$1,704,152
Moulin Rouge! The Musical	9908	$1,381,323
MJ The Musical	9754	$1,314,241
Six	8150	$1,284,951

Source: Broadway World, www.broadwayworld.com/grosses.cfm

64. Which of these shows had the smallest gross income?

65. Which show had the largest attendance?

66. Write the largest gross income number in words.

67. Which show had the smallest attendance? Write this number in words.

68. Two shows had gross income of over two million dollars. Which shows were they?

69. Four shows had attendance of fewer than 10,000 people. Which shows were they?

70. Which show had the smaller gross income: *MJ The Musical* or *Six*?

71. Which show had higher attendance: *Hamilton* or *Wicked*?

Hottest Tickets

The longest-running show on Broadway of all time is Andrew Lloyd Webber's musical *The Phantom of the Opera*, which opened on January 26, 1988, and is still running at the Majestic Theatre on Broadway. Across the Atlantic, on London's West End, the longest running show is the murder mystery play *The Mousetrap*, which opened in 1952 and is still running.

57. No; 480,105

58. Yes

59. Sam Burns

60. Scottie Scheffler; fourteen million, forty-six thousand, nine hundred ten dollars; 10,000,000 + 4,000,000 + 40,000 + 6000 + 900 + 10

61. Will Zalatoris

62. 2104 points; two thousand one hundred four

63. Seven million, twelve thousand, six hundred seventy-two dollars

64. *Six*

65. *Wicked*

66. Three million, four hundred fifty-one thousand, seven hundred twenty-four dollars

67. *Plaza Suite;* seven thousand, seven hundred ninety-one

68. *The Music Man* and *Hamilton*

69. *Plaza Suite, Moulin Rouge! The Musical, MJ The Musical,* and *Six*

70. *Six*

71. *Wicked*

72. One hundred sixty-four million, eight hundred seventy thousand, two hundred thirty-four dollars

73. 40,000,000 + 7,000,000 + 500,000 + 40,000 + 7000 + 200 + 30 + 1

74. *Spider-Man: No Way Home*

75. $183,651,655; One hundred eighty-three million, six hundred fifty-one thousand, six hundred fifty-five dollars

Writing & Thinking

76. 0 is a whole number and not a natural number. Both whole numbers and natural numbers include 1, 2, 3, 4, 5, and so on.

77. 10^{100}

78. 1: ten millions, 3: millions, 2: thousands, 7: hundreds; In the 2020 census, the four most populous states were California: 39,538,223; Texas: 29,145,505; Florida: 21,538,187; and New York: 20,201,249.

79. Hyphens are used for two-digit numbers larger than 20 that do not end in a zero.

Use the table to answer the following questions.

Top 10 Domestic Gross for Movies Opening in 2021

Rank	Movie Title	Studio	Opening	Total Gross
1	*Spider-Man: No Way Home*	Sony	$260,138,569	$804,416,709
2	*Shang-Chi and the Legend of the Ten Rings*	WDSMP	$75,388,688	$224,543,292
3	*Venom: Let There Be Carnage*	Sony	$90,033,210	$213,550,366
4	*Black Widow*	WDSMP	$80,366,312	$183,651,655
5	*F9: The Fast Saga*	Universal	$70,043,165	$173,005,945
6	*Eternals*	WDSMP	$71,297,219	$164,870,234
7	*Sing 2*	Universal	$22,326,230	$162,700,840
8	*No Time to Die*	MGM	$55,225,007	$160,891,007
9	*A Quiet Place Part II*	Paramount	$47,547,231	$169,072,261
10	*Ghostbusters: Afterlife*	Sony	$44,008,406	$129,360,575

Source: Information courtesy of Box Office Mojo. Used with permission. www.boxofficemojo.com

72. For *Eternals,* write the total domestic gross income in words.

73. Write the opening domestic income for *A Quiet Place Part II* in expanded notation.

74. Which movie had the largest gross domestic income in 2021?

75. Write the total domestic gross income for *Black Widow* in standard form and in words.

Writing & Thinking

76. How are natural numbers and whole numbers different and how are they the same?

77. A googol is the name of a very large power of ten. Look up the meaning of this term on the internet. (**Note:** Ten to the power of a googol is called a googolplex.)

78. According to the 2020 Census, Pennsylvania is the fifth most populous state in the United States with 13,002,700 people. Name the place value of each nonzero digit in this number. Use the internet to find the populations of the four most populous states.[9]

79. When are hyphens used to write numbers in English words?

9 Source: www.census.gov/data/tables/2020/dec/2020-apportionment-data.html

1.2 Addition and Subtraction with Whole Numbers

Objectives

A. Add whole numbers.

B. Recognize and use the properties of addition.

C. Calculate the perimeter of geometric figures.

D. Subtract whole numbers.

A Addition with Whole Numbers

An electronics store sold 2 cell phones one day and 3 the next day. The total number of cell phones sold in those two days is 5. The process of finding the total is called **addition**.

Figure 1

Addition is indicated either by writing the numbers horizontally separated by plus signs (+), or by writing the numbers vertically in a column. The numbers being added are called **addends**, and the result of the addition is called the **sum**.

2	+	3	=	5	or	2 addend
addend		addend		sum		+ 3 addend
						5 sum

Be sure to keep the digits aligned (in column form) by place value so you will be adding ones to ones, tens to tens, and so on. **Neatness is a necessity in mathematics.**

Adding large numbers or adding several numbers can be done by writing the numbers vertically so that the **place values are lined up** in columns. Then only the digits with the same place value are added.

Adding Whole Numbers

1. Write the numbers vertically so that the **place values are lined up** in columns.
2. Add only the digits with the same place value.

PROCEDURE

Example 1 Adding Whole Numbers

Add: 623 + 172

Solution

$$\begin{array}{r} 623 \\ +172 \\ \hline 5 \end{array}$$ ← Addend, ← Addend; Add ones.

$$\begin{array}{r} 623 \\ +172 \\ \hline 95 \end{array}$$ Add tens.

$$\begin{array}{r} 623 \\ +172 \\ \hline 795 \end{array}$$ Add hundreds.

1. Add.

$$\begin{array}{r} 461 \\ +\ 228 \\ \hline \end{array}$$

$$\begin{array}{r} 623 \\ +172 \\ \hline 795 \end{array}$$

623 ← Do not write all the steps.
+172 Write only the addends and the sum.
795 ← Sum

Now work margin exercise 1.

If the sum of the digits in one or more columns is more than 9, then the following procedure for carrying digits needs to be followed.

Carrying When Adding Whole Numbers

If the sum of the digits in one column is more than 9,

1. write the ones digit of the sum in that column and
2. **carry** the tens digit of the sum as a number to be added to the next column to the left.

PROCEDURE

Example 2 illustrates these ideas.

2. Add.

$$\begin{array}{r} 463 \\ +\ 38 \\ \hline \end{array}$$

Example 2 Adding Whole Numbers When Carrying is Required

Add: 475 + 59

Solution

Carry the 1 by writing it above the tens column. →

$$\begin{array}{r} {}^{1} \\ 47\mathbf{5} \\ +\ \ 5\mathbf{9} \\ \hline \mathbf{4} \end{array}$$

Write the 4 in the ones column. →

Add ones:
5 ones + 9 ones = 14 ones
= 1 ten + 4 ones

Carry the 1 by writing it above the hundreds column. →

$$\begin{array}{r} {}^{1\,1} \\ 4\mathbf{7}5 \\ +\ \ \mathbf{5}9 \\ \hline \mathbf{3}4 \end{array}$$

Write the 3 in the tens column. →

Add tens:
7 tens + 5 tens + 1 ten = 13 tens
= 1 hundred + 3 tens

$$\begin{array}{r} {}^{1\,1} \\ \mathbf{4}75 \\ +\ \ 59 \\ \hline 534 \end{array}$$

Write the 5 in the hundreds column. →

Add hundreds:
4 hundreds + 1 hundred = 5 hundreds

Now work margin exercise 2.

When adding more than two digits, you can sometimes add faster by looking for combinations of digits that total 10. This idea is illustrated in Example 3.

Example 3 Adding Whole Numbers Using Sums of 10

Add: 17 + 34 + 86

Solution

To add, we note the combinations that total 10 to find the sums quickly.

Carry the 1 by writing it above the tens column.

$$\begin{array}{r} {}^{1} \\ 1\,7 \\ 3\,4 \\ +\,8\,6 \\ \hline 7 \end{array}$$

Add ones: 7 + 4 + 6 = 7 + 10 = 17 ones
= 1 ten + 7 ones

Write the 7 in the ones column.

$$\begin{array}{r} {}^{1} \\ 1\,7 \\ 3\,4 \\ +\,8\,6 \\ \hline 1\,3\,7 \end{array}$$

Add tens: 1 + 1 + 3 + 8 = 10 + 3 = 13 tens
= 1 hundred and 3 tens

Write the 3 in the tens column and the 1 in the hundreds column.

Now work margin exercise 3.

3. Add.

$$\begin{array}{r} 63 \\ 75 \\ +\ 47 \\ \hline \end{array}$$

Example 4 Application: Adding Whole Numbers

Stan bought a television set for \$859, a stereo for \$697, and a computer for \$1285. What total amount did he spend?

Solution

The total amount spent is the sum.

$$\begin{array}{r} {}^{1\,2\,2} \\ 8\,5\,9 \\ 6\,9\,7 \\ +\,1\,2\,8\,5 \\ \hline 2\,8\,4\,1 \end{array}$$ Total spent

He spent a total of \$2841.

Now work margin exercise 4.

4. Diedra bought a couch for \$715, a coffee table for \$199, and a washer and dryer for \$863. What total amount did she spend?

CALCULATORS

Using a Calculator to Add Whole Numbers

(**Note:** Your instructor may allow the use of calculators. If so, you should keep in mind that a calculator is a tool to enhance, not replace, the necessary skills and abilities needed to solve problems. Your personal experience, knowledge, and general understanding of each concept will determine the ease with which you complete the problems presented.)

To add numbers on a calculator, you will need the addition key [+] and the equal sign key [=].
To add the values given in Example 4, press the keys

[8] [5] [9] [+] [6] [9] [7] [+] [1] [2] [8] [5].

Then press [=]. The display will read 2841.

To delete the problem after completing it, press CLEAR or AC, or simply begin typing the next expression. (**Note:** Your calculator may have both an AC and a C button. In that case, AC clears everything while C clears only your last entry.)

B The Properties of Addition

When two numbers are added, the order of the numbers does not matter. That is, $4 + 9 = 13$ and $9 + 4 = 13$. This fact is called the **commutative property of addition**.

Commutative Property of Addition

The **order** of the numbers in addition can be reversed without changing the sum. For example,

$$3 + 14 = 17 \quad \text{and} \quad 14 + 3 = 17.$$

PROPERTIES

To add three numbers, such as $6 + 3 + 5$, first we add two of the numbers and then add the third. Which two should we add first? The answer is that the sum is the same either way.

$$6+3+5=(6+3)+5=9+5=14$$

and

$$6+3+5=6+(3+5)=6+8=14$$

This example illustrates the **associative property of addition**.

Associative Property of Addition

The **grouping** of the numbers in addition can be changed without changing the sum. For example,

$$(8+12)+7=20+7=27 \quad \text{and} \quad 8+(12+7)=8+19=27.$$

PROPERTIES

Math Tip

When adding multiple numbers together, it is sometimes easier to add certain numbers together first. For example, consider the sum $38 + 27 + 3$. Which order is easier to add mentally: 38 plus 27 first and then add 3, or 27 plus 3 first and then add 38?

$(38+27)+3$	$38+(27+3)$
$65+3$	$38+30$
68	68

This is one advantage of using the associative property of addition.

Another property, called the **additive identity property**, involves addition with 0. Whenever 0 is added to a number, the result is the original number.

Additive Identity Property

The sum of a number and 0 is that same number. For example,

$$8 + 0 = 8 \quad \text{and} \quad 0 + 19 = 19.$$

The number 0 is called the **additive identity**.

PROPERTIES

Example 5 Recognizing the Properties of Addition

Each of the properties of addition is illustrated.

a. $40 + 3 = 3 + 40$ Commutative property of addition

As a check, we see that $40 + 3 = 43$ and $3 + 40 = 43$.

b. $2 + (5 + 9) = (2 + 5) + 9$ Associative property of addition

As a check, we see that $2 + (14) = 16$ and $(7) + 9 = 16$.

c. $86 + 0 = 86$ Additive identity property

Now work margin exercise 5.

5. For each statement, state the property of addition illustrated.

a. $(9+1)+7=9+(1+7)$

b. $25 + 4 = 4 + 25$

c. $39 + 0 = 39$

Note that Example 5a shows a change in the **order** of two numbers, while Example 5b shows a change in the **grouping** (association) of the numbers.

C Calculating the Perimeter of a Polygon

Familiar geometric figures called polygons, such as triangles, squares, and rectangles, are defined in terms of line segments (called **sides**). A **polygon** is a figure in a flat surface (a plane) with three or more sides. Each side is a line segment. Several common polygons are pictured here.

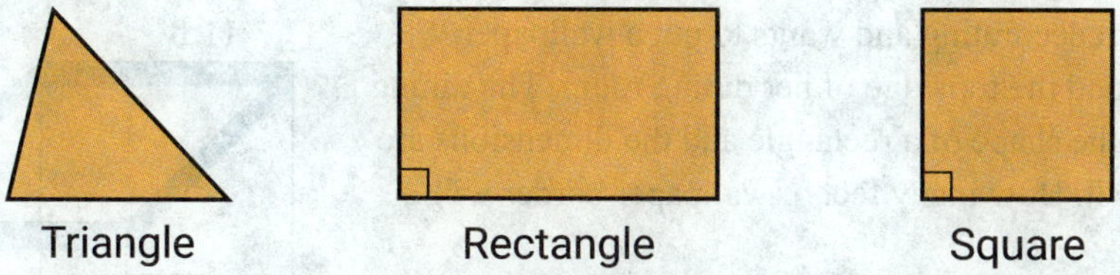

Figure 2

The **perimeter** of a geometric figure is the distance around the figure. For polygons, the perimeter is the sum of the lengths of its sides and is measured in linear units, such as inches, feet, yards, miles, centimeters, or meters. For example, suppose we are weatherproofing a house for winter by installing insulation strips around the door frame. In order to know how much insulation to buy, we measure the lengths of the sides of the door and add all four measurements together.

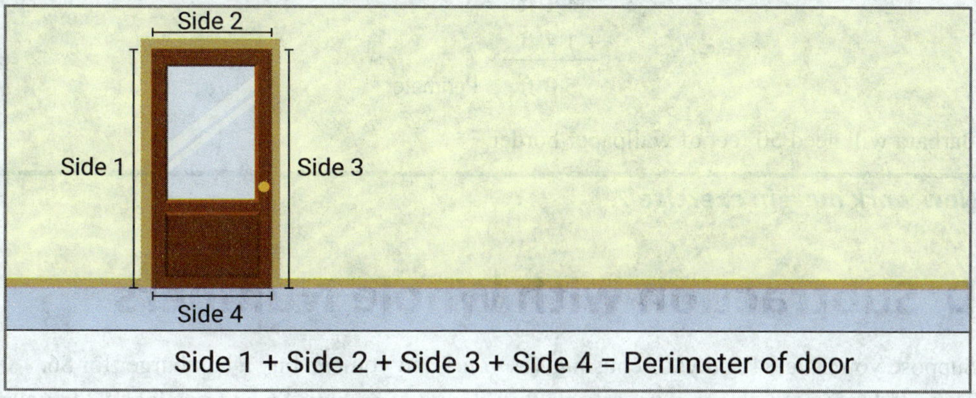

Figure 3

6. Calculate the perimeter of the polygon.

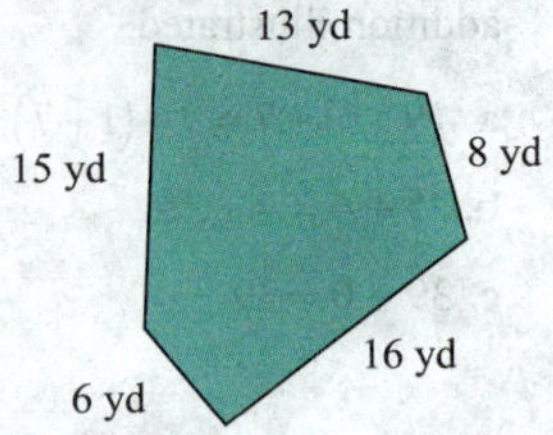

Example 6 Calculating the Perimeter of a Polygon

Calculate the perimeter of the polygon.

(**Note:** A 5-sided polygon is called a **pentagon**.)

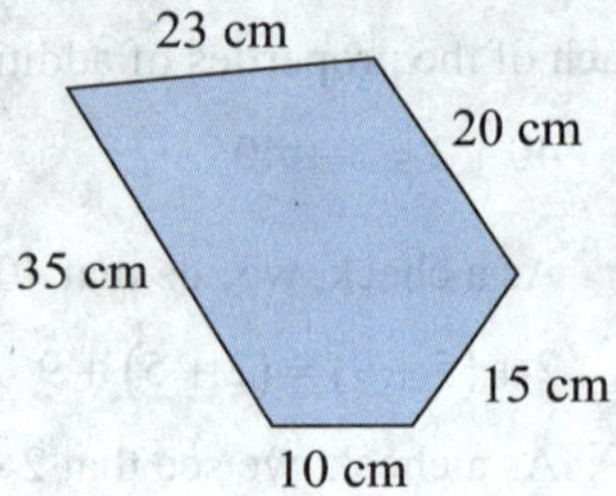

Solution

To find the perimeter, calculate the sum of the lengths of the sides.

$$\begin{array}{r} \overset{1}{1}0\text{ cm} \\ 15\text{ cm} \\ 20\text{ cm} \\ 23\text{ cm} \\ +\ 35\text{ cm} \\ \hline 103\text{ cm} \end{array} \quad \text{Perimeter}$$

The perimeter of the pentagon is 103 cm.

Now work margin exercise 6.

7. Nina is weatherproofing her house for winter and wants to install insulation strips around a door frame. The door measures 80 inches high by 36 inches wide. How many inches of weather stripping will she need?

Example 7 Application: Calculating the Perimeter of a Rectangle

Barbara is redecorating and wants to put a wallpaper border around the top edge of her dining room. The dining room is in the shape of a rectangle and the dimensions are 11 ft by 14 ft. How many feet of wallpaper border will she need?

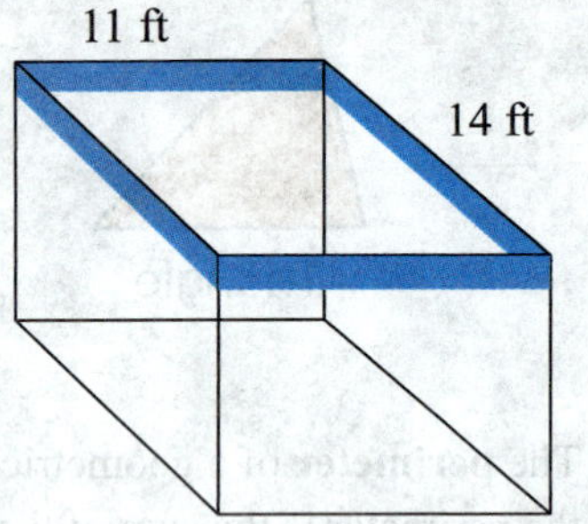

Solution

To determine the amount of wallpaper border needed, find the distance around the top edge of the room. To do this, add the lengths of each side of the room together.

$$\begin{array}{r} \overset{1}{1}1\text{ ft} \\ 14\text{ ft} \\ 11\text{ ft} \\ +\ 14\text{ ft} \\ \hline 50\text{ ft} \end{array} \quad \text{Perimeter}$$

Barbara will need 50 feet of wallpaper border.

Now work margin exercise 7.

D Subtraction with Whole Numbers

Suppose you have \$4 and your cousin gives you \$2 so you can buy a hamburger for \$6. That is \$4 + \$2 = \$6. Now, if we reverse this idea and you have \$6 and you loan \$2 to your cousin to buy a milkshake, how much will you have left? In this case \$2 is "taken away" from \$6. This is **subtraction** and the answer is the **difference**. Subtraction is indicated with a **minus sign** (–).

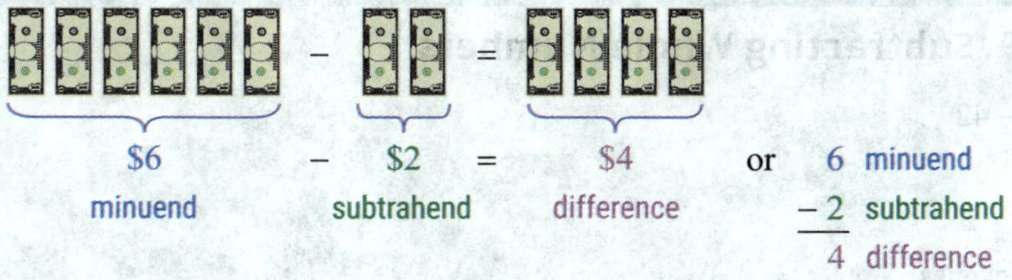

Figure 4

Subtraction

Subtraction is the operation of taking one amount, or number, away from another. (This is the opposite of adding the two amounts.)

The **difference** is the result of subtracting one number (called the **subtrahend**) from another number (called the **minuend**).

DEFINITION

Because of the relationship between addition and subtraction, we see that subtraction can be checked by addition. (See Example 8.)

Example 8 Subtracting Whole Numbers

Subtract and check by addition.

a. 10 – 7 **b.** 9 – 9 **c.** 8 – 3

Solution

a. 10 – 7 = 3 Check: 7 + 3 = 10

b. 9 – 9 = 0 Check: 9 + 0 = 9

c. 8 – 3 = 5 Check: 3 + 5 = 8

Now work margin exercise 8.

8. Subtract and check by addition.

a. 15 – 6

b. 11 – 7

c. 7 – 1

When subtracting numbers with two digits or more, we generally write the numbers in a vertical format. In this way, the digits with the same place value can be aligned.

Subtracting Whole Numbers

1. Write the numbers vertically so that the **place values are lined up** in columns.
2. Subtract only the digits with the same place value.
3. Check by adding the difference to the subtrahend. The sum must be the minuend.

PROCEDURE

9. Subtract.

$$\begin{array}{r} 65 \\ -\ 21 \\ \hline \end{array}$$

Example 9 Subtracting Whole Numbers

Subtract: 96 – 42

Solution

$$\begin{array}{r} 9\,\mathbf{6} \\ -\ 4\,\mathbf{2} \\ \hline \mathbf{4} \end{array}$$ Subtract ones.

$$\begin{array}{r} \mathbf{9}\ 6 \\ -\ \mathbf{4}\ 2 \\ \hline \mathbf{5}\ 4 \end{array}$$ Subtract tens. Difference

Check

We can check by adding.

$$\begin{array}{rl} 54 & \text{Difference} \\ +\ 42 & \text{Subtrahend} \\ \hline 96 & \text{Sum (minuend)} \end{array}$$

Now work margin exercise 9.

Teaching Note

To increase student interest, you might discuss the technique of subtraction by adding in the subtrahend rather than borrowing in the minuend. Point out that making a digit in the subtrahend larger by 1 has the same effect as borrowing 10 from a digit in the minuend. For example:

$$\begin{array}{r} 942 \\ -\ 357 \\ \hline \end{array} \qquad \begin{array}{r} 9\,\overset{14}{\cancel{4}}\,\overset{12}{\cancel{2}} \\ -\ \overset{4}{\cancel{3}}\,\overset{6}{\cancel{5}}\,7 \\ \hline 585 \end{array}$$

gives the same result as

$$\begin{array}{r} \overset{8}{\cancel{9}}\,\overset{\overset{13}{\cancel{3}}}{\cancel{4}}\,\overset{12}{\cancel{2}} \\ -\ 357 \\ \hline 585 \end{array}$$

This method is taught in many other countries, and some of your students may prefer to use it. In any case, they will probably be surprised to find that there are methods other than those they have learned for performing basic operations.

If the digit in the subtrahend (the number being subtracted) is larger than the corresponding digit in the minuend (the number being subtracted from), then we must borrow. For example, consider the following problem.

$$\begin{array}{r} 83 \\ -\ 47 \\ \hline \end{array}$$

The 7 in the ones place in 47 is larger than the 3 in the ones place in 83. So we borrow 1 ten from the tens place.

8 tens – 1 ten = 7 tens → $\begin{array}{r} \overset{7}{\cancel{8}}\,\overset{13}{\cancel{3}} \\ -\ 47 \end{array}$ ← 1 ten + 3 ones = 10 ones + 3 ones = 13 ones

Now subtract.

$$\begin{array}{r} \overset{7}{\cancel{8}}\,\overset{13}{\cancel{3}} \\ -\ 47 \\ \hline 36 \end{array}$$

Check

Add the difference to the subtrahend. The sum should be the minuend.

$$\begin{array}{rl} \overset{1}{3}6 & \text{Difference} \\ +\ 47 & \text{Subtrahend} \\ \hline 83 & \text{Sum (minuend)} \end{array}$$

The next example illustrates borrowing more than once. In finding the difference 742 – 259, we see that the 9 in the ones place is larger than 2 and then the 5 in the tens place is larger than 3.

Example 10 Subtracting Whole Numbers by Borrowing

Subtract: 742 – 259

Solution

Step 1: Since 9 is larger than 2, borrow 1 ten from the tens place and add it to the ones place.

```
   3 12
 7 4̸ 2̸     In the tens column: 4 tens – 1 ten = 3 tens
– 2 5 9     In the ones column: 1 ten + 2 ones = 12 ones
```

Step 2: Since 5 is larger than 3, borrow 1 hundred from the hundreds place and add it to the tens place.

```
    13
 6  3̸ 12
 7̸ 4̸ 2̸     In the hundreds column: 7 hundreds – 1 hundred = 6 hundreds
– 2 5 9     In the tens column: 1 hundred + 3 tens = 13 tens
```

Step 3: Now subtract.

```
    13
 6  3̸ 12
 7̸ 4̸ 2̸
– 2 5 9
  4 8 3
```

Check

Add the difference to the subtrahend. The sum should be the minuend.

```
  1 1
  4 8 3
+ 2 5 9
  7 4 2
```

Now work margin exercise 10.

10. Subtract.

$$\begin{array}{r} 341 \\ -187 \\ \hline \end{array}$$

Example 11 Subtracting Whole Numbers by Borrowing

Subtract: 800 – 65

Solution

Step 1: Since we cannot borrow from the 0 in the tens place, we end up borrowing 1 hundred from the hundreds place.

```
 7 10
 8̸ 0̸ 0     In the hundreds column: 8 hundreds – 1 hundred = 7 hundreds
–   6 5     In the tens column: 1 hundred + 0 tens = 10 tens
```

Step 2: Now, borrow 1 ten from the tens place.

```
    9
 7 1̸0̸ 10
 8̸ 0̸ 0̸     In the tens column: 10 tens – 1 ten = 9 tens
–   6 5     In the ones column: 1 ten + 0 ones = 10 ones
```

11. Subtract:

$$\begin{array}{r} 700 \\ -42 \\ \hline \end{array}$$

Step 3: Subtract.

$$\begin{array}{r} \overset{7}{\cancel{8}}\,\overset{\overset{9}{\cancel{10}}}{\cancel{0}}\,\overset{10}{\cancel{0}} \\ -\quad 6\;5 \\ \hline 7\;3\;5 \end{array}$$

Check

Add the difference to the subtrahend. The sum should be the minuend.

$$\begin{array}{r} \overset{1}{7}\,\overset{1}{3}\,5 \\ +\quad 6\;5 \\ \hline 8\;0\;0 \end{array}$$

Now work margin exercise 11.

---

12. What number should be added to 268 to get a sum of 654?

Example 12 Finding a Missing Addend

What number should be added to 546 to get a sum of 831?

Solution

We know the sum and one addend. To find the missing addend, subtract 546 from 831.

$$\begin{array}{rl} \overset{7}{\cancel{8}}\,\overset{\overset{12}{\cancel{2}}}{\cancel{3}}\,\overset{11}{\cancel{1}} & \text{Sum} \\ -\,5\;4\;6 & \text{Addend} \\ \hline 2\;8\;5 & \text{Difference} \end{array}$$

The number to be added is 285.

Now work margin exercise 12.

---

13. The cost of repairing Robert's used DVD player will be \$165. To buy a new DVD player, Robert will have to pay \$129. How much more would it cost to fix the old DVD player than to buy a new one?

Example 13 Application: Subtracting Whole Numbers

The cost of repairing Ed's used TV is \$395. To buy a new TV, he will have to pay \$447. How much more would Ed have to pay for a new TV than to have his old TV repaired?

Solution

$$\begin{array}{r} \overset{3}{\cancel{4}}\,\overset{14}{\cancel{4}}\,7 \\ -\,3\;9\;5 \\ \hline 5\;2 \end{array}$$

Ed would pay \$52 more for a new TV than to have his old TV repaired.

Now work margin exercise 13.

CALCULATORS

Using a Calculator to Subtract Whole Numbers

To subtract numbers on a calculator, you will need the subtraction key [−] and the equal sign key [=]. To subtract the values given in Example 13, press the keys

[4] [4] [7] [−] [3] [9] [5].

Then press [=]. The display will read 52.

Note: Do not confuse the [−] key and the [(−)] or [+/−] key. Subtraction always uses the [−] key. The other keys are used with negative numbers, which will be introduced in Chapter 8.

Example 14 Application: Adding and Subtracting Numbers

In pricing a new car, Jason found that he would have to pay a base price of $15,200 plus $1025 in taxes and $575 for license fees. If the bank loaned him $10,640, how much cash would Jason need to buy the car?

Solution

In this problem, experience tells us that we must add and subtract even though there are no specific directions to do so. First, we add Jason's expenses and then we subtract the amount of the bank loan.

Expenses

15,200	Base price
1025	Taxes
+ 575	License fees
16,800	Total expenses

Cash Needed

16,800	Total expenses
− 10,640	Bank loan
6160	Cash

Jason would need $6160 in cash to buy the car.

Now work margin exercise 14.

14. In pricing a new motorcycle, Alex found that he would have to pay a base price of $10,470 plus $630 in taxes and $70 for the license fee. If the bank loaned him $7530, how much cash would Alex need to buy the motorcycle?

Margin Exercise Answers

1. 689 **2.** 501 **3.** 185 **4.** $1777 **5. a.** Associative property of addition **b.** Commutative property of addition **c.** Additive identity property **6.** 58 yd **7.** 232 inches **8. a.** 9; 6 + 9 = 15 **b.** 4; 7 + 4 = 11 **c.** 6; 1 + 6 = 7 **9.** 44 **10.** 154 **11.** 658 **12.** 386 **13.** $36 **14.** $3640

1.2 Exercises

Concept Check

Fill-in-the-Blank. Complete each sentence using information found in this section.

1. When numbers are added, the result (or answer) is called the ______.
2. The numbers being added in an addition problem are called ______.
3. When the grouping of numbers is changed in an addition problem, the ______ property of addition is being utilized.
4. The ________ property of addition indicates that the order of two numbers being added can be changed.
5. Another name for the distance around a geometric figure is its _______.
6. The result in a subtraction problem is the _________.

1.2 Exercises

Concept Check

1. sum
2. addends
3. associative
4. commutative
5. perimeter
6. difference

7. False; A polygon has three or more sides.
8. True
9. False; Borrowing must occur.
10. True

Practice
1. 16
2. 12
3. 10
4. 22
5. 25
6. 18
7. Commutative property of addition
8. Commutative property of addition
9. Associative property of addition
10. Associative property of addition
11. Additive identity
12. Additive identity
13. 78
14. 58
15. 74
16. 61
17. 361
18. 1027
19. 689
20. 954
21. 16,072
22. 12,133
23. 10,500
24. 10,282
25. 53,904
26. 73,855

True/False. Determine whether each statement is true or false. If a statement is false, explain how it can be changed so the statement will be true. (**Note:** There may be more than one acceptable change.)

7. A polygon is a geometric figure in a plane with two or more sides.
8. To find the perimeter of a rectangle, add the lengths of the four sides.
9. When subtracting, sometimes the digit being subtracted is larger than the digit it is being subtracted from and so "carrying" must occur.
10. If your bank account has a balance of $743 and you want to withdraw $115, you would use subtraction to find that the new balance would be $628.

Practice

Find each sum mentally.

1. $7+(6+3)$
2. $(2+3)+7$
3. $3+2+5$
4. $8+9+5$
5. $4+9+7+5$
6. $6+2+9+1$

For each statement, state the property of addition illustrated and show that the statement is true by performing the addition. See Example 5.

7. $(2+3)+4=(3+2)+4$
8. $(8+1)+3=(1+8)+3$
9. $2+(1+6)=(2+1)+6$
10. $(8+7)+3=8+(7+3)$
11. $7+(6+0)=7+6$
12. $5+3=(5+0)+3$

Add. See Examples 1 through 3.

13. $\begin{array}{r} 24 \\ +54 \\ \hline \end{array}$
14. $\begin{array}{r} 15 \\ +43 \\ \hline \end{array}$
15. $18+56$
16. $27+34$
17. $\begin{array}{r} 268 \\ +93 \\ \hline \end{array}$
18. $\begin{array}{r} 981 \\ +46 \\ \hline \end{array}$
19. $\begin{array}{r} 128 \\ +561 \\ \hline \end{array}$
20. $\begin{array}{r} 832 \\ +122 \\ \hline \end{array}$
21. $\begin{array}{r} 6530 \\ +9542 \\ \hline \end{array}$
22. $\begin{array}{r} 5791 \\ +6342 \\ \hline \end{array}$
23. $9315+1185$
24. $8324+1958$
25. $\begin{array}{r} 21,442 \\ +32,462 \\ \hline \end{array}$
26. $\begin{array}{r} 19,213 \\ +54,642 \\ \hline \end{array}$

27. $\begin{array}{r} 213{,}116 \\ +116{,}018 \\ \hline \end{array}$

28. $\begin{array}{r} 317{,}085 \\ +146{,}413 \\ \hline \end{array}$

29. $\begin{array}{r} 165 \\ 276 \\ +394 \\ \hline \end{array}$

30. $\begin{array}{r} 876 \\ 279 \\ +143 \\ \hline \end{array}$

31. $91 + 340 + 29 + 69$

32. $86 + 32 + 142 + 75$

33. $\begin{array}{r} 124{,}564 \\ 345{,}025 \\ +1{,}671{,}000 \\ \hline \end{array}$

34. $\begin{array}{r} 123{,}456 \\ 456{,}123 \\ +1{,}879{,}282 \\ \hline \end{array}$

Subtract. See Examples 8 through 11.

35. $\begin{array}{r} 42 \\ -31 \\ \hline \end{array}$

36. $\begin{array}{r} 89 \\ -76 \\ \hline \end{array}$

37. $\begin{array}{r} 275 \\ -131 \\ \hline \end{array}$

38. $\begin{array}{r} 468 \\ -217 \\ \hline \end{array}$

39. $\begin{array}{r} 61 \\ -48 \\ \hline \end{array}$

40. $\begin{array}{r} 74 \\ -29 \\ \hline \end{array}$

41. $\begin{array}{r} 52 \\ -27 \\ \hline \end{array}$

42. $\begin{array}{r} 83 \\ -54 \\ \hline \end{array}$

43. $\begin{array}{r} 126 \\ -32 \\ \hline \end{array}$

44. $\begin{array}{r} 139 \\ -67 \\ \hline \end{array}$

45. $\begin{array}{r} 543 \\ -167 \\ \hline \end{array}$

46. $\begin{array}{r} 474 \\ -286 \\ \hline \end{array}$

47. $\begin{array}{r} 900 \\ -307 \\ \hline \end{array}$

48. $\begin{array}{r} 600 \\ -368 \\ \hline \end{array}$

49. $\begin{array}{r} 3275 \\ -1744 \\ \hline \end{array}$

50. $\begin{array}{r} 5387 \\ -2643 \\ \hline \end{array}$

27. 329,134
28. 463,498
29. 835
30. 1298
31. 529
32. 335
33. 2,140,589
34. 2,458,861
35. 11
36. 13
37. 144
38. 251
39. 13
40. 45
41. 25
42. 29
43. 94
44. 72
45. 376
46. 188
47. 593
48. 232
49. 1531
50. 2744

51. 1424
52. 5871
53. 694
54. 4617
55. 2,806,644
56. 3,800,559
57. 1,006,958
58. 7,352,439
59. 45 cm
60. 22 in.
61. 108 ft
62. 42 cm
63. 31 m
64. 54 km

51. $\begin{array}{r} 4900 \\ -3476 \\ \hline \end{array}$

52. $\begin{array}{r} 8007 \\ -2136 \\ \hline \end{array}$

53. $\begin{array}{r} 5070 \\ -4376 \\ \hline \end{array}$

54. $\begin{array}{r} 7602 \\ -2985 \\ \hline \end{array}$

55. $\begin{array}{r} 7,085,076 \\ -4,278,432 \\ \hline \end{array}$

56. $\begin{array}{r} 6,543,222 \\ -2,742,663 \\ \hline \end{array}$

57. $\begin{array}{r} 4,000,000 \\ -2,993,042 \\ \hline \end{array}$

58. $\begin{array}{r} 8,000,000 \\ -647,561 \\ \hline \end{array}$

Calculate the perimeter of each geometric figure. See Example 6.

59.

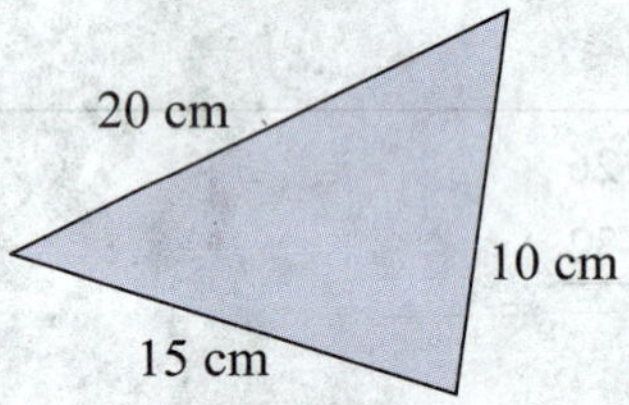

60.

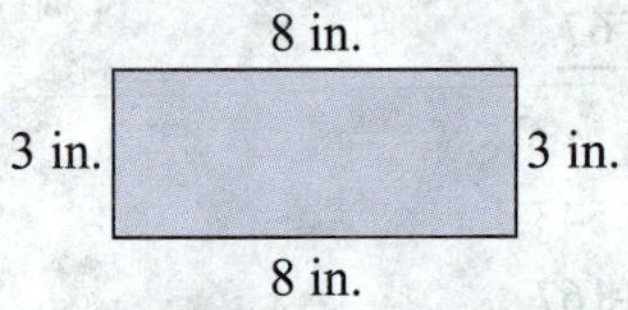

61.

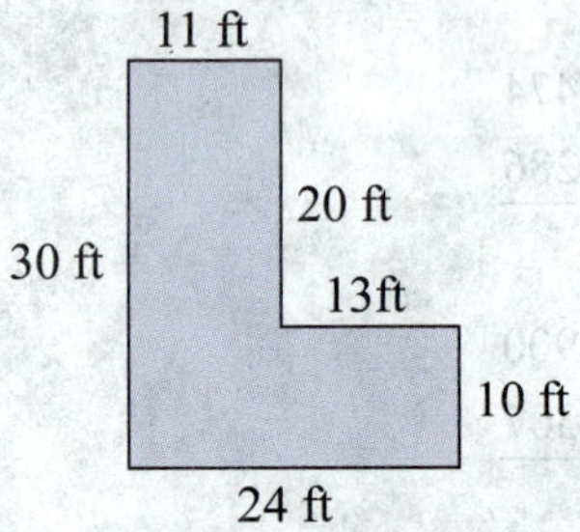

62.

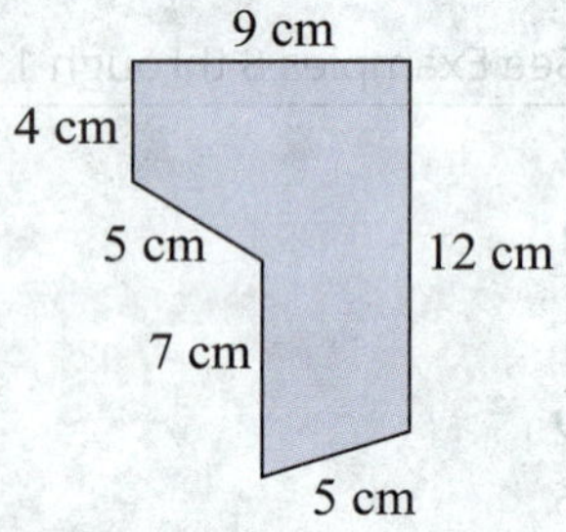

63.

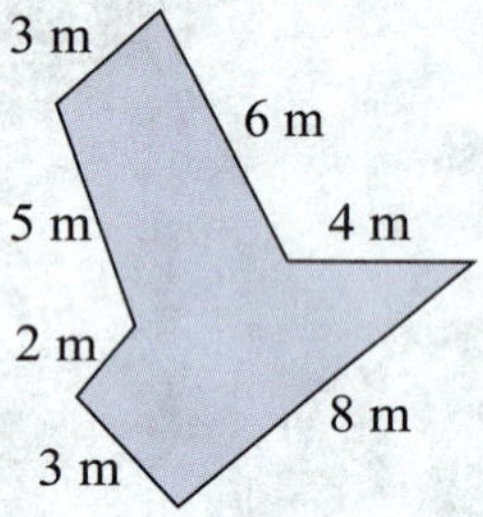

64.

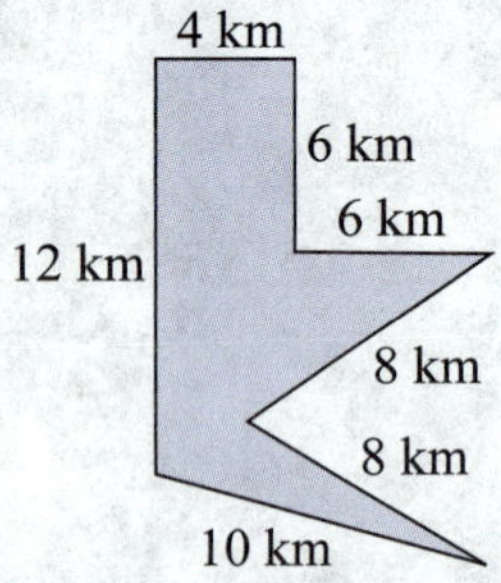

Applications

Solve.

65. Find the perimeter of a parking lot that is in the shape of a rectangle 50 yards wide and 75 yards long.

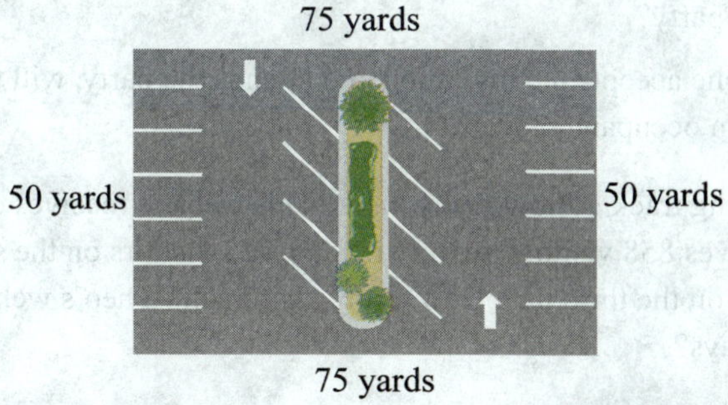

66. A window is in the shape of a triangle placed on top of a square. The length of each of two equal sides of the triangle is 24 inches and the third side is 36 inches long. The length of each side of the square is 36 inches long. Find the perimeter of the window.

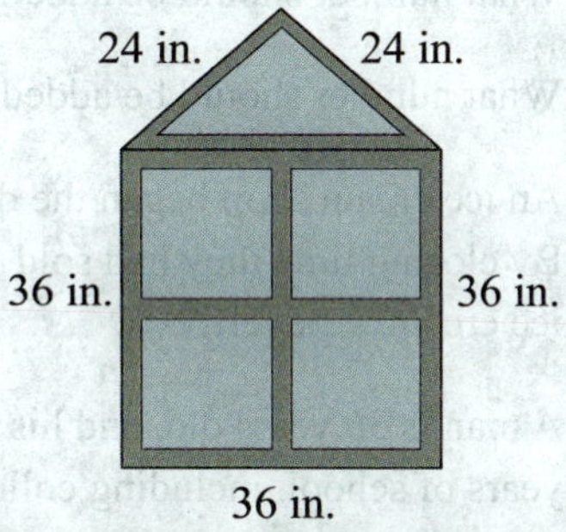

67. The Magley family has the following monthly budget: $815 mortgage; $69 electric; $47 water; and $122 phone bills (including cell phones). What is the family's budget for each month for these expenses?

68. In one year, Stream Line Appliance Co. made 4217 gas stoves; 3947 electric stoves; and 9576 toasters. What was the total number of appliances Stream Line Appliance Co. produced that year?

69. Mr. Juarez kept the mileage records indicated in the table shown here. How many miles did he drive during the six months?

Month	Mileage
Jan.	546
Feb.	378
Mar.	496
Apr.	357
May	503
June	482

70. The Modern Products Corp. showed profits as indicated in the table for the years 2020–2023. What were the company's total profits for the years 2020–2023?

Year	Profits
2020	$1,078,416
2021	$1,270,842
2022	$2,000,593
2023	$1,963,472

Applications

65. 250 yards

66. 156 in.

67. $1053

68. 17,740 appliances

69. 2762 miles

70. $6,313,323

71. **a.** 125 **b.** Yes, $125 \leq 125$
72. $2636
73. 222
74. 140
75. 18 gallons
76. 16 years
77. $480,000
78. $30,800
79. $67
80. $97
81. $39,100
82. $2096

71. Three friends rent a local party room for a holiday party. Due to the local fire code, the maximum occupancy of the party room is 125 people. Manuel invites 43 people, George invites 52 people, and Ali invites 27 people to the party. (Don't forget that the hosts are also attending!)

a. If everyone accepts the invitation and attends the party, how many people will be at the party?

b. If everyone accepts the invitation and attends the party, will they be within the maximum occupancy?

72. Theo is keeping track of how many visitors his website receives each day. The website receives 858 visitors on the first day, 723 visitors on the second day, and 1055 visitors on the third day. How many visitors did Theo's website receive during those three days?

73. What number should be added to 978 to get a sum of 1200?

74. What number should be added to 860 to get a sum of 1000?

75. An ice cream shop began the day with 32 gallons of mint chocolate chip ice cream. By closing time they had sold 14 gallons. How many gallons of mint chocolate chip ice cream were left?

76. A man is 36 years old, and his wife is 34 years old. Together, they have attended 28 years of school, including college. If the man attended 12 years of school, how many years did his wife attend?

77. The Kingston Construction Co. made a bid of $7,350,000 to build a stretch of freeway, but the Beach City Construction Co. made a lower bid of $6,870,000. How much lower was the Beach City bid?

78. Two printing companies bid on a print job for a new textbook with an initial run of 20,000 units. The first bid was for $134,400 and the second bid was $165,200. How much higher was the second bid?

79. Brad is interviewing for architecture jobs all over the country. For one job interview he has to fly to Chicago from Dallas. He found a round trip plane ticket for $250 plus $42 in taxes and fees. If the company agreed to reimburse him for $225 of his travel expenses, how much of the total ticket cost will he have to pay out of pocket?

80. Mason needs to change his train reservation. His new ticket price is $325 plus $21 in taxes. A total of $249 from his previous train ticket will be applied to the new purchase. How much will it cost Mason to switch his train reservation?

81. A couple sold their house for $135,000. They paid the realtor $8100, and other expenses of the sale came to $800. If they owed the bank $87,000 for the mortgage, what were their net proceeds from the sale?

82. Alexis decides to purchase a new bedroom set for $2150. She also decided to have it delivered to her house for an additional $90 fee. Taxes of $156 were then added. Because she purchased all the pieces together, the store is giving her a $300 discount. What was the total cost of the bedroom set?

83. Two landscaping companies made bids on the landscaping of a new apartment complex. Company A bid $550,000 for materials and plants and $225,000 for labor. Company B bid $600,000 for materials and plants and $182,000 for labor. Which company had the lower total bid? How much lower was it?

84. In pricing a four-door car, Pat found she would have to pay a base price of $19,500 plus $1170 in taxes and $250 for license fees. For a two-door of the same make, she would pay a base price of $18,700 plus $1122 in taxes and $230 for license fees. Including all expenses, how much cheaper was the two-door model?

85. The Smith family is creating a simple monthly budget to keep track of their spending habits. Each month they pay $1085 for their mortgage. Their average monthly utility bills are $104 for electric, $49 for water & sewer, and $109 for a cable TV, internet, and mobile phone bundle. They budget $550 per month for food. They also have a car payment of $295 per month, insurance payment of $65 per month, and they estimate $225 for fuel.

a. Fill in the table with the missing information.

Category	Expenses
Mortgage	$1085
Utilities	______
Food	$550
Car costs	______

b. How much money does the Smith family need to budget for these expenses each month?

c. If their combined family income is $3500 per month (after taxes), how much money do they have for other bills and expenses each month?

86. Daria has a small garden in her backyard. Lately she has had trouble with rabbits and deer eating her vegetables, so she wants to put a fence around the garden. The garden is in the shape of a rectangle with a length of 12 feet and a width of 8 feet.

a. How many feet of fencing will Daria need to buy?

b. After determining how much fencing she needs to buy, Daria realizes that she doesn't need to put any fencing on one side of the garden because it is along the side of her garage. The side next to garage has a length of 12 feet. With this new information, how many feet of fencing will Daria need to buy?

c. While at the home improvement store, a salesperson suggests that Daria install a gate in the fence so she doesn't have to climb over the fence to work in her garden. The gate that she picks out is 4 feet wide and will go in place of one section of fencing. Not including the length of the gate, many feet of fencing will Daria need to buy?

Writing & Thinking

87. List three properties of addition and give an example of each.

88. Explain when "carrying" should be used in addition with whole numbers and give an example.

89. Explain when "borrowing" would be used in subtraction and give an example.

90. Give an example when you might use subtraction (other than in a class).

83. Company A; $7000

84. $868

85. **a.** Utilities $262; Car costs $585 **b.** $2482 **c.** $1018

86. **a.** 40 ft **b.** 28 ft **c.** 24 ft

Writing & Thinking

87. Commutative property of addition, associative property of addition, and additive identity property; Examples will vary.

88. If a sum of digits is greater than 9, the tens digit of the sum should be added to the column to the left; Examples will vary.

89. Borrowing is required when a digit is smaller than the digit being subtracted; Examples will vary.

90. Answers will vary. Subtraction may be used when paying bills, buying items, losing weight, etc

Objectives

A. Recognize and use the properties of multiplication.

B. Recognize and use the distributive property.

C. Multiply whole numbers.

D. Multiply whole numbers mentally by using powers of 10.

E. Calculate the area of a rectangle.

1.3 Multiplication with Whole Numbers

If you buy five 6-packs of soda, to determine the total number of cans of soda purchased, you can add five 6s.

$$6 + 6 + 6 + 6 + 6 = 30$$

Figure 1

The process of addition with the same number is also known as **repeated addition**. Or you can **multiply** 5 times 6 and get the same result.

$$5 \cdot 6 = 30$$

The raised dot indicates multiplication.

In either case, you know that you have 30 cans of soda.

The result of multiplication is called a **product**. The two numbers being multiplied are called **factors** of the product. In the following example, the repeated addend (75) and the number of times it is used (4) are both **factors** of the **product** 300.

$$75 + 75 + 75 + 75 = 4 \cdot 75 = 300$$

factor factor product

Several notations can be used to indicate multiplication. In this text, to avoid possible confusion between the letter x (used as a variable) and the $\times$ sign, we will use the raised dot and parentheses most of the time.

Symbols for Multiplication

Symbol		**Example**
$\cdot$	raised dot	$4 \cdot 175$
$(\ )$	numbers inside or next to parentheses	$4(175)$ or $(4)175$ or $(4)(175)$
$\times$	cross sign	4×175 or $\begin{array}{r} 175 \\ \underline{\times\ 4} \end{array}$

A The Properties of Multiplication

As with addition, the operation of multiplication has several properties. Multiplication is **commutative**, **associative**, and has an identity. The number 1 is the **multiplicative identity**. Multiplication by 0 always gives a product of 0, and this fact is called the **multiplication property of 0**.

Commutative Property of Multiplication

The **order** of the numbers in multiplication can be reversed without changing the product. For example,

$$3 \cdot 4 = 12 \quad \text{and} \quad 4 \cdot 3 = 12.$$

PROPERTIES

Associative Property of Multiplication

The **grouping** of the numbers in multiplication can be changed without changing the product. For example,

$$(4 \cdot 2) \cdot 5 = 8 \cdot 5 = 40 \quad \text{and} \quad 4 \cdot (2 \cdot 5) = 4 \cdot 10 = 40.$$

PROPERTIES

Multiplicative Identity Property

The product of any number and 1 is that same number. For example,

$$6 \cdot 1 = 6 \quad \text{and} \quad 1 \cdot 14 = 14.$$

The number 1 is called the **multiplicative identity**.

PROPERTIES

Multiplication Property of 0 (or Zero-Factor Law)

The product of a number and 0 is always 0. For example,

$$63 \cdot 0 = 0 \quad \text{and} \quad 0 \cdot 7 = 0.$$

The product of a number and 0 is always 0.

PROPERTIES

Example 1 Recognizing the Properties of Multiplication

Each of the properties of multiplication is illustrated.

a. $5 \times 6 = 6 \times 5$ Commutative property of multiplication

As a check, we see that $5 \times 6 = 30$ and $6 \times 5 = 30$.

b. $2 \cdot (5 \cdot 9) = (2 \cdot 5) \cdot 9$ Associative property of multiplication

As a check, we see that $2 \cdot (45) = 90$ and $(10) \cdot 9 = 90$.

c. $8 \cdot 1 = 8$ Multiplicative identity property

d. $196 \cdot 0 = 0$ Multiplication property of 0

Now work margin exercise 1.

1. For each statement, state the property of multiplication illustrated.

a. $8(12) = 12(8)$

b. $0 \cdot 57 = 0$

c. $35 \cdot 1 = 35$

d. $7 \cdot (3 \cdot 2) = (7 \cdot 3) \cdot 2$

Teaching Note

This figure illustrates the distributive property and shows two ways to find $6(3+5)$. You might want to create more figures like this in class to help convince your students that the distributive property makes sense.

$6(3+5) = 6 \cdot 8 = 48$

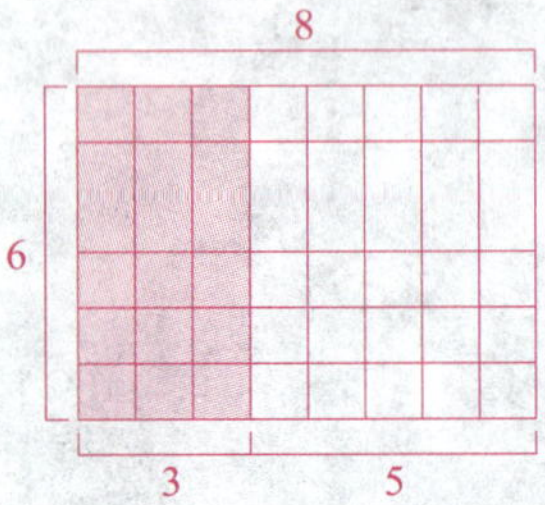

$6(3+5) = 6 \cdot 3 + 6 \cdot 5 = 18 + 30 = 48$

B The Distributive Property

Suppose there are two families attending the county fair together. One family has 5 people and the other family has 4 people. The admission for the county fair is \$7 per person. To determine how much money the families must pay all together to visit the fair, we must simplify the expression \$7(5 + 4). There are two ways to do this.

We can add first, then multiply.

$\$7(5+4) = \$7(9) = \$63$ This method finds the total number of people attending and multiplies that value by the admission price per person.

However, we can also **distribute** the multiplication over the addition as follows.

$$\begin{aligned}\$7(5+4) &= \$7 \cdot 5 + \$7 \cdot 4 \\ &= \$35 + \$28 = \$63\end{aligned}$$

This method finds the cost each family pays for admission and then adds those two costs together.

Each of the numbers inside the parentheses is multiplied by 7 and the answer is the same. The families will pay a total of \$63 to attend the county fair.

The Distributive Property

Multiplication can be distributed over addition. For example,

$$5(8+3) = 5 \cdot 8 + 5 \cdot 3.$$

PROPERTIES

2. Rewrite each expression by using the distributive property and then simplify.

a. $3(10+6)$

b. $5(7+4)$

Example 2 Using the Distributive Property

Use the distributive property to simplify each expression.

a. $6(3+5)$ **b.** $4(7+8)$

Solution

a. $6(3+5) = 6 \cdot 3 + 6 \cdot 5 = 18 + 30 = 48$

b. $4(7+8) = 4 \cdot 7 + 4 \cdot 8 = 28 + 32 = 60$

Now work margin exercise 2.

C Multiplication with Whole Numbers

In Examples 3 and 4, multiplication is explained step-by-step using the distributive property. Then, in Example 5, the steps are combined to show the multiplication using standard form.

Example 3 Multiplying Whole Numbers

Multiply: $4 \cdot 73$

Solution

By the distributive property we can write

$$4 \cdot 73 = 4(3+70) = 4 \cdot 3 + 4 \cdot 70 = 12 + 280 = 292.$$

This process is usually written in a vertical format, as the following steps indicate.

Step 1: Carry the 1 to the tens column.

$$\begin{array}{r} {}^{1} \\ 73 \\ \times\ \ 4 \\ \hline 2 \end{array}$$

Multiply:
$4 \cdot 3 = 12$ ones
$= 1$ ten $+ 2$ ones

Write 2 in the ones column.

Step 2:

$$\begin{array}{r} {}^{1} \\ 73 \\ \times\ \ 4 \\ \hline 292 \end{array}$$

Multiply, then add the carried number:
$4 \cdot 7 + 1 = 29$ tens
$= 2$ hundreds $+ 9$ tens

Write 9 in the tens column and 2 in the hundreds column.

Now work margin exercise 3.

3. Multiply.

$$\begin{array}{r} 18 \\ \times\ \ 5 \\ \hline \end{array}$$

Example 4 Multiplying Whole Numbers

Multiply: $37 \cdot 27$

Solution

The steps of multiplication are shown in finding the product $37 \cdot 27$.

Step 1: First, multiply $37 \cdot 7$.

$$\begin{array}{r} {}^{4} \\ 37 \\ \times 27 \\ \hline 259 \end{array}$$

$37 \cdot 7 = 259$

Step 2: Next, multiply $37 \cdot 20$.

$$\begin{array}{r} {}^{1} \\ 37 \\ \times 27 \\ \hline 259 \\ 740 \end{array}$$

$37 \cdot 20 = 740$

Step 3: 259 and 740 are called **partial products**. We now add them to find the product of $37 \cdot 27$.

$$\begin{array}{r} 37 \\ \times 27 \\ \hline 259 \\ 740 \\ \hline 999 \end{array}$$

Add to find the final product.

Now work margin exercise 4.

4. Multiply.

$$\begin{array}{r} 26 \\ \times\ 34 \\ \hline \end{array}$$

5. Multiply.

$$\begin{array}{r} 15 \\ \times\ 32 \\ \hline \end{array}$$

Example 5 Multiplying Whole Numbers

Multiply: $93 \cdot 46$

Solution

The standard form of multiplication is used here to find the product $93 \cdot 46$.

$$\begin{array}{rl} \overset{1}{9}3 & \\ \times\ 46 & \\ \hline 558 & 93 \cdot 6 = 558 \\ 3720 & 93 \cdot 40 = 3720 \\ \hline 4278 & \text{Product} \end{array}$$

Now work margin exercise 5.

CALCULATORS

Using a Calculator to Multiply Whole Numbers

To multiply numbers on a calculator, you will need the multiplication key [×] and the equal sign key [=]. To multiply the values given in Example 5, press the keys

[9] [3] [×] [4] [6].

Then press [=]. The display will read 4278.

D Powers of 10

Some of the powers of 10 are 1, 10, 100, 1000, 10,000, and 100,000, etc.

$$\begin{gathered} 1 \\ 10 \\ 10 \cdot 10 = 100 \\ 10 \cdot 10 \cdot 10 = 1000 \\ 10 \cdot 10 \cdot 10 \cdot 10 = 10{,}000 \\ 10 \cdot 10 \cdot 10 \cdot 10 \cdot 10 = 100{,}000 \end{gathered}$$

and so on.

Multiplication by powers of 10 is useful in explaining multiplication with whole numbers in general. Such multiplication should be done mentally and quickly. The following examples illustrate an important pattern.

$$\begin{gathered} 6 \cdot 1 = 6 \\ 6 \cdot 10 = 60 \\ 6 \cdot 100 = 600 \\ 6 \cdot 1000 = 6000 \end{gathered}$$

If one of two whole number factors is 10, the product will be the other factor with one zero (0) written to the right of it. Two zeros (00) are written to the right of the other factor when multiplying by 100, and three zeros (000) are written when multiplying by 1000. Will multiplication by one million (1,000,000) result in writing six zeros to the right of the other factor? The answer is yes.

We can summarize these ideas in the following way.

Multiplying Whole Numbers by Powers of 10

To multiply a whole number:

by 10, write 0 to the right;
by 100, write 00 to the right;
by 1000, write 000 to the right;
by 10,000, write 0000 to the right;
and so on.

PROCEDURE

Many products can be found mentally by using the properties of multiplication and the technique of multiplying by powers of 10. The processes are written out in the following examples, but they can easily be done mentally with practice.

Example 6 Multiplying Whole Numbers that End with 0s

Multiply.

a. $6 \cdot 700$ **c.** $200 \cdot 800$

b. $50 \cdot 700$ **d.** $7000 \cdot 9000$

Solution

a.
$$\begin{aligned} 6 \cdot 700 &= 6(7 \cdot 100) \\ &= (6 \cdot 7)100 \\ &= 42 \cdot 100 \\ &= 4200 \end{aligned}$$

c.
$$\begin{aligned} 200 \cdot 800 &= (2 \cdot 100)(8 \cdot 100) \\ &= (2 \cdot 8)(100 \cdot 100) \\ &= 16 \cdot 10{,}000 \\ &= 160{,}000 \end{aligned}$$

b.
$$\begin{aligned} 50 \cdot 700 &= (5 \cdot 10)(7 \cdot 100) \\ &= (5 \cdot 7)(10 \cdot 100) \\ &= 35 \cdot 1000 \\ &= 35{,}000 \end{aligned}$$

d.
$$\begin{aligned} 7000 \cdot 9000 &= (7 \cdot 1000)(9 \cdot 1000) \\ &= (7 \cdot 9)(1000 \cdot 1000) \\ &= 63 \cdot 1{,}000{,}000 \\ &= 63{,}000{,}000 \end{aligned}$$

Now work margin exercise 6.

E The Area of a Rectangle

Area is the measure of the interior, or enclosed region, of a plane surface and is measured in **square units**.

Area of a Rectangle

The **area** of a rectangle (measured in square units) is found by multiplying its length by its width.

DEFINITION

Math Tip

Powers of 10 can be used when performing mental math. Consider $9 \cdot 87$. Notice that $9 \cdot 87 = 9(80 + 7)$. After applying the distributive property, the multiplication breaks down into two steps—one involving a power of ten.

$$9(80)+9(7)$$
$$720+63$$
$$783$$

6. Multiply.

a. $5 \cdot 900$

b. $30 \cdot 600$

c. $400 \cdot 500$

d. $6000 \cdot 8000$

Math Tip

Since multiplication is commutative, a 3-inch by 2-inch rectangle has the same area as a 2-inch by 3-inch rectangle.

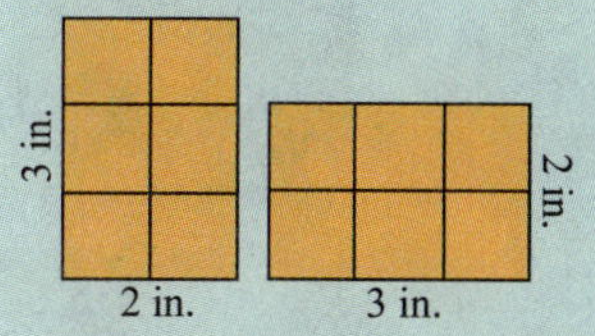

$$(3 \text{ in.})(2 \text{ in.}) = 6 \text{ sq in.}$$
$$(2 \text{ in.})(3 \text{ in.}) = 6 \text{ sq in.}$$

The concept of area is illustrated in Figure 2 in terms of square inches.

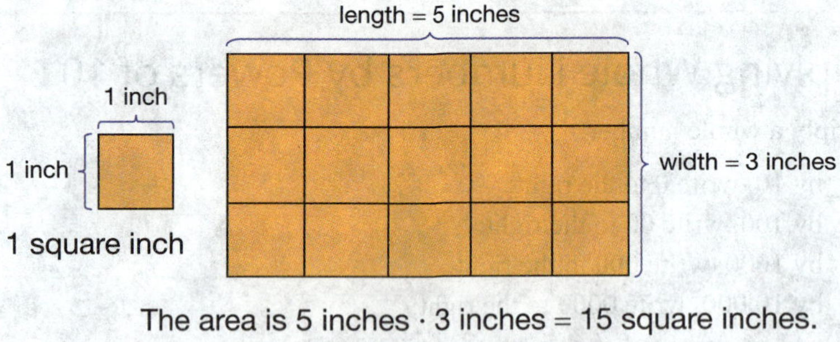

Figure 2

Some of the units of area in the metric system are square meters, square decimeters, square centimeters, and square millimeters. In the US customary system, some of the units of area are square yards (sq yd), square feet (sq ft), and square inches (sq in.).

7. Calculate the area of an elementary school playground that is in the shape of a rectangle that is 52 yd wide and 73 yd long.

Example 7 Application: Calculating the Area of a Rectangle

Calculate the area of a rectangular plot of land with the dimensions shown here.

Solution

To find the area, we multiply 186 ft · 92 ft.

$$\begin{array}{r} 186 \\ \times \quad 92 \\ \hline 372 \\ 16{,}740 \\ \hline 17{,}112 \end{array} \text{ sq ft}$$

The area of the plot of land is 17,112 sq ft.

Now work margin exercise 7.

Margin Exercise Answers

1. a. Commutative property of multiplication **b.** Multiplication property of 0 **c.** Multiplicative identity property **d.** Associative property of multiplication **2. a.** $3 \cdot 10 + 3 \cdot 6 = 30 + 18 = 48$ **b.** $5 \cdot 7 + 5 \cdot 4 = 35 + 20 = 55$ **3.** 90 **4.** 884 **5.** 480 **6. a.** 4500 **b.** 18,000 **c.** 200,000 **d.** 48,000,000 **7.** 3796 sq yd

1.3 Exercises

Concept Check

Fill-in-the-Blank. Complete each sentence using information found in this section.

1. When numbers are multiplied, the answer is called the ________.
2. Another way of looking at multiplication is as if it were repeated _________.
3. When a number is multiplied by 0, the answer is always ________.
4. $7 \cdot 12 = 12 \cdot 7$ is an example of the _______ property of multiplication.
5. According to the associative property of multiplication, the _________ of the numbers can be changed.
6. To find the area of a rectangle, multiply the _______ by the ________.

True/False. Determine whether each statement is true or false. If a statement is false, explain how it can be changed so the statement will be true. (**Note:** There may be more than one acceptable change.)

7. The numbers being multiplied are called the divisors.
8. According to the multiplicative identity, $1 \cdot 25 = 52$.
9. According to the distributive property, $4 \cdot (7+2) = 4 \cdot 7 + 4 \cdot 2$.
10. The associative property of multiplication indicates that length can be multiplied by width or width can be multiplied by length to get the same answer.

Practice

Multiply. See Examples 3 through 5.

1. $\begin{array}{r} 84 \\ \times\ 2 \\ \hline \end{array}$
2. $\begin{array}{r} 21 \\ \times\ 6 \\ \hline \end{array}$
3. $\begin{array}{r} 27 \\ \times\ 6 \\ \hline \end{array}$
4. $\begin{array}{r} 48 \\ \times\ 9 \\ \hline \end{array}$
5. $2(427)$
6. $3(108)$
7. $4 \cdot 702$
8. $3 \cdot 503$
9. $\begin{array}{r} 44 \\ \times 40 \\ \hline \end{array}$
10. $\begin{array}{r} 93 \\ \times 30 \\ \hline \end{array}$
11. $\begin{array}{r} 42 \\ \times 56 \\ \hline \end{array}$
12. $\begin{array}{r} 76 \\ \times 26 \\ \hline \end{array}$
13. $51 \cdot 83$
14. $35 \cdot 84$
15. $(17)(62)$
16. $(74)(31)$
17. $\begin{array}{r} 106 \\ \times\ 72 \\ \hline \end{array}$
18. $\begin{array}{r} 207 \\ \times 83 \\ \hline \end{array}$
19. $\begin{array}{r} 114 \\ \times\ 25 \\ \hline \end{array}$
20. $\begin{array}{r} 116 \\ \times\ 21 \\ \hline \end{array}$

1.3 Exercises

Concept Check

1. product
2. addition
3. 0 (or zero)
4. Commutative
5. grouping
6. length, width
7. False; The numbers being multiplied are called factors.
8. False; $1 \cdot 25 = 25$.
9. True
10. False; This is the commutative property of multiplication.

Practice

1. 168
2. 126
3. 162
4. 432
5. 854
6. 324
7. 2808
8. 1509
9. 1760
10. 2790
11. 2352
12. 1976
13. 4233
14. 2940
15. 1054
16. 2294
17. 7632
18. 17,181
19. 2850
20. 2436

21. $\begin{array}{r} 430 \\ \times\ 50 \\ \hline \end{array}$

22. $\begin{array}{r} 180 \\ \times\ 30 \\ \hline \end{array}$

23. $\begin{array}{r} 420 \\ \times\ 104 \\ \hline \end{array}$

24. $\begin{array}{r} 210 \\ \times\ 301 \\ \hline \end{array}$

25. $\begin{array}{r} 4321 \\ \times\ 765 \\ \hline \end{array}$

26. $\begin{array}{r} 3517 \\ \times\ 284 \\ \hline \end{array}$

27. $\begin{array}{r} 3463 \\ \times\ 1743 \\ \hline \end{array}$

28. $\begin{array}{r} 4251 \\ \times\ 2386 \\ \hline \end{array}$

29. $13 \cdot 7 \cdot 0$

30. $(96)(0)(4)$

31. $830 \cdot 10 \cdot 1$

32. $(260)(70)(1)$

For each statement, state the property of multiplication illustrated and show that the statement is true by performing the multiplication. See Example 1.

33. $8 \cdot 0 = 0$

34. $0 \cdot 17 = 0$

35. $2 \cdot (3 \cdot 6) = (2 \cdot 3) \cdot 6$

36. $3 \cdot (1 \cdot 7) = (3 \cdot 1) \cdot 7$

37. $2 \cdot (6 \cdot 8) = 2 \cdot (8 \cdot 6)$

38. $(7 \cdot 3) \cdot 4 = (3 \cdot 7) \cdot 4$

39. $5 \cdot 1 = 5$

40. $1 \cdot 19 = 19$

Multiply using the technique of multiplying by powers of ten. See Example 6.

41. $30 \cdot 30$

42. $50 \cdot 50$

43. $25 \cdot 100$

44. $47 \cdot 1000$

45. $20 \cdot 200$

46. $50 \cdot 700$

47. $20 \cdot 8000$

48. $40 \cdot 2000$

49. $\begin{array}{r} 3000 \\ \times\ 500 \\ \hline \end{array}$

50. $\begin{array}{r} 7000 \\ \times\ 800 \\ \hline \end{array}$

51. $\begin{array}{r} 80{,}000 \\ \times\ 4\,000 \\ \hline \end{array}$

52. $\begin{array}{r} 90{,}000 \\ \times\ 7\,000 \\ \hline \end{array}$

Rewrite each expression by using the distributive property and then simplify. See Example 2.

53. $3(9+7)$

54. $9(2+9)$

55. $6(3+11)$

56. $7(8+4)$

Calculate the area of each rectangle. See Example 7.

57.

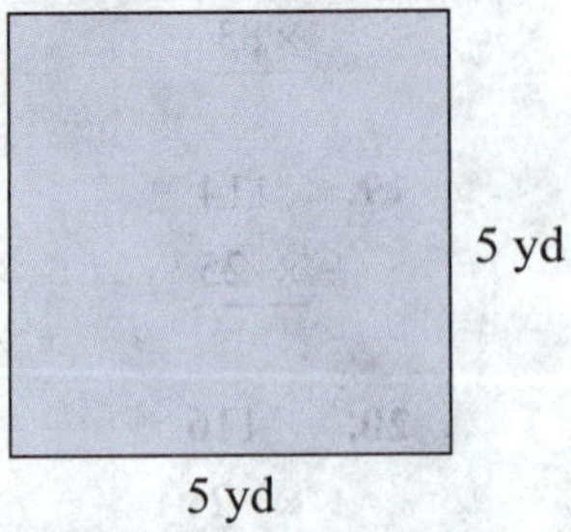

58.

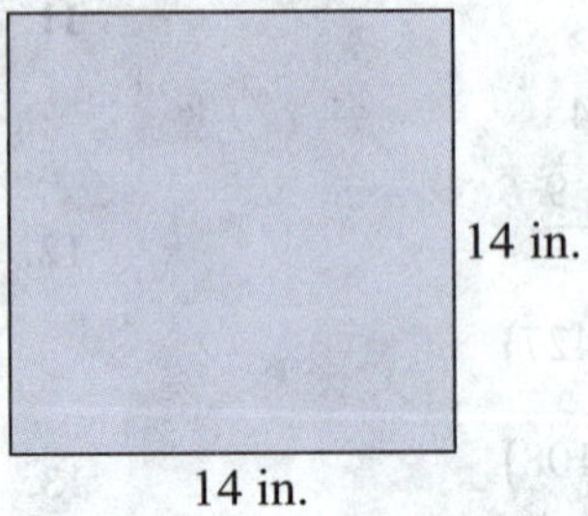

21. 21,500
22. 5400
23. 43,680
24. 63,210
25. 3,305,565
26. 998,828
27. 6,036,009
28. 10,142,886
29. 0
30. 0
31. 8300
32. 18,200
33. Zero-factor law
34. Zero-factor law
35. Associative property of multiplication
36. Associative property of multiplication
37. Commutative property of multiplication
38. Commutative property of multiplication
39. Multiplicative identity
40. Multiplicative identity
41. 900
42. 2500
43. 2500
44. 47,000
45. 4000
46. 35,000
47. 160,000
48. 80,000
49. 1,500,000
50. 5,600,000
51. 320,000,000
52. 630,000,000
53. $3 \cdot 9 + 3 \cdot 7 = 27 + 21 = 48$
54. $9 \cdot 2 + 9 \cdot 9 = 18 + 81 = 99$
55. $6 \cdot 3 + 6 \cdot 11 = 18 + 66 = 84$
56. $7 \cdot 8 + 7 \cdot 4 = 56 + 28 = 84$
57. 25 square yards

59.

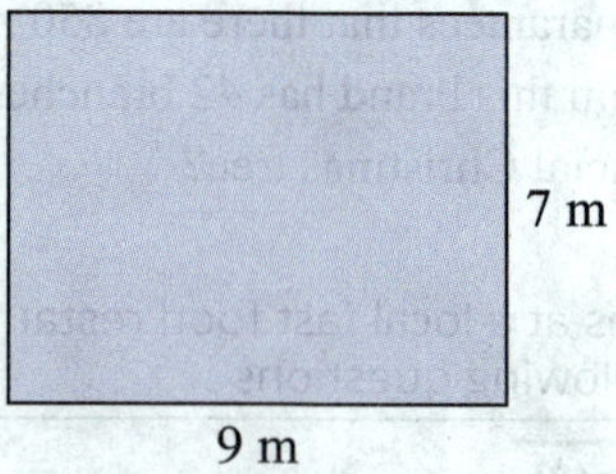

60.

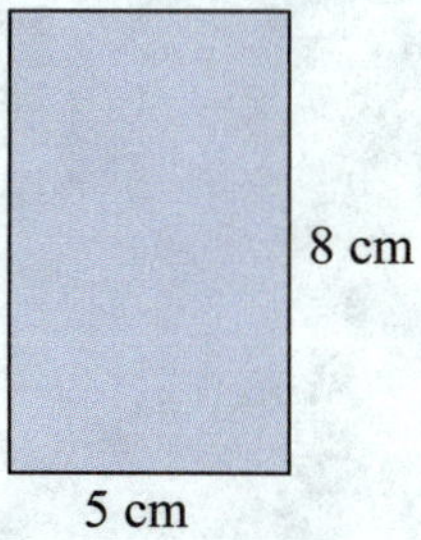

61.

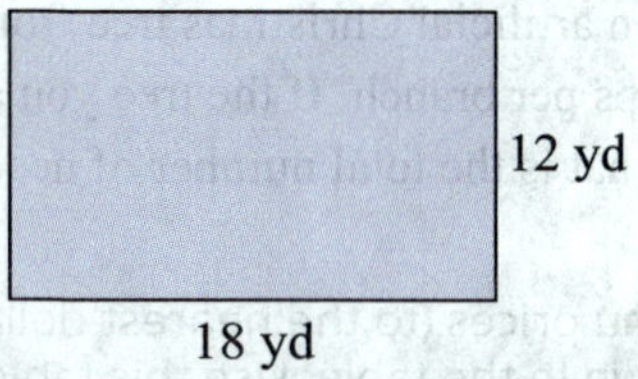

62.

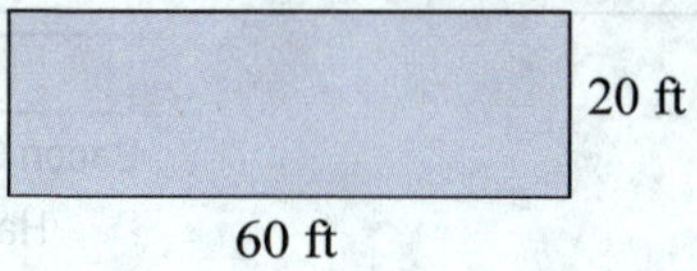

Applications

Solve.

63. The Math Club members decided to attend the national meeting of the NCTM (National Council of Teachers of Mathematics) and had a book sale to raise money for the event. Registration fees were $85 per member and the club had 35 members. How much money did the club need to raise for registration fees?

64. Students at the local community college must pay $83 for a math textbook. If there are 43 students in the class, find the total amount the class will spend on textbooks.

65. A network television station has approximately 18 minutes of commercial time in each hour. How many minutes of commercial time does the network have in a one-day programming schedule of 20 hours? In one week?

66. A group of 15 friends are gathering at a restaurant. The restaurant is having a special where each person can order a three-course meal for $35. If all 15 friends order this special, how much will the total bill be?

67. If one regular pack of candy contains 250 calories, how many calories are there in 37 packs of the same candy?

68. A sandwich shop buys 372 loaves of bread for the week. If each loaf of bread has 24 slices, how many slices of bread were purchased?

69. Your company bought 18 new cars, each with air conditioning and anti-lock brakes, at a price of $15,800 per car. How much did your company pay for these cars?

70. According to the US Fish and Wildlife Service, migratory birds are imported at a value of about $19 each. Suppose that about 800,000 live birds are imported each year. What is the total value of these imported birds?

71. It's lambing season at Fluffy Clouds Farm. There are 117 ewes expected to give birth in the next week. If each sheep is anticipated to have on average 2 lambs, what is the total expected number of lambs?

58. 196 square inches

59. 63 square meters

60. 40 square centimeters

61. 216 square yards

62. 1200 square feet

Applications

63. $2975

64. $3569

65. 360 minutes; 2520 minutes

66. $525

67. 9250 calories

68. 8928 slices of bread

69. $284,400

70. $15,200,000

71. 234

72. 35,700
73. $23
74. Order **a.** is more expensive by $3.
75. 648 square feet
76. 384 square inches
77. 10,080 square feet

72. An artificial Christmas tree from a certain brand guarantees that there are 850 needle tips per branch. If the tree you are interested in from this brand has 42 branches, what is the total number of needle tips on the artificial Christmas tree?

The menu prices (to the nearest dollar) for certain items at a local fast food restaurant are shown in the table. Use this table to answer the following questions

Item	Price($)
Bacon Cheeseburger	4
Hamburger	3
French Fries	2
Onion Rings	3
Soda	1
Milkshake	2
Cookie	1

73. Sally and her two friends ordered 2 bacon cheeseburgers, 1 hamburger, 3 orders of French fries, and 3 milkshakes. What was the total cost of their order?

74. George is hungry and can't decide what to order. He is trying to decide between

a. 2 bacon cheeseburgers, 1 order of onion rings, 1 soda, and 2 cookies; or

b. 1 bacon cheeseburger, 2 orders of French fries, 1 milkshake, and 1 cookie.

Which of the two orders is more expensive? By how much?

75. A rectangular pool measures 36 feet long by 18 feet wide. Find the area of the pool in square feet.

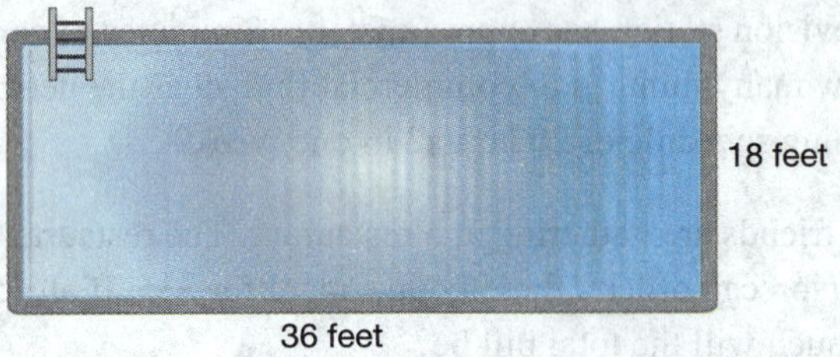

76. A painting is mounted in a rectangular frame (16 inches by 24 inches) and hung on a wall. How many square inches of wall space will the framed painting cover?

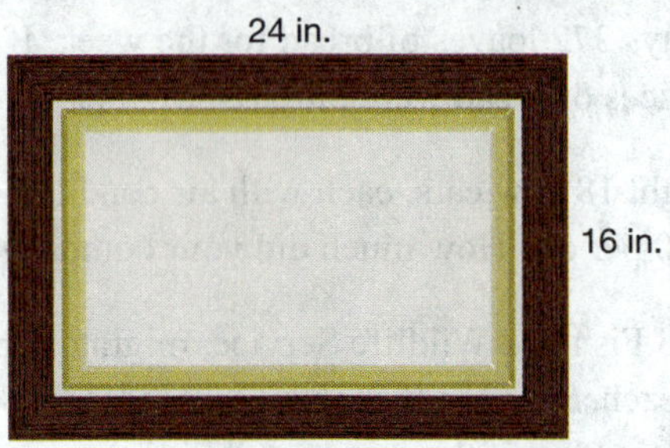

77. Cheyenne has been commissioned by her city to paint a mural on the side of a brick building and must calculate the area of the wall so she can plan and scale her mural. The side of the building measures 120 feet tall and 84 feet wide. Find the area of the mural.

78. A rectangular lot for a house measures 210 feet long by 175 feet wide. Find the area of the lot in square feet.

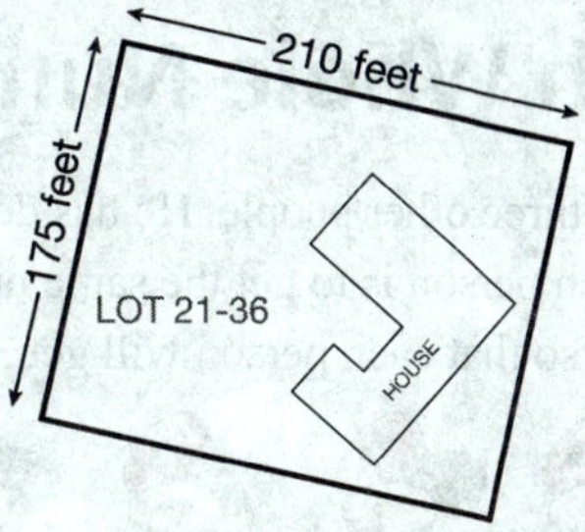

Writing & Thinking

79. Explain, in your own words, what the zero-factor law indicates.

80. Explain, in your own words, why 1 is called the multiplicative identity.

81. Name the property that uses both multiplication and addition and give an example of it.

82. Give an example of when you might use multiplication (other than in a class).

78. 36,750 square feet

Writing & Thinking

79. Zero times any number is 0.

80. Because any number multiplied by 1 results in the original number.

81. The distributive property distributes multiplication to two (or more) numbers that are being added; Examples will vary.

82. Answers will vary. Multiplication may be used to find total cost for a group, the production in a factory over a month, how much food to buy for a party, etc.

Objectives

A. Divide whole numbers.

B. Know how to divide with 0.

C. Divide by using long division.

1.4 Division with Whole Numbers

A Division with Whole Numbers

George is having a meeting with three other people. He has decided to bring a dozen donuts as part of the refreshments. If each person is to get the same number of donuts, then the 12 donuts are **divided** into 4 groups so that each person will get 3 donuts.

Figure 1

The process of separating the donuts into equally-sized groups is called **division**. We can use the division sign $(\div)$ to indicate the division procedure as follows.

$$12 \div 4 = 3$$

Read "12 divided by 4 equals 3."

Dividend ÷ Divisor = Quotient

Two other notations that indicate division are the following.

$$4\overline{)12}^{\,3}$$

Divisor → 4; Dividend → 12; Quotient → 3

$$\frac{12}{4} = 3$$

Dividend → 12; Divisor → 4; Quotient → 3

Now consider the multiplication problem $8 \cdot 10$. We know that $8 \cdot 10 = 80$ and that 8 and 10 are **factors** of 80. They are also called **divisors** of 80. In division, we want to find how many times one number is contained in another. How many 8s are in 80? There are ten 8s in 80, and we say that 80 **divided** by 8 is 10 (or $80 \div 8 = 10$). In this case, we know the results of the division because we are already familiar with the related multiplication, $8 \cdot 10$. Thus, division can be thought of as the reverse of multiplication.

Division with Whole Numbers

Division						Multiplication					
Dividend		**Divisor**		**Quotient**		**Factors**				**Product**
24	÷	6	=	4	since	6	·	4	=	24
35	÷	7	=	5	since	7	·	5	=	35

PROCEDURE

1. Divide.

a. $28 \div 7$

b. $\frac{42}{6}$

c. $\frac{57}{57}$

d. $86 \div 1$

Example 1 Understanding the Basics of Division

These examples illustrate the relationship between division and multiplication.

a. $24 \div 6 = 4$ because $4 \cdot 6 = 24$

b. $26 \div 2 = 13$ because $2 \cdot 13 = 26$

c. $\frac{72}{8} = 9$ because $8 \cdot 9 = 72$

d. $\frac{35}{1} = 35$ because $1 \cdot 35 = 35$

e. $\frac{17}{17} = 1$ because $17 \cdot 1 = 17$

Now work margin exercise 1.

Notice that Examples 1d and 1e are examples of the following two properties of division.

Division by 1

Any number divided by one is itself.

Example: $\frac{4}{1} = 4$

PROPERTIES

Division of a Number by Itself

Any nonzero number divided by itself equals 1.

Example: $\frac{3}{3} = 1$

PROPERTIES

B Division Involving 0

There are two cases to consider with division involving 0. One is when 0 is divided by a number. The other is dividing by 0. Both can be explained in terms of multiplication. For example,

$\frac{0}{7} = 0$ because $0 = 0 \cdot 7$. By the multiplication property of 0

However, if we divide by 0, we have

$$\frac{7}{0} = \underline{?} \text{ which would indicate } 7 = 0 \cdot \underline{?}.$$

But this is not possible because $0 \cdot \underline{?} = 0$ for any value of $\underline{?}$. What this means is that we cannot divide by 0. We say that

$$\frac{7}{0} \text{ is } \textbf{undefined}.$$

Similarly, consider the problem

$$0 \div 0 = \underline{?}.$$

This would mean $0 = 0 \cdot \underline{?}$, which is a true statement for any value of $\underline{?}$. Since there is not a unique value for $\underline{?}$, we agree that

$$0 \div 0 \text{ is } \textbf{undefined} \text{ also.}$$

That is, division by 0 is not possible.

Don't Divide by Zero!

Fans of science fiction time travel are familiar with the term "singularity." In the movies, it is generally understood to be something the time travelers try to avoid because it's really, really dangerous. As it turns out, "singularity" is an actual physics term that describes a situation where a physics formula divides by zero. Remember the wisdom of the time travelers and don't divide by zero. It's really, really dangerous!

Teaching Note

Technically the term $\frac{0}{0}$, as students who progress to calculus will see, is indeterminate. At this stage, it is listed as undefined to remove the temptation for a student to divide by 0.

These ideas lead to the following two rules about division involving 0.

Division Involving 0

Case 1: Any nonzero whole number can be divided into 0 and the quotient is 0.

Example: $\frac{0}{7} = 0$

Case 2: If any whole number, including 0, is divided by 0 the quotient is **undefined**.

Example: $\frac{13}{0}$ is undefined.

PROPERTIES

2. Divide.

   a. $29 \div 0$

   b. $\frac{0}{8}$

Example 2 Understanding Division with 0

These examples illustrate several situations involving division with 0.

a. $0 \div 23 = 0$ or $\frac{0}{23} = 0$ or $23\overline{)0}$ (quotient 0)

(0 can be the quotient.)

b. $14 \div 0$ is undefined or $\frac{14}{0}$ is undefined or $0\overline{)14}$ (quotient undefined)

(0 cannot be the divisor.)

Now work margin exercise 2.

C Long Division

Division does not always involve factors (or exact divisors), and we need a more general idea and method for performing division. The more general method is a process known as the **division algorithm**[1] (or the method of **long division**). This process can be written in the following format.

$$
\begin{array}{r}
4 \\
6\overline{)27} \\
-24 \\
\hline
3
\end{array}
$$

Divisor: 6
Quotient: 4
Dividend: 27
$6 \cdot 4 = 24$
$27 - 24 = 3$
Remainder: 3

The following examples illustrate long division in a step-by-step way. Study these examples carefully. Remember that the **remainder** must be less than the divisor.

Attention!

Division problems can be checked using multiplication. To do so, multiply the divisor by the quotient and add the remainder. This will give you the dividend. This is shown in the following examples.

1 An algorithm is a process or pattern of steps to be followed in solving a problem or, sometimes, only a certain part of a problem.

Example 3 Dividing Whole Numbers

Divide: $683 \div 7$

Solution

Step 1:

$$\begin{array}{r} 9 \\ 7\overline{)683} \\ \underline{-63} \\ 5 \end{array}$$

Trial divide 7 into 68. This gives 9 in the tens position. Now multiply $7 \cdot 9$.
Subtract 63.

Step 2:

$$\begin{array}{r} 97 \\ 7\overline{)683} \\ \underline{-63} \\ 53 \\ \underline{-49} \\ 4 \end{array}$$

97: Quotient

53: Bring down the 3 from the dividend, then trial divide 7 into 53.

−49: Now multiply $7 \cdot 7$ and subtract 49.

4: Remainder

Check

Divisor	·	Quotient	+	Remainder	=	Dividend
7	·	97	+	4		
679			+	4	=	683

So, $683 \div 7 = 97$ R4.

Now work margin exercise 3.

3. Divide.

$415 \div 6$

Example 4 Dividing Whole Numbers

Divide: $696 \div 30$

Solution

Step 1:

$$\begin{array}{r} 2 \\ 30\overline{)696} \\ \underline{-60} \\ 9 \end{array}$$

Trial divide 30 into 69, or 3 into 6.
This gives 2 in the tens position. Now multiply $30 \cdot 2$ and subtract 60.

Step 2:

$$\begin{array}{r} 23 \\ 30\overline{)696} \\ \underline{-60} \\ 96 \\ \underline{-90} \\ 6 \end{array}$$

23: Quotient

96: Bring down the 6 from the dividend, then trial divide 30 into 96, or 3 into 9.

−90: This gives 3 in the ones position. Now multiply $30 \cdot 3$ and subtract 90.

6: Remainder

Check

Divisor	·	Quotient	+	Remainder	=	Dividend
30	·	23	+	6		
690			+	6	=	696

So, $696 \div 30 = 23$ R6.

Now work margin exercise 4.

4. Divide.

$340 \div 16$

5. Divide.

$31\overline{)9571}$

Example 5 Dividing Whole Numbers

Divide: 9325 ÷ 45

Solution

Step 1:

$$\begin{array}{r} 2 \\ 45\overline{)9325} \\ \underline{-90} \\ 3 \end{array}$$

Trial divide 45 into 93, or 4 into 9.

This gives 2 in the hundreds position. Now multiply 45 · 2 and subtract 90.

Step 2:

$$\begin{array}{r} 20 \\ 45\overline{)9325} \\ \underline{-90} \\ 32 \\ \underline{-0} \\ 32 \end{array}$$

Bring down the 2 from the dividend. 45 does not divide into 32, so write 0 in the tens column. Now multiply 45 · 0 and subtract 0.

Step 3:

$$\begin{array}{r} 208 \\ 45\overline{)9325} \\ \underline{-90} \\ 32 \\ \underline{-0} \\ 325 \\ \underline{360} \end{array}$$

Bring down the 5 from the dividend, then trial divide 45 into 325, or 4 into 32.
The trial quotient of 8 is too large since 45 · 8 = 360 is larger than 325.

Step 4:

$$\begin{array}{r} 207 \\ 45\overline{)9325} \\ \underline{-90} \\ 32 \\ \underline{-0} \\ 325 \\ \underline{-315} \\ 10 \end{array}$$

Quotient

Now the trial quotient is 7. Since 45 · 7 = 315 and 315 is smaller than 325, 7 is the desired number.

Remainder

Check

Divisor	·	Quotient	+	Remainder	=	Dividend
207	·	45	+	10		
9315			+	10	=	9325

So 9325 ÷ 45 = 207 R10.

Now work margin exercise 5.

In Step 2 of Example 5, we wrote 0 in the quotient because 45 did not divide into 32. Many students fail to write the 0. This error obviously changes the value of the quotient.

Common Error

Consider 6936 ÷ 17 as follows.

Wrong Solution

```
     48
17 )6936
   -68
     136
    -136
       0
```

Correct Solution

```
     408
17 )6936
   -68
     13
     -0
     136
    -136
       0
```

Writing the 0 in the quotient gives the correct quotient of 408, which is considerably different from 48.

CAUTION

Completion Example 6 Dividing Whole Numbers

Finish the long division process to find the quotient and remainder.

```
    83
4 )334
  -32
    14
   -12
     2   Remainder
```

So, 334 ÷ 4 = 83 R2.

6. Divide.

$7\overline{)186}$

Now work margin exercise 6.

If the remainder is 0 in a division problem, then both the divisor and the quotient are **factors** (or **divisors**) of the dividend. We say that both factors **divide exactly** into the dividend. As a check, note that the product of the divisor and quotient will be the dividend.

Example 7 Using Division to Find Factors

Show that 19 and 43 are **factors** (or **divisors**) of 817.

Solution

In order to show that 19 and 43 are factors of 817, we divide. Both 817 ÷ 19 and 817 ÷ 43 will give a remainder of 0 if indeed these numbers are factors of 817. We only need to try one of them to find out if the remainder is 0.

7. Show that 26 and 68 are factors (or divisors) of 1768.

$$\begin{array}{r} 43 \\ 19\overline{)817} \\ \underline{-76} \\ 57 \\ \underline{-57} \\ 0 \end{array}$$

Check

$$\begin{array}{r} 43 \\ \times 19 \\ \hline 387 \\ +430 \\ \hline 817 \end{array}$$

Thus, 19 and 43 are indeed factors of 817.

Now work margin exercise 7.

8. It costs $1428 for a group of 6 friends to rent a beach house for a week. How much will each person pay?

Example 8 Application: Dividing Whole Numbers

Tasty Treat Bakery just received an order for 1296 cupcakes. When the order is shipped, 6 cupcakes will be put into 1 box. How many boxes will the bakery need to fill the order?

Solution

The number of boxes can be found by dividing the total number of cupcakes by the number of cupcakes in each box.

$$\begin{array}{r} 216 \\ 6\overline{)1296} \\ \underline{-12} \\ 09 \\ \underline{-6} \\ 36 \\ \underline{-36} \\ 0 \end{array}$$

Thus, 216 boxes will be needed.

Now work margin exercise 8.

CALCULATORS

Using a Calculator to Divide Whole Numbers

To divide numbers on a calculator, you will need the division key [÷] and the equal sign key [=].

To divide the values given in Example 8, press the keys

[1] [2] [9] [6] [÷] [6].

Then press [=]. The display will read 216.

Notice, however, that when we divide the values [1] [2] [9] [6] [÷] [1] [3], the display reads 99.692307692.... This value is not a whole number. This means the quotient has a remainder and, for now, is better calculated by hand. (The quotient in this calculation is a decimal number. Decimal numbers will be covered in Chapter 3.)

Completion Example Answers

6. 83, 12, 2; 334 ÷ 4 = 83 R2

Margin Exercise Answers

1. a. 4 **b.** 7 **c.** 1 **d.** 86 **2. a.** Undefined **b.** 0 **3.** 69 R1 **4.** 21 R4

5. 308 R23 **6.** 26 R4 **7.** $\begin{array}{r} 68 \\ 26\overline{)1768} \\ \underline{-156} \\ 208 \\ \underline{-208} \\ 0 \end{array}$ **8.** \$238

1.4 Exercises

Concept Check

Fill-in-the-Blank. Complete each sentence using information found in this section.

1. When dividing, the remainder must be less than the _______.

2. To check a division problem, multiply the quotient by the divisor and then add the ________. The result should be the _________.

3. When any nonzero number is divided by itself, the result is always _____.

4. Factors can be called ______ of the dividend.

5. When a number is divided by zero, the result is always _______.

True/False. Determine whether each statement is true or false. If a statement is false, explain how it can be changed so the statement will be true. (**Note:** There may be more than one acceptable change.)

6. If a division problem has a nonzero remainder, then the divisor and quotient are factors of the dividend.

7. $13 \div 1 = 13$

8. $12 \div 0 = 12$

9. $\frac{0}{7}$ is undefined.

Practice

Divide. See Examples 2 through 6.

1. $5\overline{)35}$

2. $8\overline{)48}$

3. $6\overline{)72}$

4. $3\overline{)51}$

5. $13\overline{)0}$

6. $42\overline{)0}$

7. $6\overline{)32}$

8. $4\overline{)25}$

9. $0\overline{)23}$

10. $0\overline{)51}$

11. $240 \div 6$

12. $120 \div 4$

13. $205 \div 5$

14. $217 \div 7$

15. $7\overline{)310}$

1.4 Exercises

Concept Check

1. Divisor

2. Remainder, dividend

3. 1

4. Divisors

5. Undefined

6. False; If a division problem has a zero remainder.

7. True

8. False; $12 \div 0$ is undefined.

9. False; $\frac{0}{7} = 0$.

Practice

1. 7

2. 6

3. 12

4. 17

5. 0

6. 0

7. 5 R2

8. 6 R1

9. Undefined

10. Undefined

11. 40

12. 30

13. 41

14. 31

15. 44 R2

16. 88 R8

17. 9

16. $9\overline{)800}$
17. $12\overline{)108}$
18. $16\overline{)128}$
19. $600 \div 25$
20. $182 \div 13$
21. $15\overline{)750}$
22. $13\overline{)260}$
23. $312 \div 20$
24. $161 \div 15$
25. $20\overline{)305}$
26. $27\overline{)356}$
27. $30\overline{)847}$
28. $13\overline{)305}$
29. $73\overline{)148}$
30. $68\overline{)207}$
31. $49\overline{)993}$
32. $27\overline{)841}$
33. $68\overline{)210}$
34. $47\overline{)237}$
35. $11\overline{)4406}$
36. $13\overline{)3917}$
37. $50\overline{)3065}$
38. $40\overline{)2163}$
39. $502\overline{)98,762}$
40. $317\overline{)70,365}$
41. $417\overline{)169,719}$
42. $201\overline{)105,123}$

Verify the given factors.

43. Show that 22 and 32 are both factors of 704 by using long division.
44. Show that 28 and 36 are both factors of 1008 by using long division.
45. Show that 35 and 45 are both factors of 1575 by using long division.
46. Show that 56 and 39 are both factors of 2184 by using long division.

Applications

Solve.

47. A mother has 12 cookies that she will divide equally among 4 children. How many cookies will each child get?
48. One pint of Ben and Jerry's Crème Brûlée Ice Cream has 64 grams of fat. If there are 4 servings per pint, how many grams of fat are in each serving?
49. Mia Scott tutors students in reading and makes $25 per student. If she made $475 in a week, how many students did she tutor?
50. A box of cereal has a net weight of 429 grams. If a serving of cereal is 33 grams, how many servings are there per box of cereal?
51. If one person can paint a small house in 48 hours, how long will it take a crew of 8 people to paint the house, assuming that all 8 work at the same speed and do not interfere with each other?
52. Seven students sold a total of 392 raffle tickets. Assuming that each student sold the same number of raffle tickets, how many tickets did each student sell?
53. The Cedarville Baseball Camp has 198 children in attendance. How many 9-member teams can this club have if each child plays on one team?

18. 8
19. 24
20. 14
21. 50
22. 20
23. 15 R12
24. 10 R11
25. 15 R5
26. 13 R6
27. 28 R7
28. 23 R6
29. 2 R2
30. 3 R3
31. 20 R3
32. 31 R4
33. 3 R6
34. 5 R2
35. 400 R6
36. 301 R4
37. 61 R15
38. 54 R3
39. 196 R370
40. 221 R308
41. 407
42. 523
43. $704 \div 22 = 32$; $704 \div 32 = 22$
44. $1008 \div 28 = 36$; $1008 \div 36 = 28$
45. $1575 \div 45 = 35$; $1575 \div 35 = 45$
46. $2184 \div 56 = 39$; $2184 \div 39 = 56$

Applications

47. 3 cookies
48. 16 grams
49. 19 students
50. 13 servings
51. 6 hours
52. 56 raffle tickets
53. 22 teams

54. US Astronaut Peggy Whitson orbited the Earth 6032 times during her space flights on the International Space Station. If the International Space Station orbits the Earth 16 times per day, how many days was Peggy Whitson in space?[2]

55. A community has 5978 square feet available for individual gardens, which will be evenly distributed among 14 people. How much space will each person get?

56. Thirteen men purchase a boat together. If the cost of the boat is $33,462, how much will each man contribute if each contributes an equal amount?

57. A large chicken farm ships 65,076 eggs during a typical week. How many dozen eggs does this represent? (**Note:** dozen = 12)

58. For the 2021–2022 academic year, the average tuition cost for four years at a public, 4-year institution was $37,396. If tuition did not increase each year, how much would you pay per year for the four years you were in college?[3]

59. Smithfield High School paid $29,022 for six upright pianos. How much did each piano cost?

60. The area of every NFL football field is 57,600 square feet. If a bag of grass seed covers 50 square feet, how many bags of grass seed will be needed to cover one football field?[4]

61. 262,800 pounds of pasta is served every year at Mama Melrose's Ristorante Italiano at Disney MGM studios. How many pounds of pasta are served every day, assuming the restaurant is open 365 days in each year?[5]

62. Sophia decides to use a store credit deal where she won't have to pay any interest on her purchase if she pays off the entire purchase amount within a year. She bought $1920 worth of furniture and plans to make equal-sized monthly payments for six months. How much will Sophia pay per month?

63. Xander has $150 to spend on video games at a summer sale. If the average price of a video game is $16 (including tax), how many games can he buy? How much money will he have left over?

64. A lantern manufacturer is preparing their newly created lanterns for storage at the warehouse. There are 7685 lanterns packaged and ready to go. Only 8 packaged lanterns can fit into a box. How many boxes are needed? How many lanterns are left over and not boxed?

Writing & Thinking

65. List the four terms used for the parts of a division problem. Then give an example of a division problem and label the parts using those terms.

66. Explain how you would check a division problem that has a nonzero remainder.

67. Discuss how division is related to multiplication.

68. Give an example of when you might use division (other than in a class).

2 Source: National Aeronautics and Space Administration
3 Source: educationdata.org/average-cost-of-college
4 Source: National Football League
5 Source: www.diningindisney.com

54. 377 days
55. 427 square feet
56. $2574 per man
57. 5423 dozen
58. $9349 per year
59. $4837 per piano
60. 1152 bags
61. 720 pounds
62. $320 a month
63. 9 games with $6 left over
64. 7680 boxes with 5 left over

Writing & Thinking

65. The main terms used in division are: dividend, divisor, quotient, and remainder. Examples will vary.

66. To check a division problem, multiply the quotient and divisor, and then add the remainder. The result should equal the original dividend.

67. Division can be thought of as the reverse of multiplication and as a result, can be used to check the result of a multiplication problem. Answers will vary.

68. Answers will vary. Division may be used to equally distribute money, to find your portion of a bill, to find the number of students per classroom, etc.

Objectives

A. Round whole numbers.

B. Estimate sums and differences by using rounded numbers.

C. Estimate products and quotients by using rounded numbers.

1.5 Rounding and Estimating with Whole Numbers

A Rounding Whole Numbers

Rounding a number, that is, approximating the number, is a procedure performed in many situations in everyday life. You may need to round your income to the nearest thousand when applying for a loan, your manager may ask for an estimate of inventory to the nearest 100 items, or a website may give a report of views per day to the nearest 100. The place value you round to depends on the situation and how you are going to use the answer.

Rounding Numbers

To **round** a given number means to find another number close to the given number. The desired place of accuracy must be stated.

DEFINITION

For example, if you were asked to round 872, you would not know what to do unless you were told the position or place of accuracy desired. The number lines in Figure 1 help to illustrate the problem.

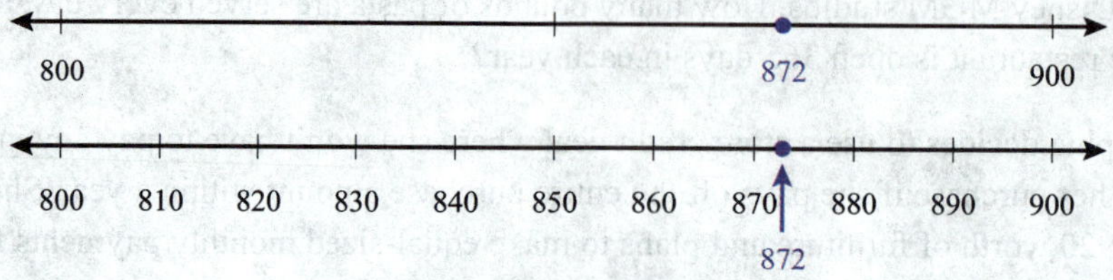

Figure 1

We can see that 872 is closer to 900 than to 800. So, **to the nearest hundred**, 872 rounds to 900. Also, 872 is closer to 870 than to 880. So, **to the nearest ten**, 872 rounds to 870.

In Example 1, we use **number lines** as visual aids in understanding the rounding process. On number lines, whole numbers are used to label equally-spaced points.

Example 1 Rounding Whole Numbers

Use a number line to round each number as indicated.

a. 43 to the nearest ten

b. 5500 to the nearest thousand

1. Use a number line to round each number as indicated.

a. 5835 to the nearest hundred

b. 85 to the nearest ten

Solution

a. To round 43 to the nearest ten:

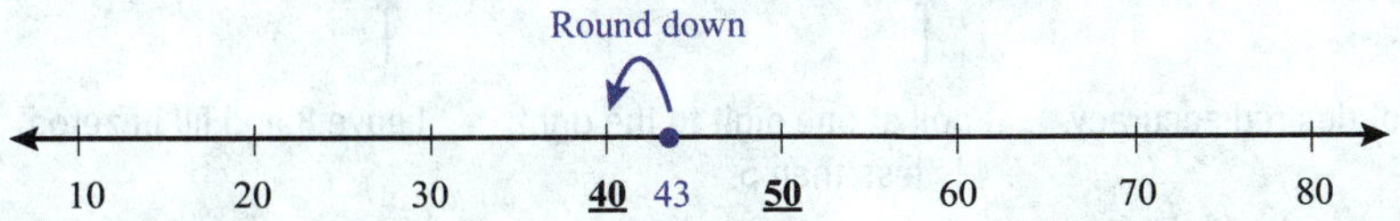

We see that 43 is closer to 40 than to 50. Thus, 43 rounds to 40 (to the **nearest ten**).

b. To round 5500 to the nearest thousand:

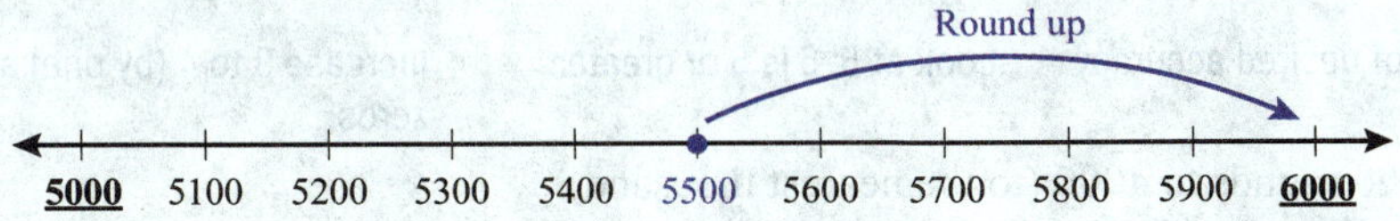

We see that 5500 is the same distance from 5000 as it is from 6000. In situations like this, we round up to the larger number. Thus, 5500 rounds to 6000 (to the **nearest thousand**).

Now work margin exercise 1.

Using number lines as aids for understanding is fine, but for practical purposes, the following procedure is more useful.

Rounding Rule for Whole Numbers

1. Look at the single digit just to the right of the digit in the place of desired accuracy.

a. **If this digit is less than 5**, leave the digit in the place of desired accuracy as it is, and replace all digits to the right with zeros. All digits to the left remain unchanged.

b. **If this digit is 5 or greater**, increase the digit in the desired place of accuracy by one and replace all digits to the right with zeros. All digits to the left remain unchanged unless a 9 is made one larger. Then the 9 is replaced by 0 and the next digit to the left is increased by 1.

PROCEDURE

Teaching Note

One or both of the following "tips" may help your students round numbers correctly.

1. Have students underline the digit that is in the place of desired accuracy.
2. Name the digit in the desired place of accuracy the "roundee" and name the digit immediately to the right the "rounder."

Example 2 Rounding Whole Numbers

Round each number as indicated.

a. 6849 to the nearest hundred

b. 3500 to the nearest thousand

c. 597 to the nearest ten

d. 20,560 to the nearest ten thousand

2. Round each number as indicated.

a. 7436 to the nearest ten

b. 25,000 to the nearest ten thousand

c. 139,800 to the nearest thousand

d. 8,079,200 to the nearest million

Solution

a. 6849 → 6849 → 6800

Place of desired accuracy. Look at one digit to the right; 4 is less than 5. Leave 8 and fill in zeros.

So 6849 rounds to 6800 (to the nearest hundred).

b. 3500 → 3500 → 4000

Place of desired accuracy. Look at 5; 5 is 5 or greater. Increase 3 to 4 (by one) and fill in zeros.

So 3500 rounds to 4000 (to the nearest thousand).

c. 597 → 597 → 600

Place of desired accuracy. Look at 7; 7 is 5 or greater. Increase 9 to 10 (this changes the 5 to a 6).

So 597 rounds to 600 (to the nearest ten).

d. 20,560 → 20,560 → 20,000

Place of desired accuracy. Look at 0; 0 is less than 5. Leave 2 and fill in zeros.

So 20,560 rounds to 20,000 (to the nearest ten thousand).

Now work margin exercise 2.

3. A redwood tree measures 284 feet in height. Round this value to the nearest ten.

Example 3 Application: Rounding Whole Numbers

A jar at a local candy store is filled with 2709 jelly beans and put on display in the store window. Round this value to the nearest thousand.

Solution

To round 2709 to the nearest thousand:

2709 → 2709 → 3000

Place of desired accuracy. Look at 7; 7 is 5 or greater. Increase 2 to 3 (by one) and fill in zeros.

So 2709 rounds to 3000 (to the nearest thousand).

Now work margin exercise 3.

B Estimating Sums and Differences

One use for rounded numbers is to **estimate** an answer (or to find an **approximate** answer) before any calculations are made with the given numbers.

To **estimate an answer** means to use rounded numbers in a calculation to form an idea of what the size of the actual answer should be. In some situations, an estimated answer may be sufficient. For example, a shopper may simply estimate the total cost of purchases to be sure that he or she has enough cash to cover the cost.

Estimating a Sum or Difference

1. Round each number to the place of the **leftmost** digit.
2. Perform the addition or subtraction with these rounded numbers.

PROCEDURE

Math Tip

When rounding to the place of the leftmost digit, single-digit numbers are considered to be already rounded.

Example 4 Estimating Sums of Whole Numbers

Estimate the sum; then find the actual sum.

$$\begin{array}{r} 68 \\ 925 \\ +487 \\ \hline \end{array}$$

Solution

Note that in this example, numbers are rounded to different places because they are of different sizes. That is, the leftmost digit does not have the same place value for all numbers.

To estimate the sum, round each number to the place of the leftmost digit and then add these rounded numbers.

$$\begin{array}{rcrl} 68 & \longrightarrow & 70 & \text{Rounded value of 68} \\ 925 & \longrightarrow & 900 & \text{Rounded value of 925} \\ +487 & \longrightarrow & +500 & \text{Rounded value of 487} \\ \hline & & 1470 & \text{Estimated sum} \end{array}$$

Now find the sum, keeping in mind that the answer should be close to 1470.

$$\begin{array}{rl} {}^{1\,2} & \\ 68 & \\ 925 & \\ +487 & \\ \hline 1480 & \text{Actual sum} \\ & \text{This sum is very close to 1470.} \end{array}$$

Now work margin exercise 4.

Note that the sum in Example 4 is very close to our estimate. If we had arrived at a sum that was far away from our estimate, like 14,000 in Example 4, we would know that we made an error and that we should check our work.

4. Estimate the sum; then find the actual sum.

$$\begin{array}{r} 176 \\ 84 \\ +75 \\ \hline \end{array}$$

Teaching Note

This method of estimation has the advantage of being fast and relatively easy because there is only one significant digit in each number. Accuracy cannot be guaranteed with any method since an estimation is just that, an estimation. Remind students that one of the purposes of estimation is to help them develop some judgment skills.

5. Estimate the sum; then find the actual sum.

```
 7824
  133
+ 904
```

Completion Example 5 Estimating Sums of Whole Numbers

Estimate the sum; then find the actual sum.

```
  5483
   232
+  657
```

Solution

First, estimate the sum by rounding each number to the place of the leftmost digit. Then add these rounded numbers.

```
  5483 ⟶    5000
   232 ⟶     200
+  657 ⟶  +  700
            5900   Estimated sum
```

Now find the sum, and compare your answer with the estimated sum. They should be "close."

```
  5483
   232
+  657
  6372   Actual sum
```

Now work margin exercise 5.

6. Estimate the difference; then find the actual difference.

```
 2685
− 847
```

Example 6 Estimating Differences of Whole Numbers

Estimate the difference; then find the actual difference.

```
  2783
−  975
```

Solution

To estimate the difference, round each number to the place of the leftmost digit and then subtract these rounded numbers.

```
  2783 ⟶    3000   Rounded value of 2783
−  975 ⟶  − 1000   Rounded value of 975
            2000   Estimated difference
```

Now find the difference, keeping in mind that the answer should be close to 2000.

```
  1 17 7 13
  2 7  8 3
−   9  7 5
  1 8  0 8   Actual difference
```

This difference is very close to 2000.

Now work margin exercise 6.

Example 7 Application: Estimating Differences of Whole Numbers

At the beginning of July, your bank account balance is \$3859. Over the course of the month, you spend \$823. Estimate the remaining balance in your account. Then find the actual balance. (Assume no deposits were made.)

Solution

In order to find the balance, we must subtract the amount of money you spent throughout the month from the starting balance. First, estimate this difference.

\$3859	→	4000 (borrow: 3 10)	Rounded value of 3859
− 823	→	− 800	Rounded value of 823
		\$3200	Estimated difference

Now we find the difference. This should be close to 3200.

$$\begin{array}{r} 3859 \\ -\ 823 \\ \hline \$3036 \end{array}$$ Actual difference

Thus, the remaining balance in your account is \$3036.

Now work margin exercise 7.

7. The World of Pets pet store orders 1598 bags of dog food in September. During the month, the store sells 597 bags. Estimate the remaining number of unsold bags of dog food at the end of the month. Then find the actual number of bags of dog food left.

The use of rounded numbers to approximate answers demands some understanding of numbers in general and some judgment as to how "close" an estimate can be to be acceptable. In particular, when all numbers are rounded up or all numbers are rounded down, the estimate might not be "close enough" to detect a large error. Still, the process is worthwhile and can give quick and useful estimates in many cases.

CAUTION

C Estimating Products and Quotients

Products and quotients can be estimated, just as sums and differences can be estimated. Once again, we will use the rule of rounding to the leftmost digit of each number before performing the operation. Example 8 illustrates the estimating procedure with multiplication.

Estimating a Product

1. Round each number to the place of the **leftmost** digit.
2. Multiply the rounded numbers.

PROCEDURE

8. Estimate the product; then find the actual product.

$$\begin{array}{r} 18 \\ \times\ 74 \\ \hline \end{array}$$

Example 8 Estimating Products of Whole Numbers

Estimate the product; then find the actual product.

$$\begin{array}{r} 62 \\ \times\ 38 \\ \hline \end{array}$$

Solution

To estimate the product, round each number to the place of the leftmost digit and then multiply these rounded numbers.

$$\begin{array}{rcrl} 62 & \longrightarrow & 60 & \text{Rounded value of 62} \\ \times\ 38 & \longrightarrow & \times\ 40 & \text{Rounded value of 38} \\ \hline & & 2400 & \text{Estimated product} \end{array}$$

Now find the product, keeping in mind that the answer should be close to 2400.

$$\begin{array}{rl} \overset{1}{6}2 & \\ \times\ 38 & \\ \hline 496 & \\ +1860 & \\ \hline 2356 & \text{Actual product} \\ & \text{The actual product is close to 2400.} \end{array}$$

Now work margin exercise 8.

9. The principal of North Valley High School wants to buy 85 computers for the computer lab. Each new computer costs $472. Estimate the total cost of the new computers. Then find the actual cost.

Example 9 Application: Estimating Products of Whole Numbers

An apartment complex is planning to furnish all of its 93 apartments with new dishwashers. The owner can purchase the dishwashers for $267 each. Estimate the total cost of buying dishwashers for every apartment. Then find the actual total cost to the apartment complex.

Solution

In order to find the total cost to the apartment complex, we must multiply the number of dishwashers purchased by the cost of each dishwasher.

We begin by estimating the product: $267 · 93.

$$\begin{array}{rcrl} \$267 & \longrightarrow & \$300 & \text{Rounded value of 267} \\ \times\ 93 & \longrightarrow & \times\ 90 & \text{Rounded value of 93} \\ \hline & & \$27,000 & \text{Estimated product} \end{array}$$

The product should be near $27,000.

$$\begin{array}{r} \$267 \\ \times\ 93 \\ \hline 801 \\ +24030 \\ \hline \$24,831 \end{array}$$

Thus, the actual total cost to the apartment complex is $24,831.

Now work margin exercise 9.

Estimating a Quotient

1. Round both the divisor and dividend to the place of the **leftmost** digit.
2. Divide with the rounded numbers. (When calculating an estimate, ignore the remainder if there is one.)

PROCEDURE

Example 10 Estimating Quotients of Whole Numbers

Estimate the quotient, then find the actual quotient.

$$8875 \div 25$$

Solution

To estimate the quotient, round each number to the place of the leftmost digit and then divide these rounded numbers.

$$8875 \div 25 \longrightarrow 9000 \div 30$$

$$\begin{array}{r} 300 \\ 30\overline{)9000} \\ \underline{-90} \\ 00 \\ \underline{-0} \\ 00 \\ \underline{-0} \\ 0 \end{array}$$

Estimated quotient (300)

Now find the quotient, keeping in mind that the answer should be near 300.

$$\begin{array}{r} 355 \\ 25\overline{)8875} \\ \underline{-75} \\ 137 \\ \underline{-125} \\ 125 \\ \underline{-125} \\ 0 \end{array}$$

Actual quotient (355)

Now work margin exercise 10.

Example 11 Estimating Quotients of Whole Numbers

Estimate the quotient, then find the actual quotient.

$$325 \div 42$$

10. Estimate the quotient, then find the actual quotient.

$3942 \div 18$

Math Tip

Recall that

quotient · divisor = dividend.

$$\begin{array}{r} \text{quotient} \\ \text{divisor}\overline{)\text{dividend}} \end{array}$$

In the division problem $\begin{array}{r} 3 \\ 2\overline{)6} \end{array}$, 3 is the quotient, 2 is the divisor, and 6 is the dividend.

11. Estimate the quotient, then find the actual quotient.

$882 \div 76$

Solution

First, estimate the quotient.

$$325 \div 42 \longrightarrow 300 \div 40$$

$$\begin{array}{r} 7 \\ 40\overline{)300} \\ -280 \\ \hline 20 \end{array}$$ Estimated quotient

Now find the quotient, keeping in mind that the answer should be near 7.

$$\begin{array}{r} 7 \\ 42\overline{)325} \\ -294 \\ \hline 31 \end{array}$$ Actual quotient; Remainder

The quotient is 7 and the remainder is 31. In this case, the quotient is the same as the estimated value. The true remainder is different.

Now work margin exercise 11.

12. Estimate the quotient, then find the actual quotient.

$36\overline{)1340}$

Completion Example 12 Estimating Quotients of Whole Numbers

Estimate the quotient, then find the actual quotient.

$$6461 \div 21$$

Solution

First, estimate the quotient.

$$\begin{array}{r} 300 \\ 20\overline{)6000} \end{array}$$ Estimate

Now find the quotient.

$$\begin{array}{r} 307 \\ 21\overline{)6461} \\ -63 \\ \hline 16 \\ -0 \\ \hline 161 \\ -147 \\ \hline 14 \end{array}$$ Actual quotient; Remainder

Is your estimate close to the actual quotient? yes

What is the difference between your estimate and the actual quotient? 7

Now work margin exercise 12.

Example 13 Application: Estimating Quotients of Whole Numbers

A group of 11 friends bought tickets to a local jazz concert. The total price of all 11 tickets was \$341. Estimate the cost of each ticket. Then calculate the actual cost of each ticket.

Solution

In order to find the cost of each person's ticket, we must divide the total cost of the tickets by the number of tickets that were purchased.

We begin by estimating the quotient.

$$341 \div 11 \longrightarrow 300 \div 10$$

$$\begin{array}{r} 30 \\ 10\overline{)300} \\ \underline{-300} \\ 0 \end{array}$$ Estimated quotient

The quotient should be near 30.

$$\begin{array}{r} 31 \\ 11\overline{)341} \\ \underline{-33} \\ 11 \\ \underline{-11} \\ 0 \end{array}$$ Actual quotient; 0 = Remainder

Thus, the actual cost of each ticket is \$31.

Now work margin exercise 13.

13. The Sunnyside Elementary School is taking a field trip to the zoo. If one school bus can seat 22 people, and there are 462 students and teachers going on the field trip, estimate the number of buses needed. Then find the actual number of buses.

Completion Example Answer

5. 700; 5900; 6372 **12.** 300 estimate; 307 actual quotient; 14 remainder; yes; 7

Margin Exercise Answers

1. a. 5800 **b.** 90 **2. a.** 7440 **b.** 30,000 **c.** 140,000 **d.** 8,000,000 **3.** 280 feet **4.** Estimate: 360; sum: 335 **5.** Estimate: 9000; sum: 8861 **6.** Estimate: 2200; difference: 1838 **7.** Estimate: 1400 bags; difference: 1001 bags **8.** Estimate: 1400; product: 1332 **9.** Estimate: \$45,000; product: \$40,120 **10.** Estimate: 200; quotient: 219 **11.** Estimate: 11; quotient: 11 R46 **12.** Estimate: 25; quotient: 37 R8 **13.** Estimate: 25 buses; quotient: 21 buses

1.5 Exercises

Concept Check

Fill-in-the-Blank. Complete each sentence using information found in this section.

1. When rounding, consider the place being rounded to and the number to its ________.

2. 37,068,155 rounded to the ten thousands place is ________, rounded to the millions place is ________.

3. When rounding, it is important that the desired ________ of accuracy be known.

1.5 Exercises

Concept Check

1. Right
2. 37,070,000; 37,000,000
3. Place

4. 5
5. 0s
6. Rounding
7. True
8. False; It becomes 440,000.
9. True
10. True

Practice

1. 80;

2. 90;

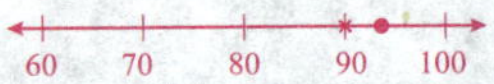

3. 660;

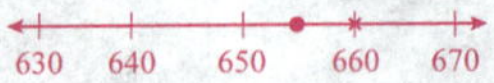

4. 380;

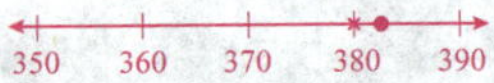

5. 500;

6. 300;

7. a. 2 b. 6 c. 6, 2, 3, 6 d. 1430, ten
8. a. 6 b. 3 c. 3, 6, 3, 1, 5 d. 96,000; thousand
9. 30
10. 80
11. 30
12. 60

4. When rounding, if a digit to the right of the place of accuracy is ________ or greater, the digit in the place of accuracy increases by 1.

5. With whole numbers, once rounding has occurred, the digits to the right of the desired place of accuracy become ________.

6. Finding a number close to a given number is called ________.

True/False. Determine whether each statement is true or false. If a statement is false, explain how it can be changed so the statement will be true. (**Note:** There may be more than one acceptable change.)

7. Rounding means finding a number close to the given number, using a specified place of accuracy.

8. When rounded to the ten thousands place, 435,613 becomes 400,000.

9. To estimate the answer for a division problem, begin by rounding both the divisor and dividend.

10. If estimated, 4250 ÷ 51 is 4000 ÷ 50 = 80.

Practice

Use a number line to round each number as indicated. (Graph the number on the line with a heavy dot and mark the rounded number with an x.) See Example 1.

1. 78 (nearest ten)
2. 94 (nearest ten)
3. 655 (nearest ten)
4. 382 (nearest ten)
5. 479 (nearest hundred)
6. 258 (nearest hundred)

Fill in the blanks to correctly complete each statement.

7. Round 1426 to the nearest ten.
   a. The digit in the tens position is ________.
   b. The next digit to the right is ________.
   c. Since ________ is greater than 5, change ________ to ________ and replace ________ with 0.
   d. So 1426 rounds to ________ to the nearest ________.

8. Round 96,315 to the nearest thousand.
   a. The digit in the thousands position is ________.
   b. The next digit to the right is ________.
   c. Since ________ is less than 5, leave ________ as it is and replace ________, ________, and ________ with 0.
   d. So 96,315 rounds to ________ to the nearest ________.

Round each number as indicated. See Example 2.

To the nearest ten:

9. 31
10. 82
11. 25
12. 55

13. 503

14. 204

15. 996

16. 998

To the nearest hundred:

17. 75

18. 63

19. 637

20. 215

21. 4163

22. 4475

23. 9954

24. 9972

To the nearest thousand:

25. 6912

26. 3790

27. 9610

28. 9900

29. 62,265

30. 13,499

31. 13,501

32. 28,559

To the nearest ten thousand:

33. 5697

34. 8787

35. 62,200

36. 44,593

37. 125,000

38. 615,000

39. 1,603,587

40. 3,527,054

Estimate each answer; then find the actual sum or difference. See Examples 4 through 6.

41. $\begin{array}{r} 83 \\ 62 \\ +\ 78 \\ \hline \end{array}$

42. $\begin{array}{r} 49 \\ 31 \\ +\ 86 \\ \hline \end{array}$

43. $\begin{array}{r} 146 \\ 259 \\ +\ 384 \\ \hline \end{array}$

44. $\begin{array}{r} 475 \\ 126 \\ +\ 572 \\ \hline \end{array}$

45. $\begin{array}{r} 5742 \\ 6271 \\ 8156 \\ +\ 972 \\ \hline \end{array}$

46. $\begin{array}{r} 483 \\ 1681 \\ 3054 \\ +\ 4006 \\ \hline \end{array}$

47. $\begin{array}{r} 854 \\ -367 \\ \hline \end{array}$

48. $\begin{array}{r} 692 \\ -427 \\ \hline \end{array}$

13. 500
14. 200
15. 1000
16. 1000
17. 100
18. 100
19. 600
20. 200
21. 4200
22. 4500
23. 10,000
24. 10,000
25. 7000
26. 4000
27. 10,000
28. 10,000
29. 62,000
30. 13,000
31. 14,000
32. 29,000
33. 10,000
34. 10,000
35. 60,000
36. 40,000
37. 130,000
38. 620,000
39. 1,600,000
40. 3,530,000
41. 220; 223
42. 170; 166
43. 800; 789
44. 1200; 1173
45. 21,000; 21,141
46. 9500; 9224
47. 500; 487
48. 300; 265

49. 20,000; 20,804
50. 40,000; 41,223
51. 210,000; 181,300
52. 400,000; 352,315
53. a. C b. D c. B d. E e. A
54. a. C b. B c. A d. D
55. a. C b. A c. E d. D e. B
56. a. B b. D c. A d. C
57. 240; 224

49. $\begin{array}{r} 63{,}504 \\ -\ 42{,}700 \\ \hline \end{array}$

50. $\begin{array}{r} 74{,}305 \\ -\ 33{,}082 \\ \hline \end{array}$

51. $\begin{array}{r} 275{,}600 \\ -\ 94{,}300 \\ \hline \end{array}$

52. $\begin{array}{r} 450{,}315 \\ -\ 98{,}000 \\ \hline \end{array}$

53. Match each product with the closest estimate of that product. Perform any calculations mentally.

	Product	**Estimate**
______	**a.** $16 \cdot 18$	**A.** 210
______	**b.** $6(78)$	**B.** 300
______	**c.** $11 \cdot 32$	**C.** 400
______	**d.** $(8)(69)$	**D.** 480
______	**e.** $25(7)$	**E.** 560

54. Match each product with the closest estimate of that product. Perform any calculations mentally.

	Product	**Estimate**
______	**a.** $37(500)$	**A.** 200
______	**b.** $37(50)$	**B.** 2000
______	**c.** $37(5)$	**C.** 20,000
______	**d.** $37(5000)$	**D.** 200,000

55. Match each quotient with the closest estimate of that quotient. Perform any calculations mentally.

	Quotient	**Estimate**
______	**a.** $9\overline{)910}$	**A.** 6
______	**b.** $34 \div 5$	**B.** 10
______	**c.** $34\overline{)12{,}000}$	**C.** 100
______	**d.** $18\overline{)3900}$	**D.** 200
______	**e.** $216 \div 18$	**E.** 300

56. Match each quotient with the closest estimate of that quotient. Perform any calculations mentally.

	Quotient	Estimate
______	**a.** $3\overline{)870}$	**A.** 30
______	**b.** $3\overline{)87{,}000}$	**B.** 300
______	**c.** $3\overline{)87}$	**C.** 3000
______	**d.** $3\overline{)8700}$	**D.** 30,000

Estimate each answer; then find the actual product or quotient. (Remember that single-digit numbers are already considered to be rounded.) See Examples 8 and 10 through 12.

57. 56×4

58. 84×3

59. 17×32

60. 16×26

61. 420×104

62. 673×186

63. $18\overline{)216}$

64. $11\overline{)99}$

65. $49\overline{)993}$

66. $18\overline{)773}$

67. $50\overline{)3065}$

68. $37\overline{)2003}$

Applications

Solve.

69. A certain manufacturer pays his workers for every ten widgets that a person produces, rounded to the nearest ten. If Lucas produced 4564 widgets, how many widgets will he be paid for?

70. Oliver decides to keep a journal of the number of calories he eats each day, rounded to the nearest ten. Today, he calculates that he ate 2348 calories. How many calories should he record in his journal?

71. As of 2021, the average household credit card debt was $8590. Round this number to the nearest hundred dollars.

72. On a typical day, 46,931 vehicles are driven through the northbound toll gates at a busy bridge. Round this number to the nearest hundred vehicles.

73. A Honda Accord EX sedan with automatic transmission has a sticker price of $26,120. Round this number to the nearest thousand dollars.

74. On a single weekday, 791,567 people rode the public transportation system in a major metropolitan area. How many people is this, rounded to the nearest thousand people?

75. A salesperson drives 96,469 miles a year. Round this number to the nearest ten thousand miles.

58. 240; 252
59. 600; 544
60. 600; 416
61. 40,000; 43,680
62. 140,000; 125,178
63. 10; 12
64. 10; 9
65. 20; 20 R13
66. 40; 42 R17
67. 60; 61 R15
68. 50; 54 R5

Applications

69. 4560 widgets
70. 2350 calories
71. $8600
72. 46,900 vehicles
73. $26,000
74. 792,000 people
75. 100,000 miles

76. 2,150,000
77. $40,000; $35,316
78. 283,000 students
79. $1100; $1076
80. $720; $738
81. 330 sets; 316 sets
82. $18,000; $21,575
83. $6000; $7634
84. $20,000; $17,446

76. In 2021, there were 2,147,070 electric vehicles in the United States. Round this number to the nearest ten thousand.[1]

77. College expenses for a private four-year college in the 2022–2023 academic year are as follows:

Tuition & Fees	$25,243
Room & Board	$8996
Books & Supplies	$1077

Estimate the total cost to attend for a year using rounded numbers to the nearest thousand. Then calculate the actual cost.

78. In 2021, the University of California campuses had the following undergraduate enrollments:

University of California 2021 Enrollment

Campus	Enrollment	Campus	Enrollment
UC Berkeley	45,036	UC Riverside	26,847
UC Davis	40,050	UC San Diego	41,885
UC Irvine	36,505	UC Santa Barbara	26,124
UC Los Angeles	46,116	UC Santa Cruz	19,841

Source: www.universityofcalifornia.edu/infocenter/fall-enrollment-glance

Approximately how many undergraduate students were enrolled in the University of California system in 2021?

79. Emilia is sending her three children to different week long daytime summer camps. The cost of the summer camp is $239 for the youngest child, $487 for the middle child, and $350 for the oldest child. Estimate how much it will cost to send the three kids to summer camp.

80. Mateo purchased the following items for his home office: one laptop computer for $483, one wireless router for $99, one printer for $138, and a pack of printer paper for $18. Estimate the total purchase price using rounded numbers. Then calculate the actual purchase price.

81. Paula is in charge of keeping inventory of the medical supplies at the hospital she works for. According to her records, the hospital began the month with 384 sets of crutches. But now, at the end of the month, there are only 68 sets remaining. Estimate the number of sets of crutches the hospital used in the last month, then calculate the actual number of sets of crutches they used.

82. Elizabeth paid $2191 to add on special paint, 18" alloy wheels, and a cargo net to the car she just purchased. If she paid a total of $23,766 for the car with the add-ons, estimate the base price of the car using rounded numbers. Then find the actual base price.

83. A total of 22 microphones are to be purchased. Each microphone costs $347. Estimate the total purchase price by using rounded numbers. Then calculate the actual purchase price.

84. A jewelry store purchased 11 fine diamond-studded watches at a wholesale cost of $1,586 each. Estimate the total cost for all of the watches by using rounded numbers. Then, calculate the actual cost for all of the watches.

1 Source: evadoption.com

85. A summer camp has 396 campers who will be going on an outing. Estimate the number of vans needed to accommodate the entire camp if each van can comfortably hold 11 passengers. Then calculate the actual number of vans that will be needed.

86. Four people decide to purchase equal shares of a beach house that costs $312,760. Estimate the contribution each person makes by using rounded numbers. Then calculate the actual contribution.

87. A technical school plans on purchasing four pieces of test equipment that cost $648 each, plus two work benches at $284 each. Estimate the total cost of the purchase using rounded numbers. Then, calculate the actual cost.

88. Ramón is running a sand volleyball tournament soon and must purchase some new equipment. He needs three new nets, which cost $159 each. He also needs five new sets of boundary lines, which cost $86 each. Estimate the total cost of the new equipment. Then calculate the actual cost.

Writing & Thinking

89. In your own words, define estimation.

90. List the steps to use in estimating a product in a multiplication problem.

91. Compare and contrast rounding and estimating.

92. Give an example when you might use rounding and/or estimation (other than in a class).

85. 40 vans; 36 vans

86. $75,000; $78,190

87. $3000; $3160

88. $1050; $907

Writing & Thinking

89. Estimation uses rounded values to find an approximate sum, difference, product, etc. Answers will vary.

90. 1. Round each number to the place of the leftmost digit. 2. Perform the multiplication on these numbers.

91. Rounding is used to find another number close to the given number. Estimation uses rounded values to find an approximate sum, difference, product, etc. Answers will vary.

92. Answers will vary. Estimation may be used when tallying up a bill, creating a budget, determining attendance at a function, etc.

Objectives

A. Solve problems by using the basic strategy for solving word problems.

B. Solve word problems involving consumer applications.

C. Solve word problems involving geometric applications.

D. Solve word problems by finding the average of a set of numbers.

1.6 Problem Solving with Whole Numbers

A Solving Word Problems

In this section, the applications are varied and sometimes involve combinations of the four operations of addition, subtraction, multiplication, and division. You must read each problem carefully and decide, using your previous experiences, just what operations to perform. Learn to enjoy the challenge of word problems, keeping the following ideas in mind.

1. At least try something.
2. If you do nothing, you learn nothing.
3. Even by making errors in technique or judgment you are learning what does not work.
4. Do not be embarrassed by mistakes.

The problems discussed in this section come under the following headings: consumer items, checking, geometry, average, and reading graphs. The steps in the basic strategy listed here will help give an organized approach regardless of the type of problem.

Basic Strategy for Solving Word Problems

1. READ: Read the problem carefully.
2. SET UP: Draw any type of figure or diagram that might be helpful, and decide what operations are needed.
3. SOLVE: Perform the operations to solve the problem.
4. CHECK: Check your work and check that your answer seems reasonable.

PROCEDURE

1. Craig is tracking how many visitors his website gets each day. It had 194 visitors on Friday and 233 visitors on Saturday. On Sunday, after posting new content, the website had 574 visitors. How many total visitors did his website have over those three days?

Example 1 Application: Adding Whole Numbers

Mary Ann is finalizing her spring break plans. First, she plans to visit her grandma, who lives 288 miles away. Then she plans to drive 145 miles to the beach and spend a couple days with her boyfriend before driving 203 miles home. How many total miles will Mary Ann drive on her spring break road trip?

Solution

Step 1: READ: Read the problem carefully. The keyword **total** indicates addition.

Step 2: SET UP: In this problem, the three amounts to be added are 288 miles, 145 miles, and 203 miles or

$$288 + 145 + 203 = \text{total miles.}$$

Step 3: SOLVE:

$$\begin{array}{r} \overset{1\,1}{288} \\ 145 \\ +203 \\ \hline 636 \end{array} \quad \text{Total miles driven}$$

Mary Ann will drive a total of 636 miles.

Step 4: CHECK: Mary Ann will drive about 300 miles to her grandma's house, 100 miles to the beach, and 200 miles home and

$$300 + 100 + 200 = 600,$$

so a total of 636 miles seems reasonable.

Now work margin exercise 1.

Example 2 Application: Multiplying Whole Numbers

The owner of a semi-pro baseball team orders baseballs from the manufacturer at the beginning of the season. The team is scheduled to play 32 games and they anticipate using 8 new baseballs each game. How many baseballs should the owner buy?

Solution

Step 1: READ: Read the problem carefully. The order for the entire season is found by multiplication.

Step 2: SET UP: To find the total number of baseballs to be ordered, multiply the number of games by the number of balls to be used in each game.

$$32 \cdot 8 = \text{total number of baseballs}$$

Step 3: SOLVE:

$$\begin{array}{rl} \overset{1}{3}2 & \text{Games} \\ \times \ \ 8 & \text{Balls for each game} \\ \hline 256 & \text{Total to be ordered} \end{array}$$

The owner should order 256 baseballs for the season.

Step 4: CHECK: The team plays about 30 games and uses about 10 balls per game and

$$30 \cdot 10 = 300,$$

so a total order of 256 balls seems reasonable.

Now work margin exercise 2.

Example 3 Application: Dividing Whole Numbers

Nineteen friends and neighbors decided to buy tickets to a basketball game with the LA Lakers playing the Boston Celtics. Because they were buying a large number of tickets they got a special group price of $1558. If they decided to divide the amount equally, what did each of them pay for a ticket?

2. The owner of a bakery is ordering ingredients for the week. The bakery uses 46 pounds of flour per day and is open every day. How much flour should be ordered for the week?

3. Orange Grove Elementary School received a donation of school supplies which included 2760 pencils. The supplies are to be evenly divided between twenty-three teachers. How many pencils will each teacher receive?

Solution

Step 1: READ: Read the problem carefully. The keyword is **divide** (in the phrase "divide the amount equally").

Step 2: SET UP: The amount to be divided equally among the 19 people is $1558. So the group price is to be divided by 19.

$$1558 \div 19 = \text{price per ticket}$$

Step 3: SOLVE:

$$\begin{array}{r} 82 \\ 19\overline{)1558} \\ \underline{-152} \\ 38 \\ \underline{-38} \\ 0 \end{array}$$

Each person paid $82 for a ticket.

Step 4: CHECK: Because there are almost 20 people in the group and $20 \cdot 80 = 1600$, the individual price of $82 seems reasonable. This can be checked directly by multiplication.

$$19 \cdot \$82 = \$1558$$

Now work margin exercise 3.

B Problem Solving: Consumer Applications

4. Mrs. Spencer bought a new car for her daughter for $27,550. The salesperson added $1600 for taxes and $475 for license fees. If she made a down payment of $4500 and financed the rest through her bank, how much did she finance?

Example 4 Application: Calculating Loan Amounts

Mr. Lukin bought a used car for $8000. Taxes of $640 and license fees of $320 were then added to the purchase price. He made a down payment of $2000 and financed the rest through his credit union. What was the amount of his loan from the credit union?

Solution

Step 1: READ: Read the problem carefully. Mr. Lukin's expenses are given and he made a down payment. One keyword is **added** (indicating the sum of the price, taxes, and license fees). Then we need to realize a down payment is subtracted from the expenses. This difference will be the loan amount.

Step 2: SET UP: In this problem, find the total of the expenses by adding the price, the taxes, and the license fees.

$$\$8000 + \$640 + \$320 = \text{total expenses}$$

Then subtract the down payment ($2000).

The result will be the amount of the loan.

Step 3: SOLVE:

Add Expenses		**Subtract Down Payment**	
\$8000	Price	\$8960	Total expenses
640	Taxes	−2000	Down payment
+ 320	License fees	\$6960	Amount of loan
\$8960	Total expenses		

After a down payment of \$2000, Mr. Lukin's loan will be \$6960.

Step 4: CHECK: With a price of \$8000 and a down payment of \$2000, the loan amount should be close to \$6000. Because of the extra fees, the actual loan amount of \$6960 seems reasonable.

Now work margin exercise 4.

Example 5 Application: Balancing a Checking Account

In January, Kathleen opened a checking account and deposited \$2500. During the month, she made another deposit of \$800 and made debit card purchases of \$132, \$425, \$196, and \$350. What was the balance in her account at the end of the month?

Solution

Step 1: READ: Read the problem carefully. To balance a checking account, the total of the debit card purchases is subtracted from the total of the deposits.

Step 2: SET UP: There are two deposits to be added (\$2500 and \$800). The total of the debit card purchases (\$132, \$425, \$196, and \$350) is subtracted from the deposits. The result will be the balance at the end of the month.

Step 3: SOLVE:

Total Deposits		**Total Debit Card Purchases**	
1		2 1	
\$2500		\$132	
+ 800		425	
\$3300	Total deposits	196	
		+ 350	
		\$1103	Total of debits

Now subtract to find the balance.

9	
2 10 10	
\$3300	
−1103	
\$2197	Balance

Kathleen's balance is \$2197.

Step 4: CHECK: The deposits are about \$3000 and the debits are about \$1000. So, a balance of \$2197 is reasonable.

Now work margin exercise 5.

5. In December, Kurt opened a checking account and deposited \$2000 into it. During that month, he wrote checks for \$57, \$120, and \$525, and used his debit card for purchases totaling \$630. He also made another deposit of \$200. What was his balance at the end of the month?

C Problem Solving: Geometric Applications

6. A square picture is mounted in a square frame. (A **square** is a rectangle with the lengths of all sides equal.) If the size of the picture in the frame is 6 inches by 6 inches and the frame is 8 inches by 8 inches, what is the area of the frame?

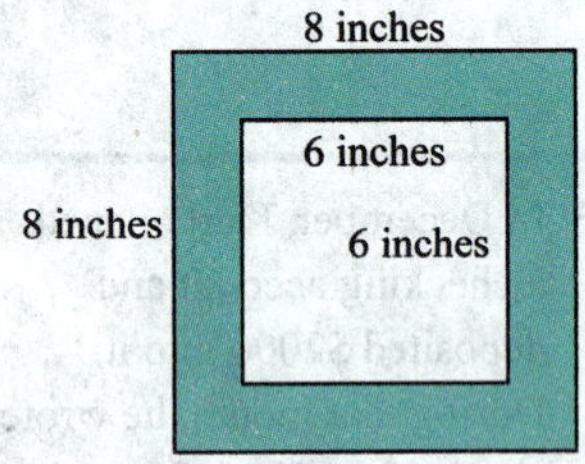

Example 6 Application: Finding the Area of Rectangles

A rectangular picture is mounted in a rectangular frame. If the size of picture (inside the frame) is 14 inches by 20 inches and the frame is 16 inches by 22 inches, what is the area of the frame?

Solution

Step 1: READ: Read the problem carefully. Remember that the area of a rectangle is found by multiplying length times width.

Step 2: SET UP: In this case, a figure is very helpful. Include the dimensions of the rectangles.

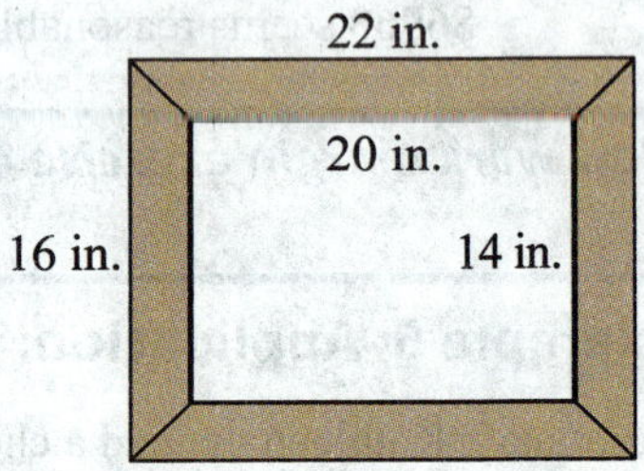

Find the area of the larger rectangle (the picture and frame) by multiplying $16 \cdot 22$. Then find the area of the smaller rectangle (just the picture) by multiplying $14 \cdot 20$. The area of the frame will be the difference between the two areas.

Step 3: SOLVE:

Larger Area	**Smaller Area**	**Area of Frame**
$\begin{array}{r} \overset{1}{1}6 \\ \times\ 22 \\ \hline 32 \\ 320 \\ \hline 352 \end{array}$ sq in.	$\begin{array}{r} 14 \\ \times\ 20 \\ \hline 280 \end{array}$ sq in.	$\begin{array}{r} \overset{2}{\cancel{3}}\overset{15}{\cancel{5}}2 \\ -\ 280 \\ \hline 72 \end{array}$ sq in.

The area of the frame is 72 sq in.

Step 4: CHECK: An area of 72 sq in. for the frame seems reasonable. The area of the picture itself is not much different from the area of the outer rectangle.

Now work margin exercise 6.

D Problem Solving: Average

A topic closely related to addition and division with whole numbers is the **average** (also called the **arithmetic average** or **mean**) of a set of numbers. Your grade in this course may be based on the average of your exam scores. The news and media report average income, average life expectancy, average sales, average attendance at sporting events, and so on. The average of a set of numbers is one representation of the concept of the "middle" of the set. It may or may not actually be one of the numbers in the set.

Finding the Average of a Set of Numbers

1. Find the sum of the given set of numbers.
2. Divide this sum by the number of numbers in the set. This quotient is called the **average** of the given set of numbers.

PROCEDURE

Example 7 Calculating an Average

Find the average of the following set of numbers: 15, 8, 90, 35, 27.

Solution

Step 1: First, find the sum of the numbers.

$$\begin{array}{r} \overset{2}{1}5 \\ 8 \\ 90 \\ 35 \\ +\ 27 \\ \hline 175 \end{array} \quad \text{Sum}$$

Step 2: Now, divide the sum by 5, since we have a list of five numbers.

$$\begin{array}{r} 35 \\ 5\overline{)175} \\ \underline{-15} \\ 25 \\ \underline{-25} \\ 0 \end{array} \quad \text{Average}$$

The average of the set of numbers is 35.

Now work margin exercise 7.

7. Find the average of the following set of numbers: 18, 29, 6, 33, 14, 26.

In Example 8, a bar graph is used to display a company's profits for each month over a 6-month period.

Example 8 Application: Calculating an Average

Top-Notch Sporting Goods recorded its profits for tennis rackets for six months. The following bar graph indicates the profits for each of the months from January to June.

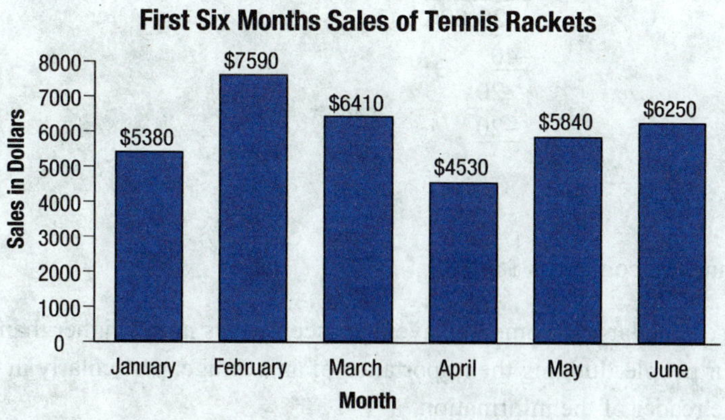

a. What month had the highest profits?

b. What month had the lowest profits?

c. What is the average monthly profit over the six months?

8. The following bar graph depicts the number of home runs a baseball player hits in five consecutive seasons.

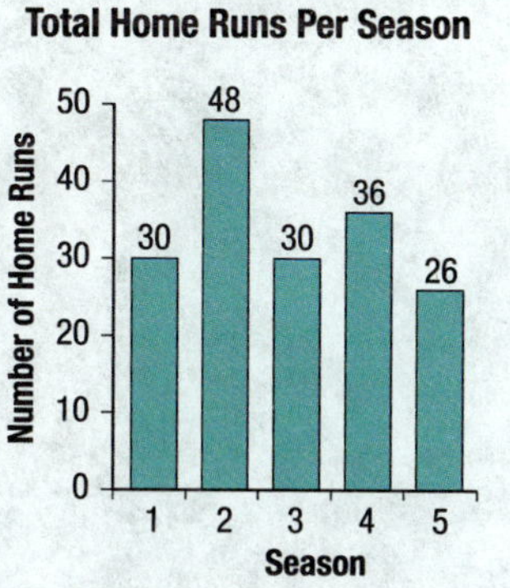

a. In which season did he hit the most home runs?

b. In which season did he hit the fewest home runs?

c. What is his average number of home runs per season?

Solution

a. From the bar graph, we can see that the month with the highest profits was February with $7590.

b. April was the month with the lowest profits with $4530.

c. The average monthly profit can be found by finding the sum of the profits for each month and dividing by 6.

Add Profits

$$\begin{array}{r} \scriptsize{3\ 3} \\ \$\,5\,3\,8\,0 \\ 7\,5\,9\,0 \\ 6\,4\,1\,0 \\ 4\,5\,3\,0 \\ 5\,8\,4\,0 \\ +\quad 6\,2\,5\,0 \\ \hline \$\,3\,6,0\,0\,0 \end{array} \quad \text{Sum}$$

Divide by 6

$$\begin{array}{r} 6000 \\ 6\overline{)36,000} \\ \underline{-36,000} \\ 0 \end{array} \quad \text{Average}$$

The average monthly profit for the six months was $6000 per month.

(In this case, we see that the average can be very useful. The store manager can use the monthly profits for planning and budgeting.)

Now work margin exercise 8.

9. Five cities have populations of 5000, 10,000, 15,000, 20,000 and 95,000. What is the average population of the cities?

Example 9 Application: Calculating an Average

Five people in a survey reported the following incomes for one year: $35,000; $41,000; $58,000; $72,000; $214,000. What was the average annual income for these five people?

Solution

Add Incomes

$$\begin{array}{r} \scriptsize{2\ 2} \\ \$\,3\,5,0\,0\,0 \\ 4\,1,0\,0\,0 \\ 5\,8,0\,0\,0 \\ 7\,2,0\,0\,0 \\ +\,2\,1\,4,0\,0\,0 \\ \hline \$\,4\,2\,0,0\,0\,0 \end{array} \quad \text{Sum}$$

Divide by 5

$$\begin{array}{r} 84,000 \\ 5\overline{)420,000} \\ \underline{-40} \\ 20 \\ \underline{-20} \\ 0 \end{array} \quad \text{Average}$$

The average annual income was $84,000.

Note: Because of one large income, the average income was much higher than the income of the other four people. Judging the importance of an average, particularly in a case like this, is up to the reader of the information.

Now work margin exercise 9.

Example 10 Application: Calculating an Average

On an English exam, two students scored 95, five scored 86, one scored 82, one scored 78, and six scored 75. What was the average score for the class?

Solution

There were fifteen students in the class. We can multiply as follows rather than add all fifteen scores.

$$\begin{array}{r} \overset{1}{9}5 \\ \times\ 2 \\ \hline 190 \end{array} \quad \begin{array}{r} \overset{3}{8}6 \\ \times\ 5 \\ \hline 430 \end{array} \quad \begin{array}{r} 82 \\ \times\ 1 \\ \hline 82 \end{array} \quad \begin{array}{r} 78 \\ \times\ 1 \\ \hline 78 \end{array} \quad \begin{array}{r} \overset{3}{7}5 \\ \times\ 6 \\ \hline 450 \end{array}$$

Next, we add the five products to find the sum of all of the scores.

$$\begin{array}{r} {}^{3\ 1} \\ 190 \\ 430 \\ 82 \\ 78 \\ +\,450 \\ \hline 1230 \end{array} \quad \text{Sum}$$

Finally, divide by 15 because the total represents 15 scores.

$$\begin{array}{r} 82 \\ 15\overline{)1230} \\ -1200 \\ \hline 30 \\ -30 \\ \hline 0 \end{array} \quad \text{Average}$$

The average score for the class was 82.

Now work margin exercise 10.

10. In a recent high school basketball game, two of the starters scored 18 points each, two scored 10 points, and one scored 4 points. What was the average number of points scored among the starters?

Margin Exercise Answers

1. 1001 visitors **2.** 322 pounds **3.** 120 pencils **4.** \$25,125 **5.** \$868 **6.** 28 square inches **7.** 21 **8. a.** Season 2 **b.** Season 5 **c.** 34 home runs **9.** 29,000 people **10.** 12 points

1.6 Exercises

Concept Check

Fill-in-the-Blank. Complete each sentence using information found in this section.

1. The words "product" or "double" indicate that the operation of __________ is involved.

2. To find the average of a set of numbers, the first step is to _______ the given numbers.

3. When solving word problems, the very first thing to do is to ________ the problem carefully.

1.6 Exercises

Concept Check

1. multiplication
2. add
3. read

4. addition
5. check, reasonable
6. subtraction
7. True
8. False; The sum is 316.
9. False; Quotient indicates division.
10. True

Applications

1. 1103 calories
2. $2372
3. 420 syringes
4. $839
5. 648 miles
6. 508 tickets
7. 64 people
8. 596 pages

4. The words "sum" and "increased by" are keywords to indicate the problem requires _________.

5. In solving word problems you should always ______ your work. Then make sure that the answer seems _________.

6. "Difference" and "less than" indicate that __________ will be involved in the problem.

True/False. Determine whether each statement is true or false. If a statement is false, explain how it can be changed so the statement will be true. (**Note:** There may be more than one acceptable change.)

7. Averages are found by performing addition and then division.

8. The sum of 312 and 4 is 1248.

9. The word "quotient" indicates multiplication.

10. After reading a problem carefully, the next step might be to make a diagram or draw a figure.

Applications

Solve.

1. Steven is calculating how many calories are in his lunch. He has a hamburger that has 354 calories, a medium fry that has 365 calories, and a chocolate milk shake that has 384 calories. How many total calories is his meal?

2. Shawna is refurnishing her living room. She buys a sofa and loveseat set for $1549, a coffee table for $245, an end table for $99, and a rug for $479. How much did she pay for all of the items to redecorate her living room?

3. Meghan is taking inventory of syringes at a clinic. She finds 240 in the first supply closet, 115 in the second supply closet, and 65 in the third supply closet. How many syringes does the clinic have in these three closets?

4. Omar is filling out tax forms and needs to total his charitable donations. During the year, he donated $234 to Goodwill, $345 to the Red Cross, and $260 to Child's Play. What is the total amount of donations that Omar will report on his taxes?

5. Marcus is moving from Cleveland, Ohio, to Houston, Texas, which is a 1299 mile drive. In one day, he drives 651 miles. How many more miles must he drive to reach Houston, Texas?

6. Sahar is at an arcade and wins 3483 tickets. She spends 2975 of the tickets on prizes from the redemption area. How many tickets does Sahar have left?

7. Tommy and Liz invited 250 people to their wedding. So far, 186 people have returned their RSVP. How many people have not sent in their RSVP?

8. Kyle is reading a book that has 978 pages. He has read 382 pages so far. How many pages does Kyle need to read to finish the book?

9. A secretary is making an inventory of office supplies. She finds 12 full boxes of blue pens which have 25 pens in each box. How many pens will she indicate on the inventory?

10. Sean has a part-time job as a data entry clerk. He earns $11 per hour and works for 22 hours each week. How much money does Sean make per week?

11. A jeweler makes a necklace that requires 128 beads. He receives an order for 6 necklaces. How many beads will he need to fulfill the order?

12. A juice bar sells fresh squeezed orange juice. To make an 8 ounce glass of orange juice, they use 4 oranges. The juice bar typically sells 86 glasses of orange juice per day. How many oranges do they typically use each day?

13. During a food drive, people donated 208 cans of baked beans. The cans of baked beans will be equally distributed to 8 food kitchens in the area. How many cans of baked beans will each food kitchen receive?

14. Isabella took out a 6 month loan for $2700 to cover car repairs. If she makes equal monthly payments, how much will she pay towards the loan each month?

15. Sixteen families want to raise money to build a playground on an empty lot for the neighborhood children. The total cost to buy the empty lot and build the playground is $18,800. How much money will each family need to raise if they all raise the same amount of money?

16. A homeless shelter is holding a fundraising event with a goal of $8250 to make repairs to their building. There are 11 companies participating in the fundraiser and each company plans to raise the same amount of money. How much money does each team need to raise to reach the goal?

17. Mark bought a new bicycle for $950. He also paid an additional $61 in sales tax. If he traded in his old bicycle for $320 and contributed that amount to the total for the new bicycle, how much does he owe?

18. To purchase a new dining room set for $1200, Mrs. Steel had to pay an additional $72 in sales tax. If she made an initial payment of $486, how much did she still owe?

19. Jorge bought a new phone for $250. The store had a deal that gave him $10 off. He also bought 3 video games for $48 each. What did he pay total for the phone and video games?

20. Vivian is shopping for groceries. She estimates that the total of the groceries will be $140. She has $8 worth of coupons to cash in. While in the checkout line, she also decides to purchase two magazines, which cost $3 each. How much will Vivian pay upon checkout?

21. To purchase a new refrigerator for $1200 including tax, Mr. Kline paid $240 down and the remainder in six equal monthly payments. What were his monthly payments?

22. To purchase a new 47-inch LED flat screen tv with a surround sound system that sells for $1300 including tax, Mr. Daley paid $200 down and the remainder in five equal monthly payments. How much were his monthly payments?

9. 300 pens
10. $242
11. 768 beads
12. 344 oranges
13. 26 cans
14. $450
15. $1175
16. $750
17. $691
18. $786
19. $384
20. $138
21. $160
22. $220

23. $255
24. $874
25. $85
26. $266
27. $150
28. $262
29. Blue car for $9112; $33 cheaper
30. Gym 1; $8 cheaper
31. 174 cm

23. For a class in statistics, Anthony bought a new graphing calculator for $95, special graphing paper for $8, a USB flash drive for $10, a textbook for $105, and a workbook for $37. How much did he spend for this class?

24. Lynn decided to take up surfing. She bought a new surfboard for $675, a wet suit for $130, a beach towel for $12, and a new swimsuit for $57. How much money did she spend? (Sales tax was included in the prices.)

25. Pat needed art supplies for a new course at the local community college. She bought a portfolio for $32, a zinc plate for $44, etching ink for $12, and three sheets of rag paper for a total of $6. She received a student discount of $9. How much did she spend on art supplies?

26. Michael is purchasing supplies for his upcoming classes. He buys a textbook for $180, a calculator for $90, a pack of pens for $5, and four notebooks for a total of $11. The bookstore is offering a discount of $20 off when students spend at least $200. How much did Michael's supplies cost?

27. Paula is training for a marathon and decided to buy some new clothes. She bought a pair of running shoes for $84, two pairs of socks for $5 a pair, one pair of shorts for $26, and two shirts for $15 each. If taxes are included in the prices, how much did she spend?

28. Miguel decided to go shopping for school clothes before college started in the fall. How much did he spend if he bought four pairs of pants for $21 per pair, five shirts for $18 each, three pairs of socks for $4 a pair, and two pairs of shoes for $38 a pair?

29. Alan wants to buy a new car. He could buy a red one for $8500 plus $510 in sales tax and $135 in fees, or he could buy a blue one for $8700 plus $522 in sales tax and $140 in fees. If the manufacturer is giving a $250 rebate on the blue model, which car would be cheaper for Alan? How much cheaper would it be?

30. Alexis is looking to join a gym with classes and personal training for one year. Gym 1 charges $90 to activate the membership, then charges $350 for the year for membership (including classes and personal training), and an additional $28 in taxes and fees. Gym 2 has no membership activation charge, but costs $380 for membership for the year, $75 for classes and personal training, and $21 in taxes and fees. Which gym membership will be cheaper for Alexis? How much cheaper will it be for the year-long membership?

31. A regular hexagon is a six-sided figure with all six sides equal and all six angles equal. Find the perimeter of a regular hexagon with one side of 29 centimeters.

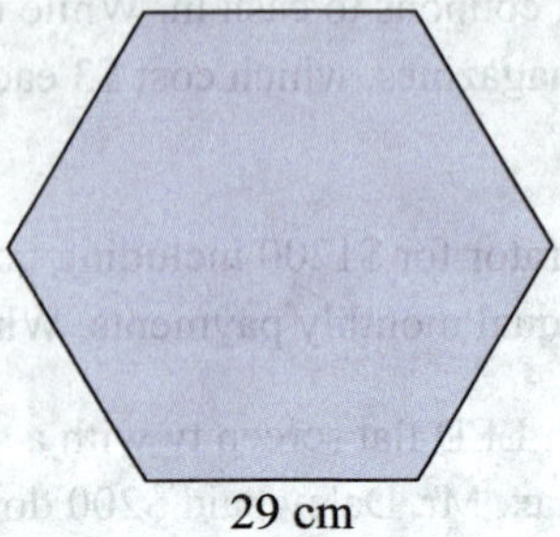

32. A regular octagon is an eight-sided figure with all eight sides equal and all eight angles equal. Find the perimeter of a regular hexagon with one side of 18 feet.

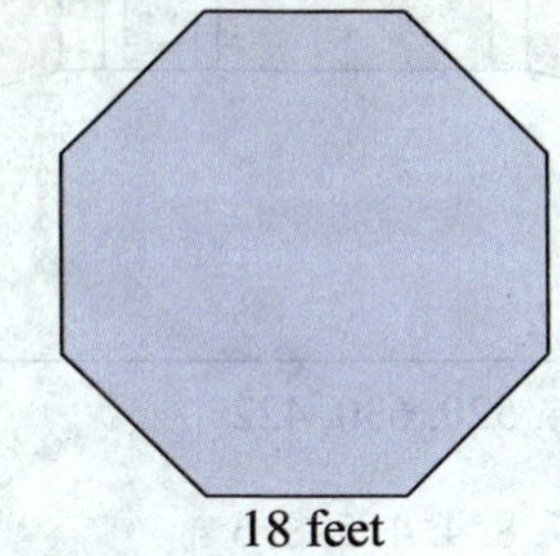

33. An isosceles triangle (two sides equal) is placed on one side of a square as shown in the figure. If each of the two equal sides of the triangle is 14 inches long and the square is 18 inches on each side, what is the perimeter of the figure?

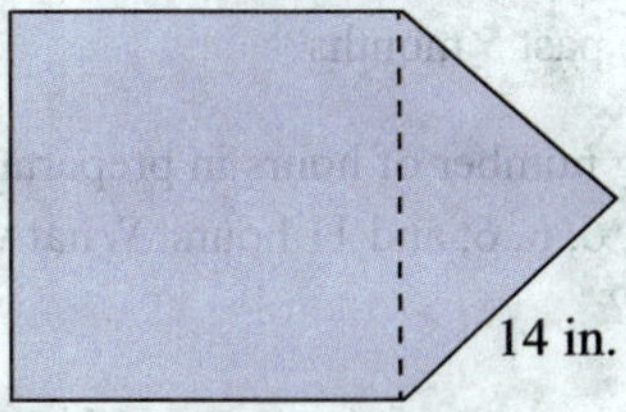

34. A scalene triangle (no sides equal) is placed on one side of a rectangle. The three sides of the triangle are 6 centimeters, 10 centimeters, and 14 centimeters, and the rectangle has a width of 6 centimeters and a length of 8 centimeters. Find the perimeter of the figure.

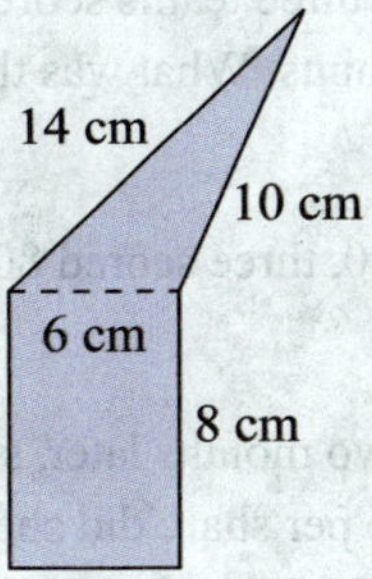

35. A square that is 10 inches on a side is placed inside a rectangle that has a width of 20 inches and a length of 24 inches. What is the area of the region inside the rectangle that surrounds the square? (Find the area of the shaded region in the figure.)

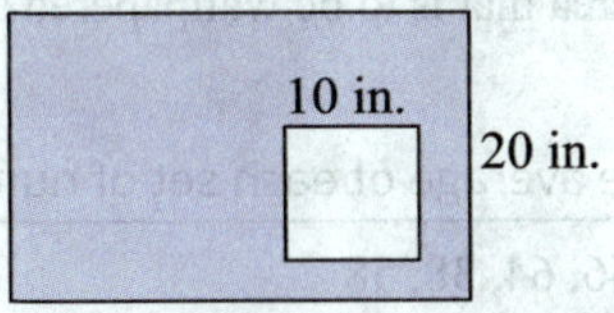

36. A rectangle that has a width of 6 inches and a height of 8 inches is placed inside of a square that has a side length of 15 inches. What is the area of the region inside the square that surrounds the rectangle? (Find the area of the shaded region in the figure.)

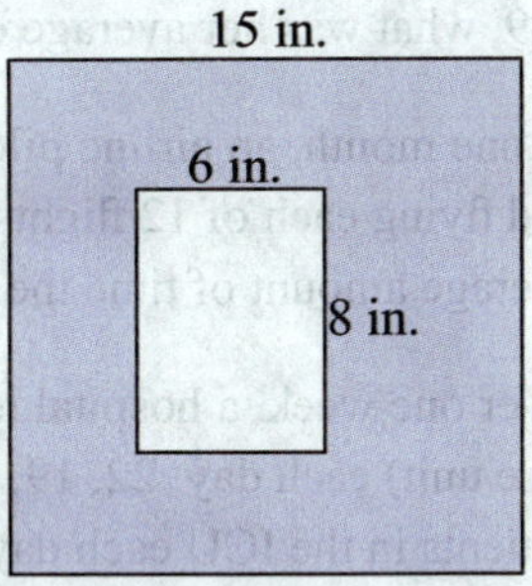

37. Except for the window space that is cut into the wall, one wall in a room is to be painted a light tan color. The wall is 8 feet by 15 feet and the window is 3 feet by 4 feet. There are two other walls that are 8 feet by 15 feet that are to be painted the same color. What is the total area of the three walls that are to be painted?

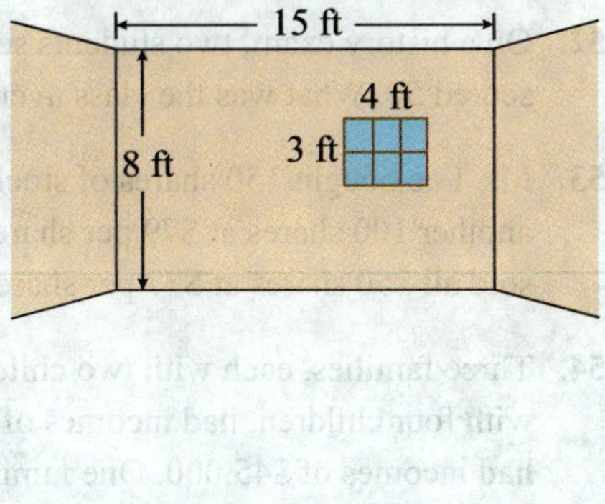

32. 144 ft
33. 82 in.
34. 46 cm
35. 380 sq in.
36. 177 sq in.
37. 348 sq ft

38. 299 sq ft
39. 54
40. 103
41. 485
42. 502
43. 6
44. 6
45. $321
46. $61
47. 7 hours
48. 20 patients
49. $125
50. $932
51. 21 points
52. 85
53. $76.60/share; $100 profit
54. $50,900

38. One wall of a room is 8 feet by 10 feet and a rectangular door 3 feet by 7 feet is to be cut in the wall. The remainder of the wall is to be wallpapered. There are three other walls in the room that are 8 feet by 10 feet and these are also to be wallpapered. What is the total area that is to be wallpapered?

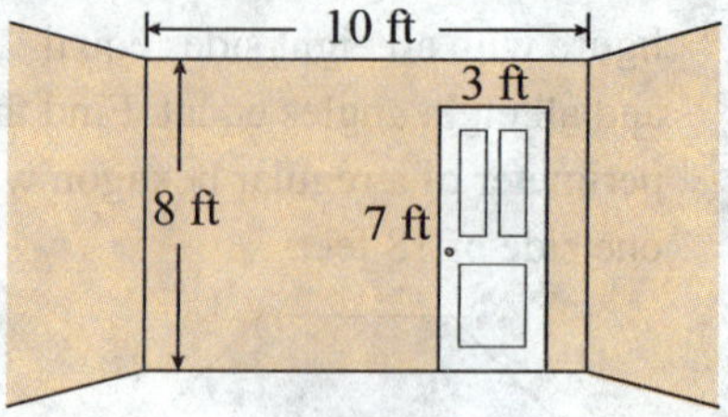

Find the average of each set of numbers. See Example 7.

39. 56, 64, 38, 58

40. 102, 113, 97, 100

41. 512, 618, 332, 478

42. 436, 520, 630, 422

43. 6, 7, 8, 4, 4, 5, 8, 6

44. 5, 4, 5, 6, 5, 8, 9, 6

Solve.

45. The Lee family spent the following amounts for groceries: $338 in June; $307 in July; $318 in August. What was the average amount they spent for groceries in these three months?

46. If Rina's cell phone bills for the past five months have been $56, $63, $52, $85, and $49, what was her average cell phone bill for the past 5 months?

47. In one month, an airline pilot spent the following number of hours in preparation for and flying each of 12 flights: 6, 8, 9, 6, 7, 7, 7, 5, 6, 6, 6, and 11 hours. What was the average amount of time the pilot spent per flight?

48. Over one week, a hospital had the following number of patients in the ICU (intensive care unit) each day: 22, 19, 23, 19, 17, 21, 19. What was the average number of patients in the ICU each day during the week?

49. Bill wanted to compare car insurance rates to find out the average amount he should pay for car insurance. If he looked at four different companies and the monthly rates were $164, $107, $131, and $98, what is the average monthly rate of car insurance?

50. A salesman sold items from his sales list for $972, $834, $1005, $1050, and $799. What was the average price per item?

51. During a sports trivia game, one team scored 35 points, three teams scored 23 points, two teams scored 18 points, and two teams scored 14 points. What was the average score of the teams?

52. On a history exam, two students scored 95, six scored 90, three scored 80, and one scored 50. What was the class average?

53. Ms. Lee bought 150 shares of stock at $75 per share. Two months later, she bought another 100 shares at $79 per share. What average price per share did she pay? If she sold all 250 shares at $77 per share, what was her profit?

54. Three families, each with two children, had incomes of $56,000. Two families, each with four children, had incomes of $62,000. Four families, each with two children, had incomes of $45,000. One family had no children and an income of $37,000. What was the average income per family?

55. Lacie owns a shoe store. During the first week of a month, Lacie sold 41 pairs of shoes and received two shipments, one with 26 pairs of shoes and one with 10 pairs of shoes. During the second week of the month, Lacie received a shipment of 20 pairs of shoes and sold 35 pairs of shoes. During the third week of the month, Lacie sold 38 pairs of shoes and received two shipments, one with 22 pairs of shoes and one with 19 pairs of shoes. During the final week of the month, Lacie received three shipments, one with 16 pairs of shoes, one with 24 pairs of shoes, and one with 14 pairs of shoes, and she sold 29 pairs of shoes.

a. What was the average weekly difference between Lacie's shipments and her sales?

b. What was Lacie's inventory at the end of the month if she began the month with an inventory of 246 pairs of shoes?

56. During July, Mr. Rodriguez made deposits in his checking account of $400 and $750 and wrote checks totaling $625. During August, his deposits were $632, $322, and $798, and his checks totaled $978. In September, his deposits were $520, $436, $200, and $376, and his checks totaled $836.

a. What was the average monthly difference between his deposits and his withdrawals?

b. What was his bank balance at the end of September if he had a balance of $500 on July 1?

57. The five longest rivers in the world are given in the following table. What is the average length of these rivers?

Five Longest Rivers in the World

River	Length (miles)
Nile	4132
Amazon	3980
Yangtze	3917
Mississippi/Missouri	3902
Yenisei	3434

58. The 10 largest cities in South Carolina have the following approximate populations. What is the average population of these cities?

10 Largest Cities in South Carolina

City	Population
Charleston	141,931
Columbia	128,386
North Charleston	122,297
Mount Pleasant	97,129
Rock Hill	76,818
Greenville	75,993
Summerville	54,993
Goose Creek	46,194
Hilton Head Island	39,930
Sumter	39,768

Source: "Cities in South Carolina by Population (2022)", World Population Review, Accessed 5 April 2022, worldpopulationreview.com.

55. **a.** 12 pairs of shoes **b.** 254 pairs of shoes

56. **a.** $665 **b.** $2495

57. 3873 miles

58. 82,344 people

Writing & Thinking

59. **1.** Read the problem carefully. **2.** Draw any type of figure or diagram that might be helpful and decide what operations are needed. **3.** Perform the operations to solve the problem. **4.** Check your work.

60. Answers will vary.

61. Answers will vary. Averages may be used when looking at attendance, money made per month, commute time, etc.

62. Answers will vary.

Writing & Thinking

59. State the basic strategy for solving word problems.

60. Make up three word problems that include keywords to indicate operations such as addition, subtraction, multiplication, and division. Underline the keywords.

61. Give an example where you might use average (other than in a class).

62. Discuss how you used mathematics to solve some problem in your life this week. (You know you did!)

1.7 Exponents and Order of Operations

Objectives

- **A.** Identify the base and exponent in an exponential expression.
- **B.** Evaluate expressions with exponents.
- **C.** Evaluate expressions with 1 and 0 as exponents.
- **D.** Use the order of operations to simplify expressions containing whole numbers.

A The Terms Base and Exponent

Repeated multiplication by the same number can be shortened by using **exponents**. For example, if 3 is used as a factor five times, we can write

5 is the exponent

$$3 \cdot 3 \cdot 3 \cdot 3 \cdot 3 = 3^5 = 243.$$

3 is the base   243 is the product

When looking at $3^5 = 243$, 3 is the **base**, 5 is the **exponent**, and 243 is the **product**. Exponents are written slightly to the right and above the base. The expression 3^5 is an **exponential expression**.

Example 1 Identifying the Base and Exponent

Identify the base and exponent in each exponential expression.

a. 6^2 **b.** 10^4

Solution

a. 6 is the base, and 2 is the exponent. **b.** 10 is the base, and 4 is the exponent.

Now work margin exercise 1.

1. Identify the base and exponent in each exponential expression.
   a. 8^3
   b. 14^6

In expressions with exponent 2, the base is said to be **squared**. In expressions with exponent 3, the base is said to be **cubed**. With other exponents, the base is said to be **"to the _____ power."**

Note

There is usually some confusion about the use of the word "power." Since 2^5 is read "two to the fifth power," it is natural to think of 5 as the power. This is not true. A power is not an exponent. A power is the product indicated by a base raised to an exponent. Thus, for the equation $2^5 = 32$, think of the phrase "two to the fifth power" in its entirety. The corresponding power is the product, 32.

Example 2 Writing Expressions using Exponents

In order to illustrate exponential notation, we show several products written with repeated multiplication and the equivalent exponential expressions.

	With Repeated Multiplication	With Exponents	
a.	$7 \cdot 7 = 49$	$7^2 = 49$	7^2 is read "7 to the second power" or "7 squared."
b.	$3 \cdot 3 = 9$	$3^2 = 9$	3^2 is read "3 to the second power" or "3 squared."
c.	$2 \cdot 2 \cdot 2 = 8$	$2^3 = 8$	2^3 is read "2 to the third power" or "2 cubed."
d.	$10 \cdot 10 \cdot 10 \cdot 10 = 10{,}000$	$10^4 = 10{,}000$	10^4 is read "10 to the fourth power."

Now work margin exercise 2.

2. Rewrite each expression using exponents.
   a. $100 \cdot 100 \cdot 100 = 1{,}000{,}000$
   b. $2 \cdot 2 \cdot 2 \cdot 2 = 16$
   c. $1 \cdot 1 \cdot 1 = 1$
   d. $8 \cdot 8 = 64$

B Evaluating Expressions with Exponents

3. Evaluate each exponential expression.
 a. 9^2
 b. 2^5
 c. 5^4

Example 3 Evaluating Exponential Expressions

Evaluate each exponential expression.

a. 6^3 **b.** 5^2 **c.** 2^6

Solution

a. $6^3 = 6 \cdot 6 \cdot 6 = 216$

b. $5^2 = 5 \cdot 5 = 25$

c. $2^6 = 2 \cdot 2 \cdot 2 \cdot 2 \cdot 2 \cdot 2$
$= 64$

Now work margin exercise 3.

Common Error

Wrong Solution

DO NOT multiply the base times the exponent.

$10^2 = 10 \cdot 2$ WRONG

$6^4 = 6 \cdot 4$ WRONG

Correct Solution

DO multiply the base times itself.

$10^2 = 10 \cdot 10$ CORRECT

$6^4 = 6 \cdot 6 \cdot 6 \cdot 6$ CORRECT

CAUTION

C The Exponents 1 and 0

If there is no exponent written with a number, then the exponent is understood to be 1. That is, any number raised to the first power is equal to itself.

The Exponent 1

Any number raised to the first power is equal to itself.

For example, $8^1 = 8$, $6^1 = 6$, and $193^1 = 193$.

DEFINITION

When the exponent 0 is used for any base except 0, the value of the expression is defined to be 1.

The Exponent 0

Any nonzero number raised to the 0 power equals 1.

For example, $2^0 = 1$, $5^0 = 1$, and $46^0 = 1$.

Note: The expression 0^0 is undefined.

DEFINITION

Note

To help in understanding the exponent 0, note the pattern in the following powers with base 3.

$3^2 = 9$, $3^1 = 3$, and $3^0 = 1$

Each value is the previous value divided by 3. In this way defining $3^0 = 1$ makes sense. (Try this idea with bases other than 3.)

Example 4 Evaluating Exponential Expressions

Evaluate each exponential expression.

a. 9^1 **b.** 8^0 **c.** 15^0 **d.** 10^1

Solution

a. $9^1 = 9$ **b.** $8^0 = 1$ **c.** $15^0 = 1$ **d.** $10^1 = 10$

Now work margin exercise 4.

4. Evaluate each exponential expression.

a. 6^0

b. 1^1

c. 813^1

d. 54^0

Teaching Note

Some of your students may question values with the exponent 0. In this case you might show more patterns with various powers of a base as follows: $2^4 = 16$, $2^3 = 8$, $2^2 = 4$, $2^1 = 2$, and $2^0 = 1$, where each value is the previous one divided by 2.

CALCULATORS

Using a Calculator to Evaluate Exponential Expressions

There are two possible ways to evaluate an exponential expression on a calculator. If the exponent is a 2, one option is to use the [x^2] key. For all other exponents (and 2 if you prefer), use the [^] or [x^y] key. For example, to evaluate the expression 11^2, press the keys

[1] [1] [^] [2] [=] OR [1] [1] [x^2] [=].

Using either method, the display will read 121.

D Order of Operations with Whole Numbers

Mathematicians have agreed on a set of rules for the order of operations in simplifying (or evaluating) any numerical expression involving addition, subtraction, multiplication, division, and exponents. These rules are used in all branches of mathematics and computer science to ensure that there is only one correct answer, regardless of how complicated an expression might be. For example, the expression $5 \cdot 2^3 - 14 \div 2$ might be simplified as follows.

$$
\begin{aligned}
5 \cdot 2^3 - 14 \div 2 &= 5 \cdot 8 - 14 \div 2 \\
&= 40 - 14 \div 2 \\
&= 40 - 7 \\
&= 33
\end{aligned}
$$

Correct

However, if we were simply to proceed from left to right as in the following steps, we would get an entirely different answer.

$$
\begin{aligned}
5 \cdot 2^3 - 14 \div 2 &= 10^3 - 14 \div 2 \\
&= 1000 - 14 \div 2 \\
&= 986 \div 2 \\
&= 493
\end{aligned}
$$

Incorrect

By the standard set of rules for order of operations, we can conclude that **the first answer (33) is the accepted correct answer** and that **the second answer (493) is incorrect**.

Teaching Note

Students will find the exercises both challenging and frustrating. These frustrations generally arise from a limited ability to follow directions and pay attention to details. Common errors occur in evaluating expressions such as $3 + 2(7 + 5)$ and $20 - 15 \div 3 + 2$, where the students will add $3 + 2$ first because it "looks right." You can probably help them a great deal by continually reminding them that, outside parentheses, addition and subtraction are the last operations to be performed.

Rules for Order of Operations

1. Simplify within grouping symbols, such as parentheses ( ), brackets [ ], or braces { }. (If there are more than one pair of grouping symbols, start with the innermost grouping symbols.)
2. Evaluate any exponential expressions.
3. Moving from **left to right**, perform any multiplication or division in the order in which it appears.
4. Moving from **left to right**, perform any addition or subtraction in the order in which it appears.

PROCEDURE

Note

Note that in Rule 3 neither multiplication nor division has priority over the other. Whichever of these operations occurs first, moving **left to right**, is done first. In Rule 4, addition and subtraction are handled in the same way. Unless they occur within grouping symbols, **addition and subtraction are the last operations to be performed.**

Teaching Note

You might want to show the mnemonic in a vertical format as

P
E
M/D
A/S

to reinforce the relationship between multiplication and division, and addition and subtraction.

A well-known mnemonic device for remembering the rules for order of operations is the following.

Please	Excuse	My	Dear	Aunt	Sally
↓	↓	↓	↓	↓	↓
Parentheses	Exponents	Multiplication	Division	Addition	Subtraction

Even though the mnemonic PEMDAS is helpful, remember that multiplication and division are performed as they appear, left to right; the same is true for addition and subtraction.

The following examples show how to apply these rules. In some cases, more than one step can be performed at the same time. (This is possible when parts are separated by + or − signs or are within separate sets of grouping symbols, as illustrated in Examples 6–9). **Work through each example step-by-step, and rewrite the examples on a separate sheet of paper.**

5. Simplify:

$6 \cdot 2 - 4 \cdot 3 + 8$

Example 5 Using the Order of Operations with Whole Numbers

Simplify: $14 \div 7 + 3 \cdot 2 - 5$

Solution

In this example there are no grouping symbols or exponents, so we begin with multiplication and division (left to right).

$$\underbrace{14 \div 7} + 3 \cdot 2 - 5$$

$= 2 + \underbrace{3 \cdot 2} - 5$ Divide before multiplying in this case.

$= \underbrace{2 + 6} - 5$ Multiply before adding or subtracting.

$= \underbrace{8 - 5}$ Add before subtracting in this case.

$= 3$ Subtract.

Now work margin exercise 5.

Example 6 Using the Order of Operations with Whole Numbers

Simplify: $2 \cdot 3^2 + 18 \div 3^2$

Solution

$$\begin{aligned} & 2 \cdot 3^2 + 18 \div 3^2 & \\ &= 2 \cdot 9 + 18 \div 9 & \text{Evaluate the exponential expressions.} \\ &= 18 + 2 & \text{Multiply and divide.} \\ &= 20 & \text{Add.} \end{aligned}$$

Now work margin exercise 6.

6. Simplify:
$6^2 \div 9 + 3 - 14 \div 7$

Teaching Note

You may want to point out to your students that + and – signs sometimes separate expressions and these expressions (called terms in algebra) can be evaluated at the same time. The + sign is used in this way in Example 6.

Example 7 Using the Order of Operations with Whole Numbers

Simplify: $(6+2)+(8+1)\div 9$

Solution

$$\begin{aligned} & (6+2)+(8+1)\div 9 & \\ &= 8 + 9 \div 9 & \text{Operate within parentheses.} \\ &= 8 + 1 & \text{Divide.} \\ &= 9 & \text{Add.} \end{aligned}$$

Now work margin exercise 7.

7. Simplify:
$(9+3)\div 4+(9-1)$

Example 8 Using the Order of Operations with Whole Numbers

Simplify: $30 \div 10 \cdot 2^3 + 3(6-2)$

Solution

$$\begin{aligned} & 30 \div 10 \cdot 2^3 + 3(6-2) & \\ &= 30 \div 10 \cdot 2^3 + 3(4) & \text{Operate within parentheses.} \\ &= 30 \div 10 \cdot 8 + 3(4) & \text{Evaluate the exponential expression.} \\ &= 3 \cdot 8 + 3(4) & \text{Divide.} \\ &= 24 + 12 & \text{Multiply each part.} \\ & 36 & \text{Add.} \end{aligned}$$

Now work margin exercise 8.

8. Simplify:
$2(5^2-15)-12+3\cdot 2^3$

Example 9 Using the Order of Operations with Whole Numbers

Simplify: $2\left[5^2+\left(2\cdot 3^2-10\right)\right]$

9. Simplify:
$4\cdot 3-\left[\left(2+3^2\right)-10\right]$

Solution

$$2[5^2 + (2 \cdot 3^2 - 10)]$$

$= 2[25 + (2 \cdot 9 - 10)]$ Evaluate the exponential expressions.

$= 2[25 + (18 - 10)]$ Multiply inside the parentheses.

$= 2(25 + 8)$ Subtract inside the parentheses.

$= 2(33)$ Add inside the parentheses.

$= 66$ Multiply.

Now work margin exercise 9.

10. Simplify:

$6^2 - 4(3^2 - 2) - 5 + 4 \cdot 2^2$

Completion Example 10 Using the Order of Operations

Simplify: $3(2+2^2)-6-3\cdot 2^2$

Solution

$$3(2 + 2^2) - 6 - 3 \cdot 2^2$$

$= 3(2 + \underline{4}) - 6 - 3 \cdot \underline{4}$ Evaluate the exponential expressions.

$= 3(\underline{6}) - 6 - 3 \cdot \underline{4}$ Add inside the parentheses.

$= \underline{18} - 6 - \underline{12}$ Multiply each part.

$= \underline{12} - \underline{12}$ Subtract.

$= \underline{0}$ Subtract.

Now work margin exercise 10.

11. Simplify:

$[20 + 15(2) \div 3 - 18](2^2 - 1)$

Teaching Note

Now might be a good time to inform the students that while calculators definitely increase the speed and accuracy of many calculations, calculators are not a substitute for understanding concepts such as order of operations. If they rely solely on their calculators, particularly without using parentheses, they will incorrectly evaluate expressions such as $(8 + 6) \div 2$ and $(8 + 6)^2$. Also, not **all** calculators follow the rules for the order of operations that are consistent with the rules used in arithmetic and algebra.

Completion Example 11 Using the Order of Operations

Simplify: $(14-10)[(5+3^2)\div 2+5]$

Solution

$$(14 - 10)[(5 + 3^2) \div 2 + 5]$$

$= (14 - 10)[(5 + \underline{9}) \div 2 + 5]$ Evaluate the exponential expression.

$= (14 - 10)[(\underline{14}) \div 2 + 5]$ Add inside the innermost parentheses.

$= \underline{4}(\underline{7} + 5)$ Subtract inside the parentheses and divide inside the brackets.

$= \underline{4}(\underline{12})$ Add inside the parentheses.

$= \underline{48}$ Multiply.

Now work margin exercise 11.

CALCULATORS

Calculators and the Order of Operations

Some calculators are programmed to follow the order of operations. To test if your calculator is programmed in this manner, evaluate $7 + 4 \cdot 2$. Press the keys [7] [+] [4] [×] [2], and then [=]. If your calculator displays 15, then it followed the order of operations. This means that most of the time, you can type in the problem as it is written and the calculator will provide the correct answer. If the calculator displays 22, it did not follow the rules for order of operations and you will need to be extra careful when evaluating expressions. You will need to break down the expression into smaller steps or add parentheses to force the calculator to perform the operations in the correct order.

To evaluate an expression with parentheses, you will use the parentheses buttons, [(] and [)]. To evaluate the expression $(6+2)+(8+1)\div 9$ (assuming your calculator follows the order of operations), press the keys

[(] [6] [+] [2] [)] [+] [(] [8] [+] [1] [)] [÷] [9].

Then press [=]. The display will read 9.

Completion Example Answers

10. $3(2+4)-6-3\cdot 4=3(6)-6-3\cdot 4=18-6-12=12-12=0$

11. $(14-10)[(5+9)\div 2+5]=(14-10)[(14)\div 2+5]=(4)(7+5)=(4)(12)=48$

Margin Exercise Answers

1. a. base: 8; exponent: 3 **b.** base: 14; exponent: 6 **2. a.** $100^3=1{,}000{,}000$ **b.** $2^4=16$ **c.** $1^3=1$ **d.** $8^2=64$ **3. a.** 81 **b.** 32 **c.** 625 **4. a.** 1 **b.** 1 **c.** 813 **d.** 1 **5.** 8 **6.** 5 **7.** 11 **8.** 32 **9.** 11 **10.** 19 **11.** 36

1.7 Exercises

Concept Check

Fill-in-the-Blank. Complete each sentence using information found in this section.

1. When an expression has an exponent of 3, the base is said to be ______.
2. Exponents are used to represent repeated ______.
3. In 2^4 the "2" is called the ______ and the "4" is called the ______.
4. 10 squared is equal to ______.
5. If any nonzero number has an exponent of 0, the value is always ______.
6. If there are multiple grouping symbols to be simplified, begin with the ______ group.

True/False. Determine whether each statement is true or false. If a statement is false, explain how it can be changed so the statement will be true. (**Note:** There may be more than one acceptable change.)

7. Nine squared is equal to eighteen.
8. $2^7=128$
9. 7^0 is undefined.
10. According to the order of operations, multiplication is always performed before division.

1.7 Exercises

Concept Check

1. Cubed
2. Multiplication
3. Base, exponent
4. 100
5. 1
6. Innermost
7. False; Equals 81
8. True
9. False; 7^0 is 1.
10. False; Multiplication and division are performed at the same time, from left to right.

Practice

1. 11^3
2. 13^3
3. 7^4
4. 6^5
5. $2^3 \cdot 3^2$
6. $2^2 \cdot 5^3$
7. $5^3 \cdot 7^2$
8. $3^3 \cdot 7^3$
9. $2 \cdot 3^2 \cdot 11^2$
10. $2^4 \cdot 11^2 \cdot 13^2$
11. a. 4 b. 2 c. 16
12. a. 6 b. 2 c. 36
13. a. 2 b. 3 c. 8
14. a. 3 b. 3 c. 27
15. a. 1 b. 6 c. 1
16. a. 1 b. 5 c. 1
17. a. 5 b. 3 c. 125
18. a. 4 b. 3 c. 64
19. a. 2 b. 4 c. 16
20. a. 2 b. 6 c. 64
21. a. 9 b. 2 c. 81
22. a. 11 b. 2 c. 121
23. a. 7 b. 2 c. 49
24. a. 7 b. 3 c. 343
25. a. 3 b. 5 c. 243
26. a. 4 b. 5 c. 1024
27. a. 30 b. 2 c. 900
28. a. 40 b. 2 c. 1600
29. a. 20 b. 3 c. 8000
30. a. 15 b. 2 c. 225
31. a. 1 b. 57 c. 1
32. a. 1 b. 99 c. 1
33. a. 4 b. 0 c. 1
34. a. 19 b. 0 c. 1
35. a. 13 b. 1 c. 13
36. a. 24 b. 1 c. 24
37. a. 22 b. 0 c. 1
38. a. 99 b. 0 c. 1
39. 21
40. 28

Practice

Rewrite each product by using exponents. See Example 2.

1. $11 \cdot 11 \cdot 11$
2. $13 \cdot 13 \cdot 13$
3. $7 \cdot 7 \cdot 7 \cdot 7$
4. $6 \cdot 6 \cdot 6 \cdot 6 \cdot 6$
5. $2 \cdot 2 \cdot 2 \cdot 3 \cdot 3$
6. $2 \cdot 2 \cdot 5 \cdot 5 \cdot 5$
7. $5 \cdot 5 \cdot 5 \cdot 7 \cdot 7$
8. $3 \cdot 3 \cdot 3 \cdot 7 \cdot 7 \cdot 7$
9. $2 \cdot 3 \cdot 3 \cdot 11 \cdot 11$
10. $2 \cdot 2 \cdot 2 \cdot 2 \; 11 \cdot 11 \cdot 13 \cdot 13$

For each exponential expression **a.** identify the base, **b.** identify the exponent, and **c.** evaluate the exponential expression. See Examples 1 through 4.

11. 4^2
12. 6^2
13. 2^3
14. 3^3
15. 1^6
16. 1^5
17. 5^3
18. 4^3
19. 2^4
20. 2^6
21. 9^2
22. 11^2
23. 7^2
24. 7^3
25. 3^5
26. 4^5
27. 30^2
28. 40^2
29. 20^3
30. 15^2
31. 1^{57}
32. 1^{99}
33. 4^0
34. 19^0
35. 13^1
36. 24^1
37. 22^0
38. 99^0

Simplify. See Examples 5 through 11.

39. $6 + 5 \cdot 3$
40. $18 + 2 \cdot 5$
41. $20 - 4 \div 4$
42. $6 - 15 \div 3$
43. $32 - 14 + 10$
44. $25 - 10 + 11$
45. $15 \div 15 + 10 \cdot 2$
46. $12 - 3 \cdot 4 \div 2$
47. $18 \div 2 - 1 - 3 \cdot 2$
48. $2 + 3 \cdot 7 - 10 \div 2$
49. $6 \cdot 2 \div 3 - 5 + 13$
50. $12 - 4 \cdot 3 \div 3 + 1$
51. $14 \cdot 2 \div 7 \div 2 + 10$
52. $3 \cdot 8 \div 4 + 1 \cdot 2$

53. $19-5(3-1)$

54. $(7-2)\cdot 6-10$

55. $(2+3\cdot 4)\div 7-2$

56. $(2+3)\cdot 4\div 5-4$

57. $15(2+3)-65-5$

58. $13(10-7)-20-19$

59. $2\cdot 5^2-8\div 2$

60. $4^2\div 2+9\div 3$

61. $4\div 2^2+3\cdot 2^2$

62. $16\div 2^4+9\div 3^2$

63. $(4+3)^2+(2+3)^2$

64. $(2+1)^2+(4+1)^2$

65. $(2^3+2)\div 5+7^2\div 7$

66. $30\div 2-11+2(5-1)^3$

67. $(10+1)\left[(5-2)^2+3(4-3)\right]$

68. $(12-2)\left[4(6-3)+(4-3)^2\right]$

69. $75+3\left[2(3+6)^2-10^2\right]$

70. $100+2\left[3(4^2-6)+2^3\right]$

71. $16+3\left[17+2^3\div 2^2-4\right]$

72. $21-(15-4^3\div 2^4+1)\div 3$

73. $10^3-2\left[(13+3)\div 2^4+18\div 3^2\right]$

74. $5^2\left[6^2\div 9\cdot(2+3)\div 2^2\right]-10^2$

Applications

Solve.

75. Neville bought 15 boxes of trading cards. Each box has 10 packs of trading cards. Each pack of trading cards contains 20 cards. He adds 132 cards that he already owns to the newly purchased cards. Then, Neville evenly distributes all of the cards to 6 of his friends. How many trading cards would each person get?

a. If you simplify the expression $15\cdot 10\cdot 20+132\div 6$ using the order of operations, will you get the correct answer? If not, explain what is wrong with the expression.

b. What is the answer? If necessary, write the corrected expression to get the correct results when following the order of operations.

76. Robert is purchasing shirts for his weekend soccer team. The shirts he wants to buy are normally \$25 each but are on sale for \$10 off. His team has a total of 11 players. How much will he spend to buy the shirts?

a. If you simplify the expression $\$25-\$10\cdot 11$ using the order of operations, will you get the correct answer? If not, explain what is wrong with the expression.

b. What is the answer? If necessary, write the corrected expression to get the correct results when following the order of operations.

77. Camila is a seamstress and is creating wedding dresses. She has 126 yards of silk fabric. For each dress, the skirt requires 4 yards of silk and the bodice requires 2 yards of silk. How many dresses can she make with the amount of silk she has?

a. If you simplify the expression $126\div 4+2$ using the order of operations, will you get the correct answer? If not, explain what is wrong with the expression.

b. What is the answer? If necessary, write the corrected expression to get the correct results when following the order of operations.

41. 19
42. 1
43. 28
44. 26
45. 21
46. 6
47. 2
48. 18
49. 12
50. 9
51. 12
52. 8
53. 9
54. 20
55. 0
56. 0
57. 5
58. 0
59. 46
60. 11
61. 13
62. 2
63. 74
64. 34
65. 9
66. 132
67. 132
68. 130
69. 261
70. 176
71. 61
72. 17
73. 994
74. 25

Applications

75. **a.** No. Here it shows that we are only dividing the old trading cards by 6 friends versus both the old and new trading cards by 6 friends.

b. 522; $\dfrac{15\cdot 10\cdot 20+132}{6}$

Writing & Thinking

78. Use your calculator to find the following values and discuss, in your own words, any pattern that you notice.

a. 86^0

b. 623^0

c. 9072^0

79. Give one example where addition should be completed before multiplication.

80. Explain how someone might think that $1 + 3^2 = 16$. Then, explain why this would not be correct.

Collaborative Learning

81. In groups of three to four students, use a calculator to evaluate 20^{10} and 10^{20}. Discuss what you think is the meaning of the notation on the display.

(**Note:** The notation is a form of a notation called scientific notation and is discussed in detail in Chapter 12. Different calculators may use slightly different forms.)

---

76. **a.** No. The formula incorrectly shows that each shirt is \$10, and the total cost would be $\$25 - \$10 \cdot 11 = \$25 - \$110 = -\$85$.

b. \$165; $(\$25 - \$10) \cdot 11$

77. **a.** No. The formula shows that we are using $\frac{126}{4}$ yards of fabric to make only skirts, not full dresses. Adding 2 yards of silk after will only be enough for 1 bodice.

b. 20; $\frac{126}{4+2}$

Writing & Thinking

78. **a.** 1 **b.** 1

c. 1; So far, it appears that any number raised to the 0th power is equal to 1.

79. If addition is within parentheses (or other grouping symbols), addition would be performed first.

80. Adding 1 + 3 first, then squaring, would give 16. But the order of operations indicates that 3 should be squared first and then 1 added. This gives a result of 10.

Collaborative Learning

81. 1.024E13, 1E20; Answers will vary.

1.8 Tests for Divisibility

Objectives

A. Check for divisibility by using the tests for 2, 3, 4, 5, 6, 9, and 10.

A Tests for Divisibility

When working with fractions (as we will be shortly), being able to divide quickly and easily by small numbers is a valuable skill. We will be looking for **factors**, so we will want to know if a whole number is **divisible by** a number before actually dividing.

Divisibility

If a number can be divided by another number so that the remainder is 0, then we say

1. the number is **divisible by** the divisor, or

2. the divisor **divides** the dividend.

DEFINITION

Math Tip

Recall that factor · factor = product.

In the multiplication problem $3 \cdot 7 = 21$, the factors are 3 and 7 and the product is 21.

For example, $46 \div 2 = 23$, so 46 is divisible by 2. Or, 2 divides 46.

There are simple tests that can be performed mentally to determine whether a number is divisible by 2, 3, 4, 5, 6, 9, or 10 **without actually dividing**.

Divisibility by 2

A number is divisible by 2 (is an **even number**) if the ones digit is 0, 2, 4, 6, or 8. (**Note:** A number is an **odd number** if it is not divisible by 2.)

DEFINITION

Example 1 Determining Divisibility by 2

Determine whether each of the following numbers is divisible by 2.

a. 674 **b.** 357

Solution

a. 674 is divisible by 2 since the ones digit is 4 (an even digit).

b. 357 is not divisible by 2 since the ones digit is not 0, 2, 4, 6, or 8.

Now work margin exercise 1.

1. Is 548 divisible by 2? Explain why or why not.

Divisibility by 3

A number is divisible by 3 if the sum of the digits is divisible by 3.

DEFINITION

2. Is 7912 divisible by 3? Explain why or why not.

Example 2 Determining Divisibility by 3

Determine whether each of the following numbers is divisible by 3.

a. 6801 **b.** 356

Solution

a. 6801 is divisible by 3 since $6 + 8 + 0 + 1 = 15$, and 15 is divisible by 3.

b. 356 is not divisible by 3 since $3 + 5 + 6 = 14$, and 14 is not divisible by 3.

Now work margin exercise 2.

Divisibility by 4

A number is divisible by 4 if the number formed by the last two digits is divisible by 4. (00 is considered to be divisible by 4.)

DEFINITION

3. Is 2476 divisible by 4? Explain why or why not.

Example 3 Determining Divisibility by 4

Determine whether each of the following numbers is divisible by 4.

a. 9036 **b.** 6700 **c.** 15,031

Solution

a. 9036 is divisible by 4 since 36 (the number formed by the last two digits) is divisible by 4.

b. 6700 is divisible by 4 since 00 is considered to be divisible by 4.

c. 15,031 is not divisible by 4 since 31 is not divisible by 4.

Now work margin exercise 3.

Divisibility by 5

A number is divisible by 5 if the ones digit is 0 or 5.

DEFINITION

4. Is 6827 divisible by 5? Explain why or why not.

Example 4 Determining Divisibility by 5

Determine whether each of the following numbers is divisible by 5.

a. 1365 **b.** 970 **c.** 1863

Solution

a. 1365 is divisible by 5 since the ones digit is 5.

b. 970 is divisible by 5 since the ones digit is 0.

c. 1863 is not divisible by 5 since the ones digit is not 0 or 5.

Now work margin exercise 4.

Divisibility by 6

A number is divisible by 6 if it is divisible by both 2 and 3.

DEFINITION

Example 5 Determining Divisibility by 6

Determine whether each of the following numbers is divisible by 6.

a. 9054 **b.** 17,000

Solution

a. 9054 is divisible by 2 since the ones digit is 4. 9054 is divisible by 3 since $9 + 0 + 5 + 4 = 18$, and 18 is divisible by 3. Therefore, 9054 is divisible by 6.

b. 17,000 is divisible by 2 since the ones digit is 0. 17,000 is not divisible by 3 since $1 + 7 + 0 + 0 + 0 = 8$, and 8 is not divisible by 3. Therefore, 17,000 is not divisible by 6.

Now work margin exercise 5.

5. Is 1576 divisible by 6? Explain why or why not.

Divisibility by 9

A number is divisible by 9 if the sum of the digits is divisible by 9.

DEFINITION

Example 6 Determining Divisibility by 9

Determine whether each of the following numbers is divisible by 9.

a. 2530 **b.** 873

Solution

a. 2530 is not divisible by 9 since $2 + 5 + 3 + 0 = 10$, and 10 is not divisible by 9.

b. 873 is divisible by 9 since $8 + 7 + 3 = 18$, and 18 is divisible by 9.

Now work margin exercise 6.

6. Is 4653 divisible by 9? Explain why or why not.

Divisibility by 10

A number is divisible by 10 if the ones digit is 0.

DEFINITION

Example 7 Determining Divisibility by 10

Determine whether each of the following numbers is divisible by 10.

a. 12,530 **b.** 841

7. Is 8510 divisible by 10? Explain why or why not.

Solution

a. 12,530 is divisible by 10 since the ones digit is 0.

b. 841 is not divisible by 10 since the ones digit is not 0.

Now work margin exercise 7.

8. Is 612 divisible by 9? Explain why or why not.

Completion Example 8 Using the Divisibility Rules

Complete each sentence.

a. 250 is divisible by 10 since the ones digit is 0.

b. 5712 is divisible by 4 since the number formed by the last 2 digits (12) is divisible by 4.

c. 5402 is not divisible by 3 since $5 + 4 + 0 + 2 = 11$, and 11 is not divisible by 3.

d. 6036 is divisible by 6 since 6036 is divisible by both 2 and 3. (It is divisible by 2 since the ones digit is 6 and it is divisible by 3 since $6 + 0 + 3 + 6 = 15$, and 15 is divisible by 3.)

Now work margin exercise 8.

Completion Example Answer

8. a. the ones digit is 0 **b.** the number formed by the last 2 digits (12) is divisible by 4 **c.** $5 + 4 + 0 + 2 = 11$, and 11 is not divisible by 3 **d.** 6036 is divisible by both 2 and 3 (It is divisible by 2 since the ones digit is 6 and it is divisible by 3 since $6 + 0 + 3 + 6 = 15$, and 15 is divisible by 3.)

Margin Exercise Answers

1. Yes, the ones digit is 8, an even digit. **2.** No, $7 + 9 + 1 + 2 = 19$, and 19 is not divisible by 3. **3.** Yes, 76 is divisible by 4. **4.** No, the ones digit is not 0 or 5. **5.** No, $1 + 5 + 7 + 6 = 19$, and 19 is not divisible by 3. **6.** Yes, $4 + 6 + 5 + 3 = 18$, and 18 is divisible by 9. **7.** Yes, the ones digit is 0. **8.** Yes, $6 + 1 + 2 = 9$, and 9 is divisible by 9.

1.8 Exercises

Concept Check

Fill-in-the-Blank. Complete each sentence using information found in this section.

1. A number is divisible by 5 if the last digit is a/an _____ or a/an _____.

2. A number is divisible by 9 if the sum of the digits is divisible by _____.

3. A number is divisible by 4 is the number formed by the last two digits is divisible by _____.

4. A number is divisible by 6 if the number is divisible by both _____ and _____.

5. Every even whole number is divisible by _____.

6. A number divisible by 10 has _____ as its ones digit.

1.8 Exercises

Concept Check

1. 0, 5
2. 9
3. 4
4. 2, 3
5. 2
6. 0
7. True
8. False; 6801 is divisible by 3.
9. False; 7605 is divisible by 5.
10. True

Practice

1. 3, 5
2. 3, 9
3. 2, 3, 4, 6, 9
4. 2, 3, 5, 6, 10
5. 3
6. 2
7. 3, 9
8. 2, 4
9. 3, 5
10. 3, 9
11. None
12. None
13. None
14. None
15. 2, 3, 4, 5, 6, 10
16. 2, 3, 5, 6, 10
17. 2, 5, 10
18. 2, 3, 6

True/False. Determine whether each statement is true or false. If a statement is false, explain how it can be changed so the statement will be true. (**Note:** There may be more than one acceptable change.)

7. A number that is divisible by 10 is also divisible by 2 and 5.
8. 6801 is divisible by 9.
9. 7605 is divisible by 10.
10. 5,187,042 is divisible by 3.

Practice

Using the tests for divisibility, determine which of 2, 3, 4, 5, 6, 9, and 10 (if any) will divide exactly into each given number. See Examples 1 through 8.

1. 105	**13.** 571	**25.** 732	**37.** 9000
2. 81	**14.** 331	**26.** 888	**38.** 5700
3. 72	**15.** 480	**27.** 441	**39.** 8805
4. 150	**16.** 510	**28.** 555	**40.** 1155
5. 471	**17.** 370	**29.** 705	**41.** 4422
6. 154	**18.** 402	**30.** 549	**42.** 5476
7. 333	**19.** 777	**31.** 576	**43.** 45,000
8. 664	**20.** 897	**32.** 792	**44.** 10,000
9. 375	**21.** 466	**33.** 1234	**45.** 12,324
10. 567	**22.** 695	**34.** 8765	**46.** 75,495
11. 443	**23.** 795	**35.** 4321	
12. 173	**24.** 885	**36.** 4391	

For each set of numbers, make up any four 3-digit numbers that you can think of that are divisible by all of the given numbers. (There are many possible answers.)

47. 2 and 3	**49.** 2, 3, and 10	**51.** 5 and 6
48. 2 and 9	**50.** 2, 5, and 9	**52.** 3 and 4

Applications

Solve.

53. A school district raises $10,125 in scholarship money for local students. The school board wants to split the money evenly between multiple students. The board is considering splitting the money between 3, 4, 6, or 9 students. Which of the options would allow the money to be split evenly among the students? How much money would each student receive for each group size that allows the money to be split evenly?

19. 3
20. 3
21. 2
22. 5
23. 3, 5
24. 3, 5
25. 2, 3, 4, 6
26. 2, 3, 4, 6
27. 3, 9
28. 3, 5
29. 3, 5
30. 3, 9
31. 2, 3, 4, 6, 9
32. 2, 3, 4, 6, 9
33. 2
34. 5
35. None
36. None
37. 2, 3, 4, 5, 6, 9, 10
38. 2, 3, 4, 5, 6, 10
39. 3, 5
40. 3, 5
41. 2, 3, 6
42. 2, 4
43. 2, 3, 4, 5, 6, 9, 10
44. 2, 4, 5, 10
45. 2, 3, 4, 6
46. 3, 5
47. 102, 108, 114, 120; Answers will vary.
48. 108, 126, 144, 162; Answers will vary.
49. 120, 150, 180, 210; Answers will vary.
50. 180, 270, 360, 450; Answers will vary.
51. 120, 150, 180, 210; Answers will vary.
52. 108, 120, 132, 144; Answers will vary.

Applications

53. 3 students would get $3375 each; 9 students would be $1125 each.

54. You are on a team that is participating in a charity walk with a goal to raise $12,400. Each team member agrees to raise the same amount of money. If the possible team sizes are 5, 6, 9, or 10 members, which team sizes allow the goal amount to be evenly split between the team members? How much money would each team member raise for each team size that can evenly split the goal amount?

55. A company is working on a project that will take 440 hours of work to complete. The manager in charge of the project has the option to have 4, 6, or 8 people work on the project. If the manager wants to evenly divide the work between the team members, which team size will evenly split the work hours? How many hours would each team member spend on the project for each team size that evenly splits the work hours?

56. Marc is moving from Charleston, SC, to Houston, TX, which is a distance of approximately 1060 miles. He plans to sightsee along the way while driving the same number of miles each day. Marc wants to make the trip in 2, 3, 4, or 5 days. Which trip length would split the miles traveled evenly between all of the days? How many miles would Marc need to drive each day for each option that evenly splits the trip length?

Writing & Thinking

57. **a.** If a number is divisible by both 3 and 5, then it will be divisible by 15. Give two examples.

b. However, a number might be divisible by 3 and not by 5. Give two examples.

c. Also, a number might be divisible by 5 and not 3. Give two examples.

58. Explain, in your own words, why a number ending in 0 is divisible by 2 and 5.

59. Create a 5 digit number that is divisible by 3, 5, 9, and 10. Explain how this number meets each of the tests for divisibility.

60. A digit is missing in the number 7☐,488

a. Explain the process that you would use to find all possible values of the missing digit that would make the resulting number divisible by 3.

b. Use the process from part a. to find all values for the missing digit so that the number will be divisible by 3. Keep in mind that there may be one value, several values, or no value that makes the number divisible by 3.

61. Think about the rules for divisibility and explain why the following statement is true: "A number may be divisible by 5 and not divisible by 10." Give two examples to illustrate your point.

62. If a number is divisible by both 2 and 4, must it also be divisible by 8? Explain your reasoning and give two examples to support your reasoning.

---

54. 5 people would raise $2480 each; 10 people would raise $1240 each.

55. 4 team members would work 110 hours each; 8 team members would work 55 hours each.

56. 2 days driving 530 miles per day; 4 days driving 265 miles per day; 5 days driving 212 miles per day

Writing & Thinking

57. **a.** 30, 45; Answers will vary.
b. 9, 12; Answers will vary.
c. 10, 25; Answers will vary.

58. A number ending in 0 is divisible by 10, and both 2 and 5 are factors of 10. Also, the number is even (and divisible by 2) and meets the divisibility test for 5. Answers will vary.

59. For example: 15,030. The number is divisible by 3 and 9 because digits add up to 9 and the number is divisible by 5 and 10 because the number ends in 0.

60. **a.** Find the sum of the known digits. Then find a digit that, when added to the sum of the known digit, results in a number that is divisible by 3.
b. 0, 3, 6, 9

61. Answers will vary. Any number that ends in 5 is divisible by 5 and not divisible by 10. Two examples are 15 and 25.

62. Answers will vary. Not always. To be divisible by 8, a number must have 2 as a factor three times. Two counterexamples are 12 and 20.

1.9 Prime Numbers and Prime Factorizations

Objectives

A. Understand the difference between prime and composite numbers.

B. Determine whether a number is prime or composite.

C. Find the prime factorization of a composite number.

D. Find all factors of a composite number.

A Prime and Composite Numbers

Every counting number (or natural number), except 1, has two or more factors (or divisors). The following is a list of some counting numbers and their factors.

Counting Numbers		Factors
3	→	1, 3
14	→	1, 2, 7, 14
17	→	1, 17
18	→	1, 2, 3, 6, 9, 18
21	→	1, 3, 7, 21
23	→	1, 23
36	→	1, 2, 3, 4, 6, 9, 12, 18, 36

Note that in this list, 3, 17, and 23 each have exactly two different factors (or divisors)—itself and 1. Such numbers are called **prime numbers**. The other numbers in the list, 14, 18, 21, and 36, have more than two different factors and are called **composite numbers**.

History

Around 300 BC, the Greek mathematician Euclid proved that there are infinitely many prime numbers. This proof is called *Euclid's theorem* and was first proven in Euclid's mathematical treatise *Elements*. It is worth noting that since there is an infinite number of primes, there is no largest prime number.

Prime Number

A **prime number** is a counting number greater than 1 that has exactly two *different* factors (or divisors)—itself and 1.

DEFINITION

Composite Number

A **composite number** is a counting number with more than two different factors (or divisors).

DEFINITION

Attention!

$1 = 1 \cdot 1$, and 1 is the only factor of 1. 1 does not have **exactly two different** factors, and it does not have more than two different factors. Thus, 1 is **neither** a prime nor a composite number.

1. Show that each number is prime.
 a. 13
 b. 19

Example 1 Determining Prime Numbers

The following numbers are prime because each has exactly two different factors, 1 and itself.

2: 2 has exactly two different factors, 1 and 2.

7: 7 has exactly two different factors, 1 and 7.

11: 11 has exactly two different factors, 1 and 11.

29: 29 has exactly two different factors, 1 and 29.

Now work margin exercise 1.

2. Show that each number is composite.
 a. 25
 b. 32

Example 2 Determining Composite Numbers

The following numbers are composite.

12: $1 \cdot 12 = 12$, $2 \cdot 6 = 12$, and $3 \cdot 4 = 12$. So 1, 2, 3, 4, 6, and 12 are all factors of 12, and 12 has more than two different factors.

33: $1 \cdot 33 = 33$ and $3 \cdot 11 = 33$. So 1, 3, 11, and 33 are all factors of 33, and 33 has more than two different factors.

Now work margin exercise 2.

B Determining Whether a Number is Prime

The following table shows prime numbers less than 50. Composite numbers are crossed out.

1	2	3	4	5	6	7	8	9	10
11	12	13	14	15	16	17	18	19	20
21	22	23	24	25	26	27	28	29	30
31	32	33	34	35	36	37	38	39	40
41	42	43	44	45	46	47	48	49	50

Figure 1

Teaching Note
Students should be encouraged to memorize the list of prime numbers less than 50. We will assume later that, when reducing fractions, the students readily recognize these primes.

We see that the prime numbers less than 50 are

2, 3, 5, 7, 11, 13, 17, 19, 23, 29, 31, 37, 41, 43, and **47.**

Two important facts about prime numbers are the following.

1. **Even Numbers:** 2 is the only even prime number.
2. **Odd Numbers:** All other prime numbers are odd numbers. But, not all odd numbers are prime.

The following procedure of dividing by prime numbers can be used to determine whether a number is prime. (Computers are used to determine whether very large numbers are prime.)

Determining Whether a Number is Prime

Divide the number by progressively larger prime numbers (2, 3, 5, 7, 11, and so forth) until one of the following is true.

1. **The remainder is 0.** This means that the prime number is a factor and **the given number is composite.**
2. **You find a quotient smaller than the prime divisor.** This means that **the given number is prime** because it has no smaller prime factors.

PROCEDURE

Example 3 Determining Whether a Number is Prime

Is 605 a prime number?

Solution

The ones digit is 5. Therefore, 605 is divisible by 5 and is not prime. The number 605 is a composite number.

Note: $605 = 5 \cdot 121 = 5 \cdot 11 \cdot 11$; that is, 5, 11, and 121 are factors of 605.

Now work margin exercise 3.

3. Is 404 a prime number? Explain your reasoning.

Teaching Note

You may want to have students divide by some composite numbers and their prime factors to better understand why it is only necessary to divide by prime numbers to determine if a number is prime.

Example 4 Determining Whether a Number is Prime

Is 103 a prime number?

Solution

Tests for 2, 3, and 5 fail. (The number 103 is not even; $1 + 0 + 3 = 4$ and 4 is not divisible by 3; and the ones digit is not 0 or 5.)

Divide by 7:

$$\begin{array}{r} 14 \\ 7\overline{)103} \\ \underline{-7} \\ 33 \\ \underline{-28} \\ 5 \end{array}$$

The quotient is greater than the divisor.

The remainder is not 0.

Divide by 11:

$$\begin{array}{r} 9 \\ 11\overline{)103} \\ \underline{-99} \\ 4 \end{array}$$

The quotient is smaller than the divisor, so we are done.

The remainder is not 0.

The number 103 is prime.

Note: There is no point in dividing by larger numbers such as 13 or 17 because the quotient would only get smaller, and if any of these larger numbers were a divisor, the smaller quotient would have been found to be a divisor earlier in the procedure.

Now work margin exercise 4.

4. Is 247 a prime number? Explain your reasoning.

5. Is 113 a prime number? Explain your reasoning.

Example 5 Determining Whether a Number is Prime

Is 221 a prime number?

Solution

Tests for 2, 3, and 5 fail. (The number 221 is not even; $2 + 2 + 1 = 5$ and 5 is not divisible by 3; and the ones digit is not 0 or 5.)

Divide by 7:

$$\begin{array}{r} 31 \\ 7\overline{)221} \\ -21 \\ \hline 11 \\ -7 \\ \hline 4 \end{array}$$

The quotient is greater than the divisor.

The remainder is not 0.

Divide by 11:

$$\begin{array}{r} 20 \\ 11\overline{)221} \\ -22 \\ \hline 01 \\ -0 \\ \hline 1 \end{array}$$

The quotient is greater than the divisor.

The remainder is not 0.

Divide by 13:

$$\begin{array}{r} 17 \\ 13\overline{)221} \\ -13 \\ \hline 91 \\ -91 \\ \hline 0 \end{array}$$

The remainder is 0.

The number 221 is composite.

Note: $221 = 13 \cdot 17$; that is, 13 and 17 are factors of 221.

Now work margin exercise 5.

6. Is 239 a prime number? Explain your reasoning.

Completion Example 6 Determining Whether a Number is Prime

Is 199 a prime number?

Solution

Tests for 2, 3, and 5 fail. (The number 199 is not __even__; $1 + 9 + 9 =$ __19__ which is not divisible by 3; and the ones digit is not 0 or 5.)

Divide 199 by 7: __28 R3__

Divide 199 by 11: __18 R1__

Divide 199 by 13: __15 R4__

Divide 199 by __17__ : __11 R12__

199 is a __prime__ number.

Now work margin exercise 6.

C Finding a Prime Factorization

Later, when we operate with fractions, we will want to factor numbers so that all of the factors are prime numbers. This is called finding the **prime factorization** of a number. For example, to find the prime factorization of 70, we start with any two factors of 70.

1. $70 = 7 \cdot 10$ 7 is prime, but 10 is not prime.

By factoring 10, we find the following.

2. $70 = 7 \cdot 10 = 7 \cdot 2 \cdot 5$ Now all of the factors are prime.

This last product $(7 \cdot 2 \cdot 5)$ is the **prime factorization** of 70.

Also, we could have started with different factors, as follows.

3. $70 = 2 \cdot 35 = 2 \cdot 5 \cdot 7$ The prime factorization is the same.

Because multiplication is commutative, the factors may be written in any order. For consistency, we will generally write the factors in ascending order, from smallest to largest.

Regardless of the factors used in the beginning, **there is only one prime factorization for any composite number**. This important fact is called the **Fundamental Theorem of Arithmetic**.

The Fundamental Theorem of Arithmetic

Every composite number has exactly one prime factorization.

THEOREM

Finding the Prime Factorization of a Composite Number

1. Factor the composite number into any two factors.
2. Factor each factor that is not prime.
3. Continue this process until all factors are prime.

PROCEDURE

The **prime factorization** of a number is a factorization of that number using only prime factors.

Many times, the beginning factors needed to start the process for finding a prime factorization can be found by using the tests for divisibility by 2, 3, 5, 6, 9, and 10, as discussed in Section 1.8. This was one purpose for developing these tests, and you should review them or write them down for easy reference.

Teaching Note

The skill of finding prime factorizations is absolutely indispensable in our work with fractions. You may want to emphasize this fact several times, and be sure students understand the distinction between the terms "prime factorization" and "factors."

Teaching Note

Another format for finding the prime factorization is to divide mentally by prime numbers, as follows.

$$\begin{array}{r} 2\underline{|140} \\ 7\underline{|70} \\ 5\underline{|10} \\ 2 \end{array}$$

The division stops when the last quotient is a prime number. The prime factorization is the product of all the prime divisors.

$$140 = 2 \cdot 2 \cdot 5 \cdot 7$$

Beginning students might prefer this style. However, they should be aware that the format used in the text is more consistent with algebraic methods they will see later.

Teaching Note

You may want to mention to your students that the prime factorization of a prime number is simply itself. For example, the prime factorization of 5 is 5.

Example 7 Finding the Prime Factorization of a Number

Find the prime factorization of 90.

7. Find the prime factorization of 140.

Solution

$90 = 9 \cdot 10$	Since the units digit is 0, we know that 10 is a factor.
$= 3 \cdot 3 \cdot 2 \cdot 5$	9 and 10 can both be factored so that each factor is a prime number. This is the prime factorization.

We can also start with a different set of factors.

$90 = 3 \cdot 30$	3 is prime, but 30 is not.
$= 3 \cdot 10 \cdot 3$	10 is not prime.
$= 3 \cdot 2 \cdot 5 \cdot 3$	All factors are prime.

Note: The final prime factorization was the same in both factor trees even though the first pair of factors was different.

Since multiplication is commutative, the order of the factors is not important. What is important is that all the factors are prime. Writing the factors in ascending order, we can write $90 = 2 \cdot 3 \cdot 3 \cdot 5$ or, with exponents, $90 = 2 \cdot 3^2 \cdot 5$.

Now work margin exercise 7.

8. Find the prime factorization of each number.

a. 74

b. 100

c. 234

Example 8 Finding the Prime Factorization of a Number

Find the prime factorization of each number.

a. 65 **b.** 72 **c.** 294

Solution

a. $65 = 5 \cdot 13$ 5 is a factor because the units digit is 5. Since both 5 and 13 are prime, $5 \cdot 13$ is the prime factorization.

b.

$72 = 8 \cdot 9$	9 is a factor because the sum of the digits is 9.
$= 2 \cdot 4 \cdot 3 \cdot 3$	
$= 2 \cdot 2 \cdot 2 \cdot 3 \cdot 3$	Prime factorization
$= 2^3 \cdot 3^2$	Using exponents

c.

$294 = 2 \cdot 147$	2 is a factor because the units digit is even.
$= 2 \cdot 3 \cdot 49$	3 is a factor of 147 because the sum of the digits is divisible by 3.
$= 2 \cdot 3 \cdot 7 \cdot 7$	Prime factorization
$= 2 \cdot 3 \cdot 7^2$	Using exponents

If we begin with the product $294 = 6 \cdot 49$, we see that the prime factorization is the same.

$$294 = 6 \cdot 49$$
$$= 2 \cdot 3 \cdot 7 \cdot 7$$
$$= 2 \cdot 3 \cdot 7^2$$

Now work margin exercise 8.

Completion Example 9 Finding the Prime Factorization of a Number

Find the prime factorization of each number.

a. 60 **b.** 308

Solution

a. $60 = 6 \cdot \underline{10}$

$= 2 \cdot 3 \cdot \underline{2} \cdot \underline{5}$ Prime factorization

$= \underline{2^2 \cdot 3 \cdot 5}$ Using exponents

b. $308 = 4 \cdot \underline{77}$

$= \underline{2} \cdot \underline{2} \cdot \underline{7} \cdot \underline{11}$ Prime factorization

$= \underline{2^2 \cdot 7 \cdot 11}$ Using exponents

Now work margin exercise 9.

9. Find the prime factorization of each number.

a. 276

b. 550

D Finding All Factors of a Composite Number

Once the prime factorization of a composite number is known, all of the factors (or divisors) of that number can be found. For a number to be a factor of a composite number, it must be either 1, the number itself, one of the prime factors, or the product of two or more of the prime factors.

Factors of a Composite Number

The only factors (or divisors) of a composite number are

1. 1 and the number itself,
2. each prime factor, and
3. products formed by all combinations of the prime factors (including repeated factors).

DEFINITION

Math Tip

Finding large prime numbers is difficult enough, but what makes prime numbers so useful is that it is even more difficult to factor large numbers back into their primes. You can factor 30 by hand $(2 \cdot 2 \cdot 3 \cdot 5)$, but it would take considerable effort (and maybe a computer) to factor 1,071,029 (which is $1021 \cdot 1049$). This is what makes large prime numbers the perfect numbers to use for encryption.

Example 10 Finding the Factors of a Composite Number

Find all the factors of 60.

Solution

The prime factorization of 60 is $2^2 \cdot 3 \cdot 5$. Thus, the factors are

1. 1 and the number itself, 60
2. Each prime factor: 2, 3, 5

10. Find all the factors of 42.

3. Products of all combinations of the prime factors:

$2 \cdot 2 = 4$, $2 \cdot 3 = 6$, $2 \cdot 5 = 10$, $3 \cdot 5 = 15$,

$2 \cdot 2 \cdot 3 = 12$, $2 \cdot 2 \cdot 5 = 20$, $2 \cdot 3 \cdot 5 = 30$

The factors are 1, 2, 3, 4, 5, 6, 10, 12, 15, 20, 30, and 60.

Now work margin exercise 10.

11. Find all the factors of 63.

Completion Example 11 Finding the Factors of a Composite Number

Find all the factors of 154.

Solution

The prime factorization of 154 is $2 \cdot 7 \cdot 11$. Thus, the factors are

1. 1 and 154.
2. Each prime factor: 2, 7, and 11.
3. Products of all combinations of the prime factors:

$2 \cdot 7 = 14$,

$2 \cdot 11 = 22$,

$7 \cdot 11 = 77$

The factors of 154 are 1, 2, 7, 11, 14, 22, 77, and 154.

Now work margin exercise 11.

One interesting application of factors of counting numbers (which is very useful in beginning algebra) involves finding two factors whose sum is some specified number. See the following example.

12. Find two factors of 84 such that their product is 84 and their sum is 25.

Example 12 Using Factors of Counting Numbers

Find two factors of 70 such that their product is 70 and their sum is 19.

Solution

Here we have listed the pairs of factors and their sums in a table format. The pair we are looking for can be found by looking at the list of sums.

Factors of 70	Sum of Factors
$1 \cdot 70 = 70$	$1 + 70 = 71$
$2 \cdot 35 = 70$	$2 + 35 = 37$
$5 \cdot 14 = 70$	$5 + 14 = 19$
$7 \cdot 10 = 70$	$7 + 10 = 17$

Thus, the numbers we are looking for are 5 and 14 because

$$5 \cdot 14 = 70 \quad \text{and} \quad 5 + 14 = 19.$$

Now work margin exercise 12.

Completion Example Answers

6. 199 is not even; $1+9+9=19$; $199 \div 7 = 28$ R3, $199 \div 11 = 18$ R1; $199 \div 13 = 15$ R4, $199 \div 17 = 11$ R12; 199 is a prime number

9. a. $6 \cdot 10 = 2 \cdot 3 \cdot 2 \cdot 5 = 2^2 \cdot 3 \cdot 5$ **b.** $4 \cdot 77 = 2 \cdot 2 \cdot 7 \cdot 11 = 2^2 \cdot 7 \cdot 11$

11. 1 and 154; 2, 7, and 11; $2 \cdot 7 = 14$, $2 \cdot 11 = 22$, $7 \cdot 11 = 77$; 1, 2, 7, 11, 14, 22, 77, and 154

Margin Exercise Answers

1. a. 13 has exactly two factors, 1 and 13. **b.** 19 has exactly two factors, 1 and 19. **2. a.** 1, 5, and 25 are all factors of 25. **b.** 1, 2, 4, 8, 16, and 32 are all factors of 32. **3.** No, the factors of 404 are 1, 2, 4, 101, 202, and 404, making 404 composite. **4.** No, the factors of 247 are 1, 13, 19, and 247, making 247 composite. **5.** Yes, tests for 2, 3, and 5 fail. Dividing by 7 and 11 fails to yield a divisor, and when dividing by 11, the quotient is less than 11. This means 113 is prime. **6.** Yes, tests for 2, 3, and 5 all fail. Dividing by 7, 11, 13, and 17 fails to yield a divisor, and when dividing by 17, the quotient is less than 17. This means 239 is prime. **7.** $2^2 \cdot 5 \cdot 7$ **8. a.** $2 \cdot 37$ **b.** $2^2 \cdot 5^2$ **c.** $2 \cdot 3^2 \cdot 13$ **9. a.** $2^2 \cdot 3 \cdot 23$ **b.** $2 \cdot 5^2 \cdot 11$ **10.** 1, 2, 3, 6, 7, 14, 21, and 42 **11.** 1, 3, 7, 9, 21, 63 **12.** 21 and 4

1.9 Exercises

Concept Check

Fill-in-the-Blank. Complete each sentence using information found in this section.

1. A composite number is a counting number with more than ______ different factors.
2. A prime number has exactly ______ different factors.
3. A prime number must be a number greater than ______.
4. Each composite number has exactly ______ prime factorization.
5. A prime number's factors are ______ and 1.
6. The number 1 is neither ______ nor ______.

True/False. Determine whether each statement is true or false. If a statement is false, explain how it can be changed so the statement will be true. (**Note:** There may be more than one acceptable change.)

7. A prime number has exactly 1 factor.
8. A composite number has 2 or more factors.
9. 231 is a prime number.
10. All the factors of 30 are 1, 2, 3, 5, 6, 10, 15 and 30.

Practice

In your own words, write the definition of each term.

1. prime number
2. composite number

1.9 Exercises

Concept Check

1. two
2. two
3. 1
4. 1
5. itself
6. prime; composite
7. False; A prime number has exactly 2 factors.
8. False; A composite number more than 2 factors.
9. False; 231 is a composite number.
10. True

Practice

1. A number whose only factors are 1 and itself. Answers will vary.
2. A number that is divisible by a number other than 1 and itself. Answers will vary.
3. Prime
4. Prime
5. Composite; 1, 3, 7, 9, 21, 63
6. Composite; 1, 3, 5, 15, 25, 75
7. Prime
8. Prime
9. Composite; 1, 5, 41, 205
10. Composite; 1, 2, 251, 502
11. Composite; 1, 11, 13, 143
12. Composite; 1, 11, 47, 517
13. 4, 6
14. 3, 4
15. 1, 12
16. 1, 7
17. 3, 8
18. 3, 12
19. 4, 4
20. 5, 5

21. $2^3 \cdot 3$
22. $2^2 \cdot 7$
23. 3^3
24. 5^3
25. 2^4
26. 3^4
27. $2^2 \cdot 3^2$
28. $2^3 \cdot 3^2$
29. $2^2 \cdot 5$
30. $2 \cdot 5^2$
31. $2 \cdot 5 \cdot 7$
32. $3 \cdot 5 \cdot 7$
33. 37 is prime
34. 43 is prime
35. $2 \cdot 3 \cdot 5^2$
36. $2^3 \cdot 3 \cdot 5$
37. $2 \cdot 3 \cdot 5 \cdot 7$
38. $2 \cdot 3 \cdot 5 \cdot 11$
39. $2^3 \cdot 3^2 \cdot 5$
40. $2^3 \cdot 3 \cdot 5^2$
41. $2^3 \cdot 5^2 \cdot 11$
42. $2^2 \cdot 5^3 \cdot 7$
43. $2^3 \cdot 5^3$
44. $2^4 \cdot 5^4$
45. a. $2 \cdot 3^2$
b. 1, 2, 3, 6, 9, 18
46. a. $2^2 \cdot 7$
b. 1, 2, 4, 7, 14, 28
47. a. $2 \cdot 3 \cdot 5$
b. 1, 2, 3, 5, 6, 10, 15, 30
48. a. $2 \cdot 3 \cdot 7$
b. 1, 2, 3, 6, 7, 14, 21, 42
49. a. $2^2 \cdot 3 \cdot 7$
b. 1, 2, 3, 4, 6, 7, 12, 14, 21, 28, 42, 84
50. a. $2 \cdot 3^2 \cdot 5$
b. 1, 2, 3, 5, 6, 9, 10, 15, 18, 30, 45, 90
51. a. $5^2 \cdot 7$
b. 1, 5, 7, 25, 35, 175
52. a. $5^2 \cdot 11$
b. 1, 5, 11, 25, 55, 275

Determine whether each number is prime or composite. If the number is composite, find at least three factors of the number. See Examples 3 through 6, 10, and 11.

3. 47
4. 59
5. 63
6. 75
7. 103
8. 107
9. 205
10. 502
11. 143
12. 517

Two numbers are given. Find two factors of the first number such that their product is the first number and their sum is the second number. See Example 12.

13. 24, 10
14. 12, 7
15. 12, 13
16. 7, 8
17. 24, 11
18. 36, 15
19. 16, 8
20. 25, 10

Find the prime factorization of each number. Use the tests for divisibility for 2, 3, 4, 5, 6, 9, and 10 whenever they help to find beginning factors. See Examples 7 through 9.

21. 24
22. 28
23. 27
24. 125
25. 16
26. 81
27. 36
28. 72
29. 20
30. 50
31. 70
32. 105
33. 37
34. 43
35. 150
36. 120
37. 210
38. 330
39. 360
40. 600
41. 2200
42. 3500
43. 1000
44. 10,000

For each number, **a.** find the prime factorization and **b.** find all the factors.

45. 18
46. 28
47. 30
48. 42
49. 84
50. 90
51. 175
52. 275
53. 300
54. 700

Applications

Solve.

55. Twenty-four pencils are to be distributed evenly between the members of a group. What are the possible group sizes if each person in the group is to receive the same number of pencils?

56. Madeline plans to travel the world for a year and spend the same number of weeks in each country she visits. How many different countries can she visit during the year? (**Hint:** There are 52 weeks in a year.)

57. A chocolatier makes 72 specialty truffles. She wants to sell packages that each have the same number of truffles. What are her options for the number of truffles that can be in a package?

58. A radio station has 120 concert tickets to use to create prize packages to give to listeners. Each prize package will have the same number of tickets. What are the possible sizes for the packages of tickets?

Writing & Thinking

59. Are all odd numbers also prime numbers? Explain your answer.

60. Are all prime numbers also odd numbers? Explain your answer.

61. Explain how the tests for divisibility are helpful in finding prime factorizations.

62. Explain the difference between factors of a number and multiples of that number.

Collaborative Learning

63. In higher level mathematics, number theorists have proven the following theorem. Write the prime factorization of a number in exponential form. Add 1 to each exponent. The product of these sums is the number of factors of the original number.

For example, $60 = 2^2 \cdot 3 \cdot 5 = 2^2 \cdot 3^1 \cdot 5^1$. Adding 1 to each exponent and forming the product gives

$$(2+1)(1+1)(1+1) = 3 \cdot 2 \cdot 2 = 12$$

and there are twelve factors of 60.

Use this theorem to find the number of factors of each of the following numbers. Find as many of these factors as you can.

a. 700 **b.** 660 **c.** 450

64. Mathematicians have been interested since ancient times in searching for perfect numbers. A perfect number is a counting number that is equal to the sum of its proper divisors (divisors not including itself). For example, the first perfect number is 6. The proper divisors of 6 are 1, 2, and 3, and $1 + 2 + 3 = 6$. With the class separated into groups of 2 to 4 students, each team is to try to find the second and third perfect numbers. (**Hint:** The second perfect number is between 20 and 30, and the third perfect number is between 450 and 500.)

Collaborative Learning

63. **a.** 700 has 18 factors. They are: 1, 2, 4, 5, 7, 10, 14, 20, 25, 28, 35, 50, 70, 100, 140, 175, 350, and 700.

b. 660 has 24 factors. They are: 1, 2, 3, 4, 5, 6, 10, 11, 12, 15, 20, 22, 30, 33, 44, 55, 60, 66, 110, 132, 165, 220, 330, and 660.

c. 450 has 18 factors. They are: 1, 2, 3, 5, 6, 9, 10, 15, 18, 25, 30, 45, 50, 75, 90, 150, 225, and 450.

64. 28, 496

53. **a.** $2^2 \cdot 3 \cdot 5^2$
b. 1, 2, 3, 4, 5, 6, 10, 12, 15, 20, 25, 30, 50, 60, 75, 100, 150, 300

54. **a.** $2^2 \cdot 5^2 \cdot 7$
b. 1, 2, 4, 5, 7, 10, 14, 20, 25, 28, 35, 50, 100, 140, 175, 250, 350, 700

Application

55. 1, 2, 3, 4, 6, 8, 12, 24

56. 1, 2, 4, 13, 26, 52

57. 1, 2, 3, 4, 6, 8, 9, 12, 18, 24, 36, 72

58. 1, 2, 3, 4, 5, 6, 8, 10, 12, 15, 20, 24, 40, 60, 120

Writing & Thinking

59. No, some odd numbers are the product of two or more odd prime factors, for example, $3 \cdot 3 = 9$, $3 \cdot 5 = 15$, $3 \cdot 7 = 21$, etc.

60. No, the number 2 is prime and even.

61. The tests for divisibility can be used to find relatively small factors quickly. Knowing these factors makes finding the prime factorization easier.

62. Factors of a number are numbers that divide evenly into that number, while multiples are found by multiplying the number by the counting numbers 1, 2, 3, and so on. For example, 2 and 5 are factors of 10 and 10, 20, 30, 40, and so on, are multiples of 10.

Teaching Note

At the end of the exercise you may want to share with your students that Euclid proved that $2^{P-1}(2^P - 1)$ is an even perfect number whenever 2^{P-1} is prime. Using this formula we can generate the 3rd and 4th perfect numbers: for $P = 7$: $2^6(2^7 - 1) = 8128$; for $P = 13$: $2^{12}(2^{13} - 1) = 33{,}550{,}336$

Chapter 1 Project

Aspiring to New Heights!

An activity to demonstrate the use of whole numbers in real life.

You may have never heard of the Willis Tower, but it once was the tallest building in the United States. This structure was originally named the Sears Tower when it was built in 1973, and it held the title of the tallest building in the world for almost 25 years. The name was changed in 2009 when Willis Group Holdings obtained the right to rename the building as part of their lease for a large portion of the office space in the building.

The Willis Tower, which is 1451 feet tall and located in Chicago, Illinois, was the tallest building in the Western Hemisphere until May 2, 2013. On this date a 408 foot spire was placed on the top of One World Trade Center in New York to bring its total height to a patriotic 1776 feet. One World Trade Center now claims the designation of being the tallest building in the United States and the Western Hemisphere.

1. The Willis Tower has an unusual construction. It is comprised of 9 square tubes of equal size, which are really separate buildings, and the tubes extend to different heights. The footprint of the Willis Tower is a 225 foot by 225 foot square. In answering the questions below, be sure to use the correct units of measurement on your answers.
   - **a.** Since the footprint of the Willis Tower is a square measuring 225 feet on each side and it is comprised of 9 square tubes of equal size, what is the side length of each tube? (It might help to draw a diagram.)
   - **b.** What is the perimeter around the footprint, or base of the Willis Tower?
   - **c.** What is the area of the base of the Willis Tower?
   - **d.** What is the area of the base for one square tube?
   - **e.** Write the values from parts c. and d. in words.
2. Suppose there are plans to alter the landscape around Willis Tower. The city engineers have proposed adding a concrete sidewalk 6 feet wide around the base of the building. A drawing of the proposal is shown. (**Note:** This drawing is not to scale.)

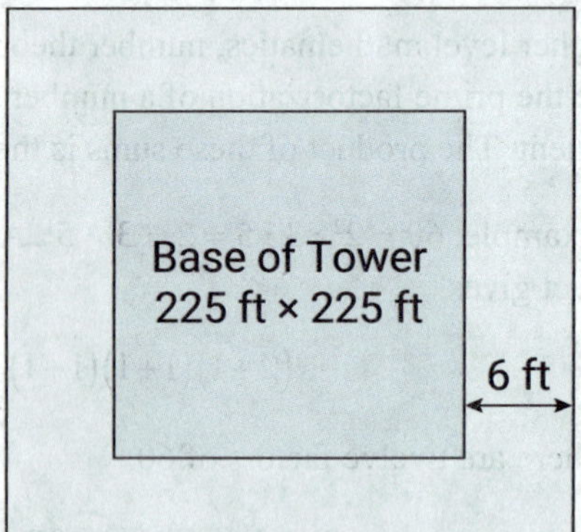

   - **a.** Determine the total area of the base of the tower including the new sidewalk.
   - **b.** Write down the area of just the base of the tower that you determined in part 1c.
   - **c.** Determine the area covered by the concrete sidewalk around the building. (**Hint:** You only want the area between the two squares.)
   - **d.** If a border were to be placed around the outside edge of the concrete sidewalk, how many feet of border would be needed?
   - **e.** If the border is only sold by the yard, how many yards of border will be needed? (**Note:** 1 yard = 3 feet.)
   - **f.** Round the value from part e. to the nearest ten.
   - **g.** Round the value from part e. to the nearest one hundred.

Chapter 1 Project: Aspiring to New Heights!

1. **a.** Show square that is 225 feet by 225 feet with 9 equal regions. Side length is $225 \div 3 = 75$ feet, so each tube is 75 feet by 75 feet.
   b. $4 \cdot 225 = 900$ feet **c.** $225 \cdot 225 = 50{,}625$ square feet
   d. $75 \cdot 75 = 5{,}625$ square feet
   e. Fifty thousand six hundred twenty-five; Five thousand six hundred twenty-five

2. **a.** $(225 + 12)(225 + 12) = 237 \cdot 237 = 56{,}169$ square feet
   b. 50,625 square feet **c.** $56{,}169 - 50{,}625 = 5{,}544$ square feet
   d. $4 \cdot 237 = 948$ feet **e.** $948 \div 3 = 316$ yards
   f. 320 yds **g.** 300 yds

Chapter 1 Project

Just Between You and Me

An activity to demonstrate the use of whole number arithmetic in real life.

Cryptography deals with turning a message from plaintext into a code known as ciphertext. Ciphertext appears to the general public as nonsense but can be decoded by the intended recipient. Over the past few hundred years, the field of cryptography has advanced greatly. Around 2000 years ago, all that was usually needed to encode (or encrypt) a message was whole number arithmetic.

To work with this classic form of encryption, often called a shift cipher, there is just one additional thing we need to know. Shift ciphers work similar to how a 12-hour clock shows the same time every 12 hours, even though it's a different time, or even a different day. Extending this concept to the alphabet, Table 1 assigns each letter a number, starting with 0, and repeats the cycle after the letter Z.

A	B	C	D	E	F	G	H	I	J	K	L	M	N	O	P	Q	R	S	T	U	V	W	X	Y	Z	A	B	C	...
0	1	2	3	4	5	6	7	8	9	10	11	12	13	14	15	16	17	18	19	20	21	22	23	24	25	26	27	28	...

Using subtraction, we can find that the numbers representing the two As are 26 apart. Similarly, the two Bs are 26 apart. Just like the digits on a clock begin to repeat after 12 hours, the letters will begin to repeat after 26 letters.

1. Notice C is represented by 28.
   - **a.** Determine the remainder when 28 is divided by 26.
   - **b.** Compare the value of the remainder found in part a. to the values in Table 1. What do you notice about this value and 28?

The idea behind a *shift* cipher is to shift the letters in the plaintext by a given number of letters to form the ciphertext. This encryption can be done with a right-shift or a left-shift.

A *right-shift* is obtained by adding the same value to each of the numbers associated with the letters. Since each letter in Table 1 is associated with a whole number, we will add the value of the right-shift, and then divide by 26, using the letter associated with the *remainder*.

For example, we will apply a right-shift by 14 to encrypt the plaintext HELLO:

Step 1: H, E, L, L, O is turned into 7, 4, 11, 11, 14.

Step 2: Adding 14 to each of these results in 21, 18, 25, 25, 28.

Step 3: When dividing by the 26, the remainders are 21, 18, 25, 25, 2, so the ciphertext is VSZZC.

2. Encrypt the word ZYGOTE using a right-shift by 3.

A *left-shift* is obtained by subtracting the same value from each of the numbers associated with the letters. We will first add 26 to the number associated with the letter, subtract the value of the left-shift, and then divide by 26, using the letter associated with the *remainder*.

For example, to decrypt the ciphertext VSZZC (which was encrypted using a right-shift by 14), we will apply a left-shift by 14:

Step 1: VSZZC becomes 21, 18, 25, 25, 2. Adding 26 to each of these results in 47, 44, 51, 51, 28.

Step 2: Subtracting 14 from each of these results in 33, 30, 37, 37, 14.

Step 3: When dividing by 26, the remainders are 7, 4, 11, 11, 14, so the plaintext is HELLO.

Notice that we added 26 to each number during Step 1. This is a necessary step when applying a left-shift to avoid negative numbers.

3. When the message has more than one word, all letters are capitalized and there are no spaces nor punctuation. This adds to the look of the text being nonsense, but for someone that knows how the plaintext was encoded, part of decoding will be properly separating things out. Suppose a colleague sends the following encrypted message, and you know the colleague used a right-shift by 8.

QABPMUMMBQVOIBVWWV

   a. Convert the letters in the ciphertext to numbers using Table 1.
   b. What would it take to undo a right-shift by 8? State this and state what numbers are obtained.
   c. Convert these back to letters using Table 1.
   d. Separate out the plain text into individual words to completely recover your colleague's original message.

Suppose you are not the intended recipient of the following ciphertext, but you have intercepted the message. You know that a right-shift by a certain amount has been used, but you do not know by how much.

EWWLEWSLLZWUGXXWWKZGHSLFAFW

For very short messages, a "brute force" approach could be used to check every possible right-shift and undo that shift. But that would mean trying 25 different possibilities. This would be too time-consuming for messages moderately long to very long.

4. Why would there 25 possibilities to check?

5. Instead of checking every possible right-shift, messages can often be decoded using the fact that E is the most commonly used letter in the English language.
   a. Count how many times each letter occurs in the intercepted ciphertext.
   b. Assume that the letter that shows up most frequently in the ciphertext was the letter E in the plaintext. What was the amount of the right-shift?
   c. Based on your answer from part b., decrypt the message. Separate out the plain text into individual words to completely recover the plaintext of the original message you intercepted.

Chapter 1 Project: Just Between You and Me

1. a. 2
   b. Both 2 and 28 represent the letter C
2. CBJRWH
3. a. 16, 0, 1, 15, 12, 20, 12, 12, 1, 16, 21, 14, 8, 1, 21, 22, 22, 21
   b. Left-shift by 8. So, add 26 to all the numbers in **a.**, then take the remainders after dividing by 26. 8, 18, 19, 7, 4, 12, 4, 4, 19, 8, 13, 6, 0, 19, 13, 14, 14, 13.
   c. ISTHEMEETINGATNOON
   d. Is the meeting at noon?
4. A shift of 0 to the right does not change anything. So, right-shifts would be 1, 2, 3 …, up to 25 to the right and then it is back to the start.
5. a. E:2, W:7, L:4, S:2, Z:2, U:1, G:2, X:2, K:1, H:1, F:2, A:1
   b. Since W is associated with 22 and E is associated with 4, the right-shift was by 18.
   c. MEETMEATTHECOFFEESHOPATNINE. Separated into individual words, "Meet me at the coffee shop at nine."

CHAPTER 2

2 Fractions and Mixed Numbers

Connections

To be a well-informed citizen, it's important to pay attention to politics and the political process since they affect everyone in some way. In local, state, and national elections, registered voters make choices about various ballot measures and who will represent them in the government. In major issues at the state and national levels, pollsters use mathematics (in particular, statistics and statistical methods) to indicate attitudes and to predict how the electorate will vote, typically accurate within certain percentage ranges. When there is an important election in your area, read news articles and listen to news reports for mathematically related statements predicting the outcomes.

Certain states, such as Ohio, are considered swing states. This means that during presidential elections no single candidate has an overwhelming support from the voters in the state. Swing state status is based on elections in recent history. Ohio is an important swing state because, since the 1960s, the candidate that won the electoral votes in Ohio has typically also won the presidential election.

Suppose that to determine funding for a political advertising campaign, a candidate's campaign staff takes a survey of voters in Ohio. They determine there are 8000 registered voters in a certain precinct, and $\frac{3}{8}$ of the voters are undecided about who they will vote for. The survey indicates that $\frac{4}{5}$ of these undecided voters agree with the candidate's platform. From this information, the campaign staff can determine how many people can be influenced to vote for their candidate with a series of campaign advertisements. How many undecided voters agree with the candidate's platform?

Objectives

A. Understand the basic concepts of fractions.

B. Graph fractions on a number line.

C. Understand the basic concepts of mixed numbers.

D. Graph mixed numbers on a number line.

E. Change mixed numbers to improper fractions.

F. Change improper fractions to mixed numbers.

2.1 Introduction to Fractions and Mixed Numbers

A Introduction to Fractions

Numbers such as $\frac{2}{3}$ (read "two-thirds") are said to be in **fraction form**. The top number, 2, is called the **numerator** and the bottom number, 3, is called the **denominator**.

$$\frac{2}{3} \begin{array}{l} \leftarrow \text{Numerator} \\ \leftarrow \text{Denominator} \end{array}$$

Fractions can be used to indicate parts of a whole. For example, if a whole candy bar has 7 equal parts, then the fraction $\frac{3}{7}$ (read "three-sevenths") indicates that we are considering 3 of those parts.

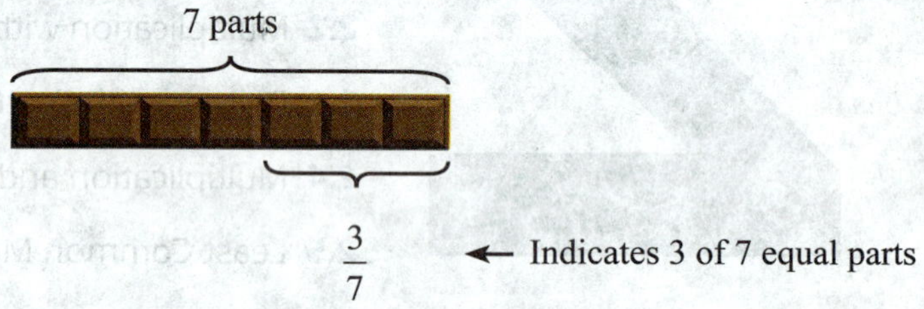

Figure 1

The whole candy bar can be represented as $\frac{7}{7}$.

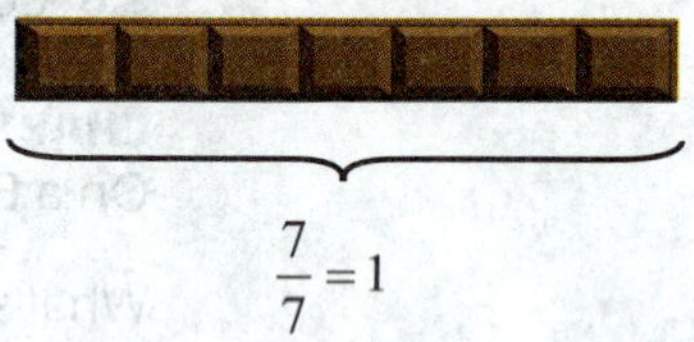

Figure 2

Example 1 shows several fractions indicating parts of a whole.

1. Write a fraction indicating

   a. the shaded part of the figure and

   b. the unshaded part of the figure.

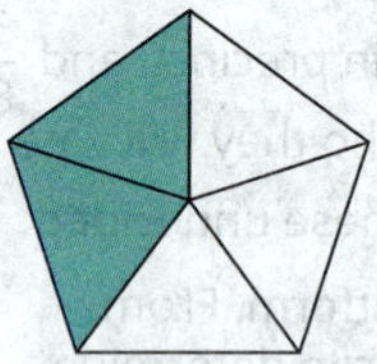

Example 1 Understanding Fractions

Write a fraction indicating

a. the shaded part of the rectangle and

b. the unshaded part of the rectangle.

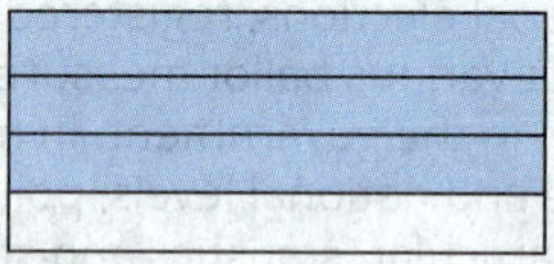

Solution

a. In the rectangle, 3 of the 4 equal parts are shaded. Thus, $\frac{3}{4}$ of the rectangle is shaded.

b. $\frac{1}{4}$ is not shaded.

Now work margin exercise 1.

Example 2 Understanding Fractions

Write a fraction indicating

a. the remaining portion of the pizza and

b. the missing portion of the pizza.

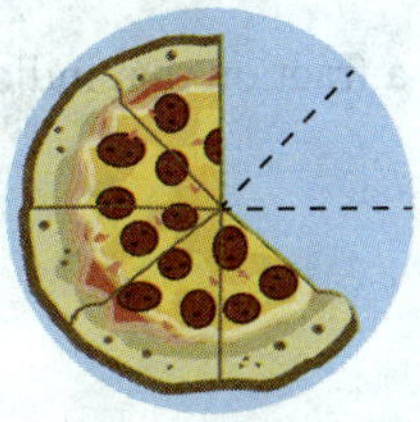

Solution

a. The pizza was cut into 8 equal pieces. The 5 pieces remaining represent $\frac{5}{8}$ of the pizza.

b. The missing portion of the pizza represents $\frac{3}{8}$ of the pizza.

Now work margin exercise 2.

2. Write a fraction indicating

a. the portion of the parking spaces that are occupied and

b. the portion of the parking spaces that are available.

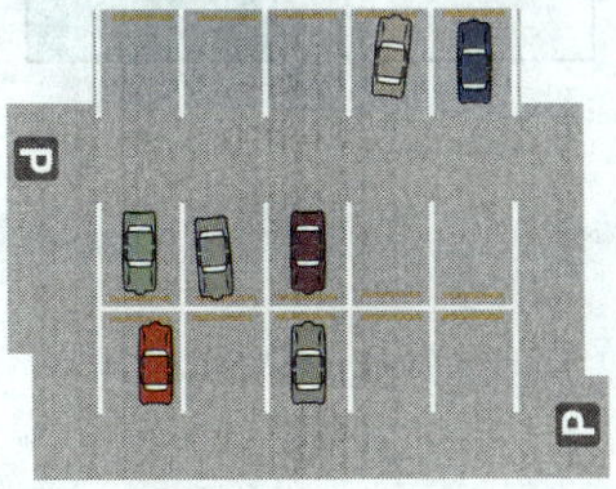

Proper Fractions and Improper Fractions

A **proper fraction** is a fraction in which the numerator is less than the denominator. (Proper fractions have values less than 1.)

Examples of proper fractions: $\frac{2}{3}$, $\frac{7}{8}$, and $\frac{32}{60}$

An **improper fraction** is a fraction in which the numerator is greater than or equal to the denominator. (Improper fractions have values greater than or equal to 1.)

Examples of improper fractions: $\frac{15}{8}$, $\frac{14}{14}$, and $\frac{250}{100}$

DEFINITION

Example 3 Understanding Proper Fractions

Draw a figure to represent the fraction $\frac{5}{6}$.

Solution

$\frac{5}{6}$ indicates 5 of 6 equal parts. Drawing a figure to represent this fraction, we divide a circle into 6 equal sections and shade 5 of them. (**Note:** Figures other than circles can be used.)

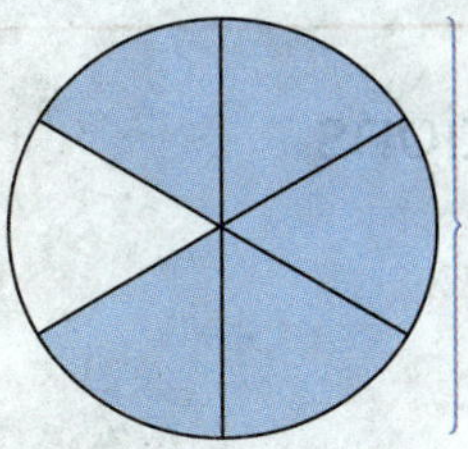

Now work margin exercise 3.

3. Draw a figure to represent the fraction $\frac{3}{8}$.

4. Write a fraction that indicates the shaded parts of the figure.

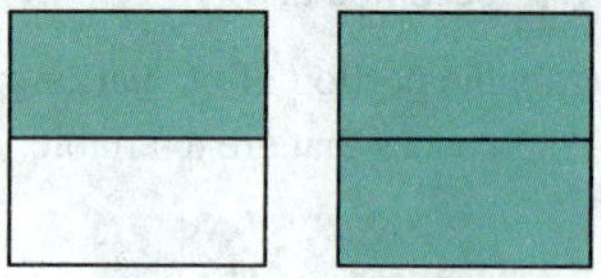

Example 4 Understanding Improper Fractions

Write a fraction that indicates the shaded parts of the figure.

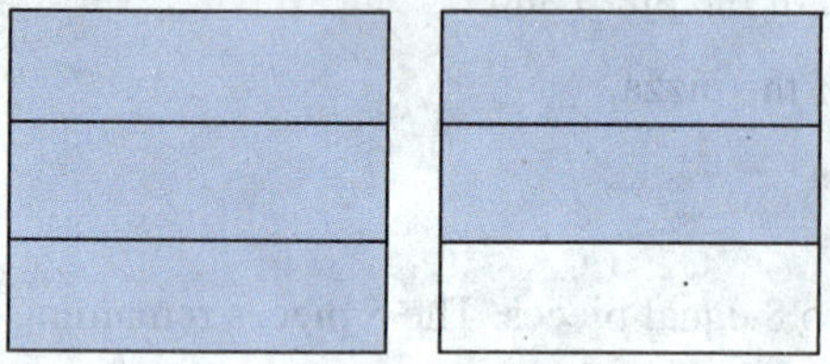

Solution

There are two squares, each separated into 3 equal parts. This means that the denominator is 3. The shading here indicates 5 of these equal parts, which means the numerator is 5. Thus, the shaded part of the figure can be represented by the improper fraction $\frac{5}{3}$.

Now work margin exercise 4.

Whole numbers can be thought of as fractions with denominator 1. Thus, in fraction form,

$$0=\frac{0}{1},\quad 1=\frac{1}{1},\quad 2=\frac{2}{1},\quad 3=\frac{3}{1},\quad \text{and so on.}$$

Fraction notation indicates division. For example, $24 \div 8$ can be written in the fraction form $\frac{24}{8}$, which indicates that the numerator is to be divided by the denominator. Thus,

$$\frac{24}{8}=3,\quad \frac{45}{5}=9,\quad \text{and} \quad \frac{0}{5}=0.$$

To discuss fractions in general we need the concept of a variable.

Variable

A **variable** is a symbol (generally a letter of the alphabet) that is used to represent an unknown number.

DEFINITION

Because we know that division by 0 is **undefined**, no denominator can be 0. Thus, in the fraction form $\frac{a}{b}$, we write $b \neq 0$ (read, "b is not equal to 0").

Math Tip

Recall from Chapter 1 that if any whole number, including 0, is divided by 0 the quotient is **undefined**.

The Number 0 in Fractions

For any nonzero value of b, $\frac{0}{b}=\mathbf{0}$.

For any value of a, $\frac{a}{0}$ is **undefined**.

DEFINITION

Example 5 Evaluating Fractions Involving 0

Find the value of each expression.

a. $\frac{0}{36}$ **b.** $\frac{0}{124}$ **c.** $\frac{17}{0}$ **d.** $\frac{1}{0}$

Solution

a. $\frac{0}{36}=0$ **b.** $\frac{0}{124}=0$ **c.** $\frac{17}{0}$ is undefined **d.** $\frac{1}{0}$ is undefined

Now work margin exercise 5.

5. Find the value of each expression.

a. $\frac{0}{45}$

b. $\frac{10}{0}$

B Graphing Fractions on a Number Line

We have seen how to "picture" fractions as parts of a whole: part of a whole candy bar, part of a pizza, or part of a shaded region of a whole figure. Another way to visualize a fraction is to mark a corresponding point on a number line. For example, to graph the fraction $\frac{2}{3}$, proceed as follows.

1. Divide the interval (distance) from 0 to 1 into 3 equal parts.
2. Graph (or shade) the second mark to the right of 0.

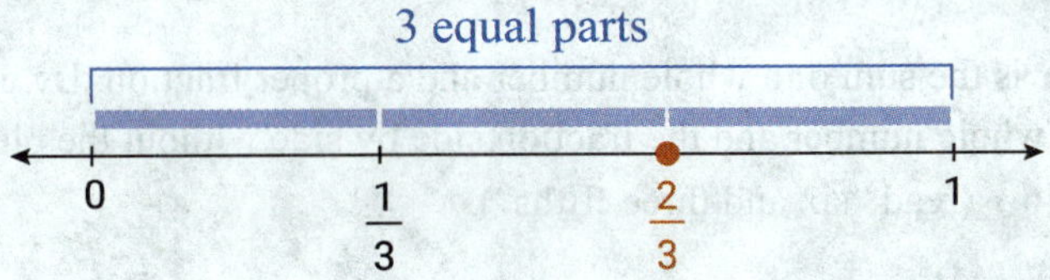

Example 6 Graphing Proper Fractions

Graph each proper fraction on a number line.

a. $\frac{4}{5}$ **b.** $\frac{2}{7}$ **c.** $\frac{5}{6}$

Solution

a.

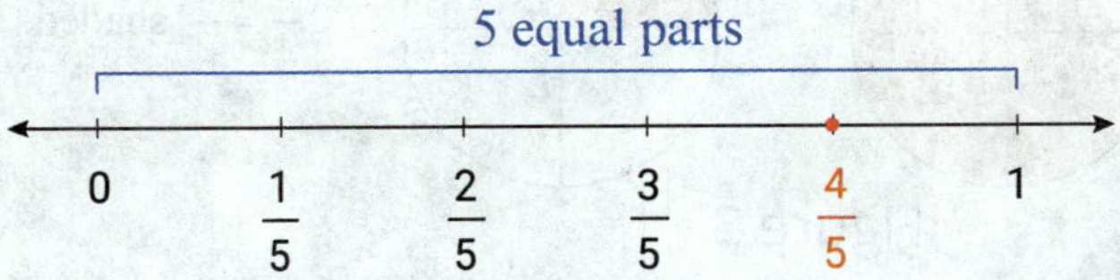

b.

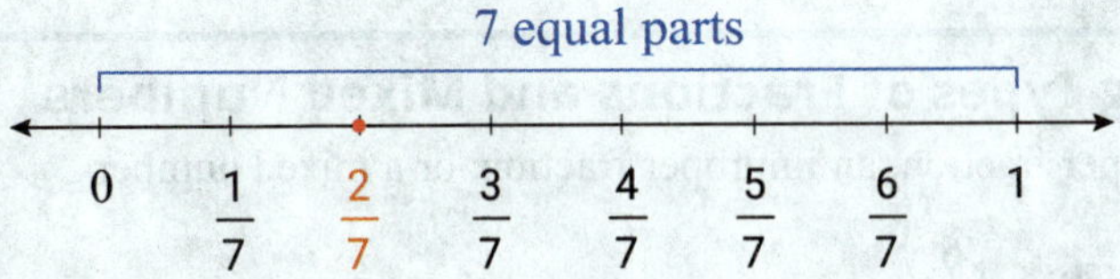

c.

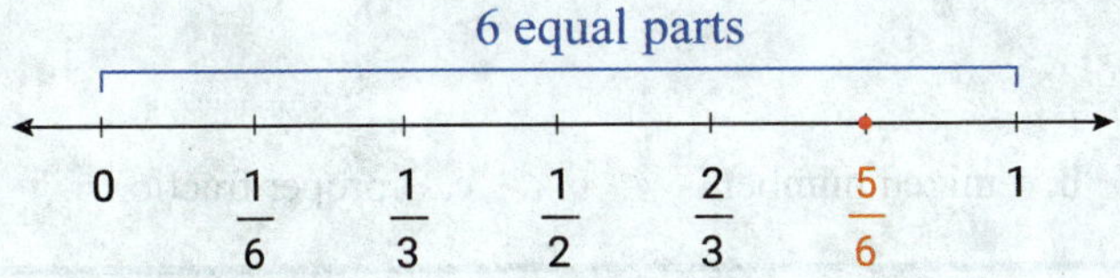

Now work margin exercise 6.

6. Graph each proper fraction on a number line.

a. $\frac{3}{4}$

b. $\frac{1}{5}$

c. $\frac{1}{2}$

We know that with improper fractions the numerator is equal to or greater than the denominator. This means that the graph of an improper fraction will be at 1 or to the right of 1 on a number line.

7. Graph each improper fraction on a number line.

a. $\frac{4}{3}$

b. $\frac{9}{4}$

Example 7 Graphing Improper Fractions

Graph each of the following improper fractions on a number line.

a. $\frac{7}{5}$ **b.** $\frac{13}{8}$

Solution

a.

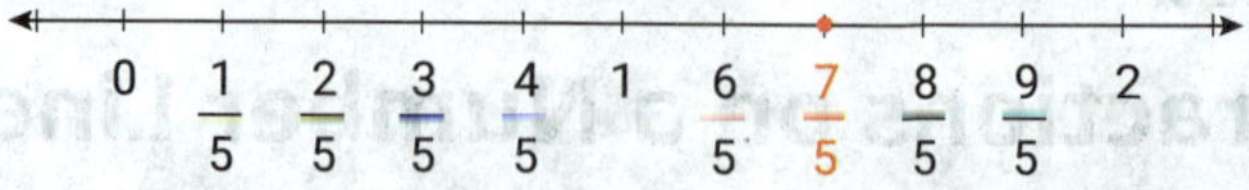

b.

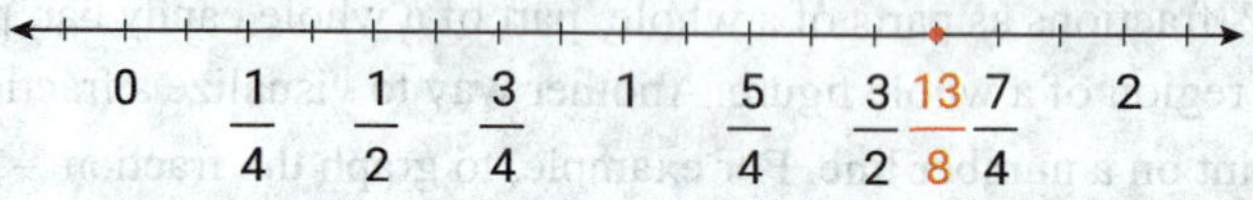

Now work margin exercise 7.

C Introduction to Mixed Numbers

A **mixed number** is the sum of a whole number and a proper fraction. By convention, we usually write the whole number and the fraction side by side without the plus sign. For example, $6+\frac{3}{5}=6\frac{3}{5}$ (read "six and three-fifths").

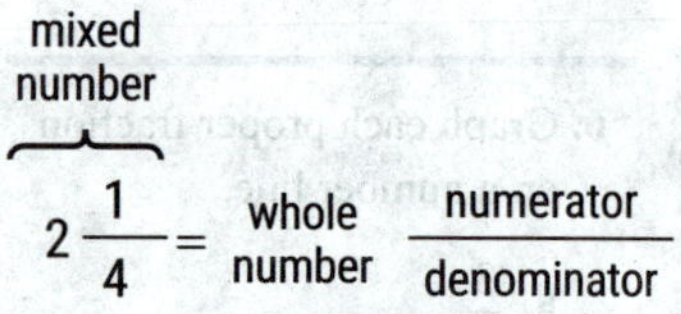

Terminology for mixed numbers

Typically, people are familiar with mixed numbers and use them frequently. For example, a carpenter might measure a board to be $2\frac{1}{4}$ feet long, or an architect might want to shade $2\frac{1}{4}$ circles in a drawing. A related question would be how many fourths (quarters) of a circle would be shaded? As shown in Figure 3, nine-fourths would be shaded. Thus, the form of an answer might be written as an improper fraction or a mixed number.

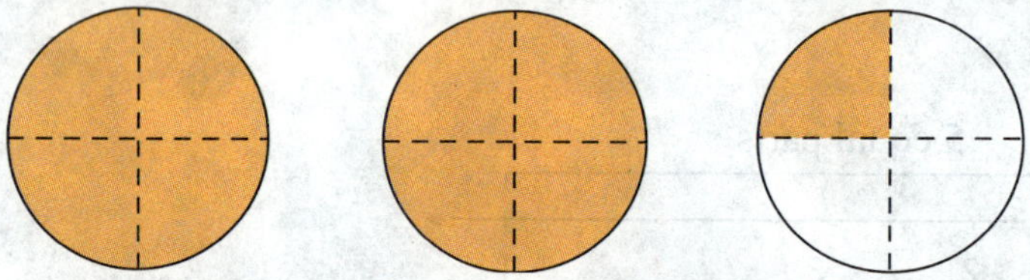

$2\frac{1}{4}=\frac{9}{4}$ shaded

Figure 3

8. Identify each number as a proper fraction, an improper fraction, or a mixed number.

a. $5\frac{2}{3}$

b. $\frac{3}{2}$

c. $\frac{2}{3}$

Example 8 Identifying Types of Fractions and Mixed Numbers

Identify each number as a proper fraction, an improper fraction, or a mixed number.

a. $\frac{8}{7}$ **b.** $1\frac{6}{7}$ **c.** $\frac{7}{8}$

Solution

a. improper fraction **b.** mixed number **c.** proper fraction

Now work margin exercise 8.

Example 9 Application: Understanding Mixed Numbers

A recipe calls for the amount of oil indicated in the figure.

a. Write a mixed number indicating the amount of oil in the measuring cups.

b. Write this amount as an improper fraction.

Solution

a. Each cup is marked in fourths and we see that there is a total of $1\frac{1}{4}$ cups.

b. As an improper fraction, $1\frac{1}{4}$ cups $= \frac{5}{4}$ cups.

Now work margin exercise 9.

9. Cassandra has the following eggs in her refrigerator.

a. Write a mixed number indicating how many cartons of eggs she has (12 eggs equals 1 carton).

b. Write this amount as an improper fraction.

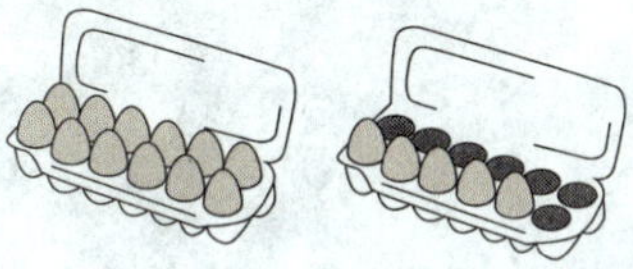

Example 10 Application: Understanding Mixed Numbers

A wooden rod is cut to the length indicated in the figure. Write the length of the rod as a mixed number.

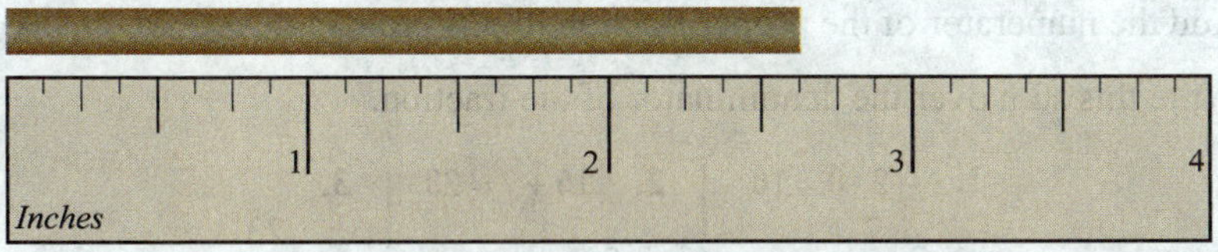

Solution

The ruler is marked in eighths of an inch. The rod measures $2\frac{5}{8}$ in.

Now work margin exercise 10.

10. A ribbon is cut to the length indicated in the figure. Write the length of the ribbon as a mixed number.

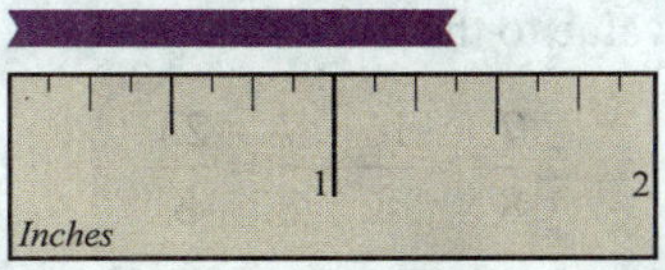

D Graphing Mixed Numbers on a Number Line

We know that mixed numbers are greater than or equal to 1 with a whole number part and a fraction part. To graph a mixed number we can proceed as with fractions by making marks on a number line that correspond to the denominator of the fraction part. For example, to graph the mixed number $1\frac{3}{4}$ proceed as follows.

1. Mark the intervals from 0 to 1 and from 1 to 2 into 4 equal parts.
2. Graph (or shade) the mark at $1\frac{3}{4}$.

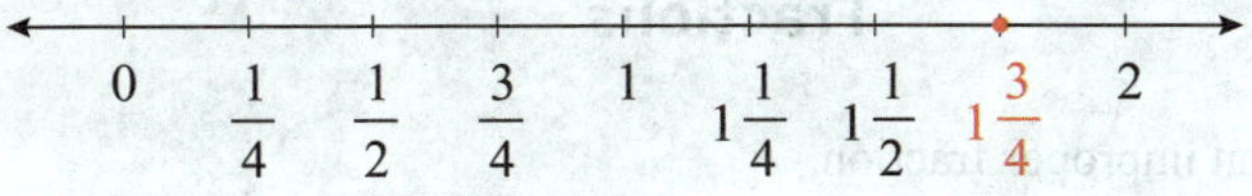

11. Graph each mixed number on a number line.

a. $1\frac{3}{4}$

b. $2\frac{2}{3}$

Example 11 Graphing Mixed Numbers

Graph each of the following mixed numbers on a number line.

a. $2\frac{1}{3}$ **b.** $3\frac{1}{2}$

Solution

a.

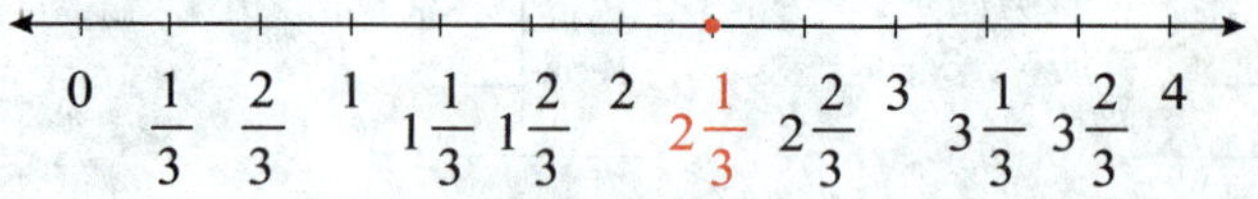

b.

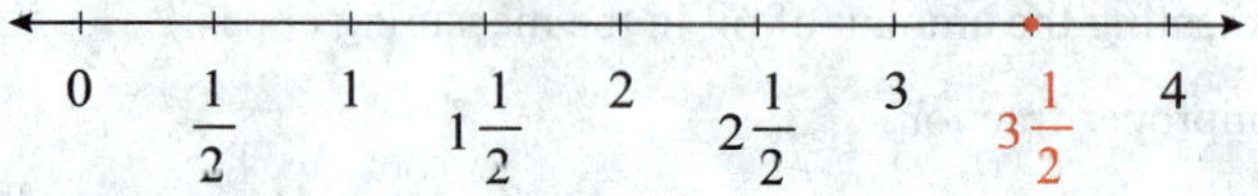

Now work margin exercise 11.

Math Tip

Writing $2\frac{7}{8}$ as an improper fraction involves changing the whole-number part into a fraction with the same denominator as the fractional part, and then adding that value to the fractional part.

$$2\frac{7}{8} = \frac{16}{8} + \frac{7}{8} = \frac{23}{8}$$

The three steps described in the procedure box show a shortcut for accomplishing this.

E Changing Mixed Numbers to Improper Fractions

The following procedure can be used to change mixed numbers to improper fractions.

Changing a Mixed Number to an Improper Fraction

1. Multiply the whole number by the denominator of the proper fraction.
2. Add the numerator of the proper fraction to this product.
3. Write this sum over the denominator of the fraction.

1. $2 \cdot 8 = 16$ $\quad 2\frac{7}{8} =$ | **2.** $16 + 7 = 23$ $\quad 2\frac{7}{8} =$ | **3.** $\frac{23}{8}$

PROCEDURE

12. Change $10\frac{4}{9}$ to an improper fraction.

Example 12 Changing Mixed Numbers to Improper Fractions

Change $8\frac{9}{10}$ to an improper fraction.

Solution

Step 1: Multiply the whole number by the denominator: $8 \cdot 10 = 80$

Step 2: Add the numerator: $80 + 9 = 89$

Step 3: Write this sum over the denominator: $8\frac{9}{10} = \frac{89}{10}$

Now work margin exercise 12.

13. Change $3\frac{5}{9}$ to an improper fraction.

Completion Example 13 Changing Mixed Numbers to Improper Fractions

Change $11\frac{2}{3}$ to an improper fraction.

Solution

Step 1: Multiply the whole number by the denominator: $11 \cdot 3 = \underline{33}$

Step 2: Add the numerator: $\underline{33} + \underline{2} = \underline{35}$

Step 3: Write this sum over the denominator: $11\frac{2}{3} = \frac{35}{3}$

Now work margin exercise 13.

F Changing Improper Fractions to Mixed Numbers

To reverse the process (that is, to change an improper fraction to a mixed number), we use the fact that a fraction can indicate division.

Changing an Improper Fraction to a Mixed Number

1. Divide the numerator by the denominator. The quotient is the whole number part of the mixed number.
2. Write the remainder over the denominator as the fraction part of the mixed number.

PROCEDURE

Example 14 Changing Improper Fractions to Mixed Numbers

Change $\frac{67}{5}$ to a mixed number.

Solution

Step 1: Divide the numerator by the denominator. The quotient is the whole number part of the mixed number.

$$\begin{array}{r} 13 \\ 5\overline{)67} \\ \underline{-50} \\ 17 \\ \underline{-15} \\ 2 \end{array}$$

13 Whole number part

2 Remainder

Step 2: Write the remainder over the denominator as the fraction part of the mixed number:

$$\frac{67}{5} = 13 + \frac{2}{5} = 13\frac{2}{5}.$$

Now work margin exercise 14.

14. Change $\frac{92}{11}$ to a mixed number.

15. Change $\frac{77}{6}$ to a mixed number.

Example 15 Changing Improper Fractions to Mixed Numbers

Change $\frac{85}{2}$ to a mixed number.

Solution

Step 1: Divide the numerator by the denominator. The quotient is the whole number part of the mixed number.

$$\begin{array}{r} 42 \\ 2\overline{)85} \\ \underline{-8} \\ 05 \\ \underline{-4} \\ 1 \end{array}$$

42 — Whole number part

1 — Remainder

Step 2: Write the remainder over the denominator as the fraction part of the mixed number:

$$\frac{85}{2} = 42 + \frac{1}{2} = 42\frac{1}{2}.$$

Now work margin exercise 15.

CALCULATORS

Using a Calculator to Convert Between Fractions and Mixed Numbers

Many scientific calculators have a fraction button, [a b/c]. To enter a mixed number or fraction in your calculator using this button, press [a b/c] between the whole number and the numerator, and again between the numerator and denominator. (To enter a fraction, simply press [a b/c] between the numerator and denominator.)

If your calculator includes this feature, then you can use it to convert between improper fractions and mixed numbers. Consider the mixed number $2\frac{1}{3}$. To enter this mixed number in your calculator, press the keys

[2] [a b/c] [1] [a b/c] [3].

The calculator will display this as 2⌟1/3, which means $2\frac{1}{3}$.

To convert this to an improper fraction press [2nd] [a b/c] (or [SHIFT] [a b/c]). This accesses the A b/c ◁▷ d/e feature that converts mixed numbers to fractions and vice versa. The display will read 7/3 which means $\frac{7}{3}$.

Similarly, to convert $\frac{5}{4}$ to a mixed number, press [5] [a b/c] [4] [=]. The display will read 1⌟1/4 which means $1\frac{1}{4}$.

Completion Example Answers

13. $11 \cdot 3 = 33$; $33 + 2 = 35$; $11\frac{2}{3} = \frac{35}{3}$

Margin Exercise Answers

1. a. $\frac{2}{5}$ **b.** $\frac{3}{5}$ **2. a.** $\frac{7}{15}$ **b.** $\frac{8}{15}$ **3.** 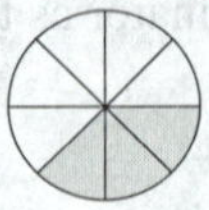 **4.** $\frac{3}{2}$ (3 of the equal parts)

5. a. 0 **b.** Undefined

6. a. 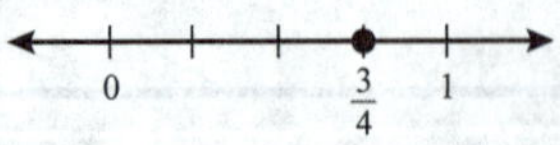

b. 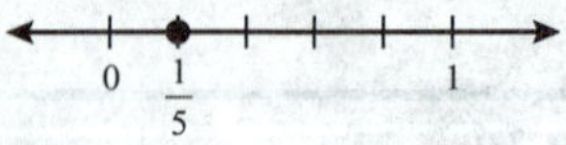

c. 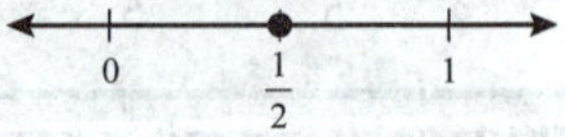

7. a. 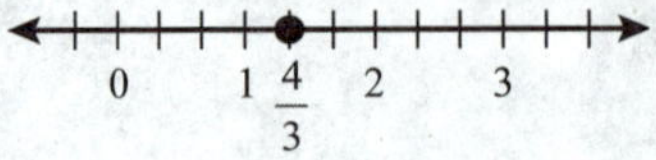

b. 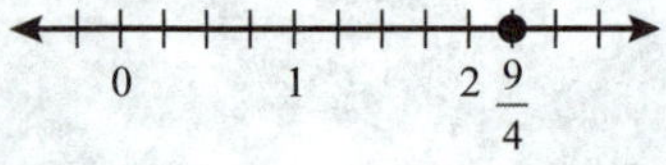

8. a. Mixed number **b.** Improper fraction **c.** Proper fraction **9. a.** $1\frac{5}{12}$ **b.** $\frac{17}{12}$ **10.** $1\frac{3}{8}$ in.

11. a. 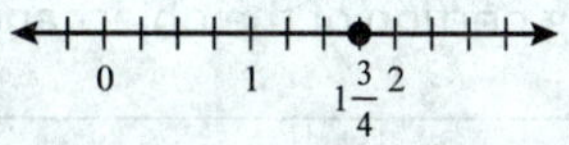

b. 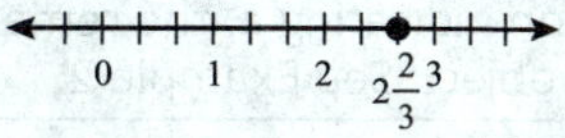

12. $\frac{94}{9}$ **13.** $\frac{32}{9}$ **14.** $8\frac{4}{11}$ **15.** $12\frac{5}{6}$

2.1 Exercises

Concept Check

Fill-in-the-Blank. Complete each sentence using information found in this section.

1. If a fraction has a numerator that is equal to or larger than the denominator, it is a/an ________ fraction.

2. A fraction that has a zero in the denominator is considered to be ___________.

3. The sum of a whole number and a proper fraction is called a/an ________ number.

4. The first step in changing an improper fraction into a mixed number is to divide the ___________ by the ___________.

True/False. Determine whether each statement is true or false. If a statement is false, explain how it can be changed so the statement will be true. (**Note:** There may be more than one acceptable change.)

5. In $\frac{11}{13}$, the denominator is 11.

6. $\frac{0}{6} = 0$

7. $\frac{17}{0}$ is undefined.

Practice

For each figure, write a fraction indicating **a.** the shaded part of the figure and **b.** the unshaded part of the figure. See Example 1.

1.

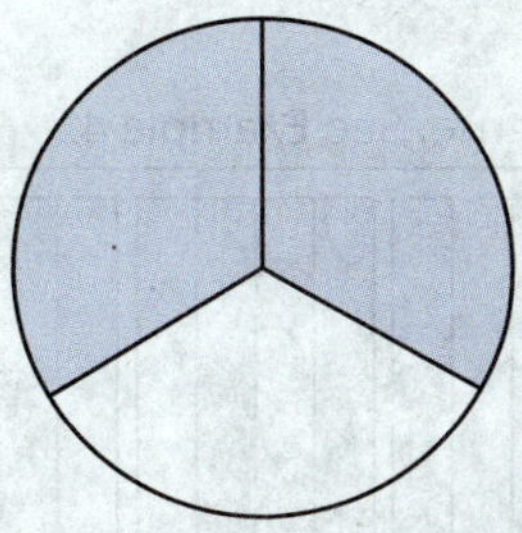

2.

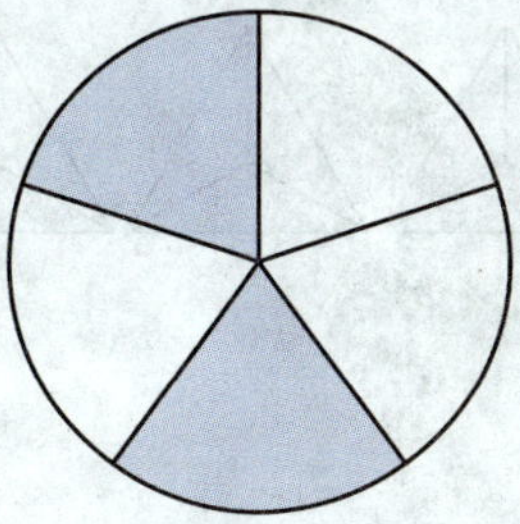

3.

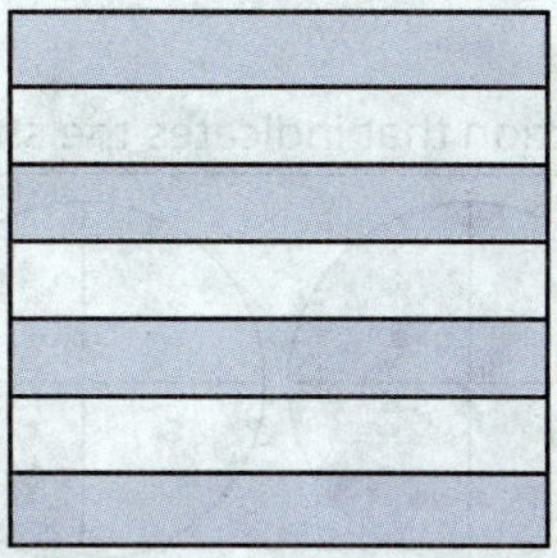

4.

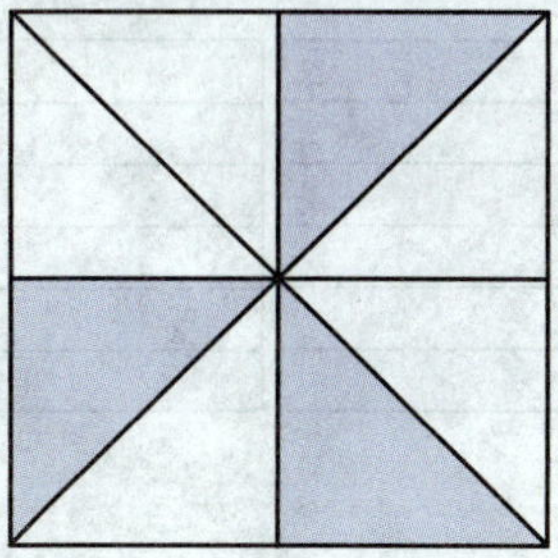

2.1 Exercises

Concept Check

1. improper

2. undefined

3. mixed

4. numerator, denominator

5. False; The numerator is 11.

6. True

7. True

Practice

1. a. $\frac{2}{3}$ **b.** $\frac{1}{3}$

2. a. $\frac{2}{5}$ **b.** $\frac{3}{5}$

3. a. $\frac{4}{7}$ **b.** $\frac{3}{7}$

4. a. $\frac{3}{8}$ **b.** $\frac{5}{8}$

5. a. $\frac{1}{6}$ b. $\frac{5}{6}$

6. a. $\frac{9}{16}$ b. $\frac{7}{16}$

7. a. $\frac{7}{12}$ b. $\frac{5}{12}$

8. a. $\frac{10}{13}$ b. $\frac{3}{13}$

9. a. $\frac{7}{30}$ b. $\frac{23}{30}$

10. a. $\frac{7}{11}$ b. $\frac{4}{11}$

11. 

12. 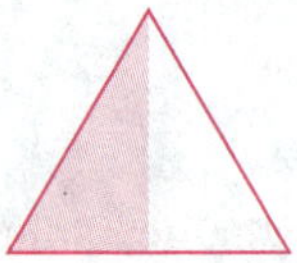

13. 

14. 

15. $\frac{5}{4}$

16. $\frac{11}{6}$

17. $\frac{10}{4}$

18. $\frac{7}{3}$

For each figure, write a fraction indicating **a.** the remaining portion of the object and **b.** the missing portion of the object. See Example 2.

5. 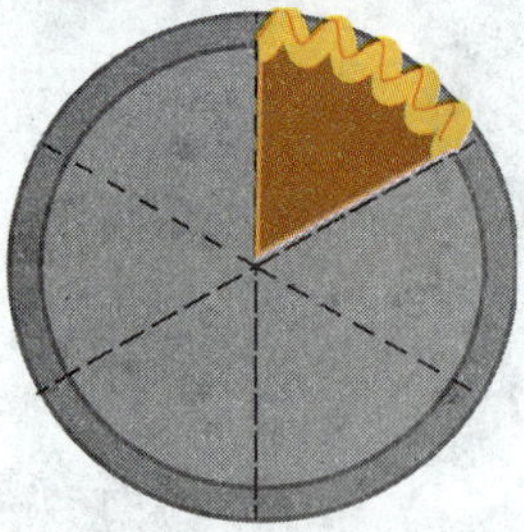

8. 

6. 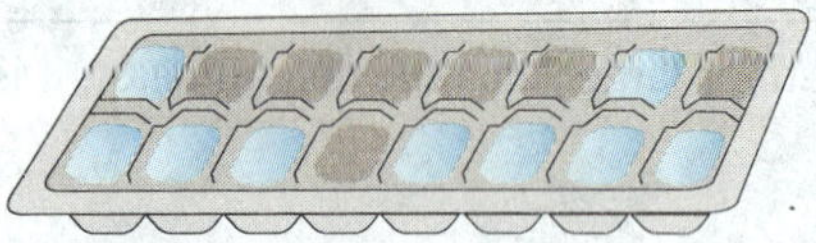

9. 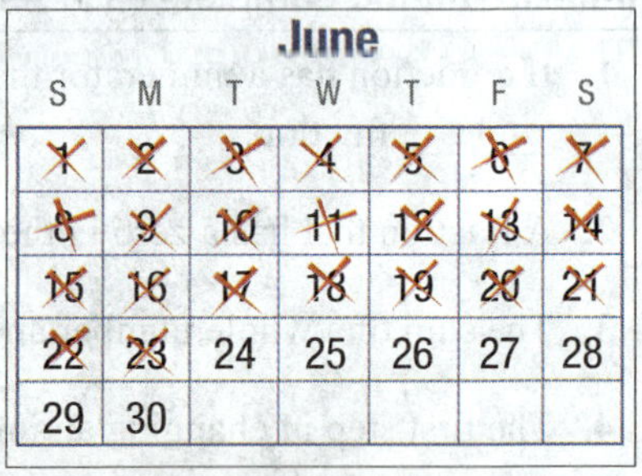

7. 

10. 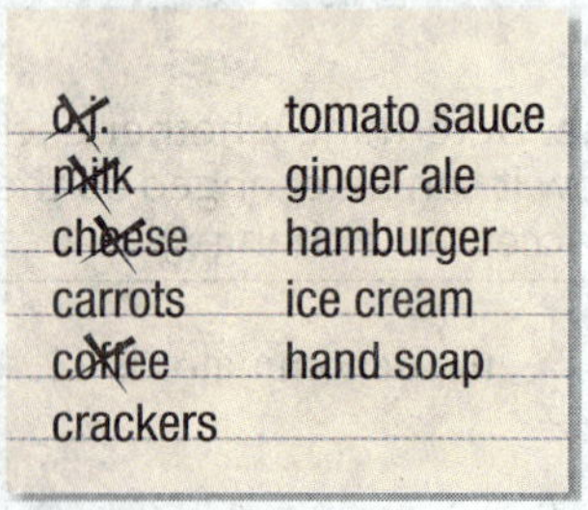

Draw a figure to represent each fraction. See Example 3.

11. $\frac{1}{3}$

12. $\frac{1}{2}$

13. $\frac{4}{5}$

14. $\frac{3}{4}$

Write a fraction that indicates the shaded parts of each figure. See Example 4.

15. 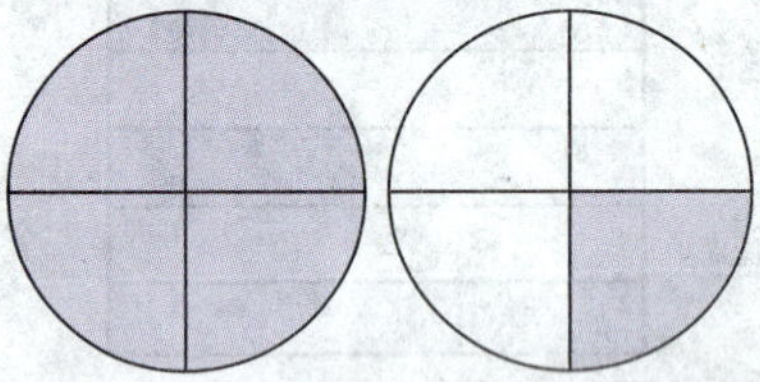

17. 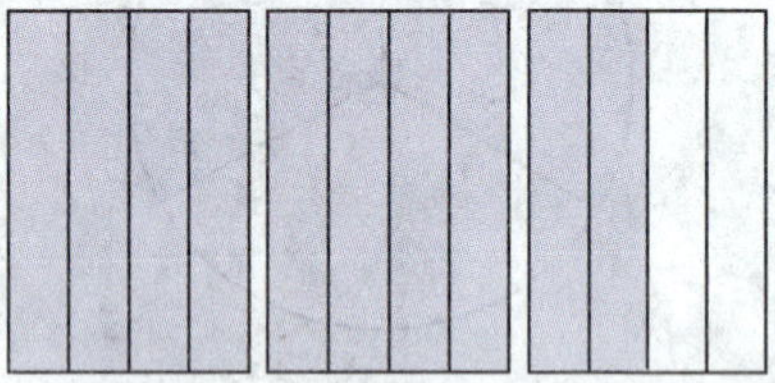

16. 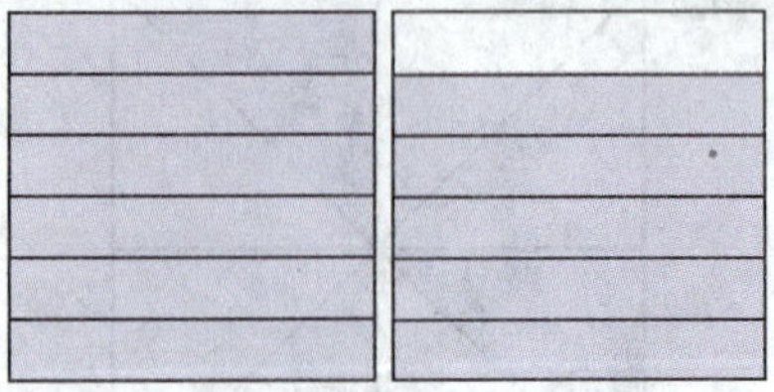

18. 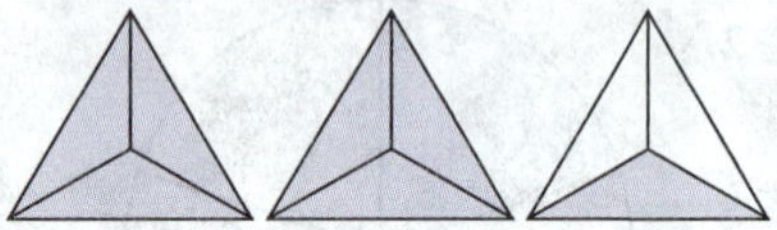

Find the value of each expression. See Example 5.

19. $\dfrac{0}{6}$

20. $\dfrac{0}{35}$

21. $\dfrac{15}{0}$

22. $\dfrac{2}{0}$

Graph each fraction on a number line. See Examples 6 and 7.

23. $\dfrac{3}{5}$

24. $\dfrac{3}{8}$

25. $\dfrac{6}{5}$

26. $\dfrac{8}{3}$

Identify each number as a proper fraction, an improper fraction, or a mixed number. See Example 8.

27. $1\dfrac{1}{2}$

28. $\dfrac{5}{3}$

29. $\dfrac{7}{8}$

30. $7\dfrac{5}{12}$

Write each amount described as **a.** a mixed number and **b.** an improper fraction. See Example 9.

31. Isabella brought 2 boxes of doughnuts to a meeting. The figure shows the remaining amount of doughnuts.

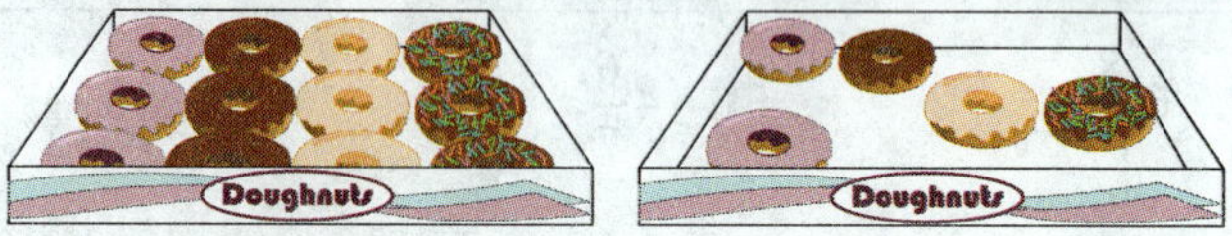

32. A recipe calls for the amount of tomato juice indicated in the figure.

33. Shane has two blister packs of gum. The figure shows the remaining amount of gum.

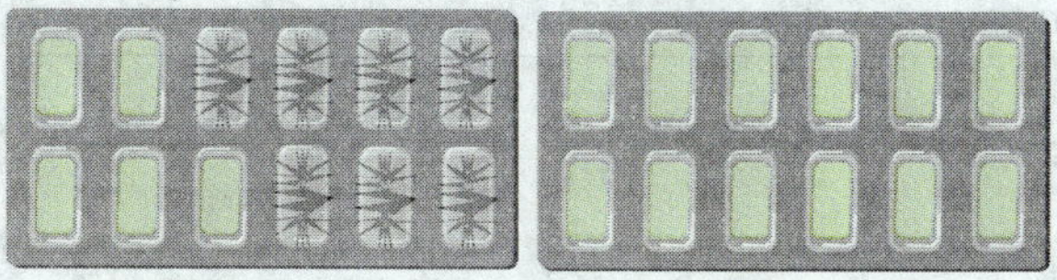

34. Cassandra has the following eggs in her refrigerator.

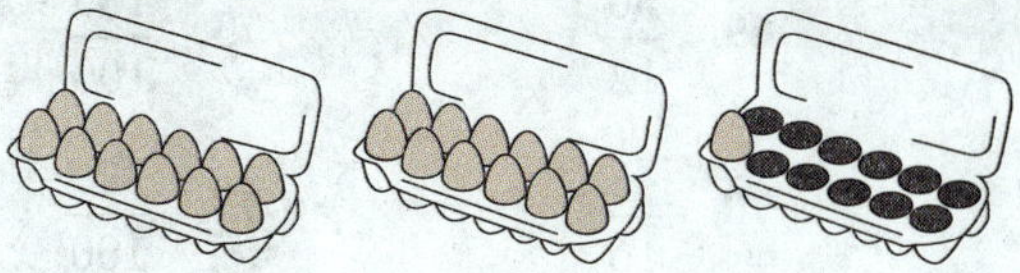

19. 0

20. 0

21. Undefined

22. Undefined

23.

24.
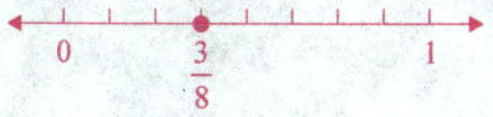

25.

26.

27. Mixed number

28. Improper fraction

29. Proper fraction

30. Mixed number

31. **a.** $1\dfrac{5}{12}$ **b.** $\dfrac{17}{12}$

32. **a.** $1\dfrac{3}{4}$ **b.** $\dfrac{7}{4}$

33. **a.** $1\dfrac{5}{12}$ **b.** $\dfrac{17}{12}$

34. **a.** $2\dfrac{1}{12}$ **b.** $\dfrac{25}{12}$

35. $2\frac{3}{8}$ in.

36. $3\frac{5}{8}$ in.

37. $1\frac{7}{8}$ in.

38. $3\frac{1}{8}$ in.

39. 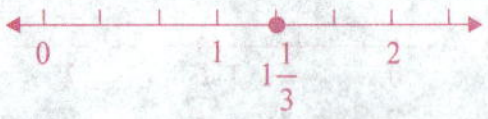

40. 

41. 

42. 

43. $\frac{8}{5}$

44. $\frac{5}{3}$

45. $\frac{9}{4}$

46. $\frac{19}{4}$

47. $\frac{68}{10}$

48. $\frac{94}{10}$

49. $\frac{37}{8}$

50. $\frac{52}{8}$

51. $\frac{34}{7}$

52. $\frac{50}{7}$

53. $\frac{17}{15}$

54. $\frac{20}{17}$

55. $\frac{128}{12}$

56. $\frac{151}{50}$

57. $\frac{701}{100}$

58. $\frac{619}{100}$

Write a mixed number to describe the length indicated in each figure. See Example 10.

35. 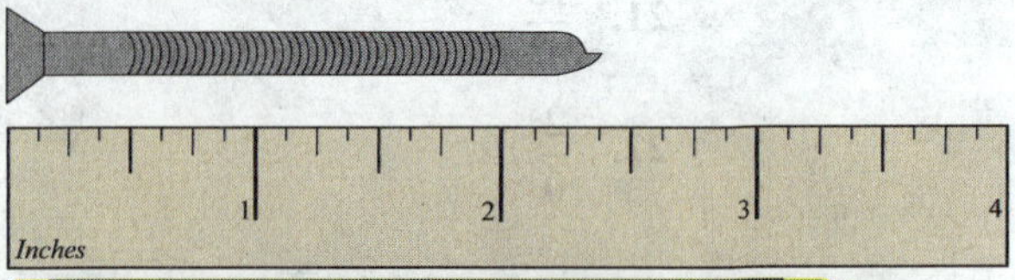

36. 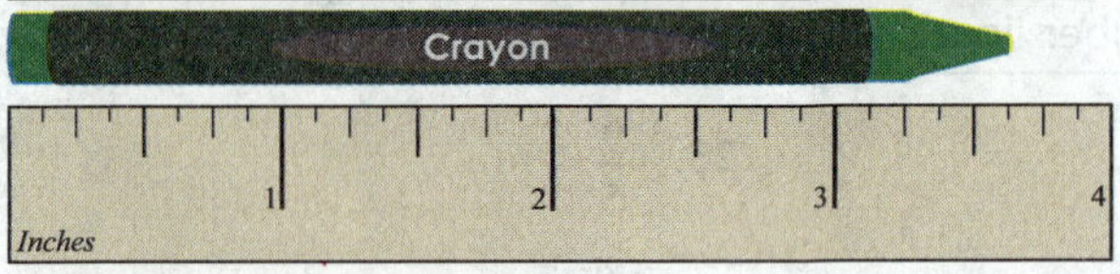

37. 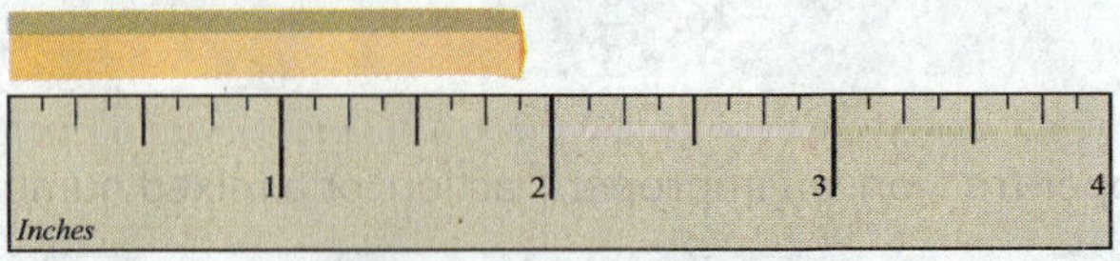

38. 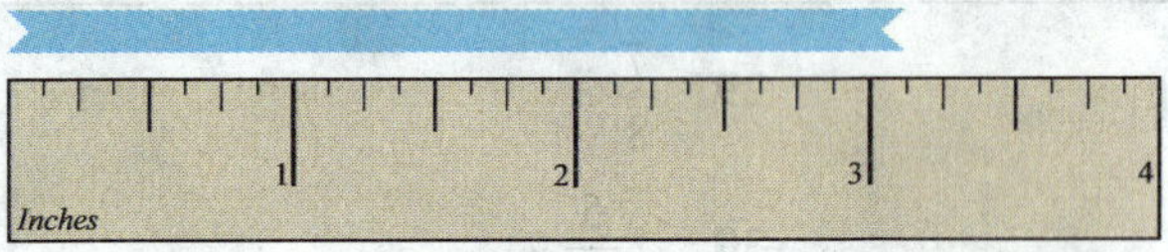

Graph each mixed number on a number line. See Example 11.

39. $1\frac{1}{3}$ **40.** $3\frac{1}{4}$ **41.** $2\frac{2}{5}$ **42.** $1\frac{3}{8}$

Change each mixed number to an improper fraction. See Examples 12 and 13.

43. $1\frac{3}{5}$ **47.** $6\frac{8}{10}$ **51.** $4\frac{6}{7}$ **55.** $10\frac{8}{12}$

44. $1\frac{2}{3}$ **48.** $9\frac{4}{10}$ **52.** $7\frac{1}{7}$ **56.** $3\frac{1}{50}$

45. $2\frac{1}{4}$ **49.** $4\frac{5}{8}$ **53.** $1\frac{2}{15}$ **57.** $7\frac{1}{100}$

46. $4\frac{3}{4}$ **50.** $6\frac{4}{8}$ **54.** $1\frac{3}{17}$ **58.** $6\frac{19}{100}$

Change each improper fraction to a mixed number. See Examples 14 and 15.

59. $\frac{4}{3}$ **63.** $\frac{5}{2}$ **67.** $\frac{37}{8}$ **71.** $\frac{35}{20}$

60. $\frac{11}{8}$ **64.** $\frac{17}{8}$ **68.** $\frac{29}{6}$ **72.** $\frac{14}{5}$

61. $\frac{13}{2}$ **65.** $\frac{27}{10}$ **69.** $\frac{36}{12}$ **73.** $\frac{185}{100}$

62. $\frac{19}{4}$ **66.** $\frac{33}{10}$ **70.** $\frac{48}{16}$ **74.** $\frac{329}{100}$

Applications

Solve.

75. If you had \$20 and you spent \$9 for a hamburger, fries, and a soft drink, what fraction of your money did you spend? What fraction would you still have?

76. In a class of 35 students, 6 students received As on a mathematics exam. What fraction of students received an A? What fraction of students did not receive an A?

77. A software company receives 45 technical support calls in one hour. Twenty-three of the calls are related to customers forgetting their passwords. What fraction of the calls was related to customers forgetting their passwords?

78. A certain brand of plain bagels has 146 calories per bagel. In each bagel, 115 calories come from carbohydrates. What fraction of the calories is from carbohydrates?

79. What fraction of a minute does 43 seconds represent? (**Hint:** There are 60 seconds in a minute.)

80. There are 5280 feet in a mile. What fraction of a mile does 923 feet represent?

81. A computer stores data on a hard drive in the form of bits, bytes, and sectors.

a. Each byte is made up of eight bits. What fraction of a byte is a bit?

b. A sector on a hard drive is traditionally 512 bytes. A byte is what fraction of a sector?

c. If a computer stores 159 bytes of data, what fraction of a sector does that amount of data take up?

82. The gas tank of a car holds 14 gallons of gas. What fraction of the tank does 9 gallons of gas take up?

83. A small box will hold 12 books. Kathleen has 35 books to pack into small boxes.

a. Write an improper fraction to describe the number of boxes that will be filled by Kathleen's books.

b. Change the improper fraction from part a. to a mixed number to describe the number of boxes that will be filled by Kathleen's books.

84. A cup holds 8 ounces of liquid. You have 29 ounces of juice to pour into cups.

a. Write an improper fraction to describe the number of cups that will be filled with juice.

b. Change the improper fraction from part a. to a mixed number to describe the number of cups that will be filled with juice.

59. $1\frac{1}{3}$

60. $1\frac{3}{8}$

61. $6\frac{1}{2}$

62. $4\frac{3}{4}$

63. $2\frac{1}{2}$

64. $2\frac{1}{8}$

65. $2\frac{7}{10}$

66. $3\frac{3}{10}$

67. $4\frac{5}{8}$

68. $4\frac{5}{6}$

69. 3

70. 3

71. $1\frac{15}{20}$

72. $2\frac{4}{5}$

73. $1\frac{85}{100}$

74. $3\frac{29}{100}$

Applications

75. $\frac{9}{20}$; $\frac{11}{20}$

76. $\frac{6}{35}$; $\frac{29}{35}$

77. $\frac{23}{45}$

78. $\frac{115}{146}$

79. $\frac{43}{60}$

80. $\frac{923}{5280}$

81. **a.** $\frac{1}{8}$ **b.** $\frac{1}{512}$ **c.** $\frac{159}{512}$

82. $\frac{9}{14}$

83. a. $\frac{35}{12}$ b. $2\frac{11}{12}$

84. a. $\frac{29}{8}$ b. $3\frac{5}{8}$

Writing & Thinking

85. In your own words, list the parts of a fraction and briefly describe the purpose of each part.

86. Give an example of a situation where you might use fractions and/or mixed numbers outside of class.

87. Show and explain, using diagrams and words, why $2\frac{3}{5}=\frac{13}{5}$.

88. Explain how to change an improper fraction into a mixed number.

Writing & Thinking

85. The two parts are the numerator and the denominator. The denominator represents the number of pieces in a whole and the numerator represents the number of these pieces being considered.

86. Answers will vary. For example, fractions and mixed numbers are often used in recipes.

87. Multiply the denominator by the whole number and add the numerator. This number is the new numerator. The denominator stays the same.

88. Divide the numerator by the denominator. The quotient is the whole number part of the mixed number. The remainder is the numerator of the fraction part of the mixed number. The denominator of the fraction part is the same as the original denominator.

2.2 Multiplication with Fractions

Objectives

A. Multiply fractions.

B. Reduce fractions to lowest terms.

C. Multiply and reduce fractions to lowest terms.

A Multiplication with Fractions

Finding the product of two fractions can be thought of as finding one fractional part of another fraction. For example, when we multiply $\frac{1}{2}$ by $\frac{1}{4}$ we are finding $\frac{1}{2}$ **of** $\frac{1}{4}$.

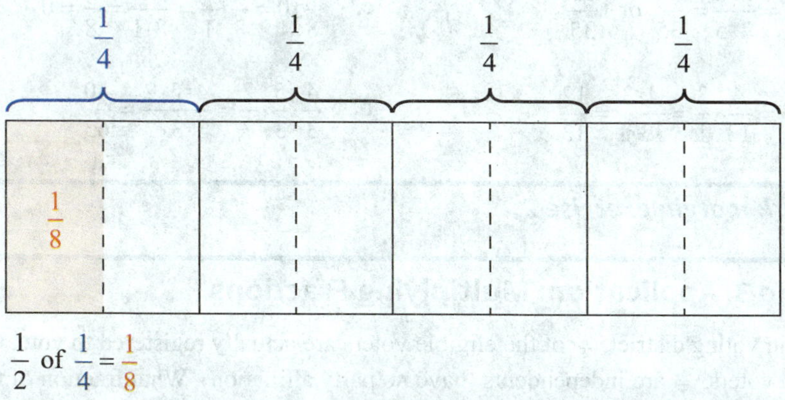

Figure 1

Now we state the rule for multiplying fractions. Remember that any whole number can be written in fraction form with a denominator of 1 and no denominator can be 0.

Multiplying Fractions

1. Multiply the numerators.
2. Multiply the denominators.

$$\frac{a}{b}\cdot\frac{c}{d}=\frac{a\cdot c}{b\cdot d}\quad (b,d\neq 0)$$

For example, $\frac{1}{2}\cdot\frac{3}{7}=\frac{1\cdot 3}{2\cdot 7}=\frac{3}{14}$.

PROCEDURE

Example 1 Multiplying Fractions

Find $\frac{2}{5}$ of $\frac{7}{3}$.

Solution

Remember, to find a fraction **of** a number means to multiply the number by the fraction.

$$\frac{2}{5}\cdot\frac{7}{3}=\frac{2\cdot 7}{5\cdot 3}=\frac{14}{15}$$

Now work margin exercise 1.

1. Find $\frac{3}{4}$ of $\frac{5}{8}$.

2. Multiply.

a. $\frac{2}{3}\cdot\frac{11}{9}$

b. $\frac{4}{5}\cdot 7$

c. $\frac{0}{4}\cdot\frac{11}{16}$

d. $\frac{1}{7}\cdot\frac{3}{5}\cdot\frac{3}{2}$

Example 2 Multiplying Fractions

Multiply.

a. $\frac{6}{7}\cdot\frac{8}{5}$ b. $\frac{4}{13}\cdot 3$ c. $\frac{9}{8}\cdot 0$ d. $\frac{4}{3}\cdot\frac{5}{3}\cdot\frac{2}{7}$

Solution

a. $\frac{6}{7}\cdot\frac{8}{5}=\frac{6\cdot 8}{7\cdot 5}=\frac{48}{35}$ or $1\frac{13}{35}$

c. $\frac{9}{8}\cdot 0=\frac{9}{8}\cdot\frac{0}{1}=\frac{9\cdot 0}{8\cdot 1}=\frac{0}{8}=0$

b. $\frac{4}{13}\cdot 3=\frac{4}{13}\cdot\frac{3}{1}=\frac{4\cdot 3}{13\cdot 1}=\frac{12}{13}$

d. $\frac{4}{3}\cdot\frac{5}{3}\cdot\frac{2}{7}=\frac{4\cdot 5\cdot 2}{3\cdot 3\cdot 7}=\frac{40}{63}$

Now work margin exercise 2.

3. For a certain piece of furniture, $\frac{1}{8}$ of the included hardware is wing nuts. Of these wing nuts, $\frac{1}{20}$ are defective. What fraction of the included hardware is defective wing nuts?

Example 3 Application: Multiplying Fractions

In a certain voting district, $\frac{3}{5}$ of the eligible voters are actually registered to vote. Of those registered voters, $\frac{2}{7}$ are independents (have no party affiliation). What fraction of the eligible voters are registered independents?

Solution

Step 1: READ: Read the problem carefully. There are two types of voters to consider: registered voters and independent voters.

Step 2: SET UP: In looking for the fraction of registered independent voters, we know that $\frac{2}{7}$ of $\frac{3}{5}$ of eligible voters are registered independents, so we will need to multiply $\frac{2}{7}\cdot\frac{3}{5}$.

Step 3: SOLVE: $\frac{2}{7}\cdot\frac{3}{5}=\frac{2\cdot 3}{7\cdot 5}=\frac{6}{35}$

Thus, $\frac{6}{35}$ of the eligible voters are registered as independents.

Step 4: CHECK: The fraction $\frac{3}{5}$ is a little more than $\frac{1}{2}$ and $\frac{1}{2}$ of $\frac{2}{7}$ is $\frac{1}{7}$, which means that approximately $\frac{1}{7}$ of eligible voters are registered as independents. Since $\frac{1}{7}=\frac{5}{35}$, the fraction $\frac{6}{35}$ seems like a reasonable answer as it is a little more than $\frac{5}{35}$.

Now work margin exercise 3.

Both the commutative property and the associative property of multiplication apply to fractions.

Commutative Property of Multiplication

The **order** of the fractions being multiplied can be reversed without changing the product. Symbolically, if $\frac{a}{b}$ and $\frac{c}{d}$ are fractions, then

$$\frac{a}{b}\cdot\frac{c}{d}=\frac{c}{d}\cdot\frac{a}{b}.\quad (b,d\neq 0)$$

For example, $\frac{2}{5}\cdot\frac{1}{3}=\frac{1}{3}\cdot\frac{2}{5}$.

PROPERTIES

Associative Property of Multiplication

The **grouping** of the fractions being multiplied can be changed without changing the product. Symbolically, if $\frac{a}{b}$, $\frac{c}{d}$, and $\frac{e}{f}$ are fractions, then

$$\left(\frac{a}{b}\cdot\frac{c}{d}\right)\cdot\frac{e}{f}=\frac{a}{b}\cdot\left(\frac{c}{d}\cdot\frac{e}{f}\right).\quad (b,d,f\neq 0)$$

For example, $\left(\frac{5}{6}\cdot\frac{2}{3}\right)\cdot\frac{1}{4}=\frac{5}{6}\cdot\left(\frac{2}{3}\cdot\frac{1}{4}\right)$.

PROPERTIES

Example 4 Recognizing the Properties of Multiplication

Each of the properties of multiplication is illustrated.

a. $\frac{3}{4}\cdot\frac{1}{7}=\frac{1}{7}\cdot\frac{3}{4}$ Commutative property of multiplication

As a check, we see that $\frac{3}{4}\cdot\frac{1}{7}=\frac{3}{28}$ and $\frac{1}{7}\cdot\frac{3}{4}=\frac{3}{28}$.

b. $\left(\frac{5}{8}\cdot\frac{3}{2}\right)\cdot\frac{3}{4}=\frac{5}{8}\cdot\left(\frac{3}{2}\cdot\frac{3}{4}\right)$ Associative property of multiplication

As a check, we see that

$$\left(\frac{5}{8}\cdot\frac{3}{2}\right)\cdot\frac{3}{4}=\frac{15}{16}\cdot\frac{3}{4}=\frac{45}{64}\quad\text{and}\quad\frac{5}{8}\cdot\left(\frac{3}{2}\cdot\frac{3}{4}\right)=\frac{5}{8}\cdot\frac{9}{8}=\frac{45}{64}.$$

Now work margin exercise 4.

4. For each statement, state the property of multiplication illustrated.

a. $\frac{1}{3}\cdot\left(\frac{1}{2}\cdot\frac{1}{4}\right)=\left(\frac{1}{3}\cdot\frac{1}{2}\right)\cdot\frac{1}{4}$

b. $\frac{1}{2}\cdot\frac{5}{8}=\frac{5}{8}\cdot\frac{1}{2}$

B Reducing Fractions to Lowest Terms

Figure 2 shows how a whole can be separated into equal parts in several ways. The shaded parts represent the same "fraction" of the whole.

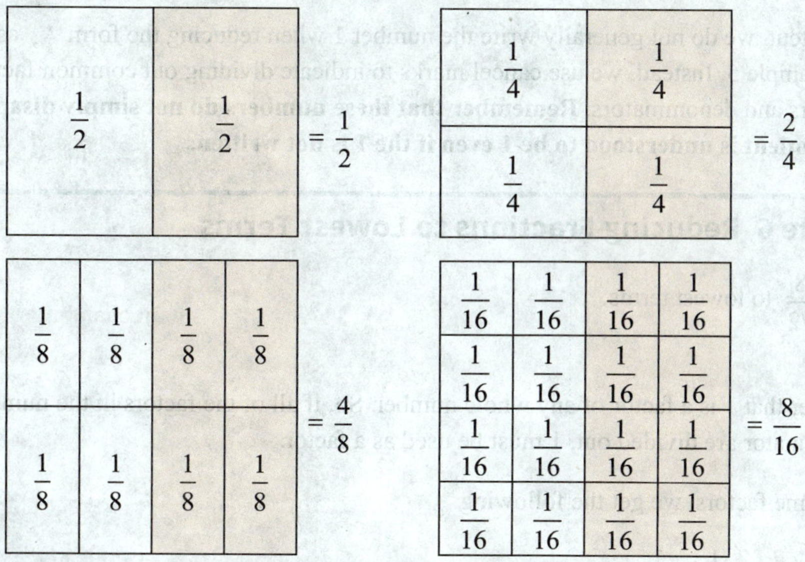

Figure 2

From the shading in Figure 2, we see that $\frac{1}{2}=\frac{2}{4}=\frac{4}{8}=\frac{8}{16}$. These fractions are **equal fractions**. Equal fractions are said to be **equivalent**.

Finding an equivalent fraction in lower terms (with a smaller denominator) is called **reducing** the fraction. **A fraction is reduced to lowest terms if the numerator and denominator have no common factors other than 1.**

The use of prime factors is very helpful in reducing to lowest terms.

Reducing a Fraction to Lowest Terms

1. Factor the numerator and denominator into prime factors.
2. Use the fact that $\frac{k}{k}=1$ and "divide out" all common factors.

Note: Reduced fractions may be improper fractions.

PROCEDURE

CALCULATORS

Using a Calculator to Reduce Fractions

To reduce fractions on a calculator, you will need the fraction button, [a b/c]. Reduce a fraction such as $\frac{8}{72}$ by pressing the keys [8] [a b/c] [7] [2]. Then press [=].

The calculator will display 1/9 or $\frac{1}{9}$, the simplified form of the fraction.

5. Reduce each fraction to lowest terms.

a. $\frac{8}{36}$

b. $\frac{49}{28}$

Example 5 Reducing Fractions to Lowest Terms

Reduce each fraction to lowest terms.

a. $\frac{15}{20}$ b. $\frac{35}{21}$

Solution

a. $\frac{15}{20}=\frac{3\cdot 5}{2\cdot 2\cdot 5}=\frac{3}{2\cdot 2}\cdot\frac{5}{5}=\frac{3}{4}\cdot 1=\frac{3}{4}$

b. $\frac{35}{21}=\frac{5\cdot 7}{3\cdot 7}=\frac{5}{3}\cdot\frac{7}{7}=\frac{5}{3}\cdot 1=\frac{5}{3}$

Note that the rules for divisibility quickly indicate that 5 is a factor of 15, 20, and 35, and 3 is a factor of 21.

Now work margin exercise 5.

As a shortcut, we do not generally write the number 1 when reducing the form $\frac{k}{k}$, as we did in Example 5. Instead, we use cancel marks to indicate dividing out common factors in numerators and denominators. **Remember that these numbers do not simply disappear. Their quotient is understood to be 1 even if the 1 is not written.**

6. Reduce $\frac{16}{64}$ to lowest terms.

Example 6 Reducing Fractions to Lowest Terms

Reduce $\frac{8}{72}$ to lowest terms.

Solution

Remember that 1 is a factor of any whole number. So, if all of the factors in the numerator or denominator are divided out, 1 must be used as a factor.

Using prime factors, we get the following.

$$\frac{8}{72}=\frac{\cancel{2}\cdot\cancel{2}\cdot\cancel{2}\cdot 1}{\cancel{2}\cdot\cancel{2}\cdot\cancel{2}\cdot 3\cdot 3}=\frac{1}{9}$$

Or, if you see that 8 is a common factor, divide it out. But remember that 1 is a factor.

$$\frac{8}{72}=\frac{\cancel{8}\cdot 1}{\cancel{8}\cdot 9}=\frac{1}{9}$$

Now work margin exercise 6.

Example 6 illustrated the fact that when reducing, common factors in the numerator and denominator need not be prime. However, the advantage of using prime factors is that you can be certain that the fraction is reduced to lowest terms.

Completion Example 7 Reducing Fractions to Lowest Terms

Reduce $\frac{52}{65}$ to lowest terms.

Solution

Finding a common factor could be difficult here. Prime factoring helps.

$$\frac{52}{65}=\frac{2\cdot 2\cdot \cancel{13}}{5\cdot \cancel{13}}=\frac{4}{5}$$

Now work margin exercise 7.

Example 8 Application: Reducing Fractions to Lowest Terms

Suppose you had \$25 and you spent \$15 to buy music.

a. What fraction of your money did you spend on music?

b. What fraction do you still have?

Solution

a. The fraction you spent is $\frac{15}{25}=\frac{3\cdot \cancel{5}}{5\cdot \cancel{5}}=\frac{3}{5}$.

b. Since you still have \$25 – \$15 = \$10, the fraction you still have is $\frac{10}{25}=\frac{2\cdot \cancel{5}}{5\cdot \cancel{5}}=\frac{2}{5}$.

Now work margin exercise 8.

C Multiplying and Reducing Fractions

Now we can multiply fractions and reduce all in one step by using prime factors (or other common factors). If you have any difficulty understanding how to multiply and reduce, use prime factors. By using prime factors, you can be sure that you have not missed a common factor and that your answer is reduced to lowest terms.

Examples 9 through 12 illustrate how to multiply and reduce at the same time by factoring the numerators and the denominators. Note that **if all the factors in the numerator or denominator divide out, then 1 must be used as a factor.** (See Examples 11 and 12.)

7. Reduce $\frac{48}{66}$ to lowest terms.

8. Suppose you had \$42 and spent \$30 to buy books.

a. What fraction of your money did you spend on books?

b. What fraction do you still have?

Math Tip

The fractional part you spent plus the fractional part you have left equals 1.

$$\frac{3}{5}+\frac{2}{5}=\frac{5}{5}=1$$

9. Multiply and reduce to lowest terms:

$\frac{4}{15}\cdot\frac{35}{48}$

Teaching Note

Encourage students to reduce as much as possible before multiplying. You can point out that time is wasted by multiplying numerators and denominators and then reducing the result, since the original numerators and denominators will factor back out again while reducing.

Example 9 Multiplying and Reducing Using Prime Factors

Multiply and reduce to lowest terms: $\frac{15}{28}\cdot\frac{7}{9}$

Solution

Do not start by multiplying the numerators and denominators. The results would simply be large numbers that would then need to be factored. Using prime factors, we have the following.

$$\frac{15}{28}\cdot\frac{7}{9}=\frac{15\cdot 7}{28\cdot 9}=\frac{\cancel{3}\cdot 5\cdot\cancel{7}}{2\cdot 2\cdot\cancel{7}\cdot\cancel{3}\cdot 3}=\frac{5}{2\cdot 2\cdot 3}=\frac{5}{12}$$

Note that we did not find the product in the numerator or denominator before factoring.

Now work margin exercise 9.

10. Multiply and reduce to lowest terms:

$\frac{5}{6}\cdot\frac{8}{7}\cdot\frac{14}{10}$

Example 10 Multiplying and Reducing Using Prime Factors

Multiply and reduce to lowest terms: $\frac{9}{10}\cdot\frac{25}{32}\cdot\frac{44}{33}$

Solution

Using prime factors, we have

$$\frac{9}{10}\cdot\frac{25}{32}\cdot\frac{44}{33}=\frac{9\cdot 25\cdot 44}{10\cdot 32\cdot 33}=\frac{\cancel{3}\cdot 3\cdot\cancel{5}\cdot 5\cdot\cancel{2}\cdot\cancel{2}\cdot\cancel{11}}{2\cdot\cancel{5}\cdot 2\cdot 2\cdot 2\cdot\cancel{2}\cdot\cancel{2}\cdot\cancel{3}\cdot\cancel{11}}=\frac{3\cdot 5}{2\cdot 2\cdot 2\cdot 2}=\frac{15}{16}.$$

Now work margin exercise 10.

11. Multiply and reduce to lowest terms:

$16\cdot\frac{12}{100}\cdot\frac{5}{36}$

Example 11 Multiplying and Reducing Using Prime Factors

Multiply and reduce to lowest terms: $\frac{17}{50}\cdot\frac{25}{34}\cdot 8$

Solution

$$\frac{17}{50}\cdot\frac{25}{34}\cdot 8=\frac{17\cdot 25\cdot 8}{50\cdot 34\cdot 1}$$

$$=\frac{\cancel{17}\cdot\cancel{25}\cdot\cancel{2}\cdot\cancel{2}\cdot 2}{\cancel{2}\cdot\cancel{25}\cdot\cancel{2}\cdot\cancel{17}\cdot 1}$$

$$=\frac{2}{1}=2$$

Note that $8=\frac{8}{1}$.

In this example, 25 is a common factor that is not a prime number.

Now work margin exercise 11.

12. Multiply and reduce to lowest terms:

$\frac{15}{6}\cdot\frac{4}{19}\cdot\frac{38}{10}$

Completion Example 12 Multiplying and Reducing Using Prime Factors

Multiply and reduce to lowest terms: $\frac{55}{26}\cdot\frac{8}{44}\cdot\frac{91}{35}$

Solution

$$\frac{55}{26}\cdot\frac{8}{44}\cdot\frac{91}{35}=\frac{55\cdot 8\cdot 91}{26\cdot 44\cdot 35}$$
$$=\frac{\cancel{5}\cdot\cancel{11}\cdot\cancel{2}\cdot\cancel{2}\cdot\cancel{2}\cdot\cancel{7}\cdot\cancel{13}}{\cancel{2}\cdot\cancel{13}\cdot\cancel{2}\cdot\cancel{2}\cdot\cancel{11}\cdot\cancel{7}\cdot\cancel{5}}$$
$$=\frac{1}{1}$$
$$=1$$

Now work margin exercise 12.

Example 13 Application: Multiplying and Reducing Fractions

A study showed that $\frac{5}{8}$ of the members of a public service organization were in favor of a new set of bylaws. If the organization had a membership of 200 people, how many were in favor of the changes in the bylaws?

Solution

Step 1: READ: Read the problem carefully. Note we need to find a fraction of the membership.

Step 2: SET UP: To find $\frac{5}{8}$ of 200, multiply.

Step 3: SOLVE: $\frac{5}{8}\cdot 200=\frac{5}{8}\cdot\frac{200}{1}=\frac{5\cdot\cancel{8}\cdot 25}{\cancel{8}\cdot 1}=\frac{125}{1}=125$

Thus, there are 125 members in favor of the bylaw changes.

Step 4: CHECK: The fraction $\frac{5}{8}$ is a little more than $\frac{1}{2}$, which gives an approximation of 100 members in favor of the law. Therefore, the answer of 125 members is reasonable as it is a little more than 100.

Now work margin exercise 13.

Another method frequently used to multiply and reduce at the same time is to divide numerators and denominators by common factors whether they are prime or not. If these factors are easily determined, then this method is probably faster. But common factors are sometimes missed with this method whereas they are not missed with the prime factorization method.

Examples 9 through 12 are shown again as Examples 14 through 17, this time using the division method.

Example 14 Multiplying and Reducing Using the Division Method

Multiply and reduce to lowest terms: $\frac{15}{28}\cdot\frac{7}{9}$

Solution

$$\frac{\overset{5}{\cancel{15}}}{\underset{4}{\cancel{28}}}\cdot\frac{\overset{1}{\cancel{7}}}{\underset{3}{\cancel{9}}}=\frac{5}{12}$$

3 divides both 15 and 9.
7 divides both 7 and 28.

Now work margin exercise 14.

13. A delivery service found that $\frac{9}{10}$ of its drivers wore their seat belts. If this company employed 500 drivers, how many drivers wore their seat belts?

14. Multiply and reduce to lowest terms:

$\frac{3}{16}\cdot\frac{40}{45}$

15. Multiply and reduce to lowest terms:

$\frac{14}{15}\cdot\frac{25}{21}\cdot\frac{3}{10}$

Example 15 Multiplying and Reducing Using the Division Method

Multiply and reduce to lowest terms: $\frac{9}{10}\cdot\frac{25}{32}\cdot\frac{44}{33}$

Solution

$$\frac{\overset{3}{\cancel{9}}}{\underset{2}{\cancel{10}}}\cdot\frac{\overset{5}{\cancel{25}}}{\underset{8}{\cancel{32}}}\cdot\frac{\overset{\overset{1}{\cancel{4}}}{\cancel{44}}}{\underset{\underset{1}{\cancel{3}}}{\cancel{33}}}=\frac{15}{16}$$

5 divides both 25 and 10.
11 divides both 44 and 33.
4 divides both 4 and 32.
3 divides both 9 and 3.

Now work margin exercise 15.

16. Multiply and reduce to lowest terms:

$\frac{8}{9}\cdot 15\cdot\frac{4}{30}$

Example 16 Multiplying and Reducing Using the Division Method

Multiply and reduce to lowest terms: $\frac{17}{50}\cdot\frac{25}{34}\cdot 8$

Solution

$$\frac{\overset{1}{\cancel{17}}}{\underset{\underset{1}{\cancel{2}}}{\cancel{50}}}\cdot\frac{\overset{1}{\cancel{25}}}{\underset{\underset{1}{\cancel{2}}}{\cancel{34}}}\cdot\frac{\overset{\overset{2}{\cancel{4}}}{\cancel{8}}}{1}=\frac{2}{1}=2$$

17 divides both 17 and 34.
25 divides both 25 and 50.
2 divides both 8 and 2.
2 divides both 4 and 2.

Now work margin exercise 16.

17. Multiply and reduce to lowest terms:

$\frac{17}{100}\cdot\frac{27}{34}\cdot\frac{25}{9}$

Example 17 Multiplying and Reducing Using the Division Method

Multiply and reduce to lowest terms: $\frac{55}{26}\cdot\frac{8}{44}\cdot\frac{91}{35}$

Solution

$$\frac{\overset{\overset{1}{\cancel{5}}}{\cancel{55}}}{\underset{\underset{1}{\cancel{2}}}{\cancel{26}}}\cdot\frac{\overset{\overset{1}{\cancel{4}}}{\cancel{8}}}{\underset{\underset{1}{\cancel{4}}}{\cancel{44}}}\cdot\frac{\overset{\overset{1}{\cancel{7}}}{\cancel{91}}}{\underset{\underset{1}{\cancel{7}}}{\cancel{35}}}=\frac{1}{1}=1$$

11 divides both 55 and 44. 13 divides both 91 and 26.
5 divides both 5 and 35. 7 divides both 7 and 7.
2 divides both 8 and 2. 4 divides both 4 and 4.

Now work margin exercise 17.

Completion Example Answers

7. $\frac{52}{65}=\frac{2\cdot 2\cdot\cancel{13}}{5\cdot\cancel{13}}=\frac{4}{5}$ **12.** $\frac{55}{26}\cdot\frac{8}{44}\cdot\frac{91}{35}=\frac{55\cdot 8\cdot 91}{26\cdot 44\cdot 35}=\frac{\cancel{5}\cdot\cancel{11}\cdot\cancel{2}\cdot\cancel{2}\cdot\cancel{2}\cdot\cancel{7}\cdot\cancel{13}}{\cancel{2}\cdot\cancel{13}\cdot\cancel{2}\cdot\cancel{2}\cdot\cancel{11}\cdot\cancel{7}\cdot\cancel{5}}=\frac{1}{1}=1$

Margin Exercise Answers

1. $\frac{15}{32}$ **2. a.** $\frac{22}{27}$ **b.** $\frac{28}{5}$ or $5\frac{3}{5}$ **c.** 0 **d.** $\frac{9}{70}$ **3.** $\frac{1}{160}$ of the hardward is defective wing nuts

4. a. Associative property of multiplication **b.** Commutative property of multiplication

5. a. $\frac{2}{9}$ **b.** $\frac{7}{4}$ **6.** $\frac{1}{4}$ **7.** $\frac{8}{11}$ **8. a.** $\frac{5}{7}$ **b.** $\frac{2}{7}$ **9.** $\frac{7}{36}$ **10.** $\frac{4}{3}$ or $1\frac{1}{3}$ **11.** $\frac{4}{15}$ **12.** 2

13. 450 drivers wore their seat belts **14.** $\frac{1}{6}$ **15.** $\frac{1}{3}$ **16.** $\frac{16}{9}$ or $1\frac{7}{9}$ **17.** $\frac{3}{8}$

2.2 Exercises

Concept Check

Fill-in-the-Blank. Complete each sentence using information found in this section.

1. Any whole number can be written in fraction form with denominator _______.

2. Finding a fraction "of" a number requires _______.

3. To reduce a fraction to lowest terms, divide out any common _______.

4. If all the factors in the numerator or denominator are divided out, then _______ must be used as a factor.

5. To multiply and reduce at the same time, divide numerators and denominators by _______ factors.

6. Finding _______ factorizations may help in multiplying and reducing at the same time.

True/False. Determine whether each statement is true or false. If a statement is false, explain how it can be changed so the statement will be true. (**Note:** There may be more than one acceptable change.)

7. Multiplication can be used to find $\frac{1}{2}$ of $\frac{2}{9}$.

8. $\frac{3}{4} \cdot \frac{9}{10} = \frac{27}{40}$

9. The statement $\frac{1}{3} \cdot \frac{2}{5} = \frac{2}{5} \cdot \frac{1}{3}$ is an example of the associative property of multiplication.

10. The number 1 is always a factor of the numerator and the denominator.

Practice

Multiply. See Examples 1 and 2.

1. $\frac{1}{4} \cdot \frac{1}{4}$
2. $\frac{1}{7} \cdot \frac{1}{7}$
3. $\frac{2}{5} \cdot \frac{2}{5}$
4. $\frac{3}{7} \cdot \frac{3}{7}$
5. $\frac{0}{3} \cdot \frac{5}{7}$
6. $\frac{0}{4} \cdot \frac{7}{6}$
7. $\frac{4}{1} \cdot \frac{3}{1}$
8. $\frac{2}{1} \cdot \frac{5}{1}$
9. $\frac{3}{5} \cdot \frac{4}{7}$
10. $\frac{2}{3} \cdot \frac{5}{11}$
11. $\frac{5}{8} \cdot \frac{3}{4}$
12. $\frac{7}{6} \cdot \frac{5}{2}$
13. $\frac{15}{1} \cdot \frac{3}{2}$
14. $\frac{6}{5} \cdot \frac{7}{1}$
15. $\frac{1}{2} \cdot \frac{3}{7} \cdot \frac{9}{2}$
16. $\frac{7}{3} \cdot \frac{2}{5} \cdot \frac{1}{9}$
17. $\frac{7}{8} \cdot \frac{7}{9} \cdot \frac{7}{3}$
18. $\frac{8}{5} \cdot \frac{8}{5} \cdot \frac{7}{1}$

2.2 Exercises

Concept Check

1. 1
2. multiplication
3. factors
4. 1
5. common
6. prime
7. True
8. True
9. False; The statement $\frac{1}{3} \cdot \frac{2}{5} = \frac{2}{5} \cdot \frac{1}{3}$ is an example of the commutative property of multiplication.
10. True

Practice

1. $\frac{1}{16}$
2. $\frac{1}{49}$
3. $\frac{4}{25}$
4. $\frac{9}{49}$
5. 0
6. 0
7. $\frac{12}{1}$ or 12
8. $\frac{10}{1}$ or 10
9. $\frac{12}{35}$
10. $\frac{10}{33}$
11. $\frac{15}{32}$
12. $\frac{35}{12}$
13. $\frac{45}{2}$
14. $\frac{42}{5}$
15. $\frac{27}{28}$

16. $\frac{14}{135}$
17. $\frac{343}{216}$
18. $\frac{448}{25}$
19. $\frac{1}{6}$
20. $\frac{1}{6}$
21. $\frac{4}{45}$
22. $\frac{12}{35}$
23. $\frac{2}{9}$
24. $\frac{3}{16}$
25. $\frac{1}{3}$
26. $\frac{1}{4}$
27. $\frac{3}{4}$
28. $\frac{3}{10}$
29. $\frac{5}{11}$
30. $\frac{7}{13}$
31. 0
32. 0
33. $\frac{2}{5}$
34. $\frac{7}{18}$
35. $\frac{12}{35}$
36. $\frac{27}{56}$
37. $\frac{1}{2}$
38. $\frac{1}{2}$
39. $\frac{2}{3}$
40. $\frac{3}{4}$

19. Find $\frac{1}{3}$ of $\frac{1}{2}$.

20. Find $\frac{1}{2}$ of $\frac{1}{3}$.

21. Find $\frac{2}{3}$ of $\frac{2}{15}$.

22. Find $\frac{4}{7}$ of $\frac{3}{5}$.

23. Find $\frac{1}{3}$ of $\frac{2}{3}$.

24. Find $\frac{1}{4}$ of $\frac{3}{4}$.

Reduce each fraction to lowest terms. If it is already in lowest terms, simply rewrite the fraction. See Examples 5 through 7.

25. $\frac{3}{9}$

26. $\frac{2}{8}$

27. $\frac{9}{12}$

28. $\frac{6}{20}$

29. $\frac{5}{11}$

30. $\frac{7}{13}$

31. $\frac{0}{25}$

32. $\frac{0}{16}$

33. $\frac{16}{40}$

34. $\frac{14}{36}$

35. $\frac{12}{35}$

36. $\frac{27}{56}$

37. $\frac{16}{32}$

38. $\frac{25}{50}$

39. $\frac{26}{39}$

40. $\frac{36}{48}$

41. $\frac{42}{63}$

42. $\frac{64}{96}$

43. $\frac{25}{76}$

44. $\frac{12}{35}$

45. $\frac{48}{12}$

46. $\frac{72}{36}$

47. $\frac{24}{100}$

48. $\frac{72}{100}$

49. $\frac{30}{70}$

50. $\frac{70}{100}$

51. $\frac{144}{156}$

52. $\frac{121}{165}$

53. $\frac{150}{135}$

54. $\frac{140}{112}$

55. $\frac{390}{260}$

56. $\frac{560}{420}$

Multiply and reduce to lowest terms. See Examples 9 through 17. (**Hint:** Factor before multiplying.)

57. $\frac{1}{3}\cdot\frac{3}{4}$

58. $\frac{3}{7}\cdot\frac{5}{3}$

59. $\frac{2}{3}\cdot\frac{4}{3}$

60. $\frac{1}{5}\cdot\frac{3}{5}$

61. $\frac{1}{5}\cdot\frac{4}{7}$

62. $\frac{3}{5}\cdot\frac{2}{7}$

63. $\frac{5}{16}\cdot\frac{16}{15}$

64. $\frac{14}{9}\cdot\frac{3}{14}$

65. $\frac{7}{8}\cdot\frac{9}{14}$

66. $\frac{6}{7}\cdot\frac{5}{12}$

67. $\frac{10}{18}\cdot\frac{9}{5}$

68. $\frac{8}{10}\cdot\frac{5}{4}$

69. $\frac{2}{21}\cdot\frac{15}{22}$

70. $\frac{3}{16}\cdot\frac{20}{21}$

71. $\frac{15}{27}\cdot\frac{9}{30}$

72. $\frac{25}{9}\cdot\frac{3}{100}$

73. $8\cdot\frac{5}{12}$

74. $9\cdot\frac{7}{24}$

75. $\frac{32}{20}\cdot\frac{13}{9}\cdot\frac{7}{26}$

76. $\frac{20}{32}\cdot\frac{9}{13}\cdot\frac{26}{7}$

77. $\frac{9}{10}\cdot\frac{35}{40}\cdot\frac{25}{15}$

78. $\frac{5}{12}\cdot\frac{56}{42}\cdot\frac{90}{54}$

79. $\frac{17}{100}\cdot\frac{27}{34}\cdot\frac{25}{9}\cdot 6$

80. $\frac{13}{28}\cdot\frac{7}{9}\cdot\frac{45}{39}\cdot 4$

Applications

Solve.

81. A pizza is to be cut into fourths. Each of these fourths is to be cut into thirds. What fraction of the pizza is each of the final pieces?

82. One of Maria's birthday presents was a box of candy. Half of the candy was chocolate covered and one-fourth of the chocolate-covered candy had cherries inside. What fraction of the candy were chocolate-covered cherries?

83. A recipe calls for $\frac{3}{4}$ cups of flour. How much flour should be used if only half of the recipe is to be made?

84. In a box of ink cartridges for a printer, $\frac{5}{6}$ of the cartridges are not black ink. Of these nonblack ink cartridges, $\frac{1}{2}$ are magenta. What fraction of the cartridges are magenta?

85. Of the books in a personal library, $\frac{3}{4}$ are fiction. Of these fiction books, $\frac{3}{5}$ are paperback. What fraction of the books in the library are fiction and paperbacks?

86. While shopping for a new smartphone, Jasmine found that $\frac{2}{3}$ of the smartphones within her budget have the ability to record high-definition videos. Of these smartphones, $\frac{2}{5}$ have 64 GB of memory. What fraction of the smartphones within her budget can take high-definition video and have 64 GB of memory?

87. A glass is 8 inches tall. If the glass is $\frac{3}{4}$ full of water, what is the height of the water in the glass?

88. A study showed that $\frac{3}{5}$ of the students in an elementary school were left-handed. If the school had an enrollment of 600 students, how many were left-handed?

89. A hexagonal dance floor has a total area of 720 square feet. If the floor separates into 6 triangles, each of which is $\frac{1}{6}$ the total area, find the area of one of the triangles.

90. If you go on a bicycle trip of 75 miles in the mountains and $\frac{1}{5}$ of the trip is downhill, what fraction of the trip is not downhill? How many miles are not downhill?

91. Suppose that a ball is dropped from a height of 40 feet, and that each time the ball bounces, it bounces back to $\frac{1}{2}$ the height it dropped. How high will the ball bounce on its third bounce?

92. You have a wire that is 40 cm long. If you take $\frac{3}{4}$ of the wire and bend it into a square, find the length of one side of the square. You may want to draw a figure to help solve the problem. (**Hint:** $s = \frac{P}{4}$ where s is the side length of one side of the square and P is the perimeter of the square.)

41. $\frac{2}{3}$
42. $\frac{2}{3}$
43. $\frac{25}{76}$
44. $\frac{12}{35}$
45. 4
46. 2
47. $\frac{6}{25}$
48. $\frac{18}{25}$
49. $\frac{3}{7}$
50. $\frac{7}{10}$
51. $\frac{12}{13}$
52. $\frac{11}{15}$
53. $\frac{10}{9}$
54. $\frac{5}{4}$
55. $\frac{3}{2}$
56. $\frac{4}{3}$
57. $\frac{1}{4}$
58. $\frac{5}{7}$
59. $\frac{8}{9}$
60. $\frac{3}{25}$
61. $\frac{4}{35}$
62. $\frac{6}{35}$
63. $\frac{1}{3}$
64. $\frac{1}{3}$
65. $\frac{9}{16}$

66. $\frac{5}{14}$

67. 1

68. 1

69. $\frac{5}{77}$

70. $\frac{5}{28}$

71. $\frac{1}{6}$

72. $\frac{1}{12}$

73. $\frac{10}{3}$ or $3\frac{1}{3}$

74. $\frac{21}{8}$ or $2\frac{5}{8}$

75. $\frac{28}{45}$

76. $\frac{45}{28}$

77. $\frac{21}{16}$ or $1\frac{5}{16}$

78. $\frac{25}{27}$

79. $\frac{9}{4}$

80. $\frac{5}{3}$

Applications

81. $\frac{1}{12}$

82. $\frac{1}{8}$

83. $\frac{3}{8}$

84. $\frac{5}{12}$

85. $\frac{9}{20}$

86. $\frac{4}{15}$

87. 6 in.

88. 360 students

89. 120 square feet

90. $\frac{4}{5}$; 60 miles

91. 5 ft

92. $\frac{15}{2}$ or $7\frac{1}{2}$ cm

93. a. 35 women
b. 21 women
c. Yes
d. Pass by 3 votes

94. 975 students

95. a. 1500 voters
b. 2500 voters
c. 1000 Democrats
d. 2500 voters

96. 90 pages

Writing & Thinking

97. No. If a fraction is less than 1 then its product with another number will be less than that other number. So, if the other number is less than 1, the product will be less than 1. Answers will vary.

98. Find the prime factorization of the numerator and denominator then divide out any common prime factors. The remaining factors represent the reduced fraction.

99. To multiply two fractions, multiply the numerators, multiply the denominators, and then reduce the product to lowest terms. Examples will vary.

93. The student senate has 75 members, and $\frac{7}{15}$ of them are women. A change in the senate constitution is being considered, and at the present time (before debating has begun), a survey shows that $\frac{3}{5}$ of the women and $\frac{4}{5}$ of the men are in favor of this change.

a. How many women are on the student senate?

b. How many women on the senate are in favor of the change?

c. If the change requires $\frac{2}{3}$ majority vote in favor to pass, would the change pass if the vote were taken today?

d. By how many votes would the change pass or fail?

94. There are 3000 students at Mountain High School, and $\frac{1}{4}$ of these students are seniors. If $\frac{3}{5}$ of the seniors are in favor of the school forming a debating team and $\frac{7}{10}$ of the remaining students (not seniors) are also in favor of forming a debating team, how many students do not favor this idea?

95. There are 4000 registered voters in Roseville and $\frac{3}{8}$ of these voters are registered Democrats. A survey indicates that $\frac{2}{3}$ of the registered Democrats are in favor of Bond Measure A and $\frac{3}{5}$ of the other registered voters are in favor of this measure.

a. How many of the voters are registered Democrats?

b. How many of the voters are not registered Democrats?

c. How many of the registered Democrats favor Measure A?

d. How many of the registered voters favor Measure A?

96. A student can read $\frac{1}{5}$ of a book in 3 hours. If the book contains 450 pages, how many pages can he read in 3 hours?

Writing & Thinking

97. If two fractions are between 0 and 1, can their product be more than 1? Explain.

98. List the steps you would use in using prime factorization to reduce a fraction.

99. Explain the process of multiplying two fractions. Give an example of a product that cannot be reduced.

2.3 Division with Fractions

Objectives

A. Find the reciprocal of a fraction.

B. Divide fractions.

A Reciprocals

If the product of two fractions is 1, then the fractions are called **reciprocals** of each other. (Remember that whole numbers can also be written in fraction form.) For example,

$$\frac{5}{8} \text{ and } \frac{8}{5} \text{ are reciprocals because } \frac{5}{8}\cdot\frac{8}{5}=\frac{\cancel{5}\cdot\cancel{8}}{\cancel{8}\cdot\cancel{5}}=1.$$

Reciprocals

The reciprocal of $\frac{a}{b}$ is $\frac{b}{a}$ ($a \neq 0$ and $b \neq 0$). The product of a nonzero number and its reciprocal is always 1.

$$\frac{a}{b}\cdot\frac{b}{a}=1$$

Note: $0=\frac{0}{1}$, but $\frac{1}{0}$ is undefined. That is, **the number 0 has no reciprocal.**

DEFINITION

Example 1 Finding Reciprocals

Find the reciprocal of $\frac{2}{3}$.

Solution

The reciprocal of $\frac{2}{3}$ is $\frac{3}{2}$. Note that $\frac{2}{3}\cdot\frac{3}{2}=\frac{\cancel{2}\cdot\cancel{3}}{\cancel{3}\cdot\cancel{2}}=1.$

Now work margin exercise 1.

1. Find the reciprocal of $\frac{7}{8}$.

Example 2 Finding Reciprocals

Find the reciprocal of 10.

Solution

The reciprocal of 10 is $\frac{1}{10}$. Note that $10\cdot\frac{1}{10}=\frac{10}{1}\cdot\frac{1}{10}=\frac{\cancel{10}\cdot 1}{1\cdot\cancel{10}}=1.$

(Remember that 10 can be written as $\frac{10}{1}$ as a fraction.)

Now work margin exercise 2.

2. Find the reciprocal of 16.

B Division with Fractions

In the division problem $14 \div 2$, we are asking how many 2s are in 14. Similarly the division problem $\frac{2}{3} \div \frac{1}{6}$ is asking how many $\frac{1}{6}$s there are in $\frac{2}{3}$. The figure below shows that there are four $\frac{1}{6}$s in $\frac{2}{3}$.

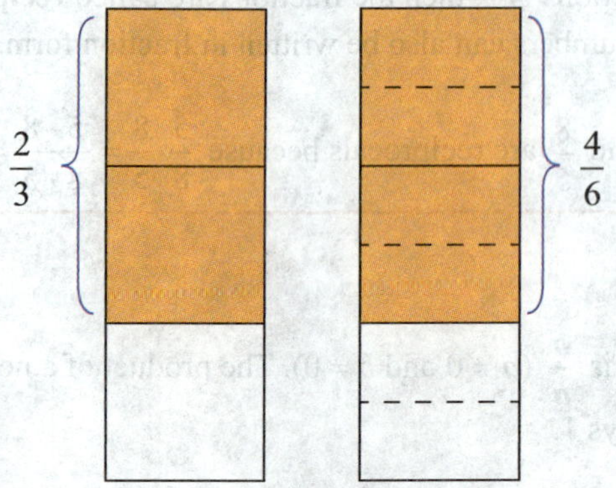

Figure 1

The division can be accomplished in the following way.

$$\frac{2}{3} \div \frac{1}{6} = \frac{2}{3} \cdot \frac{6}{1} = \frac{12}{3} = 4$$ Multiply by the reciprocal of $\frac{1}{6}$; namely, $\frac{6}{1}$.

Math Tip

The division problem $\frac{2}{3} \div \frac{1}{6}$ can be written in fraction form as $\frac{\frac{2}{3}}{\frac{1}{6}}$. This is called a *complex fraction*. We will learn more about these in Section 14.5.

Dividing Fractions

To divide by any nonzero number, multiply by its reciprocal. In general,

$$\frac{a}{b} \div \frac{c}{d} = \frac{a}{b} \cdot \frac{d}{c}, \quad \text{where } b, c, d \neq 0.$$

For example, $\frac{1}{2} \div \frac{4}{3} = \frac{1}{2} \cdot \frac{3}{4} = \frac{3}{8}$.

PROCEDURE

3. Divide: $\frac{4}{9} \div \frac{3}{5}$

Example 3 Dividing Fractions

Divide: $\frac{3}{4} \div \frac{2}{3}$

Solution

The reciprocal of $\frac{2}{3}$ is $\frac{3}{2}$, so we multiply by $\frac{3}{2}$.

$$\frac{3}{4} \div \frac{2}{3} = \frac{3}{4} \cdot \frac{3}{2} = \frac{9}{8} \text{ or } 1\frac{1}{8}$$

Now work margin exercise 3.

Example 4 Dividing Fractions

Divide: $\frac{3}{4} \div 4$

Solution

The reciprocal of 4 is $\frac{1}{4}$, so we multiply by $\frac{1}{4}$.

$$\frac{3}{4} \div 4 = \frac{3}{4} \cdot \frac{1}{4} = \frac{3}{16}$$

Now work margin exercise 4.

4. Divide: $\frac{7}{9} \div 5$

Teaching Note

You might want to mention to your class that multiplying by $\frac{1}{4}$ is the same as dividing by 4, multiplying by $\frac{1}{3}$ is the same as dividing by 3, and so on. A good example would be $\frac{3}{4} \div 3 = \frac{3}{4} \cdot \frac{1}{3}$ is the same as dividing $\frac{3}{4}$ into 3 equal parts.

As with any multiplication problem, we reduce whenever possible by factoring numerators and denominators.

Example 5 Dividing and Reducing Fractions

Divide and reduce to lowest terms: $\frac{16}{27} \div \frac{8}{9}$

Solution

The reciprocal of $\frac{8}{9}$ is $\frac{9}{8}$, so we multiply by $\frac{9}{8}$. Reduce by factoring.

$$\frac{16}{27} \div \frac{8}{9} = \frac{16}{27} \cdot \frac{9}{8} = \frac{\cancel{8} \cdot 2 \cdot \cancel{9}}{3 \cdot \cancel{9} \cdot \cancel{8}} = \frac{2}{3}$$

Now work margin exercise 5.

5. Divide and reduce to lowest terms: $\frac{10}{21} \div \frac{5}{14}$

Example 6 Dividing and Reducing Fractions

Divide and reduce to lowest terms: $\frac{9}{4} \div \frac{9}{2}$

Solution

$$\frac{9}{4} \div \frac{9}{2} = \frac{9}{4} \cdot \frac{2}{9} = \frac{\cancel{9} \cdot \cancel{2} \cdot 1}{\cancel{2} \cdot 2 \cdot \cancel{9}} = \frac{1}{2}$$

Now work margin exercise 6.

6. Divide and reduce to lowest terms: $\frac{3}{10} \div \frac{9}{5}$

Completion Example 7 Dividing and Reducing Fractions

Divide and reduce to lowest terms: $\frac{13}{4} \div \frac{39}{5}$

Solution

$$\frac{13}{4} \div \frac{39}{5} = \frac{13}{4} \cdot \frac{5}{39} = \frac{13 \cdot 5}{4 \cdot 3 \cdot 13} = \frac{5}{12}$$

Now work margin exercise 7.

7. Divide and reduce to lowest terms: $\frac{7}{12} \div \frac{9}{3}$

8. Divide and reduce to lowest terms: $\frac{8}{13} \div \frac{8}{13}$

Completion Example 8 Dividing and Reducing Fractions

Divide and reduce to lowest terms: $\frac{4}{9} \div \frac{4}{9}$

Solution

$$\frac{4}{9} \div \frac{4}{9} = \frac{4}{9} \cdot \underline{\frac{9}{4}} = \frac{4 \cdot 9}{9 \cdot \underline{4}} = \underline{1}$$

Now work margin exercise 8.

9. The result of multiplying two numbers is $\frac{5}{24}$. If one of the numbers is $\frac{7}{8}$, what is the other number?

Example 9 Finding a Missing Number

The result of multiplying two numbers is $\frac{7}{16}$. If one of the numbers is $\frac{3}{4}$, what is the other number?

Solution

$\frac{7}{16}$ is the result of multiplying two numbers. Divide $\frac{7}{16}$ by $\frac{3}{4}$ to find the second number.

$$\frac{7}{16} \div \frac{3}{4} = \frac{7}{16} \cdot \frac{4}{3} = \frac{7 \cdot \cancel{4}}{\cancel{4} \cdot 4 \cdot 3} = \frac{7}{12}$$

The other number is $\frac{7}{12}$.

Check by multiplying: $\frac{3}{4} \cdot \frac{7}{12} = \frac{\cancel{3} \cdot 7}{4 \cdot 4 \cdot \cancel{3}} = \frac{7}{16}$

Now work margin exercise 9.

10. You have 24 inches of adhesive tape and you want to cut the tape into $\frac{3}{16}$ in. strips to hang pictures. How many strips will you have?

Example 10 Application: Dividing Fractions

A baker uses $\frac{1}{8}$ cup of lemon juice for one apple pie in her special apple pie recipe. If she has $\frac{3}{4}$ cup of lemon juice on hand, how many apple pies can she make using this amount of lemon juice?

Solution

Step 1: READ: Read the problem carefully. We know the total amount of lemon juice she has and how much is used in one pie.

Step 2: SET UP: To find the number of pies she can bake, divide the amount of lemon juice on hand by the amount used in one pie.

$$\frac{3}{4} \div \frac{1}{8} = \text{number of pies}$$

Step 3: SOLVE: $\frac{3}{4} \div \frac{1}{8} = \frac{3}{4} \cdot \frac{8}{1} = \frac{3 \cdot 2 \cdot \cancel{4}}{\cancel{4} \cdot 1} = 3 \cdot 2 = 6$

She can make 6 apple pies using $\frac{3}{4}$ cup of lemon juice.

Step 4: CHECK: There are two eighths in one quarter, which means each quarter of a cup of lemon juice can make two pies. Since there are three quarters of a cup of lemon juice, $2 \cdot 3 = 6$, which means 6 pies can be made. So the answer is reasonable.

Now work margin exercise 10.

Example 11 Application: Multiplying and Dividing Fractions

A box contains 30 pieces of candy. This is $\frac{3}{5}$ of the maximum amount of this candy the box can hold.

a. Is the maximum amount of candy the box can hold more or less than 30 pieces?

b. If you want to multiply $\frac{3}{5}$ times 30, would the product be more or less than 30?

c. What is the maximum number of pieces of candy the box can hold?

Solution

a. The maximum number of pieces of candy is more than 30, because $\frac{3}{5}$ is less than the whole box.

b. Less than 30.

c. To find the maximum number of pieces, divide.

$$30 \div \frac{3}{5} = \frac{30}{1} \cdot \frac{5}{3} = \frac{\cancel{3} \cdot 10 \cdot 5}{1 \cdot \cancel{3}} = 50$$

The maximum number of pieces the box will hold is 50.

Now work margin exercise 11.

11. A truck is towing 3000 pounds. This is $\frac{2}{3}$ of the truck's towing capacity.

a. Is the truck's towing capacity more or less than 3000 pounds?

b. If you were to multiply $\frac{2}{3}$ times 3000, would the product be more or less than 3000?

c. What is the towing capacity of the truck?

Completion Example Answers

7. $\frac{13}{4} \div \frac{39}{5} = \frac{13}{4} \cdot \frac{5}{39} = \frac{\cancel{13} \cdot 5}{4 \cdot 3 \cdot \cancel{13}} = \frac{5}{12}$ **8.** $\frac{4}{9} \div \frac{4}{9} = \frac{4}{9} \cdot \frac{9}{4} = \frac{\cancel{4} \cdot \cancel{9}}{\cancel{9} \cdot \cancel{4}} = 1$

Margin Exercise Answers

1. $\frac{8}{7}$ **2.** $\frac{1}{16}$ **3.** $\frac{20}{27}$ **4.** $\frac{7}{45}$ **5.** $\frac{4}{3}$ or $1\frac{1}{3}$ **6.** $\frac{1}{6}$ **7.** $\frac{7}{36}$ **8.** 1 **9.** $\frac{5}{21}$ **10.** 128 strips of tape **11. a.** More than 3000 lb **b.** Less than 3000 **c.** 4500 pounds

2.3 Exercises

Concept Check

Fill-in-the-Blank. Complete each sentence using information found in this section.

1. The _______ of $\frac{7}{8}$ is $\frac{8}{7}$.
2. The product of any nonzero number and its reciprocal is always _______.

2.3 Exercises

Concept Check

1. reciprocal
2. 1
3. $\frac{1}{5}$
4. no
5. reciprocal
6. 9
7. False; The reciprocal of 1 is 1.
8. False; The product of a nonzero number and its reciprocal is 1.
9. False; The reciprocal of 12 is $\frac{1}{12}$.
10. True

Practice

1. $\frac{4}{3}$
2. $\frac{5}{6}$
3. 3
4. 8
5. $\frac{1}{2}$
6. $\frac{1}{7}$
7. No reciprocal
8. 1
9. $\frac{7}{12}$
10. $\frac{9}{21}$
11. $\frac{8}{9}$
12. $\frac{4}{15}$
13. $\frac{5}{7}$
14. $\frac{3}{11}$
15. 0
16. 0
17. Undefined
18. Undefined
19. $\frac{7}{5}$
20. $\frac{11}{3}$
21. $\frac{1}{3}$
22. $\frac{1}{4}$
23. $\frac{7}{6}$
24. $\frac{3}{4}$
25. $\frac{4}{9}$
26. $\frac{12}{5}$
27. $\frac{7}{16}$

3. The reciprocal of 5 is ________.

4. The number 0 has ________ reciprocal.

5. To divide by any nonzero number, multiply by its ________.

6. The result of $\frac{3}{5} \div \frac{1}{15}$ is ________.

True/False. Determine whether each statement is true or false. If a statement is false, explain how it can be changed so the statement will be true. (**Note:** There may be more than one acceptable change.)

7. The reciprocal of 1 is undefined.

8. The product of a nonzero number and its reciprocal is undefined.

9. The reciprocal of 12 is $\frac{12}{1}$.

10. The result of $\frac{1}{3} \div \frac{1}{6}$ is 2.

Practice

Find the reciprocal. See Examples 1 and 2.

1. $\frac{3}{4}$

2. $\frac{6}{5}$

3. $\frac{1}{3}$

4. $\frac{1}{8}$

5. 2

6. 7

7. 0

8. 1

9. $\frac{12}{7}$

10. $\frac{21}{9}$

Divide and reduce to lowest terms. See Examples 5 through 8.

11. $\frac{2}{3} \div \frac{3}{4}$

12. $\frac{1}{5} \div \frac{3}{4}$

13. $\frac{3}{7} \div \frac{3}{5}$

14. $\frac{2}{11} \div \frac{2}{3}$

15. $0 \div \frac{5}{6}$

16. $0 \div \frac{6}{7}$

17. $\frac{5}{6} \div 0$

18. $\frac{6}{7} \div 0$

19. $\frac{3}{5} \div \frac{3}{7}$

20. $\frac{2}{3} \div \frac{2}{11}$

21. $\frac{5}{16} \div \frac{15}{16}$

22. $\frac{4}{23} \div \frac{16}{23}$

23. $\frac{7}{18} \div \frac{3}{9}$

24. $\frac{3}{14} \div \frac{2}{7}$

25. $\frac{5}{12} \div \frac{15}{16}$

26. $\frac{8}{25} \div \frac{2}{15}$

27. $\frac{13}{40} \div \frac{26}{35}$

28. $\frac{17}{48} \div \frac{51}{90}$

29. $\frac{5}{6} \div \frac{3}{4}$

30. $\frac{3}{4} \div \frac{5}{6}$

31. $\frac{12}{27} \div \frac{10}{18}$

32. $\frac{14}{15} \div \frac{21}{25}$

33. $\frac{20}{21} \div \frac{15}{42}$

34. $\frac{16}{33} \div \frac{24}{55}$

35. $\frac{3}{7} \div \frac{3}{7}$

36. $\frac{6}{13} \div \frac{6}{13}$

37. $\frac{16}{27} \div \frac{7}{18}$

38. $\frac{17}{20} \div \frac{3}{14}$

39. $\frac{25}{36} \div \frac{5}{24}$

40. $\frac{26}{35} \div \frac{39}{40}$

41. $\frac{3}{10} \div \frac{7}{8}$

42. $\frac{5}{6} \div \frac{13}{4}$

43. $\frac{7}{8} \div \frac{15}{2}$

44. $\frac{13}{16} \div \frac{3}{2}$

45. $\frac{21}{5} \div 3$

46. $\frac{15}{8} \div 5$

47. $\frac{41}{6} \div 2$

48. $\frac{1}{7} \div 14$

49. $3 \div \frac{21}{5}$

50. $5 \div \frac{15}{8}$

51. $14 \div \frac{1}{7}$

52. $24 \div \frac{1}{4}$

Applications

Solve.

53. The result of multiplying two numbers is $\frac{2}{5}$. If one of the numbers is $\frac{5}{6}$, what is the other number?

54. The result of multiplying two numbers is $\frac{3}{8}$. If one of the numbers is $\frac{15}{7}$, what is the other number?

55. The floor of the Atlantic Ocean is spreading apart at an average rate of $\frac{3}{50}$ of a meter per year. How long will it take for the ocean floor to spread 12 meters?

56. A small private college has determined that about $\frac{11}{25}$ of the students that it accepts will actually enroll. If the college wants 550 freshmen to enroll, how many should it accept?

57. An airplane is carrying 180 passengers. This is $\frac{9}{10}$ of the capacity of the airplane.

a. Is the capacity of the airplane more or less than 180?

b. If you were to multiply 180 times $\frac{9}{10}$, would the product be more or less than 180?

c. What is the capacity of the airplane?

58. Due to environmental considerations, homeowners in a particularly dry area have been asked to use less water than usual. One home is currently using 630 gallons per day. This is $\frac{7}{10}$ of the usual amount of water used in this home.

a. Is the usual amount of water used more or less than 630 gallons?

b. If you were to multiply $\frac{7}{10}$ times 630, would the product be more or less than 630?

c. What is the usual amount of water used in this home?

28. $\frac{15}{24}$
29. $\frac{10}{9}$
30. $\frac{9}{10}$
31. $\frac{4}{5}$
32. $\frac{10}{9}$
33. $\frac{8}{3}$
34. $\frac{10}{9}$
35. 1
36. 1
37. $\frac{32}{21}$
38. $\frac{119}{30}$
39. $\frac{10}{3}$
40. $\frac{16}{21}$
41. $\frac{12}{35}$
42. $\frac{10}{39}$
43. $\frac{7}{60}$
44. $\frac{13}{24}$
45. $\frac{7}{5}$
46. $\frac{3}{8}$
47. $\frac{41}{12}$
48. $\frac{1}{98}$
49. $\frac{5}{7}$
50. $\frac{8}{3}$
51. 98
52. 96

Applications

53. $\frac{12}{25}$

54. $\frac{7}{40}$

55. 200 years

56. 1250 freshmen

57. a. More b. Less
c. 200 passengers

58. a. More b. Less
c. 900 gallons of water

59. a. More b. Less
c. 8000 steel rods per week

60. a. more b. less
c. 12,5000 tons

Writing & Thinking

61. $0=\frac{0}{1}$ and the reciprocal would be $\frac{1}{0}$ but division by 0 is undefined. So 0 has no reciprocal.

62. $15 \div 3 = 5$ and this indicates that there are 5 threes in 15. However, $15 \div \frac{1}{3} = 45$ which indicates that there are 45 one-thirds in 15.

63. $12 \div 3 = 4$ and $12 \cdot \frac{1}{3} = 4$. We see that dividing by 3 is the same as multiplying by $\frac{1}{3}$, the reciprocal of 3.

64. Yes. For example,
$\frac{2}{3} \div \frac{1}{6} = \frac{2}{3} \cdot \frac{6}{1} = \frac{2 \cdot 2 \cdot \cancel{3}}{\cancel{3} \cdot 1} = \frac{4}{1} = 4.$
Answers will vary.

65. No. For example,
$\frac{4}{5} \neq \frac{5}{4}.$
Answers will vary.

59. A manufacturing plant is currently producing 6000 steel rods per week. Because of difficulties getting materials, this number is only $\frac{3}{4}$ of the plant's potential production.

a. Is the potential production number more or less than 6000 rods?

b. If you were to multiply $\frac{3}{4}$ times 6000, would the product be more or less than 6000?

c. What is the plant's potential production?

60. A grove of orange trees was struck by an off-season frost and the result was a relatively poor harvest. This year's crop was 10,000 tons of oranges, which is about $\frac{4}{5}$ of the usual crop.

a. Is the usual crop more or less than 10,000 tons of oranges?

b. If you were to multiply 10,000 times $\frac{4}{5}$, would the product be more or less than 10,000?

c. About how many tons of oranges are usually harvested?

Writing & Thinking

61. Explain why the number 0 has no reciprocal.

62. Show that the phrases "15 divided by 3" and "15 divided by one-third" have different meanings.

63. Show that the phrases "12 divided by three" and "12 times one-third" have the same meaning.

64. If two fractions are between 0 and 1, can their quotient be more than 1? Explain.

65. Is division a commutative operation? Explain briefly and give three examples using fractions to help justify your answer.

2.4 Multiplication and Division with Mixed Numbers

Objectives

A. Multiply mixed numbers.

B. Find the area of a rectangle and triangle with fractions and mixed numbers.

C. Divide mixed numbers.

A Multiplication with Mixed Numbers

Remember that a mixed number is the sum of a whole number and a proper fraction. To avoid confusion when multiplying mixed numbers, change each mixed number to an improper fraction and then use the techniques for multiplying fractions. Also, use prime factorizations (or other factors) and reduce as you multiply.

Multiplying Mixed Numbers

1. Change each mixed number to an improper fraction.
2. Factor the numerator and denominator of each fraction, and then reduce and multiply.
3. Change the answer to a mixed number or leave it in fraction form. (The choice sometimes depends on what use is to be made of the answer.)

PROCEDURE

Example 1 Multiplying Mixed Numbers

Multiply: $\left(1\frac{1}{2}\right)\left(2\frac{1}{5}\right)$

Solution

Change each mixed number to an improper fraction, then multiply the fractions.

$$\left(1\frac{1}{2}\right)\left(2\frac{1}{5}\right)=\left(\frac{3}{2}\right)\left(\frac{11}{5}\right)=\frac{3\cdot 11}{2\cdot 5}=\frac{33}{10}\text{ or }3\frac{3}{10}$$

Now work margin exercise 1.

1. Multiply: $\left(3\frac{2}{3}\right)\left(1\frac{1}{4}\right)$

Example 2 Multiplying and Reducing Mixed Numbers

Multiply and reduce to lowest terms: $\frac{5}{6}\cdot 3\frac{3}{10}$

Solution

$$\frac{5}{6}\cdot 3\frac{3}{10}=\frac{5}{6}\cdot\frac{33}{10}=\frac{\cancel{5}\cdot\cancel{3}\cdot 11}{2\cdot\cancel{3}\cdot 2\cdot\cancel{5}}=\frac{11}{4}\text{ or }2\frac{3}{4}$$

Now work margin exercise 2.

2. Multiply and reduce to lowest terms: $\frac{1}{12}\cdot 9\frac{3}{8}$

Example 3 Multiplying and Reducing Mixed Numbers

Multiply and reduce to lowest terms: $4\frac{2}{3}\cdot 1\frac{1}{7}\cdot 2\frac{1}{16}$

3. Multiply and reduce to lowest terms: $3\frac{3}{4}\cdot 1\frac{1}{6}\cdot 4\frac{2}{5}$

Solution

$$4\frac{2}{3}\cdot 1\frac{1}{7}\cdot 2\frac{1}{16}=\frac{14}{3}\cdot\frac{8}{7}\cdot\frac{33}{16}=\frac{\cancel{2}\cdot\cancel{7}\cdot\cancel{8}\cdot\cancel{3}\cdot 11}{\cancel{3}\cdot\cancel{7}\cdot\cancel{8}\cdot\cancel{2}}=\frac{11}{1}=11$$

Now work margin exercise 3.

4. Multiply and reduce to lowest terms: $\left(\frac{7}{30}\right)\left(4\frac{2}{3}\right)\left(2\frac{1}{7}\right)$

Example 4 Multiplying and Reducing Mixed Numbers

Multiply and reduce to lowest terms: $\left(\frac{5}{33}\right)\left(3\frac{1}{7}\right)\left(6\frac{1}{8}\right)$

Solution

$$\frac{5}{33}\cdot 3\frac{1}{7}\cdot 6\frac{1}{8}=\frac{5}{33}\cdot\frac{22}{7}\cdot\frac{49}{8}=\frac{5\cdot\cancel{2}\cdot\cancel{11}\cdot\cancel{7}\cdot 7}{3\cdot\cancel{11}\cdot\cancel{7}\cdot\cancel{2}\cdot 4}$$

$$=\frac{35}{12}\text{ or }2\frac{11}{12}$$

Now work margin exercise 4.

5. Multiply and reduce to lowest terms: $2\frac{1}{2}\cdot 5\frac{3}{5}$

Completion Example 5 Multiplying and Reducing Mixed Numbers

Multiply and reduce to lowest terms: $2\frac{4}{5}\cdot 8\frac{3}{4}$

Solution

$$2\frac{4}{5}\cdot 8\frac{3}{4}=\frac{14}{5}\cdot\frac{35}{4}$$

$$=\frac{2\cdot 7\cdot 5\cdot 7}{5\cdot 2\cdot 2}=\frac{49}{2}\text{ or }24\frac{1}{2}$$

Now work margin exercise 5.

In Section 2.2, we discussed the idea that finding a fractional part of a number indicates multiplication. The key word is **of**, and we will find that this same idea is related to decimals and percents in later chapters. For example, we are familiar with expressions such as "take off 40% **of** the list price," and "three-tenths **of** our net monthly income goes to taxes." This concept is emphasized again here with mixed numbers.

6. Find $\frac{4}{7}$ of 21.

Example 6 Finding Fractional Parts of Mixed Numbers

Find $\frac{3}{5}$ of 40.

Solution

Remember, to find a fraction **of** a number means to **multiply** the number by the fraction.

$$\frac{3}{5}\cdot 40=\frac{3}{5}\cdot\frac{40}{1}=\frac{3\cdot\cancel{5}\cdot 8}{\cancel{5}\cdot 1}=24$$

Now work margin exercise 6.

Example 7 Finding Fractional Parts of Mixed Numbers

Find $\frac{2}{3}$ of $5\frac{1}{4}$.

Solution

$$\frac{2}{3}\cdot 5\frac{1}{4}=\frac{2}{3}\cdot\frac{21}{4}=\frac{\cancel{2}\cdot\cancel{3}\cdot 7}{\cancel{3}\cdot\cancel{2}\cdot 2}=\frac{7}{2}\text{ or }3\frac{1}{2}$$

Now work margin exercise 7.

7. Find $\frac{1}{3}$ of $7\frac{1}{2}$.

B Area

As we discussed in Section 1.3, **area** is the measure of the interior of a closed plane surface. For a rectangle, the area is the product of the length times the width. For a triangle, as illustrated in Figure 1, the area is half the area of the rectangle, which is half the product of the base times the height.

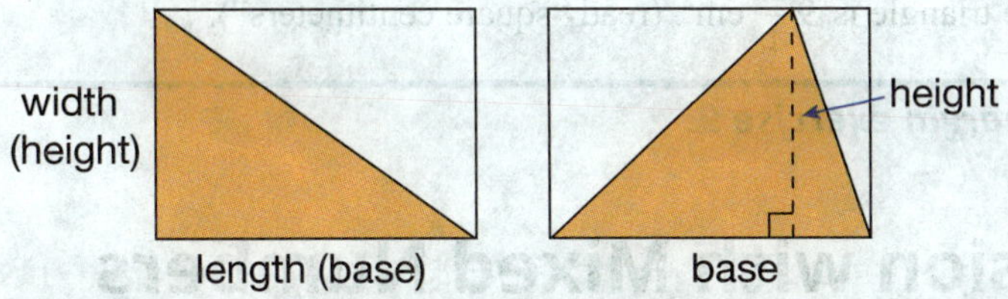

Figure 1

The area of a triangle is $\frac{1}{2}$ of the product of the base times the height. (Remember that **area is measured in square units.**)

Example 8 Application: Finding the Area of a Rectangle

Linda is framing a rectangular poster that measures $5\frac{1}{4}$ inches long by $4\frac{1}{2}$ inches wide. What is the area of the glass needed to cover the poster?

Solution

Find the area of the rectangle by multiplying the length by the width

$$5\frac{1}{4}\cdot 4\frac{1}{2}=\frac{21}{4}\cdot\frac{9}{2}=\frac{3\cdot 7\cdot 3\cdot 3}{2\cdot 2\cdot 2}=\frac{189}{8}\text{ or }23\frac{5}{8}$$

Thus, the area of the glass is $23\frac{5}{8}$ in.^{2} (read "square inches").

Now work margin exercise 8.

8. The free-throw lane on a basketball court is $18\frac{5}{6}$ feet long and 12 feet wide. What is the area inside the lane?

9. Find the area of a triangle with base $6\frac{3}{5}$ inches and height $4\frac{2}{3}$ inches.

Example 9 Finding the Area of a Triangle

Find the area of the triangle with base 6 cm and height $3\frac{1}{4}$ cm.

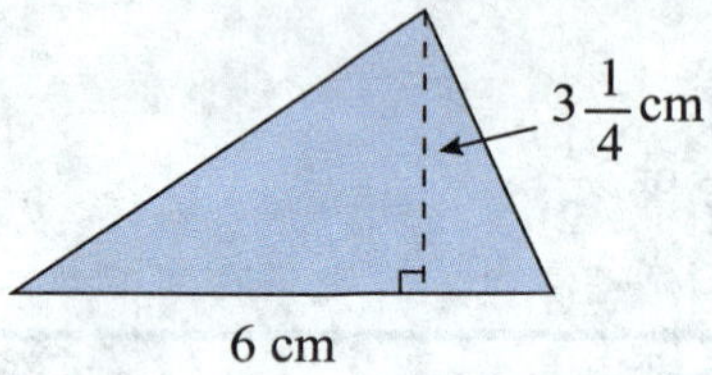

Solution

The area can be found by multiplying $\frac{1}{2}$ by the product of the base times the height.

$$\frac{1}{2}\cdot 6\cdot 3\frac{1}{4}=\frac{1}{2}\cdot\frac{6}{1}\cdot\frac{13}{4}=\frac{1\cdot \cancel{2}\cdot 3\cdot 13}{\cancel{2}\cdot 1\cdot 4}=\frac{39}{4}\text{ or }9\frac{3}{4}$$

The area of the triangle is $9\frac{3}{4}$ cm^2. (read "square centimeters").

Now work margin exercise 9.

Math Tip

To help understand how the area of a triangle is equal to base times height, consider the following two triangles. Each triangle is half of the area of the rectangle. Triangle 1 has an area of 15 and triangle 2 has an area of 25.

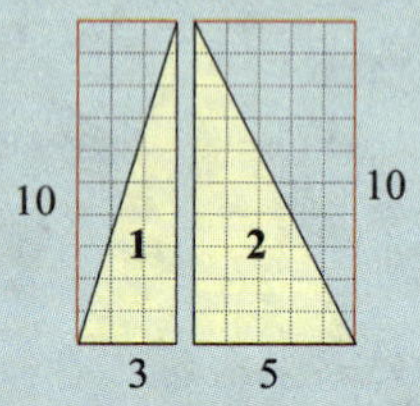

Since the triangles have the same height, we can join them. The area of the new larger triangle is still half of the area of the new larger rectangle (15 + 25 = 40).

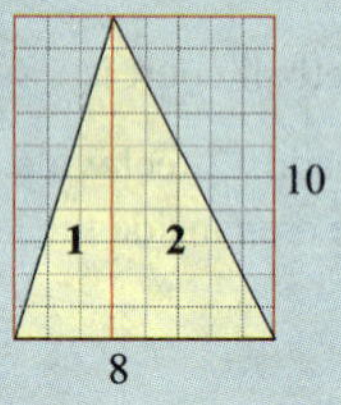

C Division with Mixed Numbers

Division with mixed numbers is the same as division with fractions, which was discussed in Section 2.3. Simply change each mixed number to an improper fraction before dividing. Recall that **to divide by any nonzero number, multiply by its reciprocal**. That is, for $c \neq 0$ and $d \neq 0$, the **reciprocal** of $\frac{c}{d}$ is $\frac{d}{c}$, and

$$\frac{a}{b}\div\frac{c}{d}=\frac{a}{b}\cdot\frac{d}{c}.$$

Dividing with Mixed Numbers

1. Change each mixed number to an improper fraction.
2. Multiply by the reciprocal of the divisor.
3. Reduce, if possible.

PROCEDURE

Example 10 Dividing and Reducing Mixed Numbers

Divide and reduce to lowest terms: $3\frac{1}{4} \div 7\frac{4}{5}$

Solution

$$3\frac{1}{4} \div 7\frac{4}{5} = \frac{13}{4} \div \frac{39}{5}$$

$$= \frac{13}{4} \cdot \frac{5}{39}$$ Note that the divisor is $\frac{39}{5}$ and we multiply by its reciprocal, $\frac{5}{39}$.

$$= \frac{\cancel{13} \cdot 5}{4 \cdot \cancel{13} \cdot 3} = \frac{5}{12}$$

Now work margin exercise 10.

10. Divide and reduce to lowest terms: $2\frac{2}{5} \div 3\frac{3}{4}$

Example 11 Dividing and Reducing Mixed Numbers

Divide and reduce to lowest terms: $6 \div 7\frac{7}{8}$

Solution

First, change the mixed number $7\frac{7}{8}$ to the improper fraction $\frac{63}{8}$ and then multiply by its reciprocal.

$$6 \div 7\frac{7}{8} = \frac{6}{1} \div \frac{63}{8}$$

$$= \frac{6}{1} \cdot \frac{8}{63}$$

$$= \frac{2 \cdot \cancel{3} \cdot 8}{1 \cdot \cancel{3} \cdot 3 \cdot 7} = \frac{16}{21}$$

Now work margin exercise 11.

11. Divide and reduce to lowest terms: $12 \div 8\frac{4}{7}$

Completion Example 12 Dividing and Reducing Mixed Numbers

Divide and reduce to lowest terms: $3\frac{1}{15} \div 1\frac{1}{5}$

Solution

$$3\frac{1}{15} \div 1\frac{1}{5} = \frac{46}{15} \div \underline{\frac{6}{5}}$$

$$= \frac{46}{15} \cdot \underline{\frac{5}{6}}$$

$$= \frac{2 \cdot 23 \cdot 5}{3 \cdot 5 \cdot \underline{2} \cdot \underline{3}} = \underline{\frac{23}{9}} \quad \text{or} \quad \underline{2\frac{5}{9}}$$

Now work margin exercise 12.

12. Divide and reduce to lowest terms: $4\frac{2}{3} \div 2\frac{1}{9}$

13. A carpenter has a board that is 21 feet long. For shelving, he wants to cut the board into pieces of $2\frac{1}{3}$ feet. How many shelves can he make?

Example 13 Application: Dividing Mixed Numbers

A landscaper has $30\frac{3}{4}$ pounds of fertilizer available to use for fertilizing trees. If he plans to use $\frac{3}{4}$ pounds of fertilizer on one tree, how many trees can he fertilize?

Solution

Step 1: READ: Read the problem carefully. We know the total amount of fertilizer he has and how much he plans to use on one tree.

Step 2: SET UP: To find the number of trees he can fertilize, divide the total amount of fertilizer by the amount to be used on one tree.

$$30\frac{3}{4} \div \frac{3}{4} = \text{number of trees.}$$

Step 3: SOLVE: $30\frac{3}{4} \div \frac{3}{4} = \frac{123}{4} \div \frac{3}{4} = \frac{123}{4} \cdot \frac{4}{3} = \frac{\cancel{3} \cdot 41 \cdot \cancel{4}}{\cancel{4} \cdot \cancel{3}} = 41$

He can fertilize 41 trees.

Step 4: CHECK: The landscaper is using less than 1 pound of fertilizer for each tree, which means he can fertilize more than 30 trees. So, 41 is a reasonable answer.

Now work margin exercise 13.

CALCULATORS

Using a Calculator to Multiply and Divide Fractions

To multiply or divide two fractions or mixed numbers, enter the expression as you did when working with whole numbers in Chapter 1, but use the [a b/c] key to enter the fractions or mixed numbers. The result will be given to you in simplified form.

For example, to multiply $\frac{5}{8} \cdot \frac{6}{11}$, press the keys

[5] [a b/c] [8] [×] [6] [a b/c] [1] [1].

Then press [=]. The display will read 15/44 which means $\frac{15}{44}$.

Similarly, to divide $2\frac{1}{2} \div 1\frac{4}{5}$, press the keys

[2] [a b/c] [1] [a b/c] [2] [÷] [1] [a b/c] [4] [a b/c] [5].

Then press [=]. The display will read 1⌟7/18 which means $1\frac{7}{18}$.

Completion Example Answers

5. $\frac{14}{5} \cdot \frac{35}{4} = \frac{\cancel{2} \cdot 7 \cdot \cancel{5} \cdot 7}{\cancel{5} \cdot \cancel{2} \cdot 2} = \frac{49}{2}$ or $24\frac{1}{2}$ **12.** $\frac{46}{15} \div \frac{6}{5} = \frac{46}{15} \cdot \frac{5}{6} = \frac{\cancel{2} \cdot 23 \cdot \cancel{5}}{3 \cdot \cancel{5} \cdot \cancel{2} \cdot 3} = \frac{23}{9}$ or $2\frac{5}{9}$

Margin Exercise Answers

1. $\frac{55}{12}$ or $4\frac{7}{12}$ **2.** $\frac{25}{32}$ **3.** $\frac{77}{4}$ or $19\frac{1}{4}$ **4.** $\frac{7}{3}$ or $2\frac{1}{3}$ **5.** 14 **6.** 12 **7.** $\frac{5}{2}$ or $2\frac{1}{2}$ **8.** 226 ft^2

9. $15\frac{2}{5}$ in.^{2} **10.** $\frac{16}{25}$ **11.** $\frac{7}{5}$ or $1\frac{2}{5}$ **12.** $\frac{42}{19}$ or $2\frac{4}{19}$ **13.** 9 shelves

2.4 Exercises

Concept Check

Fill-in-the-Blank. Complete each sentence using information found in this section.

1. A _______ number is the sum of a whole number and a proper fraction.
2. When multiplying or dividing with mixed numbers, change each mixed number into a/an _______ fraction.
3. The mixed number $5\frac{1}{3}$ is equal to the improper fraction _______.
4. The reciprocal of $2\frac{1}{3}$ is _______.
5. To divide by $1\frac{1}{2}$ multiply by _______.
6. When multiplying or dividing mixed numbers, the answer can be written as a mixed number or a/an _______.

True/False. Determine whether each statement is true or false. If a statement is false, explain how it can be changed so the statement will be true. (**Note:** There may be more than one acceptable change.)

7. When multiplying or dividing with mixed numbers, the answer should always be simplified, if possible.
8. Multiplication or division with mixed numbers can be accomplished by changing the mixed numbers to improper fractions.
9. The mixed number $4\frac{1}{5}$ is equal to $\frac{9}{5}$.
10. The reciprocal of $7\frac{2}{5}$ is $\frac{5}{37}$.

Practice

Multiply and reduce to lowest terms. Write your answer in mixed number form. See Examples 1 through 7.

1. $\frac{2}{3} \cdot 3\frac{1}{4}$
2. $\frac{3}{5} \cdot 4\frac{1}{6}$
3. $4\frac{1}{3} \cdot \frac{2}{13}$
4. $2\frac{2}{7} \cdot \frac{3}{16}$
5. $\left(2\frac{4}{5}\right)\left(1\frac{1}{7}\right)$
6. $\left(2\frac{1}{3}\right)\left(3\frac{1}{4}\right)$
7. $5\frac{1}{3}\left(2\frac{1}{2}\right)$
8. $4\frac{1}{5}\left(1\frac{1}{3}\right)$
9. $\left(1\frac{3}{5}\right)1\frac{1}{4}$
10. $2\frac{1}{4}\left(3\frac{1}{9}\right)$
11. $\left(9\frac{1}{3}\right)3\frac{3}{4}$
12. $\left(11\frac{1}{4}\right)\left(2\frac{2}{15}\right)$
13. $\left(6\frac{2}{3}\right)\left(5\frac{1}{7}\right)$
14. $12\frac{1}{2} \cdot 3\frac{1}{3}$
15. $4\frac{3}{8} \cdot 2\frac{4}{5}$

2.4 Exercises

Concept Check

1. Mixed
2. Improper
3. $\frac{16}{3}$
4. $\frac{3}{7}$
5. $\frac{2}{3}$
6. Fraction
7. True
8. True
9. False; The mixed number $4\frac{1}{5}$ is equal to $\frac{21}{5}$.
10. True

Practice

1. $2\frac{1}{6}$
2. $2\frac{1}{2}$
3. $\frac{2}{3}$
4. $\frac{3}{7}$
5. $3\frac{1}{5}$
6. $7\frac{7}{12}$
7. $13\frac{1}{3}$
8. $5\frac{3}{5}$
9. 2
10. 7
11. 35
12. 24
13. $34\frac{2}{7}$
14. $41\frac{2}{3}$
15. $12\frac{1}{4}$

16. $13\frac{1}{2}$
17. 6
18. 3
19. $11\frac{33}{35}$
20. $72\frac{9}{20}$
21. $1\frac{3}{10}$
22. $2\frac{1}{10}$
23. $7\frac{13}{21}$
24. $9\frac{38}{63}$
25. $8\frac{2}{3}$
26. $3\frac{1}{3}$
27. 126
28. 108
29. $2\frac{1}{4}$
30. $10\frac{1}{2}$
31. 18
32. 30
33. $11\frac{2}{3}$
34. $3\frac{2}{3}$
35. $3\frac{1}{9}$
36. $1\frac{5}{12}$
37. 4
38. 13
39. $5\frac{1}{5}$
40. $3\frac{5}{6}$
41. $1\frac{1}{2}$
42. $1\frac{1}{2}$

16. $6\frac{3}{8}\cdot 2\frac{2}{17}$

17. $5\frac{1}{4}\cdot 1\frac{1}{7}$

18. $2\frac{1}{6}\cdot 1\frac{5}{13}$

19. $\left(4\frac{3}{4}\right)\left(2\frac{1}{5}\right)\left(1\frac{1}{7}\right)$

20. $\left(6\frac{3}{16}\right)\left(2\frac{1}{11}\right)\left(5\frac{3}{5}\right)$

21. $1\frac{3}{32}\cdot 1\frac{1}{7}\cdot 1\frac{1}{25}$

22. $1\frac{5}{16}\cdot 1\frac{1}{3}\cdot 1\frac{1}{5}$

23. $1\frac{3}{7}\cdot 1\frac{5}{35}\cdot 4\frac{10}{15}$

24. $2\frac{9}{12}\cdot 2\frac{4}{9}\cdot 1\frac{18}{42}$

25. $\frac{5}{8}\cdot 2\frac{3}{5}\cdot 5\frac{1}{3}$

26. $4\frac{4}{24}\cdot 1\frac{12}{36}\cdot \frac{9}{15}$

27. $1\frac{1}{3}\cdot 18\cdot 3\frac{1}{2}\cdot 1\frac{18}{36}$

28. $2\frac{1}{2}\cdot 1\frac{4}{7}\cdot 21\cdot 2\frac{6}{33}$

29. $\frac{17}{100}\cdot \frac{27}{34}\cdot 2\frac{7}{9}\cdot 6$

30. $9\frac{3}{8}\cdot \frac{16}{36}\cdot 9\cdot \frac{7}{25}$

31. Find $\frac{3}{7}$ of 42.

32. Find $\frac{5}{9}$ of 54.

33. Find $\frac{7}{9}$ of 15.

34. Find $\frac{1}{6}$ of 22.

35. Find $\frac{5}{6}$ of $3\frac{11}{15}$.

36. Find $\frac{11}{12}$ of $1\frac{6}{11}$.

Divide and reduce to lowest terms. Write your answer in mixed number form. See Examples 10 through 12.

37. $3\frac{1}{2}\div\frac{7}{8}$

38. $3\frac{5}{7}\div\frac{2}{7}$

39. $4\frac{1}{3}\div\frac{5}{6}$

40. $2\frac{3}{10}\div\frac{3}{5}$

41. $5\frac{1}{4}\div 3\frac{1}{2}$

42. $6\frac{2}{3}\div 4\frac{4}{9}$

43. $6\frac{3}{5}\div 2\frac{1}{10}$

44. $\left(6\frac{3}{4}\right)\div 4\frac{1}{2}$

45. $3\frac{1}{10}\div 2\frac{1}{2}$

46. $\left(2\frac{1}{7}\right)\div 2\frac{1}{17}$

47. $7\frac{1}{5}\div 3$

48. $4\frac{1}{5}\div 3$

49. $7\frac{1}{3}\div 4$

50. $6\frac{5}{6}\div 2$

51. $3\div 4\frac{1}{5}$

52. $2\div 6\frac{5}{6}$

53. $4\frac{5}{8}\div\frac{1}{4}$

54. $\left(6\frac{3}{11}\right)\div\left(\frac{3}{4}\right)$

55. $\left(7\frac{1}{3}\right)\div\left(\frac{1}{4}\right)$

56. $4\frac{1}{5}\div 3\frac{1}{3}$

57. $1\frac{1}{32}\div 2\frac{3}{4}$

58. $2\frac{1}{17}\div 1\frac{1}{4}$

59. $1\frac{11}{14}\div 2\frac{2}{7}$

60. $13\frac{1}{7}\div 4\frac{2}{11}$

61. $1\frac{7}{8}\div 5$

62. $2\frac{4}{5}\div 7$

63. $2\frac{2}{49}\div 3\frac{3}{14}$

64. $1\frac{29}{55}\div 3\frac{5}{24}$

65. $\left(1\frac{1}{32}\right)\div 3\frac{2}{3}$

66. $\left(10\frac{2}{7}\right)\div 4\frac{1}{2}$

Applications

Solve.

67. You are planning a trip of 615 miles (round trip), and you know that your car gets an average of $27\frac{1}{3}$ miles per gallon of gas. You also know that your gas tank holds $15\frac{1}{2}$ gallons of gas.

a. How many gallons of gas will you use on this trip?

b. If the gas you buy costs $4 per gallon, how much should you plan to spend on this trip for gas?

68. You just drove your car 450 miles and used 20 gallons of gas. You know that the gas tank on your car holds $16\frac{1}{2}$ gallons of gas.

a. What is the most number of miles you can drive on one tank of gas?

b. If the gas you buy costs $4 per gallon, what would you pay to fill one-half of your tank?

69. If you drive your car to work $6\frac{3}{10}$ miles one way each weekday (5 days a week), how many miles do you drive each week going to and from work?

70. Clea can copy edit an average of $6\frac{3}{4}$ pages per hour. Approximately how many pages will she have copy edited after 36 hours?

71. A cookie recipe makes 8 dozen cookies and calls for $2\frac{1}{2}$ cups of flour. You only want to make 2 dozen cookies for yourself.

a. What fraction of the entire recipe will you be making?

b. How many cups of flour will you need to make 2 dozen cookies?

72. A right triangle is a triangle with one right angle (measure 90°). The two sides that form the right angle are called legs, and they are perpendicular to each other. The longest side is called the hypotenuse. The legs can be treated as the base and height of the triangle. Find the area of the right triangle shown here.

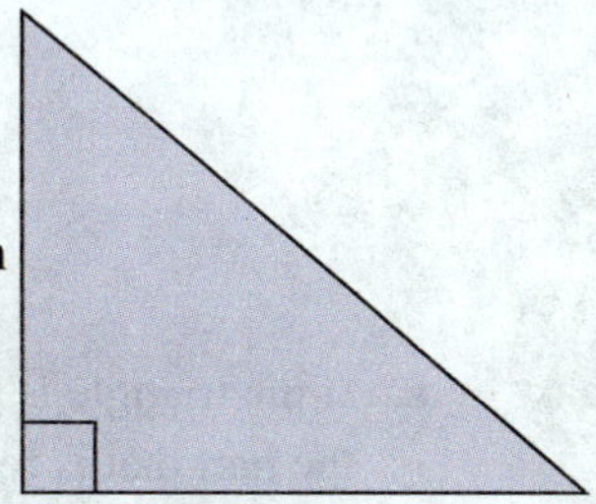

73. Find the area of the triangle in the figure shown here.

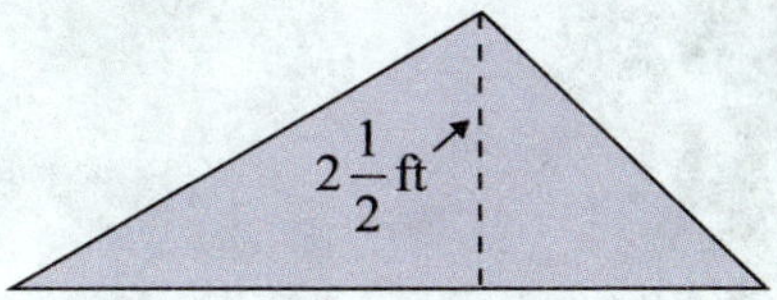

74. Find the area of a rectangle that has sides of length $5\frac{7}{8}$ meters and $4\frac{1}{2}$ meters.

43. $3\frac{1}{7}$
44. $1\frac{1}{2}$
45. $1\frac{6}{25}$
46. $1\frac{2}{49}$
47. $2\frac{2}{5}$
48. $1\frac{2}{5}$
49. $1\frac{5}{6}$
50. $3\frac{5}{12}$
51. $\frac{5}{7}$
52. $\frac{12}{41}$
53. $18\frac{1}{2}$
54. $8\frac{4}{11}$
55. $29\frac{1}{3}$
56. $1\frac{13}{50}$
57. $\frac{3}{8}$
58. $1\frac{11}{17}$
59. $\frac{25}{32}$
60. $3\frac{1}{7}$
61. $\frac{3}{8}$
62. $\frac{2}{5}$
63. $\frac{40}{63}$
64. $\frac{288}{605}$
65. $\frac{9}{32}$
66. $2\frac{2}{7}$

Applications

67. a. $22\frac{1}{2}$ gallons b. \$90

68. a. $371\frac{1}{4}$ miles
b. \$33

69. 63 miles

70. 243 pages

71. a. $\frac{1}{4}$ b. $\frac{5}{8}$ of a cup

72. 10 cm^2

73. $7\frac{1}{2}$ ft^2

74. $26\frac{7}{16}$ m^2

75. a. $34\frac{2}{3}$ in. b. $75\frac{1}{9}$ $in.^2$

76. $15\frac{7}{15}$ servings

77. $6\frac{1}{2}$ feet

78. 12 cars

79. $2\frac{12}{13}$ inches per hour

80. 15 steps

81. a. 1 sq ft b. 23 sq ft

82. a. 12 m b. 468 sq m

75. The perimeter of a square can be found by multiplying the length of one side by 4. The area can be found by squaring the length of one side. The length of one side of a square is $8\frac{2}{3}$ inches.

a. Find the perimeter of the square.

b. Find the area of the square.

76. A serving of Rice Krispies cereal is $1\frac{1}{4}$ cups. One box of Rice Krispies cereal contains approximately $19\frac{1}{3}$ cups of cereal. How many servings of cereal are in a box?

77. The area of a small garden is $35\frac{3}{4}$ square feet. The width of the garden is $5\frac{1}{2}$ feet. What is the length of the garden?

78. It takes Barry $2\frac{1}{4}$ hours to detail a car. If he worked 27 hours during the week, how many cars did he detail?

79. Silver Lake, Colorado, had a record-setting snowstorm in April of 1921. The snow fell continuously for $32\frac{1}{2}$ hours and accumulated 95 inches of snow. What was the average accumulation of snow per hour during this snow storm? [1]

80. According to building codes for the state of Washington, the maximum height of a step in a staircase is $7\frac{3}{4}$ inches. If a staircase is $116\frac{1}{4}$ inches high and is built using the maximum step height, how many steps are in the staircase?

81. A flag is in the shape of a rectangle that is 4 feet by 6 feet with a right triangle in one corner. The right triangle is 1 foot high and 2 feet long.

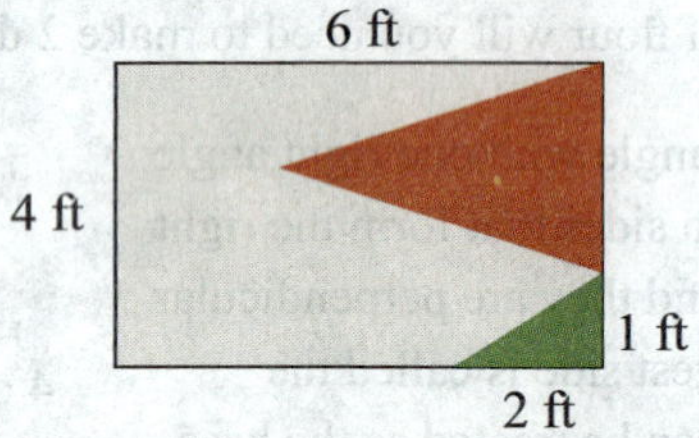

a. If the triangle is green and the rest of the flag is red and tan, what is the area of the part that is red?

b. What is the area of the flag that is red and tan?

1 Source: NOAA.gov

82. Paul's yard is in the shape of a rectangle that is 20 meters wide and 24 meters long. In one corner of his yard, Paul has a triangular area that is covered by a wooden deck. This triangular area is 4 meters high and 620 meters long. The rest of the yard is grass and flower gardens.

a. What is the area of Paul's wooden deck?

b. What is the area of the yard that is covered in grass and flower gardens?

Writing & Thinking

83. Explain how to multiply two mixed numbers. Give an example of a product that can be reduced and illustrate how you would reduce the product.

84. Suppose the product of $5\frac{7}{10}$ and some other number is $10\frac{1}{2}$. Answer the following questions without doing any calculations.

a. Do you think that this other number is more than 1 or less than 1? Why?

b. Find the other number.

85. Give an example of two mixed numbers being divided, and then explain the steps you used.

86. Compare and contrast multiplying two mixed numbers and dividing two mixed numbers.

Writing & Thinking

83. Change each mixed number to an improper fraction and then multiply the two fractions. Then, change the result to a mixed number. For example,

$$2\frac{1}{11}\cdot 5\frac{1}{2}=\frac{23}{11}\cdot\frac{11}{2}=\frac{23}{2}=11\frac{1}{2}.$$

84. a. More, since $10\frac{1}{2}>5\frac{7}{10}$

b. $\frac{35}{19}$ or $1\frac{16}{19}$

85.

$$2\frac{1}{5}\div 1\frac{4}{5}=\frac{11}{5}\div\frac{9}{5}=\frac{11}{5}\cdot\frac{5}{9}=\frac{11\cdot\cancel{5}}{\cancel{5}\cdot 9}=\frac{11}{9}$$

or $1\frac{2}{9}$; **1.** Convert the mixed numbers into improper fractions, **2.** Find the reciprocal of the divisor, **3.** Multiply the numerators and the denominators and reduce if possible, and **4.** Change the answer to mixed number form. (**Note:** The mixed number form is preferred, but the fraction form is acceptable.)

86. Whether multiplying or dividing mixed numbers, the numbers need to be converted to improper fractions first. In division, the next step requires finding the reciprocal of the divisor. The final step for both multiplication and division is to multiply the numerators and the denominators, reduce if possible, and change the result to the form of a mixed number. (The fraction form is acceptable.)

Objectives

A. Find the LCM of a set of counting numbers using prime factorizations.

B. Find equivalent fractions.

2.5 Least Common Multiple (LCM)

A Least Common Multiple (LCM)

In this section, we discuss how to find the least common multiple (LCM) of a set of counting numbers, a skill we will apply in our work with adding and subtracting fractions.

The **multiples** of a number are the products of that number with the counting numbers. Thus, the first multiple of any counting number is the number itself (the number times 1), and all other multiples are larger than that number. The multiples of 8 ($8 \cdot 1, 8 \cdot 2, 8 \cdot 3$, and so on) and 12 ($12 \cdot 1, 12 \cdot 2, 12 \cdot 3$, and so on) are shown here.

Counting Numbers:	**1,**	**2,**	**3,**	**4,**	**5,**	**6,**	**7,**	**8,**	**9,**	**10,**	**11,**	**…**
Multiples of 8:	8,	16,	24,	32,	40,	48,	56,	64,	72,	80,	88,	…
Multiples of 12:	12,	24,	36,	48,	60,	72,	84,	96,	108,	120,	132,	…

The common multiples of 8 and 12 are highlighted. They are 24, 48, 72, 96, and so on. The smallest of these, 24, is the **least common multiple (LCM)** of 8 and 12.

Least Common Multiple (LCM)

The **least common multiple (LCM)** of two (or more) counting numbers is the smallest number that is a multiple of each of these numbers.

DEFINITION

Teaching Note

There are other methods for finding the LCM, which may be even easier to use at first. However, the method in the text allows for a solid understanding that carries over to use with fractions.

You may want to also show your students this alternate method for finding the LCM of a set of counting numbers.

1. Write the numbers horizontally and find a prime number that will divide into more than one of the numbers, if possible.
2. Divide by that prime and write the quotients beneath the dividends. Rewrite any numbers not divided beneath themselves.
3. Continue the process until no two numbers have a common prime divisor.
4. The LCM is the product of all the prime divisors and the last set of quotients.

The following method, involving prime factorizations, is one way to find the least common multiple of a set of counting numbers.

Finding the LCM of a Set of Counting Numbers

1. Find the prime factorization of each number.

2. List the prime factors that appear in any one of the prime factorizations.

3. Find the product of these primes using each prime the most number of times it appears in any one of the prime factorizations.

PROCEDURE

1. Find the LCM of 12 and 42.

Example: Find the LCM of 15, 49, and 35.

5	15	49	35
7	3	49	7
3	3	7	1
	1	7	1

LCM is $3 \cdot 5 \cdot 7 \cdot 7 = 735$.

Example 1 Finding the Least Common Multiple (LCM)

Find the LCM of 20 and 45.

Solution

Step 1: Prime factorizations:

$20 = 2 \cdot 2 \cdot 5$ Two 2s, one 5

$45 = 3 \cdot 3 \cdot 5$ Two 3s, one 5

Step 2: 2, 3, and 5 are the only prime factors.

Step 3: The most number of times each prime factor appears in any one factorization:

two 2s (In 20)
two 3s (In 45)
one 5 (One in 20 or one in 45)

So, LCM = $\mathbf{2} \cdot \mathbf{2} \cdot \mathbf{3} \cdot \mathbf{3} \cdot \mathbf{5} = 180$.

180 is the smallest number divisible by both 20 and 45. (Note also that the LCM, 180, contains all the factors of the numbers 20 and 45.)

Now work margin exercise 1.

Example 2 Finding the Least Common Multiple (LCM)

Find the LCM of 12, 18, and 48.

Solution

Step 1: Prime factorizations:

$12 = 2 \cdot 2 \cdot 3$ Two 2s, one 3
$18 = 2 \cdot \mathbf{3} \cdot \mathbf{3}$ One 2, two 3s
$48 = \mathbf{2} \cdot \mathbf{2} \cdot \mathbf{2} \cdot \mathbf{2} \cdot 3$ Four 2s, one 3

Step 2: 2 and 3 are the only prime factors.

Step 3: The most number of times each prime factor appears in any one factorization:

four 2s (In 48)
two 3s (In 18)

So, LCM = $\mathbf{2} \cdot \mathbf{2} \cdot \mathbf{2} \cdot \mathbf{2} \cdot \mathbf{3} \cdot \mathbf{3} = 144$.

144 is the smallest number divisible by 12, 18, and 48.

Now work margin exercise 2.

Completion Example 3 Finding the Least Common Multiple (LCM)

Find the LCM of 36, 24, and 48.

Solution

Step 1: Prime factorizations:

$36 = 2 \cdot 2 \cdot \underline{3} \cdot \underline{3}$
$24 = 2 \cdot 2 \cdot \underline{2} \cdot \underline{3}$
$48 = 2 \cdot 2 \cdot 2 \cdot \underline{2} \cdot \underline{3}$

Step 2: __2__ and __3__ are the only prime factors.

2. Find the LCM of 18, 21, and 42.

Teaching Note
You might want to inform students that this is the technique for finding the least common denominator when adding and subtracting fractions. This knowledge may serve as a motivating factor for some of them.

3. Find the LCM of 12, 30, and 40.

Step 3: The most number of times each prime factor appears in any one factorization:

___Four___ 2s (in 48)

___Two___ 3s (in 36)

So, LCM = ___$2 \cdot 2 \cdot 2 \cdot 2 \cdot 3 \cdot 3$___ = ___144___.

___144___ is the smallest number divisible by 36, 24, and 48.

Now work margin exercise 3.

4. Find the LCM of 36, 45, and 60.

Example 4 Finding the Least Common Multiple (LCM)

Find the LCM of 27, 30, and 42.

Solution

$$\left.\begin{aligned} 27 &= 3 \cdot 3 \cdot 3 \\ 30 &= 2 \cdot 3 \cdot 5 \\ 42 &= 2 \cdot 3 \cdot 7 \end{aligned}\right\} \text{LCM} = 2 \cdot 3 \cdot 3 \cdot 3 \cdot 5 \cdot 7 = 1890$$

Now work margin exercise 4.

5. Find the LCM of 9 and 49.

Example 5 Finding the Least Common Multiple (LCM)

Find the LCM of 8 and 25.

Solution

$$\left.\begin{aligned} 8 &= 2 \cdot 2 \cdot 2 \\ 25 &= 5 \cdot 5 \end{aligned}\right\} \text{LCM} = 2 \cdot 2 \cdot 2 \cdot 5 \cdot 5 = 200$$

In this case, where the two numbers have no common prime factors, the LCM is the product of the two numbers.

Now work margin exercise 5.

Teaching Note

Understanding the relationship between each number in a set and the prime factorization of the LCM of these numbers will increase the ease and speed with which students can perform addition and subtraction with fractions. In these operations, each denominator (and corresponding numerator) is multiplied by the number of times that the denominator divides into the LCM of the denominators. You may want to reinforce this idea several times.

In Example 4, we found that 1890 is the LCM for the three numbers 27, 30, and 42. This tells us that 1890 is the smallest number that is divisible by 27, 30, and 42. The next question is how many times each of these numbers divides into 1890. In particular, how many times does 27 divide into 1890? We could simply divide by using long division; however, this is not necessary because we know the prime factorizations of 27 and 1890.

$$1890 = 2 \cdot 3 \cdot 3 \cdot 3 \cdot 5 \cdot 7 = \underbrace{(3 \cdot 3 \cdot 3)}_{27} \cdot \underbrace{(2 \cdot 5 \cdot 7)}_{70} = 27 \cdot 70$$

Thus, if we were to divide 1890 by 27, we would get a quotient of 70.

The following examples illustrate how to use prime factorizations to tell how many times each number in a set divides into the LCM for that set of numbers. **This method will be very useful when adding and subtracting fractions.**

Example 6 Finding the Least Common Multiple (LCM)

Find the LCM for 27, 30, and 42; then state how many times each number divides into the LCM.

Solution

Recall from Example 4, LCM = $2 \cdot 3 \cdot 3 \cdot 3 \cdot 5 \cdot 7 = 1890$.

Now determine how many times each number divides into the LCM.

$$1890 = 2 \cdot 3 \cdot 3 \cdot 3 \cdot 5 \cdot 7 = \underbrace{(3 \cdot 3 \cdot 3)}_{27} \cdot \underbrace{(2 \cdot 5 \cdot 7)}_{70} = 27 \cdot 70$$

So, 27 divides into 1890 70 times.

$$1890 = 2 \cdot 3 \cdot 3 \cdot 3 \cdot 5 \cdot 7 = \underbrace{(2 \cdot 3 \cdot 5)}_{30} \cdot \underbrace{(3 \cdot 3 \cdot 7)}_{63} = 30 \cdot 63$$

So, 30 divides into 1890 63 times.

$$1890 = 2 \cdot 3 \cdot 3 \cdot 3 \cdot 5 \cdot 7 = \underbrace{(2 \cdot 3 \cdot 7)}_{42} \cdot \underbrace{(3 \cdot 3 \cdot 5)}_{45} = 42 \cdot 45$$

So, 42 divides into 1890 45 times.

Now work margin exercise 6.

6. Find the LCM of 15, 35, and 42; then state how many times each number divides into the LCM.

Example 7 Finding the Least Common Multiple (LCM)

Find the LCM for 12, 18, and 66; then state how many times each number divides into the LCM.

Solution

Begin by finding the LCM.

$$\left.\begin{aligned} 12 &= 2 \cdot 2 \cdot 3 \\ 18 &= 2 \cdot 3 \cdot 3 \\ 66 &= 2 \cdot 3 \cdot 11 \end{aligned}\right\} \text{LCM} = 2 \cdot 2 \cdot 3 \cdot 3 \cdot 11 = 396$$

Now determine how many times each number divides into the LCM.

$$396 = 2 \cdot 2 \cdot 3 \cdot 3 \cdot 11 = \underbrace{(2 \cdot 2 \cdot 3)}_{12} \cdot \underbrace{(3 \cdot 11)}_{33} = 12 \cdot 33$$

So, 12 divides into 396 33 times.

$$396 = 2 \cdot 2 \cdot 3 \cdot 3 \cdot 11 = \underbrace{(2 \cdot 3 \cdot 3)}_{18} \cdot \underbrace{(2 \cdot 11)}_{22} = 18 \cdot 22$$

So, 18 divides into 396 22 times.

$$396 = 2 \cdot 2 \cdot 3 \cdot 3 \cdot 11 = \underbrace{(2 \cdot 3 \cdot 11)}_{66} \cdot \underbrace{(2 \cdot 3)}_{6} = 66 \cdot 6$$

So, 66 divides into 396 6 times.

Now work margin exercise 7.

7. Find the LCM for 10, 18, and 75; then state how many times each number divides into the LCM.

8. Find the LCM of 20, 25, and 36; then state how many times each number divides into the LCM.

Completion Example 8 Finding the Least Common Multiple (LCM)

Find the LCM of 30, 50, and 63; then state how many times each number divides into the LCM.

Solution

Begin by finding the LCM.

$30 = \underline{2 \cdot 3 \cdot 5}$

$50 = \underline{2 \cdot 5 \cdot 5}$ LCM = $\underline{2 \cdot 3 \cdot 3 \cdot 5 \cdot 5 \cdot 7} = 3150$

$63 = \underline{3 \cdot 3 \cdot 7}$

Now determine how many times each number divides into the LCM:

$3150 = \underline{2 \cdot 3 \cdot 3 \cdot 5 \cdot 5 \cdot 7} = (2 \cdot 3 \cdot 5) \cdot (\underline{3 \cdot 5 \cdot 7})$

$= 30 \cdot \underline{105}$

So, 30 divides into 3150 __105__ times.

$3150 = \underline{2 \cdot 3 \cdot 3 \cdot 5 \cdot 5 \cdot 7} = (2 \cdot 5 \cdot 5) \cdot (\underline{3 \cdot 3 \cdot 7})$

$= 50 \cdot \underline{63}$

So, 50 divides into 3150 __63__ times.

$3150 = \underline{2 \cdot 3 \cdot 3 \cdot 5 \cdot 5 \cdot 7} = (3 \cdot 3 \cdot 7) \cdot (\underline{2 \cdot 5 \cdot 5})$

$= 63 \cdot \underline{50}$

So, 63 divides into 3150 __50__ times.

Now work margin exercise 8.

Many events occur at regular intervals of time. Weather satellites may orbit the earth once every 10 hours or once every 12 hours. Delivery trucks arrive once a day or once a week at department stores. Traffic lights change once every 3 minutes or once every 2 minutes. As illustrated in Example 9, the least common multiple can be used to explain how often related events occur.

9. A walker and two joggers begin using a track at the same time and starting place. Their lap times are 6, 3, and 5 minutes, respectively.

a. In how many minutes will they be together at the starting place?

b. How many laps will each person have completed at this time?

Example 9 Application: Finding the LCM

Suppose it takes three weather satellites—**A**, **B**, and **C**—different lengths of time to orbit the earth. Satellite **A** takes 24 hours, **B** takes 18 hours, and **C** takes 12 hours. If they are directly above each other now, as shown in Part **a.** of the given figure, in how many hours will they again be directly above each other in the position shown in Part **a.**? How many orbits will each satellite have made in that time?

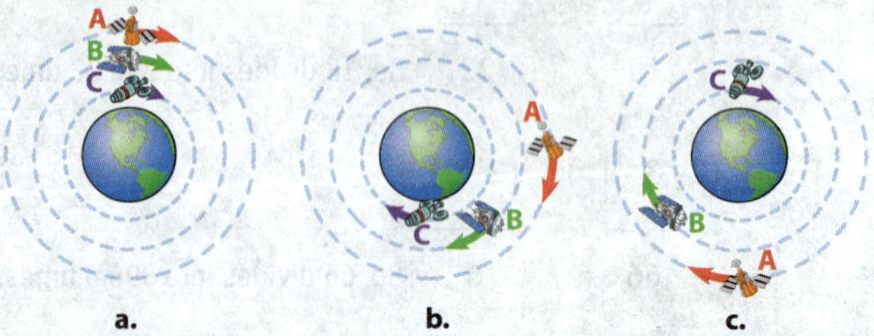

a. Beginning Positions **b.** Positions after 6 hours **c.** Positions after 12 hours

Solution

Step 1: READ: Read the problem carefully and, in this situation, study the figures. Note that there are two questions to be answered. The answer to the number of orbits depends on the number of hours it takes for the satellites to align again.

Step 2: SET UP: Since **A** takes 24 hours to make one complete orbit, the solution must be a multiple of 24. Similarly, the solution must be a multiple of 18 and a multiple of 12 to allow for complete orbits of satellites **B** and **C**.

Step 3: SOLVE: The solution is the LCM of 24, 18, and 12.

$$\left.\begin{aligned} 24 &= \mathbf{2} \cdot \mathbf{2} \cdot \mathbf{2} \cdot 3 \\ 18 &= 2 \cdot \mathbf{3} \cdot \mathbf{3} \\ 12 &= 2 \cdot 2 \cdot 3 \end{aligned}\right\} \text{LCM} = \mathbf{2} \cdot \mathbf{2} \cdot \mathbf{2} \cdot \mathbf{3} \cdot \mathbf{3} = 72$$

Thus, the satellites will align again at the position shown in Part **a.** in 72 hours (or 3 days). Note the following.

Satellite **A** will have made 3 orbits: $24 \cdot 3 = 72$

Satellite **B** will have made 4 orbits: $18 \cdot 4 = 72$

Satellite **C** will have made 6 orbits: $12 \cdot 6 = 72$

Step 4: CHECK: Because **C** goes faster than either **A** or **B**, it is reasonable that **C** makes more orbits in 72 hours than the other two satellites. Since **B** goes faster than **A**, it is also reasonable that **B** makes more orbits than **A** in 72 hours.

Now work margin exercise 9.

B Finding Equivalent Fractions

We know that any nonzero whole number divided by itself is 1. In fraction notation, this can be written $1 = \frac{1}{1} = \frac{2}{2} = \frac{3}{3} = \frac{4}{4} = \frac{19}{19} = \frac{63}{63} = \frac{k}{k}$. We also know that 1 is the multiplicative identity for whole numbers; that is, for any whole number k, $k \cdot 1 = k$. The number 1 is the multiplicative identity for fractions as well. For example,

$$\frac{3}{4} = \frac{3}{4} \cdot \mathbf{1} = \frac{3}{4} \cdot \frac{\mathbf{5}}{\mathbf{5}} = \frac{15}{20}.$$

The fractions $\frac{3}{4}$ and $\frac{15}{20}$ are equal and we say that we have found a fraction **equivalent** to $\frac{3}{4}$.

Math Tip

You may recognize that the equation $\frac{3}{4} = \frac{15}{20}$ is a *proportion*. Proportions and solving for unknown values in a proportion will be covered in Section 4.2.

Finding Equivalent Fractions

To find a fraction equivalent to $\frac{a}{b}$, multiply the numerator and denominator by the same nonzero whole number.

$$\frac{a}{b} = \frac{a}{b} \cdot \frac{k}{k} = \frac{a \cdot k}{b \cdot k}, \text{ where } b, k \neq 0 \left(\frac{k}{k} = 1\right)$$

For example, $\frac{2}{3} = \frac{2}{3} \cdot \frac{5}{5} = \frac{10}{15}$.

PROCEDURE

The following examples illustrate the use of this technique of finding equivalent fractions. Note carefully the importance of the choice of the form $\frac{k}{k}$.

10. Find the missing numerator that will make the fractions equivalent.

$$\frac{2}{3} = \frac{?}{12}$$

Example 10 Finding Equivalent Fractions

Find the missing numerator that will make the fractions equivalent.

$$\frac{3}{4} = \frac{?}{28}$$

Solution

Because $4 \cdot 7 = 28$, multiply by $\frac{7}{7}$.

$$\frac{3}{4} = \frac{3}{4} \cdot \frac{7}{7} = \frac{3 \cdot 7}{4 \cdot 7} = \frac{21}{28}$$

Thus, $\frac{3}{4} = \frac{21}{28}$.

Now work margin exercise 10.

11. Find the missing numerator that will make the fractions equivalent.

$$\frac{5}{8} = \frac{?}{24}$$

Example 11 Finding Equivalent Fractions

Find the missing numerator that will make the fractions equivalent.

$$\frac{9}{10} = \frac{?}{30}$$

Solution

Because $10 \cdot 3 = 30$, multiply by $\frac{3}{3}$.

$$\frac{9}{10} = \frac{9}{10} \cdot \frac{3}{3} = \frac{9 \cdot 3}{10 \cdot 3} = \frac{27}{30}$$

Thus, $\frac{9}{10} = \frac{27}{30}$.

Now work margin exercise 11.

Completion Example 12 Finding Equivalent Fractions

Find the missing numerator that will make the fractions equivalent.

$$\frac{3}{5} = \frac{?}{55}$$

Solution

Because $5 \cdot \underline{11} = 55$, we have $\frac{3}{5} = \frac{3}{5} \cdot \frac{\underline{11}}{\underline{11}} = \frac{\underline{33}}{\underline{55}}$. Thus, $\frac{3}{5} = \frac{\underline{33}}{55}$.

Now work margin exercise 12.

12. Find the missing numerator that will make the fractions equivalent.

$$\frac{4}{7} = \frac{?}{63}$$

Completion Example Answers

3. $2 \cdot 2 \cdot 3 \cdot 3$; $2 \cdot 2 \cdot 2 \cdot 3$; $2 \cdot 2 \cdot 2 \cdot 2 \cdot 3$; 2 and 3; Four 2s; Two 3s; LCM $= 2 \cdot 2 \cdot 2 \cdot 2 \cdot 3 \cdot 3 = 144$; 144

8. $30 = 2 \cdot 3 \cdot 5$, $50 = 2 \cdot 5 \cdot 5$, $63 = 3 \cdot 3 \cdot 7$; LCM $= 2 \cdot 3 \cdot 3 \cdot 5 \cdot 5 \cdot 7 = 3150$;

$3150 = 2 \cdot 3 \cdot 3 \cdot 5 \cdot 5 \cdot 7 = (2 \cdot 3 \cdot 5)(3 \cdot 5 \cdot 7) = 30 \cdot 105$; 105 times;

$3150 = 2 \cdot 3 \cdot 3 \cdot 5 \cdot 5 \cdot 7 = (2 \cdot 5 \cdot 5)(3 \cdot 3 \cdot 7) = 50 \cdot 63$, 63 times;

$3150 = 2 \cdot 3 \cdot 3 \cdot 5 \cdot 5 \cdot 7 = (3 \cdot 3 \cdot 7)(2 \cdot 5 \cdot 5) = 63 \cdot 50$, 50 times

12. $5 \cdot 11$; $\frac{3}{5} = \frac{3}{5} \cdot \frac{11}{11} = \frac{33}{55}$; $\frac{3}{5} = \frac{33}{55}$

Margin Exercise Answers

1. 84 **2.** 126 **3.** 120 **4.** 180 **5.** 441 **6.** LCM = 210; 210 = 15 · 14; 210 = 35 · 6; 210 = 42 · 5 **7.** LCM = 450; 450 = 10 · 45; 450 = 18 · 25; 450 = 75 · 6 **8.** LCM = 900; 900 = 20 · 45; 900 = 25 · 36; 900 = 36 · 25 **9. a.** 30 minutes **b.** 5, 10, and 6 laps **10.** 8 **11.** 15 **12.** 36

2.5 Exercises

Concept Check

Fill-in-the-Blank. Complete each sentence using information found in this section.

1. 5, 10, 15, 20, 25, … are ________ of the number 5.
2. LCM stands for _______ _________ ___________.
3. To find the LCM of a set of counting numbers, the first step is to find the _______ ____________ of each number.
4. The ___________ of a number are the products of that number with the counting numbers.
5. The LCM of 20 and 25 is _____.

2.5 Exercises

Concept Check

1. multiples
2. least common multiple
3. prime factorization
4. multiples
5. 100

6. False; The LCM of 15 and 25 is 75.
7. True
8. False; The first five multiples of 4 are 4, 8, 12, 16, and 20.
9. True

Practice

1. 5, 10, 15, 20, 25, 30, 35, 40, 45, 50, 55, 60
2. 6, 12, 18, 24, 30, 36, 42, 48, 54, 60, 66, 72
3. 12, 24, 36, 48, 60, 72, 84, 96, 108, 120, 132, 144
4. 15, 30, 45, 60, 75, 90, 105, 120, 135, 150, 165, 180
5. 30
6. 60
7. 15
8. 12
9. 30
10. 60
11. 30
12. 36
13. 200
14. 210
15. 150
16. 200
17. 40
18. 30
19. 112
20. 252
21. 196
22. 140
23. 66
24. 105
25. 10
26. 24
27. 36
28. 140
29. 60
30. 60

True/False. Determine whether each statement is true or false. If a statement is false, explain how it can be changed so the statement will be true. (**Note:** There may be more than one acceptable change.)

6. The LCM of 15 and 25 is 50.

7. The first five multiples of 9 are 9, 18, 27, 36, and 45.

8. The first five multiples of 4 are 4, 8, 12, 20, and 24.

9. When given larger numbers, the most efficient way to find the LCM is to use the prime factorization method.

Practice

List the first twelve multiples of each number.

1. 5

2. 6

3. 12

4. 15

From the lists you made in Exercises 1 through 4, find the least common multiple for the following pairs of numbers.

5. 5 and 6

6. 5 and 12

7. 5 and 15

8. 6 and 12

9. 6 and 15

10. 12 and 15

Find the LCM of each set of numbers. See Examples 1 through 5.

11. 6, 10

12. 9, 12

13. 25, 40

14. 30, 70

15. 50, 75

16. 40, 100

17. 5, 8

18. 3, 10

19. 16, 28

20. 36, 42

21. 28, 98

22. 28, 70

23. 2, 3, 11

24. 3, 5, 7

25. 2, 5, 10

26. 3, 4, 8

27. 4, 6, 9

28. 4, 14, 35

29. 6, 10, 12

30. 10, 12, 20

31. 15, 45, 90

32. 30, 60, 120

33. 10, 20, 30, 40

34. 15, 25, 30, 40

For each set of numbers, **a.** find the LCM and **b.** state how many times each number divides into the LCM. See Examples 6 through 8.

35. 10, 15, 25

36. 14, 35, 49

37. 6, 24, 30

38. 15, 30, 105

39. 12, 18, 27

40. 10, 18, 90

41. 20, 56, 45

42. 40, 56, 196

For each equation, find the missing numerator that will make the fractions equivalent. See Examples 10 through 12.

43. $\frac{5}{8} = \frac{?}{24}$

44. $\frac{2}{5} = \frac{?}{25}$

45. $\frac{1}{16} = \frac{?}{64}$

46. $\frac{1}{9} = \frac{?}{45}$

47. $\frac{6}{7} = \frac{?}{49}$

48. $\frac{6}{5} = \frac{?}{45}$

49. $\frac{3}{4} = \frac{?}{20}$

50. $\frac{5}{8} = \frac{?}{16}$

51. $\frac{10}{11} = \frac{?}{44}$

52. $\frac{5}{21} = \frac{?}{42}$

53. $\frac{5}{8} = \frac{?}{96}$

54. $\frac{5}{12} = \frac{?}{108}$

55. $\frac{11}{12} = \frac{?}{48}$

56. $\frac{4}{7} = \frac{?}{35}$

57. $\frac{2}{3} = \frac{?}{48}$

58. $\frac{2}{5} = \frac{?}{65}$

59. $\frac{9}{10} = \frac{?}{90}$

60. $\frac{7}{10} = \frac{?}{80}$

61. $\frac{9}{10} = \frac{?}{100}$

62. $\frac{3}{10} = \frac{?}{100}$

63. $\frac{7}{6} = \frac{?}{36}$

64. $\frac{2}{3} = \frac{?}{27}$

65. $\frac{10}{3} = \frac{?}{33}$

66. $\frac{9}{7} = \frac{?}{84}$

31. 90

32. 120

33. 120

34. 600

35. **a.** LCM = 150
b. $150 = 10 \cdot 15 = 15 \cdot 10 = 25 \cdot 6$

36. **a.** LCM = 490
b. $490 = 14 \cdot 35 = 35 \cdot 14 = 49 \cdot 10$

37. **a.** LCM = 120
b. $120 = 6 \cdot 20 = 24 \cdot 5 = 30 \cdot 4$

38. **a.** LCM = 210
b. $210 = 15 \cdot 14 = 30 \cdot 7 = 105 \cdot 2$

39. **a.** LCM = 108
b. $108 = 12 \cdot 9 = 18 \cdot 6 = 27 \cdot 4$

40. **a.** LCM = 90
b. $90 = 10 \cdot 9 = 18 \cdot 5 = 90 \cdot 1$

41. **a.** LCM = 2520
b. $1260 = 20 \cdot 63 = 28 \cdot 45 = 45 \cdot 28$

42. **a.** LCM = 1960
b. $1960 = 40 \cdot 49 = 56 \cdot 35 = 196 \cdot 10$

43. 15

44. 10

45. 4

46. 5

47. 42

48. 54

49. 15

50. 10

51. 40

52. 10

53. 60

54. 45

55. 44

56. 20
57. 32
58. 26
59. 81
60. 56
61. 90
62. 30
63. 42
64. 18
65. 110
66. 108

Applications

67. a. 60 minutes
b. 4, 3, and 2 trips, respectively
68. a. Every 24 days
b. Every 24 days
69. a. 90 hours
b. 6 and 5 orbits, respectively
70. a. Every 70 days
b. 14, 10, and 7 times, respectively
71. a. 840 days
b. 105, 70, 60, and 56 trips, respectively
72. a. 360 pieces
b. 15 boxes, 10 boxes, and 8 boxes, respectively
73. a. 840 seconds
b. 24 laps, 21 laps, and 20 laps, respectively

Applications

Solve.

67. Three security guards meet at the front gate for coffee before they walk around inspecting buildings at a manufacturing plant. The guards take 15, 20, and 30 minutes, respectively, for the inspection trip.

a. If they start at the same time, in how many minutes will they meet again at the front gate for coffee?

b. How many trips will each guard have made?

68. Three trucks drivers have dinner together whenever all three are at the routing station at the same time. The first driver's route takes 6 days, the second driver's route takes 8 days, and the third driver's route takes 12 days.

a. How frequently do the three drivers have dinner together?

b. How frequently do the first two drivers meet?

69. Two astronauts miss connections at their first meeting in space.

a. If one astronaut circles the earth every 15 hours and the other every 18 hours, in how many hours will they meet again at the same place?

b. How many more orbits will each astronaut have to complete between missing their first meeting and making their second meeting?

70. Three neighbors mow their lawns at different intervals during the summer months. The first one mows every 5 days, the second every 7 days, and the third every 10 days.

a. How frequently do they mow their lawns on the same day?

b. How many times does each neighbor mow in between the times when they all mow together?

71. Four women work for the same book company selling textbooks. They leave the home office on the same day and take 8 days, 12 days, 14 days, and 15 days, respectively, to visit schools in their own sales regions.

a. In how many days will they all meet again at the home office?

b. How many sales trips will each have made in this time?

72. A fruit production company has three packaging facilities, each of which uses different-sized boxes as follows: 24 pieces/box, 36 pieces/box, and 45 pieces/box.

a. Assuming that the truck provides the same quantity of uniformly-sized pieces of fruit to all three packaging facilities, what is the minimum number of pieces of fruit that will be delivered so that no fruit will be left over?

b. How many boxes will each facility package?

73. Three swimmers decide to swim laps together, and they will quit when they reach the starting end of the pool together. The first swimmer can swim a lap in 35 seconds, the second will take 40 seconds, and the third takes 42 seconds.

a. How many seconds will it take before they quit?

b. How many laps will each swimmer swim in that interval?

74. Two analog clocks are sitting next to each other. The first clock keeps perfect time, where the minute hand takes 60 minutes to travel completely around the dial. The second clock runs fast and the minute hand makes one complete revolution in 55 minutes.

a. Assuming that both clocks are started so that the minute hands are at 12, how many minutes will it take until both minute hands return to 12 at the same time?

b. How many hours does this represent?

75. Two satellites are in polar orbit around the earth. One takes 8 hours to complete its orbit and the other 15 hours.

a. Assuming that the satellites start over the North Pole, how long will it take before they are both over the North Pole again?

b. How many orbits does each satellite make during this time?

76. Three cars are being driven around the same road circuit. Car 1 can drive it in 10 minutes, car 2 takes 20 minutes, and car 3 takes 40 minutes.

a. Assuming that all the cars start at the same time, how long will it take for them to reach the starting point together?

b. How long will it take for the first two cars to meet?

c. How long will it take for cars 1 and 3 to meet?

77. Four ships leave the port on the same day. They take 12, 15, 18, and 30 days, respectively, to sail their routes and reload cargo. How frequently do these four ships leave this port on the same day?

78. A small college offers some courses at intervals greater than one year. In one department of this college, four courses are each offered at different intervals as follows. One course is offered every 2 years, a second course is held every 3 years, a third course is given every 4 years, and a special elective is offered only once every 6 years. How long will it take for these courses to be offered during the same year after the previous time this happened?

79. A certain star has four planets, all moving in the same plane. The first one takes 4 years to go around the star, the second takes 6 years, the third 10 years, and the fourth takes 15 years. Assuming that they are initially aligned, how many years will it take until they are aligned again?

80. Four tour buses depart the terminal on the same day. The itinerary of the first bus takes 6 days, it takes 12 days for the second bus, it takes 14 days for the third bus, and it takes 21 days for the fourth bus. Assuming that another tour begins the same day that the bus returns to the terminal, how many days will it take until all four buses return to the terminal on the same day?

81. A box of 6 peanut clusters is sold for $3. How much would a box of 24 peanut clusters cost if the price per peanut cluster remains the same?

82. For fieldtrips, school policy requires that 2 adults are present for every 13 students. How many adults are needed on a fieldtrip for 52 students?

74. **a.** 660 minutes
b. 11 hours

75. **a.** 120 hours
b. 15 orbits and 8 orbits, respectively

76. **a.** 40 minutes
b. 20 minutes
c. 40 minutes

77. Every 180 days

78. 12 years

79. 60 years

80. 84 days

81. $12

82. 8

Writing & Thinking

83. Since the LCM is constructed using the prime factors of each number in the set, by definition, each number will divide the LCM.

84. Since the LCM is constructed using the largest number of prime factors found in any one number in the set, the LCM will be at least as large as any number in the set.

85. Multiplying the two numbers together will give the LCM if those two numbers have no common factors. If they have any factors in common, then you would only use that common factor once. Examples will vary.

86. Answers will vary. The method of finding the prime factorization of each number first reduces the risk of determining an incorrect value for the LCM.

Writing & Thinking

83. Explain, in your own words, why each number in a set divides evenly into the LCM of that set of numbers.

84. Explain, in your own words, why the LCM of a set of numbers is greater than or equal to each number in the set.

85. Explain why simply multiplying two numbers together will not necessarily find the LCM of those numbers. Give an example of when it would find the LCM and an example when it would not.

86. There are several ways to find the least common multiple. Discuss the method you prefer and explain why.

2.6 Addition and Subtraction with Fractions

Objectives

A. Add fractions with the same denominator.

B. Add fractions with different denominators.

C. Subtract fractions with the same denominator.

D. Subtract fractions with different denominators.

A Adding Fractions with the Same Denominator

When adding two or more fractions with the same denominator, we can think of this denominator as the "name" of each fraction. The sum has this common name. Just as 3 *oranges* plus 2 *oranges* gives a total of 5 *oranges*, 3 *eighths* plus 2 *eighths* gives a total of 5 *eighths*. This denominator is called a **common denominator**. The following figure illustrates how the sum of the two fractions $\frac{3}{8}$ and $\frac{2}{8}$ might be diagrammed.

$$\frac{3}{8}+\frac{2}{8}=\frac{5}{8}$$

Figure 1

Adding Fractions with the Same Denominator

1. Add the numerators.
2. Keep the common denominator. $\qquad \dfrac{a}{b}+\dfrac{c}{b}=\dfrac{a+c}{b},\ b\neq 0$
3. Reduce, if possible.

For example, $\frac{1}{7}+\frac{4}{7}=\frac{1+4}{7}=\frac{5}{7}$.

PROCEDURE

Example 1 Adding Fractions with the Same Denominator

Add.

a. $\frac{4}{15}+\frac{6}{15}$ **b.** $\frac{1}{11}+\frac{5}{11}+\frac{3}{11}$

Solution

a. $\frac{4}{15}+\frac{6}{15}=\frac{4+6}{15}=\frac{10}{15}=\frac{2\cdot\cancel{5}}{3\cdot\cancel{5}}=\frac{2}{3}$ **b.** $\frac{1}{11}+\frac{5}{11}+\frac{3}{11}=\frac{1+5+3}{11}=\frac{9}{11}$

Now work margin exercise 1.

1. Add.

a. $\frac{3}{12}+\frac{5}{12}$

b. $\frac{1}{9}+\frac{4}{9}+\frac{3}{9}$

B Adding Fractions with Different Denominators

We know that fractions to be added will not always have the same denominator. Figure 2 illustrates how the two fractions $\frac{1}{3}$ and $\frac{2}{5}$ might be diagrammed.

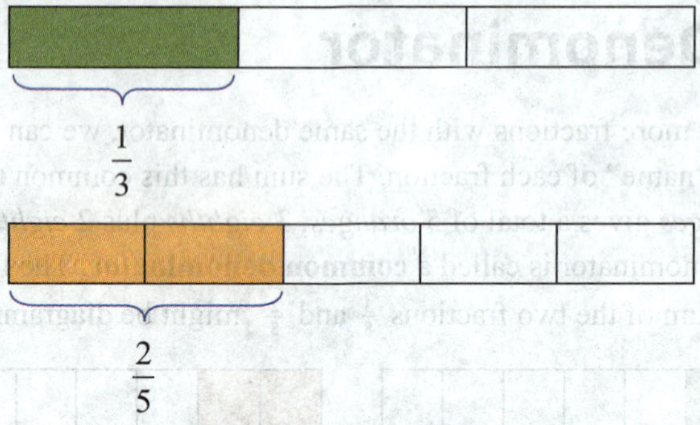

Figure 2

To find the result of adding these two fractions, a common unit (that is, a common denominator) is needed. By dividing each third into five parts and each fifth into three parts, we find that a common unit is fifteenths. Now the two fractions can be added as shown in Figure 3.

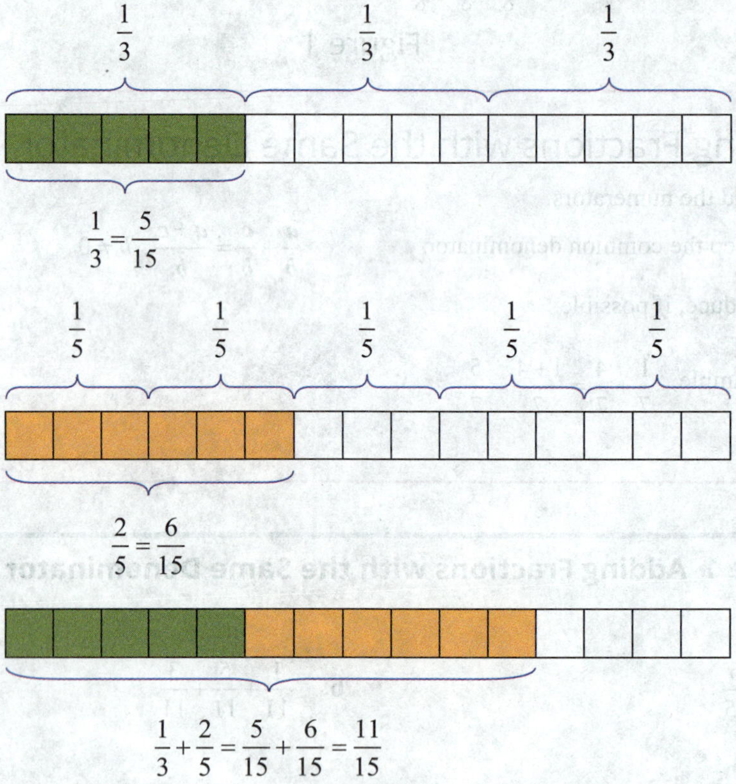

Figure 3

Of course, drawing diagrams every time two or more fractions are to be added would be difficult and time-consuming. A better way is to find the least common denominator of all the fractions, change each fraction to an equivalent fraction with this common denominator, and then add. In Figure 3, the least common denominator was the product of the denominators. In general, the **least common denominator (LCD)** is the least common multiple (LCM) of the denominators (see Section 2.5).

Example 2 Finding the Least Common Denominator (LCD)

Find the LCD for $\frac{3}{10}$ and $\frac{11}{12}$.

Solution

Using prime factorization:

$$\left.\begin{aligned} 10 &= 2 \cdot 5 \\ 12 &= 2 \cdot 2 \cdot 3 \end{aligned}\right\} \text{LCD} = 2 \cdot 2 \cdot 3 \cdot 5 = 60$$

Now work margin exercise 2.

2. Find the LCD for $\frac{5}{9}$ and $\frac{7}{18}$.

Teaching Note

You may want to emphasize the "least" from least common multiple with your students. Some students may want to just multiply the denominators. In that case, you could show with denominators 9 and 18 how much easier it is to deal with 18 (the LCM) than $9 \cdot 18 = 162$.

Adding Fractions with Different Denominators

1. Find the least common denominator (LCD).
2. Change each fraction into an equivalent fraction with that denominator.
3. Add the new fractions.
4. Reduce, if possible.

PROCEDURE

Example 3 Adding Fractions with Different Denominators

Add: $\frac{3}{8}+\frac{7}{12}$

Solution

Step 1: Find the LCD.

$$\left.\begin{aligned} 8 &= 2 \cdot 2 \cdot 2 \\ 12 &= 2 \cdot 2 \cdot 3 \end{aligned}\right\} \text{LCD} = 2 \cdot 2 \cdot 2 \cdot 3 = 24$$

$$= (2 \cdot 2 \cdot 2) \cdot 3 = 8 \cdot 3$$

$$= (2 \cdot 2 \cdot 3) \cdot 2 = 12 \cdot 2$$

Note: You might not need to use prime factorizations to find the LCD. If the denominators are numbers that are familiar to you, then you might be able to find the LCD simply by inspection.

3. Add: $\frac{1}{6}+\frac{3}{10}$

Step 2: Find fractions equivalent to $\frac{3}{8}$ and $\frac{7}{12}$ with denominator 24.

$$\frac{3}{8}=\frac{3}{8}\cdot\frac{3}{3}=\frac{9}{24}$$ Multiply by $\frac{3}{3}$ because $8\cdot 3=24$.

$$\frac{7}{12}=\frac{7}{12}\cdot\frac{2}{2}=\frac{14}{24}$$ Multiply by $\frac{2}{2}$ because $12\cdot 2=24$.

Step 3: Add.

$$\frac{3}{8}+\frac{7}{12}=\frac{9}{24}+\frac{14}{24}=\frac{9+14}{24}=\frac{23}{24}$$

Step 4: Reduce, if possible.

$$\frac{23}{24}=\frac{23}{2\cdot 2\cdot 2\cdot 3}=\frac{23}{24}$$ The fraction cannot be reduced because 23 and 24 have only 1 as a common factor.

Now work margin exercise 3.

Common Error

The following common error must be avoided.

Add: $\frac{3}{5}+\frac{1}{15}$

Wrong Solution

You **cannot** divide out across the + sign.

$$\frac{\overset{1}{\cancel{3}}}{5}+\frac{1}{\underset{5}{\cancel{15}}}=\frac{1}{5}+\frac{1}{5}=\frac{2}{5}$$

Correct Solution

$$\frac{3}{5}+\frac{1}{15}=\frac{3}{5}\cdot\frac{3}{3}+\frac{1}{15}=\frac{9}{15}+\frac{1}{15}=\frac{10}{15}$$

NOW reduce by dividing out a factor of 5.

$$\frac{10}{15}=\frac{2\cdot\cancel{5}}{3\cdot\cancel{5}}=\frac{2}{3}$$

CAUTION

4. Add: $\frac{2}{21}+\frac{9}{28}$

Example 4 Adding Fractions with Different Denominators

Add: $\frac{7}{45}+\frac{7}{36}$

Solution

Step 1: Find the LCD.

$$\left.\begin{aligned}45&=3\cdot 3\cdot 5\\36&=2\cdot 2\cdot 3\cdot 3\end{aligned}\right\}\text{LCD}=2\cdot 2\cdot 3\cdot 3\cdot 5=180$$

$$=(3\cdot 3\cdot 5)(2\cdot 2)=45\cdot 4$$

$$=(2\cdot 2\cdot 3\cdot 3)(5)=36\cdot 5$$

Step 2: Steps 2, 3, and 4 can be written in one step.

$$\begin{aligned}\frac{7}{45}+\frac{7}{36}&=\frac{7}{45}\cdot\frac{4}{4}+\frac{7}{36}\cdot\frac{5}{5}\\&=\frac{28}{180}+\frac{35}{180}\\&=\frac{63}{180}=\frac{\cancel{3}\cdot\cancel{3}\cdot7}{2\cdot2\cdot\cancel{3}\cdot\cancel{3}\cdot5}=\frac{7}{20}\end{aligned}$$

Note that in adding fractions, we also may choose to write them vertically. The process is the same.

$$\begin{array}{rcrcr}\frac{7}{45}&=&\frac{7}{45}\cdot\frac{4}{4}&=&\frac{28}{180}\\+\frac{7}{36}&=&+\frac{7}{36}\cdot\frac{5}{5}&=&+\frac{35}{180}\\\hline&&&&\frac{63}{180}=\frac{\cancel{3}\cdot\cancel{3}\cdot7}{2\cdot2\cdot\cancel{3}\cdot\cancel{3}\cdot5}=\frac{7}{20}\end{array}$$

Now work margin exercise 4.

Example 5 Adding Three Fractions with Different Denominators

Add: $\frac{2}{3}+\frac{1}{6}+\frac{5}{12}$

Solution

Step 1: LCD = 12 — You can simply observe this or use prime factorizations.

Step 2: Steps 2, 3, and 4 can be written in one step.

$$\begin{aligned}\frac{2}{3}+\frac{1}{6}+\frac{5}{12}&=\frac{2}{3}\cdot\frac{4}{4}+\frac{1}{6}\cdot\frac{2}{2}+\frac{5}{12}\\&=\frac{8}{12}+\frac{2}{12}+\frac{5}{12}\\&=\frac{15}{12}=\frac{\cancel{3}\cdot5}{2\cdot2\cdot\cancel{3}}=\frac{5}{4}\text{ or }1\frac{1}{4}\end{aligned}$$

Now work margin exercise 5.

5. Add: $\frac{1}{4}+\frac{2}{5}+\frac{7}{10}$

Example 6 Application: Adding Fractions

Rachel is mailing two letters at the post office. One letter weighs $\frac{1}{3}$ ounces and the other weighs $\frac{3}{7}$ ounces. What is the combined weight of the letters?

Solution

Step 1: READ: Read the problem carefully. We want to find the total weight of the two letters.

Step 2: SET UP: To find the combined weight of the two letters, add the weights: $\frac{1}{3}+\frac{3}{7}$

6. Catherine is at the grocery store buying fruit. She buys one grapefruit that weighs $\frac{3}{5}$ lb, one pomegranate that weighs $\frac{1}{2}$ lb, and a pear that weighs $\frac{3}{8}$ lb. What is the total weight of fruit that she will be purchasing?

Step 3: SOLVE: The LCD of 3 and 7 is 21. Using this information, we can add the two fractions.

$$\begin{aligned}\frac{1}{3}+\frac{3}{7} &= \frac{1}{3}\cdot\frac{7}{7}+\frac{3}{7}\cdot\frac{3}{3}\\ &= \frac{7}{21}+\frac{9}{21}\\ &= \frac{16}{21}\end{aligned}$$

Thus, the total weight of the two letters is $\frac{16}{21}$ ounces.

Step 4: CHECK: Since each letter weighs less than half an ounce, the total will be less than one ounce. Thus, the answer $\frac{16}{21}$ ounces seems to be a reasonable answer.

Now work margin exercise 6.

7. Fred's total income for the year was $30,000, and he spent $\frac{1}{3}$ of his income on traveling and $\frac{1}{10}$ of his income on entertainment. What portion (or fraction) of his income did he spend on these two items and what was the total amount spent?

Example 7 Application: Adding Fractions

Keith's total income for the year was $36,000, and he spent $\frac{1}{4}$ of his income on rent and $\frac{1}{12}$ of his income on his car. What portion (or fraction) of his income did he spend on these two items and what was the total amount spent?

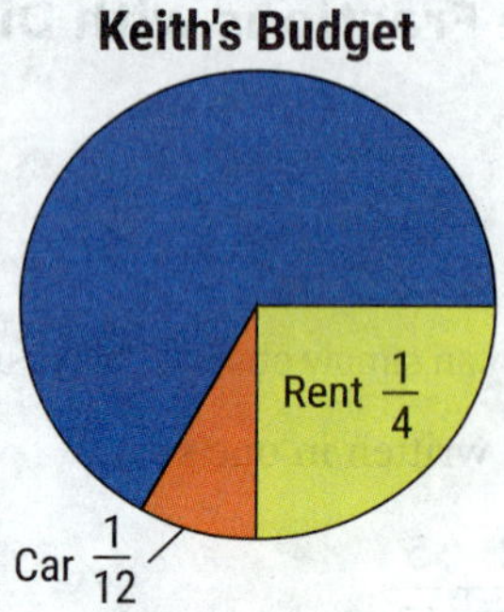

Solution

Step 1: READ: Read the problem carefully. In this case, the diagram is helpful. There are two questions here: What portion (or fraction) of his income was spent on these two items and what was the total amount spent?

Step 2: SET UP: To find the fraction spent on these items, add the fractions: $\frac{1}{4}+\frac{1}{12}$.

Then to find the amount spent, multiply $36,000 by the sum of the fractions.

Step 3: SOLVE: The LCD is 12.

$$\frac{1}{4}+\frac{1}{12} = \frac{1}{4}\cdot\frac{3}{3}+\frac{1}{12} = \frac{3}{12}+\frac{1}{12} = \frac{4}{12} = \frac{\cancel{4}\cdot 1}{\cancel{4}\cdot 3} = \frac{1}{3}$$

Now multiply $\frac{1}{3}$ times $36,000.

$$\frac{1}{3}\cdot 36{,}000 = \frac{1\cdot \overset{12{,}000}{\cancel{36{,}000}}}{\underset{1}{\cancel{3}}\cdot 1} = 12{,}000$$

Keith spent $\frac{1}{3}$ of his income for these two items for a total of $12,000.

Step 4: CHECK: Since he spent $\frac{1}{3}$ of his income on these two items, the answer of $12,000 is reasonable (and accurate).

Now work margin exercise 7.

In Section 1.2, we discussed the commutative and associative properties for addition with whole numbers. Note that these same properties apply to addition with fractions.

Commutative Property of Addition

The **order** of the fractions being added can be reversed without changing the sum.

Symbolically, if $\frac{a}{b}$ and $\frac{c}{d}$ are fractions, then

$$\frac{a}{b}+\frac{c}{d}=\frac{c}{d}+\frac{a}{b}. \quad (b,d \neq 0)$$

For example, $\frac{1}{5}+\frac{2}{3}=\frac{2}{3}+\frac{1}{5}$.

PROPERTIES

Associative Property of Addition

The **grouping** of the fractions being added can be changed without changing the sum. Symbolically, if $\frac{a}{b}$, $\frac{c}{d}$, and $\frac{e}{f}$ are fractions, then

$$\frac{a}{b}+\left(\frac{c}{d}+\frac{e}{f}\right)=\left(\frac{a}{b}+\frac{c}{d}\right)+\frac{e}{f}. \quad (b,d,f \neq 0)$$

For example, $\left(\frac{2}{7}+\frac{4}{5}\right)+\frac{1}{2}=\frac{2}{7}+\left(\frac{4}{5}+\frac{1}{2}\right)$.

PROPERTIES

C Subtracting Fractions with the Same Denominator

Finding the difference between two fractions with a common denominator is similar to finding the sum. The numerators are simply subtracted instead of added. Just as with addition, the common denominator "names" each fraction.

Figure 4 shows how the difference between $\frac{4}{5}$ and $\frac{1}{5}$ might be diagrammed.

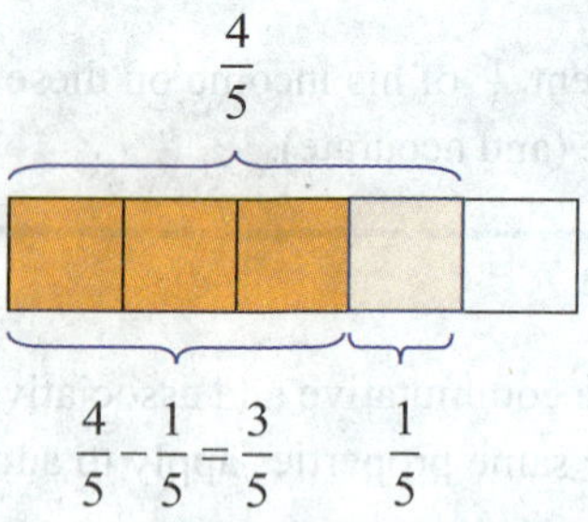

Figure 4

Subtracting Fractions with the Same Denominator

1. Subtract the numerators.
2. Keep the common denominator. $\frac{a}{b}-\frac{c}{b}=\frac{a-c}{b},\ b\neq 0$
3. Reduce, if possible.

For example, $\frac{9}{11}-\frac{4}{11}=\frac{5}{11}$.

PROCEDURE

8. Subtract: $\frac{11}{15}-\frac{8}{15}$

Example 8 Subtracting Fractions with the Same Denominator

Subtract: $\frac{9}{10}-\frac{7}{10}$

Solution

$$\frac{9}{10}-\frac{7}{10}=\frac{9-7}{10}=\frac{2}{10}=\frac{\cancel{2}\cdot 1}{\cancel{2}\cdot 5}=\frac{1}{5}$$

The difference is reduced just as any fraction is reduced.

Now work margin exercise 8.

D Subtracting Fractions with Different Denominators

To subtract fractions with different denominators, we use the least common denominator (LCD) to find equivalent fractions, just as we did when adding fractions with different denominators.

Subtracting Fractions with Different Denominators

1. Find the least common denominator (LCD).
2. Change each fraction into an equivalent fraction with that denominator.
3. Subtract the new fractions.
4. Reduce, if possible.

PROCEDURE

Example 9 Subtracting Fractions with Different Denominators

Subtract: $\frac{9}{10}-\frac{2}{15}$

Solution

Step 1: Find the LCD.

$$\left.\begin{array}{l}10=2\cdot 5\\ 15=3\cdot 5\end{array}\right\}\text{LCD}=2\cdot 3\cdot 5=30$$

Step 2: Find fractions equivalent to $\frac{9}{10}$ and $\frac{2}{15}$ with denominator 30.

$$\frac{9}{10}=\frac{9}{10}\cdot\frac{3}{3}=\frac{27}{30}$$

$$\frac{2}{15}=\frac{2}{15}\cdot\frac{2}{2}=\frac{4}{30}$$

Step 3: Subtract.

$$\frac{9}{10}-\frac{2}{15}=\frac{27}{30}-\frac{4}{30}=\frac{23}{30}$$

Step 4: Reduce, if possible.

$$\frac{23}{30}=\frac{23}{2\cdot 3\cdot 5}=\frac{23}{30}$$

The fraction cannot be reduced because 23 and 30 have only 1 as a common factor.

Now work margin exercise 9.

9. Subtract: $\frac{7}{11}-\frac{3}{5}$

Example 10 Subtracting Fractions with Different Denominators

Subtract: $\frac{7}{20}-\frac{3}{28}$

Solution

Step 1: Find the LCD.

$$\left.\begin{array}{l}20=2\cdot 2\cdot 5\\ 28=2\cdot 2\cdot 7\end{array}\right\}\text{LCD}=\text{LCD}=2\cdot 2\cdot 5\cdot 7=140$$

$$=(2\cdot 2\cdot 5)\cdot 7=20\cdot 7$$

$$=(2\cdot 2\cdot 7)\cdot 5=28\cdot 5$$

10. Subtract: $\frac{7}{12}-\frac{2}{15}$

Step 2: Steps 2, 3, and 4 can be written in one step.

$$\frac{7}{20}-\frac{3}{28}=\frac{7}{20}\cdot\frac{7}{7}-\frac{3}{28}\cdot\frac{5}{5}$$
$$=\frac{49}{140}-\frac{15}{140}$$
$$=\frac{34}{140}=\frac{\cancel{2}\cdot 17}{\cancel{2}\cdot 70}=\frac{17}{70}$$

Or, we can write the fractions vertically.

$$\frac{7}{20}=\frac{7}{20}\cdot\frac{7}{7}=\frac{49}{140}$$
$$-\frac{3}{28}=-\frac{3}{28}\cdot\frac{5}{5}=-\frac{15}{140}$$
$$\frac{34}{140}=\frac{\cancel{2}\cdot 17}{\cancel{2}\cdot 70}=\frac{17}{70}$$

Now work margin exercise 10.

11. Subtract: $3-\frac{5}{12}$

Example 11 Subtracting Fractions with Different Denominators

Subtract: $1-\frac{5}{8}$

Solution

Step 1: LCD = 8 Since 1 can be written as $\frac{1}{1}$, the common denominator is $8\cdot 1=8$.

Step 2: Steps 2, 3, and 4 can be written in one step.

$$1-\frac{5}{8}=\frac{1}{1}\cdot\frac{8}{8}-\frac{5}{8}=\frac{8}{8}-\frac{5}{8}=\frac{8-5}{8}=\frac{3}{8}$$

Now work margin exercise 11.

12. The Cavalier baseball team lost 30 games in one season. If $\frac{1}{3}$ of their losses were by more than 5 runs, and $\frac{2}{5}$ of their losses were by more than 4 runs, what fraction of their losses were by exactly 5 runs?

Example 12 Application: Subtracting Fractions with Different Denominators

The Narragansett Grays baseball team lost 90 games in one season. If $\frac{1}{5}$ of their losses were by 1 or 2 runs and $\frac{4}{9}$ of their losses were by 3 or fewer runs, what fraction of their losses were by exactly 3 runs?

Solution

Step 1: READ: Read the problem carefully. We want to find the fraction of losses that were by exactly 3 runs.

Step 2: SET UP: The losses that were by 3 or fewer runs include those that were by 1 or 2 runs. So, to find the fraction that were by exactly 3 runs, we need to subtract: $\frac{4}{9}-\frac{1}{5}$.

Step 3: SOLVE: The LCD is 45.

$$\frac{4}{9}-\frac{1}{5}=\frac{4}{9}\cdot\frac{5}{5}-\frac{1}{5}\cdot\frac{9}{9}=\frac{20}{45}-\frac{9}{45}=\frac{11}{45}$$

The Grays lost $\frac{11}{45}$ of their losses by exactly 3 runs.

Step 4: CHECK: Since $\frac{1}{5}=\frac{2}{10}$ and $\frac{4}{9}$ is close to $\frac{4}{10}$, approximately $\frac{2}{10}$ of their losses were by exactly 3 runs. Because $\frac{2}{10}=\frac{9}{45}$, the answer of $\frac{11}{45}$ seems reasonable.

Now work margin exercise 12.

Margin Exercise Answers

1. a. $\frac{2}{3}$ **b.** $\frac{8}{9}$ **2.** 18 **3.** $\frac{7}{15}$ **4.** $\frac{5}{12}$ **5.** $\frac{27}{20}$ or $1\frac{7}{20}$ **6.** $\frac{59}{40}$ or $1\frac{19}{40}$ lb **7.** $\frac{13}{30}$, \$13,000

8. $\frac{1}{5}$ **9.** $\frac{2}{55}$ **10.** $\frac{9}{20}$ **11.** $\frac{31}{12}$ or $2\frac{7}{12}$ **12.** $\frac{1}{15}$ of their losses were by exactly 5 runs

2.6 Exercises

Concept Check

Fill-in-the-Blank. Complete each sentence using information found in this section.

1. When adding fractions, it is important that the _______ are the same before adding.

2. LCD stands for _______ _______ _______.

3. The LCD is the least common multiple of the _______.

4. When subtracting fractions with the same denominator, _______ the numerators and keep the _______ _______.

5. In subtraction with fractions with different denominators, each fraction is changed to a/an _______ fraction with the LCD as its denominator.

True/False. Determine whether each statement is true or false. If a statement is false, explain how it can be changed so the statement will be true. (**Note:** There may be more than one acceptable change.)

6. The final step in adding fractions is to reduce, if possible.

7. The process for finding the LCD is the same as the process for finding the LCM.

8. LCD represents the Least Common Digit.

9. When subtracting fractions, simply subtract the numerators and the denominators.

10. Subtraction of fractions requires that the fractions have the same denominators.

2.6 Exercises

Concept Check

1. denominators
2. least common denominator
3. denominators
4. subtract, common denominator
5. equivalent
6. True
7. True
8. False; LCD stands for least common denominator.
9. False; When subtracting fractions, subtract the numerators and keep the common denominator.
10. True

Practice

1. $\frac{5}{14}$
2. $\frac{10}{11}$
3. 2
4. 1
5. $\frac{1}{5}$
6. $\frac{3}{5}$
7. $\frac{3}{4}$
8. $\frac{5}{7}$
9. $\frac{7}{9}$
10. $\frac{11}{16}$

11. $\frac{11}{8}$
12. $\frac{3}{2}$
13. $\frac{62}{45}$
14. $\frac{47}{42}$
15. $\frac{13}{12}$
16. $\frac{19}{18}$
17. $\frac{3}{4}$
18. $\frac{5}{6}$
19. $\frac{7}{10}$
20. $\frac{11}{16}$
21. 1
22. $\frac{11}{9}$
23. $\frac{17}{20}$
24. $\frac{17}{21}$
25. $\frac{5}{6}$
26. $\frac{1}{2}$
27. $\frac{11}{20}$
28. $\frac{5}{8}$
29. $\frac{17}{20}$
30. $\frac{31}{36}$
31. $\frac{317}{1000}$
32. $\frac{753}{1000}$
33. $\frac{631}{100}$
34. $\frac{583}{100}$

Practice

Add and reduce to lowest terms. See Examples 1 through 5.

1. $\frac{3}{14}+\frac{2}{14}$

2. $\frac{3}{11}+\frac{7}{11}$

3. $\frac{7}{5}+\frac{3}{5}$

4. $\frac{9}{10}+\frac{1}{10}$

5. $\frac{1}{20}+\frac{3}{20}$

6. $\frac{3}{25}+\frac{12}{25}$

7. $\frac{1}{8}+\frac{3}{8}+\frac{2}{8}$

8. $\frac{5}{14}+\frac{4}{14}+\frac{1}{14}$

9. $\frac{7}{45}+\frac{11}{45}+\frac{17}{45}$

10. $\frac{14}{32}+\frac{7}{32}+\frac{1}{32}$

11. $\frac{5}{8}+\frac{3}{4}$

12. $\frac{5}{6}+\frac{2}{3}$

13. $\frac{7}{9}+\frac{3}{5}$

14. $\frac{5}{6}+\frac{2}{7}$

15. $\frac{1}{4}+\frac{5}{6}$

16. $\frac{2}{9}+\frac{5}{6}$

17. $\frac{2}{5}+\frac{7}{20}$

18. $\frac{3}{4}+\frac{1}{12}$

19. $\frac{2}{5}+\frac{3}{10}$

20. $\frac{3}{8}+\frac{5}{16}$

21. $\frac{1}{12}+\frac{2}{3}+\frac{1}{4}$

22. $\frac{5}{18}+\frac{1}{2}+\frac{4}{9}$

23. $\frac{2}{5}+\frac{3}{10}+\frac{3}{20}$

24. $\frac{2}{7}+\frac{4}{21}+\frac{1}{3}$

25. $\frac{1}{4}+\frac{1}{20}+\frac{8}{15}$

26. $\frac{1}{5}+\frac{2}{15}+\frac{1}{6}$

27. $\frac{1}{5}+\frac{1}{10}+\frac{1}{4}$

28. $\frac{1}{5}+\frac{1}{40}+\frac{1}{4}$

29. $\frac{2}{5}+\frac{3}{10}+\frac{3}{20}$

30. $\frac{1}{3}+\frac{5}{36}+\frac{7}{18}$

31. $\frac{3}{10}+\frac{1}{100}+\frac{7}{1000}$

32. $\frac{7}{10}+\frac{5}{100}+\frac{3}{1000}$

33. $6+\frac{1}{100}+\frac{3}{10}$

34. $5+\frac{13}{100}+\frac{7}{10}$

Subtract and reduce to lowest terms. See Examples 8 through 11.

35. $\frac{9}{10}-\frac{3}{10}$

36. $\frac{7}{8}-\frac{5}{8}$

37. $\frac{11}{12}-\frac{7}{12}$

38. $\frac{21}{15}-\frac{11}{15}$

39. $\frac{12}{13}-\frac{4}{13}$

40. $\frac{5}{9}-\frac{1}{9}$

41. $\frac{5}{6}-\frac{1}{3}$

42. $\frac{5}{6}-\frac{1}{2}$

43. $\frac{11}{15}-\frac{3}{10}$

44. $\frac{9}{10}-\frac{3}{4}$

45. $\frac{8}{10}-\frac{3}{15}$

46. $\frac{9}{14}-\frac{2}{21}$

47. $\frac{3}{4}-\frac{2}{3}$

48. $\frac{2}{3}-\frac{1}{4}$

49. $\frac{3}{8}-\frac{1}{16}$

50. $\frac{7}{6}-\frac{2}{3}$

51. $\frac{5}{4}-\frac{3}{5}$

52. $\frac{2}{3}-\frac{2}{7}$

53. $\frac{25}{18}-\frac{21}{27}$

54. $\frac{21}{16}-\frac{15}{24}$

55. $1-\frac{9}{10}$

56. $1-\frac{1}{16}$

57. $2-\frac{9}{16}$

58. $6-\frac{2}{3}$

59. $\frac{14}{35}-\frac{12}{30}$

60. $\frac{20}{35}-\frac{24}{42}$

61. $\frac{76}{100}-\frac{7}{10}$

62. $\frac{54}{100}-\frac{5}{10}$

63. $\frac{1}{10}-\frac{8}{100}$

64. $\frac{3}{100}-\frac{1}{1000}$

Applications

Solve.

65. Three pieces of mail weigh $\frac{1}{2}$ ounce, $\frac{1}{5}$ ounce, and $\frac{3}{10}$ ounce. What is the total weight of the letters?

66. Using a microscope, a scientist measures the diameters of three hairs to be $\frac{1}{1000}$ inch, $\frac{3}{1000}$ inch, and $\frac{1}{100}$ inch. What is the total of these three diameters?

67. A machinist drills four holes in a straight line. Each hole has a diameter of $\frac{1}{10}$ inch and there is $\frac{1}{4}$ inch between the holes. What is the distance between the outer edges of the first and last holes?

35. $\frac{3}{5}$
36. $\frac{1}{4}$
37. $\frac{1}{3}$
38. $\frac{2}{3}$
39. $\frac{8}{13}$
40. $\frac{4}{9}$
41. $\frac{1}{2}$
42. $\frac{1}{3}$
43. $\frac{13}{30}$
44. $\frac{3}{20}$
45. $\frac{3}{5}$
46. $\frac{23}{42}$
47. $\frac{1}{12}$
48. $\frac{5}{12}$
49. $\frac{5}{16}$
50. $\frac{1}{2}$
51. $\frac{13}{20}$
52. $\frac{8}{21}$
53. $\frac{11}{18}$
54. $\frac{11}{16}$
55. $\frac{1}{10}$
56. $\frac{15}{16}$
57. $\frac{23}{16}$
58. $\frac{16}{3}$

59. 0

60. 0

61. $\frac{3}{50}$

62. $\frac{1}{25}$

63. $\frac{1}{50}$

64. $\frac{29}{1000}$

Applications

65. 1 ounce

66. $\frac{7}{500}$ inch

67. $\frac{23}{20}$ inches

68. $\frac{49}{80}$ inch

69. $\frac{5}{8}$ inch

70. $\frac{7}{8}$

71. **a.** $\frac{5}{18}$ **b.** $6000

72. **a.** $\frac{19}{60}$ **b.** $950 **c.** $11,400

73. $\frac{7}{24}$

74. $\frac{1}{5}$ ton

75. $\frac{11}{36}$

76. **a.** $\frac{3}{4}$ **b.** $\frac{1}{4}$

77. **a.** $\frac{3}{4}$ **b.** 0 (the pie is totally consumed)

78. $\frac{5}{14}$ of the truck

68. A notebook contains a piece of cardboard as a back cover that is $\frac{1}{16}$ inch thick. It has a front cover that is $\frac{1}{4}$ inch thick. All together, the sheets of paper between the front and back are $\frac{3}{10}$ inch thick. What is the total thickness of the notebook?

69. A carpenter is installing baseboard and toe molding. If the baseboard is $\frac{3}{8}$ inch thick and the toe molding (to be put in front of the baseboard) is $\frac{1}{4}$ inch thick, what is the total thickness of the two trim pieces?

70. A recipe calls for the following spices: $\frac{1}{2}$ teaspoon of turmeric, $\frac{1}{4}$ teaspoon of ginger, and $\frac{1}{8}$ teaspoon of cumin. What is the total quantity of these three spices?

71. Alyssa's investment strategy is to put $\frac{1}{6}$ of her paycheck into a savings account and another $\frac{1}{9}$ into a retirement account.

a. What fraction of her salary does Alyssa invest each month?

b. If she maintains this strategy for 24 paychecks and receives $900 per paycheck, how much money will she have saved?

72. Bryce has a monthly income of $3000. $\frac{1}{10}$ of it goes to college savings, $\frac{2}{15}$ to general savings, and $\frac{1}{12}$ to retirement.

a. What fraction of Bryce's income is being saved?

b. How much money is being saved each month?

c. What is the total amount saved in a 12-month year?

73. A $\frac{7}{8}$ inch pipe is to be shortened to $\frac{7}{12}$ inch. How much must be removed?

74. Near the end of the snow season, the road salt supply for a small college had dwindled down to $\frac{7}{10}$ ton. When the next snow storm came, $\frac{1}{2}$ ton of salt was used for the roads. How much road salt was left?

75. Xavier has driven to a national park with no gas stations and he wants to drive around some before leaving the park. He knows he can safely make it to the nearest gas station on $\frac{1}{4}$ of a tank of gas. If the tank is currently $\frac{5}{9}$ full, what fraction of a tank of gasoline does he have to use for touring the park?

76. About $\frac{1}{2}$ of all incoming solar radiation is absorbed by the earth, $\frac{1}{5}$ is absorbed by the atmosphere, and $\frac{1}{20}$ is scattered by the atmosphere. The rest is reflected by the earth and clouds.

a. What fraction of solar radiation is absorbed or scattered?

b. What fraction of solar radiation is reflected by the earth and clouds?

77. Karina has $\frac{3}{4}$ of an apple pie left over from a party last night. Her roommates found it and cut themselves three unequal sized pieces in the following amounts: $\frac{1}{3}$ of a pie, $\frac{1}{4}$ of a pie, and $\frac{1}{6}$ of a pie.

a. What fraction of a full pie did Karina's roommates take?

b. What fraction of the pie is left over?

78. A moving truck has to make stops at three different apartments to collect items for three different moves. The first apartment takes up $\frac{3}{7}$ of the truck and the second apartment takes up $\frac{3}{14}$ of the truck. What fraction of the space in the moving truck is left to fit in the items from the last apartment?

79. Jayden has a large homework assignment to do this weekend. He is able to get $\frac{4}{15}$ of the assignment done Friday after class. If he doesn't want to leave more than $\frac{2}{9}$ of the assignment to do for Sunday, what fraction of the assignment must he complete on Saturday?

80. A pentagon (a five-sided figure) has 4 sides of length $\frac{1}{3}, \frac{1}{4}, \frac{1}{5},$ and $\frac{1}{6}$ inches. If the perimeter of the pentagon is 1 inch, find the length of the fifth side.

81. At a local farmers market, $\frac{1}{2}$ of the stands sell produce only, $\frac{2}{5}$ of the stands sell produce and crafts, and the remaining stands sell only crafts. What fraction of the stands sell only crafts?

82. At a county fair, $\frac{1}{3}$ of the attendees are county residents and $\frac{3}{8}$ of them are state residents who do not live in the county. The remaining attendees are people from out of state. What fraction of the attendees are from out of state?

Writing & Thinking

83. Explain how finding the LCM relates to LCDs.

84. Explain the steps to follow when adding or subtracting fractions with unlike denominators

85. Give an example of a situation where you might add or subtract fractions (other than in class).

86. Pick one problem in this section that gave you some difficulty. Explain briefly why you had difficulty and why you think that you can better solve problems of this type in the future.

79. $\frac{23}{45}$ of the assignment

80. $\frac{1}{20}$ inch

81. $\frac{1}{10}$

82. $\frac{7}{24}$

Writing & Thinking

83. The LCM finds the least common multiple of a set of numbers. The LCD does the same thing for the set of numbers determined by the denominators.

84. **1.** Find the LCD. **2.** Change each fraction to an equivalent fraction that has the common denominator. **3.** Add or subtract the numerators and keep the common denominator. **4.** Reduce if possible.

85. Answers will vary. Adding with fractions may be used when cooking or when measuring sewing or construction materials. Subtracting with fractions may be used when cooking or when measuring sewing or construction materials.

86. Answers will vary.

Objectives

A. Add mixed numbers.

B. Subtract mixed numbers.

C. Subtract mixed numbers by borrowing.

D. Add and subtract mixed numbers by first changing them to improper fractions.

2.7 Addition and Subtraction with Mixed Numbers

A Addition with Mixed Numbers

Since a mixed number represents the sum of a whole number and a proper fraction, two or more mixed numbers can be added by adding the fraction parts and whole numbers separately.

Adding Mixed Numbers

1. Add the fraction parts.
2. Add the whole numbers.
3. Write the answer as a mixed number with the fraction part less than 1.

PROCEDURE

1. Add: $8\frac{1}{5}+3\frac{3}{5}$

Example 1 Adding Mixed Numbers with the Same Denominator

Add: $5\frac{2}{9}+8\frac{5}{9}$

Solution

We can write each mixed number as a sum and then use the commutative and associative properties of addition to treat the whole numbers and fraction parts separately.

$$\begin{aligned}5\frac{2}{9}+8\frac{5}{9}&=5+\frac{2}{9}+8+\frac{5}{9}\\&=(5+8)+\left(\frac{2}{9}+\frac{5}{9}\right)\\&=13+\frac{7}{9}=13\frac{7}{9}\end{aligned}$$

Or, vertically,

$$\begin{array}{r}5\frac{2}{9}\\+8\frac{5}{9}\\\hline 13\frac{7}{9}\end{array}$$

Now work margin exercise 1.

Example 2 Adding Mixed Numbers with Different Denominators

Add: $5\frac{1}{6}+2\frac{7}{18}$

Solution

In this case, the fractions do not have the same denominator. The LCD is 18.

$$\begin{aligned} 5\frac{1}{6} &= 5\frac{1}{6}\cdot\frac{3}{3} &= 5\frac{3}{18} \\ +2\frac{7}{18} &= +2\frac{7}{18} &= +2\frac{7}{18} \\ \hline & & 7\frac{10}{18} = 7\frac{5}{9} \end{aligned}$$

Now work margin exercise 2.

2. Add: $5\frac{1}{2}+3\frac{1}{5}$

If the sum of the proper fractions is more than 1, rewrite this sum as a mixed number and add it to the sum of the whole numbers. Example 3 illustrates this situation.

Example 3 Adding Mixed Numbers

Add: $5\frac{3}{4}+9\frac{3}{10}$

Solution

The LCD is 20.

$$\begin{aligned} 5\frac{3}{4} &= 5\frac{3}{4}\cdot\frac{5}{5} &= 5\frac{15}{20} \\ +9\frac{3}{10} &= +9\frac{3}{10}\cdot\frac{2}{2} &= +9\frac{6}{20} \\ \hline & & 14\frac{21}{20} = 14+1\frac{1}{20} = 15\frac{1}{20} \end{aligned}$$

Fraction is greater than 1. Change to a mixed number.

Now work margin exercise 3.

3. Add: $3\frac{4}{5}+7\frac{2}{7}$

Example 4 Calculating Perimeter

A triangle has sides measuring $3\frac{1}{3}$ meters, $6\frac{4}{5}$ meters, and $6\frac{14}{15}$ meters. Find the perimeter of the triangle.

Solution

We find the perimeter by adding the lengths of the three sides. The LCD is 15.

4. A triangle has sides measuring $4\frac{1}{4}$ meters, $6\frac{1}{4}$ meters, and $6\frac{1}{2}$ meters. Find the perimeter of the triangle.

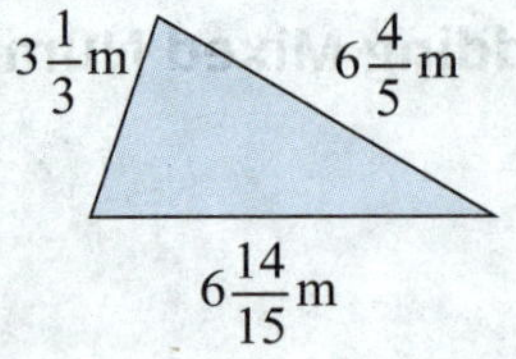

$$
\begin{array}{rcccl}
3\frac{1}{3} & = & 3\frac{1}{3}\cdot\frac{5}{5} & = & 3\frac{5}{15} \\
6\frac{4}{5} & = & 6\frac{4}{5}\cdot\frac{3}{3} & = & 6\frac{12}{15} \\
+\ 6\frac{14}{15} & = & +\ 6\frac{14}{15} & = & +\ 6\frac{14}{15} \\
\hline
 & & & & 15\frac{31}{15} = 15 + 2\frac{1}{15} = 17\frac{1}{15} \text{ meters}
\end{array}
$$

The perimeter is $17\frac{1}{15}$ meters.

Now work margin exercise 4.

B Subtraction with Mixed Numbers

Subtraction with mixed numbers is similar to addition in that we subtract the fraction parts and the whole numbers separately.

Subtracting Mixed Numbers

1. Subtract the fraction parts.
2. Subtract the whole numbers.

PROCEDURE

5. Subtract: $6\frac{4}{5} - 2\frac{1}{5}$

Example 5 Subtracting Mixed Numbers with the Same Denominator

Subtract: $10\frac{3}{7} - 6\frac{2}{7}$

Solution

$$10\frac{3}{7} - 6\frac{2}{7} = (10-6) + \left(\frac{3}{7} - \frac{2}{7}\right)$$
$$= 4 + \frac{1}{7} = 4\frac{1}{7}$$

Or, we can subtract vertically.

$$
\begin{array}{r}
10\frac{3}{7} \\
-\ 6\frac{2}{7} \\
\hline
4\frac{1}{7}
\end{array}
$$

Now work margin exercise 5.

Example 6 Subtracting Mixed Numbers with Different Denominators

Subtract: $13\frac{4}{5}-7\frac{1}{3}$

Solution

The LCD is 15.

$$\begin{aligned} 13\frac{4}{5} &= 13\frac{4}{5}\cdot\frac{3}{3} &= 13\frac{12}{15} \\ -\ 7\frac{1}{3} &= -\ 7\frac{1}{3}\cdot\frac{5}{5} &= -\ 7\frac{5}{15} \\ \hline & & 6\frac{7}{15} \end{aligned}$$

Now work margin exercise 6.

6. Subtract: $11\frac{1}{2}-4\frac{2}{7}$

C Subtracting Mixed Numbers by Borrowing

In subtracting with mixed numbers, there are cases where the fraction part of the number being subtracted is larger than the first fraction. By "borrowing" 1 from the whole number part in these cases, we can rewrite the first number as a whole number plus an improper fraction. Then the subtraction of the fraction parts can proceed as before.

Subtracting Mixed Numbers by Borrowing

When subtracting mixed numbers, if the fraction part of the number being subtracted is larger than the first fraction, then borrow using the following steps.

1. "Borrow" 1 from the first whole number.
2. Add this 1 to the first fraction. (This will always result in an improper fraction that is larger than the fraction being subtracted.)
3. Subtract.

PROCEDURE

Math Tip

Borrowing a 1 with fractions can be visualized by using circles, where each circle represents a whole 1. The following figures illustrate a situation where 1 is being borrowed from 5 and the denominator of the fraction part being subtracted is 3.

5 wholes

4 wholes + $\frac{3}{3}$

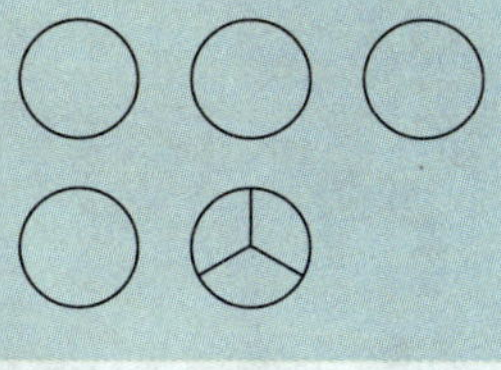

Example 7 Subtracting Mixed Numbers by Borrowing

Subtract: $7-4\frac{2}{3}$

Solution

In this case, the number 7 has no fraction part, so we must **borrow** 1 from it. We then write 1 as $\frac{3}{3}$ so that its denominator will be the same as that of the fraction part of the other number, $\frac{2}{3}$.

7. Subtract: $17-9\frac{6}{11}$

$$\begin{array}{rcl} 7 & = & 6\frac{3}{3} \\ -\,4\frac{2}{3} & = & -\,4\frac{2}{3} \\ \hline & & 2\frac{1}{3} \end{array}$$

Borrow: $7 = 6 + 1 = 6 + \frac{3}{3} = 6\frac{3}{3}$.

Now work margin exercise 7.

8. Subtract: $10\frac{1}{8} - 3\frac{5}{8}$

Teaching Note

After doing a few examples similar to Examples 7–10, you may want to let the students know that a shortcut to finding the improper fraction form when "borrowing 1" can be found by simply adding the numerator and denominator of the original fraction. For example, with $4\frac{2}{9}$, add $2 + 9 = 11$ and write $4\frac{2}{9} = 3\frac{11}{9}$. Remind them that this works because $1 = \frac{9}{9}$ and $\frac{9}{9} + \frac{2}{9} = \frac{9+2}{9}$.

Example 8 Subtracting Mixed Numbers by Borrowing

Subtract: $4\frac{2}{9} - 1\frac{5}{9}$

Solution

$\frac{5}{9}$ is larger than $\frac{2}{9}$, so **borrow** 1 from 4. Rewrite.

$$\begin{array}{rcl} 4\frac{2}{9} & = & 3\frac{11}{9} \\ -\,1\frac{5}{9} & = & -\,1\frac{5}{9} \\ \hline & & 2\frac{6}{9} = 2\frac{2}{3} \end{array}$$

Borrow: $4\frac{2}{9} = 3 + 1 + \frac{2}{9} = 3 + \frac{9}{9} + \frac{2}{9} = 3\frac{11}{9}$

Now work margin exercise 8.

9. Subtract: $16\frac{2}{25} - 8\frac{1}{10}$

Example 9 Subtracting Mixed Numbers by Borrowing

Subtract: $13\frac{5}{12} - 6\frac{13}{20}$

Solution

The LCD is 60.

$$\begin{array}{rcccccl} 13\frac{5}{12} & = & 13\frac{5}{12} \cdot \frac{5}{5} & = & 13\frac{25}{60} & = & 12\frac{85}{60} \\ -\,6\frac{13}{20} & = & -\,6\frac{13}{20} \cdot \frac{3}{3} & = & -\,6\frac{39}{60} & = & -\,6\frac{39}{60} \\ \hline & & & & & & 6\frac{46}{60} = 6\frac{23}{30} \end{array}$$

Borrow: $13\frac{25}{60} = 12 + 1 + \frac{25}{60} = 12 + \frac{60}{60} + \frac{25}{60} = 12\frac{85}{60}$.

Now work margin exercise 9.

Completion Example 10 Subtracting Mixed Numbers by Borrowing

Subtract: $12\frac{3}{4}-7\frac{9}{10}$

Solution

The LCD is 20.

$$\begin{array}{rcrcrcr} 12\frac{3}{4} & = & 12\frac{3}{4}\cdot\frac{5}{5} & = & 12\frac{15}{20} & = & 11\frac{35}{20} \\ -7\frac{9}{10} & = & -7\frac{9}{10}\cdot\frac{2}{2} & = & -7\frac{18}{20} & = & -7\frac{18}{20} \\ \hline & & & & & & 4\frac{17}{20} \end{array}$$

Difference

Now work margin exercise 10.

10. Subtract: $8\frac{3}{10}-2\frac{4}{5}$

Example 11 Application: Subtracting Mixed Numbers

When he was first riding a trail with his new bicycle, Karl found that he was taking $25\frac{1}{2}$ minutes to ride the 4-mile loop. Now he takes $18\frac{3}{5}$ minutes to ride the same 4 miles. By how many minutes has he improved in **a.** riding 4 miles? **b.** riding 1 mile?

Solution

Step 1: READ: Read the problem carefully. In this problem there are two questions. What was his improved time for the distance of **a.** 4 miles and **b.** for 1 mile.

Step 2: SET UP:

a. To find the improved time for 4 miles, subtract the two given times: $25\frac{1}{2}-18\frac{3}{5}$.

b. Then to find the improved time for 1 mile, divide the answer from part **a.** by 4.

Step 3: SOLVE:

a. The LCD is 10.

$$\begin{array}{rcrcrcr} 25\frac{1}{2} & = & 25\frac{1}{2}\cdot\frac{5}{5} & = & 25\frac{5}{10} & = & 24\frac{15}{10} \\ -18\frac{3}{5} & = & -18\frac{3}{5}\cdot\frac{2}{2} & = & -18\frac{6}{10} & = & -18\frac{6}{10} \\ \hline & & & & & & 6\frac{9}{10} \end{array}$$

So, in riding 4 miles, Karl has improved by $6\frac{9}{10}$ minutes.

11. When he first started jogging, Ted could run a 3-mile route in $24\frac{3}{4}$ minutes. Now he can run the same route in $19\frac{4}{5}$ minutes. How many minutes has he cut from **a.** his 3-mile route? **b.** each mile?

b. To find his improvement time in riding one mile, we divide the answer in part a. by 4.

$$6\frac{9}{10} \div 4 = \frac{69}{10} \div \frac{4}{1} = \frac{69}{10} \cdot \frac{1}{4} = \frac{69}{40} = 1\frac{29}{40}$$

He has improved by $1\frac{29}{40}$ minutes per mile.

Step 4: CHECK: The two times seem reasonable. With the whole number parts only, we have 25 – 18 = 7 which is very close to $6\frac{9}{10}$.

Now work margin exercise 11.

D Optional Approach to Addition and Subtraction with Mixed Numbers

An optional approach to adding and subtracting mixed numbers is to simply do all the work with improper fractions, as we did earlier with multiplication and division. You may find this approach easier. In any case, addition and subtraction can be accomplished only if the denominators are the same. So you must still be concerned about least common denominators. For comparison purposes, Examples 12 through 14 are the same problems as in Examples 1, 2, and 10.

12. Add using improper fractions:

$6\frac{5}{8} + 7\frac{3}{8}$

Example 12 Adding Mixed Numbers Using Improper Fractions

Add using improper fractions: $5\frac{2}{9} + 8\frac{5}{9}$

Solution

First, change each number into its corresponding improper fraction, then add.

$$5\frac{2}{9} + 8\frac{5}{9} = \frac{47}{9} + \frac{77}{9} = \frac{124}{9} \text{ or } 13\frac{7}{9}$$

Now work margin exercise 12.

13. Add using improper fractions:

$3\frac{7}{12} + 8\frac{2}{3}$

Example 13 Adding Mixed Numbers Using Improper Fractions

Add using improper fractions: $5\frac{1}{6} + 2\frac{7}{18}$

Solution

Change each mixed number to fraction form. Then write equivalent fractions using the LCD and add.

The LCD is 18.

$$5\frac{1}{6} + 2\frac{7}{18} = \frac{31}{6} + \frac{43}{18} = \frac{31}{6} \cdot \frac{3}{3} + \frac{43}{18} = \frac{93}{18} + \frac{43}{18} = \frac{136}{18} = \frac{\cancel{2} \cdot 68}{\cancel{2} \cdot 9} = \frac{68}{9} \text{ or } 7\frac{5}{9}$$

Now work margin exercise 13.

Example 14 Subtracting Mixed Numbers Using Improper Fractions

Subtract using improper fractions: $12\frac{3}{4}-7\frac{9}{10}$

Solution

The LCD is 20.

$$12\frac{3}{4}-7\frac{9}{10}=\frac{51}{4}-\frac{79}{10}=\frac{51}{4}\cdot\frac{5}{5}-\frac{79}{10}\cdot\frac{2}{2}=\frac{255}{20}-\frac{158}{20}=\frac{97}{20}\text{ or }4\frac{17}{20}$$

Now work margin exercise 14.

14. Subtract using improper fractions:

$10\frac{2}{13}-7\frac{4}{5}$

CALCULATORS

Using a Calculator to Add and Subtract Fractions

To add, subtract, multiply, or divide two fractions or mixed numbers, enter the expression as you did when working with whole numbers in Chapter 1, but use the [a b/c] key to enter the fractions or mixed numbers. The result will be given to you in simplified form.

For example, to subtract $\frac{5}{6}-\frac{4}{15}$, press the keys

[5] [a b/c] [6] [−] [4] [a b/c] [1] [5].

Then press [=]. The display will read 17/30 which means $\frac{17}{30}$.

Similarly, to add $4\frac{1}{3}+1\frac{5}{6}$, press the keys

[4] [a b/c] [1] [a b/c] [3] [+] [1] [a b/c] [5] [a b/c] [6]. Then press [=].

The display will read 6 ⌋ 1 / 6 which means $6\frac{1}{6}$.

Completion Example Answers

10. $12\frac{15}{20}-7\frac{18}{20}$; $11\frac{35}{20}-7\frac{18}{20}$; $4\frac{17}{20}$

Margin Exercise Answers

1. $11\frac{4}{5}$ **2.** $8\frac{7}{10}$ **3.** $11\frac{3}{35}$ **4.** 17 meters **5.** $4\frac{3}{5}$ **6.** $7\frac{3}{14}$ **7.** $7\frac{5}{11}$ **8.** $6\frac{1}{2}$ **9.** $7\frac{49}{50}$

10. $5\frac{1}{2}$ **11. a.** $4\frac{19}{20}$ minutes **b.** $1\frac{13}{20}$ minutes **12.** 14 **13.** $\frac{49}{4}$ or $12\frac{1}{4}$ **14.** $\frac{153}{65}$ or $2\frac{23}{65}$

2.7 Exercises

Concept Check

1. Mixed number, whole numbers
2. Same
3. Reduced (or simplified)
4. Fractional, whole numbers
5. Improper, 1
6. Improper
7. True
8. True
9. False; LCDs are required when adding or subtracting mixed numbers.
10. False; $12-5\frac{1}{3}=6\frac{2}{3}$

Practice

1. 10
2. 14
3. $28\frac{1}{2}$
4. $7\frac{1}{4}$
5. $14\frac{5}{7}$
6. $15\frac{3}{11}$
7. $7\frac{2}{3}$
8. $10\frac{3}{4}$
9. $10\frac{3}{8}$
10. $14\frac{1}{16}$
11. $8\frac{10}{21}$
12. $8\frac{8}{15}$
13. $5\frac{17}{26}$
14. $7\frac{19}{20}$

2.7 Exercises

Concept Check

Fill-in-the-Blank. Complete each sentence using information found in this section.

1. When adding mixed numbers, if the sum of the fractional parts is more than 1, rewrite this sum as a/an ________ ________ and add it to the sum of the ________ ________.

2. When adding (or subtracting) the fractional parts of mixed numbers, the denominators must be the ________.

3. All answers should be ________, if possible.

4. When subtracting mixed numbers, subtract the ________ parts first, then subtract the ________ ________.

5. When subtracting mixed numbers, if the fraction part being subtracted is larger than the other fraction part, the smaller fraction part must be changed to a/an ________ fraction by borrowing ________ from its whole number.

6. You can always add or subtract with mixed numbers by first changing them to ________ fractions.

True/False. Determine whether each statement is true or false. If a statement is false, explain how it can be changed so the statement will be true. (**Note:** There may be more than one acceptable change.)

7. $3\frac{1}{5}+5\frac{1}{2}=8\frac{7}{10}$

8. When adding (or subtracting) mixed numbers, the final answer should be written as a mixed number.

9. LCDs are not required when adding or subtracting mixed numbers.

10. $12-5\frac{1}{3}=7\frac{1}{3}$

Practice

Add and reduce to lowest terms. Write your answer in mixed number form. See Examples 1 through 3.

1. $5\frac{1}{3}+4\frac{2}{3}$

2. $7\frac{1}{2}+6\frac{1}{2}$

3. $21\frac{3}{4}+6\frac{3}{4}$

4. $3\frac{5}{8}+3\frac{5}{8}$

5. $10+4\frac{5}{7}$

6. $8+7\frac{3}{11}$

7. $4\frac{1}{2}+3\frac{1}{6}$

8. $6\frac{5}{12}+4\frac{1}{3}$

9. $3\frac{1}{4}+7\frac{1}{8}$

10. $11\frac{3}{4}+2\frac{5}{16}$

11. $5\frac{1}{7}+3\frac{1}{3}$

12. $2\frac{1}{5}+6\frac{1}{3}$

13. $5\frac{1}{2}+\frac{2}{13}$

14. $7\frac{3}{4}+\frac{1}{5}$

15. $5\frac{11}{12}+3\frac{5}{6}$

16. $\begin{array}{r} 9\frac{2}{3} \\ +\ \frac{5}{6} \\ \hline \end{array}$

17. $\begin{array}{r} 3\frac{1}{4} \\ +5\frac{5}{12} \\ \hline \end{array}$

18. $\begin{array}{r} 21\frac{1}{3} \\ +13\frac{1}{18} \\ \hline \end{array}$

19. $7\frac{3}{5}+2\frac{1}{8}$

20. $7\frac{2}{3}+1\frac{1}{16}$

21. $8\frac{2}{9}+4\frac{1}{27}$

22. $5\frac{1}{8}+6\frac{3}{16}$

23. $9\frac{1}{8}+3\frac{7}{12}$

24. $2\frac{5}{8}+6\frac{5}{6}$

25. $3\frac{1}{3}+4\frac{1}{4}+5\frac{1}{5}$

26. $2\frac{1}{2}+3\frac{1}{3}+5\frac{1}{5}$

27. $\frac{3}{7}+2\frac{1}{14}+2\frac{1}{6}$

28. $4\frac{1}{3}+8\frac{3}{8}+6\frac{1}{6}$

29. $\begin{array}{r} 5\frac{1}{8} \\ 1\frac{1}{5} \\ +3\frac{1}{40} \\ \hline \end{array}$

30. $\begin{array}{r} 4\frac{5}{48} \\ 2\frac{1}{8} \\ +1\frac{1}{12} \\ \hline \end{array}$

Subtract and reduce to lowest terms. Write your answer in mixed number form. See Examples 5 through 10.

31. $\begin{array}{r} 5\frac{1}{2} \\ -1 \\ \hline \end{array}$

32. $\begin{array}{r} 7\frac{3}{4} \\ -2 \\ \hline \end{array}$

33. $\begin{array}{r} 6\frac{1}{2} \\ -2\frac{1}{2} \\ \hline \end{array}$

34. $\begin{array}{r} 9\frac{1}{4} \\ -5\frac{1}{4} \\ \hline \end{array}$

35. $\begin{array}{r} 7\frac{9}{10} \\ -3\frac{3}{10} \\ \hline \end{array}$

36. $\begin{array}{r} 14\frac{5}{8} \\ -11\frac{3}{8} \\ \hline \end{array}$

37. $\begin{array}{r} 4\frac{7}{8} \\ -1\frac{1}{4} \\ \hline \end{array}$

38. $\begin{array}{r} 9\frac{5}{16} \\ -2\frac{1}{4} \\ \hline \end{array}$

39. $\begin{array}{r} 10\frac{5}{6} \\ -4\frac{2}{3} \\ \hline \end{array}$

40. $\begin{array}{r} 9\frac{13}{15} \\ -7\frac{2}{5} \\ \hline \end{array}$

41. $\begin{array}{r} 5\frac{4}{5} \\ -2\frac{1}{4} \\ \hline \end{array}$

42. $\begin{array}{r} 7\frac{9}{10} \\ -4\frac{1}{7} \\ \hline \end{array}$

43. $\begin{array}{r} 8\frac{5}{6} \\ -2\frac{1}{4} \\ \hline \end{array}$

44. $\begin{array}{r} 10\frac{7}{12} \\ -6\frac{1}{9} \\ \hline \end{array}$

45. $18\frac{7}{8}-2\frac{2}{3}$

46. $15\frac{1}{2}-2\frac{1}{3}$

47. $1-\frac{3}{4}$

48. $1-\frac{5}{16}$

49. $2-\frac{7}{8}$

50. $2-\frac{3}{5}$

51. $7-6\frac{2}{3}$

15. $9\frac{3}{4}$

16. $10\frac{1}{2}$

17. $8\frac{2}{3}$

18. $34\frac{7}{18}$

19. $9\frac{29}{40}$

20. $26\frac{35}{48}$

21. $12\frac{7}{27}$

22. $11\frac{5}{16}$

23. $12\frac{17}{24}$

24. $9\frac{11}{24}$

25. $12\frac{47}{60}$

26. $11\frac{1}{30}$

27. $4\frac{2}{3}$

28. $18\frac{7}{8}$

29. $9\frac{7}{20}$

30. $7\frac{5}{16}$

31. $4\frac{1}{2}$

32. $5\frac{3}{4}$

33. 4

34. 4

35. $4\frac{3}{5}$

36. $3\frac{1}{4}$

37. $3\frac{5}{8}$

38. $7\frac{1}{16}$

52. $5-4\frac{1}{5}$

53. $17\frac{3}{7}-5\frac{3}{10}$

54. $13\frac{1}{5}-7\frac{3}{8}$

55. $\begin{array}{r} 8\frac{3}{32} \\ -4\frac{3}{16} \\ \hline \end{array}$

56. $\begin{array}{r} 4\frac{9}{16} \\ -2\frac{7}{8} \\ \hline \end{array}$

57. $\begin{array}{r} 17\frac{3}{7} \\ -4\frac{15}{28} \\ \hline \end{array}$

58. $\begin{array}{r} 22\frac{4}{25} \\ -5\frac{3}{5} \\ \hline \end{array}$

59. $\begin{array}{r} 18\frac{3}{7} \\ -15\frac{2}{3} \\ \hline \end{array}$

60. $\begin{array}{r} 7\frac{1}{5} \\ -3\frac{1}{2} \\ \hline \end{array}$

Applications

Solve.

61. A bus trip is made in three parts. The first part takes $2\frac{1}{3}$ hours, the second part takes $2\frac{1}{2}$ hours, and the third part takes $3\frac{3}{4}$ hours. How long does the entire trip take?

62. A construction company was contracted to build three sections of highway. One section was $20\frac{7}{10}$ kilometers, the second section was $3\frac{4}{10}$ kilometers, and the third section was $11\frac{6}{10}$ kilometers. What was the total length of the highway built?

63. A quadrilateral (four-sided figure) has sides that measure $3\frac{1}{2}$ inches, $2\frac{1}{4}$ inches, $3\frac{5}{8}$ inches, and $2\frac{3}{4}$ inches. What is the total distance around the quadrilateral?

64. A pentagon (five-sided figure) has sides that measure $3\frac{3}{8}$ centimeters, $5\frac{1}{2}$ centimeters, $6\frac{1}{4}$ centimeters, $9\frac{1}{10}$ centimeters, and $4\frac{7}{8}$ centimeters. What is the perimeter in centimeters of the pentagon?

65. Among the top 20 highest grossing movies of 2021, A *Quiet Place Part II* earned \$$160\frac{1}{10}$ million, *Ghostbusters: Afterlife* earned \$$122\frac{3}{10}$ million, and *Free Guy* earned \$$121\frac{3}{5}$ million. What was the total earnings of these three films in 2021?

66. A carpenter buys boards of five widths at the lumber yard. The widths, in inches, are $3\frac{1}{2}, 9\frac{1}{4}, 11\frac{1}{14}, 7\frac{1}{4}$, and $5\frac{1}{2}$. If he bought one of each of these boards, what would be the total width of the boards if they were lined up next to each other?

67. On average, the air that we inhale includes $1\frac{1}{4}$ parts water and the air we exhale includes $5\frac{9}{10}$ parts water. How many more parts water are in the exhaled air?

68. A person who is running will burn about $14\frac{7}{10}$ calories each minute and a person who is walking will burn about $5\frac{1}{2}$ calories each minute. How many more calories does a runner burn per minute than a walker?

69. Sara can paint a room in $3\frac{3}{4}$ hours, and Emily can paint a room of the same size in $4\frac{1}{5}$ hours. How many hours are saved by having Sara paint a room of this size?

39. $6\frac{1}{6}$
40. $2\frac{7}{15}$
41. $3\frac{11}{20}$
42. $3\frac{53}{70}$
43. $6\frac{7}{12}$
44. $4\frac{17}{36}$
45. $16\frac{5}{24}$
46. $13\frac{1}{6}$
47. $\frac{1}{4}$
48. $\frac{11}{16}$
49. $1\frac{1}{8}$
50. $1\frac{2}{5}$
51. $\frac{1}{3}$
52. $\frac{4}{5}$
53. $12\frac{9}{70}$
54. $5\frac{33}{40}$
55. $3\frac{29}{32}$
56. $1\frac{11}{16}$
57. $12\frac{25}{28}$
58. $16\frac{14}{25}$
59. $2\frac{16}{21}$
60. $3\frac{7}{10}$

Applications

61. $8\frac{7}{12}$ hours

70. A teacher graded two sets of test papers. The first set took $3\frac{3}{4}$ hours to grade and the second set took $2\frac{3}{5}$ hours. How much faster did she grade the second set?

71. Mike takes $1\frac{1}{2}$ hours to clean a pool and Tom takes $2\frac{1}{3}$ hours to clean the same pool. How much longer does it take Tom to clean the pool?

72. A certain stock was selling for $43\frac{7}{8}$ dollars per share. One month later it was selling for $48\frac{1}{2}$ dollars per share. How much did the stock increase in price?

73. A salesman drove $5\frac{3}{4}$ hours one day and $6\frac{1}{2}$ hours the next day. How much more time did he spend driving on the second day?

74. A board is 16 feet long. If two pieces are cut from the board, one $6\frac{3}{4}$ feet long and the other $3\frac{1}{2}$ feet long, what is the length of the remaining piece of the original board?

75. Jade wants to lose 10 pounds and she currently weighs 180. During the first month of her weight loss routine, she loses $3\frac{1}{4}$ pounds. During the second month she loses $3\frac{1}{2}$ pounds. How much more weight does she need to lose to meet her goal?

76. Mr. Johnson originally weighed 240 pounds. During each month of six months of dieting and exercise he lost $5\frac{1}{2}$ pounds, $2\frac{3}{4}$ pounds, $4\frac{5}{16}$ pounds, $1\frac{3}{4}$ pounds, $2\frac{5}{8}$ pounds, and $3\frac{1}{4}$ pounds. What did he weigh at the end of the six months?

77. The 5 top-grossing concert tours of 2021 are shown in the table.

Band/Artist	Concert Tour Revenue (in millions of dollars)
The Rolling Stones	$115\frac{1}{2}$
Harry Styles	$86\frac{7}{10}$
Weezer, Fall Out Boy, and Green Day	$67\frac{3}{10}$
Eagles	$59\frac{1}{5}$
Dead & Company	$50\frac{1}{5}$

Source: "Rolling Stones, Harry Styles Were Top-Grossing Tours of 2021," Forbes, December 14, 2021, www.forbes.com/sites/marisadellatto/2021/12/14/rolling-stones-harry-styles-were-top-grossing-tours-of-2021.

a. What total amount did these five concert tours earn?

b. How much more did Harry Styles earn than the Eagles?

c. How much more did The Rolling Stones earn than Weezer, Fall Out Boy, and Green Day?

62. $35\frac{7}{10}$ kilometers

63. $12\frac{1}{8}$ inches

64. $29\frac{1}{10}$ centimeters

65. $404 million

66. $36\frac{4}{7}$ inches

67. $4\frac{13}{20}$ parts

68. $9\frac{1}{5}$ calories

69. $\frac{3}{5}$ hours

70. $1\frac{3}{20}$ hours

71. $\frac{5}{6}$ hour

72. $\$4\frac{5}{8}$

73. $\frac{3}{4}$ hour

74. $5\frac{3}{4}$ feet

75. $3\frac{1}{4}$ pounds

76. $219\frac{13}{16}$ pounds

77. **a.** $\$378\frac{9}{10}$ million **b.** $\$27\frac{1}{2}$ million **c.** $\$48\frac{1}{5}$ million

78. a. 2 (12-ft) pieces, $3\frac{1}{2}$ ft left over;

b. There will be no difference

Writing & Thinking

79. Fractions should be first in case the fraction being subtracted is larger than the other fraction and 1 needs to be borrowed from the whole number.

80. Answers will vary. Adding or subtracting mixed numbers may be used when cooking or when measuring sewing or construction materials.

81. a. 6 b. 600 c. 60,000

82. Part a.; In Parts b. and c., fractions are multiplied by larger numbers

83. a. 4 b. 11 c. 100

84. a. $16 \div 2 = 8$

b. Answers will vary.

c. 6

78. A carpenter wants to build a single stand-alone bookshelf that will include two shelves, a top, a bottom, and two sides. He will need three pieces of lumber each $3\frac{1}{4}$ feet long (for the two shelves and a bottom), one piece $3\frac{3}{4}$ feet long for the top and two pieces each $3\frac{1}{2}$ feet long for the sides. At the local home supply store, the lumber needed comes in lengths of 8 feet and 12 feet.

a. How many pieces of 12 feet will he need to buy to build the bookshelf?

b. Would he have been better off to buy lengths of 8 feet?

Writing & Thinking

79. In subtracting with mixed numbers explain why fractional parts should be subtracted before the whole numbers.

80. Give an example when you might add or subtract mixed numbers (other than in class).

81. Truncated mixed numbers are the whole number part of a mixed number. For example, 4 is the truncated mixed number of $4\frac{3}{5}$. Using truncated mixed numbers, estimate each of the following products mentally.

a. $2\frac{1}{7} \cdot 3\frac{5}{8}$ **b.** $20\frac{1}{7} \cdot 30\frac{5}{8}$ **c.** $200\frac{1}{7} \cdot 300\frac{5}{8}$

82. Of your three estimated answers in the previous exercise, which one do you think is closest to the actual product? Why?

83. Using truncated mixed numbers, estimate each of the following differences mentally.

a. $15\frac{6}{7} - 11\frac{3}{4}$ **b.** $136\frac{17}{40} - 125\frac{23}{30}$ **c.** $945\frac{1}{10} - 845\frac{3}{100}$

84. Consider the following problem: The product of two numbers is $16\frac{1}{2}$. If one of the numbers is $2\frac{3}{4}$, what is the other number?

a. Rewrite the problem using truncated mixed numbers and solve this new problem.

b. Does this technique make it easier for you to understand the original problem? Why or why not?

c. What is the solution to the original problem?

2.8 Comparisons and Order of Operations with Fractions

Objectives

A. Compare fractions.

B. Use the order of operations to simplify expressions containing fractions.

C. Find the average of a set of fractions.

A Comparing Fractions

For any two numbers on a number line, the smaller number is to the left of the larger number. To compare numbers, we will use the symbols < (read as "is less then") and > (read as "is greater than"). As examples,

$$2 < 5 \quad \text{and} \quad 9 > 4.$$

"two is less than 5" "nine is greater than four"

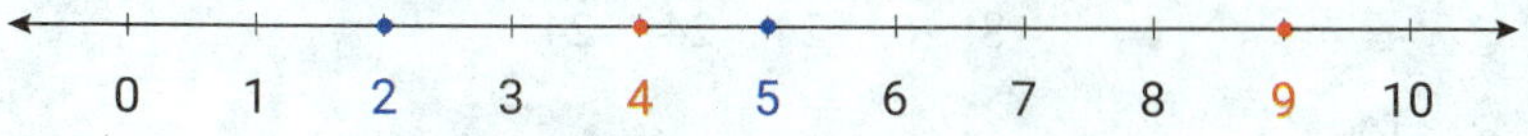

Figure 1

Fractions can be compared and graphed in the same way. For two fractions with the same denominator, a comparison is made by comparing the numerators. For example,

$$\text{because } 1 < 3 \text{, we have } \frac{1}{7} < \frac{3}{7}.$$

And graphically, $\frac{1}{7}$ is to the left of $\frac{3}{7}$. To graph the two fractions, we can divide and mark the interval from 0 to 1 into seven equal parts. Then the distance from one mark to the next is $\frac{1}{7}$.

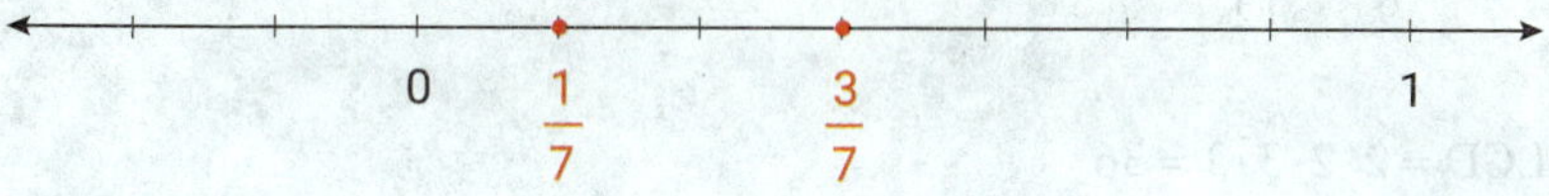

Figure 2

Teaching Note

Although not covered in the text, you may want to teach your students that when comparing negative fractions, they should place any negative signs (−) with the numerators, and then compare numerators.

To compare two fractions with different denominators, proceed as follows.

Comparing Two Fractions

1. Find the least common denominator (LCD).
2. Change each fraction into an equivalent fraction with that denominator.
3. Compare the numerators.

PROCEDURE

1. Which is larger, $\frac{9}{22}$ or $\frac{10}{33}$? How much larger?

💬 Math Tip

In Example 1, notice that the scale used on the number line is $\frac{1}{24}$, meaning there are 24 divisions from 0 to 1. When graphing two fractions with different denominators, the LCD is a good value to use to divide the whole number intervals.

Example 1 Comparing Fractions

Which is larger, $\frac{5}{6}$ or $\frac{7}{8}$? How much larger?

Solution

Step 1: $\text{LCD} = 2 \cdot 2 \cdot 2 \cdot 3 = 24$

Step 2: Find fractions equivalent to $\frac{5}{6}$ and $\frac{7}{8}$ with denominator 24.

$$\frac{5}{6} = \frac{5}{6} \cdot \frac{4}{4} = \frac{20}{24} \quad \text{and} \quad \frac{7}{8} = \frac{7}{8} \cdot \frac{3}{3} = \frac{21}{24}$$

Step 3: $\frac{7}{8} > \frac{5}{6}$, since $21 > 20$.

$$\frac{7}{8} - \frac{5}{6} = \frac{21}{24} - \frac{20}{24} = \frac{1}{24}$$

$\frac{7}{8}$ is larger by $\frac{1}{24}$.

Graphing these values on a number line, we have $\frac{7}{8} > \frac{5}{6}$ and $\frac{7}{8}$ lies to the right of $\frac{5}{6}$.

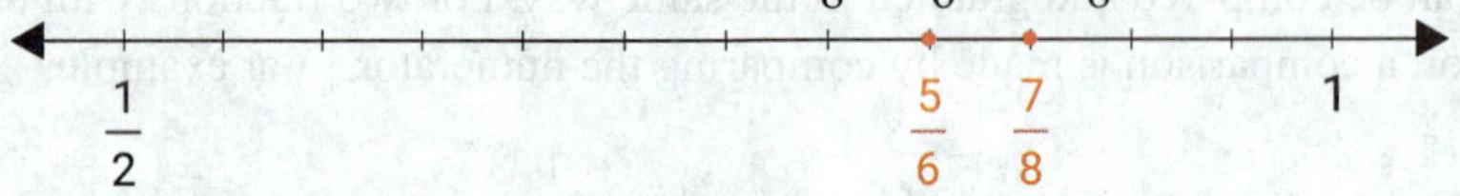

Now work margin exercise 1.

2. Which is larger, $\frac{7}{9}$ or $\frac{19}{24}$? How much larger?

Example 2 Comparing Fractions

Which is larger, $\frac{8}{9}$ or $\frac{11}{12}$? How much larger?

Solution

Step 1: $\text{LCD} = 2 \cdot 2 \cdot 3 \cdot 3 = 36$

Step 2: $\frac{8}{9} = \frac{8}{9} \cdot \frac{4}{4} = \frac{32}{36}$ and $\frac{11}{12} = \frac{11}{12} \cdot \frac{3}{3} = \frac{33}{36}$

Step 3: $\frac{11}{12} > \frac{8}{9}$, since $33 > 32$.

$$\frac{11}{12} - \frac{8}{9} = \frac{33}{36} - \frac{32}{36} = \frac{1}{36}$$

$\frac{11}{12}$ is larger by $\frac{1}{36}$.

Graphing these values on a number line, we have $\frac{11}{12} > \frac{8}{9}$ and $\frac{11}{12}$ lies to the right of $\frac{8}{9}$.

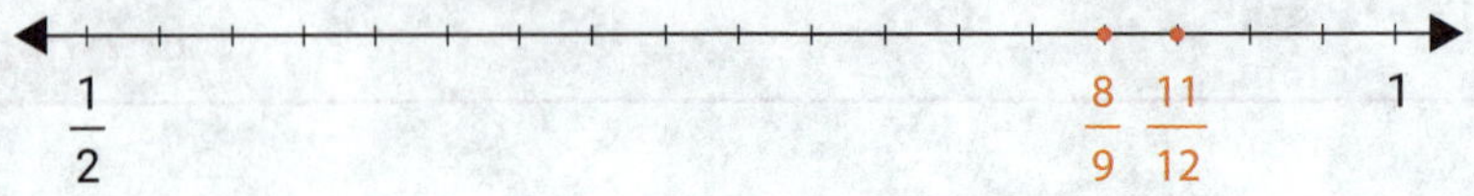

Now work margin exercise 2.

Example 3 Comparing Fractions

Arrange $\frac{2}{3}$, $\frac{7}{10}$, and $\frac{9}{15}$ in order from smallest to largest. Then find the difference between the largest and the smallest fraction.

Solution

Step 1: $\text{LCD} = 2 \cdot 3 \cdot 5 = 30$

Step 2: $\frac{2}{3} = \frac{2}{3} \cdot \frac{\mathbf{10}}{\mathbf{10}} = \frac{20}{30}$; $\frac{7}{10} = \frac{7}{10} \cdot \frac{\mathbf{3}}{\mathbf{3}} = \frac{21}{30}$; $\frac{9}{15} = \frac{9}{15} \cdot \frac{\mathbf{2}}{\mathbf{2}} = \frac{18}{30}$

Step 3: Smallest to largest: $\frac{9}{15}, \frac{2}{3}, \frac{7}{10}$ (since $18 < 20 < 21$)

Graphically, we have

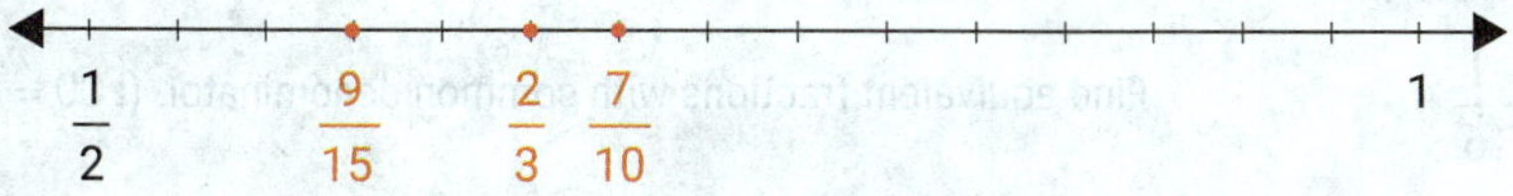

$$\frac{7}{10} - \frac{9}{15} = \frac{21}{30} - \frac{18}{30} = \frac{3}{30} = \frac{1}{10}$$

The difference between the largest and smallest fraction is $\frac{1}{10}$.

Now work margin exercise 3.

3. Arrange $\frac{7}{12}$, $\frac{5}{9}$, and $\frac{2}{3}$ in order from smallest to largest. Then find the difference between the largest and the smallest fraction.

B Order of Operations with Fractions

An expression with fractions may involve more than one operation. To evaluate such an expression, use the **rules for order of operations** just as they were discussed for whole numbers in Chapter 1. The rules are restated here for easy reference. Remember that to add or subtract fractions, you need a common denominator; to divide, multiply by the reciprocal of the divisor.

Rules for Order of Operations

1. Simplify within grouping symbols, such as parentheses ( ), brackets [ ], or braces { }. (If there are more than one pair of grouping symbols, start with the innermost grouping symbols.)
2. Evaluate any exponential expressions.
3. Moving from **left to right**, perform any multiplication or division in the order in which it appears.
4. Moving from **left to right**, perform any addition or subtraction in the order in which it appears.

PROPERTIES

4. Simplify: $\frac{2}{5}\cdot\frac{1}{4}+\frac{1}{9}\div\frac{2}{5}$

Example 4 Using the Order of Operations with Fractions

Simplify: $\frac{1}{2}\div\frac{3}{4}+\frac{5}{6}\cdot\frac{1}{5}$

Solution

$$\frac{1}{2}\div\frac{3}{4}+\frac{5}{6}\cdot\frac{1}{5}$$

$$=\frac{1}{\cancel{2}}\cdot\frac{\cancel{4}^{2}}{3}+\frac{5}{6}\cdot\frac{1}{5}$$ Divide. (Multiply by the reciprocal of the divisor.)

$$=\frac{2}{3}+\frac{\cancel{5}}{6}\cdot\frac{1}{\cancel{5}}$$ Multiply and reduce.

$$=\frac{2}{3}+\frac{1}{6}$$ Multiply and reduce.

$$=\frac{2}{3}\cdot\frac{2}{2}+\frac{1}{6}$$ Find equivalent fractions with common denominator. (LCD = 6.)

$$=\frac{4}{6}+\frac{1}{6}$$ Multiply.

$$=\frac{5}{6}$$ Add.

Now work margin exercise 4.

5. Simplify: $\frac{1}{4}+\left(\frac{1}{3}\right)^2\div 3$

Example 5 Using the Order of Operations with Fractions

Simplify: $\frac{9}{10}-\left(\frac{1}{4}\right)^2+\frac{1}{2}$

Solution

$$\frac{9}{10}-\left(\frac{1}{4}\right)^2+\frac{1}{2}$$ Evaluate the exponential expression. Remember, $\left(\frac{1}{4}\right)^2=\left(\frac{1}{4}\right)\cdot\left(\frac{1}{4}\right)$.

$$=\frac{9}{10}-\frac{1}{16}+\frac{1}{2}$$

$$=\frac{9}{10}\cdot\frac{8}{8}-\frac{1}{16}\cdot\frac{5}{5}+\frac{1}{2}\cdot\frac{40}{40}$$ Find equivalent fractions with the common denominator. (LCD = 80)

$$=\frac{72}{80}-\frac{5}{80}+\frac{40}{80}$$ Multiply.

$$=\frac{107}{80}\text{ or }1\frac{27}{80}$$ Add and subtract from left to right.

Now work margin exercise 5.

Often, a good strategy for evaluating expressions with mixed numbers that contain several operations is to change each mixed number to an improper fraction and then perform the operations.

Example 6 Using the Order of Operations with Fractions

Simplify: $3\frac{2}{5} \div \left(\frac{1}{4} + \frac{3}{5}\right)$

Solution

$$3\frac{2}{5} \div \left(\frac{1}{4} + \frac{3}{5}\right)$$

$$= \frac{17}{5} \div \left(\frac{1}{4} \cdot \frac{5}{5} + \frac{3}{5} \cdot \frac{4}{4}\right)$$ Change the mixed number to an improper fraction, and within the parentheses find equivalent fractions with the common denominator.

$$= \frac{17}{5} \div \left(\frac{5}{20} + \frac{12}{20}\right)$$ Multiply within the parentheses.

$$= \frac{17}{5} \div \frac{17}{20}$$ Add inside the parentheses.

$$= \frac{17}{5} \cdot \frac{20}{17}$$ Divide. (Multiply by the reciprocal of the divisor.)

$$= \frac{\cancel{17} \cdot 4 \cdot \cancel{5}}{\cancel{5} \cdot \cancel{17}}$$ Multiply and reduce by factoring.

$$= \frac{4}{1} = 4$$

Now work margin exercise 6.

6. Simplify: $\left(\frac{5}{12} + \frac{2}{3}\right) \div 4\frac{1}{3}$

Completion Example 7 Using the Order of Operations

Simplify: $\frac{7}{2} \cdot \frac{1}{4} + \frac{9}{10} \div \left(\frac{3}{5}\right)^2$

Solution

$$\frac{7}{2} \cdot \frac{1}{4} + \frac{9}{10} \div \left(\frac{3}{5}\right)^2$$

$$= \frac{7}{2} \cdot \frac{1}{4} + \frac{9}{10} \div \underline{\frac{9}{25}}$$

$$= \frac{7}{2} \cdot \frac{1}{4} + \frac{9}{10} \cdot \underline{\frac{25}{9}}$$

$$= \underline{\frac{7}{8}} + \underline{\frac{5}{2}}$$

$$= \underline{\frac{27}{8}} \text{ or } \underline{3\frac{3}{8}}$$

Now work margin exercise 7.

7. Simplify: $\frac{2}{3} + \frac{2}{7} \div \left(\frac{4}{5} - \frac{4}{7}\right)$

C Average

Another topic related to order of operations is that of average. As discussed in Section 1.6, the **average** of a set of numbers can be found by adding the numbers and then dividing this sum by the quantity of numbers in the set. Now, we can find the average of mixed numbers as well as whole numbers.

8. Find the average of $3\frac{1}{4}$, $3\frac{1}{12}$, and $2\frac{5}{6}$.

Example 8 Finding the Average

Find the average of $1\frac{1}{2}$, $2\frac{3}{4}$, and $3\frac{5}{8}$.

Solution

Finding the average is the same as evaluating the expression

$$\left(1\frac{1}{2}+2\frac{3}{4}+3\frac{5}{8}\right)\div 3.$$

Find the sum first. The LCD = 8.

$$\begin{aligned} 1\frac{1}{2} &= \quad 1\frac{4}{8} \\ 2\frac{3}{4} &= \quad 2\frac{6}{8} \\ +3\frac{5}{8} &= +3\frac{5}{8} \\ \hline & \quad 6\frac{15}{8} = 7\frac{7}{8} \end{aligned}$$

Now divide by 3.

$$7\frac{7}{8}\div 3=\frac{63}{8}\cdot\frac{1}{3}=\frac{\cancel{3}\cdot 3\cdot 7}{8\cdot\cancel{3}}=\frac{21}{8} \text{ or } 2\frac{5}{8}$$

The average is $2\frac{5}{8}$.

Now work margin exercise 8.

CALCULATORS

Using a Calculator to Simplify Expressions with Fractions

Many scientifc calculators have a fraction button [a b/c]. To enter a mixed number or fraction in your calculator using this button, press [a b/c] between the whole number and the numerator and again between the numerator and denominator. (To enter a fraction, simply press [a b/c] between the numerator and denominator.)

To add, subtract, multiply, or divide two fractions or mixed numbers, enter the expression as you did when working with whole numbers in Chapter 1, but use the [a b/c] key to enter the fractions or mixed numbers. The result will be given to you in simplified form.

For example, to multiply $\frac{5}{8}\cdot\frac{6}{11}$, press the keys

[5] [a b/c] [8] [×] [6] [a b/c] [1] [1]. Then press [=].

The display will read 15/44.

Similarly, to add $4\frac{1}{3}+1\frac{5}{6}$, press the keys

4 a b/c 1 a b/c 3 + 1 a b/c 5 a b/c 6. Then press =.

The display will read 6 ⌟ 1/6.

Completion Example Answers

7. $\frac{7}{2}\cdot\frac{1}{4}+\frac{9}{10}\div\frac{9}{25}=\frac{7}{2}\cdot\frac{1}{4}+\frac{9}{10}\cdot\frac{25}{9}=\frac{7}{8}+\frac{5}{2}=\frac{27}{8}$ or $3\frac{3}{8}$

Margin Exercise Answers

1. $\frac{9}{22}$ is larger by $\frac{7}{66}$. 2. $\frac{19}{24}$ is larger by $\frac{1}{72}$. 3. $\frac{5}{9},\frac{7}{12},\frac{2}{3};\ \frac{1}{9}$ 4. $\frac{17}{45}$ 5. $\frac{31}{108}$ 6. $\frac{1}{4}$

7. $\frac{23}{12}$ or $1\frac{11}{12}$ 8. $\frac{55}{18}$ or $3\frac{1}{18}$

2.8 Exercises

Concept Check

Fill-in-the-Blank. Complete each sentence using information found in this section.

1. To compare two or more fractions, change each fraction to an equivalent fraction with the LCD and then compare the _______.
2. Once fractions have common denominators, they can be compared by looking at the _______.
3. According to the rules for order of operations, after exponential expressions have been simplified, perform any _______ or _______ from left to right.
4. According to the rules for order of operations, first simplify within _______ _______.
5. To find the average of a set of numbers, add the numbers and _______ that sum by the quantity of numbers in the set.

True/False. Determine whether each statement is true or false. If a statement is false, explain how it can be changed so the statement will be true. (**Note:** There may be more than one acceptable change.)

6. The rules for order of operations are the same for fractions, mixed numbers, and whole numbers.
7. According to the rules for order of operations, multiplication occurs before division.
8. An average is found by adding all the numbers in the set and then dividing by the quantity of numbers in the set.

2.8 Exercises

Concept Check

1. Numerators
2. Numerators
3. Multiplication, division
4. Grouping symbols
5. Divide
6. True
7. False; According to the rules for order of operations, multiplication and division occur from left to right.
8. True

Practice

1. $\frac{3}{4}$ by $\frac{1}{12}$
2. $\frac{1}{2}$ by $\frac{1}{10}$
3. $\frac{7}{10}$ by $\frac{1}{6}$
4. $\frac{4}{10}$ by $\frac{1}{40}$
5. $\frac{17}{20}$ by $\frac{1}{20}$
6. $\frac{31}{36}$ by $\frac{1}{36}$
7. $\frac{13}{20}$ by $\frac{1}{40}$
8. $\frac{7}{24}$ by $\frac{1}{72}$
9. Equal
10. $\frac{37}{100}$ by $\frac{1}{20}$
11. $\frac{21}{25}$ by $\frac{11}{400}$
12. $\frac{7}{9}$ by $\frac{10}{441}$
13. $\frac{11}{48}$ by $\frac{1}{60}$
14. $\frac{17}{60}$ by $\frac{3}{140}$
15. $\frac{3}{8}, \frac{2}{5}, \frac{1}{2}; \frac{1}{8}$
16. $\frac{5}{8}, \frac{2}{3}, \frac{3}{4}; \frac{1}{8}$
17. $\frac{1}{4}, \frac{1}{3}, \frac{1}{2}; \frac{1}{4}$
18. $\frac{1}{9}, \frac{1}{7}, \frac{1}{5}; \frac{4}{45}$
19. $\frac{7}{9}, \frac{31}{36}, \frac{17}{18}; \frac{1}{6}$
20. $\frac{25}{48}, \frac{9}{16}, \frac{5}{8}; \frac{5}{48}$
21. $\frac{8}{9}, \frac{9}{10}, \frac{11}{12}; \frac{1}{36}$
22. $\frac{11}{12}, \frac{19}{20}, \frac{7}{6}; \frac{1}{4}$
23. $\frac{5}{42}, \frac{1}{3}, \frac{3}{7}; \frac{13}{42}$
24. $\frac{40}{36}, \frac{31}{24}, \frac{17}{12}; \frac{11}{36}$

Practice

For each pair of fractions, determine which fraction is larger and by how much it is larger. See Examples 1 and 2.

1. $\frac{2}{3}, \frac{3}{4}$
2. $\frac{1}{2}, \frac{2}{5}$
3. $\frac{7}{10}, \frac{8}{15}$
4. $\frac{4}{10}, \frac{3}{8}$
5. $\frac{4}{5}, \frac{17}{20}$
6. $\frac{5}{6}, \frac{31}{36}$
7. $\frac{13}{20}, \frac{5}{8}$
8. $\frac{10}{36}, \frac{7}{24}$
9. $\frac{14}{35}, \frac{12}{30}$
10. $\frac{37}{100}, \frac{24}{75}$
11. $\frac{13}{16}, \frac{21}{25}$
12. $\frac{7}{9}, \frac{37}{49}$
13. $\frac{17}{80}, \frac{11}{48}$
14. $\frac{11}{42}, \frac{17}{60}$

Arrange each set of fractions in order from smallest to largest. Then find the difference between the largest and smallest fractions. See Example 3.

15. $\frac{1}{2}, \frac{2}{5}, \frac{3}{8}$
16. $\frac{2}{3}, \frac{3}{4}, \frac{5}{8}$
17. $\frac{1}{2}, \frac{1}{3}, \frac{1}{4}$
18. $\frac{1}{5}, \frac{1}{7}, \frac{1}{9}$
19. $\frac{7}{9}, \frac{31}{36}, \frac{17}{18}$
20. $\frac{5}{8}, \frac{25}{48}, \frac{9}{16}$
21. $\frac{8}{9}, \frac{9}{10}, \frac{11}{12}$
22. $\frac{7}{6}, \frac{11}{12}, \frac{19}{20}$
23. $\frac{1}{3}, \frac{5}{42}, \frac{3}{7}$
24. $\frac{17}{12}, \frac{40}{36}, \frac{31}{24}$
25. $\frac{1}{100}, \frac{3}{1000}, \frac{20}{10{,}000}$
26. $\frac{32}{100}, \frac{298}{1000}, \frac{3}{10}$

Simplify. See Examples 4 through 7.

27. $\frac{1}{2} \div \frac{7}{8} + \frac{1}{7} \cdot \frac{2}{3}$
28. $\frac{1}{2} \div \frac{1}{2} + \frac{2}{3} \cdot \frac{2}{3}$
29. $\frac{3}{5} \cdot \frac{1}{6} + \frac{1}{5} \div 2$
30. $\frac{2}{3} \cdot \frac{1}{4} + \frac{1}{2} \div 3$
31. $5 - \frac{3}{4} \div 3$
32. $6 - \frac{5}{8} \div 4$
33. $\frac{5}{8} \cdot \frac{1}{10} \div \frac{3}{4} + \frac{1}{6}$
34. $\frac{2}{15} \cdot \frac{1}{4} \div \frac{3}{5} + \frac{1}{27}$
35. $\left(\frac{7}{15} + \frac{8}{21}\right) \div \frac{3}{35}$
36. $\left(\frac{5}{14} + \frac{3}{10}\right) \div \frac{23}{35}$
37. $\left(\frac{1}{2} - \frac{1}{3}\right) \div \left(\frac{5}{8} + \frac{3}{16}\right)$
38. $\left(\frac{7}{8} - \frac{3}{16}\right) \div \left(\frac{1}{3} - \frac{1}{4}\right)$
39. $\left(\frac{5}{6} - \frac{1}{3}\right) \div \left(\frac{1}{2} + \frac{1}{5}\right)$
40. $\left(\frac{5}{6} - \frac{2}{3}\right) \div \left(\frac{5}{8} - \frac{1}{16}\right)$
41. $\left(\frac{1}{2}\right)^2 - \left(\frac{1}{4}\right)^3$
42. $\left(\frac{1}{3}\right)^3 - \left(\frac{1}{9}\right)^2$
43. $\left(\frac{1}{3} + \frac{1}{5}\right) \cdot \left(\frac{3}{4} - \frac{1}{6}\right)$
44. $\left(\frac{2}{5} + \frac{1}{4}\right) \cdot \left(\frac{3}{4} - \frac{1}{2}\right)$
45. $\frac{2}{3} + \frac{3}{4} + \left(\frac{1}{2}\right)^2$
46. $\frac{1}{2} \div \frac{2}{3} + \left(\frac{1}{3}\right)^2$
47. $\left(\frac{3}{4} - \frac{1}{2}\right) \div \left(1 + \frac{1}{3}\right)$
48. $\left(\frac{1}{8} + \frac{1}{2}\right) \div \left(1 - \frac{2}{5}\right)$

49. $\left(\frac{1}{3}\right)^2+\left(\frac{1}{6}\right)^2+\frac{2}{3}$

50. $\left(\frac{2}{3}\right)^2+\left(\frac{1}{3}\right)^2-\frac{5}{9}$

51. $\left(2+\frac{1}{5}\right)\div 2\frac{1}{5}$

52. $\left(4+\frac{1}{2}\right)\div 4\frac{1}{2}$

53. $\left(\frac{5}{8}+\frac{5}{8}\right)\div\left(2\frac{1}{2}\right)^2$

54. $\left(\frac{7}{12}+\frac{7}{12}\right)\div\left(2\frac{1}{3}\right)^2$

55. $\left(6\frac{1}{100}+5\frac{3}{100}\right)\div\left(2\frac{1}{10}+3\frac{1}{10}\right)$

56. $\left(8\frac{31}{100}+3\frac{33}{100}\right)\div\left(1\frac{3}{10}+5\frac{3}{10}\right)$

57. $3\frac{1}{2}\cdot 5\frac{1}{3}+\frac{5}{12}\div\frac{15}{16}$

58. $2\frac{1}{2}\cdot 3\frac{1}{5}\div\frac{3}{4}+\frac{7}{10}$

59. $\left(5\frac{2}{3}-1\frac{1}{6}\right)-\left(3\frac{1}{2}+\frac{1}{6}\right)$

60. $\left(7\frac{1}{5}-2\frac{1}{3}\right)-\left(6\frac{1}{10}-2\right)$

61. $\frac{5}{8}-\frac{1}{3}\cdot\frac{2}{5}+6\frac{1}{10}$

62. $\frac{5}{6}-\frac{1}{2}\cdot\frac{4}{7}+5\frac{1}{10}$

63. $\left(2\frac{4}{9}+1\frac{1}{18}\right)\div\left(1\frac{2}{9}-\frac{1}{6}\right)$

64. $\left(2\frac{1}{6}+1\frac{5}{12}\right)\div\left(1\frac{3}{4}-\frac{2}{3}\right)$

65. $1\frac{1}{6}\cdot 1\frac{2}{19}\div\frac{7}{8}+\frac{1}{38}$

66. $2\frac{1}{13}\cdot 3\frac{1}{3}\div\frac{5}{6}+\frac{7}{39}$

Find each average. See Example 8.

67. Find the average of the numbers $\frac{5}{6}$, $\frac{1}{15}$, and $\frac{17}{30}$.

68. Find the average of the numbers $\frac{5}{8}$, $\frac{3}{10}$, and $1\frac{1}{4}$.

69. Find the average of the numbers $\frac{7}{8}$, $\frac{9}{10}$, and $1\frac{3}{4}$.

70. Find the average of the numbers $\frac{5}{6}$, $1\frac{2}{3}$, and $\frac{3}{4}$.

71. Find the average of the numbers $5\frac{1}{8}$, $7\frac{1}{2}$, $4\frac{3}{4}$, and $10\frac{1}{2}$.

72. Find the average of the numbers $4\frac{7}{10}$, $3\frac{9}{10}$, $5\frac{1}{100}$, and $11\frac{3}{20}$.

73. Find the average of the numbers $7\frac{3}{5}$, $4\frac{3}{5}$, $8\frac{1}{3}$, and $3\frac{1}{18}$.

74. Find the average of the numbers $6\frac{4}{7}$, $3\frac{1}{4}$, $1\frac{11}{12}$, and $5\frac{6}{7}$.

25. $\frac{20}{10{,}000}, \frac{3}{1000}, \frac{1}{100}; \frac{1}{125}$

26. $\frac{298}{1000}, \frac{3}{10}, \frac{32}{100}; \frac{11}{500}$

27. $\frac{2}{3}$

28. $\frac{13}{9}=1\frac{4}{9}$

29. $\frac{1}{5}$

30. $\frac{1}{3}$

31. $\frac{19}{4}=4\frac{3}{4}$

32. $\frac{187}{32}=5\frac{27}{32}$

33. $\frac{1}{4}$

34. $\frac{5}{54}$

35. $\frac{89}{9}=9\frac{8}{9}$

36. 1

37. $\frac{8}{39}$

38. $\frac{33}{4}=8\frac{1}{4}$

39. $\frac{5}{7}$

40. $\frac{8}{27}$

41. $\frac{15}{64}$

42. $\frac{2}{81}$

43. $\frac{14}{45}$

44. $\frac{13}{80}$

45. $\frac{5}{3}=1\frac{2}{3}$

46. $\frac{31}{36}$

47. $\frac{3}{16}$

48. $\frac{25}{24}=1\frac{1}{24}$

49. $\frac{29}{36}$

50. 0

51. 1

52. 1

53. $\frac{1}{5}$

54. $\frac{3}{14}$

55. $\frac{138}{65}=2\frac{8}{65}$

56. $\frac{97}{55}=1\frac{42}{55}$

57. $\frac{172}{9}=19\frac{1}{9}$

58. $\frac{341}{30}=11\frac{11}{30}$

59. $\frac{5}{6}$

60. $\frac{23}{30}$

61. $\frac{791}{120}=6\frac{71}{120}$

62. $\frac{1231}{210}=5\frac{181}{210}$

63. $\frac{63}{19}=3\frac{6}{19}$

64. $\frac{43}{13}=3\frac{4}{13}$

65. $\frac{3}{2}=1\frac{1}{2}$

66. $\frac{331}{39}=8\frac{19}{39}$

67. $\frac{22}{45}$

68. $\frac{29}{40}$

69. $\frac{47}{40}=1\frac{7}{40}$

70. $\frac{13}{12}=1\frac{1}{12}$

71. $\frac{223}{32}=6\frac{31}{32}$

72. $\frac{619}{100}=6\frac{19}{100}$

73. $\frac{2123}{360}=5\frac{323}{260}$

Applications

Solve.

75. If the product of $5\frac{1}{2}$ and $2\frac{1}{4}$ is added to the quotient of $\frac{9}{10}$ and $\frac{3}{4}$, what is the total?

76. If $\frac{9}{10}$ of 70 is divided by $\frac{3}{4}$ of 10, what is the quotient?

77. An $8\frac{1}{2}$ inch by $10\frac{3}{4}$ inch picture is placed in a frame that is $\frac{9}{16}$ inches wide. Find the perimeter of the outside edges of the frame. (Reminder: The perimeter of a rectangle is computed by adding twice the length to twice the width.)

78. A recipe calls for $2\frac{1}{5}$ cups of stewed tomatoes and $1\frac{2}{3}$ cups of fully cooked beans. If the recipe is multiplied by $2\frac{1}{2}$, how much tomato/bean mixture will there be?

79. The force between two charged particles is a product of terms divided by the square of the distance between two particles. Assume that the product of the terms is $\frac{3}{5}$ and the distance between the two particles is $\frac{1}{25}$. What is the force?

80. Two painters paint $76\frac{1}{2}$ feet of fencing in one day. The first painter works for $2\frac{1}{2}$ hours and the second works for $5\frac{3}{5}$ hours. How many feet of fencing are painted in each hour? (**Hint:** Add the number of hours then divide the length of fencing by this sum.)

81. A customer orders some items from an online store. Included in this order are five boxes of candy that weigh $1\frac{1}{3}$ pounds each. The order also contains two of each of the following items: a $\frac{2}{3}$ pound box of chewing gum, a $2\frac{1}{2}$ pound can of peanuts, and a $1\frac{1}{4}$ pound box of gourmet popping corn. If the packaging materials weigh $1\frac{1}{2}$ pounds, what is the total shipping weight of the order?

82. An art book has 40 two-sided pages of pictures, each of which is $\frac{1}{32}$ inch thick. Each two-sided page is protected by a $\frac{1}{80}$ inch thick piece of paper. Each side of the book is bound by a $\frac{1}{6}$ inch cover. What is the total thickness of the book?

83. Emma goes to a buffet where the cost of the meal is determined by the weight of the food. Among the items currently on the buffet are $\frac{8}{9}$ kg of potato salad and $\frac{3}{4}$ kg of chicken salad. Emma puts $\frac{3}{16}$ of the potato salad and $\frac{2}{15}$ of the chicken salad onto her plate. She also takes 2 slices of bread which weigh $\frac{1}{30}$ kg each.

a. What is the total weigh of the food taken? (The weight of the plate is not included.)

b. If the restaurant charges $18 for each kg of food (including taxes), how much will Emma spend?

84. A building was partially destroyed by a bad storm and it had to be torn down because it was condemned by the city. At the beginning of the demolition project, $\frac{8}{9}$ of the building was standing. After the first day of work $\frac{1}{3}$ of the building was standing. If the crew spent $6\frac{2}{3}$ hours of work on the project during the first day, what fraction of the building was torn down each hour?

85. A person has a collection of four snakes with the following lengths: $1\frac{1}{4}$ feet, $3\frac{1}{8}$ feet, $\frac{7}{8}$ feet, and $2\frac{1}{3}$ feet. What is the average length of the four snakes?

86. A town that normally does not receive significant snow experienced measurable snow for three weeks in a row. The first week it snowed $2\frac{1}{4}$ inches, the second week it snowed $9\frac{2}{3}$ inches, and the third week it snowed $1\frac{5}{6}$ inches. What was the average weekly snowfall for this three-week period?

Writing & Thinking

87. a. If two fractions are between 0 and 1, can their sum be more than 1? Explain.

b. If two fractions are between 0 and 1, can their product be more than 1? Explain.

88. If a fraction is between 0 and 1 and the fraction is squared, will the result be larger or smaller than the original fraction? Explain.

89. Consider the fraction $\frac{1}{2}$.

a. If this fraction is divided by 2, will the quotient be more or less than $\frac{1}{2}$?

b. If this fraction is divided by 3, will the quotient be more or less than the quotient in part a.?

90. Will the quotient always get smaller and smaller when a nonzero whole number is divided by larger and larger whole numbers? Can you think of a case in which this is not true? What happens when 0 is divided by larger and larger whole numbers?

74. $\frac{739}{168} = 4\frac{67}{168}$

Applications

75. $\frac{543}{40} = 13\frac{23}{40}$

76. $\frac{42}{5} = 8\frac{2}{5}$

77. 43 inches

78. $\frac{29}{3} = 9\frac{2}{3}$ cups

79. 375

80. $\frac{85}{9} = 9\frac{4}{9}$ feet per hour

81. 17 pounds

82. $\frac{25}{12} = 2\frac{1}{12}$ inches

83. a. $\frac{1}{3}$ kg **b.** \$6

84. $\frac{1}{12}$

85. $\frac{91}{48} = 1\frac{43}{48}$ feet

86. $\frac{55}{12} = 4\frac{7}{12}$ inches

Writing & Thinking

87. a. Yes; If both fractions are greater than one half, the sum will be greater than one.

b. No; Multiplying a number by a fraction between 0 and 1 results in a product that is less than the original number.

88. Smaller, because multiplying a number by a fraction between 0 and 1 results in a product that is less than the original number.

89. a. Less than **b.** Less than

90. Yes. For 0, the quotient will remain 0.

Chapter 2 Project

On a Budget

An activity to demonstrate the use of fractions in real life.

Samantha's friend Meghan is always traveling to exotic locations when she goes on vacation. One day, Samantha asked Meghan how she was able to afford it. Meghan told her it was simple: she makes a budget and sets aside a portion of her income each month for a vacation. Meghan told Samantha that they could meet at her house for dinner on Saturday and she would be glad to show her how to make a budget.

The table below shows the amount Meghan spends on each category in her budget. Assume Meghan makes $4400 a month (after taxes) and answer the following questions. Fractions should be written in reduced form.

Meghan's Budget

Category	Amount	Category	Amount
Rent	$1100	Car Loan	$528
Groceries	$352	IRA	$264
Utilities	$220	Charity	$440
Dining Out	$264	Savings	$275
Vacation	$440	Car Insurance	$176
Gas	$88	Phone	$110

1. Samantha earns a different amount per month and wants to compare her spending to Meghan's, so she decides to convert all of Meghan's spending to fractions. Find the fractional values for Meghan's categories by dividing each amount by her monthly income and reducing the fractions to lowest terms.

Meghan's Budget

Category	Amount	Category	Amount
Rent	$\frac{1}{4}$	Car Loan	$\frac{3}{25}$
Groceries	$\frac{2}{25}$	IRA	$\frac{3}{50}$
Utilities	$\frac{1}{20}$	Charity	$\frac{1}{10}$
Dining Out	$\frac{3}{50}$	Savings	$\frac{1}{16}$
Vacation	$\frac{1}{10}$	Car Insurance	$\frac{1}{25}$
Gas	$\frac{1}{50}$	Phone	$\frac{1}{40}$

2. Find the sum of the fractions in the budget. Begin by adding the fractions that have like denominators, then add the fractions by finding a common denominator.

3. Samantha tells Meghan that her budget must be incomplete. Meghan informs her that the remaining is what she spends on miscellaneous items like cleaning supplies, toiletries, and clothes. What fraction of her budget does Meghan spend on miscellaneous items?

4. Samantha determines that she spends $\frac{1}{10}$ of her income on groceries and $\frac{1}{8}$ dining out. What fraction of her income in spent on food?

5. What fraction of Megan's income is spent on food?

6. Find the difference between the fractional amount that Meghan and Samantha spend for food. Who uses the largest fractional part of their income for food?

7. Add the fractional values for the car loan, car insurance, and gas to determine the fractional amount of Meghan's income that is set aside for her car.

8. Samantha doesn't have a car payment, but she has to pay for parking at her apartment. She has determined that she spends $\frac{1}{50}$ of her monthly income on car insurance, $\frac{3}{50}$ on parking, and $\frac{1}{25}$ on gas. What fractional part of her income is used for her car?

9. Find the difference between the fractional amount that Meghan and Samantha spend on their cars. Who uses the largest fractional part of their income for car expenses?

10. Meghan plans to move into a new apartment with a roommate and travel more often. She needs to revise her budget accordingly.

Meghan's Budget

Category	Amount	Category	Amount
Rent	$\frac{1}{8}$	Car Loan	$\frac{3}{25}$
Groceries	$\frac{1}{25}$	IRA	$\frac{3}{50}$
Utilities	$\frac{1}{40}$	Charity	$\frac{1}{10}$
Dining Out	$\frac{3}{25}$	Savings	$\frac{1}{16}$
Vacation	$\frac{1}{5}$	Car Insurance	$\frac{1}{25}$
Gas	$\frac{1}{25}$	Phone	$\frac{1}{40}$

a. Halve the first three fractions in the first column. Since Megan will have a roommate and will be gone more, she expects to spend half as much on rent, groceries, and utilities. Place the new values in the new table above.

b. Double the last three fractions in the first column. Since Megan will be traveling, she expects to spend twice as much for dinning out, vacation, and gas. Place the new values in the table.

c. Find the total fractional portion budgeted for the new budget plan if the second column remains the same.

d. Would Meghan be able to afford this new budget on her current income? In other words, do the new fractions in the table add up to a fraction less than 1?

e. If the new budget in part d. is over budget, what fraction is it over? If the new budget is under budget, what fractional portion is left over for miscellaneous items?

Chapter 2 Project: On a Budget

1. See table

2. $\frac{387}{400}$

3. $\frac{13}{400}$

4. $\frac{9}{40}$

5. $\frac{7}{50}$

6. $\frac{17}{200}$; Samantha spends more on food.

7. $\frac{9}{50}$

8. $\frac{3}{25}$

9. $\frac{3}{50}$; Meghan spends more on car expenses.

10. **a.** and **b.** See Table.

c. $\frac{383}{400}$

d. Yes, the sum of the fractions is $\frac{383}{400}$ which is less than 1.

e. $\frac{17}{400}$

Chapter 2 Project

What's Cookin', Good Lookin'?

An activity to demonstrate the use of fractions in real life.

You are baking shortbread cookies for two events: your mother's birthday and your mathlete's celebration party. You have your grandfather Giuseppe's famous recipe and want to make the exact amount of cookies needed for both parties with one batch of dough, meaning you must adjust the size of the original recipe.

Giuseppe's Famous Shortbread Cookies

Ingredient	Amount Needed
Butter	1 cup
Sugar	$\frac{1}{2}$ cup
Vanilla Extract	1 teaspoon
Salt	1 teaspoon
Flour	2 cups
Makes 36 cookies	

No extra directions are given, typical of grandpa, huh?

1. Eighteen people are coming to your mother's birthday and 9 people are coming to the celebration party. How many total cookies do you need, assuming each person eats one cookie?

2. First, let's determine how to adjust the recipe to make cookies for 18 people.

a. What fraction of the recipe do you need to make 18 cookies? (**Hint**: Place the part over the whole number of cookies the recipe makes and reduce.)

b. Multiply each of the ingredients by the fraction you found in part a.

3. Next, let's determine how to adjust the recipe to make cookies for 9 people.

a. What fraction of the recipe do you need to make 9 cookies?

b. Multiply each of the ingredients by the fraction you found in part a.

4. Find the sum of the individual ingredients from Problems 2 and 3 to create one new recipe. Rewrite any improper fractions as mixed numbers.

5. Now you have an adjusted recipe to make the exact number of cookies needed. You have the following measuring cups and spoons: 1 cup, $\frac{1}{2}$ cup, $\frac{1}{4}$ cup, 1 teaspoon, $\frac{1}{2}$ teaspoon, $\frac{1}{4}$ teaspoon, and 1 tablespoon.

a. How would you measure out $\frac{3}{4}$ cup of an ingredient with the given cups?

b. Is it possible to measure out all of your ingredients with the given cups and spoons? If not, which ingredients?

c. Find at least two pairs of fractions that add to $\frac{3}{8}$.

d. If $\frac{1}{4}$ cup is equal to 4 tablespoons, how many tablespoons would be in $\frac{1}{8}$ cup?

e. How could you measure out $\frac{3}{8}$ cup of an ingredient with this new knowledge?

f. List at least two ways you could measure out the needed amount of flour.

6. Your sister gifted you with some local artisanal pecan flour, and you want substitute some of the flour in the recipe with the pecan flour with a one-to-one substitution. Using the available measuring cups, list three combinations of measurements that you could use to combine the normal flour and pecan flour and still measure out the total needed amount of flour.

7. Is it necessary to create a new adjusted recipe each time you want to make a certain amount of cookies? What are some additional strategies you could use to make a certain amount of cookies for a party?

8. Perform an internet search for recipes for your top three favorite cookies.

a. State at least three ingredients that are commonly found in these recipes that are not included in the shortbread cookie recipe.

b. Do you think any of these additional common baking ingredients are difficult to halve? Are any of these easy to double?

9. When doubling or halving a recipe, some common baking ingredients should not be doubled or halved. Perform an internet search and explain in your own words why that is true.

Chapter 2 Project: What's Cookin', Good Lookin'?

1. 27 cookies

2. a. $\frac{1}{2}$ of the recipe

b. $\frac{1}{2}$ cup butter, $\frac{1}{4}$ cup sugar, $\frac{1}{2}$ teaspoon vanilla extract, $\frac{1}{2}$ teaspoon salt, 1 cup flour

3. a. $\frac{1}{4}$ of the recipe

b. $\frac{1}{4}$ cup butter, $\frac{1}{8}$ cup sugar, $\frac{1}{4}$ teaspoon vanilla extract, $\frac{1}{4}$ teaspoon salt, $\frac{1}{2}$ cup flour

4.

Ingredient	Amount Needed
Butter	$\frac{3}{4}$ cup
Sugar	$\frac{3}{8}$ cup
Vanilla Extract	$\frac{3}{4}$ teaspoon
Salt	$\frac{3}{4}$ teaspoon
Flour	$1\frac{1}{2}$ cups
Makes 27 cookies	

5. a. Using the $\frac{1}{4}$ cup measuring cup three times will give $\frac{3}{4}$ cup.

b. No, sugar

c. Answers may vary. Some examples include $\frac{3}{4}$ or $\frac{1}{4}+\frac{1}{8}$ or $\frac{3}{16}+\frac{3}{16}$.

d. 2 tablespoons

e. $\frac{1}{4}$ cup plus 2 tablespoons or 6 tablespoons

f. Answers may vary. Some examples include 1 cup plus $\frac{1}{2}$ cup, three $\frac{1}{2}$ cups, or six $\frac{1}{4}$ cups.

6. Answers may vary. Some examples include 1 cup flour and $\frac{1}{2}$ cup pecan flour, $\frac{3}{4}$ cup flour and $\frac{3}{4}$ cup pecan flour, or 1 cup plus $\frac{1}{4}$ cup flour and $\frac{1}{4}$ cup pecan flour.

7. No, it is not necessary. Answers may vary. You could make two batches separately, or you could make more than needed and eat the leftovers.

8. a. Answers may vary. Some examples are eggs, baking powder, baking soda, other extracts, or milk.

b. Answers may vary. Eggs can be difficult to halve. Anything that is a small teaspoon measurement may be difficult to halve. Most ingredients are easily doubled using the measuring cups given.

9. Answers may vary. Doubling spices can overwhelm a recipe, such as salt, cinnamon, etc. Baking powder and baking soda follow specific ratios and may result in ruined cookies when doubled.

CHAPTER 3

3

Decimal Numbers

Connections

Many people use decimal numbers on a daily basis to discuss sports. Batting averages in baseball and save percentages in hockey are both calculated to the nearest thousandth. Quarterback ratings in football and average points per game in basketball are both calculated to the nearest tenth. Decimal numbers are also used to record many of the world's sport records. For example, for the 100-meter freestyle, the men's swimming record is 44.84 seconds by Kyle Chalmers of Russia, and the women's swimming record is 51.71 seconds by Sarah Sjöström of Sweden.

During the NBA 2021–2022 regular season, the top three players based on points per game were Giannis Antetokounmpo of the Milwaukee Bucks, Luka Doncic of the Dallas Mavericks, and Nikola Jokic of the Denver Nuggets. Antetokounmpo and Doncic tied with 31.7 points per game, followed by Jokic with 31.0 points per game. During the season, Antetokounmpo played 67 games, Doncic played 65 games, and Jokic played 74 games. How would you determine how many total points each player scored during the 2021-2022 regular season? Which player scored the most points during the season?

Source: "Season Leaders|Stats", NBA, Accessed August 9, 2022, www.nba.com/stats/players/.

Objectives

A. Read and write decimal numbers.

B. Compare decimal numbers.

C. Round decimal numbers.

3.1 Introduction to Decimal Numbers

A Reading and Writing Decimal Numbers

Decimal notation uses a place value system and a **decimal point**, with whole numbers written to the left and fractions written to the right of the decimal point. Numbers written in decimal notation are said to be **decimal numbers** (or simply **decimals**). The values of several places in the decimal system are shown in Figure 1.

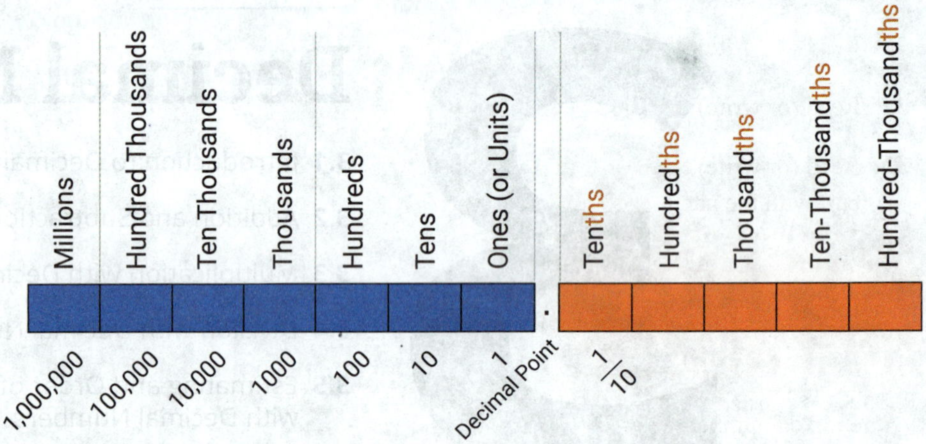

Figure 1

Fractions with denominators that are powers of 10 are called **decimal fractions**. For example, the fraction $\frac{1}{10}$ can be written in decimal form as 0.1 (read "one tenth") and can be diagrammed by shading a square as shown in Figure 2.

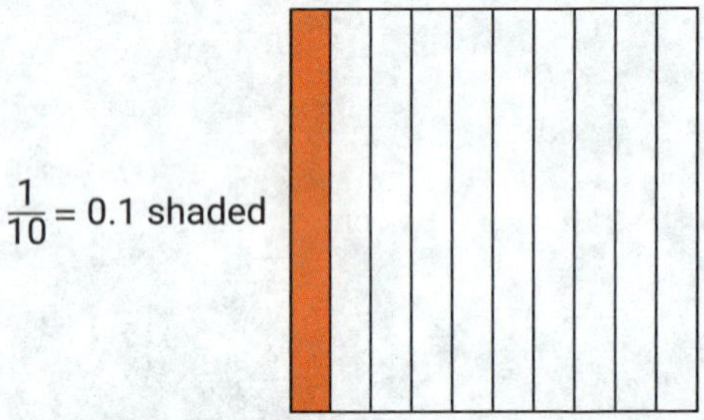

Figure 2

Similarly, the fractions $\frac{1}{100}$ (0.01 in decimal form and read "one hundredth") and $\frac{45}{100}$ (0.45 in decimal form and read "45 hundredths") can be diagrammed by shading a square that has been cut into 100 equal pieces as shown in Figure 3.

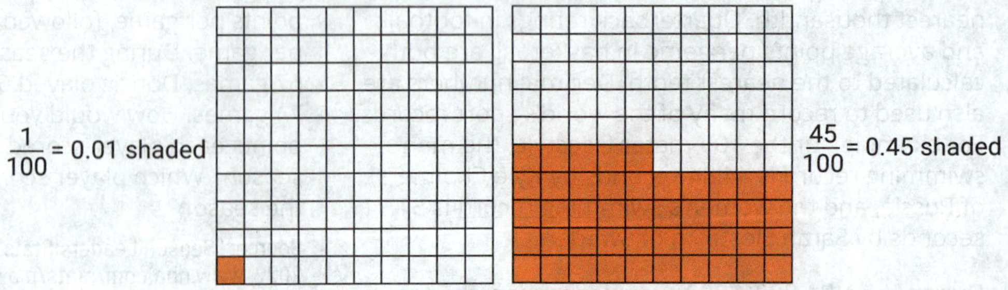

Figure 3

Familiar decimal numbers include dollars and cents. For example, you might pay \$45.25 for a new shirt or \$3.50 for a taco. These numbers are mixed numbers and can be written as follows.

$\$45\frac{25}{100} = \$45.25$ Read "forty-five dollars and twenty-five cents."

$\$3\frac{50}{100} = \$3.50$ Read "three dollars and fifty cents."

In general, decimal numbers are read (and written) according to the following convention.

Reading or Writing a Decimal Number

1. Read (or write) the whole number.
2. Read (or write) the word "**and**" in place of the decimal point.
3. Read (or write) the fraction part as a whole number. Then, name the fraction part with the name of the place value of the last digit on the right.

PROCEDURE

Note

If there is no whole number part, then 0 can be written to the left of the decimal point. For example, .6 can be written as 0.6 and in most cases 0.6 is preferred.

Example 1 Reading and Writing Decimal Numbers

Write the mixed number $48\frac{6}{10}$ in decimal notation and in words.

Solution

48 . 6 In decimal notation

↓ ↓ ↓

forty-eight and six tenths In words

And indicates the decimal point; the digit 6 is in the tenths position.

Now work margin exercise 1.

1. Write the mixed number in decimal notation and in words.

$19\frac{3}{10}$

Example 2 Reading and Writing Decimal Numbers

Write the mixed number $12\frac{75}{10,000}$ in decimal notation and in words.

Solution

Two 0s must be inserted as placeholders so the digits 7 and 5 will be in the right places.

12 . 0075 In decimal notation

↓ ↓ ↓

twelve and seventy-five ten-thousandths In words

And indicates the decimal point; the digit 5 is in the ten-thousandths position.

Now work margin exercise 2.

2. Write the mixed number in decimal notation and in words.

$39\frac{184}{10,000}$

Attention!

The letters **ths** (or **th**) at the end of a word indicate a fraction part (a part to the right of the decimal point).

six hundred = 600	two hundred thousand = 200,000
six hundred**ths** = 0.06	two hundred-thousand**ths** = 0.00002

3. Write each number in decimal notation.

   a. one thousand two hundred and five ten-thousandths

   b. one thousand two hundred five ten-thousandths

Example 3 Reading and Writing Decimal Numbers

Write each number in decimal notation. Note how **and** indicates the decimal point.

a. four hundred and two thousandths

b. four hundred two thousandths

Solution

a. four hundred and two thousandths

400 . 002

The digit 2 is in the thousandths position.

Two 0s must be inserted as placeholders.

b. four hundred two thousandths

0 . 402

Now work margin exercise 3.

B Comparing Decimal Numbers

Determining the relative size of numbers, as we did with fractions in Section 2.8, can be a very useful skill. The following technique can be used to compare two decimal numbers to see which is smaller or larger.

Comparing Two Decimal Numbers

1. Moving **left to right**, compare digits with the same place value. (Insert 0s to the right to continue comparison, if necessary.)
2. When one compared digit is larger, then the corresponding number is larger.

PROCEDURE

Example 4 Comparing Decimal Numbers

Which number is larger: 3.126 or 3.14?

Solution

Comparing digits with the same place value from left to right (until we find a mismatch) gives the following.

3. 1 2 6
↕ Mismatch
3. 1 4 0

Because 4 > 2, the number 3.14 is greater than 3.126. That is,

$$3.14 > 3.126.$$

Graphing these values on a number line, we have 3.14 > 3.126 and 3.14 lies to the right of 3.126.

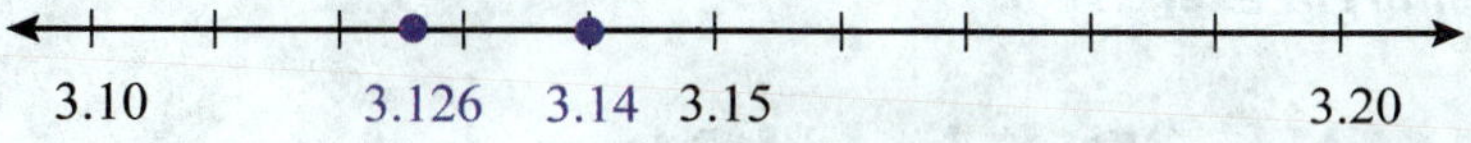

Now work margin exercise 4.

4. Which number is larger: 6.438 or 6.44?

Example 5 Comparing Decimal Numbers

Which number is larger: 0.08 or 0.085?

Solution

Comparing digits with the same place value from left to right (until we find a mismatch) gives the following.

0.0 8 0
↕ Mismatch
0.0 8 5

Because 5 > 0, we know that 0.085 > 0.08.

Graphing these values on a number line, we have 0.085 > 0.08 and 0.085 lies to the right of 0.08.

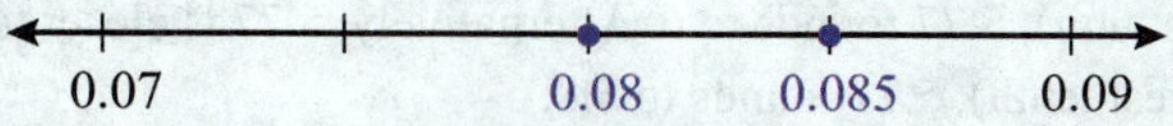

Now work margin exercise 5.

5. Which number is larger: 9.25 or 9.251?

6. Arrange the following three decimal numbers in order from smallest to largest. Then, graph them on a number line.

2.6, 2.06, 2.46

Example 6 Comparing Decimal Numbers

Arrange the following three decimal numbers in order from smallest to largest: 6.67, 5.14, 6.28. Then, graph them on a number line.

Solution

Comparing digits with the same place value in 6.67, 5.14, and 6.28 (from left to right) gives the following.

In the ones place, 5 < 6 so 5.14 is smaller than both 6.67 and 6.28.

Then, in the tenths place, 2 < 6 so 6.28 is smaller than 6.67.

Thus, from smallest to largest, the numbers are 5.14, 6.28, 6.67.

Graphing the numbers on a number line, we have the following.

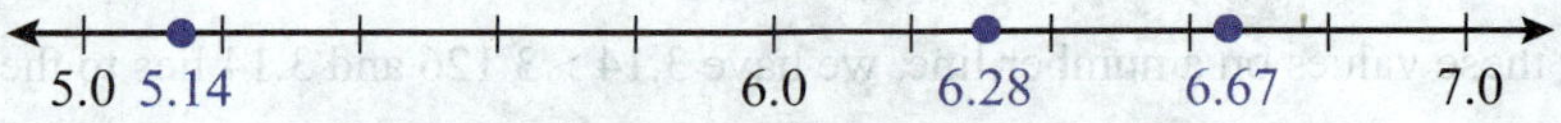

Now work margin exercise 6.

C Rounding Decimal Numbers

Measuring devices such as rulers, meter sticks, speedometers, micrometers, and surveying transits give only approximate measurements (see Figure 4). Whether the units are large (such as miles and kilometers) or small (such as inches and centimeters), there are always smaller, more accurate units (such as eighths of an inch and millimeters) that could be used to indicate a measurement. As a result, we are constantly dealing with approximate (or rounded) numbers in our daily lives.

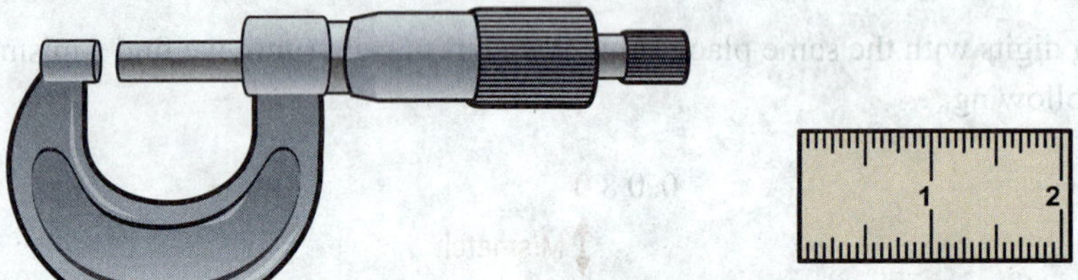

a. Micrometer **b.** Inch Ruler

Figure 4

When rounding a decimal number, the desired place of accuracy must be stated. For example, looking at the number line below, we can see that 5.77 is closer to 6.0 than it is to 5. So (to the nearest unit), 5.77 rounds to 6. Alternatively, 5.77 is closer to 5.8 than it is to 5.7. So (to the nearest tenth), 5.77 rounds to 5.8.

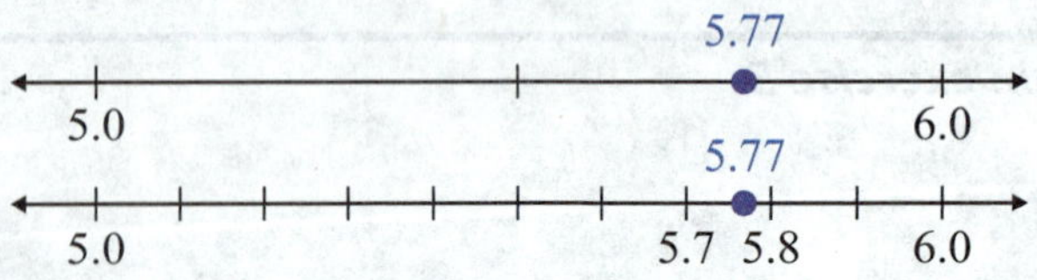

Figure 5

There are several rules for rounding decimal numbers. The rule chosen in a particular situation depends on the use of the numbers and whether there might be some sort of penalty. For example, the IRS allows rounding to the nearest dollar on income tax forms while NASA must be very exact with their calculations to make sure spacecrafts dock properly.

In this text, we will use the following rule for rounding decimal numbers.

Rounding Rule for Decimal Numbers

1. Look at the single digit one place value to the right of the digit in the place of desired accuracy.
   a. **If this digit is less than 5,** leave the digit in the desired place of accuracy as it is and replace all digits to the right with zeros. All digits to the left remain unchanged.
   b. **If this digit is 5 or greater,** increase the digit in the desired place of accuracy by one and replace all digits to the right with zeros. All digits to the left remain unchanged unless a 9 is made one larger. Then, the 9 is replaced by 0 and the next digit to the left is increased by 1.
2. Zeros to the right of the place of accuracy that are also to the right of the decimal point must be dropped. In this way, the place of accuracy is clearly understood. If a rounded number has a 0 in the desired place of accuracy, then that 0 remains.

PROCEDURE

Math Tip

When calculating sales tax, we round to the nearest hundredth. This rounding is used because the smallest unit of currency is one penny, which is represented by $0.01.

Example 7 Rounding Decimal Numbers

Round 18.649 to the nearest tenth.

Solution

6 is in the tenths place.

↓

18.649

↑

The next digit to the right is 4.

Since 4 is less than 5, leave the 6 and replace 4 and 9 with 0s.

Now, the 6 is to the right of the decimal point and the 0s to the right of the 6 in 18.600 are dropped so that the place of accuracy is indicated by the 6.

Thus, 18.649 rounds to 18.6 to the nearest tenth.

Now work margin exercise 7.

7. Round 9.337 to the nearest tenth.

8. Round 5.0983 to the nearest hundredth.

Example 8 Rounding Decimal Numbers

Round 5.83971 to the nearest thousandth.

Solution

9 is in the thousandths place.

↓

5.83971

↑

The next digit to the right is 7.

Since 7 is greater than 5, increase 9 by one and replace 7 and 1 with 0s. (Increasing 9 by one gives 10, which affects the digit 3 as well.)

Thus, 5.83971 rounds to 5.840 to the nearest thousandth, and only two 0s are dropped. Notice that the digit 3 changed to a 4 since the 9 was made one larger during rounding.

Now work margin exercise 8.

9. Round 2.3953 to the nearest thousandth.

Completion Example 9 Rounding Decimal Numbers

Round 0.6753 to the nearest hundredth.

Solution

The digit in the hundredths position is __7__.

The next digit to the right is __5__.

Since __5__ is equal to 5, change the __7__ to __8__ and replace __5__ and __3__ with 0s.

So, 0.6753 rounds to __0.68__ to the nearest hundredth (with two 0s dropped).

Now work margin exercise 9.

Completion Example Answers

9. The digit in the hundredths position is 7; The next digit to the right is 5; Since 5 is equal to 5, change the 7 to 8 and replace 5 and 3 with 0s; 0.6753 rounds to 0.68

Margin Exercise Answers

1. 19.3; nineteen and three tenths **2.** 39.0184; thirty-nine and one hundred eighty-four ten-thousandths **3. a.** 1200.0005 **b.** 0.1205 **4.** 6.44 **5.** 9.251 **6.** 2.06, 2.46, 2.6

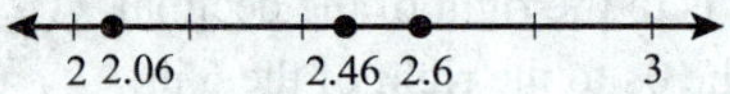

7. 9.3 **8.** 5.10 **9.** 2.395

3.1 Exercises

Concept Check

Fill-in-the-Blank. Complete each sentence using information found in this section.

1. When reading or writing decimal numbers, the decimal point is represented by the word __________.

2. The place value three places to the right of the decimal is the __________ place.

3. 2.08 would be written as __________ and __________ __________.

3.1 Exercises

Concept Check

1. and
2. thousandths
3. two, eight hundredths
4. place value
5. right
6. right
7. True
8. False; 92.586 is less than 92.6.
9. False; On a number line, any number to the right of another number is larger than that other number.

4. To compare two decimal numbers, move from left to right and compare digits with the same __________.

5. When rounding, all zeros to the __________ of the decimal point can be dropped, unless they are needed to emphasize (or show) the place of accuracy.

6. When rounding decimal numbers, look at the digit to the __________ of the place of accuracy to determine if the digit in the place of accuracy should remain the same or increase by 1.

True/False. Determine whether each statement is true or false. If a statement is false, explain how it can be changed so the statement will be true. (**Note:** There may be more than one acceptable change.)

7. Two hundred thousand, four hundred six and twelve hundredths can be written as 200,406.12.

8. 92.586 is greater than 92.6.

9. On a number line, any number to the left of another number is larger than that other number.

10. When a decimal number is rounded, all numbers to the right of the place of accuracy become zeros in the final answer.

Practice

Write each mixed number in decimal notation. See Examples 1 and 2.

1. $6\frac{5}{10}$
2. $8\frac{4}{10}$
3. $18\frac{76}{100}$
4. $2\frac{57}{100}$
5. $56\frac{3}{100}$
6. $13\frac{2}{100}$
7. $37\frac{498}{1000}$
8. $62\frac{547}{1000}$
9. $87\frac{3}{1000}$
10. $12\frac{37}{1000}$

Write each decimal number in words. See Examples 1 and 2.

11. 0.9
12. 0.5
13. 20.7
14. 96.3
15. 1.53
16. 2.79
17. 19.102
18. 18.051
19. 800.009
20. 500.005

Write each number in decimal notation. See Example 3.

21. three tenths
22. seven tenths
23. seventeen and nine tenths
24. eight hundred and three tenths
25. twenty-three hundredths
26. seventy-two hundredths
27. six and twenty-eight thousandths
28. fourteen and ninety-seven thousandths

10. False; When a decimal number is rounded, all numbers to the right of the place of accuracy become zeroes in the final answer, unless they are needed to show the place of accuracy.

Practice

1. 6.5
2. 8.4
3. 18.76
4. 2.57
5. 56.03
6. 13.02
7. 37.498
8. 62.547
9. 87.003
10. 12.037
11. Nine tenths
12. Five tenths
13. Twenty and seven tenths
14. Ninety-six and three tenths
15. One and fifty-three hundredths
16. Two and seventy-nine hundredths
17. Nineteen and one hundred two thousandths
18. Eighteen and fifty-one thousandths
19. Eight hundred and nine thousandths
20. Five hundred and five thousandths
21. 0.3
22. 0.7
23. 17.9
24. 800.3
25. 0.23
26. 0.72
27. 6.028
28. 14.097

29. 0.4502

30. 0.7165

31. 0.27

32. 0.48

33. 0.163

34. 4.537

35. 24.295

36. 110.241

37. 0.01

38. 4.002

39.

40.

41.

42.

43.

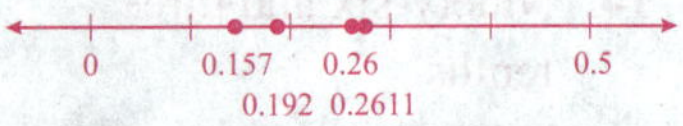

44.

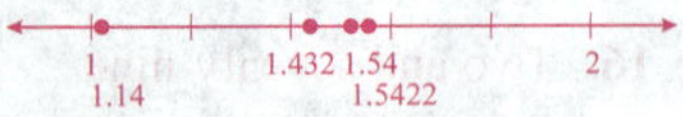

45. 7, 8, 8, 7, 8, 8, 34.8

46. 5, 3, 3, 5, 3, 70.5

47. 5, 2, 2, 5, 2, 3.0065

48. 1, 6, 6, 1, 2, 6, 8.0052

49. 8.6

50. 3.3

51. 10.0

52. 5.0

53. 1.68

54. 19.44

55. 0.08

56. 0.04

57. 0.057

58. 0.639

59. 3.003

29. four thousand five hundred two ten-thousandths

30. seven thousand one hundred sixty-five ten-thousandths

For each pair of decimal numbers, determine which number is larger. See Examples 4 and 5.

31. 0.26, 0.27

32. 0.45, 0.48

33. 0.153, 0.163

34. 4.537, 4.527

35. 23.521, 24.295

36. 110.241, 101.862

37. 0.01, 0.009

38. 4.002, 4.0008

Arrange each set of decimal numbers in order from smallest to largest. Then, graph the numbers on a number line. See Example 6.

39. 0.3, 0.03, 0.33

40. 0.55, 0.05, 0.5

41. 0.2, 0.26, 0.17

42. 1.8, 1.75, 1.86

43. 0.157, 0.2611, 0.192, 0.26

44. 1.432, 1.54, 1.14, 1.5422

Fill in the blanks to correctly complete each statement. See Examples 7 through 9.

45. Round 34.78 to the nearest tenth.

a. The digit in the tenths position is ___.

b. The next digit to the right is ___.

c. Since ___ is greater than 5, change ___ to ___ and replace ___ with 0.

d. So 34.78 rounds to _____ to the nearest tenth.

46. Round 70.53 to the nearest tenth.

a. The digit in the tenths position is ___.

b. The next digit to the right is ___.

c. Since ___ is less than 5, leave ___ as it is and replace ___ with 0.

d. So 70.53 rounds to _____ to the nearest tenth.

47. Round 3.00652 to the nearest ten-thousandth.

a. The digit in the ten-thousandths position is ___.

b. The next digit to the right is ___.

c. Since ___ is less than 5, leave ___ as it is and replace ___ with 0.

d. So 3.00652 rounds to _____ to the nearest ten-thousandth.

48. Round 8.00516 to the nearest ten-thousandth.

a. The digit in the ten-thousandths position is ___.

b. The next digit to the right is ___.

c. Since ___ is greater than 5, change ___ to ___ and replace ___ with 0.

d. So 8.00516 rounds to _____ to the nearest ten-thousandth.

Round each decimal number to the nearest tenth. See Example 7.

49. 8.555	**50.** 3.251	**51.** 9.961	**52.** 4.980

Round each decimal number to the nearest hundredth. See Example 9.

53. 1.677	**54.** 19.444	**55.** 0.0764	**56.** 0.0439

Round each decimal number to the nearest thousandth. See Example 8.

57. 0.0572	**58.** 0.6388	**59.** 3.00254	**60.** 19.01655

Round each decimal number to the nearest whole number (or nearest unit).

61. 4.41	**62.** 8.73	**63.** 29.999	**64.** 19.888

Round each decimal number to the nearest hundred.

65. 5163	**66.** 6475	**67.** 435.7	**68.** 263.9

Round each decimal number to the nearest thousand.

69. 103,499	**70.** 37,495	**71.** 50,766.4	**72.** 40,397.9

Applications

In each exercise, write the decimal numbers that are not whole numbers in words.

73. The tallest unicycle ever ridden was 114.8 feet tall, and was ridden by Sam Abrahams (with a safety wire suspended from an overhead crane) for a distance of 28 feet in Pontiac, Michigan, on January 29, 2004. [1]

74. A penny dated from 1959 through 1982 had an original weight of 3.11 grams. A penny dated 1983 or later had an original weight of 2.5 grams.

75. One yard is equal to 36 inches. One yard is also approximately equal to 0.914 meter. One meter is approximately equal to 1.09 yards. One meter is also approximately equal to 39.37 inches. (Thus, a meter is longer than a yard by about 3.37 inches.)

76. One foot is equal to 12 inches. One foot is also equal to 30.48 centimeters. One square foot is approximately 0.093 square meters.

77. One quart of water weighs approximately 2.0825 pounds.

78. The number π is approximately equal to 3.14159.

79. Euler's number e is approximately equal to 2.71828.

80. The largest state in the United States is Alaska, which covers approximately 656.4 thousand square miles. The second largest state is Texas, which covers approximately 268.6 thousand square miles. Alaska is more than 10 times the size of Wisconsin (twenty-third in size), which covers about 65.5 thousand square miles.

1 Source: semcycle.biz/record

60. 19.017
61. 4
62. 9
63. 30
64. 20
65. 5200
66. 6500
67. 400
68. 300
69. 103,000
70. 37,000
71. 51,000
72. 40,000

Applications

73. One hundred fourteen and eight tenths

74. Three and eleven hundredths; Two and five tenths

75. Nine-hundred fourteen thousandths; One and nine hundredths; Thirty-nine and thirty-seven hundredths; Three and thirty-seven hundredths

76. Thirty and forty-eight hundreds; Ninety-three thousandths

77. Two and eight-hundred twenty-five ten-thousandths

78. Three and fourteen thousand one hundred fifty-nine hundred-thousandths

79. Two and seventy-one thousand eight hundred twenty-eight hundred-thousandths

80. Six hundred fifty-six and four tenths; Two hundred sixty-eight and six tenths; Sixty-five and five tenths

81. Nine and fifty-eight hundredths; Nineteen and nineteen hundredths; Forty-three and three hundredths

82. Ninety-two and nine tenths; Sixty seven and twenty-four hundredths; Three hundred sixty-five and two tenths; Two hundred twenty-four and seven tenths

83. Thirty-five and eight tenths; Twenty-six and nine tenths; Eighteen and nine tenths; Twelve and three tenths; Seven and two tenths

Writing & Thinking

84. Answers will vary.

85. Answers will vary. Example: If the bill is $129, then that would be 129 and one hundred twenty-nine.

86. Numbers in word form are often used for official documents and/or on checks.

87. Moving left to right, compare digits with the same place value. When one compared digit is larger, the corresponding number is larger.

81. A few track world records are as follows: 9.58 seconds for 100 meters (by Usain Bolt, Jamaica, 2009); 19.19 seconds for 200 meters (by Usain Bolt, Jamaica, 2009); 43.03 seconds for 400 meters (by Wayde van Niekerk, South Africa, 2016). [2]

82. The mean distance from the Sun to Earth is about 92.9 million miles and from the Sun to Venus is 67.24 million miles. One period of revolution of the Earth about the Sun takes 365.2 days, and one period of revolution of Venus about the sun takes 224.7 days.

83. An interesting fact about aging is that the longer you live, the longer you can expect to live. A white male of age 40 can expect to live 35.8 more years; of age 50, can expect to live 26.9 more years; of age 60 can expect to live 18.9 more years; of age 70 can expect to live 12.3 more years; and of age 80 can expect to live 7.2 more years. (This same phenomenon is true of men and women of all races.)

Writing & Thinking

84. In your own words, state why the word "and" is so commonly misused when numbers are spoken and/or written. Bring an example of this from a newspaper, magazine, or television show to share with the class.

85. Suppose you are writing a check to pay your phone bill. How would you write the amount due in standard notation and in English word form?

86. Discuss situations where you think it is particularly appropriate (or necessary) to write numbers in English word form.

87. Using parts a. and b. as examples, explain in your own words how you can tell quickly when one decimal number is larger (or smaller) than another decimal number.

a. The decimal number 2.765274 is larger than the decimal number 2.763895.

b. The decimal number 17.345678 is larger than the decimal number 17.345578.

2 Source: trackandfieldnews.com/records/mens-world-records

3.2 Addition and Subtraction with Decimal Numbers

Objectives

A. Add decimal numbers.

B. Subtract decimal numbers.

A Addition with Decimal Numbers

Addition with decimal numbers can be accomplished by writing the decimal numbers one under the other and keeping the decimal points aligned vertically. In this way, whole numbers are added to whole numbers, tenths added to tenths, hundredths to hundredths, and so on. The decimal point in the sum is in line with the decimal points in the addends.

Decimal points are aligned vertically.

$$\begin{array}{rl} 6.15 & \text{Addend} \\ +3.42 & \text{Addend} \\ \hline 9.57 & \text{Sum} \end{array}$$

Any number of 0s may be written to the right of the last digit in the fractional part of a number to help keep the digits in the correct alignment. This will not change the value of any number or the value of the sum.

Adding Decimal Numbers

1. Write the numbers vertically.
2. Keep the decimal points aligned vertically.
3. Keep digits with the same place values aligned. Zeros may be written in to help keep the digits aligned properly.
4. Add, just as with whole numbers, keeping the decimal point in the sum aligned with the other decimal points.

PROCEDURE

Math Tip

When solving addition problems in your head, it is often easier to first sum any numbers that add to ten, and then add the remaining number(s). For example, 4 + 7 + 6 could be rearranged and simplified as

$$(4+6)+7=10+7=17.$$

Example 1 Adding Decimal Numbers

Add: 6.3 + 5.42 + 14.07

Solution

$$\begin{array}{r} {}^{1} \\ 6.30 \\ 5.42 \\ +14.07 \\ \hline 25.79 \end{array}$$

Decimal points are aligned vertically.

0 is written in to help keep digits aligned.

Now work margin exercise 1.

1. Add: 14.86 + 5.5 + 9.1

2. Add: 1.093 + 37 + 24.58

Example 2 Adding Decimal Numbers

Add: 56.2 + 85.75 + 29.001

Solution

$$\begin{array}{r} {}^{1\ 2} \\ 56.2\mathbf{00} \\ 85.75\mathbf{0} \\ +\ 29.001 \\ \hline 170.951 \end{array}$$

0s are written in to help keep digits aligned.

Now work margin exercise 2.

3. Add:
23.8 + 4.2567 + 11 + 3.01

Example 3 Adding Decimal Numbers

Add: 9 + 4.86 + 37.479 + 0.6

Solution

$$\begin{array}{r} {}^{2\ 1\ 1} \\ 9.\mathbf{000} \\ 4.86\mathbf{0} \\ 37.479 \\ +\ \ 0.6\mathbf{00} \\ \hline 51.939 \end{array}$$

The decimal point is understood to be to the right of 9, as in 9.0, 9.00, or 9.000.

0s are written in to help keep digits aligned.

Now work margin exercise 3.

4. Mr. Riley went to a thrift store and bought a rug for $50.79, a lamp for $33.68, and a night stand for $72.11. How much did he spend? (Tax was included in the prices.)

Example 4 Application: Adding Decimal Numbers

Mrs. Finn went to the local store and bought a pair of shoes for $42.50, a blouse for $25.60, and a skirt for $37.55. How much did she spend? (Tax was included in the prices.)

Solution

$$\begin{array}{r} {}^{1\ 1} \\ \$\ \ 42.50 \\ 25.60 \\ +\ \ 37.55 \\ \hline \$105.65 \end{array}$$

Mrs. Finn spent $105.65.

Now work margin exercise 4.

B Subtraction with Decimal Numbers

Subtracting Decimal Numbers

1. Write the numbers vertically.
2. Keep the decimal points aligned vertically.
3. Keep digits with the same place values aligned. Zeros may be written in to help keep the digits aligned properly.
4. Subtract, just as with whole numbers, keeping the decimal point in the difference aligned with the other decimal points.

PROCEDURE

Example 5 Subtracting Decimal Numbers

Subtract: 16.715 − 4.823

Solution

We subtract in a vertical format with the decimal points aligned vertically.

```
    16
  5 6̸ 11
 1 6̸.7̸ 1 5
−   4.8 2 3
 1 1.8 9 2
```

Now work margin exercise 5.

5. Subtract: 9.262 − 3.371

Example 6 Subtracting Decimal Numbers

Subtract: 21.2 − 13.716

Solution

```
   10 11 9
  1 0̸ 1̸ 10 10
 2̸ 1̸.2̸ 0̸ 0̸    0s are written in to help keep the digits aligned.
−1 3.7 1 6
   7.4 8 4
```

Now work margin exercise 6.

6. Subtract: 170.44 − 28.973

Example 7 Application: Adding and Subtracting Decimal Numbers

At the bookstore, Mrs. Gonzalez bought a textbook for $55, art supplies for $32.50, and computer supplies for $29.25. If sales tax was $9.34, how much change did she receive from a gift certificate worth $150?

7. Mr. Jones had $1000 in his checking account. He wrote checks for $280.25, $640.32, $54.19, and $16.56. How much does he have left in his checking account?

Solution

Step 1: READ: Read the problem carefully. Note that finding the amount of change she will receive will involve finding the difference between the value of the gift certificate and the total of her expenses.

Step 2: SET UP: Find the total of her expenses including sales tax.

$$\$55 + \$32.50 + \$29.25 + \$9.34$$

Then, subtract this total from the amount of the gift certificate, \$150.

Step 3: SOLVE: Add expenses.

$$\begin{array}{r} {}^{1\ 2\ 1} \\ \$\ \ 55.00 \\ 32.50 \\ 29.25 \\ +\quad 9.34 \\ \hline \$126.09 \end{array}$$ Total expenses

Now subtract the total expenses from \$150.

$$\begin{array}{r} {}^{4\ \ 9\ \ 9}_{\ \ 10\,10\,10} \\ \$150.00 \\ -\ 126.09 \\ \hline \$\ \ 23.91 \end{array}$$

She received \$23.91 in change.

Step 4: CHECK: By rounding the cost of each item (and the sales tax) to the nearest \$10, the total cost is approximately \$130. This means Mrs. Gonzalez will receive approximately \$20 in change. So the answer of \$23.91 seems reasonable.

Now work margin exercise 7.

Margin Exercise Answers

1. 29.46 **2.** 62.673 **3.** 42.0667 **4.** \$156.58 **5.** 5.891 **6.** 141.467 **7.** \$8.68

3.2 Exercises

Concept Check

Fill-in-the-Blank. Complete each sentence using information found in this section.

1. When adding decimal numbers, write the addends __________.

2. To help keep corresponding digits aligned, __________ may be written in place of missing place values.

3. Once the numbers and decimal points have been aligned, addition or subtraction can be performed the same as with __________ numbers.

4. For subtraction of decimal numbers, keep the decimal points aligned __________.

3.2 Exercises

Concept Check

1. vertically
2. zeros
3. whole
4. vertically

True/False. Determine whether each statement is true or false. If a statement is false, explain how it can be changed so the statement will be true. (**Note:** There may be more than one acceptable change.)

5. Vertical alignment is the preferred method of adding decimal numbers because digits with the same place value are easily lined up.

6. It is important to align the decimal points vertically when adding decimal numbers.

7. In subtracting decimal numbers, line up all the last digits vertically.

8. Once decimal points and corresponding digits have been aligned vertically, add or subtract from left to right.

Practice

Add. See Examples 1 through 3.

1. 42.08 + 8.005

2. 14.5 + 9.09

3. 1.3408 + 6.539

4. 5.0015 + 2.443

5. 3.07 + 596

6. 2.3 + 10.022

7. 152.3 + 4.005

8. 22.051 + 0.2006

9. 0.09 + 165.1 + 72.55

10. 3.8771 + 0.307 + 4.0086

11. 1.1 + 0.32 + 2.4 + 6.01

12. 3.2 + 0.39 + 1.004 + 4.205

13. 37.02 + 25 + 6.4 + 3.89

14. 5 + 2.37 + 436 + 10.88

15. 5 + 6.1 + 0.4

16. 8 + 0.7 + 4.2

17. 2.59 + 16.9 + 0.051

18. 2.48 + 51.22 + 10.734

19. 9 + 5.6 + 0.58 + 25.133

20. 3.766 + 9.33 + 14 + 206

5. True

6. True

7. False; In subtracting decimal numbers, line up the decimal points and corresponding digits vertically.

8. False; Once decimal points and corresponding digits have been aligned vertically, add or subtract from right to left.

Practice

1. 50.085
2. 23.59
3. 7.8798
4. 7.4445
5. 599.07
6. 12.322
7. 156.305
8. 22.2516
9. 237.74
10. 8.1927
11. 9.83
12. 8.799
13. 72.31
14. 454.25
15. 11.5
16. 12.9
17. 19.541
18. 64.434
19. 40.313
20. 233.096

21. 11.131
22. 15.216
23. 17.8
24. 35.6
25. 2.13
26. 3.52
27. 32.7
28. 43.4
29. 45.01
30. 64.06
31. 4.7974
32. 7.5903
33. 8.93
34. 11.79
35. 1.44
36. 15.89
37. 0.9757
38. 64.955
39. 4.895
40. 9.381
41. 28.475
42. 17.011
43. 2.9434
44. 2.9239
45. 17.5616
46. 20.6804
47. 10 cm
48. 12.56 in.

Subtract. See Examples 5 and 6.

21. 39.542
− 28.411

22. 46.918
− 31.702

23. 87.1
− 69.3

24. 63.4
− 27.8

25. 4.28 − 2.15

26. 8.73 − 5.21

27. 86.4 − 53.7

28. 57.2 − 13.8

29. 51.3
− 6.29

30. 72.6
− 8.54

31. 4.8
− 0.0026

32. 7.6
− 0.0097

33. 17.83 − 8.9

34. 14.59 − 2.8

35. 5.2 − 3.76

36. 29.5 − 13.61

37. 1.0057 − 0.03

38. 78.015 − 13.06

39. 22.418 − 17.523

40. 26.274 − 16.893

41. 31.009
− 2.534

42. 23.006
− 5.995

43. 4.
− 1.0566

44. 13.
− 10.0761

45. 18 − 0.4384

46. 23 − 2.3196

Find the perimeter of each figure.

47.

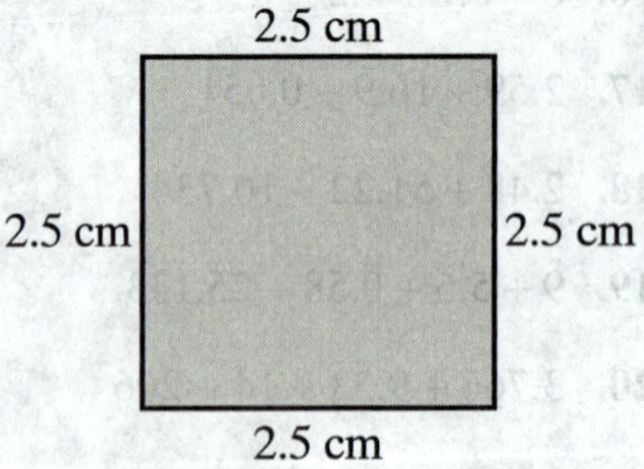

48.

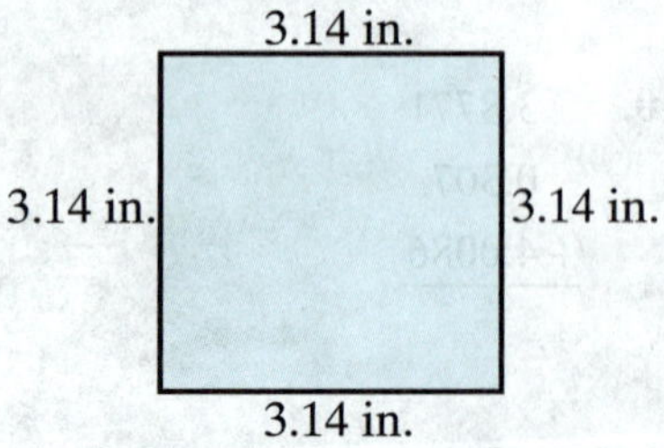

49.

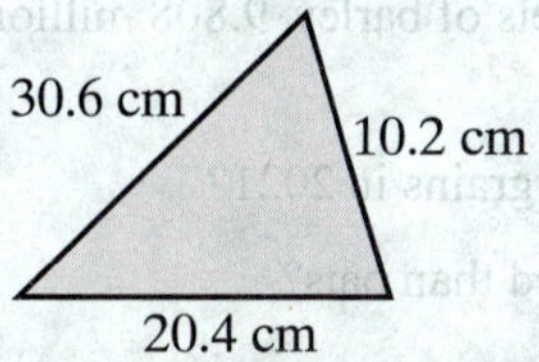

51.

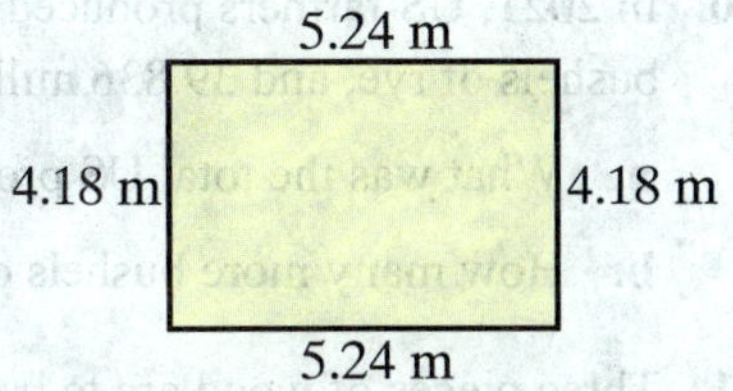

50.

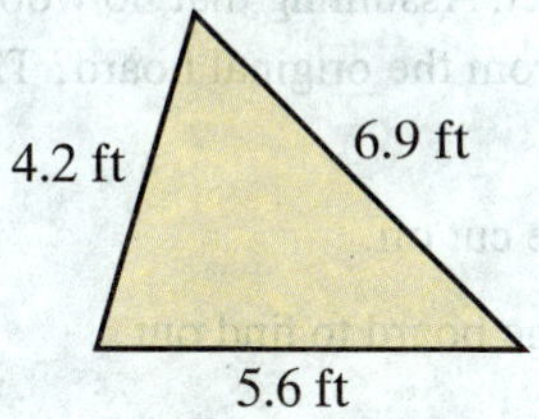

52.

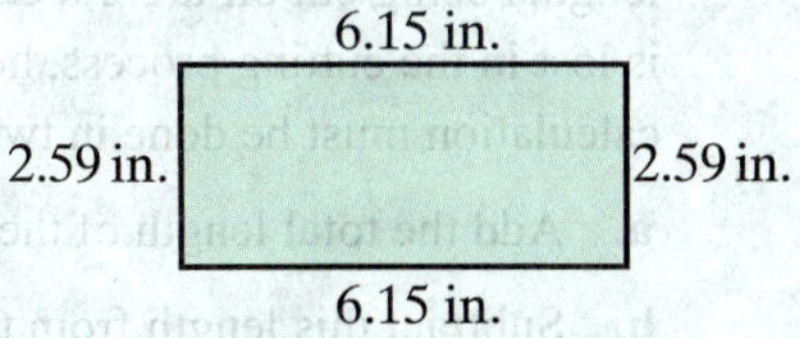

Applications

Solve.

53. Mr. Johnson bought the following items at a department store: slacks, \$32.50; shoes, \$43.75; shirt, \$18.60.

a. How much did he spend?

b. What was his change if he gave the clerk a \$100 bill? (Tax was included in the prices.)

54. Theresa got a haircut for \$38.00, a manicure for \$15.50, and left an \$8.00 tip.

a. How much did she spend?

b. How much change should she receive from \$70? (Tax was included in the prices.)

55. David is preparing a four-sided garden plot with unequal sides of 7.5 feet, 26.34 feet, 36.92 feet, and 12.07 feet. How many feet of edging material must he use? (This is the same as finding the perimeter of the plot.)

56. An architect's scale drawing shows a triangle that measures 2.25 in. on one side, 3.75 in. on the second side, and 4.5 in. on the third side. What is the perimeter of the triangle in the drawing?

57. Suppose your checking account shows a balance of \$253.70 at the beginning of the month. During the month you make deposits of \$635, \$279, and \$428, and you make debit purchases for \$124.60, \$89.55, \$680, and \$29.80. Find the end-of-the-month balance on your account.

58. Suppose your checking account shows a balance of \$382.35 at the beginning of the month. During the month, you make deposits of \$580.00, \$300.00, \$182.50, and \$45.00, and you make debit purchases for \$85.35, \$210.50, \$43.75, and \$650. Find the end-of-the-month balance in your account.

59. In 2021, US farmers raised the following amounts of livestock: 35.1 million calves, 3.2 million sheep, and 1.6 million goats.[1]

a. What was the total amount of these livestock in 2021?

b. How many more calves than sheep were there?

1 Source: US Department of Agriculture, National Agriculture Statistics Survey.

49. 61.2 cm

50. 16.7 ft

51. 18.84 m

52. 17.48 in.

Applications

53. **a.** \$94.85 **b.** \$5.15

54. **a.** \$61.50 **b.** \$8.50

55. 82.83 feet

56. 10.5 inches

57. \$671.75

58. \$500.25

59. **a.** 39.9 million

b. 31.9 million

60. a. 167.317 b. 77.837

61. a. 6.97 feet b. 2.73 feet

62. a. $1560.38

b. **$1058.03**

c. $502.35

63. a. 12.15 gallons

b. 3.25 gallons

64. a. $15.19 b. $4.81

65. $6089.90

66. 5.27 hours

60. In 2021, US farmers produced 117.673 million bushels of barley, 9.808 million bushels of rye, and 39.836 million bushels of oats.

a. What was the total US production of these three grains in 2021?

b. How many more bushels of barley were produced than oats?

61. Three pieces of wood are to be cut off of a board that is 9.7 feet long. The respective lengths being cut off are 2 feet, 3.12 feet, and 1.85 feet. Assuming that no wood is lost in the cutting process, how much is left over from the original board? This calculation must be done in two steps:

a. Add the total length of the three boards that were cut off.

b. Subtract this length from the original length of the board to find out what remains.

62. Arianna is an hourly worker who works a different number of hours each week. For a given month, she receives four paychecks as follows: $325.27, $450.83, $273.30, and $510.98. Each month she has to pay $352.78 for car payments, $650 for rent, and $55.25 for insurance. How much does Arianna have remaining for all of her other expenses? To find the answer, do the following three steps:

a. Compute her monthly pay.

b. Compute her total of the three listed monthly expenses.

c. Subtract her expenses from her monthly pay.

63. A group of people own three dairy cows to supply their own milk needs. During one milking, the first cow produced 2.79 gallons of milk, the second produced 5.34 gallons, while the third added 4.02 gallons. All of this milk was put into a 15.4 gallon holding bucket. [2]

a. How much milk did the three cows produce during this milking?

b. How much more milk can the bucket hold?

64. Kevin made the following purchases at the local grocery store: steak for $11.27, lettuce for $2.85, one tomato for $0.63, plus $0.44 for sales tax.

a. Compute the total cost.

b. If Kevin pays with a $20 bill, how much change will he receive?

65. Jazmin has saved $19,273.22 towards the purchase of a new car, and she will take out a loan for any extra cost. The agreed upon price of the car with its accessories is $23,925.50, and the taxes, registration fees, and insurance add up to $1437.62. How much must Jazmin borrow? (**Hint:** To solve this, first add up the total cost for the car and taxes, etc. Next, subtract Jazmin's savings to determine how much must be borrowed.)

66. A company has the policy of paying overtime for all hours over 37.75 worked during a week. The daily hours Marcos worked this week were, day 1: 8.33 hours; day 2: 6.2 hours; day 3: 9.87 hours; day 4: 8.5 hours; day 5: 10.12 hours. How many hours of overtime will Marcos receive this week? (**Hint:** First add the total number of hours worked during the week, and then subtract the threshold of 37.75 hours. The remainder will be the overtime hours.)

2 Source: www.raw-milk-facts.com/dairy_cow_breeds.html

67. A special steel alloy for knife blades consists of iron along with a mixture of other chemical elements. What is the total weight of the "ingredients" if 10 tons of iron is combined with 0.28 ton of carbon, 1.58 ton of chromium, 0.269 ton of vanadium, and 0.0488 ton of manganese? [3]

68. A college student is purchasing the necessary textbooks and supplies for an education course, which consist of the textbook for $144.37, the accompanying lesson plan guide book for $38.17, and a package of 4 by 6 inch index cards for $1.79. In addition, she is purchasing a candy bar for $0.71 and a piece of bubble gum for $0.06. How much will her total be?

69. A young child has a bunch of change in his pocket, and he wants to purchase a comic book which costs $2.26. If he has $1.25 in quarters, $0.40 in dimes, $0.35 in nickels, and $0.07 in pennies, does he have enough money to buy the comic book? If not, how much extra must he get so that he can buy it? (**Hint:** Add up the value of his coin collection. If it exceeds the cost, he has enough money. If not, subtract the cash on hand from the cost of the comic book.)

70. A college student is receiving assistance from two sources, which will be applied to the following expenses: $17,993.74 for tuition and fees, $7248.39 for room and board, and $1537.71 for books and supplies. If one of the sources of assistance supplies $9438.72 and the other $8300, how much must the student cover? (**Hint:** Add the total costs, add up the total assistance, and subtract this from the total cost to find how much the student must cover.)

71. A tightly-controlled timer is set at 759.99991 seconds.

a. What will its setting be if the time is decreased by 1.00082 seconds?

b. What will its setting be if the original time is increased by 1.00082 seconds?

72. An oven with very precise temperature control is 400.2 °F. If the temperature is reduced by 15.327 degrees, what will be the new temperature?

73. Four unequal lemons were squeezed for juice, and they put out 0.0043 liter, 0.00056 liter, 0.01002 liter, and 0.0097 liter respectively. Compute the total amount of juice collected.

74. Tim eats dinner at a restaurant. The appetizer costs $8.23, the entrée costs $21.78, the beverage costs $2.63, and the dessert costs $6.47 (the tax is included in these costs). If he gives a $7.50 tip, what is the total cost of the meal?

75. Raven sells her crafts at a farmers market each weekend. On Saturday, she sold items for $15.76, $17.24, and $34.22.

a. Compute her total revenue.

b. If Raven set a revenue goal of $100 for the weekend, how much revenue must she make on Sunday to reach her goal?

76. Miguel is participating in an online health challenge to walk 7.7 miles in one week. So far, he has walked 2.3 miles, 1.5 miles, 0.6 miles, and 1.2 miles.

a. How many miles has Miguel walked so far?

b. How many more miles must Miguel walk to complete the challenge?

67. 12.1778 tons

68. $185.10

69. No, he is short $0.19.

70. $9041.12

71. **a.** 758.99909 sec

b. 761.00073 sec

72. 384.873 degrees

73. 0.02458 liter

74. $46.61

75. **a.** $67.22

b. $32.78

76. **a.** 5.6 miles

b. 2.1 miles

3 Source: zknives.com/knives/steels/steelchart.ph

Writing & Thinking

77. Decimal numbers need to be aligned vertically so that numbers with the same place value are being added together. If not, then a 60 may be added to a 7 as if it were a 6 being added to a 7, giving 13, not the value of 67 that it should be.

78. Some decimal numbers have a different number of places beyond the decimal. To keep the places aligned, sometimes it becomes necessary to insert zeroes to the end of a decimal number.

79. **a.** Answers will vary.
b. Answers will vary.

80. **a.** Answers will vary.
b. Answers will vary.

81. Sophia did not align the decimal points in the addends vertically, so she was not adding corresponding place values.

Writing & Thinking

77. Why is it important that the decimal points and numbers be aligned vertically when adding or subtracting decimals?

78. Describe a problem where adding zeros might be helpful in adding or subtracting decimal numbers.

79. Suppose that you are given two decimal numbers with 0 as their whole number part.

a. Explain how the sum might be more than 1.

b. Explain why the sum cannot be more than 2.

80. Suppose you are given two decimal numbers with 0 as their whole number parts.

a. Explain why the difference would be less than 1.

b. Explain how the difference might be 0.

81. Sophia was asked to find the sum: 4.689 + 14.03. Her work is below. Explain why her answer is incorrect.

$$\begin{array}{r} \scriptstyle 1\ \ 1\ \ \ \ \\ 4.689 \\ +14.03 \\ \hline 6.092 \end{array}$$

3.3 Multiplication with Decimal Numbers

Objectives

A. Multiply decimal numbers.

B. Multiply by powers of 10.

A Multiplication with Decimal Numbers

Multiplication with decimal numbers is similar to multiplication with whole numbers. The difference is that you need to determine where to place the decimal point in the product. Two examples are shown here, in both fraction form and decimal form, to illustrate how the decimal point is to be placed in the product.

Products in Fraction Form	Products in Decimal Form	
$\frac{4}{10} \cdot \frac{6}{100} = \frac{24}{1000}$	0.4	← 1 decimal place in the factor
	× 0.0 6	← 2 decimal places in the factor
	0.0 2 4	← 3 decimal places in the product
$\frac{5}{1000} \cdot \frac{7}{100} = \frac{35}{100,000}$	0.0 0 5	← 3 decimal places in the factor
	× 0.0 7	← 2 decimal places in the factor
	0.0 0 0 3 5	← 5 decimal places in the product

As these examples indicate, there is **no need to keep the decimal points lined up for multiplication**.

The following rule states how to multiply decimal numbers and place the decimal point in the product.

Multiplying Decimal Numbers

1. Multiply the two numbers as if they were whole numbers.
2. Count the total number of digits to the right of the decimal points in both numbers being multiplied.
3. Place the decimal point in the product so that the number of digits to the right of the decimal point is the same as that found in Step 2. (0s may need to be inserted. See Example 3.)

PROCEDURE

Example 1 Multiplying Decimal Numbers

Multiply: $2.432 \cdot 5.1$

Solution

```
    2.4 3 2  ← 3 decimal places ⎫
×       5.1  ← 1 decimal place  ⎭ Total of 4 decimal places
    2 4 3 2
  1 2 1 6 0 0
  1 2.4 0 3 2  ← 4 decimal places in the product
```

Now work margin exercise 1.

1. Multiply: $3.813 \cdot 4.2$

2. Multiply: $(5.65)(1.14)$

Math Tip

Zeros at the end of a decimal number, like 54.810, can be dropped off the number because the value of the number will not change.

$$54.810 = 54.81$$

$$\frac{54{,}810}{1000} = \frac{5481}{100}$$

Example 2 Multiplying Decimal Numbers

Multiply: $4.35(12.6)$

Solution

```
      4.3 5  ←  2 decimal places ┐
 ×    1 2.6  ←  1 decimal place  ┘ Total of 3 decimal places
    -------
      2 6 1 0
      8 7 0 0
    4 3 5 0 0
    ---------
    5 4.8 1 0  ←  3 decimal places in the product
```

Now work margin exercise 2.

3. Multiply: $0.23(0.002)$

Math Tip

Zeros after the decimal at the beginning of a number, like 0.000322, are necessary because they are placeholders. The value of the number would change if we removed them.

$$0.000322 \neq 322$$

$$\frac{322}{1{,}000{,}000} \neq \frac{322}{1}$$

Example 3 Multiplying Decimal Numbers

Multiply: $(0.046)(0.007)$

Solution

```
        0.0 4 6  ←  3 decimal places ┐
 ×      0.0 0 7  ←  3 decimal place  ┘ Total of 6 decimal places
    -----------
    0.0 0 0 3 2 2  ←  6 decimal places in the product
```

Note that three 0s are inserted between the 3 and the decimal point in the product to get a total of 6 decimal places.

Now work margin exercise 3.

4. Multiply: $0.29 \cdot 11.4$

Completion Example 4 Multiplying Decimal Numbers

Multiply: $3.4 \cdot 5.8$

Solution

```
      3.4  ←  1 decimal place(s) ┐
 ×    5.8  ←  1 decimal place(s) ┘ Total of 2 decimal place(s)
    -----
    2 7 2
  1 7 0 0
  -------
  1 9.7 2     2 decimal place(s) in the product
```

Now work margin exercise 4.

5. You plan to buy fertilizer to spread on your garden. The garden is 15 feet long and 8.25 feet wide. What is the area that needs to be fertilized?

Example 5 Application: Multiplying Decimal Numbers

You are going to paint an accent wall in your living room. The wall measures 9 feet high by 12.25 feet long. What is the area you are going to paint?

Solution

Find the area of the wall by multiplying the length by the width.

$$\begin{array}{r} 12.25 \\ \times \quad 9 \\ \hline 110.25 \end{array}$$

12.25 ← 2 decimal places
× 9 ← 0 decimal places
} Total of 2 decimal places
110.25 ← 2 decimal places in the product

The area of the accent wall is 110.25 ft^2.

Now work margin exercise 5.

B Multiplying by Powers of 10

As discussed in Section 1.3, the **powers of 10** are

1, 10, 100, 1000, 10,000, 100,000, and so on.

Multiplication by powers of 10 can be accomplished mentally by using the following general guidelines. (Remember that with whole numbers the decimal point is understood to be just to the right of the ones digit.)

Multiplying by Powers of 10 (10, 100, 1000, and so on)

1. Count the number of 0s in the power of 10.
2. Move the decimal point to the right the same number of places as the number of 0s found in Step 1.

Multiplication by **10** moves the decimal point **one** place **to the right**.
Multiplication by **100** moves the decimal point **two** places **to the right**.
Multiplication by **1000** moves the decimal point **three** places **to the right**.
And so on.

PROCEDURE

Math Tip

The powers of 10 can be written using exponents.

$10^0 = 1$, $10^1 = 10$, $10^2 = 100$, $10^3 = 1000$, $10^4 = 10{,}000$

Notice that the number of zeros is the same as the exponent. Multiplying by 10^1, or 10, moves the decimal 1 place to the right. Multiplying by 10^4, or 10,000, moves the decimal 4 places to the right. So multiplying by a power of 10 moves the decimal point to the right the same number of places as the exponent.

Example 6 Multiplying by Powers of 10

The following products illustrate multiplying by powers of 10.

a. $10(1.59) = 15.9$ Move the decimal point 1 place to the right.

b. $100(2.68) = 268. = 268$ Move the decimal point 2 places to the right.

c. $1000(0.9653) = 965.3$ Move the decimal point 3 places to the right.

d. $10{,}000(7.2) = 72{,}000. = 72{,}000$ Move the decimal point 4 places to the right.

Now work margin exercise 6.

6. Multiply.

a. $10(0.13)$

b. $100(9.57)$

c. $1000(8.78)$

d. $10{,}000(5.8)$

Completion Example Answers

4. 1 decimal place, 1 decimal place, total of 2 decimal places; 19.72; 2 decimal places in the product

Margin Exercise Answers

1. 16.0146 **2.** 6.441 **3.** 0.00046 **4.** 3.306 **5.** 123.75 ft^2 **6. a.** 1.3 **b.** 957 **c.** 8780 **d.** 58,000

3.3 Exercises
Concept Check
1. whole numbers
2. 5
3. 3, right
4. 10
5. 10, 100, 1000
6. False; The decimal points do not need to be aligned vertically when multiplying decimal numbers.
7. True
8. True
9. False; Multiplying by 100 requires that the decimal point be moved 2 places to the right.

Practice
1. 0.42
2. 0.24
3. 0.04
4. 0.09
5. 21.6
6. 38.4
7. 0.42
8. 0.9
9. 0.004
10. 0.009
11. 0.108
12. 0.161
13. 0.073
14. 0.096
15. 0.0276
16. 0.0195
17. 0.0486
18. 0.0568
19. 0.0034
20. 0.0025
21. 0.3375
22. 0.4852
23. 1.500
24. 2.600
25. 1.4000

3.3 Exercises

Concept Check

Fill-in-the-Blank. Complete each sentence using information found in this section.

1. The first step in multiplying decimal numbers is to multiply the two numbers as if they were ________ ________.
2. If 7.16 is multiplied by 13.497, the answer will have ____ decimal places in it.
3. When multiplying by 1000, move the decimal point ______ place(s) to the _______.
4. Move the decimal point one place to the right when multiplying by ______.
5. The first 3 powers of 10 are ___, ___, and ___.

True/False. Determine whether each statement is true or false. If a statement is false, explain how it can be changed so the statement will be true. (**Note:** There may be more than one acceptable change.)

6. The decimal points should be aligned vertically when multiplying decimal numbers.
7. The first step in the multiplication of decimal numbers is to multiply the numbers as if they were whole numbers.
8. When multiplying decimal numbers, the answer should have the same number of decimal places as the total number of decimal places in the numbers being multiplied.
9. Multiplying by 100 requires that the decimal point be moved 100 places to the right.

Practice

Multiply. See Examples 1 through 4.

1. $(0.6)(0.7)$
2. $(0.3)(0.8)$
3. $(0.2)(0.2)$
4. $(0.3)(0.3)$
5. $8(2.7)$
6. $4(9.6)$
7. $(1.4)(0.3)$
8. $(1.5)(0.6)$
9. $(0.2)(0.02)$
10. $(0.3)(0.03)$
11. $(5.4)(0.02)$
12. $(2.3)(0.07)$
13. $(7.3)(0.01)$
14. $(3.2)(0.03)$
15. $(0.23)(0.12)$
16. $(0.15)(0.13)$
17. $(8.1)(0.006)$
18. $(7.1)(0.008)$
19. $(0.34)(0.01)$
20. $(0.25)(0.01)$
21. $3(0.1125)$
22. $4(0.1213)$
23. $4(0.375)$
24. $8(0.325)$
25. $1.6 \cdot 0.875$
26. $3.2 \cdot 0.375$
27. $5.3 \cdot 0.75$
28. $6.9 \cdot 0.25$
29. $4.8 \cdot 0.25$
30. $2.4 \cdot 0.75$
31. $0.83 \cdot 6.1$
32. $8.1 \cdot 0.72$
33. $0.27 \cdot 0.24$

34. $0.36 \cdot 0.17$

35. $0.36 \cdot 0.5$

36. $0.28 \cdot 0.5$

37. $\begin{array}{r} 6.884 \\ \times 9.5 \\ \hline \end{array}$

38. $\begin{array}{r} 5.392 \\ \times 6.5 \\ \hline \end{array}$

39. $\begin{array}{r} 0.08 \\ \times 0.542 \\ \hline \end{array}$

40. $\begin{array}{r} 0.833 \\ \times 0.04 \\ \hline \end{array}$

Multiply mentally by using your knowledge of multiplication by powers of 10. See Example 6.

41. $10(0.435)$

42. $10(0.186)$

43. $100(3.46)$

44. $100(7.82)$

45. $100(16.1)$

46. $100(38.2)$

47. $1000(4.1782)$

48. $1000(9.7521)$

49. $1000(0.38)$

50. $1000(0.47)$

51. $1000(0.0079)$

52. $1000(0.0092)$

53. $10{,}000(0.00615)$

54. $10{,}000(0.01756)$

55. $10{,}000(0.005)$

56. $10{,}000(0.0043)$

Applications

Solve.

57. Find the perimeter and area of a square with sides 3.2 ft long.

58. Find the perimeter and area of a square with sides 4.7 mm long.

59. Find the perimeter and area of a rectangle with length 10.5 cm and width 2.8 cm.

60. Find the perimeter and area of a rectangle with length of 18.04 m and width 6.25 m.

61. To buy a car, you can pay $2036.50 in cash, or you can put down $400 and make 18 monthly payments of $104.30. How much would you save by paying cash?

62. To buy a new washer and dryer, you can pay $1737.83 in cash, or you can put $350 down and make 12 monthly payments of $129.54. How much would you save by paying cash?

63. At one point, Ohio funded its public libraries with $39.87 spent for each person. How much funding would have been received that year by a library that served a town of 23,500 people?

64. In March of 2022, the average price per kilogram paid by the United States for beef imported from Australia and New Zealand was $6.25. At this price, what would be the value of 20,500 kilograms of beef?

65. The total cost of a purchase is $583.47. If the sales tax is 0.065 of the purchase price, how much will the customer have to pay? (**Hint:** To calculate the sales tax, multiply the purchase price by the tax rate.) Round your answer to the nearest cent.

26. 1.2000
27. 3.975
28. 1.725
29. 1.200
30. 1.800
31. 5.063
32. 5.832
33. 0.0648
34. 0.0612
35. 0.18
36. 0.14
37. 65.3980
38. 35.048
39. 0.04336
40. 0.03332
41. 4.35
42. 1.86
43. 346
44. 782
45. 1610
46. 3820
47. 4178.2
48. 9752.1
49. 380
50. 470
51. 7.9
52. 9.2
53. 61.5
54. 175.6
55. 50
56. 43

Applications

57. 12.8 ft; 10.24 ft^2
58. 18.8 mm; 22.09 mm^2
59. 26.6 cm; 29.40 cm^2
60. 48.58 m; 112.7500 m^2
61. $240.90
62. $166.65
63. $936,945
64. $128,125
65. $621.40

66. $29,099.34

67. 2.8926 in.

68. 25.7 millimeters

69. $932.60

70. $744.55

71. Discount: $45.30; Total: $288.67

72. 93 points

73. $1320

74. 820 grams

Writing & Thinking

75. In multiplication with decimal numbers, placement of the decimal point must be considered. Otherwise, multiplication with whole numbers and decimal numbers are the same.

76. Answers will vary.

66. The list price for a particular automobile with accessories is $35,487. If the customer will pay cash with no trade-in, the dealer will give a discount of 0.18 times the list price. What will the car cost as a straight cash transaction? This cash value price will be the list price minus the discount.

67. A hardbound textbook has 257 pages, each of which is 0.0098 inches thick. The front and back cover are each 0.187 inches thick. How thick is the book?

68. A Blu-ray disc has a thickness of 1.2 millimeters. Every stack of Blu-ray discs is wrapped in packaging that adds 3.25 millimeters to both the top and the bottom of the stack. How thick is a stack of 16 Blu-ray discs with the packages?

69. The salesperson at a clothing store receives a salary of $375.50 plus a weekly commission of 0.18 times all sales over $850 for the week. How much will that person earn if he sells $3945 that week? (**Hint:** Compute how much over the threshold of $850 the total sales were, and then multiply it by the rate of commission. Then add his salary to that value.)

70. Geoffrey works at a job with regular pay of $12.50 an hour, up to 40 hours in a week. Any work which exceeds 40 hours is paid $18.25 an hour. If he works for 53.4 hours, what will his total pay be? (**Hint:** First compute the pay for the first 40 hours; second, compute how many hours of overtime worked; third compute the overtime pay; and fourth, add the regular and overtime pay.)

71. You purchase four concert tickets that cost $75.50 each and use a code to receive a discount of 0.15 times the ticket price (before fees and taxes). The taxes and fees for the purchase total $31.97. Determine the value of the discount and the total amount paid for the tickets.

72. The price of gas is $3.279 and you add 11.3 gallons of gas to your car's gas tank. With your fuel rewards card, you earn 2.5 points for every $1 you spend on gas. How many reward points will you earn from this purchase? (Round your answer to the nearest whole number.)

73. A school district plans to purchase 1000 three-ring binders for the school year. If the binders will cost the school $1.32 each, what is the total cost of the binders?

74. A craft supply company sells glass beads in packages of 10,000. Each bead weighs approximately 0.03 grams and the packaging weighs 520 grams. What is the total weight of the package?

Writing & Thinking

75. In your own words, discuss the similarities and differences between multiplication with whole numbers and multiplication with decimal numbers.

76. Discuss, briefly, situations in which you might use multiplication with decimal numbers in your daily life.

3.4 Division with Decimal Numbers

Objectives

A. Divide decimal numbers.

B. Calculate the average amount per unit.

C. Divide by powers of 10.

A Division with Decimal Numbers

The process of division (called the **division algorithm**) with decimal numbers is, in effect, the same as division with whole numbers. With whole numbers, we are concerned with the remainder. But, with decimal numbers, we are concerned with the placement of the decimal point in the quotient. For example, consider 950 ÷ 40 as shown here.

$$\begin{array}{rrl} & 23. & \leftarrow \text{Quotient} \\ \text{Divisor} \rightarrow & 40\overline{)950.} & \leftarrow \text{Dividend} \\ & \underline{-80} & \\ & 150 & \\ & \underline{-120} & \\ & 30 & \leftarrow \text{Remainder} \end{array}$$

By adding 0s onto the dividend, we can continue to divide and get a decimal quotient rather than a whole number with a remainder.

$$\begin{array}{rrl} & 23.75 & \leftarrow \text{Quotient is a decimal number} \\ \text{Divisor} \rightarrow & 40\overline{)950.00} & \leftarrow \text{0s added on} \\ & \underline{-80} & \\ & 150 & \\ & \underline{-120} & \\ & 300 & \\ & \underline{-280} & \\ & 200 & \\ & \underline{-200} & \\ & 0 & \end{array}$$

If the divisor is a decimal number rather than a whole number, multiply both the divisor and dividend by the smallest power of 10 that will make the new divisor a whole number. For example, we can write

$$6.2\overline{)63.86} \quad \text{as} \quad \frac{63.86}{6.2} \cdot \frac{10}{10} = \frac{638.6}{62}.$$

This means that

$$6.2\overline{)63.86} \quad \text{is the same as} \quad 62\overline{)638.6}.$$

Similarly, we can write

$$1.23\overline{)4.6125} \quad \text{as} \quad \frac{4.6125}{1.23} \cdot \frac{100}{100} = \frac{461.25}{123} \quad \text{or} \quad 123\overline{)461.25}.$$

Dividing Decimal Numbers

1. Move the decimal point in the divisor to the right so that the divisor is a whole number.
2. Move the decimal point in the dividend the same number of places to the right.
3. Place the decimal point in the quotient directly above the new decimal point in the dividend.
4. Divide just as with whole numbers. (0s may be added as needed to the dividend.)

PROCEDURE

1. Divide. $9.36 \div 0.5$

Example 1 Dividing Decimal Numbers

Divide. $64.17 \div 6.2$

Solution

Step 1: Write down the numbers.

$$6.2\overline{)64.17}$$

Steps 2 and 3: Move both decimal points one place to the right so that the divisor becomes a whole number. Then place the decimal point in the quotient.

$$\begin{array}{r} . \\ 6.2.\overline{)64.1.7} \end{array}$$ ← Decimal point in quotient

Step 4: Proceed to divide as with whole numbers.

$$\begin{array}{r} 10.35 \\ 62.\overline{)641.70} \\ -62 \\ \hline 21 \\ -0 \\ \hline 217 \\ -186 \\ \hline 310 \\ -310 \\ \hline 0 \end{array}$$ ← Add 0s as needed.

Check

As with division with whole numbers, checking can be performed by multiplying the quotient by the divisor and adding the remainder. The result must be the dividend. In this example, $6.2 \cdot 10.35 + 0 = 64.17$.

Now work margin exercise 1.

Note

In moving the decimal point, you are multiplying both the divisor and dividend by the same power of 10.

Be sure to place the decimal point in the quotient **before actually dividing.**

There must be a digit in the quotient above every digit to the right of the decimal point in the dividend.

Completion Example 2 Dividing Decimal Numbers

Divide. 24.225 ÷ 4.25

Solution

$$\begin{array}{r} 5.\underline{7} \\ 4.25.\overline{)\,24.22.5} \\ -\underline{2125.0} \\ \underline{2975} \\ -\underline{2975} \\ \underline{0} \end{array}$$

← Decimal point in quotient

Now work margin exercise 2.

2. Divide. 14.872 ÷ 2.86

In each of the preceding examples, the remainder was eventually 0. This will not always be the case. We generally agree to some place of accuracy for the quotient before the division is performed. If the remainder is not 0 by the time this place of accuracy is reached in the quotient, then we divide to one more decimal place and round the quotient to the desired place of accuracy. 0s may need to be inserted as placeholders in the dividend.

Math Tip

The place of accuracy is different for different situations and measurements. Situations that deal with money like sales tax and discounts are rounded to the nearest hundredth. Measurements such as your body temperature and weight are usually rounded to the nearest tenth.

Dividing When the Remainder is Not 0

1. Decide how many decimal places are to be in the quotient.
2. Divide until the quotient has been calculated to one place to the right of the place of desired accuracy.
3. Using this last digit, round the quotient to the desired place of accuracy.

PROCEDURE

Terminating and Nonterminating Decimal Numbers

If the remainder is eventually 0, the decimal number is said to be **terminating**. For example, 14.2 and 0.3425 are terminating decimal numbers.

If the remainder is not eventually 0, the decimal number is said to be **nonterminating**. For example, 1.33333… is a nonterminating decimal number.

DEFINITION

Example 3 Dividing Decimal Numbers

Divide (to the nearest tenth). 82.3 ÷ 2.9

Solution

Divide until the quotient has been calculated to the hundredths place (place to the right of the tenths place), then round to the nearest tenth.

3. Divide (to the nearest tenth). 83.5 ÷ 5.6

```
                 28.37 ≈ 28.4      Hundredths
                                   Read "approximately equals"
                                   Round to the nearest tenth.
2.9.) 82.3.00                      Add 0s as needed.
     -58
      243
     -232
        110
        -87
         230
        -203
          27
```

Thus, $82.3 \div 2.9 \approx 28.4$. Accurate to the nearest tenth

Now work margin exercise 3.

4. Divide (to the nearest hundredth). $14.393 \div 9.3$

Example 4 Dividing Decimal Numbers

Divide (to the nearest hundredth). $1.83 \div 4.1$

Solution

Divide until the quotient has been calculated to the thousandths place (one place to the right of the hundredths place), then round to the nearest hundredth.

```
                 0.446 ≈ 0.45      Read "approximately equals"
                                   Round to the nearest hundredth.
4.1.) 1.8.300                      Add 0s as needed.
     -164
       190
      -164
        260
       -246
         14
```

Thus, $1.83 \div 4.1 \approx 0.45$. Accurate to the nearest hundredth

Now work margin exercise 4.

B Average Amount per Unit

We know from Section 1.6 that the **average** of a set of numbers can be found by adding the numbers and then dividing the sum by the number of numbers in the set. The term **average** is used in phrases such as "an average speed of 43 miles per hour" or "the average price of a pair of shoes." These averages can also be found by division. If the total amount of a quantity (distance, dollars, gallons of gas, etc.) and a number of units (time, items bought, miles, etc.) are known, then we can find the **average amount per unit** by dividing the total amount by the number of units.

Example 5 Application: Calculating Average Amount per Unit

The gas tank of a car holds 18.5 gallons of gasoline. Determine how many miles per gallon the car averages if it will go 518 miles on one tank of gas.

Solution

Divide to find the average number of miles per gallon.

$$\begin{array}{r} 28.0 \\ 18.5.\overline{)\,518.0.0} \\ -370 \\ \hline 1480 \\ -1480 \\ \hline 0 \end{array}$$

Miles per gallon

The car averages 28 miles per gallon.

Now work margin exercise 5.

If an average amount per unit is known, then a corresponding total amount can be found by multiplying by the number of units. For example, if you ride your bicycle at an average speed of 15.2 miles per hour, then the distance you travel over a certain period of time can be found by multiplying your average speed by the time you spend riding in hours.

Example 6 Application: Calculating Total Distance Traveled

If you ride your bicycle at an average speed of 15.2 miles per hour, how far will you ride in 3.5 hours?

Solution

Multiply the average speed by the number of hours.

$$\begin{array}{rl} 15.2 & \text{Miles per hour} \\ \times\quad 3.5 & \text{Hours} \\ \hline 760 & \\ 4560 & \\ \hline 53.20 & \text{Miles} \end{array}$$

You will ride 53.2 miles in 3.5 hours.

Now work margin exercise 6.

C Dividing by Powers of 10

In Section 3.3, we found that multiplication by powers of 10 can be accomplished by moving the decimal point to the right. Division by powers of 10 can be accomplished by moving the decimal point to the left.

5. If a sprinter runs 100 meters in 12.5 seconds, what is his average speed in meters per second?

Math Tip

Recall that *per* means to divide. For example, miles per gallon means miles divided by gallons.

6. On average, you can store about 277.75 songs per gigabyte of memory. How many songs could you save on an MP3 player with 8 gigabytes of memory?

Math Tip

The notation $\frac{\text{miles}}{\text{hour}}$ is read as miles per hour. Recall that *per* means divide. So $\frac{\text{miles}}{\text{hour}}$ means miles divided by hours.

If you know that you are traveling $\frac{50 \text{ miles}}{\text{hour}}$, and you want to know how many miles you will travel in 3 hours, you multiply:

$$\frac{50 \text{ miles}}{1 \cancel{\text{hour}}} \cdot 3 \cancel{\text{hours}} = 150 \text{ miles.}$$

Dividing a Decimal Number by a Power of 10 (10, 100, 1000, and so on)

1. Count the number of 0s in the power of 10.
2. Move the decimal point to the left the same number of places as the number of 0s found in Step 1.

Division by **10** moves the decimal point **one** place **to the left**.
Division by **100** moves the decimal point **two** places **to the left**.
Division by **1000** moves the decimal point **three** places **to the left**.
And so on.

PROCEDURE

Two general guidelines will help you to understand how to work with powers of 10.

1. Multiplication by a power of 10 will make a number larger, so move the decimal point to the right.
2. Division by a power of 10 will make a number smaller, so move the decimal point to the left.

7. Find each quotient.

a. $1.6 \div 10$

b. $83.46 \div 1000$

c. $732 \div 100$

Example 7 Dividing by Powers of 10

The following quotients illustrate dividing by powers of 10.

a. $5.23 \div 100 = \frac{5.23}{100} = 0.0523$ Move the decimal point 2 places to the left.

b. $817 \div 10 = \frac{817}{10} = 81.7$ Move the decimal point 1 place to the left.

c. $5.933 \div 1000 = \frac{5.933}{1000} = 0.005933$ Move the decimal point 3 places to the left.

Now work margin exercise 7.

Completion Example Answers

2.
$$\begin{array}{r} 5.7 \\ 425\overline{)2422.5} \\ \underline{-2125.0} \\ 2975 \\ \underline{-2975} \\ 0 \end{array}$$

Margin Exercise Answers

1. 18.72 **2.** 5.2 **3.** 14.9 **4.** 1.55 **5.** 8 meters per second **6.** 2222 songs **7. a.** 0.16 **b.** 0.08346 **c.** 7.32

3.4 Exercises

Concept Check

Fill-in-the-Blank. Complete each sentence using information found in this section.

1. If the divisor is a decimal number, multiply both the divisor and dividend by a power of 10 so that the divisor is a/an ________ number.
2. If the decimal point is moved 3 places to the right in the divisor, then the decimal point in the ________ must also be moved 3 places to the right.
3. Once the divisor is a whole number and the decimal point is in the proper place in the quotient, divide as with ________.
4. If a remainder is eventually 0, the decimal number is said to be a/an ________ decimal number.
5. When dividing by a power of 10, move the decimal point to the _____ the same number of places as there are zeros in the power of 10.

True/False. Determine whether each statement is true or false. If a statement is false, explain how it can be changed so the statement will be true. (**Note:** There may be more than one acceptable change.)

6. The first step in division with decimal numbers is to move the decimal point in the divisor to the right so that the divisor is a whole number.
7. Moving the decimal point in a divisor requires that the decimal point also be moved in the dividend.
8. The decimal point should be placed in the quotient before actually dividing.
9. Dividing by a power of 10 involves dividing by 10, 20, 30, 40, etc.

Practice

Divide. See Examples 1 and 2.

1. $4.68 \div 2$
2. $9.36 \div 3$
3. $1.71 \div 3$
4. $1.36 \div 4$
5. $1.62 \div 9$
6. $4.95 \div 5$
7. $0.064 \div 0.8$
8. $0.063 \div 0.7$
9. $4.484 \div 1.9$
10. $7.336 \div 1.4$
11. $82.24 \div 0.04$
12. $37.02 \div 0.03$
13. $16.35 \div 2.5$
14. $30.94 \div 6.5$
15. $48 \div 2.4$
16. $168 \div 5.6$

Divide and round to the nearest tenth. See Example 3.

17. $8\overline{)455}$
18. $8\overline{)261}$
19. $\dfrac{6.538}{9.4}$
20. $\dfrac{3.927}{5.3}$
21. $4.6 \div 5$
22. $5.7 \div 8$

3.4 Exercises

Concept Check

1. whole
2. dividend
3. whole numbers
4. terminating
5. left
6. True
7. True
8. True
9. False; Dividing by a power of 10 involves dividing by 10, 100, 1000, etc.

Practice

1. 2.34
2. 3.12
3. 0.57
4. 0.34
5. 0.18
6. 0.99
7. 0.08
8. 0.09
9. 2.36
10. 5.24
11. 2056
12. 1234
13. 6.54
14. 4.76
15. 20
16. 30
17. 56.9
18. 32.6
19. 0.7
20. 0.7
21. 0.9
22. 0.7

23. 728.7
24. 214.1
25. 21
26. 56.4
27. 0.01
28. 0.01
29. 5.78
30. 3.74
31. 12.12
32. 4.86
33. 0.56
34. 0.22
35. 3.8
36. 16.7
37. 13.81
38. 4.231
39. 0.785
40. 0.682
41. 0.794
42. 0.689
43. 0.5036
44. 0.08796

23. $43.721 \div 0.06$

24. $12.847 \div 0.06$

25. $1.62\overline{)34}$

26. $1.33\overline{)75}$

Divide and round to the nearest hundredth. See Example 4.

27. $\dfrac{0.1463}{24}$

28. $\dfrac{0.2249}{23}$

29. $0.42753 \div 0.074$

30. $0.2433 \div 0.065$

31. $1.23\overline{)14.91129}$

32. $3.14\overline{)15.25631}$

33. $9\overline{)5}$

34. $9\overline{)2}$

Divide mentally by using your knowledge of division by powers of 10. See Example 7.

35. $\dfrac{38}{10}$

36. $\dfrac{167}{10}$

37. $\dfrac{138.1}{10}$

38. $\dfrac{42.31}{10}$

39. $\dfrac{7.85}{10}$

40. $\dfrac{6.82}{10}$

41. $79.4 \div 100$

42. $68.9 \div 100$

43. $50.36 \div 100$

44. $87.96 \div 100$

45. $316.4963 \div 100$

46. $125.725 \div 100$

47. $18.6 \div 1000$

48. $96.4 \div 1000$

49. $45.621 \div 1000$

50. $10.413 \div 1000$

51. $1173.85 \div 1000$

52. $98{,}716.742 \div 1000$

53. $\dfrac{1.54}{10{,}000}$

54. $\dfrac{169.9}{10{,}000}$

Batter, Batter, Swing!

The Houston Astros team batting average of .267 was the best of the 2021 season. The Astros had a great season that year and eventually played against the Atlanta Braves in the World Series. The Astros lost to the Braves in 2021, but went on to win the World Series in 2022.

Applications

Solve.

55. If four new tires with custom rims cost \$958.24, what did each individual tire with rim cost?

56. If you bought 6 books for a total price of \$142.98, what average amount did you pay per book, including tax?

57. If the total price of a stereo was \$312.70 including a tax of 0.06 times the list price, you can find the list price by dividing the total price by 1.06. What was the list price? (**Note:** 1.06 represents the list price plus 0.06 times the list price.)

58. If the total price of a tablet PC was \$266.43 including a tax of 0.07 times the list price, you can find the list price by dividing the total price by 1.07. What was the list price? (**Note:** 1.07 represents the list price plus 0.07 times the list price.)

59. In 2021, the Houston Astros had a team batting average of .267 and 1496 hits. Find the number of team at bats, to the nearest whole number, by dividing hits by batting average.

60. In 2021, Julio Urias of the Los Angeles Dodgers had 20 wins and a 0.87 wining percentage. Find Julio Urias's total number of pitching decisions (games won or lost), to the nearest whole number, by dividing wins by winning percentage.

61. Walter Payton played football for the Chicago Bears for 13 years. In those years he carried the ball 3838 times for a total of 16,726 yards. What was his average yardage per carry (to the nearest tenth)?

62. If a car travels 330 miles on 15 gallons of gas, how many miles per gallon does it average?

63. A professor has graded a test of five students, and their scores were 76.4, 100, 84.7, 10.2, and 68.3. What is the average of these five scores?

64. Suppose that the total interest paid on a 30-year mortgage of $60,000 will be $189,570. What will be the payment each month if the payments are to pay off both the loan and the interest?

Writing & Thinking

65. In your own words, discuss the similarities and differences between division with whole numbers and division with decimal numbers.

66. List the steps you would follow in working a division problem with decimal numbers.

67. Discuss briefly, situations in which you might use division with decimal numbers in your daily life.

Collaborative Learning

68. Do you know how to find the gas mileage (miles per gallon) that your car is using? If you are not sure, proceed as follows and compare your mileage with other students in the class. (Your car might need some work if the mileage is not consistent or it is much worse than other similar sized cars.)

Step 1: Fill up your gas tank and write down the mileage indicated on the odometer.

Step 2: Drive the car for a few days.

Step 3: Fill up your gas tank again and write down the number of gallons needed to fill the tank and the new mileage indicated on the odometer.

Step 4: Find the number of miles that you drove by subtracting the new and old numbers indicated on the odometer.

Step 5: Divide the number of miles driven by the number of gallons needed to fill the tank. This number is your gas mileage (miles per gallon).

45. 3.164963
46. 1.25725
47. 0.0186
48. 0.0964
49. 0.045621
50. 0.010413
51. 1.17385
52. 98.716742
53. 0.000154
54. 0.01699

Applications

55. $239.56
56. $23.83
57. $295
58. $249
59. 5603 at bats
60. 23 games
61. 4.4 yards per carry
62. 22 mpg
63. 67.92
64. $693.25

Writing & Thinking

65. In division with decimal numbers, placement of the decimal point must be considered. Also, with whole numbers a remainder is considered while, with decimal numbers, rounding the quotient is considered. Otherwise, division with whole numbers and decimal numbers are the same.

66. To divide decimal numbers:

1. Move the decimal point in the divisor to the right so that the divisor is a whole number.
2. Move the decimal point in the dividend the same number of places to the right.
3. Place the decimal point in the quotient directly above the new decimal point in the dividend.
4. Divide just as with whole numbers.

67. Answers will vary.

Collaborative Learning

68. Answers will vary.

Objectives

A. Estimate sums and differences with rounded decimal numbers.

B. Estimate products and quotients with rounded decimal numbers.

C. Use the order of operations to simplify expressions containing decimal numbers.

3.5 Estimating and Order of Operations with Decimal Numbers

A Estimating Sums and Differences

We can estimate a sum (or difference) by rounding each number to the place of the **leftmost nonzero digit** and then adding (or subtracting) these rounded numbers. This technique of estimating answers is especially helpful when working with decimal numbers, where the placement of the decimal point is so important.

Example 1 Estimating Sums of Decimal Numbers

Estimate the sum; then find the actual sum.

$$74 + 3.529 + 52.61$$

Solution

First, estimate by adding rounded numbers.

74	rounds to →	70
3.529	rounds to →	4
52.61	rounds to →	+ 50
		124 ← Estimate

Now, find the actual sum.

$$\begin{array}{r} 74.000 \\ 3.529 \\ +\ 52.610 \\ \hline 130.139 \end{array}$$ ← Actual sum

Now work margin exercise 1.

1. Estimate the sum; then find the actual sum.

   6.68 + 103 + 21.94

In Example 1, the estimated sum and the actual sum are reasonably close (the difference is about 6). This leads us to have confidence in the answer.

If, for example, 74 had been written as 0.740 (instead of 74.000), then addition would have given a sum as shown.

Decimal in the wrong place

$$\begin{array}{r} 0.740 \\ 3.529 \\ +52.610 \\ \hline 56.879 \end{array}$$

Here, the difference between the estimate of 124 and the wrong sum of 56.879 is over 60 (a relatively large amount), and an error should be suspected.

Math Tip

Estimation requires some individual judgment, and you might get slightly different results depending on how you estimate. Regardless of the method used for estimation, there will be cases when the estimate is not particularly accurate.

Completion Example 2 Estimating Differences of Decimal Numbers

Estimate the difference; then find the actual difference.

$$132.418 - 17.526$$

Solution

First, estimate by subtracting rounded numbers.

132.418	rounds to →	100	
17.526	rounds to →	− 20	
		80	← Estimate

Now, find the actual difference.

$$\begin{array}{r} 132.418 \\ -\ \ 17.526 \\ \hline 114.892 \end{array} \leftarrow \text{Actual difference}$$

Now work margin exercise 2.

2. Estimate the difference; then find the actual difference.

1240.68 − 38.09

B Estimating Products and Quotients

Estimating products can be done by rounding each number to the place of the **leftmost nonzero digit** and multiplying these rounded numbers. This technique is particularly helpful in verifying the correct placement of the decimal point in the actual product.

Example 3 Estimating Products of Decimal Numbers

Estimate the product; then find the actual product.

$$(0.356)(6.1)$$

Solution

First, estimate by multiplying rounded numbers.

0.356	rounds to →	0.4	
6.1	rounds to →	× 6	
		2.4	← Estimate

Now, find the actual product and use the estimation to help place the decimal point.

$$\begin{array}{r} 0.356 \\ \times \quad 6.1 \\ \hline 356 \\ 21360 \\ \hline 2.1716 \end{array} \leftarrow \text{Actual product}$$

The estimated product, 2.4, helps verify that we placed the decimal point correctly in the product, 2.1716, since both numbers are close to 2.

Now work margin exercise 3.

3. Estimate the product; then find the actual product.

(11.2)(0.44)

As with addition, subtraction, and multiplication, we can use estimating with division to help in placing the decimal point in the quotient and to verify the reasonableness of the quotient. In order to estimate with division, round both the divisor and the dividend to the place of the **leftmost nonzero digit** and then divide with these rounded values.

4. Estimate the quotient; then find the actual quotient to the nearest hundredth.

13.18 ÷ 2.41

Example 4 Estimating Quotients of Decimal Numbers

Estimate the quotient; then find the actual quotient to the nearest hundredth.

$$6.2 \div 0.302$$

Solution

First, estimate by dividing rounded numbers. (6.2 rounds to 6 and 0.302 rounds to 0.3.)

```
        2 0.   ← Estimate
     ________
0.3.) 6.0.
     −6
     ___
      0 0
      −0
      ___
        0
```

Now, find the actual quotient to the nearest hundredth.

```
               2 0.5 2 9 ≈ 2 0.5 3   ← Actual quotient
         ______________
0.3 0 2.) 6.2 0 0.0 0 0
         −6 0 4
         ______
            1 6 0
          −   0
          _______
            1 6 0 0
          −1 5 1 0
          _________
                9 0 0
              −6 0 4
              _______
              2 9 6 0
             −2 7 1 8
             _________
                2 4 2
```

The estimated value, 20, is very close to the rounded quotient, 20.53, so we can be confident in the accuracy of our calculations.

Now work margin exercise 4.

Some applications may involve several operations. The problems do not usually say directly to add, subtract, or multiply. Experience and reasoning abilities are needed to decide which operations to perform with the given numbers. Example 5 illustrates a problem involving several steps and how answers can be estimated.

Example 5 Application: Estimating with Decimal Numbers

You can buy a car for $19,000 cash or you can make a down payment of $3750 and then pay $483.33 each month for 36 months. How much can you save by paying cash?

Solution

Step 1: Find the amount paid in monthly payments by multiplying the amount of each payment by 36. In this case, judgment dictates that we use 36 and do not round to 40, since we do not want to gain four full monthly payments in our estimate.

Estimate		Actual Amount	
$ 500		$ 483.33	
× 36		× 36	
3 000		2 899 98	
15 000		14 499 90	
$18,000	Estimated monthly payments	$17,399.88	Actual monthly payments

Step 2: Find the total amount paid by adding the down payment to the answer in Step 1.

Estimate		Actual Amount	
$4 000	Down payment	$3 750.00	Down payment
+18,000	Monthly payments	+17,399.88	Monthly payments
$22,000	Estimated total	$21,149.88	Actual total

Step 3: Find the savings by subtracting $19,000 (the cash price) from the answer to Step 2.

Estimate		Actual Amount	
$22,000	Estimated total	$21,149.88	Total paid
− 19,000	Cash price	− 19,000.00	Cash price
$3 000	Estimated savings	$2 149.88	Savings by paying cash

Thus, you can save $2149.88 by paying cash.

Check

The estimated $3000 saved by paying cash is reasonably close to the actual savings of $2149.88, so we can be confident in the accuracy of our calculations. The amount saved by paying cash is also a reasonable portion of the total price of the car.

Now work margin exercise 5.

Note that although estimates are close to the actual value, they do not guarantee the absolute accuracy of this actual value. The estimates help only in placing decimal points and checking the reasonableness of answers. Experience and understanding are needed to judge whether or not a particular answer is reasonably close to an estimate or vice versa.

5. How much will you save on a car by paying $456.25 a month for 12 months, as opposed to $260.35 a month for 24 months? Assume the down payment is the same.

Teaching Note

Example 5 is a good illustration of the need for judgment in any estimation. In Step 1, if 12 is rounded to 10, the resulting estimate will actually be a negative number. You might ask the students the implication of the negative result (paying cash would increase the cost) and why this does not make sense.

C Order of Operations with Decimal Numbers

Just as with whole numbers, fractions, and mixed numbers, expressions with decimal numbers may involve more than one operation. To evaluate such an expression, use the rules for order of operations just as they were discussed in Sections 1.7 and 2.8.

6. Simplify:

$2.4 \cdot 11.1 - 4.4^2$

Example 6 Using the Order of Operations with Decimal Numbers

Simplify: $3.1^2 + 7.05 \div 1.5$

Solution

$$\begin{aligned} &3.1^2 + 7.05 \div 1.5 \\ &= 9.61 + 7.05 \div 1.5 &&\text{Evaluate the exponential expression.} \\ &= 9.61 + 4.7 &&\text{Divide.} \\ &= 14.31 &&\text{Add.} \end{aligned}$$

Now work margin exercise 6.

7. Simplify:

$7.3 + 3.2\left(0.5^2 + 1.1\right)$

Example 7 Using the Order of Operations with Decimal Numbers

Simplify: $2.1(45.2 - 10.8) - 15.38$

Solution

$$\begin{aligned} &2.1(45.2 - 10.8) - 15.38 \\ &= 2.1(34.4) - 15.38 &&\text{Subtract inside the parentheses.} \\ &= 72.24 - 15.38 &&\text{Multiply.} \\ &= 56.86 &&\text{Subtract.} \end{aligned}$$

Now work margin exercise 7.

Completion Example Answers

2. $100 - 20 = 80$ (estimate); 114.892

Margin Exercise Answers

Estimated answers may vary.

1. Estimate: 127; sum: 131.62 **2.** Estimate: 960; difference: 1202.59 **3.** Estimate: 4; product: 4.928 **4.** Estimate: 5; quotient: 5.47 **5.** \$773.40 **6.** 7.28 **7.** 11.62

3.5 Exercises

Concept Check

Fill-in-the-Blank. Complete each sentence using information found in this section.

1. One way to estimate a difference with decimal numbers is to round each number to the __________ nonzero digit, and perform the subtraction.

2. In order to estimate a sum, round each __________ before performing the addition.

3. Estimating with decimal numbers is especially important because it helps ensure that the __________ was placed correctly.

4. One way to estimate a product or quotient is to round each number to the __________ nonzero digit, then perform the indicated operation.

5. When using the order of operations, the next step after removing symbols of inclusion is to evaluate any ________ ________.

True/False. Determine whether each statement is true or false. If a statement is false, explain how it can be changed so the statement will be true. (**Note:** There may be more than one acceptable change.)

6. An estimate of the sum 71.369 + 49.1 is 120.

7. One way to estimate the product of decimal numbers is to round the numbers to the rightmost nonzero digit before performing the multiplication.

8. An estimate of the quotient 16.469 ÷ 3.87 would be 4.

9. Experience and understanding are needed to decide whether or not a particular answer is reasonably close to an estimate.

10. According to the rules for order of operations, addition and subtraction should be performed before multiplication and division.

Practice

Estimate each sum; then find the actual sum. See Example 1.

1. $\begin{array}{r} 29.03 \\ +\ 3.79 \\ \hline \end{array}$

2. $\begin{array}{r} 58.2 \\ +\ 63.02 \\ \hline \end{array}$

3. 13 + 8.79

4. 9.66 + 14

5. $\begin{array}{r} 51.07 \\ 45.2 \\ +\ 6.19 \\ \hline \end{array}$

6. $\begin{array}{r} 4.22 \\ 71.6 \\ +\ 36.75 \\ \hline \end{array}$

7. 0.7 + 0.3 + 2.31

8. 6 + 5.1 + 0.8

3.5 Exercises

Concept Check

1. leftmost
2. addend
3. decimal point
4. leftmost
5. exponential expressions
6. True
7. False; When estimating the product of decimal numbers, round the numbers to the leftmost nonzero digit.
8. False; When estimating 16.469 ÷ 3.87, the answer would be 5.
9. True
10. False; According to the rules for order of operations, multiplication and division should be performed before addition and subtraction.

Practice

1. 34; 32.82
2. 120; 121.22
3. 19; 21.79
4. 20; 23.66
5. 106; 102.46
6. 114; 112.57
7. 3; 3.31
8. 12; 11.9

9. 188; 207.45
10. 49.2; 53.401
11. 20; 26.08
12. 50; 47.75
13. 1; 1.55
14. 4; 3.92
15. 10; 9.09
16. 160; 165.92
17. 2.4; 2.621
18. 0.23; 0.2759
19. 7000; 6548.06
20. 4100; 3800.63
21. 6; 4.86
22. 60; 48.6
23. 0.72; 0.7452
24. 0.12; 0.1364
25. 6; 6.2144
26. 3; 3.696
27. 5; 5.17
28. 12; 10.08
29. 2; 2.032
30. 1.5; 1.535
31. 0.06; 0.0486
32. 0.006; 0.00486
33. 5.4; 4.9077
34. 0.045; 0.046046
35. 2; 2.05
36. 0.02; 0.02
37. 10; 9.45
38. 100; 94.47
39. 10,000; 9446.67
40. 1000; 944.67
41. 2000; 2115
42. 3000; 3482

9. $\begin{array}{r} 121.6 \\ 55.9 \\ 8.32 \\ +\,21.63 \\ \hline \end{array}$

10. $\begin{array}{r} 44.4 \\ 3.211 \\ 0.19 \\ +\,5.6 \\ \hline \end{array}$

Estimate each difference; then find the actual difference. See Example 2.

11. $\begin{array}{r} 51.21 \\ -\,25.13 \\ \hline \end{array}$

12. $\begin{array}{r} 91.67 \\ -\,43.92 \\ \hline \end{array}$

13. $5.3 - 3.75$

14. $12.82 - 8.9$

15. $\begin{array}{r} 22 \\ -\,12.91 \\ \hline \end{array}$

16. $\begin{array}{r} 204 \\ -\,38.08 \\ \hline \end{array}$

17. $\begin{array}{r} 3.21 \\ -\,0.589 \\ \hline \end{array}$

18. $\begin{array}{r} 0.345 \\ -\,0.0691 \\ \hline \end{array}$

19. $7521.22 - 973.16$

20. $4732.61 - 931.98$

Estimate each product; then find the actual product. See Example 3.

21. $1.62(3)$

22. $16.2(3)$

23. $(0.92)(0.81)$

24. $(0.22)(0.62)$

25. $(0.64)(9.71)$

26. $(33.6)(0.11)$

27. $(4.7)(1.1)$

28. $(6.3)(1.6)$

29. $\begin{array}{r} 5.08 \\ \times\,0.4 \\ \hline \end{array}$

30. $\begin{array}{r} 3.07 \\ \times\,0.5 \\ \hline \end{array}$

31. $(1.62)(0.03)$

32. $1.62(0.003)$

33. $\begin{array}{r} 86.1 \\ \times\,0.057 \\ \hline \end{array}$

34. $\begin{array}{r} 0.506 \\ \times\,0.091 \\ \hline \end{array}$

Estimate each quotient; then find the actual quotient rounded to the nearest hundredth, if necessary. See Example 4.

35. $3.1\overline{)6.36}$

36. $3.1\overline{)0.0636}$

37. $28.34 \div 3$

38. $28.34 \div 0.3$

39. $0.003\overline{)28.34}$

40. $0.03\overline{)28.34}$

41. $0.1\overline{)211.5}$

42. $0.01\overline{)34.82}$

43. $282.4 \div 3.6$

44. $632.9 \div 8.2$

45. $23\overline{)71}$

46. $69\overline{)293}$

47. $6.27 \div 25.7$

48. $24.31 \div 85.3$

Simplify. See Examples 6 and 7.

49. $1.5^2 + 1.25 \div 0.5$

50. $5.4 \div 0.4 - 2.4^2$

51. $\dfrac{14.26 - 6.56}{2.5}$

52. $\dfrac{16.68 - 6.88}{2.5}$

53. $1.52 - 0.56 + 2.2 \cdot 6.5$

54. $12.6 - 5.88 + 13.9 \cdot 6.5$

55. $8.6 \div 2.15 + 3.6 \cdot 20.3$

56. $43.5 + 23.78 \div 5.8 \cdot 5.4$

57. $3.1(50 - 25.8) - 12.9$

58. $4.1(38.6 - 29.8) + 8.6$

59. $1.7(4.1 + 3.3) - 6.2$

60. $2.5(6.8 - 4.3) + 1.1$

61. $40.7 - (2.5^2 + 7.25) \div 0.5$

62. $97.5 + (30.46 - 4.6^2) \div 1.5$

63. $9.3 \cdot 1.1 + 2.4(5.7 - 8.4 \div 2.1)$

64. $5.4 \div 1.2 - 0.6(6.2 - 7.2 \div 1.5)$

65. $(1.3 + 5.9) \cdot 2.6 + 8.16$

66. $4.2 + 23.79 \div (7.4 - 3.5)$

67. $(4.2 + 4.8) \cdot (3.3^2 - 9.9) + 7.2$

68. $(5.7 + 2.9)^2 - (2.1 + 4.1) \cdot 3.4$

Applications

Solve.

69. Jim is packing three sculptures in a box for shipping. The weights of the sculptures are 5.63 pounds, 12.4 pounds, and 3 pounds. The shipping materials weigh 17.4 pounds.

a. Estimate the total weight.

b. Find the actual weight.

70. A road trip is being planned from New York City, NY, to Lawrence, KS. The first leg of the trip will be to Cleveland, OH, which will take 2.817 hours driving time; then on to Des Moines, IA, which will take 11.15 hours; followed by Kansas City, MO, which will take exactly 3 hours; and the final leg to Lawrence, KS, which will take 0.7 hours.

a. Estimate the driving time.

b. Calculate the actual driving time.

43. 75; 78.44
44. 75; 77.18
45. 3.5; 3.09
46. 4.29; 4.25
47. 0.2; 0.24
48. 0.22; 0.28
49. 4.75
50. 7.74
51. 3.08
52. 3.92
53. 15.26
54. 97.07
55. 77.08
56. 65.64
57. 62.12
58. 44.68
59. 6.38
60. 7.35
61. 13.7
62. 103.7
63. 14.31
64. 3.66
65. 26.88
66. 10.3
67. 16.11
68. 52.88

Applications

69. **a.** 39 pounds
b. 35.43 pounds

70. **a.** 17 hours
b. 17.667 hours

71. a. 15 feet b. 15.03 feet

72. a. $600 b. $13.35

73. a. $5000 b. $5959.80

74. a. $250 b. $294.30

75. a. $900 b. $891.50

c. $10,698.00

76. a. $756 b. $737.64

77. a. 600 miles b. 480.7

78. a. 20 mph b. 23.38 mph

71. Five boxes of unequal size are placed side-by-side along a wall, where the first box is 2.36 feet wide, the second is 1.76 feet wide, the third is 3.8 feet wide, the fourth is 0.94 feet wide, and the fifth is 6.17 feet wide.

a. Estimate the length of all the boxes together.

b. Calculate the actual length of all the boxes together.

72. Suppose the price to buy and install granite countertops is $64.85 per square foot.

a. Estimate the price of a new 11 square foot granite countertop that was installed.

b. What is the exact amount paid for the countertop?

73. Suppose Ted is ordering more refrigerators for his stockroom after an appliance sale. He orders 12 new refrigerators at the whole sale price of $496.65.

a. Estimate the total price Ted paid.

b. What is the exact price paid?

74. A sales person at an appliance store receives a 0.12 commission on all items she sells. Assume a customer purchases a front loading washing machine for $996.45, a matching dryer for $891.58, and two matching pedestals for $282.17 each.

a. Estimate what the commission will be. (**Hint:** Add all the prices before computing the commission. Commission = total sales times commission rate.)

b. What is the actual commission on the sale? Round your answer to the nearest cent.

75. Each month, Sydney is setting aside 0.26 times her take-home pay of $3428.84 for a down payment on a condominium.

a. Estimate how much is being placed in savings each month.

b. Compute the exact value of the savings each month. Round your answer to the nearest hundredth.

c. How much money is being saved each year? Use the monthly savings found in part **b.** for the computation.

76. Suppose the sale price of an LED Smart TV is $683 and sales tax is figured at 0.08 times the price.

a. Approximately what total amount is paid for the TV?

b. What is the exact amount paid for the TV? Round your answer to the nearest cent.

77. Suppose a car averages 25.3 miles per gallon.

a. Approximately how far will it go on 19 gallons of gas?

b. What is the actual distance the car will go on 19 gallons of gas?

78. Bicycle racer Peter Sagan rode 125.09 miles in 5.35 hours.

a. Estimate how fast he was riding per hour.

b. What was his average speed per hour (to the nearest hundredth)?

79. A quarter section of beef can be bought cheaper than the same amount of meat purchased a few pounds at a time. Suppose it costs $765.80 to purchase a quarter section which weighs 150 pounds.

a. Estimate the cost per pound for this quarter section.

b. What is the actual cost per pound for this quarter section?

80. According to the latest measurements, Mt. Everest is 29,035 feet above sea level. There are 5280 feet in a mile. [1]

a. Estimate how high Mt. Everest is in miles.

b. What is the actual elevation in miles? Round your answer to the nearest tenth of a mile.

81. Suppose a car travels 330 miles on 15 gallons of gas.

a. Approximately how many miles does the car travel per gallon?

b. Exactly how many miles does the car travel per gallon?

82. Kristin bought three pens that cost $1.25 each, two binders that cost $2.55, and had a coupon for $1.55 off of her total purchase.

a. Approximately how much did Kristin owe, not including tax?

b. Determine the exact amount that Kristin paid for the purchase, not including tax.

83. Stephen is taking 7 credit hours of courses at his local community college. Each credit hour costs $105.50. He also has to pay two technology fees of $16.75 each.

a. Approximately how much will Stephen's tuition be for the semester?

b. Determine the exact amount that Stephen will pay for tuition.

84. Last week, Sam worked 6 days driving a delivery truck. He drove 437.8 miles each day the first two days. Then the next three days he drove 562.6 miles each day. On the final day, he drove 521.4 miles.

a. Approximately how many miles did Sam drive during the week?

b. Determine the exact mileage Sam drove during the week.

c. If Sam earns $0.28 per mile, how much money did he earn during the week? (Round to the nearest cent, if necessary.)

85. Maya and Michelle run a small online business after school. This week, they fulfilled online orders that brought in profits of $11.20, $26.89, $8.90, and $51.33.

a. Estimate the total profit from the orders.

b. Find the exact total profit from the orders.

c. If Maya and Michelle split the profits evenly, how much will each of them get?

86. Kayla purchased four shirts for a total of $69.52, including a tax of $5.72.

a. Estimate the price of each shirt before tax.

b. Find the exact price of each shirt before tax.

79. a. $5 per pound **b.** $5.11 per pound

80. a. 6 miles **b.** 5.5 miles

81. a. 15 mpg **b.** 22 mpg

82. a. $7 **b.** $7.30

83. a. $740 **b.** $772

84. a. 3100 miles **b.** 3084.8 miles **c.** $863.74

85. a. $99 **b.** $98.32 **c.** $49.16

86. a. $16 **b.** $15.95

1 Source: www.mnteverest.net/history.html

Writing & Thinking

87. Answers will vary. The estimate says you are paying $4000 more with the payment plan than with cash. This is a significant overestimate since the payment and the number of months were both rounded up.

88. The two procedures would produce different results. Method **a.** is more accurate since you only round one figure. In method **b.** you multiply two rounded amounts which increases the difference between the approximated amount and the actual amount. Answers will vary.

Writing & Thinking

87. In Example 5, we stated the following problem:

> You can buy a car for $19,000 cash or you can make a down payment of $3750 and pay $483.33 each month for 36 months. How much can you save by paying cash?

Estimate the savings by rounding all values. (This includes rounding 36 months to 40 months.) Explain why this estimated savings does not seem reasonable, and explain why we must be careful about using rounded numbers in practical applications.

88. Suppose you are only interested in an approximate answer for a product. Would there be any difference in the products produced by the following two procedures?

a. First multiply the two numbers as they are and then round the product to the desired place of accuracy.

b. First round each number to the desired place of accuracy and then multiply the rounded numbers.

Explain why you think these two procedures would produce the same result or different results.

3.6 Decimal Numbers and Fractions

Objectives

A. Change decimal numbers to fractions.

B. Change fractions to decimal numbers.

C. Simplify expressions with both decimal numbers and fractions.

A Changing Decimal Numbers to Fractions

To understand how decimal numbers and fractions are related, we show how to change from one form to the other. From Section 3.1, we know that terminating decimal numbers can be written in fraction form with denominators that are powers of 10. For example,

$$0.35 = \frac{35}{100} \quad \text{and} \quad 0.017 = \frac{17}{1000}$$

The rightmost digit, 5, is in the hundredths position, so the fraction has 100 in the denominator.

The rightmost digit, 7, is in the thousandths position, so the fraction has 1000 in the denominator.

In each case, the denominator is the power of 10 that corresponds to the rightmost digit. Thus, we proceed as follows to change a decimal number to fraction form.

Changing from Decimal Numbers to Fractions

A decimal number less than 1 (digits are to the right of the decimal point) can be written in fraction form by writing a fraction with

1. A numerator that consists of the whole number formed by all the digits of the decimal number to the right of the decimal point, and
2. A denominator that is the power of 10 that corresponds to the rightmost digit. (For example, a denominator of 100 corresponds to hundredths.)

PROCEDURE

Math Tip

When changing a decimal number to a fraction, you may notice that the number of 0s in the denominator will be the same as the number of digits to the right of the decimal in the original number. For example, a value that ends in the hundredths place has 2 digits to the right of the decimal, which means the denominator will be 100 (notice 100 has 2 zeros).

In Examples 1 through 3, each decimal number is changed to fraction form and then reduced, if possible, by using the factoring techniques discussed in Chapter 2 for reducing fractions.

Example 1 Changing Decimal Numbers to Fractions

Change each decimal number to a fraction in lowest terms.

a. 0.25 **b.** 0.32

Solution

a. $0.25 = \frac{25}{100} = \frac{\cancel{25} \cdot 1}{4 \cdot \cancel{25}} = \frac{1}{4}$

Hundredths

b. $0.32 = \frac{32}{100} = \frac{\cancel{4} \cdot 8}{\cancel{4} \cdot 25} = \frac{8}{25}$

Hundredths

Now work margin exercise 1.

1. Change each decimal number to a fraction in lowest terms.

a. 0.48

b. 0.95

2. Change each decimal number to a fraction in lowest terms.

a. 0.763

b. 0.032

Example 2 Changing Decimal Numbers to Fractions

Change each decimal number to a fraction in lowest terms.

a. 0.131 **b.** 0.075

Solution

a. $0.131 = \frac{131}{1000}$ (Thousandths)

b. $0.075 = \frac{75}{1000} = \frac{\not{25} \cdot 3}{\not{25} \cdot 40} = \frac{3}{40}$ (Thousandths)

Now work margin exercise 2.

3. Change 15.8 to a mixed number with the fraction in lowest terms.

Example 3 Changing Decimal Numbers to Fractions

Change 2.6 to a mixed number with the fraction in lowest terms.

Solution

As a mixed number,

$$2.6 = 2\frac{6}{10} = 2\frac{\not{2} \cdot 3}{\not{2} \cdot 5} = 2\frac{3}{5}$$

(Tenths)

Now work margin exercise 3.

B Changing Fractions to Decimal Numbers

To change a fraction to decimal form, divide the numerator by the denominator. As we noted in Section 3.4:

1. If the remainder is eventually 0, the decimal number is said to be **terminating**.
2. If the remainder is not eventually 0, the decimal number is said to be **nonterminating**.

The following examples illustrate fractions that convert to terminating decimal numbers.

Math Tip

The following are examples of terminating numbers.

$\frac{7}{10} = 0.7, \quad \frac{5}{8} = 0.625$

The following are examples of nonterminating numbers.

$\frac{2}{3} = 0.66666\ldots$ or $0.\overline{6}$

$\frac{3}{7} = 0.\overline{428571}$

Example 4 Changing Fractions to Decimal Numbers

Change $\frac{3}{8}$ to a decimal number.

Solution

$$\frac{3}{8} \rightarrow 8\overline{)3.000}$$

$$\begin{array}{r} 0.375 \\ 8\overline{)3.000} \\ \underline{-24} \\ 60 \\ \underline{-56} \\ 40 \\ \underline{-40} \\ 0 \end{array}$$

Thus, $\frac{3}{8} = 0.375.$

Now work margin exercise 4.

4. Change $\frac{2}{5}$ to a decimal number.

Example 5 Changing Fractions to Decimal Numbers

Change $\frac{5}{4}$ to a decimal number.

Solution

$$\frac{5}{4} \rightarrow 4\overline{)5.00}$$

$$\begin{array}{r} 1.25 \\ 4\overline{)5.00} \\ \underline{-4} \\ 10 \\ \underline{-8} \\ 20 \\ \underline{-20} \\ 0 \end{array}$$

Thus, $\frac{5}{4} = 1.25.$

Now work margin exercise 5.

5. Change $\frac{33}{20}$ to a decimal number.

Nonterminating decimal numbers can be **repeating** or **nonrepeating**. A **repeating decimal** has a repeating pattern to its digits. Every fraction with a whole number numerator and nonzero denominator is either terminating or repeating. Such numbers are called **rational numbers**.

The following examples illustrate how some fractions (using division) convert to repeating decimal numbers.

Math Tip

The values $\frac{2}{3} = 0.\overline{6}$ and $\frac{3}{7} = 0.\overline{428571}$ are examples of repeating nonterminating decimals. The value $\pi = 3.14159\ldots$ is an example of a nonrepeating nonterminating decimal.

6. Change $\frac{2}{15}$ to a decimal number.

Teaching Note

The following discussion will help to develop an understanding of the concept of infinity as used in mathematics and will capture the imagination and interest of many students. Consider the following fact.

$$\frac{1}{9} = 0.11111...$$
$$\frac{2}{9} = 2 \cdot \frac{1}{9} = 0.22222...$$

And so on until

$$\frac{9}{9} = 9 \cdot \frac{1}{9} = 0.99999...$$

Thus, we can show 0.99999.... is equal to 1.

Example 6 Changing Fractions to Decimal Numbers

Change $\frac{7}{12}$ to a decimal number.

Solution

$$\frac{7}{12} \rightarrow 12\overline{)7.0000}$$

0.5833	← The 3 will repeat without end.
12)7.0000	
−6 0	
1 00	
−96	
40	
−36	
40	
−36	
4	← Continuing to divide will give a remainder of 4 each time.

We write $\frac{7}{12} = 0.58333...$ where the three dots (called an ellipsis) mean "and so on" or to continue the pattern without stopping.

Or, if we agree to round, we can write $\frac{7}{12} \approx 0.583$ (to the nearest thousandth).

Now work margin exercise 6.

7. Change $\frac{3}{13}$ to a decimal number.

Teaching Note (cont.)

Also, as shown by subtraction, the difference between these two numbers equals 0.

$1.00000... - 0.99999... = 0.00000...$

Algebraically, you can show that if $x = 0.99999...$, then

$$10x = 9.99999...$$
$$10x = 9 + 0.99999...$$
$$10x = 9 + x$$
$$9x = 9$$
$$x = 1.$$

Example 7 Changing Fractions to Decimal Numbers

Change $\frac{1}{7}$ to a decimal number.

Solution

$$\frac{1}{7} \rightarrow 7\overline{)1.000000}$$

0.142857	← The six digits will repeat in the same pattern without end.
7)1.000000	
−7	
30	
−28	
20	
−14	
60	
−56	
40	
−35	
50	
−49	
1	← The remainder will repeat in sequence 3, 2, 6, 4, 5, 1, 3 and so on. Therefore, the digits in the sequence will also repeat in sequence without end.

We write $\frac{1}{7} = 0.142857142857...$

Now work margin exercise 7.

Another way of writing repeating decimal numbers is to write a **bar** over the repeating digits. Thus, we can write

$$\frac{7}{12} = 0.58\overline{3} \quad \text{and} \quad \frac{1}{7} = 0.\overline{142857}.$$

CALCULATORS

Using a Calculator to Change Fractions to Decimal Numbers

To change a fraction to a decimal number on a calculator, simply divide the numerator by the denominator. For example, to change the fraction $\frac{5}{8}$ to a decimal number, press the keys

[5] [÷] [8]. Then press [=].

The display will read 0.625.

C Simplifying Expressions with Both Decimal Numbers and Fractions

As the following examples illustrate, we can perform operations and comparisons with both fractions and decimal numbers by changing the fractions to decimal form.

Example 8 Simplifying Expressions with Decimals and Fractions

Find the sum $10\frac{1}{2} + 7.32 + 5\frac{3}{5}$ in decimal form.

Solution

$$\begin{aligned} 10\tfrac{1}{2} &= 10.50 && \tfrac{1}{2} = 0.50 \\ 7.32 &= 7.32 \\ +\,5\tfrac{3}{5} &= 5.60 && \tfrac{3}{5} = 0.60 \\ \hline &\;\; 23.42 \end{aligned}$$

Now work margin exercise 8.

8. Find the sum $2.88 + \frac{1}{4} + 13\frac{9}{10}$ in decimal form.

Example 9 Comparing Decimal Numbers and Fractions

Determine which is larger, $\frac{3}{16}$ or 0.18. How much larger is the larger number?

9. Which is larger: $\frac{9}{16}$ or 0.52? How much larger?

Solution

To compare the two numbers, begin by changing $\frac{3}{16}$ to decimal form.

$$\begin{array}{r} 0.1875 \\ 16\overline{)3.0000} \\ \underline{16} \\ 140 \\ \underline{128} \\ 120 \\ \underline{112} \\ 80 \\ \underline{80} \\ 0 \end{array}$$

Thus, $\frac{3}{16} = 0.1875$.

Comparing, we see that 0.1875 is larger than 0.18. Now subtract the smaller number from the larger number.

$$\begin{array}{r} 0.1875 \\ -0.1800 \\ \hline 0.0075 \end{array}$$ Difference

Thus, $\frac{3}{16}$ is larger than 0.18 by 0.0075.

Now work margin exercise 9.

10. Apples cost $2.10 per pound and bananas cost $0.42 per pound. How much will you pay for 5 pounds of apples and $3\frac{1}{2}$ pounds of bananas?

Example 10 Application: Decimal and Fraction Expressions

An auto mechanic charges $35.60 per hour of labor. How much would he charge for a job which would take $2\frac{1}{2}$ hours of labor and required $273.49 for parts?

Solution

Step 1: READ: Read the problem carefully. In this case, we need to combine his charge for labor with the cost of the parts.

Step 2: SET UP: To find the charge for labor, multiply the cost per hour times the number of hours: $\$35.60 \cdot 2\frac{1}{2}$

Then add the cost of the parts: $273.49

Step 3: SOLVE: Change $2\frac{1}{2}$ to decimal form $\left(2\frac{1}{2} = 2.5\right)$ and multiply by $35.60.

$$\begin{array}{r} \$35.60 \\ \times \quad 2.5 \\ \hline 17800 \\ 71200 \\ \hline \$89.000 \end{array}$$ Charge for labor

Now add the cost of the parts to the cost of labor.

$$\begin{array}{rl} \overset{1\ 1}{\$89.00} & \text{Cost of labor} \\ +\$273.49 & \text{Cost of parts} \\ \hline \$\ \ 362.49 & \text{Total charge for job} \end{array}$$

The total charge for the job would be $362.49.

Step 4: CHECK: The cost of labor per hour is approximately $40. This means that $2\frac{1}{2}$ hours of labor will cost approximately $100. Considering that the parts alone cost $273.49, the answer of $362.49 seems reasonable.

Now work margin exercise 10.

Margin Exercise Answers

1. a. $\frac{12}{25}$ **b.** $\frac{19}{20}$ **2. a.** $\frac{763}{1000}$ **b.** $\frac{4}{125}$ **3.** $15\frac{4}{5}$ **4.** 0.4 **5.** 1.65 **6.** 0.13333...

7. 0.230769230769... **8.** 17.03 **9.** $\frac{9}{16}$ is larger than 0.52 by 0.0425 **10.** $11.97

3.6 Exercises

Concept Check

Fill-in-the-Blank. Complete each sentence using information found in this section.

1. To change a decimal number to a fraction, use the digits of the decimal number as the ________ and a power of 10 as the ________.
2. To change a decimal number to a fraction, find the power of 10 that names the position of the rightmost digit as the ________.
3. Nonterminating decimal numbers can be categorized as either ________ or ________ decimal numbers.
4. To change a fraction to a decimal number, divide the ________ by the ________.
5. In decimal form, $\frac{3}{5}$ is a ________ decimal number.
6. When computing problems with both fractions and decimal numbers, sometimes all numbers need to be in ________ form to preserve accuracy.

True/False. Determine whether each statement is true or false. If a statement is false, explain how it can be changed so the statement will be true. (**Note:** There may be more than one acceptable change.)

7. When a decimal number is changed to a fraction, the denominator will be the power of 10 that names the rightmost digit of the decimal number.
8. When a decimal number is changed to a fraction, the numerator can be determined by using the whole number that is formed by all the digits of the decimal number.
9. Fractions can always be converted to decimal form without losing accuracy.
10. In decimal form, $\frac{1}{3}$ is repeating and nonterminating.

3.6 Exercises

Concept Check

1. numerator, denominator
2. denominator
3. repeating, nonrepeating
4. numerator, denominator
5. terminating
6. fraction
7. True
8. True
9. False; In some cases, fractions can be converted to decimal form without losing accuracy.
10. True

Practice

1. $\frac{9}{10}$
2. $\frac{3}{10}$
3. $\frac{526}{1000}$
4. $\frac{625}{1000}$
5. $\frac{51}{10}$
6. $\frac{29}{10}$
7. $\frac{815}{100}$
8. $\frac{635}{100}$
9. $\frac{1}{5}$
10. $\frac{4}{5}$

11. $\frac{9}{50}$
12. $\frac{9}{25}$
13. $\frac{9}{40}$
14. $\frac{91}{200}$
15. $\frac{25}{4}$ or $6\frac{1}{4}$
16. $\frac{11}{4}$ or $2\frac{3}{4}$
17. 0.05
18. 0.04
19. 0.625
20. 0.375
21. $0.\overline{6}$
22. $0.1\overline{6}$
23. $0.2\overline{7}$
24. $0.41\overline{6}$
25. 6.67
26. 4.44
27. 0.48
28. 0.68
29. 0.03
30. 0.07
31. 1.43
32. 1.23
33. 0.7
34. 4.4
35. 11.59
36. 14.63
37. 1.13
38. 1.05
39. 1
40. 13
41. 6.50
42. 9.92
43. 23.41
44. 15.50
45. 120.31
46. 36.68

Practice

Change each decimal number to a fraction. Do not reduce.

1. 0.9 **2.** 0.3 **3.** 0.526 **4.** 0.625 **5.** 5.1 **6.** 2.9 **7.** 8.15 **8.** 6.35

Change each decimal number to a fraction or mixed number in lowest terms. See Examples 1 through 3.

9. 0.2 **10.** 0.8 **11.** 0.18 **12.** 0.36 **13.** 0.225 **14.** 0.455 **15.** 6.25 **16.** 2.75

Change each fraction to a decimal number. If the decimal number is nonterminating, write it using a bar over the repeating pattern of digits. See Examples 4 through 7.

17. $\frac{1}{20}$ **18.** $\frac{1}{25}$ **19.** $\frac{5}{8}$ **20.** $\frac{3}{8}$ **21.** $\frac{2}{3}$ **22.** $\frac{1}{6}$ **23.** $\frac{5}{18}$ **24.** $\frac{5}{12}$

Change each fraction to a decimal number rounded to the nearest hundredth. See Example 6.

25. $\frac{20}{3}$ **26.** $\frac{40}{9}$ **27.** $\frac{16}{33}$ **28.** $\frac{15}{22}$ **29.** $\frac{1}{32}$ **30.** $\frac{1}{14}$ **31.** $\frac{30}{21}$ **32.** $\frac{16}{13}$

Simplify each expression by writing all of the numbers in decimal form and performing the indicated operations. Round to the nearest hundredth, if necessary. See Example 8.

33. $\frac{1}{4}+0.25+\frac{1}{5}$

34. $\frac{3}{4}+\frac{1}{10}+3.55$

35. $6+2\frac{37}{100}+3\frac{11}{50}$

36. $2\frac{53}{100}+5\frac{1}{10}+7$

37. $1\frac{1}{4}-0.125$

38. $2\frac{4}{5}-1.75$

39. $3.1-2\frac{1}{10}$

40. $36.2-23\frac{1}{5}$

41. $\left(5\frac{1}{10}\right)^2(0.25)$

42. $\left(2\frac{1}{10}\right)^2(1.5)^2$

43. $\left(1\frac{3}{4}\right)\left(2\frac{1}{2}\right)(5.35)$

44. $\left(1\frac{4}{5}\right)\left(3\frac{1}{4}\right)(2.65)$

45. $72.186\div\frac{3}{5}$

46. $9.17\div\frac{1}{4}$

47. $5\frac{54}{100}\div2.1$

48. $3\frac{26}{100}\div1.5$

49. $\frac{1}{7}+0.355+\frac{2}{3}$

50. $\frac{1}{3}+\frac{3}{10}+3.452$

51. $\frac{7}{9}+\frac{3}{5}+0.418$

52. $0.5+8\frac{1}{6}+2\frac{11}{50}$

For each pair of numbers, determine which number is larger. Then, determine the difference between the two numbers. (If the difference is a nonterminating number, write it using a bar over the repeating pattern of digits.) See Example 9.

53. $0.28, \frac{3}{11}$

54. $\frac{1}{3}, 0.3$

55. $\frac{22}{7}, 3.3$

56. $\frac{4}{9}, 0.5$

57. $3.5, 3\frac{2}{3}$

58. $5\frac{3}{4}, 5.5$

Arrange each set of numbers in order from smallest to largest.

59. $0.76, \frac{3}{4}, \frac{7}{10}$

60. $0.63, \frac{5}{8}, 0.64$

61. $\frac{5}{16}, 0.3126, 0.314$

62. $0.083, \frac{41}{500}, \frac{2}{25}$

Applications

Change any decimal number (that is not a whole number) to a fraction or mixed number.

63. By 2020 census estimates, there were 93.8 people per square mile in the United States.

64. A gallon of milk weighs 8.6 lb.

65. The median age for men at the beginning of their first marriage is 30.3 years. The median age for women at the beginning of their first marriage is 28.4 years.

66. The maximum speed of a giant tortoise on land is about 0.17 mph.

67. There are about 21.5 students per teacher in a public school.

68. The surface gravity on Mars is about 0.38 times the gravity on Earth. The atmospheric pressure on Mars is about 0.01 times the atmospheric pressure on Earth.

Change any fraction or mixed number (that is not a whole number) to a decimal number. Round each number to the nearest hundredth.

69. In 2022, it was estimated that the average American adult spent $\frac{7}{12}$ of the time spent on their phone on social media.

70. In a recent year about $\frac{8}{57}$ of the advertising budget in the automotive industry was spent on newspaper ads.

71. In 2016, the average price of unleaded gasoline in California was $2\frac{9}{10}$ times the price it was in 1970.

72. The expected winner of a student senate presidential election received $\frac{21}{46}$ of the votes.

47. 2.64

48. 2.17

49. 1.16

50. 5.31

51. 1.80

52. 10.89

53. 0.28 is larger; $0.00\overline{72}$

54. $\frac{1}{3}$ is larger; $0.0\overline{3}$

55. 3.3 is larger; $0.\overline{1571428}$

56. 0.5 is larger; $0.0\overline{5}$

57. $3\frac{2}{3}$ is larger; $0.0\overline{16}$

58. $5\frac{3}{4}$ is larger; 0.25

59. $\frac{7}{10}, \frac{3}{4}, 0.76$

60. $\frac{5}{8}, 0.63, 0.64$

61. $\frac{5}{16}, 0.3126, 0.314$

62. $\frac{2}{25}, \frac{41}{500}, 0.083$

Applications

63. $93\frac{4}{5}$

64. $8\frac{3}{5}$

65. $30\frac{3}{10}$; $28\frac{2}{5}$

66. $\frac{17}{100}$

67. $21\frac{1}{2}$

68. $\frac{38}{100}$ or $\frac{19}{50}$; $\frac{1}{100}$

69. 0.58

70. 0.14

71. 2.9

72. 0.46

73. 17.92 inches
74. 121.8 mg
75. 20.2 miles
76. 5.1 seconds
77. 60 students that drive to school and have a job are taking AP English
78. $14.22
79. 0.9 ounces
80. $25.10
81. 25 slides
82. $40.75

Writing & Thinking

83. For the numerator, write the whole number formed by all the digits of the decimal number, and for the denominator, write the power of 10 that corresponds to the rightmost digit. Reduce the fraction, if possible.
84. Answers will vary.
85. This problem could be solved by converting all numbers to fractions or all numbers to decimal numbers. Answers will vary.

Solve.

73. A rectangle has a length of 6.4 inches and a width that is $\frac{2}{5}$ times the length. Find the perimeter of the rectangle.

74. A coffee cup holds $14\frac{1}{3}$ ounces of coffee. If each ounce of a certain brewed coffee contains 8.5 mg of caffeine, how much caffeine is in one of these cups of coffee (round to the nearest tenth of a mg)?

75. During five days this week, Mark ran 3.2 miles, $2\frac{3}{4}$ miles, 5.1 miles, $7\frac{1}{3}$ miles, and 1.8 miles. Find the total distance that Mark ran this week, to the nearest tenth of a mile.

76. Sarah and Iygen are working on a physics lab together. They each take measurements of three different timed experiments. Sarah has an analog stop watch and clocks experimental times of $5\frac{1}{8}$, $4\frac{3}{4}$, and $5\frac{1}{2}$ seconds. Iygen has a digital stop watch and clocks experimental times of 5.2, 4.6, and 5.3 seconds. Use all 6 times to determine the average time for all of the experiments, to the nearest tenth.

77. Summerville High School has 3375 students, $\frac{2}{3}$ of whom drive to school. Of the students who drive to school, 0.24 of them have an after-school job. Exactly $\frac{1}{9}$ of the students who drive to school and have a job are taking AP English. How many students who drive to school and have a job are taking AP English?

78. A preschool provides peanut butter sandwiches for each of its 45 children once per week. If each child receives $\frac{8}{5}$ ounces of peanut butter on a sandwich, and peanut butter costs $3.16 per pound (16 ounces), how much does the preschool spend per week on peanut butter?

79. A loaf of bread weighs 21.6 ounces. Mauricio cut off a third of the loaf to save for later and then cut the remaining portion into 16 equal slices. What was the weight of each slice of the 16 slices he cut?

80. A husband and wife join three single friends for dinner at a Mexican restaurant. Because they shared some appetizers and many entrees are the same price, the five of them decided to simply split the bill five ways. The husband and wife will pay for theirs together. If the total bill was $62.75, determine how much the husband and wife owe towards the total bill.

81. Janet and Marion are performing an experiment. They each examine 50 slides of viruses for signs of activity. After the examination, Janet reported that $\frac{1}{5}$ of her slides were active and Marion found that 0.3 of her slides were active. How many total slides showed signs of active viruses?

82. A pound of roasted cashews costs $13.50 and a pound of roasted peanuts costs $5.25. How much will it cost to purchase $2\frac{1}{2}$ pounds of cashews and $1\frac{1}{3}$ pounds of peanuts?

Writing & Thinking

83. Describe the process used to change a terminating decimal number to a fraction.

84. Do you find it is easier to convert decimal numbers into fractions or fractions into decimal numbers? Explain why.

85. List 2 different ways to solve this problem: $\frac{1}{2}+3.67-\frac{1}{8}$. State which method you prefer and why.

Chapter 3 Project

What Would You Weigh on the Moon?

An activity to demonstrate the use of decimal numbers in real life.

The following table contains the surface gravity of each planet in the same solar system as the Earth, as wells Earth's moon, and the sun. The acceleration due to gravity g at the surface of a planet is given by the formula

$$g = \frac{GM}{R^2},$$

where M is the mass of the planet, R is its radius, and G is the gravitational constant. From the formula you can see that a planet with a larger mass M will have a greater value for surface gravity. Also the larger the radius R of the planet, the smaller the surface gravity.

If you look at different sources, you may find that surface gravity varies slightly from one source to another due to different values for the radius of some planets, especially the gas giants: Jupiter, Saturn, Uranus, and Neptune.

Planet	Surface Gravity (m/s²)	Relative Surface Gravity	Fractional Equivalent
Earth	9.78	1.00	1
Jupiter	23.10	2.36	$2\frac{9}{25}$
Mars	3.72	0.38	$\frac{19}{50}$
Mercury	3.78	0.39	$\frac{39}{100}$
Moon	1.62	0.17	$\frac{17}{100}$
Neptune	11.15	1.14	$1\frac{7}{50}$
Saturn	9.05	0.93	$\frac{93}{100}$
Sun	274.00	28.02	$28\frac{1}{50}$
Uranus	8.69	0.89	$\frac{89}{100}$
Venus	9.07	0.93	$\frac{93}{100}$

1. Compare the surface gravity of each planet or celestial body to the surface gravity of the Earth by dividing each planet's surface gravity by that of the Earth's, as listed in the table above. (This is referred to as **relative surface gravity**.) Round your answer to the nearest hundredth and place your results in the third column of the table. The values for Earth and Jupiter have been done for you. (**Note:** Comparing Earth to itself results in a value of 1.)

2. For Jupiter, the relative surface gravity value of 2.36 means that the gravity on Jupiter is 2.36 times that of Earth; therefore, your weight on Jupiter would be approximately 2.36 times your weight on Earth. (Although mass is a constant and doesn't change regardless of what planet you are on, your weight depends on the pull of gravity). Explain what the relative surface gravity value means for Mars.

3. Calculate your weight on the moon by taking your present weight (in kg or pounds) and multiplying it by the moon's relative surface gravity.

4. Approximately how many times larger is the surface gravity of the sun compared to that of Mars? Round to the nearest whole number.

5. Convert each value in column three to a mixed number and place the result in column four. Be sure to reduce all fractions to lowest terms.

   a. Which is larger, the relative surface gravity of Mercury or $\frac{2}{5}$?

   b. Which is smaller, the relative surface gravity of the moon or $\frac{4}{25}$?

   c. Write the fractional equivalent of Jupiter's relative surface gravity as an improper fraction in lowest terms.

Chapter 3 Project: What Would You Weigh on the Moon?

2. The gravity on Mars is 0.38 times that of Earth or a little more than $\frac{1}{3}$ of Earth's gravity.
3. Answers will vary. (Weight on Earth) × 0.17 = (Weight on the moon)
4. 74 times larger.
5. **a.** $\frac{2}{5}$

   b. $\frac{4}{25}$

   c. $\frac{59}{25}$

Chapter 3 Project: A Trip to The Grocery Store

1. $9.56.
2. The total cost is $8.40. One way this answer can be found is by subtracting $0.50 from the cost of one can of beans to obtain $0.84. By multiplying this by 10, we obtain $8.40.
3. **a.** $0.09

   b. $0.11

   c. Bread Brand A has the better deal per slice. Brand A is the better deal overall, because even though the total price is higher, the price per slice is less.
4. **a.** Brand C: $0.03, Brand D: $0.02, Brand E: $0.01

   b. No, it is not cheaper by unit than Brand E. The unit price, even on sale, is $0.03, which is more expensive than Brand E. It is not a better deal to buy Brand C's paper towels, even on sale.

   c. Using the rounded unit price of $0.01, the total Brand D would need to be is $5.00.
5. Student answers will vary. As long as their total is under $20.00 and above $18.00, this is acceptable. One possible answer is to pick 1 of everything plus 1 additional bag of potatoes.

Chapter 3 Project

A Trip to The Grocery Store

An activity to demonstrate the use of decimal numbers in real life.

Decimals can be encountered in many areas of our daily lives. When we order food at a restaurant or go to the gas station, we see numbers written and expressed to the nearest hundredth. The grocery store is the same; we see prices for items, prices per unit, sale stickers, and so on, and we need to be able to find the best prices to get the most for our money.

For your upcoming trip to the grocery store, and suppose you have made a full list of items that you will need to purchase. In each question, you will be evaluating the prices of each item on your list using various operations with decimals.

1. The first three items you need to buy are a gallon of milk, 2 cartons of eggs, and a stick of butter. You have found that $3.49 is the best price for milk, $2.57 is the best price for one carton of eggs, and $0.93 is the best price for a stick of butter. What is the total cost of these items?

2. For canned beans, the grocery store is running a promotional deal where you will receive 50 cents off each canned good as long as you buy 10 of them. If the price of one can of beans is $1.34, what is the total price for 10 cans of beans? Explain how you found your answer.

3. When it comes to buying a loaf of bread, you have several options. To find the best deal, in this case, we need to find the unit price. The unit price is the cost per one item in a larger group. In other words, the unit price of the bread would be the cost per each slice of bread as opposed to the cost for the entire loaf. Assume each slice is the same size.

Bread Brand A has 25 slices and costs $2.35.
Bread Brand B has 18 slices and costs $1.95.

a. What is the unit price for a slice of Brand A's bread? Round to the nearest hundredth.

b. What is the unit price for a slice of Brand B's bread? Round to the nearest hundredth.

c. Which brand has the better deal per slice of bread? How do you know?

4. The store has three brands of paper towels for sale, with varying prices and amounts of individual sheets. Assume each sheet has the same dimensions.

Brand C is $12.90 for 400 sheets.
Brand D is $8.36 for 500 sheets.
Brand E is $9.30 for 775 sheets.

a. Calculate the unit price for each brand per sheet. Round each to the nearest hundredth.

b. During the weekend, the grocery store is running a sale on paper towels. If Brand C is on sale for a discount of 20%, is it now cheaper than Brand E, by unit price? Explain your answer.

c. What total price would Brand D need to be to have the same unit price as Brand E? (Hint: Use the rounded value you found in part a.)

5. To finish your grocery trip, you have to make choices with your remaining money. What is one possible combination of items you could purchase assuming you have $20.00 left to spend, you must purchase at least one of each item listed, and your total must be greater than $18.00. Justify your answer by showing your total spent and explain why you chose the items you did.

One pound of bananas for $0.64
A 5-lb bag of apples for $5.63
A 3-lb bag of potatoes for $3.68
A 2-lb bag of baby carrots for $1.99
A container of guacamole for $2.67

CHAPTER 4

4 Ratios, Proportions, and Percents

Connections

The concepts of ratios, proportions, and percents are used often in everyday life. Unit rates, a special type of ratio, are used to give the price per ounce of products in a grocery store so you can decide which is the better buy between two options. Proportions are used in cooking and baking to get the correct balance of ingredients in a recipe. And percents are used from the amount of active ingredient in a health product to the amount of interest you pay per month on your credit card. No matter your career or stage in life, ratios, proportions, and percents play an active role in your day-to-day activities. Homeowners often need to purchase home-improvement supplies such as paint, fertilizer, and grass seed for repairs and maintenance. In many cases, a homeowner must purchase supplies in quantities greater than what is needed for a particular job because the manufacturer's or distributor's packaging sizes do not fit the homeowner's exact needs. When this happens, the store will typically not sell part of a can of paint or a part of a bag of fertilizer. Fortunately, a little application of mathematics can help stretch any homeowner's dollar by minimizing any excess amounts that must be bought.

Suppose that during the late spring, you decide to treat your lawn with a fertilizer and weed killer combination. One bag contains 18 pounds with a recommended coverage of 5000 square feet. If your lawn is in the shape of a rectangle that is 150 feet long by 220 feet wide, how many pounds of the fertilizer and weed killer combination do you need to cover the lawn? How many bags will you need to purchase?

Objectives

A. Use ratios to compare two quantities.

B. Find rates.

C. Find unit rates.

D. Compare unit prices.

4.1 Ratios and Unit Rates

A Introduction to Ratios

We know two meanings for fractions.

1. To indicate a part of a whole.

$$\frac{3}{8} \text{ means } \frac{3 \text{ pieces of pie}}{8 \text{ pieces in the whole pie}}$$

2. To indicate division.

$$\frac{3}{8} \text{ means } 3 \div 8 \text{ or } 8\overline{)3.000} = 0.375$$

$$\begin{array}{r} 0.375 \\ 8\overline{)3.000} \\ \underline{-2\,4} \\ 60 \\ \underline{-56} \\ 40 \\ \underline{-40} \\ 0 \end{array}$$

A third use for fractions is to compare two quantities. Such a comparison is called a **ratio**. For example, the ratio $\frac{3}{4}$ (read "the ratio of 3 to 4") might mean $\frac{3 \text{ dollars}}{4 \text{ dollars}}$ or $\frac{3 \text{ hours}}{4 \text{ hours}}$. Note that the numerator is read first and the denominator second. Also, the units in the numerator and denominator are the same. (If the units are not the same, the ratio is called a **rate**.)

Math Tip

Understanding how to set up and interpret ratios and rates will help you in both your personal and professional life. Ratios and rates are used in everyday activities such as in cooking, interacting with social media, watching television, photography, and driving, just to name a few. Even health indicators such as your heart rate and blood pressure are expressed as ratios.

Ratios

A **ratio** is a comparison of two quantities by division. The ratio of *a* to *b* can be written as

$$\frac{a}{b} \quad \text{or} \quad a:b \quad \text{or} \quad a \text{ to } b.$$

DEFINITION

Attention!

The most common notation for ratios is fraction notation, and that is the notation we will use in this section.

Ratios have the following characteristics.

1. Ratios can be reduced, just as fractions can be reduced.

2. The common units in a ratio can be canceled, just as factors can be canceled in fractions.

3. Generally, ratios are written with whole numbers in the numerator and denominator, and this fraction is reduced to lowest terms.

Example 1 Application: Writing Ratios

The floor of a room measures 14 feet long and 10 feet wide. An architect drawing a scale model of the room needs to determine the ratio of the room's length to its width. Write this ratio as a fraction in lowest terms.

Solution

The ratio of length to width (in that order) in fraction form is

$$\frac{\text{length}}{\text{width}}=\frac{14\ \cancel{\text{feet}}}{10\ \cancel{\text{feet}}}=\frac{\cancel{2}\cdot 7}{\cancel{2}\cdot 5}=\frac{7}{5}.$$

Note that, in addition to reducing the fraction to lowest terms, the common units (feet) were canceled.

Now work margin exercise 1.

1. A picture window in the room being drawn by the architect is 36 in. tall and 30 in. wide. Write the ratio of height to width as a fraction in lowest terms.

Remember that when numbers are written in fraction form, the fraction bar indicates division. As illustrated in Example 2, if the numerator and/or denominator in a ratio contains mixed numbers, we change the mixed numbers to improper fractions and then divide.

Example 2 Writing Ratios that Compare Mixed Numbers

Write each ratio as a fraction in lowest terms.

a. $2\frac{1}{2}$ to $5\frac{1}{2}$ **b.** 3 days to $1\frac{1}{2}$ days

Solution

a. $$\frac{2\frac{1}{2}}{5\frac{1}{2}}=\frac{\frac{5}{2}}{\frac{11}{2}}=\frac{5}{2}\div\frac{11}{2}=\frac{5}{\cancel{2}}\cdot\frac{\cancel{2}}{11}=\frac{5}{11}$$

b. $$\frac{3\ \cancel{\text{days}}}{1\frac{1}{2}\ \cancel{\text{days}}}=\frac{3}{\frac{3}{2}}=\frac{3}{1}\div\frac{3}{2}=\frac{\cancel{3}}{1}\cdot\frac{2}{\cancel{3}}=\frac{2}{1}$$

Note that in ratios two numbers are being compared, so the ratio is 2 to 1, not just 2.

Now work margin exercise 2.

2. Write each ratio as a fraction in lowest terms.

a. $1\frac{1}{3}$ to $4\frac{2}{3}$

b. 7 months to $1\frac{3}{4}$ months

Example 3 Writing Ratios that Compare Decimal Numbers

Two lengths of pipe are shown. Write the ratio of the length of the longer pipe to the length of the shorter pipe by comparing the lengths in meters.

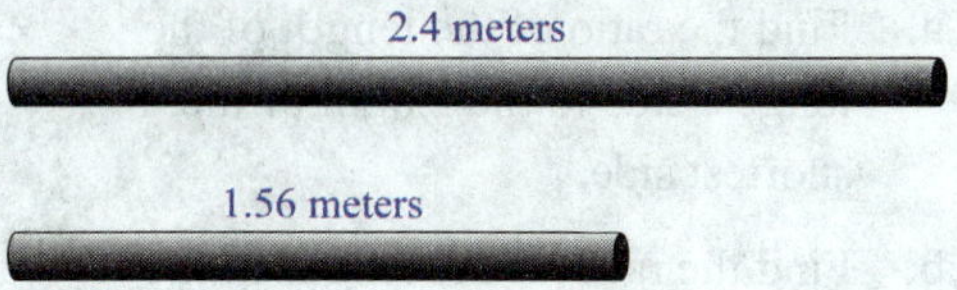

3. The lengths of two screws are shown. Write the ratio of the length of the longer screw to the length of the shorter screw by comparing the lengths in centimeters.

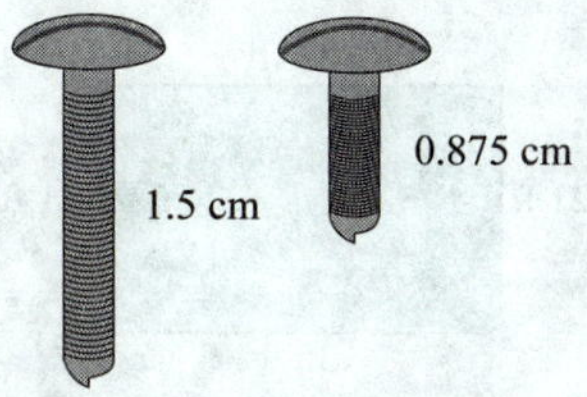

Solution

$$\frac{\text{length of longer pipe}}{\text{length of shorter pipe}} = \frac{2.4 \cancel{\text{ meters}}}{1.56 \cancel{\text{ meters}}} = \frac{2.4}{1.56} \cdot \frac{100}{100} = \frac{240}{156} = \frac{\cancel{12} \cdot 20}{\cancel{12} \cdot 13} = \frac{20}{13}$$

The ratio is $\frac{20}{13}$.

Now work margin exercise 3.

4. Use the information from the circle graph in Example 4 to
   a. find the ratio of money budgeted for utilities to money budgeted for the mortgage payment, and
   b. find the ratio of money budgeted for food to the total budget.

Example 4 Application: Writing Ratios from Graphs

The circle graph shown here illustrates a monthly budget with categories for food, mortgage payment, utilities, taxes, and other.

Monthly Household Budget

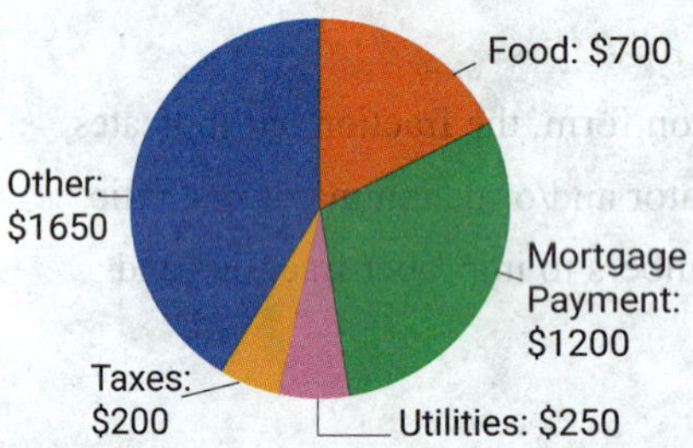

Use the information from the graph to

a. find the ratio of money budgeted for food to money budgeted for taxes and

b. find the ratio of the mortgage payment to the total budget.

Solution

a. The ratio of money budgeted for food to money budgeted for taxes is

$$\frac{\$700}{\$200} = \frac{7 \cdot \cancel{100}}{2 \cdot \cancel{100}} = \frac{7}{2}.$$

b. The total budget is found by summing the amount spent in each category.

$$\$700 + \$1200 + \$250 + \$200 + 1650 = \$4000$$

So the ratio of the mortgage payment to the total budget is

$$\frac{\$1200}{\$4000} = \frac{\cancel{400} \cdot 3}{\cancel{400} \cdot 10} = \frac{3}{10}.$$

Now work margin exercise 4.

5. The length of a rectangle is 35 ft and its width is 21 ft.
   a. Find the ratio of the length to the width.
   b. Find the ratio of the width to the length.

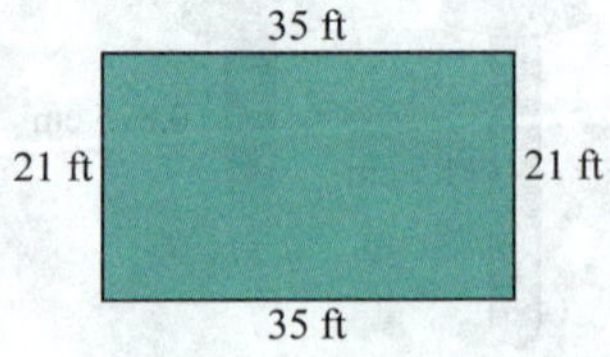

Example 5 Writing Ratios in Geometry

The lengths of the sides of a triangle are 10 in., 24 in., and 26 in.

a. Find the ratio of the length of the longest side to the length of the shortest side.

b. Find the ratio of the length of the shortest side to the length of the longest side.

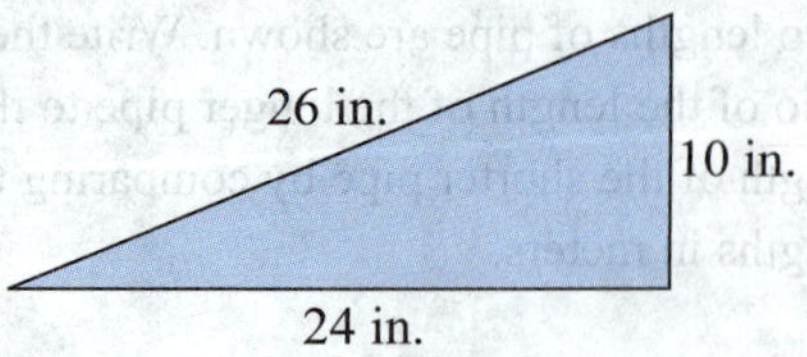

Solution

a. The ratio of the length of the longest side to the length of the shortest side is

$$\frac{26\text{ in.}}{10\text{ in.}} = \frac{2 \cdot 13}{2 \cdot 5} = \frac{13}{5}.$$

b. The ratio of the length of the shortest side to the length of the longest side is

$$\frac{10\text{ in.}}{26\text{ in.}} = \frac{2 \cdot 5}{2 \cdot 13} = \frac{5}{13}.$$

Now work margin exercise 5.

> **Note**
>
> Notice that the answers to parts a. and b. in Example 5 illustrate how the order of the comparison in the ratio is critical.

When measurements of the same type are compared, the measurements should be in the same units. For example *feet* to *feet*, *minutes* to *minutes*, *ounces* to *ounces*. When the units are different, but of the same type, the units should be changed so the smaller unit is used in both the numerator and denominator. This will guarantee whole numbers in the ratio. For example, as illustrated in Example 6, if units of time are used (in this case, hours and minutes), the numbers should be changed so all units are minutes.

The following table of equivalent measurements will be helpful in setting up ratios that compare measurements. (These measurements and others will be discussed later in Chapter 5.)

Length	Weight
12 inches = 1 foot 3 feet = 1 yard 5280 feet = 1 mile	16 ounces = 1 pound 2000 pounds = 1 ton
Capacity	**Time**
2 cups = 1 pint 2 pints = 1 quart 4 quarts = 1 gallon	60 seconds = 1 minute 60 minutes = 1 hour 24 hours = 1 day 7 days = 1 week

Table 1: Equivalent Measures

> **Math Tip**
>
> It is good practice to show units in your math work to help ensure that your answer is in terms of the units that you are expecting.

Example 6 Writing Ratios that Compare Measurements

Write the ratio of 30 minutes to 2 hours as a fraction in lowest terms.

Solution

The smaller unit is minutes. Because 1 hour = 60 minutes, we have 2 hours = 120 minutes. The ratio can be written as

$$\frac{30\text{ minutes}}{2\text{ hours}} = \frac{30\text{ minutes}}{120\text{ minutes}} = \frac{30 \cdot 1}{30 \cdot 4} = \frac{1}{4}.$$

Now work margin exercise 6.

6. Write the ratio of 4 weeks to 12 days as a fraction in lowest terms.

B Rates

A **rate** is a ratio with different units in the numerator and denominator. The units do not divide out (or cancel). For example, if a person walks 20 feet in 8 seconds the rate is

$$\frac{20 \text{ feet}}{8 \text{ seconds}} = \frac{\cancel{4} \cdot 5 \text{ feet}}{\cancel{4} \cdot 2 \text{ seconds}} = \frac{5 \text{ feet}}{2 \text{ seconds}}.$$

7. Write the following rate as a fraction in lowest terms: 165 students to 6 sections of biology.

Example 7 Writing a Rate

Write the following rate as a fraction in lowest terms: 150 kilometers (km) to 8 gallons (gal) of gas.

Solution

$$\frac{150 \text{ km}}{8 \text{ gal}} = \frac{\cancel{2} \cdot 75 \text{ km}}{\cancel{2} \cdot 4 \text{ gal}} = \frac{75 \text{ km}}{4 \text{ gal}}$$

Now work margin exercise 7.

8. Suppose a player has a batting average of .200. What does this indicate?

Example 8 Application: Batting Average

During baseball season, major league players' batting averages are published online. Suppose a player has a batting average of .250. What does this indicate?

Solution

A batting average is a ratio (or rate) of hits to times at bat. Thus, a batting average of .250 means

$$.250 = \frac{250 \text{ hits}}{1000 \text{ times at bat}}.$$

Reducing gives

$$.250 = \frac{250}{1000} = \frac{\cancel{250} \cdot 1}{\cancel{250} \cdot 4} = \frac{1}{4} = \frac{1 \text{ hit}}{4 \text{ times at bat}}.$$

This means that we can expect this player to get 1 hit for every 4 times he comes to bat. (Note that batting averages are always given to the nearest thousandth.)

Now work margin exercise 8.

C Unit Rates

A rate with a 1 in the denominator is called a **unit rate**. For example, you may attend a concert that costs \$20 per person. This is equivalent to \$20/person or $\frac{\$20}{1 \text{ person}}$. Other examples of units rates are 70 miles per hour or \$52 per month. (Note that *per* means *divided by*.)

Changing Rates to Unit Rates

To make a rate a unit rate, divide the numerator by the denominator so the denominator is 1 unit.

PROCEDURE

Example 9 Application: Writing a Unit Rate

A bicyclist rides over level ground at a steady speed of 27 miles (mi) in 2.25 hours (hr). Find her speed in miles per hour.

Solution

The rate is indicated by the ratio $\frac{27 \text{ mi}}{2.25 \text{ hr}}$.

To find the unit rate divide the numerator by the denominator.

$$\frac{27 \text{ mi}}{2.25 \text{ hr}} = \frac{2700 \text{ mi}}{225 \text{ hr}} = \frac{12 \cdot \cancel{225} \text{ mi}}{\cancel{225} \cdot 1 \text{ hr}} = 12 \text{ mi/hr}$$

Thus, her speed is 12 miles per hour, or 12 mph.

Now work margin exercise 9.

9. A swimmer swims laps at a steady pace of 6 laps in $\frac{1}{4}$ hour. Find his speed in laps per hour.

Example 10 Application: Writing a Unit Rate

Find the rate of a frog hopping across a road if the road is 20 ft wide and the frog hops 10 times to get across the road. (Assume each hop is the same distance.)

Solution

The rate can be found by comparing the number of feet covered to the number of hops needed.

$$\frac{20 \text{ ft}}{10 \text{ hops}} = \frac{\cancel{10} \cdot 2 \text{ ft}}{\cancel{10} \cdot 1 \text{ hops}} = 2 \text{ ft/hop}$$

Thus, the frog's rate is 2 feet per hop, or 2 ft/hop.

Now work margin exercise 10.

10. Find the rate of cupcake consumption at an open house if the guests consume 196 cupcakes in 4 hours.

Example 11 Application: Writing a Unit Rate

You find yourself reading a particularly exciting detective thriller novel. In one afternoon in the library, you read 300 pages of this book in 2.5 hours. What was your reading rate? That is, how many pages did you read in one hour?

Solution

$$\frac{300 \text{ pages}}{2.5 \text{ hours}} = \frac{3000 \text{ pages}}{25 \text{ hours}} = \frac{\cancel{25} \cdot 120 \text{ pages}}{\cancel{25} \cdot 1 \text{ hours}} = 120 \text{ pages/hour}$$

Thus, your reading rate was 120 pages per hour, or 120 pages/hour.

Now work margin exercise 11.

11. An office's new copier printed 630 pages in 15 minutes. How many pages did it print per minute?

D Comparing Unit Prices

> **Note**
>
> In the past, many consumers did not understand the concept of unit price or know how to determine such a number. Now, most states have a law that grocery stores must display the unit price for certain goods they sell so that consumers can be fully informed.

One common example of unit rate is the **unit price** (or **price per unit**). When you buy groceries, the same item may be packaged in two (or more) sizes. Since you want to get the most for your money, you want the better (or best) buy and, generally, want to buy the item with the lowest price per unit. To find the price per unit, divide the price of the item by the number of units.

In the following examples, we show the division process as you would write it on paper. However, you may choose to use a calculator to perform these operations. In either case, your first step should be to write each ratio so you can clearly see the values you are dividing and that all ratios are comparing the same types of units.

Notice that the comparisons being made in the examples and exercises are with different amounts of the same item.

12. Which is a better buy: a 2-liter bottle of soda for \$1.09 or a 3-liter bottle for \$1.49?

Example 12 Application: Comparing Unit Prices

A 16-ounce jar of grape jam is priced at \$3.99 and a 9.5-ounce jar of the same jam is \$2.69. Which is the better buy?

Solution

To solve this problem, we must divide to find the price per ounce of each jar of jam. Then we will compare the results to decide which is the better buy. In this problem, we convert dollars to cents (\$3.99 = 399¢ and \$2.69 = 269¢) so that the results will be ratios of cents per ounce. (Answers are rounded to the nearest tenth of a cent.)

16-ounce jar

$$\frac{399¢}{16\text{ oz}} \rightarrow \begin{array}{r} 24.93 \\ 16\overline{)399.00} \\ \underline{-32} \\ 79 \\ \underline{-64} \\ 15\ 0 \\ \underline{-14\ 4} \\ 60 \\ \underline{-48} \\ 12 \end{array} \quad \text{Or } 24.9¢ \text{ per ounce}$$

The 16-ounce jar costs 24.9¢ per ounce.

9.5-ounce jar

$\frac{269¢}{9.5\text{ oz}} \rightarrow 9.5.\overline{)269.0.00}$ = 28.31 Or 28.3¢ per ounce

```
       28.31
9.5.)269.0.00
    -190
      79 0
     -76 0
       3 0 0
      -2 8 5
         1 50
         -95
          55
```

The 9.5-ounce jar costs 28.3¢ per ounce.

Thus, the larger jar (16 ounces for \$3.99) is the better buy because the price per ounce is less.

Now work margin exercise 12.

Example 13 Application: Comparing Unit Prices

At the local Shop and Save grocery store, bottles of pancake syrup come in three different sizes.

36 fluid ounces for \$5.29

24 fluid ounces for \$4.69

12 fluid ounces for \$3.99

Find the price per fluid ounce (to the nearest tenth of a cent) for each size of bottle and tell which is the best buy.

Solution

After each ratio is set up, the numerator is changed from dollars and cents to just cents so that the division will yield cents per fluid ounce.

36 fluid ounces for \$5.29

$\frac{\$5.29}{36\text{ oz}} = \frac{529¢}{36\text{ oz}} \rightarrow 36\overline{)529.00}$ = 14.69 Or 14.7¢ per fluid ounce

```
     14.69
36)529.00
  -36
   169
  -144
    25 0
   -21 6
     3 40
    -3 24
       16
```

Therefore, the largest bottle costs 14.7¢ per fluid ounce, or 14.7¢/fl oz.

Note: "14.7¢/fl oz." is read "14.7¢ per fluid ounce."

13. Which is a better buy: a 12.5-ounce box of cereal for \$2.00, a 16-ounce box for \$3.00, or a 24-ounce box for \$3.50? What is the price per ounce (to the nearest tenth of a cent) for each?

24 fluid ounces for $4.69

$$\frac{\$4.69}{24\text{ oz}} = \frac{469¢}{24\text{ oz}} \rightarrow \begin{array}{r} 19.54 \\ 24\overline{)469.00} \\ \underline{-24} \\ 229 \\ \underline{-216} \\ 13\ 0 \\ \underline{-12\ 0} \\ 1\ 00 \\ \underline{-96} \\ 4 \end{array}$$

Or 19.5¢ per fluid ounce

The medium-sized bottle costs 19.5¢/fl oz.

12 fluid ounces for $3.99

$$\frac{\$3.99}{12\text{ oz}} = \frac{399¢}{12\text{ oz}} \rightarrow \begin{array}{r} 33.25 \\ 12\overline{)399.00} \\ \underline{-36} \\ 39 \\ \underline{-36} \\ 3\ 0 \\ \underline{-2\ 4} \\ 60 \\ \underline{-60} \\ 0 \end{array}$$

Or 33.3¢ per fluid ounce

The smallest bottle costs 33.3¢/fl oz.

Looking at the price per fluid ounce for each bottle, the largest container (36 fluid ounces) has the lowest price per fluid ounce and is therefore the best buy.

Now work margin exercise 13.

In each of the examples, the larger (or largest) amount of units was the better (or best) buy. Often, larger amounts are less expensive because packaging costs less and the manufacturer wants you to buy more of the product. However; the larger amount is not always the better buy because of other considerations such as storage space or the consumer's individual needs. For example, people who do not use much pancake syrup may want to buy a smaller bottle. Even though they pay more per unit, it's more economical in the long run not to have to throw any away.

Special Comment on the Term **per**

Be aware that the term **per** can be interpreted to mean **divided by**. For example,

cents **per** ounce	means	cents **divided by** ounces
dollars **per** pound	means	dollars **divided by** pounds
miles **per** hour	means	miles **divided by** hours
miles **per** gallon	means	miles **divided by** gallons

Margin Exercise Answers

1. $\frac{6}{5}$ **2. a.** $\frac{2}{7}$ **b.** $\frac{4}{1}$ **3.** $\frac{12}{7}$ **4. a.** $\frac{5}{24}$ **b.** $\frac{7}{40}$ **5. a.** $\frac{5}{3}$ **b.** $\frac{3}{5}$ **6.** $\frac{7}{3}$

7. $\frac{55 \text{ students}}{2 \text{ sections}}$ **8.** We can expect the player to get 1 hit for every 5 times at bat. **9.** 24 laps/hr **10.** 49 cupcakes/hour **11.** 42 pgs/min **12.** 3-liter bottle for $1.49 **13.** 24-ounce box for $2.00; 16¢/oz (12.5-ounce box); 18.8¢/oz (16-ounce box); 14.6¢/oz (24-ounce box)

4.1 Exercises

Concept Check

Fill-in-the-Blank. Complete each sentence using information found in this section.

1. When two quantities are compared by division, the comparison is called a/an ________.

2. The ratio $\frac{3}{8}$ is read "the ratio of ________ to ________."

3. Ratios can be ____________ to lowest terms.

4. When numbers in a ratio have different units, the ratio is said to be a/an ______.

5. A unit rate is a rate with denominator __________.

6. To find a unit price (or price per unit), divide the __________ by the number of units.

True/False. Determine whether each statement is true or false. If a statement is false, explain how it can be changed so the statement will be true. (**Note:** There may be more than one acceptable change.)

7. The units in the numerator and denominator of a ratio must be the same, or need to be able to be converted to the same units.

8. The order of the numbers in a ratio or a rate is irrelevant as long as the numbers are reduced.

9. The ratio 8:2 can be reduced to the ratio 4.

10. To make a unit rate, divide the numerator by the denominator.

Practice

Write each ratio as a fraction in lowest terms. See Examples 1 through 3.

1. 18 to 28

2. 25 to 30

3. 3.2 to 2.4

4. 2.1 to 3.5

5. $\frac{2}{3}$ to $\frac{5}{6}$

6. $\frac{10}{9}$ to $\frac{5}{12}$

7. $5.50 to $8.00

8. $12.75 to $18.50

4.1 Exercises

Concept Check

1. ratio
2. three; eight
3. reduced
4. rate
5. 1
6. price
7. True
8. False; The order of the numbers in a ratio or a rate is important.
9. False; The ratio 8:2 can be reduced to the ratio 4:1.
10. True

Practice

1. $\frac{9}{14}$
2. $\frac{5}{6}$
3. $\frac{4}{3}$
4. $\frac{3}{5}$
5. $\frac{4}{5}$
6. $\frac{8}{3}$
7. $\frac{11}{16}$
8. $\frac{51}{74}$

9. $\frac{7}{24}$

10. $\frac{6}{19}$

11. $\frac{1}{2}$

12. $\frac{1}{3}$

13. $\frac{4}{1}$ or 4

14. $\frac{6}{5}$

15. $\frac{1\text{ minute}}{40\text{ seconds}}$ or $\frac{3}{2}$

16. $\frac{2\text{ hours}}{15\text{ minutes}}$ or $\frac{8}{1}=8$

17. $\frac{3}{4}$

18. $\frac{3}{2}$

19. $\frac{4}{7}$

20. $\frac{3}{4}$

21. $\frac{\$2\text{ profit}}{\$5\text{ invested}}$

22. $\frac{\$1\text{ profit}}{\$5\text{ invested}}$

23. $\frac{1\text{ teacher}}{12\text{ students}}$

24. $\frac{1\text{ teacher}}{18\text{ students}}$

25. $\frac{1\text{ hit}}{5\text{ times at bat}}$

26. $\frac{1\text{ hit}}{4\text{ times at bat}}$

27. $\frac{10\text{ children}}{3\text{ families}}$

28. $\frac{15\text{ children}}{8\text{ families}}$

29. $\frac{5\text{ scholarships}}{42\text{ students}}$

30. $\frac{3\text{ scholarships}}{40\text{ students}}$

31. 60 miles per hour

32. 70 miles per hour

9. $3\frac{1}{2}$ donuts to 12 donuts

10. 3 months to $9\frac{1}{2}$ months

For each ratio, change to common units in the numerator and denominator and then write each ratio as a fraction in lowest terms. See Example 6.

11. 1 dime to 4 nickels

12. 5 nickels to 3 quarters

13. 5 dollars to 5 quarters

14. 6 dollars to 50 dimes

15. 5 minutes to 200 seconds

16. 4 hours to 30 minutes

17. 18 inches to 2 feet

18. 36 inches to 2 feet

19. 8 days to 2 weeks

20. 21 days to 4 weeks

Write each rate as a fraction in lowest terms. See Example 7.

21. $200 in profit to $500 invested

22. $200 in profit to $1000 invested

23. 2 teachers to 24 students

24. 3 teachers to 54 students

25. 100 hits to 500 times at bat

26. 125 hits to 500 times at bat

27. 1500 children to 450 families

28. 900 children to 480 families

29. 25 scholarships to 210 students

30. 12 scholarships to 160 students

Write each rate as a unit rate. See Examples 9 through 11.

31. 270 miles to 4.5 hours

32. 245 miles to 3.5 hours

33. 50 miles to 2 gallons of gas

34. 60 miles to 5 gallons of gas

35. 1260 words to 30 minutes

36. 840 words to 15 minutes

37. 2250 gallons of water to 25 people

38. 3600 gallons of water to 16 people

39. $612 to 48 hours

40. $399 to 28 hours

Find the unit price (to the nearest tenth of a cent) of each set of items and tell which one is the better (or best) purchase. See Examples 12 and 13.

41. ***Coffee beans:*** 1.75 oz at $1.99, 12 oz at $7.99

42. ***Coffee:*** 11.5 oz at $3.99, 39 oz at $8.99

43. ***Socks:*** 5 pairs for $12.50, 3 pairs for $8.25

44. ***Tennis balls:*** 3-pack for $2.99, 12-pack for $10.98

45. ***Sliced ham:*** 12 oz at $3.99, 8 oz at $2.89

46. ***Frozen orange juice:*** 16 fl oz at $2.99, 12 fl oz at $2.19

47. ***Mayonnaise:*** 8 oz at $1.59, 16 oz at $2.69, 32 oz at $3.89, 64 oz at $6.79

48. ***Aluminum foil:*** 200 sq ft at $7.39, 75 sq ft at $3.69, 50 sq ft at $3.19, 25 sq ft at $1.49

Applications

Solve.

49. A serving of three home-baked chocolate chip cookies weighs 40 grams and contains 12 grams of fat. What is the rate, in lowest terms, of fat grams to total grams?

50. In recent years, 18 out of every 100 students taking the SAT (Scholastic Aptitude Test) at a local school have scored 600 or above on the mathematics portion of the test. Write the ratio, in lowest terms, of the number of scores 600 or above to the number of scores below 600.

51. In a recent year, Albany, NY reported a total of 60 clear days, the rest being cloudy or partly cloudy. For a 365-day year, write the rate, in lowest terms, of clear days to cloudy or partly cloudy days.

52. Different sellers for an online book broker sell the same book title for a wide variety of prices. For a certain title, the highest selling price (mint condition) is \$35, while the lowest selling price is \$10. What is the ratio of the highest price to the lowest?

53. One truck can carry 9.2 tons of cargo, while a second truck can haul 11.5 tons. What is the ratio of their capacities? (**Hint:** Multiply both the numerator and denominator by 10 to remove the decimal point.)

54. If the CEO of a major corporation makes \$4,000,000 a year, and the average annual salary of the workers is \$30,000 a year, what is the ratio of the CEO's salary to that of the regular workers?

55. John has collected 7 quarters, while Walt has collected 25 dimes. How much money does Walt have compared with John's accumulation? (**Hint:** When converting the coin collections to monetary values, express them in cents.)

56. A recipe calls for 6 cups of milk and 4 cups of bread crumbs. What is the ratio of milk to bread crumbs?

57. A farmer has a long narrow rectangular piece of land. If the length is 9504 feet and the width is 1320 feet, what is the ratio of the length to the width?

58. Lemonade can be made by mixing lemon juice, sugar, and water in the ratio 1:1:6. Find the amount of each ingredient in 12 quarts of lemonade.

The circle graph shown here illustrates how Reece spent his time the day before his final exams. Use the information from the graph to answer the following questions.

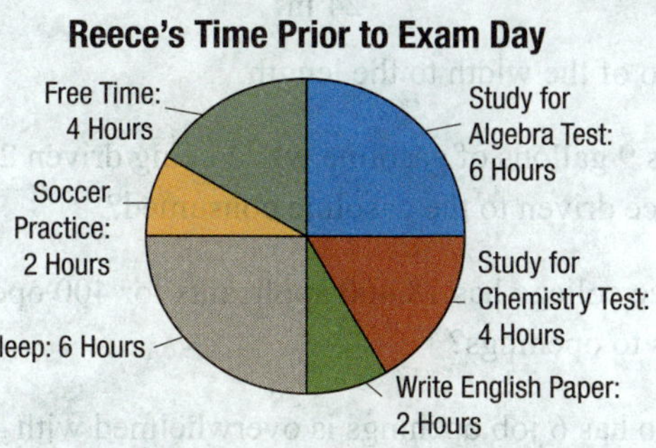

59. What is the ratio of time Reece spent studying for his algebra exam to time spent studying for his chemistry exam?

33. 25 miles per gallon
34. 12 miles per gallon
35. 42 words per minute
36. 56 words per minute
37. 90 gallons per person
38. 225 gallons per person
39. \$12.75 per hour
40. \$14.25 per hour
41. 113.7¢/oz; 66.6¢/oz; 12 oz at \$7.99
42. 34.7¢/oz; 23.1¢/oz; 39 oz at \$9.39
43. 250¢/pair; 275¢/pair; 5 pairs at \$12.50
44. 99.7¢/ball; 91.5¢/ball; 3-pack at \$10.98
45. 33.3¢/oz; 36.1¢/oz; 12 oz at \$3.99
46. 18.7¢/fl oz; 18.3¢/fl oz; 12 fl oz at \$2.19
47. 19.9¢/oz; 16.8¢/oz; 12.2¢/oz; 10.6¢/oz; 64 oz at \$6.79
48. 3.7¢/sq ft; 4.9¢/sq ft; 6.4¢/sq ft; 6¢/sq ft; 200 sq ft at \$7.39

Applications

49. $\frac{3 \text{ fat grams}}{10 \text{ grams}}$
50. $\frac{9}{41}$
51. $\frac{12 \text{ clear days}}{61 \text{ cloudy days}}$
52. $\frac{7}{2}$
53. $\frac{4}{5}$
54. $\frac{400}{3}$
55. $\frac{10}{7}$
56. $\frac{3}{2}$
57. $\frac{36}{5}$

58. 1.5 quarts lemon juice, 1.5 quarts sugar, 9 quarts water

59. $\frac{3}{2}$

60. $\frac{1}{3}$

61. $\frac{1}{12}$

62. $\frac{1}{6}$

63. $\frac{4}{5}$

64. $\frac{3}{5}$

65. $\frac{4}{15}$

66. $\frac{4}{25}$

67. a. $\frac{7}{3}$ b. $\frac{3}{7}$

68. a. $\frac{8}{3}$ b. $\frac{3}{8}$

69. 25 miles per gallon

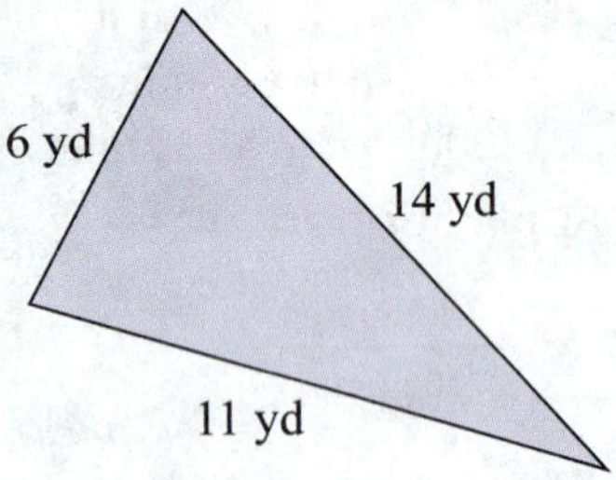

70. 46 applicants per opening

71. 74 applicants per opening

72. 6 outlets per office

73. 5 ipads per classroom

60. What is the ratio of time Reece spent at soccer practice to time spent sleeping?

61. What is the ratio of time Reece spent writing his English paper to the total number of hours in the day?

62. What is the ratio of time Reece spent in free time to the total number of hours in the day?

Iliana is organizing an after school program for kids in her school. The circle graph shown here illustrates the number of kids that have signed up for each session. Use the information from the graph to answer the following questions.

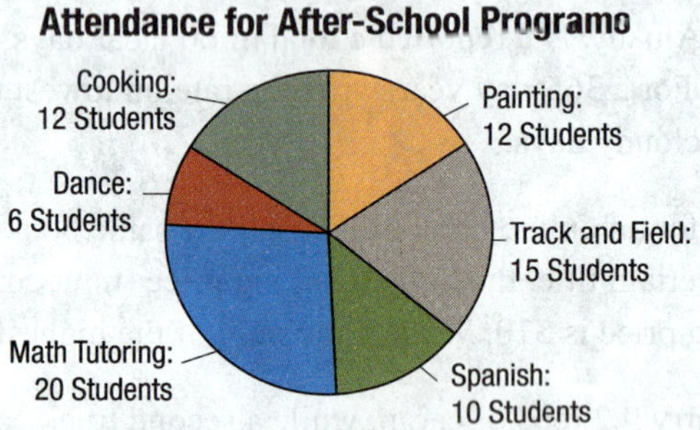

63. What is the ratio of children in Cooking to children in Track and Field?

64. What is the ratio of children in Dance to children in Spanish?

65. What is the ratio of children in Math Tutoring to total children in the program?

66. What is the ratio of children in Painting to total children in the program?

Solve.

67. The lengths of the sides of a triangle are 6 yd, 11 yd, and 14 yd.

a. Find the ratio of the length of the longest side to the length of the shortest side.

b. Find the ratio of the length of the shortest side to the length of the longest side.

68. The length of a rectangle is 24 in. and its width is 9 in.

a. Find the ratio of the length to the width.

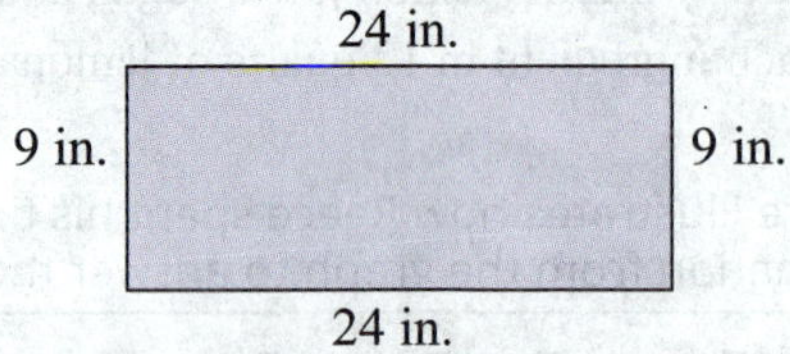

b. Find the ratio of the width to the length.

69. If a car consumes 9 gallons of gasoline while being driven 225 miles, what is the unit rate of the distance driven to the gasoline consumed?

70. A certain selective college has 18,400 applicants for 400 openings. What is the unit rate of applicants to openings?

71. An employer who has 6 job openings is overwhelmed with 444 applicants. What is the rate of applicants per job opening?

72. In the new office building he is designing, Steve is including 216 electrical outlets in the offices. If the building will have 36 similar sized offices, what is the unit rate of outlets to offices?

73. A benefactor donated new iPads to his local elementary school. If he gave the school 210 ipads and the school has 42 classrooms, what is the unit rate of ipads per classroom?

74. An automobile manufacturer produces 6000 cars in a week. Find their production rate of cars per day. (Assume a 5 day workweek.)

75. An office entered a local contest for best Christmas decorations and won. If the grand prize was $1500 and the money was split evenly between all 24 employees, what were the dollar per person winnings?

76. A restaurant charges a flat fee of $6.50 to fill up a salad bowl with whatever you want from their salad bar. If the salad Marianita made weighed 12.5 ounces, what was its cost per ounce?

77. Suppose a baseball player has a batting average of .300. What does this indicate?

78. Suppose a baseball player has a batting average of .280. What does this indicate?

79. A store sells 42-gallon trash bags in two different amounts. If you can buy a box of 50 trash bags for $25.98 or a box of 24 trash bags for $14.98, which option is the better deal per trash bag?

80. A landscaping company needs to purchase grass seed to plant a new lawn. Brand A sells 225 square feet worth of seed for $29.97. Brand B sells 400 square feet worth of seed for $55.97. Which option should they go with?

81. A shopper paid $2.12 for 4 pounds of bananas. What was the price per pound of bananas?

82. A carton of 18 free-range organic eggs costs $5.94. What is the cost per egg?

83. A 24-pack of 12-ounce cans of cola costs $6.00. What is the price per can?

84. A 24-pack of 16.9-ounce bottles of water is sold for $3.84. What is the price per bottle of water?

85. A shopper buys a 14-ounce bag of candy for $5.99 and uses a coupon for $1 off. What is the price per ounce after the discount?

86. A shopper buys a 24-pack of 3-ounce cans of cat food for $16.18 and uses a coupon for $3.25 off. What is the price per can after the discount?

87. A car travels 441 miles and uses 14 gallons of gas. The fuel efficiency rating for the car is advertised to be 33 miles per gallon.

   a. Find the rate of miles per gallon for the car.

   b. Does the car's mileage per gallon match the advertised mileage per gallon? If not, is it higher or lower?

88. A receptionist at a luxury spa sold 32 memberships during her 5-day work week. She is required to sell an average of 6 memberships per work day.

   a. Find the rate of memberships per day that the receptionist sold.

   b. Does the average number of membership sales by the receptionist match her daily quota? If not, is it higher or lower?

74. 1200 cars per day

75. $62.50 per person

76. 52¢ per ounce

77. We can expect the player to get 3 hits for every 10 times at bat.

78. We can expect the player to get 7 hits for every 25 times at bat.

79. 50 trash bags for $25.98

80. Brand A; 225 sq ft for $29.97

81. $0.53 per pound

82. $0.33 per egg

Highest Batting Average

As of 2019, Ty Cobb holds the record for the highest batting average in Major League Baseball, at .366 or .367, depending on the source. He achieved this over 11,434 total career at bats.

83. $0.25 per can

84. $0.16 per bottle

85. $0.36 per ounce

86. $0.54 per can

87. **a.** 31.5 miles per gallon
b. No; 31.5 miles per gallon is lower than the advertised amount.

88. **a.** 6.4 memberships per day
b. No; 6.4 memberships per day is higher than the quota.

89. 63 miles per hour is lower than the posted speed limit.

90. 150 milliliters per hour matches the prescribed rate.

Writing & Thinking

91. The ratio can be written as $\frac{5}{3}$, 5:3, or 5 to 3. The preferred method is as a fraction because it is easier to simplify and manipulate in mathematical operations.

92. Examples will vary. Rates may be used to compare costs when shopping, driving speeds, power usage, etc.

93. a. A ratio with different units in the numerator and denominator (the units do not cancel out); Answers will vary.

b. A rate with a denominator value of 1 unit; Answers will vary.

c. The average price for each unit of an item containing multiple units; Answers will vary.

94. Numerator

89. A motorist drives along an interstate highway for 3 hours and travels a total of 189 miles. The posted speed limit is 65 miles per hour. Does the motorist's average speed (in miles per hour) match the posted speed limit? If not, is it higher or lower?

90. A nurse sets up a 750 milliliter bag of intravenous solution. The bag empties in 5 hours. The patient's doctor prescribed the solution to be administered at a rate of 150 milliliters per hour. Does the rate that the nurse set the bag at match the prescribed rate? If not, is it higher or lower?

Writing & Thinking

91. Demonstrate three different ways the ratio comparing 5 apples to 3 apples can be written. Choose one form and explain why it is the preferred form when using ratios in a math course.

92. Give an example of when the use of ratios or rates might be relevant in a real life situation.

93. In your own words, define each of the following terms.

a. Rate

b. Unit rate

c. Price per unit

94. When finding price per unit, will monetary units be located in the numerator or the denominator of the rate?

4.2 Proportions

Objectives

A. Verify proportions.

B. Solve proportions.

C. Solve application problems using proportions.

A Introduction to Proportions

One way to compare the two fractions $\frac{3}{6}$ and $\frac{4}{8}$ is to find a common denominator and then compare the numerators. For example,

$$\frac{3}{6}=\frac{3}{6}\cdot\frac{4}{4}=\frac{12}{24} \quad\text{and}\quad \frac{4}{8}=\frac{4}{8}\cdot\frac{3}{3}=\frac{12}{24}.$$

We see that, because the resulting numerators are equal, the original fractions are equal.

Another comparison technique is to reduce both fractions.

$$\frac{3}{6}=\frac{\cancel{3}\cdot 1}{\cancel{3}\cdot 2}=\frac{1}{2} \quad\text{and}\quad \frac{4}{8}=\frac{\cancel{4}\cdot 1}{\cancel{4}\cdot 2}=\frac{1}{2}$$

Because the original fractions reduce to the same fraction, we see that the original fractions are equal.

A third method is to set up an equation, called a **proportion**, which states that the two ratios are equal. If this statement is true, then the **cross products** must be equal. Cross products are found by cross multiplying, as illustrated here.

$\frac{3}{6}=\frac{4}{8}$ is true because $3\cdot 8=6\cdot 4$. ← Cross products

Proportions

A **proportion** is a statement that two ratios are equal.

In symbols, $\frac{\boldsymbol{a}}{\boldsymbol{b}}=\frac{\boldsymbol{c}}{\boldsymbol{d}}$ $(b\neq 0, d\neq 0)$ is a proportion.

A proportion is true if the **cross products**, $a\cdot d$ and $b\cdot c$, are equal.

DEFINITION

Math Tip

The cross product idea is a result of multiplying both sides of the equation by the product of the denominators as in the following examples.

$$\frac{3}{6}=\frac{4}{8}$$

$$\overset{8}{\cancel{48}}\cdot\frac{3}{\cancel{6}}=\frac{4}{\cancel{8}}\cdot\overset{6}{\cancel{48}}$$

$$8\cdot 3=4\cdot 6$$

or

$$\frac{a}{b}=\frac{c}{d}$$

$$\cancel{b}d\cdot\frac{a}{\cancel{b}}=\frac{c}{\cancel{d}}\cdot b\cancel{d}$$

$$d\cdot a=c\cdot b$$

Example 1 Verifying Proportions

Compare the cross products to determine whether each proportion is true or false.

a. $\frac{6}{8}=\frac{15}{20}$ **b.** $\frac{5}{8}=\frac{7}{10}$ **c.** $\frac{9}{13}=\frac{4.5}{6.5}$

Solution

a. $6\cdot 20=120$ and $8\cdot 15=120$

Therefore, the cross products are equal and the proportion $\frac{6}{8}=\frac{15}{20}$ is true.

b. $5\cdot 10=50$ and $8\cdot 7=56$

Because the cross products are not equal $(50\neq 56)$, the proportion $\frac{5}{8}=\frac{7}{10}$ is false.

1. Determine whether each proportion is true or false.

a. $\frac{8}{28}=\frac{15}{35}$

b. $\frac{5}{6}=\frac{7}{9}$

c. $\frac{3.5}{4}=\frac{5.25}{6}$

c.

$$\begin{array}{r} 6.5 \\ \times\ \ 9 \\ \hline 58.5 \end{array} \quad \text{and} \quad \begin{array}{r} 4.5 \\ \times\ 13 \\ \hline 135 \\ 450 \\ \hline 58.5 \end{array}$$

Because the cross products are equal, the proportion $\frac{9}{13}=\frac{4.5}{6.5}$ is true.

Now work margin exercise 1.

2. Determine whether the given proportion is true or false.

$$\frac{\frac{4}{5}}{\frac{1}{3}}=\frac{4}{3}$$

Completion Example 2 Verifying Proportions

Determine whether the given proportion is true or false. (**Hint:** Change each mixed number to an improper fraction.)

$$\frac{2\frac{1}{3}}{7}=\frac{3\frac{1}{4}}{9\frac{3}{4}}$$

Solution

$$2\frac{1}{3}\cdot 9\frac{3}{4}=\frac{7}{\cancel{3}}\cdot\frac{\overset{13}{\cancel{39}}}{4}=\underline{\frac{91}{4}} \quad \text{and} \quad 7\cdot 3\frac{1}{4}=\frac{7}{1}\cdot\frac{13}{4}=\underline{\frac{91}{4}}$$

Because the cross products are __equal__, the proportion is __true__.

Now work margin exercise 2.

B Solving Proportions

If one of the numbers in a proportion is unknown, it can be replaced with a variable, such as x. A **variable** is a letter that represents an unknown number. For example,

$$\frac{x}{12}=\frac{2}{3}.$$

We want to find the value of this unknown number that makes the proportion true. That is, we want to **solve the proportion**. This can be done as follows.

$$\frac{x}{12}=\frac{2}{3}$$

$3\cdot x=12\cdot 2$ Find the cross products and set them to equal each other.

$\frac{\cancel{3}\cdot x}{\cancel{3}}=\frac{24}{3}$ Divide both sides by 3, the number that multiplies the variable.

$x=8$ Simplify.

Solving a Proposition

1. Find the **cross products** (or **cross multiply**) and then set the cross products equal to each other.
2. Divide both sides of the equation by the number that multiplies the variable.
3. Simplify.

PROCEDURE

Example 3 Solving Proportions

Find the value of x if $\frac{4}{8}=\frac{5}{x}$.

Solution

$\frac{4}{8}=\frac{5}{x}$ Write the proportion.

$4\cdot x=8\cdot 5$ Find the cross products and set them equal to each other.

$\frac{\cancel{4}\cdot x}{\cancel{4}}=\frac{40}{4}$ Divide both sides by 4, the number that multiplies the variable.

$x=10$ Simplify.

Now work margin exercise 3.

3. Find the value of x if $\frac{12}{x}=\frac{9}{15}$.

Example 4 Solving Proportions

Find the value of y if $\frac{6}{16}=\frac{y}{24}$.

Solution

$\frac{6}{16}=\frac{y}{24}$ Write the proportion.

$6\cdot 24=16\cdot y$ Find the cross products and set them equal to each other.

$\frac{144}{16}=\frac{\cancel{16}\cdot y}{\cancel{16}}$ Divide both sides by 16, the number that multiplies the variable.

$9=y$ Simplify.

Note that the variable may appear on the right side of the equation as well as on the left side of the equation. In either case, **we divide both sides of the equation by the number that multiplies the variable.**

Alternative Solution

Reduce the fraction $\frac{6}{16}$ before solving the proportion to keep the numbers smaller and easier to work with.

4. Find the value of y if $\frac{14}{18}=\frac{y}{63}$.

$\frac{6}{16}=\frac{y}{24}$ Write the proportion.

$\frac{3}{8}=\frac{y}{24}$ Reduce the fraction: $\frac{6}{16}=\frac{\cancel{2}\cdot 3}{\cancel{2}\cdot 8}=\frac{3}{8}$

$3\cdot 24=8\cdot y$ Proceed to solve as before.

$\frac{72}{8}=\frac{\cancel{8}\cdot y}{\cancel{8}}$

$9=y$

Now work margin exercise 4.

5. Find the value of y if $\frac{4}{7}=\frac{y}{2.8}$.

Example 5 Solving Proportions

Find the value of y if $\frac{5}{9}=\frac{y}{4.5}$.

Solution

$\frac{5}{9}=\frac{y}{4.5}$ Write the proportion.

$5\cdot 4.5=9\cdot y$ Find the cross products and set them equal to each other.

$\frac{22.5}{9}=\frac{\cancel{9}\cdot y}{\cancel{9}}$ Divide both sides by 9, the number that multiplies the variable.

$2.5=y$ Simplify.

Now work margin exercise 5.

6. Find the value of x if $\frac{\frac{1}{2}}{6}=\frac{5}{x}$.

Example 6 Solving Proportions

Find the value of n if $\frac{n}{7}=\frac{20}{\frac{2}{3}}$.

Solution

$\frac{n}{7}=\frac{20}{\frac{2}{3}}$ Write the proportion.

$\frac{2}{3}\cdot n=7\cdot 20$ Find the cross products and set them equal to each other.

$\frac{\cancel{\frac{2}{3}}\cdot n}{\cancel{\frac{2}{3}}}=\frac{140}{\frac{2}{3}}$ Divide both sides by $\frac{2}{3}$, the number that multiplies the variable.

$n=140\cdot\frac{3}{2}$ Multiply by the reciprocal.

$n=210$ Simplify.

Now work margin exercise 6.

Completion Example 7 Solving Proportions

Find the value of y if $\frac{2\frac{1}{2}}{6} = \frac{3}{y}$.

Solution

First change the mixed number $2\frac{1}{2}$ to the improper fraction $\frac{5}{2}$.

$\frac{2\frac{1}{2}}{6} = \frac{3}{y}$ Write the proportion.

$\frac{\frac{5}{2}}{6} = \frac{3}{y}$ Change the mixed number to an improper fraction.

$\frac{5}{2} \cdot y = 3 \cdot 6$ Find the cross products and set them equal to each other.

$\frac{\frac{5}{2} \cdot y}{\frac{5}{2}} = \frac{18}{\frac{5}{2}}$ Divide both sides by $\frac{5}{2}$, the number that multiplies the variable.

$y = 18 \cdot \frac{2}{5}$ Multiply by the reciprocal.

$y = \frac{36}{5}$ Simplify.

Now work margin exercise 7.

7. Find the value of z if $\frac{2\frac{2}{3}}{5} = \frac{\frac{1}{5}}{z}$.

C Applications of Proportions

Problems that involve two ratios are the type that can be solved using proportions.

Solving an Application Using a Proportion

1. Identify the unknown quantity and use a variable to represent this quantity.
2. Set up a proportion in which the units are compared in the same order. (Make sure that the units are labeled so they can be seen to be in the right order.)
3. Solve the proportion.

PROCEDURE

Example 8 Application: Solving Proportions

A motorcycle will travel 352 miles on 11 gallons of gas. How many miles will this motorcycle travel on 15 gallons of gas?

8. Two pounds of candy costs \$4.75. How many pounds of candy can be bought for \$28.50?

Math Tip

Although the way you set up the proportions may vary, the units that result from the cross products should be the same. In Example 8, the cross products have the units miles · gallons.

So you may not have the following set ups.

$$\frac{\text{miles}}{\text{gallons}} = \frac{\text{gallons}}{\text{miles}}$$

$$\frac{\text{gallons}}{\text{miles}} = \frac{\text{miles}}{\text{gallons}}$$

Verify this for yourself.

Solution

Assign the variable: Let x = unknown number of miles.

Set up the proportion: $\frac{352 \text{ miles}}{11 \text{ gallons}} = \frac{x \text{ miles}}{15 \text{ gallons}}$ The units are in the same order (miles to gallons) in each ratio.

Solve the proportion:

$$352 \cdot 15 = 11 \cdot x$$

$$\frac{5280}{11} = \frac{\cancel{11} \cdot x}{\cancel{11}}$$

$$480 = x$$

The motorcycle will travel 480 miles on 15 gallons of gas.

Now work margin exercise 8.

In Example 8, *any* of the following proportions will yield the same answer.

$\frac{11 \text{ gallons}}{352 \text{ miles}} = \frac{15 \text{ gallons}}{x \text{ miles}}$ For each ratio, the units in the numerator are different than the units in the denominator. However, both numerators have the same units and both denominators have the same units

$\frac{352 \text{ miles}}{x \text{ miles}} = \frac{11 \text{ gallons}}{15 \text{ gallons}}$ For each ratio, the units in the numerator and denominator are the same. Additionally, the numerators correspond and the denominators correspond.

$\frac{x \text{ miles}}{352 \text{ miles}} = \frac{15 \text{ gallons}}{11 \text{ gallons}}$ For each ratio, the units in the numerator and denominator are the same. Additionally, the numerators correspond and the denominators correspond.

9. Precinct 1 has 520 registered voters and Precinct 2 has 630 registered voters. In the last election, the ratio of actual voters to registered voters was the same in both precincts. If there were 104 actual voters in Precint 1, how many actual voters were there in Precinct 2?

Example 9 Application: Solving Proportions

An architect draws the plans for a building by using a scale of $\frac{1}{2}$ inch to represent 10 feet. How many feet are represented by 6 inches?

Solution

Assign the variable: Let y = unknown number of feet.

Set up the proportion:

$\frac{\frac{1}{2} \text{ inch}}{6 \text{ inches}} = \frac{10 \text{ feet}}{y \text{ feet}}$ For each ratio in the proportion, the units in the numerator and denominator are the same. Additionally, the numerators correspond and the denominators correspond.

Solve the proportion:

$$\frac{1}{2} \cdot y = 6 \cdot 10$$

$$\frac{\cancel{\frac{1}{2}} \cdot y}{\cancel{\frac{1}{2}}} = \frac{60}{\frac{1}{2}}$$

$$y = 60 \cdot \frac{2}{1}$$

$$y = 120$$

On these plans, 6 inches represents 120 feet.

Now work margin exercise 9.

Example 10 Application: Solving Proportions

A tire on a car makes 250 revolutions per minute. How many revolutions will the tire make in one hour?

Solution

Assign the variable: Let x = the number of revolutions made in one hour.

Set up the proportion: Since 1 hr = 60 min, we have

$$\frac{250 \text{ rev}}{1 \text{ min}} = \frac{x \text{ rev}}{60 \text{ min}}$$

Solve the proportion:

$$250 \cdot 60 = 1 \cdot x$$
$$15{,}000 = x$$

The tire will make 15,000 revolutions in one hour.

Now work margin exercise 10.

10. An investor expects to make \$8 for every \$125 he invests. If he invests \$1000, how much money should he expect to make?

Completion Example 11 Application: Solving Proportions

A recommended method of diluting weed killer is 3 capfuls of weed killer to 2 gallons of water. How many capfuls of weed killer should be mixed with 5 gallons of water?

Solution

Assign the variable: Let x = unknown number of capfuls of weed killer.

Set up the proportion: $\frac{3 \text{ capfuls}}{2 \text{ gallons}} = \frac{x \text{ capfuls}}{5 \text{ gallons}}$

Solve the proportion:

$$5 \cdot \underline{\ 3\ } = x \cdot \underline{\ 2\ }$$
$$\frac{15}{2} = \frac{x \cdot \cancel{2}}{\cancel{2}}$$
$$\frac{15}{2} = x$$

$\underline{\frac{15}{2} \text{ or } 7\frac{1}{2}}$ capfuls of weed killer should be mixed with 5 gallons of water.

Now work margin exercise 11.

11. In June, it took Freddie 39 hours to paint six apartments in an apartment building. How long should he expect it to take him to paint another four apartments in July, assuming they are all the same size?

12. 4 out of 10 students in a certain college class are female. If there are a total of 35 students in the class, how many are female?

Completion Example 12 Application: Solving Proportions

A jelly manufacturer puts 2.5 ounces of sugar into every 6-ounce jar of jelly. How many ounces of jelly can be made with 300 ounces of sugar?

Solution

Assign the variable: Let x = unknown amount of jelly.

Set up the proportion: $\frac{2.5 \text{ oz sugar}}{6 \text{ oz jelly}} = \frac{300 \text{ oz sugar}}{x \text{ oz jelly}}$

Solve the proportion:

$$2.5 \cdot x = 6 \cdot 300$$

$$\frac{\cancel{2.5} \cdot x}{\cancel{2.5}} = \frac{1800}{2.5}$$

$$x = \underline{720}$$

300 ounces of sugar will make ___720___ ounces of jelly.

Now work margin exercise 12.

Nurses and doctors work with proportions when prescribing medicine and giving injections. Medical texts write proportions in the form

$$5:60::x:72 \qquad \text{instead of} \qquad \frac{5}{60} = \frac{x}{72}.$$

With either notation, the solution is found by finding the cross products, setting them equal to each other, and solving the equation.

13. A nurse is instructed to run an IV drip. The nurse uses a bag which holds 1000 mL of solution. How many hours will the IV run before the bag needs to be replaced given the proportion below?

80 mL : 1 hr :: 1000 mL : x hr

Example 13 Application: Solving Proportions Written in Medical Notation

A certain pain medicine is administered using a solution in the ratio of 2 grams of pain medicine to 32 ounces of solution. A nurse is told to give a patient 10 grams of the pain medicine. Given the following proportion, how many ounces of solution should the nurse give the patient?

2 grams : 32 ounces :: 10 grams : x ounces

Solution

$$2:32::10:x$$

$$\frac{2}{32} = \frac{10}{x}$$

$$2 \cdot x = 32 \cdot 10$$

$$\frac{\cancel{2} \cdot x}{\cancel{2}} = \frac{320}{2}$$

$$x = 160$$

The nurse should give the patient 160 ounces of solution.

Now work margin exercise 13.

Completion Example Answers

2. $\frac{91}{4}, \frac{91}{4}$; equal, true **7.** $3 \cdot 6$; 18; 18; $\frac{36}{5}$ **11.** 2, 3; $\frac{15}{2} = \frac{x \cdot \cancel{2}}{\cancel{2}}$; $\frac{15}{2}$; $\frac{15}{2}$ or $7\frac{1}{2}$ **12.** 2.5, 6;

$\frac{\cancel{2.5} \cdot x}{\cancel{2.5}} = \frac{1800}{2.5}$; 720; 720

Margin Exercise Answers

1. a. False **b.** False **c.** True **2.** False **3.** $x = 20$ **4.** $y = 49$ **5.** $y = 1.6$ **6.** $x = 60$ **7.** $z = \frac{3}{8}$
8. 12 pounds **9.** 126 voted in Precinct 2 **10.** $64 **11.** 26 hours **12.** 14 students **13.** 12.5 hr

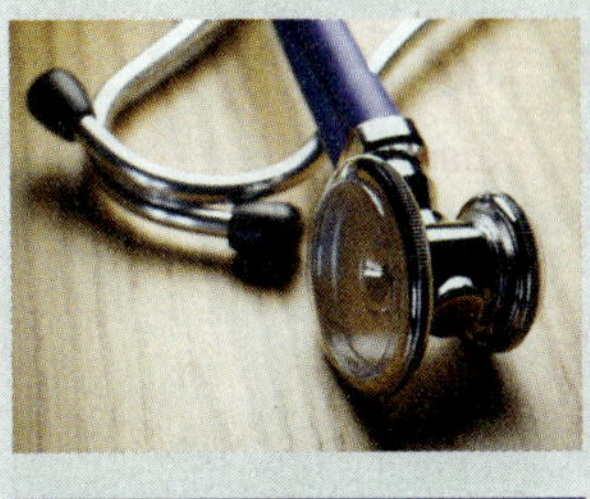

The First IV

The first use of IV treatment was by a Scottish physician, Dr. Thomas Latta (1796–1833). During a cholera epidemic, Dr. Latta observed that his patients were losing water in their blood. His great intuition was that salt water would be the best imitation of blood, and he pioneered the use of a pump to get the saline solution directly into the circulatory system. His patients recovered so quickly that the editor of the prestigious journal *The Lancet* called it "more like the workings of a miraculous and supernatural agent."

4.2 Exercises

Concept Check

Fill-in-the-Blank. Complete each sentence using information found in this section.

1. A proportion is true if the ________ ________ are equal.
2. When two fractions are equal to each other, a/an ________ is created.
3. The cross products of $\frac{r}{k} = \frac{t}{f}$ are ________ and ________.
4. To solve a word problem using a proportion, the first step is to __________ the unknown quantity.
5. When setting up a proportion to solve a word problem, the same ______ should be compared in the same order.

True/False. Determine whether each statement is true or false. If a statement is false, explain how it can be changed so the statement will be true. (**Note:** There may be more than one acceptable change.)

6. A proportion is a statement that two ratios are being multiplied.
7. Cross canceling is used to determine if a proportion is true.
8. In order to solve the proportion $\frac{16}{36.8} = \frac{x}{27.6}$ we construct the equation $36.8x = 441.6$.
9. When using proportions to solve a word problem, there is only one correct way to set up the proportion.
10. The proportions $\frac{36 \text{ tickets}}{\$540} = \frac{x \text{ tickets}}{\$75}$ and $\frac{x \text{ tickets}}{36 \text{ tickets}} = \frac{\$75}{\$540}$ will yield the same answer.

4.2 Exercises

Concept Check

1. cross products or two ratios
2. proportion
3. *rf* and *kt*
4. identify
5. units
6. False; A proportion is a statement that two ratios are equal.
7. False; Cross products are used to determine if a proportion is true.
8. True
9. False; When using proportions to solve a word problem, there are many correct ways to set up the proportion.
10. True

Practice

1. True
2. False
3. True
4. True
5. True
6. False
7. True
8. True
9. False
10. False
11. True
12. True
13. True
14. True
15. False
16. False
17. True
18. True
19. False
20. True
21. $x = 10$
22. $x = 12$
23. $x = 50$
24. $R = 60$
25. $D = 100$
26. $B = 40$
27. $A = \frac{21}{2}$ or 10.5
28. $x = \frac{42}{5}$ or 8.4
29. $x = 1$
30. $x = 1$
31. $y = 6$
32. $x = 20$
33. $x = \frac{1}{4}$ or 0.25
34. $x = \frac{3}{2}$ or 1.5
35. $R = 50$
36. $A = 135$
37. $A = 27.3$

Practice

Determine whether each proportion is true or false. See Examples 1 and 2.

1. $\frac{3}{6} = \frac{4}{8}$
2. $\frac{9}{8} = \frac{7}{6}$
3. $\frac{2}{5} = \frac{4}{10}$
4. $\frac{3}{5} = \frac{6}{10}$
5. $\frac{12}{15} = \frac{20}{25}$
6. $\frac{19}{16} = \frac{20}{17}$
7. $\frac{1}{4} = \frac{25}{100}$
8. $\frac{7}{8} = \frac{875}{1000}$
9. $\frac{2}{3} = \frac{66}{100}$
10. $\frac{1}{3} = \frac{33}{100}$
11. $\frac{12}{18} = \frac{14}{21}$
12. $\frac{11}{22} = \frac{17}{34}$
13. $\frac{7}{16} = \frac{3\frac{1}{2}}{8}$
14. $\frac{10}{17} = \frac{5}{8\frac{1}{2}}$
15. $\frac{3\frac{1}{5}}{\frac{2}{5}} = \frac{3\frac{3}{5}}{\frac{3}{5}}$
16. $\frac{1\frac{1}{4}}{1\frac{1}{2}} = \frac{\frac{1}{4}}{\frac{1}{2}}$
17. $\frac{6}{1.56} = \frac{2}{0.52}$
18. $\frac{3.75}{3} = \frac{7.5}{6}$
19. $\frac{8.5}{6.5} = \frac{4.5}{3.5}$
20. $\frac{12}{1.09} = \frac{36}{3.27}$

Solve each proportion. See Examples 3 through 7.

21. $\frac{5}{4} = \frac{x}{8}$
22. $\frac{3}{6} = \frac{6}{x}$
23. $\frac{1}{2} = \frac{x}{100}$
24. $\frac{3}{5} = \frac{R}{100}$
25. $\frac{3}{5} = \frac{60}{D}$
26. $\frac{8}{B} = \frac{6}{30}$
27. $\frac{A}{3} = \frac{7}{2}$
28. $\frac{6}{x} = \frac{5}{7}$
29. $\frac{\frac{1}{2}}{x} = \frac{5}{10}$
30. $\frac{\frac{1}{3}}{x} = \frac{3}{9}$
31. $\frac{1}{4} = \frac{1\frac{1}{2}}{y}$
32. $\frac{1}{12} = \frac{1\frac{2}{3}}{x}$
33. $\frac{x}{1} = \frac{1\frac{1}{4}}{5}$
34. $\frac{1}{5} = \frac{x}{7\frac{1}{2}}$
35. $\frac{150}{300} = \frac{R}{100}$
36. $\frac{A}{450} = \frac{30}{100}$
37. $\frac{A}{42} = \frac{65}{100}$
38. $\frac{98}{100} = \frac{B}{35}$
39. $\frac{3}{2} = \frac{B}{4.5}$
40. $\frac{7}{x} = \frac{2}{6.5}$
41. $\frac{3.5}{2.6} = \frac{10.5}{B}$
42. $\frac{7.2}{y} = \frac{4.8}{14.4}$
43. $\frac{7.8}{1.3} = \frac{x}{0.25}$
44. $\frac{4.1}{3.2} = \frac{x}{6.4}$

45. $\frac{13.5}{B} = \frac{15}{100}$ **46.** $\frac{19.2}{96} = \frac{R}{100}$

47. 120 mL : 2 hr :: 1500 mL : x hr

48. x mL : 3 hr :: 1800 mL : 12 hr

49. 3 grams : x ounces :: 1.2 grams : 8 ounces

50. 2.5 grams : 18 ounces :: x grams : 54 ounces

51. 50 mg : 30 min :: 225 mg : x min

52. x mg : 45 min :: 240 mg : 120 min

Applications

Solve.

53. The quality of concrete is based on the ratio of bags of cement to cubic yards of gravel. One batch of concrete consists of 27 bags of cement mixed into 9 cubic yards of gravel, while a second has 15 bags of cement mixed with 5 cubic yards of gravel. Determine whether the ratio of cement to gravel is the same for both batches.

54. Two children were selling lemonade at different stands. At the first stand the child mixed 2 cups of lemon juice with 8 cups of water, while at the second stand 3 cups of lemon juice were mixed with 13 cups of water. Determine if the two lemonades are of the same strength.

55. One propane heater will operate for $2\frac{2}{5}$ hours on $4\frac{2}{3}$ pounds of propane, while a second heater will operate for $4\frac{4}{5}$ hours on $9\frac{1}{3}$ pounds of propane. Determine if they operate at the same efficiency.

56. One olive press will extract 3.2 cups of oil from 15.6 pounds of olives, while the second press will extract 9.5 cups of oil from 48.8 pounds of olives. Assuming that the olives came from the same batch, determine if the two presses operate at the same efficiency.

57. A florist sells a bouquet of 12 red roses for \$46 and a bouquet of 18 red roses for \$62. Determine if the cost per rose is the same for each bouquet.

58. To color 1 cup of frosting the color of a granny smith apple, 10 drops of green and 25 drops of yellow food coloring are used. To color a larger amount of frosting, 28 drops of green and 65 drops of yellow food coloring are used. Determine if the ratio of green to yellow food coloring is the same for each batch of frosting.

59. When making a small "Energizing" scented candle, 5 drops of peppermint scent and 3 drops of lemongrass scent are added to the wax. When making a large "Energizing" scented candle 35 drops of peppermint scent and 21 drops of lemongrass scent are added to the wax. Determine if the ratio of peppermint scent to lemongrass scent is the same for each size candle.

38. $B = 34.3$
39. $B = 6.75$
40. $x = 22.75$
41. $B = 7.8$
42. $y = 21.6$
43. $x = 1.5$
44. $x = 8.2$
45. $B = 90$
46. $R = 20$
47. $x = 25$ hr
48. $x = 450$ mL
49. $x = 20$ ounces
50. $x = 7.5$ grams
51. $x = 135$ min
52. $x = 90$ mg

Applications

53. They are the same.
54. They are different.
55. They are the same.
56. They are different.
57. They are different.
58. They are different.
59. They are different.

60. They are different.
61. \$25
62. 21 mice
63. 180 minutes or 3 hours
64. 3360 miles
65. \$4.17
66. 154 miles
67. $1\frac{1}{2}$ cups or 1.5 cups
68. \$120
69. 33 hits
70. \$1275
71. 437.5 grams
72. 2100 minutes or 35 hours
73. 20 yards
74. \$34
75. \$400
76. \$648
77. 9 hours
78. 200 feet

60. While competing in a speed run to complete the first level of a video game, Alex scores a total of 2352 points in 98 seconds. Shannon scored a total of 2314 points in 89 seconds. Determine if the ratio of points per second is the same for Alex and Shannon.

61. If the odds on a horse are that a \$1 bet wins \$5 if the horse wins, how much can be won with a \$5 bet?

62. If a cat normally catches 3 mice in a day, how many would you expect it to catch in a week?

63. An English teacher must read and grade 27 essays. If the teacher takes 20 minutes to read and grade 3 essays, how much time will he need to grade all 27 essays?

64. A salesman determined that he drove 560 miles every two weeks. How far would he drive in three months (12 weeks)?

65. If one dozen (12) eggs cost \$1.39, what would three dozen eggs cost?

66. If 1 inch represents 35 miles on a map, how many miles do 4.4 inches represent?

67. If 2 cups of flour are needed to make 12 biscuits, how much flour is needed to make 9 of the same kind of biscuits?

68. A baseball team bought 8 bats for \$96. What would they pay for 10 bats?

69. A baseball player gets 22 hits in 4 weeks. How many hits would you expect him to get in 6 weeks?

70. If property taxes are figured at \$1.50 for every \$100 in evaluation, what taxes will be paid on a home valued at \$85,000?

71. Two units of a certain gas weigh 175 grams. What is the weight of 5 units of this gas?

72. A typist can type 8 pages of a manuscript in 56 minutes. How long will this typist take to type 300 pages?

73. The price of a certain fabric is \$1.75 per yard. How many yards can be bought with \$35 (not including tax)?

74. Sales tax is figured at 6¢ for every \$1.00 of merchandise purchased. What is the purchase price on an item that had sales tax of \$2.04?

75. A sales woman makes \$8 for every \$100 worth of product that she sells. What will she make if she sells \$5000 worth of product?

76. An investor made \$144 in one year on a \$1000 investment. What would she have earned if her investment had been \$4500?

77. If you can drive 286 miles in $5\frac{1}{2}$ hours, how long will it take you to drive 468 miles at the same rate of speed?

78. An architect drew plans for a city park using a scale of $\frac{1}{4}$ inch to represent 25 feet. How many feet would 2 inches represent?

79. One bag of Weed Killer & Fertilizer contains 18 pounds of fertilizer and weed treatment with a recommended coverage of 5000 square feet. If your lawn is in the shape of a rectangle 150 feet by 220 feet, how many pounds of Weed Killer & Fertilizer do you need to cover the lawn?

80. One bag of dichondra lawn food contains 20 pounds of fertilizer and its recommended coverage is 4000 square feet. If you want to cover a lawn that is in the shape of a rectangle 120 feet by 160 feet, how many pounds of lawn food do you need?

81. A condominium owner pays property taxes of $2000 per year. If taxes are figured at a rate of $1.25 for every $100 in value, what is the value of the condominium?

82. You can buy 3 onions for $2.89 at the grocery store. What would be the cost of 9 onions?

83. If a 4-year-old laughs 300 times per day on average, how many times does the child laugh in a year (assuming 365 days in a year)?

84. An electric fan makes 180 revolutions per minute. How many revolutions will the fan make if it runs for 24 hours?

Writing & Thinking

85. In your own words, clarify how you can know that a proportion is set up correctly or not.

86. List the steps to solve a word problem using a proportion.

87. A problem states:

"On a map from Detroit to Chicago, the distance is 4 inches. The actual distance is 300 miles. What is the distance on the map for an actual distance of 750 miles?"

When Traci set up this problem, her proportion was: $\frac{4 \text{ inches}}{300 \text{ miles}} = \frac{750 \text{ miles}}{x \text{ inches}}$.

Explain and correct the problem(s) with Traci's proportion.

88. Give an example when the use of proportions would be helpful (outside of a class).

79. 118.8 pounds

80. 96 pounds

81. $160,000

82. $8.67

83. 109,500 times

84. 259,200 revolutions

Writing & Thinking

85. A proportion has been set up correctly if the same units are in the same location in both ratios.

86. **1.** Identify the unknown quantity and assign it a variable, **2.** Set up a proportion in which the units are compared in the same order, and **3.** Solve the proportion.

87. Traci's proportion has inches in the numerator in one ratio and in the denominator in the other. The units should be the same in the numerators and the same in the denominators. This means she should have used the proportion $\frac{4 \text{ inches}}{300 \text{ miles}} = \frac{x}{750 \text{ miles}}$ or $\frac{300 \text{ miles}}{4 \text{ inches}} = \frac{750 \text{ miles}}{x}$.

88. Examples will vary. Proportions may be used to determine prices for more or less of an item, travel time remaining, actual length from a scale drawing, etc.

Objectives

A. Understand percents.

B. Change decimal numbers to percents.

C. Change percents to decimal numbers.

4.3 Decimals and Percents

A Percents

Percent is an important part of your daily life. For example, consider the following common uses of percent.

- Income tax is a **percent** of your income.
- Property tax is a **percent** of the value of your home.
- Sales tax is a **percent** of the price of goods sold by retailers.
- Your take-home income is a **percent** of your actual earnings.
- Interest is a **percent** earned on the amount invested.
- You will spend a **percent** of your income on food each month.

The word percent comes from the Latin *per centum*, meaning per hundred. So **percent means hundredths**, or **the ratio of a number to 100**. The symbol % is called the **percent sign**. This sign has the same meaning as the fraction $\frac{1}{100}$. For example,

$$\frac{35}{100} = 35\left(\frac{1}{100}\right) = 35\% \quad \text{and} \quad \frac{60}{100} = 60\left(\frac{1}{100}\right) = 60\%.$$

In Figure 1, the large square is partitioned into 100 small squares. Each small square represents 1%, or $\frac{1}{100}$, of the large square. Thus, the shaded portion is $\frac{40}{100} = 40 \cdot \frac{1}{100} = 40\%$ of the large square.

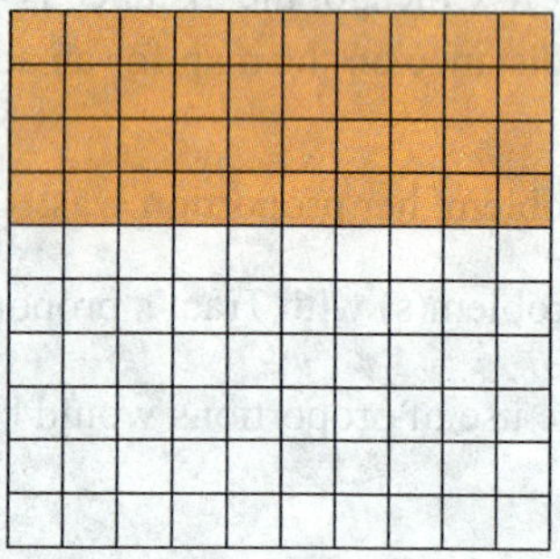

Figure 1

As will be illustrated in Example 1, a fraction with denominator 100 can be easily written as a percent. The numerator is not changed, regardless of whether it is a whole number, decimal number, or mixed number. The fraction $\frac{1}{100}$ is replaced by the % sign.

Example 1 Changing Fractions with Denominators of 100 to Percents

Change each fraction to a percent.

a. $\frac{7}{100} = 7 \cdot \frac{1}{100} = 7\%$ Remember that percent means hundredths.

b. $\frac{83}{100} = 83 \cdot \frac{1}{100} = 83\%$

c. $\frac{6.4}{100} = 6.4 \cdot \frac{1}{100} = 6.4\%$ Notice that the decimal point is not moved. The numerator is unchanged.

d. $\frac{100}{100} = 100 \cdot \frac{1}{100} = 100\%$ All of something is 100% of that thing.

e. $\frac{240}{100} = 240 \cdot \frac{1}{100} = 240\%$ If the numerator is larger than 100, then the number is larger than 1 and it is more than 100%.

Now work margin exercise 1.

1. Change each fraction to a percent.

a. $\frac{9}{100}$

b. $\frac{1.25}{100}$

c. $\frac{125}{100}$

d. $\frac{62}{100}$

e. $\frac{0}{100}$

B Changing Decimal Numbers to Percents

One way to change a decimal number to a percent is to

1. change the decimal number to fraction form with denominator 100, then
2. change the fraction to percent form.

Consider the examples in Table 1.

Decimal Form		Fraction Form		Percent Form
0.47	=	$\frac{47}{100} = 47 \cdot \frac{1}{100}$	=	47%
0.9	=	$\frac{9}{10} = \frac{90}{100} = 90 \cdot \frac{1}{100}$	=	90%
0.325	=	$\frac{325}{1000} = \frac{32.5}{100} = 32.5 \cdot \frac{1}{100}$	=	32.5%

Table 1

By studying these examples carefully, we can see a pattern emerge. We see that we can go directly to the percent form from the decimal form by moving the decimal point two places to the right and writing the % sign.

Changing a Decimal Number to a Percent

1. Move the decimal point two places to the right.
2. Write the % sign.

PROCEDURE

2. Change each decimal number to a percent.
   a. 0.0035
   b. 2.17
   c. 0.871
   d. 0.1

Example 2 Changing Decimal Numbers to Percents

Change each decimal number to a percent.

a. 0.254 **b.** 0.005 **c.** 1.5 **d.** 0.2

Solution

a. 0.25.4 = 25.4% ← % symbol added.
Decimal point moved two places to the right

b. 0.00.5 = 0.5% ← % symbol added. Note that this is less than 1%.
Decimal point moved two places to the right

c. 1.50. = 150% ← % symbol added. Note that this is more than 100%.
Decimal point moved two places to the right (a 0 is inserted)

d. 0.20. = 20% ← % symbol added.
Decimal point moved two places to the right (a 0 is inserted)

Now work margin exercise 2.

C Changing Percents to Decimal Numbers

To change percents to decimal numbers, we reverse the procedure of changing decimal numbers to percents. For example,

$$38.\% = 38 \cdot \frac{1}{100} = \frac{38}{100} = 0.38$$

↑ Understood decimal point

As indicated in the following procedure, the same result can be found by moving the decimal point two places to the left and deleting the % sign.

Changing a Percent to a Decimal Number

1. Move the decimal point two places to the left.
2. Delete the % sign.

PROCEDURE

Example 3 Changing Percents to Decimal Numbers

Change each percent to a decimal number.

a. 76.% = 0.76 ← % symbol deleted

↑ Understood decimal point ↑ Decimal point moved two places to the left

b. 18.5% = 0.185

c. 50% = 0.50

d. 100% = 1.00

e. 0.25% = 0.0025 Notice that when moving the decimal point two places to the left, two zeros were added as placeholders.

Now work margin exercise 3.

3. Change each percent to a decimal number.

a. 40%

b. 211%

c. 0.6%

d. 29.37%

e. 102%

The following relationships between decimal numbers and percents provide helpful guidelines in changing from one form to the other.

Relationships Between Decimal Numbers and Percents

A decimal number that is

a. less than 0.01 is less than 1%,

b. between 0.01 and 0.10 is between 1% and 10%,

c. between 0.10 and 1.00 is between 10% and 100%, and

d. more than 1 is more than 100%.

PROPERTIES

Margin Example Answers

1. a. 9% **b.** 1.25% **c.** 125% **d.** 62% **e.** 0% **2. a.** 0.35% **b.** 217% **c.** 87.1% **d.** 10%
3. a. 0.40 **b.** 2.11 **c.** 0.006 **d.** 0.2937 **e.** 1.02

4.3 Exercises

Concept Check

Fill-in-the-Blank. Complete each sentence using information found in this section.

1. If 83 out of 100 pieces of paper are shaded, then the shaded pieces of paper make up _______ % of the total.

2. Percent is the ratio of a number to _______.

3. To change a decimal number to a percent, move the decimal point _______ places to the _______.

4. To change a percent to a decimal number, move the decimal point _______ places to the _______.

True/False. Determine whether each statement is true or false. If a statement is false, explain how it can be changed so the statement will be true. (**Note:** There may be more than one acceptable change.)

5. If a decimal number is less than 1, then the equivalent percent will be less than 100%.

6. It is not possible to have a percent greater than 100%.

7. A decimal number that is between 0.01 and 0.10 is between 10% and 100%.

8. To change from a percent to a decimal, simply omit the percent sign.

Practice

Find the percent of each square that is shaded.

1.

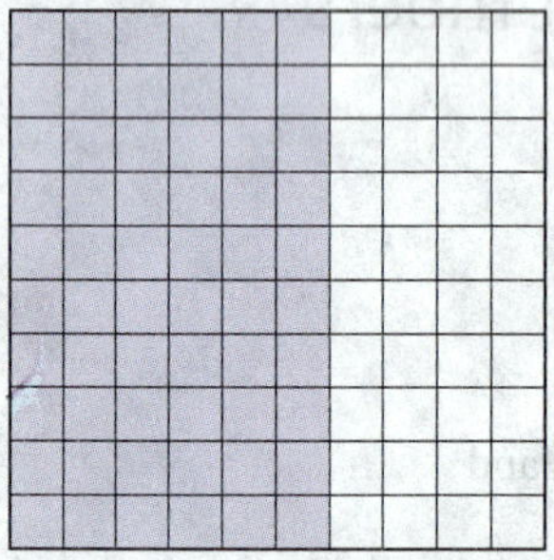

3.

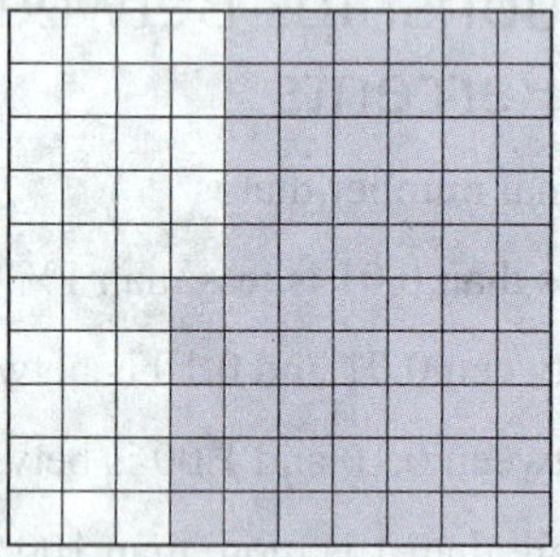

2.

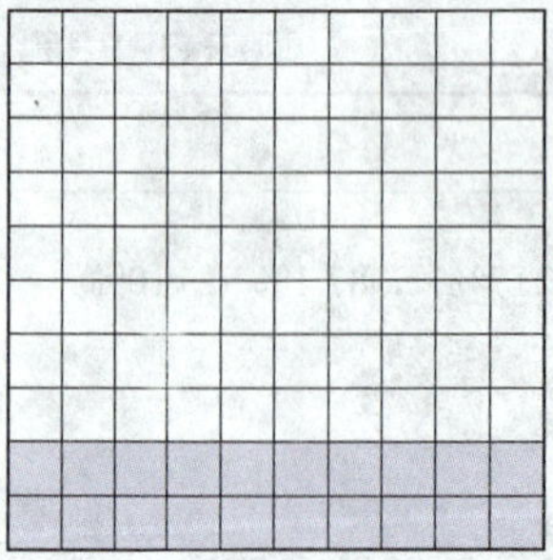

4.

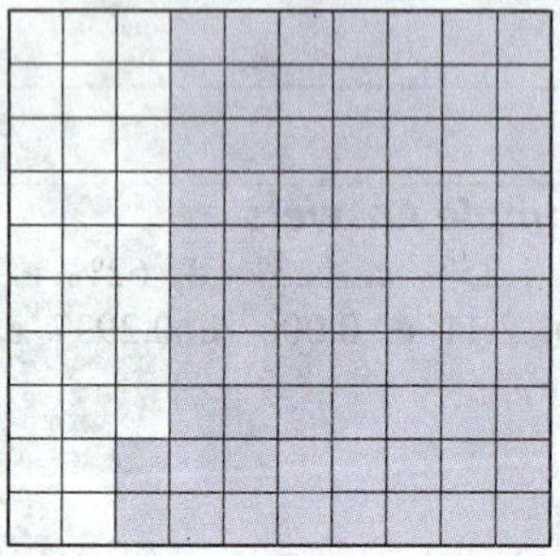

4.3 Exercises

Concept Check

1. 83
2. 100
3. 2; right
4. 2; left
5. True
6. False; It is possible to have a percent greater than 100%.
7. False; A decimal number that is between 0.01 and 0.10 is between 1% and 10%.
8. False; To change from a percent to a decimal, move the decimal point two places to the left and omit the percent sign.

Practice

1. 60%
2. 20%
3. 65%
4. 72%
5. 12.5%
6. 0.5%
7. 100%
8. 0%
9. 20%
10. 80%
11. 9%
12. 7%
13. 15%
14. 53%
15. 125%
16. 336%
17. 13.4%
18. 99.9%
19. 0.48%
20. 0.53%
21. 0.5%
22. 0.2%
23. 2.14%
24. 1.62%

5. 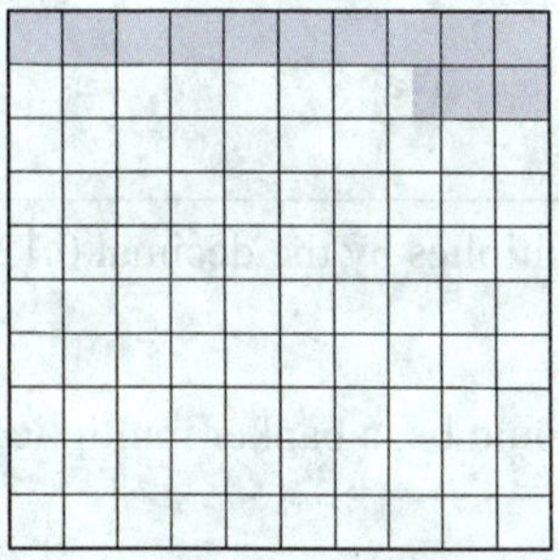

6. 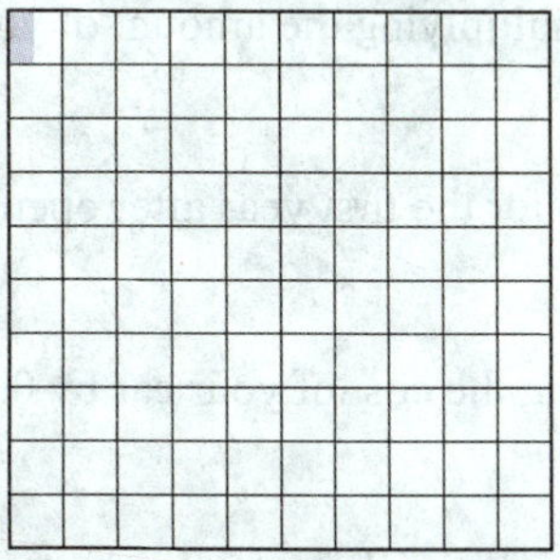

7. 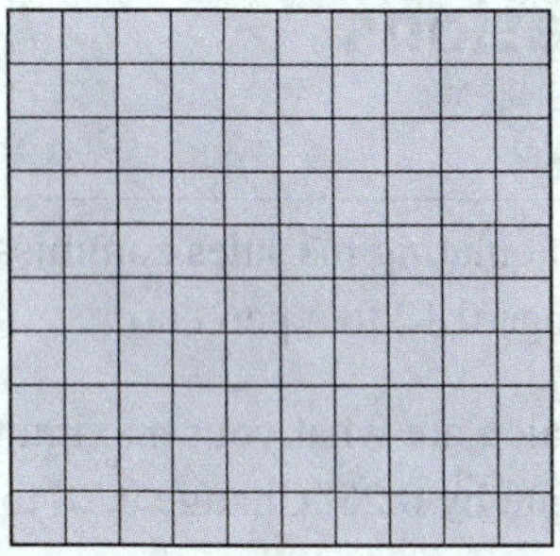

8. 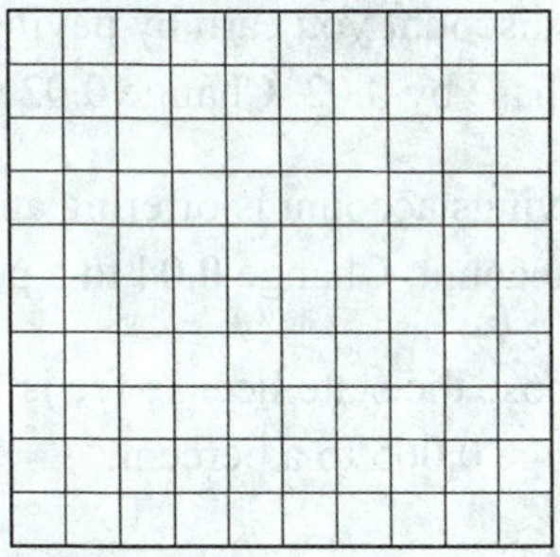

Change each fraction to a percent. See Example 1.

9. $\frac{20}{100}$

10. $\frac{80}{100}$

11. $\frac{9}{100}$

12. $\frac{7}{100}$

13. $\frac{15}{100}$

14. $\frac{53}{100}$

15. $\frac{125}{100}$

16. $\frac{336}{100}$

17. $\frac{13.4}{100}$

18. $\frac{99.9}{100}$

19. $\frac{0.48}{100}$

20. $\frac{0.53}{100}$

21. $\frac{0.5}{100}$

22. $\frac{0.2}{100}$

23. $\frac{2.14}{100}$

24. $\frac{1.62}{100}$

Change each decimal number to a percent. See Example 2.

25. 0.02
26. 0.09
27. 0.1
28. 0.7
29. 0.36
30. 0.52
31. 0.40
32. 0.60
33. 0.025
34. 0.035
35. 0.005
36. 0.004
37. 0.128
38. 0.368
39. 1.12
40. 1.75
41. 2.3
42. 6.7
43. 2
44. 25

Change each percent to a decimal number. See Example 3.

45. 2%
46. 7%
47. 18%
48. 42%
49. 25%
50. 75%
51. 20%
52. 30%
53. 125%
54. 179%
55. 120%
56. 150%
57. 6.5%
58. 2.5%
59. 17.3%
60. 10.1%
61. 0.26%
62. 0.52%
63. 1.37%
64. 2.44%

25. 2%
26. 9%
27. 10%
28. 70%
29. 36%
30. 52%
31. 40%
32. 60%
33. 2.5%
34. 3.5%
35. 0.5%
36. 0.4%
37. 12.8%
38. 36.8%
39. 112%
40. 175%
41. 230%
42. 670%
43. 200%
44. 2500%
45. 0.02
46. 0.07
47. 0.18
48. 0.42
49. 0.25
50. 0.75
51. 0.2
52. 0.3
53. 1.25
54. 1.79
55. 1.2
56. 1.5
57. 0.065
58. 0.025
59. 0.173
60. 0.101
61. 0.0026
62. 0.0052
63. 0.0137
64. 0.0244

Applications

Solve.

65. In calculating his sales commission, Mr. Howard multiplies by the decimal 0.12. Change 0.12 to a percent.

66. To calculate what your maximum house payment should be, a banker multiplied your income by 0.28. Change 0.28 to a percent.

67. The discount you earn by paying cash is found by multiplying the amount of your purchase by 0.02. Change 0.02 to a percent.

68. A savings account is offering an interest rate of 0.04 for the first year after opening the account. Change 0.04 to a percent.

69. Suppose the state license fee is figured by multiplying the cost of your car by 0.065. Change 0.065 to a percent.

70. The interest rate on a loan is 6.4%. Change 6.4% to a decimal.

71. The sales commission for the clerk at a retail store is figured at 8.5%. Change 8.5% to a decimal.

72. Suppose that sales tax is figured at 7.25%. Change 7.25% to a decimal.

73. At Mr. Jones' sporting goods store, the rate of profit based on sales price is 45%. Change 45% to a decimal.

74. The discount during a special sale on dresses is 30%. Change 30% to a decimal.

75. According to the last US Census, per capita property taxes in New Jersey were 523% higher than those in Alabama. Write 523% as a decimal.

76. According to the Aluminum Association, the percentage of aluminum cans recycled in 2015 was 64.3%, and this percentage fell to 63.6% in 2018. Write 64.3% and 63.6% as decimals.

Writing & Thinking

77. Describe the relationship between percent and the number 100.

78. Describe a situation where more than 100% is possible. Describe a situation where it is impossible to have more than 100%.

79. Compare and contrast the process for converting percents to decimal numbers and changing decimal numbers to percents.

79. Both conversions involve moving the decimal point two places. The difference is whether the decimal point moves left or right. When changing a percent to decimal, the decimal point moves two places to the left and when changing a decimal to a percent, the decimal point moves two places to the right.

Applications

65. 12%

66. 28%

67. 2%

68. 4%

69. 6.5%

70. 0.064

71. 0.085

72. 0.0725

73. 0.45

74. 0.3

75. 5.23

76. 0.643 and 0.636

Writing & Thinking

77. Percent means per centum or per 100. For example, fifty-eight percent means 58 out of 100. Percent can be written as a fraction with 100 in the denominator as in 58/100. The decimal equivalent, 0.58 is read as "fifty-eight hundredths," indicating percent can be written using the hundredths place, another connection.

78. Sometimes it is possible to have percentages over 100%. For example, if a bank account had \$50 and someone put in \$75, the amount deposited would have been 150% (starting \$50 + \$25 more) of the original amount. However, sometimes more than 100% is not possible. For example, if a gas tank can hold exactly 16 gallons of gas, 110% of the capacity could not be put into the tank.

4.4 Fractions and Percents

Objectives

A. Change fractions to percents.

B. Change percents to fractions.

A Changing Fractions to Percents

As discussed earlier, if a fraction has denominator 100, it can be changed to a percent by writing the numerator and adding the % sign. If the denominator is a factor of 100,

1, 2, 4, 5, 10, 20, 25, or 50, (factors of 100)

the fraction can be changed to an equivalent fraction with denominator 100 and then changed to a percent.

$$\frac{1}{4}=\frac{1}{4}\cdot\frac{25}{25}=\frac{25}{100}=25\%$$
$$\frac{3}{5}=\frac{3}{5}\cdot\frac{20}{20}=\frac{60}{100}=60\%$$
$$\frac{17}{50}=\frac{17}{50}\cdot\frac{2}{2}=\frac{34}{100}=34\%$$

However, most fractions do not have factors of 100 as denominators. So, another approach (easily applied with calculators) is to change the fraction to decimal form by dividing the numerator by the denominator. The resulting quotient will be a decimal number. The decimal number can then be changed to a percent by moving the decimal point two places to the right and writing the % sign.

Math Tip

Percents make it easier to compare results, such as your grades on tests. Suppose you got 40 questions out of 50 correct on your first test and 12 out of 15 on your second test. Which test did you perform better on? Since 40 out of 50 is 80% and 12 out of 15 is 80%, you performed equally well on both tests.

Changing a Fraction to a Percent

1. Change the fraction to a decimal number. (Divide the numerator by the denominator.)
2. Change the decimal number to a percent.

PROCEDURE

Example 1 Changing Fractions to Percents

Change $\frac{5}{8}$ to a percent.

Solution

Note that 8 is not a factor of 100, so we divide using long division (or using a calculator).

$$\begin{array}{r} 0.625 \\ 8\overline{)5.000} \\ -4\,8 \\ \hline 20 \\ -16 \\ \hline 40 \\ -40 \\ \hline 0 \end{array}$$

This can be done with a calculator.

Now change 0.625 to a percent. Move the decimal point two places to the right and write the % sign.

$$\frac{5}{8}=0.625=62.5\%$$

Now work margin exercise 1.

1. Change $\frac{13}{16}$ to a percent.

2. Change $\frac{27}{50}$ to a percent.

Example 2 Changing Fractions to Percents

Change $\frac{11}{20}$ to a percent.

Solution

Divide.

$$\begin{array}{r} 0.55 \\ 20\overline{)11.00} \\ -10\,00 \\ \hline 1\,00 \\ -1\,00 \\ \hline 0 \end{array}$$

Thus, $\frac{11}{20} = 0.55 = 55\%$.

Or, we can notice that 20 is a factor of 100 and write

$$\frac{11}{20} = \frac{11}{20} \cdot \frac{5}{5} = \frac{55}{100} = 55\%.$$

Now work margin exercise 2.

3. Change $1\frac{1}{2}$ to a percent.

Example 3 Changing Mixed Numbers to Percents

Change $2\frac{1}{4}$ to a percent.

Solution

Change $2\frac{1}{4}$ to decimal form, then change the decimal number to a percent as follows.

$$2\frac{1}{4} = 2.25 = 225\%$$

Now work margin exercise 3.

4. Change $2\frac{4}{5}$ to a percent.

Completion Example 4 Changing Mixed Numbers to Percents

Change $3\frac{3}{5}$ to a percent.

Solution

Since $3\frac{3}{5}$ is larger than 1, the percent will be more than $\underline{100}\%$.

Change $3\frac{3}{5}$ to decimal form, then change the decimal number to a percent.

$$3\frac{3}{5} = 3.\underline{60} = \underline{360}\%$$

Now work margin exercise 4.

Example 5 Changing Fractions to Percents

Change $\frac{2}{3}$ to a percent (rounded to the nearest tenth of a percent).

Solution

Divide. To find the solution to the nearest tenth of a percent, we must round the decimal quotient to the nearest thousandth.

$$\begin{array}{r} 0.6666 \approx 0.667 \quad \text{To the nearest thousandth} \\ 3\,\overline{)\,2.0000} \\ \underline{-1\,8} \\ 20 \\ \underline{-18} \\ 20 \\ \underline{-18} \\ 20 \\ \underline{-18} \\ 2 \end{array}$$

Thus, $\frac{2}{3} \approx 0.667 = 66.7\%$.

Now work margin exercise 5.

5. Change $\frac{3}{11}$ to a percent (rounded to the nearest tenth of a percent).

Example 6 Application: Changing Fractions to Percents

During the years 1921 to 2012, the Los Angeles Dodgers baseball team played in 18 World Series Championships and won 6 of them. What percent of these championships did the Dodgers win (rounded to the nearest tenth of a percent)?

Solution

The percent won can be found by changing the fraction $\frac{6}{18}$ to decimal form and then changing the decimal number to a percent.

$$\begin{array}{r} 0.3333 \approx 0.333 \quad \text{To the nearest thousandth} \\ 18\,\overline{)\,6.0000} \\ \underline{-5\,4} \\ 60 \\ \underline{-54} \\ 60 \\ \underline{-54} \\ 60 \\ \underline{-54} \\ 6 \end{array}$$

Thus, $\frac{6}{18} \approx 0.333 = 33.3\%$.

The Dodgers won 33.3% of these championships.

Now work margin exercise 6.

6. The Braves little league baseball team won 14 of their 15 games this season. Find the percent of games won this season (rounded to the nearest tenth of a percent).

B Changing Percents to Fractions

To change a percent to fraction form or mixed number form we can proceed as follows. Note that if the fraction is greater than 1 you may want to change it to mixed number form.

Changing a Percent to a Fraction or a Mixed Number

1. Write the percent as a fraction with 100 as the denominator and delete the % sign.
2. Reduce the fraction, if possible.

PROCEDURE

7. Change 80% to a fraction and reduce, if possible.

Example 7 Changing Percents to Fractions

Change 60% to a fraction and reduce, if possible.

Solution

$$60\% = \frac{60}{100} = \frac{3 \cdot \cancel{20}}{5 \cdot \cancel{20}} = \frac{3}{5}$$

Now work margin exercise 7.

8. Change 235% to a mixed number and reduce, if possible.

Example 8 Changing Percents to Mixed Numbers

Change 130% to a mixed number and reduce, if possible.

Solution

$$130\% = \frac{130}{100} = \frac{13 \cdot \cancel{10}}{10 \cdot \cancel{10}} = \frac{13}{10} = 1\frac{3}{10}$$

Now work margin exercise 8.

Common Misunderstanding Concerning Percents

The fractions $\frac{1}{4}$ and $\frac{1}{2}$ are often confused with the percents $\frac{1}{4}\%$ and $\frac{1}{2}\%$. The differences can be clarified by using decimal numbers.

Percent	Decimal Number	Fraction
$\frac{1}{4}\%$ (or 0.25%)	0.0025	$\frac{1}{400}$
$\frac{1}{2}\%$ (or 0.5%)	0.005	$\frac{1}{200}$
25%	0.25	$\frac{1}{4}$
50%	0.50	$\frac{1}{2}$

Thus, $\frac{1}{4} = 0.25$ and $\frac{1}{4}\% = 0.0025$

$$0.25 \neq 0.0025$$

Similarly, $\frac{1}{2} = 0.50$ and $\frac{1}{2}\% = 0.005$

$$0.50 \neq 0.005$$

You can think of $\frac{1}{4}$ as being one-fourth of a dollar (a quarter) and $\frac{1}{4}\%$ as being one-fourth of a penny. Similarly, $\frac{1}{2}$ can be thought of as one-half of a dollar and $\frac{1}{2}\%$ as one-half of a penny.

CAUTION

You should memorize the equivalent percent, decimal number, and fraction values in the following box. Look for patterns related to the fractions. Many calculations with these expressions can be done mentally.

Common Equivalent Percent, Decimal Number, and Fraction Values

$1\% = 0.01 = \frac{1}{100}$

$25\% = 0.25 = \frac{1}{4}$

$50\% = 0.50 = \frac{1}{2}$

$75\% = 0.75 = \frac{3}{4}$

$100\% = 1.00 = 1$

$33\frac{1}{3}\% = 0.33\overline{3} = \frac{1}{3}$

$66\frac{2}{3}\% = 0.66\overline{6} = \frac{2}{3}$

$12\frac{1}{2}\% = 0.125 = \frac{1}{8}$

$37\frac{1}{2}\% = 0.375 = \frac{3}{8}$

$62\frac{1}{2}\% = 0.625 = \frac{5}{8}$

$87\frac{1}{2}\% = 0.875 = \frac{7}{8}$

PROPERTIES

4.4 Exercises

Concept Check

1. 100
2. decimal
3. 2; right
4. 100
5. False; Fractions that have denominators other than 100 can be changed to a percent.
6. False; The fraction $\frac{1}{5}$ is equivalent to 20%.
7. False; When changing from a percent to a mixed number, the fraction should be reduced.
8. True

Practice

1. 3%
2. 7%
3. 16%
4. 29%
5. 50%
6. 10%
7. 75%
8. 25%
9. 5%
10. 55%
11. 40%
12. 80%
13. 26%
14. 86%
15. 87.5%
16. 62.5%
17. 4%
18. 24%
19. 12.5%
20. 62.5%
21. 58.3%
22. 8.3%
23. 116.7%

Completion Example Answers

4. 100%; 3.60, 360%

Margin Example Answers

1. 81.25% **2.** 54% **3.** 150% **4.** 280% **5.** 27.3% **6.** 93.3% **7.** $\frac{4}{5}$ **8.** $2\frac{7}{20}$

4.4 Exercises

Concept Check

Fill-in-the-Blank. Complete each sentence using information found in this section.

1. The easiest fractions to convert to a percent are those that have a denominator of 100 or those with denominators that are a factor of _______.
2. If a fraction does not have a denominator that is a factor of 100, it can still be changed to a percent by finding its equivalent _______ number form.
3. If a fraction has been converted to a decimal number, it then can to be changed to a percent by moving the decimal point _______ places to the _______.
4. To change a percent to a fraction, begin by removing the percent sign and writing the percent as a numerator over _______.

True/False. Determine whether each statement is true or false. If a statement is false, explain how it can be changed so the statement will be true. (**Note:** There may be more than one acceptable change.)

5. Fractions that have denominators other than 100 cannot be changed to a percent.
6. The fraction $\frac{1}{5}$ is equivalent to $\frac{1}{5}$%.
7. When changing from a percent to a mixed number, the fraction does not need to be reduced.
8. $75\% = 0.75 = \frac{3}{4}$

Practice

Change each fraction or mixed number to a percent. If necessary, round to the nearest tenth of a percent. See Examples 1 through 5.

1. $\frac{3}{100}$	**6.** $\frac{1}{10}$	**11.** $\frac{2}{5}$	**16.** $\frac{35}{56}$
2. $\frac{7}{100}$	**7.** $\frac{3}{4}$	**12.** $\frac{4}{5}$	**17.** $\frac{1}{25}$
3. $\frac{16}{100}$	**8.** $\frac{1}{4}$	**13.** $\frac{13}{50}$	**18.** $\frac{6}{25}$
4. $\frac{29}{100}$	**9.** $\frac{1}{20}$	**14.** $\frac{43}{50}$	**19.** $\frac{1}{8}$
5. $\frac{1}{2}$	**10.** $\frac{11}{20}$	**15.** $\frac{56}{64}$	**20.** $\frac{5}{8}$

21. $\frac{7}{12}$ **24.** $2\frac{5}{6}$ **27.** $1\frac{1}{20}$ **30.** $5\frac{3}{10}$

22. $\frac{1}{12}$ **25.** $1\frac{1}{4}$ **28.** $1\frac{17}{20}$ **31.** $2\frac{1}{15}$

23. $1\frac{1}{6}$ **26.** $1\frac{3}{4}$ **29.** $2\frac{1}{10}$ **32.** $4\frac{1}{18}$

Change each percent to a fraction or mixed number and reduce, if possible. See Examples 7 and 8.

33. 5% **34.** 4% **35.** 15% **36.** 48% **37.** 17% **38.** 43% **39.** 25% **40.** 50%

41. 33% **42.** 66% **43.** 100% **44.** 300% **45.** 120% **46.** 150% **47.** 125% **48.** 536%

49. 12.5% **50.** 62.5% **51.** 0.75% **52.** 0.25% **53.** $\frac{1}{2}$% **54.** $\frac{1}{4}$% **55.** $12\frac{1}{2}$%

56. $37\frac{1}{2}$% **57.** $16\frac{2}{3}$% **58.** $83\frac{1}{3}$% **59.** $33\frac{1}{3}$% **60.** $66\frac{2}{3}$%

Find the missing forms of each number.

	Fraction form	Decimal form	Percent form
61.	$\frac{5}{8}$	______	______
62.	$\frac{11}{20}$	______	______
63.	______	0.09	______
64.	______	1.75	______
65.	______	______	36%
66.	______	______	10.5%

Applications

Solve.

67. A department store offers a 35% discount during a special sale on dresses. Change 35% to a fraction reduced to lowest terms.

68. A department store offers a 30% discount during a special sale on men's suits. Change 30% to a fraction reduced to lowest terms.

69. Betty pays income tax of 24% on her pay from working as a labor and delivery nurse. Express 24% as a fraction reduced to lowest terms.

24. 283.3%
25. 125%
26. 175%
27. 105%
28. 185%
29. 210%
30. 530%
31. 206.7%
32. 405.6%
33. $\frac{1}{20}$
34. $\frac{1}{25}$
35. $\frac{3}{20}$
36. $\frac{12}{25}$
37. $\frac{17}{100}$
38. $\frac{43}{100}$
39. $\frac{1}{4}$
40. $\frac{1}{2}$
41. $\frac{33}{100}$
42. $\frac{33}{50}$
43. 1
44. 3
45. $1\frac{1}{5}$
46. $1\frac{1}{2}$
47. $1\frac{1}{4}$
48. $5\frac{9}{25}$
49. $\frac{1}{8}$
50. $\frac{5}{8}$
51. $\frac{3}{400}$

52. $\frac{1}{400}$

53. $\frac{1}{200}$

54. $\frac{1}{400}$

55. $\frac{1}{8}$

56. $\frac{3}{8}$

57. $\frac{1}{6}$

58. $\frac{5}{6}$

59. $\frac{1}{3}$

60. $\frac{2}{3}$

61. a. 0.625 b. 62.5%

62. a. 0.55 b. 55%

63. a. $\frac{9}{100}$ b. 9%

64. a. $\frac{7}{4}$ or $1\frac{3}{4}$ b. 175%

65. a. $\frac{9}{25}$ b. 0.36

66. a. $\frac{21}{200}$ b. 0.105

Applications

67. $\frac{7}{20}$

68. $\frac{3}{10}$

69. $\frac{6}{25}$

70. $3\frac{17}{20}$

71. $\frac{16}{125}$

72. $\frac{3}{8}$

73. $\frac{7}{20{,}000}$

74. a. $8\frac{1}{3}$% b. 25%; c. $6\frac{1}{4}$% or 6.25%

70. The portfolio of an investor increased by 385%. Express this as a mixed number reduced to lowest terms.

71. Creaky Cowboy Café is unable to sell 12.8% of their pastries on a given day. Express 12.8% as a fraction reduced to lowest terms.

72. A cat gives birth to a litter of kittens with different coat patterns. 37.5% of the kittens are orange. Express 37.5% as a fraction reduced to lowest terms.

73. A certain very precise voltmeter will measure the voltage with no greater than 0.035% error. This means that the difference between the actual voltage and the indicated voltage is no greater than 0.035% of the measured voltage; this is expressed as ±0.035%. Express this maximum error as a fraction reduced to lowest terms.

74. Consider each measurement.

a. There are 12 inches in a foot. What percent of a foot is an inch?

b. There are 4 quarts in a gallon. What percent of a gallon is a quart?

c. There are 16 ounces in a pound. What percent of a pound is an ounce?

75. In a sophomore class of 250 students, 10 students represent the sophomore class on the student council. What percent of the class is on the student council?

76. Out of a possible total of 240 points on an exam, David received 204 points. What percent of the exam did David get correct?

77. Three gallons of fluoride are in a 4,000,000 gallon supply of drinking water. What percent of the drinking water is fluoride?

78. To receive a Bachelor of Science (BS) degree at a certain college, the student must complete a total of 128 credit hours, of which 41 of these credits must be general education Core Skills courses. What percent of the total curriculum is dedicated to general education courses?

79. According to the laws in the United States, a person can be arrested for driving under the influence of alcohol if the blood alcohol concentration (BAC) is 0.08% or greater, where BAC in decimal form is defined as the number of grams of alcohol in 1 milliliter of blood.[1]

a. Assume that 2 milliliters of blood has 0.0022 grams of alcohol, what is the BAC in decimal units? (**Hint:** Divide the amount of alcohol by the amount of blood.)

b. Express this as a percent.

c. Does this exceed the legal limit of 0.08%?

80. In 2020, approximately 88,000,000 Americans were between the ages of 25 and 44. If the total US population was approximately 329,000,000 people, what percentage (to the nearest percent) of the population is between the ages of 25 and 44?[2]

1 Source: dui.drivinglaws.org/drink-table.php
2 Source: www.statista.com/statistics/241488/population-of-the-us-by-sex-and-age/

81. A recipe to create self-rising flour says to mix 120 grams of all-purpose flour with 5 grams of salt and 8 grams of baking powder. The salt and baking powder mixture make up what percent of the self-rising flour? Round your answer to the nearest tenth of a percent.

82. A sample of perfume is made with 21 drops of peach oil, 35 drops of rosemary oil, and 14 drops of sandalwood oil. The oil mixture is then diluted with 100 drops of perfumer's alcohol. Round your answers to the nearest tenth of a percent. (**Hint:** The denominator is equal to the total number of drops that make up the perfume.)

a. What percent of the perfume sample does the rosemary make up?

b. The perfumer's alcohol makes up what percent of the perfume sample?

c. The oil mixture makes up what percent of the perfume sample?

Writing & Thinking

83. Justify why mixed numbers are a larger percentage than proper fractions alone. (Consider the value of 100%.)

84. Describe the process to change a percent to a fraction or mixed number.

75. 4%

76. 85%

77. 0.000075%

78. 32.03%

79. **a.** 0.0011 **b.** 0.11% **c.** Yes

80. 27%

81. 9.8%

82. **a.** 20.6% **b.** 58.8% **c.** 41.2%

Writing & Thinking

83. 100% = 1 so anytime there is a mixed number, which has a value greater than 1, the percentage will be greater than 100%. Proper fractions (numerator is smaller than denominator) have a value less than 1 and therefore the percentage will be less than 100%.

84. To change a percent to a fraction or mixed number, **1.** Write the percent as a fraction with 100 as the denominator and drop the % sign and **2.** Reduce the fraction if possible.

Objectives

A. Solve problems using the percent proportion $\frac{P}{100}=\frac{A}{B}$.

4.5 Solving Percent Problems Using Proportions

A Solving Problems Using the Percent Proportion $\frac{P}{100}=\frac{A}{B}$

We know that percent means hundredths and percent is the ratio of a number to 100. For example, 25% can be written in the ratio form $\frac{25}{100}$. Using this ratio concept,

"**25%** of **60** is **15**" can be written as the proportion $\frac{25}{100}=\frac{15}{60}$.

"25%	of	60	is	15"
↓		↓		↓
Percent		**Base**		**Amount**

In general,

"P**%** of B is A" can be written as the proportion $\frac{P}{100}=\frac{A}{B}$.

"P%	of	B	is	A"
↓		↓		↓
Percent		**Base**		**Amount**

Math Tip

When you find a percent of a given number, the amount will either be

a. smaller than the given number if the percent is less than 100% or

b. larger than the given number if the percent is more than 100%.

The Percent Proportion $\frac{P}{100}=\frac{A}{B}$

For the proportion $\frac{P}{100}=\frac{A}{B}$

$P\% =$ **percent** (written as the ratio $\frac{P}{100}$).

$B =$ **base** (number that we are finding the percent of).

$A =$ **amount** (a part of the base).

FORMULA

There are three basic types of percent problems. Each of these three types can be solved by using a proportion of the form $\frac{P}{100}=\frac{A}{B}$. If any two of the values P, A, and B are known, then the third can be found by solving the resulting proportion.

Three Basic Types of Percent Problems and the Proportion $\frac{P}{100}=\frac{A}{B}$

Type 1: Find the **amount** given the base and the percent.

What is $P\%$ of B?

What is 65% of 500? $\frac{65}{100}=\frac{A}{500}$

Type 2: Find the **base** given the percent and the amount.

$P\%$ of what number is A?

57% of what number is 51.3? $\frac{57}{100}=\frac{51.3}{B}$

Type 3: Find the **percent** given the base and the amount.

What percent of B is A?

What percent of 170 is 204? $\frac{P}{100}=\frac{204}{170}$

FORMULA

Note

The operations in these examples can be performed with a calculator or by hand. In either case, the equations should be written so that the = signs are aligned one under the other. Also, **writing the equations and the calculated values will help you remember whether you are multiplying or dividing.**

The following examples illustrate how to substitute into the proportion $\frac{P}{100}=\frac{A}{B}$ and to solve the resulting proportion for the unknown number.

Example 1 Finding the Amount

What is 65% of 500?

Solution

$P\% = 65\%$ and $B = 500$. We want to find the **amount** A.

$$\frac{65}{100}=\frac{A}{500} \quad \text{Substitute } P=65 \text{ and } B=500.$$

$$65\cdot 500=100\cdot A \quad \text{Find the cross products and set them equal to each other.}$$

$$\frac{32{,}500}{100}=\frac{\cancel{100}\cdot A}{\cancel{100}} \quad \text{Divide both sides by 100.}$$

$$325=A \quad \text{Simplify.}$$

So 65% of 500 is **325**.

Now work margin exercise 1.

1. What is 15% of 80?

Math Tip

You can check for the reasonableness of the answer in Example 1 using percents that are easy to calculate. 50% of 500 is 250 and 10% of 500 is 50. So 325 seems reasonable for 65% of 500.

2. 86% of ________ is 430.

Math Tip

To check for the reasonableness of Example 2, we can use the facts that 50% of 90 is 45 and 10% of 90 is 9. This means that 60% of 90 is 45 + 9 = 54. Since 57% < 60% and 51.3 < 54, 90 seems reasonable.

3. What percent of 90 is 19.8?

4. Dietary experts recommend that adults on a 2000-calorie-per-day diet should eat 300 g of carbohydrates per day. If a granola bar claims to provide 6% of the recommended daily carbohydrates, how many grams of carbohydrates does it contain?

Example 2 Finding the Base

57% of ________ is 51.3.

Solution

$P\% = 57\%$ and $A = 51.3$. We want to find the **base** $\boldsymbol{B}$.

$$\frac{57}{100} = \frac{51.3}{B}$$ Substitute $P = 57$ and $A = 51.3$.

$$57 \cdot B = 100 \cdot 51.3$$ Find the cross products and set them equal to each other.

$$\frac{\cancel{57} \cdot B}{\cancel{57}} = \frac{5130}{57}$$ Divide both sides by 57.

$$B = 90$$ Simplify.

So 57% of **90** is 51.3.

Now work margin exercise 2.

Example 3 Finding the Percent

What percent of 170 is 204?

Solution

$B = 170$ and $A = 204$. We want to find the **percent** $\boldsymbol{P}$.

$$\frac{P}{100} = \frac{204}{170}$$ Substitute $B = 170$ and $A = 204$.

$$P \cdot 170 = 100 \cdot 204$$ Find the cross products and set them equal to each other.

$$\frac{P \cdot \cancel{170}}{\cancel{170}} = \frac{20,400}{170}$$ Divide both sides by 170.

$$P = 120$$ Simplify.

So **120%** of 170 is 204.

Now work margin exercise 3.

Example 4 Application: Finding the Amount

Many food product labels now list total calories and number of calories from fat. Dietary experts believe that a healthy diet has at most 30% of its calories derived from fat. Following this guideline, if an adult consumes 2500 calories per day, at most how many calories should be from fat?

Solution

In this problem, we want to find 30% of the total diet of 2500 calories. So, with $P\% = 30\%$ and $B = 2500$, we want to find the value of A.

Substitution in the proportion $\frac{P}{100} = \frac{A}{B}$ gives

$$\frac{30}{100} = \frac{A}{2500}$$
$$30 \cdot 2500 = 100 \cdot A$$
$$\frac{75{,}000}{100} = \frac{\cancel{100} \cdot A}{\cancel{100}}$$
$$750 = A.$$

So in a healthy diet of 2500 calories per day, no more than 750 calories should be derived from fat.

Now work margin exercise 4.

Example 5 Application: Finding the Percent

Greg is hoping to raise $750 for a charity bicycle ride. Currently, his friends and family have donated $330 toward the cause. What percent of his goal has he reached?

Solution

Greg has raised $330 towards his goal. We want to find what percent this amount is of $750. So, with $B = \$750$ and $A = \$330$, We want to find the value of P. Substitution in the proportion $\frac{P}{100} = \frac{A}{B}$ gives

$$\frac{P}{100} = \frac{330}{750}$$
$$P \cdot 750 = 100 \cdot 330$$
$$\frac{P \cdot \cancel{750}}{\cancel{750}} = \frac{33{,}000}{750}$$
$$P = 44$$

Greg has reached 44% of his goal.

Now work margin exercise 5

5. Angela is shopping for a new tablet. Of the 15 different tablets she is considering, only 6 have a rear-facing 8-megapixel camera. What percent of the tablets have this type of camera?

Margin Exercise Answers

1. 12 **2.** 500 **3.** 22% **4.** 18 g **5.** 40%

4.5 Exercises

Concept Check

Fill-in-the-Blank. Complete each sentence using information found in this section.

1. Using a proportion equation, $P\%$ represents ________ and is written as the ratio $\frac{P}{\_\_\_\_\_\_}$.

2. In the proportion equation $\frac{P}{100} = \frac{A}{B}$, the letter A represents ________, or part of the base.

4.5 Exercises

Concept Check

1. percent, 100

2. amount

3. 172.3
4. 56
5. $\frac{95}{100} = \frac{A}{60}$
6. True
7. False; In the proportion $\frac{P}{100} = \frac{65}{200}$, the base is 200.
8. False; In the problem "What is 26% of 720?" the missing number is the amount.
9. False; The base is not always larger than the amount.

Practice

1. 7.5
2. 51
3. 15
4. 18
5. 64
6. 9
7. 47
8. 83
9. 900
10. 250
11. 150
12. 70
13. 62
14. 46
15. 20%
16. 25%
17. 150%
18. 300%
19. 150
20. 250
21. 30
22. 80
23. 24
24. 16
25. 33.3%
26. 33.3%

3. In the proportion $\frac{P}{100} = \frac{172.3}{150}$, ________ represents the amount.
4. In the proportion $\frac{56}{100} = \frac{A}{542}$, ________ represents the percent.
5. For the question "What is 95% of 60?" the corresponding proportion would be ________.

True/False. Determine whether each statement is true or false. If a statement is false, explain how it can be changed so the statement will be true. (**Note:** There may be more than one acceptable change.)

6. Percent problems can be solved with a proportion if two of the three parts P, A, and B are known.
7. In the proportion $\frac{P}{100} = \frac{65}{200}$, the base is 65.
8. In the problem "What is 26% of 720?" the missing number is the base.
9. Because the base represents the whole, it is always larger than the amount.

Practice

Use the proportion $\frac{P}{100} = \frac{A}{B}$ to find each unknown quantity. Round percents to the nearest tenth of a percent. All other answers should be rounded to the nearest hundredth, if necessary. See Examples 1 through 3.

1. Find 15% of 50.
2. What is 85% of 60?
3. 25% of 60 is _____.
4. 40% of 45 is _____.
5. 80% of 80 is _____.
6. 10% of 90 is _____.
7. Find 100% of 47.
8. Find 100% of 83.
9. 27 is 3% of what number?
10. 5 is 2% of what number?
11. 30% of _____ is 45.
12. 50% of _____ is 35.
13. 100% of _____ is 62.
14. 150% of _____ is 69.
15. What percent of 50 is 10?
16. What percent of 48 is 12?
17. What percent of 50 is 75?
18. What percent of 14 is 42?
19. _____% of 44 is 66.
20. _____% of 56 is 140.
21. _____% of 70 is 21.
22. _____% of 90 is 72.
23. 75% of 32 is _____.
24. 25% of 64 is _____.
25. What percent of 15 is 5?
26. What percent of 54 is 18?
27. _____% of 120 is 48.
28. _____% of 100 is 38.
29. 14 is 20% of what number?
30. 32 is 20% of what number?

31. 23 is ____% of 10.

32. 18 is ____% of 10.

33. Find 5% of 72.

34. Find 60% of 42.

35. What number is 50% of 35?

36. What number is 30% of 85?

37. 76.8 is 15% of what number?

38. 142.6 is 23% of what number?

39. 78 is 120% of what number?

40. 200 is 125% of what number?

41. 20.25 is 25% of _____.

42. 43.68 is 84% of _____.

43. What percent of 100 is 66.5?

44. What percent of 100 is 24.9?

45. Find 13.5% of 95.

46. Find 18.5% of 345.

47. What percent of 160 is 200?

48. What percent of 120 is 258?

49. ____ is 96% of 35.

50. ____ is 17% of 425.

51. 150% of 70 is _____.

52. 250% of 52 is _____.

53. 86.5% of what number is 100?

54. 68.5% of what number is 27?

55. 440 is 110% of what number?

56. 450 is 150% of what number?

57. 48 is ____% of 24.

58. 92 is _____% of 23.

59. 179.2 is ____% of 640.

60. 132.6 is _____% of 390

61. 40 is $33\frac{1}{3}$% of _____.

62. 8 is $66\frac{2}{3}$% of _____.

Applications

Solve.

63. In 2016, the Los Angeles Dodgers led the major leagues in home attendance, drawing an average of 45,720 fans to their home games. This figure represented 81.64% of the capacity of Dodger Stadium. Estimate how many fans the stadium can hold (to the nearest ten) when it is filled to capacity. [1]

64. Baldwin-Wallace College, a small liberal arts college in Ohio, had a total enrollment of 3332 students in the 2021–2022 school year. Of that number, 83.1% of the students were undergraduates. How many undergraduates were enrolled at Baldwin-Wallace College in 2021–2022? Round your answer to the nearest whole number. [2]

65. Within the borders of West Virginia, the length of Interstate 77 is 186 miles. Of this, 88 miles is a toll road. What percentage of Interstate 77 in West Virginia is a toll road? Round your answer to the nearest hundredth of a percent.

66. While Ashley Johnston was on the TV show "The Biggest Loser," she dropped her weight to 191 pounds, which was 51.45% of her starting weight. What was her starting weight? Round your answer to the nearest pound. [3]

1 Source: espn.go.com/mlb/attendance
2 Source: "About," Baldwin Wallace University, www.bw.edu/about
3 Source: msnbc.msn.com/id/35487357/ns/health-fitness

27. 40
28. 38
29. 70
30. 160
31. 230
32. 180
33. 3.6
34. 25.2
35. 17.5
36. 25.5
37. 512
38. 620
39. 65
40. 160
41. 81
42. 52
43. 66.5%
44. 24.9%
45. 12.82
46. 63.83
47. 125%
48. 215%
49. 33.6
50. 72.25
51. 105
52. 130
53. 115.61
54. 39.42
55. 400
56. 300
57. 200
58. 400
59. 28
60. 34
61. 120
62. 12

Applications

63. 56,000 fans
64. 2769 undergraduates
65. 47.31%
66. 371 pounds

67. $97,600
68. 11.0%
69. **a.** 93.75%
b. 6.25%; Keep this player
70. 160 free throws
71. 27 aces
72. Non-Resident Alien: 1453; Native American: 135; Hispanic: 2197; Asian or Pacific Islander: 3633; Black: 1081; White: 7579
73. Army: 13.44%; Marine Corps: 6.66%; Navy: 15.90%; Air Force: 19.21%; Coast Guard: 16.43%

Writing & Thinking

74. The four parts of the proportion equation are percent, the number 100, amount, and base. The percent is the number of "pieces" out of 100. (The 100 is necessary to denote the per cent/per 100.) The amount is the number that is part of the whole (or "pieces" of the whole) and the base is the whole.
75. Proportions would work for mixed numbers because a mixed number can be rewritten as a fraction. The only additional step required would be to change the mixed number to an improper fraction and then solve the proportion as normal.

67. You want to purchase a new home for $122,000. The bank will loan you 80% of the purchase price. How much will the bank loan you? (This amount is called your mortgage and you will pay it off over several years with interest. For example, a 30-year loan will probably cost you a total of more than 3 times the original loan amount.)

68. Two people dined at the La Fiesta Mexican Grill where the food cost was $31.18. To that, a tax of $3.43 was added. What is the percentage of this tax? Round your answer to the nearest tenth of a percent.

69. A kicker on a professional football team made 45 of 48 field goal attempts.

a. What percent of his attempts did he make?

b. What percent did he miss? Would you keep this player on your team or trade for a new kicker?

70. In one season, a basketball player missed 15% of her free throws. How many free throws did she attempt if she made 136 free throws?

71. In one tennis match, Andy served an "ace" on 15% of his serves. If he served 180 times, how many aces did he serve?

72. The University of Southern California has a total of 16,897 undergraduate students. Find the number of students in each of six groups using the percents shown in the graph below. Round each answer to the nearest whole number.[4]

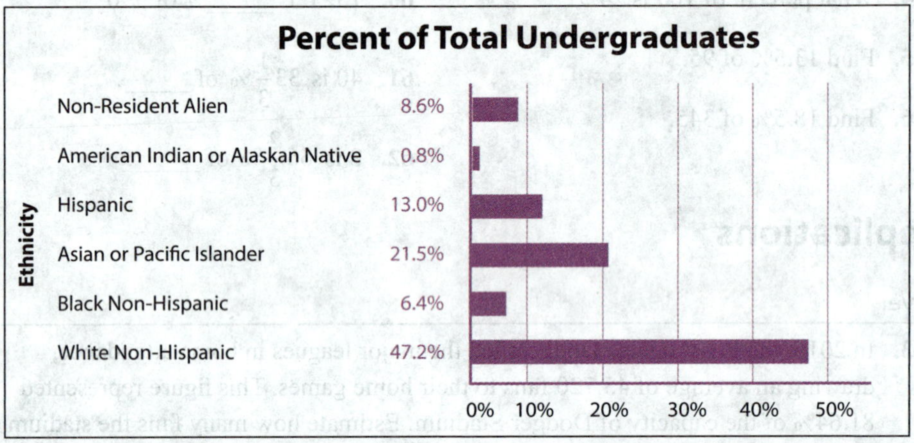

73. Find the percent of women in each of the 5 branches of the military from the data shown in the following table. Round each answer to the nearest hundredth of a percent.

Component	Military	Female
Army	561,979	75,507
Marine Corps	202,612	13,493
Navy	323,139	51,385
Air Force	329,640	63,310
Coast Guard	41,327	6,790

Writing & Thinking

74. List the four parts of the proportion equation and give a brief definition of each one.

75. Can a mixed number be used in a proportion? Justify your answer.

4 Source: education-portal.com/directory/school/University_of_Southern_California.html

4.6 Solving Percent Problems Using Equations

Objectives

A. Solve percent problems using the equation $R \cdot B = A$.

A Solving Percent Problems Using the Equation $R \cdot B = A$

The equation $R \cdot B = A$ (where R represents percent in decimal form or fraction form) is called the basic equation for solving percent problems. To see how this basic equation is used, consider the following statement:

16% of 50 is 8.

This statement can be translated into an equation in the following way.

Statement →	16%	of	50	is	8
	↑ Percent changed to decimal number ↓	↑ "of" changed to times sign ↓	↕	↑ "is" changed to equal sign ↓	↕
Equation →	0.16	$\cdot$	50	=	8
	↓	↓	↓	↓	↓
Basic equation →	RATE	$\cdot$	BASE	=	AMOUNT

The terms represented in the basic equation are the same as those discussed in the previous section and are explained in detail in the following box.

Terms Related to the Basic Equation $R \cdot B = A$

For the basic equation $R \cdot B = A$,

R = **rate** or percent (as a decimal number or fraction).

B = **base** (number that we are finding the percent of).

A = **amount** (a part of the base).

DEFINITION

Note

The basic equation can also be written in the form $A = R \cdot B$. This form is convenient when solving for the amount A.

Many people have a difficult time with percent problems when trying to decide whether to multiply or divide. By using the basic equation $R \cdot B = A$ (or the proportion discussed earlier), these difficulties can be avoided. The decision to multiply or divide to find the unknown quantity is determined when the given numbers are substituted into the equation.

Three Basic Types of Percent Problems and the Formula $R \cdot B = A$

Type 1: Find the **amount** given the base and the percent (rate).

What is 65% of 800?

$$\begin{array}{ccccc} A & = & R & \cdot & B \\ \downarrow & & \downarrow & & \downarrow \\ A & = & 0.65 & \cdot & 800 \end{array}$$

Type 2: Find the **base** given the percent (rate) and the amount.

42% of what number is 157.5?

$$\begin{array}{ccccc} R & \cdot & B & = & A \\ \downarrow & & \downarrow & & \downarrow \\ 0.42 & \cdot & B & = & 157.5 \end{array}$$

Type 3: Find the **percent** (rate) given the base and the amount.

What percent of 92 is 115?

$$\begin{array}{ccccc} R & \cdot & B & = & A \\ \downarrow & & \downarrow & & \downarrow \\ R & \cdot & 92 & = & 115 \end{array}$$

FORMULA

Note

of means to multiply. (The raised dot, ·, is used in the percent formula.)

is means equals (=).

1. What is 10% of 137?

Example 1 Finding the Amount

What is 65% of 800?

Solution

In this case, $R = 65\% = 0.65$ and $B = 800$. We want to find the **amount** A.

$$A = R \cdot B$$
$$A = 0.65 \cdot 800$$
$$A = 520$$

So 65% of 800 is **520**.

Now work margin exercise 1.

2. 8% of what number is 2?

Note

The operations in these examples can be performed with a calculator or by hand. In either case, the equations should be written so that the = signs are aligned one under the other. Also, **writing the equations and the calculated values will help you remember whether you are multiplying or dividing.**

Example 2 Finding the Base

42% of what number is 157.5?

Solution

Here, $R = 42\% = 0.42$ and $A = 157.5$. We want to find the **base** B.

$$R \cdot B = A$$
$$0.42 \cdot B = 157.5$$
$$\frac{\cancel{0.42} \cdot B}{\cancel{0.42}} = \frac{157.5}{0.42} \quad \text{Divide both sides by 0.42.}$$
$$B = 375$$

So 42% of **375** is 157.5.

Now work margin exercise 2.

Example 3 Finding the Percent

________% of 92 is 115.

Solution

For this problem, $B = 92$ and $A = 115$. We want to find the **percent** $\boldsymbol{R}$.

$$R \cdot B = A$$
$$R \cdot 92 = 115$$
$$\frac{R \cdot \cancel{92}}{\cancel{92}} = \frac{115}{92} \quad \text{Divide both sides by 92.}$$
$$R = 1.25$$
$$R = 125\%$$

We have **125**% of 92 is 115.

Now work margin exercise 3.

3. ________% of 56 is 67.2.

Math Tip

In Example 3, we can verify that the answer is reasonable by noticing that $92 < 115$. This relationship means that the percent should be greater than 100%.

Percents can be changed to fraction form, and in some cases, the fraction form will simplify the work. For example, we know that

$$75\% = \frac{3}{4}, \qquad 33\frac{1}{3}\% = \frac{1}{3}, \qquad \text{and} \qquad 12\frac{1}{2}\% = \frac{1}{8}.$$

The following examples illustrate the use of fractions. (See the table of values at the end of Section 4.4 for other percent and fraction equivalents.)

Example 4 Finding the Amount

Find 75% of 56.

Solution

Here, $R = 75\% = \frac{3}{4}$ and $B = 56$. We want to find the **amount** A.

$$A = R \cdot B$$
$$A = \frac{3}{\cancel{4}} \cdot \overset{14}{\cancel{56}}$$
$$A = 3 \cdot 14$$
$$A = 42$$

So 75% of 56 is **42**.

Now work margin exercise 4.

4. Find 150% of 60.

Example 5 Finding the Base

250 is 62.5% of ________.

Solution

For this problem, $R = 62.5\% = \frac{5}{8}$ and $A = 250$. We want to find the **base** $\boldsymbol{B}$.

5. 14 is $87\frac{1}{2}\%$ of ________.

Math Tip

We can check the answer in Example 5 for reasonableness by noticing that 200 is 50% of 400 and 40 is 10% of 400. This means that 60% of 400 is $200 + 40 = 240$. So 250 seems reasonable for 62.5% of 400.

$$R \cdot B = A$$

$$\frac{5}{8} \cdot B = 250$$

$$\frac{\frac{5}{8} \cdot B}{\frac{5}{8}} = \frac{250}{\frac{5}{8}}$$ Divide both sides by $\frac{5}{8}$.

$$B = \overset{50}{250} \cdot \frac{8}{5}$$ Multiply by the reciprocal.

$$B = 50 \cdot 8$$

$$B = 400$$

Thus, 250 is 62.5% of **400**.

Now work margin exercise 5.

Tips for Working with Percents

The following two comments are helpful in understanding percents and the relative sizes of the bases and the amounts.

1. A percent is just another form of a fraction: $R = \frac{A}{B}$.
2. When you find a percent of a given number, the amount will be
   a. smaller than the given number if the percent is less than 100% or
   b. larger than the given number if the percent is more than 100%.

6. Bradley is the manager at a cafe. He has observed that, typically, 80% of the drinks served are coffee drinks. At the end of the day, the coffee shop served 232 coffee drinks. How many drinks did the cafe serve that day?

Example 6 Application: Finding the Base

Maggie is taking an online survey. Currently, she has answered 27 questions, and the progress bar at the bottom of her screen tells her she has completed 60% of the survey. How many questions are on the survey?

Solution

For this problem, $R = 60\% = 0.60$ and $A = 27$. To find the base, substitute into the equation and solve for B.

$$R \cdot B = A$$

$$0.60 \cdot B = 27$$

$$\frac{0.60 \cdot B}{0.60} = \frac{27}{0.60}$$ Divide both sides by 0.60.

$$B = 45$$

There are 45 questions on the survey.

Now work margin exercise 6.

Example 7 Application: Finding the Amount

Marcus feels he needs to get 80% of the questions in his homework assignment correct to be confident that he has mastered the material. If today's assignment contains 15 questions, how many questions does Marcus need to answer correctly to feel he has achieved mastery of this material?

Solution

For this problem, $R = 80\% = 0.80$ and $B = 15$. To find the amount, substitute into the equation and solve for A.

$$R \cdot B = A$$
$$0.80 \cdot 15 = A$$
$$12 = A$$

Marcus needs to answer **12** questions correctly to feel he has mastered the material.

Now work margin exercise 7.

7. Jack works at a nature preserve observing the bald eagle population. Over the last 5 years, the population has been growing at a rate of 4% each year (each year the population is 104% of the previous year). Last year, the nature preserve was home to 50 pairs of bald eagles. How many pairs of bald eagles can Jack expect to observe this year?

Margin Exercise Answers
1. 13.7 **2.** 25 **3.** 120% **4.** 90 **5.** 16 **6.** 290 drinks **7.** 52 pairs

4.6 Exercises

Concept Check

Fill-in-the-Blank. Complete each sentence using information found in this section.

1. In the equation $R \cdot B = A$, B represents the ________.
2. In the equation $R \cdot B = A$, R is written in ________ form.
3. The word "of" implies ________.
4. In the problem: "63% of what number is 792?" the number being sought is the ________.
5. The basic equation can be written in the form $A = R \cdot B$. This form is most convenient when the unknown value is the ________.

True/False. Determine whether each statement is true or false. If a statement is false, explain how it can be changed so the statement will be true. (**Note:** There may be more than one acceptable change.)

6. In order to solve the equation $0.56 \cdot B = 12$ for the base B one would multiply 12 by 0.56.
7. In the problem "126% of 720 is what number?" the missing number is the amount.
8. The solution to the problem "50% of what number is 352?" could be found by solving the equation $50 \cdot B = 352$.
9. If the base is 120 and the rate is greater than 100%, then the amount will be greater than 120.

4.6 Exercises

Concept Check

1. base
2. decimal number
3. multiplication or times
4. base
5. amount, A
6. False; In order to solve the equation $0.56 \cdot B = 12$ for the base, B, one would divide 12 by 0.56.
7. True
8. False; The solution to the problem "50% of what number is 352?" could be found by solving the equation $0.5 \cdot B = 352$.
9. True

Practice

1. 7
2. 18
3. 3.1
4. 4.5
5. 9
6. 21
7. 42
8. 300
9. 150
10. 250
11. 20%
12. 75%
13. 150
14. 85
15. 50
16. 20
17. 70
18. 75
19. 36
20. 1.48
21. 33.3%
22. 62.5%
23. 35%
24. 67%
25. 12.5
26. 21.7
27. 18
28. 41
29. 110
30. 86
31. 180%
32. 150%
33. 614
34. 520
35. 38
36. 73
37. 80
38. 25
39. 200%
40. 150%

Practice

Use the equation $R \cdot B = A$ to find each unknown quantity. Round percents to the nearest tenth of a percent. All other answers should be rounded to the nearest hundredth, if necessary. See Examples 1 through 5.

1. 10% of 70 is what number?

2. 20% of 90 is what number?

3. 5% of 62 is what number?

4. 6% of 75 is what number?

5. Find 75% of 12.

6. Find 60% of 35.

7. 150% of ____ is 63.

8. 110% of ____ is 330.

9. ____% of 60 is 90.

10. ____% of 60 is 150.

11. What percent of 75 is 15?

12. What percent of 120 is 90?

13. 3 is 2% of what number?

14. 17 is 20% of what number?

15. ____% of 34 is 17.

16. ____% of 30 is 6.

17. 21 is 30% of what number?

18. 75 is 100% of what number?

19. 100% of 36 is ____.

20. 1% of 148 is ____.

21. What percent of 48 is 16?

22. What percent of 88 is 55?

23. What percent of 100 is 35?

24. What percent of 100 is 67?

25. What number is 50% of 25?

26. What number is 31% of 70?

27. 25% of 72 is ____.

28. 82% of 50 is ____.

29. 22 is 20% of ____.

30. 86 is 100% of ____.

31. 18 is what percent of 10?

32. 15 is what percent of 10?

33. 92.1 is 15% of what number?

34. 119.6 is 23% of what number?

35. 9.5 is 25% of ____.

36. 29.2 is 40% of ____.

37. 100 is 125% of ____.

38. 30 is 120% of ____.

39. 36 is ____% of 18.

40. 60 is ____% of 40.

41. 45 is 60% of what number?

42. 12% of ____ is 15.

43. ____ is 96% of 17.

44. ____ is 84% of 32.

45. 18% of 325 is what number?

46. 28% of 460 is what number?

47. Find 15.2% of 75.

48. Find 67.5% of 140.

49. 87.5% of what number is 21?

50. 37.5% of what number is 12?

51. Find 120% of 60.

52. Find 150% of 126.

53. What percent of 32 is 30.4?

54. What percent of 240 is 76.8?

55. 17.4 is ____% of 60.

56. 64.8 is ____% of 180.

57. ____% of 4 is 7.

58. ____% of 20 is 28.

59. 24 is $33\frac{1}{3}\%$ of what number?

60. 60 is $66\frac{2}{3}\%$ of what number?

Use fractions to solve each percent problem. See Examples 4 and 5.

61. Find 50% of 32.

62. 25% of 150 is what number?

63. What is $12\frac{1}{2}\%$ of 80?

64. What is $62\frac{1}{2}\%$ of 16?

65. $33\frac{1}{3}\%$ of 75 is what number?

66. Find $66\frac{2}{3}\%$ of 60.

67. 75% of what number is 21?

68. 50% of what number is 35?

69. $37\frac{1}{2}\%$ of what number is 62.1?

70. $12\frac{1}{2}\%$ of what number is 76.3?

Applications

Solve.

71. Only 27% of American deaths in the Revolutionary War occurred in battle. All other mortalities were the result of things such as exposure, disease, and starvation. Of the estimated 25,300 deaths, how many men died in battle (to the nearest hundred)?

72. In 2016, Expedia paid its CEO $94.6 million in compensation (salary, bonus, and stock options). This was 2252% of his total compensation in 2020. What was his total compensation package in 2020? (Round to the nearest tenth of a million.)

73. During his presidency, from 1945 to 1953, Harry S. Truman vetoed 250 congressional bills, and 12 of those vetoes were overridden. What percent of Truman's vetoes were overridden?

74. William O. Douglas served as an associate justice on the Supreme Court for 36 years, from 1939 to 1975, the longest tenure of any Supreme Court justice in history. He lived to be 82. What percent of his life (to the nearest one percent) did he spend on the Supreme Court?

75. 14 karat gold is an alloy which consists of 58.33% gold by weight, while the remainder is alloy material. If a goldsmith starts with 20 grams of pure gold, how much will the 14 karat gold alloy weigh? Round your answer to the nearest hundredth of a gram.

76. The minimum down payment to obtain the best financing rate on a house is 20%. Assuming that John has set aside $35,000 and wants to take advantage of the best financing rate, what is the most expensive house he can purchase?

77. A student missed 6 problems on a mathematics test and received a grade of 85%. If all the problems were of equal value, how many problems were on the test?

78. Magazine publishers will often give significant discounts to subscribers. The cover price of *TV Guide* is $145.01 for 12 months, but a 12 month subscription costs only $20.00! What is the percent savings (to the nearest tenth) for someone who subscribes to TV Guide?

41. 75
42. 125
43. 16.32
44. 26.88
45. 58.5
46. 128.8
47. 11.4
48. 94.5
49. 24
50. 32
51. 72
52. 789
53. 95%
54. 32%
55. 29%
56. 36%
57. 175%
58. 140%
59. 72
60. 90
61. 16
62. 37.5
63. 10
64. 10
65. 25
66. 40
67. 28
68. 70
69. 165.6
70. 610.4

Applications

71. 6800
72. $4.2 million
73. 4.8%
74. 44%
75. 34.29 g
76. $175,000
77. 40
78. 86.2%

79. In 2021, the final standings of the top two teams in each division of the American and National Leagues in baseball were as listed here. With this information, find the percentage (Pct.) of wins for each team on this list (accurate to three decimal places).

American League	Won	Lost	Pct.	National League	Won	Lost	Pct.
East				**East**			
Tampa Bay	100	62	**a.**	Atlanta	88	73	**g.**
Boston	92	70	**b.**	Philadelphia	82	80	**h.**
Central				**Central**			
Chicago White Sox	93	69	**c.**	Milwaukee	95	67	**i.**
Cleveland	80	82	**d.**	St. Louis	90	72	**j.**
West				**West**			
Houston	95	67	**e.**	San Francisco	107	55	**k.**
Seattle	90	72	**f.**	LA Dodgers	106	56	**l.**

Source: www.mlb.com/standings/2021

80. A political candidate hired a survey company to ask 15,000 constituents what issues matter the most to them for the upcoming election. The bar graph shows the results broken down into six main issues. Find the number of constituents that say each issue matters the most.

What Matters Most to Voters?

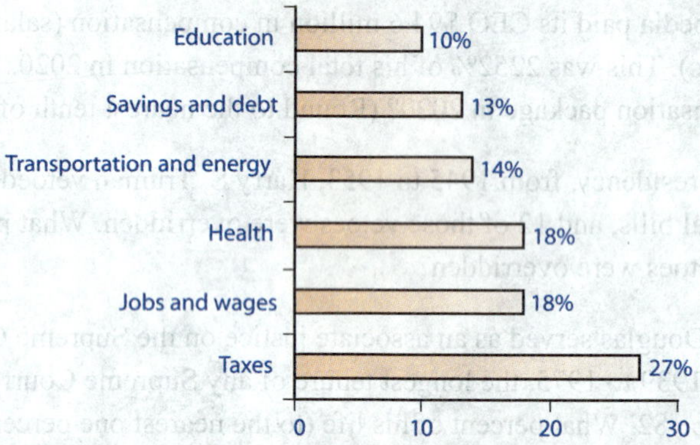

Writing & Thinking

81. List the three parts of the equation $R \cdot B = A$ and define each one in your own words.

82. Explain the connection between the proportion $\frac{P}{100} = \frac{A}{B}$ and the equation $R \cdot B = A$.

83. Explain how to determine which number is the rate (percent), which one is the amount, and which one is the base.

84. Discuss when, in your opinion, it might be better to use the proportion $\frac{P}{100} = \frac{A}{B}$ and when it might be better to use the equation $R \cdot B = A$ in solving a percent problem.

79. **a.** 61.728% **b.** 56.790% **c.** 57.407% **d.** 49.383% **e.** 58.642% **f.** 55.556% **g.** 54.658% **h.** 50.617% **i.** 58.642% **j.** 55.556% **k.** 66.049% **l.** 65.432%

80. Education: 1500; Savings and debt: 1950; Transportation and energy: 2100; Health: 2700; Jobs and wages: 2700; Taxes: 4050

Writing & Thinking

81. The three parts are the rate, base, and amount. The rate is the percent but should be written in decimal number form. The amount is the number that is part of the whole and the base is the whole.

82. P is a whole number percentage and R is the percentage written as a decimal number, so $\frac{P}{100} = R$. Thus $R = \frac{A}{B}$, and we can multiply both sides of the equation by B to obtain $R \cdot B = A$.

83. The amount is the number that is often near the word "is." The base is the number that often follows the word "of." The rate is the number written either as a fraction or as a decimal number that has not been identified as the amount or the base, and usually appears before the word "of."

84. If the amount is the missing number, the equation $R \cdot B = A$ is usually a more efficient choice. If the percentage is missing, or if you struggle with remembering how to write a decimal number as a percentage, the proportion $\frac{P}{100} = \frac{A}{B}$ eliminates that uncertainty and easily sets up a solution for the percentage. If the base is the missing number, either proportion or equation can be used, whichever you prefer.

4.7 Applications of Percent

Objectives

- **A.** Know and use Pólya's four-step process for problem solving.
- **B.** Calculate a discount.
- **C.** Calculate sales tax.
- **D.** Calculate commission.
- **E.** Calculate percent increase and percent decrease.
- **F.** Calculate percent of profit.

A Pólya's Four-Step Process for Problem Solving

George Pólya (1887–1985), a famous professor at Stanford University, studied the process of discovery learning. Among his many accomplishments, he developed the following four-step process as an approach to problem solving.

1. Understand the problem.
2. Devise a plan.
3. Carry out the plan.
4. Look back over the results.

There are a variety of types of applications discussed throughout this course and in subsequent courses in mathematics, and you will find these four steps helpful as guidelines for understanding and solving all of them. Applying the necessary skills to solve exercises, such as adding fractions or solving equations, is not the same as accumulating the knowledge to solve problems. **Problem solving can involve careful reading, reflection, and some original or independent thought.**

History

George Pólya is one of the most influential mathematicians of the 20th century, not only for his contributions to math but also to education. His father, Jakab Pollák, changed his name from a Jewish sounding name to Pólya, a more Hungarian sounding one, to improve his chances of a university post which he received shortly before he died. George Pólya was a professor of mathematics at ETH Zürich from 1914 to 1940 and at Stanford University from 1940 to 1953. Stanford University named Pólya Hall in his honor. The hall was built while Pólya was still teaching, which caused him to complain that it made people think he was dead!

Basic Steps for Solving Word Problems

1. Understand the problem. For example,
   a. Read the problem.
   b. Understand all the words.
   c. If it helps, restate the problem in your own words.
2. Devise a plan. For example, use one or all of the following.
   a. Guess, estimate, or make a list of possibilities.
   b. Draw a picture or diagram.
   c. Use a variable and form an equation.
3. Carry out the plan. For example,
   a. Try all the possibilities you have listed.
   b. Study your picture or diagram for insight into the solution.
   c. Solve any equation that you may have set up.
4. Look back over the results. For example,
   a. Can you see an easier way to solve the problem?
   b. Does your solution actually work? Does it make sense in terms of the wording of the problem? Is it reasonable?
   c. If there is an equation, check your answer in the equation.

PROCEDURE

In this section, we will discuss applications that involve percent and the types of percent problems discussed in Sections 4.5 and 4.6. Your instructor may allow the use of calculators. If so, you should keep in mind that **a calculator is a tool to enhance, not replace, the necessary skills and abilities related to problem solving**. Your personal experience, knowledge, and general understanding of problem-solving situations will determine the ease with which you grasp the problems presented. Study the following examples carefully.

B Discount

To attract customers or to sell goods that have been in stock for some time, retailers and manufacturers offer a **discount**, a reduction in the original price (or marked price) of an item. The new, reduced price is called the **sale price**. The **discount** is the difference between the original price and the sale price. The **rate of discount** (or **percent of discount**) is the percent of the original price to be discounted.

Terms Related to Discount

Discount: difference between the original price and the sale price

Sale price: original price minus the discount

Rate of discount: percent of original price to be discounted

DEFINITION

As you study the examples and work through the exercises, be sure to **label each amount of money as to what it represents**. This will help in organizing the results and determining what operations to perform.

1. A pair of shoes originally priced at $52 is now discounted 25%.
   a. What is the amount of the discount?
   b. What is the sale price?

Example 1 Application: Solving Discount Problems

A refrigerator that regularly sells for $1200 is on sale at a 20% discount.

a. What is the amount of the discount?

b. What is the sale price?

Solution

a. To find the discount, we find 20% of $1200.

20% of $1200 is ______.

$$\frac{P}{100} = \frac{A}{B}$$
$$\frac{20}{100} = \frac{A}{1200}$$
$$20 \cdot 1200 = 100 \cdot A$$
$$\frac{24{,}000}{100} = \frac{\cancel{100} \cdot A}{\cancel{100}}$$
$$240 = A$$

OR

$$A = R \cdot B$$
$$A = 0.20 \cdot 1200$$
$$A = 240$$

The discount is $240.

Math Tip

The order of the parts in Example 1 could have been reversed. In that case, we would first find 80% (100% − 20% = 80%) of $1200 to find the sale price, and then subtract this result from $1200 to find the discount.

b. Find the sale price by subtracting the discount from the original price.

$1200	Original price
− 240	Discount
$960	Sale price

The sale price is $960.

Now work margin exercise 1.

Example 2 Application: Solving Discount Problems

Large, fluffy towels were on sale at a discount of 30%. If the sale price was $8.40, what was the original price?

Solution

In this case, we already know the sale price. Now we need to reason that because the discount was 30%, the sale price represents 70% (100% − 30% = 70%) of the original price. Also, note that the original price will be more than the sale price of $8.40.

70% of ______ is $8.40.

$$\frac{P}{100} = \frac{A}{B}$$
$$\frac{70}{100} = \frac{8.40}{B}$$
$$70 \cdot B = 100 \cdot 8.40$$
$$\frac{\cancel{70} \cdot B}{\cancel{70}} = \frac{840}{70}$$
$$B = 12$$

OR

$$R \cdot B = A$$
$$0.70 \cdot B = 8.40$$
$$\frac{\cancel{0.70} \cdot B}{\cancel{0.70}} = \frac{8.40}{0.70}$$
$$B = 12$$

The original price of the towels was $12.00 each.

Now work margin exercise 2.

2. Dylan buys a watch for $90. This price was after a 40% discount off the original price. What did the watch cost originally?

C Sales Tax

Sales tax is a tax charged on the selling price of goods sold by retailers. The **rate of the sales tax** (a percent of the selling price) varies from state to state (or even city to city in some cases). States and cities use sales tax to pay for public services.

Terms Related to Sales Tax

Sales tax: tax based on selling price

Rate of sales tax: percent of selling price

DEFINITION

3. Assuming a 7% sales tax rate, what would be the final cost of a discounted pair of shoes priced at $39?

Example 3 Application: Solving Sales Tax Problems

If the rate of sales tax is 6%, what would be the final cost of a laptop priced at $899?

Solution

First, find the sales tax by taking 6% of $899.

6% of 899 is _____.

$$\frac{P}{100} = \frac{A}{B} \qquad \text{OR} \qquad A = R \cdot B$$

$$\frac{6}{100} = \frac{A}{899} \qquad\qquad A = 0.06 \cdot 899$$

$$6 \cdot 899 = 100 \cdot A \qquad\qquad A = 53.94$$

$$\frac{5394}{100} = \frac{\cancel{100} \cdot A}{\cancel{100}}$$

$$53.94 = A$$

Next, find the final cost by adding the sales tax to the original price.

	$899.00	Original price
+	53.94	Sales tax
	$952.94	Final cost

The final cost of the laptop would be $952.94.

Now work margin exercise 3.

D Commission

A **commission** is a fee paid to an agent or salesperson for a service. Commissions are usually a percent of a negotiated contract or a percent of sales. In some cases, a salesperson's entire salary is earned through commission. In other cases, the salesperson earns a base salary plus a commission on sales above a certain amount.

4. Lynsay earns a salary of $1250 a month plus a commission of 5% on all electronics she sells at her job at the local computer store. What did she earn the month she sold $28,640 in electronics?

Example 4 Application: Solving Commission Problems

Susan sells women's shoes. She earns a salary of $2000 a month plus a commission of 8% on what she sells over $8500. What did Susan earn the month she sold $22,500 worth of shoes?

Solution

First, find the base for her commission by subtracting $8500 from her sales.

	$22,500	Total sales
−	8 500	
	$14,000	Base for commission

Next, find the amount of the commission by taking 8% of $14,000.

$$\frac{P}{100} = \frac{A}{B} \qquad \text{OR} \qquad A = R \cdot B$$

$$\frac{8}{100} = \frac{A}{14{,}000} \qquad A = 0.08 \cdot 14{,}000$$

$$8 \cdot 14{,}000 = 100 \cdot A \qquad A = 1120 \quad \text{Amount of commission}$$

$$\frac{112{,}000}{100} = \frac{\cancel{100} \cdot A}{\cancel{100}}$$

$$1120 = A \quad \text{Amount of commission}$$

Finally, add her salary and the amount of the commission to find her earnings for the month.

	\$2000	Salary
+	1120	Commission
	\$3120	Total pay for the month

Susan earned \$3120 for the month.

Now work margin exercise 4.

E Percent Increase and Percent Decrease

The value of property, such as a home, can increase or decrease over time. Similarly, the value of stocks in the stock market, populations of cities and countries, and quantities of products sold in a year can, and do, increase or decrease. At times, it is helpful to know by what percent the value changed. This is called finding the **percent increase** (or the **percent decrease**). Examples 5 and 6 illustrate these ideas. (**Note:** In many cases, the increase in value is called **appreciation**, and the decrease in value is called **depreciation**.)

Is Appreciation Always Good for You?

If something appreciates in value, we generally think it's a good thing. For a country's currency, it's often the opposite. Suppose 1 US dollar can trade for 4 Turkish liras. If we sell a computer worth \$1000, then Turkish people pay 4000 liras for that computer. But if the dollar appreciates to 6 liras, then that same computer now costs 6000 liras, and Turkish people may be less able to buy it. As a result, currency appreciation may make it harder to export goods.

Example 5 Application: Calculating Percent Increase

Akshi sells homemade greeting cards on the Internet. Last year she sold 160 cards. This year she sold 176 cards. What was her percent increase in sales from last year to this year?

Solution

First, find the increase in sales from last year to this year.

	176	This year's sales
−	160	Last year's sales
	16	Amount of change

Now determine the percent increase by finding what percent 16 is of 160. Use 160 as the base because we want to find the percent increase from last year's sales.

5. Last year, the local community college had 425 students declare a STEM (Science, Technology, Engineering, and Mathematics) major. This year, that number increased to 442 students. What was the percent increase in students declaring STEM majors between last year and this year?

$$\frac{P}{100}=\frac{A}{B} \qquad \text{OR} \qquad R\cdot B=A$$

$$\frac{P}{100}=\frac{16}{160} \qquad\qquad R\cdot 160=16$$

$$P\cdot 160=100\cdot 16 \qquad\qquad \frac{R\cdot \cancel{160}}{\cancel{160}}=\frac{16}{160}$$

$$\frac{P\cdot \cancel{160}}{\cancel{160}}=\frac{1600}{160} \qquad\qquad R=0.1$$

$$P=10$$

Akshi's percent increase in sales was 10%.

Now work margin exercise 5.

6. Tom is looking to sell his house. Unfortunately, the housing market in his area is not doing well, and he has to sell his house for less than he bought it. If Tom's purchasing price for the house was \$200,000 and he plans to sell it for \$194,000, what is the percent decrease in the value of Tom's house?

Example 6 Application: Calculating Percent Decrease

Three years ago, you bought a new car for \$25,000. Your business is doing well and you are now looking for another new car and want to trade in your first car. The dealer has told you that the trade in value of your car is now \$14,500. What is the percent decrease in the value of your car?

Solution

First find the actual decrease in the value.

\$25,000	Original price
− 14,500	Current price
\$10,500	Decrease in value

Now we find the percent decrease by finding what percent \$10,500 (the decrease in value) is of \$25,000 (the original price).

$$\frac{P}{100}=\frac{A}{B} \qquad \text{OR} \qquad R\cdot B=A$$

$$\frac{P}{100}=\frac{10,500}{25,000} \qquad\qquad R\cdot 25,000=10,500$$

$$P\cdot 25,000=100\cdot 10,500 \qquad\qquad \frac{R\cdot \cancel{25,000}}{\cancel{25,000}}=\frac{10,500}{25,000}$$

$$\frac{P\cdot \cancel{25,000}}{\cancel{25,000}}=\frac{1,050,000}{25,000} \qquad\qquad R=0.42$$

$$P=42$$

So the percent decrease in the value of your car is 42%.

Now work margin exercise 6.

F Percent of Profit

Manufacturers and retailers are concerned with the profit on each item produced or sold. In this sense, the profit made on an item is the difference between the selling price to the customer and the cost to the company. For example, suppose that a department store can buy a certain light fixture from a manufacturer for \$80 (the store's cost) and sell the fixture for \$100 (the selling price). The profit for the store is the difference between the selling price and the cost of the fixture. That is, in this example,

$$\text{Profit} = \$100 - \$80 = \$20.$$

The **percent of profit** is a ratio. And, as the following discussion indicates, this ratio can be based either on cost or on selling price.

Profit and Percent of Profit

Profit: the difference between selling price and cost

Profit = Selling price − Cost Profit = \$100 − \$80 = \$20

Percent of Profit: There are two types; both are ratios with **profit in the numerator.**

1. Percent of profit **based on cost** (cost is the denominator):

$$\frac{\textbf{Profit}}{\textbf{Cost}} = \textbf{\% of profit based on cost} \qquad \frac{20}{80} = \frac{1}{4} = 25\%$$

2. Percent of profit **based on selling price** (selling price is the denominator):

$$\frac{\textbf{Profit}}{\textbf{Selling Price}} = \textbf{\% of profit based on selling price} \qquad \frac{20}{100} = \frac{1}{5} = 20\%$$

FORMULA

Math Tip

A percent is a ratio. In fact, if the basic equation $R \cdot B = A$ is solved for R, we get the rate (or percent) $R = \frac{A}{B}$. That is, the rate (or percent) is a ratio.

Example 7 Application: Calculating Percent of Profit

A retail store markets calculators that cost the store \$45 each and are sold to customers for \$72 each.

a. What is the profit on each calculator?

b. What is the percent of profit based on cost?

c. What is the percent of profit based on selling price?

Solution

a. First find the profit.

\$72	Selling price
− \$45	Cost
\$27	Profit

The profit is \$27 per calculator.

For parts b. and c., use a ratio and then change the fraction to a percent to find each percent of profit.

b. For percent of profit based on cost, remember that cost is in the denominator.

$$\frac{\text{Profit}}{\text{Cost}} = \frac{\$27}{\$45} = \frac{3}{5} = 60\%$$

The percent of profit based on cost is 60%.

7. A shoe store sells certain shoes for \$110 each when the shoes actually cost the store \$50.

a. What is the profit on each pair of shoes?

b. What is the percent of profit based on cost?

c. What is the percent of profit based on selling price?

Note

Percent of profit **based on cost** is normally higher than percent of profit **based on selling price** because the selling price is usually larger than the cost. The business community reports whichever percent serves its purpose better. Your responsibility as an investor or consumer is to know which percent is reported and what it means to you.

c. For percent of profit based on selling price, remember that selling price is in the denominator.

$$\frac{\text{Profit}}{\text{Selling Price}} = \frac{\$27}{\$72} = \frac{3}{8} = 37.5\%$$

The percent of profit based on selling price is 37.5%.

Now work margin exercise 7.

Margin Exercise Answers

1. a. \$13 **b.** \$39 **2.** \$150 **3.** \$41.73 **4.** \$2682 **5.** 4% **6.** 3% **7. a.** \$60 **b.** 120% **c.** ≈51.55%

4.7 Exercises

Concept Check

Fill-in-the-Blank. Complete each sentence using information found in this section.

1. The final step in solving word problems is to ________ over the result to see if the result is reasonable.
2. The amount of reduction in the original selling price is called a/an ________. The reduced price is the ________ price.
3. Sales tax is a percentage of the ________ price. This tax is added to the buyer's cost.
4. The fee paid to an agent or salesperson for a service in called a/an ________.
5. If the value of an item increases, the increase is value can be called ________.
6. Percent of profit can be based on either ________ or ________ price.

True/False. Determine whether each statement is true or false. If a statement is false, explain how it can be changed so the statement will be true. (**Note:** There may be more than one acceptable change.)

7. If an item is selling for a 35% discount, the customer will pay 65% of the original price.
8. If you must pay 7% sales tax on a purchase, the total cost you will pay is 170% of the total before tax.
9. A car was purchased in 1965 for \$3800. It sold for \$1200 in 2011. This is an example of depreciation.
10. Profit is determined by subtracting selling price from the cost.

4.7 Exercises

Concept Check

1. look back or check
2. discount; sale
3. (total) sale or selling
4. commission
5. appreciation
6. cost; selling
7. True
8. False; If you must pay 7% sales tax on a purchase, the total cost you will pay is 107% of the total before tax.
9. True
10. False; Profit is determined by subtracting cost from selling price.

Applications

Solve.

1. A television which normally sells for $300 is priced at a 10% discount.
   a. Find the amount of the discount.
   b. Find the sale price.

2. A new briefcase was priced at $275. Suppose that it were marked down 30%.
   a. What would be the amount of the discount?
   b. What would be the new price?

3. A store owner received a 3% discount from the manufacturer when she bought $15,500 worth of dresses.
   a. What was the amount of the discount?
   b. What did she pay for the dresses?

4. In order to get more subscribers, a book club offered three books for a total price of $7.02. The total selling price was originally $17.55 for all three books.
   a. What was the amount of the discount?
   b. Based on the original selling price, what was the rate of the discount on these three books?

5. Sheets are marked $22.50 and pillowcases $7.50. What is the sale price of each item if each item is discounted 25% off the marked price?

6. Towels were on sale at a discount of 30%. If the sale price was $3.01, what was the original price?

7. Headphones were on sale for $49.00. What was the original price if the sale price represents a discount of 20%?

8. ▦ An auto supply store received a shipment of auto parts and a bill for $845.30. Some of the parts were not as ordered, and they were returned immediately. The value of the parts returned was $175.50. The terms of the billing provided the store with a 2% discount if it paid with cash (for the parts it kept) within two weeks. What did the store pay for the parts it kept if it paid cash within two weeks?

9. In the roofing business, shingles are sold by the "square," which is enough material to cover a 10 ft by 10 ft square (or 100 square feet). A roofing supplier has a closeout on shingles at a 30% discount.
   a. If the original price was $230 per square, what is the sale price per square?
   b. How much would a roofer pay for 34 squares?

10. An auto dealer paid $8730 for a large order of special parts. This was not the original price. The amount paid reflects a 3% discount off the original price because the dealer paid cash. What was the original price of the parts?

11. If the sales tax rate is 6.5%, what is the tax on an $800 purchase?

Applications

1. a. $30 b. $270
2. a. $82.50 b. $192.50
3. a. $465 b. $15,035
4. a. $10.53 b. 60%
5. $16.88; $5.63
6. $4.30
7. $61.25
8. $656.40
9. a. $161 b. $5474
10. $9000
11. $52

Why Do Bostonians Shop in New Hampshire?

Sales taxes are not the same everywhere. Different states have different sales tax rates. Within a state, different localities may have different rates. In California, for example, the 2019 minimum sales tax rate is 7.25%, but within Los Angeles County, the rate is at least 9.5% and as high as 10.5% in the city of Santa Fe Springs. At the other extreme, five states in the United States don't have sales tax: Alaska, Delaware, Montana, New Hampshire, and Oregon. And that's why people from Boston go shopping in New Hampshire.

12. a. $1.81 b. $32.01
13. a. $6.93 b. $122.38
14. $8.12
15. 1.5%
16. a. $750 b. $600 c. $636
17. a. $250 b. 20% c. $216
18. a. The paperback b. $4.13
19. $11,700
20. $35,000
21. $875
22. $7500
23. $28,000

12. The sales tax in a certain state is figured at 6%.

a. How much tax is there on a purchase of $30.20?

b. What is the total amount paid for the purchase?

13. Suppose sales tax is figured at 6%.

a. How much tax would be paid on the purchase of three textbooks priced at $55.00, $25.50, and $43.95?

b. What would be the total cost of all three books?

14. If sales tax is figured at 7.25%, how much tax will be added to the total purchase price of three textbooks priced at $25.00, $35.00, and $52.00?

15. The property taxes on a house were $1050. What was the tax rate if the house was valued at $70,000?

16. The discount on a computer was $150. This was a 20% discount.

a. What was the original selling price of the computer?

b. What was the sale price?

c. What was the total amount paid for the computer if a 6% sales tax was added to the sale price?

17. The discount on men's suits was $50, and they were on sale for $200.

a. What was the original selling price?

b. What was the rate of discount?

c. What was the total amount paid for the suit if an 8% sales tax was added to the sale price?

18. Taylor is enrolled in a calculus course. She has the choice of buying the text in hardback form for $60.00 or in paperback form for $46.50. Tax is figured at 5% of the selling price. The bookstore buys back hardback books for 40% of the selling price and paperback books for 30% of the selling price.

a. Which book is the more economical buy for Taylor if she sells her book back to the bookstore at the end of the semester?

b. How much does she save?

19. A realtor works on 6% commission. What is his commission on a house he sold for $195,000?

20. A realtor selling commercial property works on a 4% commission. What is her commission on a building she sold for $875,000?

21. A car saleswoman earns a commission of 7% on each car she sells. How much did she earn on the sale of a car for $12,500?

22. A realtor works on 6% commission. What is his commission on a house he sold for $125,000?

23. If a salesman works on a 10% commission (no monthly salary), how much merchandise will he have to sell to earn $2800 in one month?

24. A realtor works on a 5% commission. How much would she need to sell a house for in order to earn $24,250 in commission?

25. A sales clerk receives a monthly salary of $500 plus a commission of 6% on all sales over $3500. What did the clerk earn the month she sold $8000 in merchandise?

26. A sales clerk receives a monthly salary of $1295 plus a commission of 7% on all sales over $2500. What did the clerk earn the month she sold $16,000 in merchandise?

27. A shoe saleswoman works on a fixed salary of $1440 per month plus an 8% commission. How much did she make during the month in which she sold $13,500 worth of shoes and accessories?

28. ▦ Suppose you sell your home for $180,000 and you owe the bank $60,000 on the mortgage. You pay a real estate agent a commission of 6% of the selling price, and other fees and taxes totaling $1200. How much money do you make from the sale?

29. ▦ Suppose you sell your three-bedroom home for $180,000 and you owe the balance of the mortgage of $55,000 to the bank. You pay a real estate agent a fee of 6% of the selling price and other fees and taxes that total $1500. How much money do you make from the sale?

30. ▦ Scott is buying a beach house off the coast of South Carolina for $260,000. The bank will loan him $225,000. If he must also pay a 4% commission to the real estate agent and $3800 in taxes and other fees, how much cash does he need?

31. ▦ A computer programmer was told that he would be given a bonus of 5% of any money his programs could save the company. How much would he have to save the company to earn a bonus of $500?

32. Central Valley Community College had 48 teams compete at their 3rd annual corn hole tournament. The following year they had 54 teams compete. What was the percent increase in competing teams?

33. Due to the increasing cost of breakfast cereals, more and more people are buying private-label brands rather than national brands. In a recent year, the sale of private-label cereals rose from 170 million boxes to 180 million boxes. What was the percent increase in sales (to the nearest tenth of a percent)?

34. ▦ The decade from 1960 to 1970 saw the largest 10-year increase in the number of male elementary and high school teachers in our nation's history. There was an increase from 402,000 to 691,000. What was the percent increase from 1960 to 1970 (to the nearest one percent)?

35. ▦ The average attendance to a Yankees game in 2009 was 45,364 fans. In 2010 the average attendance grew to 46,491 fans. Find the percent increase in attendance. Round your answer to the nearest thousandth.

36. ▦ In 1966, the student enrollment at California Polytechnic State University in San Luis Obispo, CA, was 7740. In 1977, the university had 15,502 students. Since that time the enrollment growth has slowed. What was the percent increase of student enrollment during that eleven year period? Round your answer to the nearest tenth of a percent.[1]

1 Source: www.lib.calpoly.edu/universityarchives/history/timeline

24. $485,000
25. $770
26. $2240
27. $2520
28. $108,000
29. $112,700
30. $49,200
31. $10,000
32. 12.5%
33. 5.9%
34. 72%
35. 2.484%
36. 100.3%

The 28/36 Rule

Nobody wants to lend money to someone who cannot pay it back. Banks use multiple measures to decide whether a borrower would qualify for a mortgage. The most important factor is how much you earn; the more money you make, the more banks are willing to lend to you. According to Investopedia, a general rule of thumb is that banks want their borrowers to spend no more than 28% of gross income on housing expenses (mortgage, insurance, condo fees) and no more than 36% on all debts (mortgage, car loans, student loans).

37. 214,926 people
38. 7.4 billion people
39. a. $200.00 b. 11.11%
40. 1.0%
41. 38%
42. 49.3%
43. a. $7 b. $33\frac{1}{3}\%$
c. 25%
44. a. $50 b. 20%
c. $16\frac{2}{3}\%$ or $16.\overline{6}\%$
45. a. $180 b. 40%
c. $28\frac{4}{7}\%$ or 28.6%

37. Frisco, Texas, had the largest population growth among large cities in 2022. In 2021, the population of Frisco was 208,262. The percent increase between 2021 and 2022 was 3.2%. What was the population of Frisco in 2022 (to the nearest whole number)?

38. The population of the world was 7.9 billion in 2022 with an expected percent increase of 1% per year. At this rate of increase, what was the expected population of the world for the year 2024? Round your answer to the nearest tenth of a billion. (**Hint:** first find the expected population for 2023. Then use this answer to find the expected population for 2024.) [2]

39. In April 2022, Best Buy sold a 12.9″ iPad Pro for the reduced price of $1599.99. The original price was $1799.99.

a. How much was the price reduced in terms of dollars?

b. Find the percent decrease or reduction. Round your answer to the nearest hundredth of a percent.

40. The 2010 population of Wheeling, WV, was 43,002 while the 2015 population dropped to 42,573. What was the percent decrease of population in that five year period? Round your answer to the nearest tenth of a percent. [3]

41. The circulation of the Washington Post newspaper was approximately 633,100 in 2009, and it dropped to 395,234 in 2015. What was the percent decrease in circulation? Round your answer to the nearest percent. [4]

42. The Dow Jones Industrial Index had a peak of 13,930 in October of 2007, but dropped to a minimum of 7,063 in February 2009. Fortunately this dip was short lived, and the market started increasing again. What was the percent decrease in the stock market drop according to the Dow Jones Industrial Index during this sixteen month interval? Round your answer to the nearest tenth of a percent. [5]

43. A company manufactures and sells plastic boxes that cost $21 each to produce, and that sell for $28 each.

a. How much profit does the company make on each box?

b. What is the percent of profit based on cost?

c. What is the percent of profit based on selling price?

44. Men's suits were on sale for $300. Each one cost the store owner $250.

a. What was the profit for the store?

b. What as the store's percent profit based on cost?

c. What was the store's percent profit based on selling price?

45. The cost of a 65″ smart TV to a store owner was $450, and she sold the TV for $630.

a. What was her profit?

b. What was her percent of profit based on cost?

c. What as her percent of profit based on selling price?

2 Source: Worldometer, www.worldometers.info/world-population/
3 Source: www.city-data.com/zips/26003.html
4 Source: www.capitolcommunicator.com/washington-post-circulation-drops-37-percent-since-2009-states-dcrtv/
5 Source: www.stockcharts.com/charts/historical/djia2000.html

46. A set of golf clubs cost a golf pro \$400, and she sold them in the pro shop for \$550.
   a. What was her profit?
   b. What was her percent of profit based on cost?
   c. What was her percent of profit based on selling price?

47. An art gallery sells paintings by a local artist for \$2500 each. The gallery owner has agreed to pay the artist \$2000 for each painting of a certain size.
   a. What is the profit on each painting?
   b. What is the percent of profit based on cost?
   c. What is the percent of profit based on selling price?

48. The cost of an 85″ smart TV to a store owner was \$3300 and he sold the TV for \$4500.
   a. What was his profit?
   b. What was his percent of profit based on cost?
   c. What was his percent of profit based on selling price?

49. The Golf Pro Shop had a set of 10 golf clubs that were marked on sale for \$860. This was a discount of 20% off the original selling price.
   a. What was the original selling price?
   b. If the clubs cost the Golf Pro Shop \$602, what was its profit?
   c. What was the shop's percent of profit based on the original selling price?
   d. What was the percent of profit based on the sale price?

50. A car dealer bought a 10-year-old used car for \$2500. He marked up the price so that he would make a profit of 25% based on his cost.
   a. What was the selling price?
   b. If the customer paid 8% of the selling price in taxes and fees, what was the customer's total cost for the car?

Another common use of percent is to calculate the tip on a bill. Tips are most commonly left in restaurants, but also apply to any service-based industry. Solve each problem, using the guidelines outlined in the problem to find the tip.

51. You invited your friend to lunch at the local coffeehouse, and the bill totaled \$12.60. Your friend offered to leave the tip. What amount did he leave if he tipped 15% on the total bill?

52. Mrs. Chung had two large pizzas delivered to her home for \$26.00. If she tipped the delivery person 18%, what total amount did she pay the driver?

53. On a recent business trip, Ken and Joe had breakfast at the restaurant next to the motel where they were staying. Ken's breakfast bill was \$6.95 and Joe's was \$8.75. How much should each man have left as a tip if each tipped 20%?

54. The bill for a family dining at a restaurant was \$38.40. What would a 15% tip be? What total amount should they leave on the table if they want to go before the waiter comes to pick up the money?

46. a. \$150 b. 37.5% c. $27\frac{3}{11}\%$ or $27.\overline{27}\%$
47. a. \$500 b. 25% c. 20%
48. a. \$1200 b. 36.4% c. $26\frac{2}{3}\%$ or $26.\overline{6}\%$
49. a. \$1075 b. \$258 c. 24% d. 30%
50. a. \$3125 b. \$3375
51. \$1.89
52. \$30.68
53. \$1.39; \$1.75
54. \$5.76; \$44.16

55. $8.64
56. $1.76; $13.51
57. $31.80
58. **a.** $13 **b.** $81.90
59. **a.** $3.60 **b.** $22.50
60. $28.18
61. $111.60

Writing & Thinking

62. The four steps to solving word problems are:
1. Understand the problem,
2. Devise a plan using various resources such as diagrams/pictures, estimates, and/or equations,
3. Carry out the plan, and
4. Look back over the results or check the answer to determine if it is reasonable. Answers will vary.

Possibilities include:
1. Understanding is central so that one begins looking for the correct item,
2. The plan is the most important as it sets the stage to find the unknown,
3. The calculations are most important because if not done properly, the answer will be wrong, or
4. Checking the results will find any errors and determine if the result is reasonable. Any variation of these are correct.

55. Juan took his date out for dinner before the senior prom. The total bill, including tax, was $48.00. How much did he tip the waitress if he left 18% (after tax)? What were his total expenses for the meal?

56. Walt decided to treat himself to lunch, and the bill was $11.75, including tax. What amount did he leave as a 15% tip (after tax)? What was the total amount he paid for lunch?

57. A lawyer took four clients to dinner. The total bill for dinner and drinks was $150.00 plus a 6% sales tax. What amount did she leave as a tip if she calculated the tip after tax at 20%?

58. A math teacher took two of her colleagues to dinner, and the bill was $65.00 plus a 6% sales tax.

a. What amount did she leave as a 20% tip (before tax)?

b. What was the total amount, including tip, she paid for the dinner?

59. Ann paid for lunch at the local pizza parlor for three of her study partners because they were celebrating having passed a statistics exam. The bill was $18.00 plus tax at 5%.

a. What amount did she leave as a 20% tip (before tax)?

b. What was the total amount of her bill, including tip and taxes?

60. On Monday night, Alondra and Parker decided to stay home, watch a football game, and have Chinese food delivered. If the bill was $23.00 plus tax at 7.5% and they tipped the driver 15% of the bill (before tax), what total amount did they pay?

61. The coach invited the basketball team to his home after the last game of the season and had six large pizzas delivered. If the bill was $90.00 plus tax at 6% and he gave the driver an 18% tip (before tax), what total amount did he pay?

Writing & Thinking

62. List the four basic steps for solving word problems. Of the four basic steps for solving word problems, which step do you think is the most important and why?

63. Determine how to calculate sales tax when eating out and relate this process to either a proportion and/or using the amount/base/rate equation. Give an example.

64. A man weighed 200 pounds. He lost 20 pounds in 3 months. Then he gained back 20 pounds 2 months later.

a. What percent of his weight did he lose in the first 3 months?

b. What percent of his weight did he gain back?

c. The loss and gain are the same, but the two percentages are different. Explain why.

65. Explain the process to determine how to find percent of profit based on

a. cost.

b. selling price.

Collaborative Learning

With the class separated into teams of two to four students, each team is to analyze the following problem and decide how to answer the related questions. Then each team leader is to present the team's answers and related ideas to the class for general discussion.

66. Jerry works in a bookstore and gets a salary of $500 per month plus a commission of 3% on whatever he sells. Wilma works in the same store, but she decided to work on a straight 8% commission.

a. At what amount of sales will Jerry and Wilma make the same amount of money?

b. Up to that point, who will be making more?

c. After that point, who would be making more? Explain briefly. (If you were offered a job at this bookstore, which method of payment would you choose?)

63. Sales tax and tips are percentages of some item or service. The percent is the rate, while the cost of the item being purchased is the base. The amount is then the sales tax itself, which is being compared to the base. A sales tax might be 8%, as in "what is 8% of the cost of the item purchased?"

64. **a.** He lost 10% of his original weight.

b. He gains back $11\frac{1}{9}\%$ of his weight

c. The percentages are different because the gain of 20 pounds is based on his new weight of 180 pounds, not his original weight of 200 pounds.

65. For both types of percent of profit, the profit (the difference between selling price and cost) must be determined first. The profit is then used as a numerator. In profit based on cost, the denominator is the cost. In profit based on selling price, the selling price is the denominator. The fraction is then divided and the result is a percent of profit, either based on cost or selling price.

Collaborative Learning

66. **a.** $10,000 **b.** Jerry **c.** Wilma. Answers will vary.

4.8 Simple and Compound Interest

Objectives

A. Calculate simple interest.

B. Calculate compound interest by repeated addition.

C. Calculate compound interest using the compound interest formula.

D. Calculate the new value of an object after inflation or depreciation.

A Simple Interest

Interest is money paid for the use of money. The amount of money that is borrowed or invested is called the **principal**. The charge for borrowing or lending, usually a percent of the principal, is called the **interest rate** or **rate of interest**. Unless otherwise stated, the rate of interest is assumed to be a yearly (or annual) rate. The time of the loan is expressed in years or a fraction of a year. Eventually the borrower must pay back the original loan amount (the principal) plus the interest.

The following formula is used to calculate **simple interest**. (As a reminder, a **formula** is a general statement, usually an equation, that relates two or more variables.)

Simple Interest Formula

Interest = Principal · rate · time

Writing the formula using letters, we have $I = P \cdot r \cdot t$, where

I = interest (earned or paid),
P = principal (the amount invested or borrowed),
r = rate of interest (stated as an annual rate) in decimal or fraction form, and
t = time (years or fraction of a year).

FORMULA

Although the rate of interest is generally given in the form of a percent, the calculations in the formula are made by changing the percent into decimal form or fraction form. (**Note:** When dealing with money, we round to the hundredths place.)

1. If you were to borrow \$1500 at 8.5% for one year, how much interest would you pay?

Example 1 Application: Calculating Simple Interest

You want to borrow \$2000 from your bank for one year. If the interest rate is 5.5%, how much interest would you pay?

Solution

The interest is found by using the formula $I = P \cdot r \cdot t$, with

$$P = \$2000, \quad r = 5.5\% = 0.055, \quad \text{and} \quad t = 1 \text{ year}.$$

$$I = 2000 \cdot 0.055 \cdot 1 = 110$$

You would pay \$110 in interest.

Now work margin exercise 1.

Completion Example 2 Application: Calculating Simple Interest

Sylvia borrowed \$2400 at 5% interest for 3 months. How much interest did she have to pay?

Solution

The interest is found by using the formula $I = P \cdot r \cdot t$, with

$$P = \$2400, \quad r = 5\% = \underline{\ 0.05\ }, \quad \text{and} \quad t = 3 \text{ months} = \frac{3}{12} \text{ year} = \underline{\ \frac{1}{4}\ } \text{ year.}$$

$$I = \underline{\ 2400\ } \cdot \underline{\ 0.05\ } \cdot \underline{\ \frac{1}{4}\ } = \underline{\ 30\ }$$

Sylvia had to pay _\$30_ in interest.

Now work margin exercise 2.

2. Ralph borrowed \$3000 at 7% for two months. How much interest did he pay?

Example 3 Application: Calculating Total Amount Due

Carmen loaned \$500 to a friend for 6 months at an interest rate of 8%. How much will her friend pay her at the end of the 6 months?

Solution

The total amount paid to Carmen at the end of the six months will be the original amount loaned plus the interest.

The interest is found by using the formula $I = P \cdot r \cdot t$, with

$$P = \$500, \quad r = 8\% = 0.08, \quad \text{and} \quad t = 6 \text{ months} = \frac{6}{12} \text{ year} = \frac{1}{2} \text{ year.}$$

$$I = 500 \cdot \overset{0.04}{\cancel{0.08}} \cdot \frac{1}{\cancel{2}} = 500 \cdot 0.04 = 20$$

The interest is \$20 and the total amount to be paid at the end of 6 months is

$$\text{principal} + \text{interest} = \$500 + \$20 = \$520.$$

Now work margin exercise 3.

3. Stacey loaned her aunt \$2000 at 10% interest. How much will Stacey's aunt pay her, if she repays the entire loan at the end of 9 months?

If you know the values of any three of the variables in the formula $I = P \cdot r \cdot t$, you can find the unknown value by substituting into the formula and solving for the unknown, as illustrated in Examples 4 and 5.

Example 4 Application: Calculating Principal using Simple Interest

What principal would you need to invest at a rate of 6% to earn \$450 in 6 months?

4. What principal would you need to invest if your investment will return 8% interest and you want to make \$500 in interest in 3 months?

Solution

Here the principal P is unknown. We do know

$$I = \$450, \quad r = 6\% = 0.06, \quad \text{and} \quad t = \frac{6}{12} = \frac{1}{2} \text{ year.}$$

Substituting and solving for P, we have

$$450 = P \cdot 0.06 \cdot \frac{1}{2}$$
$$450 = P \cdot 0.03$$
$$\frac{450}{0.03} = \frac{P \cdot \cancel{0.03}}{\cancel{0.03}} \qquad \text{Divide both sides by 0.03.}$$
$$15{,}000 = P$$

You would need to invest a principal amount of \$15,000 to earn \$450 in 6 months at a rate of 6%.

Now work margin exercise 4.

5. How long will it take for \$1000 at 6% to earn \$50 in simple interest?

Example 5 Application: Calculating Time using Simple Interest

Stuart is investing \$1500 at an interest rate of 4%. How long will it take for his investment to earn \$15 in simple interest?

Solution

In this case, the unknown is time t. We do know

$$I = \$15, \quad R = 4\% = 0.04, \quad \text{and} \quad P = \$1500.$$

Substituting and solving for t gives

$$15 = 1500 \cdot 0.04 \cdot t$$
$$15 = 60 \cdot t$$
$$\frac{15}{60} = \frac{\cancel{60} \cdot t}{\cancel{60}} \qquad \text{Divide both sides by 60.}$$
$$\frac{1}{4} = t$$

Stuart will earn \$15 in interest after $\frac{1}{4}$ year (or 3 months).

Now work margin exercise 5.

B Compound Interest

Interest paid on interest earned is called **compound interest.** To calculate compound interest, we can calculate the simple interest for each period of time that interest is compounded. **A new principal is used for each calculation.** The new principal is the previous principal plus the earned interest. The calculations can be performed in a step-by-step manner, as indicated in the following procedure.

Calculating Compound Interest

1. Use the formula $I = P \cdot r \cdot t$, to calculate simple interest.

   Let $t = \frac{1}{n}$ where n is the number of periods per year for compounding.

   For example:

   for compounding annually, $n = 1$ and $t = \frac{1}{1} = 1$.

   for compounding semiannually, $n = 2$ and $t = \frac{1}{2}$

   for compounding quarterly, $n = 4$ and $t = \frac{1}{4}$

   for compounding bimonthly, $n = 6$ and $t = \frac{1}{6}$

   for compounding monthly, $n = 12$ and $t = \frac{1}{12}$

   for compounding daily, $n = 365$ and $t = \frac{1}{365}$

2. Add this interest to the principal to create a new value for the principal.
3. Repeat steps 1 and 2 however many times the interest is to be compounded.

PROCEDURE

Math Tip

If you had \$1,000,000 and you invested it at 5% interest compounded *yearly*, you would have \$2,653,297.71 in 20 years. If you had invested that same \$1,000,000 at 5% interest compounded *daily*, you would have \$2,718,095.67 after 20 years. The difference in compounding changed the interest by \$64,797.96. That is the power of compounding.

As Benjamin Franklin said, "Money makes money. And the money that makes money makes more money."

In Examples 6 and 7, compound interest is calculated in the step-by-step manner just outlined. This process is, to say the least, somewhat laborious and time-consuming. However, it does serve to develop a basic understanding of the concept of compound interest. After these examples, we will introduce a formula that can be used to calculate compound interest.

Example 6 Application: Calculating Compound Interest

If a savings account of \$1200 is compounded annually (once a year) at 5%, how much interest will be earned in three years?

Solution

The account is compounded annually, so $n = 1$ and $t = \frac{1}{1} = 1$.

Use the formula for simple interest, $I = P \cdot r \cdot t$, with $r = 5\% = 0.05$ and $t = 1$.

The principal will change each year.

1. **First year:** the principal is $P = \$1200$.

   $I = \$1200 \cdot 0.05 \cdot 1 = \$60.00$ Interest for the first year

2. **Second year:** the new principal is $P = \$1200 + \$60 = \$1260$.

   $I = \$1260 \cdot 0.05 \cdot 1 = \$63.00$ Interest for the second year

3. **Third year:** the new principal is $P = \$1260 + \$63 = \$1323$.

   $I = \$1323 \cdot 0.05 \cdot 1 = \$66.15$ Interest for the third year

6. George deposits \$500 in a savings account that pays 6% interest compounded annually. How much interest will he earn in three years?

The total interest earned in three years will be

$$\begin{array}{r} \$60.00 \\ \$63.00 \\ + \quad \$66.15 \\ \hline \$189.15 \end{array}$$

Note: Because the principal is larger each year, the interest earned increases each year.

Now work margin exercise 6.

7. If a principal of $3000 is compounded quarterly (4 times a year) at 7%, what will be the balance in the account at the end of one year?

Make Compound Interest Work for You

With compounding, interest on a loan can grow exponentially over time. A $10,000 loan at a 5% interest rate would incur $2840 in interest over 5 years. Most loans require monthly payments, which help slow down the interest accumulated. With a required minimum monthly payment of $188.71, the same loan would incur $1322.74 in interest. You can make compound interest work for you by paying a little bit more each month. For the same example, by paying an extra $20 a month, the loan would be paid off 7 months early.

Example 7 Application: Calculating Compound Interest

If an account of $5000 is compounded monthly (12 times a year) at 6%, what will be the balance in the account at the end of four months?

Solution

The account is compounded monthly, so $n = 12$ and $t = \frac{1}{12}$.

Use the formula for simple interest, $I = P \cdot r \cdot t$, with $r = 6\% = 0.06$ and $t = \frac{1}{12}$.

The principal will change each month.

1. **First month:** the principal is $P = \$5000$.

   $I = \$5000 \cdot 0.06 \cdot \frac{1}{12} = \$25.00$ Interest for the first month

2. **Second month:** the new principal is $P = \$5000 + \$25 = \$5025$.

   $I = \$5025 \cdot 0.06 \cdot \frac{1}{12} = \$25.125 \approx \$25.13$ Interest for the second month

3. **Third month:** the new principal is $P = \$5025 + \$25.13 = \$5050.13$.

   $I = \$5050.13 \cdot 0.06 \cdot \frac{1}{12} \approx \$25.25$ Interest for the third month

4. **Fourth month:** the new principal is $P = \$5050.13 + \$25.25 = \$5075.38$.

   $I = \$5075.38 \cdot 0.06 \cdot \frac{1}{12} \approx \$25.38$ Interest for the fourth month

The total interest earned in four months will be

$$\begin{array}{r} \$25.00 \\ \$25.13 \\ \$25.25 \\ + \quad \$25.38 \\ \hline \$100.76 \end{array}$$

The balance in the account will be $\$5000.00 + \$100.76 = \$5100.76$.

Now work margin exercise 7.

The steps outlined in Examples 6 and 7 illustrate how to adjust the principal for each new time period of the compounding process. The interest is greater each time period because the new adjusted principal (the old principal plus the interest over the previous time period) is greater. In this process, we have calculated the actual interest for each time period. This process is simplified by using the compound interest formula.

C The Compound Interest Formula

The following compound interest formula does not calculate the actual interest for each time period. In fact, this formula does not calculate the interest directly. This formula calculates the total accumulated **amount** (also called the **future value** of the principal).

Compound Interest Formula

When interest is compounded, the total **amount** A accumulated (including principal and interest) is given by the formula

$$A = P\left(1+\frac{r}{n}\right)^{nt},$$

where

P = principal (the amount invested or borrowed),
r = annual interest rate in decimal or fraction form,
t = time (in years), and
n = number of compounding periods in 1 year.

FORMULA

Example 8 Application: Using the Compound Interest Formula

Max invested \$4500 at 9% to be compounded monthly. What will be the amount in his account in 5 years?

Solution

$P = \$4500,\quad r = 9\% = 0.09,\quad n = 12$ times per year, $\quad t = 5$ years

Substituting into the formula gives the following.

$$\begin{aligned} A &= \$4500\left(1+\frac{0.09}{12}\right)^{12(5)} \\ &= \$4500(1+0.0075)^{60} \\ &= \$4500(1.0075)^{60} \\ &\approx \$7045.56 \quad \text{Rounded to the nearest cent} \end{aligned}$$

The amount in Max's account in 5 years will be \$7045.56

Now work margin exercise 8.

8. Dana invested \$3000 in a special account at her bank at 4.5% interest to be compounded monthly. What will be the amount in her account after 4 years?

Teaching Note

You might want to discuss the reasonableness of the formula with the following algebraic development using $n = 1$ for three years.

First year: $A = P(1+r)$

For the second year, we replace P in the original formula with the new principal $P(1+r)$ which leads to the expression

$$A = [P(1+r)](1+r) = P(1+r)^2.$$

For the third year, we replace P in the original formula with the new principal $P(1+r)^2$ and arrive at

$$A = [P(1+r)^2](1+r) = P(1+r)^3.$$

CALCULATORS

Using a Calculator to Calculate Compound Interest

Most scientific calculators automatically follow the rules for order of operations. (See Section 1.7 to learn how to test whether your calculator does or not.) Therefore, you can enter all of the numbers and parentheses just as you see them placed in the formula. The one exception is that you will need to put the exponent nt in parentheses, or simply enter its product.

Looking at Example 8, you can enter the exponent as 60 or enter it in parentheses as (12 × 5). To enter the full calculation in the calculator, press the keys

[4] [5] [0] [0] [×] [(] [1] [+] [0] [.] [0] [9] [÷] [1] [2] [)] [^] [(] [1] [2] [×] [5] [)].

Then press [=]. The display will read 7045.564621.

Thus, we see that whether we do the calculations step-by-step or let the calculator do all the steps internally, Max's account will hold $7045.56 in 5 years. (**Note:** As you work the problems in this section, there may be problems where your answers vary slightly from the answers provided. Discrepancies like this can occur due to rounding in intemediate steps, but these answers are still considered correct.)

Total Interest Earned

To find the total interest earned on an investment that has earned interest by compounding, subtract the initial principal from the accumulated amount.

$$I = A - P$$

FORMULA

9. How much interest did Dana earn in the investment described in margin exercise 8?

Example 9 Application: Calculating Total Interest Earned

How much interest did Max earn on the investment described in Example 8?

Solution

$$\begin{aligned} I &= A - P \\ &= 7045.56 - 4500.00 \\ &= 2545.56 \end{aligned}$$

Max earned $2545.56 in interest.

Now work margin exercise 9.

10. a. Use the compound interest formula to find the value of $10,000 invested for 5 years if it is compounded every 6 months at 9% interest.

b. Find the amount of interest earned.

Completion Example 10 Application: Compound Interest

a. Use the compound interest formula to find the value of $6000 invested for 4 years if it is compounded quarterly at 12%.

b. Find the amount of interest earned.

Solution

a. With a scientific calculator, follow the steps outlined in Example 8 with

$$P = \$6000, \quad r = 12\% = 0.12, \quad n = 4, \quad \text{and} \quad t = 4.$$

Substituting into the formula gives the following.

$$A = \$6000\left(1+\frac{\underline{0.12}}{\underline{4}}\right)^{(4\cdot\underline{4})}$$

$$= \$6000\left(1+\underline{0.03}\right)^{\underline{16}}$$

$$= \$6000\left(\underline{1.03}\right)^{\underline{16}}$$

$$\approx \$\underline{9628.24} \quad \text{Rounded to the nearest cent}$$

The value (or amount) will be \$ 9628.24 .

b. The interest earned will be

$I = \$\underline{9628.24} - \$6000.00 = \$\underline{3628.24}$.

Now work margin exercise 10.

D Inflation and Depreciation

Inflation measures the general increase in prices of goods and services from one year to the next. For example, if inflation is at 6% annually, then your income will need to increase by 6% each year to be able to afford to buy the same items (or to live the same lifestyle) that you did the year before. The inflation rate is most often calculated using the US Consumer Price Index (CPI) a measure, put out by the Bureau of Labor Statistics, of the average change over time in the prices of goods and services. Many worker bargaining committees tie their salary requests to inflation and the government's Consumer Price Index each year.

The effect of inflation on the cost of a particular item can be calculated in the same manner we calculated interest compounded annually. That is, we use the compound interest formula with $n = 1$. For example, consider a \$3.00 loaf of bread. What will be the cost of this loaf in 20 years, if annual inflation remains constant at 5%? We have,

$$A = \$3.00\left(1+\frac{0.05}{1}\right)^{1\cdot 20} = \$3.00(1+0.05)^{20} \approx \$7.96 \quad \text{Rounded to the nearest cent}$$

Thus, the projected cost of bread in 20 years is more than double the original cost. The cost has gone from \$3.00 a loaf to \$7.96 a loaf. Note that this is not a super-improved loaf of bread; it is the same bread you are eating today.

Relatedly, if the annual percent increases in your salary are less than the inflation rate of 5%, each dollar you make purchases less. For example, assume you receive a 3% cost of living raise each year. This means that in 20 years, based on the rate of increase of your salary, the comparable price for a loaf of bread is

$$A = \$3.00(1+0.03)^{20} \approx \$5.42 \quad \text{Rounded to the nearest cent}$$

Assuming you could only afford to pay \$3.00 for a loaf of bread in the present year, 20 years later you can only afford \$5.42 per loaf, while the actual price is \$7.96. In other words, your spending power is reduced.

Social Security and Inflation

Social Security benefits keep pace with inflation by adding a cost-of-living adjustment, or COLA, effective each December. Social Security's COLA is based on a specific portion of CPI called the Consumer Price Index for Urban Wage Earners and Clerical Workers (CPI-W). In 2019, CPI-W was about 2.5 times the cost of living of the 1982 base year.

Inflation

The adjusted amount A due to **inflation** is

$$A = P(1+r)^t,$$

where

P = principal (the original value),
r = annual rate of inflation in decimal notation, and
t = time (in years).

FORMULA

11. Suppose that your income is $21,600 per year and that each year you can expect to receive a cost-of-living raise that matches the inflation rate. If inflation is steady at 4% per year, find your yearly income in 15 years.

Example 11 Application: Calculating Inflation

Suppose that your income is $28,800 per year and that each year you can expect to receive a cost-of-living raise that matches the inflation rate. If inflation is steady at 3% per year, find your yearly income in:

a. 5 years

b. 10 years

c. 20 years

Solution

To calculate your yearly income in 5, 10, and 20 years, evaluate $A = \$28{,}800(1+0.03)^t$ at $t = 5$, $t = 10$, and $t = 20$.

a. for $t = 5$: $A = \$28{,}800(1+0.03)^5 \approx \$33{,}387.09$ Rounded to the nearest cent

b. for $t = 10$: $A = \$28{,}800(1+0.03)^{10} \approx \$38{,}704.79$ Rounded to the nearest cent

c. for $t = 20$: $A = \$28{,}800(1+0.03)^{20} \approx \$52{,}016.00$ Rounded to the nearest cent

Thus, if you keep the same job for 20 years with your salary matching inflation, you will be making $33,387.09 in 5 years, $38,704.79 in 10 years, and $52,016.00 in 20 years.

Now work margin exercise 11.

Math Tip

We all know that new vehicles depreciate in value, but what factors influence this depreciation? A strong fan base or a good reputation can help vehicles retain their value, while the high cost of repairs or outdated technology may cause them to lose value quicker. The average 5-year depreciation values for vehicles in 2021 ranged from 9.2% (for a Jeep Wrangler) to 65.1% (for a BMW X5). So if retained value is important to you, be sure to do your homework before you buy a new car.

Depreciation is a decrease in the value of an item over time. For income tax purposes, it is also a method for distributing the cost of an item over its lifetime. If we know the original value and the rate of depreciation of an item, we can calculate its current market value. For example, if a car was purchased for $15,000 and its value depreciates 20% each year, then its value each year is 80% of its value the previous year.

Value after 1 year: $\$15{,}000(0.80) = \$12{,}000$

Value after 2 years: $\underbrace{[\$15{,}000(0.80)]}_{\text{First year}}(0.80) = \$12{,}000(0.80) = \$9600$

or

$\$15{,}000(0.80)^2 = \$15{,}000(0.64) = \$9600.$

Depreciation

The current value V of an item due to **depreciation** is

$$V = P(1-r)^t,$$

where

P = principal (the original value),
r = annual rate of depreciation in decimal notation, and
t = time (in years).

FORMULA

Example 12 Application: Calculating the Current Value Due to Depreciation

Suppose that a certain make of automobile depreciates 15% each year. Find the current market value of one of these automobiles if it is 5 years old and its original value was $40,000.

Solution

$P = \$40{,}000, \quad r = 15\% = 0.15, \quad \text{and} \quad t = 5 \text{ years}$

Using the formula for the current value due to depreciation, we have the following.

$$V = P(1-r)^t = \$40{,}000(1-0.15)^5 \approx \$17{,}748.21 \quad \text{Rounded to the nearest cent}$$

The current market value of the automobile is $17,748.21.

Now work margin exercise 12.

12. Suppose that a certain boat depreciates at a rate of 18% each year. Find the current market value of the boat if it is 7 years old and its original value was $13,000.

Completion Example Answers

2. 0.05; $\frac{1}{4}$ year; $I = 2400 \cdot 0.05 \cdot \frac{1}{4} = 30$; $30

10. a. $\left(1+\frac{0.12}{4}\right)^{(4\cdot 4)}$; $(1+0.03)^{16}$; $(1.03)^{16}$; 9628.24; $9628.24 **b.** $9628.24, $3628.24

Margin Exercise Answers

1. $127.50 **2.** $35 **3.** $2150 **4.** $25,000 **5.** $\frac{5}{6}$ year (or 10 months) **6.** $95.51 **7.** $3215.57 **8.** $3590.44 **9.** $590.44 **10. a.** $15,529.69 **b.** $5529.69 **11.** $38,900.38 **12.** $3240.71

4.8 Exercises

Concept Check

1. interest
2. principal
3. simple
4. years
5. quarterly
6. compound
7. True
8. False; Simple interest is not compounded.
9. False; Compound interest is earned on the principal and interest earned.
10. False; Inflation can be treated in the same manner as interest compounded annually.

Applications

1. $30
2. $160
3. $32
4. $108
5. $200
6. $112.50
7. $100
8. $32.67
9. $37.50
10. $833.33

4.8 Exercises

Concept Check

Fill-in-the-Blank. Complete each sentence using information found in this section.

1. Money paid for the use of money is ________.
2. The amount of money being invested or borrowed is known as the ________.
3. Money borrowed and paid back in one payment is calculated with ________ interest.
4. In the formula $I = P \cdot r \cdot t$, the units for time must be in ________
5. If interest is compounded four times per year, it is said to be compounded ________.
6. Interest paid on interest earned is known as ________ interest.

True/False. Determine whether each statement is true or false. If a statement is false, explain how it can be changed so the statement will be true. (**Note:** There may be more than one acceptable change.)

7. In the simple interest formula, the rate can be written as a decimal number or a fraction.
8. Simple interest can be compounded monthly or quarterly.
9. Interest cannot be earned on interest, only the principal.
10. Inflation can be treated in the same manner as simple interest.

Applications

Solve each problem using the formula for simple interest. Round your answer to the nearest cent, if necessary. Assume 1 year equals 360 days. See Examples 1 through 5.

1. What is the simple interest paid on $500 at 6% for one year?
2. What is the simple interest paid on $2000 at 8% for one year?
3. What will be the interest earned in one year on a savings account of $800 if the bank pays 4% interest?
4. If interest is paid at 6% for one year, what will a principal of $1800 earn?
5. How much interest would be paid on a loan of $5000 at 8% for 6 months?
6. How much interest would be paid on a loan of $3000 at 5% for 9 months?
7. Stacey loaned her brother $1500 for 8 months at 10% interest. How much interest did she earn?
8. Find the simple interest paid on a savings account of $2800 for 120 days at 3.5%.
9. If you were to borrow $1000 at 5% for nine months, how much interest would you pay?
10. What principal will earn $50 in interest if it is invested at 6% for one year?

11. What principal will earn $50 in interest if it is invested at 8% for 90 days?

12. What principal will earn $75 in interest if it is invested for 60 days at 9%?

13. What principal will earn $75 in interest if it is invested at 5% for 6 months?

14. What principal would have to be invested at 8% for 60 days to earn interest of $500?

15. How long will it take for $1000 invested at 5% to earn $50 in simple interest?

16. What length of time will it take to earn $70 in simple interest if $2000 is invested at 7%?

17. How many days must you leave $1000 in a savings account at 5.5% to have a balance of $1011?

18. If interest is paid at 6% for one year, what will a principal of $1800 earn?

19. If a principal of $900 is invested at a rate of 4% for 90 days, what will be the interest earned?

20. If you borrow $750 for 30 days at 9%, how much interest will you pay?

21. How much interest would be paid on a 60-day loan of $500 at 4%?

22. What interest rate would you be paying if you borrowed $1000 for 6 months and paid $60 in interest?

23. What rate of interest is charged if a loan of $2500 for 90 days is paid off with $2562.50?

24. A friend wants to borrow $500 from you for 8 months and is willing to pay you interest at 6%. How much would he owe you at the end of the 8 months?

25. If you charge $1000 worth of merchandise at a local department store at 18% interest, how much will you owe at the end of 60 days?

26. A bank decides to loan $5 million to a contractor to build new homes. How much interest will the bank earn in one year if the interest rate is 9.2%?

27. Every 6 months a stock pays 10% in dividends (interest on investment). What will be the earnings of $14,600 invested for 6 months? (Remember, the rates of interest are given as annual rates.)

28. A credit card company has $120 million loaned to its customers at 18.9%. How much interest will it earn in one month?

29. A department store keeps $15 million in merchandise in stock. If the store pays interest at 9% on a bank loan for this stock, how much interest will the store pay in 3 months' time?

30. You buy an oven on sale from $500 to $450, but you don't make a payment for 60 days and are charged interest at a rate of 18%.

    a. How much do you pay for the oven by waiting 60 days to pay?

    b. How much do you save by buying the oven on sale? (Sales tax is not included here.)

11. $2500
12. $5000
13. $3000
14. $37,500
15. 1 year
16. 6 months or $\frac{1}{2}$ year
17. 72 days
18. $108
19. $9
20. $5.63
21. $3.33
22. 12%
23. 10%
24. $520
25. $1030
26. $460,000
27. $730
28. $1,890,000
29. $337,500
30. a. $463.50 **b.** $36.50

31. a. $1000
b. 9 months or $\frac{3}{4}$ year

32. a. $66.25 b. 5366.25

33. $1030

34. 10%

35. 240 days or $\frac{2}{3}$ year

36. 1 year

37. a. $16 b. $100
c. 30 days or $\frac{1}{12}$ year
d. 8.5%

38. a. $7.50
b. 60 days or $\frac{1}{6}$ year
c. 20% d. $180

39. a. $100 b. $2100, $105
c. $2205, $110.25
d. $315.25

31. Carlos has a savings account with a balance of $25,000.

a. If the savings account is drawing interest at 8%, how much interest will he earn in 6 months?

b. How long must he leave the money in the account to earn $1500?

32. A savings account of $5300 is left for 90 days drawing interest at a rate of 5%.

a. How much interest is earned?

b. What is the amount in the account at the end of 90 days?

33. If you charge $1000 worth of merchandise at a local department store at 18% interest, how much will you owe at the end of 60 days?

34. Ms. Lee accumulated $240,000 and she wants to live on the interest each year. If she needs $2000 a month to live on, what interest rate must she earn on her money?

35. Mr. Smith has a savings account of $2500 that draws 4.5% interest. How many days will it take for him to earn $75?

36. A small airline company borrowed $7.5 million to buy some new airplanes. The loan rate was 7.5%, and the airline paid $562,500 in interest. What was the length of time of the loan?

37. Determine the missing item in each row.

Principal	Rate	Time	Interest
$400	16%	90 days	**a.**
b.	15%	120 days	$5.00
$560	12%	**c.**	$5.60
$2700	**d.**	40 days	$25.50

38. Determine the missing item in each row.

Principal	Rate	Time	Interest
$600	15%	30 days	**a.**
$500	18%	**b.**	$15.00
$450	**c.**	90 days	$22.50
d.	10%	30 days	$1.50

Solve each problem by repeatedly using the formula for calculating simple interest. Round your answer to the nearest cent, if necessary. See Examples 6 and 7.

39. You loan your cousin $2000 at 5% compounded annually for 3 years. How much interest will your cousin owe you?

a. First year: $I = 2000 \cdot 0.05 \cdot 1 =$ ________

b. Second year: $I =$ ________ $\cdot\, 0.05 \cdot 1 =$ ________

c. Third year: $I =$ ________ $\cdot\, 0.05 \cdot 1 =$ ________

d. The total interest is ________.

40. John borrowed \$5000 from his uncle at 6% compounded annually for 4 years. How much interest will he owe his uncle at the end of 4 years?

a. First year: $I = 5000 \cdot 0.06 \cdot 1 =$ ______

b. Second year: $I =$ ______ $\cdot\, 0.06 \cdot 1 =$ ______

c. Third year: $I =$ ______ $\cdot\, 0.06 \cdot 1 =$ ______

d. Fourth year: $I =$ ______ $\cdot\, 0.06 \cdot 1 =$ ______

e. The total interest is ______.

41. If \$9000 is deposited in a savings account at 4% compounded monthly, what will be the balance in the account in 4 months?

a. First month: $I = 9000 \cdot 0.04 \cdot \frac{1}{12} =$ ______

b. Second month: $I =$ ______ $\cdot\, 0.04 \cdot \frac{1}{12} =$ ______

c. Third month: $I =$ ______ $\cdot\, 0.04 \cdot \frac{1}{12} =$ ______

d. Fourth month: $I =$ ______ $\cdot\, 0.04 \cdot \frac{1}{12} =$ ______

e. The total interest earned is ______.

f. The balance in the account is ______.

42. Jeremy put \$3500 in a savings account at 5.5% compounded quarterly for 6 months. What will be the balance on the account at the end of 6 months?

a. First quarter: $I = 3500 \cdot 0.055 \cdot \frac{1}{4} =$ ______

b. Second quarter: $I =$ ______ $\cdot\, 0.055 \cdot \frac{1}{4} =$ ______

c. The total interest earned is ______.

d. The balance in the account is ______.

43. Your cousin loans you \$3000 compounded annually at 4% for 4 years. How much interest will your cousin owe you?

44. Keri borrowed \$4000 from her aunt compounded annually at 5% for 6 years. How much interest will she owe her aunt at the end of 6 years?

45. If \$9000 is deposited in a savings account compounded monthly at 4%, what will be the balance in the account in 3 months?

46. If interest is calculated at 10% compounded quarterly, what will be the value of \$15,000 in 9 months?

Solve each problem by using the compound interest formula. Round your answer to the nearest cent, if necessary. See Examples 8 through 10.

47. You deposit \$1500 at 4% to be compounded semiannually. How much interest will you earn in 3 years?

48. A principal of \$2500 is deposited at 6% to be compounded monthly. How much will the account be worth in 6 months?

40. a. \$300
b. \$5300, \$318
c. \$5618, \$337.08
d. \$5955.08, \$357.30
e. \$1312.38

41. a. \$30
b. \$9030, \$30.10
c. \$9060.10, \$30.20
d. \$9090.30, \$30.30
e. \$120.60
f. \$9120.60

42. a. \$48.13
b. \$3548.13, \$48.79
c. \$96.92 d. \$3596.92

43. \$509.58

44. \$360.38

45. \$9090.30

46. \$16,153.36

47. \$189.24

48. \$2575.94

49. a. $1051.56 b. $51.56

50. a. $6863.93
b. About $20 c. $21.47

51. a. $28,051.03
b. $329.40

52. a. $3153.49
b. No; $2234.08

53. a. $634.13 b. No
c. Monthly compounding allows interest earned the previous month to gain interest the following month, but semiannual compounding waits six months for this process to begin.

54. a. $26,620
b. $26,801.92; No
c. Explanations will vary. By compounding semiannually, interest is posted twice as frequently, so more interest is earned on interest accrued.

55. a. $1493.42 b. $93.42

56. a. $24,492.29 b. $95.63

49. Suppose you can invest $10,000 at an interest rate of 10%.

a. What will be the interest on $10,000 compounded daily at 10% for one year?

b. What is the difference between this and simple interest at 10% for one year?

50. Suppose you can invest $5000 at an interest rate of 8%.

a. Find the value of $5000 compounded quarterly at 8% for 4 years.

b. What do you think the difference in interest would be if the money were compounded daily: about $5, $20, or over $100?

c. Find the exact difference in interest.

51. Suppose $20,000 is invested in a savings account.

a. What would the value of the savings account be at the end of 5 years if interest were calculated at 7% compounded annually?

b. How much more would be earned if the interest were compounded daily?

52. Suppose $3000 is invested in a savings account.

a. Suppose the money is invested at 5% and compounded monthly for one year. Find the accumulated value.

b. Is the accumulated amount the same if the original principal of $3000 is compounded annually for 12 years? If not, what is the difference?

53. Suppose $5000 is invested at an interest rate of 12%.

a. Calculate the interest in one year on $5000 compounded monthly at 12%.

b. Suppose the interest is compounded semiannually. Is the accumulated value the same?

c. If not, explain why not in your own words.

54. Suppose $20,000 is invested in a savings account.

a. What will be the value of a $20,000 savings account at the end of 3 years if interest is calculated at 10% compounded annually?

b. Suppose the interest is compounded semiannually. What is the value? Is the value the same?

c. If not, explain why not in your own words.

55. Suppose $10,000 is invested into a high-yield savings account.

a. Calculate the interest earned in one year on $10,000 compounded monthly at 14%.

b. What is the difference between this and simple interest at 14% for 1 year?

56. Suppose that $50,000 is invested in a certificate of deposit for 5 years and the interest rate is 8%.

a. What will be the interest earned if it is compounded monthly?

b. How much more interest would be earned if it were compounded daily?

57. Suppose you have \$25,000 to invest.

a. Find the value of \$25,000 compounded daily at 5% for 20 years.

b. Do you think that the amount will be doubled or more than doubled if the rate is doubled to 10%?

c. Find the amount if the rate is 10%.

Find the amount (A) and the interest (I) for the given information.

	Compounding Period	Principal	Annual Rate	Time	A	$I = A - P$
58.	Quarterly	\$1000	10%	5 yr	**a.**	**b.**
59.	Monthly	\$1000	10%	5 yr	**a.**	**b.**
60.	Daily	\$1000	10%	5 yr	**a.**	**b.**
61.	Monthly	\$5000	7.5%	10 yr	**a.**	**b.**
62.	Daily	\$25,000	8%	20 yr	**a.**	**b.**
63.	Daily	\$25,000	12%	20 yr	**a.**	**b.**

Solve. Round your answer to the nearest cent, if necessary. See Examples 11 and 12.

64. Kevin is currently spending \$1300 per month on rent and utilities. Assuming an annual inflation rate of 3%, how much should he plan to spend on rent and utilities per month 2 years from now?

65. In 2013, the average price for a gallon of milk was \$3.00, and the average price for a loaf of bread was \$2.50. If the inflation rate was 6% per year, how much did these items cost in 2016?

66. Sam receives a cost-of-living raise equal to inflation each year. If inflation was steady at 5% annually, and his current yearly income is \$56,800, what was Sam's yearly income 4 years ago?

67. The current market value of Kayla's car is \$24,000. She plans to trade in her current car and buy a new one in 5 years. If the car depreciates at 12% per year, what will be the market value of her car when she is ready to trade it in?

68. Stephen has a fishing boat that he bought 3 years ago. He has decided to sell it, and the boat is valued at \$8500. If the yearly rate of depreciation for his boat is 13.2%, how much did he originally pay for the boat?

69. Gavin bought a new truck last year for \$29,900. This year, he decided that he wants to trade it in for a smaller car. He can resell the truck for 26,500. What was the rate of depreciation for the year?

70. A house is appraised at \$125,000. Assuming 3% constant inflation, what will be its value, to the nearest thousand dollars, in 30 years?

71. If a new pickup truck is valued at \$18,000, what will be its value in 3 years if it depreciates 22% each year?

57. a. \$67,952.39 **b.** More **c.** \$184,675.81
58. a. \$1638.62 **b.** \$638.62
59. a. \$1645.31 **b.** 645.31
60. a. \$1648.61 **b.** \$647.61
61. a. \$10,560.33 **b.** \$5560.33
62. a. \$123,804.10 **b.** \$98,804.10
63. a. \$275,470.73 **b.** \$250,470.73
64. \$1379.17
65. Milk: \$4.01; Bread: \$3.35
66. \$46,729.50
67. \$12,665.57
68. \$12,997.50
69. 11.4%
70. \$303,000
71. \$8541.94

72. Suppose that an apartment complex is purchased for \$1,500,000. For property tax purposes, the land is considered to be 30% of the value of the property. For income tax purposes, the owners are allowed to depreciate the value of the buildings by 5% per year. What will be the value of the apartment complex (buildings and land) in 10 years? (**Note:** This will not be the market value, but it will form the basis for capital gains taxes when the property is sold.)

Writing & Thinking

73. List the four parts involved in the simple interest formula. In your own words, define each one.

74. Compare and contrast simple interest with compound interest.

75. a. What will be the value of \$10,000 compounded weekly at 10% for 3 years?

b. Use your calculator to choose values for t to use in the formula until you find approximately how many years of daily compounding are needed for the value to accumulate to \$20,000.

t	$A=P\left(1+\frac{0.10}{365}\right)^{365t}$	A

Collaborative Learning

With the class separated into teams of two to four students, each team is to analyze the following problem related to compound interest. Each team leader is to discuss the results found by the team and how the team arrived at these results. A general classroom discussion should follow with the class coming to an understanding of the concepts of present value and future value.

76. Suppose that you would like to set aside some money today for your child's college education. Your child is 3 years old and would be starting college at the age of 18. What amount should you invest today (called the present value) at 8% compounded daily to accumulate \$40,000 (called the future value) for your child's education?

72. \$1,078,673.79

Writing & Thinking

73. The simple interest formula is $I = P \cdot r \cdot t$ where I is interest, P is principal, r is rate, and t is time. Interest is the amount of money paid for the use of money. The principal is the starting amount invested. Rate is the interest rate and should be written as a decimal or fraction. Time is the amount of time, in years, that interest is being earned on the principal. Time can be written as a decimal or fraction. When a decimal is used, it should only be when it is a terminating decimal so that no rounding is required, which could change the value calculated.

74. Interest is money paid for the use of money, whether it is simple or compounded. Simple interest is derived using a formula that calculates the interest based on the starting principal one time over any amount of time. Compound interest not only earns interest on the principal but also on the interest itself. It can be a series of simple interest calculations where the principal changes each time to include the interest earned.

75. Answers will vary.

a. \$13,498.03

b. About 7 years.

In groups of three to four students, work through the following problem. Discuss your answers in class.

77. Three monthly incomes are listed in the table and each receives a yearly cost-of-living raise.

a. Find each monthly income after 5 years for annual inflation rates of 4%, 6%, and 8%. Round to the nearest cent.

Monthly Income	4%	6%	8%
$2000			
$2500			
$4000			

b. Discuss the difference between starting with a $2000 monthly income with an 8% yearly pay raise and starting with a $4000 monthly income with a 4% yearly pay raise. (**Hint:** Compare each starting monthly income with the monthly income after 5 years.)

c. Suppose that the cost-of-living increases at a rate of 6% each year and you only received a 4% raise each year. Discuss how this might affect your way of living and what actions you might take.

Collaborative Learning

76. Answers will vary. $12,049.37. Some students may use the formula $A = P\left(1+\frac{r}{n}\right)^{nt}$ and solve for P.

77. a.

Monthly Income	4%	6%	8%
$2000	$2433.31	$2676.45	$2938.66
$2500	$3041.63	$3345.56	$3673.32
$4000	$4866.61	$5352.90	$5877.31

b. Answers will vary. The total increase in income after 5 years is similar for the lower income with an 8% yearly pay raise and the higher income with the 4% yearly pay raise.

c. Answers will vary.

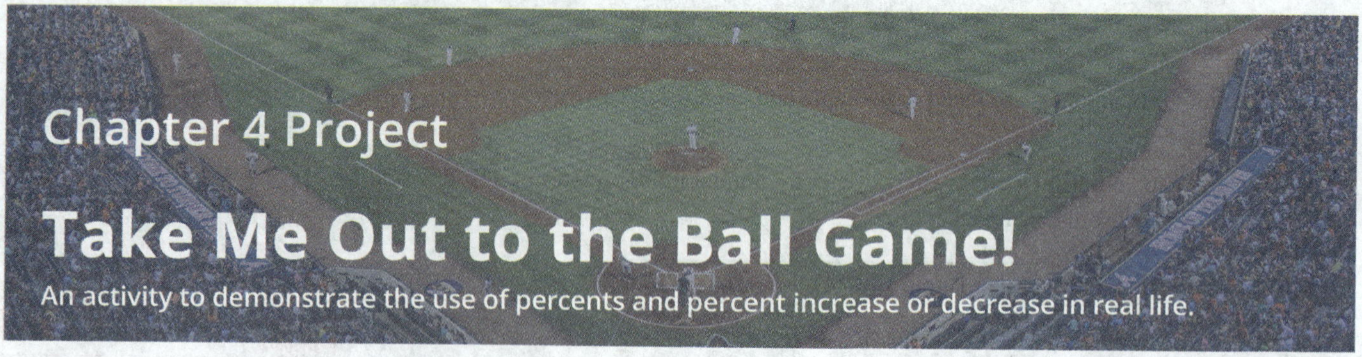

Chapter 4 Project

Take Me Out to the Ball Game!

An activity to demonstrate the use of percents and percent increase or decrease in real life.

The Atlanta Braves baseball team has been one of the most popular baseball teams for fans, not only from Georgia, but throughout the Carolinas and the southeastern United States. The Braves franchise started playing at the Atlanta-Fulton County Stadium in 1966 and this continued to be their home field for 30 years. In 1996, the Centennial Olympic Stadium that was built for the 1996 Summer Olympics was converted to a new ballpark for the Atlanta Braves. The ballpark was named Turner Field and was opened for play in 1997. In 2017, the Braves moved to a new stadium named SunTrust Park, which was renamed as Truist Park in 2020 when SunTrust and BB&T merged.

Round all percents to the nearest whole percent.

1. The Atlanta-Fulton County Stadium had a seating capacity of 52,769 fans. Turner Field had a seating capacity of 50,096. Truist Park has a seating capacity of 41,149.
   a. Determine the decrease in seating capacity between Turner Field and the original Braves stadium. 5%
   b. Determine the percent decrease in seating capacity between Truist Park and Turner field. 18%
2. The Centennial Olympic Stadium had approximately 85,000 seats. Some of the seating was removed in order to convert it to the Turner Field ballpark. Rounding the number of seats in Turner Field to the nearest thousand, what is the approximate percent decrease in seating capacity from the original Olympic stadium? 41%
3. The Braves were ranked 2nd out of 16 teams in 1999 and the average attendance was 40,554. Again in 2021, they were ranked 2nd out of 15 teams and the average attendance was 28,746. What was the percent decrease in attendance from 1999 to 2021? [1] 29%
4. The highest average attendance for the Braves was 47,960 in 1993 at the Atlanta-Fulton County Stadium. The lowest average attendance was 6642 in 1975 at the Atlanta-Fulton County Stadium. What is the percent increase from the lowest attendance to the highest? [2] 622%
5. Chipper Jones, a popular Braves third baseman, retired in July 2013. He started his career with the Braves in 1993 at the age of 21. [3]
   a. In 2001, Chipper had 189 hits in 572 at-bats. Calculate Chipper's batting average for the season by dividing the number of hits by the number of at-bats. Round to the nearest thousandth. .330
   b. In 2008, Chipper had 160 hits in 439 at-bats. Calculate Chipper's batting average for the season by dividing the number of hits by the number of at-bats. Round to the nearest thousandth. .364
   c. Calculate the percent change in Chipper's batting average from 2001 to 2008. 110%
   d. Does this represent a percent increase or decrease? Percent increase
6. In 2001, Chipper had 102 RBIs (runs batted in). In 2008, Chipper had only 75 RBIs.
   a. Calculate the percent change in RBIs from 2001 to 2008. 26%
   b. Does this represent a percent increase or decrease? Percent decrease

1 Source: "Atlanta Braves Attendance, Stadiums, and Park Factors," Baseball Reference, Accessed May 3, 2022, www.baseball-reference.com/teams/ATL/attend.shtml.
2 Source: "Atlanta Braves Attendance Data," Baseball Almanac, Accessed May 3, 2022, baseball-almanac.com/teams/bravatte.shtml.
3 Source: "Chipper Jones Stats," Baseball Almanac, Accessed May 3, 2022, www.baseball-almanac.com/players/player.php?p=jonesch06.

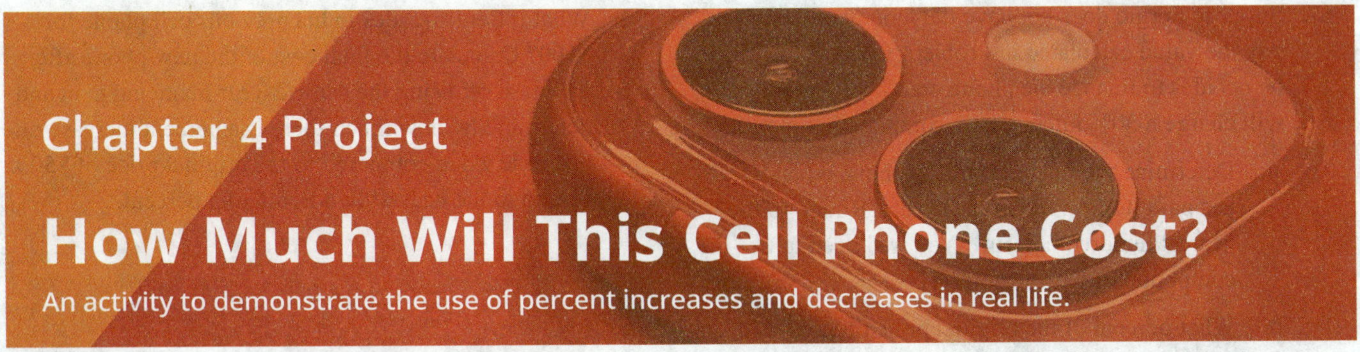

Chapter 4 Project

How Much Will This Cell Phone Cost?

An activity to demonstrate the use of percent increases and decreases in real life.

Sergio works in electronics sales. He wants to purchase a new cell phone from his place of employment, which offers a 10% employee discount. Sales tax is 7% in the city where the store is located. The original price of the cell phone Sergio wants to buy is $500.

For the following problems, round monetary amounts to the nearest cent. Otherwise, do not round unless indicated.

1. Suppose the store calculates the discount before applying sales tax.

- **a.** What is the price of the phone following the 10% discount, before tax?
- **b.** What is the total after applying tax to the answer in part a.?

2. Let's compare the price Sergio pays to the original price of the phone.

- **a.** Does the price Sergio pays represent an increase or a decrease from the original price? By how much?
- **b.** What percent of the original price for the phone is the price that Sergio pays in Problem 1 part a.?
- **c.** Does the answer to part b. represent a percent increase or a percent decrease from the original price?
- **d.** Calculate the percent increase or percent decrease of the price of the phone.

3. Consider whether the order of applying the discount and applying the tax matters.

- **a.** Find the cost of the phone if the tax is applied first, followed by the discount.
- **b.** How does the cost found in part a. compare to the cost found in Problem 1 part b.?

4. Sergio's coworker Haruto is looking at a cell phone with an original price of $450. Consider whether this change in original price will affect the *percent* of the original price he will pay after the employee discount.

- **a.** Repeat Problem 1 parts a. and b. and Problem 2 part b. for the phone Haruto is interested in.
- **b.** Does the value from part a. represent a percent increase or a percent decrease from the original price? By how much?
- **c.** How does the value in part b. compare to the value from Problem 2 part d.?

5. Zendaya works at the same store as Sergio and Haruto. The final price she pays for a cell phone, including both the discount and tax, is $311.53. What was the original price for her cell phone?

6. Marya works at another electronics store in the same city (which means the sales tax is also 7%), but her employer only offers a 5% employee discount. Like Sergio, she is looking at a cell phone with an original price of $500.

- **a.** Repeat Problem 1 parts a. and b. and Problem 2 part b. for the phone Marya is interested in.
- **b.** Does this represent a percent increase or a percent decrease? By how much?
- **c.** How do the percents for the employee discount and tax determine if there is a percent increase or a percent decrease?

7. Suppose Sergio does not have the money on hand to pay the value calculated in Problem 1 part b. in full, so he decides to finance the phone on a 2-year plan that will charge a 30% interest rate, compounded monthly.

   a. Determine the final cost of the phone when Sergio has paid it off in full after 2 years. Round to the nearest hundredth. (**Hint:** You will need to use the compound interest formula.)

   b. What percent of the original price of $500 will Sergio have paid?

   c. Does this represent a percent decrease or a percent increase? By how much?

8. What if Sergio trades in his previous cell phone for $100? Calculate the final cost of the new phone after paying it off using the same 2-year financing plan and determine by how much there is a percent increase or decrease when compared to the original price of $500. Round to the nearest hundredth of a percent.

Chapter 4 Project: How Much Will This Cell Phone Cost?

1. a. $(1-0.10)\cdot 500 = \$450$ will be the price after tax.

   b. $(1+0.07)\cdot 450 = \$481.50$ will be the total cost after applying tax.

2. a. Sergio's payment is a decrease of $18.50 from the original price.

   b. $481.50 is 96.3% of $500.

   c. This represents a percent decrease.

   d. The percent decrease is 3.7%.

3. The following shows that the order of applying the discount and applying the tax does not matter.

   a. If tax is applied first, the cost before the discount is $(1+0.07)\cdot 500 = \$535$. Applying the discount to this results in $(1-0.10)\cdot 535 = \$481.50$.

   b. The answer in part **a.** is the same as the answer found in Problem 1 part **b.**

4. a. Haruto pays $(1-0.10)\cdot 450 = \$405$ after the discount but before tax. After tax, he pays $433.35. $433.35 is 96.3% of $450.

   b. This represents a percent decrease by 3.7%.

   c. This is the same as the answer to Problem 2 part **d.**

5. After rounding to the nearest hundredth of a dollar, $311.53 is 96.3% of $323.50. $323.50 is the original price of her cell phone.

6. a. Before tax, Marya has $(1-0.05)\cdot 500 = \$475$ to pay. After tax, she has $(1+0.07)\cdot 475 = \$508.25$ to pay. $508.25 is 101.65% of the original price.

   b. This represents a percent increase by 1.65%.

   c. Answers may vary. Students should mention that if the employee discount is less than the sales tax, then there is a percent increase in price. If the employee discount is greater than the sales tax, then there is a percent decrease in price.

7. a. $A = P\left(1+\frac{r}{n}\right)^{nt} = 481.50\left(1+\frac{0.30}{12}\right)^{12(2)} \approx \$870.90$ is what he will pay after financing.

   b. $870.90 is 174.18% of $500.

   c. This represents a percent increase of 74.18%.

8. With a trade-in, Sergio will have $381.50 left to pay after the employee discount and the tax but before applying the financing.
$A = P\left(1+\frac{r}{n}\right)^{nt} = 381.50\left(1+\frac{0.30}{12}\right)^{12(2)} \approx \$690.03$ is what he will pay after financing. This is approximately a 38.01% percent increase.

CHAPTER 5

5 Measurement

Connections

Measurements surround us as we go about our daily lives. When you drive to school, work, or a friend's house, you must obey the speed limits that are posted in miles per hour. When you fill your car with gasoline, the fuel pump keeps track of how many gallons it dispenses. When you cook, the ingredients are measured by weight or volume. All of these measurements play a role in how you interact with the world around you. Knowing how different units of measurement work together will help you be more successful in your day-to-day life.

Suppose you purchase a French cookbook to expand your culinary skills. The book you purchased has recipes that are a direct translation from French, meaning that the ingredient measurements and cooking temperatures are given in metric units. For example, the recipe for *Pain au Chocolat* calls for 120 milliliters of milk and needs the oven to be preheated to 190 °C. If you are living in the United States, you might only have measuring cups that measure fluid ounces and an oven that has a temperature dial in degrees Fahrenheit. How can you determine how many cups of milk you need for the recipe and to what temperature to preheat your oven?

Objectives

A. Recognize the basic units of measure in the US customary system.

B. Use multiplication and division to convert between US units of measure.

C. Use unit fractions to convert between US units of measure.

5.1 US Measurements

A The US System of Measurement

You measure things every day: the distance you drive to work, the time it takes you to get to school, how much gas you use in a week, the area you mow in your yard, the weight of the apples you just bought. Possible measurements you use are miles, gallons, square feet, and pounds. These are units of measure in the **US customary system**. Most of the rest of the world uses the **metric system** with meters, liters, and grams. (See the next sections for more information on the metric system.)

In the **US customary system**, the units are not systematically related. Historically, some of the units were associated with parts of the body, which would vary from person to person. For example, a foot was the length of a person's foot and a yard was the distance from the tip of one's nose to the tip of one's fingers with the arm outstretched.

There is considerably more consistency now because the official weights and measures are monitored by the government. Table 1 below shows relationships between basic units used for length, weight, capacity, and time in the US customary system. (**Note:** The time equivalents are used in both the US customary and metric systems.)

Math Tip

It's important to keep in mind that *ounces* and *fluid ounces* are two very different measurements: ounces measure weight and fluid ounces measure capacity. The values of the two measurements are rarely interchangeable. Consider the following measurements.

1 fluid ounce of cooked oats weighs 0.43 ounces.

1 fluid ounce of water weighs 1.04 ounces.

1 fluid ounce of honey weighs 1.44 ounces.

Measurements Used in the US Customary System

US Units of Length			US Units of Weight		
12 inches (in.)	=	1 foot (ft)	16 ounces (oz)	=	1 pound (lb)
36 inches	=	1 yard (yd)	2000 pounds	=	1 ton (T)
3 feet	=	1 yard			
5280 feet	=	1 mile (mi)			
US Units of Capacity			**Units of Time**		
8 fluid ounces (fl oz)	=	1 cup (c)	60 seconds (sec)	=	1 minute (min)
2 cups	=	1 pint (pt) = 16 fluid ounces	60 minutes	=	1 hour (hr)
2 pints	=	1 quart (qt)	24 hours	=	1 day
4 quarts	=	1 gallon (gal)	7 days	=	1 week

Table 1

There is no simple way to convert from one unit of measure to another. You must simply memorize the information in Table 1. You are probably familiar with many of the conversions.

1. Use Table 1 to convert each measurement.

a. 16 oz = ________ lb

b. 1 qt = ________ pt

c. 1 day = ________ hr

d. 1 mile = ________ ft

Example 1 Basic Conversions in the US Customary System

Use Table 1 to convert each measurement.

a. 1 gal = _____ qt

b. 3 ft = _____ yd

c. 60 min = _____ hr

d. 1 T = _____ lb

Solution

a. 1 gal = 4 qt

b. 3 ft = 1 yd

c. 60 min = 1 hr

d. 1 T = 2000 lb

Now work margin exercise 1.

B Using Multiplication and Division to Convert US Units of Measure

When converting measurements that are not given directly in Table 1, you must decide to either multiply or divide. The following two statements will help you decide. In either case, you need to memorize the conversions in Table 1.

Using Multiplication and Division to Convert Measurements

1. **Multiply** to convert to smaller units. (There will be more smaller units.)
2. **Divide** to convert to larger units. (There will be fewer larger units.)

PROCEDURE

Teaching Note

The multiplication and division method and the unit fraction method are both presented here. You may choose to use one method or both.

Example 2 Converting US Units of Measure Using Multiplication/Division

Use multiplication or division to convert each measurement.

a. 3 c = _____ fl oz

b. 5 gal = _____ qt

c. 150 min = _____ hr

d. 39 in. = _____ ft

Solution

a. You are converting from a *larger* unit to a *smaller* unit (a *cup* is larger than a *fluid ounce*), so multiply. Because there are 8 fluid ounces in 1 cup, multiply by 8.

$$3 \text{ c} = 3 \cdot 8 \text{ fl oz} = 24 \text{ fl oz}$$

b. You are converting from a *larger* unit to a *smaller* unit (a *gallon* is larger than a *quart*), so multiply. Because there are 4 quarts in 1 gallon, multiply by 4.

$$5 \text{ gal} = 5 \cdot 4 \text{ qt} = 20 \text{ qt}$$

c. You are converting from a *smaller* unit to a *larger* unit (a *minute* is smaller than an *hour*), so divide. Because there are 60 minutes in 1 hour, divide by 60.

$$150 \text{ min} = \frac{150}{60} \text{ hr} = \frac{5}{2} \text{ hr} = 2\frac{1}{2} \text{ hr or } 2.5 \text{ hr}$$

d. You are converting from a *smaller* unit to a *larger* unit (an *inch* is smaller than a *foot*), so divide. Because there are 12 inches in 1 foot, divide by 12.

$$39 \text{ in.} = \frac{39}{12} \text{ ft} = \frac{13}{4} \text{ ft} = 3\frac{1}{4} \text{ ft or } 3.25 \text{ ft}$$

Now work margin exercise 2.

2. Use multiplication or division to convert each measurement.

a. 3 yd = _______ in.

b. 4 c = _______ fl oz

c. 28 days = _______ weeks

d. 7500 lb = _______ T

3. A bag of assorted Halloween candy weighs 72 ounces. How many pounds of candy are in the bag?

Example 3 Application: Converting US Units of Measure

A bottle of water contains 16.8 fluid ounces. How many cups of water are in the bottle?

Solution

There are 8 fluid ounces in 1 cup and we want to change from fluid ounces to cups. To change from a smaller unit to a larger unit, divide.

$$16.8 \text{ fl oz} = \frac{16.8}{8} \text{ c} = 2.1 \text{ c}$$

Thus, the bottle contains 2.1 cups of water.

Now work margin exercise 3.

C Using Unit Fractions to Convert US Units of Measure

A second method of conversion involves the use of unit fractions. A **unit fraction** is a fraction equivalent to 1. For example,

$$\frac{3 \text{ ft}}{3 \text{ ft}} = \frac{\cancel{3} \cancel{\text{ft}}}{\cancel{3} \cancel{\text{ft}}} = 1.$$

But from Table 1, we know that 3 ft = 1 yd. With this fact, we can write two unit fractions with numerator or denominator 1 as follows.

$$\frac{1 \text{ yd}}{3 \text{ ft}} = 1 \quad \text{and} \quad \frac{3 \text{ ft}}{1 \text{ yd}} = 1$$

Similarly,

$$\frac{1 \text{ min}}{60 \text{ sec}} = 1 \quad \text{and} \quad \frac{60 \text{ sec}}{1 \text{ min}} = 1.$$

This type of fraction is useful in making a conversion. Because the value of the fraction is 1, **multiplication by a unit fraction** does not change the value of the expression being converted. Use the following guidelines in determining which type of fraction to use.

Using Unit Fractions to Convert Measurements

1. The numerator should be in the units of measure of the desired result.
2. The denominator should be in the original units of measure.

PROCEDURE

Example 4 Using Unit Fractions to Convert US Units of Measure

Use unit fractions to convert each measurement.

a. 21 ft = ______ yd

b. 15 hr = ______ min

c. $6\frac{1}{2}$ qt = ______ pt

d. 40 oz = ______ lb

Solution

a. Choose the unit fraction with yards in the numerator and feet in the denominator.

$$\frac{1\text{ yd}}{3\text{ ft}} = 1$$

Now multiply by this fraction as follows.

$$21\text{ ft} = 21\cancel{\text{ft}} \cdot \frac{1\text{ yd}}{3\cancel{\text{ft}}} = \frac{21}{3}\text{ yd} = 7\text{ yd}$$

Note that the measure label of feet (ft) divides out and the result is in yards (yd).

b. Choose the unit fraction with minutes in the numerator and hours in the denominator.

$$\frac{60\text{ min}}{1\text{ hr}} = 1$$

Now multiply by this fraction as follows.

$$15\text{ hr} = 15\cancel{\text{hr}} \cdot \frac{60\text{ min}}{1\cancel{\text{hr}}} = 15 \cdot 60\text{ min} = 900\text{ min}$$

Note that, as with the multiplication/division method, the number became larger because minutes are a smaller unit than hours.

c. Choose the unit fraction with pints in the numerator and quarts in the denominator.

$$\frac{2\text{ pt}}{1\text{ qt}} = 1$$

Now multiply by this fraction as follows.

$$6\frac{1}{2}\text{ qt} = 6\frac{1}{2}\cancel{\text{qt}} \cdot \frac{2\text{ pt}}{1\cancel{\text{qt}}} = \frac{13}{\cancel{2}} \cdot \cancel{2}\text{pt} = 13\text{ pt}$$

d. Choose the unit fraction with pounds in the numerator and ounces in the denominator.

$$\frac{1\text{ lb}}{16\text{ oz}} = 1$$

Now multiply by this fraction as follows.

$$40\text{ oz} = 40\cancel{\text{oz}} \cdot \frac{1\text{ lb}}{16\cancel{\text{oz}}} = \frac{40}{16}\text{ lb} = \frac{5}{2}\text{ lb} = 2\frac{1}{2}\text{ lb or } 2.5\text{ lb}$$

Now work margin exercise 4.

4. Use unit fractions to convert each measurement.

a. 10,000 lb = ______ T

b. $5\frac{2}{3}$ days = ______ hr

c. 128 fl oz = ______ c

d. 3 mi = ________ ft

5. The average movie run time is around 1.8 hours. How many minutes long is the average movie?

Example 5 Application: Converting US Units of Measure

A fully grown African elephant can weigh as much as 7.5 tons. How many pounds is this?

Solution

There are 2000 pounds in 1 ton. Using a unit fraction to convert from tons to pounds gives the following.

$$7.5\text{ T} = 7.5\ \cancel{\text{T}} \cdot \frac{2000\text{ lb}}{1\ \cancel{\text{T}}} = 7.5 \cdot 2000\text{ lb} = 15{,}000\text{ lb}$$

Thus, a fully grown African elephant can weigh as much as 15,000 pounds.

Now work margin exercise 5.

6. How many fluid ounces are in 8 gallons of apple juice?

Example 6 Application: Converting US Units of Measure

Determine how many seconds are in a 5-day work week assuming an 8 hr work day.

Solution

This number can be found as follows.

$$5\text{ days} = 5\ \cancel{\text{days}} \cdot \frac{8\ \cancel{\text{hr}}}{1\ \cancel{\text{day}}} \cdot \frac{60\ \cancel{\text{min}}}{1\ \cancel{\text{hr}}} \cdot \frac{60\text{ sec}}{1\ \cancel{\text{min}}} = 5 \cdot 8 \cdot 60 \cdot 60\text{ sec} = 144{,}000\text{ sec}$$

Thus, there are 144,000 seconds in a 5-day work week.

Now work margin exercise 6.

Margin Exercise Answers

1. a. 1 **b.** 2 **c.** 24 **d.** 5280 **2. a.** 108 **b.** 32 **c.** 4 **d.** $3\frac{3}{4}$ or 3.75 **3.** $4\frac{1}{2}$ or 4.5 pounds **4. a.** 5 **b.** 136 **c.** 16 **d.** 15,840 **5.** 108 minutes **6.** 1024 fluid ounces

5.1 Exercises

Concept Check

Fill-in-the-Blank. Complete each sentence using information found in this section.

1. There are two cups in 1 ____, which is equivalent to ___ fluid ounces.
2. Sixteen ounces equals 1 _____ and 2000 ______ equals 1 ton.
3. Both 3 feet and 36 inches equal 1 ____.
4. When converting from one unit of measure to another smaller unit, ____________ is necessary.
5. When using a unit fraction for conversions, the numerator should have the same units as the result and the denominator should be in the units to be _________.
6. A fraction equivalent to 1 is called a ____ fraction.

5.1 Exercises

Concept Check

1. pint, 16
2. pound, pounds
3. yard
4. multiplication
5. converted
6. unit
7. True
8. False; One mile is equivalent to 5280 feet.
9. True
10. True

Practice

1. 12
2. 7
3. 1

True/False. Determine whether each statement is true or false. If a statement is false, explain how it can be changed so the statement will be true. (**Note:** There may be more than one acceptable change.)

7. Capacity can be measured using ounces, quarts, and gallons.

8. One mile is equivalent to 2000 feet.

9. To convert from smaller units to larger units, division will be required.

10. Multiplication by a unit fraction does not change the value of the expressions being converted.

Practice

Use the units you have memorized from Table 1 to convert each measurement. See Example 1.

1. 1 ft = ___ in.

2. 1 week = ___ days

3. 2000 lb = ___ T

4. 2 c = ___ pt

5. 4 qt = ___ gal

6. 36 in. = ___ yd

7. 1 mi = ___ ft

8. 1 lb = ___ oz

9. 24 hr = ___ day

10. 12 in. = ___ ft

11. 1 c = ___ fl oz

12. 1 qt = ___ pt

Use multiplication or division to convert each measurement. See Example 2.

13. 3 ft = ___ in.

14. 2 lb = ___ oz

15. 5 min = ___ sec

16. 3 yd = ___ ft

17. 10,560 ft = ___ mi

18. 6000 lb = ___ T

19. 3 weeks = ___ days

20. 5 yd = ___ ft

21. 72 in. = ___ yd

22. 48 in. = ___ ft

23. 90 min = ___ hr

24. 9 qt = ___ gal

Use unit fractions to convert each measurement. First write the unit fraction you are going to use, then perform the conversion. See Example 4.

25. 7 yd = ___ ft

26. 5 qt = ___ pt

27. 6 pt = ___ qt

28. 32 fl oz = ___ c

29. 13 qt = ___ gal

30. 3 pt = ___ fl oz

31. 18 in. = ___ ft

32. 24 oz = ___ lb

33. 3 mi = ___ ft

34. 5 T = ___ lb

35. 7920 ft = ____ mi

36. 7000 lb = ___ T

4. 1
5. 1
6. 1
7. 5280
8. 16
9. 1
10. 1
11. 8
12. 2
13. 36
14. 32
15. 300
16. 9
17. 2
18. 3
19. 21
20. 15
21. 2
22. 4
23. 1.5
24. 2.25
25. $\frac{3\text{ ft}}{1\text{ yd}}$; 21
26. $\frac{2\text{ pt}}{1\text{ qt}}$; 10
27. $\frac{1\text{ qt}}{2\text{ pt}}$; 3
28. $\frac{1\text{ c}}{8\text{ fl oz}}$; 4
29. $\frac{1\text{ gal}}{4\text{ qt}}$; 3.25
30. $\frac{16\text{ fl oz}}{1\text{ pt}}$; 48
31. $\frac{1\text{ ft}}{12\text{ in.}}$; 1.5
32. $\frac{1\text{ lb}}{16\text{ oz}}$; 1.5
33. $\frac{5280\text{ ft}}{1\text{ mi}}$; 15,840
34. $\frac{2000\text{ lb}}{1\text{ T}}$; 10,000
35. $\frac{1\text{ mi}}{5280\text{ ft}}$; 1.5
36. $\frac{1\text{ T}}{2000\text{ lb}}$; 3.5

37. 8
38. 180
39. 32,000
40. 52,800
41. 4
42. 13
43. 88
44. 42
45. 1.5
46. 2.5
47. 150
48. 54

Applications

49. $\frac{7}{8}$ or 0.875 square feet
50. The large jar; $0.005/oz
51. $93.22
52. 76 fluid ounces
53. $\frac{1}{2}$ or 0.5 miles
54. $4\frac{1}{4}$ or 4.25 scoops
55. 15 miles per hour
56. 45,842,400
57. The small bag; $0.005¢
58. 3 hours
59. 126,720 inches

Convert each measurement. See Examples 1, 2, and 4.

37. 4 pt = ___ c

38. 3 hr = ___ min

39. 16 T = ___ lb

40. 10 mi = ___ ft

41. 96 hr = ___ days

42. 39 ft = ___ yd

43. 5.5 lb = ___ oz

44. 3.5 ft = ___ in.

45. 6 qt = ___ gal

46. 150 min = ___ hr

47. 2.5 min = ___ sec

48. 1.5 yd = ___ in.

Applications

Solve.

49. Find the area (in sqaure feet) of a rectangle that is $1\frac{1}{2}$ ft by 7 in.

50. A small jar of peanut butter sells for $0.08 per ounce. A large jar of peanut butter sells for $1.20 per pound. Which is the better buy and by how much (in cents per ounce)?

51. Sheer fabric costs $7.99 per yard. If it will take 35 feet of fabric to make drapes for the entire house, how much must you spend on fabric for the drapes, to the nearest cent?

52. While cleaning out the garage, Nelson discovers some containers of oil that need to be taken in for recycling. The containers hold 20 fluid ounces, 3 cups, and 1 quart, respectively. Find the total amount of oil ready to be recycled (in fluid ounces).

53. Joel runs track for Northside High School. He runs the 220-yard sprint, the $\frac{1}{4}$-mile hurdles, and the $\frac{1}{2}$-mile relay (of which he runs one leg, which is $\frac{1}{8}$ of a mile). How far does Joel run in total (in miles)?

54. How many scoops will it take a 2-ton crane (it can scoop 2 tons of material at one time) to move 17,000 pounds of dirt?

55. A ball dropped from the top of a building hits the ground at a speed of 22 feet per second. How fast does the ball hit the ground in miles per hour?

56. The author of this textbook spent 1 year, 23 weeks, 5 days, and 14 hours writing it. How many seconds is this? (**Hint:** There are 52 weeks in a year.)

57. A small bag of cookies sells for $0.12 per ounce. A large bag of cookies sells for $2.00 per pound. Which is the better buy and by how much (in cents per ounce)?

58. Adrian exercised three days during the past week. He exercised 50 minutes and 15 seconds during the first day, 84 minutes and 25 seconds on the second day, and 45 minutes and 20 seconds on the third day. How many hours did he spend exercising during the week?

59. Determine how many inches are in 2 miles. (**Hint:** You will need to use more than one conversion factor.)

60. Each load of laundry washed uses 2 fluid ounces of a certain laundry detergent. Determine how many loads of laundry can be washed with a one-gallon jug of laundry detergent. (**Hint:** First determine how many fluid ounces are in one gallon.)

61. Ashleigh works at a coffee shop and is paid $12 per hour. How much is she paid for working for 5 hours and 25 minutes?

62. A baby weighed 7 pounds and 9 ounces at birth. At the last checkup, the baby weighed 12 pounds and 2 ounces. How much weight did the baby gain, in ounces?

63. Marco has a goal to drink 10 cups of water each day. How many 16-ounce water bottles must Marco drink each day to reach his goal?

64. Blake had 2.5 feet of ribbon left over from his last project and purchased a new 5-yard package of the same ribbon. How many inches of the ribbon does Blake have?

Writing & Thinking

65. Colby needs to find out how many yards are in one mile. What two sets of equivalent units would he need to make that determination?

66. In your own words, explain when you would multiply and when you would divide when converting between units.

67. Briefly describe a unit fraction and explain when and how it would be used.

68. Give at least two examples of when you might want to convert between units of measure (outside of a class).

60. 64 loads of laundry

61. $65

62. 73 oz

63. 5

64. 210 in.

Writing & Thinking

65. Colby would need to know that there are 3 feet in a yard and 5280 feet in a mile.

66. When converting from smaller to larger units, division is used because fewer larger units are needed to make the same measurement. When converting from larger to smaller units, multiplication is used because more of the smaller units are needed to make the same measurement.

67. A unit fraction is a fraction with different units that is equivalent to 1. It can be used to convert between units. When used, the numerator should contain the same units as the desired outcome and the denominator should have the same unit as the unit being converted.

68. Answers will vary but may include measuring for carpeting, sod, material for clothing, baking, timing an event or sale, or weighing items.

Objectives

A. Convert between metric units of length.

B. Be familiar with the metric prefixes mega-, giga-, and tera-.

C. Convert between metric units of area.

5.2 The Metric System: Length and Area

A Units of Length in the Metric System

The **metric system** of measurement is used by about 90% of the people in the world. The United States is the only major industrialized country still committed to the US customary system (formerly called the English system). Even in the United States, the metric system is used in many fields of study and business, such as medicine, science, the computer industry, and the military. Industries involved in international trade must be familiar with the metric system.

The **meter** is the basic unit of length in the metric system. It is a little longer than a yard. Smaller and larger units are named by putting a prefix in front of the basic unit—for example, **centi**meter (about the width of a dime) and **kilo**meter (used in place of miles to measure distance). The prefixes we will use are shown in boldface print in Table 1. Other prefixes that indicate extremely small units are micro-, nano-, pico-, femto-, and atto-. Prefixes that indicate extremely large units are mega-, giga-, and tera-. You are probably familiar with computer terms such as megabytes, gigabytes, and terabytes.

Metric Measures of Length

1 **milli**meter (mm)	=	0.001 meter	1 m = 1000 mm	1 cm = 10 mm
1 **centi**meter (cm)	=	0.01 meter	1 m = 100 cm	
1 **deci**meter (dm)	=	0.1 meter	1 m = 10 dm	
1 meter (m)	=	1.0 meter (the basic unit)		
1 **deka**meter (dam)	=	10 meters		
1 **hecto**meter (hm)	=	100 meters		
1 **kilo**meter (km)	=	1000 meters		

Table 1

Writing Metric Units of Measure

In the metric system,

1. A 0 is written to the left of the decimal point if there is no whole number part (0.287 m).
2. No commas are used in writing numbers. If a number has more than four digits (to the left or right of the decimal point), the digits are grouped in threes from the decimal point with a space between the groups (25 000 m or 0.000 34 m).

PROCEDURE

As indicated in Table 1, the metric units of length are related to each other by powers of 10. That is, simply multiply by 10 to get the equivalent measure expressed in the next smaller unit. Thus, you will have **more** of a **smaller** unit.

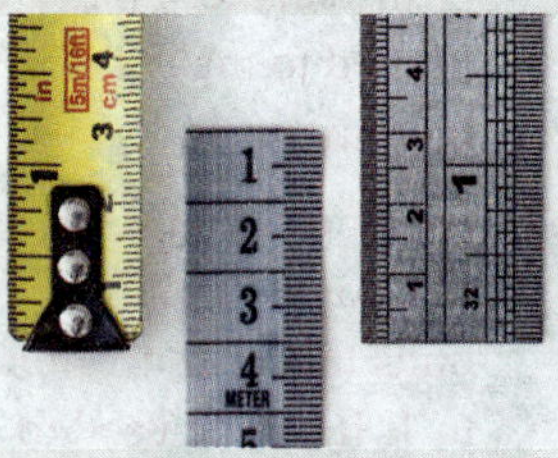

Metric Prefixes

In the metric system, there are three base units: meter (length), gram (mass), and liter (volume). Various prefixes are used to help clarify the measure of an object. Kilo- (thousands), hecto- (hundreds), and deka- (tens) are used when describing quantities larger than the base unit. These large prefixes are derived from Greek words. When quantities are smaller than the base unit, prefixes such as deci- (tenths), centi- (hundredths), and milli- (thousandths) are used. These small prefixes are derived from Latin words.

For example,

$$1\text{ m} = 10\text{ dm} = 100\text{ cm} = 1000\text{ mm}.$$

Conversely, divide by 10 to get the equivalent measure expressed in the next larger unit. Thus, you will have **less** of a **larger** unit. For example,

$$1\text{ mm} = 0.1\text{ cm} = 0.01\text{ dm} = 0.001\text{ m}.$$

There are two basic methods of converting units of measurement in the metric system:

1. multiplying by unit fractions and
2. moving the decimal point as guided by prefixes on a conversion line.

Both methods are discussed here.

From Section 5.1, we know that a **unit fraction** is a fraction that compares equivalent measures and is equal to 1. In a unit fraction, either the numerator or the denominator is 1. For example, in the metric system,

$$\frac{1\text{ cm}}{10\text{ mm}} = 1 \quad\text{and}\quad \frac{100\text{ cm}}{1\text{ m}} = 1.$$

Multiplication by a unit fraction can be used to convert units of measurement in the metric system. Remember the following guidelines in determining which type of fraction to use.

Using Unit Fractions to Convert Measurements

1. The numerator should be in the units of measure of the desired result.
2. The denominator should be in the original units of measure.

PROCEDURE

Example 1 Converting Metric Units of Length

Convert each measurement using unit fractions.

a. 5.6 m = ______ cm

b. 23.5 cm = ______ mm

c. 375 mm = ______ m

d. 1055 m = ______ km

Solution

a. $5.6\text{ m} = 5.6\ \cancel{\text{m}} \cdot \frac{100\text{ cm}}{1\ \cancel{\text{m}}} = 5.6 \cdot 100\text{ cm} = 560\text{ cm}$

b. $23.5\text{ cm} = 23.5\ \cancel{\text{cm}} \cdot \frac{10\text{ mm}}{1\ \cancel{\text{cm}}} = 23.5 \cdot 10\text{ mm} = 235\text{ mm}$

c. $375\text{ mm} = 375\ \cancel{\text{mm}} \cdot \frac{1\text{ m}}{1000\ \cancel{\text{mm}}} = \frac{375}{1000}\text{ m} = 0.375\text{ m}$

d. $1055\text{ m} = 1055\ \cancel{\text{m}} \cdot \frac{1\text{ km}}{1000\ \cancel{\text{m}}} = \frac{1055}{1000}\text{ km} = 1.055\text{ km}$

Now work margin exercise 1.

1. Convert each measurement using unit fractions.

a. 6.4 cm = ______ mm

b. 35 m = ________ cm

c. 5.9 m = ________ km

d. 320 mm = ________ m

2. A red wood ant can grow to have a length of 0.009 m. How many millimeters is this?

Example 2 Application: Converting Metric Units of Length

The highest peak of Mount Everest has a height of 8848 meters. What is this height in kilometers?

Solution

There are 1000 meters in 1 kilometer. Converting from meters to kilometers, we have

$$8848 \text{ m} = 8848 \cancel{\text{m}} \cdot \frac{1 \text{ km}}{1000 \cancel{\text{m}}} = \frac{8848}{1000} \text{ km} = 8.848 \text{ km}.$$

Thus, Mt. Everest has a height of 8.848 km.

Now work margin exercise 2.

Another technique for changing units in the metric system can be illustrated with the concept of a number line with the metric units of measure listed in order from largest to smallest (from left to right), as shown in Figure 1.

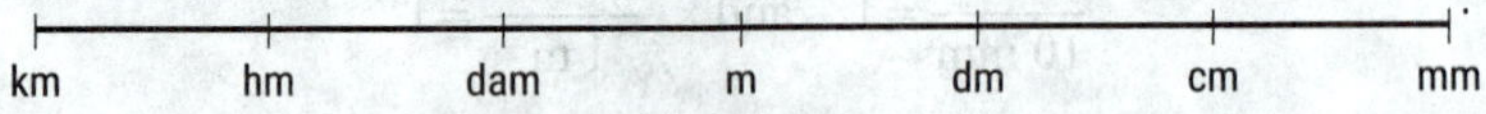

Figure 1

Simply move the decimal point in the direction of change using one digit per space.

3. Convert 65 m to hectometers using a metric conversion line.

Example 3 Converting Metric Units of Length

Convert 56 cm to meters using a metric conversion line.

Solution

Note that the decimal point is aligned over the original unit of metric length.

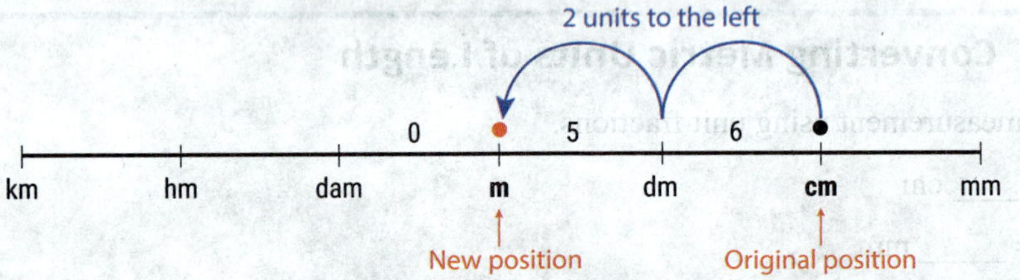

Thus, 56 cm = 0.56 m.

Now work margin exercise 3.

Example 4 Converting Metric Units of Length

Convert 13.5 m to millimeters using a metric conversion line.

Solution

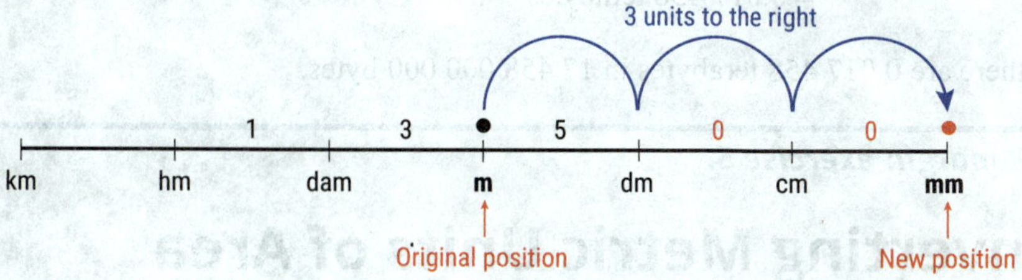

Thus, 13.5 m = 13 500 mm.

Now work margin exercise 4.

4. Convert 43 m to centimeters using a metric conversion line.

B The Prefixes Mega-, Giga-, and Tera-

The metric prefixes mega-, giga-, and tera- often appear in today's technological world. For example, you may be familiar with megapixels on your camera or gigahertz for computing speed. In a computer's memory, a byte is used to store characters, punctuation marks, and numbers. Computer memory may be measured in units as follows.

Large Metric Prefixes

Prefix	Multiplier	Example
kilo-	1 thousand	1 kilobyte (KB) = 1000 bytes
mega-	1 million	1 megabyte (MB) = 1 000 000 bytes
giga-	1 billion	1 gigabyte (GB) = 1 000 000 000 bytes
tera-	1 trillion	1 terabyte (TB) = 1 000 000 000 000 bytes

Table 2

Example 5 The Prefixes Mega-, Giga-, and Tera-

a. How many pixels are in 12 megapixels?

b. How many terabytes are in 17 458 000 000 bytes?

Solution

a. Using unit fractions: $12 \text{ megapixels} = 12 \text{ megapixels} \cdot \frac{1\,000\,000 \text{ pixels}}{1 \text{ megapixel}}$

$$= 12 \cdot 1\,000\,000 \text{ pixels}$$

$$= 12\,000\,000 \text{ pixels}$$

Thus, there are 12 000 000 pixels in 12 megapixels.

5. a. How many hertz are in 4.3 gigahertz?

b. How many megatons are in 2 125 000 metric tons?

b. $17\ 458\ 000\ 000 \text{ bytes} = 17\ 458\ 000\ 000 \cancel{\text{ bytes}} \cdot \dfrac{1 \text{ terabyte}}{1\ 000\ 000\ 000\ 000 \cancel{\text{ bytes}}}$

$$= \frac{17\ 458\ 000\ 000}{1\ 000\ 000\ 000\ 000} \text{ terabyte}$$

$$= 0.017\ 458 \text{ terabytes}$$

Thus, there are 0.017 458 terabytes in 17 458 000 000 bytes.

Now work margin exercise 5.

C Converting Metric Units of Area

Recall that area is measured in **square units**. In metric units, a square that is 1 centimeter long on each side is said to have an area of 1 square centimeter (or 1 cm^2). Figure 2 illustrates the area concept with a rectangle of area 28 cm^2.

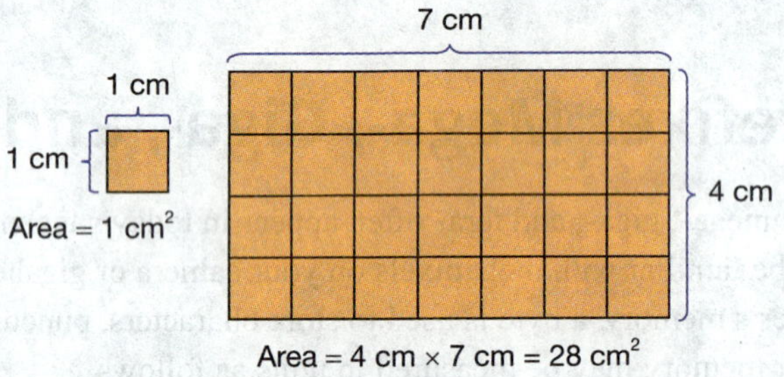

There are 28 squares that are each 1 cm^2 in the large rectangle.

Figure 2

Math Tip

Since each metric unit is 10 more than the previous one, each square metric unit is $10 \cdot 10 = 100$ more than the previous one.

For example,

$$\begin{aligned} 1 \text{ cm}^2 &= 1 \text{ cm} \cdot 1 \text{ cm} \\ &= 10 \text{ mm} \cdot 10 \text{ mm} \\ &= 100 \text{ mm}^2 \end{aligned}$$

and

$$\begin{aligned} 1 \text{ dm}^2 &= 1 \text{ dm} \cdot 1 \text{ dm} \\ &= 10 \text{ cm} \cdot 10 \text{ cm} \\ &= 100 \text{ cm}^2. \end{aligned}$$

Table 3 shows metric measures of areas. For example, the area of a tennis court might be measured in square meters, while the area of this page of paper might be measured in square centimeters. Other measures, listed in Table 4, are used for measuring land.

Metric Measures of Area	
$1 \text{ cm}^2 =$	100 mm^2
$1 \text{ dm}^2 = 100 \text{ cm}^2 =$	$10\ 000 \text{ mm}^2$
$1 \text{ m}^2 = 100 \text{ dm}^2 = 10\ 000 \text{ cm}^2 =$	$1\ 000\ 000 \text{ mm}^2$
$1 \text{ dam}^2 =$	100 m^2
$1 \text{ hm}^2 = 100 \text{ dam}^2 =$	$10\ 000 \text{ m}^2$
$1 \text{ km}^2 = 100 \text{ hm}^2 = 10\ 000 \text{ dam}^2 =$	$1\ 000\ 000 \text{ m}^2$

Table 3

Just as unit fractions were used to convert metric measures of length, they can be used to convert measures of area. For example,

$$\frac{1 \text{ cm}^2}{100 \text{ mm}^2} = 1 \quad \text{and} \quad \frac{10\ 000 \text{ cm}^2}{1 \text{ m}^2} = 1.$$

Example 6 Converting Metric Units of Area

Convert each measurement using unit fractions.

a. $5\ cm^2 =$ _____ mm^2

b. $4600\ mm^2 =$ _____ m^2

Solution

a. $5\ cm^2 = 5\ \cancel{cm^2} \cdot \frac{100\ mm^2}{1\ \cancel{cm^2}} = 5 \cdot 100\ mm^2 = 500\ mm^2$

Note that the decimal point is moved 2 places to the right.

b. $4600\ mm^2 = 4600\ \cancel{mm^2} \cdot \frac{1\ m^2}{1\ 000\ 000\ \cancel{mm^2}} = 0.0046\ m^2$

Note that the decimal point is moved 6 places to the left.

Now work margin exercise 6.

6. Convert each measurement using unit fractions.

a. $86\ m^2 =$ ________ cm^2

b. $0.06\ mm^2 =$ ________ dm^2

The number line concept can also be used to change units of area. Figure 3 shows a metric conversion line with metric units of area listed in order from largest to smallest (from left to right). Notice in Table 3 that the metric units of area are related to each other by powers of 100. This means that two digits are used in each space on the line.

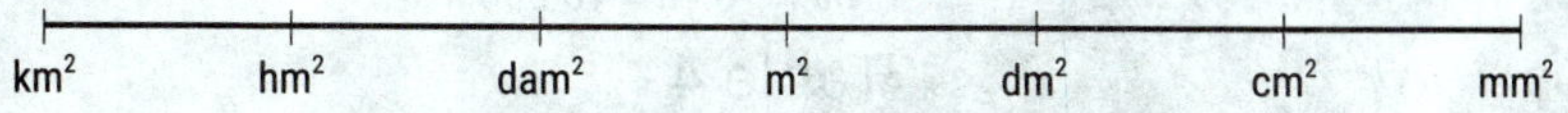

Figure 3

Example 7 Converting Metric Units of Area

Convert 253 mm^2 to square centimeters using a metric conversion line.

Solution

Note that the decimal point is aligned over the original unit of metric area and that two digits are used in each space on the line.

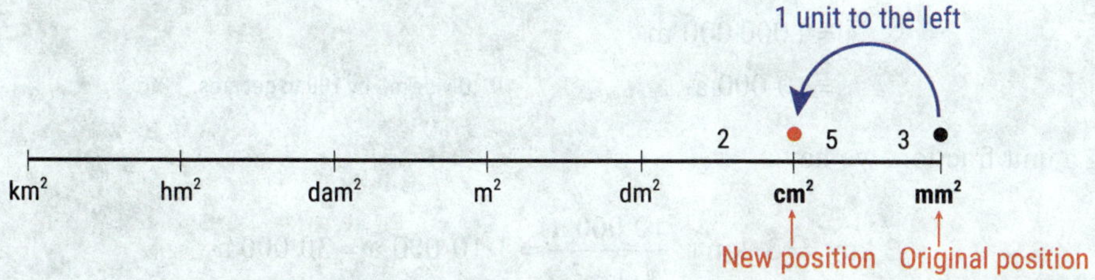

Thus, $253\ mm^2 = 2.53\ cm^2$.

Now work margin exercise 7.

7. Convert 365 mm^2 to square meters using a metric conversion line.

8. Convert 4.261 km^2 to square meters using a metric conversion line.

Example 8 Converting Metric Units of Area

Convert 6.1 m^2 to square centimeters using a metric conversion line.

Solution

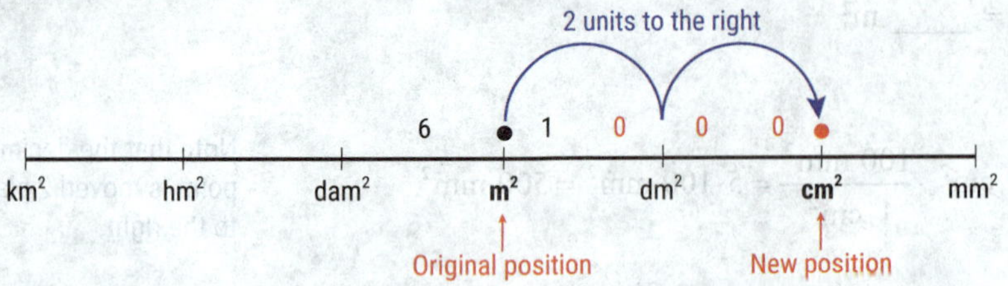

Thus, 6.1 m^2 = 61 000 cm^2.

Now work margin exercise 8.

A square with each side 10 meters long encloses an area of 1 **are (a)**. A **hectare (ha)** is 100 ares. The are and hectare are used to measure land area in the metric system.

Metric Measures of Land Area
1 a = 100 m^2
1 ha = 100 a = 10 000 m^2

Table 4

9. Convert 49 500 km^2 to hectares.

Example 9 Converting Metric Units of Land Area

Convert 3 km^2 to ares.

Solution

To make this conversion, we need to know how many ares are in 1 km^2. Because 1 km = 1000 m we have

$$\begin{aligned} 1 \text{ km}^2 &= 1 \text{ km} \cdot 1 \text{ km} \\ &= 1000 \text{ m} \cdot 1000 \text{ m} \\ &= 1\,000\,000 \text{ m}^2 \\ &= 10\,000 \text{ a} \end{aligned}$$

Divide m^2 by 100 to get ares.

Using a unit fraction, we have

$$3 \text{ km}^2 = 3 \cancel{\text{km}^2} \cdot \frac{10\,000 \text{ a}}{1 \cancel{\text{km}^2}} = 3 \cdot 10\,000 \text{ a} = 30\,000 \text{ a}$$

Thus, 3 km^2 = 30 000 a.

Now work margin exercise 9.

Margin Exercise Answers

1. a. 64 **b.** 3500 **c.** 0.0059 **d.** 0.320 **2.** 9 mm **3.** 0.65 hm **4.** 4300 cm **5. a.** 4 300 000 000 hertz **b.** 2.125 megatons **6. a.** 860 000 **b.** 0.000 006 **7.** 0.000 365 m^2 **8.** 4 261 000 m^2 **9.** 4 950 000 hectares

5.2 Exercises

Concept Check

Fill-in-the-Blank. Complete each sentence using information found in this section.

1. In the metric system, commas are not used. Digits are instead grouped in _______ with spaces between the groups.

2. If there is no whole number of units in the metric system, a/an _____ is written to the left of the decimal point.

3. The basic unit of length in the metric system is the _____.

4. In the metric system, conversions from smaller to larger units require _______ by a power of 10.

5. To change from larger to smaller metric units, _______ by a power of 10.

True/False. Determine whether each statement is true or false. If a statement is false, explain how it can be changed so the statement will be true. (**Note:** There may be more than one acceptable change.)

6. To change from smaller units to larger units, multiplication must be used.

7. Units of length in the metric system are named by putting a prefix in front of the basic unit meter, for example, centimeter.

8. In metric units, a square that is 1 centimeter long on each side is said to have an area of 1 centimeter.

Practice

Use this ruler to help measure (or approximate) each length. (Answers will vary.) The ruler shown here is marked with millimeters and centimeters.

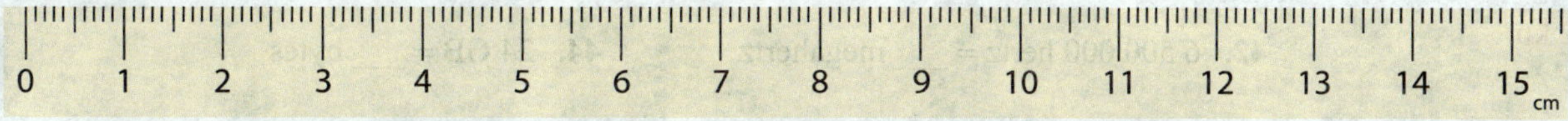

1. The width of a pencil
2. The diameter (distance across) of a penny
3. The width of your thumb
4. The width of your little finger
5. The thickness of a watch band
6. The height of a cell phone
7. The width of a paper clip
8. The thickness of a dime
9. The length and width of a dollar bill
10. The length and width of your school ID

What metric unit of length would you use to measure each item? (Choose from millimeters, meters, and kilometers.)

11. A sprint distance on a track

12. The length of a room

5.2 Exercises

Concept Check

1. threes
2. 0
3. meter
4. division
5. multiply
6. False; To change from smaller units to larger units, division must be used.
7. True
8. False; In metric units, a square that is 1 centimeter long on each side is said to have an area of 1 square centimeter.

Practice

1.-10. Answers will vary.

11. Meters

12. Meters

13. Meters
14. Kilometers
15. Millimeters
16. Millimeters
17. 300
18. 600
19. 80
20. 19
21. 1500
22. 13 600
23. 3.6
24. 14
25. 0.82
26. 0.48
27. 0.0525
28. 0.01977
29. 0.75
30. 0.185
31. 0.245 m
32. 0.087 m
33. 0.23 m
34. 0.32 cm
35. 10 km
36. 11 m
37. 200 m
38. 140 km
39. 0.00679 km
40. 0.003872 km
41. 150.3
42. 6.5
43. 30 000 000
44. 24 000 000 000
45. 960
46. 452
47. 5
48. 0.39
49. 500 000
50. 3 000 000
51. 1300; 130 000
52. 640; 64 000
53. 115 000; 11 500 000

13. The height of a building

14. The distance between Houston, TX, and San Antonio, TX

15. The width of the head of a straight pin

16. The thickness of a stick of gum

Convert each measurement. See Examples 1, 3, and 4.

17. 3 m = ___ cm

18. 60 m = ___ dm

19. 0.8 m = ___ cm

20. 1.9 cm = ___ mm

21. 1.5 m = ___ mm

22. 13.6 km = ___ m

23. 36 mm = ___ cm

24. 140 cm = ___ dm

25. 82 cm = ___ m

26. 4.8 mm = ___ cm

27. 5.25 cm = ___ m

28. 19.77 m = ___ km

29. 750 mm = ___ m

30. 185 m = ___ km

31. Change 245 mm to meters.

32. Change 87 mm to meters.

33. Convert 23 cm to meters.

34. Convert 3.2 mm to centimeters.

35. How many kilometers are in 10 000 m?

36. How many meters are in 1100 cm?

37. What number of meters is equivalent to 20 000 cm?

38. What number of kilometers is equivalent to 140 000 m?

39. Express 679 cm in kilometers.

40. Express 3872 mm in kilometers.

Convert each measurement. See Example 5.

41. 150 300 000 000 bytes = ___ GB

42. 6 500 000 hertz = ___ megahertz

43. 30 MB = ___ bytes

44. 24 GB = ___ bytes

Convert each measurement. See Examples 6 through 9.

45. 9.6 cm^2 = ________ mm^2

46. 4.52 cm^2 = ________ mm^2

47. 500 mm^2 = ________ cm^2

48. 39 mm^2 = ________ cm^2

49. 0.5 m^2 = ________ mm^2

50. 3 m^2 = ________ mm^2

51. 13 dm^2 = ________ cm^2 = ________ mm^2

52. 6.4 dm^2 = ________ cm^2 = ________ mm^2

53. 11.5 m^2 = ________ cm^2 = ________ mm^2

54. 3.6 m^2 = ________ cm^2 = ________ mm^2

55. 0.04 m^2 = ________ cm^2 = ________ mm^2

56. 0.6 m^2 = ________ cm^2 = ________ mm^2

57. 6.7 a = __________ m^2

58. 0.45 a = __________ m^2

59. 200 a = __________ m^2

60. 0.8 a = __________ m^2

61. Change 5.75 km^2 to hectares.

62. Change 0.4 km^2 to hectares.

63. 9.56 ha = __________ a = __________ m^2

64. 0.27 ha = __________ a = __________ m^2

65. 6.25 m^2 = __________ a = __________ ha

66. 35 m^2 = __________ a = __________ ha

Applications

Solve.

67. A triangle has a base measuring 4 cm and a height measuring 16 mm. Determine the area of the triangle in cm^2.

68. For the 10-km-long Bridge Run, officials set up water tables every 635 meters, starting at the start line. How many water tables are set up for the Bridge Run?

69. The current world record for the men's long jump is 8.95 m, set by Mike Powell in 1991. Bob Beamon held the previous record of 8.90 m for 23 years. By how many mm did Mike Powell beat Bob Beamon's record?

70. DeQuan has four pieces of metal that must be welded end-to-end in order to form a spill tray. If the pieces measure 35 cm, 112 mm, 4 decimeters, and 1.2 meters, how long will the tray be (in meters)?

71. A rectangular backyard measures 4 hectometers by 11 dekameters. How many square meters is the backyard?

72. A section of railroad track measuring 2.1 km in length needs to be replaced. Each railroad tie is 4 decimeters wide and they are to be spaced 0.8 m apart. How many railroad ties will be needed to complete this section of track?

73. A certain type of computer processor has a speed of 6 000 000 000 hertz. What is the speed of this computer processor in gigahertz?

74. A demolition crew used 5 megatons of dynamite to clear away a rock formation. How many metric tons of dynamite did they use?

75. Two hard drives contain 3.75 terabytes and 4.25 terabytes of information, respectively. What is the difference in the amount of information stored on the two hard drives in gigabytes? (**Hint:** Convert terabytes to bytes, then to gigabytes.)

76. A camera with 14 megapixels has how many more pixels than a camera with 12 megapixels?

77. A farm has an area of 7500 ares. What is the area of the farm in hectares?

78. Arches National Park has an area of 310 300 000 m^2. What is the area of Arches National Park in square kilometers?

54. 36 000; 3 600 000
55. 400; 40 000
56. 6000; 600 000
57. 670
58. 45
59. 20 000
60. 80
61. 575 ha
62. 40 ha
63. 956; 95 600
64. 27; 2700
65. 0.0625; 0.000625
66. 0.35; 0.0035

Applications

67. 3.2 cm^2
68. 16 tables
69. 50 mm
70. 2.062 m
71. 44 000 square meters
72. 1750 railroad ties
73. 6 gigahertz
74. 5 000 000 metric tons
75. 500 gigabytes
76. 2 000 000 more pixels
77. 75 hectares
78. 310.3 square kilometers

79. 70 000 square meters

80. 3.5332 square centimeters

Writing & Thinking

81. Converting among the U.S. customary systems requires knowledge of the equivalencies and there is no consistency among them so they must be memorized. Within the metric system, typically all that is requred for conversions is to simply move the decimal point the desired number of places according to the prefixes.

82. Conversion of metric units requires merely moving the decimal. If converting from larger units to smaller units, multiply by the appropriate power of 10 to achieve the proper conversion. If converting from smaller units to larger units, divide by the appropriate power of 10 to achieve the proper conversion.

83. Each category of metric units has a base unit. The prefixes determine how many or what fraction of the base unit is being used. For example, the basic unit of length is meter and a millimeter is 1/1000 of a meter, a centimeter is 1/100 of a meter, and a kilometer is 1000 meters.

84. One use is in computer science where symbols and words are measured in bytes. For example, a kilobyte is 1000 bytes, and a megabyte is one million bytes.

79. Two lakes have areas of 1.25 km^2 and 1.18 km^2, respectively. What is the difference in area of the two lakes in square meters?

80. A US nickel has an area of approximately 353.32 mm^2. What is the area of a nickel in square centimeters?

Writing & Thinking

81. Compare and contrast ease of converting units in the US customary system and the metric system.

82. Discuss how to convert units within the metric system.

83. Discuss the meaning of prefixes like milli-, centi-, and kilo- in metric units. Give examples.

84. Discuss the meaning of the prefixes kilo-, mega-, giga-, and tera- in technology.

5.3 The Metric System: Capacity and Weight

Objectives

A. Convert between metric units of capacity.

B. Convert between metric units of weight.

A Units of Capacity (Liquid Volume) in the Metric System

In the metric system, capacity (liquid volume) is measured in **liters** (abbreviated L). You are probably familiar with 1-L and 2-L bottles of soda on your grocer's shelf. Other liquids commonly measured in liters are gasoline, milk, and water. A liter is the volume enclosed in a cube that is 10 cm on each edge. So a capacity of 1 liter is equal to

$$10 \text{ cm} \times 10 \text{ cm} \times 10 \text{ cm} = 1000 \text{ cm}^3 = 1 \text{ L}.$$

That is, the cubic box shown in Figure 1 would hold 1 liter of liquid.

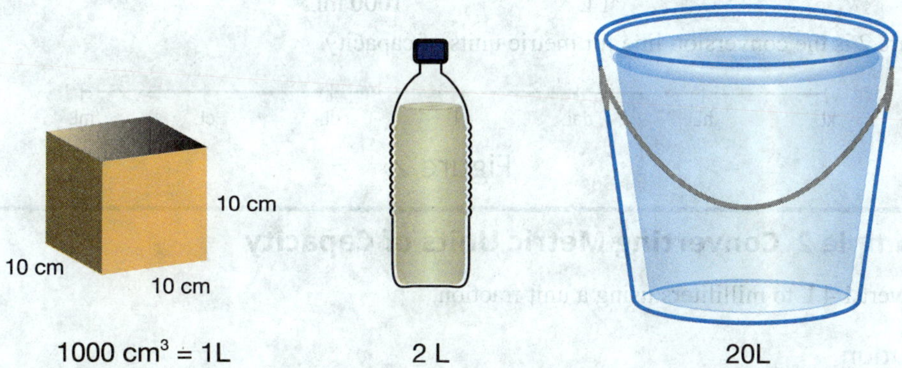

Figure 1

The prefixes kilo-, deka-, deci-, centi-, and milli- all indicate the same part of the liter as they do of the meter. The units of capacity you will see most often are liters (L) and milliliters (mL). A milliliter is equal to 1 cm^3 and is commonly used in dosages of medicine.

Metric Measures of Capacity (Liquid Volume)		
1 **milli**liter (mL)	=	0.001 liter
1 liter (L)	=	1.0 liter
1 **hecto**liter (hL)	=	100 liters
1 **kilo**liter (kL)	=	1000 liters

Table 1

Equivalent Measures of Capacity					
1000 mL	=	1 L	1 mL	=	1 cm^3
1000 L	=	1 kL	1 L	=	1000 cm^3
10 hL	=	1 kL			

Table 2

1. Choose the best metric unit of capacity to complete each statement.

a. You may get 3 __________ of blood drawn.

b. A swimming pool may contain 60 __________ of water.

Example 1 Common Metric Units of Capacity

Choose the best metric unit of capacity to complete each statement.

a. A jug of orange juice might hold 2 _______.

b. Flushing a dog's eyes with saline solution requires 10 _______ of solution.

Solution

a. A jug of orange juice might hold 2 liters (or 2 L).

b. Flushing a dog's eyes with saline solution requires 10 milliliters of solution (or 10 mL).

Now work margin exercise 1.

In converting metric units of capacity, we can use either unit fractions or a metric conversion line just as we did with length. For example, two unit fractions are

$$\frac{1000 \text{ mL}}{1 \text{ L}} = 1 \quad \text{and} \quad \frac{1 \text{ L}}{1000 \text{ mL}} = 1.$$

Figure 2 is the conversion line for metric units of capacity.

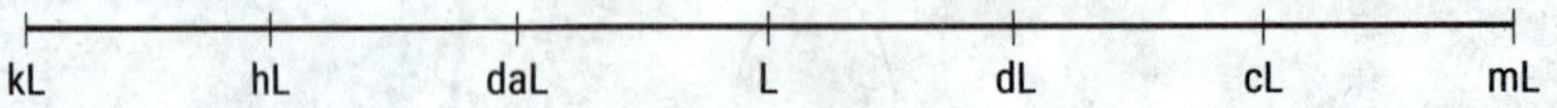

Figure 2

2. Convert 2.81 kL to liters using a unit fraction.

Example 2 Converting Metric Units of Capacity

Convert 1.4 L to milliliters using a unit fraction.

Solution

$$1.4 \text{ L} = 1.4 \not{\text{L}} \cdot \frac{1000 \text{ mL}}{1 \not{\text{L}}} = 1.4 \cdot 1000 \text{ mL} = 1400 \text{ mL}$$

Now work margin exercise 2.

3. Convert 13.3 L to kiloliters using a metric conversion line.

Example 3 Converting Metric Units of Capacity

Convert 2.5 mL to liters using a metric conversion line.

Solution

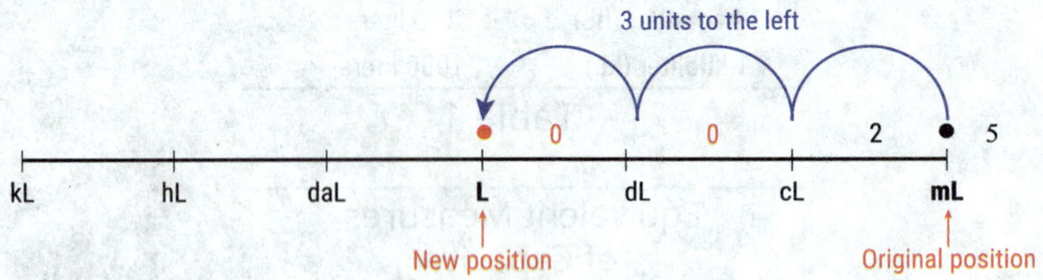

Thus, 2.5 mL = 0.0025 L.

Now work margin exercise 3.

Example 4 Converting Metric Units of Capacity

Convert 3.65 L to milliliters **a.** using a unit fraction and **b.** using a metric conversion line.

Solution

There are 1000 milliliters in 1 liter. We can convert 3.65 L as follows.

a. Using a unit fraction:

$$3.65\text{ L} = 3.65\ \cancel{\text{L}} \cdot \frac{1000\text{ mL}}{1\ \cancel{\text{L}}} = 3.65 \cdot 1000\text{ mL} = 3650\text{ mL}$$

b. Using a metric conversion line:

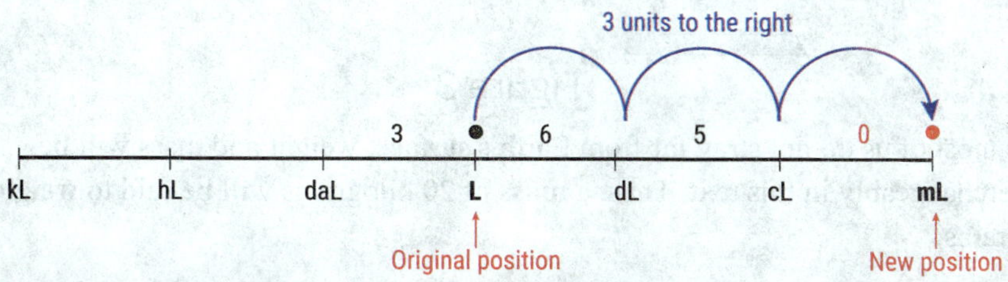

Thus, with either method, we have 3.65 L = 3650 mL.

Now work margin exercise 4.

4. Convert 49 kL to liters using a unit fraction or a metric conversion line.

Example 5 Application: Converting Metric Units of Capacity

To measure the volume of an object, we can measure the amount of water it displaces. During an experiment, a ball bearing displaces 6 mL of water. Convert this amount to liters **a.** using a unit fraction and **b.** using a metric conversion line.

Solution

There are 1000 milliliters in 1 liter. So, the conversion of 6 mL to L can be accomplished as follows.

a. Using a unit fraction:

$$6\text{ mL} = 6\ \cancel{\text{mL}} \cdot \frac{1\text{ L}}{1000\ \cancel{\text{mL}}} = \frac{6}{1000}\text{ L} = 0.006\text{ L}$$

b. Using a metric conversion line:

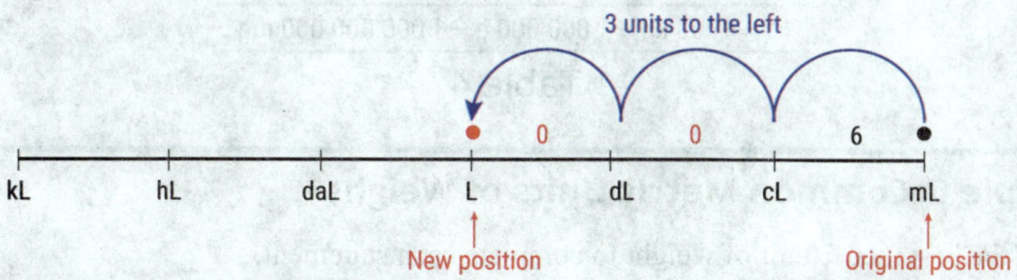

With either method, we see that 6 mL of water is equivalent to 0.006 L of water.

Now work margin exercise 5.

5. The average American consumes 169 208 milliliters of carbonated soft drinks per year. Convert this amount to liters using a unit fraction or a metric conversion line.

Precise Definitions

In November 2018, at the 26th meeting of the General Conference on Weights and Measures in Vienna, the new official definition of the kilogram was unanimously approved. The new definition for the kilogram is expressed in terms of Planck's constant and the speed of light. The previous standard was based on a cylinder made from a platinum alloy, copies of which were held in environmentally monitored safes around the world.

Photo credit: National Institute of Standards and Technology

B Units of Weight (Mass) in the Metric System

Mass is the amount of material in an object. Regardless of where the object is in space, its mass remains the same. (See Figure 3.) **Weight** is the force of Earth's gravitational pull on an object. The farther an object is from Earth, the less the gravitational pull of the Earth. Thus, astronauts experience weightlessness in space, but their mass is unchanged.

The two objects have the same mass and balance on an equal arm balance, regardless of their location in space.

Figure 3

Because most of us do not stray far from Earth's surface, weight and mass will be used interchangeably in this text. Thus, a **mass** of 20 kilograms will be said to **weigh** 20 kilograms.

The basic unit of mass in the metric system is the **kilogram**, which is about 2.2 pounds. In some fields, such as medicine, the **gram** (about the mass of a paper clip) is more convenient as a basic unit than the kilogram. Large masses, such as loaded trucks and railroad cars, are measured by the **metric ton** (1000 kilograms, or about 2200 pounds). (See Tables 3 and 4.)

Metric Measures of Mass		
1 **milli**gram (mg)	=	0.001 gram
1 **centi**gram (cg)	=	0.01 gram
1 **deci**gram (dg)	=	0.1 gram
1 gram (g)	=	1.0 gram
1 **deka**gram (dag)	=	10 grams
1 **hecto**gram (hg)	=	100 grams
1 **kilo**gram (kg)	=	1000 grams
1 metric ton (t)	=	1000 kilograms

Table 3

Equivalent Measures of Mass					
1000 mg	=	1 g	0.001 g	=	1 mg
1000 g	=	1 kg	0.001 kg	=	1 g
1000 kg	=	1 t	0.001 t	=	1 kg
1 t = 1000 kg = 1 000 000 g = 1 000 000 000 mg					

Table 4

Example 6 Common Metric Units of Weight

Choose the best metric unit of weight to complete each statement.

a. A piece of paper might weigh 4.5 ____________.

b. A container of cheese might weigh 0.5 ____________.

6. Choose the best metric unit of weight to complete each statement.

a. An apple may weigh 160 ____________.

b. A textbook may weigh 2.5 ____________.

Solution

a. A piece of paper might weigh 4.5 <u>grams</u>.

b. A container of cheese might weigh 0.5 <u>kilograms</u> (or 0.5 kg).

Now work margin exercise 6.

The most common metric units of weight are grams, kilograms, and milligrams. As with length and capacity, conversions among units of weight can be accomplished using unit fractions and using a metric conversion line. Examples of two unit fractions are

$$\frac{1000\text{ mg}}{1\text{ g}}=1 \quad\text{and}\quad \frac{1\text{ g}}{1000\text{ mg}}=1.$$

Figure 4 is the conversion line for metric units of weight.

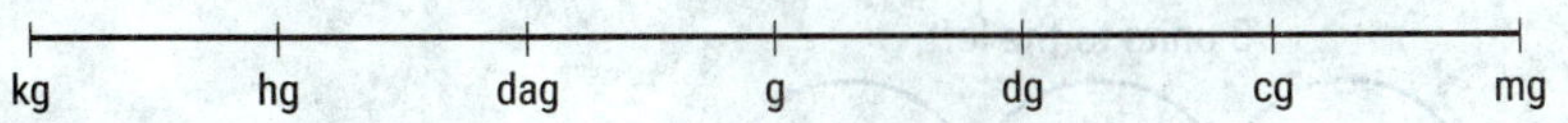

Figure 4

Example 7 Converting Metric Units of Weight

Convert 34 g to milligrams **a.** using a unit fraction and **b.** using a metric conversion line.

Solution

There are 1000 milligrams in 1 gram. So, the conversion of 34 g to milligrams can be accomplished as follows.

a. Using a unit fraction:

$$34\text{ g}=34\ \cancel{\text{g}}\cdot\frac{1000\text{ mg}}{1\ \cancel{\text{g}}}=34\cdot 1000\text{ mg}=34\ 000\text{ mg}$$

b. Using a metric conversion line:

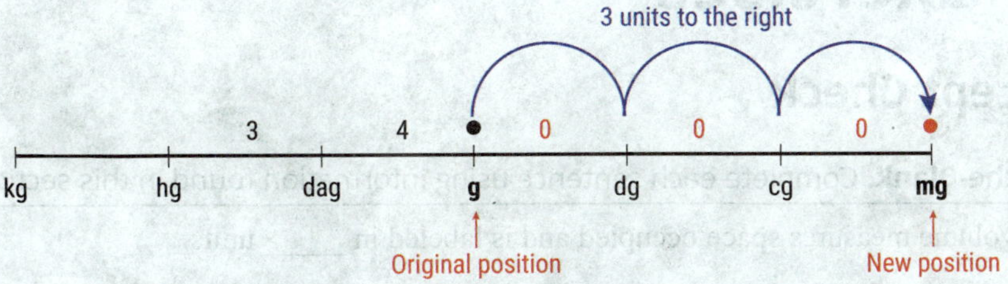

Thus, with either method, we have 34 g = 34 000 mg.

Now work margin exercise 7.

7. Convert 14.9 kg to grams using a unit fraction or a metric conversion line.

8. 825 milligrams of table salt are added to a meal. Convert this mass to grams using a unit fraction or a metric conversion line.

Example 8 Application: Converting Metric Units of Weight

A box of detergent weighs 475 grams. Convert this mass to kilograms **a.** using a unit fraction and **b.** using a metric conversion line.

Solution

There are 1000 grams in 1 kilogram. We can convert 475 g to kg as follows.

a. Using a unit fraction:

$$475\text{ g} = 475\ \cancel{\text{g}} \cdot \frac{1\text{ kg}}{1000\ \cancel{\text{g}}} = \frac{475}{1000}\text{ kg} = 0.475\text{ kg}$$

b. Using a metric conversion line:

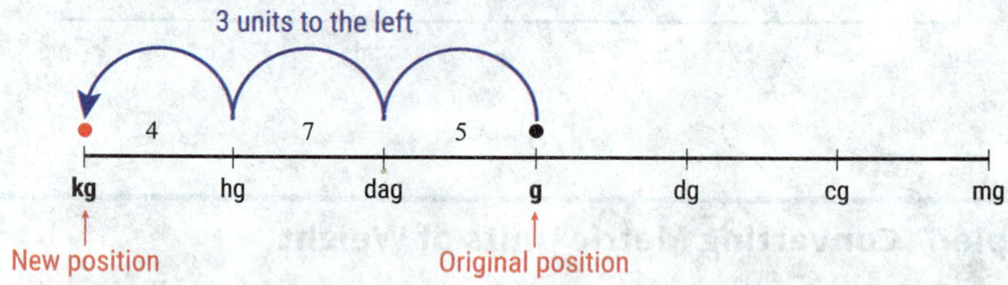

Each method shows that 475 g of detergent is equivalent to 0.475 kg of detergent.

Now work margin exercise 8.

Margin Exercise Answers

1. a. milliliters **b.** kiloliters **2.** 2810 L **3.** 0.0133 kL **4.** 49 000 L **5.** 169.208 L
6. a. grams **b.** kilograms **7.** 14 900 g **8.** 0.825 g

5.3 Exercises

Concept Check

Fill-in-the-Blank. Complete each sentence using information found in this section.

1. Volume measures space occupied and is labeled in _____ units.
2. Liquid volume in the metric system is measured in ______ (L).
3. The force of the Earth's gravitational pull on an object is the object's ______.
4. The amount of material in an object is its _______.
5. The basic unit of mass in the metric system is the _______ (kg).

True/False. Determine whether each statement is true or false. If a statement is false, explain how it can be changed so the statement will be true. (**Note:** There may be more than one acceptable change.)

6. One milliliter is equivalent to one cubic centimeter.
7. Volume is measured in square units.

5.3 Exercises

Concept Check

1. cubic
2. liters
3. weight
4. mass
5. kilogram
6. True
7. False; Volume is measured in cubic units.
8. False; In 1 liter there are 1000 milliliters.
9. False; A metric ton and a US customary ton are not equal (a metric weighs about 2200 US pounds).
10. True

8. In 1 liter there are 100 milliliters.

9. A metric ton and a US customary ton are equal (a metric ton weighs about 2000 US pounds).

10. A dekagram contains 10 grams.

Practice

What metric unit of capacity (liquid volume) would you use to measure each item? (Choose from milliliters, liters, and kiloliters.) See Example 1.

1. A bottle of perfume
2. A bottle of milk
3. A can of motor oil
4. A large fish tank at an aquarium
5. A drop of water
6. A flu shot

Convert each measurement. See Examples 2 through 4.

7. 2 L = ___ mL
8. 25 L = ___ mL
9. 19 mL = ___ L
10. 75 mL = ___ L
11. 13 L = ___ mL
12. 90 L = ___ mL
13. 500 mL = ___ L
14. 600 mL = ___ L
15. 6.3 kL = ___ L
16. 4.27 kL = ___ L
17. 76.4 L = ___ kL
18. 89.2 L = ___ kL
19. 950 mL = ___ L
20. 125 mL = ___ L
21. 1.25 L = ___ mL
22. 5.45 L = ___ mL
23. Change 5.3 L to milliliters.
24. Change 5.3 mL to liters.

What metric unit of weight (mass) would you use to measure each item? (Choose from milligrams, grams, and kilograms.) See Example 6.

25. A suitcase
26. Your weight
27. A cell phone
28. An apple
29. An aspirin
30. A ladybug

Convert each measurement. See Example 7.

31. 2 g = ___ mg
32. 7 kg = ___ g
33. 7.58 t = ___ kg
34. 5.6 t = ___ kg
35. 0.54 g = ___ mg
36. 3.94 g = ___ mg
37. 2000 g = ___ kg
38. 600 mg = ___ g
39. 34.5 mg = ___ g
40. 92.3 g = ___ kg

Practice

1. Milliliters
2. Liters
3. Liters
4. Kiloliters
5. Milliliters
6. Milliliters
7. 2000
8. 25,000
9. 0.019
10. 0.075
11. 13 000
12. 90 000
13. 0.5
14. 0.6
15. 6300
16. 4270
17. 0.0764
18. 0.0892
19. 0.95
20. 0.125
21. 1250
22. 5450
23. 5300 mL
24. 0.0053 L
25. Kilograms
26. Kilograms
27. Grams
28. Grams
29. Milligrams
30. Milligrams
31. 2000
32. 7000
33. 7580
34. 5600
35. 540
36. 3940
37. 2
38. 0.6
39. 0.0345
40. 0.0923

41. 0.091
42. 0.042
43. 4 600 000
44. 19 800 000
45. 2.963
46. 3.547
47. 5000 kg
48. 17 000 kg
49. 96 000 mg
50. 342 000 g
51. 75 kg
52. 3g
53. 0.0016 g
54. 1600 mg
55. 0.000 34 kg
56. 0.00896 g
57. 7 000 000 g
58. 400 000 g
59. m
60. kL
61. L
62. mg
63. km
64. mm
65. L
66. mL
67. g
68. kg
69. 0.006
70. 40
71. 0.4
72. 0.058
73. 9 000 000 cm^3
74. 730 cm^3

Applications

75. 60 doses
76. 600 minutes

41. 91 kg = ___ t

42. 42 kg = ___ t

43. 4.6 kg = ___ mg

44. 19.8 kg = ___ mg

45. 2963 kg = ___ t

46. 3547 kg = ___ t

47. How many kilograms are there in 5 metric tons?

48. How many kilograms are there in 17 metric tons?

49. Express 96 g in milligrams.

50. Express 342 kg in grams.

51. Convert 75 000 g to kilograms.

52. Convert 3000 mg to grams.

53. How many grams are in 1.6 mg?

54. How many milligrams are in 1.6 g?

55. Change 0.34 g to kilograms.

56. Change 8.96 mg to grams.

57. Convert 7 metric tons to grams.

58. Convert 0.4 t to grams.

What metric unit of measurement would you use to measure each item? (Choose from mm, m, km, mL, L, kL, mg, g, and kg.)

59. The depth of a swimming pool

60. The volume of a swimming pool

61. The amount of water in a glass

62. How heavy a microchip is

63. How far you drive to work

64. The diameter of a quarter

65. The volume of toothpaste in a tube

66. The amount of medicine in an injection

67. How heavy an iPhone is

68. The weight of a whale

Convert each measurement. (**Hint:** Since each metric unit is 10 times the next smaller unit, each cubic metric unit is 10 · 10 · 10 = 1000 times the next smaller unit.)

69. 6 dm^3 = ___ m^3

70. 0.04 cm^3 = ___ mm^3

71. 400 cm^3 = ___ L

72. 5800 cm^3 = ___ hL

73. Convert 9 kL to cubic centimeters.

74. Convert 0.73 L to cubic centimeters.

Applications

Solve.

75. How many 5-mL doses of liquid medication can be given from a vial containing 3 deciliters?

76. Salvador's backyard has a swimming pool that holds 960 hectoliters of water. He has a water pump that will pump 16 dekaliters of water per minute into the pool. How long will it take for Salvador to fill his pool?

77. Thomas cleans up after Chemistry lab and part of this job is to empty all of the beakers into a bucket that can be removed for proper disposal. The bucket holds 4 L. There are 5 beakers each containing 50 cL of solution, 3 beakers each with 200 mL of solution, and 2 more beakers with 3 deciliters of solution. Will Thomas be able to dispose of all the leftover solution using just this one bucket? Why or why not?

78. One cup of flour is approximately 120 grams. How many cups of flour can you get out of a bag of flour weighing 2.4 kg?

79. A small dump truck can haul 3 metric tons of sand in one load. If each grain of sand weighs 0.05 mg, how many grains of sand can the dump truck hold in one load?

80. A metric weigh set comes with 10 weights: 2 of them weighing 5 g each, 4 of them weighing 1 g each, and one each weighing 5 mg, 10 mg, 2 cg, and 5 cg. What is the heaviest object that can be weighed using this set (in grams)?

81. Three small fish tanks hold 13 860 cm^3, 28 290 cm^3, and 15 624 cm^3 of water each. How many liters of water would be needed to fill the fish tanks? Round your answer to the nearest tenth of a liter.

82. An apple orchard has plastic bags capable of holding 5 kilograms of weight. If each apple weighs approximately 200 grams, how many bags are needed to package 450 apples?

Writing & Thinking

83. In the metric system, the common unit of capacity is the liter. Discuss how you would change from a measure of liters to milliliters.

84. Conlin said that his mass would change if he lived on Jupiter. Is he correct? Explain.

85. If you were to tell someone what your weight is in metric units, what unit of measure would you use, grams or kilograms? Explain briefly.

77. Yes. The total amount of solution to be disposed is 3.7 L.

78. 20 cups

79. 60 000 000 000 grains

80. 14.085 g

81. 57.8 L

82. 18

Writing & Thinking

83. You could change to milliliters by multiplying by 1000 or by using a unit fraction where the numerator is 1000 mL and the denominator is the given measure in liters.

84. Mass is consistent. Weight is determined by the gravitational pull on an object. Conlin's weight would change if he were any place other than on Earth's surface. His mass would stay the same regardless of where he is located.

85. You would probably use kilograms because a gram is approximately the weight of a paperclip and a kilogram is about 2.2 pounds.

Objectives

A. Convert between US and metric units of temperature.

B. Convert between US and metric units of length.

C. Convert between US and metric units of area.

D. Convert between US and metric units of capacity (liquid volume).

E. Convert between US and metric units of weight (mass).

History

In 1724, Dutch physicist and inventor Daniel Gabriel Fahrenheit proposed a temperature scale that was created by dividing the range from the freezing point of salted water (0 °F) to the temperature of the human body (96 °F) into 96 points. In 1742, Swedish astronomer and physicist Anders Celsius proposed a temperature scale in which water boils at 0 degrees and the temperature at which water freezes is 100 degrees. Two years later, that proposed scale was reversed to become the Celsius scale we know today.

1. Use a straight edge across the two thermometers to convert each temperature.

a. 0 °C

b. 50 °F

5.4 US and Metric Equivalents

A Temperature

We begin the following discussion of equivalent measures between the US customary and metric systems with measures of temperature.

Temperature

US customary measure is in **degrees Fahrenheit** (°F).

Metric measure is in **degrees Celsius** (°C).

DEFINITION

The two scales are shown on the thermometers in Figure 1. Approximate conversions can be found by reading along a ruler or the edge of a piece of paper held horizontally across the page.

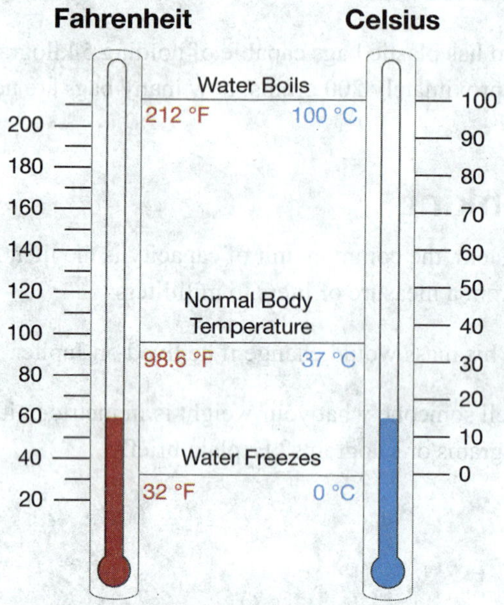

Figure 1

Example 1 Equivalent Measures of Temperature

Hold a straight edge horizontally across the thermometers in Figure 1 and you will read the following.

100 °C = 212 °F	Water boils at sea level
40 °C = 104 °F	Hot day in the desert
20 °C = 68 °F	Comfortable room temperature

Now work margin exercise 1.

The two thermometers in Figure 1 illustrate various relationships between the Celsius and Fahrenheit scales. However, the following two formulas serve to provide exact conversions between the two scales.

Temperature Formulas

F = Fahrenheit temperature and C = Celsius temperature

$$F = \frac{9 \cdot C}{5} + 32 \qquad C = \frac{5(F - 32)}{9}$$

FORMULA

Example 2 Converting Units of Temperature

Convert $F = 86°$ to its equivalent measurement in degrees Celsius.

Solution

$$C = \frac{5(F-32)}{9} = \frac{5((86)-32)}{9} = \frac{5(54)}{9} = 30$$

Thus, 86 °F = 30 °C.

Now work margin exercise 2.

2. Convert $F = 77°$ to its equivalent measurement in degrees Celsius.

Example 3 Converting Units of Temperature

Convert $C = 35°$ to its equivalent measurement in degrees Fahrenheit.

Solution

$$F = \frac{9 \cdot C}{5} + 32 = \frac{9 \cdot (35)}{5} + 32 = 63 + 32 = 95$$

Thus, 35 °C = 95 °F.

Now work margin exercise 3.

3. Convert $C = 45°$ to its equivalent measurement in degrees Fahrenheit.

B Length

Units of length in the US customary system were discussed in Section 5.1 and units of length in the metric system were discussed in Section 5.2. In this section, we discuss conversion techniques between the two systems. Approximate conversions (rounded) are given in Table 1. (The one exact conversion is 1 in. = 2.54 cm.)

Note

Most of the conversion units are not exact and slightly different answers are possible, depending on the conversion units used. Generally, we will round conversions to the nearest hundredth.

Measurement Conversions

US to Metric	Metric to US
1 in. = 2.54 cm (exact)	1 cm ≈ 0.394 in.
1 ft ≈ 0.305 m	1 m ≈ 3.28 ft
1 yd ≈ 0.914 m	1 m ≈ 1.09 yd
1 mi ≈ 1.61 km	1 km ≈ 0.62 mi

Table 1

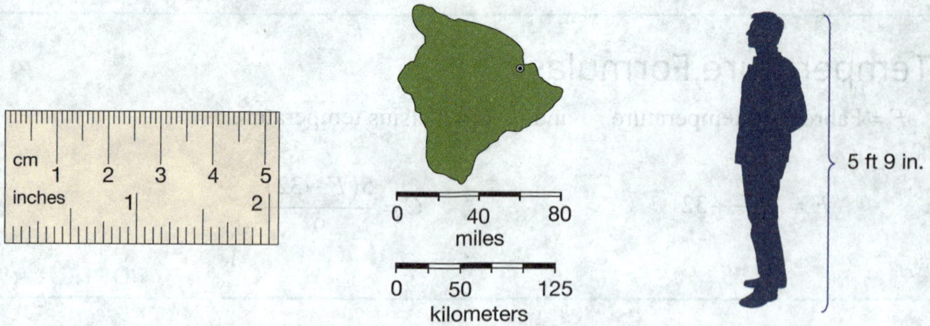

Ruler — 1 in. = 2.54 cm

Hawaii — 80 mi ≈ 128.8 km

Man — 5 ft 9 in. = 175.26 cm

Figure 2

In converting from one system of measure to the other, we will use unit fractions based on the equivalent measures given in the tables. For example,

$$\frac{2.54\text{ cm}}{1\text{ in.}} = 1 \quad \text{and} \quad \frac{1\text{ mi}}{1.61\text{ km}} \approx 1.$$

4. Convert each measurement, rounding to the nearest hundredth.

a. 84 ft = ________ m

b. 100 yd = ________ m

c. 75 cm = ________ in.

d. 500 m = ________ ft

Example 4 Converting Units of Length

Convert each measurement, rounding to the nearest hundredth.

a. 6 in. = ______ cm

b. 25 mi = ______ km

c. 10 km = ______ mi

d. 30 m = ______ ft

e. 100 m = ______ yd

Solution

a. $6\text{ in.} = 6\ \cancel{\text{in.}} \cdot \frac{2.54\text{ cm}}{1\ \cancel{\text{in.}}} = 6 \cdot 2.54\text{ cm} = 15.24\text{ cm}$

b. $25\text{ mi} \approx 25\ \cancel{\text{mi}} \cdot \frac{1.61\text{ km}}{1\ \cancel{\text{mi}}} = 25 \cdot 1.61\text{ km} = 40.25\text{ km}$

c. $10\text{ km} \approx 10\ \cancel{\text{km}} \cdot \frac{0.62\text{ mi}}{1\ \cancel{\text{km}}} = 10 \cdot 0.62\text{ mi} = 6.2\text{ mi}$

d. $30\text{ m} \approx 30\ \cancel{\text{m}} \cdot \frac{1\text{ ft}}{0.305\ \cancel{\text{m}}} = \frac{30}{0.305}\text{ ft} \approx 98.36\text{ ft}$

e. $100\text{ m} \approx 100\ \cancel{\text{m}} \cdot \frac{1\text{ yd}}{0.914\ \cancel{\text{m}}} = \frac{100}{0.914}\text{ yd} \approx 109.41\text{ yd}$

With this information, do you think that a 100 m dash is longer (or shorter) than a 100 yd dash?

Now work margin exercise 4.

Math Tip

Multiplying by a conversion factor greater than 1 will give you a bigger number. This goes along with the intuition that converting from a larger unit to a smaller unit will give more of that smaller unit. Multiplying by a number less than 1 gives a smaller number. This goes along with the intuition that converting from a smaller unit to a larger unit will give less of that larger unit.

C Area

As with units of length, units of area in the US customary system are not systematically related as they are in the metric system. Table 2 shows some of the basic units of area in the US customary system.

US Customary Units of Area
1 ft² = 144 in.²
1 yd² = 9 ft²
1 acre = 4840 yd² = 43,560 ft²

Table 2

We will use the equivalent measures of length (rounded) in Table 1 to convert units of area from the US customary system to metric units of area, and vice versa.

Measurement Conversions

US to Metric	Metric to US
1 in.² ≈ 6.45 cm²	1 cm² ≈ 0.155 in.²
1 ft² ≈ 0.093 m²	1 m² ≈ 10.764 ft²
1 yd² ≈ 0.836 m²	1 m² ≈ 1.196 yd²
1 acre ≈ 0.405 ha	1 ha ≈ 2.47 acres

Table 3

Figure 3

Example 5 Converting Units of Area

Convert each measurement, rounding to the nearest hundredth.

a. 40 yd² = ______ m²

b. 100 cm² = ______ in.²

c. 6 acres = ______ ha

d. 5 ha = ______ acres

Solution

a. $40\text{ yd}^2 \approx 40\ \cancel{\text{yd}^2} \cdot \dfrac{0.836\text{ m}^2}{1\ \cancel{\text{yd}^2}} = 40 \cdot 0.836\text{ m}^2 = 33.44\text{ m}^2$

b. $100\text{ cm}^2 \approx 100\ \cancel{\text{cm}^2} \cdot \dfrac{0.155\text{ in.}^2}{1\ \cancel{\text{cm}^2}} = 100 \cdot 0.155\text{ in.}^2 = 15.5\text{ in.}^2$

c. $6\text{ acres} \approx 6\ \cancel{\text{acres}} \cdot \dfrac{0.405\text{ ha}}{1\ \cancel{\text{acre}}} = 6 \cdot 0.405\text{ ha} = 2.43\text{ ha}$

5. Convert each measurement, rounding to the nearest hundredth.

a. 53 in.² = ________ cm²

b. 50 m² = ________ ft²

c. 16 acres = ______ ha

d. 3 ha = ________ acres

d. $5 \text{ ha} \approx 5 \text{ ha} \cdot \dfrac{2.47 \text{ acres}}{1 \text{ ha}} = 5 \cdot 2.47 \text{ acres} = 12.35 \text{ acres}$

Now work margin exercise 5.

D Capacity (Liquid Volume)

Just as when converting units of length, we use a table of conversions (rounded) between the two systems. See Table 4.

Measurement Conversions

US to Metric	Metric to US
1 qt ≈ 0.946 L	1 L ≈ 1.06 qt
1 gal ≈ 3.785 L	1 L ≈ 0.264 gal

1 mL is equal to 1 cubic centimeter (cc or cm^3).
These units are commonly used in medicine.

Table 4

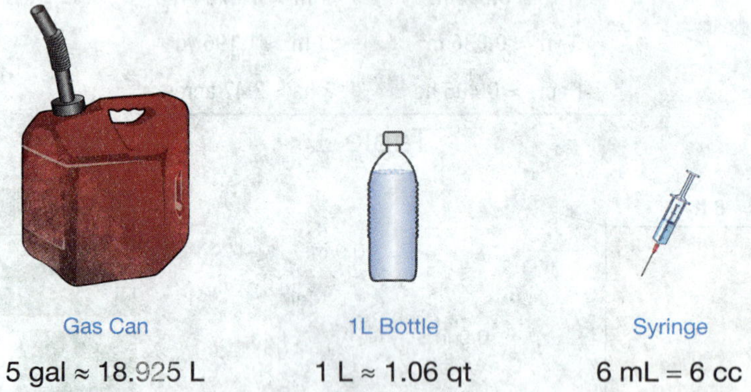

Figure 4

6. Convert each measurement, rounding to the nearest hundredth.

a. 4 gal = ________ L

b. 16 L = ________ qt

c. 28 qt = ________ L

Example 6 Converting Units of Capacity (Liquid Volume)

Convert each measurement, rounding to the nearest hundredth.

a. 20 gal = _____ L **b.** 42 L = _____ gal **c.** 6 qt = _____ L

Solution

a. $20 \text{ gal} \approx 20 \text{ gal} \cdot \dfrac{3.785 \text{ L}}{1 \text{ gal}} = 20 \cdot 3.785 \text{ L} = 75.7 \text{ L}$

b. $42 \text{ L} \approx 42 \text{ L} \cdot \dfrac{0.264 \text{ gal}}{1 \text{ L}} = 42 \cdot 0.264 \text{ gal} \approx 11.09 \text{ gal}$

c. $6 \text{ qt} \approx 6 \text{ qt} \cdot \dfrac{0.946 \text{ L}}{1 \text{ qt}} = 6 \cdot 0.946 \text{ L} \approx 5.68 \text{ L}$

Now work margin exercise 6.

E Weight (Mass)

Table 5 gives approximate conversions of weight (mass) in the US Customary and Metric systems.

Measurement Conversions

US to Metric	Metric to US
1 oz ≈ 28.35 g	1 g ≈ 0.035 oz
1 lb ≈ 0.454 kg	1 kg ≈ 2.205 lb

Table 5

Bag of Fertilizer — Road Bike

25 lb ≈ 11.35 kg — 9 kg ≈ 19.845 lb

Figure 5

Example 7 Converting Units of Weight (Mass)

Convert each measurement, rounding to the nearest hundredth.

a. 5 lb = _____ kg **b.** 15 kg = _____ lb **c.** 35 oz = _____ g

Solution

a. $5 \text{ lb} \approx 5 \cancel{\text{lb}} \cdot \frac{0.454 \text{ kg}}{1 \cancel{\text{lb}}} = 5 \cdot 0.454 \text{ kg} = 2.27 \text{ kg}$

b. $15 \text{ kg} \approx 15 \cancel{\text{kg}} \cdot \frac{2.205 \text{ lb}}{1 \cancel{\text{kg}}} = 15 \cdot 2.205 \text{ lb} \approx 33.08 \text{ lb}$

c. $35 \text{ oz} \approx 35 \cancel{\text{oz}} \cdot \frac{28.35 \text{ g}}{1 \cancel{\text{oz}}} = 35 \cdot 28.35 \text{ g} = 992.25 \text{ g}$

Now work margin exercise 7.

7. Convert each measurement, rounding to the nearest hundredth.

a. 3 oz = ________ g

b. 42 g = ________ oz

c. 22 kg = ________ lb

Margin Exercise Answers

1. a. 32 °F **b.** 10 °C **2.** 25 °C **3.** 113 °F **4. a.** 25.62 **b.** 91.4 **c.** 29.55 **d.** 1640 **5. a.** 341.85 **b.** 538.2 **c.** 6.48 **d.** 7.41 **6. a.** 15.14 **b.** 16.96 **c.** 26.49 **7. a.** 85.05 **b.** 1.47 **c.** 48.51

5.4 Exercises

Concept Check

1. Fahrenheit
2. Celsius
3. 2.54
4. 0.093
5. 0.946
6. 2.205
7. False; Water freezes at 32 degrees Fahrenheit.
8. True
9. False; A 5K (km) run is shorter than a 5 mile run.
10. True

Practice

1. 77
2. 176
3. 50
4. 32
5. 45
6. 200
7. 122
8. 95
9. 0 °C
10. 5 °C
11. 59 °F
12. 86 °F
13. 2.74
14. 4.57
15. 11.99
16. 8.72
17. 83.82
18. 53.34
19. 27.88
20. 131.23
21. 32.19
22. 56.33
23. 2.74 m
24. 4.57 m
25. 96.6 km
26. 62.14 km
27. 124.22 mi

5.4 Exercises

Concept Check

Fill-in-the-Blank. Complete each sentence using information found in this section.

1. In the US customary system, temperature is measured in degrees _______.
2. In the metric system, temperature is measured in degrees _______.
3. One inch equals exactly ____ centimeters.
4. One square foot is about _____ square meters.
5. One quart is approximately _____ liters.
6. One kilogram is approximately _____ pounds.

True/False. Determine whether each statement is true or false. If a statement is false, explain how it can be changed so the statement will be true. (**Note:** There may be more than one acceptable change.)

7. Water freezes at 32 degrees Celsius.
8. When converting between US customary and metric units, often the results will be approximations.
9. A 5K (km) run is longer than a 5 mile run.
10. One square meter covers more area than one square yard.

Practice

Convert each measurement. Round to the nearest hundredth, if necessary. See Examples 2 and 3.

1. 25 °C = ___ °F
2. 80 °C = ___ °F
3. 10 °C = ___ °F
4. 0 °C = ___ °F
5. 113 °F = ___ °C
6. 392 °F = ___ °C
7. 50 °C = ___ °F
8. 35 °C = ___ °F
9. Change 32 °F to degrees Celsius.
10. Change 41 °F to degrees Celsius.
11. Change 15 °C to degrees Fahrenheit.
12. Change 30 °C to degrees Fahrenheit.

Convert each measurement. Round to the nearest hundredth, if necessary. See Example 4.

13. 9 ft = ___ m
14. 15 ft = ___ m
15. 11 m = ___ yd
16. 8 m = ___ yd
17. 33 in. = ___ cm
18. 21 in. = ___ cm
19. 8.5 m = ___ ft
20. 40 m = ___ ft
21. 20 mi = ___ km
22. 35 mi = ___ km
23. How many meters are in 3 yd?
24. How many meters are in 5 yd?
25. Change 60 miles to kilometers.

26. Change 100 kilometers to miles.

27. Convert 200 kilometers to miles.

28. Convert 100 miles to kilometers.

29. How many inches are in 50 cm?

30. How many inches are in 100 cm?

31. Convert 14 inches to centimeters.

32. Convert 14 centimeters to inches.

Convert each measurement. Round to the nearest hundredth, if necessary. See Example 5.

33. 3 in.^{2} = ________ cm^2

34. 16 in.^{2} = ________ cm^2

35. 600 ft^2 = ________ m^2

36. 300 ft^2 = ________ m^2

37. 100 yd^2 = ________ m^2

38. 250 yd^2 = ________ m^2

39. 1000 acres = ________ ha

40. 250 acres = ________ ha

41. How many acres are in 300 ha?

42. How many acres are in 400 ha?

43. Change 5 m^2 to square feet.

44. Change 10 m^2 to square feet.

45. Change 30 cm^2 to square inches.

46. Change 50 cm^2 to square inches.

Convert each measurement. Round to the nearest hundredth, if necessary. See Example 6.

47. 4 qt = ___ L

48. 7 qt = ___ L

49. 4 L = ___ gal

50. 9 L = ___ gal

51. 10 qt = ___ L

52. 20 qt = ___ L

53. 78 mL = ___ cc

54. 91 cc = ___ mL

55. 42 L = ___ gal

56. 50 L = ___ gal

57. How many quarts are in 10 L?

58. How many quarts are in 25 L?

59. How many liters are in 50 gal?

60. How many liters are in 36 gal?

Convert each measurement. Round to the nearest hundredth, if necessary. See Example 7.

61. 33 kg = ___ lb

62. 95 kg = ___ lb

63. 35 oz = ___ g

64. 55 oz = ___ g

65. 10 lb = ___ kg

66. 70 lb = ___ kg

67. 100 g = ___ oz

68. 53 g = ___ oz

69. Convert 16 oz to grams.

70. Convert 64 oz to grams.

71. How many pounds are in 120 kg?

72. How many pounds are in 500 kg?

28. 160.93 km
29. 19.69 in.
30. 39.37 in.
31. 35.56 cm
32. 5.51 in.
33. 19.35
34. 103.23
35. 55.74
36. 27.87
37. 83.61
38. 209
39. 405
40. 101.25
41. 741.32 acres
42. 988.42 acres
43. 53.82 ft^2
44. 107.64 ft^2
45. 4.65 in.^{2}
46. 7.75 in.^{2}
47. 3.79
48. 6.62
49. 1.06
50. 9.54
51. 9.46
52. 18.92
53. 78
54. 91
55. 11.10
56. 13.21
57. 10.57 qt
58. 26.42 qt
59. 189.25 L
60. 136.28 L
61. 72.75
62. 209.44
63. 992.23
64. 0.43
65. 4.56
66. 31.75
67. 3.53
68. 1.87

69. 453.59 g
70. 1814.37 g
71. 264.55 lb
72. 1102.31 lb

Applications

73. 177 °C
74. 75 degrees Fahrenheit
75. 9.7 miles per hour
76. 109 meters
77. 226.3 km
78. 1.9 cm wide, 0.8 cm deep, 3.6 cm long
79. 3035.14 m²
80. 1.215 hectares
81. a. 11 145.6 cm²
b. 1.116 m²
82. 51 times

Applications

Solve.

73. While visiting her aunt in Germany, Helga wants to surprise her aunt with a cake. She brought her mom's cake recipe with her from Georgia. The recipe says to bake the cake at 350 degrees Fahrenheit but the temperature gauge on her aunt's oven is in degrees Celsius. What temperature should Helga set her aunt's oven to in order to bake the cake at the correct temperature? Round the temperature to the nearest degree.

74. Michael, who lives in Fargo, North Dakota, is packing for his trip to France. He looks up the weather for the duration of his trip to help him determine what type of clothing to pack. The average temperature for the week he will be in France is 24 degrees Celsius. What will the average temperature be in degrees Fahrenheit? Round the temperature to the nearest degree.

75. The Palace at Versailles is 23.5 km from the center of the city of Paris, France. If it takes Pierre 1.5 hours to bike between the two places, what is his speed in miles per hour (to the nearest tenth)?

76. Roger Bannister is famous for being the first man to break the 4-minute mile. He ran the mile in 3 minutes and 59 seconds. Today, most track meets do not have a one-mile race. Instead, they have the "metric mile" race, which is 1500 meters. How much further is the actual mile race than the 1500-meter race (rounded to the nearest whole meter)?

77. The Ironman Triathlon championship in Hawaii consists of a swim of 3.86 km, a bike ride of 180.25 km, and finishes with a run equal to the length of a standard marathon. A marathon is typically 26.2 miles. What is the total length of the Ironman Triathlon in kilometers? Round the length to the nearest tenth of a km.

78. Darren needs to retool his widget maker so that it takes measurements in centimeters instead of inches. The widget maker currently makes widgets that are $\frac{3}{4}$ of an inch wide, $\frac{1}{3}$ of an inch deep, and $1\frac{3}{7}$ inches long. Determine the measurements of the widgets in centimeters (to the nearest tenth of a centimeter).

79. Suppose that the home you are buying sits on a rectangular shaped lot that is 270 feet by 121 feet. Convert this area to square meters.

80. A new manufacturing building covers an area of 3 acres. How many hectares of ground does the new building cover?

81. A painting of a landscape is on a rectangular canvas that measures 3 feet by 4 feet.

a. How many square centimeters of wall space will the painting cover when it is hanging?

b. How many square meters?

82. A turkey baster holds 150 mL of liquid. How many times would the turkey baster need to be filled in order to empty a 2-gallon bin of water from a dehumidifier?

83. You are making a large pot of soup and the recipe calls for 2 liters of vegetable stock. (Stock is similar to broth.) If the only cans of vegetable stock available at the super market hold 10 fluid ounces, how many cans will you need? (**Hint:** See Section 5.1 for conversions within the US Customary system.)

84. Rachel has taken up French cooking and is making chocolate *éclairs*. Since the recipe is in French, the measurements are given using metric units. The *éclair* recipe calls for 150 g butter, 225 mL water, 225 g flour, and 300 mL heavy cream. Convert the measurements given in grams into ounces and the measurements given in milliliters into fluid ounces. Round each measurement to the nearest whole number.

85. Kristy is shopping for gourmet chocolates online. The website of her favorite chocolatier only lists the mass of each box of chocolate in grams. A small box of chocolates has a mass of 153 grams, a medium box has a mass of 309 grams, and a large box has a mass of 595.4 grams. Find the weight in ounces of each box of chocolates. Round each weight to the nearest tenth.

86. How many liters of water are needed to fill a 20-gallon fish tank?

87. Biologists inject a solution of hormones into female fish to encourage egg development. They inject 0.1 milliliters of solution per kilogram of the fish's body weight. How many milliliters of solution should they use for a 60-pound fish? Round your answer to the nearest tenth of a milliliter.

Writing & Thinking

88. Abby said that water boils at 100 degrees. Joel said it boils at 212 degrees. Who is correct and why?

89. Paola mistakenly thought that a meter stick was the same as a yard stick. Explain her mistake.

90. Most conversions between the US customary system of measure and metric system are not exact. Explain why this is true and give any exceptions.

91. Kai and Kristen were converting between US customary measures and metric measures. Their answers were close but not the same. The teacher said they were both right. How could this be?

83. 7 cans

84. 5 oz butter, 8 fl oz water, 8 oz flour, 10 fl oz heavy cream

85. Answers will vary. Small: 5.4 ounces; Medium: 10.8 ounces or 10.9 ounces; Large: 1 pound 5 ounces or (21 ounces)

86. 75.7 L

87. 2.7 mL

Writing & Thinking

88. Both would be correct if Abby was referring to degrees Celsius and Joel was referring to degrees Fahrenheit.

89. One meter is equivalent to 1.09 yards so they are close in length.

90. Most US and metric measurement conversions are not exact numbers. Therefore, most of the conversions are approximations. The exception to this "rule" is that 1 inch = 2.54 cm

91. Because conversions between the metric system and the US system require rounding, it is possible that two answers could be right although they might be slightly different. If Kai used the US to Metric equivalent and Kristen used the Metric to US equivalent, they would both have a correct answer but their answers will be slightly different answers because they used different approximations.

Chapter 5 Project

Metric Cooking

An activity to demonstrate the use of metric to US conversions in real life.

Your grandma lives outside of the United States and has emailed you her famous apple pie recipe to use at an upcoming party. As you start to bake the pie on the day of the party, you realize grandma's recipe is written using only metric units. Looking through the supplies in your kitchen, you find a scale that measures weight in ounces and some measuring spoons that measure volume in $\frac{1}{8}$, $\frac{1}{4}$, and $\frac{1}{2}$ of a teaspoon and 1 whole tablespoon. In order to successfully bake the pie in time for the party, you must quickly convert the metric measurements to US measurements. You start with the pie crust ingredients.

Pie Crust

Ingredient	Metric Measurement	US Measurement
Flour	280 g	9.8 oz
Vegetable Shortening	90 g	3.15 oz
Unsalted Butter	50 g	1.75 oz
Cold Water	90 mL	6.12 tbsp
Salt	5 mL	1.015 tsp

1. Fill in the third column of the pie crust table by converting the measurements of each ingredient using the correct conversion factors. Use 1 mL = 0.068 tbsp and 1 mL = 0.203 tsp. (**Note:** tbsp stands for tablespoon and tsp stands for teaspoon.)

2. The recipe requires the oven to be preheated to 230 °C, but your oven measures degrees in Fahrenheit. What temperature should you preheat your oven to? 446 °F

   While the oven is preheating, you begin to prepare the ingredients for the apple filling.

Apple Filling

Ingredient	Metric Measurement	US Measurement
Apples	1 kg	35 oz
Sugar	100 g	3.5 oz
Cornstarch	15 mL	3.045 tsp
Cinnamon	2.5 mL	0.5075 tsp
Salt	0.5 mL	0.1015 tsp
Nutmeg	0.5 mL	0.1015 tsp
Butter	25 g	0.875 oz

3. Fill in the third column of the apple filling table by converting the measurements of each ingredient using the correct conversion factors. Use 1 mL = 0.068 tbsp and 1 mL = 0.203 tsp.

4. Since the measuring spoons can only measure $\frac{1}{8}$, $\frac{1}{4}$, and $\frac{1}{2}$ of a teaspoon and 1 whole tablespoon, what is the most reasonable way to round each US volume measurement in each of the tables? Answers will vary.

In addition to the usual recipe, your grandma has listed several variations on the recipe that depend on the taste of the apples. For bland apples, you should add 20 mL of lemon juice to the filling. For sour apples, you should increase the sugar to 140 g.

5. You taste the apples and decide you need to increase the sugar to 140 g. You have already added 100 g. How many more ounces of sugar do you need to add to reach 140 g? 1.4 oz

6. You notice that the recipe requires a pie pan with a diameter of 23 cm. After measuring the pie pan you discover it is 9 inches in diameter. Is the pie pan the right size for the apple pie? What is its diameter in inches? Round to the nearest whole number. Yes, 9 inches

Chapter 5 Project

Confused Conversions

An activity to investigate the proper use of conversions between systems of measurements in real life.

Imagine talking to your European friend and you remark that the daily high was 30 degrees. Things can get a little confusing: 30 °F is a cold snap while 30 °C is a very warm summer day. This confusion could be avoided by making sure that you are clear about which system's units you are using.

Could confusing measurements between the metric system and the US system cause more harm than just a misunderstanding? We will explore some real-life situations where things didn't go very well.

1. In 1983, Air Canada Flight 143, the infamous Gimli Glider made an emergency landing in Gimli, Manitoba, when it ran out of fuel midair. The mistake can be traced to a refueling procedure: the plane was supposed to be fueled using kilograms, but it was instead refueled using pounds.
   - **a.** Considering that there are approximately 0.454 kilograms in one pound, determine the number of kilograms in 100 pounds of fuel.
   - **b.** Suppose an empty tank can hold 400 kilograms of fuel. Instead, 400 pounds of fuel are added to the tank. How many more kilograms of fuel would be needed to get a full tank?
2. In 1998, NASA had a slight miscommunication with the Mars Climate Orbiter. This was a $125-million spacecraft designed to study Mars's atmosphere. It was forever lost in space once it was made to accelerate too quickly.
   - **a.** The standard metric unit for impulse is the Newton-second (N · s) while the customary English unit is the pound-second (lb · s). We know that one Newton is approximately equal to 4.45 pounds. What is the value of 250 N · s in lb · s?
   - **b.** The Mars Climate Orbiter was calibrated to receive impulse information in N · s but NASA inadvertently converted the number to lb · s. How many times larger was the impulse value that was relayed to the orbiter compared to the correct value.
3. A grain is a unit of measure equal to about 0.065 grams. In 1999, the Institute for Safe Medication Practices reported a case of a patient who received 0.5 *grams* of phenobarbital (a sedative) instead of the prescribed 0.5 *grains*.
   - **a.** Determine the dose in grains of 0.5 grams of phenobarbital. Round to the nearest tenth.
   - **b.** What is a possible consequence of giving a patient 0.5 grams instead of 0.5 grains?
4. Perform an internet search for "Verizon cents versus dollars." This is a video of a customer disputing their data charges with a cellphone carrier. Explain how his billing issue is related to our confused conversions problems.
5. Have you ever found yourself in a situation where a misunderstanding about units and conversions created a problem? Either describe your own situation or research one on the internet.

Chapter 5 Project: Confused Conversions

1. **a.** Since $1\text{ lb} = 0.454\text{ kg}$, $100\text{ lb} \cdot \dfrac{0.454\text{ kg}}{1\text{ lb}} = 45.4\text{ kg}.$
   b. 400 pounds of fuel are equal to $4(45.4\text{ kg}) = 181.6\text{ kg}$ of fuel. So, we would still need another 218.4 kg to fill up the tank.
2. **a.** $1\text{ N} = 4.45\text{ lb}$ so $(250\text{ N}\cdot\text{s}) \cdot \dfrac{4.45\text{ lb}}{1\text{ N}} = 1112.5\text{ lb}\cdot\text{s}$
   b. 4.45 times larger.
3. **a.** $1\text{ grain} = 0.065\text{ g}$ so $0.5\text{ g} \cdot \dfrac{1\text{ grain}}{0.065\text{ g}} \approx 7.7\text{ grains}$
   b. The amount received by the patient is over 10 times the amount prescribed which can lead to a fatal overdose.
4. The customer representatives are confusing 0.002 cents with 0.002 dollars. These values are 100 times apart.
5. Answers will vary.

Chapter 5 Project

Confused Conversions

[illegible] the [illegible] your friend [illegible] that the [illegible] was 80 degrees. Things can get a bit confusing [illegible] could be avoided by making sure that you [illegible]

[illegible] conversions [illegible] between the metric system and the U.S. system [illegible] a better understanding [illegible] some real-life situations [illegible] as well.

1. In 1999, [illegible] landing on [illegible] Mars [illegible] the [illegible]

   a. Considering that there are approximately 0.453 kilograms in a pound, [illegible] determine the number of [illegible]

   b. Suppose [illegible] 400 [illegible] How many [illegible] would [illegible]

2. In 1999, NASA [illegible] with the Mars Climate Orbiter [illegible]

   a. [illegible] equal to 4.45 [illegible] What is the [illegible]

   b. The Mars Climate Orbiter was [illegible]

5. A unit of measure [illegible] 0.06 grams. In 1999, [illegible]

   a. [illegible]

   b. [illegible]

6. [illegible]

CHAPTER 6

6 Geometry

Connections

Sometimes geometry can feel a little removed from our day to day lives. You might catch yourself asking questions such as "When am I ever going to use the formula for the area of a rectangle?" or "Why do I need to know the characteristics of right triangles?" Upon closer inspection of our surroundings, it is very hard to deny that geometry is present in many aspects of modern life. This topic is present in a variety of situations, such as a family deciding how much paint to purchase to repaint their living room and a town government budgeting for the construction of a new school building.

The next time you walk through a building, look around and notice the geometric shapes that were used in its construction. The walls, ceiling, and floors are likely rectangles. Many stairways are supported by triangular structures. And windows can be squares, rectangles, or circles.

Suppose your college decides to paint all four walls of a classroom that is 25 ft by 20 ft with walls that are 10 ft high. The administration has selected a paint with a coverage of 350 ft^2 per gallon. How would you determine many gallons of paint will be needed to cover the classroom walls if you ignore all openings such as windows and doors?

Objectives

A. Recognize points, lines, planes, rays, and angles.

B. Classify an angle by its measure.

C. Recognize complementary and supplementary angles.

D. Recognize congruent, vertical, and adjacent angles.

E. Know when lines are parallel and perpendicular.

F. Classify a triangle by its sides and angles.

6.1 Angles and Triangles

A Introduction to Geometry

Plane geometry is the study of the properties of figures in a plane. The three most basic ideas in plane geometry are **point**, **line**, and **plane**. These terms are considered so fundamental that they are simply called **undefined terms**. These undefined terms provide the foundation for the study of geometry and the definitions of other geometric figures such as **line segment**, **ray**, **angle**, **triangle**, **polygon**, and so on.

Point, Line, Plane

Undefined Term	Representation	Discussion
Point	A • Point A	A point is represented by a dot. Points are labeled with capital letters.
Line	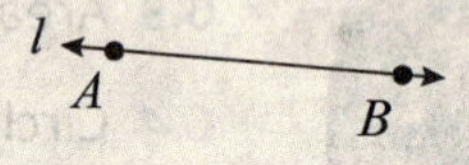 Line l or line $\overleftrightarrow{AB}$	A line has no beginning or end. Lines are labeled with lowercase letters or by two points on the line.
Plane	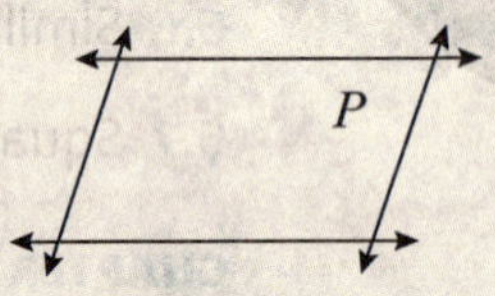 Plane P	Flat surfaces, such as a table top or wall, represent portions of planes. Planes are labeled with capital letters.

DEFINITION

History

As early as the 7th century BC, the Greeks used practical knowledge to formulate the abstract subject we know as Geometry. The word *geometry* comes from the combination of the Greek words *ge* (Earth) and *métron* (measure).

We begin the discussion of angles with the definitions of a **ray** and an **angle** by using the undefined terms **point** and **line**.

Ray and Angle

Term	Definition	Illustrations with Notation
Ray	A **ray** consists of a point (called the **endpoint**) and all the points on a line on one side of that point.	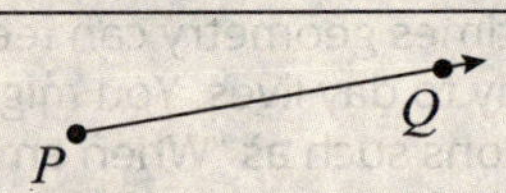 Ray $\overrightarrow{PQ}$ with endpoint P
Angle	An **angle** consists of two rays with a common endpoint. (The two rays are called the **sides** of the angle and the endpoint is called the **vertex**.)	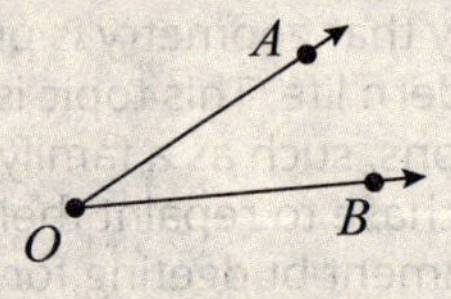 $\angle AOB$ with vertex O

DEFINITION

Labeling Angles

There are three common ways of labeling angles:

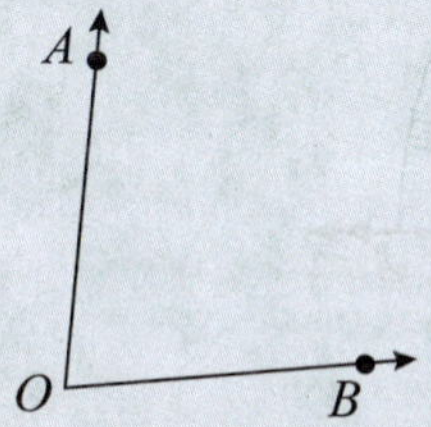

A. $\angle AOB$

Using three capital letters with the vertex as the middle letter.

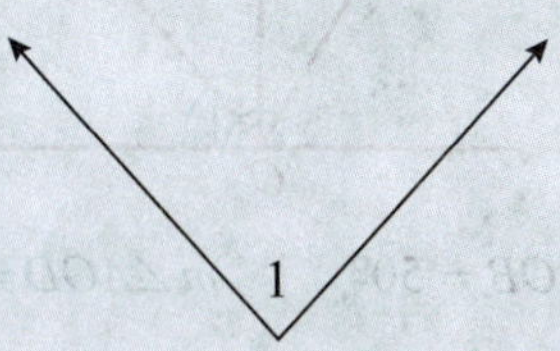

B. $\angle 1$

Using single numbers such as 1, 2, 3.

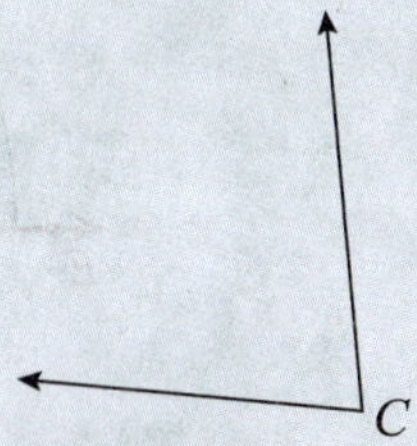

C. $\angle C$

Using the single capital letter at the vertex when the meaning is clear.

PROCEDURE

Math Tip

When using three letters to name an angle, the two outside letters can be reversed as long as the vertex is the middle letter. That is, $\angle AOB$ and $\angle BOA$ are two names for the same angle.

B Classifying Angles by Their Measures

Every angle has a measurement, or **measure**, associated with it. The base unit when measuring angles is **one degree** (symbolized 1°). There are 360 degrees in a complete circle (one full revolution). In Figure 1, a device called a protractor shows that the measure of $\angle AOB$ is 5 degrees. We write $m\angle AOB = 5°$.

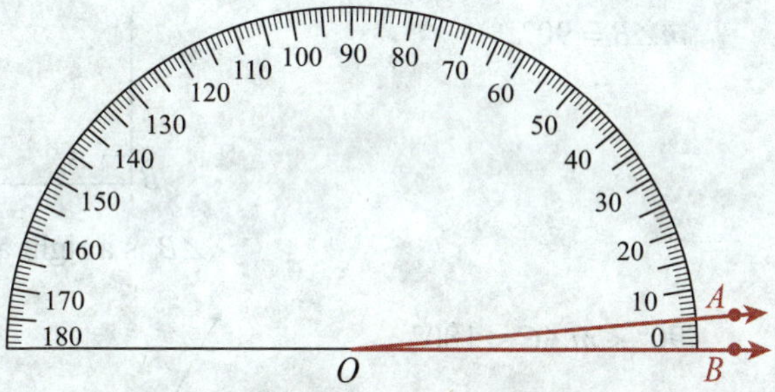

The protractor shows $m\angle AOB = 5°$.

Figure 1

To measure an angle with a protractor, lay the bottom edge of the protractor along one side of the angle with the vertex at the marked center point. Then, read the measure from the protractor where the other side of the angle crosses it.

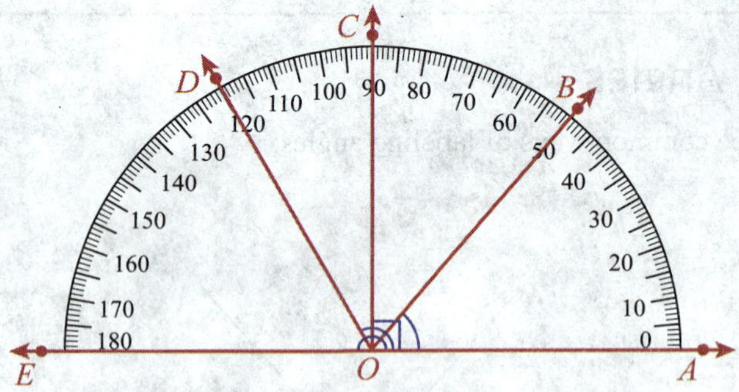

$m\angle AOB = 50^\circ$ $m\angle AOD = 120^\circ$

$m\angle AOC = 90^\circ$ $m\angle AOE = 180^\circ$

Figure 2

Angles can be classified according to their measures.

Angles Classified by Measure

Name	Measure	Illustrations with Notation
Acute	$0^\circ < m\angle A < 90^\circ$	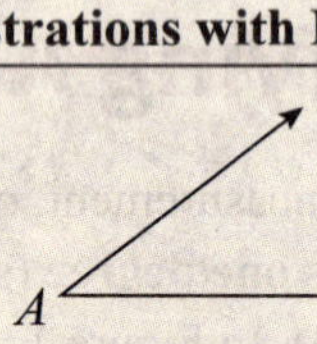 $\angle A$ is an acute angle.
Right	$m\angle B = 90^\circ$	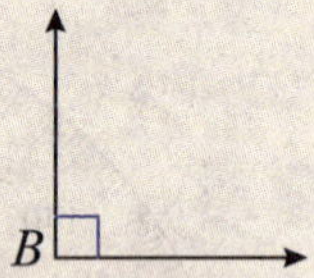 $\angle B$ is a right angle.
Obtuse	$90^\circ < m\angle C < 180^\circ$	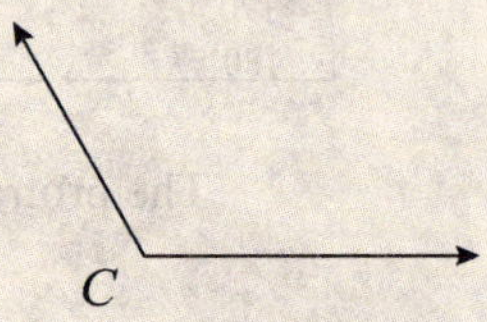 $\angle C$ is an obtuse angle.
Straight	$m\angle D = 180^\circ$	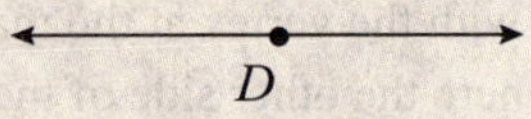 $\angle D$ is a straight angle. (The rays are in opposite directions.)

DEFINITION

Figure 3 is used for Examples 1 and 2.

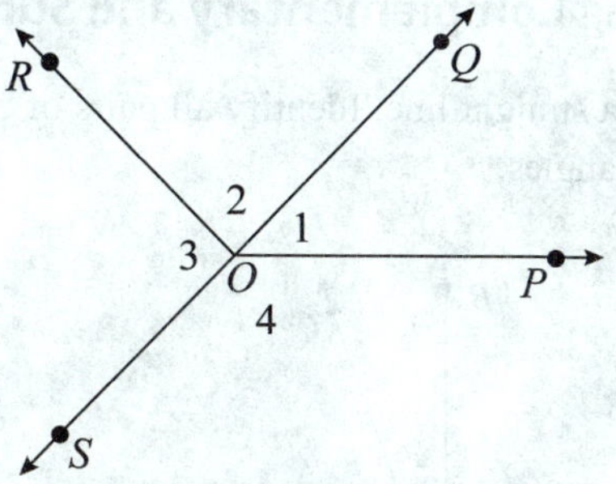

Figure 3

Example 1 Measuring Angles

Check the measures of **a.** $\angle 1$, **b.** $\angle 2$, **c.** $\angle 3$, and **d.** $\angle 4$ (from Figure 3) using a protractor. (You may need to extend the rays to be able to read the numbers on your protractor.)

Solution

a. $m\angle 1 = 45°$

b. $m\angle 2 = 90°$

c. $m\angle 3 = 90°$

d. $m\angle 4 = 135°$

Now work margin exercise 1.

1. Find the measures of $\angle 1$ and $\angle 2$ using a protractor.

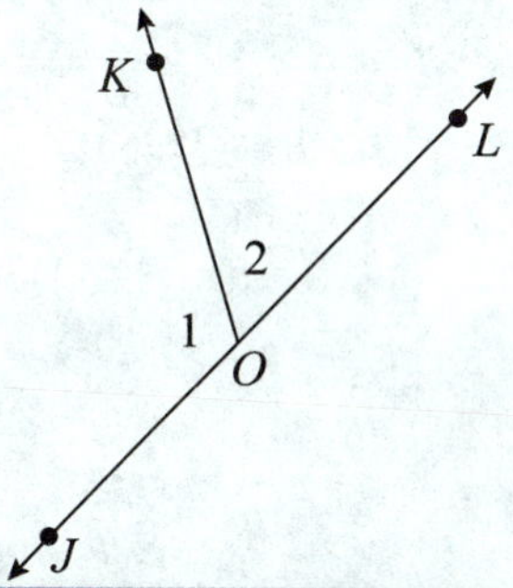

Example 2 Classifying Angles by Their Measure

Classify each angle (from Figure 3) as acute, right, obtuse, or straight.

a. $\angle 1$ **b.** $\angle 2$ **c.** $\angle POR$

Solution

a. $\angle 1$ is acute since $0° < m\angle 1 < 90°$.

b. $\angle 2$ is a right angle since $m\angle 2 = 90°$.

c. $\angle POR$ is obtuse since $m\angle POR = m\angle 1 + m\angle 2 = 45° + 90° = 135°$ and $135° > 90°$.

Now work margin exercise 2.

2. Classify each angle (from Figure 3) as acute, right, obtuse, or straight.

a. $\angle 3$

b. $\angle 4$

c. $\angle SOQ$

C Complementary and Supplementary Angles

Complementary and Supplementary Angles

1. Two angles are **complementary** if the sum of their measures is 90°.
2. Two angles are **supplementary** if the sum of their measures is 180°.

DEFINITION

3. In the figure shown $\overleftrightarrow{MQ}$ is a straight line. Identify all pairs of complementary and supplementary angles.

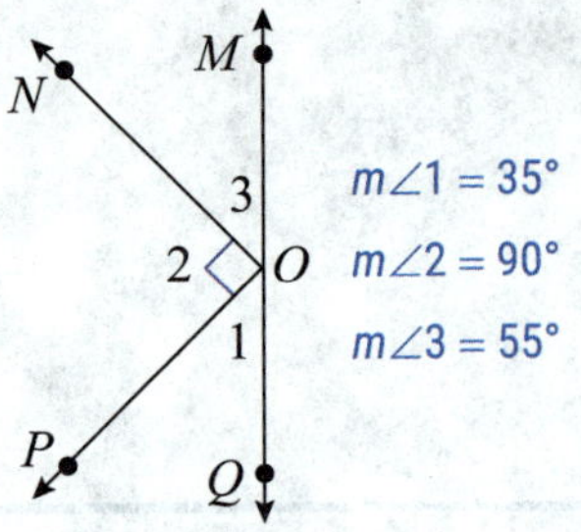

4. In the figure shown in Example 4, assume $m\angle ROQ = 40°$.

a. What is the measure of $\angle SOR$?

b. Are $\angle SOR$ and $\angle ROP$ equal?

Example 3 Identifying Complementary and Supplementary Angles

In the figure shown, $\overleftrightarrow{AD}$ is a straight line. Identify all pairs of complementary, supplementary, and straight angles.

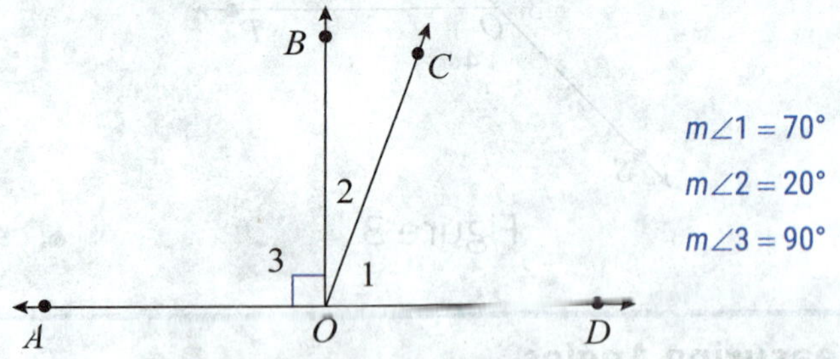

Solution

a. $\angle 1$ and $\angle 2$ are complementary since $m\angle 1 + m\angle 2 = 90°$.

b. $\angle COD$ and $\angle COA$ are supplementary.
$m\angle COA = m\angle 3 + m\angle 2 = 90° + 20° = 110°$,
so $m\angle COD + m\angle COA = 70° + 110° = 180°$.

c. $\angle AOD$ is a straight angle since $m\angle AOD = 180°$.

d. $\angle BOA$ and $\angle BOD$ are supplementary; and in this case $m\angle BOD = 90°$.

Now work margin exercise 3.

Example 4 Calculating Measures of Angles

In the figure shown, $\overleftrightarrow{PS}$ is a straight line and $m\angle QOP = 30°$.

a. Find the measure of $\angle QOS$.

b. Find the measure of $\angle SOP$.

c. Are any pairs of angles supplementary?

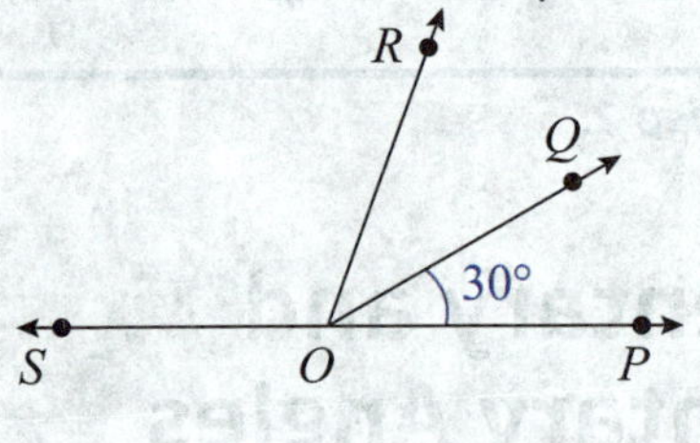

Solution

a. $m\angle QOS = 180° - 30° = 150°$

b. $m\angle SOP = 180°$

c. Yes, $\angle QOP$ and $\angle QOS$ are supplementary and $\angle ROP$ and $\angle ROS$ are supplementary.

Now work margin exercise 4.

D Congruent, Vertical, and Adjacent Angles

If two angles have the same measure, they are said to be **congruent angles** (symbolized as ≅). As shown in Figure 4, $m\angle A = m\angle B = 30°$ and $\angle A \cong \angle B$ (read "angle *A* **is congruent** to angle *B*").

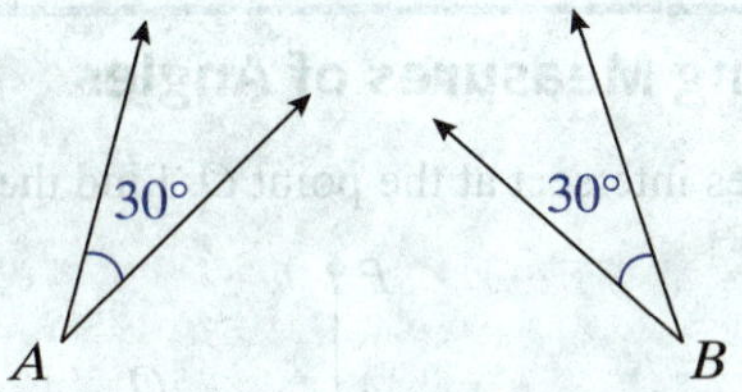

$\angle A \cong \angle B$

Figure 4

Example 5 Identifying Congruent Angles

Identify the congruent angles in the figure.

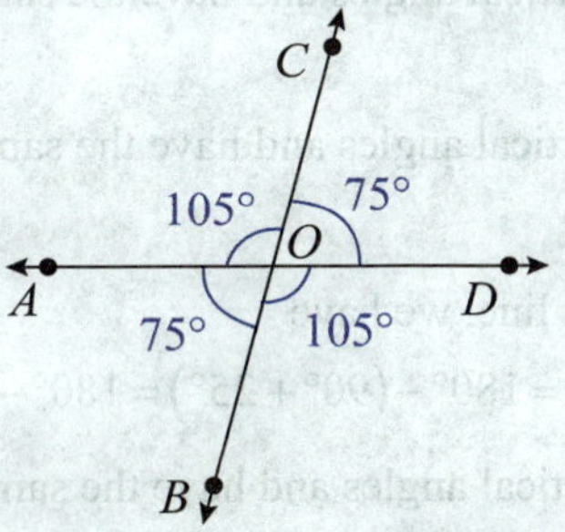

Solution

$\angle AOB \cong \angle COD$ and $\angle COA \cong \angle DOB$

Now work margin exercise 5.

5. Identify the congruent angles in the figure.

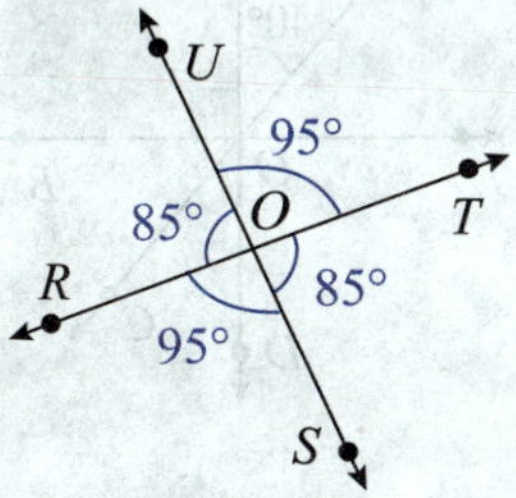

Two lines **intersect** if they have exactly one point in common. If two lines intersect, then two pairs of **vertical angles** are formed. Vertical angles are opposite each other. (See Figure 5.)

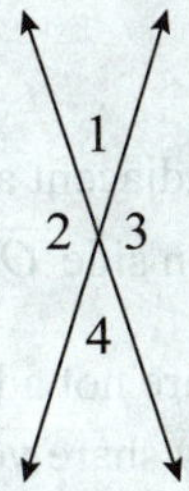

$\angle 1$ and $\angle 4$ are vertical angles.

$\angle 2$ and $\angle 3$ are vertical angles.

Figure 5

Vertical Angles

Vertical angles are congruent. That is, vertical angles have the same measure. (**Note:** In Figure 5, $\angle 1 \cong \angle 4$ and $\angle 2 \cong \angle 3$.)

DEFINITION

6. In the figure shown, three lines intersect at the point O. Find the measure of each angle.

a. $\angle COD$

b. $\angle DOE$

c. $\angle EOF$

d. $\angle BOC$

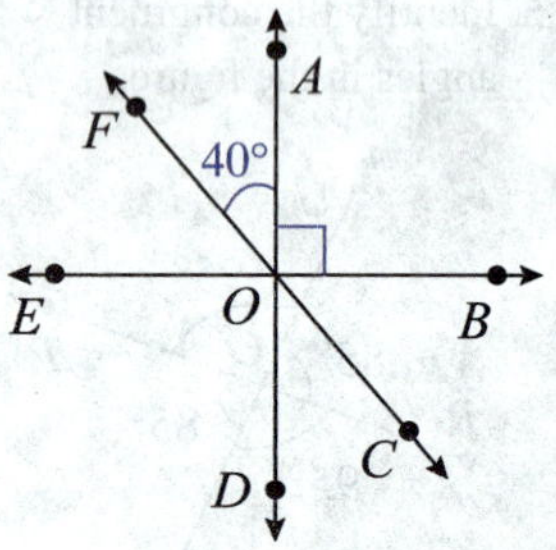

Example 6 Calculating Measures of Angles

In the given figure, three lines intersect at the point O. Find the measure of each angle.

a. $\angle TOU$

b. $\angle ROS$

c. $\angle POQ$

d. $\angle SOT$

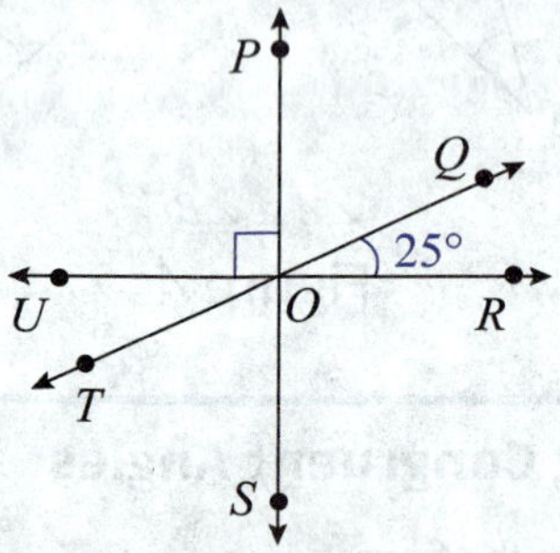

Solution

a. $\angle QOR$ and $\angle TOU$ are vertical angles and have the same measure. This means that $m\angle TOU = 25°$.

b. $\angle POU$ and $\angle ROS$ are vertical angles and have the same measure. This means that $m\angle ROS = 90°$.

c. Because $\overleftrightarrow{RU}$ is a straight line, we have

$$m\angle POQ = 180° - (90° + 25°) = 180° - 115° = 65°.$$

d. $\angle POQ$ and $\angle SOT$ are vertical angles and have the same measure. This means $m\angle SOT = 65°$.

Now work margin exercise 6.

Adjacent Angles

Two angles are **adjacent** if they share a common vertex and a common side but do not overlap. (See Figure 6.)

DEFINITION

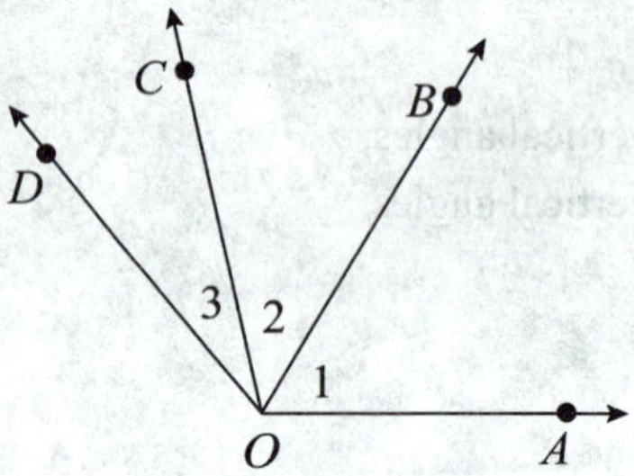

$\angle 1$ and $\angle 2$ are adjacent angles. They have the common side $\overrightarrow{OB}$.

$\angle 2$ and $\angle DOB$ are not adjacent even though they share vertex O and common side $\overrightarrow{OB}$.

Figure 6

Example 7 Finding Adjacent Angles

In the figure shown, $\overleftrightarrow{AC}$ and $\overleftrightarrow{BD}$ are straight lines. Name an angle adjacent to $\angle EOD$.

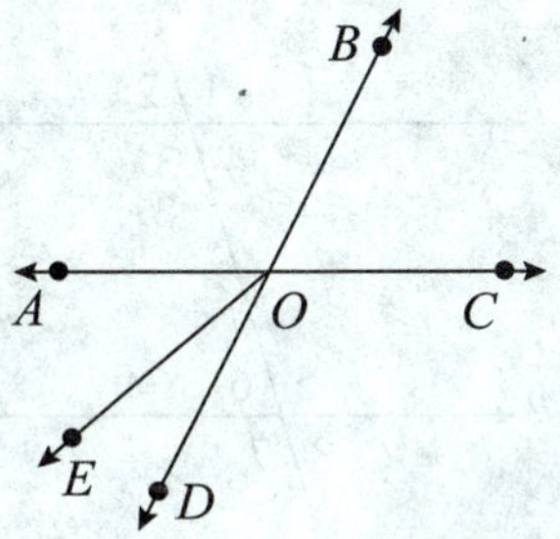

Solution

One angle adjacent to $\angle EOD$ is $\angle AOE$. Two other angles adjacent to $\angle EOD$ are $\angle BOE$ and $\angle COD$.

Now work margin exercise 7.

7. In the figure shown, $\overleftrightarrow{WY}$ and $\overleftrightarrow{VX}$ are straight lines. Name an angle adjacent to $\angle XQZ$.

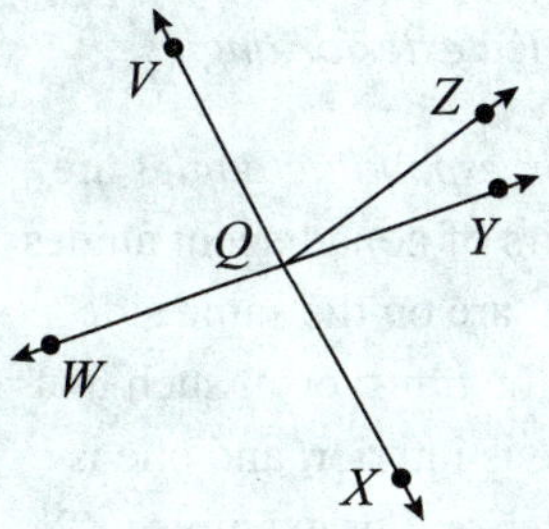

E Parallel and Perpendicular Lines

Parallel Lines and Perpendicular Lines

Term	Definition	Illustrations with Notation
Parallel Lines	Two lines are **parallel** (symbolized $\parallel$) if they are in the same plane and do not intersect.	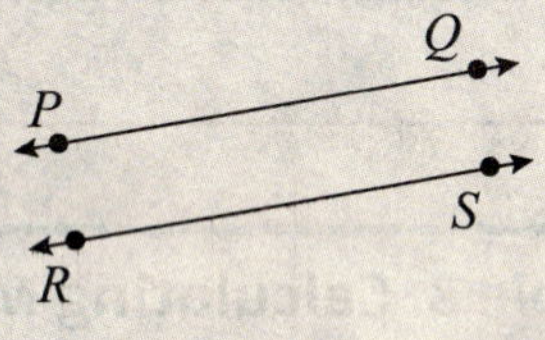 $\overleftrightarrow{PQ}$ is parallel to $\overleftrightarrow{RS}$ ($\overleftrightarrow{PQ} \parallel \overleftrightarrow{RS}$)
Perpendicular Lines	Two lines are **perpendicular** (symbolized $\perp$) if they intersect to form right angles.	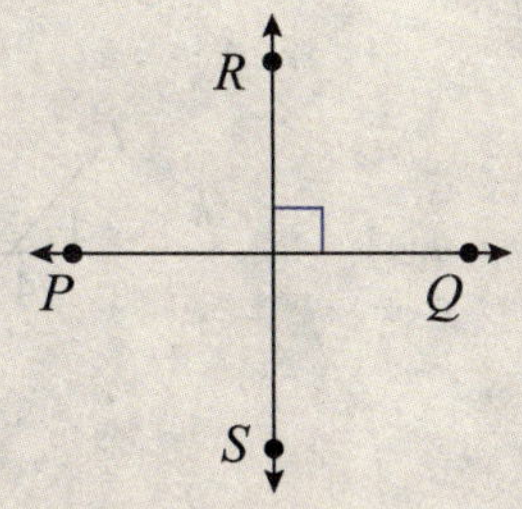 $\overleftrightarrow{PQ}$ is perpendicular to $\overleftrightarrow{RS}$ ($\overleftrightarrow{PQ} \perp \overleftrightarrow{RS}$)

DEFINITION

A **transversal** is a line in a plane that intersects two or more lines in that plane at different points. As shown in Figure 7, eight angles are formed when a transversal intersects two lines.

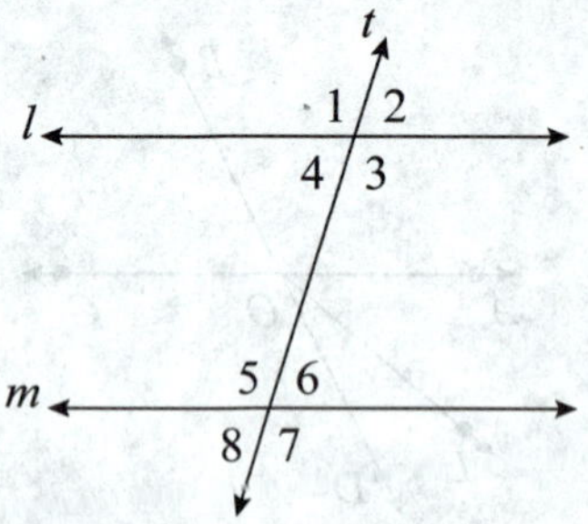

Figure 7

Math Tip

In Figure 7, ∠3, ∠4, ∠5, and ∠6 are called *interior angles* and ∠1, ∠2, ∠7, and ∠8 are called *exterior angles*.

Corresponding angles are pairs of nonadjacent angles that are on the same side of the transversal such that one is interior, and one is exterior. For example, ∠2 and ∠6.

Alternate interior angles are nonadjacent interior angles that are on opposite sides of the transversal. For example, ∠3 and ∠5.

Some of the eight angles formed are named as follows.

- Corresponding angles: ∠1 and ∠5, ∠2 and ∠6, ∠3 and ∠7, ∠4 and ∠8
- Alternate interior angles: ∠3 and ∠5, ∠4 and ∠6

Parallel Lines and a Transversal

If two parallel lines are cut by a transversal, then the following two statements are true.

1. **Corresponding angles are congruent.**
2. **Alternate interior angles are congruent.**

PROPERTIES

Example 8 Calculating Measures of Angles

In the figure shown, lines *k* and *l* are parallel, *t* is a transversal, and $m\angle 1 = 50°$. Find the measures of the other 7 angles.

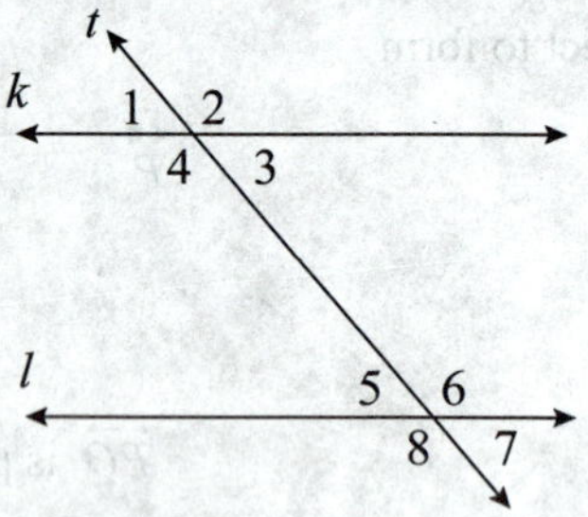

Solution

One way of reasoning (among several) is as follows.

∠1 and ∠3 are vertical angles so $m\angle 1 = m\angle 3 = 50°$.

∠1 and ∠2 are supplementary angles so $m\angle 2 = 180° - 50° = 130°$.

∠2 and ∠4 are vertical angles so $m\angle 4 = 130°$.

8. In the figure shown, lines *l* and *m* are parallel, *t* is a transversal, and $m\angle 2 = 80°$. Find $m\angle 4$, $m\angle 5$, and $m\angle 6$.

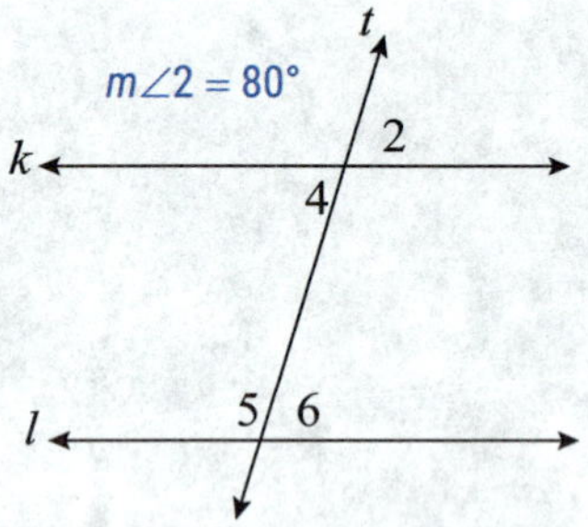

Now, because $\angle 1$ and $\angle 5$ are corresponding angles, they have the same measures which means $m\angle 5 = 50°$.

Because $\angle 4$ and $\angle 6$ are alternate interior angles, they have the same measures which means $m\angle 6 = 130°$.

Again, using vertical angles, $m\angle 5 = m\angle 7 = 50°$ and $m\angle 6 = m\angle 8 = 130°$.

Now work margin exercise 8.

F Classifying Triangles

A **line segment** consists of two points on a line and all the points between them.

line segment $\overline{AB}$

Figure 8

A **triangle** consists of three line segments that join three points that are not collinear (do not lie on a straight line). The line segments are called the **sides** of the triangle and the points are called the **vertices** of the triangle. If the points are labeled *A*, *B*, and *C*, the triangle is symbolized ΔABC. (See Figure 9.)

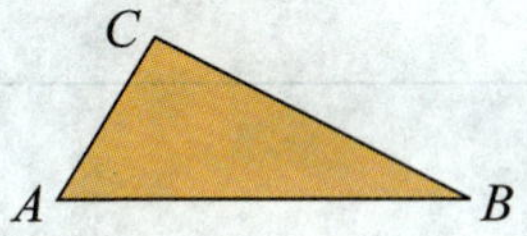

ΔABC with vertices *A*, *B*, and *C* and sides $\overline{AB}$, $\overline{BC}$, and $\overline{AC}$.

Figure 9

The sides of a triangle are said to determine three angles, and these angles are labeled by the vertices. Thus, the angles of ΔABC are $\angle A$, $\angle B$, and $\angle C$.

Triangles are classified in two ways:

1. According to the lengths of their sides.
2. According to the measures of their angles.

The corresponding names and properties are listed in the following tables.

Note

The line segment with endpoints *A* and *B* is indicated by placing a bar over the letters, as in $\overline{AB}$. The length of the segment is indicated by writing only the letters, as in *AB*.

Triangles Classified by Sides

(**Note:** In the figures, sides with equal length are indicated by the same number of tic marks.)

Name	Property	Example	
Scalene	No two sides have equal lengths.	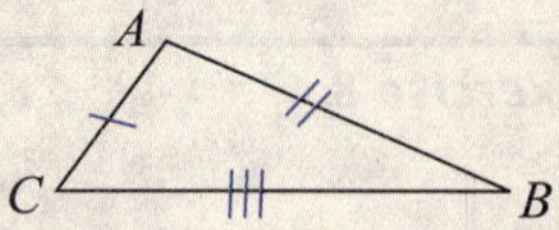	ΔABC is scalene since no two sides have equal lengths.
Isosceles	At least two sides have equal lengths.	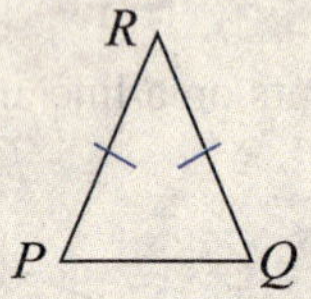	ΔPQR is isosceles since $PR = QR$.
Equilateral	All three sides have equal lengths.	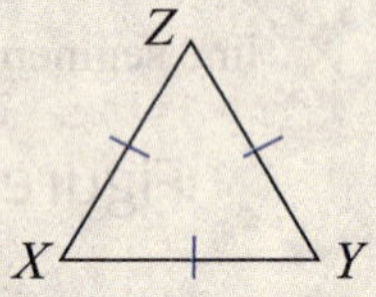	ΔXYZ is equilateral since $XY = XZ = YZ$

DEFINITION

Triangles Classified by Angles

Name	Property	Example	
Acute	All three angles are acute.	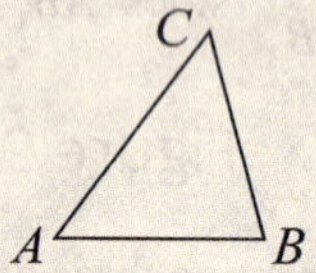	$\angle A$, $\angle B$, and $\angle C$ are all acute so ΔABC is acute.
Right	One angle is a right angle.	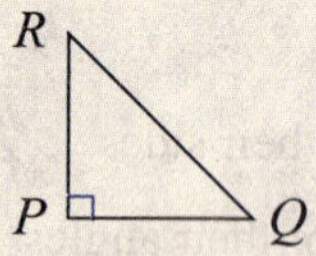	$m\angle P = 90°$, so ΔPRQ is a right triangle.
Obtuse	One angle is obtuse.	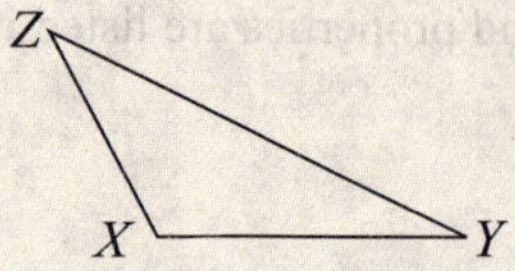	$\angle X$ is obtuse, so ΔXYZ is an obtuse triangle.

DEFINITION

Every triangle is said to have six parts—namely, three angles and three sides. Two sides of a triangle are said to **include** the angle at their common endpoint or vertex. The third side is said to be **opposite** this angle.

The sides in a right triangle have special names. The longest side, opposite the right angle, is called the **hypotenuse**, and the other two sides are called **legs**. (See Figure 10.)

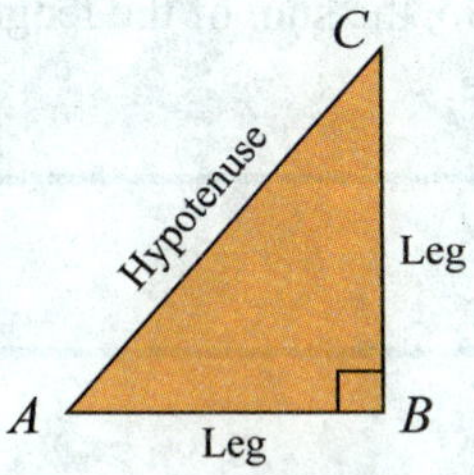

ΔABC is a right triangle.

$m\angle B = 90$

$\overline{AC}$ is opposite $\angle B$.

$\overline{AB}$ and $\overline{BC}$ include $\angle B$ (the right angle).

Figure 10

Three Properties of Triangles

In a triangle, the following statements are true:

1. The sum of the measures of the angles is 180°.
2. The sum of the lengths of any two sides must be greater than the length of the third side.
3. Longer sides are opposite angles with larger measures.

PROPERTIES

Example 9 Classifying a Triangle by Its Sides

In ΔABC, $AB = AC$. What kind of triangle is ΔABC?

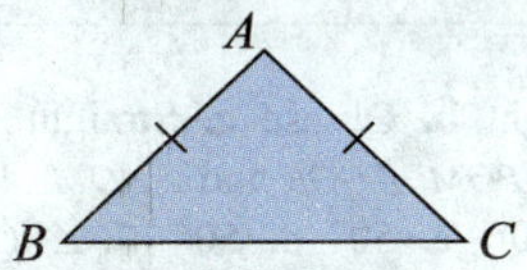

Solution

ΔABC is isosceles because two sides have equal lengths.

Now work margin exercise 9.

9. In ΔDEF, $DE = 5$, $EF = 7$, and $DF = 8$. What kind of triangle is ΔDEF?

Example 10 Determining Whether a Triangle Exists

ΔPQR was drawn and then the lengths of the sides were labeled as shown in the figure. Explain why the labels cannot be correct.

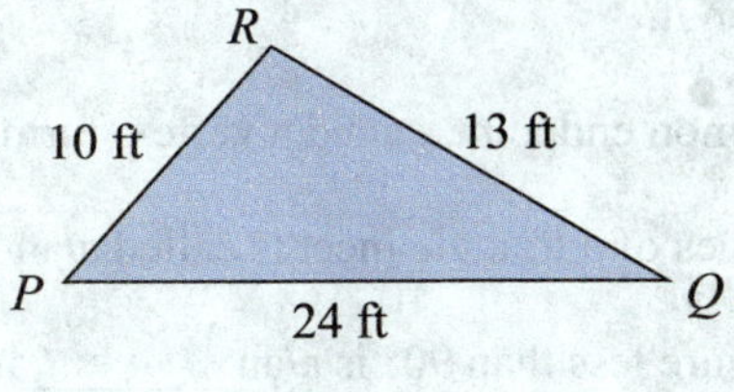

10. Is it possible to have a triangle with sides of length 15 in., 37 in. and 51 in.?

Solution

The labels cannot be correct because $PR + QR = 10$ ft + 13 ft = 23 ft and $PQ = 24$ ft, which is greater than the sum of the other two sides. In a triangle, the sum of the lengths of any two sides must be greater than the length of the third side.

Now work margin exercise 10.

11. Now assume that in ΔBOR, $m\angle B = 35°$ and $m\angle O = 55°$. Answer each question.

a. What is $m\angle R$?

b. Which side is opposite $\angle B$?

c. Which sides include $\angle B$?

d. Is ΔBOR a right triangle? Why or why not?

Example 11 Analyzing Triangles

In ΔBOR, $m\angle B = 50°$ and $m\angle O = 70°$.

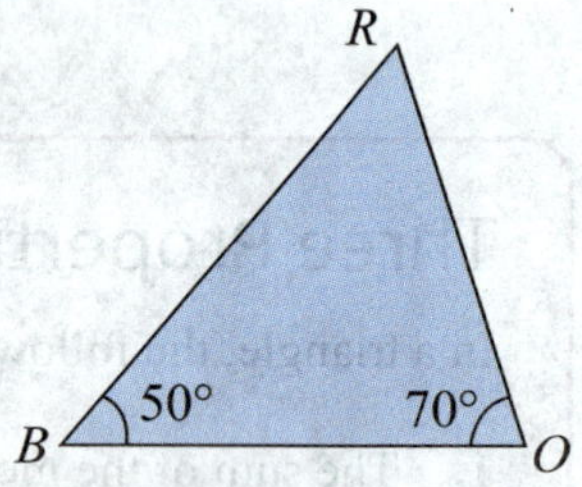

a. What is $m\angle R$?

b. Which side is opposite $\angle R$?

c. Which sides include $\angle R$?

d. Is ΔBOR a right triangle? Why or why not?

Solution

The sum of the measures of the angles must be 180°.

a. Since $50° + 70° = 120°$, $m\angle R = 180° - 120° = 60°$.

b. $\overline{BO}$ is opposite $\angle R$.

c. $\overline{RB}$ and $\overline{RO}$ include $\angle R$.

d. ΔBOR is not a right triangle because none of the angles is a right angle.

Now work margin exercise 11.

Margin Exercise Answers

1. $m\angle 1 = 120°$, $m\angle 2 = 60°$ **2. a.** Right **b.** Obtuse **c.** Straight **3.** Complementary: $\angle MON$ and $\angle POQ$ Supplementary: $\angle QOP$ and $\angle POM$, $\angle QON$ and $\angle NOM$ **4. a.** 110° **b.** No **5.** $\angle ROS \cong \angle TOU$ and $\angle ROU \cong \angle SOT$ **6. a.** 40° **b.** 90° **c.** 50° **d.** 50° **7.** $\angle VQZ$, $\angle ZQW$ or $\angle WQX$ **8.** $m\angle 4 = 80°$, $m\angle 5 = 100°$, $m\angle 6 = 80°$ **9.** Scalene **10.** Yes **11. a.** 90° **b.** $\overline{RO}$ **c.** $\overline{BR}$ and $\overline{BO}$ **d.** Yes, because $m\angle R = 90°$.

6.1 Exercises

Concept Check

Fill-in-the-Blank. Complete each sentence using information found in this section.

1. A/An _____ has no beginning or end and is labeled with a lowercase letter or by the labels of two points on it.

2. Two rays with a common endpoint, called a vertex, form a/an _______.

3. A point where two sides of a triangle meet is called a/an _________.

4. An angle with a measure less than 90° is a/an _______ angle.

5. An angle that measures 180° is a/an _________ angle.

6.1 Exercises

Concept Check

1. line
2. angle
3. vertex
4. acute
5. straight

6. If the sum of the measures of two angles is 180°, they are said to be ______________ angles.

7. Two lines are ______________ if they intersect and form right angles.

8. A triangle with no equal sides is a/an _______ triangle.

9. A triangle with three equal sides is a/an _________ triangle.

10. When the measures of the three angles in a triangle are added, the sum is _____ degrees.

True/False. Determine whether each statement is true or false. If a statement is false, explain how it can be changed so the statement will be true. (**Note:** There may be more than one acceptable change.)

11. The sum of the measures of two complementary angles is equal to the measure of one right angle.

12. The sum of the measures of complementary angles is greater than the sum of the measures of supplementary angles.

13. Adjacent angles are two angles that share a vertex and a common side but do not overlap.

14. If two lines in a plane are not parallel, then they are perpendicular.

15. A triangle with sides of 4 inches, 4 inches, and 3 inches is an isosceles triangle.

16. A triangle with three angles that each measure less than 90 degrees is an acute triangle.

Practice

Use a protractor to find the measure of each angle. (**Note:** You may need to extend the rays to be able to read the numbers on your protractor.) See Example 1.

1.

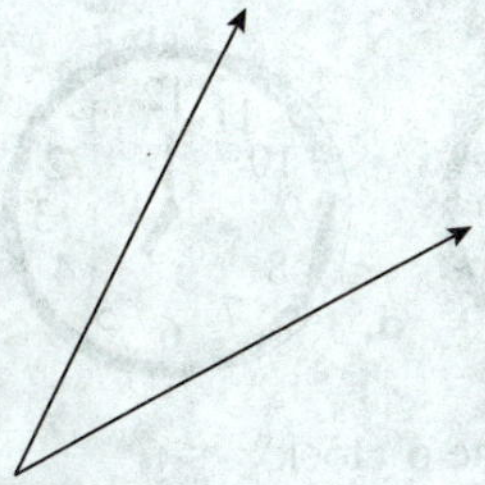

2.

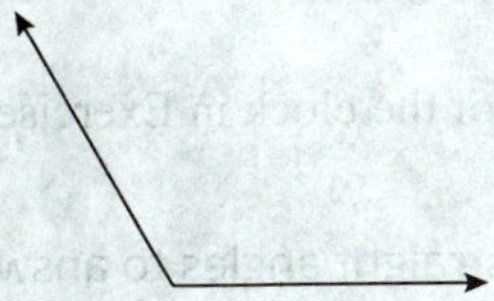

3.

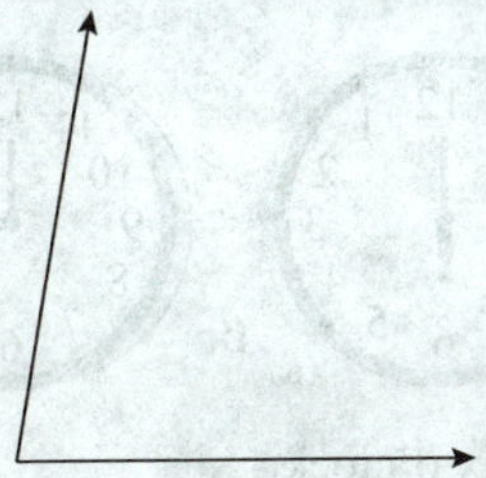

4.

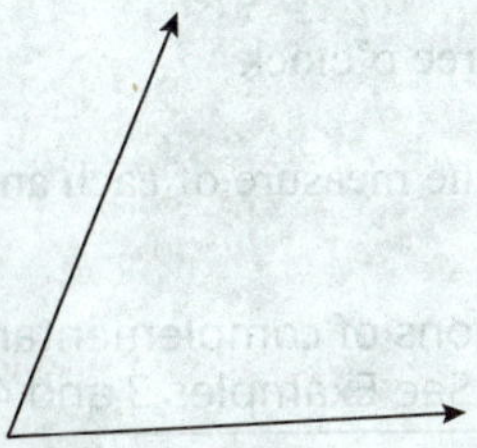

Classify each angle as acute, right, obtuse, or straight. See Example 2.

5.

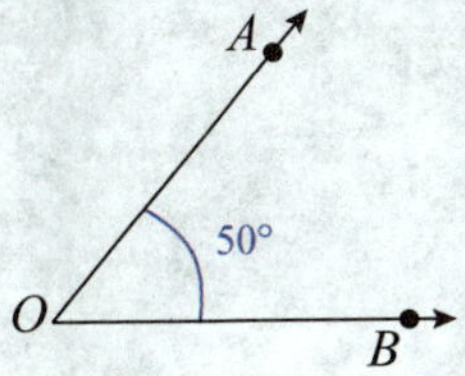

6.

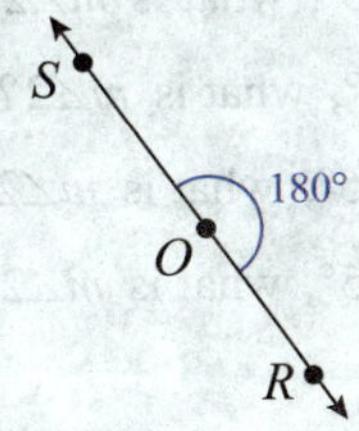

6. supplementary
7. perpendicular
8. scalene
9. equilateral
10. 180
11. True
12. False; The sum of the measures of complementary angles is less than the sum of measures of supplementary angles.
13. True
14. False; If two lines in a plane are not parallel, then they intersect.
15. True
16. True

Practice

1. 35°
2. 120°
3. 80°
4. 65°
5. Acute
6. Straight

7. Obtuse
8. Obtuse
9. Acute
10. Right
11. a. Obtuse b. Acute
c. Right
12. a. Straight b. Right
c. Acute
d. Obtuse
13. a. 180° b. 90° c. 30°
d. 150°
14. a. 75° b. 87° c. 45°
d. 15°

7.

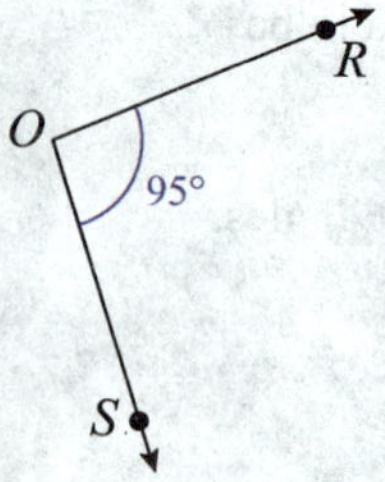

8.

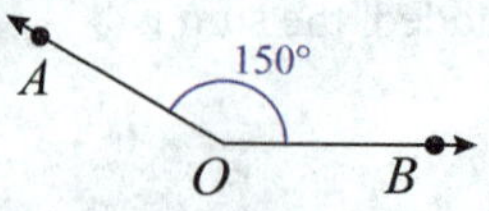

9.

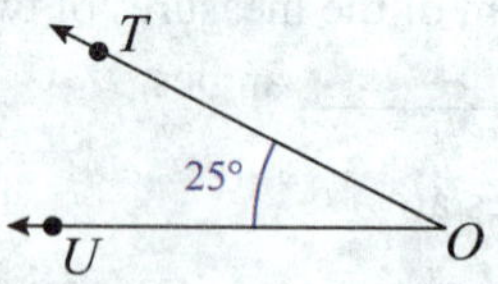

10.

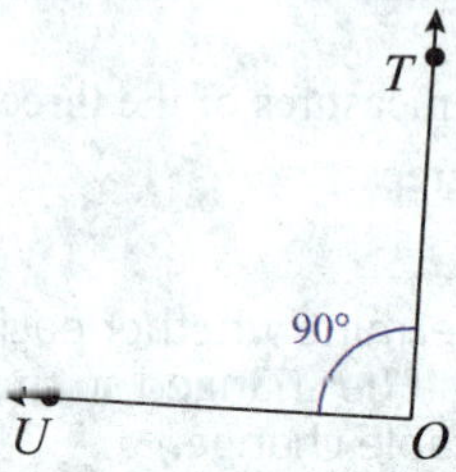

Use the definitions of acute, right, obtuse, and straight angles to answer the questions. See Example 2.

11. In the figure shown, $\overleftrightarrow{DC}$ is a straight line and m$\angle BOA = 90°$.

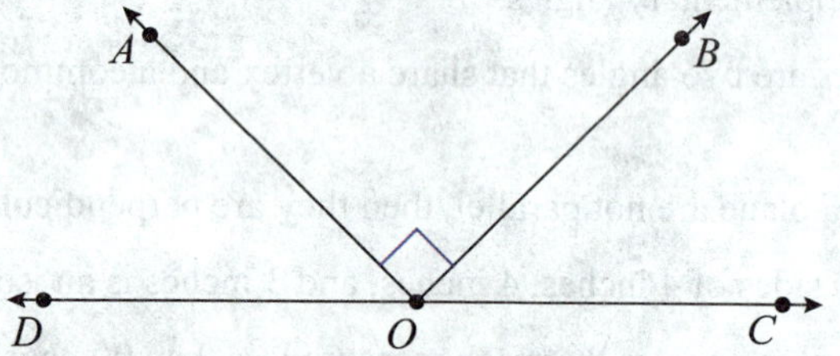

a. What type of angle is $\angle AOC$?

b. What type of angle is $\angle BOC$?

c. What type of angle is $\angle BOA$?

12. Name the type of angle formed by the hands on a clock.

a. at six o'clock

b. at three o'clock

c. at one o'clock

d. at five o'clock

13. What is the measure of each angle formed by the hands of the clock in Exercise 12?

Use the definitions of complementary, supplementary, and straight angles to answer each question. See Examples 3 and 4.

14. Assume that $\angle 1$ and $\angle 2$ are complementary.

a. If $m\angle 1 = 15°$, what is $m\angle 2$?

b. If $m\angle 1 = 3°$, what is $m\angle 2$?

c. If $m\angle 1 = 45°$, what is $m\angle 2$?

d. If $m\angle 1 = 75°$, what is $m\angle 2$?

15. Assume $\angle 3$ and $\angle 4$ are supplementary.

a. If $m\angle 3 = 45°$, what is $m\angle 4$?

b. If $m\angle 3 = 90°$, what is $m\angle 4$?

c. If $m\angle 3 = 110°$, what is $m\angle 4$?

d. If $m\angle 3 = 135°$, what is $m\angle 4$?

16. In the figure shown,

a. Name all of the pairs of supplementary angles.

b. Name all the pairs of complementary angles.

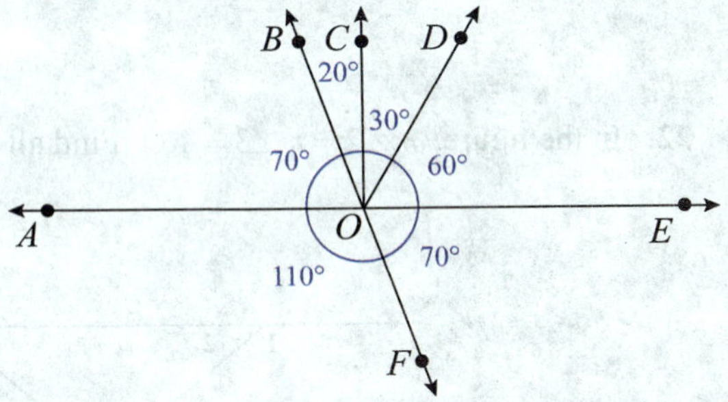

Use the definitions of adjacent and vertical angles to answer each question. See Examples 6 through 8.

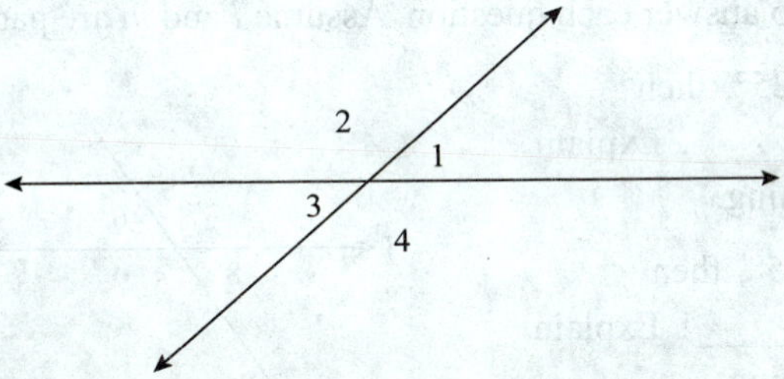

17. The figure shows two intersecting lines.

a. If $m\angle 1 = 30°$, what is $m\angle 2$?

b. Is $m\angle 3 = 30°$? Give a reason for your answer other than the fact that $\angle 1$ and $\angle 3$ are vertical angles.

c. Name two pairs of congruent angles.

d. Name four pairs of adjacent angles.

18. The figure shows two intersecting lines where $m\angle 1 = 30°$. Find the measures of the other three angles.

19. Given that $m\angle 1 = 42°$ in the figure, find the measures of the other three angles.

20. In the figure shown, $\overleftrightarrow{AB}$ is a straight line.

a. Name two pairs of adjacent angles.

b. Name two vertical angles, if there are any.

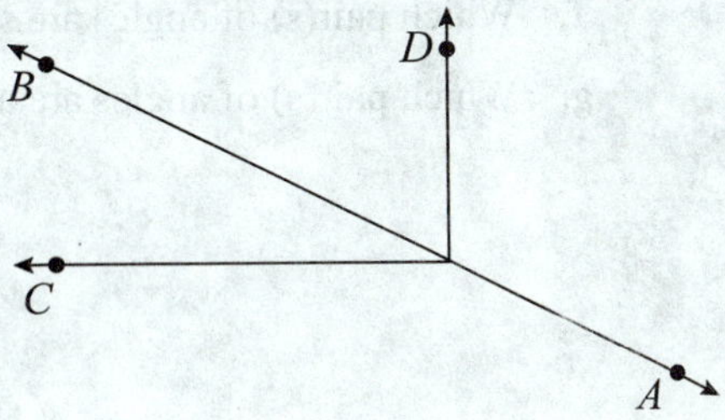

15. a. 135° **b.** 90° **c.** 70° **d.** 45°

16. a. $\angle AOB$ and $\angle BOE$; $\angle AOB$ and $\angle AOF$; $\angle AOC$ and $\angle COE$; $\angle AOD$ and $\angle DOE$; $\angle AOF$ and $\angle FOE$; $\angle BOC$ and $\angle COF$; $\angle BOD$ and $\angle DOF$; $\angle BOE$ and $\angle EOF$

b. $\angle AOB$ and $\angle BOC$; $\angle COD$ and $\angle DOE$; $\angle BOC$ and $\angle EOF$

17. a. 150°

b. Yes; $\angle 2$ and $\angle 3$ are supplementary.

c. $\angle 1$ and $\angle 3$; $\angle 2$ and $\angle 4$

d. $\angle 1$ and $\angle 2$; $\angle 2$ and $\angle 3$; $\angle 3$ and $\angle 4$; $\angle 1$ and $\angle 4$

18. $m\angle 2 = 150°$; $m\angle 3 = 30°$; $m\angle 4 = 150°$

19. $m\angle 2 = 138°$; $m\angle 3 = 42°$ $m\angle 4 = 138°$

20. a. $\angle CPB$ and $\angle BPD$; $\angle BPD$ and $\angle APD$; $\angle CPA$ and $\angle CPB$; $\angle BPD$ and $\angle DPA$

b. There are none.

21. **a.** $m\angle 2 = 70°$; $m\angle 3 = 90°$; $m\angle 4 = 20°$; $m\angle 5 = 70°$

b. $\angle 3$

c. $\angle 2$ and $\angle 5$

22. $m\angle 3 = m\angle 5$; $m\angle 2 = m\angle 5$; $m\angle 1 = m\angle 4$; $m\angle 1 = m\angle 6$; $m\angle 4 = m\angle 6$

23. **a.** 125°; $\angle 1$ and $\angle 3$ are vertical angles.

b. 55°; $\angle 8$ and $\angle 6$ are vertical angles.

c. $m\angle 7 = 125°$; $\angle 6$ and $\angle 7$ are supplementary angles.

d. Yes; $\angle 2$ and $\angle 6$ are corresponding angles.

24. **a.** $\angle QPR$ and $\angle RPB$

b. $\angle APR$

c. $\angle APQ$ and $\angle QPB$

d. None

e. $\angle QPR$ and $\angle RPB$

f. $\angle APQ$ and $\angle QPB$; $\angle APR$ and $\angle RPB$

g. $\angle APQ$ and $\angle QPB$; $\angle APQ$ and $\angle QPR$; $\angle APR$ and $\angle RPB$; $\angle QPR$ and $\angle RPB$

21. In the figure shown, *l*, *m*, and *n* are straight lines with $m\angle 1 = 20°$ and $m\angle 6 = 90°$.

a. Find the measures of the other four angles.

b. Which angle is supplementary to $\angle 6$?

c. Which angles are complementary to $\angle 1$?

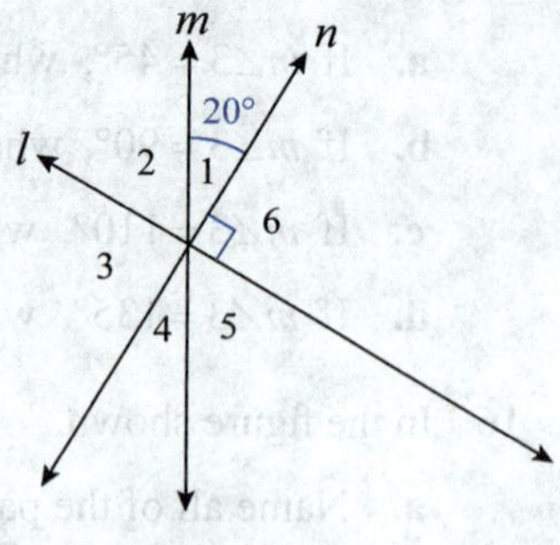

22. In the figure, $m\angle 2 = m\angle 3 = 40°$. Find all other pairs of angles that are congruent.

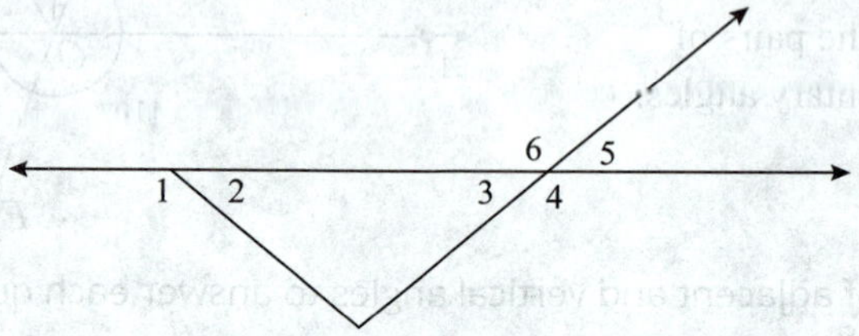

23. Use the figure to answer each question. Assume *l* and *m* are parallel.

a. If $m\angle 1 = 125°$, then $m\angle 3 =$ _______. Explain your reasoning.

b. If $m\angle 8 = 55°$, then $m\angle 6 =$ _______. Explain your reasoning.

c. What is $m\angle 7$? Explain your reasoning.

d. Does $m\angle 2 = m\angle 6$? Explain your reasoning.

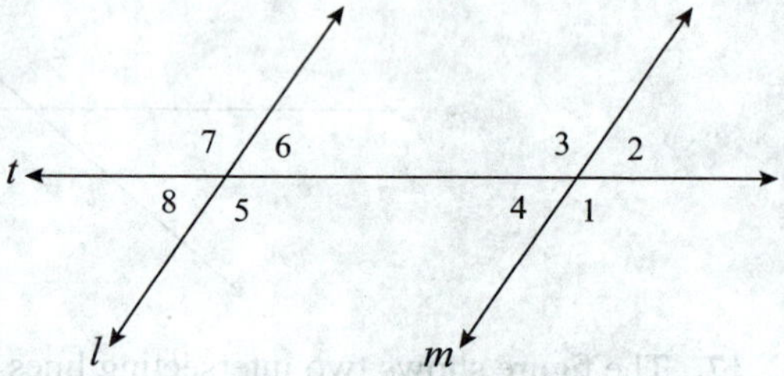

24. Lines $\overline{AB}$ and $\overline{PQ}$ are perpendicular.

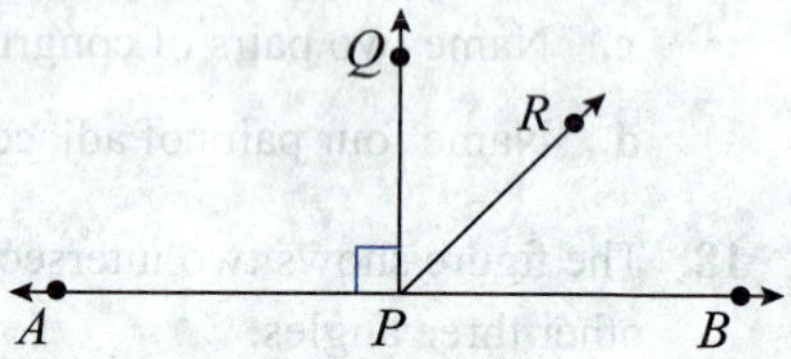

a. Which angle(s) are acute?

b. Which angle(s) are obtuse?

c. Which angle(s) are right angles?

d. Which pair(s) of angles are vertical angles?

e. Which pair(s) of angles are complementary?

f. Which pair(s) of angles are supplementary?

g. Which pair(s) of angles are adjacent?

Classify each triangle in the most precise way possible, given the indicated lengths of its sides and/or measures of its angles. See Example 9.

25.

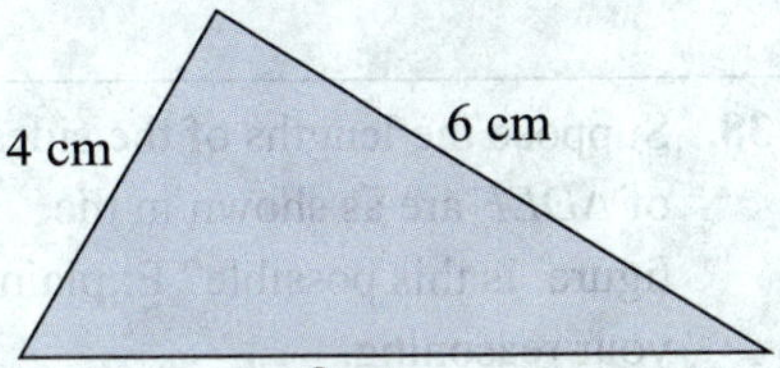

26.

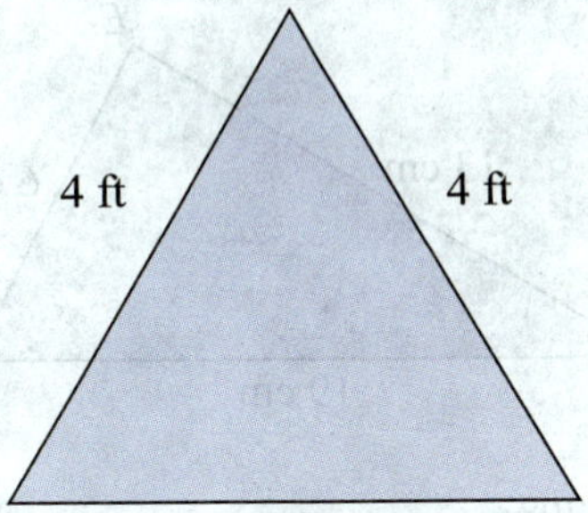

27.

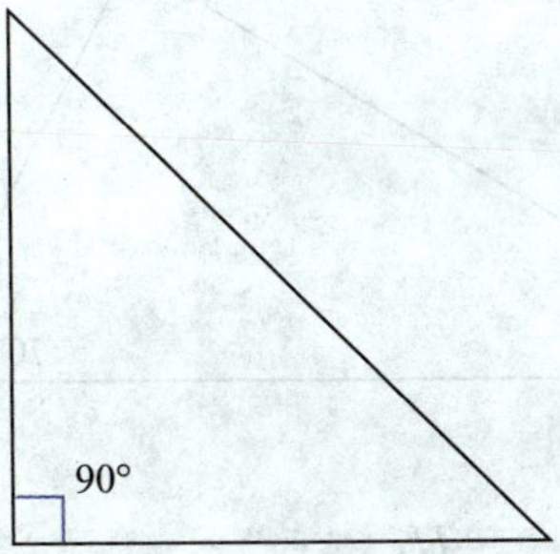

28.

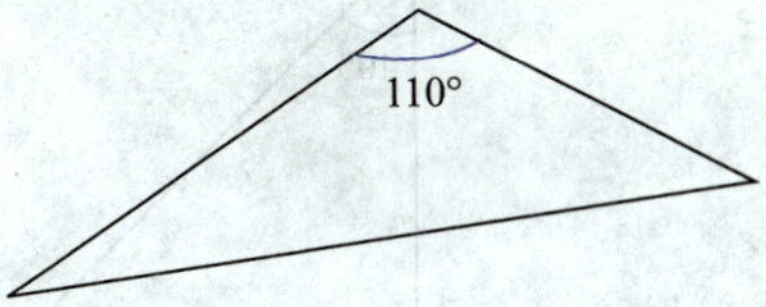

29.

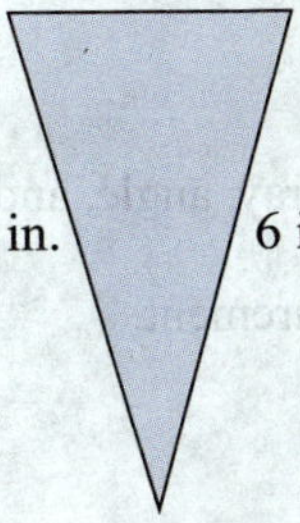

30.

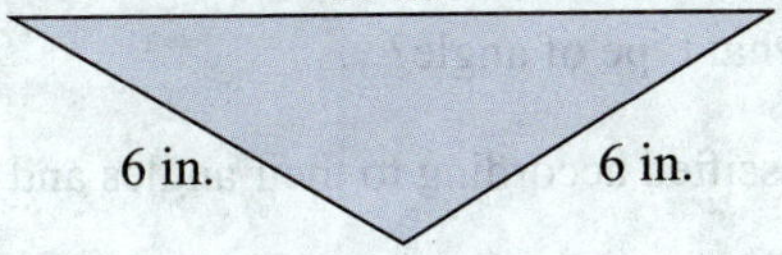

31.

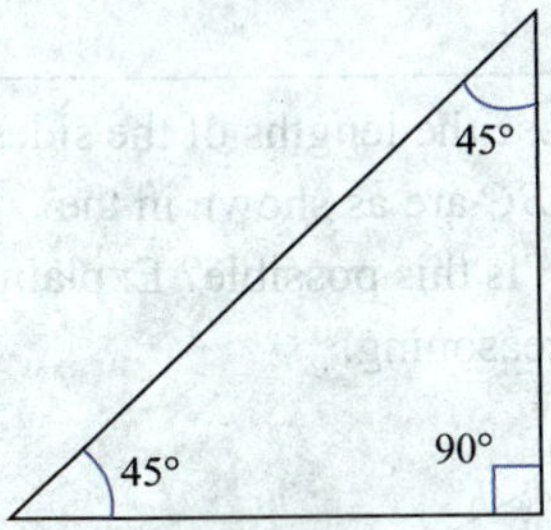

32.

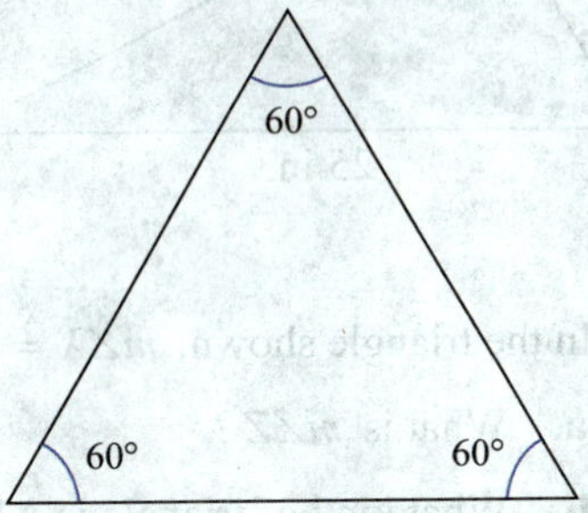

33.

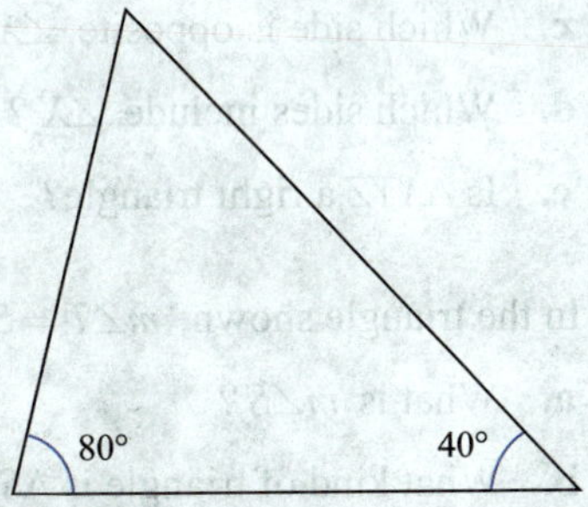

34.

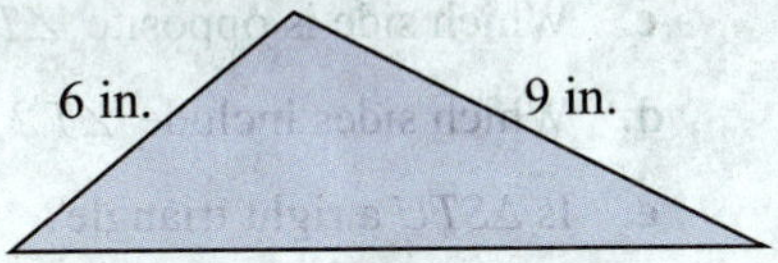

35.

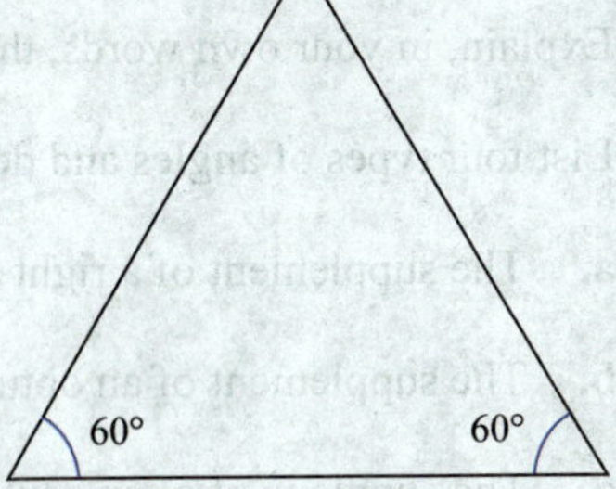

36.

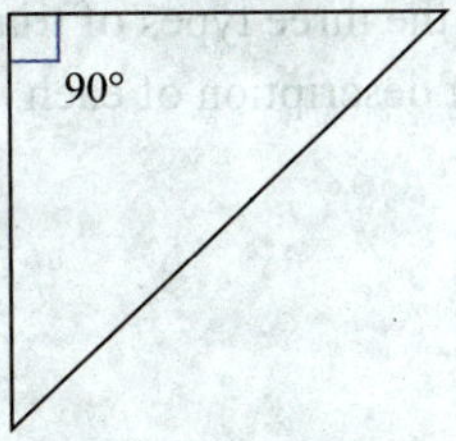

25. Scalene
26. Equilateral
27. Right
28. Obtuse
29. Isosceles
30. Isosceles
31. Isosceles and right
32. Acute
33. Acute
34. Scalene
35. Acute
36. Isosceles and right

Applications

37. Yes, since $25 < 12 + 15$.

38. No, since $19 = 13 + 6$.

39. a. $m\angle Z = 80°$

b. Acute

c. $\overline{YZ}$

d. $\overline{XZ}$ and $\overline{XY}$

e. No, no angle is 90°

40. a. $m\angle S = 90°$

b. Right

c. $\overline{SU}$

d. $\overline{ST}$ and $\overline{TU}$

e. Yes, $m\angle S = 90°$

Writing & Thinking

41. A ray is similar to a line in that it has at least one end that continues infinitely. An angle is formed by two rays that have a common endpoint, called a vertex. A line has no endpoint and continues indefinitely in opposite directions in the same plane.

42. Four types of angles are acute, right, obtuse, and straight. An acute angle measures between 0 and 90 degrees. A right angle measures 90 degrees. An obtuse angle measures between 90 and 180 degrees. A straight angle measures 180 degrees.

43. a. A right angle

b. An acute angle

c. An obtuse angle

44. Acute, right, and obtuse triangles are classified according to their angles. In an acute triangle, all three angles measure less than 90 degrees each. A right triangle has one right angle. An obtuse angle has one angle that is greater than 90 degrees.

Applications

Solve.

37. Suppose the lengths of the sides of ΔABC are as shown in the figure. Is this possible? Explain your reasoning.

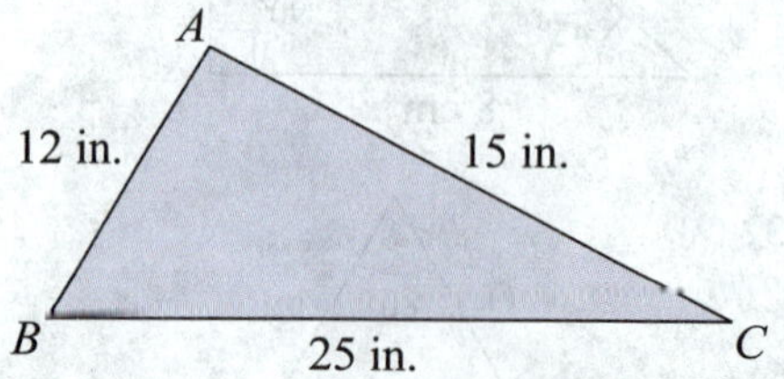

38. Suppose the lengths of the sides of ΔDEF are as shown in the figure. Is this possible? Explain your reasoning.

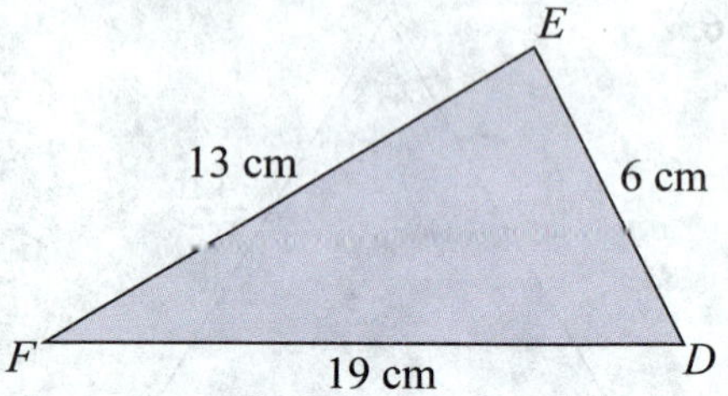

39. In the triangle shown, $m\angle X = 30°$ and $m\angle Y = 70°$.

a. What is $m\angle Z$?

b. What kind of triangle is ΔXYZ?

c. Which side is opposite $\angle X$?

d. Which sides include $\angle X$?

e. Is ΔXYZ a right triangle?

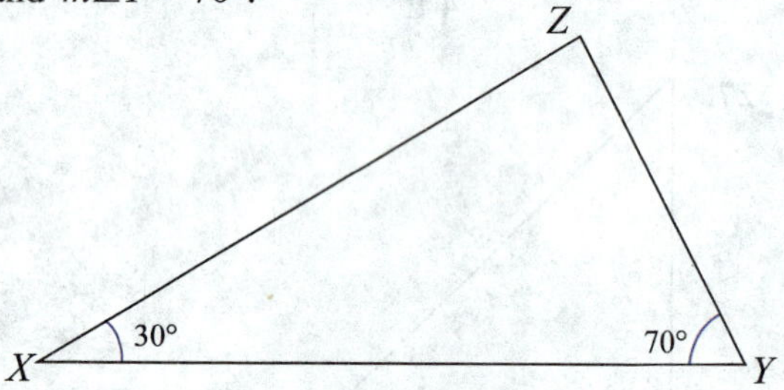

40. In the triangle shown, $m\angle T = 50°$ and $m\angle U = 40°$.

a. What is $m\angle S$?

b. What kind of triangle is ΔSTU?

c. Which side is opposite $\angle T$?

d. Which sides include $\angle T$?

e. Is ΔSTU a right triangle?

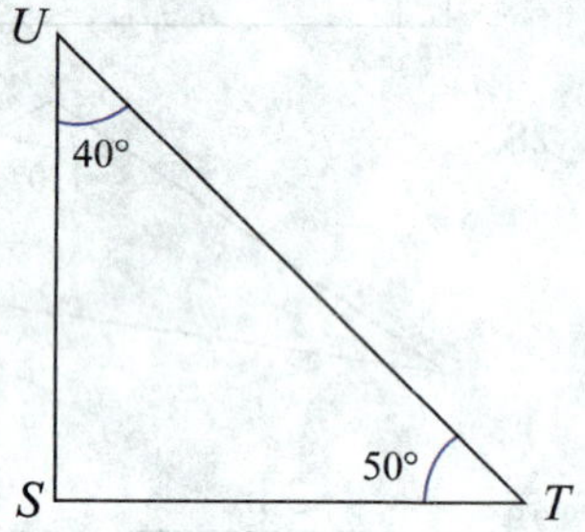

Writing & Thinking

41. Explain, in your own words, the relationships between vertex, ray, angle, and line.

42. List four types of angles and define each one in terms of measurement.

43. a. The supplement of a right angle is what type of angle?

b. The supplement of an obtuse angle is what type of angle?

c. The supplement of an acute angle is what type of angle?

44. List the three types of triangles that are classified according to their angles and give a brief description of each.

6.2 Perimeter

Objectives

A. Recognize various types of polygons.

B. Calculate the perimeters of polygons.

A Polygons

Familiar geometric figures called **polygons** (such as triangles, rectangles and squares) are defined in terms of line segments.

Polygon

A **polygon** is a closed plane figure, with three or more sides, in which each side is a line segment.

Each point where two sides meet is called a **vertex**.

Note: A **closed figure** begins and ends at the same point.

DEFINITION

A **triangle** is a polygon with three sides.

A **parallelogram** is a four-sided polygon with both pairs of opposite sides parallel and equal in length.

A **rectangle** is a parallelogram with four right angles.

A **square** is a rectangle in which all four sides are the same length.

A **trapezoid** is a four-sided polygon with exactly one pair of opposite sides that are parallel.

An illustration of each of these polygons is shown in Figure 1.

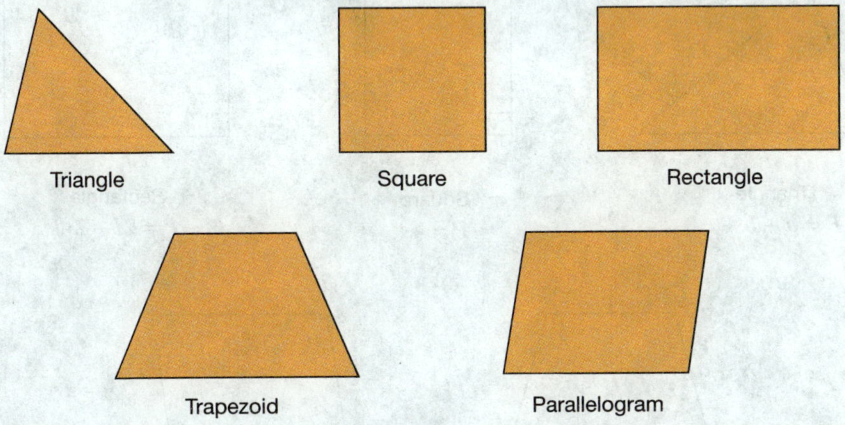

Figure 1

B Finding the Perimeter of a Polygon

As stated in Chapter 4, a **formula** is a general statement (usually an equation) that relates two or more variables. The formulas given here are used to calculate the **perimeters** of polygons.

Perimeter

The **perimeter** P of a polygon is the sum of the lengths of its sides.

DEFINITION

The formulas for the perimeters of several polygons are given here. For a rectangle, l represents length and w represents width. (Customarily, the width is the shorter of the two.) For a square, s is the length of one side, and all four sides have the same length. When using these formulas, you should be familiar with the following basic units of length and the corresponding abbreviations. **Be sure to label your answers with the correct units.**

From the Metric System	From the US Customary System
millimeter (mm)	inch (in.)
centimeter (cm)	foot (ft)
meter (m)	yard (yd)
kilometer (km)	mile (mi)

Table 1

Perimeter Formulas for Five Polygons

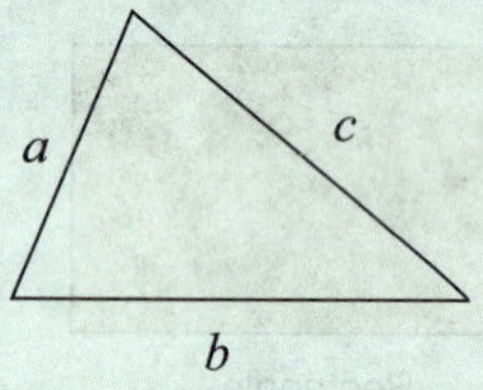

Triangle
$P = a + b + c$

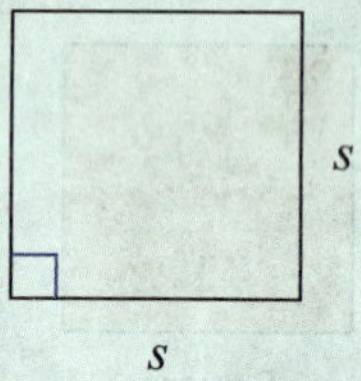

Square
$P = 4s$

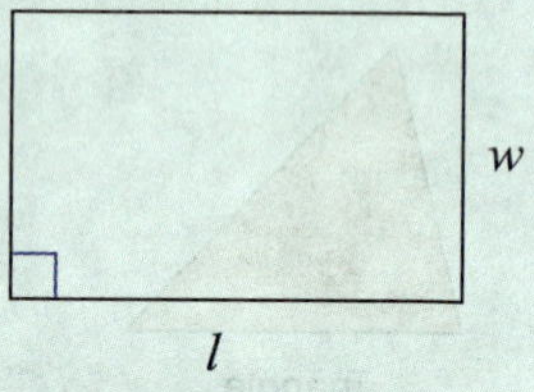

Rectangle
$P = 2l + 2w$

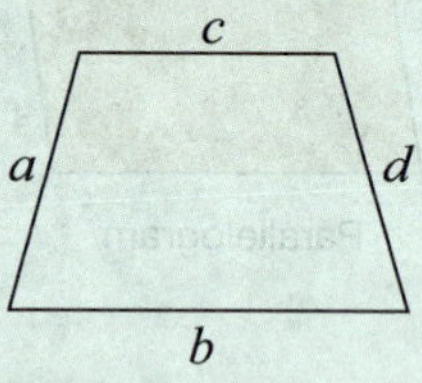

Trapezoid
$P = a + b + c + d$

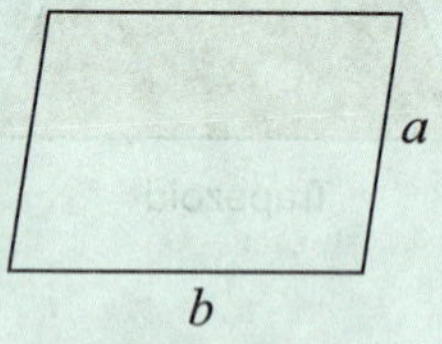

Parallelogram
$P = 2a + 2b$

Note that each formula represents the sum of the lengths of the sides.

FORMULA

Teaching Note

The examples in this section illustrate how figures can help students better understand problems, particularly when they involve geometry. Discuss the appropriateness of all visual aids regardless of how amateurish they may seem to the student.

Example 1 Calculating the Perimeter of a Square

Calculate the perimeter of a square with sides of length 16 in.

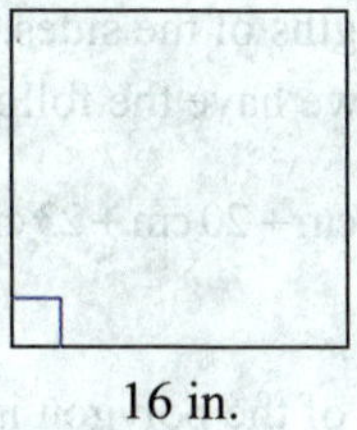

Solution

Using the formula for the perimeter of a square, we have the following.

$$P = 4s$$
$$P = 4 \cdot 16 \text{ in.}$$
$$= 64 \text{ in.}$$

The perimeter of the square is 64 in.

Now work margin exercise 1.

1. Calculate the perimeter of a square with sides of length 8 ft.

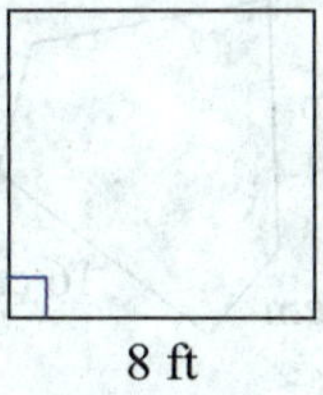

Example 2 Calculating the Perimeter of a Triangle

Calculate the perimeter of a triangle with sides of length 40 mm, 70 mm, and 80 mm.

Solution

First draw a figure and label the lengths of the sides.

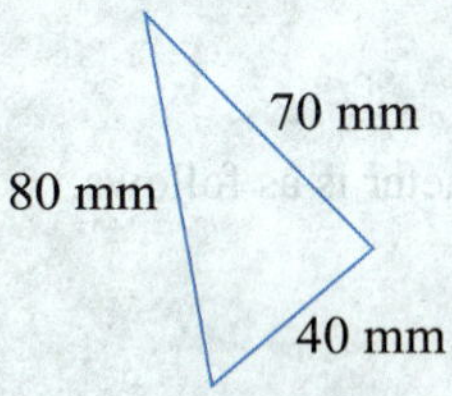

Now, using the formula for the perimeter of a triangle, we have the following.

$$P = a + b + c$$
$$P = 40 \text{ mm} + 70 \text{ mm} + 80 \text{ mm}$$
$$= 190 \text{ mm}$$

The perimeter of the triangle is 190 mm.

Now work margin exercise 2.

2. Calculate the perimeter of a triangle with sides of length 15 ft, 40 ft, and 30 ft. It may be helpful to begin by drawing the figure.

Example 3 Calculating the Perimeter of a Rectangle

Calculate the perimeter of the rectangle.

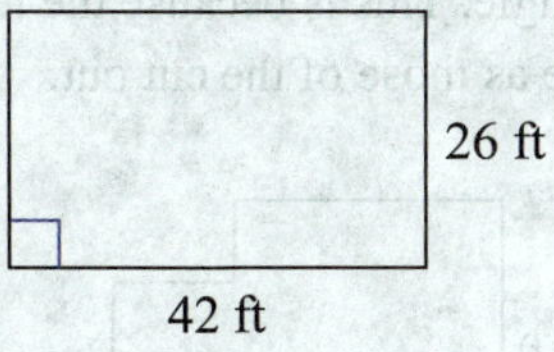

Solution

Using the formula for the perimeter of a rectangle, we have the following.

$$P = 2l + 2w$$
$$P = 2 \cdot 42 \text{ ft} + 2 \cdot 26 \text{ ft}$$
$$= 84 \text{ ft} + 52 \text{ ft}$$
$$= 136 \text{ ft}$$

The perimeter of the rectangle is 136 ft.

Now work margin exercise 3.

3. Calculate the perimeter of the rectangle.

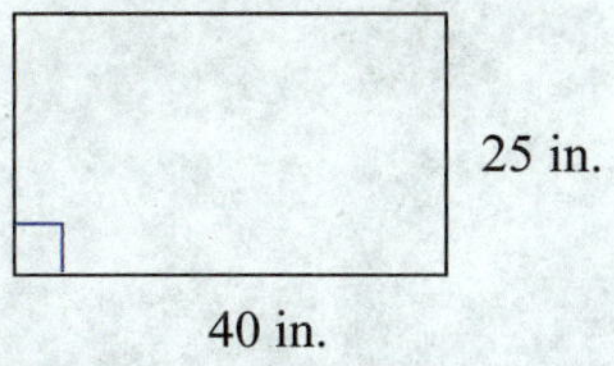

4. Calculate the perimeter of the polygon.

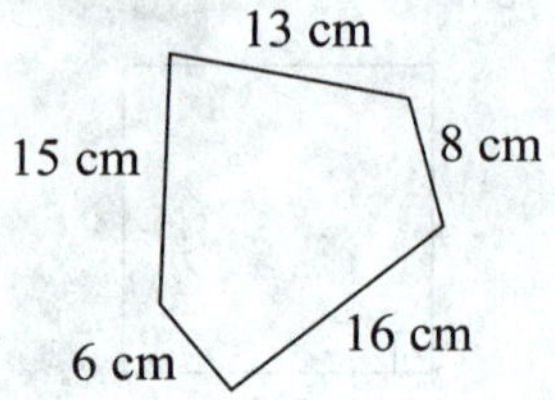

Example 4 Calculating the Perimeter of a Polygon

Calculate the perimeter of the polygon. (**Note:** A 5-sided polygon is called a **pentagon**.)

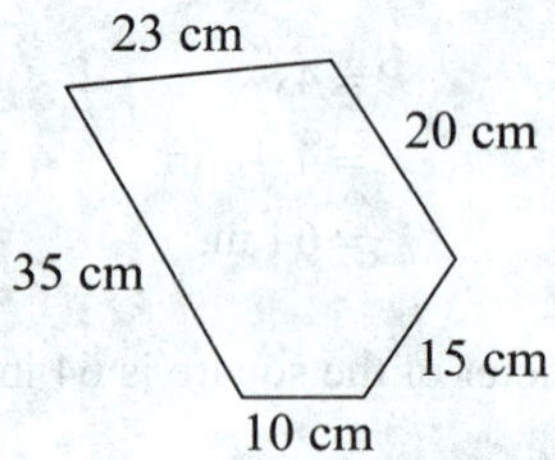

Solution

The perimeter of any polygon is found by adding the lengths of the sides. Thus, for this pentagon we have the following.

$$\begin{aligned} P &= 10\text{ cm} + 15\text{ cm} + 20\text{ cm} + 23\text{ cm} + 35\text{ cm} \\ &= 103\text{ cm} \end{aligned}$$

The perimeter of the polygon is 103 cm.

Now work margin exercise 4.

5. Calculate the perimeter of the polygon.

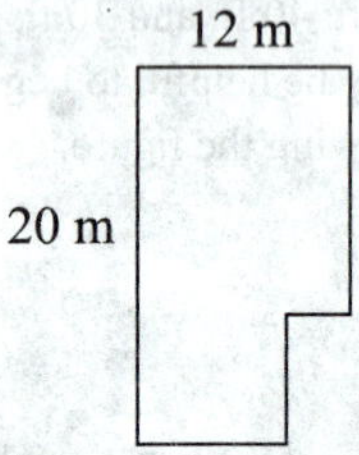

Example 5 Calculating the Perimeter of a Polygon

Calculate the perimeter of the polygon.

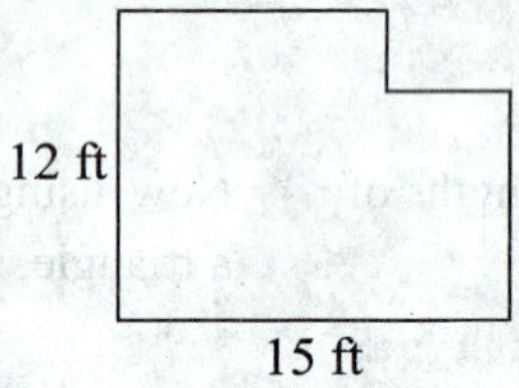

Solution

If a rectangle has length 15 ft and width 12 ft, then its perimeter is as follows.

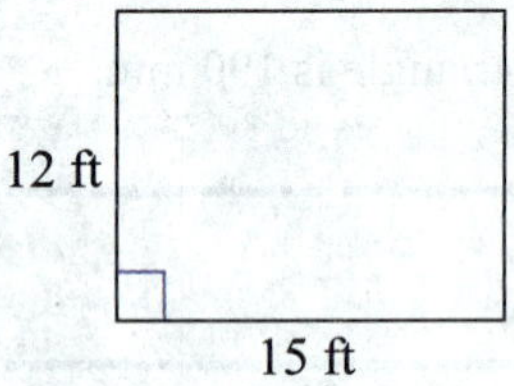

$$\begin{aligned} P &= 2 \cdot 15\text{ ft} + 2 \cdot 12\text{ ft} \\ &= 30\text{ ft} + 24\text{ ft} \\ &= 54\text{ ft} \end{aligned}$$

Now, if a small rectangle is cut from one corner of the original rectangle, then a new shape is formed. An interesting fact is that, regardless of the size of the cut out rectangle, this new shape will have the **same perimeter** as the original rectangle. This is because the length and width of the segments of the indented corner are the same as those of the cut out.

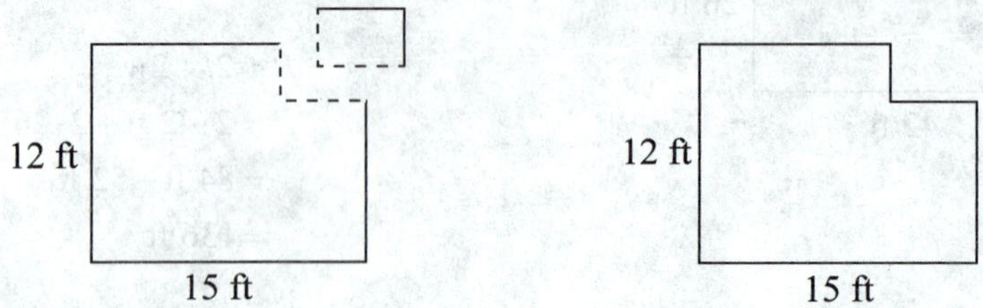

Thus, the perimeter of the polygon is 54 ft.

Now work margin exercise 5.

Example 6 Application: Calculating the Perimeter of a Polygon

For security, a chain link fence is to be built on the edge of a property surrounding a new warehouse. The property is L-shaped as shown.

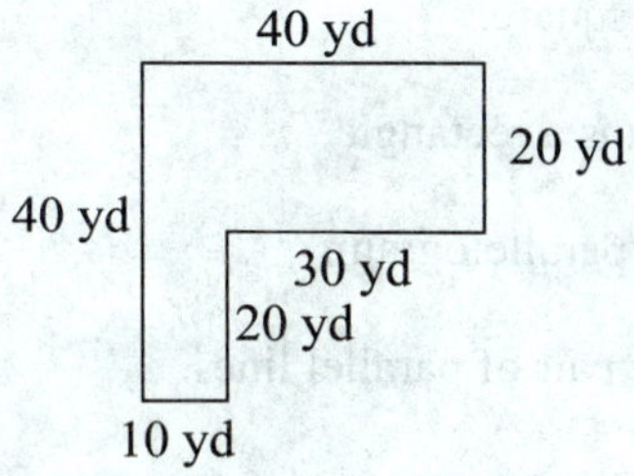

a. How many yards of fencing will be needed?

b. How much will be spent on fencing if the fencing that was chosen costs \$12.50 per yard?

Solution

a. As the fence is to be put up along the edge of the property, the amount of fencing needed is equivalent to the perimeter of the property.

$$P = 40\text{ yd} + 40\text{ yd} + 20\text{ yd} + 30\text{ yd} + 20\text{ yd} + 10\text{ yd} = 160\text{ yd}$$

They will need 160 yd of fencing.

b. $\text{Cost} = \$12.50 \cdot 160 = \$2000$

The cost of the fencing will be \$2000.

Now work margin exercise 6.

6. Brandy is adding plastic trimming around the edge of her new swimming pool, which has the following shape.

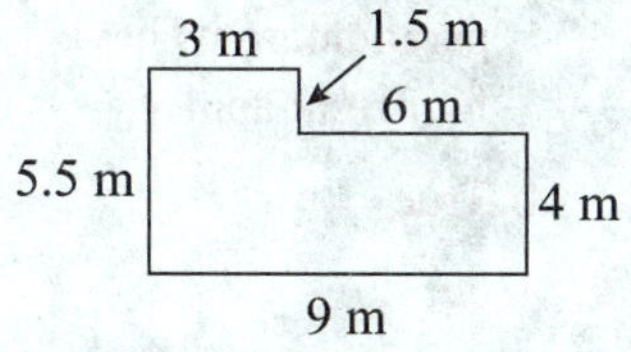

a. How many meters of trimming will she need?

b. What will be the total cost of the trimming if she can buy it for \$7.50 a meter?

Margin Exercise Answers

1. 32 ft **2.** 85 ft **3.** 130 in. **4.** 58 cm **5.** 64 m **6. a.** 29 m **b.** \$217.50

6.2 Exercises

Concept Check

Fill-in-the-Blank. Complete each sentence using information found in this section.

1. A closed plane figure with at least three sides in which each side is a line segment is a/an ________.
2. A four-sided polygon with both pairs of opposite sides parallel is a/an ________.
3. A polygon with four sides in which adjacent sides form right angles is a/an ________.
4. A figure that begins and ends at the same point is known as a/an ________ figure.
5. A polygon with three sides is known as a/an ________.
6. The sum of the lengths of the sides of a polygon is its ________.

6.2 Exercises

Concept Check

1. polygon
2. parallelogram
3. rectangle
4. closed
5. triangle
6. perimeter

7. a. True
b. False; Not all rectangles have four equal sides.
8. a. False; Not all parallelograms have four right angles.
b. True
9. True
10. True
11. a. B b. C c. A d. E e. D

Practice
1. 44 cm
2. 152 mm
3. 116 cm
4. $12\frac{1}{6}$ ft
5. 18 km
6. 44 m
7. 143 in.
8. 38.1 in.
9. 48.1 yd
10. 166 ft

True/False. Determine whether each statement is true or false. If a statement is false, explain how it can be changed so the statement will be true. (**Note:** There may be more than one acceptable change.)

7. a. Every square is a rectangle.
b. Every rectangle is a square.
8. a. Every parallelogram is a rectangle.
b. Every rectangle is a parallelogram.
9. A trapezoid has only one pair of parallel lines.
10. The formula to calculate the perimeter of a square is $P = 4s$ where s is the length of one side.

Match each formula for perimeter to its corresponding geometric figure.

11. a. Square — A. $P = 2l + 2w$
b. Parallelogram — B. $P = 4s$
c. Rectangle — C. $P = 2b + 2a$
d. Trapezoid — D. $P = a + b + c$
e. Triangle — E. $P = a + b + c + d$

Practice

Calculate the perimeter of each figure described. See Examples 1 through 3.

1. A parallelogram with sides of length 15 cm and 7 cm.
2. A parallelogram with sides of length 42 mm and 34 mm.
3. A rectangle with sides of length 24 cm and 34 cm.
4. A rectangle with sides of length $3\frac{1}{4}$ ft and $2\frac{5}{6}$ ft.
5. A square with sides of length $4\frac{1}{2}$ km.
6. A square with sides of length 11 m.
7. A triangle with sides of length 21 in., 67 in., and 55 in.
8. A triangle with sides of length 7.5 in., 17 in., and 13.6 in.
9. A trapezoid with sides of length 14.2 yd, 10.1 yd, 8 yd, and 15.8 yd.
10. A trapezoid with sides of length 31 ft, 39 ft, 45 ft, and 51 ft.

Calculate the perimeter of each figure.

11.

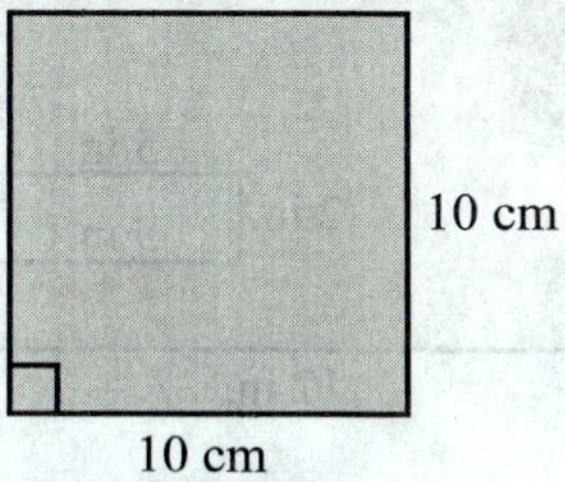

12.

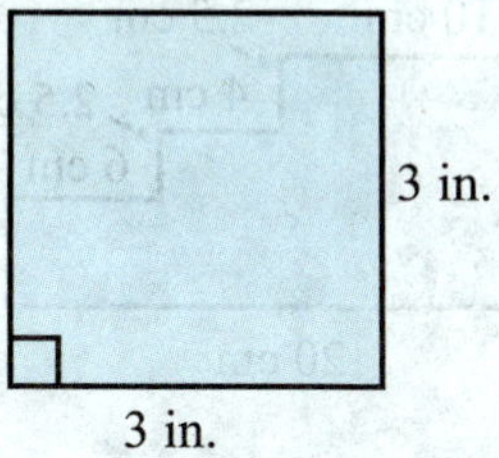

13.

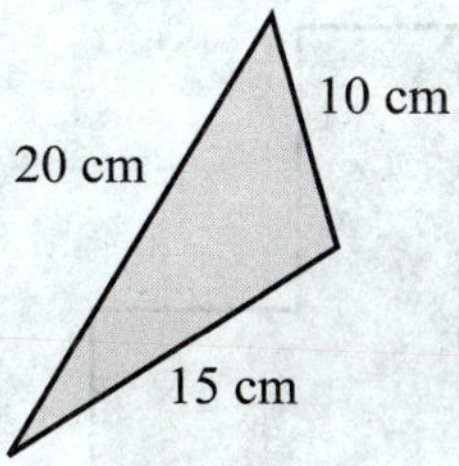

14.

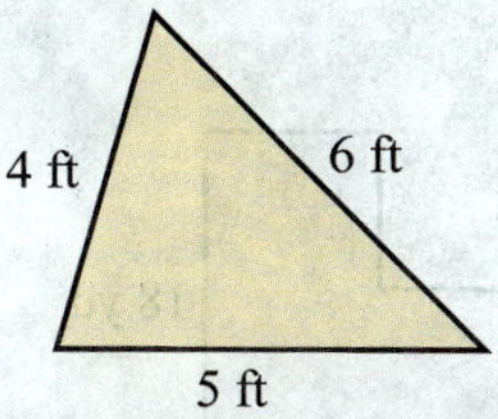

15.

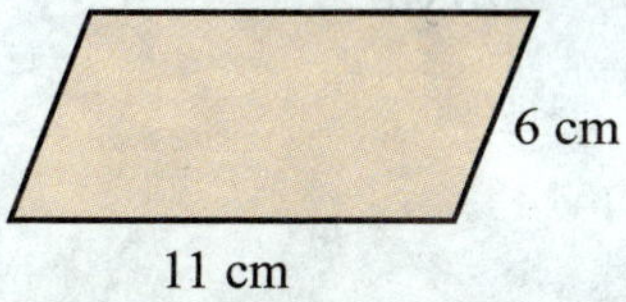

16.

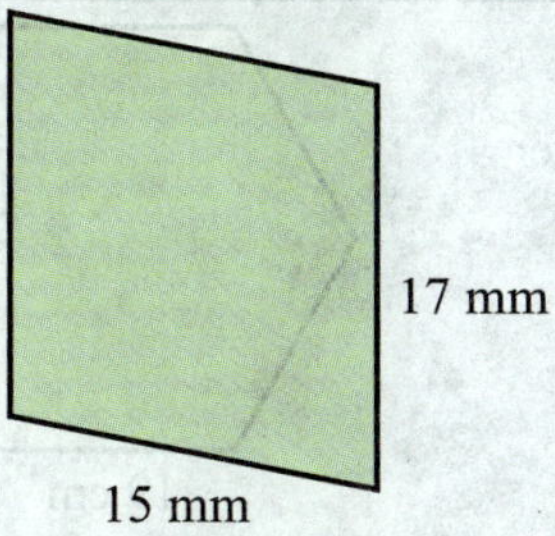

17.

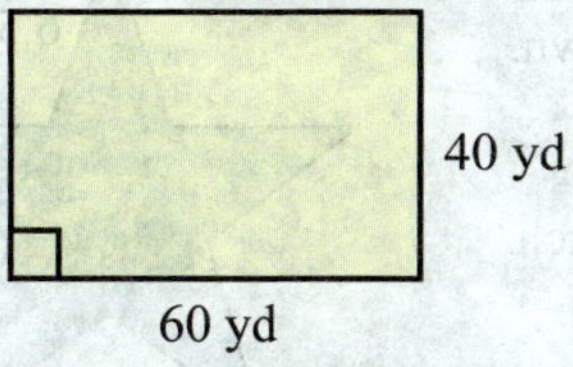

18.

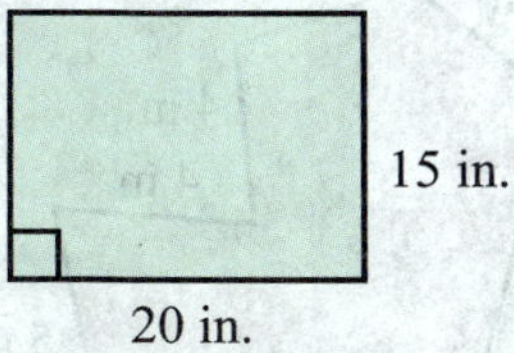

19.

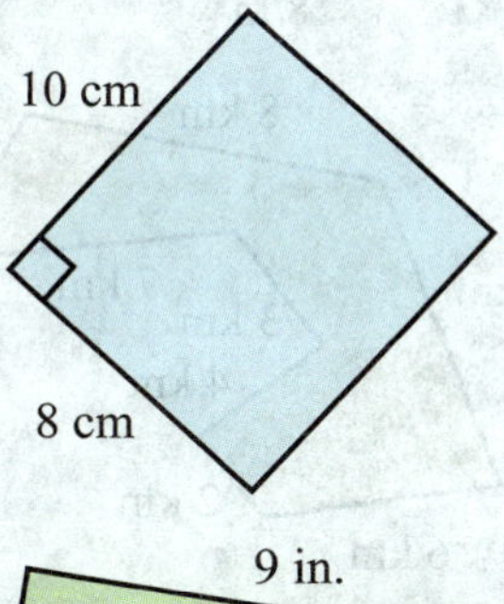

20.

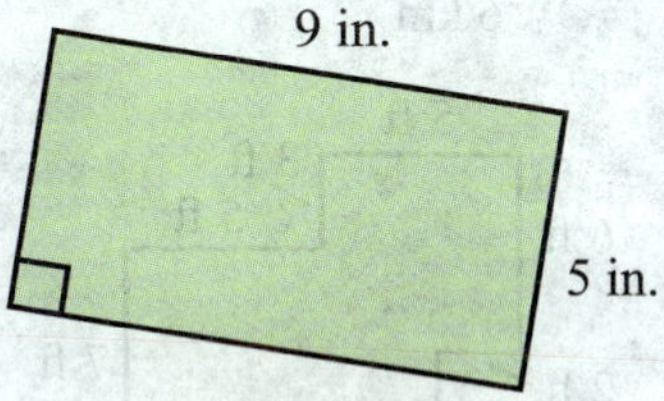

21.

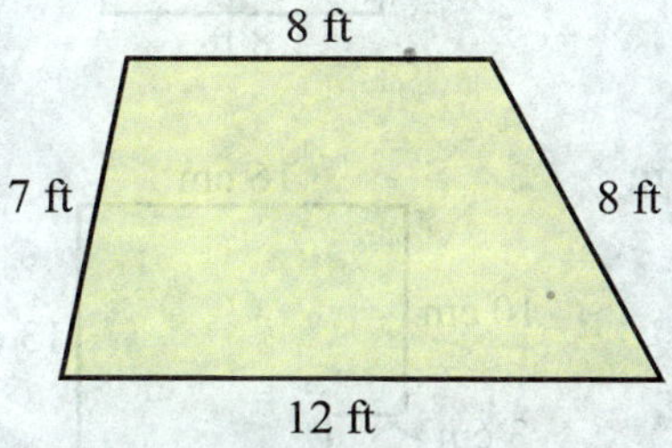

22.

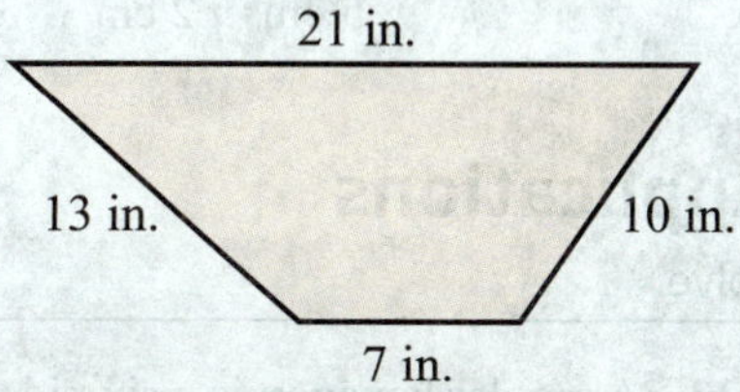

23.

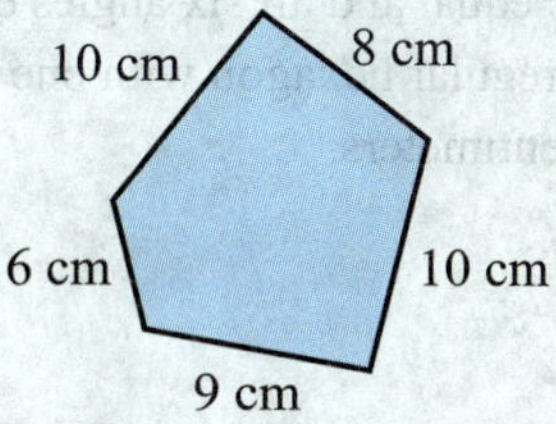

24.

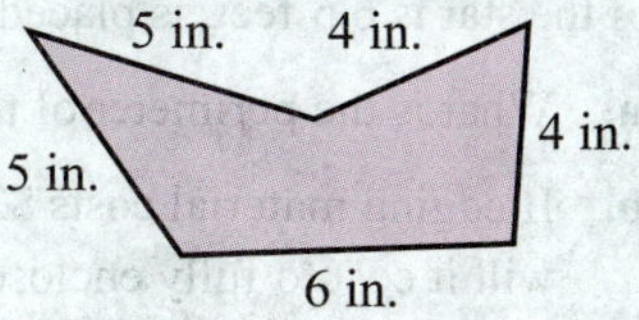

11. 40 cm
12. 12 in.
13. 45 cm
14. 15 ft
15. 34 cm
16. 64 mm
17. 200 yd
18. 70 in.
19. 36 cm
20. 28 in.
21. 35 ft
22. 51 in.
23. 43 cm
24. 24 in.

25. 36 m
26. 41 km
27. 40 ft
28. 62 cm
29. 50 in.
30. 60 cm
31. 108 ft
32. 76 yd

Applications

33. 114 cm
34. a. 65 ft b. $156

25.

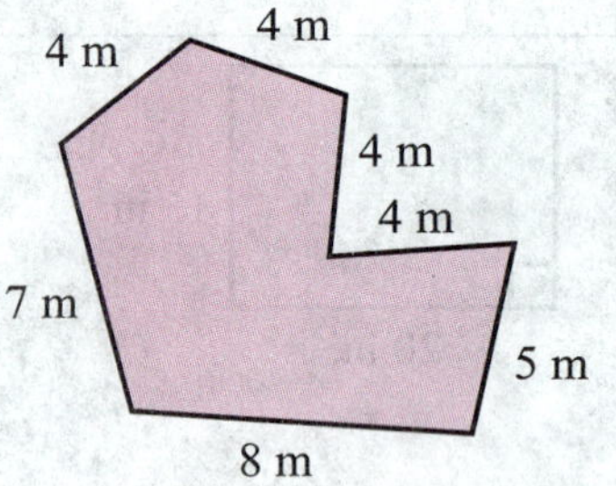

26.

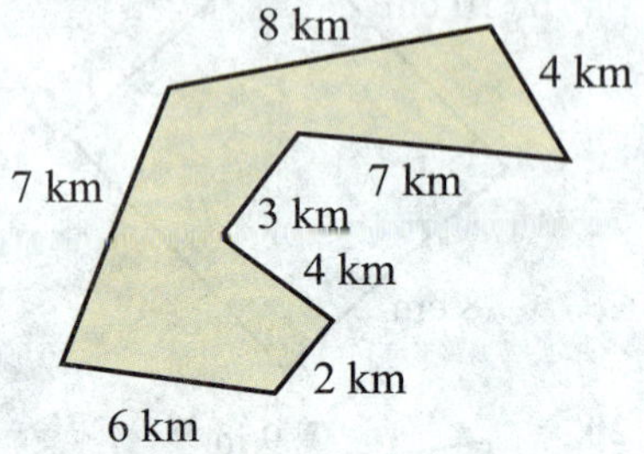

27.

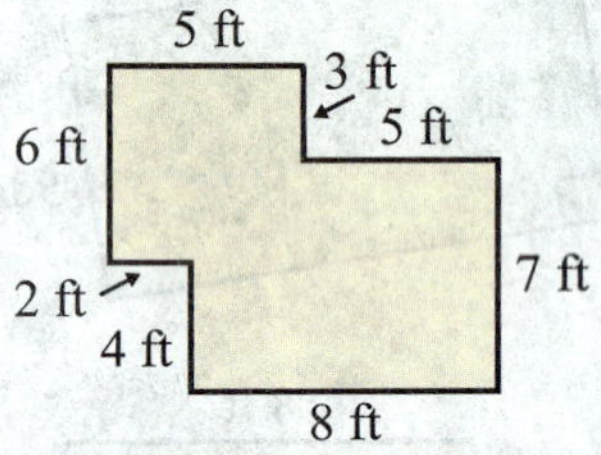

28.

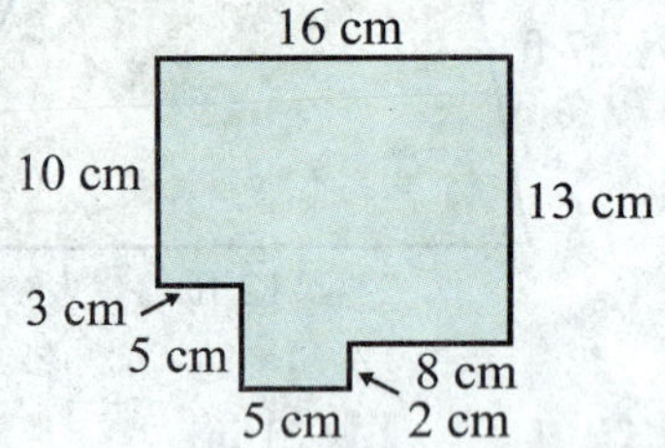

29.

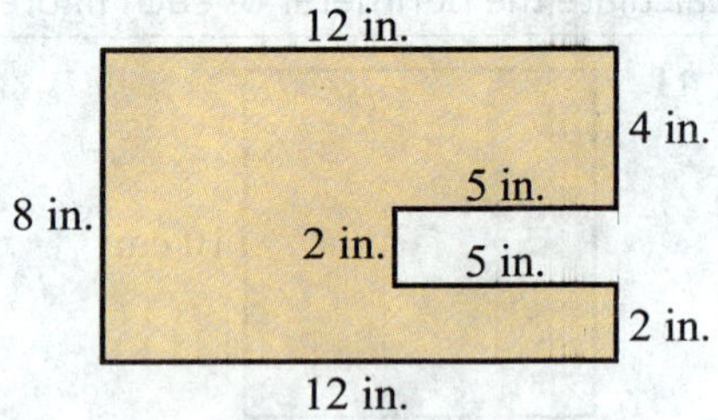

30.

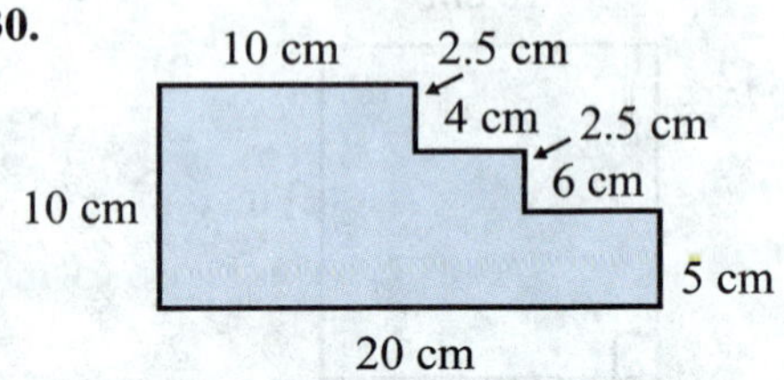

31.

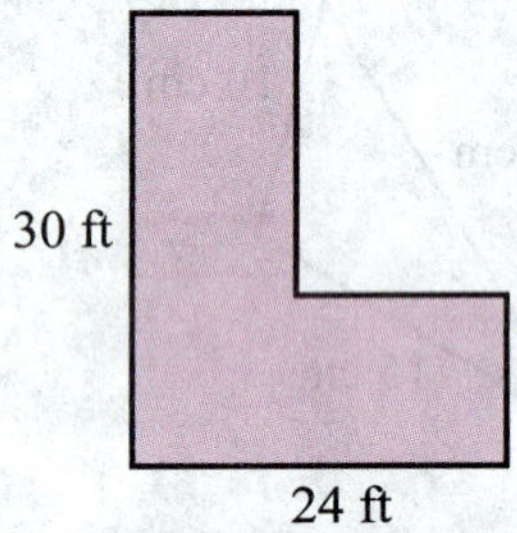

32.

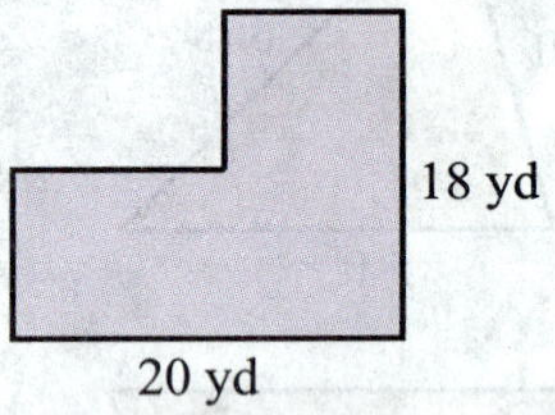

Applications

Solve.

33. A regular hexagon is a six-sided figure with all six sides equal and all six angles equal. Find the perimeter of a regular hexagon with one side measuring 19 centimeters.

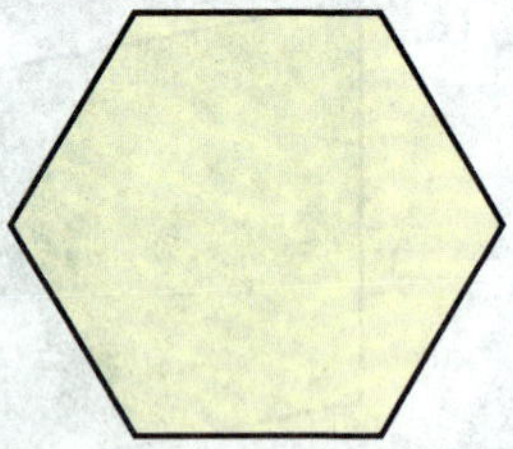

34. A five-pointed star-shaped flower plot, where each edge of the star is 6.5 feet, is placed in the middle of a lawn.

a. What is the perimeter of the plot?

b. If edging material costs $2.40 per foot, how much will it cost to fully enclose the star?

35. A rectangular picture frame is 10.5 inches high and 8.7 inches wide. How much picture framing material must be used to frame the picture? (**Hint:** This is the same as the perimeter of the outer edge.)

36. A regular octagon is an eight-sided figure with all eight sides equal and all eight angles equal. Find the perimeter of a regular octagon if one side measures 12 inches. (**Note:** Where do you see regular octagons an a regular basis?)

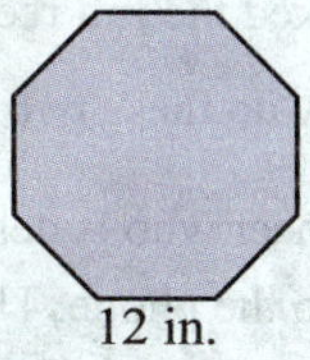

37. The Pentagon near Washington, D.C., is a five-sided building where each outside wall is 921 feet. [1]

a. What is the perimeter of the building?

b. If it takes a person 0.00341 minutes to walk 1 foot, how long will it take the person to walk completely around the building? Round your answer to the nearest tenth of a minute.

38. A public building consists of a rectangular portion used for office space that is 95 feet long and 27 feet wide connected to an octagonal shaped auditorium where each of the eight sides is the same length as the width of the rectangular portion. The roof is constructed so that there are rain gutters on all sides of the building. How much guttering must be purchased? (**Hint:** One side of the rectangle and one side of the octagon will not be included in computing the perimeter.)

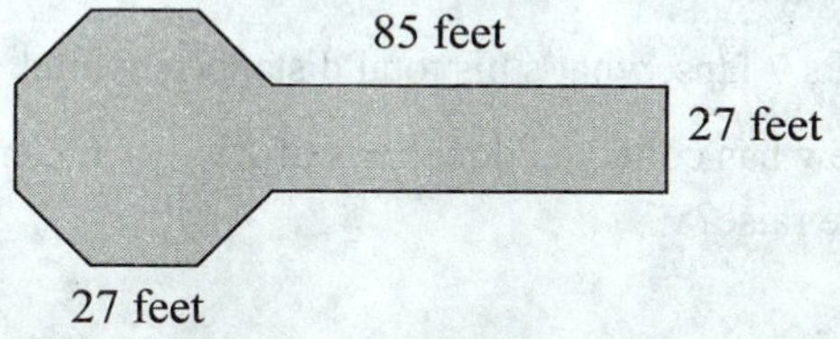

39. A property owner has an odd-shaped lot as shown.

a. What is the perimeter of the lot?

b. If the cost of constructing a fence is $15.00 per foot, how much will it cost to construct a fence around the perimeter of the lot?

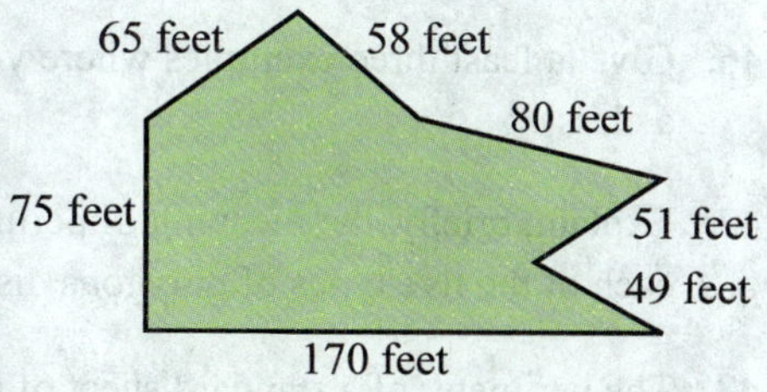

35. 38.4 in.

36. 96 in.; stop signs

37. a. 4605 ft

b. 15.7 minutes

38. 386 ft

39. a. 548 feet **b.** $8220

1 Source: www.infoplease.com/spot/pentagon1.html

40. **a.** 4180 ft **b.** 4180 ft
c. They are identical.

41. $13\frac{4}{5}$ inches

42. 27 feet

43. $17\frac{1}{2}$ km

44. **a.** 2800 m
b. $700

Writing & Thinking

45. Some of the polygons are: triangle (3 sides), square (4 sides), rectangle (4 sides), parallelogram (4 sides), and trapezoid (4 sides).

46. Examples will vary. Road signs can be square, rectangular, or triangular. TVs and many appliances like stoves and refrigerators are rectangular. When a cardboard box is broken down it can form a rectangle or a parallelogram. An athletic field or court can be square or rectangular. Yard fencing might be in the shape of a trapezoid

47. Perimeter is the distance around a figure. Formulas for the perimeter of: triangle $(P = a + b + c)$, square $(P = 4s)$, rectangle $(P = 2l + 2w)$, trapezoid $(P = a + b + c + d)$, and parallelogram $(P = 2a + 2b)$.

48. In each case, the perimeter is the same.

40. For exercise, John will walk along the path that is indicated by the solid line in the drawing. Note that he cuts a corner where both sides are the same length.

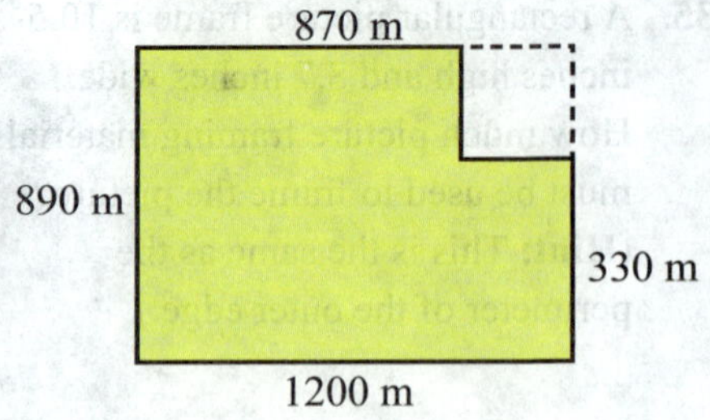

a. How many meters did John walk?

b. How long would the walk have been if John didn't cut the corner, but rather walked the full rectangle, as indicated by the dotted line?

c. How do these two distances compare?

41. An engineer who is designing a new smartphone decides to add a soft neoprene edging to the phone. The phone itself is $4\frac{1}{2}$ inches tall and $2\frac{2}{5}$ inches wide. How much neoprene edging is needed to go along the outside edge of each smartphone?

42. Jessica wants to add a decorative fringe to a throw rug. The rug is a rectangle with length 8 feet and width 5 feet. If Jessica wants to buy 1 foot more than the perimeter of the rug, how many feet of fringe must she buy?

43. While training for a triathlon, Carly maps out a route with one side cutting through a lake, one side to bike, a third side to run, plus a final side to cool down and return to her starting point. She starts by swimming a length of $\frac{1}{2}$ km, then bikes a length of 8 km, runs a length of 5 km, and has a cool-down jog of 4 km back to her starting place. What is the total distance of this practice route?

44. Matthew is participating in a walk-a-thon to raise money for his school. His class is walking a rectangular path on the school property. The path is 150 m long and 50 m wide.

a. If Matthew walks 7 laps, what's his total distance covered

b. Suppose Matthew has collected donations of 25¢ per meter walked. How much money would he raise?

Writing & Thinking

45. Name as many polygons as you can and include the number of sides for each one.

46. Give at least three examples where you might see specific polygons (outside of a class).

47. Explain, briefly, the meaning of perimeter. Write the formula for the perimeter of each of the five types of polygons discussed in this section.

48. The perimeter of a standard sheet of paper ($8\frac{1}{2}$ in. by 11 in.) is $P = 2 \cdot 8\frac{1}{2} + 2 \cdot 11 = 17 + 22 = 39$ inches. Use a pair of scissors to cut a rectangle from one corner of a standard sheet of paper and measure the perimeter of the new figure. Repeat this process several times (in some cases, cut more than one rectangle from a different corner and measure the perimeter each time). Give a brief explanation of the results in each case.

6.3 Area

Objectives

A. Calculate the areas of polygons.

A Finding the Area of a Polygon

Area is a measure of the interior of (or surface enclosed by) a plane figure and is measured in **square units**. The concept of area was discussed in Chapter 1 and is illustrated again in Figure 1. (**Note:** in.^{2} is read "inches squared" or "square inches.")

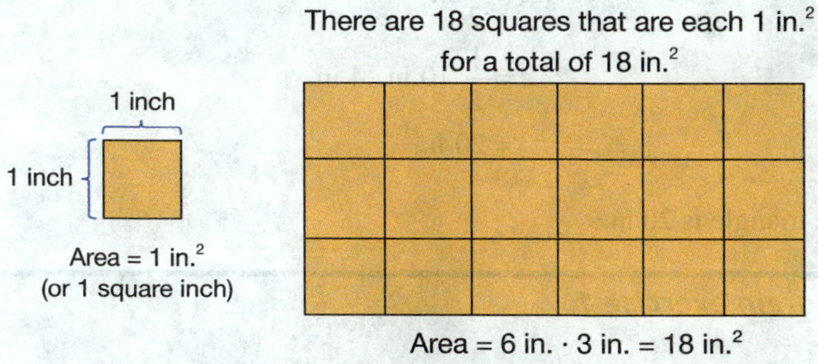

Figure 1

You should be familiar with the following basic units of area and the corresponding abbreviations.

From the Metric System	From the US Customary System
square millimeters (mm^2)	square inches (in.^{2})
square centimeters (cm^2)	square feet (ft^2)
square meters (m^2)	square yards (yd^2)
square kilometers (km^2)	square miles (mi^2)

Table 1

The formulas for the areas of the geometric figures shown here are independent of the units of measure used. **Be sure to label your answers with the correct units.**

Area Formulas for Five Polygons

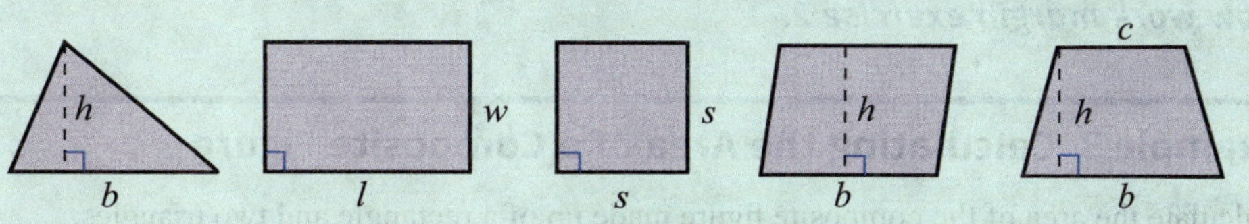

Triangle	Rectangle	Square	Parallelogram	Trapezoid
$A=\frac{1}{2}bh$	$A=lw$	$A=s^2$	$A=bh$	$A=\frac{1}{2}h(b+c)$

Note: The letter h is used to represent the **height** of the figure. The height is also called the **altitude** and is perpendicular to the base.

FORMULA

1. Calculate the area of a triangle with height 7 mm and base 8 mm.

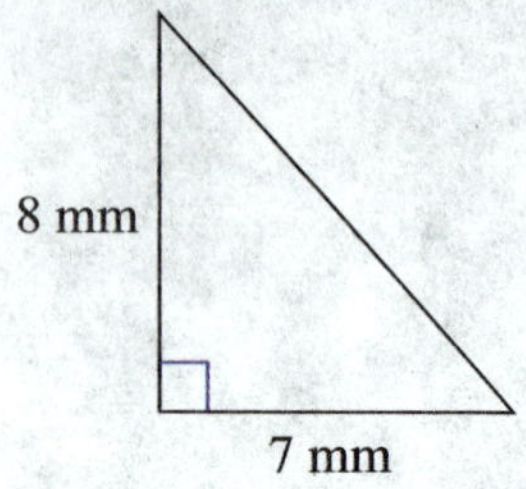

Example 1 Calculating the Area of a Triangle Using a Formula

Calculate the area of a triangle with height 4 in. and base 10 in. (Be sure to label the answer in square inches.)

Solution

Now, using the formula for the area of a triangle, we have the following.

$$A = \frac{1}{2}bh$$

$$A = \frac{1}{2} \cdot 10 \text{ in.} \cdot 4 \text{ in.}$$

$$= 20 \text{ in.}^2$$

The area of the triangle is 20 in.^{2}

Now work margin exercise 1.

2. Calculate the area of a trapezoid with altitude 3 cm and parallel sides of length 9 cm and 15 cm. It may be helpful to begin by drawing the figure.

Example 2 Calculating the Area of a Trapezoid Using a Formula

Calculate the area of a trapezoid with altitude 6 in. and parallel sides of length 12 in. and 24 in.

Solution

First draw a figure and label the lengths of the known parts.

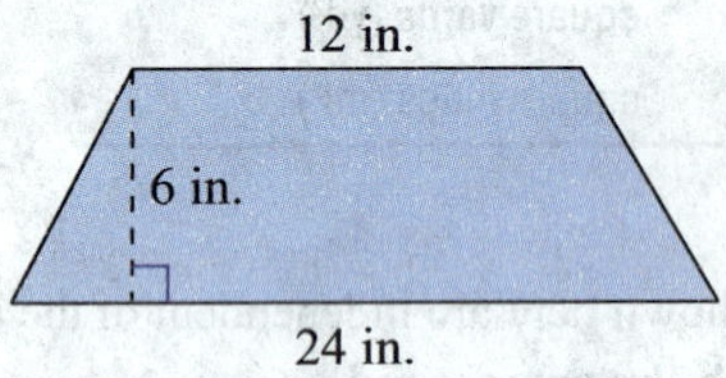

Now, using the formula for the area of a trapezoid, we have the following.

$$A = \frac{1}{2}h(b+c)$$

$$A = \frac{1}{2} \cdot 6 \text{ in.} \cdot (24 \text{ in.} + 12 \text{ in.})$$

$$= 3 \text{ in.} \cdot 36 \text{ in.}$$

$$= 108 \text{ in.}^2$$

The area of the trapezoid is 108 in.^{2}

Now work margin exercise 2.

3. Calculate the area of the composite figure.

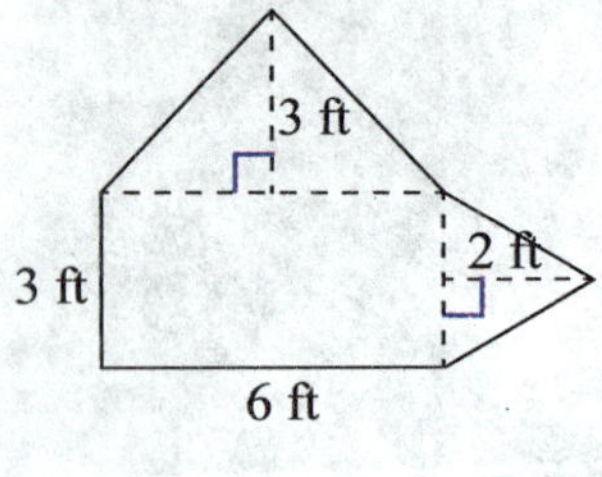

Example 3 Calculating the Area of a Composite Figure

Calculate the area of the composite figure made up of a rectangle and two triangles.

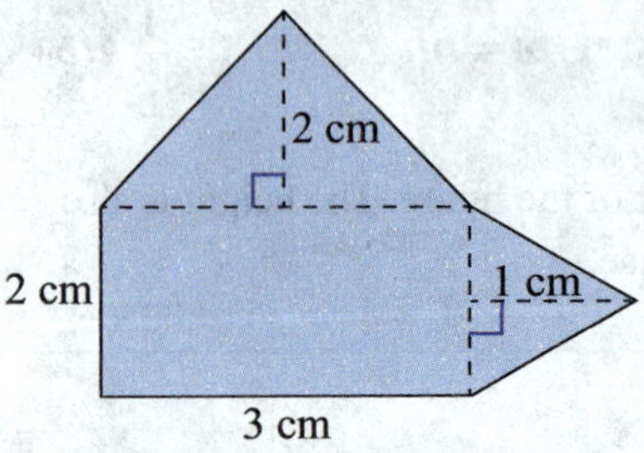

Solution

To find the area of this figure, calculate the area of each part and then add the three areas. The figure is made up of two triangles and one rectangle.

Rectangle	**Larger Triangle**	**Smaller Triangle**
$A = lw$	$A = \frac{1}{2}bh$	$A = \frac{1}{2}bh$
$A = 2 \text{ cm} \cdot 3 \text{ cm} = 6 \text{ cm}^2$	$A = \frac{1}{2} \cdot 3 \text{ cm} \cdot 2 \text{ cm} = 3 \text{ cm}^2$	$A = \frac{1}{2} \cdot 2 \text{ cm} \cdot 1 \text{ cm} = 1 \text{ cm}^2$

$$\begin{aligned}\text{Total area} &= 6 \text{ cm}^2 + 3 \text{ cm}^2 + 1 \text{ cm}^2 \\ &= 10 \text{ cm}^2\end{aligned}$$

The area of the composite figure is 10 cm².

Now work margin exercise 3.

Example 4 Calculating Area

A square is cut out of a rectangle as shown. Calculate the area of the shaded region.

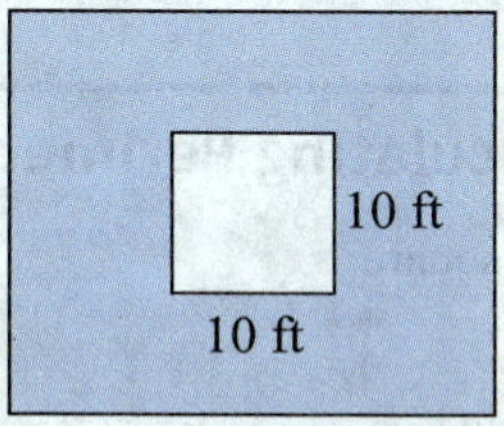

4. A rectangle is cut out of a rectangle as shown. Calculate the area of the shaded region.

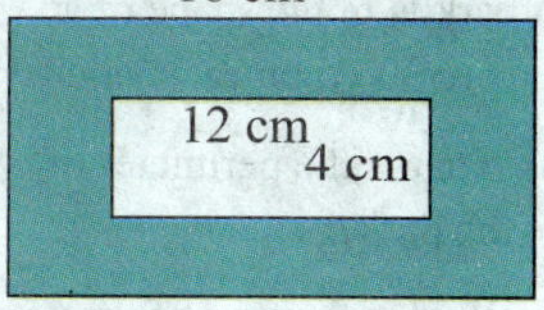

Solution

There are three steps in calculating the area of the shaded region. First, calculate the area of the outer figure. Then, calculate the area of the inner figure. Finally, calculate the difference between the areas.

Step 1: Calculate the area of the rectangle.

$$\begin{aligned}A &= lw \\ A &= 30 \text{ ft} \cdot 25 \text{ ft} = 750 \text{ ft}^2\end{aligned}$$

Step 2: Calculate the area of the square.

$$\begin{aligned}A &= s^2 \\ A &= (10 \text{ ft})^2 = 100 \text{ ft}^2\end{aligned}$$

Step 3: Calculate the difference between the two areas.

$$\text{Area of shaded region} = 750 \text{ ft}^2 - 100 \text{ ft}^2 = 650 \text{ ft}^2$$

The area of the shaded region is 650 ft².

Now work margin exercise 4.

5. Again, this polygon is a rectangle with a rectangular piece missing. Calculate the area of the polygon.

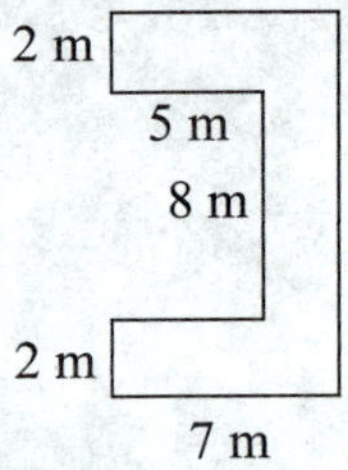

Math Tip

Another way to solve this type of problem is to find the area of the large rectangle and subtract the area of the small missing rectangle.

Example 5 Calculating Area

The polygon shown here is a rectangle with a rectangular piece missing. Calculate the area of the polygon.

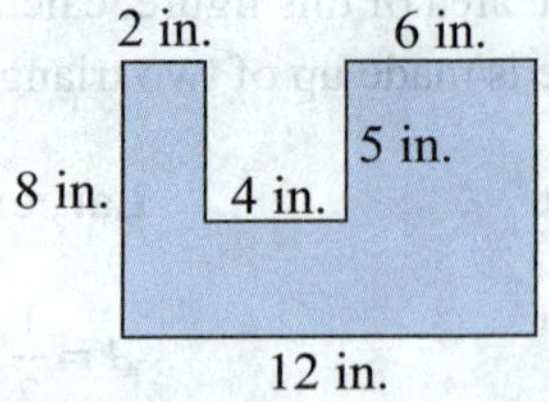

Solution

There are several ways to find the area of this figure. One way is to calculate the area of each of the three parts as illustrated here and then add the three areas.

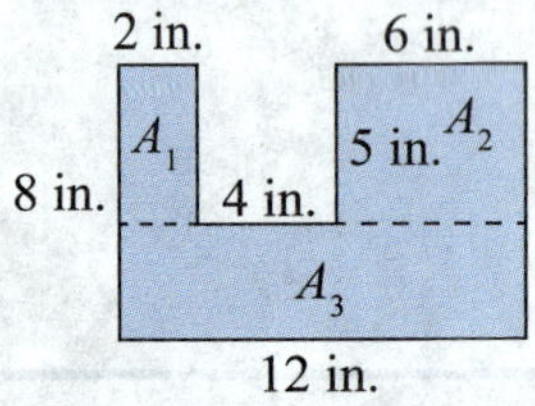

$A_1 = 2 \text{ in.} \cdot 5 \text{ in.} = 10 \text{ in.}^2$

$A_2 = 5 \text{ in.} \cdot 6 \text{ in.} = 30 \text{ in.}^2$

$A_3 = 12 \text{ in.} \cdot 3 \text{ in.} = 36 \text{ in.}^2$

Notice that the height of the bottom rectangle is $8 - 5 = 3$.

$A_{\text{Total}} = 10 \text{ in.}^2 + 30 \text{ in.}^2 + 36 \text{ in.}^2 = 76 \text{ in.}^2$

Note: A_1 is read "A sub 1", A_2 is read "A sub 2", and so on.

The area of the polygon is 76 in.^{2}

Now work margin exercise 5.

6. A park is in the shape of a square 30 yd on each side.

a. What is the perimeter of the park?

b. What is the area of the park?

Example 6 Application: Calculating Perimeter and Area

A baseball infield is in the shape of a square 90 ft on each side.

a. What is the perimeter of the infield?

b. What is the area of the infield?

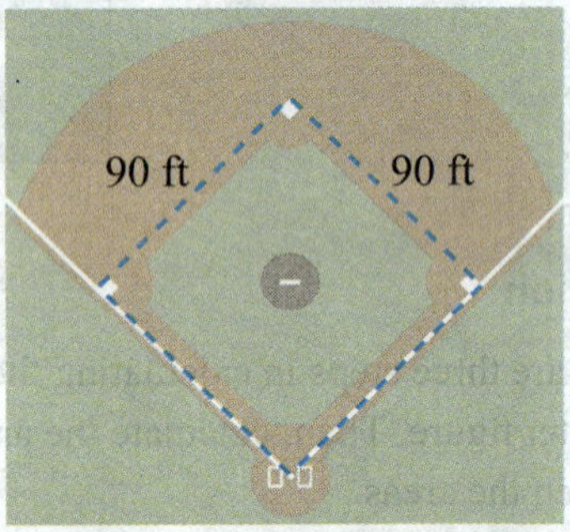

Solution

a. $P = 4s$

$P = 4 \cdot 90 \text{ ft} = 360 \text{ ft}$

b. $A = s^2$

$A = (90 \text{ ft})^2 = 8100 \text{ ft}^2$

The perimeter of the infield is 360 ft and the area is 8100 ft^2.

Now work margin exercise 6.

Margin Exercise Answers

1. 28 mm^2 **2.** 36 cm^2 **3.** 30 ft^2 **4.** 80 cm^2 **5.** 44 m^2 **6. a.** 120 yd **b.** 900 yd^2

6.3 Exercises

Concept Check

Fill-in-the-Blank. Complete each sentence using information found in this section.

1. When measuring area use ________ units.

2. The measure of the interior of a plane figure is the ______ of the figure.

3. $A = bh$ is the formula for the area of a/an ____________.

4. The b in area formulas represents the figure's _______.

5. $A = s^2$ is the formula for the area of a/an __________.

6. $A = \frac{1}{2}bh$ is the formula for the area of a/an _________.

True/False. Determine whether each statement is true of false. If a statement is false, explain how it can be changed so the statement will be true. (**Note:** There may be more than one acceptable change.)

7. The $(b+c)$ in the trapezoid area formula represents the sum of the lengths of the base and the corners.

8. The height of a triangle is the distance between the base and the vertex opposite the base.

9. The area formula for a triangle is $A = a + b + c$.

10. The area formula for a trapezoid is $A = \frac{1}{2}h(b+c)$.

Practice

Calculate the area of each figure described. See Examples 1 and 2.

1. A square with sides of length 9 ft.

2. A square with sides of length 6 in.

3. A rectangle with length 21 km and width 25 km.

4. A rectangle with length $1\frac{1}{4}$ mi and width $2\frac{1}{2}$ mi.

5. A parallelogram with height 2.3 ft and base 11.9 ft.

6. A parallelogram with height 5 m and base 12 m.

7. A triangle with height $\frac{8}{9}$ in. and base $\frac{5}{12}$ in.

8. A triangle with height 16.4 cm and base 8.2 cm.

9. A trapezoid with height 10 cm and parallel sides of length 15 cm and 18 cm.

10. A trapezoid with height 30 mm and parallel sides of length 45 mm and 50 mm.

6.3 Exercises

Concept Check

1. square
2. area
3. parallelogram
4. base
5. square
6. triangle
7. False; The $(b+c)$ in the trapezoid area formula represents the sum of the lengths of the two parallel bases.
8. True
9. False; The area formula for a triangle is $A = \frac{1}{2}bh$.
10. True

Practice

1. 81 ft^2
2. 36 in.^2
3. 525 km^2
4. $\frac{25}{8}$ mi^2
5. 27.37 ft^2
6. 60 m^2
7. $\frac{5}{27}$ in.^2
8. 67.24 cm^2
9. 165 cm^2
10. 1425 mm^2

11. 48 in.^{2}
12. 360 cm^2
13. 99 ft^2
14. 176 yd^2
15. 162 yd^2
16. 1200 ft^2
17. 1925 cm^2
18. 25 yd^2
19. 196 in.^{2}
20. 1 ft^2
21. 48 cm^2
22. 96 ft^2
23. 160 in.^{2}
24. 162 yd^2
25. 60 in.^{2}

Calculate the area of each figure. See Examples 1 through 5.

11.
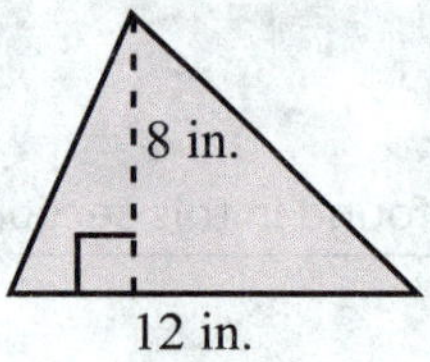

12.
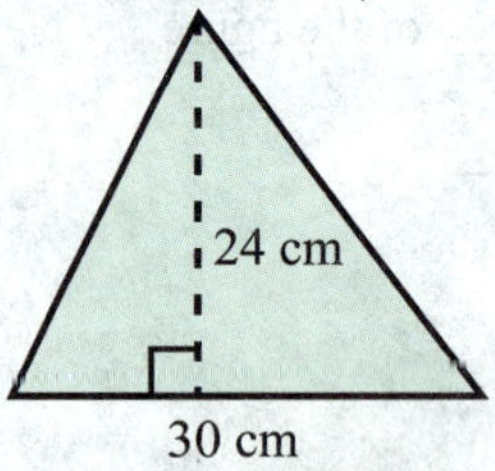

13.
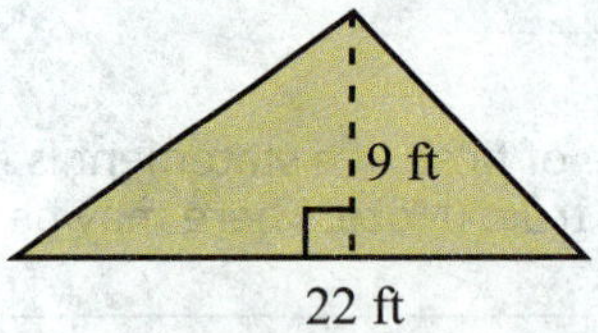

14.
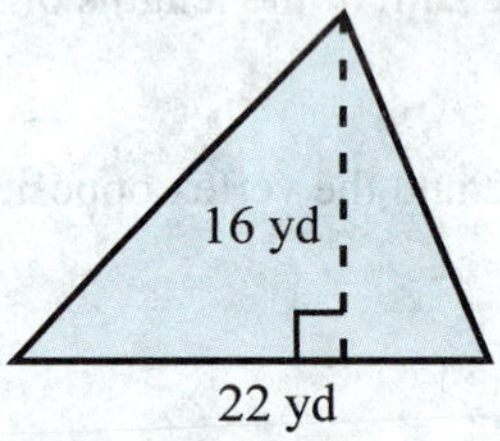

15.
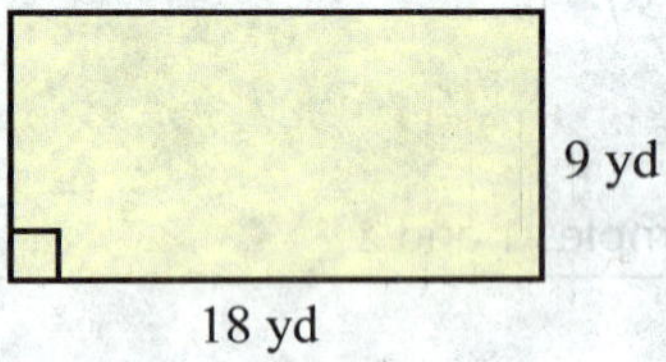

16.
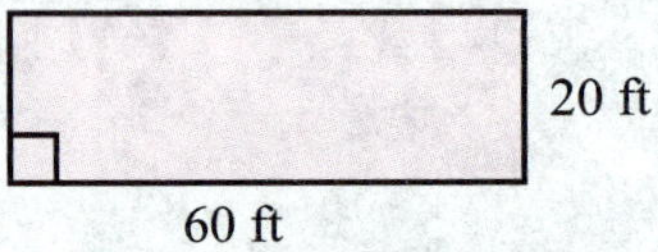

17.
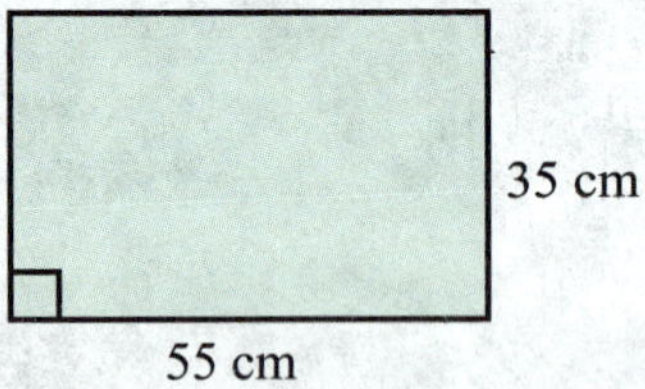

18.
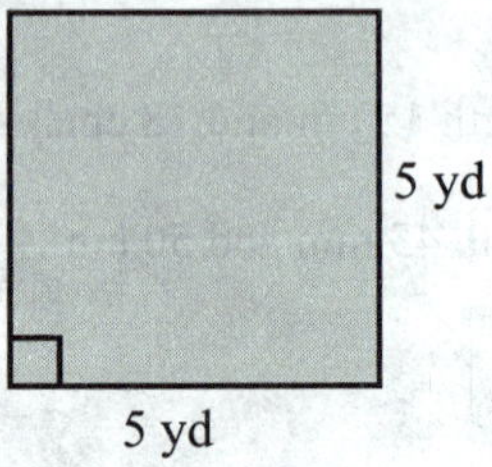

19.
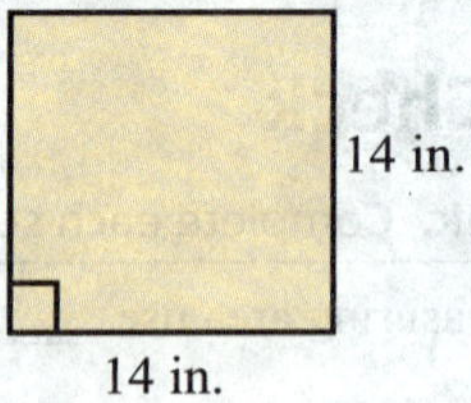

20.
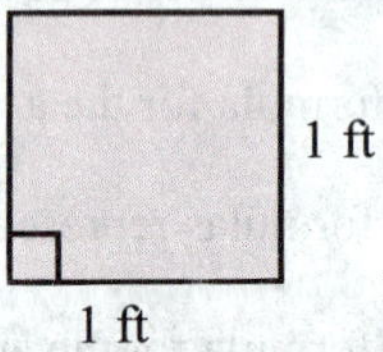

21.
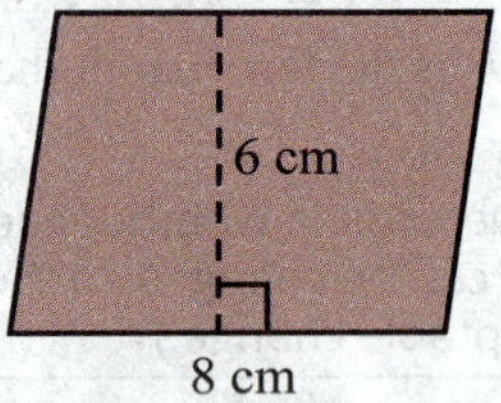

22.
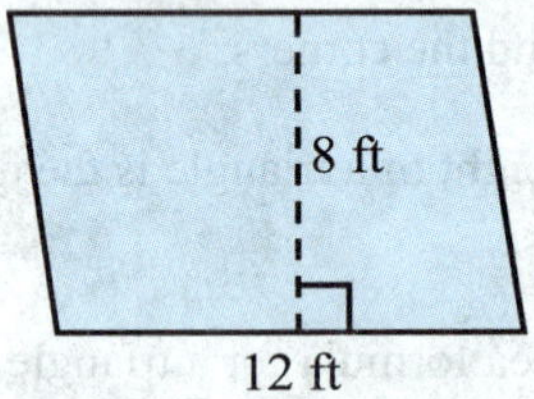

23.
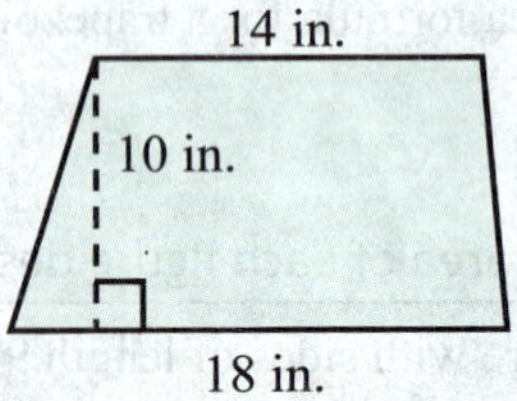

24.
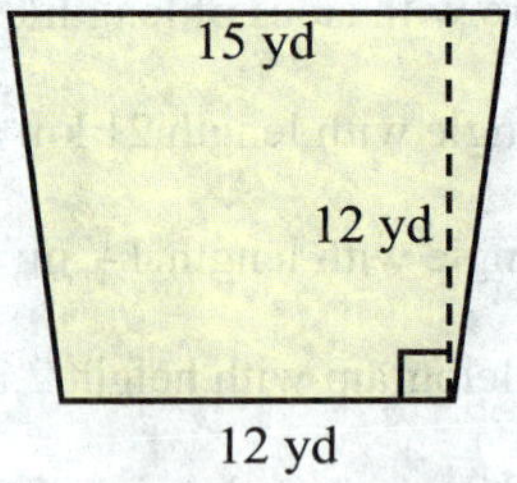

25.
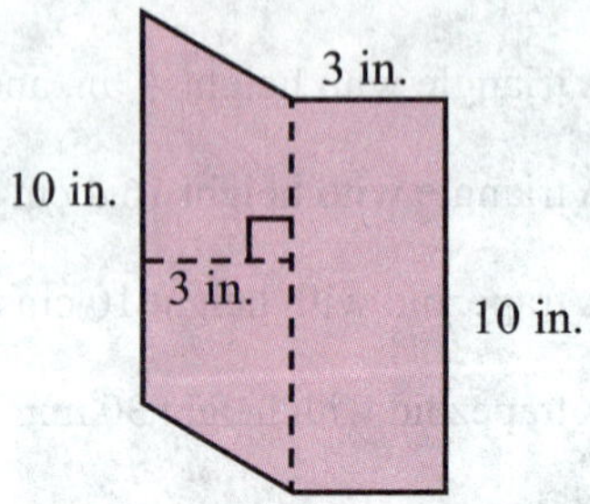

26.

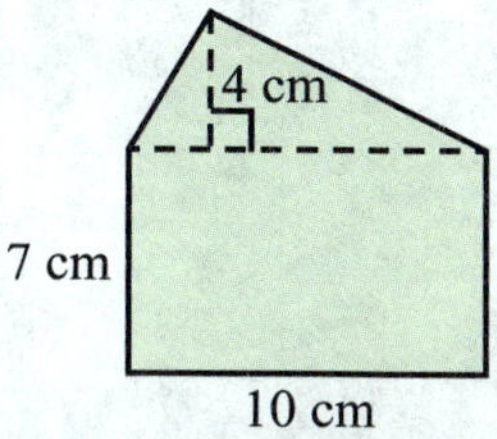

27.

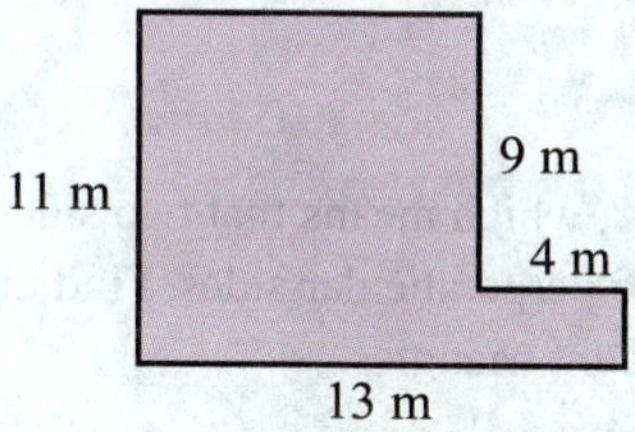

28.

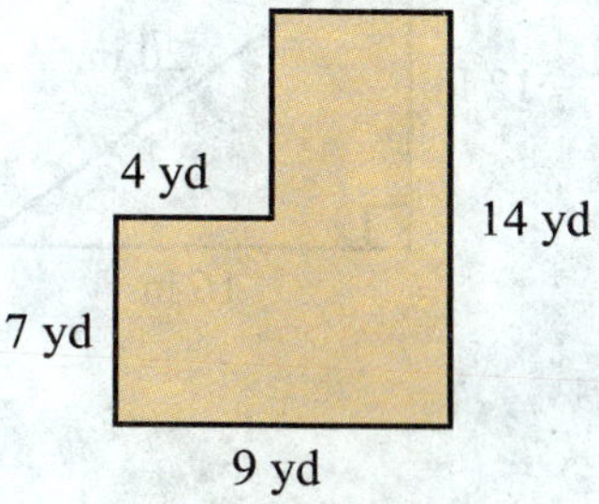

29.

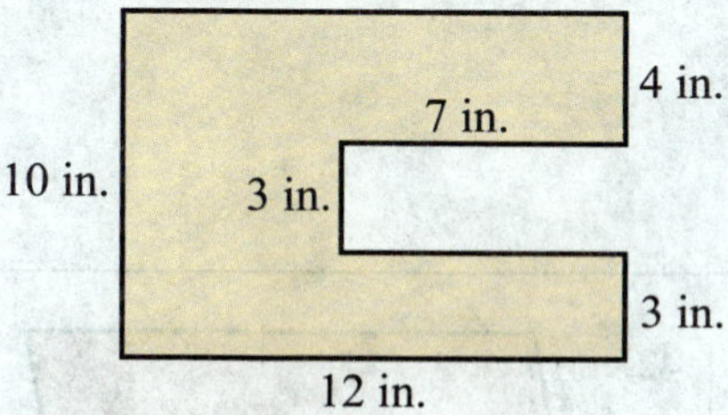

30.

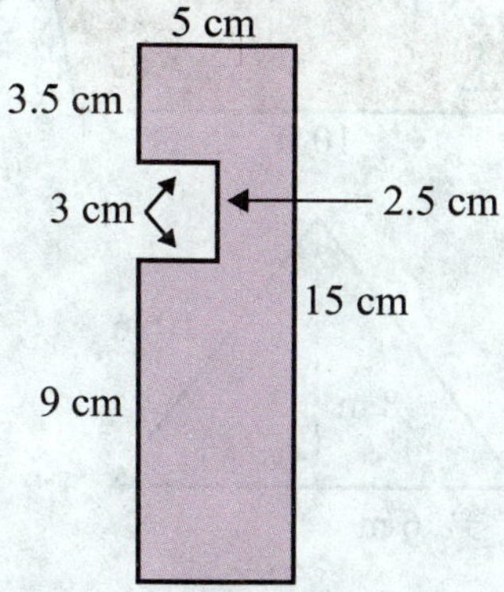

31.

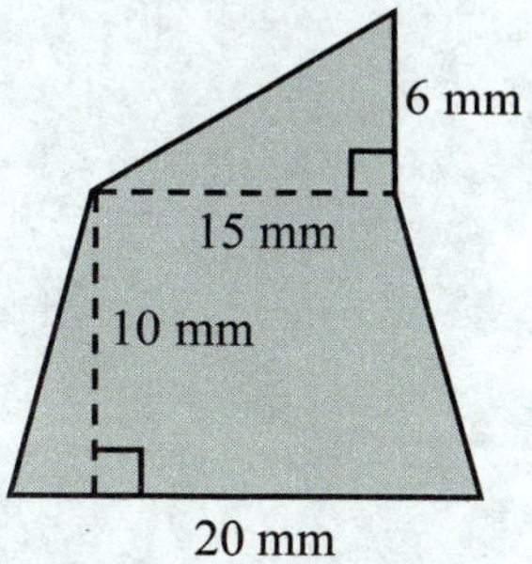

32.

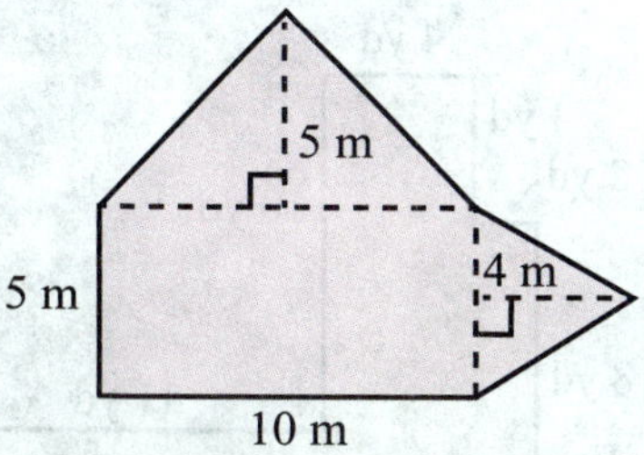

33.

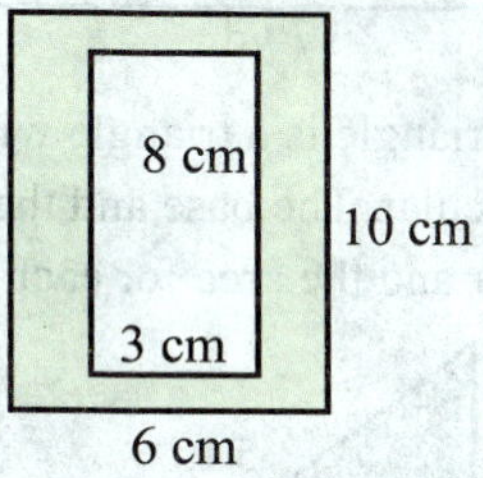

34.

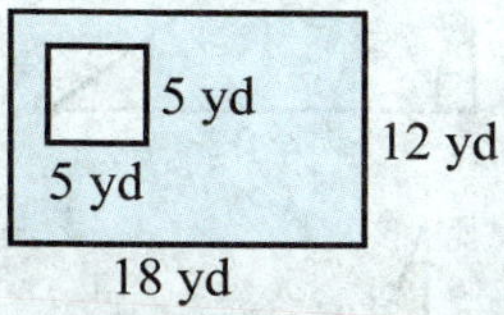

35.

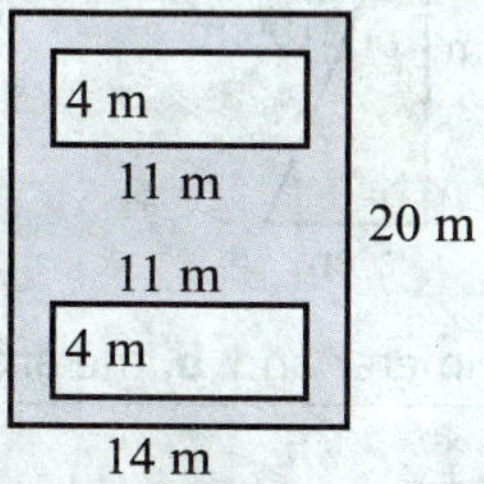

36.

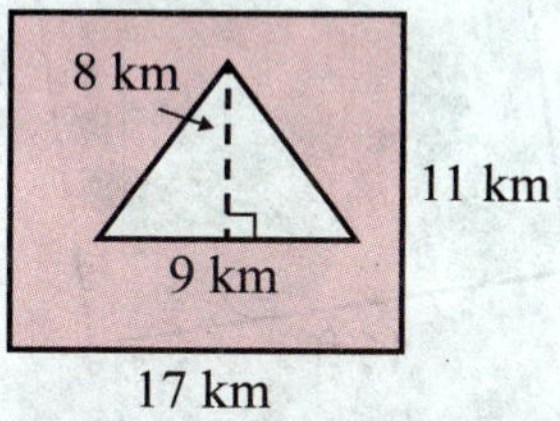

37.

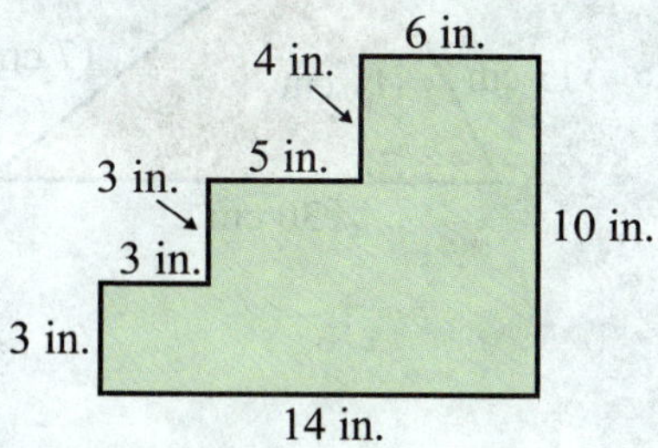

26. 90 cm^2
27. 107 m^2
28. 98 yd^2
29. 99 in.^{2}
30. 67.5 cm^2
31. 220 mm^2
32. 85 m^2
33. 36 cm^2
34. 191 yd^2
35. 192 m^2
36. 151 km^2
37. 99 in.^{2}

38. 86 yd^2

39. a. 12 ft; 6 ft^2

b. 30 cm; 30 cm^2

c. 48 in.; 96 in.^{2}

40. a. 56 ft b. 192 ft^2

41. a. 70 cm b. 220 cm^2

42. b. 32 ft b. 50 ft^2

43. a. 32 m b. 48 m^2

38.

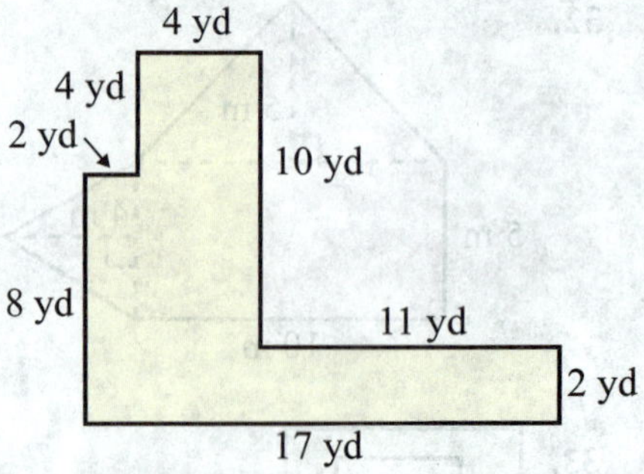

39. A **right** triangle is a triangle with one angle of 90°, which means that two sides are perpendicular. The base and the height are the two perpendicular sides. Find the perimeter and the area for each of the following right triangles.

a.

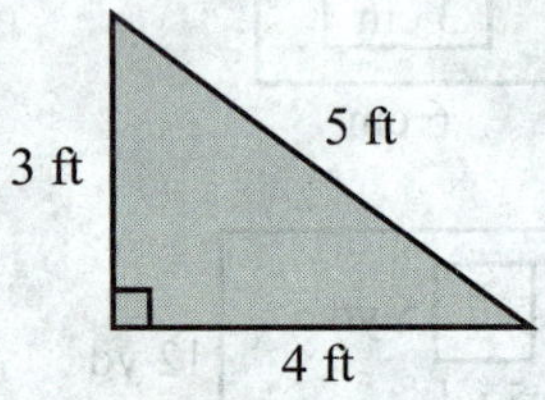

c.

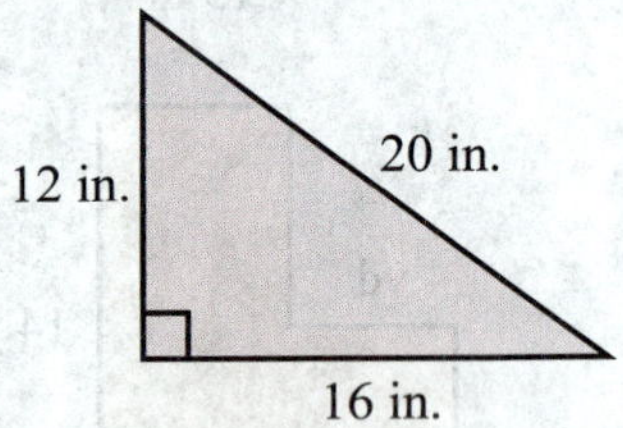

b.

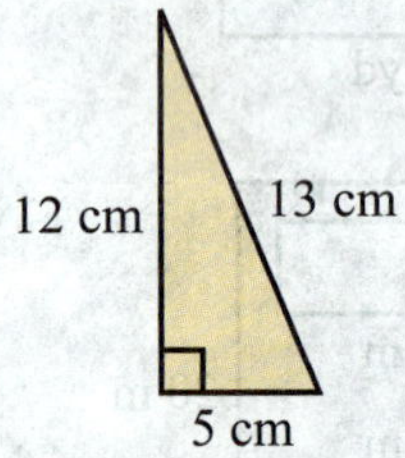

Find **a.** the perimeter and **b.** the area.

40.

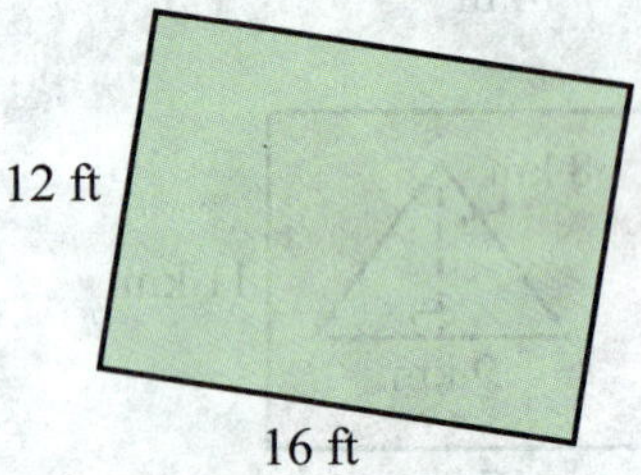

41.

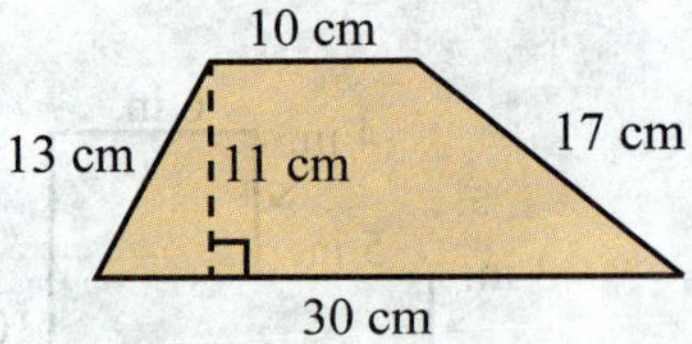

42.

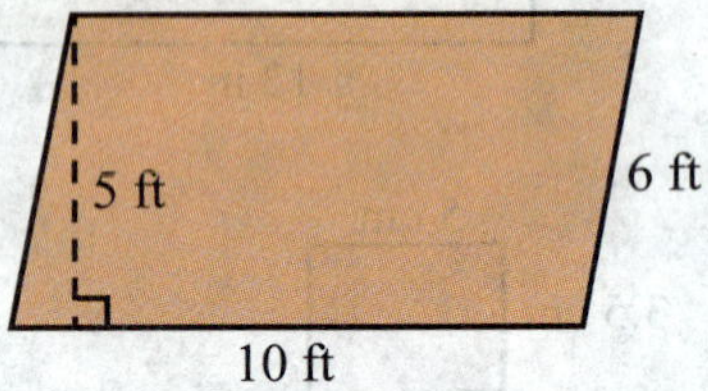

43.

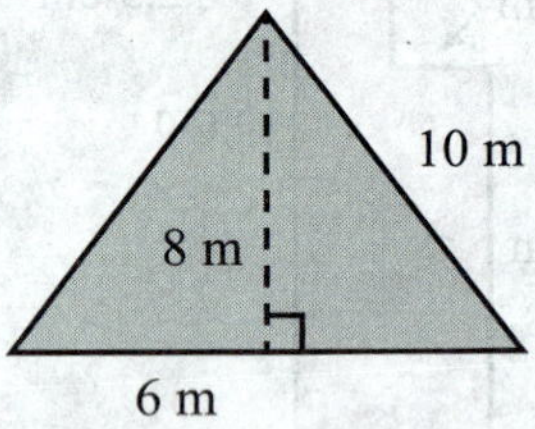

Applications

Solve.

44. The boundaries of a certain small town form a parallelogram with a length of 4.5 miles and a height of 2.6 miles. What is the area within the town limits?

45. Vinyl tile is to be laid on the floor of a rectangular room which is 17 feet long and 12 feet wide. How many square feet of tile must be put down?

46. A homeowner has a triangular section of her yard that needs fresh sod. If the base of the triangle is 8 feet and the height is 12 feet, what is the area of this section of yard?

47. Taylor is putting a quilt together using square pieces of fabric. If each square has a side length of 7 inches, and she wants the quilt to be 8 squares by 8 squares, what will the area of the quilt be in square inches?

48. The main stage at a theater is in the shape of a trapezoid. The owner of the theater is planning to install a new specially designed flooring system on the stage. The stage is 12 feet wide in the front and 15 feet wide in the back. The stage is 10 feet deep. How much flooring will the manager need?

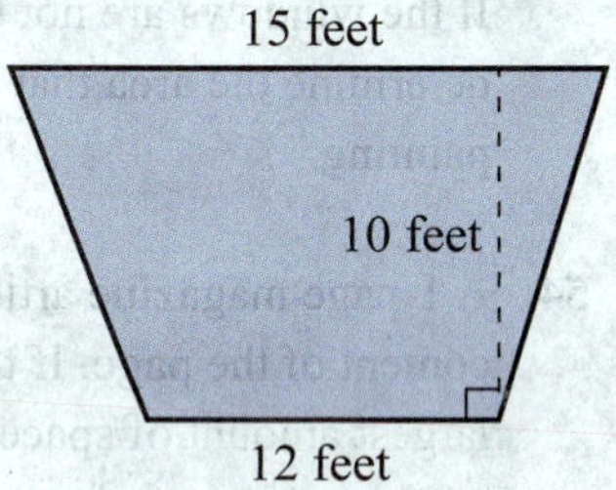

49. A sailboat has a triangular sail with the dimensions as shown in the drawing. Note that the 12 foot measurement is the height of the triangle.

a. What is the area of the sail?

b. What is the perimeter of the sail?

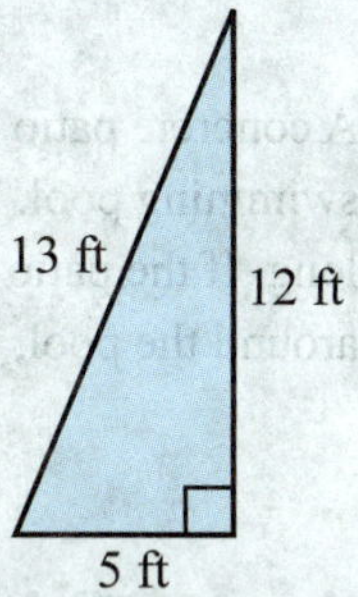

50. A square electronics circuit board is 18 centimeters on each side. On the center of one of the edges is an 8 by 1.5 centimeter rectangular lip for plugging in.

a. What is the total perimeter of the circuit board, including the lip?

b. What is the area of the circuit board?

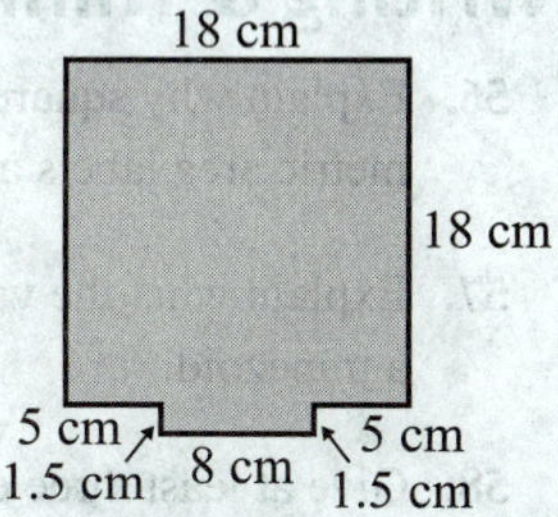

51. David is planting a five-sided lawn as shown in the figure. The lawn consists of a 50 foot by 40 foot rectangle and an attached 14 foot high triangle.

a. What is the area of the lawn to be planted?

b. If one pound of grass seed will cover 200 square feet, how many pounds will be necessary to cover the entire lawn? (**Hint:** Divide the area by the number of square feet that one pound of seed will cover.)

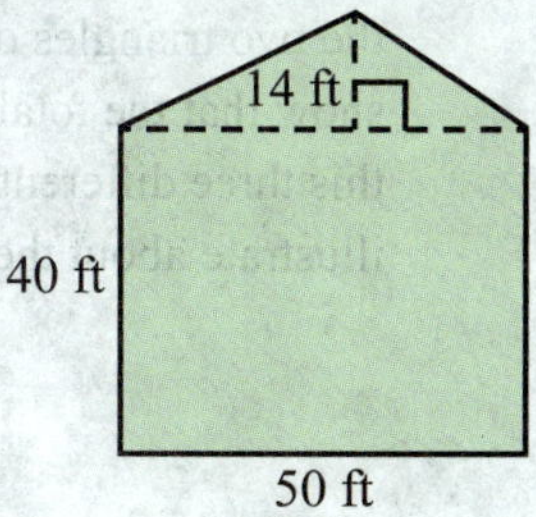

Applications

44. 11.7 square miles

45. 204 square feet

46. 48 square feet

47. 3136 square inches

48. 135 square feet

49. **a.** 30 ft^2 **b.** 30 ft

50. **a.** 75 cm **b.** 336 cm^2

51. **a.** 2350 square feet

b. 11.75 pounds

52. a. 148.5 ft^2 b. \$1410.75

53. 123 square feet

54. 108 square inches

55. 426 square meters

Writing & Thinking

56. Square units are used for area because area is measuring how many "squares" can fit in a particular area. Examples of metric area labels are km^2 and m^2. US customary area labels are in.^{2} and yd^2.

57. The expression $(b+c)$ represents the sum of the two bases in a trapezoid. The bases are the two sides that are parallel in a trapezoid.

58. Examples will vary. Area is required when purchasing carpet, laying sod, putting on a roof, painting walls, building a deck, and with construction projects in general.

59. The area of the interior triangle is equal to half of the area of the rectangle no matter where the point is chosen.

52. A trapezoidal patio with a square opening for flowers is to be constructed. As shown on the drawing, the ends of the trapezoid are 12 ft and 9 ft respectively, with a height of 15 ft. Each side of the square cutout is 3 ft.

a. What is the area of the concrete surface?

b. If the charge to pour and finish the concrete is \$9.50 a square foot, what will it cost?

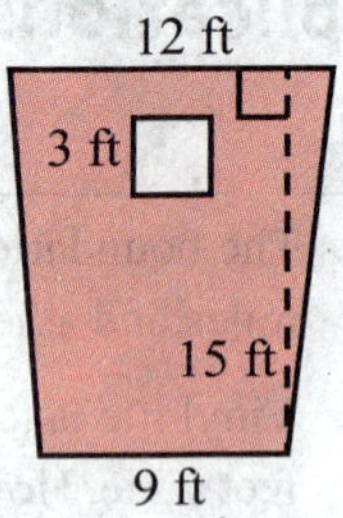

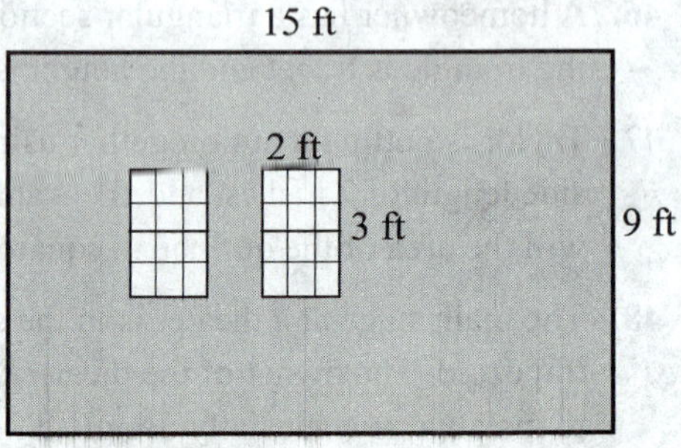

53. Bill is painting a wall in his living room. The wall is 9 feet high and 15 feet wide. The wall has two rectangular windows that measure 2 feet by 3 feet. If the windows are not to be painted, determine the area that Bill will be painting.

54. A 1-page magazine article must have 1-inch margins of blank space surrounding the content of the page. If the magazine pages are 11 inches by 14 inches, determine the largest amount of space that will contain print on this page.

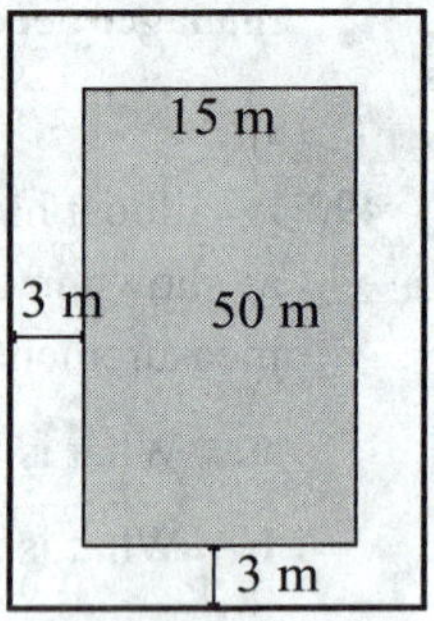

55. A concrete patio is being poured to surround a rectangular swimming pool. The pool is 15 meters wide by 50 meters long. If the patio is to be a uniform 3 meters width all around the pool, find the area of the concrete patio.

Writing & Thinking

56. Explain why square units are used for labeling areas. Give two examples each of metric area labels and US customary area labels.

57. Explain what the value of $(b+c)$ represents in the formula for the area of a trapezoid.

58. Give at least three examples where finding the area of a figure would be helpful (outside of a class).

59. Draw a rectangle and choose any point on one side of the rectangle. Draw line segments to the vertices on the opposite side (forming three triangles). Now cut out the two triangles on each end. Place these triangles inside the remaining triangle to show that the total of the two areas is equal to the area of the remaining triangle. Do this three different times choosing a different point each time. What fact does this illustrate about the area of a triangle?

6.4 Circles

Objectives

A. Know the terms circle, center, radius, and diameter.

B. Calculate the circumference and area of a circle.

A Circles

The concepts of perimeter and area were discussed in Sections 6.2 and 6.3. In this section, we discuss circles and the formulas for circumference (perimeter) and area that are related to circles.

Circles

Circle: The set of all points in a plane that are some fixed distance from a fixed point called the **center** of the circle.

Radius: The distance from the center of a circle to any point on the circle. (The letter r is used to represent the radius of a circle.)

Diameter: The distance from one point on a circle to another point on the circle measured through the center. (The letter d is used to represent the diameter of a circle and $d = 2r$.)

Circumference: Perimeter of (or distance around) a circle.

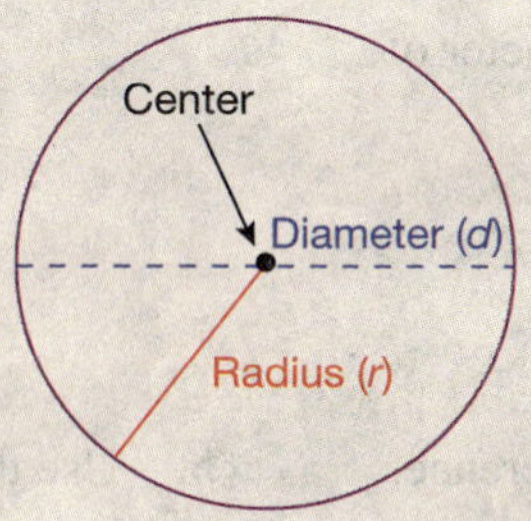

DEFINITION

History

The circle has been intriguing man since before recorded history and is used to symbolize many concepts such as infinity, wholeness, stability, and perfection. For example, wedding bands are considered symbols of everlasting love because they are circles and as such have no beginning and no end. Similarly, the individual rings in the symbol of the Olympic games represent continuity and the human being, while the five rings interlocked represent the five continents of the world coming together to compete.

B Formulas for Circumference and Area of a Circle

Note carefully that a diameter is twice as long as a radius. That is, $d = 2r$. This relationship leads to two formulas for circumference, one with the diameter and the other with the radius.

Formulas for Circles

Circumference: $C = 2\pi r$ and $C = \pi d$

Area: $A = \pi r^2$

FORMULA

1. Calculate **a.** the circumference and **b.** the area of a circle with a radius of 11 m.

Note

The Greek letter π (pi) is the symbol used for the constant 3.1415926535.... This number is an infinite, nonrepeating decimal number. For our purposes, we will use $\pi \approx 3.14$ (accurate to hundredths). However, you should always be aware that 3.14 is only an approximation for π and that related answers are only approximations.

Example 1 Calculating the Circumference and Area of a Circle

Calculate **a.** the circumference and **b.** the area of a circle with a radius of 6 ft.

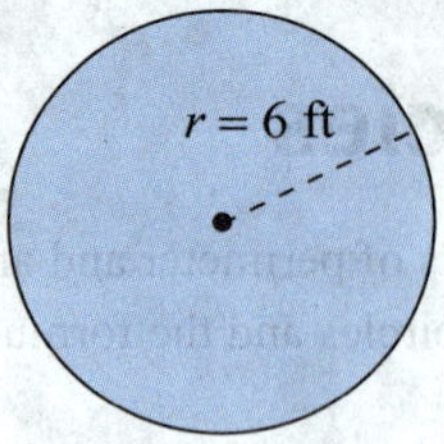

Solution

a. Use the formula for circumference:

$$C = 2\pi r$$
$$C \approx 2 \cdot 3.14 \cdot 6 \text{ ft}$$
$$= 37.68 \text{ ft}$$

The circumference is 37.68 ft.

b. Use the formula for area:

$$A = \pi r^2$$
$$A \approx 3.14 \cdot (6 \text{ ft})^2$$
$$= 3.14 \cdot 36 \text{ ft}^2$$
$$= 113.04 \text{ ft}^2$$

The area is 113.04 ft^2.

Now work margin exercise 1.

2. Calculate **a.** the circumference and **b.** the area of a circle with a diameter of 15 ft.

Example 2 Calculating the Circumference and Area of a Circle

Calculate **a.** the circumference and **b.** the area of a circle with a diameter of 5.2 in.

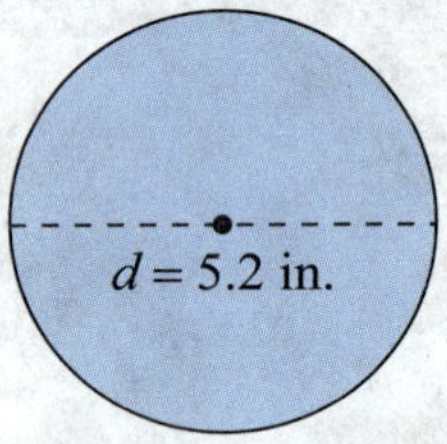

Solution

a. Use the formula for circumference:

$$C = \pi d$$
$$C \approx 3.14 \cdot 5.2 \text{ in.}$$
$$= 16.328 \text{ in.}$$

The circumference is 16.328 in.

b. Use the formula for area:

$$A = \pi r^2$$

In this case, $r = \frac{1}{2} \cdot 5.2 \text{ in.} = 2.6 \text{ in.}$

$$A \approx 3.14 \cdot (2.6 \text{ in.})^2$$
$$= 3.14 \cdot 6.76 \text{ in.}^2$$
$$= 21.2264 \text{ in.}^2$$

The area is 21.2264 in.^2

Now work margin exercise 2.

3. Calculate the perimeter of the following figure.

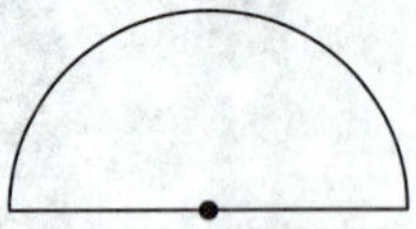

Example 3 Calculating the Perimeter

Calculate the perimeter of the figure shown: a semicircle (half of a circle) and a diameter. The radius is 4 ft long.

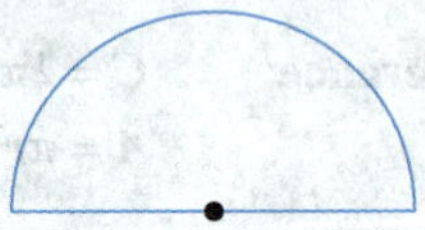

Solution

In order to find the perimeter of the figure, find the perimeter of the semicircle and then add the diameter.

Length of semicircle $= \frac{1}{2}C = \frac{1}{2}(2\pi r) \approx \frac{1}{2}(2 \cdot 3.14 \cdot 4 \text{ cm}) = 12.56 \text{ cm}$

Diameter $= 2r = 2 \cdot 4 \text{ cm} = 8 \text{ cm}$

Perimeter = semicircle + diameter = 12.56 cm + 8 cm = 20.56 cm

The perimeter is 20.56 cm.

Now work margin exercise 3.

Example 4 Calculating the Area of a Washer

Calculate the area of the washer (shaded portion) with dimensions as shown in the figure.

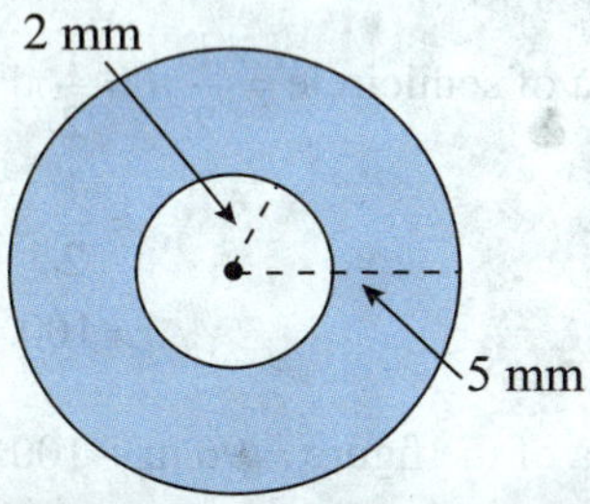

Solution

Subtract the area of the inside (smaller) circle from the area of the outside (larger) circle.

Area of Larger Circle

$A = \pi r^2$

$A \approx 3.14 \cdot (5 \text{ mm})^2$

$= 3.14 \cdot 25 \text{ mm}^2$

$= 78.50 \text{ mm}^2$

Area of Smaller Circle

$A = \pi r^2$

$A \approx 3.14 \cdot (2 \text{ mm})^2$

$= 3.14 \cdot 4 \text{ mm}^2$

$= 12.56 \text{ mm}^2$

Area of Washer

$$\begin{array}{r} 78.50 \text{ mm}^2 \\ -\ 12.56 \text{ mm}^2 \\ \hline 65.94 \text{ mm}^2 \end{array}$$

The area of the washer is 65.94 mm^2.

Now work margin exercise 4.

Example 5 Calculating Perimeter and Area

Calculate **a.** the perimeter and **b.** the area of the figure shown here with a rectangular base and a semicircle attached to the top. The rectangle has a length of 16 m and a width of 6 m.

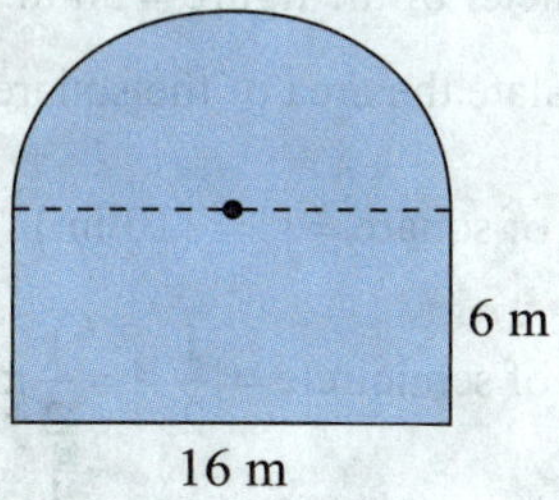

4. Calculate the area of the shaded portion of the figure shown.

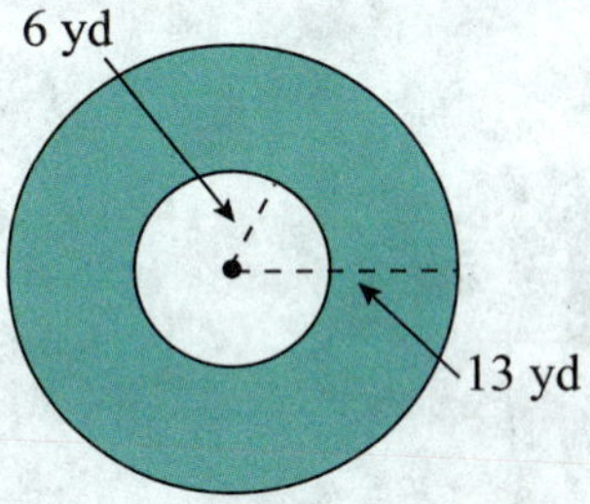

5. Calculate **a.** the perimeter and **b.** the area of the figure shown here with a rectangular base and a semicircle attached to the top. The rectangle has a length of 4 ft and a width of 7 ft.

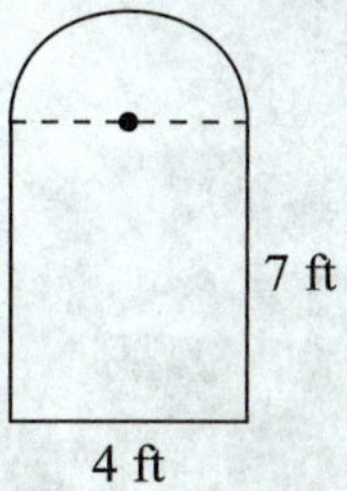

Solution

a. The perimeter of the figure is the sum of the lengths of three sides of the rectangle and the length of the semicircle. (**Note:** The diameter of the semicircle is 16 m.)

Sum of the lengths of three sides of the rectangle = 6 m + 16 m + 6 m = 28 m

$$\text{Length of semicircle} = \frac{1}{2}C = \frac{1}{2}\pi d \approx \frac{1}{2}\cdot 3.14\cdot 16\text{ m} = 25.12\text{ m}$$

Perimeter of the figure = 28 m + 25.12 m = 53.12 m

b. Calculate the area of the rectangle then add the area of the semicircle.

$$\text{Area of rectangle} = lw = 16\text{m}\cdot 6\text{m} = 96\text{ m}^2$$

$$\text{Area of semicircle} = \frac{1}{2}A = \frac{1}{2}\pi r^2$$

$$\approx \frac{1}{2}\cdot 3.14\cdot(8\text{ m})^2$$

In this case, $r = \frac{1}{2}d = \frac{1}{2}\times 16\text{m} = 8\text{m}$.

$$= 100.48\text{ m}^2$$

Area of the figure = 96 m + 100.48 m = 196.48 m²

The perimeter of the figure is 53.12 m and the area of the figure is 196.48 m².

Now work margin exercise 5.

6. Calculate **a.** the perimeter and **b.** the area of the figure shown here with a square base and a semicircle cut out of one side. One side of the square is 6 cm long.

Example 6 Calculating Perimeter and Area

Calculate **a.** the perimeter and **b.** the area of the figure shown here with a square base and a semicircle cut out of one side. One side of the square is 10 in. long.

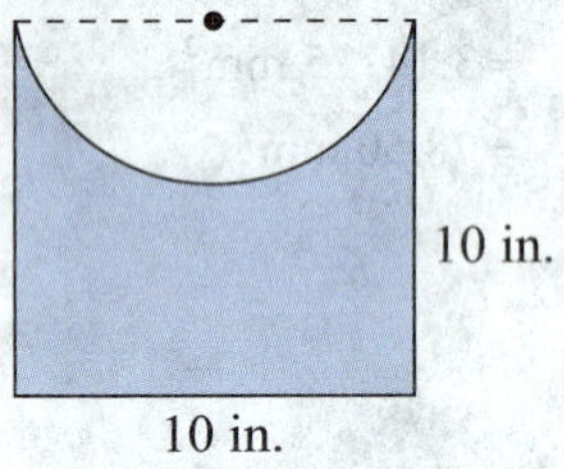

Solution

a. The perimeter of the figure is the sum of the lengths of three sides of the square and the length of the semicircle. (**Note:** The diameter of the semicircle is 10 in.)

Sum of the lengths of three sides of the square = 10 in. + 10 in. + 10 in. = 30 in.

$$\text{Length of semicircle} = \frac{1}{2}C = \frac{1}{2}\pi d \approx \frac{1}{2}\cdot 3.14\cdot 10\text{ in.} = 15.7\text{ in.}$$

Perimeter of the figure = 30 in. + 15.7 in. = 45.7 in.

b. Calculate the area of the square then subtract the area of the semicircle.

$$\text{Area of square} = s^2 = (10\text{ in.})^2 = 100\text{ in.}^2$$

$$\text{Area of semicircle} = \frac{1}{2}A = \frac{1}{2}\pi r^2$$

$$\approx \frac{1}{2}\cdot 3.14\cdot(5\text{ in.})^2$$

In this case, $r = \frac{1}{2}d = \frac{1}{2}\times 10\text{in.} = 5\text{in.}$

$$= 39.25\text{ in.}^2$$

Area of the figure = 100 in.^{2} – 39.25 in.^{2} = 60.75 in.^{2}

The perimeter of the figure is 45.7 in. and the area of the figure is 60.75 in.^{2}

Now work margin exercise 6.

CALCULATOR

Entering π on a Calculator

Many calculators will have a [π] key (may also be accessed using [SHIFT] [EXP] or [SHIFT] [10^x]) which can be used in calculating circumferences and areas involving π. Be aware, though, that the value of π inserted by a calculator will be rounded to a minimum of 10 decimal places. As the problems in this section were calculated using $\pi = 3.14$, you may see slight differences in the answers due to rounding. For example, a circle with radius 6 has a circumference of 37.68 ft when calculated using $\pi = 3.14$. Alternatively, to find the circumference using a calculator, press the keys

[2] [×] [π] [×] [6]. Then press [=].

The display will read 37.699111843... (your calculator may display more or fewer digits depending on its settings). Rounding this value to two decimal places gives us 37.70 ft. Even though the two circumferences differ by 0.02, both answers are considered correct.

Margin Exercise Answers

1. a. 69.08 m **b.** 379.94 m^2 **2. a.** 47.1 ft **b.** 176.625 ft^2 **3.** 46.26 in. **4.** 417.62 yd^2
5. a. 24.28 ft **b.** 34.28 ft^2 **6. a.** 27.42 cm **b.** 21.87 cm^2

6.4 Exercises

Concept Check

Fill-in-the-Blank. Complete each sentence using information found in this section.

1. The line segment from one point on a circle through the center to the point directly opposite this point is a/an __________.

2. The set of all points in a plane that are equidistant from a fixed point is a/an _______.

3. The line segment from the center of a circle to any point on the circle is a/an _________.

4. The perimeter of a circle is called the ______________.

5. 3.14151926535… represents pi, which is represented by the Greek letter _____.

6. $C = \pi d$ is the formula to find the ______________ of a circle.

True/False. Determine whether each statement is true of false. If a statement is false, explain how it can be changed so the statement will be true. (**Note:** There may be more than one acceptable change.)

7. Every radius on a circle has the same length.

8. Exact answers can be achieved when substituting 3.14 in place of π.

6.4 Exercises

Concept Check

1. diameter
2. circle
3. radius
4. circumference
5. π
6. circumference
7. True
8. False; Approximations can be achieved when substituting 3.14 in place of π.

9. False; The length of the radius of a circle is half of the length of the diameter.
10. False; $A = \pi r^2$

Practice

1. a. 31.4 ft b. 78.5 ft^2
2. a. 6.28 ft b. 3. 14 ft^2
3. a. 19.468 yd b. 30.1754 yd^2
4. a. 7.536 in b. 4.5216 $in.^2$
5. a. 3.14 m b. 0.785 m^2
6. a. 9.42 ft b. 7.065 ft^2
7. a. 188.4 cm b. 2826 cm^2
8. a. 43.96 m b. 153.86 m^2
9. a. 4.71 ft b. 1.76625 ft^2
10. a. 76.616 mi b. 467.3576 mi^2
11. a. 37.68 cm b. 113.04 cm^2
12. a. 47.1 cm b. 176.625 cm^2
13. a. 94.2 in. b. 706.5 $in.^2$
14. a. 75.36 mm b. 452.16 mm^2
15. a. 35.98 in. b. 76.93 $in.^2$
16. a. 25.7 ft b. 39.25 ft^2
17. a. 10.71 cm b. 7.065 cm^2
18. a. 53.55 yd b. 176.625 yd^2

9. The length of the diameter of a circle is half of the length of the radius.
10. The area of a circle is found by using the formula $A = \pi d$.

Practice

Calculate **a.** the circumference and **b.** the area of each circle described. Use $\pi \approx 3.14$. See Examples 1 and 2.

1. A circle with radius 5 ft
2. A circle with radius 1 ft
3. A circle with diameter 6.2 yd
4. A circle with diameter 2.4 in.
5. A circle with radius 0.5 m
6. A circle with radius 1.5 ft
7. A circle with diameter 60 cm
8. A circle with diameter 14 m
9. A circle with radius $\frac{3}{4}$ ft
10. A circle with radius $12\frac{1}{5}$ mi

Calculate **a.** the perimeter and **b.** the area of each figure. Use $\pi \approx 3.14$. See Examples 1 through 3, 5, and 6.

11.
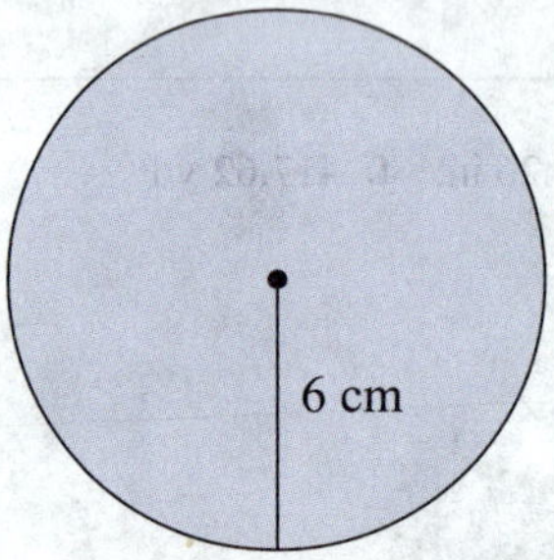

12.
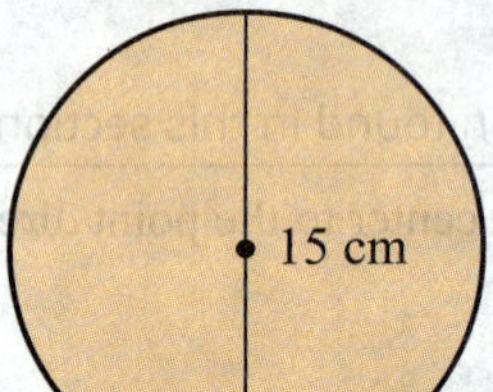

13.
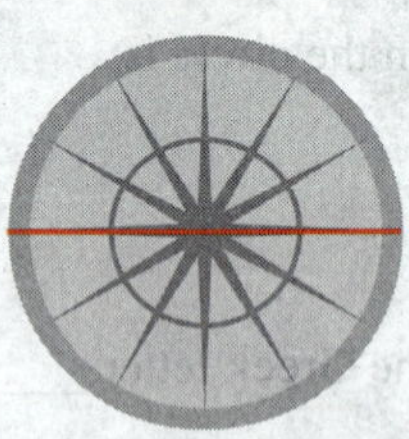

14.
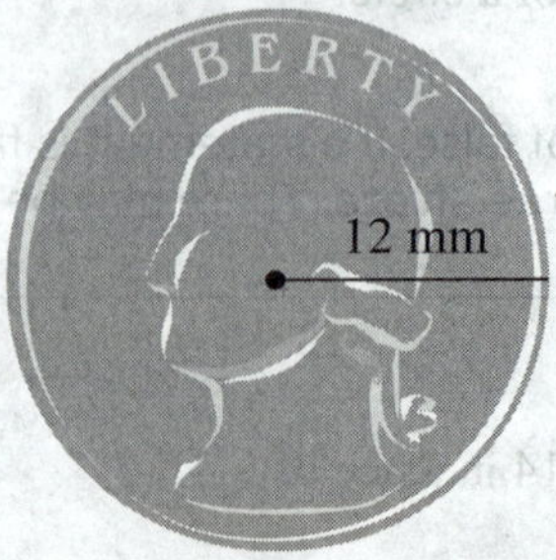

15.
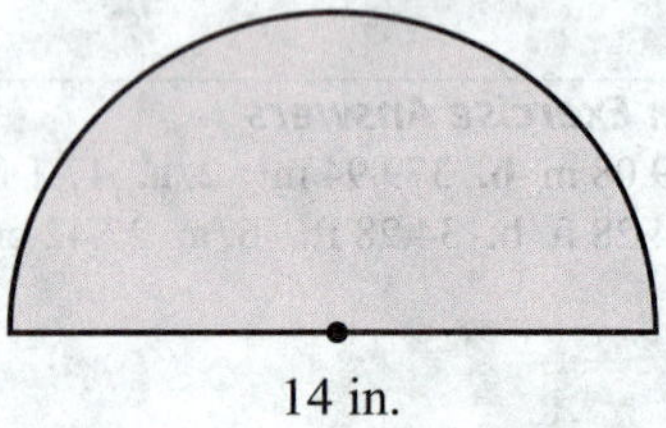

16.
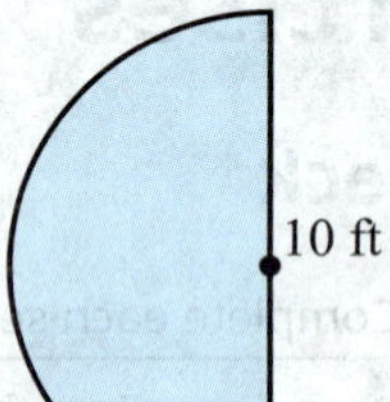

17.
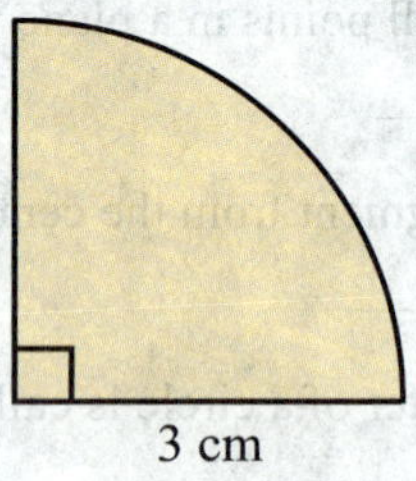

18.
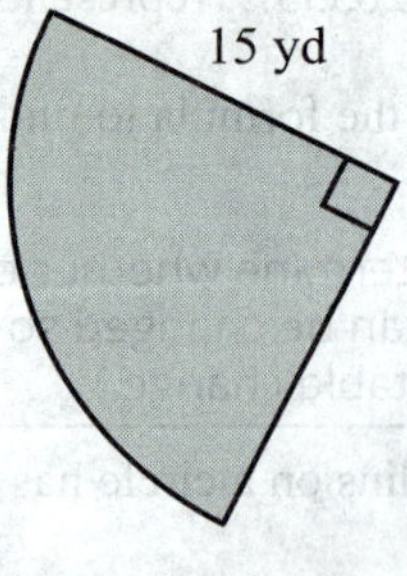

19.

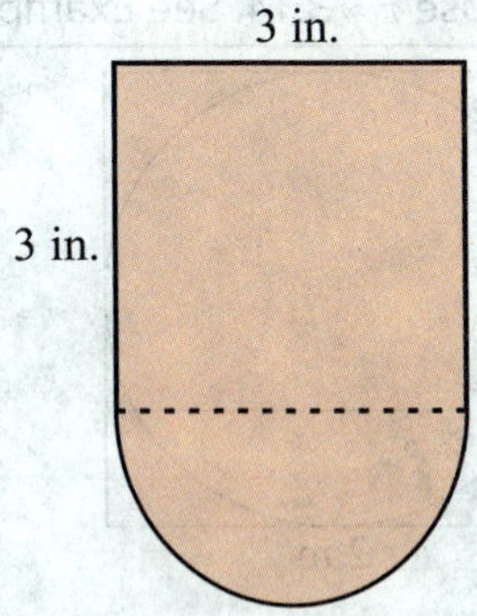

20.

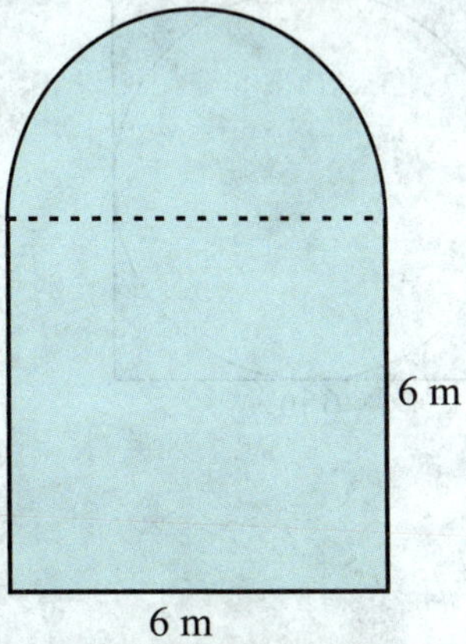

21.

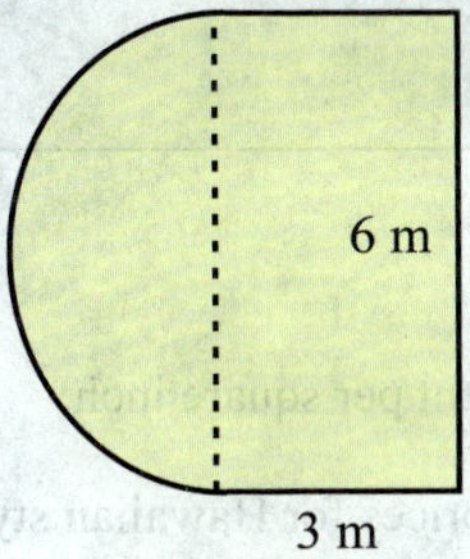

22.

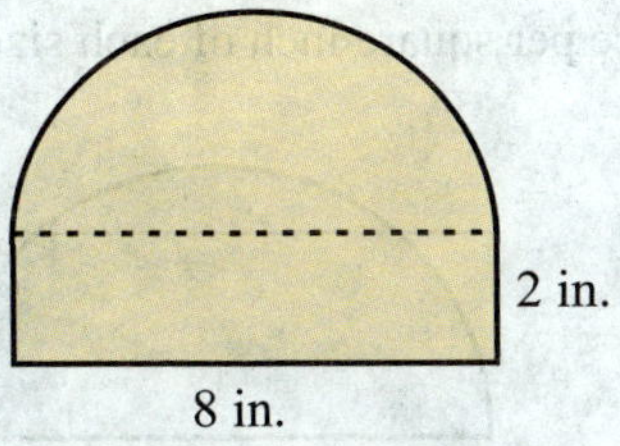

23.

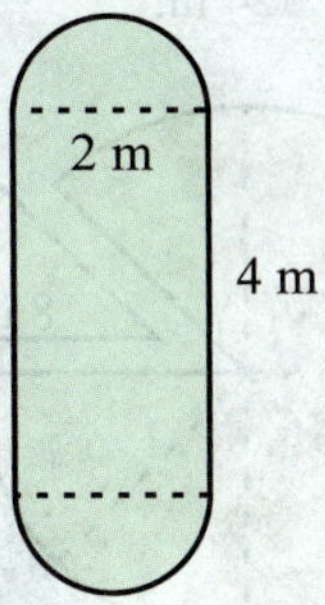

24.

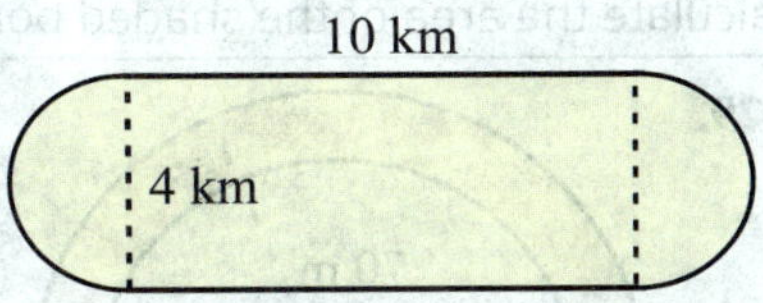

25.

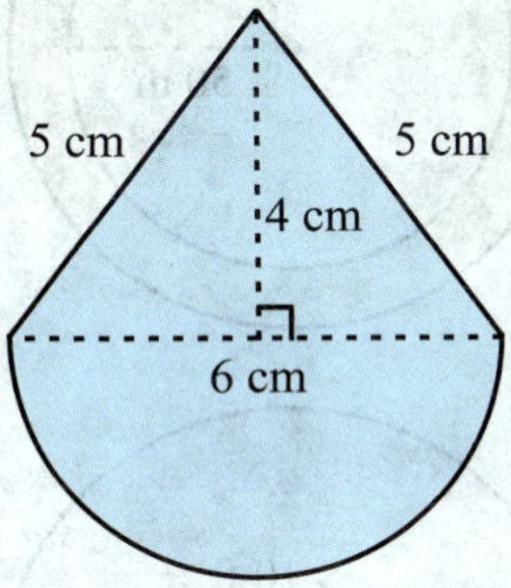

26.

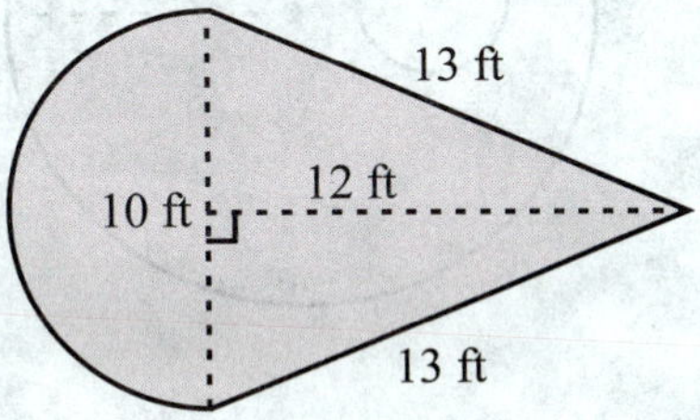

27.

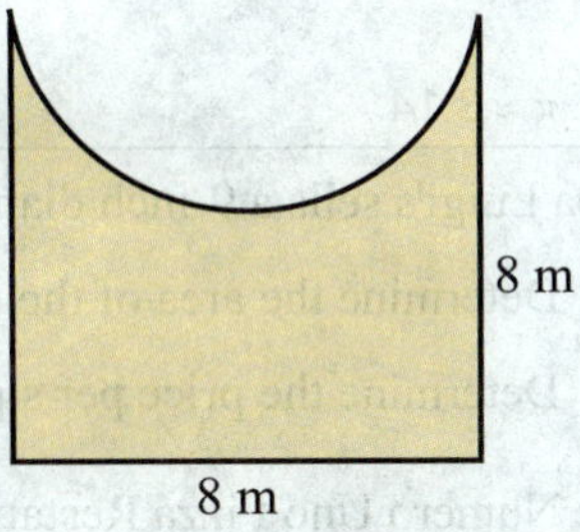

28.

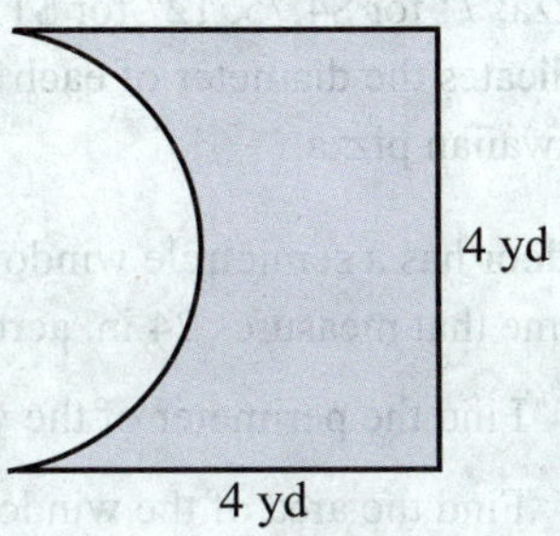

19. **a.** 13.71 in.
b. 12.5325 in.^{2}

20. **a.** 27.42m **b.** 50.13 m^2

21. **a.** 21.42 m **b.** 32.13 m^2

22. **a.** 24.56 in. **b.** 41.12 in.^{2}

23. **a.** 14.28 m **b.** 11.14 m^2

24. **a.** 32.56 km
b. 52.56 km^2

25. **a.** 19.42 cm
b. 26.13 cm^2

26. **a.** 41.7 ft **b.** 99.25 ft^2

27. **a.** 36.56 m **b.** 38.88 m^2

28. **a.** 18.28 yd **b.** 9.72 yd^2

29. 7536 m^2

30. 2512 ft^2

31. 0.86 m^2

32. 7.74 in.^{2}

Applications

33. a. 63.6 square inches

 b. $0.13 per square inch

34. $0.12; $0.10; $0.09

35. a. 61.68 in.

 b. 226.08 in.^{2}

36. a. 22.28 in

 b. 25.12 in.^{2}

Calculate the area of the shaded portion of each figure. Use $\pi \approx 3.14$. See Example 4.

29.

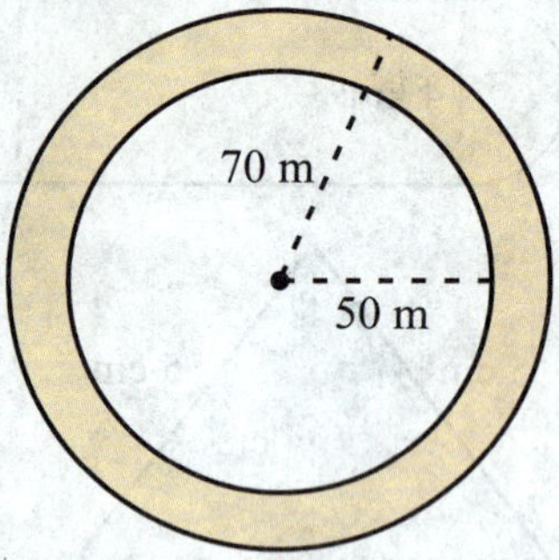

31.

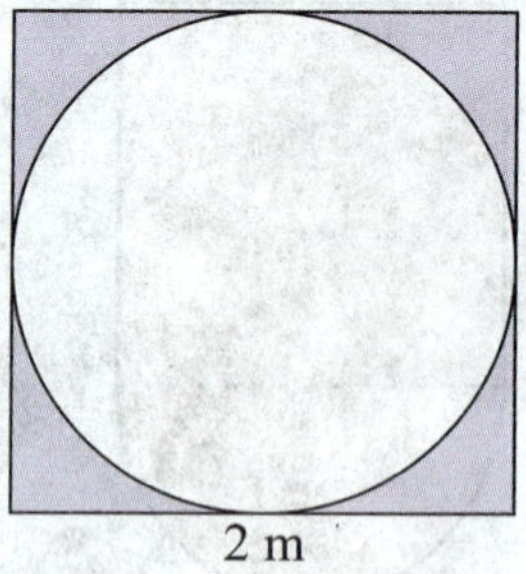

32.

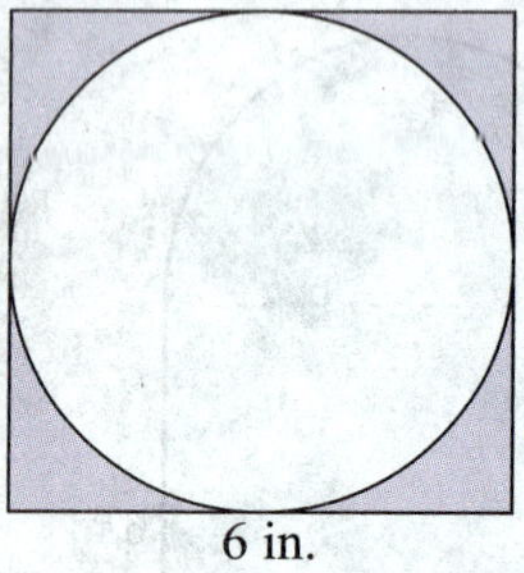

30.

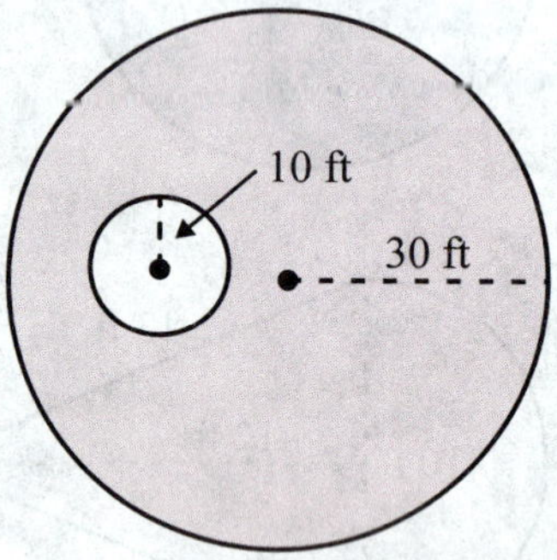

Applications

Solve. Use $\pi \approx 3.14$.

33. Papa Luigi's sells a 9-inch diameter pizza for $8.

a. Determine the area of the pizza to the nearest tenth.

b. Determine the price per square inch to the nearest cent per square inch.

34. The Numero Uno Pizza Restaurant listed the following prices for Hawaiian style pizza: 7″ for $4.75, 12″ for $11.45, and 14″ for $13.95 where the number of inches indicates the diameter of each pizza. Find the price per square inch of each size of Hawaiian pizza.

35. A door has a semicircle window above the door frame that measures 24 in. across the bottom.

a. Find the perimeter of the window.

b. Find the area of the window.

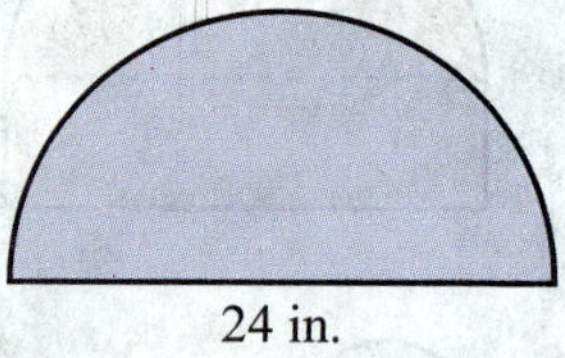

36. A large 16 in. pizza is cut into eight pieces.

a. What is the perimeter of a single piece?

b. What is the area of this piece of pizza?

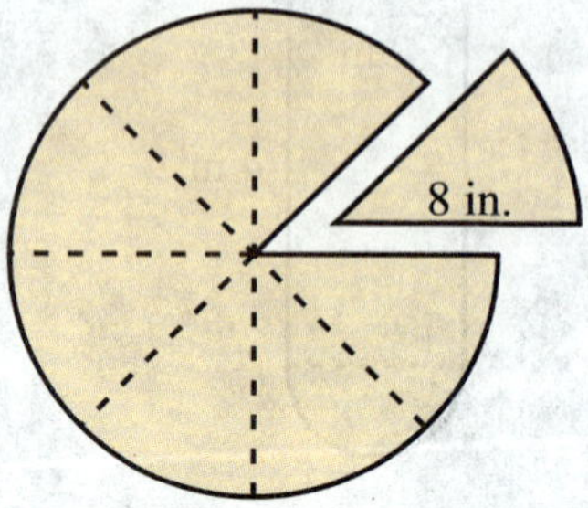

37. Some cities use traffic circles rather than traffic lights at major intersections. A traffic circle is a circular road at the intersection, where the driver will enter, drive counterclockwise, and then leave on the desired road. If the outer diameter of the circle is 220 feet, and the road is 25 feet wide, what is the surface area of the traffic circle?

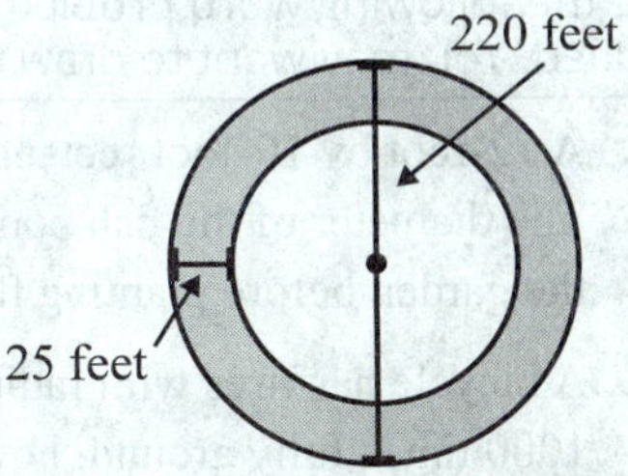

38. An entry door is in the shape of a rectangle with the top part being a semicircle, as shown. Compute its area. (**Hint:** To find the height of the rectangular portion of the door, subtract the height (radius) of the semicircle from the total height of the door.)

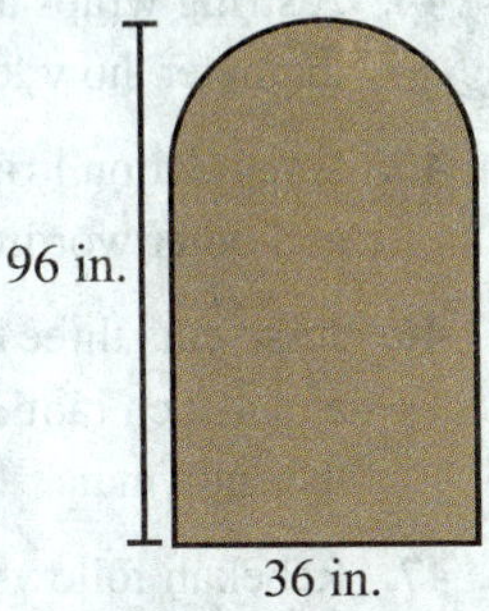

39. One common design for speaker cabinets in high quality stereo systems is the bass reflex cabinet, where there is a port (either rectangular or circular) in the front of the cabinet in addition to the speaker itself. If the front panel is 26 inches high and 18 inches wide, what will the area of the panel be if a 12-inch hole is cut out for the speaker, and a 10 inch by 4 inch rectangular port is cutout as shown?

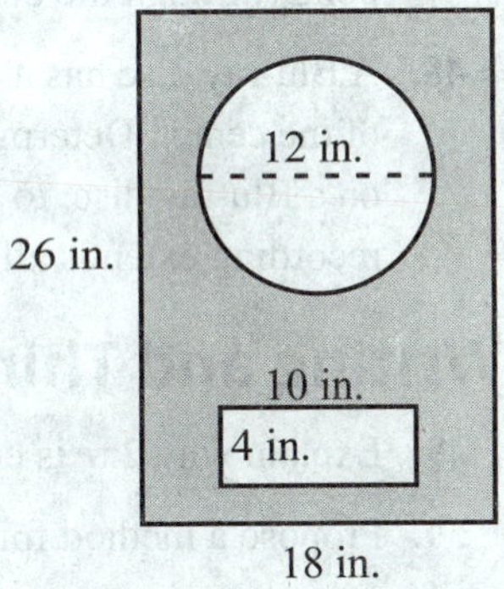

40. A 10-inch flywheel has a square opening in the center to connect it to the driving shaft. Assuming that this square hole is 1 inch on the side, what is the area of the flywheel?

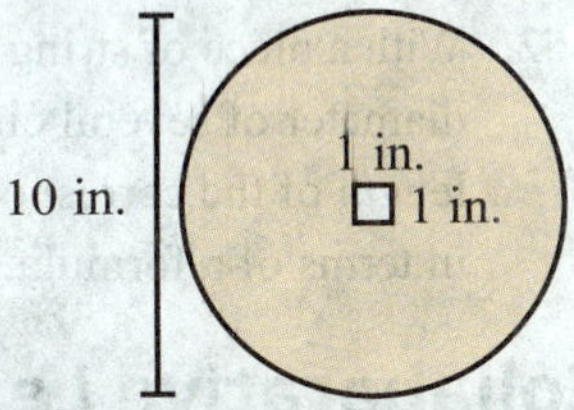

41. Brain Freeze Ice Cream is creating a new wooden sign in the shape of an ice cream cone to display their flavors of the day. The outline of the cone is shown.

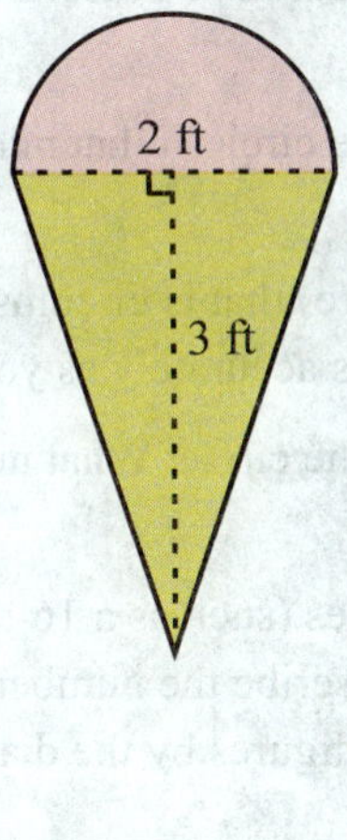

a. The sign will be painted to look like an ice cream cone. What is the area of the sign to be painted? Round to the nearest hundredth.

b. The sign will have a metal border to protect it from damage. How much metal border will be needed? Round to the nearest hundredth.

37. 15,307.5 ft^2
38. 3316.68 in.^{2}
39. 314.96 in.^{2}
40. 77.5 in.^{2}
41. **a.** 4.57 square feet
b. 10.34 feet

Solve the following word problems by finding either the circumference or the area specified. You may want to draw figures to help solve the problems. (Use $\pi \approx 3.14$.)

42. A 12-foot by 10-foot rectangular flower garden has a circular fish pond in the center. The diameter of the fish pond is 6 feet. The gardener plans to place new top soil in the garden before planting flowers. What is the area that will be covered by top soil?

43. A bicycle has tires with radius of 10 in. If John pedals this bike so that the tires rotate 1000 times fully around, how far does John ride his bike?

44. A string wraps around the center of a coffee can 6 times. If the can is 8 cm in diameter, how long is the string, to the nearest whole number?

45. A regulation Frisbee measures 10.75 inches in diameter. How much room for logos and other wording is available on the top of the Frisbee, rounded to the nearest tenth?

46. Jesse cuts three disks out of cardboard to make a snowman. If the disks are of 2-, 3-, and 4-inch radii, respectively, determine how much total cardboard was used to make the snowman.

47. Rebekah follows a nature trail once around a circular pond. If the distance across the pond (through the center) is 1500 ft, how far did Rebekah walk?

48. A Blu-ray disc has a radius of 6 cm, with a hole measuring 0.5 cm in radius cut out of the center. Determine the total amount of physical space available for recording on a Blu-ray disc, to the nearest whole number. (Assume the space available for recording extends all the way to the edges.)

Writing and Thinking

49. Explain why $2\pi r$ is equivalent to d.

50. Propose a method for calculating the area of a semicircle and justify your method.

51. Explain why answers that are calculated using π (approximately 3.14) are not precise, but are close approximations.

52. With a piece of string and a ruler, carefully measure the circumference and the diameter of several circular figures in your home. Divide each circumference by the length of the corresponding diameter. What result do you observe? Explain this result in terms of a formula.

Collaborative Learning

53. First, draw a circle with a diameter of 10 cm (radius 5 cm). Next, draw a square outside the circle so that each side of the square just touches the circle. (The square is said to be circumscribed about the circle, and the circle is said to be inscribed in the square.)

a. Divide the perimeter of the square by the diameter of the circle. What number did you get?

Next, on the same figure, draw an octagon (8-sided figure) that is circumscribed about the circle. Measure the perimeter of the octagon as accurately as you can.

b. Divide the perimeter of the octagon by the diameter of the circle. What number did you get?

c. Now try drawing circumscribed polygons with more sides (such as a 16-sided figure, a 32-sided figure, and so on) about the circle. Describe the numbers you think you would get by dividing the perimeters of these figures by the diameter of the circle.θ

---

42. 91.74 square feet

43. 62,800 in. (a little less than one mile)

44. 151 cm

45. 90.7 in.^{2}

46. 91.06 in.^{2}

47. 4710 ft

48. 112 cm^2

Writing & Thinking

49. One diameter is equal to two radii. Thus, $d = 2r$ and $C = 2\pi r = \pi d$.

50. Because a semicircle is one-half of a circle, the area of a semicircle will be one-half of the area of the same-sized circle. **1.** Find the area of the circle. **2.** Take half of that area. **3.** Label with square units.

51. Pi is a nonrepeating non-terminating decimal and therefore, when using π, a rounded number must be used for calculations. Anytime a rounded number is used in calculations, the result will be an approximation.

52. The result should be close to π in each case. $\frac{C}{d} = \pi$

Collaborative Learning

53. a. 4

b. Answers will vary somewhat, but they should be about 3.3.

c. Answers will vary. The numbers will be come closer to π as the number of sides increases.

6.5 Volume and Surface Area

Objectives

A. Calculate the volumes of solids.

B. Calculate the surface areas of solids.

A Volume

Volume is a measure of the space enclosed by a three-dimensional figure and is measured in **cubic units**. The volume or space contained within a cube that is 1 centimeter on each edge is **one cubic centimeter**, or 1 cm^3, as shown in Figure 1. A cubic centimeter is about the size of a sugar cube.

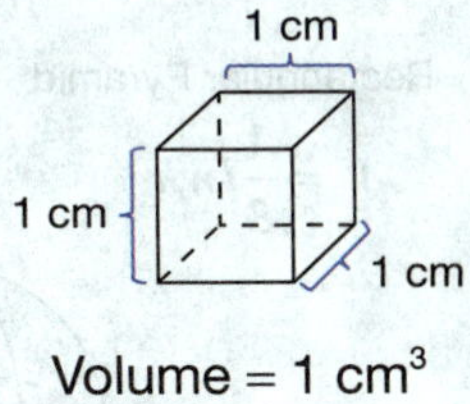

Figure 1

A rectangular solid that has edges of 3 cm, 2 cm, and 5 cm has a volume of

$$3 \text{ cm} \times 2 \text{ cm} \times 5 \text{ cm} = 30 \text{ cm}^3.$$

We can think of the rectangular solid as being three layers of ten cubic centimeters, as shown in Figure 2.

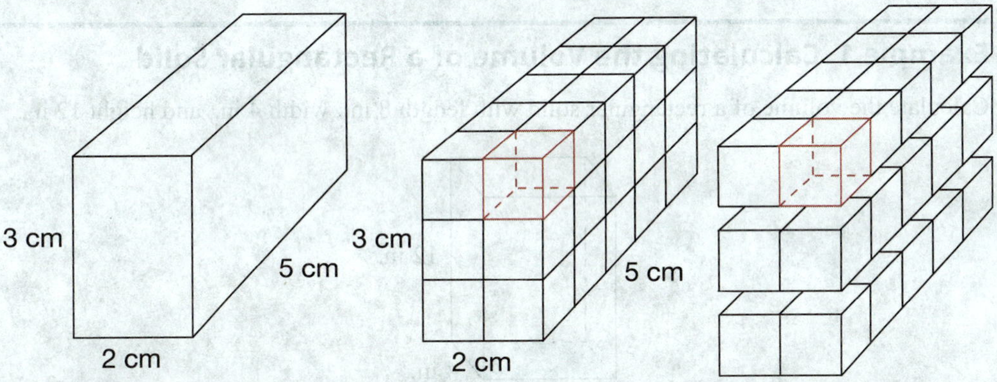

Figure 2

Some of the basic units of volume and their corresponding abbreviations are as follows. **Be sure to label your answers with the correct units.**

From the Metric System	From the US Customary System
cubic millimeters (mm^3)	cubic inches ($in.^3$)
cubic centimeters (cm^3)	cubic feet (ft^3)
cubic meters (m^3)	cubic yards (yd^3)

Table 1

Five common geometric solids and the corresponding formulas for calculating the volume of each type of solid are shown here.

Teaching Note

To further students' conceptual understanding of volume, you may want to show students that as long as they can find the area of the base of a prism or cylinder, they can find the volume of the object by multiplying the area of the base by the height. Similarly, you can explain to students that any pyramid or cone has $\frac{1}{3}$ the volume of its corresponding prism or cylinder.

Volume Formulas for Five Geometric Solids

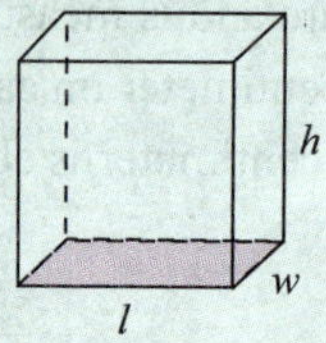

Rectangular Solid

$$V = lwh$$

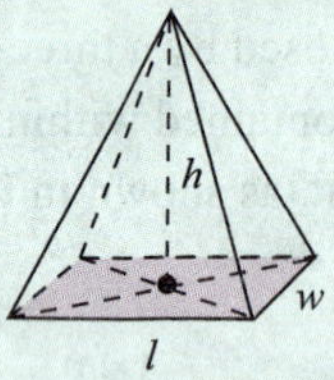

Rectangular Pyramid

$$V = \frac{1}{3}lwh$$

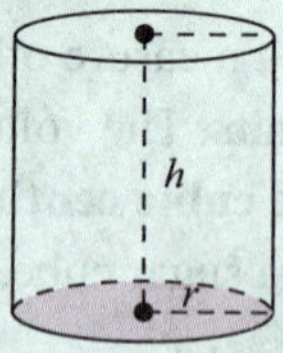

Right Circular Cylinder

$$V = \pi r^2 h$$

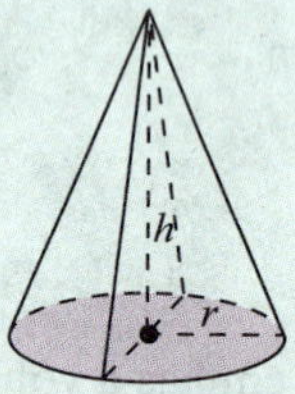

Right Circular Cone

$$V = \frac{1}{3}\pi r^2 h$$

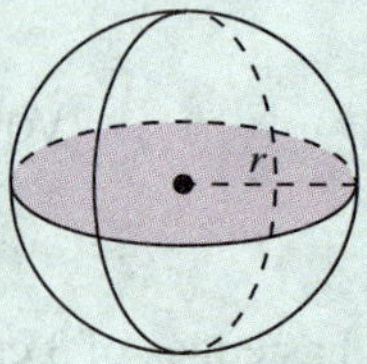

Sphere

$$V = \frac{4}{3}\pi r^3$$

FORMULA

1. Calculate the volume of a rectangular solid with length 15 in., width 6 in., and height 9 in.

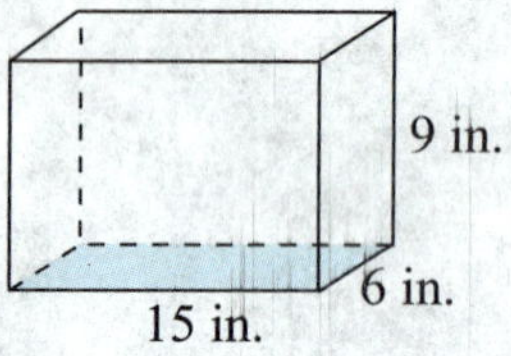

Example 1 Calculating the Volume of a Rectangular Solid

Calculate the volume of a rectangular solid with length 8 in., width 4 in., and height 12 in.

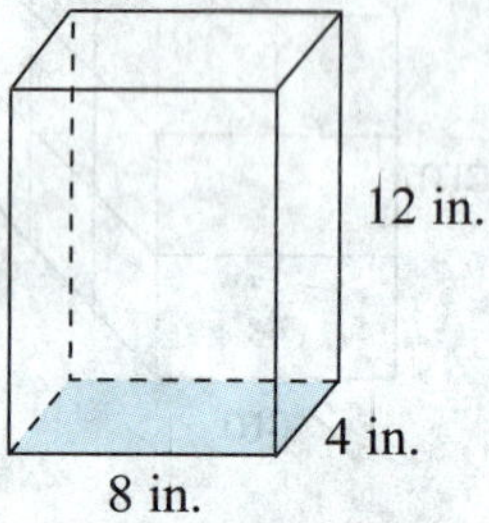

Solution

Using the formula for the volume of a rectangular solid, we have the following.

$$V = lwh$$
$$V = 8 \text{ in.} \cdot 4 \text{ in.} \cdot 12 \text{ in.}$$
$$= 384 \text{ in.}^3$$

The volume of the rectangular solid is 384 in.^{3}

Now work margin exercise 1.

Example 2 Calculating the Volume of a Sphere

Calculate the volume of a sphere with radius 9 cm. Use $\pi = 3.14$.

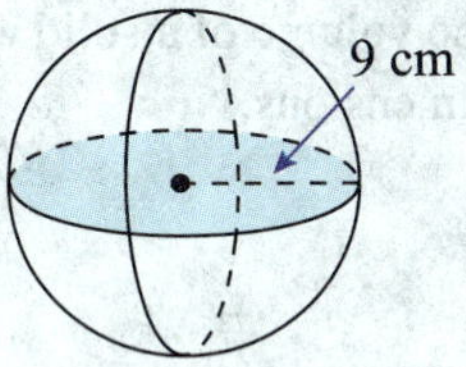

Solution

Using the formula for the volume of a sphere, we have the following.

$$V = \frac{4}{3}\pi r^3$$
$$V \approx \frac{4}{3}\cdot 3.14\cdot(9\text{ cm})^3$$
$$= \frac{4}{3}\cdot 3.14\cdot 729\text{ cm}^3$$
$$= 3052.08\text{ cm}^3$$

The volume of the sphere is 3052.08 cm^3.

Now work margin exercise 2.

2. Calculate the volume of a sphere with radius 3 ft.

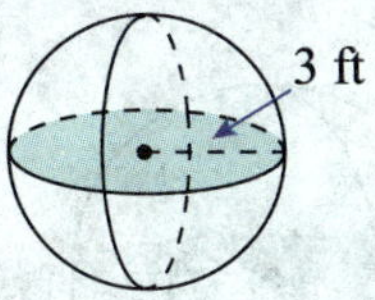

Example 3 Calculating the Volume of a Cone

What is the volume of a cone with a height of 12 mm and a circular base with a diameter of 8 mm?

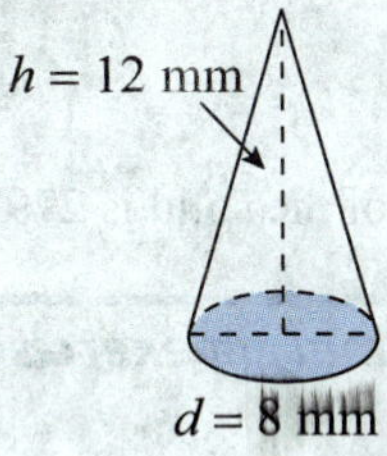

Solution

We know the diameter of 8 mm, but we need the radius.

The radius is half of the diameter. $r = \frac{1}{2}\cdot 8\text{ mm} = 4\text{ mm}$

So, applying the formula for the volume of a cone, we have the following.

$$V = \frac{1}{3}\pi r^2 h$$
$$V \approx \frac{1}{3}\cdot 3.14\cdot(4\text{ mm})^2\cdot 12\text{ mm}$$
$$= 3.14\cdot 64\text{ mm}^3$$
$$= 200.96\text{ mm}^3$$

The volume of the cone is 200.96 mm^3.

Now work margin exercise 3.

3. What is the volume of a cone with a height of 3 m and a circular base with a diameter of 6 m?

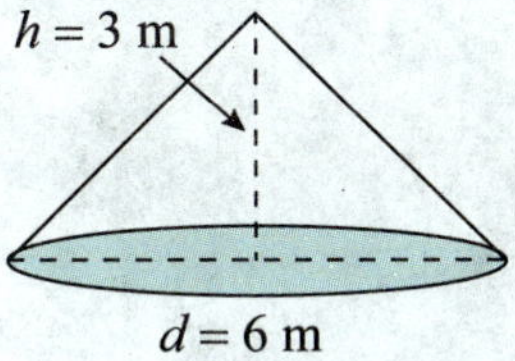

4. Calculate the volume of a solid with the indicated dimensions.

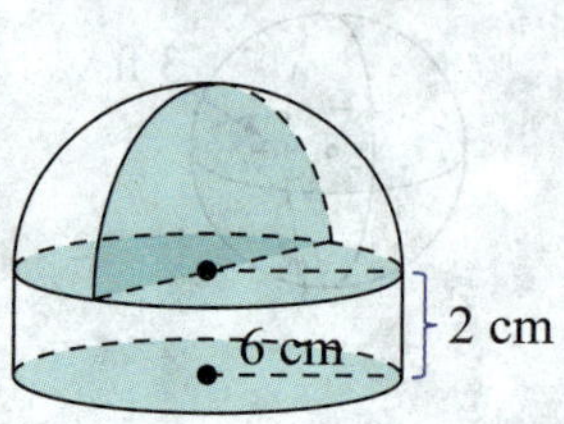

Example 4 Calculating the Volume of a Solid

Calculate the volume of a solid with the indicated dimensions.

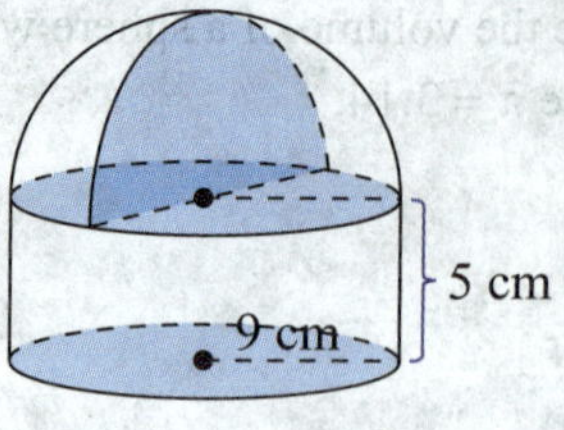

Solution

From the figure, we see that the bottom portion of the solid is a cylinder and on top of the cylinder is a **hemisphere** (one-half of a sphere). Thus, the volume of the solid will be the sum of the volumes of the cylinder and the hemisphere.

Volume of Cylinder

$$V = \pi r^2 h$$
$$V \approx 3.14(9 \text{ cm})^2 (5 \text{ cm})$$
$$= 1271.7 \text{ cm}^3$$

Volume of Hemisphere

$$V = \frac{1}{2} \cdot \frac{4}{3} \pi r^3 \quad \text{Half the volume of a sphere}$$
$$V \approx \frac{2}{3}(3.14)(9 \text{ cm})^3$$
$$= 1526.04 \text{ cm}^3$$

Total Volume

$$\begin{array}{r} 1271.70 \text{ cm}^3 \\ +1526.04 \text{ cm}^3 \\ \hline 2797.74 \text{ cm}^3 \end{array}$$

The volume of the solid is 2797.74 cm^3.

Now work margin exercise 4.

A **cube** is a rectangular solid in which the length, width, and height are all equal. If s is the length of one edge, the formula for the volume of the cube is $V = s^3$.

5. Calculate the volume of the cube in both cubic inches and cubic feet. (Remember, 1 ft = 12 in.)

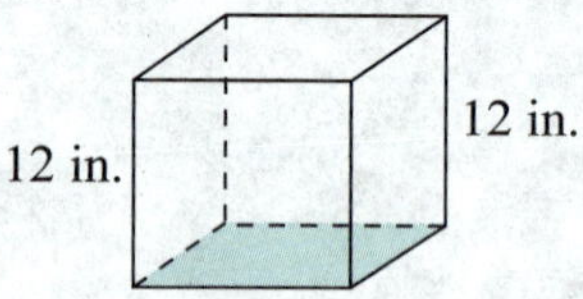

Example 5 Calculating the Volume of a Cube

Calculate the volume of the cube in both cubic yards and cubic feet. (Remember, 1 yd = 3 ft.)

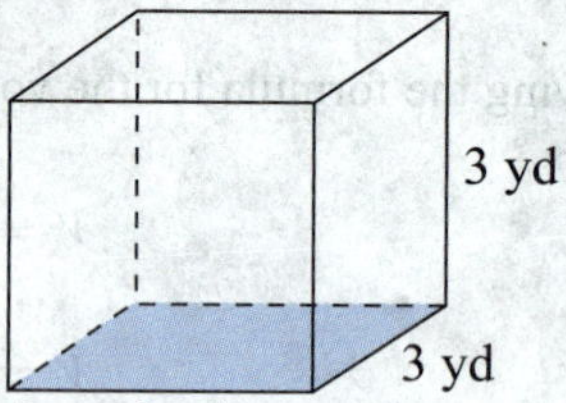

Solution

We apply the formula $V = s^3$ twice; once by using $s = 3$ yards and once by using $s = 9$ feet.

$$V = s^3 \qquad V = s^3$$
$$V = (3 \text{ yd})^3 \qquad V = (9 \text{ ft})^3$$
$$= 27 \text{ yd}^3 \qquad = 729 \text{ ft}^3$$

The volume of the cube is 27 yd^3 or 729 ft^3.

Notice that because feet are smaller than yards, there are many more cubic feet than cubic yards in the cube.

Now work margin exercise 5.

B Surface Area

The **surface area** (*SA*) of a geometric solid is a measure of the area of the outside surface in square units. Surface area is particularly important in fields related to building and construction.

As an example, the surface area of a rectangular solid is found by adding the areas of the six sides. Each side is a rectangle. The front and back are the same. The left and right sides are the same. The top and bottom are the same. If the edges are labeled as l, w, and h, we see that the total surface area can be calculated by using the following formula.

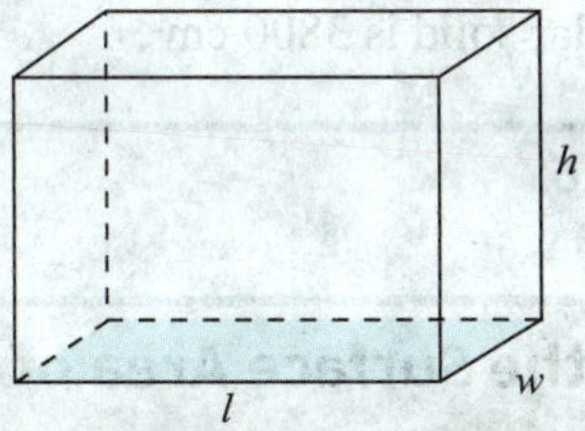

$$SA = 2lw + 2wh + 2lh$$

Figure 3

Cylinders and spheres are two other solids where surface area can be an important consideration. The cost of building and maintaining the surface area of such figures can be a determining factor for an architect in using a particular shape in designing structures. The formulas for surface areas of three solids are given here.

Be sure to label your answers with square units.

Surface Area Formulas for Three Geometric Solids

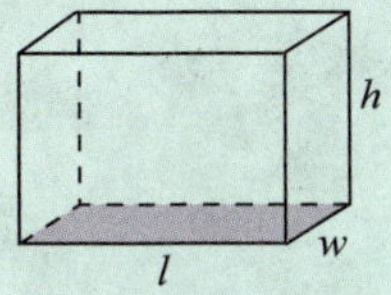

Rectangular Solid

$$SA = 2lw + 2wh + 2lh$$

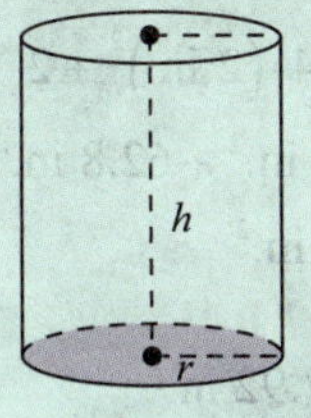

Right Circular Cylinder

$$SA = 2\pi r^2 + 2\pi rh$$

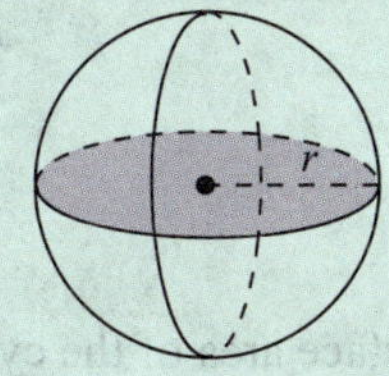

Sphere

$$SA = 4\pi r^2$$

FORMULA

Math Tip

Consider the construction of the following rectangular solid.

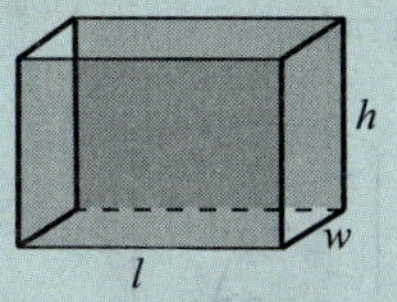

If you unfold and flatten the shape, it can help you visualize the different areas that are to be added together to find the total surface area.

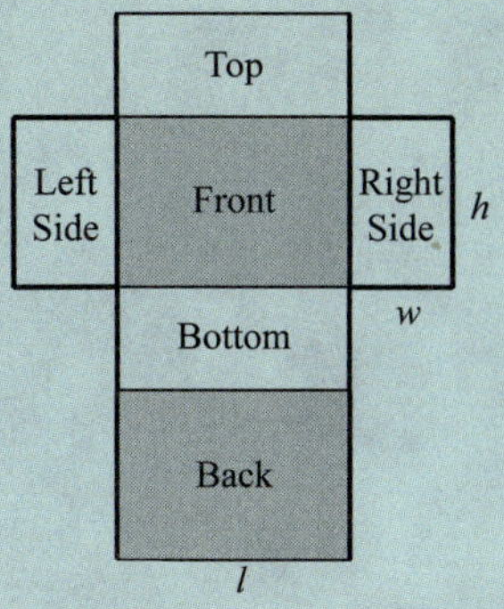

6. Calculate the surface area of a rectangular solid with length 15 ft, width 4 ft, and height 12 ft.

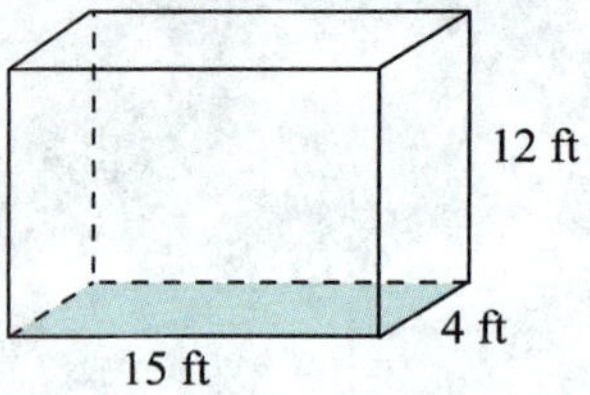

Example 6 Calculating the Surface Area of a Rectangular Solid

Calculate the surface area of a rectangular solid with length 30 cm, width 10 cm, and height 40 cm.

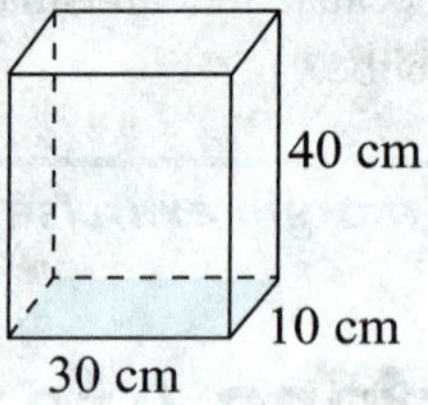

Solution

Using the formula for the surface area of a rectangular solid, we have the following

$$SA = 2lw + 2wh + 2lh$$

$$\begin{aligned} SA &= 2 \cdot 30 \text{ cm} \cdot 10 \text{ cm} + 2 \cdot 10 \text{ cm} \cdot 40 \text{ cm} + 2 \cdot 30 \text{ cm} \cdot 40 \text{ cm} \\ &= 600 \text{ cm}^2 + 800 \text{ cm}^2 + 2400 \text{ cm}^2 \\ &= 3800 \text{ cm}^2 \end{aligned}$$

The surface area of the rectangular solid is 3800 cm^2.

Now work margin exercise 6.

7. Calculate the surface area of a cylinder with a height of 7 yd and a circular base with a radius of 4 yd.

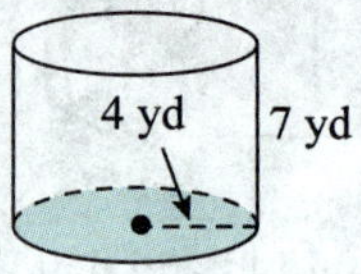

Example 7 Calculating the Surface Area of a Cylinder

Calculate the surface area of a coffee can in the shape of a cylinder with a height of 5 in. and a circular base with a radius of 2 in.

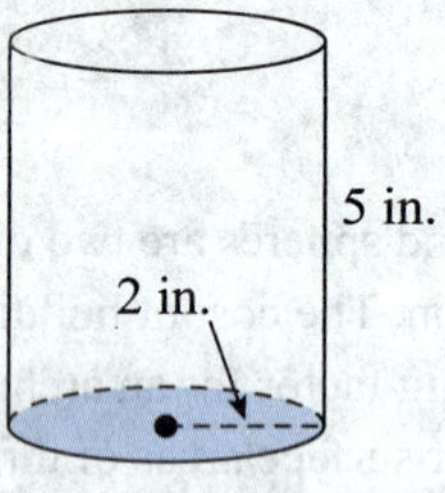

Solution

Using the formula for the surface area of a cylinder, we have the following.

$$SA = 2\pi r^2 + 2\pi rh$$

$$\begin{aligned} SA &\approx 2 \cdot 3.14 \cdot (2 \text{ in.})^2 + 2 \cdot 3.14 \cdot 2 \text{ in.} \cdot 5 \text{ in.} \\ &= 25.12 \text{ in.}^2 + 62.8 \text{ in.}^2 \\ &= 87.92 \text{ in.}^2 \end{aligned}$$

The surface area of the cylinder is 87.92 in.^2

Now work margin exercise 7.

Margin Exercise Answers

1. 810 in.^3 **2.** 113.04 ft^3 **3.** 28.26 m^3 **4.** 678.24 cm^3 **5.** 1728 in.^3 or 1 ft^3 **6.** 576 ft^2 **7.** 276.32 yd^2

6.5 Exercises

Concept Check

Fill-in-the-Blank. Complete each sentence using information found in this section.

1. The measure of the space enclosed by a three-dimensional figure is its __________.
2. Volume is measured in _______ units.
3. The measure of the area of the outside surface of a geometric solid is the solid's ________ _____.
4. Surface area is measured in _______ units.
5. $SA = 2\pi r^2 + 2\pi rh$ is the formula for the surface area of a/an _______ _______ ________.
6. $SA = 4\pi r^2$ is the formula for the surface area of a/an _________.

True/False. Determine whether each statement is true or false. If a statement is false, explain how it can be changed so the statement will be true. (**Note:** There may be more than one acceptable change.)

7. To find the volume of a can of corn, the formula $V = \pi r^2 h$ would be used.
8. $V = lwh$ is the formula for the surface area of a rectangular solid.
9. The area of the paper label on a can of peaches is an example of surface area.
10. To find the volume of a rectangular solid, the areas of each surface are added together.

Match each formula for volume to its corresponding geometric figure.

11. a. Rectangular solid **A.** $V = \frac{4}{3}\pi r^3$

b. Rectangular pyramid **B.** $V = \frac{1}{3}\pi r^2 h$

c. Right circular cylinder **C.** $V = lwh$

d. Right circular cone **D.** $V = \pi r^2 h$

e. Sphere **E.** $V = \frac{1}{3}lwh$

6.5 Exercises

Concept Check

1. volume
2. cubic
3. surface area
4. square
5. right circular cylinder
6. sphere
7. True
8. False; $V = lwh$ is the formula for the volume of a rectangular solid.
9. True
10. False; To find the surface area of a rectangular solid, the areas of each surface are added together.
11. **a.** C **b.** E **c.** D **d.** B **e.** A

Practice

1. 70 in.^{3}
2. 1696.6 in.^{3}
3. 381.51 cm^3
4. 904.32 ft^3
5. 12.56 mm^3
6. 80 cm^3
7. 60 in.^{3}
8. 96 in.^{3}
9. 14.13 ft^3
10. 381.51 cm^3
11. 401.92 m^3
12. 282.6 km^3
13. 376.8 ft^3
14. 12.56 dm^3

Practice

Calculate the volume of each solid. See Examples 1 through 5. Use $\pi \approx 3.14$.

1. A rectangular solid with length 5 in., width 2 in., and height 7 in.
2. A right circular cylinder 15 in. high and 1 ft in diameter.
3. A sphere with radius 4.5 cm.
4. A sphere with diameter 12 ft.
5. A right circular cone 3 mm high with a 2 mm radius.
6. A rectangular pyramid with length 8 cm, width 1 cm, and height 30 cm.

Calculate the volume of each solid. See Examples 1 through 5. Use $\pi \approx 3.14$.

7.

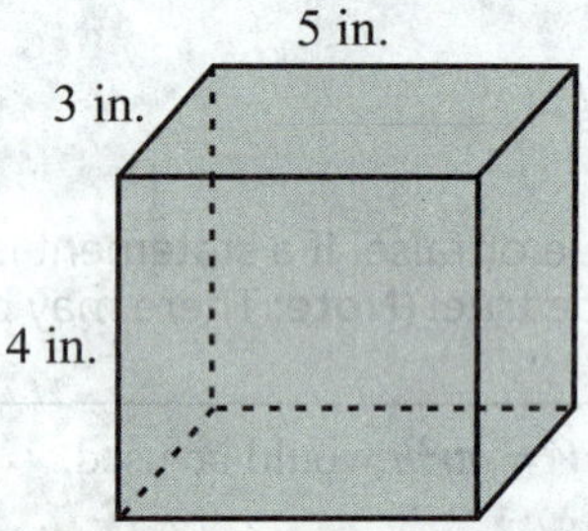

8.

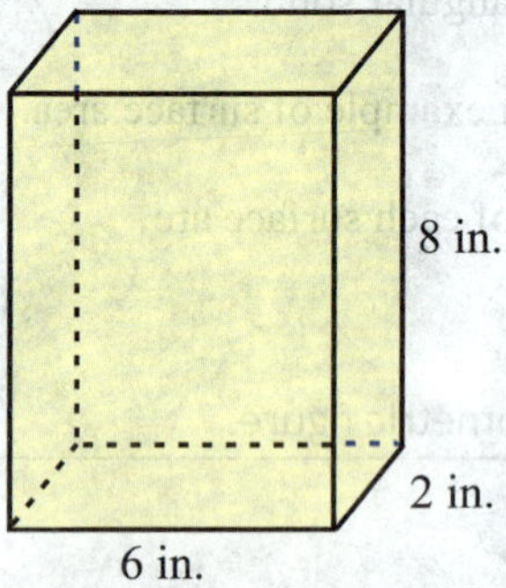

9.

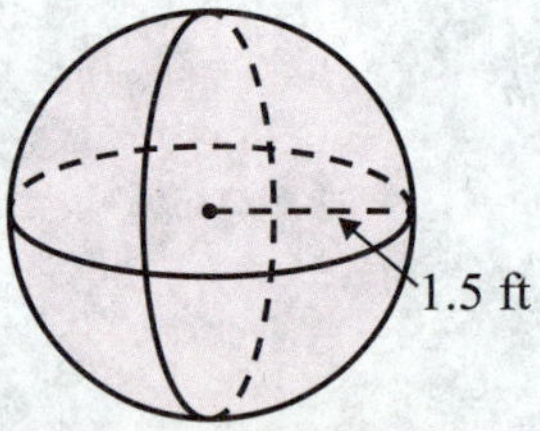

10.

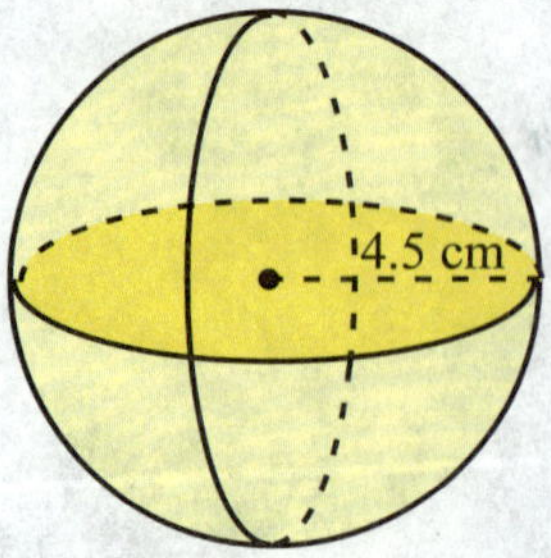

11.

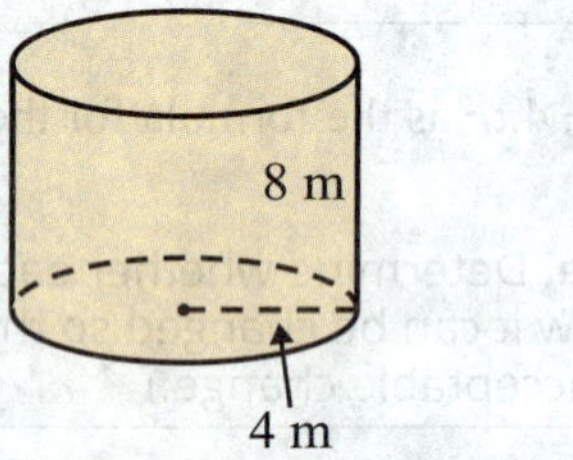

12.

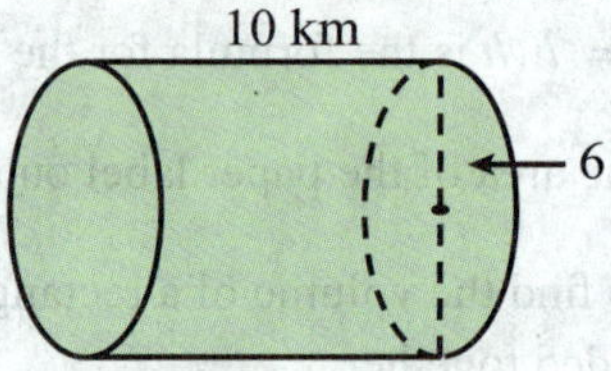

13.

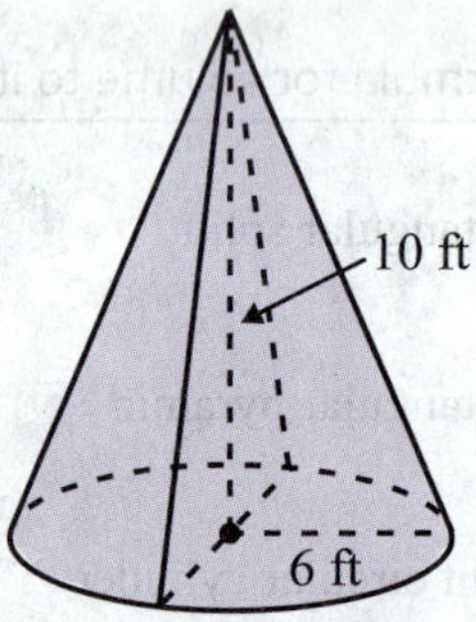

14.

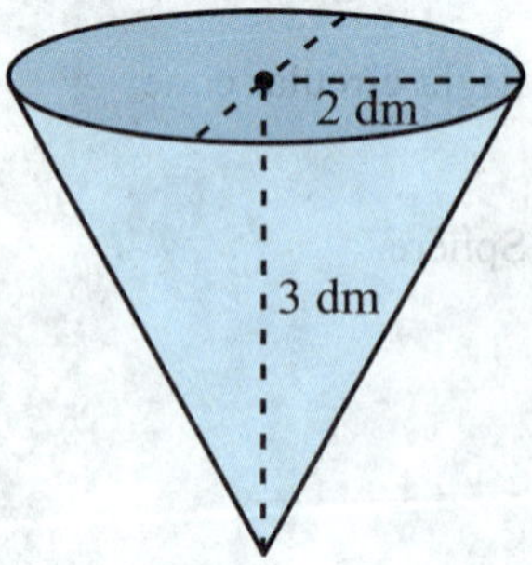

15.

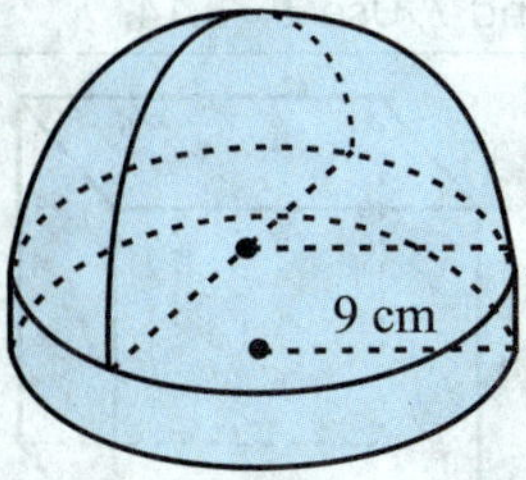

16.

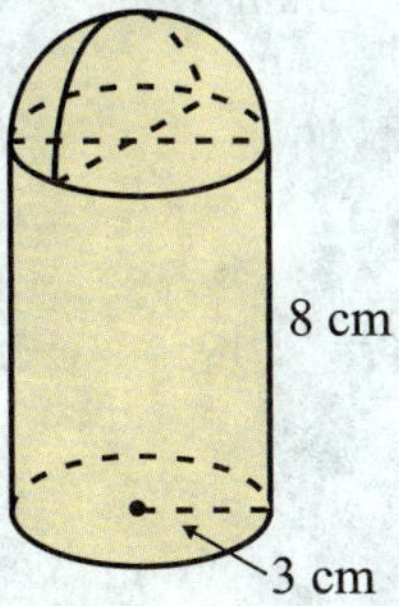

17.

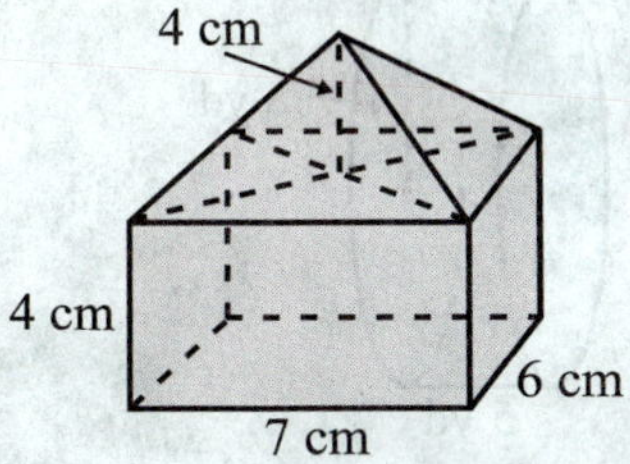

18.

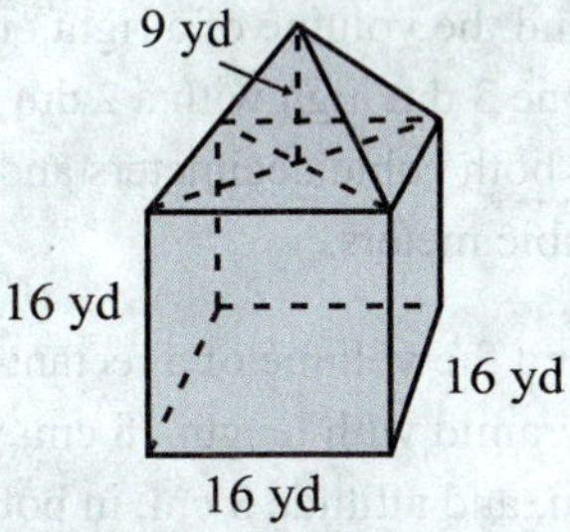

19.

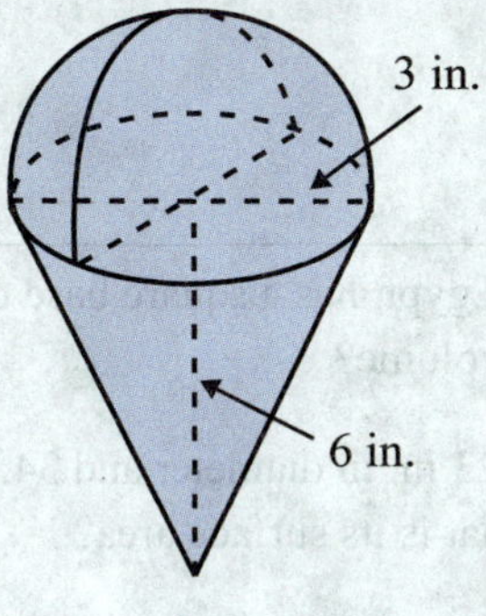

20.

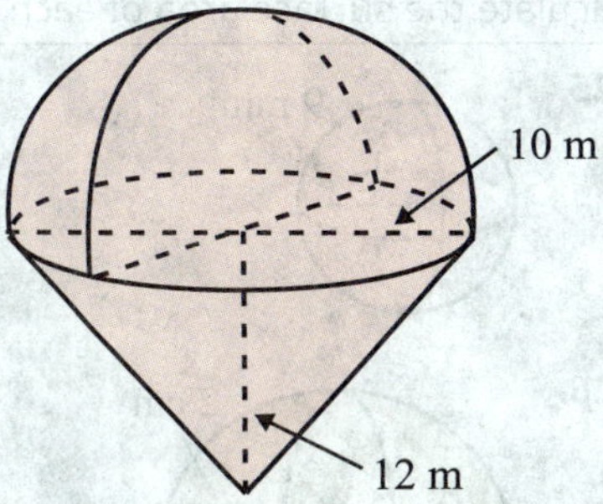

21.

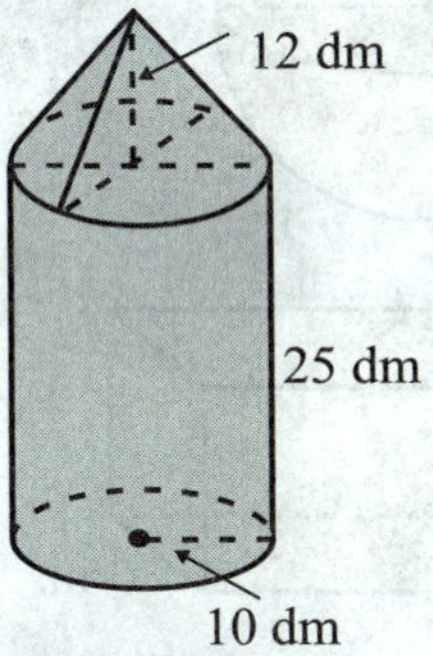

22.

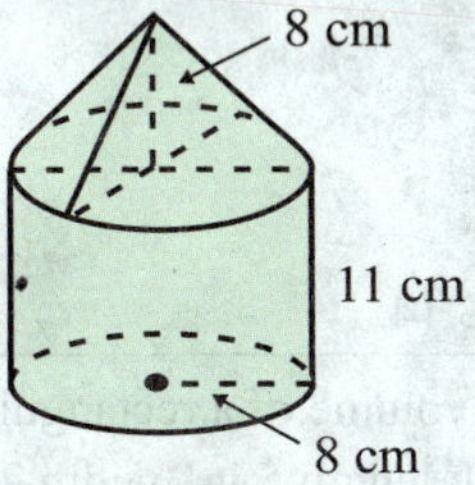

23.

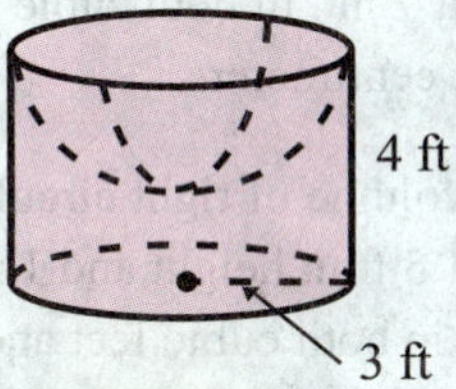

24.

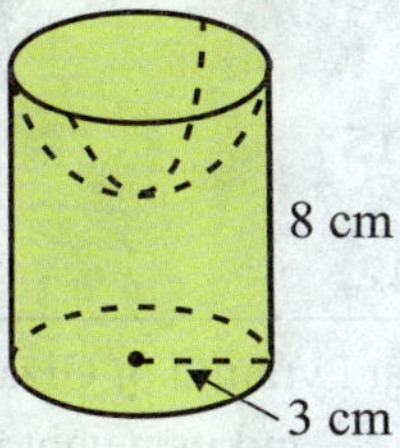

15. 2289.06 cm^3
16. 282.6 cm^2
17. 224 cm^3
18. 4864 yd^3
19. 113.04 $in.^3$
20. 3349.33 m^3
21. 9106 dm^3
22. 2746.45 cm^3
23. 56.52 ft^3
24. 169.56 cm^3

25. 1017.36 mm^2

26. 50.24 cm^2

27. 122 $in.^2$

28. 150 km^2

29. 226.08 m^2

30. 653.12 yd^2

31. $V = 70$ $in.^3$; $V = 1147.09$ cm^3

32. $V = 1.1775$ ft^3; $V = 0.032\,97$ m^3

33. $V = 12.56$ dm^3; $V = 0.012\,56$ m^3

34. $V = 180$ cm^3; $V = 180\,000$ mm^3

Applications

35. 2,596,902 m^3

36. 3322.12 $in.^2$

Calculate the surface area of each solid. See Examples 6 and 7. Use $\pi \approx 3.14$.

25.

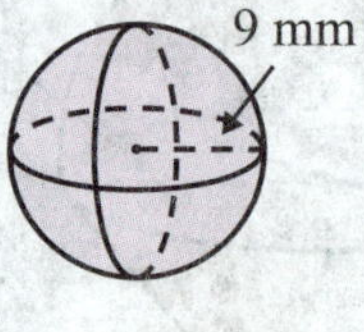

26.

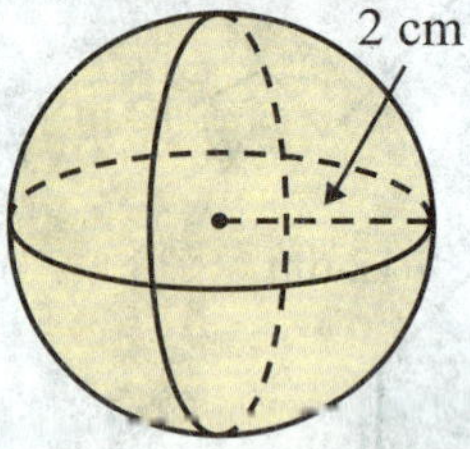

27.

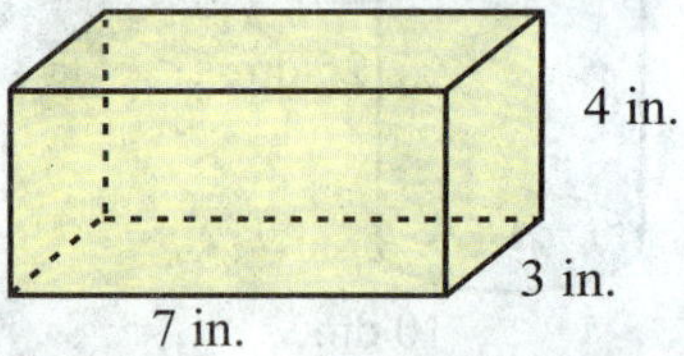

28.

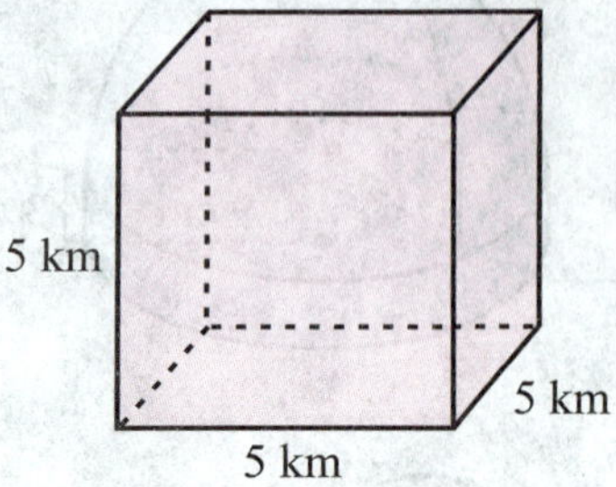

29.

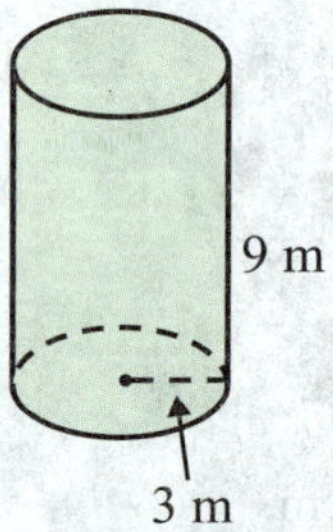

30.

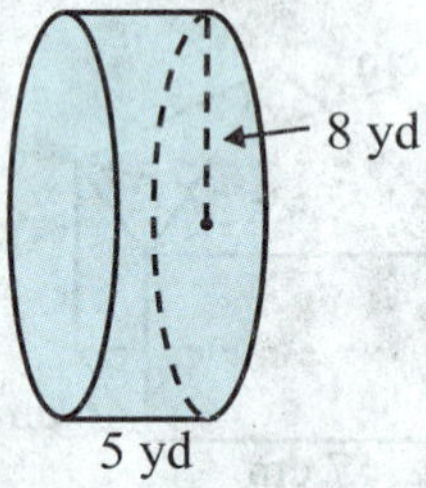

Solve. Use $\pi \approx 3.14$.

31. Find the volume of a rectangular solid with length 5 in., width 2 in., and height 7 in., in both cubic inches and cubic centimeters.

32. Find the volume of right circular cylinder 1.5 ft in height and 1 ft in diameter, in both cubic feet and cubic meters.

33. Find the volume of a right circular cone 3 dm high with a 2 dm radius, in both cubic decimeters and cubic meters.

34. Find the volume of a rectangular pyramid with length 18 cm, width 10 cm, and altitude 3 cm, in both cubic centimeters and cubic millimeters.

Applications

Solve. Use $\pi \approx 3.14$.

35. The Great Pyramid of Giza, which is located in Egypt, has a square base of 231 m on each side, and its height is 146 m. What is its volume?

36. A standard 55 gallon round steel drum is about 23 in. in diameter and 34.5 in. high. Assuming that the drum is totally enclosed, what is its surface area?

37. A 6 in. tall ice cream cone is filled solid with ice cream where the final scoop of ice cream forms a perfect hemisphere above the top of the cone. What is the total volume of ice cream in the cone if the top of the cone has a 2.2 in. opening? Round your answer to the nearest hundredth.

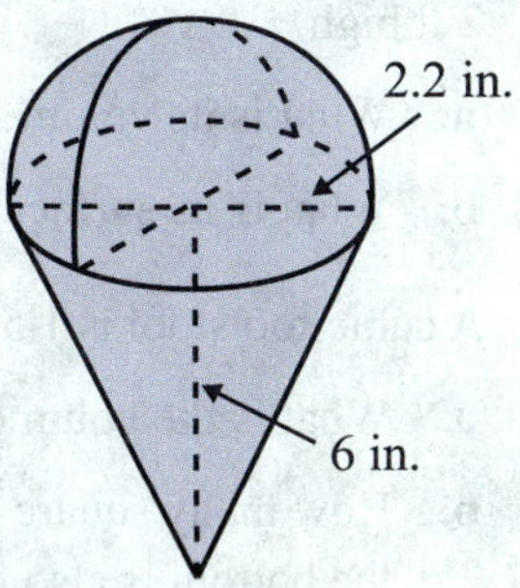

38. A cylindrical trash can has a hemispherical top (with a trap door for the trash). If the diameter of the can is 16 in. and its total height is 38 in., find its volume. (**Hint:** Begin by finding the height of the straight part of the can.)

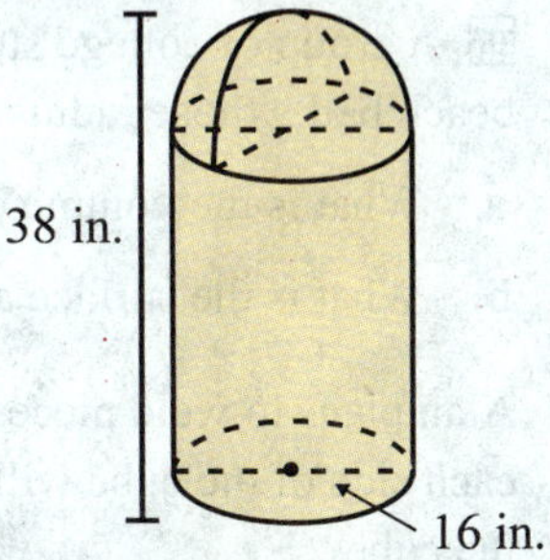

39. A rectangular tent with straight sides has a pyramidal shaped roof. The dimensions of the rectangular portion are 12 ft long, 10 ft wide, and 6 ft high. The peak of the pyramid is 2 ft above the top edge of the walls. What is the volume of the inside of the tent?

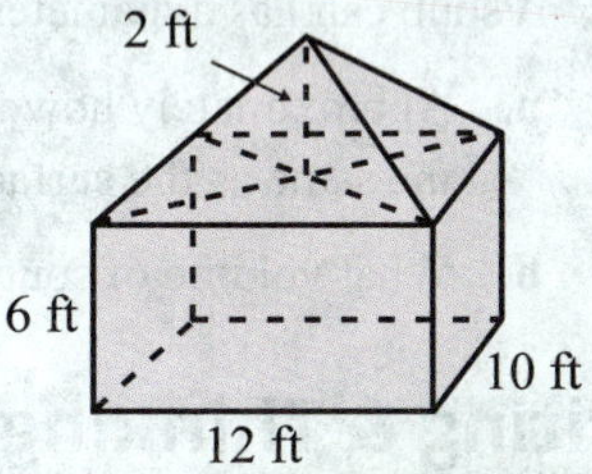

40. Disposable paper drinking cups, like those used at water coolers, are often cone-shaped. Find the volume of such a cup that is 9 cm high with a 3.2 cm radius. Express the answer to the nearest milliliter.

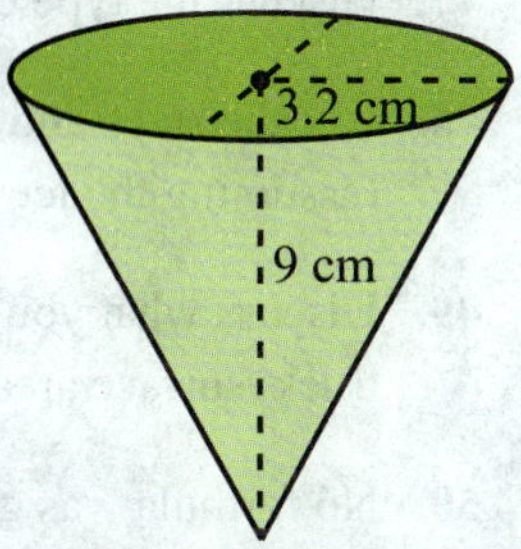

41. A manufacturer is to design a can in the shape of a right circular cylinder to hold 0.5 L of juice concentrate. If the can must have a diameter of 7 cm, how tall will the can be (to the nearest centimeter)?

37. 10.39 in.^{3}

38. 7100.6 in.^{3}

39. 800 ft^3

40. 96 mL

41. 13 cm

42. **a.** 30 ft^3 **b.** 62 ft^2

43. **a.** 3.375 ft^3 **b.** 13.5 ft^2

44. **a.** 4.19 ft^3 **b.** 12.56 ft^2

45. 1536 square inches

46. **a.** 61.23 square inches **b.** 35.33 cubic inches

Writing & Thinking

47. Volume is measured in cubic units. Volume takes up a three-dimensional space and the units can be thought of as small cubes which leads to the concept of cubic units.

48. To find the volume of the ice cream itself, the volume is half the volume of a sphere and the formula used would be $V = \frac{1}{2} \cdot \frac{4}{3}\pi r^3 = \frac{2}{3}\pi r^3$. The result is the volume of the ice cream that is exposed. Next find the volume of the cone by using the formula $V = \frac{1}{3}\pi r^2 h$. The volume of the ice cream cone will be the sum of these two volumes.

49. Volume is more important because it determines how many packages or what sized packages the driver can fit into the truck. Three-dimensional space (volume) is more important for this job.

42. A specialty lined storage chest has the inside dimensions of 5 ft long, 3 ft wide, and 2 ft high.

a. What is its volume?

b. What is its surface area?

43. A cubic footstool is 1.5 ft long in each direction.

a. What is the volume of the footstool?

b. How many square feet of material are necessary to cover the footstool? Assume the bottom is also being covered by the material.

44. A group of college students went to the beach and inflated their 2-ft spherical beach ball, whose radius is 1 ft.

a. What is the volume of the ball? Round your answer to the nearest hundredth.

b. What is the surface area of the ball?

45. Alan plans to weld pieces of metal together to create a metal cube. The length of each side of the cube will be 16 inches. How much metal will he need to create the cube?

46. A soup can has a diameter of 3 inches and a height of 5 inches.

a. Approximately how much material is needed to create the soup can? (**Hint:** It is the same as the surface area.)

b. What volume of soup can fit inside of the soup can? Round to the nearest tenth.

Writing & Thinking

47. Discuss the type of units used for volume and explain why.

48. List the steps and formulas you would use to find the volume of an ice cream cone (assuming the ice cream itself forms a perfect half sphere).

49. Discuss what you think would be more important to a UPS driver when loading his truck, surface area or volume.

50. No formula was given in the text for the surface area of a rectangular pyramid. Create a plan to find the surface area of this type of figure. Include formulas and operations you would use.

50. A rectangular pyramid has five surfaces as sides (4 triangles) and base (rectangle or square). To find the surface area, find the area of each of the 4 triangles by using the formula $A = \frac{1}{2}bh$. Multiply this result by 4. Find the area of the rectangle by using the formula $A = lw$. Add these results to find the surface area of the rectangular pyramid.

6.6 Similar and Congruent Triangles

Objectives

A. Know the properties of similar triangles.

B. Know the properties of congruent triangles.

A Similar Triangles

Two triangles are said to be **similar triangles** if they have the same "shape." They may or may not have the same "size." More formally, if two triangles are similar, they have the following two properties.

Similar Triangles

1. In similar triangles, the **corresponding angles have the same measure**. (We say that the corresponding angles are congruent.)
2. In similar triangles, the **lengths of the corresponding sides are proportional**.

DEFINITION

In similar triangles, **corresponding sides** are those sides opposite the congruent angles (angles with the same measure) in the respective triangles. In Figure 1, triangles *ABC* and *DEF* are similar.

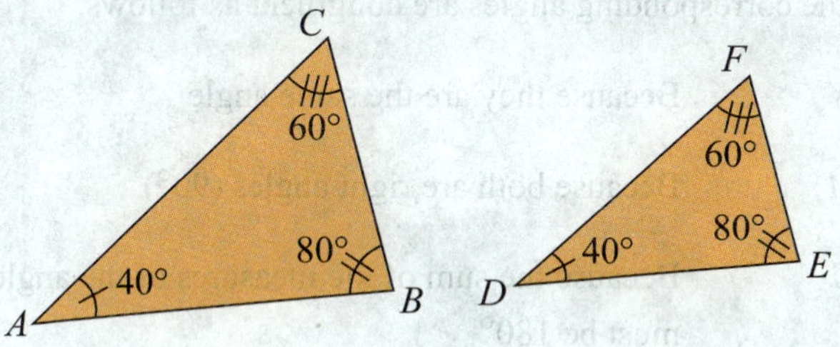

Figure 1

For the similar triangles in Figure 1, the following relationships are true.

$\angle A$ corresponds to $\angle D$; $m\angle A = m\angle D$. $\overline{BC}$ corresponds to $\overline{EF}$.

$\angle B$ corresponds to $\angle E$; $m\angle B = m\angle E$. $\overline{AC}$ corresponds to $\overline{DF}$.

$\angle C$ corresponds to $\angle F$; $m\angle C = m\angle F$. $\overline{AB}$ corresponds to $\overline{DE}$.

$$\frac{BC}{EF} = \frac{AC}{DF} = \frac{AB}{DE}$$

The corresponding angles are congruent. **The lengths of corresponding sides are proportional.**

We write $\Delta ABC \sim \Delta DEF$. (~ is read "is similar to.")

Note: The notation for similar triangles indicates the respective correspondences of angles and sides by the order in which the vertices of each triangle are identified. For example, we could have written $\Delta BCA \sim \Delta EFD$ or $\Delta CAB \sim \Delta FDE$. Be sure to match up corresponding angles and sides when indicating similar triangles.

Math Tip

In Section 6.1, we learned that if two parallel lines are cut by a transversal, the corresponding angles are congruent. This lesson uses the term *corresponding* in a different way; it means *having a relationship*. Corresponding parts match or go together. Just because the parts correspond does not mean that they are congruent. Recall in our previous lesson, the lines had to be parallel for the corresponding angles to be congruent and in this lesson the triangles must be similar if the corresponding angles are congruent.

1. Which of the following statements is correct in reference to the figure in Example 1?

a. $\Delta BAC \sim \Delta XYA$

b. $\Delta CBA \sim \Delta YXA$

Example 1 Determining Whether Triangles are Similar

Given the two triangles ΔABC and ΔAXY with

$$m\angle ABC = m\angle AXY = 90° \text{ (as shown in the figure)}$$

determine whether or not $\Delta ABC \sim \Delta AXY$.

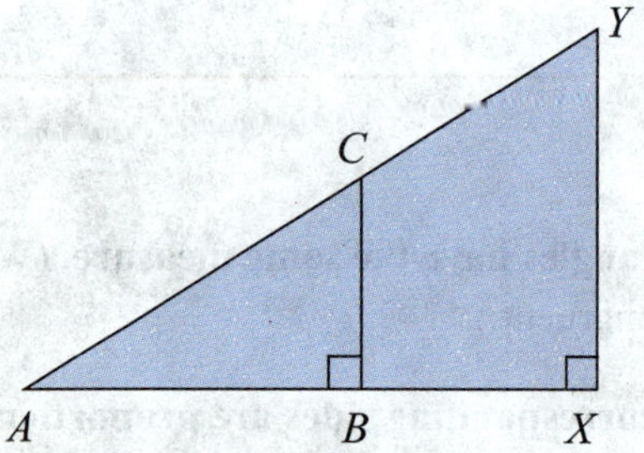

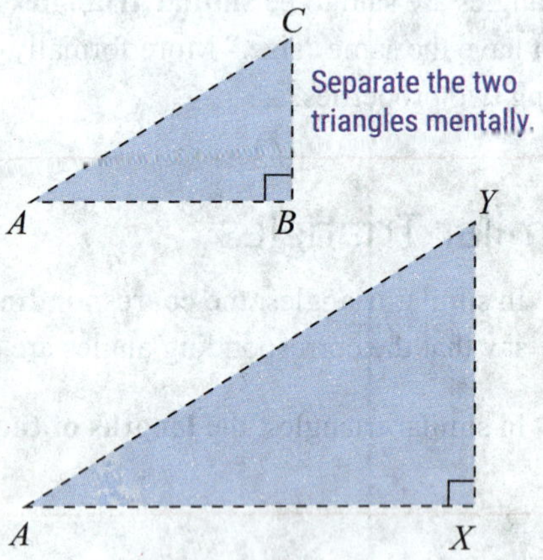

Solution

We can show that the corresponding angles are congruent as follows.

$m\angle CAB = m\angle YAX$	Because they are the same angle
$m\angle CBA = m\angle YXA$	Because both are right angles (90°)
$m\angle BCA = m\angle XYA$	Because the sum of the measures of the angles in each triangle must be 180°

Therefore, the corresponding angles are congruent, and the triangles are similar.

Now work margin exercise 1.

2. Given that $AC = 5$ cm in the figure in Example 2, find CY. (**Hint:** $CY = AY - AC$)

Example 2 Finding Unknown Values Using Similar Triangles

Refer to the figure shown. If $AB = 4$ cm, $BX = 2$ cm, and $BC = 3$ cm, find XY.

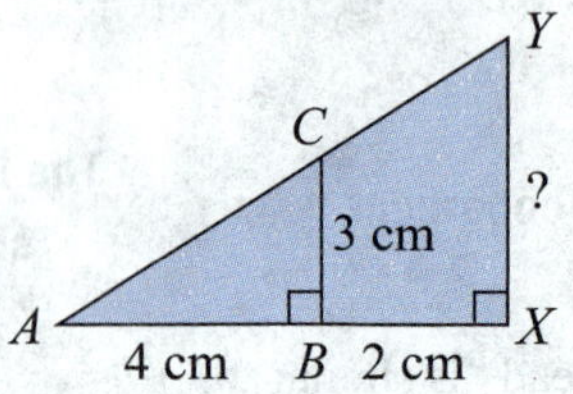

Solution

From Example 1, we know that $\Delta ABC \sim \Delta AXY$; therefore, the lengths of their corresponding sides are proportional. Since $\overline{AB}$ and $\overline{AX}$ are corresponding sides (they are opposite congruent angles) and $\overline{BC}$ and $\overline{XY}$ are corresponding sides (they are opposite congruent angles), the following proportion is true.

$$\frac{AB}{AX}=\frac{BC}{XY}$$

Note that $AX = AB + BX = 4 + 2 = 6$ cm.

Thus,

$$\frac{4\text{ cm}}{6\text{ cm}}=\frac{3\text{ cm}}{XY}$$
$$4\cdot XY = 3\cdot 6$$
$$\frac{\cancel{4}\cdot XY}{\cancel{4}}=\frac{18}{4}$$
$$XY = 4.5\text{ cm.}$$

Now work margin exercise 2.

Example 3 Application: Finding Unknown Values Using Similar Triangles

A very tall, magnificently decorated tree was on display in an outside patio area at the mall during the month before Christmas. Joann knew that the height of a nearby lamppost was 10 feet. At 2 p.m. on a Tuesday afternoon, she measured the shadow of the lamppost to be $3\frac{1}{2}$ feet long and the shadow of the tree to be 21 feet long. With her understanding of similar triangles, Joann was able to calculate the height of the tree. What was the height of the tree?

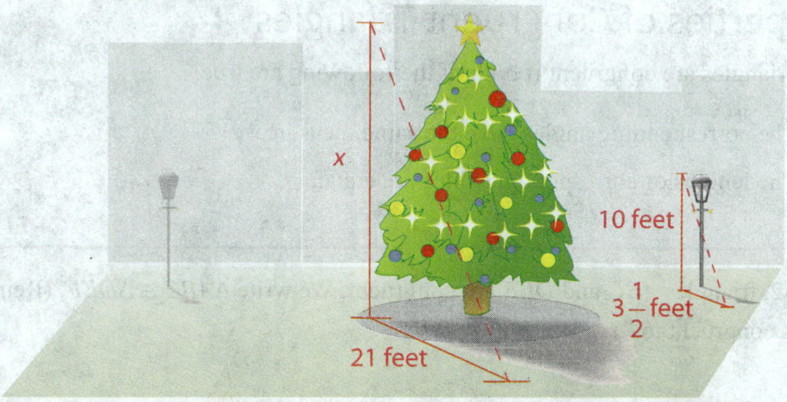

3. To determine the height of an abandoned building, an architect measures the length of the shadow the building casts. At the same time, another architect determines that the length of a shadow cast by a 3-foot stick is 2 feet. If the building casts a shadow that is 18 feet long, what is the height of the building?

Teaching Note

You might comment to your students that light rays travel in parallel lines and this makes the two triangles in the figure similar triangles.

Solution

By letting x represent the height of the tree, Joann set up and solved the following proportion.

$$\frac{x \text{ ft (height of tree)}}{10 \text{ ft (height of lamppost)}} = \frac{21 \text{ ft (length of tree shadow)}}{3\frac{1}{2} \text{ ft (length of post shadow)}}$$

$$\frac{7}{2} \cdot x = 21 \cdot 10$$

$$\frac{\frac{7}{2} \cdot x}{\frac{7}{2}} = \frac{210}{\frac{7}{2}}$$

$$x = \overset{30}{\cancel{210}} \cdot \frac{2}{\cancel{7}}$$

$$x = 60$$

The tree was 60 feet tall.

Now work margin exercise 3.

B Congruent Triangles

If two triangles have the same "shape" and the same "size", they are congruent. Two congruent triangles have the following two properties.

Properties of Congruent Triangles

Two triangles are congruent if both of the following are true.

1. The corresponding angles have the same measure.
2. The lengths of corresponding sides are equal.

PROPERTIES

In Figure 2, triangles ABC and DEF are congruent. We write $\Delta ABC \cong \Delta DEF$. (Remember $\cong$ is read "is congruent to.")

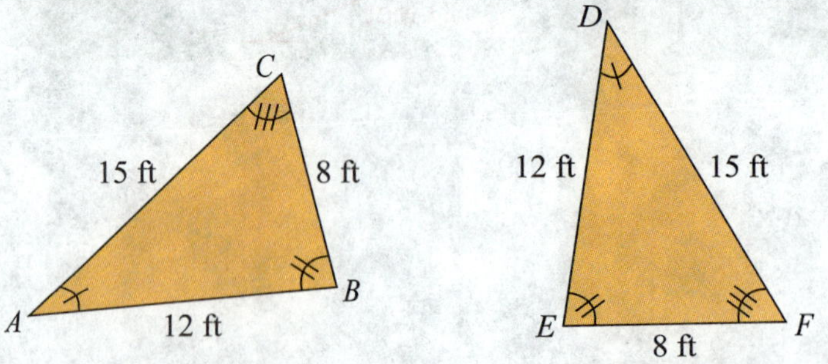

Figure 2

Note that the corresponding angles are congruent. That is, $m\angle A = m\angle D$, $m\angle B = m\angle E$, and $m\angle C = m\angle F$. And, corresponding sides have the same lengths.

The following three properties can be used to determine whether two triangles are congruent.

Determining Congruent Triangles

1. Side-Side-Side (SSS)

If two triangles are such that the lengths of the three sides of one triangle are equal to the lengths of corresponding sides of the other triangle, then the two triangles are congruent.

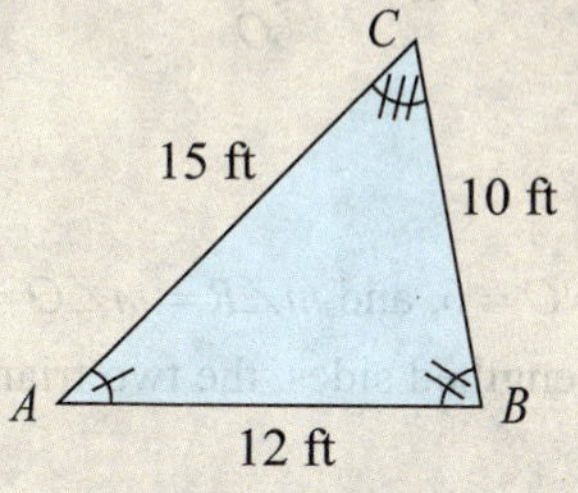

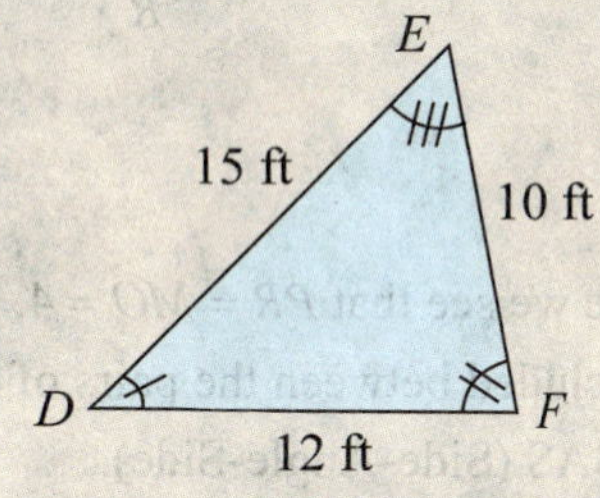

Triangle *ABC* is congruent to triangle *DFE* by SSS.

2. Side-Angle-Side (SAS)

If two triangles are such that the lengths of two sides of one triangle equal the lengths of corresponding sides of the other triangle and the angles included between the sides are congruent, then the two triangles are congruent.

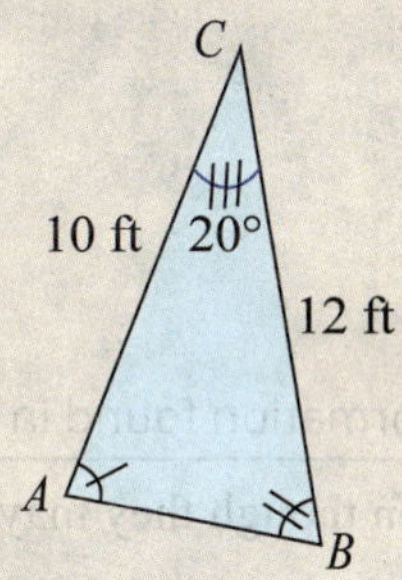

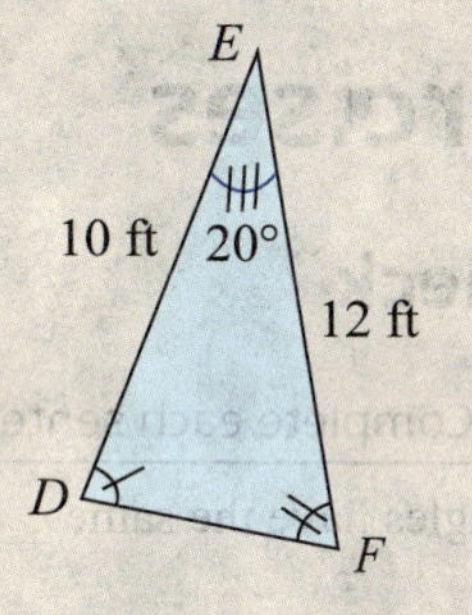

Triangle *ABC* is congruent to triangle *DFE* by SAS.

3. Angle-Side-Angle (ASA)

If two triangles are such that two angles in one triangle are congruent to two angles in the other triangle and the lengths of the sides included between the angles are equal, then the two triangles are congruent.

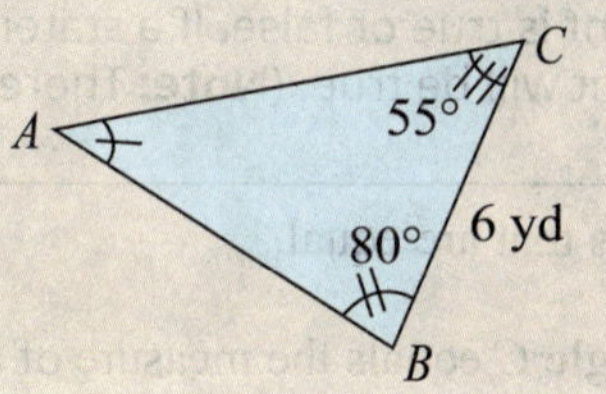

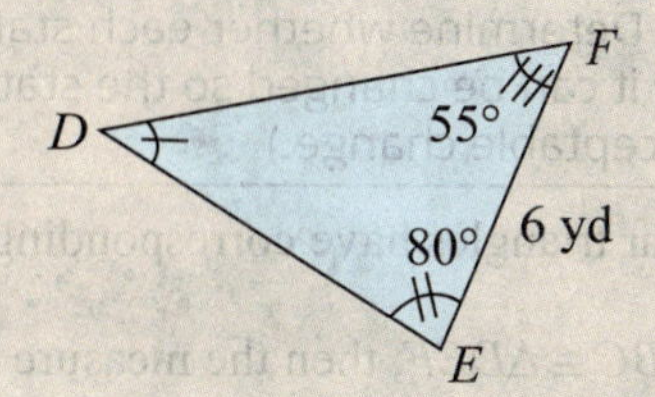

Triangle *ABC* is congruent to triangle *DEF* by ASA.

DEFINITION

4. Determine whether triangles *JKL* and *MNO* are congruent. If they are congruent, state the property that confirms that they are congruent.

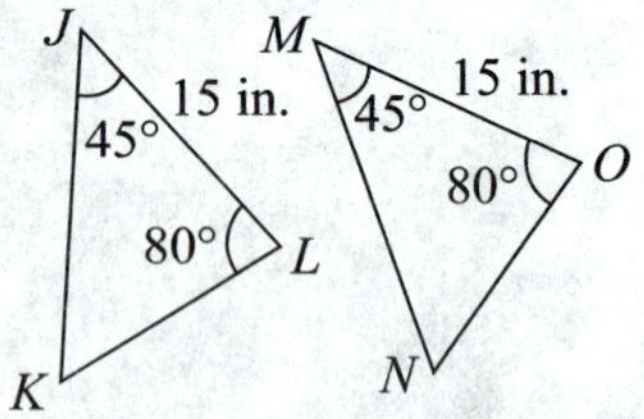

Example 4 Determining Whether Triangles are Congruent

Determine whether triangles *PQR* and *MNO* are congruent.

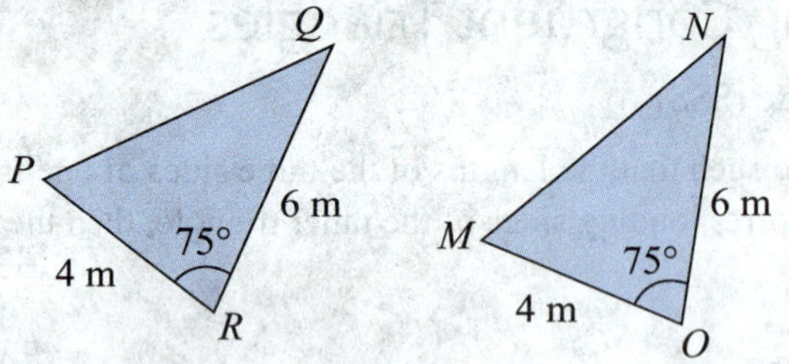

Solution

From the figure we see that $PR = MO = 4$, $QR = NO = 6$, and $m\angle R = m\angle O = 75$. As $\angle R$ and $\angle O$ are included between the pairs of equal-lengthed sides, the two triangles are congruent by SAS (Side-Angle-Side).

Now work margin exercise 4.

Margin Exercise Answers

1. b. **2.** 2.5 cm **3.** 27 feet **4.** Yes, by ASA

6.6 Exercises

Concept Check

Fill-in-the-Blank. Complete each sentence using information found in this section.

1. Similar triangles have the same __________ even though they may not have the same size.
2. In similar triangles, corresponding angles are ____________.
3. In similar triangles, the lengths of corresponding sides are ____________.
4. In congruent triangles, the lengths of corresponding sides are ____________.

True/False. Determine whether each statement is true or false. If a statement is false, explain how it can be changed so the statement will be true. (**Note:** There may be more than one acceptable change.)

5. Similar triangles have corresponding sides that are equal.
6. If $\Delta ABC \cong \Delta DEF$, then the measure of angle C equals the measure of angle D.
7. If $\Delta ABC \sim \Delta DEF$, then $AC = DF$.
8. Congruent triangles have corresponding angles that are equal.

6.6 Exercises

Concept Check

1. shape
2. congruent
3. proportional
4. equal
5. False; Similar triangles have corresponding sides that are proportional.
6. False; If $\Delta ABC \cong \Delta DEF$, then the measure of angle C equals the measure of angle F.
7. False; If $\Delta ABC \cong \Delta DE$ then $AC = DF$.
8. True

Practice

Determine whether each pair of triangles is similar. If the pair of triangles is similar, explain why and indicate the similarity by using the ~ symbol. See Example 1.

1.

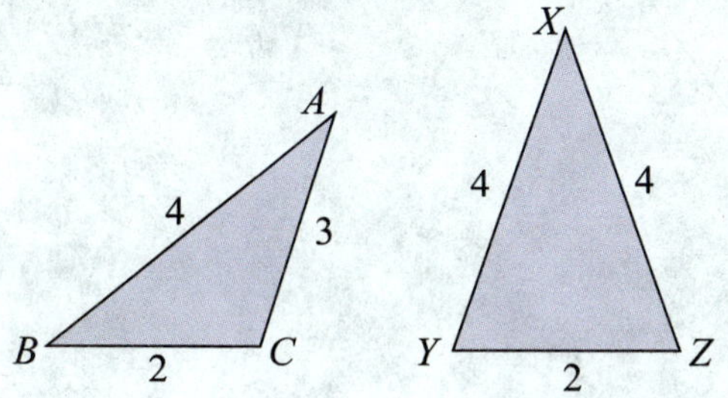

2.

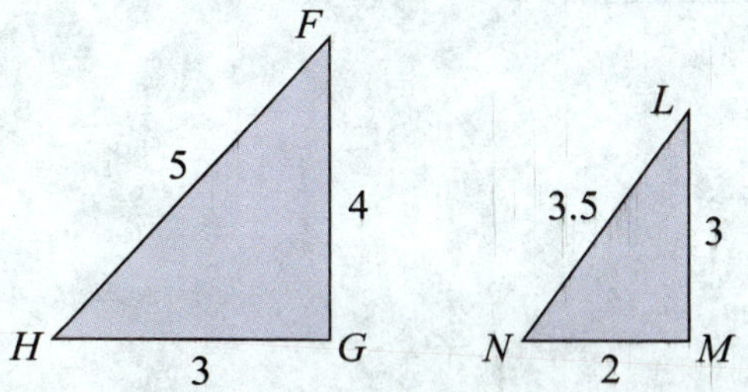

3.

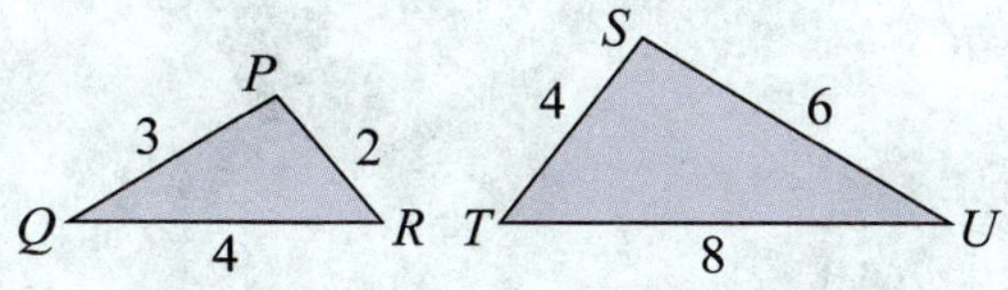

4.

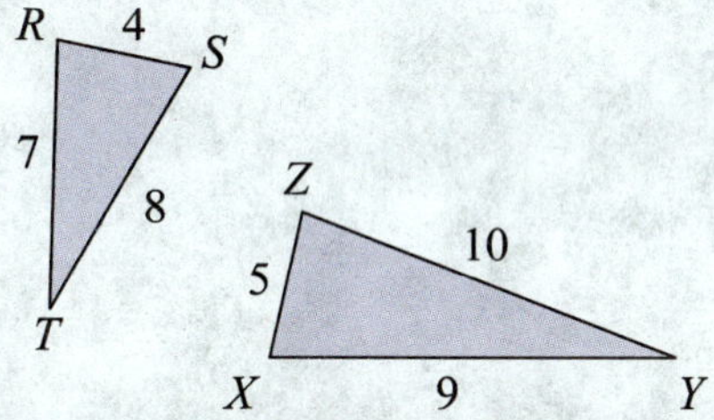

5.

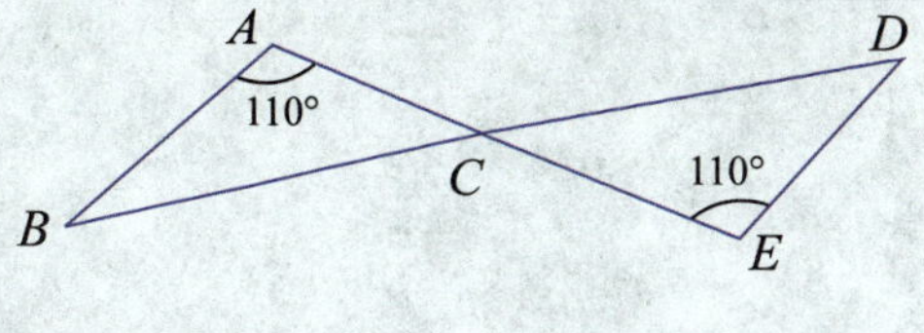

6.

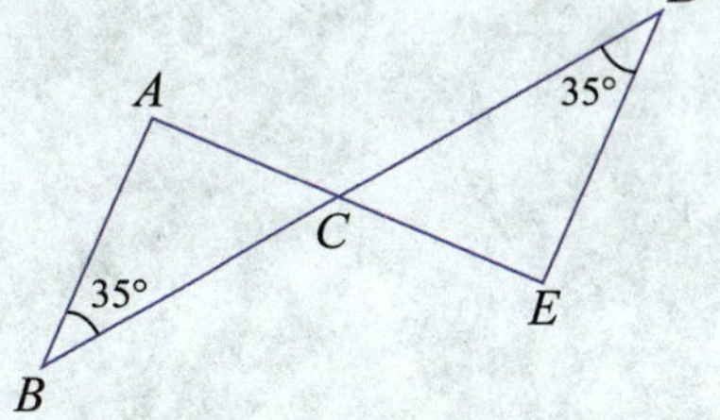

Practice

1. The triangles are not similar. The corresponding sides are not proportional.
2. The triangles are not similar. The corresponding sides are not proportional.
3. $\Delta PQR \sim \Delta SUT$. All pairs of corresponding sides are proportional to the ratio 1:2.
4. The triangles are not similar. The corresponding sides are not proportional.
5. $\Delta ABC \sim \Delta EDC$. The corresponding angles have the same measure.
6. $\Delta ABC \sim \Delta EDC$. The corresponding angles have the same measure.

7. $x = 50°; y = 70°$
8. $x = 55°; y = 100°$
9. $x = 50°; y = 50°$
10. $x = 10°; y = 50°$
11. $x = 20°; y = 100°$
12. $x = 35°; y = 90°$
13. $x = 4.8; y = 7.2$

Find the values for *x* and *y*. See Example 2.

7. $\Delta ABC \sim \Delta XYZ$

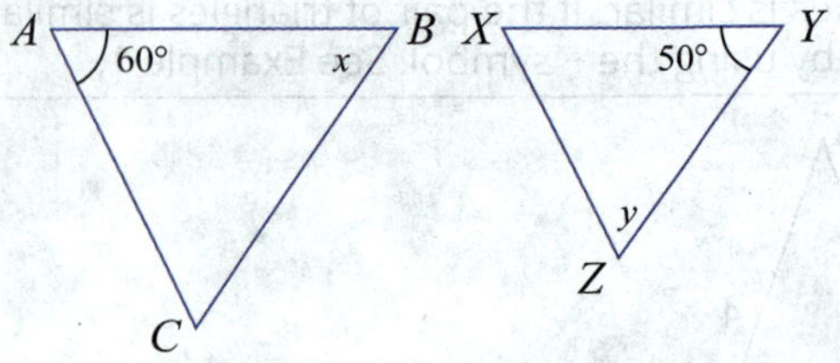

8. $\Delta DEF \sim \Delta TRS$

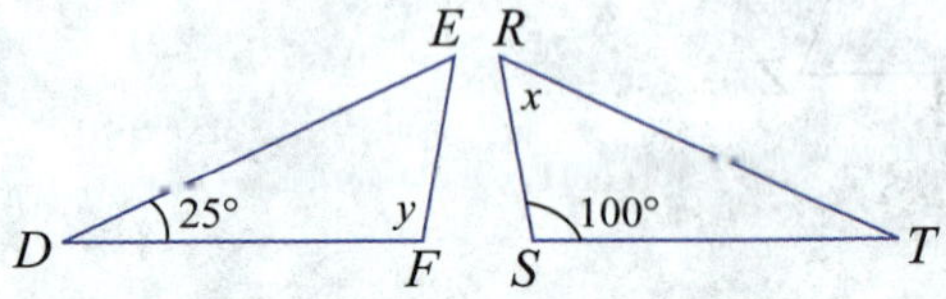

9. $\Delta ABC \sim \Delta ADE$

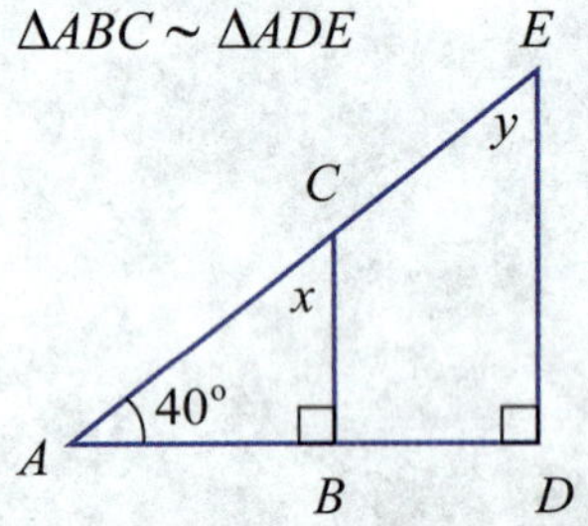

10. $\Delta WVS \sim \Delta UTS$

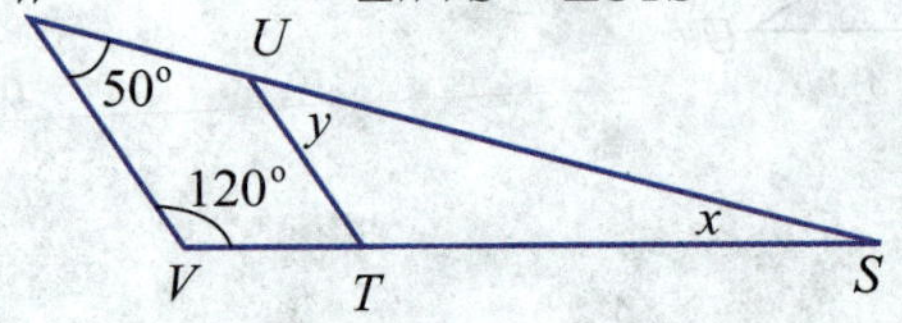

11. $\Delta ABE \sim \Delta CDE$

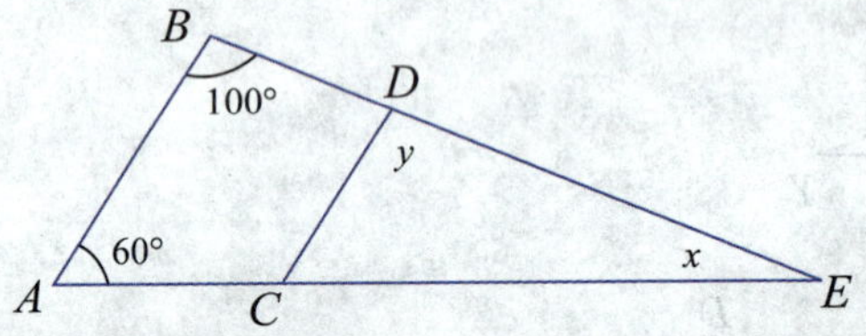

12. $\Delta MPR \sim \Delta NPQ$

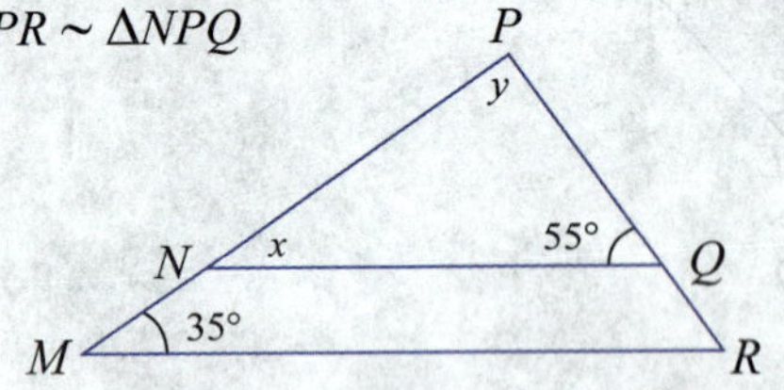

13. $\Delta RST \sim \Delta UVW$

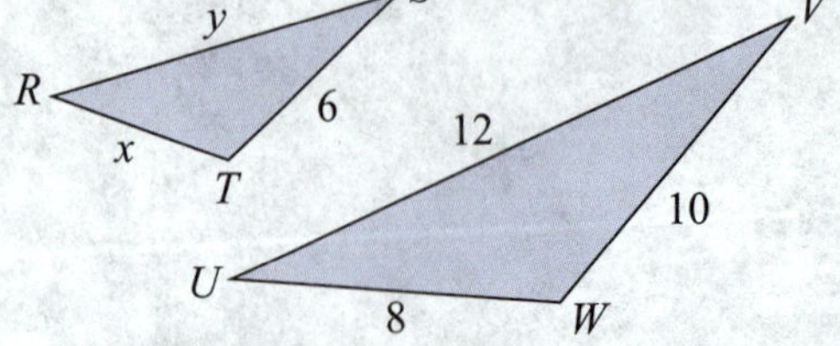

14. $\Delta LMN \sim \Delta OPQ$

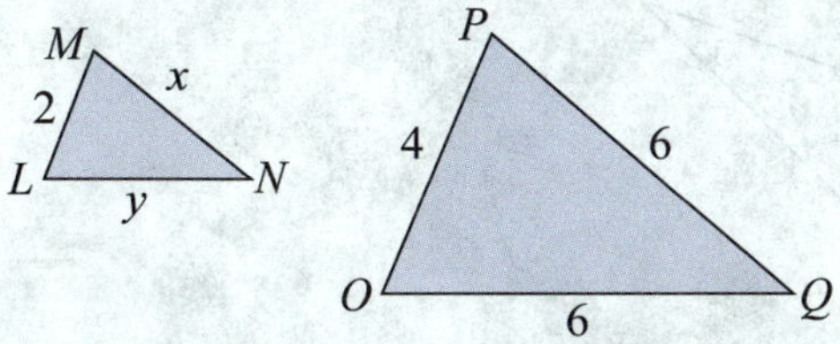

15. $\Delta PQR \sim \Delta STV$

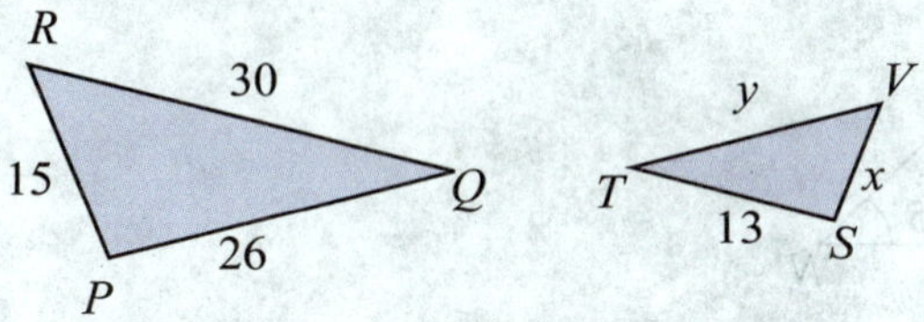

16. $\Delta ABC \sim \Delta EDC$

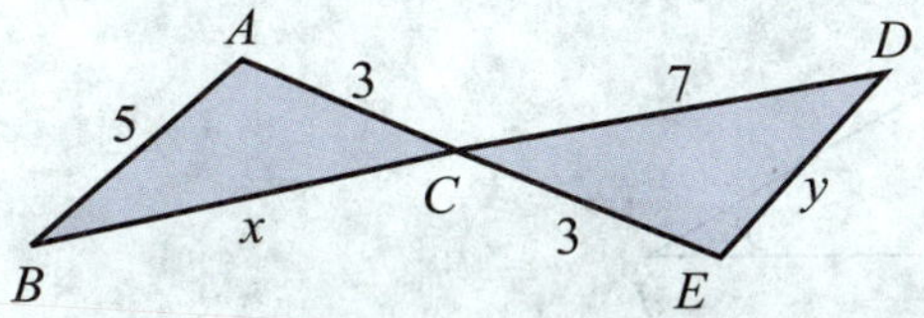

17. $\Delta PQR \sim \Delta SUT$

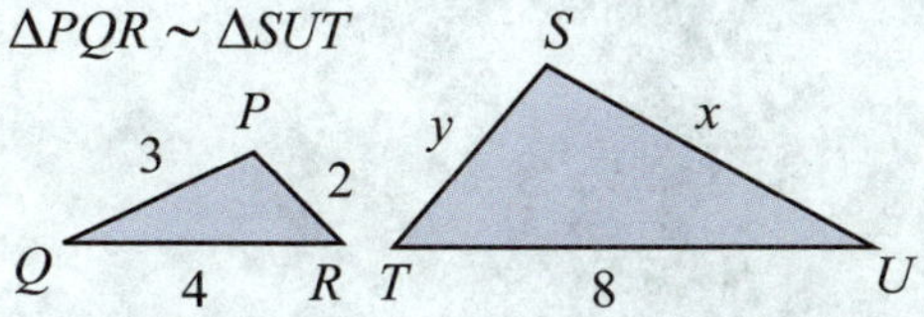

18. $\Delta FGH \sim \Delta LMN$

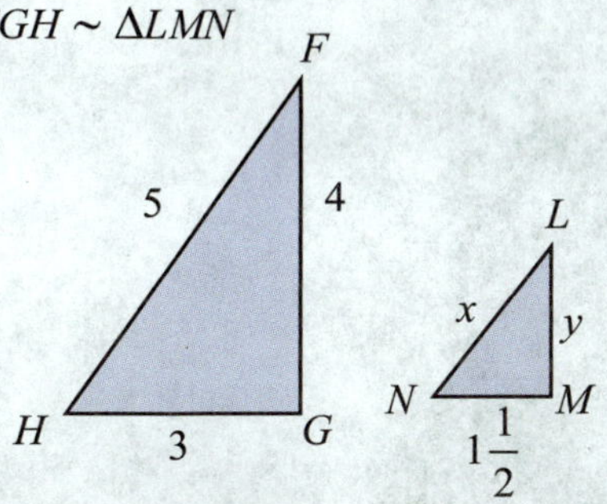

19. $\Delta ABC \sim \Delta EDC$

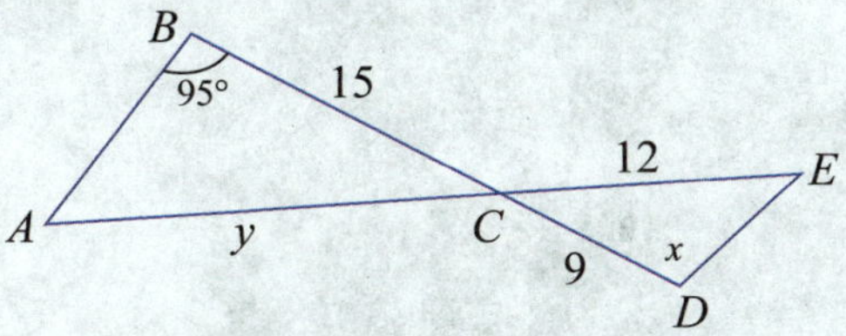

20. $\Delta ABC \sim \Delta EDC$

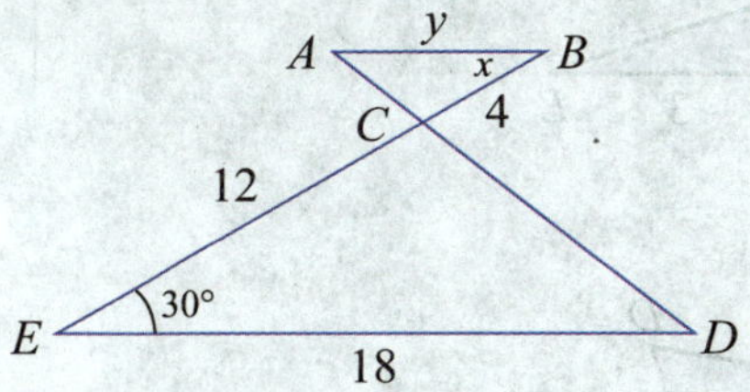

14. $x = 3; y = 3$

15. $x = 7.5; y = 15$

16. $x = 7; y = 5$

17. $x = 6; y = 4$

18. $x = 2.5; y = 2$

19. $x = 95°; y = 20$

20. $x = 30°; y = 6$

21. $x = 25°; y = 20°$
22. $x = 45°; y = 100°$
23. $x = 25°; y = 40°$
24. $x = 95°; y = 43°$
25. $x = 12; y = 10$
26. $x = 8; y = 2$
27. $x = 4.8; y = 2.5$

21. $\Delta QRS \sim \Delta QTU$

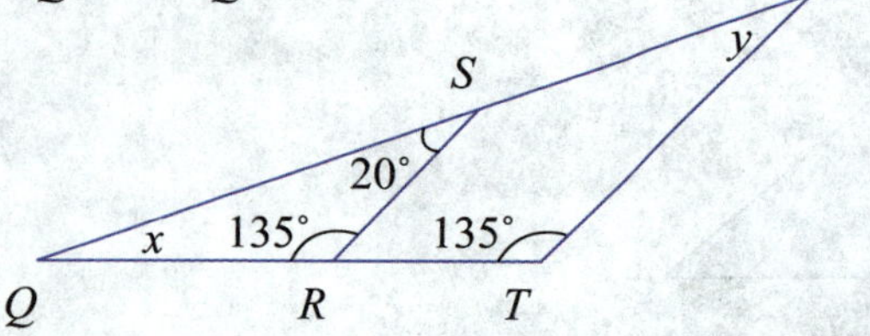

22. $\Delta KLM \sim \Delta KNO$

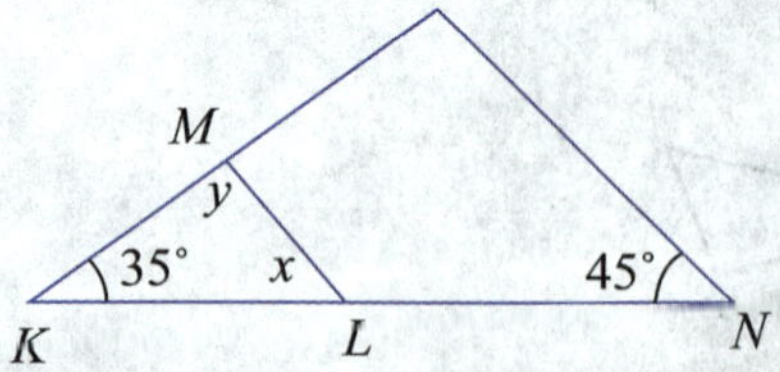

23. $\Delta GHI \sim \Delta GJK$

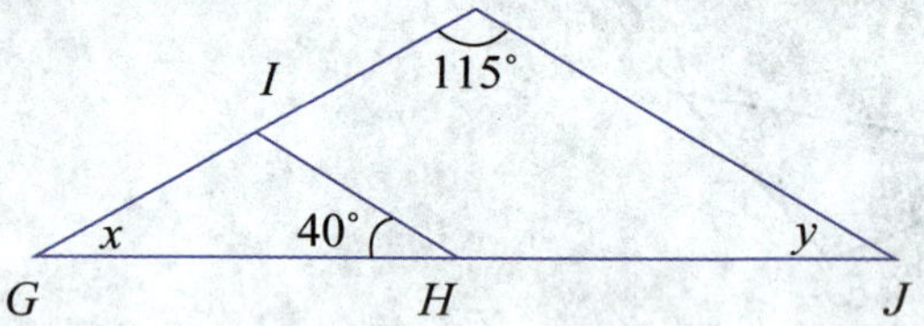

24. $\Delta ABC \sim \Delta ADE$

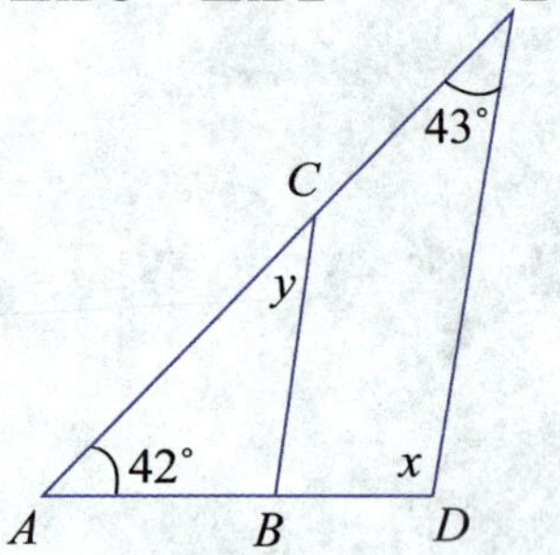

25. $\Delta XYZ \sim \Delta LJK$

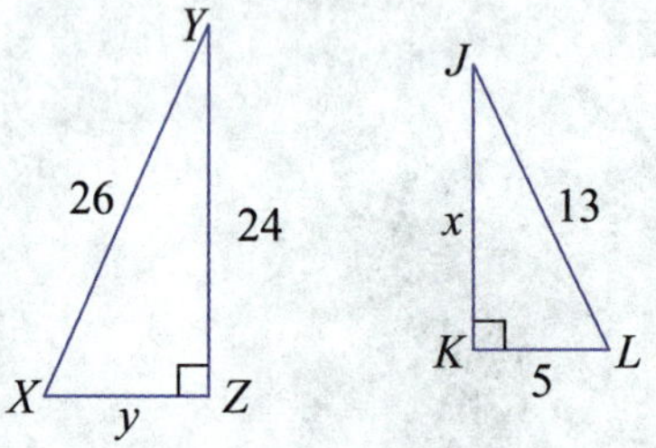

26. $\Delta ABC \sim \Delta DEF$

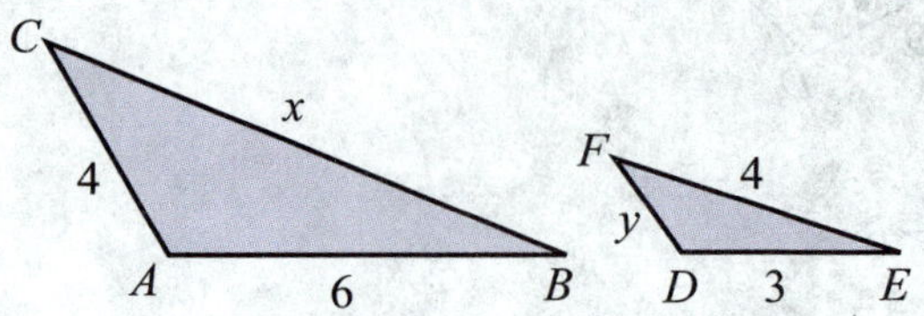

27. $\Delta ABC \sim \Delta EDC$

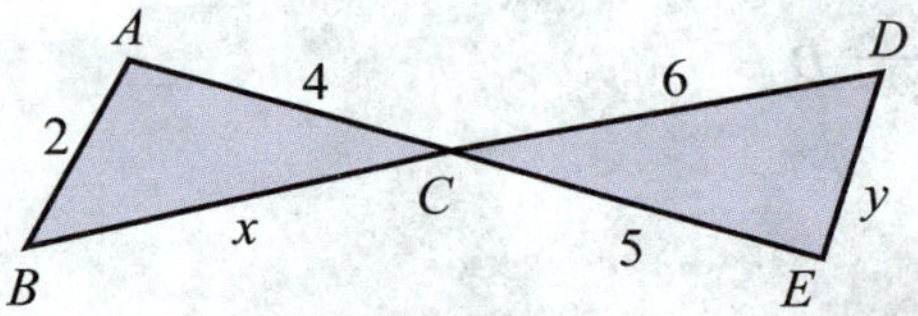

28. $\Delta ABC \sim \Delta PQR$

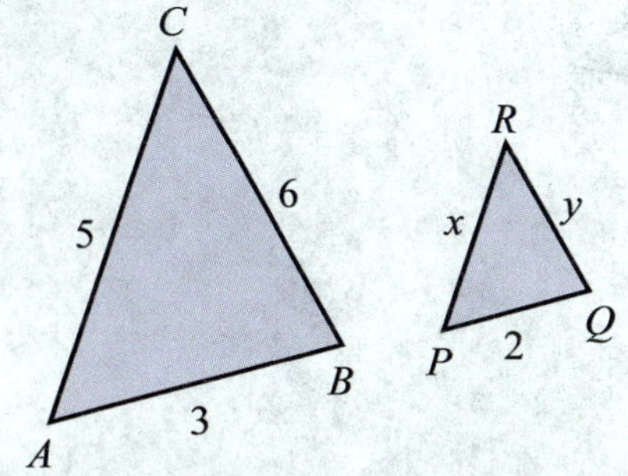

Determine whether each pair of triangles is congruent. If the pair of triangles is congruent, state the property that confirms that they are congruent. See Example 4.

29.

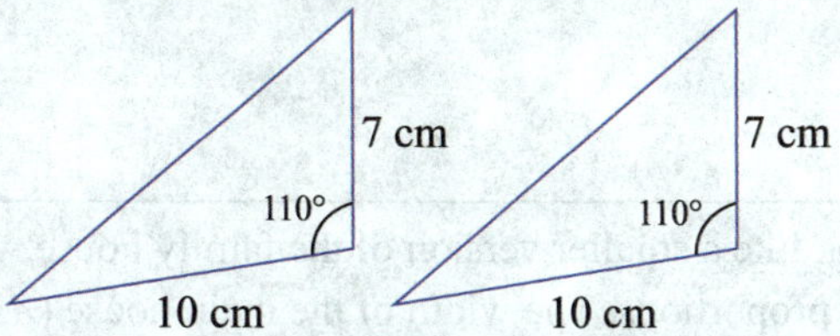

30.

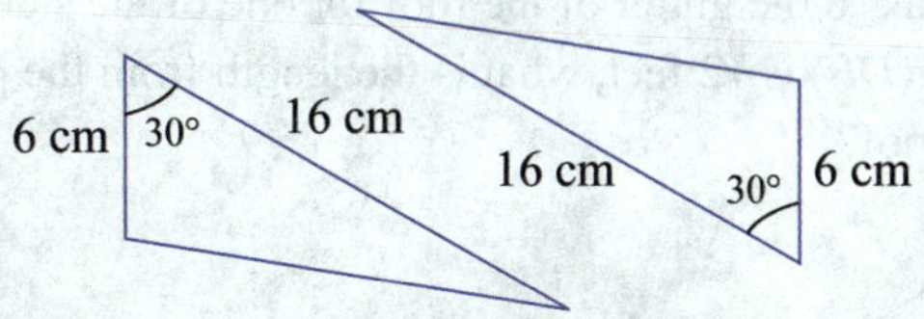

31.

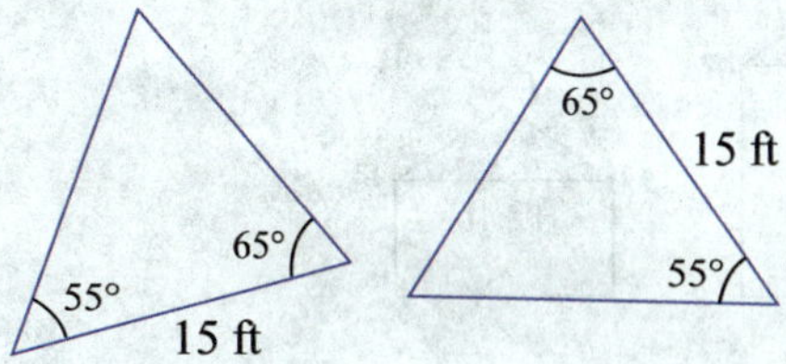

32.

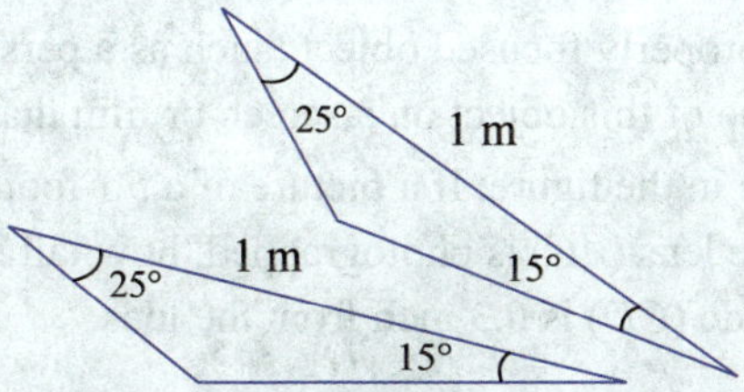

33.

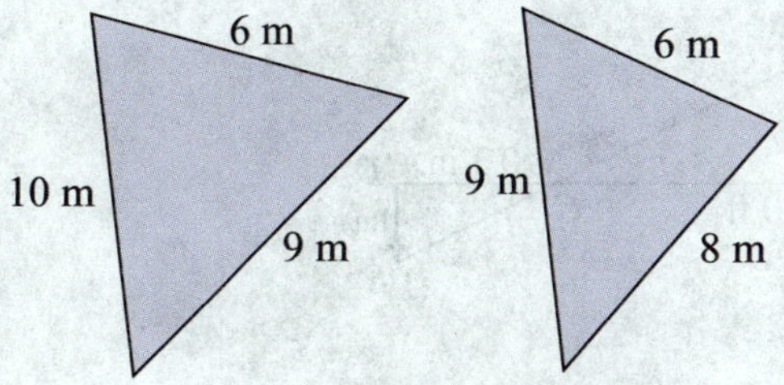

34.

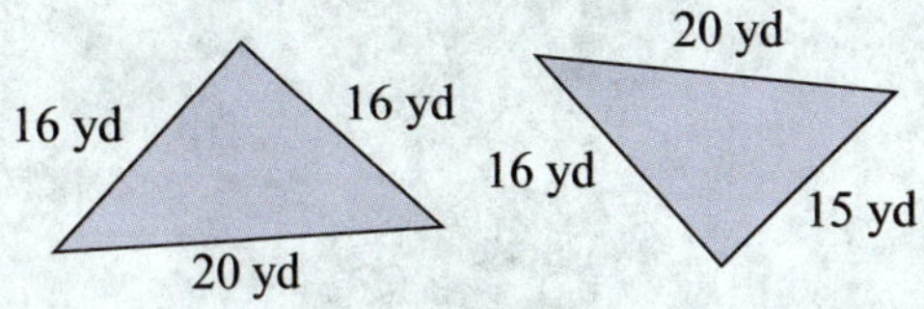

28. $x = \frac{10}{3}$ or $3\frac{1}{3}$; $y = 4$

29. Congruent by SAS

30. Congruent by SAS

31. Congruent by ASA

32. Congruent by ASA

33. Not congruent

34. Not congruent

35. Congruent by SSS

36. Congruent by SSS

Applications

37. 7.5 feet

38. 0.1 in.

35.

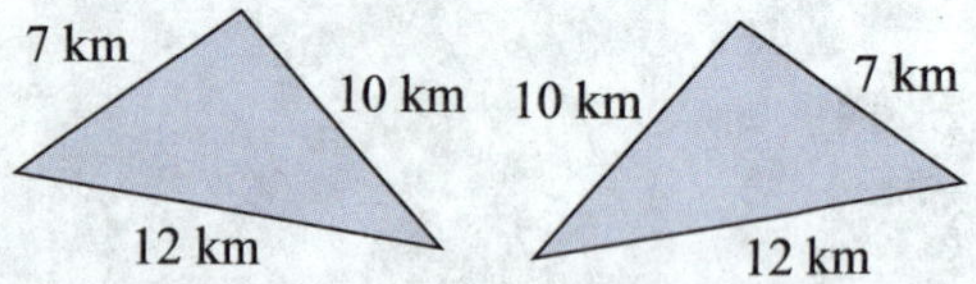

36.

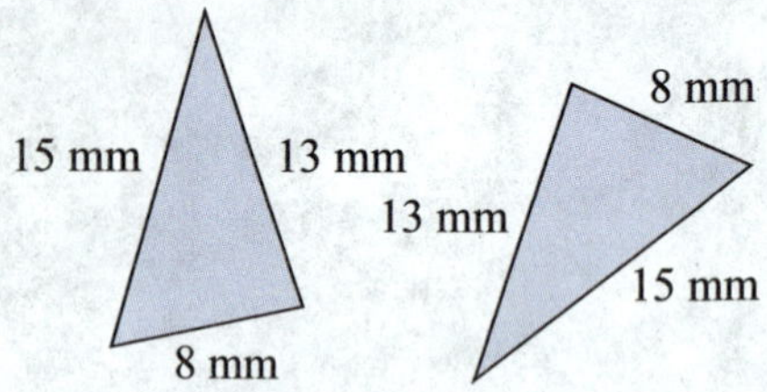

Applications

Solve.

37. A child's playhouse is built to look like a smaller version of the family house, where the ends of the roofs have similar proportions. The width of the main house (*AB*) is 32 feet and the length from the peak to the gutter of the roof for one of the sides is 20 feet. If the width of the playhouse (*DF*) is 12 feet, what is the length from the peak to the gutter (*DE*) of the playhouse roof?

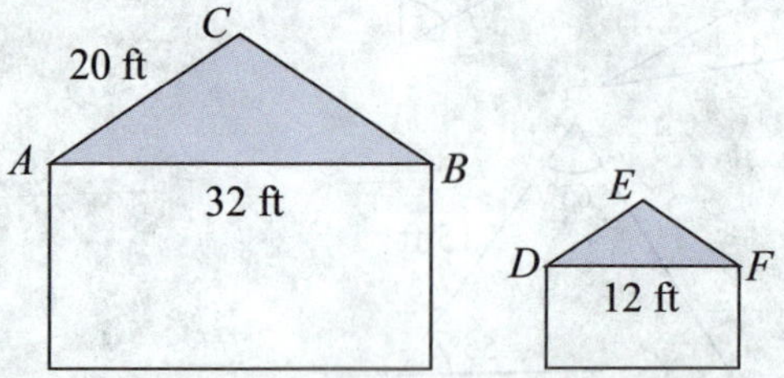

38. A camera uses a lens that will look at a properly focused object (such as a person or a tree) and then display an inverted image of this object on a screen or film that is on the opposite side of the lens as shown in the figure. If a picture of a 50-foot tall building (*AB*), which is 150 feet from the lens (*AC*) is photographed, how tall is the image (*DE*) if the film on the opposite side (*CD*) is 0.3 inch from the lens.

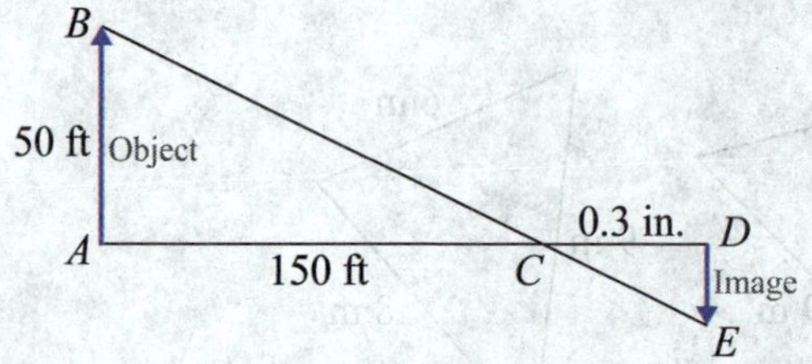

39. A surveyor is trying to figure out the length of a large quarry. The surveyor sees two trees at the edge of the quarry that are close together, and notices that the line between the two trees is parallel to the line that connects the edges of the quarry at its widest. We will call the two trees points *R* and *S* and the points *X* and *Y* will be the endpoints of the line connecting the two edges of the quarry. Drawing an imaginary line from *R* to *Y* and from *S* to *X* creates two similar triangles, ΔRST and ΔYXT, with a common point *T*. If the length of *RS* is 25 yd, the length of *ST* is 16 yd, and the length of *XT* is 80 yd, what is the length of the quarry (the distance between *X* and *Y*)? (See the figure.)

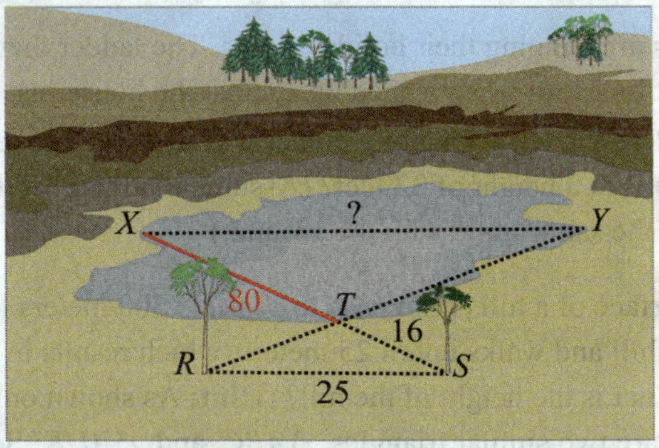

40. You and a friend are walking to class and want to figure out the height of the tree next to your building. Your friend is exactly 6 ft tall and casts 4 ft shadow. The tree casts a 12 ft shadow. How tall is the tree?

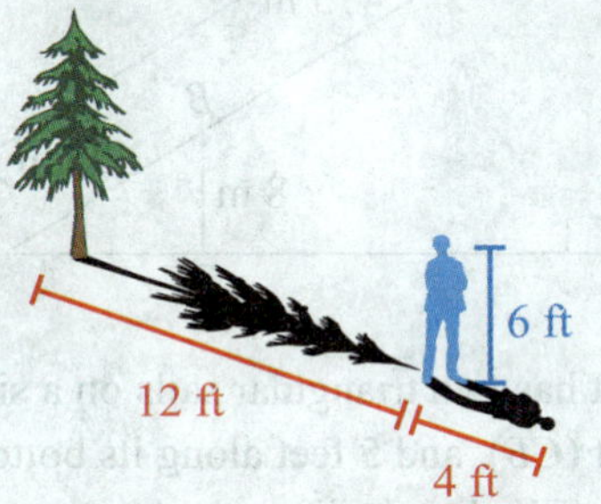

41. Thales was a mathematician circa 500 B.C. who wanted to know the height of the Great Pyramid. He discovered he could calculate the height of the pyramid using similar triangles. He stuck a rod in the ground that rose 5 ft into the air and cast an 8 ft shadow. If, at the same time, the length of the shadow from the Great Pyramid was 768 ft, determine the height of the Great Pyramid.

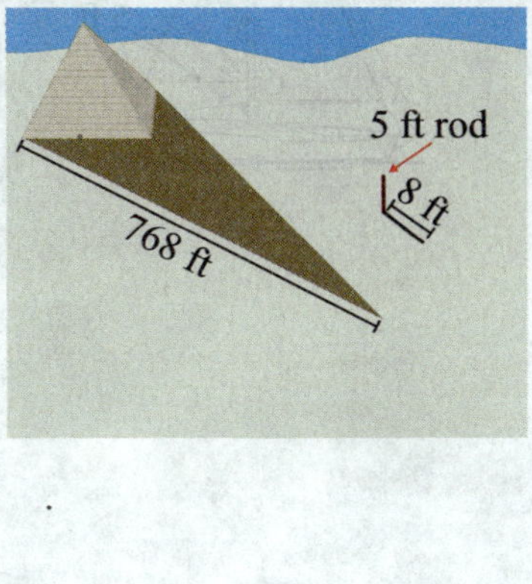

39. 125 yd
40. 18 ft
41. 480 ft

History

Thales was a Greek mathematician who lived approximately 624 BC to 540 BC. While he was an engineer by trade, he was considered by Aristotle to be the first philosopher in the Greek tradition. Thales has also been referred to as the Father of Science. While it is difficult to give Thales credit for measuring the Great Pyramid as described in Exercise 41 (due to a lack of written records), there are stories of him doing this as well as using this method to measure the distances of ships at sea.

42. 1 ft
43. 48 ft
44. 160 ft
45. 6.9 ft

42. At what height off the table would you need to hit a ping pong ball for it to skim the net (0.5 feet tall) and hit on the edge of the opposite side of the table 9 feet away from you? Assume that the net is in the exact center of the table (with 4.5 feet on either side) and that your paddle is directly above the edge of the table.

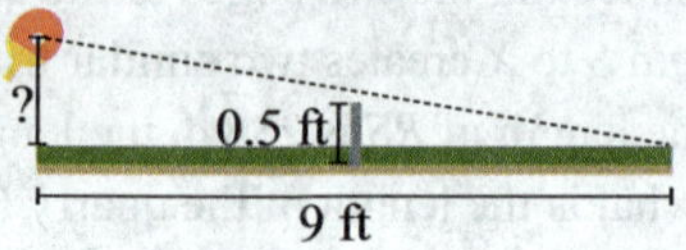

43. Your neighbors are hanging their holiday lights. The ladder they are currently using is 12 feet long and when leaned up against the house just reaches the top of their 8-foot tall porch. How long of a ladder will they need to reach the top of their chimney which is at a height of 32 feet? (Assume that both ladders are placed such that they make the same angle with the ground.)

44. The sloping surface of a hill (from base to peak) is 500 meters long. Paul starts at the base of the hill and walks uphill 25 meters, which results in a gain of 8 meters in elevation. What is the height of the hill? (**Hint:** As shown on the accompanying drawing, there are two similar triangles ΔABC and ΔADE. Solve for DE.)

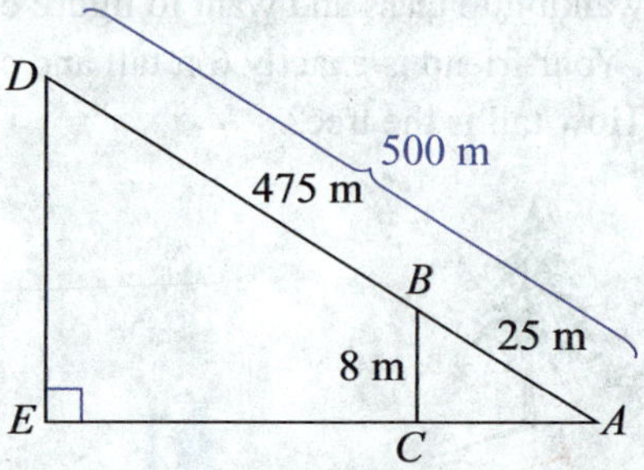

45. A sloop is a sailboat that has two triangular sails on a single mast. If the smaller sail is 12 feet along the mast (CB), and 5 feet along its bottom (AC), and the larger sail is 16.5 feet along the mast (ZY), how wide is the larger sail at the bottom (XZ) if ΔABC and ΔXYZ are similar triangles? Round your answer to the nearest tenth.

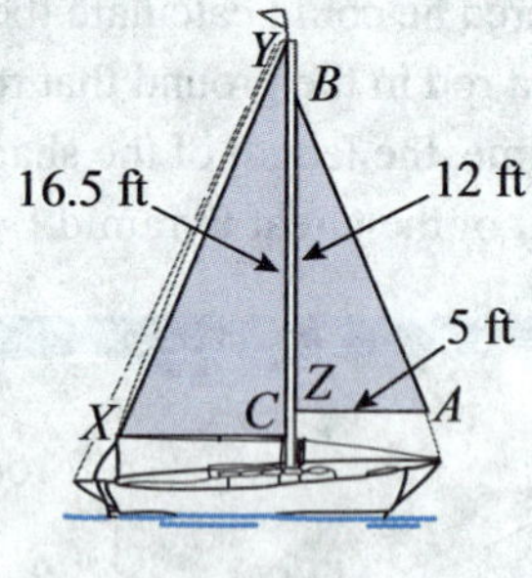

Writing & Thinking

46. In ΔXYZ and ΔUVW, $m\angle Z = 30°$, and $m\angle W = 30°$.

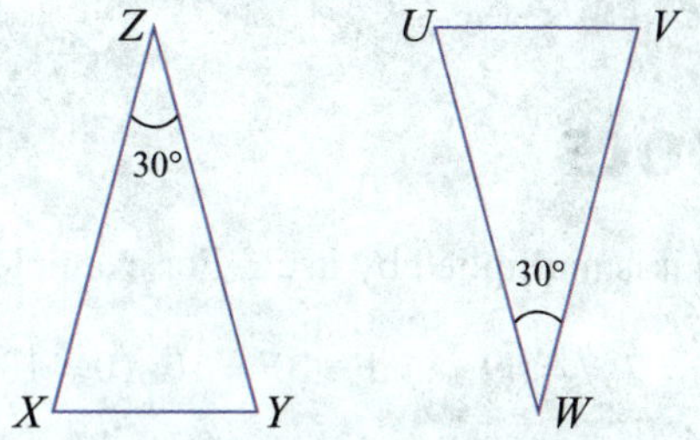

a. If both triangles are isosceles, what are the measures of the other four angles? (In an isosceles triangle, the angles opposite the equal side must be congruent.)

b. Are the triangles similar? Explain why.

47. Consider ΔABC and ΔDEF.

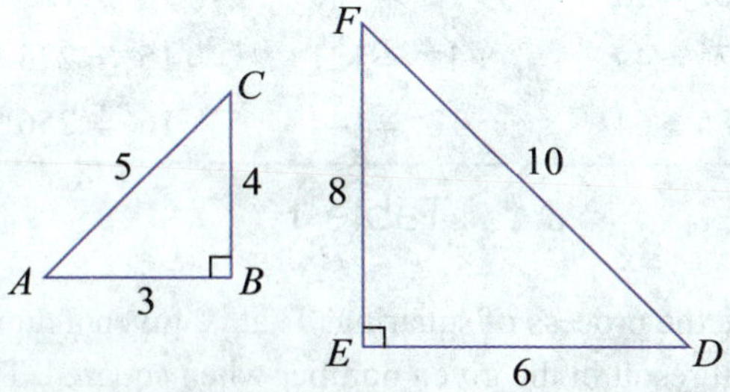

a. Are the triangles similar? Explain why.

b. If so, which angles are congruent?

48. Determine the errors in the following statement. Assume $\Delta ABC \sim \Delta DEF$.

a. Corresponding angles are congruent. This means $m\angle A = m\angle D$, $m\angle B = m\angle F$, and $m\angle C = m\angle E$.

b. Corresponding sides are the same length.

49. Kelly needs to determine whether two triangles are similar. She was given the following information.

For ΔABC and ΔDEF, $AB = 3.6$, $AC = 2.4$, $BC = 2$ and $DE = 9$, $DF = 6$, $EF = 5$.

What should be her first step?

50. Given the following information, $AB = 5.5$, $AC = 36$, $BC = 14$ and $DE = 1.1$, $DF = 7.2$, $EF = 4.2$. Jordan has begun calculating whether or not ΔABC and ΔDEF are similar. Find the error in his calculations.

$$\frac{5.5}{36} \stackrel{?}{=} \frac{7.2}{1.1}$$
$$0.15 \neq 0.65$$

Writing & Thinking

46. a. Each of the other four angles measure 75°.

b. The triangles are similar since the three pairs of corresponding angles are congruent.

47. a. The triangles are similar since the three pairs of corresponding sides are proportional.

b. $m\angle A = m\angle D$, $m\angle B = m\angle E$, $m\angle C = m\angle F$

48. a. $\angle B$ should be equal to $m\angle E$ and $m\angle C$ should be equal to $m\angle F$.

b. Corresponding sides are proportional, not that they have the same length.

49. She needs to calculate the ratios of each pair of corresponding sides. Answers will vary.

50. The proportion has been set up incorrectly. He has set $\frac{AB}{AC} = \frac{DF}{DE}$ when he should have set $\frac{AB}{AC} = \frac{DE}{DF}$.

Objectives

A. Find the square root of a perfect square.

B. Approximate square roots using a calculator.

C. Use the Pythagorean Theorem to solve application problems.

6.7 Square Roots and the Pythagorean Theorem

A Square Roots

A number is **squared** when it is multiplied by itself. For example,

$$7^2 = 7 \cdot 7 = 49 \quad \text{and} \quad 10^2 = 10 \cdot 10 = 100.$$

The result is called a **perfect square**. Thus, 49 and 100 are perfect squares. Table 1 shows the perfect squares found by squaring the whole numbers from 1 to 20.

Squares of Whole Numbers from 1 to 20

$1^2 = 1$	$5^2 = 25$	$9^2 = 81$	$13^2 = 169$	$17^2 = 289$
$2^2 = 4$	$6^2 = 36$	$10^2 = 100$	$14^2 = 196$	$18^2 = 324$
$3^2 = 9$	$7^2 = 49$	$11^2 = 121$	$15^2 = 225$	$19^2 = 361$
$4^2 = 16$	$8^2 = 64$	$12^2 = 144$	$16^2 = 256$	$20^2 = 400$

Table 1

Now we want to reverse the process of squaring. That is, given a number, we want to find another number that will result in the given number when squared. This is called **finding the square root** of the given number. For example, find the **square root** of 49. What number squared will result in 49?

Since $7^2 = 49$, the number 7 is called the **square root** of 49,

and since $10^2 = 100$, the number 10 is called the **square root** of 100.

We write

$$\sqrt{49} = 7 \quad \text{and} \quad \sqrt{100} = 10.$$

Similarly,

$$\sqrt{25} = 5 \quad \text{since} \quad 5^2 = 25,$$
$$\sqrt{121} = 11 \quad \text{since} \quad 11^2 = 121, \text{ and}$$
$$\sqrt{1} = 1 \quad \text{since} \quad 1^2 = 1.$$

Terminology of Radicals

The symbol $\sqrt{}$ is called a **radical sign**.

The number under the radical sign is called the **radicand**.

The complete expression, such as $\sqrt{49}$, is called a **radical** or **radical expression**.

DEFINITION

Table 2 contains the square roots of the perfect square numbers from 1 to 400 (the squares of the numbers from 1 to 20). Note that this table is just another way of looking at Table 1.

Square Roots of Perfect Squares from 1 to 400

$\sqrt{1}=1$	$\sqrt{25}=5$	$\sqrt{81}=9$	$\sqrt{169}=13$	$\sqrt{289}=17$
$\sqrt{4}=2$	$\sqrt{36}=6$	$\sqrt{100}=10$	$\sqrt{196}=14$	$\sqrt{324}=18$
$\sqrt{9}=3$	$\sqrt{49}=7$	$\sqrt{121}=11$	$\sqrt{225}=15$	$\sqrt{361}=19$
$\sqrt{16}=4$	$\sqrt{64}=8$	$\sqrt{144}=12$	$\sqrt{256}=16$	$\sqrt{400}=20$

Table 2

Example 1 Evaluating Perfect Squares

Use your memory of the values in Table 1 to evaluate each expression.

a. 15^2 **b.** 11^2

Solution

a. $15^2 = 225$ **b.** $11^2 = 121$

Now work margin exercise 1.

1. Evaluate each expression.

a. 18^2

b. 9^2

Example 2 Evaluating Square Roots

Use your memory of the values in Table 2 to evaluate each expression.

a. $\sqrt{256}$ **b.** $\sqrt{81}$

Solution

a. $\sqrt{256} = 16$ **b.** $\sqrt{81} = 9$

Now work margin exercise 2.

2. Evaluate each expression.

a. $\sqrt{36}$

b. $\sqrt{169}$

Most numbers are not perfect squares and the square roots of these numbers are not found as easily as those in Example 2 and in Table 2. In fact, most square roots are **irrational numbers (infinite nonrepeating decimals)**; that is, most square roots can be only approximated with decimals.

B Approximating Square Roots

Consider the fact that 5 is between 4 and 9. Using inequality signs we can write the following.

$4 < 5 < 9$

So, it seems reasonable (and it is true) that

$$\sqrt{4} < \sqrt{5} < \sqrt{9},$$

and we have

$$2 < \sqrt{5} < 3.$$

Therefore, even though we may not know the exact value of $\sqrt{5}$, we do know that it is between 2 and 3.

Decimal number approximations to $\sqrt{5}$ are shown here to illustrate more accurate estimates.

$$(2.2)^2 = 4.84 < 5$$
$$(2.23)^2 = 4.9729 < 5$$
$$(2.236)^2 = 4.999696 < 5$$
$$(2.237)^2 = 5.004169 > 5$$

Thus, we can now say that $\sqrt{5}$ is between 2.236 and 2.237.

3. Use a calculator to approximate each square root to the nearest ten-thousandth.

a. $\sqrt{5}$

b. $\sqrt{15}$

Example 3 Calculating Square Roots Using a Calculator

Use a calculator to approximate each square root to the nearest ten-thousandth.

a. $\sqrt{2}$ **b.** $\sqrt{18}$

Solution

a. $\sqrt{2} \approx 1.4142$ **b.** $\sqrt{18} \approx 4.2426$

Now work margin exercise 3.

CALCULATOR

Using a Calculator to Find Square Roots

To find the square root of a number, you will need the square root key, [√x]. If you do not see this key on your calculator, check the function above the [x²] key. The square root function is often accessed by pressing [2nd] [x²]. When entering a square root, calculators vary as to whether to press the [√x] key before or after entering the radicand. You may need to consult your calculator's manual or your instructor to find the correct sequence of keys.

Thus, to approximate the values asked for in Example 3, press the following keys.

a. [√x] [2] [=] OR [2] [√x]

The display will read 1.414213562. Rounding to the nearest ten-thousandth, we have 1.4142.

b. [√x] [1] [8] [=] OR [1] [8] [√x]

The display will read 4.242640687. Rounding to the nearest ten-thousandth, we have 4.2426.

C The Pythagorean Theorem

The following discussion involving right triangles serves as an application of squares and square roots and uses the terms defined as follows.

Terms Related to Right Triangles

Right triangle: A triangle containing a right (90°) angle

Hypotenuse: The longest side of a right triangle; the side opposite the right angle

Leg: Each of the other two sides of a right triangle (the sides that are not the hypotenuse)

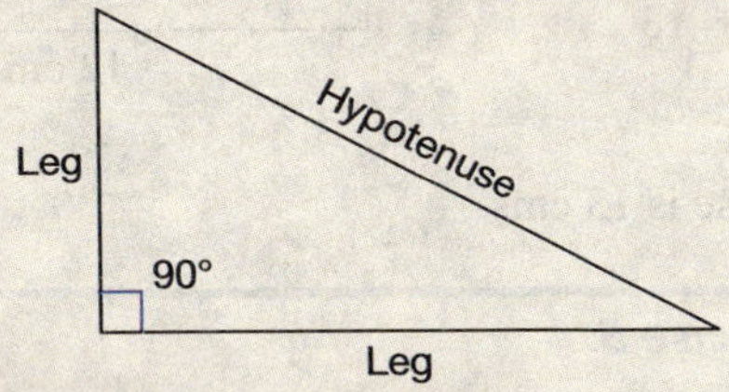

DEFINITION

Pythagoras (c. 570 to c. 495 BC), a famous Greek mathematician, is given credit for discovering the following theorem (although historians have found that the facts of the theorem were known before the time of Pythagoras).

The Pythagorean Theorem

In a right triangle, the square of the length of the hypotenuse is equal to the sum of the squares of the lengths of the two legs.

$$c^2 = a^2 + b^2$$

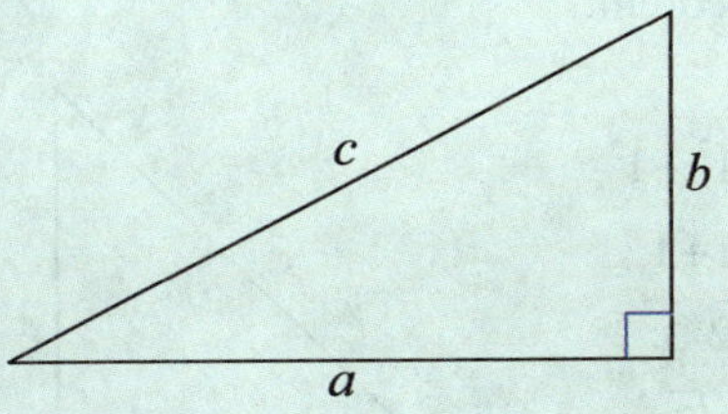

THEOREM

History

Pythagoras was a Greek philosopher and mathematician who lived from approximately 570 BC to 490 BC. While Pythagoras wrote no books during his lifetime, he is credited with establishing the theory of numbers in the world and in music. Although we don't really know how Pythagoras died, there are several interesting accounts of his death ranging from starvation to suicide to murder. Who knew that being a mathematician could be so dangerous?

Example 4 Verifying Right Triangles

Show that a triangle with sides of lengths 3 inches, 4 inches, and 5 inches must be a right triangle.

Solution

If the triangle is a right triangle, then its three sides must satisfy the property stated in the Pythagorean Theorem: $c^2 = a^2 + b^2$. Or, in this case, $5^2 = 3^2 + 4^2$. Since $25 = 9 + 16$ is a true statement, the triangle is a right triangle.

Now work margin exercise 4.

4. Is a triangle with sides of lengths 4 cm, 7 cm, and 8 cm a right triangle?

5. Find the length of the hypotenuse of a right triangle with legs of length 8 cm and 15 cm.

Example 5 Finding the Length of the Hypotenuse

Find the length of the hypotenuse of a right triangle with legs of length 12 cm and 5 cm.

Solution

Let $c =$ the length of the hypotenuse.

Now, by the Pythagorean Theorem:

$$c^2 = 12^2 + 5^2$$
$$c^2 = 144 + 25$$
$$c^2 = 169$$
$$c = \sqrt{169} = 13$$

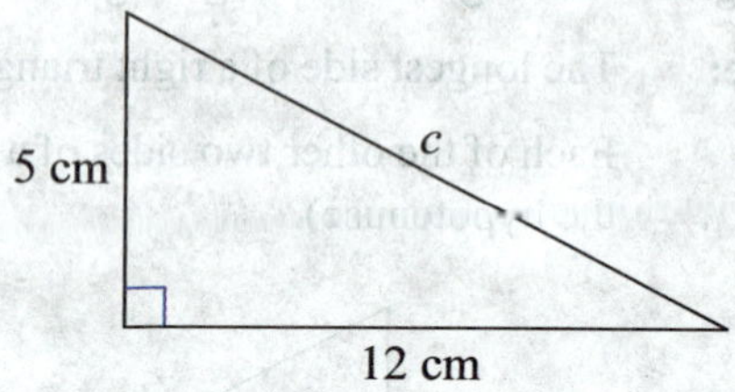

The length of the hypotenuse is 13 cm.

Now work margin exercise 5.

Rarely do all three sides of a right triangle have whole number values. Examples 6 and 7 illustrate right triangles with irrational numbers as the lengths of the hypotenuses.

6. Find the length of the hypotenuse (to the nearest hundredth) of a right triangle in which both legs have a length of 4 yd.

Example 6 Finding the Length of the Hypotenuse

Find the length of the hypotenuse (to the nearest hundredth) of a right triangle in which both legs have a length of 1 meter.

Solution

Let $c =$ the length of the hypotenuse

Now, by the Pythagorean Theorem:

$$c^2 = 1^2 + 1^2$$
$$c^2 = 1 + 1$$
$$c^2 = 2$$
$$c = \sqrt{2}$$
$$c \approx 1.41$$

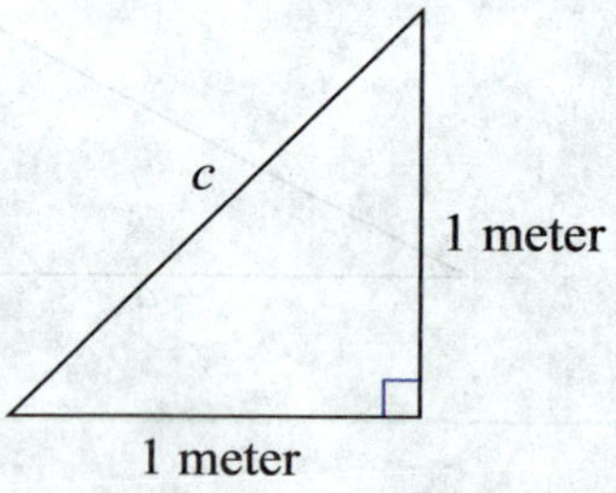

The length of the hypotenuse is $\sqrt{2}$ meters (or about 1.41 meters).

Now work margin exercise 6.

Math Tip

There are many different proofs of the Pythagorean Theorem. The following image is Euclid's geometric proof of the theorem. Although it is not a formal proof, it is an interesting visualization of the Pythagorean Theorem using a 3-4-5 right triangle.

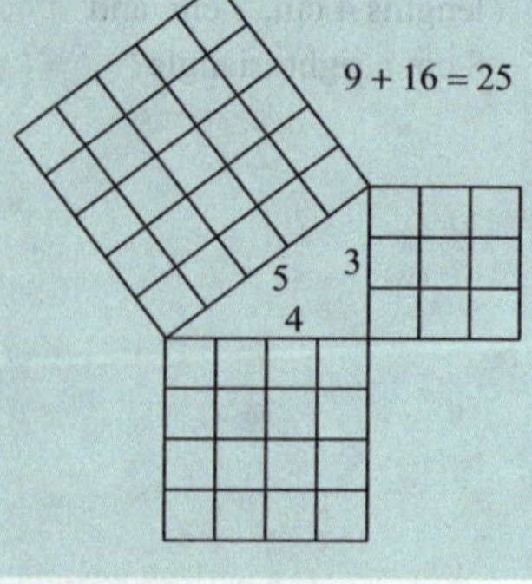

Example 7 Application: Finding the Length of the Hypotenuse

A guy wire is attached to the top of a telephone pole and anchored to the ground 10 feet from the base of the pole. If the pole is 20 feet high, what is the length of the guy wire to the nearest hundredth?

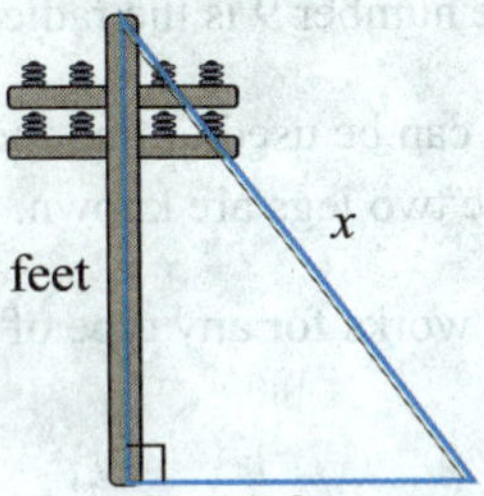

Solution

Let $x =$ the length of the guy wire.

Then, by the Pythagorean Theorem,

$$x^2 = 10^2 + 20^2$$
$$x^2 = 100 + 400$$
$$x^2 = 500$$
$$x = \sqrt{500} \approx 22.36$$

The guy wire is about 22.36 feet long.

Now work margin exercise 7.

7. A 35-foot-tall tree casts a shadow which is 30 feet long. Find the distance from the top of the tree to the end of its shadow to the nearest hundredth.

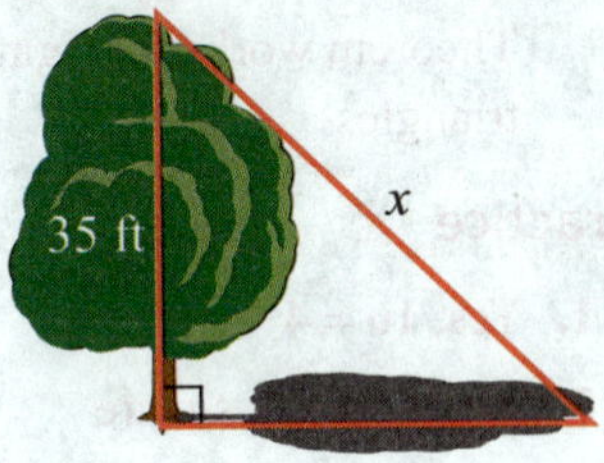

Margin Exercise Answers

1. a. 324 **b.** 81 **2. a.** 6 **b.** 13 **3. a.** 2.2361 **b.** 3.8730 **4.** No, $8^2 \neq 7^2 + 4^2$ **5.** 17 cm
6. $\sqrt{32}$ yd or about 5.66 yd **7.** 46.10 ft

6.7 Exercises

Concept Check

Fill-in-the-Blank. Complete each sentence using information found in this section.

1. The complete expression $\sqrt{81}$ is a/an ____________ or ____________ expression.

2. When a number is multiplied by itself, the result is a/an _________ _________.

3. The number under the radical sign is the ________.

4. $\sqrt{\ }$ is called a/an __________ sign.

5. The Pythagorean Theorem can be helpful in finding the lengths of sides of _______ triangles.

6. $c^2 = a^2 + b^2$ is the equation part of the ____________ Theorem.

6.7 Exercises

Concept Check

1. radical, radical
2. perfect square
3. radicand
4. radical
5. right
6. Pythagorean

7. True
8. False; In the expression $\sqrt{81}$, the number 81 is the radicand.
9. True
10. False; The Pythagorean Theorem works for right triangles.

Practice

1. Yes, $16 = 4^2$
2. Not a perfect square
3. Not a perfect square
4. Not a perfect square
5. Yes, $400 = 20^2$
6. Yes, $256 = 16^2$
7. Yes, $121 = 11^2$
8. Yes, $144 = 12^2$
9. 144
10. 289
11. 400
12. 64
13. 6
14. 9
15. 15
16. 13
17. 14
18. 19
19. 30
20. 60
21. 2.8284
22. 2.2361
23. 4.8990
24. 5.6569
25. 5.8310
26. 3.1623
27. 28.2843
28. 22.3607
29. 15.5563
30. 22.1133
31. 0.5
32. 0.6
33. 1.23

True/False. Determine whether each statement is true or false. If a statement is false, explain how it can be changed so the statement will be true. (**Note:** There may be more than one acceptable change.)

7. 49 is a perfect square.
8. In the expression $\sqrt{81}$, the number 9 is the radicand.
9. The Pythagorean Theorem can be used to find the length of the longest side of a right triangle if the lengths of the two legs are known.
10. The Pythagorean Theorem works for any type of triangle.

Practice

State whether or not each number is a perfect square.

1. 16	**4.** 40	**7.** 121
2. 48	**5.** 400	**8.** 144
3. 45	**6.** 256	

Evaluate each expression. See Examples 1 and 2.

9. 12^2	**13.** $\sqrt{36}$	**17.** $\sqrt{196}$
10. 17^2	**14.** $\sqrt{81}$	**18.** $\sqrt{361}$
11. 20^2	**15.** $\sqrt{225}$	**19.** $\sqrt{900}$
12. 8^2	**16.** $\sqrt{169}$	**20.** $\sqrt{3600}$

Use a calculator to approximate each square root to the nearest ten-thousandth. See Example 3.

21. $\sqrt{8}$	**28.** $\sqrt{500}$	**35.** $\sqrt{1.96}$
22. $\sqrt{5}$	**29.** $\sqrt{242}$	**36.** $\sqrt{2.89}$
23. $\sqrt{24}$	**30.** $\sqrt{489}$	**37.** $\sqrt{0.0025}$
24. $\sqrt{32}$	**31.** $\sqrt{0.25}$	**38.** $\sqrt{0.0016}$
25. $\sqrt{34}$	**32.** $\sqrt{0.36}$	**39.** $\sqrt{0.003}$
26. $\sqrt{10}$	**33.** $\sqrt{1.5129}$	**40.** $\sqrt{0.004}$
27. $\sqrt{800}$	**34.** $\sqrt{4.6225}$	

Answer each question.

41. a. Do you think that $\sqrt{39}$ is more than 6 or less than 6? Explain your reasoning.

    b. Do you think that $\sqrt{39}$ is more than 7 or less than 7? Explain your reasoning.

42. a. Do you think that $\sqrt{18}$ is more than 4 or less than 4? Explain your reasoning.

b. Do you think that $\sqrt{18}$ is more than 5 or less than 5? Explain your reasoning.

For each problem, **a.** use your understanding of square roots to estimate the value of each square root. Then, **b.** use your calculator to find the value of each square root accurate to the nearest ten-thousandth.

43. a. The nearest whole numbers to $\sqrt{50}$ are ____ and ____.

b. ▦ Find the value of $\sqrt{50}$.

44. a. The nearest whole numbers to $\sqrt{23}$ are ____ and ____.

b. ▦ Find the value of $\sqrt{23}$.

45. a. The nearest whole numbers to $\sqrt{11}$ are ____ and ____.

b. ▦ Find the value of $\sqrt{11}$.

46. a. The nearest whole numbers to $\sqrt{72}$ are ____ and ____.

b. ▦ Find the value of $\sqrt{72}$.

47. a. The nearest whole numbers to $\sqrt{60}$ are ____ and ____.

b. ▦ Find the value of $\sqrt{60}$.

48. a. The nearest whole numbers to $\sqrt{80}$ are ____ and ____.

b. ▦ Find the value of $\sqrt{80}$.

Use the Pythagorean Theorem to determine whether or not each triangle is a right triangle. See Example 4.

49.

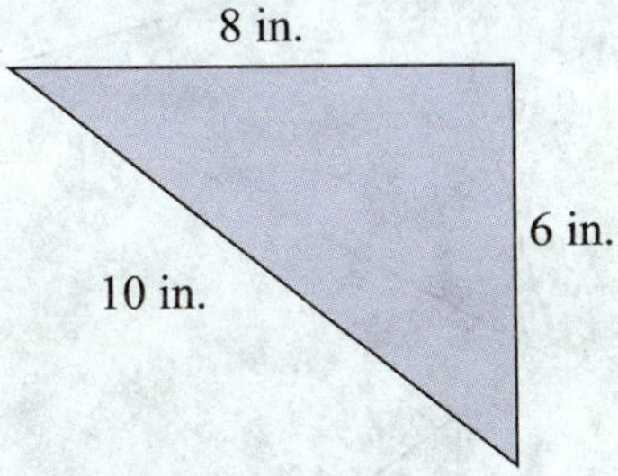

50.

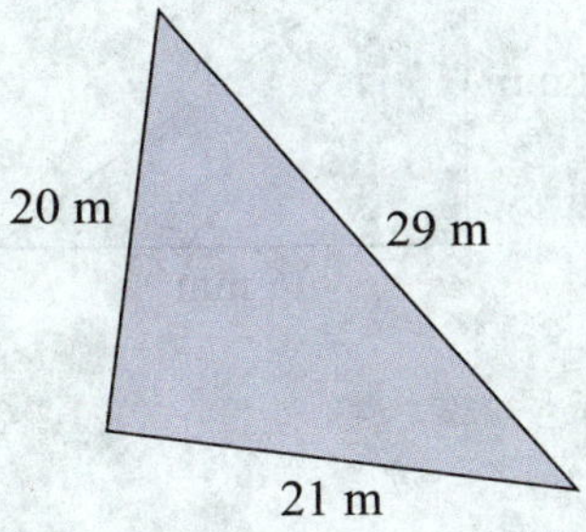

51.

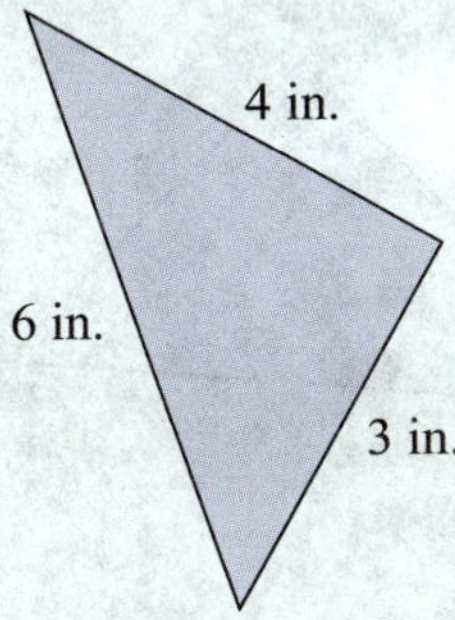

52.

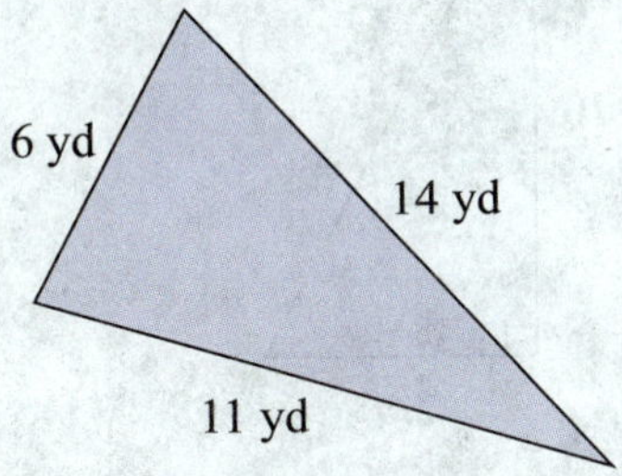

34. 2.15

35. 1.4

36. 1.7

37. 0.05

38. 0.04

39. 0.0548

40. 0.0632

41. a. Answers will vary. $\sqrt{39} > \sqrt{36} = 6$, so $\sqrt{39} > 6$

b. Answers will vary. $\sqrt{39} < \sqrt{49} = 7$, so $\sqrt{39} < 7$

42. a. Answers will vary. $\sqrt{18} > \sqrt{16} = 4$, so $\sqrt{18} > 4$

b. Answers will vary. $\sqrt{18} < \sqrt{25} = 5$, so $\sqrt{25} < 5$

43. a. 7 and 8 **b.** 7.0711

44. a. 4 and 5 **b.** 4.7958

45. a. 3 and 4 **b.** 3.3166

46. a. 8 and 9 **b.** 8.4853

47. a. 7 and 8 **b.** 7.7460

48. a. 8 and 9 **b.** 8.9443

49. Yes, $6^2 + 8^2 = 10^2$

50. Yes, $21^2 + 20^2 = 29^2$

51. No, $3^2 + 4^2 \neq 6^2$

52. No, $11^2 + 6^2 \neq 14^2$

53. Yes, $5^2 + 12^2 = 13^2$
54. No, $2^2 + 3^2 \neq 4^2$
55. $c = 2.23$
56. $c = 4.24$
57. $c = 5$
58. $c = 10.77$
59. $c = 20.62$
60. $c = 5.66$ cm
61. $c = 14.14$ cm
62. $c = 20$ mm

53.

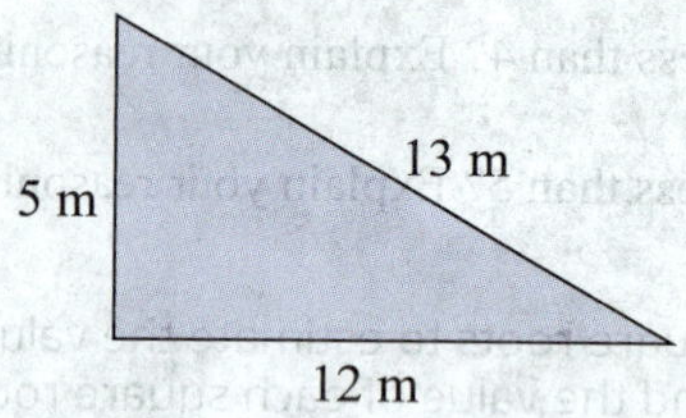

54.

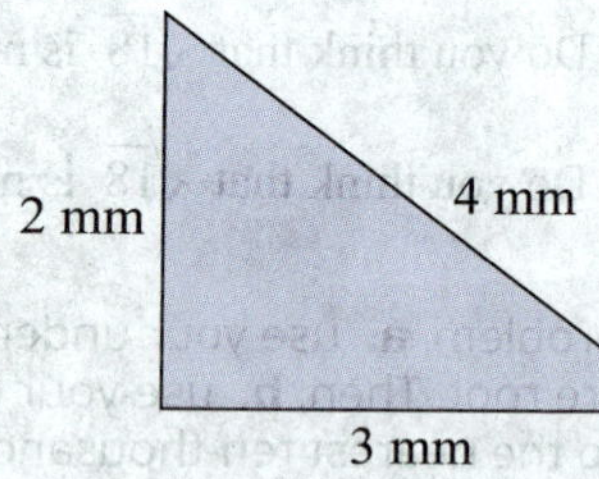

Find the hypotenuse for each right triangle accurate to the nearest hundredth. See Examples 5 and 6.

55.

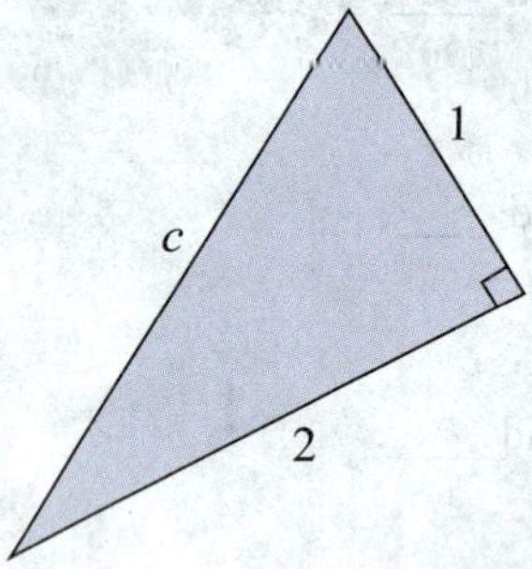

56.

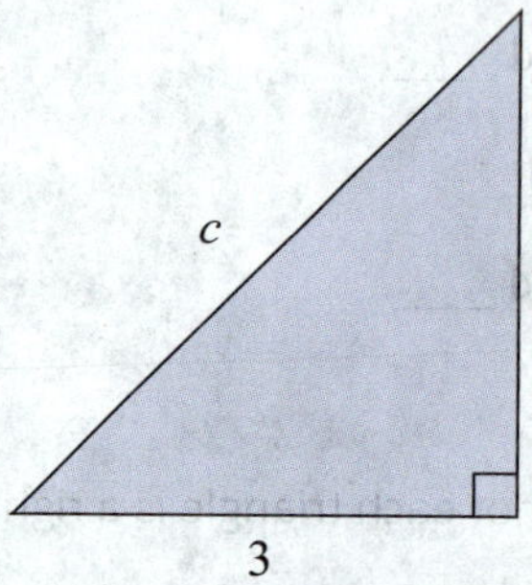

57.

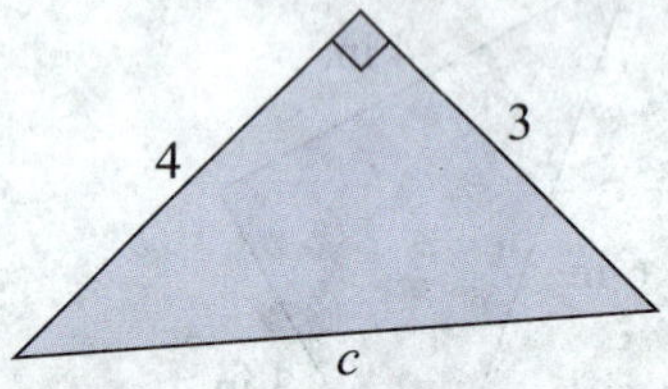

58.

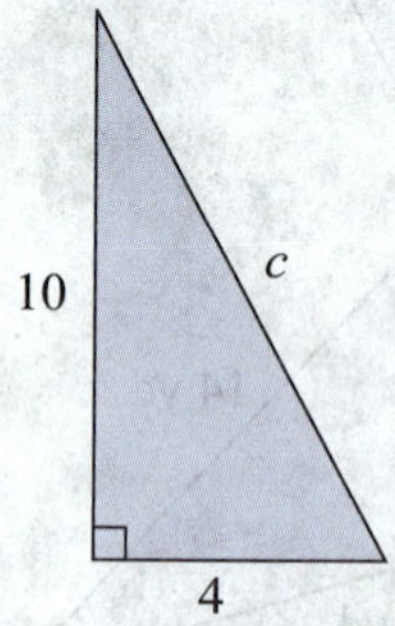

59.

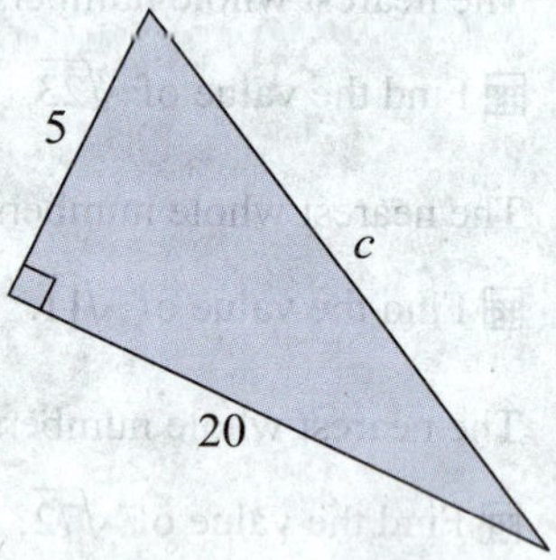

60.

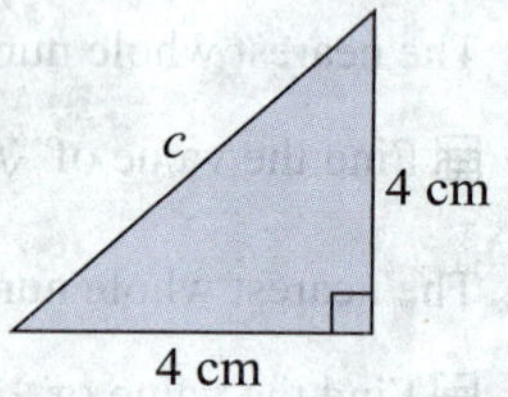

61.

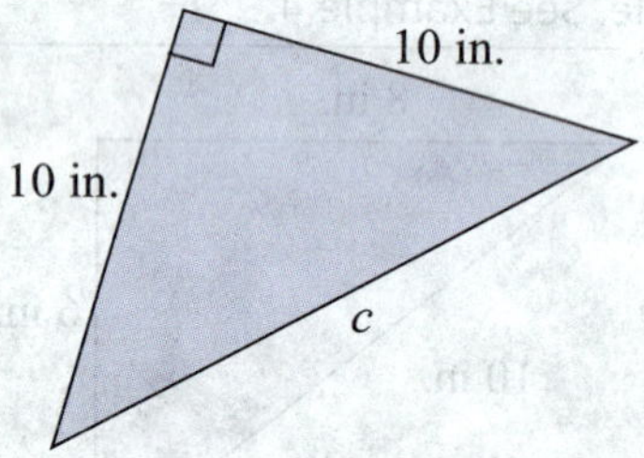

62.

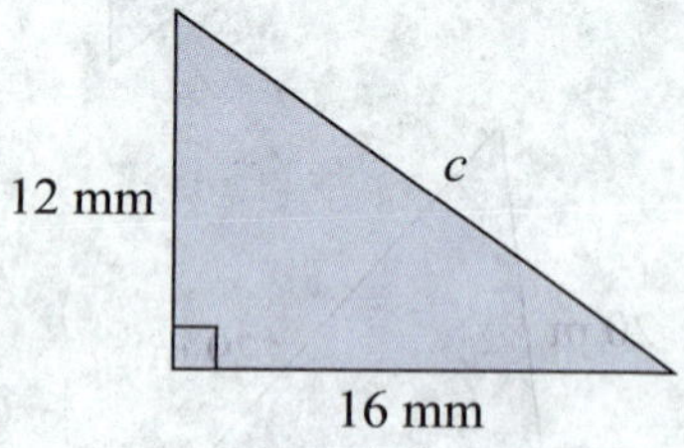

Applications

Solve.

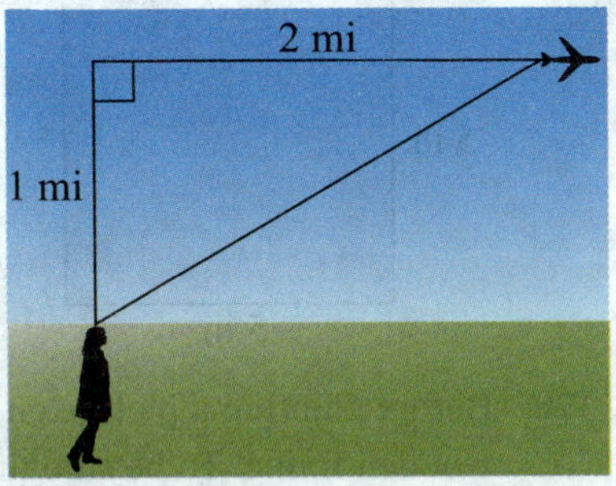

63. If an airplane passes directly over your head at an altitude of 1 mile, how far (to the nearest hundredth of a mile) is the airplane from your position after it has flown 2 miles farther at the same altitude?

64. The GE Building in New York is 850 feet tall (70 stories). At a certain time of day, the building casts a shadow 100 feet long. Find the distance from the top of the building to the tip of the shadow (to the nearest tenth of a foot).

65. The base of a fire engine ladder is 30 feet from a building and reaches to a third floor window 50 feet above ground level. Find the length of the ladder to the nearest hundredth of a foot.

66. A forestay that helps support a ship's mast reaches from the top of the mast, which is 20 meters high, to a point on the deck 10 meters from the base of the mast. What is the length of the forestay (to the nearest tenth of a meter)?

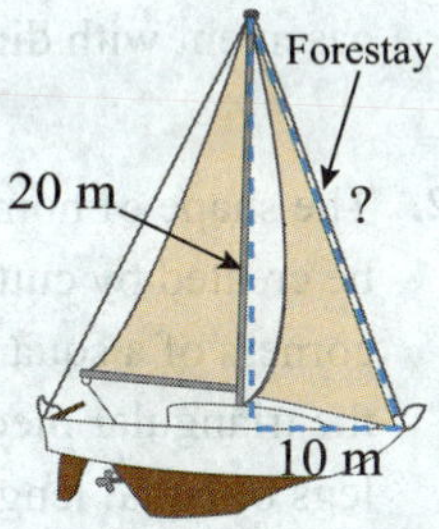

67. The Xerox Center building in Chicago is 500 feet tall. At a certain time of day, it casts a shadow that is 150 feet long. At that time of day, what is the distance (to the nearest tenth of a foot) from the tip of the shadow to the top of the Xerox building?

68. The shape of a baseball infield is a square with sides 90 feet long.

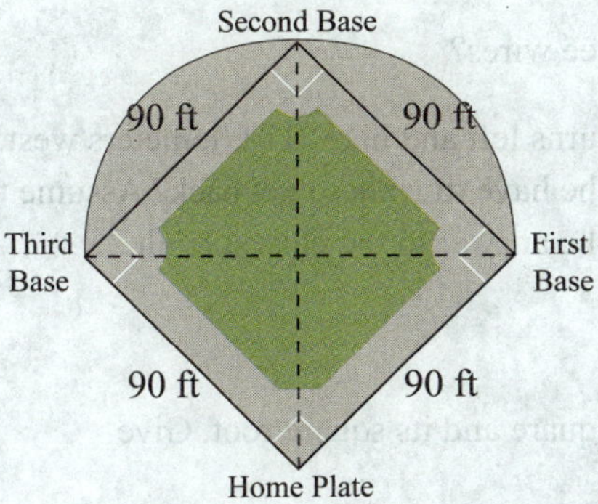

a. Find the distance (to the nearest tenth of a foot) from home plate to second base.

b. The diagonals of the square intersect halfway between home plate and second base. If the pitcher's mound is 60.5 feet from home plate, is the pitcher's mound closer to home plate or to second base?

Applications

63. 2.24 miles

64. 8.559 feet long

65. 58.31 ft

66. 22.4 meters

67. 522.0 feet

68. a. 126.3 feet

b. Closer to home plate

69. **a.** 20 in. **b.** 25 in.^{2} **c.** 7.0711 in.

70. 44.9 inches

71. 64.9 inches

72. 17.0 inches

73. **a.** 17 feet **b.** 51 feet

74. 14.2 km

Writing & Thinking

75. A perfect square is the result when a number is squared. For example, 5 squared is 25, therefore 25 is the perfect square of 5. Then if one finds the square root of 25, the answer is 5 because the square root is the original number that was squared to get the result of 25.

76. The radical sign is $\sqrt{\ }$. The radicand is the number under the radical sign. A radical expression includes the radical sign and its radicand. For example, in the expression $\sqrt{36}$ the entire expression is called a radical expression, 36 is the radicand, and the symbol $\sqrt{\ }$ is called the radical sign.

69. A diagonal of a square is the line segment from one corner to an opposite corner of a square. A square has side length of 5 in.

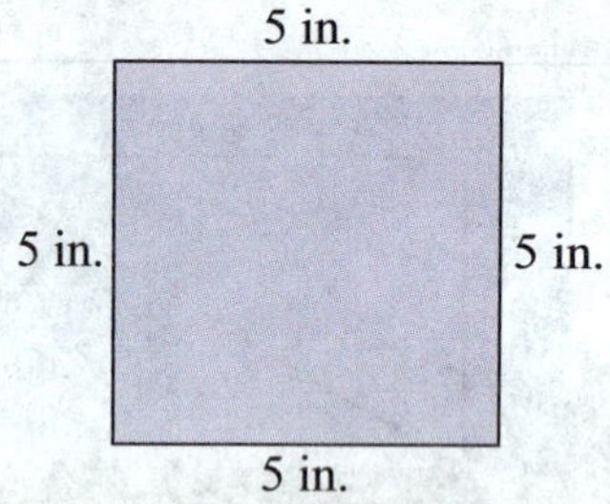

a. Find the perimeter of the square.

b. Find the area of the square.

c. Find the length of a diagonal of the square.

70. Before painting a picture on canvas, an artist must stretch the canvas on a rectangular wooden frame. To be sure that the corners of the canvas are true right angles, the artist can measure the diagonals of the stretched canvas. What should be the diagonal measure, to the nearest tenth of an inch, of a canvas whose sides are 24 inches and 30 inches in length?

71. While installing windows in a new home, a builder measures the diagonals of rectangular window casements to verify that their corners are true right angles. What should be the diagonal measure, to the nearest tenth of an inch, of a window casement with dimensions 36 inches and 54 inches.

72. The shape of home plate in the game of baseball can be created by cutting off two triangular pieces at the corners of a square, as shown in the figure. If each of the triangular pieces has a hypotenuse of 12 inches and legs of equal length, what is the length of one side of the original square, to the nearest tenth of an inch?

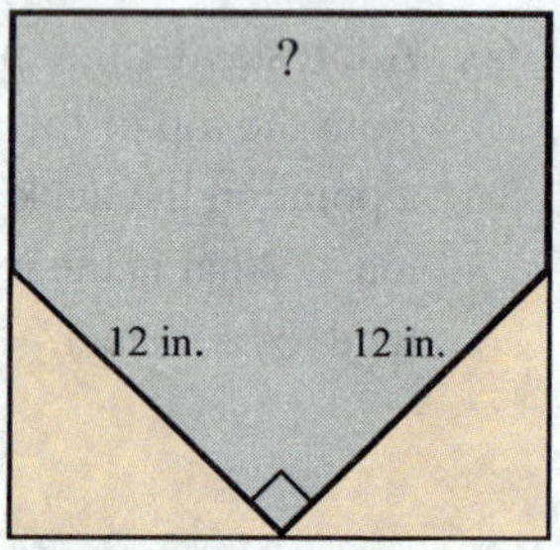

73. The city needs to replant an 18-foot tree. To ensure the tree does not fall over, three wires are to be attached to the tree 3 feet from the top. The wires will extend 8 feet from the base of the tree.

a. How long will each wire be? Round to the nearest hundredth.

b. What will be the total length of the three wires?

74. A hiker hikes 9 kilometers north and then turns left and hikes 11 kilometers west. If she takes the shortest path, how long will she have to walk to get back? Assume the terrain is flat with no obstructions. Round the answer to the closest tenth.

Writing & Thinking

75. Explain the connection between a perfect square and its square root. Give an example.

76. Determine the difference between a radical sign, radicand, and radical expression.

77. Find the length of one side of the square in the figure shown with the square inscribed in a circle of radius 20 meters. (Each corner of the square lies on the circle.) Use $\pi \approx 3.14$.

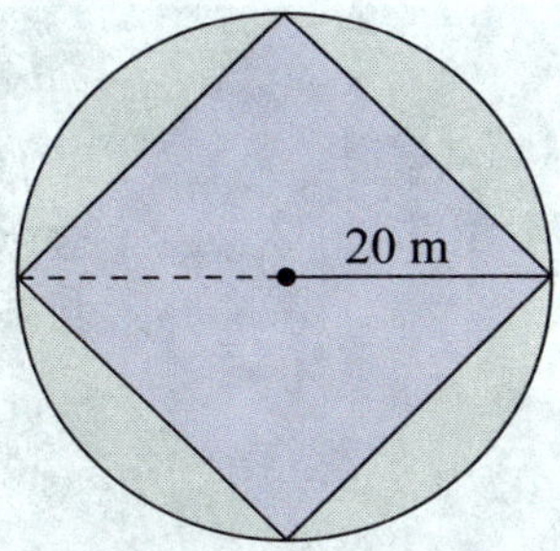

78. To create a square inside a square, a quilting pattern requires four triangular pieces like the one shaded in the figure shown here. If the square in the center measures 12 centimeters on a side and two legs of each triangle are of equal length, how long are the legs of each triangle to the nearest tenth of a centimeter?

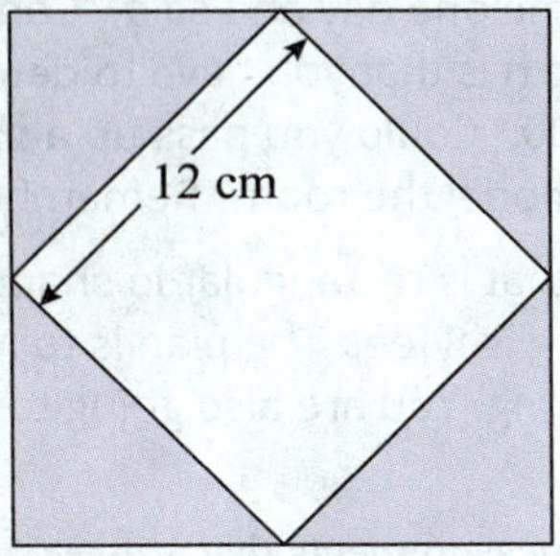

79. If three whole numbers satisfy the Pythagorean Theorem, these three numbers are called a Pythagorean triple. For example, 3, 4, and 5 are a Pythagorean triple because $3^2 + 4^2 = 5^2$ (or $9 + 16 = 25$). Another Pythagorean triple is 5, 12, and 13 because $5^2 + 12^2 = 13^2$ (or $25 + 144 = 169$). Complete the following table by finding a, b, and c, and telling which sets of these three numbers (if any) constitute a Pythagorean triple. The first one is done for you.

m	n	$a = 2nm$	$b = m^2 - n^2$	$c = m^2 + n^2$	Pythagorean Triple?
5	1	10	24	26	yes: $10^2 + 24^2 = 26^2$
7	1				
3	2				
7	2				
5	3				
11	3				
13	7				

Extension: Choose some of your own numbers for m and n. Are your results Pythagorean triples? (**Note:** m must be larger than n so $m^2 - n^2$ will be positive.)

77. 28.28 m

78. 8.5 cm

79.

m	n	$a = 2nm$	$b = m^2 - n^2$	$c = m^2 + n^2$	Pythagorean Triple?
5	1	10	24	26	Yes: $10^2 + 24^2 = 26^2$
7	1	14	48	50	Yes: $14^2 + 48^2 = 50^2$
3	2	12	5	13	Yes: $12^2 + 5^2 = 13^2$
7	2	28	45	53	Yes: $28^2 + 45^2 = 53^2$
5	3	30	16	34	Yes: $30^2 + 16^2 = 34^2$
11	3	66	112	130	Yes: $66^2 + 112^2 = 130^2$
13	7	182	120	218	Yes: $182^2 + 120^2 = 218^2$

Chapter 6 Project: This Mixtape Is Fire!

1. a. 16.22 in. **b.** 16.44 in.^{2}

2. a. 31.775 in.^{2} **b.** 7.01 in.^{3}

3. a. 50.47 in.^{2} **b.** 10.74 in.^{3}

c. The less common CD case will require less plastic film to wrap. The surface area of the first case is about 63.30 in.^{2} and the surface area of the less common case is about 62.77 in.^{2} A smaller surface area means less wrapping.

4. a. Cutting out 4 in. along the top results in a length of 8.375 in. Cutting 3 in. out from the right results in a length of 9.375 in. The length of the hypotenuse of the triangle is 5 in. by using the Pythagorean Theorem. Adding all of the sides results in a perimeter of 47.5 in.

b. 147.14 in.^{2}

5. a. 37.68 in.

b. First, find the area of the entire record. Next, find the area of the inner circle. Subtract the area of the smaller circle from the larger circle. 105.98 in.^{2}

c. No. Explanations will vary. Students should point out that the records are not uniform cylinders and that the stack will not be a solid cylinder. The formula for the volume of a cylinder can be used as an approximation for the volume of the stack of 500 records.

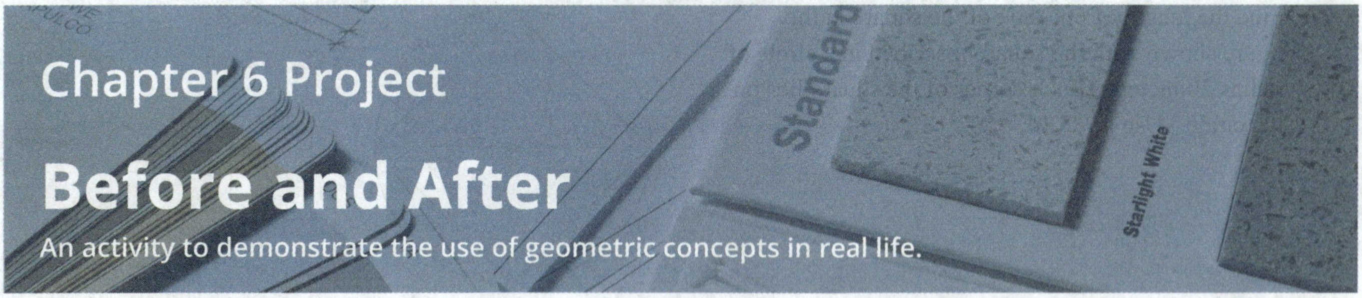

Chapter 6 Project

Before and After

An activity to demonstrate the use of geometric concepts in real life.

Suppose HGTV came to your home one day and said, "Congratulations, you have just won a FREE makeover for any room in your home! The only catch is that you have to determine the amount of materials needed to do the renovations and keep the budget under \$2000." Could you pass up a deal like that? Would you be able to calculate the amount of flooring and paint needed to remodel the room? Remember it's a FREE makeover if you can!

Let's take an average size room that is rectangular in shape and measures 16 feet 3 inches in width by 18 feet 9 inches in length. The height of the ceiling is 8 feet. The plan is to repaint all the walls and the ceiling and to replace the carpet on the floor with hardwood flooring. You are also going to put crown molding around the top of the walls for a more sophisticated look.

1. Take the length and width measurements that are in feet and inches and convert them to a fractional number of feet and reduce to lowest terms. (Remember that there are 12 inches in a foot. For example, 12 feet 1 inch is $12\frac{1}{12}$ feet.)
2. Now convert these same measurements to decimal numbers.
3. Determine the number of square feet of flooring needed to redo the floor. (Express your answer in terms of a decimal and do not round the number.) 304.6875 square feet
4. If the flooring comes in boxes that contain 24 square feet, how many boxes of flooring will be needed? (Remember that the store only sells whole boxes of flooring.) 13 boxes
5. If the flooring you chose costs \$74.50 per box, how much will the hardwood flooring for the room cost (before sales tax)? \$968.50
6. Figure out the surface area of the four walls and the ceiling that need to be painted, based on the room's dimensions. (We will ignore any windows, doors, or closets since this is an estimate.) 864.6875 square feet
7. Assume that a gallon of paint covers 350 square feet and you are going to have to paint the walls and the ceilings **twice** to cover the current paint color. Determine how many gallons of paint you need to paint the room. (Again, assume that you can only buy whole gallons of paint. Any leftover paint can be used for touch-ups.) 5 gallons
8. If the paint you have chosen costs \$18.95 per gallon, calculate the cost of the paint (before sales tax). \$94.75
9. Determine how many feet of crown molding will be needed to go around the top of the room. 70 feet
10. The molding comes in 12-foot sections only. How many sections will you need to buy?
11. If the molding costs \$2.49 per linear foot, determine the cost of the molding (before sales tax). \$179.28
12. Calculate the cost of all the materials for the room makeover (before sales tax).
    a. Were you able to stay within budget for the project?
    b. If so, then what extras could you add? If not, what could you adjust in this renovation to stay within budget?
    c. Using sales tax in your area, calculate the final price of the room makeover with sales tax included.

Chapter 6 Project: Before and After

1. Length = $18\frac{3}{4}$ feet, Width = $16\frac{1}{4}$ feet
2. Length = 18.75 feet, Width = 16.25 feet
10. Will need to buy 6 sections of molding.
12. a. Yes, with a total cost of materials equal to \$1,242.53
    b. Answers will vary. Extras that could be added: window treatments, rugs, furnishings, etc.
    c. Answers will vary, but will be well under budget of \$2000.

Chapter 6 Project

This Mixtape Is Fire!

An activity to demonstrate the use of geometric formulas and properties in real life.

While most music listening today happens via online streaming, music is still distributed in a variety of physical formats, such as cassette tapes, CDs, and vinyl records.

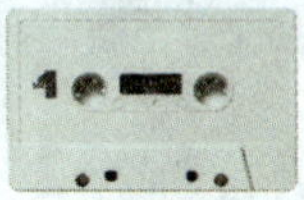
Cassette Tape

Compact Disc (CD)

Vinyl Record

Imagine that you are a musical artist who wants to produce and distribute your own music using cassette tapes, CDs, and vinyl records. To save money, you plan to design the album artwork yourself and send the art files to the manufacturer along with the music files. The album artwork contains the cover art, information about each song, and all the credits. The dimensions of the album artwork for each type of physical media varies.

1. For the cassette tape, the album artwork (called a J-card) is commonly printed on a 4-inch by 4.11-inch rectangle that is folded into a J shape and placed into a plastic case.
   a. What is the perimeter of this sleeve?
   b. What is the area of this sleeve?

2. The plastic case for a cassette tape commonly measures 4.25 inches by 2.75 inches by 0.6 inches. The case is in the shape of a rectangular prism.
   a. What is the surface area of a cassette case?
   b. What is the volume of a cassette case? Round to the nearest hundredth.

3. For the CD, the front artwork is commonly printed onto a 4.724-inch square and the back artwork is printed onto a 4.724-inch by 5.96-inch rectangle. Both the front and back artwork is inserted into a plastic case with dimensions 4.91 inches by 5.61 inches by 0.39 inches. The case is in the shape of a rectangular prism. The entire case is then wrapped in plastic film.
   a. What is the total area of the front and back artwork? Round to the nearest hundredth.
   b. What is the volume of the CD case? Round to the nearest hundredth.
   c. A less common CD case has dimensions 4.92 inches by 5.51 inches by 0.41 inches. Which CD case size would require less plastic film to wrap it? How do you know?

4. The vinyl record sleeve has a 12.375-inch square for the cover art. Suppose you want to designate a triangle at the top right of the cover art for the artist's name. This triangle has one leg along the right side of the sleeve measuring 3 inches and another leg along the top side of the sleeve measuring 4 inches.
   a. What is the perimeter of the cover art excluding the triangle? Explain how you found this perimeter, using the Pythagorean Theorem in your explanation.
   b. What is the area of the cover art excluding the triangle? Round to the nearest hundredth.

5. Each vinyl record is circular with a diameter of 12 inches. Use the approximation $\pi \approx 3.14$.
   a. What is the circumference of a 12-inch vinyl record?
   b. In the middle of each record, there is a circle with a diameter of 3 inches with no music on it where a paper label is placed to identify the record. What is the area of the playable surface of this vinyl record that is not the center label? Explain how you found your answer.
   c. You plan to have 500 vinyl records made for your album. At their center, each record has a height of 0.3 inches. The height along the outside of the record is 0.2 inches with a rounded edge. Can you use the formula for a volume of a cylinder to calculate the exact volume of a stack of 500 records? Explain your answer.

Chapter 6 Project
This Mixtape is [illegible]

While most music listening today happens via online streaming, music is still distributed in a variety of other formats, such as cassette tapes, CDs, and vinyl records.

Cassette Tape — Compact Disc (CD) — Vinyl Record

Imagine that you are a musician/artist who wants to produce and distribute your own music using cassette tapes, CDs, and vinyl records. To save money, you plan on designing the album artwork yourself and sending the art files to the manufacturer along with the music files. The album artwork contains the cover art, the artist's name, the song, and other [illegible] fits. The dimensions of the album artwork for each type of physical media varies.

1. For the cassette tape, the album artwork is called a [illegible] and is commonly printed on a [illegible] inch by [illegible] inch rectangle, then folded into a [illegible] and placed into the plastic case.

   a. What is the perimeter of this sleeve?

   b. What is the area of this sleeve?

2. The plastic case for a cassette tape commonly measures [illegible] inches by [illegible] inches by [illegible] inches. The case is in the shape of a rectangular prism.

   a. What is the surface area of a cassette case?

   b. What is the volume of one cassette case? Round to the nearest tenth.

3. For the CD, the front artwork is commonly printed onto a [illegible] inch square and the back artwork [illegible] onto a [illegible] inch by [illegible] inch rectangle. Both the front and back artwork is inserted into a plastic case with dimensions [illegible] inches by [illegible] inches by [illegible] inches. The case is in the shape of a rectangular prism. The entire case is then wrapped in plastic film.

   a. What is the total area of the front and back artwork? Round to the nearest hundredth.

   b. What is the volume of the CD case? Round to the nearest hundredth.

   c. A less common CD case has dimensions [illegible] inches by [illegible] inches by [illegible] inches. Which CD case design would require less plastic film to wrap it? How do you know?

4. The [illegible] sleeve for the cover art [illegible] is a [illegible] square. Suppose you want to use a [illegible] triangle as the [illegible] of the cover art for the artist's name. The triangle has one leg along the right side of the sleeve measuring [illegible] inches and the other leg along the bottom of the sleeve [illegible] inches.

   a. What is the perimeter of the cover art's equilateral triangle? Explain how you found the perimeter using the Pythagorean Theorem in your explanation.

   b. What is the area of the cover art section of the triangle? Round to the nearest hundredth.

5. Each vinyl record is circular with a diameter of 12 inches. Use the approximation [illegible].

   a. What is the circumference of a 12-inch vinyl record?

   b. In the middle of each record, there is a circle with a circumference of [illegible] inches, which contains a [illegible] label [illegible] placed to identify the record. What is the area of the playable surface of this vinyl record (that is, not the center label)? Explain how you found your answer.

6. You plan to have [illegible] made for your album. At the [illegible] cost, before the [illegible] are made, [illegible] on the outside of the [illegible] would be [illegible]. Use the formula for the volume of a cylinder to calculate the exact volume of a stack of [illegible] records. Explain your answer.

CHAPTER 7

7 Statistics, Graphs, and Probability

7.1 Statistics: Mean, Median, Mode, and Range

7.2 Reading Graphs

7.3 Constructing Graphs from Databases

7.4 Probability

CHAPTER 7 PROJECTS

What's My Average?

Misleading Graphs

Connections

From data in reports at work to news from media outlets, people come across statistics and probability daily. A lot of the statistics calculated about the US population are the result of the data collected in the US Census, which is updated every 10 years. Besides being used in news and research, these statistics can be used by businesses to determine the best way to make money long term.

The field of actuary science uses statistics to help insurance companies determine how much to charge customers. For example, your insurance premium is based on the likelihood of you filing a claim based on personal information, such as your age, your gender, and where you live. These companies base their rates on statistics gathered over many years. Your age affects insurance rates because drivers under the age of 25 on average have a higher rate of fatal accidents. Similarly, your location can affect your insurance rate since accident rates vary by location. For example, Fort Collins, CO, has a lower accident rate per driver than Portland, OR. Knowing how to interpret statistics and probability will help you make and understand decisions that affect your everyday life.

Suppose you are employed at the local city planning department. Over the past three-week period, or 15 business days, the department issued the following number of building permits to residents and contractors.

17, 19, 18, 35, 30, 29, 23, 14, 18, 16, 20, 18, 18, 25, 30

For the monthly review meeting, you need to present this data in a meaningful way. How would you calculate the mean, median, mode, and range of this data? What do these measurements tell you about the number of building permits issued?

Objectives

A. Recognize common statistical terms.

B. Calculate mean, median, mode, and range.

7.1 Statistics: Mean, Median, Mode, and Range

History

The origins of statistics may be traced to the 18th century, but the modern field of statistics was developed in the late 19th and early 20th centuries. The English mathematician Karl Pearson (1857–1936) is known as the founder of mathematical statistics. Around the same time, the contributions of English polymath Francis Galton (1822–1911) helped found statistical theory. The Royal Statistical Society was founded in 1834 and has always been an inclusive society: it gained Florence Nightingale as its first female member in 1858 and Stella Cunliffe was the first female president in 1975.

A Statistical Terms

A January 21, 2022, article in the Los Angeles Times reported that the **median** price of a home in Southern California was $697,500. By counties, the median price of homes were as follows.

San Bernardino: $485,750	Los Angeles: $805,000
Orange County: $935,000	Riverside County: $550,000
Ventura County: $751,000	San Diego County: $743,000

So, if you are moving to Southern California or already live there, these are very interesting numbers; but what do they mean? (Do you know the median price of homes in your area?) Just what is a median, and is it different from a mean? Both the mean and the median are statistics that we are going to study in this section.

Statistics is the study of how to gather, organize, analyze, and interpret numerical information. A **statistic** is a particular measure or characteristic of a part, or **sample**, of a larger collection of items called the **population**. The population can be a collection of people, animals, objects, or numbers related to information of interest.

In this section, we will study only four numerical statistics that are easily found or calculated: **mean**, **median**, **mode**, and **range**. The mean, median, and mode, also called **measures of central tendency**, are measures that describe the "average" or "middle" of a set of data. (**Note:** The concept of average was introduced in Section 1.6.) The range is a measure that describes how "spread out" the data are. Other measures that you might read about in newspapers and magazines and would certainly study in a course in statistics are

standard deviation and **variance** (both measures of how "spread out" data are),

z-**score** (a measure that compares numbers that are expressed in different units; for example, your test score on a mathematics exam and your performance in a physical exercise), and

correlation coefficient (a measure of how two different types of data might be related; for example, the relationship between a person's income and the number of years of schooling they have).

You will need a semester or two of algebra in order to be able to study and understand these and other statistics, so keep working hard. The following terms and their definitions are necessary for understanding the ideas and problems in this section.

Terms Used in the Study of Statistics

Data: Value(s) measuring some characteristic of interest such as income, height, weight, grade point averages, scores on tests, and so on. (We will consider only numerical data.)

Statistic: A single number describing some characteristic of the data.

Mean: The sum of all the data divided by the number of data items. (The arithmetic average of the data.)

Median: The middle of the data after the data have been arranged in order (smallest to largest or vice versa). (The median may or may not be one of the data items.)

Mode: The data item(s) that appears the most number of times. (A set of data may have more than one mode.)

Range: The difference between the largest and smallest data items.

DEFINITION

B Calculating Mean, Median, Mode, and Range

The following two sets of data are used to answer the questions in Examples 1, 2, and 3.

Group A

Annual Income for 8 Individuals

\$28,000	\$22,000	\$25,000	\$27,000
\$45,000	\$80,000	\$25,000	\$30,000

Table 1

Group B

The Time (in Minutes) of 11 Movies

100 min	90 min	113 min	110 min	88 min	90 min
155 min	88 min	105 min	93 min	90 min	

Table 2

Example 1 Application: Finding the Mean

Find the mean annual income for the individuals in Group A.

Solution

The mean is the average of the data. Therefore, we add the salaries and divide the sum by 8.

Math Tip

The average of a set of numbers is the value that represents the "middle" of the set and can be calculated in different ways. Mean, median, and mode are three types of averages and each one provides a different view of the data. The average used needs to be considered carefully or it can cause you to jump to the wrong conclusion. Consider the following salaries: \$12,000, \$12,100, \$12,200, \$12,300, \$12,400, \$80,000, and \$80,000. The averages for these salaries are \$80,000 (mode), \$31,157.42 (mean), and \$12,300 (median). Consider the different pictures painted by these different averages.

1. Find the mean movie time for the movies in Group B, rounded to the nearest minute.

Add:	Divide:	
$ 28,000 22,000 25,000 27,000 45,000 80,000 25,000 + 30,000 $282,000	$$\begin{array}{r} 35,250 \\ 8\overline{)282,000} \\ \underline{-24} \\ 42 \\ \underline{-40} \\ 2\,0 \\ \underline{-1\,6} \\ 40 \\ \underline{-40} \\ 00 \\ \underline{-0} \\ 0 \end{array}$$	Mean annual income

The mean annual income is $35,250.

Now work margin exercise 1.

Finding the Median

1. Arrange the data in order, either from smallest to largest or largest to smallest.
2. a. If there is an **odd** number of items, the median is the middle item.
   b. If there is an **even** number of items, the median is found by calculating the average of the two middle items. (**Note:** This value may or may not be in the data.)

PROCEDURE

2. 12 students took a make-up exam for a history class, and their scores are listed below. Find the median score of the 12 students.

Scores	
86	77
79	67
92	84
98	91
93	88
95	77

Example 2 Application: Finding the Median

Find the median annual income for the 8 individuals in Group A and the median time for the movies in Group B.

Solution

First, we arrange both sets of data in order from smallest to largest.

Group A (Incomes)	
1. $22,000	
2. $25,000	
3. $25,000	
4. $27,000	The median is between
5. $28,000	$27,000 and $28,000.
6. $30,000	
7. $45,000	
8. $80,000	

Group B (Movie Times)	
1. 88 min	
2. 88 min	
3. 90 min	
4. 90 min	
5. 90 min	
6. 93 min	The median is 93 min.
7. 100 min	
8. 105 min	
9. 110 min	
10. 113 min	
11. 155 min	

Group A:

There are 8 items (an **even** number) so we find the middle two items and average them. For this set of data, the middle two items are the 4th and 5th items or \$27,000 and \$28,000. (Count 4 from the top and 4 from the bottom.)

$$\text{median income} = \frac{27{,}000 + 28{,}000}{2} = \frac{55{,}000}{2} = \$27{,}500$$

Thus, the median annual income is \$27,500.

Group B:

Group B has 11 items (an **odd** number) and the median is the 6th item.

Thus, the median movie time is 93 minutes.

(Notice that in Group A the median is not one of the data items; while in Group B the median is one of the data items.)

Now work margin exercise 2.

Once the data have been ranked (as we did in Example 2), the mode (if there is one) and the range are easily determined. Remember that the mode is the item that occurs most frequently, and the range is the difference between the largest and smallest items in each set of data.

Example 3 Application: Finding the Mode and Range

For Group A and Group B, find **a.** the mode and **b.** the range.

Solution

a. From the arranged data in Example 2, we can find the mode by determining the most frequent item in each group.

Group A: The mode is \$25,000. (\$25,000 occurs twice and no other item occurs more than once.)

Group B: The mode is 90 minutes. (90 min occurs three times and no other item occurs more than twice.)

b. Again referring to the arranged data in Example 2, we can calculate each range as follows.

range = (largest value) – (smallest value)

Group A: Range = \$80,000 – \$22,000 = \$58,000

Group B: Range = 155 – 88 = 67 minutes

Now work margin exercise 3.

Completion Example 4 Application: Using the Mean to Find a Missing Value

Suppose that your grade in this class is the mean of 5 exam scores: 4 sectional exams and 1 comprehensive final exam, with each exam scored on an equal basis of 100 points. On the first 4 exams, you have scores of 85, 78, 82, and 70. What is the lowest score you can get on the final exam and still earn a grade of B in the course? (Assume that to get a B your mean score must be between 80 and 89.)

3. Find **a.** the mode and **b.** the range for the following set of data.

Ages of 8 People			
35	14	37	51
28	19	35	45

4. Suppose that your grade in this class is based on 4 exam scores: 3 sectional exams and 1 comprehensive final exam, with each exam scored on a basis of 100 points. On the first 3 exams you have scores of 93, 95, and 75. What is the lowest score you can get on the final exam and still earn a grade of A in the course? (Assume that to get an A, you must average 90 or above.)

Solution

Solve the problem by first finding the total number of points needed to obtain a mean of 80 on 5 exams. Then calculate the number of points accumulated on the first four exams. Finally, subtract the number of points accumulated from the total needed. This will give the number of points needed on the final exam.

a. Find the total number of points needed for a B. Since the 5 exams are to have a mean score of 80 (or more), then the total number of points must be at least the following product.

$$\begin{array}{r} 80 \\ \times \;\; 5 \\ \hline 400 \end{array} \quad \text{Minimum total points needed for a B}$$

b. Calculate the number of points you have accumulated on the first 4 exams.

$$\begin{array}{r} 85 \\ 78 \\ 82 \\ +70 \\ \hline 315 \end{array} \quad \text{Points accumulated}$$

c. To obtain a mean of 80 (or more) on your 5 exams, you need the following score on the final exam.

$$\begin{array}{rl} 400 & \text{Total points needed} \\ -\,315 & \text{Points accumulated} \\ \hline 85 & \text{Points needed on the final exam} \end{array}$$

Thus, you need at least a score of 85 on your final exam to earn a grade of B for the course. (**Note:** We will see an algebraic approach to solving problems of this type in Chapter 9.)

Note

Of the four statistics mentioned in this section, the mean and median are most commonly used. Many people feel that the mean (or arithmetic average) is relied on too much in reporting central tendencies. A few very large (or very small) data items can distort the mean as a picture of a central tendency. As you can see in the Group A data, the median of $27,500 is probably more representative of the data than the mean of $35,250. Note how the one income of $80,000 raises the mean considerably.

When you read an article in a magazine or newspaper that reports means or medians, you should now have a better understanding of the implications.

Now work margin exercise 4.

Completion Example Answers

4. a. 800 × 5 = 400 **b.** 85 + 78 + 82 + 70 = 315 **c.** 400 − 315 = 85; 85

Margin Exercise Answers

1. 102 minutes **2.** 87 **3. a.** Mode: 35 **b.** Range: 37 **4.** 97

7.1 Exercises

Concept Check

Fill-in-the-Blank. Complete each sentence using information found in this section.

1. The study of how to gather, organize, analyze, and interpret numerical information is called ________.

2. A particular measure or characteristic of a sample is called a/an ________.

3. The value(s) measuring some characteristic of interest is considered ________.

7.1 Exercises

Concept Check

1. statistics
2. statistic
3. data

4. The sum of all the numerical data divided by the number of data items is known as the _______ of the data.

5. The median is the middle term if there is a/an _____ number of items.

6. The median is the average of the two middle terms if there is a/an _______ number of items.

True/False. Determine whether each statement is true or false. If a statement is false, explain how it can be changed so the statement will be true. (**Note:** There may be more than one acceptable change.)

7. Data is the value(s) of a particular characteristic of interest such as number of people, weight, temperatures, or innings pitched.

8. The range is the difference between the first number listed in the data and the last number listed.

9. The number that appears the greatest number of times in a set of data is the sample.

10. The mean and median are never the same number.

4. mean

5. odd

6. even

7. True

8. False; The range is the difference between the largest and smallest numbers in a data set.

9. False; The number that appears the greatest number of times in a set of data is the mode.

10. False; The mean and median can be the same number.

Practice

For each set of data, find **a.** the mean, **b.** the median, **c.** the mode (if any), and **d.** the range. See Examples 1 through 3.

1. The ages of the first five US presidents on the date of their inaugurations were as follows. (The presidents were Washington, Adams, Jefferson, Madison, and Monroe.)

57, 61, 57, 57, 58

2. Dr. Wright recorded the following nine test scores for students in his statistics course.

95, 82, 85, 71, 65, 85, 62, 77, 98

3. Family incomes in a survey of eight students are as follows.

$35,000, $28,000, $42,000, $71,000,
$63,000, $36,000, $51,000, $63,000

4. Stacey went to six different auto repair shops to get the following estimates to repair her car.

$425, $525, $325, $300, $500, $325

5. Mike kept track of his golf scores for twelve rounds of eighteen holes each. His scores were as follows. Round to the nearest tenth, if necessary.

85, 90, 82, 85, 87, 80,
78, 82, 88, 82, 86, 81

Practice

1. **a.** 58 **b.** 57
c. 57 **d.** 4

2. **a.** 80 **b.** 82
c. 85 **d.** 36

3. **a.** $48,625 **b.** $46,500
c. $63,000 **d.** $43,000

4. **a.** $400 **b.** $375
c. $325 **d.** $225

5. **a.** 83.833 **b.** 83.5
c. 82 **d.** 12

6. a. 22 b. 19
c. 18 d. 21

7. a. 17.6 in. b. 14.9 in.
c. None d. 19.7 in.

8. a. 6.2 hours b. 6 hours
c. 6 hours d. 6 hours

9. a. 96,360 m³
b. 84,500 m³
c. None d. 82,700 m³

10. a. 12,000 mi
b. 12,000 mi
c. 14,000 mi d. 6000 mi

11. a. $4892.60 b. $5100
c. None d. $2890

12. a. $18,901.50
b. $18,731.5
c. None d. $3156

13. a. 22.3 b. 21
c. 23 d. 19

6. The city planning department issued the following numbers of building permits over a three-week period (15 business days).

17, 19, 18, 35, 30, 29, 23, 14,
18, 16, 20, 18, 18, 25, 30

7. On a one-day fishing trip, Mr. and Mrs. Milster recorded the following lengths of the ten fish they caught (measured in inches).

14.3, 13.6, 10.5, 15.5, 20.1,
10.9, 12.4, 25.0, 30.2, 23.5

8. Fifteen college students reported the following hours of sleep the night before an exam.

4, 6, 6, 7, 6.5, 6.5, 7.5, 8.5,
5, 6, 4.5, 5.5, 9, 3, 8

9. The volume (in thousands of cubic meters) of the world's 5 largest dams are as follows.

Atatürk Dam:	84,500
Fort Peck Dam:	96,000
Hourtribijk:	78,000
Oahe Dam:	70,300
Tarbela Dam:	153,000

10. The Big City fire department reported the following mileage for tires used on their nine fire trucks.

14,000, 14,000, 11,000, 15,000, 9000,
14,000, 12,000, 10,000, 9000

11. Resident tuition charged by 10 colleges are as follows. Round to the nearest tenth, if necessary.

$4088, $5175, $5076, $3639, $3756,
$6529, $4479, $5922, $5138, $5124

12. Nonresident tuition charged by 10 colleges are as follows.

$19,934, $19,000, $19,948, $18,236, $17,890,
$18,950, $17,764, $20,920, $17,860, $18,513

13. The following list contains the ages of 20 students surveyed in a college chemistry class.

18, 23, 23, 23, 22, 18, 21, 20, 18, 20,
19, 20, 21, 19, 23, 36, 35, 26, 17, 24

14. The local high school basketball team scored the following points per game during their 20-game season.

85,	60,	62,	70,	75,	52,	88,	50,	80,	72,
90,	85,	85,	93,	70,	75,	68,	73,	65,	82

15. The winning margin for each of the first 40 Super Bowls are as follows.

25,	19,	9,	16,	3,	21,	7,	17,
10,	4,	18,	17,	4,	12,	17,	5,
10,	29,	22,	36,	19,	32,	4,	45,
1,	13,	35,	17,	23,	10,	14,	7,
15,	7,	27,	3,	27,	3,	3,	11

16. The local weather station recorded the following daily high temperatures (in degrees Fahrenheit) for one month.

75°,	76°,	76°,	78°,	85°,	82°,
85°,	88°,	90°,	90°,	88°,	95°,
96°,	92°,	88°,	88°,	80°,	80°,
78°,	80°,	78°,	76°,	77°,	75°,
75°,	74°,	70°,	70°,	72°,	73°

17. The distances from Chicago to selected cities are as follows.

Boston:	980 miles
Cleveland:	345 miles
Dallas:	930 miles
Denver:	1050 miles
Detroit:	280 miles
Indianapolis:	190 miles
Los Angeles:	2110 miles
Miami:	1390 miles
New Orleans:	950 miles
San Francisco:	2210 miles
Seattle:	2050 miles

18. The numbers of passengers (to the nearest thousand) at 10 airports are as follows.

ATL (Atlanta):	101,489,000
PEK (Beijing):	89,938,000
DXB (Dubai):	78,010,000
ORD (Chicago):	76,942,000
HND (Tokyo):	75,316,000
LHR (London):	74,989,000
LAX (Los Angeles):	74,704,000
HKG (Hong Kong):	68,342,000
CDG (Charles de Gualle):	65,771,000
DFW (Dallas Fort Worth)	64,072,000

Source: "World Top 30 Airports", www.world-airport-codes.com

Super Bowl History

The first Super Bowl was held on January 15, 1967, and Roman numerals are used to refer to this annual event instead of the year. Two teams are tied with the most wins, with 6 wins each—the Pittsburgh Steelers (IX, X, XIII, XIV, XL, and XLIII) and New England Patriots (XXXVI, XXXVIII, XXXIX, XLIX, LI, and LIII). The New England Patriots also have the most appearances in the big game (11 total). There are four teams that have never appeared: the Cleveland Browns, the Detroit Lions, the Jacksonville Jaguars, and the Houston Texans.

14. a. 74 **b.** 74 **c.** 85 **d.** 43

15. a. 15.425 **b.** 14.5 **c.** 17 and 3 **d.** 44

16. a. 81° **b.** 79° **c.** 88° **d.** 26°

17. a. 1135 miles **b.** 980 miles **c.** None **d.** 2020 miles

18. a. 76,957,300 **b.** 75,152,500 **c.** None **d.** 37,417,000

19. a. 12,371,400
b. 11,872,500
c. None d. 6,955,000
20. a. 171.53 b. 56
c. 2, 4 d. 1522

Application

21. 79
22. 72

19. The number of volumes (to the nearest thousand) in college libraries in the United States is as follows.

Harvard University Library:	16,832,000
Yale University Library:	15,200,000
University of Michigan Library:	13,829,000
University of Illinois at Urbana-Champaign:	13,158,000
Columbia University:	12,200,000
University of California, Berkeley Libraries:	11,545,000
Kings College Libraries:	11,189,000
University of Texas Libraries:	9,990,000
Indiana University Library:	9,934,000
University of Chicago Library:	9,837,000

20. The Medal of Honor is the nation's highest military awarded for uncommon valor by men and women in battle. The numbers of medals awarded by war are as follows.

War	Medals	War	Medals
Civil War:	1523	World War I:	126
Korean Expedition (1871):	15	Occupation of Nicaragua:	2
Indian Wars:	426	World War II:	417
Spanish-American War:	110	Korean War:	145
Second Samoan Civil War:	4	Vietnam War:	258
Philippine-American War:	86	USS *Liberty* Incident:	1
Boxer Rebellion:	59	Battle of Mogadishu:	2
Occupation of Veracruz:	56	Iraq War:	4
United States Occupation of Haiti:	8	War in Afghanistan:	14
Dominican Republic Occupation:	3		

Source: Congressional Medal of Honor Society

Applications

Solve.

21. Suppose that you are to take four hourly exams and a final exam in your chemistry class. Each exam has a maximum of 100 points and you must average between 75 and 82 points to receive a passing grade of C. If you have scores of 83, 65, 70, and 78 on the hourly exams, what is the minimum score you can make on the final exam and receive a grade of C? (First, explain your strategy in solving this problem. Then, solve the problem.)

22. Suppose that the instructor in the class in Exercise 21 has informed the class that the lowest of your hourly exam scores will be replaced by your score on the final exam, provided that your final exam score is higher. (That is, the final exam score may be counted twice.) Now, what is the minimum score that you can make on the final exam and still receive a grade of C? (First, explain your strategy in solving this problem. Then, solve the problem.)

23. The number of farms in the United States is decreasing. The following graph shows the number of farms (in millions) for each decade since 1940. As you can tell from the graph, the number of farms seems to be leveling off somewhat.

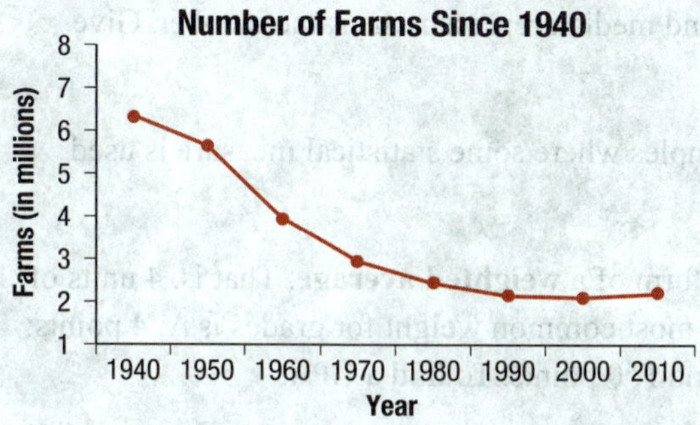

1940: 6.35	**1950:** 5.65
1960: 3.96	**1970:** 2.95
1980: 2.44	**1990:** 2.15
2000: 2.11	**2010:** 2.2

a. Find the mean number of farms from 1940 to 2010.

b. Find the mean number of farms from 1980 to 2010.

24. Financial institutions sometimes fail, as they did during the Great Recession in the US (which lasted from December 2007 through June 2009). The effects of this recession was felt for several years, as indicated by the number of bank failures from 2007 through 2016 (as reported by the Federal Deposit Insurance Company). Find the **a.** mean, **b.** median, **c.** mode, and **d.** range over these ten years.

2007: 3	**2008:** 25	**2009:** 140	**2010:** 157	**2011:** 92
2012: 51	**2013:** 24	**2014:** 18	**2015:** 8	**2016:** 5

25. The armed forces of countries are measured in several ways: active troops, reserve troops, tanks, navy (carriers, cruisers, frigates, destroyers, submarines), and combat aircraft. In 2017, the following countries had active troops in the following numbers according to the International Institute for Strategic Studies.

Afghanistan:	171,200
Egypt:	438,500
China:	2,183,000
France:	202,950
United States:	1,347,300
Chile:	64,750

a. What was the mean number of active troops for these countries? Round your answer to the nearest tenth.

b. What was the median number of active troops for these countries?

26. Each state pays its governor a salary. During 2020, the governors' salaries from ten states were as follows. Find **a.** the mean, **b.** median, **c.** mode, and **d.** range for these ten salaries.

Pennsylvania:	$201,729
Georgia:	$175,000
Vermont:	$184,100
New York:	$225,000
California:	$209,747
Illinois:	$181,670
Tennessee:	$198,780
Massachusetts:	$185,000
New Jersey:	$175,000
Washington:	$182,179

23. a. 3.48 million farms
b. 2.23 million farms

24. a. 52.3 **b.** 24.5
c. None **d.** 154

25. a. 734,616.7
b. 320,725

26. a. $168,594
b. $173,449
c. $175,000 **d.** $37,251

Writing & Thinking

27. State how to determine the median of a set of data.

28. Determine whether or not the mean and median represent the same number. Give examples to justify your answer.

29. Give three different and specific examples where some statistical measure is used (outside of a class).

30. Your grade point average (GPA) is a form of a **weighted average**. That is, 4 units of A counts more than 4 units of B. The most common weight for grades is A, 4 points; B, 3 points; C, 2 points; D, 1 point; and F, 0 points. To find a GPA,

1. Multiply the points for each grade by the number of units for the course.
2. Find the sum of these products.
3. Divide this sum by the total number of units taken.

Find the GPA (to the nearest tenth) for each of the following situations:

a. 3 units of A in astronomy, 4 units of B in geometry, 3 units of C in sociology, and 4 units of A in biology.

b. 5 units of B in history, 4 units of C in calculus, 3 units of A in computer science, and 4 units of D in geology.

c. Your own GPA for the last semester (or your anticipated GPA for this semester).

Collaborative Learning

31. With the class separated into teams of 2 to 4 students, each team is to go on campus and survey 50 students and ask each student how many minutes it takes him or her to drive to school from home.

a. Each team is to find the mean, median, mode, and range for the 50 responses.

b. Each team is to bring all the data to class, and the class is to pool the information and find the mean, median, mode, and range for the pooled data.

c. The class is to discuss the results of the individual teams and the pooled data and what use such information might have for the administration of the college.

Writing & Thinking

27. The first step to finding the median is always to arrange the data in order. The median is the middle number. If there is an even number of item, average the two middle numbers to find the median.

28. The mean and median may be the same number as in the set of data: 12, 16, 20. However, more generally, the mean and median are two different numbers as in the set of data: 22, 23, 36. In this set, the median is 23, while the mean is 27.

29. Answers will vary. Examples include: median income or home value for a particular area, mean temperature at a certain level in the ocean, mode of days having a particular heat index in a particular geographic location, or the range of SAT or ACT scores for a graduating class.

30. **a.** 3.3 **b.** 2.4

c. Answers will vary.

Collaborative Learning

31. Answers will vary.

7.2 Reading Graphs

A Introduction to Graphs

Graphs are pictures of numerical information. Graphs appear almost daily in newspapers and magazines and frequently in textbooks and corporate reports. Well-drawn graphs can organize and communicate information accurately, effectively, and fast. Most computers can be programmed to draw graphs, and anyone whose work involves a computer in any way will probably be expected to understand graphs and even to create graphs.

There are many different types of graphs, each type particularly well-suited to the display and clarification of certain types of information. There have been various graphs used throughout the text. In this section, we will discuss in more detail the uses of bar graphs, circle graphs, line graphs, and histograms.

Four Types of Graphs and Their Purposes

1. **Bar Graphs:** To emphasize comparative amounts
2. **Circle Graphs:** To help in understanding percents or parts of a whole (Circle graphs are also called **pie charts**.)
3. **Line Graphs:** To indicate tendencies or trends over a period of time
4. **Histograms:** To indicate data in **classes** (a range or interval of numbers)

DEFINITION

A common characteristic of all graphs is that they are intended to communicate information about numerical data quickly and easily. With this in mind, note the following three properties of all graphs.

Properties of Graphs

Every graph should

1. be clearly labeled,
2. be easy to read, and
3. have an appropriate title.

PROPERTIES

Objectives

A. Learn the purposes and properties of graphs.

B. Read bar graphs.

C. Read circle graphs.

D. Read line graphs.

E. Read histograms.

Math Tip

Not all graphs are created equal. Since the purpose of a graph is to communicate information in a concise manner, be sure that the type of graph chosen presents the information clearly. For example, if a bar graph or a circle graph might work, consider whether you want to compare total amounts (bar graph) or if you want to compare how the data is made up as a whole (circle graph).

B Bar Graphs

1. Use the bar graph in Example 1 to answer the following questions.

a. What were the sales in March?

b. What were the sales during the lowest-sales month?

c. What was the amount of decrease in sales between January and February?

d. What was the percent decrease in sales between January and February (to the nearest tenth of a percent)?

Example 1 Reading a Bar Graph

Examine the bar graph and answer the questions given. Note that the scale on the left (sales) and the categories at the bottom (months) are clearly labeled and the graph itself has a title.

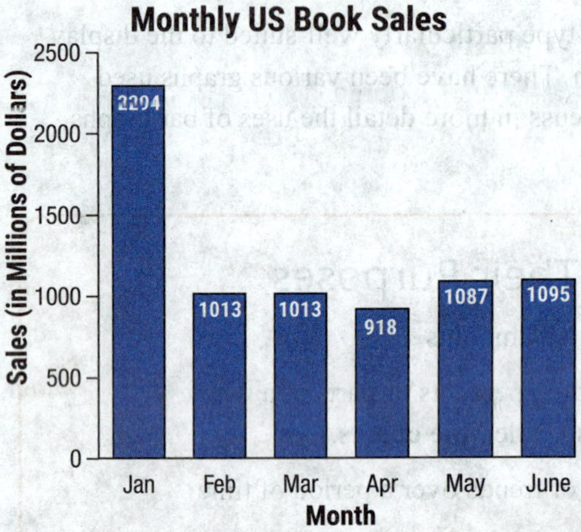

a. What were the sales in February?

b. During what month were sales lowest?

c. During what month were sales highest?

d. What were the sales during the highest-sales month?

e. What were the sales in June?

f. What was the amount of increase in sales between April and May?

g. What was the percent increase in sales from April to May (to the nearest tenth of a percent)?

Solution

a. Look at the bar for February and read to the left to find the amount in sales (or read the number at the top of the bar). Note that the scale on the left of the graph says that sales are in millions of dollars. This gives an answer of $1013 million.

b. To determine which month had the lowest amount of sales, look for the shortest bar. The month that sales were lowest is April.

c. To determine which month had the highest amount of sales, look for the tallest bar. The month that sales were highest is January.

d. Sales were highest in January. Looking at the top of the bar and reading to the left gives sales of $2294 million.

e. To find the sales in June, look at the bar for June and read over to the left. Doing this gives an answer of $1095 million.

f. To determine the amount of increase in sales from April to May, first read the sales for April ($918 million). Then read the sales for May ($1087 million). Subtracting April sales from May sales gives a difference of $169 million.

	$1087	May sales
−	918	April sales
	$ 169	Increase in sales

g. To find the percent of increase in sales from April to May take the amount of increase in sales (found in part f. above) and divide by the sales in April.

$$\frac{\text{Increase in April sales}}{\text{April sales}} \quad \frac{\$169 \text{ million}}{\$918 \text{ million}} = 0.184095... \approx 18.4\% \text{ increase}$$

Now work margin exercise 1.

C Circle Graphs

Example 2 Reading a Circle Graph

Examine the circle graph. This graph shows the percent of a household's annual income they plan to budget for various expenses. Suppose the household has an annual income of \$45,000. Use the information in the graph to calculate how much money will be budgeted for each expense.

Household Budget for One Year

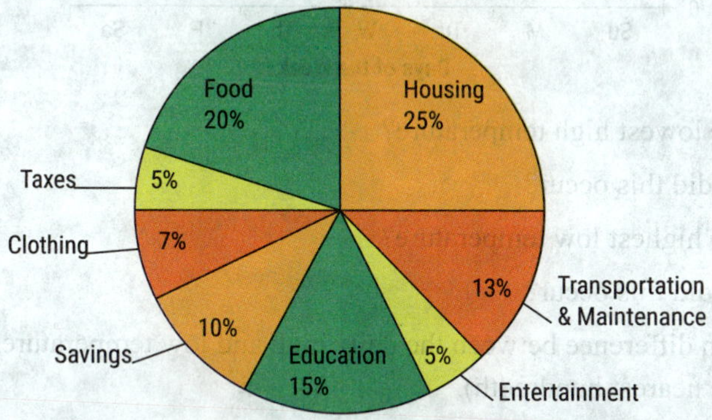

Solution

To calculate the budget amount for each item multiply the corresponding percent by the total household income.

Item		Amount
Housing	$0.25 \times \$45{,}000 =$	\$11,250
Food	$0.20 \times \$45{,}000 =$	\$9000
Taxes	$0.05 \times \$45{,}000 =$	\$2250
Clothing	$0.07 \times \$45{,}000 =$	\$3150
Savings	$0.10 \times \$45{,}000 =$	\$4500
Education	$0.15 \times \$45{,}000 =$	\$6750
Entertainment	$0.05 \times \$45{,}000 =$	\$2250
Transportation & Maintenance	$0.13 \times \$45{,}000 =$	\$5850
	What is the total of all amounts?	\$45,000

Now work margin exercise 2.

2. Using the circle graph in Example 2, how much will the family spend on each of the following expenses if the family income increases to \$55,000 and the percent budgeted for each expense does not change?

a. Housing

b. Savings

c. Clothing

d. Food

D Line Graphs

Example 3 Reading a Line Graph

Examine the line graph. This graph shows the relationships between daily high and low temperatures. You can see that temperatures tended to rise during the week but fell sharply on Saturday.

3. Use the line graph in Example 3 to answer the following questions.

a. What was the high temperature for the week (highest daily high)?

b. What was the low temperature for the week (lowest daily low)?

c. Find the mean low temperature for the week (to the nearest hundredth).

d. Find the maximum daily difference between daily high and low temperatures in a single day for the week.

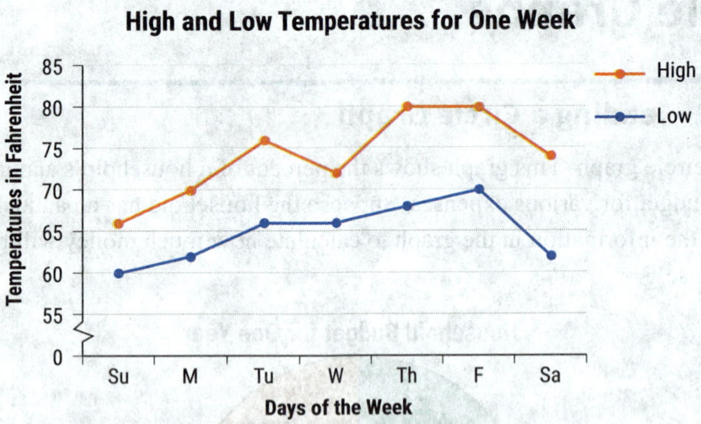

Note

In the line graph, notice that the temperature scale on the left does not start at 0 °F. There is a break indicated in that scale.

a. What was the lowest high temperature?

b. On what day did this occur?

c. What was the highest low temperature?

d. On what day did this occur?

e. Find the mean difference between the daily high and low temperatures for the week shown (to the nearest hundredth).

Solution

a. To find the lowest high temperature, look at the line labeled "High" and find the lowest point on that line graph. This shows that the lowest high temperature is <u>66 °F</u>.

b. Looking directly down from the point located in part a., we see that the lowest high temperature occurred on <u>Sunday</u>.

c. To find the highest low temperature, look at the line labeled "Low" and find the highest point on that line graph. This shows that the highest low temperature is <u>70 °F</u>.

d. Looking directly down from the point located in part c., we see that the highest low temperature occurred on Friday.

e. First, find the sum of the differences:

Sunday:	66 − 60	=	6 °F
Monday:	70 − 62	=	8 °F
Tuesday:	76 − 66	=	10 °F
Wednesday:	72 − 66	=	6 °F
Thursday:	80 − 68	=	12 °F
Friday:	80 − 70	=	10 °F
Saturday:	74 − 62	=	12 °F
			64 °F Sum of differences

Then divide the sum of the differences by the number of days to find the mean of these differences.

$$64 \div 7 = 9.142\ldots \approx 9.14$$

Thus, the mean difference between the daily high and low temperatures is approximately 9.14 °F.

Now work margin exercise 3.

E Histograms

For bar graphs (see Example 1), we label the base line for each bar with individual names for people, months, days of the week, or other categories. Another type of bar graph is called a **histogram**. In a histogram, the base line is labeled with numbers that indicate the boundaries of ranges (or intervals) of numbers. This range of numbers is called a **class** (each bar on a histogram represents a class). The bars are placed next to each other with no space between them.

Terms Related to Histograms

Class: A range (or interval) of numbers that contains data items.

Lower class limit: The smallest number that belongs to a class.

Upper class limit: The largest number that belongs to a class.

Class boundaries: Numbers that are halfway between the upper limit of one class and the lower limit of the next class.

Class width: The difference between the class boundaries of a class (the width of each bar).

Frequency: The number of data items in a class.

DEFINITION

Math Tip

Some advantages of a histogram over a bar graph are that histograms show the frequency of occurrences of different data values, and the overall shape of the graph can provide additional insight into the data set. Some advantages of a bar graph over a histogram are that a bar graph can display categorical or numerical data and the order of the categories can be rearranged.

Example 4 Reading a Histogram

Examine the following histogram. This histogram summarizes the scores of 50 students on an English placement test. Refer to the graph to answer the given questions. Assume the scores are whole numbers.

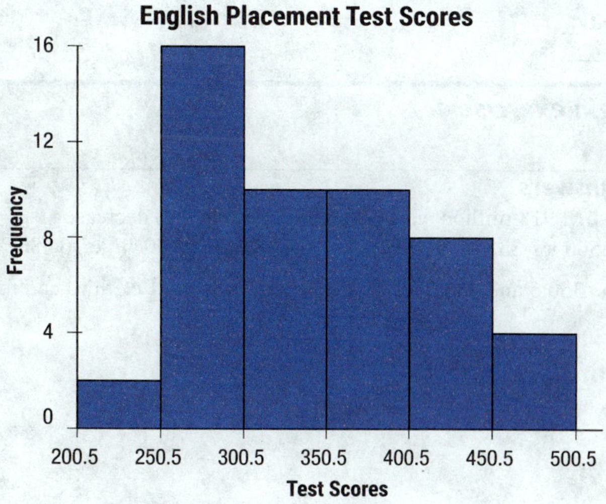

a. How many classes are represented?

b. What are the class limits of the first class?

c. What are the class boundaries of the second class?

d. What is the width of each class?

4. Use the histogram in Example 4 to answer the following questions.

a. What are the class limits of the fifth class?

b. What are the class boundaries of the fourth class?

c. Which class has the least frequency?

d. What percent of the scores are below 300.5?

e. Which two classes appear to be equal?

e. Which class has the greatest frequency?

f. What is this frequency?

g. What percent of the scores are between 200.5 and 250.5?

h. What percent of the scores are above 400.5?

Solution

a. Each bar on the histogram represents a class. Thus, in this histogram, there are 6 classes.

b. The class limits consist of the smallest number and the largest number that belong to a class. The first class has a lower class limit of 201 and an upper class limit of 250.

c. The class boundaries are listed on the horizontal axis of the graph. Looking at the second class, or second bar, we see that the lower class boundary is 250.5 and the upper class boundary is 300.5.

d. Each class has the same width and this is found by subtracting the lower class boundary from the upper class boundary of any class. Using the class boundaries just found for the second class gives 300.5 – 250.5 = 50. Thus, the class width of each class is 50.

e. To find the class with the greatest frequency, look for the tallest bar. We see that the second class has the greatest frequency.

f. Looking at the bar for the second class and then to the left, we see that the frequency of the second class is 16.

g. The graph shows that there are 2 scores between 200.5 and 250.5. We want to divide this number by the total number of scores: 2 + 16 + 10 + 10 + 8 + 4 = 50.

Now dividing 2 by 50 gives $\frac{2}{50} = 0.04 = 4\%$.

h. Looking at the graph we see that the fifth and sixth classes represent scores above 400.5. The frequencies of these classes are 8 and 4. Now dividing the total of these classes by the total number of scores gives the percent of scores greater than 400.5: $\frac{12}{50} = 0.24 = 24\%$.

Now work margin exercise 4.

Margin Exercise Answers

1. a. \$1013 million **b.** \$918 million **c.** \$1281 million **d.** 55.8% decrease
2. a. \$13,750 **b.** \$5500 **c.** \$3850 **d.** \$11,000 **3. a.** 80 °F **b.** 60 °F **c.** 64.86 °F **d.** 12 °F
4. a. 401 and 450 **b.** 350.5 and 400.5 **c.** first class **d.** $\frac{18}{50} = 36\%$ **e.** third and fourth classes

7.2 Exercises

Concept Check

Fill-in-the-Blank. Complete each sentence using information found in this section.

1. Pictures of numerical information are _______.
2. To emphasize comparative amounts, the best way to present the information is to use ______ graphs.
3. To help understand percentages or parts of a whole, the preferred method is to use _______ graphs.
4. A/An _______ graph is used to indicate tendencies or trends over time.
5. The largest whole number that belongs to a class in a histogram is the ________ class limit.
6. The number of data items in a class is the ____________.

True/False. Determine whether each statement is true or false. If a statement is false, explain how it can be changed so the statement will be true. (**Note:** There may be more than one acceptable change.)

7. Graphs should always be clearly labeled, easy to read, and have appropriate titles.
8. Circle graphs show trends over a period of time.
9. The frequency is the number of data items in a class.
10. Numbers that are halfway between the upper limit of one class and the lower limit of the next class are the class boundaries.

Applications

Answer the questions using the given graphs.

1. The following bar graph shows the number of students in five fields of study at a university.

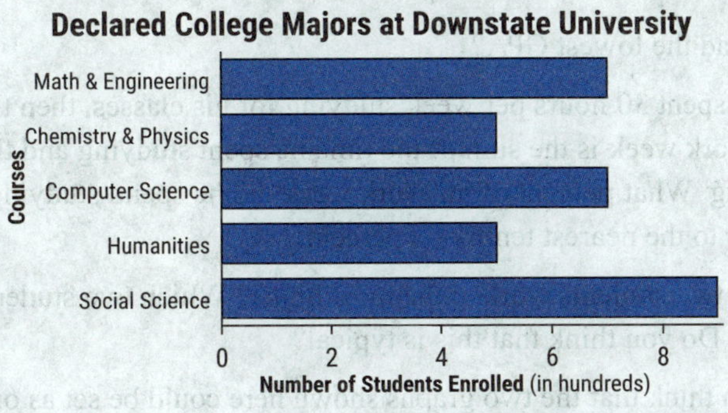

   a. Which field(s) of study has the largest number of declared majors?

   b. Which field(s) of study has the smallest number of declared majors?

   c. How many declared majors are indicated in the entire graph?

   d. What percent are computer science majors? Round your answer to the nearest tenth of a percent.

7.2 Exercises

Concept Check

1. graphs
2. bar
3. circle
4. line
5. upper
6. frequency
7. True
8. False; Line graphs show trends over a period of time.
9. True
10. True

Applications

1. **a.** Social Science
   b. Chemistry & Physics, Humanities
   c. About 3300
   d. About 21.2%

2. a. Second Monday, 800
b. 150 c. 6300
d. About 12.7%

3. a. Sue b. Bob and Sue
c. 85.7%
d. Bob and Sue, Bob and Sue, Yes, in most cases
e. No, the vertical scales represent two different types of quantities

2. The following bar graph shows the number of vehicles that crossed one intersection during a two-week period.

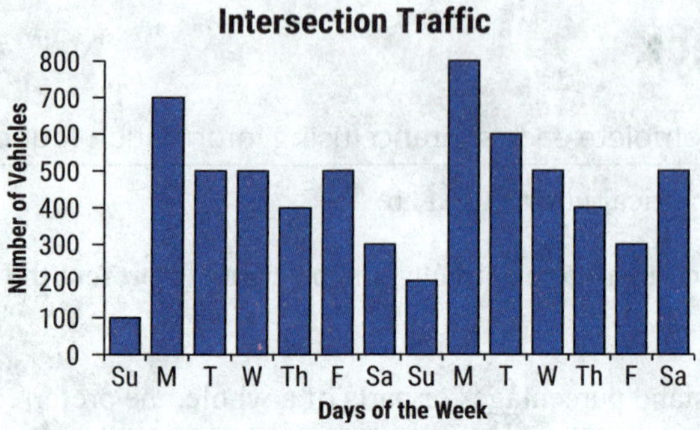

a. On which day did the highest number of vehicles cross the intersection? How many crossed that day?

b. What was the mean number of vehicles that crossed the intersection on the two Sundays?

c. What was the total number of vehicles that crossed the intersection during the two weeks?

d. About what percent of the total traffic was counted on Saturdays? Round your answer to the nearest tenth of a percent.

3. The following bar graphs show the number of hours worked each week and the GPA's of five college students. When comparing the following two graphs, assume that all five students graduated with comparable grades from the same high school.

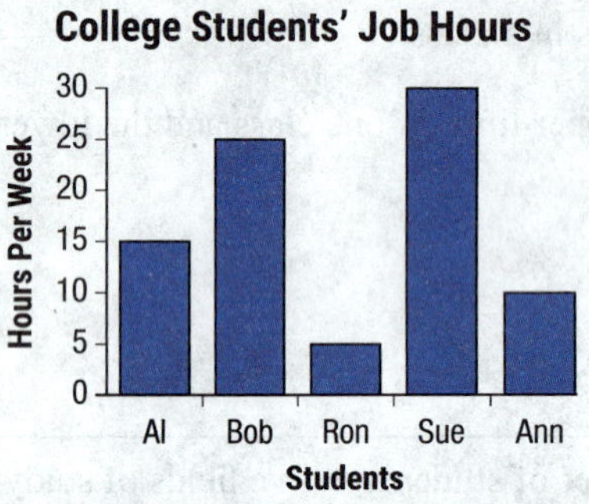

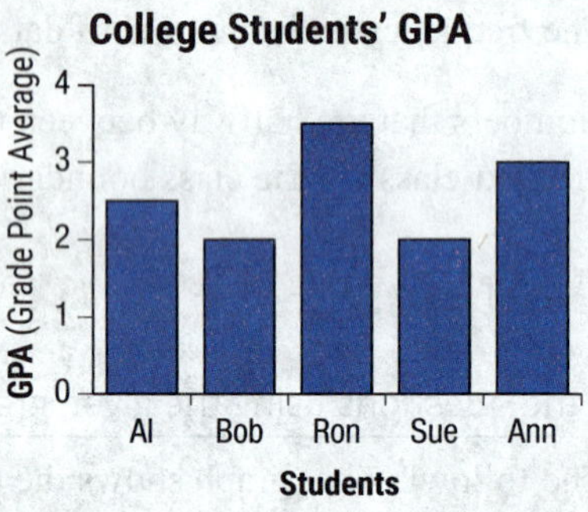

a. Who worked the most hours per week?

b. Who had the lowest GPA?

c. If Ron spent 30 hours per week studying for his classes, then the length of his total work week is the sum of the time he spent studying and the time he spent working. What percent of his work week did he spend studying? Round your answer to the nearest tenth of a percent.

d. Which two students worked the most hours? Which two students had the lowest GPAs? Do you think that this is typical?

e. Do you think that the two graphs shown here could be set as one graph? If so, show how you might do this.

4. The following circle graph represents the various areas of spending for a school with a total budget of $34,500,000.

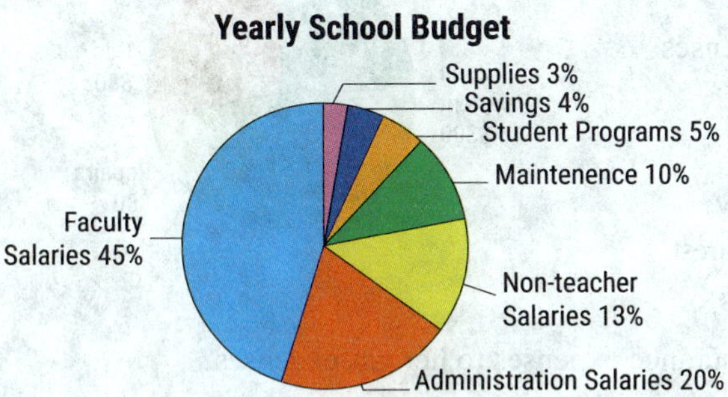

a. What amount will be allocated to each category?

b. What percent will be for expenditures other than salaries?

c. How much will be spent on maintenance and supplies?

d. How much more will be spent on teachers' salaries than on administration salaries?

5. The following circle graph represents the types of shows broadcast on television station KCBA. The station is off the air from 2 a.m. to 6 a.m., so they have only 20 hours of daily programming. Sports are not shown in the graph because they are considered special events.

a. In the 20-hour period shown, how much time (in minutes) is devoted daily to each category?

b. What category has the most time devoted to it?

c. How much total time (in minutes) is devoted to drama, reality TV, and sitcoms?

6. Mike just graduated from college and decided that he should try to live within a budget. The circle graph shows the categories he chose and the percents he allowed. His beginning take home salary is $24,000.

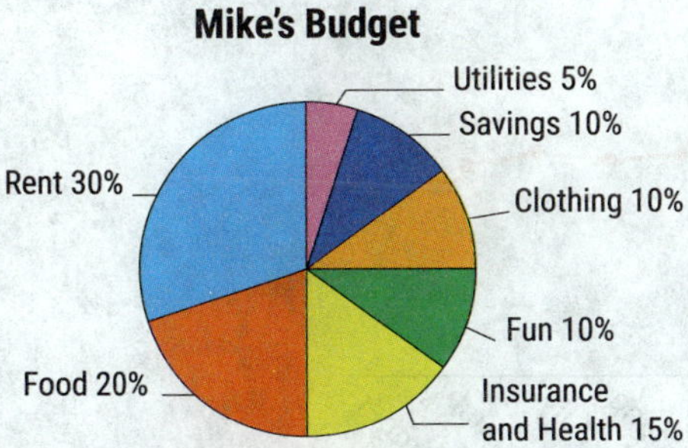

a. How much did he budget for each category?

b. What category was smallest in his budget?

c. What total amount did he budget for food, clothing, and rent?

7. The following circle graph represents the various sources of income for a city government with a total income of $100,000,000.

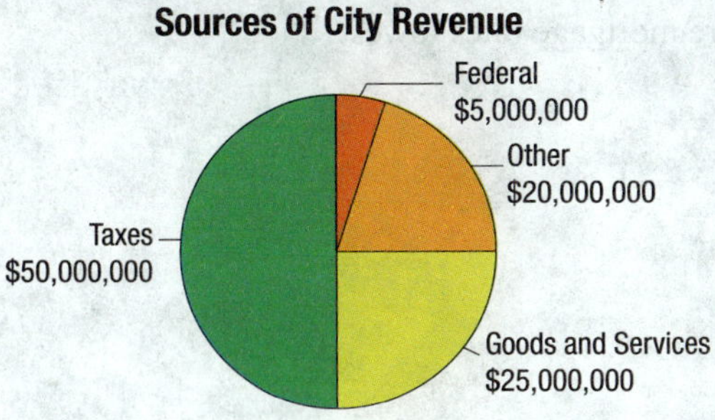

a. What is the city's largest source of income?

b. What percent of income comes from good and services?

c. What is the ratio of income from taxes to the total income?

4. **a.** Faculty salaries: $15,525,000; Administration salaries: $6,900,000; Student programs: $1,725,000; Savings: $1,380,000; Non-teacher salaries: $4,485,000; Maintenance: $3,450,000; Supplies: $1,035,000
b. 22% **c.** $4,485,000
d. $8,625,000

5. **a.** News: 300 min; Movies: 120 min; Sitcoms: 156 min; Reality TV: 180 min; Drama: 144 min; Children's shows: 120 min; Commercials: 180 min
b. News **c.** 480 min

6. **a.** Rent: $7200; Food: $4800; Insurance and Health: $3600; Clothing: $2400; Utilities: $1200; Fun: $2400; Savings: $2400
b. Utilities **c.** $14,400

7. **a.** Taxes **b.** 25% **c.** 1:2

8. a. $530
b. Repairs: 15.1%; Gas: 15.1%; Insurance 13.2%; Loan Payment: 56.6%
c. 7:8

9. a. February and May
b. 6 inches c. March
d. 3.58 inches

10. a. April b. September

8. The following circle graph shows Sally's car expenses for the month of June.

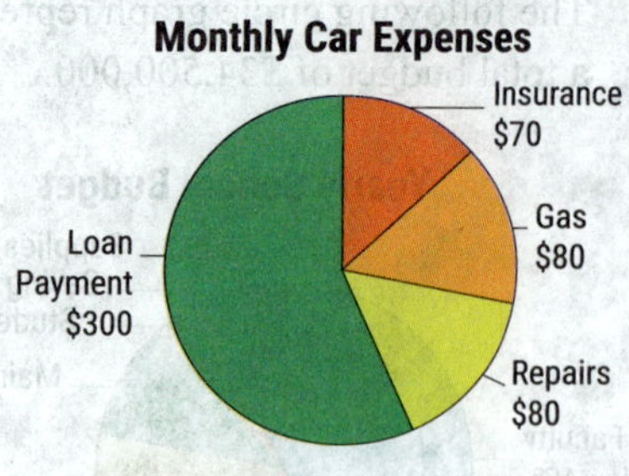

a. What were her total car expenses for the month?

b. What percent of her expenses did she spend on each category? Round your answers to the nearest tenth of a percent.

c. What was the ratio of her insurance expenses to her gas expenses?

9. The following line graph shows the total monthly rainfall in a certain city over five months.

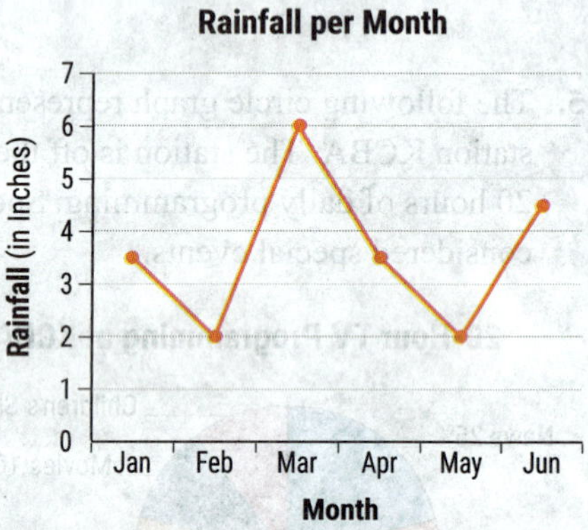

a. Which months had the least rainfall?

b. What was the most rainfall in a month?

c. What month had the most rainfall?

d. What was the mean rainfall over the six-month period (to the nearest hundredth)?

10. The following line graph shows the average percent of product returns experienced at a new shoe store during the first nine months of business.

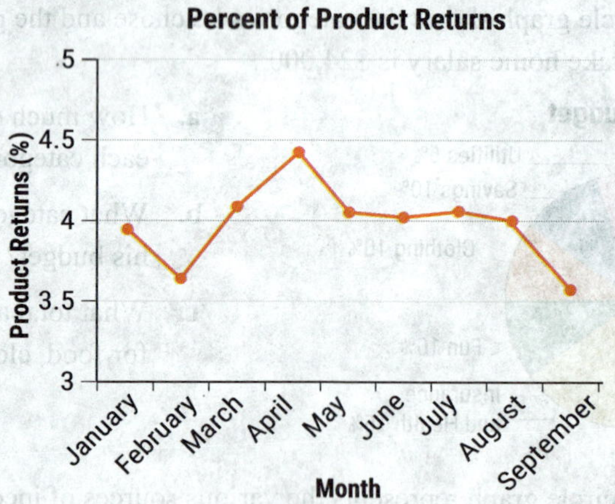

a. During what month or months were mortgage rates highest?

b. During what month or months were mortgage rates lowest?

11. The following line graph shows the number of home runs hit each month of a baseball season by two players, Mark and Nelson.

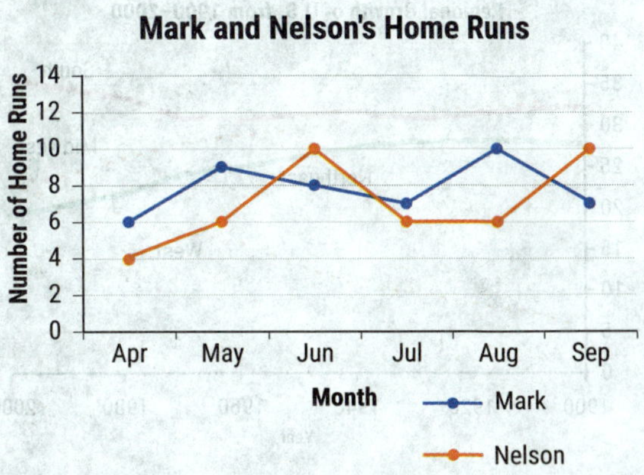

a. During which month did Mark hit the most home runs?

b. How much higher was Mark's total for that month than for his lowest month?

c. In what months did Nelson hit fewer home runs than Mark?

d. What was the difference between Nelson and Mark's home runs in July?

e. What percent of Nelson's total home runs did he hit in May?

f. What percent of Mark's home runs did he hit in April? Round your answer to the nearest percent.

12. The following line graphs show the stock market prices for oil, steel, and wheat over the course of a week. Assume that on Monday morning you had 100 shares of each of the three stocks shown.

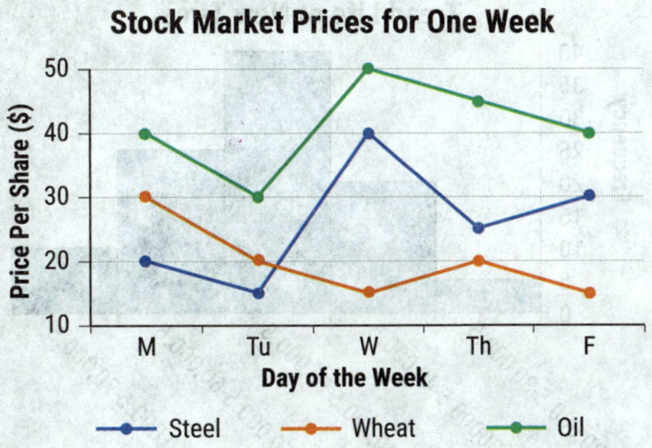

a. If you held the stock all week, on which stock would you have lost money?

b. How much money would you have lost on that stock?

c. On which stock would you have gained money?

d. How much money would you have gained on that stock?

e. On which stock could you have made the most money if you had sold it at the best time?

f. How much money could you have made had you sold that stock at the best time?

11. a. August **b.** 4
c. April, May, July, August
d. 1 **e.** 14% **f.** 13%

12. a. Wheat **b.** $1500
c. Steel **d.** $1000
e. Steel **f.** $2000

13. a. West: 5%; Northeast: 28%; Midwest: 35%; South: 32%
b. West: 22%; Northeast: 19%; Midwest: 23%; South: 36%
c. South d. 5%
e. West f. 5%, 1900
g. 36%, 2000
h. Midwest

14. a. 5 b. 10,000
c. Third class d. 40
e. 30,000.5 and 40,000.5
f. 100 g. 5% h. 35%

13. The following line graph shows the change in the percent of the US population living in each of four major regions during the twentieth century. [1]

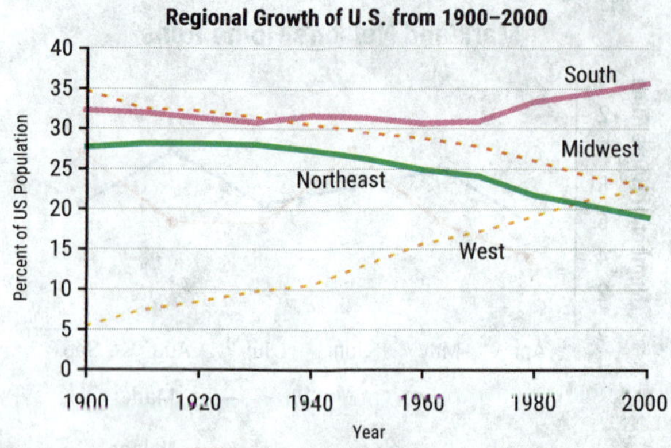

a. Approximately what percent of the population was in each of the four regions in 1900?

b. Approximately what percent of the population was in each of the four regions in 2000?

c. Which region seems to have had the most stable percent of the population between 1900 and 2000?

d. What is the difference between the highest and lowest percent for this region?

e. Which region has had the most growth?

f. What was its lowest percent and when?

g. What was its highest percent and when?

h. Which region has had the most decline?

14. The following histogram summarizes the tread life for 100 types of new tires.

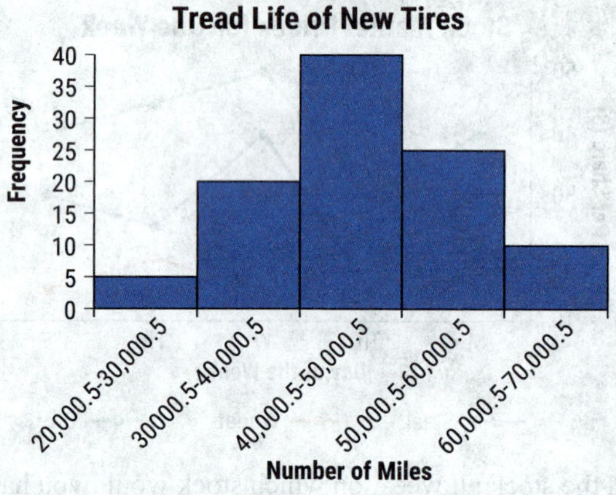

a. How many classes are represented?

b. What is the width of each class?

c. Which class has the highest frequency?

d. What is this frequency?

e. What are the class boundaries of the second class?

f. How many tires were tested?

g. What percent of tires were in the first class?

h. What percent of the tires lasted more than 50,000 miles?

1 Source: U.S. Census Bureau, decennial census of population, 1900 to 2000

15. A certain number of new cars were evaluated to find how many miles per gallon could be driven with a gallon of gas. The data is summarized in the following histogram. Assume the miles per gallon are whole numbers.

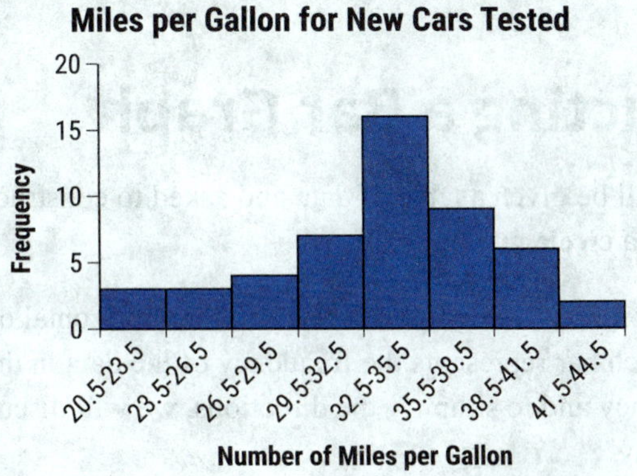

   a. How many classes are represented?
   b. What is the class width?
   c. Which class has the smallest frequency?
   d. What is this frequency?
   e. What are the class limits for the third class?
   f. How many cars were tested?
   g. How many cars tested below 30 miles per gallon?
   h. What percent of the cars tested about 38 miles per gallon?

15. **a.** 8 **b.** 3 **c.** Eight class **d.** 2 **e.** 27, 29 **f.** 50 **g.** 10 **h.** 16%

Writing & Thinking

16. State three properties or characteristics that should be true of all graphs so that they can communicate numerical data quickly and easily.

17. List the four types of graphs discussed in the text and briefly give the purpose of each.

18. Give three different and specific examples where graphs are used (outside of a class).

19. Compare and contrast a bar graph and a histogram.

Writing & Thinking

16. All graphs should be **1.** clearly labeled, **2.** easy to read, and **3.** have appropriate titles.

17. The four types of graphs are **1.** bar graphs, used for comparative amounts; **2.** circle graphs, also known as pie charts, to help understand percents or parts of a whole; **3.** line graphs, to indicate tendencies or trends over time; and **4.** histograms to indicate data in a range or interval of numbers (called classes).

18. Examples will vary. Pie charts/circle graphs are used when indicating what percent of the population uses public transportation versus driving a car or walking. Bar graphs are used to indicate how many people polled intend to vote for a particular political candidate. Line graphs are used to show the change in a person's financial portfolio. Histograms can show numbers of people making a particular range of income.

19. Both bar graphs and histograms use bars to indicate information. A bar graph uses categories and has spaces between the bars to separate the categories. A histogram uses boundaries of intervals instead of categories and the bars have no space between the classes.

Objectives

A. Construct a bar graph from a set of data.

B. Construct a circle graph from a set of data.

7.3 Constructing Graphs from Databases

A Constructing a Bar Graph

In this section, you will be given a table of data and asked to construct your own graph, either a **bar graph** or a **circle graph**.

Note

All of the graphs discussed here can be created with a computer and a spreadsheet program such as Microsoft Excel. If you have access to a computer, your instructor may choose to have you work the problems in this section with a spreadsheet program.

As illustrated in Section 7.2, a bar graph may have either horizontal or vertical bars. In both cases, the length of each bar represents the frequency of the data in the category being graphed. For consistency and to simplify the directions, we will discuss the construction of vertical bar graphs only.

Steps to Follow in Constructing a Vertical Bar Graph

1. Draw a vertical axis and a horizontal axis. Label these axes appropriately.
2. Mark an appropriate scale on the vertical axis to represent the frequency of each category. (The scale must be uniform. That is, the distance between consecutive marks must represent the same amount.)
3. Mark the categories of data along the horizontal axis.
4. Draw the vertical bar for each category so that the height of the bar reaches the frequency of the data in that category. (**Note:** Make sure the bars have the same width.)
5. Give the graph a title.

PROCEDURE

1. Construct a bar graph that represents the following data.

GPA of 5 College Students at the End of the Semester

Student	GPA
Ian	2.50
Marcus	3.25
Luke	3.00
Jenna	1.75
Denise	3.75

Example 1 Constructing a Bar Graph

Construct a bar graph that represents the following population data.

Population of the 8 Largest US Cities in the US 2020 Census (to the nearest thousand)

	City	Population
1.	New York, NY	8,253,000
2.	Los Angeles, CA	3,970,000
3.	Chicago, IL	2,678,000
4.	Houston, TX	2,316,000
5.	Phoenix, AZ	1,708,000
6.	Philadelphia, PA	1,578,000
7.	San Antonio, TX	1,567,000
8.	San Diego, CA	1,422,000

Source: www.census.gov

Solution

Step 1 and 2: Draw the vertical axis and horizontal axis and mark a scale on the vertical axis that will encompass the numbers from 0 to 8.253 million people. (On this graph, we have chosen to mark the numbers from 0 to 9.0 in a scale of 1 unit.)

Step 3 and 4: The horizontal axis marks are labeled with the names of the cities represented in alphabetical order. The height of each vertical bar corresponds to the population (in millions) of each city as given.

Step 5: Give the graph a title.

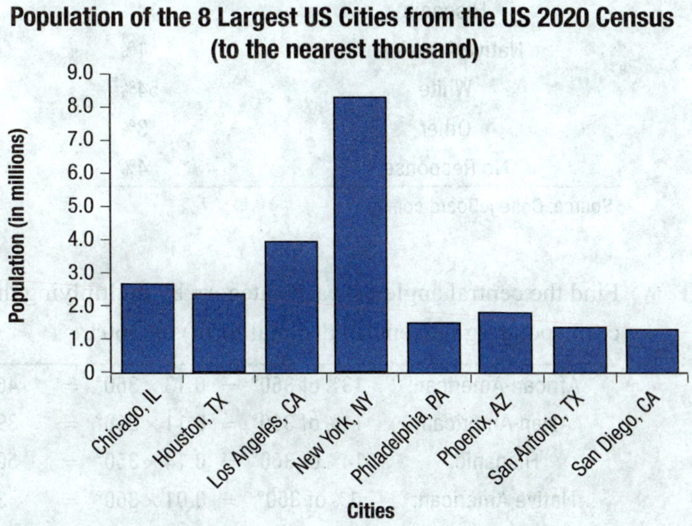

Now work margin exercise 1.

B Constructing a Circle Graph

A **circle graph** (or **pie chart**) is a circle that is marked in **sectors** (pie-shaped wedges) that correspond to percentages of data in each category represented. Refer to Section 6.1 for information on angles and protractors and Section 6.4 for information on circles.

If a radius of a circle is rotated completely around the circle (like the motion of the minute hand of a clock in 1 hour), the radius is said to rotate 360°. Therefore, to find the central angle needed in the graph to represent a percent of the data, multiply that percent (in decimal form) by 360°.

Steps to Follow in Constructing a Circle Graph

1. Find the central angle (angle at the center of the circle) for each category by multiplying the corresponding percent (in decimal form) by 360°.
2. Draw a circle.
3. Draw each central angle (use a protractor).
4. Label each sector with the name and corresponding percent of each category.
5. Give the graph a title.

PROCEDURE

2. Construct a circle graph that represents the following data.

2018 School Budget for Richland County

Budget Item	%
Faculty Salaries	40%
Administration	20%
Student Programs	15%
Non-teacher Salaries	12%
Maintenance	10%
Supplies	3%

Example 2 Constructing a Circle Graph

Construct a circle graph that represents the following data.

Ethnic Breakdown of Students Who Took the SAT (Scholastic Assessment Test), Nationwide, 2010

Ethnicity	% SAT Taken
African-American	13%
Asian-American	11%
Hispanic	14%
Native American	1%
White	54%
Other	3%
No Response	4%

Source: CollegeBoard.com

Solution

Step 1: Find the central angle for each category by multiplying the corresponding percent (in decimal form) by 360°.

African-American:	13% of 360°	=	$0.13 \times 360°$	=	46.8°
Asian-American:	11% of 360°	=	$0.11 \times 360°$	=	39.6°
Hispanic:	14% of 360°	=	$0.14 \times 360°$	=	50.4°
Native American:	1% of 360°	=	$0.01 \times 360°$	=	3.6°
White:	54% of 360°	=	$0.54 \times 360°$	=	194.4°
Other:	3% of 360°	=	$0.03 \times 360°$	=	10.8°
No response:	4% of 360°	=	$0.04 \times 360°$	=	14.4°

Math Tip

To check your work, add up the degrees from Step 1. The sum should be 360 degrees.

Step 2, 3, and 4: Draw a circle, mark the central angles as close to the actual degrees as is practical, and label each sector. Note that the order of the sectors (pie slices) is not important.

Step 5: Give the graph a title.

Ethnic Breakdown for SAT, Nationwide, 2010

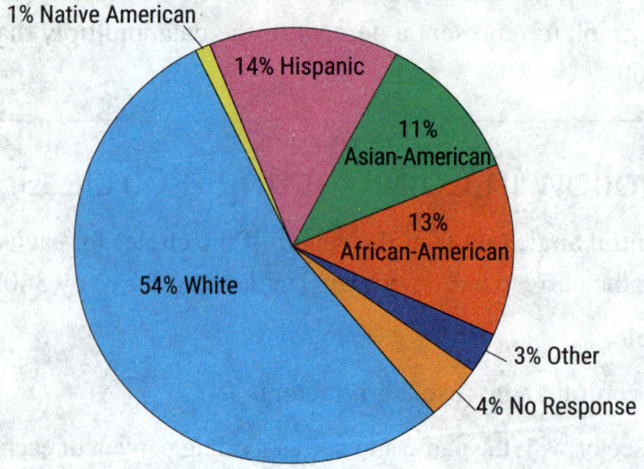

Now work margin exercise 2.

Margin Exercise Answers

1. **College Students' GPA**

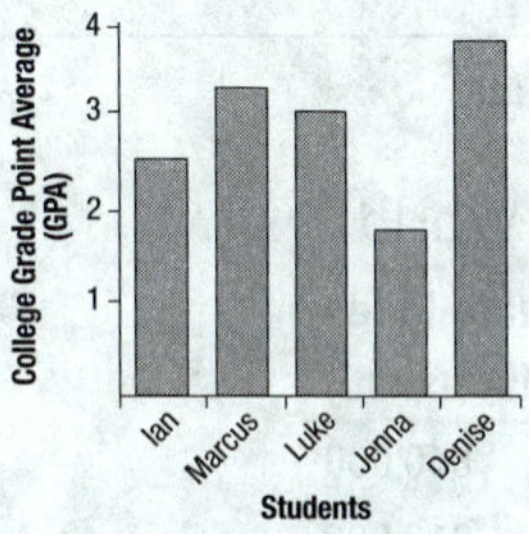

2. **2013 Richland County School Budget**

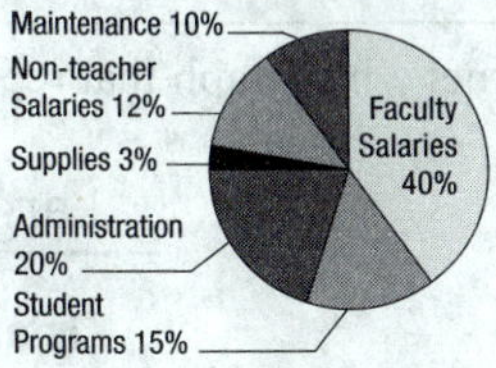

7.3 Exercises

Concept Check

Fill-in-the-Blank. Complete each sentence using information found in this section.

1. In a vertical bar graph, the categories of data are written along the _________ axis.
2. In a vertical bar graph, the bar for each category is to have a height that represents the _________ of the data for that category.
3. The final step in creating a bar graph is to make sure the bars have the same _______.
4. The first step in constructing a circle graph is to find the central ______ for each category.
5. In the creation of a circle graph, the central angle is found by multiplying the corresponding percent by _______ degrees.
6. The final step in constructing a circle graph is to label each sector with the ______ and corresponding _______ of each category.

True/False. Determine whether each statement is true or false. If a statement is false, explain how it can be changed so the statement will be true. (**Note:** There may be more than one acceptable change.)

7. In creating a vertical bar graph, a bar's width should vary based on the number it represents.
8. The first step in constructing a vertical bar graph is to draw a vertical and a horizontal axis.
9. Bar graphs can have either vertical or horizontal bars.
10. Titles are unnecessary for circle graphs.

7.3 Exercises

Concept Check

1. horizontal
2. frequency
3. width
4. angle
5. 360
6. name, percent
7. False; In creating a vertical bar graph, all bar widths should be the same.
8. True
9. True
10. False; Titles are necessary for understanding circle graphs.

Applications

1.

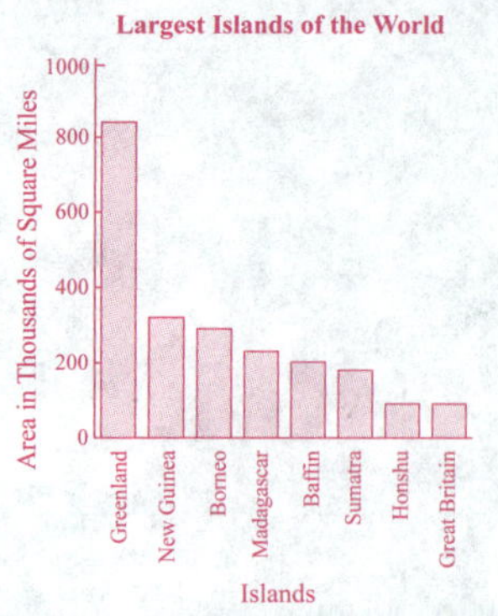

2.

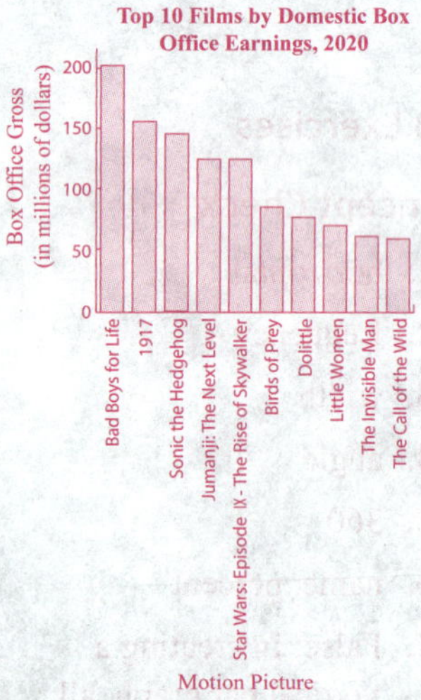

3.

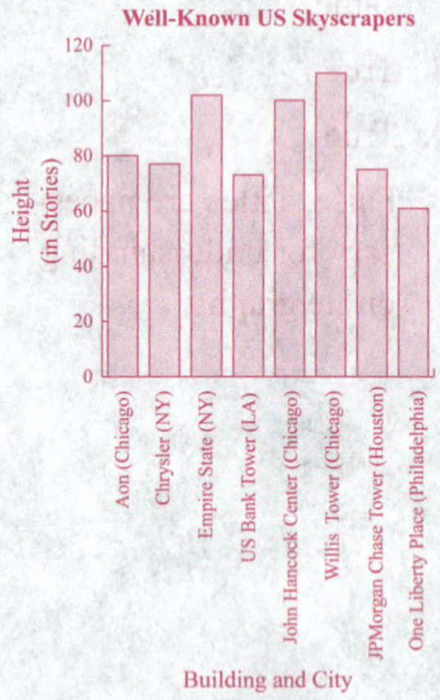

Applications

For each set of data, construct the specified graph.

1. Construct a bar graph that represents the following data.

Largest Islands of the World

Island	Area in Square Miles (nearest ten thousand)
Greenland	840,000
New Guinea	310,000
Borneo	290,000
Madagascar	230,000
Baffin	200,000
Sumatra	180,000
Honshu	90,000
Great Britain	90,000

2. Construct a bar graph that represents the following data.

10 Top Films by Domestic Box Office Earnings, 2020

Motion Picture	Box Office Gross (in millions of dollars)
Bad Boys for Life	204.4
1917	157.9
Sonic the Hedgehog	146.1
Jumanji: The Next Level	124.7
Star Wars: Episode IX - The Rise of Skywalker	124.5
Birds of Prey	84.2
Dolittle	77.0
Little Women	70.5
The Invisible Man	64.9
The Call of the Wild	62.3

Source: Information courtesy of Box Office Mojo. Used with permission. www.boxofficemojo.com

3. Construct a bar graph that represents the following data.

Well-Known United States Skyscrapers

Building and City	Height (in stories)
Aon Center (Chicago)	80
Chrysler (NY)	77
Empire State (NY)	102
U.S. Bank Tower(LA)	73
John Hancock Center (Chicago)	100
Willis Tower (Chicago)	110
JPMorgan Chase Tower (Houston)	75
One Liberty Place (Philadelphia)	61

4. Construct a bar graph that represents the following data.

Population of Countries that Begin with the Letter L (2016 Estimates, to nearest thousand)

Country	Population
Laos	7,019,000
Latvia	1,966,000
Lebanon	6,238,000
Lesotho	1,953,000
Liberia	4,300,000
Libya	6,542,000
Liechtenstein	38,000
Lithuania	2,854,000
Luxembourg	582,000

Source: CIA World Facebook

5. Construct a bar graph that represents the following data.

Notable Modern Bridges

Bridge and Location	Span (in meters)
Brooklyn (New York)	500
Bosporus (Istanbul)	1100
Fourth Road (Scotland)	1000
George Washington (New York)	1100
Golden Gate (California)	1300
Humber (England)	1400
Mackinac Straits (Michigan)	1200
Newport (Rhode Island)	500
Ponte 25 de Abril (Portugal)	1000
Verazano-Narrows (New York)	1300

6. Construct a bar graph that represents the following data.

World Population by Decade

Year	Total World Population
1960	3,039,451,023
1970	3,706,618,163
1980	4,453,831,714
1990	5,278,639,789
2000	6,082,966,429
2010	6,852,472,823
2020	7,756,873,419

Source: US Census Bureau

4.

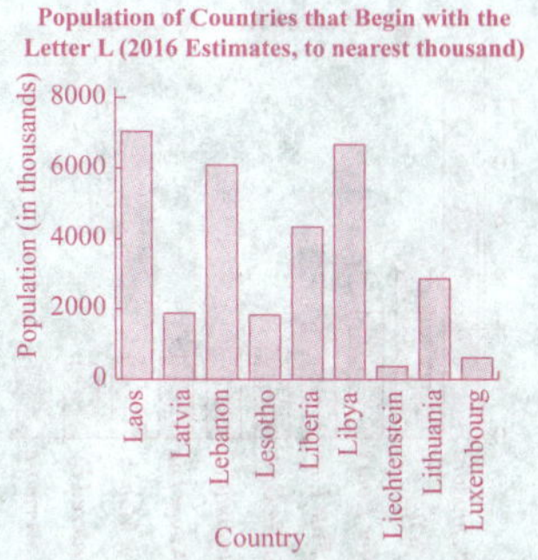

5.

6.

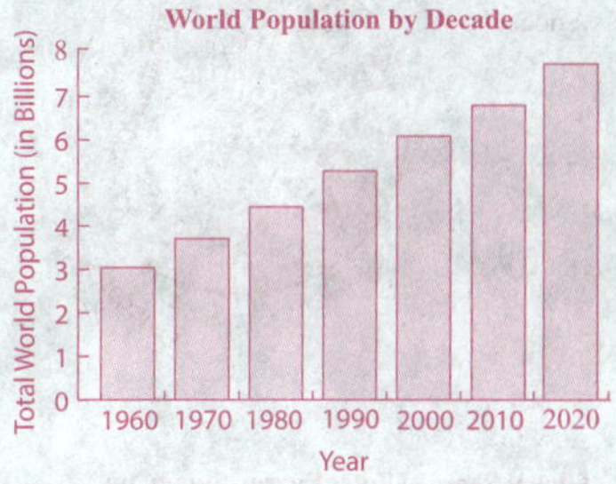

7. Construct a bar graph that represents the following data.

Top 10 Selling Albums in 2020

Album	Sales
Taylor Swift, *folklore*	1,276,000
BTS, *MAP OF THE SOUL : 7*	674,000
The Weekend, *After Hours*	480,000
Harry Styles, *Fine Line*	420,000
Billie Elish, *When We All Fall Asleep, Where Do We Go?*	348,000
Lady Gaga, *Chromatica*	331,000
Juice WRLD, *Legends Never Die*	301,000
Halsey, *Maniac*	301,000
Eminem, *Music To Be Murdered By*	287,000
Taylor Swift, *evermore*	283,000

Source: "Year-End Report U.S. 2020," MRC Data

8. Construct a circle graph that represents the following data.

Percent of Population with Particular Blood Types

Type of Blood	Percent of Population
O positive (O^+)	38%
O negative (O^-)	7%
A positive (A^+)	34%
A negative (A^-)	6%
B positive (B^+)	9%
B negative (B^-)	2%
AB positive (AB^+)	3%
AB negative (AB^-)	1%

Source: AABB.org

9. Construct a circle graph that represents the following data.

Global Share of Total Energy Supply by Source, 2019

Source of Energy	Percent
Oil	30.9%
Coal	26.8%
Natural Gas	23.2%
Biofuels	9.4%
Nuclear	5.0%
Hydro	2.5%
Other	2.2%

Source: Key World Energy Statistics, 2021. International Energy Agency.

7.

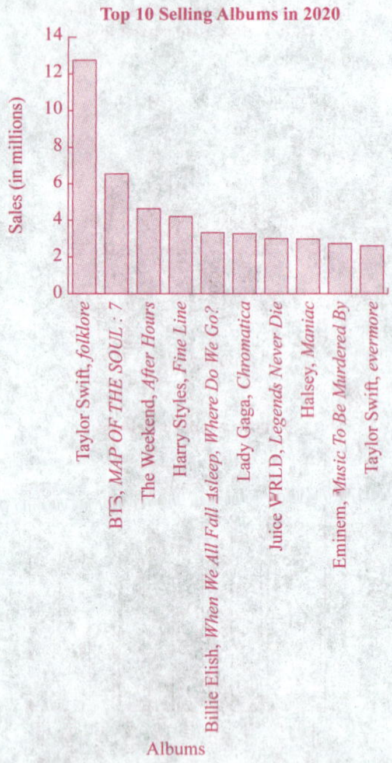

8.

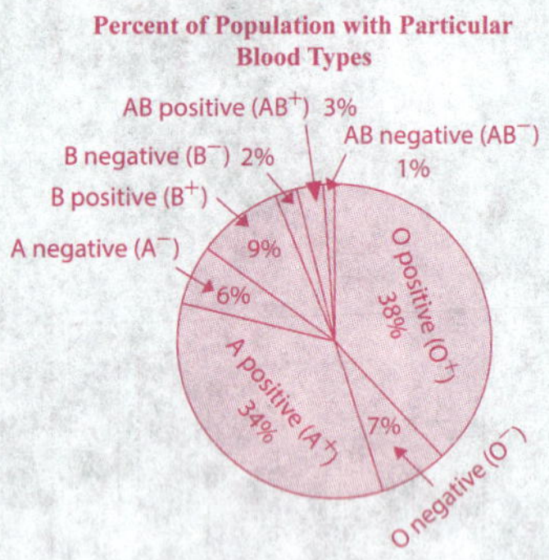

9.

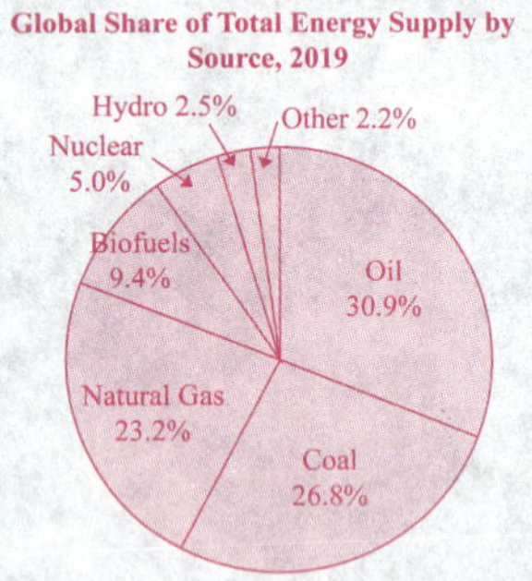

10.

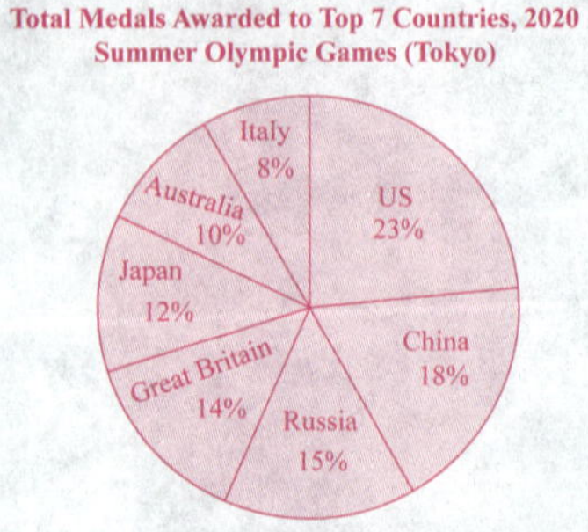

10. Construct a circle graph that represents the following data. Divide each number of medals by the total number to get the corresponding percent for each country.

Total Medals Awarded to Top 7 Countries, 2020 Summer Olympic Games (Tokyo)

Country	Number of Medals
United States	113
China	88
Russia	71
Great Britain	65
Japan	58
Australia	46
Italy	40
Total	481

Source: Olympic.org

11. Construct a circle graph that represents the following data.

2021 Distribution of Employment by Age

Age Range	Percent
16 to 19	3%
20 to 24	9%
25 to 34	23%
35 to 44	21%
45 to 54	20%
55 to 64	17%
65 and over	7%

Source: Bureau of Labor Statistics

12. Construct a circle graph that represents the following data.

Cholesterol Levels of 100 Students

Cholesterol Level	Number of Students
Recommended	35
Borderline	15
Moderate Risk	40
High Risk	10

11.

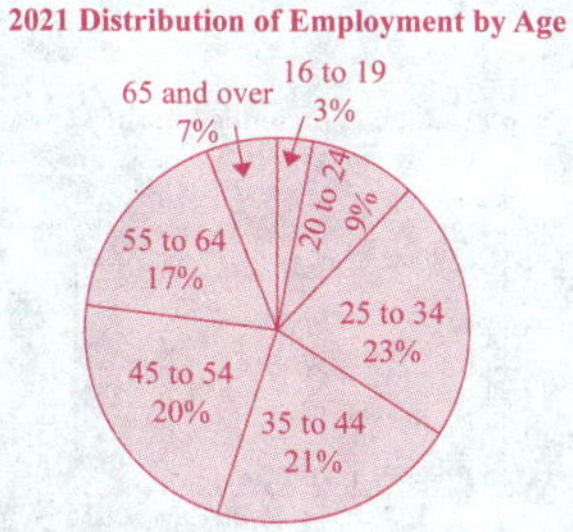

12.

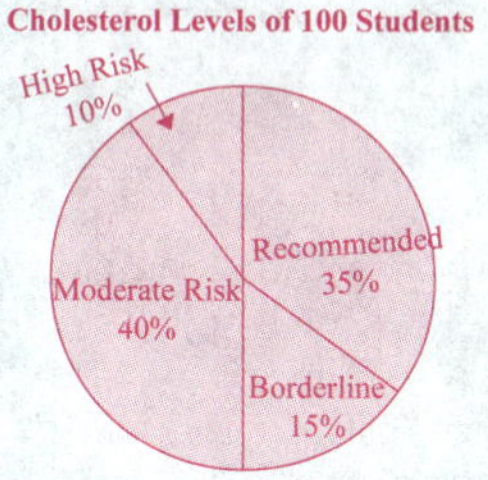

13.

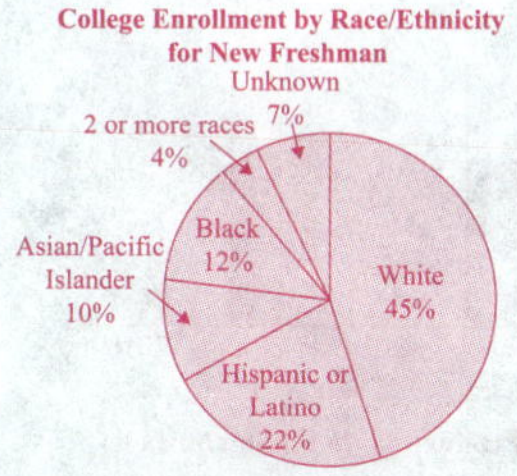

Writing & Thinking

14. Constructing a graph would require a thorough understanding of the data and concepts represented in the graph as well as the proper type of graph to best communicate the information.

15. Answers will vary. A person could mislabel or not label the graph at all. This mistake will make it difficult to properly interpret or understand the information presented. If the person does not pay attention to neatness and organization, the graph may be difficult to read. A person might forget to use appropriate titles which can decrease the understanding or purpose of the graph.

Collaborative Learning

16.

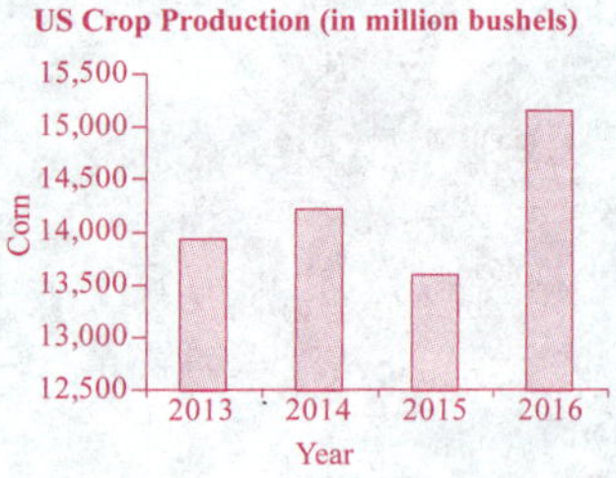

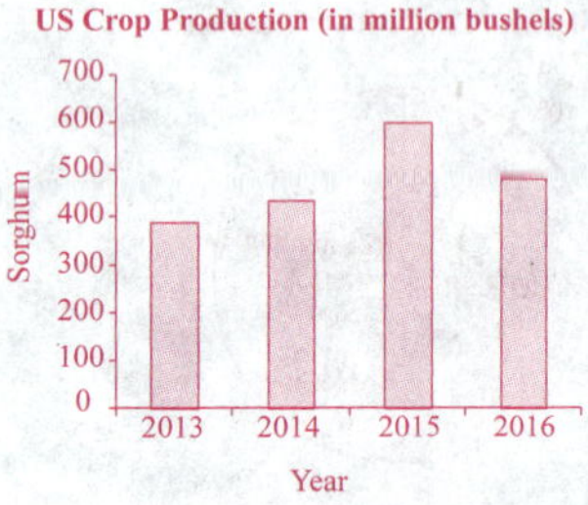

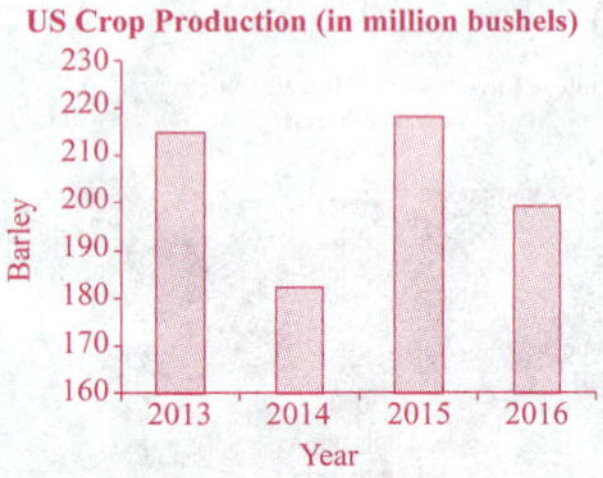

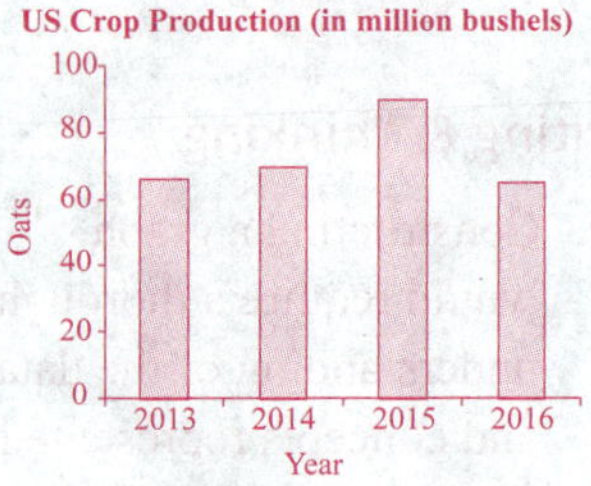

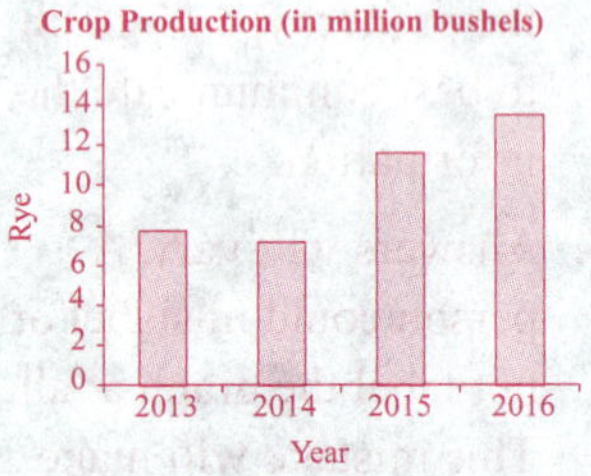

13. Construct a circle graph that represents the following data.

College Enrollment by Race/Ethnicity for New Freshman

Ethnicity	Percent
White	45
Hispanic or Latino	22
Asian/Pacific Islander	10
Black	12
2 or more races	4
Unknown	7

Writing & Thinking

14. While most graphs can be created through the use of a computer, give at least one benefit from constructing a graph yourself.

15. List three mistakes a person might make when constructing a graph.

Collaborative Learning

16. With the class separated into teams of two to four students, find access to a computer and a computer spreadsheet program, such as Excel, and enter the following data related to crop production on farms. Have the program find the mean and median for each category of crop for the years listed here. Also, have the program draw a bar graph for each category of crop by using the years listed here on the base line. Each team leader is to discuss how the team learned to use the program and bring printouts of the graphs. A general classroom discussion related to the power and efficiency of the computer should follow.

Crop Production in the U.S. (in Million Bushels)

Year	Corn	Sorghum	Barley	Oats	Rye
2013	13,925	389	215	65.88	7.67
2014	14,215	433	182	69.68	7.19
2015	13,601	597	218	89.54	11.62
2016	15,148	480	199	64.77	13.45

Source: usda.mannlib.cornell.edu/MannUsda/homepage.do

7.4 Probability

A Determining Sample Spaces

Activities involving chance such as tossing a coin, rolling a die, spinning a wheel in a game, and predicting weather are called **experiments**. The likelihood of a particular result is called its **probability**.

A weather person might predict rain with a probability of 30%.

In tossing a fair coin, the probability of heads (H) is $\frac{1}{2}$.

Terms Related to Probability

Outcome: An individual result of an experiment.

Sample Space: The set of all possible outcomes of an experiment.

Event: Some (or all) of the outcomes from the sample space.

DEFINITION

A **tree diagram** can be used to "picture" the possible outcomes of an experiment. Each branch of the tree diagram shows a separate outcome. For example, the tree diagram shown here illustrates the two possible outcomes of tossing a coin, heads (H) or tails (T).

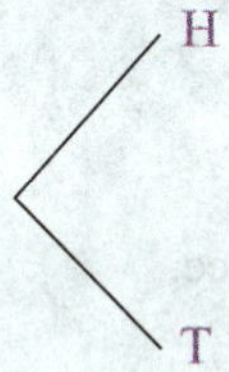

Figure 1: Tree Diagram

Example 1 Finding All Outcomes Using a Tree Diagram

A 6-sided die is rolled. Draw a tree diagram illustrating the possible outcomes of this experiment.

Solution

Since the die has 6 sides, we will label the possible outcomes as 1, 2, 3, 4, 5, and 6.

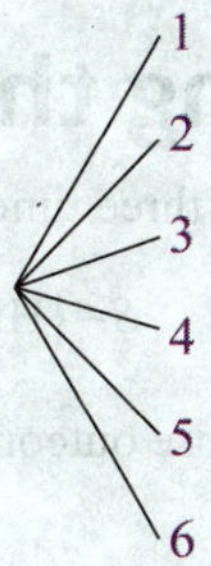

Now work margin exercise 1.

Objectives

A. Use a tree diagram to determine the sample space of an experiment.

B. Calculate the probability of an event.

7.4 Exercises

Concept Check

1. tree
2. sample space
3. probability
4. experiment
5. sample space
6. 0
7. False; The individual result of an experiment is an outcome.
8. True
9. True
10. True

Applications

1.

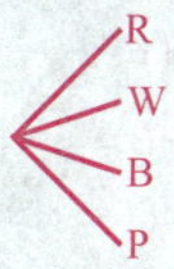

$S = \{R, W, B, P\}$
R = red, W = white,
B = blue, P = purple

2.

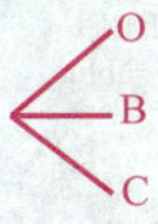

$S = \{O, B, C\}$
O = original, B = BBQ,
C = cheddar

1. One red, one white, one blue, one green, and one yellow marble are put in a bag and then a single marble is drawn at random. Draw a tree diagram illustrating the possible outcomes of this experiment.

2. A 4-sided die is rolled twice. Draw a tree diagram illustrating the outcomes of the experiment and list the outcomes in the sample space.

3.

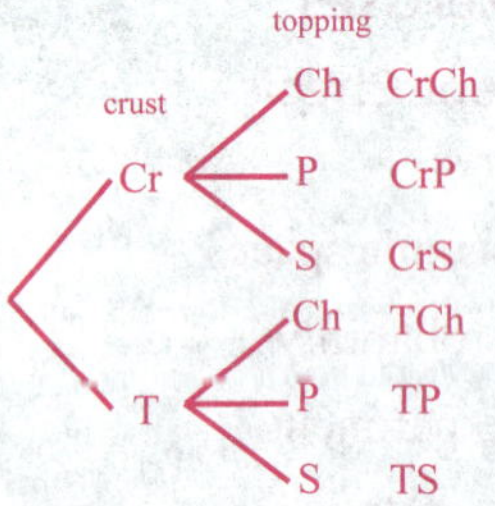

S = {CrCh, CrP, CrS, TCh, TP, TS}
Cr = crispy, T = thick,
Ch = cheese, P = pepperoni,
S = sausage

3. A coin is tossed and then a six-sided die is rolled. Draw a tree diagram illustrating the outcomes of the experiment and list the outcomes in the sample space.

4.

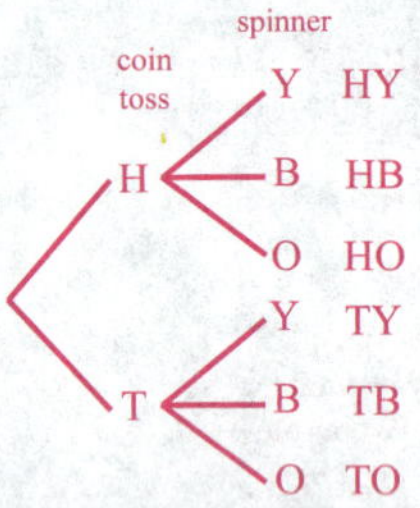

S = {HY, HB, HO, TY, TB, TO}
Y = yellow, B = blue, O = orange

5.

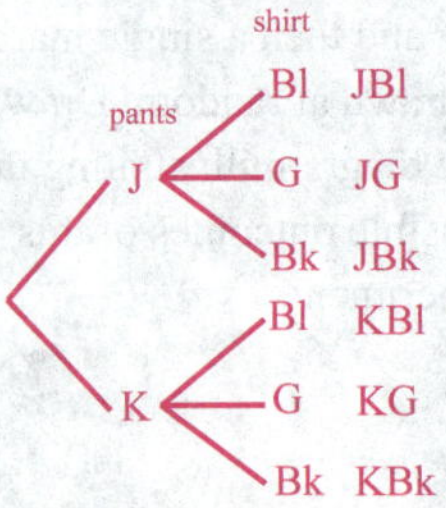

S = {JBl, JG, JBk, KBl, KG, KBk}
J = jeans, K = khaki pants, Bl = blue,
G = green, Bk = black

Example 2 Finding the Sample Space Using a Tree Diagram

A coin is tossed twice. Draw a tree diagram illustrating the possible outcomes of the experiment and list the outcomes in the sample space.

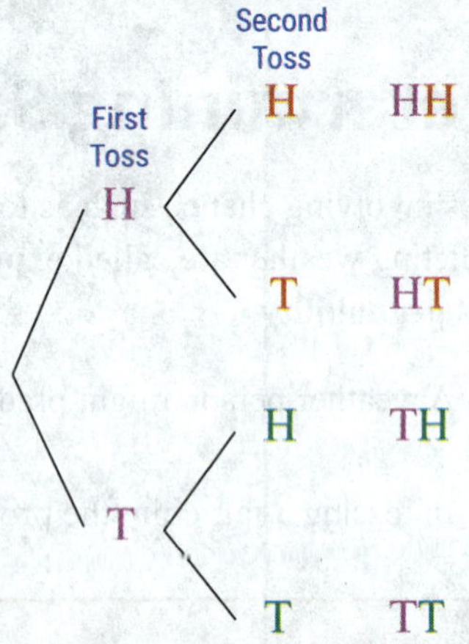

Outcomes: HH, HT, TH, TT

Solution

Following the branches (left to right) shows the four outcomes in the sample space.

$$S = \{HH, HT, TH, TT\}$$

Now work margin exercise 2.

Example 3 Finding the Sample Space Using a Tree Diagram

A coin is tossed and then one of the numbers (1, 2, and 3) is chosen at random from a box. Draw a tree diagram illustrating the possible outcomes of the experiment and list the outcomes in the sample space.

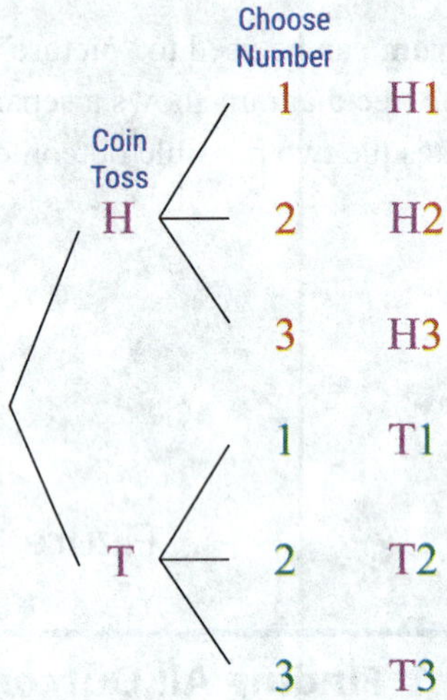

Outcomes: H1, H2, H3, T1, T2, T3

Solution

There are six outcomes in the sample space.

$$S = \{H1, H2, H3, T1, T2, T3\}$$

Now work margin exercise 3.

B Finding the Probability of an Event

If a coin is tossed three times, there are eight outcomes.

$$S=\{HHH, HHT, HTH, HTT, THH, THT, TTH, TTT\}$$

Choosing only those outcomes with exactly two heads is an event. If we label this event A, then

$$A = \{HHT, HTH, THH\}.$$

The probability of an event is the likelihood of it occurring. Because there are eight possible outcomes in the sample space and three in event A, we have probability of $A = \frac{3}{8}$.

6.

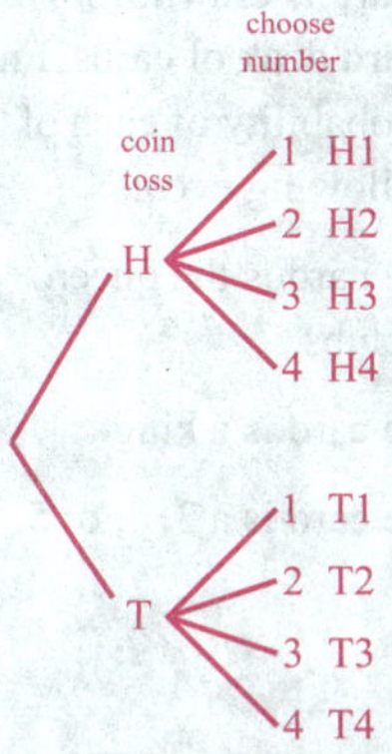

$S = \{H1, H2, H3, H4, T1, T2, T3, T4\}$

Probability of an Event

$$\text{probability of an event} = \frac{\text{number of outcomes in event}}{\text{number of outcomes in sample space}}$$

DEFINITION

Basic Characteristics of Probabilities

1. Probabilities are between 0 and 1, inclusive.

a. If an event can never occur, its probability is 0.

b. If an event will always occur, its probability is 1.

2. The sum of the probabilities of the outcomes in a sample space is 1.

PROPERTIES

Example 4 Calculating a Probability

A coin is tossed twice. Find the probability that both tosses are tails, TT.

Solution

There are four outcomes in the sample space: $S = \{\text{HH}, \text{HT}, \text{TH}, \text{TT}\}$.

One of these outcomes is the event TT. So,

$$\text{probability of TT} = \frac{1}{4}.$$

Now work margin exercise 4.

4. If a coin is tossed three times, what is the probability that the tosses are all heads or all tails?

Example 5 Calculating a Probability

A six-sided die is rolled once. Find the probability of rolling an even number.

Solution

There are six possible outcomes in rolling a die. The sample space is

$$S = \{1, 2, 3, 4, 5, 6\}.$$

Of these six, three are even numbers: 2, 4, 6. So,

$$\text{probability of an even number} = \frac{3}{6} = \frac{1}{2}.$$

Now work margin exercise 5.

5. There are three M&M candies in a box: one red, one yellow, and one green. Two M&Ms are chosen from the box. What is the probability that these M&Ms are the red one and the green one?

6. One card is drawn from a standard deck of cards. Find the probability of each of the following events.

a. The card is the queen of clubs.

b. The card is a king.

c. The card is a 3, 4, or 5.

7.

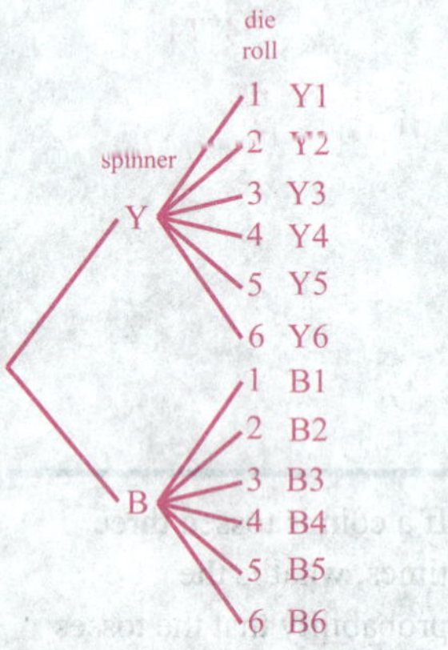

S = {Y1, Y2, Y3, Y4, Y5, Y6, B1, B2, B3, B4, B5, B6}
Y = yellow, B = blue

8.

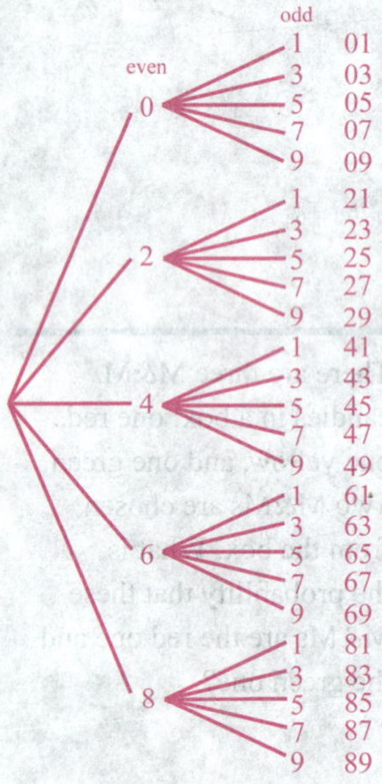

S = {01, 03, 05, 07, 09, 21, 23, 25, 27, 29, 41, 43, 45, 47, 49, 61, 63, 65, 67, 69, 81, 83, 85, 87, 89}

Example 6 Calculating a Probability

A standard deck of cards is 52 cards with four suits (hearts, diamonds, spades, and clubs) and 13 cards in each suit. The cards are ace, king, queen, jack, 10, 9, 8, 7, 6, 5, 4, 3, and 2. If one card is drawn from a deck of cards, find the probability of each of the following events.

a. The card is a 10.

b. The card is a diamond.

c. The card is a 2 or a 3.

Solution

a. There are four 10s in a deck (one 10 in each suit). So,

$$\text{probability of a 10} = \frac{4}{52} = \frac{\cancel{4} \cdot 1}{\cancel{4} \cdot 13} = \frac{1}{13}.$$

b. There are 13 diamonds in a deck (diamonds is one of the suits). So,

$$\text{probability of a diamond} = \frac{13}{52} = \frac{\cancel{13} \cdot 1}{4 \cdot \cancel{13}} = \frac{1}{4}.$$

c. There are four 2s and four 3s in a deck (one of each in each suit) for a total of 8. So,

$$\text{probability of a 2 or a 3} = \frac{8}{52} = \frac{\cancel{4} \cdot 2}{\cancel{4} \cdot 13} = \frac{2}{13}.$$

Now work margin exercise 6.

Margin Exercise Answers

1.

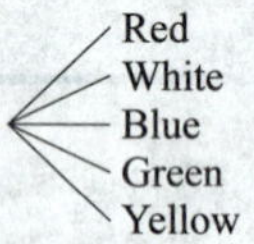

2.

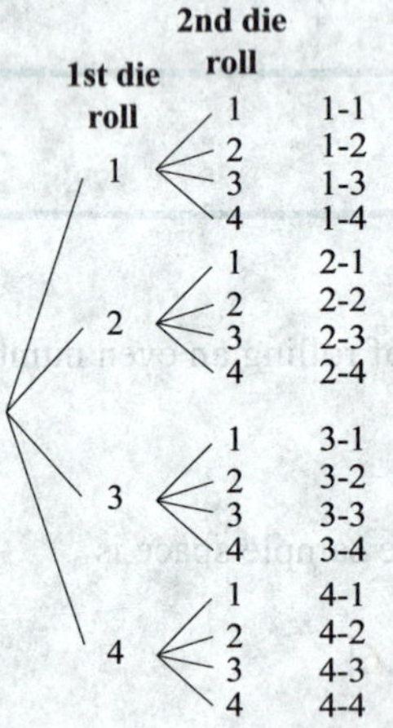

S = {1-1, 1-2, 1-3, 1-4, 2-1, 2-2, 2-3, 2-4, 3-1, 3-2, 3-3, 3-4, 4-1, 4-2, 4-3, 4-4}

3.

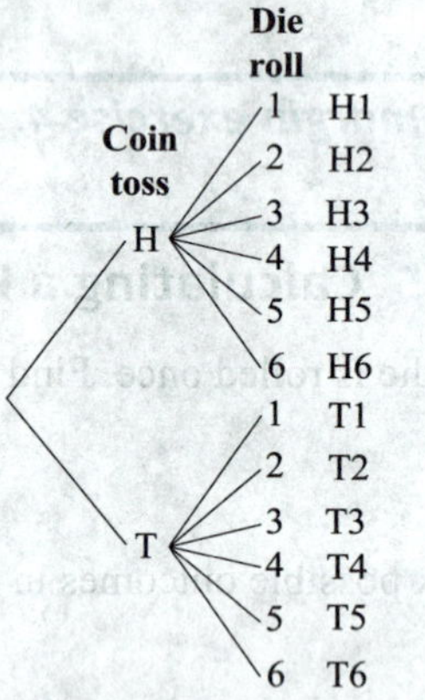

S = {H1, H2, H3, H4, H5, H6, T1, T2, T3, T4, T5, T6}

4. $\frac{1}{4}$

5. $\frac{1}{3}$ **6. a.** $\frac{1}{52}$ **b.** $\frac{1}{13}$ **c.** $\frac{3}{13}$

7.4 Exercises

Concept Check

Fill-in-the-Blank. Complete each sentence using information found in this section.

1. A/An __________ diagram can be used to view possible outcomes of an experiment.
2. The set of all possible outcomes of an experiment is its __________ __________.
3. The likelihood of a particular result is the __________ of the result.
4. An activity in which the result is random is a/an __________.
5. The probability of an event is the number of outcomes in the event divided by the number of outcomes in the __________ __________.
6. The probability of an event that can never occur is __________.

True/False. Determine whether each statement is true or false. If a statement is false, explain how it can be changed so the statement will be true. (**Note:** There may be more than one acceptable change.)

7. The individual result of an experiment is a probability.
8. An event is some or all of the outcomes from the sample space.
9. A single result of an experiment is an outcome.
10. The probability of a tossed coin showing either heads or tails is 1.

Applications

For each experiment, draw a tree diagram illustrating the possible outcomes and list the outcomes in the sample space. See Examples 1 through 3.

1. Four marbles are in a box: one red, one white, one blue, and one purple. One marble is chosen.
2. There are three flavors of potato chips to choose from: original, BBQ, and cheddar. One flavor is chosen.
3. When ordering pizza, a crust is chosen (crispy or thick) followed by choosing a topping (cheese, pepperoni, or sausage).
4. A game is played by tossing a coin followed by spinning a spinner with three colors: yellow, blue, and orange.
5. You have two pairs of pants: jeans and khaki pants. You also have three shirts: a blue one, a green one, and a black one. How many outfits can you make?
6. A coin is tossed followed by choosing a number from 1 to 4.
7. A spinner with two colors (yellow and blue) is spun followed by rolling a six-sided die.

9.

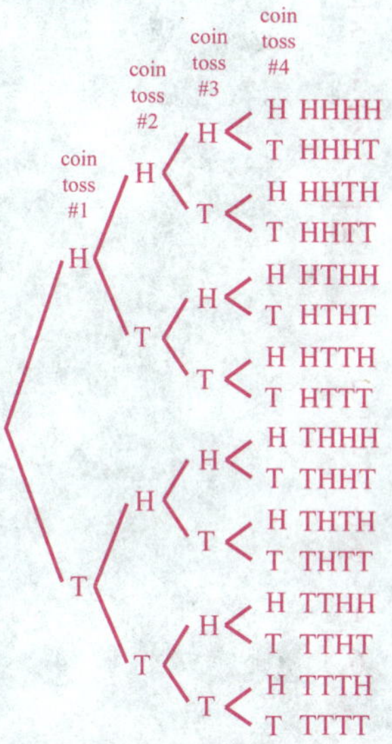

S = {HHHH, HHHT, HHTH, HHTT, HTHH, HTHT, HTTH, HTTT, THHH THHT, THTH, THTT, TTHH, TTHT, TTTH, TTTT}

10.

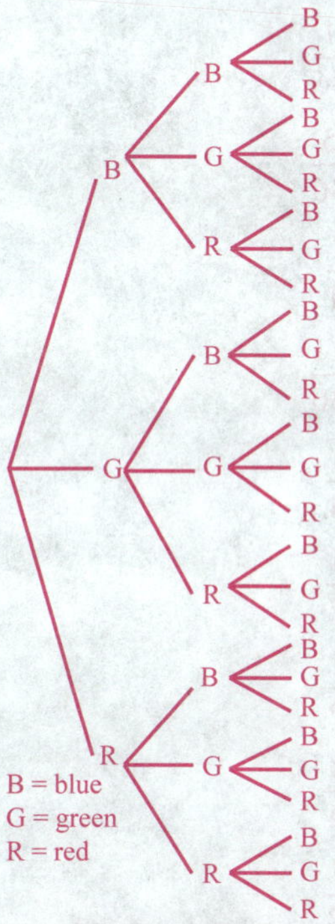

S = {BBB, BBG, BBR, BGB, BGG, BGR, BRB, BRG, BRR, GBB, GBG, GBR, GGB, GGG, GGR, GRB, GRG, GRR, RBB, RBG, RBR, RGB, RGG, RGR, RRB, RRG, RRR}

11. $\frac{2}{5}$
12. $\frac{3}{5}$
13. $\frac{1}{20}$
14. $\frac{1}{3}$
15. $\frac{3}{4}$
16. $\frac{1}{6}$
17. $\frac{1}{6}$
18. $\frac{5}{36}$
19. 0
20. $\frac{1}{12}$
21. $\frac{1}{36}$
22. $\frac{5}{36}$
23. $\frac{1}{4}$
24. $\frac{11}{36}$
25. $\frac{1}{4}$
26. $\frac{1}{2}$
27. $\frac{1}{2}$
28. $\frac{3}{4}$
29. $\frac{1}{2}$
30. $\frac{5}{12}$
31. 1
32. $\frac{1}{6}$
33. $\frac{1}{13}$
34. 0

8. Two digits from 0 to 9 are chosen at random. The first one is even and the second one is odd.

9. A coin is tossed 4 times.

10. A spinner has three colors: blue, green, and red. The spinner is spun 3 times.

For each problem, calculate the probability described. See Examples 4 through 6.

11. A box contains 5 marbles: two red, one white, two blue. What is the probability of choosing a blue marble from the box?

12. A machine contains only 5 gumballs: three yellow, one white, one green. What is the probability of getting a yellow gumball when you put a coin in the machine?

13. Your English professor chooses students randomly to answer questions. If the class has 20 students, what is the probability that you will be selected to answer the next question?

14. There are four socks in a drawer (two black and two blue). Two socks are chosen at random. What is the probability that the chosen socks are a matching pair?

15. In a survey of 124 dentists, 93 dentists said they would recommend Healthy White brand toothpaste, 23 dentists said they would recommend Super White brand toothpaste, and 8 dentists said they would recommend Extreme White brand toothpaste. What is the probability that one of the dentists will recommend Healthy White brand toothpaste?

16. A type of candy has six flavors: cherry, strawberry, lemon, green apple, grape, and orange. You open a new bag of candy that has three of each flavor. You reach in and grab a piece of candy without looking. What is the probability that the candy you pick is lemon-flavored?

Not all experiments are easily pictured with tree diagrams. If two six-sided dice are rolled, there are 36 possible outcomes and these outcomes can be represented in table form as shown here. In this table, we have illustrated the two dice in different colors, red and blue. Use this table to find the probabilities of the events described.

	1	2	3	4	5	6
1	1, 1	1, 2	1, 3	1, 4	1, 5	1, 6
2	2, 1	2, 2	2, 3	2, 4	2, 5	2, 6
3	3, 1	3, 2	3, 3	3, 4	3, 5	3, 6
4	4, 1	4, 2	4, 3	4, 4	4, 5	4, 6
5	5, 1	5, 2	5, 3	5, 4	5, 5	5, 6
6	6, 1	6, 2	6, 3	6, 4	6, 5	6, 6

17. a sum of 7

18. a sum of 6

19. a sum of 15

20. a sum of 4

21. a sum of 2

22. a sum of 8

23. neither die is even

24. at least one die is 6

A box contains four pieces of paper with the numbers 1, 2, 3, and 4 written on them. A piece of paper is drawn at random. Find the probability of each event.

25. The number is 4.

26. The number is odd.

27. The number is less than 3.

28. The number is not 3.

A spinner (shown) with the numbers from 1 to 12 is spun once. Find the probability of each event.

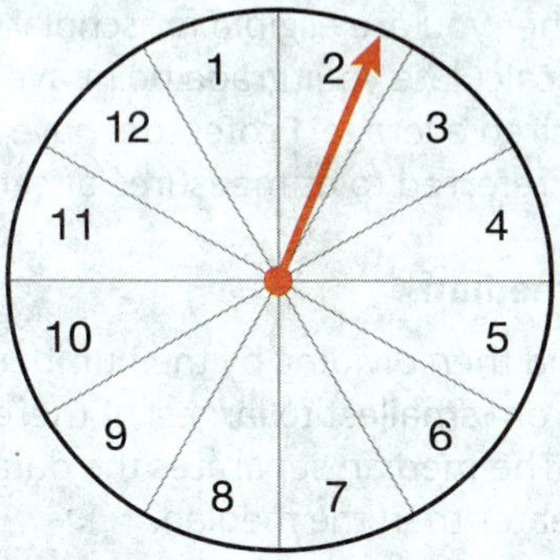

29. The number is an even number.

30. The number is a prime number.

31. The number is less than 13.

32. The number is a multiple of 5.

A standard deck of cards has 52 cards with four suits (hearts, diamonds, spades, and clubs) and 13 cards in each suit. The hearts and diamonds are red cards and the spades and clubs cards are black cards. The cards are ace, king, queen, jack, 10, 9, 8, 7, 6, 5, 4, 3, and 2. If one card is drawn from the deck of cards, find the probability of each event.

33. The card is an ace.

34. The card is a 1.

35. The card is a club.

36. The card is red.

37. The card is the king of hearts.

38. The card is the 3 of diamonds.

39. The card is a queen or jack.

40. The card is 10, 9, or 8.

Writing & Thinking

41. List at least three activities that are experiments of chance.

42. Explain the benefit(s) of using a tree diagram to determine probability.

43. State the two basic characteristics of probabilities and discuss what events will have

a. a probability of 0 and **b.** a probability of 1.

44. Explain, in your own words, why the probability of an event cannot be greater than 1.

35. $\frac{1}{4}$

36. $\frac{1}{2}$

37. $\frac{1}{52}$

38. $\frac{1}{52}$

39. $\frac{2}{13}$

40. $\frac{3}{13}$

Writing & Thinking

41. Chance experiments include, but are not limited to, tossing a coin, spinning a bottle, drawing a card from a standard deck of cards, picking numbers in the lottery, choosing straws, and picking colored marbles.

42. A tree diagram gives a "picture" of possible outcomes of an experiment and illustrates the possible outcomes.

43. Probabilities are between 0 and 1, inclusive. The sum of the probabilities of the outcomes in a sample space is 1. An event has probability 0 if it can never occur, such as rolling a 7 on a six-sided die. An event has probability 1 if it will always occur, such as rolling one of the numbers 1, 2, 3, 4, 5, or 6 on a six-sided die.

44. Answers will vary. Essentially, the number of outcomes in an event cannot be greater than the number of outcomes in the sample space. Thus, the probability of an event cannot be greater than one.

Chapter 7 Project

What's My Average?

An activity to investigate the relationships among different measures of central tendency.

If you are a college student, then grades are important to you. They determine whether you are eligible for scholarships or getting into a particular college or program of choice. It is important to be able to calculate your grade point average in a class and to be able to determine the score you need on a test to reach your desired average. Professors have many different ways of calculating your average for a class. Measures of average are often referred to as measures of central tendency.

For this project, you will be working with two of these measures, the **mean** and the **median**.

Recall that the **mean** of a set of data is found by adding all the numbers in the set and then dividing by the number of data values. The **median** is the middle number once you arrange the data in order from smallest to largest. If there is an even number of data values, then the median is the mean of the two middle values. The median separates the data into two parts such that 50% of the data values are less than the median and 50% are greater than the median.

Jonathan and Tristen are two students in Dr. Hawkes Math 230 class. So far, Dr. Hawkes has given 5 tests and the students' scores are listed below.

Jonathan	Tristen
24	80
98	84
86	88
96	72
96	81

1. Calculate the mean and median of Jonathan's grades.

2. Calculate the mean and median of Tristen's grades.

3. Compare the two measures of *average* for each student.
   a. Are the mean and median similar for Jonathan?
   b. Are the mean and median similar for Tristen?
   c. Based on the **mean**, who has the best *average* in the class?
   d. Based on the **median**, who has the best *average* in the class?

4. In your opinion, which student has the most consistent test scores? Explain your reasoning.

5. If each student had scored 2 points higher on each test, how would this affect
   a. The mean of their grades?
   b. The median?

6. Dr. Hawkes is planning on giving one more test in the class. His grading scale is as follows.

A	93–100
B	85–92
C	74–84
D	69–73
F	Below 69

   a. What is the lowest score each student can make on the test and still end up with a grade of C for the class (based on the **mean** of all test scores)?
   b. Who has to make the higher grade on the last test to get a C, Jonathan or Tristen?
   c. If the last test counts double (equivalent to two test grades) what is the lowest score each student can make on the test in order to make a B in the class (based on the **mean** of all test scores)? (Do not round the mean.)
   d. If the last test counts double, who has to make the higher grade on the last test to get a B, Jonathan or Tristen?

7. Based on the work you have done in Problems 1 through 6, which measure do you think is the *best* measure of a student's *average* grade, the mean or the median? (Explain your reasoning by looking at this question from both Jonathan and Tristen's point of view.)

Chapter 7 Project

Misleading Graphs

An activity to demonstrate the use of bar graphs in real life.

Unfortunately, we can be intentionally or unintentionally misled by statistics, particularly when graphs are used to convey findings. In this project, you'll learn how to spot when this occurs with a bar graph and how to fix the representation.

Suppose the following table reported education levels among all young adults (18–24 years old) within the United States for a specific year.

Highest Education Level Attained	Percent (%)
Neither a High School Diploma nor a G.E.D.	5
High School Diploma or G.E.D.	65
Bachelor's Degree	20
Beyond Bachelor's Degree	10

1. The percent value in the first category is lower than all the others. For the other three categories, calculate how many times bigger the percents are compared to the percent value in the first category.

2. Suppose a graphic designer presents the information in the following bar graph. For each of the four categories, calculate the area of the bar shown. (Recall that the area of a square is $A = s^2$.)

Highest Education Level Attained

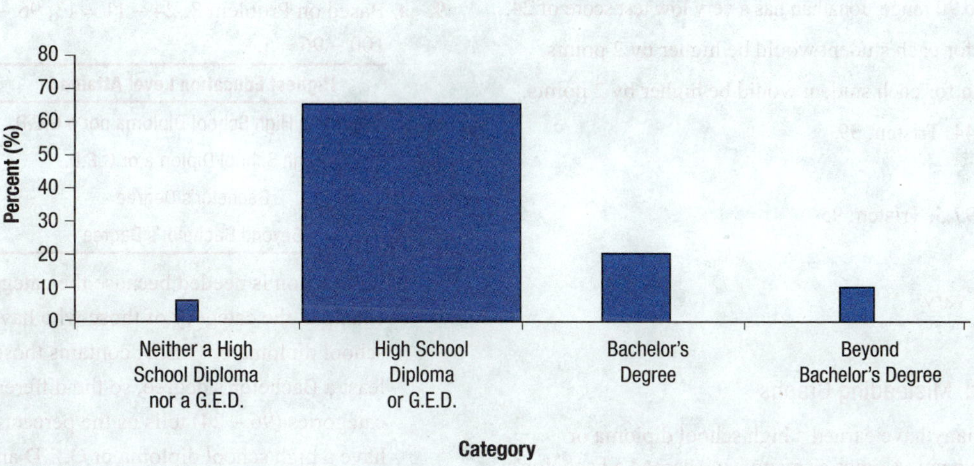

3. Similar to Problem 1, the area in the first category is smaller than the areas for all the other categories. Calculate how many times bigger the area of each of the other three are compared to the area in the first category.

4. There is a relationship between the area comparisons in Problem 3 to the percent comparisons in Problem 1. What is it? (**Hint:** What type of geometric figures are shown in the graph and what calculation does this suggest?)

5. Review Section 7.3. Which step of constructing a vertical bar graph was skipped, whether intentional or not? How could this alter the perceptions of those reading the graph in Problem 2?

6. Construct a vertical bar graph that follows all the steps shown in Section 7.3.

7. Although we are able to do the calculations and make the comparisons to spot the misrepresentation, why do you think the skipped step mentioned in Problem 5 is important? (**Note:** Whether for short reading or a deep dive, look into the works of psychologist Jean Piaget.)

8. Suppose that one year later the information was updated and presented in the following table. While it may be understandable that the information was presented this way, it cannot be used to construct a bar graph for educational levels of young adults in the US. Why?

Educational Level	Percent (%)
At Least a high School Diploma or G.E.D.	96
At Least a Bachelor's Degree	24
Beyond a Bachelor's Degree	11

9. Use the information from the table in Problem 8 to do the following.

   a. Construct a new table like the one from the beginning of the project. (**Hint:** Subtraction is required.)

   b. Explain why subtraction was required to construct the table in part a.

   c. Construct a vertical bar graph to go along with the table in part a.

Chapter 7 Project: What's My Average?

1. mean = 80, median = 96
2. mean = 81, median = 81
3. **a.** No, the median is much higher than the mean.
   b. Yes, they are the same.
   c. Tristen
   d. Jonathan
4. Answers will vary. For example: Tristen, because all his scores are in the 70 to 90 range. Jonathan has a very low test score of 24.
5. **a.** The mean for each student would be higher by 2 points.
   b. The median for each student would be higher by 2 points.
6. **a.** Jonathan: 44, Tristen: 39
   b. Jonathan
   c. Jonathan: 97.5, Tristen: 95
   d. Jonathan
7. Answers will vary.

Chapter 7 Project: Misleading Graphs

1. 13 times as many have earned a high school diploma or G.E.D. as have not. 4 times as many have earned a bachelor's degree as those that have not earned a high school diploma nor G.E.D. 2 times as many have earned a degree beyond a bachelor's degree as those that have not earned a high school diploma nor a G.E.D.
2. The areas are, from left to right, 25, 4225, 400, and 100.
3. The area of the second square is 169 times bigger than the area of the first. The area of the third square is 16 times bigger than the area of the first. The area of the fourth square is 4 times bigger than the area of the first.
4. $13^2 = 169$, $4^2 = 16$, and $2^2 = 4$. The area comparisons are larger than the percent comparisons. In particular, squaring the percent comparisons yields the area comparisons.
5. The widths are not the same. Thus, when comparing any pair of categories, the size ratios are greatly exaggerated.
6. 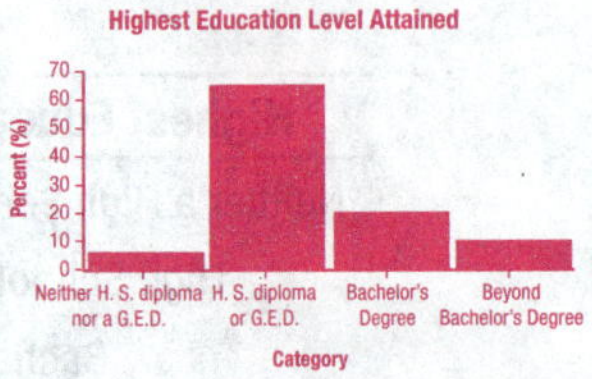

7. The step is included for at least two possible reasons. One is that younger readers may only be able to focus on one dimension. Even with older graph readers, though the calculations and comparisons can be done, if reading quickly, they will believe the greatly exaggerated size ratios.
8. The categories overlap. (For example, someone that has a degree beyond the bachelor's level already has a bachelor's degree.)
9. **a.** Based on Problem 8; 24 − 11 = 13, 96 − 24 = 72, and 100 − 96 = 4.

Highest Education Level Attained	Percent (%)
Neither a High School Diploma nor a G.E.D.	4
High School Diploma or G.E.D.	72
Bachelor's Degree	13
Beyond Bachelor's Degree	11

   b. Subtraction is needed because the categories overlap. For example, the category of those who have at least a high school diploma or G.E.D. contains those who have at least a Bachelor's degree, so the difference between those categories (96 − 24) tells us the percent of people who have a high school diploma or G.E.D and do not have a Bachelor's degree (72).

   c. 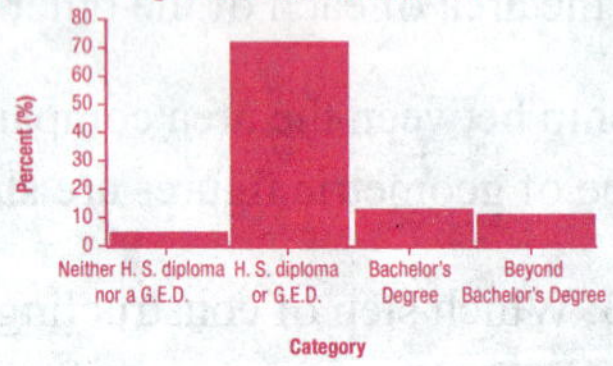

CHAPTER 8

8 Introduction to Algebra

Connections

The Swiss psychologist Jean Piaget once said,

> "Logic and mathematics are nothing other than specialized linguistic structures."

Variables are used as place holders for numerical values in algebraic expressions and equations. Expressions and equations can be seen as tools that allow us to record, reuse, and communicate important relationships using a universally understood language.

For example, in the United States, we commonly measure temperature in degrees Fahrenheit (°F). In this temperature scale, 32 °F is the freezing temperature of water while 212 °F is the boiling point. Most of the rest of the world uses the Celsius scale (°C), where water freezes at 0 °C and boils at 100 °C. The two scales are related by the following formula.

$$C = \frac{5}{9}(F - 32)$$

According to the National Oceanic and Atmospheric Administration, the ocean water temperature at Conimicut Lighthouse in Rhode Island was 39.9 °F on February 2, 2020. How can this formula be used to determine the water temperature in degree Celsius?

8.1 The Real Number Line and Absolute Value

Objectives

A. Graph real numbers on a number line.

B. Identify types of numbers.

C. Recognize the inequality symbols and use them to compare real numbers.

D. Find the absolute value of a real number.

Numbers and number concepts form the foundation for the study of algebra. In this chapter, you will learn about positive and negative numbers and how to operate with these numbers. The following list indicates ways that positive and negative numbers occur frequently in our daily lives.

	Negative	Zero	Positive
Temperatures are recorded as:	below zero	zero	above zero
The stock market will show:	a loss	no change	a gain
Altitude can be measured as:	below sea level	sea level	above sea level
Businesses will report:	losses	break even	profits

A Number Lines

The study of algebra requires that we know a variety of types of numbers and their names. The set of numbers

$$\mathbb{N} = \{1,2,3,4,5,6,7,8,9,10,11,\ldots\}$$

is called the **counting numbers** or **natural numbers**. The three dots (an ellipsis) indicate that the pattern is to continue without end. Putting 0 with the set of natural numbers gives the set of **whole numbers**.

$$\mathbb{W} = \{0,1,2,3,4,5,6,7,8,9,10,11,\ldots\}$$

Thus, mathematicians make the distinction that 0 is a whole number but not a natural number.

Math Tip

When creating a number line, you want to determine a starting point, a scale, and the values that you wish to display. Choose where you want to start and place your first value. (Zero is often a good value to begin with.) Your placement of the next value on the number line will determine the scale. The spacing between the tick marks needs to be consistent and should represent the same distance.

To help in understanding different types of numbers and their relationships to each other, we begin with a "picture" called a **number line**. For example, choose some point on a horizontal line and label it with the number 0 (see Figure 1).

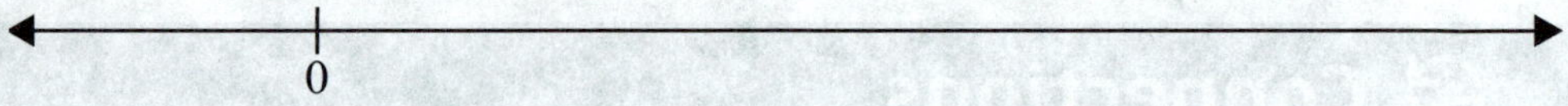

Figure 1

Now choose another point on the line to the right of 0 and label it with the number 1 (see Figure 2). This point is arbitrary and once it is chosen, a scale is determined for the remainder of the line.

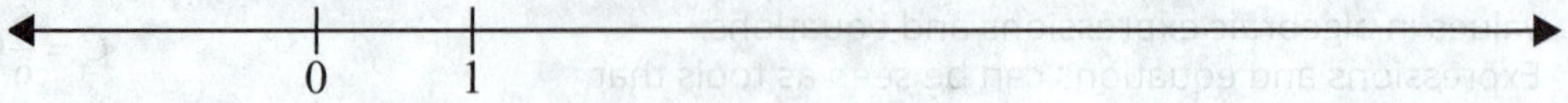

Figure 2

We now have a number line. Points corresponding to all the whole numbers are determined. The point corresponding to 2 is the same distance from 1 as 1 is from 0. The point 3 is the same distance from 2, and so on (see Figure 3).

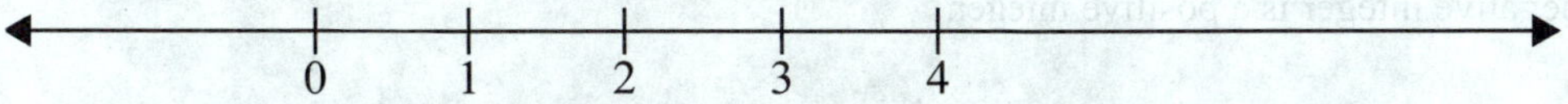

Figure 3

The **graph** of a number is the point that corresponds to the number, and the number is called the **coordinate** of the point. We will follow the convention of using the terms "number" and "point" interchangeably. For example, one point might be "seven" and another point "two." The graph of 7 is shown by the large red dot in Figure 4.

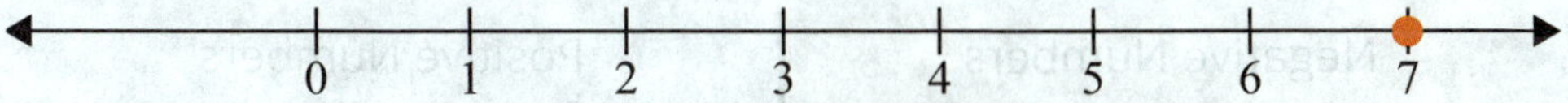

Figure 4

The graph of the set $A = \{2, 4, 6\}$ is shown in Figure 5.

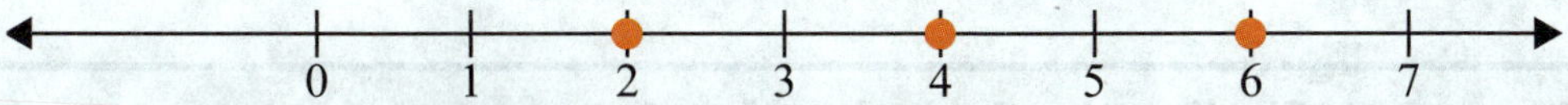

Figure 5

On a horizontal number line, the point one unit to the left of 0 is the **opposite** of 1. It is called **negative** 1 and is symbolized –1. Similarly, the point two units to the left of 0 is the opposite of 2, called negative 2, and symbolized –2, and so on (see Figure 6).

The opposite of 1 is –1;	The opposite of –1 is $-(-1) = +1$;
The opposite of 2 is –2;	The opposite of –2 is $-(-2) = +2$;
The opposite of 3 is –3; and so on.	The opposite of –3 is $-(-3) = +3$; and so on.

Numbers and Their Opposites

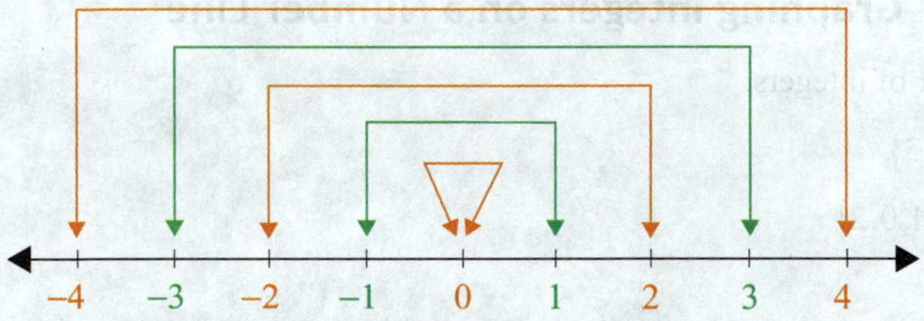

Figure 6

Note

The negative sign $(-)$ indicates the opposite of a number as well as a negative number. It is also used, as we will see in Section 8.3, to indicate subtraction. (Note on your calculator the subtraction key on the right side and the negative key $(-)$ at the bottom of the keypad.)

Integers

The set of numbers consisting of the whole numbers and their opposites is called the set of **integers**: $\mathbb{Z} = \{\ldots, -3, -2, -1, 0, 1, 2, 3, \ldots\}$.

DEFINITION

The natural numbers are also called **positive integers**. Their opposites are called **negative integers. Zero is its own opposite and is neither positive nor negative** (see Figure 7). Note that the opposite of a positive integer is a negative integer, and the opposite of a negative integer is a positive integer.

Integers: $\{\ldots, -3, -2, -1, 0, 1, 2, 3, \ldots\}$

Positive integers: $\{1, 2, 3, 4, 5, \ldots\}$

Negative integers: $\{\ldots -4, -3, -2, -1\}$

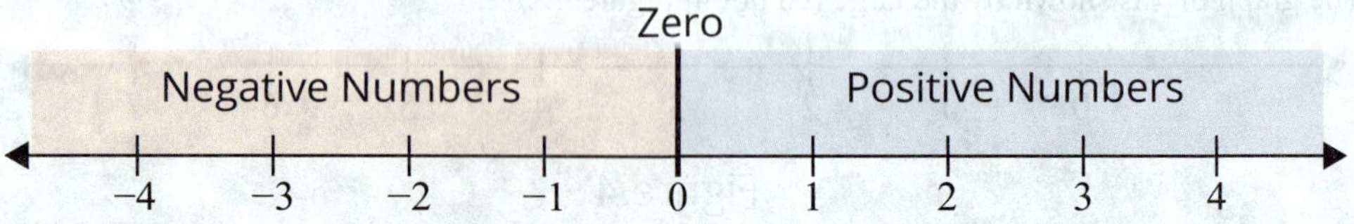

Figure 7

1. Find the opposite of each integer.

a. +10

b. −8

Example 1 Finding the Opposite of an Integer

Find the opposite of each integer.

a. +7

b. −4

Solution

a. −7

b. $-(-4)$ or $+4$ In words, the opposite of −4 is +4.

Now work margin exercise 1.

2. Graph each set of integers.

a. $\{-2, -1, 0\}$

b. $\{-7, -4, -2, 0, 1\}$

Example 2 Graphing Integers on a Number Line

Graph each set of integers.

a. $\{-3, -1, 1, 3\}$

b. $\{-4, -2, -1, 0, 2\}$

Solution

a.

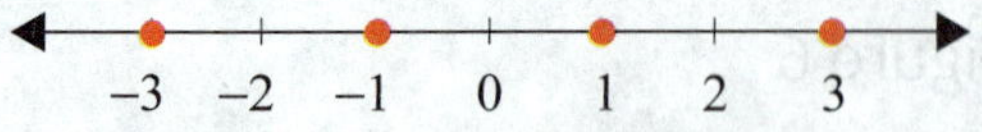

b.

Now work margin exercise 2.

B Types of Numbers

Numbers that can be written as fractions and whose numerators and denominators are integers have the technical name **rational numbers** ($\mathbb{Q}$). Positive and negative decimal numbers and the integers themselves can also be classified as rational numbers. For example, the following numbers are all rational numbers.

$$1.3 = \frac{13}{10},\quad 5 = \frac{5}{1},\quad -4 = \frac{-4}{1},\quad \frac{3}{8},\quad \text{and}\quad \frac{17}{6}$$

Variables are needed to be able to state rules and definitions in general forms. The definition is restated here for easy reference and for emphasis.

Variables

A **variable** is a symbol (generally a letter of the alphabet) that is used to represent an unknown number or any one of several numbers.

DEFINITION

The variables a and b are used to represent integers in the following definition of a **rational number**.

Teaching Note
You might mention that the word "rational," in this case, comes from "ratio".

Rational Numbers

A **rational number** is a number that can be written in the form of $\frac{a}{b}$, where a and b are integers and $b \neq 0$. ($\neq$ is read "is not equal to.")

OR

A **rational number** is a number that can be written in decimal form as a terminating decimal or as an infinite repeating decimal.

DEFINITION

Numbers that cannot be written as fractions with integer numerators and denominators are called **irrational numbers**. This includes numbers like $\sqrt{2}$, $\sqrt{3}$, π, and $\sqrt[3]{5}$. These numbers can be written as **infinite nonrepeating decimal numbers**.

All rational numbers and irrational numbers are classified as **real numbers** ($\mathbb{R}$) and can be written in some decimal form. They can also be represented on a number line (see Figure 8). The number line is called the **real number line**.

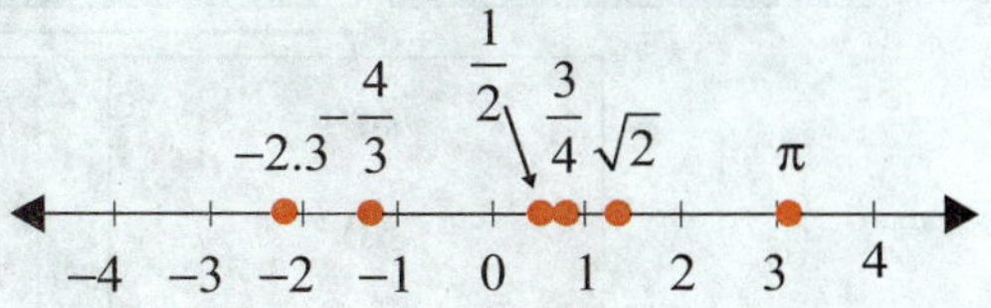

Figure 8

The following decimal values and approximations can be found with a calculator.

Examples of Rational Numbers

$\frac{3}{4} = 0.75$ This decimal number is terminating.

$\frac{1}{3} = 0.33333333\ldots = 0.\overline{3}$ There is an infinite number of 3s in this repeating pattern. (The bar above the 3 indicates the pattern repeats without end.)

$\frac{3}{11} = 0.27272727\ldots = 0.\overline{27}$ This repeating pattern shows that there may be more than one digit in the pattern.

Examples of Irrational Numbers

$\sqrt{2} = 1.414213562\ldots$ There is an infinite number of digits with no repeating pattern.

$\pi = 3.141592653\ldots$ There is an infinite number of digits with no repeating pattern.

1.41441444144441... There is an infinite number of digits and a pattern of sorts. However, the pattern is nonrepeating.

Figure 9 illustrates the relationships among the various categories of real numbers.

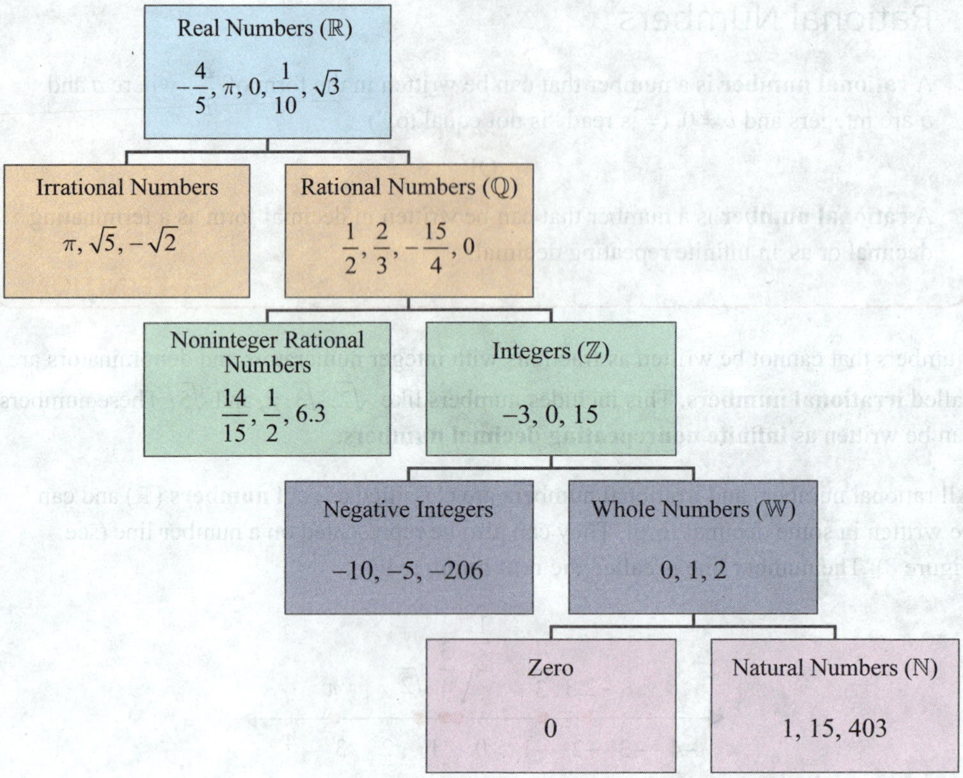

Figure 9

Math Tip

When a fraction such as $\frac{1}{3}$ is written in decimal form as $0.\overline{3}$, the bar indicates repeating digits, and the two forms represent the exact same value. Writing the number as a truncated decimal, such as 0.3, results in an approximation. The exact value or fractional form is usually preferred.

Math Tip

Another way to illustrate the relationships between the various categories of real numbers is with a Venn diagram, as shown here.

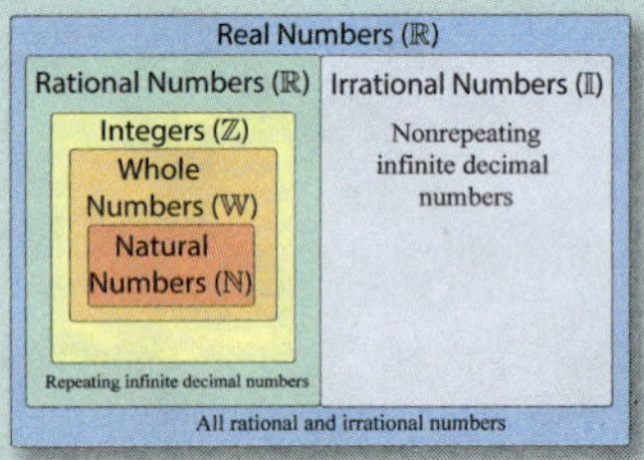

Example 3 Identifying Types of Numbers

List the numbers in the set $S = \left\{-5, -\frac{3}{4}, 0, \sqrt{2}, 17\right\}$ that are

a. whole numbers

b. integers

c. rational numbers

d. real numbers

Solution

a. 0 and 17 are whole numbers.

b. -5, 0, and 17 are integers.

c. $-5, -\frac{3}{4}, 0$, and 17 are rational numbers.

d. All numbers in S are real numbers.

Now work margin exercise 3.

3. List the numbers in the set

$S = \left\{-6, -\frac{1}{7}, 1, \sqrt{11}, 20\right\}$

that are:

a. whole numbers

b. integers

c. rational numbers

d. real numbers

Example 4 Graphing Sets of Numbers

Graph the set of **real numbers** $\left\{-\frac{3}{4}, 0, 1, 1.5, 3\right\}$.

Solution

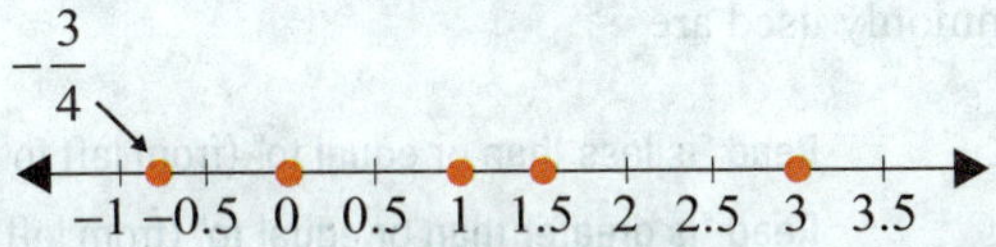

Now work margin exercise 4.

4. Graph the set of real numbers

$\left\{-2.5, -1, 0, \frac{5}{4}, 4\right\}$.

Example 5 Graphing Sets of Numbers

Graph all **natural numbers** less than or equal to 3.

Solution

Remember that the natural numbers are 1, 2, 3, 4, ...

Now work margin exercise 5.

5. Graph all natural numbers less than 6.

C Inequality Symbols

On a horizontal number line, **smaller numbers are always to the left of larger numbers.** Two symbols used to indicate order are

	$<$	Read "is less than" (from left to right)
and	$>$	Read "is greater than" (from left to right)

Using the real number line in Figure 10, you can see the following relationships.

Using <		Using >	
$0 < 3$	0 is less than 3	$3 > 0$	3 is greater than 0
$-2 < 1$	−2 is less than 1	$1 > -2$	1 is greater than −2
$-7 < -4$	−7 is less than −4	$-4 > -7$	−4 is greater than −7
$\frac{3}{8} < \frac{9}{8}$	$\frac{3}{8}$ is less than $\frac{9}{8}$	$\frac{9}{8} > \frac{3}{8}$	$\frac{9}{8}$ is greater than $\frac{3}{8}$

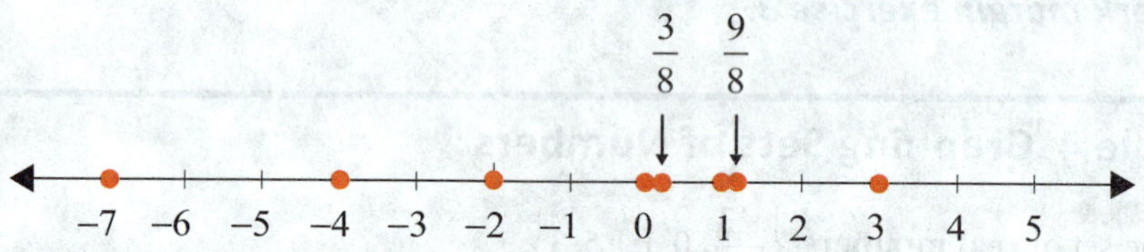

Figure 10

Two other symbols commonly used are

	$\leq$	Read "is less than or equal to" (from left to right)
and	$\geq$	Read "is greater than or equal to" (from left to right)

For example, $5 \geq -10$ is true since 5 is greater than −10. Also, $5 \geq 5$ is true since 5 does equal 5.

Note

Each symbol can be read from left to right as is indicated in "Symbols of Equality and Inequality." However, each symbol can also be read from right to left. Thus, any inequality can be read in two ways. For example, 6 < 10 can be read from left to right as "6 is less than 10," but also from right to left as "10 is greater than 6." We will see that this flexibility is particularly useful when reading expressions with variables in Section 9.7.

Symbols of Equality and Inequality

Reading from left to right:

$=$	is equal to	$\neq$	is not equal to
$<$	is less than	$>$	is greater than
$\leq$	is less than or equal to	$\geq$	is greater than or equal to

Example 6 Verifying Inequalities

Determine whether each statement is true or false.

a. $7 < 15$

b. $3 > -1$

c. $-\frac{1}{2} \leq -\frac{1}{4}$

d. $2.7 \geq 2.7$

e. $-5 > -6$

Solution

a. $7 < 15$ is true, since 7 is less than 15. 7 is to the left of 15 on the number line.

b. $3 > -1$ is true, since 3 is greater than -1. 3 is to the right of -1 on the number line.

c. $-\frac{1}{2} \leq -\frac{1}{4}$ is true, since $-\frac{1}{2}$ is less than $-\frac{1}{4}$. $-\frac{1}{2}$ is to the left of $-\frac{1}{4}$ on the number line. (Reminder, to compare two fractions find equivalent fractions using the LCD as the denominator. Then, compare the numerators.)

d. $2.7 \geq 2.7$ is true, since 2.7 is equal to 2.7.

e. $-5 > -6$ is true, since -5 is greater than -6.

Now work margin exercise 6.

6. Determine whether each of the following statements is true or false.

a. $2 < 10$

b. $6 < -5$

c. $9 \geq -9$

d. $8.5 \leq 8.5$

e. $-3 < -7$

Note

$7 < 15$ can be read as "7 is less than 15" or as "15 is greater than 7."

$3 > -1$ can be read as "3 is greater than -1" or as "-1 is less than 3."

D Absolute Value

In working with the real number line, you may have noticed that any integer and its opposite lie the same number of units from 0 on the number line. For example, both +7 and -7 are seven units from 0 (see Figure 11). The + and − signs indicate direction and the 7 indicates distance.

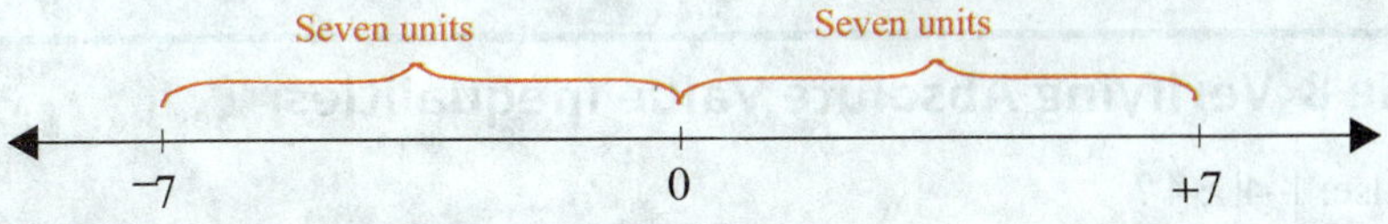

Figure 11

The **distance a number is from 0 on a number line is called its absolute value** and is symbolized by two vertical bars, $|\ \ |$. Thus, $|+7| = 7$ and $|-7| = 7$. Similarly,

$|3| = 3$

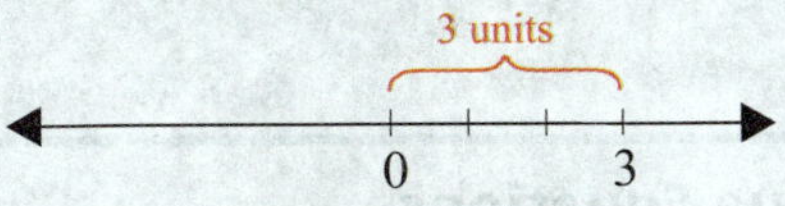

3 is 3 units from 0.

$|-4| = 4$

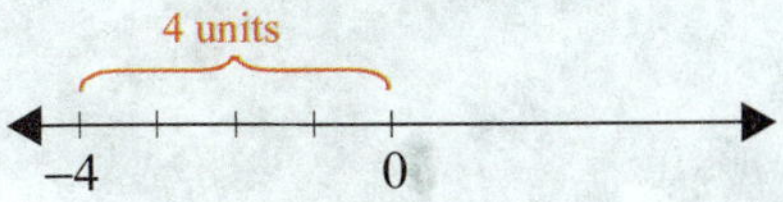

−4 is 4 units from 0.

$|-2| = 2$

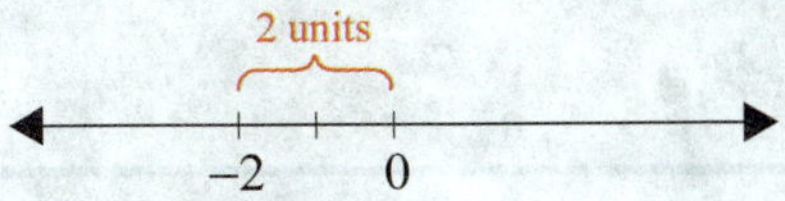

−2 is 2 units from 0.

$|0| = 0$

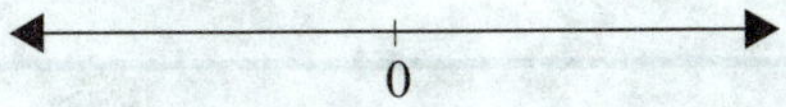

$\left|-\frac{4}{3}\right| = \frac{4}{3}$

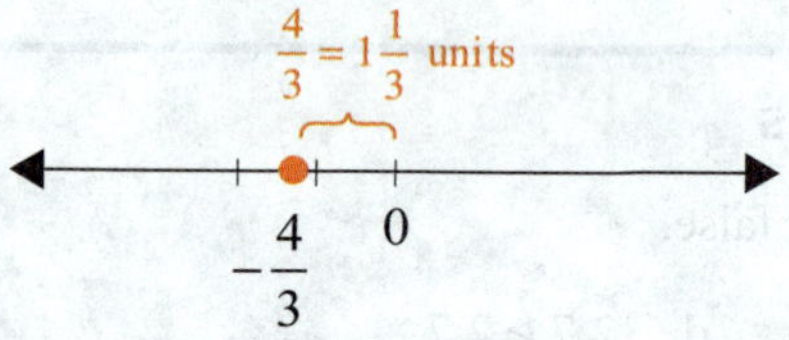

$-\frac{4}{3}$ is $\frac{4}{3} = 1\frac{1}{3}$ units from 0.

Teaching Note

You may want to show your students how to use the absolute value key and the negative key on the graphing calculator at this time. These ideas are included in the exercises.

Since distance (similar to length) is never negative, the absolute value of a number is never negative. Or, the absolute value of a nonzero number is always positive.

Absolute Value

The **absolute value** of a real number is its distance from 0 on a number line. Note that the absolute value of a real number is never negative.

DEFINITION

7. Simplify each absolute value expression.

a. $|+4|$

b. $|-3.3|$

c. $-|-7.4|$

Example 7 Finding Absolute Values

Simplify each absolute value expression.

a. $|6.3|$ **b.** $|-5.1|$ **c.** $-|-2.9|$

Solution

a. $|6.3| = 6.3$ The number 6.3 is 6.3 units from 0.

b. $|-5.1| = 5.1$ The number −5.1 is 5.1 units from 0.

c. $-|-2.9| = -(2.9) = -2.9$ The opposite of the absolute value of −2.9.

Now work margin exercise 7.

8. True or false: $|-20| \le 20$?

Example 8 Verifying Absolute Value Inequalities

True or false: $|-4| \ge 4$?

Solution

True, since $|-4| = 4$ and $4 \ge 4$.

Now work margin exercise 8.

9. If $|z| = 3$, what are the possible values of z?

Example 9 Solving Absolute Value Equations

If $|x| = 7$, what are the possible values for x?

Solution

$x = 7$ or $x = -7$ since $|7| = 7$ and $|-7| = 7$.

Now work margin exercise 9.

10. If $|y| = -19$, what are the possible values of y?

Example 10 Solving Absolute Value Equations

If $|x| = -3$, what are the possible values for x?

Solution

There are no values of x for which $|x| = -3$. The absolute value can never be negative. There is **no solution**.

Now work margin exercise 10.

Example 11 Application: Solving Absolute Value Equations

During the manufacturing of machine parts, certain measurements of the part must stay within a tolerance range or the part will be defective. The quality control manager determines that the maximum amount the length of a screw can vary from its targeted length of 12 millimeters and not be defective is 2 millimeters. If x represents the maximum acceptable difference in length from the target, either positive or negative, then $|x| = 2$. What are the maximum amounts that the screw length can vary from its target?

Solution

To find the maximum the screw length can vary from its target, solve the equation $|x| = 2$ millimeters. This gives the following two values for x: $x = -2$ millimeters and $x = 2$ millimeters.

Thus, the maximum amounts that the screw length can vary are -2 millimeters and 2 millimeters.

Now work margin exercise 11.

11. A quality control manager determines that the maximum amount the width of a latch can vary from its targeted length and not be defective is 3 millimeters. If y represents the maximum acceptable error, either positive or negative, then what are the maximum amounts that the hatch width can vary from its target?

Margin Exercise Answers

1. a. -10 **b.** $+8$ **2. a.** 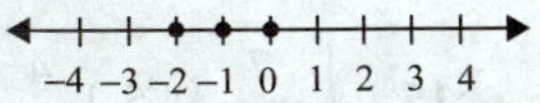

b. 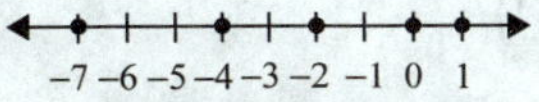

3. a. 1, 20 **b.** -6, 1, 20 **c.** $-6, -\frac{1}{7}, 1, 20$ **d.** All numbers in S are real numbers.

4. 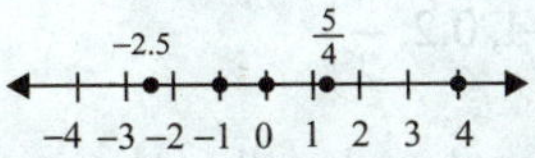

5. 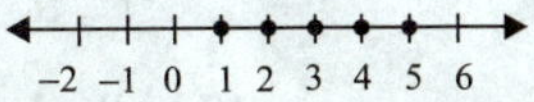

6. a. True **b.** False **c.** True **d.** True **e.** False **7. a.** 4 **b.** 3.3 **c.** -7.4 **8.** True **9.** $z = 3, -3$ **10.** No solution **11.** -3 millimeters and 3 millimeters

8.1 Exercises

Concept Check

Fill-in-the-Blank. Complete each sentence using information found in this section.

1. The set of numbers that includes the whole numbers and their opposites is the set of __________
2. A number that can be written as a fraction is a/an __________ number.
3. Infinite nonrepeating decimal numbers are __________ numbers.
4. A number's distance from 0 on a number line is the number's __________.

8.1 Exercises

Concept Check

1. integers
2. rational
3. irrational
4. absolute value
5. graph
6. inequality

7. True
8. True
9. True
10. False; Zero is neither positive nor negative.

Practice

1. 3
2. 7
3. 0
4. 0
5. −2
6. −6
7. 

8. 

9. 

10. 

11. 

12. 

13. 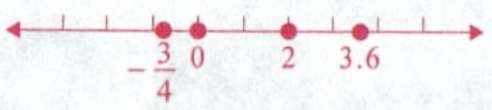

14. 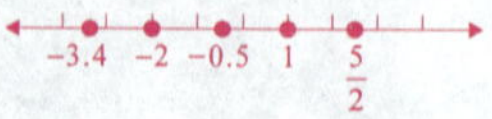

15. 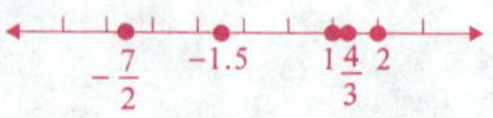

16. 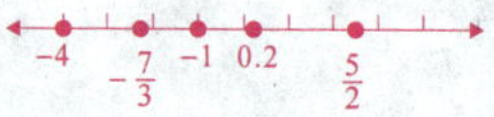

17. 

18. 

19. 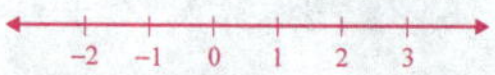

20. 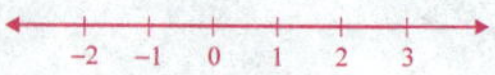

21. 4, 8
22. 0, 4, 8
23. −7, −2, 0, 4, 8
24. $-\sqrt{6}, \sqrt{5}, \sqrt{11}$
25. $-7, -2, -\frac{5}{3}, -1.4, 0, \frac{3}{5}, 4, 5.9, 8$
26. All numbers in A are real numbers.
27. <
28. <

5. The ________ of a number is the point that corresponds to the number on a number line.

6. The symbols < and > are known as ________ symbols.

True/False. Determine whether each statement is true or false. If a statement is false, explain how it can be changed so the statement will be true. (**Note:** There may be more than one acceptable change.)

7. On a number line, smaller numbers are always to the left of larger numbers.

8. The absolute value of a negative number is a positive number.

9. All whole numbers are also integers.

10. Zero is a positive number.

Practice

Find the opposite of each integer. See Example 1.

1. −3
2. −7
3. 0
4. −0
5. +2
6. +6

Graph each set of real numbers on a real number line. See Examples 2, 4, and 5.

7. $\{1, 2, 5, 6\}$
8. $\{-3, -2, 0, 1\}$
9. $\{2, -3, 0, -1\}$
10. $\{-2, -1, 4, -3\}$
11. $\left\{0, -1, \frac{7}{4}, 3, 1\right\}$
12. $\left\{-2, -1, -\frac{1}{3}, 2\right\}$
13. $\left\{-\frac{3}{4}, 0, 2, 3.6\right\}$
14. $\left\{-3.4, -2, -0.5, 1, \frac{5}{2}\right\}$
15. $\left\{-\frac{7}{2}, -1.5, 1, \frac{4}{3}, 2\right\}$
16. $\left\{-4, -\frac{7}{3}, -1, 0.2, \frac{5}{2}\right\}$
17. all whole numbers less than 4
18. all negative integers greater than −4
19. all whole numbers less than 0
20. all natural numbers less than or equal to −1

List the numbers in the set $A = \left\{-7, -\sqrt{6}, -2, -\frac{5}{3}, -1.4, 0, \frac{3}{5}, \sqrt{5}, \sqrt{11}, 4, 5.9, 8\right\}$ that are described in each exercise. See Example 3.

21. Natural numbers
22. Whole numbers
23. Integers
24. Irrational numbers
25. Rational numbers
26. Real numbers

Fill in each blank with the appropriate symbol that will make the statement true: <, >, or =. See Examples 6 through 8.

27. 4 ___ 6

28. -3 ___ 1

29. -2 ___ -4

30. -8 ___ 0

31. -20 ___ -19

32. -67 ___ -50

33. $-(-4.3)$ ___ 4.3

34. 5.6 ___ $-(-8.7)$

35. $-\frac{3}{4}$ ___ -1

36. -2.3 ___ $-2\frac{3}{10}$

37. $\frac{1}{3}$ ___ $\frac{1}{2}$

38. $-\frac{1}{2}$ ___ $-\frac{1}{3}$

39. $|-4|$ ___ 4

40. $|7|$ ___ -7

41. $|-8|$ ___ -8

42. -15 ___ $|-15|$

Determine whether each statement is true or false. If a statement is false, rewrite it in a form that is a true statement. (There may be more than one way to correct a statement.) See Examples 6 and 8.

43. $0 = -0$

44. $-22 < -16$

45. $-9 > -8.5$

46. $-17 \le 17$

47. $4.7 \ge 3.5$

48. $|-5| = 5$

49. $-|-7| \ge -|7|$

50. $|-8| \ge 4$

51. $-|-3| < -|4|$

52. $\left|\frac{5}{2}\right| < 2$

List the possible values for x for each statement. See Examples 9 and 10.

53. $|x| = 5$

54. $|x| = 8$

55. $|x| = 2$

56. $|x| = 0$

57. $|x| = -6$

58. $|x| = -1$

59. $|x| = 23$

60. $|x| = 105$

Choose the response that correctly completes each sentence. Assume that the variables represent integers. In each problem, give two examples that illustrate your reasoning.

61. $|a|$ is (never, sometimes, always) equal to a.

62. $|x|$ is (never, sometimes, always) equal to $-x$.

63. $|y|$ is (never, sometimes, always) equal to a positive integer.

64. $|x|$ is (never, sometimes, always) greater than x.

29. >
30. <
31. <
32. <
33. =
34. <
35. >
36. =
37. <
38. <
39. =
40. >
41. >
42. <
43. True
44. True
45. False; $-9 < -8.5$
46. True
47. True
48. True
49. True
50. True
51. False; $-|-3| > -|4|$
52. False; $\left|-\frac{5}{2}\right| > 2$
53. $x = -5, 5$
54. $x = -8, 8$
55. $x = -2, 2$
56. $x = 0$
57. No solution
58. No solution
59. $x = -23, 23$
60. $x = -105, 105$
61. Sometimes. Examples will vary.
62. Sometimes. Examples will vary.
63. Always. Examples will vary.
64. Sometimes. Examples will vary.

Applications

Solve. Represent each quantity with a signed integer.

65. The Alvin is a manned deep-ocean research submersible that has explored the wreck of the Titanic. The operating depth of the Alvin is 4500 meters below sea level.

66. The Mariana trench is the deepest known location on the Earth's ocean floor. The deepest known part of the Mariana Trench is approximately 11 kilometers below sea level.

67. Mount Everest is considered to be the highest mountain on Earth. Its peak reaches to a height of approximately 8844 meters.

68. The lowest temperature ever recorded was at the Vostok Station on Antarctica. On July 21, 1983, the temperature was approximately 128 degrees Fahrenheit below zero.

69. Terrence placed a drop of colored water on the center of a white strand of yarn and measured how much the color spread. Before placing the drop, he predicts that the color will spread no more than 3 inches away from the initial drop.

a. Write an absolute value inequality using the variable x to represent the predicted spread.

b. Graph the solution set of integers for the absolute value inequality from part a. on the given number line, placing the initial drop at the point 0.

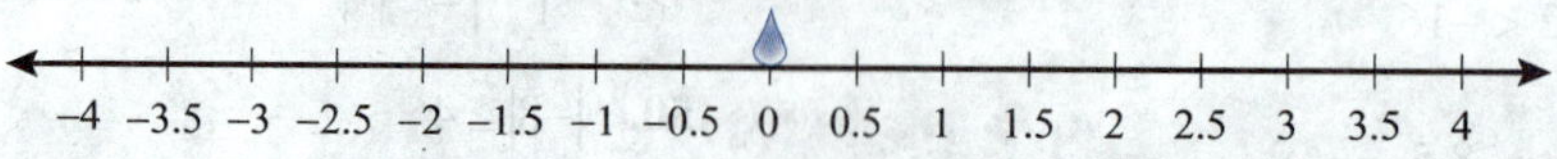

70. A ready-to-assemble bookcase contains wooden boards that have predrilled holes along with the screws and washers needed for assembly. The screws used to assemble the bookcase need to have a length of 38 mm with a tolerance of 2 mm. If the screw is too short, it won't be able to hold the pieces of wood together. If the screw is too long, it might stick out of the other end of the board.

a. What is the largest length the screws can have before they are too long?

b. What is the smallest length the screws can have before they are too short?

c. Graph the tolerance of the screw. (Graph only the integers in the tolerance range.)

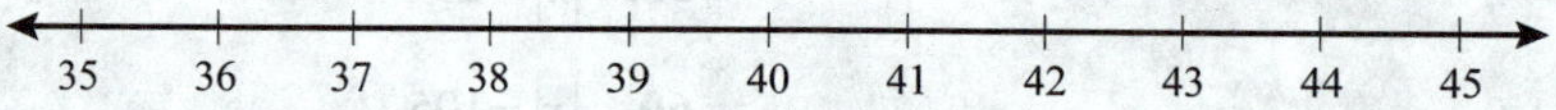

71. A freezer in a biology lab is supposed to be kept at 0 °C. A lab assistant places a thermometer in the freezer and marks down the temperature every half hour. She records the following temperatures in degrees Celsius: $\left\{2, -\frac{1}{2}, \frac{1}{2}, 1\frac{1}{2}, -\frac{5}{2}\right\}$.

a. Graph the set of temperatures on a number line.

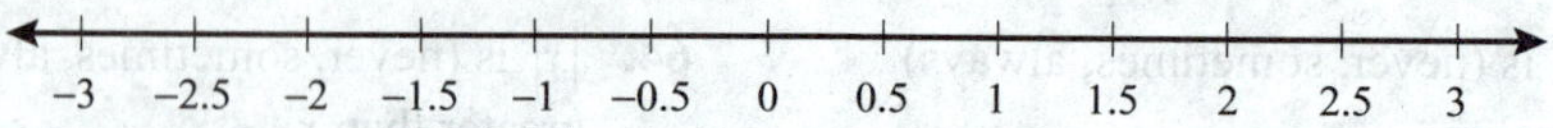

b. Which value is the furthest away from 0?

Applications

65. −4500 meters

66. −11 kilometers

67. +8844 meters

68. −128 degrees Fahrenheit

69. **a.** $|x| \leq 3$

b.

70. **a.** 40 mm **b.** 36 mm

c.

71. **a.**

b. $-\frac{5}{2}$ has the furthest distance from zero.

72. **a.** $x = -3$ mm, 3 mm

b. The length of the screw can be 3 mm longer or 3 mm shorter than 15 mm.

c. 12 mm and 18 mm

73. **a.** $x = -5$ millimeters, 5 millimeters

b. The length of the board can be 5 mm longer or 5 mm shorter than 1 meter.

c. 1.05 m or 0.95 m

74. **a.** 70 °F

b. 84 °F

75. **a.** Fresno

b. El Centro

72. During the manufacture of machine parts, certain measurements of the part must stay within a tolerance range or the part will be defective. The quality control manager determines that the maximum amount the length of a screw can vary from its targeted length of 15 millimeters and not be defective is 3 millimeters. If x represents the maximum acceptable difference in length from the target, either positive or negative, then $|x| = 3$.

   a. Solve the equation $|x| = 3$ to determine the maximum amounts that the screw length can vary from its target length.

   b. What do the answers from part a. mean?

   c. Determine the maximum and minimum lengths of a screw with the given tolerance range.

73. A carpenter needs to cut a board 1 meter long to create the backboard for a bookshelf. He has determined that the maximum amount the length of the board can vary from its required length and not cause structural issues is 5 mm. If x represents the maximum acceptable difference in length from the target, either positive or negative, then $|x| = 5$.

   a. Solve the equation $|x| = 5$ to determine the maximum amounts that the board length can vary from its target length.

   b. What do the answers from part a. mean?

   c. Determine the maximum and minimum lengths of a board with the given tolerance range.

74. Cynthia believes that the ideal daytime temperature for growing tomato plants is 77 °F, and the plants will be okay as long as the temperature does not vary more than 7 °F from that ideal temperature during the day.

   a. How low can the daytime temperature be to accommodate the tomato plant growth?

   b. How high can the daytime temperature be to accommodate the tomato plant growth?

75. The following table shows the elevation of six California cities. (Sea level is defined to have 0 feet of elevation.)

City	Elevation (in feet)
Alameda	50
Death Valley	–282
El Centro	–39
Fresno	296
Salton City	–125
Windsor	118

   a. Which of these cities has an elevation farthest away from sea level?

   b. Which of these cities has an elevation closest to sea level?

Writing & Thinking

76. Give one example each of the use of a positive number, a negative number, and the number zero (outside of a class).

77. Explain, in your own words, how an expression such as $-y$ might represent a positive number.

78. Compare and contrast absolute value with opposites.

Writing & Thinking

76. Examples will vary. Temperature can be negative, zero, or positive. Altitude/elevation can be below sea level (a negative number), at sea level (0), or above sea level (a positive number). A person's banking/credit accounts can have balances that are negative, 0, or positive. A particular stock can trade lower (negative), higher (positive), or at the same price (0).

77. If y is a negative number then $-y$ represents a positive number. For example, if $y = -2$, then $-y = -(-2) = 2$.

78. Absolute value is always positive or zero because it is measuring a distance. Opposites always change signs because they are located the same distance from zero but in the opposite direction so that their sign is also opposite.

Objectives

A. Add real numbers.

8.2 Addition with Real Numbers

A Addition with Real Numbers

Picture a straight line in an open field and numbers marked on a number line. An archer stands at 0 and shoots an arrow to +3, then stands at 3 and shoots the arrow 5 more units in the positive direction (to the right). Where will the arrow land? (See Figure 1.)

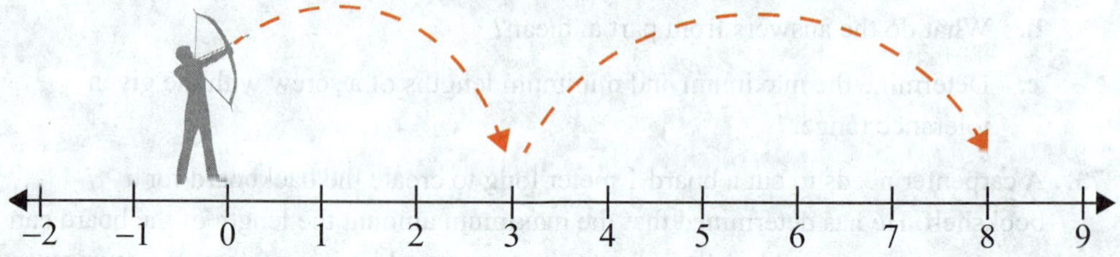

Figure 1

Naturally, you have figured out that the answer is +8. What you have done is added the two positive integers, +3 and +5.

$$(+3)+(+5)=+8 \quad \text{or} \quad 3+5=8$$

Suppose another archer shoots an arrow in the same manner as the first but in the opposite direction. Where would his arrow land? The arrow lands at −8. You have just added −3 and −5. (See Figure 2.)

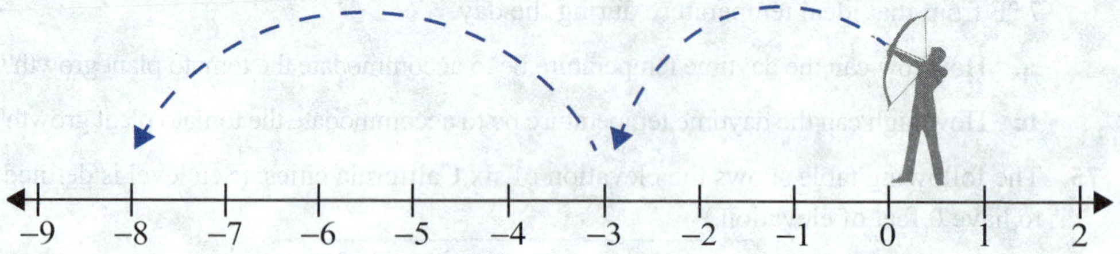

$$(-3)+(-5)=-8$$

Figure 2

If the archer stands at 0 and shoots an arrow to +3 and then the archer goes to +3 and turns around and shoots an arrow 5 units in the opposite direction, where will the arrow stick? Would you believe at −2? (See Figure 3.)

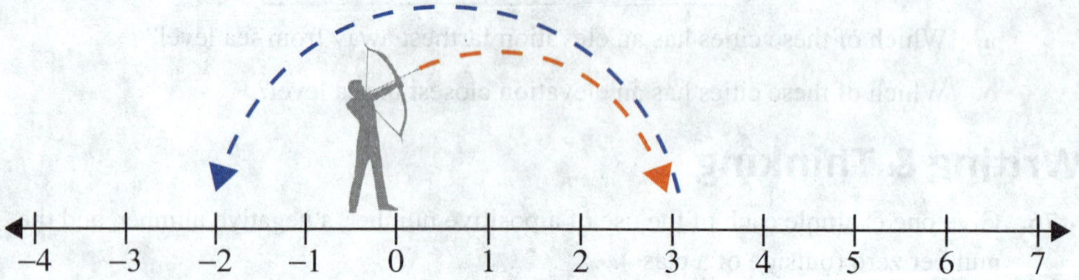

$$(+3)+(-5)=-2$$

Figure 3

For our final archer, the first shot is to –3. Then after going to –3, he turns around and shoots 5 units in the opposite direction. Where is the arrow? It is at +2. (See Figure 4.)

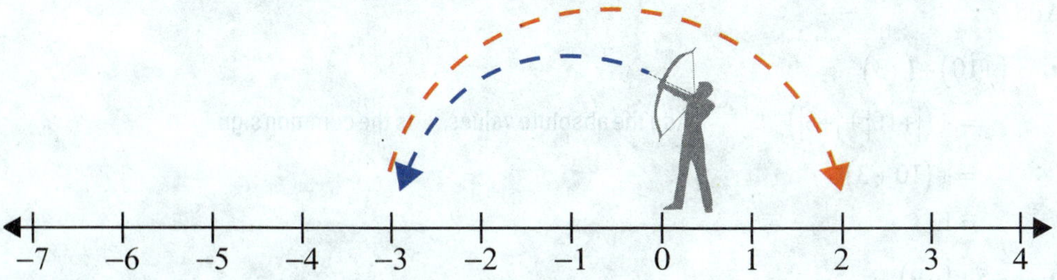

$$(-3)+(+5)=+2$$

Figure 4

In summary:

1. The sum of two positive real numbers is positive.

(+3)	+	(+5)	=	+8
↑	↑	↑	↑	↑
positive	plus	positive	is	**positive**

2. The sum of two negative real numbers is negative.

(–3)	+	(–5)	=	–8
↑	↑	↑	↑	↑
negative	plus	negative	is	**negative**

3. The sum of a positive real number and a negative real number may be negative or positive (or zero) depending on which number is farther from 0.

(+3)	+	(–5)	=	–2	(+5)	+	(–3)	=	+2
↑	↑	↑	↑	↑	↑	↑	↑	↑	↑
positive	plus	negative	is	**negative**	positive	plus	negative	is	**positive**

The rules for addition with real numbers can be summarized in the following manner.

Rules for Addition with Real Numbers

1. To add two real numbers with **like signs**,
   a. add their absolute values and
   b. use the common sign.
2. To add two real numbers with **unlike signs**,
   a. subtract their absolute values (the smaller from the larger) and
   b. use the sign of the number with the larger absolute value.

PROCEDURE

Study the following examples carefully to help in understanding the rules for addition.

1. Add.

a. $4 + 11$

b. $(-15)+(-5)$

c. $3.1 + 2.8$

d. $\left(-\frac{2}{5}\right)+\left(-\frac{1}{3}\right)$

Note

The positive sign (+) may be omitted when writing positive numbers, but the negative sign (–) must always be written for negative numbers. Thus, if there is no sign in front of a real number, the real number is understood to be positive.

Example 1 Adding Real Numbers with Like Signs

Add.

a. $(+10)+(+3)$

$= +(|+10|+|+3|)$ Add the absolute values, + is the common sign.

$= +(10+3)$

$= 13$

b. $-5+(-9)$

$= -(|-5|+|-9|)$ Add the absolute values, – is the common sign.

$= -(5+9)$

$= -14$

c. $-2.5+(-1.4)$

$= -(|-2.5|+|-1.4|)$ Add the absolute values, – is the common sign.

$= -(2.5+1.4)$

$= -3.9$

d. $\frac{3}{4}+\frac{1}{3}$

$= +\left(\left|\frac{3}{4}\right|+\left|\frac{1}{3}\right|\right)$ Add the absolute values, + is the common sign.

$= +\left(\frac{3}{4}+\frac{1}{3}\right)$

$= +\left(\frac{3}{4}\cdot\frac{3}{3}+\frac{1}{3}\cdot\frac{4}{4}\right)$ (Reminder, to add fractions use the LCD to write each fraction over a common denominator. Then, add the numerators.)

$= +\left(\frac{9}{12}+\frac{4}{12}\right)$

$= \frac{13}{12}$

Now work margin exercise 1.

2. Add.

a. $10+(-2)$

b. $-12 +5$

c. $1.9+(-4.2)$

d. $-\frac{1}{3}+\frac{7}{9}$

Example 2 Adding Real Numbers with Unlike Signs

Add.

a. $7+(-6)$

$= +(|7|-|-6|)$ Subtract the absolute values. Use + because $|7| > |-6|$.

$= +(7-6)$

$= 1$

b. $-3+8$

$= +(|8|-|-3|)$ Subtract the absolute values. Use + because $|8| > |-3|$.

$= +(8-3)$

$= 5$

c. $8.7+(-9.5)$

$=-(|-9.5|-|8.7|)$ Subtract the absolute values. Use – because $|-9.5|>|8.7|$.

$=-(9.5-8.7)$

$=-0.8$

d. $-\frac{5}{8}+\frac{1}{2}$

$=-\left(\left|-\frac{5}{8}\right|-\left|\frac{1}{2}\right|\right)$ Subtract the absolute values. Use – because $\left|-\frac{5}{8}\right|>\left|\frac{1}{2}\right|$.

$=-\left(\frac{5}{8}-\frac{1}{2}\right)$

$=-\left(\frac{5}{8}-\frac{1}{2}\cdot\frac{4}{4}\right)$ (Reminder, to subtract fractions use the LCD to write each fraction over a common denominator. Then, subtract the numerators.)

$=-\left(\frac{5}{8}-\frac{4}{8}\right)$

$=-\frac{1}{8}$

Now work margin exercise 2.

One technique to add more than two real numbers is to add left to right.

Example 3 Adding Three or More Real Numbers

Add.

a. $-3+2+(-5)$

b. $6.0+(-4.3)+(-1.5)$

Solution

a. $-3+2+(-5)=-1+(-5)=-6$

b. $6.0+(-4.3)+(-1.5)=1.7+(-1.5)=0.2$

Now work margin exercise 3.

Another technique for adding several real numbers is to add the positive and negative real numbers separately and then add the results. (We are, in effect, using the commutative and associative properties of addition.)

Example 4 Adding Three or More Real Numbers

Add.

a. $-9+13+(-11)$

b. $7+(-20)+(-30)+8$

Solution

a. $-9+13+(-11)=-9+(-11)+13=-20+13=-7$

b. $7+(-20)+(-30)+8=7+8+(-20)+(-30)=15+(-50)=-35$

Now work margin exercise 4.

Math Tip

Recall that when adding numbers with unlike signs, the answer will have the sign of the number that has the largest absolute value.

3. Add.

a. $-7+5+(-3)$

b. $-3.2+(-6.1)+5.7$

4. Add.

a. $-15+6+(-25)$

b. $16+(-12)+(-5)+8$

8.2 Exercises

Concept Check

1. positive
2. positive
3. negative
4. absolute values
5. equation
6. False; The sum of a positive and negative number can be positive, negative, or zero.
7. True
8. False; The sum of two positive numbers is always positive, zero is neither positive nor negative.

Practice

1. 13
2. 5
3. −4
4. −10
5. 0
6. −6
7. 5
8. −9
9. −13
10. 0
11. −8
12. 9
13. −10
14. 7
15. 0
16. 13
17. 17
18. −2.5
19. −4.6
20. −18
21. −29
22. 8
23. −9
24. −10.1

Margin Exercise Answers

1. a. 15 **b.** −20 **c.** 5.9 **d.** $-\frac{11}{15}$ **2. a.** 8 **b.** −7 **c.** −2.3 **d.** $\frac{4}{9}$ **3. a.** −5 **b.** −3.6 **4. a.** −34 **b.** 7

8.2 Exercises

Concept Check

Fill-in-the-Blank. Complete each sentence using information found in this section.

1. If there is no sign in front of a number, it is understood to be a __________ number.
2. The sum of two positive real numbers is always __________.
3. The sum of two negative real numbers is always __________.
4. To find the sum of numbers with unlike signs, subtract their __________.
5. A statement that two expressions are equal is called a/an __________.

True/False. Determine whether each statement is true or false. If a statement is false, explain how it can be changed so the statement will be true. (**Note:** There may be more than one acceptable change.)

6. The sum of a positive number and a negative number is always positive.
7. When adding numbers with unlike signs, the results uses the sign of the number with the larger absolute value.
8. The sum of two positive numbers can equal zero.

Practice

Add. Reduce any fractions to lowest terms.

1. $4+9$
2. $8+(-3)$
3. $(-9)+5$
4. $(-7)+(-3)$
5. $(-9)+9$
6. $2+(-8)$
7. $11+(-6)$
8. $(-12)+3$
9. $-18+5$
10. $26+(-26)$
11. $-5+(-3)$
12. $11+(-2)$
13. $(-2)+(-8)$
14. $10+(-3)$
15. $17+(-17)$
16. $(-7)+20$
17. $21+(-4)$
18. $2.1+(-4.6)$
19. $-1.5+(-3.1)$
20. $(-15)+(-3)$

21. $(-12)+(-17)$

22. $24+(-16)$

23. $-4+(-5)$

24. $-4.3+(-5.8)$

25. $-6.9+(-8.5)$

26. $(-6)+(-8)$

27. $9+(-12)$

28. $-12+9$

29. $(-33)+(-21)$

30. $(-21)+18$

31. $9.7+(-12.2)$

32. $-19.6+4.1$

33. $\frac{3}{14}+\frac{3}{14}$

34. $\frac{1}{10}+\frac{3}{10}$

35. $\frac{3}{4}+\left(-\frac{1}{8}\right)$

36. $\frac{5}{17}+\left(-\frac{15}{34}\right)$

37. $-\frac{5}{2}+\frac{3}{4}$

38. $-\frac{1}{6}+\frac{7}{15}$

39. $-3+4+(-8)$

40. $(-9)+(-6)+5$

41. $3.2+(-1.2)+(-2.5)$

42. $-5.3+1.9+(-0.7)$

43. $102+(-21)+(-5)$

44. $130+(-45)+(-32)$

45. $-210+(-200)+100$

46. $-18+(-15)+(-30)$

47. $35+2+(-5)+(-5)$

48. $-56+(-3)+(-1)+3$

49. $3.7+(-0.6)+1.4+(-2.2)$

50. $-7.5+(-2.4)+3.5+6.1$

Add. Be sure to find the absolute values first.

51. $13+|-5|$

52. $|-2|+(-5)$

53. $|-10|+|-4|$

54. $|-7|+(+7)$

55. $|-18|+|+17|$

56. $|-14|+|-6|$

Add using a calculator.

57. $47+(-29)+66$

58. $56+(-41)+(-28)$

59. $2932+4751+(-3876)$

60. $(-8154)+2147+(-136)$

61. $(-16{,}945)+(-27{,}302)+(-53{,}467)$

62. $(-12{,}299)+15{,}631+(-47{,}558)$

25. −15.4
26. −14
27. −3
28. −3
29. −54
30. −3
31. −2.5
32. −15.5
33. $\frac{3}{7}$
34. $\frac{2}{5}$
35. $\frac{5}{8}$
36. $-\frac{5}{34}$
37. $-\frac{7}{4}$
38. $\frac{3}{10}$
39. −7
40. −10
41. −0.5
42. −4.1
43. 76
44. 53
45. −310
46. −63
47. 27
48. −57
49. 2.3
50. −0.3
51. 18
52. −3
53. 14
54. 14
55. 35
56. 20
57. 84
58. −13
59. 3807
60. −6143
61. −97,714
62. −44,226

Applications

63. Profit: \$6000

64. a. \$45,000 + (−\$8000) + (−\$2000) + \$15,000

b. \$50,000

65. a. −0.3 cm, −0.1 cm

b. 0.02 cm

c. Increase

66. Floor 3

67. 275 feet

68. −2 °C

69. 8 feet

70. −17

Applications

Solve.

63. The table shows the reported profit or loss per quarter as reported by a business. Did the business have a total positive or negative profit for the year?

Quarter	Profit/Loss
1	\$15,000
2	−\$8000
3	−\$2000
4	\$1000

64. For 2024, a business reports a profit of \$45,000 during the first quarter, a loss of \$8000 during the second quarter, a loss of \$2000 during the third quarter, and a profit of \$15,000 during the fourth quarter.

a. Write an addition expression to represent the total profit made by the company in 2024. Do not simplify.

b. Simplify the expression from part a.

65. A climatologist takes weekly measurements of the height of a glacier near the North Pole. She keeps track of how much the glacier's height either increased or decreased during the week. Her results are presented in the table.

Week	Increase	Decrease
1	0.25 cm	
2		0.3 cm
3		0.1 cm
4	0.17 cm	

a. Which measurements in the table would have a negative value?

b. Calculate the total change in height of the glacier over the four weeks. (**Hint:** Find the sum.)

c. Did the total height of the glacier increase or decrease by the end of the four weeks?

66. A passenger boards an elevator five floors below the ground floor. In this building, the ground floor is floor 0 and the floor above the ground floor is floor 1. The elevator goes up 8 floors before the passenger exits the elevator. At which floor did the passenger exit the elevator?

67. A submarine dives to a depth of 250 feet below the surface. It rises 75 feet before diving an additional 100 feet. What is the final depth of the submarine?

68. The temperature at 2 a.m. was −17 °C. By 2 p.m., the temperature increased a total of 15 °C. What was the temperature at 2 p.m.?

69. The tallest hill of a roller coaster is 282 feet above the ground. The hill descends 290 feet before leveling out. What is the lowest point of this hill of the roller coaster?

70. From the noon weather report to the evening weather report, the temperature changed from 72 °F to 55 °F. This situation can be represented by the equation $72 + t = 55$, where t represents the change in temperature. Determine which of the following values satisfies the equation: −13, 13, −17, 17.

71. At the end of the first inning of a baseball game, the home team had a score of 3 points. At the end of the ninth inning, the home team had a score of 11 runs. This situation can be represented by the equation $3 + x = 11$, where x represents the change in score. Determine which of the following values satisfies the equation: −8, 8, −4, 4.

72. Charlotte is a zoologist and part of her job is to keep track of the growth rate of a recently born koala. She writes in her report that the koala weighs 5.6 ounces more than it did when it was born a month ago, and the current weight is 28.4 ounces. This can be translated into a mathematical equation as $w + 5.6 = 28.4$, where w is the weight in ounces of the koala at birth. Determine the birth weight of the koala by substituting each of the following values into the equation to find the solution: 22.7 ounces, 23.2 ounces, 22.8 ounces, 23.8 ounces.

73. Part of Noam's job as an accountant is to keep track of the amount of money in the reserve fund. Last week, the fund started with \$1253.75 and only one transaction was made. This week, the fund started with \$1155.89. This change in value can be written in equation form as \$1253.75 + t = \$1155.89, where t is the amount of the transaction. Determine the amount of the transaction by substituting each of the following values into the equation to find the solution: −\$150.14, −\$97.86, −\$89.86, −\$97.14.

74. Trevor is installing a hardwood floor in a customer's living room. The length of the room is $17\frac{3}{8}$ feet. Since the flooring comes in 12-foot pieces, Trevor needs to determine the length he must cut off of one of the boards to make it fit. The amount he needs to trim off of one of the boards can be represented by the equation $12 + (12 - r) = 17\frac{3}{8}$, where r is the amount to be removed from one of the boards. Determine the amount that needs to be trimmed off of one of the boards by substituting each of the following values into the equation to find the solution: $5\frac{5}{8}$ feet, $5\frac{3}{8}$ feet, $6\frac{5}{8}$ feet, $6\frac{3}{8}$ feet.

Writing & Thinking

75. Describe, in your own words, how the sum of the absolute values of two numbers might be 0. (Is this even possible?)

76. Describe in your own words the conditions under which the sum of two integers will be 0.

Choose the response that correctly completes each statement. In each problem, give two examples that illustrate your reasoning.

77. If x and y are real numbers, then $x + y$ is (never, sometimes, always) equal to 0.

78. If x and y are real numbers, then $x + y$ is (never, sometimes, always) negative.

79. If x and y are real numbers, then $x + y$ is (never, sometimes, always) positive.

80. If x is a positive real number and y is a negative real number, then $x + y$ is (never, sometimes, always) equal to 0.

81. If x and y are positive real numbers, then $x + y$ is (never, sometimes, always) equal to 0.

82. If x and y are both negative real numbers, then $x + y$ is (never, sometimes, always) equal to 0.

83. If x is a negative real number, then $-x$ is (never, sometimes, always) negative.

84. If x is a positive real number, then $-x$ is (never, sometimes, always) negative.

71. 8
72. The solution is 22.8 ounces.
73. The solution is −\$97.86.
74. The solution is $6\frac{5}{8}$ feet.

Writing & Thinking

75. $|0| + |0| = 0$
76. If they are opposites.
77. Sometimes. Examples will vary.
78. Sometimes. Examples will vary.
79. Sometimes. Examples will vary.
80. Sometimes. Examples will vary.
81. Never. Examples will vary.
82. Never. Examples will vary.
83. Never. Examples will vary.
84. Always. Examples will vary.

Objectives

A. Find the additive inverse (opposite) of a given number.

B. Subtract real numbers.

C. Find the change in value between two real numbers.

8.3 Subtraction with Real Numbers

A Additive Inverses

In basic arithmetic, subtraction is defined in terms of addition. For example, we know that the difference 32 – 25 is equal to 7 because 25 + 7 = 32. A beginning student in arithmetic does not know how to find a difference such as 15 – 20, where a larger number is subtracted from a smaller number, because negative numbers are not yet defined and there is no way to add a positive number to 20 and get 15. Now, with our knowledge of negative numbers, we will define subtraction in such a way that larger numbers may be subtracted from smaller numbers. We will still define subtraction in terms of addition, but we will apply our new rules for addition with real numbers.

Before developing methods of **subtraction** with real numbers, we state and illustrate an important relationship between any real number and its opposite.

Additive Inverse

The **opposite** of a real number is called its **additive inverse**. The sum of a number and its additive inverse is zero. Symbolically, for any real number a,

$$a+(-a)=0.$$

DEFINITION

1. Find the additive inverse (opposite) of each number.

a. 14

b. –9.2

Example 1 Finding Additive Inverses

Find the additive inverse (opposite) of each number.

a. 3 **b.** –7.3 **c.** 0

Solution

a. The additive inverse of 3 is –3, because $3+(-3)=0$.

b. The additive inverse of –7.3 is $-(-7.3)=+7.3,$ because $(-7.3)+7.3=0.$

c. The additive inverse of 0 is –0 = 0. That is, 0 is its own opposite.

Now work margin exercise 1

B Subtraction with Real Numbers

Intuitively, we can think of addition of numbers (positive or negative or both) as "piling on" or "accumulating" numbers. For example, when we add positive numbers (or negative numbers), the sum is a number more positive (or more negative). Figures 1 and 2 illustrate these ideas.

a. Adding positive numbers piles on or accumulates the numbers in a positive direction.

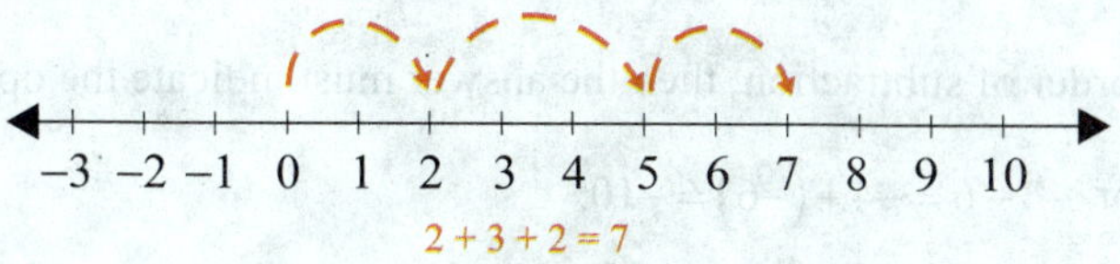

Figure 1

b. Adding negative numbers piles on or accumulates the numbers in a negative direction.

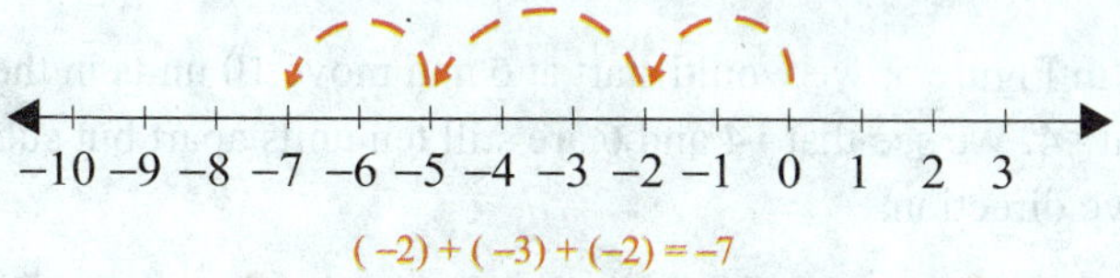

Figure 2

In **subtraction**, we want to find the "difference between" two numbers. On a number line, this translates as the "distance between" the two numbers **with direction considered.** As illustrated in Figure 3, the distance between 6 and 1 is five units. To find this distance, we subtract: $6-1=6+(-1)=5$. Note that to subtract 1, we add the opposite of 1 (which is –1).

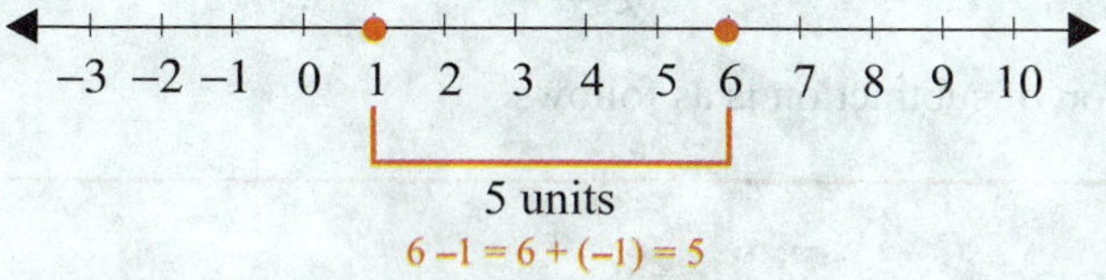

Figure 3

In Figure 4, we see that the distance between 6 and –4 is ten units. Again, to find this distance, we subtract $6-(-4)$. But we know that we must have 10 as an answer. To get 10, we add the opposite of –4 (which is +4) as follows: $6-(-4)=6+(+4)=10$.

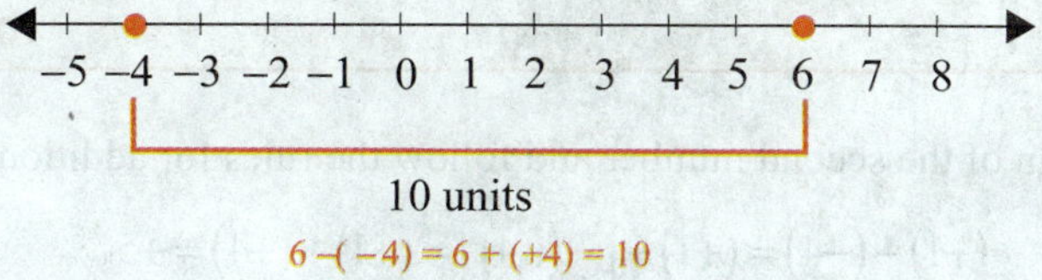

Figure 4

To understand the "direction" in subtraction, think of subtraction on the number line as

(end value) – (beginning value).

This means that for $6-(-4)=6+(+4)=10$,

we have 6 – (–4) = 6 + (+4) = 10

↑ (end value) – ↑ (beginning value)

and, as illustrated in Figure 4, we would start at –4 and move 10 units in the **positive direction** to end at 6.

As illustrated in Figures 1 through 4, **subtraction is defined in terms of addition**.

To subtract, add the opposite of the number being subtracted.

If we reverse the order of subtraction, then the answer must indicate the opposite direction.

This means that for $-4-6=-4+(-6)=-10$,

we have $-4 \quad - \quad 6 \quad = \quad -4+(-6)=-10$

(end value) − (beginning value)

and, as illustrated in Figure 5, we would start at 6 and move 10 units in the **negative direction** to end at −4. We see that −4 and 6 are still ten units apart but subtraction now indicates a negative direction.

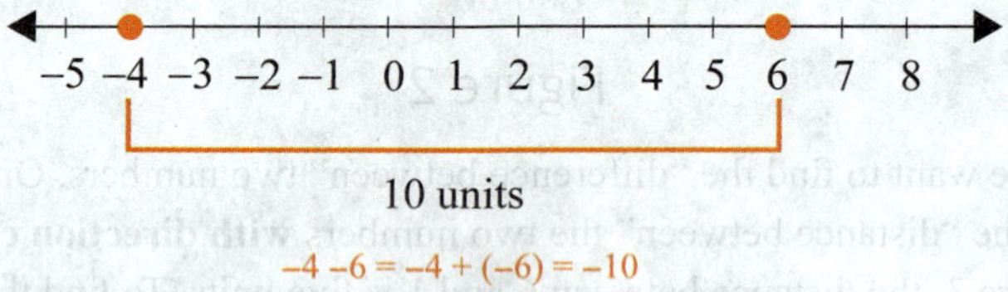

Figure 5

The formal definition of subtraction is as follows.

Subtraction

For any real numbers a and b,

$$a-b=a+(-b).$$

In words, to subtract b from a, **add** the **opposite** of b to a.

DEFINITION

We change the sign of the second number and follow the rules for addition.

$$(+1)-(-4)=(+1)+[-(-4)]=(+1)+(+4)=+5$$

add opposite

$$(-10)-(-3)=(-10)+[-(-3)]=(-10)+(+3)=-7$$

add opposite

2. Subtract.

a. $-3-8$

b. $-4-(-4)$

c. $-1.5-(-5.3)$

d. $-\frac{1}{8}-\frac{5}{8}$

e. $\frac{4}{5}-\left(-\frac{1}{2}\right)$

Example 2 Subtracting Real Numbers

Subtract.

a. $-4-1=(-4)+(-1)=-5$

b. $-1-(-8)=(-1)+(+8)=7$

c. $10.3-(-2.3)=10.3+(+2.3)=10.3+2.3=12.6$

d. $-\frac{3}{16}-\frac{1}{16}=-\frac{3}{16}+\left(-\frac{1}{16}\right)=-\frac{4}{16}=-\frac{\not{4}\cdot 1}{\not{4}\cdot 4}=-\frac{1}{4}$

e. $-\frac{2}{3}-\left(-\frac{1}{6}\right)=-\frac{2}{3}+\left(+\frac{1}{6}\right)=\left(-\frac{2}{3}\cdot\frac{2}{2}\right)+\frac{1}{6}=-\frac{4}{6}+\frac{1}{6}=-\frac{3}{6}=-\frac{1}{2}$

Now work margin exercise 2.

In practice, the notation $a - b$ is thought of as addition of signed numbers. That is, because $a-b=a+(-b)$, we think of the plus sign, +, as being present in $a - b$. In fact, an expression such as 4 – 19 can be thought of as "four plus negative nineteen." We have the following.

$$4-19=4+(-19)=-15$$
$$-25-30=-25+(-30)=-55$$
$$-3-(-17)=-3+(+17)=14$$
$$24-11-6=24+(-11)+(-6)=7$$

Generally, the second step is omitted and we go directly to the answer by computing the sum mentally.

$$4-19=-15$$
$$-25-30=-55$$
$$-3-(-17)=14$$
$$24-11-6=7$$

Math Tip

Notice that the symbol – is used in two different ways. One way is a verb meaning *subtract*, as in $a - b$. The other way is as an adjective meaning *negative*, as in $a+(-b)$. We tie the two uses together with $a-b=a+(-b)$.

C Change in Value

To find the **change in value** between two numbers, take the end value and subtract the beginning value. Symbolically,

change in value = (end value) – (beginning value).

Example 3 Application: Calculating Change in Value

At noon on Tuesday the temperature was 34 °F. By noon on Thursday the temperature had changed to −5 °F. How much did the temperature change between Tuesday and Thursday?

3. At 3 p.m. on Friday the temperature was 63 °F. By midnight the temperature had changed to 44 °F. How much did the temperature change between 3 p.m. and midnight?

Solution

For change in value:

$$(\text{end value}) - (\text{beginning value}) = (-5) - (+34)$$
$$= -5 + (-34)$$
$$= -39$$

Between Tuesday and Thursday the temperature changed −39 °F (or dropped 39 °F).

Now work margin exercise 3.

4. A drone plane flew from an altitude of 25,000 ft to an altitude of 14,000 ft. What was the change in altitude?

Example 4 Application: Calculating Change in Value

A jet pilot flew her plane from an altitude of 30,000 ft to an altitude of 12,000 ft. What was the change in altitude?

Solution

end altitude	−	beginning altitude	=	change in altitude
12,000	−	30,000	=	−18,000 ft

This means that the plane *descended* 18,000 ft.

Now work margin exercise 4.

The **net change** in a measure is the algebraic sum of several numbers. Example 5 illustrates how positive and negative numbers can be used to find the net change of weight (gain or loss) and sales over a period of time.

5. If Andrea's sales for last week were as follows, what were her total net sales?

Day	Net Sales
Monday	3
Tuesday	8
Wednesday	3
Thursday	6
Friday	−1

Example 5 Application: Calculating Net Change

Susan is a salesperson for a shoe store. Last week her sales of pairs of shoes were as follows.

Day	Sales	Returns	Daily Net Sales
Monday	7	1	6
Tuesday	3	0	3
Wednesday	2	4	−2
Thursday	6	1	5
Friday	8	3	5

What were Susan's net sales for last week?

Solution

$$6 + 3 + (-2) + 5 + 5 = 17$$

Susan's net sales for the week were 17 pairs of shoes.

Now work margin exercise 5.

Example 6 Application: Calculating Net Change

Robert weighed 230 lb when he started to diet. The first month he lost 7 lb, the second month he gained 2 lb, and the third month he lost 5 lb. What was his weight after 3 months of dieting? What was his net change in weight?

Solution

His weight after 3 months can be calculated as follows.

$$\begin{aligned} &230+(-7)+(+2)+(-5)\\ &=223+(+2)+(-5)\\ &=225+(-5)\\ &=220\text{ lb}\end{aligned}$$

His net change in weight was

$$230-220=10\text{ lb.}$$

Thus, Robert's weight after 3 months of dieting was 220 lb which was a net change of 10 lb.

Now work margin exercise 6.

6. Shawn weighed 236 lb when he started to diet. The first month he lost 8 lb, the second month he gained 3 lb, and the third month he lost 10 lb. What was his weight after 3 months of dieting? What was his net change in weight?

Margin Exercise Answers

1. a. −14 **b.** +9.2 **2. a.** −11 **b.** 0 **c.** 3.8 **d.** $-\frac{3}{4}$ **e.** $\frac{13}{10}$ **3.** −19 °F **4.** −11,000 ft **5.** 19 sales **6.** 221 lb; 15 lb

8.3 Exercises

Concept Check

Fill-in-the-Blank. Complete each sentence using information found in this section.

1. The ______ ______ is the opposite of a real number.
2. On a horizontal number line, to move in the positive direction is to move to the ______.
3. On a horizontal number line, to move in the negative direction is to move to the ______.
4. To subtract b from a, add the ______ of b to a.
5. To find the change in value between two numbers, take the end value and ______ the beginning value.
6. The algebraic sum of several numbers is their ______ ______.

True/False. Determine whether each statement is true or false. If a statement is false, explain how it can be changed so the statement will be true. (**Note:** There may be more than one acceptable change.)

7. The sum of a number and its additive inverse is the number itself.
8. The additive inverse of negative seven is seven.

8.3 Exercises

Concept Check

1. additive inverse
2. right
3. left
4. opposite
5. subtract
6. net change
7. False; The sum of a number and its additive inverse is zero.
8. True

9. True
10. True

Practice

1. -11
2. -17
3. 6
4. 2
5. -4.7
6. 3.4
7. 0
8. $-\frac{9}{16}$
9. $\frac{5}{7}$
10. 257
11. 5
12. -2
13. -10
14. -35
15. 7
16. 12
17. 3
18. -16
19. 12
20. -15
21. 24
22. -8
23. 5.9
24. 7
25. -4
26. -15.6
27. 10
28. -0.8
29. $-\frac{7}{20}$
30. $\frac{1}{3}$
31. $-\frac{1}{4}$
32. $\frac{2}{15}$
33. $\frac{3}{40}$

9. We can think of addition of numbers as accumulating numbers.
10. The expression "15 – 7" can be thought of as "fifteen plus negative seven."

Practice

Find the additive inverse (opposite) of each real number. See Example 1.

1. 11
2. 17
3. -6
4. -2
5. 4.7
6. -3.4
7. 0
8. $\frac{9}{16}$
9. $-\frac{5}{7}$
10. -257

Subtract. Reduce fractions to lowest terms. See Example 2.

11. $8-3$
12. $5-7$
13. $-4-6$
14. $-18-17$
15. $3-(-4)$
16. $5-(-7)$
17. $-8-(-11)$
18. $-14-2$
19. $0-(-12)$
20. $-8-7$
21. $16-(-8)$
22. $15-23$
23. $2.8-(-3.1)$
24. $5.3-(-1.7)$
25. $-1.4-2.6$
26. $-8.5-7.1$
27. $1.6-(-8.4)$
28. $1.5-2.3$
29. $\frac{2}{5}-\frac{3}{4}$
30. $\frac{7}{15}-\frac{2}{15}$
31. $\frac{5}{16}-\frac{9}{16}$
32. $\frac{5}{6}-\frac{7}{10}$
33. $\frac{9}{20}-\frac{3}{8}$
34. $\frac{3}{14}-\frac{5}{6}$

Perform the indicated operation to find the net change in value. See Examples 5 and 6.

35. $-6+(-4)-5$
36. $-2-2+11$
37. $6+(-3)+(-4)$
38. $-3+(-7)+2$
39. $-5-2-(-4)$
40. $-8-5-(-3)$
41. $-7-(-2)+6$
42. $-3-(-3)+(-6)$
43. $9.7-1.6-(8.1)$
44. $-11.3+5.3-7.9$

Simplify the expression on each side of the blank and then fill in the blank with the proper symbol: <, >, or =.

45. $-6+(-2)$ ______ $3+(-8)$

46. $-4-(-3)$ ______ $-4+(-3)$

47. $5.1-8.2$ ______ $8.2-5.1$

48. $7-(-3)$ ______ $-3-7$

49. $\frac{11}{2}+\left(-\frac{3}{4}\right)$ ______ $\frac{11}{2}-\frac{3}{4}$

50. $0-6$ ______ $0-(-6)$

51. $-8-(-8)$ ______ $-14-13$

52. $-\frac{7}{4}-\left(-\frac{3}{8}\right)$ ______ $\frac{1}{2}-\frac{7}{3}$

53. $-151-86$ ______ $-107-141$

54. $2.5-6.2$ ______ $-1.1-2.3$

▦ Use your calculator to simplify the expression on each side of the blank and then fill in the blank with the proper symbol: <, >, or =.

55. $648-(-396)$ ______ $124-163$

56. $-19{,}824-23{,}417$ ______ $12{,}793-(-14{,}387)$

57. $-43{,}931-(-28{,}677)$ ______ $-(13{,}665+21{,}425)$

58. $-(24{,}295+13{,}107)$ ______ $-48{,}261-(-16{,}276)$

Solve.

59. Find the difference between −5 and −6. (**Hint:** Subtract the numbers in the order given.)

60. Find the difference between 30 and −12. (**Hint:** Subtract the numbers in the order given.)

61. Subtract −3 from −10.

62. Subtract −2 from 6.

63. Subtract 13 from −13.

64. Subtract 20 from −20.

65. Find the sum of −12 and 6. Then subtract −17.

66. Find the sum of 11 and −13. Then subtract 25.

Applications

Solve.

67. At 2 p.m., the temperature was 76 °F. At 8 p.m., the temperature was 58 °F. What was the change in temperature?

68. According to NASA, the temperature on the Earth's moon during the day is 260 °F and during the night it is −280 °F. What is the daily temperature change on the moon? [1]

69. On July 1, a certain stock opened at \$53 per share. One month later, on August 1, the stock opened at \$60 per share. Find the change in price of the stock.

1 Source: NASA.gov

34. $-\frac{13}{21}$
35. −15
36. 7
37. −1
38. −8
39. −3
40. −10
41. 1
42. −6
43. 0
44. −13.9
45. $-8<-5$
46. $-1>-7$
47. $-3.1<3.1$
48. $10>-10$
49. $\frac{19}{4}=\frac{19}{4}$
50. $-6<6$
51. $0>-27$
52. $-\frac{11}{8}>-\frac{11}{6}$
53. $-237>-248$
54. $-3.7<-3.4$
55. $1044>-39$
56. $-43{,}241<27{,}180$
57. $-15{,}254>-35{,}090$
58. $-37{,}402<-31{,}985$
59. 1
60. 42
61. −7
62. 8
63. −26
64. −40
65. 11
66. −27

Applications

67. −18°F (a decrease of 18 degrees Fahrenheit)
68. −540°F (a decrease of 540 degrees Fahrenheit)

69. \$7 (an increase of 7 dollars)

70. −\$750 (a decrease of 750 dollars)

71. −14,787 ft (a decrease of 14,787 feet)

72. 30,280 ft (an increase of 30,280 feet)

73. a. Negative b. Positive

c. May and June

d. Answer will vary. March and April or April and May

74. 2152 miles

75. \$250,404

76. \$39,100

77. \$847

70. Mr. Meade is having a hard time selling his old car. He just slashed the price from \$3500 to \$2750. By how much did he change the price?

71. If you travel from the top of Mt. Whitney, elevation 14,505 ft, to the floor of Death Valley, elevation 282 ft below sea level, what is the change in elevation?

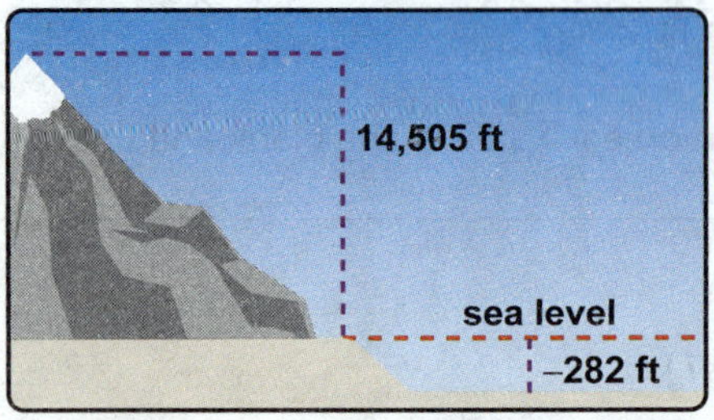

72. A submarine submerged 280 ft below the surface of the sea and fired a rocket that reached an altitude of 30,000 ft. What was the change in altitude of the rocket?

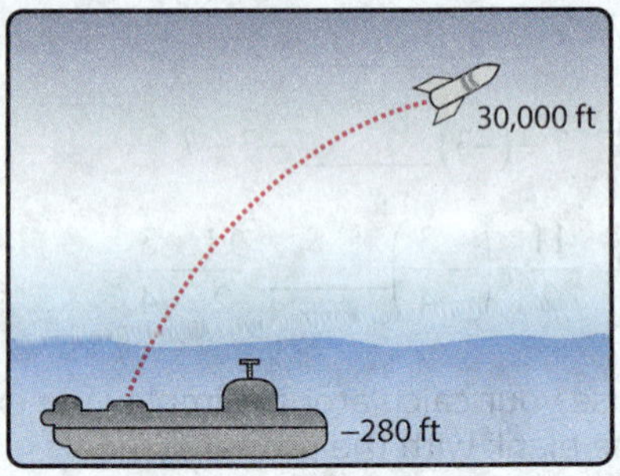

73. The highest temperatures recorded per month from July 2012 through June 2013 in Marquette, MI, are shown. Use the graph to answer the questions. [2]

Highest Recorded Temperatures, July 2012-June 2013, Marquette, MI

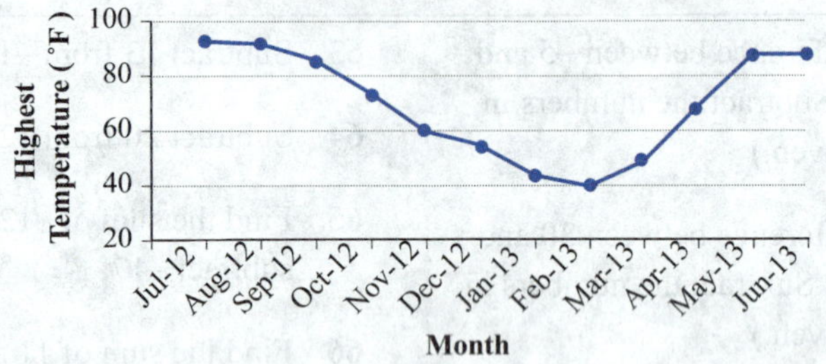

a. Was the change in temperature negative or positive from October 2012 to November 2012?

b. Was the change in temperature negative or positive from April 2013 to May 2013?

c. Between which two months did the smallest change in high temperature occur?

d. Between which two months did the largest change take place?

74. James decided to take a road trip from Wilmington, NC, to Los Angeles, CA. The total distance of the trip is 2590 miles. He stops for a break after he has traveled 438 miles. How many more miles must James travel before he reaches Los Angeles?

75. The Kingston Construction Co. made a bid of \$7,043,272 to build a stretch of freeway, but the Beach City Construction Co. made a lower bid of \$6,792,868. How much lower was the Beach City bid?

76. A couple sold their house for \$175,000. They paid the realtor \$9100, and other expenses of the sale came to \$800. If they owed the bank \$126,000 for the mortgage, what were their net proceeds from the sale?

77. If you had \$980 in your checking account and you wrote a check for \$358 and made a deposit of \$225, what would be the balance in the account?

2 Source: weather.gov

78. In pricing a four-door car, Pat found she would have to pay a base price of $30,500 plus $2135 in taxes and $1006 for license fees. For a two-door sedan car of the same make, she would pay a base price of $25,000 plus $1750 in taxes and $825 for license fees. Including all expenses, how much cheaper was the two door model?

79. The cost of repairing Sylvia's dishwasher would be $350 for parts (including tax) plus $105 for labor. To buy a new dishwasher, she would pay $670 plus $40 in sales tax, and a friend of hers would pay her $90 for her old dishwasher. How much more would Sylvia have to pay to buy a new dishwasher than to have her old one repaired?

80. During a football game, the home team gained 28 yards, gained 9 yards, lost 2 yards, gained 5 yards, gained 1 yard, gained 10 yards, lost 5 yards on a penalty, and lost 11 yards. What was the team's net yardage on this possession of the football?

81. Beginning at a temperature of 10 °C, the temperature in a scientific experiment was measured hourly for four hours. It dropped 5 °C, dropped 8 °C, dropped 6 °C, then rose 3 °C. What was the final temperature recorded?

82. Harry went on a diet for 5 weeks. During those 5 weeks, Harry lost 5 pounds, gained 3 pounds, lost 2 pounds, lost 4 pounds, and gained 1 pound. What was his total loss (or gain) for the 5 weeks? If he weighed 210 pounds when he started the diet plan, what did he weigh at the end of the 5-week period?

83. In a 5-day week the NASDAQ posted a gain of 38 points, a loss of 65 points, a loss of 32 points, a gain of 10 points, and a gain of 15 points. If the NASDAQ started the week at 2350 points, what was the market at the end of the week?

84. Many personal securities accounts have a cash fund to receive dividends. The investor is free to withdrawal money from that account, as well as make deposits. Unlike regular checking accounts, there is no overdraft penalty when the balance is less than zero dollars; instead, a nominal interest rate is charged until the balance becomes positive. In addition, this type of account will accrue interest on positive balances. At the beginning of an account period the balance is $3520, and five transactions are made before the next statement is posted as follows: $300 deposit, $2500 withdrawal, $1800 withdrawal, $253 deposit, and $450 withdrawal. What is the balance at the close of this accounting period?

85. A family has a rain cistern to collect rainwater runoff from the roof to water their lawn and garden. When it rains, the garden does not need to be watered, and the cistern just collects additional water for future use. When it doesn't rain, the stored water is needed to water the plants. At the beginning of a five-week period, the cistern has 873 gallons of water, and the net intake/outflow was noted at the end of each week. First it rose by 461 gallons, then it dropped by 349 gallons, then it dropped by 217 gallons, followed by a 275 gallon increase, and a 177 gallon increase. How many gallons does the cistern contain after this five week period?

86. John goes to a casino in Las Vegas and plays six games. His wins and losses for each respective game was $15 gained, $50 lost, $20 lost, $80 gained, $100 lost, and $40 lost. What was John's net gain or loss?

78. $6066
79. $165
80. Gain of 35 yards
81. −6 °C (or 6 °C below 0 °C)
82. Lost 7 pounds; 203 pounds
83. 2316 points
84. −$677
85. 1220 gallons
86. −$115 (a loss of 115 dollars)

87. a. 1454 people
b. 198,580 people
88. a. 19 trumpets
b. −6 trumpets or decreased by 6
89. The solution is 90.85 g
90. a. 92.7% b. −0.8%
91. 3 °F
92. 10 lb loss; 215 lb
93. 4238 points

87. The population of San Bernardino County, California, in 2003 was 197,126. The population shift for the next five years is shown on the chart below. [3]

Year	Population Shift
2004	+2696
2005	−554
2006	+162
2007	−955
2008	+105

a. What was the change in population in that five year period?

b. What is the population in 2008?

88. The manager of a music store is required to keep track of the inventory for each instrument that the store rents out. The store had twenty-five trumpets in stock at the beginning of the school year before twelve trumpets were rented out. Five trumpets were returned by the end of the first semester. Seven additional trumpets were rented at the beginning of the second semester. Eight trumpets were returned at the end of the second semester.

a. How many trumpets did the store have in stock at the end of the second semester?

b. What was the change in the trumpet inventory between the beginning of the school year and the end of the second semester?

89. A chemist is determining the water content of different foods in his lab by completely dehydrating food samples (that is, he is removing all of the water). He started an experiment with a fresh white potato that had a mass of 115 g. The mass of the potato after being dehydrated was 24.15 g. This situation can be described by the equation $115 - w = 24.15$, where w is the amount of water mass the potato lost in grams. Determine the water mass of the fresh potato by substituting each of the following values into the equation to find the solution: 90.15 g, −90.15 g, 90.85 g, −90.85 g

90. Grace keeps track of her grade average in Introduction to Anthropology throughout the semester. After the first exam, she had a 93.5%. A project decreased her average grade by 1.5%. The next exam decreased her grade by 4.7%. A term paper increased her grade by 2.3%. The final increased her grade by 3.1%.

a. What was Grace's grade average at the end of the semester?

b. What was the change in Grace's grade average during the semester?

91. Beginning with a temperature of 8 °F above zero, the temperature was measured hourly for 4 hours. It rose 3 °F, dropped 7 °F, dropped 2 °F, and rose 1 °F. What was the final temperature recorded?

92. George and his wife went on a diet plan for 5 weeks. During these 5 weeks, George lost 5 pounds, gained 2 pounds, lost 4 pounds, lost 6 pounds, and gained 3 pounds. What was his total loss or gain for these 5 weeks? If he weighed 225 pounds when he started the diet, what did he weigh at the end of the 5-week period?

93. In a 5-day week, the NASDAQ stock market posted a gain of 145 points, a loss of 100 points, a loss of 82 points, a gain of 50 points, and a gain of 25 points. If the NASDAQ started the week at 4200 points, what was the market at the end of the week?

3 Source: www.idcide.com/citydata/ca/san-bernardino.htm

94. In 10 running plays in a football game, the halfback gained 2 yards, gained 12 yards, lost 5 yards, lost 3 yards, gained 22 yards, gained 3 yards, gained 7 yards, lost 2 yards, gained 4 yards, and gained 45 yards. What was his net yardage for the game? This was a good game for him, but not his best.

Writing & Thinking

95. Explain, in your own words, how to find the difference between a positive and a negative number.

96. What is the additive inverse of 0? Why?

97. Under what conditions can the difference between two negative numbers be a positive number?

98. Explain how a subtraction problem is changed to an addition problem. Give an example.

94. 85 yd

Writing & Thinking

95. Add the opposite of the second number to the first number.

96. 0; because $-0 = 0$.

97. Answers will vary. When the absolute value of the number being subtracted is greater than the absolute value of the other number, the difference will be positive.

98. By definition, subtraction is accomplished by adding the opposite of the number being subtracted. For example: $-18 - 12$ becomes $-18 + (-12) = -36$. Examples will vary.

Objectives

A. Multiply real numbers.

B. Divide real numbers.

C. Find the average (mean) of a set of real numbers.

8.4 Multiplication and Division with Real Numbers

A Multiplication with Real Numbers

Multiplication is shorthand for repeated addition. That is,

$$7+7+7+7+7=5\cdot 7=35$$

and

$$(-6)+(-6)+(-6)=3(-6)=-18.$$

Similarly,

$$(-2)+(-2)+(-2)+(-2)+(-2)=5(-2)=-10.$$

Repeated addition with a negative number results in a product of a positive number and a negative number. Since the sum of negative numbers is negative, we have the following general rule.

The product of a positive real number and a negative real number is negative.

1. Multiply. Reduce fractions to lowest terms.

a. $4(-7)$

b. $7(-4)$

c. $8(-5)$

d. $35(-1)$

e. $4(-1.6)$

f. $\frac{5}{3}\cdot\left(-\frac{21}{10}\right)$

Example 1 Multiplying Positive and Negative Real Numbers

Multiply. Reduce fractions to lowest terms.

a. $5(-3)=(-3)+(-3)+(-3)+(-3)+(-3)=-15$

b. $3(-5)=(-5)+(-5)+(-5)=-15$

c. $7(-10)=-70$

d. $42(-1)=-42$

e. $4(-6.2)=-24.8$

f. $\frac{3}{7}\cdot\left(-\frac{14}{9}\right)=-\frac{\cancel{3}\cdot 2\cdot \cancel{7}}{\cancel{7}\cdot \cancel{3}\cdot 3}=-\frac{2}{3}$

Now work margin exercise 1.

Math Tip

When you see two sets of parentheses such as $(3)(-6)$, it means to multiply the numbers that are inside the parentheses. So $(3)(-6)=-18$.

The product of two negative real numbers can be explained in terms of opposites. We also need the fact that for any real number a, we can think of the **opposite of *a***, $(-a)$, as the product of -1 and a. That is, we have

$$-a=-1\cdot a.$$

Thus,

$$-4=-1\cdot 4=-1(4).$$

Now we use this idea to consider the product of two negative numbers. For the product $-4(-7)$ we have,

$$-4(-7)=-1(4)(-7)=-1[(4)(-7)]=-1(-28)=-(-28)=28.$$

This reasoning leads to the following rule.

The product of two negative real numbers is positive.

Example 2 Multiplying Two Negative Real Numbers

Multiply. Reduce fractions to lowest terms.

a. $(-4)(-9) = +36$

b. $-\frac{3}{2}\left(-\frac{2}{9}\right) = +\frac{\cancel{3} \cdot \cancel{2}}{\cancel{2} \cdot \cancel{3} \cdot 3} = +\frac{1}{3}$

c. $-2(-6.7) = +13.4$

d. $(-1)(-5)(-3)(-2) = 5(-3)(-2) = -15(-2) = +30$

Now work margin exercise 2.

2. Multiply. Reduce fractions to lowest terms.

a. $-6(-4)$

b. $-\frac{4}{5}\left(-\frac{11}{6}\right)$

c. $-5(-3.5)$

d. $(-2)(-3)(-4)(-2)$

What happens if a number is multiplied by 0? For example, $3(0) = 0 + 0 + 0 = 0$. In fact, **multiplication by 0 always gives a product of 0**.

Example 3 Multiplying by 0

Multiply.

a. $0 \cdot 6 = 0$

b. $-13 \cdot 0 = 0$

Now work margin exercise 3.

3. Multiply.

a. $0(9)$

b. $-19(0)$

The rules for multiplication with real numbers can be summarized as follows. Remember that these rules apply to integers, decimals, and fractions.

Rules for Multiplication with Real Numbers

If a and b are positive real numbers, then

1. The product of two positive numbers is positive: $a \cdot b = +ab$.
2. The product of two negative numbers is positive: $(-a)(-b) = +ab$.
3. The product of a positive number and a negative number is negative: $a(-b) = -ab$.
4. The product of 0 and any number is 0: $a \cdot 0 = 0$ and $(-a) \cdot 0 = 0$.

PROCEDURE

B Division with Real Numbers

The rules for multiplication lead directly to the rules for division because division is defined in terms of multiplication. For convenience, division is indicated in fraction form.

Division with Real Numbers

For real numbers a, b, and x (where $b \neq 0$),

$$\boldsymbol{\frac{a}{b} = x} \text{ means that } \boldsymbol{b \cdot x = a}.$$

For real numbers a and b (where $b \neq 0$),

$$\boldsymbol{\frac{a}{0}} \text{ is } \textbf{undefined}, \text{ but } \boldsymbol{\frac{0}{b} = 0}.$$

DEFINITION

The following discussion explains why division by 0 is undefined.

Division by 0 is Undefined

1. Suppose that $a \neq 0$ and $\frac{a}{0} = x$. Then, since division is related to multiplication, we must have $a = 0 \cdot x$, but this is not possible because $0 \cdot x = 0$ for any value of x and we stated that $a \neq 0$.

2. Suppose that $\frac{0}{0} = x$. Then, $0 = 0 \cdot x$, which is true for all values of x. But, for the division to be defined we must have a unique answer for x.

Therefore, in any case, we conclude that division by 0 is undefined.

DEFINITION

4. Divide.

a. $\frac{42}{7}$

b. $\frac{-42}{7}$

c. $\frac{42}{-7}$

d. $\frac{-42}{-7}$

e. $\frac{-3}{0}$

f. $\frac{0}{5}$

Example 4 Dividing Real Numbers

Divide.

a. $\frac{36}{9} = 4$ because $9 \cdot 4 = 36$.

b. $\frac{-36}{9} = -4$ because $9(-4) = -36$.

c. $\frac{36}{-9} = -4$ because $-9(-4) = 36$.

d. $\frac{-36}{-9} = 4$ because $-9(4) = -36$.

e. $\frac{8}{0}$ is undefined because division by 0 is undefined.

f. $\frac{0}{-6} = 0$ because $-6 \cdot 0 = 0$.

Now work margin exercise 4.

The division problems in Example 4 illustrated the following rules for division with real numbers.

Rules for Division with Real Numbers

If a and b are positive real numbers (where $b \neq 0$),

1. The quotient of two positive numbers is positive: $\frac{a}{b} = +\frac{a}{b}$.
2. The quotient of two negative numbers is positive: $\frac{-a}{-b} = +\frac{a}{b}$.
3. The quotient of a positive number and a negative number is negative $\frac{-a}{b} = -\frac{a}{b}$ and $\frac{a}{-b} = -\frac{a}{b}$.

PROCEDURE

Note

The following common rules about multiplication and division with two nonzero real numbers are helpful in remembering the signs of answers.

1. If the numbers have the same sign, both the product and quotient will be positive.
2. If the numbers have different signs, both the product and quotient will be negative.

Example 5 Dividing Fractions and Decimals

Divide. Reduce fractions to lowest terms. Round answers with decimals to the nearest tenth.

a. $-\frac{16}{7} \div \left(-\frac{2}{21}\right) = -\frac{16}{7} \cdot \left(-\frac{21}{2}\right) = +\frac{\cancel{2} \cdot 8 \cdot 3 \cdot \cancel{7}}{\cancel{7} \cdot \cancel{2}} = \frac{24}{1} = +24$ (Reminder: To divide by a fraction, multiply by its reciprocal.)

b. $-\frac{20}{12} \div \frac{15}{2} = -\frac{20}{12} \cdot \frac{2}{15} = -\frac{\cancel{4} \cdot \cancel{5} \cdot 2}{\cancel{4} \cdot 3 \cdot 3 \cdot \cancel{5}} = -\frac{2}{9}$

c. $-5.7 \div 4.2 \approx -1.4$ (To the nearest tenth)

d. $-16.54 \div (-5.1) \approx +3.2$ (To the nearest tenth)

Now work margin exercise 5.

5. Divide. Reduce fractions to lowest terms. Round answers with decimals to the nearest tenth.

a. $\frac{14}{5} \div \left(-\frac{7}{30}\right)$

b. $-\frac{21}{10} \div \frac{6}{5}$

c. $-3.4 \div 1.6$

d. $-21.32 \div (-7.4)$

C Average (or Mean)

Average (or **mean**) has already been discussed with whole numbers, fractions, and decimal numbers. Now we apply this discussion to negative numbers. Newspapers and magazines report average income, average sales, average attendance at sporting events, and so on. The mean of a set of numbers is particularly important in the study of statistics. For example, scientists might be interested in the mean IQ of the students attending a certain university or the mean height of students in the fourth grade.

Average

The **average** (or **mean**) of a set of numbers is the value found by adding the numbers in the set and then dividing the sum by the number of numbers in the set.

DEFINITION

6. At noon on five consecutive days in Mears, Michigan, the temperatures were $-3°$, $5°$, $8°$, $-4°$, and $14°$ (in degrees Fahrenheit.) (Negative numbers represent temperatures below zero.) Find the average of these noonday temperatures.

Example 6 Application: Calculating an Average

At noon on five consecutive days in Aspen, Colorado, the temperatures were $-5°$, $7°$, $6°$, $-7°$, and $14°$ (in degrees Fahrenheit). (Negative numbers represent temperatures below zero). Find the average of these noonday temperatures.

Solution

First, add the five temperatures. $(-5)+7+6+(-7)+14=15$

Now divide the sum, 15, by the number of temperatures, 5.

$$\frac{15}{5}=3$$

The average noon temperature was 3 °F.

Now work margin exercise 6.

7. In a placement exam for biology, a group of ten students had the following scores: 4 students scored 72, 2 students scored 78, 3 students scored 85, and 1 student scored 91. What was the mean score for this group of students?

Example 7 Application: Calculating an Average

In a placement exam for mathematics, a group of ten students had the following scores: 3 students scored 75, 2 students scored 80, 1 student scored 82, 3 students scored 85, and 1 student scored 88. What was the mean score for this group of students?

Solution

To find the total of all the scores, we multiply and then add. This is more efficient than adding all ten scores.

$$\left.\begin{aligned}75\cdot 3&=225\\80\cdot 2&=160\\82\cdot 1&=82\\85\cdot 3&=255\\88\cdot 1&=88\end{aligned}\right\}\text{ Multiply.}$$

$$225+160+82+255+88=810 \quad \text{Add.}$$

$$810\div 10=81 \quad \text{Divide by the number of scores.}$$

The mean score on the placement test for this group of students was 81.

Now work margin exercise 7.

8. The following speeds (in miles per hour) of twenty cars were recorded at a certain point on a busy road.

45	55	43
47	48	51
54	52	44
46	41	49
50	48	44
45	43	47
53	49	

Find the average speed of these cars. Round your answer to the nearest tenth.

Example 8 Application: Calculating an Average

The following speeds (in miles per hour) of fifteen cars were recorded at a certain point on a freeway.

70	75	65	60	61	64	68	72
59	68	82	76	70	68	50	

Find the average speed of these cars.

Solution

Using a calculator, the sum of the speeds is 1008 mph.

Dividing by 15 gives the average speed.

$$1008 \div 15 = 67.2 \text{ mph}$$

The average speed was 67.2 mph.

Now work margin exercise 8.

Margin Exercise Answers

1. a. −28 **b.** −28 **c.** −40 **d.** −35 **e.** −6.4 **f.** $-\frac{7}{2}$ **2. a.** +24 **b.** $+\frac{22}{15}$ **c.** +17.5 **d.** +48

3. a. 0 **b.** 0 **4. a.** 6 **b.** −6 **c.** −6 **d.** 6 **e.** undefined **f.** 0 **5. a.** 12 **b.** $-\frac{7}{4}$ **c.** −2.1 **d.** 2.9
6. 4 °F **7.** 79 **8.** 47.7 mph

8.4 Exercises

Concept Check

Fill-in-the-Blank. Complete each sentence using information found in this section.

1. The value found by adding numbers and then dividing the sum by the number of items in the set is the ________ (or ____) of the numbers.

2. When a positive number and negative number are multiplied, the result is a ________ number.

3. If two positive numbers are being multiplied, the product is a __________ number.

4. If two negative numbers are being divided, the quotient is a _________ number.

5. When zero is divided by a nonzero number, the answer is always ___.

6. The quotient of a number divided by zero is __________.

True/False. Determine whether each statement is true or false. If a statement is false, explain how it can be changed so the statement will be true. (**Note:** There may be more than one acceptable change.)

7. If a negative number is divided by a positive number, the result will be a negative number.

8. The product of zero and a number is zero.

9. If two numbers have the same sign, both the product and the quotient of the two numbers will be negative.

10. The mean of a set of numbers is always positive.

8.4 Exercises

Concept Check

1. average, mean
2. negative
3. positive
4. positive
5. 0
6. undefined
7. True
8. True
9. False; The product and quotient will be positive.
10. False; The mean of a set of numbers can be positive, negative, or zero.

Practice

1. -12
2. -30
3. 48
4. 57
5. 56
6. 22
7. -21
8. -35
9. 56
10. 66
11. 26
12. 72
13. -70
14. -60
15. -24
16. -162
17. -288
18. 18
19. 0
20. 0
21. -9
22. 19
23. -7.31
24. 0.52
25. $-\frac{1}{6}$
26. $-\frac{6}{35}$
27. $\frac{18}{5}$
28. $\frac{9}{14}$
29. $-\frac{12}{5}$
30. $-\frac{35}{6}$
31. $-\frac{9}{2}$
32. $\frac{16}{455}$
33. 4
34. 2
35. -6

Practice

Multiply. Reduce fractions to lowest terms. See Examples 1 through 3.

1. $4(-3)$
2. $6(-5)$
3. $12 \cdot 4$
4. $19 \cdot 3$
5. $(-8)(-7)$
6. $(-11)(-2)$
7. $-3 \cdot 7$
8. $-7 \cdot 5$
9. $(-14)(-4)$
10. $(-11)(-6)$
11. $(-13)(-2)$
12. $(-8)(-9)$
13. $10(-7)$
14. $(-5)(12)$
15. $(-2)(-3)(-4)$
16. $(-6)(-3)(-9)$
17. $-8 \cdot 4 \cdot 9$
18. $(-3)(2)(-3)$
19. $(-7)(-16)(0)$
20. $-9 \cdot 0 \cdot 4$
21. $(-2)(4.5)$
22. $(-5)(-3.8)$
23. $4.3(-1.7)$
24. $(-2.6)(-0.2)$
25. $-\frac{3}{8} \cdot \frac{4}{9}$
26. $\frac{4}{5} \cdot \frac{-3}{14}$
27. $\frac{-4}{5} \cdot \frac{-9}{2}$
28. $-\frac{3}{4} \cdot -\frac{6}{7}$
29. $-4 \cdot \frac{3}{5}$
30. $-7 \cdot \frac{5}{6}$
31. $\frac{5}{2} \cdot \frac{-15}{10} \cdot \frac{6}{5}$
32. $\frac{-4}{7} \cdot \frac{2}{5} \cdot \frac{-2}{13}$

Divide. Reduce fractions to lowest terms. Round answers with decimals to the nearest tenth. See Examples 4 and 5.

33. $\frac{-8}{-2}$
34. $\frac{-20}{-10}$
35. $\frac{-30}{5}$
36. $\frac{-51}{3}$
37. $\frac{-26}{-13}$
38. $\frac{-91}{-7}$
39. $\frac{0}{6}$
40. $\frac{16}{0}$
41. $\frac{39}{-13}$
42. $\frac{44}{-4}$
43. $\frac{-34}{2}$
44. $\frac{-36}{9}$
45. $\frac{-3}{0}$
46. $\frac{0}{-7}$
47. $\frac{-60}{-12}$
48. $\frac{-48}{-16}$
49. $\frac{-4.8}{8}$
50. $\frac{-5.6}{7}$
51. $\frac{-4}{-0.2}$
52. $\frac{-3}{-8}$
53. $\frac{2.99}{-1.3}$
54. $\frac{2.8}{-1.4}$
55. $\frac{-2}{15} \div \frac{8}{5}$
56. $\frac{-3}{5} \div \frac{-9}{10}$
57. $\frac{6}{11} \div \frac{4}{3}$
58. $\frac{-10}{3} \div \frac{-7}{5}$
59. $\frac{-9}{14} \div \frac{54}{35}$
60. $\frac{45}{8} \div \frac{35}{12}$

Determine whether each statement is true or false. If a statement is false, rewrite it in a form that is true. (There may be more than one correct new form.)

61. $(-4)(6) \geq 3 \cdot 8$

62. $(-7)(-9) \leq 3 \cdot 21$

63. $-\frac{3}{4} \cdot \frac{5}{8} = \frac{15}{16}\left(-\frac{1}{2}\right)$

64. $(-6.0)(9.1) = (6.1)(-9.0)$

65. $6(-3) > (-14) + (-4)$

66. $7 + 8 > (-10) + (-5)$

67. $-\frac{2}{3} + \frac{1}{2} \leq \frac{1}{2} - \frac{1}{4}$

68. $1.7 + (-3.9) < (-1.4) + (-4.2)$

69. $(-4)(9) = (-24) + (-12)$

70. $14 + 6 \leq (-2)(-10)$

Use a graphing calculator to find the value of each expression. Round quotients to the nearest hundredth, if necessary. Remember the negative sign [(-)] is next to [ENTER]. For example, $-14.8 \div (-5)$ would appear as follows.

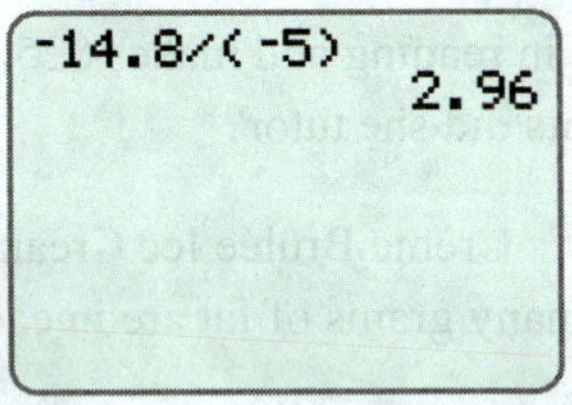

71. $(27)(-24)(-180)$

72. $(-461)(-45)(-17)$

73. $(54)(-17)(-24)$

74. $(-77{,}459) \div 29$

75. $(-62{,}234) \div (-37)$

76. $(-35 - 45 - 56) \div 3$

77. $(-52 - 30 - 40 - 60) \div 4$

78. $72 \div (15 - 22)$

79. $95 \div (-3 - 7)$

80. $-15.3 \div (-5.4)$

81. $(-13.4)(-2.5)(-1.63)$

82. $(-2.5)(-3.41)(-10.6)$

Applications

Solve.

83. Find the mean of the following set of integers: −10, 15, 16, −17, −34, and −42.

84. Find the mean of the following set of integers: −72, −100, −54, 82, and −96.

85. The Math Club members decided to attend the national meeting of the NCTM (National Council of Teachers of Mathematics) and had a book sale to raise money for the event. Registration fees were \$85 per member and the club had 35 members. How much money did the club need to raise for registration fees?

86. If one regular pack of candy contains 250 calories, how many calories are there in 37 packs of the same candy?

87. A sandwich shop buys 372 loaves of bread for the week. If each loaf of bread has 24 slices, how many slices of bread were purchased?

88. Students at the local community college must pay \$83 for a math textbook. If there are 43 students in the class, find the total amount the class will spend on textbooks.

36. −17
37. 2
38. 13
39. 0
40. Undefined
41. −3
42. −11
43. −17
44. −4
45. Undefined
46. 0
47. 5
48. 3
49. −0.6
50. −0.8
51. 20
52. 0.4
53. −2.3
54. −2
55. $-\frac{1}{12}$
56. $\frac{2}{3}$
57. $\frac{9}{22}$
58. $\frac{50}{21}$
59. $-\frac{5}{12}$
60. $\frac{27}{14}$
61. False; $(-4)(6) < 3 \cdot 8$
62. True
63. True
64. False; $(-6.0)(9.1) > (6.1)(-9.0)$
65. False; $6(-3) = (-14) + (-4)$
66. True
67. True
68. False; $1.7 + (-3.9) > (-1.4) + (-4.2)$
69. True
70. True

71. 116,640
72. −352,665
73. 22,032
74. −2671
75. 1682
76. −45.33
77. −45.5
78. −10.29
79. −9.5
80. 2.83
81. −54.605
82. −90.365

Applications

83. −12
84. −48
85. $2975
86. 9250 calories
87. 8928 slices of bread
88. $3569
89. $442,800
90. $15,200,000
91. 3 cookies
92. 56 raffle tickets
93. 22 teams
94. 19 students
95. 17 grams
96. 6 hours
97. $10,740 per year
98. $4837
99. 1152 bags
100. 377 days
101. 427 sq ft
102. $2574
103. 5423 dozen

89. Your company bought 18 new cars, each with blind-spot monitoring and backup cameras, at a price of $24,600 per car. How much did your company pay for these cars?

90. According to the US Fish and Wildlife Service, migratory birds are imported at a value of about $19 each. Suppose that about 800,000 live birds are imported each year. What is the total value of these imported birds?

91. A mother has 12 cookies that she will equally divide among 4 children. How many cookies will each child get?

92. Seven students sold a total of 392 raffle tickets. Assuming that each student sold the same number of raffle tickets, how many tickets did each student sell?

93. The Cedarville Baseball Camp has 198 campers. How many 9-member teams can this camp have?

94. Jane Scott tutors students in reading and makes $25 per student. If she makes $475 in a week, how many students did she tutor?

95. One pint of Ben and Jerry's Crème Brûlée Ice Cream has 68 grams of fat. If there are 4 servings per pint, how many grams of fat are in each serving? [1]

96. If one person can paint a small house in 48 hours, how long will it take a crew of 8 people to paint the house, assuming that all 8 work at the same speed, and do not interfere with each other?

97. For 2021–2022, the average tuition cost for four years at a public 4-year institution was $42,960. If tuition did not increase each year, how much would you pay per year for the four years you were in college? [2]

98. Smithfield High School paid $29,022 for six pianos. How much did each piano cost?

99. The area of every NFL football field is 57,600 square feet. If a bag of grass seed covers 50 square feet, how many bags of grass seed will be needed to cover one grass football field? [3]

100. US Astronaut Peggy Whitson orbited the Earth 6032 times during her space flights on the International Space Station. If the International Space Station orbits the Earth 16 times a day, how many days was Peggy Whitson in space? [4]

101. A community has 5978 square feet available for individual gardens that will be evenly distributed among 14 people. How much space will each person get?

102. Thirteen men purchase a boat together. If the cost of the boat is $33,462, how much will each man contribute if each contributes an equal amount?

103. A large chicken farm produces 65,076 eggs during a typical week. How many dozen eggs does this represent? (**Note:** 1 dozen = 12 eggs)

1 Source: Ben and Jerry's
2 Source: www.collegeboard.com
3 Source: National Football League
4 Source: National Aeronautics and Space Administration

104. 262,800 pounds of pasta is served every year at Mama Melrose's Ristorante Italiano at Disney MGM Studios. How many pounds of pasta are served every day, assuming there are 365 days in each year? [5]

105. Katelyn is having a birthday party and has invited 11 friends. For lunch, Katelyn's parents bought 3 large pizzas that have 8 slices per pizza. How many slices of pizza will each child get? (Assume only Katelyn and her 11 friends are eating the pizza.)

106. The costs of a one-way flight from Baltimore, MD, to Orlando, FL, for seven different flight times are as follows: $73, $171, $147, $87, $89, $111, $100. What is the average cost of a flight from Baltimore to Orlando? [6]

107. The numbers of fatal motor vehicle accidents in the United States each year from 2005 to 2015 are as follows: 39,252; 38,648; 37,435; 34,172; 30,862; 30,296; 29,867; 31,006; 30,202; 30,056; 32,166. What was the average number of fatal accidents in the United States per year from 2005 to 2015? (Round your answer to the nearest tenth.) [7]

108. During one day, Sebastian made several transactions with his checking account. He deposited $150 at the beginning of the day, bought groceries for $45.50, filled his car with gas for $39, bought tutoring supplies for $15, and deposited $120 at the end of the day.

a. Which amounts are credits to Sebastian's account? Be sure to include the sign.

b. Which amounts are debits to his account? Be sure to include the sign.

c. What was the average transaction amount that Sebastian made during the day? (**Hint:** Be sure to use the amounts from parts a. and b.)

109. An auto tire manufacturer recommends using 35 psi of air pressure in their standard tires. A tire has a leak that causes the air pressure of the tire to change at a rate of –2 psi per hour.

a. How much will the tire's air pressure change after 4 hours?

b. A tire with a standard air pressure of 35 psi is considered to be flat when it has only 24.5 psi of pressure. What change in air pressure will cause the tire to be considered flat?

c. How long will it take the tire to lose the amount of air pressure determined in part b.?

104. 720 pounds
105. 2 slices
106. $111.14
107. 33,087.5 accidents
108. **a.** $150, $120
b. −$45.50, −$39, −$15
c. $34.10
109. **a.** −8 psi
b. −10.5 psi
c. 5.25 hours (or 5 hours 15 minutes)

5 Source: www.diningindisney.com
6 Source: Expedia.com
7 Source: NHTSA

110. **a.** −10.83 °F
b. Increase
c. −9.14 °F
d. The average temperature from part c. is higher.
e. 26 °F
111. 26.6 calls
112. 9.27 milligrams per deciliter
113. 80
114. 78.6

110. ▦ In King Salmon, Alaska, the lowest monthly temperature was recorded for several months as shown in the table. Use the data to answer the following questions. Round your answers to the nearest hundredth.[8]

Lowest Monthly Temperature in King Salmon, Alaska

Month	Temperature
October 2012	4 °F
November 2012	−4 °F
December 2012	−22 °F
January 2013	−11 °F
February 2013	−14 °F
March 2013	−18 °F

a. What was the average of the low temperatures over these months?

b. The lowest temperature recorded for April 2013 was 1 °F. Will the average low temperature increase or decrease if this data value is now used in calculating the average? Do not calculate the average.

c. Find the average lowest temperature from October 2012 through April 2013. (**Hint:** part b. gives the lowest temperature for April 2013.)

d. How do the average temperatures from parts a. and c. compare?

e. What is the range of the lowest monthly temperatures for the months given in the table?

111. Twenty business executives made the following numbers of telephone calls during one week. Find the mean number of calls (to the nearest tenth) made by these executives.

20	16	14	11	51
40	36	28	52	25
18	16	42	49	12
18	22	33	9	19

112. The blood calcium level (in milligrams per deciliter) for 20 patients was reported as follows.

8.2	10.2	9.3	8.5	7.3
9.7	9.6	8.3	9.8	9.1
9.4	11.1	10.0	8.5	9.9
8.6	10.2	9.4	9.1	9.2

Find the mean blood calcium level (to the nearest hundredth) for these patients.

113. Fifteen students scored the following scores on an exam in accounting: 1 scored 67, 4 scored 73, 3 scored 77, 2 scored 80, 3 scored 88, and 2 scored 93. What was the average score for these students?

114. On an exam in history, a class of twenty-one students had the following test scores: 4 scored 65, 3 scored 70, 6 scored 78, 2 scored 82, 1 scored 85, 3 scored 91, and 2 scored 95. What was the mean score (to the nearest tenth) on this test for the class?

8 Source: weather.gov

The frequency of a number is a count of how many times that number appears. In statistics, data is commonly given in the table form of a frequency distribution as illustrated. To find the mean, multiply each number by its frequency, add these products, and divide the sum by the sum of the frequencies.

115. The heights of the top 30 NBA scorers for the 2021–2022 season are listed in the frequency table. Find the mean height for these men.[9]

Height (in inches)	73	74	75	76	77	78	79	80	81	82	83	84	85
Frequency	3	3	1	2	3	2	3	4	1	2	4	1	1

116. The students in a psychology class were asked the number of books that they had read in the last month. The following frequency distribution indicates the results. Find the mean number of books read (to the nearest tenth) by these students

Number of Books	0	1	2	3	4	5
Frequency	3	2	6	4	2	1

117. The following bar graph shows the approximate amounts of time per week spent by people watching broadcast TV or a streaming service for six groups (by age and sex) of people 18 years of age and older. What is the average amount of time per week people over the age of 18 spend watching broadcast TV or a streaming service? (Assume each group has the same number of people.)

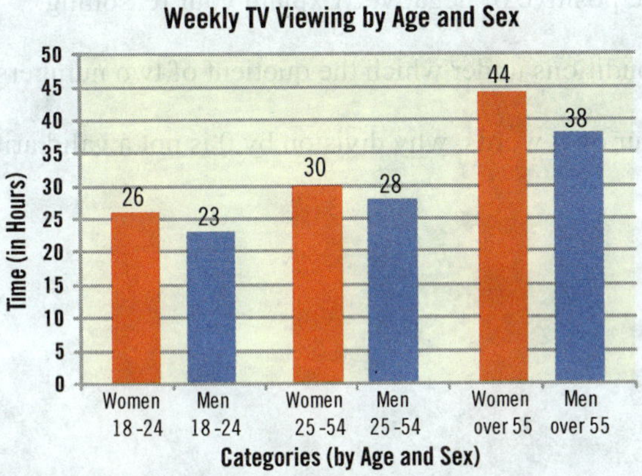

118. The following pictograph shows the number of phone calls received by a 1-hour radio talk show in Seattle during one week. (Not all calls actually get on the air.) What was the mean number of calls per show received that week?

115. 78.6 inches
116. 2.2 books
117. 31.5 hours
118. 42 calls

9 Source: NBA

119. 18,631.8 square miles

Writing & Thinking

120. Negative; The product of every two negative numbers will be positive and this result multiplied by the remaining negative will give a negative answer.

121. Positive; The product of two negative numbers will be positive so an even number of negatives will give a positive product.

122. Dividing 0 by any number other than 0 gives a quotient of 0.

123. Answers will vary.

119. The following bar graph shows the area of each of the five Great Lakes. Lake Superior, with an area of about 31,700 square miles, is the world's largest fresh water lake. What is the mean size of these lakes?

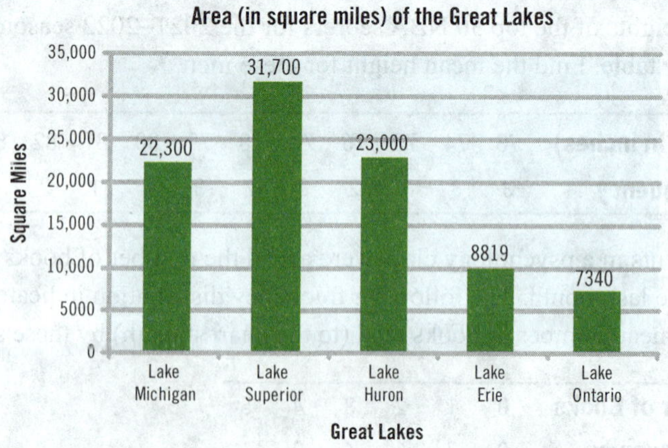

Writing & Thinking

120. If you multiply an odd number of negative numbers together, do you think that the product will be positive or negative? Explain your reasoning.

121. If you multiply an even number of negative numbers together, do you think that the product will be positive of negative? Explain your reasoning.

122. Explain the conditions under which the quotient of two numbers is 0.

123. Explain, in your own words, why division by 0 is not a valid arithmetic operation.

8.5 Order of Operations with Real Numbers

Objectives

A. Use the rules for order of operations

A Using the Order of Operations

Consider the following expression.

$$45 \div 9 + 2 \cdot 3^2$$

Try evaluating the expression before you read the discussion below. Did you get 63 or 23 or 441?

As discussed in Chapters 1, 2, and 3, mathematicians have agreed on a set of rules for the order of performing operations when evaluating numerical expressions. These expressions may contain grouping symbols, exponents, and the operations of addition, subtraction, multiplication, and division. The rules are called the **rules for order of operations** and are used throughout all levels of mathematics and science (as well as by calculators) so that there is only one correct answer for the value of an expression.

Now, using these rules, the expression $45 \div 9 + 2 \cdot 3^2$ can be evaluated as follows.

$$\begin{aligned} 45 \div 9 + 2 \cdot 3^2 &= 45 \div 9 + 2 \cdot \mathbf{9} && \text{Evaluate the exponent.} \\ &= \mathbf{5} + 2 \cdot 9 && \text{Divide.} \\ &= 5 + \mathbf{18} && \text{Multiply.} \\ &= \mathbf{23} && \text{Add.} \end{aligned}$$

In this section, we will see how to apply these rules with all real numbers, including positive and negative integers, fractions, and decimal numbers.

Rules for Order of Operations

1. Simplify within grouping symbols, such as parentheses ( ), brackets [ ], and braces { }, working from the innermost grouping outward.
2. Evaluate any exponential expressions.
3. Moving from **left to right,** perform any multiplications **or** divisions **in the order they appear**.
4. Moving from **left to right**, perform any additions **or** subtractions **in the order they appear**.

PROCEDURE

Note

Other grouping symbols are the absolute value bars (such as $|3+5|$), the fraction bar (as in $\frac{4+7}{10+1}$), and the square root symbol (such as $\sqrt{5+11}$).

These rules are very explicit and should be studied carefully. Note that in rule 3, neither multiplication nor division has priority over the other. Whichever of these operations occurs first, moving left to right, is done first. In rule 4, addition and subtraction are handled in the same way. Unless they occur within grouping symbols, **addition and subtraction are the last operations to be performed**.

A well-known mnemonic device for remembering the rules for order of operations is Please Excuse My Dear Aunt Sally (PEMDAS).

Teaching Note

You might want to show the mnemonic in a vertical format as

P
E
M/D
A/S

to reinforce the relationship between multiplication and division and addition and subtraction.

Math Tip

Remember to follow the order of operations and simplify within grouping symbols first. The expression $(12 \div 3) \cdot 4$ is not the same as $12 \div (3 \cdot 4)$.

Notice that

$$(12 \div 3) \cdot 4 = 4 \cdot 4 = 16$$

and

$$12 \div (3 \cdot 4) = 12 \div 12 = 1.$$

Please	Excuse	My	Dear	Aunt	Sally
Parentheses	Exponents	Multiplication	Division	Addition	Subtraction

Caution

Even though the mnemonic **PEMDAS** is helpful, remember that multiplication and division are performed as they appear, left to right. Also, addition and subtraction are performed as they appear, left to right.

For example,

$$12 \div 3 \cdot 4 = 4 \cdot 4 - 16, \quad \text{Division first}$$

but

$$12 \cdot 3 \div 4 = 36 \div 4 = 9. \quad \text{Multiplication first}$$

CAUTION

A negative sign in front of a variable expression indicates that the expression is to be multiplied by -1. For example,

$$-x^2 = -1 \cdot x^2.$$

This is consistent with the rules for order of operations (indicating that exponents come before multiplication) and is particularly useful in determining the values of expressions involving negative numbers and exponents. For example, each of the expressions

$$-7^2 \text{ and } (-7)^2$$

has a different value. By the order of operations,

$$-7^2 = -1 \cdot 7^2 = -1 \cdot 49 = -49,$$

but

$$(-7)^2 = (-7)(-7) = 49.$$

In the second expression, the base is a negative number. Remember that if the base is a negative number, then the negative number must be placed in parentheses.

The following examples show how to apply the rules. In some cases, more than one step can be performed at the same time. This is possible when parts are separated by + or – signs or are within separate grouping symbols. **Work through each of the following examples step by step, and rewrite the examples on a separate sheet of paper.**

Example 1 Using the Order of Operations with Real Numbers

Simplify: $36 \div 4 - 6 \cdot 2^2$

Solution

$36 \div 4 - 6 \cdot 2^2$

$= 36 \div 4 - 6 \cdot 4$	Evaluate exponents.
$= 9 - 24$	Divide and multiply, left to right.
$= -15$	Subtract.

Now work margin exercise 1.

1. Simplify: $56 \div 7 - 5 \cdot 3^2$

Example 2 Using the Order of Operations with Real Numbers

Simplify: $2(3^2 - 1) - 3 \cdot 2^3$

Solution

$2(3^2 - 1) - 3 \cdot 2^3$

$= 2(9 - 1) - 3 \cdot 8$	Evaluate exponents.
$= 2(8) - 3 \cdot 8$	Simplify within groupings.
$= 16 - 24$	Multiply.
$= -8$	Subtract (or add algebraically).

Now work margin exercise 2.

2. Simplify: $2(5^2 - 9) - 2 \cdot 3^3$

Example 3 Using the Order of Operations with Real Numbers

Simplify: $(2-5)^2 + |2 - 5^2| - 2^3$

Solution

$(2-5)^2 + |2 - 5^2| - 2^3$

$= (-3)^2 + |2 - 5^2| - 2^3$	Subtract inside the parentheses.
$= 9 + |2 - 25| - 8$	Evaluate exponents.
$= 9 + |-23| - 8$	Subtract inside the absolute value bars.
$= 9 + 23 - 8$	Evaluate the absolute value.
$= 24$	Add and subtract.

Now work margin exercise 3.

3. Simplify:
$(1-3)^2 + |9 - 4^2| - 1^3$

4. Simplify:

$7-3\left[\left(4\cdot 6-3^2\right)\div 5+2^2\right]$

Example 4 Using the Order of Operations with Real Numbers

Simplify: $9-2\left[\left(3\cdot 5-7^2\right)\div 2+2^2\right]$

Solution

$9-2\left[\left(3\cdot 5-7^2\right)\div 2+2^2\right]$

$=9-2\left[\left(3\cdot 5-49\right)\div 2+4\right]$ Evaluate exponents.

$=9-2\left[\left(15-49\right)\div 2+4\right]$ Multiply inside the parentheses.

$=9-2\left[\left(-34\right)\div 2+4\right]$ Subtract inside the parentheses.

$=9-2\left[-17+4\right]$ Divide inside the brackets.

$=9-2\left[-13\right]$ Add inside the brackets.

$=9+26$ Multiply.

$=35$ Add.

Note: Because of the rules for order of operations at no time did we even subtract 9 – 2.

Now work margin exercise 4.

5. Simplify:

$5\left(4^2-3^2\right)-4\cdot 2^3$

Completion Example 5 Using the Order of Operations

Simplify: $6\left(5^2-4^2\right)-2\cdot 3^3$.

Solution

$6\left(5^2-4^2\right)-2\cdot 3^3$

$=6\left(\underline{25}-\underline{16}\right)-2\cdot\underline{27}$ Evaluate exponents.

$=6\left(\underline{9}\right)-2\cdot 27$ Simplify within parentheses.

$=\underline{54}-\underline{54}$ Multiply.

$=\underline{0}$ Subtract.

Now work margin exercise 5.

6. Simplify:

$3.5-4\left[\left(-2\right)^2-3.2\right]+12.3$

Example 6 Using the Order of Operations with Real Numbers

Simplify: $5.2-3\left[\left(-3\right)^2-2.4\right]+14.1$

Solution

$5.2-3\left[\left(-3\right)^2-2.4\right]+14.1$

$=5.2-3\left[9-2.4\right]+14.1$ Evaluate exponents.

$=5.2-3\left[6.6\right]+14.1$ Simplify within groupings.

$=5.2-19.8+14.1$ Multiply.

$=-0.5$ Add and subtract.

Now work margin exercise 6.

Example 7 Using the Order of Operations with Real Numbers

Simplify: $\left|-\frac{3}{5}\cdot\frac{5}{6}\right|-\frac{1}{4}\div\left(\frac{5}{2}\right)^2$

Solution

$$\left|-\frac{3}{5}\cdot\frac{5}{6}\right|-\frac{1}{4}\div\left(\frac{5}{2}\right)^2$$

$$=\left|-\frac{3}{5}\cdot\frac{5}{6}\right|-\frac{1}{4}\div\left(\frac{25}{4}\right)$$ Evaluate exponents.

$$=\left|-\frac{\cancel{3}}{\cancel{5}}\cdot\frac{\cancel{5}}{\underset{2}{\cancel{6}}}\right|-\frac{1}{\cancel{4}}\cdot\frac{\cancel{4}}{25}$$ Divide, then multiply and reduce.

$$=\left|-\frac{1}{2}\right|-\frac{1}{25}$$

$$=\frac{1}{2}-\frac{2}{50}$$ Evaluate the absolute value.

$$=\frac{25}{50}-\frac{2}{50}$$ Find the LCD.

$$=\frac{23}{50}$$ Subtract.

Now work margin exercise 7.

7. Simplify:

$$\left|-\frac{5}{3}\cdot\frac{3}{10}\right|-\frac{1}{9}\div\left(\frac{4}{3}\right)^2$$

CALCULATOR

Order of Operations with a Calculator

Some calculators are programmed to follow the rules for order of operations. To test whether your calculator is programmed in this manner, let's evaluate $7+4\cdot 2$. Press the keys

[7] [+] [4] [×] [2] and then [=].

If the answer on your calculator screen shows 15, then you do have a built in order of operations. This means that most of the time, you can type in the problem as it is written with the calculator performing the order of operations in a proper manner. If the calculator gave you an answer of 22, then you will need to be careful when entering problems.

If your calculator has parentheses buttons, [(] and [)], you can evaluate the expression $(6+2)+(8+1)\div 9$ by pressing the keys

[(] [6] [+] [2] [)] [+] [(] [8] [+] [1] [)] [÷] [9].

Then press [=].

The display will read 9.

Completion Example Answers

5. $6(25-16)-2\cdot 27=6(9)-2\cdot 27=54-54=0$

Margin Exercise Answers

1. −37 **2.** −22 **3.** 10 **4.** −14 **5.** 3 **6.** 12.6 **7.** $\frac{7}{16}$

8.5 Exercises

Concept Check

1. grouping symbols
2. innermost
3. left, right
4. appear
5. −1
6. grouping
7. True
8. False; Addition and subtraction are the last operations to be performed.
9. True
10. False; The mnemonic is PEMDAS.

Practice

1. **a.** 36 **b.** 16
2. **a.** 8 **b.** 50
3. −25
4. 0
5. −10
6. −59
7. −45
8. 45
9. −137
10. −48
11. 152
12. −3
13. −189
14. −530
15. 143
16. 426

8.5 Exercises

Concept Check

Fill-in-the-Blank. Complete each sentence using information found in this section.

1. When following the rules for order of operations, simplify within ______ ______ first.
2. Start by simplifying the ______ grouping symbol and working outward.
3. When performing multiplication and division, move from _____ to ______.
4. When performing addition and subtraction, perform the operations in the order they _______, moving left to right.
5. A negative sign in front of a variable means the variable is being multiplied by ____.
6. Parentheses, brackets, and braces are known as ________ symbols.

True/False. Determine whether each statement is true or false. If a statement is false, explain how it can be changed so the statement will be true. (**Note:** There may be more than one acceptable change.)

7. If there are no grouping symbols, multiplication should always be performed before addition.
8. When following the rules for order of operations, powers indicated by exponents should be evaluated last.
9. The square root symbol is a grouping symbol.
10. A well-known mnemonic device for remembering the rules for order of operations is SADMEP.

Practice

Simplify.

1. **a.** $24 \div 4 \cdot 6$

   b. $24 \cdot 4 \div 6$
2. **a.** $20 \div 5 \cdot 2$

   b. $20 \cdot 5 \div 2$
3. $15 \div (-3) \cdot 3 - 10$
4. $20 \cdot 2 \div 2^2 + 5(-2)$
5. $3^2 \div (-9) \cdot (4 - 2^2) + 5(-2)$
6. $4^2 \div (-8)(-2) + 3(2^2 - 5^2)$
7. $14 \cdot 3 \div (-2) - 6(4)$

8. $6(13-15)^2\cdot 8\div 2^2+3(-1)$

9. $-10+15\div(-5)\cdot 3^2-10^2$

10. $16\cdot 3\div\left(2^2-5\right)$

11. $2-5\left[(-20)\div(-4)\cdot 2-40\right]$

12. $9-6\left[(-21)\div 7\cdot 2-(-8)\right]$

13. $(7-10)\left[49\div(-7)+20\cdot 3-(-10)\right]$

14. $(9-11)\left[(-10)^2\cdot 2+6(-5)^2-10^2+3\cdot 5\right]$

15. $8-9\left[(-39)\div(-13)+7(-2)-(-2)^2\right]$

16. $6-20\left[(-15)\div 3\cdot 5+6\cdot 2\div 3\right]$

17. $|16-20|\left[32\div|3-5|-5^2\right]$

18. $|10-30|\left[4^2\cdot|5-8|\div(-2)^2+|17-18|\right]$

19. $(-10)+(-2)+|2-4|$

20. $|16-20|+(-10)^2+5^2$

21. $\frac{3}{8}\cdot\frac{4}{5}+\frac{1}{15}$

22. $\frac{1}{4}\cdot\frac{12}{15}+\frac{2}{7}$

23. $\frac{1}{3}\div\frac{1}{2}-\frac{5}{6}\cdot\frac{3}{4}$

24. $\frac{2}{9}\div\frac{14}{3}-\frac{1}{6}\cdot\frac{4}{7}$

25. $\left(\frac{5}{6}\right)^2\div\frac{5}{12}-\frac{3}{8}$

26. $\left(\frac{2}{5}\right)^2\cdot\frac{5}{8}+\frac{1}{5}\div\frac{3}{4}$

27. $\frac{7}{6}\cdot 2^2-\frac{2}{3}\div\frac{1}{2}$

28. $\frac{3}{4}\div 3^2-4\left(\frac{1}{2}\right)^2$

29. $\left(\frac{3}{4}\right)\div\left(\frac{3}{5}\right)\cdot\frac{7}{8}+\frac{3}{16}$

30. $\left(\frac{2}{3}\right)\div\frac{7}{12}\cdot\frac{2}{7}+\left(\frac{1}{2}\right)^2$

31. $\left(-\frac{9}{10}\right)+\frac{5}{8}\cdot\frac{4}{5}\div\frac{6}{10}+\frac{2}{3}$

32. $\frac{5}{8}\div\frac{5}{2}+\left(-\frac{1}{2}\right)^2\cdot\frac{2}{5}$

33. $-0.7-8.5\div 1.7$

34. $-0.4-2.6\cdot 1.5$

35. $(3.1+1.1)\div(5.7-6.9)$

36. $(3.2-6.5)\cdot 2^2$

37. $-15\div\left(\frac{1}{4}-\frac{7}{8}\right)$

38. $-12\div\left(\frac{1}{2}+\frac{1}{10}\right)$

39. $(-5-7)\div-4-8$

40. $4(-2)^2-10\div 5+1$

Solve.

41. Find the average of the five numbers: –7, 8, –3, 5, and 2.

42. Find the average of the six numbers: –1, –2, –3, 3, 2, and 1.

43. If the square of $\frac{7}{8}$ is subtracted from the square of $\frac{3}{4}$, what is the difference?

44. Find the quotient if the sum of $\frac{1}{5}$ and $\frac{2}{15}$ is divided by the difference between $\frac{7}{8}$ and $\frac{3}{4}$.

Use a graphing calculator to evaluate each expression.

45. $3.4\div 4+5\cdot 8.32$

46. $8.1\div 5+16.3\cdot 7$

47. $0.75\div 1.5+7\cdot 3.1^2$

48. $1.05\div(-3)\cdot 3.7-1.1^2$

17. −36
18. 260
19. −10
20. 129
21. $\frac{11}{30}$
22. $\frac{17}{35}$
23. $\frac{1}{24}$
24. $-\frac{1}{21}$
25. $\frac{31}{24}$
26. $\frac{11}{30}$
27. $\frac{10}{3}$
28. $-\frac{11}{12}$
29. $\frac{41}{32}$
30. $-\frac{33}{28}$
31. $\frac{3}{5}$
32. $\frac{7}{20}$
33. −5.7
34. −4.3
35. −3.5
36. −13.2
37. 24
38. −20
39. −5
40. 15
41. 1
42. 0
43. $-\frac{13}{64}$
44. $\frac{8}{3}$
45. 42.45
46. 115.72
47. 67.77
48. −2.505

49. 15.41
50. 3.956

Applications

51. a. $\$11 \cdot \left(1\frac{1}{4}\right) + \$8.50$

b. \$22.25

c. \$31

52. a. $0.10 \cdot 4 \cdot (\$102 + 7.50)$

b. \$394.20

53. a. $-\$42 - \$35 - (3 \cdot \$5)$

b. −\$92

54. a. $115 - 20(3 + 1.5)$

b. 25 yards

c. Yes

49. $6.32 \cdot 8.4 \div 16.8 + 3.5^2$

50. $(82.7 + 16.2) \div (14.83 - 19.83)^2$

Applications

Solve.

51. Madeline sells homemade aprons online and needs to determine how to charge for each apron. To create each apron, she spends \$8.50 on supplies and it takes her $1\frac{1}{4}$ hours to cut and sew each one. Madeline wants to charge \$11 per hour of work plus the cost of supplies.

a. Write an expression to describe how much each apron will cost.

b. Evaluate the expression to determine the selling cost of each apron.

c. Madeline will sew a name or initials onto the apron for an additional charge of \$1.75 per letter. If Kathy orders an apron and wants her name sewn onto it, how much will the apron cost?

52. The Matthews family, a family of 4, is planning a trip to New York City. During their visit, they want to see the Broadway play *Beetlejuice*. The tickets cost \$102 each. The Matthews purchase the tickets online and the website charges a service fee of \$7.50 per ticket. The website is running a sale where the Matthews can get 10% off of their entire purchase.

a. Write an expression to describe how much of a discount the Matthews will receive on their purchase.

b. What is the final purchase price of the tickets?

53. Dennis overdrew his checking account and ended up with a balance of −\$42. The bank charged a \$35 overdraft fee and an additional \$5 fee for every day the account was overdrawn. Dennis left his account overdrawn for 3 days.

a. Write an expression to show the balance of Dennis's checking account after 3 days.

b. Simplify the expression in part a. to find the balance of Dennis's checking account after 3 days.

54. Camila is a seamstress and is creating bridesmaid dresses. She has 115 yards of satin fabric. For each dress, the skirt requires 3 yards of satin and the bodice requires 1.5 yards of satin. She plans to make 20 dresses.

a. Write an expression to show how much fabric Camila will have left over after making the dresses.

b. Simplify the expression in part a. to determine how much fabric Camila will have left over.

c. Camila wants to make shawls from the leftover fabric. Each shawl requires 1.25 yards of satin. Can she make 15 shawls?

55. During harvest season, farmers donate fresh food to a local food kitchen. To make sure the food doesn't spoil, the food kitchen distributes the food between themselves and 5 other food kitchens in the area. One farmer donates $12\frac{1}{2}$ pounds of potatoes, another farmer donates $15\frac{3}{4}$ pounds of potatoes, and a third farmer donates $11\frac{3}{4}$ pounds of potatoes. The food kitchen finds that $1\frac{1}{4}$ pounds of the donated potatoes are rotten.

a. Write an expression to show how many pounds of potatoes each food kitchen will receive.

b. Simplify the expression from part a. to determine how many pounds of potatoes each food kitchen will receive.

56. Casey wants to put together some back-to-school gifts for local families in need. She has contacted companies directly and worked out deals to get backpacks for \$10 each, headphones for \$4 each, a pack of crayons for \$0.50 each, and a combo pack consisting of a notebook, a folder, and a pencil for \$1 per combo pack.

a. Write an expression to describe how much each gift will cost Casey, assuming each gift consists of one backpack, one pair of headphones, one pack of crayons, and 2 combo packs.

b. How much will Casey spend in total if she is able to give 5 back-to-school gifts?

57. You and three friends are planning a weekend trip. You plan to share a hotel room that is \$225 a night, go on a city tour that costs \$20 per person, and go to a baseball game that is \$15 per person.

a. Write an expression to describe the total cost of the trip, assuming you are all staying for two nights.

b. If you split the cost equally, how much will each person pay for the trip, not counting additional expenses?

Writing & Thinking

58. Explain, in your own words, why the following expression cannot be evaluated.

$$\left(24-2^4\right)+6(3-5)\div\left(3^2-9\right)$$

59. Consider any number between 0 and 1. If you square this number, will the result be larger or smaller than the original number? Is this always the case? Explain.

60. Consider any number between −1 and 0. If you square this number, will the result be larger or smaller than the original number? Is this always the case? Explain.

55. a. $\left(12\frac{1}{2}+15\frac{3}{4}+11\frac{3}{4}-1\frac{1}{4}\right)\div 6$

b. $6\frac{11}{24}$ pounds of potatoes

56. a. \$10 + \$4 + \$0.5 + 2 · \$1

b. \$82.50

57. a. 2 · \$225 + 4 · \$20 + 4 · \$15

b. \$147.50

Writing & Thinking

58. $\left(3^2-9\right)=0$ and division by 0 is undefined.

59. Smaller; When any positive number is multiplied by a fraction (or decimal) between 0 and 1, the result will be smaller. This is what is happening when a number between 0 and 1 is squared. Answers will vary.

60. It will always be larger. If you square a negative number you get a positive number. All positive numbers are larger than all negative numbers.

Objectives

A. Identify properties of real numbers that justify given statements.

8.6 Properties of Real Numbers

A Identifying Properties of Real Numbers

At this point, we have covered operations with whole numbers, fractions, decimals, and negative numbers. As we continue to work with all types of real numbers and algebraic expressions, we will need to understand that certain properties are true for addition and multiplication. However, these same properties are not true for subtraction and division. For example,

the order of the numbers in addition ***does not*** change the result:

$$17 + 5 = 5 + 17 = 22;$$

while the order of the numbers in subtraction ***does*** change the result:

$$17 - 5 = 12 \quad \text{but} \quad 5 - 17 = -12.$$

We say that addition is **commutative** while subtraction is **not commutative**. The various properties of real numbers under the operations of addition and multiplication are summarized here. These properties are used throughout algebra and mathematics in developing formulas and general concepts.

Properties of Addition and Multiplication

In this table a, b, and c are real numbers.

Name of Property	For Addition	For Multiplication
Commutative property	$\boldsymbol{a+b=b+a}$ $3+6=6+3$	$\boldsymbol{ab=ba}$ $4\cdot 9=9\cdot 4$
Associative property	$\boldsymbol{(a+b)+c=a+(b+c)}$ $(2+5)+4=2+(5+4)$	$\boldsymbol{a(bc)=(ab)c}$ $6\cdot(2\cdot 7)=(6\cdot 2)\cdot 7$
Identity	$\boldsymbol{a+0=0+a=a}$ $20+0=0+20=20$	$\boldsymbol{a\cdot 1=1\cdot a=a}$ $-2\cdot 1=1\cdot(-2)=-2$
Inverse	$\boldsymbol{a+(-a)=0}$ $10+(-10)=0$	$\boldsymbol{a\cdot\frac{1}{a}=1\ (\text{for } a\neq 0)}$ $3\cdot\frac{1}{3}=1$

PROPERTIES

Note

The number 0 is called the **additive identity** because when 0 is added to a number, the result is the same number. Likewise, the number 1 is called the **multiplicative identity** because when a number is multiplied by 1, the result is the same number. Also, the **additive inverse** of a number is its **opposite** and the **multiplicative inverse** of a number is its **reciprocal**.

Properties of Addition and Multiplication (cont.)

In this table a, b, and c are real numbers.

Zero-Factor Law

$\boldsymbol{a \cdot 0 = 0 \cdot a = 0}$ $\qquad -5 \cdot 0 = 0 \cdot (-5) = 0$

Distributive Property of Multiplication over Addition

$\boldsymbol{a(b+c) = ab + ac}$ $\qquad 3(x+5) = 3 \cdot x + 3 \cdot 5 = 3x + 15$

PROPERTIES

Math Tip

The zero-factor law goes by many different names. A few other names are

the zero-product law,

the zero property of multiplication, and

the multiplication property of zero.

The raised dot is optional when indicating multiplication between two variables or a number and a variable. For example,

$$x \cdot y = xy \quad \text{and} \quad 5 \cdot x = 5x.$$

In the case of a number and a variable, the number is called the coefficient of the variable. So,

in $5x$, the number 5 is the **coefficient** of x.

Thus, in an illustration of the distributive property involving a variable, we can write

$$5(x-6) = 5 \cdot x - 5 \cdot 6 = 5x - 30.$$

Example 1 Identifying Properties of Addition and Multiplication

State the name of each property being illustrated.

a. $(-7) + 13 = 13 + (-7)$

b. $8 + (9+1) = (8+9) + 1$

c. $(-25) \cdot 1 = -25$

d. $3(x+y) = 3x + 3y$

e. $0 \cdot 14 = 0$

f. $4 \cdot (3 \cdot 2) = (4 \cdot 3) \cdot 2$

g. $-(5+y) = -1(5+y)$
$= -1 \cdot 5 + (-1)y$
$= -5 - y$

Solution

a. Commutative property of addition

b. Associative property of addition

c. Multiplicative identity

d. Distributive property

e. Zero-factor law

f. Associative property of multiplication

g. Distributive property

Now work margin exercise 1.

1. State the name of each property being illustrated.

a. $-2 \cdot (4 \cdot 16) = (-2 \cdot 4) \cdot 16$

b. $5(z+w) = 5z + 5w$

c. $(-37) \cdot 0 = 0$

d. $14 + (2+5) = (14+2) + 5$

e. $-7(-15) = -15(-7)$

f. $0 + 28 = 28$

g. $5 + (-5) = 0$

2. State the property illustrated and show that the statement is true for the value given for the variable.

a. $x+21=21+x$ given that $x=-7$

b. $(5\cdot 4)x=5(4x)$ given that $x=2$

c. $11(y+3)=11y+33$ given that $y=-4$

Example 2 Identifying Properties of Addition and Multiplication

For each of the following equations, state the property illustrated, and show that the statement is true for the value given for the variable by substituting the value in the equation and evaluating.

a. $x+14=14+x$ given that $x=-4$

b. $(3\cdot 6)x=3(6x)$ given that $x=5$

c. $12(y+3)=12y+36$ given that $y=-2$

Solution

a. The commutative property of addition is illustrated.

$$(-4)+14=10 \text{ and } 14+(-4)=10$$

b. The associative property of multiplication is illustrated.

$$(3\cdot 6)\cdot 5=18\cdot 5=90 \text{ and } 3\cdot(6\cdot 5)=3\cdot 30=90$$

c. The distributive property is illustrated.

$$12(-2+3)=12(1)=12 \text{ and } 12(-2)+36=-24+36=12$$

Now work margin exercise 2.

Margin Exercise Answers

1. a. Associative property of multiplication **b.** Distributive property **c.** Zero-factor law **d.** Associative property of addition **e.** Commutative property of multiplication **f.** Additive identity **g.** Additive inverse **2. a.** Commutative property of addition $(-7)+21=14$ and $21+(-7)=14$ **b.** Associative property of multiplication $(5\cdot 4)\cdot 2=40$ and $5\cdot(4\cdot 2)=40$ **c.** Distributive property $11(-4+3)=-11$ and $11(-4)+33=-11$

8.6 Exercises

Concept Check

Fill-in-the-Blank. Complete each sentence using information found in this section.

1. The multiplicative inverse of a number is its ________.
2. The ______ ________ of all numbers is 1.
3. Zero multiplied by a number or variable is an example of the _____-______ law.
4. The distributive property involves two operations, ______ and ______.
5. The additive inverse of a number is the _______ of that number.
6. In the term $8x$, the 8 is the ___________ of the variable.

8.6 Exercises

Concept Check

1. reciprocal
2. multiplicative identity
3. zero-factor
4. addition, multiplication
5. opposite
6. coefficient
7. False; The commutative property of addition allows the order to change.
8. True
9. False; The additive identity of all numbers is 0.
10. False; The commutative property only works for addition and multiplication.

Practice

1. $3+7$
2. $6\cdot(9\cdot 3)$
3. $4\cdot 19$
4. $5+18$
5. $30+48$
6. $(16+9)+11$
7. $(2\cdot 3)\cdot x$
8. $3x+15$
9. $(3+x)+7$
10. $9x+45$
11. 0

True/False. Determine whether each statement is true or false. If a statement is false, explain how it can be changed so the statement will be true. (**Note:** There may be more than one acceptable change.)

7. Changing the order of the numbers in an addition problem is allowed because of the associative property of addition.
8. The equation $(8\cdot2)\cdot5=8\cdot(2\cdot5)$ is an example of the associative property of multiplication.
9. The additive identity of all numbers is 1.
10. The commutative property works for division and subtraction.

Practice

Complete the expressions using the given property. Do not simplify.

1.	$7+3=$ ______	commutative property of addition
2.	$(6\cdot9)\cdot3=$ ______	associative property of multiplication
3.	$19\cdot4=$ ______	commutative property of multiplication
4.	$18+5=$ ______	commutative property of addition
5.	$6(5+8)=$ ______	distributive property
6.	$16+(9+11)=$ ______	associative property of addition
7.	$2\cdot(3x)=$ ______	associative property of multiplication
8.	$3(x+5)=$ ______	distributive property
9.	$3+(x+7)=$ ______	associative property of addition
10.	$9(x+5)=$ ______	distributive property
11.	$6\cdot0=$ ______	zero-factor law
12.	$6\cdot1=$ ______	multiplicative identity
13.	$0+(x+7)=$ ______	additive identity
14.	$0\cdot(-13)=$ ______	zero-factor law
15.	$2(x-12)=$ ______	distributive property
16.	$(-5)+5=$ ______	additive inverse
17.	$6.3+(-6.3)=$ ______	additive inverse
18.	$3\cdot\frac{1}{3}=$ ______	multiplicative inverse

12. 6
13. $x+7$
14. 0
15. $2x-24$
16. 0
17. 0
18. 1
19. Commutative property of addition
20. Commutative property of multiplication
21. Multiplicative identity
22. Additive identity
23. Associative property of addition
24. Commutative property of addition
25. Commutative property of multiplication
26. Distributive property
27. Commutative property of multiplication
28. Associative property of multiplication
29. Multiplicative inverse
30. Multiplicative inverse
31. Additive inverse
32. Additive inverse
33. Multiplicative identity
34. Associative property of multiplication
35. Zero-factor law
36. Additive identity
37. Associative property of addition
38. Distributive property
39. $6(11)=66$ and $6\cdot3+6\cdot8=66$
40. $7(3)=21$ and $7\cdot8-7\cdot5=21$
41. $10(-7)=-70$ and $10\cdot2-10\cdot9=-70$

42. $13(8)=104$ and $13\cdot 5+13\cdot 3=104$

43. Commutative property of multiplication; $6\cdot 4=4\cdot 6=24$

44. Commutative property of addition; $19+3=3+19=22$

45. Associative property of addition; $8+(5+(-2)) = (8+5)+(-2)=11$

46. Associative property of multiplication; $(2\cdot 7)\cdot 4=2\cdot(7\cdot 4)=56$

47. Distributive property; $5(4+18)=5(4)+90=110$

48. Associative property of addition; $(2(3)+14)+3 = 2(3)+(14+3)=23$

49. Associative property of multiplication; $(6\cdot(-2))\cdot 9=6\cdot(-2\cdot 9) = -108$

50. Commutative property of multiplication; $11\cdot 4=4\cdot 11=44$

51. Commutative property of addition; $3+(-34)=(-34)+3=-31$

52. Distributive property; $3(-2+15)=3(-2)+45=39$

53. Commutative property of addition; $2(3+4)=2(4+3)=14$

54. Commutative property of multiplication; $(-2+2)(-2-4) =(-2-4)(-2+2)=0$

55. Commutative property of addition; $5+(4-15)=(4-15)+5 =-6$

State the name of each property being illustrated. See Example 1.

19. $5+16=16+5$

20. $5\cdot 16=16\cdot 5$

21. $32\cdot 1=32$

22. $32+0=32$

23. $5+(3+1)=(5+3)+1$

24. $5+(3+1)=(3+1)+5$

25. $13(y+2)=(y+2)\cdot 13$

26. $13(y+2)=13y+26$

27. $6(2\cdot 9)=(2\cdot 9)\cdot 6$

28. $6(2\cdot 9)=(6\cdot 2)\cdot 9$

29. $5\cdot\frac{1}{5}=1$

30. $14\cdot\frac{1}{14}=1$

31. $7.1+(-7.1)=0$

32. $(-9)+9=0$

33. $1\cdot 14.2=14.2$

34. $(5\cdot 3)\cdot(-7)=5(3\cdot(-7))$

35. $5.68\cdot 0=0\cdot 5.68=0$

36. $0+5.68=5.68$

37. $2+(x+6)=(2+x)+6$

38. $2(x+6)=2x+12$

First evaluate each expression using the rules for order of operations and then use the distributive property to evaluate the same expression. The value must be the same.

39. $6(3+8)$

40. $7(8-5)$

41. $10(2-9)$

42. $13(5+3)$

For each of the following equations, state the property illustrated, and show that the statement is true for the value of $x = 4$, $y = -2$, or $z = 3$ by substituting the corresponding value in the equation and evaluating. See Example 2.

43. $6\cdot x=x\cdot 6$

44. $19+z=z+19$

45. $8+(5+y)=(8+5)+y$

46. $(2\cdot 7)\cdot x=2\cdot(7x)$

47. $5(x+18)=5x+90$

48. $(2z+14)+3=2z+(14+3)$

49. $(6\cdot y)\cdot 9=6\cdot(y\cdot 9)$

50. $11\cdot x=x\cdot 11$

51. $z+(-34)=-34+z$

52. $3(y+15)=3y+45$

53. $2(3+x)=2(x+3)$

54. $(y+2)(y-4)=(y-4)(y+2)$

55. $5+(x-15)=(x-15)+5$

56. $z+(4+x)=(4+x)+z$

57. $(3x)\cdot 5=3\cdot(x\cdot 5)$

58. $(x+y)+z=x+(y+z)$

Applications

Solve

59. Jessica works part-time at a retail store and makes $11 an hour. During one week, she worked $6\frac{1}{2}$ hours on Monday and $4\frac{1}{4}$ hours on Thursday.

a. Determine the amount of money she earned during the week by evaluating the expression $\$11\cdot\left(6\frac{1}{2}+4\frac{1}{4}\right)$.

b. Rewrite this expression to remove the parentheses using one of the properties talked about in this section.

c. What property did you use in part b. to rewrite the expression?

60. Robin went to the grocery store to buy a few items she needed in order to cook dinner. She bought milk for $3.99, rolls for $2.25, a package of steaks for $12.01, and some marinade for $1.75. Before getting to the checkout line, Robin remembered that she only had $20 in her purse. Did she have enough money to buy the food items if the store does not charge sales tax on food?

a. Write an expression to find the total of Robin's food purchases. Do not simplify.

b. Robin doesn't have a calculator to determine the total cost of her items. She wants to make sure that she has enough money to buy them. Rearrange the expression from part a. so that she could quickly find the total using mental math.

c. What properties did you use in part b. to rewrite the expression?

d. Did Robin have enough money to purchase all of the items?

61. Jordan didn't balance his checking account during the week and ended up overdrawing his account. He had a starting balance of $85.04 and wrote checks for two bills for the amounts of $28.79 and $50.00. He also used his debit card to purchase lunch for $12.16. In order to avoid an overdraft fee, Jordan must deposit enough money today to bring his balance back to a minimum of zero.

a. Write an expression to find the current balance of Jordan's checking account. Do not simplify.

b. Evaluate the expression from part a. to determine the current balance of Jordan's checking account.

c. Write an equation to show Jordan's current checking account balance plus the amount he must deposit today to bring the balance to zero.

d. What property is illustrated in part c.?

Writing & Thinking

62. a. The distributive property illustrated as $a(b+c)=ab+ac$ is said to "distribute multiplication over addition." Explain, in your own words, the meaning of this phrase.

b. What would an expression that "distributes addition over multiplication" look like? Explain why this would or would not make sense.

56. Commutative property of addition; $3+(4+4)=(4+4)+3=11$

57. Associative property of multiplication; $(3\cdot4)\cdot5=3\cdot(4\cdot5)=60$

58. Associative property of addition; $(4+(-2))+3$ $=4+(-2+3)=5$

Applications

59. a. $118.25

b. $\$11\cdot\left(6\frac{1}{2}\right)+\$11\cdot\left(4\frac{1}{4}\right)$

c. Distributive property

60. a. $3.99 + $2.25 + $12.01 + $1.75

b. ($3.99 + $12.01) + ($2.25 + $1.75) = $16 + $4 = $20

c. Commutative and associative properties of addition

d. Yes

61. a. $85.04 – $28.79 – $50.00 – $12.16 or $85.04 – ($28.79 + $50.0 + $12.16)

b. –$5.91

c. –$5.91 + $5.91 = 0

d. The additive inverse property

Writing & Thinking

62. a. Multiply each term that is part of the sum (in the parentheses) by a.

b. An expression that "distributes addition over multiplication" would look like $a+(bc)=(a+b)(a+c)$, however this is not a true statement. Answers will vary.

Objectives

A. Identify like terms.

B. Simplify algebraic expressions by combining like terms.

C. Evaluate algebraic expressions for given values of the variables.

8.7 Simplifying and Evaluating Algebraic Expressions

A Identifying Like Terms

A number is called a **constant**. Any constant, power of a variable, or the products of a number and powers of variables is called a **term**. Examples of terms are

$$16, \quad \frac{3}{4}x, \quad -5.2, \quad 1.3xy, \quad -5x^2, \quad 14b^3, \quad \text{and} \quad -\frac{x}{y}.$$

A number written next to a variable (as in $-10x$) or a variable written next to another variable (as in xy) indicates multiplication. In the term $-4x^2$, the constant -4 is called the **numerical coefficient** of x^2 (or simply the **coefficient** of x^2). If a term has only one variable, then the exponent of the variable is called the **degree** of the term. Constants are said to be of zero degree.

Coefficient	Coefficient	Coefficient	Constant
$5x^3$	$\frac{1}{2}y^2$	$-7.1x$	-3
Third-degree term	Second-degree term	First-degree term	Zero-degree term

If no number is written next to a variable, then the coefficient is understood to be 1. For example,

$$x = 1 \cdot x, \quad a^3 = 1 \cdot a^3, \quad \text{and} \quad xy = 1 \cdot xy.$$

If a negative sign (–) is written next to a variable, then the coefficient is understood to be –1. For example,

$$-y = -1 \cdot y, \quad -x^2 = -1 \cdot x^2, \quad \text{and} \quad -xy = -1 \cdot xy.$$

Like Terms

Like terms (or **similar terms**) are terms that contain the same variables (if any) raised to the same powers.

DEFINITION

Like Terms

$-6, 1.84, 132,$ and $\frac{2}{5}$	are like terms because each term is a constant.
$\frac{1}{2}a, 15a, -2.6a, 3a$	are like terms because each term contains the same variable a raised to the same exponent, 1. (Remember that $a = a^1$.) These terms are first-degree in a.
$5xy^2$ and $-3.2xy^2$	are like terms because each term contains the same two variables, x and y, where x is first-degree in both terms and y is second-degree in both terms.

Unlike Terms

$8x$ and $-9x^2$	are unlike terms (**not** like terms) because the variable x is not of the same degree in both terms. $8x$ is first-degree in x and $-9x^2$ is second-degree in x.
$\frac{1}{3}b^2$ and $2a^2b$	are not like terms because the variables are not of the same degree in both terms.

Example 1 Identifying Like Terms

Identify the like terms in the following list of terms.

$$6, \quad 2.2x, \quad -10.1, \quad -x, \quad 3x^2z, \quad -5x, \quad \frac{3}{4}x^2z, \quad 0$$

Solution

$6, -10.1,$ and 0 are like terms; all are constants.

$2.2x, -x,$ and $-5x$ are like terms; all have the variable x.

$3x^2z$ and $\frac{3}{4}x^2z$ are like terms; all have the variables x^2z.

Now work margin exercise 1.

1. Identify the like terms in the following list of terms.

$\frac{5}{3}, 10b, -1, 4a^2b, -3b,$
$2a^2b, -5.65b, 0$

Math Tip

Pay careful attention to the exponents of like terms. Terms may appear similar to each other but not be like terms due to differences in the exponents. For example,

$$x^2z \quad \text{and} \quad xz^2$$

are not like terms since the exponent is on different variables.

B Simplifying Algebraic Expressions by Combining Like Terms

An **algebraic expression** is a combination of variables and numbers using any of the operations of addition, subtraction, multiplication, or division, as well as exponents. Examples of algebraic expressions are

$$x^2 - 14, \quad \frac{2xy}{3z^3} + y^2, \quad \frac{12x-9}{3x-4}, \quad \text{and} \quad 10x^3 - 4x^2 - 11x + 1.$$

To simplify algebraic expressions, we **combine like terms** whenever possible.

The procedure for combining like terms uses an alternate form of the distributive property as shown below. (The distributive property was first introduced in Section 1.3.)

$$ba + ca = (b + c)a$$

$$7x + 2x = (7 + 2)x = 9x$$

Thus, by using the distributive property in this sense, we can combine the like terms $7x$ and $2x$. (Intuitively, we can think of x as a label or name. Thus, just as 7 oranges + 2 oranges = 9 oranges, $7x + 2x = 9x$.)

Similarly, we can write

$$17y - 14y = (17 - 14)y - 3y$$

and

$$4n - n + 2n = (4 - 1 + 2)n = 5n.$$ (Notice that -1 is the coefficient of $-n$.)

Combining Like Terms

To **combine like terms**, add (or subtract) the coefficients and keep the common variable expression.

DEFINITION

Example 2 Combining Like Terms

Simplify each expression by combining like terms.

a. $8x + 10x$

b. $6.5y - 2.3y + 3$

c. $2x^2 + 3x + x^2 - x$

d. $4(n - 7) + 5(n + 1)$

e. $3a - 52a^2 + 9$

f. $\frac{x + 3x}{2} + 5x$

Solution

a. $8x + 10x = (8 + 10)x$ — Distributive property

$= 18x$

b. $6.5y - 2.3y + 3 = (6.5 - 2.3)y + 3$ — Distributive property

$= 4.2y + 3$

c. $2x^2 + 3x + x^2 - x$

$= 2x^2 + x^2 + 3x - x$ — Commutative property of addition

$= (2 + 1)x^2 + (3 - 1)x$ — Distributive property. (**Note:** $+x^2 = 1x^2$ and $-x = -1x$.)

$= 3x^2 + 2x$

2. Simplify each expression by combining like terms.

a. $3x + 9x$

b. $4.2z + 5 - 3.1z$

c. $4x^2 + 7y + 3x^2 - 2y$

d. $6(a - 2) + 4(a + 5)$

e. $-2x^2 + 5y^2 + 9 - y$

f. $\frac{x + 5x}{3} + 6x$

d. $4(n-7)+5(n+1)$

$= 4n-28+5n+5$ Simplify by using the distributive property twice.

$= 4n+5n-28+5$ Commutative property of addition

$= (4+5)n+(-28+5)$ Distributive property

$= 9n-23$ Combine like terms.

e. $3a-52a^2+9$ This expression is already simplified since it has no like terms.

f. A fraction bar represents division and is a grouping symbol, similar to parentheses. So, combine like terms in the numerator first.

$$\frac{x+3x}{2}+5x=\frac{4x}{2}+5x$$

$$=\frac{4}{2}\cdot x+5x$$

$= 2x+5x$ Reduce the fraction.

$= 7x$ Combine like terms.

Now work margin exercise 2.

Math Tip

It is often easier to first group like terms using the commutative property before combining the like terms together.

C Evaluating Algebraic Expressions

Algebraic expressions can be evaluated for given values of the variables by substituting one value for each variable into the expression and then following the rules for order of operations. When like terms are present, the process of evaluation can be made easier by first combining like terms.

Parentheses must be used around negative numbers when substituting.

Without parentheses, an evaluation can be dramatically changed and lead to wrong answers, particularly when even exponents are involved. We analyze with the exponent 2 as follows.

In general, except for $x = 0$, the following is true.

1. $-x^2$ **is negative**. ($-6^2 = -1\cdot 6^2 = -1\cdot 36 = -36$, this is -1 times the square of 6.)
2. $(-x)^2$ **is positive**. $[(-6)^2 = (-6)(-6) = 36$, this is the square of $-6.]$
3. $-x^2 \neq (-x)^2$ $(-36 \neq 36)$

Evaluating an Algebraic Expression

1. Combine like terms, if possible.
2. Substitute the values given for any variables.
3. Follow the rules for order of operations.

(**Note:** Terms separated by + and – signs may be evaluated at the same time. Then the value of the expression can be found by adding and subtracting from left to right.)

PROCEDURE

3. a. Evaluate x^2 for $x = 10$ and for $x = -5$.

b. Evaluate $-x^2$ for $x = 10$ and for $x = -5$.

Example 3 Evaluating Algebraic Expressions

a. Evaluate x^2 for $x = 3$ and for $x = -4$.

b. Evaluate $-x^2$ for $x = 3$ and for $x = -4$.

Solution

a. For $x = 3,\ x^2 = (3)^2 = 9.$

For $x = -4,\ x^2 = (-4)^2 = 16.$

b. For $x = 3,\ -x^2 = -(3)^2 = -1(9) = -9.$

For $x = -4,\ -x^2 = -(-4)^2 = -1(16) = -16.$

Now work margin exercise 3.

4. Simplify and evaluate $4x + 6 + 3x$ for $x = -4$.

Example 4 Simplifying and Evaluating Algebraic Expressions

Simplify and evaluate $2x + 5 + 7x$ for $x = -3$.

Solution

First, simplify the expression by combining like terms.

$$\begin{aligned} 2x+5+7x &= 2x+7x+5 \\ &= 9x+5 \end{aligned}$$

Now, substitute −3 for x (using parentheses around −3 to be sure the signs are correct), and evaluate this simplified expression by following the rules for order of operations.

$$\begin{aligned} 9x+5 &= 9(-3)+5 \\ &= -27+5 \\ &= -22 \end{aligned}$$

Now work margin exercise 4.

5. Simplify and evaluate $5ab - 8ab + 2a - 3a$ for $a = -3,\ b = 1$

Example 5 Simplifying and Evaluating Algebraic Expressions

Simplify and evaluate $3ab - 4ab + 6a - a$ for $a = 2,\ b = -1$.

Solution

Simplify first.

$$3ab - 4ab + 6a - a = -ab + 5a$$

Now evaluate.

$$\begin{aligned} -ab+5a &= -1(2)(-1)+5(2) \\ &= 2+10 \\ &= 12 \end{aligned}$$

Note: $-ab = -1ab$

Now work margin exercise 5.

Example 6 Simplifying and Evaluating Algebraic Expressions

Simplify and evaluate $\frac{7x+5x}{4}-2(x+1)$ for $x=5$.

Solution

Simplify first.

$$\begin{aligned}\frac{7x+5x}{4}-2(x+1)&=\frac{12x}{4}-2x-2\\&=3x-2x-2\\&=x-2\end{aligned}$$

Now evaluate.

$$\begin{aligned}x-2&=(5)-2\\&=3\end{aligned}$$

Now work margin exercise 6.

6. Simplify and evaluate $\frac{7y+2y}{3}+2(y-8)$ for $y=4$.

Completion Example 7 Simplifying and Evaluating Expressions

Simplify and evaluate $5x^2-2x^2+x+3x^2-2x$ for $x=-1$.

Solution

First, simplify the expression by combining like terms.

$$\begin{aligned}5x^2-2x^2+x+3x^2-2x&=(\underline{5-2+3})x^2+(\underline{1-2})x\\&=\underline{6}\,x^2-\underline{1}\,x\end{aligned}$$

Now, substitute -1 for x and evaluate.

$$\begin{aligned}6x^2-x&=6(\underline{-1})^2-(\underline{-1})\\&=6(\underline{1})-(\underline{-1})\\&=\underline{6}+\underline{1}\\&=\underline{7}\end{aligned}$$

Now work margin exercise 7.

7. Simplify and evaluate $4x^2-3x+2x^2+4x$ for $x=-2$.

Completion Example Answer

7. $(5-2+3)x^2+(1-2)x=6x^2-1x$; $6(-1)^2-(-1)=6(1)-(-1)=6+1=7$

Margin Exercise Answers

1. $\frac{5}{3}$, -1, and 0 are like terms; $10b$, $-3b$, and $-5.65b$ are like terms; and $4a^2b$ and $2a^2b$ are like terms. **2. a.** $12x$ **b.** $1.1z+5$ **c.** $7x^2+5y$ **d.** $10a+8$ **e.** Already simplified **f.** $8x$ **3. a.** For $x=10$, $x^2=100$; For $x=-5$, $x^2=25$ **b.** For $x=10$, $-x^2=-100$; For $x=-5$, $-x^2=-25$ **4.** $7x+6$; -22 **5.** $-3ab-a$; 12 **6.** $5y-16$; 4 **7.** $6x^2+x$; 22

8.7 Exercises

Concept Check

Fill-in-the-Blank. Complete each sentence using information found in this section.

1. If no number is written next to a variable, the coefficient is understood to be the number __________.
2. Any constant, variable, or product or quotient of a constant and/or variable is a/an __________ .
3. A single number is a/an __________ .
4. In a term, the number being multiplied by the variable is the numerical __________ of that term.
5. To combine like terms, add or subtract the __________ and keep the common variable expression.
6. When substituting, __________ must be used around negative numbers.

True/False. Determine whether each statement is true or false. If a statement is false, explain how it can be changed so the statement will be true. (**Note:** There may be more than one acceptable change.)

7. A variable that does not appear to have an exponent has an exponent of 1.
8. In the term $-9x$, nine is being subtracted from x.
9. In the term "$12a$," 12 is the constant.
10. Like terms have the same coefficients.

Practice

Identify the like terms in each list of terms. See Example 1.

1. $-5, 3, 7x, 8, 9x, 3y$
2. $-2x^2, -13x^3, 5x^2, 14x^2, 10x^3$
3. $5xy, -x^2, -6xy, 3x^2y, 5x^2y, 2x^2$
4. $3ab^2, -ab^2, 8ab, 9a^2b, -10a^2b, ab, 12a^2$
5. $24, 8.3, 1.5xyz, -1.4xyz, -6, xyz, 5xy^2z, 2xyz^2$
6. $-35y, 1.62, -y^2, -y, 3y^2, \frac{1}{2}, 75y, 2.5y^2$

Simplify each expression by combining like terms. See Example 2.

7. $8x + 7x$
8. $3y + 8y$
9. $5x + (-2x)$

8.7 Exercises

Concept Check

1. 1
2. term
3. constant
4. coefficient
5. coefficients
6. parentheses
7. True
8. False; In the term $-9x$, -9 is being multiplied by x.
9. False; In the term $12a$, 12 is the coefficient.
10. False; Like terms have the same variables with same exponents.

Practice

1. -5, 3, and 8 are like terms; $7x$ and $9x$ are like terms.
2. $-2x^2$, $5x^2$, and $14x^2$ are like terms; $10x^3$ and $-13x^3$ are like terms.
3. $-x^2$ and $2x^2$ are like terms; $-6xy$ and $5xy$ are like terms; $3x^2y$ and $5x^2y$ are like terms.
4. $8ab$ and ab are like terms; $3ab^2$ and $-ab^2$ are like terms; $9a^2b$ and $-10a^2b$ are like terms.
5. 24, -6, and 8.3 are like terms; $1.5xyz$, $-1.4xyz$, and xyz are like terms.
6. $\frac{1}{2}$ and 1.62 are like terms; $-35y$, $-y$, and $75y$ are like terms; $-y^2$, $3y^2$ and $2.5y^2$ are like terms.
7. $15x$
8. $11y$
9. $3x$
10. $4x$
11. $5y^2$
12. $11z^2$

10. $7x+(-3x)$

11. $6y^2-y^2$

12. $16z^2-5z^2$

13. $3x-5x+12x$

14. $2a+14a-25a$

15. $6c-13c+5c$

16. $40x-30x-10x$

17. $4x+2+3x$

18. $3x-1+x$

19. $5x^2-3x^2+2x$

20. $-2x^2-x^2-x$

21. $7x^2-4x^2+20$

22. $14y^3-25+8y^3$

23. $2x^2-2y+5x^2+6x^2$

24. $4a+2a-3b-a$

25. $4x+7-8+3x$

26. $-5x-1+8+9x$

27. $2n^2-6n+1-4n^2+8n-4$

28. $3n^2+2n-5-n^2+n-4$

29. $3-5x^2+4x^2+20x+42-17x$

30. $13x+12x^2+15x-35-41-2x^2$

31. $3(n+1)+n$

32. $2(n-4)+n+1$

33. $5(a-b)+2a-3b$

34. $4a-3b+2(a+2b)$

35. $3(2x+y)+2(x-y)$

36. $4(x+5y)+3(2x-7y)$

37. $y-\frac{4y+5y}{3}$

38. $z-\frac{3z+5z}{4}$

39. $\frac{2x+3x}{3}+x$

40. $\frac{2y+4y}{5}-2y$

Evaluate the expression for $x=2$ and $y=-3$. See Example 3.

41. $-x^2$

42. $-y^2$

43. $(-x)^2$

44. $(-y)^2$

45. $-x$

46. $-y$

Simplify each expression and then evaluate the expression for $x=4$, $y=3$, $a=-2$, and $b=-1$. See Examples 4 through 7.

47. $5y+4-2y$

48. $7b-17-b$

49. $3(y-1)+2(y+2)$

50. $4(y+3)+5(y-2)$

51. $3.1a^2-0.9a^2+4a-5.3a^2$

52. $8.3x^2-5.7x^2+x^2+2$

53. $2.4(x+1)+1.3(x-1)$

54. $1.3(y+2)-2.6(8-y)$

55. $\frac{3a+5a}{-2}+12a$

56. $8a+\frac{5a+4a}{9}$

57. $\frac{-4b-2b}{-3}+\frac{2b+5b}{7}$

58. $\frac{5b+3b}{4}+\frac{-4b-b}{-5}$

13. $10x$
14. $-9a$
15. $-2c$
16. 0
17. $7x+2$
18. $4x-1$
19. $2x^2+2x$
20. $-3x^2-x$
21. $3x^2+20$
22. $22y^3-25$
23. $13x^2-2y$
24. $5a-3b$
25. $7x-1$
26. $4x+7$
27. $-2n^2+2n-3$
28. $2n^2+3n-9$
29. $-x^2+3x+45$
30. $10x^2+28x-76$
31. $4n+3$
32. $3n-7$
33. $7a-8b$
34. $6a+b$
35. $8x+y$
36. $10x-y$
37. $-2y$
38. $-z$
39. $\frac{8x}{3}$
40. $-\frac{4y}{5}$
41. -4
42. -9
43. 4
44. 9
45. -2
46. 3
47. $3x+4$; 13
48. $6b-17$; -23
49. $5y+1$; 16
50. $9y+2$; 29
51. $-3.1a^2+4a$; -20.4
52. $3.6x^2+2$; 59.6

53. $3.7x + 1.1$; 15.9
54. $3.9y - 18.2$; -6.5
55. $8a$; -16
56. $9a$; -18
57. $3b$; -3
58. $3b$; -3
59. $-x^2 + 5x - 7$; -21
60. $4x^2 - 4x + 5$; 29
61. $3y^2 - y$; 4
62. $2y^2 - 6y - 5$; 3
63. $4y^3 - 4y^2 + 5y + 1$; -12
64. $7y^3 + 5y^2 - 6$; -8
65. $5x^2 + 9x - 22$; -20
66. $3y^2 - 16y - 5$; 14
67. $2a^2 - 2a$; 4
68. $-a^3 - 2a - 7$; -4
69. $9ab - 7a + 2b$; 21
70. $3ab - 3a + 3b$; 3
71. $14a - 15b + 2c + 2$; 24
72. $12a + 8b - 3c - 64$; -101
73. $10a + 10b + 10c$; 0
74. 0; 0

Applications

75. \$50,000
76. 0 feet
77. 375 pounds
78. 7000 square yards
79. **a.** $47s + 143m$
b. 6170 boxes

Simplify each expression and then evaluate the expression for $x = -2$ and $y = -1$. See Examples 4 and 7.

59. $2x^2 - 3x^2 + 5x - 8 + 1$

60. $5x^2 - 4x + 2 - x^2 + 3$

61. $y^2 + 2y^2 + 2y - 3y$

62. $y^2 + y^2 - 8y + 2y - 5$

63. $y^3 + 3y^3 + 5y - 4y^2 + 1$

64. $7y^3 + 4y^2 + 6 + y^2 - 12$

65. $2(x^2 - 3x - 5) + 3(x^2 + 5x - 4)$

66. $5(y^2 - 4y + 3) - 2(y^2 - 2y + 10)$

Simplify each expression and then evaluate the expression for $a = -1$, $b = -2$, and $c = 3$. See Example 5.

67. $a^2 - a + a^2 - a$

68. $a^3 - 2a^3 - 3a + a - 7$

69. $5ab - 7a + 4ab + 2b$

70. $2ab + 4b - 3a + ab - b$

71. $14(a+7) - 15(b+6) + 2(c-3)$

72. $12(a-3) + 8(b-2) - 3(c+4)$

73. $20(a+b+c) - 10(a+b+c)$

74. $16(a-b+c) + 16(-a+b-c)$

Applications

Solve.

75. An apartment management company owns a property with 100 units. The company has determined that the profit made per month from the property can be calculated using the equation $P = -10x^2 + 1500x - 6000$, where x is the number of units rented per month. How much profit does the company make when 80 units are rented?

76. A ball is thrown upward from an initial height of 96 feet with an initial velocity of 16 feet per second. After t seconds, the height of the ball can be described by the expression $-16t^2 + 16t + 96$. What is the height of the ball after 3 seconds?

77. You have a 5-pound dumbbell that you use for wrist curls and a 10-pound dumbbell that you use for bicep curls. Your little brother likes to brag that if he lifts your 5-pound dumbbell two times, he's actually lifting 10 pounds. Using that logic, you could claim that the amount of weight you lift in a day can be modeled by the expression $5x + 10y$, where x is the number of wrist curls you do and y is the number of bicep curls you do. How much could you claim to have lifted on a day you did 15 wrist curls and 30 bicep curls?

78. Dan has 340 yards of fencing available to enclose a rectangular field. The area of the enclosure using this fencing can be modeled by $A = 170x - x^2$, where x is the width of the field. If the field ends up being 70 yards wide, how much will the area be?

79. A moving company starts the week with 72 bundles of small boxes and 50 bundles of medium boxes. During the week, they use 25 bundles of small boxes and 32 bundles of medium boxes. At the end of the week, they buy 125 bundles of medium boxes. The total number of boxes at the end of the week can be modeled by the expression $72s + 50m - 25s - 32m + 125m$, where s represents the number of boxes in a bundle of small boxes and m represents the number of boxes in a bundle of medium boxes.

a. Simplify the expression by combining like terms.

b. How many boxes are in stock at the end of the week if there are 40 boxes in a small bundle of boxes and 30 boxes in a medium bundle of boxes?

80. During a sale, all newly released video games are priced the same and all Blu-ray discs are priced the same. During the first day of the sale, Mitchell buys 4 video games and 6 Blu-ray discs. The next day he buys 2 more video games and returns 2 of the Blu-ray discs. The amount of money Mitchell spends can be modeled by $4v + 6d + 2v - 2d$, where v represents the cost of each video game and d represents the cost of each Blu-ray disc.

a. Simplify the expression by combining like terms.

b. How much did Mitchell spend if each video game costs \$35 and each Blu-ray disc costs \$19?

Writing & Thinking

81. Define constant and variable. Explain why those particular words are used.

82. Discuss like and unlike terms and give an example of each.

83. The text recommends simplifying an expression (combining like terms) before evaluating. Do you think this is necessary?
Evaluate the expression $4x^2 - 5(x+2) + 3x + 10 + 2x$ for $x = 3$:

a. by substituting and then evaluating.

b. by first simplifying and then evaluating.

Which method would you recommend? Why?

84. Explain the difference between -13^2 and $(-13)^2$.

80. **a.** $6v + 4d$ **b.** \$286

Writing & Thinking

81. A number is a constant and its value does not change regardless of where it appears in an expression or equation. A variable, usually represented by a letter, represents an unknown number or numbers.

82. Like terms have the same variables with the same exponents. For example, $4a^2bc^3$ and $-3a^2bc^3$ are like terms. Unlike terms either have different variables or possibly the same variables with different exponents. For example $6ab$ and $-9a^2b$ are unlike terms and $5xy$ and $13ax$ are unlike terms.

83. **a.** For $x = 3$,

$$4x^2 - 5(x+2) + 3x + 10 + 2x$$
$$= 4(3)^2 - 5(3+2) + 3(3) + 10 + 2(3)$$
$$= 4(9) - 5(5) + 9 + 10 + 6$$
$$= 36 - 25 + 9 + 10 + 6$$
$$= 36$$

b.

$$4x^2 - 5(x+2) + 3x + 10 + 2x$$
$$= 4x^2 - 5x - 10 + 3x + 10 + 2x$$
$$= 4x^2$$

For $x = 3$,

$$4x^2 = 4(3)^2$$
$$= 4(9) = 36$$

The second method, as it requires fewer calculations.

84. -13^2 is the square of 13 multiplied by -1 while $(-13)^2$ is the square of -13. This means that $-13^2 = -169$ and $(-13)^2 = 169$.

Objectives

A. Translate English phrases into algebraic expressions.

B. Translate algebraic expressions into English phrases.

C. Create word problems that fit a given equation.

8.8 Translating English Phrases and Algebraic Expressions

A Translating English Phrases into Algebraic Expressions

Algebra is a language of mathematicians, and to understand mathematics, you must understand the language. We want to be able to change English phrases into their "algebraic" equivalents and vice versa. So if a problem is stated in English, we can translate the phrases into algebraic symbols and proceed to solve the problem according to the rules and methods developed for algebra.

Certain words are the keys to the basic operations. Some of these words are listed here and emphasized in boldface in Example 1.

Math Tip

You must be careful when translating the phrase *less than*. The phrase "x is less than 3" is translated as $x < 3$, while the phrase "x less than 3" is translated as $3 - x$. Knowing whether you are comparing numbers or writing an algebraic expression will help determine how to translate the phrase.

Key Words To Look For When Translating Phrases

Addition	Subtraction	Multiplication	Division	Exponent (Powers)
add	subtract (from)	multiply	divide	square of
sum	difference	product	quotient	cube of
plus	minus	times	ratio	
more than	less than	twice, double		
increased by	decreased by	triple		
total	less			

The following examples illustrate how these key words used in English phrases can be translated into algebraic expressions. **Note that in each case "a number" or "the number" implies the use of a variable (an unknown quantity).**

Example 1 Translating English Phrases into Algebraic Expressions

Change each phrase into its equivalent algebraic expression.

English Phrase	Algebraic Expression
a. a number **plus** three	$z + 3$
three **added to** z	
the **sum** of z and three	
three **more than** a number	
z **increased by** three	
b. the **product** of three and x	$3x$
three **times** a number	
three **multiplied** by the number represented by x	

1. Change each phrase into its equivalent algebraic expression.

   a. the product of seven and x

   b. five increased by n

   c. four times the quantity found by adding a number to two

   d. three more than the product of two and a number

   e. nine multiplied by a number, less four

   f. the quotient of three and a number

English Phrase	Algebraic Expression
c. **twice** the **sum** of x and one the **product** of two with the **sum** of x and one two **times** the quantity found by **adding** a number to one	$2(x+1)$
d. **twice** x, **plus** one the **sum** of **twice** x and one two **times** x, **increased** by one one **more than** the **product** of two and a number	$2x+1$
e. the **difference** between five **times** a number and three three **less than** the **product** of a number and five five **times** a number, **minus** three three **subtracted from** $5n$ five **multiplied by** a number, **less** three	$5n-3$
f. the **quotient of** a number and six n **divided by** six the **ratio** of a number and six	$\frac{n}{6}$
g. the **square of** a number a number **squared**	x^2
h. the **cube of** a number a number **cubed**	n^3

Now work margin exercise 1.

Math Tip

The use of a comma can help clarify ambiguity in English phrases with multiple operations.

Attention!

In Example 1a, the phrase "the sum of z and three" was translated as $z + 3$. If the expression had been translated as $3 + z$, there would have been no mathematical error because addition is commutative. That is, $z + 3 = 3 + z$. However, in Part **e**, the phrase "three less than the product of a number and five" must be translated as it was because subtraction is **not** commutative. Thus,

"three less than five times a number" means $5n - 3$

while "three less five times a number" means $3 - 5n$.

Therefore, be very careful when writing and/or interpreting expressions indicating subtraction. The same is true with expressions involving division.

2. Change each phrase into its equivalent algebraic expression.

a. the number of inches in f feet

b. the cost of renting a trailer for one day and driving x miles if the rate is \$25 per day plus \$0.33 per mile

Example 2 Application: Translating English Phrases

Change each phrase into its equivalent algebraic expression.

English Phrase	Algebraic Expression
a. the number of minutes in h hours	$60h$
b. the cost of renting a truck for one day and driving x miles if the rate is \$30 per day plus \$0.25 per mile	$30 + 0.25x$

Now work margin exercise 2.

The words **quotient** and **difference** deserve special mention because their use implies that the numbers given are to be operated on **in the order given**. That is, division and subtraction are done with the values in the same order that they are given in the problem.

the quotient of y and 5 ⟶ $\frac{y}{5}$

the quotient of 5 and y ⟶ $\frac{5}{y}$

the difference between 6 and x ⟶ $6 - x$

the difference between x and 6 ⟶ $x - 6$

If we did not have these agreements concerning subtraction and division, then the phrases just illustrated might have more than one interpretation and be considered **ambiguous**.

An **ambiguous phrase** is a phrase whose meaning is not clear or that may have two or more interpretations. This is a common occurrence in ordinary everyday language, and misunderstandings occur frequently. Imagine the difficulties diplomats have in communicating ideas from one language to another trying to avoid ambiguities. Even the order of subjects, verbs, and adjectives may not be the same from one language to another. Translating grammatical phrases in any language into mathematical expressions is quite similar. To avoid ambiguous phrases in mathematics, we try to be precise in the use of terminology, to be careful with grammatical construction, and to follow the rules for order of operations.

B Translating Algebraic Expressions into English Phrases

Consider the three expressions to be translated into English:

$$7(n+1),\ 6(n-3),\ \text{and}\ 7n+1.$$

In the first two expressions, we indicate the parentheses with a phrase such as "the quantity" or "the sum of" or "the difference between." **Without the parentheses, we agree that the operations used in the expression are to be indicated in the order given.** Thus,

$7(n+1)$ can be translated as "seven times the sum of a number and one,"

$6(n-3)$ can be translated as "six times the difference between a number and three,"

while $7n+1$ can be translated as "seven times a number, plus one."

Example 3 Translating Algebraic Expressions into English Phrases

Change each algebraic expression into an equivalent English phrase. In each case translate the variable as "a number."

a. $5x$

b. $2n+8$

c. $3(a-2)$

Solution

	Algebraic Expression	Possible English Phrase
a.	$5x$	the product of five and a number
b.	$2n+8$	twice a number, increased by eight
c.	$3(a-2)$	three times the difference between a number and two

Now work margin exercise 3.

3. Change each algebraic expression into an equivalent English phrase.

a. $10x$

b. $4a+7$

c. $7(n-5)$

C Translating Equations into Word Problems

Have you ever said or heard someone say something like,

"I can do the mathematics. I just have trouble solving word problems."

Maybe you are one of these people. Basically, what they mean is that they can perform simple arithmetic or algebraic manipulations. However, reasoning and translating ideas into symbols that represent the information given in a "word problem" is where they are having trouble. Well, the bad news is that the reasoning, translating, and solving are the most important parts of mathematics. The skills needed for operating with numbers in arithmetic

What's the Difference?

Have you heard this joke?

Q: What's the difference between old and vintage?

A: About $300.

In real life, our everyday language tends to be less precise in the use of *difference*. When we say *difference* we tend to mean the positive difference. In math, we subtract in the exact order stated, so the math difference between old and vintage is actually a negative number, −$300.

and solving equations in algebra are just that—skills. For sure, without these skills, you would not be able to solve word problems. However, they are not sufficient. That is, you need much more in the way of understanding abstract concepts to be able to solve word problems and to solve problems in your daily life. The good news is that with experience and practice all of this can be learned.

Soon you will be given a chance to solve word problems by translating English phrases into algebraic expressions and sentences into equations. The solutions to the equations are the solutions to the corresponding word problems. In this section, to help in understanding the abstract relationship between word problems and algebraic equations, we are going to reverse the process. You are going to be given an equation and asked to "make up" your own related word problem that might use this equation to find a solution. Consider the following example.

4. Make up a word problem which might use the equation $x + 3x = 19$ in its solution

Example 4 Translating Equations into Word Problems

For each equation, make up your own word problem that might use the equation in its solution. Remember that the variable can be translated into something like "a number" or "some number."

a. $5x + 10 = -10$ **b.** $3y + 25 = 2(y + 6)$

Solution

a. Some number is multiplied by 5 and the product is increased by 10. If the result is equal to -10, what is the number?

b. If 25 is added to the product of 3 and a number, the result will be equal to twice the sum of the same number and 6. What is the number?

Now work margin exercise 4.

Note

In Example 4, the translations are not unique. In fact, there are many ways to make up a problem for each equation. However, all word problems should result in the same equation. You should be able to show your word problem to your classmates and have them agree that the related equation will give the solution to the problem.

Margin Exercise Answers

1. a. $7x$ **b.** $5 + n$ **c.** $4(y+2)$ **d.** $2x + 3$ **e.** $9x - 4$ **f.** $\frac{3}{n}$ **2. a.** $12f$ **b.** $25 + 0.33x$ **3.** Answers will vary. **a.** the product of ten and a number **b.** four times a number increased by seven **c.** seven times the difference between a number and five **4.** Answers will vary. For example, a number plus three times that number is equal to nineteen. What is the number?

8.8 Exercises

Concept Check

Fill-in-the-Blank. Complete each sentence using information found in this section.

1. A phrase is considered ____________ if its meaning is not clear or if it has two or more possible interpretations.

2. Phrases such as "a number" or "the number" imply the use of a/an ____________.

3. Key words such as "decreased by" and "minus" indicate the operation of ____________.

4. The key words "cube of" and "square of " mean ____________ are involved.

8.8 Exercises

Concept Check

1. ambiguous
2. variable
3. subtraction
4. exponents

5. "Twice" and "three times" indicate the operation of ____________.

6. "Divide" and "quotient" specify that ____________ should be used.

True/False. Determine whether each statement is true or false. If a statement is false, explain how it can be changed so the statement will be true. (**Note:** There may be more than one acceptable change.)

7. The order in which the values are given is particularly important when working with subtraction and division problems.

8. "More than" and "increased by" are key phrases specifying the operation of subtraction.

9. Division is indicated by the phrase "five less than a number."

10. Key phrases for parentheses can be used to limit ambiguity in English phrases.

Practice

Write the algebraic expressions described by the English phrases. Choose your own variable. See Example 1.

1. six added to a number
2. seven more than a number
3. four less than a number
4. a number decreased by thirteen
5. the quotient of twice a number and ten
6. the difference between a number and three, all divided by seven
7. four subtracted from the product of six and a number
8. eight minus twice a number
9. the sum of four times a number and twice the same number
10. the sum of nine times a number and the same number
11. fifteen decreased by twice a number
12. twenty decreased by the product of four and a number
13. three times a number, less five times the same number
14. seven times a number, decreased by twice the number
15. nine times the sum of a number and two
16. three times the difference between a number and eight
17. thirteen less than the product of four and the sum of a number and one
18. four more than the product of eight and the difference between a number and six
19. eight more than the product of three and the sum of a number and six
20. six less than twice the difference between a number and seven

5. multiplication
6. division
7. True
8. False; "More than" and "increased by" are key phrases specifying the operation of addition.
9. False; Subtraction is indicated by the phrase "five less than a number."
10. True

Practice

1. $x+6$
2. $n+7$
3. $y-4$
4. $z-13$
5. $\frac{2m}{10}$
6. $\frac{x-3}{7}$
7. $6y-4$
8. $8-2x$
9. $4x+2x$
10. $9y+y$
11. $15-2y$
12. $20-4n$
13. $3x-5x$
14. $7x-2x$
15. $9(x+2)$
16. $3(x-8)$
17. $4(x+1)-13$
18. $8(x-6)+4$
19. $3(x+6)+8$
20. $2(x-7)-6$

21. four less than the product of three and the difference between seven and a number

22. nine more than twice the sum of seventeen and a number

23. eighteen less than the quotient of a number and two

24. seven increased by the quotient of a number and five

Translate each pair of English phrases into algebraic expressions. Notice the differences between the algebraic expressions and the corresponding English phrases.

25. a. six less than a number

b. six less a number

26. a. twenty less than a number

b. twenty less a number

27. a. five less than three times a number

b. five less three times a number

28. a. six less than four times a number

b. six less four times a number

Write the algebraic expression described by the English phrases using the given variables. See Example 2.

29. the number of hours in d days

30. the cost of x graphing calculators if one calculator costs $115

31. the cost of x gallons of gasoline if the cost of one gallon is $3.15

32. the number of seconds in m minutes

33. the number of days in y years (Assume 365 days in a year.)

34. the cost of x pounds of candy priced at $4.95 a pound

35. the number of days in t weeks and three days

36. the number of minutes in h hours and twenty minutes

37. the points scored by a football team on t touchdowns and one field goal (a touchdown is 7 points and a field goal is 3 points)

38. the amount of vacation days an employee has after w weeks if she gets 0.2 vacation days for every week she works

39. the cost of renting a car for one day and driving m miles if the rate is $20 per day plus 15 cents per mile

40. the cost of purchasing a fishing rod and reel if the rod costs x dollars and the reel costs $8 more than twice the cost of the rod

41. the perimeter of a rectangle if the width is w centimeters and the length is three centimeters less than twice the width

42. the area of a square with side length of c centimeters

Translate each algebraic expression into an equivalent English phrase. (There may be more than one correct translation.) See Examples 3 and 4.

43. $4x$

44. $-9x$

45. $x + 5$

46. $x - 12$

47. $4x - 7$

48. $3x + 5$

49. $7(x+1)$

50. $3(x+2)$

21. $3(7-x)-4$
22. $2(17+x)+9$
23. $\frac{x}{2}-18$
24. $7+\frac{y}{5}$
25. a. $x-6$ b. $6-x$
26. a. $y-20$ b. $20-y$
27. a. $3x-5$ b. $5-3x$
28. a. $4n-6$ b. $6-4n$
29. $24d$
30. $115x$
31. $\$3.15x$
32. $60m$
33. $365y$
34. $\$4.95x$
35. $7t+3$
36. $60h+20$
37. $7t+3$
38. $0.2w$
39. $\$20+\$0.15m$
40. $\$x+(\$2x+\$8)=\$3x+\$8$
41. $2w+2(2w-3)=6w-6$
42. c^2
43. Four times a number
44. The product of a number and negative nine
45. A number increased by five
46. A number decreased by twelve
47. Four times a number decreased by seven
48. Five more than three times a number
49. Seven times the sum of a number and one
50. Three times the sum of a number and two

51. $-2(x-8)$

52. $10(x+4)$

53. $5(2x+3)$

54. $3(4x-5)$

55. $\frac{6}{x-1}$

56. $\frac{9}{x+3}$

57. $6x+x-1$

58. $5x-x+2$

59. $8+2(x-1)$

60. $5-3(x+1)$

Translate each pair of expressions into equivalent English phrases. (There may be more than one correct translation.) Notice the differences between the algebraic expressions and the corresponding English phrases.

61. $3x+7;\ 3(x+7)$

62. $4x-1;\ 4(x-1)$

63. $7x-3;\ 7(x-3)$

64. $5(x+6);\ 5x+6$

Writing & Thinking

65. Explain why translating addition and multiplication problems from English into algebra may be easier than changing subtraction or division problems from English into algebra. (Consider the properties previously studied.)

66. Explain the difference between $5(n+3)$ and $5n+3$ when converting from algebra to English.

67. Make up your own word problem that might use the given equation in its solution. Be creative! Translate the variable into something like "a strange number," or "the age of a dog," or "an amount invested."

a. $2x+3=-4$

b. $3x-2=-5$

c. $n+(n+1)=25$

d. $n+(n+2)=135$

e. $2x+3x=x$

f. $x=5x-6x$

51. Negative two times the difference between a number and eight

52. Ten times the sum of a number and four

53. Five times the sum of twice a number and three

54. Three times the difference between four times a number and five

55. The quotient of six and the difference between a number and one

56. Nine divided by the sum of a number and three

57. Six times a number plus the number minus one

58. The product of five and a number minus the number plus two

59. Eight increased by twice the difference between a number and one

60. Five decreased by three times the sum of a number and one

61. The sum of three times a number and seven; three times the sum of a number and seven

62. One less than four times a number; four times the difference between a number and one

63. The product of seven and a number minus three; seven times the difference between a number and three

64. Five times the sum of a number and six; five times a number plus six

Writing & Thinking

65. The Commutative Property of Addition and Multiplication permits the order of items being added or multiplied to change and still have the same result. This property does not hold true for subtraction or division. Therefore, order is important for subtraction and division problems or the answer will change or be incorrect.

66. For the first expression, a phrase such as "the quantity" or "the sum of" should be used since 5 is being multiplied by both n and 3. That is, the first expression would be "five times the sum of a number and three," while the second expression would be "the sum of five times a number and three."

67. Answers will vary.

Chapter 8 Project

Going to Extremes!

An activity to demonstrate the use of signed numbers in real life.

When asked what the highest mountain peak in the world is, most people would say Mount Everest. This answer may be correct, depending on what you mean by highest. According to Geology.com, there may be other contenders for this important distinction.

The peak of Mount Everest is 8850 meters (or 29,035 feet) above sea level, giving it the distinction of being the mountain with the highest altitude in the world. However, Mauna Kea is a volcano on the big island of Hawaii whose peak is over 10,000 meters above the nearby ocean floor, which makes it taller than Mount Everest. A third contender for the highest mountain peak is Chimborazo, an inactive volcano in Ecuador. Although Chimborazo only has an altitude of 6263 meters (20,549 feet) above sea level, it is the highest mountain above Earth's center. How could a mountain that is only 6263 meters tall be higher than a mountain that is 8850 meters tall? Because the Earth is not really a sphere but an "oblate spheroid," which means that the Earth is widest at the equator. Chimborazo is 1° south of the equator which makes it about 2000 meters farther from the Earth's center than Mount Everest.

What about the other extreme? What is the lowest point on Earth? As you might have guessed, there is also more than one candidate for that distinction. The lowest exposed area of land on Earth's surface is on the Dead Sea shore at 413 meters below sea level. The Bentley Subglacial Trench in Antarctica is the lowest point on Earth that is not covered by ocean but is covered by ice. This trench reaches 2555 meters below sea level. The deepest point on the ocean floor occurs 10,916 meters below sea level in the Mariana Trench in the Pacific Ocean.

1. Calculate the **difference** in elevation between Mount Everest and Chimborazo in both meters and feet. What operation does the word **difference** imply? 2587 meters, 8486 feet; subtraction

2. Write an expression to calculate the **difference** in elevation between the peak of Mount Everest and the lowest point on the Dead Sea shore in meters and then simplify. $8850 \text{ m} - (-413 \text{ m}) = 9263 \text{ m}$

3. If you were to travel from the bottom of the Mariana Trench to the top of Mount Everest, how many meters would you travel? 19,766 meters

4. If Mount Everest were magically moved and placed at the bottom of the Mariana Trench, how many meters of water would lie above Mount Everest's peak? 2066 meters

5. How much farther below sea level (in meters) is the Mariana Trench as compared to the Dead Sea shore? 10,503 meters

6. Add the elevations (in meters) together for Mount Everest, Chimborazo, the Dead Sea Shore, the Bentley Subglacial Trench, and the Mariana Trench and show your result. Is this number positive or negative? Would this value represent an elevation above or below sea level?

7. Convert the results in Problems 2 through 4 above to feet using the conversion factor 1 meter = 3.28 feet. Do not round your answers.

8. Convert the results in Problem 7 from feet to miles using the conversion factor 1 mile = 5280 feet. (Round your answers to the nearest thousandth.)

9. Using the height of Mount Everest in meters as an example, the conversions in Problems 7 and 8 could have been combined to do the conversion from meters directly to miles by using the following sequence of conversion factors:

$$8850 \text{ m} \cdot \frac{3.28 \not{\text{ft}}}{1 \not{\text{m}}} \cdot \frac{1 \not{\text{m}}}{5280 \not{\text{ft}}} \approx 5.498 \text{ miles}$$

(A mountain peak over 5 miles high!)

Now notice that in doing the conversions, the units for meters and feet cancel out since they appear in both the numerator and denominator, leaving only the unit of miles in the numerator of your result. This is called dimensional analysis and is extremely helpful in converting measurements to make sure you end up with the correct answer and the correct units on your result.

Now verify that this sequence of conversions works by taking the results from Problems 2 through 4 and applying both conversion factors above. How do your results compare to the results from Problem 8? (Round your answers to the nearest thousandth.)

10. There is more than one way that this conversion could have been performed. Using the conversion factors 1 km = 1000 m and 1 mile = 1.61 km, convert the results in Problems 2 through 4 from meters to miles by using these factors in sequence similar to Problem 9 and performing a dimensional analysis. (Round your answer to the nearest thousandth.) Do you get exactly the same results? Why do you think this is so?

Chapter 8 Project: Going to Extremes!

6. 1229 meters; positive, above sea level

7. 30,382.64 feet; 64,832.48 feet; 6776.48 feet

8. 5.754 miles; 12.279 miles; 1.283 miles

9. Results match Question 8.

10. 5.753 miles; 12.277 miles; 1.283 miles. The answers do not match exactly because the conversion factors are rounded.

Chapter 8 Project: Ordering Operations

1. **a.** 3

b. The computer computed the operations from left to right

c. ((8 + 0) * 3 – 10) / 2

d. 8 + (0 * 3) – (10 / 2)

e. Answers will vary. Grouping symbols are necessary for clarity.

2. **a.** –17

b. The computer calculated (–4)^2 instead of –(4^2)

c. (–4)^2 – 10 / 2 + 4

3. **a.** $-6x - 10y$

b. The computer combined $3x + 5y$ into $8xy$

c. Yes, the error is still in combining the unlike terms.

4. **a.** The issue is with phrasing. You should have typed "five times the sum of two and three"

b. The issue is with the computer program since "less" and "less than" do not mean the same thing.

Chapter 8 Project

Ordering Operations

An activity to explore the orders of operations.

Did you know that the order of operations you learn in math courses is a relatively new set of rules? The rules were created in the early 1900s and solidified into the current form along with the creation of computers and computer languages. Before the 1600s, mathematical notation was not commonly used, and mathematical expressions and equations were written out in words. Any phrasing that was ambiguous (that is, could be understood in more than one way) was avoided. When simplifying expressions in mathematics, care must be taken to properly follow the order of operations. If you stray from the order, you will reach a different conclusion than intended.

Suppose you are creating a computer program to follow the rules for the order of operations. The computer uses * for multiplication, / for division, and ^ to indicate exponents.

1. As your first test, you enter "8 + 0 * 3 – 10 / 2" into the computer program and it returns a value of 7.
   a. Following the order of operations, what would you expect as the result?
   b. Explain the error(s) made by the computer program.
   c. Insert grouping symbols into the expression so that it will simplify to the value returned by the computer.
   d. Rewrite the expression (using parentheses if needed) so that the computer program will return the expected value from part a. if the computer program is not modified.
   e. Compare the answers from parts c. and d. What do you notice?

2. After modifying your code, you verify that the previous expression returns the correct value. For the next test, you enter "–4^2 – 10 / 2 + 4" and the computer returns a value of 15.
   a. Following the order of operations, what would you expect as the result?
   b. Explain the error(s) made by the computer program.
   c. Rewrite the original expression so that it will simplify to the value returned by the computer.

3. After further modification, the computer program can now properly follow the order of operations. You next add code to allow the program to simplify algebraic expressions. You test your code by entering "$-2(3x + 5y)$" and it returns $-16xy$.
   a. What would you expect the program to return?
   b. Explain the error(s) made by the computer program.
   c. Is this the same error as the computer returning a result of $30xy$ when you enter "$3(4x + 6y)$"?

4. As a final step, you write code to allow the program to translate English phrases into algebraic expressions. You enter the phrase "five times two plus three" and the program returns "5 * 2 + 3".
   a. If you expected "5(2 + 3)" in return, is the issue with the computer program or your phrasing? Explain what the issue is.
   b. Next, you enter the phrase "twelve less two times a number" and the program returns "$2x - 12$". Is the issue the computer program or phrasing?

CHAPTER 9

Solving Linear Equations and Inequalities

Connections

You have probably heard that mathematics is the language of science, engineering, and business. While that is considered to be true, it can be difficult to see where some of the skills you develop in a math course are applied in those areas.

For example, the ability to solve linear equations is particularly useful in the study of free markets in economics. Economists want to be able to understand how changes in the price of goods affect consumer behavior. One way to gain this understanding is to analyze the demand for a specific product.

The demand of a product is the number of units q that the market is willing to absorb at a price p. In other words, it's how many items consumers are willing and able to purchase. Typically, the lower the price of a product, the more units consumers will be willing to purchase. To illustrate this relationship, consider the US demand for wheat in 1981, which can be illustrated by the following equation.

$$q = 3550 - 266p$$

In this equation, q is measured in millions of bushels and p is the price per bushel in dollars.

How can you determine the price per bushel when the demand hit 2630 million bushels?

Objectives

A. Determine whether a given real number is a solution to an equation.

B. Solve equations of the form $x + b = c$.

C. Solve equations of the form $ax = c$.

9.1 Solving Linear Equations: $x + b = c$ and $ax = c$

In this section, we will discuss solving linear equations in the following two forms:

$$x + b = c \quad \text{and} \quad ax = c.$$

In these equations, we treat a, b, and c as constants and x as the unknown quantity.

In the following sections, we will combine the techniques developed here and discuss solving linear equations in the forms:

$$ax + b = c \quad \text{and} \quad ax + b = cx + d.$$

A Checking Solutions in Equations

An **equation** is a statement that two algebraic expressions are equal. That is, both expressions represent the same number. If an equation contains a variable, any number that gives a true statement when substituted for the variable is called a **solution** to the equation. The solutions to an equation form a **solution set**. The process of finding the solution set is called **solving the equation**.

Note

A linear equation in x is also called a **first-degree equation in x** because the variable x can be written with the exponent 1. That is, $x = x^1$.

Linear Equation in x

If a, b, and c are **constants** and $a \neq 0$ then a **linear equation in x** is an equation that can be written in the form

$$ax + b = c.$$

DEFINITION

We begin by determining whether a particular number satisfies an equation. A number is said to be a **solution** (or to **satisfy an equation**) if it gives a true statement when substituted for the variable.

1. Determine whether the given number is a solution to the equation.
   a. $x + 3 = 12$ given that $x = 4$
   b. $y + 3.7 = 4.2$ given that $y = 0.5$
   c. $4.3 - z = 2.1$ given that $z = 2.2$
   d. $|y| + 2 = 9$ given that $y = -7$

Example 1 Checking Given Solutions in Equations

Determine whether the given real number is a solution to the given equation by substituting for the variable and checking to see if the resulting equation is true or false.

a. $x + 5 = -2$ given that $x = -7$

b. $1.4 + z = 0.5$ given that $z = -1.1$

c. $5.6 - y = 2.9$ given that $y = 2.7$

d. $|z| - 14 = -3$ given that $z = -10$

Solution

a. $(-7) + 5 = -2$ is true, so -7 **is** a solution.

b. $1.4 + (-1.1) = 0.5$ is false because $1.4 + (-1.1) = 0.3 \neq 0.5$. So, -1.1 **is not** a solution.

c. $5.6 - (2.7) = 2.9$ is true, so 2.7 **is** a solution.

d. $|-10| - 14 = -3$ is false because $|-10| - 14 = 10 - 14 = -4 \neq -3$. So, -10 **is not** a solution.

Now work margin exercise 1.

Note

Notice that parentheses were used around negative numbers in the substitutions. This should be done to keep operations properly separated, particularly when negative numbers are involved.

B Solving Equations of the Form $x + b = c$

To begin, we need the **addition principle of equality**.

Addition Principle of Equality

If the same algebraic expression is added to both sides of an equation, the new equation has the same solutions as the original equation. Symbolically, if A, B, and C are algebraic expressions, then the equations

$$A = B$$

and

$$A + C = B + C$$

have the same solutions.

PROPERTIES

Math Tip

In the form $x + b = c$, the variables b and c are constants. This means that b and c stand for real numbers and can appear as either negative or positive numbers.

Equations with the same solutions are said to be **equivalent equations**.

The objective of solving linear (or first-degree) equations is to find an equivalent equation with the variable by itself (with a coefficient of +1) on one side of the equation and any constants on the other side. The following procedure will help in solving linear equations such as

$$x - 3 = 7, \quad -11 = y + 5, \quad 3z - 2z + 2.5 = 3.2 + 0.8, \quad \text{and} \quad x - \frac{2}{5} = \frac{3}{10}.$$

Solving Linear Equations that Simplify to the Form $x + b = c$

1. Combine like terms on both sides of the equation.
2. Use the **addition principle of equality** and add the opposite of the constant b to both sides of the equation. The objective is to isolate the variable on one side of the equation (either the left side or the right side) with a coefficient of +1.
3. Check your answer by substituting it for the variable in the original equation.

PROCEDURE

Every linear equation has exactly one solution. This means that once a solution has been found, there is no need to search for another solution.

Note that, as illustrated in the examples in this section, variables other than x may be used.

2. Solve the equation:

$x - 5 = 12$

Example 2 Solving Linear Equations of the Form $x + b = c$

Solve the equation: $x - 3 = 7$

Solution

$x - 3 = 7$	Write the equation.
$x - 3 + 3 = 7 + 3$	Add 3 (the opposite of -3) to both sides.
$x = 10$	Simplify.

Check

$x - 3 = 7$	
$(10) - 3 \overset{?}{=} 7$	Substitute $x = 10$.
$7 = 7$	True statement

Now work margin exercise 2.

3. Solve the equation:

$-8 = x + 4$

Example 3 Solving Linear Equations of the Form $x + b = c$

Solve the equation: $-11 = y + 5$

Solution

$-11 = y + 5$	Write the equation. Note that the variable can be on the right side.
$-11 - 5 = y + 5 - 5$	Add -5 (the opposite of $+5$) to both sides. This is the same as subtracting 5 from both sides.
$-16 = y$	Simplify.

Check

$-11 = y + 5$	
$-11 \overset{?}{=} (-16) + 5$	Substitute $y = -16$.
$-11 = -11$	True statement

Now work margin exercise 3.

4. Solve the equation:

$x - 3.7 = 0$

Example 4 Solving Linear Equations of the Form $x + b = c$

Solve the equation: $-4.1 + x = 0$

Solution

$-4.1 + x = 0$	Write the equation.
$-4.1 + x + 4.1 = 0 + 4.1$	Add 4.1 (the opposite of -4.1) to both sides.
$x = 4.1$	Simplify.

Check

$-4.1 + x = 0$	
$-4.1 + (4.1) \overset{?}{=} 0$	Substitute $x = 4.1$.
$0 = 0$	True statement

Now work margin exercise 4.

Example 5 Solving Linear Equations of the Form $x + b = c$

Solve the equation: $x - \frac{2}{5} = \frac{3}{10}$

Solution

$x - \frac{2}{5} = \frac{3}{10}$	Write the equation.
$x - \frac{2}{5} + \frac{2}{5} = \frac{3}{10} + \frac{2}{5}$	Add $\frac{2}{5}$ $\left(\text{the opposite of } -\frac{2}{5}\right)$ to both sides.
$x = \frac{3}{10} + \frac{4}{10}$	Simplify. (The common denominator is 10.)
$x = \frac{7}{10}$	Simplify.

Check

$x - \frac{2}{5} = \frac{3}{10}$	
$\left(\frac{7}{10}\right) - \frac{2}{5} \stackrel{?}{=} \frac{3}{10}$	Substitute $x = \frac{7}{10}$.
$\frac{7}{10} - \frac{4}{10} \stackrel{?}{=} \frac{3}{10}$	The common denominator is 10.
$\frac{3}{10} = \frac{3}{10}$	True statement

Now work margin exercise 5.

5. Solve the equation:

$x - \frac{3}{8} = \frac{3}{4}$

Example 6 Simplifying and Solving Linear Equations

Simplify and solve: $3z - 2z + 2 = 3 + 8$

Solution

$3z - 2z + 2 = 3 + 8$	Write the equation.
$z + 2 = 11$	Combine like terms on both sides of the equation.
$z + 2 - 2 = 11 - 2$	Add -2 (the opposite of $+2$) to both sides. This is the same as subtracting 2 from both sides.
$z = 9$	Simplify.

Check

$3z - 2z + 2 = 3 + 8$	
$3(9) - 2(9) + 2 \stackrel{?}{=} 3 + 8$	Substitute $z = 9$.
$27 - 18 + 2 \stackrel{?}{=} 3 + 8$	Simplify.
$11 = 11$	True statement

Now work margin exercise 6.

6. Simplify and solve:

$5z - 4z + 3 = 5 - 7$

Completion Example 7 Simplifying and Solving Linear Equations

Supply the reasons for each step in solving the equation.

$5x - 4x - 1.5 = 6.3 + 4.0$

7. Simplify and solve:

$4z + 0.8 - 3z = 3.1 + 1.9$

Solution

$5x-4x-1.5=6.3+4.0$	Write the equation.
$x-1.5=10.3$	Combine like terms on both sides of the equation.
$x-1.5+1.5=10.3+1.5$	Add 1.5 (the opposite of −1.5) to both sides of the equation.
$x=11.8$	Simplify.

Now work margin exercise 7.

C Solving Equations of the Form $ax = c$

To solve equations of the form $ax = c$, where $a \neq 0$, we can use the idea of the reciprocal of the coefficient a. For example, as we studied earlier, the reciprocal of $\frac{3}{4}$ is $\frac{4}{3}$ and $\frac{3}{4} \cdot \frac{4}{3} = 1$. Also, we need the **multiplication** (or **division**) **principle of equality** as stated in the following box.

Multiplication (or Division) Principle of Equality

If both sides of an equation are multiplied by (or divided by) the same nonzero constant, the new equation has the same solutions as the original equation. Symbolically, if A and B are algebraic expressions and C is any nonzero constant, then the equations

$$A = B,$$

$$AC = BC \text{ where } C \neq 0,$$

$$\text{and } \frac{A}{C} = \frac{B}{C} \text{ where } C \neq 0$$

have the same solutions.

PROPERTIES

Remember that the objective is to get the variable by itself on one side of the equation. That is, we want the variable to have +1 as its coefficient. The following procedure will accomplish this.

Solving Linear Equations that Simplify to the Form $ax = c$

1. Combine like terms on both sides of the equation.
2. Use the **multiplication** (or **division**) **principle of equality** and multiply both sides of the equation by the reciprocal of the coefficient of the variable (**or divide both sides by the coefficient itself**). The coefficient of the variable will become +1.
3. Check your answer by substituting it for the variable in the original equation.

PROCEDURE

Example 8 Solving Linear Equations of the Form $ax = c$

Solve the linear equation: $5x = 20$

Solution

$5x = 20$	Write the equation.
$\frac{1}{5}\cdot(5x) = \frac{1}{5}\cdot 20$	Multiply both sides by $\frac{1}{5}$, the reciprocal of 5.
$\left(\frac{1}{5}\cdot 5\right)x = \frac{1}{5}\cdot\frac{20}{1}$	Use the associative property of multiplication.
$1\cdot x = 4$	Simplify.
$x = 4$	

Check

$5x = 20$	
$5\cdot(4) \stackrel{?}{=} 20$	Substitute $x = 4$.
$20 = 20$	True statement

Multiplying by the reciprocal of the coefficient is the same as **dividing** by the coefficient itself. So, we can multiply both sides by $\frac{1}{5}$, as we did, or we can divide both sides by 5. In either case, the coefficient of x becomes +1.

$5x = 20$	
$\frac{5x}{5} = \frac{20}{5}$	Divide both sides by 5.
$x = 4$	Simplify.

Now work margin exercise 8.

8. Solve the equation:

$3x = 33$

Example 9 Solving Linear Equations of the Form $ax = c$

Solve the linear equation: $1.1x + 0.2x = 12.2 - 3.1$

Solution

When decimal coefficients or constants are involved, you might want to use a calculator to perform some of the arithmetic.

$1.1x + 0.2x = 12.2 - 3.1$	Write the equation.
$1.3x = 9.1$	Combine like terms.
$\frac{1.3x}{1.3} = \frac{9.1}{1.3}$	Divide both sides by 1.3 so that the coefficient will become +1.
$x = 7.0$	Use a calculator or pencil and paper to divide.

9. Solve the equation:

$2.5x + 0.4x = 15.2 - 3.6$

Check

$$1.1x + 0.2x = 12.2 - 3.1$$

$$1.1(7) + 0.2(7) \stackrel{?}{=} 12.2 - 3.1$$ Substitute $x = 7$.

$$7.7 + 1.4 \stackrel{?}{=} 9.1$$ Simplify.

$$9.1 = 9.1$$ True statement

Now work margin exercise 9.

10. Solve the equation:

$-x = -15$

Example 10 Solving Linear Equations of the Form $ax = c$

Solve the linear equation: $-x = 4$

Solution

$$-x = 4$$ Write the equation.

$$-1x = 4$$ -1 is the coefficient of x.

$$\frac{-1x}{-1} = \frac{4}{-1}$$ Divide by -1 so that the coefficient will become $+1$.

$$x = -4$$

Check

$$-x = 4$$

$$-(-4) \stackrel{?}{=} 4$$ Substitute $x = -4$.

$$4 = 4$$ True statement

Now work margin exercise 10.

11. Solve the equation:

$\frac{2x}{3} = \frac{4}{5}$

Completion Example 11 Solving Equations of the Form $ax = c$

Supply the reasons for each step in solving the equation.

$$\frac{4x}{5} = \frac{3}{10}$$ (This could also be written $\frac{4}{5}x = \frac{3}{10}$ because $\frac{4}{5}x$ is the same as $\frac{4x}{5}$.)

Solution

$$\frac{4x}{5} = \frac{3}{10}$$ Write the equation.

$$\frac{5}{4} \cdot \frac{4}{5}x = \frac{5}{4} \cdot \frac{3}{10}$$ Multiply both sides by $\frac{5}{4}$.

$$1 \cdot x = \frac{1 \cdot \cancel{5}}{4} \cdot \frac{3}{2 \cdot \cancel{5}}$$

$$x = \frac{3}{8}$$ Simplify.

Now work margin exercise 11.

Example 12 Application: Solving Linear Equations

The original price of a Blu-Ray player was reduced by \$45.50. The sale price was \$165.90. Solve the equation $y - 45.50 = 165.90$ to determine the original price of the Blu-Ray player.

Solution

$$y - 45.50 = 165.90$$

$$y - 45.50 + 45.50 = 165.90 + 45.50$$ Use the addition principle of equality by adding 45.50 to both sides.

$$y = 211.40$$ Simplify.

The original price of the Blu-Ray player was \$211.40.

Now work margin exercise 12.

12. The original price of a wool coat was reduced by \$15.80. The reduced price was \$84.79. Solve the equation $y - 15.80 = 84.79$ to determine the original price of the coat.

Completion Example Answers

7. $5x - 4x - 1.5 = 6.3 + 4.0$ Write the equation.

$x - 1.5 = 10.3$ Combine like terms on both sides of the equation.

$x - 1.5 + 1.5 = 10.3 + 1.5$ Add 1.5 (the opposite of −1.5) to both sides of the equation.

$x = 11.8$ Simplify.

11. $\frac{4x}{5} = \frac{3}{10}$ Write the equation.

$\frac{5}{4} \cdot \frac{4}{5} x = \frac{5}{4} \cdot \frac{3}{10}$ Multiply both sides by $\frac{5}{4}$.

$1 \cdot x = \frac{1 \cdot \cancel{5}}{4} \cdot \frac{3}{2 \cdot \cancel{5}}$ Simplify.

$x = \frac{3}{8}$

Margin Exercise Answers

1. a. $x = 4$ is not a solution **b.** $y = 0.5$ is a solution **c.** $z = 2.2$ is a solution **d.** $y = -7$ is a solution

2. $x = 17$ **3.** $x = -12$ **4.** $x = 3.7$ **5.** $x = \frac{9}{8}$ **6.** $z = -5$ **7.** $z = 4.2$ **8.** $x = 11$ **9.** $x = 4$ **10.** $x = 15$

11. $x = \frac{6}{5}$ **12.** The original price of the wool coat was \$100.59.

9.1 Exercises

Concept Check

Fill-in-the-Blank. Complete each sentence using information found in this section.

1. A/An ________ is a statement that two algebraic expressions are equal.
2. If an equation contains a variable, any number that gives a true statement when substituted for the variable is a/an ________ of the equation.
3. The ________ principle of ________ involves adding the same algebraic expression to both sides of an equation.

9.1 Exercises

Concept Check

1. equation
2. solution
3. addition, equality

4. constants
5. dividing
6. multiplication
7. True
8. False; Solving the equation
9. True
10. True

Practice

1. $x = -2$ is a solution
2. $z = 18$ is a solution
3. $x = 4$ is not a solution
4. $x = 1$ is not a solution
5. $x = -4$ is a solution
6. $x = 5$ is a solution
7. $x = -18$ is a solution
8. $z = -72$ is not a solution
9. $x = -28$ is a solution
10. $x = -11$ is a solution
11. $x = 7$
12. $x = 19$
13. $y = -4$
14. $y = -7$
15. $x = -19$
16. $x = -27$
17. $n = 37$
18. $n = 56$
19. $z = -6$
20. $z = 17$
21. $x = 5$
22. $x = -1$
23. $y = -5.9$
24. $y = -5.3$
25. $x = -1.2$
26. $x = 1.1$
27. $x = \frac{11}{20}$
28. $n = \frac{1}{2}$
29. $x = 9$
30. $x = 12$
31. $y = 8$
32. $y = 3$

4. The objective of solving linear equations is to get the variable (with a coefficient of +1) on one side of the equation and any ________ on the other side.

5. Multiplying by the reciprocal of the coefficient of the variable is the same as ________ by the coefficient.

6. If both sides of an equation are multiplied by the same nonzero constant, the ________ principle of equality can be used.

True/False. Determine whether each statement is true or false. If a statement is false, explain how it can be changed so the statement will be true. (**Note:** There may be more than one acceptable change.)

7. When an algebraic expression is added to both sides of an equation, the new equation has the same solutions as the original equation.

8. The process of finding the solution set to an equation is called simplifying the equation.

9. A linear equation in x is also called a first-degree equation in x.

10. Equations with the same solutions are said to be equivalent equations.

Practice

Determine whether the given number is a solution to the given equation by substituting and then evaluating. See Example 1.

1. $x + 4 = 2$ given that $x = -2$
2. $z + (-12) = 6$ given that $z = 18$
3. $x - 3 = -7$ given that $x = 4$
4. $x - 2 = -3$ given that $x = 1$
5. $-10 + x = -14$ given that $x = -4$
6. $-9 - x = -14$ given that $x = 5$
7. $-26 + |x| = -8$ given that $x = -18$
8. $42 + |z| = -30$ given that $z = -72$
9. $|x| - |-3| = 25$ given that $x = -28$
10. $|-2| + |x| = 13$ given that $x = -11$

Solve each equation. See Examples 2 through 11.

11. $x - 6 = 1$
12. $x - 10 = 9$
13. $y + 7 = 3$
14. $y + 12 = 5$
15. $x + 15 = -4$
16. $x + 17 = -10$
17. $22 = n - 15$
18. $36 = n - 20$
19. $6 = z + 12$
20. $18 = z + 1$
21. $x - 20 = -15$
22. $x - 10 = -11$
23. $y + 3.4 = -2.5$
24. $y + 1.6 = -3.7$
25. $x + 3.6 = 2.4$
26. $x + 2.7 = 3.8$
27. $x + \frac{1}{20} = \frac{3}{5}$
28. $n - \frac{2}{7} = \frac{3}{14}$

29. $5x = 45$

30. $9x = 108$

31. $32 = 4y$

32. $51 = 17y$

33. $\frac{3x}{4} = 15$

34. $\frac{5x}{7} = 65$

35. $\frac{y}{5} = 2$

36. $\frac{x}{3} = -4$

37. $-1 = \frac{x}{8}$

38. $0 = \frac{x}{15}$

39. $7x - 8x = 13 - 25$

40. $10n - 11n = 20 - 14$

41. $3n - 2n + 6 = 14$

42. $7n - 6n + 13 = 22$

43. $1.7y + 1.3y = 6.3$

44. $2.5y + 7.5y = 4.2$

45. $\frac{3}{4}x = \frac{5}{3}$

46. $\frac{5}{6}x = \frac{5}{3}$

47. $7.5x = -99.75$

48. $-14 = 0.7x$

49. $1.5y - 0.5y + 6.7 = -5.3$

50. $2.6y - 1.6y - 5.1 = -2.9$

51. $10x - 9x - \frac{1}{2} = -\frac{9}{10}$

52. $6x - 5x + \frac{3}{4} = -\frac{1}{12}$

53. $1.4x - 0.4x + 2.7 = -1.3$

54. $3.5y - 2.5y - 6.3 = -1.0 - 2.5$

55. $\frac{7x}{4} - \frac{3x}{4} + \frac{7}{8} = \frac{3}{2}$

56. $\frac{5n}{2} - \frac{3n}{2} + \frac{4}{5} = \frac{7}{5} - \frac{1}{10}$

57. $6.2 = -3.5 + 7n - 6n$

58. $-7.2 = 1.3n - 0.3n - 1.0$

59. $1.7x = -5.1 - 1.7$

60. $3.2x = 2.8 - 9.2$

Use a calculator to help solve the following equations.

61. $y + 32.861 = -17.892$

62. $x - 41.625 = 59.354$

63. $17.61x - 16.61x + 27.059 = 9.845$

64. $14.83y - 8.65 - 13.83y = 17.437 + 1.0$

65. $2.637x = 648.702$

66. $-0.3057y = 316.7052$

67. $-x = 145.6 + 17.89 - 10.32$

68. $-y = 143.5 + 178.462 - 200$

Applications

Solve.

69. The Japanese writing system consists of three sets of characters, two with 81 characters (which all Japanese students must know), and a third, *kanji*, with over 50,000 characters (of which only some are used in everyday writing). If a Japanese student knows 2107 total characters, solve the equation $x + 2(81) = 2107$ to determine the number of *kanji* characters the student knows.

33. $x = 20$
34. $x = 91$
35. $y = 10$
36. $y = -12$
37. $x = -8$
38. $x = 0$
39. $x = 12$
40. $n = -6$
41. $n = 8$
42. $n = 9$
43. $y = 2.1$
44. $y = 0.42$
45. $x = \frac{20}{9}$
46. $x = 2$
47. $x = -13.3$
48. $x = -20$
49. $y = -12$
50. $y = 2.2$
51. $x = -\frac{2}{5}$
52. $x = -\frac{5}{6}$
53. $x = -4$
54. $y = 2.8$
55. $x = \frac{5}{8}$
56. $n = \frac{1}{2}$
57. $n = 9.7$
58. $n = -6.2$
59. $x = -4$
60. $x = -2$
61. $y = -50.753$
62. $x = 100.979$
63. $x = -17.214$
64. $y = 27.087$
65. $x = 246$
66. $y = -1036$
67. $x = -153.17$
68. $y = -121.962$

Applications

69. 1945 *kanji* characters

70. **a.** Answers will vary.
b. $x = 200$
c. The patient will receive 200 milliliters of intravenous solution per hour.

71. **a.** Answers will vary.
b. $x = 12.5$
c. The garden should be 12.5 feet wide.

72. 10,500 graduate students

73. 4.24 light years

74. 1500 likes

75. 11,500 words

76. 50 pizzas

77. 2250 students

70. A nurse must give a patient 800 milliliters of intravenous solution over 4 hours. This can be represented by the equation $4x = 800$, where x represents the amount of solution the patient receives per hour in milliliters.

a. Why was multiplication chosen in the equation?

b. Solve the equation to determine the value of x.

c. What does the answer to part b. mean? Write a complete sentence.

71. John is making a garden in his backyard. He buys enough topsoil to cover 300 square feet. John wants the garden to go along the side of his garage, which is 24 feet in length. To determine how wide the garden needs to be, John uses the equation $24x = 300$, where x is the width of the garden in feet.

a. Why was multiplication chosen in this equation?

b. Solve the equation to determine the value of x.

c. What does the answer to part b. mean? Write a complete sentence.

72. A university enrolls both undergraduate and graduate students in all programs of study. There are a total of 28,000 students enrolled. Of this total, 17,500 students are undergraduates. Solve the equation $17{,}500 + x = 28{,}000$ to determine how many graduate students are enrolled in the university.

73. ▦ The diameter of the Milky Way is approximately 23,585 times the distance from the sun to the nearest star, Proxima Centauri. Considering that the Milky Way is roughly 100,000 light years across, solve the equation, $23{,}585x = 100{,}000$ to find the number of light years from the sun to this star. (Round your answer to the nearest hundredth.)

74. A group of students at Homestate University decide to start a math club. They create a Facebook page for their club, and their goal is get 5000 "likes" for their page. Three months after they launch their club and Facebook page, they have received a total of 3500 likes. Solve the equation $3500 + x = 5000$ to determine how many more likes they need to get to reach their goal.

75. An author is determined to have his first novel published by the publisher of George Orwell's *1984*, his favorite book. However, his contract with the publisher requires his novel to be at least 75,000 words, and he has only written 63,500 words. Solve the equation, $63{,}500 + x = 75{,}000$ to determine how many more words he must write.

76. The best pizza parlor in town slices their large pizzas so that each pizza contains 8 slices. Joe's fraternity hosts a pizza party for its members and guests, and the fraternity orders large pizzas from the best pizza parlor in town. By the end of the party, 400 slices of pizza had been eaten and all of the pizza boxes were empty. Solve the equation $8x = 400$ to determine how many pizzas were ordered for the party.

77. During rush week at Homestate University, the fraternities and sororities pledge a combined total of 450 freshmen. These 450 freshmen represent $\frac{1}{5}$ of the school's total enrollment. Solve the equation $\frac{1}{5}x = 450$ to determine the total number of students enrolled at Homestate University.

78. The inventory manager's computer crashed and he did not have a backup of his data. The company manager is requesting an inventory report for the week for a specific item. The inventory manager knows that there are currently 1472 of that item in stock. During the week, a shipment arrived with 1500 of the item. The company also shipped out 975 of the item during the week. This situation can be represented by $x + 1500 - 975 = 1472$, where x is the number of items in the inventory at the beginning of the week.

a. Why were the operations of addition and subtraction chosen in this equation?

b. Solve the equation to determine the value of x.

c. What does the answer to part b. mean? Write a complete sentence.

79. Clara has \$4200 saved to use as a down payment on the new car she is buying that costs \$15,750. She will have to get a loan to pay for the rest of the cost. This situation can be modeled by $4200 + x = 15,750$, where x is the amount of the loan in dollars.

a. Why was the operation of addition chosen in this equation?

b. Solve the equation to determine the value of x.

c. What does the answer to part b. mean? Write a complete sentence.

80. A sculptor has decided to begin a project to make scale models of famous landmarks out of stone. His first model will be of one of the moai, giant human figures carved from stone on Easter Island. If his model is to be $\frac{1}{12}$ scale, and the original moai weighs 75 tons, solve the equation $12x = 75$ to determine how many tons his completed sculpture will weigh.

Writing & Thinking

81. a. Is the expression $6 + 3 = 9$ an equation? Explain.

b. Is 4 a solution to the equation $5 + x = 10$? Explain.

78. a. Answers will vary.
b. $x = 947$
c. There were 947 of that item in the inventory at the beginning of the week.

79. a. Answers will vary.
b. $x = 11,550$
c. Clara will need a loan for \$11,550.

80. 6.25 tons

Writing & Thinking

81. a. Yes. It is stating that $6 + 3$ is equal to 9.
b. No. If we substitute 4 for x, we get the statement $9 = 10$, which is not true.

Objectives

A. Solve equations of the form $ax + b = c$.

9.2 Solving Linear Equations: $ax + b = c$

A Solving Equations of the Form $ax + b = c$

In Section 9.1, the equations to be solved were in one of two forms: $x + b = c$ or $ax = c$. In the first type, the coefficient of the variable x was always +1 and we used the addition principle to add or subtract constants to solve the equation. In the second type, we used the multiplication principle (or division principle) to multiply both sides by the reciprocal of the coefficient of the variable (or divide both sides by the coefficient itself). Now we will apply both of these techniques in solving equations of the form $ax + b = c$ where the coefficient a is a number other than 1.

Solving Linear Equations that Simplify to the Form $ax + b = c$

1. Combine like terms on both sides of the equation.
2. Use the **addition principle of equality** and add the opposite of the constant b to both sides of the equation.
3. Use the **multiplication** (or **division**) **principle of equality** and multiply both sides of the equation by the reciprocal of the coefficient of the variable (**or divide both sides by the coefficient itself**). The coefficient of the variable will become +1.
4. Check your answer by substituting it for the variable in the original equation.

PROCEDURE

1. Solve the equation:

$2x + 5 = -21$

Example 1 Solving Linear Equations of the Form $ax + b = c$

Solve the equation: $3x + 3 = -18$

Solution

$3x+3=-18$	Write the equation.
$3x+3-3=-18-3$	Add −3 to both sides.
$3x=-21$	Simplify.
$\frac{3x}{3}=\frac{-21}{3}$	Divide both sides by 3.
$x=-7$	Simplify.

Check

$3x+3=-18$	
$3(-7)+3 \stackrel{?}{=} -18$	Substitute $x = -7$.
$-21+3 \stackrel{?}{=} -18$	Simplify.
$-18=-18$	True statement

Now work margin exercise 1.

Example 2 Solving Linear Equations of the Form $ax + b = c$

Solve the equation: $-26 = 2y - 14 - 4y$

Solution

$-26 = 2y - 14 - 4y$	Write the equation.
$-26 = -2y - 14$	Combine like terms.
$-26 + 14 = -2y - 14 + 14$	Add 14 to both sides.
$-12 = -2y$	Simplify.
$\frac{-12}{-2} = \frac{-2y}{-2}$	Divide both sides by -2.
$6 = y$	Simplify.

Check

$-26 = 2y - 14 - 4y$	
$-26 \stackrel{?}{=} 2(6) - 14 - 4(6)$	Substitute $y = 6$.
$-26 \stackrel{?}{=} 12 - 14 - 24$	Simplify.
$-26 = -26$	True statement

Now work margin exercise 2.

2. Solve the equation:

$-18 = 2y - 8 - 7y$

Examples 3 and 4 illustrate solving equations with decimal coefficients. You may choose to work with the decimal coefficients as they are. However, another approach, shown here, is to multiply both sides in such a way to give integer coefficients. Generally, integers are easier to work with than decimals.

Example 3 Solving Linear Equations Involving Decimals

Solve the equation: $16.53 - 18.2z - 6.23 = 1.2$

Solution

$16.53 - 18.2z - 6.23 = 1.2$	Write the equation.
$100(16.53 - 18.2z - 6.23) = 100(1.2)$	Multiply both sides by 100. (This results in integer coefficients.)
$1653 - 1820z - 623 = 120$	Simplify.
$1030 - 1820z = 120$	Combine like terms.
$1030 - 1820z - 1030 = 120 - 1030$	Add -1030 to both sides.
$-1820z = -910$	Simplify.
$\frac{-1820z}{-1820} = \frac{-910}{-1820}$	Divide both sides by -1820.
$z = 0.5 \left(\text{or } z = \frac{1}{2} \right)$	Simplify.

Check

$16.53 - 18.2z - 6.23 = 1.2$	
$16.53 - 18.2(0.5) - 6.23 \stackrel{?}{=} 1.2$	Substitute $z = 0.5$.
$16.53 - 9.1 - 6.23 \stackrel{?}{=} 1.2$	Simplify.
$1.2 = 1.2$	True statement

Now work margin exercise 3.

3. Solve the equation:

$3.69 + 12.8z - 6.25 = 0$

4. Solve the equation:

$0.25x + 0.5x - 2.3 = 1.45$

Example 4 Solving Linear Equations Involving Decimals

Solve the equation: $5.1x + 7.4 - 1.8x = -9.1$

Solution

Because the decimal numbers are accurate to tenths, we could multiply both sides by 10 to get integer coefficients. However, we can also work with the decimal numbers directly as shown here.

$5.1x + 7.4 - 1.8x = -9.1$	Write the equation.
$3.3x + 7.4 = -9.1$	Combine like terms.
$3.3x + 7.4 - 7.4 = -9.1 - 7.4$	Add -7.4 to both sides.
$3.3x = -16.5$	Simplify.
$\frac{3.3x}{3.3} = \frac{-16.5}{3.3}$	Divide both sides by the coefficient 3.3.
$x = -5$	Simplify.

Check

$5.1x + 7.4 - 1.8x = -9.1$	
$5.1(-5) + 7.4 - 1.8(-5) \stackrel{?}{=} -9.1$	Substitute $x = -5$.
$-25.5 + 7.4 + 9 \stackrel{?}{=} -9.1$	Simplify.
$-9.1 = -9.1$	True statement

Now work margin exercise 4.

> **Math Tip**
>
> Recall that the least common multiple (LCM) of two (or more) counting numbers is the smallest number that is a multiple of each of these numbers.

Examples 5 and 6 illustrate solving equations with coefficients that are fractions. You may choose to work with the coefficients as they are. However, another approach, shown here, is to multiply both sides by the LCM of the denominators, which will give integer coefficients. Generally, integers are easier to work with than fractions.

5. Solve the equation:

$\frac{3}{5}x - \frac{7}{3} = -\frac{17}{15}$

Example 5 Solving Linear Equations Involving Fractions

Solve the equation: $\frac{5}{6}x - \frac{5}{2} = -\frac{10}{9}$

Solution

$\frac{5}{6}x - \frac{5}{2} = -\frac{10}{9}$	Write the equation.
$18\left(\frac{5}{6}x - \frac{5}{2}\right) = 18\left(-\frac{10}{9}\right)$	Multiply both sides by 18 (the LCM of the denominators).
$18\left(\frac{5}{6}x\right) - 18\left(\frac{5}{2}\right) = 18\left(-\frac{10}{9}\right)$	Apply the distributive property.
$\frac{\overset{3}{\cancel{18}}}{1} \cdot \frac{5}{\cancel{6}}x - \frac{\overset{9}{\cancel{18}}}{1} \cdot \frac{5}{\cancel{2}} = \frac{\overset{2}{\cancel{18}}}{1} \cdot \left(-\frac{10}{\cancel{9}}\right)$	Multiply.
$15x - 45 = -20$	Simplify.

$15x - 45 + 45 = -20 + 45$	Add 45 to both sides.
$15x = 25$	Simplify.
$\frac{15x}{15} = \frac{25}{15}$	Divide both sides by 15.
$x = \frac{5}{3}$	Simplify.

Check

$\frac{5}{6}x - \frac{5}{2} = -\frac{10}{9}$	
$\frac{5}{6}\left(\frac{5}{3}\right) - \frac{5}{2} \stackrel{?}{=} -\frac{10}{9}$	Substitute $x = \frac{5}{3}$.
$\frac{25}{18} - \frac{45}{18} \stackrel{?}{=} -\frac{20}{18}$	Simplify.
$-\frac{20}{18} = -\frac{20}{18}$	True statement

Now work margin exercise 5.

Note

About Checking:
Checking can be quite time-consuming and doesn't need to be done for every problem. This is particularly important on exams. You should check only if you have time after the entire exam is completed.

Example 6 Solving Linear Equations Involving Fractions

Solve the equation: $\frac{1}{2}x + \frac{3}{4}x + \frac{5}{2} - \frac{2}{3}x = -1$

Solution

$\frac{1}{2}x + \frac{3}{4}x + \frac{5}{2} - \frac{2}{3}x = -1$	Write the equation.
$12\left(\frac{1}{2}x + \frac{3}{4}x + \frac{5}{2} - \frac{2}{3}x\right) = 12(-1)$	Multiply both sides by 12 (the LCM of the denominators).
$12\left(\frac{1}{2}x\right) + 12\left(\frac{3}{4}x\right) + 12\left(\frac{5}{2}\right) - 12\left(\frac{2}{3}x\right) = 12(-1)$	Apply the distributive property.
$6x + 9x + 30 - 8x = -12$	Simplify.
$7x + 30 = -12$	Combine like terms.
$7x + 30 - 30 = -12 - 30$	Add −30 to both sides.
$7x = -42$	Simplify.
$\frac{7x}{7} = \frac{-42}{7}$	Divide both sides by 7.
$x = -6$	Simplify. Checking will show that −6 is the solution.

Now work margin exercise 6.

6. Solve the equation:

$$\frac{1}{3}x + \frac{3}{8}x + \frac{3}{2} - \frac{3}{4}x = -3$$

7. Solve the equation:

$$\frac{2}{3}x-\frac{5}{9}=\frac{1}{18}$$

Completion Example 7 Solving Equations Involving Fractions

Solve the equation: $\frac{3}{4}y+\frac{1}{2}=\frac{5}{8}$

Solution

$\frac{3}{4}y+\frac{1}{2}=\frac{5}{8}$	Write the equation.
$\underline{8}\left(\frac{3}{4}y+\frac{1}{2}\right)=\underline{8}\left(\frac{5}{8}\right)$	Multiply both sides by $\underline{8}$.
$\underline{8}\left(\frac{3}{4}y\right)+\underline{8}\left(\frac{1}{2}\right)=\underline{8}\left(\frac{5}{8}\right)$	Apply the distributive property.
$\underline{6}y+\underline{4}=\underline{5}$	Simplify.
$\underline{6}y+4-\underline{4}=5-\underline{4}$	Subtract $\underline{4}$ from both sides.
$6y=\underline{1}$	Simplify.
$\frac{6y}{\underline{6}}=\frac{1}{\underline{6}}$	Divide both sides by $\underline{6}$.
$y=\frac{1}{\underline{6}}$	Simplify.

Now work margin exercise 7.

Completion Example Answers

7.

$\frac{3}{4}y+\frac{1}{2}=\frac{5}{8}$	Write the equation.
$8\left(\frac{3}{4}y+\frac{1}{2}\right)=8\left(\frac{5}{8}\right)$	Multiply both sides by 8.
$8\left(\frac{3}{4}y\right)+8\left(\frac{1}{2}\right)=8\left(\frac{5}{8}\right)$	Apply the distributive property.
$6y+4=5$	Simplify.
$6y+4-4=5-4$	Subtract 4 from both sides.
$6y=1$	Simplify.
$\frac{6y}{6}=\frac{1}{6}$	Divide both sides by 6.
$y=\frac{1}{6}$	Simplify.

Margin Exercise Answers

1. $x=-13$ **2.** $y=2$ **3.** $z=0.2$ **4.** $x=5$ **5.** $x=2$ **6.** $x=108$ **7.** $x=\frac{11}{12}$

9.2 Exercises

Concept Check

Fill-in-the-Blank. Complete each sentence using information found in this section.

1. The first step in solving linear equations that simplify to the form $ax + b = c$ is to combine _______ terms on both sides of the equation.
2. When solving a linear equation that has been simplified to the form $ax + b = c$, use the __________ principle of equality and add the __________ of the constant b to both sides of the equation.
3. Once you have a variable term on one side of the equation and a constant term on the other, use the __________ principle of equality and multiply both sides of the equation by the reciprocal of the coefficient of the variable.
4. When you multiply both sides of the equation by the reciprocal of the coefficient of the variable, the coefficient of the variable will become ______.
5. Check your answer by __________ it in for the variable in the original equation.

True/False. Determine whether each statement is true or false. If a statement is false, explain how it can be changed so the statement will be true. (**Note:** There may be more than one acceptable change.)

6. If an equation of the form $ax + b = c$ uses decimal or fractional coefficients, the addition and multiplication principles of equality cannot be used.
7. The first step in solving $2x + 3 = 9$ is to add 3 to both sides.
8. To solve an equation that has been simplified to $4x = 12$, you need to multiply both sides by $\frac{1}{4}$, or divide both sides by 4.
9. When solving a linear equation with decimal coefficients, one approach is to multiply both sides in such a way to give integer coefficients before solving.

Practice

Solve each equation. See Examples 1 through 7.

1. $3x+11=2$
2. $3x+10=-5$
3. $5x-4=6$
4. $4y-8=-12$
5. $6x+10=22$
6. $3n+7=19$
7. $9x-5=13$
8. $2x-4=12$
9. $1-3y=4$
10. $5-2x=9$
11. $14+9t=5$
12. $5+2x=-7$
13. $-5x+2.9=3.5$
14. $3x+2.7=-2.7$
15. $10+3x-4=18$
16. $5+5x-6=9$

9.2 Exercises

Concept Check

1. like
2. addition, opposite
3. multiplication
4. +1
5. substituting
6. False; The addition and multiplication principles of equality can be used with decimal or fractional coefficients.
7. False; Subtract 3 from both sides.
8. True
9. True

Practice

1. $x=-3$
2. $x=-5$
3. $x=2$
4. $y=-1$
5. $x=2$
6. $n=4$
7. $x=2$
8. $x=8$
9. $y=-1$
10. $x=-2$
11. $t=-1$
12. $x=-6$
13. $x=-0.12$
14. $x=-1.8$
15. $x=4$
16. $x=2$

17. $x=0$
18. $x=\frac{1}{9}$
19. $y=0$
20. $y=0$
21. $x=-2$
22. $y=2$
23. $y=-6$
24. $y=-6$
25. $n=6$
26. $n=-4$
27. $n=8$
28. $n=-4$
29. $x=0$
30. $x=0$
31. $x=-7$
32. $x=-\frac{11}{3}$
33. $x=-\frac{1}{8}$
34. $x=-\frac{27}{10}$
35. $x=-\frac{13}{2}$
36. $x=\frac{23}{8}$
37. $x=-\frac{21}{5}$
38. $x=-\frac{4}{7}$
39. $x=-\frac{8}{15}$
40. $x=2$
41. $y=\frac{28}{5}$
42. $y=23$
43. $x=2$
44. $x=-\frac{1}{4}$
45. $y=\frac{7}{5}$
46. $y=-10$
47. $x=-4.5$
48. $x=2.2$
49. $x=-44$

17. $15=7x+7+8$

18. $14=9x+5+8$

19. $5y-3y+2=2$

20. $6y+8y-7=-7$

21. $x-4x+25=31$

22. $3y+9y-13=11$

23. $-20=7y-3y+4$

24. $-20=5y+y+16$

25. $4n-10n+35=1-2$

26. $-5n-3n+2=34$

27. $3n-15-n=1$

28. $2n+12+n=0$

29. $5.4x-0.2x=0$

30. $0=5.1x+0.3x$

31. $\frac{1}{2}x+7=\frac{7}{2}$

32. $\frac{3}{5}x+4=\frac{9}{5}$

33. $\frac{1}{2}-\frac{8}{3}x=\frac{5}{6}$

34. $\frac{2}{5}-\frac{1}{2}x=\frac{7}{4}$

35. $\frac{3}{2}=\frac{1}{3}x+\frac{11}{3}$

36. $\frac{11}{8}=\frac{1}{5}x+\frac{4}{5}$

37. $\frac{7}{2}-5-\frac{5}{2}x=9$

38. $\frac{8}{3}+2-\frac{7}{3}x=6$

39. $\frac{5}{8}x-\frac{1}{4}x+\frac{1}{2}=\frac{3}{10}$

40. $\frac{1}{2}x+\frac{3}{4}x-\frac{5}{3}=\frac{5}{6}$

41. $\frac{y}{2}+\frac{1}{5}=3$

42. $\frac{y}{3}-\frac{2}{3}=7$

43. $\frac{7}{8}=\frac{3}{4}x-\frac{5}{8}$

44. $\frac{1}{10}=\frac{4}{5}x+\frac{3}{10}$

45. $\frac{y}{7}+\frac{y}{28}+\frac{1}{2}=\frac{3}{4}$

46. $\frac{5y}{6}-\frac{7y}{8}-\frac{1}{12}=\frac{1}{3}$

47. $x+1.2x+6.9=-3.0$

48. $3x-0.75x-1.72=3.23$

49. $10=x-0.5x+32$

50. $33=y+3-0.4y$

51. $2.5x+0.5x-3.5=2.5$

52. $4.7-0.5x-0.3x=-0.1$

53. $6.4+1.2x+0.3x=0.4$

54. $5.2-1.3x-1.5x=-0.4$

55. $-12.13=2.42y+0.6y-13.64$

56. $-7.01=1.75x+3.05x-8.45$

57. $-0.4x+x+17.2=18.1$

58. $y-0.75y+13.76=14.66$

59. $0=17.3x-15.02x-0.456$

60. $0=20.5x-16.35x+0.1245$

Use a calculator to help solve the following equations.

61. $0.15x+5.23x-17.815=15.003$

62. $15.97y-12.34y+16.95=8.601$

63. $13.45x-20x-17.36=-24.696$

64. $26.75y-30y+23.28=4.4625$

Applications

Solve.

65. The tickets for a concert featuring the new hit band, Flying Sailor, sold out in 2.5 hours. If there were 35,000 tickets sold, solve the equation $35,000 - 2.5x = 0$ to find the number of tickets sold per hour.

66. Katie's nutritionist recommends that she follows a diet of 2000 calories per day. Katie eats 4 times a day, eating the same number of calories at each sitting. However, every morning she stops at the local coffee shop and treats herself to a large flavored coffee that contains 240 calories. Solve the equation $4x + 240 = 2000$ to determine how many calories Katie should eat at each sitting.

67. Salim is a student in a course on the modern British novel. He is given an assignment to read a 350 page novel in 7 days. He reads the first 38 pages of the novel on the day he receives his assignment and decides to finish the novel by reading the same amount of pages of each day until the assignment is due. Solve the equation $6x + 38 = 350$ to determine how many pages Salim should read each day.

68. All snacks (candy, popcorn, and soda) cost \$3.50 each at the local movie theater. Admission tickets cost \$7.50 each. After a long week, Carlos treats himself to a night at the movies. His movie night budget is \$25 and he spends all his movie money. Solve the equation \$3.50x + \$7.50 = \$25.00 to determine how many snacks Carlos can buy.

69. The Political Science Club at Homestate University is planning to host an election night party for members and guests. The club plans to serve cookies and estimates it will need a total of 1500 cookies in 6 varieties for the party. The club orders 300 chocolate chip cookies and an equal number of cookies in each of the remaining 5 varieties. Solve the equation $5x + 300 = 1500$ to determine how many cookies of each remaining variety will be ordered.

70. All courses in the Homestate University graduate school are worth 3 credits. To earn a master's degree, a student must earn a total of 36 credits. The student's thesis work counts as 6 credits. Solve the equation $3x + 6 = 36$ to determine how many courses a student must take to earn a master's degree.

71. A rectangular-shaped parking lot is to have a perimeter of 450 yards. If the width must be 90 yards because of a building code, solve the equation $2l + 2(90) = 450$ to determine the length of the parking lot.

72. Jeff, who lives in England, is reading a letter from his pen pal in the United States. His pen pal says that the temperature was 97.7 degrees Fahrenheit that day, making it too hot to play soccer outside. Jeff doesn't know how hot this is, because he is used to temperatures in Celsius. Help Jeff solve the equation, $1.8C + 32 = 97.7$ to determine the temperature in degrees Celsius.

73. The tallest man-made structure in the world is the Burj Khalifa in Dubai, which stands at 2717 feet tall. The tallest tree in the world is a redwood tree in California. If 7 of these trees were stacked on top of each other, they would still be 59.1 feet shorter than the Burj Khalifa. Solve the equation, $7x + 59.1 = 2717$ to determine the height of the tree.

50. $y = 50$
51. $x = 2$
52. $x = 6$
53. $x = -4$
54. $x = 2$
55. $y = 0.5$
56. $x = 0.3$
57. $x = 1.5$
58. $y = 3.6$
59. $x = 0.2$
60. $x = -0.03$
61. $x = 6.1$
62. $y = -2.3$
63. $x = 1.12$
64. $y = 5.79$

Applications

65. 14,000 tickets per hour
66. 440 calories
67. 52 pages
68. 5 snacks
69. 240 cookies of each remaining variety
70. 10 courses
71. 135 yards
72. 36.5 °C
73. 379.7 feet

74. a. Amount in stock; amount of individual cake pops sold; unknown number of packages of 4 sold; amount of cake pops left
b. $x = 18$
c. 18 packages of cake pops were sold.

75. a. Lowest temperature; Change in temperature per hour; Current temperature
b. $x = 4$
c. The low temperature occurred 4 hours ago.

76. a. $x = 200$
b. There are 200 syringes per box.

Writing & Thinking

77. a. The 4 should have been multiplied by 3 so that the 3 was distributed over the entire left-hand side of the equation; Correct answer is $x = 15$.
b. 3 should be subtracted from each side, not from each term, and $5x - 3$ doesn't simplify to $2x$; Correct answer is $x = \frac{8}{5}$.

74. A bakery sells cake pops individually and in packages of 4. At the beginning of the day, the bakery had 114 cake pops in stock. They sold 34 individual cake pops and several packages of cake pops. At the end of the day, there were 8 cake pops left. This situation can be modeled by the equation $114 - 34 - 4x = 8$, where x is the number of packages of cake pops sold.

a. Explain what each term in the equation $114 - 34 - 4x = 8$ represents in the situation.

b. Solve the equation to determine the value of x.

c. What does the answer to part b. mean? Write a complete sentence.

75. The lowest temperature of the night was reported to be 24 °F. The weather report mentioned that the temperature has steadily risen 1.5 degrees per hour since the lowest temperature of the day and it is currently 30 °F. This situation can be modeled by the equation $24 + 1.5x = 30$, where x is the time in hours since the lowest temperature was recorded.

a. Explain what each term in the equation $24 + 1.5x = 30$ represents in the situation.

b. Solve the equation to determine the value of x.

c. What does the answer to part b. mean? Write a complete sentence.

76. While taking inventory, a nurse records that there are $\frac{3}{5}$ of a box of syringes in one closet, two boxes that are $\frac{1}{8}$ full in another closet, and 24 syringes in the supply cart. He calculates the total to be 194 syringes. The staff member who reorders supplies is new and doesn't know how many syringes are in the boxes that the clinic uses, so she sets up the equation $\frac{3}{5}x + 2\left(\frac{1}{8}x\right) + 24 = 194$, where x is the number of syringes in a box.

a. Solve the equation to determine the value of x.

b. What does the answer to part a. mean? Write a complete sentence.

Writing & Thinking

77. Find the error(s) made in solving each equation and give the correct solution.

a.
$$\frac{1}{3}x + 4 = 9$$
$$3 \cdot \frac{1}{3}x + 4 = 3 \cdot 9$$
$$x + 4 = 27$$
$$x + 4 - 4 = 27 - 4$$
$$x = 23$$

b.
$$5x + 3 = 11$$
$$(5x - 3) + (3 - 3) = 11 - 3$$
$$2x + 0 = 8$$
$$\frac{2x}{2} = \frac{8}{2}$$
$$x = 4$$

9.3 Solving Linear Equations: $ax + b = cx + d$

Objectives

A. Solve equations of the form $ax + b = cx + d$.

B. Determine whether an equation is a conditional equation, an identity, or a contradiction.

A Solving Equations of the Form $ax + b = cx + d$

Now we are ready to solve linear equations of the most general form: $ax + b = cx + d$, where constants and variables may be on both sides. There may also be parentheses or other symbols of inclusion. **Remember that the objective is to get the variable on one side of the equation by itself with a coefficient of +1.**

Solving Linear Equations that Simplify to the Form $ax + b = cx + d$

1. Simplify each side of the equation by removing any grouping symbols and combining like terms on both sides of the equation.
2. Use the **addition principle of equality** and add the opposite of a constant term and/or variable term to both sides of the equation so that variables are on one side and constants are on the other side.
3. Use the **multiplication** (or **division**) **principle of equality** and multiply both sides of the equation by the reciprocal of the coefficient of the variable (**or divide both sides by the coefficient itself**). The coefficient of the variable will become +1.
4. Check your answer by substituting it for the variable in the original equation.

PROCEDURE

Note that after using the addition principle of equality it does not matter if the variables are on the left side or the right side of the equation. The important thing is that all variables are on one side and all constants are on the other.

Example 1 Solving Linear Equations of the Form $ax + b = cx + d$

Solve the equation: $5x + 3 = 2x - 18$

Solution

$5x + 3 = 2x - 18$	Write the equation.
$5x + 3 - 3 = 2x - 18 - 3$	Add -3 to both sides.
$5x = 2x - 21$	Simplify.
$5x - 2x = 2x - 21 - 2x$	Add $-2x$ to both sides.
$3x = -21$	Simplify.
$\frac{3x}{3} = \frac{-21}{3}$	Divide both sides by 3.
$x = -7$	Simplify.

1. Solve the equation:

$7x + 4 = 3x - 20$

Check

$$5x+3=2x-18$$

$$5(-7)+3\stackrel{?}{=}2(-7)-18 \quad \text{Substitute } x=-7.$$

$$-35+3\stackrel{?}{=}-14-18 \quad \text{Simplify.}$$

$$-32=-32 \quad \text{True statement}$$

Now work margin exercise 1.

2. Solve the equation:

$6x+3-2x=2x-15+4$

Example 2 Solving Linear Equations of the Form $ax+b=cx+d$

Solve the equation: $4x+1-x=2x-13+5$

Solution

$4x+1-x=2x-13+5$	Write the equation.
$3x+1=2x-8$	Combine like terms.
$3x+1-1=2x-8-1$	Add -1 to both sides.
$3x=2x-9$	Simplify.
$3x-2x=2x-9-2x$	Add $-2x$ to both sides.
$x=-9$	Simplify.

Check

$4x+1-x=2x-13+5$	
$4(-9)+1-(-9)\stackrel{?}{=}2(-9)-13+5$	Substitute $x=-9$.
$-36+1+9\stackrel{?}{=}-18-13+5$	Simplify.
$-26=-26$	True statement

Now work margin exercise 2.

3. Solve the equation:

$5y+1.4=7y-2.8$

Example 3 Solving Linear Equations Involving Decimals

Solve the equation: $6y+2.5=7y-3.6$

Solution

$6y+2.5=7y-3.6$	Write the equation.
$6y+2.5+3.6=7y-3.6+3.6$	Add 3.6 to both sides.
$6y+6.1=7y$	Simplify.
$6y+6.1-6y=7y-6y$	Add $-6y$ to both sides.
$6.1=y$	Simplify.

Check

$6y+2.5=7y-3.6$	
$6(6.1)+2.5\stackrel{?}{=}7(6.1)-3.6$	Substitute $y=6.1$.
$36.6+2.5\stackrel{?}{=}42.7-3.6$	Simplify.
$39.1=39.1$	True statement

Now work margin exercise 3.

Math Tip

Another good approach to Example 3 is to first multiply each term by 10 to clear the decimals.

Example 4 Solving Linear Equations Involving Fractions

Solve the equation: $\frac{1}{3}x+\frac{13}{15}=\frac{3}{5}x-1$

Solution

$$\frac{1}{3}x+\frac{13}{15}=\frac{3}{5}x-1$$ Write the equation.

$$15\left(\frac{1}{3}x+\frac{13}{15}\right)=15\left(\frac{3}{5}x-1\right)$$ Multiply both sides by 15 (the LCM of the denominators.)

$$15\left(\frac{1}{3}x\right)+15\left(\frac{13}{15}\right)=15\left(\frac{3}{5}x\right)-15(1)$$ Apply the distributive property.

$$5x+13=9x-15$$ Simplify.

$$5x+13-13=9x-15-13$$ Add –13 to both sides.

$$5x=9x-28$$ Simplify.

$$5x-9x=9x-28-9x$$ Add $-9x$ to both sides.

$$-4x=-28$$ Simplify.

$$\frac{-4x}{-4}=\frac{-28}{-4}$$ Divide both sides by – 4.

$$x=7$$ Simplify. Checking will show that 7 is the solution.

Now work margin exercise 4.

4. Solve the equation:
$\frac{1}{4}x+\frac{3}{4}=\frac{1}{2}x+2$

Teaching Note

Example 4 shows adding constants to both sides at the stage

$5x + 13 = 9x - 15.$

You might want to show adding variables at this stage as well.

$$5x+13-5x=9x-15-5x$$
$$13=4x-15$$
$$13+15=4x-15+15$$
$$28=4x$$
$$\frac{28}{4}=\frac{4x}{4}$$
$$7=x$$

In this case, the variable coefficient is positive and on the right side. This can be an interesting point for discussion.

Example 5 Solving Linear Equations Involving Parentheses

Solve the equation: $2(y-7)=4(y+1)-26$

Solution

$$2(y-7)=4(y+1)-26$$ Write the equation.

$$2y-14=4y+4-26$$ Use the distributive property.

$$2y-14=4y-22$$ Combine like terms.

$$2y-14+22=4y-22+22$$ Add 22 to both sides. Here we will put the variable on the right side to get a positive coefficient of y.

$$2y+8=4y$$ Simplify.

$$2y+8-2y=4y-2y$$ Add $-2y$ to both sides.

$$8=2y$$ Simplify.

$$\frac{8}{2}=\frac{2y}{2}$$ Divide both sides by 2.

$$4=y$$ Simplify. Checking will show that 4 is the solution.

Now work margin exercise 5.

5. Solve the equation:

$3(y-5)=8(y+1)-3$

6. Solve the equation:
$-5(3x+7)-11=-7(x-1)-5$

Example 6 Solving Linear Equations Involving Parentheses

Solve the equation: $-2(5x+13)-2=-6(3x-2)-41$

Solution

$-2(5x+13)-2=-6(3x-2)-41$	Write the equation.
$-10x-26-2=-18x+12-41$	Use the distributive property. Be careful with the signs.
$-10x-28=-18x-29$	Combine like terms.
$-10x-28+18x=-18x-29+18x$	Add $18x$ to both sides.
$8x-28=-29$	Simplify.
$8x-28+28=-29+28$	Add 28 to both sides.
$8x=-1$	Simplify.
$\frac{8x}{8}=\frac{-1}{8}$	Divide both sides by 8.
$x=-\frac{1}{8}$	Simplify. Checking will show that $-\frac{1}{8}$ is the solution.

Now work margin exercise 6.

7. Solve the equation:
$2-2(x+5)=3(x-6)$

Completion Example 7 Solving Equations Involving Parentheses

Solve the equation: $4(x+3)=2(3x-1)+6$

Solution

$4(x+3)=2(3x-1)+6$	Write the equation.
$4x+\underline{12}=6x-\underline{2}+6$	Use the distributive property.
$4x+12=6x+\underline{4}$	Combine like terms.
$4x+12-\underline{12}=6x+4-\underline{12}$	Subtract $\underline{12}$ from both sides.
$4x=6x-\underline{8}$	Simplify.
$4x-\underline{6x}=6x-8-\underline{6x}$	Subtract $\underline{6x}$ from both sides.
$\underline{-2}x=-8$	Simplify.
$\frac{-2x}{\underline{-2}}=\frac{-8}{\underline{-2}}$	Divide both sides by $\underline{-2}$.
$x=\underline{4}$	Simplify.

Now work margin exercise 7.

B Conditional Equations, Identities, and Contradictions

When solving equations, there are times when we are concerned with the number of solutions that an equation has. If an equation has a finite number of solutions (the number of solutions is a countable number), the equation is said to be a **conditional equation**. As stated earlier, every linear equation has exactly one solution. Thus, **every linear equation is a conditional equation**. However, in some cases, simplifying an equation will lead to a statement that is always true, such as $0 = 0$. In these cases, the original equation is called an **identity** and has an infinite number of solutions, which can be written as all real numbers or $\mathbb{R}$. (Recall that the **real numbers** consist of all rational and irrational numbers.) If the equation simplifies to a statement that is never true, such as $0 = 2$, then the original equation is called a **contradiction** and there is no solution. Table 1 summarizes these ideas.

Type of Equation	Number of Solutions
Conditional	Finite number of solutions
Identity	Infinite number of solutions
Contradiction	No solution

Table 1

Math Tip

If an equation has an *infinite number of solutions*, then there is an endless number of solutions that would satisfy the equation.

Example 8 Determining Types of Equations

Determine whether the equation $3(x+5)+1=-11$ is a conditional equation, an identity, or a contradiction.

Solution

$3(x+5)+1=-11$	Write the equation.
$3x+15+1=-11$	Use the distributive property.
$3x+16=-11$	Combine like terms.
$3x+16-16=-11-16$	Add -16 to both sides.
$3x=-27$	Simplify.
$\frac{3x}{3}=\frac{-27}{3}$	Divide both sides by 3.
$x=-9$	Simplify.

The equation has one solution. Therefore, it is a conditional equation.

Now work margin exercise 8.

8. Determine whether the equation $4(x-17)+3x=7x+49$ is a conditional equation, an identity, or a contradiction.

Example 9 Determining Types of Equations

Determine whether the equation $3(x-25)+3x=6(x+10)$ is a conditional equation, an identity, or a contradiction.

Solution

$3(x-25)+3x=6(x+10)$	Write the equation.
$3x-75+3x=6x+60$	Use the distributive property.
$6x-75=6x+60$	Combine like terms.
$6x-75-6x=6x+60-6x$	Add $-6x$ to both sides.
$-75=60$	Simplify.

The last equation is never true. Therefore, the original equation is a contradiction and has no solution.

Now work margin exercise 9.

9. Determine whether the equation $-3(x-4)=12-2x-x$ is a conditional equation, an identity, or a contradiction.

10. Determine whether the equation $5x + 9 = -13 + 3x$ is a conditional equation, an identity, or a contradiction.

Example 10 Determining Types of Equations

Determine whether the equation $-2(x-7)+x=14-x$ is a conditional equation, an identity, or a contradiction.

Solution

$-2(x-7)+x=14-x$	Write the equation.
$-2x+14+x=14-x$	Use the distributive property.
$14-x=14-x$	Combine like terms.
$14-x-14=14-x-14$	Add -14 to both sides.
$-x=-x$	Simplify.
$-x+x=-x+x$	Add x to both sides.
$0=0$	Simplify.

The last equation is always true. Therefore, the original equation is an identity, and has an infinite number of solutions. Every real number is a solution.

Now work margin exercise 10.

Completion Example Answers

7.

$4(x+3)=2(3x-1)+6$	Write the equation.
$4x+\underline{12}=6x-\underline{2}+6$	Use the distributive property.
$4x+12=6x+\underline{4}$	Combine like terms.
$4x+12-\underline{12}=6x+4-\underline{12}$	Subtract $\underline{12}$ from both sides.
$4x=6x-\underline{8}$	Simplify.
$4x-\underline{6x}=6x-8-\underline{6x}$	Subtract $\underline{6x}$ from both sides.
$\underline{-2}x=-8$	Simplify.
$\dfrac{-2x}{\underline{-2}}=\dfrac{-8}{\underline{-2}}$	Divide both sides by $\underline{-2}$.
$x=\underline{4}$	Simplify.

Margin Exercise Answers

1. $x=-6$ **2.** $x=-7$ **3.** $y=2.1$ **4.** $x=-5$ **5.** $y=-4$ **6.** $x=-6$ **7.** $x=2$ **8.** contradiction **9.** identity **10.** conditional

9.3 Exercises

Concept Check

Fill-in-the-Blank. Complete each sentence using information found in this section.

1. A/An __________ is an equation that has an infinite number of solutions.

2. If an equation has a finite number of solutions, it is a/an __________ equation.

3. Every linear equation is a/an ____________ equation.

9.3 Exercises

Concept Check

1. identity
2. conditional
3. conditional
4. contradiction
5. real, $\mathbb{R}$
6. True
7. True

4. If a linear equation simplifies to a statement that is never true, then the original equation is called a/an ________.

5. The solution set of an identity can be written as all ______ numbers or ______.

True/False. Determine whether each statement is true or false. If a statement is false, explain how it can be changed so the statement will be true. (**Note:** There may be more than one acceptable change.)

6. Every linear equation has exactly one solution.

7. If a linear equation simplifies to a statement that is always true, then the original equation is called an identity.

8. If an equation has no solution, it is called an identity.

9. The most general form of a linear equation is $ax + b = cx + d$.

Practice

Solve each equation. See Examples 1 through 7.

1. $3x+2=x-8$
2. $5x+1=2x-5$
3. $4n-3=n+6$
4. $6y+3=y-7$
5. $3y+18=7y-6$
6. $2y+5=8y+10$
7. $3x+11=8x-4$
8. $9x+3=5x-9$
9. $14n=3n$
10. $1.6x=0.8x$
11. $6y-2.1=y-2.1$
12. $13x+5=2x+5$
13. $2(z+1)=3z+3$
14. $6x-3=3(x+2)$
15. $16y+23y-3=16y-2y+2$
16. $5x-2x+4=3x+x-1$
17. $0.25+3x+6.5=0.75x$
18. $0.9y+3=0.4y+1.5$
19. $6.5+1.2x=0.5-0.3x$
20. $x-0.1x+0.8=0.2x+0.1$
21. $\frac{2}{3}x+1=\frac{1}{3}x-6$
22. $\frac{4}{5}n+2=\frac{2}{5}n-4$
23. $\frac{y}{5}+\frac{3}{4}=\frac{y}{2}+\frac{3}{4}$
24. $\frac{5n}{6}+\frac{1}{9}=\frac{3n}{2}+\frac{1}{9}$
25. $\frac{3}{8}\left(y-\frac{1}{2}\right)=\frac{1}{8}\left(y+\frac{1}{2}\right)$
26. $\frac{1}{2}\left(\frac{x}{2}+1\right)=\frac{1}{3}\left(\frac{x}{2}-1\right)$
27. $\frac{2x}{3}+\frac{x}{3}=-\frac{3}{4}+\frac{x}{2}$
28. $\frac{3}{4}x+\frac{1}{5}x=\frac{1}{2}x-\frac{3}{10}$
29. $x+\frac{2}{3}x-2x=\frac{x}{6}-\frac{1}{8}$
30. $3x+\frac{1}{2}x-\frac{2}{5}x=\frac{x}{10}+\frac{7}{20}$
31. $3(1+9x)=6(2-4x)$
32. $4(5-x)=8(3x+10)$

8. False; It is called a contradiction.

9. True

Practice

1. $x=-5$
2. $x=-2$
3. $n=3$
4. $y=-2$
5. $y=6$
6. $y=-\frac{5}{6}$
7. $x=3$
8. $x=-3$
9. $n=0$
10. $x=0$
11. $y=0$
12. $x=0$
13. $z=-1$
14. $x=3$
15. $y=\frac{1}{5}$
16. $x=5$
17. $x=-3$
18. $y=-3$
19. $x=-4$
20. $x=-1$
21. $x=-21$
22. $n=-15$
23. $y=0$
24. $n=0$
25. $y=1$
26. $x=-10$
27. $x=-\frac{3}{2}$
28. $x=-\frac{2}{3}$
29. $x=\frac{1}{4}$
30. $x=\frac{7}{60}$
31. $x=\frac{3}{17}$
32. $x=-\frac{15}{7}$

33. $3(4x-1)=4(2x-3)+8$

34. $7(2x-1)=5(x+6)-13$

35. $5-3(2x+1)=4(x-5)+6$

36. $-2(y+5)-4=6(y-2)+2$

37. $8+4(2x-3)=5-(x+3)$

38. $8(3x+5)-9=9(x-2)+14$

39. $4.7-0.3x=0.5x-0.1$

40. $5.8-0.1x=0.2x-0.2$

41. $0.2(x+3)=0.1(x-5)$

42. $0.4(x+3)=0.3(x-6)$

43. $\frac{1}{2}(4-8x)=\frac{1}{3}(4x+7)-3$

44. $3+\frac{1}{4}(x-4)=\frac{2}{5}(2+3x)$

45. $0.6x-22.9=1.5x-18.4$

46. $0.1y+3.8=5.72-0.3y$

47. $0.12n+0.25n-5.895=4.3n$

48. $0.15n+32n-21.0005=10.5n$

49. $0.7(x+14.1)=0.3(x+32.9)$

50. $0.8(x-6.21)=0.2(x-24.84)$

Determine whether each equation is a conditional equation, an identity, or a contradiction. See Examples 8 through 10.

51. $2(3x-1)+5=3$

52. $-2x+13=-2(x-7)$

53. $5x+13=-2(x-7)+3$

54. $3x+9=-3(x-3)+6x$

55. $7(x-1)=-3(3-x)+4x$

56. $3(x-2)+4x=6(x-1)+x$

57. $5(x+1)=3(x+1)+2(x+1)$

58. $8x-20+x=-3(5-2x)+3(x-4)$

59. $2x+3x=5.2(3-x)$

60. $5.2x+3.4x=0.2(x-0.42)$

Use a calculator to help solve each equation.

61. $0.17x-23.0138=1.35x+36.234$

62. $48.512-1.63x=2.58x+87.63553$

63. $0.32(x+14.1)=2.47x+2.21795$

64. $1.6(9.3+2x)=0.2(3x+133.94)$

Applications

Solve.

65. Caitlyn and Steve are planning their wedding reception and must decide between two catering halls. The first site, A Wedding Space, rents for \$800 for one day and charges \$50 per person for dinner. The second venue, A Wedding Place, costs \$1000 to rent for one day and charges \$40 per person for the same dinner. Solve the equation $800 + 50x = 1000 + 40x$ to determine how many guests they can invite so that the cost they pay will be the same at both wedding catering halls.

66. The value of a new car depreciates at a rate of about \$250 per month. Suppose a car originally costs \$30,000. The car was bought with a \$1000 down payment and a loan with 0% financing for 60 months with payments of \$200 a month. Solve the equation $30{,}000 - 250t = 29{,}000 - 200t$ to determine how many months it will take for the value of the vehicle to equal the amount owed on the loan?

33. $x=-\frac{1}{4}$
34. $x=\frac{8}{3}$
35. $x=\frac{8}{5}$
36. $y=-\frac{1}{2}$
37. $x=\frac{2}{3}$
38. $x=-\frac{7}{3}$
39. $x=6$
40. $x=20$
41. $x=-11$
42. $x=-30$
43. $x=\frac{1}{2}$
44. $x=\frac{24}{19}$
45. $x=-5$
46. $y=4.8$
47. $n=-1.5$
48. $n=0.97$
49. $x=0$
50. $x=0$
51. Conditional
52. Contradiction
53. Conditional
54. Identity
55. Contradiction
56. Identity
57. Identity
58. Contradiction
59. Conditional
60. Conditional
61. $x=-50.21$
62. $x=-9.293$
63. $x=1.067$
64. $x=4.58$

Applications

65. 20 guests
66. 20 months

67. Heidrick has won a chance to "Eat the Elite" at the Zombie Dash 7-mile race. He will receive a 2.25-mile head start. The elite runners run at a pace of 12.6 miles per hour and Heidrick runs at a pace of 6 miles per hour. Solve the equation $12.6t = 6t + 2.25$ to determine how many hours it will take the elite runners catch up to Heidrick? Enter your answer in hours rounded to the nearest hundredth.

68. A company has two packaging options for shipping quantities of a certain inventory item. Option A uses 20 boxes and there are 5 items unpacked. Option B requires more filler and uses 23 boxes, where each box holds 2 fewer items than Option A and there are only 3 items unpacked. This situation can be represented by $20x + 5 = 23(x - 2) + 3$, where x is the number of items that can fit in the box used for Option A.

   a. What does $20x + 5$ represent in the equation?

   b. What does $x - 2$ represent?

   c. Solve the equation for x.

   d. Check the solution.

   e. What does the answer from part c. mean? Write a complete sentence.

69. Two advertisement flyers have the same area. The first flyer has a length of 12 inches and a width of x inches. The second flyer has a length of 4 inches and a width that is 10 inches more than x. This situation can be represented by $12x = 4(10 + x)$, where x is the width of the first flyer.

   a. What does $12x$ represent in the equation?

   b. What does $10 + x$ represent?

   c. Solve the equation for x.

   d. Check the solution.

   e. What does the answer from part c. mean? Write a complete sentence.

70. The manager of a café wants to list a price for the weekly featured combo that includes tax. He wants to sell a medium house-blend coffee with a pastry for a total of \$5.45. He doesn't know which pastry to sell with the coffee to avoid losing money on the combo. The medium coffee costs \$2.75 and the tax is 9%. He uses the equation $1.09(2.75 + x) = 5.45$ to determine the price of the pastry, which is represented by the variable x.

   a. What does the sum $2.75 + x$ represent?

   b. Solve the equation for x.

   c. Which of the following pastries would you choose to be a part of the combo? Explain why you made your choice.

   cherry pie for \$2.50, coffee cake for \$2.25, bagel for \$2.00

71. A farmer is putting a shed on his property. He has two designs. One uses wood and would cost \$2 per square foot plus an extra \$8400 in materials. The other design is metal and would cost \$4 per square foot plus an additional \$8800. Both sheds are the same size, and the wood shed costs $\frac{3}{4}$ what the metal shed costs. Solve the equation $2x + 8400 = \frac{3}{4}(4x + 8800)$ to determine how many square feet the shed will be.

67. 0.34 hours

68. **a.** The number of items packaged with Option A
b. The number of items that each box in Option B can hold
c. $x = 16$
d. Solution should be correct if part c. is correct.
e. The boxes used in Option A can hold 16 items.

69. **a.** The area of the first flyer
b. The width of the second flyer
c. $x = 5$
d. Solution should be correct if part c. is correct.
e. The first flyer has a width of 5 inches.

70. **a.** The cost of the special before tax
b. $x = 2.25$
c. The coffee cake or bagel will be cost effective. Answers will vary.

71. 1800 square feet

72. 150 boxes

73. 240 sundaes

74. 3 hours

Writing & Thinking

75. a. $5x+1$ b. $x=6$
c. Answers will vary.

76. a. $x+6=20$; conditional, since there is one age that will make the statement true
b. $x+6=x+8$; contradiction, since he can't be both 6 years older and 8 years older
c. $x+6=(x+3)+3$; identity, since the statement will be true regardless of Ryan's current age

72. Two rival shoe companies want to rent the same building for their office and shipping space. Schulster's Shoes would use 1000 square feet for offices, 600 for shipping, and another 6 square feet for every packaged box of shoes. Shoes, Shoes, Shoes! would use 750 square feet for offices, 400 for shipping, and 9 square feet per packaged box of shoes. If only one company may occupy the building and the maximum amount of inventory kept by both companies is the same, solve the equation, $1000+600+6x=750+400+9x$ to determine the maximum number of boxes of shoes each company plans to have.

73. An ice cream shop is having a special "Ice Cream Sunday" event in which they are giving away giant mixed sundaes of 3 scoops of vanilla ice cream and 2 scoops of chocolate. If they have 24 gallons of chocolate and 36 gallons of vanilla to start with, solve the equation, $36-\frac{1}{20}(3x)=24-\frac{1}{20}(2x)$ to determine how many sundaes they will have made when they run out of ice cream. (For this problem, we assume a gallon equals 20 scoops.)

74. A guitarist and a drummer are getting ready for a gig. The length of the gig will depend on how much material they have prepared. For every hour of the show, the guitarist must practice for 5 days and the drummer for 3 days. Since the guitarist already knows some of the songs, he saves 3 days of practice time. The drummer hurts his hand and loses 3 days of practice. If they plan to start and finish practicing at the same time, solve the equation, $5x-3=3x+3$ to determine how long the show will be.

Writing & Thinking

75. Answer each question.

a. Simplify the expression $3(x+5)+2(x-7)$.

b. Solve the equation $3(x+5)+2(x-7)=31$.

c. How are the methods you used to answer parts a. and b. similar? How are they different?

76. Write an equation to represent each situation, using x to represent Ryan's current age. Determine whether each equation is a conditional equation, an identity, or a contradiction, and explain why that makes sense for the situation represented.

a. In 6 years, Ryan will be 20 years old.

b. In 6 years, Ryan will be 8 years older than he is now.

c. In 6 years, Ryan will be 3 years older than he will be 3 years from now.

9.4 Working with Formulas

Objectives

A. Solve applied problems by evaluating known formulas at given values of the variables.

B. Solve formulas for specified variables in terms of the other variables.

A Evaluating Formulas

Formulas are general rules or principles stated mathematically. There are many formulas in such fields of study as business, economics, medicine, physics, and chemistry as well as mathematics.

Listed here are a few formulas with a description of the meaning of each formula. There are, of course many more formulas, some of which you will see in the exercises.

Formula	Meaning
$I = Prt$	The **simple interest** (I) earned by investing money is equal to the product of the principal P times the rate of interest r times the time t in years.
$C = \frac{5}{9}(F - 32)$	**Temperature** in degrees Celsius (C) equals $\frac{5}{9}$ times the difference between the Fahrenheit temperature (F) and 32.
$d = rt$	The **distance traveled** (d) equals the product of the rate of speed r and the time t.
$P = 2l + 2w$	The **perimeter of a rectangle** (P) is equal to twice the length l plus twice the width w.
$L = 2\pi rh$	The **lateral surface area** (L) (top and bottom not included) of a cylinder is equal to 2π times the radius r of the base times the height h.
$F = ma$	In physics, the **force** (F) acting on an object is equal to its mass m times its acceleration a.
$\alpha + \beta + \gamma = 180°$	The **sum of the angles of a triangle** (α, β, and γ) is 180°.**Note:** α, β, and γ are the Greek lowercase letters alpha, beta, and gamma, respectively.
$P = a + b + c$	The **perimeter of a triangle** (P) is equal to the sum of the lengths of the three sides a, b, and c.

Table 1

Note

Be sure to use the letters just as they are given in the formulas. In mathematics, there is little or no flexibility between capital and small letters as they are used in formulas. In general, capital letters have special meanings that are different from corresponding small letters. For example, capital ***A*** may mean the area of a triangle and small ***a*** may mean the length of one side, two completely different ideas.

If you know values for all but one variable in a formula, you can substitute those values and find the value of the unknown variable using the techniques for solving equations discussed earlier in this chapter. This section presents a variety of formulas from real-life situations, with complete descriptions of the meanings of these formulas. Working with these formulas will help you become familiar with a wide range of applications of algebra and provide practice in solving equations.

1. You get a loan for \$2000 for 60 days at an annual interest rate of 6%. How much do you have to pay at the end of the loan period, 60 days?

Example 1 Application: Evaluating Formulas

The formula for calculating simple interest is

$I = Prt$, where

I = **interest** (earned or paid)

P = **principal** (the amount invested or borrowed)

r = **rate of interest** (stated as an annual or yearly rate in percent form)

t = **time** (in years)

The rate of interest is usually given in percent form and converted to decimal or fraction form for calculations.

Maribel loaned \$5000 to a friend for 3 months at an annual interest rate of 8%. How much will her friend pay her at the end of the 3 months?

Solution

Here, $P = \$5000,$

$r = 8\% = 0.08,$

$t = 3 \text{ months} = \frac{3}{12} \text{ year} = \frac{1}{4} \text{ year}.$

Find the interest by substituting in the formula $I = Prt$ and evaluating.

$$\begin{aligned} I &= 5000 \cdot \overset{0.02}{\cancel{0.08}} \cdot \frac{1}{\cancel{4}} \\ &= 5000 \cdot 0.02 \\ &= 100.00 \end{aligned}$$

The interest is \$100 and the amount to be paid at the end of 3 months is principal + interest = \$5000 + \$100 = \$5100.

Now work margin exercise 1.

2. Given $C = \frac{5}{9}(F-32)$, find F if $C = 50°$.

Example 2 Evaluating Formulas

Given the formula $C = \frac{5}{9}(F-32)$, first find C if $F = 212°$ (212 degrees Fahrenheit) and then find F if $C = 20°$ (20 degrees Celsius).

Solution

$F = 212°$, so substitute 212 for F in the formula.

$$\begin{aligned} C &= \frac{5}{9}(212-32) \\ &= \frac{5}{9}(180) \\ &= 100 \end{aligned}$$

That is, 212 °F is the same as 100 °C. Water will boil at 212 °F at sea level. This means that, if the temperature is measured in degrees Celsius instead of degrees Fahrenheit, water will boil at 100 °C at sea level.

$C = 20°$, so substitute 20 for C in the formula.

$20 = \frac{5}{9}(F-32)$	Now solve for F.
$\frac{9}{5} \cdot 20 = \frac{9}{5} \cdot \frac{5}{9}(F-32)$	Multiply both sides by $\frac{9}{5}$.
$36 = F - 32$	Simplify.
$36 + 32 = F - 32 + 32$	Add 32 to both sides.
$68 = F$	Simplify.

That is, a temperature of 20 °C is the same as a comfortable spring day temperature of 68 °F.

Now work margin exercise 2.

Example 3 Application: Evaluating Formulas

The lifting force F exerted on an airplane wing is found by multiplying some constant k by the area A of the wing's surface and by the square of the plane's velocity v. The formula is $F = kAv^2$. Find the force on a plane's wing during takeoff if the area of the wing is 120 ft², k is $\frac{4}{3}$, and the plane is traveling 80 miles per hour during takeoff.

Solution

We know that $k = \frac{4}{3}$, $A = 120$, and $v = 80$. Substitution gives

$$F = \frac{4}{3} \cdot 120 \cdot 80^2$$
$$= \frac{4}{3} \cdot 120 \cdot 6400$$
$$= 160 \cdot 6400$$
$$= 1{,}024{,}000$$

The force is measured in pounds.

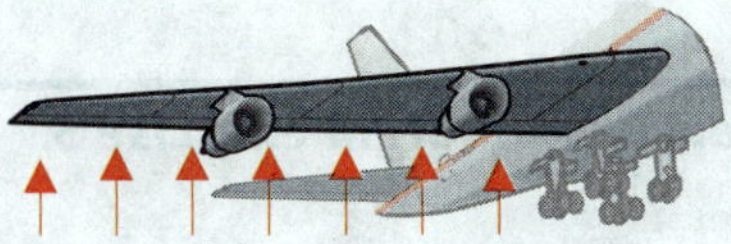

Thus the force on the plane's wing during takeoff is 1,024,000 lb.

Now work margin exercise 3.

3. Use the formula $F = kAv^2$ to find the force exerted on a plane's wing during takeoff if the area of the wing is 100 ft², $k = \frac{5}{4}$, and $v = 90$ mph.

Example 4 Evaluating Formulas

The perimeter of a triangle is 38 feet. One side is 5 feet long and a second side is 18 feet long. How long is the third side?

Solution

Using the formula $P = a + b + c$, substitute $P = 38$, $a = 5$, and $b = 18$. Then solve for the third side.

$$P = a + b + c$$
$$38 = 5 + 18 + c$$
$$38 = 23 + c$$
$$38 - 23 = 23 + c - 23$$
$$15 = c$$

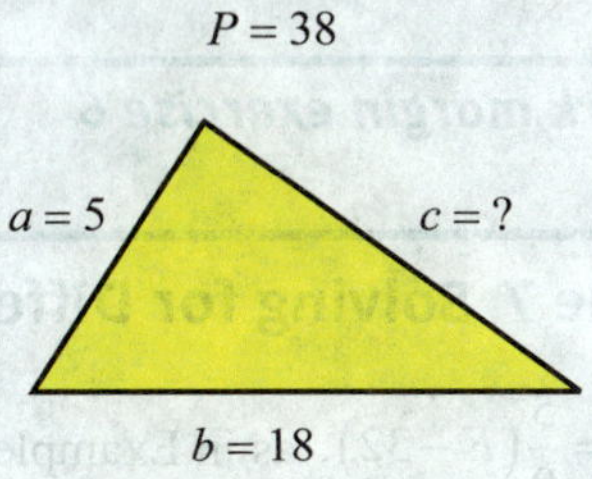

The third side is 15 feet long.

Now work margin exercise 4.

4. Two angles of a triangle measure 65° and 85°. What is the measure of the third angle of the triangle?

B Solving Formulas for Different Variables

We say that the formula $d = rt$ is "solved for" d in terms of r and t. Similarly, the formula $A = \frac{1}{2}bh$ is solved for A in terms of b and h, and the formula $P = R - C$ (profit is equal to revenue minus cost) is solved for P in terms of R and C. Many times, we want to use a certain formula in another form. We want the formula "solved for" some variable other than the one given in terms of the remaining variables. **Treat the variables just as you would constants in solving linear equations.** Study the following examples carefully.

5. Given $P = IV$, solve for I in terms of P and V.

Example 5 Solving for Different Variables

Given $d = rt$, solve for t in terms of d and r. We want to represent the time in terms of distance and rate. We will use this concept later in word problems.

Solution

$d = rt$ — Treat r and d as if they were constants.

$\frac{d}{r} = \frac{rt}{r}$ — Divide both sides by r.

$\frac{d}{r} = t$ — Simplify.

Now work margin exercise 5.

6. Given $P = \frac{I}{rt}$, solve for t in terms of I, r, and P.

Example 6 Solving for Different Variables

Given $V = \frac{k}{P}$, solve for P in terms of V and k.

Solution

$V = \frac{k}{P}$

$P \cdot V = P \cdot \frac{k}{P}$ — Multiply both sides by P.

$PV = k$ — Simplify.

$\frac{PV}{V} = \frac{k}{V}$ — Divide both sides by V.

$P = \frac{k}{V}$ — Simplify.

Now work margin exercise 6

7. Given $y = \frac{2}{5}(x+3)$, solve for x.

Example 7 Solving for Different Variables

Given $C = \frac{5}{9}(F - 32)$, as in Example 2, solve for F in terms of C. This would give a formula for finding Fahrenheit temperature given a Celsius temperature value.

Solution

$$C = \frac{5}{9}(F-32)$$ Treat C as a constant.

$$\frac{9}{5}\cdot C = \frac{9}{5}\cdot\frac{5}{9}(F-32)$$ Multiply both sides by $\frac{9}{5}$.

$$\frac{9}{5}C = F-32$$ Simplify.

$$\frac{9}{5}C+32 = F-32+32$$ Add 32 to both sides.

$$\frac{9}{5}C+32 = F$$ Simplify.

Thus, $F = \frac{9}{5}C+32$ is solved for F, and $C = \frac{5}{9}(F-32)$ is solved for C.

These are two forms of the same formula.

Now work margin exercise 7.

Teaching Note

You might want to explain that multiplication by $\frac{9}{5}$ in Example 7 makes the next step easier than it would be if we used the distributive property and removed the parentheses that way. Some judgment must be used in a case like this. You could show that

$$C = \frac{5}{9}(F-32)$$

$$C = \frac{5}{9}\cdot F - \frac{5}{9}\cdot 32$$

is more difficult to manipulate and solve for F.

Example 8 Solving for Different Variables

Given the equation $2x + 4y = 10$,

a. solve for x in terms of y, and then

b. solve for y in terms of x.

This equation is typical of the algebraic equations that we will discuss in Chapter 10.

Solution

a. Solving for x yields the following.

$$2x+4y=10$$ Treat $4y$ as a constant.

$$2x+4y-4y=10-4y$$ Subtract $4y$ from both sides. (This is the same as adding $-4y$.)

$$2x=10-4y$$ Simplify.

$$\frac{2x}{2}=\frac{10}{2}-\frac{4y}{2}$$ Divide each term on both sides by 2.

$$x=5-2y$$ Simplify.

b. Solving for y yields the following.

$$2x+4y=10$$ Treat $2x$ as a constant.

$$2x+4y-2x=10-2x$$ Subtract $2x$ from both sides.

$$4y=10-2x$$ Simplify.

$$\frac{4y}{4}=\frac{10}{4}-\frac{2x}{4}$$ Divide each term on both sides by 4.

$$y=\frac{5}{2}-\frac{x}{2}$$ Simplify.

Alternatively, we can write the following.

$$y=\frac{5-x}{2} \quad \text{or} \quad y=-\frac{1}{2}x+\frac{5}{2}$$ All forms are correct.

Now work margin exercise 8

8. Given the equation $16y + 25z = 400$,

a. solve for y in terms of z, and then

b. solve for z in terms of y.

9. Given $4x + 8y + 12z = 16$, solve for x.

Completion Example 9 Solving for Different Variables

Given $3x - y = 15$, solve for y in terms of x.

Solution

Supply the reasons for each step in the following solution.

$$3x - y = 15$$

$$3x - y - 3x = 15 - 3x \quad \text{Subtract } 3x \text{ from both sides.}$$

$$-y = 15 - 3x$$

$$-1(-y) = -1(15 - 3x) \quad \text{Multiply both sides by } -1 \text{ (or divide both sides by } -1\text{).}$$

$$y = -15 + 3x \quad \text{Simplify using the distributive property.}$$

or $$y = 3x - 15$$

Now work margin exercise 9.

Completion Example Answers

9.

$$3x - y = 15$$

$$3x - y - 3x = 15 - 3x \quad \text{Subtract } 3x \text{ from both sides.}$$

$$-y = 15 - 3x$$

$$-1(-y) = -1(15 - 3x) \quad \text{Multiply both sides by } -1 \text{ (or divide both sides by } -1\text{).}$$

$$y = -15 + 3x \quad \text{Simplify using the distributive property.}$$

or $$y = 3x - 15$$

Margin Exercise Answers

1. \$2020 **2.** $F = 122\ °F$ **3.** 1,012,500 lb **4.** 30° **5.** $I = \frac{P}{V}$ **6.** $t = \frac{I}{Pr}$ **7.** $x = \frac{5}{2}y - 3$
8. a. $y = \frac{400 - 25z}{16}$ or $y = -\frac{25}{16}z + 25$ **b.** $z = \frac{400 - 16y}{25}$ or $z = -\frac{16}{25}y + 16$ **9.** $x = 4 - 2y - 3z$

9.4 Exercises

Concept Check

Fill-in-the-Blank. Complete each sentence using information found in this section.

1. Formulas are general rules or principles stated __________.

2. The ______ ________ earned by investing money is equal to the product of the principle times the rate of interest times the time in one year or part of a year.

3. The distance traveled equals the product of the rate of speed and the _____.

4. The _______ of a rectangle is equal to twice the length plus twice the width.

5. If you know values for all but one variable in a formula, you can __________ those values and find the value of the unknown variable by solving the equation.

6. If you want to use a formula in another form, treat the variables just as you would ________ in solving linear equations.

9.4 Exercises

Concept Check

1. mathematically
2. simple interest
3. time
4. perimeter
5. substitute
6. constants

True/False. Determine whether each statement is true or false. If a statement is false, explain how it can be changed so the statement will be true. (**Note:** There may be more than one acceptable change.)

7. When using formulas, typically it does not matter if capital or lower case letters are used: $A = a$, $C = c$, etc.

8. If the perimeter and length are known, $P = 2l + 2w$ can be used to find the width of a rectangle.

9. Rate of interest is stated as an annual rate in percent form.

Applications

Refer to Example 1 for information concerning simple interest and the related formula $I = Prt$. (**Note:** Use 365 days in a year and 12 months in each year.)

Simple Interest

1. You want to borrow $4000 at 12% for only 146 days. How much interest would you pay?

2. For how many days must you leave $1000 in a savings account at 5.5% to earn $11.00 in interest?

3. What principal would you need to invest to earn $450 in simple interest in 6 months if the interest rate was 9%?

4. After one month, Gustav received $25 in simple interest on his savings account of $12,000. What was the interest rate?

5. A savings account of $3500 is left for 9 months and draws simple interest at a rate of 7%.
   a. How much interest is earned?
   b. What is the balance in the account at the end of the 9 months?

6. Tim just deposited $2562.50 to pay off a 3 month loan of $2500.
   a. How much of what he deposited was interest on the loan?
   b. What rate of interest was he charged?

In the following application problems, read the descriptions carefully and then substitute the values given in the problem for the corresponding variables in the formulas. Evaluate the resulting expression for the unknown variable. See Examples 1 through 4.

Velocity

If an object is shot upward with an initial velocity v_0 in feet per second, the velocity v in feet per second is given by the formula $v = v_0 - 32t$, where t is time in seconds. (v_0 is read "v sub zero." The $_0$ is called a subscript.)

7. An object projected upward with an initial velocity of 106 feet per second has a velocity of 42 feet per second. How many seconds have passed?

8. Find the initial velocity of an object if the velocity after 4 seconds is 48 feet per second.

7. False; Case matters in formulas.
8. True
9. True

Applications

1. $192
2. 73 days
3. $10,000
4. 2.5%
5. a. $183.75 b. $3683.75
6. a. $62.50 b. 10%
7. 2 seconds
8. 176 feet per second

9. 4 milliliters
10. 100 milligrams
11. $1030
12. 0.5 years
13. 14 rafters
14. 16 studs
15. 336 in. or 28 ft
16. 3 ft

Medicine

In nursing, one procedure for determining the dosage for a child is

$$\text{child's dosage} = \frac{\text{age of child in years}}{\text{age of child} + 12} \cdot \text{adult dosage.}$$

9. If the adult dosage of a drug is 20 milliliters, how much should a 3-year-old child receive?

10. If the adult dosage of a drug is 340 milligrams, how much should a 5-year-old child receive?

Investments

The total amount of money in an account with P dollars invested in it is given by the formula $A = P + Prt$, where r is the rate expressed as a decimal and t is time (one year or part of a year).

11. If $1000 is invested at 6% interest, find the total amount in the account after 6 months.

12. How long will it take an investment of $600, at an annual rate of 5%, to be worth $615?

Construction

The number N of rafters in a roof or studs in a wall can be found by the formula $N = \frac{L}{d} + 1$, where L is the length of the roof or wall and d is the center-to-center distance from one rafter or stud to the next. Note that L and d must be in the same units.

13. How many rafters will be needed to build a roof 26 ft long if they are placed 2 ft apart center-to-center?

14. A wall has studs placed 16 in. apart center-to-center. If the wall is 20 ft long, how many studs are in the wall?

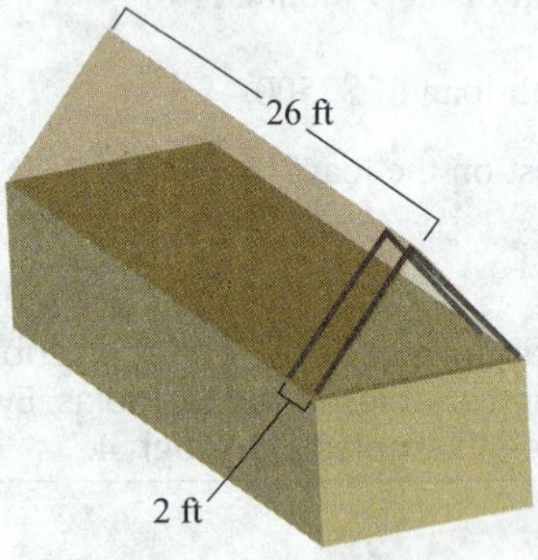

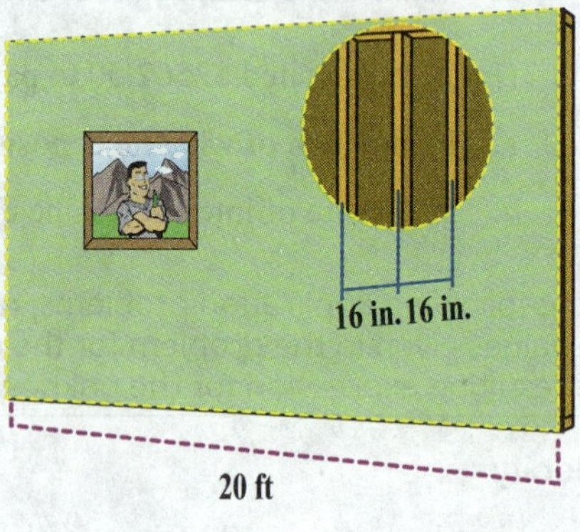

15. How long is a wall if it requires 22 studs placed 16 in. apart center-to-center?

16. What should the center-to-center distance be if you are building a 33 ft long roof using 12 rafters?

Cost

The total cost C of producing x items can be found by the formula $C = ax + k$, where a is the cost per item and k is the fixed costs (rent, utilities, and so on).

17. Find the total cost of producing 30 items if each costs \$15 and the fixed costs are \$580.

18. The total cost to produce 80 dolls is \$1097.50. If each doll costs \$9.50 to produce, find the fixed costs.

19. It costs a company \$3.60 to produce a calculator. Last week the total costs were \$1308. If the fixed costs are \$480 weekly, how many calculators were produced last week?

20. Each week a carpentry shop builds 60 end tables for a total cost of \$5340. If the fixed costs for a week are \$750, what is the cost to produce each end table?

Profit

The profit P is given by the formula $P = R - C$, where R is the revenue and C is the cost.

21. Find the revenue (income) of a company that shows a profit of \$3.2 million and costs of \$1.8 million.

22. Find the revenue of a company that shows a profit of \$3.2 million and costs of \$5.7 million.

Depreciation

Many items decrease in value as time passes. This decrease in value is called depreciation. One type of depreciation is called linear depreciation. The value V of an item after t years is given by $V = C - Crt$, where C is the original cost and r is the rate of depreciation expressed as a decimal.

23. If you buy a car for \$6000 and it depreciates linearly at a rate of 10% per year, what will be its value after 6 years?

24. A contractor buys a 4-year-old piece of heavy equipment valued at \$20,000. If the original cost of this equipment was \$25,000, find the rate of depreciation.

Distance, Rate, Time

The distance traveled d is given by the formula $d = rt$, where r is the rate of speed and t is the time it takes.

25. How long will a truck driver take to travel 350 miles if he averages 50 mph?

26. What is the average rate of speed of a biker who bikes 21.92 miles in 68.5 minutes?

27. What is Jonathan's average rate of speed if he hikes 10.4 miles in 6.4 hours?

28. How long will it take a train traveling at 40 mph to go 140 miles?

17. \$1030
18. \$337.50
19. 230 calculators
20. \$76.50
21. \$5 million
22. \$8.9 million
23. \$2400
24. 5%
25. 7 hours
26. 0.32 miles per minute
27. 1.625 mph
28. 3.5 hours

29. $b = P - a - c$
30. $s = \frac{P}{3}$
31. $m = \frac{F}{a}$
32. $d = \frac{C}{\pi}$
33. $w = \frac{A}{l}$
34. $C = R - P$
35. $n = \frac{R}{p}$
36. $k = v - gt$
37. $P = A - I$
38. $h = \frac{L}{2\pi r}$
39. $m = 2A - n$
40. $a = P - 2b$
41. $t = \frac{I}{Pr}$
42. $E = RI$
43. $b = \frac{P - a}{2}$
44. $b^2 = c^2 - a^2$
45. $\beta = 180° - \alpha - \gamma$
46. $x = \frac{y - b}{m}$
47. $h = \frac{V}{lw}$
48. $t = \frac{v_0 - v}{g}$
49. $b = \frac{2A}{h}$
50. $I = \frac{E}{R}$
51. $h = \frac{V}{\pi r^2}$
52. $L = \frac{R}{2A}$
53. $g = \frac{mv^2}{2K}$
54. $y = \frac{4 - x}{4}$
55. $y = \frac{6 - 2x}{3}$
56. $y = 3x - 14$

Solve each formula for the indicated variable. See Examples 5 through 9.

29. $P = a + b + c$; solve for b.

30. $P = 3s$; solve for s.

31. $F = ma$; solve for m.

32. $C = \pi d$; solve for d.

33. $A = lw$; solve for w.

34. $P = R - C$; solve for C.

35. $R = np$; solve for n.

36. $v = k + gt$; solve for k.

37. $I = A - P$; solve for P.

38. $L = 2\pi rh$; solve for h.

39. $A = \frac{m + n}{2}$; solve for m.

40. $P = a + 2b$; solve for a.

41. $I = Prt$; solve for t.

42. $R = \frac{E}{I}$; solve for E.

43. $P = a + 2b$; solve for b.

44. $c^2 = a^2 + b^2$; solve for b^2.

45. $\alpha + \beta + \gamma = 180°$; solve for β.

46. $y = mx + b$; solve for x.

47. $V = lwh$; solve for h.

48. $v = -gt + v_0$; solve for t.

49. $A = \frac{1}{2}bh$; solve for b.

50. $R = \frac{E}{I}$; solve for I.

51. $V = \pi r^2 h$; solve for h.

52. $A = \frac{R}{2L}$; solve for L.

53. $K = \frac{mv^2}{2g}$; solve for g.

54. $x + 4y = 4$; solve for y.

55. $2x + 3y = 6$; solve for y.

56. $3x - y = 14$; solve for y.

57. $5x + 2y = 11$; solve for x.

58. $-2x + 2y = 5$; solve for x.

59. $A = \frac{1}{2}h(b + c)$; solve for b.

60. $A = P(1 + rt)$; solve for r.

61. $R = \frac{3(x - 12)}{8}$; solve for x.

62. $-2x - 5 = -3(x + y)$; solve for x.

63. $3y - 2 = x + 4y + 10$; solve for y.

64. $V = \frac{1}{3}\pi r^2 h$; solve for h.

Determine a formula for each of the following situations.

65. Each ticket for a concert costs $\$t$ per person and parking costs \$9.00. What is the total cost per car C if there are n people in a car?

66. A-to-Z Truck Rentals charges \$25 per day plus \$0.75 per mile for a 10-foot rental truck. What would you pay per day for renting the truck from A-to-Z if you were to drive the truck x miles in one day?

67. Top-of-the-Line computer company knows that the cost (labor and materials) of producing a computer is \$325 per computer per week and the fixed overhead costs (lighting, rent, etc.) are \$5400 per week. What are the company's weekly costs of producing n computers per week?

68. If the Top-of-the-Line computer company (see Exercise 67) sells its computers for \$683 each, what is its profit per week if it sells the same number n that it produces? (Remember that profit is equal to revenue minus costs, or $P = R - C$.)

Solve.

69. Samantha uses a credit promotion at a home improvement store where she doesn't have to pay any interest on her purchase as long as she pays off the entire balance within 6 months. She purchases \$8000 in merchandise. If she fails to pay off the balance within 6 months, then she will be charged \$600 in interest. Samantha lost the paper work and wants to determine the interest rate on her purchase.

a. Which formula from Table 1 fits this situation?

b. Match the variables in the formula from part a. to the information provided.

c. The formula from part a. needs to be solved for which variable?

d. What is the interest rate on her purchase? (Remember to convert to a percent.)

70. In a physics lab, a ball is rolled down an incline that has a machine at the bottom that calculates the force of impact. The ball has a mass of 1.5 kilograms. After several trials, the average force of impact is calculated to be 12.75 kg · m/s². The researchers need to determine the average acceleration of the ball at the moment it struck the machine.

a. Which formula from Table 1 fits this situation?

b. Match the variables in the formula from part a. to the information provided.

c. The formula from part a. needs to be solved for which variable?

d. What was the average acceleration of the ball in m/s²?

71. Charles is experimenting with a new sail design for his sailboat and needs to keep the total area of the triangular sail to 150 square feet. The base of the sail must be exactly 3 times the height of the sail.

a. What geometric formula for area should be used?

b. Write an expression for the base of the formula using the variable *h* for height.

c. Substitute the expression from part b. into the area formula for the base.

d. Solve this formula for the height squared.

e. What would you have to do to both sides of the equation in part d. to solve the formula for the height?

f. Substitute 150 for the area of the sail in the formula from part d. and solve for the height of the sail.

g. What is the length of the base of the sail?

57. $x = \frac{11-2y}{5}$

58. $x = \frac{2y-5}{2}$

59. $b = \frac{2A-hc}{h}$ or $b = \frac{2A}{h} - c$

60. $r = \frac{A-P}{Pt}$

61. $x = \frac{8R+36}{3}$ or $x = \frac{8R}{3} + 12$

62. $x = 5 - 3y$

63. $y = -x - 12$

64. $h = \frac{3V}{\pi r^2}$

65. $C = nt + 9$

66. $C = 0.75x + 25$

67. $C = 325n + 5400$

68. $P = 358n - 5400$

69. **a.** $I = Prt$ **b.** $P = \$8000$; $I = \$600$; $t = 0.5$ **c.** r **d.** $r = 15\%$

70. **a.** $F = ma$ **b.** $F = 12.75\text{ kg} \cdot \text{m/s}^2$; $m = 1.5$ **c.** a **d.** $a = 8.5\text{ m/s}^2$

71. **a.** $A = \frac{1}{2}bh$ **b.** $b = 3h$ **c.** $A = \frac{1}{2}(3h)h = \frac{3}{2}h^2$ **d.** $h^2 = \frac{2}{3}A$ **e.** Take the square root of both sides **f.** 10 **g.** 30 feet

Writing & Thinking

72. The formula $z = \dfrac{x - \bar{x}}{s}$ is used extensively in statistics. In this formula, x represents one value in a set of data, $\bar{x}$ represents the average (or mean) of those numbers in the set, and s represents a value called the standard deviation of the numbers. (The standard deviation is a positive number and is a measure of how "spread out" the numbers are.) The values for z are called z-scores, and they measure the number of standard deviation units a number x is from the average $\bar{x}$.

a. If $\bar{x} = 70$, what will be the z-score for $x = 70$? Does this z-score depend on the value of s? Explain.

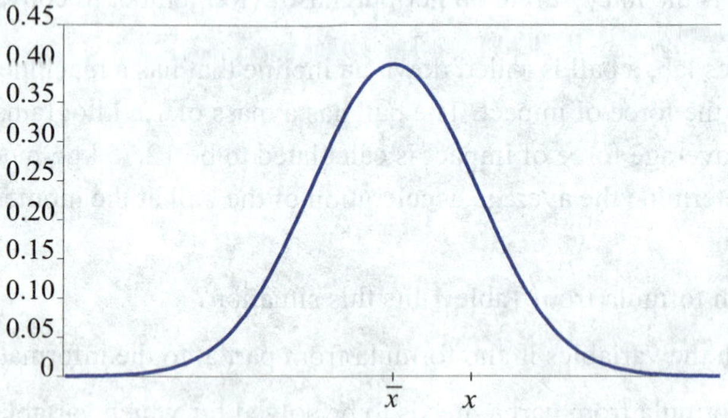

b. For what values of x will the corresponding z-scores be negative?

c. Calculate your z-score on each of the last two test scores in this class. (Your instructor will give you the average and standard deviation for each test.) What do these scores tell you about your performance on the two exams?

73. Suppose that, for a particular set of exam scores, $\bar{x} = 72$ and $s = 6$. Find the z-score that corresponds to each of the following scores.

a. 78 **b.** 66 **c.** 81 **d.** 0

Standard Normal (z-scores)

It's often hard to compare values on different scales, but the z-score can help. Suppose Connie scored a 63 on her math exam (where the class average was 60 with a standard deviation of 5) and Mike scored a 78 on his history exam (where the class average was 75 with a standard deviation of 10). Who scored relatively better? By converting both test scores to z-scores, it's easier to compare the values.

$$z_{Connie} = \frac{63 - 60}{5} = 0.6$$

$$z_{Mike} = \frac{78 - 75}{10} = 0.3$$

Since Connie's z-score is larger than Mike's, Connie did relatively better on her math exam than Mike did on his history exam.

Writing & Thinking

72. a. 0; No, because the numerator will be zero and thus the whole fraction will be equal to zero for all values of s.
b. $x < 70$
c. Answers will vary.

73. a. 1 **b.** −1 **c.** 1.5
d. −12

9.5 Applications: Number Problems and Consecutive Integers

Objectives

A. Use linear equations to solve number problems.

B. Use linear equations to solve word problems involving consecutive integers.

C. Use linear equations to solve applications.

A Number Problems

In Section 8.8, we discussed translating English phrases into algebraic expressions, and in Section 4.7, we learned the Pólya four-step process as an approach to problem solving, stated here.

1. Understand the problem.
2. Devise a plan.
3. Carry out the plan.
4. Look back over the results.

Now we will use those skills to read number problems and translate the sentences and phrases in the problem into a related equation. The solution of this equation will be the solution to the problem.

Example 1 Solving Number Problems

If a number is decreased by 36 and the result is 76 less than twice the number, what is the number?

Solution

Let n = the unknown number.

A number is decreased by 36	The result is	76 less than twice the number
$n - 36$	$=$	$2n - 76$

$$
\begin{aligned}
n - 36 - n &= 2n - 76 - n \\
-36 &= n - 76 \\
-36 + 76 &= n - 76 + 76 \\
40 &= n
\end{aligned}
$$

The number is 40.

Now work margin exercise 1.

1. If a number is decreased by 28 and the result is 92 less than twice the number, what is the number?

Example 2 Solving Number Problems

Three times the sum of a number and 5 is equal to twice the number plus 5. Find the number.

2. Six times the sum of a number and 3 is equal to 3 times the number plus 3. Find the number.

Solution

Let x = the unknown number.

3 times the sum of a number and 5	Is equal to	Twice the number plus 5
$3(x+5)$	$=$	$2x+5$

$$3x+15=2x+5$$
$$3x+15-2x=2x+5-2x$$
$$x+15=5$$
$$x+15-15=5-15$$
$$x=-10$$

The number is –10.

Now work margin exercise 2.

3. One integer is 3 more than 4 times a second integer. Their sum is 28. What are the two integers?

Example 3 Solving Number Problems

One integer is 4 more than three times a second integer. Their sum is 24. What are the two integers?

Solution

Let n = the second integer, then $3n + 4$ = the first integer.

$$(1^{\text{st}}\text{ integer})+(2^{\text{nd}}\text{ integer})=24 \quad \text{Their sum is 24.}$$
$$(3n+4)+n=24$$
$$4n+4=24$$
$$4n+4-4=24-4$$
$$4n=20$$
$$\frac{4n}{4}=\frac{20}{4}$$
$$n=5$$
$$3n+4=19$$

The two integers are 5 and 19.

Now work margin exercise 3.

B Consecutive Integers

Remember that the set of **integers** consists of the whole numbers and their opposites.

$$\mathbb{Z}=\{\ldots,-4,-3,-2,-1,0,1,2,3,4,\ldots\}$$

Even integers are integers that are divisible by 2. The even integers are

$$\{\ldots,-6,-4,-2,0,2,4,6,\ldots\}.$$

Odd integers are integers that are not even. If an odd integer is divided by 2 the remainder will be 1. The odd integers are

$$\{\ldots,-5,-3,-1,1,3,5,\ldots\}.$$

The following terms and ways of representing the integers must be understood before attempting the problems.

Consecutive Integers

Integers are **consecutive** if each is 1 more than the previous integer. Three consecutive integers can be represented as

$$n, \quad n+1, \quad \text{and} \quad n+2,$$

where n is an integer.

DEFINITION

Math Tip

There are several additional ways to represent three consecutive integers. For example, $n-2$, $n-1$, n and $n-1$, n, $n+1$ are other valid ways to represent three consecutive integers.

Consecutive Even Integers

Even integers are **consecutive** if each is 2 more than the previous even integer. Three consecutive even integers can be represented as

$$n, \quad n+2, \quad \text{and} \quad n+4,$$

where n is an **even** integer.

DEFINITION

Note

In this section, we will be dealing only with integers. Therefore, if you get a result that has a fraction or decimal number (not an integer), you will know that an error has been made and you should correct some part of your work.

Consecutive Odd Integers

Odd integers are **consecutive** if each is 2 more than the previous odd integer. Three consecutive odd integers can be represented as

$$n, \quad n+2, \quad \text{and} \quad n+4,$$

where n is an **odd** integer.

DEFINITION

Note that consecutive even and consecutive odd integers are represented in the same way.

$$n, \quad n+2, \quad \text{and} \quad n+4$$

The value of the first integer, n, determines whether the remaining integers are odd or even. For example,

<u>*n* is odd</u>		or	<u>*n* is even</u>	
If	$n=11$,		If	$n=36$,
then	$n+2=13$		then	$n+2=38$
and	$n+4=15.$		and	$n+4=40.$

4. The sum of three consecutive odd integers is −9. What are the integers?

Example 4 Finding Consecutive Integers

The sum of three consecutive odd integers is −3. What are the integers?

Solution

Let n = the first odd integer,
then $n + 2$ = the second odd integer,
and $n + 4$ = the third odd integer.

Set up and solve the related equation.

$$\left(1^{\text{st}} \text{ integer}\right)+\left(2^{\text{nd}} \text{ integer}\right)+\left(3^{\text{rd}} \text{ integer}\right)=-3$$
$$n+(n+2)+(n+4)=-3$$
$$3n+6=-3$$
$$3n+6-6=-3-6$$
$$3n=-9$$
$$\frac{3n}{3}=\frac{-9}{3}$$
$$n=-3$$
$$n+2=-1$$
$$n+4=1$$

The three consecutive odd integers are −3, −1, and 1.

Now work margin exercise 4.

5. Find three consecutive even integers such that 6 less than three times the first is equal to the sum of the other two.

Example 5 Finding Consecutive Integers

Find three consecutive integers such that the sum of the first and third is 76 less than three times the second.

Solution

Let n = the first integer,
then $n + 1$ = the second integer,
and $n + 2$ = the third integer.

Set up and solve the related equation.

$$\left(1^{\text{st}} \text{ integer}\right)+\left(3^{\text{rd}} \text{ integer}\right)=3\left(2^{\text{nd}} \text{ integer}\right)-76$$
$$n+(n+2)=3(n+1)-76$$
$$2n+2=3n+3-76$$
$$2n+2=3n-73$$
$$2n+2+73=3n-73+73$$
$$2n+75=3n$$
$$2n+75-2n=3n-2n$$
$$75=n$$
$$76=n+1$$
$$77=n+2$$

The three consecutive integers are 75, 76, and 77.

Check

$75+77=152$ and $3(76)-76=228-76=152$

Now work margin exercise 5.

C Applications of Linear Equations

As you learn more abstract mathematical ideas, you will find that you will use these ideas and the related processes to solve a variety of everyday problems as well as problems in specialized fields of study. Generally, you may not even be aware of the fact that you are using your mathematical knowledge. However, these skills and ideas will be part of your thinking and problem solving techniques for the rest of your life.

Example 6 Application: Calculating Living Expenses

Joe wants to budget $\frac{2}{5}$ of his monthly income for rent. He found an apartment he likes for \$800 a month. What monthly income does he need to be able to afford this apartment?

Solution

Let $x =$ Joe's monthly income, then $\frac{2}{5}x =$ rent.

$$\frac{2}{5}x = 800$$

$$\frac{5}{2} \cdot \frac{2}{5}x = \frac{5}{2} \cdot \frac{800}{1}$$

$$x = 2000$$

Joe's monthly income should be \$2000.

Now work margin exercise 6.

6. Jim plans to budget $\frac{3}{7}$ of his monthly income to send his son Taylor to private school. If the school he'd like Taylor to attend costs \$1200 a month, what monthly income does Jim need to be able to afford the school?

Example 7 Application: Calculating Costs of Purchases

A student bought a calculator and a textbook for a total of \$200.80 (including tax). The textbook cost \$20.50 more than the calculator. He then challenged a friend to calculate the cost of each item.

Solution

His friend proceeded as follows.
Let $x =$ cost of the calculator,
then $x + 20.50 =$ cost of the textbook.
The equation to be solved is:

$$\underbrace{x + 20.50}_{\text{Cost of textbook}} + \underbrace{x}_{\text{Cost of calculator}} = \underbrace{200.80}_{\text{Total spent}}$$

$$2x + 20.50 = 200.80$$

$$2x + 20.50 - 20.50 = 200.80 - 20.50$$

$$2x = 180.30$$

$$\frac{2x}{2} = \frac{180.30}{2}$$

$$x = 90.15 \quad \text{Cost of calculator}$$

$$x + 20.50 = 110.65 \quad \text{Cost of textbook}$$

The calculator costs \$90.15 and the textbook costs \$90.15 + \$20.50 = \$110.65, with tax included in each price.

Now work margin exercise 7.

7. Graeme bought bagpipes and a kilt for a total of \$507.30 (including tax). If the bagpipes cost \$70.40 more than the kilt, what was the cost of each item?

Margin Exercise Answers

1. 64 **2.** –5 **3.** 5 and 23 **4.** –5, –3, and –1 **5.** 12, 14, and 16 **6.** \$2800 **7.** Bagpipes: \$288.85, Kilt: \$218.45

9.5 Exercises

Concept Check

Fill-in-the-Blank. Complete each sentence using information found in this section.

1. Integers are ________ if each is 1 more than the previous integer.

2. Three consecutive integers can be represented as n, ________, and ______.

3. Three consecutive even integers can be represented as n, ________, and ______.

4. The whole numbers and their opposites form the set of ________.

5. Even integers are integers that are divisible by ___.

6. Odd integers are integers that are not _____.

True/False. Determine whether each statement is true or false. If a statement is false, explain how it can be changed so the statement will be true. (**Note:** There may be more than one acceptable change.)

7. If an odd integer is divided by 2, the remainder will be 1.

8. To find 3 consecutive odd integers, you could use n, $n+1$, and $n+3$.

9. Odd integers are integers that are divisible by 1.

10. Even integers are consecutive if each is 2 more than the previous even integer.

Practice

For each statement, define a variable to represent the unknown quantity. (Be sure that your variable definition includes what quantity is being measured and the units used to measure it.) Then, write an expression, equation, or inequality to represent it.

1. Three-fifths of the savings account balance

2. The radius of the circle increased by 6 centimeters

3. Four times the length of a side of the square is 20 feet.

4. Fifteen percent of your annual salary is 3000 dollars.

Read each problem carefully, translate the various phrases into algebraic expressions, set up an equation, and solve the equation. See Examples 1 through 5.

5. Five less than a number is equal to 13 decreased by the number. Find the number.

6. Three less than twice a number is equal to the number. What is the number?

7. Thirty-six is 4 more than twice a certain number. Find the number.

9.5 Exercises

Concept Check

1. consecutive

2. $n+1, n+2$

3. $n+2, n+4$

4. integers

5. 2

6. even

7. True

8. False; $n, n+2, n+4$

9. False; Integers that are not even

10. True

Practice

1. x = savings account balance in dollars (or other unit of money); $\frac{3}{5}x$

2. r = the radius of the circle in cm; $r+6$

3. s = the length of a side of the square; $4s = 20$

4. s = your annual salary in dollars; $0.15s = 3000$

5. $x - 5 = 13 - x$; 9

6. $2x - 3 = x$; 3

7. $36 = 2x + 4$; 16

8. $15 - 2x = 27$; -6

9. $7x = 2x + 35$; 7

10. $2x - 3 = 6 - x$; 3

11. $3x + 14 = 6 - x$; -2

12. $\frac{x}{7} + 2 = -3$; -35

8. Fifteen decreased by twice a number is 27. Find the number.

9. Seven times a certain number is equal to the sum of twice the number and 35. What is the number?

10. The difference between twice a number and 3 is equal to 6 decreased by the number. Find the number.

11. Fourteen more than 3 times a number is equal to 6 decreased by the number. Find the number.

12. Two added to the quotient of a number and 7 is equal to −3. What is the number?

13. The quotient of twice a number and 5 is equal to the number increased by 6. What is the number?

14. Three times the sum of a number and 4 is equal to −9. Find the number.

15. Four times the difference between a number and 5 is equal to the number increased by 4. What is the number?

16. When 17 is added to 6 times a number, the result is equal to 1 plus twice the number. What is the number?

17. If the sum of twice a number and 5 is divided by 11, the result is equal to the difference between 4 and the number. Find the number.

18. If the difference between a number and 21 is divided by 2, the result is 4 times the number. What is the number?

19. Twice a number increased by 3 times the number is equal to 4 times the sum of the number and 3. Find the number.

20. Twice the difference between a number and 10 is equal to 6 times the number plus 16. What is the number?

21. The sum of two consecutive odd integers is 60. What are the integers?

22. The sum of two consecutive even integers is 78. What are the integers?

23. Find three consecutive integers whose sum is 69.

24. Find three consecutive integers whose sum is 93.

25. The sum of four consecutive integers is 74. What are the integers?

26. Find four consecutive integers whose sum is 90.

27. 171 minus the first of three consecutive integers is equal to the sum of the second and third. What are the integers?

28. If the first of three consecutive integers is subtracted from 120, the result is the sum of the second and third. What are the integers?

29. Four consecutive integers are such that if 3 times the first is subtracted from 208, the result is 50 less than the sum of the other three. What are the integers?

30. Find two consecutive integers such that twice the first plus three times the second equals 83.

31. Find three consecutive even integers such that the first plus twice the second is 54 less than four times the third.

32. Find three consecutive odd integers such that 4 times the first is 44 more than the sum of the second and third.

33. Find three consecutive even integers such that if the first is subtracted from the sum of the second and third, the result is 66.

34. Find three consecutive even integers such that their sum is 168 more than the second.

13. $\frac{2x}{5} = x + 6$; -10

14. $3(x+4) = -9$; -7

15. $4(x-5) = x + 4$; 8

16. $6x + 17 = 1 + 2x$; -4

17. $\frac{2x+5}{11} = 4 - x$; 3

18. $\frac{x-21}{2} = 4x$; -3

19. $2x + 3x = 4(x+3)$; 12

20. $2(x-10) = 6x + 16$; -9

21. $n + (n+2) = 60$; $29, 31$

22. $n + (n+2) = 78$; $38, 40$

23. $n + (n+1) + (n+2) = 69$; $22, 23, 24$

24. $n + (n+1) + (n+2) = 93$; $30, 31, 32$

25. $n + (n+1) + (n+2) + (n+3) = 74$; $17, 18, 19, 20$

26. $n + (n+1) + (n+2) + (n+3) = 90$; $21, 22, 23, 24$

27. $171 - n = (n+1) + (n+2)$; $56, 57, 58$

28. $120 - n = (n+1) + (n+2)$; $39, 40, 41$

29. $208 - 3n = (n+1) + (n+2) + (n+3) - 50$; $42, 43, 44, 45$

30. $2n + 3(n+1) = 83$; $16, 17$

31. $n + 2(n+2) = 4(n+4) - 54$; $42, 44, 46$

32. $4n = (n+2) + (n+4) + 44$; $25, 27, 29$

33. $(n+2) + (n+4) - n = 66$; $60, 62, 64$

34. $n + (n+2) + (n+4) = (n+2) + 168$; $82, 84, 86$

35. $2n + 3(n+2) = 2(n+4) + 7$; $3, 5, 7$

36. $3n+2(n+4)=6(n+2)$
-20; 16, 18, 20

Applications

37. a. The unknown value is the number of postcards purchased.
b. $p=9$
c. Brooke purchased 9 postcards.

38. a. The unknown value is the number of boxes of cupcakes that can be purchased.
b. $19.50+8.50c=45$
c. $c=3$
d. Robin can purchase 3 boxes of cupcakes.

39. a. The number of slices each pizza was cut into
b. $\frac{1}{4}p+\frac{1}{2}p+\frac{3}{8}p=9$
c. $p=8$
d. Each pizza was cut into 8 slices.

40. $c+(c+67.51)=329.49$; calculator: \$130.99, textbook: \$198.50

41. $(c-58.96)+c=96.94$; Flash drive: \$18.99; Printer: \$77.95

42. $2x+90{,}000=310{,}000$; \$110,000

35. Find three consecutive odd integers such that the sum of twice the first and three times the second is 7 more than twice the third.

36. Find three consecutive even integers such that the sum of three times the first and twice the third is twenty less than six times the second.

Applications

Solve.

37. An art show charges \$12.50 for admission and sells 4-inch by 6-inch postcards of works by the featured artists for \$2.75 each. Brooke attends the art show and spends a total of \$37.25 on admission and postcards. The situation can be modeled by $\$37.25 = \$12.50 + \$2.75p$.

a. The unknown value is represented by the variable p in the equation. What is the unknown value in this situation?

b. Solve the equation for the variable.

c. What does the answer to part b. mean? Write a complete sentence.

38. Robin is in charge of purchasing desserts for a dinner party that her nonprofit organization is throwing. She decides to buy a cake and several specialty cupcakes from Barbara's Bombtastic Bakery. She needs to buy one 8-inch round cake, which costs \$19.50. She has \$45 to spend and will spend the leftover amount on cupcakes, which are \$8.50 for a box of 4. How many boxes of cupcakes can Robin purchase?

a. What is the unknown value in this problem? Let the variable c represent this unknown value.

b. Write an equation to represent this situation.

c. Solve the equation for the variable.

d. What does the answer to part c. mean? Write a complete sentence.

39. For his Superbowl party, John bought 3 large pizzas: a pepperoni, a sausage and mushroom, and a Hawaiian pizza with ham and pineapple. Each pizza was cut into the same number of slices. After the party was over, there was $\frac{1}{4}$ of the pepperoni pizza left, $\frac{1}{2}$ of the sausage and mushroom pizza left, and $\frac{3}{8}$ of the Hawaiian pizza left. There were a total of 9 pieces of pizza leftover. How many slices was each pizza cut into?

a. What is the unknown value in this problem? Let the variable p represent this unknown value.

b. Write an equation to represent this situation.

c. Solve the equation for the variable.

d. What does the answer to part c. mean? Write a complete sentence.

40. A mathematics student bought a calculator and a textbook for a course in statistics. If the textbook costs \$67.51 more than the calculator, and the total cost for both was \$329.49, what was the cost of each item?

41. The total cost of a computer flash drive and an all-in-one printer was \$96.94, including tax. If the cost of the flash drive was \$58.96 less than the printer, what was the cost of each item?

42. A real estate agent says that the current value of a 25-year old home is \$90,000 more than twice its value when it was new. If the current value is \$310,000, what was the value of the home when it was new?

43. On average, the number of electric guitars sold in Texas each year is 91,399, which is seven times the average number of guitars sold each year in Wyoming. How many electric guitars, on average, are sold each year in Wyoming?

44. A classic car is now selling for \$1500 more than three times its original price. If the selling price is now \$12,000, what was the car's original price?

45. On August 24, the Fernandez family received 19 pieces of mail, consisting of magazines, bills, letters, and ads. If they received the same number of magazines as letters, three more bills than letters, and five more ads than bills, how many magazines did they receive?

46. A pet boarding facility cared for a total of 542 dogs and cats in a given year. If the facility cared for 56 more dogs than cats, how many cats did the facility care for that year?

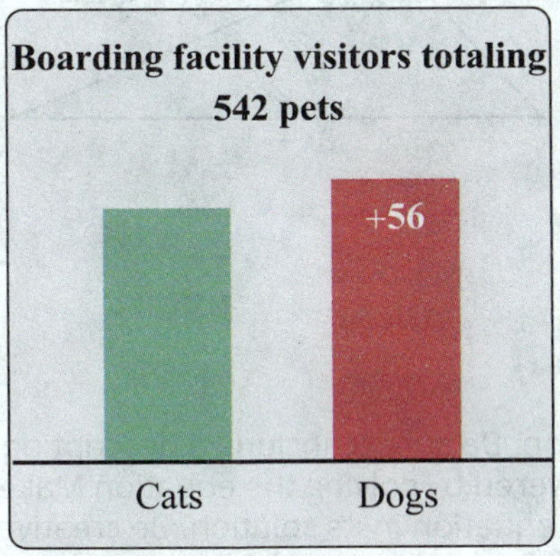

47. Lucinda bought two buckets of golf balls for the driving range. She gave the pro-shop clerk a 50-dollar bill and received \$10.50 in change. What was the cost of one bucket of golf balls? (Tax was included.)

48. A guitar manufacturer spent \$158 million on the production of acoustic and electric guitars last year. If the amount the company spent producing acoustic guitars was \$68 million more than it spent on producing electric guitars, how much did the company spend producing electric guitars?

49. The cost to rent a ballroom at a convention center for one day is \$800 for the first 2 hours. Every additional hour that the ballroom is in use costs an additional \$50 per hour, which is used to cover security fees. If you owe \$1450 for a one-day rental of the ballroom, how many hours did you use the ballroom?

50. The cost to rent a party room at an arcade is a fixed price per hour plus \$15 per child. If a 3-hour room rental with 20 children costs \$330, what is the fixed price per hour for the room rental?

51. A-to-Z Truck Rentals charges \$19.99 per day plus 65¢ per mile driven to rent a pick-up truck. For a one day trip, Louis paid a rental fee of \$127.24. How many miles did he drive?

43. $7x = 91{,}399$; 13,057

44. $3x + 1500 = 12{,}000$; \$3500

45. $x + x + (x+3) + (x+3+5) = 19$; 2 magazines

46. $(x+56) + x = 542$; 243 cats

47. $50 - 2x = 10.50$; \$19.75

48. $(x+68) + x = 158$; \$45 million on electric guitars

49. $800 + 50(x-2) = 1450$; 15 hours

50. $3x + 15(20) = 390$; \$30

51. $19.99 + 0.65x = 127.24$; 165 miles

52. $x + x + 2 + x + 6 = 29$; 7 feet, 9 feet, 13 feet

53. $n + 4n - 4 + 2n + 7 = 59$; 8 cm, 28 cm, 23 cm

54. $x + x - 50{,}000 = 500{,}000$; Lot: \$275,000, House: \$225,000

55. $x + 3x - 1 + 2x + 5 = 64$; 10 inches, 29 inches, 25 inches

Note: Answers for 56 through 61 will vary.

56. The difference between five times a number and the number is equal to 8; $x = 2$

57. Twice a number increased by 3 equals 9; $x = 3$

58. Find two consecutive integers whose sum is 33; $n = 16, 17$

59. Find 3 consecutive even integers such that the sum of the first and the third is 3 times the second; $n = -2, 0, 2$

60. Find two consecutive integers such that 3 times the second is 53 more than the first; $n = 25, 26$

61. Find a number such that twice the sum of the number and 2, decreased by 6 is equal to the sum of the number and 4 minus the number; $n = 3$

Writing & Thinking

62. **a.** $n, n+2, n+4, n+6$

b. $n, n+2, n+4, n+6$

c. Yes. Answers will vary.

52. A 29-foot board is cut into three pieces at a sawmill. The second piece is 2 feet longer than the first and the third piece is 4 feet longer than the second. What are the lengths of the three pieces?

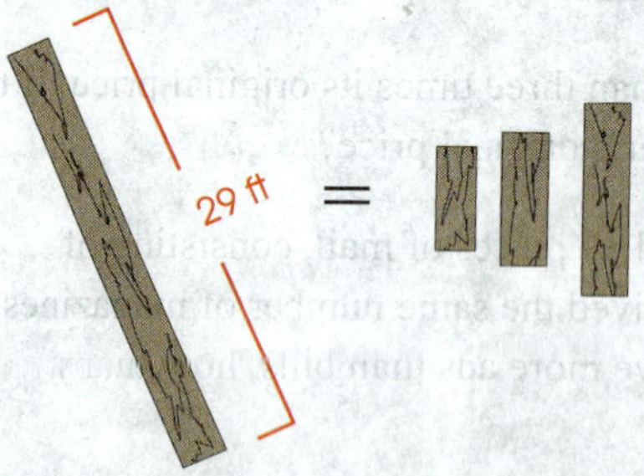

53. The three sides of a triangle are n, $4n - 4$, $2n + 7$ (as shown in the figure). If the perimeter of the triangle is 59 cm, what is the length of each side? (**Reminder:** The perimeter of a triangle is the sum of the lengths of the sides.)

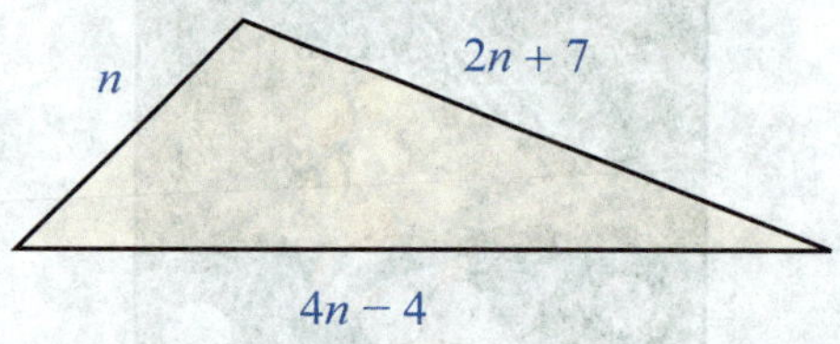

54. Joe Johnson decided to buy a lot and build a house on the lot. He knew that the cost of constructing the house was going to be \$50,000 less than the cost of the lot. He told a friend that the total cost was going to be \$500,000. As a test to see if his friend remembered the algebra they had together in school, he challenged his friend to calculate what he paid for the lot and what he was going to pay for the house. What was the cost of the lot and the cost of the house?

55. The three sides of a triangle are x, $3x - 1$, and $2x + 5$ (as shown in the figure). If the perimeter of the triangle is 64 inches, what is the length of each side?

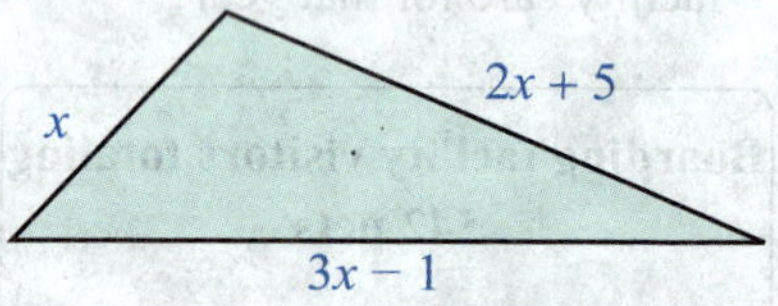

Write a word problem which leads to each equation. Be sure to include a description of the situation and a question which would be answered by solving the equation Make up your own word problem that might use the given equation in its solution. Be creative! Then solve the equation and check to see that the answer is reasonable.

56. $5x - x = 8$

57. $2x + 3 = 9$

58. $n + (n+1) = 33$

59. $n + (n+4) = 3(n+2)$

60. $3(n+1) = n + 53$

61. $2(n+2) - 6 = n + 4 - n$

Writing & Thinking

62. **a.** How would you represent four consecutive odd integers?

b. How would you represent four consecutive even integers?

c. Are these representations the same? Explain.

9.6 Applications: Distance-Rate-Time, Interest, Average, and Cost

Objectives

A. Use linear equations to solve distance-rate-time problems.

B. Use linear equations to solve simple interest problems.

C. Use linear equations to solve average problems.

D. Use linear equations to solve cost problems.

A Distance-Rate-Time

Word problems (or applications) are designed to teach you to read carefully, to organize information, and to think clearly. Whether a particular problem is easy for you depends a great deal on your personal experiences and general reasoning abilities. The problems generally do not give specific directions to add, subtract, multiply, or divide. You must decide what relationships are indicated through careful analysis of the problem.

Problems involving distance usually make use of the relationship indicated by the formula $d = rt$, where d = distance, r = rate, and t = time. A chart or table showing the known and unknown values is quite helpful and is illustrated in the next example.

Example 1 Application: Solving Distance-Rate-Time Problems

A brother and sister leave a family reunion at the same time and drive their cars in opposite directions. The brother's speed is 50 mph and the sister's speed is 65 mph. When will they be 460 miles apart?

Solution

Let t = travel time.

	Rate	•	Time	=	Distance
Brother's trip	50	·	t	=	$50t$
Sister's trip	65	·	t	=	$65t$

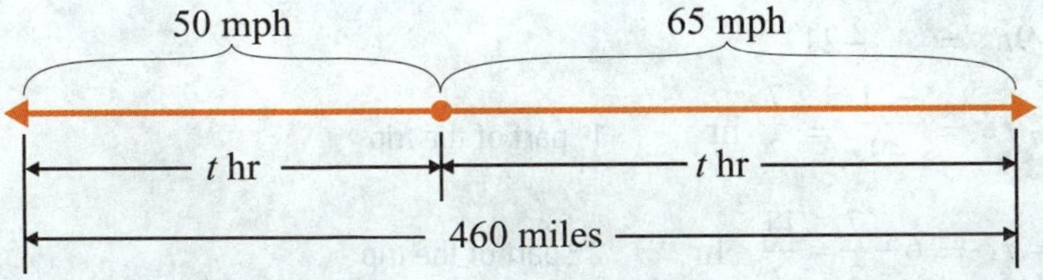

Brother's distance + Sister's distance = Total Distance — Form an equation relating the given information.

$$50t + 65t = 460$$ Solve the equation.

$$115t = 460$$

$$t = 4$$ Travel time

The brother and sister will be 460 miles apart in 4 hours.

Check

$50(4) = 200$ miles (1st car)

$65(4) = 260$ miles (2nd car)

$200 + 260 = 460$ miles in total

Now work margin exercise 1.

1. A husband and wife leave their home at the same time and drive in opposite directions. The husband's speed is 55 mph and the wife's speed is 60 mph. When will they be 230 miles apart?

2. Liang drove 200 miles in 4 hours. His average speed for the first part of the trip was 60 mph and his average speed for the second part of the trip was 45 mph. How long did each part of the trip take him?

Example 2 Application: Solving Distance-Rate-Time Problems

A motorist averaged 45 mph for the first part of a trip and 54 mph for the last part of the trip. If the total trip of 303 miles took 6 hours, what was the time for each part?

Solution

We can think of the equation as the total time minus the time for the 1st part of the trip is equal to the time for the 2nd part.

Let t = time for 1st part of trip
$6 - t$ = time for 2nd part of trip

	Rate	•	Time	=	Distance
1st part	45	·	t	=	$45t$
2nd part	54	·	$6-t$	=	$54(6-t)$

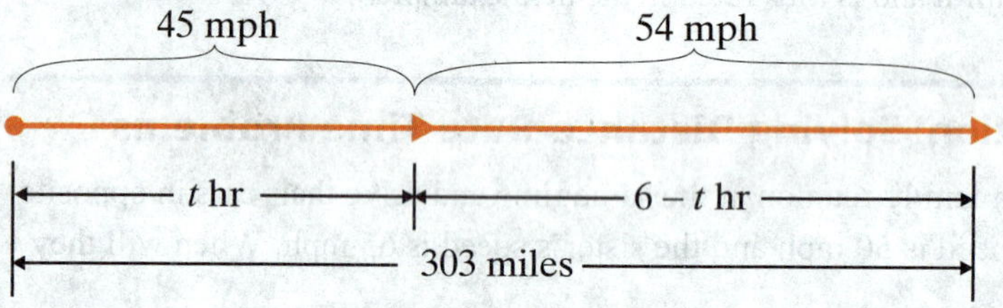

$$\underbrace{\text{1st part distance}} + \underbrace{\text{2nd part distance}} = \underbrace{\text{Total distance}}$$ Form an equation relating the given information.

$$45t + 54(6-t) = 303$$ Solve the equation.

$$45t + 324 - 54t = 303$$

$$324 - 9t = 303$$

$$-9t = -21$$

$$t = \frac{-21}{-9} = \frac{7}{3} \text{ hr}$$ 1st part of the trip

$$6 - t = 6 - \frac{7}{3} = \frac{11}{3} \text{ hr}$$ 2nd part of the trip

The first part took $\frac{7}{3}$ hr or $2\frac{1}{3}$ hr.

The second part took $\frac{11}{3}$ hr or $3\frac{2}{3}$ hr.

Check

$45 \cdot \frac{7}{3} = 15 \cdot 7 = 105$ miles (1st part)

$54 \cdot \frac{11}{3} = 18 \cdot 11 = 198$ miles (2nd part)

$105 + 198 = 303$ miles in total

Now work margin exercise 2.

B Simple Interest

To solve problems related to interest on money invested for one year, you need to know the basic relationship among the principal P (amount invested), the annual rate of interest r, and the amount of interest I (money earned). This relationship is described in the formula $Pr = I$. (This is the formula for simple interest, $I = Prt$, with $t = 1$.) We use this relationship in Example 3.

Example 3 Application: Solving Interest Problems

Kara has had $40,000 invested for one year, some in a savings account which paid 7% and the rest in a high-risk stock which yielded 12% for the year. If her interest income last year was $3550, how much did she have in the savings account and how much did she invest in the stock?

Solution

Let $x =$ amount invested at 7%

$40,000 - x =$ amount invested at 12%   Total amount invested minus amount invested at 7% represents the amount invested at 12%.

	Principal	•	Rate	=	Interest
Savings Account	x	·	0.07	=	$0.07(x)$
Stock	$40,000 - x$	·	0.12	=	$0.12(40,000 - x)$

Interest at 7% + Interest at 12% = Total interest

$$\begin{aligned} 0.07(x) + 0.12(40,000 - x) &= 3550 \\ 7x + 12(40,000 - x) &= 355,000 \\ 7x + 480,000 - 12x &= 355,000 \\ -5x &= -125,000 \\ x &= 25,000 \\ 40,000 - x &= 15,000 \end{aligned}$$

Multiply both sides of the equation by 100 to eliminate the decimal numbers.

Amount invested at 7%

Amount invested at 12%

Kara had $25,000 in the savings account at 7% interest and $15,000 in the high-risk stock at 12% interest.

Check

$25,000(0.07) = 1750$ and $15,000(0.12) = 1800$

and $\$1750 + \$1800 = \$3550$

Now work margin exercise 3.

3. Mark has had $15,000 invested for 1 year, some in a low-risk stock yielding 4% and the rest in a high-risk stock yielding 15%. If his investment has earned $1997 in interest, how much did he invest in the low-risk stock and how much in the high-risk stock?

C Average (or Mean)

The **average** (or mean) of a set of numbers is the sum of the numbers divided by the number of numbers in the set. In this section, we use the concept of average to find unknown numbers.

4. Suppose you earned the following four scores on the first four biology tests of the semester, 90, 95, 97, and 90, respectively. You can exempt the final if your average on the five tests is 85 or higher. What is the minimum score you can earn on the fifth test and still exempt the final?

Example 4 Application: Finding The Average (or Mean)

Suppose that you have scores of 85, 92, 82, and 88 on four exams in your English class. What score will you need on the fifth exam to have an average of 90?

Solution

Let x = your score on the fifth exam.

The sum of all the scores, including the unknown fifth exam, divided by 5 must equal 90.

$$\frac{85+92+82+88+x}{5}=90$$

$$\frac{347+x}{5}=90$$

$$5\cdot\frac{347+x}{5}=5\cdot 90$$

$$347+x=450$$

$$x=103$$

Assuming that each exam is worth 100 points, you cannot attain an average of 90 on the five exams.

Now work margin exercise 4.

5. Using the graph in Example 5, find the difference in enrollment between the largest university and the smallest university listed.

Example 5 Application: Using Bar Graphs

The bar graph shows the undergraduate enrollment at the main campuses at six Big Ten universities during a certain year. Use the graph to find the following. (Note that the units on the graph are in thousands.)

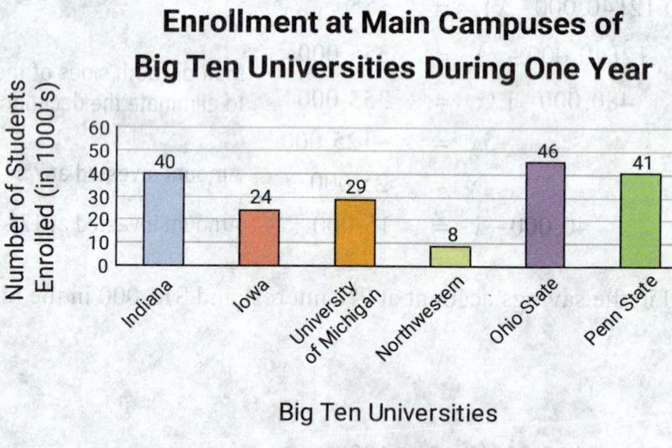

a. Find the average enrollment over the six schools. (Round to the nearest thousand.)

b. Which university had the lowest enrollment?

c. Find the difference in enrollment between Ohio State and Penn State.

Solution

a. Find the sum: $40+24+29+8+46+41=188.$

Divide by 6: $188\div 6\approx 31.$

The average enrollment was approximately 31,000 students.

b. Northwestern had the lowest enrollment with 8000 students.

c. The difference was 46,000 – 41,000 = 5000 students.

Now work margin exercise 5.

D Cost

Example 6 Application: Calculating Cost

A jeweler paid \$350 for a ring. He wants to price the ring for sale so that he can give a 30% discount on the marked selling price and still make a profit of 20% on his cost. What should be the marked selling price of the ring?

Solution

We make use of the relationship

$S - C = P$ (Selling price – Cost = Profit).

Let x = marked selling price.
Then $x - 0.30x$ = actual selling price
and 350 = cost.

$$\underbrace{\text{Actual selling price}} - \underbrace{\text{Cost}} = \underbrace{\text{Profit}}$$

$$\begin{aligned} x - 0.30x - 350 &= 0.20(350) \\ 0.70x - 350 &= 70 \\ 0.70x &= 420 \\ x &= 600 \end{aligned}$$

The actual selling price is the marked selling price minus the 30% discount on the ring.

The profit is 20% of what he paid originally.

The jeweler should set the selling price at \$600.

Check

Step 1:

\$600	marked selling price
× 0.30	discount %
\$180	discount

Step 2:

\$600	marked selling price
– \$180	discount
\$420	actual selling price

Step 3:

\$420	actual selling price
– \$350	cost
\$70	profit

As a double check,

\$350	cost
× 0.20	profit %
\$70	profit

Now work margin exercise 6.

6. A used car salesman paid \$1500 for a car. He plans to sell the car at a 25% discount on the marked selling price, but still wants to make a profit of 15% on his cost. What should be the marked selling price of the car?

Margin Exercise Answers

1. 2 hours **2.** First part took $\frac{4}{3}$ or $1\frac{1}{3}$ hours; Second part took $\frac{8}{3}$ or $2\frac{2}{3}$ hours. **3.** \$2300 in the low-risk stock; \$12,700 in the high-risk stock. **4.** 53 **5.** 38,000 students **6.** \$2300

9.6 Exercises

Concept Check

Fill-in-the-Blank. Complete each sentence using information found in this section.

1. To calculate distance, find the product of the _____ and _____.
2. In the formula $Pr = I$, I represents _____, P represents _______, and r represents _______.
3. In a simple interest problem, if r is 0.12, this represents an interest rate of _______.
4. Given cost x, a discount of ____% can be represented as $0.25x$.
5. In the formula, $Pr = I$, the time period is ____ year(s).
6. When solving distance-rate-time problems, a/an ______ or _______ showing the known and unknown values is helpful.

True/False. Determine whether each statement is true or false. If a statement is false, explain how it can be changed so the statement will be true. (**Note:** There may be more than one acceptable change.)

7. When using the formula $I = Pr$, the value of r should be written as a percent.
8. In the distance-rate-time formula $d = r \cdot t$, the value t stands for the time spent traveling.
9. Profit can be determined by subtracting the cost from the selling price.
10. The concept of average can be used to find unknown numbers.

Applications

Solve.

1. Dana finds the perfect dress for the Freshman Dance on sale at Belk. If she paid \$95.96 for the dress on sale (before tax) and the dress was marked 20% off, find the original price of the dress.
   a. Is the 20% marked off the price she paid, or the original price?
   b. How do you represent the original price; the amount of the discount?
   c. Set up an equation and solve for the original price of the dress.

9.6 Exercises

Concept Check

1. rate, time
2. interest, principal, rate
3. 12%
4. 25
5. one
6. chart, table
7. False; The value of r should be written as a decimal number.
8. True
9. True
10. True

Applications

1. a. Original price
b. x; $0.20x$ **c.** \$119.95

2. Tommy inherits $5000 from his grandmother. He decides to invest some of the money in a CD paying a 4% return and the rest in a money market account (MMA) paying a 3% return. How much did he invest in each if the total amount of return was $187.50?

   a. How do you represent the money invested in the CD? The money invested in the MMA?

   b. How do you represent the total amount of return from the CD; from the MMA?

   c. Set up an equation by equating the values of the total amount of the return.

   d. Solve the equation and answer the question.

3. Mary Kate is preparing the Chemistry lab for an experiment that calls for a solution containing 20% ethyl alcohol. However, she only has 10% solution and 25% solution.

   a. How much alcohol is there in 20 liters of 10% alcohol solution?

   b. Write an algebraic expression for the amount of alcohol contained in x liters of the 25% solution.

   c. Write an algebraic expression for the total amount of solution created when x liters of the 25% solution is added to 20 liters of the 10% solution.

   d. Take 20% of the amount in part c. and set it equal to the total from parts a. and b. to equate the amount of alcohol in the mixed solution to a 20% solution.

   e. Solve the equation found in part d. to determine how much 25% solution Mary Kate used to make a 20% solution.

4. The local grocery store offers two varieties of salami at the deli counter. The price per pound of the second, more expensive, salami is $3 more than twice the price per pound of the first variety. If Johannes buys 2 pounds of each type of salami, and his total price is $29.94, find the price per pound for each variety of salami.

   a. How do you represent the price of the first variety of salami; the second variety?

   b. How do you represent the amount spent on the first variety; the second variety?

   c. Set up an equation and find the price per pound of the first variety.

   d. What is the price per pound of the second variety? Verify the two prices are correct.

5. Brian has a collection of dimes and quarters in his pocket. He has twice as many dimes as quarters, and the total value of the coins is $2.70. How many of each type of coin does he have?

   a. Solve this problem by defining a variable, writing an equation, and solving that equation.

   b. Explain how you can check your answer without using your equation from part a.

2. a. x; $5000 - x$
   b. $0.04x$; $0.03(5000-x)$
   c. $0.04x+0.03(5000-x)=187.50$
   d. $3750 in the CD and $1250 in the MMA

3. a. 2 liters b. $0.25x$
   c. $x + 20$
   d. $0.20(x+20)=2+0.25x$
   e. 40 liters

4. a. x; $2x + 3$
   b. $2x$; $2(2x+3)$
   c. $2x+2(x+3)=29.94$; $3.99
   d. $10.98

5. a. x is the number of quarters in Brian's pocket; $0.10(2q)+0.25q=2.70$; $q=6$, so Brian has 6 quarters and 12 dimes.
   b. Check against the language of the problem: 12 is twice as much as 6, and 6 quarters ($1.50) and 12 dimes ($1.20) add up to $2.70.

6. a. Answers will vary.

Cyclist	Rate	Time	Distance
Chantelle	x	0.5	$0.5x$
Taylor	$x+6$	0.5	$0.5(x+6)$

b. $0.5x+0.5(x+6)=19$

c. $x=16$;
Chantelle's rate: 16 mph;
Taylor's rate: 22 mph

7. 1.68 mph
8. 18 mph
9. 8.75 hours
10. 4.75 hours
11. 3 hours
12. 375 mph; 450 mph
13. 50 mph; 300 miles
14. 4.8 hours
15. 7 hours
16. 54 mph
17. 36 mph; 60 mph
18. 12 miles

6. Two bicyclists, Chantelle and Taylor, start from opposite ends of a 19-mile-long bike path. Taylor rides her bike 6 mph faster than Chantelle, and the cyclists meet in 30 minutes. How fast was each of them riding? Draw a picture to represent the described problem.
   a. Use a table to organize the information given in the problem.
   b. Write an equation based on your diagram and/or table.
   c. Solve the equation and relate your answer to the original problem.
7. What is Nathan's average rate of speed if he hikes 12.6 miles in 7.5 hours?
8. What is Amy's average rate of speed if she bikes 39.6 miles in 2.2 hours?
9. Jamie plans to take the scenic route from Los Angeles to San Francisco. Her GPS tells her it is a 420-mile trip. If she figures her average speed will be 48 mph, how long will the trip take her?
10. Scott's average speed on his drive from Memphis, TN, is 60 mph. If the total trip is 285 miles, how long should he expect the drive to take?
11. Jane rides her bike to Lake Junaluska. Going to the lake, her average speed is 12 mph. On the return trip, her average speed is 10 mph. If the round trip takes a total of 5.5 hours, how long does the return trip take?
12. Two planes which are 2475 miles apart fly toward each other. Their speeds differ by 75 mph. If they pass each other in 3 hours, what is the speed of each?

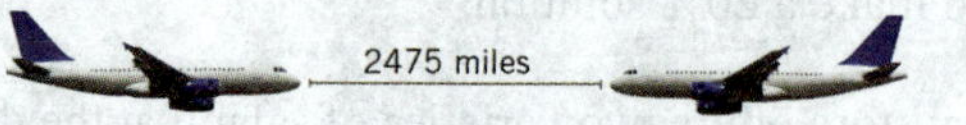

13. Marcus drives from Chicago to Detroit in 6 hours. On the return trip, his speed is increased by 10 mph and the trip takes 5 hours. Find his rate on the return trip. How far apart are the towns?
14. Tim and Barb have 8 hours to spend on a mountain hike. They can walk up the trail at an average rate of 2 mph and can walk down at an average rate of 3 mph. How long should they plan to hike uphill before turning around?
15. The Reeds are moving across Texas. Mr. Reed leaves $3\frac{1}{2}$ hours before Mrs. Reed. If his average speed is 40 mph and her average speed is 60 mph, how long will Mrs. Reed have to drive before she overtakes Mr. Reed?
16. After traveling for 40 minutes, Mr. Koole had to slow to $\frac{2}{3}$ his original speed for the rest of the trip due to heavy traffic. The total trip of 84 miles took 2 hours. Find his original speed.
17. A train leaves Cincinnati at 2:00 p.m. A second train leaves the same station in the same direction at 4:00 p.m. The second train travels 24 mph faster than the first. If the second train overtakes the first at 7:00 p.m., what is the speed of each of the two trains?
18. Maria runs through the countryside at a rate of 10 mph. She returns along the same route at 6 mph. If the total trip took 1 hour 36 minutes, how far did she run in total?

19. The distance from Atlanta, Georgia, to Washington, DC, is 620 miles. Driving in the middle of the night, it takes about 9 hours to get to Washington. Due to higher traffic volume, it takes 2 more hours to travel there during the day. What is the average rate of the driver during the day and during the night? (Round to the nearest whole number.)

20. Mr. Kent drove to a conference. The first half of the trip took 3 hours due to traffic. Traffic let up for the second half of the trip, and he was able to increase his speed by 20 mph to make sure he got there on time. Find his rates of speed if he traveled 2 hours at the second rate.

21. Jayden walked to his friend's house at a rate of 4 mph to borrow his friend's bicycle. Coming back home, he rode the bicycle at an average rate of 12 mph. The total time for the round trip was 1 hour 30 minutes. How far away does Jayden's friend live?

22. Once a week, Felicia walks/runs for a total of 6 miles. Felicia spends twice as much time walking as she does running. If she walks at a rate of 4 mph and runs three times faster than she walks, what is the time for each part?

23. Achilles is racing a tortoise and gives him a 2-hour head start. The tortoise runs at a pace of 10 miles per hour and Achilles runs at a pace of 25 miles per hour. How long will it take Achilles to catch up to the tortoise?

a. Fill out the $d = r \cdot t$ table. Let the variable t represent the amount of time that the tortoise has traveled.

	Rate (mph)	·	Time (min)	=	Distance (miles)
Tortoise					
Achilles					

b. When Achilles catches up to the tortoise, they will have traveled the same distance. Set up a linear equation using the information in the table.

c. Solve the equation from part b. for the variable.

d. How long will it take Achilles to catch up to the tortoise?

e. If the race is 35 miles long, will Achilles pass the tortoise before crossing the finish line? Show work to support your answer.

24. Amanda invests $25,000, part at 5% and the rest at 6%. The annual return on the 5% investment exceeds the annual return on the 6% investment by $40. How much did she invest at each rate?

25. Mr. Hill invests $10,000, part at 5.5% and part at 6%. The annual interest from the 5.5% investment exceeds the annual interest from the 6% investment by $251. How much did he invest at each rate?

26. The annual interest earned on a $6000 investment was $120 less than the interest earned on $10,000 invested at 1% less interest per year. What was the rate of interest on each amount?

27. Two investments totaling $16,000 produce an annual income of $1140. One investment yields 6% a year, while the other yields 8% per year. How much is invested at each rate?

19. Day: 56 mph; Night: 69 mph

20. 1st half: 40 mph; 2nd half: 60 mph

21. 4.5 miles

22. Walks for 0.6 hours or 36 minutes; runs for 0.3 hours or 18 minutes

23. a.

	Rate (mph)	·	Time (min)	=	Distance (miles)
Tortoise	10		t		$10t$
Achilles	25		$t-2$		$25(t-2)$

b. $10t = 25(t-2)$

c. $3\frac{1}{3}$ hours

d. $1\frac{1}{3}$ hours

e. Yes

24. $14,000 at 5%; $11,000 at 6%

25. $7400 at 5.5%; $2600 at 6%

26. 4.5% on $10,000; 5.5% on $6000

27. $7000 at 6%; $9000 at 8%

28. 6.5% on $4000; 6% on $3000

29. $5500 at 8%; $6500 at 10%

30. $24,000 at 4.5%; $18,000 at 6%

31. a.

	Principal ($)	· Rate =	Interest ($)
High-Risk Fund	P	0.08	$0.08P$
Low-Risk Fund	$3600 - P$	0.04	$0.04(3600 - P)$

b. $0.08P + 0.04(3600 - P) = 198$

c. $P = \$1350$

d. High-risk fund: $1350; Low-risk fund: $2250

e. Answer will vary.

32. $112.50

33. $6720 at 5.5%; $5280 at 7%

34. $5700

35. $600 at 2.5%; $800 at 4%

36. $12,000 at 1.5%; $6000 at 4%

28. The annual interest on a $4000 investment exceeds the interest earned on a $3000 investment by $80. The $4000 is invested at a 0.5% higher rate of interest than the $3000. What is the interest rate of each investment?

29. D'Andra makes two investments that total $12,000. One investment yields 8% per year and the other 10% per year. The total interest for one year is $1090. Find the amount invested at each rate.

30. A company is planning to invest $42,000 into two simple interest accounts. The annual interest rate on one of the accounts is 4.5% while the rate on the other is 6%. How much should the company invest in each account so that the two accounts will produce an equal annual interest income?

31. Savannah invests $3600 per year into her retirement account, a portion of which is a contribution match from her employer. Savannah invests the employer match in a high-risk fund that averages a return of 8% and invests the rest in a low-risk account that averages a return of 4%. She wants to earn a total of $198 in interest for the year. How much should be invested in each fund?

a. Fill out the $I = P \cdot r$ table. Let the variable P represent the amount of money invested in the high-risk fund.

	Principal ($)	·	Rate	=	Interest ($)
High-Risk Fund					
Low-Risk Fund					

b. Write an equation to represent the total interest earned for the year by using the information in the table.

c. Solve the equation from part b. for the variable.

d. How much should be invested in each fund?

e. Verify that investing the amounts from part d. at the given rates yields $198 total interest in Savannah's retirement account.

32. A particular style of shoe costs a shoe store $81 per pair. What should the selling price of the shoes be so a 10% discount results in a 25% profit?

33. Sebastian would like both of his investments, a total of $12,000, to bring him the same annual interest income. One of his investments is at 5.5% annual interest rate and the other at 7%. Find the amount of money that Sebastian should invest in each account.

34. A car dealer paid $3800 for a used car. For his upcoming Labor Day sale, he wants to offer a 10% discount off the posted selling price, but would still like to make a 35% profit. What price should he advertise for that car?

35. Gabriella got some money from her grandparents as a graduation present. She decided to invest all of it. Part of the money was invested at a 2.5% interest rate, and the rest at a 4% interest rate. She invested $200 more in the 4% account than the 2.5% account. If her annual interest income was $47, how much did she invest at each rate?

36. Jordan is earning 1.5% interest from money invested in a savings account and 4% interest on a mutual bond fund. If the total of his investments is $18,000 and the annual interest from the savings account is less than the annual interest from the bond by $60, how much has Jordan invested at each rate?

37. A small company invested $20,500 such that a part of the money is in an account with a 4% interest rate and the rest at a 5% rate. The annual interest from the 5% account is $35 more than the interest earned from the 4% account. Find the amount of money the company invested at each rate.

38. During the month of January, a department store would like to have a sale of 40% off of women's knee-high boots. The store purchased the boots for $33 per pair. How much should the selling price be, if the manager of the store wants to make a 10% profit per pair?

39. Carla plans on buying a new pair of sandals for the summer. They are on sale for 20% off of the original price. What was the original price of the sandals, if she pays $34.83, with 7.5% sales tax?

40. Last year, an individual invested some money at a 5% interest rate and $2200 less than that amount at a 6% interest rate. If his interest income was $880, how much did he invest at each rate?

41. A store purchased a certain style of leather jacket at $70 per jacket. If the store wants to sell the jackets at a 20% discount and still make a profit of 30%, what should be the marked selling price for each jacket?

42. After receiving a 10% off coupon in the mail, Mark decided that it was time to buy new headphones. Using the coupon and paying 8% sales tax, the final price came to $213.84. What was the listed price of the headphones?

43. Robin's Refurbished Wrecks purchased a used car for $2850. For the upcoming Labor Day sale, the car dealership would like to offer a 5% discount off the posted selling price of the car, but would still like to make a 40% profit. What price should the car dealership advertise for the car?

a. Use the purchase price of the car to determine how much a 40% profit will be.

b. Use the variable x to represent the actual selling price of the car. Write an expression to represent the selling price of the car after the 5% discount.

c. Write an equation the represents the situation by using the answers from parts a. and b. along with the equation *selling price – cost = profit*.

d. Solve the equation from part c.

e. What does the answer from part d. mean? Write a complete sentence.

44. Marissa has five exam scores of 75, 82, 90, 85, and 77 in her chemistry class. What score does she need on the final exam to have an average grade of 80 (and thus earn a grade of B)? (All exams have a maximum of 100 points.)

45. Gerald had scores of 80, 92, 89, and 95 on four exams in his algebra class. What score will he need on his fifth exam to have an overall average grade of 90? (All exams have a maximum of 100 points.)

46. While riding her bike to the park and back home five times, Stacey timed herself at 60 min, 62 min, 55 min (the wind was helping), 58 min, and 63 min. She had set a goal of having an average time of 60 minutes for her rides. How many minutes will she need on her sixth ride to attain her goal?

37. $11,000 at 4%; $9500 at 5%

38. $60.50

39. $40.50

40. $9200 at 5%; $7000 at 6%

41. $113.75

42. $220

43. **a.** $1140

b. $x - 0.05x$ or $0.95x$

c. $x - 0.05x - 2850 = 1140$ or $0.95x - 2850 = 1140$

d. $4200

e. Robin's Refurbished Wrecks should sell the used car for $4200.

44. 71

45. 94

46. 62 min

47. 4 phone calls
48. 6 hours
49. $20
50. 67
51. a. Not possible
b. 192
52. a. 53.8°F b. 44°F
c. −17°F

TV Watching Trend

In recent years, there has been a decline in the watching of network television. With the expansion of alternative programing options such as Netflix and Amazon Prime Video, more people are switching away from the traditional networks. This can also be seen in the nominations (and winners) of the television awards. Lately, the list of award nominees barely includes any of the traditional networks' shows or actors. The graph shows the decline in television viewing among millennials from 2014 through 2018.

Source: www.usatoday.com/story/life/tv/2018/11/12/cord-cutting-and-streaming-younger-viewers-cut-into-traditional-tv-viewing/1924404002

47. For every 4-week period, Lauren wants to make an average of 6 phone calls per week. The first week she made 9 phone calls; the second week she made 6 phone calls, and the third week 5 phone calls. How many phone calls does Lauren need to make in the fourth week to make sure she stays on track with her goal?

48. While growing up, Jason was allowed to watch TV an average of 3 hours a day over a one-week period. One particular week he watched 1 hour, 2 hours, 1 hour, 3 hours, 3 hours, and 5 hours. How many hours could Jason watch the seventh and last day of the week and still obey his parents?

49. A college student realized that he was spending too much money on video games. For the remaining 5 months of the year, his goal is to spend an average of $50 a month towards his hobby. How much can he spend in December, taking into consideration that the other 4 months he spent $70, $25, $105, $30, respectively?

50. Wade has scores of 59, 68, 76, 84, and 69 on the first five tests in his social studies class. He knows that the final exam counts as two tests. What score will he need on the final to have an average of 70? (All tests and exams have a maximum of 100 points.)

51. A statistics student has grades of 86, 91, 95, and 76 on four hour-long exams. What score must he receive on the final exam to have an average grade of 90 if

a. the final is equivalent to a single hour-long exam (100 points maximum)?

b. the final is equivalent to two hour-long exams (200 points maximum)?

52. Consider the monthly temperatures over a year for a city in New Zealand.

a. Find the average temperature for the year. (Round to the nearest tenth.)

b. Find the minimum average monthly temperature for the year.

c. Find the difference in average monthly temperature between the months of June and December.

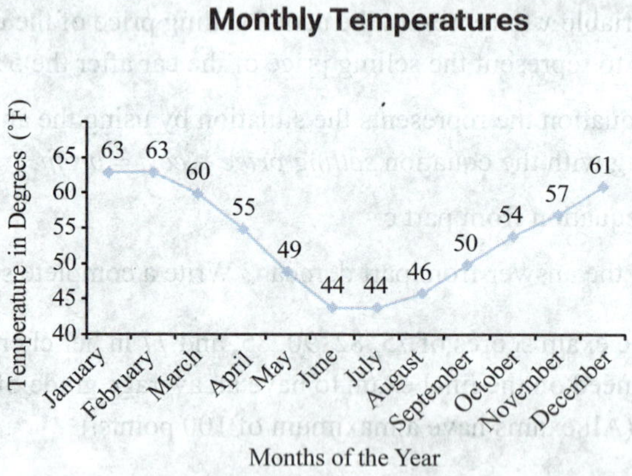

53. Consider the averages of monthly rainfall over a year for Visakhapatnam, India. [1]

a. Find the average rainfall for the year. (Round to the nearest tenth.)

b. Find the maximum average monthly rainfall for the year.

c. Find the difference in average monthly rainfall between the months of October and December.

Average Monthly Rainfall in Visakhapatnam, India

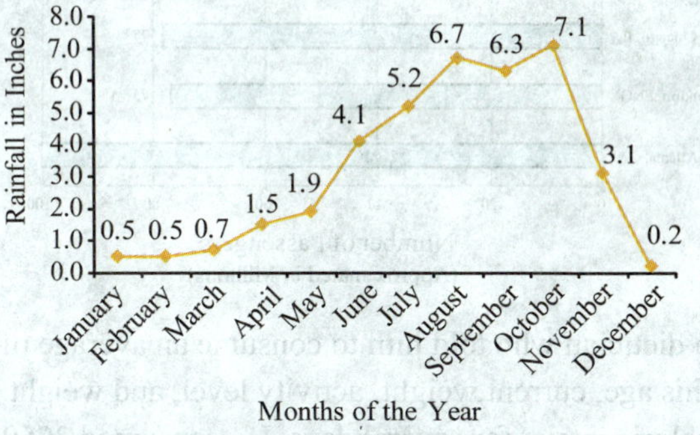

54. Consider the yearly profits of the world's largest companies in 2020. [2]

a. Find the average profits of the companies in the year 2020. (Round to the nearest tenth of a billion.)

b. Find the yearly profits of SoftBank Group in the year 2020.

c. What was the difference in profits between ICBC and Berkshire Hathaway?

Profits of World's Largest Companies in 2020

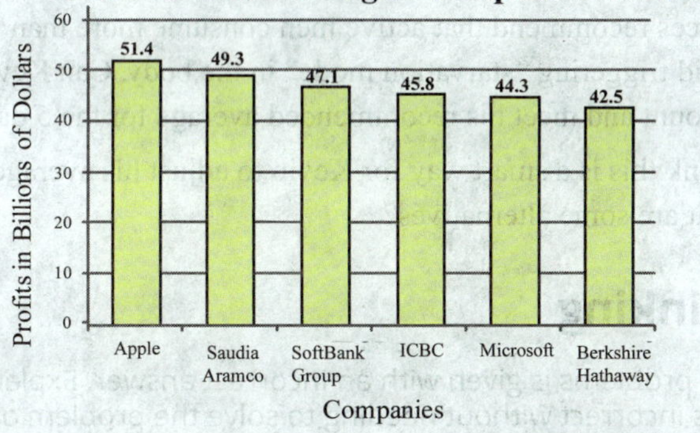

53. a. 3.2 inches
b. 7.1 inches
c. 6.9 inches

54. a. $46.7 billion
b. $47.1 billion
c. $3.3 billion

1 Source: www.weather.com

2 Source: "Leading companies in the world in 2020, by net income," Statistica, August 2021, www.statista.com/statistics/269857/most-profitable-companies-worldwide/.

55. a. 80.4 million
b. 26 million
c. 75 million

56. a. $\frac{2050 + 2200 + 2300 + 2400 + x}{5} = 2100$
b. $x = 1550$ calories
c. Yes
d. Answers will vary.

Writing & Thinking

57. If each equal side is 9 cm long, that would make the perimeter more than 18 cm; Correct answer: 6 cm

58. A $90 textbook marked down 50% would cost $45, not $60; Correct answer: $120

59. He would not be able to paddle faster than the current, so he would not be able to return upriver at all; Correct answer: $\frac{2}{3}$ mph

55. Consider the number of passengers at the following airports.[3]

a. Find the average number of passengers.

b. Find the difference in passengers between Atlanta, GA, and Tokyo.

c. What was the total number of passengers to go through London?

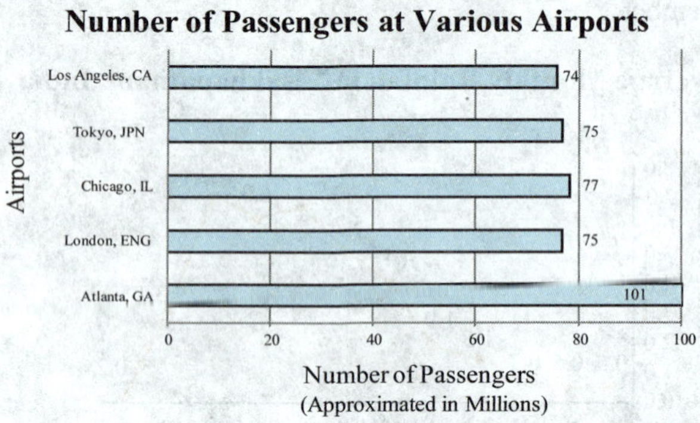

56. Kevin consulted a dietician who told him to consume an average of 2100 calories per day based on his age, current weight, activity level, and weight goals. Kevin kept track of his calorie intake for several days. He consumed 2050 calories on Monday, 2200 calories on Tuesday, 2300 calories on Wednesday, and 2400 calories on Thursday. How many calories would he need to consume on Friday to have an average calorie intake of 2100 for the five days?

a. Set up an equation to solve for the amount of calories Kevin would need to consume on Friday. Use the variable x to represent the number of calories needed.

b. Solve the equation from part a. for the variable.

c. Some sources recommend that active men consume more than 1500 calories per day to avoid triggering "starvation mode" in the body. Can Kevin stay above this calorie amount and meet his recommended average for the 5 days?

d. Do you think this is a smart way for Kevin to adjust his average calorie intake? If not, what are some alternatives?

Writing & Thinking

Each of the following problems is given with an incorrect answer. Explain how you can tell that the answer is incorrect without needing to solve the problem or do any algebra; then, solve the problem correctly.

57. The perimeter of an isosceles triangle is 16 cm. Since the triangle is isosceles, two sides have the same length; the third side is 2 cm shorter than one of the two equal sides. Find the length of one of the two equal sides. **Incorrect answer: 9 cm**

58. Leela found a used textbook, which was marked down 50% from the price of the new textbook. If the used textbook cost $60, how much did the new textbook cost? **Incorrect answer: $90**

59. Kareem can paddle his kayak at 6 mph in still water. He decides to go kayaking on the local river. He paddles downriver (with the current) for 2 hours; then he turns around and paddles upriver (against the current) for 2.5 hours, returning to his starting point. How fast is the current in the river? **Incorrect answer: 27 mph**

3 Source: Airports Council International – North America

9.7 Solving Linear Inequalities in One Variable

Objectives

A. Graph intervals of real numbers.

B. Solve linear inequalities of the form $ax < c$ and $x + b < c$.

C. Solve linear inequalities of the form $ax + b < c$ and $ax + b < cx + d$.

D. Use linear inequalities to solve applications.

A Intervals of Real Numbers

Suppose that a and b are two real numbers and that $a < b$. The set of all real numbers between a and b is called an **interval of real numbers**. Various types of intervals and their corresponding interval notations are listed in Table 1. In this notation, brackets [and] indicate an endpoint is included and parentheses (and) indicate an endpoint is not included.

Type of Interval	Algebraic Notation	Interval Notation	Graph
Open Interval	$a < x < b$	(a, b)	
Closed Interval	$a \le x \le b$	$[a, b]$	
Half-Open Interval	$a \le x < b$ $a < x \le b$	$[a, b)$ $(a, b]$	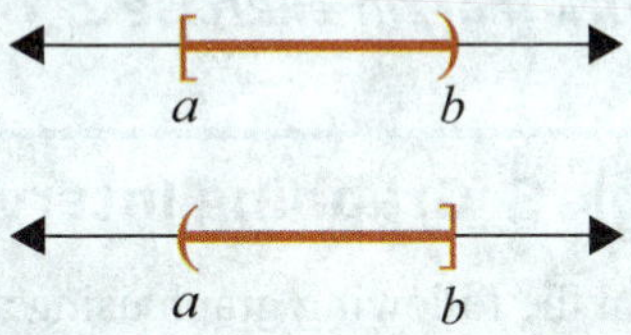
Open Interval	$x > a$ $x < b$	(a, ∞) $(-\infty, b)$	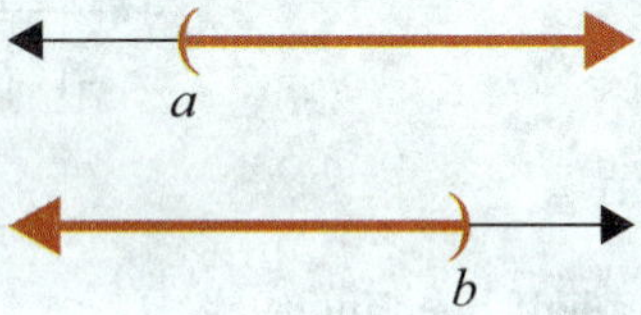
Half-Open Interval	$x \ge a$ $x \le b$	$[a, \infty)$ $(-\infty, b]$	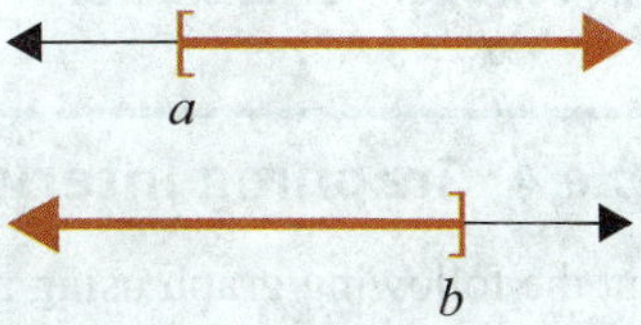

Table 1

Teaching Note

The alternate notation of open and closed circles may be more familiar to your students, however we chose to graph using parentheses and brackets in order to correspond to interval notation. Also, this notation is common in higher level mathematics.

Note

The symbol for infinity ∞ (or $-\infty$) does not represent a number. The symbol is used to indicate that the interval is to include all real numbers from some point on (either in the positive direction or the negative direction) without end.

The three types of intervals are defined as follows.

- In an **open interval**, neither endpoint is included.
- In a **closed interval**, both endpoints are included.
- In a **half-open interval**, only one endpoint is included.

1. Graph the half-open interval $(-\infty, 6]$.

Example 1 Graphing Intervals

Graph the open interval $(3, \infty)$.

Solution

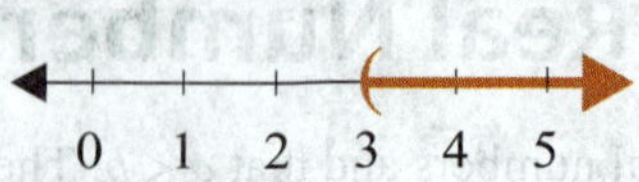

Now work margin exercise 1.

2. Graph the closed interval $-1 \le x \le 1$.

Example 2 Graphing Intervals

Graph the half-open interval $0 < x \le 4$.

Solution

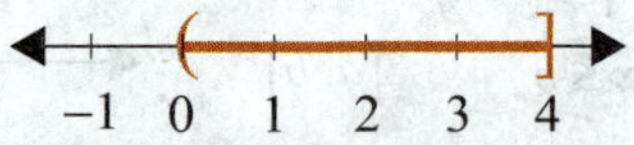

Now work margin exercise 2.

3. Represent the following graph using algebraic notation, and state what kind of interval it is.

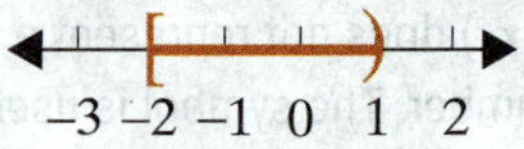

Example 3 Graphing Intervals

Represent the following graph using algebraic notation, and state what kind of interval it is.

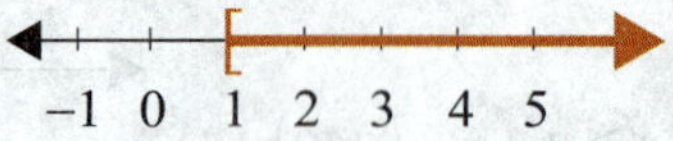

Solution

$x \ge 1$ is a half-open interval.

Now work margin exercise 3.

4. Represent the following graph using interval notation, and state what kind of interval it is.

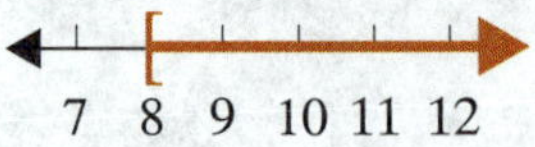

Example 4 Graphing Intervals

Represent the following graph using interval notation, and state what kind of interval it is.

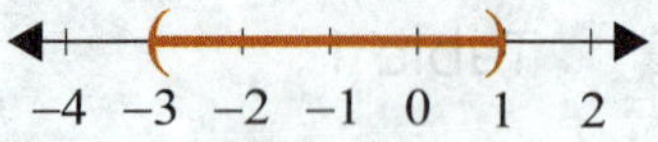

Solution

$(-3, 1)$ is an open interval.

Now work margin exercise 4.

B Solving Linear Inequalities: $ax < c$ and $x + b < c$

Linear inequalities are inequalities that relate two linear expressions. For example,

$$x < 3, \quad y + 3.2 \geq 5.6, \quad 2t \leq \frac{5}{16}, \quad 5x - 8 < 7, \quad \text{and} \quad 3x - 14 > 2x + 1$$

are linear inequalities. A **solution** to an inequality is any number that gives a true statement when substituted for the variable. Note that, unless otherwise stated, the variables represent real numbers.

For example, 0 is a solution to the inequality $x < 3$ since $0 < 3$ is a true statement. The number 4 is not a solution since $4 < 3$ is a false statement.

The set of all solutions to an inequality is called the **solution set**. Because the solutions to inequalities generally involve an infinite set of real numbers, we illustrate the solution set with a graph on the number line. In this case, the solution set for the inequality $x < 3$ is the open interval shown here.

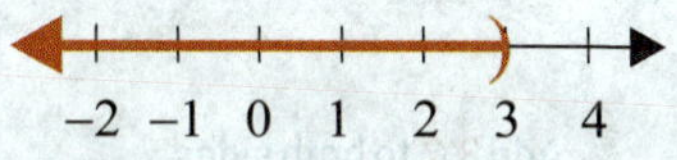

To solve an inequality such as $x + 3 > 5$, use the following steps.

$$x + 3 > 5$$

$$x + 3 - 3 > 5 - 3 \qquad \text{Add } -3 \text{ to both sides.}$$

$$x > 2 \qquad \text{Simplify.}$$

The solution set can be graphed as shown.

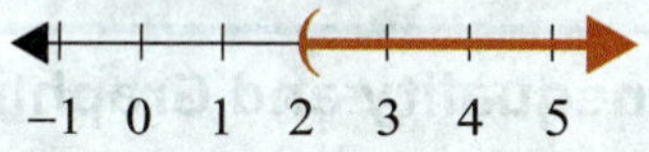

(Notice how similar this process is to solving equations.)

Two inequalities that have the same solution set are said to be **equivalent**. The following **addition principle** gives the needed tool to solve inequalities in the form $x + b < c$.

> **Attention!**
>
> Any principle or rule stated for inequalities with the < symbol holds true for ≤, >, and ≥ as well.

Addition Principle for Solving Linear Inequalities

If A and B are algebraic expressions and C is a real number, then the inequalities

$$A < B$$

and

$$A + C < B + C$$

are equivalent.

(If a real number is added to both sides of an inequality, the new inequality is equivalent to the original.)

PROPERTIES

5. Solve the inequality $y - 1.5 < 3.7$ and graph the solution set.

Example 5 Solving an Inequality and Graphing the Solution Set

Solve the inequality $y - 3.2 \geq 5.6$ and graph the solution set. Write the solution set using interval notation.

Solution

$$y - 3.2 \geq 5.6$$

$$y - 3.2 + 3.2 \geq 5.6 + 3.2 \qquad \text{Add } 3.2 \text{ to both sides.}$$

$$y \geq 8.8 \qquad \text{Simplify.}$$

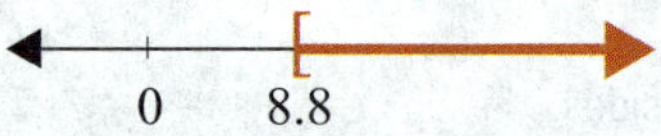

y is in $[8.8, \infty)$. Note that the interval is a half-open interval.

Now work margin exercise 5.

6. Solve the inequality $x + \frac{2}{3} > \frac{1}{3}$ and graph the solution set.

Example 6 Solving an Inequality and Graphing the Solution Set

Solve the inequality $t + \frac{4}{5} \leq \frac{1}{5}$ and graph the solution set. Write the solution set using interval notation.

Solution

$$t + \frac{4}{5} \leq \frac{1}{5}$$

$$t + \frac{4}{5} - \frac{4}{5} \leq \frac{1}{5} - \frac{4}{5} \qquad \text{Add } -\frac{4}{5} \text{ to both sides.}$$

$$t \leq -\frac{3}{5} \qquad \text{Simplify.}$$

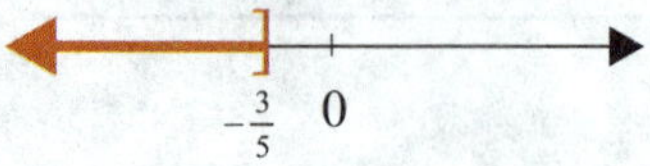

t is in $\left(-\infty, -\frac{3}{5}\right]$. Note that the interval is a half-open

Now work margin exercise 6.

Before solving inequalities of the form $ax < c$ where the coefficient a is not 1, we need to consider the effects of multiplying, or dividing, both sides of the inequality by a positive or a negative real number. For example, we know that $6 < 10$.

Multiply both sides by 3	**Multiply both sides by –3**	**Divide both sides by –2**
$6 < 10$	$6 < 10$	$6 < 10$
$3\cdot 6 \ ? \ 3\cdot 10$	$-3\cdot 6 \ ? \ -3\cdot 10$	$\frac{6}{-2} \ ? \ \frac{10}{-2}$
$18 < 30$	$-18 > -30$	$-3 > -5$

Multiplying or dividing both sides of an inequality by a negative real number causes the "sense" of the inequality to be reversed. The sense of the inequality means "less than" or "greater than."

Multiplication Principle for Solving Linear Inequalities

If A and B are algebraic expressions and C is a positive real number, then the inequalities

$$A < B$$

and

$$AC < BC$$

are equivalent.

If A and B are algebraic expressions and C is a negative real number, then the inequalities

$$A < B$$

and

$$AC > BC$$

are equivalent.

(In other words, if both sides are multiplied by a negative number, then the sense of the inequality is reversed.)

PROPERTIES

Example 7 Solving an Inequality and Graphing the Solution Set

Solve the inequality $5x > 15$ and graph the solution set. Write the solution set using interval notation.

Solution

$5x > 15$

$\frac{5x}{5} > \frac{15}{5}$ Divide both sides by 5.

$x > 3$ Simplify.

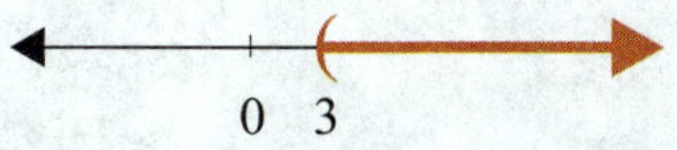

x is in $(3, \infty)$. Note that the interval is an open interval.

Now work margin exercise 7.

7. Solve the inequality $3x < 18$ and graph the solution set.

8. Solve the inequality $-4x \le 24.8$ and graph the solution set.

Example 8 Solving an Inequality and Graphing the Solution Set

Solve the inequality $-3x \le 12.3$ and graph the solution set. Write the solution set using interval notation.

Solution

$$-3x \le 12.3$$

$$\frac{-3x}{-3} \ge \frac{12.3}{-3} \qquad \text{Divide both sides by } -3 \text{ and reverse the inequality.}$$

$$x \ge -4.1 \qquad \text{Simplify.}$$

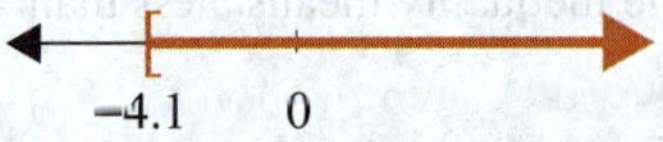

x is in $[-4.1, \infty)$. Note that the interval is a half-open interval.

Now work margin exercise 8.

C Solving Linear Inequalities: $ax + b < c$ and $ax + b < cx + d$

As with solving linear equations, solving linear inequalities involves finding equivalent inequalities of simpler forms. We want the variable with a coefficient of +1 on one side of the inequality and any constants on the other. To accomplish this we use the following steps.

Solving Linear Inequalities

1. Combine like terms on each side of the inequality symbol.
2. Use the addition principle of inequality to add the opposites of constants or variable expressions to both sides so that constants are on one side and variables on the other.
3. Use the multiplication (or division) principle of inequality to multiply both sides by the reciprocal of the coefficient of the variable (or divide both sides by the coefficient itself). The coefficient of the variable will become +1. **If this coefficient is negative, be sure to reverse the sense of the inequality.**
4. A quick (and generally satisfactory) check is to select any one number in your solution set and substitute it into the original inequality. If the statement is false, you need to look for an error in your solution.

PROCEDURE

Example 9 Solving Linear Inequalities

Solve the inequality $6x+5\le -1$ and graph the solution set. Write the solution set using interval notation.

Solution

$$6x+5\le -1 \quad \text{Write the inequality.}$$
$$6x+5-5\le -1-5 \quad \text{Add }-5\text{ to both sides.}$$
$$6x\le -6 \quad \text{Simplify.}$$
$$\frac{6x}{6}\le\frac{-6}{6} \quad \text{Divide both sides by 6.}$$
$$x\le -1 \quad \text{Simplify.}$$

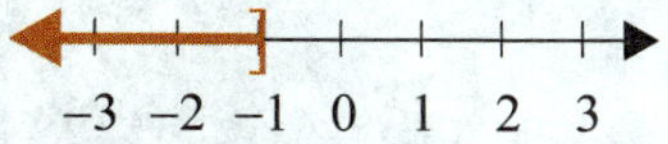

x is in $(-\infty,-1]$.

Check

As a check, pick a number in the solution interval and substitute it into the original inequality. For example, choose -3 and substitute it for x.

$$6(-3)+5\overset{?}{\le}-1$$
$$-18+5\overset{?}{\le}-1$$
$$-13\le -1 \quad \text{True statement}$$

Note that it is not possible to check every number in an interval with infinitely many solutions. But a true result, as shown here, gives confidence in the solution interval.

Now work margin exercise 9.

9. Solve the inequality $4x+8>-16$ and graph the solution set. Write the solution set using interval notation.

Example 10 Solving Linear Inequalities

Solve the linear inequality $x-3>3x+4$ and graph the solution set. Write the solution set using interval notation.

Solution

$$x-3>3x+4 \quad \text{Write the inequality.}$$
$$x-3-x>3x+4-x \quad \text{Add }-x\text{ to both sides.}$$
$$-3>2x+4 \quad \text{Simplify.}$$
$$-3-4>2x+4-4 \quad \text{Add }-2\text{ to both sides.}$$
$$-7>2x \quad \text{Simplify.}$$
$$\frac{-7}{2}>\frac{2x}{2} \quad \text{Divide both sides by 2.}$$
$$\frac{-7}{2}>x \quad \text{Simplify.}$$
$$\text{or} \quad x<-\frac{7}{2}$$

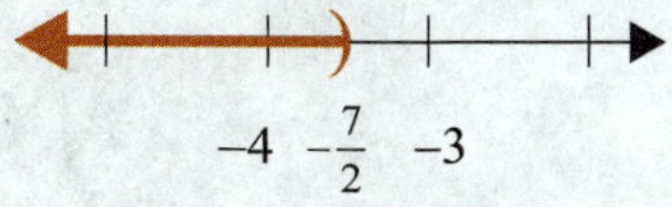

x is in $\left(-\infty,-\frac{7}{2}\right)$.

Now work margin exercise 10.

10. Solve the inequality $x-2\le 6x-6$ and graph the solution set. Write the solution set using interval notation.

11. Solve the inequality $3-8x \le x+4$ and graph the solution set. Write the solution set using interval notation.

Example 11 Solving Linear Inequalities

Solve the linear inequality $6-4x \le x+1$ and graph the solution set. Write the solution set using interval notation.

Solution

$6-4x \le x+1$	Write the inequality.
$6-4x-x \le x+1-x$	Add $-x$ to both sides.
$6-5x \le 1$	Simplify.
$6-5x-6 \le 1-6$	Add -6 to both sides.
$-5x \le -5$	Simplify.
$\frac{-5x}{-5} \ge \frac{-5}{-5}$	Divide both sides by -5. Note that the inequality sign is reversed because of division by a negative number!
$x \ge 1$	Simplify.

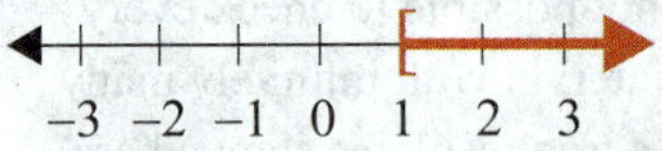

x is in $[1, \infty)$.

Now work margin exercise 11.

12. Solve the inequality $4x+3 \ge 7x-(5-x)$ and graph the solution set. Write the solution set using interval notation.

Completion Example 12 Solving Linear Inequalities

Supply the reasons for each step in solving the linear inequality and graph the solution set. Write the solution set using interval notation.

$2x+5 < 3x-(7-x)$

Solution

$2x+5 < 3x-(7-x)$	Write the inequality.
$2x+5 < 3x-7+x$	Distribute the negative sign.
$2x+5 < 4x-7$	Combine like terms.
$2x+5-2x < 4x-7-2x$	Add $-2x$ to both sides.
$5 < 2x-7$	Simplify.
$5+7 < 2x-7+7$	Add 7 to both sides.
$12 < 2x$	Simplify.
$\frac{12}{2} < \frac{2x}{2}$	Divide both sides by 2.
$6 < x$	Simplify.

Graph the interval on a number line.

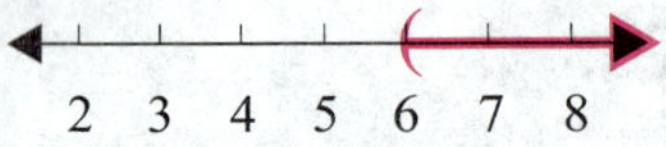

x is in $(6, \infty)$.

Now work margin exercise 12.

D Applications of Linear Inequalities

In the following two examples, we show how inequalities can be related to real-world problems.

Example 13 Application: Using Inequalities

A math student has grades of 85, 98, 93, and 90 on four examinations. If he must average 90 or better to receive an A for the course, what scores can he receive on the final exam and earn an A? (Assume that the final exam counts the same as the other exams.)

Solution

Let x = the score on final exam.

The average is found by adding the scores and dividing by 5.

$$\frac{85+98+93+90+x}{5} \geq 90$$

$$\frac{366+x}{5} \geq 90 \quad \text{Simplify the numerator.}$$

$$5\left(\frac{366+x}{5}\right) \geq 5 \cdot 90 \quad \text{Multiply both sides by 5.}$$

$$366+x \geq 450 \quad \text{Simplify.}$$

$$366+x-366 \geq 450-366 \quad \text{Add } -366 \text{ to each side.}$$

$$x \geq 84$$

If the student scores 84 or more on the final exam, he will average 90 or more and receive an A in math.

Now work margin exercise 13.

13. A certain anesthetic must be administered in a way that requires the average dosage to remain under 450 milligrams per hour in order to guarantee patient safety. This anesthetic is being used during a 4-hour surgery. For the first three hours, the dosages were as follows: 500 milligrams the first hour, 380 milligrams during the second hour, and 520 milligrams the third hour. What must the final dosage remain under to be safely administered?

Example 14 Application: Using Inequalities

Taylor is at the local fair and purchases 30 ride tickets. She wants to get her money's worth and she only has time to go on 8 rides. The rollercoaster requires 5 tickets to board and the merry-go-round requires 3 tickets to board. What is the maximum number of times she can ride the rollercoaster?

Solution

Let x = number of rollercoaster rides,
then $8 - x$ = number of merry-go-round rides.

Taylor cannot use more than 30 tickets.

$$5x+3(8-x) \leq 30$$

$$5x+24-3x \leq 30$$

$$2x+24 \leq 30$$

$$2x+24-24 \leq 30-24$$

$$2x \leq 6$$

$$\frac{2x}{2} \leq \frac{6}{2}$$

$$x \leq 3$$

Taylor can go on the rollercoaster ride at most 3 times.

Now work margin exercise 14.

14. Ashley is planning her wedding and is ordering flowered centerpieces. She wants some rose centerpieces and some lily centerpieces. The rose centerpieces cost \$225, and the lily centerpieces cost \$175. If she needs a total of 20 centerpieces, and she has a budget of \$3900 to spend, what is the maximum number of rose centerpieces she can buy?

9.7 Exercises

Concept Check

1. interval
2. open
3. closed
4. linear expressions
5. addition
6. multiplication
7. True
8. False; The sense of the inequality should reverse.
9. False; Only one value in the solution set needs to be checked.
10. True

Practice

1. $(-1, \infty)$

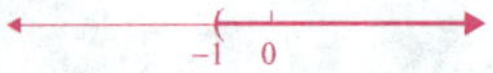

2. $[-2, 4)$

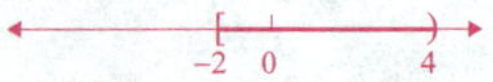

3. $(-\infty, 5]$

4. $[0, 3]$

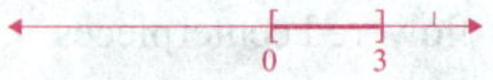

5. $[-5, -1]$

6. $(3, 8)$

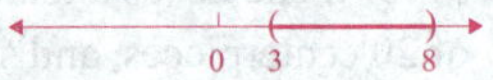

7. $[-7, -4)$

8. $(-\infty, -6]$

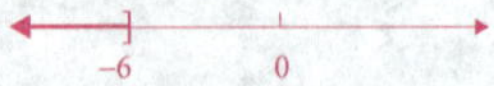

9. 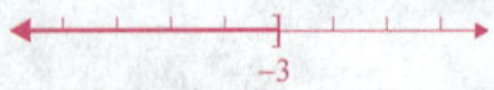

Half-open interval

10. 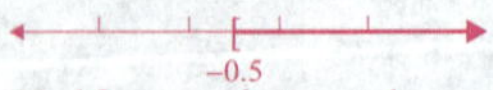

Half-open interval

11. 

Open interval

Completion Example Answers

12.

$2x+5<3x-(7-x)$	Write the inequality.
$2x+5<3x-7+x$	Distribute the negative sign.
$2x+5<4x-7$	Combine like terms.
$2x+5-2x<4x-7-2x$	Add $-2x$ to both sides.
$5<2x-7$	Simplify.
$5+7<2x-7+7$	Add 7 to both sides.
$12<2x$	Simplify.
$\frac{12}{2}<\frac{2x}{2}$	Divide both sides by 2.
$6<x$	Simplify.

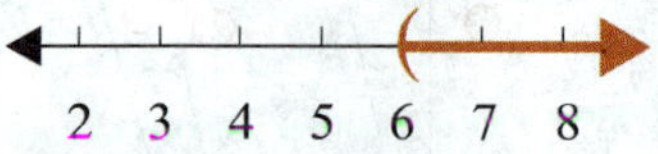

x is in $(6, \infty)$

Margin Exercise Answers

1. 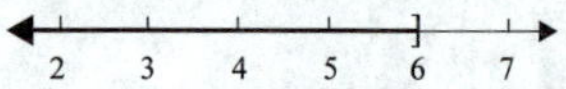 **2.** 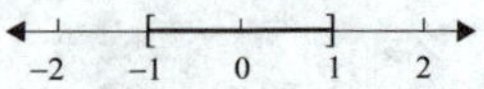 **3.** $-2 \le x < 1$ is a half-open interval.

4. $[8, \infty)$ is a half-open interval. **5.** $y < 5.2$ 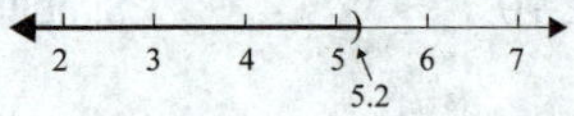

6. $x > -\frac{1}{3}$ 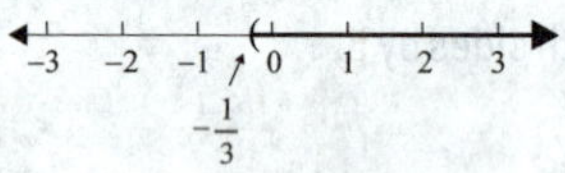 **7.** $x < 6$ 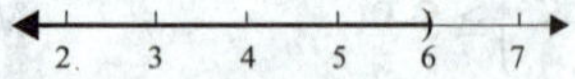

8. $x \ge -6.2$ 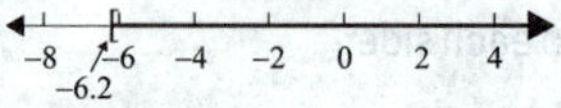 **9.** $(-6, \infty)$ 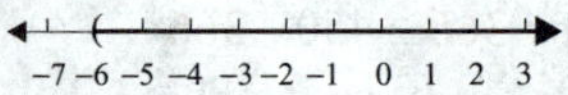

10. $\left[\frac{4}{5}, \infty\right)$ 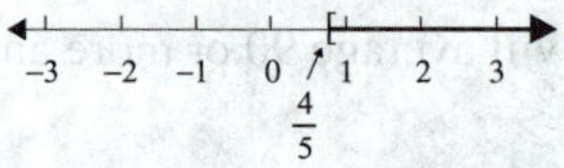 **11.** $\left[-\frac{1}{9}, \infty\right)$ 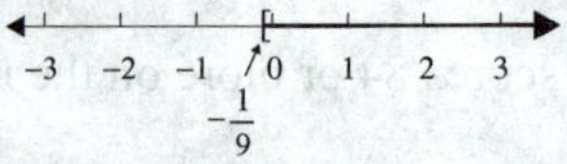 **12.** $(-\infty, 2]$

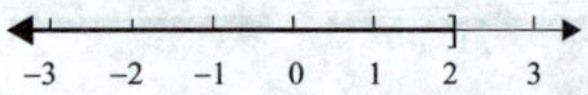

13. The maximum final dosage that can be administered must be less than 400 milligrams. **14.** Ashley can buy at most 8 rose centerpieces.

9.7 Exercises

Concept Check

Fill-in-the-Blank. Complete each sentence using information found in this section.

1. If a and b are real numbers where $a < b$, the set of all real numbers between a and b is called a/an ______ of real numbers.

2. In a/an ______ interval, neither endpoint is included.

3. In a/an ______ interval, both end points are included.

4. Linear inequalities are inequalities that relate two ______ ______.

5. If A and B are algebraic expressions and C is a real number, then the ________ principle for solving linear inequalities states that $A < B$ and $A + C < B + C$ are equivalent.

6. If A and B are algebraic expressions and C is a real number, then the ________ principle for solving linear inequalities states that $A < B$ and $AC < BC$ are equivalent.

True/False. Determine whether each statement is true or false. If a statement is false, explain how it can be changed so the statement will be true. (**Note:** There may be more than one acceptable change.)

7. If only one endpoint is included in an interval, it is called a half-open interval.
8. When both sides of a linear inequality are multiplied by a negative constant, the sense of the inequality should stay the same.
9. To check the solution set of a linear inequality, every solution in the solution set must be checked in the original inequality.
10. The infinity symbol ∞ does not represent a specific number.

Practice

Graph each interval on a real number line. See Example 1.

1. $(-1,\infty)$
2. $[-2,4)$
3. $(-\infty,5]$
4. $[0,3]$
5. $[-5,-1]$
6. $(3,8)$
7. $[-7,-4)$
8. $(-\infty,-6]$

Graph each interval on a real number line and tell what type of interval it is. See Examples 2 through 4.

9. $x \leq -3$
10. $x \geq -0.5$
11. $x > 4$
12. $x < -\frac{1}{10}$
13. $0 < x \leq 2.5$
14. $-1.5 \leq x < 3.2$
15. $-2 \leq x \leq 0$
16. $-1 \leq x \leq 1$
17. $4 > x \geq 2$
18. $0 > x \geq -5$

Solve each inequality and graph the solution set. Write each solution set using interval notation. See Examples 5 through 12.

19. $x + 1 > 5$
20. $x - 3 < 2$
21. $3 + x \leq 7$
22. $5 + x \geq 11$
23. $3 < 4 + x$
24. $9 > 6 + x$
25. $4 \geq x - 3$
26. $12 \leq x + 8$
27. $4x > 16$
28. $3x < 27$
29. $5x \leq 15$
30. $-2x \geq 6$
31. $10 > -5x$
32. $12 < 8x$
33. $14 \geq 2x$
34. $9 \leq -3x$

12.
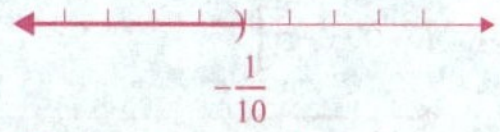
Open interval

13.
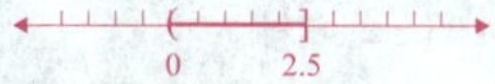
Half-open interval

14.

Half-open interval

15.

Closed interval

16.

Closed interval

17.

Half-open interval

18.
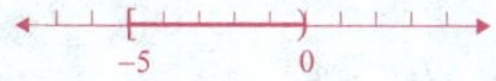
Half-open interval

19. $(4,\infty)$
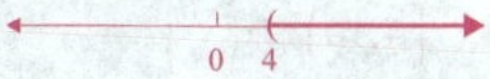

20. $(-\infty,5)$
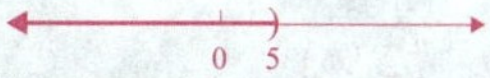

21. $(-\infty,4]$
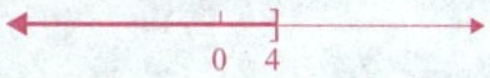

22. $[6,\infty)$
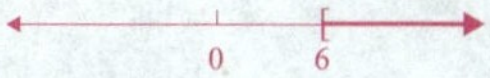

23. $(-1,\infty)$
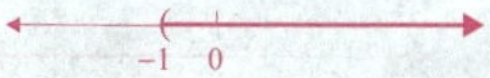

24. $(-\infty,3)$
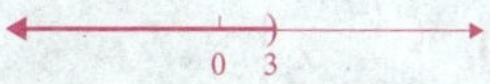

25. $(-\infty,7]$
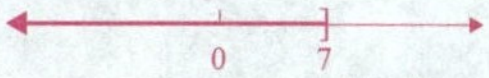

26. $[4,\infty)$
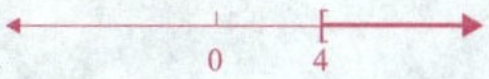

27. $(4,\infty)$
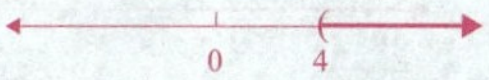

28. $(-\infty,9)$
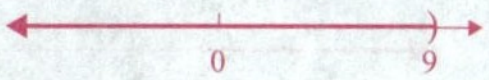

29. $(-\infty,3]$
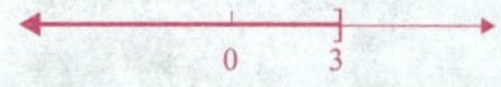

30. $(-\infty,-3]$

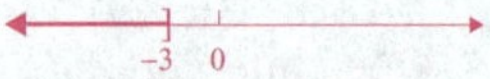

31. $(-2,\infty)$

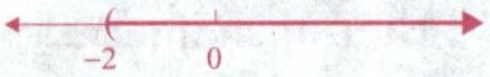

32. $\left(\frac{3}{2},\infty\right)$

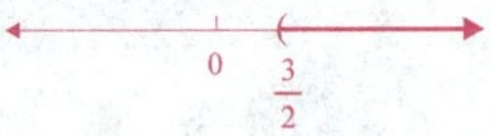

33. $(-\infty,7]$

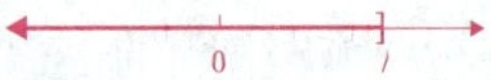

34. $(-\infty,-3]$

35. $(-\infty,1)$

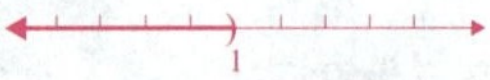

36. $[4,\infty)$

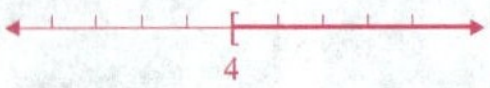

37. $(2,\infty)$

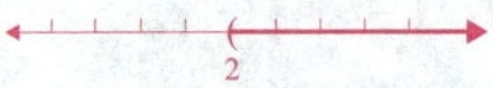

38. $(4,\infty)$

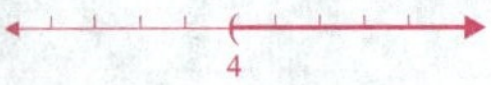

39. $\left(\frac{8}{3},\infty\right)$

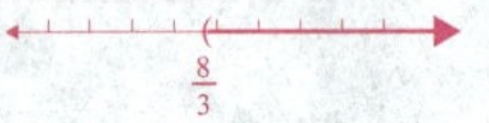

40. $\left(\frac{1}{2},\infty\right)$

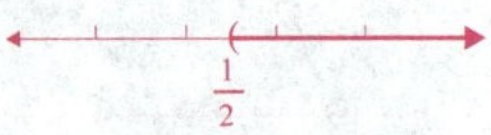

41. $[-0.4,\infty)$

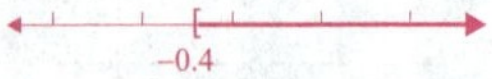

42. $(-\infty,1.2)$

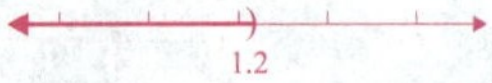

43. $(-\infty,2)$

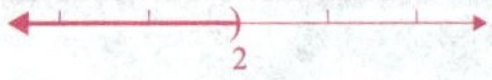

35. $2x+3<5$

36. $4x-7\geq 9$

37. $14-5x<4$

38. $23<7x-5$

39. $6x-15>1$

40. $9-2x<8$

41. $5.6+3x\geq 4.4$

42. $12x-8.3<6.1$

43. $1.5x+9.6<12.6$

44. $0.8x-2.1\geq 1.1$

45. $2+3x\geq x+8$

46. $x-6\leq 4-x$

47. $3x-1\leq 11-3x$

48. $5x+6\geq 2x-2$

49. $4-2x<5+x$

50. $4+x>1-x$

51. $x-6>3x+5$

52. $4+7x\leq 4x-8$

53. $\frac{x}{2}-1\leq\frac{5x}{2}-3$

54. $\frac{x}{4}+1\leq 5-\frac{x}{4}$

55. $\frac{x}{3}-2>1-\frac{x}{3}$

56. $\frac{5x}{3}+2>\frac{x}{3}-1$

57. $6x+5.91<1.11-2x$

58. $4.3x+21.5\geq 1.7x+0.7$

59. $6.2x-5.9>4.8x+3.2$

60. $0.9x-11.3<3.1-0.7x$

61. $4(6-x)<-2(3x+1)$

62. $-3(2x-5)\leq 3(x-1)$

63. $-(3x+8)\geq 2(3x+1)$

64. $6(3x+1)<5(1-2x)$

65. $11x+8-5x\geq 2x-(4-x)$

66. $1-(2x+8)<(9+x)-4x$

67. $5-3(4-x)+x\leq -2(3-2x)-x$

68. $x-2(x+3)\geq 7-(4-x)+11$

69. $\frac{2(x-1)}{3}<\frac{3(x+1)}{4}$

70. $\frac{3(x-2)}{2}\geq\frac{4(x-1)}{3}$

71. $\frac{x-2}{4}>\frac{x+2}{2}+6$

72. $\frac{x+4}{9}\leq\frac{x}{3}-2$

73. $\frac{2x+7}{4}\leq\frac{x+1}{3}-1$

74. $\frac{4x}{7}-3>\frac{x-6}{2}-4$

Represent each of the following statements as an inequality involving a variable *x*, and graph its solution set on a number line.

75. You must be at least 58 inches in height to ride this roller coaster.

76. There are fewer than 12 days left before final exams.

77. Gifts worth \$5 or less do not need to be declared.

78. Arsenic levels over 10 parts per billion may be dangerous.

Applications

Solve.

79. A statistics student has grades of 82, 95, 93, and 78 on four hour-long exams. He must average 90 or higher to receive an *A* for the course. What scores can he receive on the final exam and earn an *A* if:

a. The final is equivalent to a single hour-long exam (100 points maximum)?

b. The final is equivalent to two hourly exams (200 points maximum)?

80. To receive a grade of B in a chemistry class, Melissa must average 80 or more but less than 90. If her five hour-long exam scores were 75, 82, 90, 85, and 77, what score does she need on the final exam (100 points maximum) to earn a grade of B?

81. A car salesman makes \$1000 each week that he works and makes approximately \$250 commission for each car he sells. If a car salesman wants to make at least \$3500 in one week, how many cars does he need to sell?

82. Allison is ordering boxes of 24 tea bags from a website. The website is having a promotion where each box of tea comes with 2 free sample packs, and each sample pack contains 3 tea bags. If Allison has an empty container that holds 150 tea bags, what is the largest number of boxes of tea Allison can order and not overfill the container?

83. WildLily Florist is creating arrangements for a wedding this weekend. The large arrangements use 8 flowers and the small arrangements use 5 flowers.

a. Let x represent the number of large arrangements. Write an algebraic expression for the number of small arrangements, if there are 15 tables that need an arrangement.

b. Write an algebraic expression representing the total number of flowers used in the 15 arrangements.

c. If the bride has paid for 100 flowers, use an inequality to determine the maximum number of the 15 arrangements that can be large.

84. John's algebra test consists of 19 questions, 13 equations and 6 word problems. Each equation is worth 4 points, and each word problem is worth 8 points. Assume there is no partial credit on this test.

a. Let w be the number of word problems John gets correct. Write an expression for the number of points John will get from the word problem part of his test.

b. Assuming John gets every equation correct, write an inequality that will help determine the fewest number of word problems he can get correct and still make an 80 on the test. What is the fewest number he can get correct?

c. Let x be the number of equations John gets correct. Write an expression for the number of points John will get from the equation part of his test.

d. Assuming John gets every word problem correct, write an inequality that will help determine the fewest number of equations he can get correct and still get an 80 on the test. What is the fewest number of equations he can get correct?

44. $[4, \infty)$

45. $[3, \infty)$

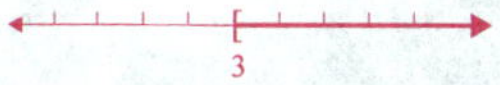

46. $(-\infty, 5]$

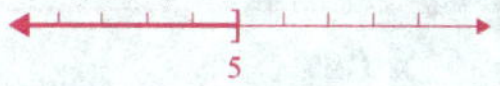

47. $(-\infty, 2]$

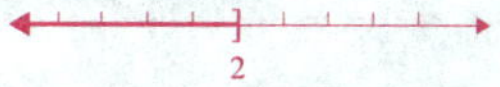

48. $\left[-\frac{8}{3}, \infty\right)$

49. $\left(-\frac{1}{3}, \infty\right)$

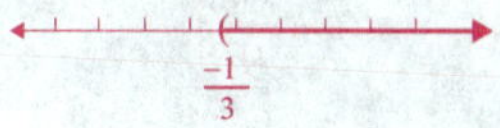

50. $\left(-\frac{3}{2}, \infty\right)$

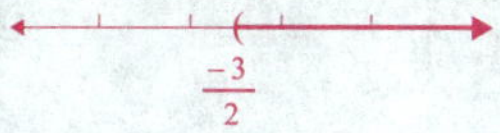

51. $\left(-\infty, -\frac{11}{2}\right)$

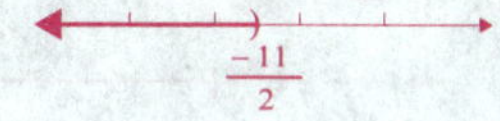

52. $(-\infty, -4]$

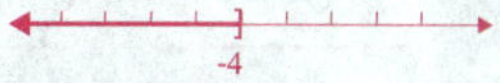

53. $[1, \infty)$

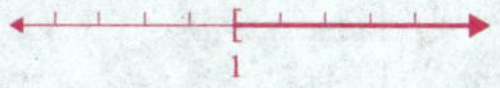

54. $(-\infty, 8]$

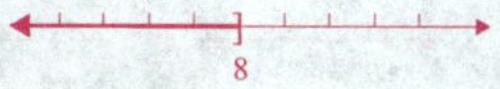

55. $\left(\frac{9}{2}, \infty\right)$

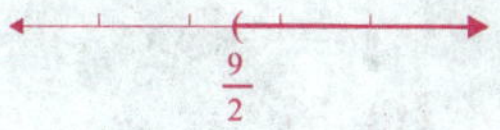

56. $\left(-\frac{9}{4}, \infty\right)$

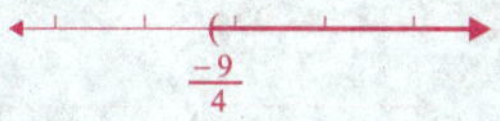

57. $(-\infty, -0.6)$

-0.6

58. $[-8, \infty)$

-8

59. $(6.5, \infty)$

6.5

60. $(-\infty, 9)$

9

61. $(-\infty, -13)$

-13

62. $[2, \infty)$

2

63. $\left(-\infty, -\frac{10}{9}\right]$

$-\frac{10}{9}$

64. $\left(-\infty, -\frac{1}{28}\right)$

$\frac{-1}{28}$

65. $[-4, \infty)$

-4

66. $(-\infty, 16)$

16

67. $(-\infty, 1]$

1

68. $(-\infty, -10]$

-10

69. $(-17, \infty)$

-17

70. $[10, \infty)$

10

71. $(-\infty, -30)$

-30

85. Dr. Smyth has an attendance clause in his course syllabus that a student loses 5 points on his or her final grade average for every unexcused absence the student has after his or her first three unexcused absences. If Kara must have a 70 to pass the course, determine the largest number of unexcused absences Kara can have and still have any chance to pass the course.

86. Tracy needs to purchase 25 pastries for the PTA Teachers' Breakfast. Bear Claws cost \$1.75 each and Apple Turnovers cost \$2.15 each. If Tracy's budget is \$50, find the maximum number of Apple Turnovers that Tracy can purchase.

87. The maximum occupancy for a concert in Thompson-Boling Arena is 24,000 people. However, for every 15 tickets sold, there must be one worker present (security, food service, admissions, etc.). Determine the maximum number of tickets than can be sold.

88. ▦ Phineas wants to build a nuclear-powered submarine to take his friends on a tour of the Arctic Circle. At least twice as much titanium must be used in the construction of the shell of the sub as the amount of stainless steel used in its construction. The cost of titanium is \$500 per pound and the cost of steel is \$300 per pound. If Phineas has only \$1,000,000 to spend on metal for the sub, determine the greatest number of pounds of metal (both together) that can be used to construct his submarine. (Round your answer to the nearest pound.)

89. Nicole has just moved to Orlando and discovered that Florida residents can purchase 4-day tickets to Disney World for \$55 per day. Annual passes (with certain restrictions) for Florida residents are \$390. Nicole is trying to decide if she thinks she will go to the park enough times to make it worth buying an annual pass. Use the formula $55x \leq 390$, where x is the number of days spent visiting at Disney World, to determine how many times she would have to go in order for the annual pass to be the better deal.

90. Fernando has already consumed 270 grams of carbohydrates and is on a diet that restricts his carbohydrate consumption to no more than 300 grams of carbohydrates per day. A serving of 6 crackers has 21 grams of carbohydrates. Solve the inequality $\frac{21}{6}c + 270 \leq 300$ to determine how many crackers (c) Fernando can eat without going over his goal. Write your answer as a whole number.

91. Jeph is in charge of buying office supplies for the nonprofit organization he works for. He has \$400 to spend. He needs to buy a printer that costs \$150, a box of printer paper for \$60, and some ink cartridges for \$12.50 each. What is the maximum number of ink cartridges that Jeph can buy? (**Note:** Tax is not included in the sales price.)

a. Set up the linear inequality. Use the variable c to represent the number of ink cartridges.

b. Solve the equation from part a. for the variable.

c. What does the answer from part b. mean? Write a complete sentence.

92. Sarah is participating in National Novel Writers Month where she has to write a rough draft of a novel with at least 50,000 words during the month of November. At the end of the day on November 20th, she has a total of 32,500 words. What is the minimum number of words that Sarah needs to write each day for the rest of the month to make the goal of 50,000 words?

a. Set up the linear inequality. Use the variable w to represent the number of words per day.

b. Solve the equation from part a. for the variable.

c. What does the answer from part b. mean? Write a complete sentence.

93. Andrew needs to earn at least a B in each class to keep his scholarship. The grade in his economics class is based on five exams that are equally weighted. On the first four exams, Andrew received the following scores: 92, 74, 80, 72. Andrew needs an average of at least 80 to earn a B for the class. What range of scores does he need on the fifth exam to keep his scholarship?

a. Set up the linear inequality. Use the variable E to represent the fifth exam score.

b. Solve the equation from part a. for the variable.

c. What does the answer from part b. mean? Write a complete sentence.

Writing & Thinking

94. a. Write a list of three situations where inequalities might be used in daily life.

b. Illustrate theses situations with algebraic inequalities and appropriate numbers.

72. $[11, \infty)$

73. $\left(-\infty, -\frac{29}{2}\right]$

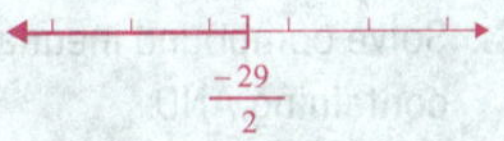

74. $(-56, \infty)$

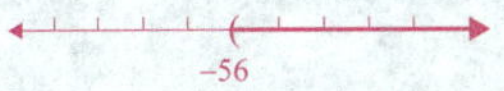

75. $x \geq 58$

76. $x < 12$

77. $x \leq 5$

78. $x > 10$

Applications

79. a. The student would need a score higher than 102 points, which is not possible. Thus he cannot earn an A in the course.
b. The student must score at least 192 points to earn an A in the course.

80. Melissa must score at least a 71 to earn a B for the course.

81. He must sell at least 10 cars.

82. Allison can buy at most 5 boxes of tea.

83. a. $15 - x$
b. $8x + 5(15 - x)$
c. 8 large arrangements

84. a. $8w$
b. 4 word problems
c. $4x$
d. $4x + 8(6) \geq 80$; 8 equations

85. $100 - 5(x - 3) \geq 70$; 9 unexcused absences

86. $1.75(25 - x) + 2.15x \leq 50$; 15 Apple Turnovers

87. 22,500 tickets

88. $500(2x) + 300x \leq 1{,}000{,}000$; 2307 lb

89. 8 times

90. 8 crackers

91. a. $150 + 60 + 12.50c \leq 400$
b. If you solve the inequality you get $c \leq 15.2$, but you can't buy 0.2 ink cartridges, so $c \leq 15$.
c. Jeph can buy at most 15 ink cartridges.

92. a. $32{,}500 + 10w \geq 50{,}000$
b. $w \geq 1750$
c. Sarah will need to write at least 1750 words per day to meet her goal.

93. a. $\frac{92 + 74 + 80 + 72 + E}{5} \geq 80$
b. $E \geq 82$
c. Andrew needs to earn at least an 82 on the fifth exam.

Writing & Thinking

94. a. Answers will vary.
b. Answers will vary.

Objectives

A. Find the union and intersection of two sets.
B. Graph compound inequalities on a real number line.
C. Solve compound inequalities containing AND.
D. Solve compound inequalities containing OR.

9.8 Compound Inequalities

A Union and Intersection of Sets

A **set** is a collection of objects or numbers. The items in the set are called **elements**. Sets are indicated with braces $\{\ \}$ and named with capital letters. If the elements are listed within the braces, the set is said to be in **roster form**. For example, the following sets are in roster form.

$$A = \left\{\frac{3}{4}, 1.7, 2, 4, 6, \sqrt{89}\right\}$$

$$\mathbb{W} = \{0, 1, 2, 3, 4, 5 \ldots\}$$

$$\mathbb{Z} = \{\ldots, -4, -3, -2, -1, 0, 1, 2, 3, 4, \ldots\}$$

The symbol $\in$ is read "is an element of" and is used to indicate that a particular number belongs to a set. For example, in the previous illustrations, $1.7 \in A$, $0 \in \mathbb{W}$, and $-3 \in \mathbb{Z}$.

The following discussion highlights the two important set concepts **union** and **intersection**. The concepts of union and intersection are part of set theory, which is very useful in a variety of courses including abstract algebra, probability, and statistics. These concepts are also used in analyzing inequalities and analyzing relationships among sets in general.

Math Tip

The *or* we use in this course is known as an *inclusive or*. If you have the sets A and B, then A or B means you would include the elements that appear in at least one of the sets A or B. This includes the elements that appear in only set A, the elements that appear in only set B, and the elements that appear in both sets A and B.

Union and Intersection

The **union** (symbolized $\cup$, as in $A \cup B$) of two (or more) sets is the set of all elements that belong to either one set or the other set or to both sets.

The **intersection** (symbolized $\cap$, as in $A \cap B$) of two (or more) sets is the set of all elements that belong to both sets.

The word **or** is used to indicate union and the word **and** is used to indicate intersection.

DEFINITION

Math Tip

Any elements in an intersection or union of sets are only listed once.

For example, if $A = \{1, 2, 3\}$ and $B = \{2, 3, 4\}$, then the numbers that belong to A **or** B is the set $A \cup B = \{1, 2, 3, 4\}$. The set of numbers that belong to A **and** B is the set $A \cap B = \{2, 3\}$. These relationships can be illustrated using the following Venn diagram.

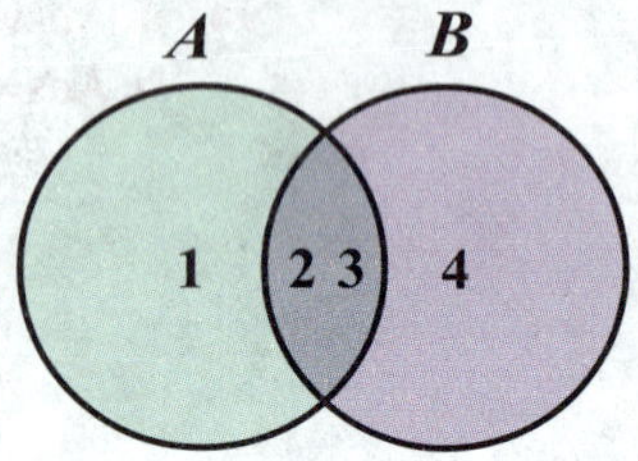

Example 1 Finding the Union and Intersection of Two Sets

Consider the two sets $A=\{0,1,5,7\}$ and $B=\{-3,0,5,6,7\}$. Then determine the sets **a.** $A\cup B$ and **b.** $A\cap B$.

Solution

a. $A\cup B=\{-3,0,1,5,6,7\}$ These elements belong to A ***or*** to B ***or*** to both A and B.

b. $A\cap B=\{0,5,7\}$ These elements belong to both A ***and*** B.

Now work margin exercise 1.

1. Consider the two sets $A=\{2,4,6,8\}$ and $B=\{1,2,4,8\}$.

Determine:

a. $A\cup B$

b. $A\cap B$

Example 2 Finding the Union and Intersection of Two Sets

Find **a.** the union and **b.** the intersection of the two sets $P=\left\{-6,-3.5,-\frac{1}{2},0,2,4,6\right\}$ and $Q=\{-6,-4,-2,0,2\}$.

Solution

a. $P\cup Q=\left\{-6,-4,-3.5,-2,-\frac{1}{2},0,2,4,6\right\}$ These elements belong to P ***or*** to Q ***or*** to both P ***and*** Q.

b. $P\cap Q=\{-6,0,2\}$ These elements belong to both P ***and*** Q.

Now work margin exercise 2.

2. Find **a.** the union and **b.** the intersection of the two sets $M=\left\{-4,-1,\frac{1}{3},5,6.4,9\right\}$ and $N=\{-4,-3,-2,-1,0\}$.

If the elements in a set can be counted, as in A above, the set is said to be **finite**. If the elements cannot be counted, the set is said to be **infinite**. If a set has absolutely no elements, it is called the **empty set** or **null set** and is written in the form $\{\ \}$ or with the special symbol $\varnothing$. For example, the set of all people over 15 feet tall is the empty set, $\varnothing$.

The notation $\{x\mid \quad\}$ is read "the set of all x such that…" and is called **set-builder notation**. The vertical bar (|) is read "such that." For example,

$$\{x\mid x \text{ is a positive even integer less than } 10\} \text{ is read}$$

"the set of all x such that x is a positive even integer less than 10."

This particular set could also have been written in the roster form $\{2,4,6,8\}$. In a case such as $\{x\mid x<7\}$, roster form is not possible.

Example 3 Finding the Union and Intersection of Two Sets

Find the union and the intersection of the two sets.

$$F=\{y\mid y \text{ is a positive even integer less than } 6\}$$

and

$$G=\{y\mid y \text{ is a positive odd integer less than } 6\}$$

3. Find the union and intersection of the two sets

$F=\left\{\begin{matrix} y\mid y \text{ is a negative} \\ \text{integer greater than } -4\end{matrix}\right\}$

and

$G=\left\{\begin{matrix} y\mid y \text{ is a positive} \\ \text{integer less than } 4\end{matrix}\right\}$.

What's Nothing?

There's a popular logic joke.

"Nothing is better than eternal happiness. A sandwich is better than nothing. Therefore, a sandwich is better than eternal happiness."

This is a play on the ambiguous meaning of nothing. Mathematically, the first sentence translates to $\{x|x \text{ is better than eternal happiness}\} = \varnothing$. The second means something different: Let y be the value (nutritional, monetary, etc.) of a sandwich, then $y > 0$. In the first case, *nothing* refers to an empty set. In the second, *nothing* refers to zero value.

Solution

Although these sets are given in set-builder form they are easier to relate if they are both in roster form as $F = \{2,4\}$ and $G = \{1,3,5\}$. Now we have the following.

Union: $F \cup G = \{1,2,3,4,5\}$ These are the elements that belong to F ***or*** G or to both F ***and*** G.

Intersection: $F \cap G = \varnothing$ *The solution is the empty set since there are no elements in both F and G.*

Now work margin exercise 3.

Similarly, union and intersection notation can be used for sets with inequalities.

For example, $\{x|x < a \textbf{ or } x > b\}$ can be written in the form

$$\{x|x < a\} \cup \{x|x > b\}.$$

Also, $\{x|x > a \textbf{ and } x < b\}$ can be written in the form

$$\{x|x > a\} \cap \{x|x < b\} \text{ or } \{x|a < x < b\}.$$

Example 4 and Example 5 illustrate how the concepts of **union** (indicated by **or**) and **intersection** (indicated by **and**) are related to the graphs of intervals of real numbers on a real number line.

B Compound Inequalities

Two inequalities related by and (indicating intersection) or or (indicating union) are called **compound inequalities**. Note that, unless otherwise stated, the variables represent real numbers.

For example,

$x \le 2$ and $x \ge 0$

and $x > 5$ or $x \le 4$

are compound inequalities.

4. Graph the set $\{y|y \ge -1 \textbf{ and } y < 3\}$.

Example 4 Graphing Compound Inequalities

Graph the set $\{x|x \le 2 \textbf{ and } x \ge 0\}$. The word **and** implies those values of x that satisfy both inequalities.

Solution

$x \le 2$

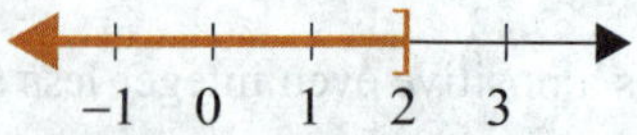

$x \ge 0$

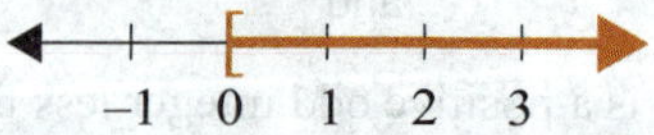

$x \le 2$ **and** $x \ge 0$

The solution graph shows the intersection $\cap$ of the first two graphs. In other words, the third graph shows the points in common between the first two graphs in this example.

This set can also be indicated as $\{x|0 \le x \le 2\}$.

Now work margin exercise 4.

Example 5 Graphing Compound Inequalities

Graph the set $\{x|x > 5 \textbf{ or } x \le 4\}$. The word **or** implies those values of x that satisfy **at least one** of the inequalities.

Solution

$x > 5$

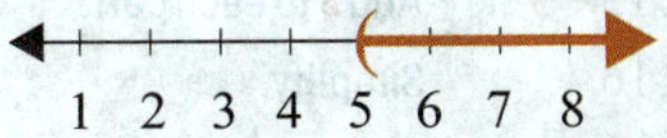

$x \le 4$

$x > 5$ **or** $x \le 4$

The solution graph shows the union $\cup$ of the first two graphs.

Now work margin exercise 5.

5. Graph the set $\{t|t \le 8 \textbf{ or } t > 10\}$.

C Solving Compound Inequalities Containing AND

The solution set of $\{x\,|-5 \le 4x-1 \text{ and } 4x-1 < 11\}$ is the **intersection** of the solution sets of the two inequalities. This compound inequality can be indicated in the abbreviated notation $\{x\,|-5 \le 4x-1 < 11\}$. To find the solution set we follow the steps for solving inequalities given in Section 9.7. As stated there, the objective is to isolate the variable with coefficient +1.

Example 6 Solving Compound Inequalities Containing AND

Solve the compound inequality $-5 \le 4x - 1 < 11$ and graph the solution set. Write the solution set using interval notation.

Solution

$-5 \le 4x-1 < 11$	Write the inequality.
$-5+1 \le 4x-1+1 < 11+1$	Add 1 to each part.
$-4 \le 4x < 12$	Simplify.
$\frac{-4}{4} \le \frac{4x}{4} < \frac{12}{4}$	Divide each part by 4.
$-1 \le x < 3$	Simplify.

The solution set is the half-open interval $[-1, 3)$.

Now work margin exercise 6.

6. Solve the inequality $-12 < 5x + 3 \le 8$ and graph the solution set. Write the solution set using interval notation.

7. Solve the inequality $-3 < 5 - 4x < 17$ and graph the solution set. Write the solution set using interval notation.

Example 7 Solving Compound Inequalities Containing AND

Solve the compound inequality $0.5 \le -0.3 - 0.2x$ and $-0.3 - 0.2x \le 1.3$ and graph the solution set. Write the solution set using interval notation.

Solution

To begin, we write this inequality in abbreviated notation as $0.5 \le -0.3 - 0.2x \le 1.3$.

$$\begin{array}{rcll}
0.5 \le & -0.3-0.2x & \le 1.3 & \text{Write the inequality.} \\
10(0.5) \le & 10(-0.3-0.2x) & \le 10(1.3) & \text{Multiply each part by 10 to get integer coefficients and constants.} \\
5 \le & -3-2x & \le 13 & \text{Simplify.} \\
5+3 \le & -3-2x+3 & \le 13+3 & \text{Add 3 to each part.} \\
8 \le & -2x & \le 16 & \text{Simplify.} \\
\frac{8}{-2} \ge & \frac{-2x}{-2} & \ge \frac{16}{-2} & \text{Divide each part by } -2\text{. Note that the inequalities change sense.} \\
-4 \ge & x & \ge -8 & \text{Simplify.} \\
(\text{or } -8 \le & x & \le -4) &
\end{array}$$

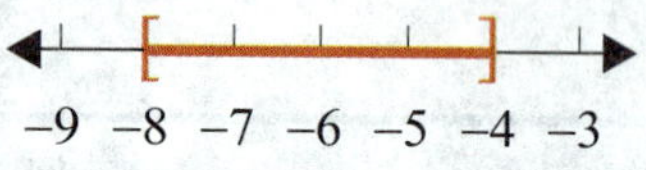

The solution set is the closed interval $[-8, -4]$.

Now work margin exercise 7.

8. Solve the inequality $7 \le \frac{5x-1}{2} \le 13$ and graph the solution set. Write the solution set using interval notation.

Example 8 Solving Compound Inequalities Containing AND

Solve the compound inequality $0 < \frac{3x-5}{4} < 3$ and graph the solution set. Write the solution set using interval notation.

Solution

$$\begin{array}{rcll}
0 < & \frac{3x-5}{4} & < 3 & \text{Write the inequality.} \\
0 \cdot 4 < & \frac{3x-5}{4} \cdot 4 & < 3 \cdot 4 & \text{Multiply each part by 4 to clear the fraction.} \\
0 < & 3x-5 & < 12 & \text{Simplify each part.} \\
0+5 < & 3x-5+5 & < 12+5 & \text{Add 5 to each part.} \\
5 < & 3x & < 17 & \text{Simplify.} \\
\frac{5}{3} < & \frac{3x}{3} & < \frac{17}{3} & \text{Divide each part by 3.} \\
\frac{5}{3} < & x & < \frac{17}{3} & \text{Simplify.}
\end{array}$$

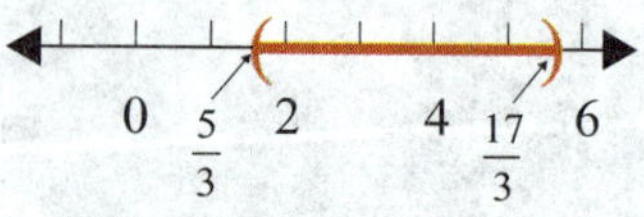

The solution set is the open interval $\left(\frac{5}{3}, \frac{17}{3}\right)$.

Now work margin exercise 8.

D Solving Compound Inequalities Containing OR

The solution set of the compound inequality $\{x \mid x \le 4 \text{ or } x > 5\}$ is the **union** of the solutions sets of the two inequalities. There is no abbreviated notation for the union of the two solution sets.

To solve for a solution set of the union of two inequalities, we solve each inequality separately. Example 9 illustrates this method.

Example 9 Solving Compound Inequalities Containing OR

Solve the compound inequality $2x-3>-2$ or $-5x+7\ge 22$ and graph the solution set. Write the solution set using interval notation.

Solution

Solve the inequalities separately.

$$\begin{aligned} 2x-3&>-2 \\ 2x-3+3&>-2+3 \\ 2x&>1 \\ \frac{2x}{2}&>\frac{1}{2} \\ x&>\frac{1}{2} \end{aligned} \qquad \text{or} \qquad \begin{aligned} -5x+7&\ge 22 \\ -5x+7-7&\ge 22-7 \\ -5x&\ge 15 \\ \frac{-5x}{-5}&\le\frac{15}{-5} \\ x&\le -3 \end{aligned}$$

The graph of the solution set is

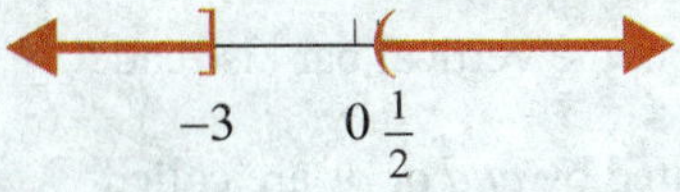

In interval notation, the solution set is $(-\infty,-3]\cup\left(\frac{1}{2},\infty\right)$.

Now work margin exercise 9.

9. Solve the inequality and graph the solution set. Write the solution set using interval notation.

$-3x+5>8$ or $2x-9\ge -1$

Example 10 Solving Compound Inequalities Containing OR

Solve the compound inequality $3x+4\ge 1$ or $2x-7\le -3$ and graph the solution set. Write the solution set using interval notation.

Solution

Solve the inequalities separately.

$$\begin{aligned} 3x+4&\ge 1 \\ 3x+4-4&\ge 1-4 \\ 3x&\ge -3 \\ \frac{3x}{3}&\ge\frac{-3}{3} \\ x&\ge -1 \end{aligned} \qquad \text{or} \qquad \begin{aligned} 2x-7&\le -3 \\ 2x-7+7&\le -3+7 \\ 2x&\le 4 \\ \frac{2x}{2}&\le\frac{4}{2} \\ x&\le 2 \end{aligned}$$

The graph of the solution set is

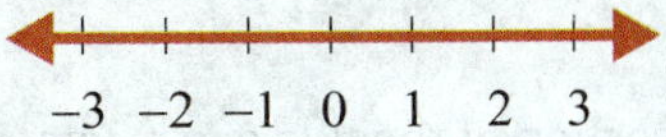

In interval notation, the solution set is $(-\infty,\infty)$.

Now work margin exercise 10.

10. Solve the inequality $4x-3\ge 1$ or $5x+2<12$ and graph the solution set. Write the solution set using interval notation.

Margin Exercise Answers

1. a. $A \cup B = \{1,2,4,6,8\}$ **b.** $A \cap B = \{2,4,8\}$ **2. a.** $M \cup N = \left\{-4,-3,-2,-1,0,\frac{1}{3},5,6,4,9\right\};$

b. $M \cap N = \{-4,-1\}$ **3.** $F \cup G = \{-3,-2,-1,1,2,3\};$ $F \cap G = \varnothing$

4. 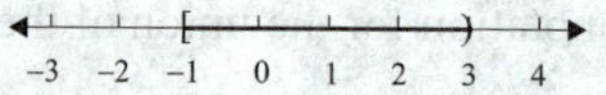

5. 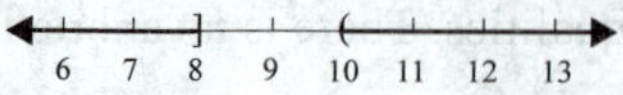

6. 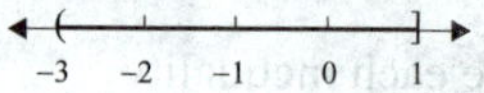

$(-3,1]$ **7.** 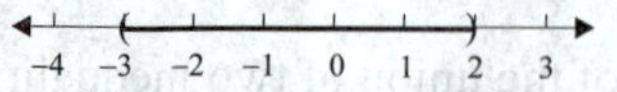

$(-3,2)$

8. 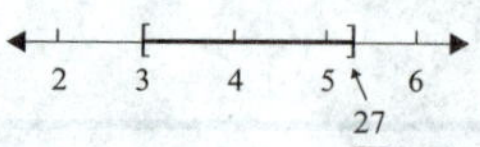

$\left[3,\frac{27}{5}\right]$ **9.** 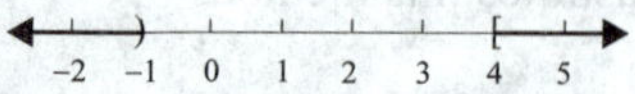

$(-\infty,-1) \cup [4,\infty)$

10. 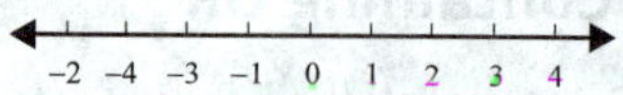

$(-\infty,\infty)$

9.8 Exercises

Concept Check

Fill-in-the-Blank. Complete each sentence using information found in this section.

1. The items in a set are called ______.
2. The word ______ is used to indicate union and the word ______ is used to indicate intersection.
3. In set-builder notation, the vertical bar | is read "______ ______."
4. Two inequalities related by *and* or *or* are called ______ inequalities.
5. A set is in ______ form if the elements are listed within braces, such as $A = \{1,2,3,5,8,\ldots\}$.
6. If the elements in a set cannot be counted, the set is said to be ______.

True/False. Determine whether each statement is true or false. If a statement is false, explain how it can be changed so the statement will be true. (**Note:** There may be more than one acceptable change.)

7. The union of two sets is the set of all elements that belong to both sets.
8. The intersection of two sets is the set of elements that belong to just one set or the other, but not both.
9. A null set contains no elements.
10. The solution set of a compound inequality containing *and* is the union of the solution sets of the two inequalities.

Practice

Find the union and the intersection of the two given sets. See Examples 1 through 3.

1. $A = \{2,4,6,8\}$, $B = \{1,2,3,4\}$ **2.** $R = \{-1,0,1\}$, $S = \{-2,-1,1,2\}$

9.8 Exercises

Concept Check

1. elements
2. or, and
3. such that
4. compound
5. roster
6. infinite
7. False; The union of two sets contains elements that belong to either one set, the other set, or both sets.
8. False; The intersection of two sets contains elements that belong to both sets.
9. True
10. False; The solution is the intersection of the solution sets of the two inequalities.

Practice

1. Union: $\{1, 2, 3, 4, 6, 8\}$; Intersection: $\{2, 4\}$
2. Union: $\{-2, -1, 0, 1, 2\}$; Intersection: $\{-1, 1\}$
3. Union: $\{-4, -3, -2, -1, 0, 1, 2, 4, 6\}$; Intersection: $\{0\}$
4. Union: $\{6, 8, 10, 12, 15, 20, 25\}$; Intersection: $\{10\}$
5. Union: $\{1, 2, 4, 8, 16, 32, 64\}$; Intersection: $\{1, 4, 16\}$
6. Union: $\{-10, -6, -2, 0, 1, 3, 5, 7\}$; Intersection: $\{-10, -2, 0, 5\}$
7. $(3,\infty) \cup (-\infty,2]$

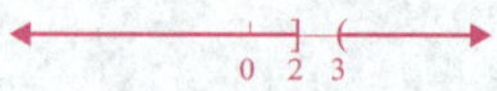

8. $[-5,-1]$

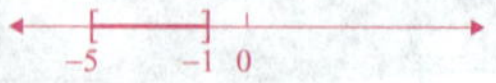

9. $(3, 8)$

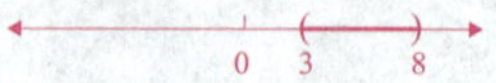

3. $P=\{-4,-2,0,2,4,6\}$, $Q=\{-3,-1,0,1\}$

4. $A=\{10,15,20,25\}$, $B=\{6,8,10,12\}$

5. $E=\{1,2,4,8,16,32\}$, $F=\{1,4,16,64\}$

6. $C=\{-10,-6,-2,0,5,7\}$, $D=\{-10,-2,0,1,3,5\}$

Graph each set of numbers on a real number line. See Examples 4 and 5.

7. $\{x \mid x>3 \text{ or } x\le 2\}$

8. $[-5,-1]$

9. $(3,8)$

10. $[-7,-4)$

11. $(-\infty,-6]$

12. $(4,\infty)$

13. $(-\infty,5)$

14. $(-\infty,4]$

Graph each set of numbers on a real number line.

15. $\{x \mid x \text{ is a whole number less than } 3\}$

16. $\{x \mid x \text{ is an integer with } |x|\le 3\}$

17. $\{x \mid x \text{ is a prime number less than } 20\}$

18. $\{x \mid x \text{ is a positive whole number divisible by } 0\}$

Use set-builder notation to indicate each set of numbers as described.

19. The set of all real numbers between 3 and 5, including 3

20. The set of all real numbers between −4 and 4

21. The set of all real numbers greater than or equal to −2.5

22. The set of all real numbers between −1.8 and 5, including both of these numbers

The graphs of intervals of real numbers are given. **a.** Use set-builder notation to indicate the interval of numbers shown in each graph. **b.** Use interval notation to represent the graph.

23. 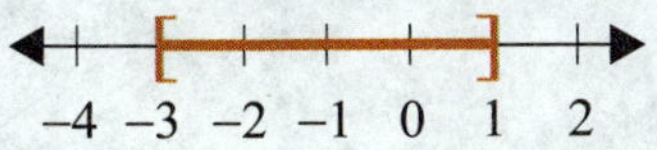

24. 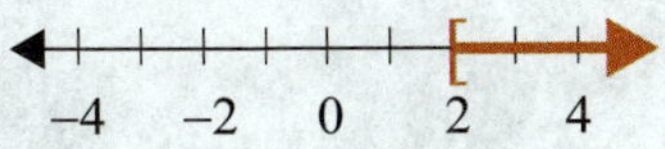

25. 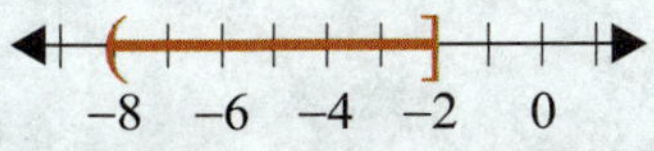

26. 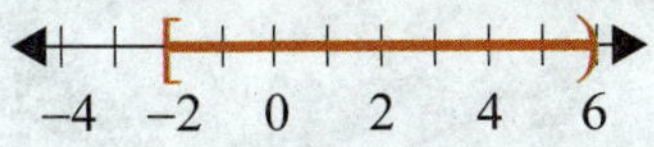

27. 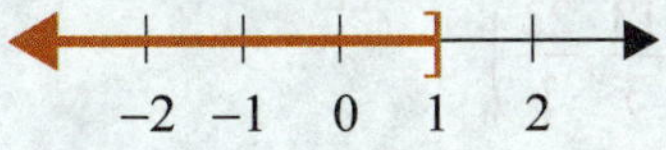

28. 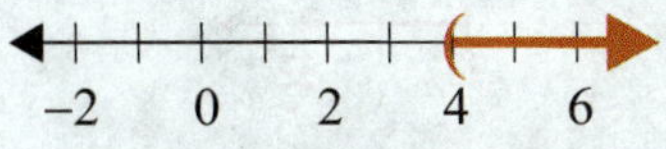

29. 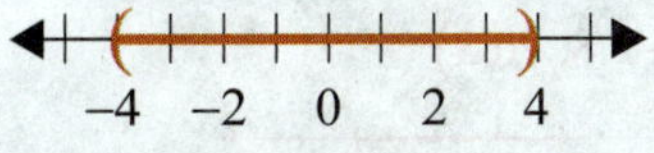

30. 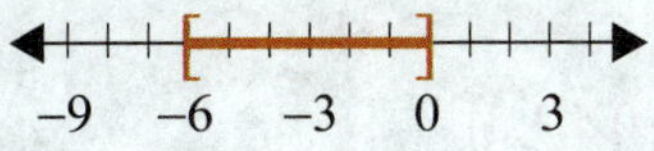

Solve each compound inequality and graph its solution set. Write each solution set using interval notation. See Examples 6 through 10.

31. $\{x \mid x+3>2 \text{ and } x-1<5\}$

32. $\{x \mid x<-1 \text{ or } 2x+1\ge 3\}$

33. $\{x \mid 3x-1\le 5 \text{ or } x+2\ge 8\}$

34. $\{x \mid 5x>15 \text{ and } x+3<10\}$

10. $[-7,-4)$

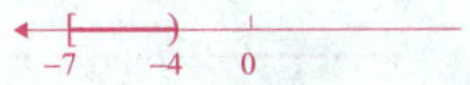

11. $(-\infty,-6]$

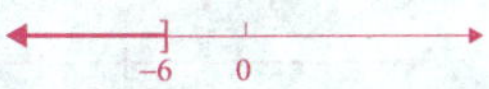

12. $(4,\infty)$

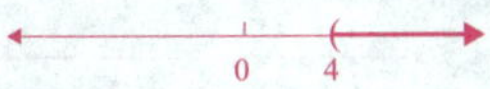

13. $(-\infty,5)$

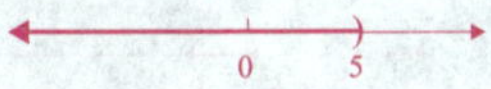

14. $(-\infty,4]$

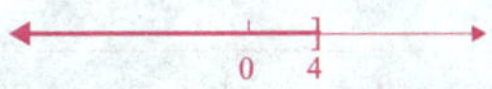

15. 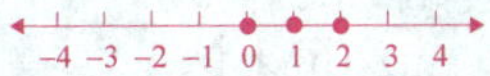

16. 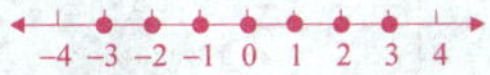

17. 

18. No solution

19. $\{x \mid 3\le x<5\}$

20. $\{x \mid -4<x<4\}$

21. $\{x \mid x\ge -2.5\}$

22. $\{x \mid -1.8\le x\le 5\}$

23. **a.** $\{x \mid -3\le x\le 1\}$ **b.** $[-3,1]$

24. **a.** $\{x \mid x\ge 2\}$ **b.** $[2,\infty)$

25. **a.** $\{x \mid -8<x\le -2\}$ **b.** $(-8,-2]$

26. **a.** $\{x \mid -2\le x<6\}$ **b.** $[-2,6)$

27. **a.** $\{x \mid x\le 1\}$ **b.** $(-\infty,1]$

28. **a.** $\{x \mid x>4\}$ **b.** $(4,\infty)$ **c.** Open interval

29. **a.** $\{x \mid -4<x<4\}$ **b.** $(-4,4)$

30. **a.** $\{x \mid -6\le x\le 0\}$ **b.** $[-6,0]$ **c.** Closed interval

31. $(-1,6)$

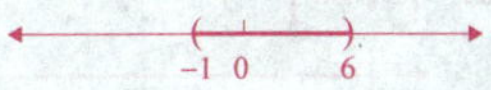

32. $(-\infty,-1)\cup[1,\infty)$

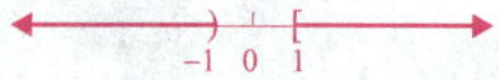

33. $(-\infty,2]\cup[6,\infty)$

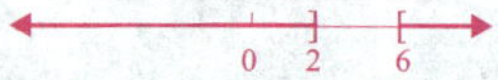

34. $(3, 7)$

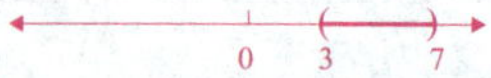

35. $(-1,0)$

36. $(-9,\ 1]$

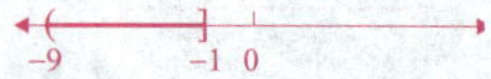

37. $\left[-\dfrac{16}{3},\dfrac{1}{2}\right]$

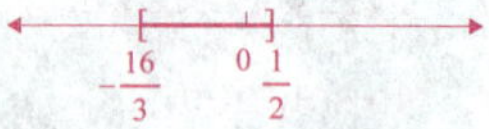

38. $\left[-\dfrac{1}{2},\dfrac{1}{2}\right)$

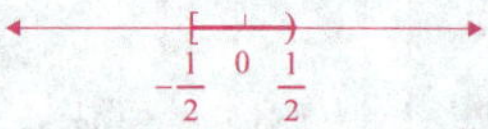

39. $(-\infty,0.4]\cup[5,\infty)$

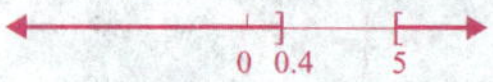

40. $[-6, 9)$

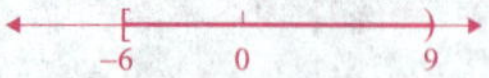

41. $(-9,1)$

42. $[-4, 0]$

43. $\left[\dfrac{1}{2},\dfrac{3}{2}\right]$

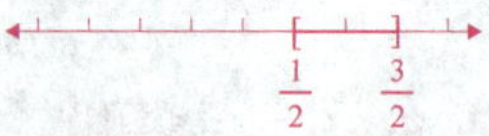

44. $(-2, 3)$

45. $[3, 15]$

35. $\{x\mid 4x+3>-1 \text{ and } -2x+5>5\}$

36. $\{x\mid 12>-2x-6 \text{ or } x+3\le 2\}$

37. $\{x\mid -13\le 3x+3 \text{ and } 0\ge 2x-1\}$

38. $\{x\mid -6x+2\le 5 \text{ and } 2x+1<2\}$

39. $\{x\mid -0.8\le -4x+0.8 \text{ or } 0.3x\ge 1.5\}$

40. $\left\{x\mid -4\le \dfrac{1}{2}x-1 \text{ and } \dfrac{2}{3}x-1<5\right\}$

41. $-4<x+5<6$

42. $2\le -x+2\le 6$

43. $3\ge 4x-3\ge -1$

44. $13>3x+4>-2$

45. $1\le \dfrac{2}{3}x-1\le 9$

46. $-2\le \dfrac{1}{2}x-5\le -1$

47. $14>-2x-6>4$

48. $-11\ge -3x+2>-20$

49. $-1.5<2x+4.1<3.5$

50. $0.9<3x+2.4<6.9$

Writing & Thinking

51. Explain what is meant by the terms *and* and *or* when describing compound inequalities.

46. $[6, 8]$

47. $(-10, -5)$

48. $\left[\dfrac{13}{3},\dfrac{22}{3}\right)$

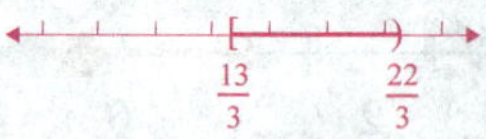

49. $(-2.8, -0.3)$

50. $(-0.5, 1.5)$

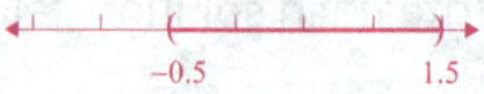

Writing & Thinking

51. Answers will vary. Answers should mention that *and* indicates the intersection and *or* indicates the union.

9.9 Absolute Value Equations

Objectives

A. Solve absolute value equations.

B. Solve equations with two absolute value expressions.

A Solving Absolute Value Equations

The definition of **absolute value** was given in Section 8.1 and is stated again here for easy reference.

> **Absolute Value**
>
> The **absolute value** of a number is its distance from 0 on a number line.
>
> DEFINITION

Equations involving absolute value may have more than one solution (all of which must be included when giving an answer). For example, suppose that $|x| = 3$. Since both 3 and -3 are three units from 0, we have either $x = 3$ or $x = -3$. We can say that the solution set is $\{3, -3\}$. In general, **any number and its opposite have the same absolute value.**

> **Solving Absolute Value Equations**
>
> For $c > 0$:
>
> **a.** If $|\boldsymbol{x}| = \boldsymbol{c}$, then $\boldsymbol{x} = \boldsymbol{c}$ or $\boldsymbol{x} = -\boldsymbol{c}$.
>
> **b.** If $|ax + b| = c$, then $\boldsymbol{ax + b = c}$ or $\boldsymbol{ax + b = -c}$.
>
> DEFINITION

Note

If the absolute value expression is isolated on one side of the equation, we say that the equation is in **standard form**. You may need to manipulate the absolute value equation to get it into standard form before you can solve it. (See Example 1d.)

Example 1 Solving Absolute Value Equations

Solve each absolute value equation.

a. $|x| = 5$

b. $|3x - 4| = 5$

c. $|4x - 1| = -8$

d. $5|3x + 17| - 4 = 51$

Solution

a. $x = 5$ or $x = -5$

b.

$$\begin{aligned} 3x - 4 &= 5 \\ 3x - 4 + 4 &= 5 + 4 \\ 3x &= 9 \\ x &= 3 \end{aligned} \qquad \text{or} \qquad \begin{aligned} 3x - 4 &= -5 \\ 3x - 4 + 4 &= -5 + 4 \\ 3x &= -1 \\ x &= -\frac{1}{3} \end{aligned}$$

c. There is no number that has a negative absolute value. Therefore, this equation has no solution. (The solution is $\varnothing$ and the equation is a contradiction.)

1. Solve each absolute value equation.

a. $|x| = 8$

b. $|5x - 2| = 8$

c. $|2x + 5| = -6$

d. $7|4x + 1| + 5 = 54$

Math Tip

Absolute value equations will always involve solving two equations. One equation is for the positive value of the expression inside the absolute value bars. The other equation is for the negative value of the expression inside the absolute value bars.

d.

$$5|3x+17|-4=51$$ Write the equation.

$$5|3x+17|-4+4=51+4$$ Add 4 to both sides.

$$5|3x+17|=55$$ Simplify.

$$\frac{5|3x+17|}{5}=\frac{55}{5}$$ Divide both sides by 5 to put the equation in standard form.

$$|3x+17|=11$$ Simplify.

$$3x+17=11 \quad \text{or} \quad 3x+17=-11$$

$$3x=-6 \qquad 3x=-28$$

$$x=-2 \qquad x=-\frac{28}{3}$$

Now work margin exercise 1.

B Solving Equations with Two Absolute Value Expressions

If two numbers have the same absolute value, then either they are equal or they are opposites of each other. This fact can be used to solve equations that involve two absolute value expressions.

Solving Equations with Two Absolute Value Expressions

If $|a|=|b|$, then either $\boldsymbol{a=b}$ or $\boldsymbol{a=-b}$.

More generally,

if $|ax+b|=|cx+d|$, then either $\boldsymbol{ax+b=cx+d}$ or $\boldsymbol{ax+b=-(cx+d)}$.

DEFINITION

2. Solve: $|4x-5|=|3x-16|$

Example 2 Solving Equations with Two Absolute Value Expressions

Solve: $|x+5|=|2x+1|$

Solution

In this case, the two expressions $(x+5)$ and $(2x+1)$ are equal to each other or are opposites of each other.

$$|x+5|=|2x+1|$$

$$x+5=2x+1 \quad \text{or} \quad x+5=-(2x+1)$$

$$x+5-x=2x+1-x \qquad x+5=-2x-1$$

$$5=x+1 \qquad x+5+2x=-2x-1+2x$$

$$5-1=x+1-1 \qquad 3x+5=-1$$

$$4=x \qquad 3x+5-5=-1-5$$

$$3x=-6$$

$$\frac{3x}{3}=\frac{-6}{3}$$

$$x=-2$$

Note the use of parentheses. We want the opposite of the entire expression $(2x+1)$.

Make sure to check that both 4 and −2 satisfy the original equation.

Now work margin exercise 2.

Margin Exercise Answers

1. a. $x=-8,\ 8$ **b.** $x=-\frac{6}{5},\ 2$ **c.** no solution **d.** $x=-2,\ \frac{3}{2}$ **2.** $x=-11,\ 3$

9.9 Exercises

Concept Check

Fill-in-the-Blank. Complete each sentence using information found in this section.

1. If an absolute value expression is isolated on one side of an equation, the equation is in ________ form.
2. If two numbers have the same absolute value, then either they are ______ or they are ______ of each other.
3. The absolute value of a number is its ________ from 0 on the number line.
4. The absolute value of any number must be _______ or 0.

True/False. Determine whether each statement is true or false. If a statement is false, explain how it can be changed so the statement will be true. (**Note:** There may be more than one acceptable change.)

5. Equations involving absolute value can only have one solution.
6. If two numbers have the same absolute value, they must be equal to each other.
7. There is no number that has a negative absolute value.
8. If $|a|=|b|$, we can only rewrite it as $a=b$.

9.9 Exercises

Concept Check

1. standard
2. equal, opposites
3. distance
4. positive
5. False; Equations involving absolute value can have more than one solution.
6. False; Must be equal OR the opposite of
7. True
8. False; The equation can also be rewritten as $a=-b$.

Practice

Solve each absolute value equation. See Examples 1 and 2.

1. $|x| = 8$
2. $|x| = 6$
3. $|z| = -\frac{1}{5}$
4. $|z| = \frac{1}{5}$
5. $|x+3| = 2$
6. $|y+5| = -7$
7. $|6x-1| = 9$
8. $|3x+1| = 8$
9. $|6n+4| = 8$
10. $|3x-5| = 10$
11. $|3x+4| = -9$
12. $|-2x+1| = -3$
13. $|-5x+10| = 0$
14. $|6y+4| = 0$
15. $|-4x+1| = 7$
16. $|-3x+4| = 7$
17. $|5x-2|+4 = 7$
18. $|2x-7|-1 = 0$
19. $|-3x+4|-2 = 3$
20. $|-x+5|+1 = 9$
21. $\left|\frac{1}{4}x - \frac{1}{2}\right| = 6$
22. $\left|\frac{1}{5}y - \frac{2}{3}\right| = \frac{2}{3}$
23. $5\left|\frac{x}{2}+1\right| - 7 = 8$
24. $6\left|\frac{x}{5}-2\right| + 5 = 11$
25. $3\left|\frac{x}{3}+1\right| - 5 = -2$
26. $2\left|\frac{x}{4}-3\right| + 6 = 10$
27. $|2x-1| = |x+2|$
28. $|2x-5| = |x-3|$
29. $|x+3| = |x-5|$
30. $|x-8| = |x+4|$
31. $|3x+1| = |4-x|$
32. $|5x+4| = |1-3x|$
33. $\left|\frac{3x}{2}+2\right| = \left|\frac{x}{4}+3\right|$
34. $\left|\frac{x}{3}-4\right| = \left|\frac{5x}{6}+1\right|$
35. $\left|\frac{2x}{5}-3\right| = \left|\frac{x}{2}-1\right|$
36. $\left|\frac{4x}{3}+7\right| = \left|\frac{x}{4}+2\right|$

Practice

1. $x = -8, 8$
2. $x = -6, 6$
3. No solution
4. $z = -\frac{1}{5}, \frac{1}{5}$
5. $x = -5, -1$
6. No solution
7. $x = -\frac{4}{3}, \frac{5}{3}$
8. $x = -3, \frac{7}{3}$
9. $n = -2, \frac{2}{3}$
10. $x = -\frac{5}{3}, 5$
11. No solution
12. No solution
13. $x = 2$
14. $y = -\frac{2}{3}$
15. $x = -\frac{3}{2}, 2$
16. $x = -1, \frac{11}{3}$
17. $x = -\frac{1}{5}, 1$
18. $x = 3, 4$
19. $x = -\frac{1}{3}, 3$
20. $x = -3, 13$
21. $x = -22, 26$
22. $y = 0, \frac{20}{3}$
23. $x = -8, 4$
24. $x = 5, 15$
25. $x = -6, 0$
26. $x = 4, 20$
27. $x = -\frac{1}{3}, 3$
28. $x = 2, \frac{8}{3}$
29. $x = 1$
30. $x = 2$
31. $x = -\frac{5}{2}, \frac{3}{4}$
32. $x = -\frac{5}{2}, -\frac{3}{8}$
33. $x = -\frac{20}{7}, \frac{4}{5}$
34. $x = -10, \frac{18}{7}$
35. $x = -20, \frac{40}{9}$
36. $x = -\frac{108}{19}, -\frac{60}{13}$

9.10 Absolute Value Inequalities

A Solving Absolute Value Inequalities

Now consider an inequality with an absolute value, such as $|x| < 3$. For a number to have an absolute value less than 3, it must be within 3 units of 0. That is, the numbers between −3 and 3 have their absolute values less than 3 because they are within 3 units of 0. Thus, we have the following for $|x| < 3$.

Algebraic Notation	Graph	Interval Notation
$|x| < 3$ $-3 < x < 3$ **(the intersection)**	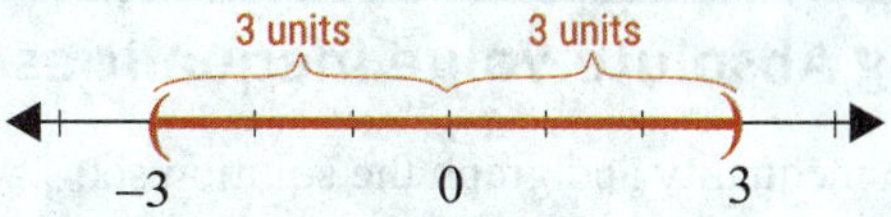	$(-3, 3)$

Table 1

The inequality $|x-5| < 3$ means that the distance between x and 5 is less than 3. That is, we want all the values of x that are within 3 units of 5. The inequality is solved algebraically as follows.

$$|x-5| < 3$$

$$-3 < x-5 < 3 \qquad x-5 \text{ is between } -3 \text{ and } 3.$$

$$-3+5 < x-5+5 < 3+5 \qquad \text{Add 5 to each part of the expression, just as in solving linear inequalities.}$$

$$2 < x < 8 \qquad \text{Simplify each expression.}$$

$$x \text{ is in } (2, 8) \qquad \text{Use interval notation.}$$

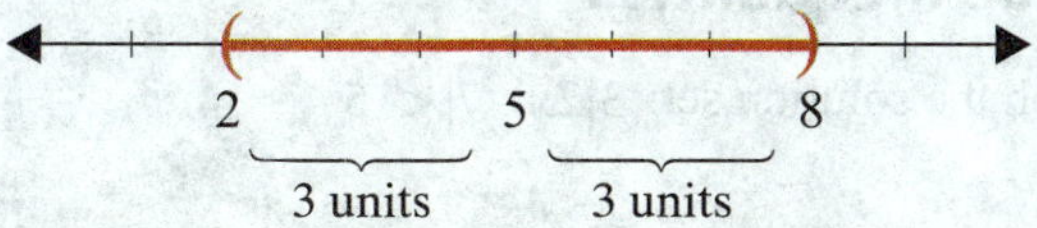

The values for x are between 2 and 8 and are within 3 units of 5.

> **Solving Absolute Value Inequalities with < (or ≤)**
>
> For $c > 0$:
>
> **a.** If $|x| < c$, then $-c < x < c$.
>
> **b.** If $|ax+b| < c$, then $-c < ax+b < c$.
>
> The inequalities in **a.** and **b.** are also true if $<$ is replaced by $\leq$.
>
> DEFINITION

Objectives

A. Solve absolute value inequalities.

Math Tip

The intersection of $-3 < x$ and $x < 3$ can be combined into a single inequality $-3 < x < 3$. Recall that the word **and** represents the intersection of two sets, and any number that is a solution must satisfy **both** inequalities.

Note

If the absolute value expression is isolated on one side of the inequality, we say that the inequality is in **standard form**. You may need to manipulate the absolute value inequality to get it into standard form before you can solve it. (See Examples 3 and 5.)

1. Solve the absolute value inequality $|x| < 3$ and graph the solution set. Write the solution set using interval notation.

Example 1 Solving Absolute Value Inequalities

Solve the absolute value inequality and graph the solution set: $|x| \leq 6$

Solution

$|x| \leq 6$

$-6 \leq x \leq 6$

So, x is in $[-6, 6]$.

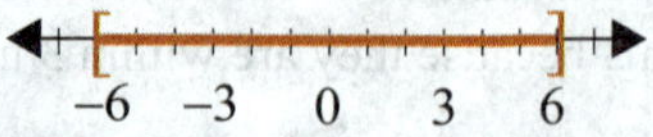

Now work margin exercise 1.

2. Solve the absolute value inequality $|x+1| \leq 4$ and graph the solution set. Write the solution set using interval notation.

Example 2 Solving Absolute Value Inequalities

Solve the absolute value inequality and graph the solution set: $|x+3| < 2$

Solution

$|x+3| < 2$

$-2 < x+3 < 2$

$-2-3 < x+3-3 < 2-3$

$-5 < x < -1$

So, x is in $(-5, -1)$.

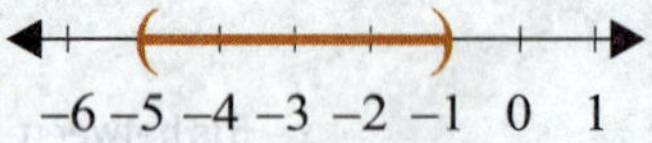

Now work margin exercise 2.

3. Solve the absolute value inequality $4|2x+3| < 20$ and graph the solution set. Write the solution set using interval notation.

Example 3 Solving Absolute Value Inequalities

Solve the absolute value inequality and graph the solution set: $3|2x-7| < 15$

Solution

$3|2x-7| < 15$

$|2x-7| < 5$

Divide both sides by 3 in order to get the inequality in standard form. Remember, we must get the expression in **standard form** before solving an absolute value inequality.

$-5 < 2x-7 < 5$

$-5+7 < 2x-7+7 < 5+7$

$2 < 2x < 12$

$1 < x < 6$

So, x is in $(1, 6)$.

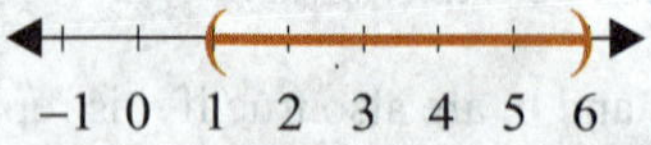

Now work margin exercise 3.

Example 4 Solving Absolute Value Inequalities

Solve the absolute value inequality and graph the solution set: $|x+9| < -\frac{1}{2}$

Solution

Since absolute value is always nonnegative (greater than or equal to 0), no number has an absolute value less than $-\frac{1}{2}$. Thus, there is **no solution**, $\varnothing$.

Now work margin exercise 4.

4. Solve the absolute value inequality $|4x+3| \le -3$ and graph the solution set. Write the solution set using interval notation.

Example 5 Solving Absolute Value Inequalities

Solve the absolute value inequality and graph the solution set: $|2x+4|+4 < 7$

Solution

$|2x+4|+4 < 7$

$|2x+4| < 3$ — Add −4 to both sides in order to get the expression in standard form.

$-3 < 2x+4 < 3$

$-7 < 2x < -1$

$-\frac{7}{2} < x < -\frac{1}{2}$

So, x is in $\left(-\frac{7}{2}, -\frac{1}{2}\right)$.

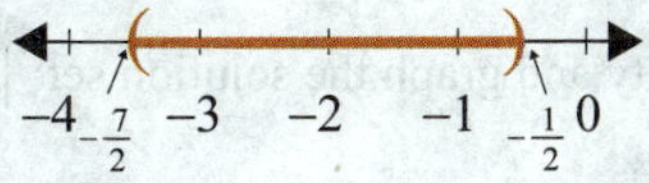

Now work margin exercise 5.

5. Solve the absolute value inequality $|3x-4|+1 \le 9$ and graph the solution set. Write the solution set using interval notation.

We have been discussing inequalities in which the absolute value is less than some positive number. Now consider an inequality where the absolute value is greater than some positive number, such as $|x| > 3$. For a number to have an absolute value greater than 3, its distance from 0 must be greater than 3. That is, numbers that are greater than 3 **or** less than −3 will have absolute values greater than 3. Thus, we have the following for $|x| > 3$.

Algebraic Notation	Graph	Interval Notation
$\lvert x\rvert > 3$ $x > 3$ **or** $x < -3$ **(the union)**	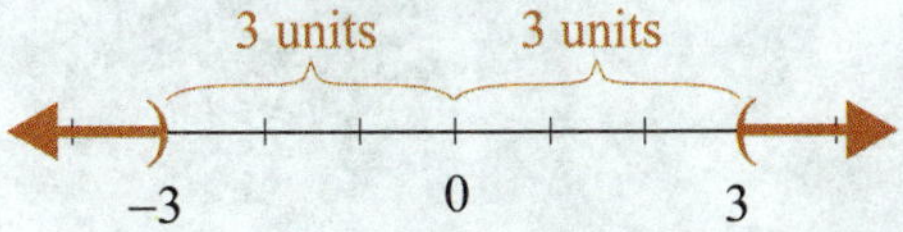	$(-\infty, -3) \cup (3, \infty)$

Table 2

Math Tip

The expression "$x > 3$ or $x < -3$" cannot be combined into one inequality expression without using the absolute value notation. The word *or* must separate the inequalities since any number that satisfies one or the other is a solution to the absolute value inequality. There are no numbers that satisfy both inequalities.

The inequality $|x-5| > 6$ means that the distance between x and 5 is more than 6. That is, we want all values of x that are more than 6 units from 5. The inequality is solved algebraically as follows.

$|x-5| > 6$ indicates that

$x-5 < -6$ **or** $x-5 > 6$. — $x-5$ is less than −6 or greater than 6.

Solving both inequalities gives the following.

$$x - 5 + 5 < -6 + 5 \quad \textbf{or} \quad x - 5 + 5 > 6 + 5$$ Add 5 to each side, just as in solving linear inequalities.

$$x < -1 \qquad\qquad x > 11$$ Simplify.

So, x is in $(-\infty, -1) \cup (11, \infty)$.

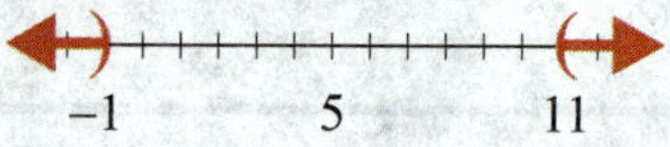

Note: The values for x less than -1 or greater than 11 are more than 6 units from 5. Thus, we can interpret the inequality $|x - 5| > 6$ to mean that the distance from x to 5 is greater than 6.

Solving Absolute Value Inequalities with > (or ≥)

For $c > 0$:

a. If $|x| > c$, then $x < -c$ **or** $x > c$.

b. If $|ax + b| > c$, then $ax + b < -c$ **or** $ax + b > c$.

The inequalities in **a.** and **b.** are true if $>$ is replaced by $\geq$.

DEFINITION

6. Solve the absolute value inequality $|x| > 2$ and graph the solution set. Write the solution set using interval notation.

Example 6 Solving Absolute Value Inequalities

Solve the absolute value inequality and graph the solution set: $|x| \geq 5$

Solution

$|x| \geq 5$

$x \leq -5 \quad \text{or} \quad x \geq 5$

So, x is in $(-\infty, -5] \cup [5, \infty)$.

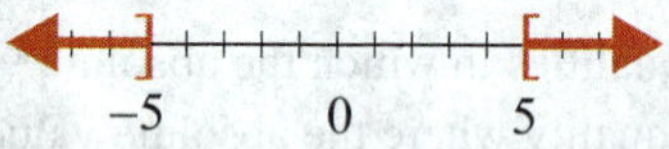

Now work margin exercise 6.

7. Solve the absolute value inequality $|2x + 7| \geq 3$ and graph the solution set. Write the solution set using interval notation.

Example 7 Solving Absolute Value Inequalities

Solve the absolute value inequality and graph the solution set: $|4x - 3| > 2$

Solution

$|4x - 3| > 2$

$4x - 3 < -2 \quad \text{or} \quad 4x - 3 > 2$

$4x < 1 \qquad\qquad 4x > 5$

$x < \dfrac{1}{4} \qquad\qquad x > \dfrac{5}{4}$

So, x is in $\left(-\infty, \dfrac{1}{4}\right) \cup \left(\dfrac{5}{4}, \infty\right)$.

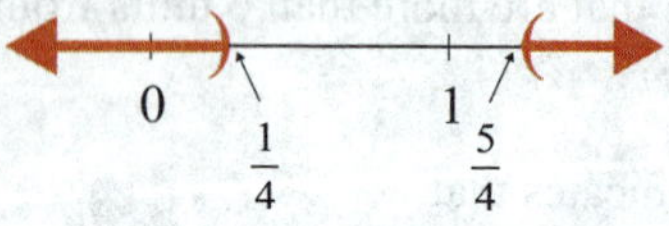

Now work margin exercise 7.

Example 8 Solving Absolute Value Inequalities

Solve the absolute value inequality and graph the solution set: $|3x-8| > -6$

Solution

There is nothing to do here except observe that no matter what is substituted for x, the absolute value will be greater than -6. Absolute value is always nonnegative (greater than or equal to 0). The solution to the inequality is **all real numbers**, so shade the entire number line. In interval notation, x is in $(-\infty, \infty)$.

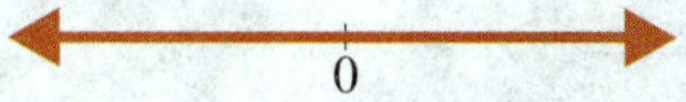

Now work margin exercise 8.

8. Solve the absolute value inequality $|5x-1| \geq -1$ and graph the solution set. Write the solution set using interval notation.

Example 9 Solving Absolute Value Inequalities

Solve the absolute value inequality and graph the solution set: $|2x-5|-5 \geq 4$

Solution

$$|2x-5|-5 \geq 4$$

$$|2x-5| \geq 9$$

Add 5 to both sides in order to get the inequality in **standard form**.

$$2x-5 \leq -9 \quad \text{or} \quad 2x-5 \geq 9$$

$$2x \leq -4 \qquad\qquad 2x \geq 14$$

$$x \leq -2 \qquad\qquad x \geq 7$$

So, x is in $(-\infty, -2] \cup [7, \infty)$.

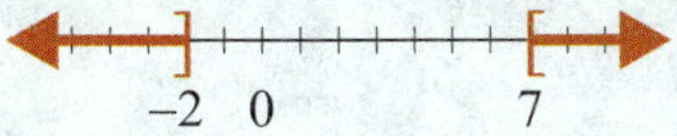

Now work margin exercise 9.

9. Solve the absolute value inequality $|3x+2|+1 \geq 8$ and graph the solution set. Write the solution set using interval notation.

Margin Exercise Answers

1. 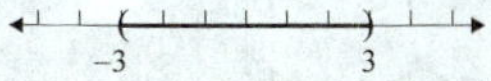

$(-3, 3)$

2. 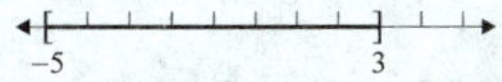

$[-5, 3]$

3. 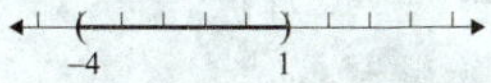

$(-4, 1)$

4. 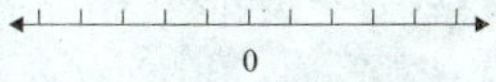

no solution

5. 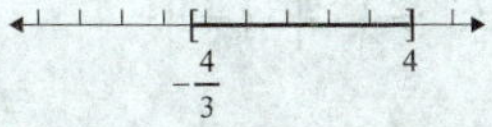

$\left[-\frac{4}{3}, 4\right]$

6. 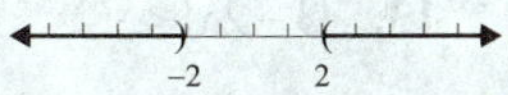

$(-\infty, -2) \cup (2, \infty)$

7. 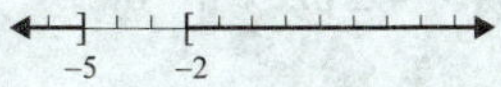

$(-\infty, -5] \cup [-2, \infty)$

8. 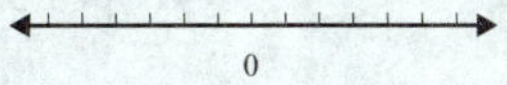

$(-\infty, \infty)$

9. 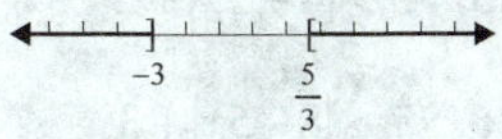

$(-\infty, -3] \cup \left[\frac{5}{3}, \infty\right)$

9.10 Exercises

Concept Check

1. standard
2. distance, greater
3. or
4. real
5. False; Only one statement/inequality must be true.
6. False; The solution is an intersection.
7. False; Must be greater than 2
8. True

Practice

1. $(-\infty,\infty)$

2. $(-\infty,-3]\cup[3,\infty)$

3. $\left(-\frac{4}{5},\frac{4}{5}\right)$

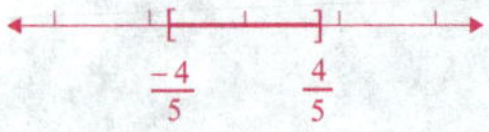

4. $\left(-\infty,-\frac{7}{2}\right]\cup\left[\frac{7}{2},\infty\right)$

5. $(-\infty,1)\cup(5,\infty)$

6. $[-1, 9]$

7. $[-10, -2]$

8. 

No solution

9. $(-\infty,-8]\cup[-2,\infty)$

10. $(-5,7)$

9.10 Exercises

Concept Check

Fill-in-the-Blank. Complete each sentence using information found in this section.

1. If an absolute value expression is isolated on one side of an inequality, the inequality is in ________ form.

2. The inequality $|x-6|>5$ means that the ______ between x and 6 is ____ than 5.

3. If $|x|>c$ then $x<-c$ ______ $x>c$.

4. If an inequality is always true, such as $|3x-8|>-6$, then the solution is all ______ numbers.

True/False. Determine whether each statement is true or false. If a statement is false, explain how it can be changed so the statement will be true. (**Note:** There may be more than one acceptable change.)

5. If the solution is a union, there are two statements or inequalities, both of which must be true.

6. If the solution to a compound inequality is $-4<x<6$, then the solution is a union.

7. For a number to have absolute value greater than 2, its distance from 0 must be less than 2.

8. The inequality $|2x+9|<-2$ has no solution.

Practice

Solve each of the absolute value inequalities and graph the solution sets. Write each solution using interval notation. See Examples 1 through 9.

1. $|x|\ge -2$
2. $|x|\ge 3$
3. $|x|\le \frac{4}{5}$
4. $|x|\ge \frac{7}{2}$
5. $|x-3|>2$
6. $|y-4|\le 5$
7. $|x+6|\le 4$
8. $|x+2|\le -4$
9. $|x+5|\ge 3$
10. $|x-1|<6$
11. $|2x-1|\ge 2$
12. $|3x+4|>-8$
13. $|3-2x|<-2$
14. $|4+3x|>5$
15. $|5+4x|\le 3$
16. $|5x-2|<8$
17. $|3x+4|-1<0$
18. $|2x-3|-3\le 0$
19. $\left|\frac{3x}{2}-4\right|\ge 5$
20. $\left|\frac{3}{7}y+\frac{1}{2}\right|>2$
21. $|2x-9|-7\le 4$
22. $|3x-7|+4\le 4$
23. $-4<|6x-1|+4$
24. $4\le|3x+1|-6$
25. $5>|4-2x|+2$
26. $7>|8-5x|+3$
27. $3|4x+5|-5>10$
28. $6|4x-7|+7>19$
29. $4|7x+9|-3<17$
30. $2|7x-3|+4\ge 12$

Writing & Thinking

A set of real numbers is described. **a**. Sketch a graph of the set on a real number line. **b**. Represent each set using absolute value notation. **c**. Represent each set using interval notation. If the set is one interval, state what type of interval it is.

31. The set of real numbers between -10 and 10, inclusive

32. The set of real numbers within 7 units of 4

33. The set of real numbers more than 6 units from 8

34. The set of real numbers greater than or equal to 3 units from -1

35. The set of real numbers within 2 units of -5

11. $\left(-\infty,-\frac{1}{2}\right]\cup\left[\frac{3}{2},\infty\right)$

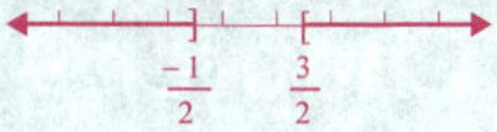

12. $(-\infty,\infty)$

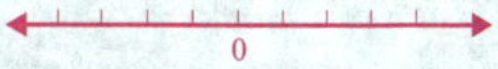

13. 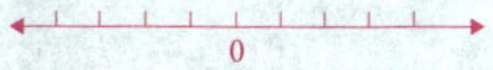

No solution

14. $(-\infty,-3)\cup\left(\frac{1}{3},\infty\right)$

15. $\left[-2,\frac{1}{2}\right]$

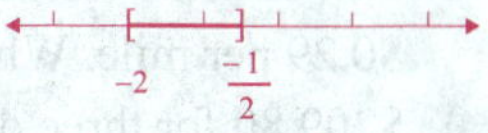

16. $\left(-\frac{6}{5},2\right)$

17. $\left(-\frac{5}{3},-1\right)$

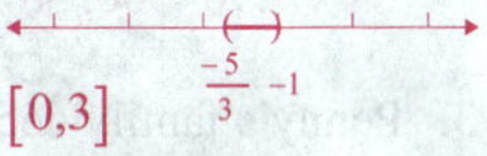

18. $[0,3]$

19. $\left(-\infty,-\frac{2}{3}\right]\cup[6,\infty)$

20. $\left(-\infty,-\frac{35}{6}\right)\cup\left(\frac{7}{2},\infty\right)$

21. $[-1,10]$

22. $\left\{\frac{7}{3}\right\}$

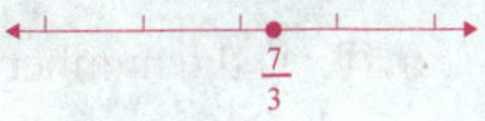

23. $(-\infty,\infty)$

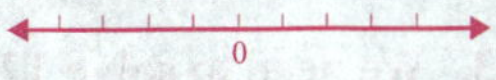

24. $\left(-\infty,-\frac{11}{3}\right]\cup[3,\infty)$

25. $\left(\frac{1}{2},\frac{7}{2}\right)$

26. $\left(\frac{4}{5},\frac{12}{5}\right)$

27. $\left(-\infty,-\frac{5}{2}\right)\cup(0,\infty)$

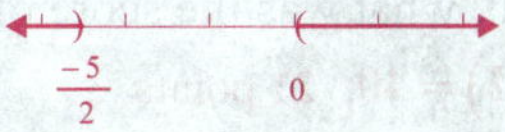

28. $\left(-\infty,\frac{5}{4}\right)\cup\left(\frac{9}{4},\infty\right)$

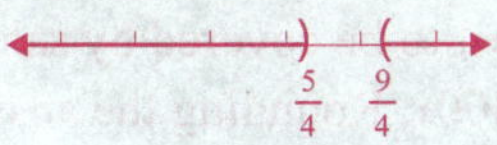

29. $\left(-2,-\frac{4}{7}\right)$

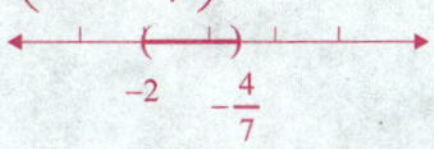

30. $\left(-\infty,-\frac{1}{7}\right]\cup[1,\infty)$

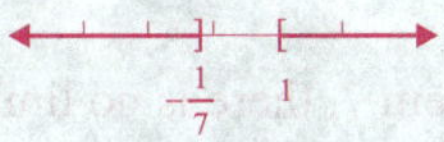

Writing & Thinking

31. a.

b. $|x|\le 10$

c. $[-10,10]$, Closed interval

32. a.

b. $|x-4|<7$

c. $(-3,11)$, Open interval

33. a.

b. $|x-8|>6$

c. $(-\infty,2)\cup(14,\infty)$

34. a.

b. $|x+1|\ge 3$

c. $(-\infty,-4]\cup[2,\infty)$

35. a.

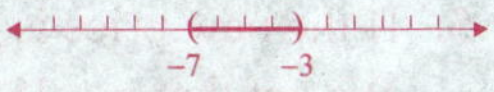

b. $|x+5|<2$

c. $(-7,-3)$, Open interval

Chapter 9 Project

A Linear Vacation

An activity to demonstrate the importance of solving linear equations in real life.

The process of finding ways to use math to solve real-life scenarios is called mathematical modeling. In the following activity you will be using linear equations to model some real-life scenarios that arise during a family vacation.

For each question, write a linear equation in one variable and then solve.

1. Penny and her family went on vacation to Florida and decided to rent a car to do some sightseeing. The cost of the rental car was a fixed price per day plus \$0.29 per mile. When she returned the car, the bill was \$209.80 for three days and they had driven 320 miles. What was the fixed price per day to rent the car?
   $0.29(320)+3x=209.80$; \$39 per day
2. Penny's son Chase wanted to go to the driving range to hit some golf balls. Penny gave the pro-shop clerk \$60 for three buckets of golf balls and received \$7.50 in change. What was the cost of each bucket?
   $3x+7.50=60$; \$17.50 per bucket
3. Penny's family decided to go to the Splash Park. They purchased two adult tickets and two child tickets. The adult tickets were $1\frac{1}{2}$ times the price of the child tickets and the total cost for all four tickets was \$130. What was the cost of each type of ticket?
   $2\left(1\frac{1}{2}x\right)+2x=130$; Child ticket price: \$26.00; Adult ticket price: \$39.00
4. Penny's family went shopping at a nearby souvenir shop where they decided to buy matching T-shirts. If they bought four T-shirts and a \$9.99 bottle of sunscreen for a total cost of \$89.51, before tax, how much did each T-shirt cost? $4x+9.99=89.51$; \$19.88 per T-Shirt
5. Penny and her family went out to eat at a local restaurant. Three of them ordered a shrimp basket, but her daughter Meghan ordered a basket of chicken tenders, which was \$4.95 less than the shrimp basket. If the total order before tax was \$46.85, what was the price of a shrimp basket? $3x+(x-4.95)=46.85$; \$12.95 per shrimp basket
6. While on the beach, Penny and her family decided to play a game of volleyball. Penny and her son beat her husband and daughter by two points. If the combined score of both teams was 40, what was the score of the winning team? $x+(x+2)=40$; 21 points

Chapter 9 Project: Breaking Even

1. **a.** $12{,}000+10(230)=14{,}300$ dollars.
   b. $x=43$ guitars.
   c. We need to solve the inequality $12{,}000 + 230x > 50{,}000$. This gives us that the minimum number is 166 (we round up because the number of guitars must be an integer).
2. **a.** $400(10)=4000$ dollars.
   b. If x is the number of guitars produced, we need to solve the inequality $400x \geq 50{,}000$. This gives us that the minimum number of guitars is 125.
3. We need to solve $12{,}000 + 230x = 400x$. Rounding the answer, we get 71 guitars.
4. A lower fixed cost should lower the break-even point since there would be less cost to be covered by sales. We solve $10{,}000 + 230x = 400x$. Rounding the answer, we get 59 guitars.
5. If the new price is p, we solve $12{,}000+230(80)=80p$. We get that $p=380$ dollars.
6. $400x-(12{,}000+230x)>0$
7. Rounding up, we get $(71,\infty)$. This represents the possible number of guitars that can be manufactured and sold to return a positive profit.
8. Based on the solution to Problem 7, there is no limit to the number of guitars that can be manufactured and sold to return a positive profit. This is not realistic. Explanations will vary.

Chapter 9 Project

Breaking Even

An activity to demonstrate the use of linear equations and inequalities in business.

In manufacturing, the production cost for an item usually has two components: a fixed cost and a variable cost. You can think of the fixed cost as money that must be spent to operate the business regardless of production level. Examples of fixed cost include paying the rent or mortgage for the manufacturing facility or insurance on the property. The variable cost reflects the funds that must be spent to produce one unit of the product. Variable cost should account for things like raw materials and labor costs.

1. A guitar manufacturer has daily fixed cost of $12,000 and each guitar costs $230 to build.
   a. Determine the total cost of producing 10 guitars in one day.
   b. How many guitars were produced in a day when the total cost was $21,890?
   c. What is the minimum number of guitars produced in a day that will make the total cost exceed $50,000?
2. The revenue for a manufacturer is the income generated from selling products. Revenue is defined as the price per unit times the number of units sold. The guitar manufacturer from our previous problem can sell each guitar for $400.
   a. Determine the revenue when 10 guitars are sold.
   b. What is the minimum number of guitars that the manufacturer must sell in a day so that revenue is at least $50,000?
3. Profit is defined as revenue minus cost. The break-even point is the production level at which the profit is exactly zero. Above that level, the manufacturer returns a profit. Determine the break-even point for this guitar manufacturer. In other words, find the number of guitars that must be manufactured and sold in a day to cover all the manufacturing cost.
4. If the manufacturer could decrease fixed cost from $12,000 to $10,000, would you expect the break-even point to go up or down? Explain. Use a linear equation to verify your prediction.
5. Assume that the manufacturer keeps fixed cost at $12,000 with a production cost of $230 per guitar. What should the sale price be to keep a break-even point of 80 units?
6. Write a linear inequality that represents the company returning a positive profit if the fixed cost is $12,000 with a production cost of $230 per guitar and a sale price of $400 per guitar.
7. Solve the inequality found in Problem 6. Round to the nearest integer and write the solution set in interval notation. Explain what this solution set represents.
8. Based on the solution set found in Problem 7, is there a limit to the number of guitars that can be manufactured and sold and still return a profit? Do you think this is realistic? Explain your answer.

10

CHAPTER 10

Graphing Linear Equations and Inequalities

Connections

Rideshare companies such as Uber and Lyft are becoming increasingly more popular in the United States. The website carsurance.net reported that one in every four people in America used a rideshare service at least once per month in 2020. With these numbers, the industry is projected to surpass $220 billion dollars in value by the year 2025.

Prices for the rideshare services vary slightly from company to company, and different markets have their own pricing formulas. However, the general idea is the same: a customer pays a fixed fee to request a ride and then each mile driven costs a certain amount.

Suppose that in Columbus, Ohio, the approximate pricing formulas are as follows: Uber charges a flat fee of $2.80 plus $1.60 per mile and Lyft charges a flat fee of $3.80 plus $1.30 per mile. For very short rides, Uber seems to cost less than Lyft. But as the ride gets longer, Lyft becomes a better deal. The cost of each service as a function of miles is graphed here.

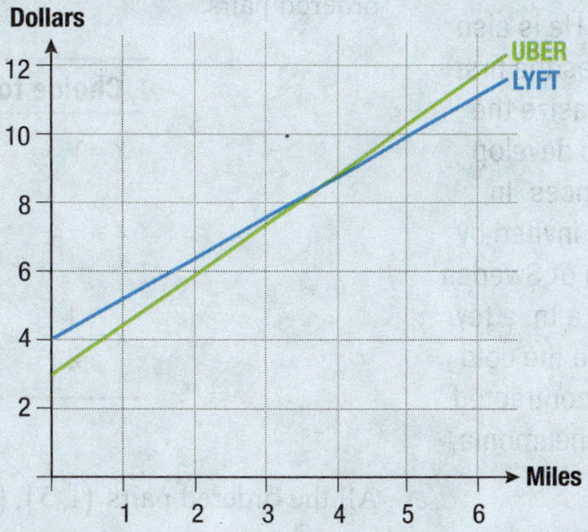

Given this information, could you determine what length a trip must be for a ride with Lyft to be less expensive than a ride with Uber?

Objectives

A. Relate ordered pairs of real numbers using an equation in two variables.

B. Plot ordered pairs of real numbers in the Cartesian coordinate system.

C. Find ordered pairs that satisfy a given linear equation.

D. Identify points on a graph.

10.1 The Cartesian Coordinate System

René Descartes (1596–1650), a famous French mathematician, developed a system for solving geometric problems using algebra. This system is called the **Cartesian coordinate system** or the **rectangular coordinate system**. Descartes based his system on a relationship between points in a plane and **ordered pairs** of real numbers. This section begins by relating algebraic formulas with ordered pairs and then shows how these ideas can be related to geometry.

A Man of Many Talents

In addition to his contributions to mathematics, René Descartes was also one of the founders of modern philosophy. His best known philosophical statement was "I think, therefore I am." He is also often regarded as the first thinker to emphasize the use of reason to develop the natural sciences. In 1649, René was invited by Queen Christina of Sweden to tutor her at 5 a.m. a few times a week. In the cold morning air, he contracted and died from pneumonia.

A Equations in Two Variables

The equation $d = 60t$ represents a relationship between the pair of variables t and d, where t is time and d is distance. For example, if a car is driven at 60 mph (the average speed) for $t = 3$ hours, then $d = 60 \cdot 3 = 180$ miles. With the understanding that t is first and d is second, we can represent $t = 3$ and $d = 180$ in the form of an **ordered pair** $(t, d) = (3, 180)$. We say that $(3, 180)$ **is a solution of** (or **satisfies**) the equation $d = 60t$. **The order of the numbers in an ordered pair is critical.** Thus, we see that the ordered pair $(3, 180)$ is different from $(180, 3)$.

In a similar way, the interest I earned on a principal P invested at 5% can be calculated with the equation $I = 0.05P$. Solutions are of the form (P, I), and one solution is $(100, 5)$, indicating that an investment of \$100 would earn \$5 in interest for one year.

For the equation $y = 2x + 3$, ordered pairs are in the form (x, y), and $(2, 7)$ satisfies the equation. If $x = 2$, then substituting in the equation gives $y = 2(2) + 3 = 7$. In the ordered pair (x, y), x is called the **first coordinate** and y is called the **second coordinate**. To find ordered pairs that satisfy an equation in two variables, we can **choose any value** for one variable and find the corresponding value for the other variable by substituting into the equation. For example, **for the equation** $\boldsymbol{y = 2x + 3}$, we can find the following ordered pairs.

Choice for x	Substitution	Ordered Pair
$x = 1$	$y = 2(1) + 3 = 5$	$(1, 5)$
$x = -2$	$y = 2(-2) + 3 = -1$	$(-2, -1)$
$x = \frac{1}{2}$	$y = 2\left(\frac{1}{2}\right) + 3 = 4$	$\left(\frac{1}{2}, 4\right)$

Table 1

All the ordered pairs $(1, 5)$, $(-2, -1)$, and $\left(\frac{1}{2}, 4\right)$ satisfy the equation $y = 2x + 3$.

There are an infinite number of such ordered pairs. Any real number could have been chosen for x and the corresponding value for y calculated.

Since the equation $y = 2x + 3$ is solved for y, we say that the value of y "depends" on the choice of x. Thus, in an ordered pair of the form (x, y), the first coordinate x is called the **independent variable** and the second coordinate y is called the **dependent variable**.

In the following table, the first variable in each case is the independent variable and the second variable is the dependent variable. Corresponding ordered pairs would be of the form (t, d), (P, I), and (x, y). The choices for the values of the independent variables are arbitrary. There are an infinite number of other values that could have just as easily been chosen.

$d = 60t$

t	d	(t, d)
5	60(5)	(5, 300)
10	60(10)	(10, 600)
12	60(12)	(12, 720)
15	60(15)	(15, 900)

$I = 0.05P$

P	I	(P, I)
100	0.05(100)	(100, 5)
200	0.05(200)	(200, 10)
500	0.05(500)	(500, 25)
1000	0.05(1000)	(1000, 50)

$y = 2x + 3$

x	y	(x, y)
−2	2(−2) + 3	(−2, −1)
−1	2(−1) + 3	(−1, 1)
0	2(0) + 3	(0, 3)
3	2(3) + 3	(3, 9)

Table 2

Math Tip

Independent variables are named such since they do **not** depend on the value of the other (dependent) variable. This means we can choose any convenient value (such as 0 or 1) for the independent variable when it comes to solving or simplifying.

B Plotting Ordered Pairs

The Cartesian coordinate system relates algebraic equations and ordered pairs to geometry. In this system, two number lines intersect at right angles and separate the plane into four **quadrants**. The **origin**, designated by the ordered pair $(0, 0)$, is the point of intersection of the two lines. The horizontal number line is called the **horizontal axis** or ***x*-axis**. The vertical number line is called the **vertical axis** or ***y*-axis**. Points that lie on either axis are not in any quadrant. They are simply on an axis. (See Figure 1.)

Quadrant I
x is positive
y is postive

Quadrant II
x is negative
y is postive

Quadrant III
x is negative
y is negative

Quadrant IV
x is positive
y is negative

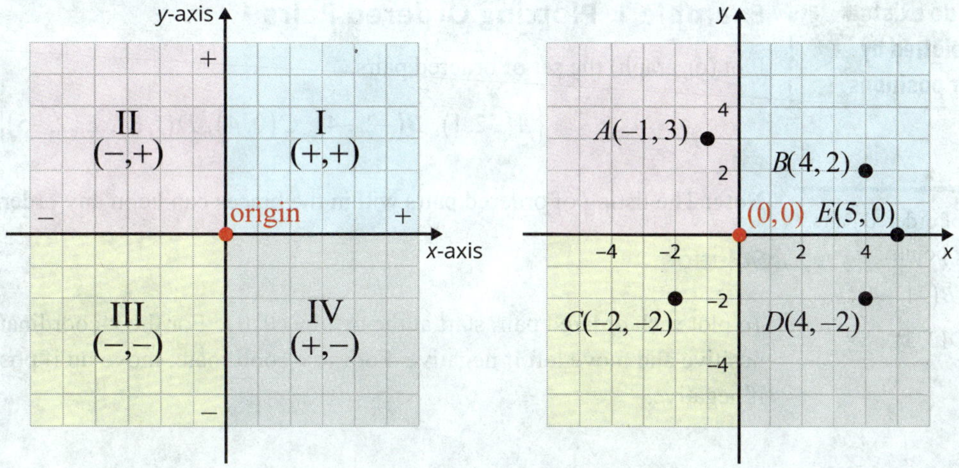

Figure 1

The following important relationship between ordered pairs of real numbers and points in a plane is the cornerstone of the Cartesian coordinate system.

Teaching Note

For students unfamiliar with the concept of a plane, you can explain that a table top, or similar flat surface, can be used to represent a part of a plane.

One-to-One Correspondence

There is a **one-to-one correspondence** between points in a plane and ordered pairs of real numbers.

DEFINITION

In other words, for each point in a plane there is one and only one corresponding ordered pair of real numbers, and for each ordered pair of real numbers there is one and only one corresponding point in the plane.

> **Note**
>
> Although this discussion is related to ordered pairs of real numbers, most of the examples use ordered pairs of **integers**. This is because ordered pairs of integers are relatively easy to plot and read on a graph. Ordered pairs with fractions, decimals, or radicals must be plotted by estimating the positions of the points. The precise coordinates intended for such points can be difficult or impossible to read because large dots must be used so the points can be seen. **Even with these difficulties, you should understand that we are discussing ordered pairs of real numbers and that points with fractions, decimals, and radicals as coordinates do exist and should be plotted by estimating their positions.**

The **points** $A(2,1)$, $B(-2,3)$, $C(-3,-2)$, $D(1,-2)$, and $E(3,0)$ are shown on the graph in Figure 2. (**Note:** An ordered pair of real numbers and the corresponding point on the graph are frequently used to refer to each other. Thus, the ordered pair $(2,1)$ and the point $(2,1)$ are interchangeable ideas.)

Point	Quadrant
$A(2,1)$	I
$B(-2,3)$	II
$C(-3,-2)$	III
$D(1,-2)$	IV
$E(3,0)$	x-axis

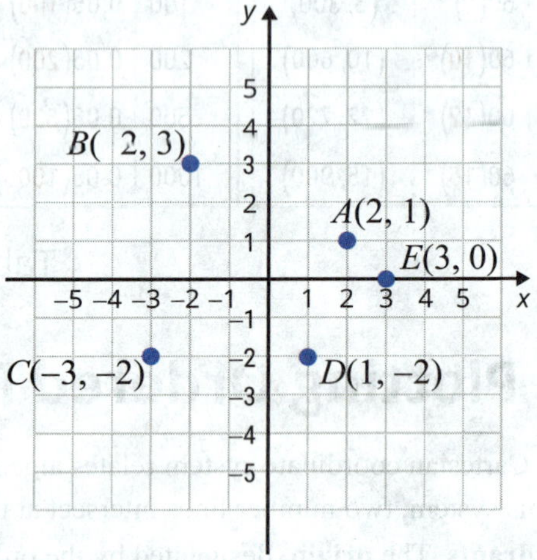

Figure 2

Example 1 Plotting Ordered Pairs

Plot (or graph) the set of ordered pairs.

$$\{A(-2,1),\ B(-2,-4),\ C(0,4),\ D(1,3),\ E(2,-5)\}$$

Note: The listing of ordered pairs within the braces can be in any order.

Solution

To plot each ordered pair, start at the **origin** $(0,0)$. For the x-coordinate, move right if positive and move left if negative. For the y-coordinate, move up if positive and move down if negative.

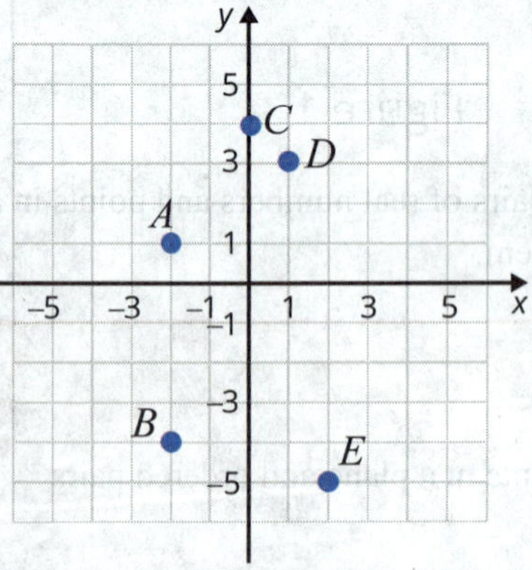

For $A(-2,1)$, move 2 units left and 1 unit up.

For $B(-2,-4)$, move 2 units left and 4 units down.

For $C(0,4)$, move no units left or right and 4 units up.

For $D(1,3)$, move 1 unit right and 3 units up.

For $E(2,-5)$, move 2 units right and 5 units down.

Now work margin exercise 1.

1. Plot the set of ordered pairs, and label the points.
$\{A(-3,2), B(-1,-4), C(3,-1), D(3,3), E(4,0)\}$

Example 2 Plotting Ordered Pairs

Plot (or graph) the set of ordered pairs.

$$\{A(-1,3),\ B(0,1),\ C(1,-1),\ D(2,-3),\ E(3,-5)\}$$

Solution

To plot each ordered pair, start at the **origin**, and move as follows.

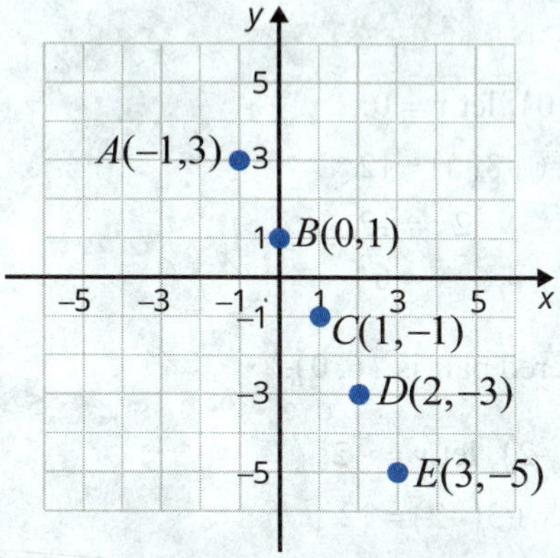

For $A(-1,3)$, move 1 unit left and 3 units up.

For $B(0,1)$, move no units left or right and 1 unit up.

For $C(1,-1)$, move 1 unit right and 1 unit down.

For $D(2,-3)$, move 2 units right and 3 units down.

For $E(3,-5)$, move 3 units right and 5 units down.

Now work margin exercise 2.

2. Plot the set of ordered pairs, and label the points.

$\{A(-3,-1), B(-2,0), C(0,2), D(3,1), E(4,-3)\}$

Teaching Note

You may need to remind your students that there is no significance in the order of the elements in a set. However, the order of values in the ordered pair is important.

C Finding Ordered Pairs That Satisfy Linear Equations

Math Tip

You can choose values for either x or y to find ordered pairs that satisfy the equation.

The points (ordered pairs) in Example 2 can be shown to satisfy the equation $y=-2x+1$. For example, using $x=-1$ in the equation yields

$$\begin{aligned} y &= -2(-1)+1 \\ &= 2+1=3 \end{aligned}$$

and the ordered pair $(-1,3)$ satisfies the equation. Similarly, letting $y=1$ gives

$$\begin{aligned} (1) &= -2x+1 \\ 0 &= -2x \\ 0 &= x, \end{aligned}$$

and the ordered pair $(0,1)$ also satisfies the equation.

We can write all the ordered pairs in Example 2 in table form.

x	$-2x+1=y$	(x,y)
−1	$-2(-1)+1=3$	$(-1,3)$
0	$-2(0)+1=1$	$(0,1)$
1	$-2(1)+1=-1$	$(1,-1)$
2	$-2(2)+1=-3$	$(2,-3)$
3	$-2(3)+1=-5$	$(3,-5)$

Table 3

3. Determine the missing coordinates in the ordered pairs so that each point will satisfy the equation $x+5y=15$.

$(0,\ ), (\ ,2), (15,\ ), (30,\ )$

Example 3 Finding Ordered Pairs

Find the missing coordinates in the ordered pairs so that each point will satisfy the equation $2x + 3y = 12$.

$$(0,\ ),\ (3,\ ),\ (\ ,0),\ (\ ,-2)$$

Solution

The missing values can be found by substituting the given values for x (or for y) into the equation $2x + 3y = 12$ and solving for the other variable.

For $(0,\ )$, let $x = 0$:

$$2(0)+3y=12$$
$$3y=12$$
$$y=4$$

The ordered pair is $(0, 4)$.

For $(3,\ )$, let $x = 3$:

$$2(3)+3y=12$$
$$6+3y=12$$
$$3y=6$$
$$y=2$$

The ordered pair is $(3, 2)$.

For $(\ ,0)$, let $y = 0$:

$$2x+3(0)=12$$
$$2x=12$$
$$x=6$$

The ordered pair is $(6, 0)$.

For $(\ ,-2)$, let $y = -2$:

$$2x+3(-2)=12$$
$$2x-6=12$$
$$2x=18$$
$$x=9$$

The ordered pair is $(9, -2)$.

Now work margin exercise 3.

4. Complete the table so that each ordered pair will satisfy the equation $y=-3x+2$.

x	y	(x, y)
0		
	1	
–2		
	0	

Example 4 Finding Ordered Pairs

Complete the table so that each ordered pair will satisfy the equation $y = -3x + 1$.

x	y	(x, y)
0		
	4	
$\frac{1}{3}$		
3		

Solution

Substitute each given value for x or y into the equation $y = -3x + 1$ to find the ordered pairs and complete the table.

For $x = 0$:

$$y=-3(0)+1$$
$$y=0+1$$
$$y=1$$

For $y = 4$:

$$(4)=-3x+1$$
$$3=-3x$$
$$-1=x$$

For $x = \frac{1}{3}$:

$$y=-3\left(\frac{1}{3}\right)+1$$
$$y=-1+1$$
$$y=0$$

For $x = 3$:

$$y=-3(3)+1$$
$$y=-9+1$$
$$y=-8$$

The completed table is as follows.

x	*y*	(*x*, *y*)
0	1	(0, 1)
–1	4	(–1, 4)
$\frac{1}{3}$	0	$\left(\frac{1}{3}, 0\right)$
3	–8	(3, –8)

Now work margin exercise 4.

Example 5 Determining Ordered Pairs

Determine which, if any, of the ordered pairs $(0, -2)$, $\left(\frac{2}{3}, 0\right)$, and $(2, 5)$ satisfy the equation $y = 3x - 2$.

Solution

To determine whether an ordered pair is a solution to an equation, substitute the *x*-coordinate for *x* and the *y*-coordinate for *y* in the equation $y = 3x - 2$ to see if the result is a true statement.

For the ordered pair $(0, -2)$, let $x = 0$ and $y = -2$.

$$(-2) \stackrel{?}{=} 3(0) - 2$$
$$-2 \stackrel{?}{=} 0 - 2$$
$$-2 = -2 \quad \text{True statement}$$

For the ordered pair $\left(\frac{2}{3}, 0\right)$, let $x = \frac{2}{3}$ and $y = 0$.

$$(0) \stackrel{?}{=} 3\left(\frac{2}{3}\right) - 2$$
$$0 \stackrel{?}{=} 2 - 2$$
$$0 = 0 \quad \text{True statement}$$

For the ordered pair $(2, 5)$, let $x = 2$ and $y = 5$.

$$(5) \stackrel{?}{=} 3(2) - 2$$
$$5 \stackrel{?}{=} 6 - 2$$
$$5 \neq 4 \quad \text{False statement}$$

So the ordered pairs $(0, -2)$ and $\left(\frac{2}{3}, 0\right)$ satisfy the equation, and the ordered pair $(2, 5)$ does not.

Now work margin exercise 5.

5. Determine which, if any, of the ordered pairs $(0, -3)$, $(3, 6)$, and $\left(\frac{1}{4}, -2\right)$ satisfy the equation $y = 4x - 3$.

D Identifying Points on a Graph

6. The graphs of two lines are given. Use the grid to locate three points on each line.

a.

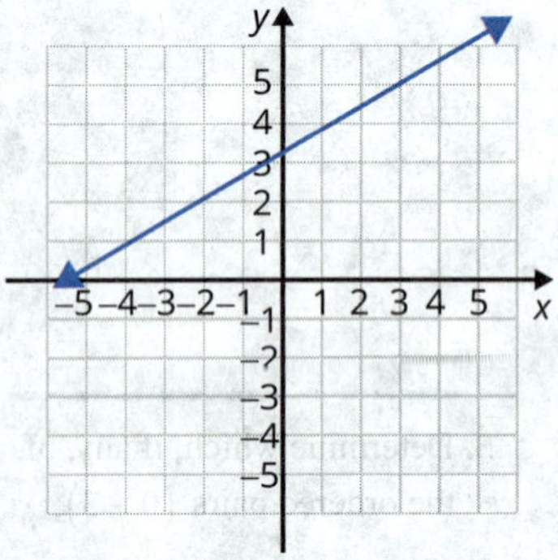

b.

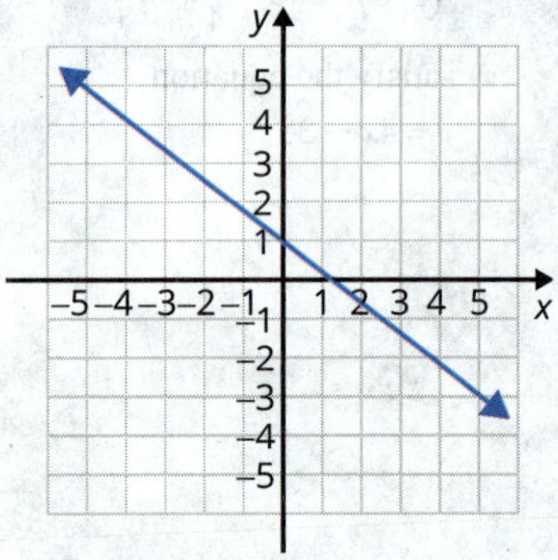

Example 6 Locating Points on the Graph of a Line

The graphs of two lines are given. Each line contains an infinite number of points. Use the grid to help you locate (or estimate) three points on each line.

a.

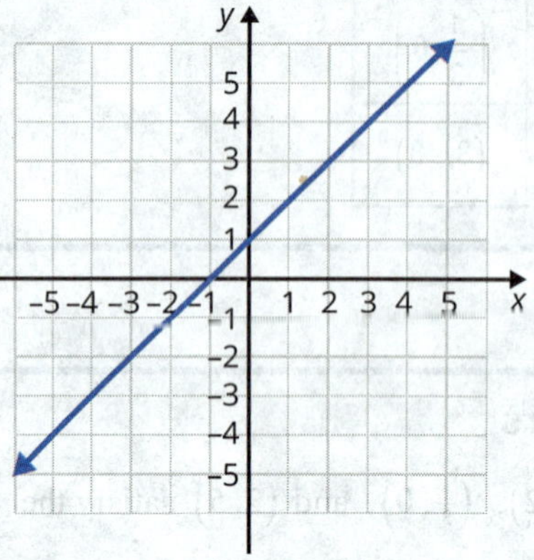

b.

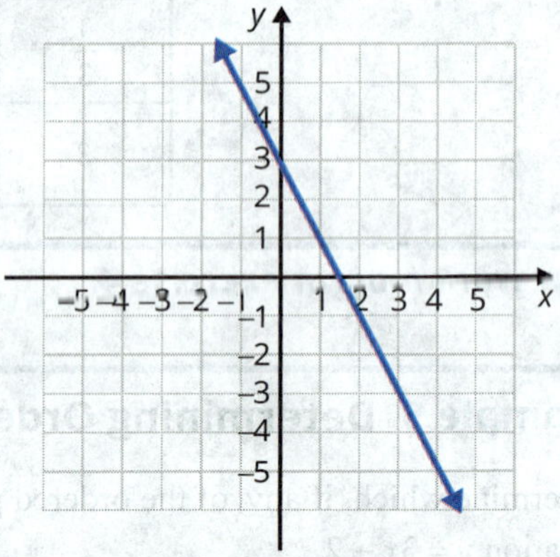

Solution

a. Three points on this graph are $(-2, -1), (1, 2)$, and $(3, 4)$. (Of course, there is more than one correct answer to this type of question. Use your own judgment.)

b. Three points on this graph are $(0, 3), (1, 1)$, and $(2, -1)$. (You may also estimate with fractions. For example, one point appears to be approximately $\left(\frac{1}{2}, 2\right)$.)

Now work margin exercise 6.

Margin Exercise Answers

1.

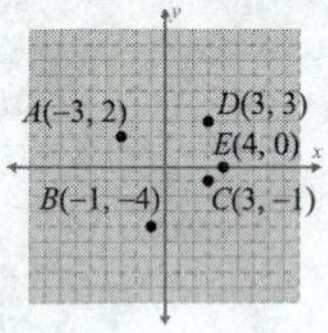

2.

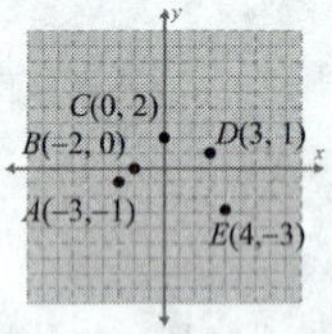

3. $(0, 3), (5, 2), (15, 0), (30, -3)$

4.

x	y	(x, y)
0	2	$(0, 2)$
$\frac{1}{3}$	1	$\left(\frac{1}{3}, 1\right)$
–2	8	$(-2, 8)$
$\frac{2}{3}$	0	$\left(\frac{2}{3}, 0\right)$

5. $(0, -3)$ and $\left(\frac{1}{4}, -2\right)$ satisfy the equation $y = 4x - 3$ **6. a.** $(-4, 1), (-2, 2), (3, 5)$

b. $(-5, 5), (0, 1), (5, -3)$

10.1 Exercises

Concept Check

Fill-in-the-Blank. Complete each sentence using information found in this section.

1. The Cartesian coordinate system has a vertical and a horizontal line that separate a plane into four _________.

2. In an ordered pair, x represents the _________ (first/second) coordinate and y represents the _________ (first/second) coordinate.

3. If an ordered pair has two negative coordinates, the graph of the corresponding point is in Quadrant _________.

4. If an ordered pair satisfies an equation, it is a/an _________ of the equation.

5. The point of intersection of the x-axis and y-axis is called the _________.

6. Linear equations have a/an _________ number of solutions.

True/False. Determine whether each statement is true or false. If a statement is false, explain how it can be changed so the statement will be true. (**Note:** There may be more than one acceptable change.)

7. The graph of every ordered pair that has a positive x-coordinate and a negative y-coordinate can be found in Quadrant IV.

8. To find the y-value that corresponds with $x = 2$, substitute 2 for x into the given equation and solve for y.

9. If $(-7, 3)$ is a solution of $y = 3x + 24$, then $(-7, 3)$ satisfies $y = 3x + 24$.

10. If point $A = (0, 4)$, then point A lies on the x-axis.

Practice

For each graph, list the set of ordered pairs corresponding to the points on the graph.

1.

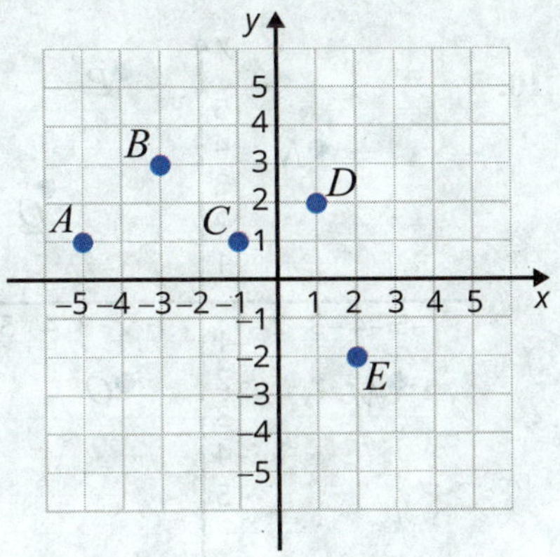

2.

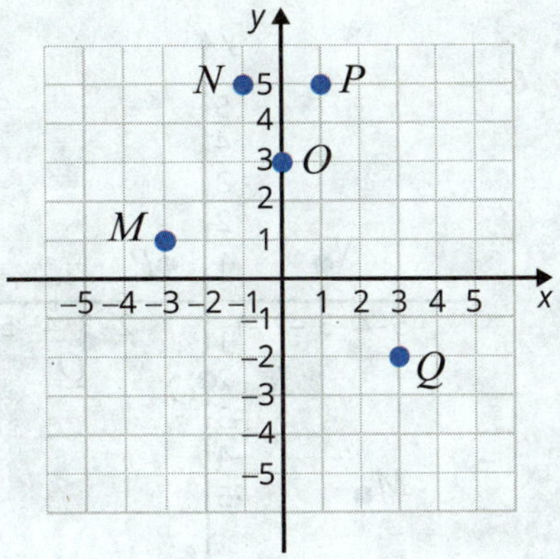

10.1 Exercises

Concept Check

1. quadrants
2. first, second
3. III
4. solution
5. origin
6. infinite
7. True
8. True
9. True
10. False; y-axis

Practice

1. $\{A(-5,1), B(-3,3), C(-1,1), D(1,2), E(2,-2)\}$
2. $\{M(-3,1), N(-1,5), O(0,3), P(1,5), Q(3,-2)\}$

3. $\left\{A(-3,-2), B(-1,-3), C(-1,3), D(0,0), E(2,1)\right\}$

4. $\left\{M(-3,2), N(-1,-1), O(1,5), P(3,-2), Q(5,5)\right\}$

5. $\left\{A(-4,4), B(-3,-4), C(0,-4), D(0,3), E(4,1)\right\}$

6. $\left\{M(-4,-5), N(-3,1), O(0,-2), P(1,1), Q(3,-1)\right\}$

7. $\left\{A(-3,-5), B(-1,4), C(0,-1), D(3,1), E(6,0)\right\}$

8. $\left\{M(-6,2), N(-1,6), O(0,0), P(1,-6), Q(6,3)\right\}$

9. $\left\{A(-5,0), B(-2,2), C(-1,-4), D(0,6), E(2,0)\right\}$

10. $\left\{M(-5,-2), N(-3,4), O(2,-2), P(2,6), Q(4,3)\right\}$

11.

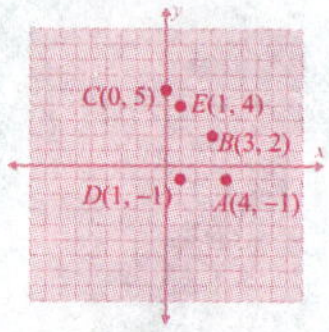

12.

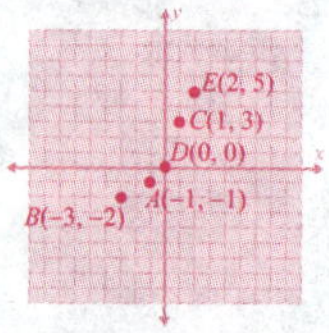

13.

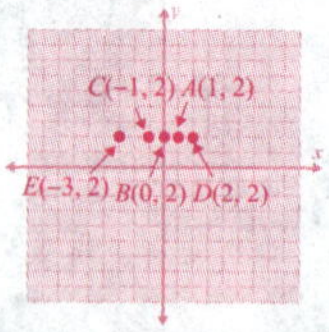

3.

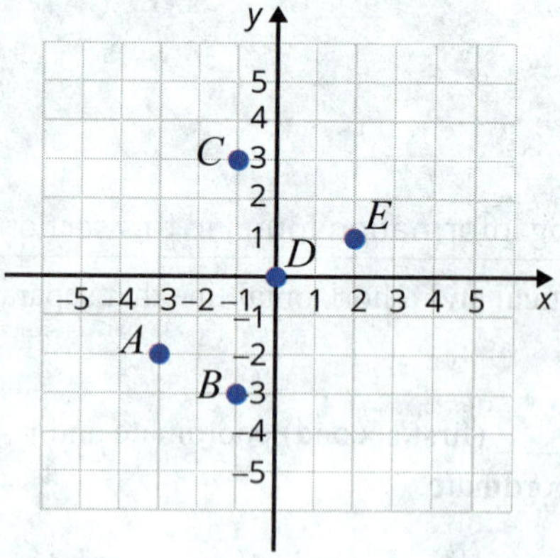

4.

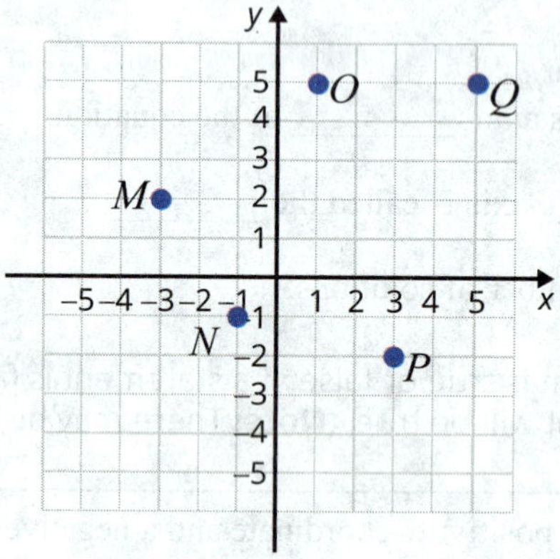

5.

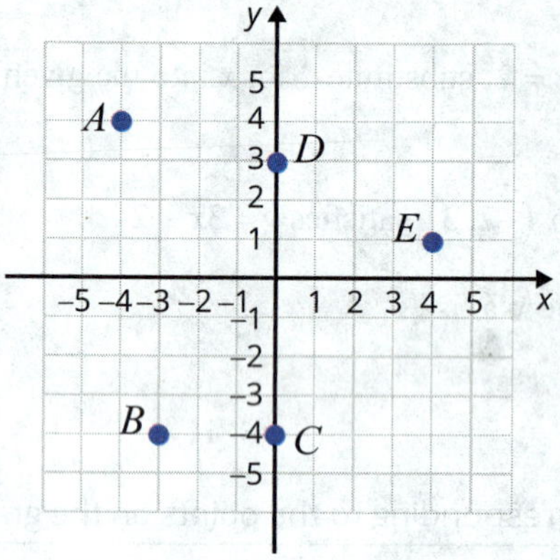

6.

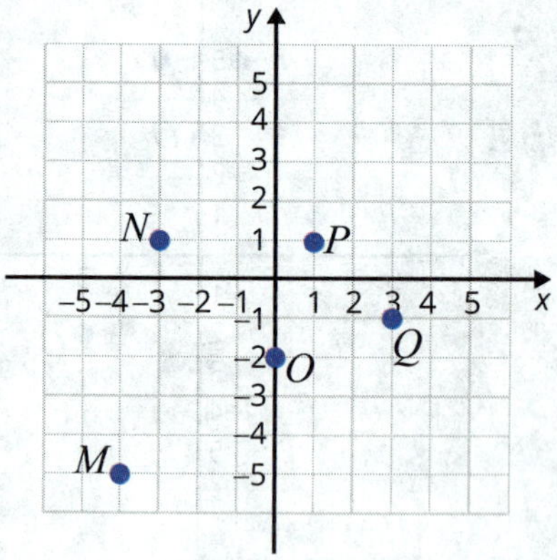

7.

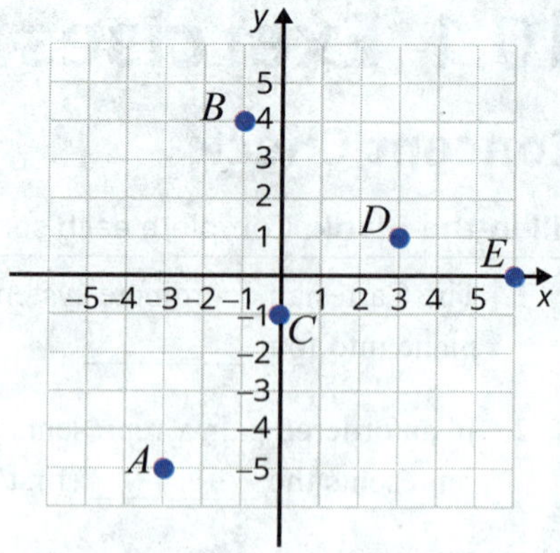

8.

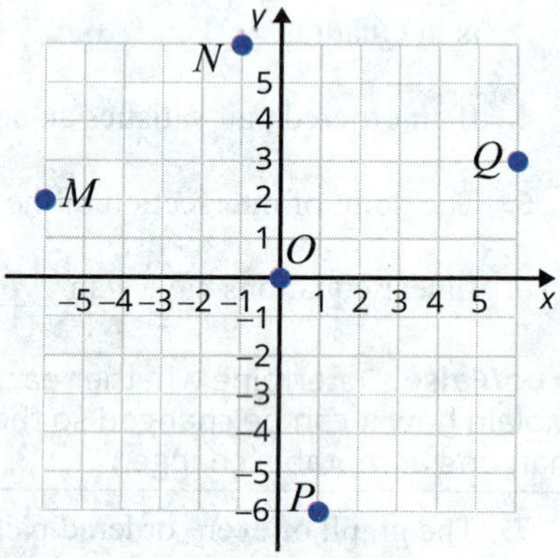

9.

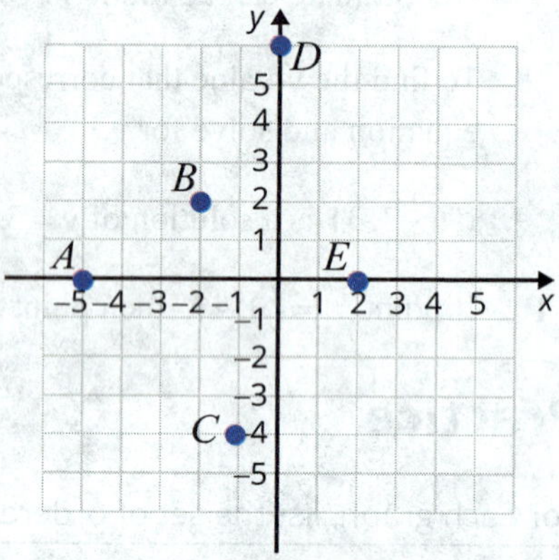

10.

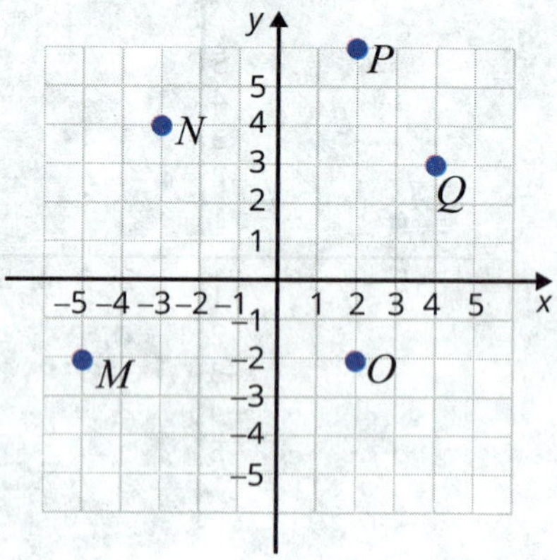

Plot each set of ordered pairs and label the points. See Examples 1 and 2.

11. $\{A(4,-1), B(3,2), C(0,5), D(1,-1), E(1,4)\}$

12. $\{A(-1,-1), B(-3,-2), C(1,3), D(0,0), E(2,5)\}$

13. $\{A(1,2), B(0,2), C(-1,2), D(2,2), E(-3,2)\}$

14. $\{A(-1,4), B(0,-3), C(2,-1), D(4,1), E(-1,-1)\}$

15. $\{A(1,0), B(3,0), C(-2,1), D(-1,1), E(0,0)\}$

16. $\{A(-1,-1), B(0,1), C(1,3), D(2,5), E(3,10)\}$

17. $\{A(4,1), B(0,-3), C(1,-2), D(2,-1), E(-4,2)\}$

18. $\{A(0,1), B(1,0), C(2,-1), D(3,-2), E(4,-3)\}$

19. $\{A(1,4), B(-1,-2), C(0,1), D(2,7), E(-2,-5)\}$

20. $\{A(0,0), B(-1,3), C(3,-2), D(0,4), E(-7,0)\}$

21. $\left\{A(1,-3), B\left(-4,\frac{3}{4}\right), C\left(2,-2\frac{1}{2}\right), D\left(\frac{1}{2},4\right)\right\}$

22. $\left\{A\left(\frac{3}{4},\frac{1}{2}\right), B\left(2,-\frac{5}{4}\right), C\left(\frac{1}{3},-2\right), D\left(-\frac{5}{3},2\right)\right\}$

23. $\{A(1.6,-2), B(3,2.5), C(-1,1.5), D(0,-2.3)\}$

24. $\{A(-2,2), B(-3,1.6), C(3,0.5), D(1.4,0)\}$

Determine the missing coordinate in each of the ordered pairs so that the point will satisfy the equation given. See Example 3.

25. $x - y = 4$

a. (0,__)

b. (2,__)

c. (__,0)

d. (__,−3)

26. $x + y = 7$

a. (0, __)

b. (−1,__)

c. (__,0)

d. (__,3)

27. $x + 2y = 6$

a. (0, __)

b. (2, __)

c. (__,0)

d. (__, 4)

28. $3x + y = 9$

a. (0, __)

b. (4,__)

c. (__,0)

d. (__,3)

14.

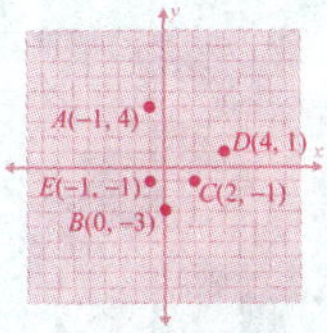

15.

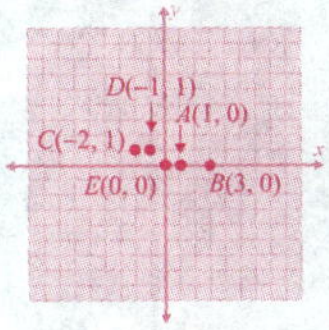

16.

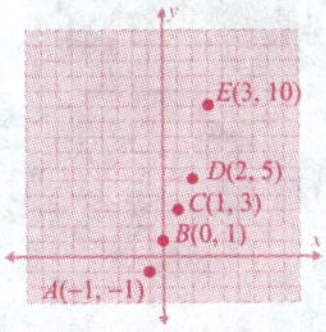

17.

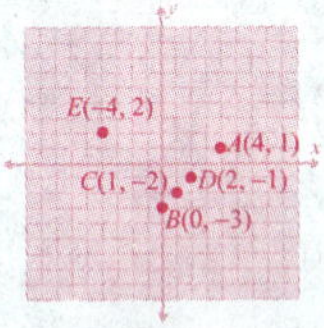

18.

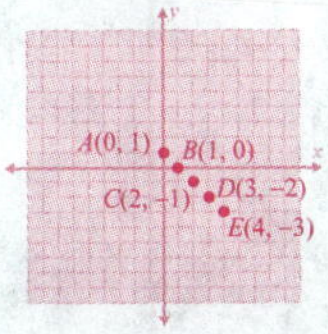

19.

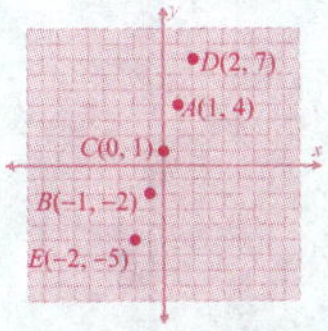

20.

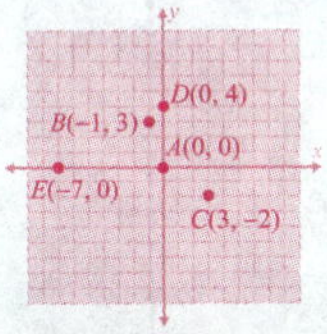

21.

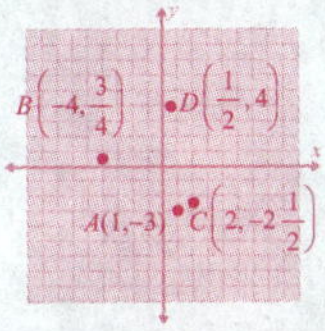

22.

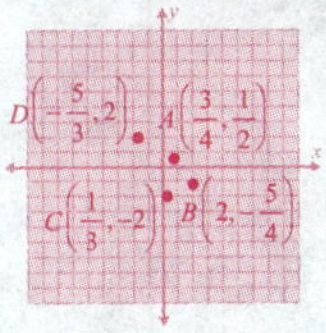

23.

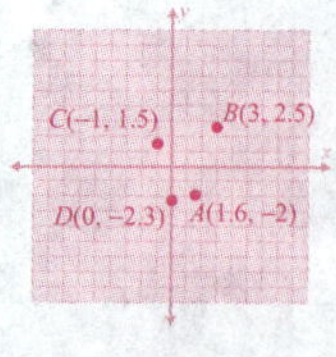

24.

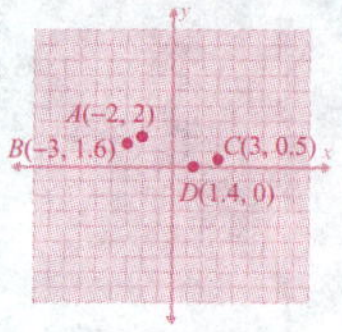

25. $(0, -4), (2, -2), (4, 0), (1, -3)$

26. $(0, 7), (-1, 8), (7, 0), (4, 3)$

27. $(0, 3), (2, 2), (6, 0), (-2, 4)$

28. $(0, 9), (4, -3), (3, 0), (2, 3)$

29. $(0, -8), (1, -4), (2, 0), (3, 4)$

30. $(0, -1), (4, 1), (2, 0), (8, 3)$

31. $(0, 2), \left(-1, \frac{8}{3}\right), (3, 0), (6, -2)$

32. $(0, 5), \left(2, \frac{5}{3}\right), (3, 0), \left(\frac{3}{5}, 4\right)$

33. $\left(0, \frac{-7}{4}\right), (1, -1), \left(\frac{7}{3}, 0\right), \left(3, \frac{1}{2}\right)$

34. $\left(0, \frac{6}{5}\right), \left(\frac{1}{2}, 1\right), (3, 0), (-2, 2)$

35.

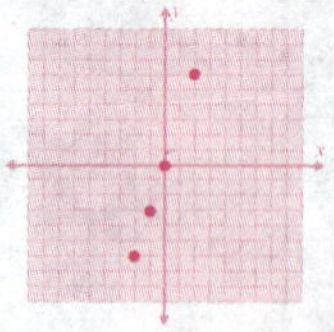

$(0, 0), (-1, -3), (-2, -6), (2, 6)$

29. $4x - y = 8$

a. $(0, \_\_)$

b. $(1, \_\_)$

c. $(\_\_, 0)$

d. $(\_\_, 4)$

30. $x - 2y = 2$

a. $(0, \_\_)$

b. $(4, \_\_)$

c. $(\_\_, 0)$

d. $(\_\_, 3)$

31. $2x + 3y = 6$

a. $(0, \_\_)$

b. $(-1, \_\_)$

c. $(\_\_, 0)$

d. $(\_\_, -2)$

32. $5x + 3y = 15$

a. $(0, \_\_)$

b. $(2, \_\_)$

c. $(\_\_, 0)$

d. $(\_\_, 4)$

33. $3x - 4y = 7$

a. $(0, \_\_)$

b. $(1, \_\_)$

c. $(\_\_, 0)$

d. $\left(\_\_, \frac{1}{2}\right)$

34. $2x + 5y = 6$

a. $(0, \_\_)$

b. $\left(\frac{1}{2}, \_\_\right)$

c. $(\_\_, 0)$

d. $(\_\_, 2)$

Complete the tables so that each ordered pair will satisfy the given equation. Plot the resulting sets of ordered pairs. See Example 4.

35. $y = 3x$

x	y
0	
	−3
−2	
	6

36. $y = -2x$

x	y
0	
	4
3	
	−2

37. $y = 2x - 3$

x	y
0	
	−1
−2	
	3

38. $y = 3x + 5$

x	y
0	
	−4
−2	
	2

39. $y = 9 - 3x$

x	y
0	
	0
1	
	–3

40. $y = 6 - 2x$

x	y
0	
	0
–2	
	–2

41. $y = \frac{3}{4}x + 2$

x	y
0	
	5
–4	
	$\frac{5}{4}$

42. $y = \frac{3}{2}x - 1$

x	y
0	
	2
–2	
	$-\frac{5}{2}$

43. $3x - 5y = 9$

x	y
0	
	0
–2	
	–1

44. $4x + 3y = 6$

x	y
0	
	0
3	
	–1

45. $5x - 2y = 10$

x	y
0	
	0
–1	
	5

46. $3x - 2y = 12$

x	y
	0
0	
	–3
6	

47. $2x + 3.2y = 6.4$

x	y
0	
3.2	
	0.8
	–0.2

48. $3x + y = -2.4$

x	y
	0
0	
	0.6
1.6	

36.

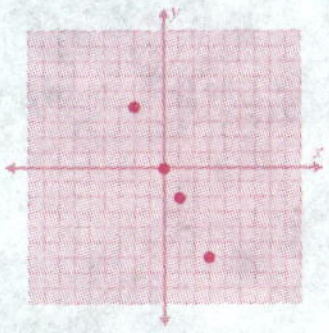

$(0, 0), (-2, 4), (3, -6), (1, -2)$

37.

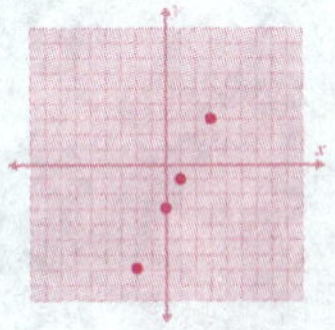

$(0, -3), (1, -1), (-2, -7), (3, 3)$

38.

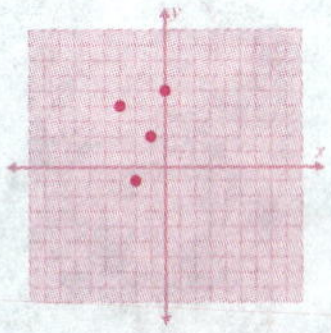

$(0, 5), (-3, 4), (-2, -1), (-1, 2)$

39.

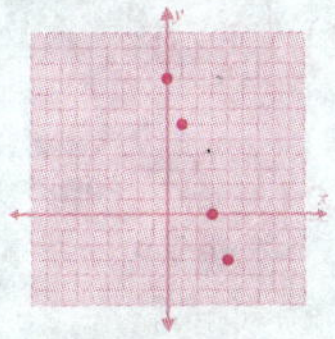

$(0, 9), (3, 0), (1, 6), (4, -3)$

40.

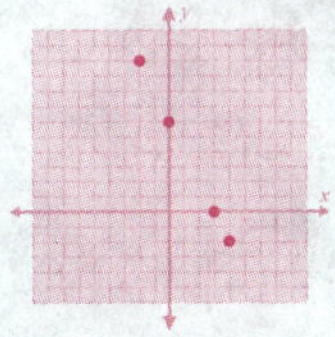

$(0, 6), (3, 0), (-2, 10), (4, -2)$

41.

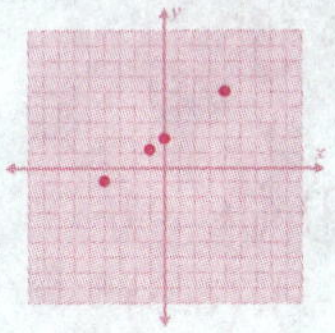

$(0, 2), (4, 5), (-4, -1), \left(-1, \frac{5}{4}\right)$

42. 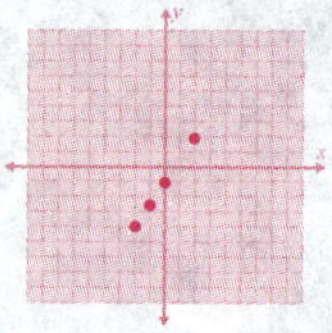

$(0,-1),(2,2),$
$(-2,-4),\left(-1,-\frac{5}{2}\right)$

43. 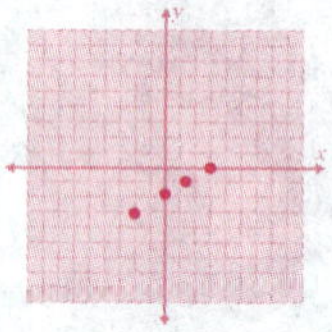

$\left(0,-\frac{9}{5}\right),(3,0),$
$(-2,-3),\left(\frac{4}{3},-1\right)$

44. 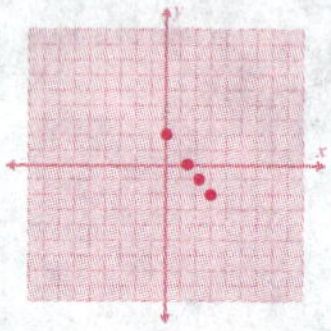

$(0,2),\left(\frac{3}{2},0\right),$
$(3,-2),\left(\frac{9}{4},-1\right)$

45. 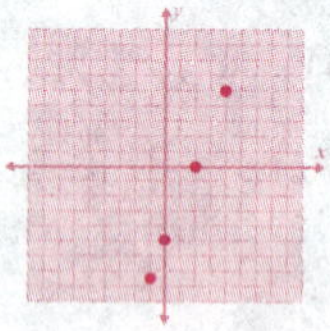

$(0,-5),(2,0),$
$\left(-1,-\frac{15}{2}\right),(4,5)$

46. 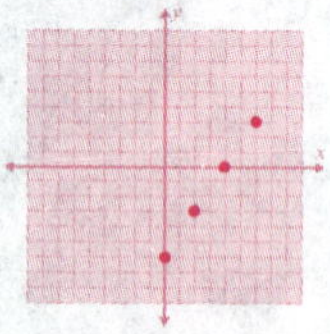

$(4,0),(0,-6),$
$(2,-3),(6,3)$

Determine which, if any, of the ordered pairs satisfy the given equation. See Example 5.

49. $y=2x-4$

a. $(1,1)$
b. $(2,0)$
c. $(1,-2)$
d. $(3,2)$

50. $y=-4x+5$

a. $\left(\frac{3}{4},2\right)$
b. $(4,0)$
c. $(1,1)$
d. $(0,3)$

51. $x+2y=-1$

a. $(1,-1)$
b. $(1,0)$
c. $(2,1)$
d. $(3,-2)$

52. $2x-3y=7$

a. $(1,3)$
b. $\left(\frac{1}{2},-2\right)$
c. $\left(\frac{7}{2},0\right)$
d. $(2,1)$

53. $2x+5y=8$

a. $(4,0)$
b. $(2,1)$
c. $(1,1.2)$
d. $(1.5,1)$

54. $3x+4y=10$

a. $(-2,3)$
b. $(0,2.5)$
c. $(4,-2)$
d. $(1.2,1.6)$

The graph of a line is shown. List any three points on each line. (There is more than one correct answer.) See Example 6.

55. 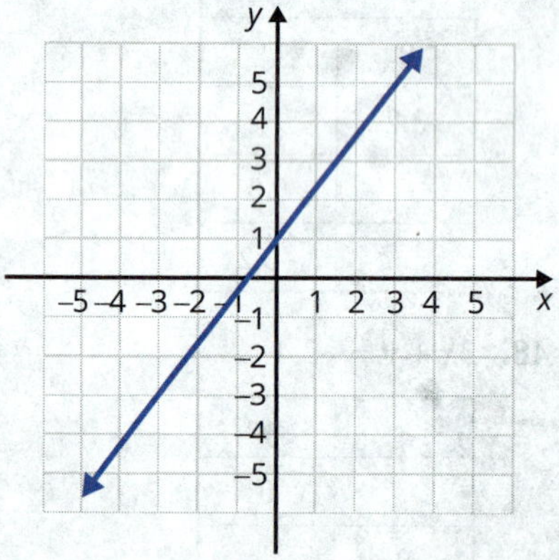

56. 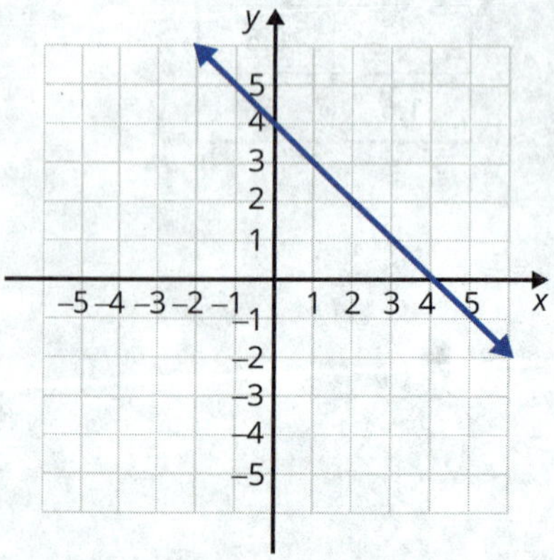

57. 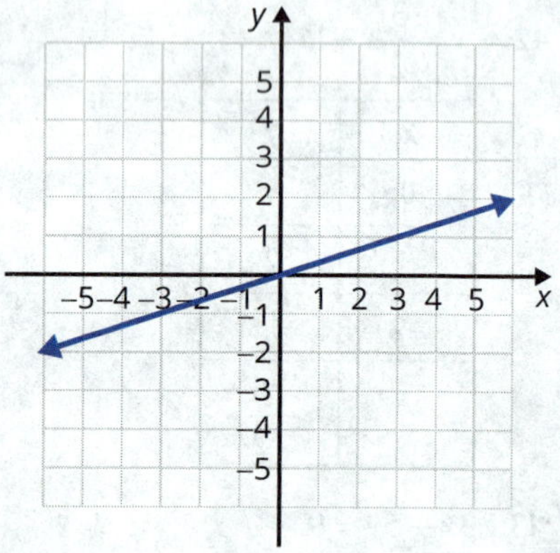

58. 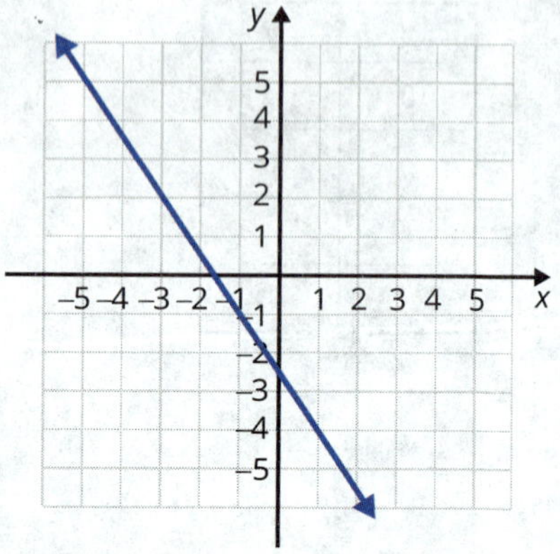

59.

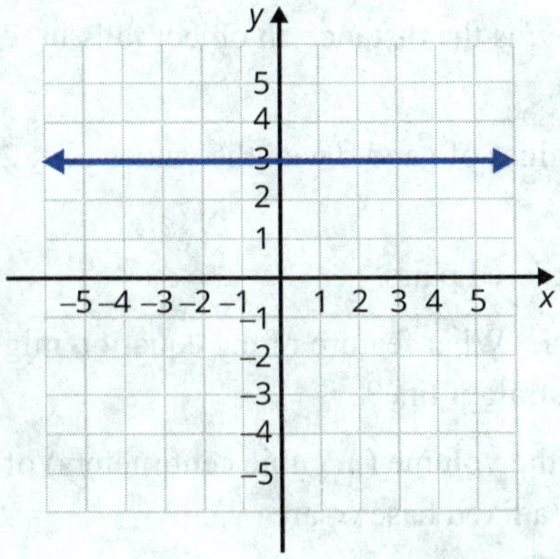

62.

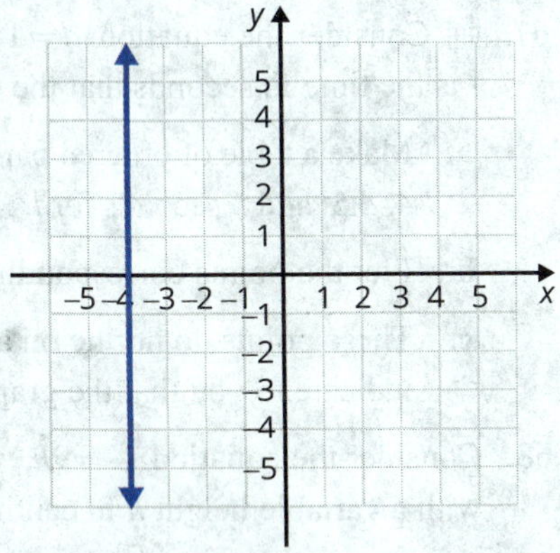

60.

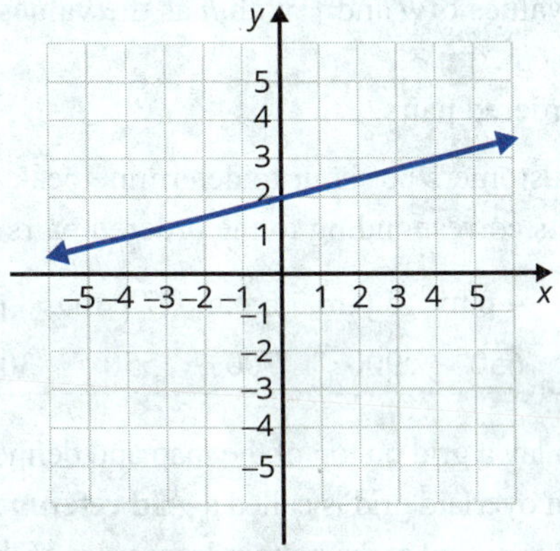

63.

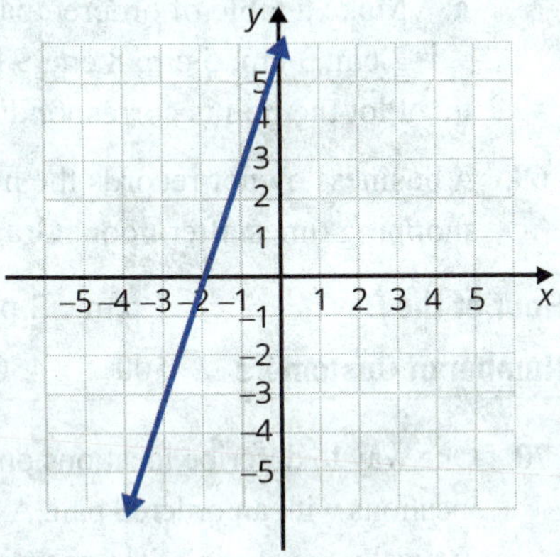

61.

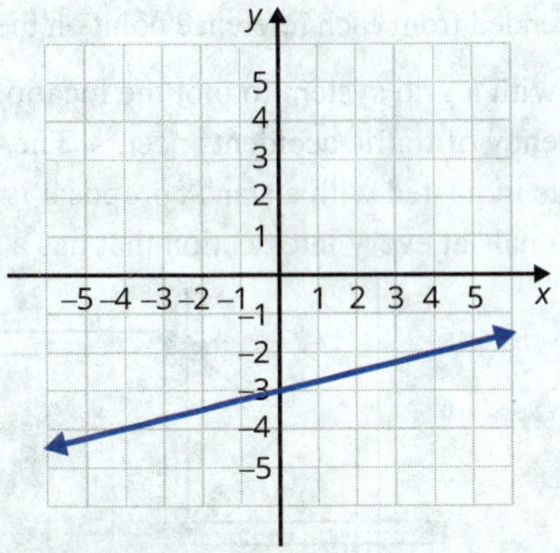

64.

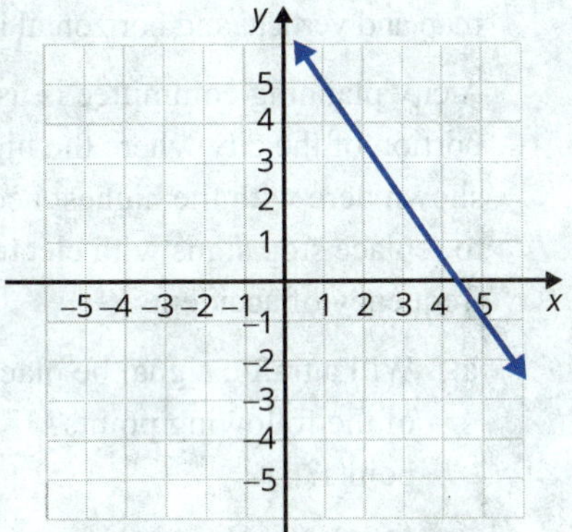

Applications

Solve.

65. At one point in 2017, the exchange rate from US dollars to Euros was $E = 0.85D$ where E is Euros and D is dollars.

a. Make a table of ordered pairs for the values of D and E if D has the values \$100, \$200, \$300, \$400, and \$500.

b. Plot the points corresponding to the ordered pairs.

66. Consider the equation $F = \frac{9}{5}C + 32$, where C is temperature in degrees Celsius and F is the corresponding temperature in degrees Fahrenheit.

a. Make a table of ordered pairs for the values of C and F if C has the values −20°, −10°, −5°, 0°, 5°, 10°, and 15°.

b. Plot the points corresponding to the ordered pairs.

47.

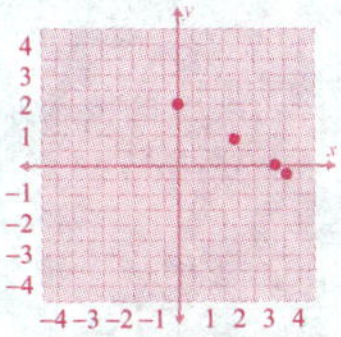

$(0, 2), (3.2, 0), (1.92, 0.8), (3.52, -0.2)$

48.

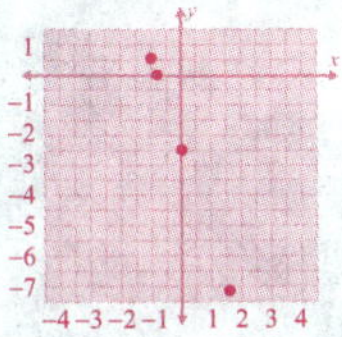

$(-0.8, 0), (0, -2.4), (-1, 0.6), (1.6, -7.2)$

49. b, c, d

50. a, c

51. a, d

52. b, c

53. a, c, d

54. b, d

55. For example, $(-3, -3)$, $(0, 1)$, and $(3, 5)$.

56. For example, $(0, 4)$, $(2, 2)$, and $(4, 0)$.

57. For example, $(-3, -1)$, $(0, 0)$, and $(3, 1)$.

58. For example, $(-5, 5)$, $(-3, 2)$, and $(1, -4)$.

59. For example, $(-2, 3)$, $(0, 3)$, and $(4, 3)$.

60. For example, $(-4, 1)$, $(0, 2)$, and $(4, 3)$.

61. For example, $(-4, -4)$, $(0, -3)$, and $(4, -2)$.

62. For example, $(-4, -3)$, $(-4, 0)$, and $(-4, 2)$.

63. For example, $(-3, -3)$, $(-2, 0)$, and $(-1, 3)$.

64. For example, $(1, 5)$, $(3, 2)$, and $(5, -1)$.

Applications

67. Consider the equation $d = 16t^2$, where d is the distance an object falls in feet and t is the time in seconds that the object falls.

a. Make a table of ordered pairs for the values of t and d with the values of 1, 2, 3.5, 4, 4.5, and 5 seconds for t.

b. Plot the points corresponding to the ordered pairs.

c. These points do not lie on a straight line. What feature of the equation might indicate to you that the graph is not a straight line?

68. Consider the equation $V = 9h$, where V is the volume (in cubic centimeters) of a box with a variable height h in centimeters and a fixed base of area 9 cm².

a. Make a table of ordered pairs for the values of h and V with h as the values 2 cm, 3 cm, 5 cm, 8 cm, 9 cm, and 10 cm.

b. Plot the points corresponding to the ordered pairs.

69. A business owner records the number of customers per hour to determine peak shopping times after noon. Graph the points corresponding to the ordered pairs.

Hour of the Day	1 p.m.	2 p.m.	3 p.m.	4 p.m.	5 p.m.	6 p.m.	7 p.m.	8 p.m.
Number of Customers	500	450	200	650	900	700	550	300

70. One way to describe locations on a map is to lay a grid on top of the map and define locations with an ordered pair. A map with an overlaid grid is called a **grid reference**. Typically the origin of the coordinate system is placed at the bottom left corner of the map and vertical and horizontal lines are extended from each reference point on the axes.

A city planning committee is using a map with a grid system to plot the locations in a portion of the city where the highest frequency of traffic accidents occurs. The map is shown here with the highest accident points indicated with a star. A proposal is made to replace stop signs with electric traffic signals at every intersection that has a high frequency of accidents.

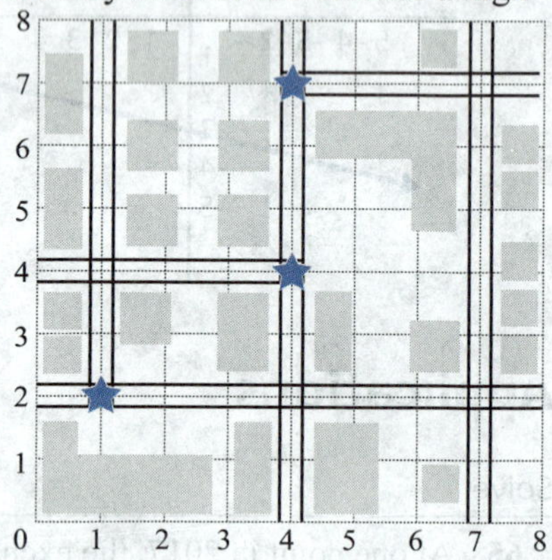

a. Will a traffic signal be placed at any of the following points? If so, which points?

(2, 1), (4, 4), (0, 5)

b. Which points not listed in part a. will have traffic signals?

65. a.

D	E
100	85
200	170
300	255
400	340
500	425

b.

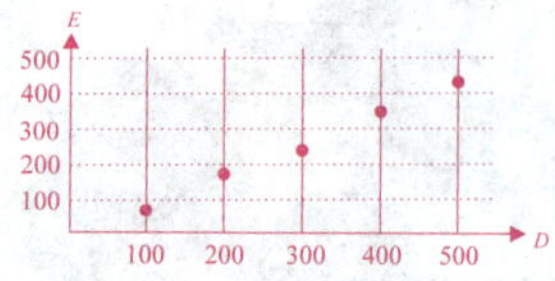

66. a.

C	F
−20	−4
−10	14
−5	23
0	32
5	41
10	50
15	59

b.

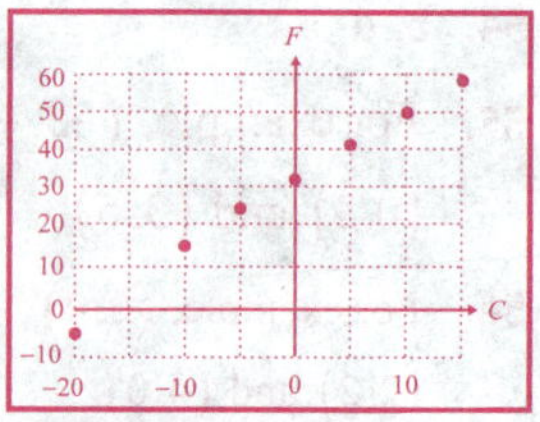

67. a.

t	d
1	16
2	64
3.5	196
4	256
4.5	324
5	400

b.

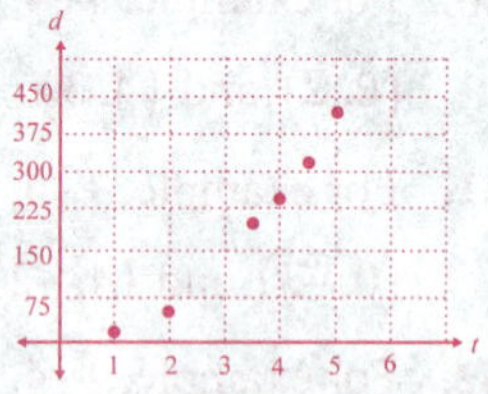

c. The time t is squared in the equation.

71. The equation $t = \frac{d}{r}$ can be used to determine how long it will take you to travel a certain distance while traveling at a specified rate, or speed. Suppose you have to travel a distance of 150 miles. (**Hint:** Rate is in miles per hour, time is in hours, and distance is in miles.)

a. Complete the table of values.

r (mph)	*t* (hours)
25	
50	
60	

b. Graph the ordered pairs from the table. Try using a scale of 10 for each increment on the *r*-axis.

c. What happens to the value of time as the rate increases?

d. Complete the table of values. You may want to solve the given equation, $t = \frac{d}{r}$, for *r* first. Remember that $d = 150$ miles.

r (mph)	*t* (hours)
	5
	2.5
	1
	0.5
	0.1

e. As the value of *t* gets smaller (and closer to zero), how does the rate change?

72. Suppose you deposit \$1000 into an account that earns simple interest at a rate of 4%. Use the simple interest formula $I = Prt$ to solve the following problems.

a. Fill in the given table to determine the amount of interest earned if you keep the deposit in the account for 3 months, 6 months, 1 year, 2 years, and 3 years.

t (years)	*I* (\$)
$\frac{1}{4}$	
$\frac{1}{2}$	
1	
2	
3	

b. Plot the ordered pairs from the table in part a. on the coordinate plane and draw a line through the points.

c. Since graphs are models of situations, they can be used to predict values. Use the graph to predict the amount of interest that will be earned after 9 months. (Remember to convert the time to years.)

d. Use the graph to predict the amount of interest that will be earned after one and a half years.

68. a.

h	*V*
2	18
3	27
5	45
8	72
9	81
10	90

b.

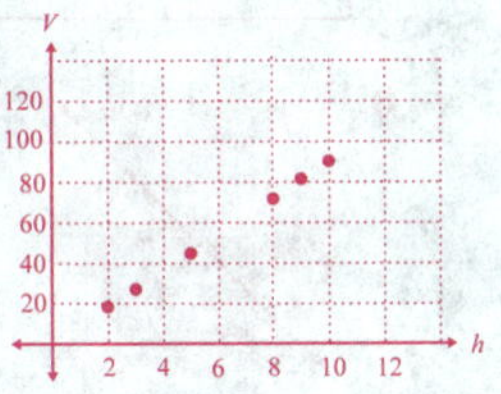

69.

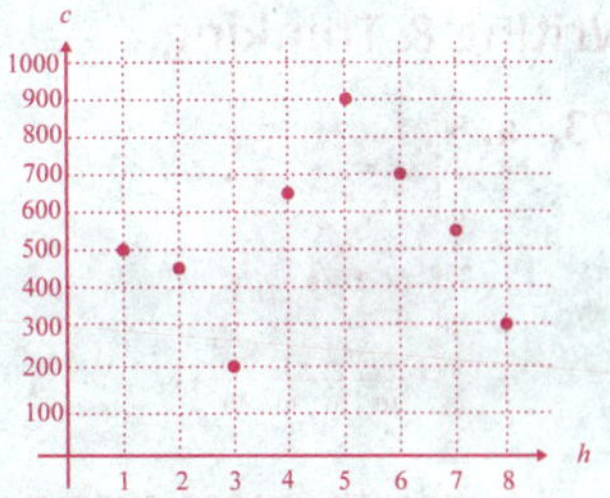

70. a. Yes, at (4, 4)

b. (1, 2), (4, 7)

71. a.

r (mph)	*t* (hours)
25	6
50	3
60	2.5

b.

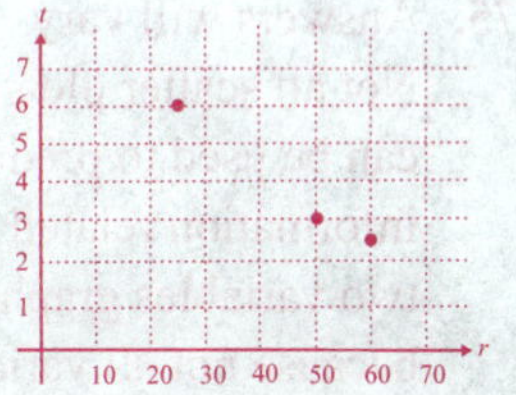

c. The time decreases as the rate increases.

d.

r (mph)	*t* (hours)
30	5
60	2.5
150	1
300	0.5
1500	0.1

e. Answers will vary. Rate increases as the time decreases.

72. a.

t (years)	I ($)
$\frac{1}{4}$	10
$\frac{1}{2}$	20
1	40
2	80
3	120

b.

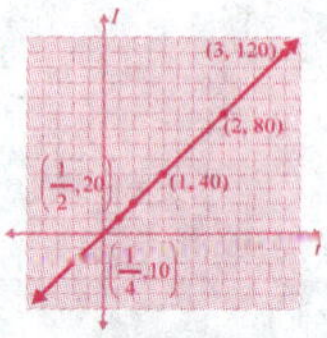

c. $30 **d.** $60

Writing & Thinking

73. a.

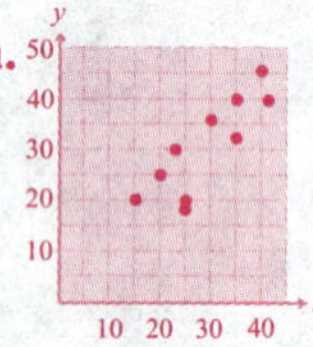

b. Yes, the more push-ups a person can do it appears the more sit-ups he/she can do.

c. 24 sit-ups, 33 sit-ups, 36 sit-ups, 45 sit-ups. Answers will vary.

74. Answers will vary.

75. Answers will vary. Not all scatter plots can be used to predict information related to the two variables graphed because not all variables are related.

Writing & Thinking

In statistics, data is sometimes given in the form of ordered pairs where each ordered pair represents two pieces of information about one person. For example, ordered pairs might represent the height and weight of a person or the person's number of years of education and that person's annual income. The ordered pairs are plotted on a graph and the graph is called a **scatter diagram (or scatter plot)**. Such scatter diagrams are used to see if there is any pattern to the data and, if there is, then the diagram is used to predict the value for one of the variables if the value of the other is known. For example, if you know that a person's height is 5 ft 6 in., then his or her weight might be predicted from information indicated in a scatter diagram that has several points of known information about height and weight.

73. a. The following table of values indicates the number of push-ups and the number of sit-ups that ten students did in a physical education class. Plot these points in a scatter diagram.

Person	#1	#2	#3	#4	#5	#6	#7	#8	#9	#10
x (push-ups)	20	15	25	23	35	30	42	40	25	35
y (sit-ups)	25	20	20	30	32	36	40	45	18	40

b. Does there seem to be a pattern in the relationship between push-ups and sit-ups? What is this pattern?

c. Using the scatter diagram in part a., predict the number of sit-ups that a student might be able to do if he or she has just done each of the following numbers of push-ups: 22, 32, 35, and 45. (**Note:** In each case, there is no one correct answer. The answers are only estimates based on the diagram.)

74. Ask ten friends or fellow students what their heights and shoe sizes are. (You may want to ask all men or all women since the scales for men's and women's shoe sizes are different.) Organize the data in table form and then plot the corresponding scatter diagram. Knowing your own height, does the pattern indicated in the scatter diagram seem to predict your shoe size?

75. Ask ten friends or fellow students what their heights and ages are. Organize the data in table form and then plot the corresponding scatter diagram. Knowing your own height, does the pattern indicated in the scatter diagram seem to predict your age? Do you think that all scatter diagrams can be used to predict information related to the two variables graphed? Explain.

10.2 Graphing Linear Equations in Two Variables

Objectives

A. Graph linear equations by plotting points on the line.

B. Graph linear equations by finding and plotting their x- and y-intercepts.

C. Graph horizontal and vertical lines.

A Graphing Linear Equations by Plotting Points

In Section 10.1, we discussed ordered pairs of real numbers and graphed a few points (ordered pairs) that satisfied particular equations. Now, suppose we want to graph all the points that satisfy an equation such as

$$y = -3x + 3.$$

The **solution set** for equations of this type (in the two variables x and y) consists of an infinite set of ordered pairs in the form (x, y) that satisfy the equation.

To find some of the solutions of the equation $y = -3x + 3$, we form a table (as we did in Section 10.1) by

1. choosing arbitrary values for x and
2. finding the corresponding values for y by substituting into the equation.

In Table 1, we have found five ordered pairs that satisfy the equation and graphed the corresponding points in Figure 1.

Choices x	Substitutions $-3x + 3 = y$	Results (x, y)
-1	$-3(-1)+3=6$	$(-1, 6)$
0	$-3(0)+3=3$	$(0, 3)$
$\frac{2}{3}$	$-3\left(\frac{2}{3}\right)+3=1$	$\left(\frac{2}{3}, 1\right)$
2	$-3(2)+3=-3$	$(2, -3)$
3	$-3(3)+3=-6$	$(3, -6)$

Table 1

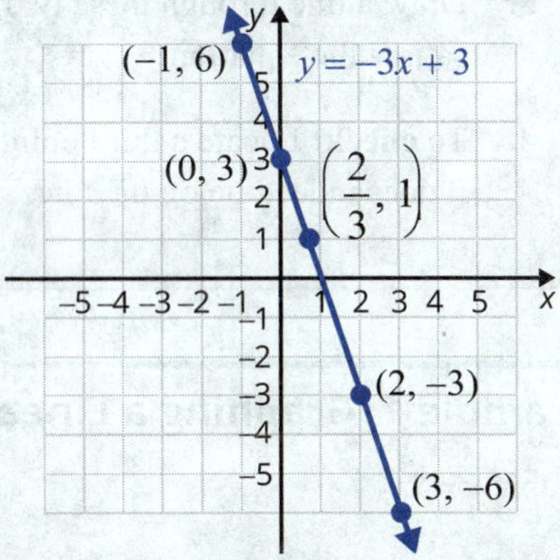

Figure 1

The five points in Figure 1 appear to lie on a line. They do in fact lie on a line, and any ordered pair that satisfies the equation $y = -3x + 3$ will also lie on that same line.

Just as we use the terms **ordered pair** and **point** (the graph of an ordered pair) interchangeably, we use the terms **equation** and **graph of an equation** interchangeably. The equations

$$2x + 3y = 4, \quad y = -5, \quad x = 1.4, \quad \text{and} \quad y = 3x + 2$$

are called **linear equations**, and their graphs are lines on the Cartesian coordinate system.

Note

Note that in the standard form $Ax + By = C$, A and B may be positive, negative, or 0, but A and B cannot **both** equal 0.

Standard Form of a Linear Equation

Any equation of the form

$$Ax + By = C,$$

where A, B, and C are real numbers and A and B are not both equal to 0, is called the **standard form of a linear equation.**

DEFINITION

Every line corresponds to some linear equation, and the graph of every linear equation is a line. We know from geometry that **two points determine a line**. This means that the graph of a linear equation can be found by locating any two points that satisfy the equation.

Graphing a Linear Equation in Two Variables

1. Locate any two points that satisfy the equation. (Choose values for x and y that lead to values that are easily calculated for the other variable. Remember that there are an infinite number of choices for either x or y. Once a value for x or y is chosen, the corresponding value for the other variable is found by substituting into the equation.)
2. Plot these two points on a Cartesian coordinate system.
3. Draw a line through these two points. (**Note:** Every point on that line will satisfy the equation.)
4. **To check:** Locate a third point that satisfies the equation, and check to see that it does indeed lie on the line.

PROCEDURE

1. Graph: $y = 3x$

Example 1 Graphing a Linear Equation in Two Variables

Graph: $y = 2x$

Solution

Substitute −1, 0, and 1 for x.

x	$y = 2x$	y
−1	$y = 2(-1)$	−2
0	$y = 2(0)$	0
1	$y = 2(1)$	2

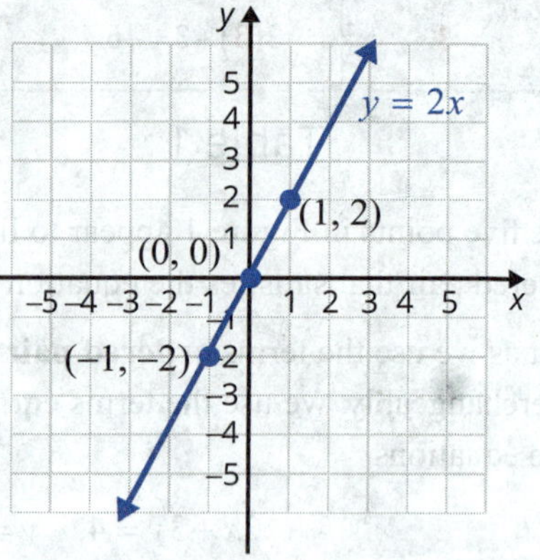

Now work margin exercise 1.

Example 2 Graphing a Linear Equation in Two Variables

Graph: $2x + 3y = 6$

Solution

Make a table with headings x and y and, whenever possible, **choose values for x or y that lead to values that are easily calculated for the other variable.** (Values chosen for x and y are red and bolded.)

x	$2x + 3y = 6$	y
0	$2(\mathbf{0}) + 3y = 6$	2
−3	$2(\mathbf{-3}) + 3y = 6$	4
3	$2x + 3(\mathbf{0}) = 6$	**0**
$\frac{5}{2}$	$2x + 3\left(\mathbf{\frac{1}{3}}\right) = 6$	$\mathbf{\frac{1}{3}}$

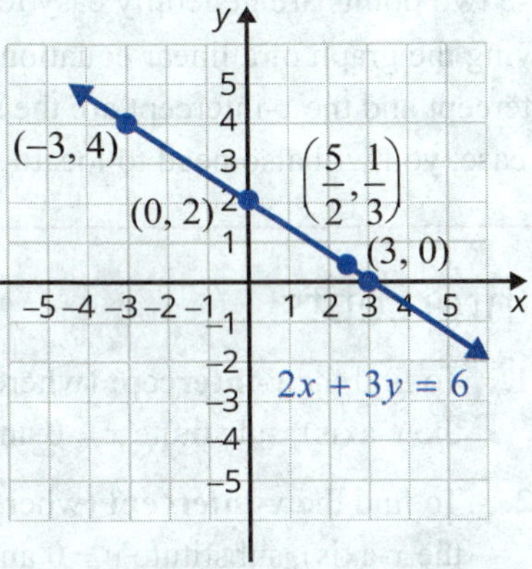

Now work margin exercise 2.

2. Graph: $3x + 2y = 6$

Example 3 Graphing a Linear Equation in Two Variables

Graph: $x - 2y = 1$

Solution

Solving the equation for x gives $x = 2y + 1$. Substitute 0, 1, and 2 for y.

x	$x = 2y + 1$	y
1	$x = 2(\mathbf{0}) + 1$	**0**
3	$x = 2(\mathbf{1}) + 1$	**1**
5	$x = 2(\mathbf{2}) + 1$	**2**

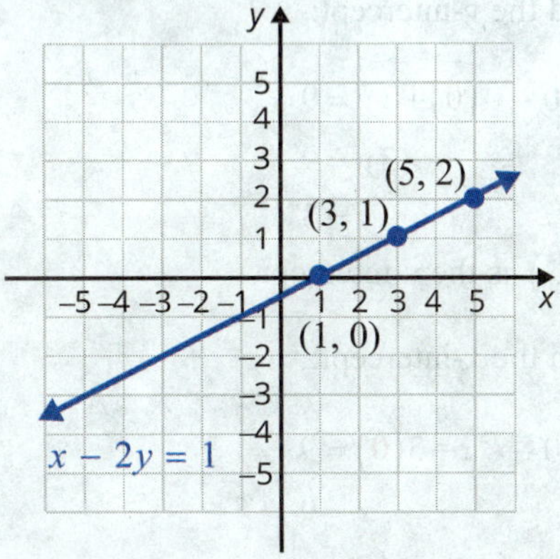

Now work margin exercise 3.

3. Graph: $2x - y = 2$

Math Tip

Choosing values of 0 for x and y usually leads to simpler calculations.

B Using *x*- and *y*-Intercepts to Graph Linear Equations

While the choice of the values for x or y can be arbitrary, letting $x = 0$ will locate the point on the graph where the line crosses (or intercepts) the y-axis. This point is called the ***y*-intercept** and is of the form $(0, y)$. The ***x*-intercept** is the point found by letting $y = 0$. This is the point where the line crosses (or intercepts) the x-axis and is of the form $(x, 0)$. These two points are generally easy to locate and are frequently used as the two points for drawing the graph of a linear equation. If a line passes through the point $(0, 0)$, then the y-intercept and the x-intercept are the same point; namely, the origin (see Example 1). In this case, you will also need to locate some other point to draw the graph.

Note

In general, the intercepts are easy to find because substituting 0 for *x* or *y* leads to an easy solution for the other variable. However, when the intercepts result in a point with fractional (or decimal) coordinates and estimation is involved, then a third point that satisfies the equation should be found to verify that the line is graphed correctly.

Intercepts

1. To find the ***y*-intercept** (where the line crosses the y-axis), substitute $x = 0$ and solve for y.
2. To find the ***x*-intercept** (where the line crosses the x-axis), substitute $y = 0$ and solve for x.

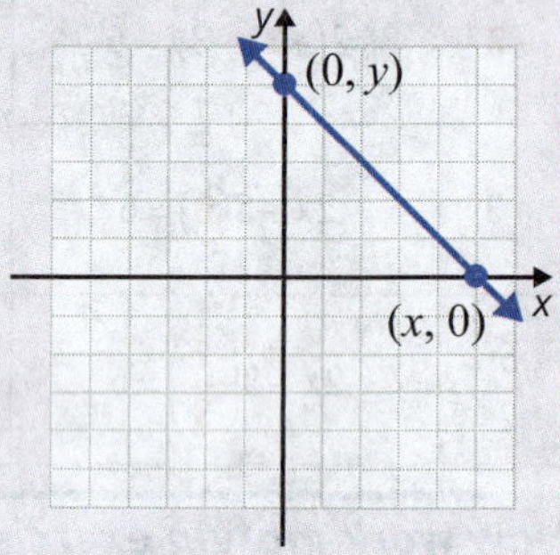

PROCEDURE

4. Graph $x + 2y = 6$ by locating the x-intercept and the y-intercept.

Example 4 Using Intercepts to Graph Linear Equations

Graph $x + 3y = 9$ by locating the y-intercept and the x-intercept.

Solution

Find the y-intercept:

$$x = 0 \rightarrow (0) + 3y = 9$$
$$3y = 9$$
$$y = 3$$

$(0, 3)$ is the y-intercept.

Find the x-intercept:

$$y = 0 \rightarrow x + 3(0) = 9$$
$$x = 9$$

$(9, 0)$ is the x-intercept.

Plot the two intercepts and draw the line that contains them.

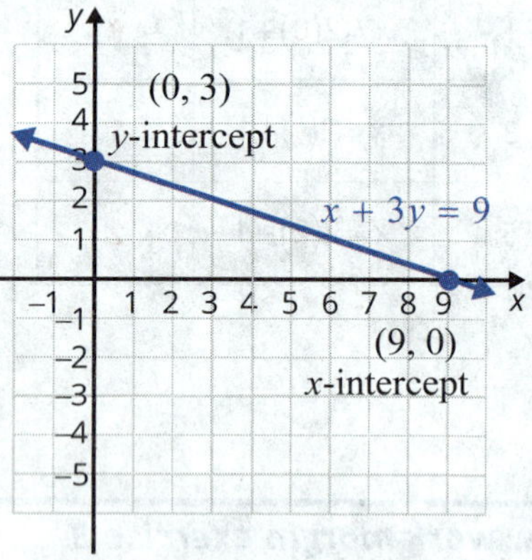

Math Tip

It is a good idea to use a third point to check your work when graphing using intercepts. In Example 4, we can verify visually and algebraically that the line also contains the point (3, 2).

Now work margin exercise 4.

Example 5 Using Intercepts to Graph Linear Equations

Graph $3x - 2y = 12$ by locating the y-intercept and the x-intercept.

Solution

Find the y-intercept:

$$x = 0 \rightarrow 3(0) - 2y = 12$$
$$-2y = 12$$
$$y = -6$$

$(0, -6)$ is the y-intercept.

Find the x-intercept:

$$y = 0 \rightarrow 3x - 2(0) = 12$$
$$3x = 12$$
$$x = 4$$

$(4, 0)$ is the x-intercept.

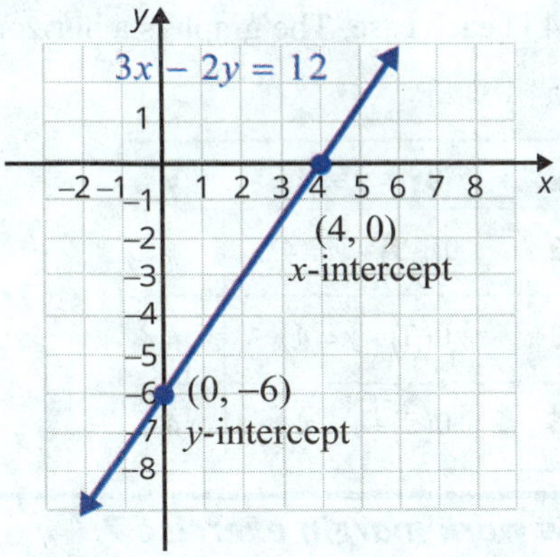

Plot the two intercepts and draw the line that contains them.

Now work margin exercise 5.

5. Graph $5x - 3y = 15$ by locating the x-intercept and the y-intercept.

Completion Example 6 Using Intercepts to Graph Equations

Graph $x - 5y = 5$ by locating the y-intercept and the x-intercept.

Solution

Find the y-intercept:

$x = 0 \rightarrow$ __0__ $- 5y = 5$

__$-5y = 5$__

__$y = -1$__

__(0, −1)__ is the y-intercept.

Find the x-intercept:

$y = 0 \rightarrow x - 5 \cdot ($__0__$) = 5$

__$x - 0 = 5$__

__$x = 5$__

__(5, 0)__ is the x-intercept.

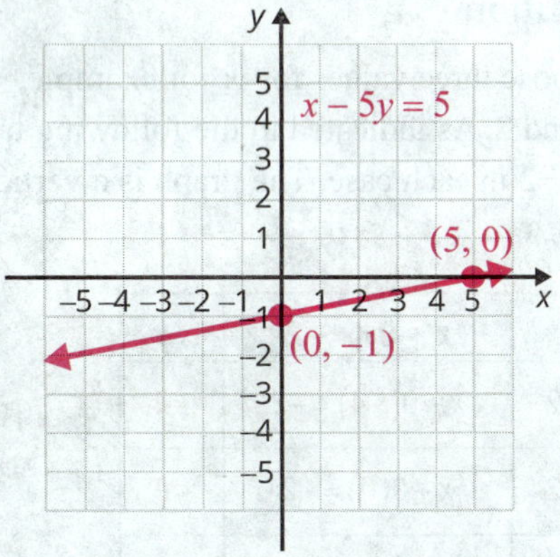

Plot the two intercepts and draw the line that contains them.

Now work margin exercise 6.

6. Graph $4x - y = 8$ by locating the x-intercept and the y-intercept.

C Graphing Horizontal and Vertical Lines

Now consider the linear equation $y = 4$ (which can be written in standard form as $0x + y = 4$). Regardless of the value substituted for x, the corresponding y-value will be 4. The graph of this equation is a **horizontal line** with y-intercept $(0, 4)$, as illustrated in Example 7.

10.2 Exercises

Concept Check

1. y
2. x
3. infinite
4. $Ax + By = C$
5. line
6. 2
7. True
8. True
9. False; Horizontal lines have y-intercepts.
10. False; $(x, 0)$

7. Graph the line $y = -1$.

Practice

1. a.
2. e.
3. d.
4. c.
5. f.
6. b.
7. 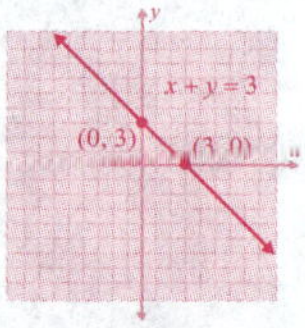

8. 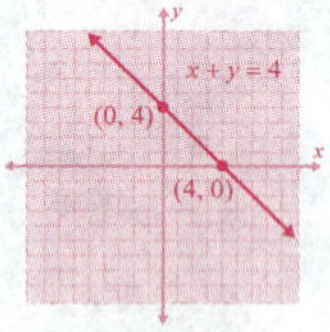

Example 7 Graphing Horizontal Lines

Graph the line $y = 4$ (or $0x + y = 4$).

Solution

Choose three values for x; for example, -3, 3, and 5. As indicated in the following table, $y = 4$ in each case. The graph is a horizontal line.

x	$0x + y = 4$	y
-3	$0(-3) + y = 4$	4
3	$0(3) + y = 4$	4
5	$0(5) + y = 4$	4

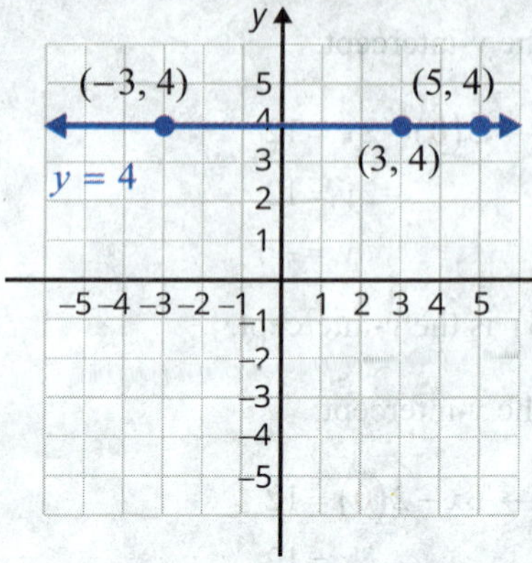

Now work margin exercise 7.

Next, consider the linear equation $x = -2$ (which can be written in standard form as $x + 0y = -2$). Regardless of the value substituted for y, the corresponding x-value will be -2. The graph of this equation is a **vertical line** with x-intercept $(-2, 0)$ as illustrated in Example 8.

8. Graph the line $x = 4$.

9. 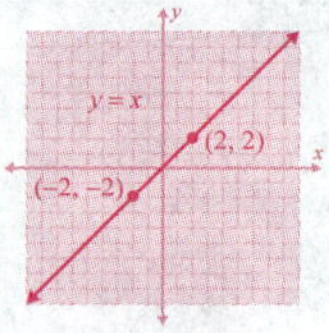

10. 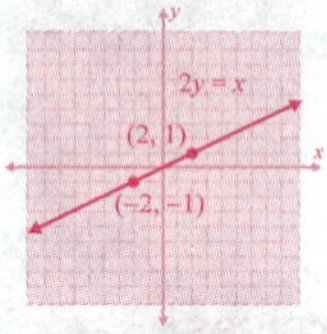

11. 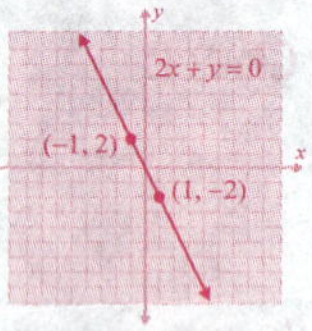

12. 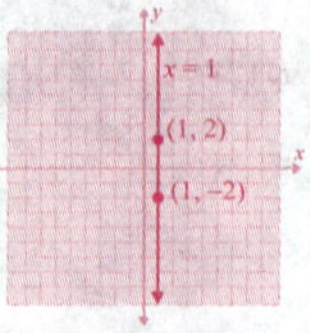

13. 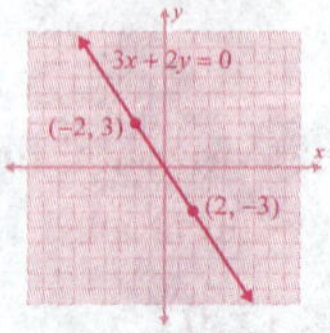

Example 8 Graphing Vertical Lines

Graph the line $x = -2$ (or $x + 0y = -2$).

Solution

Choose three values for y; for example, -4, 0, and 2. As indicated in the following table, $x = -2$ in each case. The graph is a vertical line.

x	$x + 0y = -2$	y
-2	$x + 0(-4) = -2$	-4
-2	$x + 0(0) = -2$	0
-2	$x + 0(2) = -2$	2

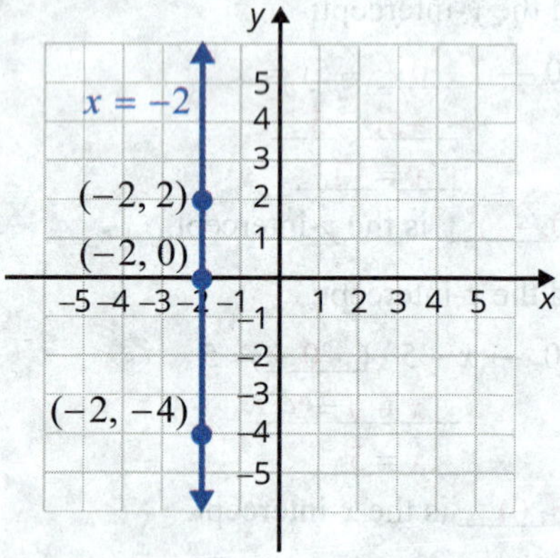

Now work margin exercise 8.

Horizontal and Vertical Lines

For real numbers a and b, the graph of

$y = b$ is a **horizontal line** and $x = a$ is a **vertical line.**

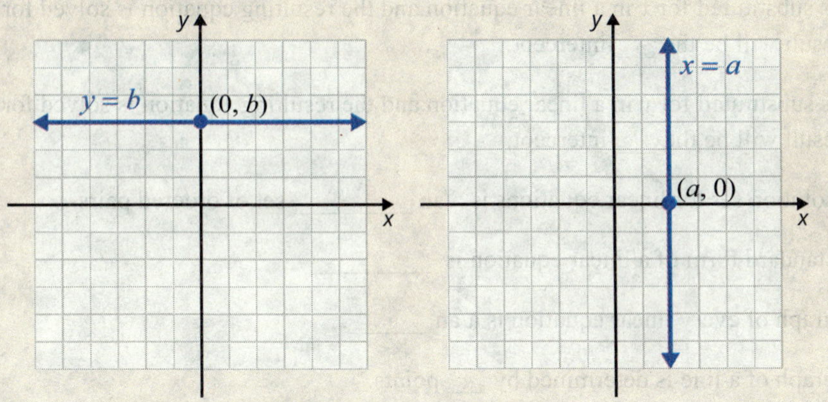

DEFINITION

Completion Example Answers

6. $x = 0 \rightarrow 0 - 5y = 5;\ -5y = 5;\ y = -1;\ (0, -1)$ is the y-intercept;
$y = 0 \rightarrow x - 5 \cdot 0 = 5;\ x - 0 = 5;\ x = 5;\ (5, 0)$ is the x-intercept.

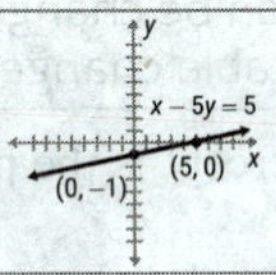

Margin Exercise Answers

1.

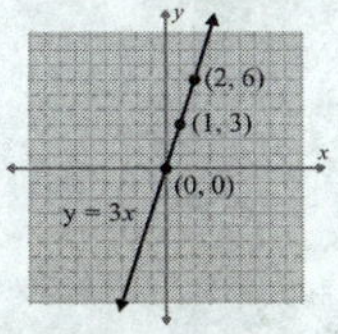

2.

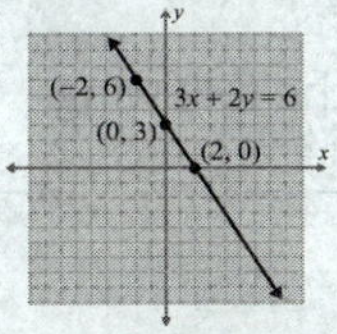

3.

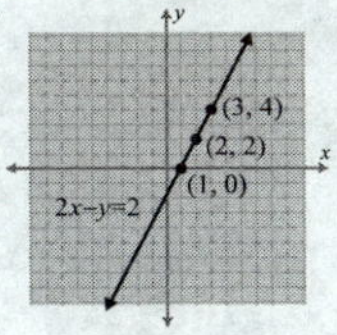

4.

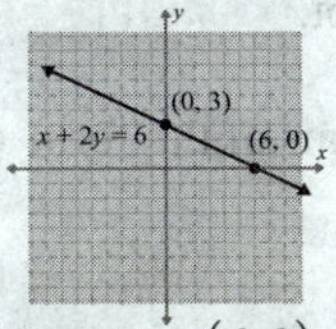

x-intercept $= (6, 0)$,

y-intercept $= (0, 3)$

5.

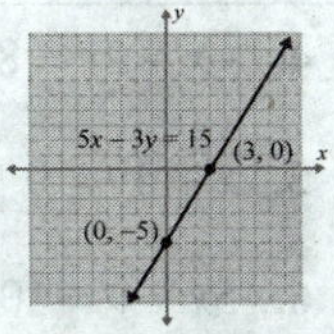

x-intercept $= (3, 0)$,

y-intercept $= (0, -5)$

6.

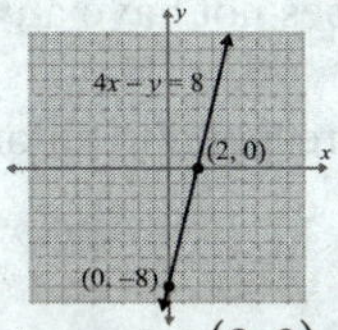

x-intercept $= (2, 0)$,

y-intercept $= (0, -8)$

7.

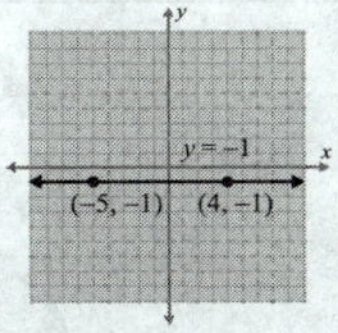

8.

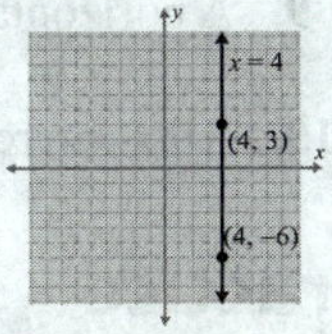

14.

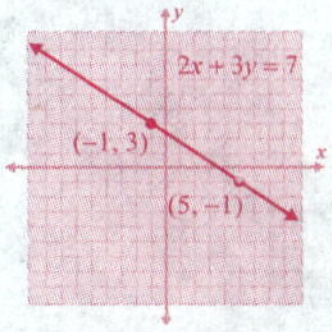

15.

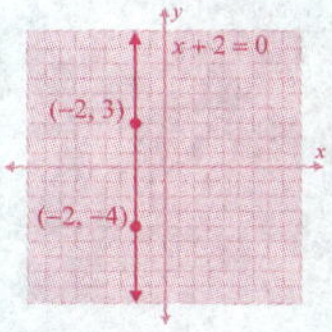

16.

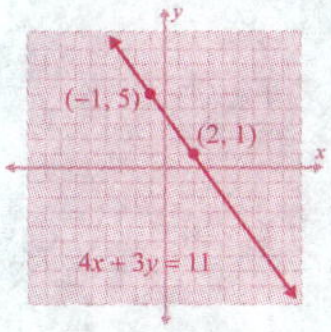

17.

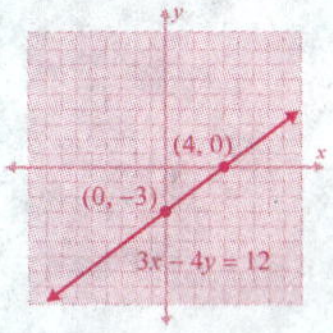

18.

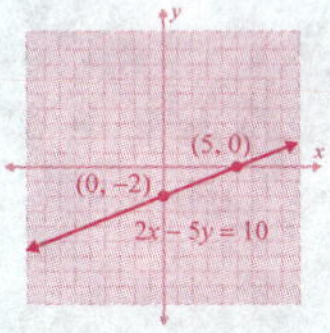

19.

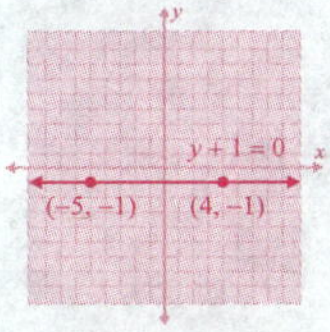

20.

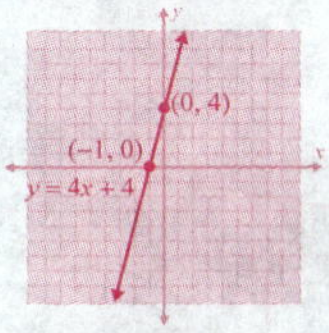

21.

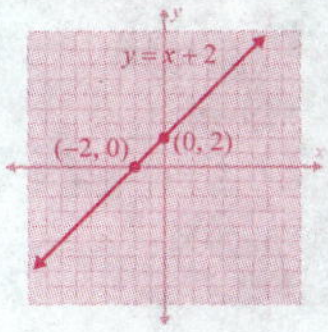

22.

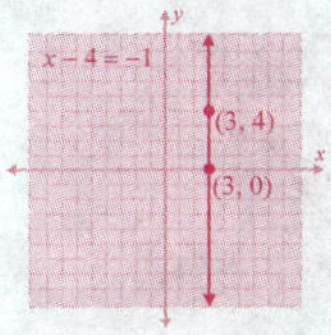

23.

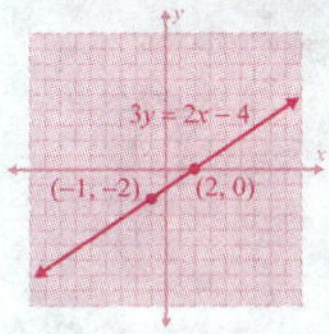

24.

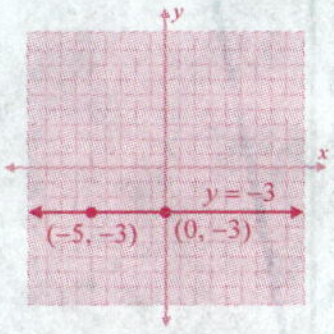

25.

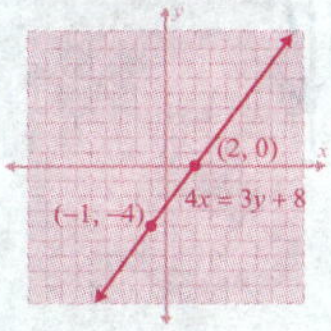

26.

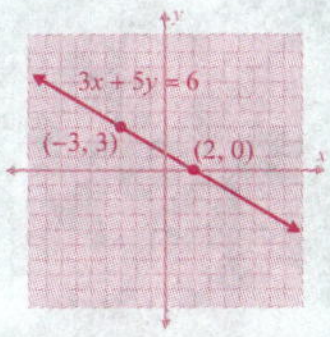

27.

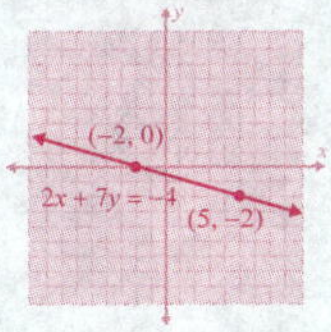

28.

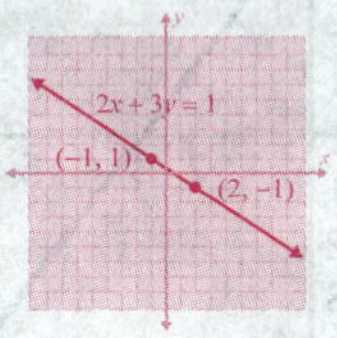

29.

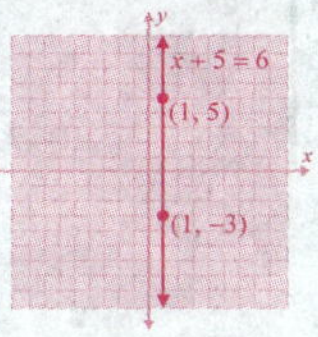

30.

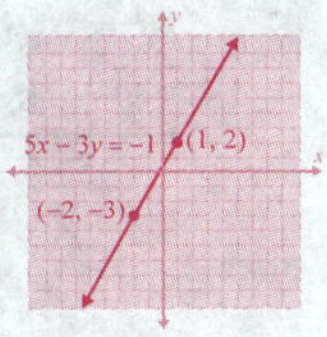

31.

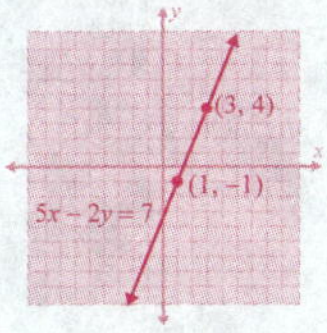

32.
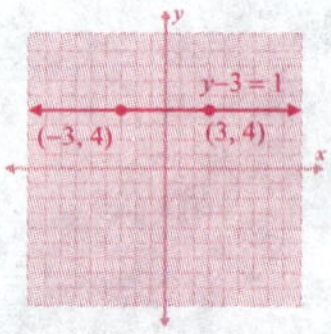

33.
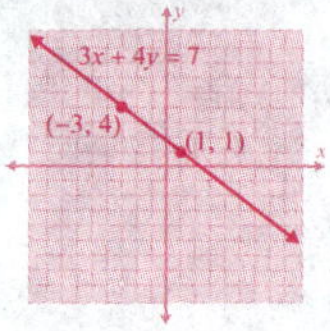

34.
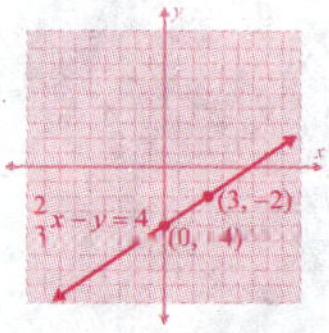

35.
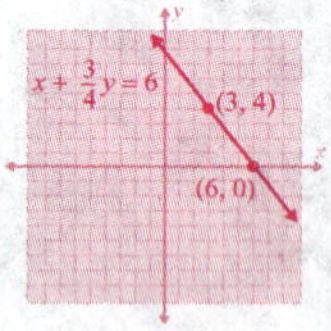

36.
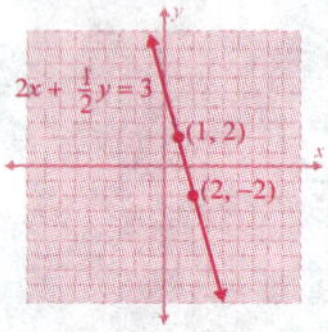

37.
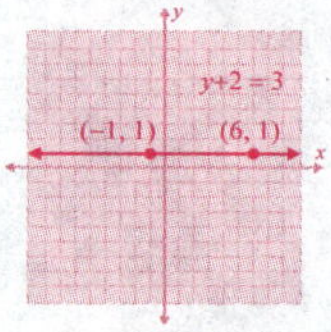

38.
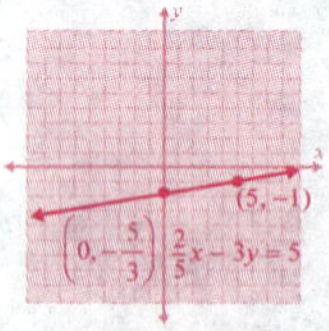

39.
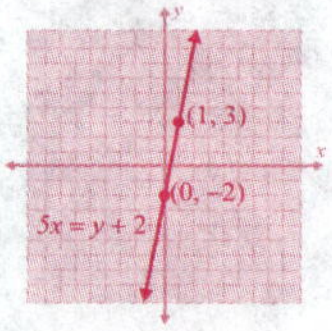

40.
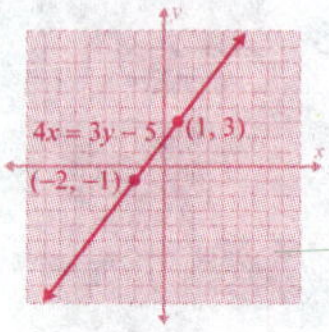

41.
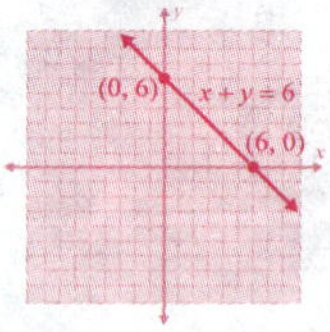

10.2 Exercises

Concept Check

Fill-in-the-Blank. Complete each sentence using information found in this section.

1. If 0 is substituted for x in a linear equation and the resulting equation is solved for y, the result will be the ___-intercept.
2. If 0 is substituted for y in a linear equation and the resulting equation is solved for x, the result will be the ___-intercept.
3. The solution set for linear equations is a/an _________ set of ordered pairs.
4. The standard form of a linear equation is __________.
5. The graph of every linear equation is a/an _______.
6. The graph of a line is determined by ___ points.

True/False. Determine whether each statement is true or false. If a statement is false, explain how it can be changed so the statement will be true. (**Note:** There may be more than one acceptable change.)

7. The y-intercept is the point where a line crosses the y-axis.
8. The terms ordered pair and point are used interchangeably.
9. A horizontal line does not have a y-intercept.
10. All x-intercepts correspond to an ordered pair of the form $(0, y)$.

Practice

Use your knowledge of y-intercepts and x-intercepts to match each of the following equations with its graph.

1. $4x + 3y = 12$
2. $4x - 3y = 12$
3. $x + 2y = 8$
4. $-x + 2y = 8$
5. $x + 4y = 0$
6. $5x - y = 10$

a.
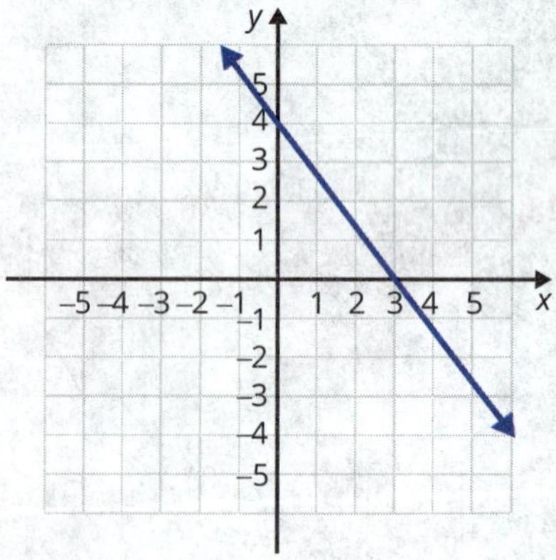

b.
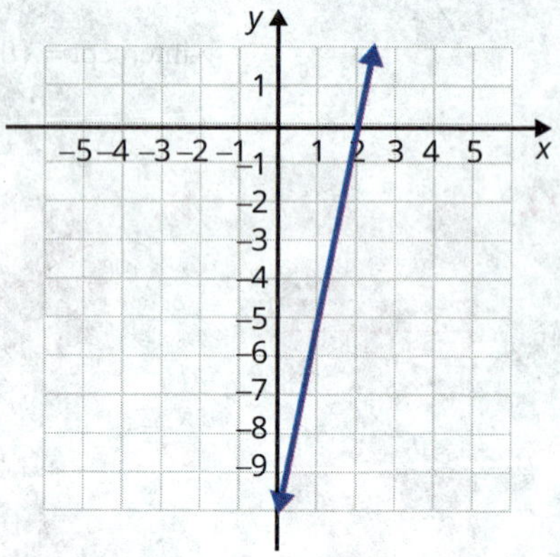

c.

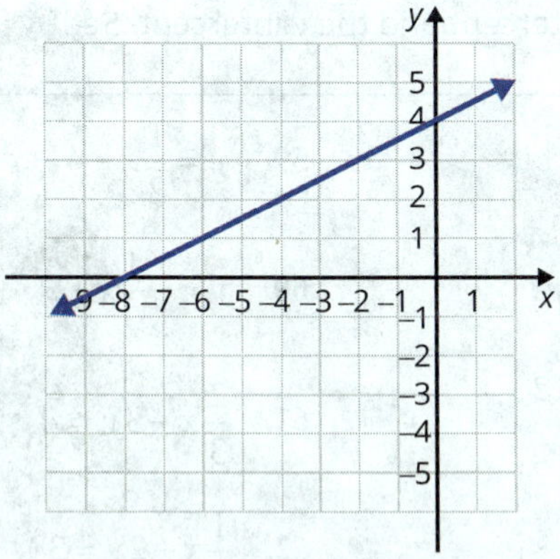

e.

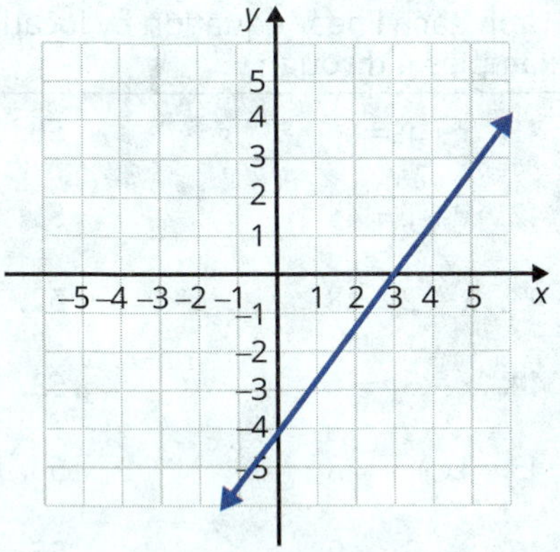

d.

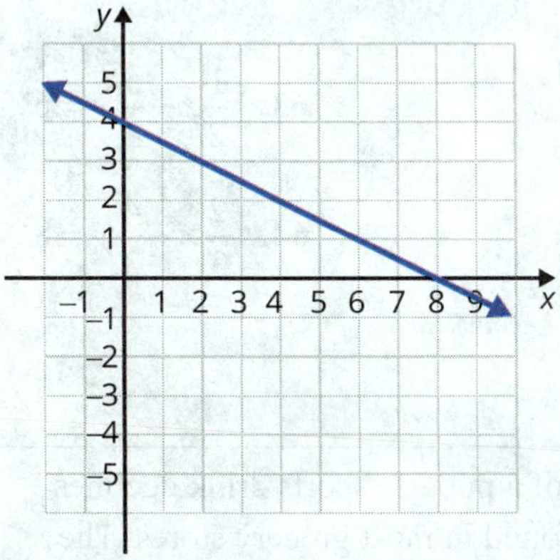

f.

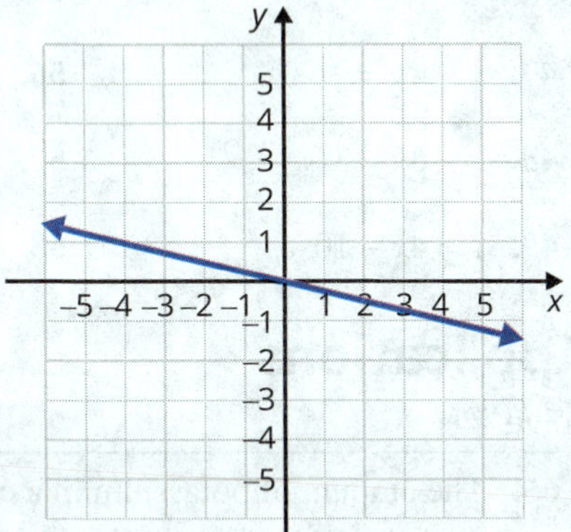

Graph each linear equation by locating at least two ordered pairs that satisfy the given equation. See Examples 1 through 3.

7. $x + y = 3$

8. $x + y = 4$

9. $y = x$

10. $2y = x$

11. $2x + y = 0$

12. $x = 1$

13. $3x + 2y = 0$

14. $2x + 3y = 7$

15. $x + 2 = 0$

16. $4x + 3y = 11$

17. $3x - 4y = 12$

18. $2x - 5y = 10$

19. $y + 1 = 0$

20. $y = 4x + 4$

21. $y = x + 2$

22. $x - 4 = -1$

23. $3y = 2x - 4$

24. $y = -3$

25. $4x = 3y + 8$

26. $3x + 5y = 6$

27. $2x + 7y = -4$

28. $2x + 3y = 1$

29. $x + 5 = 6$

30. $5x - 3y = -1$

31. $5x - 2y = 7$

32. $y - 3 = 1$

33. $3x + 4y = 7$

34. $\frac{2}{3}x - y = 4$

35. $x + \frac{3}{4}y = 6$

36. $2x + \frac{1}{2}y = 3$

37. $y + 2 = 3$

38. $\frac{2}{5}x - 3y = 5$

39. $5x = y + 2$

40. $4x = 3y - 5$

42.

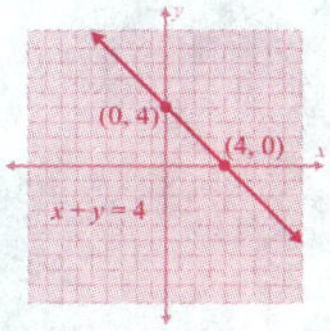

43.

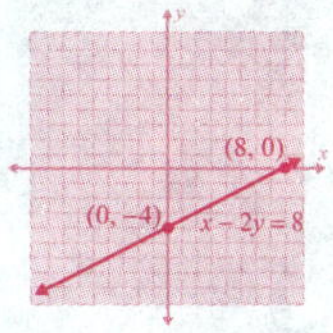

44.

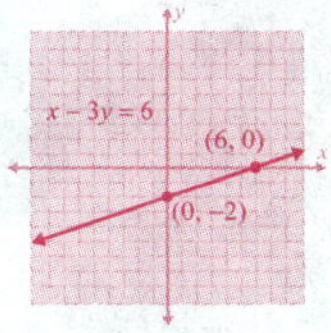

45.

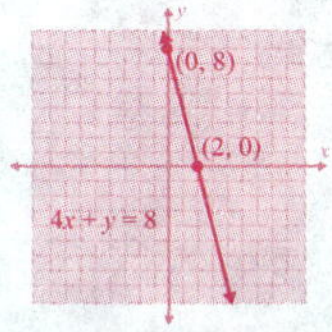

46.

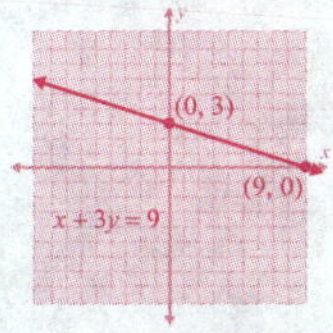

47.

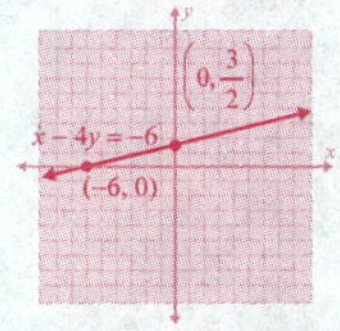

48.

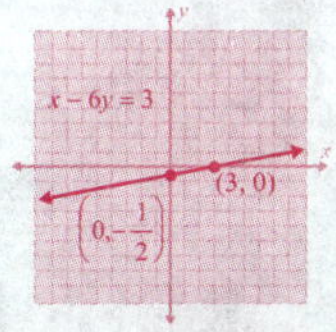

49.

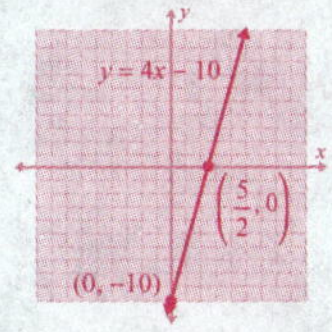

50.

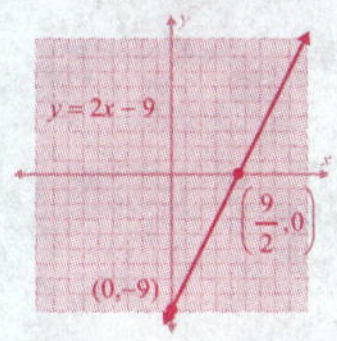

51.

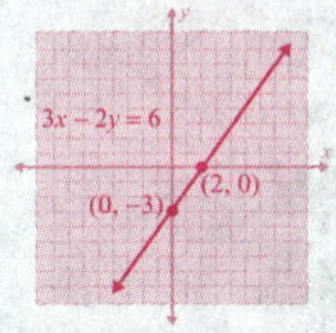

Graph each linear equation by locating the *x*-intercept and the *y*-intercept. See Examples 4 through 6.

41. $x+y=6$

42. $x+y=4$

43. $x-2y=8$

44. $x-3y=6$

45. $4x+y=8$

46. $x+3y=9$

47. $x-4y=-6$

48. $x-6y=3$

49. $y=4x-10$

50. $y=2x-9$

51. $3x-2y=6$

52. $5x+2y=10$

53. $2x+3y=12$

54. $3x+7y=-21$

55. $3x-7y=-21$

56. $3x+2y=15$

57. $5x+3y=7$

58. $2x+3y=5$

59. $y=\frac{1}{2}x-4$

60. $y=-\frac{1}{3}x+3$

61. $\frac{2}{3}x-3y=4$

62. $\frac{1}{2}x+2y=3$

63. $\frac{1}{2}x-\frac{3}{4}y=6$

64. $\frac{2}{3}x+\frac{4}{3}y=8$

Applications

Solve.

65. The amount of potassium in a clear bottle of a popular sports drink declines over time when exposed to the UV lights found in most grocery stores. The amount of potassium in a container of this sports drink is given by the equation $y=-30x+360$, where *y* represents the mg of potassium remaining after *x* days on the shelf. Find both the *x*-intercept and *y*-intercept, and interpret the meaning of each in the context of this problem.

66. Mr. Adler has found that the grade each student gets in his Introductory Algebra course directly correlates with the amount of time spent doing homework and is represented by the equation $y=7x+30$, where *y* represents the numerical score the student receives on an exam (out of 100 points) after spending *x* hours per week doing homework. Find the *y*-intercept and interpret its meaning in this context.

67. Barbara's Bombtastic Bakery is donating cookies to a charity bake sale. The bakery decides to donate chocolate chip cookies and peanut butter cookies by the dozen, and they want to donate a total of 30 dozen cookies. To determine the possible combinations, the bakery uses the equation $C+P=30$, where *C* is the number of dozens of chocolate chip cookies and *P* is the number of dozens of peanut butter cookies.

a. Find the intercepts of the given equation.

b. Graph the equation using the intercepts.

c. Are there any solutions to the equation that do not make sense in the context of the problem? Explain why.

d. If Barbara's Bombtastic Bakery decides to donate 16 dozen peanut butter cookies, how many dozen chocolate chip cookies will be donated?

52.
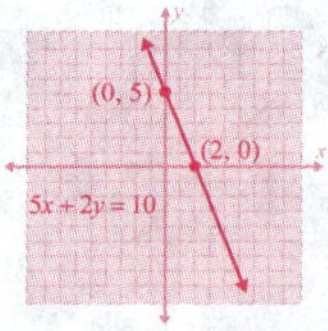

53.
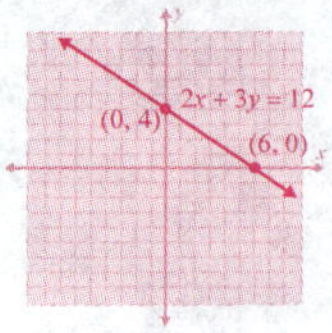

54.
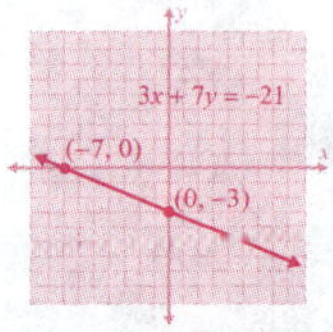

55.
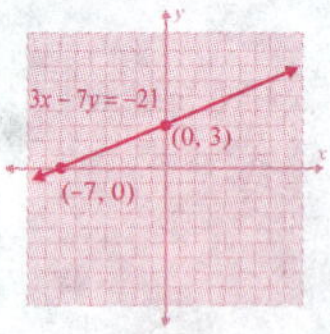

56.
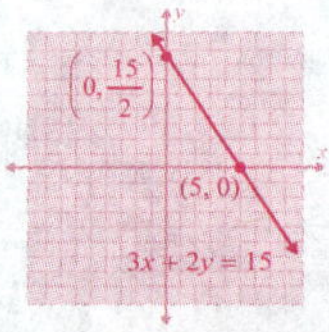

57.
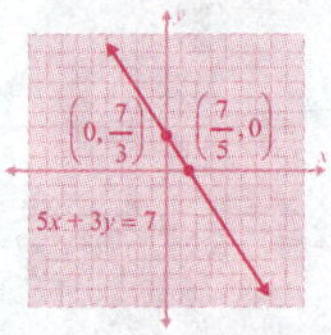

58.
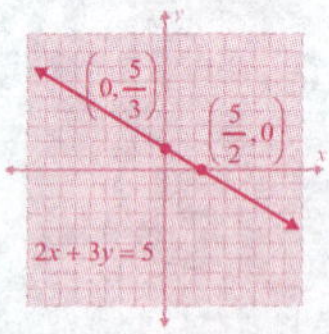

59.
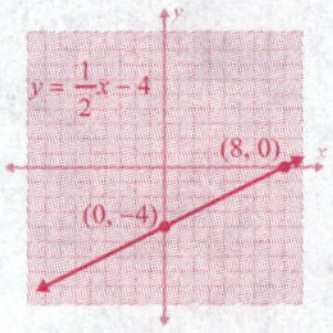

60.
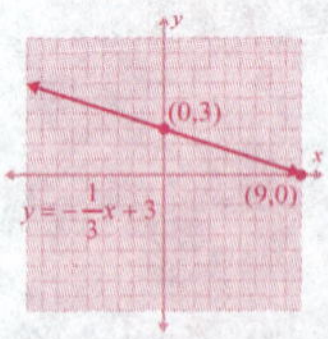

61.
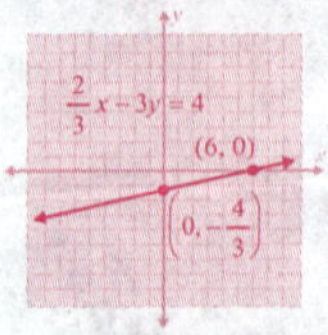

68. Ibuprofen can be given to a child to treat a fever. For a fever lower than 102.5 °F, the recommended dosage is 2.2 milligrams per pound of body weight. This can be modeled by the equation $D = 2.2w$, where D is the dosage of ibuprofen in milligrams and w is the weight of the child in pounds.

a. Create a table to determine several coordinate pairs that satisfy the equation.

b. Graph the equation using the coordinate pairs from part a.

c. Are there any solutions from the graph that do not make sense in the context of the problem? Explain why.

d. How much ibuprofen should be given to a child that weighs 45 pounds?

69. Mason purchased a card game for \$20. The 5-card expansion packs cost \$1 per pack. Mason wants to keep track of how much money he spends on the game, so he creates the equation $C = 20 + 1n$, where C is the total cost in dollars and n is the number of expansion packs purchased.

a. Create a table to determine several coordinate pairs that satisfy the equation.

b. Graph the equation using the coordinate pairs from part a.

c. Are there any solutions from the graph that do not make sense in the context of the problem? Explain why.

d. If Mason buys 37 expansion packs, what will the total cost be?

Writing & Thinking

70. Explain, in your own words, why it is sufficient to find the x-intercept and y-intercept to graph a line (assuming that they are not the same point).

71. Explain, in your own words, how you can determine if an ordered pair is a solution to an equation.

62.

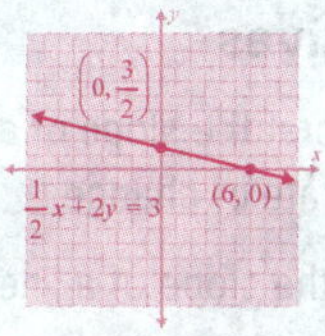

63.

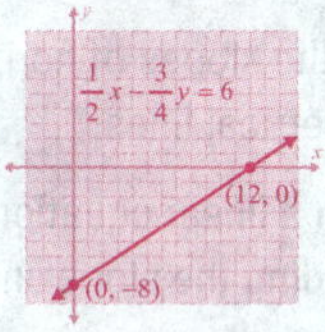

64.

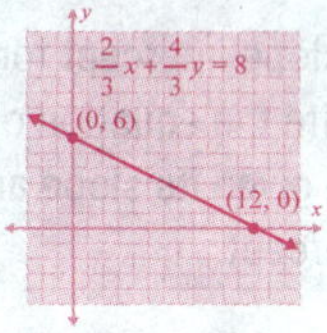

Applications

65. The x-intercept is $(12, 0)$ meaning that after 12 days, there will be 0 mg of potassium remaining in the bottle of sports drink. The y-intercept is $(0, 360)$ meaning that the original amount of potassium in the sports drink is 360 mg.

66. The y-intercept is $(0, 30)$, meaning that if a student does no homework at all, the student will get a score of 30 points on the exam.

67. a. C-intercept: $(30, 0)$
P-intercept: $(0, 30)$

b.

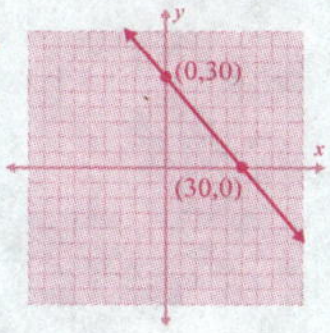

c. Answers will vary. Negative number of cookies;Total amount of cookies is over 30 dozen.

d. 14 dozen chocolate chip cookies

68. a.

w	15	20	45
D	33	44	99

b.

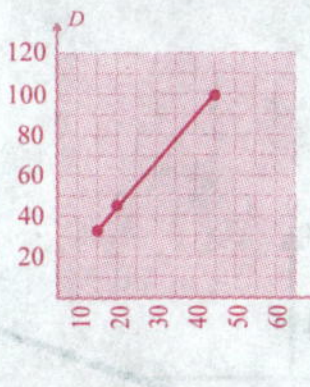

c. Answers will vary. Negative weights; Negative amounts of medicine.

d. 99 milligrams

69. a.

n	5	10	37
C	25	30	57

b.

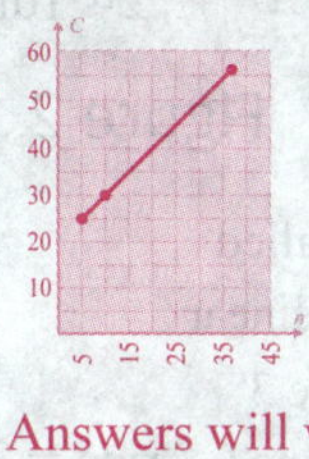

c. Answers will vary. Negative number of songs; Negative cost.

d. \$57

Writing & Thinking

70. Two (unique) points determine a line.

71. Substitute the x and y values into the equation. Then evaluate both sides to see if the equation is true.

Objectives

A. Interpret the slope of a line as a rate of change.

B. Find the slope of a line given two points on the line.

C. Find the slopes of horizontal and vertical lines.

D. Graph a linear equation by finding the slope and *y*-intercept.

E. Use slope-intercept form to write the equation of a line given its slope and *y*-intercept.

10.3 Slope-Intercept Form

A The Meaning of Slope

If you ride a bicycle up a mountain road, you certainly know when the **slope** (a measure of steepness called the **grade** for roads) increases because you have to pedal harder. The contractor who built the road was aware of the slope because trucks traveling the road must be able to control their downhill speed and be able to stop in a safe manner. A carpenter given a set of house plans calling for a roof with a pitch of 7 : 12 knows that for every 7 feet of rise (vertical distance) there are 12 feet of run (horizontal distance). That is, the ratio of rise to run is $\frac{\text{rise}}{\text{run}} = \frac{7}{12}$.

Figure 1

Note that this ratio can be in units other than feet, such as inches or meters. (See Figure 2.)

$$\frac{\text{rise}}{\text{run}} = \frac{7 \text{ inches}}{12 \text{ inches}} = \frac{3.5 \text{ feet}}{6 \text{ feet}} = \frac{14 \text{ feet}}{24 \text{ feet}}$$

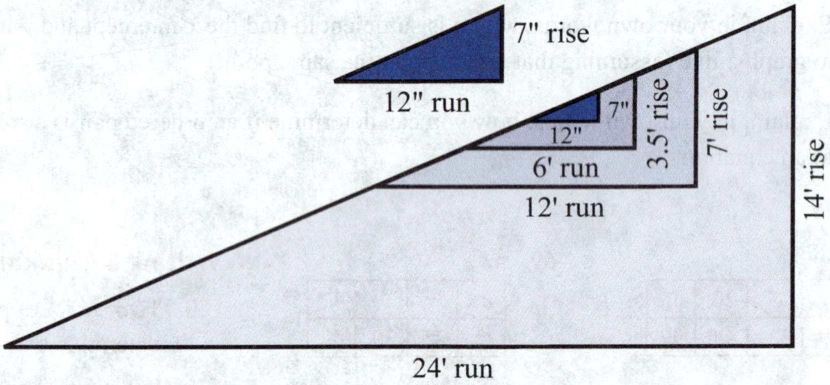

Figure 2

For a line, the **ratio of rise to run** is called the **slope of the line**. The graph of the linear equation $y = \frac{1}{3}x + 2$ is shown in Figure 3. What do you think is the slope of the line? Do you think that the slope is positive or negative? Do you think the slope might be $\frac{1}{3}$? $\frac{3}{1}$? 2?

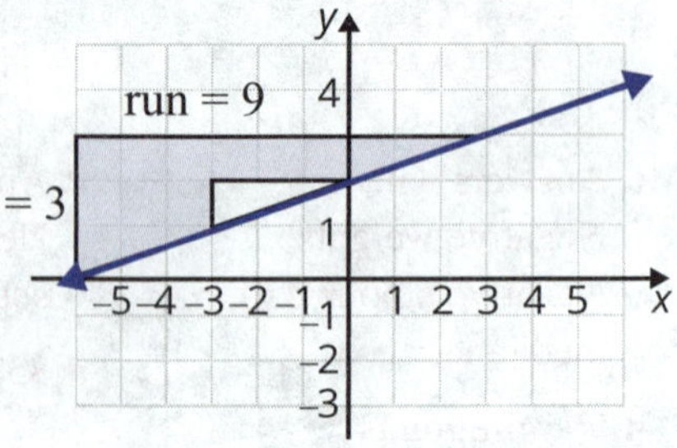

$$\text{slope} = \frac{\text{rise}}{\text{run}} = \frac{3}{9} = \frac{1}{3}$$

Figure 3

The concept of slope also relates to situations that involve **rate of change**. For example, the graphs in Figure 4 illustrate slope as miles per hour that a car travels and as pages per minute that a printer prints. (**Note:** Be sure to look at the scales on the axes when reading a graph.)

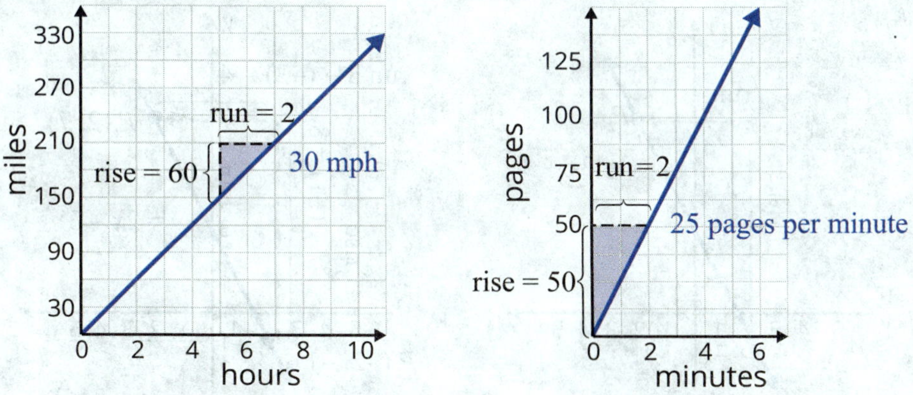

slope = rate of change

$= \frac{60 \text{ miles}}{2 \text{ hours}}$

= 30 mph

slope = rate of change

$= \frac{50 \text{ pages}}{2 \text{ minutes}}$

= 25 pages per minute

Figure 4

In general, the ratio of a change in one variable (say y) to a change in another variable (say x) is called the **rate of change of y with respect to x**. Figure 5 shows how the rate of change (the slope) can change over periods of time.

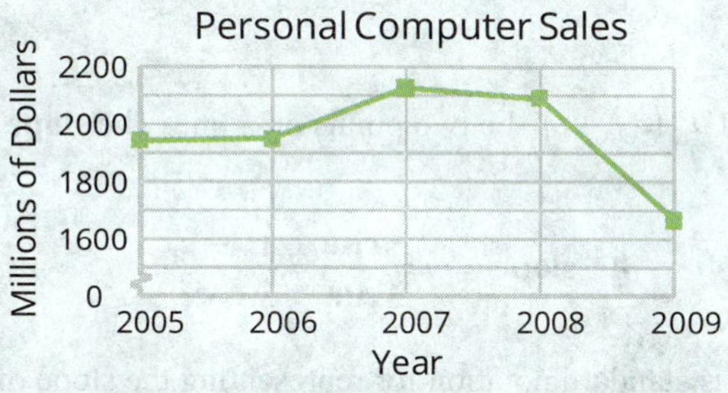

Source: Consumer Electronics Association

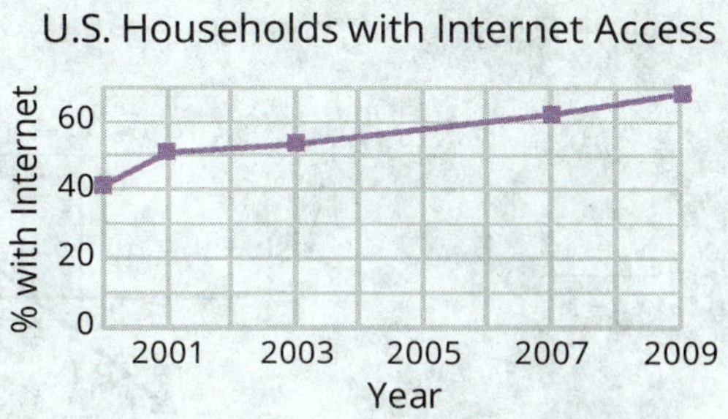

Source: U.S. Dept. of Commerce

Figure 5

B Calculating the Slope

> **Note**
>
> In the notation P_1, 1 is called a **subscript** and P_1 is read "P sub 1." Similarly, P_2 is read "P sub 2." Subscripts are used in labeling and are not used in calculations.

Consider the line $y = 2x + 3$ and two points on the line, $P_1(-2, -1)$ and $P_2(2, 7)$, as shown in Figure 6.

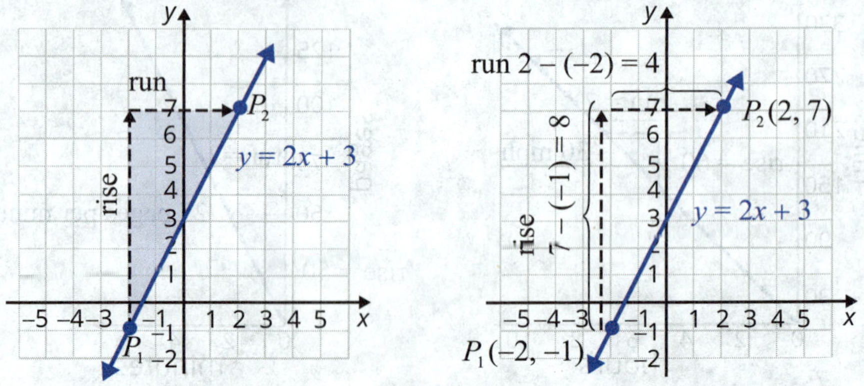

Figure 6

For the line $y = 2x + 3$ and using the points $(-2, -1)$ and $(2, 7)$ that are on the line,

$$\textbf{slope} = \frac{\text{rise}}{\text{run}} = \frac{\text{difference in } y\text{-values}}{\text{difference in } x\text{-values}} = \frac{7-(-1)}{2-(-2)} = \frac{8}{4} = 2.$$

From similar illustrations and the use of subscript notation, we can develop the following formula for the slope of any line.

> ### Slope
>
> Let $P_1(x_1, y_1)$ and $P_2(x_2, y_2)$ be two points on a line. The **slope** can be calculated as follows.
>
> $$\textbf{slope} = \boldsymbol{m} = \frac{\textbf{rise}}{\textbf{run}} = \frac{\boldsymbol{y_2 - y_1}}{\boldsymbol{x_2 - x_1}}$$
>
> **Note:** The letter m is standard notation for representing the slope of a line.
>
> FORMULA

$$\text{slope} = m = \frac{\text{rise}}{\text{run}} = \frac{y_2 - y_1}{x_2 - x_1}$$

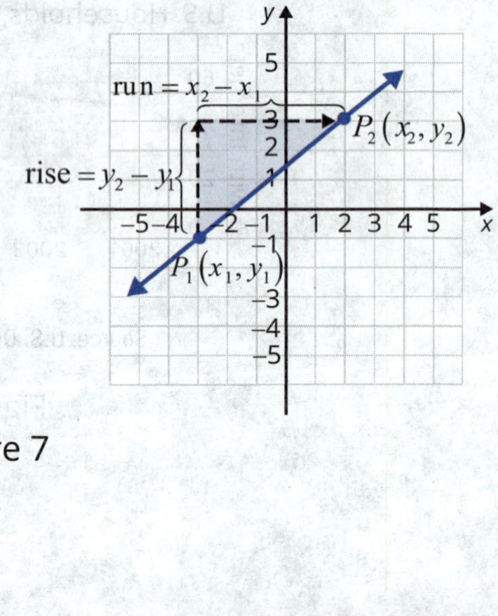

Figure 7

Example 1 Finding the Slope of a Line

Find the slope of the line that contains the points $(-1, 2)$ and $(3, 5)$, then graph the line.

Solution

For (x_1, y_1), use $(-1, 2)$ and for (x_2, y_2), use $(3, 5)$.

$$\text{slope} = m = \frac{y_2 - y_1}{x_2 - x_1} = \frac{5-2}{3-(-1)} = \frac{3}{4}$$

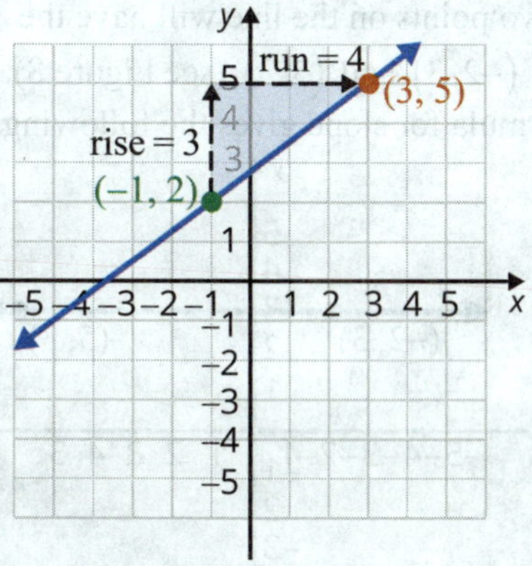

Or, for (x_1, y_1), use $(3, 5)$ and for (x_2, y_2), use $(-1, 2)$.

$$\text{slope} = m = \frac{y_2 - y_1}{x_2 - x_1} = \frac{2-5}{-1-3} = \frac{-3}{-4} = \frac{3}{4}$$

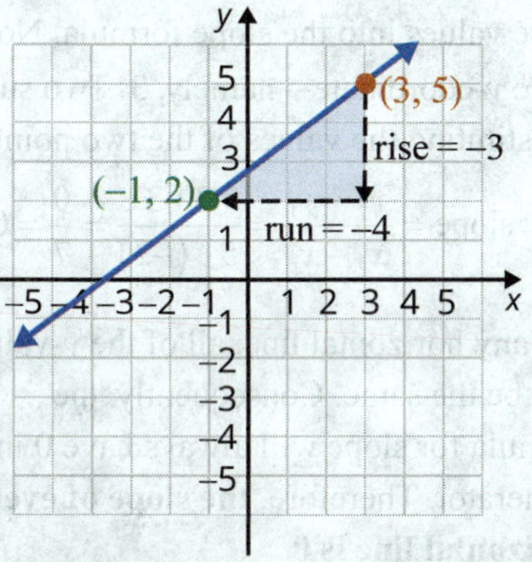

Now work margin exercise 1.

1. Find the slope of the line that contains the points $(3, -1)$ and $(4, 2)$, and then graph the line.

As we see in Example 1, **the slope is the same even if the order of the points is reversed.** The important part of the procedure is that **the coordinates must be subtracted in the same order in both the numerator and the denominator.**

In general,

$$\text{slope} = \frac{y_2 - y_1}{x_2 - x_1} = \frac{y_1 - y_2}{x_1 - x_2}.$$

Example 2 Finding the Slope of a Line

Find the slope of the line that contains the points $(1, 3)$ and $(5, 1)$, then graph the line.

Solution

For (x_1, y_1), use $(1, 3)$ and for (x_2, y_2), use $(5, 1)$.

$$\text{slope} = m = \frac{1-3}{5-1} = \frac{-2}{4} = -\frac{1}{2}$$

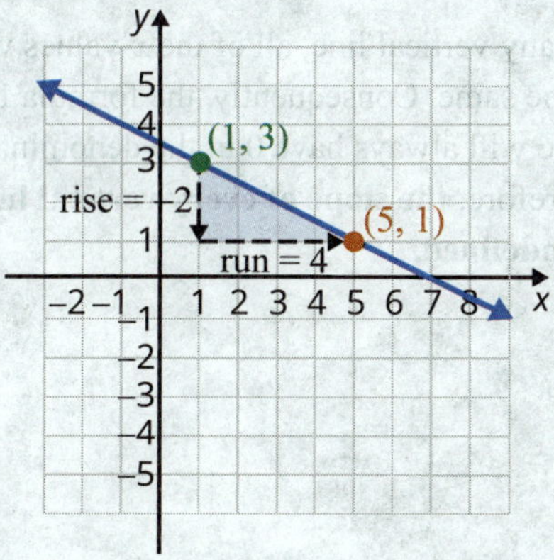

Now work margin exercise 2.

2. Find the slope of the line that contains the points $(0, 5)$ and $(4, 2)$, and then graph the line.

Positive and Negative Slope

Lines with **positive slope go up** (increase) as we move along the line from left to right.

Lines with **negative slope go down** (decrease) as we move along the line from left to right.

DEFINITION

C Slopes of Horizontal and Vertical Lines

In Section 10.2, we discussed the graphs of horizontal lines ($y = b$) and vertical lines ($x = a$), but the slopes of lines of these types were not discussed.

To find the slope of a horizontal line, such as $y = 3$, find two points on the line and substitute these values into the slope formula. Note that any two points on the line will have the same y-coordinate—namely, 3. Two such points are $(-2, 3)$ and $(5, 3)$ (see Figure 8). Substituting the values of the two points into the formula for slope gives the following.

$$\text{slope} = \frac{y_2 - y_1}{x_2 - x_1} = \frac{3-3}{5-(-2)} = \frac{0}{7} = 0$$

For any horizontal line, all of the y-values will be the same. Consequently, the formula for slope will always have 0 in the numerator. Therefore, **the slope of every horizontal line is 0.**

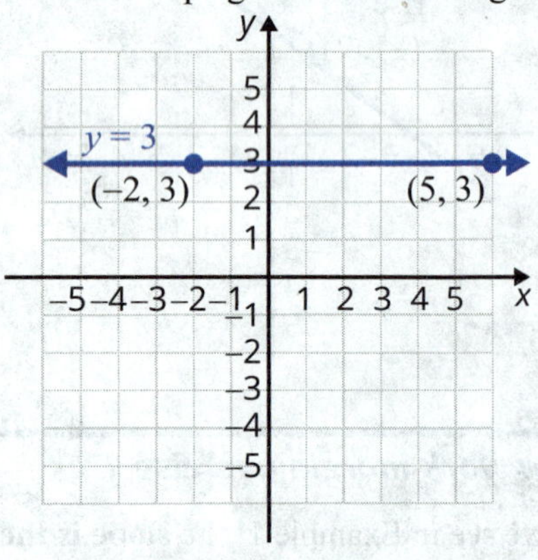

Figure 8

To find the slope of a vertical line, such as $x = 4$, find two points on the line and substitute these values into the slope formula. Note that any two points on the line will have the same x-coordinate—namely, 4. Two such points are $(4, 1)$ and $(4, 6)$ (see Figure 9). The slope formula gives the following.

$$\text{Slope} = \frac{y_2 - y_1}{x_2 - x_1} = \frac{6-1}{4-4} = \frac{5}{0}, \textbf{ which is undefined.}$$

(Remember, division by 0 is undefined; thus, the slope is undefined.)

For any vertical line, all of the x-values will be the same. Consequently, the formula for slope will always have 0 in the denominator. Therefore, **the slope of every vertical line is undefined.**

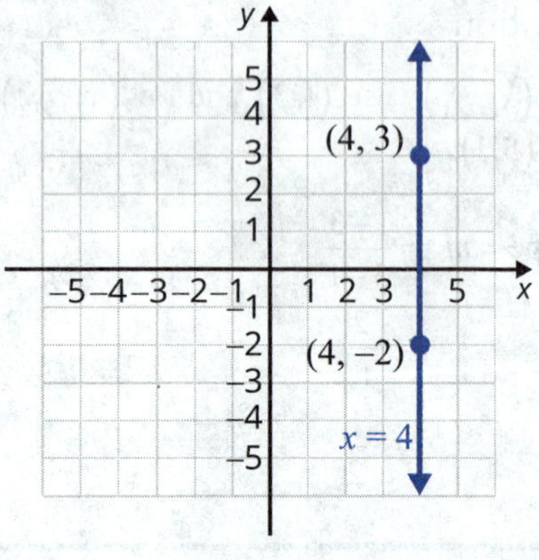

Figure 9

Horizontal and Vertical Lines

The following two general statements are true for horizontal and vertical lines.

1. For **horizontal lines** (of the form $y = b$), the **slope is 0**.
2. For **vertical lines** (of the form $x = a$), the **slope is undefined**.

DEFINITION

Example 3 Finding the Slope of a Horizontal Line

Find the equation and slope of the horizontal line through the point $(-2, 5)$.

Solution

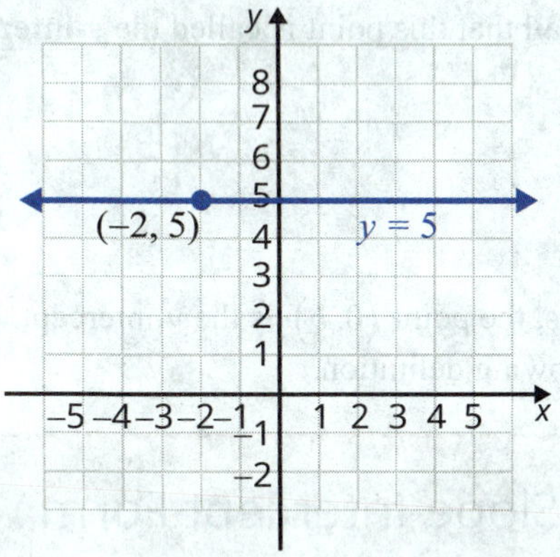

The equation is $y = 5$ and the slope is 0.

Now work margin exercise 3.

3. Find the equation and slope of the horizonal line through the point $(3,-2)$.

Example 4 Finding the Slope of a Vertical Line

Find the equation and slope of the vertical line through the point $(3, 2)$.

Solution

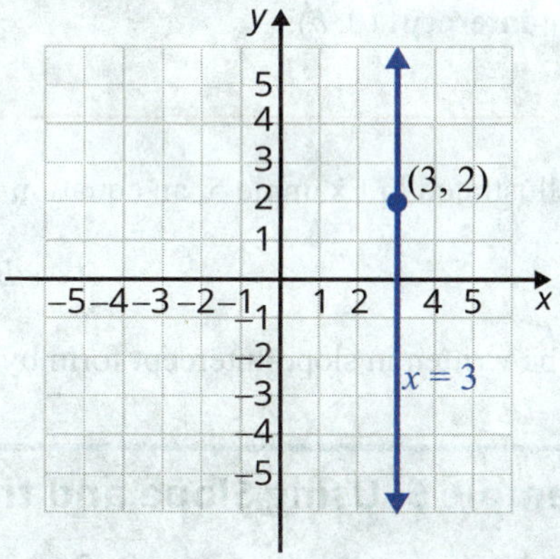

The equation is $x = 3$ and the slope is undefined.

Now work margin exercise 4.

4. Find the equation and slope of the vertical line through the point $(2,4)$.

D Slope-Intercept Form

There are certain relationships between the coefficients in the equation of a line and the graph of that line. For example, consider the equation

$$y = 5x - 7.$$

First, find two points on the line and calculate the slope. The points $(0, -7)$ and $(2, 3)$ both satisfy the equation.

$$\text{slope} = m = \frac{3-(-7)}{2-0} = \frac{10}{2} = 5$$

Observe that the slope, $m = 5$, is the same as the coefficient of x in the equation $y = 5x - 7$. This is not just a coincidence. In fact, if a linear equation is solved for y, then the coefficient of x will always be the slope of the line.

The Slope m

For an equation in the form $\mathbf{y = mx + b}$, the slope of the line is m.

DEFINITION

For the line $y = mx + b$, the point where $x = 0$ is the point where the line crosses the y-axis. Recall that this point is called the ***y*-intercept**. By letting $x = 0$, we get

$$y = mx + b$$
$$y = m \cdot 0 + b$$
$$y = b.$$

Thus, the point $(0, b)$ is the y-intercept. The concepts of slope and y-intercept lead to the following definition.

Slope-Intercept Form

An equation of the form

$$y = mx + b$$

is called the **slope-intercept** form for the equation of a line that has **slope** m and ***y*-intercept** $(0, b)$.

DEFINITION

As illustrated in Example 5, an equation in **standard form**

$$Ax + By = C \text{ with } B \neq 0$$

can be written in slope-intercept form by solving for y.

Example 5 Using Slope and the *y*-Intercept to Graph a Line

Find the slope and y-intercept of $-2x + 3y = 6$, and graph the line.

Solution

Solve for y.

$$-2x + 3y = 6$$
$$3y = 2x + 6$$
$$\frac{3y}{3} = \frac{2x}{3} + \frac{6}{3}$$
$$y = \frac{2}{3}x + 2$$

Thus, $m = \frac{2}{3}$, which is the slope, and b is 2, making the y-intercept $(0, 2)$.

5. Find the slope and y-intercept of $-4x + 2y = 12$, and graph the line.

10.3 Exercises

Concept Check

1. run 2. change
3. positive 4. undefined
5. 0
6. slope, y-intercept
7. True
8. False; Slope is the same regardless of the order of the points.
9. True 10. True

Practice

1. $m = 5$
2. m is undefined
3. $m = -\frac{1}{7}$ 4. $m = -\frac{8}{7}$
5. $m = 0$

As shown in the graph, if we "rise" 2 units up and "run" 3 units to the right **from the *y*-intercept** $(0, 2)$, we locate another point, $(3, 4)$. The line can be drawn through these two points.

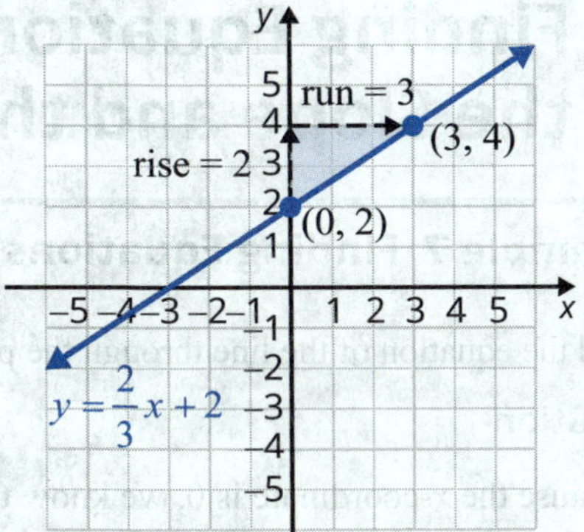

As shown in the second graph, we could also first "run" 3 units right and "rise" 2 units up from the *y*-intercept to locate the point $(3, 4)$ on the graph.

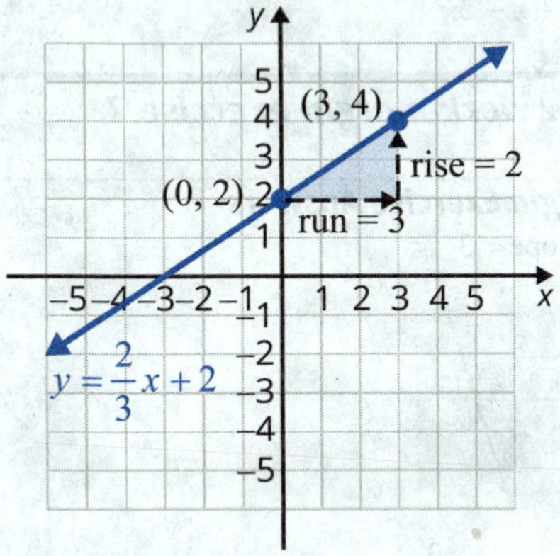

Now work margin exercise 5.

Example 6 Using Slope and the *y*-Intercept to Graph a Line

Find the slope and *y*-intercept of $x + 2y = -6$, and graph the line.

Solution

Solve for *y*.

$$x + 2y = -6$$
$$2y = -x - 6$$
$$\frac{2y}{2} = \frac{-x}{2} - \frac{6}{2}$$
$$y = -\frac{1}{2}x - 3$$

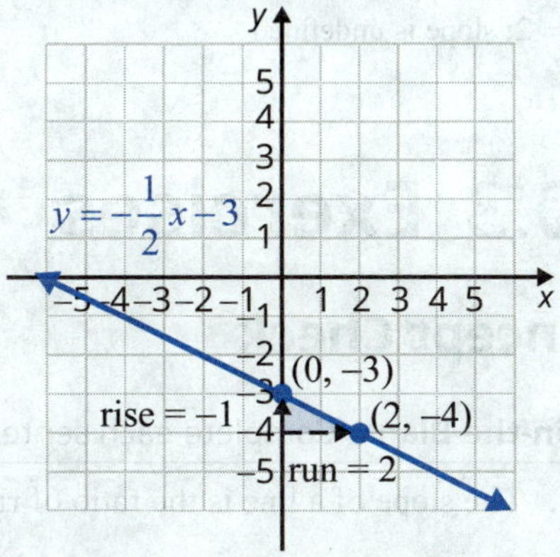

Thus, $m = -\frac{1}{2}$, which is the slope, and b is -3, making the *y*-intercept $(0, -3)$.

We can treat $m = -\frac{1}{2}$ as $m = \frac{-1}{2}$ and the "rise" as -1 and the "run" as 2. Moving from $(0, -3)$ as shown in the graph, we locate another point $(2, -4)$ on the graph and draw the line.

Now work margin exercise 6.

6. m is undefined

7. $m = \frac{1}{2}$ 8. $m = \frac{3}{2}$

9. $m = 2$ 10. $m = -\frac{3}{10}$

11. $m = \frac{1}{5}$ 12. $m = \frac{5}{4}$

13.

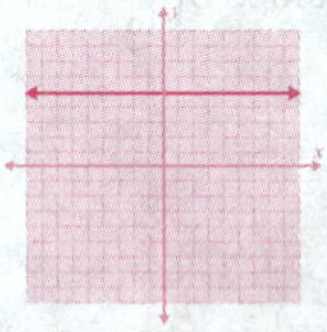

Horizontal line; $m = 0$

14.

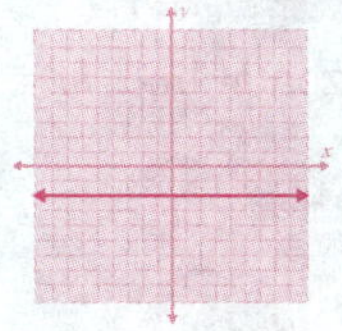

Horizontal line; $m = 0$

15.

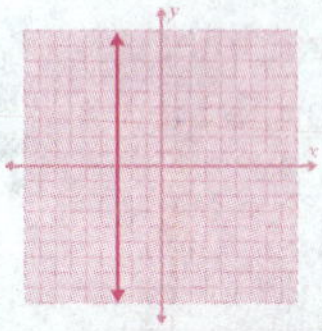

Vertical line; m is undefined

6. Find the slope and *y*-intercept of $3x + 2y = -10$, and graph the line.

Math Tip

In Example 6, we also could have written $m = -\frac{1}{2}$ as $m = \frac{1}{-2}$. We interpret this as a "rise" of 1 and a "run" of -2. Moving from the *y*-intercept $(0,-3)$, we locate the alternate point $(-2,-2)$ on the line.

16.

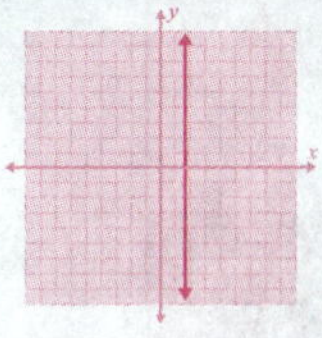

Vertical line; m is undefined

E Finding Equations of Lines Given the Slope and the *y*-Intercept

7. Find the equation of the line through the point $(0, -3)$ with a slope of $\frac{2}{3}$.

Example 7 Finding Equations Given the Slope and the *y*-Intercept

Find the equation of the line through the point $(0, -2)$ with slope $\frac{1}{2}$.

Solution

Because the x-coordinate is 0, we know that the point $(0, -2)$ is the y-intercept. So $b = -2$. The slope is $\frac{1}{2}$. So $m = \frac{1}{2}$. Substituting in slope-intercept form, $y = mx + b$, gives the result $y = \frac{1}{2}x - 2$.

Now work margin exercise 7.

Margin Exercise Answers

1. slope = 3

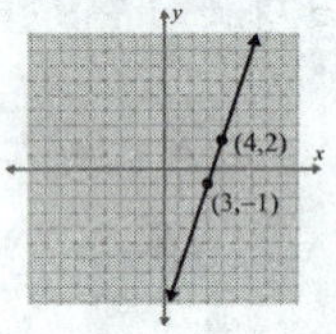

2. slope = $-\frac{3}{4}$

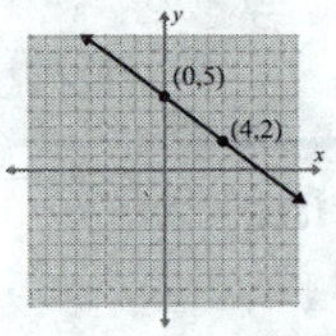

3. $y = -2$; slope is 0

4. $x = 2$; slope is undefined

5. $m = 2$; y-intercept = $(0, 6)$

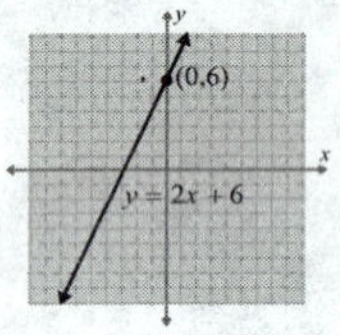

6. $m = -\frac{3}{2}$; y-intercept = $(0, -5)$

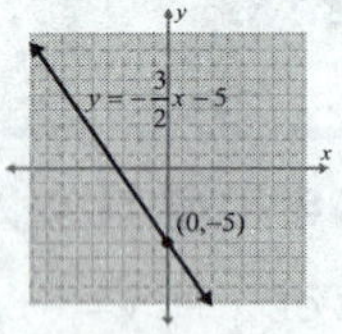

7. $y = \frac{2}{3}x - 3$

10.3 Exercises

Concept Check

Fill-in-the-Blank. Complete each sentence using information found in this section.

1. The slope of a line is the ratio of rise to _________.
2. Another name for slope is the rate of _________.
3. A line that rises (increases) from left to right has a/an _________ slope.
4. The slope of every vertical line is _________.
5. The slope of every horizontal line is _________.
6. In the equation $y = mx + b$, m represents the _________ and $(0, b)$ represents the _________.

17.

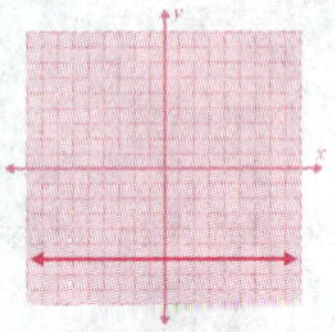

Horizontal line; $m = 0$

18.

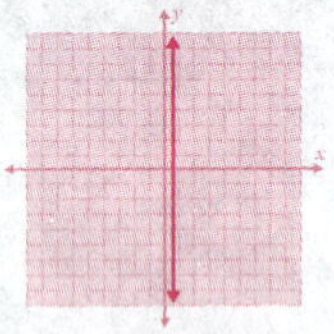

Vertical line; m is undefined

19.

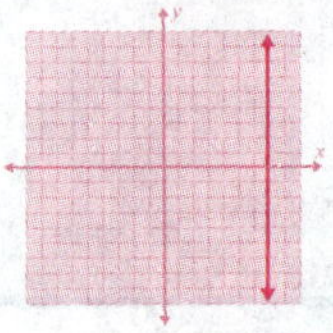

Vertical line; m is undefined

20.

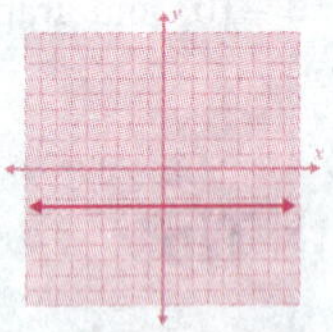

Horizontal line; $m = 0$

21.

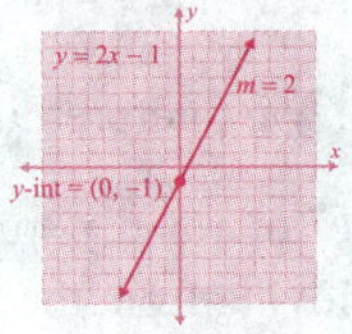

22.

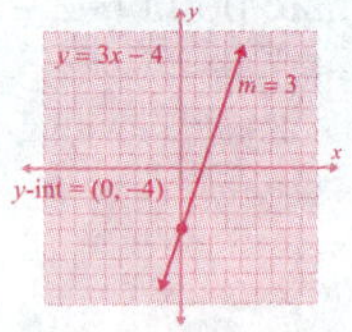

23.

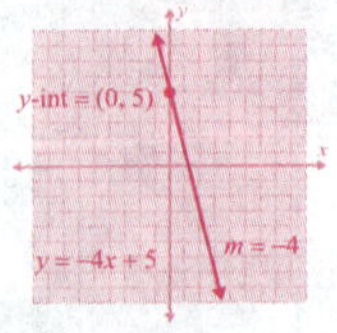

True/False. Determine whether each statement is true or false. If a statement is false, explain how it can be changed so the statement will be true. (**Note:** There may be more than one acceptable change.)

7. If the y-intercept and the slope of a line are given, there is enough information to write the equation of the line.

8. When using the slope formula, the slope of a line changes if the order of the points is reversed.

9. A line that falls (decreases) from left to right has a negative slope.

10. The line that represents the equation $y = 2x + 4$ has a y-intercept of $(0, 4)$.

Practice

Find the slope of the line determined by each pair of points. See Examples 1 and 2.

1. $(2, 4); (1, -1)$
2. $(1, -2); (1, 4)$
3. $(-6, 3); (1, 2)$
4. $(-3, 7); (4, -1)$
5. $(-5, 8); (3, 8)$
6. $(-2, 3); (-2, -1)$
7. $(5, 1); (3, 0)$
8. $(0, 0); (-2, -3)$
9. $\left(\frac{3}{4}, \frac{3}{2}\right); (1, 2)$
10. $\left(4, \frac{1}{2}\right); (-1, 2)$
11. $\left(\frac{3}{2}, \frac{4}{5}\right); \left(-2, \frac{1}{10}\right)$
12. $\left(\frac{7}{2}, \frac{3}{4}\right); \left(\frac{1}{2}, -3\right)$

Determine whether each equation represents a horizontal line or vertical line and give its slope. Graph the line. See Examples 3 and 4.

13. $y = 5$
14. $y = -2$
15. $x = -3$
16. $x = 1.7$
17. $3y = -18$
18. $4x = 2.4$
19. $-3x + 21 = 0$
20. $2y + 5 = 0$

Write each equation in slope-intercept form. Find the slope and y-intercept, and then use them to draw the graph. See Examples 5 and 6.

21. $y = 2x - 1$
22. $y = 3x - 4$
23. $y = 5 - 4x$
24. $y = 4 - x$
25. $y = \frac{2}{3}x - 3$
26. $y = \frac{2}{5}x + 2$
27. $x + y = 5$
28. $x - 2y = 6$
29. $x + 5y = 10$
30. $4x + y = 0$
31. $4x + y + 3 = 0$
32. $2x + 7y + 7 = 0$
33. $2y - 8 = 0$

24.
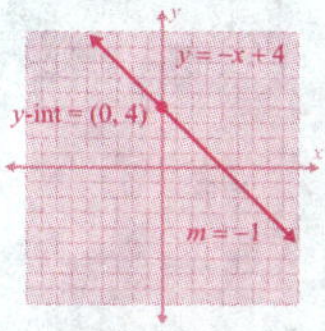

25.
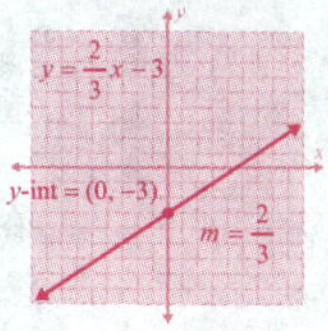

26.
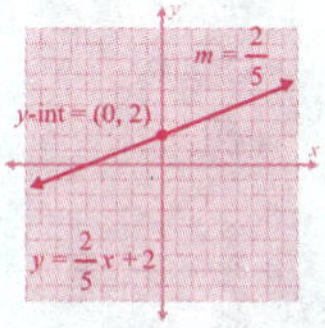

27.
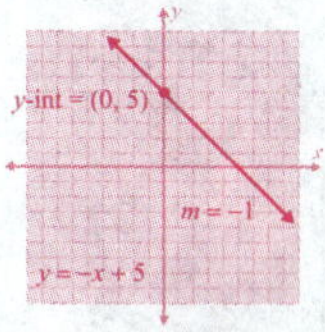

28.
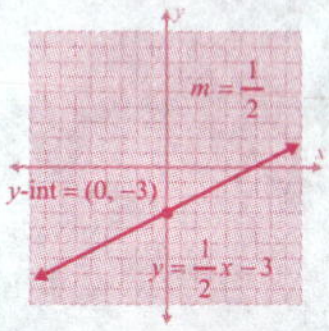

29.
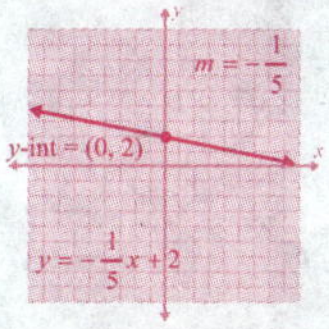

30.
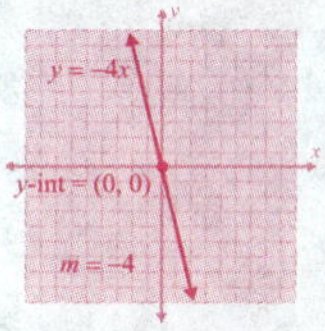

31.
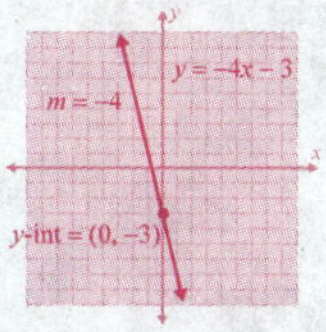

32.
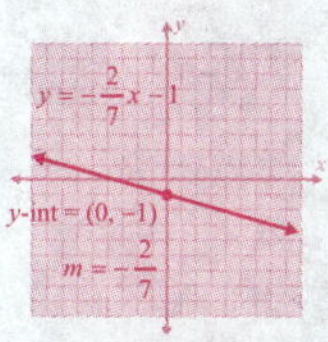

33.
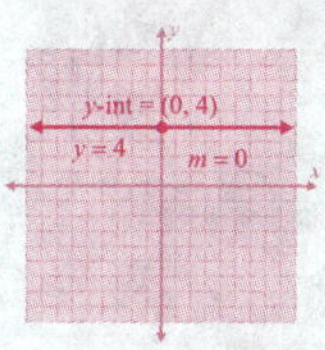

34. $3y - 9 = 0$ **39.** $5x - 6y = 18$ **44.** $7x + 2y = 4$

35. $2x = 3y$ **40.** $3x + 6 = 6y$ **45.** $6y = -6 + 3x$

36. $4x = y$ **41.** $5 - 3x = 4y$ **46.** $4x = 3y - 7$

37. $3x + 9 = 0$ **42.** $5x = 11 - 2y$ **47.** $5x - 2y + 5 = 0$

38. $4x + 7 = 0$ **43.** $6x + 4y = -8$ **48.** $6x + 5y = -15$

In reference to the equation $y = mx + b$, sketch the graphs of three lines for each of the two characteristics listed below.

49. $m > 0$ and $b > 0$ **51.** $m > 0$ and $b < 0$

50. $m < 0$ and $b > 0$ **52.** $m < 0$ and $b < 0$

Find an equation in slope-intercept form for the line passing through the given point with the given slope. See Example 7.

53. $(0, 3); m = -\frac{1}{2}$ **57.** $(0, -5); m = 4$ **61.** $(0, -3); m = -\frac{5}{6}$

54. $(0, 2); m = \frac{1}{3}$ **58.** $(0, 9); m = -1$ **62.** $(0, -1); m = -\frac{3}{2}$

55. $(0, -3); m = \frac{2}{5}$ **59.** $(0, -4); m = 1$

56. $(0, -6); m = \frac{4}{3}$ **60.** $(0, 6); m = -5$

The graph of a line is shown with two points labeled. Find **a.** the slope, **b.** the *y*-intercept (if there is one), and **c.** the equation of the line in slope-intercept form.

63.

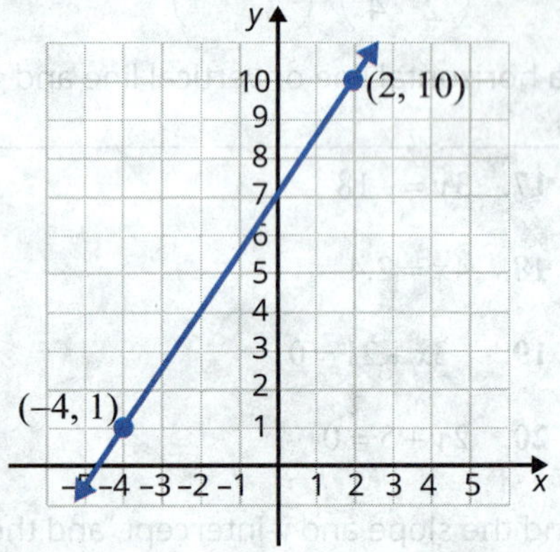

64.

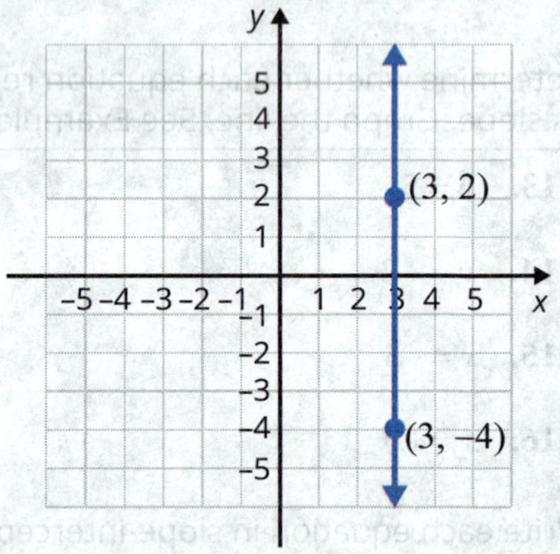

65.

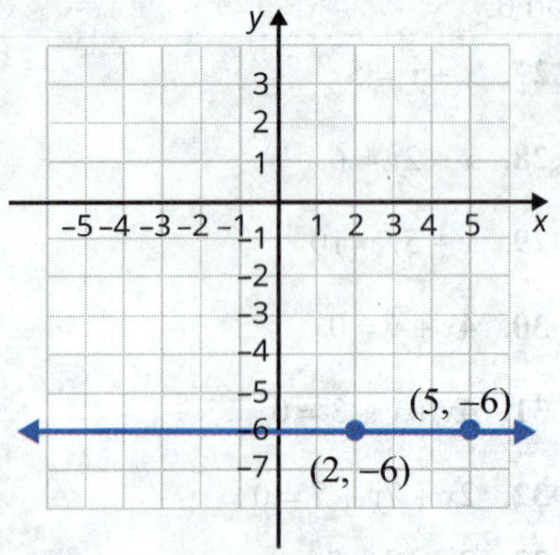

66.

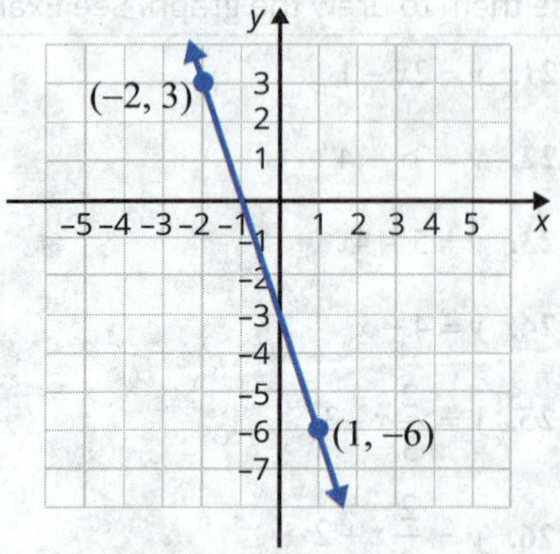

34.

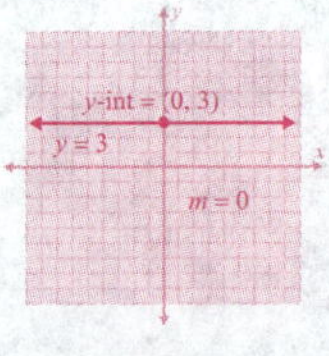

35.

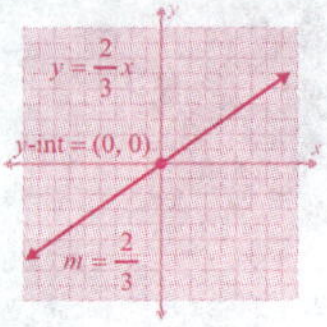

36.

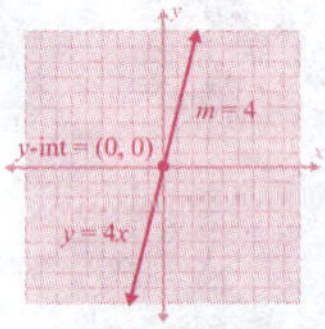

37.

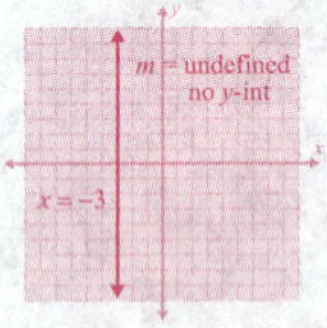

38.

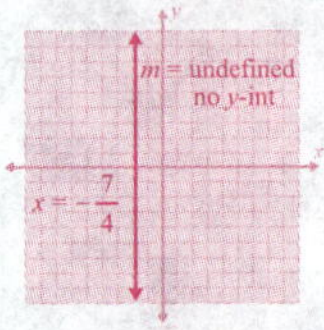

39.

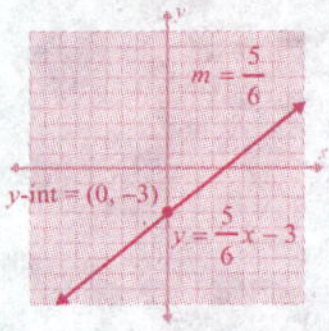

40.

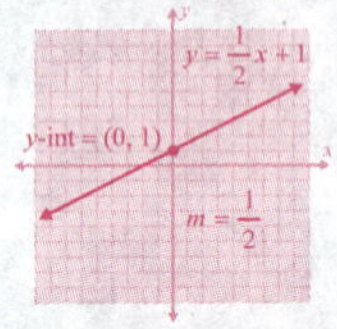

41.

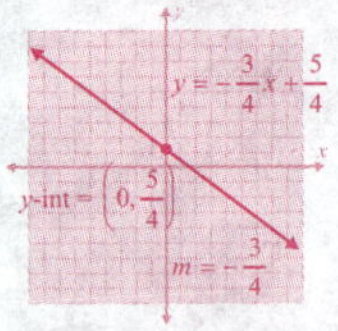

42.

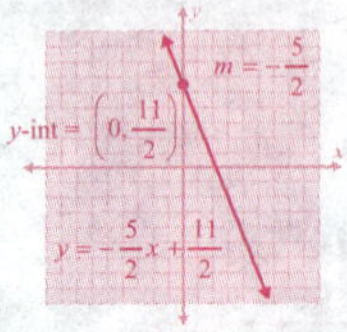

43.

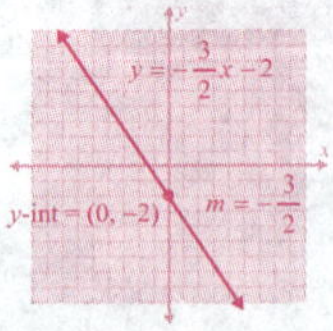

67.

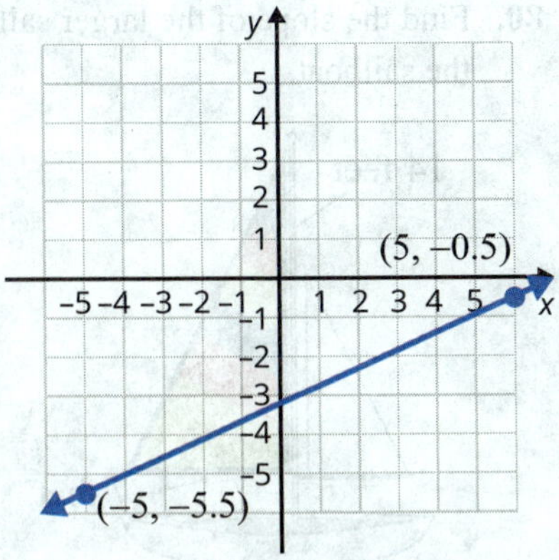

68.

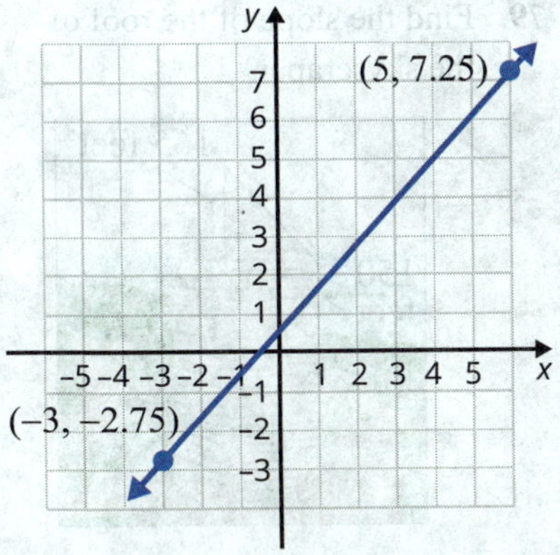

69.

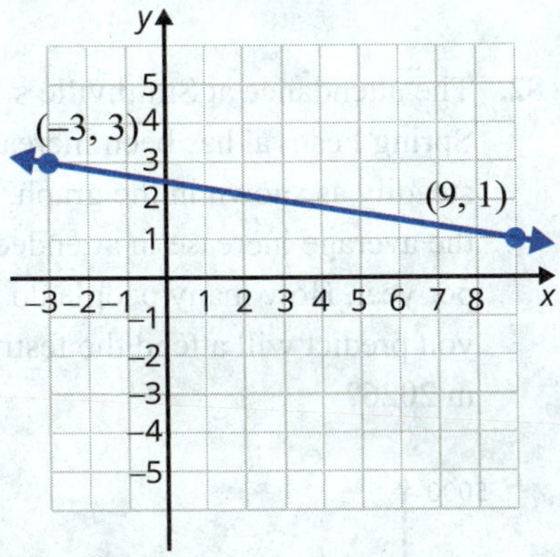

70.

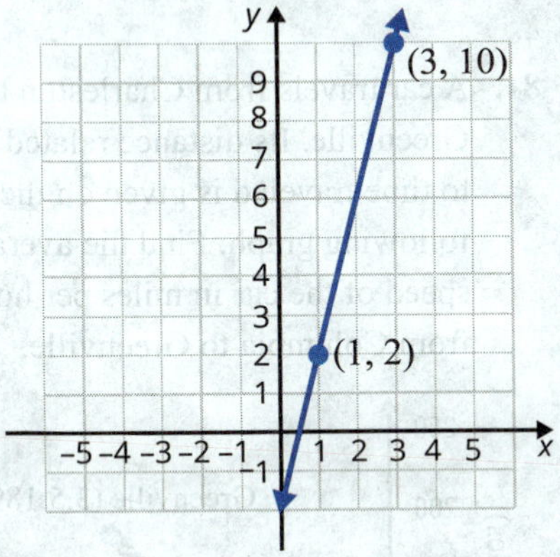

Points are said to be **collinear** if they lie on a straight line. If points are collinear, then the slope of the line through any two of them must be the same (because the line is the same line). Use this idea to determine whether the three points in each of the sets are collinear.

71. $\{(-1, 3), (0, 1), (5, -9)\}$

72. $\{(-2, -4), (0, 2), (3, 11)\}$

73. $\{(-2, 0), (0, 30), (1.5, 5.25)\}$

74. $\{(-1, -7), (1, 1), (2.5, 7)\}$

75. $\left\{\left(\frac{2}{3}, \frac{1}{2}\right), \left(0, \frac{5}{6}\right), \left(-\frac{3}{4}, \frac{29}{24}\right)\right\}$

76. $\left\{\left(\frac{3}{2}, -\frac{1}{3}\right), \left(0, \frac{1}{6}\right), \left(-\frac{1}{2}, \frac{3}{4}\right)\right\}$

Applications

Solve.

77. Find the slope of the ski slope.

78. Find the slope of the road.

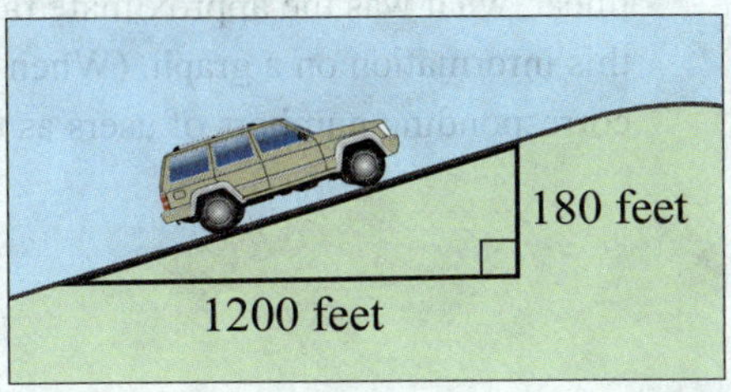

44.

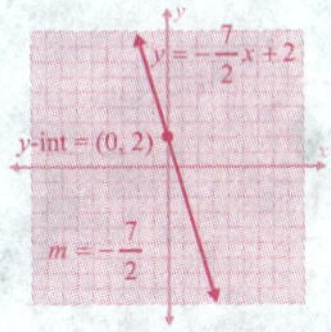

45.

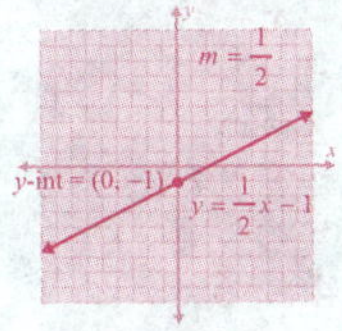

46.

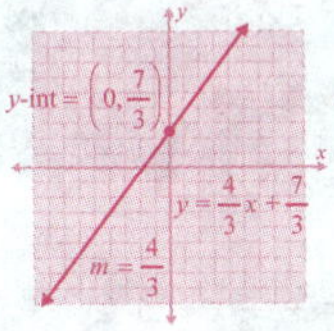

47.

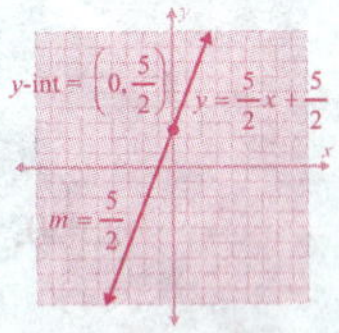

48.

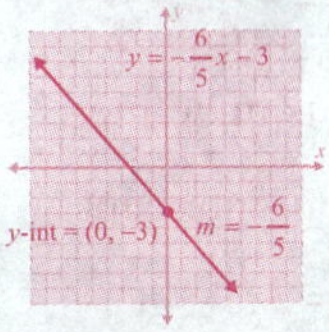

49. Answers will vary.

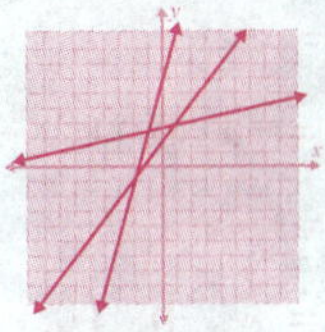

50. Answers will vary.

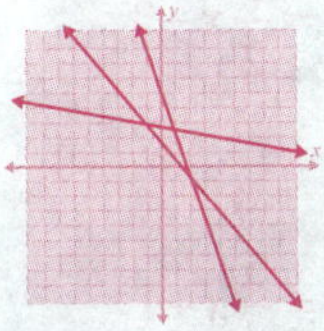

51. Answers will vary.

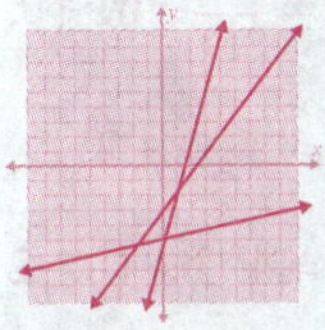

52. Answers will vary.

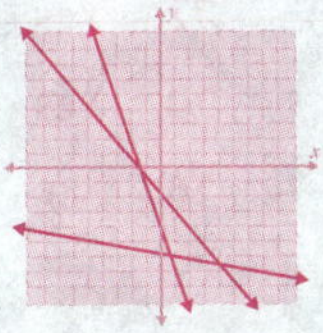

53. $y=-\frac{1}{2}x+3$

54. $y=\frac{1}{3}x+2$

55. $y=\frac{2}{5}x-3$

56. $y=\frac{4}{3}x-6$

57. $y=4x-5$

58. $y=-x+9$

59. $y=x-4$

60. $y=-5x+6$

61. $y=-\frac{5}{6}x-3$

62. $y=-\frac{3}{2}x-1$

63. **a.** $m=-\frac{1}{6}$ **b.** $\left(0,\frac{5}{2}\right)$

c. $y=-\frac{1}{6}x+\frac{5}{2}$

64. **a.** m is undefined

b. no y-intercept

c. $x=3$

65. **a.** $m=0$ **b.** $(0,-6)$

c. $y=-6$

66. **a.** $m=-3$ **b.** $(0,-3)$

c. $y=-3x-3$

67. **a.** $m=\frac{1}{2}$ **b.** $(0,-3)$

c. $y=\frac{1}{2}x-3$

68. **a.** $m=\frac{5}{4}$ **b.** $(0,1)$

c. $y=\frac{5}{4}x+1$

69. **a.** $m=-\frac{1}{3}$ **b.** $(0,2)$

c. $y=-\frac{1}{3}x+2$

70. **a.** $m=4$ **b.** $(0,-2)$

c. $y=4x-2$

71. Yes **72.** Yes

73. No **74.** Yes

75. Yes **76.** No

79. Find the slope of the roof of the skyscraper.

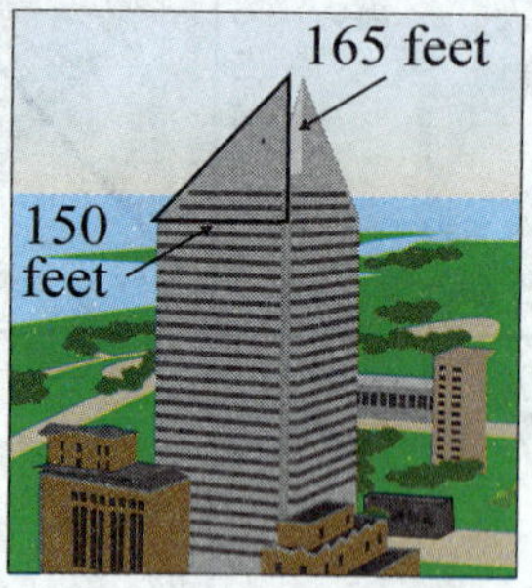

80. Find the slope of the larger sail on the sailboat.

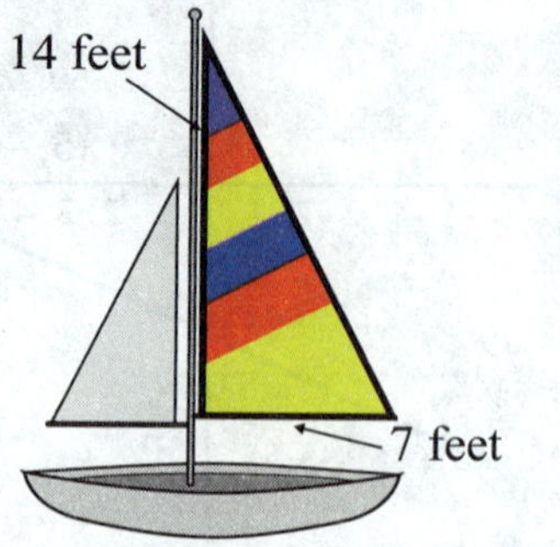

81. A car travels from Charleston to Greenville. Its distance related to time traveled is given on the following graph. Find the average speed of the car in miles per hour from Columbia to Greenville.

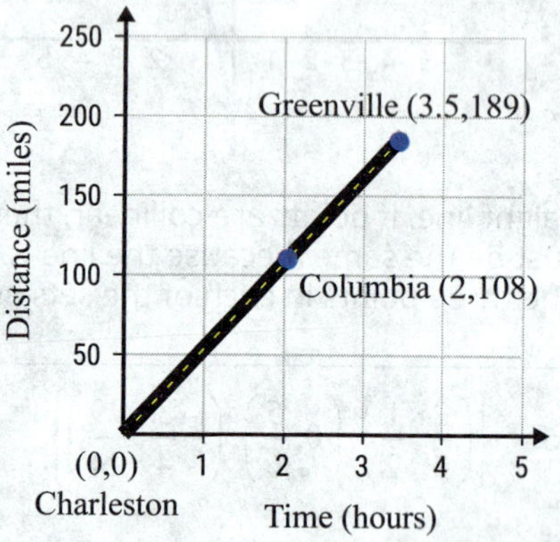

82. The attendance at Smithville's Spring Festival has been increasing steadily as shown in the graph. Find the average increase in attendees per year. How many people do you predict will attend the festival in 2026?

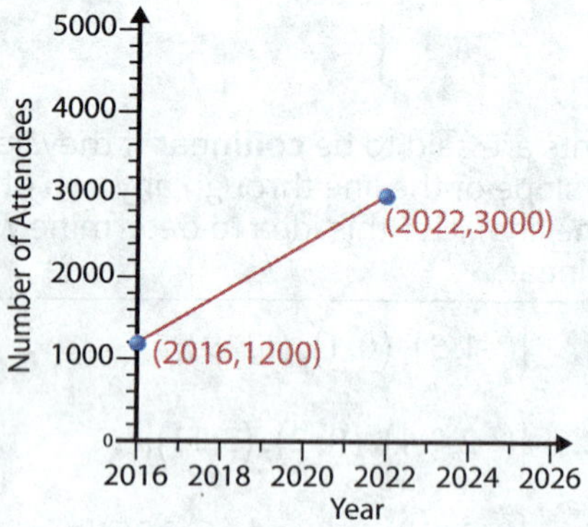

83. John bought his new car for \$35,000 in the year 2022. He knows that the value of his car has depreciated linearly. If the value of the car in 2025 was \$23,000, what was the annual rate of depreciation of his car? Show this information on a graph. (When graphing, use years as the x-coordinates and the corresponding values of the car as the y-coordinates.)

84. The number of people in the United States with cell phones was about 198 million in 2011 and about 232 million in 2016. If the growth in the usage of cell phones was linear, what was the approximate rate of growth per year from 2011 to 2016. Show this information on a graph. (When graphing, use years as the x-coordinates and the corresponding numbers of users as the y-coordinates.) [1]

1 Source: www.statista.com/statistics/231612/number-of-cell-phone-users-usa/

85. The Millennium Force roller coaster at Cedar Point in Sandusky, Ohio, has been voted the Best Steel Coaster by Golden Ticket eleven times since 2001. The Millennium Force is known for its steep slope on the first hill and speeds up to 93 miles per hour. The descent from the first hill starts at 310 feet above the ground and ends 10 feet above the ground. The descent runs approximately 53 feet. Find the slope of this hill to the nearest hundredth. (**Hint:** The slope is running downhill. Consider how that affects the slope.)

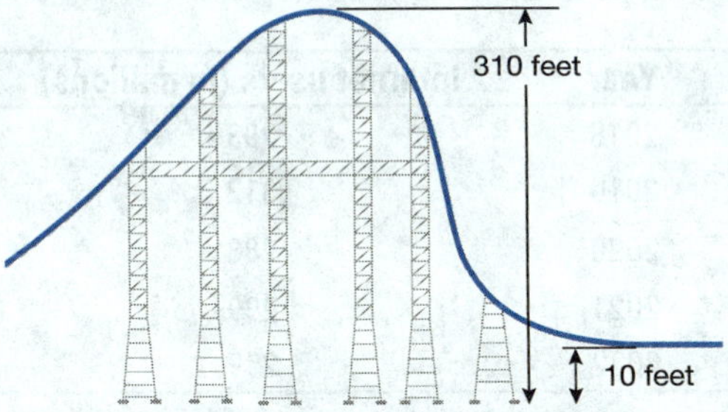

86. The grade, or slope, of a road is commonly given as a percentage. The grade can be determined by multiplying the slope by 100. Calculate the slope of each road or track and then determine its grade. Round each percent to the nearest hundredth if necessary.

a. A road increases in height 5 feet for every 120 feet of run.

b. The railway line with the steepest grade that does not run on a track system is the Lisbon tramway network in Portugal which has a section that increases 5 feet in height for every 37 feet of run.

c. A road on the Route des Crêtes (Route of the Ridges) in France has an elevation that increases in height 450 feet over 1500 feet.

87. Jared sells paintings at an open-air market. He starts his work day with \$30 and sells each painting for \$15. Jared wants to create a linear equation to model this situation where y is the amount of money Jared has at the end of the work day and x is the number of paintings sold.

a. The slope, or rate of change, is the increase in the amount of money Jared makes when he sells a painting. Determine the value of the slope and list the units for both variables.

b. The y-coordinate of the y-intercept of this equation is the amount of money Jared has before he sells any paintings. What is the y-intercept?

c. Write a linear equation in slope-intercept form to model this situation using the answers from parts a. and b.

d. Graph the equation from part c.

e. Are there any solutions to the equation which do not make sense in the context of the problem? Explain why.

f. Use the graph to determine the amount of money Jared will have after selling 4 paintings.

Applications

77. $\frac{3}{4}$ **78.** $\frac{3}{20}$

79. $\frac{11}{10}$ **80.** 2

81. 54 mph

82. 300 people per year; 4200 people

83. \$4000/year

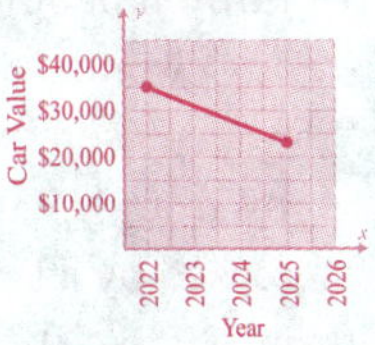

84. 6.8 million/year

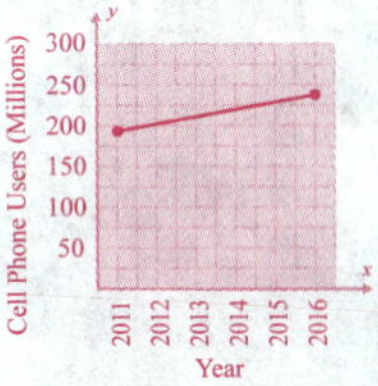

85. −5.66

86. **a.** 4.17% **b.** 13.51%
c. 30%

87. **a.** Slope is \$15; Units are dollars and paintings
b. $(0, 30)$
c. $y = 15x + 30$
d.

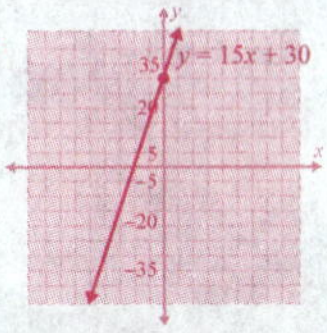

e. Answers will vary. x cannot be negative since you can't sell a negative number of paintings.
f. \$90

88. **a.** and **b.**

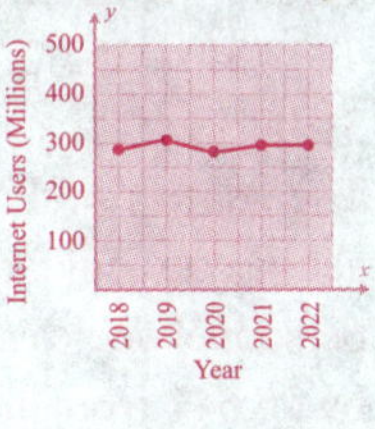

c. 19, −24, 11, 0

88. a. and **b.**

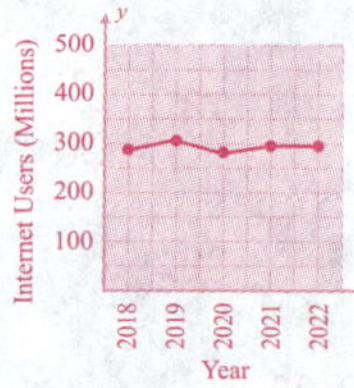

c. 19, –24, 11, 0

d. The number of internet users increased by 19 million people/year (mppy) from '18 to '19, decreased by 24 mppy from '19 to '20, increased by 11 mppy from '20 to '21, and stayed the same from '21 to '22.

88. The given table shows the estimated number of internet users from 2018 through 2022. The number of users for each year is shown in millions.

a. Plot these points on a graph.

b. Connect the points with line segments.

c. Find the slope of each line segment.

d. Interpret each slope as a rate of change.

Year	Internet users (in millions)
2018	293
2019	312
2020	288
2021	299
2022	299

Source: www.statista.com/statistics/276445/number-of-internet-users-in-the-united-states

89. a. and **b.**

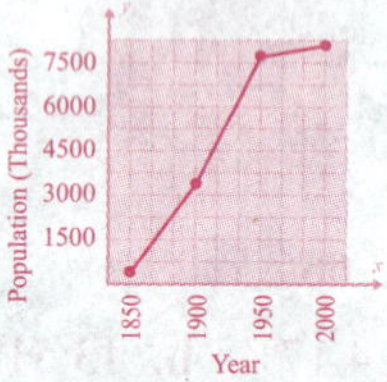

c. 58,433.1; 89,095.1; 2326.42

d. The population of New York increased by 58,433 people/year from 1850–1900; 89,095 ppy from 1900–1950; and 2326 ppy from 1950–2000.

89. The following table shows the urban growth from 1850 to 2000 in New York, NY.

a. Plot these points on a graph.

b. Connect the points with line segments.

c. Find the slope of each line segment.

d. Interpret each slope as a rate of change.

Year	Population
1850	515,547
1900	3,437,202
1950	7,891,957
2000	8,008,278

Source: U.S. Census Bureau

90. a. and **b.**

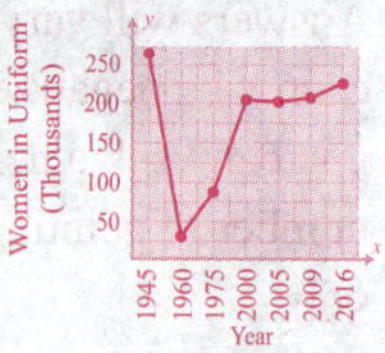

c. 15,647.07; 4354.53; 4229.32; –205.2; 450; 3,209.14

d. The number of female active duty military personnel decreased by 15,647 women/year from 1945–1960; increased by 4355 wpy from '60–'75; increased by 4229 wpy from '5–'00; decreased by 205 wpy from '00–'05; increased by 450 wpy from '05–'09; and increased by 3,209.14 wpy from '09–'16.

90. The following graph shows the number of female active duty military personnel over a span from 1945 to 2016. The number of women listed includes both officers and enlisted personnel from the Army, the Navy, the Marine Corps, and the Air Force.

a. Plot these points on a graph.

b. Connect the points with line segments.

c. Find the slope of each line segment.

d. Interpret each slope as a rate of change.

Women in the Department of Defense

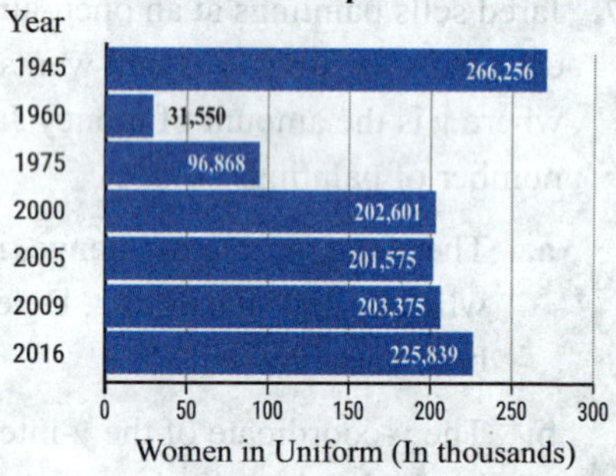

Source: U.S. Dept. of Defense

91. a. and **b.**

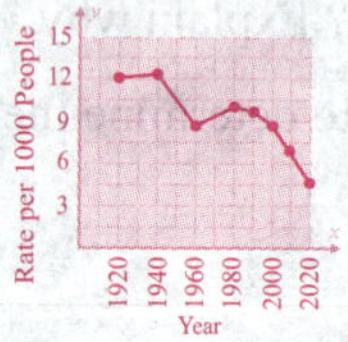

c. –0.18, 0.075, –0.02, –0.13, –0.17, –0.17

d. # of marriages decreased 0.18 marriages/1000 people people from

91. The following graph shows the rates of marriage per 1000 people in the US over a span from 1940 to 2020.

a. Plot these points on a graph.

b. Connect the points with line segments.

c. Find the slope of each line segment.

d. Interpret each slope as a rate of change.

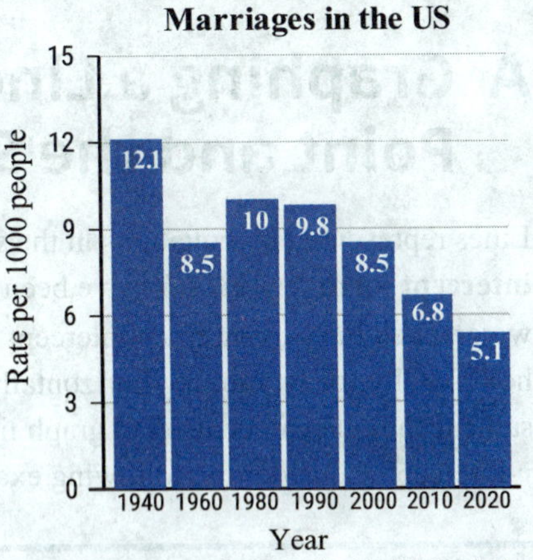

Source: U.S. National Center for Health Statistics

Writing & Thinking

92. a. Explain in your own words why the slope of a horizontal line must be 0.

b. Explain in your own words why the slope of a vertical line must be undefined.

93. a. Describe the graph of the line $y = 0$.

b. Describe the graph of the line $x = 0$.

94. In the formula $y = mx + b$, explain the meaning of m and the meaning of b.

95. The slope of a road is called a **grade**. A steep grade is cause for truck drivers to have slow speed limits in mountains. What do you think that a "grade of 12%" means? Draw a picture of a right triangle that would indicate a grade of 12%.

Collaborative Learning

96. The class should be divided into teams of 2 or 3 students. Each team will need access to a digital camera, a printer, and a ruler.

a. Take pictures of 8 things with a defined slope. (**Suggestions:** A roof, a stair railing, a beach umbrella, a crooked tree, etc. Be creative!)

b. Print each picture.

c. Use a ruler to draw a coordinate system on top of each picture. You will probably want to use increments of in. or cm, depending on the size of your picture.

d. Identify the line in each picture whose slope you are calculating and then use the coordinate systems you created to identify the coordinates of two points on each line.

e. Use the points you just found to calculate the slope of the line in each picture.

f. Share your findings with the class.

'40–'60, increased 0.075 marriages/1000 people from '60–'80, decreased 0.02 marriages/1000 people from '80–'90, decreased 0.13 marriages/1000 people from '90–'00, decreased 0.17 marriages/1000 people from '00–'10, and decreased 0.17 marriages/1000 people per year from '10–'20.

Writing & Thinking

92. a. For any horizontal line, all of the y values will be the same. Thus the formula for slope will always have 0 in the numerator making the slope of every horizontal line 0.

b. For any vertical line, all of the x values will be the same. Thus the formula for slope will always have 0 in the denominator making the slope of every vertical line undefined.

93. a. The x-axis

b. The y-axis

94. $m =$ slope of the line, $(0, b) = y$-intercept.

95. A grade of 12% means that the slope of the road is 0.12. For every 100 feet of horizontal distance (run) there is a vertical distance (rise) of 12 feet.

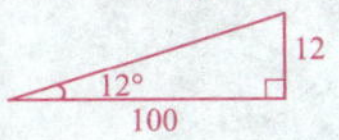

Collaborative Learning

96. Answers will vary.

Objectives

A. Graph a linear equation given its slope and a point on the line.

B. Use point-slope form to write the equation of a line given its slope and a point on the line.

C. Use point-slope form to write the equation of a line given two points on the line.

D. Write equations of parallel and perpendicular lines.

10.4 Point-Slope Form

A Graphing a Line Given a Point and the Slope

Lines represented by equations in the **standard form** $Ax + By = C$ and in the **slope-intercept form** $y = mx + b$ have been discussed in Sections 10.2 and 10.3. In Section 10.3, we graphed lines using the y-intercept $(0, b)$ and the slope by moving vertically and then horizontally (or by moving horizontally and then vertically) from the y-intercept. This same technique can be used to graph lines if the given point is on the line but is not the y-intercept. Consider the following example.

1. Graph the line with slope $m = -\frac{1}{3}$ and which passes through the point $(3, 4)$.

Example 1 Graphing a Line Given a Point and the Slope

Graph the line with slope $m = -\frac{3}{4}$ and which passes through the point $(2, 5)$.

Solution

Start from the point $(2, 5)$ and locate another point on the line using the slope as $\frac{\text{rise}}{\text{run}} = \frac{-3}{4}$ or $\frac{3}{-4}$. There are many ways to proceed.

Here are two possible ways:

1. Move 4 units right and 3 units down, or
2. Move 3 units down and 4 units right.

Either way, you arrive at the same point $(6, 2)$.

This means that we can move from the given point either with the rise first or the run first.

Note: Any numbers in the ratio of -3 to 4 can be used for the moves, such as -6 to 8 or 9 to -12.

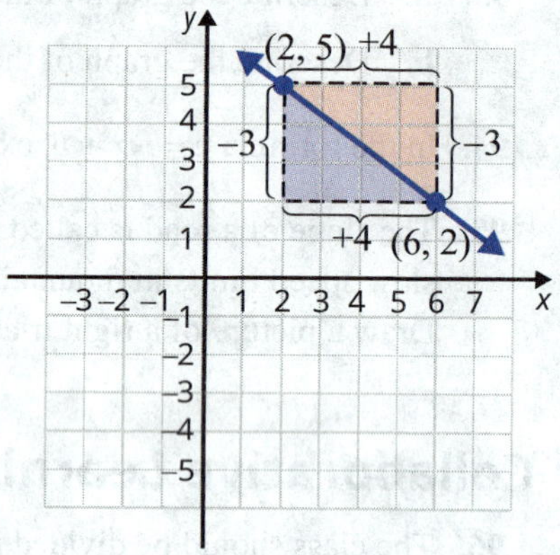

Now work margin exercise 1.

B Point-Slope Form

Now consider finding the equation of the line given the point (x_1, y_1) on the line and the slope m. If (x, y) is **any other point** on the line, then the slope formula gives the following equation.

$$\frac{y - y_1}{x - x_1} = m$$

Multiplying both sides of this equation by the denominator (assuming the denominator is not 0) gives

$$y - y_1 = m(x - x_1),$$

which is called **point-slope form.**

For example, suppose that a point $(x_1, y_1) = (8, 3)$ and the slope $m = -\frac{3}{4}$ are given.

If (x, y) represents any point on the line other than $(8, 3)$, then substituting into the formula for slope gives the following.

$$\frac{y-y_1}{x-x_1}=m \quad \text{Formula for slope}$$

$$\frac{y-3}{x-8}=-\frac{3}{4} \quad \text{Substitute the given information.}$$

$$(x-8)\left(\frac{y-3}{x-8}\right)=-\frac{3}{4}(x-8) \quad \text{Multiply both sides by } (x-8).$$

$$y-3=-\frac{3}{4}(x-8) \quad \text{Point-slope form: } y-y_1=m(x-x_1)$$

From this point-slope form, we can manipulate the equation to get the other two forms.

$$y-3=-\frac{3}{4}(x-8)$$

$$y-3=-\frac{3}{4}x+6$$

$$y=-\frac{3}{4}x+9 \quad \text{Slope-intercept form: } y=mx+b$$

$$\text{or} \quad 4(y)=4\left(-\frac{3}{4}x+9\right)$$

$$4y=-3x+36$$

$$3x+4y=36 \quad \text{Standard form: } Ax+By=C$$

Point-Slope Form

An equation of the form

$$\mathbf{y-y_1=m(x-x_1)}$$

is called the **point-slope form** for the equation of a line that contains the point (x_1, y_1) and has slope m.

DEFINITION

Example 2 Finding Equations of Lines Using the Slope and a Point

Find the equation of the line with a slope of $-\frac{1}{2}$ and which passes through the point $(2, 3)$. Graph the line using the point and slope.

Solution

Substitute the values into the point-slope form.

$$y-y_1=m(x-x_1)$$

$$y-3=-\frac{1}{2}(x-2) \quad \text{Point-slope form}$$

$$y-3=-\frac{1}{2}x+1$$

$$y=-\frac{1}{2}x+4 \quad \text{Slope-intercept form}$$

$$\text{or} \quad 2y=-x+8$$

$$x+2y=8 \quad \text{Standard form}$$

2. Find the equation of the line with a slope of $-\frac{2}{3}$ and which passes through the point $(3,1)$. Graph the line using the point and slope.

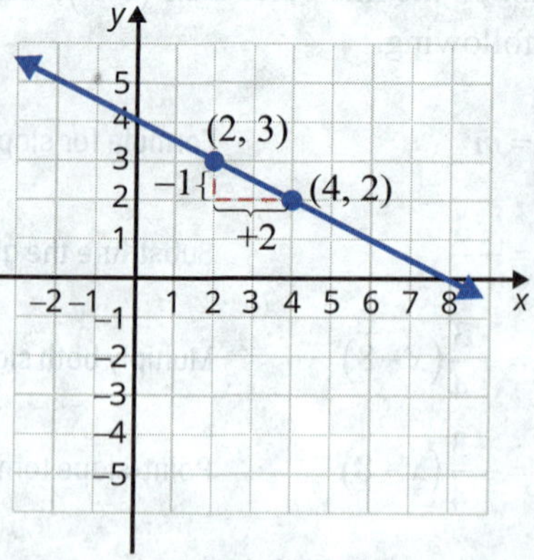

The point one unit down and two units right from $(2, 3)$ will be on the line because the slope is $m = \frac{\text{rise}}{\text{run}} = \frac{-1}{2} = -\frac{1}{2}$.

With a negative slope, either the rise is negative and the run is positive, or the rise is positive and the run is negative. In either case, as the previous figure and the following figure illustrate, the line is the same.

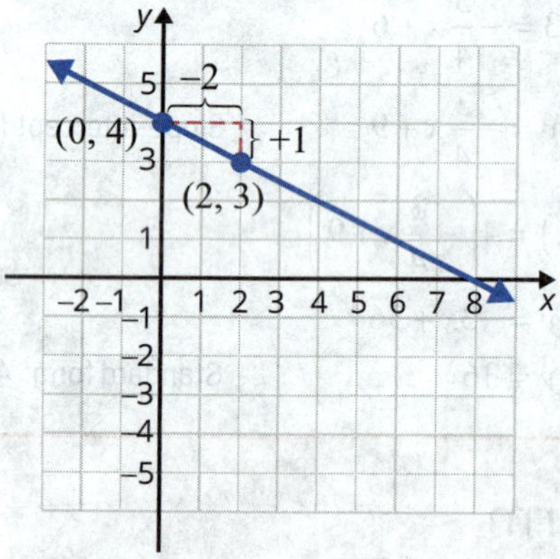

The point one unit up and two units to the left from $(2, 3)$ is on the line because the slope is $m = \frac{\text{rise}}{\text{run}} = \frac{1}{-2} = -\frac{1}{2}$.

Now work margin exercise 2.

In Example 2, the equation of the line is written in all three forms: point-slope form, slope-intercept form, and standard form. Generally, any one of these forms is sufficient. However, there are situations in which one form is preferred over the others. Therefore, manipulation among the forms is an important skill. Also, if the answer in the text is in one form and your answer is in another form, you should be able to recognize that the answers are equivalent.

C Finding Equations of Lines Given Two Points

Given two points that lie on a line, the equation of the line can be found using the following method.

Finding the Equation of a Line Given Two Points

To find the equation of a line given two points on the line:

1. Use the formula $m = \frac{y_2 - y_1}{x_2 - x_1}$ (where $x_2 \neq x_1$) to find the slope.
2. Use this slope, m, and either point in the point-slope formula $y - y_1 = m(x - x_1)$ to find the equation.

PROCEDURE

Example 3 Finding Equations Given Two Points

Find the equation of the line containing the two points $(-1, 2)$ and $(4, -2)$.

Solution

First, find the slope.

$$m = \frac{y_2 - y_1}{x_2 - x_1} = \frac{-2-2}{4-(-1)} = \frac{-4}{5} = -\frac{4}{5}$$

Now, use one of the given points and the point-slope form for the equation of a line.

(**Note:** $(-1, 2)$ and $(4, -2)$ are shown to illustrate that either point may be used.)

Using $(-1, 2)$		Using $(4, -2)$
$y - y_1 = m(x - x_1)$	Point-slope form	$y - y_2 = m(x - x_2)$
$y - 2 = -\frac{4}{5}[x-(-1)]$	Substitute.	$y-(-2) = -\frac{4}{5}(x-4)$
$y - 2 = -\frac{4}{5}x - \frac{4}{5}$	Distributive property	$y + 2 = -\frac{4}{5}x + \frac{16}{5}$
$y = -\frac{4}{5}x - \frac{4}{5} + 2$	Solve for *y*.	$y = -\frac{4}{5}x + \frac{16}{5} - 2$
$y = -\frac{4}{5}x + \frac{6}{5}$	Slope-intercept form	$y = -\frac{4}{5}x + \frac{6}{5}$
or $4x + 5y = 6$	Standard form	or $4x + 5y = 6$

Now work margin exercise 3.

3. Find the equation of the line containing the two points $(-2, 4)$ and $(0, -1)$.

Teaching Note

Emphasize, using several examples, that either point may be used after the slope is determined, and have students demonstrate to their own satisfaction that either point produces the same, or an equivalent, equation for the line.

Example 4 Finding Equations of Lines Using a Graph

Write an equation in standard form for the following line.

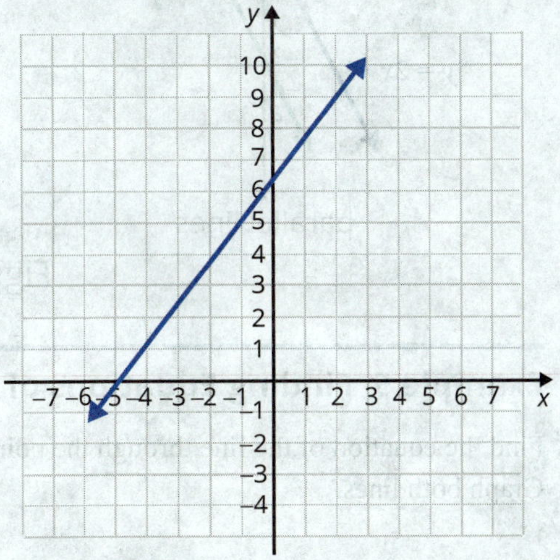

Solution

First, we must identify two points on the line. From the graph, we can see that $(2, 9)$ and $(-1, 5)$ are points on the line.

Next, we find the slope of the line as follows.

$$m = \frac{9-5}{2-(-1)} = \frac{4}{3}$$

4. Write an equation in standard form for the following line.

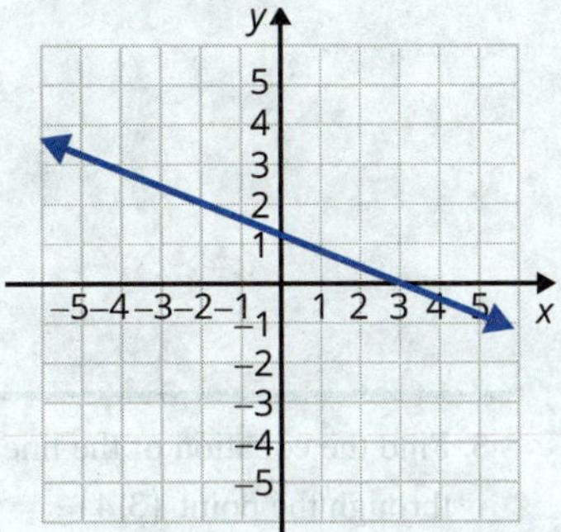

Now, we substitute the values into point-slope form and then rewrite the equation in standard form.

$$y - y_1 = m(x - x_1)$$
$$y - 5 = \frac{4}{3}(x - (-1))$$
$$y - 5 = \frac{4}{3}x + \frac{4}{3}$$
$$y = \frac{4}{3}x + \frac{19}{3}$$
$$3y = 4x + 19$$
$$-4x + 3y = 19$$

Now work margin exercise 4.

D Parallel and Perpendicular Lines

Note

All vertical lines (undefined slopes) are parallel to one another. All horizontal lines (zero slopes) are also parallel to one another. Vertical and horizontal lines are perpendicular to each other.

Parallel and Perpendicular Lines

Parallel lines are lines that never intersect and that have the **same slope**.

Perpendicular lines are lines that intersect at 90° (right) angles and have slopes that are **negative reciprocals** of each other. Horizontal lines are perpendicular to vertical lines.

DEFINITION

As illustrated in Figure 1, the lines $y = 2x + 1$ and $y = 2x - 3$ are **parallel**. They have the same slope, 2. The lines $y = \frac{2}{3}x + 1$ and $y = -\frac{3}{2}x - 2$ are **perpendicular**. Their slopes, $\frac{2}{3}$ and $-\frac{3}{2}$, are negative reciprocals of each other.

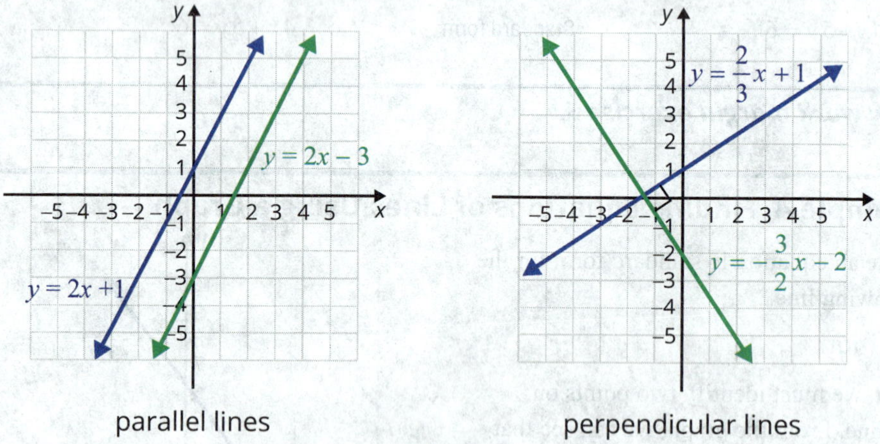

Figure 1

5. Find the equation of the line through the point $(3,4)$ and parallel to the line $3x + 2y = -2$. Graph both lines.

Example 5 Finding Equations of Parallel Lines

Find the equation of the line through the point $(3, 2)$ and parallel to the line $5x + 3y = 1$. Graph both lines.

Solution

First, solve for y to find the slope of the given line.

$$5x+3y=1$$
$$3y=-5x+1$$
$$y=-\frac{5}{3}x+\frac{1}{3} \qquad \text{Thus, any parallel to this line has slope } -\frac{5}{3}.$$

Now, use the point-slope form $y-y_1=m(x-x_1)$ with $m=-\frac{5}{3}$ and $(x_1, y_1)=(3, 2)$.

$$y-2=-\frac{5}{3}(x-3) \qquad \text{Point-slope form}$$
$$3(y-2)=-5(x-3) \qquad \text{Multiply both sides by the LCD, 3.}$$
$$3y-6=-5x+15 \qquad \text{Simplify.}$$
$$5x+3y=21 \qquad \text{Standard form}$$
$$\text{or} \quad y=-\frac{5}{3}x+7 \qquad \text{Slope-intercept form}$$

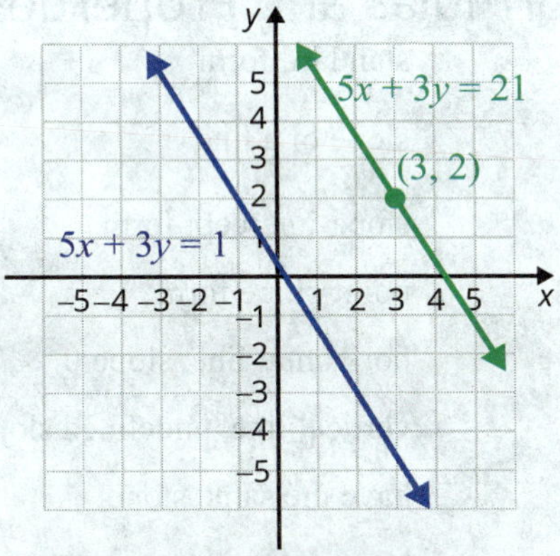

Now work margin exercise 5

Example 6 Finding Equations of Perpendicular Lines

Find the equation of the line through the point $(3, 2)$ and perpendicular to the line $5x + 3y = 1$. Graph both lines.

Solution

We know from Example 5 that the slope of the line $5x + 3y = 1$ is $-\frac{5}{3}$. Thus, any line perpendicular to this line must have slope $m=\frac{3}{5}$ (the negative reciprocal of $-\frac{5}{3}$).

Now, using the point-slope form $y-y_1=m(x-x_1)$ with $m=\frac{3}{5}$, and $(x_1, y_1)=(3, 2)$, we have the following.

$$y-2=\frac{3}{5}(x-3) \qquad \text{Point-slope form}$$
$$5(y-2)=3(x-3) \qquad \text{Multiply both sides by the LCD, 5.}$$
$$5y-10=3x-9 \qquad \text{Simplify.}$$
$$-3x+5y=1 \qquad \text{Standard form}$$
$$\text{or} \quad y=\frac{3}{5}x+\frac{1}{5} \qquad \text{Slope-intercept form}$$

6. Find the equation of the line through the point $(3,4)$ and perpendicular to the line $3x+2y=-2$. Graph both lines.

10.4 Exercises

Concept Check

1. negative reciprocals
2. $y - y_1 = m(x - x_1)$
3. same
4. standard
5. slope
6. Vertical
7. True
8. True
9. True
10. False; Pairs of lines can be parallel, perpendicular, or neither.

Practice

1. **a.** $m = 2$ **b.** $(3,1)$

c.

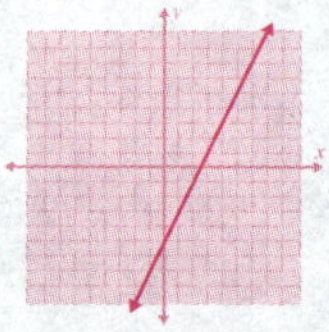

2. **a.** $m = \frac{1}{2}$ **b.** $(1,4)$

c.

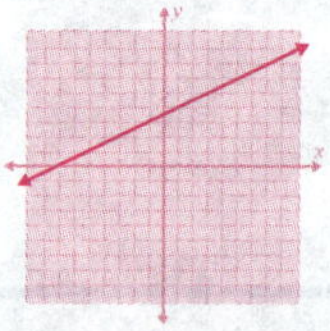

3. **a.** $m = -5$ **b.** $(0,-2)$

c.

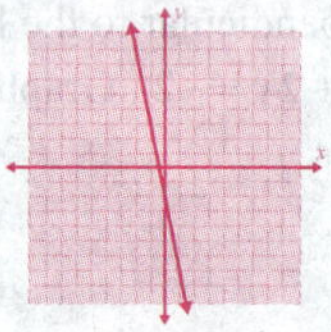

4. **a.** $m = -1$ **b.** $(-8,0)$

c.

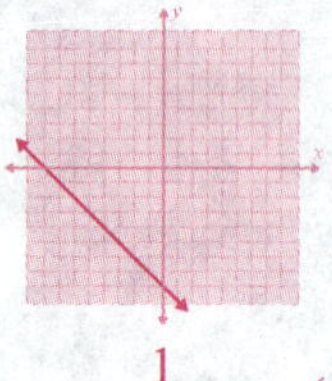

5. **a.** $m = -\frac{1}{4}$ **b.** $(-2,3)$

c.

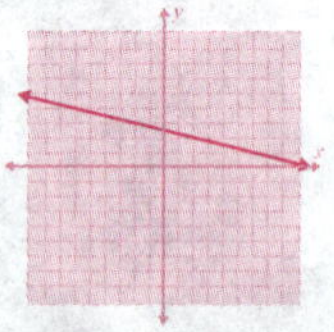

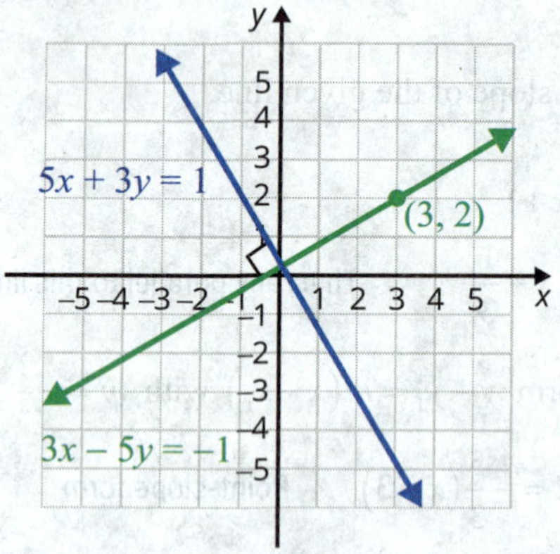

Now work margin exercise 6.

For easy reference, the following table summarizes what we know about lines.

Summary of Formulas and Properties of Lines

1.	$Ax + By = C$	standard form
2.	$m = \frac{y_2 - y_1}{x_2 - x_1}$	slope of a line
3.	$y = mx + b$	slope-intercept form
4.	$y - y_1 = m(x - x_1)$	point-slope form
5.	$y = b$	horizontal line, slope 0
6.	$x = a$	vertical line, undefined slope
7.	parallel lines	have the same slope
8.	perpendicular lines	have slopes that are negative reciprocals of each other

PROPERTIES

Margin Exercise Answers

1.

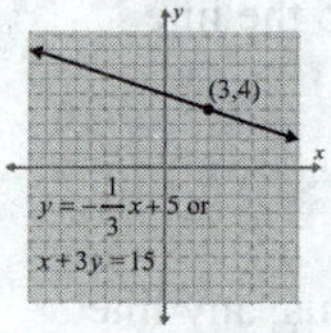

2.

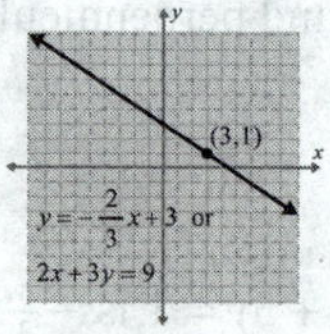

$y = -\frac{2}{3}x + 3$ or $2x + 3y = 9$

3. $y = -\frac{5}{2}x - 1$ or $5x + 2y = -2$

4. $2x + 5y = 6$

5.

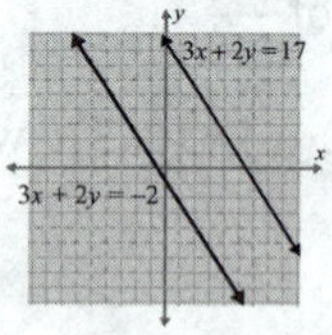

$y = -\frac{3}{2}x + \frac{17}{2}$ or $3x + 2y = 17$

6.

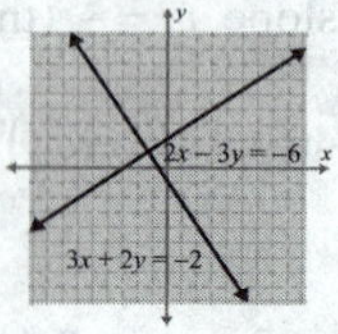

$y = \frac{2}{3}x + 2$ or $2x - 3y = -6$

10.4 Exercises

Concept Check

Fill-in-the-Blank. Complete each sentence using information found in this section.

1. Perpendicular lines have slopes that are ________ ________ of each other.
2. The point-slope form for an equation is ____________.
3. Parallel lines have the _______ slope.
4. Lines represented by the equation $Ax + By = C$ are in _________ form.
5. In the equation $y - y_1 = m(x - x_1)$, m represents the _______.
6. __________ lines are of the form $x = a$.

True/False. Determine whether each statement is true or false. If a statement is false, explain how it can be changed so the statement will be true. (**Note:** There may be more than one acceptable change.)

7. Given two perpendicular lines (neither of which have slope 0), we know that one has a positive slope and the other has a negative slope.
8. If Line 2 is parallel to Line 3, then the slope of Line 2 equals the slope of Line 3.
9. A line perpendicular to a horizontal line has a slope that is undefined.
10. All pairs of lines are either parallel or perpendicular.

Practice

Find **a.** the slope, **b.** a point on the line, and **c.** the graph of the line for the following equations in point-slope form.

1. $y - 1 = 2(x - 3)$
2. $y - 4 = \frac{1}{2}(x - 1)$
3. $y + 2 = -5(x)$
4. $y = -(x + 8)$
5. $y - 3 = -\frac{1}{4}(x + 2)$
6. $y + 6 = \frac{1}{3}(x - 7)$

Find an equation in standard form for the line passing through the given point with the given slope. Graph the line. See Examples 1 and 2.

7. $(-2,1)$; $m = -2$
8. $(3,4)$; $m = 3$
9. $(5,-2)$; $m = 0$
10. $(0,0)$; $m = -3$
11. $(-3,6)$; $m = \frac{1}{2}$
12. $(-3,-1)$; m is undefined
13. $(7,10)$; $m = \frac{3}{5}$
14. $(-1,-1)$; $m = -\frac{1}{4}$
15. $\left(-2, \frac{1}{3}\right)$; $m = \frac{2}{3}$
16. $\left(\frac{5}{2}, \frac{1}{2}\right)$; $m = -\frac{4}{3}$

6. a. $m = \frac{1}{3}$ **b.** $(7,-6)$

c.

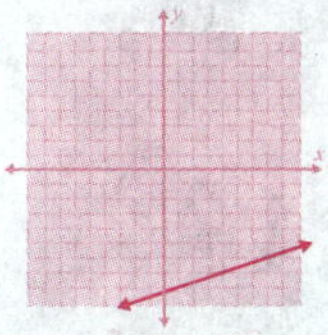

7. $2x + y = -3$

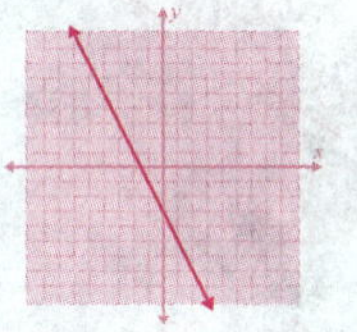

8. $3x - y = 5$

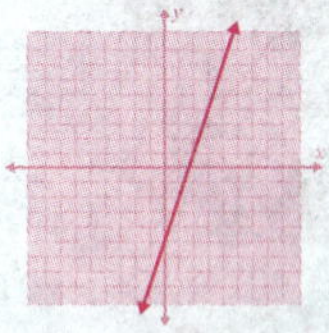

9. $y = -2$

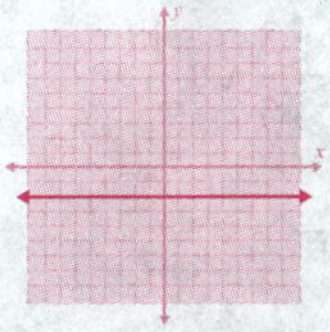

10. $3x + y = 0$

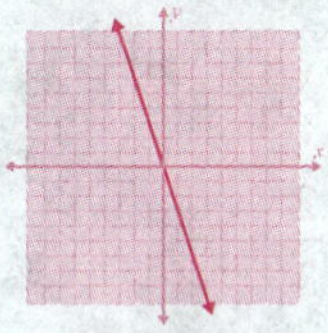

11. $x - 2y = -15$

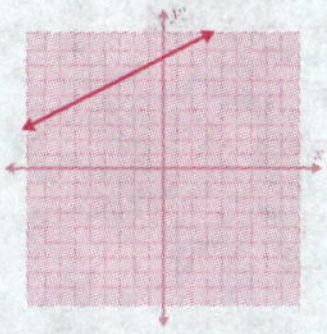

12. $x = -3$

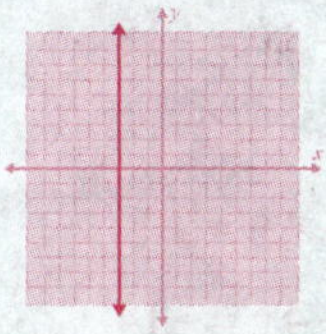

13. $3x - 5y = -29$

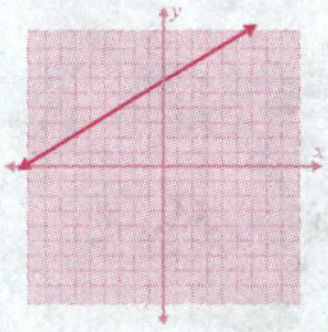

14. $x+4y=-5$

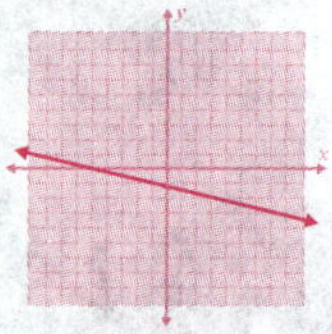

15. $2x-3y=-5$

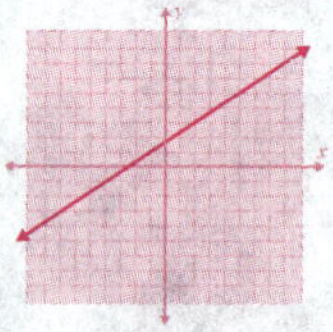

16. $8x+6y=23$

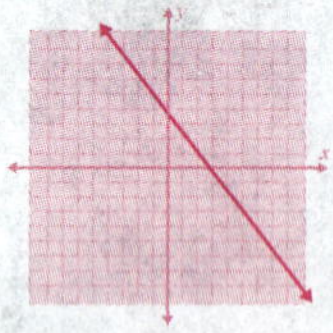

17. $y=\frac{1}{2}x+\frac{9}{2}$

18. $y=-\frac{3}{5}x+\frac{11}{5}$

19. $y=-\frac{1}{7}x+\frac{2}{7}$

20. $y=\frac{5}{7}x-\frac{8}{7}$

21. $y=-\frac{5}{4}x+2$

22. $y=\frac{2}{3}x-\frac{5}{3}$

23. $y=-5$

24. $y=-\frac{7}{2}x+4$

25. $y=-x+4$

26. $y=\frac{1}{4}x$

27. $x-y=1$

28. $4x-3y=-18$

29. $3x+y=2$

30. $5x+4y=20$

31. $3x-4y=-14$

32. $2x-y=-7$

33. $7x+3y=-5$

34. $2x+3y=-7$

35. $y=6$ **36.** $x=-1$

Find an equation in slope-intercept form for the line passing through the two given points. See Example 3.

17. $(-5,2);\ (3,6)$

18. $(-3,4);\ (2,1)$

19. $(-5,1);\ (2,0)$

20. $(-4,-4);\ (3,1)$

21. $(0,2);\ \left(1,\frac{3}{4}\right)$

22. $\left(\frac{5}{2},0\right);\ \left(2,-\frac{1}{3}\right)$

23. $(2,-5);\ (4,-5)$

24. $(0,4);\ \left(1,\frac{1}{2}\right)$

25. $(-2,6);\ (3,1)$

26. $(8,2);\ (0,0)$

Find an equation in standard form for each line shown. See Example 4.

27.

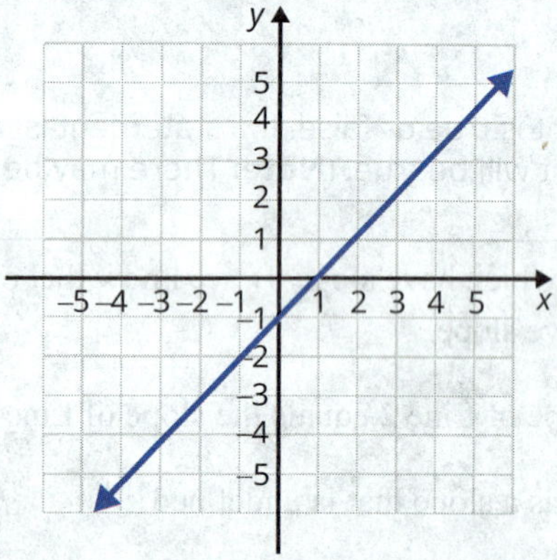

30.

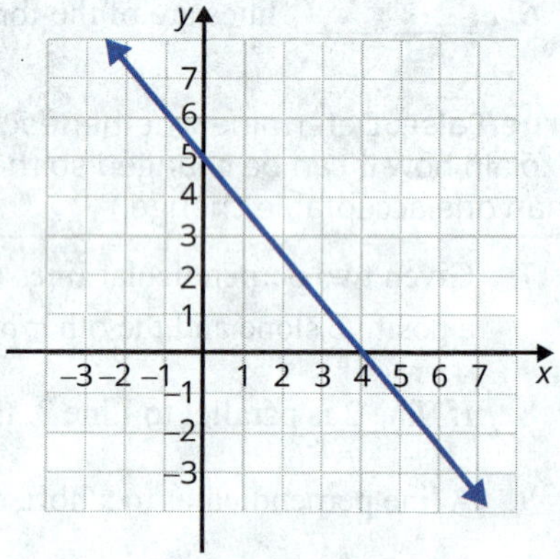

28.

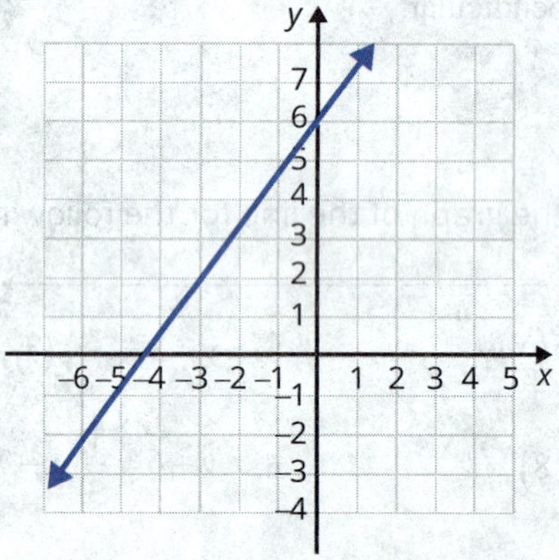

31.

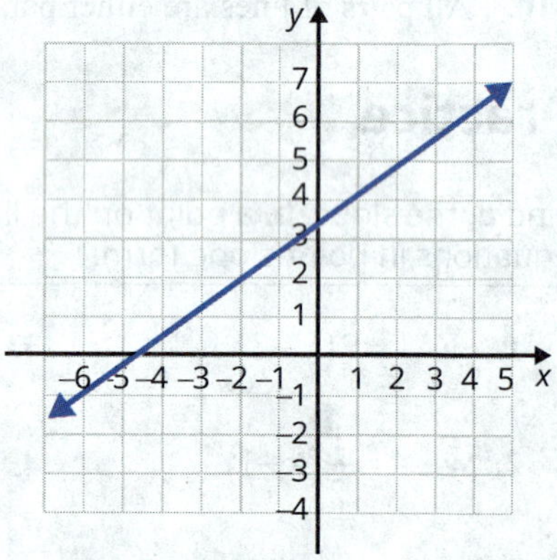

29.

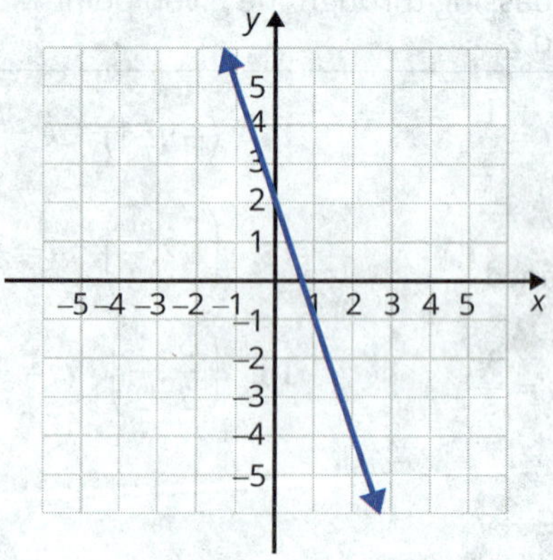

32.

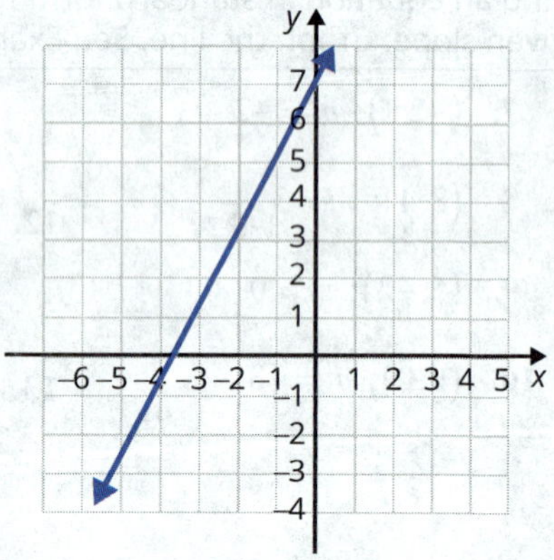

33.

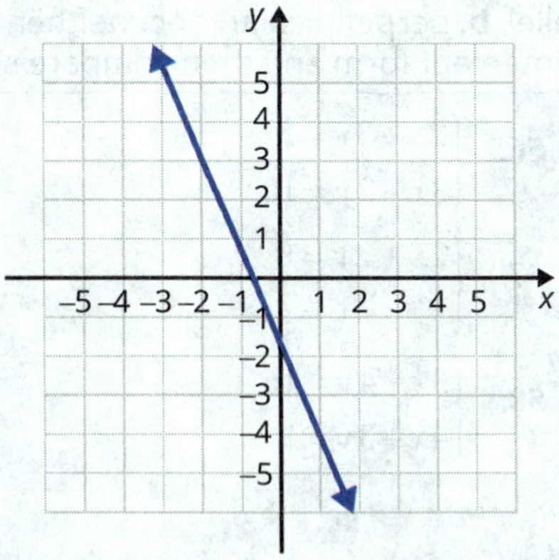

34.

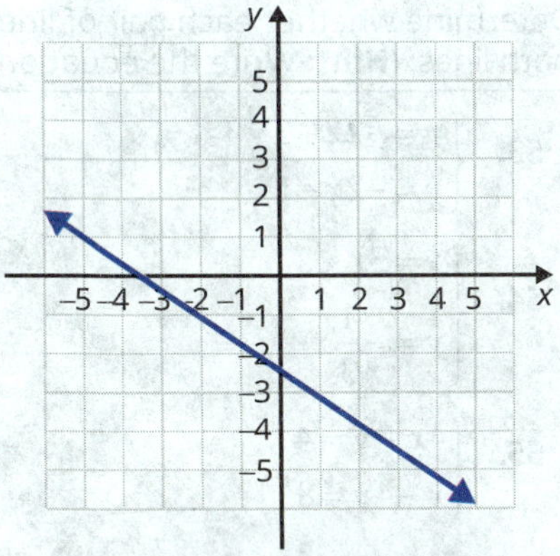

Find an equation in slope-intercept form that satisfies each set of conditions. See Examples 5 and 6.

35. Find an equation for the horizontal line through the point $(-2, 6)$.

36. Find an equation for the vertical line through the point $(-1, -4)$.

37. Write an equation for the line parallel to the x-axis and containing the point $(2, 7)$.

38. Find an equation for the line parallel to the y-axis and containing the point $(2, -4)$.

39. Find an equation for the line perpendicular to $x = 4$ and that passes through $(-1, 7)$.

40. Find an equation for the line parallel to the line $-6y = 1$ and containing the point $(-3, 2)$.

41. Write an equation for the line parallel to the line $2x - y = 4$ and containing the origin. Graph both lines.

42. Find an equation for the line parallel to $7x - 3y = 1$ and containing the point $(1, 0)$. Graph both lines.

43. Write an equation for the line parallel to $5x = 7 + y$ and through the point $(-1, -3)$. Graph both lines.

44. Write an equation for the line that contains the point $(2, 2)$ and is perpendicular to the line $4x + 3y = 4$. Graph both lines.

45. Find an equation for the line that passes through the point $(4, -1)$ and is perpendicular to the line $5x - 3y + 4 = 0$. Graph both lines.

46. Write an equation for the line that is perpendicular to $8 - 3x - 2y = 0$ and passes through the point $(-4, -2)$.

47. Write an equation for the line through the origin that is perpendicular to $3x - y = 4$.

48. Find an equation for the line that is perpendicular to $2x + y = 5$ and that passes through $(6, -1)$.

49. Write an equation for the line that is perpendicular to $2x - y = 7$ and has the same y-intercept as $x - 3y = 6$.

50. Find an equation for the line with the same y-intercept as $5x + 4y = 12$ and that is perpendicular to $3x - 2y = 4$.

51. Show that the points $A(-2, 4)$, $B(0, 0)$, $C(6, 3)$, and $D(4, 7)$ are the vertices of a rectangle. (Plot the points and show that opposite sides are parallel and that adjacent sides are perpendicular.)

52. Show that the points $A(0, -1)$, $B(3, -4)$, $C(6, 3)$, and $D(9, 0)$ are the vertices of a parallelogram. (Plot the points and show that opposite sides are parallel.)

37. $y = 7$ **38.** $x = 2$

39. $y = 7$ **40.** $y = 2$

41. $y = 2x$

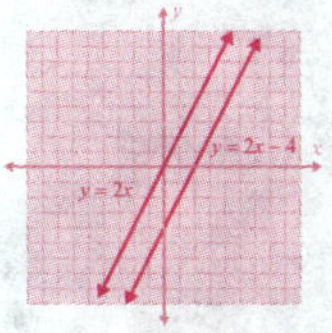

42. $y = \frac{7}{3}x - \frac{7}{3}$

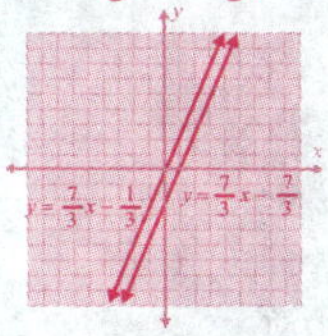

43. $y = 5x + 2$

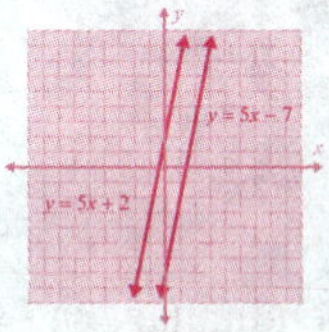

44. $y = \frac{3}{4}x + \frac{1}{2}$

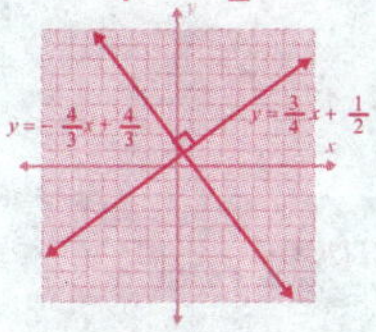

45. $y = -\frac{3}{5}x + \frac{7}{5}$

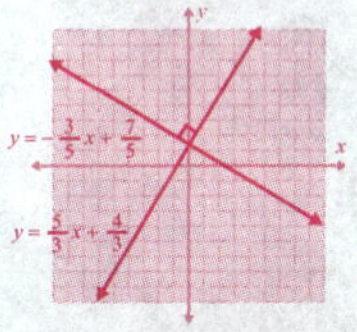

46. $y = \frac{2}{3}x + \frac{2}{3}$

47. $y = -\frac{1}{3}x$

48. $y = \frac{1}{2}x - 4$

49. $y = -\frac{1}{2}x - 2$

50. $y = -\frac{2}{3}x + 3$

Determine whether each pair of lines is **a.** parallel, **b.** perpendicular, or **c.** neither. Graph both lines. (**Hint:** Write the equations in slope-intercept form and then compare slopes.)

53. $\begin{cases} y=-2x+3 \\ y=-2x-1 \end{cases}$

54. $\begin{cases} y=3x+2 \\ y=-\dfrac{1}{3}x+6 \end{cases}$

55. $\begin{cases} 4x+y=4 \\ x-4y=8 \end{cases}$

56. $\begin{cases} 2x+3y=5 \\ 3x+2y=10 \end{cases}$

57. $\begin{cases} 2x+2y=9 \\ 2x-y=6 \end{cases}$

58. $\begin{cases} 3x-4y=16 \\ 4x+3y=15 \end{cases}$

Applications

Solve.

59. The cost for an airline to fly from Raleigh, NC, to Nashville, TN, is \$5000. The airline charges \$100 for the one-way ticket from Raleigh to Nashville.

a. Find an equation for the profit P made by the airline on this one-way flight if they sell t tickets.

b. Use the equation found in part a. to determine the number of tickets that must be sold for the airline to "break even;" that is, for the profit to be equal to 0?

60. A scrapbooking club offers a monthly sticker subscription that is \$9.95 for any selection of 100 stickers, then charges \$0.10 for each additional sticker over 100.

a. Write an equation for the total bill, b, in a month in which you order s stickers (where s is at least 100).

b. Jenny's bill last month was \$27.85. How many stickers did she order?

61. Betsy earns 10 hours of paid time off (PTO) per month and the accumulated hours rollover each month. Two months into the current year, she has accumulated a total of 80 hours of PTO. Let y be the number of PTO hours accumulated and x be the number of months.

a. Graph the line that represents her projected PTO accumulation for the current calendar year if she does not use any PTO. (Graph should specify that x-axis is months and y-axis is hours of PTO.)

b. Write the equation in point-slope form that represents Betsy's projected PTO accumulation.

62. The price p of a college textbook increases as the number of pages n increases. In fact, the price increases \$20 for every 100 pages that are added to the textbook.

a. Assuming there are no "fixed" costs, find an equation for the price of a textbook in terms of the number of pages.

b. Use the equation found in part a. to approximate the price of a 560-page textbook.

63. A taxi charges a fare of \$5.00 plus \$0.25 per eighth of a mile for a ride.

a. Find an equation for the fare f in terms of the number of miles m.

b. Use the equation found in part a. to determine the cost for a 15-mile ride.

51.
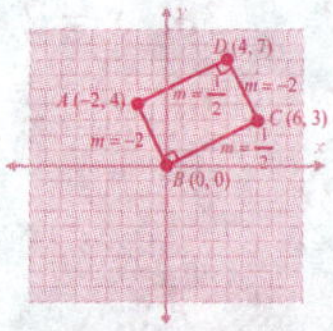

52.
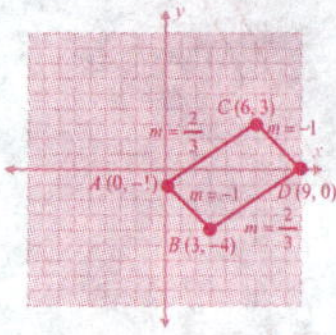

53.
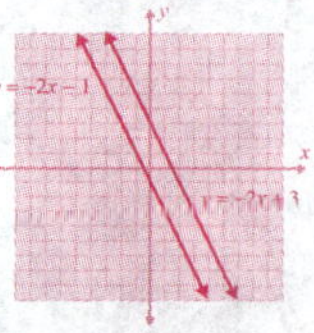

Parallel

54.
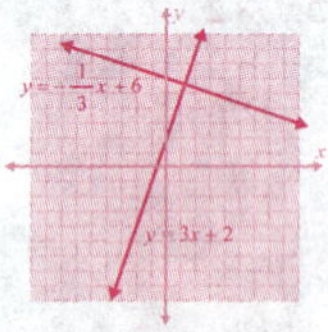

Perpendicular

55.
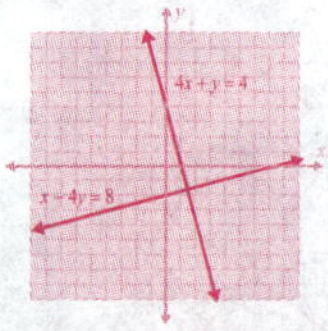

Perpendicular

56.
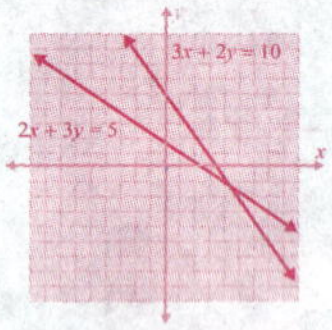

Neither

57.
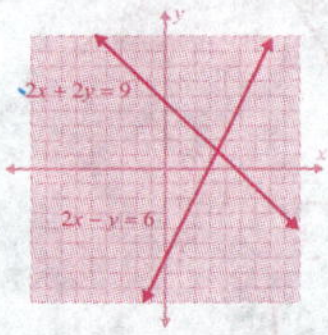

Neither

58.
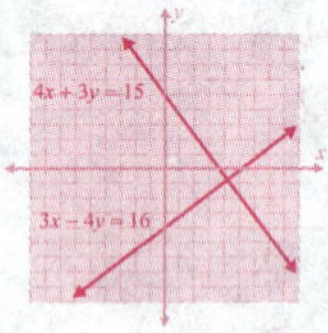

Perpendicular

Applications

59. a. $P=100t-5000$

b. 50 tickets

64. Natalie invested some money in a simple interest savings fund. After 2 years, she earned \$120 in interest. After 5 years, she earned \$300 in interest.

a. Write two ordered pairs from the information given where x represents the time in years and y represents the amount of interest earned.

b. Find the slope of the line which contains the two ordered pairs from part a.

c. Write the point-slope equation that models the situation.

d. Rewrite this equation in $y = mx + b$ form.

65. An archaeology crew finds the foundation of a house during a dig. The corners of the foundations are plotted on their grid map at the following points: $(1, 7)$, $(3, 2)$, $(9, 4)$, and $(7, 9)$.

a. Plot the points on the coordinate plane.

b. Find the slope of each side of the foundation.

c. Are any of the sides parallel? If so, which sides?

d. Are any of the sides perpendicular to each other? If so, which sides?

e. Is the foundation in the form of a geometric shape? If so, which shape

Writing & Thinking

66. Ramps for persons with disabilities are now built into most buildings and walkways. (If ramps are not present in a building, then there must be elevators.) What do you think that the slope of a ramp should be in order to be considered accessible? Look in your library or contact your local building permit office to find the recommended slope for such ramps.

60. a. $b = 9.95 + 0.10(s - 100)$

b. 279 stickers

61. a.

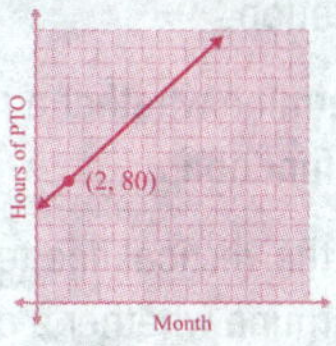

b. $y - 80 = 10(x - 2)$

62. a. $p = \frac{1}{5}n$ **b.** \$112

63. a. $f = 5 + 2m$ **b.** \$35

64. a. $(2, 120)$; $(5, 300)$

b. $m = 60$

c. $y - 120 = 60(x - 2)$ or $y - 300 = 60(x - 5)$

d. $y = 60x$

65. a.

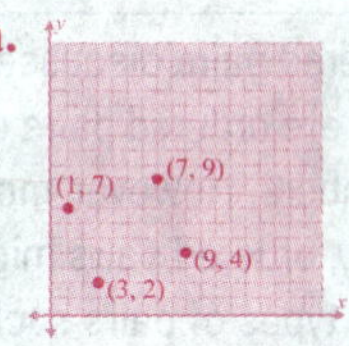

b. Top: $m = \frac{1}{3}$;

Bottom: $m = \frac{1}{3}$;

Left: $m = -\frac{5}{2}$

Right: $m = -\frac{5}{2}$

c. Top is parallel with bottom, left is parallel with right.

d. No

e. Parallelogram

Writing & Thinking

66. The Americans with Disabilities Act recommends a slope of $\frac{1}{12}$ for business use and a slope of $\frac{1}{6}$ for residential use. Answers will vary.

Objectives

A. Find the domain and range of a relation.

B. Determine whether a relation is a function.

C. Use the vertical line test to determine whether a graph is the graph of a function.

D. Determine the domains of linear and nonlinear functions.

E. Evaluate functions written in function notation.

F. Use a graphing calculator to graph functions.

10.5 Introduction to Functions and Function Notation

Everyday use of the term **function** is not far from the technical use in mathematics. For example, distance traveled is a function of time, profit is a function of sales, heart rate is a function of exertion, and interest earned is a function of principal invested. In this sense, one variable depends on (or is a function of) another.

Mathematicians distinguish between graphs of ordered pairs of real numbers as those that represent **functions** and those that do not. For example, every equation of the form $y = mx + b$ represents a function and we say that y is a function of x. Thus, lines that are not vertical are the graphs of functions. As the following discussion will show, vertical lines do not represent functions.

Note

The ordered pairs discussed in this text are ordered pairs of real numbers. However, more generally, ordered pairs might be other types of pairs such as (child, mother), (city, state), or (name, batting average).

A Finding Domain and Range

Relation, Domain, and Range

A **relation** is a set of ordered pairs of real numbers.

The **domain, *D***, of a relation is the set of all first coordinates in the relation.

The **range, *R***, of a relation is the set of all second coordinates in the relation.

DEFINITION

In the graph of a relation, the horizontal axis (the x-axis) is called the **domain axis**, and the vertical axis (the y-axis) is called the **range axis**.

1. Find the domain and range of each of the following relations.

a. $g = \left\{\begin{matrix}(4,5),(7,3),\\(3,6),(7,5)\end{matrix}\right\}$

b. $f = \left\{\begin{matrix}(-2,3),(-4,-3),\\(0,0)\end{matrix}\right\}$

Example 1 Finding the Domain and Range

Find the domain and range for each of the following relations.

a. $g = \{(5,7),(6,2),(6,3),(-1,2)\}$ **b.** $f = \{(-1,1),(1,5),(0,3)\}$

Solution

a. $D = \{5, 6, -1\}$ The set of all the first coordinates in g

$R = \{7, 2, 3\}$ The set of all the second coordinatesin g

Notice that 6 is written only once in the domain and 2 is written only once in the range, even though each appears more than once in the relation.

b. $D = \{-1, 1, 0\}$ The set of all the first coordinates in f

$R = \{1, 5, 3\}$ The set of all the second coordinates in f

Now work margin exercise 1

Example 2 Reading Domain and Range from the Graph of a Relation

Identify the domain and range from the graph of each relation.

a.

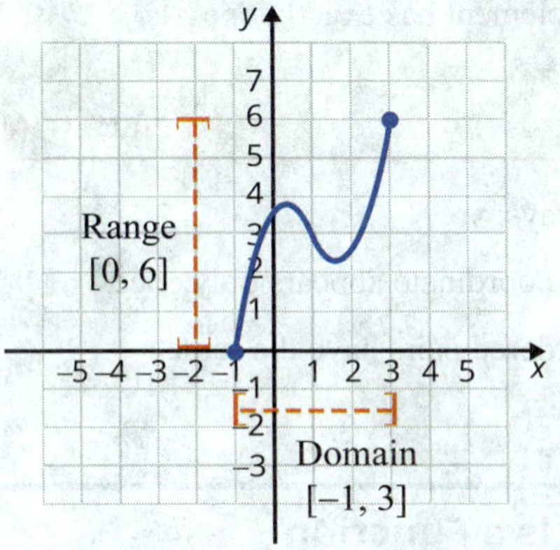

The domain consists of the set of x-values for all points on the graph. In this case, the domain is the interval $[-1, 3]$. The range consists of the set of y-values for all points on the graph. In this case, the range is the interval $[0, 6]$.

b.

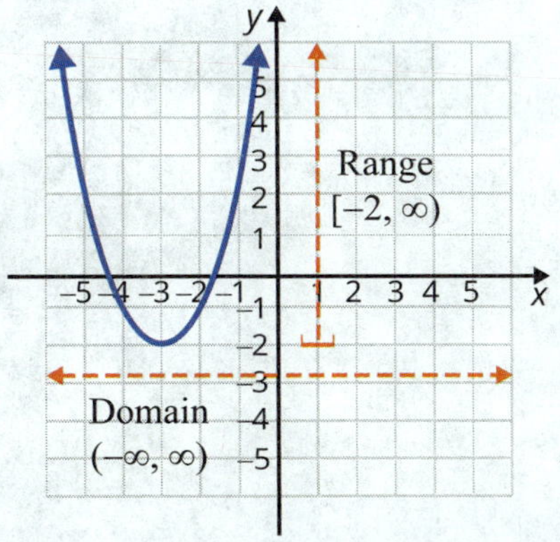

There is no restriction on the x-values which means that for every real number there is a point on the graph with that number as its x-value. Thus, the domain is the set of all real numbers: $(-\infty, \infty)$. The y-values begin at –2 and then increase to infinity. The range is the interval $[-2, \infty)$.

Now work margin exercise 2.

2. Identify the domain and range from the graph of each relation.

a.

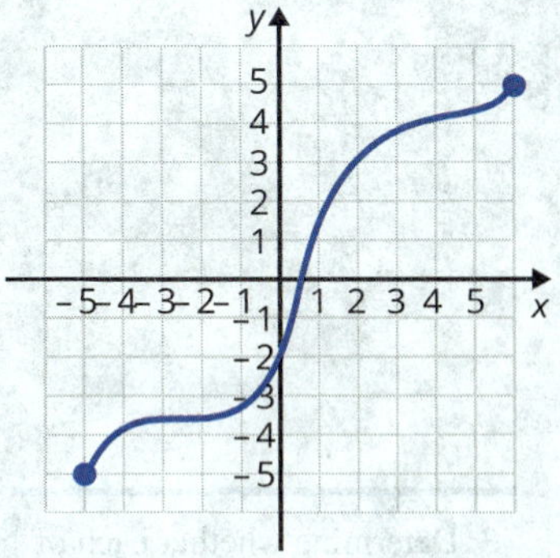

b.

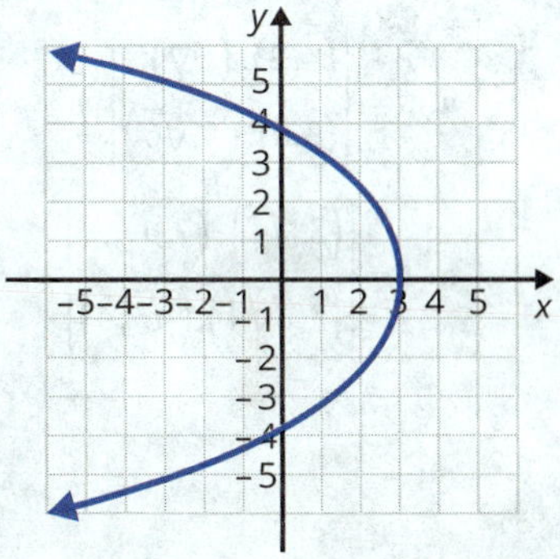

B Identifying Functions

The relation $f = \{(-1, 1), (1, 5), (0, 3)\}$, used in Example 1b, meets a particular condition in that each first coordinate has a unique corresponding second coordinate. Such a relation is called a **function**. Notice that g in Example 1a is **not** a function because the first coordinate 6 has two different corresponding second coordinates, 2 and 3. Also, for ease in discussion and understanding, the relations illustrated in Examples 1 and 3 have only a finite number of ordered pairs. The graphs of these relations are isolated dots or points. As we will see, the graphs of most relations and functions have an infinite number of points, and their graphs are smooth curves.

Functions

A **function** is a relation in which each domain element has exactly one corresponding range element.

DEFINITION

The definition can also be stated in the following ways:

1. A function is a relation in which each first coordinate appears only once.
2. A function is a relation in which no two ordered pairs have the same first coordinate.

3. Determine whether each of the following relations is a function.

a. $s = \left\{ \begin{matrix} (4,2), (5,7), \\ (3,8), (4,\sqrt{6}) \end{matrix} \right\}$

b. $t = \left\{ \begin{matrix} (3,\sqrt{4}), (7,4), \\ (-3,\sqrt{4}), (\sqrt{3},4) \end{matrix} \right\}$

Example 3 Determining If a Relation Is a Function

Determine whether each of the following relations is a function.

a. $s = \left\{ (2,3), (1,6), (2,\sqrt{5}), (0,-1) \right\}$

b. $t = \left\{ (1,5), (3,5), (\sqrt{2},5), (-1,5), (-4,5) \right\}$

Solution

a.

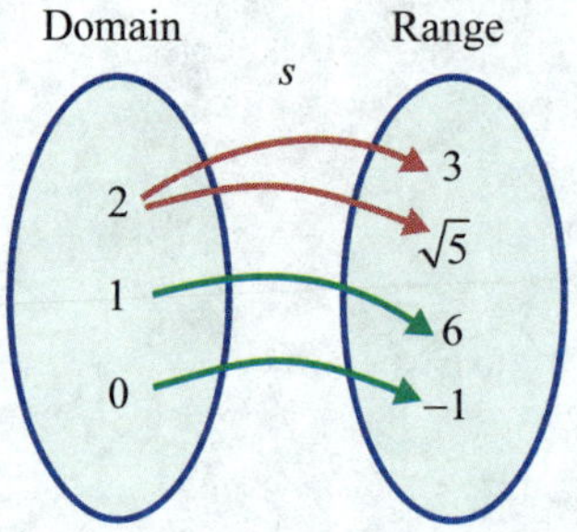

s is not a function. The number 2 appears as a first coordinate more than once.

b.

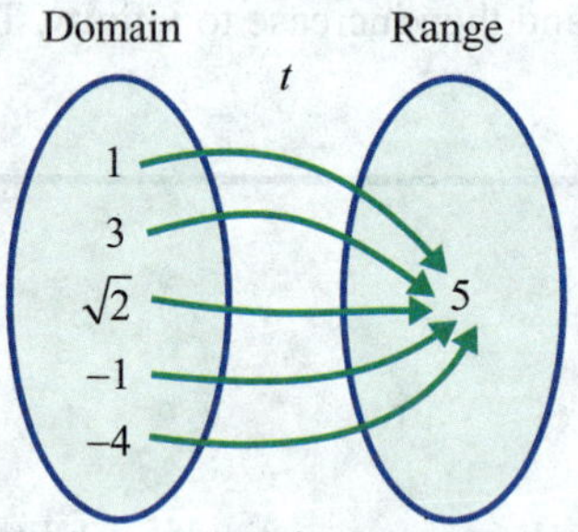

t is a function. Each first coordinate appears only once. The fact that the second coordinates are all the same has no effect on the concept of a function.

Now work margin exercise 3.

C Vertical Line Test

If one point on the graph of a relation is directly above or below another point on the graph, then these points have the same first coordinate (or x-coordinate). Such a relation is **not** a function. Therefore, the **vertical line test** can be used to tell whether a graph represents a function or not.

Vertical Line Test

If **any** vertical line intersects the graph of a relation at more than one point, then the relation is **not** a function.

PROCEDURE

Math Tip

The vertical line test verifies whether the graph satisfies the definition of a function: each domain element has exactly one corresponding range element. For each domain element, a vertical line can be drawn. If any of these vertical lines intersects the graph at more than one point, then the domain element represented by the vertical line has more than one corresponding range element, which means the graph does not represent a function.

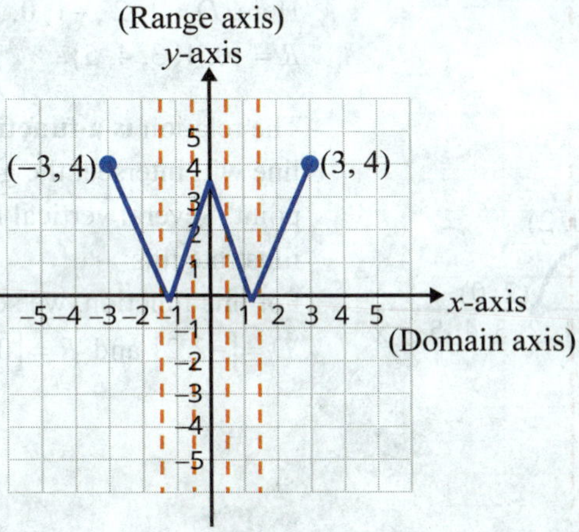

Figure 1

The vertical lines in Figure 1 indicate that this graph represents a function. From the graph, we see that the domain of the function is the interval of real numbers $[-3, 3]$ and the range of the function is the interval of real numbers $[0, 4]$.

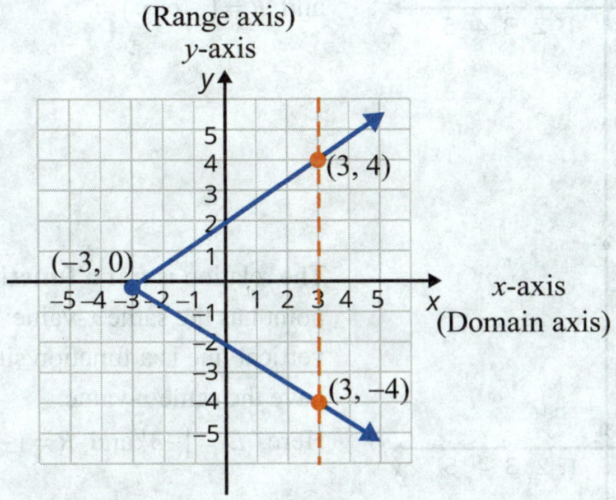

Figure 2

The relation in Figure 2 is **not** a function because the vertical line drawn intersects the graph at more than one point, $(3, 4)$ and $(3, -4)$. Thus, for that x-value, there is more than one corresponding y-value. Here $D = [-3, \infty)$ and $R = (-\infty, \infty)$.

4. Use the vertical line test to determine whether each of the following graphs represents a function. Then list the domain and range of each graph.

a.

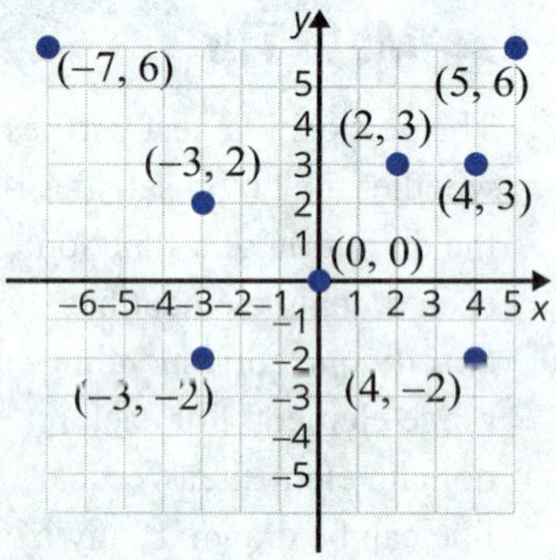

b.

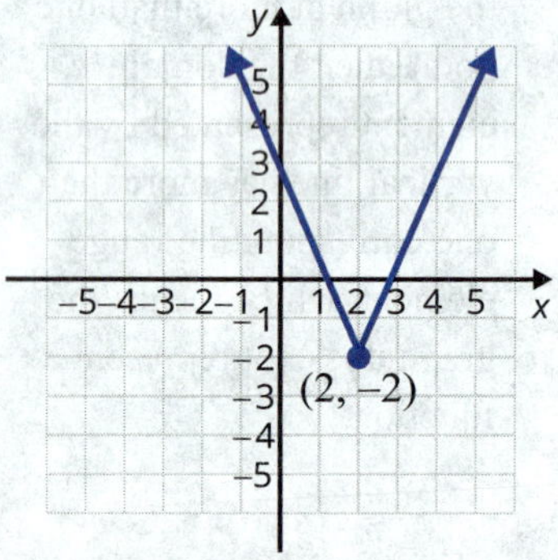

c.

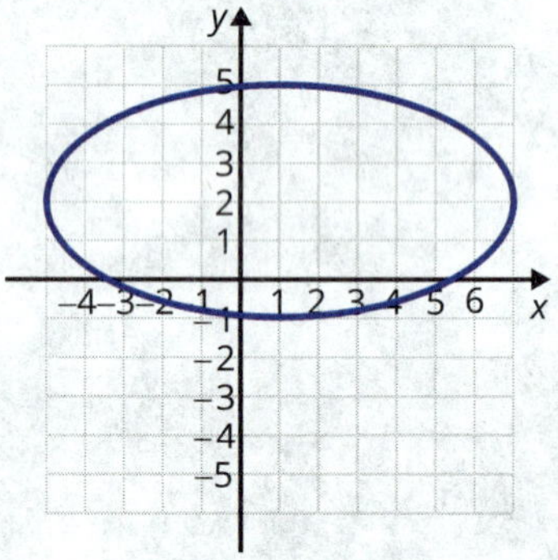

Example 4 Using the Vertical Line Test

Use the vertical line test to determine whether each graph represents a function. Then list the domain and range of each graph.

a.

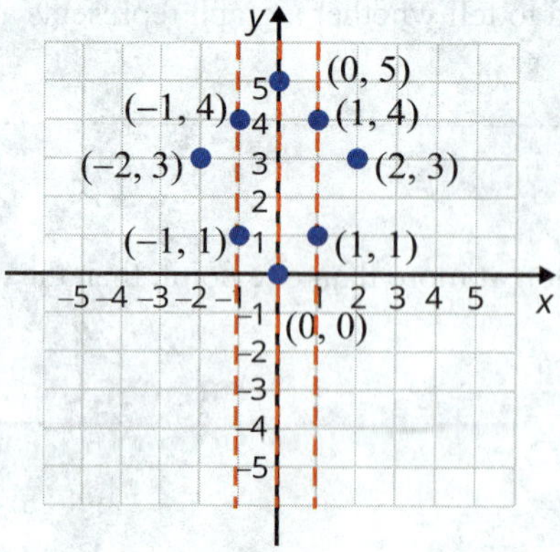

The relation is **not a function** since a vertical line can be drawn that intersects the graph at more than one point. Listing the ordered pairs shows that several x-coordinates appear more than once.

$$r = \begin{Bmatrix} (-2,3),(-1,1),(-1,4), \\ (0,0),(0,5),(1,1),(1,4),(2,3) \end{Bmatrix}$$

Here, $D = \{-2, -1, 0, 1, 2\}$ and $R = \{0, 1, 3, 4, 5\}$.

b.

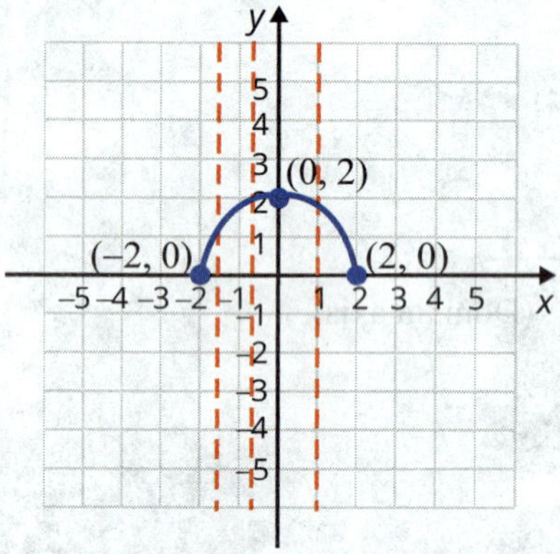

The relation **is a function**. No vertical line will intersect the graph at more than one point. Several vertical lines are drawn to illustrate this.

For this function, we see from the graph that $D = [-2, 2]$ and $R = [0, 2]$.

c.

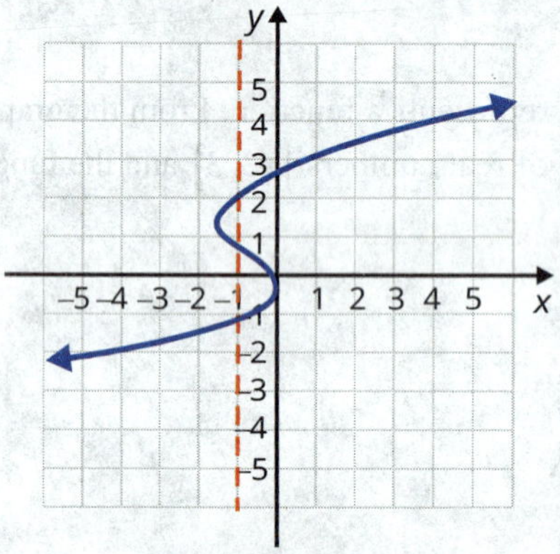

The relation is **not a function**.

At least one vertical line (drawn) intersects the graph at more than one point.

Here, $D = (-\infty, \infty)$ and $R = (-\infty, \infty)$.

d.

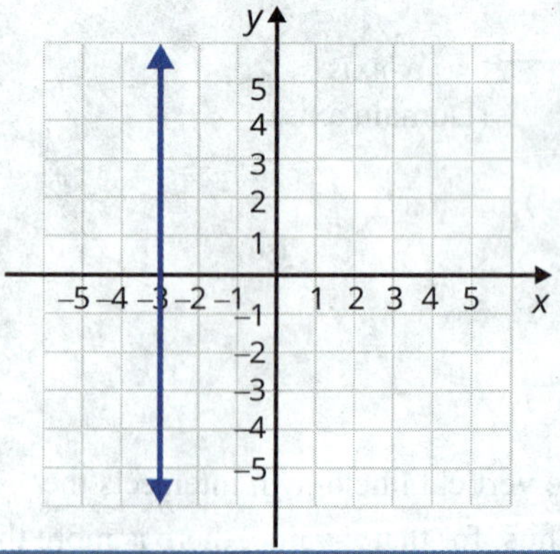

The relation is **not a function**. Every point has the same x-value: $x = -3$. In fact, no vertical line is a function since every point will have the same x-value.

Here, $D = \{-3\}$ and $R = (-\infty, \infty)$.

Now work margin exercise 4.

D Finding Domains of Functions

All nonvertical lines represent functions. Thus we have the following definition for a linear function.

Linear Function

A **linear function** is a function represented by an equation of the form

$$y = mx + b.$$

The domain of a linear function is the set of all real numbers:

$$D = (-\infty, \infty).$$

DEFINITION

If the graph of a linear function is not a horizontal line, then the range is also the set of all real numbers. If the line is horizontal, then the domain is still the set of all real numbers; however, the range is a set containing just a single number. For example, the graph of the linear equation $y = 5$ is a horizontal line. The domain of the function is the set of all real numbers and the range is the set containing only the number 5 (written $\{5\}$). Figures 3 and 4 show two linear functions and the domain and range of each function.

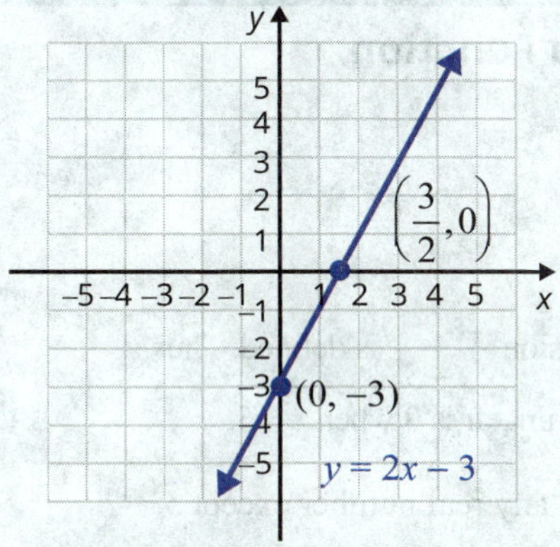

The graph of the linear function $y = 2x - 3$ is shown here.

$D = \{\text{all real numbers}\} = (-\infty, \infty)$

$R = \{\text{all real numbers}\} = (-\infty, \infty)$

Figure 3

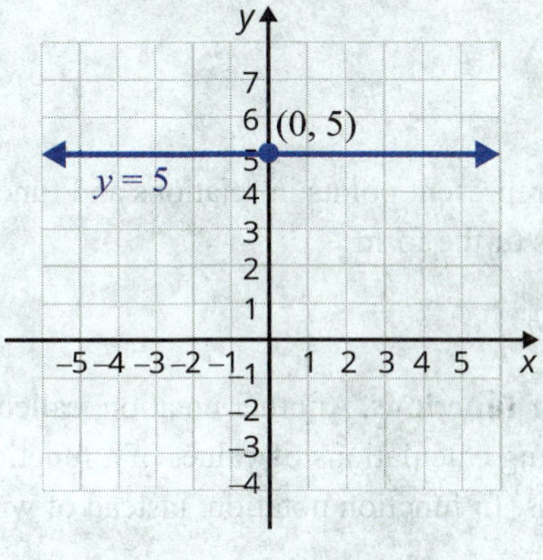

The graph of the linear function $y = 5$ is shown here.

$D = \{\text{all real numbers}\} = (-\infty, \infty)$

$R = \{5\}$

Figure 4

As we have seen, the domain for a linear function is the set of all real numbers. Now for the nonlinear function

$$y = \frac{2}{x-1}$$

we say that the domain (all possible values for x) is every real number for which the expression $\frac{2}{x-1}$ is defined. Because the denominator cannot be 0, the domain consists of all real numbers except 1. That is, $D = (-\infty, 1) \cup (1, \infty)$, or simply $x \neq 1$. We adopt the following rule concerning equations and domains:

Unless a finite domain is explicitly stated, the domain will be implied to be the set of all real x-values for which the given function is defined. That is, the domain consists of all values of x that give real values for y.

Attention!

In determining the domain of a function, one fact to remember at this stage is that **no denominator can equal 0**. In future chapters, we will discuss other nonlinear functions with limited domains.

5. Find the domain for the function $y = \frac{3x-2}{x+3}$.

Example 5 Finding the Domain of a Function

Find the domain for the function $y = \frac{2x+1}{x-5}$.

Solution

The domain is all real numbers for which the expression $\frac{2x+1}{x-5}$ is defined. Thus, $D = (-\infty, 5) \cup (5, \infty)$, or $x \neq 5$, because the denominator is 0 when $x = 5$.

Note: Here interval notation tells us that x can be any real number except 5.

Now work margin exercise 5.

E Function Notation

We have used the ordered pair notation (x, y) to represent points in relations and functions. As the vertical line test will show, linear equations of the form

$$y = mx + b,$$

where the equation is solved for y, represent **linear functions**. Another notation, called **function notation**, is more convenient for indicating calculations of values of a function and indicating operations performed with functions. In function notation, instead of writing y, write $f(x)$, which is read "***f* of *x***."

The letter f is the name of the function. The letters f, g, h, F, G, and H are commonly used in mathematics, but any letter other than x will do. We have used r, s, and t in previous examples.

The linear equation $y = -3x + 2$ represents a linear function, and we can replace y with $f(x)$ as follows.

$$f(x) = -3x + 2$$

Now, in function notation, $f(4)$ means to replace x with 4 in the function.

$$f(4) = -3 \cdot (4) + 2 = -12 + 2 = -10$$

Thus, the ordered pair $(4, -10)$ can be written as $(4, f(4))$.

Example 6 Evaluating Functions

For the function $g(x) = 4x + 5$, find:

a. $g(2)$ **b.** $g(-1)$ **c.** $g(0)$

Solution

a. $g(2) = 4(2) + 5 = 13$

b. $g(-1) = 4(-1) + 5 = 1$

c. $g(0) = 4(0) + 5 = 5$

Now work margin exercise 6.

6. For the function $g(x) = 3x - 2$, find:

a. $g(3)$

b. $g(-2)$

c. $g(0)$

Function notation is valid for a wide variety of types of functions. Example 7 illustrates the use of function notation with a nonlinear function.

Example 7 Evaluating Nonlinear Functions

For the function $h(x) = x^2 - 3x + 2$, find:

a. $h(4)$ **b.** $h(0)$ **c.** $h(-3)$

Solution

a. $h(4) = (4)^2 - 3(4) + 2 = 16 - 12 + 2 = 6$

b. $h(0) = (0)^2 - 3(0) + 2 = 0 - 0 + 2 = 2$

c. $h(-3) = (-3)^2 - 3(-3) + 2 = 9 + 9 + 2 = 20$

Now work margin exercise 7.

7. For the function $f(y) = y^3 - 2y + 5$, find:

a. $f(1)$

b. $f(0)$

c. $f(-3)$

Given the graph of $y = f(x)$, we can find points on the graph for values of x by realizing that the notation $f(x)$ is another way of referring to the corresponding y-values. By looking at the graph of $y = f(x)$ given here, we can see that $f(-2) = 6$, $f(1) = -3$, and $f(2) = -2$.

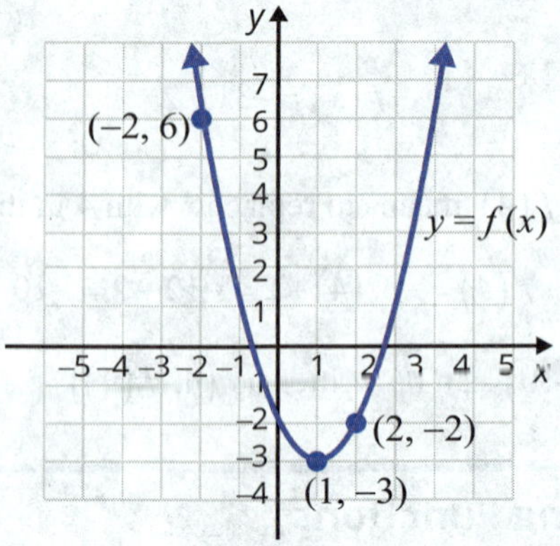

Figure 5

8. Use the graph of $f(x)$ to find each of the values.

a. $f(4)$

b. $f(-4)$

c. $f(-2)$

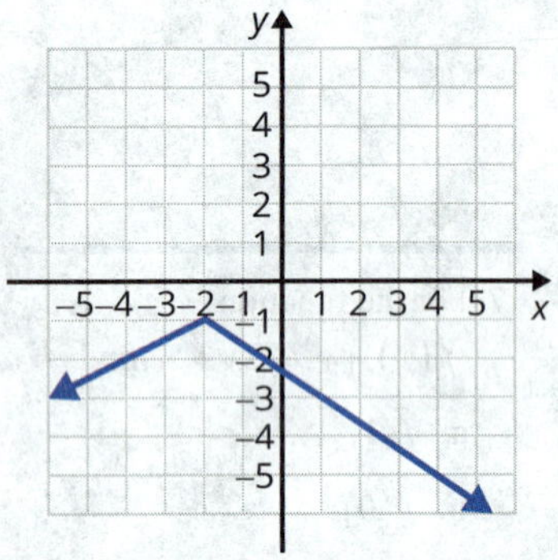

Example 8 Evaluating Functions From a Graph

Using the graph of $g(x)$, find each of the following values.

a. $g(-2)$ **b.** $g(0)$ **c.** $g(3)$

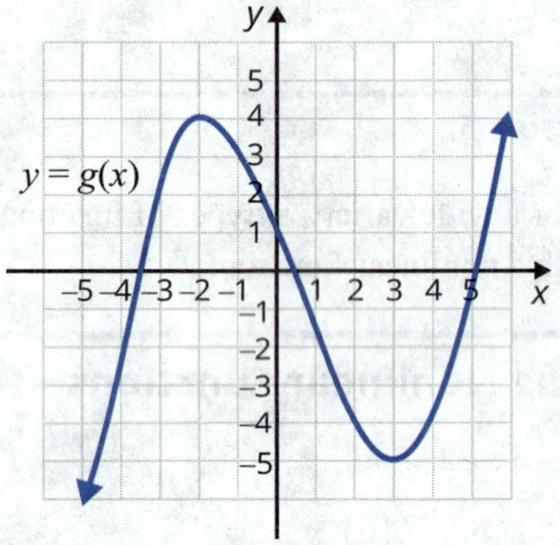

Solution

Using the graph, we need to find the y-value that corresponds to each x-value.

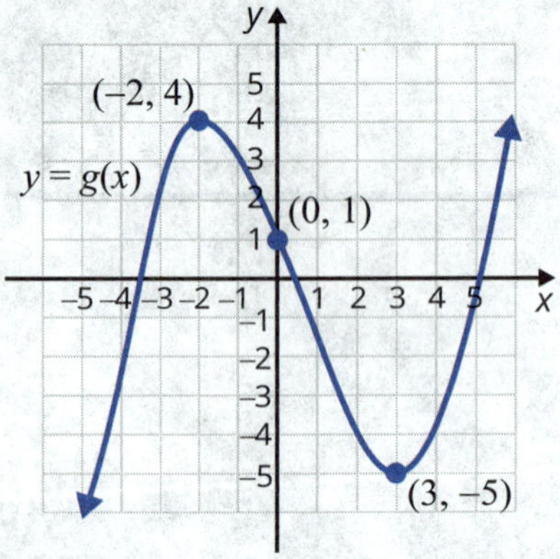

Thus, the function values are as follows.

a. $g(-2) = 4$

b. $g(0) = 1$

c. $g(3) = -5$

Now work margin exercise 8.

F Using a Graphing Calculator to Graph Functions

There are many types and brands of graphing calculators available. For convenience and so that directions can be specific, only the TI-84 Plus graphing calculator is used in the related discussions in this text. Other graphing calculators may be used, but the steps required may be different from those indicated in the text. If you choose to use another calculator, be sure to read the manual for your calculator and follow the relevant directions.

In any case, remember that a calculator is just a tool to allow for fast calculations and to help in understanding some abstract concepts. A calculator does not replace your ability to think and reason or the need for algebraic knowledge and skills.

You should practice and experiment with your calculator until you feel comfortable with the results. **Do not be afraid of making mistakes. Note that [CLEAR] or [2nd] QUIT will get you out of most trouble and allow you to start over.**

Some Basics about the TI-84 Plus

1. [MODE]: Turn the calculator [ON] and press the [MODE] key. The screen should be highlighted as shown below. If it is not, use the arrow keys in the upper right corner of the keyboard to highlight the correct words and press [ENTER]. It is particularly important that Func is highlighted. This stands for function. See the manual for the meanings of the rest of the terms.

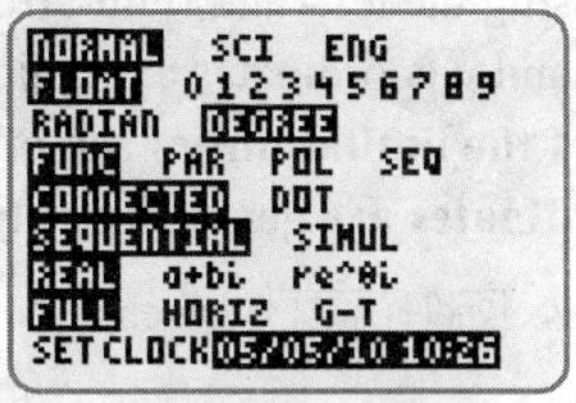

Figure 6

2. [WINDOW]: Press the [WINDOW] key and the standard window will be displayed. By default, the standard window displays a graph with x-values and y-values ranging from –10 to 10 with tick marks on the axes every 1 unit.

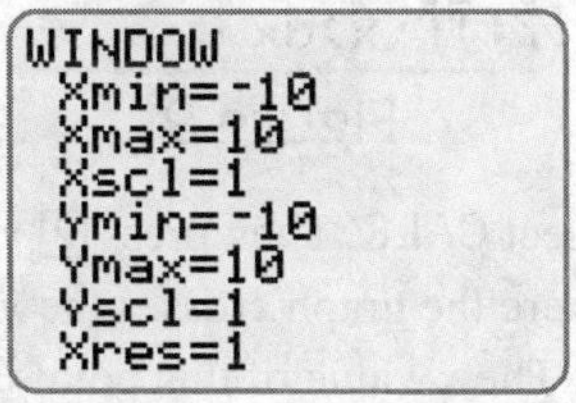

Figure 7

This window can be changed at any time by changing the individual numbers or pressing the [ZOOM] key and selecting an option from the menu displayed. Because of the shape of the display screen, the standard screen is not a square screen (one unit along the x-axis looks longer than one unit along the y-axis). Be aware that the slopes of lines are not truly depicted unless the screen is in a scale of about 3 : 2. A square screen can be attained by pressing the [ZOOM] key and selecting ZSquare or by pressing the [WINDOW] key and setting Xmin = −15 and Xmax = 15 to give the x-axis a length of 30 and the y-axis a length of 20 (a ratio of 3:2).

3. [Y=]: The [Y=] key is in the upper left corner of the keyboard. This key will allow up to ten functions to be entered. These functions are labeled as Y1,..., Y9,YØ. The variable x may be entered using the [X,T,Θ,n] key. The [^] key is used to indicate exponents. (Also note that the negative sign [(-)] is next to the [ENTER] key.) For example, the equation $y = x^2 + 3x$ would be entered as follows.

Y1=X^2+3X

To change an entry, practice with the keys [DEL] (delete), [CLEAR], and [2nd] INS (insert).

Figure 8

> **Note**
>
> When a function is first entered in the [Y=] list, the = sign will be highlighted. (See Figure 8.) This indicates that the function is selected to be graphed. To change the selection, use the arrow keys to highlight the = sign and press [ENTER].

4. [GRAPH]: If this key is pressed, then the screen will display the graph of the functions that are entered in the [Y=] list and selected with the = sign highlighted. The graph is displayed using the minimum and maximum values indicated in the current window. In many cases, the window must be changed to accommodate the domain and range of the function or to show a point where two functions intersect.

5. [TRACE]: The [TRACE] key will display the current graph even if it is not already displayed and give the x- and y-coordinates of a point highlighted on the graph. The curve may be traced by pressing the left- and right-arrow keys. At each point on the graph, the corresponding x- and y-coordinates are indicated at the bottom of the screen. **(Remember that because of the limitations of the pixels (lighted dots) on the screen, these x- and y-coordinates are generally only approximations.)**

6. CALC: The CALC key (press [2nd] [TRACE]) gives a menu with seven items. Items 1–5 are used with graphs.

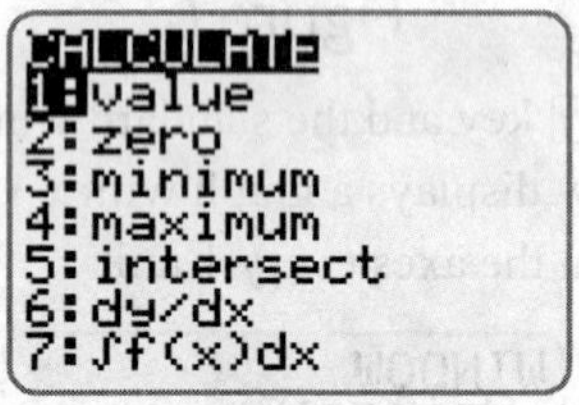

Figure 9

After displaying a graph, select CALC. Then press [2] and follow the steps outlined below to locate the point where the graph crosses the x-axis (the x-intercept). The graph must actually cross the axis. The x-value of this point is called a zero of the function because the corresponding y-value will be 0.

Step 1: With the left arrow, move the cursor to the left of the x-intercept on the graph. Press [ENTER] in response to the question "LeftBound?".

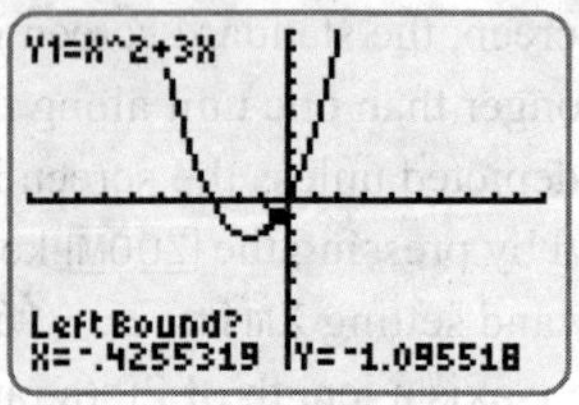

Figure 10

Step 2: With the right arrow, move the cursor to the right of the x-intercept on the graph. Press [ENTER] in response to the question "`RightBound?`".

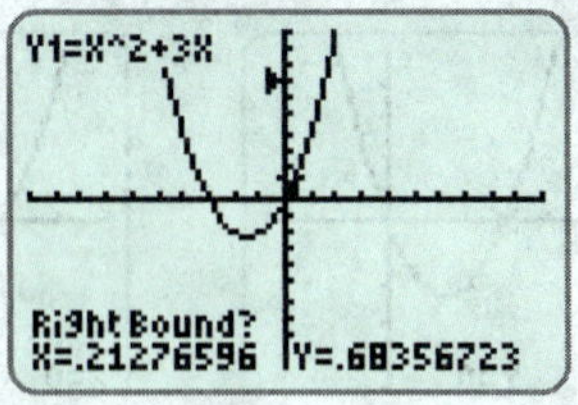

Figure 11

Teaching Note

Tell your students that if the message ERR: SYNTAX
`1: Quit`
`2: Goto`
appears on the display, the error will be highlighted if they press "`2: Goto`".

Step 3: With the left arrow, move the cursor near the x-intercept. Press [ENTER] in response to the question "`Guess?`". The calculator's estimate of the zero will appear at the bottom of the display.

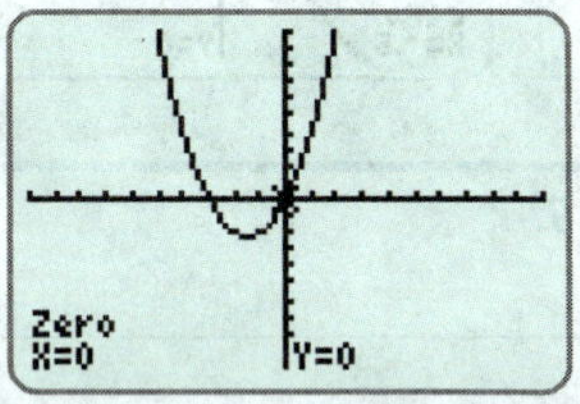

Figure 12

Example 9 Graphing Functions with a TI-84 Plus

Use a TI-84 Plus graphing calculator to find the graphs of each of the following functions. Use the CALC key to find the point(s) where each graph intersects the x-axis. Changing the window may help you get a better or more complete picture of the function. This is a judgement call on your part.

a. $3x + y = -1$

b. $y = x^2 + 3x$

c. $y = 2x - 1$; $y = 2x + 1$; $y = 2x + 3$

9. Use a graphing calculator to find the graphs of each of the following functions. Then use the CALC key to find the point(s) where each graph intersects the x-axis.

a. $2x + 3y = -8$

b. $y = x^3 + 2x$

c. $y = 4x - 2$;
$y = 4x + 3$;
$y = 4x + 7$

Solution

a. To have the calculator graph a nonvertical straight line, you must first solve the equation for y. Solving for y gives

$$y = -3x - 1.$$

(It is important that the [(-)] key be used to indicate the negative sign in front of $3x$. This is not the same as the subtraction key.)

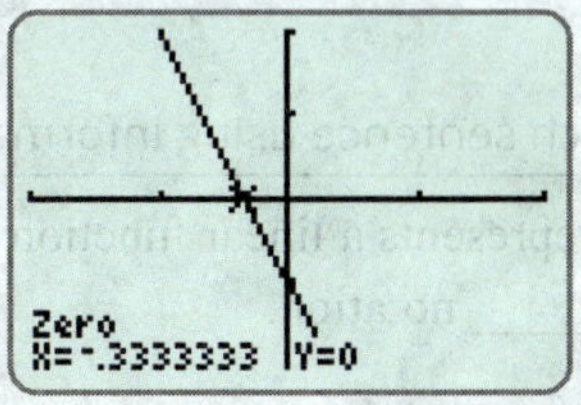

Note

Vertical lines are not functions and cannot be graphed by the calculator in function mode.

10.5 Exercises

Concept Check

1. function
2. vertical
3. domain
4. range
5. domain
6. range
7. True

Note

The standard window shows 96 pixels across the window and 64 pixels up and down the window. This gives a ratio of 3 to 2 and can give a slightly distorted view of the actual graph because the vertical pixels are squeezed into a smaller space. For Example 9c, the graphs of all three functions are in the standard window. Experiment by changing the window to a square window, say −9 to 9 for x and −6 to 6 for y. Then graph the functions and notice the slight differences (and better representation) in the appearances on the display.

8. False; A function is a relation in which…
9. True
10. False; Not a function

Practice

1. $\{(-4,0),(-1,4),(1,2),(2,5),(6,-3)\}$; $D=\{-4,-1,1,2,6\}$; $R=\{-3,0,2,4,5\}$; Function
2. $\{(-4,-1),(0,0),(0,3),(2,-2),(3,1)\}$; $D=\{-4,0,2,3\}$; $R=\{-2,-1,0,1,3\}$; Not a function
3. $\{(-5,-4),(-4,-2),(-2,-2),(1,-2),(2,1)\}$; $D=\{-5,-4,-2,1,2\}$; $R=\{-4,-2,1\}$; Function

b. Since the graph of this function has two x-intercepts, we have shown the graph twice. Each graph shows the coordinates of a distinct x-intercept.

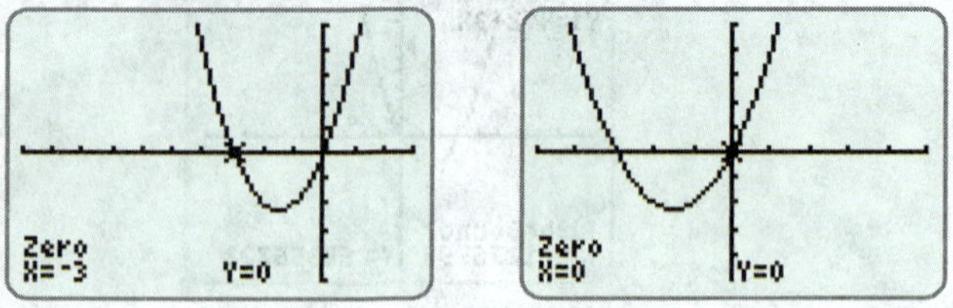

c. $y = 2x - 1$ $\quad$ $y = 2x + 1$ $\quad$ $y = 2x + 3$

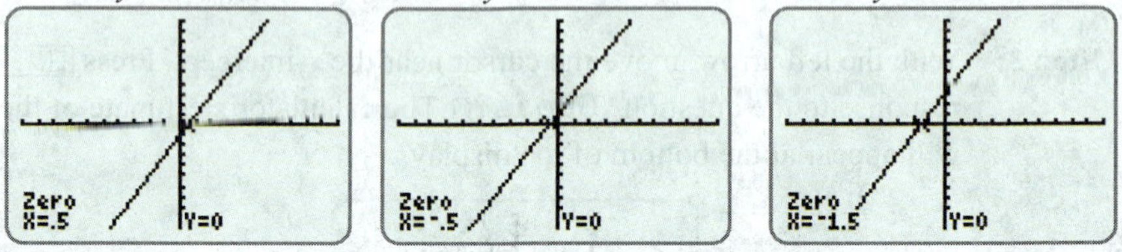

Now work margin exercise 9.

Margin Exercise Answers

1. a. $D=\{4,7,3\}$; $R=\{5,3,6\}$ **b.** $D=\{-2,-4,0\}$; $R=\{3,-3,0\}$ **2. a.** $D=[-5,6]$; $R=[-5,5]$ **b.** $D=(-\infty,3]$; $R=(-\infty,\infty)$

3. a. not a function **b.** function **4. a.** not a function; $D=\{-7,-3,0,2,4,5\}$, $R=\{-2,0,2,3,6\}$

b. function; $D=(-\infty,\infty)$, $R=[-2,\infty)$ **c.** not a function; $D=[-5,7]$, $R=[-1,5]$ **5.** $D=(-\infty,-3)\cup(-3,\infty)$ or $x \neq -3$

6. a. $g(3)=7$ **b.** $g(-2)=-8$ **c.** $g(0)=-2$ **7. a.** $f(1)=4$ **b.** $f(0)=5$ **c.** $f(-3)=-16$ **8. a.** $f(4)=-5$ **b.** $f(-4)=-2$ **c.** $f(-2)=-1$

9. a.

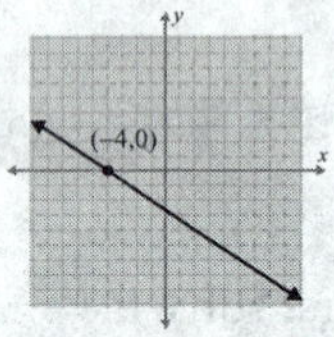

b.

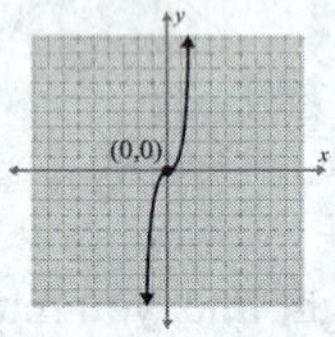

c.

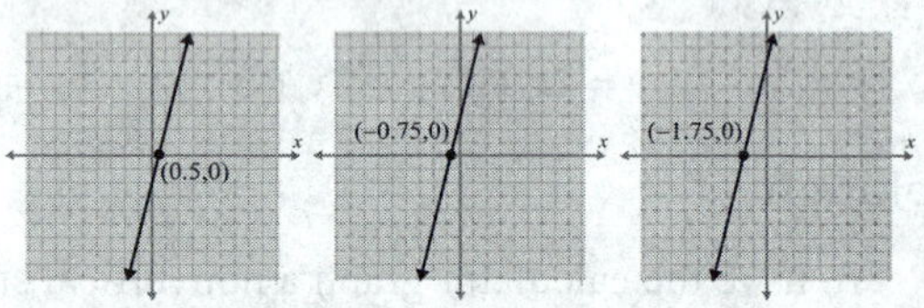

10.5 Exercises

Concept Check

Fill-in-the-Blank. Complete each sentence using information found in this section.

1. The equation $y = mx + b$ represents a linear function and $f(x) = mx + b$ is the same equation written in ________ notation.
2. The ________ line test can be used to determine if a relation is a function.
3. The set of all first coordinates in a relation is the ________, D.
4. The set of all second coordinates in a relation is the ________, R.

5. In the graph of a relation, the x-axis is called the __________ axis.

6. In the graph of a relation, the y-axis is called the __________ axis.

True/False. Determine whether each statement is true or false. If a statement is false, explain how it can be changed so the statement will be true. (**Note:** There may be more than one acceptable change.)

7. If the domain of a linear function is not explicitly stated, the implied domain is the set of all values of x that produce real values for y.

8. A relation is a function in which each domain element has exactly one corresponding range element.

9. In a function, the range elements can have more than one corresponding domain element.

10. If $s = \{(1,-6),(3,5),(4,0),(1,2)\}$, then s is a function.

Practice

List the sets of ordered pairs that correspond to the points. State the domain and range and indicate which of the relations are also functions. See Examples 1 through 3.

1.
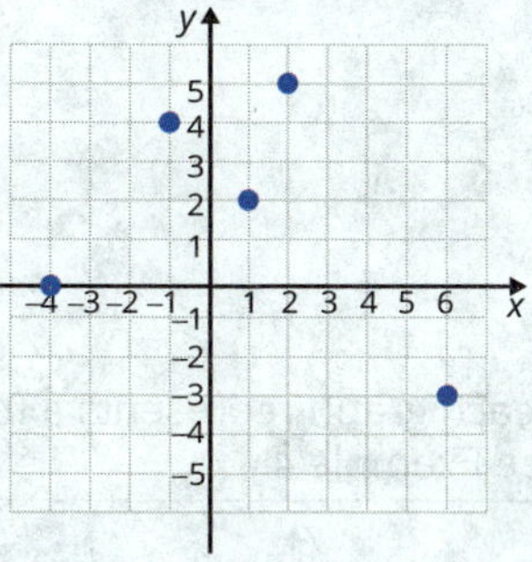

4.
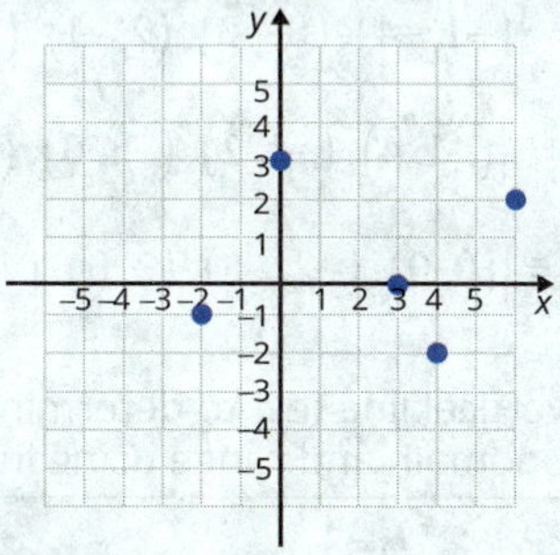

2.
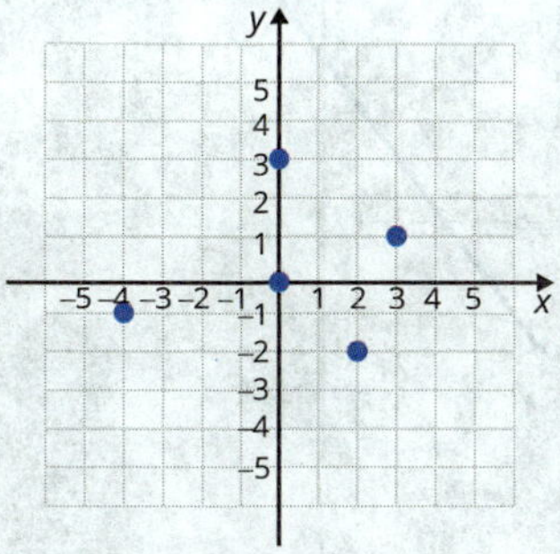

5.
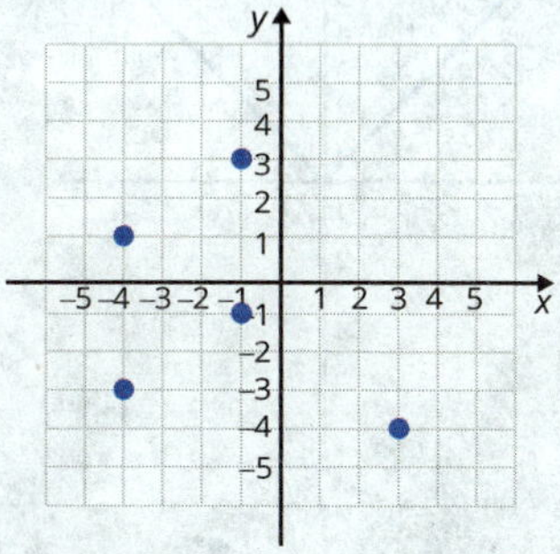

3.
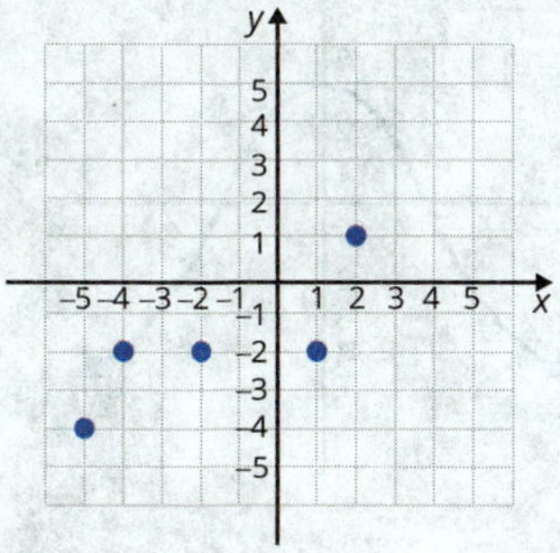

6.
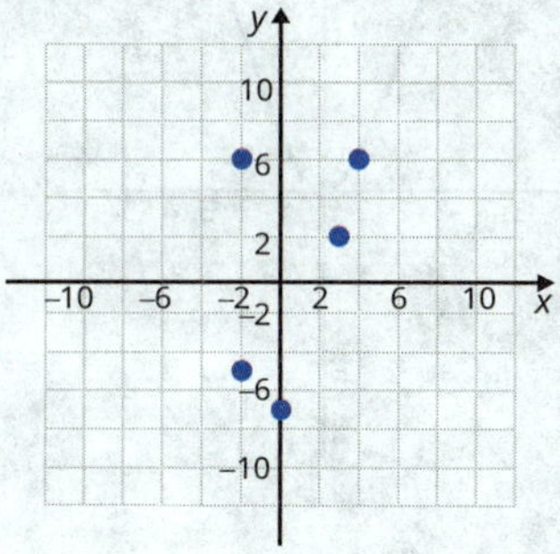

4. $\{(-2,-1),(0,3),(3,0),(4,-2),(6,2)\}$; $D=\{-2,0,3,4,6\}$; $R=\{-2,-1,0,2,3\}$; Function

5. $\{(-4,-3),(-4,1),(-1,-1),(-1,3),(3,-4)\}$; $D=\{-4,-1,3\}$; $R=\{-4,-3,-1,1,3\}$; Not a function

6. $\{(-2,6),(-2,-5),(0,-7),(3,2),(4,6)\}$; $D=\{-2,0,3,4\}$; $R=\{-7,-5,2,6\}$; Not a function

7. $\{(-5,-5),(-5,3),(0,5),(1,-2),(1,2)\}$; $D=\{-5,0,1\}$; $R=\{-5,-2,2,3,5\}$; Not a function

8. $\{(-3,3),(-1,2),(1,1),(3,0),(5,-1)\}$; $D=\{-3,-1,1,3,5\}$; $R=\{-1,0,1,2,3\}$; Function

9.
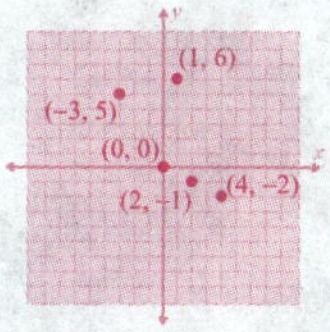

$D=\{-3,0,1,2,4\}$; $R=\{-2,-1,0,5,6\}$; Function

10.
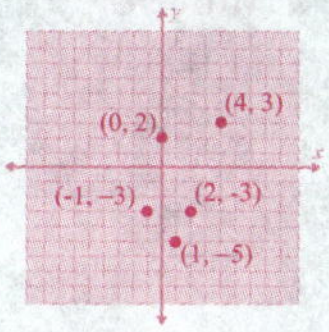

$D=\{-1,0,1,2,4\}$; $R=\{-5,-3,2,3\}$; Function

11.

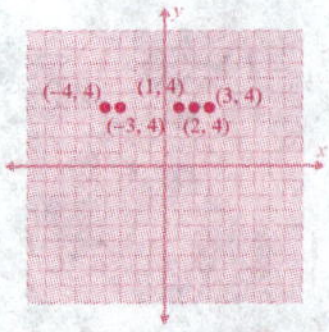

$D=\{-4,-3,1,2,3\}$;
$R=\{4\}$;
Function

12.

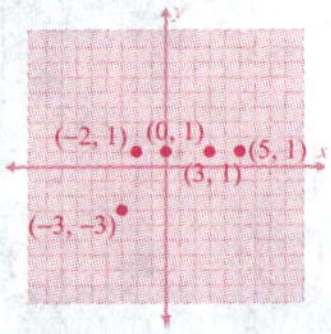

$D=\{-3,-2,0,3,5\}$;
$R=\{-3,1\}$;
Function

13.

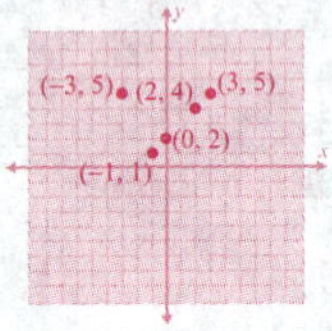

$D=\{-3,-1,0,2,3\}$;
$R=\{1,2,4,5\}$;
Function

14.

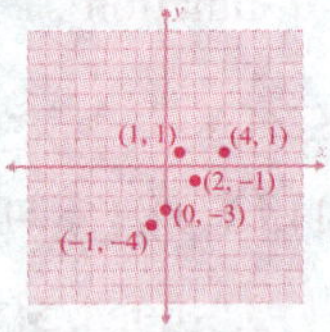

$D=\{-1,0,1,2,4\}$;
$R=\{-4,-3,-1,1\}$;
Function

15.

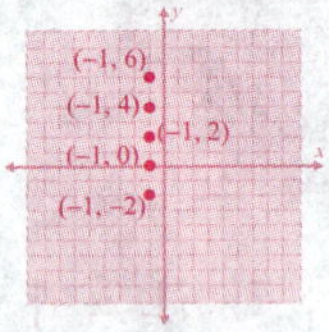

$D=\{-1\}$;
$R=\{-2,0,2,4,6\}$;
Not a function

7.

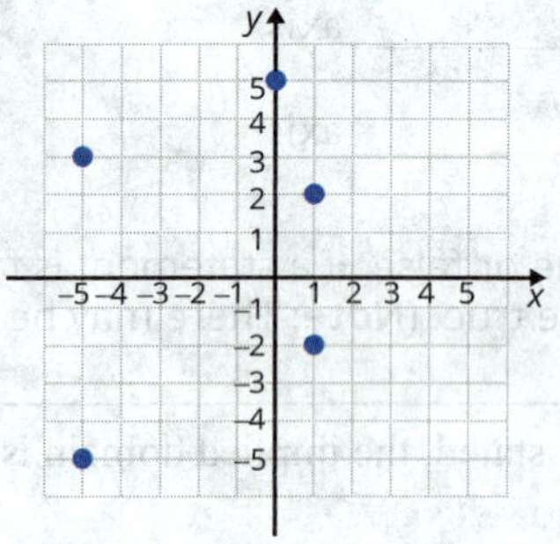

8.

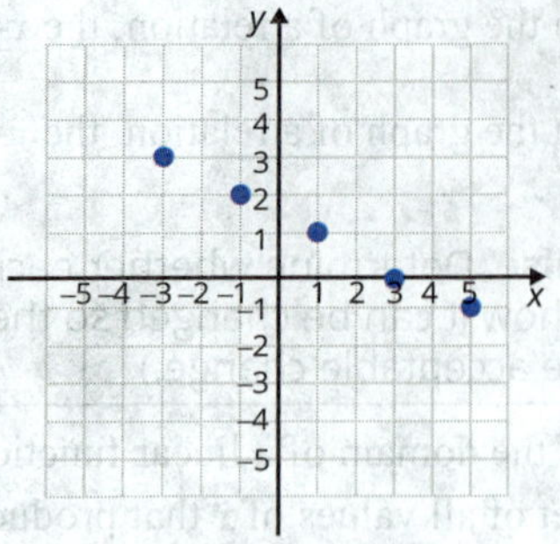

Graph the relations. State the domain and range and indicate which of the relations are functions. See Examples 1 through 3.

9. $f=\{(0,0),(1,6),(4,-2),(-3,5),(2,-1)\}$

10. $h=\{(1,-5),(2,-3),(-1,-3),(0,2),(4,3)\}$

11. $g=\{(-4,4),(-3,4),(1,4),(2,4),(3,4)\}$

12. $f=\{(-3,-3),(0,1),(-2,1),(3,1),(5,1)\}$

13. $s=\{(0,2),(-1,1),(2,4),(3,5),(-3,5)\}$

14. $t=\{(-1,-4),(0,-3),(2,-1),(4,1),(1,1)\}$

15. $f=\{(-1,4),(-1,2),(-1,0),(-1,6),(-1,-2)\}$

16. $g=\{(0,0),(-2,-5),(2,0),(4,-6),(5,2)\}$

Use the vertical line test to determine whether or not each graph represents a function. State the domain and range using interval notation. See Example 4.

17.

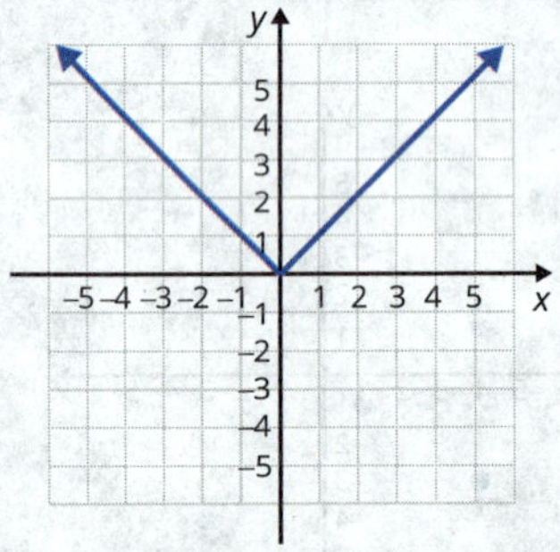

19.

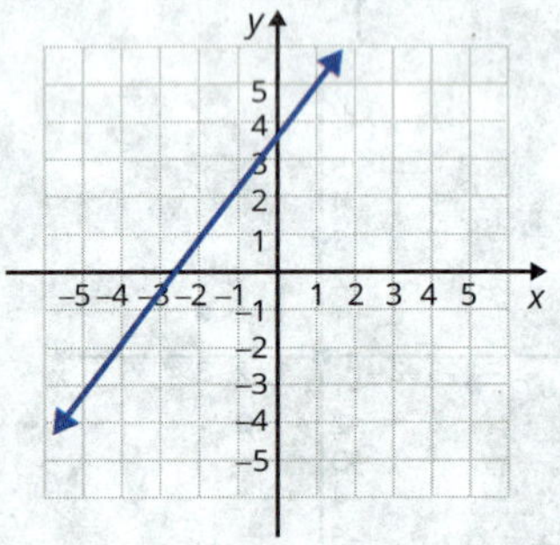

18.

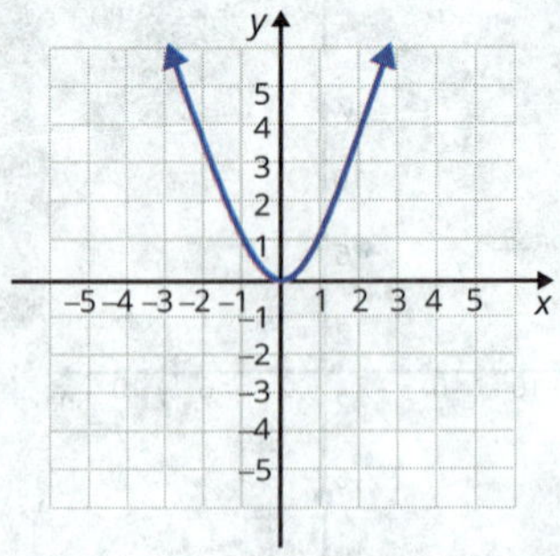

20.

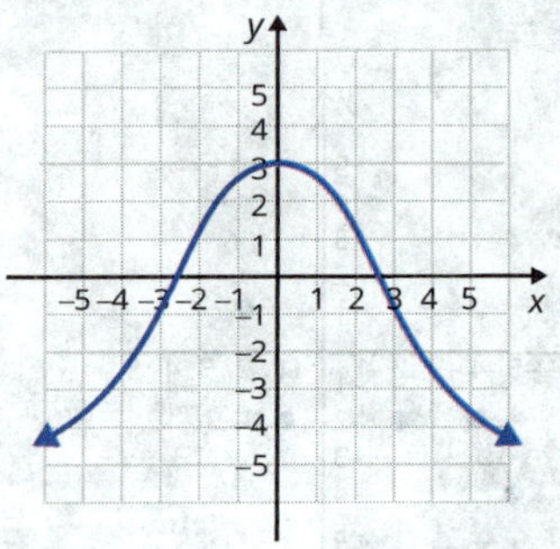

21.

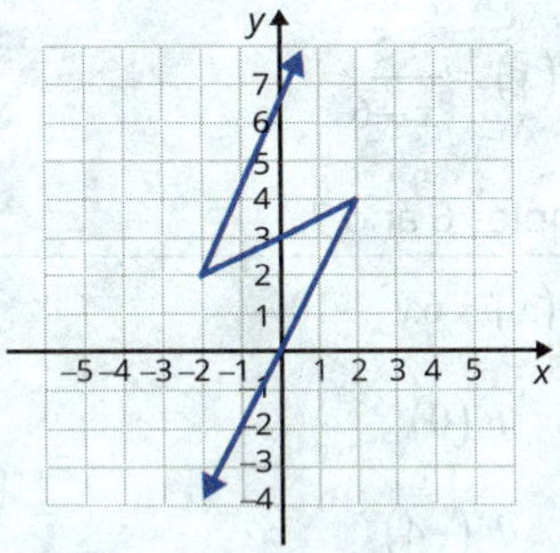

25.

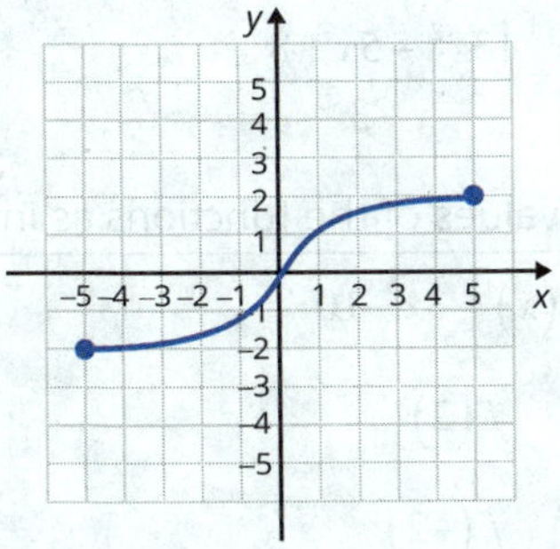

22.

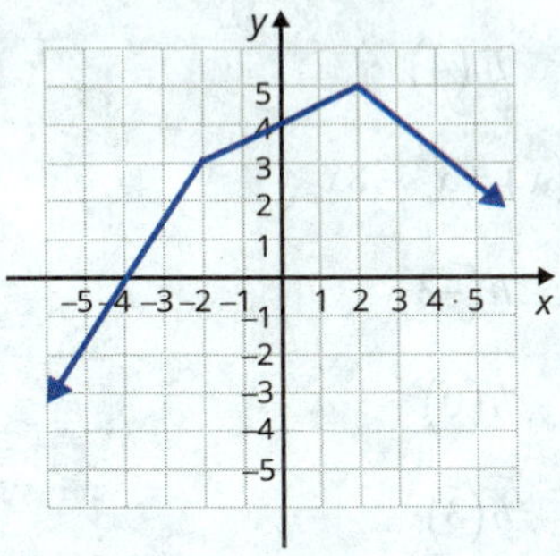

26.

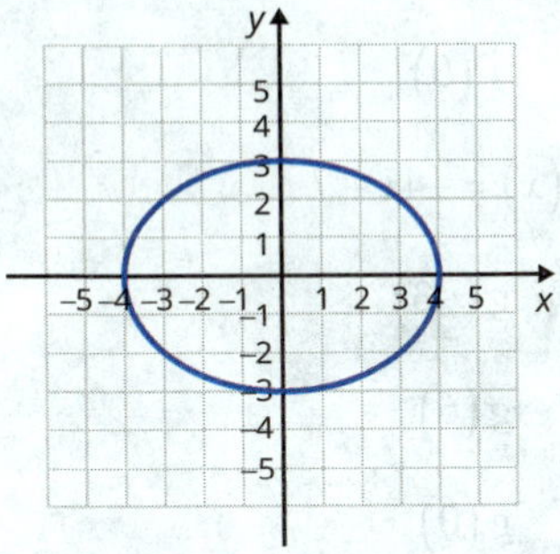

23.

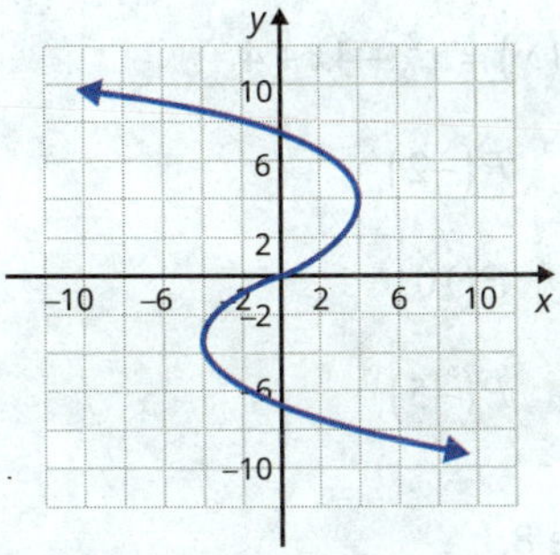

27.

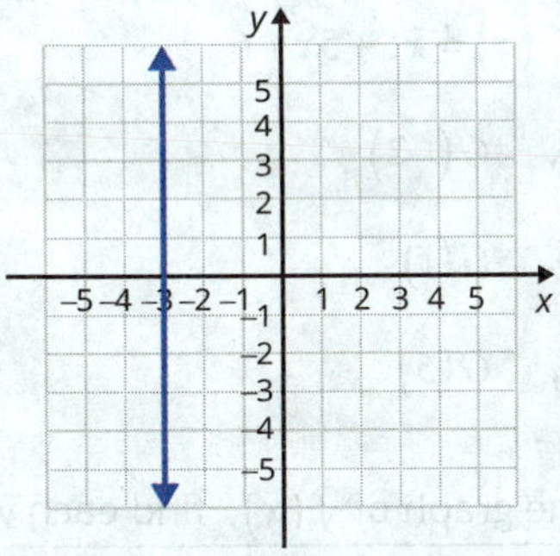

24.

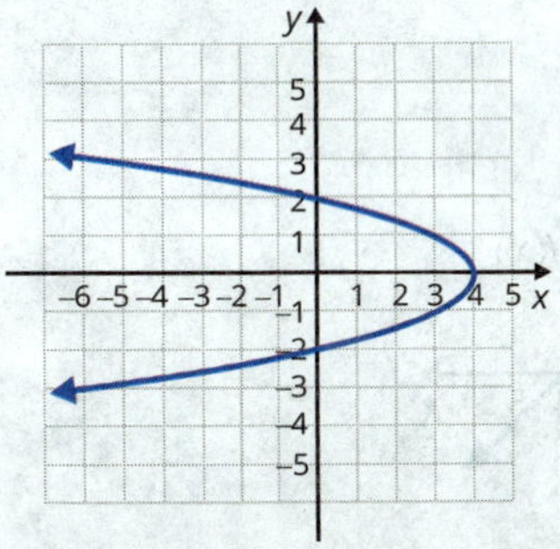

28.

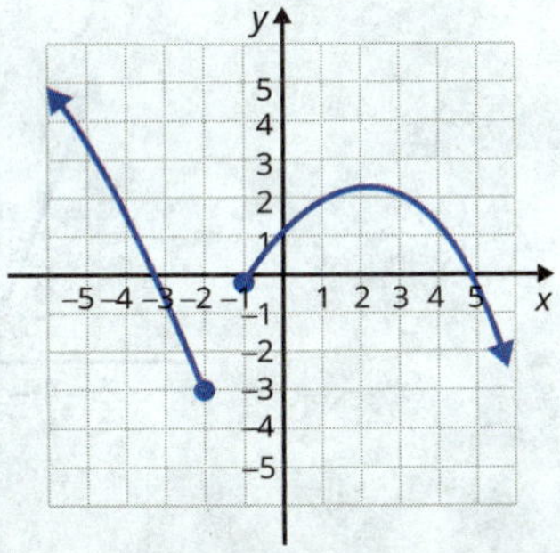

Express the function as a set of ordered pairs for the given equation and given domain. (**Hint:** Substitute each domain element for *x* and find the corresponding *y*-coordinate.)

29. $y = 3x + 1;\ D = \left\{-9, -\frac{1}{3}, 0, \frac{4}{3}, 2\right\}$

30. $y = -\frac{3}{4}x + 2;\ D = \{-4, -2, 0, 3, 4\}$

31. $y = 1 - 3x^2;\ D = \{-2, -1, 0, 1, 2\}$

32. $y = x^3 - 4x;\ D = \left\{-1, 0, \frac{1}{2}, 1, 2\right\}$

State the domains of the functions. See Example 5.

33. $y = -5x + 10$

34. $2x + y = 14$

35. $g(x) = \frac{8}{x}$

36. $h(x) = \frac{7}{3x}$

16.

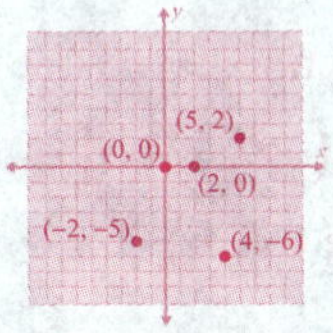

$D = \{-2, 0, 2, 4, 5\}$;
$R = \{-6, -5, 0, 2\}$;
Function

17. Function; $D = (-\infty, \infty)$; $R = [0, \infty)$

18. Function; $D = (-\infty, \infty)$; $R = [0, \infty)$

19. Function; $D = (-\infty, \infty)$; $R = (-\infty, \infty)$

20. Function; $D = (-\infty, \infty)$; $R = (-\infty, 3]$

21. Not a function; $D = (-\infty, \infty)$; $R = (-\infty, \infty)$

22. Function; $D = (-\infty, \infty)$; $R = (-\infty, 5]$

23. Not a function; $D = (-\infty, \infty)$; $R = (-\infty, \infty)$

24. Not a function; $D = (-\infty, 4]$; $R = (-\infty, \infty)$

25. Function; $D = [-5, 5]$; $R = [-2, 2]$

26. Not a function; $D = [-4, 4]$; $R = [-3, 3]$

27. Not a function; $D = \{-3\}$; $R = (-\infty, \infty)$

28. Function; $D=(-\infty,-2]\cup[-1,\infty)$; $R=(-\infty,\infty)$

29. $\left\{(-9,-26),\left(-\frac{1}{3},0\right),(0,1),\left(\frac{4}{3},5\right),(2,7)\right\}$

30. $\left\{(-4,5),\left(-2,\frac{7}{2}\right),(0,2),\left(3,-\frac{1}{4}\right),(4,-1)\right\}$

31. $\{(-2,-11),(-1,-2),(0,1),(1,-2),(2,-11)\}$

32. $\left\{(-1,3),(0,0),\left(\frac{1}{2},-\frac{15}{8}\right),(1,-3),(2,0)\right\}$

33. $D=(-\infty,\infty)$

34. $D=(-\infty,\infty)$

35. $D=(-\infty,0)\cup(0,\infty)$ or $x\neq 0$

36. $D=(-\infty,0)\cup(0,\infty)$ or $x\neq 0$

37. $D=(-\infty,3)\cup(3,\infty)$ or $x\neq 3$

38. $D=(-\infty,6)\cup(6,\infty)$ or $x\neq 6$

39. **a.** −4 **b.** −16 **c.** −10

40. **a.** 19 **b.** −17 **c.** 7

41. **a.** 0 **b.** 12 **c.** 56

42. **a.** −10 **b.** 86 **c.** 86

43. **a.** −3 **b.** 0 **c.** 3

44. **a.** 0 **b.** 144 **c.** 9

45. $f(1)=3$

46. $f(-3)=1$

47. $f(4)=0$

48. $f(-1)=2$

37. $y=\dfrac{13x^2-5x+8}{x-3}$

38. $f(x)=\dfrac{35}{x-6}$

Find the values of the functions as indicated. See Examples 6 and 7.

39. $f(x)=3x-10$

a. $f(2)$

b. $f(-2)$

c. $f(0)$

40. $g(x)=-4x+7$

a. $g(-3)$

b. $g(6)$

c. $g(0)$

41. $G(x)=x^2+5x+6$

a. $G(-2)$

b. $G(1)$

c. $G(5)$

42. $F(x)=6x^2-10$

a. $F(0)$

b. $F(-4)$

c. $F(4)$

43. $h(x)=x^3-8x$

a. $h(-3)$

b. $h(0)$

c. $h(3)$

44. $P(x)=x^2+4x+4$

a. $P(-2)$

b. $P(10)$

c. $P(-5)$

Using the graph of $f(x)$, find each value. See Example 8.

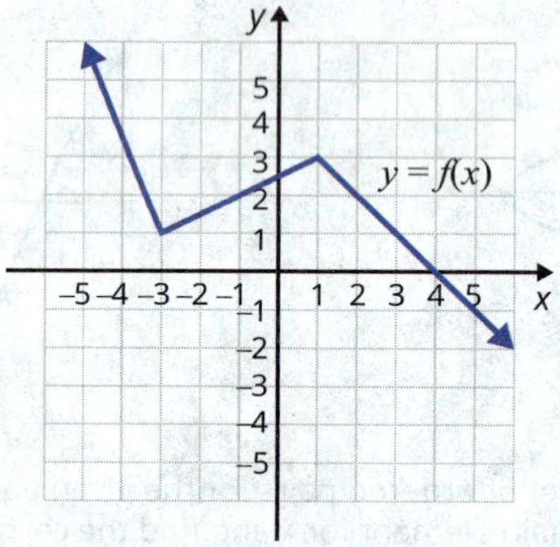

45. $f(1)$

46. $f(-3)$

47. $f(4)$

48. $f(-1)$

Use a graphing calculator to graph the functions. Use the CALC features to find *x*-intercepts, if any. (The value of *y* will be 0 at those points.) For absolute value functions, select the `MATH` menu, then the `NUM` menu, and then abs (. Remember to press) after entering the absolute value. See Example 9.

49. $y=6$

50. $y=4x$

51. $y=x+5$

52. $y=-2x+3$

53. $y=x^2-4x$

54. $y=1+2x-x^2$

55. $y = -|3x|$

56. $y = |x+2|$

57. $y = |x^2 - 3x|$

58. $y = 2x^3 - 5x^2 + 1$

59. $y = -x^3 + 3x - 1$

60. $y = x^4 - 10x^2 + 9$

Use the CALC features of the calculator to find the coordinates of any points of intersection of the graphs. (**Hint:** The `intersect` item on the CALC menu will help in finding the point (or points) of intersection of two functions, if there is one.) In the Y = menu use both Y1 = and Y2 = to be able to graph both functions at the same time.

61. $y = 3x + 2$
$y = 4 - x$

62. $y = 2 - x$
$y = x$

63. $y = 2x - 1$
$y = x^2$

64. $y = x + 3$
$y = -x^2 + x + 7$

The calculator display shows an incorrect graph for the corresponding equation. Explain how you know, by just looking at the graph, that a mistake has been made.

65. $y = 2x + 5$

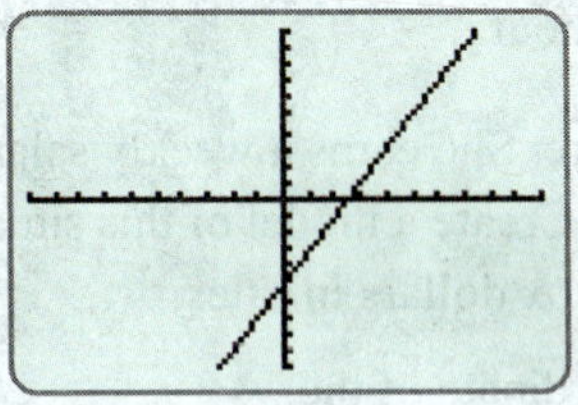

66. $y = -3x + 4$

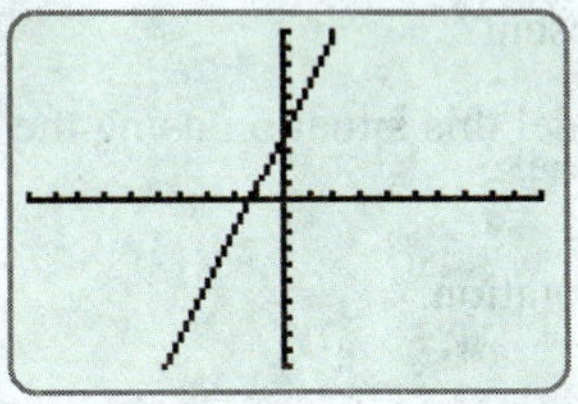

67. $y = \frac{2}{3}x - 2$

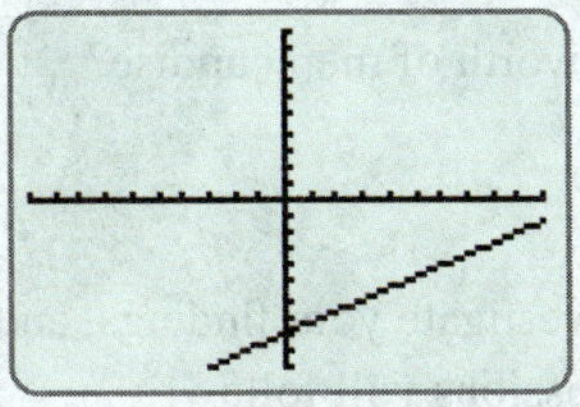

68. $y = -4x$

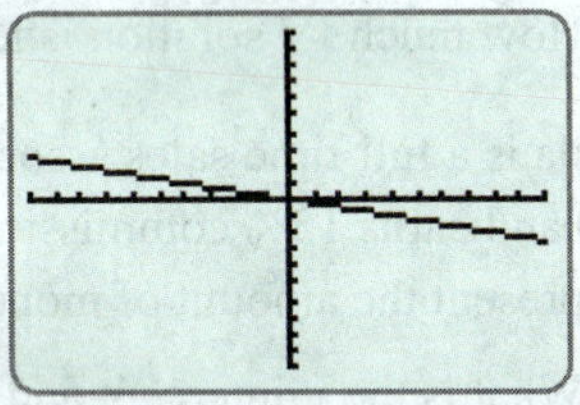

69. $y = -\frac{1}{3}x$

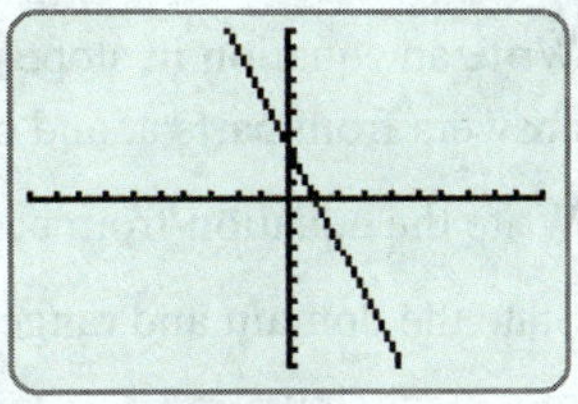

49.

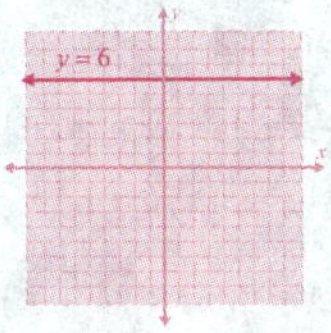

50.

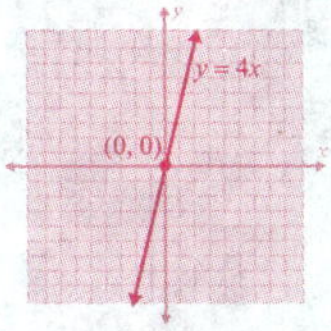

51.

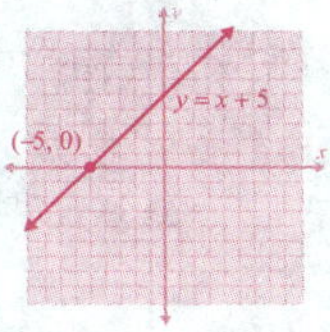

52.

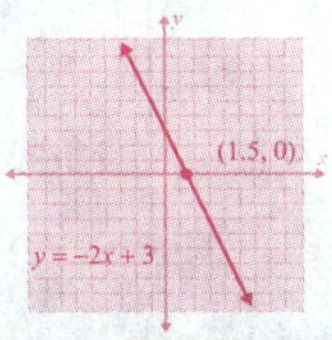

53.

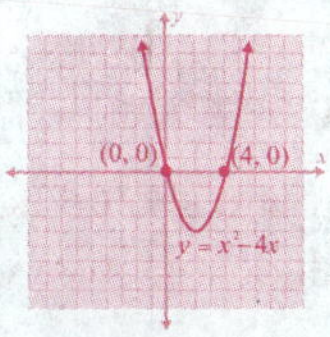

54.

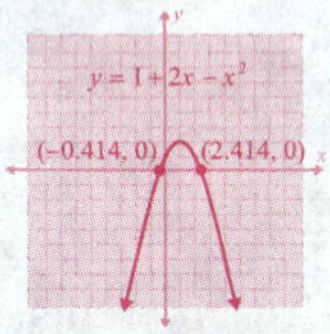

55.

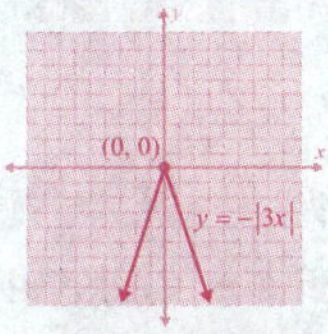

56.

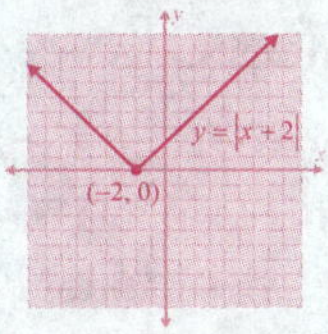

57.

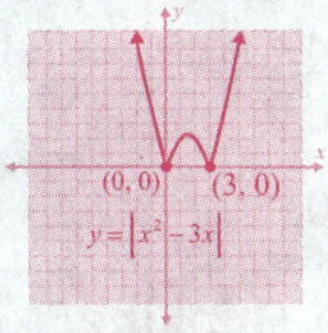

58.

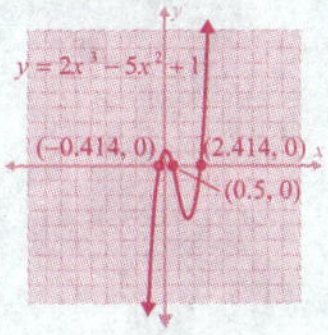

59.

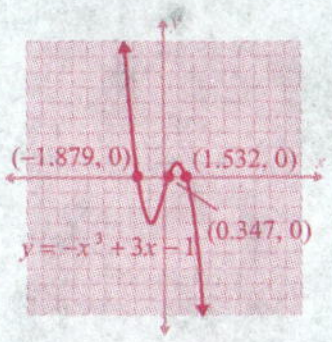

60.

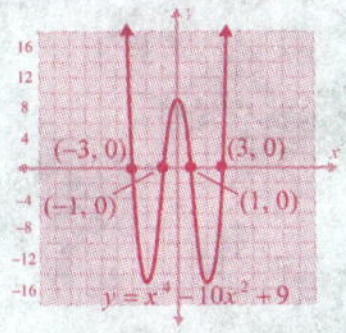

61.

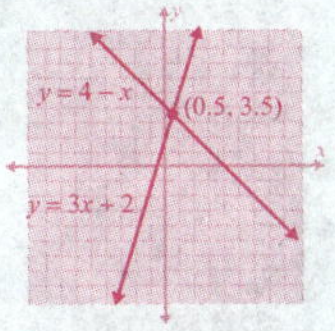

62.

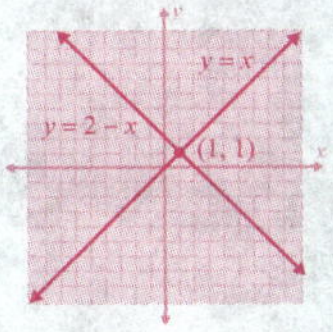

63.

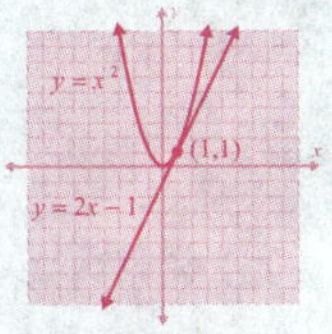

64.

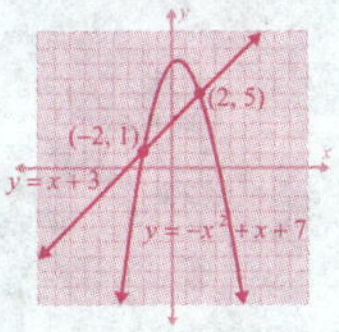

Applications

Solve.

70. A nurse hangs a 1000-milliliter IV bag that is set to drip at 120 milliliters per hour. Create a model of this situation to represent the amount of IV solution left in the bag after x hours.

a. The y-intercept is the amount of IV solution in the bag initially (time = 0). What is the y-intercept?

b. The slope is equal to the rate that the IV solution is dispensed per hour. What is the slope? (**Hint:** Consider whether the amount of IV solution in the bag is increasing or decreasing and how this would affect the slope.)

c. Write an equation in slope-intercept form to model this situation.

d. Write the equation from part c. using function notation.

e. State the domain and range of the function.

f. State any additional restrictions that should be made on the domain for it to make sense in the context of this problem.

g. How much IV solution is left in the bag after 5 hours?

71. Ariella is a full-time sales associate at a clothing store. She earns a weekly salary of \$250 and earns 15% commission on all of her sales. Create a model of this situation to represent the amount of money Ariella makes after x dollars in sales.

a. What is the y-intercept and what does the y-coordinate of the y-intercept represent?

b. What is the slope and what does this value represent?

c. Write an equation in slope-intercept form to model this situation using the answers from parts a. and b.

d. Write the equation from part c. using function notation.

e. State the domain and range of the function.

f. State any additional restrictions that should be made on the domain for it to make sense in the context of this problem.

g. How much will Ariella make if she sells \$5000 worth of merchandise?

Writing & Thinking

72. Enter a variety of functions in your calculator, investigate your findings, and report these to your class. Certainly, interesting discussions will follow!

---

65. y-intercept $= (0,-5)$ (should be $(0,5)$)

66. Slope $= 4$ (should be -3)

67. y-intercept $= (0,-8)$ (should be $(0,-2)$)

68. Slope $= -\frac{1}{4}$ (should be the reciprocal, -4)

69. Slope $= -3$ (should be the reciprocal, $-\frac{1}{3}$) and y-intercept $= (0,2)$ (should be $(0,0)$)

Applications

70. **a.** $(0,1000)$ **b.** $m=-120$

c. $y=-120x+1000$

d. $f(x)=-120x+1000$

e. Domain is all real numbers; Range is all real numbers

f. Domain should not include negative values

g. 400 milliliters

71. **a.** $(0, 250)$, represents Ariella's base weekly salary

b. 0.15; Represents the commission rate

c. $y = 0.15x + 250$

d. $f(x) = 0.15x + 250$

e. Domain is all real numbers; Range is all real numbers

f. Answers will vary. Some students may say sales cannot be negative, some will allow negative sales due to returns.

g. \$1000

Writing & Thinking

72. Answers will vary.

10.6 Graphing Linear Inequalities in Two Variables

Objectives

A. Graph linear inequalities.

B. Use a graphing calculator to graph linear inequalities.

A Graphing Linear Inequalities

In this section, we will develop techniques for analyzing and graphing linear inequalities. We need the following terminology.

Half-plane	A straight line separates a plane into two **half-planes**. The points on one side of the line are in one of the half-planes, and the points on the other side of the line are in the other half-plane.
Boundary line	The line itself is called the **boundary line**.
Closed half-plane	If the boundary line is included in the solution set, then the half-plane is said to be **closed**.
Open half-plane	If the boundary line is not included in the solution set, then the half-plane is said to be **open**.

Figure 1 shows both an open half-plane and a closed half-plane with the line $2x - 3y = 10$ as the boundary line. The solution set is the shaded region.

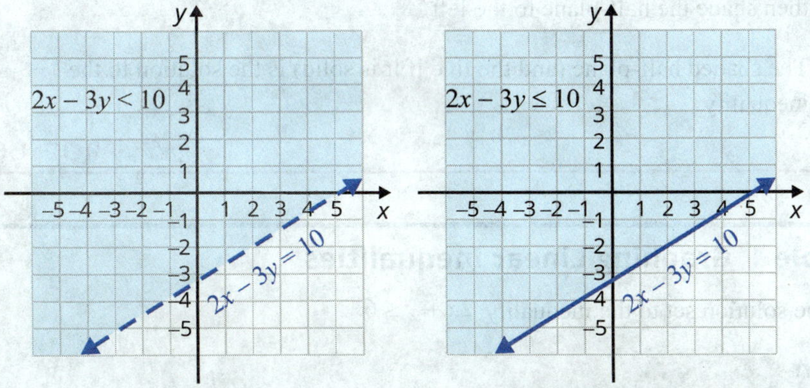

The points on the line $2x - 3y = 10$ are not included in the solution set, so the line is dashed. The half-plane is open.

The points on the line $2x - 3y = 10$ are included in the solution set, so the line is solid. The half-plane is closed.

Figure 1

Graphing Linear Inequalities

1. First, graph the boundary line (dashed if the inequality is $<$ or $>$, solid if the inequality is $\leq$ or $\geq$).

2. Next, determine which side of the line to shade using one of the following methods.

Method 1

a. Test any one point obviously on one side of the line.

b. If the test-point satisfies the inequality, shade the half-plane on that side of the line. Otherwise, shade the other half-plane.

Note: The point $(0, 0)$, if it is not on the boundary line, is usually the easiest point to test.

Method 2

a. Solve the inequality for y (assuming that the line is not vertical).

b. If the solution shows $y <$ or $y \leq$, then shade the half-plane below the line.

c. If the solution shows $y >$ or $y \geq$, then shade the half-plane above the line.

Note: If the boundary line is vertical, then solve for x. If the solution shows $x >$ or $x \geq$, then shade the half-plane to the right. If the solution shows $x <$ or $x \leq$, then shade the half-plane to the left.

3. The shaded half-plane (and the line if it is solid) is the solution to the inequality.

PROCEDURE

1. Graph the solution set to the inequality $9x - 3y \leq 21$.

Example 1 Graphing Linear Inequalities

Graph the solution set to the inequality $2x + y \leq 6$.

Solution

Method 1 is used in this example.

Step 1: Graph the boundary line $2x + y = 6$ as a solid line because the inequality is $\leq$ (less than or **equal to**).

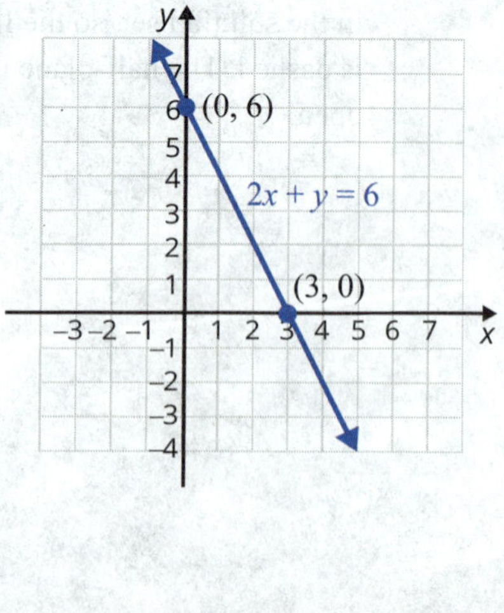

Step 2: Test any point on one side of the line. In this example, we have chosen $(0, 0)$.

$$2(0)+(0)\le 6$$
$$0\le 6$$

This is a true statement.

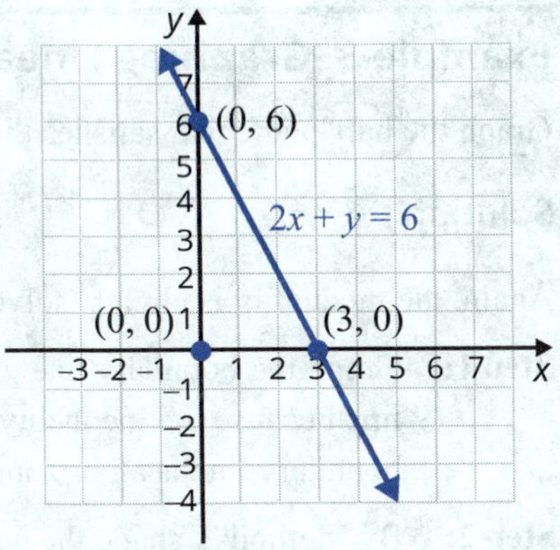

Teaching Note

While some students will recognize that the origin $(0, 0)$ is the easiest point to use as a test point, it is a good idea to have students "test" some points on both sides of the boundary line upon first introduction, so that they develop confidence in the test (and solution set).

Step 3: Shade the half-plane on the same side of the line as the point $(0, 0)$. The shaded half-plane and the boundary line is the solution set to the inequality.

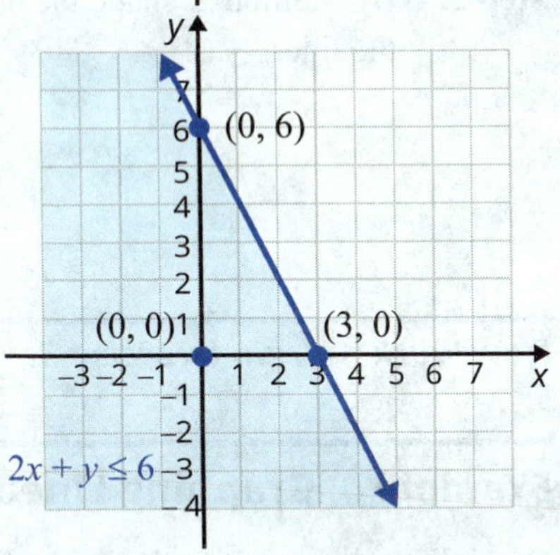

Now work margin exercise 1.

Example 2 Graphing Linear Inequalities

Graph the solution set to the inequality $y > 2x$.

Solution

Since the inequality is already solved for y, Method 2 is easy to apply.

Step 1: Graph the boundary line $y = 2x$ as a dashed line because the inequality is >.

Step 2: The solution shows >, so by Method 2, the graph consists of those points above the line. Shade the half-plane above the line.

Note: As a check, we see that the point $(-3,0)$ gives $0>-6,$ a true statement. Thus, we know we have shaded the correct half-plane.

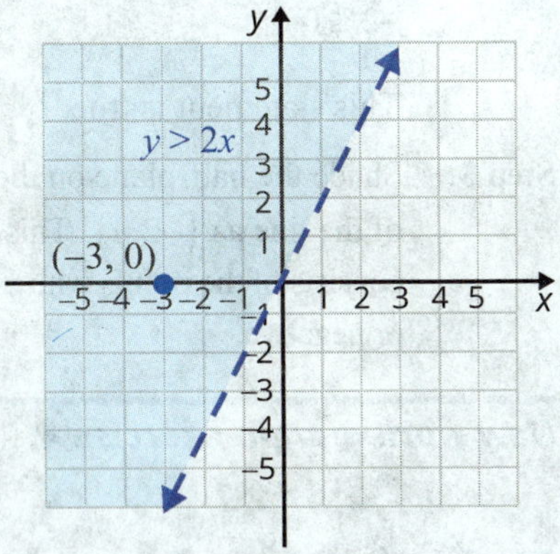

Now work margin exercise 2.

2. Graph the solution set to the inequality $y < 3x$.

3. Graph the solution set to the inequality $y < 4$.

Example 3 Graphing Linear Inequalities

Graph the half-plane that satisfies the inequality $y > 1$.

Solution

Again, the inequality is already solved for y and Method 2 is used.

Step 1: Graph the boundary line $y = 1$ as a dashed line because the inequality is >. (The boundary line is a horizontal line.)

Step 2: By Method 2, shade the half-plane above the line.

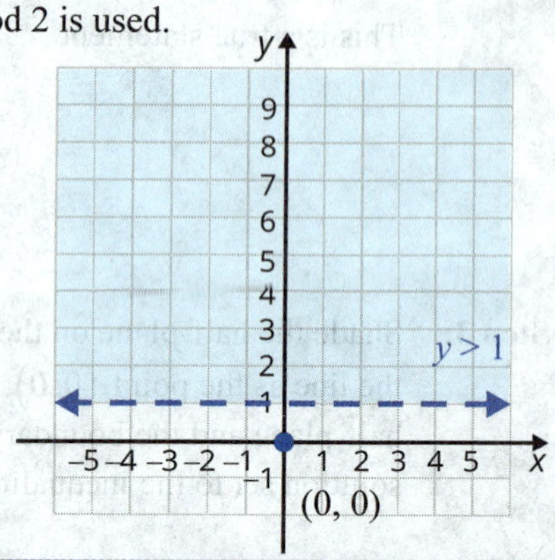

Now work margin exercise 3.

4. Graph the solution set to the inequality $x > -2$.

Example 4 Graphing Linear Inequalities

Graph the solution set to the inequality $x \le 0$.

Solution

The boundary line is a vertical line and Method 1 is used.

Step 1: Graph the boundary line $x = 0$ as a solid line because the inequality is ≤ (less than or **equal to**). Note that this is the y-axis.

Step 2: Test the point $(-2, 1)$.

$-2 \le 0$

This statement is true.

Step 3: Shade the half-plane on the same side of the line as $(-2, 1)$. This half-plane consists of the points with x-coordinates 0 or negative.

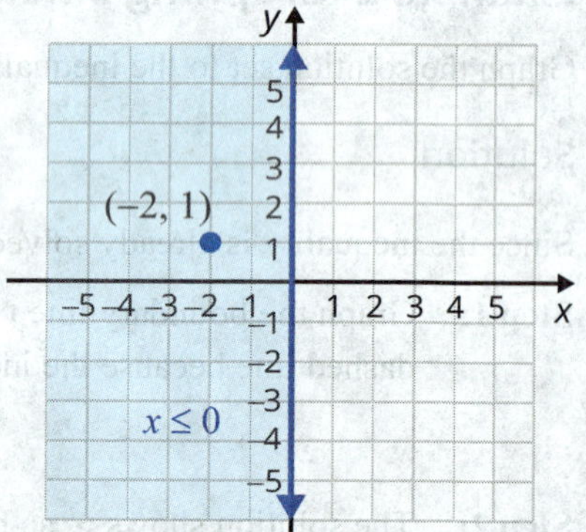

Now work margin exercise 4.

B Using a Graphing Calculator to Graph Linear Inequalities

The first step in using the TI-84 Plus (or any other graphing calculator) to graph a linear inequality is to solve the inequality for y. This step is necessary because calculators require the equations to be entered in this form. Thus, Method 2 for graphing the correct half-plane is appropriate.

Notice that when you press the Y= key on the calculator, a slash (\ ) appears to the left of the Y expression as in \Y1 =. This slash is actually a command to the calculator to graph the corresponding function as a solid line or curve. If you move the cursor to position it over the slash and hit ENTER repeatedly, the following options will appear.

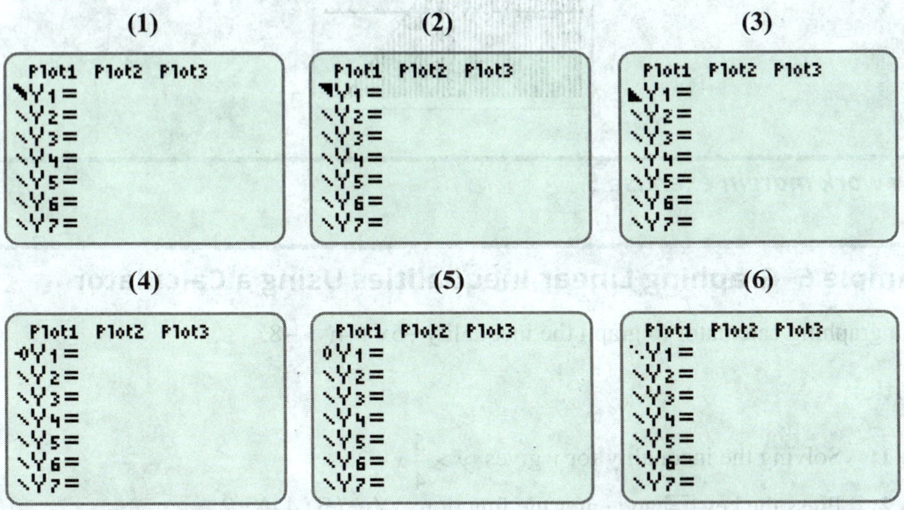

Figure 2

If the slash (which is actually four dots if you look closely) becomes a set of three dots **(6)**, then the corresponding graph of the function will be dotted. By setting the shading above the slash **(2)**, the corresponding graph on the display will show shading above the line or curve. By setting the shading below the slash **(3)**, the corresponding graph on the display will show shading below the line or curve. (**Note:** When shading above or below the curve, the boundary curve will always appear to be solid, so the calculator is not good for determining whether the boundary curve is included or not.) Options 1, 4, and 5 are not used when graphing linear inequalities. The following examples illustrate two situations.

Example 5 Graphing Linear Inequalities Using a Calculator

Use a graphing calculator to graph the inequality $2x + y \leq 7$.

Solution

Step 1: Solving the inequality for y gives $y \leq -2x + 7$.

Step 2: Press the key Y= and enter the function: \Y1=-2X+7.

Step 3: Go to the \ and hit ENTER three times so that the display appears as follows.

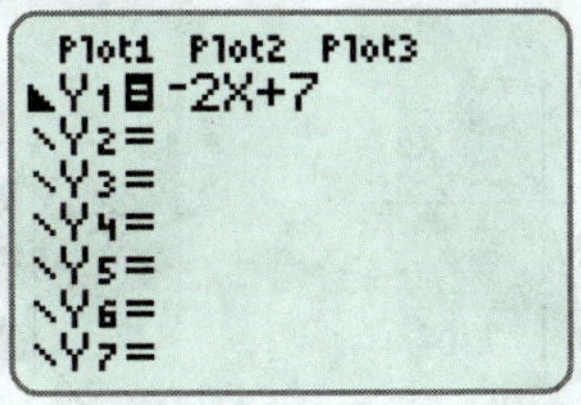

5. Use a graphing calculator to graph the inequality $5x - 8 > y - 6$.

Step 4: Press GRAPH and (using the standard window settings) the following graph should appear on the display.

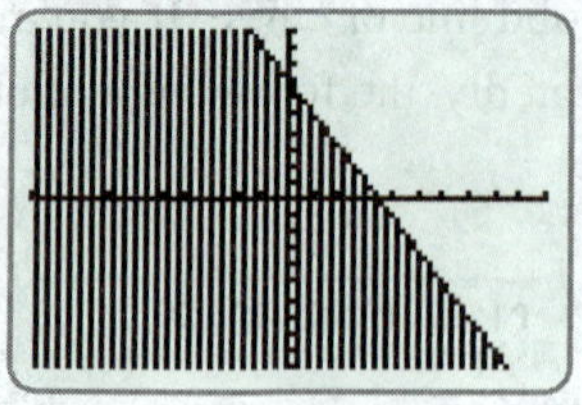

Now work margin exercise 5.

Example 6 Graphing Linear Inequalities Using a Calculator

Use a graphing calculator to graph the inequality $-5x + 4y > -8$.

Solution

Step 1: Solving the inequality for y gives $y > \frac{5}{4}x - 2$.

Step 2: Press the key Y= and enter the function: \Y1=(5/4)X-2.

Step 3: Go to the \ and hit ENTER two times so that the display appears as follows.

Step 4: Press GRAPH and (using the standard window settings) the following graph should appear on the display. (**Note:** The inequality is >, so the boundary line should actually be dashed.)

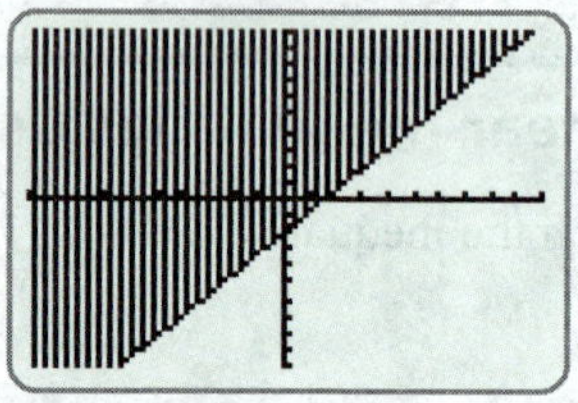

Now work margin exercise 6.

6. Use a graphing calculator to graph the inequality $6x - 3y \le 7$.

Margin Exercise Answers

1. 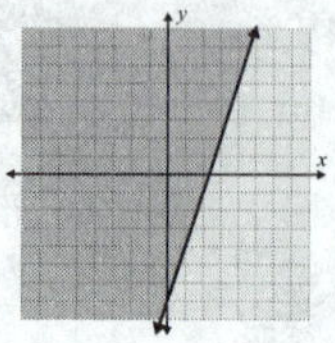
2. 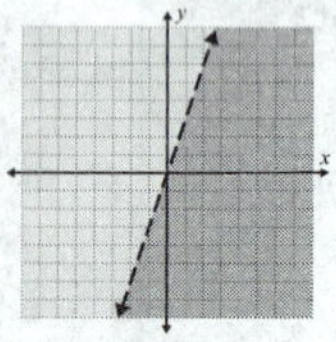
3. 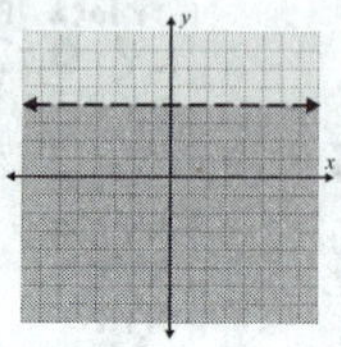
4. 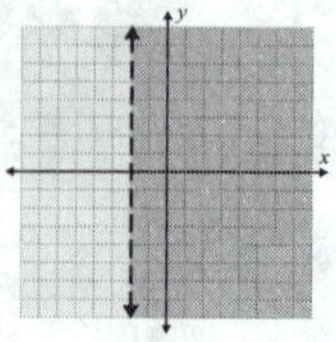
5. 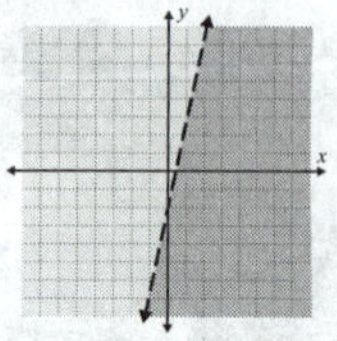
6. 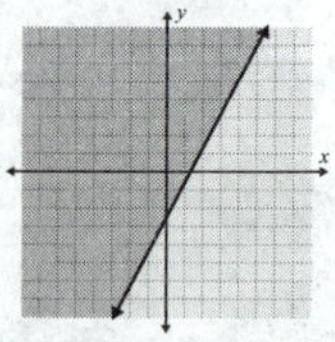

10.6 Exercises

Concept Check

1. test
2. satisfies
3. open
4. boundary
5. closed
6. dashed
7. True
8. False; The boundary line is included if the solution set is a closed half-plane.
9. False; The boundary line is solid if the inequality uses ≤ or ≥.
10. False; Slope has no effect on whether the boundary line is part of the solution.

Practice

1. 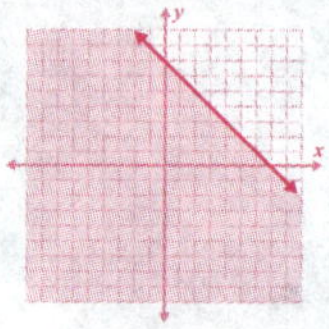
2. 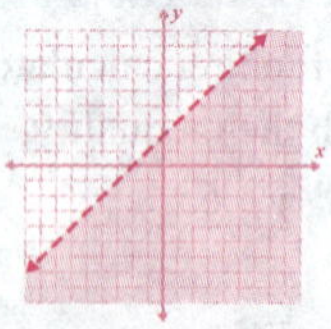
3. 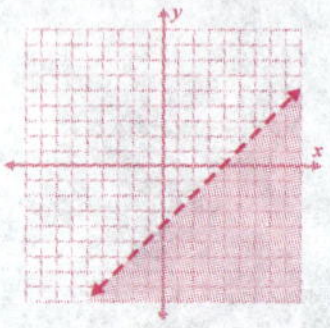
4. 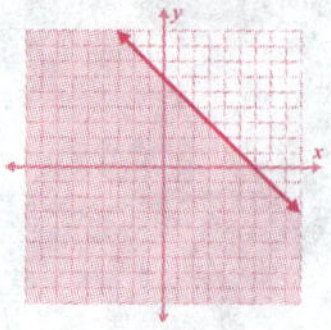
5. 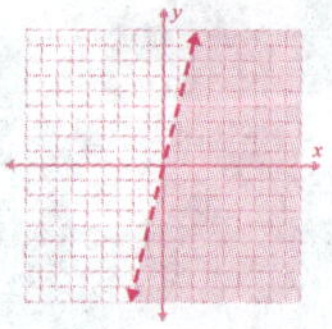

10.6 Exercises

Concept Check

Fill-in-the-Blank. Complete each sentence using information found in this section.

1. To determine which half-plane is a solution of a linear inequality (and therefore should be shaded), _________ any point clearly on one side of the boundary line.

2. If a point is tested on one side of the boundary line and it _________ the inequality, shade that side of the boundary line. The shaded region is the solution set.

3. If a boundary line is not included in the solution set, the solution is a/an _________ half-plane.

4. A straight line that separates two half-planes is called a/an _________ line.

5. If a boundary line is part of the solution set, the graph of the solution set is a/an _________ half-plane.

6. The boundary line should be _________ if the inequality is $<$ or $>$.

True/False. Determine whether each statement is true or false. If a statement is false, explain how it can be changed so the statement will be true. (**Note:** There may be more than one acceptable change.)

7. A solid boundary line indicates that the points on that line are included in the solution.

8. If the solution set is an open half-plane, then the boundary line is included in the solution.

9. The boundary line is solid when the inequality uses a $<$ or $>$ symbol.

10. The slope of an inequality is used to determine whether the boundary line is included in the solution.

Practice

Graph the solution set of each of the linear inequalities. See Examples 1 through 4.

1. $x+y\le 7$
2. $x-y>-2$
3. $x-y>4$
4. $x+y\le 6$
5. $y<4x$
6. $y<-2x$
7. $y\ge -3x$
8. $y>x$
9. $x-2y>8$
10. $x+3y\le 3$
11. $4x+y\ge 2$
12. $5x-y<4$
13. $y\le 5-3x$
14. $y\ge 8-2x$
15. $2y-x\le 0$
16. $x+y>0$
17. $x+4\ge 0$
18. $x-5\le 0$
19. $y\ge -2$
20. $y+3<0$
21. $4x<-3y+9$
22. $3x<2y-4$
23. $3y>4x+6$
24. $5x<2y-6$
25. $x+3y<7$
26. $3x+4y>11$
27. $\frac{1}{2}x-y>1$
28. $\frac{1}{3}x+y\ge 3$
29. $\frac{2}{3}x+y\ge 4$
30. $2x-\frac{4}{3}y>8$

6.
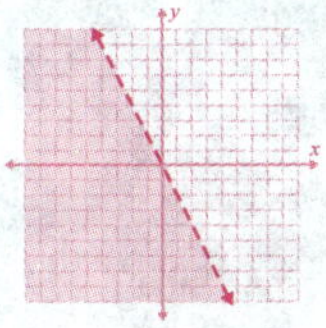

7.
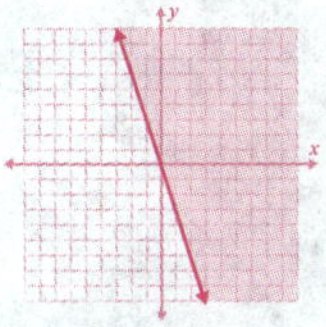

8.
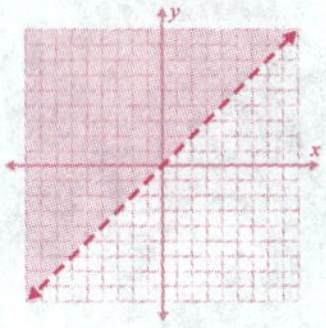

9.
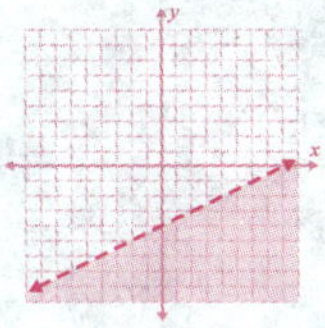

10.
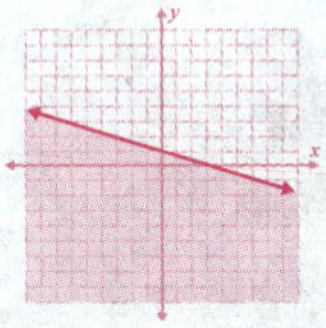

11.
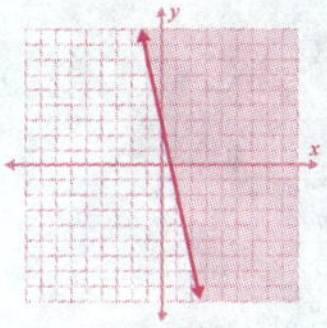

12.
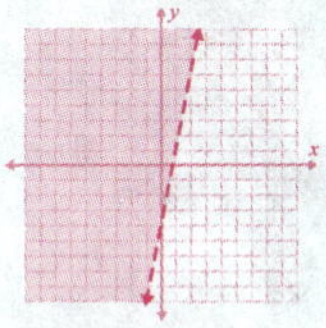

13.
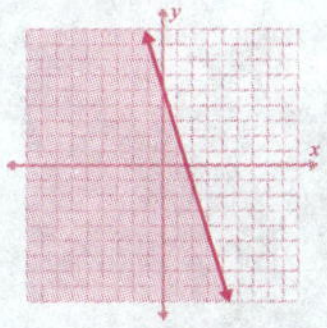

14.
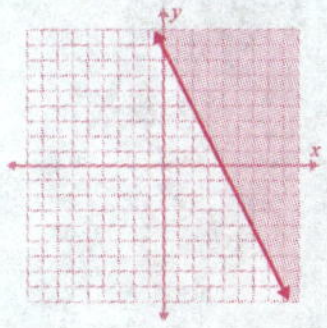

15.
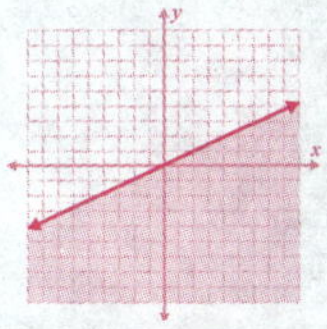

16.
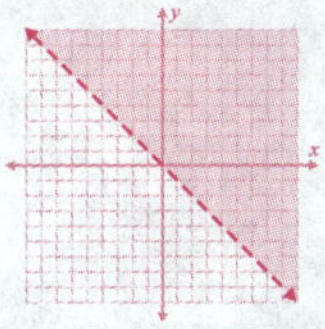

17.
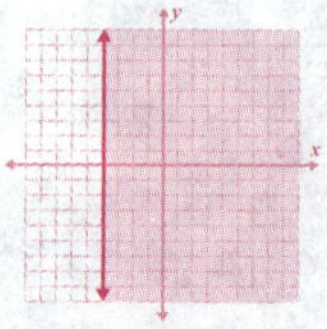

18.
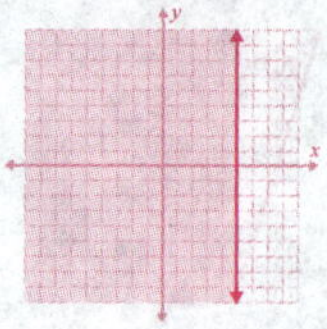

19.
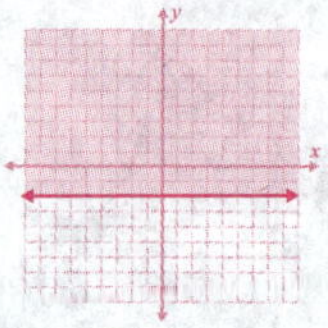

20.
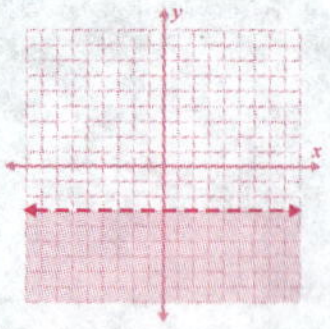

21.
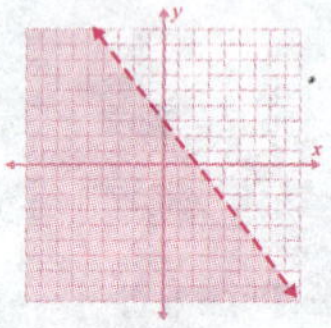

22.
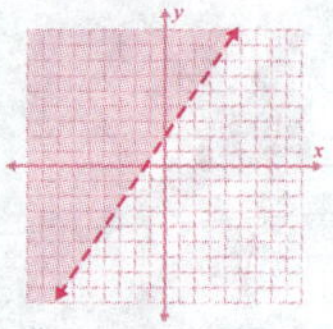

23.
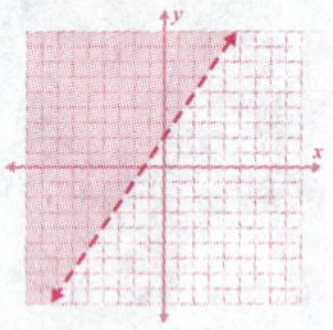

24.
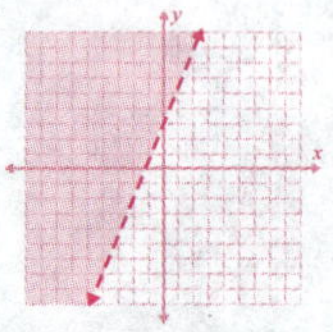

25.
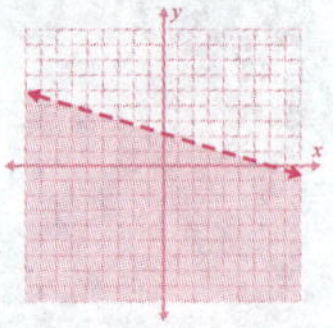

26.
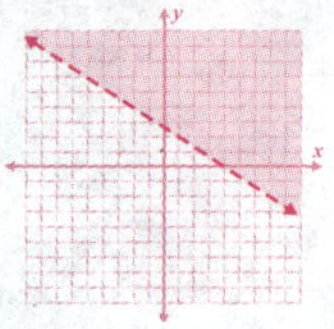

Use a graphing calculator to graph each of the linear inequalities. See Examples 5 and 6.

31. $y > \frac{1}{2}x$

32. $2x \geq -6y$

33. $x - y \leq 5$

34. $x + 2y > 8$

35. $y \geq -3$

36. $y \leq -4$

37. $2x + y \leq 6$

38. $x - 3y \geq 9$

39. $3x + 2y \geq 12$

40. $3x - 4y > 15$

Applications

Solve.

41. The grade for a 1-credit-hour survey class is based on an exam and a project, which are worth a maximum of 50 points each. The sum of the two scores must be at least 75 points for a student to earn a passing grade.

   a. Let the amount of points earned on the exam be represented by the variable x and the amount of points earned on the project be represented by the variable y. Create a linear inequality to describe the solution set for a passing grade.

   b. Graph the linear inequality from part a.

   c. A student earns 45 points on their final exam and 22 points on their project. Plot this point on the graph. Did this student earn a passing grade?

   d. Are there any points in the solution set that do not make sense for this situation?

42. A fail-safe is installed on a device with two electrical inputs. If the sum of the inputs is greater than 250 kilowatts, the fail-safe will activate and cause the machine to switch off.

   a. Let one electrical input be represented by the variable x and the other be represented by the variable y. Create a linear inequality to describe the values that will activate the fail-safe.

   b. Graph the linear inequality.

   c. The device has electrical inputs of 95 kilowatts and 145 kilowatts. Plot this point on the graph. Will the fail-safe activate and switch off the device? Explain why.

43. Janessa has been trying to live a heathier life, so she bought a wrist fitness tracker. At the end of the first week, she connected it to the computer to download the data. The line, given by $m = \frac{3}{10}d + 2$, best represents her first week's data, where m represents miles walked in a day and d represents the day number (for example, on day 1, $d = 1$). The given graph is meant to represent the best fit line for the data taken once a day at the same time each day.

   a. Fill out the following table using $m = \frac{3}{10}d + 2$, where the input is the day number.

Domain, *d*	**1**	**2**	**3**	**4**	**5**	**6**	**7**
Range, *m*							

   b. Assuming the best fit line continues, how many miles would you expect Janessa to walk by the end of the second week (day 14)? Round your answer to the nearest tenth.

Writing & Thinking

44. Explain in your own words how to test to determine which side of the graph of an inequality should be shaded.

45. Describe the difference between a closed and an open half-plane.

27. 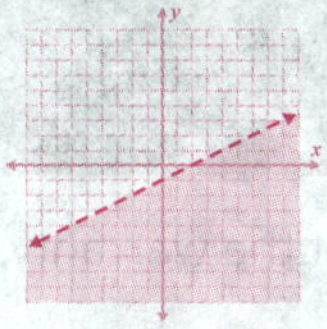

28. 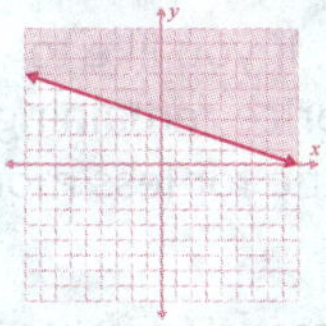

29. 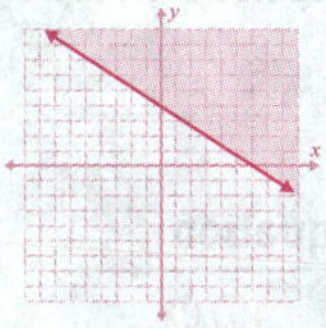

30. 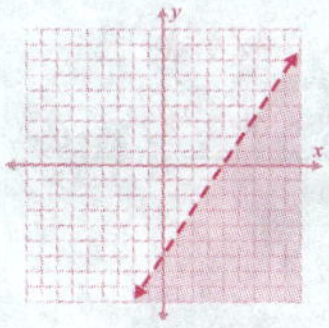

31. 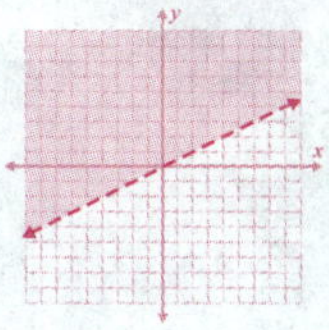

32. 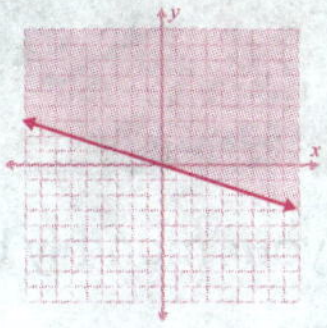

33. 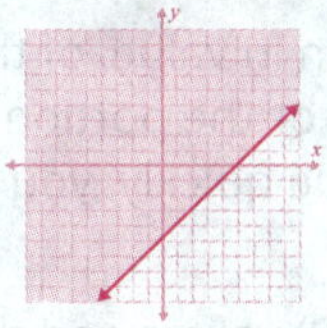

34. 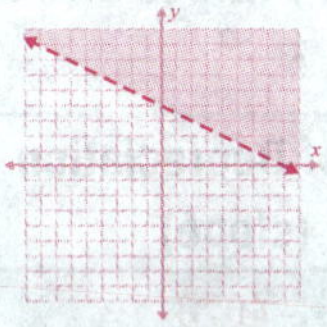

35. 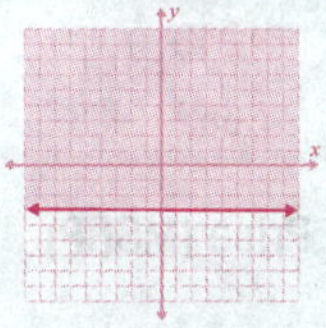

36. 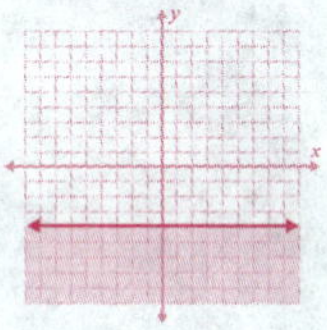

37. 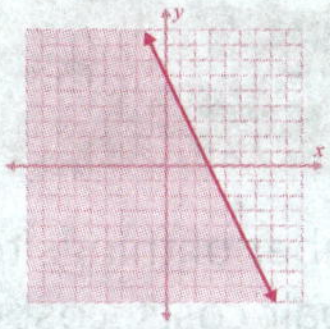

38. 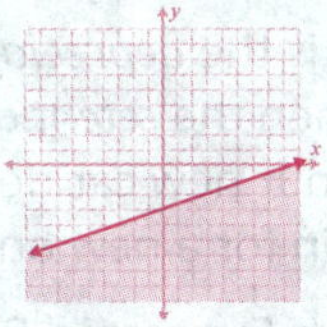

39. 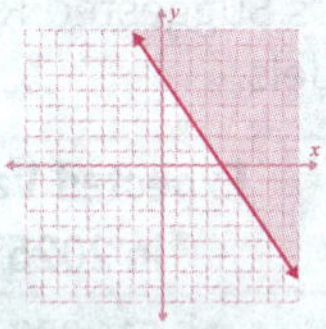

40. 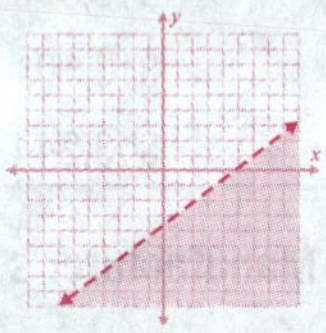

Chapter 10 Project: Demand and It Shall Be Supplied

1. a. As price increases, consumers are less likely to purchase an item.

b. We have $m = -10$ T-shirts per dollar. This means that an increase of one dollar in price causes the company to sell 10 fewer shirts.

c. We have $b = 500$. This means that at a price of zero dollars, the market would absorb 500 T-shirts. That is, this is the maximum number of shirts the company has the material to make without making any profit from sales.

d. $d = 500 - 10p$

2. a. As price increases, the manufacturer has more money to produce more T-shirts.

b. We have $m = 5$ T-shirts per dollar. This means that an increase of one dollar in price allows the company to produce an extra 5 T-shirts.

c. We have $b = 175$. This means that at a price of zero dollars, the company can produce 175 T-shirts.

d. $s = 175 + 5p$

3. a. 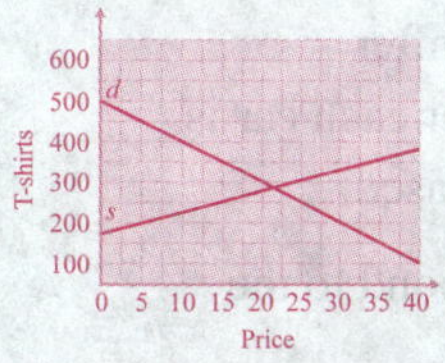

b. $21.67. If the T-shirts are priced at $21.67, the number of T-shirts manufactured will match the demand, which is approximately 283 T-Shirts.

4. Answers may vary. A common example is gasoline. The demand is inelastic because there is no substitute for it and consumers are forced to purchase gas to drive to work and other activities.

Applications

41. a. $x + y \geq 75$

b. 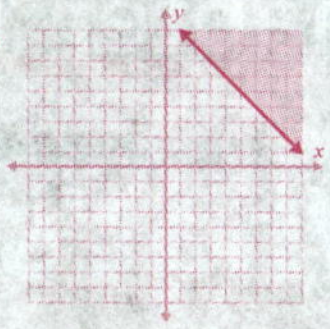

c. No; $67 < 75$

d. Answers will vary; Any points with x or y < 0 or > 50.

42. a. $x + y > 250$

b. 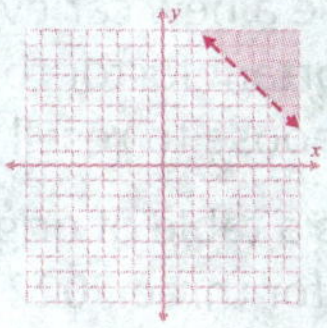

c. No; $145 + 95 < 250$

43. a.

Domain, d	Range, m
1	$\frac{23}{10}$
2	$\frac{13}{5}$
3	$\frac{29}{10}$
4	$\frac{16}{5}$
5	$\frac{7}{2}$
6	$\frac{19}{5}$
7	$\frac{41}{10}$

b. 6.2

Writing & Thinking

44. Test any point not on the line. If the test point satisfies the inequality, shade the half-plane on that side of the line. Otherwise, shade the other half-plane.

45. A closed half-plane includes points on the line and is symbolized by a solid boundary line. An open half-plane does not include points on the line and is symbolized by a dashed boundary line.

Chapter 10 Project

What's Your Car Worth?

An activity to demonstrate the use of linear models in real life.

When buying a new car, there are a number of things to keep in mind: your monthly budget, length of the warranty, routine maintenance costs, potential repair costs, cost of insurance, etc.

One thing you may not have considered is the *depreciation*, or reduction in value, of the car over time. If you like to purchase a new car every 3 to 5 years, then the *retention value* of a car, or the portion of the original price remaining, is an important factor to keep in mind. If your new car depreciates in value quickly, you may have to settle for less money if you choose to resell it later or trade it in for a new one.

Below is a table of original Manufacturer's Suggested Retail Price (MSRP) values and the anticipated retention value after 3 years for three 2022 mid-price car models.

Car Model	2022 MSRP	Expected Value in 2025	Rate of Depreciation (slope)	Linear Equation
Mini Cooper	\$28,600	\$16,390	−4070	$y = 28{,}600 - 4070x$
Toyota Camry	\$25,845	\$13,101	−4248	$y = 25{,}845 - 4248x$
Ford Taurus	\$30,230	\$13,244	−5662	$y = 30{,}230 - 5662x$

Manufacturer's Suggested Retail Price (MSRP)

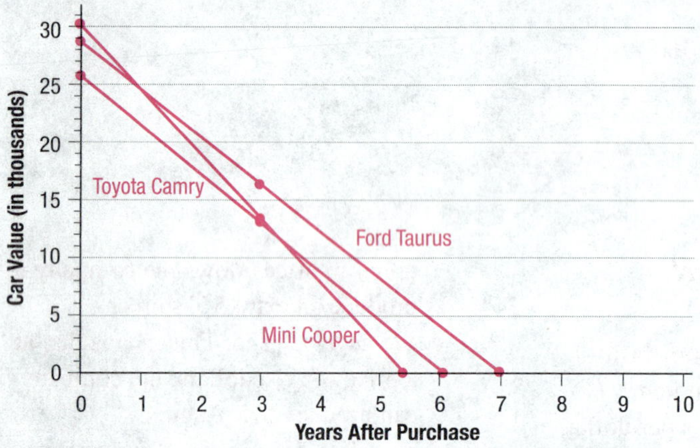

1. The x-axis of the graph is labeled "Years after Purchase." Recall that the MSRP value for each car is for the year 2022 when the car was purchased.
   a. What value on the x-axis will correspond to the year 2022? 0
   b. Using the value from part a. as the x-coordinate and the MSRP values in column two as the y-coordinates, plot three points on the graph corresponding to the value of the three cars at time of purchase. See graph.
   c. What value on the x-axis will correspond to the year 2025? 3
   d. Using the value from part c. as the x-coordinate and the expected car values in column three as the y-coordinates, plot three points on the graph corresponding to the value of the three cars in 2025. See graph.
2. Draw a line segment on the graph connecting the pair of points for each car model. Label each line segment after the car model it represents and label each point with a coordinate pair, (x, y). Consider using a different color when plotting each line segment to help you identify the three models. See graph.

3. Use the slope formula, $m = \frac{y_2 - y_1}{x_2 - x_1}$, to answer the following questions.
   a. Calculate the rate of depreciation for each model by calculating the slope (or rate of change) between each pair of corresponding points using the slope formula and enter it into the appropriate row of column four of the table.
   b. Are the slopes calculated above positive or negative? Explain why.
   c. Interpret the meaning of the slope for the Toyota Camry making sure to include the units for the variables.
   d. Which car model depreciates in value the fastest? Explain how you determined this.
4. Use the slope-intercept form of an equation, $y = mx + b$, for the following problems.
   a. Write an equation to model the depreciation in value over time of each car (in years). Place these in column five of the table.
   b. What does the y-intercept represent for each car?
5. Use the equations from Problem 4 for the following problems.
   a. Predict the value of the Mini Cooper 4 years after purchase. $12,320
   b. Predict the value of the Ford Taurus $2\frac{1}{2}$ years after purchase. $16,075
6. Determine from the graph how long it takes from the time of purchase until the Ford Taurus and the Toyota Camry have the same value. (It may be difficult to read the coordinates for the point of intersection, but you can get a rough idea of the value from the graph. You can find the exact point of intersection by setting the two equations equal to one another and solving for x.)
   a. After how many years are the car values for the Ford Taurus and the Toyota Camry the same? Round to the nearest tenth. 3.1 years
   b. What is the approximate value of both cars at this point in time? Round to the nearest 100 dollars. $12,700
7. How long will it take for the Toyota Camry to fully depreciate (reach a value of zero)?
   a. For the first method, extend the line segment between the two points plotted for the Toyota Camry until it intersects the horizontal axis. The x-intercept is the time at which the value of the car is zero.
   b. Substitute 0 for y in the equation you developed for the Toyota Camry and solve for x. Round to the nearest year.
   c. Compare the results from parts a. and b. Are the results similar? Why or why not?
8. How long will it take for the Ford Taurus to fully depreciate? (Repeat Problem 7 for the Ford Taurus.) Round to the nearest year. 5 years
9. Why is there such a difference in depreciation for the Camry and the Taurus? Do some research on a reliable Internet site and list two reasons why cars depreciate at different rates.
10. Based on what you have learned from this activity, do you think retention value will be a significant factor when you purchase your next car? Why or why not? Answers will vary.

Chapter 10 Project: What's Your Car Worth?

3. **a.** See column four of table.
   b. Slopes are negative because car value is declining over time.
   c. The Toyota Camry is depreciating at a rate of $4248 per year.
   d. The Ford Taurus depreciates the fastest because it has the largest negative slope.
4. **a.** See column five of table.
   b. The original purchase price of MSRP in 2022.
7. **a.** Answers will vary.
   b. 6 years
   c. Answers should be similar. It depends on the accuracy of the graph drawn.
9. Answers will vary. Possibilities include: better quality, customer loyalty, requires less maintenance and repair.

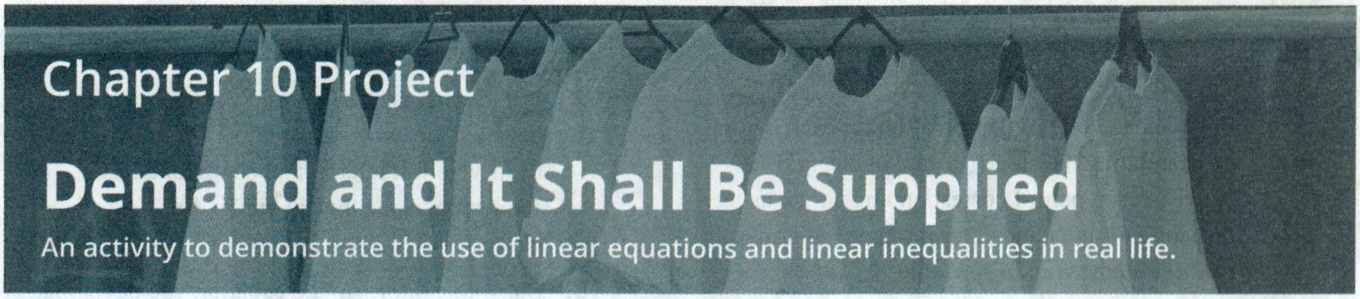

Chapter 10 Project

Demand and It Shall Be Supplied

An activity to demonstrate the use of linear equations and linear inequalities in real life.

In economics, the demand for a product is the number of units of the product that the market is willing to absorb at a certain price. That is, the demand is the number of units of the product that sells at any given time. The most basic model for the demand d as a function of the price p is given by a linear equation

$$d = b + mp,$$

where m and b are real numbers.

1. The owner of a T-shirt company believes that the demand d (in units) for their Stranger Things T-shirt follows the linear demand model from the introduction, where p is the price of one T-shirt and m and b are real numbers. The company owner knows that they can sell 300 Stranger Things T-shirts for \$20 each but only 250 T-shirts at \$25 each.
   - **a.** Explain why it is reasonable to believe that the demand of T-shirts decreases as the price increases.
   - **b.** Compute the value of the real number m. How would you interpret the value you have found?
   - **c.** Compute the value of the real number b. How would you interpret the value you have found?
   - **d.** Write the demand equation using your answers from parts b. and c.

The supply for a product is the number of units of a product that the manufacturer can make available at a certain price. The most basic model for the supply s as a function of the price p is given by a linear equation

$$s = b + mp,$$

where m and b are real numbers.

2. The T-shirt company from Problem 1 can produce 275 shirts when the price is \$20. If the price is raised to \$25, they can produce 300 shirts.
   - **a.** Explain why it is reasonable to believe that the supply of T-shirts increases as the price increases.
   - **b.** Compute the value of the real number m. How would you interpret the value you have found?
   - **c.** Compute the value of the real number b. How would you interpret the value you have found?
   - **d.** Write the supply equation using your answers from parts b. and c.
3. The equilibrium price is the price for which the demand is equal to the supply. At this price, the number of units produced is exactly the number of units absorbed by the market.
   - **a.** Graph the demand and supply equations for the T-shirt company on the same coordinate plane. Recall that the x-axis should represent price.
   - **b.** Find the equilibrium price rounded to the nearest cent and explain what it means in words.
4. Some products have a demand that is not sensitive to changes in price. That is, a large variation in price will not produce a corresponding large variation in demand. This phenomenon is defined as inelastic demand. Perform an internet search to find an example of a product with an inelastic demand. Explain why the demand for the product is inelastic.

CHAPTER 11

11 Systems of Linear Equations

Connections

Sometimes a single equation does not provide enough information to find a solution. When this occurs, we must construct more than one equation and use more than one variable, then use these equations to find an appropriate solution, if one exists. This is referred to as solving a system of linear equations.

Chinese mathematicians recorded methods for solving systems of two equations sometime between the 10th and 2nd century BC, as described in the book *The Nine Chapters on the Mathematical Art*. However, it wasn't until the 18th and 19th centuries that methods for solving systems of equations were introduced to the Western world by the European mathematicians Gabriel Cramer, Carl Frederick Gauss, and Wilhelm Jordan. Thanks in large part to their work, algorithms have been derived for calculating solutions to systems with large numbers of equations and variables. With the invention of computers, the time it takes to solve these systems has been drastically reduced, and solving equations now plays a prominent role in the fields of engineering, physics, chemistry, computer science, and economics.

Suppose a farmer is building a fenced-in pasture for his cows and has 180 meters of fencing to use. The farmer wants the pasture to be in the shape of a rectangle where the length is two times the width. The farmer can use this information, along with the fact that the perimeter of a rectangle is $P = 2l + 2w$, to determine the dimensions of the pasture. What are the dimensions of the desired pasture?

Objectives

A. Determine whether a given ordered pair is a solution to a specified system of linear equations.

B. Solve systems of linear equations by graphing.

C. Use a graphing calculator to solve systems of linear equations.

Systems of Equations

The study of systems of equations involves multiple equations with multiple variables. A solution of a particular system must satisfy all of the equations at the same time. Around 200 BC, Chinese mathematicians devised a method to solve a system of two equations in two variables. Later, the Swiss mathematician Gabriel Cramer (in 1750) and the German mathematician Wilhelm Jordon (in 1888) expanded the algorithms for systems of equations.

11.1 Systems of Linear Equations: Solutions by Graphing

Many applications involve two (or more) quantities, and by using two (or more) variables, we can form linear equations using the information given. Such a set of equations is called a **system**, and in this chapter we will develop techniques for solving systems of linear equations.

Graphing systems of two equations in two variables is helpful in visualizing the relationships between the equations. However, this approach is somewhat limited in finding solutions since numbers might be quite large or solutions might involve fractions or decimals that must be estimated on the graph. Therefore, other algebraic techniques are necessary to accurately solve systems of linear equations. We will discuss two of these methods in later sections: substitution and addition. These algebraic techniques are also necessary when we try to solve more than two equations with more than two variables. Graphing in three dimensions will be left to later courses.

A Checking Solutions to Systems of Equations

Many applications, as we will see in Sections 11.4 and 11.5, involve solving pairs of linear equations. Such pairs are said to form a **system of linear equations** or **a set of simultaneous equations**. For example,

$$\begin{cases} x-2y=0 \\ 3x+y=7 \end{cases} \quad \text{and} \quad \begin{cases} y=-x+4 \\ y=2x+1 \end{cases}$$

are two systems of linear equations.

Each individual equation has an infinite number of solutions. That is, there are an infinite number of ordered pairs that satisfy each equation. However, we are interested in finding ordered pairs that satisfy **both** equations.

The **solution of a system** of linear equations is the set of ordered pairs (or points) that satisfy **both** equations. To determine whether a particular ordered pair is a solution to a system, substitute the values for x and y into **both** equations. If the results for both equations are true statements, then the ordered pair is a solution to the system.

1. Show that $(6, 4)$ is a solution to the system
$$\begin{cases} x-3y=-6 \\ 3x+2y=26 \end{cases}$$

Example 1 Checking Solutions to Systems (Solution)

Show that $(2,1)$ is a solution to the system $\begin{cases} x-2y=0 \\ 3x+y=7 \end{cases}$

Solution

Substitute $x = 2$ and $y = 1$ into **both** equations.

In the first equation:

$$(2)-2(1) \stackrel{?}{=} 0$$
$$2-2=0 \quad \text{True statement}$$

In the second equation:

$$3(2)+(1) \stackrel{?}{=} 7$$
$$6+1=7 \quad \text{True statement}$$

Because $(2, 1)$ satisfies both equations, $(2, 1)$ **is a solution to the system**.

Now work margin exercise 1.

Example 2 Checking Solutions to Systems (Not a Solution)

Show that $(0, 4)$ is not a solution to the system $\begin{cases} y = -x + 4 \\ y = 2x + 1 \end{cases}$

Solution

Substitute $x = 0$ and $y = 4$ into **both** equations.

In the first equation:

$$(4) \stackrel{?}{=} -(0) + 4$$
$$4 = 0 + 4 \quad \text{True statement}$$

In the second equation:

$$(4) \stackrel{?}{=} 2(0) + 1$$
$$4 = 0 + 1 \quad \text{False statement}$$

Because $(0, 4)$ does not satisfy **both** equations, $(0, 4)$ **is *not* a solution to the system**.

Now work margin exercise 2.

2. Show that $(3, 2)$ is not a solution to the system

$$\begin{cases} y = -2x + 8 \\ y = 3x - 8 \end{cases}$$

Teaching Note

Consider giving some practical sounding applications in which scientists, business analysts, or engineers might want two conditions to depend on two variables and a simultaneous solution is needed. For example, an architect might need a rectangular window that has a perimeter of 140 inches with a length that is 4 more inches than the width. (While this situation may be set up as a single equation, it is also useful for the illustration of a system.)

B Solving Systems of Linear Equations by Graphing

Solving a System of Linear Equations by Graphing

1. Graph both linear equations on the same set of axes.
2. Observe the point of intersection (if there is one).
   a. If the slopes of the two lines are different, then the lines intersect at one and only one point. The system has a single point as its solution.
   b. If the lines have the same slope and different y-intercepts, then the lines are parallel. The system has no solution.
   c. If the lines are the same line, then all the points on the line form the solution. There are an infinite number of solutions.
3. Check the solution (if there is one) in both of the original equations.

PROCEDURE

Consistent and Inconsistent Systems of Linear Equations

1. A system is **consistent** if it has one or more solutions.
2. A system is **inconsistent** if it has no solutions.

DEFINITION

3. Solve the system of equations by graphing.

$$\begin{cases} y = 2x + 6 \\ y = -x - 3 \end{cases}$$

Note

Before working the examples in this section, you might want to review graphing lines in Chapter 10. Recall that lines can be graphed by plotting intercepts, finding a point on the line and using the slope to find another point, or finding two points.

Example 3 Solving Systems (One Solution/A Consistent System)

Solve the system of equations by graphing.

$$\begin{cases} y = -x + 4 \\ y = 2x + 1 \end{cases}$$

Solution

In Example 2, we showed that $(0,4)$ is not a solution to the given system.

However, by graphing both equations, we see that the two lines appear to **intersect** at the point $(1,3)$.

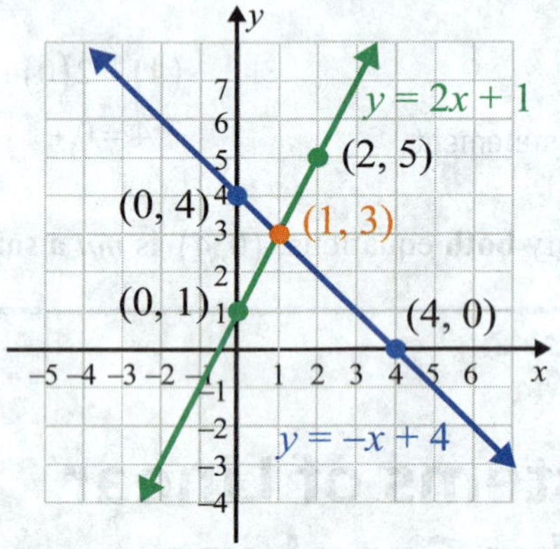

We can check that $(1,3)$ satisfies both equations by substituting $x = 1$ and $y = 3$ into both equations.

$$\begin{aligned} y &= 2x + 1 \\ (3) &\stackrel{?}{=} 2(1) + 1 \\ 3 &= 3 \end{aligned} \qquad \begin{aligned} y &= -x + 4 \\ (3) &\stackrel{?}{=} -(1) + 4 \\ 3 &= 3 \end{aligned}$$

Since both substitutions give true statements, the point $(1,3)$ **is the solution to the system** and the system is **consistent**.

Now work margin exercise 3.

4. Solve the system of equations by graphing.

$$\begin{cases} -x + 2 = 2y \\ x + 2y = 8 \end{cases}$$

Example 4 Solving Systems (No Solution/An Inconsistent System)

Solve the system of equations by graphing.

$$\begin{cases} y = -x + 4 \\ x + y = 2 \end{cases}$$

Solution

The graphs of the lines of the equations in the system are shown.

The lines are **parallel**. This can be shown by writing the equations in slope-intercept form and noting that they have the same slope and different y-intercepts.

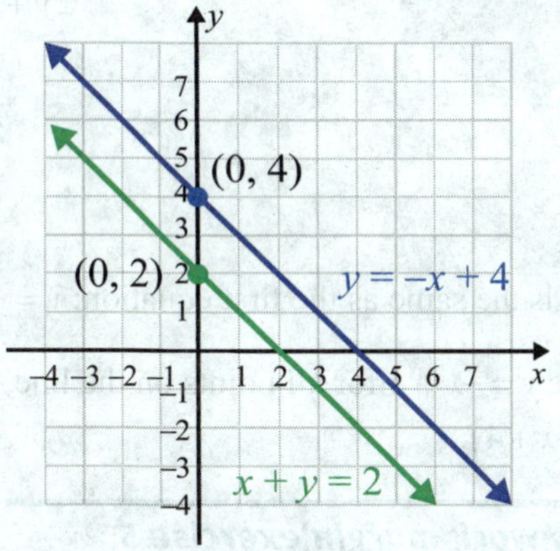

First equation: $y = -x + 4$ with $m = -1$ and y-intercept 4

Second equation: $y = -x + 2$ with $m = -1$ and y-intercept 2

Thus the lines are **parallel** and do not intersect, and there is **no solution** to the system. The system is **inconsistent**.

Now work margin exercise 4.

Dependent and Independent Systems of Linear Equations

If the graphs of two linear equations are

a. the same line, then the equations are **dependent**.

b. different lines, then the equations are **independent**.

DEFINITION

Example 5 Solving Systems (Infinite Solutions/A Dependent System)

Solve the system of equations by graphing.

$$\begin{cases} y = -x + 4 \\ 2x + 2y = 8 \end{cases}$$

Solution

The graphs of both equations are as follows.

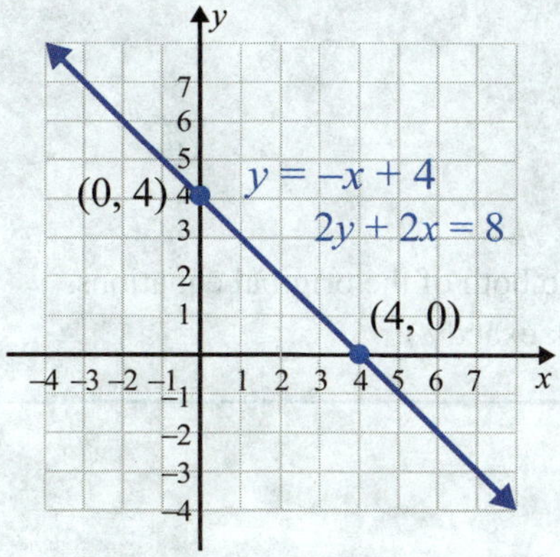

There is just one line because both equations represent the same line. That is, they **coincide**. Any point that satisfies one of the equations will also satisfy the other. The system has an **infinite number of solutions** and the equations are **dependent**. Putting both equations in the slope-intercept form shows that they are identical.

5. Solve the system of equations by graphing.

$$\begin{cases} 3x = y + 4 \\ 6x - 8 = 2y \end{cases}$$

Solving $2y + 2x = 8$ for y gives the following.

$$\begin{aligned} 2y+2x&=8 \\ 2y&=-2x+8 \\ \frac{2y}{2}&=\frac{-2x}{2}+\frac{8}{2} \\ y&=-x+4 \end{aligned}$$

Teaching Note

To help your students understand the notation for the solutions in a dependent system, you might have them substitute several numbers for y and for x.

This is the same as the first equation: $y = -x + 4$.

Since $y = -x + 4$ for all points on the line, we can write the **set of all solutions in the form** $(x, -x+4)$.

Now work margin exercise 5.

<table>
<tr><th colspan="3">System</th></tr>
<tr><td>$\begin{cases} y=-x+4 \\ y=2x+1 \end{cases}$</td><td>$\begin{cases} y=-x+4 \\ x+y=2 \end{cases}$</td><td>$\begin{cases} y=-x+4 \\ 2x+2y=8 \end{cases}$</td></tr>
<tr><th colspan="3">Graph</th></tr>
<tr><td>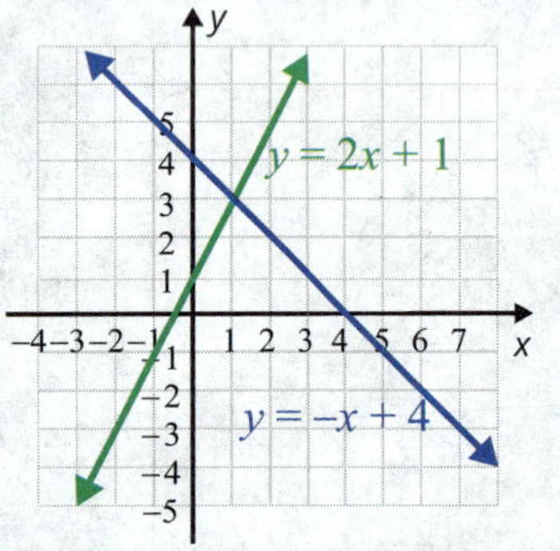
</td><td>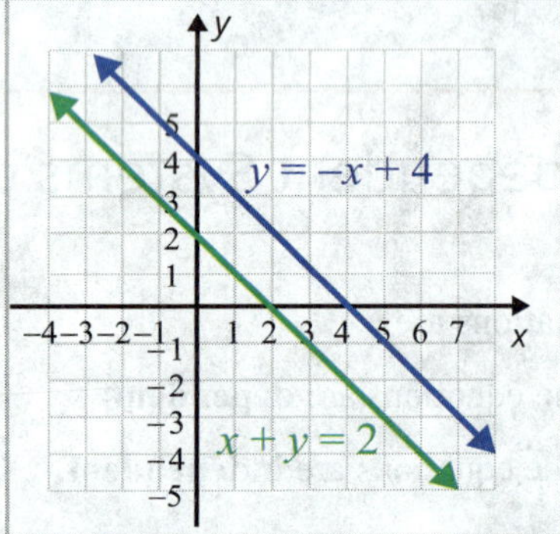
</td><td>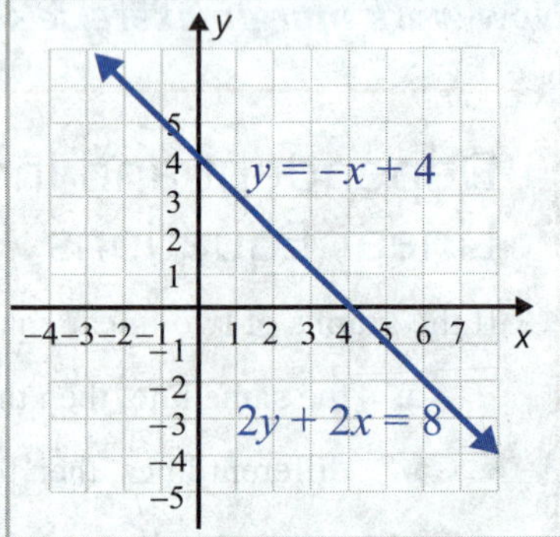
</td></tr>
<tr><th colspan="3">Classification</th></tr>
<tr><td>The system $\begin{cases} y=-x+4 \\ y=2x+1 \end{cases}$ is consistent and the equations are independent.</td><td>The system $\begin{cases} y=-x+4 \\ x+y=2 \end{cases}$ is inconsistent and the equations are independent.</td><td>The system $\begin{cases} y=-x+4 \\ 2x+2y=8 \end{cases}$ is consistent and the equations are dependent.</td></tr>
</table>

Table 1

Attention!

When using the graphing method, be sure to

1. graph the lines as accurately as you can, and
2. check your solution by substituting it back into both of the original equations. (Of course, fractional estimates may not check exactly.)

Example 6 illustrates estimating the solution.

Example 6 Solving a System that Requires Estimation

Solve the system of equations by graphing.

$$\begin{cases} x - 3y = 4 \\ 2x + y = 3 \end{cases}$$

Solution

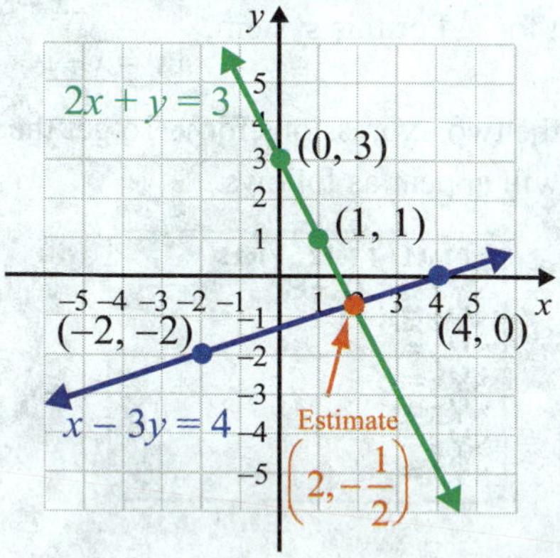

The two lines intersect at one point. However, we can only estimate the point of intersection as $\left(2, -\frac{1}{2}\right)$. In this situation be aware that, although graphing gives a good "estimate," finding exact solutions to the system is not likely.

Check

Substitute $x = 2$ and $y = -\frac{1}{2}$.

$$(2) - 3\left(-\frac{1}{2}\right) \stackrel{?}{=} 4 \qquad \text{and} \qquad 2(2) + \left(-\frac{1}{2}\right) \stackrel{?}{=} 3$$

$$\frac{7}{2} \neq 4 \qquad\qquad \frac{7}{2} \neq 3$$

Thus, checking shows that the estimated solution $\left(2, -\frac{1}{2}\right)$ does not satisfy either equation. The estimated point of intersection is just that, an estimate. The following sections will provide algebraic techniques for solving systems of equations that would give the exact solution as $\left(\frac{13}{7}, -\frac{5}{7}\right)$.

Now work margin exercise 6.

6. Solve the system of equations by graphing.

$$\begin{cases} y = 4x - 3 \\ y - 2x = 0 \end{cases}$$

C Using a Graphing Calculator to Solve Systems of Linear Equations

A graphing calculator can be used to locate (or estimate) the point of intersection of two lines (and therefore the solution to the system).

7. Use the graphing calculator's intersect function to solve the system of equations.

$$\begin{cases} x+y=-5 \\ x-3y=-1 \end{cases}$$

Example 7 Using a Graphing Calculator To Solve a System

Use a graphing calculator's intersect function to solve the system of equations.

$$\begin{cases} 2x+y=8 \\ x-y=1 \end{cases}$$

Solution

Step 1: Solve each equation for y. For this system, $\begin{cases} y=-2x+8 \\ y=x-1 \end{cases}$

Step 2: Press [Y=] and enter the two expressions for y. To get the variable X, press [X,T,Θ,n]. The display screen will appear as follows.

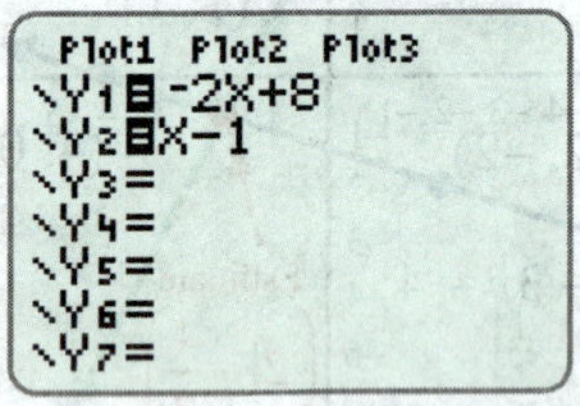

Step 3: Press [GRAPH]. (Both lines and their intersection should appear. If not, you may need to adjust the window.)

Step 4: Press [2nd] and CALC. Select intersect. The cursor will appear on one of the lines. Use the right- or left-arrow keys to get near the point of intersection and press [ENTER]. Then move the up- or down-arrow keys to get to the other line. Now, use the right- or left-arrow keys to move closer to the point of intersection on this line and press [ENTER]. Follow the directions for Guess? by moving the cursor to the point of intersection and pressing [ENTER].

Step 5: The answer $x = 3$ and $y = 2$ will appear at the bottom of the display screen.

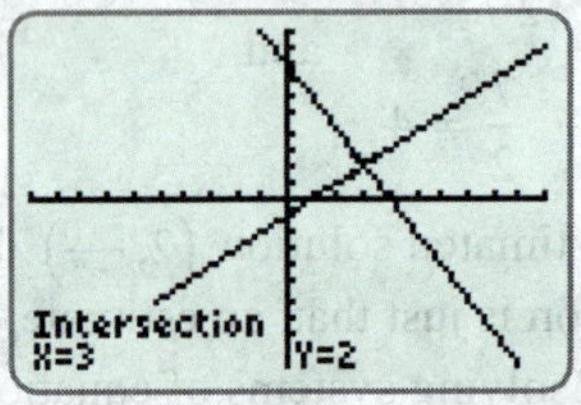

Note: Step 4 may seem somewhat complicated, but give it a try. It is accurate and can be fun! (If the lines are parallel (an inconsistent system) the calculator will give an error message.)

Now work margin exercise 7.

Teaching Note

You may want to show your students how "TRACE" works in locating the intersection. This is basically a one-step operation, but the calculator will generally give decimal approximations only.

Margin Exercise Answers

1. $6-12=-6$ True statement; $18+8=26$ True statement **2.** $2=-6+8$ True statement; $2=9-8$ False statement **3.**

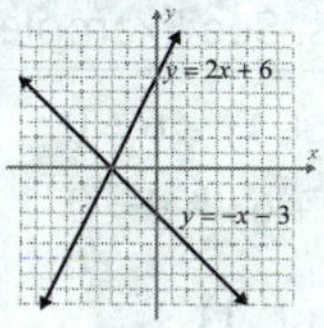

$(-3,0)$ **4.**

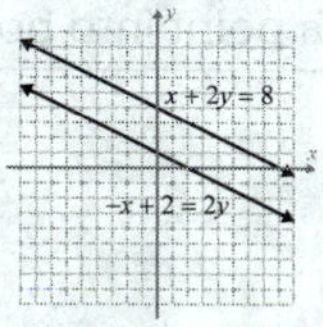

No solution

5.

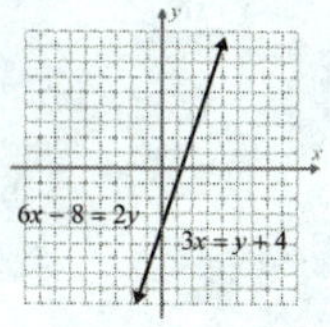

$(x, 3x-4)$ **6.**

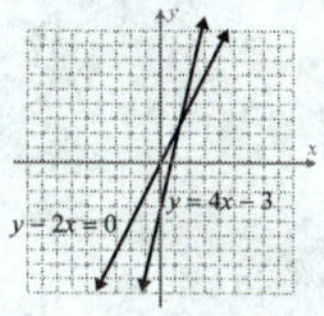

$\left(\frac{3}{2}, 3\right)$ **7.**

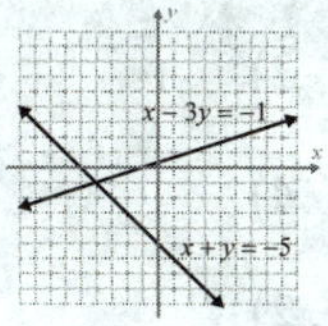

$(-4,-1)$

11.1 Exercises

Concept Check

Fill-in-the-Blank. Complete each sentence using information found in this section.

1. A system of equations is also called a set of ______ equations.
2. When solving a system of two linear equations, the goal is to find ordered pairs that satisfy _____ equations.
3. To solve a system of linear equations by graphing, you should graph both equations on the ______ set of axes.
4. A dependent system has _________ solution(s).
5. A consistent system has exactly ___ solution(s).
6. An inconsistent system has ___ solution(s).

True/False. Determine whether each statement is true or false. If a statement is false, explain how it can be changed so the statement will be true. (**Note:** There may be more than one acceptable change.)

7. To check a solution, substitute it into one of the equations. If the solution satisfies one equation it will satisfy all of the equations.
8. A system of equations with graphs that are parallel lines has exactly one solution.
9. A system of equations with graphs that intersect at one point has exactly one solution.
10. A system of equations with graphs that are the same line has infinitely many solutions.

11.1 Exercises

Concept Check

1. simultaneous
2. both
3. same
4. infinite
5. one
6. no
7. False; The solution must be checked in all equations.
8. False; Graphs that are parallel indicate no solutions.
9. True
10. True

Practice

1. c
2. None
3. a, c
4. b
5. $(0,2)$
6. $(-1,1)$
7. $(4,2)$
8. $(1,-3)$
9. $m_1=-2, b_1=3,$ $m_2=-2, b_2=\frac{5}{2}$
10. $m_1=\frac{3}{5}, b_1=-\frac{1}{5},$ $m_2=\frac{3}{5}, b_2=-\frac{2}{5}$
11. $m_1=\frac{1}{2}, b_1=3,$ $m_2=\frac{1}{2}, b_2=-\frac{1}{2}$
12. $m_1=3, b_1=-8,$ $m_2=3, b_2=-6$
13.

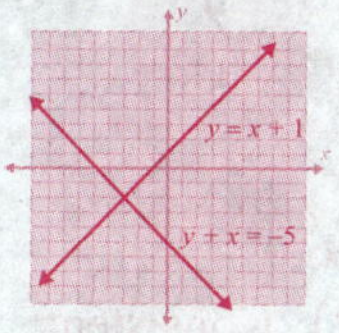

$(-3,-2)$; Consistent; Independent

Practice

Determine which of the given points, if any, lie on both of the lines in the systems of equations by substituting each point into both equations. See Examples 1 and 2.

1. $\begin{cases} x - y = 6 \\ 2x + y = 0 \end{cases}$

a. $(1, -2)$

b. $(4, -2)$

c. $(2, -4)$

d. $(-1, 2)$

2. $\begin{cases} x + 3y = 5 \\ 3y = 4 - x \end{cases}$

a. $(2, 1)$

b. $(2, -2)$

c. $(-1, 2)$

d. $(4, 0)$

3. $\begin{cases} 2x + 4y - 6 = 0 \\ 3x + 6y - 9 = 0 \end{cases}$

a. $(1, 1)$

b. $(2, 0)$

c. $\left(0, \frac{3}{2}\right)$

d. $(-1, 3)$

4. $\begin{cases} 5x - 2y - 5 = 0 \\ 5x = -3y \end{cases}$

a. $(1, 0)$

b. $\left(\frac{3}{5}, -1\right)$

c. $(0, 0)$

d. $(1, 4)$

The graphs of the lines represented by each system of equations are given. Determine the solution of the system by looking at the graph. Check your solution by substituting into both equations. See Example 3.

5. $\begin{cases} x + 2y = 4 \\ x - y = -2 \end{cases}$

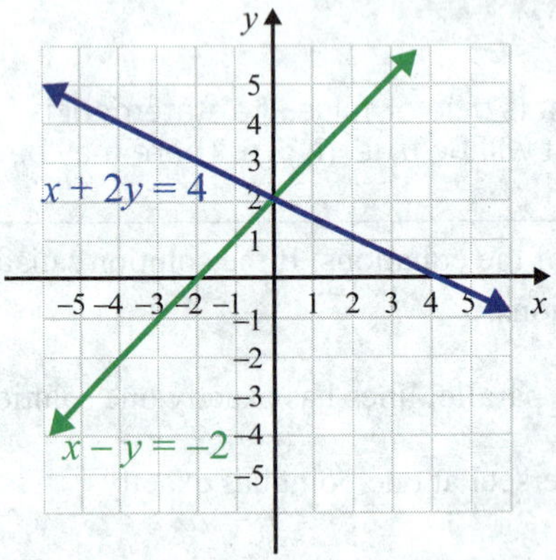

6. $\begin{cases} x + 2y = 1 \\ 2x + y = -1 \end{cases}$

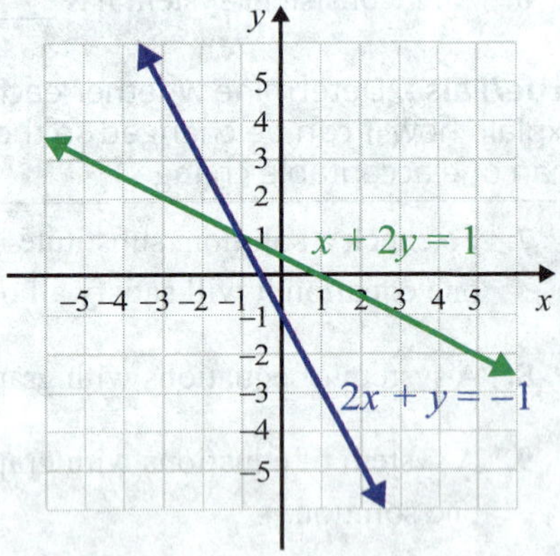

14.

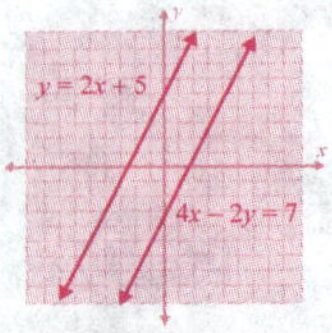

No solution; Inconsistent; Independent

15.

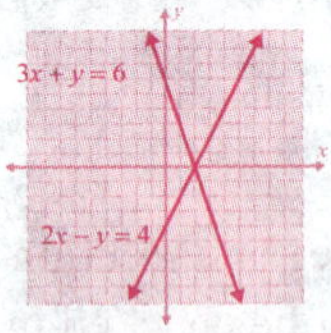

$(2, 0)$; Consistent; Independent

16.

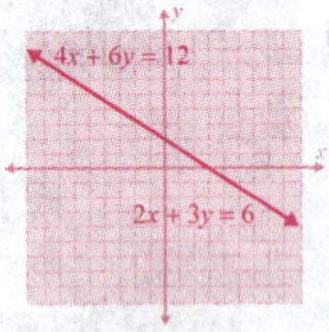

$\left(x, -\frac{2}{3}x + 2\right)$; Consistent; Dependent

17.

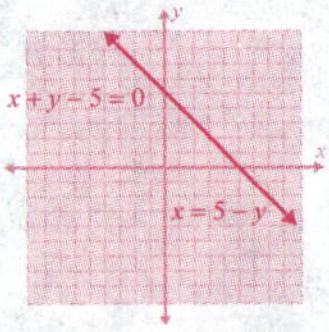

$(x, -x + 5)$; Consistent; Dependent

18.

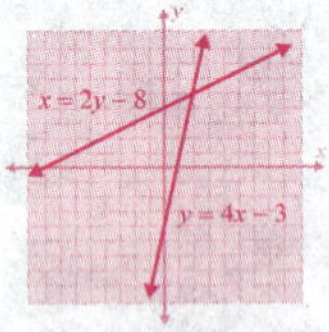

$(2, 5)$; Consistent; Independent

19.

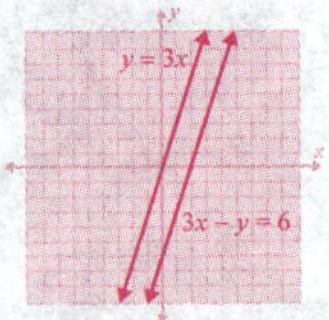

No solution; Inconsistent; Independent

20.

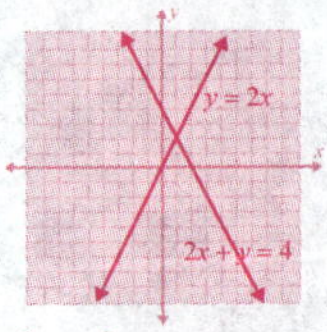

$(1, 2)$; Consistent; Independent

7. $\begin{cases} 2x - y = 6 \\ 3x + y = 14 \end{cases}$

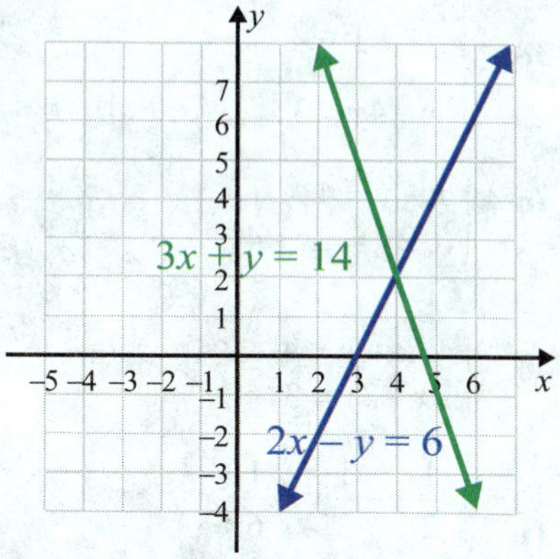

8. $\begin{cases} x - y = 4 \\ 3x - y = 6 \end{cases}$

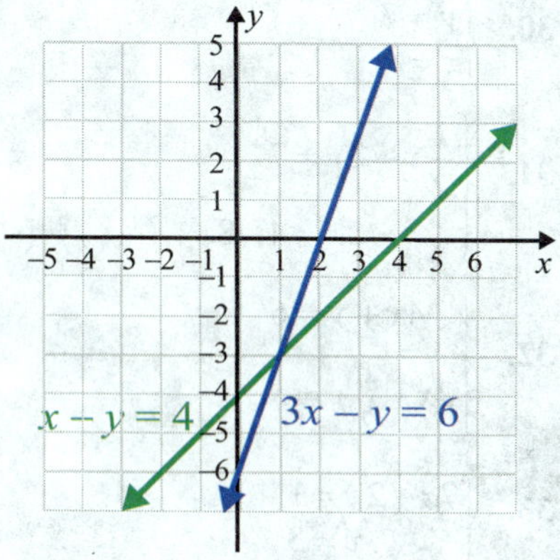

Show that each system of equations is inconsistent by determining the slope of each line and the *y*-intercept. (That is, show that the lines are parallel and do not intersect.) See Example 4.

9. $\begin{cases} 2x + y = 3 \\ 4x + 2y = 5 \end{cases}$

10. $\begin{cases} 3x - 5y = 1 \\ 6x - 10y = 4 \end{cases}$

11. $\begin{cases} y = \dfrac{1}{2}x + 3 \\ x - 2y = 1 \end{cases}$

12. $\begin{cases} 3x - y = 8 \\ x - \dfrac{1}{3}y = 2 \end{cases}$

Solve each system of equations by graphing. Then, state whether the system is consistent or inconsistent and whether the equations are dependent or independent. See Examples 3 through 6.

13. $\begin{cases} y = x + 1 \\ y + x = -5 \end{cases}$

14. $\begin{cases} y = 2x + 5 \\ 4x - 2y = 7 \end{cases}$

15. $\begin{cases} 2x - y = 4 \\ 3x + y = 6 \end{cases}$

16. $\begin{cases} 2x + 3y = 6 \\ 4x + 6y = 12 \end{cases}$

17. $\begin{cases} x + y - 5 = 0 \\ x = 5 - y \end{cases}$

18. $\begin{cases} y = 4x - 3 \\ x = 2y - 8 \end{cases}$

19. $\begin{cases} 3x - y = 6 \\ y = 3x \end{cases}$

20. $\begin{cases} y = 2x \\ 2x + y = 4 \end{cases}$

Solve each system of equations by graphing. See Examples 3 through 6.

21. $\begin{cases} x - y = 5 \\ x = -3 \end{cases}$

22. $\begin{cases} x - 2y = 4 \\ x = 4 \end{cases}$

23. $\begin{cases} x + 2y = 8 \\ 3x - 2y = 0 \end{cases}$

24. $\begin{cases} x + y = 8 \\ 5y = 2x + 5 \end{cases}$

25. $\begin{cases} 5x - 4y = 5 \\ 8y = 10x - 10 \end{cases}$

26. $\begin{cases} 2x + y = 0 \\ 4x + 2y = -8 \end{cases}$

27. $\begin{cases} 4x + 3y + 7 = 0 \\ 5x - 2y + 3 = 0 \end{cases}$

28. $\begin{cases} 4x - 2y = 10 \\ -6x + 3y = -15 \end{cases}$

21. $(-3, -8)$

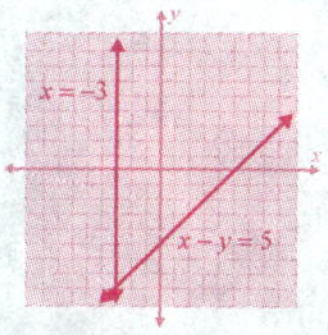

22. $(4, 0)$

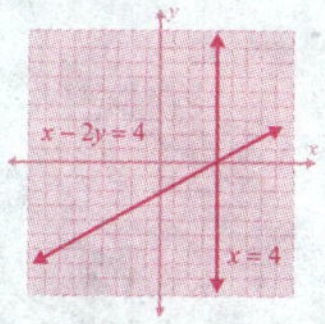

23. $(2, 3)$

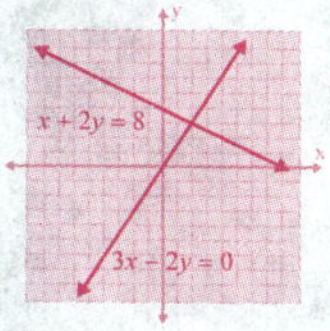

24. $(5, 3)$

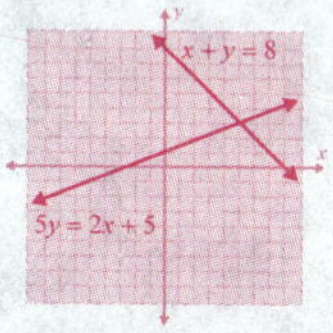

25. $\left(x, \dfrac{5}{4}x - \dfrac{5}{4}\right)$

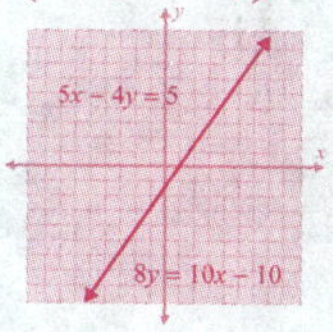

26. No solution

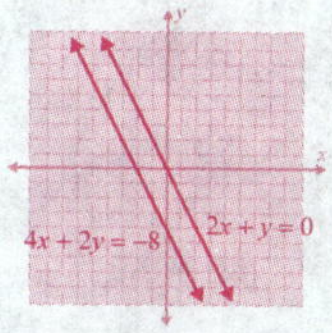

27. $(-1, -1)$

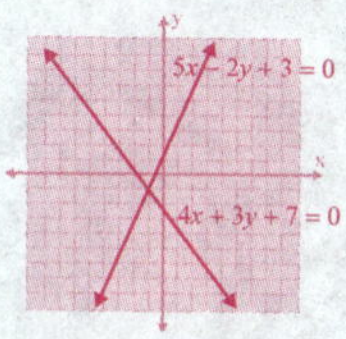

28. $(x, 2x - 5)$

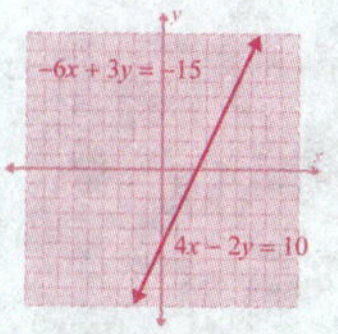

29. $(5,-1)$

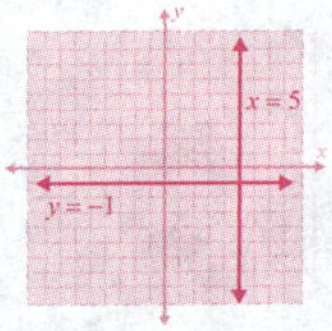

30. $(8,7)$

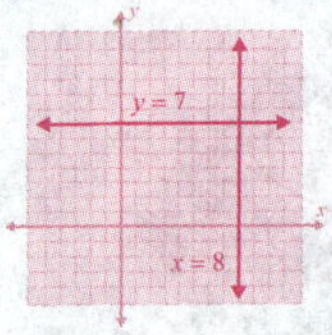

31. No solution

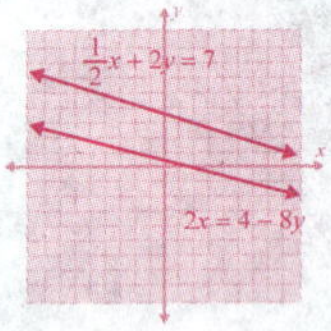

32. $(x,-4x+6)$

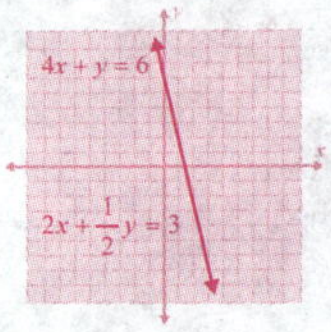

33. $(1,3)$

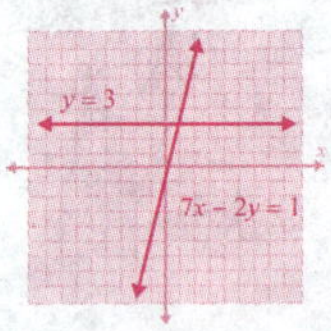

34. $(1.5,-2.5)$

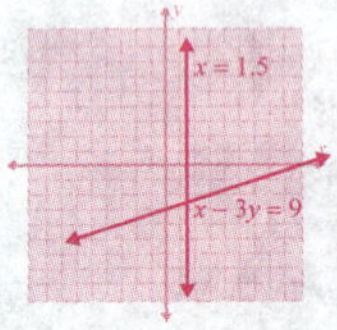

35. $\left(x,\dfrac{1}{2}x+2\right)$

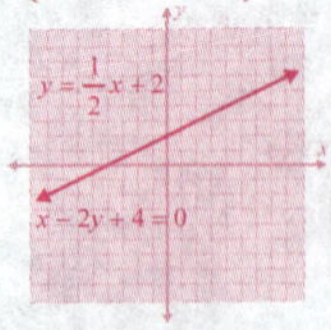

36. No solution

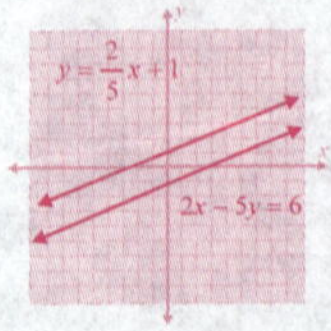

29. $\begin{cases} x=5 \\ y=-1 \end{cases}$

30. $\begin{cases} y=7 \\ x=8 \end{cases}$

31. $\begin{cases} \dfrac{1}{2}x+2y=7 \\ 2x=4-8y \end{cases}$

32. $\begin{cases} 4x+y=6 \\ 2x+\dfrac{1}{2}y=3 \end{cases}$

33. $\begin{cases} 7x-2y=1 \\ y=3 \end{cases}$

34. $\begin{cases} x=1.5 \\ x-3y=9 \end{cases}$

35. $\begin{cases} y=\dfrac{1}{2}x+2 \\ x-2y+4=0 \end{cases}$

36. $\begin{cases} 2x-5y=6 \\ y=\dfrac{2}{5}x+1 \end{cases}$

37. $\begin{cases} 2x+3y=5 \\ 3x-2y=1 \end{cases}$

38. $\begin{cases} \dfrac{2}{3}x+y=2 \\ x-4y=3 \end{cases}$

39. $\begin{cases} x-y=4 \\ 2y=2x-4 \end{cases}$

40. $\begin{cases} x+y=4 \\ 2x-3y=3 \end{cases}$

41. $\begin{cases} \dfrac{1}{2}x+\dfrac{1}{3}y=\dfrac{1}{6} \\ \dfrac{1}{4}x+\dfrac{1}{4}y=0 \end{cases}$

42. $\begin{cases} \dfrac{1}{4}x-y=\dfrac{13}{4} \\ \dfrac{1}{3}x+\dfrac{1}{6}y=-\dfrac{1}{6} \end{cases}$

Use a graphing calculator's intersect function to solve each system of equations. If necessary, round values to the nearest ten-thousandth. (Remember to solve each equation for y and then enter them as Y1 and Y2 in the Y= menu.) See Example 7.

43. $\begin{cases} x+2y=9 \\ x-2y=-7 \end{cases}$

44. $\begin{cases} x-3y=0 \\ 2x+y=7 \end{cases}$

45. $\begin{cases} y=2 \\ 2x-3y=-3 \end{cases}$

46. $\begin{cases} 2x-3y=0 \\ 3x+3y=\dfrac{5}{2} \end{cases}$

47. $\begin{cases} y=-3 \\ 2x+y=0 \end{cases}$

48. $\begin{cases} 2x-3y=1.25 \\ x+2y=5 \end{cases}$

49. $\begin{cases} x+y=3.5 \\ -2x+5y=7.7 \end{cases}$

50. $\begin{cases} 4x+y=-0.5 \\ x+2y=-8 \end{cases}$

Applications

Each of the following applications has been modeled using a system of equations. Solve the system graphically.

51. The sum of two numbers is 25 and their difference is 15. What are the two numbers? Let x be one number and y be the other number.

The corresponding modeling system is $\begin{cases} x+y=25 \\ x-y=15 \end{cases}$

52. The perimeter of a rectangle is 50 m and the length is 5 m longer than the width. Find the dimensions of the rectangle.
Let x be the length and y be the width.

The corresponding modeling system is $\begin{cases} 2x+2y=50 \\ x-y=5 \end{cases}$

53. OSHA recommends that swimming pool owners clean their pool decks with a solvent composed of a 12% chlorine solution and a 3% chlorine solution. Fifteen gallons of the solvent consists of 6% chlorine. How much of each of the mixing solutions were used?
Let x be the number of gallons of the 12% solution and y be the number of gallons of the 3% solution.

The corresponding modeling system is $\begin{cases} x+y=15 \\ 0.12x+0.03y=0.06(15) \end{cases}$

54. A student bought a calculator and a textbook for a course in algebra. He told his friend that the total cost was \$170 (without tax) and that the calculator cost \$20 more than twice the cost of the textbook. What was the cost of each item?
Let x be the cost of the calculator and y be the cost of the textbook.

The corresponding modeling system is $\begin{cases} x+y=170 \\ x=2y+20 \end{cases}$

55. Felix is trying to decide between two job offers. Company A is offering \$15 per hour with a \$500 signing bonus and Company B is offering \$18 per hour with a \$100 signing bonus. Graph the system of linear equations to estimate how many hours he must work to make Company B the best financial choice. Round up to the nearest hour.

The corresponding modeling system is $\begin{cases} y=15x+500 \\ y=18x+100 \end{cases}$

56. Silas is comparing two credit card offers. Credit Card A has no annual fee and offers 1.5% cash back on purchases. Credit Card B has a \$95 annual fee and offers 6% back on purchases. Let x be the amount spent and y be the amount of cash back, minus annual fees.

The corresponding modeling system is $\begin{cases} y=0.015x \\ y=0.06x-95 \end{cases}$

Find the ordered pair that represents the amount Silas needs to spend to earn the same amount of cash back minus his annual fees, and the amount of cash back he earns after spending that amount.

57. Mackenzie plans to join a gym and wants to decide between the two gyms that are close to her home, Fit4Life and Workout Nation. At Fit4Life, she would pay a \$20 per month membership fee and \$10 per class. At Workout Nation, she would pay a \$45 per month membership fee and \$5 per class. Mackenzie wants to determine how many classes she would have to take per month for both gyms to have the same cost.

a. Write two equations to represent the situation. Use the variable y to represent the total cost per month and the variable x to represent the number of classes.

b. Graph the two equations on the same coordinate plane.

c. Find the point of intersection.

d. What does the point of intersection mean? Write a complete sentence.

e. If Mackenzie plans to take 10 classes per month, which gym would be the better deal?

37. $(1,1)$

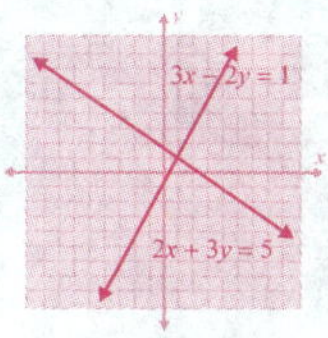

38. $(3,0)$

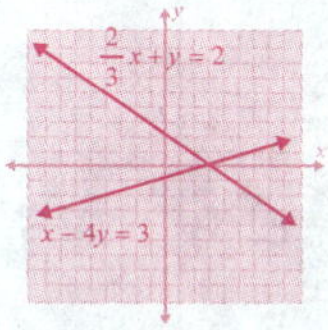

39. No solution

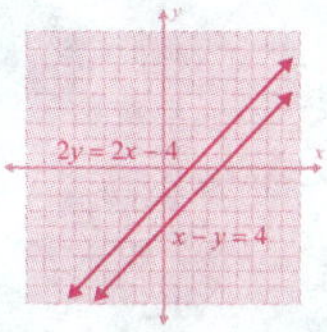

40. $(3,1)$

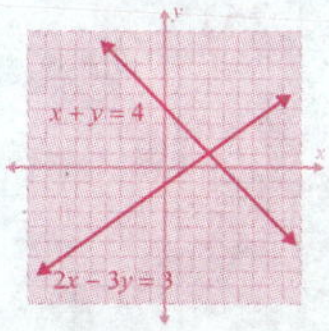

41. $(1,-1)$

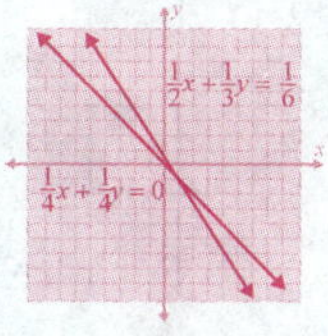

42. $(1,-3)$

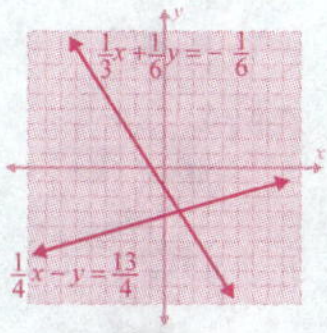

43. $(1,4)$

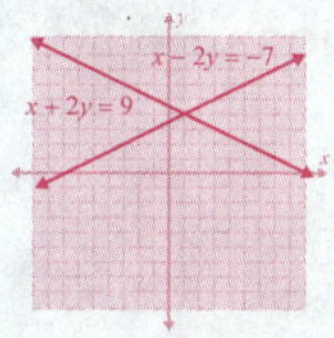

44. $(3,1)$

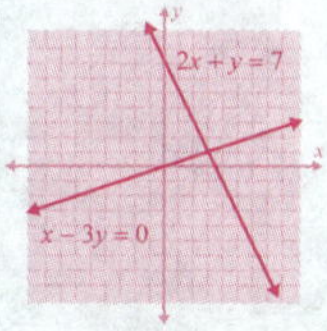

58. You are planning a vacation and would like to spend your money wisely. You decide to fly to your destination and then rent a car when you get there. Discount Car Rentals charges \$10 per day for the economy class car and \$0.10 per mile. Cars For Hire charges \$15 per day and \$0.05 per mile. You need to determine which car rental service offers the best deal.

a. Write two equations to represent the situation. Use the variable y to represent the total cost per day and the variable x to represent the number of miles driven per day.

b. Graph the two equations on the same coordinate plane.

c. Find the point of intersection.

d. What does the point of intersection mean? Write a complete sentence.

e. You expect to drive at most 75 miles per day. Which car rental company should you rent the car from?

Writing & Thinking

59. Explain, in your own words, why the answer to a consistent system of linear equations can be written as an ordered pair.

45. (1.5, 2)

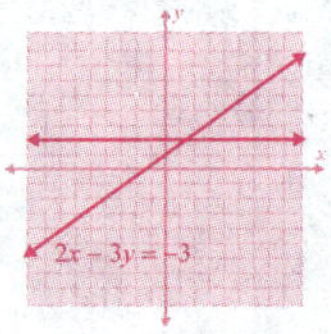

46. (0.5, 0.3333)

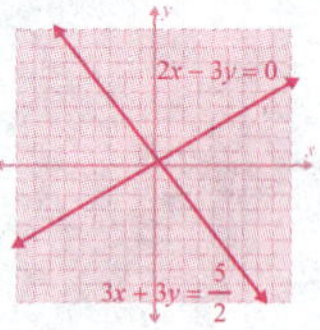

47. (1.5, −3)

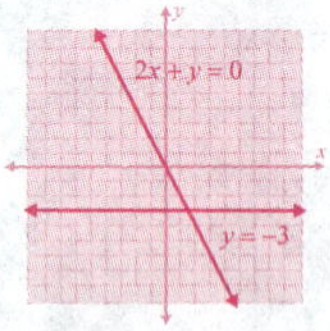

48. (2.5, 1.25)

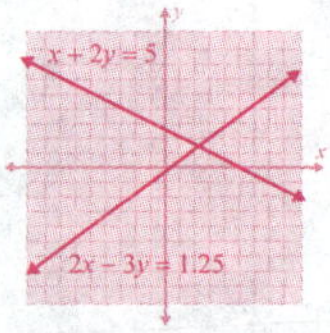

49. (1.4, 2.1)

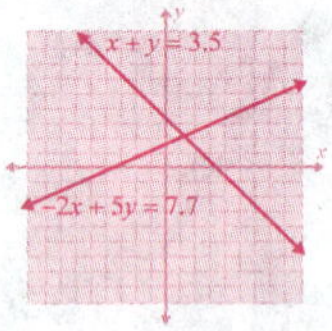

50. (1, −4.5)

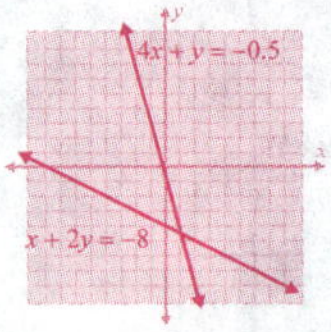

Applications

51.

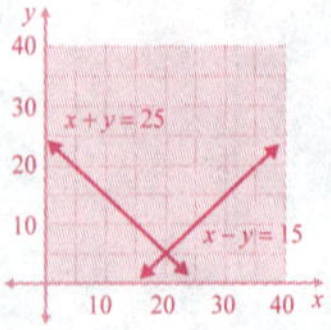

(20, 5)

52.

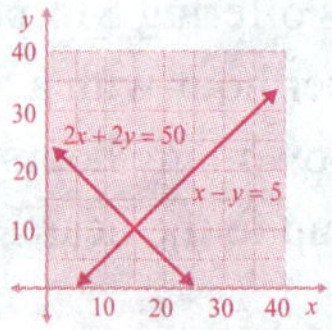

15 m × 10 m

53.

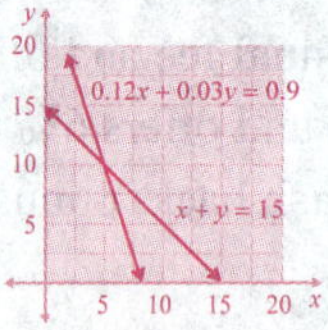

5 gallons of 12%; 10 gallons of 3%

54. Calc. = \$120; Book = \$50

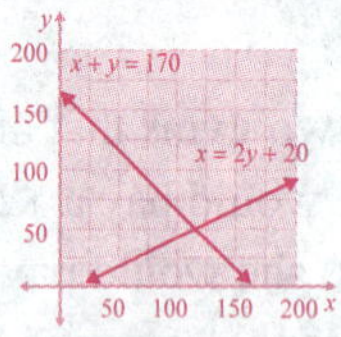

55. 134 hours

56. (\$2111.11, \$31.67)

57. **a.** Fit4life: $y = 20 + 10x$; Workout Nation: $y = 45 + 5x$

b.

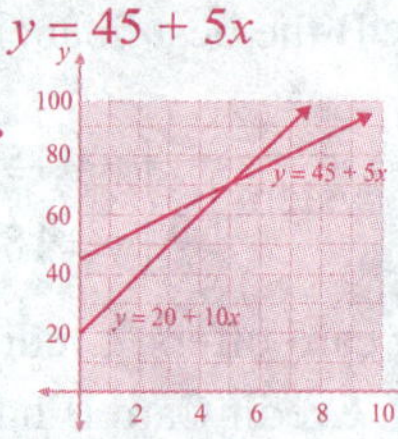

c. (5, 70)

d. The gym memberships will cost the same amount per month if 5 classes are taken.

e. Workout Nation

58. **a.** Discount Car Rentals: $y = 15 + 0.05x$; Cars for Hire: $y = 10 + 0.10x$

b.

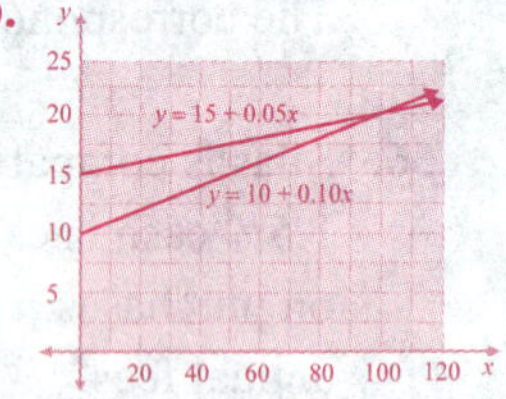

c. (100, 20)

d. The car rentals will cost the same amount per day when 100 miles are driven.

e. Cars for Hire

Writing & Thinking

59. The solution to a consistent system of linear equations is a single point, which is easily written as an ordered pair.

11.2 Systems of Linear Equations: Solutions by Substitution

Objectives

A. Use the method of substitution to solve systems of linear equations.

A Solving Systems of Linear Equations by Substitution

As we discussed in Section 11.1, solving systems of linear equations by graphing is somewhat limited in accuracy. The graphs must be drawn very carefully and even then the points of intersection (if there are any) can be difficult to estimate accurately.

In this section, we will develop an algebraic method called the **method of substitution**. The objective in the substitution method is to eliminate one of the variables so that a new equation is formed with just one variable. If this new equation has one solution, then the solution to the system is a single point. If this new equation is never true, then the system has no solution. If this new equation is always true, then the system has an infinite number of solutions.

Solving a System of Linear Equations by Substitution

1. Solve one of the equations for one of the variables.
2. Substitute the resulting expression into the other equation.
3. Solve this new equation, if possible, and then substitute back into one of the original equations to find the value of the other variable. (This is known as **back substitution**.)
4. Check the solution in both of the original equations.

PROCEDURE

To illustrate the substitution method, consider the following system.

$$\begin{cases} y = -2x + 5 \\ x + 2y = 1 \end{cases}$$

How would you substitute? Since the first equation is already solved for y, a reasonable substitution would be to replace y with $-2x + 5$ in the second equation. Try this and see what happens.

First equation already solved for y: $y = -2x + 5$

Substitute into second equation: $x + 2y = 1$

$$x + 2(-2x + 5) = 1$$

We now have one equation in only one variable, namely x. The problem has been reduced from one of solving two equations in two variables to solving one equation in one variable. Solve this equation for x. Then, find the corresponding y-value by substituting this x-value into **either of the two original equations**.

$$
\begin{aligned}
x+2(-2x+5)&=1\\
x-4x+10&=1\\
-3x&=1-10\\
-3x&=-9\\
x&=3
\end{aligned}
$$

Back substitute $x = 3$ into $y = -2x + 5$.

$$
\begin{aligned}
y&=-2(3)+5\\
y&=-6+5\\
y&=-1
\end{aligned}
$$

Thus, the solution to the system is the point $(3, -1)$.

Substitution is not the only algebraic technique for solving a system of linear equations. It does work in all cases but is most often used when one of the equations is easily solved for one variable.

In the following examples, note how the results in Example 3 indicate that the system has no solution and the results in Example 4 indicate that the system has an infinite number of solutions.

1. Solve the system:

$$\begin{cases} x=3 \\ y=7-4x \end{cases}$$

Example 1 Solving Systems by Substitution (One Solution)

Use the method of substitution to solve the following system of linear equations.

$$\begin{cases} x=-5 \\ y=2x+9 \end{cases}$$

Solution

The first equation is already solved for x. Substituting -5 for x in the second equation gives the corresponding value for y.

$$y=2(-5)+9=-10+9=-1$$

The solution to the system is $(-5,-1)$.

Check

Substitution shows that $(-5, -1)$ satisfies **both** of the equations in the system.

$$
\begin{aligned}
x&=-5 & \qquad y&=2x+9\\
(-5)&=-5 & (-1)&\stackrel{?}{=}2(-5)+9\\
-5&=-5 & -1&=-1
\end{aligned}
$$

Now work margin exercise 1.

Example 2 Solving Systems by Substitution (One Solution)

Use the method of substitution to solve the following system of linear equations.

$$\begin{cases} y = \frac{5}{6}x + 2 \\ \frac{1}{6}x + y = 8 \end{cases}$$

Solution

The first equation is already solved for y. Substituting $\frac{5}{6}x + 2$ for y in the second equation gives the following.

$$\frac{1}{6}x + \left(\frac{5}{6}x + 2\right) = 8$$

$$6 \cdot \frac{1}{6}x + 6 \cdot \left(\frac{5}{6}x + 2\right) = 6 \cdot 8$$ Multiply both sides by 6, the LCD.

$$x + 5x + 12 = 48$$
$$6x + 12 = 48$$
$$6x = 48 - 12$$
$$6x = 36$$
$$x = 6$$

Substituting 6 for x in the first equation gives the corresponding value for y.

$$y = \frac{5}{6}(6) + 2 = 5 + 2 = 7$$

The solution to the system is $(6, 7)$.

Check

Substitution shows that $(6, 7)$ satisfies **both** of the equations in the system.

$$y = \frac{5}{6}x + 2 \qquad \frac{1}{6}x + y = 8$$

$$(7) \overset{?}{=} \frac{5}{6}(6) + 2 \qquad \frac{1}{6}(6) + (7) \overset{?}{=} 8$$

$$7 = 7 \qquad 8 = 8$$

Now work margin exercise 2.

2. Solve the system:

$$\begin{cases} y = \frac{2}{5}x - 3 \\ \frac{3}{5}x + y = 7 \end{cases}$$

Example 3 Solving Systems by Substitution (No Solution)

Use the method of substitution to solve the following system of linear equations.

$$\begin{cases} 3x + y = 1 \\ 6x + 2y = 3 \end{cases}$$

3. Solve the system:

$$\begin{cases} 2x + y = 1 \\ 10x + 5y = 4 \end{cases}$$

Solution

Solving the first equation for y gives $y=1-3x$. Substituting $1-3x$ for y in the second equation gives the following.

$$\begin{aligned}6x+2(1-3x)&=3\\6x+2-6x&=3\\2&=3\end{aligned}$$

This last equation $(2=3)$ is **false**. This tells us that the system is inconsistent and has **no solution**. Graphically, the lines are parallel and there is no intersection.

Now work margin exercise 3.

4. Solve the system:

$$\begin{cases}-2x+y=1\\10x-5y=-5\end{cases}$$

Example 4 Solving Systems by Substitution (Infinite Solutions)

Use the method of substitution to solve the following system of linear equations.

$$\begin{cases}x-2y=1\\3x-6y=3\end{cases}$$

Solution

Solving the first equation for x gives $x=1+2y$. Substituting $1+2y$ for x in the second equation gives the following.

$$\begin{aligned}3(1+2y)-6y&=3\\3+6y-6y&=3\\3&=3\end{aligned}$$

This last equation $(3=3)$ is **always true**. This tells us that the equations are dependent and that the system has an **infinite number of solutions.** The solutions are of the form $(1+2y,\ y)$ for all values of y. (Or, solving one of the equations for y we have $\left(x,\ \frac{1}{2}x-\frac{1}{2}\right)$ for all values of x.)

Now work margin exercise 4.

Math Tip

In Example 4, the reason we started by solving the first equation for x is that x is the only variable in the system that does not have a coefficient. This means we did not have to perform division to isolate the variable, only addition (and therefore saved some time). Solving for any of the other three variables first would still give the same final answer.

As illustrated in Example 4, the solution can take two forms for a system with an infinite number of solutions. In one form, we can solve for x and then use this expression for x in the ordered pair format $(x,\ y)$. For example, in Example 4, solving either equation for x gives $x=1+2y$ and we can write $(1+2y,\ y)$ to represent all solutions.

Alternatively, we can solve for y, which gives $y=\frac{1}{2}x-\frac{1}{2}$, and we have the form $\left(x,\ \frac{1}{2}x-\frac{1}{2}\right)$ for all solutions. Try substituting various values for x and y in these expressions and you will see that all results satisfy both equations.

Example 5 Solving Systems by Substitution (Decimal Numbers)

Use the method of substitution to solve the following system of linear equations.

$$\begin{cases} x+y=5 \\ 0.2x+0.3y=0.9 \end{cases}$$

Solution

Solving the first equation for y gives $y=5-x$. Substituting $5-x$ for y in the second equation gives the following.

$$\begin{aligned} 0.2x+0.3y&=0.9 \\ 10(0.2x+0.3y)&=10(0.9) \\ 2x+3(5-x)&=9 \\ 2x+15-3x&=9 \\ -x&=9-15 \\ -x&=-6 \\ \frac{-x}{-1}&=\frac{-6}{-1} \\ x&=6 \\ y=5-x&=5-(6)=-1 \end{aligned}$$

Multiply by 10 so that the corresponding coefficients and constants will be integers.

The solution to the system is $(6,-1)$.

To check, substitute $x=6$ and $y=-1$ in both of the original equations.

Now work margin exercise 5.

5. Solve the system:

$$\begin{cases} x+y=6 \\ 0.9x+0.4y=0.9 \end{cases}$$

Completion Example 6 Solving Systems by Substitution

Use the method of substitution to solve the following system of linear equations.

$$\begin{cases} x+y=3 \\ 2x-y=12 \end{cases}$$

Solution

Solving the first equation for x gives $x=$ ___3 − y___. Substituting ___3 − y___ for x in the second equation gives the following.

$$\begin{aligned} 2(\underline{3-y})-y&=12 \\ \underline{6-2y}-y&=12 \\ 6-\underline{3y}&=12 \\ -3y&=\underline{6} \\ y&=\underline{-2} \\ x=3-(\underline{-2})&=\underline{5} \end{aligned}$$

The solution to the system is ___(5, −2)___.

Now work margin exercise 6.

6. Solve the system:

$$\begin{cases} x-y=-4 \\ 4x+3y=5 \end{cases}$$

Completion Example Answers

6. $x = 3 - y$; $3 - y$;

$$
\begin{aligned}
2(3-y)-y&=12\\
6-2y-y&=12\\
6-3y&=12\\
-3y&=6\\
y&=-2\\
x=3-(-2)&=5
\end{aligned}
$$

The solution to the system is $(5,-2)$.

Margin Exercise Answers

1. $(3,-5)$ **2.** $(10,1)$ **3.** No solution **4.** $(x,1+2x)$ or $\left(\frac{y-1}{2},y\right)$ **5.** $(-3,9)$ **6.** $(-1,3)$

11.2 Exercises

Concept Check

Fill-in-the-Blank. Complete each sentence using information found in this section.

1. The first step when solving a system of equations using the method of substitution is to solve for one of the ______ in one of the equations.
2. The second step is to substitute the resulting expression into the ______ expression.
3. If the equation formed after substitution is never true, then the system has ______ solution(s).
4. If the equation formed after substitution is always true, then the system has a/an ______ number of solutions.
5. After solving the equation formed after substitution, the value of the variable is substituted into one of the original expressions to find the value of the other variable. This is known as ______ ______.
6. The solution should be checked in ______ equations.

True/False. Determine whether each statement is true or false. If a statement is false, explain how it can be changed so the statement will be true. (**Note:** There may be more than one acceptable change.)

7. The method of substitution reduces the problem from one of solving two equations in two variables to solving one equation in one variable.
8. The method of substitution is most often used when one of the equations is impossible to graph.
9. The method of substitution is more accurate than the graphing method.
10. When using the method of substitution, you should always solve the first equation for x.

11.2 Exercises

Concept Check

1. variables
2. other
3. no
4. infinite
5. back substitution
6. both
7. True
8. False; The method is used most often used when one of the equations is easily solved for a variable.
9. True
10. False; You should find the easiest variable to solve for in either equation.

Practice

1. $(2,4)$
2. $(3,3)$
3. $(x,3x-7)$
4. $(-1,1)$
5. $(-6,-2)$
6. No solution
7. $(4,1)$
8. $(0,3)$
9. No solution
10. $(4,-5)$
11. $(3,2)$
12. No solution

Practice

Use the method of substitution to solve each system. See Examples 1 through 6.

1. $\begin{cases} x+y=6 \\ y=2x \end{cases}$

2. $\begin{cases} 5x+2y=21 \\ x=y \end{cases}$

3. $\begin{cases} 3x-7=y \\ 2y=6x-14 \end{cases}$

4. $\begin{cases} y=3x+4 \\ 2y=3x+5 \end{cases}$

5. $\begin{cases} x=3y \\ 3y-2x=6 \end{cases}$

6. $\begin{cases} 4x=y \\ 4x-y=7 \end{cases}$

7. $\begin{cases} x-5y+1=0 \\ x=7-3y \end{cases}$

8. $\begin{cases} 2x+5y=15 \\ x=y-3 \end{cases}$

9. $\begin{cases} 7x+y=9 \\ y=4-7x \end{cases}$

10. $\begin{cases} 3y+5x=5 \\ y=3-2x \end{cases}$

11. $\begin{cases} 3x-y=7 \\ x+y=5 \end{cases}$

12. $\begin{cases} 4x-2y=5 \\ y=2x+3 \end{cases}$

13. $\begin{cases} 3x+5y=-13 \\ y=3-2x \end{cases}$

14. $\begin{cases} 15x+5y=20 \\ y=-3x+4 \end{cases}$

15. $\begin{cases} x-y=5 \\ 2x+3y=0 \end{cases}$

16. $\begin{cases} 4x=8 \\ 3x+y=8 \end{cases}$

17. $\begin{cases} 2y=5 \\ 3x-4y=-4 \end{cases}$

18. $\begin{cases} x+y=8 \\ 3x+2y=8 \end{cases}$

19. $\begin{cases} y=2x-5 \\ 2x+y=-3 \end{cases}$

20. $\begin{cases} 2x+3y=5 \\ x-6y=0 \end{cases}$

21. $\begin{cases} x+5y=1 \\ x-3y=5 \end{cases}$

22. $\begin{cases} 3x+8y=-2 \\ x+2y=-1 \end{cases}$

23. $\begin{cases} 9x+3y=6 \\ 3x=2-y \end{cases}$

24. $\begin{cases} 5x+2y=-10 \\ 10x=-3-4y \end{cases}$

25. $\begin{cases} x-2y=-4 \\ 3x+y=-5 \end{cases}$

26. $\begin{cases} x+4y=3 \\ 3x-4y=7 \end{cases}$

27. $\begin{cases} 3x-y=-1 \\ 7x-4y=0 \end{cases}$

28. $\begin{cases} x+5y=-1 \\ 2x+7y=1 \end{cases}$

29. $\begin{cases} x+3y=5 \\ 3x+2y=7 \end{cases}$

30. $\begin{cases} 3x-4y-39=0 \\ 2x-y-13=0 \end{cases}$

31. $\begin{cases} \frac{1}{4}x-\frac{3}{2}y=-5 \\ -x+6y=20 \end{cases}$

32. $\begin{cases} \frac{-4}{3}x+2y=7 \\ \frac{8}{3}x-4y=-5 \end{cases}$

33. $\begin{cases} 6x-y=15 \\ 0.2x+0.5y=2.1 \end{cases}$

34. $\begin{cases} x+2y=3 \\ 0.4x+y=0.6 \end{cases}$

35. $\begin{cases} 0.2x-0.1y=0 \\ y=x+10 \end{cases}$

36. $\begin{cases} 0.1x-0.2y=1.4 \\ 3x+y=14 \end{cases}$

37. $\begin{cases} 3x-2y=5 \\ y=1.5x+2 \end{cases}$

38. $\begin{cases} x=2y-7.5 \\ 2x+4y=-15 \end{cases}$

39. $\begin{cases} \frac{1}{2}x+\frac{1}{3}y=4 \\ 3x+2y=24 \end{cases}$

40. $\begin{cases} \frac{1}{3}x+\frac{1}{7}y=2 \\ 7x+3y=42 \end{cases}$

41. $\begin{cases} \frac{x}{3}+\frac{y}{5}=1 \\ x+6y=12 \end{cases}$

42. $\begin{cases} \frac{x}{5}+\frac{y}{4}-3=0 \\ \frac{x}{10}-\frac{y}{2}+1=0 \end{cases}$

13. $(4,-5)$
14. $(x,-3x+4)$
15. $(3,-2)$
16. $(2,2)$
17. $\left(2,\frac{5}{2}\right)$
18. $(-8,16)$
19. $\left(\frac{1}{2},-4\right)$
20. $\left(2,\frac{1}{3}\right)$
21. $\left(\frac{7}{2},-\frac{1}{2}\right)$
22. $\left(-2,\frac{1}{2}\right)$
23. $(x,2-3x)$
24. No solution
25. $(-2,1)$
26. $\left(\frac{5}{2},\frac{1}{8}\right)$
27. $\left(-\frac{4}{5},-\frac{7}{5}\right)$
28. $(4,-1)$
29. $\left(\frac{11}{7},\frac{8}{7}\right)$
30. $\left(\frac{13}{5},-\frac{39}{5}\right)$
31. $\left(x,\frac{1}{6}x+\frac{10}{3}\right)$
32. No solution
33. $(3,3)$
34. $(9,-3)$
35. $(10,20)$
36. $(6,-4)$
37. No solution
38. $(-7.5,0)$
39. $\left(x,-\frac{3}{2}x+12\right)$
40. $\left(x,-\frac{7}{3}x+14\right)$
41. $\left(2,\frac{5}{3}\right)$
42. $(10,4)$

Applications

43. $(20, 5)$

44. 15 m × 10 m

45. 5 gallons of 12%; 10 gallons of 3%

46. Calculator = $120; Book = $50

47. 3 months

48. a. Square tile: 4 in.^{2}; Rectangular tile: 8 in.^{2}; Backsplash: 360 in.^{2}

Applications

Each of the following applications has been modeled using a system of equations. Use the method of substitution to solve each system. (**Note:** Some of these exercises were also given in Section 11.1. Check to see that you arrived at the same answers by both methods.)

43. The sum of two numbers is 25 and their difference is 15. What are the two numbers?
Let x be one number
and y be the other number.

The corresponding modeling system is $\begin{cases} x+y=25 \\ x-y=15 \end{cases}$

44. The perimeter of a rectangle is 50 meters and the length is 5 meters longer than the width. Find the dimensions of the rectangle.
Let x be the length and y be the width.

The corresponding modeling system is $\begin{cases} 2x+2y=50 \\ x-y=5 \end{cases}$

45. OSHA recommends that swimming pool owners clean their pool decks with a solvent composed of a 12% chlorine solution and a 3% chlorine solution. Fifteen gallons of the solvent consists of 6% chlorine. How much of each of the mixing solutions were used?
Let x be the number of gallons of the 12% solution
and y be the number of gallons of the 3% solution.
The corresponding modeling system is

$$\begin{cases} x+y=15 \\ 0.12x+0.03y=0.06(15) \end{cases}$$

46. A student bought a calculator and a textbook for a course in algebra. He told his friend that the total cost was $170 (without tax) and that the calculator cost $20 more than twice the cost of the textbook. What was the cost of each item?
Let x be the cost of the calculator
and y be the cost of the textbook.
The corresponding modeling system is

$$\begin{cases} x+y=170 \\ x=2y+20 \end{cases}$$

47. A fitness center manager is trying to decide whether to charge an enrollment fee of $25 with a monthly rate of $50 or an enrollment fee of $100 with a monthly rate of $25. After how many months would it be more profitable for the manager to choose the lower enrollment fee and the higher monthly rate? Round up to the nearest month.

The corresponding modeling system is $\begin{cases} y=50x+25 \\ y=25x+100 \end{cases}$

48. Connor is retiling the backsplash of his kitchen counter. He plans on using square tiles that measure 2 inches by 2 inches and rectangular tiles that measure 2 inches by 4 inches. The backsplash measures 6 inches by 60 inches. He wants to use an equal number of rectangular tiles and square tiles. How many of each tile will he need to buy?

a. Find the area of each size tile and the area of the backsplash.

b. Write two equations to represent the situation. Use the variable s to represent the number of square tiles and the variable r to represent the number of rectangular tiles.

c. Solve the system of equations by substitution.

d. What does the solution mean? Write a complete sentence.

e. If the square tiles come in boxes of 30 and the rectangular tiles come in boxes of 18, how many boxes of each type of tile will Connor need to buy to retile the backsplash?

49. Harper is buying a gift for her friend's wedding. She found two gifts at two different stores to choose between. She plans on personalizing the selected gift with an engraved message. The first gift costs \$45 and the store charges \$0.15 per letter engraved. The second gift costs \$55 and the store charges \$0.10 per letter engraved. How many letters would the message need to contain for the two gifts to be the same cost?

a. Write two equations to represent the situation. Use the variable c to represent the total cost of the gift with engraving and the variable t to represent the number of letters in the engraved message.

b. Solve the system of equations by substitution.

c. What does the solution mean? Write a complete sentence.

d. How many letters must the engraved message contain for the second gift to be the less expensive option?

50. Carlos is buying supplies for the tutoring center where he works. He was given a budget of \$230 to spend. Calculators cost \$15 each and packs of paper are \$2.50 each. He would like to buy twice as many packs of paper as calculators. How many calculators and notebooks can Carlos buy and stay in budget? (**Note:** The tutoring center is non-profit organization and therefore does not have to pay sales tax.)

a. Write two equations to represent the situation. Use the variable c to represent the number of calculators and the variable p to represent the number of packs of paper.

b. Solve the system of equations by substitution. Write the solutions in decimal form.

c. What does the solution mean? Write a complete sentence.

d. Does the answer to part c. make sense? Explain why.

e. Based on your answer from part d., how many calculators should Carlos buy to stay within budget and to buy twice as many packs of paper as calculators?

f. If the tutoring center has at most 12 students at a time, will each student have a new calculator to work with?

Writing & Thinking

51. Explain the advantages of solving a system of linear equations

a. by graphing,

b. by substitution.

---

b. $\begin{cases} r = s \\ 4s + 8r = 360 \end{cases}$

c. $r = 30,\ s = 30$

d. Connor needs to buy 30 square tiles and 30 rectangular tiles.

e. 1 box of square tiles, 2 boxes of rectangular tiles

49. a. $\begin{cases} c = 45 + 0.15t \\ c = 55 + 0.10t \end{cases}$

b. $t = 200,\ c = 75$

c. The gifts will both cost \$75 when the engraved message contains 200 letters.

d. More than 200 letters

50. a. $\begin{cases} p = 2c \\ 15c + 2.50p = 230 \end{cases}$

b. $p = 23;\ c = 11.5$

c. To spend \$230 and have twice as many packs of paper as calculators, Carlos will need to buy 23 packs of paper and 11.5 calculators.

d. No. You cannot buy half of a calculator.

e. 11 calculators

f. No

Writing & Thinking

51. Answers will vary.

Objectives

A. Use the method of addition to solve systems of linear equations.

B. Use a system of equations to find the equation of a line through two points.

11.3 Systems of Linear Equations: Solutions by Addition

A Solving Systems of Linear Equations by Addition

We have discussed two methods for solving systems of linear equations:

1. **Graphing**
2. **Substitution**

We know that solutions found by graphing are not necessarily exact (estimation may be involved); and, in some cases, the method of substitution can lead to complicated algebraic steps. In this section, we consider a third method:

3. **Addition** (or **method of elimination**)

In the **method of addition**, as with the method of substitution, the objective is to eliminate one of the variables so that a new equation is found with just one variable, if possible.

1. If this new equation has one solution, the solution to the system is a **single point**.
2. If this new equation is never true, the system has **no solution**.
3. If this new equation is always true, the system has an **infinite number of solutions**.

Consider solving the following system where both equations are written in standard form.

$$\begin{cases} x-2y=-9 \\ x+2y=11 \end{cases}$$

In the **method of addition**, we write one equation under the other so that like terms are aligned vertically. (Note that, in this example, the coefficients of y are opposites, namely −2 and +2.)

Then, add like terms as follows.

$$\begin{array}{r} x-2y=-9 \\ \underline{x+2y=\ 11} \\ 2x\quad\ =\ \ 2 \\ x\quad\ =\ \ 1 \end{array}$$

The y terms are eliminated because the coefficients are opposites.

Now substitute $x = 1$ into either of the original equations and solve for y.

$$\begin{aligned} (1)-2y&=-9 \\ -2y&=-10 \\ y&=5 \end{aligned} \qquad \textbf{OR} \qquad \begin{aligned} (1)+2y&=11 \\ 2y&=10 \\ y&=5 \end{aligned}$$

Therefore, the **solution** to the system is $(1, 5)$.

This example was relatively easy because the coefficients for y were opposites. The general procedure can be outlined as follows.

Solving a System of Linear Equations by Addition

1. Write the equations in **standard form,** one under the other, so that **like terms are aligned**.
2. Multiply all terms of one equation by a constant (and possibly all terms of the other equation by another constant) so that **two like terms have opposite coefficients.**
3. Add the two equations by **combining like terms** and solve the resulting equation, if possible.
4. **Back substitute into one of the original equations** to find the value of the other variable.
5. Check the solution (if there is one) in both of the **original** equations.

PROCEDURE

Math Tip

When solving a system of equations, two equations may be interchanged. In other words, the order in which the equations are written has no impact on the solution.

Example 1 Solving Systems by Addition (One Solution)

Use the method of addition to solve the following system of linear equations.

$$\begin{cases} x-y=-1 \\ 2x+3y=33 \end{cases}$$

Solution

Multiply each term in the first equation by 3 and leave each term in the second equation as is. Now the y-coefficients will be opposites. Add the two equations by combining like terms. This will eliminate y. Solve for x.

$$\begin{cases} [3](x-y=-1) \longrightarrow & 3x-3y=-3 \\ 2x+3y=33 \longrightarrow & 2x+3y=33 \end{cases}$$

$$\begin{aligned} 5x &= 30 \\ x &= 6 \end{aligned}$$

Substitute $x=6$ into either one of the original equations.

$$\begin{aligned} x-y&=-1 \\ (6)-y&=-1 \\ -y&=-7 \\ y&=7 \end{aligned} \quad \textbf{OR} \quad \begin{aligned} 2x+3y&=33 \\ 2(6)+3y&=33 \\ 12+3y&=33 \\ 3y&=21 \\ y&=7 \end{aligned}$$

The solution is $(6,7)$.

Check

Substitution shows that $(6,7)$ satisfies **both** of the equations in the system.

$$\begin{aligned} x-y&=-1 \\ (6)-(7)&\overset{?}{=}-1 \\ -1&=-1 \end{aligned} \qquad \begin{aligned} 2x+3y&=33 \\ 2(6)+3(7)&\overset{?}{=}33 \\ 12+21&\overset{?}{=}33 \\ 33&=33 \end{aligned}$$

Now work margin exercise 1.

1. Solve the system:

$$\begin{cases} 6x+y=-7 \\ -3x+2y=16 \end{cases}$$

2. Solve the system:

$$\begin{cases} 6x+3y=15 \\ 2x+y=5 \end{cases}$$

Example 2 Solving Systems by Addition (Infinite Solutions)

Use the method of addition to solve the following system of linear equations.

$$\begin{cases} 3x-\frac{1}{2}y=6 \\ 6x-y=12 \end{cases}$$

Solution

Multiply the first equation by −2 so that the y-coefficients will be opposites.

$$\begin{cases} [-2]\left(3x-\frac{1}{2}y=6\right) \longrightarrow & -6x+y=-12 \\ (6x-y=12) \longrightarrow & 6x-y=12 \\ & \overline{0=0} \end{cases}$$

Because this last equation, $0=0$, is **always true**, the system has infinitely many solutions. The equations are dependent and the solution set consists of all points that satisfy the equation $6x-y=12$. Solving for y gives $y=6x-12$, and we can write the solution in the general form $(x, 6x-12)$. Or, solving for x, $x=\frac{1}{6}y+2$, and the solution can be written in the form $\left(\frac{1}{6}y+2, y\right)$.

Now work margin exercise 2.

3. Solve the system:

$$\begin{cases} 4x-y=-1 \\ -12x+3y=4 \end{cases}$$

Math Tip

There is more than one set of numbers you can multiply the pair of equations by to get opposite coefficients. For instance, in Example 3, you could have multiplied the first equation by 6 and the second equation by 4 to result in the y-coefficient being opposites.

Example 3 Solving Systems by Addition (No Solution)

Use the method of addition to solve the following system of linear equations.

$$\begin{cases} 2x-4y=1 \\ -3x+6y=2 \end{cases}$$

Solution

Multiply each term in the first equation by 3 and each term in the second equation by 2. This will result in the x-coefficients being opposites.

$$\begin{cases} [3](2x-4y=1) \longrightarrow & 6x-12y=3 \\ [2](-3x+6y=2) \longrightarrow & -6x+12y=4 \\ & \overline{0=7} \end{cases}$$

Because this last equation, $0=7$, is **false**, the system is inconsistent and has no solution.

Now work margin exercise 3.

4. Solve the system:

$$\begin{cases} x+0.5y=5.3 \\ 0.5x-2y=1.3 \end{cases}$$

Example 4 Solving Systems by Addition (Decimal Numbers)

Use the method of addition to solve the following system of linear equations.

$$\begin{cases} x+0.4y=3.08 \\ 0.1x-y=0.1 \end{cases}$$

Solution

Multiply the second equation by -10 so that the x-coefficients will be opposites.

$$\begin{cases} (x+0.4y=3.08) & \longrightarrow & 1.0x+\ 0.4y=3.08 \\ [-10](0.1x-\ \ y=0.1) & \longrightarrow & \underline{-1.0x+10.0y=-1.0} \end{cases}$$

$$10.4y=2.08 \quad \text{x is eliminated.}$$

$$y=0.2$$

Substitute $y=0.2$ into one of the original equations.

$$\begin{aligned} x+0.4y&=3.08 \\ x+0.4(0.2)&=3.08 \\ x+0.08&=3.08 \\ x&=3 \end{aligned}$$

The solution is $(3, 0.2)$.

To check, substitute $x=3$ and $y=0.2$ in both of the original equations.

Now work margin exercise 4.

Completion Example 5 Solving Systems by Addition (One Solution)

Use the method of addition to solve the following system of linear equations.

$$\begin{cases} 3x+5y=-3 \\ -7x+2y=7 \end{cases}$$

Solution

Multiply each term in the first equation by 2 and each term in the second equation by -5. This will result in the y-coefficients being opposites. Add the two equations by combining like terms which will eliminate y. Solve for x.

$$\begin{cases} [2](3x+5y=-3) & \longrightarrow & \underline{6}\ x+\underline{10}\ y=\underline{-6} \\ [-5](-7x+2y=7) & \longrightarrow & \underline{35}\ x-\underline{10}\ y=\underline{-35} \end{cases}$$

$$\underline{41}\ x \quad =\underline{-41} \quad \text{y is eliminated.}$$

$$x \quad = \underline{-1}$$

Substitute $x=\underline{-1}$ into either one of the original equations. Choosing the first equation gives the following.

$$\begin{aligned} 3(\underline{-1})+5y&=-3 \\ \underline{-3}+5y&=-3 \\ 5y&=\underline{0} \\ y&=\underline{0} \end{aligned}$$

The solution is $\underline{(-1,0)}$.

Now work margin exercise 5.

5. Solve the system:

$$\begin{cases} 3x+2y=0 \\ 5x-2y=16 \end{cases}$$

Note

In Example 5, we could eliminate x instead of y by multiplying the terms in the first equation by 7 and the terms in the second equation by 3. The solution will be the same. Try this yourself to confirm this method.

B Using Systems of Equations to Find the Equation of a Line

6. Using the formula $y = mx + b$, find the equation of the line determined by the two points $(4, -2)$ and $(6, 3)$.

Example 6 Using a System to Find the Equation of a Line

Using the formula $y = mx + b$, find the equation of the line determined by the two points $(3, 5)$ and $(-6, 2)$.

Solution

Write two equations in m and b by substituting the cooıdlnates of the points for x and y. This gives the system,

$$\begin{cases} 5 = 3m + b \\ 2 = -6m + b \end{cases}$$

Multiply the second equation by -1 so that the b-values will be opposites.

$$\begin{cases} (5 = 3m + b) & \longrightarrow & 5 = 3m + b \\ [-1]\,(2 = -6m + b) & \longrightarrow & \underline{-2 = 6m - b} \end{cases}$$

$$3 = 9m$$

$$\frac{1}{3} = m$$

Substitute $m = \frac{1}{3}$ into one of the original equations.

$$5 = 3m + b$$
$$5 = 3\left(\frac{1}{3}\right) + b$$
$$5 = 1 + b$$
$$4 = b$$

The equation of the line is $y = \frac{1}{3}x + 4$.

Now work margin exercise 6.

Now that you know three methods for solving a system of linear equations (graphing, substitution, and addition), which method should you use? Consider the following guidelines when making your decision.

Guidelines for Deciding which Method to Use when Solving a System of Linear Equations

1. The graphing method is helpful in "seeing" the geometric relationship between the lines and finding approximate solutions. A calculator can be very helpful here.
2. Both the substitution method and the addition method give exact solutions.
3. The substitution method may be reasonable and efficient if one of the coefficients of one of the variables is 1.
4. The addition method is particularly efficient if the coefficients for one of the variables are opposites.

PROCEDURE

Completion Example Answers

5. $6x + 10y = -6$ Substitute $x = -1$; $3(-1) + 5y = -3$ The solution is $(-1, 0)$.

$$\begin{array}{rl} 6x+10y &= -6 \\ 35x-10y &= -35 \\ \hline 41x &= -41 \\ x &= -1 \end{array} \qquad \begin{aligned} 3(-1)+5y &= -3 \\ -3+5y &= -3 \\ 5y &= 0 \\ y &= 0 \end{aligned}$$

Margin Exercise Answers

1. $(-2,5)$ **2.** $(x,-2x+5)$ or $\left(\frac{-1}{2}y+\frac{5}{2},y\right)$ **3.** No solution **4.** $(5,0.6)$ **5.** $(2,-3)$

6. $y=\frac{5}{2}x-12$

11.3 Exercises

Concept Check

Fill-in-the-Blank. Complete each sentence using information found in this section.

1. When using the method of addition, the objective is to ______ one of the variables so that a new equation is found with just one variable.

2. The first step of the method of addition is to write one equation under the other so that the __________ are aligned vertically.

3. To make it easier to align terms, the equations should be written in ______ form.

4. Multiply the terms of one equation by a constant so that two like terms have ______ coefficients. You may need to multiply both equations by different constants.

5. Next, add the two equations by ______ like terms and solve the resulting equation for the other variable.

6. The addition method is particularly efficient if the ______ for one of the variables are opposites.

True/False. Determine whether each statement is true or false. If a statement is false, explain how it can be changed so the statement will be true. (**Note:** There may be more than one acceptable change.)

7. When using the method of addition, the solution only needs to be checked in one of the original equations.

8. It's possible for a system of equations to have no solutions.

9. Both the addition method and the substitution method gives approximate solutions.

10. The graphing method is helpful in "seeing" the geometric relationship between the lines and finding approximate solutions.

11.3 Exercises

Concept Check

1. eliminate
2. like terms
3. standard
4. opposite
5. combining
6. coefficients
7. False; The solution always needs to be checked in both original equations.
8. True
9. False; Both methods give exact solutions.
10. True

Practice

Use the method of addition to solve each system. See Examples 1 through 5.

1. $\begin{cases} 8x - y = 29 \\ 2x + y = 11 \end{cases}$

2. $\begin{cases} x + 3y = 9 \\ x - 7y = -1 \end{cases}$

3. $\begin{cases} 3x + 2y = 0 \\ 5x - 2y = 8 \end{cases}$

4. $\begin{cases} 12x - 3y = 21 \\ 4x - y = 7 \end{cases}$

5. $\begin{cases} 2x + 2y = 5 \\ x + y = 3 \end{cases}$

6. $\begin{cases} 2x - y = 7 \\ x + y = 2 \end{cases}$

7. $\begin{cases} 3x + 3y = 9 \\ x + y = 3 \end{cases}$

8. $\begin{cases} 9x + 2y = -42 \\ 5x - 6y = -2 \end{cases}$

9. $\begin{cases} \frac{1}{2}x + y = -4 \\ 3x - 4y = 6 \end{cases}$

10. $\begin{cases} x + y = 1 \\ x - \frac{1}{3}y = \frac{11}{3} \end{cases}$

11. $\begin{cases} x + y = 12 \\ 0.05x + 0.25y = 1.6 \end{cases}$

12. $\begin{cases} x + 0.1y = 8 \\ 0.1x + 0.01y = 0.64 \end{cases}$

Solve each system of linear equations. See Examples 1 through 5.

13. $\begin{cases} x = 11 + 2y \\ 2x - 3y = 17 \end{cases}$

14. $\begin{cases} 6x - 3y = 6 \\ y = 2x - 2 \end{cases}$

15. $\begin{cases} x - 2y = 4 \\ y = \frac{1}{2}x - 2 \end{cases}$

16. $\begin{cases} 2x + y = 3 \\ 4x + 2y = 7 \end{cases}$

17. $\begin{cases} x = 3y + 4 \\ y = 6 - 2x \end{cases}$

18. $\begin{cases} y = 2x + 14 \\ x = 14 - 3y \end{cases}$

19. $\begin{cases} 7x - y = 16 \\ 2y = 2 - 3x \end{cases}$

20. $\begin{cases} 3x + y = -10 \\ 2y - 1 = x \end{cases}$

21. $\begin{cases} 4x - 2y = 8 \\ 2x - y = 4 \end{cases}$

22. $\begin{cases} x + y = 6 \\ 2x + y = 16 \end{cases}$

23. $\begin{cases} 3x + 2y = 4 \\ x + 5y = -3 \end{cases}$

24. $\begin{cases} x + 2y = 0 \\ 2x = 4y \end{cases}$

25. $\begin{cases} 4x + 3y = 2 \\ 3x + 2y = 3 \end{cases}$

26. $\begin{cases} x - 3y = 4 \\ 3x - 9y = 10 \end{cases}$

27. $\begin{cases} 5x - 2y = 17 \\ 2x - 3y = 9 \end{cases}$

28. $\begin{cases} \frac{1}{2}x + 2y = 9 \\ 2x - 3y = 14 \end{cases}$

29. $\begin{cases} 3x + 2y = 14 \\ 7x + 3y = 26 \end{cases}$

30. $\begin{cases} 4x + 3y = 28 \\ 5x + 2y = 35 \end{cases}$

Practice

1. $(4, 3)$
2. $(6, 1)$
3. $\left(1, -\frac{3}{2}\right)$
4. $(x, 4x - 7)$
5. No solution
6. $(3, -1)$
7. $(x, 3 - x)$
8. $(-4, -3)$
9. $(-2, -3)$
10. $(3, -2)$
11. $(7, 5)$
12. No solution
13. $(1, -5)$
14. $(x, 2x - 2)$
15. $\left(x, \frac{1}{2}x - 2\right)$
16. No solution
17. $\left(\frac{22}{7}, -\frac{2}{7}\right)$
18. $(-4, 6)$
19. $(2, -2)$
20. $(-3, -1)$
21. $(x, 2x - 4)$
22. $(10, -4)$
23. $(2, -1)$
24. $(0, 0)$
25. $(5, -6)$
26. No solution
27. $(3, -1)$
28. $(10, 2)$
29. $(2, 4)$
30. $(7, 0)$

31. $\begin{cases} 2x+7y=2 \\ 5x+3y=-24 \end{cases}$

32. $\begin{cases} 7x-6y=-1 \\ 5x+2y=37 \end{cases}$

33. $\begin{cases} 10x+4y=7 \\ 5x+2y=15 \end{cases}$

34. $\begin{cases} 6x-5y=-40 \\ 8x-7y=-54 \end{cases}$

35. $\begin{cases} 0.5x-0.3y=7 \\ 0.3x-0.4y=2 \end{cases}$

36. $\begin{cases} 0.6x+0.5y=5.9 \\ 0.8x+0.4y=6 \end{cases}$

37. $\begin{cases} 2.5x+1.8y=7 \\ 3.5x-2.7y=4 \end{cases}$

38. $\begin{cases} 0.75x-0.5y=2 \\ 1.5x-0.75y=7.5 \end{cases}$

39. $\begin{cases} \frac{2}{3}x-\frac{1}{2}y=\frac{2}{3} \\ \frac{8}{3}x-2y=\frac{17}{6} \end{cases}$

40. $\begin{cases} \frac{3}{4}x+\frac{1}{4}y=\frac{3}{8} \\ \frac{3}{2}x+\frac{1}{2}y=\frac{3}{4} \end{cases}$

41. $\begin{cases} \frac{1}{6}x-\frac{1}{12}y=-\frac{13}{6} \\ \frac{1}{5}x+\frac{1}{4}y=2 \end{cases}$

42. $\begin{cases} \frac{5}{3}x-\frac{2}{3}y=-\frac{29}{30} \\ 2x+5y=0 \end{cases}$

Write an equation for the line determined by the two given points by using the formula $y = mx + b$ to set up a system of equations with m and b as the unknowns. See Example 6.

43. $(2,3),(1,-2)$

44. $(0,6),(-3,-3)$

45. $(1,-3),(5,-3)$

46. $(5,3),(5,-4)$

47. $(1,2),(-3,0)$

48. $(-4,2),(5,-1)$

Applications

Each of the following applications has been modeled using a system of equations. Use the method of substitution or the method of addition to solve each system.

49. For two months, Martin used the same snow removal company to help clear his property. The company has a fixed reservation rate per month, plus an hourly rate for the amount of time that is spent clearing snow. Martin received two bills from the snow removal company. The first bill was \$150 for 4 hours of snow removal and the second bill was \$200 for 6 hours of snow removal. Find the equation of the line that represents the snow removal company's fixed reservation rate and charge per hour. Round to the nearest cent if necessary.

50. For two months, Milan used the same landscaping maintenance company. The company has a fixed reservation rate per month, plus an hourly rate for the amount of time that is spent on landscaping work. Milan received two bills from the landscaping company. The first bill was \$109.50 for 3.5 hours of landscaping work and the second bill was \$147.75 for 5.75 hours of landscaping work. Find the equation of the line that represents the landscaping company's fixed reservation rate and charge per hour. Round to the nearest cent if necessary.

31. $(-6,2)$
32. $(5,6)$
33. No solution
34. $(-5,2)$
35. $(20,10)$
36. $(4,7)$
37. $\left(2,\frac{10}{9}\right)$
38. $(12,14)$
39. No solution
40. $\left(x,-3x+\frac{3}{2}\right)$
41. $\left(-\frac{45}{7},\frac{92}{7}\right)$
42. $\left(-\frac{1}{2},\frac{1}{5}\right)$
43. $y=5x-7$; $m=5$, $b=-7$
44. $y=3x+6$; $m=3$, $b=6$
45. $y=-3$; $m=0$, $b=-3$
46. $x=5$; $m=$ undefined, $b=$ none
47. $y=\frac{1}{2}x+\frac{3}{2}$; $m=\frac{1}{2}$, $b=\frac{3}{2}$
48. $y=-\frac{1}{3}x+\frac{2}{3}$; $m=-\frac{1}{3}$, $b=\frac{2}{3}$

Applications

49. $y=25x+50$
50. $y=17x+50$

51. \$4000 at 10%; \$6000 at 6%

52. 1200 adults; 3300 children

53. 40 liters of 30% solution (x), 60 liters of 40% solution (y)

54. $\frac{7}{2}$ hours

55. a. $\begin{cases} y = 3500 + 1.5x \\ y = 1000 + 2x \end{cases}$

b. $x = 5000,\ y = 11{,}000$

c. The cards will earn the same amount of rewards points if you spend \$5000.

d. City

51. Georgia had \$10,000 to invest, and she put the money into two accounts. One of the accounts will pay 6% interest and the other will pay 10%. How much did she put in each account if the interest from the 10% account exceeded the interest from the 6% account by \$40?
Let x be the amount in the 10% account and y be the amount in the 6% account.

The system that models the problem is $\begin{cases} x + y = 10{,}000 \\ 0.10x - 0.06y = 40 \end{cases}$

52. A minor league baseball team has a game attendance of 4500 people. Tickets cost \$5 for children and \$8 for adults. The total revenue made at this game was \$26,100. How many adults and how many children attended the game?
Let x be the number of adults and y be the number of children.

The system that models the problem is $\begin{cases} x + y = 4500 \\ 8x + 5y = 26{,}100 \end{cases}$

53. How many liters each of a 30% acid solution and a 40% acid solution must be used to produce 100 liters of a 36% acid solution?
Let x be the amount of 30% solution and y be the amount of 40% solution.

The system that models the problem is $\begin{cases} x + y = 100 \\ 0.30x + 0.40y = 0.36(100) \end{cases}$

54. Two cars leave Denver at the same time traveling in opposite directions. One travels at an average speed of 55 mph and the other at 65 mph. In how many hours will they be 420 miles apart?
Let x be the time of travel for first car and y be the time of travel for second car.

The system that models the problem is $\begin{cases} x = y \\ 55x + 65y = 420 \end{cases}$

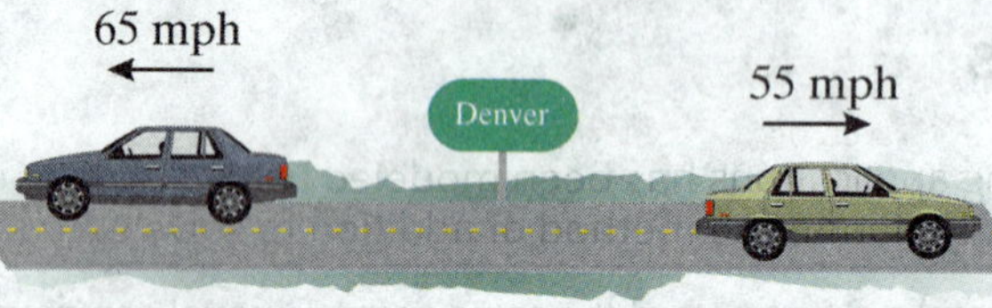

55. You are deciding between two credit cards with similar rewards programs. The City credit card will give you 3500 points as a sign-up bonus and 1.5 points for every dollar you spend. The International credit card gives you 1000 points as a sign-up bonus and gives you 2 points for every dollar you spend. How much would you have to spend to earn the same amount of rewards on each credit card?

a. Write two equations to represent the situation. Use the variable x to represent the number of dollars spent and the variable y to represent the total number of points earned.

b. Solve the system of equations by addition.

c. What does the solution mean? Write a complete sentence.

d. If you only plan to purchase \$4000 in merchandise, which credit card will give you the most points?

56. Barbara's Bombtastic Bakery uses chocolate chips in one type of cookie and in one type of muffin. The cookie recipe calls for 5 cups of chocolate chips and the muffin recipe calls for 2 cups of chocolate chips. The cookie recipe makes 30 large cookies and the muffin recipe makes 18 giant muffins. The bakery currently has 50 cups of chocolate chips and only has room in the display case for a combination of 360 cookies and muffins. The manager wants to determine how many of each item to bake to use all of the chocolate chips.

a. Write two equations to represent the situation. Use the variable x to represent the number of batches of chocolate chip cookies and the variable y to represent the number of batches of muffins.

b. Solve the system of equations by addition.

c. What does the solution mean? Write a complete sentence.

d. If the manager makes the amount of chocolate chip cookies and muffins described in part c., will the bakery be able to fulfill an order for 150 chocolate chip cookies and 160 chocolate chip muffins?

Writing & Thinking

57. Explain, in your own words, why the answer to a system with infinite solutions is written as an ordered pair with variables.

56. a. $\begin{cases} 5x+2y=50 \\ 30x+18y=360 \end{cases}$

b. $x=6,\ y=10$

c. The bakery should make 6 batches of chocolate chip cookies and 10 batches of muffins.

d. Yes

Writing & Thinking

57. Answers will vary.

Objectives

A. Use systems of linear equations to solve distance-rate-time application problems.

B. Use systems of linear equations to solve number problems.

C. Use systems of linear equations to solve amounts and cost problems.

11.4 Applications: Distance-Rate-Time, Number Problems, Amounts, and Cost

A Distance-Rate-Time Problems

Systems of equations occur in many practical situations such as

- supply and demand in business,
- velocity and acceleration in engineering,
- money and interest in investments, and
- mixture in physics and chemistry.

Many of the problems in earlier sections illustrated these ideas. However, in those exercises, the system of equations was given. As you study the applications in Sections 11.4 and 11.5, you may want to refer to some of those exercises, as well as the examples, as guides in solving the given applications. **For these applications, you will need to create your own systems of equations.**

Remember that the emphasis in all application problems is to develop your reasoning as well as your reading skills. You must learn how to translate English phrases into algebraic expressions. This is the *thinking part*.

1. Bob ran 5 miles in 1 hour running with the wind and then turned around. Over the next hour, running against the wind, he ran 4 miles. What were the wind speed and the speed Bob was running? Assume Bob maintained the same speed for his entire run.

Example 1 Application: Distance-Rate-Time (Rates Unknown)

A small plane flew 300 miles in 2 hours flying with the wind. On the return trip, flying against the wind, it traveled only 200 miles in 2 hours. What were the wind speed and the speed of the plane? (**Note:** The "speed of the plane" means how fast the plane would be flying with no wind.)

Solution

Let s = speed of plane

and w = wind speed.

When flying with the wind, the plane's actual rate will increase to $s + w$. When flying against the wind, the plane's actual rate will decrease to $s - w$. (**Note:** If the wind had been strong enough, the plane could actually have been flying backward, away from its destination.)

	Rate	×	Time	=	Distance
With the Wind	$s+w$		2		$2(s+w)$
Against the Wind	$s-w$		2		$2(s-w)$

The system of linear equations is as follows.

$$\begin{cases} 2(s+w)=300 & \text{With the wind} \\ 2(s-w)=200 & \text{Against the wind} \end{cases}$$

Apply the distributive property to the left-hand side of each equation to find an equivalent system of equations. Then use the addition method to solve the system.

$$\begin{aligned} 2s+2w &= 300 && \text{With the wind} \\ \underline{2s-2w} &\underline{= 200} && \text{Against the wind} \\ 4s \quad &= 500 \\ s \quad &= 125 && \text{Speed of plane} \end{aligned}$$

Back substitute $s = 125$ into one of the original equations.

$$\begin{aligned} 2(125+w) &= 300 \\ 125+w &= 150 \\ w &= 25 && \text{Wind speed} \end{aligned}$$

The speed of the plane was 125 mph, and the wind speed was 25 mph. (Checking will show that these numbers do give distances of 300 mi and 200 mi.)

Now work margin exercise 1.

Teaching Note

Without getting too mathematical or formal, it might be helpful to have students think of applied situations in which two or more conditions need to be considered when designing or doing something. (For example, the weight of a car for safety versus fuel economy.)

Example 2 Application: Distance-Rate-Time (Times Unknown)

Two buses leave a bus station traveling in opposite directions. One leaves at noon and the second leaves at 1 p.m. The first one travels at an average speed of 55 mph and the second one at an average speed of 59 mph. At what time will the buses be 226 miles apart?

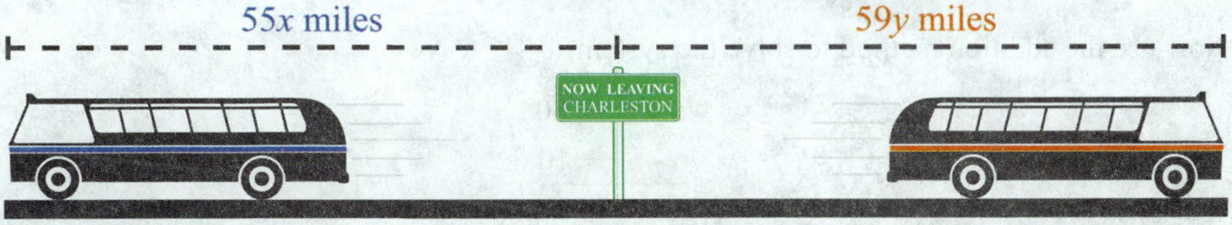

Solution

Let x = time of travel for first bus

and y = time of travel for second bus.

	Rate	×	Time	=	Distance
Bus 1	55		x		$55x$
Bus 2	59		y		$59y$

The system of linear equations is as follows.

$$\begin{cases} x = y+1 & \text{The first bus travels 1 hour longer than the second.} \\ 55x+59y=226 & \text{The sum of the distances is 226 miles.} \end{cases}$$

2. One man takes the eastbound line at 9 a.m., and his wife takes the westbound line at 10:00 a.m. The husband's train averages 25 mph, while his wife's train averages 30 mph. At what time will the husband and wife be 135 miles apart?

Note

You should consider making tables similar to those illustrated in Examples 1, 2, and 4 when working with applications. These tables can help you organize the information in a more understandable form.

Now use the substitution method to solve the system.

$$55(y+1)+59y=226$$
$$55y+55+59y=226$$
$$114y=171$$
$$y=1.5$$

Back substituting in the first equation gives $x=(1.5)+1=2.5$.

Thus, the first bus travels 2.5 hours and the second bus travels 1.5 hours. The buses will be 226 miles apart at 2:30 p.m.

Now work margin exercise 2.

B Number Problems

3. The sum of two numbers is 150 and their difference is 36. What are the two numbers?

Example 3 Solving a Number Problem

The sum of two numbers is 80 and their difference is 10. What are the two numbers?

Solution

Let x = one number

and y = the other number.

The system of linear equations is as follows.

$$\begin{cases} x+y=80 & \text{The sum is 80.} \\ x-y=10 & \text{The difference is 10.} \end{cases}$$

Now use the addition method to solve the system.

$$\begin{aligned} x+y &= 80 \\ x-y &= 10 \\ \hline 2x \quad &= 90 \\ x \quad &= 45 \end{aligned}$$

Back substitute 45 for x in the first equation.

$$(45)+y=80$$
$$y=80-45=35$$

The two numbers are 45 and 35.

Check

$45+35=80$ and $45-35=10$

Now work margin exercise 3.

Teaching Note

Anticipate that some students will recognize many of these problems as solvable with one variable rather than two. Discuss with them that these simple examples can be considered as practice for the method of substitution or addition for more difficult problems. Also, thinking in terms of two or more variables becomes a way of thought and can be used to solve many practical problems.

C Amounts and Cost Problems

Example 4 Application: Counting Coins

Mike has \$1.05 worth of change in nickels and quarters. If he has twice as many nickels as quarters, how many of each type of coin does he have?

Solution

We use two equations—one relating the number of coins and the other relating the value of the coins. (The value of each nickel is 5 cents and the value of each quarter is 25 cents.)

Let n = number of nickels

and q = number of quarters.

Coins	# of Coins	Value	Total Value
Nickels	n	0.05	$0.05n$
Quarters	q	0.25	$0.25q$

The system of linear equations is as follows.

$$\begin{cases} n = 2q \\ 0.05n + 0.25q = 1.05 \end{cases}$$

There are two times as many nickels as quarters.
The total amount of money is \$1.05.

Note carefully that in the first equation q is multiplied by 2 because the number of nickels is twice the number of quarters. Therefore, n is bigger.

The first equation is already solved for n, so substituting $2q$ for n in the second equation gives the following.

$$0.05(2q) + 0.25q = 1.05$$
$$0.10q + 0.25q = 1.05$$
$$10q + 25q = 105$$

Multiply the equation by 100 so that the coefficients and constants are integers.

$$35q = 105$$
$$q = 3$$ Number of quarters
$$n = 2q = 2(3) = 6$$ Number of nickels

Mike has 3 quarters and 6 nickels.

Now work margin exercise 4.

4. Seth has \$1.40 worth of change in nickels and dimes. If he has 3 times as many dimes as nickels, how many of each type of coin does he have?

5. Enrique is 9 years older than his sister Maria. In 5 years, he will be twice as old as Maria. How old are Enrique and Maria now?

Example 5 Application: Calculating Age

Kathy is 6 years older than her sister, Sue. In 3 years, she will be twice as old as Sue. How old is each girl now?

Solution

We use two equations—one relating their ages now and the other relating their ages in 3 years.

Let K = Kathy's age now

and S = Sue's age now.

Then the system of linear equations is as follows.

$$\begin{cases} K - S = 6 & \text{The difference in their ages is 6 years.} \\ K + 3 = 2(S+3) & \text{In 3 years each age is increased by 3 and Kathy is twice as old as Sue.} \end{cases}$$

Rewrite the second equation in standard form.

$$\begin{aligned} K + 3 &= 2(S+3) \\ K + 3 &= 2S + 6 \\ K - 2S &= 3 \end{aligned}$$

Now, solve by addition.

$$\begin{cases} (K - S = 6) & \longrightarrow & K - S = 6 \\ [-1](K - 2S = 3) & \longrightarrow & -K + 2S = -3 \end{cases}$$

$$S = 3 \quad \text{Sue's current age}$$

Back substitute $S = 3$ into one of the original equations.

$$\begin{aligned} K - (3) &= 6 \\ K &= 6 + 3 \\ K &= 9 \quad \text{Kathy's current age} \end{aligned}$$

Kathy is 9 years old and Sue is 3 years old.

Now work margin exercise 5.

6. Seven sodas and six water bottles cost \$13.55. Three sodas and four water bottles cost \$6.95. What is the cost of a soda? What is the cost of a water bottle?

Example 6 Application: Finding Amounts and Cost

Three hot dogs and two orders of french fries cost \$10.30. Four hot dogs and four orders of french fries cost \$15.60. What is the cost of a hot dog? What is the cost of an order of french fries?

Solution

Let x = cost of one hot dog

and y = cost of one order of french fries.

The system of linear equations is as follows.

$$\begin{cases} 3x+2y=10.30 & \text{Three hot dogs and two orders of french fries cost \$10.30.} \\ 4x+4y=15.60 & \text{Four hot dogs and four orders of french fries cost \$15.60.} \end{cases}$$

Both equations are in standard form. Use the addtion method to solve the system.

$$\begin{cases} [-2](3x+2y=10.30) & \longrightarrow & -6x-4y=-20.60 \\ (4x+4y=15.60) & \longrightarrow & 4x+4y=\ \ 15.60 \end{cases}$$

$$\begin{aligned} -2x &= -5.00 \\ x &= 2.50 \quad \text{Cost of one hot dog} \end{aligned}$$

Back substitute $x = 2.50$ into one of the original equations.

$$\begin{aligned} 3(2.50)+2y &= 10.30 \\ 7.50+2y &= 10.30 \\ 2y &= 2.80 \\ y &= 1.40 \quad \text{Cost of one order of fries} \end{aligned}$$

One hot dog costs \$2.50 and one order of french fries costs \$1.40.

Now work margin exercise 6.

Margin Exercise Answers

1. The wind speed was 0.5 miles per hour and Bob was running 4.5 miles per hour.
2. 12:00 p.m. **3.** 93 and 57 **4.** 12 dimes and 4 nickels **5.** Enrique is 13 and Maria is 4.
6. A soda costs \$1.25 and a water bottle costs \$0.80

11.4 Exercises

Practice

Solve each problem by setting up a system of two equations in two unknowns and solve. See Examples 1 through 6.

1. The sum of two numbers is 56. Their difference is 10. Find the numbers.

2. The sum of two numbers is 40. The sum of twice the larger and 4 times the smaller is 108. Find the numbers.

3. The sum of two numbers is 36. Three times the smaller plus twice the larger is 87. Find the two numbers.

4. The sum of two integers is 102, and the larger number is 10 more than three times the smaller. Find the two integers.

5. The difference between two integers is 13, and their sum is 87. What are the two integers?

6. The difference between two numbers is 17. Four times the smaller is equal to 7 more than the larger. What are the numbers?

11.4 Exercises

Practice

1. 23, 33
2. 14, 26
3. 15, 21
4. 23, 79
5. 37, 50
6. 8, 25

Applications

7. 80°, 100°
8. 75°, 15°
9. 55°, 55°, 70°
10. 65°, 65°, 50°
11. The rate of the boat is 10 mph and the rate of the current is 2 mph.
12. The speed of the plane is 165 mph and the speed of the wind is 33 mph.
13. Bolt's speed was 10.32 meters per second and the wind speed was 0.12 meters per second.
14. Boat, 38 mph; Current, 2 mph
15. He traveled $1\frac{1}{2}$ hours at 52 mph and 2 hours at 56 mph.
16. She traveled $3\frac{3}{4}$ hours at 60 mph and 2 hours at 55 mph.

Applications

Solve

7. Two angles are supplementary if the sum of their measures is 180°. Find two supplementary angles such that the smaller is 30° more than one half of the larger.

8. Two angles are complementary if the sum of their measures is 90°. Find two complementary angles such that one is 15° less than six times the other.

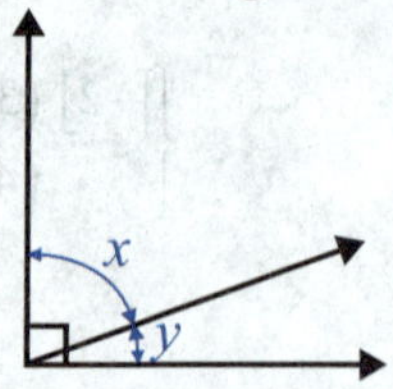

9. The sum of the measures of the three angles of a triangle is 180°. In an isosceles triangle, two of the angles have the same measure. What are the measures of the angles of an isosceles triangle in which one angle measures 15° more than each of the other two equal angles?

10. The sum of the measures of the three angles of a triangle is 180°. In an isosceles triangle, two of the angles have the same measure. What are the measures of the angles of an isosceles triangle in which each of the two equal angles measures 15° more than the third angle?

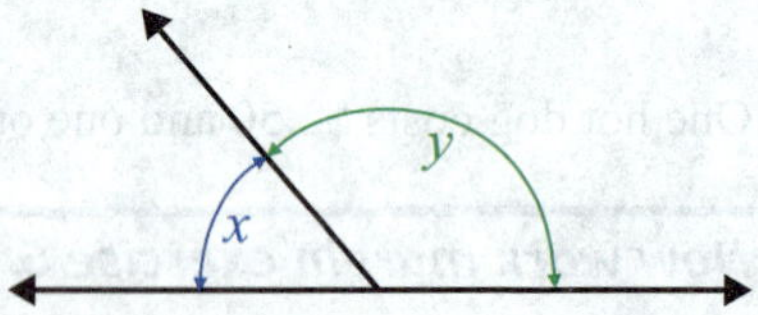

11. Liam makes a 4-mile motorboat trip downstream in 20 minutes $\left(\frac{1}{3}\text{ hr}\right)$. The return trip takes 30 minutes $\left(\frac{1}{2}\text{ hr}\right)$. Find the rate of the boat in still water and the rate of the current.

12. Mr. McKelvey finds that he can travel 1188 miles in 6 hours when flying with the wind. However, when flying against the wind, he travels only $\frac{2}{3}$ of the distance in the same amount of time. Find the speed of the plane in still air and the wind speed.

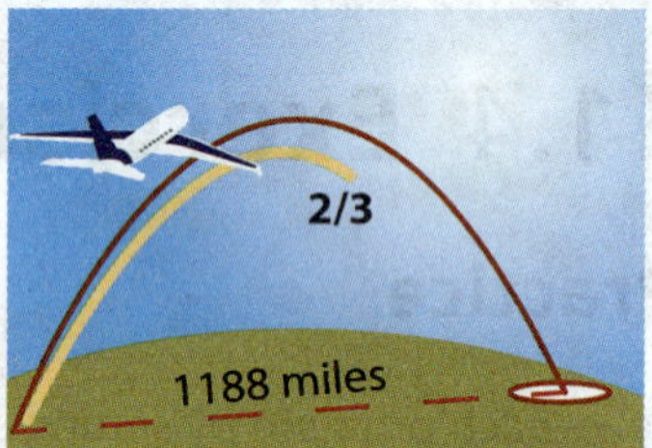

13. Usain Bolt, the world-record holder in the 100 meter dash, ran 100 meters in 9.69 seconds with no wind. He later ran the same distance in 9.58 seconds with the wind. What was his speed and what was the wind speed?

14. Jessica drove her speedboat upriver this morning. It took her 1 hour going upriver and 54 minutes going down river. If she traveled 36 miles each way, what would have been the rate of the boat in still water and what was the rate of the current (in miles per hour)?

15. Dominic went on a 190-mile business trip. He averaged 52 mph for the first part of the trip and 56 mph for the second part. If the total trip took $3\frac{1}{2}$ hours, how long did he travel at each rate?

16. Marian drove to a resort 335 miles from her home. She averaged 60 mph for the first part of her trip and 55 mph for the second part. If her total driving time was $5\frac{3}{4}$ hours, how long did she travel at each rate?

17. Marcos lives 364 miles away from his cousin Cana. They start driving at the same time and travel toward each other. Cana's speed is 11 mph faster than Marcos' speed. If they meet in 4 hrs, find their speeds.

18. Naomi and Linda live 324 miles apart. They start at the same time and travel toward each other. Naomi's speed is 8 mph greater than Linda's. If they meet in 3 hours, find their speeds.

19. Steve travels 4 times as fast as Tim. Starting at the same point, but traveling in opposite directions, they are 105 miles apart after 3 hours. Find their rates of travel.

20. Bella travels 5 mph less than twice as fast as June. Starting at the same point and traveling in the same direction, they are 80 miles apart after 4 hours. Find their speeds.

21. Two trains leave Dallas at the same time. One train travels east and the other travels west. The speed of the westbound train is 5 mph greater than the speed of the eastbound train. After 6 hours, they are 510 miles apart. Find the rate of each train. Assume the trains travel in a straight line in opposite directions.

22. A boat left Dana Point Marina at 11:00 a.m. traveling at 10 knots (nautical miles per hour). Two hours later, a Coast Guard boat left the same marina traveling at 14 knots trying to catch the first boat. If both boats traveled the same course, at what time did the Coast Guard captain anticipate overtaking the first boat?

23. A jogger runs into the countryside at a rate of 10 mph. He returns along the same route at 6 mph. If the total trip took 1 hour 36 minutes, how far did he jog?

24. A cyclist traveled to her destination at an average rate of 15 mph. By traveling 3 mph faster, she took 30 minutes less to return. What distance did she travel each way?

25. Sonja has some nickels and dimes. If she has 30 coins worth a total of $2.00, how many of each type of coin does she have?

26. Conner has a total of 27 coins consisting of quarters and dimes. The total value of the coins is $5.40. How many of each type of coin does he have?

27. A bag contains pennies and nickels only. If there are 182 coins in all and their value is $3.90, how many pennies and how many nickels are in the bag?

28. Your friend challenges you to figure out how many dimes and quarters are in a cash register. He tells you that there are 65 coins and that their value is $11.90. How many dimes and how many quarters are in the register?

29. The entry fee for a county carnival was $3.50 for adults and $2.50 for children. If the income for one day was $9950 and the attendance was 3500 people, how many adults and how many children attended the carnival that day?

30. The width of a rectangle is $\frac{3}{4}$ of its length. If the perimeter of the rectangle is 140 feet, what are the dimensions of the rectangle?

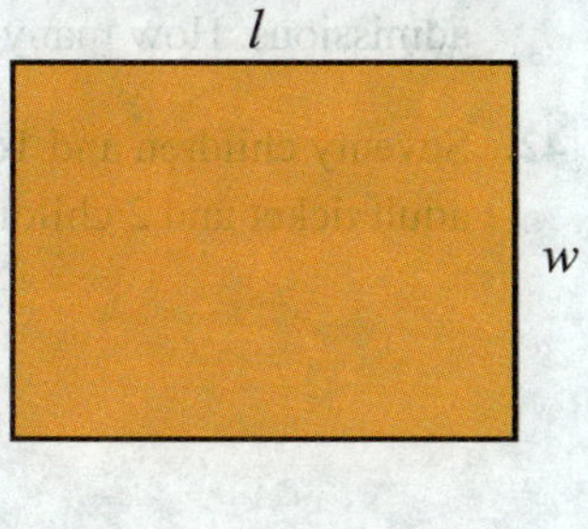

17. Marcos traveled at 40 mph and Cana traveled at 51 mph.

18. Naomi's speed was 58 mph and Linda's speed was 50 mph.

19. Steve traveled at 28 mph and Tim traveled at 7 mph.

20. Bella's speed was 45 mph and June's speed was 25 mph.

21. The westbound train was traveling 45 mph and the eastbound train was traveling 40 mph.

22. 6:00 p.m.

23. He jogged 12 miles.

24. She traveled 45 miles each way.

25. 20 nickels and 10 dimes

26. 18 quarters and 9 dimes

27. 52 nickels and 130 pennies

28. 36 quarters and 29 dimes

29. 800 adults and 2700 children attended.

30. $l = 40$ feet; $w = 30$ feet

31. $l = 14$ meters; $w = 8$ meters

32. $l = 97.5$ meters; $w = 32.5$ meters

33. 100 yards × 45 yards

34. Side of square is 15 ft; Side of hexagon is 10 ft

35. $l = 33$ meters; $w = 17$ meters

36. Ava is 12 years old; Curt is 4 years old.

37. Priscilla was 21 years old and Elvis was 32 years old.

38. 250 adult tickets and 70 children's tickets were sold.

39. She bought 10 paperbacks and 5 hardbacks.

40. Morton has 140 pounds of batteries and 40 pounds of cans.

41. 5000 general admission and 7500 reserved tickets were sold.

42. Adult ticket: \$3; Children's ticket: \$2.

31. The length of a rectangle is 10 meters more than one half of the width. If the perimeter is 44 meters, what are the length and width?

32. A farmer has 260 meters of fencing to build a rectangular corral. He wants the length to be 3 times as long as the width. What dimensions should he make his corral?

33. At present, the length of a rectangular soccer field is 55 yards longer than the width. The city wants to rearrange the area containing the soccer field into two square playing fields. A math teacher on the council told them that if the width of the current field were to be increased by 5 yards and the length cut in half, the resulting field would be a square. What are the dimensions of the field currently?

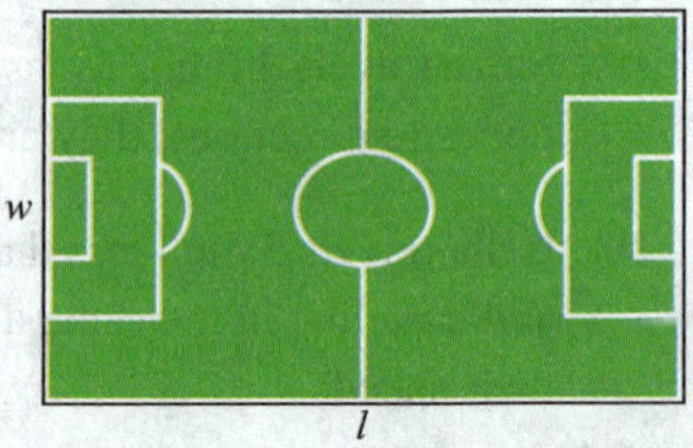

34. Consider a square and a regular hexagon (a six-sided figure with sides of equal length). One side of the square is 5 feet longer than a side of the hexagon, and the two figures have the same perimeter. What are the lengths of the sides of each figure?

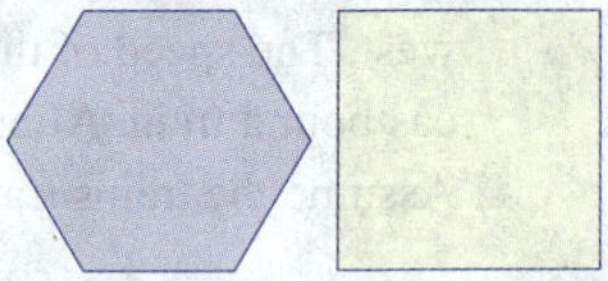

35. The length of a rectangle is 1 meter less than twice the width. If each side is increased by 4 meters, the perimeter will be 116 meters. Find the length and the width of the original rectangle.

36. Ava is 8 years older than her brother Curt. Four years from now, Ava will be twice as old as Curt. How old is each at the present time?

37. When they got married, Elvis Presley was 11 years older than his wife Priscilla. One year later, Priscilla was two-thirds of Elvis' age. How old was each of them when they got married?

38. A Christmas charity party sold tickets for \$45.00 for adults and \$25.00 for children. The total number of tickets sold was 320 and the total for the ticket sales was \$13,000. How many adult and how many children's tickets were sold?

39. Joan went to a book sale on campus and bought paperback books for \$0.25 each and hardback books for \$1.75 each. If she bought a total of 15 books for \$11.25, how many of each type of book did she buy?

40. Morton took some backup batteries and aluminum cans to the recycling center. Their total weight was 180 pounds. He received 1.5¢ per pound for the batteries and 30¢ per pound for the cans. The total received was \$14.10. How many pounds of each did Morton have?

41. Admission to a high-school baseball game is \$2.00 for general admission and \$3.50 for reserved seats. The receipts for the season were \$36,250 for 12,500 paid admissions. How many of each ticket, general and reserved, were sold?

42. Seventy children and 160 adults attended a play. The total receipts were \$620. One adult ticket and 2 children's tickets cost \$7. Find the price of each type of ticket.

43. Last summer, Ernie sold surfboards. One style sold for $625 and the other sold for $550. He sold a total of 47 surfboards. How many of each style did he sell if the sales from each style were equal?

44. The Candy Shack sells a particular candy in two different size packages. One size sells for $1.25 and the other sells for $1.75. If the store received $65.50 for 42 packages of candy, how many of each size were sold?

45. The pro shop at the Divots Country Club ordered two brands of golf balls. Titleless balls cost $1.80 each and the Done Lob balls cost $1.50 each. The total cost of Titleless balls exceeded the total cost of the Done Lob balls by $108. If equal numbers of each brand were ordered, how many dozen of each brand were ordered?

46. Sellit Realty Company gets a 6% fee for selling improved properties and 10% for selling unimproved land. Last week, the total sales were $220,000 and their total fees were $16,400. What were the sales from each of the two types of properties?

47. A men's clothing store sells two styles of sports jackets, one selling for $95 and one selling for $120. Last month, the store sold 40 jackets, with receipts totaling $4250. How many of each style did the store sell?

48. Frank bought 2 shirts and 1 pair of dress pants for a total of $55. If he had bought 1 shirt and 2 pairs of dress pants, he would have paid $68. What was the price of each shirt and each pair of dress pants?

49. At McDonalds, 3 Big Macs and 5 orders of medium French fries cost $21.53. Six Big Macs and 2 orders of medium French fries cost $28.34. What is the price of a Big Mac? What is the price of one order of medium French fries?

50. A bakery sells burnt and broken cookies for a discounted price. Burnt cookies are $0.49 and broken cookies are $0.70. If Spencer spends $49.70 to buy 80 cookies for a game night with his friends, how many of each type of cookie did Spencer buy?

51. A small manufacturer produces two kinds of radios, model *X* and model *Y*. Model *X* takes 4 hours to produce and costs $8 each to make. Model *Y* takes 3 hours to produce and costs $7 each to make. If the manufacturer decides to allot a total of 58 hours and $126 each week, how many of each model will be produced?

52. A furniture shop makes dining room chairs. Employees can build two styles of chairs. Style I takes 1 day and the materials cost $60. Style II takes $1\frac{1}{2}$ days but the materials only costs $30. If, during the last two months, they spent 36 days and $1200 building chairs, how many chairs of each style did they build?

43. 22 at $625 and 25 at $550

44. 16 packages at $1.25 and 26 packages at $1.75

45. 30 dozen (360 balls)

46. $140,000 in sales for the improved properties and $80,000 in sales for the unimproved properties

47. The store sold 22 of the $95 jackets and 18 of the $120 jackets.

48. Shirt: $14; Pair of dress pants: $27

49. One Big Mac costs $4.11 and one order of medium French fries costs $1.84.

50. 30 burnt cookies and 50 broken cookies.

51. They will produce 7 of Model X and 10 of Model Y.

52. They built 12 using Style I and 16 using Style II.

53. A petting zoo charges \$5 for children and \$10 for adults. On Tuesday, the petting zoo made \$1400 from ticket sales and sold a total of 200 tickets. How many adults and how many children visited the petting zoo on Tuesday?

a. Write two equations to describe the situation. Use the variable c to represent the number of children and the variable a to represent the number of adults.

b. Solve the system of linear equations.

c. Use the solution from part b. to write a complete sentence to answer the question from the problem.

54. A boat tour travels 4 miles downstream in 20 minutes and the return trip upstream takes 30 minutes. Find the rate of the boat and the rate of the current.

a. Change each time from minutes to hours. Write each time value as a fraction.

b. Use a table to set up a system of linear equations. Use the variable b to represent the rate of the boat and the variable c to represent the rate of the current.

c. Solve the system of linear equations.

d. Use the solution from part c. to write a complete sentence to answer the question from the problem.

Writing & Thinking

55. A two digit number can be written as ab, where a and b are the digits. We do not mean that the digits are multiplied, but the value of the number is $10a + b$. For example, the two digit number 34 has a value of $10 \cdot 3 + 4$. Set up and solve a system of equations for the following problem.
The sum of the digits of a two digit number is 13. If the digits are reversed, then the value of the number is increased by 45. What is the number?

53. a. $\begin{cases} 5c+10a=1400 \\ c+a=200 \end{cases}$

b. $c=120,\ a=80$

c. 120 children and 80 adults visited the petting zoo on Tuesday.

54. a. $\frac{1}{3}$ hour; $\frac{1}{2}$ hour

b. $\begin{cases} \frac{1}{3}(b+c)=4 \\ \frac{1}{2}(b-c)=4 \end{cases}$

	Rate	·	Time	=	Distance
Downstream	$b+c$		$\frac{1}{3}$		$\frac{1}{3}(b+c)$
Upstream	$b-c$		$\frac{1}{2}$		$\frac{1}{2}(b-c)$

c. $b=10,\ c=2$

d. The boat is traveling at a rate of 10 mph and the current is moving at a rate of 2 mph.

Writing & Thinking

55. The number is 49.

11.5 Applications: Interest and Mixture

Objectives

A. Use systems of linear equations to solve interest problems.

B. Use systems of linear equations to solve mixture problems.

In this section, we will study two more types of applications that can be solved using systems of linear equations: interest on money invested and mixture. These applications can be "wordy," and you will need to read carefully and analyze the information thoroughly to be able to translate it into a system involving two variables.

A Interest Problems

People in business and banking know several formulas for calculating interest. The formula used depends on the frequency of payment (monthly or yearly) and the type of interest (simple or compound). Penalties for late payments, and sometimes penalties for paying off a loan in full, may be involved. In any case, standard notation is the following:

P → the principal (amount of money invested or borrowed)

r → the rate of interest (an annual rate)

t → the time (in one year or part of a year)

I → the interest (paid or earned)

In this section, we will use only the basic formula for simple interest

$$I = Prt$$

with interest calculated on an annual basis. In this special case, we have $t = 1$ and the formula becomes

$$I = Pr.$$

Example 1 Application: Calculating Interest and Balances

James has two investment accounts, one pays 6% interest and the other pays 10% interest. He has \$1000 more in the 10% account than he has in the 6% account. In one year, the interest from the 10% account is \$260 more than the interest from the 6% account. How much does he have in each account?

Solution

Careful reading indicates two types of information:

1. He has two accounts.
2. He earns two amounts of interest.

Let $x =$ amount (principal) invested at 6%

and $y =$ amount (principal) invested at 10%.

Then,

$0.06x =$ interest earned on first account, and

$0.10y =$ interest earned on second account.

1. Fergus has two investment accounts for his toupee company. One pays 9% interest and the other pays 12% interest. He has \$800 more in the 12% account than he has in the 9% account. In one year, the interest from the 12% account is \$246 more than the interest from the 9% account. How much does Fergus have in each account?

Math Tip

You can use tables like those in Section 11.4 to help organize the information in a more understandable form.

Now set up two equations.

$$\begin{cases} y = x + 1000 & \text{The 10\% account has \$1000 more than the 6\% account.} \\ 0.10y - 0.06x = 260 & \text{Interest from the 10\% account is \$260 more than interest from the 6\% account.} \end{cases}$$

Because the first equation is already solved for y, we use the substitution method and substitute for y in the second equation.

$$0.10(x+1000) - 0.06x = 260$$

$$10(x+1000) - 6x = 26{,}000 \qquad \text{Multiply by 100 to get integer coefficients and constants.}$$

$$10x + 10{,}000 - 6x = 26{,}000$$

$$4x = 16{,}000$$

$$x = 4000 \qquad \text{Amount at 6\%}$$

Back substitute $x = 4000$ into one of the original equations to find y.

$$y = x + 1000 = (4000) + 1000 = 5000 \qquad \text{Amount at 10\%}$$

James has \$4000 invested at 6% and \$5000 invested at 10%.

Now work margin exercise 1.

2. Darnell has \$9000 to invest. He decides to separate his money into two accounts. One yields interest at the rate of 5% and the other at 9%. If he wants a total annual income from both accounts to be \$550, how should he split up the money?

Example 2 Application: Calculating Interest and Investments

Lila has \$7000 to invest. She decides to separate her funds into two accounts. One yields interest at the rate of 2% and the other at 5%. (The higher interest account is considered more risky. Otherwise, she would put the entire \$7000 into that account.) If she wants a total annual income from both accounts to be \$260, how should she split the money?

Solution

Again, careful reading indicates two types of information:

1. She has two accounts.
2. She earns two amounts of interest.

Let x = amount (principal) invested at 2%

and y = amount (principal) invested at 5%.

Then,

$0.02x$ = interest earned on first account

$0.05y$ = interest earned on second account.

Now set up two equations.

$$\begin{cases} x + y = 7000 & \text{The total amount invested is \$7000.} \\ 0.02x + 0.05y = 260 & \text{The total interest from both accounts is \$260.} \end{cases}$$

Both equations are in standard form. Solve by addition. Multiply the first equation by -2 and the second by 100 to get opposite coefficients for x as follows.

$$\begin{cases}[-2] & (x+\quad y=7000) \\ [100] & (0.02x+0.05y=260)\end{cases} \begin{array}{l}\longrightarrow \\ \longrightarrow\end{array} \begin{array}{rl}-2x-2y= & -14,000 \\ 2x+5y= & 26,000 \\ \hline 3y= & 12,000 \\ y= & 4000\end{array}$$

Amount at 5%

Substitute $y = 4000$ into one of the original equations.

$$x+(4000)=7000$$

$$x=3000 \quad \text{Amount at 2\%}$$

She should invest \$3000 at 2% and \$4000 at 5%.

Now work margin exercise 2.

B Mixture Problems

Problems involving mixtures occur in the fields of physics and chemistry and in such places as candy stores or coffee shops. In a mixture problem, two or more solutions are mixed together to form a final solution that satisfies certain conditions of percentage of concentration. For example, two solutions with different concentrations of salt might be mixed to create a final solution with a specific concentration of salt.

The basic plan is to write an equation that deals with only one part of the mixture (such as the salt in the mixture). The following examples explain how this can be accomplished.

Example 3 Application: Solving a Mixture Problem

How many ounces each of a 10% salt solution and a 15% salt solution must be used to produce 50 ounces of a 12% salt solution?

Solution

Let x = amount of 10% solution

and y = amount of 15% solution.

	Amount of Solution	×	Percent of Salt	=	Amount of Salt
10% Solution	x		0.10		$0.10x$
15% Solution	y		0.15		$0.15y$
12% Solution	50		0.12		$0.12(50)$

Then the system of linear equations is as follows.

$$\begin{cases} x+y=50 \\ 0.10x+0.15y=0.12(50)\end{cases}$$

The sum of the amounts of salt from the two solutions equals the total amount of salt in the final solution.

The sum of the two amounts must be 50 ounces.

3. How many ounces each of a 12% chlorine solution and a 18% chlorine solution must be used to produce 150 ounces of a 14% chlorine solution?

Teaching Note

Because "mixture" and "interest" appear to represent different contexts, students may not recognize that the structures of the equations are the same, or that the interest context is a mixture. You might emphasize the fact that both types of problems involve percent.

Now, multiply the first equation by -10 and the second by 100 to get opposite coefficients for x.

$$\begin{cases} [-10]\quad (x + \quad y = 50) & \longrightarrow & -10x - 10y = -500 \\ [100](0.10x + 0.15y = 0.12(50)) & \longrightarrow & 10x + 15y = 600 \end{cases}$$

$$5y = 100$$

$$y = 20 \quad \text{Amount of 15\%}$$

Substitute $y = 20$ into one of the original equations.

$$x + (20) = 50$$

$$x = 30 \quad \text{Amount of 10\%}$$

Use 30 ounces of the 10% solution and 20 ounces of the 15% solution.

Now work margin exercise 3.

4. How many gallons of a 17% soap solution should be mixed with a 22% soap solution to produce 120 gallons of a 21% solution?

Example 4 Application: Solving a Mixture Problem

How many gallons of a 20% acid solution should be mixed with a 30% acid solution to produce 100 gallons of a 23% solution?

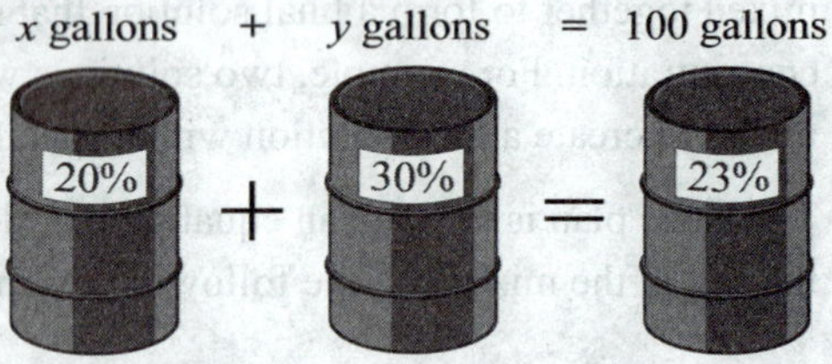

Solution

Let x = amount of 20% solution

and y = amount of 30% solution.

	Amount of Solution	×	Percent of Acid	=	Amount of Acid
20% Solution	x		0.20		$0.20x$
30% Solution	y		0.30		$0.30y$
23% Solution	100		0.23		0.23(100)

Then the system of linear equations is as follows.

$$\begin{cases} x + y = 100 \\ 0.20x + 0.30y = 0.23(100) \end{cases}$$

The sum of the two amounts must be 100 gallons.

The sum of the amounts of acid from the two solutions equals the total amount of acid in the final solution.

Multiply the first equation by -20 and the second by 100 to get opposite coefficients for x.

$$\begin{cases} [-20]\quad (x + \quad y = 100) & \longrightarrow & -20x - 20y = -2000 \\ [100](0.20x + 0.30y = 0.23(100)) & \longrightarrow & 20x + 30y = 2300 \end{cases}$$

$$10y = 300$$

$$y = 30 \quad \text{Amount of 30\%}$$

Back substitute $y = 30$ into one of the original equations.

$$x + (30) = 100$$
$$x = 70 \quad \text{Amount of 20\%}$$

70 gallons of the 20% solution should be added to 30 gallons of the 30% solution. This will produce 100 gallons of a 23% solution.

Now work margin exercise 4.

Margin Exercise Answers

1. Fergus has invested $5000 at 9% and $5800 at 12%.
2. He should invest $2500 at 9% and $6500 at 5%.
3. 50 ounces of the 18% solution and 100 ounces of the 12% solution.
4. 96 gallons of the 22% solution and 24 gallons of the 17% solution.

11.5 Exercises

Concept Check

Fill-in-the-Blank. Complete each sentence using information found in this section.

1. The formula used to calculate interest depends on the frequency of ______ and the type of ______.
2. In the interest formula $I = Prt$, P stands for the ______.
3. In the interest formula $I = Prt$, I stands for the interest ______ or ______.
4. In the interest formula $I = Prt$, r stands for the ______ of interest.
5. In the interest formula $I = Prt$, t stands for the ______ in years.
6. When solving a mixture problem, the basic plan is to write a/an ______ that deals with only one part of the mixture.

True/False. Determine whether each statement is true or false. If a statement is false, explain how it can be changed so that the statement will be true. (**Note:** There may be more than one acceptable change.)

7. When interest is calculated on an annual basis, we have $t = 0$ and the formula becomes $I = Pr$.
8. Problems involving mixture occur in the sciences such as physics and chemistry.
9. In an interest problem, time can be given in parts of a year.
10. When two or more items are mixed, the final mixture should satisfy certain conditions of percentage of concentration.

Applications

Solve each problem by setting up a system of two equations in two unknowns and solving the system. See Examples 1 through 4.

1. Carmen invested $9000. She invested part in a 6% passbook account and the rest in a 10% certificate account. If her annual interest was $680, how much did she invest at each rate?

11.5 Exercises

Concept Check

1. payment, interest
2. principal
3. paid or earned
4. rate
5. time
6. equation
7. False; When interest is calculated on an annual basis, $t = 1$.
8. True
9. True
10. True

Applications

1. $5500 at 6%; $3500 at 10%

2. $8500 at 6%; $3500 at 8%
3. $7400 at 5.5%; $2600 at 6%
4. $3200 at −3%; $6300 at 6%
5. $450
6. $1500 at 9%; $800 at 13%
7. $3500 in each or $7000 total
8. $900 at 12%; $2100 at 10%
9. $20,000 at 24%; $11,000 at 18%
10. $750 at 7%; $2250 at 8%
11. $800 at 5%; $2100 at 7%
12. $2500 at 8%; $3500 at 12%
13. $8500 at 9%; $3500 at 11%
14. $3800 at 15%; $4200 at 12%
15. $87,000 in bonds; $37,000 in certificates

2. Mr. Brown has $12,000 invested. Part is invested at 6% and the remainder at 8%. If the annual interest from the 6% investment is $230 more than the annual interest from the 8% investment, how much is invested at each rate?

3. Ten thousand dollars is invested, part at 5.5% and part at 6%. The annual interest from the 5.5% investment is $251 more than the annual interest from the 6% investment. How much is invested at each rate?

4. On two investments totaling $9500, Darius lost 3% on one and earned 6% on the other. If his net annual receipts were $282, how much was each investment?

5. Merideth has money in two savings accounts. One rate is 8% and the other is 10%. If she has $200 more in the 10% account, how much is invested at 8% if the total annual interest is $101?

6. Money is invested at two rates. One rate is 9% and the other is 13%. If there is $700 more invested at 9%, find the amount invested at each rate if the total annual interest is $239.

7. Ethan has half of his investments in stock paying an 11% dividend and the other half in a debentured stock paying 13% interest. If his total annual interest is $840, how much does he have invested?

8. Betty invested some of her money at 12% interest. She invested $300 more than twice that amount at 10%. How much is invested at each rate if her interest income is $318 annually?

9. A company invested some money in a development yielding 24% and $9000 less in a development yielding 18%. If the first investment produces $2820 more per year than the second, how much is invested in each development?

10. Victoria invests a certain amount of money at 7% annual interest and three times that amount at 8%. If her annual interest income is $232.50, how much does she have invested at each rate?

11. Jamal has a certain amount of money invested at 5% annual interest and $500 more than twice that amount invested in bonds yielding 7%. His total annual income from interest is $187. How much does he have invested at each rate?

12. A total of $6000 is invested, part at 8% and the remainder at 12%. How much is invested at each rate if the annual interest is $620?

13. Ms. Merriman has invested $12,000. Part is invested at 9% and the remainder at 11%. If the interest from the 9% investment is $380 more than the interest from the 11% investment, how much is invested at each rate?

14. Eight thousand dollars is invested, part at 15% and the remainder at 12%. If the annual interest income from the 15% investment is $66 more than the annual interest income from the 12% investment, how much is invested at each rate?

15. Morgan inherited $124,000 from her uncle. She invested a portion in bonds and the remainder in a long-term certificate account. The amount invested in bonds was $24,000 less than 3 times the amount invested in certificates. How much was invested in bonds and how much in certificates?

16. Sang has invested \$48,000, part at 6% and the rest in a higher risk investment at 10%. How much did she invest at each rate to receive \$4000 in interest after one year?

17. A metallurgist has one alloy containing 20% copper and another containing 70% copper. How many pounds of each alloy must he use to make 50 pounds of a third alloy containing 50% copper?

18. A manufacturer has received an order for 24 tons of a 60% copper alloy. His stock contains only alloys of 80% copper and 50% copper. How much of each will he need to melt together to fill the order?

19. A tobacco shop wants 50 ounces of tobacco that is 24% rare Turkish blend. How much each of a 30% Turkish blend and a 20% Turkish blend will be needed?

20. How many liters each of a 40% acid solution and a 55% acid solution must be used to produce 60 liters of a 45% acid solution?

21. A dairy farmer wants to mix a 35% protein supplement and a standard 15% protein supplement to make 1800 pounds of a high-grade 20% protein supplement. How many pounds of each should he use?

22. To meet the government's specifications, a certain alloy must be 65% aluminum. How many pounds each of a 70% aluminum alloy and a 54% aluminum alloy will be needed to produce 640 pounds of the 65% aluminum alloy?

23. A butcher shop has ground beef that is 40% fat and extra lean ground beef that is only 15% fat. How many pounds of each will be needed to obtain 50 pounds of lean ground beef that is 25% fat?

24. George decides to mix grades of gasoline in his truck. He puts in 8 gallons of regular and 12 gallons of premium for a total cost of \$55.80. If premium gasoline costs \$0.15 more per gallon than regular, what was the price of each grade of gasoline?

25. How many grams of pure acid (100% acid) and how many grams of a 40% solution should be mixed together to get a total of 30 grams of a 60% solution?

26. Dark chocolate made with 100% cacao by weight is mixed with dark chocolate that is 60% cacao. How much of each (100% cacao and 60% cacao) should be used to get a mixture of 50 pounds of chocolate that is 70% cacao by weight?

27. Pure salt is to be added to a 4% salt solution. How many ounces of salt and how many ounces of the 4% solution should be mixed together to get 60 ounces of a 20% salt solution?

28. How many liters each of a 12% iodine solution and a 30% iodine solution must be used to produce a total mixture of 90 liters of a 22% iodine solution?

29. A candymaker is making truffles using a mixture of a melted dark chocolate that is 72% cocoa and milk chocolate that is 42% cocoa. If she wants 6 pounds of melted chocolate that is 52% cocoa, how much of each type of chocolate does she need?

30. A dairy supplier needs 360 gallons of milk containing 4% butterfat. How many gallons each of milk containing 5% butterfat and milk containing 2% butterfat must be used to obtain the desired 360 gallons?

16. \$20,000 at 6%; \$28,000 at 10%
17. 20 pounds of 20%; 30 pounds of 70%
18. 8 tons of 80%; 16 tons of 50%
19. 20 ounces of 30%; 30 ounces of 20%
20. 40 liters of 40%; 20 liters of 55%
21. 450 pounds of 35%; 1350 pounds of 15%
22. 440 pounds of 70%; 200 pounds of 54%
23. 20 pounds of 40%; 30 pounds of 15%
24. Regular costs \$2.70 per gallon; Premium costs \$2.85 per gallon.
25. 10 g of pure acid; 20 g of the 40% solution
26. 12.5 lb of 100% cacao; 37.5 lb of 60% cacao
27. 10 oz of salt; 50 oz of the 4% solution
28. 40 liters of 12%; 50 liters of 30%
29. 2 lb of 72%; 4 lb of 42%
30. 240 gallons of 5%; 120 gallons of 2%

31. 7.11 ounces of 0.5% solution; 0.89 ounces of the 5% solution

32. 27 oz of 25%; 9 oz of 45%

33. a. $\begin{cases} x + y = 5000 \\ 0.04x + 0.09y = 350 \end{cases}$

Account	Principal	·	Interest Rate	=	Interest
Account A	x		0.04		$0.04x$
Account B	y		0.09		$0.09y$
Totals	$5000		N/A		$350

b. $x = 2000$, $y = 3000$

c. Sanjay should invest $2000 in Account A and $3000 in Account B to earn $350 in interest.

d. Yes; Invest more in Account B.

34. a. $\begin{cases} x + y = 20 \\ 0.03x + 0.12y = 1.2 \end{cases}$

% Solution	Amount of Solution	·	Percent of Peroxide	=	Amount of Peroxide
3%	x		0.03		$0.03x$
12%	y		0.12		$0.12y$
6%	20		0.06		$0.06(20) = 1.2$

b. $x = 13\frac{1}{3}$, $y = 6\frac{2}{3}$

c. You should mix $13\frac{1}{3}$ ounces of 3% hydrogen peroxide with $6\frac{2}{3}$ ounces of 12% hydrogen peroxide to create 20 ounces of 6% hydrogen peroxide.

35. a. $5x + 7.50y = 6.50$; $x + y = 1$

b. $x = 0.4$, $y = 0.6$

c. The manager should mix 0.4 pounds of cashews with 0.6 pounds of almonds to create a mixture which will cost $6.50 for a one-pound bag.

Writing & Thinking

36. a. Answers will vary.

b. Answers will vary.

31. 🖩 Care guidelines recommend that new body piercings are clearned with a 1% salt solution for the first few weeks with the new piercing. You have a 0.5% solution and a 5% solution, and you need to make 8 ounces of the 1% solution. How much of the 0.5% and 5% solutions will you need? (Round your answers to the nearest hundredth.)

32. A pharmacist has two solutions of alcohol. One is 25% alcohol. The other is 45% alcohol. He wants to mix these two solutions to get 36 ounces that is 30% alcohol. How many ounces of each of these two solutions should he mix together?

33. Sanjay has $5000 to invest, and he has an option to split the amount between two simple interest accounts. Account A is expected to earn 4% interest and Account B is expected to earn 9% interest. If he has a goal to make $350 in interest after one year, how much should he invest in each account?

a. Use a table to set up a system of linear equations to describe the situation. Use the variable x to represent the amount invested in Account A and the variable y to represent the amount invested in Account B.

b. Solve the system of linear equations from part a.

c. Write a complete sentence to answer the question from the problem.

d. Is it possible for Sanjay to earn more than $350 in interest? If yes, explain how.

34. You have 3% hydrogen peroxide and 12% hydrogen peroxide. You need 20 ounces of 6% hydrogen peroxide. How many ounces of each grade of hydrogen peroxide do you need to mix together to obtain the required amount of 6% hydrogen peroxide?

a. Set up a system of linear equations to describe the situation. Use the variable x to represent the amount of 3% solution and the variable y to represent the amount of 12% solution.

b. Solve the system of linear equations.

c. Write a complete sentence to answer the question from the problem.

35. King Nut Company sells freshly roasted nuts. A one-pound bag of broken fancy cashews costs $5. A one-pound bag of almonds costs $7.50. The manager wants to sell a mixture of the nuts that will cost $6.50 for a one-pound bag. How much of each type of nut should he combine to make a one-pound bag of mixed nuts?

a. Write two equations to describe the situation. Use the variable x to represent the amount of cashews and the variable y to represent the amount of almonds.

b. Solve the system of linear equations.

c. Write a complete sentence to answer the question from the problem.

Writing & Thinking

36. Your friend has $20,000 to invest and decided to invest part at 4% interest and the rest at 10% interest. Why might you advise him (or her) to invest all of it

a. at 4%?

b. at 10%?

11.6 Systems of Linear Equations: Three Variables

Objectives

A. Solve systems of linear equations in three variables.

B. Use systems of linear equations in three variables to solve applications.

A Solving Systems in Three Variables

The equation $2x + 3y - z = 16$ is called a **linear equation in three variables**. The standard form (or general form) is

$$Ax + By + Cz = D,$$

where A, B, and C are not all equal to 0. The solutions to such equations are called **ordered triples** and are of the form (x_0, y_0, z_0). One ordered triple that satisfies the equation $2x + 3y - z = 16$ is $(1, 4, -2)$. To check this, substitute $x = 1$, $y = 4$, and $z = -2$ into the equation to see if the result is 16.

$$\begin{aligned} 2(1)+3(4)-(-2) &= 2+12+2 \\ &= 16 \end{aligned}$$

There are an infinite number of ordered triples that satisfy any linear equation in three variables in which at least two of the coefficients are nonzero. Any two values may be substituted for two of the variables, and then the value for the third variable can be calculated. For example, by letting $x = -1$ and $y = 5$ in the equation $2x + 3y - z = 16$, we find the following:

$$\begin{aligned} 2(-1)+3(5)-z &= 16 \\ -2+15-z &= 16 \\ -z &= 3 \\ z &= -3 \end{aligned}$$

Hence, the ordered triple $(-1, 5, -3)$ also satisfies the equation $2x + 3y - z = 16$.

Graphs can be drawn in three dimensions by using a coordinate system involving three mutually perpendicular number lines labeled as the x-axis, y-axis, and z-axis. Three planes are formed: the xy-plane, the xz-plane, and the yz-plane. The three axes separate space into eight regions called **octants**. You can picture the first octant as the region bounded by the floor of a room and two walls with the axes meeting in a corner. The floor is the xy-plane. The axes can be ordered in a "right-hand" or "left-hand" format. Figure 1 shows the point represented by the ordered triple $(2, 3, 1)$ in a right-hand system, where the positive z-axis points up.

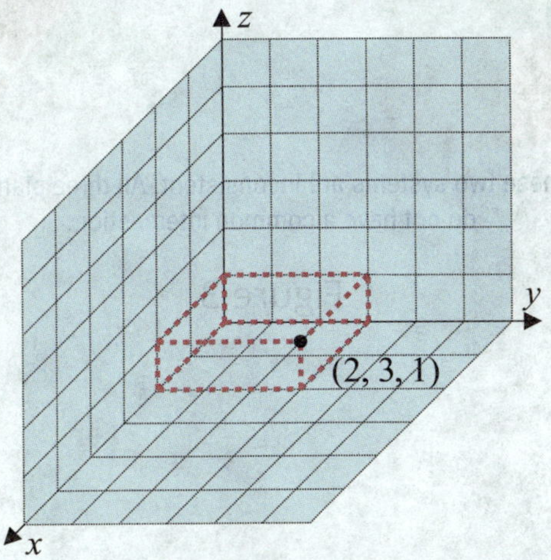

Figure 1

Teaching Note

To help students "see" the three-dimensional aspect, hold three meter sticks perpendicular to each other and discuss coordinates in class.

Math Tip

The difference between a right-hand system and a left-hand system is the direction we define the positive axes to point. In the right-hand system shown in Figure 1, the positive x-, y-, and z-axes point forward, *right*, and up, respectively. If we changed it to a left-hand system the positive x-, y-, and z-axes would point forward, *left*, and up, respectively.

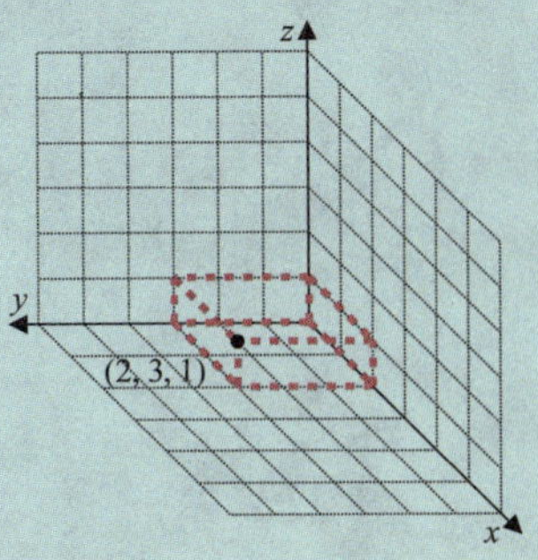

Teaching Note

Have students notice commonly created instances of two planes coming together to form a line, (for example, a wall and floor) and three planes coming together to form a point (for example, a wall, floor, and adjacent wall).

The graphs of linear equations in three variables are planes in three dimensions. A portion of the graph of $2x + 3y - z = 16$ appears in Figure 2.

Two distinct planes will either be parallel or they will intersect. If they intersect, their intersection will be a line. If three distinct planes intersect, they will intersect in a line or in a single point represented by an ordered triple.

The graphs of systems of three linear equations in three variables can be both interesting and informative, but they can be difficult to sketch and points of intersection difficult to estimate. Also, most graphing calculators are limited to graphs in two dimensions, so they are not useful in graphically analyzing systems of linear equations in three variables.

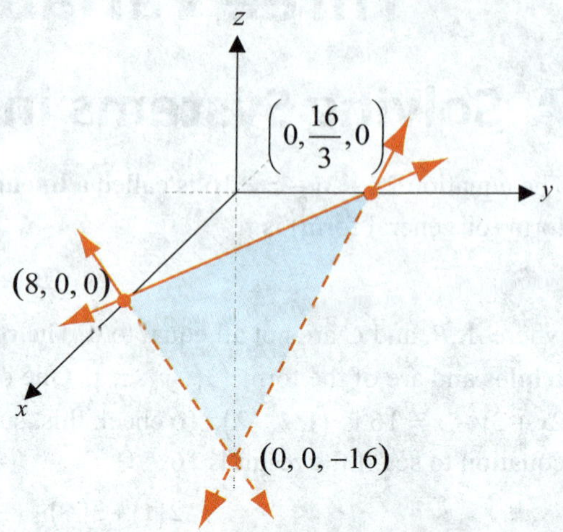

Figure 2

Therefore, in this text, only algebraic techniques for solving these systems will be discussed. Figure 3 illustrates four different possibilities for the relative positions of three planes.

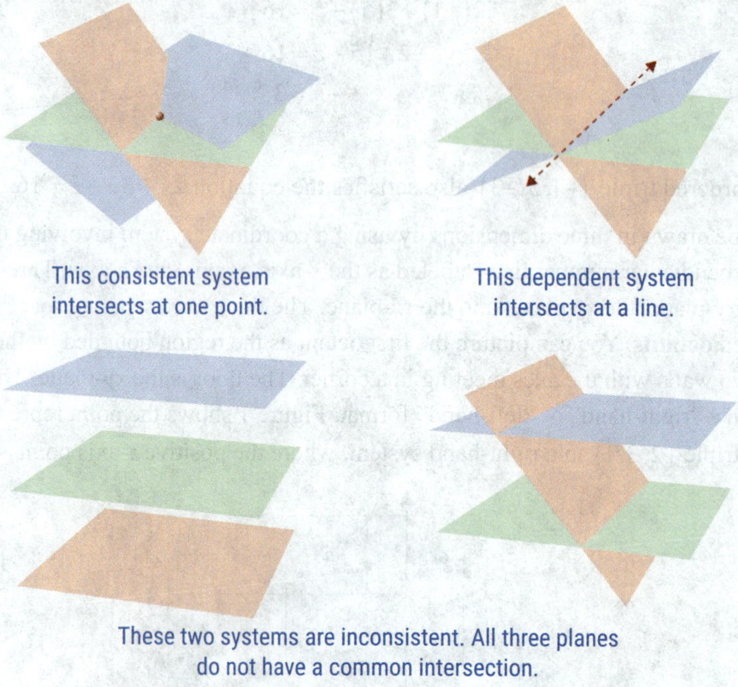

Figure 3

Solving a System of Three Linear Equations in Three Variables

1. Select two equations and eliminate one variable by using the addition method.
2. Select a different pair of equations and eliminate the **same** variable.
3. Steps 1 and 2 give **two** linear equations in **two** variables. Solve these equations by either addition or substitution, as discussed in Sections 11.2 and 11.3.
4. Back substitute the values found in Step 3 into any one of the original equations to find the value of the third variable.
5. Check the solution (if one exists) in all three of the original equations.

PROCEDURE

The solution possibilities for a system of three equations in three variables are as follows.

1. There will be exactly one ordered triple solution. (Graphically, the three planes intersect at one point.)
2. There will be an infinite number of solutions. (Graphically, the three planes intersect in a line or are the same plane.)
3. There will be no solutions. (Graphically, there are no points common to all three planes.)

The technique is illustrated with the following system.

$$\begin{cases} 2x + 3y - z = 16 & \text{(I)} \\ x - y + 3z = -9 & \text{(II)} \\ 5x + 2y - z = 15 & \text{(III)} \end{cases}$$

Step 1: Using equations (I) and (II), eliminate y.

$$\begin{array}{l} \text{(I)} \\ \text{(II)} \end{array}\begin{cases} \ (2x + 3y - z = 16) \longrightarrow \\ [3]\ (x - y + 3z = -9) \longrightarrow \end{cases} \begin{array}{r} 2x + 3y - z = 16 \\ 3x - 3y + 9z = -27 \\ \hline 5x + 8z = -11 \ \text{(IV)} \end{array}$$

Step 2: Using a different pair of equations, (II) and (III), eliminate the **same** variable, y.

$$\begin{array}{l} \text{(II)} \\ \text{(III)} \end{array}\begin{cases} [2]\ (x - y + 3z = -9) \longrightarrow \\ \ (5x + 2y - z = 15) \longrightarrow \end{cases} \begin{array}{r} 2x - 2y + 6z = -18 \\ 5x + 2y - z = 15 \\ \hline 7x + 5z = -3 \ \text{(V)} \end{array}$$

Using the results of Steps 1 and 2, solve the two equations for x and z.

$$\begin{array}{l} \text{(IV)} \\ \text{(V)} \end{array}\begin{cases} [-7]\ (5x + 8z = -11) \longrightarrow \\ [5]\ (7x + 5z = -3) \longrightarrow \end{cases} \begin{array}{r} -35x - 56z = 77 \\ 35x + 25z = -15 \\ \hline -31z = 62 \\ z = -2 \end{array}$$

Note

In Step 1, we could have chosen to begin by eliminating x or z. To be sure that you understand the process, you might want to solve the system by first eliminating x and then again by first eliminating z. In all cases, the answer will be the same.

Back substitute $z = -2$ into equation (IV) to find x.

$$\begin{aligned} 5x + 8(-2) &= -11 && \text{Use equation (IV).} \\ 5x - 16 &= -11 \\ 5x &= 5 \\ x &= 1 \end{aligned}$$

Step 3: Using $x = 1$ and $z = -2$, back substitute to find y.

$$\begin{aligned} (1) - y + 3(-2) &= -9 && \text{Use equation (II).} \\ -y - 5 &= -9 \\ -y &= -4 \\ y &= 4 \end{aligned}$$

The solution is $(1, 4, -2)$. The solution can be checked by substituting the results into **all three** of the original equations.

$$\begin{cases} 2(1) + 3(4) - (-2) = 16 & \text{(I)} \\ (1) - (4) + 3(-2) = -9 & \text{(II)} \\ 5(1) + 2(4) - (-2) = 15 & \text{(III)} \end{cases}$$

1. Solve.

$$\begin{cases} 2x + y + z = 4 \\ x + 2y + z = 1 \\ 3x + y - z = -3 \end{cases}$$

Example 1 Solving a System with Three Variables (One Solution)

Solve the system of equations.

$$\begin{cases} x - y + 2z = -4 & \text{(I)} \\ 2x + 3y + z = \frac{1}{2} & \text{(II)} \\ x + 4y - 2z = 4 & \text{(III)} \end{cases}$$

Solution

Using equations (I) and (III), eliminate z.

$$\begin{array}{lrcrcrcr} \text{(I)} & x & - & y & + & 2z & = & -4 \\ \text{(III)} & x & + & 4y & - & 2z & = & 4 \\ \hline & 2x & + & 3y & & & = & 0 \ \text{(IV)} \end{array}$$

Using equations (I) and (II), eliminate z.

$$\begin{array}{l} \text{(I)} \quad (x - y + 2z = -4) \longrightarrow x - y + 2z = -4 \\ \text{(II)} \quad [-2]\left(2x + 3y + z = \frac{1}{2}\right) \longrightarrow -4x - 6y - 2z = -1 \\ \hline \qquad\qquad\qquad -3x - 7y = -5 \ \text{(V)} \end{array}$$

Eliminate the variable x using the two equations in x and y.

$$\begin{array}{l} \text{(IV)} \quad [3] \ (2x + 3y = 0) \longrightarrow 6x + 9y = 0 \\ \text{(V)} \quad [2] \ (-3x - 7y = -5) \longrightarrow -6x - 14y = -10 \\ \hline \qquad\qquad\qquad -5y = -10 \\ \qquad\qquad\qquad y = 2 \end{array}$$

Back substituting into equation (IV) to find x yields the following.

$$\begin{aligned} 2x+3(2)&=0 \\ 2x+6&=0 \\ 2x&=-6 \\ x&=-3 \end{aligned}$$

Finally, using $x=-3$ and $y=2$, back substitute into (I).

$$\begin{aligned} (-3)-(2)+2z&=-4 \\ -5+2z&=-4 \\ 2z&=1 \\ z&=\frac{1}{2} \end{aligned}$$

The solution is $\left(-3, 2, \frac{1}{2}\right)$.

Check

The solution can be checked by substituting $\left(-3, 2, \frac{1}{2}\right)$ into all three of the original equations.

$$\begin{cases} (-3)-(2)+2\left(\frac{1}{2}\right)=-4 & \text{(I)} \\ 2(-3)+3(2)+\left(\frac{1}{2}\right)=\frac{1}{2} & \text{(II)} \\ (-3)+4(2)-2\left(\frac{1}{2}\right)=4 & \text{(III)} \end{cases}$$

Now work margin exercise 1.

Example 2 Solving a System with Three Variables (No Solution)

Solve the system of equations.

$$\begin{cases} 3x-5y+z=6 & \text{(I)} \\ x-y+3z=-1 & \text{(II)} \\ 2x-2y+6z=5 & \text{(III)} \end{cases}$$

Solution

Using equations (I) and (II), eliminate z.

$$\begin{array}{ll} \text{(I)} & [-3]\ (3x - 5y + z = 6) \longrightarrow -9x + 15y - 3z = -18 \\ \text{(II)} & \ (x - y + 3z = -1) \longrightarrow \underline{x - y + 3z = -1} \\ & \qquad\qquad\qquad\qquad\qquad\qquad -8x + 14y = -19 \ \text{(IV)} \end{array}$$

2. Solve.

$$\begin{cases} 7x+2y-z=7 \\ 3x-y-2z=1 \\ 9x-3y-6z=-3 \end{cases}$$

Using equations (II) and (III), eliminate z.

$$\begin{aligned} &\text{(II)} \quad [-2](x - y + 3z = -1) \rightarrow -2x + 2y - 6z = 2 \\ &\text{(III)} \quad (2x - 2y + 6z = 5) \rightarrow \underline{2x - 2y + 6z = 5} \\ &\qquad\qquad\qquad\qquad\qquad\qquad\qquad 0 = 7 \text{ (V)} \end{aligned}$$

This last equation is **false**. Thus, the system has **no solution**.

Now work margin exercise 2.

3. Solve.

$$\begin{cases} x + 2y + 3z = 4 \\ 2x + 3y + 3z = 7 \\ x + 3y + 6z = 5 \end{cases}$$

Example 3 Three Variables (A System with Infinitely Many Solutions)

Solve the system of equations.

$$\begin{cases} 3x - 2y + 4z = 1 & \text{(I)} \\ 3x - y + 7z = -1 & \text{(II)} \\ 6x - 5y + 5z = 4 & \text{(III)} \end{cases}$$

Solution

Using equations (I) and (II), eliminate x.

$$\begin{aligned} &\text{(I)} \quad (3x - 2y + 4z = 1) \rightarrow 3x - 2y + 4z = 1 \\ &\text{(II)} \quad [-1](3x - y + 7z = -1) \rightarrow \underline{-3x + y - 7z = 1} \\ &\qquad\qquad\qquad\qquad\qquad\qquad\qquad -y - 3z = 2 \quad \text{(IV)} \end{aligned}$$

Using equations (I) and (III), eliminate x.

$$\begin{aligned} &\text{(I)} \quad [-2](3x - 2y + 4z = 1) \rightarrow -6x + 4y - 8z = -2 \\ &\text{(III)} \quad (6x - 5y + 5z = 4) \rightarrow \underline{6x - 5y + 5z = 4} \\ &\qquad\qquad\qquad\qquad\qquad\qquad\qquad -y - 3z = 2 \quad \text{(V)} \end{aligned}$$

Using equations (IV) and (V), eliminate y.

$$\begin{aligned} &\text{(IV)} \quad (-y - 3z = 2) \rightarrow -y - 3z = 2 \\ &\text{(V)} \quad [-1](-y - 3z = 2) \rightarrow \underline{y + 3z = -2} \\ &\qquad\qquad\qquad\qquad\qquad\qquad 0 = 0 \end{aligned}$$

Because this last equation, $0 = 0$, is **always true**, the system has an **infinite number of solutions**.

Now work margin exercise 3.

B Solving Applications Using Systems in Three Variables

Example 4 Application: Using a System with Three Variables

A cash register contains $341 in $20, $5, and $2 bills. There are twenty-eight bills in all and three more $2 bills than $5 bills. How many bills of each kind are there?

Solution

Let x = number of $20 bills,

y = number of $5 bills, and

z = number of $2 bills.

$$\begin{cases} x+y+z=28 & \text{(I)} \quad \text{There are twenty-eight total bills.} \\ 20x+5y+2z=341 & \text{(II)} \quad \text{The total value is \$341.} \\ z=y+3 & \text{(III)} \quad \text{There are three more \$2 bills than \$5 bills.} \end{cases}$$

Using equations (I) and (II), eliminate x.

$$\begin{array}{ll} \text{(I)} & [-20] \quad (x + y + z = 28) \longrightarrow -20x - 20y - 20z = -560 \\ \text{(II)} & \quad (20x + 5y + 2z = 341) \longrightarrow \underline{20x + 5y + 2z = 341} \\ & \qquad\qquad - 15y - 18z = -219 \text{ (IV)} \end{array}$$

As equation (III) already has no x term, we rewrite it in the form $y - z = -3$ and use this equation along with the equation just found.

$$\begin{array}{ll} \text{(III)} & [15] \quad (y - z = -3) \longrightarrow 15y - 15z = -45 \\ \text{(IV)} & \quad (-15y - 18z = -219) \longrightarrow \underline{-15y - 18z = -219} \\ & \qquad\qquad -33z = -264 \\ & \qquad\qquad z = 8 \end{array}$$

Back substituting into equation (III) to solve for y gives the following.

$$\begin{aligned} (8) &= y+3 \\ 5 &= y \end{aligned}$$

Now we can substitute the values $z = 8$ and $y = 5$ into equation (I).

$$\begin{aligned} x+(5)+(8) &= 28 \\ x+13 &= 28 \\ x &= 15 \end{aligned}$$

There are fifteen $20 bills, five $5 bills, and eight $2 bills.

Now work margin exercise 4.

4. A piggy bank contains $21.70 in $1 coins, quarters and dimes. There are 37 coins in all and 5 more quarters than dimes. How many coins of each kind are there?

Margin Exercise Answers

1. $(1, -2, 4)$ **2.** The system has no solution. **3.** The system has an infinite number of solutions.
4. 18 dollar coins, 12 quarters, and 7 dimes

11.6 Exercises

Concept Check

1. $Ax + By + Cz = D$
2. ordered triples
3. infinite
4. x, y, z
5. octants
6. point
7. False; Choose 2 equations, eliminate 1 variable.
8. True
9. True
10. False; May be intersecting or parallel

Practice

1. $(1, 0, 1)$
2. $(2, 7, 4)$
3. $(1, 2, -1)$
4. $(1, 3, 3)$
5. Infinite number of solutions
6. $(-2, 9, 1)$

11.6 Exercises

Concept Check

Fill-in-the-Blank. Complete each sentence using information found in this section.

1. The standard form of a linear equation in three variables is ________.
2. Solutions to a linear equation in three variables are called ________ ________.
3. There can be 0, 1, or a/an ________ number of solutions to a system of three linear equations in three variables.
4. Graphs can be created in three dimensions by using a coordinate system involving three mutually perpendicular number lines—the ___-axis, ___-axis, and ___-axis.
5. The three coordinate axes separate space into eight regions called ______.
6. If three distinct planes intersect, they will do so in a straight line or in a single ______ represented by an ordered triple.

True/False. Determine whether each statement is true or false. If a statement is false, explain how it can be changed so the statement will be true. (**Note:** There may be more than one acceptable change.)

7. To find the solution for a system of three linear equations in three variables, start by choosing two variables and eliminating one equation.
8. An ordered triple believed to be a solution to a system of three linear equations in three variables should be checked in all three original equations.
9. If two distinct planes intersect, that intersection forms a straight line.
10. Two distinct planes will always intersect.

Practice

Solve each system of equations. State which systems, if any, have no solution or an infinite number of solutions. See Examples 1 through 3.

1. $$\begin{cases} x + y - z = 0 \\ 3x + 2y + z = 4 \\ x - 3y + 4z = 5 \end{cases}$$

2. $$\begin{cases} x - y + 2z = 3 \\ -6x + y + 3z = 7 \\ x + 2y - 5z = -4 \end{cases}$$

3. $$\begin{cases} 2x - y - z = 1 \\ 2x - 3y - 4z = 0 \\ x + y - z = 4 \end{cases}$$

4. $$\begin{cases} y + z = 6 \\ x + 5y - 4z = 4 \\ x - 3y + 5z = 7 \end{cases}$$

5. $$\begin{cases} x + y - 2z = 4 \\ 2x + y = 1 \\ 5x + 3y - 2z = 6 \end{cases}$$

6. $$\begin{cases} x - y + 5z = -6 \\ x + 2z = 0 \\ 6x + y + 3z = 0 \end{cases}$$

7. $\begin{cases} y+z=2 \\ x+z=5 \\ x+y=5 \end{cases}$

8. $\begin{cases} 2y+z=-4 \\ 3x+4z=11 \\ x+y=-2 \end{cases}$

9. $\begin{cases} x-y+2z=-3 \\ 2x+y-z=5 \\ 3x-2y+2z=-3 \end{cases}$

10. $\begin{cases} x-y-2z=3 \\ x+2y+z=1 \\ 3y+3z=-2 \end{cases}$

11. $\begin{cases} 2x-y+5z=-2 \\ x+3y-z=6 \\ 4x+y+3z=-2 \end{cases}$

12. $\begin{cases} 2x-y+5z=5 \\ x-2y+3z=0 \\ x+y+4z=7 \end{cases}$

13. $\begin{cases} 3x+y+4z=-6 \\ 2x+3y-z=2 \\ 5x+4y+3z=2 \end{cases}$

14. $\begin{cases} 2x+y-z=-3 \\ -x+2y+z=5 \\ 2x+3y-2z=-3 \end{cases}$

15. $\begin{cases} x-2y+z=7 \\ x-y-4z=-4 \\ x+4y-2z=-5 \end{cases}$

16. $\begin{cases} 2x-2y+3z=4 \\ x-3y+2z=2 \\ x+y+z=1 \end{cases}$

17. $\begin{cases} 2x+3y+z=4 \\ 3x-5y+2z=-5 \\ 4x-6y+3z=-7 \end{cases}$

18. $\begin{cases} x+y+z=3 \\ 2x-y-2z=-3 \\ 3x+2y+z=4 \end{cases}$

19. $\begin{cases} 2x-3y+z=-1 \\ 6x-9y-4z=4 \\ 4x+6y-z=5 \end{cases}$

20. $\begin{cases} x+6y+z=6 \\ 2x+3y-2z=8 \\ 2x+4z=3 \end{cases}$

Applications

Solve.

21. A florist is creating the bridesmaids' bouquets for a wedding. Each bouquet will cost \$92 and have a mixture of 16 flowers consisting of tiger lilies, cream roses, and white daisies. The lilies cost \$10 each, the roses cost \$6 each, and the daisies cost \$4 each. If each bouquet will have as many daisies as it has roses and lilies combined, how many of each type of flower will be in the bouquet?
    - **a.** Write a system of linear equations in three variables to describe the situation. Remember that you will have three equations. Use the variable *x* to represent the number of lilies, the variable *y* to represent the number of roses, and the variable *z* to represent the number of daisies.
    - **b.** Solve the system of equations from part a.
    - **c.** Use the solution from part b. to answer the question from the problem statement. Write a complete sentence.

7. $(4,1,1)$
8. $(1,-3,2)$
9. $(1,2,-1)$
10. Infinite number of solutions
11. $(-2,3,1)$
12. $(1,2,1)$
13. No solution
14. $(-1,1,2)$
15. $(3,-1,2)$
16. No solution
17. $(2,1,-3)$
18. $(1,-1,3)$
19. $\left(\frac{1}{2},\frac{1}{3},-1\right)$
20. $\left(\frac{5}{2},\frac{2}{3},-\frac{1}{2}\right)$

Applications

21. a. $\begin{cases} x+y+z=16 \\ 10x+6y+4z=92 \\ x+y=z \end{cases}$

b. $(3, 5, 8)$

c. Each bouquet will have 3 lilies, 5 roses, and 8 daisies.

22. a.
$$\begin{cases} 60x+45y+30z=27600 \\ x-y-z=20 \\ 2y-z=-40 \end{cases}$$

b. $(300, 80, 200)$

c. The box office sold 300 main floor tickets, 80 balcony tickets, and 200 mezzanine tickets.

23. 34, 6, 27

24. 56, 84, 49

25. 18 ones, 16 fives, 12 tens

26. 11 nickels, 7 dimes, 5 quarters

27. 19 cm, 24 cm, 30 cm

28. 100°, 30°, 50°

29. Bananas: \$0.60/lb; Apples: \$1.60/lb; Grapes: \$3.60/lb

30. House: \$115,000; Lot: \$33,000; Improvements: \$11,000

22. A theater has seats for a theatrical production located on the main floor, the balcony, and the mezzanine. For an upcoming musical, main floor tickets cost \$60 each, balcony tickets cost \$45 each, and mezzanine tickets cost \$30 each. On opening night, the ticket sales totaled \$27,600. The box office sold 20 more tickets for the main floor than they did for the balcony and mezzanine combined. The number of tickets sold for the mezzanine was 40 more than twice the number of tickets sold for the balcony. How many of each type of ticket did the box office sell?

a. Write a system of linear equations in three variables to describe the situation. Remember that you will have three equations. Use the variable x to represent the number of main floor tickets sold, the variable y to represent the number of balcony tickets sold, and the variable z to represent the number of mezzanine tickets sold.

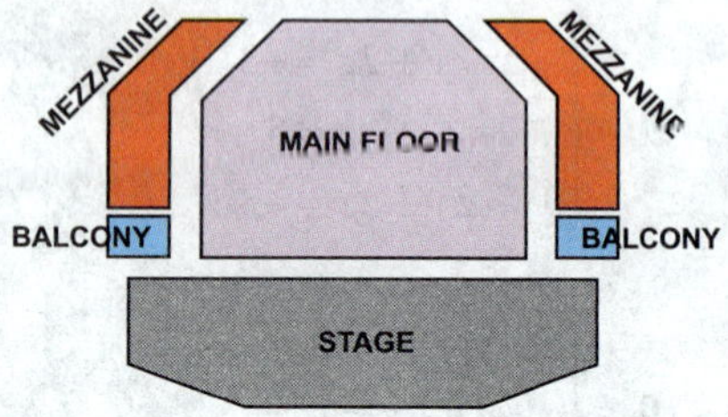

b. Solve the system of equations from part a.

c. Use the solution from part b. to answer the question from the problem statement. Write a complete sentence.

23. The sum of three integers is 67. The sum of the first and second integers is 13 more than the third integer. The third integer is 7 less than the first. Find the three integers.

24. The sum of three integers is 189. The first integer is 28 less than the second. The second integer is 21 less than the sum of the first and third integers. Find the three integers.

25. A wallet contains \$218 in \$10, \$5, and \$1 bills. There are forty-six bills in all and four more fives than tens. How many bills of each kind are there?

26. Ava has 23 coins in her purse, including nickels, dimes, and quarters. She has two more dimes than quarters, and the total value of the coins is \$2.50. How many of each kind of coin does she have?

27. The perimeter of a triangle is 73 cm. The longest side is 13 cm less than the sum of the other two sides. The shortest side is 11 cm less than the longest side. Find the lengths of the three sides.

28. The sum of the measures of the three angles of a triangle is 180°. In one particular triangle, the largest angle is 10° more than three times the smallest angle, and the third angle is one-half the largest angle. What are the measures of the three angles?

29. At Steve's Fruit Stand, 4 pounds of bananas, 2 pounds of apples, and 3 pounds of grapes cost \$16.40. Five pounds of bananas, 4 pounds of apples, and 2 pounds of grapes cost \$16.60. Two pounds of bananas, 3 pounds of apples, and 1 pound of grapes cost \$9.60. Find the price per pound of each kind of fruit.

30. The Tates are having a house built. The cost is split into three parts: the house, the lot, and the improvements. The cost of building the house is \$16,000 more than three times the cost of the lot. The cost of the improvements (the landscaping, sidewalks, and upgrades) is one-third the cost of the lot. If the total cost is \$159,000, what is the cost of each part of the construction?

31. Kai inherited \$100,000 from his aunt and decided to invest in three different accounts: savings, bonds, and stocks. The amount in his bond account was \$10,000 more than three times the amount in his stock account. At the end of the first year, the savings account returned 5%, the bond 8%, and the stocks 10% for total interest of \$7400. How much did he invest in each account?

32. Summer has saved a total of \$30,000 and wants to invest in three different stocks: Apple, Netflix, and Amazon.com, Inc. She wants the Apple amount to be \$1000 less than twice the Netflix amount and the Amazon.com, Inc. amount to be \$2000 more than the total in the other two stocks. How much should she invest in each stock?

33. A chemist wants to mix 9 liters of a 25% acid solution. Because of limited amounts on hand, the mixture needs to be created from three different solutions, one with 10% acid, another with 30% acid, and a third with 40% acid. The amount of the 10% solution must be twice the amount of the 40% solution, and the amount of the 30% solution must equal the total amount of the other two solutions. How much of each solution must be used?

34. An appliance company makes three versions of their popular stand mixer. The standard model has production costs of \$100, the deluxe model of \$175, and the premium model of \$250. On any given day the factory line makes 140 mixers and spends \$21,500 on production costs. If the number of standard mixers made equals the sum of the premium and deluxe models made, how many of each kind of mixer are made each day?

Writing & Thinking

35. Is it possible for three linear equations in three unknowns to have exactly two solutions? Explain your reasoning in detail.

36. In geometry, three non-collinear points determine a plane. (That is, if three points are not on a line, then there is a unique plane that contains all three points.). Find the values of A, B, and C (and therefore the equation of the plane) given $Ax + By + Cz = 3$ and the three points on the plane $(0, 3, 2)$, $(0, 0, 1)$ and $(-3, 0, 3)$. Sketch the plane in three dimensions as best you can by locating the three given points.

37. As stated in Exercise 36, three non-collinear points determine a plane. Find the values of A, B, and C (and therefore the equation of the plane) given $Ax + By + Cz = 10$ and the three points on the plane $(2, 0, -2)$, $(3, -1, 0)$ and $(-1, 5, -4)$. Sketch the plane in three dimensions as best you can by locating the three given points.

31. Savings: \$30,000; Bonds: \$55,000; Stocks: \$15,000

32. Apple: \$9000; Netflix: \$5000; Amazon.com, Inc.: \$16,000

33. 3 liters of 10%, 4.5 liters of 30%, 1.5 liters of 40%

34. Standard: 70 mixers; Deluxe: 40 mixers; Premium: 30 mixers

Writing & Thinking

35. No. Graphically, the three planes intersect at one point (one solution), or in a line (infinitely many solutions) or, they do not have a common intersection (no solution).

36. $A = 2, B = -1, C = 3$

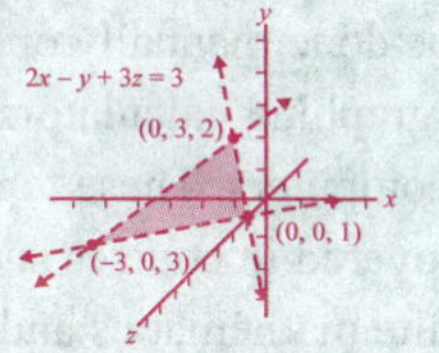

37. $A = 4, B = 2, C = -1$

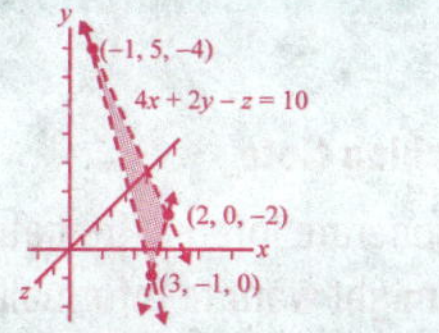

11.7 Matrices and Gaussian Elimination

Objectives

- **A.** Write systems of linear equations as matrices and use elementary row operations to manipulate them.
- **B.** Use the Gaussian elimination method to solve systems of linear equations.
- **C.** Use a graphing calculator and matrices to solve systems of linear equations.

A Introduction to Matrices

A rectangular array of numbers is called a **matrix** (plural **matrices**). Matrices are usually named with capital letters, and each number in the matrix is called an **entry**. Entries written horizontally are said to form a **row**, and entries written vertically are said to form a **column**. The matrix A has two rows and three columns and is a **2 × 3 matrix** (read "two by three matrix").

$$A = \begin{bmatrix} 3 & 4 & 0 \\ 7 & -2 & 5 \end{bmatrix} \qquad \begin{bmatrix} 3 & 4 & 0 \\ 7 & -2 & 5 \end{bmatrix} \begin{matrix} \leftarrow \text{Row 1} \\ \leftarrow \text{Row 2} \end{matrix} \qquad \begin{matrix} \text{Column 1} & \text{Column 2} & \text{Column 3} \\ \downarrow & \downarrow & \downarrow \end{matrix} \begin{bmatrix} 3 & 4 & 0 \\ 7 & -2 & 5 \end{bmatrix}$$

We say that the **dimension** of the matrix is the number of rows by the number of columns, which for this matrix is two by three (or 2 × 3). Similarly, if a matrix has three rows and two columns then its dimension is 3 × 2.

Three more examples are as follows.

$$B = \begin{bmatrix} 5 & -1 \\ 2 & 3 \end{bmatrix} \qquad C = \begin{bmatrix} 5 & -1 & 0 & 7 \\ 2 & 3 & 2 & 8 \\ 1 & -3 & 0 & 6 \end{bmatrix} \qquad D = \begin{bmatrix} 0 & 4 \\ 1 & 6 \\ -1 & 3 \end{bmatrix}$$

2 × 2 matrix 3 × 4 matrix 3 × 2 matrix

A matrix with the same number of rows as columns is called a **square matrix**. Matrix B is a square 2 × 2 matrix.

Matrices have many uses and are generated from various types of problems because they allow data to be presented in a systematic and orderly manner. (Business majors may want to look up a topic called Markov chains.)

Matrices can be used to solve systems of linear equations in which the equations are written in standard form. The two matrices derived from a system of linear equations are the **coefficient matrix** (made up of the coefficients of the variables) and the **augmented matrix** (which includes the coefficients and the constant terms).

System **Coefficient Matrix** **Augmented Matrix**

$$\begin{cases} x - y + z = -6 \\ 2x + 3y = 17 \\ x + 2y + 2z = 7 \end{cases} \qquad \begin{bmatrix} 1 & -1 & 1 \\ 2 & 3 & 0 \\ 1 & 2 & 2 \end{bmatrix} \qquad \underbrace{\left[\begin{array}{ccc|c} 1 & -1 & 1 & -6 \\ 2 & 3 & 0 & 17 \\ 1 & 2 & 2 & 7 \end{array}\right]}_{\text{Coefficients} \quad \uparrow \text{Constants}}$$

Note that 0 is the entry in the second row, third column of both matrices. This 0 corresponds to the missing z-variable in the second equation of the system. This is because the second equation could have been written as $2x + 3y + 0z = 17$.

Math Tip

Under certain conditions, matrices can be added, subtracted, and multiplied. Matrix methods of solving systems of linear equations can be done manually or with graphing calculators and computers. These topics are covered in courses such as finite mathematics and linear algebra.

Teaching Note

To generate interest in matrices, you might want to informally discuss and give simple examples of adding, subtracting, and multiplying with matrices and writing entries in the general form. Also, let students know that using row or column matrices to represent vectors is helpful in many physics and calculus applications. Most students find abstract material more worth their time and effort if they can see uses in future courses. Some of your students may even have seen or used Markov chains in a business or economics course and will want to discuss these applications.

As discussed in Section 10.3, notation with a small number to the right and below a variable is called **subscript** notation. For example, a_1 is read "*a* sub one," b_3 is read "*b* sub three," and R_2 is read "*R* sub two."

a_1 ← Subscript; ↑ Variable

b_3 ← Subscript; ↑ Variable

R_2 ← Subscript; ↑ Variable

When working with matrices, R_1 represents row 1, R_2 represents row 2, and so on.

In solving these systems, we can make any of the following three manipulations **without changing the solution set of the system**.

The system $\begin{cases} x - y + z = -6 \\ 2x + 3y + 0z = 17 \\ x + 2y + 2z = 7 \end{cases}$ is used here to illustrate some possibilities.

1. Any two equations may be interchanged.

$$\begin{cases} x - y + z = -6 \\ 2x + 3y + 0z = 17 \\ x + 2y + 2z = 7 \end{cases} \xrightarrow{R_1 \leftrightarrow R_2} \begin{cases} 2x + 3y + 0z = 17 \\ x - y + z = -6 \\ x + 2y + 2z = 7 \end{cases}$$

Here we have interchanged the first two equations.

2. All terms of any equation may be multiplied by a nonzero constant.

$$\begin{cases} x - y + z = -6 \\ 2x + 3y + 0z = 17 \\ x + 2y + 2z = 7 \end{cases} \xrightarrow{-2R_1} \begin{cases} -2x + 2y - 2z = 12 \\ 2x + 3y + 0z = 17 \\ x + 2y + 2z = 7 \end{cases}$$

Here we have multiplied each term of the first equation by −2.

3. All terms of any equation may be multiplied by a nonzero constant and these new terms may be added to like terms of another equation. **The original equation remains unchanged.**

$$\begin{cases} x - y + z = -6 \\ 2x + 3y + 0z = 17 \\ x + 2y + 2z = 7 \end{cases} \xrightarrow{R_3 - R_1} \begin{cases} x - y + z = -6 \\ 2x + 3y + 0z = 17 \\ 0x + 3y + z = 13 \end{cases}$$

Here we have multiplied the first equation by −1 (mentally) and added the results to the third equation.

When dealing with matrices, the three operations mentioned above are called **elementary row operations**. These operations are listed below and illustrated in Example 1. Follow the steps outlined in Example 1 carefully, and note how the row operations are indicated, such as $\frac{1}{2}R_3$ to indicate that all numbers in row 3 are multiplied by $\frac{1}{2}$. (Reasons for using these row operations are discussed under **Gaussian elimination** in Objective B.)

Elementary Row Operations

1. Interchange two rows.
2. Multiply a row by a nonzero constant.
3. Multiply a row by a nonzero constant and add it to another row.

If any elementary row operation is applied to a matrix, the new matrix is said to be **row-equivalent** to the original matrix.

PROPERTIES

1. a. Write the coefficient matrix and the augmented matrix for the following system.

$$\begin{cases} 2x+\ \ y+z=4 \\ x+2y+z=1 \\ 3x+\ \ y-z=-3 \end{cases}$$

b. In the augmented matrix you found in part a., interchange rows 1 and 3 and multiply row 2 by 3.

Example 1 Creating Coefficient and Augmented Matrices

For the system $\begin{cases} y+\ \ z=6 \\ x+5y-\ 4z=4 \\ 2x-6y+10z=14 \end{cases}$

a. Write the corresponding coefficient matrix and the corresponding augmented matrix.

b. In the augmented matrix found in part a., interchange rows 1 and 2 and multiply row 3 by $\frac{1}{2}$.

Solution

a. **Coefficient Matrix** **Augmented Matrix**

$$\begin{bmatrix} 0 & 1 & 1 \\ 1 & 5 & -4 \\ 2 & -6 & 10 \end{bmatrix} \qquad \left[\begin{array}{ccc:c} 0 & 1 & 1 & 6 \\ 1 & 5 & -4 & 4 \\ 2 & -6 & 10 & 14 \end{array}\right]$$

b.

$$\left[\begin{array}{ccc:c} 0 & 1 & 1 & 6 \\ 1 & 5 & -4 & 4 \\ 2 & -6 & 10 & 14 \end{array}\right] \begin{array}{c} \xrightarrow{R_1 \leftrightarrow R_2} \\ \xrightarrow{\frac{1}{2}R_3} \end{array} \left[\begin{array}{ccc:c} 1 & 5 & -4 & 4 \\ 0 & 1 & 1 & 6 \\ 1 & -3 & 5 & 7 \end{array}\right]$$

Now work margin exercise 1.

2. a. Write the coefficient matrix and the augmented matrix for the following system.

$$\begin{cases} x+\ \ y=9 \\ 4x-2y=6 \end{cases}$$

b. In the augmented matrix you found in part a., add −4 times row 1 to row 2.

Example 2 Creating Coefficient and Augmented Matrices

For the system $\begin{cases} x-\ \ y=5 \\ 3x+4y=29 \end{cases}$

a. Write the corresponding coefficient matrix and the corresponding augmented matrix.

b. In the augmented matrix found in part a., add −3 times row 1 to row 2. (**Note:** Row 1 is unchanged in the resulting matrix. Only row 2 is changed.)

Solution

a. **Coefficient Matrix** **Augmented Matrix**

$$\begin{bmatrix} 1 & -1 \\ 3 & 4 \end{bmatrix} \qquad \left[\begin{array}{cc:c} 1 & -1 & 5 \\ 3 & 4 & 29 \end{array}\right]$$

b. $\left[\begin{array}{cc|c} 1 & -1 & 5 \\ 3 & 4 & 29 \end{array}\right] \xrightarrow{R_2-3R_1} \left[\begin{array}{cc|c} 1 & -1 & 5 \\ 0 & 7 & 14 \end{array}\right]$ $3R_1$ is equal to $[3 \quad -3|15]$

Now work margin exercise 2.

With matrices, we use capital letters to name a matrix and **double subscript** notation with corresponding lower case letters to indicate both the row and column location of an entry. For example, in a matrix A, the entries will be designated as follows.

a_{11} is read "a sub one one" and indicates the entry in the first row and first column;
a_{12} is read "a sub one two" and indicates the entry in the first row and second column;
a_{13} is read "a sub one three" and indicates the entry in the first row and third column;
a_{21} is read "a sub two one" and indicates the entry in the second row and first column; and so on.

With double subscript notation, we can write the general form of a 2×3 matrix A and a 3×3 matrix B.

$$A = \begin{bmatrix} a_{11} & a_{12} & a_{13} \\ a_{21} & a_{22} & a_{23} \end{bmatrix} \qquad B = \begin{bmatrix} b_{11} & b_{12} & b_{13} \\ b_{21} & b_{22} & b_{23} \\ b_{31} & b_{32} & b_{33} \end{bmatrix}$$

This notation is used in Objective C to discuss the use of calculators to define matrices and to operate with matrices. The general form of a system of three linear equations might use this notation in the following way.

$$\begin{cases} a_{11}x + a_{12}y + a_{13}z = k_1 \\ a_{21}x + a_{22}y + a_{23}z = k_2 \\ a_{31}x + a_{32}y + a_{33}z = k_3 \end{cases}$$

A matrix is in **upper triangular form** (or just **triangular form** for our purposes) if its entries in the lower left triangular region are all 0s. The **main diagonal** consists of the entries in the positions of b_{11}, b_{22}, and b_{33}. Note that in the main diagonal, the column and row indices are equal. Thus, if all the entries below the main diagonal of a matrix are all 0s, the matrix is in triangular form, as shown.

$$B = \begin{bmatrix} b_{11} & b_{12} & b_{13} \\ 0 & b_{22} & b_{23} \\ 0 & 0 & b_{33} \end{bmatrix}$$

The upper triangular form of a matrix with all 1s in the main diagonal is called the **row echelon form** (or **ref**). We will see later that a graphing calculator can be used to change a matrix into the row echelon form.

B Gaussian Elimination

Another method of solving a system of linear equations is the **Gaussian elimination** method (named after the famous German mathematician Carl Friedrich Gauss, 1777–1855). This method makes use of augmented matrices and elementary row operations. The objective is to transform an augmented matrix into row echelon form and then use back substitution to find the values of the variables. The method is outlined as follows.

Note

We will see in dealing with polynomials later that a_{11} can be read as "a sub eleven." However, with matrices, we need to indicate the row and column corresponding to the entry. If there are more than nine rows or columns, then commas are used to separate the numbers as $a_{10,10}$. You will see the commas in use on your calculator.

Teaching Note

You might want to illustrate a matrix in lower triangular form to satisfy students' curiosity.

History

Born in 1777, Johann Carl Friedrich Gauss contributed greatly to the sciences and the study of mathematics in his lifetime. But his contributions were not limited to mathematics: he also held a strong interest in astronomy. In 1802, he held the position of director of the Göttingen observatory. One year prior, the dwarf planet Ceres was observed in the sky on January 1st, 1801. Gauss gave a prediction for its next appearance that greatly varied from his peers. When Ceres appeared again in December it was almost exactly where Gauss had predicted. How do you think he predicted it? Through mathematics of course!

Strategy for Gaussian Elimination

1. Write the augmented matrix for the system.
2. Use elementary row operations to transform the matrix into row echelon form.
3. Solve the corresponding system of equations by using back substitution.

PROCEDURE

The following examples illustrate the method. Study the steps and the corresponding comments carefully

3. Use the Gaussian elimination method with back substitution to solve the system of linear equations.

$$\begin{cases} 4x-4y=20 \\ 3x+6y=-3 \end{cases}$$

Example 3 Using Gaussian Elimination (Two Variables)

Use the Gaussian elimination method with back substitution to solve the system of linear equations.

$$\begin{cases} 2x+4y=-6 \\ 5x-\ \ y=7 \end{cases}$$

Solution

Step 1: Write the augmented matrix.

$$\left[\begin{array}{cc|c} 2 & 4 & -6 \\ 5 & -1 & 7 \end{array}\right]$$

(The following steps show how to use elementary row operations to get the matrix in row echelon form.)

Step 2: Multiply row 1 by $\frac{1}{2}$ so that the entry in the upper left corner will be 1. This will help to get 0 below the 1 in the next step.

$$\left[\begin{array}{cc|c} 2 & 4 & -6 \\ 5 & -1 & 7 \end{array}\right] \xrightarrow{\frac{1}{2}R_1} \left[\begin{array}{cc|c} 1 & 2 & -3 \\ 5 & -1 & 7 \end{array}\right]$$

Step 3: To get 0 in the lower left corner, add –5 times row 1 to row 2.

$$\left[\begin{array}{cc|c} 1 & 2 & -3 \\ 5 & -1 & 7 \end{array}\right] \xrightarrow{R_2-5R_1} \left[\begin{array}{cc|c} 1 & 2 & -3 \\ 0 & -11 & 22 \end{array}\right]$$

Step 4: Now multiply row 2 by $-\frac{1}{11}$.

$$\left[\begin{array}{cc|c} 1 & 2 & -3 \\ 0 & -11 & 22 \end{array}\right] \xrightarrow{-\frac{1}{11}R_2} \left[\begin{array}{cc|c} 1 & 2 & -3 \\ 0 & 1 & -2 \end{array}\right]$$

Step 5: The last matrix (in row echelon form) in Step 4 represents the following system of linear equations.

$$\begin{cases} x+2y=-3 \\ 0x+y=-2 \end{cases}$$

The second equation gives $y=-2$.

Back substitute to find the value for x.

$$x+2(-2)=-3$$
$$x-4=-3$$
$$x=1$$

Thus, the solution is $(1,-2)$.

Now work margin exercise 3.

Example 4 Using Gaussian Elimination (Three Variables)

Solve the following system of linear equations by using the Gaussian elimination method with back substitution.

$$\begin{cases} 2x-3y-z=-4 \\ -x+2y+z=6 \\ x-y+2z=14 \end{cases}$$

Solution

Step 1: Write the augmented matrix.

$$\left[\begin{array}{ccc|c} 2 & -3 & -1 & -4 \\ -1 & 2 & 1 & 6 \\ 1 & -1 & 2 & 14 \end{array}\right]$$

Step 2: Exchange row 1 and row 3 so that the entry in the upper left corner will be 1.

$$\left[\begin{array}{ccc|c} 2 & -3 & -1 & -4 \\ -1 & 2 & 1 & 6 \\ 1 & -1 & 2 & 14 \end{array}\right] \xrightarrow{R_1 \leftrightarrow R_3} \left[\begin{array}{ccc|c} 1 & -1 & 2 & 14 \\ -1 & 2 & 1 & 6 \\ 2 & -3 & -1 & -4 \end{array}\right]$$

Step 3: To get a 0 under the 1 in Column 1, add row 1 to row 2.

$$\left[\begin{array}{ccc|c} 1 & -1 & 2 & 14 \\ -1 & 2 & 1 & 6 \\ 2 & -3 & -1 & -4 \end{array}\right] \xrightarrow{R_2 + R_1} \left[\begin{array}{ccc|c} 1 & -1 & 2 & 14 \\ 0 & 1 & 3 & 20 \\ 2 & -3 & -1 & -4 \end{array}\right]$$

Step 4: To get a_{31} to be 0, add −2 times row 1 to row 3.

$$\left[\begin{array}{ccc|c} 1 & -1 & 2 & 14 \\ 0 & 1 & 3 & 20 \\ 2 & -3 & -1 & -4 \end{array}\right] \xrightarrow{R_3 - 2R_1} \left[\begin{array}{ccc|c} 1 & -1 & 2 & 14 \\ 0 & 1 & 3 & 20 \\ 0 & -1 & -5 & -32 \end{array}\right]$$

Step 5: Add row 2 to row 3 to arrive at triangular form.

$$\left[\begin{array}{ccc|c} 1 & -1 & 2 & 14 \\ 0 & 1 & 3 & 20 \\ 0 & -1 & -5 & -32 \end{array}\right] \xrightarrow{R_3 + R_2} \left[\begin{array}{ccc|c} 1 & -1 & 2 & 14 \\ 0 & 1 & 3 & 20 \\ 0 & 0 & -2 & -12 \end{array}\right]$$

Step 6: Then multiply row 3 by $-\frac{1}{2}$.

$$\left[\begin{array}{ccc|c} 1 & -1 & 2 & 14 \\ 0 & 1 & 3 & 20 \\ 0 & 0 & -2 & -12 \end{array}\right] \xrightarrow{-\frac{1}{2}R_3} \left[\begin{array}{ccc|c} 1 & -1 & 2 & 14 \\ 0 & 1 & 3 & 20 \\ 0 & 0 & 1 & 6 \end{array}\right]$$

Math Tip

You can check your solution by substituting the values into the original equations and confirming that you get true statements.

4. Solve the following system of linear equations using the Gaussian elimination method with back substitution.

$$\begin{cases} x+3y-2z=-3 \\ -2x-y-3z=-2 \\ 3x+y+z=8 \end{cases}$$

Step 7: The row echelon form matrix in Step 6 represents the following system of linear equations.

$$\begin{cases} x - y + 2z = 14 \\ \quad\; y + 3z = 20 \\ \qquad\quad z = 6 \end{cases}$$

From the third equation, we have $z = 6$.

Back substitution into the equation $y + 3z = 20$ gives

$$y + 3(6) = 20$$
$$y = 2.$$

Back substitution into the equation $x - y + 2z = 14$ gives

$$x - 2 + 2(6) = 14$$
$$x = 4.$$

Thus, the solution is $(4, 2, 6)$.

Now work margin exercise 4.

If the final matrix, in triangular form, has a row with all entries 0, then the system has an infinite number of solutions.

For example, solving the system $\begin{cases} x + 3y = 8 \\ 2x + 6y = 16 \end{cases}$ will result in the matrix $\left[\begin{array}{cc|c} 1 & 3 & 8 \\ 0 & 0 & 0 \end{array}\right]$.
The last line indicates that $0x + 0y = 0$, which is always true. Therefore, the solution to the system is the set of all solutions of the equation $x + 3y = 8$.

If the triangular form of the augmented matrix shows the coefficient entries in one or more rows to be all 0s and the constant not 0, then the system has no solution.

For example, the last row of the augmented matrix $\left[\begin{array}{ccc|c} 1 & 2 & 2 & 7 \\ 0 & 1 & 3 & 6 \\ 0 & 0 & 0 & 15 \end{array}\right]$ indicates that $0x + 0y + 0z = 15$. Since this is never true, the system has no solution.

Teaching Note

You may want to allow your students to explore the reduced row echelon form (rref) option in the MATRIX / MATH menu on the TI-84 Plus graphing calculator. This will lead directly to answers but may take away from their understanding of elementary row operations. You can explain how programmers must understand how to use elementary row operations to arrive at the ref and the rref used in calculators and in computers.

C Using a Graphing Calculator and Matrices to Solve Systems of Equations

The TI-84 Plus calculator can be used to define a matrix and perform matrix operations. (Matrices can be added, subtracted, and multiplied under special restrictions. These operations are saved for another course.) In particular, the Gaussian elimination method is used by the calculator to solve a system of linear equations by reducing the corresponding augmented matrix to **row echelon form (ref)**.

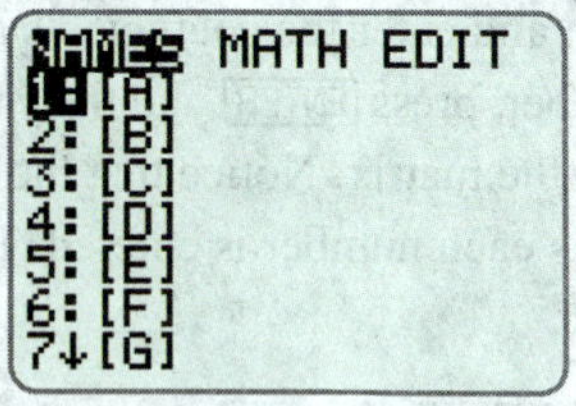

Pressing the MATRIX key (found by pressing [2nd] [x^{-1}]) will give the menu shown here. This menu will allow you to choose existing matrices to operate with them.

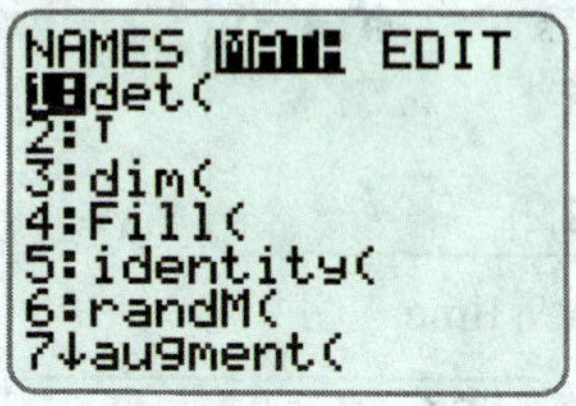

Pressing the right arrow and moving to MATH will give the choices shown. This menu will give you a list of operations you can do on matrices. We will use the reduced echelon form (ref) option here.

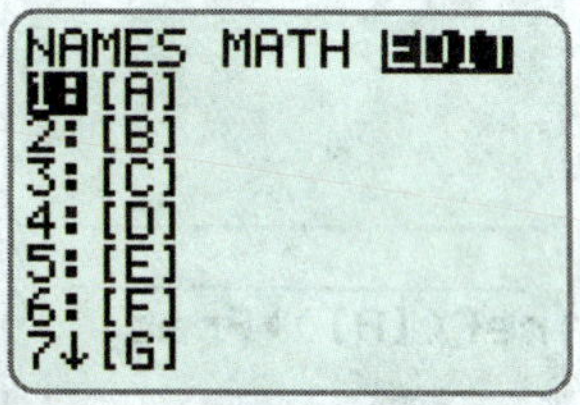

Pressing the right arrow again and moving to EDIT will give the choices shown. This menu allows you to edit existing matrices and create new ones.

Example 5 shows how to use the calculator to solve a system of three linear equations in three variables. Study each step carefully.

Example 5 Using a Graphing Calculator to Solve Systems of Equations

Use a TI-84 Plus calculator to solve the following system of linear equations.

$$\begin{cases} x+2y+\ z=1 \\ -x+\ y+\ z=-6 \\ 4x-\ y+3z=-1 \end{cases}$$

Solution

Step 1: Press [2nd] MATRIX and move to the EDIT menu. Press [ENTER]. The following display will appear.

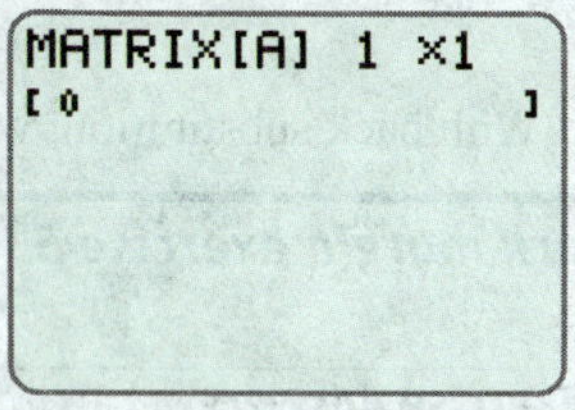

Step 2: The augmented matrix is a 3×4 matrix. So, in the top line enter [3], press [ENTER], enter [4], press [ENTER], and the display will appear as shown.

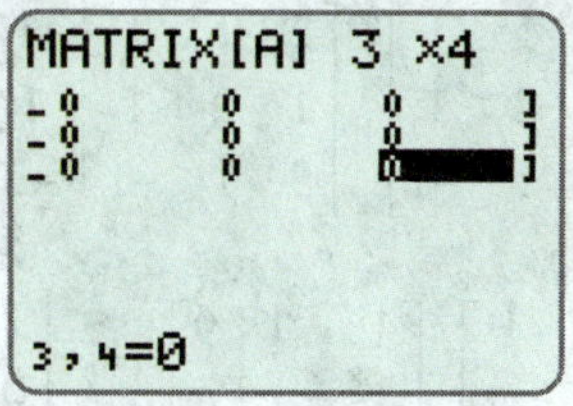

Note: If other numbers are already present on the display, just type over them. The calculator will adjust automatically.

5. Use a TI-84 Plus calculator to solve the following system of linear equations.

$$\begin{cases} x-3y+2z=10 \\ 2x+5y-2z=-9 \\ -x+6y+3z=2 \end{cases}$$

Step 3: Move the cursor to the upper left entry position and enter the coefficients and constants in the matrix. As you enter each number, press ENTER and the cursor will automatically move to the next position in the matrix. Notice that the double subscripts appear at the bottom of the display as each number is entered. The final display for matrix [A] should appear as shown.

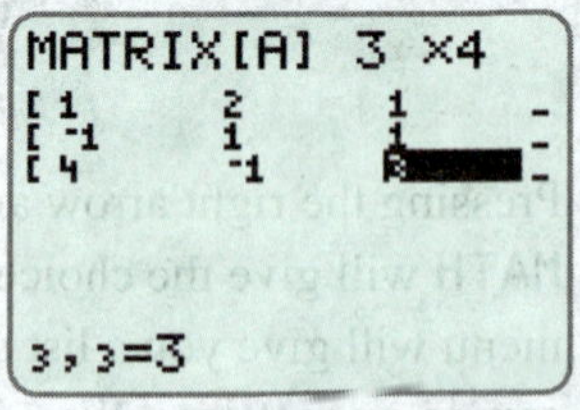

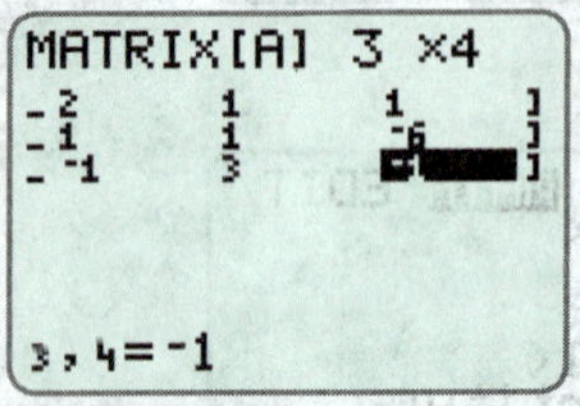

Note: The display only shows three columns at a time.

Step 4: Press 2nd `QUIT`, press 2nd `MATRIX` again, and go to `MATH`. Move the cursor down to `A:ref(` and press ENTER. The display will appear as shown.

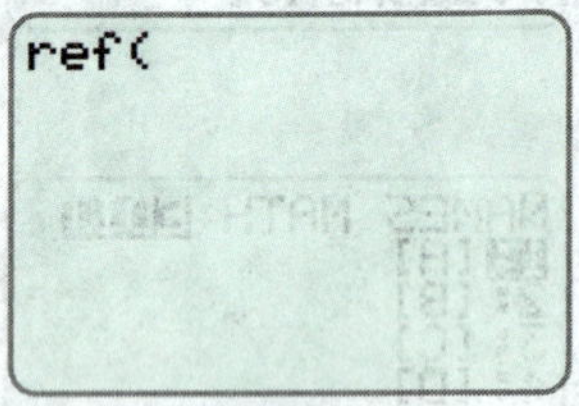

Step 5: Press 2nd `MATRIX` again and press ENTER (this selects the matrix `[A]`). Enter a right parenthesis), press the MATH key, and choose `Frac` by pressing ENTER. The display will appear as shown.

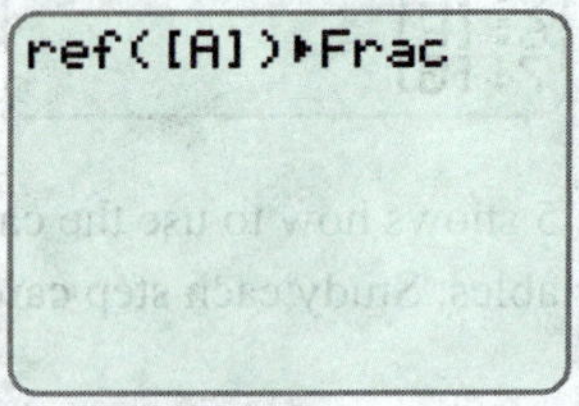

Note: You must select the matrix from the matrix menu. The calculator will not recognize the matrix if you manually type in `[A]`.

Step 6: Press ENTER and the row echelon form of matrix will appear as follows.

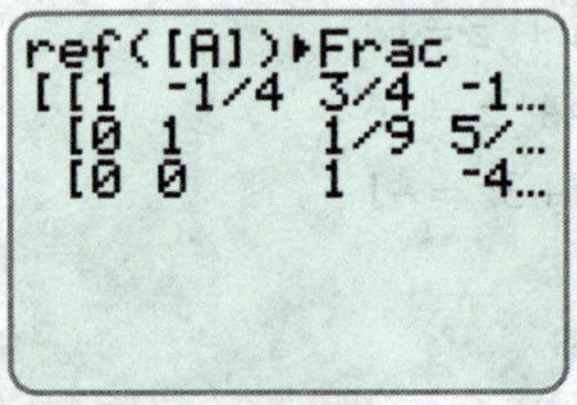

Step 7: With back substitution, we get the solution $(3,1,-4)$.

Now work margin exercise 5.

Margin Exercise Answers

1. a. $\begin{bmatrix} 2 & 1 & 1 \\ 1 & 2 & 1 \\ 3 & 1 & -1 \end{bmatrix}, \left[\begin{array}{ccc|c} 2 & 1 & 1 & 4 \\ 1 & 2 & 1 & 1 \\ 3 & 1 & -1 & -3 \end{array}\right]$ **b.** $\left[\begin{array}{ccc|c} 3 & 1 & -1 & -3 \\ 3 & 6 & 3 & 3 \\ 2 & 1 & 1 & 4 \end{array}\right]$

2. a. $\begin{bmatrix} 1 & 1 \\ 4 & -2 \end{bmatrix}, \left[\begin{array}{cc|c} 1 & 1 & 9 \\ 4 & -2 & 6 \end{array}\right]$ **b.** $\left[\begin{array}{cc|c} 1 & 1 & 9 \\ 0 & -6 & -30 \end{array}\right]$ **3.** $(3,-2)$ **4.** $(4,-3,-1)$ **5.** $(1,-1,3)$

11.7 Exercises

Concept Check

Fill-in-the-Blank. Complete each sentence using information found in this section.

1. A rectangular array of numbers is called a/an ______ (plural ______).
2. In a matrix, entries written horizontally form a/an ______ and entries written vertically form a/an ______.
3. The ______ of a matrix is the number of rows by the number of columns.
4. A matrix that is made up of the coefficients of the variables of a system of linear equations is called a/an ________ matrix.
5. A matrix that is made up of the coefficients of the variables and the constant terms of a system of linear equations is called a/an ______ matrix.
6. Interchanging equations, multiplying an equation by a constant, and adding like terms of two equations are known as ____________ operations.

True/False. Determine whether each statement is true or false. If a statement is false, explain how it can be changed so that the statement will be true. (**Note:** There may be more than one acceptable change.)

7. A matrix that has 3 rows and 5 columns is a 5×3 matrix.
8. A matrix with the same number of rows as columns, such as a 3×3 matrix, is called a square matrix.
9. Interchanging two equations in a system of linear equations will change the solution of the system.
10. In a system of linear equations, adding like terms of one equation to another equation will not change the solution of the system.

Practice

Write the coefficient matrix and the augmented matrix for the given systems of linear equations. See Examples 1 and 2.

1. $\begin{cases} 2x+2y=13 \\ 5x-y=10 \end{cases}$

2. $\begin{cases} x+4y=-1 \\ 2x-3y=7 \end{cases}$

3. $\begin{cases} 7x-2y+7z=2 \\ -5x+3y=2 \\ 4y+11z=8 \end{cases}$

4. $\begin{cases} -8x+2y-z=6 \\ 2x+3z=-3 \\ -4x-2y+5z=13 \end{cases}$

5. $\begin{cases} 3w+x-y+2z=6 \\ w-x+2y-z=-8 \\ 2x+5y+z=2 \\ w+3x+3z=14 \end{cases}$

6. $\begin{cases} 4w+x+3y-2z=13 \\ w-2x+y-4z=-3 \\ w+x+4y+2z=12 \\ -2w+3x-y-3z=5 \end{cases}$

11.7 Exercises

Concept Check

1. matrix, matrices
2. row, column
3. dimension
4. coefficient
5. augmented
6. elementary row
7. False; A matrix that has 3 rows and 5 columns is a 3 x 5 matrix.
8. True
9. False; Interchanging two equations in a system of linear equations will not change the solution.
10. True

Practice

1. $\begin{bmatrix} 2 & 2 \\ 5 & -1 \end{bmatrix}, \left[\begin{array}{cc|c} 2 & 2 & 13 \\ 5 & -1 & 10 \end{array}\right]$

2. $\begin{bmatrix} 1 & 4 \\ 2 & -3 \end{bmatrix}, \left[\begin{array}{cc|c} 1 & 4 & -1 \\ 2 & -3 & 7 \end{array}\right]$

3. $\begin{bmatrix} 7 & -2 & 7 \\ -5 & 3 & 0 \\ 0 & 4 & 11 \end{bmatrix}, \left[\begin{array}{ccc|c} 7 & -2 & 7 & 2 \\ -5 & 3 & 0 & 2 \\ 0 & 4 & 11 & 8 \end{array}\right]$

4. $\begin{bmatrix} -8 & 2 & -1 \\ 2 & 0 & 3 \\ -4 & -2 & 5 \end{bmatrix}, \left[\begin{array}{ccc|c} -8 & 2 & -1 & 6 \\ 2 & 0 & 3 & -3 \\ -4 & -2 & 5 & 13 \end{array}\right]$

5. $\begin{bmatrix} 3 & 1 & -1 & 2 \\ 1 & -1 & 2 & -1 \\ 0 & 2 & 5 & 1 \\ 1 & 3 & 0 & 3 \end{bmatrix}, \left[\begin{array}{cccc|c} 3 & 1 & -1 & 2 & 6 \\ 1 & -1 & 2 & -1 & -8 \\ 0 & 2 & 5 & 1 & 2 \\ 1 & 3 & 0 & 3 & 14 \end{array}\right]$

6. $\begin{bmatrix} 4 & 1 & 3 & -2 \\ 1 & -2 & 1 & -4 \\ 1 & 1 & 4 & 2 \\ -2 & 3 & -1 & -3 \end{bmatrix}$, $\left[\begin{array}{cccc|c} 4 & 1 & 3 & -2 & 13 \\ 1 & -2 & 1 & -4 & -3 \\ 1 & 1 & 4 & 2 & 12 \\ -2 & 3 & -1 & -3 & 5 \end{array}\right]$

7. $\begin{cases} -3x+5y=1 \\ -x+3y=2 \end{cases}$

8. $\begin{cases} 3x-y=5 \\ -2x+10y=9 \end{cases}$

9. $\begin{cases} x+3y+4z=1 \\ 2x-3y-2z=0 \\ x+y=-4 \end{cases}$

10. $\begin{cases} 2x-9y+14z=0 \\ -3x-8z=5 \\ 2x-6y+z=3 \end{cases}$

11. a. $\begin{bmatrix} -1 & 7 \\ 1 & 4 \end{bmatrix}$

b. $\begin{bmatrix} 1 & 4 \\ 2 & -14 \end{bmatrix}$

12. a. $\left[\begin{array}{cc|c} \frac{3}{2} & -1 & 4 \\ 1 & 5 & 9 \end{array}\right]$

b. $\left[\begin{array}{cc|c} 0 & -17 & -19 \\ 1 & 5 & 9 \end{array}\right]$

13. a. $\begin{bmatrix} 1 & 3 & 7 \\ 4 & -1 & 6 \\ -8 & -2 & 5 \end{bmatrix}$

b. $\begin{bmatrix} 1 & 3 & 7 \\ -16 & 0 & -7 \\ 4 & -1 & 6 \end{bmatrix}$

14. a. $\left[\begin{array}{ccc|c} 1 & 2 & -3 & 7 \\ 0 & 1 & 1 & 4 \\ 0 & -2 & -1 & 9 \end{array}\right]$

b. $\left[\begin{array}{ccc|c} 1 & 2 & -3 & 7 \\ 0 & 1 & 1 & 4 \\ 0 & 0 & 1 & 17 \end{array}\right]$

15. $(-1, 2)$

16. $(2, -1)$

17. $(-1, -1)$

Write the system of linear equations represented by each of the augmented matrices. Use *x*, *y*, and *z* as the variables.

7. $\left[\begin{array}{cc|c} -3 & 5 & 1 \\ -1 & 3 & 2 \end{array}\right]$

8. $\left[\begin{array}{cc|c} 3 & -1 & 5 \\ -2 & 10 & 9 \end{array}\right]$

9. $\left[\begin{array}{ccc|c} 1 & 3 & 4 & 1 \\ 2 & -3 & -2 & 0 \\ 1 & 1 & 0 & -4 \end{array}\right]$

10. $\left[\begin{array}{ccc|c} 2 & -9 & 14 & 0 \\ -3 & 0 & -8 & 5 \\ 2 & -6 & 1 & 3 \end{array}\right]$

Perform the indicated row operations on the given matrix. See Examples 1 and 2.

11. $\begin{bmatrix} 1 & 4 \\ -1 & 7 \end{bmatrix}$

a. Interchange rows 1 and 2

b. Multiply row 2 by −2

12. $\left[\begin{array}{cc|c} 3 & -2 & 8 \\ 1 & 5 & 9 \end{array}\right]$

a. Multiply row 1 by $\frac{1}{2}$

b. Add −3 times row 2 to row 1

13. $\begin{bmatrix} 1 & 3 & 7 \\ -8 & -2 & 5 \\ 4 & -1 & 6 \end{bmatrix}$

a. Interchange rows 2 and 3

b. Add −2 times row 3 to row 2

14. $\left[\begin{array}{ccc|c} 1 & 2 & -3 & 7 \\ 0 & -2 & -1 & 9 \\ 0 & 1 & 1 & 4 \end{array}\right]$

a. Interchange rows 2 and 3

b. Using your answer from part a., add 2 times row 2 to row 3.

Use the Gaussian elimination method to solve the given system of linear equations. See Examples 3 and 4.

15. $\begin{cases} x+2y=3 \\ 2x-y=-4 \end{cases}$

16. $\begin{cases} 4x+3y=5 \\ -x-2y=0 \end{cases}$

17. $\begin{cases} -8x+2y=6 \\ x-2y=1 \end{cases}$

18. $\begin{cases} 2x+y=-2 \\ 4x+3y=-2 \end{cases}$

19. $\begin{cases} x-3y+2z=11 \\ -2x+4y+z=-3 \\ x-2y+3z=12 \end{cases}$

20. $\begin{cases} x+2y-z=6 \\ x+3y-3z=3 \\ x+y+z=6 \end{cases}$

21. $\begin{cases} x+2y+3z=4 \\ x-y-z=0 \\ 4x-3y+z=5 \end{cases}$

22. $\begin{cases} x+y-2z=-1 \\ 3x+4y-2z=0 \\ x-y+z=4 \end{cases}$

23. $\begin{cases} x-y-2z=3 \\ x+2y-z=5 \\ 2x-3y-2z=3 \end{cases}$

24. $\begin{cases} x+y+3z=2 \\ 2x-y+z=1 \\ 4x+y+7z=5 \end{cases}$

25. $\begin{cases} x-y+5z=-6 \\ x+2z=0 \\ 6x+y+3z=0 \end{cases}$

26. $\begin{cases} x-3y-z=-4 \\ 3x-2y+z=1 \\ -2x+y+2z=13 \end{cases}$

27. $\begin{cases} x-y+2z=5 \\ 2x-2y+4z=5 \\ 3x-3y+6z=8 \end{cases}$

28. $\begin{cases} 2x-y-5z=-9 \\ x-3y+2z=0 \\ 3x+2y+10z=4 \end{cases}$

29. $\begin{cases} x-5y+2z=-7 \\ x+y+4z=7 \\ 2x-y+7z=7 \end{cases}$

30. $\begin{cases} 2x-y+5z=-2 \\ 4x+y+3z=-2 \\ x+3y-z=6 \end{cases}$

31. $\begin{cases} 3x+4z=11 \\ x+y=-2 \\ 2y+z=-4 \end{cases}$

32. $\begin{cases} y+z=2 \\ x+y=5 \\ x+z=5 \end{cases}$

Use your graphing calculator to solve the systems of linear equations. See Example 5.

33. $\begin{cases} x+y=-4 \\ 2x+3y=-12 \end{cases}$

34. $\begin{cases} 2x+3y=1 \\ x-5y=-19 \end{cases}$

35. $\begin{cases} x+y+z=10 \\ 2x-y+z=10 \\ -x+2y+2z=14 \end{cases}$

36. $\begin{cases} 2x+y+2z=2 \\ x-y+4z=1 \\ 3x-y+z=-4 \end{cases}$

37. $\begin{cases} 2y-3z=-18 \\ 4x+5y=-7 \\ 6x-z=8 \end{cases}$

38. $\begin{cases} w+x+y+z=0 \\ w+x-y-z=-2 \\ -w+3x+3y-z=11 \\ y-2z=6 \end{cases}$

39. $\begin{cases} x-3y+z=0 \\ 2x+2y-z=2 \\ x+y+z=5 \end{cases}$

40. $\begin{cases} x-2y-2z=-13 \\ 2x+y-z=-5 \\ x+y+z=6 \end{cases}$

Applications

Set up a system of linear equations that represents the information and solve the system using Gaussian elimination. See Example 4.

41. The sum of three integers is 169. The first integer is twelve more than the second integer. The third integer is fifteen less than the sum of the first and second integers. What are the integers?

42. A pizzeria sells three sizes of pizzas: small, medium, and large. The pizzas sell for \$6.00, \$8.00, and \$9.50, respectively. One evening they sold 68 pizzas for a total of \$528.00. If they sold twice as many medium-sized pizzas as large-sized pizzas, how many of each size did they sell?

18. $(-2, 2)$
19. $(-1, -2, 3)$
20. No solution
21. $(1, 0, 1)$
22. $(2, -1, 1)$
23. $(2, 1, -1)$
24. Infinite solutions
25. $(-2, 9, 1)$
26. $(-2, -1, 5)$
27. No solution
28. $(-2, 0, 1)$
29. Infinite number of solutions
30. $(-2, 3, 1)$
31. $(1, -3, 2)$
32. $(4, 1, 1)$
33. $(0, -4)$
34. $(-4, 3)$
35. $(2, 1, 7)$
36. $(-1, 2, 1)$
37. $(2, -3, 4)$
38. $\left(-\frac{13}{12}, \frac{1}{12}, \frac{8}{3}, -\frac{5}{3}\right)$
39. $\left(\frac{13}{12}, \frac{5}{4}, \frac{8}{3}\right)$
40. $\left(-\frac{1}{3}, 1, \frac{16}{3}\right)$

Applications

41. 52, 40, 77
42. 20 small, 32 medium, 16 large

43. Bacon: \$3.09/lb; Eggs: \$4.03/doz; Bread: \$1.40/loaf

44. \$127,500 at 6%; \$62,500 at 8%; \$60,000 at 11%

45. a. $\begin{cases} u + c = 12 \\ 2500u + 625c = 13{,}125 \end{cases}$

b. $\left[\begin{array}{cc|c} 1 & 1 & 12 \\ 2500 & 625 & 13{,}125 \end{array}\right]$

c. University: 3, Community College: 9

46. a. 3×4

b. $\left[\begin{array}{ccc|c} 1 & 1 & 1 & 15 \\ 46 & 35 & 40 & 617 \\ 15 & 12 & 10 & 195 \end{array}\right]$

c. $x = 7, y = 5, z = 3$

Writing & Thinking

47. Solving the second equation for z, we can back substitute into the first equation, eliminating z. The result is the equation $x + 5y = 6$, which means the system has an infinite number of solutions.

43. Caroline bought a pound of bacon, a dozen eggs, and a loaf of bread. The total cost was \$8.52. The eggs cost \$0.94 more than the bacon. The combined cost of the bread and eggs was \$2.34 more than the cost of the bacon. Find the cost of each item.

44. An investment firm is responsible for investing \$250,000 from an estate according to three conditions in the will of the deceased. The money is to be invested in three accounts paying 6%, 8%, and 11% interest. The amount invested in the 6% account is to be \$5000 more than the total invested in the other two accounts, and the total annual interest for the first year is to be \$19,250. How much is the firm supposed to invest in each account?

45. Maurice is looking at his options for colleges. He can only borrow \$13,125 a year for tuition. The community college in town does not offer every class he needs, and he cannot afford to attend the university full time. The university charges \$2500 per credit hour (u) and the community college charges \$625 per credit hour (c). In order to finish his degree on time, Maurice must take 12 credit hours per semester for both fall and spring semesters, but not for the summer semester.

a. Write the system of equations that represents this situation.

b. Write the system as a matrix.

c. How many hours at each school can he take?

46. You and 14 of your friends are planning a spring-break trip. There are three concerts happening in different locations all on the same night, and everyone wants to attend one concert. The group has \$617 to spend on the 15 tickets and \$195 to spend on transportation to the concerts. For concert x, tickets cost \$46 and transportation per person is \$15. For concert y, tickets cost \$35 and transportation per person is \$12. For concert z, tickets are \$40 and transportation per person is \$10.
Consider the following system of equations.

$$\begin{aligned} x + y + z &= 15 \\ 46x + 35y + 40z &= 617 \\ 15x + 12y + 10z &= 195 \end{aligned}$$

a. What is the dimension of the augmented matrix?

b. Set up augmented matrix.

c. Solve the system of equations using Gaussian elimination.

Writing & Thinking

47. Suppose that Gaussian elimination with a system of three linear equations in three unknowns results in the following triangular matrix. Discuss how you can use back substitution to find that the system has an infinite number of solutions and these solutions satisfy the equation $x + 5y = 6$. (**Hint:** Solve the second equation for z.)

$$\left[\begin{array}{ccc|c} 1 & 2 & -1 & 4 \\ 0 & 3 & 1 & 2 \\ 0 & 0 & 0 & 0 \end{array}\right]$$

11.8 Systems of Linear Inequalities

Objectives

A. Solve systems of linear inequalities graphically.

B. Use a graphing calculator to graph systems of linear inequalities.

A Solving Systems of Linear Inequalities

In some branches of mathematics, in particular a topic called game theory, the solution to a very sophisticated problem can involve the set of points that satisfies a system of several **linear inequalities**. In business, these ideas relate to problems such as minimizing the cost of shipping goods from several warehouses to distribution outlets. In this section, we will consider graphing the solution sets to only two inequalities. We will leave the problem solving techniques to another course.

In Section 10.6, we graphed linear inequalities of the form $y < mx + b$ (or $y \le mx + b$ or $y > mx + b$ or $y \ge mx + b$). These graphs are called **half-planes.** The line $y = mx + b$ is called the **boundary line**, and the half-planes are **open** (the boundary line is not included) or **closed** (the boundary line is included).

In this section, we will develop techniques for graphing (and therefore solving) **systems of two linear inequalities**. The **solution set** (if there are any solutions) to such a system of linear inequalities consists of the points in the **intersection** of two half-planes and portions of boundary lines indicated by the inequalities. The following procedure may be used to solve a system of linear inequalities.

Math Tip

Linear programming is the mathematical method of determining the optimal values of a system of linear equalities in regard to real-life constraints. This could mean anything from maximizing the profit of a business to minimizing the distance traveled for food delivery drivers. As you can imagine, linear programming is extremely useful for business applications in our modern-day world.

Solving a System of Two Linear Inequalities

1. For each inequality, graph the boundary line and shade the appropriate half-plane.
2. Determine the region of the graph that is common to both half-planes (the region where the shading overlaps). This region is called the **intersection** of the two half-planes and is the **solution set** of the system.
3. To check the solution, pick one test-point in the intersection and verify that it satisfies both inequalities.

Note: If there is no intersection, then the system has no solution.

PROCEDURE

Example 1 Solving Systems of Linear Inequalities

Solve the system of linear inequalities graphically: $\begin{cases} x \le 2 \\ y \ge -x + 1 \end{cases}$

Solution

Step 1: For $x \le 2$, the points are to the left of and on the line $x = 2$. For $y \ge -x + 1$, the points are above and on the line $y = -x + 1$.

Step 2: Determine the region that is common to both half-planes.

Step 3: In this case, we test the point $(0, 3)$. In the following graph, the solution is the region where the shading overlaps, along with the boundary lines of that region.

1. Solve the system of linear inequalities graphically:

$\begin{cases} y \ge 2 \\ x - y < 4 \end{cases}$

$0 \le 2$ A true statement

$3 \ge -0 + 1$ A true statement

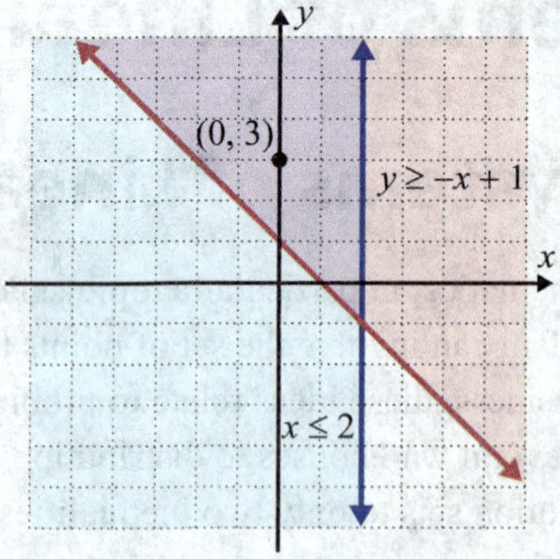

Now work margin exercise 1.

Example 2 Solving Systems of Linear Inequalities

Solve the system of linear inequalities graphically: $\begin{cases} 2x + y \le 6 \\ x + y < 4 \end{cases}$

Solution

First, solve each inequality for y: $\begin{cases} y \le -2x + 6 \\ y < -x + 4 \end{cases}$

Step 1: For $y \le -2x + 6$, the points are below and on the line $y = -2x + 6$. For $y < -x + 4$, the points are below but not on the line $y = -x + 4$.

Step 2: Determine the region that is common to both half-planes. Note that the line $y = -x + 4$ is dashed to indicate that the points on the line are not included.

Step 3: In this case, we test the point $(0, 0)$. In the following graph, the solution is the region where the shading overlaps and the portion of the solid boundary line, $2x + y = 6$, where $x > 2$.

$2 \cdot 0 + 0 \le 6$ A true statement

$0 + 0 < 4$ A true statement

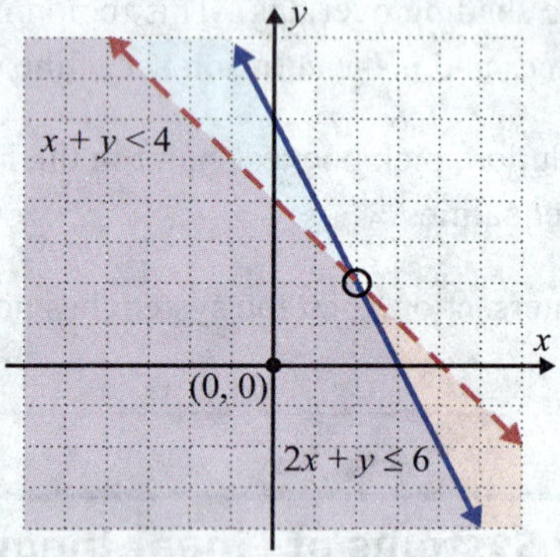

Now work margin exercise 2.

2. Solve the system of linear inequalities graphically:

$\begin{cases} x + 2y < 3 \\ 3x + y \ge -6 \end{cases}$

11.8 Exercises

Concept Check

1. test-point
2. boundary
3. closed
4. open
5. three
6. intersection
7. False; When boundary lines are parallel, the solution is either the strip between the boundary lines, a half-plane, or there is no solution.
8. False; The solution is the intersection of the graphs.
9. True
10. True

Practice

1.

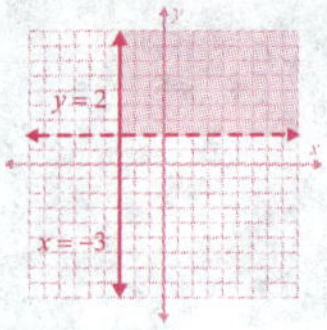

Example 3 Solving Systems of Linear Inequalities

Solve the system of linear inequalities graphically: $\begin{cases} y \ge x \\ y \le x+2 \end{cases}$

Solution

Step 1: For $y \ge x$, the points are above and on the line $y = x$. For $y \le x + 2$, the points are below and on the line $y = x + 2$.

Step 2: Determine the region that is common to both half-planes.

Step 3: We test point $(-1, 0)$. The solution set consists of the boundary lines and the region between them.

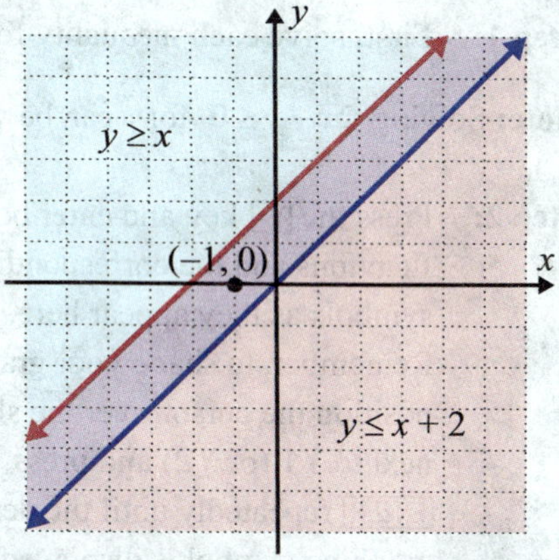

Now work margin exercise 3.

3. Solve the system of linear inequalities graphically:

$$\begin{cases} y < -3x \\ y > -3x - 4 \end{cases}$$

Possible Solutions to Parallel Systems of Linear Inequalities

When the boundary lines are parallel, there are three possibilities:

1. The common region will be in the form of a strip between two lines (as in Example 3).
2. The common region will be a half-plane, as the solution to one inequality will be entirely contained within the solution of the other inequality.
3. There will be no common region, which means there is no solution.

PROPERTIES

B Using a Graphing Calculator to Graph Systems of Linear Inequalities

To graph (and therefore solve) a system of linear inequalities (with no vertical boundary lines) with a TI-84 Plus graphing calculator, proceed as follows.

1. Solve each inequality for y.
2. Press the [Y=] key and enter the two expressions for Y1 and Y2.
3. Move to the left of Y1 and Y2 and press [ENTER] until the desired graphing symbol appears.
4. Press [GRAPH] and the desired region will be graphed as a cross-hatched area. (You may need to adjust the window so that both regions appear. If you need assistance with this, refer back to Section 10.5.)

Example 4 illustrates how this can be done.

2.

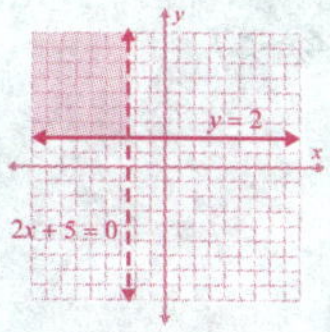

3.

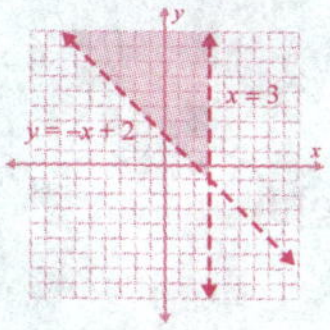

4.

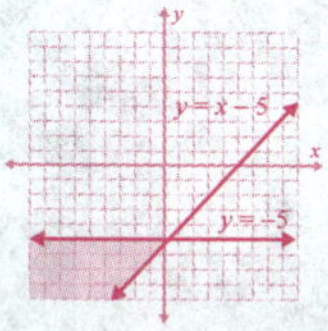

5.

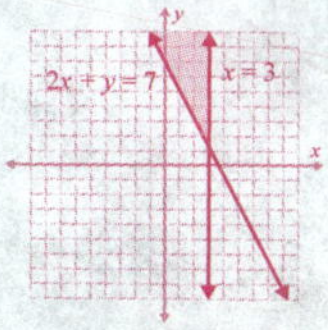

6.

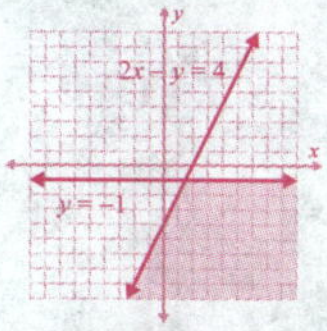

7.

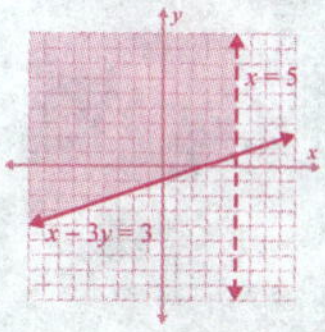

8.

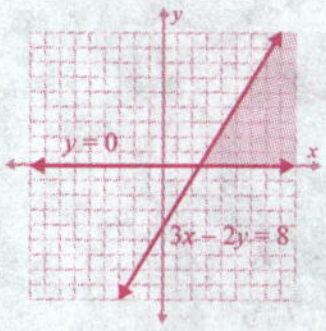

9.

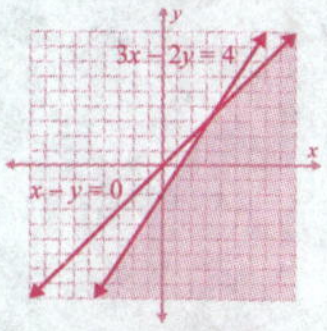

10.

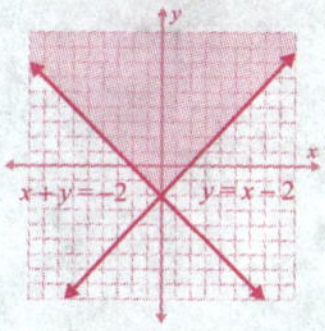

4. Use a TI-84 Plus graphing calculator to graph the following system of linear inequalities.

$$\begin{cases} y-2x \geq -5 \\ y+4x < 7 \end{cases}$$

11.

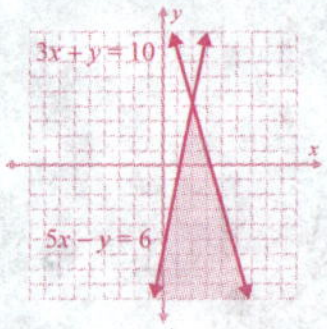

12. No solution

13.

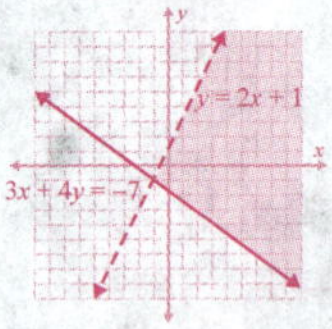

14.

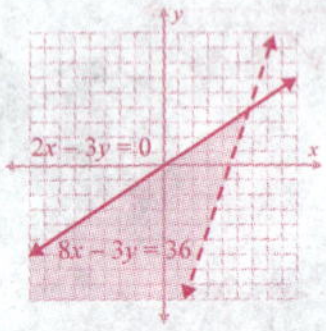

15.

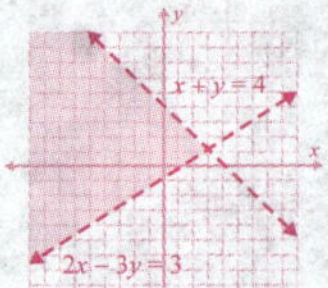

16.

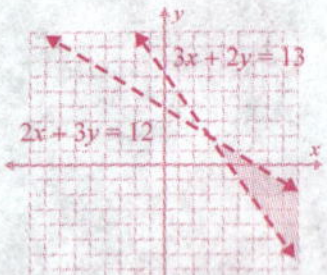

17.

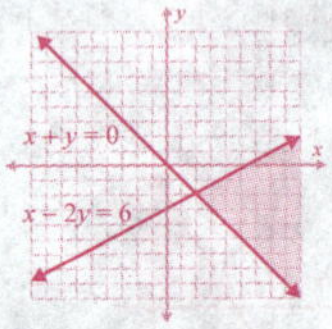

18.

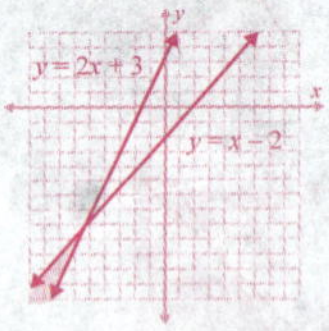

19.

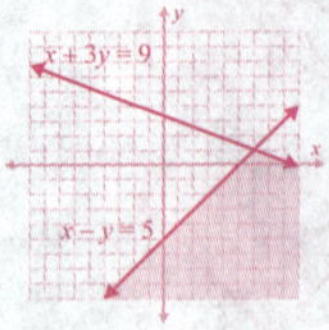

Example 4 Graphing Systems of Linear Inequalities

Use a graphing calculator to graph the system of linear inequalities: $\begin{cases} 2x+y<4 \\ 2x-y\leq 0 \end{cases}$

Solution

Step 1: First, solve each inequality for y: $\begin{cases} y<-2x+4 \\ y\geq 2x \end{cases}$

Note: Solving $2x-y\leq 0$ for y can be written as $2x\leq y$ and then as $y\geq 2x$

Step 2: Press the [Y=] key and enter both functions and the corresponding symbols as they appear here. Remember, to shade your graphs, position the cursor over the slash next to Y1 (or Y2) and press [ENTER] repeatedly until the desired graphing symbol is displayed.

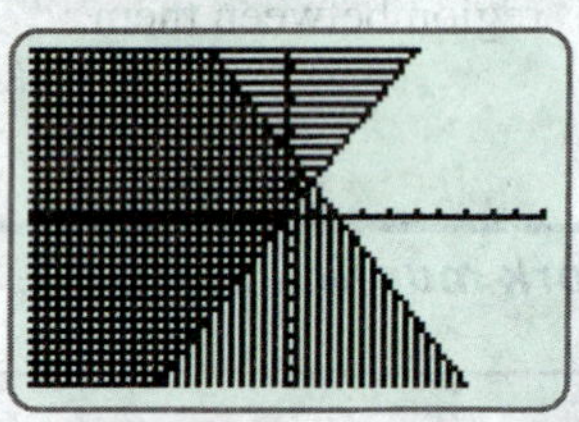

Step 3: Press [GRAPH]. The graph should appear as shown. The solution is the cross-hatched region and the points on the line $2x - y = 0$, where $x < 1$.

Now work margin exercise 4.

Margin Exercise Answers

1.

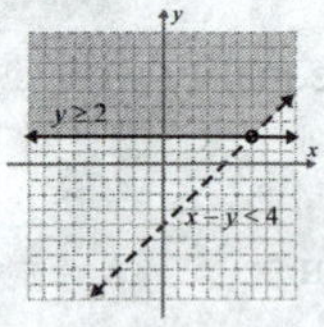

2.

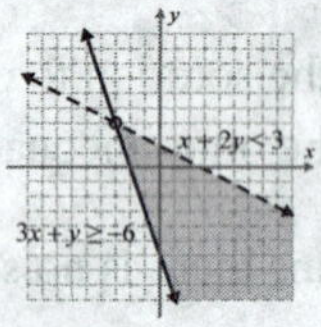

3.

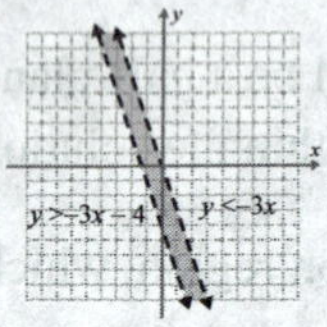

4.

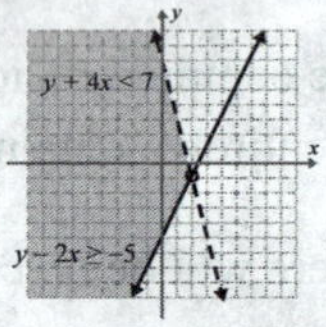

11.8 Exercises

Concept Check

Fill-in-the-Blank. Complete each sentence using information found in this section.

1. To check the graph of a system of linear inequalities, a/an ____-____ should be chosen to determine whether or not it satisfies both inequalities.

2. When graphing an inequality such as $y > mx + b$, the line $y = mx + b$ is called the ______ line.

3. If a boundary line is included in the solution, the half-plane is ______.

4. When a boundary line is not included in the solution, the half-plane is ____.

5. If the boundary lines are parallel, there are ___ possible types of solutions.

6. The solution set of a system of two linear inequalities consists of the points in the ______ of the two half-planes and portions of the boundary lines.

True/False. Determine whether each statement is true or false. If a statement is false, explain how it can be changed so the statement will be true. (**Note:** There may be more than one acceptable change.)

7. When boundary lines are parallel, the system of linear inequalities has no solution.

8. If two half-planes overlap, that region is the union of the graphs.

9. Half-planes are the graphs of linear inequalities.

10. If the graphs of two linear inequalities have no intersection, then the system has no solution.

Practice

Solve the systems of two linear inequalities graphically. See Example 1 through 3.

1. $\begin{cases} y>2 \\ x\ge -3 \end{cases}$

2. $\begin{cases} 2x+5<0 \\ y\ge 2 \end{cases}$

3. $\begin{cases} x<3 \\ y>-x+2 \end{cases}$

4. $\begin{cases} y\le -5 \\ y\ge x-5 \end{cases}$

5. $\begin{cases} x\le 3 \\ 2x+y>7 \end{cases}$

6. $\begin{cases} 2x-y>4 \\ y<-1 \end{cases}$

7. $\begin{cases} x-3y\le 3 \\ x<5 \end{cases}$

8. $\begin{cases} 3x-2y\ge 8 \\ y\ge 0 \end{cases}$

9. $\begin{cases} x-y\ge 0 \\ 3x-2y\ge 4 \end{cases}$

10. $\begin{cases} y\ge x-2 \\ x+y\ge -2 \end{cases}$

11. $\begin{cases} 3x+y\le 10 \\ 5x-y\ge 6 \end{cases}$

12. $\begin{cases} y>3x+1 \\ -3x+y<-1 \end{cases}$

13. $\begin{cases} 3x+4y\ge -7 \\ y<2x+1 \end{cases}$

14. $\begin{cases} 2x-3y\ge 0 \\ 8x-3y<36 \end{cases}$

15. $\begin{cases} x+y<4 \\ 2x-3y<3 \end{cases}$

16. $\begin{cases} 2x+3y<12 \\ 3x+2y>13 \end{cases}$

17. $\begin{cases} x+y\ge 0 \\ x-2y\ge 6 \end{cases}$

18. $\begin{cases} y\ge 2x+3 \\ y\le x-2 \end{cases}$

19. $\begin{cases} x+3y\le 9 \\ x-y\ge 5 \end{cases}$

20. $\begin{cases} x-y\ge -2 \\ x+2y<-1 \end{cases}$

21. $\begin{cases} y\le x+3 \\ x-y\le -5 \end{cases}$

22. $\begin{cases} y\ge 2x-5 \\ 3x+2y>-3 \end{cases}$

23. $\begin{cases} y\le -2x \\ y>-2x-6 \end{cases}$

24. $\begin{cases} y>x-4 \\ y<x+2 \end{cases}$

Use a graphing calculator to solve the systems of linear inequalities. See Example 4.

25. $\begin{cases} y\ge 0 \\ 3x-5y\le 10 \end{cases}$

26. $\begin{cases} y\le 0 \\ 3x+y\le 11 \end{cases}$

27. $\begin{cases} 4x-3y\ge 6 \\ 3x-y\le 3 \end{cases}$

28. $\begin{cases} 3x+2y\le 15 \\ 2x+5y\ge 10 \end{cases}$

29. $\begin{cases} 3x-4y\ge -6 \\ 3x+2y\le 12 \end{cases}$

30. $\begin{cases} 3y\le 2x+2 \\ x+2y\le 11 \end{cases}$

31. $\begin{cases} x+y\le 8 \\ 3x-2y\ge -6 \end{cases}$

32. $\begin{cases} x+y\le 7 \\ 2x-y\le 8 \end{cases}$

33. $\begin{cases} y\le x \\ y<2x+1 \end{cases}$

34. $\begin{cases} x-y\ge -2 \\ 4x-y<16 \end{cases}$

20.
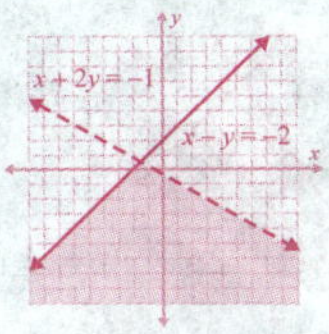

21. No solution

22.
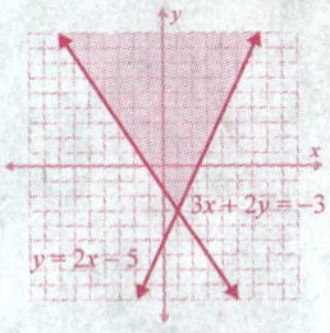

23.
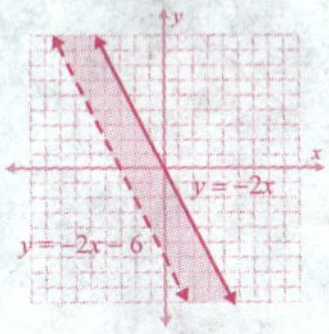

24.
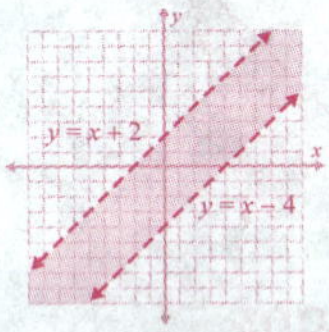

25.
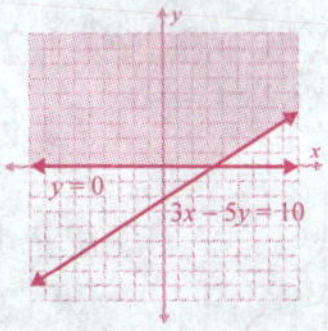

26.
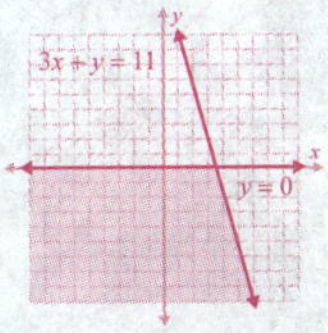

27.
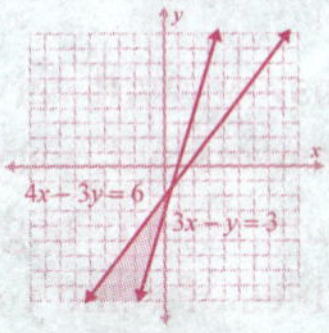

28.
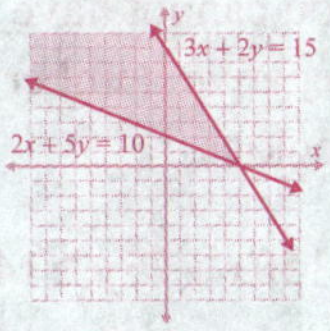

29.
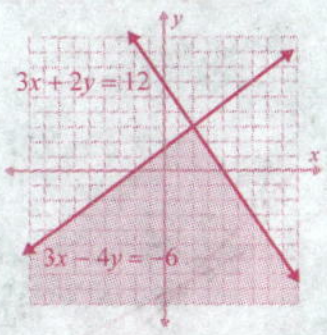

30.
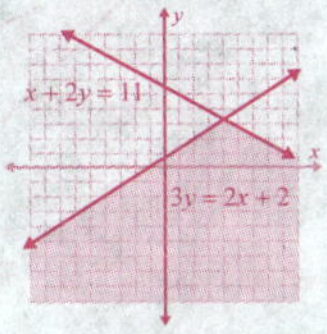

31. 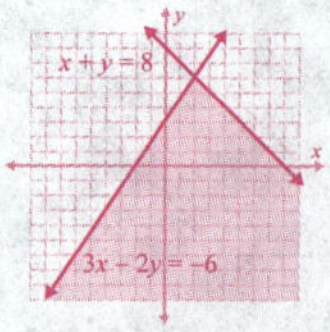

32. 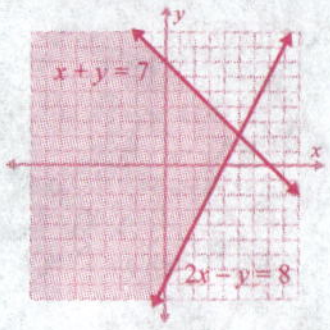

33. 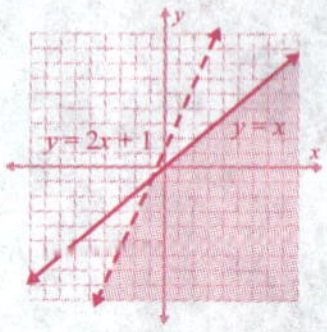

34. 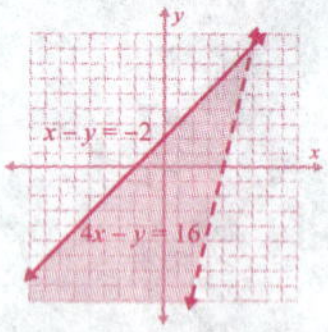

Applications

35. a. $\begin{cases} 0.1x + 0.3y + 1.2 < 5 \\ x + 2y < 30 \end{cases}$

b. 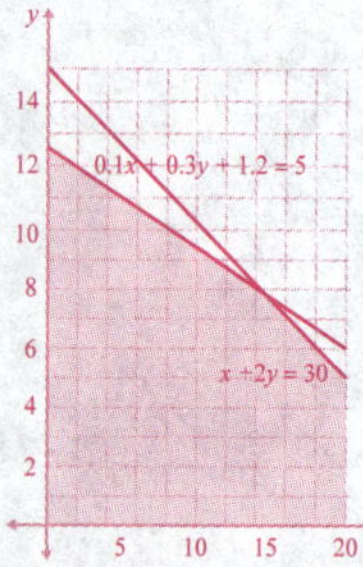

c. Answers will vary.

d. Answers will vary. For example, any negative values of x and y since you cannot have a negative number of cookies.

36. a. $\begin{cases} 150x + 75y \geq 14{,}000 \\ x + y \leq 150 \end{cases}$

b. 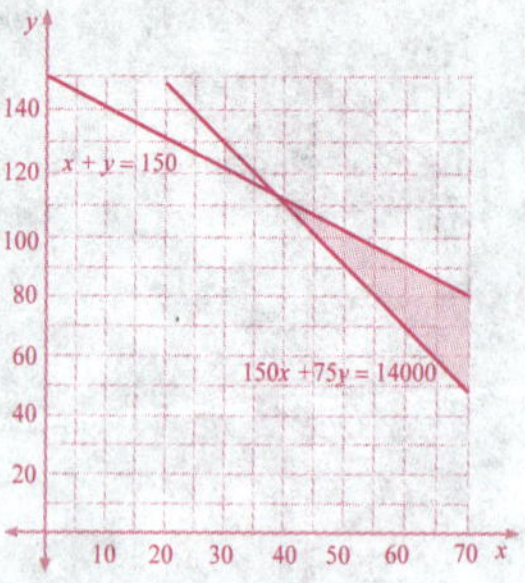

Applications

Solve.

35. Barbara's Bombtastic Bakery sells cookie bouquets where the price depends on the arrangement. Each completed bouquet arrangement needs to weigh less than 5 pounds for shipping purposes. The small cookies weigh 0.1 pounds and the large cookies weigh 0.3 pounds. The flower pot and Styrofoam weigh 1.2 pounds. The cost of each arrangement needs to be less than \$30. The small cookies cost \$1 each and the large cookies cost \$2 each. (The cost of the flower pot and foam are included in the cookie prices.)

a. Write two linear inequalities to describe the situation. Use the variable x to represent the number of small cookies and the variable y to represent the number of large cookies in a bouquet.

b. Graph the two linear inequalities on the same coordinate plane.

c. Describe the solution set for the situation.

d. Do any of the values in the solution set not make sense in the context of the problem? Explain why or why not.

36. Robin is planning a charity ball to raise money for her favorite charity. There are two different ticket options. The VIP option includes dinner, dancing, and cocktails for \$150 per ticket. The regular option includes dancing and cocktails for \$75 per ticket. Robin wants to make at least \$14,000 in ticket sales. The ballroom that is being used for the charity event has a maximum capacity of 150 people.

a. Write two linear inequalities to describe the situation. Let the variable x represent the number of VIP tickets sold and let the variable y represent the number of regular tickets sold.

b. Graph the two linear inequalities on the same coordinate plane.

c. Describe the solution set for the situation.

d. Can Robin reach her sales goal if she only sells tickets for the regular option? Explain why or why not.

Writing & Thinking

37. Graph the inequalities and explain how you can tell that there is no solution.

$$\begin{cases} y \leq 2x - 5 \\ y \geq 2x + 3 \end{cases}$$

c. Answers will vary.

d. No. $150 \cdot \$75 = \$11{,}250$

Writing & Thinking

37. The solutions of the two inequalities do not overlap.

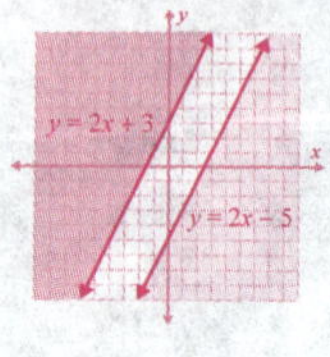

Chapter 11 Project

Don't Put All Your Eggs in One Basket!

An activity to demonstrate the use of linear systems in real life.

Have you ever heard the phrase "Don't put all your eggs in one basket"? This is a common saying that is often quoted in the investment world—and it's true. In an ever-changing economy it is important to diversify your investments. Splitting your money up into two or more funds may keep you from losing it all if one of the funds performs poorly. You may be thinking that you are too young to consider investments and saving money for retirement, but it is never too soon—especially in an economy where interest rates are extremely low. Low rates means it takes even longer to build up your nest egg. So start saving now and be sure to have more than one basket to put your eggs in! For this activity, if you need help understanding some of the investment terms, use the following website as a resource: www.investopedia.com

Let's suppose that you received a total of $5000 in cash as a graduation present from your relatives. You also have an additional $2500 that you saved from your summer job. You are thinking about investing the $7500 in two investment funds that have been recommended to you. One is currently earning 4% interest annually (conservative fund) and the other is earning 8% annually (aggressive fund). Keep in mind that interest rates fluctuate as the economy changes and there are few guarantees on the amount you will actually earn from any investment. Also, note that higher rates of interest typically indicate a higher risk on your investment.

1. If you want to earn $400 total in interest on your investments this year, how much money would you have to invest in each fund? Let the variable x be the amount invested in Fund 1 and the variable y be the amount invested in Fund 2. Recall that to calculate the interest on an investment, use the formula $I = Prt$, where P is the principal or amount invested, r is the annual interest rate, and t is the amount of time invested, which for our problem will be 1 year ($t = 1$). Use the table below to help you organize the information. Note that interest rates have to be converted to decimals before using them in an equation.

	Principal	Interest Rate	Interest
Fund 1	x	0.04	$0.04x$
Fund 2	y	0.08	$0.08y$
Total	**a.**		**b.**

a. Fill in the total amount available for investment in the bottom row of the table.

b. Fill in the total amount of interest desired in the bottom row of the table.

c. What does $0.04x$ represent in the context of this problem? hours. You'll need to convert to make units consistent throughout.)

d. What does $0.08y$ represent in the context of this problem?

e. Using the principal column of the table, write an equation in standard form involving the variables x and y to represent the total amount available for investment.

f. Using the interest column of the table, write an equation in standard form involving the variables x and y to represent the total amount of interest desired.

g. Solve the linear system of two equations derived in parts e. and f. to determine the amount to invest in each fund to earn $400 in interest. (You may use any method you choose: substitution, addition/elimination, or graphing).

h. Check to make sure that your solution to the system is correct by substituting the values from part g. for x and y into both equations and verify that the equations are true statements.

2. Suppose you decide you want to earn more interest on your investment. You now want to earn $500 in interest next year instead of $400. sing a table similar to the one in Problem 1, organize the information and follow a similar format to determine the amounts to invest in each of the funds that will earn $500 in interest in a year. $x = \$2500, y = \$5000$

3. Compare the results you obtained from Problems 1 and 2. How did the amounts in each investment change when your desired interest increased by $100?

4. Suppose you decide that $500 is not enough interest and you want to earn an additional $100 on your investments for a total of $600 in interest. Using a table similar to the one in Problem 1, organize the information and follow a similar format to determine the amounts to invest in each of the funds that will earn $600 in interest in a year. $x = \$0, y = \$7500$

5. Compare the results from Problem 4 to the results from Problems 1 and 2.
   - **a.** How much are you investing in Fund 1 to earn $600 in interest?
   - **b.** How much are you investing in Fund 2 to earn $600 in interest?
   - **c.** How do your results contradict the advice provided to you at the start of this activity?
   - **d.** Is it possible to make more than $600 in interest on your $7500 investment using these two funds? Explain why or why not?
   - **e.** What is the smallest amount of interest you can earn on your investment using these two funds? How did you determine this?

6. How much interest would you earn if you split the initial principal of $7500 equally between the two funds? $450

7. If you actually had $7500 to invest in these two funds earning 4% and 8% respectively, how would you invest the money? Explain your reasoning. Answers will vary.

Chapter 11 Project: Don't Put All Your Eggs in One Basket!

1. **a.** 7500
   b. 400
   c. The interest earned on the amount invested in Fund 1 after 1 year
   d. The interest earned on the amount invested in Fund 2 after 1 year
   e. $x + y = 7500$
   f. $0.04x + 0.08y = 400$
   g. $x = \$5000, y = \$2500$

3. The amount invested in Fund 1 was cut in half and the amount invested in Fund 2 was doubled.

5. **a.** $0 or no principal
   b. $7500 or all of the principal
   c. The advice was not to put all your eggs in one basket or to not invest all your money in one type of investment.
   d. No, it isn't possible to earn more than $600 in interest with these two funds since investing the entire principal into Fund 2 gives you $600 in interest.
   e. $300 by investing the entire principal into Fund 1.

Chapter 11 Project

Limitations of Money-Making Opportunities

An activity to demonstrate the use of solving systems of linear equations and inequalities in real life.

In this project, you will investigate how limitations (known as *constraints*) on the production of goods can lead to limitations on how much profit a company can make.

Suppose the Soaring Eagle Book Store wishes to produce two types of coffee mugs: Type A and Type B. Each Type A mug will result in a profit of \$3, and each Type B mug will result in a profit of \$2.75. Manufacturing one Type A mug requires 6 minutes on Machine I and 3 minutes on Machine II. Manufacturing one Type B mug requires 4 minutes on Machine I and 6 minutes on Machine II. According to the schedule, Machine I has 12 hours available and Machine II has 10 hours available to make the mugs. How many of each type of mug should the bookstore make to maximize its profit?

1. Using the variables x and y, identify the two unknown quantities.
2. Consider the information provided about the time available on each machine. This information allows us to write two linear inequalities in the standard form $Ax + By \leq C$. These are called constraints.
   a. Explain what each constraint represents.
   b. Explain why the constraints should be of the less-than-or-equal-to type.
   c. Write the two constraints. (**Note:** Be mindful of how the times are expressed, minutes versus hours. You'll need to convert to make units consistent throughout.)
3. Regardless of maximizing profit, there are two more "common sense" constraints. Consider whether there is a minimum or maximum number of mugs. Using this information, write two more constraints and explain why they are constraints.
4. Graph the system of linear inequalities resulting from the four constraints. (**Hint:** You may wish to first graph the two inequalities from Problem 3, shading lightly. Once you've identified the intersection, erase extra shading. Then add the other two constraints found in Problem 2 to the graph and, once you've identified the new (smaller) intersection, erase extra shading again.)
5. The final result of graphing the four constraints should be a four-cornered region. Use systems of equations to solve for the (x, y) coordinates of the four corners.
6. Recall the profit that the bookstore can make from each type of mug. Write an equation of the form $P = Ax + By$, with A and B filled in but not P.
7. Consider the following six values of P: –400, –200, 0, 200, 400, and 600. Using your answer to Problem 6, lightly graph six lines for these six different values of P over the shaded four-cornered region from Problem 5. (**Hint:** These six lines will be parallel.)
8. Would all of the lines created in Problem 7 indicate a possible value of P? Explain why or why not.
9. Is there a maximum value of P or can it keep growing? Why or why not?
10. Let's determine the maximum profit that Soaring Eagle Book Store can make from selling mugs.
    a. If there is a maximum value of P, at what point does it seem to occur?
    b. Is there anything special about the point found in part a.?
    c. Substitute this point into your $P = Ax + By$ equation from Problem 6 to calculate the maximum profit. Express your final answer in words; that is, state how many mugs of each type should be made to produce the maximum profit and what that maximum profit is.

Chapter 11 Project: Limitations of Money-Making Opportunities

1. x is the number of Type A mugs and y is the number of Type B mugs.

2. a. Each constraint represents the time available on each machine to create the two types of mugs.

   b. Machine I cannot operate more than 12 hours and Machine II cannot operate more than 10 hours.

   c. 12 hours = 720 minutes and 10 hours = 600 minutes. $6x + 4y \le 720$ (for Machine I, all in minutes) and $3x + 6y \le 600$ (for Machine II, all in minutes.)

3. The two other constraints are $x \ge 0$ and $y \ge 0$ because you cannot create a negative number of mugs.

4.

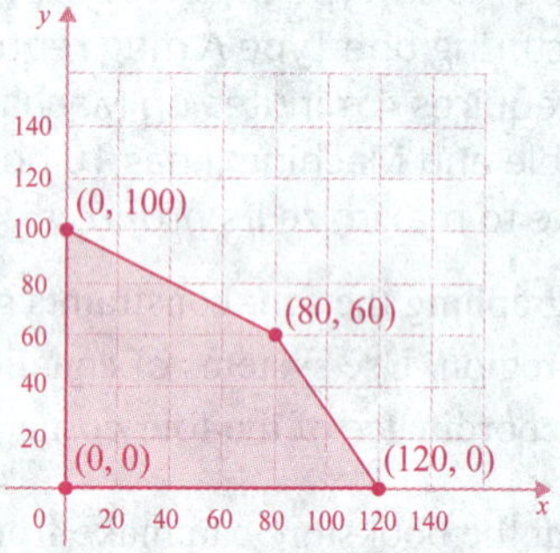

5. From $x = 0$ and $y = 0$, (0, 0). From $6x + 4y = 720$ and $y = 0$, (120, 0). From $6x + 4y = 720$ and $3x + 6y = 600$, (80, 60). From $3x + 6y = 600$ and $x = 0$, (0, 100). Each corner point is labeled in the graph from Problem 4.

6. $P = 3x + 2.75y$

7.

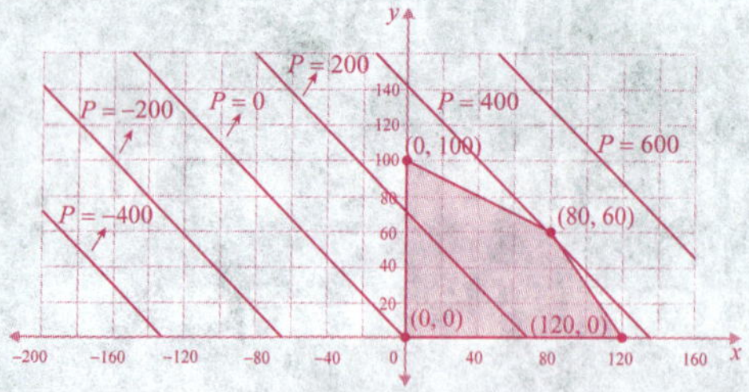

8. The first two lines ($P = -400$ and $P = -200$) do not represent possible values of P because profit should be nonnegative. The value $P = 600$ does not represent a possible value because it would require making a number of mugs that results in running the two machines for longer time than they are available. The remaining lines do represent possible values of P because the profit $P = Ax + By$ is always non-negative when x and y are nonnegative.

9. There is a maximum value of P. The value is limited by the number of mugs that can be made on the machines while they are available.

10. a. (80, 60)

    b. It is a corner point.

    c. $3 \cdot 80 + 2.75 \cdot 60 = 405$. By making 80 Type A mugs and 60 Type B mugs, the bookstore will obtain a maximum profit of \$405.

CHAPTER 12

12 Exponents and Polynomials

Connections

Physics is a natural science that deals with matter and its motion through space. Within physics, scientists often use polynomials to define the rules that govern nature. Consider the formula for force, $F = ma$ (which means force equals mass times acceleration). This formula defines why a basketball bounces higher when slammed into the ground than when given a simple dribble. It also explains why a baseball may dent a car, but a ball of paper thrown at the same speed will not. Knowing how to perform operations with polynomials and how to factor polynomials is necessary to understand the information provided by these rules of nature.

Suppose you are bowling with friends and using a 5 kg bowling ball (which is approximately 11 pounds). After you roll the ball down the lane, the ball's acceleration can be represented by the polynomial $a = 4t^2 - 12t + 7$, where a is measured in m/s^2 and t represents time in seconds. (We can assume that this is the ball's acceleration only until it strikes the pins.) How would you use the formula $F = ma$ to find a polynomial that represents the force of the ball at time t? (Note that the force of the bowling ball is measured in Newtons (N) which is $kg \cdot m/s^2$.)

Objectives

A. Use the product rule for exponents.

B. Simplify expressions that contain the exponent 0.

C. Use the quotient rule for exponents.

D. Simplify expressions that contain negative exponents.

E. Use a graphing calculator to evaluate exponential expressions.

12.1 Rules for Exponents

A The Product Rule

Previously, an **exponent** was defined as a number that tells how many times a number (called the **base**) is used in multiplication. This definition is limited because it is valid only if the exponents are positive integers. In this section, we will develop four properties of exponents that will help in simplifying algebraic expressions and expand your understanding of exponents to include variable bases and negative exponents. (In a later chapter, you will study fractional exponents.)

From earlier, we know that

Exponent

$$6^2 = 6 \cdot 6 = 36$$

Base

and

$$6^3 = 6 \cdot 6 \cdot 6 = 216.$$

Teaching Note

Be sure to emphasize that the exponent **counts** the number of factors in these expressions. This is especially important when students are dealing with expressions such as $2^3 \cdot 2^4$, because there is a tendency for them to multiply the base numbers and erroneously write 4^7.

Also, the base may be a variable so that

$$x^3 = x \cdot x \cdot x \quad \text{and} \quad x^5 = x \cdot x \cdot x \cdot x \cdot x.$$

Now to find the products of expressions such as $6^2 \cdot 6^3$ or $x^3 \cdot x^5$ and to simplify these products, we can write down all the factors as follows.

$$6^2 \cdot 6^3 = (6 \cdot 6) \cdot (6 \cdot 6 \cdot 6) = 6^5$$

$$x^3 \cdot x^5 = (x \cdot x \cdot x) \cdot (x \cdot x \cdot x \cdot x \cdot x) = x^8$$

With these examples in mind, what do you think would be a simplified form for the product $3^4 \cdot 3^3$? You were right if you thought 3^7. That is, $3^4 \cdot 3^3 = 3^7$. Notice that in each case, **the base stays the same**.

The preceding discussion, along with the basic concept of whole-number exponents, leads to the following **product rule for exponents**.

Note

Reminder about the **exponent 1**: if a variable or constant has no exponent written, the exponent is understood to be 1.

For example, $y = y^1$ and $7 = 7^1$.

In general, for any real number a, $\mathbf{a = a^1}$.

The Product Rule for Exponents

If a is a nonzero real number and m and n are integers, then

$$a^m \cdot a^n = a^{m+n}.$$

In words, to multiply powers with the same base, keep the base and add the exponents.

PROPERTIES

Example 1 Using the Product Rule for Exponents

Use the product rule for exponents to simplify the following expressions.

a. $x^2 \cdot x^4$ **b.** $a \cdot b^6$ **c.** $4^2 \cdot 4$ **d.** $2^3 \cdot 2^2$ **e.** $(-2)^4(-2)^3$

Solution

a. $x^2 \cdot x^4 = x^{2+4} = x^6$ Keep the base and add the exponents.

b. $a \cdot b^6 = ab^6$ This expression cannot be simplifed. The bases are not the same.

c. $4^2 \cdot 4 = 4^{2+1} = 4^3 = 64$ Note that the base stays 4. That is, the bases are not multiplied.

d. $2^3 \cdot 2^2 = 2^{3+2} = 2^5 = 32$ Note that the base stays 2. That is, the bases are not multiplied.

e. $(-2)^4(-2)^3 = (-2)^{4+3} = (-2)^7 = -128$

Now work margin exercise 1.

1. Use the product rule for exponents when simplifying the following expressions.

a. $x^3 \cdot x^2$

b. $y^5 \cdot y^4$

c. $5^2 \cdot 5$

d. $3^2 \cdot 3^3$

e. $(-3)^5(-3)^2$

To multiply terms that have numerical coefficients and the same variables with exponents, **the coefficients are multiplied** as usual and **the exponents are added by using the product rule**. You may want to apply the commutative property of multiplication to rearrange the variables and constants so they are grouped together. This will make it easier to perform the multiplication and find the sum of the exponents. Example 2 illustrates these concepts.

Example 2 Using the Product Rule for Exponents

Use the product rule for exponents when simplifying the following expressions.

a. $2y^2 \cdot 3y^9$ **b.** $(-3x^3)(-4x^3)$ **c.** $(-6ab^2)(8ab^3)$

Solution

a. $$2y^2 \cdot 3y^9 = 2 \cdot 3 \cdot y^2 \cdot y^9 = 6y^{2+9} = 6y^{11}$$

Apply the commutative property of multiplication. Then, multiply the coefficients 2 and 3 and add the exponents 2 and 9.

b. $$(-3x^3)(-4x^3) = (-3)(-4) \cdot x^3 \cdot x^3 = 12x^{3+3} = 12x^6$$

Coefficients –3 and –4 are multiplied and exponents 3 and 3 are added.

c. $$(-6ab^2)(8ab^3) = (-6) \cdot 8 \cdot a^1 \cdot a^1 \cdot b^2 \cdot b^3 = -48 \cdot a^{1+1} \cdot b^{2+3} = -48a^2b^5$$

Coefficients –6 and 8 are multiplied and exponents on each variable are added.

Now work margin exercise 2.

2. Use the product rule for exponents when simplifying the following expressions.

a. $4y^3 \cdot 5y^6$

b. $(-2x^4)(5x^3)$

c. $(-5ab^3)(7ab^4)$

B The Exponent 0

The product rule is stated for m and n as **integer** exponents. This means that the rule is also valid for 0 and for negative exponents. As an aid for understanding 0 as an exponent, consider the following patterns of exponents for powers of 2, 3, and 10.

Teaching Note

You might want to discuss division and subtraction of exponents as in

$\frac{2^3}{2^3} = 2^{(3-3)} = 2^0 = 1$

because $\frac{2^3}{2^3} = \frac{8}{8} = 1.$

Powers of 2	Powers of 3	Powers of 10
$2^5 = 32$	$3^5 = 243$	$10^5 = 100{,}000$
$2^4 = 16$	$3^4 = 81$	$10^4 = 10{,}000$
$2^3 = 8$	$3^3 = 27$	$10^3 = 1000$
$2^2 = 4$	$3^2 = 9$	$10^2 = 100$
$2^1 = 2$	$3^1 = 3$	$10^1 = 10$
$2^0 = ?$	$3^0 = ?$	$10^0 = ?$

Note that as the exponent decreases by one, each new power is the result of dividing the previous value by the base. As a result, we see that $2^0 = 1$, $3^0 = 1$, and $10^0 = 1$.

Another approach to understanding 0 as an exponent is to consider the product rule. Applying this rule with the exponent 0 and the fact that 1 is the multiplicative identity gives results such as the following equations.

$5^0 \cdot 5^2 = 5^{0+2} = 5^2$ This implies that $5^0 = 1$.

$4^0 \cdot 4^3 = 4^{0+3} = 4^3$ This implies that $4^0 = 1$.

This discussion leads directly to the **rule for 0 as an exponent.**

The Exponent 0

If a is a nonzero real number, then $\boldsymbol{a^0 = 1}$.

The expression 0^0 is undefined.

PROPERTIES

Attention!

Throughout this text, unless specifically stated otherwise, we will assume that the bases of exponents are nonzero.

3. Simplify the following expressions using the rule for 0 as an exponent.

a. 1587^0

b. $x^0 \cdot x^2$

c. $(-17)^0$

Example 3 Evaluating The Exponent 0

Simplify the following expressions using the rule for 0 as an exponent.

a. 10^0 **b.** $x^0 \cdot x^3$ **c.** $(-6)^0$

Solution

a. $10^0 = 1$

b. $x^0 \cdot x^3 = x^{0+3} = x^3$ or $x^0 \cdot x^3 = 1 \cdot x^3 = x^3$

c. $(-6)^0 = 1$

Now work margin exercise 3.

C The Quotient Rule

Now consider a fraction in which the numerator and denominator are powers with the same base, such as $\frac{5^4}{5^2}$ or $\frac{x^5}{x^2}$. We can write the following.

$$\frac{5^4}{5^2} = \frac{\cancel{5} \cdot \cancel{5} \cdot 5 \cdot 5}{\cancel{5} \cdot \cancel{5} \cdot 1} = \frac{5^2}{1} = 25 \quad \text{or} \quad \frac{5^4}{5^2} = 5^{4-2} = 5^2 = 25$$

and

$$\frac{x^5}{x^2} = \frac{\cancel{x} \cdot \cancel{x} \cdot x \cdot x \cdot x}{\cancel{x} \cdot \cancel{x} \cdot 1} = \frac{x^3}{1} = x^3 \quad \text{or} \quad \frac{x^5}{x^2} = x^{5-2} = x^3$$

In fractions, as just illustrated, the exponents can be subtracted. Again, the base remains the same. We now have the following **quotient rule for exponents**.

Quotient Rule for Exponents

If a is a nonzero real number and m and n are integers, then

$$\frac{a^m}{a^n} = a^{m-n}.$$

In words, to divide two powers with the same base, keep the base and subtract the exponents. (Subtract the denominator exponent from the numerator exponent.)

PROPERTIES

Teaching Note

While we are emphasizing the quotient rule for exponents here, you might want to discuss the fact that $\frac{5^4}{5^2} = \frac{625}{25} = 25$, and, as long as the base is numerical, we can always check this way.

Example 4 Using the Quotient Rule for Exponents

Use the quotient rule for exponents to simplify the following expressions.

a. $\frac{x^6}{x}$ **b.** $\frac{y^8}{y^2}$ **c.** $\frac{x^2}{x^2}$

Solution

a. $\frac{x^6}{x} = x^{6-1} = x^5$ Keep the base and subtract the exponents.

b. $\frac{y^8}{y^2} = y^{8-2} = y^6$

c. $\frac{x^2}{x^2} = x^{2-2} = x^0 = 1$ Note how this example shows another way to justify the idea that $a^0 = 1$. Since the numerator and denominator are the same and not 0, it makes sense that the fraction is equal to 1.

Now work margin exercise 4.

4. Use the quotient rule for exponents to simplify the following expressions.

a. $\frac{x^5}{x^3}$

b. $\frac{y^7}{y^4}$

c. $\frac{x^5}{x^5}$

In division, with terms that have numerical coefficients, the **coefficients are divided** as usual and **exponents are subtracted** using the quotient rule. These ideas are illustrated in Example 5.

5. Use the quotient rule for exponents when simplifying the following expressions.

a. $\frac{12x^{12}}{4x^4}$

b. $\frac{21x^8y^7}{3x^2y^3}$

Example 5 Dividing Terms with Coefficients

Use the quotient rule for exponents when simplifying the following expressions.

a. $\frac{15x^{15}}{3x^3}$ **b.** $\frac{20x^{10}y^6}{2x^2y^3}$

Solution

a.
$$\frac{15x^{15}}{3x^3} = \frac{15}{3}\cdot\frac{x^{15}}{x^3}$$
$$= 5\cdot x^{15-3}$$
$$= 5x^{12}$$

Coefficients 15 and 3 are divided and exponents 15 and 3 are subtracted.

b.
$$\frac{20x^{10}y^6}{2x^2y^3} = \frac{20}{2}\cdot\frac{x^{10}}{x^2}\cdot\frac{y^6}{y^3}$$
$$= 10\cdot x^{10-2}\cdot y^{6-3}$$
$$= 10x^8y^3$$

Coefficients 20 and 2 are divided and exponents on each variable are subtracted.

Now work margin exercise 5.

D Negative Exponents

The quotient rule for exponents leads directly to the development of an understanding of negative exponents. In Examples 4 and 5, for each base, the exponent in the numerator was larger than or equal to the exponent in the denominator. Therefore, when the exponents were subtracted using the quotient rule, the result was either a positive exponent or the exponent 0. But, what if the larger exponent is in the denominator and we still apply the quotient rule?

For example, applying the quotient rule to $\frac{4^3}{4^5}$ gives

$$\frac{4^3}{4^5} = 4^{3-5} = 4^{-2}$$

which results in a negative exponent.

But, simply reducing $\frac{4^3}{4^5}$ gives

$$\frac{4^3}{4^5} = \frac{\cancel{4}\cdot\cancel{4}\cdot\cancel{4}\cdot 1}{\cancel{4}\cdot\cancel{4}\cdot\cancel{4}\cdot 4\cdot 4} = \frac{1}{4\cdot 4} = \frac{1}{4^2}.$$

This means that $4^{-2} = \frac{1}{4^2}$.

Similar investigations will show that $2^{-1} = \frac{1}{2}$, $5^{-2} = \frac{1}{5^2}$, and $x^{-3} = \frac{1}{x^3}$.

Teaching Note

You might want to refer back to the patterns discussed when introducing the exponent 0 and help the students to logically decide that the next steps in the respective columns are $2^{-1} = \frac{1}{2}$, $3^{-1} = \frac{1}{3}$, and $10^{-1} = \frac{1}{10}$.

Rule for Negative Exponents

If a is a nonzero real number and n is an integer, then

$$a^{-n} = \frac{1}{a^n}.$$

PROPERTIES

Example 6 Negative Exponents

Use the rule for negative exponents to simplify each expression so that it contains only positive exponents.

a. 5^{-1} **b.** x^{-3} **c.** $x^{-9} \cdot x^{7}$

Solution

a. $5^{-1} = \frac{1}{5^1} = \frac{1}{5}$ Use the rule for negative exponents.

b. $x^{-3} = \frac{1}{x^3}$ Use the rule for negative exponents.

c. Here we use the product rule first and then the rule for negative exponents.

$$x^{-9} \cdot x^{7} = x^{-9+7} = x^{-2} = \frac{1}{x^2}$$

Now work margin exercise 6.

6. Use the rule for negative exponents to simplify each expression so that it contains only positive exponents.

a. 7^{-1}

b. x^{-7}

c. $x^{-11} \cdot x^{6}$

Note

There is nothing wrong with negative exponents. In fact, negative exponents are preferred in some courses in mathematics and science. However, so that all answers are the same, in this course we will consider expressions to be simplified if

1. all exponents are positive and

2. each base appears only once.

Each of the expressions in Example 7 is simplified by using the appropriate rule for exponents. Study each example carefully. In each case, **the expression is considered simplified if each base appears only once and each base has only positive exponents.**

Example 7 Combining Rules for Exponents

Simplify each expression so that it contains only positive exponents.

a. $2^{-5} \cdot 2^{8}$ **c.** $\frac{10^{-5}}{10^{-2}}$ **e.** $\frac{15x^{10} \cdot 2x^{2}}{3x^{15}}$

b. $\frac{x^6}{x^{-1}}$ **d.** $\frac{x^6 y^3}{x^2 y^5}$

Solution

a. $2^{-5} \cdot 2^{8} = 2^{-5+8}$ Use the product rule with positive and negative exponents.

$= 2^3$

$= 8$

b. $\frac{x^6}{x^{-1}} = x^{6-(-1)}$ Use the quotient rule with positive and negative exponents.

$= x^{6+1}$

$= x^7$

7. Simplify each expression so that it contains only positive exponents.

a. $3^{-7} \cdot 3^{10}$

b. $\frac{x^5}{x^{-1}}$

c. $\frac{8^{-4}}{8^{-2}}$

d. $\frac{x^7 y^3}{x^4 y^6}$

e. $\frac{8x^9 \cdot 3x^4}{6x^{16}}$

c. $\frac{10^{-5}}{10^{-2}} = 10^{-5-(-2)}$ Use the quotient rule with negative exponents.

$= 10^{-5+2}$ Add the exponents.

$= 10^{-3}$

$= \frac{1}{10^3}$ or $\frac{1}{1000}$ Use the rule for negative exponents.

d. $\frac{x^6 y^3}{x^2 y^5} = x^{6-2} y^{3-5}$ Use the quotient rule with two variables.

$= x^4 y^{-2}$

$= \frac{x^4}{y^2}$ Use the rule for negative exponents.

e. $\frac{15x^{10} \cdot 2x^2}{3x^{15}} = \frac{(15 \cdot 2)x^{10+2}}{3x^{15}}$ Use the product rule.

$= \frac{30x^{12}}{3x^{15}}$

$= 10x^{12-15}$ Use the quotient rule.

$= 10x^{-3}$

$= \frac{10}{x^3}$ Use the rule for negative exponents.

(**Note:** Notice that the negative exponent relates only to the variable x, **not** to the number 10.)

Now work margin exercise 7.

Attention!

Using the Quotient Rule

Regardless of the size of the exponents or whether they are positive or negative, the following single subtraction rule can be used with the quotient rule.

(numerator exponent – denominator exponent)

This subtraction will always lead to the correct answer.

Summary of the Rules for Exponents

The following rules are true for any nonzero real number a and integers m and n.

1. The exponent 1: $a = a^1$
2. The exponent 0: $a^0 = 1$
3. The product rule: $a^m \cdot a^n = a^{m+n}$
4. The quotient rule: $\dfrac{a^m}{a^n} = a^{m-n}$
5. Negative exponents: $a^{-n} = \dfrac{1}{a^n}$

PROPERTIES

E Using a Graphing Calculator to Evaluate Exponential Expressions

On a TI-84 Plus graphing calculator, the caret key [^] is used to indicate an exponent. Example 8 illustrates the use of a graphing calculator in evaluating expressions with exponents.

Example 8 Evaluating Exponential Expressions

Use a graphing calculator to evaluate each expression.

a. 2^{-3} **b.** 23.18^0 **c.** $(-3.2)^3(1.5)^2$

Solution

The following solutions show how the caret key [^] is used to indicate exponents. Be careful to use the negative sign key [(-)] (and not the minus sign key) for negative numbers and negative exponents. Use parentheses as they are given in the expression being evaluated.

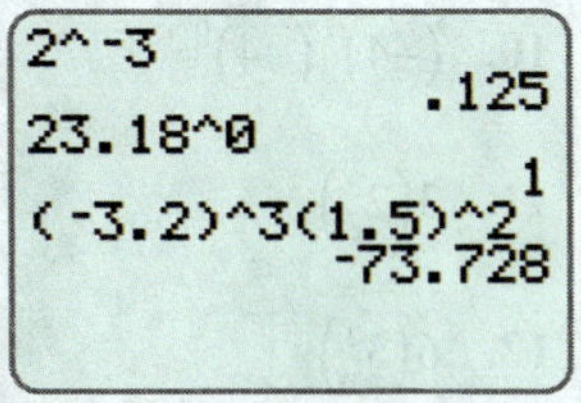

Now work margin exercise 8.

8. Use a graphing calculator to evaluate each expression.

a. 5^{-2}

b. 17.06^0

c. $(-4.7)^2(1.2)^3$

Margin Exercise Answers

1. a. x^5 **b.** y^9 **c.** 125 **d.** 243 **e.** -2187 **2. a.** $20y^9$ **b.** $-10x^7$ **c.** $-35a^2b^7$ **3. a.** 1 **b.** x^2 **c.** 1 **4. a.** x^2 **b.** y^3 **c.** 1 **5. a.** $3x^8$ **b.** $7x^6y^4$ **6. a.** $\frac{1}{7}$ **b.** $\frac{1}{x^7}$ **c.** $\frac{1}{x^5}$ **7. a.** 3^3 or 27 **b.** x^6 **c.** $\frac{1}{8^2}$ or $\frac{1}{64}$ **d.** $\frac{x^3}{y^3}$ **e.** $\frac{4}{x^3}$ **8. a.** 0.04 **b.** 1 **c.** 38.17152

12.1 Exercises

Concept Check

1. subtract
2. add
3. positive
4. undefined
5. a
6. 1
7. False; If there is no exponent written, the exponent is assumed to be 1.
8. False; Under these conditions, $a^{-n}=\frac{1}{a^n}$.
9. True
10. False; You should subtract the denominator exponent from the numerator exponent.

Practice

1. 27
2. 16,807
3. 512
4. 25
5. $\frac{1}{3}$
6. $\frac{1}{16}$
7. $\frac{1}{25}$
8. $\frac{1}{216}$
9. 16
10. −64
11. 24
12. 54
13. −500
14. −54
15. $\frac{3}{8}$
16. $\frac{4}{9}$
17. $-\frac{3}{25}$

12.1 Exercises

Concept Check

Fill-in-the-Blank. Complete each sentence using information found in this section.

1. The quotient rule for exponents says that when dividing two powers with the same base, keep the base and ________ the exponents.
2. The product rule for exponents says that when multiplying two powers with the same base, keep the base and ________ the exponents.
3. An expression is considered simplified if each base appears only once and each base has only ________ exponents.
4. The expression 0^0 is ________.
5. For all real values of a, $a^1 =$ ___.
6. For all real values of a, $a^0 =$ ___.

True/False. Determine whether each statement is true or false. If a statement is false, explain how it can be changed so the statement will be true. (**Note:** There may be more than one acceptable change.)

7. If a constant does not have an exponent written, it is assumed that the exponent is 0.
8. If a is a nonzero real number and n is an integer, then $a^{-n} = -a^n$.
9. Since the product rule is stated for integer exponents, the rule is also valid for 0 and negative exponents.
10. When using the quotient rule, you should subtract the smaller exponent from the larger exponent.

Practice

Simplify each expression. The final form of the expressions with variables should contain only positive exponents. Assume that all variables represent nonzero numbers. See Examples 1 though 7.

1. $3^2 \cdot 3$
2. $7^2 \cdot 7^3$
3. $8^3 \cdot 8^0$
4. $5^0 \cdot 5^2$
5. 3^{-1}
6. 4^{-2}
7. 5^{-2}
8. 6^{-3}
9. $(-2)^4(-2)^0$
10. $(-4)^3(-4)^0$
11. $3(2^3)$
12. $6(3^2)$
13. $-4(5^3)$
14. $-2(3^3)$
15. $3(2^{-3})$
16. $4(3^{-2})$
17. $-3(5^{-2})$
18. $-5(2^{-2})$
19. $x^2 \cdot x^3$
20. $x^3 \cdot x$
21. $y^2 \cdot y^0$
22. $y^3 \cdot y^8$
23. x^{-3}
24. y^{-2}

25. $2x^{-1}$

26. $5y^{-4}$

27. $-8y^{-2}$

28. $-10x^{-3}$

29. $5x^6y^{-4}$

30. x^0y^{-2}

31. $3x^0+y^0$

32. $5y^0-3x^0$

33. $\dfrac{7^3}{7}$

34. $\dfrac{9^5}{9^2}$

35. $\dfrac{10^3}{10^4}$

36. $\dfrac{10}{10^5}$

37. $\dfrac{2^3}{2^6}$

38. $\dfrac{5^7}{5^4}$

39. $\dfrac{x^4}{x^2}$

40. $\dfrac{x^6}{x^3}$

41. $\dfrac{x^3}{x}$

42. $\dfrac{y^7}{y^2}$

43. $\dfrac{x^7}{x^3}$

44. $\dfrac{x^8}{x^3}$

45. $\dfrac{x^{-2}}{x^2}$

46. $\dfrac{x^{-3}}{x}$

47. $\dfrac{x^4}{x^{-2}}$

48. $\dfrac{x^5}{x^{-1}}$

49. $\dfrac{x^{-3}}{x^{-5}}$

50. $\dfrac{x^{-4}}{x^{-1}}$

51. $\dfrac{y^{-2}}{y^{-4}}$

52. $\dfrac{y^3}{y^{-3}}$

53. $3x^3\cdot x^0$

54. $3y\cdot y^4$

55. $x^3\cdot x^2\cdot x^{-1}$

56. $x^{-3}\cdot x^0\cdot x^2$

57. $(4x^3)(9x^0)$

58. $(5x^2)(3x^4)$

59. $(-2x^2)(7x^3)$

60. $(3y^3)(-6y^2)$

61. $(-4x^5)(3x)$

62. $(6y^4)(5y^5)$

63. $\dfrac{8y^3}{2y^2}$

64. $\dfrac{12x^4}{3x}$

65. $\dfrac{9y^5}{3y^3}$

66. $\dfrac{-10x^5}{2x}$

67. $\dfrac{-8y^4}{4y^2}$

68. $\dfrac{12x^6}{-3x^3}$

69. $\dfrac{x^{-1}x^2}{x^3}$

70. $\dfrac{x\cdot x^3}{x^{-3}}$

71. $\dfrac{10^4\cdot 10^{-3}}{10^{-2}}$

72. $\dfrac{10\cdot 10^{-1}}{10^2}$

73. $(9x^2)^0$

74. $(-2x^{-3}y^5)^0$

75. $(9x^2y^3)(-2x^3y^4)$

76. $(-3xy)(-5x^2y^{-3})$

77. $\dfrac{-8x^2y^4}{4x^3y^2}$

78. $\dfrac{-8x^{-2}y^4}{4x^2y^{-2}}$

79. $(3a^2b^4)(4ab^5c)$

80. $(-6a^3b^4)(4a^{-2}b^8)$

81. $\dfrac{36a^5b^0c}{-9a^{-5}b^{-3}}$

82. $\dfrac{7x^2y^{-2}}{28x^0yz^{-2}}$

83. $\dfrac{25y^6\cdot 3y^{-2}}{15xy^4}$

84. $\dfrac{12a^{-2}\cdot 18a^4}{36a^2b^{-5}}$

Use a graphing calculator to evaluate each expression. Round quotients to the nearest ten-thousandth, if necessary. See Example 8.

85. $(2.16)^0$

86. $(-5.06)^2$

87. $(1.6)^{-2}$

88. $(2.5)^{-4}$

89. $(6.4)^4(2.3)^2$

90. $(-14.8)^2(21.3)^2$

18. $-\dfrac{5}{4}$

19. x^5

20. x^4

21. y^2

22. y^{11}

23. $\dfrac{1}{x^3}$

24. $\dfrac{1}{y^2}$

25. $\dfrac{2}{x}$

26. $\dfrac{5}{y^4}$

27. $-\dfrac{8}{y^2}$

28. $-\dfrac{10}{x^3}$

29. $\dfrac{5x^6}{y^4}$

30. $\dfrac{1}{y^2}$

31. 4

32. 2

33. 49

34. 729

35. $\dfrac{1}{10}$

36. $\dfrac{1}{10{,}000}$

37. $\dfrac{1}{8}$

38. 125

39. x^2

40. x^3

41. x^2

42. y^5

43. x^4

44. x^5

45. $\dfrac{1}{x^4}$

46. $\dfrac{1}{x^4}$

47. x^6

48. x^6

49. x^2

50. $\frac{1}{x^3}$

51. y^2

52. y^6

53. $3x^3$

54. $3y^5$

55. x^4

56. $\frac{1}{x}$

57. $36x^3$

58. $15x^6$

59. $-14x^5$

60. $-18y^5$

61. $-12x^6$

62. $30y^9$

63. $4y$

64. $4x^3$

65. $3y^2$

66. $-5x^4$

67. $-2y^2$

68. $-4x^3$

69. $\frac{1}{x^2}$

70. x^7

71. 1000

72. $\frac{1}{100}$

73. 1

74. 1

75. $-18x^5y^7$

76. $\frac{15x^3}{y^2}$

77. $-\frac{2y^2}{x}$

78. $-\frac{2y^6}{x^4}$

79. $12a^3b^9c$

80. $-24ab^{12}$

81. $-4a^{10}b^3c$

Applications

Solve.

91. Rylee wants to move all her files to a new hard drive that has 2^{12} GB of storage on it. She wants to designate the same amount of storage for each of 2^4 projects. How much storage should be assigned to each project? Write your answer as a power of two.

92. Trey is studying patterns in bacteria. For a positive test result in his experiment, bacteria must grow in population at a minimum rate of 3^2 in 24 hours. If the initial population of the bacteria is 3^5 and his final measurement after 24 hours is 3^8, should he mark the test as positive or negative?

93. A molecule being studied under a powerful microscope is cubic in shape. What is the volume of the molecule if the length of one side is 10^{-8} cm?

94. A hurricane caused flooding in a home at the rate of 2^3 ft^3 per hour. If that home has a storage closet that is 2^1 feet wide, 2^3 feet long, and 2^4 feet high, how long will it take the storage closet to fill with water? Write your answer as a power of 2.

95. A conference center needs an array of gift bags set up for a meeting. There will be 2^5 gift bags per row and 2^4 rows of gift bags. The delivery truck can hold 2^9 gift bags per load. How many deliveries will the truck need to make in order to supply the gift bags needed?

96. A local children's convention receives donations of 2^8 bags of candy for use as gifts for attendees. The convention has 2^7 children attending. How many bags of candy will each child receive?

97. Molly buys land that is 3^4 yards wide and 3^5 yards long. What is the area of the land? Write your answer as a power of 3.

98. Samuel wants to buy grass seed to plant in his yard. His lawn is 2^6 feet wide and 27 feet long. Each bag of grass seed will cover 2^{10} square feet. How many bags of seed should he purchase? Write your answer as a power of 2.

99. Barbara's Bombtastic Bakery makes *petit four glaces,* which are small bite-sized cakes. Each cake is in the shape of a cube that has a side length of $2x$, where x is a positive length which varies depending on the cake flavor.

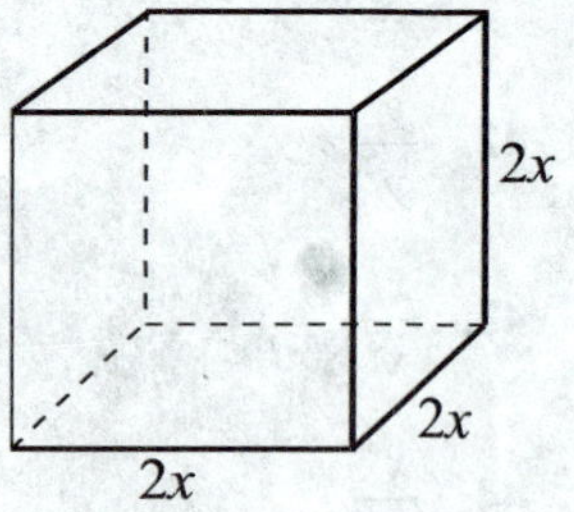

a. Write an expression using exponents to find the volume of the *petit four glaces*. Do not simplify.

b. Which exponential rule will you need to use to simplify the expression from part a.?

c. Simplify the expression from part a.

d. If $x = 2$ cm, determine the volume of the *petit four glaces* using the expression from part c.

100. A strain of the influenza virus is spreading throughout a community and the number of confirmed cases of the flu doubles every day. On day 0 (the initial day) of the outbreak, 1 person has the virus. On day 1 of the outbreak, $1 \cdot 2 = 2$ people will have the virus. On day 2 of the outbreak, $1 \cdot 2 \cdot 2 = 1 \cdot 2^2 = 4$ people will have the virus.

a. Write an exponential expression to describe how many people will have influenza virus on day 5. Write as a power of 2 and simplify.

b. Write an exponential expression to describe how many people would have the virus on day n if 3 people had the virus on day 0 of the outbreak. Write the expression in exponential form and simplify.

c. Use the expression from part b. to determine the number of people that will have the virus on day 5 of the outbreak if 3 people had the virus on day 0?

101. A standard hard drive has 2^{38} bytes of data. 1 gigabyte is equivalent to 230 bytes.

a. Write an exponential expression to determine how many gigabytes are equivalent to 2^{38} bytes?

b. Simplify the expression from part a. to determine how many gigabytes are in 2^{38} bytes.

c. What rule of exponents did you use to simplify part b.?

82. $\frac{x^2z^2}{4y^3}$

83. $\frac{5}{x}$

84. $6b^5$

85. 1

86. 25.6036

87. 0.3906

88. 0.0256

89. 8875.147264

90. 99,376.2576

Applications

91. 2^8 GB

92. Positive

93. 10^{-24} cm^3

94. 2^5 hours

95. 1 delivery

96. 2 bags of candy

97. 3^9 square yards

98. 2^1 bags of seed

99. **a.** $(2x)^3$

b. Product rule for exponents

c. $8x^3$ **d.** 64 cm^3

100. **a.** $1 \cdot 2^5 = 32$ **b.** $3 \cdot 2^n$

c. 96 people

101. **a.** $\frac{2^{38}}{230} = 28$ gigabytes

b. 256 gigabytes

c. Quotient rule

Objectives

A. Use the power rule for exponents to simplify expressions.

B. Use the rule for a power of a product to simplify expressions.

C. Use the rule for a power of a quotient to simplify expressions.

D. Use combinations of rules for exponents to simplify expressions.

12.2 Power Rules for Exponents

The summary of the rules for exponents given in Section 12.1 is repeated here for easy reference.

Summary of the Rules for Exponents

The following rules are true for any nonzero real number a and integers m and n.

1. The exponent 1: $a = a^1$
2. The exponent 0: $a^0 = 1$
3. The product rule: $a^m \cdot a^n = a^{m+n}$
4. The quotient rule: $\frac{a^m}{a^n} = a^{m-n}$
5. Negative exponents: $a^{-n} = \frac{1}{a^n}$

PROPERTIES

A Power Rule

Consider what happens when a power is raised to a power. For example, to simplify the expressions $(x^2)^3$ and $(2^5)^2$, we can write

$$(x^2)^3 = x^2 \cdot x^2 \cdot x^2 = x^{2+2+2} = x^6 \quad \text{and} \quad (2^5)^2 = 2^5 \cdot 2^5 = 2^{5+5} = 2^{10}.$$

However, this technique can be quite time-consuming when the exponent is large, such as in $(3y^3)^{17}$. The **power rule for exponents** gives a convenient way to handle powers raised to powers.

Math Tip

The previous properties of multiplication you learned still apply when multiplying the exponents and keeping the base. For example, consider the commutative property:

$$(a^m)^n = a^{mn} = a^{nm} = (a^n)^m$$

Power Rule for Exponents

If a is a nonzero real number and m and n are integers, then

$$(a^m)^n = a^{mn}.$$

In other words, the value of a power raised to a power can be found by multiplying the exponents and keeping the base.

PROPERTIES

1. Simplify each expression by using the power rule for exponents.

a. $(x^3)^5$ **c.** $(y^{-5})^3$

b. $(x^4)^{-3}$ **d.** $(3^2)^{-3}$

Example 1 Using the Power Rule for Exponents

Simplify each expression by using the power rule for exponents.

a. $(x^2)^4$ **b.** $(x^5)^{-2}$ **c.** $(y^{-7})^2$ **d.** $(2^3)^{-2}$

Solution

a. $\left(x^2\right)^4 = x^{2(4)} = x^8$ Multiply the exponents and keep the base.

b. $\left(x^5\right)^{-2} = x^{5(-2)} = x^{-10} = \frac{1}{x^{10}}$ or $\left(x^5\right)^{-2} = \frac{1}{\left(x^5\right)^2} = \frac{1}{x^{5(2)}} = \frac{1}{x^{10}}$

c. $\left(y^{-7}\right)^2 = y^{(-7)(2)} = y^{-14} = \frac{1}{y^{14}}$

d. $\left(2^3\right)^{-2} = 2^{3(-2)} = 2^{-6} = \frac{1}{2^6}$ Evaluating gives $\frac{1}{2^6} = \frac{1}{64}$.

Another approach, because we have a numerical base, would be

$\left(2^3\right)^{-2} = (8)^{-2} = \frac{1}{8^2}$. Evaluating gives $\frac{1}{8^2} = \frac{1}{64}$.

We see that while the base and exponent may be different, the value is the same.

Now work margin exercise 1.

B Rule for Power of a Product

If the base of an exponent is a product, we will see that each factor in the product is to be raised to the power indicated by the exponent. For example, $(10x)^3$ indicates that the product of 10 and x is to be raised to the 3rd power and $\left(-2x^2y\right)^5$ indicates that the product of -2, x^2, and y is to be raised to the 5th power. We can simplify these expressions as follows.

$$\begin{aligned}(10x)^3 &= 10x \cdot 10x \cdot 10x \\ &= 10 \cdot 10 \cdot 10 \cdot x \cdot x \cdot x \\ &= 10^3 \cdot x^3 \\ &= 1000x^3\end{aligned}$$

and

$$\begin{aligned}\left(-2x^2y\right)^5 &= \left(-2x^2y\right)\left(-2x^2y\right)\left(-2x^2y\right)\left(-2x^2y\right)\left(-2x^2y\right) \\ &= (-2)(-2)(-2)(-2)(-2) \cdot x^2 \cdot x^2 \cdot x^2 \cdot x^2 \cdot x^2 \cdot y \cdot y \cdot y \cdot y \cdot y \\ &= (-2)^5\left(x^2\right)^5 \cdot y^5 \\ &= -32x^{10}y^5\end{aligned}$$

We can simplify expressions such as these in a much easier fashion by using the following **rule for power of a product**.

Rule for Power of a Product

If a and b are nonzero real numbers and n is an integer, then

$$(ab)^n = a^n b^n.$$

In words, a power of a product is found by raising each factor to that power.

PROPERTIES

2. Simplify each expression by using the rule for power of a product.

a. $(4x)^2$

b. $(xy)^7$

c. $(-9ab)^2$

d. $(ab)^{-3}$

e. $(x^3y^{-4})^2$

Example 2 Using the Rule for Power of a Product

Simplify each expression by using the rule for power of a product.

a. $(5x)^2$ **b.** $(xy)^3$ **c.** $(-7ab)^2$ **d.** $(ab)^{-5}$ **e.** $(x^2y^{-3})^4$

Solution

a. $(5x)^2 = 5^2 \cdot x^2 = 25x^2$ Raise each factor to the given power.

b. $(xy)^3 = x^3 \cdot y^3 = x^3y^3$

c. $(-7ab)^2 = (-7)^2 a^2b^2 = 49a^2b^2$

d. $(ab)^{-5} = a^{-5} \cdot b^{-5} = \frac{1}{a^5} \cdot \frac{1}{b^5} = \frac{1}{a^5b^5}$ or, using the rule of negative exponents first and then the rule for the power of a product, $(ab)^{-5} = \frac{1}{(ab)^5} = \frac{1}{a^5b^5}$

e. $(x^2y^{-3})^4 = (x^2)^4(y^{-3})^4 = x^8 \cdot y^{-12} = \frac{x^8}{y^{12}}$

Now work margin exercise 2.

Negative Numbers and Exponents

In an expression such as $-x^2$, we know that -1 is understood to be the coefficient of x^2. That is, $-x^2 = -1 \cdot x^2$.

The same is true for expressions with numbers such as -7^2. That is,

$$-7^2 = -1 \cdot 7^2 = -1 \cdot 49 = -49.$$

We see that the exponent refers to 7 and not to -7. For the exponent to refer to -7 as the base, -7 **must be in parentheses** as follows.

$$(-7)^2 = (-7)(-7) = +49$$

As another example, $-2^0 = -1 \cdot 2^0 = -1 \cdot 1 = -1$ and $(-2)^0 = 1$.

CAUTION

C Rule for Power of a Quotient

If the base of an exponent is a quotient (in fraction form), we will see that both the numerator and denominator are raised to the power indicated by the exponent. For example, in the expression $\left(\frac{2}{x}\right)^3$ the quotient (or fraction) $\frac{2}{x}$ is raised to the 3rd power.

We can simplify this expression as follows.

$$\left(\frac{2}{x}\right)^3 = \frac{2}{x} \cdot \frac{2}{x} \cdot \frac{2}{x} = \frac{2 \cdot 2 \cdot 2}{x \cdot x \cdot x} = \frac{2^3}{x^3} = \frac{8}{x^3}$$

Or we can simplify the expression in a much easier manner by applying the following **rule for the power of a quotient**.

Rule for Power of a Quotient

If a and b are nonzero real numbers and n is an integer, then

$$\left(\frac{a}{b}\right)^n = \frac{a^n}{b^n}.$$

In words, a power of a quotient (in fraction form) is found by raising both the numerator and the denominator to that power.

PROPERTIES

Example 3 Using the Rule for Power of a Quotient

Simplify each expression by using the rule for the power of a quotient.

a. $\left(\frac{y}{x}\right)^5$ **b.** $\left(\frac{2}{a}\right)^4$ **c.** $\left(\frac{3}{4}\right)^3$ **d.** $\left(\frac{x}{7}\right)^2$

Solution

a. $\left(\frac{y}{x}\right)^5 = \frac{y^5}{x^5}$ Raise the numerator and denominator to the given power.

b. $\left(\frac{2}{a}\right)^4 = \frac{2^4}{a^4} = \frac{16}{a^4}$

c. $\left(\frac{3}{4}\right)^3 = \frac{3^3}{4^3} = \frac{27}{64}$

d. $\left(\frac{x}{7}\right)^2 = \frac{x^2}{7^2} = \frac{x^2}{49}$

Now work margin exercise 3.

3. Simplify each expression by using the rule for the power of a quotient.

a. $\left(\frac{x}{y}\right)^7$

b. $\left(\frac{5}{6}\right)^2$

c. $\left(\frac{3}{a}\right)^3$

d. $\left(\frac{x}{6}\right)^3$

D Using Combinations of Rules for Exponents

There may be more than one way to apply the various rules for exponents. As illustrated in the following examples, if you apply the rules correctly (even in a different sequence), the answer will be the same in every case.

Example 4 Using Combinations of Rules for Exponents

Simplify each expression by using the appropriate rules for exponents.

a. $\left(\frac{-2x}{y^2}\right)^3$ **b.** $\left(\frac{3a^2b}{a^3b^2}\right)^2$

Solution

a. $\left(\frac{-2x}{y^2}\right)^3 = \frac{(-2x)^3}{(y^2)^3} = \frac{(-2)^3x^3}{y^6} = \frac{-8x^3}{y^6}$

4. Simplify each expression by using the appropriate rules for exponents.

a. $\left(\frac{-3x}{y^3}\right)^3$

b. $\left(\frac{4a^3b^2}{a^4b}\right)^2$

Teaching Note

Example 4b illustrates two procedures for simplifying an expression. You should emphasize the idea that all of these rules for exponents allow for flexibility. Of course, the challenge is for the student to use the rule correctly regardless of the order of usage.

b. Method 1: Simplify inside the parentheses first.

$$\left(\frac{3a^2b}{a^3b^2}\right)^2 = \left(3a^{2-3}b^{1-2}\right)^2 = \left(3a^{-1}b^{-1}\right)^2 = 3^2a^{-2}b^{-2} = \frac{9}{a^2b^2}$$

Method 2: Apply the power of a quotient rule first.

$$\left(\frac{3a^2b}{a^3b^2}\right)^2 = \frac{\left(3a^2b\right)^2}{\left(a^3b^2\right)^2} = \frac{3^2a^{2(2)}b^2}{a^{3(2)}b^{2(2)}} = \frac{9a^4b^2}{a^6b^4} = 9a^{4-6}b^{2-4} = 9a^{-2}b^{-2} = \frac{9}{a^2b^2}$$

Notice that the answer is the same even though the rules were applied in a different order.

Now work margin exercise 4.

Another general approach with fractions involving negative exponents is to note that

$$\left(\frac{a}{b}\right)^{-n} = \frac{a^{-n}}{b^{-n}} = \frac{b^n}{a^n} = \left(\frac{b}{a}\right)^n.$$

In effect, there are two basic shortcuts with negative exponents and fractions:

1. Taking the reciprocal of a fraction changes the sign of any exponent in the fraction.
2. Moving any factor from numerator to denominator, or vice versa, changes the sign of the corresponding exponent.

5. Simplify:

$$\left(\frac{x^6}{y^3}\right)^{-5}$$

Example 5 Using Two Approaches with Fractional Expressions and Negative Exponents

Simplify: $\left(\frac{x^3}{y^5}\right)^{-4}$

Solution

Method 1: Use the ideas of reciprocals first.

$$\left(\frac{x^3}{y^5}\right)^{-4} = \left(\frac{y^5}{x^3}\right)^4 = \frac{y^{5(4)}}{x^{3(4)}} = \frac{y^{20}}{x^{12}}$$

Method 2: Apply the power of a quotient rule first.

$$\left(\frac{x^3}{y^5}\right)^{-4} = \frac{\left(x^3\right)^{-4}}{\left(y^5\right)^{-4}} = \frac{x^{3(-4)}}{y^{5(-4)}} = \frac{x^{-12}}{y^{-20}} = \frac{y^{20}}{x^{12}}$$

Now work margin exercise 5.

Example 6 Simplifying A More Complex Problem

This example involves the application of a variety of steps. Study it carefully and see if you can get the same result by following a different sequence of steps.

Simplify: $\left(\dfrac{2x^2y^3}{3xy^{-2}}\right)^{-2}\left(\dfrac{4x^2y^{-1}}{3x^{-5}y^3}\right)^{-1}$

Solution

Method 1: Simplify inside the parentheses first.

$$\left(\frac{2x^2y^3}{3xy^{-2}}\right)^{-2}\left(\frac{4x^2y^{-1}}{3x^{-5}y^3}\right)^{-1}=\left(\frac{2}{3}x^{2-1}y^{3-(-2)}\right)^{-2}\left(\frac{4}{3}x^{2-(-5)}y^{-1-3}\right)^{-1}$$

$$=\left(\frac{2}{3}x^1y^5\right)^{-2}\left(\frac{4}{3}x^7y^{-4}\right)^{-1}$$

$$=\left(\frac{2}{3}\right)^{-2}\cdot x^{-2}\cdot y^{5(-2)}\cdot\left(\frac{4}{3}\right)^{-1}\cdot x^{7(-1)}\cdot y^{-4(-1)}$$

$$=\frac{3^2}{2^2}\cdot\frac{3}{4}\cdot x^{-2+(-7)}\cdot y^{-10+4}$$

$$=\frac{27}{16}x^{-9}y^{-6}$$

$$=\frac{27}{16x^9y^6}$$

Method 2: Apply the power of a quotient rule first.

$$\left(\frac{2x^2y^3}{3xy^{-2}}\right)^{-2}\left(\frac{4x^2y^{-1}}{3x^{-5}y^3}\right)^{-1}=\frac{2^{-2}\cdot x^{2(-2)}\cdot y^{3(-2)}}{3^{-2}\cdot x^{-2}y^{-2(-2)}}\cdot\frac{4^{-1}\cdot x^{2(-1)}\cdot y^{-1(-1)}}{3^{-1}\cdot x^{-5(-1)}\cdot y^{3(-1)}}$$

$$=\frac{3^2}{2^2}\cdot\frac{x^{-4}}{x^{-2}}\cdot\frac{y^{-6}}{y^4}\cdot\frac{3}{4}\cdot\frac{x^{-2}}{x^5}\cdot\frac{y^1}{y^{-3}}$$

$$=\frac{9\cdot3\cdot x^{-4-2}y^{-6+1}}{4\cdot4\cdot x^{-2+5}y^{4-3}}$$

$$=\frac{27x^{-6}y^{-5}}{16x^3y^1}$$

$$=\frac{27x^{-6-3}y^{-5-1}}{16}$$

$$=\frac{27x^{-9}y^{-6}}{16}$$

$$=\frac{27}{16x^9y^6}$$

Now work margin exercise 6.

6. Simplify:

$$\left(\frac{3x^3y^2}{4xy^{-3}}\right)^{-2}\left(\frac{5x^3y^{-2}}{2x^{-4}y^2}\right)^{-2}$$

A complete summary of the rules for exponents includes the following eight rules.

Summary of the Rules for Exponents

The following rules are true for any nonzero real numbers a and b and integers m and n.

1. The exponent 1: $a = a^1$
2. The exponent 0: $a^0 = 1$
3. The product rule: $a^m \cdot a^n = a^{m+n}$
4. The quotient rule: $\frac{a^m}{a^n} = a^{m-n}$
5. Negative exponents: $a^{-n} = \frac{1}{a^n}$.
6. Power rule: $\left(a^m\right)^n = a^{mn}$.
7. Power of a product: $(ab)^n = a^n b^n$.
8. Power of a quotient: $\left(\frac{a}{b}\right)^n = \frac{a^n}{b^n}$.

PROPERTIES

Margin Exercise Answers

1. a. x^{15} **b.** $\frac{1}{x^{12}}$ **c.** $\frac{1}{y^{15}}$ **d.** $\frac{1}{3^6}$ or $\frac{1}{9^3}$ or $\frac{1}{729}$ **2. a.** $16x^2$ **b.** x^7y^7 **c.** $81a^2b^2$ **d.** $\frac{1}{a^3b^3}$ **e.** $\frac{x^6}{y^8}$ **3. a.** $\frac{x^7}{y^7}$ **b.** $\frac{25}{36}$ **c.** $\frac{27}{a^3}$ **d.** $\frac{x^3}{216}$ **4. a.** $\frac{-27x^3}{y^9}$ **b.** $\frac{16b^2}{a^2}$ **5.** $\frac{y^{15}}{x^{30}}$ **6.** $\frac{64}{225x^{18}y^2}$

12.2 Exercises

Concept Check

Fill-in-the-Blank. Complete each sentence using information found in this section.

1. Moving any term from the numerator to the denominator, or vice versa, changes the sign of the corresponding __________.
2. A power of a quotient (in fraction form) is found by raising both the __________ and the __________ to that power.
3. To find the value of a power raised to a power, __________ the exponents and ______ the base.
4. A power of a product can be found by ______ each factor to that power.
5. In an expression such as $-x^2$, we know that -1 is understood to be the ______ of x^2.

12.2 Exercises

Concept Check

1. exponent(s)
2. numerator, denominator
3. multiply, keep
4. raising
5. coefficient
6. True
7. True
8. False; The rules for exponents can be applied in any order, resulting in the same answer.
9. True

Practice

1. -81
2. -25
3. -16
4. -400
5. 1,000,000
6. 4096
7. a^6
8. $\frac{1}{b^8}$
9. $\frac{1}{x^{10}}$
10. x^6
11. $\frac{1}{256}$
12. 64
13. $9y^2$
14. a^4b^4
15. $16x^2y^2$
16. $\frac{9}{x^4}$
17. $\frac{1}{x^6y^6}$
18. $\frac{a^9}{b^6}$
19. $36x^6$
20. $9x^8$
21. $-108x^6$

True/False. Determine whether each statement is true or false. If a statement is false, explain how it can be changed so the statement will be true. (**Note:** There may be more than one acceptable change.)

6. Taking the reciprocal of a fraction changes the sign of any exponent in the fraction.

7. For an exponent to refer to −7 as the base, −7 must be in parentheses.

8. When simplifying an expression with exponents, the rules for exponents must be used in a specific order or the answer will vary.

9. The expression -8^2 simplifies to −64.

Practice

Use the rules for exponents to simplify each of the expressions. Assume that all variables represent nonzero real numbers. See Examples 1 through 6.

1. -3^4
2. -5^2
3. -2^4
4. -20^2
5. $(-10)^6$
6. $(-4)^6$
7. $(a^3)^2$
8. $(b^2)^{-4}$
9. $(x^{-5})^2$
10. $(x^{-2})^{-3}$
11. $(2^4)^{-2}$
12. $(2^{-3})^{-2}$
13. $(3y)^2$
14. $(ab)^4$
15. $(-4xy)^2$
16. $(3x^{-2})^2$
17. $(xy)^{-6}$
18. $(a^3b^{-2})^3$
19. $(6x^3)^2$
20. $(-3x^4)^2$
21. $4(-3x^2)^3$
22. $7(2y^{-2})^4$
23. $5(x^2y^{-1})$
24. $-3(7xy^2)^0$
25. $-2(3x^5y^{-2})^{-3}$
26. $-4(5x^{-3}y)^{-1}$
27. $\left(\frac{a}{b}\right)^4$
28. $\left(\frac{x}{2}\right)^3$
29. $\left(\frac{2}{3}\right)^2$
30. $\left(\frac{a}{4}\right)^3$
31. $\left(\frac{x}{y}\right)^6$
32. $\left(\frac{2}{5}\right)^2$
33. $\left(\frac{3x}{y}\right)^3$
34. $\left(\frac{-4x}{y^2}\right)^2$
35. $\left(\frac{6m^3}{n^5}\right)^0$
36. $\left(\frac{3x^2}{y^3}\right)^2$
37. $\left(\frac{-2x^2}{y^{-2}}\right)^2$
38. $\left(\frac{2x}{y^5}\right)^{-2}$
39. $\left(\frac{x}{y}\right)^{-2}$
40. $\left(\frac{2a}{b}\right)^{-1}$
41. $\left(\frac{3x}{y^{-2}}\right)^{-1}$
42. $\left(\frac{4a^2}{b^{-3}}\right)^{-3}$
43. $\left(\frac{-3}{xy^2}\right)^{-3}$
44. $\left(\frac{5xy^3}{y}\right)^2$
45. $\left(\frac{m^2n^3}{mn}\right)^2$

22. $\frac{112}{y^8}$
23. $\frac{5x^2}{y}$
24. -3
25. $-\frac{2y^6}{27x^{15}}$
26. $-\frac{4x^3}{5y}$
27. $\frac{a^4}{b^4}$
28. $\frac{x^3}{8}$
29. $\frac{4}{9}$
30. $\frac{a^3}{64}$
31. $\frac{x^6}{y^6}$
32. $\frac{4}{25}$
33. $\frac{27x^3}{y^3}$
34. $\frac{16x^2}{y^4}$
35. 1
36. $\frac{9x^4}{y^6}$
37. $4x^4y^4$
38. $\frac{y^{10}}{4x^2}$
39. $\frac{y^2}{x^2}$
40. $\frac{b}{2a}$
41. $\frac{1}{3xy^2}$
42. $\frac{1}{64a^6b^9}$
43. $-\frac{x^3y^6}{27}$
44. $25x^2y^4$
45. m^2n^4

46. $16a^4b^4$

47. $-\frac{y^2}{49x^2}$

48. $\frac{1}{8a^3b^6}$

49. $\frac{25x^6}{y^2}$

50. $\frac{y^8}{16x^8}$

51. $\frac{x^6}{y^6}$

52. $\frac{8a^6}{b^9}$

53. $\frac{36y^{14}}{x^4}$

54. $27x^{18}y^{12}$

55. $\frac{49y^4}{x^6}$

56. $\frac{25x^{12}y^6}{9}$

57. $\frac{24y^5}{x^7}$

58. $\frac{25x^9}{8y^4}$

59. $\frac{4x^3}{3}$

60. $\frac{3}{10y}$

61. $\frac{16x^{13}}{243y^6}$

62. $\frac{11{,}520b^{28}}{a}$

63. $96x^2y^4z^4$

64. $\frac{729}{49a^6b^4c^4}$

65. 19.4481

66. 0.0346

67. $208.5136x^4$

68. $140.608x^6$

69. 22.7024

70. 0.1335

46. $\left(\frac{2ab^3}{b^2}\right)^4$

47. $\left(\frac{-7^2x^2y}{y^3}\right)^{-1}$

48. $\left(\frac{2ab^4}{b^2}\right)^{-3}$

49. $\left(\frac{5x^3y}{y^2}\right)^2$

50. $\left(\frac{2x^2y}{y^3}\right)^{-4}$

51. $\left(\frac{x^3y^{-1}}{y^2}\right)^2$

52. $\left(\frac{2a^2b^{-1}}{b^2}\right)^3$

53. $\left(\frac{6y^5}{x^2y^{-2}}\right)^2$

54. $\left(\frac{3x^4}{x^{-2}y^{-4}}\right)^3$

55. $\frac{(7x^{-2}y)^2}{(xy^{-1})^2}$

56. $\frac{(-5x^3y^4)^2}{(3x^{-3}y)^2}$

57. $\frac{(3x^2y^{-1})^{-2}}{(6x^{-1}y)^{-3}}$

58. $\frac{(2x^{-3})^{-3}}{(5y^{-2})^{-2}}$

59. $\frac{(4x^{-2})(6x^5)}{(9y)(2y^{-1})}$

60. $\frac{(5x^2)(3x^{-1})^2}{(25y^3)(6y^{-2})}$

61. $\left(\frac{3xy^3}{4x^2y^{-3}}\right)^{-1}\left(\frac{2x^3y^{-1}}{9x^{-3}y^{-1}}\right)^2$

62. $\left(\frac{5a^4b^{-2}}{6a^{-4}b^3}\right)^{-2}\left(\frac{5a^3b^4}{2^{-2}a^{-2}b^{-2}}\right)^3$

63. $\left(\frac{6x^{-4}yz^{-2}}{4^{-1}x^{-4}y^3z^{-2}}\right)^{-1}\left(\frac{2^{-2}xyz^{-3}}{12x^2y^2z^{-1}}\right)^{-2}$

64. $\left(\frac{3^{-5}a^5b^3c^{-1}}{3^{-2}abc}\right)^{-2}\left(\frac{7^{-1}a^{-4}bc^2}{7^{-2}a^{-3}bc^{-2}}\right)^{-2}$

Use a graphing calculator to evaluate each expression. Round quotients to the nearest ten-thousandth, if necessary.

65. $(2.1^2)^2$

66. $(1.4^{-2})^5$

67. $(3.8x)^4$

68. $(5.2x^2)^3$

69. $\left(\frac{8.1}{1.7}\right)^2$

70. $\left(\frac{2.3}{4.5}\right)^3$

12.3 Applications: Scientific Notation

A Writing Numbers in Scientific Notation

A basic application of integer exponents occurs in scientific disciplines, such as astronomy and biology, when very large and very small numbers are involved. For example, the distance from the earth to the sun is approximately 93,000,000 miles, and the approximate radius of a carbon atom is 0.0000000077 centimeters.

In **scientific notation** (an option in all scientific and graphing calculators), decimal numbers are written as the product of a number that is greater than or equal to 1 and less than 10, and an integer power of 10. In scientific notation, there is just one digit to the left of the decimal point.

$$250{,}000 = 2.5\times10^{5}$$ This is the same as multiplying by 100,000.

$$0.000000345 = 3.45\times10^{-7}$$ This is the same as dividing by 10,000,000.

The exponent tells how many places the decimal point is to be moved and in which direction. If the exponent is positive, the decimal point is moved to the right.

$$5.6\times10^{4} = 5.6000.$$ 4 places to the right

A negative exponent indicates that the decimal point should move to the left.

$$4.9\times10^{-3} = 0.004.9$$ 3 places to the left

> ### Scientific Notation
> If N is a decimal number, then in **scientific notation**
>
> $N = a\times\mathbf{10}^{n}$ where $1\le|a|<10$ and n is an integer.
>
> DEFINITION

Example 1 Writing Decimals in Scientific Notation

Write the following decimal numbers in scientific notation.

a. 8,720,000 **b.** 0.000000376

Solution

a. $8{,}720{,}000 = 8.72\times10^{6}$ 8.72 is between 1 and 10.

To check, move the decimal point 6 places to the right and get the original number:

$$8.72\times10^{6} = 8.7\,2\,0\,0\,0\,0. = 8{,}720{,}000$$
1 2 3 4 5 6

Objectives

A. Write decimal numbers in scientific notation.

B. Use scientific notation to help simplify decimal expressions.

Scientific Notation

When writing very large and very small numbers, scientific notation is sometimes used to save space. Many calculators will automatically convert to this notation when necessary. It is believed that the concept of scientific notation was first introduced when Archimedes was asked how many gains of sand were in the universe. This concept was later advanced by René Descartes when he introduced the standard notation that uses superscripts to show exponents. Scientific notation also helps in the understanding of significant figures in science classes. An example from chemistry is Avogadro's number, the number of molecules per mole of a substance: 6.02×10^{23}.

1. Write the following decimal numbers in scientific notation.

a. 63,900,000

b. 0.00000245

b. $0.000000376 = 3.76 \times 10^{-7}$ 3.76 is between 1 and 10.

To check, move the decimal point 7 places to the left and get the original number:

$$3.76 \times 10^{-7} = 0.0000003.76 = 0.000000376$$

7 6 5 4 3 2 1

Now work margin exercise 1.

B Scientific Notation and Simplifying Expressions

2. Simplify the following expressions by first writing the decimal numbers in scientific notation and then using the properties of exponents.

a. $\dfrac{(2600)(0.0036)}{520{,}000}$

b. $\dfrac{(108{,}000)(3400)}{(0.006)(510)}$

Example 2 Using Scientific Notation while Simplifying

Simplify the following expressions by first writing the decimal numbers in scientific notation and then using the properties of exponents.

a. $\dfrac{(0.085)(41{,}000)}{0.00017}$ **b.** $\dfrac{(11{,}100)(0.064)}{(8{,}000{,}000)(370)}$

Solution

a.
$$\frac{(0.085)(41{,}000)}{0.00017} = \frac{(8.5\times10^{-2})(4.1\times10^{4})}{1.7\times10^{-4}}$$ Write each number in scientific notation.

$$= \frac{(\overset{5}{\cancel{8.5}})(4.1)}{\cancel{1.7}} \times \frac{(10^{-2})(10^{4})}{10^{-4}}$$ Reduce and simplify by using the appropriate rules for exponents.

$$= 20.5 \times \frac{10^{2}}{10^{-4}}$$
$$= 2.05\times10^{1}\times\frac{10^{2}}{10^{-4}}$$
$$= 2.05\times10^{1+2-(-4)}$$
$$= 2.05\times10^{7}$$

b.
$$\frac{(11{,}100)(0.064)}{(8{,}000{,}000)(370)} = \frac{(1.11\times10^{4})(6.4\times10^{-2})}{(8.0\times10^{6})(3.7\times10^{2})}$$ Write each number in scientific notation.

$$= \frac{(\overset{0.3}{\cancel{1.11}})(\overset{0.8}{\cancel{6.4}})}{(\cancel{8.0})(\cancel{3.7})} \times \frac{10^{2}}{10^{8}}$$ Reduce and simplify by using the appropriate rules for exponents.

$$= 0.24\times10^{2-8}$$
$$= 2.4\times10^{-1}\times10^{-6}$$
$$= 2.4\times10^{-1-6}$$
$$= 2.4\times10^{-7}$$

Now work margin exercise 2.

Example 3 Application: Scientific Notation

Light travels approximately 3×10^{8} meters per second. How many meters per minute does light travel?

Solution

Since there are 60 seconds in one minute, multiply by 60.

$$\begin{aligned}3\times10^{8}\times60&=180\times10^{8}\\&=1.8\times10^{2}\times10^{8}\\&=1.8\times10^{10}\end{aligned}$$

Thus, light travels 1.8×10^{10} meters per minute.

(**Note:** In decimal form, this is 18,000,000,000 meters per minute. So, you can see why scientists prefer scientific notation.)

Now work margin exercise 3.

3. One mole, a value often used in physics and chemistry, equals 6.02×10^{23} particles for all substances. How many particles would be in 8 moles of carbon?

Example 4 Scientific Notation and Calculators

Use a graphing calculator to evaluate each expression. Leave the answer in scientific notation.

a. $\dfrac{0.0042\cdot3{,}000{,}000}{0.21}$

b. $\dfrac{8600\left(3.0\times10^{5}\right)}{1.5\times10^{-6}}$

Solution

a. With a graphing calculator (set in scientific notation mode) the display should appear as shown here. Note that the E in the display indicates an exponent with base 10.

b. With a graphing calculator (set in scientific notation mode) the display should appear as shown here.
Note: Remember, the caret key [^] is used to indicate an exponent and that the numerator and denominator must be set in parentheses.

Now work margin exercise 4.

4. Use a graphing calculator to evaluate each expression. Leave the answer in scientific notation.

a. $\dfrac{(4{,}000{,}000)(0.00056)}{0.032}$

b. $\dfrac{(54{,}000)\left(5\times10^{3}\right)}{4.5\times10^{-5}}$

Note

You can press the [MODE] key and select SCI to have all calculations in scientific notation.

5. If light travels 3×10^8 meters per second, how many meters does light travel in one day?

Example 5 Application: Scientific Notation and Calculators

A light-year is the distance light travels in one year. Use a graphing calculator to find the length of a light-year in scientific notation if light travels 186,000 miles per second.

Solution

60 seconds = 1 minute
60 minutes = 1 hour
24 hours = 1 day
365 days = 1 year

Multiplication gives the following display on your calculator.

Thus, a light-year is 5.865696×10^{12}, or 5,865,696,000,000 miles (5 trillion, 865 billion, 696 million miles).

Now work margin exercise 5.

Margin Exercise Answers

1. a. 6.39×10^7 **b.** 2.45×10^{-6} **2. a.** 1.8×10^{-5} **b.** 1.2×10^8 **3.** 4.816×10^{24} particles
4. a. 7E4 **b.** 6E12 **5.** 2.592E13 meters

12.3 Exercises

Concept Check

Fill-in-the-Blank. Complete each sentence using information found in this section.

1. In scientific notation, decimal numbers are written as a product of a number greater than or equal to ______ and less than ______, and an integer power of 10.

2. In scientific notation, there is/are ______ digit(s) to the left of the decimal point.

3. The exponent of a number written in scientific notation tells how many places the ______ is to be moved and in what direction.

True/False. Determine whether each statement is true or false. If a statement is false, explain how it can be changed so the statement will be true. (**Note:** There may be more than one acceptable change.)

4. The exponent in the number 1.4×10^4 indicates that the decimal point should be moved 4 places to the right.

5. The exponent in the number 2.5×10^{-3} indicates that the decimal point should be moved 3 places to the right.

6. The number 3.53×10^5 is less than 8.72×10^{-4}.

7. The number 4000 written in scientific notation is 0.4×10^4.

12.3 Exercises

Concept Check

1. 1, 10
2. one
3. decimal point
4. True
5. False; The decimal point should be moved 3 places to the left.
6. False; 3.53×10^5 is greater than 8.72×10^{-4}.
7. False; 4000 written in scientific notation is 4.0×10^3.

Practice

1. 8.6×10^4
2. 9.27×10^5
3. 3.62×10^{-2}
4. 6.1×10^{-3}
5. 1.83×10^7
6. 3.76×10^8
7. 2.368×10^{-10}
8. 1.03×10^9
9. 9×10^{-7}
10. 5.71×10^{-8}
11. 3.28×10^{-11}
12. 8.45×10^8
13. 0.042
14. 0.00835
15. 7,560,000
16. 0.0000001002
17. 0.00006132
18. 851,500,000
19. 30,670,000,000
20. 93,740,000
21. 7,205,000,000
22. 400,000,000,000

Practice

Write the following numbers in scientific notation. See Example 1.

1. 86,000
2. 927,000
3. 0.0362
4. 0.0061
5. 18,300,000
6. 376,000,000
7. 0.0000000002368
8. 1,030,000,000
9. 0.0000009
10. 0.0000000571
11. 0.000000000328
12. 845,300,000

Write the following numbers in decimal form.

13. 4.2×10^{-2}
14. 8.35×10^{-3}
15. 7.56×10^{6}
16. 1.002×10^{-7}
17. 6.132×10^{-5}
18. 8.515×10^{8}
19. 3.067×10^{10}
20. 9.374×10^{7}
21. 7.205×10^{9}
22. 4×10^{11}
23. 6.91×10^{-6}
24. 7.408×10^{-9}

First write each of the numbers in scientific notation. Then perform the indicated operations and leave your answer in scientific notation. See Example 2.

25. $300\cdot0.00015$
26. $0.000024\cdot40{,}000$
27. $0.0003\cdot0.0000025$
28. $0.00005\cdot0.00013$
29. $23{,}400{,}000{,}000\cdot5{,}500{,}000{,}000$
30. $7{,}800{,}000{,}000\cdot0.00000081$
31. $\dfrac{3900}{0.003}$
32. $\dfrac{4800}{12{,}000}$
33. $\dfrac{125}{50{,}000}$
34. $\dfrac{0.0046}{230}$
35. $\dfrac{0.0000000000013}{0.000000026}$
36. $\dfrac{0.02\cdot3900}{0.013}$
37. $\dfrac{0.0084\cdot0.003}{0.21\cdot60}$
38. $\dfrac{0.005\cdot650\cdot3.3}{0.0011\cdot2500}$
39. $\dfrac{5.4\cdot0.003\cdot50}{15\cdot0.0027\cdot200}$
40. $\dfrac{0.0000000039\cdot15{,}000{,}000{,}000}{8{,}000{,}000\cdot0.0000000013}$
41. $\dfrac{(1.4\times10^{-2})(922)}{(3.5\times10^{3})(2.0\times10^{6})}$
42. $\dfrac{(4300)(3.0\times10^{2})}{(1.5\times10^{-3})(860\times10^{-2})}$
43. $\dfrac{(25)(3.75\times10^{-5})}{(0.4\times10^{11})(75\times10^{-7})}$

23. 0.00000691

24. 0.000000007408

25. $(3\times10^{2})(1.5\times10^{-4})$; 4.5×10^{-2}

26. $(2.4\times10^{-5})(4\times10^{4})$; 9.6×10^{-1}

27. $(3\times10^{-4})(2.5\times10^{-6})$; 7.5×10^{-10}

28. $(5\times10^{-5})(1.3\times10^{-4})$; 6.5×10^{-9}

29. $(2.34\times10^{10})(5.5\times10^{9})$; 1.287×10^{20}

30. $(7.8\times10^{9})(8.1\times10^{-7})$; 6.318×10^{3}

31. $\dfrac{3.9\times10^{3}}{3\times10^{-3}}$; 1.3×10^{6}

32. $\dfrac{4.8\times10^{3}}{1.2\times10^{4}}$; 4×10^{-1}

33. $\dfrac{1.25\times10^{2}}{5\times10^{4}}$; 2.5×10^{-3}

34. $\dfrac{4.6\times10^{-3}}{2.3\times10^{2}}$; 2×10^{-5}

35. $\dfrac{1.3\times10^{-12}}{2.6\times10^{-8}}$; 5×10^{-5}

36. $\dfrac{(2\times10^{-2})(3.9\times10^{3})}{1.3\times10^{-2}}$; 6×10^{3}

37. $\dfrac{(8.4\times10^{-3})(3\times10^{-3})}{(2.1\times10^{-1})(6\times10)}$; 2×10^{-6}

38. $\dfrac{(5\times10^{-3})(6.5\times10^{2})(3.3\times10^{0})}{(1.1\times10^{-3})(2.5\times10^{3})}$; 3.9×10^{0}

39. $\dfrac{(5.4\times10^{0})(3\times10^{-3})(5\times10)}{(1.5\times10)(2.7\times10^{-3})(2\times10^{2})}$; 1×10^{-1}

40. $\dfrac{(3.9\times10^{-10})(1.5\times10^{10})}{(8\times10^{6})(1.3\times10^{-9})}$; 5.625×10^{2}

41. $\dfrac{(1.4\times10^{-2})(9.22\times10^{2})}{(3.5\times10^{3})(2.0\times10^{6})}$; 1.844×10^{-9}

42. $\dfrac{(4.3\times10^{3})(3.0\times10^{2})}{(1.5\times10^{-3})(8.60\times10^{0})}$; 1×10^{8}

43. $\dfrac{(2.5\times10)(3.75\times10^{-5})}{(4\times10^{10})(7.5\times10^{-6})}$; 3.125×10^{-9}

44. $\dfrac{(1.1\times10^{4})(3.42\times10^{3})}{(1.71\times10^{-10})(5.5\times10^{-14})}$; 4×10^{30}

45. $\dfrac{(1.4\times10^{-7})(7\times10^{13})}{4\times10}$; 2.45×10^{5}

46. $\dfrac{(1.16\times10^{8})(7.2\times10^{12})}{(5.8\times10^{14})(2.4\times10^{-15})}$; 6×10^{20}

47. $\dfrac{(1.95\times10^{-5})(2.65\times10^{3})(7.56\times10^{10})}{(1.5\times10^{7})(1.3\times10^{-13})}$; 2.0034×10^{15}

48. $\dfrac{(8.8\times10)(1.048\times10^{-5})}{(3.2\times10^{7})(1.375\times10^{5})(2\times10)}$; 1.048×10^{-17}

49. 3E8

50. 9E−7

51. 8.5E7

52. 1.32E8

53. 3E22

54. 3.25E−12

55. 1.6051E15

56. 1.673E−15

57. 1E2

58. 4.04E1

59. 1.02E22

60. 4.04E12

44. $\dfrac{(1.1\times10^{4})(342\times10)}{(17.1\times10^{-11})(5.5\times10^{-14})}$

45. $\dfrac{(1.4\times10^{-7})(7\times10^{13})}{40}$

46. $\dfrac{(1.16\times10^{8})(7.2\times10^{12})}{(58\times10^{13})(2.4\times10^{-15})}$

47. $\dfrac{(1.95\times10^{-5})(2650)(7.56\times10^{10})}{(15\times10^{6})(1.3\times10^{-13})}$

48. $\dfrac{(88)(1.048\times10^{-5})}{(3.2\times10^{7})(13.75\times10^{4})(2\times10)}$

Use your calculator (set in scientific notation mode) to evaluate each expression. Leave the answer in scientific notation. See Example 4.

49. $90{,}000 \div 0.0003$

50. $0.0081\div9000$

51. $400\times175{,}000+5000\times3000$

52. $7000\times6000+200\times450{,}000$

53. $9.12\times10^{13}\div3.04\times10^{-9}$

54. $1.989\times10^{-6}\div6.12\times10^{5}$

55. $(4\times10^{6})(1.75\times10^{7})+(5.1\times10^{8})(3.01\times10^{6})$

56. $(2.37\times10^{-7})(4\times10^{-9})+(1.45\times10^{-8})(5\times10^{-8})$

57. $\dfrac{5.6\cdot0.003\cdot5000}{15\cdot0.0028\cdot20}$

58. $\dfrac{0.0006\cdot660\cdot40.4}{0.00011\cdot3600}$

59. $\dfrac{(5.6\times10^{7})(3\times10^{13})(5.1\times10^{-11})}{(1.5\times10^{-10})(2.8\times10^{-8})(2\times10^{6})}$

60. $\dfrac{(6\times10^{11})(6.6\times10^{-6})(4.04\times10^{7})}{(11\times10^{-6})(3.6\times10^{6})}$

Applications

Solve.

61. The mass of a hydrogen atom is approximately 0.00000000000000000000000167 grams. Write this number in scientific notation.

62. The circumference of the earth is approximately 1,580,000,000 inches. Express this circumference in scientific notation.

63. There are approximately 6×10^{13} cells in an adult human body. Express this number in decimal form.

64. The world population is approximately 7.5×10^9 people. Write this number in decimal form.

65. The mass of the earth is about 5,980,000,000,000,000,000,000,000,000 grams. Write this number in scientific notation.

66. One year is approximately 31,500,000 seconds. Express this time in scientific notation.

67. One light-year is approximately 9.46×10^{15} meters. The distance to a certain star is 4.3 light-years. How many meters is this?

68. Light travels approximately 3×10^{10} centimeters per second. How many centimeters would this be per minute? Per hour? Express your answers in scientific notation.

69. The mass of an atom of gold is approximately 3.27×10^{-22} grams. What would be the mass of 2000 atoms of gold? Express your answer in scientific notation.

70. An ounce of gold contains 5×10^{22} atoms. All the gold ever taken out of the earth is estimated to be 3.0×10^{31} atoms. How many ounces of gold is this?

71. There are 8.64×10^4 seconds in a day. How many seconds are in 30 days? Express this time in scientific notation.

72. A scientist measured that his sample weighed 0.0000023 grams. He wrote the value as 2.3×10^6 g on a report. Did he write the value correctly in scientific notation? If not, what should it be?

73. A scientist calculated that her experiment consumed 520,000 joules (J) of energy. She wrote the value as 52×10^4 J on a report of the experiment. Did she write the value correctly in scientific notation? If not, what should it be?

74. A molecule of table salt weighs approximately 9.704×10^{-23} grams. What would be the weight of 4,000,000 molecules of table salt?

a. Write 4,000,000 in scientific notation.

b. Write an expression to find the weight of 4,000,000 molecules of table salt.

c. Simplify the expression from part b.

d. What does the answer from part c. mean? Write a complete sentence.

Applications

61. 1.67×10^{-24} grams

62. 1.58×10^9 inches

63. 60,000,000,000,000 cells

64. 7,500,000,000 people

65. 5.98×10^{27} grams

66. 3.15×10^7 seconds

67. 4.0678×10^{16} m

68. 1.8×10^{12} cm/min; 1.08×10^{14} cm/hr

69. 6.5×10^{-19} grams

70. 6×10^8 ounces

71. 2.592×10^6 seconds

72. No. Should be 2.3×10^{-6}

73. No. Should be 5.2×10^5

74. **a.** 4.0×10^6

b. $(4.0 \times 10^6) \cdot (9.704 \times 10^{-23})$

c. 3.8816×10^{-16}

d. The weight of 4,000,000 molecules of table salt is 3.8816×10^{-16} grams.

Objectives

A. Classify polynomials by degree and number of terms.

B. Evaluate polynomial functions written using function notation.

12.4 Introduction to Polynomials

A Classifying Polynomials

A **term** is an expression that involves only multiplication and/or division with constants and/or variables. Note that a number written next to a variable or a variable written next to another variable indicates multiplication. The number written next to a variable is called the **coefficient** of the variable. For example,

$$3x, \quad -5y^2, \quad 17, \quad \text{and} \quad \frac{x^2}{y}$$

are all algebraic terms.

In the term

$3x$, 3 is the coefficient of x,

$-5y^2$, -5 is the coefficient of y^2, and

$\frac{x}{y}$, 1 is the coefficient of $\frac{x}{y}$.

A term that consists of only a number, such as 17, is also called a **constant** or a **constant term.**

Polynomials

This word *polynomial* was coined in the 17th century and joins two diverse roots: the Greek word *poly*, meaning "many," and the Latin word *nomen*, meaning "name." A polynomial can be classified by the number of terms it has (for example, monomial, binomial, and trinomial) or by its degree (for example, constant, linear, and quadratic).

Monomial

A **monomial in** x is a term of the form

$$kx^n$$

where k is a real number and n is a whole number.

n is called the **degree** of the monomial, and k is called the **coefficient**.

DEFINITION

A **monomial** may have more than one variable, and the degree of such a monomial is the sum of the degrees of its variables. For example, $4x^2y^3$ is a 5th degree monomial in x and y.

However, in this chapter, only monomials of one variable will be discussed. (**Note:** Any variable may be used in place of x.)

Monomials may have fractional or negative coefficients; however, monomials may not have fractional or negative *exponents*. These facts are part of the definition of a monomial kx^n since k (the coefficient) can be any real number, but n (the exponent) must be a whole number.

Expressions that **are not** monomials: $3\sqrt{x}, \ -15x^{\frac{2}{3}}, \ 4a^{-2}$

Expressions that **are** monomials: $17, \ 3x, \ 5y^2, \ \frac{2}{7}a^4, \ \pi x^2, \ \sqrt{2}x^3$

Since $x^0 = 1$, a nonzero constant can be multiplied by x^0 without changing its value. Thus, we say that a nonzero constant is a monomial of degree 0. For example,

$$17 = 17x^0 \text{ and } -6 = -6x^0,$$

which means that the constants 17 and −6 are monomials of degree 0. However, for the special number 0, we can write

$$0 = 0x^2 = 0x^5 = 0x^{13}$$

and we say that the constant 0 is a monomial of no degree.

A **polynomial** is a monomial or the indicated sum and/or difference of monomials. The following are all examples of polynomials.

$$3x, \quad y+5, \quad 4x^2 - 7x + 1, \quad \text{and} \quad a^{10} + 5a^3 - 2a^2 + 6$$

Polynomial

A **polynomial** is a monomial or the indicated sum and/or difference of monomials.

The **degree of a polynomial** is the largest of the degrees of its terms after like terms have been combined.

The coefficient of the term of largest degree is called the **leading coefficient**.

DEFINITION

Teaching Note

You might want to encourage students to notice that the degree of a polynomial has no relationship to the number of terms of the polynomial.

Special Terminology for Polynomials

Term	Definition	Examples
Monomial:	polynomial with one term	$-2x^3$ and $4a^5$
Binomial:	polynomial with two terms	$3x+5$ and a^2+3a
Trinomial:	polynomial with three terms	x^2+6x-7 and a^3-8a^2+12a

Polynomials with four or more terms are simply referred to as **polynomials**.

DEFINITION

Math Tip

Polynomials can also be named according to their degree. For example, a second-degree polynomial can also be called a *quadratic* polynomial. Some of the most common names you will see are as follows.

Degree 0: Zero or constant

Degree 1: Linear

Degree 2: Quadratic

Degree 3: Cubic

Knowing the degree of a polynomial function is helpful since polynomial functions of the same degree share the same general graph shape.

No special name is given to polynomials with more than three terms. They are referred to generally as polynomials. Of course, monomials, binomials, and trinomials can be referred to as polynomials as well.

Examples of polynomials are as follows.

$-1.4x^5$	a fifth-degree monomial in x (leading coefficient is −1.4)
$4z^3 - 7.5z^2 - 5z$	a third-degree trinomial in z (leading coefficient is 4)
$\frac{3}{4}y^4 - 2y^3 + 4y - 6$	a fourth-degree polynomial in y (leading coefficient is $\frac{3}{4}$)

In each of these examples, the terms have been written so that the exponents on the variables decrease in order from left to right. We say that the terms are written in **descending order**. If the exponents on the terms increase in order from left to right, we say that the terms are written in **ascending order**. As a general rule, for consistency and style in operating with polynomials, the polynomials in this chapter will be written in descending order.

1. Simplify each of the following polynomials, then state the degree and type of each polynomial.

a. $4x^2 + 5x^2$

b. $4x^7 + 5x^2 - 5x$

c. $\frac{1}{3}y^2 + 4y^3 - 3y^2 + 8$

d. $3x^3 + 4x^2 - 8 - 3x^3$

e. $-4y^{-3} + 2y^{-1} + 3y^2$

Example 1 Simplifying Polynomials

Simplify each polynomial by combining **like terms**. Write the polynomials in descending order and state the degree and type of each polynomial.

a. $5x^3 + 7x^3$

b. $6x^3 + 4x^3 - 2x$

c. $\frac{1}{2}y + 3y - \frac{2}{3}y^2 - 7$

d. $x^2 + 8x - 15 - x^2$

e. $-3y^4 + 2y^2 + y^{-1}$

Solution

a. $5x^3 + 7x^3 = (5+7)x^3 = 12x^3$ Third-degree monomial

b. $6x^3 + 4x^3 - 2x = 10x^3 - 2x$ Third-degree binomial

c. $\frac{1}{2}y + 3y - \frac{2}{3}y^2 - 7 = -\frac{2}{3}y^2 + \frac{7}{2}y - 7$ Second-degree trinomial

d. $x^2 + 8x - 15 - x^2 = 8x - 15$ First-degree binomial

e. This expression is not a polynomial since y has a negative exponent.

Now work margin exercise 1.

2. Simplify the following polynomial, then state the degree and type of the polynomial.
$x^3 + 6x^2 - 5 + 12x^3 + 9 - 8x^2$

Completion Example 2 Simplifying Polynomials

Simplify the polynomial, then state the degree and type of the polynomial.

$$x^2 + 7x - 3x + 6 - 3x^2$$

Solution

$x^2 + 7x - 3x + 6 - 3x^2 =$ ___-2___ $x^2 +$ ___4___ $x + 6$; ___second___ -degree trinomial

Now work margin exercise 2.

B Evaluating Polynomial Functions

Teaching Note

Some students may be perplexed by the notation "$p(x)$" and continue confusing it as meaning multiplication. Remind the students of the context for functions and be aware of the dual interpretations. Also, mention that the use of "p" is purely arbitrary and other letters may be used.

To evaluate a polynomial for a given value of the variable,

1. substitute that value for the variable wherever it occurs in the polynomial and
2. follow the rules for order of operations.

A convenient notation for evaluating polynomials is the function notation, $p(x)$ (read "p of x"), which was previously discussed in Section 10.5.

For example,

if $p(x) = x^2 - 4x + 13$,

then $p(5) = (5)^2 - 4(5) + 13 = 25 - 20 + 13 = 18$.

Example 3 Evaluating Polynomials

Given the polynomial $p(x) = 4x^2 + 5x - 15$, find $p(3)$.

Solution

Substitute 3 for x throughout the polynomial.

$$\begin{aligned} \text{For } p(x) &= 4x^2 + 5x - 15, \\ p(3) &= 4(3)^2 + 5(3) - 15 \\ &= 4(9) + 15 - 15 \\ &= 36 + 15 - 15 \\ &= 36 \end{aligned}$$

Now work margin exercise 3.

Example 4 Evaluating Polynomials

Given the polynomial $p(y) = 5y^3 + y^2 - 3y + 8$, find $p(-2)$.

Solution

Substitute -2 for y in $p(y) = 5y^3 + y^2 - 3y + 8$, then simplify.

$$\begin{aligned} p(-2) &= 5(-2)^3 + (-2)^2 - 3(-2) + 8 \\ &= 5(-8) + 4 - 3(-2) + 8 \\ &= -40 + 4 + 6 + 8 \\ &= -22 \end{aligned}$$

Note the use of parentheses around −2.

Now work margin exercise 4.

Completion Example 5 Evaluating Polynomials

Given the polynomial $p(x) = 5x^2 + 6x - 10$, find $p(3)$.

Solution

$p(3) = 5(\underline{\ 3\ })^2 + 6(\underline{\ 3\ }) - 10 = \underline{\ 45\ } + \underline{\ 18\ } - 10 = \underline{\ 53\ }$

Now work margin exercise 5.

3. Given the polynomial $p(x) = 3x^3 - 5x^2 - 13$, find $p(3)$.

4. Given the polynomial $p(y) = 4y^3 + 2y^2 - 8y + 2$, find $p(-2)$.

5. Given the polynomial $p(y) = 8y^3 - 9y + 22$, find $p(2)$.

6. Rewrite the polynomial expression $f(x) = 8x - 27$ by substituting for x as indicated by the function notation $f(3a-1)$.

Example 6 Evaluating Polynomials

Rewrite the polynomial expression $f(x) = 6x + 13$ by substituting for x as indicated by the function notation $f(2a+1)$.

Solution

Substitute $2a + 1$ for x throughout the polynomial.

$$f(2a+1) = 6(2a+1) + 13$$
$$= 12a + 6 + 13$$
$$= 12a + 19$$

Now work margin exercise 6.

Completion Example Answers

2. $-2x^2 + 4x + 6$; second-degree polynomial **5.** $5(3)^2 + 6(3) - 10 = 45 + 18 - 10 = 53$

Margin Exercise Answers

1. a. $9x^2$; second-degree monomial **b.** $9x^2 - 5x$; second-degree binomial **c.** $4y^3 - \frac{8}{3}y^2 + 8$; third-degree trinomial **d.** $4x^2 - 8$; second-degree binomial **e.** not a polynomial
2. $13x^3 - 2x^2 + 4$; third-degree trinomial **3.** 23 **4.** −6 **5.** 68 **6.** $24a - 35$

12.4 Exercises

Concept Check

Fill-in-the-Blank. Complete each sentence using information found in this section.

1. If the terms in a polynomial are written so that the exponents on the variable decrease in order from left to right, it is said that the expression is in ____________ order.
2. A monomial or an indicated sum or difference of monomials is known as a/an __________.
3. In a polynomial, the coefficient of the term of the largest degree is called the _________ coefficient.
4. Monomials may not have fractional or _________ exponents.
5. A polynomial with two terms is called a/an ________.
6. A trinomial is a polynomial with ___ terms.

True/False. Determine whether each statement is true or false. If a statement is false, explain how it can be changed so the statement will be true. (**Note:** There may be more than one acceptable change.)

7. A nonzero constant is a monomial with no degree.
8. A polynomial with four terms is a quadrinomial.
9. A monomial is a polynomial with one term.
10. A polynomial must have at least two terms.

12.4 Exercises

Concept Check

1. descending
2. polynomial
3. leading
4. negative
5. binomial
6. 3
7. False; Degree of 0
8. False; Four or more terms is a polynomial.
9. True
10. False; A polynomial can have one term.

Practice

1. Monomial
2. Trinomial
3. Binomial
4. Not a polynomial
5. Trinomial
6. Not a polynomial
7. Not a polynomial
8. Trinomial
9. Binomial
10. Binomial
11. $4y$; First-degree monomial; Leading coefficient 4
12. $5x^2 - x$; Second-degree binomial; Leading coefficient 5
13. $x^3 + 3x^2 - 2x$; Third-degree trinomial; Leading coefficient 1
14. $3x^2$; Second-degree monomial; Leading coefficient 3

Practice

Identify each expression as a monomial, binomial, trinomial, or not a polynomial.

1. $3x^4$
2. $5y^2 - 2y + 1$
3. $8x^3 - 7$
4. $-2x^{-2}$
5. $14a^7 - 2a - 6$
6. $17x^{\frac{2}{3}} + 5x^2$
7. $6a^3 + 5a^2 - a^{-3}$
8. $-3y^4 + 2y^2 - 9$
9. $\frac{1}{2}x^3 - \frac{2}{5}x$
10. $\frac{5}{8}x^5 + \frac{2}{3}x^4$

Simplify the polynomials. Write the polynomials in descending order and state the degree and type of each polynomial. Then, state the leading coefficient. See Examples 1 and 2.

11. $y + 3y$
12. $4x^2 - x + x^2$
13. $x^3 + 3x^2 - 2x$
14. $3x^2 - 8x + 8x$
15. $x^4 - 4x^2 + 2x^2 - x^4$
16. $2 - 6y + 5y - 2$
17. $-x^3 + 6x + x^3 - 6x$
18. $11x^2 - 3x + 2 - 7x^2$
19. $6a^5 + 2a^2 - 7a^3 - 3a^2$
20. $2x^2 - 3x^2 + 2 - 4x^2 - 2 + 5x^2$
21. $4y - 8y^2 + 2y^3 + 8y^2$
22. $2x + 9 - x + 1 - 2x$
23. $5y^2 + 3 - 2y^2 + 1 - 3y^2$
24. $13x^2 - 6x - 9x^2 - 4x$
25. $7x^3 + 3x^2 - 2x + x - 5x^3 + 1$
26. $-3y^5 + 7y - 2y^3 - 5 + 4y^2 + y^2$
27. $x^4 + 3x^4 - 2x + 5x - 10 - x^2 + x$
28. $a^3 + 2a^2 - 6a + 3a^3 + 2a^2 + 7a + 3$
29. $2x + 4x^2 + 6x + 9x^3$
30. $15y - y^3 + 2y^2 - 10y^2 + 2y - 16$

Find the values of the functions as indicated. See Examples 3 through 5.

31. Given $f(x) = 3x - 10$, find
 a. $f(2)$
 b. $f(-2)$
 c. $f(0)$
32. Given $g(x) = -4x + 7$, find
 a. $g(-3)$
 b. $g(6)$
 c. $g(0)$
33. Given $p(x) = x^2 + 14x - 3$, find $p(-1)$.
34. Given $h(x) = -5x^2 - 8x + 7$, find $h(-3)$.
35. Given $f(x) = 3x^3 - 9x^2 - 10x - 11$, find $f(3)$.
36. Given $f(y) = y^3 - 5y^2 + 6y + 2$, find $f(2)$.
37. Given $p(y) = -4y^3 + 5y^2 + 12y - 1$, find $p(-10)$.

15. $-2x^2$; Second-degree monomial; Leading coefficient -2
16. $-y$; First-degree monomial; Leading coefficient -1
17. 0; Monomial of no degree; Leading coefficient 0
18. $4x^2 - 3x + 2$; Second-degree trinomial; Leading coefficient 4
19. $6a^5 - 7a^3 - a^2$; Fifth-degree trinomial; Leading coefficient 6
20. 0; Monomial of no degree; Leading coefficient 0
21. $2y^3 + 4y$; Third-degree binomial; Leading coefficient 2
22. $-x + 10$; First-degree binomial; Leading coefficient -1
23. 4; Monomial of degree 0; Leading coefficient 4
24. $4x^2 - 10x$; Second-degree binomial; Leading coefficient 4
25. $2x^3 + 3x^2 - x + 1$; Third-degree polynomial; Leading coefficient 2
26. $-3y^5 - 2y^3 + 5y^2 + 7y - 5$; Fifth-degree polynomial; Leading coefficient -3
27. $4x^4 - x^2 + 4x - 10$; Fourth-degree polynomial; Leading coefficient 4
28. $4a^3 + 4a^2 + a + 3$; Third-degree polynomial; Leading coefficient 4
29. $9x^3 + 4x^2 + 8x$; Third-degree trinomial; Leading coefficient 9
30. $-y^3 - 8y^2 + 17y - 16$; Third-degree polynomial; Leading coefficient -1

31. a. −4 b. −16 c. −10
32. a. 19 b. −17 c. 7
33. −16
34. −14
35. −41
36. 2
37. 4379
38. −28
39. 13
40. 81
41. −28
42. −1
43. $3a^4+5a^3-8a^2-9a$
44. $6c^5+5c^2-10c+3$
45. $3a+11$
46. $-4a+14$
47. $10a+25$
48. $-12a-12$

Applications

49. a. First degree polynomial
 b. 5 feet per second
 c. 50 feet per second
 d. 95 feet per second
50. a. Second degree polynomial
 b. 44 feet c. 152 feet
 d. 324 feet

38. Given $h(a)=a^3+4a^2+a+2$, find $h(-5)$.

39. Given $p(a)=2a^4+3a^2-8a$, find $p(-1)$.

40. Given $g(x)=8x^4+2x^3-6x^2-7$, find $g(-2)$.

41. Given $g(x)=x^5-x^3+x-2$, find $g(-2)$.

42. Given $p(x)=3x^6-2x^5+x^4-x^3-3x^2+2x-1$, find $p(1)$.

A polynomial function is given. Rewrite the polynomial function by substituting for the variable as indicated in the given function notation.

43. Given $p(x)=3x^4+5x^3-8x^2-9x$, find $p(a)$.

44. Given $p(x)=6x^5+5x^2-10x+3$, find $p(c)$.

45. Given $f(x)=3x+5$, find $f(a+2)$.

46. Given $f(x)=-4x+6$, find $f(a-2)$.

47. Given $g(x)=5x-10$, find $g(2a+7)$.

48. Given $g(x)=-4x-8$, find $g(3a+1)$.

Applications

Solve.

49. A car on an amusement park ride starts with an initial velocity (or initial speed) of 5 feet per second and accelerates at a rate of 15 feet per second squared. The speed of the car during the ride can be modeled by the polynomial $p(x) = 15x + 5$, where x is the time, in seconds, since the car started moving.

a. Identify the degree of the polynomial.

b. Determine how fast the car is moving after 0 seconds.

c. Determine how fast the car is moving after 3 seconds.

d. Determine how fast the car is moving after 6 seconds.

50. A ball is thrown from the top of a building towards the ground with an initial velocity (or initial speed) of 12 feet per second. The force of gravity causes the ball to accelerate at a rate of 32 feet per second squared. The distance of the ball from the top of the building can be modeled by the polynomial $p(x) = 12x + 32x^2$ where x is the time, in seconds, since the ball was thrown.

a. Identify the degree of the polynomial.

b. Determine how far the ball is from the top of the building at 1 second.

c. Determine how far the ball is from the top of the building at 2 seconds.

d. Determine how far the ball is from the top of the building at 3 seconds.

51. Chelsea is competing in a double–elimination softball tournament, which means that a team is eliminated once they lose two games. If the winning team goes undefeated, the total number of games played will be $G(t) = 2t - 1$, where t is the number of teams participating in the tournament.

a. Identify the degree of the polynomial.

b. Determine how many games will be played if 5 teams are participating.

c. Determine how many games will be played if 6 teams are participating.

d. Determine how many games will be played if 15 teams are participating.

52. Max runs a no-kill dog shelter. To help manage finances for the upcoming year, Max uses data from the previous years to construct a mathematical model that can predict the number of dogs the shelter will have during each month of the upcoming year. The model he constructs is $D(x) = x^2 - 12x + 80$, where x is the number of the month of the year (this means that January = 1, February = 2, etc.).

a. Identify the degree of the polynomial.

b. How many animals are predicted to be in the shelter in January?

c. How many animals are predicted to be in the shelter in July?

d. How many animals are predicted to be in the shelter in December?

53. The value of a car starts to depreciate the moment you drive it off of the car dealership's lot. After two years, the approximate value of a car that originally costs \$25,000 is calculated by $V = \$25{,}000x^2$, where x is the average percent that the car retains its value per year. Cars that are well taken care of and driven sparingly will retain more value than cars that are poorly maintained and driven frequently.

a. Suppose that a car that is well maintained and rarely driven will retain an average of 90% of its value each year. What will be the approximate value of the car after 2 years?

b. Suppose that a car that is not well maintained will retain an average of 80% of its value each year. What will be the approximate value of the car after 2 years?

54. Forensic scientists can use a simple formula to approximate the time of death. This formula is based on the average body temperature of humans being 37 °C and the fact that a deceased body will lose an average of 1.5 °C per hour until the body temperature matches the temperature of the surrounding environment. The formula is $f(t) = 37 - 1.5t$, where t is the time in hours since death.

a. What is the approximate body temperature of a person that died 4 hours ago?

b. What is the approximate body temperature of a person that died 14 hours ago?

c. If the temperature of the environment in which the body was found was 20 °C, would it be reasonable for the body temperature to be the temperature from part b.? Explain why or why not.

51. **a.** First degree polynomial **b.** 9 games **c.** 11 games **d.** 29 games

52. **a.** Second degree polynomial **b.** 69 dogs **c.** 45 dogs **d.** 80 dogs

53. **a.** \$20,250 **b.** \$16,000

54. **a.** 31 °C **b.** 16 °C **c.** No, 16 °C is less than 20 °C and the body only cools until it reaches the temperature of the environment. Answers will vary.

55. a. 12 feet b. 28 feet c. 72 feet d. No. Answers will vary.

56. a. \$37.20 b. \$51.80

57. \$45,950 in 2020; \$24,950 in 2026

58. −\$1000 for 100 pairs; \$200 for 500 pairs; \$15,200 for 1000 pairs; A negative profit means the company lost that amount of money

59. 61 after 4 hours; 250,122 after 2 days (48 hours)

60. 20%

Writing & Thinking

61. He was correct. The expression is a sum/difference of monomials.

62. She ordered them by their coefficients rather than their degrees (the exponents are not in descending order); $6a^2 + 10a - 9$ is in descending order.

55. A sled going down a hill has an initial speed of 5 feet per second and a constant acceleration of 1 foot per second squared. The distance of the sled in feet from the top of the hill can be modeled by the polynomial $d(t) = 5t + \frac{1}{2}t^2$, where t is the time in seconds after the sled leaves the top of the hill.

a. Determine the distance the sled is from the top of the hill after 2 seconds.

b. Determine the distance the sled is from the top of the hill after 4 seconds.

c. Determine the distance the sled is from the top of the hill after 8 seconds.

d. Does the distance that the sled travels double when the time doubles? Explain why or why not.

56. Camilla is creating square baby quilts. She determines that the sale price of each quilt should be $p(x) = \$1.80x^2 + 4(\$0.50)x + \$15$, where x is the side length of each square blanket in feet, \$1.80 is the cost per square foot of material, \$0.50 is the cost per foot of border material, and \$15 is the amount of profit Camilla wants to make on each blanket.

a. How much will a blanket that has a side length of 3 feet cost?

b. How much will a blanket that has a side length of 4 feet cost?

57. The value of an automobile depreciates linearly over time. A new 2018 Mercedes E300 sells for \$52,950 and its value, V, after t years is given by the equation $V = 52{,}950 - 3500t$. Find the value of the 2018 Mercedes in 2020 (e.g. after 2 years), and then in 2026.

58. PDQ Tennis Shoe Co. follows a profit model of $P = 3x^2 - 15x + 2$, where P is the profit in hundreds of dollars after selling x hundred pairs of tennis shoes. Find each of the profits for PDQ after selling 100 pairs, 500 pairs, and 1000 pairs of tennis shoes. What does a negative value for profit represent?

59. The number of protozoa in a biology laboratory experiment is given by the polynomial function $p(t) = 0.04t^4 + 0.3t^3 + 2t^2$, where p is the number of protozoa after t hours. Determine the number of protozoa after 4 hours. What is the number of protozoa after 2 days? (Round both values to the nearest whole number.)

60. The percent p of material retained by a student x days after hearing a lecture is given by the polynomial function $p(x) = 100 - 5x^2$. What percent of the material is still remembered by a student 4 days after hearing the lecture?

Writing & Thinking

61. Tony was classifying expressions for a homework assignment. He said that $7y^2 + 12y - 3$ was a polynomial. Was he correct or not? Justify your answer.

62. Jeanne thought that $10a - 9 + 6a^2$ was in descending order. Explain Jeanne's error and what the correct descending order should be.

First-degree polynomials are also called linear polynomials, second-degree polynomials are called quadratic polynomials, and third-degree polynomials are called cubic polynomials. The related functions are called linear functions, quadratic functions, and cubic functions, respectively.

63. Use a graphing calculator to graph the following linear functions. (See Section 10.5 to review graphing functions on a graphing calculator.)

a. $p(x)=2x+3$

b. $p(x)=-3x+1$

c. $p(x)=\frac{1}{2}x$

64. Use a graphing calculator to graph the following quadratic functions.

a. $p(x)=x^2$

b. $p(x)=x^2+6x+9$

c. $p(x)=-x^2+2$

65. Use a graphing calculator to graph the following cubic functions.

a. $p(x)=x^3$

b. $p(x)=x^3-4x$

c. $p(x)=x^3+2x^2-5$

66. Make up a few of your own linear, quadratic, and cubic functions and graph these functions with your calculator. Using the results from Exercises 63, 64, and 65, and your own functions, describe in your own words:

a. the general shape of the graphs of linear functions.

b. the general shape of the graphs of quadratic functions.

c. the general shape of the graphs of cubic functions.

63. a.

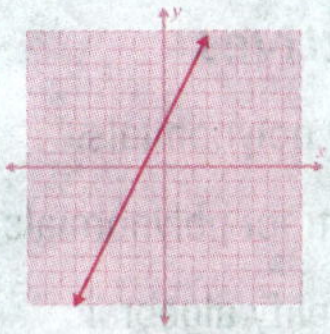

b.

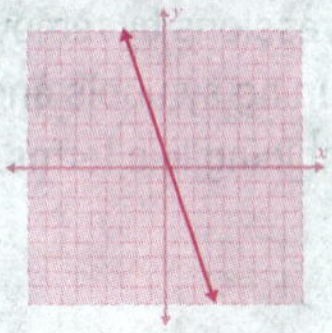

c.

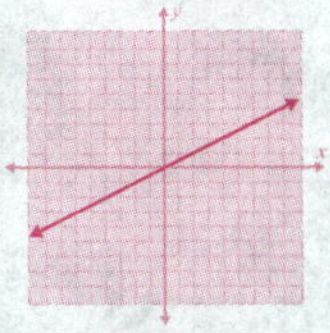

64. a.

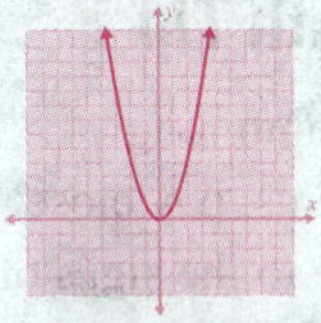

b.

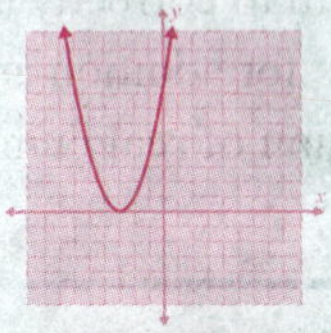

c.

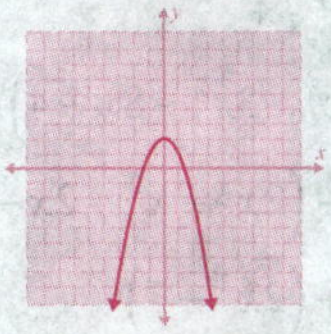

65. a.

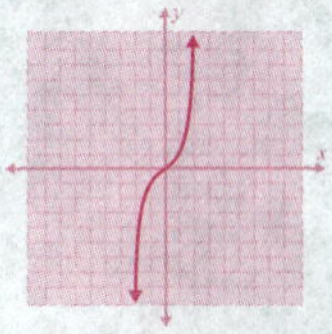

b.

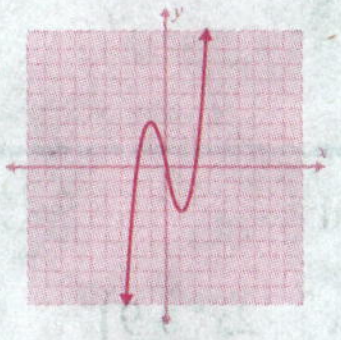

c.

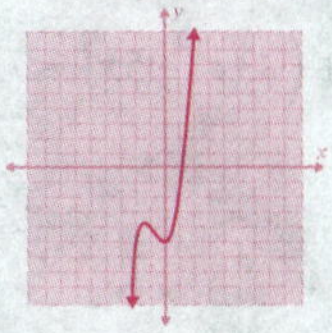

66. Answers will vary.

Objectives

A. Add polynomials.

B. Subtract polynomials.

C. Simplify algebraic expressions by removing grouping symbols and combining like terms.

12.5 Addition and Subtraction with Polynomials

A Addition with Polynomials

The **sum** of two or more polynomials is found by combining **like terms.** When adding polynomials, the polynomials may be written horizontally or vertically. For example, the polynomials written horizontally can be simplified by first grouping like terms.

$$\begin{aligned}&\left(x^2-5x+3\right)+\left(2x^2-8x-4\right)+\left(3x^3+x^2-5\right)\\&=3x^3+\left(x^2+2x^2+x^2\right)+\left(-5x-8x\right)+\left(3-4-5\right)\\&=3x^3+4x^2-13x-6\end{aligned}$$

Math Tip

When combining coefficients, pay close attention to the signs on each term to correctly apply the rules for addition and subtraction of real numbers.

If the polynomials are written vertically, we align like terms, one beneath the other, in a column format and combine like terms in each column.

$$\begin{array}{r}x^2-5x+3\\2x^2-8x-4\\3x^3+x^2\qquad-5\\\hline 3x^3+4x^2-13x-6\end{array}$$

1. Add:

$\left(6x^3+3x^2-8x+4\right)$
$+\left(2x^3-x^2+2x-8\right)$
$+\left(2x^2-7\right)$

Example 1 Adding Polynomials

Add: $\left(5x^3-8x^2+12x+13\right)+\left(-2x^2-8\right)+\left(4x^3-5x+14\right)$

Solution

$$\begin{aligned}&\left(5x^3-8x^2+12x+13\right)+\left(-2x^2-8\right)+\left(4x^3-5x+14\right)\\&=\left(5x^3+4x^3\right)+\left(-8x^2-2x^2\right)+\left(12x-5x\right)+\left(13-8+14\right)\\&=9x^3-10x^2+7x+19\end{aligned}$$

Now work margin exercise 1.

2. Add:

$\left(7x^3+2x^2+9\right)$
$+\left(-3x^3-5x^2+7x-7\right)$

Example 2 Adding Polynomials

Add: $\left(x^3-x^2+5x\right)+\left(4x^3+5x^2-8x+9\right)$

Solution

$$\begin{array}{l}x^3-\;x^2+5x\\4x^3+5x^2-8x+9\\\hline 5x^3+4x^2-3x+9\end{array}$$

Now work margin exercise 2.

Completion Example 3 Adding Polynomials

Add: $(4x^3+15x-7)+(x^3-4x^2+6x+2)$

Solution

$$\begin{aligned}&(4x^3+15x-7)+(x^3-4x^2+6x+2)\\&=(4x^3+\underline{\quad x^3\quad})+(-4x^2)+(15x+\underline{\quad 6x\quad})+(\underline{\quad -7\quad}+\underline{\quad 2\quad})\\&=\underline{\quad 5\quad}x^3-4x^2+\underline{\quad 21\quad}x-\underline{\quad 5\quad}\end{aligned}$$

Now work margin exercise 3.

3. Add:

$(6x^2-8x+3)+(x^2+4x+9)$

B Subtraction with Polynomials

A negative sign written in front of a polynomial in parentheses indicates the **opposite of the entire polynomial.** The opposite can be found by changing the sign of every term in the polynomial.

$$-(2x^2+3x-7)=-2x^2-3x+7$$

We can also think of the opposite of a polynomial as −1 times the polynomial, applying the distributive property as follows.

$$\begin{aligned}-(2x^2+3x-7)&=-1(2x^2+3x-7)\\&=-1(2x^2)-1(3x)-1(-7)\\&=-2x^2-3x+7\end{aligned}$$

The result is the same with either approach. So the **difference** between two polynomials can be found by changing the sign of each term of the second polynomial and then combining like terms.

Teaching Note

Some students may need to be reminded that "combining" like terms is just a use of the distributive property.

$$\begin{aligned}(5x^2-3x-7)-(2x^2+5x-8)&=5x^2-3x-7-2x^2-5x+8\\&=(5x^2-2x^2)+(-3x-5x)+(-7+8)\\&=3x^2-8x+1\end{aligned}$$

If the polynomials are written vertically, one beneath the other, we change the signs of the terms of the polynomial being subtracted and then combine like terms.

$$\begin{array}{r}5x^2-3x-7\\ \underline{-(2x^2+5x-8)}\end{array}\quad\longrightarrow\quad\begin{array}{r}5x^2-3x-7\\ \underline{-2x^2-5x+8}\\ 3x^2-8x+1\end{array}$$

4. Subtract:

$\left(7x^4 - 18x^3 + 8x^2 - 11\right)$
$-\left(-8x^4 - 15x^3 + 3x^2 - 12\right)$

Example 4 Subtracting Polynomials

Subtract: $\left(9x^4 - 22x^3 + 3x^2 + 10\right) - \left(5x^4 - 2x^3 - 5x^2 + x\right)$

Solution

$$\left(9x^4 - 22x^3 + 3x^2 + 10\right) - \left(5x^4 - 2x^3 - 5x^2 + x\right)$$

Apply the distributive property.

$$= 9x^4 - 22x^3 + 3x^2 + 10 - 5x^4 + 2x^3 + 5x^2 - x$$
$$= \left(9x^4 - 5x^4\right) + \left(-22x^3 + 2x^3\right) + \left(3x^2 + 5x^2\right) - x + 10$$
$$= 4x^4 - 20x^3 + 8x^2 - x + 10$$

Now work margin exercise 4.

5. Subtract:

$9x^3 - 6x^2 + 1$
$-\left(3x^3 + 3x^2 + 8x\right)$

Example 5 Subtracting Polynomials

Subtract: $8x^3 + 5x^2 - 14$
$-\left(-2x^3 + x^2 + 6x\right)$

Solution

$$\begin{array}{r} 8x^3 + 5x^2 - 14 \\ -\left(-2x^3 + x^2 + 6x\right) \\ \hline \end{array} \qquad \begin{array}{r} 8x^3 + 5x^2 + 0x - 14 \\ 2x^3 - x^2 - 6x + 0 \\ \hline 10x^3 + 4x^2 - 6x - 14 \end{array}$$

Write in 0s for missing terms to help with alignment of like terms.

Now work margin exercise 5.

6. Subtract:

$\left(6x^4 - 5x + 11\right)$
$-\left(2x^4 + 3x^3 - 4x + 8\right)$

Completion Example 6 Subtracting Polynomials

Subtract: $\left(6x^3 - 4x^2 + 8x + 14\right) - \left(2x^3 - x^2 + 3x - 9\right)$

Solution

$$\left(6x^3 - 4x^2 + 8x + 14\right) - \left(2x^3 - x^2 + 3x - 9\right)$$
$$= 6x^3 - 4x^2 + 8x + 14 - 2x^3 \underline{\ +\ } x^2 \underline{\ -\ } 3x \underline{\ +\ } 9$$
$$= \left(6x^3 \underline{\ -2x^3\ }\right) + \left(-4x^2 \underline{\ +x^2\ }\right) + \left(8x \underline{\ -3x\ }\right) + \left(14 \underline{\ +9\ }\right)$$
$$= \underline{\ 4\ }x^3 - \underline{\ 3\ }x^2 + \underline{\ 5\ }x + \underline{\ 23\ }$$

Now work margin exercise 6.

C Simplifying Algebraic Expressions

If an algebraic expression contains more than one pair of grouping symbols, such as parentheses ( ), brackets [ ], braces { }, radicals signs $\sqrt{\ }$, absolute value bars | |, or fraction bars, simplify by working to remove the innermost pair of symbols first. Apply the rules for order of operations just as if the variables were numbers and proceed to combine like terms.

Example 7 Simplifying Algebraic Expressions

Simplify the expression: $5x-\left[2x+3(4-x)+1\right]-9$

Solution

$$\begin{aligned}&5x-\left[2x+3(4-x)+1\right]-9\\&=5x-\left[2x+12-3x+1\right]-9\\&=5x-\left[-x+13\right]-9\\&=5x+x-13-9\\&=6x-22\end{aligned}$$

Work with the parentheses first since they are included inside the brackets.

Now work margin exercise 7.

7. Simplify the expression:

$3x-\left[1+4(5-2x)+3x\right]$

Example 8 Simplifying Algebraic Expressions

Simplify the expression: $-3(x-4)+2\left[x+3(x-3)\right]$

Solution

$$\begin{aligned}&-3(x-4)+2\left[x+3(x-3)\right]\\&=-3(x-4)+2\left[x+3x-9\right]\\&=-3(x-4)+2\left[4x-9\right]\\&=-3x+12+8x-18\\&=5x-6\end{aligned}$$

Work with the parentheses first since they are included inside the brackets.

Now work margin exercise 8.

8. Simplify the expression:

$-5(x-3)+3\left[x+2(x-3)\right]$

Completion Example Answers

3. $\left(4x^3+x^3\right)+\left(-4x^2\right)+\left(15x+6x\right)+\left(-7+2\right)=5x^3-4x^2+21x-5$

6. $6x^3-4x^2+8x+14-2x^3+x^2-3x+9$

$=\left(6x^3-2x^3\right)+\left(-4x^2+x^2\right)+\left(8x-3x\right)+\left(14+9\right)$

$=4x^3-3x^2+5x+23$

Margin Exercise Answers

1. $8x^3+4x^2-6x-11$ **2.** $4x^3-3x^2+7x+2$ **3.** $7x^2-4x+12$ **4.** $15x^4-3x^3+5x^2+1$
5. $6x^3-9x^2-8x+1$ **6.** $4x^4-3x^3-x+3$ **7.** $8x-21$ **8.** $4x-3$

12.5 Exercises

Concept Check

1. sign
2. innermost
3. like
4. opposite
5. variables
6. variable, exponent
7. False; All terms are subtracted.
8. True
9. False; They are like terms because they have the same variable raised to the same exponent.
10. False; Absolute value bars and radical signs are grouping symbols.

Practice

1. $3x^2+7x+2$
2. $4x^2+x-4$
3. $2x^2+11x-7$
4. $2x^2+2x+2$
5. $3x^2$
6. $5x^2+6x-10$
7. $x^2-5x+17$
8. $4x^2+7x-8$
9. $-x^2+2x-7$
10. $7x^2+14x+4$
11. $2x^3+4x^2-3x-7$
12. $2x^3-3x^2-3x-6$

12.5 Exercises

Concept Check

Fill-in-the-Blank. Complete each sentence using information found in this section.

1. To find the difference between two polynomials, first change the ______ of each term in the second polynomial.
2. If there are multiple grouping symbols within an algebraic expression, begin working on the ________ pair of symbols first.
3. To add two or more polynomials, combine ____ terms.
4. A negative sign written in front of a polynomial in parentheses indicates the ______ of the entire polynomial.
5. To simplify algebraic expressions, apply the rules for order of operations just as if the ______ were numbers and proceed to combine like terms.
6. Like terms have the same ______ raised to the same ______.

True/False. Determine whether each statement is true or false. If a statement is false, explain how it can be changed so the statement will be true. (**Note:** There may be more than one acceptable change.)

7. When subtracting one polynomial from another polynomial, only the first term of the polynomial is subtracted.
8. To simplify polynomials that are being added and subtracted, combine like terms.
9. The terms $6a^2$ and $7a^2$ are not like terms because they don't have the same coefficient.
10. Absolute value bars and radical signs are not considered grouping symbols.

Practice

Find the indicated sums. See Examples 1 through 3.

1. $(2x^2+5x-1)+(x^2+2x+3)$
2. $(x^2+3x-8)+(3x^2-2x+4)$
3. $(x^2+7x-7)+(x^2+4x)$
4. $(x^2+2x-3)+(x^2+5)$
5. $(2x^2-x-1)+(x^2+x+1)$
6. $(3x^2+5x-4)+(2x^2+x-6)$
7. $(-2x^2-3x+9)+(3x^2-2x+8)$
8. $(x^2+6x-7)+(3x^2+x-1)$
9. $(-4x^2+2x-1)+(3x^2-x+2)+(x-8)$
10. $(8x^2+5x+2)+(-3x^2+9x-4)+(2x^2+6)$
11. $(x^2+2x-1)+(3x^2-x+2)+(2x^3-4x-8)$
12. $(x^3+2x-9)+(x^2-5x+2)+(x^3-4x^2+1)$

13. $$\begin{array}{r} x^2+4x-4 \\ \underline{-2x^2+3x+1} \end{array}$$

14. $$\begin{array}{r} 2x^2+4x-3 \\ \underline{3x^2-9x+2} \end{array}$$

15. $$\begin{array}{r} x^3+3x^2\quad +x \\ \underline{-2x^3-\ x^2\ +2x-4} \end{array}$$

16. $$\begin{array}{r} 4x^3+5x^2\qquad +11 \\ \underline{2x^3-2x^2-3x-\ 6} \end{array}$$

17. $$\begin{array}{r} 7x^3+5x^2+\ x-\ 6 \\ -3x^2+4x+11 \\ \underline{-3x^3-2x^2-5x+\ 2} \end{array}$$

18. $$\begin{array}{r} x^3+5x^2+7x-3 \\ 4x^2+3x-9 \\ \underline{4x^3+2x^2\qquad -2} \end{array}$$

19. $$\begin{array}{r} x^3+\ 3x^2\qquad -4 \\ 7x^2+2x+1 \\ \underline{x^3+\ \ x^2-6x\quad} \end{array}$$

20. $$\begin{array}{r} x^3+2x^2\qquad -\ 5 \\ -2x^3\qquad +x-\ 9 \\ \underline{x^3-2x^2\qquad +14} \end{array}$$

Find the indicated differences. See Examples 4 through 6.

21. $(2x^2+4x+8)-(x^2+3x+2)$

22. $(3x^2+7x-6)-(x^2+2x+5)$

23. $(x^2-9x+2)-(4x^2-3x+4)$

24. $(6x^2+11x+2)-(4x^2-2x-7)$

25. $(2x^2-x-10)-(-x^2+3x-2)$

26. $(7x^2+4x-9)-(-2x^2+x-9)$

27. $(x^4+8x^3-2x^2-5)-(2x^4+10x^3-2x^2+11)$

28. $(x^3+4x^2-3x-7)-(3x^3+x^2+2x+1)$

29. $(-3x^4+2x^3-7x^2+6x+12)-(x^4+9x^3+4x^2+x-1)$

30. $(2x^5+3x^3-2x^2+x-5)-(3x^5-2x^3+5x^2+6x-1)$

31. $(9x^2-5)-(13x^2-6x-6)$

32. $(8x^2+9)-(4x^2-3x-2)$

33. $(3x^4-2x^3-8x-1)-(5x^3-3x^2-3x-10)$

34. $(x^5+6x^3-3x^2-5)-(2x^5+8x^3+5x+17)$

35. $$\begin{array}{r} 14x^2-6x+9 \\ \underline{-(\ 8x^2+\ \ x-9\ )} \end{array}$$

36. $$\begin{array}{r} 9x^2-3x+2 \\ \underline{-(4x^2-5x-1)} \end{array}$$

37. $$\begin{array}{r} 5x^4+8x^2+11 \\ \underline{-(-3x^4+2x^2-\ 4)} \end{array}$$

38. $$\begin{array}{r} 11x^2+5x-13 \\ \underline{-(-3x^2+5x+\ 2)} \end{array}$$

39. $$\begin{array}{r} x^3+6x^2\qquad -3 \\ \underline{-(-x^3+2x^2-3x+7)} \end{array}$$

40. $$\begin{array}{r} 3x^3\qquad +9x-17 \\ \underline{-(x^3+5x^2-2x-\ 6)} \end{array}$$

13. $-x^2+7x-3$
14. $5x^2-5x-1$
15. $-x^3+2x^2+3x-4$
16. $6x^3+3x^2-3x+5$
17. $4x^3+7$
18. $5x^3+11x^2+10x-14$
19. $2x^3+11x^2-4x-3$
20. x
21. x^2+x+6
22. $2x^2+5x-11$
23. $-3x^2-6x-2$
24. $2x^2+13x+9$
25. $3x^2-4x-8$
26. $9x^2+3x$
27. $-x^4-2x^3-16$
28. $-2x^3+3x^2-5x-8$
29. $-4x^4-7x^3-11x^2+5x+13$
30. $-x^5+5x^3-7x^2-5x-4$
31. $-4x^2+6x+1$
32. $4x^2+3x+11$
33. $3x^4-7x^3+3x^2-5x+9$
34. $-x^5-2x^3-3x^2-5x-22$
35. $6x^2-7x+18$
36. $5x^2+2x+3$
37. $8x^4+6x^2+15$
38. $14x^2-15$
39. $2x^3+4x^2+3x-10$
40. $2x^3-5x^2+11x-11$

41. $4x-13$
42. $-15x+22$
43. $-7x+9$
44. $-3x-6$
45. $2x+17$
46. $-9x+26$
47. $3x^3+13x^2+9$
48. $10x^3-3x^2+7$
49. $8x^2-x-2$
50. $-7x^2+6x-2$
51. $3x-19$
52. $5x-5$
53. $x^2-2x+13$
54. $-x^2+6x-20$
55. $7x-3$
56. $x+5$
57. $4x^2+3x+5$
58. $7x^3-2x^2+9x$
59. $7x-18$
60. $7x-6$
61. $x^2+11x-8$
62. $5x^2-16x-11$
63. $7x^3-x^2-7$
64. x^3+x^2+6x-2

Applications

65. **a.** $3.50x^3+4x^2-30$ dollars

b. $6.50x^3+4x^2-20$ dollars

c. $460

Simplify each of the following expressions. See Examples 7 and 8.

41. $5x+2(x-3)-(3x+7)$

42. $-4(x-6)-(8x+2)-3x$

43. $11+[3x-2(1+5x)]$

44. $2+[9x-4(3x+2)]$

45. $8x-[2x+4(x-3)-5]$

46. $17-[-3x+6(2x-3)+9]$

47. $3x^3-[5-7(x^2+2)-6x^2]$

48. $10x^3-[8-5(3-2x^2)-7x^2]$

49. $(2x^2+4)-[-8+2(7-3x^2)+x]$

50. $-[6x^2-3(4+2x)+9]-(x^2+5)$

51. $2[3x+(x-8)-(2x+5)]-(x-7)$

52. $3[x+(10-3x)-(8-3x)]+(2x-1)$

53. $(x^2-1)+2[4+(3-x)]$

54. $(4\ \ x^2)+3[(2x-3)-5]$

55. $-(x-5)+[6x-2(4-x)]$

56. $2(2x+1)-[5x-(2x+3)]$

Complete the following word problems.

57. Find the sum of $4x^2-3x$ and $6x+5$.

58. Subtract $2x^2-4x$ from $7x^3+5x$.

59. Subtract $3(x+1)$ from $5(2x-3)$.

60. Find the sum of $10x-2(3x+5)$ and $3(x-4)+16$.

61. Subtract $3x^2-4x+2$ from the sum of $4x^2+x-1$ and $6x-5$.

62. Subtract $-2x^2+6x+12$ from the sum of $2x^2+3x-1$ and $x^2-13x+2$.

63. Add $5x^3-8x+1$ to the difference between $2x^3+14x-3$ and x^2+6x+5.

64. Add $2x^3+4x^2+1$ to the difference between $-x^2+10x-3$ and x^3+2x^2+4x.

Applications

Solve.

65. A manufacturer estimates that it costs $2x^3+4x^2-35$ dollars to create the amount of items it would take to fill a box which has a side length of x feet. The warehouse manager determines it will cost $1.50x^3+5$ dollars to store each box for one month.

a. Add the two polynomials to determine the manufacturing and storage costs for each box of items for one month.

b. The warehouse manager knows that each box will be stored for an average of 3 months. Determine the cost to produce a box of items and store it for 3 months.

c. If the box has a side length of 4 feet, use the expression from part b. to determine how much will it cost to create and store a box of items for 3 months.

66. Carson has two loans, a loan for his car and a home equity loan that he used for home improvements. The car loan is for $15,000 and Carson plans to make monthly payments of $500 per month. The home improvement loan is for $9000 and Carson plans to make monthly payments of $300.

  a. Write an algebraic expression to describe the value of the car loan after x months.

  b. Write an algebraic expression to describe the value of the home equity loan after x months.

  c. Add together the two algebraic expressions to determine the remaining loan amount to be paid after x months for both loans combined.

  d. How much will Carson still owe on both loans after 10 months?

67. A company estimates that the revenue from selling x items is $58x$ dollars and the cost of producing x items is $31x + 40$ dollars.

  a. The company's profit is defined as revenue minus cost. Find the expression that models the profit of producing and selling x items.

  b. What is the profit from producing and selling 12 items?

68. Keri's gym membership costs $30 a month plus $6 for every class she takes. To cut back on her spending, she is thinking of switching to a gym that costs $20 a month plus $4 for every class she takes.

  a. Write an algebraic expression that describes how much Keri is currently paying per month if she takes x classes.

  b. Write an algebraic expression that describes how much Keri would pay per month at the new gym if she took x classes.

  c. Write an algebraic expression that describes Keri's savings per month if she switches gyms.

  d. How much would Keri save the first month if she switches gyms and takes 10 classes?

Writing & Thinking

69. Explain, in your own words, how to subtract one polynomial from another.

70. Give two examples that show how the sum of two binomials might not be a binomial.

66. a. $15,000 – 500x$
b. $9000 – 300x$
c. $24,000 – 800x$
d. $16,000

67. a. $27x - 40$ dollars
b. $284

68. a. $30 + 6x$ dollars
b. $20 + 4x$ dollars
c. $10 + 2x$ dollars
d. $30

Writing & Thinking

69. Add the opposite of each term in the polynomial being subtracted to the original polynomial and combine like terms.

70. $(2x+4)+(x^2+2x)$
$= x^2+4x+4;$
$(x^9+x^3)+(x^7+x^5)$
$= x^9+x^7+x^5+x^3.$
Answers will vary.

Objectives

A. Multiply a polynomial by a monomial.

B. Multiply two polynomials.

C. Use the FOIL method to multiply two binomials.

12.6 Multiplication with Polynomials

Up to this point, we have multiplied terms such as $5x^2 \cdot 3x^4 = 15x^6$ by using the product rule for exponents. Also, we have applied the distributive property to expressions such as $5(2x+3) = 10x+15$.

Now, we will use both the product rule for exponents and the distributive property to multiply polynomials. We discuss three cases here:

1. The product of a monomial with a polynomial of two or more terms
2. The product of two binomials
3. The product of a binomial with a polynomial of more than two terms

A Multiplying a Polynomial by a Monomial

Teaching Note

While we continue to emphasize the use of the distributive property in so many instances of simplifying, it is also important to point out our continued use of the commutative and associative properties toward this end. For example, in this demonstration we reach a point of having "$5x \cdot 3$" and we simplify this as "$15x$" (by the commutative and associative properties of multiplication).

Using the distributive property $a(b+c) = ab+ac$ with multiplication indicated on the left, we can find the product of a monomial with a polynomial of two or more terms as follows.

$$5x(2x+3) = 5x \cdot 2x + 5x \cdot 3 = 10x^2 + 15x$$

$$3x^2(4x-1) = 3x^2 \cdot 4x + 3x^2(-1) = 12x^3 - 3x^2$$

$$-4a^5(a^2-8a+5) = -4a^5 \cdot a^2 - 4a^5(-8a) - 4a^5 \cdot 5 = -4a^7 + 32a^6 - 20a^5$$

B Multiplying Two Polynomials

Now suppose that we want to multiply two binomials, say, $(x+3)(x+7)$. We will apply the distributive property in the following way with multiplication indicated on the right of the parentheses.

Teaching Note

You might want to show your students a rectangular format such as the following for finding the product $(x+5)(x+3)$ Some students will find this arrangement helpful in keeping track of all the terms.

	$x + 5$	
x	x^2	$5x$
+		
3	$3x$	15

$= x^2 + 8x + 15$

Compare $(x+3)(x+7)$ to $(a+b)c = ac+bc$.

Think of $(x+7)$ as taking the place of c. Thus, it takes the following form.

$$(a+b)c \quad = \quad ac \quad + \quad bc$$

$$(x+3)(x+7) \quad = \quad x(x+7) \quad + \quad 3(x+7)$$

Complete the products on the right by using the distributive property twice.

$$\begin{aligned} (x+3)(x+7) &= x(x+7)+3(x+7) \\ &= x \cdot x + x \cdot 7 + 3 \cdot x + 3 \cdot 7 \\ &= x^2 + 7x + 3x + 21 \\ &= x^2 + 10x + 21 \end{aligned}$$

In the same manner,

$$\begin{aligned}(x+2)(3x+4) &= x(3x+4)+2(3x+4)\\ &= x\cdot 3x + x\cdot 4 + 2\cdot 3x + 2\cdot 4\\ &= 3x^2+4x+6x+8\\ &= 3x^2+10x+8\end{aligned}$$

Similarly,

$$\begin{aligned}(2x-1)(x^2+x-5) &= 2x(x^2+x-5)-1(x^2+x-5)\\ &= 2x\cdot x^2+2x\cdot x+2x\cdot(-5)-1\cdot x^2-1\cdot x-1(-5)\\ &= 2x^3+2x^2-10x-x^2-x+5\\ &= 2x^3+x^2-11x+5\end{aligned}$$

The product of two polynomials can also be found by writing one polynomial under the other. **The distributive property is applied by multiplying each term of one polynomial by each term of the other.** Consider the product $(2x^2+3x-4)(3x+7)$. Now writing one polynomial under the other and applying the distributive property, we obtain the following.

Multiply by $+7$:

$$\begin{array}{rrr} 2x^2 + & 3x & -4\\ & 3x & +7\\ \hline 14x^2 + & 21x & -28\end{array}$$

Multiply by $3x$:

$$\begin{array}{rrrr} & 2x^2 + & 3x & - 4\\ & & 3x & + 7\\ \hline & 14x^2 & +21x & -28\\ 6x^3 & +9x^2 & -12x & \\ \hline \end{array}$$

Align the like terms so that they can be easily combined.

Finally, combine like terms:

$$\begin{array}{rrrr} & 2x^2 & +3x & -4\\ & & 3x & +7\\ \hline & 14x^2 & +21x & -28\\ 6x^3 + & 9x^2 & -12x & \\ \hline 6x^3 + & 23x^2 & +9x & -28\end{array}$$

Combine like terms.

Example 1 Multiplying Polynomials and Monomials

1. Multiply: $-6x(3x^2-x-3)$

Multiply: $-4x(x^2-3x+12)$

Solution

$$\begin{aligned}-4x(x^2-3x+12) &= -4x\cdot x^2-4x(-3x)-4x\cdot 12\\ &= -4x^3+12x^2-48x\end{aligned}$$

Now work margin exercise 1.

2. Multiply:
$5x^3\left(6x^4-3x^2+12\right)$

Completion Example 2 Multiplying Polynomials and Monomials

Multiply: $3x^2\left(x^2+12x-5\right)$

Solution

$$3x^2\left(x^2+12x-5\right)=\underline{\ 3x^2\ }\cdot x^2+\underline{\ 3x^2\ }\cdot 12x+\underline{\ 3x^2\ }\cdot(-5)$$
$$=\underline{\ 3\ }x^4+\underline{\ 36\ }x^3-\underline{\ 15\ }x^2$$

Now work margin exercise 2.

3. Multiply: $(3x+3)(4x-3)$

Example 3 Multiplying Two Binomials

Multiply: $(2x-4)(5x+3)$

Solution

$$(2x-4)(5x+3)=2x(5x+3)-4(5x+3)$$
$$=2x\cdot 5x+2x\cdot 3-4\cdot 5x-4\cdot 3$$
$$=10x^2+6x-20x-12$$
$$=10x^2-14x-12$$

Now work margin exercise 3.

4. Multiply: $(x-4)(7x+5)$

Completion Example 4 Multiplying Two Binomials

Multiply: $(3x+8)(5x+4)$

Solution

$$(3x+8)(5x+4)=\underline{\ 3x\ }\cdot(5x+4)+\underline{\ 8\ }\cdot(5x+4)$$
$$=\underline{\ 3x\ }\cdot 5x+\underline{\ 3x\ }\cdot 4+\underline{\ 8\ }\cdot 5x+\underline{\ 8\ }\cdot 4$$
$$=\underline{\ 15\ }x^2+\underline{\ 12\ }x+\underline{\ 40\ }x+\underline{\ 32\ }$$
$$=\underline{\ 15\ }x^2+\underline{\ 52\ }x+\underline{\ 32\ }$$

Now work margin exercise 4.

5. Multiply:
$$\begin{array}{r}4x^2+8x-2\\ \underline{3x+1}\end{array}$$

Example 5 Multiplying Polynomials

Multiply:
$$\begin{array}{r}7y^2-3y+2\\ \underline{2y+3}\end{array}$$

Solution

$$\begin{array}{rrrrl}
 & 7y^2 & -3y & +2 & \\
 & & 2y & +3 & \\
\hline
 & 21y^2 & -9y & +6 & \text{Multiply by 3.}\\
14y^3 & -6y^2 & +4y & & \text{Multiply by } 2y.\\
\hline
14y^3 & +15y^2 & -5y & +6 & \text{Combine like terms.}
\end{array}$$

Now work margin exercise 5.

Example 6 Multiplying Polynomials

Multiply: $\begin{array}{r} x^2+3x-1 \\ \underline{x^2-3x+1} \end{array}$

Solution

$$\begin{array}{rl} x^2+3x-1 & \\ \underline{x^2-3x+1} & \\ x^2+3x-1 & \text{Multiply by 1.} \\ -3x^3-9x^2+3x & \text{Multiply by } -3x. \\ \underline{x^4+3x^3-x^2} & \text{Multiply by } x^2. \\ x^4-9x^2+6x-1 & \text{Combine like terms.} \end{array}$$

Now work margin exercise 6.

6. Multiply: $\begin{array}{r} x^2+4x-2 \\ \underline{x^2-4x+2} \end{array}$

Example 7 Multiplying Polynomials

Multiply: $2x(x-3)(2x+1)$

Solution

First multiply $2x(x-3)$, then multiply this result by $(2x+1)$.

$$\begin{aligned} 2x(x-3) &= 2x\cdot x+2x\cdot(-3) \\ &= 2x^2-6x \\ (2x^2-6x)(2x+1) &= 2x^2(2x+1)-6x(2x+1) \\ &= 2x^2\cdot 2x+2x^2\cdot 1-6x\cdot 2x-6x\cdot 1 \\ &= 4x^3+2x^2-12x^2-6x \\ &= 4x^3-10x^2-6x \end{aligned}$$

Now work margin exercise 7.

7. Multiply: $3x(x+2)(4x+1)$

Example 8 Multiplying Polynomials

Multiply: $(x-5)(x+2)(x-1)$

Solution

First multiply $(x-5)(x+2)$, then multiply this result by $(x-1)$.

$$\begin{aligned} (x-5)(x+2) &= x(x+2)-5(x+2) \\ &= x^2+2x-5x-10 \\ &= x^2-3x-10 \\ (x^2-3x-10)(x-1) &= x^2(x-1)-3x(x-1)-10(x-1) \\ &= x^3-x^2-3x^2+3x-10x+10 \\ &= x^3-4x^2-7x+10 \end{aligned}$$

Now work margin exercise 8.

8. Multiply: $(x-2)(x+3)(x-6)$

C The FOIL Method

In the case of the **product of two binomials** such as $(2x+5)(3x-7)$, the **FOIL** method is useful. **F-O-I-L** is a mnemonic device (memory aid) to help in remembering which terms of the binomials to multiply together. First, by using the distributive property we can see how the terms are multiplied.

$$
\begin{aligned}
(2x+5)(3x-7) &= 2x(3x-7)+5(3x-7) \\
&= 2x \cdot 3x + 2x \cdot (-7) + 5 \cdot 3x + 5 \cdot (-7)
\end{aligned}
$$

First terms	Outside terms	Inside terms	Last terms
F	O	I	L

Now, we can use the FOIL method and then combine like terms to go directly to the answer.

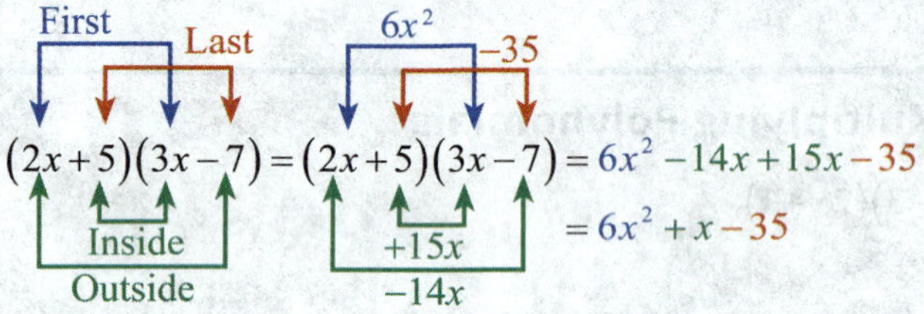

9. Use the FOIL method to multiply the binomials:

$(x+7)(3x+4)$

Example 9 Using the FOIL Method to Multiply Binomials

Use the FOIL method to multiply the binomials: $(x+3)(2x+8)$

Solution

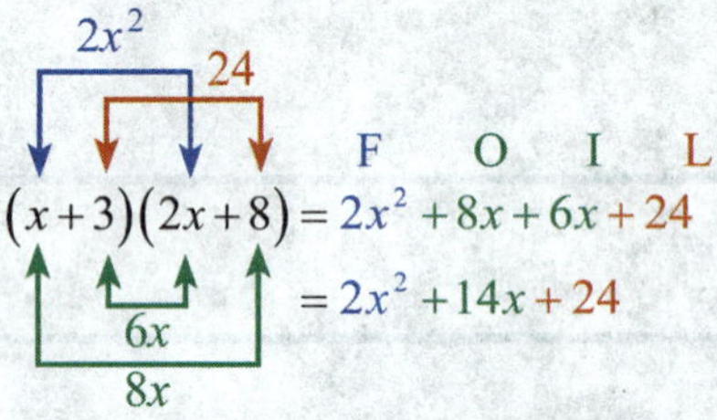

Now work margin exercise 9.

10. Use the FOIL method to multiply the binomials:

$(4x-5)(2x-2)$

Example 10 Using the FOIL Method to Multiply Binomials

Use the FOIL method to multiply the binomials: $(2x-3)(3x-5)$

Solution

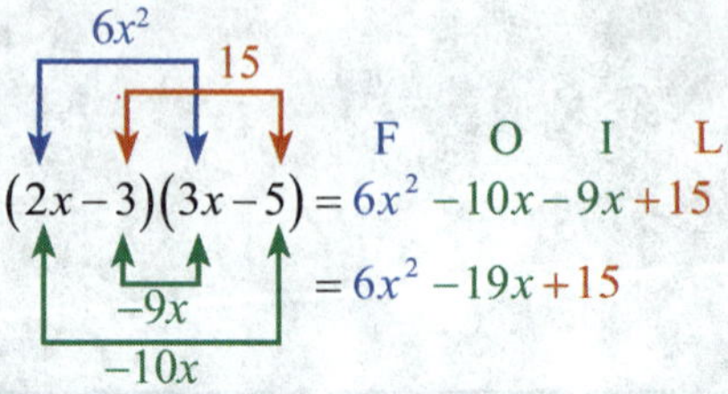

Now work margin exercise 10.

Completion Example 11 Using the FOIL Method

Use the FOIL method to multiply the binomials: $(x+11)(3x-2)$

Solution

$$(x+11)(3x-2) = x\cdot\underline{\ 3x\ } + x\cdot(\underline{\ -2\ }) + 11\cdot\underline{\ 3x\ } + 11\cdot(\underline{\ -2\ })$$
$$= \underline{\ 3\ }x^2 + \underline{\ 31\ }x - \underline{\ 22\ }$$

Now work margin exercise 11.

11. Use the FOIL method to multiply the binomials:

$(2x-8)(5x-9)$

Completion Example Answers

2. $3x^2\cdot x^2 + 3x^2\cdot 12x + 3x^2(-5) = 3x^4 + 36x^3 - 15x^2$

4. $3x\cdot(5x+4) + 8\cdot(5x+4) = 3x\cdot 5x + 3x\cdot 4 + 8\cdot 5x + 8\cdot 4$
$= 15x^2 + 12x + 40x + 32 = 15x^2 + 52x + 32$

11. $x\cdot 3x + x\cdot(-2) + 11\cdot 3x + 11\cdot(-2) = 3x^2 + 31x - 22$

Margin Exercise Answers

1. $-18x^3 + 6x^2 + 18x$ **2.** $30x^7 - 15x^5 + 60x^3$ **3.** $12x^2 + 3x - 9$ **4.** $7x^2 - 23x - 20$

5. $12x^3 + 28x^2 + 2x - 2$ **6.** $x^4 - 16x^2 + 16x - 4$ **7.** $12x^3 + 27x^2 + 6x$ **8.** $x^3 - 5x^2 - 12x + 36$

9. $3x^2 + 25x + 28$ **10.** $8x^2 - 18x + 10$ **11.** $10x^2 - 58x + 72$

12.6 Exercises

Concept Check

Fill-in-the-Blank. Complete each sentence using information found in this section.

1. The ____________ property is used to find the product of a monomial with a polynomial.
2. When multiplying two polynomials, the distributive property is applied by multiplying each ___ of one polynomial by each ___ of the other.
3. In the case of the product of two binomials, a mnemonic device called the ____ method is useful.
4. The FOIL method stands for First, _______, Inside, ______.
5. When multiplying polynomials, the last step is to combine ____ terms, if possible.

True/False. Determine whether each statement is true or false. If a statement is false, explain how it can be changed so the statement will be true. (**Note:** There may be more than one acceptable change.)

6. The distributive property can only be used to multiply a monomial and a polynomial.
7. The product of $(a+b)$ and $(c+d)$ is $ac+bd$.
8. The FOIL method is a way to remember one specific order that the distributive property can be applied.

12.6 Exercises

Concept Check

1. distributive
2. term, term
3. FOIL
4. Outside, Last
5. like
6. False; The distributive property can be used when multiplying any types of polynomials.
7. False; The product of $(a+b)$ and $(c+d)$ is $ac+ad+bc+bd$.
8. True

Practice

Multiply and simplify, if necessary.

1. $-3x^2\left(2x^3+5x\right)$
2. $5x^2\left(-4x^2+6\right)$
3. $4x^5\left(x^2-3x+1\right)$
4. $9x^3\left(2x^3-x^2+5x\right)$
5. $-1\left(y^5-8y+2\right)$
6. $-7\left(2y^4+3y^2+1\right)$
7. $-4x^3\left(x^5-2x^4+3x\right)$
8. $-2x^4\left(x^3-x^2+2x\right)$
9. $5x^3\left(5x^2-x+2\right)$
10. $-2x^2\left(x^3+5x-4\right)$
11. $a^2\left(a^5+2a^4-5a+1\right)$
12. $7t^3\left(-t^3+5t^2+2t+1\right)$
13. $(x+4)(x-3)$
14. $(x+7)(x-5)$
15. $(a+6)(a-8)$
16. $(x+2)(x-4)$
17. $(x-2)(x-1)$
18. $(x-7)(x-8)$
19. $3(t+4)(t-5)$
20. $-4(x+6)(x-7)$
21. $x(x+3)(x+8)$
22. $t(t-4)(t-7)$
23. $(2x+1)(x-4)$
24. $(3x-1)(x+4)$
25. $(6x-1)(x+3)$
26. $(8x+15)(x+1)$
27. $(2x+3)(2x-3)$
28. $(3t+5)(3t-5)$
29. $(4x+1)(4x+1)$
30. $(5x-2)(5x-2)$
31. $(y+3)\left(y^2-y+4\right)$
32. $(2x+1)\left(x^2-7x+2\right)$
33. $$\begin{array}{r} 3x+7 \\ \underline{x-5} \end{array}$$
34. $$\begin{array}{r} 2x+6 \\ \underline{x+3} \end{array}$$
35. $$\begin{array}{r} x^2+3x+1 \\ \underline{5x-9} \end{array}$$
36. $$\begin{array}{r} 8x^2+3x-2 \\ \underline{-2x+7} \end{array}$$
37. $$\begin{array}{r} 2x^2+3x+5 \\ \underline{x^2+2x-3} \end{array}$$
38. $$\begin{array}{r} 6x^2-x+8 \\ \underline{2x^2+5x+6} \end{array}$$
39. $(3x-4)(x+2)$
40. $(t+6)(4t-7)$
41. $(2x+5)(x-1)$
42. $(5a-3)(a+4)$
43. $(7x+1)(x-2)$
44. $(x-2)(3x+8)$
45. $(2x+1)(3x-8)$
46. $(3x+7)(2x-5)$

Practice

1. $-6x^5-15x^3$
2. $-20x^4+30x^2$
3. $4x^7-12x^6+4x^5$
4. $18x^6-9x^5+45x^4$
5. $-y^5+8y-2$
6. $-14y^4-21y^2-7$
7. $-4x^8+8x^7-12x^4$
8. $-2x^7+2x^6-4x^5$
9. $25x^5-5x^4+10x^3$
10. $-2x^5-10x^3+8x^2$
11. $a^7+2a^6-5a^3+a^2$
12. $-7t^6+35t^5+14t^4+7t^3$
13. x^2+x-12
14. $x^2+2x-35$
15. $a^2-2a-48$
16. x^2-2x-8
17. x^2-3x+2
18. $x^2-15x+56$
19. $3t^2-3t-60$
20. $-4x^2+4x+168$
21. x^3+11x^2+24x
22. t^3-11t^2+28t
23. $2x^2-7x-4$
24. $3x^2+11x-4$
25. $6x^2+17x-3$
26. $8x^2+23x+15$
27. $4x^2-9$
28. $9t^2-25$
29. $16x^2+8x+1$
30. $25x^2-20x+4$
31. y^3+2y^2+y+12
32. $2x^3-13x^2-3x+2$
33. $3x^2-8x-35$
34. $2x^2+12x+18$
35. $5x^3+6x^2-22x-9$
36. $-16x^3+50x^2+25x-14$
37. $2x^4+7x^3+5x^2+x-15$
38. $12x^4+28x^3+47x^2$ $+34x+48$

47. $(2x+3)(2x+3)$

48. $(5y+2)(5y+2)$

49. $(x+3)(x^2-4)$

50. $(y^2+2)(y-4)$

51. $(2x+7)(2x-7)$

52. $(3x-4)(3x+4)$

53. $(x+1)(x^2-x+1)$

54. $(x-2)(x^2+2x+4)$

55. $(7a-2)(7a-2)$

56. $(5a-6)(5a-6)$

57. $x(x+3)(x+5)$

58. $x(x-8)(x+2)$

59. $2x(x-1)(3x+2)$

60. $3x(x+1)(2x+3)$

61. $(2x+3)(x^2-x-1)$

62. $(3x+1)(x^2-x+9)$

63. $(x+1)(x+2)(x+3)$

64. $(t-1)(t-2)(t-3)$

65. $(a^2+a-1)(a^2-a+1)$

66. $(y^2+y+2)(y^2+y-2)$

67. $(t^2+3t+2)^2$

68. $(a^2-4a+1)^2$

Simplify.

69. $3x(2x+1)-2(2x+1)$

70. $x(3x+4)+7(3x+4)$

71. $3a(3a-5)+5(3a-5)$

72. $6x(x-1)+5(x-1)$

73. $5x(-2x+7)-2(-2x+7)$

74. $y(y^2+1)-1(y^2+1)$

75. $x(x^2+3x+2)+2(x^2+3x+2)$

76. $4x(x^2-x+1)+3(x^2-x+1)$

77. $(y+6)(y-6)+(y+5)(y-5)$

78. $(y-2)(y+2)+(y-1)(y+1)$

79. $(2a+1)(a-5)+(a-4)(a-4)$

80. $(x+4)(2x+1)+(x-3)(x-2)$

81. $(x-3)(x+5)-(x+3)(x+2)$

82. $(t+3)(t+3)-(t-2)(t-2)$

83. $(2a+3)(a+1)-(a-2)(a-2)$

84. $(4t-3)(t+4)-(t-2)(3t+1)$

Applications

Solve.

85. A graphic artist is designing a poster to advertise an upcoming event. The only restrictions regarding the poster size is that it must have a length of $3x$ inches and a width of $2x + 5$ inches. Find a simplified expression for the area of the poster.

86. Armon works for a company that ships artwork worldwide. The size of each item varies, but all of the art is on square canvases. Armon's job is to make the wooden shipping crates for each piece of art. In order to protect the artwork, each crate must be 10 inches deep. The crate must also be 10 inches wider and 12 inches taller than the artwork. Letting x represent the length of one side of the artwork, find the volume of the rectangular shipping crate.

39. $3x^2+2x-8$
40. $4t^2+17t-42$
41. $2x^2+3x-5$
42. $5a^2+17a-12$
43. $7x^2-13x-2$
44. $3x^2+2x-16$
45. $6x^2-13x-8$
46. $6x^2-x-35$
47. $4x^2+12x+9$
48. $25y^2+20y+4$
49. $x^3+3x^2-4x-12$
50. y^3-4y^2+2y-8
51. $4x^2-49$
52. $9x^2-16$
53. x^3+1
54. x^3-8
55. $49a^2-28a+4$
56. $25a^2-60a+36$
57. x^3+8x^2+15x
58. x^3-6x^2-16x
59. $6x^3-2x^2-4x$
60. $6x^3+15x^2+9x$
61. $2x^3+x^2-5x-3$
62. $3x^3-2x^2+26x+9$
63. $x^3+6x^2+11x+6$
64. $t^3-6t^2+11t-6$
65. a^4-a^2+2a-1
66. $y^4+2y^3+y^2-4$
67. $t^4+6t^3+13t^2+12t+4$
68. $a^4-8a^3+18a^2-8a+1$
69. $6x^2-x-2$
70. $3x^2+25x+28$
71. $9a^2-25$
72. $6x^2-x-5$
73. $-10x^2+39x-14$
74. y^3-y^2+y-1
75. x^3+5x^2+8x+4
76. $4x^3-x^2+x+3$
77. $2y^2-61$

78. $2y^2 - 5$

79. $3a^2 - 17a + 11$

80. $3x^2 + 4x + 10$

81. $-3x - 21$

82. $10t + 5$

83. $a^2 + 9a - 1$

84. $t^2 + 18t - 10$

Applications

85. $6x^2 + 15x$ square inches

86. $V = 10x^2 + 220x + 1200$ cubic inches

87. $R(x) = -8x^2 + 70x + 1500$

88. **a.** $12 + 4x$ square feet

b. $12 + 3x$ square feet

c. $x^2 + 7x + 12$ square feet

89. **a.** $A(x) = 280 - 4x^2$

b. $14 - 2x$ and $20 - 2x$

c. $B(x) = 4x^2 - 68x + 280$

d. $V(x) = 4x^3 - 68x^2 + 280x$

90. **a.** $x + 8$ **b.** $2x + 10$

c. $2x^2 + 26x + 80$

87. Theodore and Sarah have a small business selling custom-made T-shirts. They currently sell 150 shirts per month and charge $10 per T-shirt. Sarah thinks they can increase their revenue by increasing their selling price, but Theodore knows from experience that each $1 increase in price will decrease the number of T-shirts they sell by 8 shirts per month. Find an expression for the total monthly revenue Theodore and Sarah's business can generate if they change the selling price of their T-shirts. Let x represent number of $1 increases in the price of the T-shirts.

88. The smallest plot of land that you can rent at a community garden is 3 feet long by 4 feet wide.

a. Suppose you want to rent a plot of land that is x feet longer than the smallest available plot. What would the area of this plot of land be?

b. Suppose you want to rent a plot of land that is x feet wider than the smallest plot with a length of 3 feet. What would the area of this plot of land be?

c. Suppose you want to rent a plot of land that is x feet longer and x feet wider than the smallest plot of land. What would the area of this plot of land be?

89. Lee is making a box. He starts with a piece of cardboard that is 14 inches by 20 inches. He cuts a square with side length x from each corner of the box.

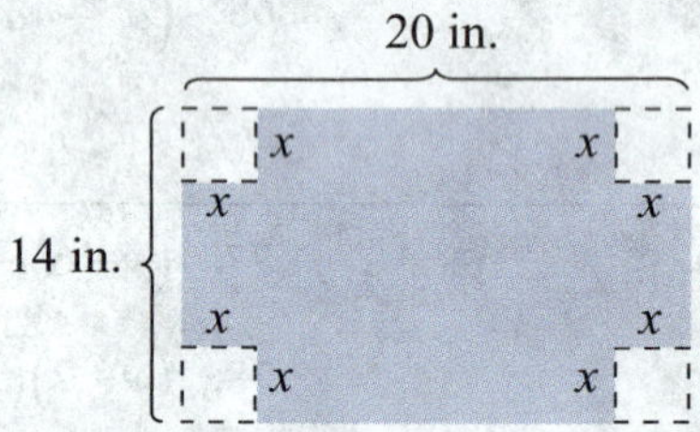

a. Write a polynomial function $A(x)$ to represent the area of the cardboard that remains after the corners are cut out.

b. When the sides of the box are folded up, what will be the side lengths of the base of the box?

c. Write a polynomial function $B(x)$ to represent the area of the base of the box when the sides are folded up.

d. The height of the box will be x inches. Write a polynomial function $V(x)$ to determine the volume of the box.

90. The glass portion of a sliding glass door has a ratio of height to width of 2 : 1. The framework around the window adds 8 inches to the width of the door and 10 inches to the height.

a. Write a polynomial expression to represent the width of the door, including the framework. Use the variable x to represent the width of the door.

b. Write a polynomial expression to represent the height of the door, including the framework.

c. Write a polynomial expression for the total area of the window, including the framework.

Writing & Thinking

91. We have seen how the distributive property is used to multiply polynomials.

 a. Show how the distributive property can be used to find the following product.

$$\begin{array}{r} 75 \\ \times\, 93 \\ \hline \end{array}$$

 (**Hint:** 75 = 70 + 5 and 93 = 90 + 3)

 b. In the multiplication algorithm for multiplying whole numbers (as in the product above), we are told to "move to the left" when multiplying. For example:

$$\begin{array}{r} 75 \\ \times\, 93 \\ \hline 15 \\ 21 \\ 45 \\ 63 \\ \hline \end{array}$$

 Why are the 21 and 45 moved one place to the left in the alignment?

 When 9 and 7 are multiplied, we move the 63 two places left. Why?

Writing & Thinking

91. a. Answers will vary.

b. Answers will vary.

Objectives

A. Multiply binomials, finding products that are the difference of two squares.

B. Square binomials, finding products that are perfect square trinomials.

12.7 Special Products of Binomials

A The Difference of Two Squares

There are a few special cases that deserve attention when multiplying polynomials. For example, consider the following product.

$$\begin{aligned}(x+7)(x-7) &= x^2 - 7x + 7x - 49 \\ &= x^2 - 49\end{aligned}$$

In this example, the middle terms, $-7x$ and $+7x$, are opposites of each other and their sum is 0. Therefore, the resulting product has only two terms and each term is a square.

$$(x+7)(x-7) = x^2 - 49$$

In fact, when two binomials are in the form of the sum and difference of the same two terms, the product will always be the difference of the squares of the terms. The product is called the **difference of two squares**.

Math Tip

The methods of multiplying polynomials learned in previous sections will give the same answer as using the difference of two squares definition. Consider the FOIL method to multiply $(x + a)(x - a)$. Notice that combining like terms causes the middle terms to cancel each other out since they sum to zero.

$$\begin{aligned}(x+a)(x-a) &= x^2 - ax + ax - a^2 \\ &= x^2 - a^2\end{aligned}$$

After the middle terms cancel, we are left with the difference of two squares. Knowing this definition allows us to save time when we recognize this special product of binomials.

Difference of Two Squares

$$(x+a)(x-a) = x^2 - a^2$$

DEFINITION

In order to quickly recognize the difference of two squares, you should know the following squares of the positive integers from 1 to 20. The squares of integers are called **perfect squares**.

Perfect Squares from 1 to 400

Number (n)	1	2	3	4	5	6	7	8	9	10
Square (n^2)	1	4	9	16	25	36	49	64	81	100

Number (n)	11	12	13	14	15	16	17	18	19	20
Square (n^2)	121	144	169	196	225	256	289	324	361	400

Table 1

Example 1 Difference of Two Squares

Find the following products.

a. $(x+4)(x-4)$ **b.** $(3y+7)(3y-7)$ **c.** $(x^3-6)(x^3+6)$

Solution

a. The two binomials represent the sum and difference of x and 4. So, the product is the difference of their squares.

$$\begin{aligned}(x+4)(x-4) &= x^2 - 4^2 \\ &= x^2 - 16 \qquad \text{Difference of two squares}\end{aligned}$$

1. Find the following products.

a. $(x+6)(x-6)$

b. $(4y+3)(4y-3)$

c. $(x^4-3)(x^4+3)$

b. $(3y+7)(3y-7)=(3y)^2-7^2$ — **Note:** $(3y)^2 = 3y \cdot 3y = 9y^2$

$=9y^2-49$ — Difference of two squares

c. $(x^3-6)(x^3+6)=(x^3)^2-6^2$ — **Note:** $(x^3)^2 = x^3 \cdot x^3 = x^6$

$=x^6-36$ — Difference of two squares

Now work margin exercise 1.

2. Find the product:

$(2x^2+6)(2x^2-6)$

Completion Example 2 Difference of Two Squares

Find the product: $(4x+9)(4x-9)$

Solution

$$(4x+9)(4x-9)=(\underline{\ 4x\ })^2-(\underline{\ 9\ })^2$$

$$=\underline{\ 16x^2\ }-\underline{\ 81\ }$$

Now work margin exercise 2.

B Squares of Binomials

Now, we consider the case where the two binomials being multiplied are the same. That is, we want to consider the **square of a binomial**. The following examples, using the distributive property, illustrate two patterns that, after some practice, allow us to go directly to the products.

$$(x+3)^2=(x+3)(x+3)=x^2+3x+3x+9$$

$$=x^2+2\cdot 3x+9$$ The middle term is doubled. $3x+3x=2\cdot 3x$

$$=x^2+6x+9$$ Perfect square trinomial

$$(x-11)^2=(x-11)(x-11)=x^2-11x-11x+121$$

$$=x^2-2\cdot 11x+121$$ The middle term is doubled. $-11x-11x=2(-11x)=-2\cdot 11x$

$$=x^2-22x+121$$ Perfect square trinomial

Note that in each case the result of squaring the binomial is a trinomial. These trinomials are called **perfect square trinomials**.

Squares of Binomials (Perfect Square Trinomials)

$(x+a)^2=x^2+2ax+a^2$ — **Square of a Binomial Sum**

$(x-a)^2=x^2-2ax+a^2$ — **Square of a Binomial Difference**

DEFINITION

3. Find the following products.

a. $(3x+5)^2$

b. $(4x-2)^2$

c. $(8-2x)^2$

d. $(2y^3-2)^2$

Example 3 Squares of Binomials

Find the following products.

a. $(2x+3)^2$ **b.** $(5x-1)^2$ **c.** $(9-x)^2$ **d.** $(y^3+1)^2$

Solution

a. The pattern for squaring a binomial gives the following

$$(2x+3)^2 = (2x)^2 + 2(3)(2x) + (3)^2$$ **Note:** $(2x)^2 = 2x \cdot 2x = 4x^2$

$$= 4x^2 + 12x + 9$$ Perfect square trinomial

b. $$(5x-1)^2 = (5x)^2 - 2(1)(5x) + (1)^2$$

$$= 25x^2 - 10x + 1$$ Perfect square trinomial

c. $$(9-x)^2 = (9)^2 - 2(x)(9) + x^2$$

$$= 81 - 18x + x^2$$

$$= x^2 - 18x + 81$$ Perfect square trinomial

d. $$(y^3+1)^2 = (y^3)^2 + 2(1)(y^3) + 1^2$$ **Note:** $(y^3)^2 = y^3 \cdot y^3 = y^6$

$$= y^6 + 2y^3 + 1$$ Perfect square trinomial

Now work margin exercise 3.

4. Find the product: $(5x-2)^2$

Completion Example 4 Squares of Binomials

Find the product: $(3x+10)^2$

Solution

$$(3x+10)^2 = (\underline{\ 3x\ })^2 + 2 \cdot \underline{\ 10\ } \cdot \underline{\ 3x\ } + (\underline{\ 10\ })^2$$

$$= \underline{\ 9\ }x^2 + \underline{\ 60\ }x + \underline{\ 100\ }$$

Now work margin exercise 4.

Common Error

Wrong Solution

Many algebra students make the following error.

$$(x+a)^2 = x^2 + a^2$$

$$(x+6)^2 = x^2 + 36$$

Correct Solution

Avoid this error by remembering that **the square of a binomial is a trinomial.**

$$(x+6)^2 = x^2 + 2 \cdot 6x + 36$$

$$= x^2 + 12x + 36$$

CAUTION

Completion Example Answers

2. $(4x)^2-(9)^2=16x^2-81$ **4.** $(3x)^2+2\cdot10\cdot3x+(10)^2=9x^2+60x+100$

Margin Exercise Answers

1. a. x^2-36 **b.** $16y^2-9$ **c.** x^8-9 **2.** $4x^4-36$ **3. a.** $9x^2+30x+25$ **b.** $16x^2-16x+4$ **c.** $4x^2-32x+64$ **d.** $4y^6-8y^3+4$ **4.** $25x^2-20x+4$

12.7 Exercises

Concept Check

Fill-in-the-Blank. Complete each sentence using information found in this section.

1. When two binomials are in the form of the sum and difference of the same term, the product is called the __________ of two squares.
2. When the two binomials being multiplied together are the same, that product is called the ______ of a binomial.
3. The result of squaring a binomial is a/an ________.
4. Trinomials that are the result of squaring binomials are called ________ square trinomials.

True/False. Determine whether each statement is true or false. If a statement is false, explain how it can be changed so the statement will be true. (**Note:** There may be more than one acceptable change.)

5. When two binomials are in the form of the sum and difference of the same term, the product will be a trinomial.
6. When the two binomials being multiplied together are the same, the product will be a trinomial.
7. Perfect square trinomials result from squaring a binomial sum or a binomial difference.
8. When finding the product of two binomials that are in the form of the sum and difference of the same two terms, the FOIL method and the difference of two squares formula will produce different results.

Practice

Find each product and identify any that are either the difference of two squares or a perfect square trinomial. See Examples 1 through 4.

1. $(x-7)^2$
2. $(x-5)^2$
3. $(x+4)(x+4)$
4. $(x+8)(x+8)$
5. $(x+3)(x-3)$
6. $(x-6)(x+6)$
7. $(x+9)(x-9)$
8. $(x+12)(x-12)$
9. $(2x+3)(x-1)$
10. $(3x+1)(2x+5)$
11. $(3x-4)^2$
12. $(3x+1)^2$
13. $(5x+2)(5x-2)$
14. $(2x+1)(2x-1)$
15. $(3x-2)(3x-2)$

12.7 Exercises

Concept Check

1. difference
2. square
3. trinomial
4. perfect
5. False; The product will be a binomial.
6. True
7. True
8. False; The two methods will have the same result.

Practice

1. $x^2-14x+49$; Perfect square trinomial
2. $x^2-10x+25$; Perfect square trinomial
3. $x^2+8x+16$; Perfect square trinomial
4. $x^2+16x+64$; Perfect square trinomial
5. x^2-9; Difference of two squares
6. x^2-36; Difference of two squares
7. x^2-81; Difference of two squares
8. x^2-144; Difference of two squares
9. $2x^2+x-3$
10. $6x^2+17x+5$
11. $9x^2-24x+16$; Perfect square trinomial
12. $9x^2+6x+1$; Perfect square trinomial
13. $25x^2-4$; Difference of two squares
14. $4x^2-1$; Difference of two squares
15. $9x^2-12x+4$; Perfect square trinomial

16. x^2+6x+9; Perfect square trinomial
17. $x^2-16x+64$; Perfect square trinomial
18. $x^2-10x+25$; Perfect square trinomial
19. $16x^2-25$; Difference of two squares
20. $-x^2+121$; Difference of two squares
21. $25x^2-81$; Difference of two squares
22. $81x^2-4$; Difference of two squares
23. $x^2-8x+16$; Perfect square trinomial
24. $9x^2+42x+49$; Perfect square trinomial
25. $4x^2-49$; Difference of two squares
26. $36x^2-25$; Difference of two squares
27. $10x^4-11x^2-6$
28. $8x^4+18x^2+7$
29. $49x^2+14x+1$; Perfect square trinomial
30. $25x^2-20x+4$; Perfect square trinomial
31. $x^2+10x+25$
32. x^2-6x+9
33. x^4-1
34. x^4-25
35. x^4+6x^2+9
36. x^6+16x^3+64
37. x^6-4x^3+4
38. x^4-8x^2+16
39. $9x^2-4$
40. $9x^2-1$
41. $16x^2-9$
42. $64x^2-25$
43. $25x^2+30x+9$
44. $9x^2+24x+16$

16. $(3+x)^2$

17. $(8-x)(8-x)$

18. $(5-x)(5-x)$

19. $(4x+5)(4x-5)$

20. $(11-x)(11+x)$

21. $(5x-9)(5x+9)$

22. $(9x+2)(9x-2)$

23. $(4-x)^2$

24. $(3x+7)^2$

25. $(2x+7)(2x-7)$

26. $(6x+5)(6x-5)$

27. $(5x^2+2)(2x^2-3)$

28. $(4x^2+7)(2x^2+1)$

29. $(1+7x)^2$

30. $(2-5x)^2$

Find each product

31. $(5+x)(5+x)$

32. $(3-x)(3-x)$

33. $(x^2+1)(x^2-1)$

34. $(x^2+5)(x^2-5)$

35. $(x^2+3)(x^2+3)$

36. $(x^3+8)(x^3+8)$

37. $(x^3-2)^2$

38. $(x^2-4)^2$

39. $(3x+2)(3x-2)$

40. $(3x-1)(3x+1)$

41. $(4x+3)(4x-3)$

42. $(8x+5)(8x-5)$

43. $(5x+3)(5x+3)$

44. $(3x+4)(3x+4)$

45. $(6x-5)^2$

46. $(7x-2)^2$

Use a calculator as an aid in multiplying the binomials.

47. $(x+1.4)(x-1.4)$

48. $(x-2.1)(x+2.1)$

49. $(x-2.5)^2$

50. $(x+1.7)^2$

51. $(x+2.15)(x-2.15)$

52. $(x+1.36)(x-1.36)$

53. $(x+1.24)^2$

54. $(x-1.45)^2$

55. $(1.42x+9.6)^2$

56. $(0.46x-0.71)^2$

57. $(11.4x+3.5)(11.4x-3.5)$

58. $(2.5x+11.4)(1.3x-16.9)$

59. $(12.6x-6.8)(7.4x+15.3)$

60. $(3.4x+6)(3.4x-6)$

Applications

Solve.

61. A square is 30 inches on each side. A square that is x inches on each side is cut from each corner of the 30-inch square.

a. Represent the area of the remaining portion of the square in the form of a polynomial function $A(x)$.

b. Represent the perimeter of the remaining portion of the square in the form of a polynomial function $P(x)$.

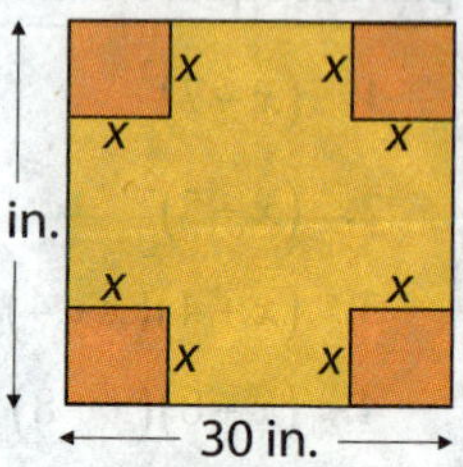

62. A rectangle has sides $(x+3)$ feet and $(x+5)$ feet. If a square that is x feet on each side is cut from the rectangle, represent the remaining area in the form of a polynomial function $A(x)$.

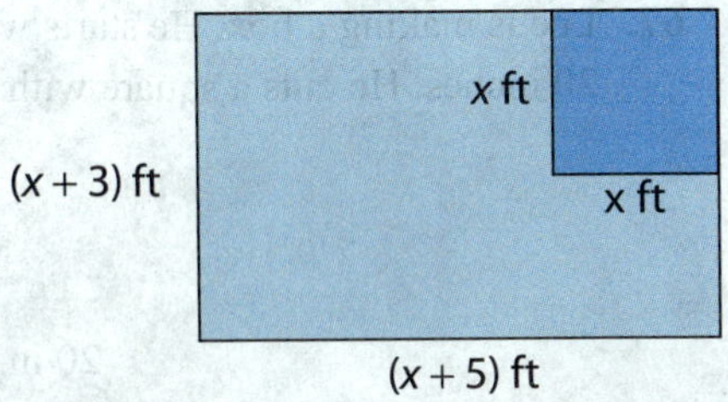

63. In the case of binomial probabilities, if x is the probability of success in one trial of an event, then the expression $f(x)=15x^4(1-x)^2$ is the probability of 4 successes in 6 trials where $0 \le x \le 1$.

a. Represent the expression $f(x)$ as a single polynomial by multiplying the polynomials.

b. If a fair coin is tossed, the probability of heads occurring is $\frac{1}{2}$. That is, $x=\frac{1}{2}$. Find the probability of 4 heads occurring in 6 tosses.

64. The Americans with Disabilities Act requires sidewalks to be x feet wide in order for wheelchairs to fit on them. At the bottom, the Empire State Building is 425 feet long and 190 feet wide and a regulation sidewalk surrounds the building.

a. Represent the area covered by the building and the sidewalk in the form of a polynomial function.

b. Represent the area covered by just the sidewalk in the form of a polynomial function.

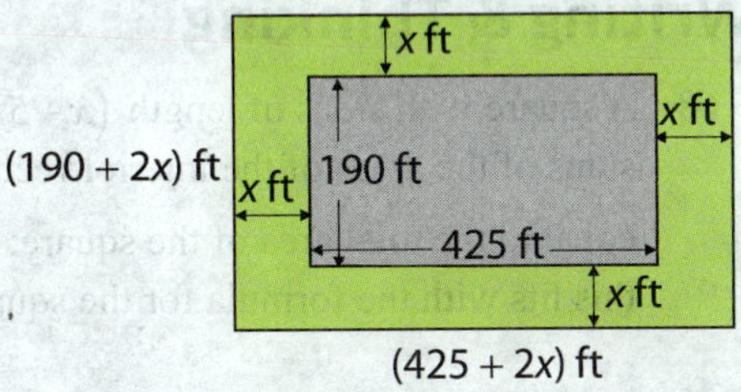

65. A rectangular piece of cardboard that is 10 inches by 15 inches has squares of length x inches on a side cut from each corner. (Assume that $0 < x < 5$.)

a. Represent the remaining area in the form of a polynomial function $A(x)$.

b. Represent the perimeter of the remaining figure in the form of a polynomial function $P(x)$.

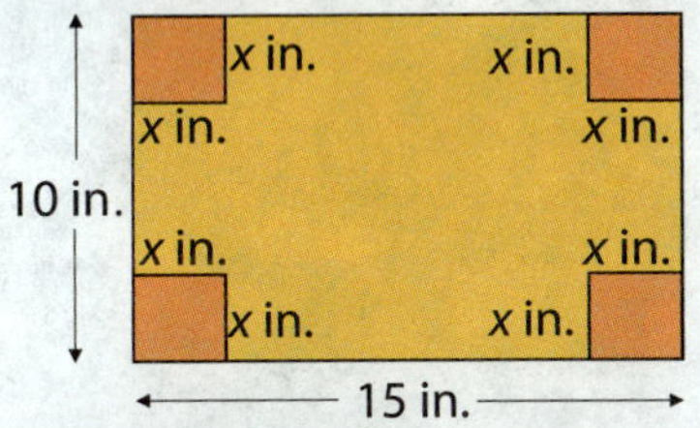

c. If the flaps of the cardboard are folded up, an open box is formed. Represent the volume of this box in the form of a polynomial function $V(x)$. (**Note:** Volume = length × width × height.)

66. The world's largest single aquarium habitat, located at the Georgia Aquarium, is 284 feet long, 126 feet wide, and 30 feet deep. Another aquarium is attempting to make a tank that is x feet longer, wider, and deeper. Represent the volume of the new tank as a polynomial function $V(x)$. (**Note:** The volume of the aquarium is the product of its length, width, and height, $V=lwh$.)

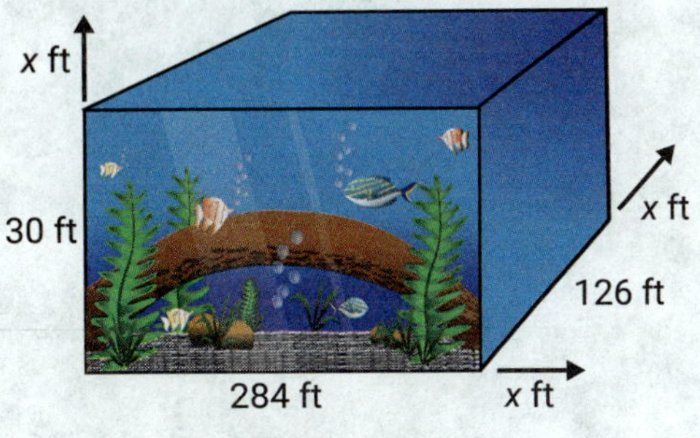

45. $36x^2-60x+25$

46. $49x^2-28x+4$

47. $x^2-1.96$

48. $x^2-4.41$

49. $x^2-5x+6.25$

50. $x^2+3.4x+2.89$

51. $x^2-4.6225$

52. $x^2-1.8496$

53. $x^2+2.48x+1.5376$

54. $x^2-2.9x+2.1025$

55. $2.0164x^2+27.264x+92.16$

56. $0.2116x^2-0.6532x+0.5041$

57. $129.96x^2-12.25$

58. $3.25x^2-27.43x-192.66$

59. $93.24x^2+142.46x-104.04$

60. $11.56x^2-36$

Applications

61. a. $A(x)=900-4x^2$

b. $P(x)=4(30-2x)+8x=120$

62. $A(x)=8x+15$

63. a. $f(x)=15x^6-30x^5+15x^4$

b. 0.234375

64. a. $A(x)=4x^2+1230x+80,750$

b. $A(x)=4x^2+1230x$

65. a. $A(x)=150-4x^2$

b. $P(x)=2(15-2x)+2(10-2x)+8x=50$

c. $V(x)=(10-2x)(15-2x)x=4x^3-50x^2+150x$

66. $V(x)=x^3+440x^2+48,084x+1,073,520$

67. a. $A(x) = 400 - 4x^2$

b. $20 - 2x$ and $20 - 2x$

c. $B(x) = 400 - 80x + 4x^2$

d. $V(I) = 400x - 80x^2 + 4x^3$

Writing & Thinking

68. $(x+5)^2 = x^2 + 2(5x) + 5^2$

Answers will vary.

67. Lee is making a box. He starts with a piece of cardboard that is 20 inches by 20 inches. He cuts a square with side length x from each corner of the box.

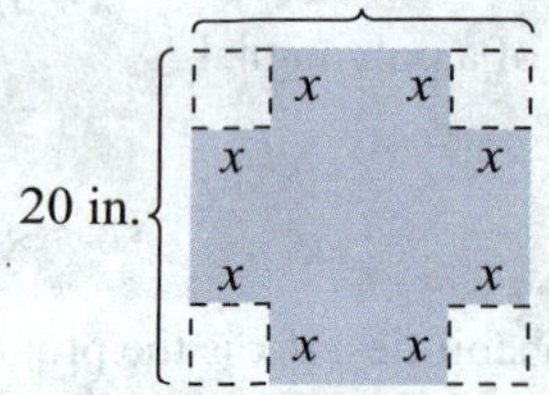

a. Write a polynomial function $A(x)$ to represent the area of the cardboard that remains after the corners are cut out.

b. When the sides of the box are folded up, what will be the side lengths of the base of the box?

c. Write a polynomial function $B(x)$ to represent the area of the base of the box when the sides are folded up.

d. The height of the box will be x inches. Write a polynomial function $V(x)$ to determine the volume of the box.

Writing & Thinking

68. A square with sides of length $(x+5)$ can be broken up as shown in the diagram. The sums of the areas of the interior rectangles and squares is equal to the total area of the square: $(x+5)^2$. Show how this fits with the formula for the square of a sum.

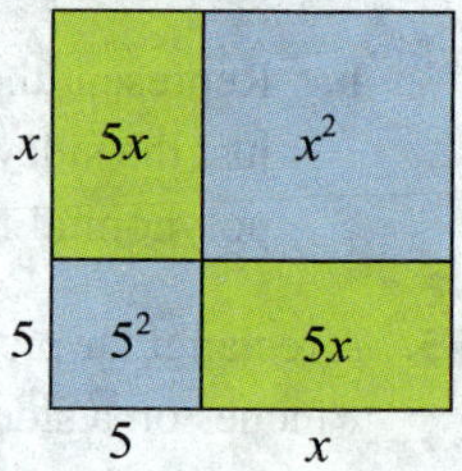

12.8 Division with Polynomials

Objectives

A. Divide a polynomial by a monomial.

B. Divide polynomials by using the division algorithm.

Fractions, such as $\frac{127}{2}$ and $\frac{1}{8}$, in which the numerator and denominator are integers are called rational numbers. Fractions in which the numerator and denominator are polynomials are called **rational expressions**. (No denominator can be 0.)

Rational Expressions: $\dfrac{x^3-6x^2+2x}{3x}, \dfrac{6x^2-7x-2}{2x-1}, \dfrac{5x^3-3x+1}{x+3},$ and $\dfrac{1}{10x}$

In this section, we will treat a rational expression as a division problem. With this basis, there are two situations to consider:

1. The denominator (divisor) is a monomial.
2. The denominator (divisor) is not a monomial.

A Dividing by a Monomial

We know that the sum of fractions with the same denominator can be written as a single fraction by adding the numerators and using the common denominator. For example,

$$\frac{5}{a}+\frac{3b}{a}+\frac{2c}{a}=\frac{5+3b+2c}{a}.$$

If instead of adding the fractions, we start with the sum and want to divide the numerator by the denominator (with a monomial in the denominator), we divide each term in the numerator by the monomial denominator and simplify each fraction.

$$\frac{4x^3+8x^2-12x}{4x}=\frac{4x^3}{4x}+\frac{8x^2}{4x}-\frac{12x}{4x}=x^2+2x-3$$

Example 1 Dividing by a Monomial

Divide each polynomial by the monomial denominator by writing each fraction as the sum (or difference) of fractions. Simplify each fraction, if possible.

a. $\dfrac{x^3-6x^2+2x}{3x}$

b. $\dfrac{15y^4-20y^3+5y^2}{5y^2}$

Solution

a. $\dfrac{x^3-6x^2+2x}{3x}=\dfrac{x^3}{3x}-\dfrac{6x^2}{3x}+\dfrac{2x}{3x}=\dfrac{x^2}{3}-2x+\dfrac{2}{3}$

b. $\dfrac{15y^4-20y^3+5y^2}{5y^2}=\dfrac{15y^4}{5y^2}-\dfrac{20y^3}{5y^2}+\dfrac{5y^2}{5y^2}=3y^2-4y+1$

Now work margin exercise 1.

1. Divide each polynomial by the monomial denominator by writing each fraction as the sum (or difference) of fractions. Simplify if possible.

a. $\dfrac{18x^5-6x^4+9x}{3x}$

b. $\dfrac{5y^4-6y^3-8y}{2y}$

B The Division Algorithm

Teaching Note

It is recommended that you remind students of the long division process so that they can see the similarities. Expect many students to have forgotten the process.

In arithmetic, the **division algorithm** (called **long division**) is the process (or series of steps) that we follow when dividing two numbers. By this division algorithm, we can find $64 \div 5$ as follows.

$$\begin{array}{r} 12 \\ 5\overline{)64} \\ \underline{-5} \\ 14 \\ \underline{-10} \\ 4 \end{array}$$

- 12 ← Quotient
- Divisor → 5; 64 ← Dividend
- −5 ← Subtract
- −10 ← Subtract
- 4 ← 4 is the remainder. (The remainder is always smaller than the divisor.)

Check: $5 \cdot 12 + 4 = 60 + 4 = 64$ (Multiply the divisor times the quotient and add the remainder. The result should be the original dividend.)

We can also write the division in fraction form with the remainder over the divisor, giving a mixed number.

$$64 \div 5 = \frac{64}{5} = 12 + \frac{4}{5} = 12\frac{4}{5}$$

In algebra, **the division algorithm with polynomials** is quite similar. In dividing one polynomial by another, with the degree of the divisor smaller than the degree of the dividend, the quotient will be another polynomial with a remainder.

For $64 \div 5$	For $P \div D$	
$64 = 12 \cdot 5 + 4$	$P = Q \cdot D + R$	where P is the dividend,
or		D is the divisor,
		Q is the quotient, and
$\frac{64}{5} = 12 + \frac{4}{5}$	$\frac{P}{D} = Q + \frac{R}{D}$	R is the remainder.

The remainder must be of smaller degree than the divisor. If the remainder is 0, then the divisor and quotient are factors of the dividend.

The Division Algorithm

For polynomials P and D, the division algorithm gives

$$\boldsymbol{\frac{P}{D} = Q + \frac{R}{D}}, \qquad \boldsymbol{D \neq 0}$$

where Q and R are polynomials and the **degree of** $\boldsymbol{R}$ **< degree of** $\boldsymbol{D}$.

THEOREM

The actual process of long division is not clear from this abstract definition. Although you are familiar with the process of long division with decimal numbers, this same procedure may seem more complicated with polynomials. The **division algorithm (long division)** is illustrated in a step-by-step form in the following example. Study it carefully.

Example 2 Using The Division Algorithm

Simplify $\dfrac{6x^2-7x-2}{2x-1}$ using the division algorithm.

Solution

Calculation	Explanation
Step 1: $2x-1\overline{)6x^2-7x-2}$	Write both polynomials in order of descending powers. **If any powers are missing, fill in with 0s.**
Step 2: $\begin{array}{r} 3x \\ 2x-1\overline{)\,6x^2-7x-2} \end{array}$	Mentally divide $6x^2$ by $2x$: $\dfrac{6x^2}{2x}=3x$. Write $3x$ above $-7x$.
Step 3: $\begin{array}{r} 3x \\ 2x-1\overline{)\,6x^2-7x-2} \\ -(6x^2-3x) \end{array}$	Multiply $3x$ times $(2x-1)$ and write the terms of the product, $6x^2-3x$, under the like terms in the dividend. Use a negative sign to indicate that the product is to be subtracted.
Step 4: $\begin{array}{r} 3x \\ 2x-1\overline{)\,6x^2-7x-2} \\ \underline{-6x^2+3x} \\ -4x \end{array}$	Subtract $6x^2-3x$ by changing signs and adding.
Step 5: $\begin{array}{r} 3x \\ 2x-1\overline{)\,6x^2-7x-2} \\ \underline{-6x^2+3x}\downarrow \\ -4x-2 \end{array}$	Bring down the -2.
Step 6: $\begin{array}{r} 3x-2 \\ 2x-1\overline{)\,6x^2-7x-2} \\ \underline{-6x^2+3x} \\ -4x-2 \end{array}$	Mentally divide $-4x$ by $2x$: $\dfrac{-4x}{2x}=-2$. Write -2 in the quotient.
Step 7: $\begin{array}{r} 3x-2 \\ 2x-1\overline{)\,6x^2-7x-2} \\ \underline{-6x^2+3x} \\ -4x-2 \\ \underline{-(-4x+2)} \end{array}$	Multiply -2 times $(2x-1)$ and write the terms of the product, $-4x+2$, under the like terms in the expression $-4x-2$. Use a negative sign to indicate that the product is to be subtracted.

2. Simplify $\dfrac{18x^3+6x^2-10x-11}{6x+4}$ using the division algorithm.

Math Tip

Another approach to Step 2 is to ask yourself, "What times the first term in the divisor will give us the first term in the dividend?" For example, "What times $2x$ will give $6x^2$?" The answer is $3x$ since $2x\cdot 3x=6x^2$. This reasoning can be used at every step of division.

Step 8:
$$\begin{array}{r} 3x-2 \\ 2x-1\overline{)\,6x^2-7x-2} \\ \underline{-6x^2+3x} \\ -4x-2 \\ \underline{+4x-2} \\ -4 \end{array}$$

Subtract $-4x+2$ by changing signs and adding.

Thus, the quotient is $3x-2$ and the remainder is -4.

In the form $Q+\frac{R}{D}$ we can write $\frac{6x^2-7x-2}{2x-1}=3x-2-\frac{4}{2x-1}$.

Check

Show $Q\cdot D+R=P$.

$$(3x-2)(2x-1)-4=6x^2-3x-4x+2-4=6x^2-7x-2$$

Now work margin exercise 2.

3. Divide $(8x^3-2x^2+21x+45)\div(4x+5)$ using the division algorithm.

Example 3 Using Long Division (Remainder 0)

Divide $(25x^3-5x^2+3x+1)\div(5x+1)$ using the division algorithm.

Solution

$$\begin{array}{r} 5x^2-2x+1 \\ 5x+1\overline{)\,25x^3-5x^2+3x+1} \\ \underline{-(25x^3+5x^2)} \\ -10x^2+3x \\ \underline{-(-10x^2-2x)} \\ 5x+1 \\ \underline{-(5x+1)} \\ 0 \end{array}$$

There is no remainder, thus the quotient is $5x^2-2x+1$.

Now work margin exercise 3.

In Example 3, because the remainder is 0, both $(5x+1)$ and $(5x^2-2x+1)$ are **factors** of $25x^3-5x^2+3x+1$. That is

$$(5x^2-2x+1)(5x+1)=25x^3-5x^2+3x+1,\text{ or } Q\cdot D=P.$$

Factoring polynomials will be discussed in detail in Chapter 13.

Example 4 Using Long Division (Terms Missing)

Simplify $\dfrac{x^4+9x^2-3x+5}{x^2-x+2}$ using the division algorithm.

Solution

Note that 0 is written as a placeholder for any missing powers of the variable. In this way, like terms are easily aligned vertically.

$$
\begin{array}{r}
x^2+\ x+8 \\
x^2-x+2\overline{\smash{)}x^4+0x^3+9x^2-3x+5} \\
\underline{-\left(x^4-\ x^3+2x^2\right)} \\
x^3+7x^2-3x \\
\underline{-\left(x^3-\ x^2+2x\right)} \\
8x^2-5x+5 \\
\underline{-\left(8x^2-8x+16\right)} \\
3x-11
\end{array}
$$

Note that the remainder is of smaller degree than the divisor.

Thus, the quotient is x^2+x+8 and the remainder is $3x-11$.

In the form $Q+\dfrac{R}{D}$, we can write $x^2+x+8+\dfrac{3x-11}{x^2-x+2}$.

Now work margin exercise 4.

4. Simplify $\dfrac{7x^3+4x-9}{x-2}$ using the division algorithm.

Margin Exercise Answers

1. a. $6x^4-2x^3+3$ **b.** $\dfrac{5y^3}{2}-3y^2-4$ **2.** $3x^2-x-1-\dfrac{7}{6x+4}$ **3.** $2x^2-3x+9$

4. $7x^2+14x+32+\dfrac{55}{x-2}$

12.8 Exercises

Concept Check

Fill-in-the-Blank. Complete each sentence using information found in this section.

1. Fractions in which the numerator and denominator are polynomials are called __________ expressions.

2. When polynomials are divided, if the remainder is 0, then both the divisor and quotient are ________ of the dividend.

3. To divide a polynomial by a monomial, divide each term in the __________ by the monomial denominator and then _______ each fraction.

12.8 Exercises

Concept Check

1. rational

2. factors

3. numerator, simplify

4. dividend, divisor
5. 0
6. True
7. False; Descending order
8. False; Missing powers should be filled in with zeros.
9. True

Practice

1. y^2-2y+3
2. $3x^2+4x+6$
3. $2x^2-3x+1$
4. $7y^2+14y+6$
5. $10x^3-11x^2+x$
6. x^4+2x^3-3
7. $-4x^2+7x-\frac{5}{2}$
8. $6x^2-4x+\frac{7}{2}$
9. $y^3-\frac{7}{2}y^2-\frac{15}{2}y+4$
10. $5y^3-\frac{7}{2}y^2+\frac{21}{4}y+\frac{21}{2}$
11. $x-6+\frac{4}{x+4}$
12. $x+10+\frac{5}{x-1}$
13. $3x-4-\frac{7}{2x-1}$
14. $2x+2-\frac{1}{5x+3}$
15. $3x+4+\frac{1}{7x-1}$
16. $5x+2-\frac{3}{3x-4}$
17. $x-9$
18. $x-7$
19. $x^2-x-\frac{3}{x-8}$
20. $x^2-4x-\frac{5}{x-2}$
21. $4x^2-6x+9-\frac{17}{x+2}$
22. $3x^2+3x+5-\frac{10}{x+1}$

4. The division algorithm with polynomials tells us that for $\frac{P}{D}=Q+\frac{R}{D}$, where P is the ________, D is the ________, Q is the quotient, and R is the remainder.

5. When dividing polynomials, the denominator (or divisor) can never equal _____.

True/False. Determine whether each statement is true or false. If a statement is false, explain how it can be changed so the statement will be true. (**Note:** There may be more than one acceptable change.)

6. When dividing polynomials, any remainder must be of smaller degree than the divisor.

7. The first step in the division algorithm is to align the polynomials in ascending order.

8. To aid in organization and clarity when dividing polynomials, it is best to fill in any missing powers with ones.

9. The process followed when dividing two polynomials is called the division algorithm with polynomials.

Practice

Express each quotient as a sum (or difference) of fractions and simplify, if possible. See Example 1.

1. $\frac{8y^3-16y^2+24y}{8y}$
2. $\frac{18x^4+24x^3+36x^2}{6x^2}$
3. $\frac{34x^5-51x^4+17x^3}{17x^3}$
4. $\frac{14y^4+28y^3+12y^2}{2y^2}$
5. $\frac{110x^4-121x^3+11x^2}{11x}$
6. $\frac{15x^7+30x^6-45x^3}{15x^3}$
7. $\frac{-56x^4+98x^3-35x^2}{14x^2}$
8. $\frac{108x^6-72x^5+63x^4}{18x^4}$
9. $\frac{16y^6-56y^5-120y^4+64y^3}{16y^3}$
10. $\frac{20y^5-14y^4+21y^3+42y^2}{4y^2}$

Divide by using the division algorithm. Write the answers in the form $Q+\frac{R}{D}$, where the degree of R is less than the degree of D. See Examples 2 through 4.

11. $\frac{x^2-2x-20}{x+4}$
12. $\frac{x^2+9x-5}{x-1}$
13. $\frac{6x^2-11x-3}{2x-1}$
14. $\frac{10x^2+16x+5}{5x+3}$
15. $\frac{21x^2+25x-3}{7x-1}$
16. $\frac{15x^2-14x-11}{3x-4}$

17. $\dfrac{x^2-12x+27}{x-3}$

18. $\dfrac{x^2-12x+35}{x-5}$

19. $\dfrac{x^3-9x^2+8x-3}{x-8}$

20. $\dfrac{x^3-6x^2+8x-5}{x-2}$

21. $\dfrac{4x^3+2x^2-3x+1}{x+2}$

22. $\dfrac{3x^3+6x^2+8x-5}{x+1}$

23. $\dfrac{x^3+6x+3}{x-7}$

24. $\dfrac{2x^3+3x-2}{x-1}$

25. $\dfrac{2x^3-5x^2+6}{x+2}$

26. $\dfrac{4x^3-x^2+13}{x-1}$

27. $\dfrac{21x^3+41x^2+13x+5}{3x+5}$

28. $\dfrac{6x^3-7x^2+14x-8}{3x-2}$

29. $\dfrac{2x^3+7x^2+10x-6}{2x+3}$

30. $\dfrac{6x^3-4x^2+5x-7}{x-2}$

31. $\dfrac{x^3-x^2-10x-10}{x-4}$

32. $\dfrac{2x^3-3x^2+7x+4}{2x-1}$

33. $\dfrac{10x^3+11x^2-12x+9}{5x+3}$

34. $\dfrac{6x^3+19x^2-3x-7}{6x+1}$

35. $\dfrac{2x^3-7x+2}{x+4}$

36. $\dfrac{2x^3+4x^2-9}{x+3}$

37. $\dfrac{9x^3-19x+9}{3x-2}$

38. $\dfrac{4x^3-8x^2-9x}{2x-3}$

39. $\dfrac{6x^3+11x^2+25}{2x+5}$

40. $\dfrac{16x^3+7x+12}{4x+3}$

41. $\dfrac{x^4-3x^3+2x^2-x+2}{x-3}$

42. $\dfrac{x^4+x^3-4x^2+x-3}{x+6}$

43. $\dfrac{x^4+2x^2-3x+5}{x-2}$

44. $\dfrac{3x^4+2x^3-2x^2-1}{x+1}$

45. $\dfrac{x^4-x^2+3}{x-\frac{1}{2}}$

46. $\dfrac{x^3+2x^2+1}{x-\frac{2}{3}}$

47. $\dfrac{3x^3+5x^2+7x+9}{x^2+2}$

48. $\dfrac{2x^4+2x^3+3x^2+6x-1}{2x^2+3}$

49. $\dfrac{x^4+x^3-4x+1}{x^2+4}$

50. $\dfrac{2x^4+x^3-8x^2+3x-2}{x^2-5}$

23. $x^2+7x+55+\dfrac{388}{x-7}$

24. $2x^2+2x+5+\dfrac{3}{x-1}$

25. $2x^2-9x+18-\dfrac{30}{x+2}$

26. $4x^2+3x+3+\dfrac{16}{x-1}$

27. $7x^2+2x+1$

28. $2x^2-x+4$

29. $x^2+2x+2-\dfrac{12}{2x+3}$

30. $6x^2+8x+21+\dfrac{35}{x-2}$

31. $x^2+3x+2-\dfrac{2}{x-4}$

32. $x^2-x+3+\dfrac{7}{2x-1}$

33. $2x^2+x-3+\dfrac{18}{5x+3}$

34. $x^2+3x-1-\dfrac{6}{6x+1}$

35. $2x^2-8x+25-\dfrac{98}{x+4}$

36. $2x^2-2x+6-\dfrac{27}{x+3}$

37. $3x^2+2x-5-\dfrac{1}{3x-2}$

38. $2x^2-x-6-\dfrac{18}{2x-3}$

39. $3x^2-2x+5$

40. $4x^2-3x+4$

41. $x^3+2x+5+\dfrac{17}{x-3}$

42. $x^3-5x^2+26x-155$
$+\dfrac{927}{x+6}$

43. $x^3+2x^2+6x+9+\dfrac{23}{x-2}$

44. $3x^3-x^2-x+1-\dfrac{2}{x+1}$

45. $x^3+\dfrac{1}{2}x^2-\dfrac{3}{4}x-\dfrac{3}{8}$
$+\dfrac{45}{16\left(x-\dfrac{1}{2}\right)}$

46. $x^2+\frac{8}{3}x+\frac{16}{9}+\frac{59}{27\left(x-\frac{2}{3}\right)}$

47. $3x+5+\frac{x-1}{x^2+2}$

48. $x^2+x+\frac{3x-1}{2x^2+3}$

49. $x^2+x-4+\frac{-8x+17}{x^2+4}$

50. $2x^2+x+2+\frac{8x+8}{x^2-5}$

51. $2x+3+\frac{6}{3x^2-2x-1}$

52. $x-6+\frac{-3x-8}{x^2-3x+5}$

53. $3x^2-10x+12+\frac{-x-14}{x^2+x+1}$

54. $2x^2+5x+3+\frac{x-4}{x^2-3x+1}$

55. $x^2-2x+7+\frac{-17x+14}{x^2+2x-3}$

56. $3x^2-3x+2+\frac{-5x+5}{x^2+x-1}$

57. x^2+3x+9

58. $x^2-5x+25$

59. $x^5-x^4+x^3-x^2+x-1$

60. $x^5+x^4+x^3+x^2+x+1$

61. $x^4+x^3+x^2+x+1+\frac{2}{x-1}$

62. $x^5-x^4+x^3-x^2+x-1+\frac{2}{x+1}$

63. $x^4-\frac{1}{2}x^3-\frac{3}{4}x^2+\frac{3}{8}x+\frac{13}{16}-\frac{13}{32\left(x+\frac{1}{2}\right)}$

64. $x^3-\frac{14}{5}x^2+\frac{56}{25}x-\frac{224}{125}+\frac{3396}{625\left(x+\frac{4}{5}\right)}$

51. $\frac{6x^3+5x^2-8x+3}{3x^2-2x-1}$

52. $\frac{x^3-9x^2+20x-38}{x^2-3x+5}$

53. $\frac{3x^4-7x^3+5x^2+x-2}{x^2+x+1}$

54. $\frac{2x^4-x^3-10x^2-3x-1}{x^2-3x+1}$

55. $\frac{x^4+3x-7}{x^2+2x-3}$

56. $\frac{3x^4-4x^2+3}{x^2+x-1}$

57. $\frac{x^3-27}{x-3}$

58. $\frac{x^3+125}{x+5}$

59. $\frac{x^6-1}{x+1}$

60. $\frac{x^6-1}{x-1}$

61. $\frac{x^5+1}{x-1}$

62. $\frac{x^6+1}{x+1}$

63. $\frac{x^5-x^3+x}{x+\frac{1}{2}}$

64. $\frac{x^4-2x^3+4}{x+\frac{4}{5}}$

Applications

Solve.

65. A moving company uses a box that has a volume of $x^3-2x^2-13x-10$ cubic inches.

a. If the height of the box is $x+2$, what is the area of the base of the box?

b. If the height of the box is $x+1$, what is the area of the base of the box?

66. A rectangular garden requires a volume of top soil modeled by the equation $200x^3+350x^2+150x$, where x is the depth in inches of top soil needed.

a. Determine an expression for the area of the garden. (**Hint:** Volume = length · width · height and Area = length · width.)

b. If the width of the garden is $10x+10$ inches, use the expression from part a. to find an expression for the length of the garden.

c. If the depth of soil needed is 3 inches, find the volume of top soil that needs to be purchased for the garden.

d. Determine how many cubic feet of top soil is needed by dividing the answer from part c. by 1728 in.^{3}. Round your answer to the nearest tenth. (**Note:** 1 cubic foot = 12 in. · 12 in. · 12 in. = 1728 in.^{3})

e. If the top soil comes in bags that contain 0.75 cubic feet of soil, how many bags will need to be purchased?

f. If the cost of the top soil is \$2.10 per bag (including tax), what will be the total cost of the top soil needed for the garden?

Writing & Thinking

67. Suppose that a polynomial is divided by $(3x-2)$ and the answer is given as

$x^2+2x+4+\dfrac{20}{3x-2}$.

What was the original polynomial? Explain how you arrived at this conclusion.

68. Suppose that a polynomial is divided by $(x+5)$ and the answer is given as

$x^2-3x+2-\dfrac{6}{x+5}$.

What was the original polynomial? Explain how you arrived at this conclusion.

69. Given that $P(x)=2x^3-8x^2+10x+15$.

a. Find $P(2)$ then divide $P(x)$ by $x-2$.

b. Find $P(-1)$ then divide $P(x)$ by $x+1$.

c. Find $P(4)$ then divide $P(x)$ by $x-4$.

Do you see any pattern in the values of $P(a)$ for $x=a$ and the remainders you found in the division process?

Applications

65. a. x^2-4x-5 square inches

b. $x^2-3x-10$ square inches

66. a. $200x^2+350x+150$

b. $20x+15$ **c.** 9000 in.^{2}

d. 5.2 cubic feet

e. 7 bags **f.** \$14.70

Writing & Thinking

67. $3x^3+4x^2+8x+12$; Multiply the divisor $(3x-2)$ by the quotient, $x^2+2x+4+\dfrac{20}{3x-2}$, to get the original polynomial.

68. $x^3+2x^2-13x+4$; Multiply the divisor $(x+5)$ by the quotient, $x^2-3x+2-\dfrac{6}{x+5}$, to get the original

polynomial.

69. a. 19;

$2x^2-4x+2+\dfrac{19}{x-2}$

b. -5;

$2x^2-10x+20-\dfrac{5}{x+1}$

c. 55;

$2x^2+10+\dfrac{55}{x-4}$

Yes; $R=P(a)$ when $P(x)$ is divided by x.

Objectives

A. Use synthetic division to divide polynomials.

B. Use the remainder theorem to find the value of a polynomial at a specific value of x.

12.9 Synthetic Division and the Remainder Theorem

A Synthetic Division

In the previous section, polynomials were divided using the division algorithm (long division). In the special case when the divisor of a rational expression is a first-degree binomial with leading coefficient 1, long division can be simplified by omitting the variables entirely and writing only certain coefficients. The procedure is called **synthetic division.** The following analysis describes how the procedure works for $\frac{5x^3+11x^2-3x+1}{x+3}$.

(Notice that $x + 3$ is first-degree with leading coefficient 1.)

a. With variables

$$
\begin{array}{r|rrrr}
 & 5x^2 & -\ 4x & +\ 9 & \\ \hline
x+3 & 5x^3 & +\ 11x^2 & -\ 3x & +\ 1 \\
 & 5x^3 & +\ 15x^2 & & \\ \hline
 & & -\ 4x^2 & -\ 3x & \\
 & & -\ 4x^2 & -\ 12x & \\ \hline
 & & & 9x & +\ 1 \\
 & & & 9x & +\ 27 \\ \hline
 & & & & -26
\end{array}
$$

b. Without variables

$$
\begin{array}{r|rrrr}
 & 5 & -\ 4 & +\ 9 & \\ \hline
1+3 & 5 & +\ 11 & -\ 3 & +\ 1 \\
 & \boxed{5} & +\ 15 & & \\ \hline
 & & -\ 4 & \boxed{-\ 3} & \\
 & & \boxed{-\ 4} & -12 & \\ \hline
 & & & 9 & \boxed{+\ 1} \\
 & & & \boxed{9} & +27 \\ \hline
 & & & & -26
\end{array}
$$

The boxed numbers in Step b. can be omitted since they are repetitions of the numbers directly above them.

c. Boxed numbers omitted

$$
\begin{array}{r|rrrr}
 & 5 & -\ 4 & +9 & \\ \hline
1+3 & 5 & +\ 11 & -\ 3 & +1 \\
 & & \underline{+\ 15} & & \\
 & & -\ 4 & & \\
 & & & \underline{-12} & \\
 & & & 9 & \\
 & & & & \underline{+27} \\
 & & & & -26
\end{array}
$$

d. Numbers moved up to fill in spaces

$$
\begin{array}{r|rrrr}
 & 5 & -\ 4 & +\ 9 & \\ \hline
1+3 & 5 & +11 & -\ 3 & +\ 1 \\
 & & +15 & -12 & +\ 27 \\ \hline
 & & -4 & +9 & -26
\end{array}
$$

Next, omit the 1 in the divisor, change +3 to –3, and write the opposites of the boxed numbers (because the quotient coefficient will now be multiplied by –3 instead of +3), as shown in Steps e. and f. This allows the numbers to be added instead of subtracted. The number 5 is written on the bottom line, and the top line is omitted. The quotient and remainder can now be read from the bottom line.

e.

$$\begin{array}{r|rrrr} & 5 & -4 & +9 & \\ \hline 1+3 & 5 & +11 & -3 & +1 \\ & & \boxed{+15} & \boxed{-12} & \boxed{+27} \\ \hline & & -4 & +9 & -26 \end{array}$$

f.

$$\begin{array}{r|rrrr} -3 & 5 & +11 & -3 & +1 \\ & \downarrow & -15 & +12 & -27 \\ \hline & 5 & -4 & +9 & -26 \end{array}$$

This represents

$$5x^2-4x+9+\frac{-26}{x+3}.$$

The numbers on the bottom now represent the coefficients of a polynomial of *one degree less than the dividend,* along with the remainder. The last number to the right is the remainder.

In summary, synthetic division can be accomplished as follows.

1. Write only the coefficients of the dividend and the opposite of the constant in the divisor.

$$\begin{array}{r|rrrr} -3 & 5 & 11 & -3 & 1 \\ & & & & \\ \hline & & & & \end{array}$$

2. Rewrite the first coefficient (5) as the first coefficient in the quotient.

$$\begin{array}{r|rrrr} -3 & 5 & 11 & -3 & 1 \\ & \downarrow & & & \\ \hline & 5 & & & \end{array}$$

3. Multiply the coefficient (5) by the constant divisor (–3) and add this product (–15) to the second coefficient.

$$\begin{array}{r|rrrr} -3 & 5 & 11 & -3 & 1 \\ & \downarrow & -15 & & \\ \hline & 5 \nearrow & -4 & & \end{array}$$

4. Continue to multiply each new coefficient by the constant divisor and add this product to the next coefficient in the dividend.

$$\begin{array}{r|rrrr} -3 & 5 & 11 & -3 & 1 \\ & \downarrow & -15 & 12 & -27 \\ \hline & 5 \nearrow & -4 \nearrow & 9 \nearrow & -26 \end{array}$$

5. The constants on the bottom line are the coefficients of the quotient and the remainder.

$$\frac{5x^3+11x^2-3x+1}{x+3}=5x^2-4x+9+\frac{-26}{x+3}$$
$$=5x^2-4x+9-\frac{26}{x+3}$$

1. Use synthetic division to write each expression in the form $Q+\dfrac{R}{D}$.

a. $\dfrac{3x^3-5x^2+2x+1}{x-1}$

b. $\dfrac{x^4+2x^3+5x-9}{x+3}$

Example 1 Using Synthetic Division

Use synthetic division to write each expression in the form $Q+\dfrac{R}{D}$.

a. $\dfrac{4x^3+10x^2+11}{x+5}$

b. $\dfrac{2x^4-x^3-5x^2-2x+7}{x-2}$

Solution

a.

$$\begin{array}{r|rrrr} -5 & 4 & 10 & 0 & 11 \\ & \downarrow & -20 & 50 & -250 \\ \hline & 4 & -10 & 50 & -239 \end{array}$$

Since there is no *x*-term, 0 is the coefficient. The coefficient is 0 for any missing term.

$$\frac{4x^3+10x^2+11}{x+5}=4x^2-10x+50+\frac{-239}{x+5}$$

$$=4x^2-10x+50-\frac{239}{x+5}$$

Math Tip

Since synthetic division only works when the divisor is of the form $(x-c)$, it is important to continue practicing the division algorithm learned in the previous section.

b.

$$\begin{array}{r|rrrrr} 2 & 2 & -1 & -5 & -2 & 7 \\ & \downarrow & 4 & 6 & 2 & 0 \\ \hline & 2 & 3 & 1 & 0 & 7 \end{array}$$

$$\frac{2x^4-x^3-5x^2-2x+7}{x-2}=2x^3+3x^2+x+\frac{7}{x-2}$$

Now work margin exercise 1.

B The Remainder Theorem

Synthetic division can be used for several purposes, one of which is to find the value of a polynomial for a particular value of x. For example, we know (from Section 12.4) that if

$$P(x)=x^3-5x^2+7x-10,$$

then

$$P(2)=2^3-5\cdot2^2+7\cdot2-10=-8.$$

Using synthetic division to divide $x^3-5x^2+7x-10$ by $x-2$, we have

$$\begin{array}{r|rrrr} 2 & 1 & -5 & 7 & -10 \\ & & 2 & -6 & 2 \\ \hline & 1 & -3 & 1 & \boxed{-8} \end{array} \longleftarrow \text{Remainder}$$

The fact that the remainder is the same as $P(2)$ is not an accident. In fact, as the following theorem states, the remainder when a polynomial is divided by a first-degree factor of the form $(x-c)$ will always be $P(c)$.

The Remainder Theorem

If a polynomial $P(x)$ is divided by $(x-c)$, then the remainder will be $P(c)$.

THEOREM

Proof:

By the division algorithm, we know that $\frac{P(x)}{x-c}=Q(x)+\frac{R}{x-c}$, where R is a constant.

(Remember that the degree of the remainder must be less than the degree of the divisor.)

Multiplying through by $(x-c)$, we have

$$P(x)=(x-c)\cdot Q(x)+R,$$

and substituting $x=c$ gives

$$\begin{aligned} P(c)&=(c-c)\cdot Q(c)+R \\ &=0\cdot Q(c)+R \\ &=0+R \\ &=R. \end{aligned}$$

The proof is complete.

Example 2 Using the Remainder Theorem and Synthetic Division

Use synthetic division to find $P(5)$ given $P(x)=-2x^2+15x-50$.

Solution

$$\begin{array}{r|rrr} 5 & -2 & 15 & -50 \\ & & -10 & 25 \\ \hline & -2 & 5 & \boxed{-25} \end{array} \quad \leftarrow \text{Remainder} = P(5)$$

Thus, $P(5)=-25$.

(Checking shows $P(5)=-2\cdot 5^2+15\cdot 5-50=-50+75-50=-25$.)

Now work margin exercise 2.

2. Use synthetic division to find $P(3)$ given $P(x)=-x^3+4x^2-2x+5$.

Example 3 Using the Remainder Theorem and Synthetic Division

Use synthetic division to find $P(-3)$, given $P(x)=3x^4+10x^3-5x^2+125$.

Note: To evaluate $P(-3)$, think of the divisor in the form $(x+3)=(x-(-3))$. That is, in the form $(x-c)$, $c=-3$.

3. Use synthetic division to find $P(-2)$ given $P(x)=4x^2+3x-7$.

Solution

$-3\rfloor$	3	10	−5	0	125
		−9	−3	24	−72
	3	1	−8	24	(53) ← Remainder = $P(-3)$

Thus, $P(-3)=53$.

Now work margin exercise 3.

Example 4 Using the Remainder Theorem and Synthetic Division

Use synthetic division to show that $(x-6)$ is a factor of $P(x)=x^3-14x^2+53x-30$.

Solution

$6\rfloor$	1	−14	53	−30
		6	−48	30
	1	−8	5	(0) ← Remainder = $P(6)$

Thus, the remainder is $P(6)=0$ and $\boldsymbol{(x-6)}$ **is a factor of** $\boldsymbol{P(x)}$.

Note: The coefficients in the quotient tell us that x^2-8x+5 is also a factor of $P(x)$.

Now work margin exercise 4.

Margin Exercise Answers

1. a. $3x^2-2x+\frac{1}{x-1}$ **b.** $x^3-x^2+3x-4+\frac{3}{x+3}$ **2.** $P(3)=8$ 3. $P(-2)=3$

4.

$2\rfloor$	1	1	2	−16
		2	6	16
	1	3	8	0

12.9 Exercises

Concept Check

Fill-in-the-Blank. Complete each sentence using information found in this section

1. Synthetic division can be used to divide polynomials when the divisor of a rational expression is a first-degree __________ with leading coefficient 1.

2. Synthetic division involves omitting the __________ entirely and writing only certain coefficients.

3. If a polynomial $P(x)$ is divided by $(x-c)$ then the __________ will be $P(c)$.

4. When performing synthetic division, you first write only the __________ of the dividend and the ______ of the constant in the divisor.

---

4. Use synthetic division to show that $(x-2)$ is a factor of $P(x)=x^3+x^2+2x-16$.

12.9 Exercises

Concept Check

1. binomial
2. variables
3. remainder
4. coefficients, opposite
5. one
6. False; The divisor must have a leading coefficient of 1.
7. True
8. True
9. True

Practice

1. **a.** $x-9$
   b. $c=3;\ P(3)=0$
2. **a.** $x-7$
   b. $c=5;\ P(5)=0$
3. **a.** $x^2-4x+33-\frac{265}{x+8}$
   b. $c=-8;\ P(-8)=-265$
4. **a.** $x^2-4x-\frac{5}{x-2}$
   b. $c=2;\ P(2)=-5$
5. **a.** $4x^2-6x+9-\frac{17}{x+2}$
   b. $c=-2;\ P(-2)=-17$
6. **a.** $3x^2+3x+5-\frac{10}{x+1}$
   b. $c=-1;\ P(-1)=-10$
7. **a.** $x^2+7x+55+\frac{388}{x-7}$
   b. $c=7;\ P(7)=388$
8. **a.** $2x^2-8x+25-\frac{98}{x+4}$
   b. $c=-4;\ P(-4)=-98$
9. **a.** $2x^2-2x+6-\frac{27}{x+3}$
   b. $c=-3;\ P(-3)=-27$

5. Synthetic division results in a quotient that is a polynomial of ____ degree less than the dividend, along with the remainder.

True/False. Determine whether each statement is true or false. If a statement is false, explain how it can be changed so the statement will be true. (**Note:** There may be more than one acceptable change.)

6. Synthetic division can be used to divide a polynomial by $2x + 3$.

7. At the end of the synthetic division process, the constants on the bottom line are the coefficients of the quotient and the remainder.

8. Synthetic division can be used to find the value of a polynomial for a particular value of x.

9. Synthetic division is only used when the divisor is a first-degree polynomial of the form $(x + c)$ or $(x - c)$.

Practice

Divide the following expressions using synthetic division. **a.** Write the answer in the form $Q + \frac{R}{D}$ where R is a constant. **b.** In each exercise, $D = (x - c)$. State the value of c and the value of $P(c)$. (Assume $P(x)$ is the numerator of the fraction.) See Examples 1 through 3.

1. $\dfrac{x^2 - 12x + 27}{x - 3}$

2. $\dfrac{x^2 - 12x + 35}{x - 5}$

3. $\dfrac{x^3 + 4x^2 + x - 1}{x + 8}$

4. $\dfrac{x^3 - 6x^2 + 8x - 5}{x - 2}$

5. $\dfrac{4x^3 + 2x^2 - 3x + 1}{x + 2}$

6. $\dfrac{3x^3 + 6x^2 + 8x - 5}{x + 1}$

7. $\dfrac{x^3 + 6x + 3}{x - 7}$

8. $\dfrac{2x^3 - 7x + 2}{x + 4}$

9. $\dfrac{2x^3 + 4x^2 - 9}{x + 3}$

10. $\dfrac{4x^3 - x^2 + 13}{x - 1}$

11. $\dfrac{x^4 - 3x^3 + 2x^2 - x + 2}{x - 3}$

12. $\dfrac{x^4 + x^3 - 4x^2 + x - 3}{x + 6}$

13. $\dfrac{x^4 + 2x^2 - 3x + 5}{x - 2}$

14. $\dfrac{3x^4 + 2x^3 + 2x^2 + x - 1}{x + 1}$

15. $\dfrac{x^4 - x^2 + 3}{x - \frac{1}{2}}$

16. $\dfrac{x^3 + 2x^2 + 1}{x - \frac{2}{3}}$

17. $\dfrac{x^5 - 1}{x - 1}$

18. $\dfrac{x^5 - x^3 + x}{x + \frac{1}{2}}$

19. $\dfrac{x^4 - 2x^3 + 4}{x + \frac{4}{5}}$

20. $\dfrac{x^6 + 1}{x + 1}$

10. a. $4x^2 + 3x + 3 + \dfrac{16}{x - 1}$
b. $c = 1$; $P(1) = 16$

11. a. $x^3 + 2x + 5 + \dfrac{17}{x - 3}$
b. $c = 3$; $P(3) = 17$

12. a. $x^3 - 5x^2 + 26x - 155 + \dfrac{927}{x + 6}$
b. $c = -6$; $P(-6) = 927$

13. a. $x^3 + 2x^2 + 6x + 9 + \dfrac{23}{x - 2}$
b. $c = 2$; $P(2) = 23$

14. a. $3x^3 - x^2 + 3x - 2 + \dfrac{1}{x + 1}$
b. $c = -1$; $P(-1) = 1$

15. a. $x^3 + \dfrac{1}{2}x^2 - \dfrac{3}{4}x - \dfrac{3}{8} + \dfrac{45}{16\left(x - \frac{1}{2}\right)}$
b. $c = \dfrac{1}{2}$; $P\left(\dfrac{1}{2}\right) = \dfrac{45}{16}$

16. a. $x^2 + \dfrac{8}{3}x + \dfrac{16}{9} + \dfrac{59}{27\left(x - \frac{2}{3}\right)}$
b. $c = \dfrac{2}{3}$; $P\left(\dfrac{2}{3}\right) = \dfrac{59}{27}$

17. a. $x^4 + x^3 + x^2 + x + 1$
b. $c = 1$; $P(1) = 0$

18. a. $x^4 - \dfrac{1}{2}x^3 - \dfrac{3}{4}x^2 + \dfrac{3}{8}x + \dfrac{13}{16} - \dfrac{13}{32\left(x + \frac{1}{2}\right)}$
b. $c = -\dfrac{1}{2}$; $P\left(-\dfrac{1}{2}\right) = -\dfrac{13}{32}$

Applications

Solve.

21. The minimum temperature for plants to grow in a certain model greenhouse can be modeled by the function $T(x) = 0.000027x^3 - 0.004144x^2 + 0.145x + 39.757$, where T is the degrees in Fahrenheit and x is the number of days since seeds were planted. Use synthetic division to find the minimum temperature after 15 days. Round your answer to the nearest tenth.

22. A moving company uses a box that has a volume of $x^3 + 7x^2 - 6x - 72$ cubic inches.

a. If the height of the box is $x + 4$, what is the area of the base of the box?

b. If the height of the box is $x - 3$, what is the area of the base of the box?

Collaborative Learning

23. With the class divided into teams of 3 or 4 students, each team should develop answers to the following questions and be prepared to discuss the answers in class.

a. First, use long division to divide the polynomial $P(x) = 2x^3 - 8x^2 + 10x + 15$ by $2x - 1$. Then, use synthetic division to divide the same polynomial by $x - \frac{1}{2}$. Do the same process with two or three other polynomials and divisors. Next compare the corresponding long and synthetic division answers and explain how the answers are related.

b. Use the results from part a. and explain algebraically the relationship of the answers when a polynomial is divided (using long division) by $ax - b$ and (using synthetic division) by $x - \frac{b}{a}$.

c. Show how the remainder theorem should be restated if $x - c$ is replaced by $ax - b$.

---

19. a. $x^3 - \frac{14}{5}x^2 + \frac{56}{25}x - \frac{224}{125} + \frac{3396}{625\left(x + \frac{4}{5}\right)}$

b. $c = -\frac{4}{5}; P\left(-\frac{4}{5}\right) = \frac{3396}{625}$

20. a. $x^5 - x^4 + x^3 - x^2 + x - 1 + \frac{2}{x+1}$

b. $c = -1; P(-1) = 2$

Applications

21. 41 °F

22. a. $x^2 + 3x - 18$ square inches

b. $x^2 + 10x + 24$ square inches

Collaborative Learning

23. a. Answers will vary.

$x^2 - \frac{7}{2}x + \frac{13}{4} + \frac{73}{4(2x-1)}$;

$2x^2 - 7x + \frac{13}{2} + \frac{73}{4\left(x - \frac{1}{2}\right)}$

b. $\frac{P(x)}{x - \frac{b}{a}} = \frac{aP(x)}{ax - b}$

c. If a polynomial $P(x)$ is divided by $(ax - b)$, then the remainder will be $P\left(\frac{b}{a}\right)$.

Answers will vary.

Chapter 12 Project

Math in a Box

An activity to demonstrate the use of polynomials in real life.

Suppose you have a piece of cardboard with a length of 32 inches and a width of 20 inches and you want to use it to create a box. You would need to cut a square out of each corner of the cardboard so that you can fold the edges up. But what size squares should you cut? Cutting four small sqaures will make a shorter box. Cutting four large squares will make a taller box.

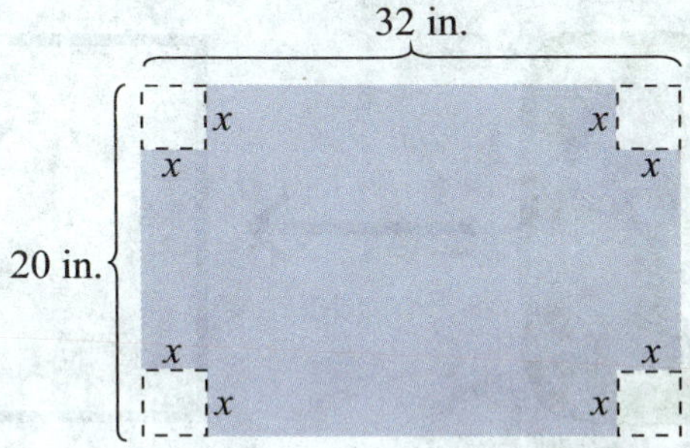

1. Since we haven't determined the size of the square to cut from each corner, let the side length of the square be represented by the variable x. Write a simplified polynomial expression in x and note the degree of the polynomial for each of the following geometric concepts.
   a. The length of the base of the box once the corners are cut out
   b. The width of the base of the box once the corners are cut out
   c. The height of the box
   d. The perimeter of the base of the box
   e. The area of the base of the box
   f. The volume of the box
2. Evaluate the volume expression for the following values of x. (Be sure to include the units of measurement.)
   a. $x = 1$ in. 540 in.^{3}
   b. $x = 2$ in. 896 in.^{3}
   c. $x = 3$ in. 1092 in.^{3}
   d. $x = 3.5$ in. 1137.5 in.^{3}
   e. $x = 6$ in. 960 in.^{3}
   f. $x = 7$ in. 756 in.^{3}
3. Based on your volume calculations for the different values of x in Problem 2, if you were trying to maximize the volume of the box, between what two values of x do you think the maximum will be? Between 3 and 6 inches
4. Using trial and error, see if you can determine the side length x of the square that maximizes the volume of the box. (**Hint:** It will be a value in the interval from Problem 3.) $x = 4$ inches
5. Using the value you found for x in Problem 4, determine the dimensions of the box that maximize its volume.
6. Calculate the volume of the box in Problem 5.

Chapter 12 Project: Math in a Box

1. a. $32 - 2x$; first degree or linear
   b. $20 - 2x$; first degree or linear
   c. x; first degree or linear
   d. $2(32 - 2x) + 2(20 - 2x) = 104 - 8x$; first degree or linear
   e. $640 - 104x + 4x^2$; second degree
   f. $640x - 104x^2 + 4x^3$; third degree
5. Length = 24 inches; width = 12 inches; height = 4 inches
6. 1152 in.^{3}

Chapter 12 Project

Small but Mighty

An activity to investigate the use of polynomials in biology.

In biology, we define a cell as the basic unit that contains the essential molecules of life and that all living things are composed of. In this activity, we will investigate why cells are usually very small.

Using a very simplified model, we can consider a cell as a small square. The sides of the square model the cell's membrane, which is the way the cell interacts with the environment, and the inside of the square models the cell's cytoplasm and nucleus, which are the parts that make the cell function.

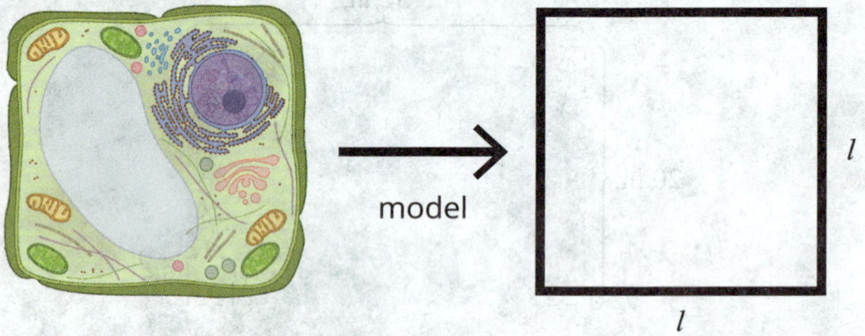

1. What is the perimeter of a cell whose side length is equal to 10 μm (micrometers)?
2. What is the area of this same cell?
3. Compute the perimeter and the area for cells whose side lengths are 20 μm, 40 μm, and 80 μm. Summarize your findings in the table below

Side Length (μm)	Perimeter (μm)	Area (μm^2)
10		
20		
40		
80		

4. If a cell has a side length given by l, write a polynomial that represents its perimeter using the variable l. What is the degree of the polynomial you found?
5. If a cell has a side length given by l, write a polynomial that represents its area using the variable l. What is the degree of the polynomial you found?

We know that the ability of a cell to obtain its nutrients is proportional to its perimeter. That is, when a cell doubles the length of its side, it will be able to obtain twice as many nutrients. We also know that the amount of nutrients needed by a cell to function is proportional to its area. If a cell doubles in area, its need for nutrients also doubles.

6. Looking at the table you completed in Problem 3, determine how many times the need for nutrients increases every time the length of the side of the cell doubled.
7. Use your answer from Problem 6 to explain why there might be a limit to a cell's size. Explain why the need for nutrients grows much faster than the growth in the ability to obtain these nutrients.
8. Do a quick internet search and find out what the smallest and largest cells found in nature are. Does the answer surprise you? Why?

Chapter 12 Project: Small but Mighty

1. $4(10\ \mu\text{m}) = 40\ \mu\text{m}$
2. $(10\ \mu\text{m})^2 = 100\ \mu\text{m}^2$
3.

Side Length (μm)	Perimeter (μm)	Area (μm^2)
10	40	100
20	80	400
40	160	1600
80	320	6400

4. Perimeter: $P(l) = 4l$, degree one
5. Area: $S(l) = l^2$, degree two
6. Every time the side doubles, the volume quadruples. This means that the need for nutrients also quadruples.
7. As the cell doubles its side, it needs 4 times the nutrients as it did before doubling in size. However, the cell can only obtain twice as many nutrients at the new size. This will make it so the cells will eventually not be able to obtain enough nutrients as they grow larger.
8. Smallest cell: small bacterium; Largest cell: ostrich egg. Answers will vary.

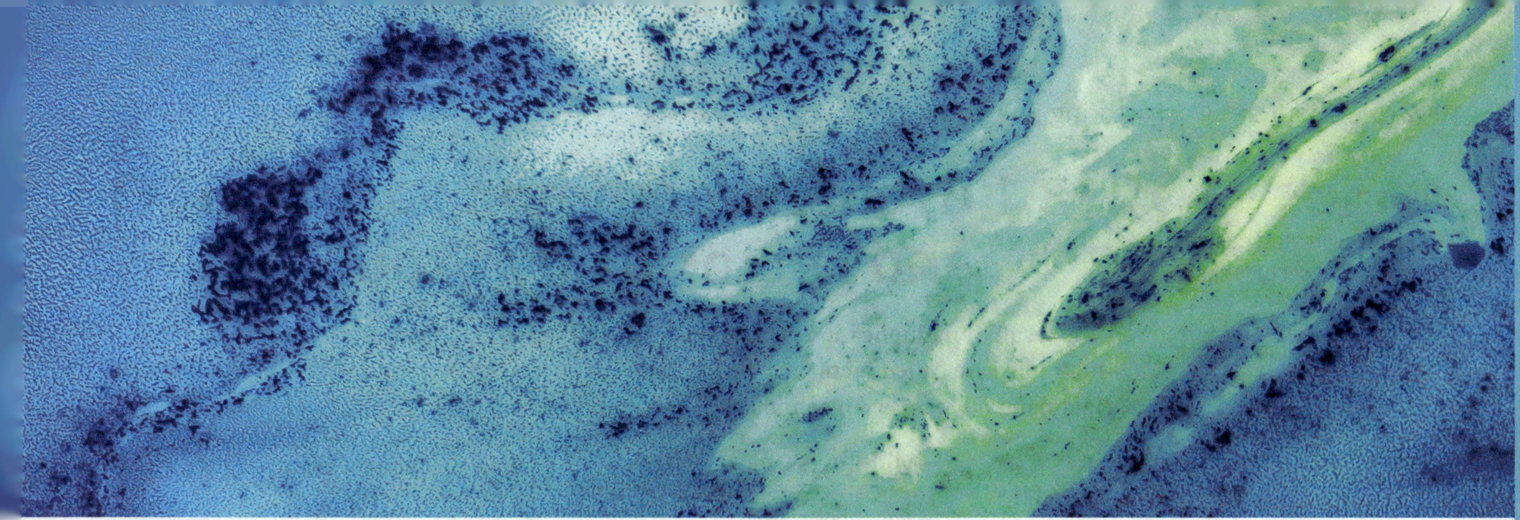

CHAPTER 13

13

Factoring Polynomials

Connections

Quadratic polynomials can be used to describe a variety of different real-life situations. For example, in physics, the path of an object that is affected by gravity after being thrown can be modeled with a quadratic polynomial. Quadratic polynomials also show up in economics in situations related to maximizing profit or minimizing cost. Geometry is another area where quadratic polynomials can be useful.

Sometimes we know how two measurements relate to each other, but we don't know the exact values of these measurements. Perhaps we know that the length of an object should be four inches longer than the width, or three times as long as the width, no matter what the width is. In situations like this, we can use a variable to represent the width and an expression related to that variable to represent the length. Both of those representations are linear expressions, but if you multiply them together, the result is a quadratic polynomial that represents the area of the rectangle.

Suppose you are planning a painting to exhibit at an art show where the only requirement is that the canvas size must be 98 square inches. You decide that your painting would look best if the canvas is twice as tall as it is wide. How would you find the dimensions of the canvas?

Objectives

A. Find the greatest common factor of a set of terms.

B. Factor polynomials by factoring out the greatest common monomial factor.

C. Factor polynomials by grouping.

13.1 Greatest Common Factor (GCF) and Factoring by Grouping

The result of multiplication is called the **product** and the numbers or expressions being multiplied are called **factors** of the product. The reverse of multiplication is called **factoring**. That is, given a product, we want to find the factors.

Multiplying Polynomials	**Factoring Polynomials**
$3x(x+5) = 3x^2 + 15x$	$3x^2 + 15x = 3x(x+5)$
factors ↑↑ ↑ ↑ product	product ↑ ↑↑ ↑ factors

Factoring polynomials relies heavily on the multiplication techniques developed in Sections 12.6 and 12.7. You must remember how to multiply in order to be able to factor. Furthermore, you will find that the skills used in factoring polynomials are necessary when simplifying rational expressions (Chapter 14) and when solving equations. In other words, study this section and Sections 13.2, 13.3, and 13.4, with extra care.

A Greatest Common Factor of a Set of Terms

The **greatest common factor** (**GCF**) of two or more integers is the largest integer that is a factor (or divisor) of all of the integers. For example, the GCF of 30 and 40 is 10. Note that 5 is also a common factor of 30 and 40, but 5 is not the *greatest* common factor. The number 10 is the largest number that will divide into both 30 and 40.

One way of finding the GCF is to use the prime factorization of each number. For example, to find the GCF for 36 and 60, we can first find the prime factorizations.

$$36 = 4 \cdot 9 = 2 \cdot 2 \cdot 3 \cdot 3$$
$$60 = 4 \cdot 15 = 2 \cdot 2 \cdot 3 \cdot 5$$

The common factors are 2, 2, and 3 and their product is the GCF of 36 and 60.

$$\text{GCF} = 2 \cdot 2 \cdot 3 = 12$$

Writing the prime factorizations using exponents gives the following.

$$36 = 2^2 \cdot 3^2 \quad \text{and} \quad 60 = 2^2 \cdot 3 \cdot 5$$

We can see that the GCF is the product of the greatest power of each prime factor that is **common to both numbers**.

$$\text{GCF} = 2^2 \cdot 3 = 12$$

This procedure can be used to find the GCF for any set of integers or algebraic terms with integer exponents.

Finding the GCF of a Set of Terms

1. Find the prime factorization of all integers and integer coefficients.
2. List all the factors that are common to all terms, including variables.
3. Raise each common factor to the smallest exponent on that factor in the list.
4. Multiply these powers to find the GCF.

Note: If there is no common prime factor or variable, then the GCF is 1.

PROCEDURE

Example 1 Finding the GCF

Find the GCF of each set of algebraic terms.

a. $\{30, 45, 75\}$

b. $\{20x^4y, 15x^3y, 10x^5y^2\}$

Solution

a. Find the prime factorization of each number.

$$30 = 2 \cdot 3 \cdot 5,\ 45 = 3^2 \cdot 5, \text{ and } 75 = 3 \cdot 5^2$$

The common factors are 3 and 5 and the greatest power of each that is **common to all the numbers** is 3^1 and 5^1.

Thus, the $\text{GCF} = 3^1 \cdot 5^1 = 15.$

b. Writing each integer coefficient in prime factored form gives the following.

$$\begin{aligned} 20x^4y &= 2^2 \cdot 5 \cdot x^4 \cdot y \\ 15x^3y &= 3 \cdot 5 \cdot x^3 \cdot y \\ 10x^5y^2 &= 2 \cdot 5 \cdot x^5 \cdot y^2 \end{aligned}$$

The common factors are 5 , x, and y and after finding the greatest power of each that is **common to all three terms**, we have 5^1, x^3, and y^1.

Thus, the $\text{GCF} = 5^1 \cdot x^3 \cdot y^1 = 5x^3y.$

Now work margin exercise 1.

1. Find the GCF of each set of algebraic terms.

a. $\{5, 10, 20\}$

b. $\{150xy, 250x^3y^2, 100x^2y^2\}$

B Factoring Out the Greatest Common Monomial Factor

Now consider the polynomial $3n + 15$. We want to write this polynomial as a product of two factors. Since

$$3n + 15 = 3 \cdot n + 3 \cdot 5,$$

Teaching Note

Emphasize the use of the distributive property as the springboard to factoring the common monomial factor from binomials and other polynomials. This will help students understand the need to factor out common binomial factors, as well.

we see that 3 is a common factor of the two terms in the polynomial. By using the distributive property, we can write the following.

$$3n+15=3\cdot n+3\cdot 5=3(n+5)$$

In this way, the polynomial $3n + 15$ has been **factored** into the product of 3 and $(n+5)$.

For a more general approach to finding factors of a polynomial, we can use the quotient rule for exponents.

$$\frac{a^m}{a^n}=a^{m-n}$$

This property is used when dividing terms. For example,

$$\frac{35x^8}{5x^2}=7x^6 \text{ and } \frac{16a^5}{-8a}=-2a^4.$$

To divide a polynomial by a monomial (a procedure that was discussed in Chapter 12), each term in the polynomial is divided by the monomial.

$$\frac{8x^3-14x^2+10x}{2x}=\frac{8x^3}{2x}-\frac{14x^2}{2x}+\frac{10x}{2x}=4x^2-7x+5$$

With practice, this division can be done mentally.

This concept of dividing each term by a monomial is part of finding a monomial factor of a polynomial. Finding the greatest common monomial factor of a polynomial means to **find the GCF of the terms of the polynomial.** This monomial will be one factor, and the sum of the various quotients found by dividing each term by the GCF will be the other factor. Thus, using the quotients just found and the fact that $2x$ is the GCF of the terms,

$$8x^3-14x^2+10x=2x(4x^2-7x+5).$$

We say that $2x$ is **factored out** and that $2x$ and $(4x^2-7x+5)$ are the factors of $8x^3-14x^2+10x$.

Monomials

Recall that a monomial is a one-term polynomial. This term can consist of a number, variables, or a product of numbers and variables. Also recall that terms in a polynomial are separated by addition and subtraction. For example, each of the following is a monomial: 7, p, $3xy^2$, $-15z^4$, $x^3y^4z^5$, and 215.

Teaching Note

You may want to remind your students that they can check their factoring by multiplying the factors. While they may not have the greatest common factor, the product must be the expression being factored.

Factoring Out the GCF

1. Find the GCF of the terms in the polynomial.
2. Divide this monomial factor into each term of the polynomial, resulting in another polynomial factor.

The product of the GCF and this new polynomial factor is a factored form of the original polynomial.

PROCEDURE

Now, let's factor $24x^6-12x^4-18x^3$. The GCF is $6x^3$ and is factored out as follows.

$$\begin{aligned}24x^6-12x^4-18x^3&=6x^3\cdot 4x^3+6x^3(-2x)+6x^3(-3)\\&=6x^3(4x^3-2x-3)\end{aligned}$$

The factoring can be checked by multiplying $6x^3$ and $4x^3 - 2x - 3$. The product should be the expression we started with.

$$6x^3\left(4x^3-2x-3\right)=24x^6-12x^4-18x^3$$

By definition, the GCF of a polynomial will have a positive coefficient. However, if the leading coefficient is negative, we may choose to factor out the negative of the GCF (or $-1 \cdot$ GCF). This technique will leave a positive coefficient for the first term of the other polynomial factor. For example, the GCF for $-10a^5+15a^4$ is $5a^4$ and we can factor as follows.

$$-10a^2+15a=5a^4\left(-2a+3\right)$$

or

$$-10a^2+15a=-5a^4\left(2a-3\right)$$

Both factorizations are correct.

Suppose we factor $8n + 16$ as follows.

$$8n+16=4\left(2n+4\right)$$

Notice that $2n + 4$ has 2 as a common factor. Therefore, we have not factored completely. An expression is **factored completely** if none of its factors can be factored.

Factoring polynomials means finding factors that are integers (other than $+1$) or polynomials with integer coefficients. If this cannot be done, we say that the polynomial is **not factorable**. For example, the polynomials x^2+36 and $3x + 17$ are not factorable.

Example 2 Factoring Out the GCF of a Polynomial

Factor each polynomial by factoring out the greatest common monomial factor.

a. $6n + 30$

b. $x^3 + x$

c. $5x^3-15x^2$

Solution

a. $6n+30=6\cdot n+6\cdot 5=6\left(n+5\right)$

Checking by multiplying gives $6\left(n+5\right)=6n+30,$ the original expression.

b. $x^3+x=x\cdot x^2+x\left(+1\right)=x\left(x^2+1\right)$ $+1$ is the coefficient of x.

Note: It is important to write the coefficient $+1$, otherwise the implication would be that 0 was the coefficient. The product must be equal to the original polynomial.

c. $5x^3-15x^2=5x^2\cdot x+5x^2\left(-3\right)=5x^2\left(x-3\right)$

Now work margin exercise 2.

2. Factor each polynomial by factoring out the greatest common monomial factor.

a. $7n + 21$

b. $3y^3 + y^4$

c. $9x + 54x^2$

3. Factor $6x^2+2x-9$ by factoring out the GCF.

Example 3 Factoring Out the GCF of a Polynomial

Factor $2x^4-3x^2+2$ by factoring out the GCF.

Solution

$2x^4-3x^2+2$

This polynomial has no common monomial factor other than 1. In fact, this polynomial is **not factorable.**

Now work margin exercise 3.

4. Factor $-9a^3-18a+9a^2$ by factoring out the GCF.

Example 4 Factoring Out the GCF of a Polynomial

Factor $-4a^5+2a^3-6a^2$ by factoring out the GCF.

Solution

The GCF is $2a^2$ and we can factor as follows.

$$-4a^5+2a^3-6a^2=2a^2\left(-2a^3+a-3\right)$$

However, the leading coefficient is negative and we can also use a GCF of $-2a^2$.

$$-4a^5+2a^3-6a^2=-2a^2\left(2a^3-a+3\right)$$

Both answers are correct. However, we will see later that having a positive leading coefficient for the polynomial in parentheses may make that polynomial easier to factor.

Now work margin exercise 4.

A polynomial may be in more than one variable. For example, $5x^2y+10xy^2$ is in the two variables x and y. Thus, the GCF may have more than one variable.

$$\begin{aligned}5x^2y+10xy^2&=5xy\cdot x+5xy\cdot 2y\\&=5xy(x+2y)\end{aligned}$$

Similarly,

$$\begin{aligned}4xy^3-2x^2y^2+8xy^2&=2xy^2\cdot 2y+2xy^2(-x)+2xy^2\cdot 4\\&=2xy^2(2y-x+4).\end{aligned}$$

5. Factor each polynomial by factoring out the GCF (or $-1\cdot$GCF).
   a. $10xy+30xy^2$
   b. $2a^2b^2-16ab^2$
   c. $5xz^2+15xz^3-20x^2z$
   d. $-3b^2d+12b^2-15b^3d^2$

Example 5 Factoring Out the GCF of a Multi-Variable Polynomial

Factor each polynomial by factoring out the GCF (or $-1\cdot$GCF).

a. $4ax^3+4ax$

b. $3x^2y^2-6xy^2$

c. $-14by^3-7b^2y+21by^2$

d. $13a^4b^5-3a^2b^9-4a^6b^3$

Solution

a. $4ax^3 + 4ax = 4ax(x^2 + 1)$ Note that $4ax = 1 \cdot 4ax$.

Checking by multiplying gives $4ax(x^2 + 1) = 4ax^3 + 4ax$, the original expression.

b. $3x^2y^2 - 6xy^2 = 3xy^2(x - 2)$

c. $-14by^3 - 7b^2y + 21by^2 = -7by(2y^2 + b - 3y)$

Note that by factoring out a negative term, in this case $-7by$, the leading coefficient in parentheses is positive.

d. $13a^4b^5 - 3a^2b^9 - 4a^6b^3 = a^2b^3(13a^2b^2 - 3b^6 - 4a^4)$

Now work margin exercise 5.

C Factoring by Grouping

Consider the expression

$$y(x+4) + 2(x+4)$$

as the sum of two terms, $y(x+4)$ and $2(x+4)$. Each of these "terms" has the **common binomial factor** $(x+4)$. Factoring out this common binomial factor by using the distributive property gives the following.

$$y(x+4) + 2(x+4) = (x+4)(y+2)$$

Similarly,

$$3(x-2) - a(x-2) = (x-2)(3-a).$$

Now consider the product of $(x+3)(y+5)$.

$$\begin{aligned}(x+3)(y+5) &= x(y+5) + 3(y+5)\\ &= xy + 5x + 3y + 15\end{aligned}$$

While this product has four terms and no like terms, we know the product has two factors, namely $(x+3)$ and $(y+5)$. Factoring polynomials with four or more terms can sometimes be accomplished by grouping the terms and using the distributive property, as in the above discussion and the following examples. Keep in mind that the common factor can be a binomial or other polynomial.

Example 6 Factoring Out a Common Binomial Factor

Factor each polynomial.

a. $3x^2(5x+1) - 2(5x+1)$ **b.** $7a(2x-3) + (2x-3)$

Solution

a. $3x^2(5x+1) - 2(5x+1) = (5x+1)(3x^2 - 2)$

6. Factor each polynomial.

a. $x^2(2x-y) + 2(2x-y)$

b. $6y(x-u) + (x-u)$

b. $7a(2x-3)+(2x-3)=7a(2x-3)+1\cdot(2x-3)$ — 1 is the understood coefficient of $(2x-3)$.

$=(2x-3)(7a+1)$

Now work margin exercise 6.

In Example 6, the common binomial factor was in parentheses. However, many times the expression to be factored is in a form with four or more terms. For example, by multiplying in Example 6a, we get the following expression.

$$3x^2(5x+1)-2(5x+1)=15x^3+3x^2-10x-2$$

The expression $15x^3+3x^2-10x-2$ has four terms with no common monomial factor; yet, we know that it has the two binomial factors $(5x+1)$ and $(3x^2-2)$.

We can find the binomial factors by **grouping**. This means looking for common factors in each group and then looking for common binomial factors. The process is illustrated in the following examples.

7. Factor $xy+2y+6x+12$ by grouping.

Example 7 Factoring Polynomials by Grouping

Factor $xy + 5x + 3y + 15$ by grouping.

Solution

$xy+5x+3y+15=(xy+5x)+(3y+15)$ — Group terms that have a common monomial factor.

$=x(y+5)+3(y+5)$ — Factoring out the monomial factors shows $(y + 5)$ is a common binomial factor

$=(y+5)(x+3)$ — Use the distributive property with $(y + 5)$ as a common factor.

Checking gives $(y+5)(x+3)=xy+5x+3y+15.$

Now work margin exercise 7.

Math Tip

In Example 7, we can use the properties of addition to rearrange and group the terms first:

$$\begin{aligned}&xy+5x+3y+15\\&=(xy+3y)+(5x+15)\\&=y(x+3)+5(x+3)\\&=(x+3)(y+5)\end{aligned}$$

By the commutative property of multiplication, the solutions are both correct.

8. Factor $xy-3y-2x+6$ by grouping.

Example 8 Factoring Polynomials by Grouping

Factor $x^2-xy-5x+5y$ by grouping.

Solution

$$\begin{aligned}x^2-xy-5x+5y&=(x^2-xy)+(-5x+5y)\\&=x(x-y)+5(-x+y)\end{aligned}$$

This does not give a common binomial factor because $(x-y)\neq(-x+y)$. However, these two expressions are **opposites**. Thus we can find a common binomial factor by factoring −5 instead of +5 from the last two terms.

$$x^2 - xy - 5x + 5y = (x^2 - xy) + (-5x + 5y)$$
$$= x(x - y) - 5(x - y)$$
$$= (x - y)(x - 5) \quad \text{Success!}$$

Now work margin exercise 8.

Example 9 Factoring Polynomials by Grouping

Factor $x^2 + ax + 3x + 3y$ by grouping.

Solution

$$x^2 + ax + 3x + 3y = (x^2 + ax) + (3x + 3y)$$
$$= x(x + a) + 3(x + y)$$

But $(x + a) \neq (x + y)$ and there is no common factor. $x^2 + ax + 3x + 3y$ is **not factorable.**

Now work margin exercise 9.

9. Factor $2x + 6 + xy - x^2$ by grouping.

Example 10 Factoring Polynomials by Grouping

Factor $xy + 5x + y + 5$ by grouping.

Solution

$$xy + 5x + y + 5 = (xy + 5x) + (y + 5)$$
$$= x(y + 5) + 1 \cdot (y + 5) \quad \text{1 is the understood coefficient of } (y+5).$$
$$= (y + 5)(x + 1)$$

Now work margin exercise 10.

10. Factor $x^2 + xy + x + y$ by grouping.

Example 11 Factoring Polynomials by Grouping

Factor $5xy + 6uv - 3vy - 10ux$ by grouping.

Solution

This example illustrates that there are times where factoring is easier if the terms are reordered. In the expression $5xy + 6uv - 3vy - 10ux$ there is no common factor in the first two terms. However, the first and third terms have a common factor so we rearrange the terms as follows.

$$5xy + 6uv - 3vy - 10ux = (5xy - 3vy) + (6uv - 10ux)$$
$$= y(5x - 3v) + 2u(3v - 5x)$$

Now we see that $5x - 3v$ and $3v - 5x$ are opposites and we factor out $-2u$ from the last two terms. The result is as follows.

11. Factor $8xy - 6wy - 12xz + 9wz$ by grouping.

$$\begin{aligned}5xy+6uv-3vy-10ux &= (5xy-3vy)+(6uv-10ux)\\ &= y(5x-3v)-2u(-3v+5x)\\ &= y(5x-3v)-2u(5x-3v) \quad \textbf{Note: } 5x-3v=-3v+5x\\ &= (5x-3v)(y-2u)\end{aligned}$$

Now work margin exercise 11.

Margin Exercise Answers

1. a. 5 **b.** $50xy$ **2. a.** $7(n+3)$ **b.** $y^3(3+y)$ **c.** $9x(1+6x)$ **3.** Not factorable **4.** $-9a(a^2+2-a)$ **5. a.** $10xy(1+3y)$ **b.** $2ab^2(a-8)$ **c.** $5xz(z+3z^2-4x)$ **d.** $-3h^2(d-4+5bd^2)$ **6. a.** $(2x-y)(x^2+2)$ **b.** $(x-u)(6y+1)$ **7.** $(x+2)(y+6)$ **8.** $(x-3)(y-2)$ **9.** Not factorable **10.** $(x+y)(x+1)$ **11.** $(4x-3w)(2y-3z)$

13.1 Exercises

Concept Check

Fill-in-the-Blank. Complete each sentence using information found in this section.

1. The result of multiplication is called the ______ and the numbers or expressions being multiplied are called ________ of the product.

2. The reverse of multiplication with polynomials is called ________.

3. GCF stands for ______________. The GCF of a set of numbers is the ________ positive integer that is a factor of all numbers in the set.

4. Factoring polynomials with four or more terms can sometimes be accomplished by _______ terms and using the distributive property.

5. If the leading coefficient in a polynomial is a negative number, you may choose to factor out the _______ of the GCF.

True/False. Determine whether each statement is true or false. If a statement is false, explain how it can be changed so the statement will be true. (**Note:** There may be more than one acceptable change.)

6. When finding the GCF of a polynomial, you need to consider only the coefficients.

7. An expression is factored completely if none of its factors can be factored.

8. One way to find the GCF of a set of numbers is to use the prime factorization of each number.

9. Binomials cannot be factored out of algebraic expressions.

13.1 Exercises

Concept Check

1. product, factors
2. factoring
3. greatest common factor, largest
4. grouping
5. negative / opposite
6. False; Variables need to be considered as well.
7. True
8. True
9. False; Binomials can be common factors.

Practice

1. 5
2. 5
3. 8

Practice

Find the GCF for each set of terms. See Example 1.

1. $\{10, 15, 20\}$
2. $\{25, 30, 75\}$
3. $\{16, 40, 56\}$
4. $\{30, 42, 54\}$
5. $\{9, 14, 22\}$
6. $\{44, 66, 88\}$
7. $\{30x^3, 40x^5\}$
8. $\{15y^4, 25y\}$
9. $\{8a^3, 16a^4, 20a^2\}$
10. $\{36xy, 48xy, 60xy\}$
11. $\{26ab^2, 39a^2b, 52a^2b^2\}$
12. $\{28c^2d^3, 14c^3d^2, 42cd^2\}$
13. $\{45x^2y^2z^2, 75xy^2z^3\}$
14. $\{21a^5b^4c^3, 28a^3b^4c^3, 35a^3b^4c^2\}$

Simplify each expression.

15. $\frac{x^7}{x^3}$
16. $\frac{x^8}{x^3}$
17. $\frac{-8y^3}{2y^2}$
18. $\frac{12x^2}{2x}$
19. $\frac{9x^5}{3x^2}$
20. $\frac{-10x^5}{2x}$
21. $\frac{4x^3y^2}{2xy}$
22. $\frac{21x^4y^3}{-3xy^2}$

Complete the factoring of the polynomial as indicated.

23. $3m + 27 = 3(\quad)$
24. $2x + 18 = 2(\quad)$
25. $5x^2 - 30x = 5x(\quad)$
26. $6y^3 - 24y^2 = 6y^2(\quad)$
27. $13ab^2 + 13ab = 13ab(\quad)$
28. $8x^2y - 4xy = 4xy(\quad)$
29. $-15xy^2 - 20x^2y - 5xy = -5xy(\quad)$
30. $-9m^3 - 3m^2 - 6m = -3m(\quad)$

Factor each polynomial by finding the GCF (or –1 · GCF). See Examples 2 through 5.

31. $11x - 121$
32. $14x + 21$
33. $16y^3 + 12y$
34. $-3x^2 + 6x$
35. $-6ax + 9ay$
36. $4ax - 8ay$
37. $10x^2y - 25xy$
38. $16x^4y - 14x^2y$
39. $-18y^2z^2 + 2yz$
40. $-14x^2y^3 - 14x^2y$
41. $8y^2 - 32y + 8$
42. $5x^2 - 15x - 5$
43. $2xy^2 - 3xy - x$
44. $ad^2 + 10ad + 25a$

4. 6
5. 1
6. 22
7. $10x^3$
8. $5y$
9. $4a^2$
10. $12xy$
11. $13ab$
12. $14cd^2$
13. $15xy^2z^2$
14. $7a^3b^4c^2$
15. x^4
16. x^5
17. $-4y$
18. $6x$
19. $3x^3$
20. $-5x^4$
21. $2x^2y$
22. $-7x^3y$
23. $m + 9$
24. $x + 9$
25. $x - 6$
26. $y - 4$
27. $b + 1$
28. $2x - 1$
29. $3y + 4x + 1$
30. $3m^2 + m + 2$
31. $11(x - 11)$
32. $7(2x + 3)$
33. $4y(4y^2 + 3)$
34. $-3x(x - 2)$
35. $-3a(2x - 3y)$
36. $4a(x - 2y)$
37. $5xy(2x - 5)$
38. $2x^2y(8x^2 - 7)$
39. $-2yz(9yz - 1)$
40. $-14x^2y(y^2 + 1)$
41. $8(y^2 - 4y + 1)$
42. $5(x^2 - 3x - 1)$

43. $x(2y^2-3y-1)$
44. $a(d^2+10d+25)$
45. $4m^2(2x^3-3y+z)$
46. $3t^2x^2(12x^2-15x+8)$
47. $-7x^2z^3(8x^2+14xz+5z^2)$
48. $17x^3y^4(2xy^2-3y+x^2)$
49. $x^4y^2(15+24x^2y^4-32x^3y)$
50. $-3x^2y^3(y+2xy+3)$
51. $(y+3)(7y^2+2)$
52. $(a-7)(6a-5)$
53. $(x-4)(3x+1)$
54. $(x+5)(2x^2+1)$
55. $(x-2)(4x^3-1)$
56. $(x+1)(9a-1)$
57. $(2y+3)(10y-7)$
58. $(x+5)(a+b)$
59. $(x-2)(a-b)$
60. $(x-10)(3a+5b)$
61. $(b+c)(x+1)$
62. $(3+a)(x+y)$
63. $(x^2+6)(x+3)$
64. $(2z^2+3)(z-7)$
65. Not factorable
66. $(x+6y)(x-4)$
67. $(3-b)(x+y)$
68. $(a+3)(x+5y)$
69. $(y-4)(5x+z)$
70. $(x+2z)(1-3y)$
71. $(z^2+3)(a+1)$
72. Not factorable
73. $(6x+1)(a+2)$
74. $(4y+3)(x-1)$
75. $(x+1)(y+1)$
76. $(x-1)(y+1)$
77. $(2y-7z)(5x-y)$
78. Not factorable
79. $(3x-4u)(y-2v)$

45. $8m^2x^3-12m^2y+4m^2z$

46. $36t^2x^4-45t^2x^3+24t^2x^2$

47. $-56x^4z^3-98x^3z^4-35x^2z^5$

48. $34x^4y^6-51x^3y^5+17x^5y^4$

49. $15x^4y^2+24x^6y^6-32x^7y^3$

50. $-3x^2y^4-6x^3y^4-9x^2y^3$

Factor each expression by factoring out the common binomial factor. See Example 6.

51. $7y^2(y+3)+2(y+3)$

52. $6a(a-7)-5(a-7)$

53. $3x(x-4)+(x-4)$

54. $2x^2(x+5)+(x+5)$

55. $4x^3(x-2)-(x-2)$

56. $9a(x+1)-(x+1)$

57. $10y(2y+3)-7(2y+3)$

58. $a(x+5)+b(x+5)$

59. $a(x-2)-b(x-2)$

60. $3a(x-10)+5b(x-10)$

Factor each of the polynomials by grouping. If a polynomial cannot be factored, write "not factorable." See Examples 7 through 11.

61. $bx+b+cx+c$

62. $3x+3y+ax+ay$

63. $x^3+3x^2+6x+18$

64. $2z^3-14z^2+3z-21$

65. $10a^2-5az+2a+z$

66. $x^2-4x+6xy-24y$

67. $3x+3y-bx-by$

68. $ax+5ay+3x+15y$

69. $5xy+yz-20x-4z$

70. $x-3xy+2z-6zy$

71. z^2+3+az^2+3a

72. x^2-5+x^2y+5y

73. $6ax+12x+a+2$

74. $4xy+3x-4y-3$

75. $xy+x+y+1$

76. $xy+x-y-1$

77. $10xy-2y^2+7yz-35xz$

78. $7xy-3y+2x^2-3x$

79. $3xy-4uy-6vx+8uv$

80. $xy+5vy+6ux+30uv$

81. $3ab+4ac+2b+6c$

82. $24y-3yz+2xz-16x$

83. $6ac-9ad+2bc-3bd$

84. $2ac-3bc+6ad-9bd$

Applications

Solve.

85. Bonnie volunteers to bring bags of candy to her child's class for the Halloween party this year. She buys one bag of candy A containing 150 pieces of candy, one bag of candy B containing 180 pieces of candy, and one bag of candy C containing 330 pieces of candy. She needs to use all the candy to create identical treat bags. How many treat bags can Bonnie make so that each one has the same number and variety of candy? How many of each type of candy will be in each bag?

86. The area of a rectangular photo can be represented by the polynomial $15x^2 + 5x$.

a. If $x = 2$ inches, find the area of the photo.

b. Factor the polynomial to find a variable expression for the length and width of the photo.

c. If $x = 2$ inches, use the answer from part b. to find the length and the width of the photo.

d. Find the area of the photo by multiplying the length and width values from part c.

e. Are the answers from parts a. and d. the same? Explain why or why not.

87. A circus performer is shot vertically into the air with an initial velocity of 48 feet per second. The height of the performer above the ground in feet can be described by the polynomial $48x - 16x^2$ after x seconds.

a. Find the height of the circus performer after 2 seconds.

b. Factor the polynomial $48x - 16x^2$.

c. Use the factored form of the polynomial from part b. to find the height of the circus performer after 2 seconds.

d. Are the answers from parts a. and c. the same? Explain why or why not.

Writing & Thinking

88. Explain why the GCF of $-3x^2 + 3$ is 3 and not -3.

80. $(x+5v)(y+6u)$

81. Not factorable

82. $(2x-3y)(z-8)$

83. $(2c-3d)(3a+b)$

84. $(2a-3b)(3d+c)$

Applications

85. She can make 30 identical treat bags, each containing 5 pieces of candy A, 6 pieces of candy B, and 11 pieces of candy C.

86. **a.** 70 square inches

b. $5x(3x+1)$; $5x$ and $3x+1$ are the length and width (or the width and length)

c. 10 inches by 7 inches

d. 70 square inches

e. Yes. They are equivalent expressions.

87. **a.** 32 feet **b.** $16x(3-x)$

c. 32 feet

d. Yes. They are equivalent expressions.

Writing & Thinking

88. Although both can be factored out of $-3x^2 + 3$, 3 is greater than -3 making it the greatest common factor.

Objectives

A. Factor trinomials with leading coefficients of 1 (of the form $x^2 + bx + c$).

B. Factor trinomials by first factoring out a common monomial factor.

13.2 Factoring Trinomials: $x^2 + bx + c$

In Section 12.6, we learned to use the FOIL method to multiply two binomials. In many cases, the simplified form of the product was a trinomial. In this section, we will learn to factor trinomials by reversing the FOIL method. In particular, we will focus on factoring trinomials in one variable with leading coefficient 1.

A Factoring Trinomials with Leading Coefficients of 1

Using the FOIL method to multiply $(x+5)$ and $(x+3)$, we find the following.

$$(x+5)(x+3) = \overset{F}{x^2} + \overset{O}{3x} + \overset{I}{5x} + \overset{L}{3\cdot 5} = x^2 + \underset{3+5}{8x} + \underset{3\cdot 5}{15}$$

We see that the leading coefficient (the coefficient of x^2) is 1, the coefficient 8 is the sum of 5 and 3, and the constant term 15 is the product of 5 and 3.

More generally, we can write the following.

$$\begin{aligned}(x+a)(x+b) &= x^2 + bx + ax + ab \\ &= x^2 + \underset{\text{sum of constants } a \text{ and } b}{(b+a)}x + \underset{\text{product of constants } a \text{ and } b}{ab}\end{aligned}$$

Now given a trinomial with leading coefficient 1, we want to find the binomial factors, if any. Reversing the relationship between a and b, as shown above, we can proceed as follows.

To factor a trinomial with leading coefficient 1, find two factors of the constant term whose sum is the coefficient of the middle term. (If these factors do not exist, the trinomial is **not factorable**.)

For example, to factor $x^2 + 11x + 30$, we need positive factors of +30 whose sum is +11.

Positive Factors of 30				Sums of These Factors
1	·	30	→	1 + 30 = 31
2	·	15	→	2 + 15 = 17
3	·	10	→	3 + 10 = 13
5	·	6	→	5 + 6 = 11

Now, because $5 \cdot 6 = 30$ and $5 + 6 = 11$, we have

$$x^2 + 11x + 30 = (x+5)(x+6).$$

Example 1 Factoring Trinomials with Leading Coefficients of 1

Factor: $x^2 + 8x + 12$

Solution

12 has three pairs of positive integer factors. Of these 3 pairs, only 2 + 6 is equal to 8.

Factors of 12				
1	·	12		1 + 12 = 13
2	·	6	→	2 + 6 = 8
3	·	4		3 + 4 = 7

Thus, $x^2 + 8x + 12 = (x+2)(x+6)$.

Note: If the middle term had been $-8x$, then we would have wanted pairs of negative integer factors to find a sum of -8. In other words, we would have found that $x^2 - 8x + 12 = (x-2)(x-6)$.

Now work margin exercise 1.

1. Factor: $x^2 + 10x + 21$

Example 2 Factoring Trinomials with Leading Coefficients of 1

Factor: $y^2 - 8y - 20$

Solution

We want a pair of integer factors of -20 whose sum is -8. In this case, because the product is negative, one of the factors must be positive and the other negative.

Factors of −20				
−1	·	20		$-1+20=19$
1	·	−20		$1+(-20)=-19$
−2	·	10		$-2+10=8$
2	·	−10	→	$2+(-10)=-8$
−4	·	5		$-4+5=1$
4	·	−5		$4+(-5)=-1$

We have listed all the pairs of integer factors of -20. You can see that 2 and -10 are the only two whose sum is -8. Thus, listing all the pairs is not necessary. This stage is called the **trial-and-error stage**. That is, you can **try** different pairs (mentally or by making a list) until you find the correct pair. If such a pair does not exist, the polynomial is **not factorable**.

In this case, we have $y^2 - 8y - 20 = (y+2)(y-10)$.

Note that, by the commutative property of multiplication, the order of the factors does not matter. That is, we can also write

$$y^2 - 8y - 20 = (y-10)(y+2).$$

Now work margin exercise 2.

2. Factor: $x^2 - x - 20$

3. Factor: $x^2 - 5x + 6$

Completion Example 3 Factoring Trinomials

Factor: $y^2 + 10y + 16$

Solution

$y^2 + 10y + 16 = (y + \underline{\quad 8 \quad})(y + \underline{\quad 2 \quad})$

Now work margin exercise 3.

Math Tip

If c is positive, it can be helpful to ask yourself "What factors multiply to c and *add* to b?"

If c is negative, it can be helpful to ask yourself "What factors multiply to c and *subtract* to b?"

Once you determine the factors, you then need to ask yourself "Which of the factors should be negative and which should be the positive?"

Factoring Trinomials of the Form $x^2 + bx + c$

To factor $x^2 + bx + c$, if possible, find a pair of integer factors of c whose sum is b.

1. If c is positive, then both factors must have the same sign.
   a. Both will be positive if b is positive.

      Example: $x^2 + 5x + 4 = (x+4)(x+1)$

   b. Both will be negative if b is negative.

      Example: $x^2 - 5x + 4 = (x-4)(x-1)$

2. If c is negative, then one factor must be positive and the other negative.

   Examples: $x^2 + 6x - 7 = (x+7)(x-1)$ and $x^2 - 6x - 7 = (x-7)(x+1)$

PROCEDURE

B Finding a Common Monomial Factor and Then Factoring

If a trinomial does not have a leading coefficient of 1, then we look for a common monomial factor. If there is a common monomial factor, factor out this common monomial factor first and then factor the remaining trinomial factor, if possible. (We will discuss factoring trinomials with leading coefficients other than 1 in the next section.) A polynomial is **completely factored** if none of its factors can be factored further.

4. Completely factor each trinomial.

   a. $7y^3 + 35y^2 - 42y$

   b. $11x^3y + 22x^2y - 33xy$

Example 4 Finding a Common Monomial Factor

Completely factor each trinomial by first factoring out the GCF in the form of a common monomial factor.

a. $5x^3 - 15x^2 + 10x$

b. $10y^5 - 20y^4 - 80y^3$

Solution

a. First factor out the GCF, $5x$.

$5x^3 - 15x^2 + 10x = 5x(x^2 - 3x + 2)$ Factored, but not completely factored

Now factor the trinomial $x^2 - 3x + 2$. Look for factors of +2 that add up to –3. Because $(-1)(-2) = +2$ and $(-1)+(-2) = -3,$ we have

$$\begin{aligned} 5x^3 - 15x^2 + 10x &= 5x\left(x^2 - 3x + 2\right) \\ &= 5x(x-1)(x-2). \quad \text{Completely factored} \end{aligned}$$

Check

The factoring can be checked by multiplying the factors.

$$\begin{aligned} 5x(x-1)(x-2) &= 5x\left(x^2 - 2x - x + 2\right) \\ &= 5x\left(x^2 - 3x + 2\right) \\ &= 5x^3 - 15x^2 + 10x \quad \text{The original expression} \end{aligned}$$

b. First factor out the GCF, $10y^3$.

$$10y^5 - 20y^4 - 80y^3 = 10y^3\left(y^2 - 2y - 8\right)$$

Now factor the trinomial $y^2 - 2y - 8$. Look for factors of –8 that add up to –2. Because $(-4)(+2) = -8$ and $(-4)+(+2) = -2,$ we have

$$10y^5 - 20y^4 - 80y^3 = 10y^3\left(y^2 - 2y - 8\right) = 10y^3(y-4)(y+2).$$

The factoring can be checked by multiplying the factors.

Now work margin exercise 4.

> **Math Tip**
>
> It is a common error to accidentally discard the common monomial factor when moving onto the next step of factoring. The best way to prevent this is to make sure you continue to write the common monomial factor on every step of the equation (as shown in Example 4). Be sure to check your work to make sure the common monomial factor wasn't discarded.

Completion Example 5 Finding a Common Monomial Factor

Completely factor $5a^2 + 25a - 180$ by first factoring out the GCF in the form of a common monomial factor.

Solution

$$\begin{aligned} 5a^2 + 25a - 180 &= 5\left(a^2 + \underline{\ 5\ }\,a - \underline{\ 36\ }\right) \\ &= 5\left(a + \underline{\ 9\ }\right)\left(a - \underline{\ 4\ }\right) \end{aligned}$$

Now work margin exercise 5.

5. Completely factor $6x^2 + 12x - 48$ by first factoring out the GCF in the form of a common monomial factor.

Identifying Polynomials That Are Not Factorable

When factoring polynomials, always look for a common monomial factor first. Then, if there is one, remember to include this common monomial factor as part of the answer. Not all polynomials are factorable. For example, no matter which combinations are tried, $x^2 + 3x + 4$ does not have two binomial factors with integer coefficients. (There are no factors of +4 that will add to +3.) We say that the polynomial is **not factorable**. A polynomial is not factorable if it cannot be factored as a product of polynomials with integer coefficients.

13.2 Exercises

Concept Check

1. constant, middle
2. trial-and-error
3. positive
4. common
5. different
6. True
7. False; One factor is negative, one is positive.
8. True
9. True

Practice

1. $\{1, 15\}, \{-1, -15\}, \{3, 5\}, \{-3, -5\}$
2. $\{1, 12\}, \{-1, -12\}, \{2, 6\}, \{-2, -6\}, \{4, 3\}, \{-4, -3\}$
3. $\{1, 20\}, \{-1, -20\}, \{4, 5\}, \{-4, -5\}, \{2, 10\}, \{-2, -10\}$
4. $\{1, 30\}, \{-1, -30\}, \{6, 5\}, \{-6, -5\}, \{15, 2\}, \{-15, -2\}, \{10, 3\}, \{-10, -3\}$
5. $\{1, -6\}, \{6, -1\}, \{2, -3\}, \{3, -2\}$
6. $\{1, -7\}, \{7, -1\}$
7. $\{1, 16\}, \{-1, -16\}, \{4, 4\}, \{-4, -4\}, \{8, 2\}, \{-8, -2\}$
8. $\{1, 18\}, \{-1, -18\}, \{2, 9\}, \{-2, -9\}, \{3, 6\}, \{-3, -6\}$
9. $\{1, -10\}, \{10, -1\}, \{5, -2\}, \{2, -5\}$
10. $\{1, -25\}, \{-1, 25\}, \{5, -5\}$
11. 4, 3
12. 25, 1
13. −7, 2
14. −6, 5
15. 8, −1

Completion Example Answers

3. $(y+8)(y+2)$ **5.** $5(a^2+5a-36)=5(a+9)(a-4)$

Margin Exercise Answers

1. $(x+3)(x+7)$ **2.** $(x-5)(x+4)$ **3.** $(x-3)(x-2)$ **4. a.** $7y(y-1)(y+6)$ **b.** $11xy(x+3)(x-1)$ **5.** $6(x+4)(x-2)$

13.2 Exercises

Concept Check

Fill-in-the-Blank. Complete each sentence using information found in this section.

1. To factor a trinomial that has 1 as its leading coefficient, find two factors of the ________ term whose sum is the coefficient of the ________ term.
2. When listing all the pairs of factors for a particular term, the ______-______-______ method is being used.
3. When factoring trinomials with leading coefficient 1, if the constant is __________, then both factors have the same sign.
4. If the leading coefficient of a trinomial is not one, the first step in factoring is to look for a/an ______ monomial factor to factor out.
5. When factoring trinomials with leading coefficient 1, if the constant term is negative, then the factors of that constant have __________ sign(s).

True/False. Determine whether each statement is true or false. If a statement is false, explain how it can be changed so the statement will be true. (**Note:** There may be more than one acceptable change.)

6. In a trinomial such as $x^2 - 5x + 4$, one would need to find two factors of 4 whose sum is negative 5.
7. In factoring a trinomial with leading coefficient 1, if the constant term is negative, then both factors must be negative.
8. The first step in factoring a trinomial is to look for a common monomial factor.
9. For a trinomial with leading coefficient 1, if no pair exists whose product is the constant and whose sum is the middle term's coefficient, then the trinomial is not factorable.

Practice

List all pairs of integer factors for each given integer. Remember to include negative integers as well as positive integers.

1. 15

2. 12

3. 20

4. 30

5. −6

6. −7

7. 16

8. 18

9. −10

10. −25

Find the pair of integers whose product is the first integer and whose sum is the second integer.

11. 12, 7

12. 25, 26

13. −14, −5

14. −30, −1

15. −8, 7

16. −40, 6

17. 36, −12

18. 16, −10

19. 20, −9

20. 4, −5

Complete each factorization as indicated.

21. $x^2+6x+5=(x+5)(\quad)$

22. $y^2-7y+6=(y-1)(\quad)$

23. $p^2-9p-10=(p+1)(\quad)$

24. $m^2+4m-45=(m-5)(\quad)$

25. $a^2+12a+36=(a+6)(\quad)$

26. $n^2-2n-3=(n-3)(\quad)$

Completely factor each trinomial. If a trinomial cannot be factored, write "not factorable." See Examples 1 through 5.

27. x^2-x-12

28. $x^2-6x-27$

29. y^2+y-30

30. $x^2+6x-36$

31. m^2+3m-1

32. $x^2+3x-18$

33. $x^2-8x+16$

34. $a^2+10a+25$

35. $x^2+7x+12$

36. a^2+a+2

37. y^2-3y+2

38. $y^2-14y+24$

39. x^2+3x+5

40. $y^2+12y+35$

41. x^2-x-72

42. y^2+8y+7

43. $z^2-15z+54$

44. $a^2+4a-21$

45. x^3+10x^2+21x

46. x^3+8x^2+15x

47. $5x^2-5x-60$

48. $6x^2+24x+18$

49. $10y^3-10y^2-60y$

50. $7y^3-70y^2+168y$

51. $4p^4+36p^3+32p^2$

52. $15m^5-30m^4+15m^3$

53. $2x^4-14x^3-36x^2$

54. $3y^6+33y^5+90y^4$

55. $2x^2-2x-72$

56. $3x^2-18x+30$

16. 10, −4

17. −6, −6

18. −8, −2

19. −5, −4

20. −1, −4

21. $x+1$

22. $y-6$

23. $p-10$

24. $m+9$

25. $a+6$

26. $n+1$

27. $(x-4)(x+3)$

28. $(x-9)(x+3)$

29. $(y+6)(y-5)$

30. Not factorable

31. Not factorable

32. $(x+6)(x-3)$

33. $(x-4)(x-4)$

34. $(a+5)(a+5)$

35. $(x+4)(x+3)$

36. Not factorable

37. $(y-1)(y-2)$

38. $(y-12)(y-2)$

39. Not factorable

40. $(y+7)(y+5)$

41. $(x+8)(x-9)$

42. $(y+7)(y+1)$

43. $(z-6)(z-9)$

44. $(a-3)(a+7)$

45. $x(x+7)(x+3)$

46. $x(x+5)(x+3)$

47. $5(x-4)(x+3)$

48. $6(x+3)(x+1)$

49. $10y(y-3)(y+2)$

50. $7y(y-6)(y-4)$

51. $4p^2(p+1)(p+8)$

52. $15m^3(m-1)(m-1)$

53. $2x^2(x-9)(x+2)$

54. $3y^4(y+6)(y+5)$

55. $2(x^2 - x - 36)$
56. $3(x^2 - 6x + 10)$
57. $2a^2(a-10)(a+6)$
58. $2a^2(a+9)(a+3)$
59. $3y^3(y-8)(y+1)$
60. $4y^3(y+6)(y+1)$
61. $x(x-2)(x-8)$
62. $x(x-3)(x+1)$
63. $5(a^2+2a-6)$
64. $6(a^2+4a+2)$
65. $20a^2(a+1)(a+1)$
66. $6x^2(x-1)(x-$

Applications

67. Base = $x + 48$; Height = x
68. Base = x; Height = $x + 28$
69. $x + 5$
70. $x + 3$
71. a. 0 feet
b. $-16(t-3)(t+2)$
c. 0 feet
d. Yes. They are equivalent expressions.

57. $2a^4 - 8a^3 - 120a^2$

58. $2a^4 + 24a^3 + 54a^2$

59. $3y^5 - 21y^4 - 24y^3$

60. $4y^5 + 28y^4 + 24y^3$

61. $x^3 - 10x^2 + 16x$

62. $x^3 - 2x^2 - 3x$

63. $5a^2 + 10a - 30$

64. $6a^2 + 24a + 12$

65. $20a^4 + 40a^3 + 20a^2$

66. $6x^4 - 12x^3 + 6x^2$

Applications

Solve.

67. The area of a triangle is $\frac{1}{2}$ the product of its base and its height. If the area of the triangle shown is given by the function $A(x) = \frac{1}{2}x^2 + 24x$, find representations for the lengths of its base and its height (where the base is longer than the height).

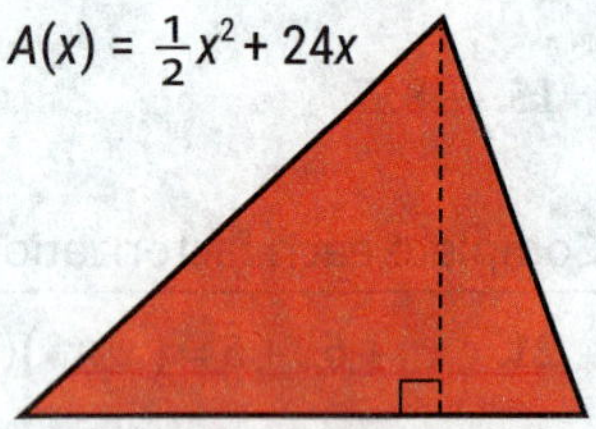

68. The area of a triangle is $\frac{1}{2}$ the product of its base and its height. If the area of the triangle shown is given by the function $A(x) = \frac{1}{2}x^2 + 14x$, find representations for the lengths of its base and its height (where the height is longer than the base).

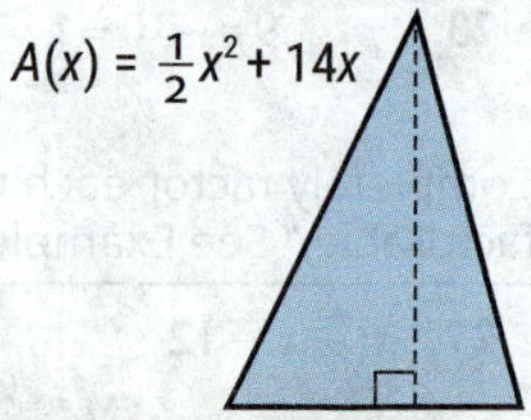

69. The area of the rectangle shown is given by the polynomial function $A(x) = 4x^2 + 20x$. If the width of the rectangle is $4x$, what is the length?

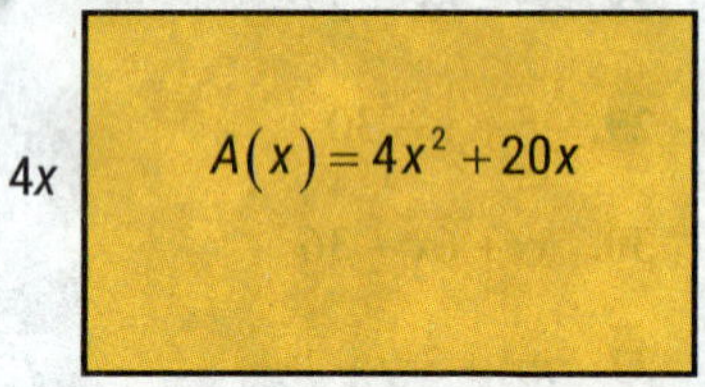

70. The area of the rectangle shown is given by the polynomial function $A(x) = x^2 + 11x + 24$. If the length of the rectangle is $(x+8)$, what is the width?

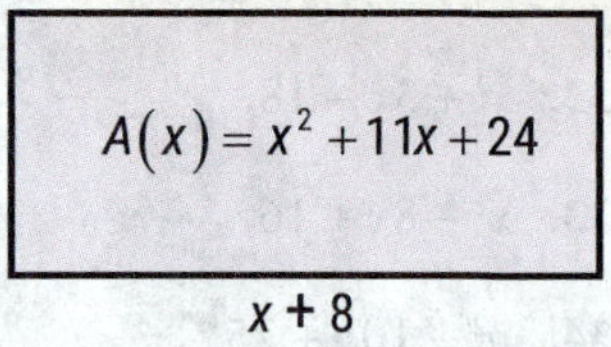

71. A ball is thrown upward from an initial height of 96 feet with an initial velocity of 16 feet per second. After t seconds, the height of the ball can be described by the polynomial $-16t^2 + 16t + 96$.

a. What is the height of the ball after 3 seconds?

b. Completely factor the polynomial $-16t^2 + 16t + 96$.

c. Use the factored form of the polynomial from part b. to find the height of the ball after 3 seconds.

d. Are the answers from parts a. and c. the same? Why do you think this is?

72. A large call center determines that the average number of calls they receive per hour of the day can be modeled by the polynomial $-x^2 + 25x - 100$, where x is the hour of the day, 1 through 24.

a. Factor the polynomial completely.

b. If the average number of calls at a certain time of day equals 26, write an equation using the polynomial given to demonstrate this fact.

c. Rewrite the equation in part b. so that all terms are on the left side of the equation and zero is on the right.

d. Factor the expression on the left side of the equation from part c.

Writing & Thinking

73. Discuss, in your own words, how the sign of the constant term determines what signs will be used in the factors when factoring trinomials.

72. a. $-1(x-5)(x-20)$

b. $-x^2 + 25x - 100 = 26$

c. $-x^2 + 25x - 126 = 0$

d. $-1(x-7)(x-18)$

Writing & Thinking

73. If the sign of the constant term is positive, the signs in the factors will both be positive or both be negative. If the sign of the constant term is negative, the sign in one factor will be positive and the sign in the other factor will be negative.

Objectives

A. Use the trial-and-error method to factor trinomials.

B. Use the *ac*-method to factor trinomials.

13.3 Factoring Trinomials: $ax^2 + bx + c$

Recall that **second-degree trinomials** in the variable x have the general form

$$ax^2 + bx + c,$$

where the coefficients a, b, and c are real numbers and $a \neq 0$. In this section, we will discuss two methods of factoring trinomials of the form $ax^2 + bx + c$ in which the coefficients are restricted to integers. These two methods are the **trial-and-error method** (a reverse of the **FOIL** method of multiplication) and the ***ac*-method** (a form of **grouping**).

A The Trial-and-Error Method of Factoring

To help in understanding this method, we need to review the FOIL method of multiplying two binomials.

$$(2x+5)(3x+1) = 6x^2 + 17x + 5$$

F L — F O I L

$$(2x+5)(3x+1) = 2x\cdot 3x + 2x\cdot 1 + 5\cdot 3x + 5\cdot 1$$

$$= 6x^2 + 17x + 5$$

F The product of the **first** two terms is $6x^2$.

O, **I** The sum of the **inner** and **outer** products is $17x$.

L The product of the **last** two terms is 5.

Now, consider the problem of factoring $6x^2 + 23x + 7$ as the product of two binomials.

$\mathbf{F} = 6x^2$ $\quad \mathbf{L} = +7$

$$6x^2 + 23x + 7 = (\quad\quad)(\quad\quad)$$

Now we use various combinations for **F** and **L** in the **trial-and-error method** as follows.

1. List all the possible combinations of factors of $6x^2$ and $+7$ in their respective **F** and **L** positions. (See the following list.)
2. Check the sum of the products in the **O** and **I** positions until you find the sum to be $+23x$.
3. If none of these sums is $+23x$, the trinomial is not factorable.

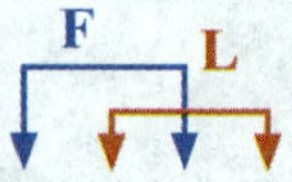

a. $(6x+1)(x+7)$

b. $(6x+7)(x+1)$

c. $(3x+1)(2x+7)$

d. $(3x+7)(2x+1)$

e. $(6x-1)(x-7)$

f. $(6x-7)(x-1)$

g. $(3x-1)(2x-7)$

h. $(3x-7)(2x-1)$

We really don't need to check these last four because the **O** and **I** would be negative, and we are looking for $+23x$. In this manner, the trial-and-error method is more efficient than it first appears to be.

Investigating only the possibilities in the list with positive constants, we need to check the sums of the outer (O) and inner (I) products to find $+23x$.

a. $(6x+1)(x+7): \mathbf{O}+\mathbf{I} = 42x + x = 43x$

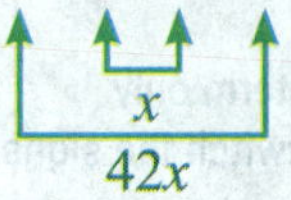

b. $(6x+7)(x+1)\ : \mathbf{O}+\mathbf{I} = 6x + 7x = 13x$

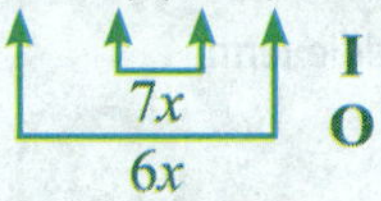

c. $(3x+1)(2x+7): \mathbf{O}+\mathbf{I} = 21x + 2x = 23x$ ← We found $23x$! With a little luck we could have found this first.

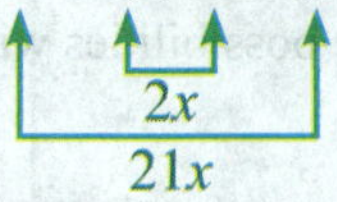

The correct factors, $(3x+1)$ and $(2x+7)$, have been found, so we don't need to take the time to try the next product in the list, $(3x+7)(2x+1)$. Thus, even though the list of possibilities of factors may be long, the actual time involved may be quite short if the correct factors are found early in the trial-and-error method. So we have

$$6x^2 + 23x + 7 = (3x+1)(2x+7).$$

Look at the constant term to determine which signs to use for the constants in the factors. The following guidelines will help limit the trial-and-error search.

Guidelines for the Trial-and-Error Method

1. If the sign of the constant term is positive (+), the signs in both factors will be the same, either both positive or both negative.
2. If the sign of the constant term is negative (–), the signs in the factors will be different, one positive and one negative.

1. Use the trial-and-error method to factor each polynomial.

   a. $x^2+8x+12$

   b. $8u^2-2u-21$

Example 1 Using the Trial-and-Error Method

Use the trial-and-error method to factor each polynomial.

a. x^2+6x+5 **b.** $4x^2-4x-15$ **c.** $6a^2-31a+5$

Solution

a. Since the middle term is $+6x$ and the constant is 5, we know that the two factors of 5 must both be positive, +5 and +1.

$$x^2+6x+5=(x+5)(x+1)$$

5x

x

b. For **F**: $4x^2=4x\cdot x$ and $4x^2=2x\cdot 2x$

For **L**: $-15=-15\cdot 1$, $-15=-1\cdot 15$, $-15=-3\cdot 5$, and $-15=-5\cdot 3$

Trials:

$$(2x-15)(2x+1)$$

−30x

2x

$2x-30x=-28x$ is the wrong middle term.

$$(2x-3)(2x+5)$$

−6x

10x

$10x-6x=+4x$ is the wrong middle term only because the sign is wrong. So just switch the signs and the factors will be right.

$$(2x+3)(2x-5)$$

6x

−10x

$-10x+6x=-4x$ is the right middle term.

Now that we have the answer, there is no need to try all the possibilities with $(4x\quad)(x\quad)$.

c. Since the middle term is $-31a$ and the constant is +5, we know that the two factors of 5 must both be negative, −5 and −1. We try **F** $=6a^2=6a\cdot a$.

$$6a^2-31a+5=(6a-1)(a-5) \qquad -30a-a=-31a$$

−a

−30a

We found the correct factors on the first try.

Now work margin exercise 1.

2. Completely factor each polynomial.

   a. $8x^3-12x^2+4x$

   b. $21x^3+49x^2-7x$

Example 2 Factoring Trinomials

Completely factor each polynomial. Be sure to look first for the greatest common monomial factor.

a. $6x^3-8x^2+2x$ **b.** $-2x^2-x+6$ **c.** $10x^3+5x^2+5x$

Solution

a. To begin, factor out the common monomial $2x$.

$$6x^3 - 8x^2 + 2x = 2x\left(3x^2 - 4x + 1\right)$$

Now, factor the trinomial $3x^2 - 4x + 1$. Since the middle term is $-4x$ and the constant is $+1$, we know that the two factors of 1 must both be negative, -1 and -1. We also know that $\mathbf{F} = 3x^2 = 3x \cdot x$. This gives

$$3x^2 - 4x + 1 = (3x - 1)(x - 1).$$ $-3x - x = -4x$

$-x$

$-3x$

Thus, $6x^3 - 8x^2 + 2x = 2x\left(3x^2 - 4x + 1\right) = 2x(3x - 1)(x - 1)$. Be careful to include the monomial term in the answer.

Check

Check the factorization by multiplying.

$$\begin{aligned} 2x(3x-1)(x-1) &= 2x\left(3x^2 - 3x - x + 1\right) \\ &= 2x\left(3x^2 - 4x + 1\right) \\ &= 6x^3 - 8x^2 + 2x \end{aligned}$$

The original polynomial

b. $$\begin{aligned} -2x^2 - x + 6 &= -1\left(2x^2 + x - 6\right) \\ &= -1(2x - 3)(x + 2) \end{aligned}$$

Notice that factoring out -1 gives a positive leading coefficient for the trinomial.

c. $10x^3 + 5x^2 + 5x = 5x\left(2x^2 + x + 1\right)$

Now consider the trinomial: $2x^2 + x + 1 = (2x + ?)(x + ?)$.

The factors of $+1$ need to be $+1$ and $+1$, but $(2x + 1)(x + 1) = 2x^2 + \underline{3x} + 1$.

So there is no way to factor and get a middle term of $+x$ for the product. This trinomial, $2x^2 + x + 1$, is **not factorable**.

We have

$$10x^3 + 5x^2 + 5x = 5x\left(2x^2 + x + 1\right).$$

Factored completely

Now work margin exercise 2.

> **Note**
>
> Remember, *to factor completely* means to find factors of the polynomial, none of which are themselves factorable. Thus,
>
> $$2x^2 + 12x + 10 = (2x + 10)(x + 1)$$
>
> is not factored completely because $2x + 10 = 2(x + 5)$.
>
> We could write
>
> $$\begin{aligned} 2x^2 + 12x + 10 &= (2x + 10)(x + 1) \\ &= 2(x + 5)(x + 1). \end{aligned}$$
>
> This problem can be avoided by first factoring out the GCF (in this case, 2).

B The *ac*-Method of Factoring

The ***ac*-method** of factoring is in reference to the coefficients a and c in the general form $ax^2 + bx + c$ and involves the method of factoring by grouping, which was discussed in Section 13.1. The method is best explained by analyzing an example and explaining each step as follows.

We consider the problem of factoring the trinomial $2x^2 + 9x + 10$, where $a = 2$, $b = 9$, and $c = 10$.

Analysis of Factoring by the *ac*-Method

	General Method	Example
	ax^2+bx+c	$2x^2+9x+10$
Step 1:	Multiply $a \cdot c$.	Multiply $2 \cdot 10 = 20$.
Step 2:	Find two integers whose product is *ac* and whose sum is *b*. If this is not possible, then the trinomial is **not factorable**.	Find two integers whose product is 20 and whose sum is 9. In this case, $4 \cdot 5 = 20$ and $4+5=9$.
Step 3:	Rewrite the middle term (*bx*) using the two numbers found in Step 2 as coefficients.	Rewrite the middle term $(+9x)$ using +4 and +5 as coefficients. $2x^2+9x+10=2x^2+4x+5x+10$
Step 4:	Factor by grouping the first two terms and the last two terms.	Factor by grouping the first two terms and the last two terms. $2x^2+4x+5x+10=(2x^2+4x)+(5x+10)$ $=2x(x+2)+5(x+2)$
Step 5:	Factor out the common binomial factor. This will give two binomial factors of the trinomial ax^2+bx+c.	Factor out the common binomial factor $(x+2)$. Thus, $2x^2+9x+10=2x^2+4x+5x+10$ $=(2x^2+4x)+(5x+10)$ $=2x(x+2)+5(x+2)$ $=(x+2)(2x+5)$.

PROCEDURE

3. Use the *ac*-method to factor $3a^2+14a+8$.

Example 3 Using the *ac*-Method

Use the *ac*-method to factor $3x^2+19x+6$.

Solution

$a=3,\ b=19,\ c=6$

Step 1: Find the product *ac*: $3 \cdot 6 = 18$.

Step 2: Find two integers whose product is 18 and whose sum is 19.

$$(+1)(+18)=18 \text{ and } (+1)+(+18)=+19$$

1 and 18 are the desired coefficients.

Step 3: Rewrite the middle term $(+19x)$ as $+\ 1x + 18x$.

$$3x^2+19x+6=3x^2+1x+18x+6$$

Step 4: Factor by grouping.

$$3x^2 + 19x + 6 = 3x^2 + 1x + 18x + 6$$
$$= x(3x+1) + 6(3x+1)$$

Step 5: Factor out the common binomial factor $(3x+1)$.

$$3x^2 + 19x + 6 = 3x^2 + 1x + 18x + 6$$
$$= x(3x+1) + 6(3x+1)$$
$$= (3x+1)(x+6)$$

Note that in Step 3, we could have written $+19x$ as $+18x + 1x$. Try this to convince yourself that the result will be the same two factors.

Now work margin exercise 3.

Example 4 Using the *ac*-Method

Use the *ac*-method to factor $12y^3 - 26y^2 + 12y$.

Solution

First factor out the greatest common factor $2y$.

$$12y^3 - 26y^2 + 12y = 2y(6y^2 - 13y + 6)$$

Now factor the trinomial $6y^2 - 13y + 6$ with $a = 6$, $b = -13$, and $c = 6$.

Step 1: Find the product *ac*: $6(6) = 36$.

Step 2: Find two integers whose product is 36 and whose sum is -13.

$$(-4)(-9) = +36 \text{ and } -4 + (-9) = -13$$

Note: This may take some time and experimentation. We do know that both numbers must be negative because the product is positive and the sum is negative. You might try prime factoring.

For example: $36 = 2 \cdot 2 \cdot 3 \cdot 3$

With combinations of these prime factors, we can write the following.

Negative Factors of 36			Sum	
-1	·	-36	$-1 + (-36) = -37$	
-2	·	-18	$-2 + (-18) = -20$	
-3	·	-12	$-3 + (-12) = -15$	
-4	·	-9	$-4 + (-9) = -13$	We can stop here!

-4 and -9 are the desired coefficients.

Step 3: Rewrite the middle term $(-13y)$ as $-9y - 4y$.

$$6y^2 - 13y + 6 = 6y^2 - 9y - 4y + 6$$

Note: -2 is factored from the last two terms so that there will be a common binomial factor $(2y-3)$.

4. Use the *ac*-method to factor $12b^3 - 9b^2 - 30b$.

Steps 4 and 5: Factor out the common binomial factor $(2y-3)$.

$$\begin{aligned} 6y^2-13y+6 &= 6y^2-9y-4y+6 \\ &= 3y(2y-3)-2(2y-3) \\ &= (2y-3)(3y-2) \end{aligned}$$

Thus, for the original expression,

$$\begin{aligned} 12y^3-26y^2+12y &= 2y(6y^2-13y+6) \\ &= 2y(2y-3)(3y-2) \end{aligned}$$

Do not forget to write the common monomial factor, $2y$, in the answer.

Now work margin exercise 4.

5. Use the *ac*-method to factor $7x^2+19x-6$.

Example 5 Using the *ac*-Method

Use the *ac*-method to factor $4x^2-5x-6$.

Solution

$a=4, b=-5, c=-6$

Step 1: Find the product *ac*: $4(-6)=-24$.

Step 2: Find two integers whose product is –24 and whose sum is –5.

Note: We know that one number must be positive and the other negative because the product is negative.

$(+3)(-8)=-24$ and $(+3)+(-8)=-5$.

Step 3: Rewrite the middle term $(-5x)$ as $+3x-8x$.

$$\begin{aligned} 4x^2-5x-6 &= 4x^2+3x-8x-6 \\ &= x(4x+3)-2(4x+3) \end{aligned}$$

Steps 4 and 5: Factor out the common binomial factor $(4x+3)$.

$$\begin{aligned} 4x^2-5x-6 &= 4x^2+3x-8x-6 \\ &= x(4x+3)-2(4x+3) \\ &= (4x+3)(x-2) \end{aligned}$$

Now work margin exercise 5.

Tips to Keep in Mind While Factoring

1. When factoring polynomials, always look for the greatest common factor first. Then, if there is one, remember to include this common factor as part of the answer.
2. **To factor completely** means to find factors of the polynomial such that none of the factors are themselves factorable.
3. Not all polynomials are factorable. (See $2x^2+x+1$ in Example 2c.) Any polynomial that cannot be factored as the product of polynomials with integer coefficients is not factorable.
4. Factoring can be checked by multiplying the factors. The product should be the original expression.

Completion Example 6 Factoring Trinomials

Completely factor each trinomial. Be sure to begin by looking for the greatest common factor.

a. $15x^2+38x+7$

b. $4y^2+6y-108$

Solution

a. $15x^2+38x+7=(5x+\underline{\quad 1\quad})(3x+\underline{\quad 7\quad})$

b. $4y^2+6y-108=2(\underline{\quad 2\quad}y^2+\underline{\quad 3\quad}y-\underline{\quad 54\quad})$
$=2(2y-\underline{\quad 9\quad})(y+\underline{\quad 6\quad})$

Now work margin exercise 6.

6. Completely factor each trinomial. Be sure to begin by looking for the greatest common factor.

a. $5x^2-32x-21$

b. $24x^2+87x-36$

Completion Example Answers

2. a. $(5x+1)(3x+7)$ **b.** $2(2y^2+3y-54)=2(2y-9)(y+6)$

Margin Exercise Answers

1. a. $(x+6)(x+2)$ **b.** $(4u-7)(2u+3)$ **2. a.** $4x(2x-1)(x-1)$ **b.** $7x(3x^2+7x-1)$
3. $(3a+2)(a+4)$ **4.** $3(b-2)(4b+5)$ **5.** $(7x-2)(x+3)$ **6. a.** $(5x+3)(x-7)$
b. $3(x+4)(8x-3)$

Note

No matter which method you use (the *ac*-method or the trial-and-error method), factoring trinomials takes time. With practice, you will become more efficient with either method. Make sure to be patient and observant.

13.3 Exercises

Concept Check

Fill-in-the-Blank. Complete each sentence using information found in this section.

1. When using the trial-and-error method to factor a trinomial of the form $ax^2 + bx + c$, you first need to list all possible combinations of ____________ of *a* and *c*, in their respective "First" and "Last" positions, according to the FOIL method.

2. The second step is to check the sums of the ______ in the O and I positions in the list until you find the sum to be *c*.

3. If none of these sums is *c*, the trinomial is not ______.

4. Look at the ___________ term to determine what signs to use for the constants in the factors.

5. When using the *ac*-method of factoring, you need to find two integers whose __________ is *ac* and whose _____ is *b*.

6. The *ac*-method of factoring uses the __________ method.

13.3 Exercises

Concept Check

1. factors
2. products
3. factorable
4. constant
5. product, sum
6. grouping
7. False; The middle term should be the sum of the inner and outer products.
8. False; The trial-and-error method of factoring is the opposite of the FOIL method.
9. False; The first step is to multiply *a* and *c*.
10. True

Practice

1. $(x+2)(x+3)$
2. $(x-2)(x-4)$
3. $(2x-5)(x+1)$
4. $(3x-7)(x+1)$
5. $(6x+5)(x+1)$
6. $(4x-3)(x-2)$

7. $-(x-2)(x-1)$
8. $-(x+2)(x+3)$
9. $(x-5)(x+2)$
10. $(x-10)(x-1)$
11. $-(x-14)(x+1)$
12. $-(x-6)(x-6)$
13. Not factorable
14. Not factorable
15. $-x(2x+1)(x-1)$
16. $-y(2y+1)(y+1)$
17. $(t-1)(4t+1)$
18. $(2x+1)(x-2)$
19. $(5a-6)(a+1)$
20. $(3a+1)(a+1)$
21. $(7x-2)(x+1)$
22. $(4x+3)(x+5)$
23. $(2x-3)(4x+1)$
24. $(2x+3)(3x+7)$
25. $(3x+4)(3x-5)$
26. Not factorable
27. $2(2x-5)(3x-2)$
28. $3(2b-1)(2b-1)$
29. $(3x-1)(x-2)$
30. $(7x+3)(x-2)$
31. $(3x-1)(3x-1)$
32. $(2x+1)(2x+1)$
33. $(3y+2)(2y+1)$
34. $(3y-4)(4y+3)$
35. $(x-1)(x-45)$
36. $(x-2)(x+8)$
37. Not factorable
38. $(5a-2)(a-1)$
39. $2b(4a-3)(a-2)$
40. $2mn(6m-1)(m-4)$
41. Not factorable
42. Not factorable
43. $(4x-1)(4x-1)$
44. $(3x+1)(x-4)$
45. $(8x-3)(8x-3)$

True/False. Determine whether each statement is true or false. If a statement is false, explain how it can be changed so the statement will be true. (**Note:** There may be more than one acceptable change.)

7. A trinomial is factorable if the middle term is the difference of the inner and outer products of two binomials.
8. The trial-and-error method of factoring a trinomial follows the same steps as the FOIL method of multiplication.
9. The first step in the *ac*-method of factoring is to rewrite the middle term.
10. Factoring can be checked by multiplying the factors and verifying that the product matches the original polynomial.

Practice

Completely factor each polynomial. If a polynomial cannot be factored, write "not factorable." See Examples 1 through 6.

1. x^2+5x+6
2. x^2-6x+8
3. $2x^2-3x-5$
4. $3x^2-4x-7$
5. $6x^2+11x+5$
6. $4x^2-11x+6$
7. $-x^2+3x-2$
8. $-x^2-5x-6$
9. $x^2-3x-10$
10. $x^2-11x+10$
11. $-x^2+13x+14$
12. $-x^2+12x-36$
13. $x^2+8x+64$
14. x^2+2x+3
15. $-2x^3+x^2+x$
16. $-2y^3-3y^2-y$
17. $4t^2-3t-1$
18. $2x^2-3x-2$
19. $5a^2-a-6$
20. $3a^2+4a+1$
21. $7x^2+5x-2$
22. $4x^2+23x+15$
23. $8x^2-10x-3$
24. $6x^2+23x+21$
25. $9x^2-3x-20$
26. $4x^2+40x+25$
27. $12x^2-38x+20$
28. $12b^2-12b+3$
29. $3x^2-7x+2$
30. $7x^2-11x-6$
31. $9x^2-6x+1$
32. $4x^2+4x+1$
33. $6y^2+7y+2$
34. $12y^2-7y-12$
35. $x^2-46x+45$
36. $x^2+6x-16$
37. $3x^2+9x+5$
38. $5a^2-7a+2$
39. $8a^2b-22ab+12b$
40. $12m^3n-50m^2n+8mn$

41. x^2+x+1

42. x^2+2x+2

43. $16x^2-8x+1$

44. $3x^2-11x-4$

45. $64x^2-48x+9$

46. $9x^2-12x+4$

47. $6x^2+2x-20$

48. $12y^2-15y+3$

49. $10x^2+35x+30$

50. $24y^2+4y-4$

51. $-18x^2+72x-8$

52. $7x^4-5x^3+3x^2$

53. $-45y^2+30y+120$

54. $-12m^2+22m+4$

55. $12x^2-60x+75$

56. $32y^2+50$

57. $6x^3+9x^2-6x$

58. $-5y^2+40y-60$

59. $12x^3-108x^2+243x$

60. $30a^3+51a^2+9a$

61. $9x^3y^3+9x^2y^3+9xy^3$

62. $48x^2y-354xy+126y$

63. $48xy^3-100xy^2+48xy$

64. $24a^2x^2+72a^2x+243x$

65. $21y^4-98y^3+56y^2$

66. $72a^3-306a^2+189a$

Writing & Thinking

67. It is true that $2x^2+10x+12=(2x+6)(x+2)=(2x+4)(x+3)$. Explain how the trinomial can be factored in two ways. Is there some kind of error?

68. It is true that $5x^2-5x-30=(5x-15)(x+2)$. Explain why this is not the completely factored form of the trinomial.

69. The volume of an open box is found by cutting equal squares (x inches on a side) from a sheet of cardboard that is 5 inches by 25 inches. The function representing this volume is $V(x)=4x^3-60x^2+125x$, where $0<x<2.5$. Factor this function and use the factors to explain, in your own words, how the function represents the volume.
(**Note:** Volume of a box = length × width × height.)

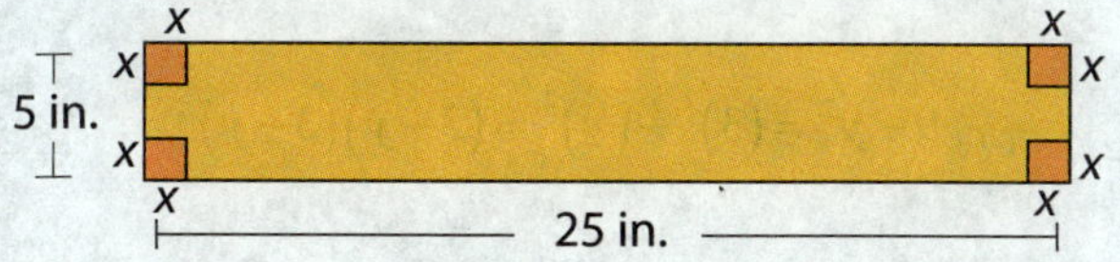

46. $(3x-2)(3x-2)$

47. $2(3x-5)(x+2)$

48. $3(4y-1)(y-1)$

49. $5(2x+3)(x+2)$

50. $4(3y-1)(2y+1)$

51. $-2(9x^2-36x+4)$

52. $x^2(7x^2-5x+3)$

53. $-15(3y+4)(y-2)$

54. $-2(m-2)(6m+1)$

55. $3(2x-5)(2x-5)$

56. $2(16y^2+25)$

57. $3x(2x-1)(x+2)$

58. $-5(y-2)(y-6)$

59. $3x(2x-9)(2x-9)$

60. $3a(5a+1)(2a+3)$

61. $9xy^3(x^2+x+1)$

62. $6y(8x-3)(x-7)$

63. $4xy(3y-4)(4y-3)$

64. $3x(8a^2x+24a^2+81)$

65. $7y^2(y-4)(3y-2)$

66. $9a(4a-3)(2a-7)$

Writing & Thinking

67. This is not an error, but the trinomial is not completely factored. The completely factored form of this trinomial is $2(x+2)(x+3)$.

68. This is not the completely factored form because $(5x-15)$ can be factored into $5(x-3)$.

69. $x(5-2x)(25-2x)$; The height is x, the width is $(5-2x)$, and the length is $(25-2x)$.

Objectives

A. Factor the difference of two squares.

B. Factor perfect square trinomials.

C. Factor the sums and differences of two cubes.

13.4 Special Factoring Techniques

In Section 12.7, we discussed the following three products.

- $(x+a)(x-a)=x^2-a^2$ Difference of two squares
- $(x+a)^2=x^2+2ax+a^2$ Square of a binomial sum
- $(x-a)^2=x^2-2ax+a^2$ Square of a binomial difference

Two objectives in this section are to learn to factor products of these types (difference of two squares and squares of binomials) without using to the trial-and-error method. That is, with practice you will learn to recognize the "form" of these products and go directly to the factors. You should memorize these three products and their names.

Teaching Note

In addition to memorizing the special factorizations, you may want to encourage students to memorize that x^2+a^2 is not factorable.

For easy reference to squares, Table 1 lists the squares of the integers from 1 to 20.

Perfect Squares from 1 to 400

Integer (n)	1	2	3	4	5	6	7	8	9	10
Square (n^2)	1	4	9	16	25	36	49	64	81	100

Integer (n)	11	12	13	14	15	16	17	18	19	20
Square (n^2)	121	144	169	196	225	256	289	324	361	400

Table 1

A Factoring the Difference of Two Squares

Consider the polynomial x^2-25. By recognizing this expression as the **difference of two squares**, we can go directly to the factors.

$$x^2-25=(x)^2-(5)^2=(x+5)(x-5)$$

Similarly, we have

$$9-y^2=(3)^2-(y)^2=(3+y)(3-y)$$

and

$$49x^2-36=(7x)^2-(6)^2=(7x+6)(7x-6).$$

Remember to *look for a common monomial factor first*.

$$6x^2y-24y=6y(x^2-4)=6y(x+2)(x-2)$$

Example 1 Factoring the Difference of Two Squares

Factor completely.

a. $3a^2b-3b$

b. x^6-400

Solution

a. $3a^2b-3b=3b(a^2-1)$ Factor out the GCF, $3b$.

$=3b(a+1)(a-1)$ Difference of two squares; Don't forget that $3b$ is a factor.

b. Even powers, such as x^6, can always be treated as squares: $x^6=(x^3)^2$.

$x^6-400=(x^3)^2-20^2$

$=(x^3+20)(x^3-20)$ Difference of two squares

Now work margin exercise 1.

1. Factor completely.

a. $7ax^2-343a$

b. y^6-100

Sum of Two Squares

The **sum of two squares** is an expression of the form x^2+a^2 and is **not factorable**. For example, x^2+36 is the sum of two squares and is not factorable. There are no factors with integer coefficients whose product is x^2+36. To understand this situation, write

$$x^2+36=x^2+0x+36$$

and note that there are no factors of 36 that will add to 0.

DEFINITION

Example 2 Factoring the Sum of Two Squares (Not Factorable)

Factor completely. Be sure to begin by looking for the greatest common factor.

a. y^2+64

b. $4x^2+100$

Solution

a. y^2+64 is the **sum of two squares and is not factorable.**

b. $4x^2+100=4(x^2+25)$ Factored completely

We see that 4 is the greatest common monomial factor and x^2+25 is the **sum of two squares** and is **not factorable**.

Now work margin exercise 2.

2. Factor completely.

a. x^2+25

b. $45x^2+20$

B Factoring Perfect Square Trinomials

We know that **squaring a binomial** leads to a **perfect square trinomial**. Therefore, factoring a perfect square trinomial gives the square of a binomial. We simply need to recognize whether a trinomial fits the "form."

In a perfect square trinomial, both the first and last terms of the trinomial must be perfect squares. If the first term is of the form x^2 and the last term is of the form a^2, then the middle term must be of the form $2ax$ or $-2ax$, as shown in the following examples.

$$x^2+8x+16=(x+4)^2 \text{ Here, } 8x=2\cdot 4\cdot x=2ax \text{ and } 16=4^2=a^2.$$

$$x^2-6x+9=(x-3)^2 \text{ Here, } -6x=-2\cdot 3\cdot x=-2ax \text{ and } 9=3^2=a^2$$

3. Factor completely.

a. $z^2+40z+400$

b. $y^2-14y+49$

c. $3x^2z-18xyz+27y^2z$

d. $(y^2+8y+16)-z^2$

Example 3 Factoring Perfect Square Trinomials

Factor completely.

a. $z^2-12z+36$

b. $4y^2+12y+9$

c. $2x^3-8x^2y+8xy^2$

d. $(x^2+6x+9)-y^2$

Solution

a. In the form $x^2-2ax+a^2$, we have $x=z$ and $a=6$.

$$\begin{aligned} z^2-12z+36 &= z^2-2\cdot 6z+6^2 \\ &= (z-6)^2 \end{aligned}$$

b. In the form $x^2+2ax+a^2$, we have $x=2y$ and $a=3$.

$$\begin{aligned} 4y^2+12y+9 &= (2y)^2+2\cdot 3\cdot 2y+3^2 \\ &= (2y+3)^2 \end{aligned}$$

c. Factor out the GCF first. Then factor the **perfect square trinomial**.

$$\begin{aligned} 2x^3-8x^2y+8xy^2 &= 2x(x^2-4xy+4y^2) \\ &= 2x\left[x^2-2\cdot 2y\cdot x+(2y)^2\right] \\ &= 2x(x-2y)^2 \end{aligned}$$

d. Treat x^2+6x+9 as a perfect square trinomial, then factor the **difference of two squares**.

$$\begin{aligned} (x^2+6x+9)-y^2 &= (x+3)^2-y^2 \\ &= (x+3+y)(x+3-y) \end{aligned}$$

Now work margin exercise 3.

C Factoring the Sums and Differences of Two Cubes

The formulas for the sums and differences of two cubes are new, and we can proceed to show that they are indeed true as follows.

$$\begin{aligned}(x+a)(x^2-ax+a^2) &= x\cdot x^2 - x\cdot ax + x\cdot a^2 + a\cdot x^2 - a\cdot ax + a\cdot a^2\\ &= x^3 - ax^2 + a^2x + ax^2 - a^2x + a^3\\ &= x^3 + a^3 \quad \text{Sum of two cubes}\end{aligned}$$

$$\begin{aligned}(x-a)(x^2+ax+a^2) &= x\cdot x^2 + x\cdot ax + x\cdot a^2 - a\cdot x^2 - a\cdot ax - a\cdot a^2\\ &= x^3 + ax^2 + a^2x - ax^2 - a^2x - a^3\\ &= x^3 - a^3 \quad \text{Difference of two cubes}\end{aligned}$$

Because we know that factoring is the reverse of multiplication, we use these formulas to factor the sums and differences of two cubes. For example,

$$\begin{aligned}x^3+27 &= (x)^3+(3)^3\\ &= (x+3)\left((x)^2-(3)(x)+(3)^2\right)\\ &= (x+3)(x^2-3x+9)\end{aligned}$$

and

$$\begin{aligned}x^6-125 &= (x^2)^3-(5)^3\\ &= (x^2-5)\left[(x^2)^2+(5)(x^2)+(5)^2\right]\\ &= (x^2-5)(x^4+5x^2+25).\end{aligned}$$

As an aid in factoring sums and differences of two cubes, Table 2 contains the cubes of the integers from 1 to 10. These cubes are called perfect cubes.

Perfect Cubes from 1 to 1000

Integer (n)	1	2	3	4	5	6	7	8	9	10
Cube (n^3)	1	8	27	64	125	216	343	512	729	1000

Table 2

Note

Tips for Working with Sums and Differences of Two Cubes

1. In each case, after multiplying the factors, the middle terms drop out and only two terms are left.
2. The trinomials in parentheses are not perfect square trinomials. These trinomials are not factorable.
3. The sign in the binomial factor agrees with the sign between the two cubes.

Example 4 Factoring Sums and Differences of Two Cubes

Factor completely.

a. x^3-8 **b.** x^6+64y^3 **c.** $16y^{12}-250$

4. Factor completely.

a. y^3-27

b. $8y^3-x^6$

c. $48x^{12}-750$

13.4 Exercises

Concept Check

1. binomial
2. squares
3. $2ax$, $-2ax$
4. $(x-a)(x^2+ax+a^2)$
5. $(x+a)(x^2-ax+a^2)$
6. 27, 64, 125, 216
7. True
8. True
9. False; The sum of two squares is not factorable.
10. True

Practice

1. $(x-5)(x+5)$
2. $(y-11)(y+11)$
3. $(9-y)(9+y)$
4. $(5-z)(5+z)$
5. $2(x-8)(x+8)$
6. $3(x-7)(x+7)$
7. $4(x-2)(x+2)(x^2+4)$
8. $5(x^2-5)(x^2+5)$
9. Not factorable
10. Not factorable
11. $(y-8)^2$
12. $(z+9)^2$
13. $-4(x-5)(x+5)$
14. $-3(2x^2-1)(2x^2+1)$
15. $(3x-5)(3x+5)$
16. $(2x-7)(2x+7)$
17. $(y-5)^2$
18. $(x+6)^2$
19. $(2x-1)^2$
20. $(7x-1)^2$
21. $(5x+3)^2$
22. $(3y+2)^2$
23. $(4x-5)^2$
24. $(3x-2)^2$

Solution

a.
$$\begin{aligned} x^3-8 &= x^3-2^3 \\ &= (x-2)(x^2+2\cdot x+2^2) \\ &= (x-2)(x^2+2x+4) \end{aligned}$$

Note: Remember that the second polynomial is not a perfect square trinomial and cannot be factored.

b.
$$\begin{aligned} x^6+64y^3 &= (x^2)^3+(4y)^3 \\ &= (x^2+4y)\left[(x^2)^2-4y\cdot x^2+(4y)^2\right] \\ &= (x^2+4y)(x^4-4x^2y+16y^2) \end{aligned}$$

c. Factor out the GCF first. Then factor the **difference of two cubes.**

$$\begin{aligned} 16y^{12}-250 &= 2(8y^{12}-125) \\ &= 2\left[(2y^4)^3-5^3\right] \\ &= 2(2y^4-5)\left[(2y^4)^2+(5)(2y^4)+5^2\right] \\ &= 2(2y^4-5)(4y^8+10y^4+25) \end{aligned}$$

Now work margin exercise 4.

Margin Exercise Answers

1. a. $7a(x-7)(x+7)$ **b.** $(y^3+10)(y^3-10)$ **2. a.** Not factorable **b.** $5(9x^2+4)$

3. a. $(z+20)^2$ **b.** $(y-7)^2$ **c.** $3z(x-3y)^2$ **d.** $(y+4-z)(y+4+z)$

4. a. $(y-3)(y^2+3y+9)$ **b.** $(2y-x^2)(4y^2+2x^2y+x^4)$ **c.** $6(2x^4-5)(4x^8+10x^4+25)$

13.4 Exercises

Concept Check

Fill-in-the-Blank. Complete each sentence using information found in this section.

1. Factoring a perfect square trinomial gives a square __________.
2. In a perfect square trinomial, both the first and last terms must be perfect ______.
3. If the first term of a perfect square trinomial is x^2, and the last term is of the form a^2, then the middle term must be of the form _____ or _____.
4. The formula for factoring the difference of cubes is $x^3 - a^3 =$ __________.
5. The formula for factoring the sum of two cubes is $x^3 + a^3 =$ __________.
6. The first 6 perfect cubes are 1, 8, ___, ___, ___, and ___.

True/False. Determine whether each statement is true or false. If a statement is false, explain how it can be changed so the statement will be true. (**Note:** There may be more than one acceptable change.)

7. The expression $x^2 + 20x + 100$ is a perfect square trinomial.

8. When factoring polynomials, always look for a common monomial factor first.

9. The sum of two squares, $(x^2 + a^2)$, is factorable.

10. Sixty-four is a perfect square and a perfect cube.

Practice

Completely factor each of the given polynomials. If a polynomial cannot be factored, write "not factorable." See Examples 1 through 4.

1. $x^2 - 25$

2. $y^2 - 121$

3. $81 - y^2$

4. $25 - z^2$

5. $2x^2 - 128$

6. $3x^2 - 147$

7. $4x^4 - 64$

8. $5x^4 - 125$

9. $y^2 + 100$

10. $4x^2 + 49$

11. $y^2 - 16y + 64$

12. $z^2 + 18z + 81$

13. $-4x^2 + 100$

14. $-12x^4 + 3$

15. $9x^2 - 25$

16. $4x^2 - 49$

17. $y^2 - 10y + 25$

18. $x^2 + 12x + 36$

19. $4x^2 - 4x + 1$

20. $49x^2 - 14x + 1$

21. $25x^2 + 30x + 9$

22. $9y^2 + 12y + 4$

23. $16x^2 - 40x + 25$

24. $9x^2 - 12x + 4$

25. $4x^3 - 64x$

26. $50x^3 - 8x$

27. $2x^3y + 32x^2y + 128xy$

28. $3x^2y - 30xy + 75y$

29. $y^2 + 6y + 9$

30. $y^2 + 4y + 4$

31. $x^2 - 20x + 100$

32. $25x^2 - 10x + 1$

33. $x^4 + 10x^2y + 25y^2$

34. $16x^4 + 8x^2y + y^2$

35. $x^3 - 125$

36. $x^3 - 64$

37. $y^3 + 216$

38. $y^3 + 1$

39. $x^3 + 27y^3$

40. $8x^3 + 1$

41. $x^2 + 64y^2$

42. $3x^3 + 81$

43. $4x^3 - 32$

44. $64x^3 + 27y^3$

45. $54x^3 - 2y^3$

46. $3x^4 + 375xy^3$

47. $x^3y + y^4$

48. $x^4y^3 - x$

49. $x^2y^2 - x^2y^5$

50. $2x^2 - 16x^2y^3$

51. $24x^4y + 81xy^4$

52. $x^6 - 64y^3$

53. $x^6 - y^9$

54. $64x^2 + 1$

55. $27x^3 + y^6$

56. $x^3 + 64z^3$

57. $8x^3 + y^3$

58. $x^3 + 125y^3$

59. $8y^3 - 8$

60. $36x^3 + 36$

61. $9x^2 - y^2$

62. $x^2 - 4y^2$

63. $x^4 - 16y^4$

25. $4x(x-4)(x+4)$

26. $2x(5x-2)(5x+2)$

27. $2xy(x+8)^2$

28. $3y(x-5)^2$

29. $(y+3)^2$

30. $(y+2)^2$

31. $(x-10)^2$

32. $(5x-1)^2$

33. $(x^2+5y)^2$

34. $(4x^2+y)^2$

35. $(x-5)(x^2+5x+25)$

36. $(x-4)(x^2+4x+16)$

37. $(y+6)(y^2-6y+36)$

38. $(y+1)(y^2-y+1)$

39. $(x+3y)(x^2-3xy+9y$

40. $(2x+1)(4x^2-2x+1)$

41. Not factorable

42. $3(x+3)(x^2-3x+9)$

43. $4(x-2)(x^2+2x+4)$

44. $(4x+3y)$
$(16x^2-12xy+9y^2)$

45. $2(3x-y)$
$(9x^2+3xy+y^2)$

46. $3x(x+5y)$
$(x^2-5xy+25y^2)$

47. $y(x+y)(x^2-xy+y^2)$

48. $x(xy-1)(x^2y^2+xy+1)$

49. $x^2y^2(1-y)(1+y+y^2)$

50. $2x^2(1-2y)$
$(1+2y+4y^2)$

51. $3xy(2x+3y)$
$(4x^2-6xy+9y^2)$

52. (x^2-4y)
$(x^4+4x^2y+16y^2)$

53. (x^2-y^3)
$(x^4+x^2y^3+y^6)$

54. Not factorable

55. $(3x+y^2)$
$(9x^2-3xy^2+y^4)$

56. $(x+4z)(x^2-4xz+16z^2)$

57. $(2x+y)(4x^2-2xy+y^2)$

58. $(x+5y)$
$(x^2-5xy+25y^2)$

59. $8(y-1)(y^2+y+1)$

60. $36(x+1)(x^2-x+1)$

61. $(3x-y)(3x+y)$

62. $(x-2y)(x+2y)$

63. $(x-2y)(x+2y)$
(x^2+4y^2)

64. $(3x-1)(3x+1)(9x^2+1)$

65. $(x-y-9)(x-y+9)$

66. $(x+2y-5)(x+2y+5)$

67. $(x-y-6)(x-y+6)$

68. $(x+2y-5)(x+2y+5)$

69. $(4x+1-y)(4x+1+y)$

70. $(x+y+3)(x-y-3)$

71. a. x^2-16

b. 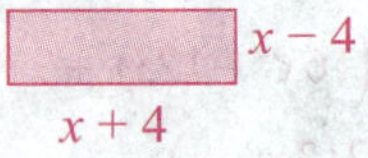

72. a. x^2+12x

b. $4x+24$

Writing & Thinking

73. a. $xy+xy+x^2+y^2$
$=x^2+2xy+y^2$
$=(x+y)^2$

b. $(x+y)(x+y)$
$=(x+y)^2$

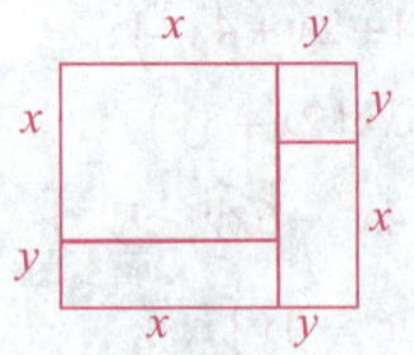

64. $81x^4-1$

65. $(x-y)^2-81$

66. $(x+2y)^2-25$

67. $(x^2-2xy+y^2)-36$

68. $(x^2+4xy+4y^2)-25$

69. $(16x^2+8x+1)-y^2$

70. $x^2-(y^2+6y+9)$

Solve.

71. a. Represent the area of the shaded region of the square shown below as the difference of two squares.

b. Use the factors of the expression in part a. to draw (and label the sides of) a rectangle that has the same area as the shaded region.

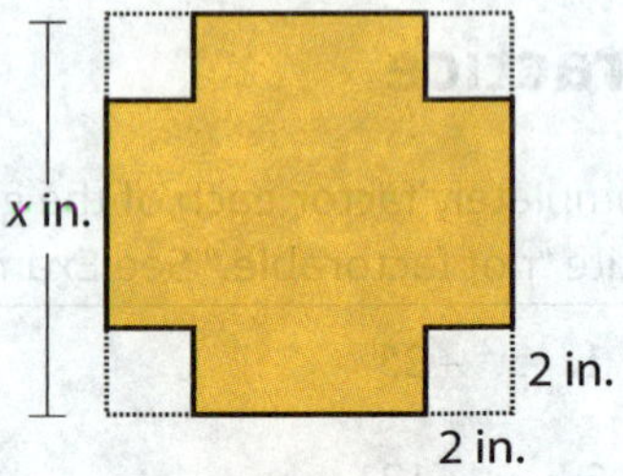

72. a. Use a polynomial function to represent the area of the shaded region of the square.

b. Use a polynomial function to represent the perimeter of the shaded figure.

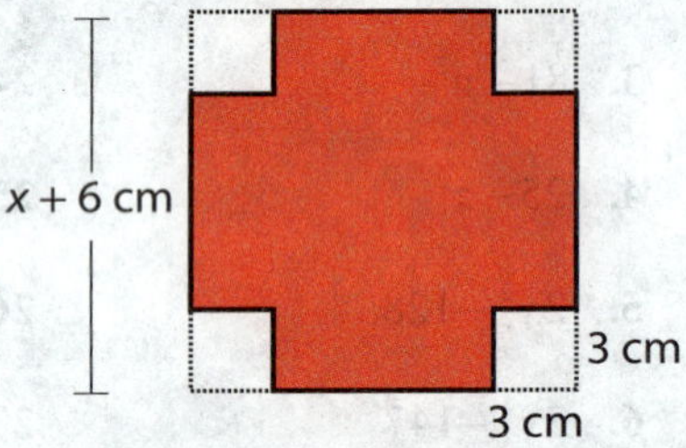

Writing & Thinking

73. a. Show that the sum of the areas of the rectangles and squares in the figure is a perfect square trinomial.

b. Rearrange the rectangles and squares in the form of a square and represent its area as the square of a binomial.

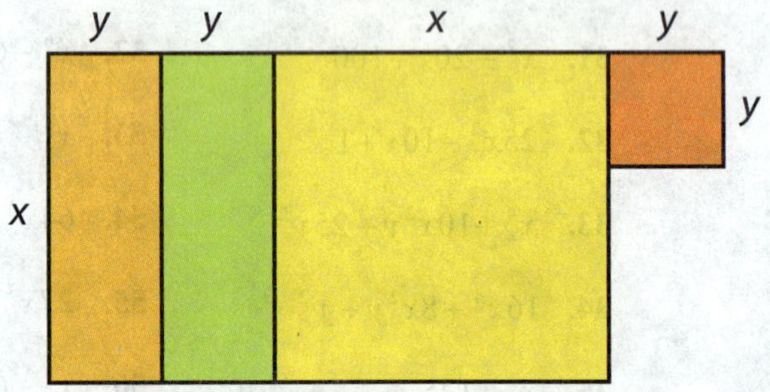

74. Compound interest is interest earned on interest. If a principal P is invested and compounded annually (once a year) at a rate of r, then the amount, A_1 accumulated in one year is $A_1 = P + Pr$.

In factored form, we have $A_1 = P + Pr = P(1+r)$.

At the end of the second year the amount accumulated is $A_2 = (P + Pr) + (P + Pr)r$.

a. Write the expression for A_2 in factored form similar to that for A_1.

b. Write an expression for the amount accumulated in three years, A_3, in factored form.

c. Write an expression for A_n the amount accumulated in n years.

d. Use the formula you developed in part c. and your calculator to find the amount accumulated if \$10,000 is invested at 6% and compounded annually for 20 years.

75. You may have heard of (or studied) the following rules for division of an integer by 3 and 9:

1. An integer is divisible by 3 if the sum of its digits is divisible by 3.

2. An integer is divisible by 9 if the sum of its digits is divisible by 9.

The proofs of both **1.** and **2.** can be started as follows.

Let *abc* represent a three-digit integer.

Then $abc = 100a + 10b + c$

$= (99+1)a + (9+1)b + c$

$=$ (now you finish the proofs)

Use the pattern just shown and prove both **1.** and **2.** for a four-digit integer.

74. a. $A_2 = P(1+r)^2$

b. $A_3 = P(1+r)^3$

c. $A_n = P(1+r)^n$

d. \$32,071.35

75. For a 3-digit integer:

abc

$= 100a + 10b + c$

$= (99+1)a + (9+1)b + c$

$= 9(11a + b) + a + b + c$

So, if the sum $(a+b+c)$ is divisible by 3 (or 9), then the number *abc* will be divisible by 3 (or 9).

For a 4-digit integer:

$abcd$

$= 1000a + 100b + 10c + d$

$= (999+1)a + (99+1)b$

$+ (9+1)c + d$

$= 9(111a + 11b + c) + a$

$+ b + c + d$

So, if the sum $(a+b+c+d)$ is divisible by 3 (or 9), then the number *abcd* will be divisible by 3 (or 9).

Objectives

A. Determine the best method to use when factoring a given polynomial.

13.5 Review of Factoring Techniques

A General Guidelines for Factoring

As we have seen so far in Chapter 13, factoring is a basic algebraic skill. Factoring is useful in solving equations and, as we will see in Chapter 14, necessary for simplifying algebraic fractions. This section is provided as extra practice in factoring. The following guidelines are provided for easy reference.

General Guidelines for Factoring Polynomials

1. **Always look for a common monomial factor first.** If the leading coefficient is negative, factor out a negative monomial, even if it is just -1.
2. **Check the number of terms.**
   - **a. Two terms:**
     1. Difference of two squares? – factorable
     2. Sum of two squares? – not factorable
     3. Difference of two cubes? – factorable
     4. Sum of two cubes? – factorable
   - **b. Three terms:**
     1. Perfect square trinomial?
     2. Use trial-and-error method?

        Guidelines for the trial-and-error method
        - **a.** If the sign of the constant term is positive, the signs in both factors will be the same, either both positive or both negative.
        - **b.** If the sign of the constant term is negative, the signs in the factors will be different, one positive and one negative.
     3. Use *ac*-method?

        Guidelines for the *ac*-method
        - **a.** Multiply $a \cdot c$.
        - **b.** Find two integers whose product is ac and whose sum is b. If this is not possible, the trinomial is not factorable.
        - **c.** Rewrite the middle term (bx) using the two numbers found in Step **b** as coefficients.
        - **d.** Factor by grouping.
   - **c. Four terms:**

     Group terms with a common factor and factor out any common binomial factor.
3. **Check the possibility of factoring any of the factors.**

Checking: Factoring can be checked by multiplying the factors. The product should be the original expression.

PROCEDURE

13.5 Exercises

Concept Check

1. trial-and-error, *ac*
2. squares, cubes
3. original
4. grouping
5. monomial
6. False; The first step is to check for a common monomial factor.

13.5 Exercises

Concept Check

Fill-in-the-Blank. Complete each sentence using information found in this section.

1. If a trinomial is not a perfect square trinomial, you can try factoring it by using the ______________ method or the ___-method.
2. When factoring a polynomial with two terms, check to see if it is of the form of the sum or difference of two _________ or two _________.
3. Factoring can be checked by multiplying the factors. The product should be the ________ expression.
4. If there are four terms when factoring, consider factoring by __________.
5. When factoring, always look for a common ______ factor first.

True/False. Determine whether each statement is true or false. If a statement is false, explain how it can be changed so the statement will be true. (**Note:** There may be more than one acceptable change.)

6. You should always start by checking the number of terms when factoring a polynomial.
7. If a trinomial is to be factored, the trial-and-error or *ac*-methods can be used.
8. If there are four terms in a polynomial, it cannot be factored.

Practice

Completely factor each of the given polynomials. If a polynomial cannot be factored, write "not factorable."

1. m^2+7m+6
2. a^2-4a+3
3. $x^2+11x+18$
4. $y^2+8y+15$
5. x^2-100
6. $n^2-8n+12$
7. m^2-m-6
8. y^2-49
9. $a^2+2a+24$
10. $-x^2-12x-35$
11. $64a^2-1$
12. $49x^2+4$
13. $x^2+10x+25$
14. $x^2+3x-10$
15. $x^2+9x-36$
16. $x^2+16x+64$
17. $3a^2+12a-36$
18. $-2y^2+24y-70$
19. $-5x^2+70x-240$
20. $7t^2+14t-168$
21. $64+49t^2$
22. $3x^2-147$
23. x^3-4x^2-12x
24. $3n^3+15n^2+18n$
25. $112a-2a^2-2a^3$
26. $200x+20x^2-4x^3$
27. $16x^3-100x$
28. $48x^3-27x$
29. $-3x^2+17x-10$
30. $2x^2+7x+3$
31. $6x^2-11x+4$
32. $12x^2-32x+5$
33. $12m^2+m-6$

7. True
8. False; It might be factorable by the grouping method.

Practice

1. $(m+6)(m+1)$
2. $(a-3)(a-1)$
3. $(x+9)(x+2)$
4. $(y+5)(y+3)$
5. $(x-10)(x+10)$
6. $(n-6)(n-2)$
7. $(m-3)(m+2)$
8. $(y-7)(y+7)$
9. Not factorable
10. $-(x+7)(x+5)$
11. $(8a-1)(8a+1)$
12. Not factorable
13. $(x+5)^2$
14. $(x+5)(x-2)$
15. $(x+12)(x-3)$
16. $(x+8)^2$
17. $3(a+6)(a-2)$
18. $-2(y-5)(y-7)$
19. $-5(x-6)(x-8)$
20. $7(t-4)(t+6)$
21. Not factorable
22. $3(x-7)(x+7)$
23. $x(x-6)(x+2)$
24. $3n(n+3)(n+2)$
25. $-2a(a+8)(a-7)$
26. $-4x(x-10)(x+5)$
27. $4x(2x-5)(2x+5)$
28. $3x(4x-3)(4x+3)$
29. $-(x-5)(3x-2)$
30. $(2x+1)(x+3)$
31. $(2x-1)(3x-4)$
32. $(2x-5)(6x-1)$
33. $(4m+3)(3m-2)$

34. $(3t-7)(2t+5)$
35. $2(2x-1)(x-3)$
36. $-2(2x-5)(x-2)$
37. $(4x-7)(2x+5)$
38. $(4x+3)(3x-1)$
39. $(5x+6)(4x-9)$
40. $(7x-5)(3x+2)$
41. $-(5x-7)(3x+2)$
42. $-(x-3)(3x+8)$
43. $-(2a-3)(4a-5)$
44. $(7x-6)(9x+2)$
45. $(4y+5)(5y-4)$
46. $(7x-3)(5x+2)$
47. $(6x-1)(3x-2)$
48. $(3x-11)(4x-1)$
49. $-6(5x-4)(5x+4)$
50. $-7x(5x-6)(5x+6)$
51. $3(4n^2-20n-25)$
52. $-2x(6x^2+x+35)$
53. $a(21a^2-13a-2)$
54. $x(13x^2+120x+100)$
55. $3x(3x-2)(4x+5)$
56. $-3x(2x+3)(5x-7)$
57. $2x(2x-1)(4x-11)$
58. $4y(2y+5)(3y-8)$
59. $5(24m^2+2m+15)$
60. $2x(9x+5)(8x-5)$
61. $(y-4)(x+3)$
62. $(2x+1)(z+5)$
63. $(x+2y)(x-6)$
64. $(2y+5)(y+3z)$
65. $-(x^2-5)(x-8)$
66. $(2x^2-3)(x-7)$
67. $(x+5)(x^2-5x+25)$
68. $(y-10)(y^2+10y+100)$
69. $x^4(y-1)(y^2+y+1)$
70. $x^3(xy-1)(x^2y^2+xy+1)$

34. $6t^2+t-35$
35. $4x^2-14x+6$
36. $-4x^2+18x-20$
37. $8x^2+6x-35$
38. $12x^2+5x-3$
39. $20x^2-21x-54$
40. $21x^2-x-10$
41. $14+11x-15x^2$
42. $24+x-3x^2$
43. $-8a^2+22a-15$
44. $63x^2-40x-12$
45. $20y^2+9y-20$
46. $35x^2-x-6$
47. $18x^2-15x+2$
48. $12x^2-47x+11$
49. $-150x^2+96$
50. $252x-175x^3$
51. $12n^2-60n-75$
52. $-12x^3-2x^2-70x$
53. $21a^3-13a^2-2a$
54. $13x^3+120x^2+100x$
55. $36x^3+21x^2-30x$
56. $63x-3x^2-30x^3$
57. $16x^3-52x^2+22x$
58. $24y^3-4y^2-160y$
59. $75+10m+120m^2$
60. $144x^3-10x^2-50x$
61. $xy+3y-4x-12$
62. $2xz+10x+z+5$
63. $x^2+2xy-6x-12y$
64. $2y^2+6yz+5y+15z$
65. $-x^3+8x^2+5x-40$
66. $2x^3-14x^2-3x+21$
67. x^3+125
68. y^3-1000
69. $x^4y^3-x^4$
70. $x^6y^3-x^3$
71. $8a^6+27b^6$
72. a^9+64b^3
73. x^6y^3-125
74. x^3y^3+216
75. $x^3+7x^2-9x-63$
76. $x^5+5x^4-4x-20$
77. $9x^2-(y+6)^2$
78. $(x+2)^2-25a^2$
79. $(y^2+20y+100)-49x^2$
80. $(t^2+22t+121)-16s^2$

71. $(2a^2+3b^2)(4a^4-6a^2b^2+9b^4)$
72. $(a^3+4b)(a^6-4a^3b+16b^2)$
73. $(x^2y-5)(x^4y^2+5x^2y+25)$
74. $(xy+6)(x^2y^2-6xy+36)$
75. $(x-3)(x+3)(x+7)$
76. $(x+5)(x^2-2)(x^2+2)$
77. $(3x+y+6)(3x-y-6)$
78. $(x+2+5a)(x+2-5a)$
79. $(y+10+7x)(y+10-7x)$
80. $(t+11+4s)(t+11-4s)$

13.6 Solving Quadratic Equations by Factoring

Objectives

A. Solve quadratic equations by factoring.

B. Write equations given the roots.

In this chapter, the emphasis has been on second-degree polynomials and functions. The methods of factoring we have studied (*ac*-method, trial-and-error method, difference of two squares, and perfect square trinomials) have been related, in general, to second-degree polynomials. Second-degree polynomials are called **quadratic polynomials** (or **quadratics**) and are particularly evident in physics with the path of thrown objects (or projectiles) affected by gravity, in mathematics involving area (area of a circle, rectangle, square, and triangle), and in many situations in higher-level mathematics.

For example, consider a ball being dropped from the top of a 720-foot-tall building. The height h of the ball above ground level at a time t is given by the polynomial function $h(t) = -16t^2 + 720$. Using this equation and skills we already possess, we could determine the height of the ball at any given time by evaluating the function at that value. We could also graph these values to model the path of the ball as it falls to the ground. However, in order to determine the exact moment when the ball strikes the ground, we need to know when the height is zero. To find this piece of information, we must solve the quadratic equation $0 = -16t^2 + 720$. This section uses the factoring skills we have gained to solve quadratic equations like this.

A Solving Quadratic Equations by Factoring

Quadratic Equations

Quadratic equations are equations that can be written in the form

$$\boldsymbol{ax^2 + bx + c = 0},$$

where a, b, and c are real numbers and $a \neq 0$.

DEFINITION

The form $ax^2 + bx + c = 0$ is called the **standard form** (or **general form**) of a **quadratic equation**. In the standard form, a quadratic polynomial is on one side of the equation and 0 is on the other side. For example, $x^2 - 8x + 12 = 0$ is a quadratic equation in standard form, while $3x^2 - 2x = 27$ and $x^2 = 25x$ are quadratic equations, just not in standard form.

These last two equations can be manipulated algebraically so that 0 is on one side. When solving quadratic equations by factoring, having 0 on one side before we factor is necessary because of the following **zero-factor law**.

Zero-Factor Law

If the product of two (or more) factors is 0, then at least one of the factors must be 0. That is, for real numbers a and b,

$$\textbf{if } \boldsymbol{a \cdot b = 0}, \textbf{ then } \boldsymbol{a = 0} \textbf{ and/or } \boldsymbol{b = 0}.$$

DEFINITION

1. Solve: $(y-7)(3y-5)=0$

Example 1 Solving Factored Quadratic Equations

Solve: $(x-5)(2x-7)=0$

Solution

Since the quadratic is already factored and the other side of the equation is 0, we use the zero-factor law and set each factor equal to 0. This process yields two linear equations, which can then be solved.

$$x-5=0 \quad \text{or} \quad 2x-7=0$$
$$x=5 \qquad\qquad 2x=7$$
$$x=\frac{7}{2}$$

Thus, the **two solutions** (or **roots**) to the original equation are $x=5$ and $x=\frac{7}{2}$. Or we can say that the solution set is $\left\{\frac{7}{2},5\right\}$.

The solutions can be **checked** by substituting them one at a time for x in the equation. That is, there will be two "checks."

Substituting $x=5$ gives the following.

$$\begin{aligned}(x-5)(2x-7)&=(5-5)(2\cdot 5-7)\\&=(0)(3)\\&=0\end{aligned}$$

Substituting $x=\frac{7}{2}$ gives the following.

$$\begin{aligned}(x-5)(2x-7)&=\left(\frac{7}{2}-5\right)\left(2\cdot\frac{7}{2}-7\right)\\&=\left(-\frac{3}{2}\right)(7-7)\\&=\left(-\frac{3}{2}\right)(0)\\&=0\end{aligned}$$

Therefore, both 5 and $\frac{7}{2}$ are solutions to the original equation.

Now work margin exercise 1.

Math Tip

Recall that when solving linear equations, we start by isolating the variable as soon as possible. When solving quadratic equations, however, we gather all terms to one side so that the other side is 0.

In Example 1, the polynomial was factored and the other side of the equation was 0. The equation was solved by setting each factor, in turn, equal to 0 and solving the resulting linear equations. These solutions tell us that the original equation has two solutions. In general, **a quadratic equation has two solutions**. In the special cases where the two factors are the same, there is only one solution and it is called a **double solution** (or **double root**). To solve Example 2, we will complete the following steps.

1. Write the quadratic equation in standard form with one side 0.
2. Factor the polynomial.
3. Set each factor equal to 0.
4. Solve the resulting linear equations.

Study these examples carefully. Checking the solution is left as an exercise for the student.

Example 2 Solving Quadratic Equations by Factoring

Solve by factoring: $3x^2 = 6x$

Solution

$3x^2 = 6x$

$3x^2 - 6x = 0$ — Write the equation in standard form with 0 on one side by subtracting $6x$ from both sides.

$3x(x-2) = 0$ — Factor out the common monomial, $3x$.

$3x = 0$ or $x - 2 = 0$ — Set each factor equal to 0.

$x = 0$ $\quad x = 2$ — Solve each linear equation.

The solutions are 0 and 2.

Now work margin exercise 2.

2. Solve by factoring:
$4v^2 = 8v$

> **Note**
>
> To make factoring easier, the terms should be arranged so that the leading coefficient is a positive number.

Example 3 Solving Quadratic Equations by Factoring

Solve by factoring: $x^2 - 8x + 16 = 0$

Solution

$x^2 - 8x + 16 = 0$

$(x-4)^2 = 0$ — The trinomial is a perfect square.

$x - 4 = 0$ or $x - 4 = 0$ — Both factors are the same, so there is only one distinct solution.

$x = 4$ $\quad x = 4$

The only solution is 4, and it is called a **double solution** (or **double root**).

Now work margin exercise 3.

3. Solve by factoring:
$x^2 - 6x + 9 = 0$

Example 4 Solving Quadratic Equations by Factoring

Solve by factoring: $4x^2 - 4x = 24$

Solution

$4x^2 - 4x = 24$ — Subtract 24 from both sides so that one side is 0.

$4x^2 - 4x - 24 = 0$ — Factor out the common monomial, 4.

$4(x^2 - x - 6) = 0$

$4(x-3)(x+2) = 0$ — The constant factor 4 can never be 0 and does not affect the solution.

$x - 3 = 0$ or $x + 2 = 0$

$x = 3$ $\quad x = -2$

The solutions are 3 and –2.

Now work margin exercise 4.

4. Solve by factoring:
$9x^2 - 27x = 36$

5. Solve by factoring:

$(x-8)^2=16$

Example 5 Solving Quadratic Equations by Factoring

Solve by factoring: $(x+5)^2=36$

Solution

$$
\begin{aligned}
(x+5)^2 &= 36 \\
x^2+10x+25 &= 36 && \text{Multiply } (x+5)^2. \\
x^2+10x-11 &= 0 && \text{Add } -36 \text{ to both sides so that one side is } 0. \\
(x+11)(x-1) &= 0 && \text{Factor.} \\
x+11=0 \quad &\text{or} \quad x-1=0 \\
x=-11 \quad & \quad\;\; x=1
\end{aligned}
$$

The solutions are −11 and 1.

Now work margin exercise 5.

6. Solve by factoring:

$2x(x-4)=3(4-x)$

Example 6 Solving Quadratic Equations by Factoring

Solve by factoring: $3x(x-1)=2(5-x)$

Solution

$$
\begin{aligned}
3x(x-1) &= 2(5-x) \\
3x^2-3x &= 10-2x && \text{Use the distributive property twice.} \\
3x^2-3x+2x-10 &= 0 && \text{Arrange terms so that 0 is on one side.} \\
3x^2-x-10 &= 0 && \text{Simplify.} \\
(3x+5)(x-2) &= 0 && \text{Factor.} \\
3x+5=0 \quad &\text{or} \quad x-2=0 \\
3x=-5 \quad & \quad\;\; x=2 \\
x=-\frac{5}{3} \quad &
\end{aligned}
$$

The solutions are $-\frac{5}{3}$ and 2.

Now work margin exercise 6.

7. Solve by factoring:

$\frac{x^2}{6}-\frac{x}{3}=\frac{1}{2}$

Example 7 Solving Quadratic Equations by Factoring

Solve by factoring: $\frac{2x^2}{15}-\frac{x}{3}=-\frac{1}{5}$

Solution

$$
\begin{aligned}
\frac{2x^2}{15}-\frac{x}{3} &= -\frac{1}{5} \\
15\cdot\frac{2x^2}{15}-\frac{x}{3}\cdot 15 &= -\frac{1}{5}\cdot 15 && \text{Multiply each term by 15, the LCM of the denominators, to get integer coefficients.} \\
2x^2-5x &= -3 && \text{Simplify.} \\
2x^2-5x+3 &= 0 && \text{Add 3 to both sides so that one side is 0.} \\
(2x-3)(x-1) &= 0 && \text{Factor.}
\end{aligned}
$$

$$2x - 3 = 0 \quad \text{or} \quad x - 1 = 0$$
$$2x = 3 \qquad x = 1$$
$$x = \frac{3}{2}$$

The solutions are $\frac{3}{2}$ and 1.

Now work margin exercise 7.

Completion Example 8 Solving Quadratic Equations by Factoring

Solve by factoring: $(x-2)^2 = 64$

Solution

$$(x-2)^2 = 64$$
$$x^2 - \underline{\quad 4 \quad}x + \underline{\quad 4 \quad} = 64$$
$$x^2 - \underline{\quad 4 \quad}x - \underline{\quad 60 \quad} = 0$$
$$(x + \underline{\quad 6 \quad})(x - \underline{\quad 10 \quad}) = 0$$
$$x + \underline{\quad 6 \quad} = 0 \quad \text{or} \quad x - \underline{\quad 10 \quad} = 0$$
$$x = \underline{\quad -6 \quad} \qquad x = \underline{\quad 10 \quad}$$

The solutions are $\underline{\quad -6 \quad}$ and $\underline{\quad 10 \quad}$.

Now work margin exercise 8.

In the next example, we show that factoring can be used to solve equations of degrees higher than second-degree. There will be more than two factors, but just as with quadratics, if the product is 0, then the solutions are found by setting each factor equal to 0.

Example 9 Solving Higher Degree Equations

Solve by factoring: $2x^3 - 4x^2 - 6x = 0$

Solution

$$2x^3 - 4x^2 - 6x = 0$$
$$2x(x^2 - 2x - 3) = 0 \qquad \text{Factor out the common monomial, } 2x.$$
$$2x(x-3)(x+1) = 0 \qquad \text{Factor the trinomial.}$$
$$2x = 0 \quad \text{or} \quad x - 3 = 0 \quad \text{or} \quad x + 1 = 0 \qquad \text{Set each factor equal to 0 and solve.}$$
$$x = 0 \qquad x = 3 \qquad x = -1$$

The solutions are 0, 3, and -1.
Or we can say that the solution set is $\{-1, 0, 3\}$.

Now work margin exercise 9.

8. Solve by factoring:

$(x+4)^2 = 9$

9. Solve by factoring:

$4x^3 - 12x^2 - 40x = 0$

Solving a Quadratic Equation by Factoring

1. Add or subtract terms as necessary so that 0 is on one side of the equation and the equation is in the standard form $ax^2 + bx + c = 0$, where a, b, and c are real numbers and $a \neq 0$.

2. Factor completely. (If there are any fractional coefficients, multiply each term by the least common denominator first so that all coefficients will be integers. Then, factor.)

3. Set each nonconstant factor equal to 0 and solve each linear equation for the unknown.

4. Check each solution, one at a time, in the original equation.

PROCEDURE

Special Comment: All of the quadratic equations in this section can be solved by factoring. That is, all of these quadratic polynomials are factorable. However, as we have seen in some of the previous sections, not all polynomials are factorable. In Chapter 16, we will develop techniques (other than factoring) for solving quadratic equations, whether the quadratic polynomial is factorable or not.

Common Error

A common error is to divide both sides of an equation by the variable x. This error can be illustrated by using the equation in Example 2.

$$3x^2 = 6x$$
$$\frac{3x^2}{x} = \frac{6x}{x}$$
$$3x = 6$$
$$x = 2$$

Do not divide by x, because you lose the solution $x = 0$.

INCORRECT

Factoring is the method to use. By factoring, you will find all solutions as shown in the previous examples.

CAUTION

B Finding an Equation Given the Roots

To help develop a complete understanding of the relationships between factors, factoring, and solving equations, we reverse the process of finding solutions. That is, we want to **find an equation that has certain given solutions (or roots)**. For example, to find an equation that has the roots

$$x = 4 \quad \text{and} \quad x = -7,$$

we proceed as follows.

1. Rewrite the linear equations with 0 on one side.
   $x-4=0$ and $x+7=0$
2. Form the product of the factors and set this product equal to 0. $(x-4)(x+7)=0$
3. Multiply the factors. The resulting quadratic equation must have the two given roots (because we know the factors).
   $x^2+3x-28=0$

The formal reasoning is based on the following theorem called the **factor theorem**.

Factor Theorem

If $x=c$ is a root of a polynomial equation in the form $P(x)=0$, then $x-c$ is a factor of the polynomial $P(x)$.

THEOREM

Note

The equation we just found, $x^2+3x-28=0$, can be multiplied by any nonzero constant and a new equation will be formed, but it will still have the same solutions, namely 4 and −7. Thus, technically, there are many equations with these two roots.

Example 10 Finding Equations with Given Roots

Find a polynomial equation with integer coefficients that has the given roots.

$x=3$ and $x=-\frac{2}{3}$

Solution

Form the linear equations and then find the product of the factors.

$x=3 \qquad x=-\frac{2}{3}$

$x-3=0 \qquad x+\frac{2}{3}=0$ Rewrite the equations with 0 on one side.

$3x+2=0$ Multiply by 3 to get integer coefficients.

Form the equation by setting the product of the factors equal to 0 and multiplying.

$(x-3)(3x+2)=0$ The equation has integer coefficients.

$3x^2-7x-6=0$ This quadratic equation has the two given roots.

Now work margin exercise 10.

10. Find a polynomial equation with integer coefficients that has the given roots.
    $x=4$ and $x=\frac{3}{2}$

Completion Example Answers

8. $x^2-4x+4=64$; $x^2-4x-60=0$; $(x+6)(x-10)=0$; $x+6=0$ or $x-10=0$;
$x=-6$ or $x=10$;
The solutions are -6 and 10.

Margin Exercise Answers

1. $y=7,\ \frac{5}{3}$ **2.** $v=0,2$ **3.** $x=3$ (double root) **4.** $x=-1,4$ **5.** $x=4,12$ **6.** $x=-\frac{3}{4},4$
7. $x=-1,3$ **8.** $x=-7,-1$ **9.** $x=-2,0,5$ **10.** $2x^2-11x+12=0$

13.6 Exercises

Concept Check

Fill-in-the-Blank. Complete each sentence using information found in this section.

1. When solving quadratic equations by factoring, it is necessary to have one side of the equation equal to ______.
2. The zero-factor law states that if the product of two or more factors equals zero, then at least one of the factors must be ______.
3. The factor theorem states that if $x = c$ is a root of a polynomial equation in the form $P(x) = 0$, then $x - c$ is a/an ______ of the polynomial $P(x)$.
4. In general, a quadratic equation has two solutions. If the two solutions are the same number, the equation is said to have a/an ________ solution or root.
5. Solutions can be checked by ____________ them one at a time for x in the equation.
6. Second-degree polynomials are called __________ polynomials.

True/False. Determine whether each statement is true or false. If a statement is false, explain how it can be changed so the statement will be true. (**Note:** There may be more than one acceptable change.)

7. When solving quadratic equations by factoring, it is important that all of the coefficients are integers.
8. The standard form for a quadratic equation is $ax^2 + bx = c$.
9. Not all quadratic equations can be solved by factoring.
10. All quadratic equations have two distinct solutions.

Practice

Solve each equation. See Example 1.

1. $(x-3)(x-2)=0$
2. $(x+5)(x-2)=0$
3. $(2x-9)(x+2)=0$
4. $(x+7)(3x-4)=0$
5. $0=(x+3)(x+3)$
6. $0=(x+10)(x-10)$
7. $(x+5)(x+5)=0$
8. $(x+5)(x-5)=0$
9. $2x(x-2)=0$
10. $3x(x+3)=0$
11. $(x+6)^2=0$
12. $5(x-9)^2=0$

Solve each equation by factoring. See Examples 2 through 9.

13. $x^2-3x-4=0$
14. $x^2+7x+12=0$

13.6 Exercises

Concept Check

1. zero
2. zero
3. factor
4. double
5. substituting
6. quadratic
7. True
8. False; Standard form is $ax^2 + bx + c = 0$.
9. True
10. False; Some quadratic equations have one double solution.

Practice

1. $x = 2, 3$
2. $x = -5, 2$
3. $x = -2, \frac{9}{2}$
4. $x = -7, \frac{4}{3}$
5. $x = -3$
6. $x = -10, 10$
7. $x = -5$
8. $x = -5, 5$
9. $x = 0, 2$
10. $x = -3, 0$
11. $x = -6$
12. $x = 9$
13. $x = -1, 4$
14. $x = -4, -3$
15. $x = -3, 4$
16. $x = 2, 9$
17. $x = -3, 0$
18. $x = 0, 3$
19. $x = 2, 4$
20. $x = -5, 6$
21. $x = -4, 3$
22. $x = -5, -2$
23. $x = -\frac{1}{2}, 3$

15. $x^2-x-12=0$

16. $x^2-11x+18=0$

17. $0=x^2+3x$

18. $0=x^2-3x$

19. $x^2+8=6x$

20. $x^2=x+30$

21. $2x^2+2x-24=0$

22. $9x^2+63x+90=0$

23. $0=2x^2-5x-3$

24. $0=2x^2-x-3$

25. $3x^2-4x-4=0$

26. $3x^2-8x+5=0$

27. $2x^2-7x=4$

28. $4x^2+8x=-3$

29. $-2x=3x^2-8$

30. $6x^2+2=-7x$

31. $4x^2-12x+9=0$

32. $25x^2-60x+36=0$

33. $8x=5x^2$

34. $15x=3x^2$

35. $9x^2-36=0$

36. $4x^2-16=0$

37. $5x^2=10x-5$

38. $2x^2=4x+6$

39. $8x^2+32=32x$

40. $6x^2=18x+24$

41. $\dfrac{x^2}{9}=1$

42. $\dfrac{x^2}{2}=8$

43. $\dfrac{x^2}{5}-x-10=0$

44. $\dfrac{2}{3}x^2+2x-\dfrac{20}{3}=0$

45. $\dfrac{x^2}{8}+x+\dfrac{3}{2}=0$

46. $\dfrac{x^2}{6}-\dfrac{1}{2}x-3=0$

47. $x^2-x+\dfrac{1}{4}=0$

48. $\dfrac{x^2}{3}-2x+3=0$

49. $x^3+8x=6x^2$

50. $x^3=x^2+30x$

51. $6x^3+7x^2=-2x$

52. $3x^3=8x-2x^2$

53. $0=x^2-100$

54. $0=x^2-121$

55. $3x^2-75=0$

56. $5x^2-45=0$

57. $x^2+8x+16=0$

58. $x^2+14x+49=0$

59. $3x^2=18x-27$

60. $5x^2=10x-5$

61. $(x-1)^2=4$

62. $(x-3)^2=1$

63. $(x+5)^2=9$

64. $(x+4)^2=16$

65. $(x+4)(x-1)=6$

66. $(x-5)(x+3)=9$

67. $27=(x+2)(x-4)$

68. $-1=(x+2)(x+4)$

69. $x(x+7)=3(x+4)$

24. $x=-1, \dfrac{3}{2}$

25. $x=-\dfrac{2}{3}, 2$

26. $x=1, \dfrac{5}{3}$

27. $x=-\dfrac{1}{2}, 4$

28. $x=-\dfrac{3}{2}, -\dfrac{1}{2}$

29. $x=-2, \dfrac{4}{3}$

30. $x=-\dfrac{2}{3}, -\dfrac{1}{2}$

31. $x=\dfrac{3}{2}$

32. $x=\dfrac{6}{5}$

33. $x=0, \dfrac{8}{5}$

34. $x=0, 5$

35. $x=-2, 2$

36. $x=-2, 2$

37. $x=1$

38. $x=-1, 3$

39. $x=2$

40. $x=-1, 4$

41. $x=-3, 3$

42. $x=-4, 4$

43. $x=-5, 10$

44. $x=-5, 2$

45. $x=-6, -2$

46. $x=-3, 6$

47. $x=\dfrac{1}{2}$

48. $x=3$

49. $x=0, 2, 4$

50. $x=-5, 0, 6$

51. $x=-\dfrac{2}{3}, -\dfrac{1}{2}, 0$

52. $x=-2, 0, \dfrac{4}{3}$

53. $x=-10, 10$

54. $x=-11, 11$

55. $x=-5, 5$

56. $x=-3, 3$

57. $x=-4$

58. $x=-7$

59. $x=3$

60. $x=1$

61. $x=-1, 3$

62. $x=2, 4$

63. $x=-8, -2$

64. $x=-8, 0$

65. $x=-5, 2$

66. $x=-4, 6$

67. $x=-5, 7$

68. $x=-3$

69. $x=-6, 2$

70. $x=-6, 3$

71. $x=-1, \frac{2}{3}$

72. $x=1, \frac{3}{2}$

73. $x=-\frac{3}{2}, 4$

74. $x=-1, -\frac{2}{3}$

75. $y^2-y-6=0$

76. $x^2-12x+35=0$

77. $2x^2+11x+5=0$

78. $4x^2+3x-1=0$

79. $8x^2-10x+3=0$

80. $18y^2-15y+2=0$

81. $x^3-x^2-6x=0$

82. $y^3+3y^2-4y=0$

83. $y^3-4y^2-3y+18=0$

84. $x^3+3x^2+3x+1=0$

Applications

85. **a.** 640 ft; 384 ft

b. 144 ft; 400 ft

c. 7 seconds; $0=-16(t+7)(t-7)$

86. **a.** $s(0)=490$; This means that the ball is dropped from a height of 490 meters.

b. 470.4 meters

c. 10 seconds

70. $x(x+9)=6(x+3)$

71. $3x(x+1)=2(x+1)$

72. $2x(x-1)=3(x-1)$

73. $x(2x+1)=6(x+2)$

74. $3x(x+3)=2(2x-1)$

Find a polynomial equation with integer coefficients that has the given roots. See Example 10.

75. $y=3, y=-2$

76. $x=5, x=7$

77. $x=-5, x=-\frac{1}{2}$

78. $x=\frac{1}{4}, x=-1$

79. $x=\frac{1}{2}, x=\frac{3}{4}$

80. $y=\frac{2}{3}, y=\frac{1}{6}$

81. $x=0, x=3, x=-2$

82. $y=0, y=-4, y=1$

83. $y=-2, y=3, y=3$ (3 is a double root.)

84. $x=-1, x=-1, x=-1$ (-1 is a triple root.)

Applications

Solve.

85. A ball is dropped from the top of a building that is 784 feet high. The height of the ball above ground level is given by the polynomial function $h(t)=-16t^2+784$, where t is measured in seconds.

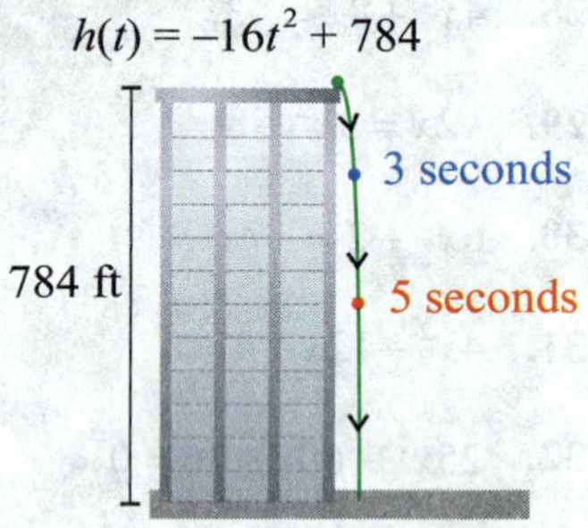

a. How high is the ball after 3 seconds? 5 seconds?

b. How far has the ball traveled in 3 seconds? 5 seconds?

c. When will the ball hit the ground? Explain your reasoning in terms of factors.

86. A tennis ball is dropped from a building. The position of the ball after t seconds is given by the polynomial function $s(t)=-4.9t^2+490$, where s is the height in meters of the ball.

a. Find $s(0)$. What does this value represent in the context of this problem?

b. How high is the tennis ball 2 seconds after it has been dropped?

c. How long before the tennis ball hits the ground?

87. A ball is thrown upward from an initial height of 96 feet with an initial velocity of 16 feet per second. After t seconds, the height of the ball can be described by the equation $h=-16t^2+16t+96$.

a. What happens when $h=0$?

b. Rewrite the equation with $h=0$.

c. Solve the equation by factoring.

d. What does the answer to part c. mean?

e. Do both solutions from part c. make sense in the context of the problem? Explain why or why not.

88. Robin is putting the finishing touches on a quilt. The quilt is currently 80 inches long by 60 inches wide, and she plans to add a border around the quilt. The width of the border on the sides will be twice the width of the border on the top and bottom of the quilt.

  a. If x is the width of the border in inches that will be added to the top and bottom of the quilt, write an expression for the length and width of the quilt with the border added.

  b. Write a simplified expression to find the area of the quilt with the border added.

  c. Robin has a total of 5712 square inches of fabric to use for the back of the quilt. Use the expression from part b. to write an equation to describe the total area of the back of the quilt.

  d. Solve the quadratic equation from part c. by factoring.

  e. Do both of the solutions from part d. make sense in the context of the problem?

  f. What is the total length and width of the quilt?

89. We know that the area of a circle is proportional to the square of the radius. In fact, if the radius of a circle is r, then the area of the circle is $A = \pi r^2$. Let's determine how the area is changed when we double the radius.

  a. Find the area of the circle for a radius of 1 in.

  b. Find the area of the circle for a radius of 2 in.

  c. Find the area of the circle for a radius of 4 in.

  d. Find the area of the circle for a radius of 8 in.

  e. Do you see the pattern? If you double the radius, by how many times does the area increase?

90. The St. Louis Arch is not quite in the shape of a parabola, but it can be closely modeled with the polynomial function $h(x) = -0.007x^2 + 0.003x + 625$, where x and $h(x)$ are both measured in feet and the center of the arch lies along the y-axis.

  a. Find $h(0)$. What does this mean in this context?

  b. Find $h(100)$. What does this mean in this context?

  c. Find $h(300)$. Use this value to approximate the total distance between the two points at which the arch hits the ground.

  d. Use your graphing calculator to sketch the graph and find the x-intercepts (to the nearest integer). What is the actual distance between the two points at which the arch hits the ground? (rounded to the nearest foot) (**Note:** See Section 10.5 to review finding x-intercepts by using a graphing calculator.)

Writing & Thinking

91. When solving equations by factoring, one side of the equation must be 0. Explain why this is so.

92. In solving the equation $(x+5)(x-4) = 6$, why can't we just put one factor equal to 3 and the other equal to 2? Certainly $3 \cdot 2 = 6$.

Writing & Thinking

91. This allows for use of the zero factor law, which says that for the product to equal zero one of the factors must equal zero. Answers will vary.

92. Answers will vary.

87. **a.** The ball will hit the ground.

b. $0 = -16t^2 + 16t + 96$

c. $t = -2, 3$

d. The ball will hit the ground after -2 seconds and after 3 seconds.

e. No, time cannot be negative.

88. **a.** Length is $80 + 2x$; width is $60 + 4x$.

b. $(80 + 2x)(60 + 4x) = 8x^2 + 440x + 4800$

c. $8x^2 + 440x + 4800 = 5712$

d. $x = 2, x = -57$

e. The negative length does not make sense in the context of the problem.

f. 84 inches by 68 inches

89. **a.** π in.^{2}

b. 4π in.^{2}

c. 16π in.^{2}

d. 64π in.^{2}

e. The area gets 4 times larger.

90. **a.** $h(0) = 625$; This means the arch is 625 feet tall at its center.

b. $h(100) = 555.3$; This means that 100 feet away from the center axis of the arch, the arch is 555.3 feet tall.

c. $h(300) = -4.1$; This means that the arch hits the ground a little less than 300 feet from the center axis of the arch.

d. x-intercepts: $(-299,0), (299,0)$; 598 feet

Objectives

A. Use quadratic equations to solve application problems.

B. Use quadratic equations to solve problems involving consecutive integers.

C. Use quadratic equations to solve problems related to the Pythagorean Theorem.

13.7 Applications: Quadratic Equations

Whether application problems cause you difficulty depends a great deal on your personal experiences and general reasoning abilities. These abilities are developed over a long period of time. A problem that is easy for you, possibly because you have had experience in a particular situation, might be quite difficult for a friend and vice versa.

Most problems do not say specifically to add, subtract, multiply, or divide. You are to know from the nature of the problem what to do. You are to ask yourself, "What information is given? What am I trying to find? What tools, skills, and abilities do I need to use?"

Word problems should be approached in an orderly manner. You should have an "attack plan."

Attack Plan for Application Problems

1. Read the problem carefully at least twice.
2. Decide what is asked for and assign a variable or variable expression to the unknown quantities. It may help to organize a chart, table, or diagram relating all the information provided.
3. Form and then solve an equation that relates the information provided. (A formula of some type may be necessary.)
4. Check your solution with the wording of the problem to be sure it makes sense.

PROCEDURE

Several types of problems lead to quadratic equations. The problems in this section are set up so that the equations can be solved by factoring. More general problems and approaches to solving quadratic equations are discussed in Chapter 16.

A Using Quadratic Equations to Solve Applications

1. One number is five less than another and the sum of their squares is 325. What are the numbers?

Example 1 Solving Quadratic Number Problems

One number is four more than another and the sum of their squares is 296. What are the numbers?

Solution

Let $x =$ smaller number

and $x + 4 =$ larger number.

Set the sum of their squares equal to 296 and solve the equation.

$$x^2 + (x+4)^2 = 296$$ Add the squares.

$$x^2 + x^2 + 8x + 16 = 296$$ Expand $(x+4)^2$.

$$2x^2 + 8x - 280 = 0$$ Write the equation in standard form.

$$2(x^2 + 4x - 140) = 0$$ Factor out the GCF.

$$2(x+14)(x-10) = 0$$ Factor the trinomial.

$x + 14 = 0$ or $x - 10 = 0$

$x = -14$ $\quad$ $x = 10$

$x + 4 = -10$ $\quad$ $x + 4 = 14$

There are two sets of answers to the problem: 10 and 14 or –14 and –10.

Check

$10^2 + 14^2 = 100 + 196 = 296$

and $(-14)^2 + (-10)^2 = 196 + 100 = 296$

Now work margin exercise 1.

Example 2 Application: Solving Quadratic Equations

In an orange grove, there are 10 more trees in each row than there are rows. How many rows are there if there are 96 trees in the grove?

Solution

Let r = number of rows

and $r + 10$ = number of trees per row.

Set up the equation and solve.

$$r(r+10) = 96$$

$$r^2 + 10r = 96$$

$$r^2 + 10r - 96 = 0$$

$$(r-6)(r+16) = 0$$

$r - 6 = 0$ or $r + 16 = 0$

$r = 6$ $\quad$ $r = -16$

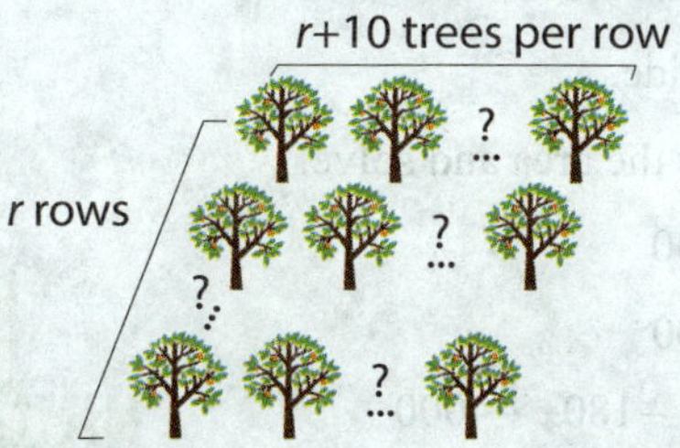

While –16 is a solution to the equation, –16 does not fit the conditions of the problem and is discarded. You cannot have –16 rows.

Thus, there are 6 rows in the grove (6 · 16 = 96 trees).

Now work margin exercise 2.

2. In a theater, there are 4 fewer seats in a row than there are rows. How many rows are there if there are 165 seats in the theater?

Example 3 Application: Solving Quadratic Equations

A rectangle has an area of 135 square meters and a perimeter of 48 meters. What are the dimensions of the rectangle?

3. A rectangle has an area of 98 square feet and a perimeter of 42 feet. What are the dimensions of the rectangle?

Math Tip

Applications involving quadratic equations can also be set up as a system of nonlinear equations. In Example 3, the system would include one equation for the perimeter and one equation for the area.

$$\begin{cases} lw = 135 \\ 2l + 2w = 48 \end{cases}$$

We will learn more about these types of systems later on.

Solution

The area of a rectangle is the product of its length and width ($A = lw$). The perimeter of a rectangle is given by $P = 2l + 2w$. Since the perimeter is 48 meters, then the length plus the width must be 24 meters (one-half of the perimeter).

Let w = width

and $24 - w$ = length.

Set up an equation for the area and solve.

$$(24 - w) \cdot w = 135 \qquad \text{Area} = lw = \text{length times width}$$
$$24w - w^2 = 135$$
$$0 = w^2 - 24w + 135$$
$$0 = (w - 9)(w - 15)$$

$$w - 9 = 0 \quad \text{or} \quad w - 15 = 0$$
$$w = 9 \qquad\qquad w = 15$$
$$l = 24 - 9 = 15 \qquad\qquad l = 24 - 15 = 9$$

The dimensions are 9 meters by 15 meters ($9 \cdot 15 = 135$).

Now work margin exercise 3.

4. A husband and wife want to screen in their back patio that is in the shape of a rectangle. If the perimeter of the three sides is 38 feet and the area of the patio is 168 square feet, what are the dimensions of the patio?

Example 4 Applications: Solving Quadratic Equations

A man wants to build a fence on three sides of a rectangular-shaped lot he owns. If 180 feet of fencing is needed and the area of the lot is 4000 square feet, what are the dimensions of the lot?

Solution

Let x = one of two equal sides

and $180 - 2x$ = third side.

Set up an equation for the area and solve.

$$x(180 - 2x) = 4000$$
$$180x - 2x^2 = 4000$$
$$0 = 2x^2 - 180x + 4000$$
$$0 = 2(x^2 - 90x + 2000)$$
$$0 = 2(x - 50)(x - 40)$$

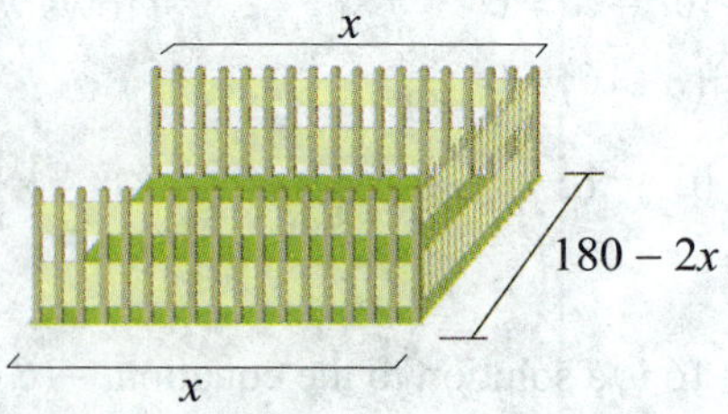

$$x - 50 = 0 \quad \text{or} \quad x - 40 = 0$$
$$x = 50 \qquad\qquad x = 40$$
$$180 - 2x = 180 - 2(50) = 80 \qquad 180 - 2x = 180 - 2(40) = 100$$

Thus, there are two possible answers: the lot is 50 feet by 80 feet or the lot is 40 feet by 100 feet.

Now work margin exercise 4.

B Consecutive Integers

In Section 9.5, we discussed applications with **consecutive integers**, **consecutive even integers**, and **consecutive odd integers**. Because the applications involved only addition and subtraction, the related equations were first degree. In this section, the applications involve squaring expressions and solving quadratic equations. For convenience, the definitions and representations of the various types of consecutive integers are repeated here.

Consecutive Integers

Integers are **consecutive** if each is 1 more than the previous integer. Three consecutive integers can be represented as $\boldsymbol{n}$, $\boldsymbol{n+1}$, and $\boldsymbol{n+2}$.

For example, 5, 6, and 7.

DEFINITION

Consecutive Even Integers

Even integers are consecutive if each is 2 more than the previous even integer. Three consecutive even integers can be represented as $\boldsymbol{n}$, $\boldsymbol{n+2}$, and $\boldsymbol{n+4}$, where n is an even integer.

For example, 24, 26, and 28.

DEFINITION

Consecutive Odd Integers

Odd integers are consecutive if each is 2 more than the previous odd integer. Three consecutive odd integers can be represented as $\boldsymbol{n}$, $\boldsymbol{n+2}$, and $\boldsymbol{n+4}$, where n is an odd integer.

For example, 41, 43, and 45.

DEFINITION

Note

Note that consecutive even and consecutive odd integers are represented in the same way. The value of the first integer n determines whether the remaining integers are even or odd.

Example 5 Consecutive Integers

Find two consecutive positive integers such that the sum of their squares is 265.

Solution

Let $n =$ first integer.

Then, $n + 1 =$ next consecutive integer.

Set up and solve the related equation.

5. Find two consecutive negative odd integers such that the sum of their squares is 202.

$$n^2+(n+1)^2=265$$
$$n^2+n^2+2n+1=265$$
$$2n^2+2n-264=0$$
$$2(n^2+n-132)=0$$
$$2(n+12)(n-11)=0$$
$$n+12=0 \quad \text{or} \quad n-11=0$$
$$n=-12 \qquad n=11$$
$$n+1=-11 \qquad n+1=12$$

Consider the solution $n = -12$. The next consecutive integer, $n + 1$, is -11. While it is true that the sum of their squares is 265, we must remember that the problem calls for **positive** consecutive integers. Therefore, we can only consider positive solutions. Hence, the two integers are 11 and 12.

Now work margin exercise 5.

6. Find three consecutive even integers such that the product of the second and third is 112 more than the first.

Example 6 Consecutive Integers

Find three consecutive odd integers such that the product of the first and second is 68 more than the third.

Solution

Let n = first odd integer

and $n + 2$ = second consecutive odd integer

and $n + 4$ = third consecutive odd integer.

Set up and solve the related equation.

$$n(n+2)=(n+4)+68$$
$$n^2+2n=n+72$$
$$n^2+n-72=0$$
$$(n-8)(n+9)=0$$
$$n-8=0 \quad \text{or} \quad n+9=0$$
$$n=8 \qquad n=-9$$
$$n+2=10 \qquad n+2=-7$$
$$n+4=12 \qquad n+4=-5$$

The three consecutive odd integers are -9, -7, and -5. Note that 8, 10, and 12 are even, and therefore cannot be considered a solution to the problem.

Now work margin exercise 6.

C The Pythagorean Theorem

A geometric topic that often generates quadratic equations is right triangles and the Pythagorean Theorem. Recall from Chapter 6 that in a **right triangle**, one of the angles is a right angle (measures 90°), and the side opposite this angle (the longest side) is called the **hypotenuse**. The other two sides are called **legs**.

The Pythagorean Theorem

In a right triangle, if c is the length of the hypotenuse and a and b are the lengths of the legs, then

$$c^2 = a^2 + b^2.$$

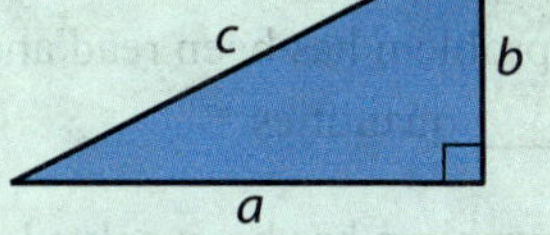

THEOREM

Example 7 Application: The Pythagorean Theorem

A support wire is 25 feet long and stretches from a tree to a point on the ground. The point of attachment on the tree is 5 feet higher than the distance from the base of the tree to the point of attachment on the ground. How far up the tree is the point of attachment?

Solution

Let x = distance from base of tree to point of attachment on ground.

Then, $x + 5$ = height of point of attachment on tree.

By the Pythagorean Theorem, we have the following.

$$(x+5)^2 + x^2 = 25^2$$
$$x^2 + 10x + 25 + x^2 = 625$$
$$2x^2 + 10x - 600 = 0$$
$$2(x^2 + 5x - 300) = 0$$
$$2(x-15)(x+20) = 0$$
$$x - 15 = 0 \quad \text{or} \quad x + 20 = 0$$
$$x = 15 \qquad\qquad x = -20$$

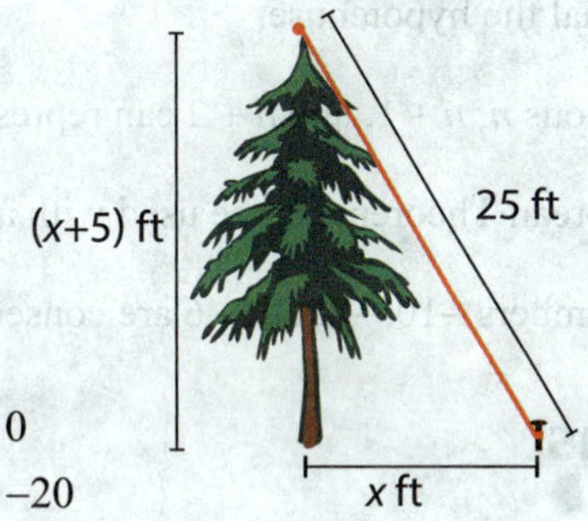

Because distance must be positive, -20 is not a possible solution.

The solution is

$$x = 15$$
$$\text{and } x + 5 = 20.$$

Thus, the point of attachment is 20 feet up the tree.

Now work margin exercise 7.

7. A guyline for a portable beach volleyball set is 10 feet long from the top of the net to where it is staked to the ground. The distance from the base of the pole to the point of attachment on the beach is 2 feet less than the height of the volleyball net. What is the height of the net?

Margin Exercise Answers

1. 15, 10 or −15, −10 **2.** 15 rows **3.** 7 feet by 14 feet **4.** 12 feet by 14 feet or 7 feet by 24 feet **5.** −11, −9 **6.** 8, 10, 12 **7.** 8 feet

13.7 Exercises

Concept Check

1. variable, unknown
2. check
3. consecutive
4. $c^2 = a^2 + b^2$
5. 1
6. $n + 2$
7. False; The sum of the squares of the lengths of the legs is equal to the hypotenuse squared.
8. True
9. False: Only with right triangles
10. True

Applications

1. $x(x+8) = -16$; $x = -4$, so the numbers are -4 and 4.
2. $x(x+10) = -25$; $x = -5$, so the numbers are -5 and 5.
3. $x^2 = 7x$; $x = 0, 7$
4. $x^2 = 2x$; $x = 0, 2$
5. $x^2 + 3x = 28$; $x = 4$
6. $x^2 + 3x = 54$; $x = 6$
7. $x(x+3) = 40$ $x = -8, 5$, so the numbers are -8 and -5 or 5 and 8.
8. $x(2x+3) = 27$; $x = 3$, so the numbers are 3 and 9.
9. $(x+5)^2 + x^2 = 53$; $x = 2$, so the numbers are 2 and 7.
10. $x^2 + (x-5)^2 = 97$; $x = -4, 9$, so the numbers are -4 and -9 or 9 and 4.
11. $x + (x+4)^2 = 38$; $x = 2$, so the integers are 2 and 6.
12. $x^2 + (3+2x) = 51$; $x = 6$, so the numbers are 6 and 15.

13.7 Exercises

Concept Check

Fill-in-the-Blank. Complete each sentence using information found in this section.

1. Once an application problem has been read and understood, you should assign a ________ to the ________ quantities.

2. After an application problem has been solved, it is important to ______ the solution with the problem to make sure the answer makes sense.

3. Even integers are ___________ if each is 2 more than the previous even integer.

4. The formula (or equation) related to the Pythagorean Theorem is ________.

5. Integers are consecutive if each is ___ more than the previous integer.

6. Two consecutive odd integers can be represented by n and ____.

True/False. Determine whether each statement is true or false. If a statement is false, explain how it can be changed so the statement will be true. (**Note:** There may be more than one acceptable change.)

7. The Pythagorean Theorem states that if the two legs of a right triangle are added, the sum will equal the hypotenuse.

8. The expressions n, $n + 1$, and $n + 2$ can represent three consecutive integers.

9. The Pythagorean Theorem can be used with any triangle.

10. The three numbers -10, -8, and -6 are consecutive even integers.

Applications

Write a quadratic equation for each of the following word problems. Then solve the word problem. Remember to check each solution with the wording of the original problem to make sure it is reasonable. See Examples 1 through 7.

1. One number is eight more than another. Their product is -16. What are the numbers?

2. One number is 10 more than another. If their product is -25, find the numbers.

3. The square of an integer is equal to seven times the integer. Find the integer.

4. The square of an integer is equal to twice the integer. Find the integer.

5. If the square of a positive integer is added to three times the integer, the result is 28. Find the integer.

6. If the square of a positive integer is added to three times the integer, the result is 54. Find the integer.

7. One number is three more than another. Their product is 40. Find the numbers.

8. One positive number is three more than twice another. If the product is 27, find the numbers.

9. One positive number is five more than another. The sum of their squares is 53. What are the numbers?

10. One number is five less than another. The sum of their squares is 97. Find the numbers.

11. The difference between two positive integers is 4. If the smaller is added to the square of the larger, the sum is 38. Find the integers.

12. One positive number is 3 more than twice another. If the square of the smaller is added to the larger, the sum is 51. Find the numbers.

13. The product of a negative integer and 5 less than twice the integer equals the integer plus 56. Find the integer.

14. Find a positive integer such that the product of the integer with a number three less than the integer is equal to the integer increased by 32.

15. The product of two consecutive positive integers is 72. Find the integers.

16. Find two consecutive integers whose product is 110.

17. Find two consecutive positive integers such that the sum of their squares is 85.

18. Find two consecutive positive integers such that the sum of their squares is 145.

19. The product of two consecutive odd integers is 63. Find the integers.

20. The product of two consecutive even integers is 80. Find the integers.

21. Find two consecutive positive integers such that the square of the second integer added to four times the first is equal to 41.

22. Find two consecutive negative integers such that 6 times the first plus the square of the second equals -14.

23. Find three consecutive positive integers such that twice the product of the two smaller integers is 88 more than the product of the two larger integers.

24. Find three consecutive odd integers such that the product of the first and third is 1 more than 4 times the second.

25. Four consecutive integers are such that, if the product of the first and third is multiplied by 6, the result is equal to the sum of the second and the square of the fourth. What are the integers?

26. Find four consecutive even integers such that the square of the sum of the first and second is equal to 60 less than twice the product of the third and fourth.

27. The length of a rectangle is twice the width. The area is 72 square inches. Find the length and width of the rectangle.

28. The length of a rectangle is three times the width. If the area is 147 square centimeters, find the length and width of the rectangle.

29. The length of a rectangle is four times the width. If the area is 64 square feet, find the length and width of the rectangle.

13. $x(2x-5)=x+56$; $x=-4$

14. $x(x-3)=x+32$; $x=8$

15. $x(x+1)=72$; $x=8$, so the integers are 8 and 9.

16. $x(x+1)=110$; $x=-11, 10$, so the integers are -11 and -10 or 10 and 11.

17. $x^2+(x+1)^2=85$; $x=6$, so the integers are 6 and 7.

18. $(x+1)^2+x^2=145$; $x=8$, so the integers are 8 and 9.

19. $x(x+2)=63$; $x=-9, 7$, so the integers are -9 and -7 or 7 and 9.

20. $x(x+2)=80$; $x=-10$, 8, so the integers are -10 and -8 or 8 and 10.

21. $4x+(x+1)^2=41$; $x=4$, so the integers are 4 and 5.

22. $6x+(x+1)^2=-14$; $x=-5, -3$, so the integers are -5 and -4 or -3 and -2.

23. $2x(x+1)=(x+1)(x+2)+88$; $x=10$, so the integers are 10, 11, and 12.

24. $x(x+4)=4(x+2)+1$; $x=-3, 3$, so the integers are -3, -1, and 1 or 3, 5, and 7.

25. $6x(x+2)=(x+1)+(x+3)^2$; $x=-2, 1$, so the integers are -2, -1, 0, and 1 or 1, 2, 3, and 4.

26. $[x+(x+2)]^2=2(x+4)(x+6)-60$; $x=2, 4$, so the integers are 2, 4, 6, and 8 or 4, 6, 8, and 10.

27. $w(2w)=72$; $w=6$, so width is 6 in. and length is 12 in.

28. $w(3w)=147$; $w=7$, so width is 7 cm and length is 21 cm.

29. $w(4w)=64$; $w=4$, so width is 4 ft and length is 16 ft.

30. $w(5w)=180$; $w=6$, so width is 6 in. and length is 30 in.

31. $l(l-4)=45$; $l=9$, so width is 5 ft and length is 9 ft.

32. $w(w+3)=54$; $w=6$, so width is 6 m and length is 9 m.

33. $\frac{1}{2}b(b-4)=16$; $b=8$, so base is 8 ft and height is 4 ft.

34. $\frac{1}{2}(h+5)h=12$; $h=3$, so height is 3 m and base is 8 m.

35. $\frac{1}{2}(h+6)h=20$; $h=4$, so the base is 10 in.

36. $\frac{1}{2}(h-3)h=9$; $h=6$, so the height is 6 ft.

37. $w(16-w)=48$; $w=4$, 12, so the rectangle is 4 in. by 12 in.

38. $w(10-w)=24$; $w=4$, 6, so the rectangle is 4 cm by 6 cm.

39. $r(r+13)=140$; $r=$ 7, so there are 7 trees in each row.

40. $n(n+3)=108$; $n=9$, so there are 9 rows.

41. $r(r+7)=144$; $r=9$, so there are 9 rows.

42. $r(r+6)=187$; $r=11$, so there are 11 rows.

30. The length of a rectangle is five times the width. If the area is 180 square inches, find the length and width of the rectangle.

31. The width of a rectangle is 4 feet less than the length. The area is 45 square feet. Find the length and width of the rectangle.

32. The length of a rectangular yard is 3 meters greater than the width. If the area of the yard is 54 square meters, find the length and width of the yard.

33. The height of a triangle is 4 feet less than the base. The area of the triangle is 16 square feet. Find the length of the base and the height of the triangle.

34. The base of a triangle exceeds the height by 5 meters. If the area is 12 square meters, find the length of the base and the height of the triangle.

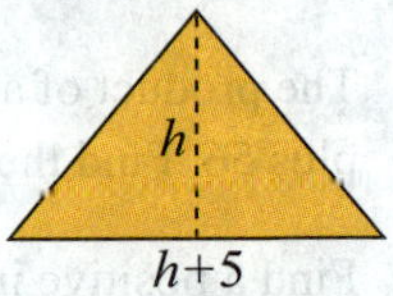

35. The base of a triangle is 6 inches greater than the height. If the area is 20 square inches, find the length of the base.

36. The base of a triangle is 3 feet less than the height. The area is 9 square feet. Find the height.

37. The perimeter of a rectangle is 32 inches. The area of the rectangle is 48 square inches. Find the dimensions of the rectangle.

38. The area of a rectangle is 24 square centimeters. If the perimeter is 20 centimeters, find the length and width of the rectangle.

39. An orchard has 140 apple trees. The number of rows exceeds the number of trees per row by 13. How many trees are there in each row?

40. One formation for an army drill team is rectangular. The number of members in each row exceeds the number of rows by 3. If there is a total of 108 members in the formation, how many rows are there?

41. A theater can seat 144 people. The number of rows is 7 less than the number of seats in each row. How many rows of seats are there?

42. An empty field on a college campus is being used for overflow parking for a football game. It currently has 187 cars in it. If the number of rows of cars is six less than the number of cars in each row, how many rows are there?

43. The parking garage at Baltimore-Washington International Airport contains 8400 parking spaces. The number of cars that can be parked on each floor exceeds the number of floors by 1675. How many floors are there in the parking garage?

44. One bookshelf in the public library can hold 175 books. The number of books on each shelf exceeds the number of shelves by 18. How many books are on each shelf?

45. The length of a rectangle is 7 centimeters greater than the width. If 4 centimeters are added to both the length and width, the new area would be 98 square centimeters. Find the dimensions of the original rectangle.

46. The width of a rectangle is 5 meters less than the length. If 6 meters are added to both the length and width, the new area will be 300 square meters. Find the dimensions of the original rectangle.

47. Susan is going to fence in a rectangular flower garden in her back yard. She has 50 feet of fencing, and she plans to use the house as the fence on one side of the garden. If the area is 300 square feet, what are the dimensions of the flower garden?

48. A rancher is going to build a corral with 52 yards of fencing. He is planning to use the barn as one side of the corral. If the area is 320 square yards, what are the dimensions?

49. A telephone pole is to have a guy wire attached to its top and anchored to the ground at a point that is at a distance 34 feet less than the height of the pole from the base. If the wire is to be 2 feet longer than the height of the pole, what is the height of the pole?

50. Lucy is standing next to the the General Sherman tree in Sequoia National Park, home of some of the largest trees in the world. The distance from Lucy to the base of the tree is 71 meters less than the height of the tree. If the distance from Lucy to the top of the tree is 1 meter more than the height of the tree, how tall is the General Sherman?

51. A Christmas tree is supported by a wire that is 1 foot longer than the height of the tree. The wire is anchored at a point whose distance from the base of the tree is 49 feet shorter than the height of the tree. What is the height of the tree?

52. An architect wants to draw a rectangle with a diagonal of 13 inches. The length of the rectangle is to be 2 inches more than twice the width. What dimensions should she make the rectangle?

53. Incline mats, or triangle mats, are offered with different levels of incline to help gymnasts learn basic moves. As the name may suggest, two sides of the mat are right triangles. If the height of the mat is 28 inches shorter than the length of the mat and the hypotenuse is 8 inches longer than the length of the mat, what is the length of the mat?

54. Bill uses mirrors to augment the "laser experience" at a laser show. At one show, he places three mirrors, A, B, C, in a right triangular form. If the distance between A and B is 15 m more than the distance between A and C, and the distance between B and C is 15 m less than the distance between A and C, what is the distance between mirror A and mirror C?

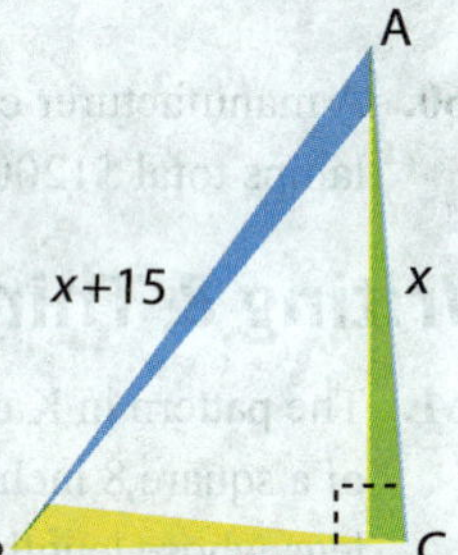

55. A support wire is attached x feet from the top of a 17-foot pole to protect the pole during a blizzard. The other end of each wire is attached to a stake x feet from the base of the pole. The wire used is 13 feet long.

a. Draw a diagram to describe the situation. Be sure to label the figure with the known information.

b. Use the Pythagorean Theorem to write an equation that describes the situation. Do not simplify.

c. Simplify the equation from part b. and solve for x.

d. Do both solutions from part b. make sense in this situation? That is, do they both result in a positive distance on the pole and a positive distance from the pole?

e. What do the answers from part b. mean? (You should have two answers.)

f. Which answer from part d. seems like the better option? Write an explanation for your choice.

43. $n(n+1675)=8400$; $n=5$, so there are 5 floors.

44. $b(b-18)=175$; $b=25$, so there are 25 books on each shelf.

45. $(w+11)(w+4)=98$; $w=3$, so the rectangle was 3 cm by 10 cm.

46. $(l+6)(l+1)=300$; $l=14$, so the rectangle was 14 m by 9 m.

47. $w=10, 15$, so width is 10 ft and length is 30 ft or width is 15 ft and length is 20 ft.

48. $w(52-2w)=320$; $w=10, 16$, so width is 10 yd and length is 32 yd or width is 16 yd and length is 20 yd.

49. $h^2+(h-34)^2=(h+2)^2$; $h=48$, so the height of the pole is 48 ft.

50. $h^2+(h-71)^2=(h+1)^2$; $h=84$, so the height of the General Sherman is 84 meters.

51. $x^2+(x-49)^2=(x+1)^2$; $x=60$, so height is 60 ft.

52. $w^2+(2w+2)^2=13^2$; $w=5$, so width is 5 in. and length is 12 in.

53. $l^2+(l-28)^2=(l+8)^2$; $l=60$, so the length of the mat is 60 inches.

54. $x^2+(x-15)^2=(x+15)^2$; $x=60$, so the distance between A and C is 60 m.

55. **a.** Figures will vary.

b. $(17-x)^2+x^2=13^2$

c. $x=12, x=5$

d. Yes

e. The wire is attached to the pole 5 feet from the ground (12 feet from the top) and attached to the stake which is 12 feet from the pole. Or the wire is attached to the pole 12 feet from the ground (5 feet from the top) and attached to the stake which is 5 feet from the pole.

f. Answers will vary.

56. a. Figures will vary.

b. $180 - 2x$

c. $180x - 2x^2 = 4000$

d. 50, 40

e. Yes

f. 50 feet by 80 feet and 40 feet by 100 feet

57. $1.50 per pound

58. $3.50 or $8.75 per drink

59. $16 or $20 per reel

60. $20 or $30 per lamp

Writing & Thinking

61. 128 in.^{2}; We can use the Pythagorean Theorem to find the length and the width (by finding the diagonals of the interior square) instead.

56. A family wants to fence in a rectangular area of their yard next to the house so their dog can play outside without being on a leash. One side of the fenced-in area will be along the side of the house, so they will only need to fence in three sides. The family decides to fence in an area of 4000 square feet and they purchase 180 feet of fencing. What are the dimensions of the fenced in area?

a. Draw a diagram to represent the situation. Use the variable x to label the two sides of the fence which will have the same length.

b. Write an expression involving x to represent the length of the third side of the fence.

c. Write an equation to represent the area of the fenced-in yard.

d. Solve the equation from part c.

e. Do both solutions make sense in the context of the problem?

f. What are the possible dimensions of the fenced-in yard?

The **demand** for a product is the number of units of the product x that consumers are willing to buy when the market price is p dollars. The consumers' **total expenditure** for the product S is found by multiplying the price times the demand. $(S = px)$ Solve the following consumer demand questions.

57. During the summer at a local market, a farmer will sell $8p + 588$ pounds of peaches at p dollars per pound. If he sold $900 worth of peaches this summer, what was the price per pound of the peaches?

58. On a hot afternoon, fans at a stadium will buy $490 - 40p$ drinks for p dollars each. If the total sales after a game were $1225, what was the price per drink?

59. When fishing reels are priced at p dollars, local consumers will buy $36 - p$ fishing reels. What is the price if total sales were $320?

60. A manufacturer can sell $100 - 2p$ lamps at p dollars each. If the receipts from the lamps total $1200, what is the price of the lamps?

Writing & Thinking

61. The pattern in Kara's linoleum flooring is in the shape of a square 8 inches on a side with right triangles (with legs whose lengths are x inches) placed on each side of the original square so that a new larger square is formed. What is the area of the new square? Explain why you do not need to find the value of x.

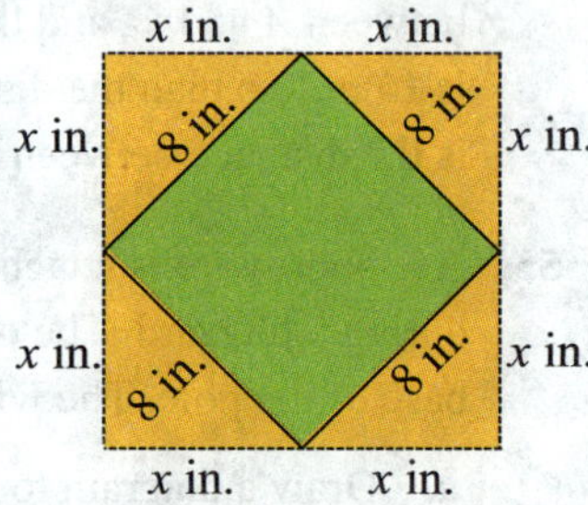

Chapter 13 Project

Building a Dog Pen

An activity to demonstrate the use of the Pythagorean Theorem and quadratic equations in real life.

Justin's house sits on a lot that is shaped like a trapezoid. He decides to make the back part of the lot usable by building a triangular dog pen for his dog Blackjack. On his lunch break at work, he decides to order the materials but realizes that he forgot to write down the actual dimensions of that area of the lot. He wants to get the pen done this weekend because his friends are coming over to help him, so he has to order the materials today so they arrive on time. Justin remembers that one of the sides is 5 feet more than the length of the shortest side and the longest side is 10 feet more than the shortest side. Can he use this information to figure out the dimensions of the area so that he can order the materials today?

1. Using the diagram of the lot provided and the variable x for the length of the shortest side of the dog pen, write an expression for the other two sides of the triangular pen and label them on the diagram.

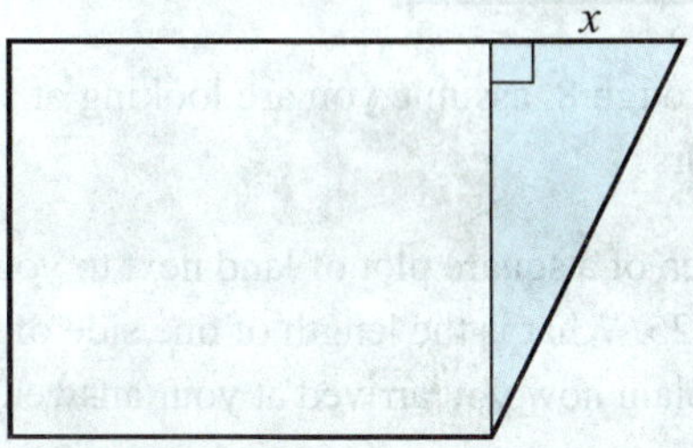

2. Using the Pythagorean Theorem, substitute the three expressions for the sides of the triangle into the formula and simplify the resulting polynomial. Be sure to move all terms to one side of the equation with the other side equal to zero. (Remember that the longest side is the hypotenuse in the formula. Make sure your leading term has a positive coefficient and that your squared binomials result in a trinomial.)

3. Use the equation you found in Problem 2 to answer the following.

a. Factor the resulting quadratic equation into two linear binomial factors.

b. Find the two solutions to the equation from part a. using the zero-factor law.

c. Do both of these solutions make sense? Explain your reasoning.

d. Using the solution (or solutions) that makes sense, substitute this value for x and determine the dimensions of the dog pen.

4. To fence in the dog pen, Justin plans to purchase chain-link fencing at a cost of \$2.58 per foot.

a. How much fencing will he need?

b. How much will the fencing cost?

5. How much area will the dog pen enclose? 150 ft^2

6. Justin decides to also put a dog house in the pen to protect Blackjack in bad weather. The dog house is rectangular in shape and measures 2.5 feet by 3 feet. Once the dog house is in the pen, how much area will Blackjack have to run in?

7. The sides of the pen form a right triangle, and the measurements of the sides of the pen were found using the Pythagorean Theorem. Any three positive integers that satisfy the Pythagorean Theorem are called a **Pythagorean triple**. There are an infinite number of these triples and numerous formulas that can be used to generate them. Do some research on the internet to find one of these formulas and use the formula to generate three more sets of Pythagorean triples. Verify that they are Pythagorean triples by substituting them into the Pythagorean Theorem. Answers will vary.

8. Another way to generate a Pythagorean triple is to take an existing triple and multiply each integer by a constant. Take the Pythagorean triple $(3, 4, 5)$ and multiply each integer in the triple by the factors below and verify that the result is also a Pythagorean triple by substituting into the Pythagorean Theorem.

a. Multiply by 3: $(9, 12, 15)$

b. Multiply by 5: $(15, 20, 25)$

c. Multiply by 8: $(24, 32, 40)$

Chapter 13 Project

Planting Seeds

An activity to demonstrate factoring polynomials in real life.

Imagine you live within walking distance from a community garden where anyone in the neighborhood can plant, grow, and harvest various vegetables and plants. You have been given a plot of land in this garden that is in the shape of a rectangle. You also have been given seeds to plant tomatoes, peppers, cucumbers, and squash.

When planting seeds in a garden, it is important to know how much space you have for the crops, and to adjust your seeds accordingly. One possible way the plot of land could be divided is shown. For Problems 1 through 5, assume this is the layout of the vegetables.

Tomatoes	Cucumbers	Squash	Tomatoes
Tomatoes	Cucumbers	Squash	Tomatoes
Peppers	Cucumbers	Squash	Peppers

Width (left side); Length (bottom)

1. The squash sections each have a width of $2x - 1$ feet. What is the total width of the column of squash?

2. If the length of the column of cucumbers is $x - 4$, and the width of each section of cucumbers is $4x + 1$, what is the total area of the three cucumber sections?

3. The section of tomatoes on the left side of the plot of land has an area of $x^2 + 6x + 9$ ft². If this plot of tomatoes were in the shape of a square, what would be the length of its side?

4. One squash section has a width of $2x - 1$ feet and a length of $4x + 2$ feet.
   a. What is the area of this section of squash?
   b. Is there any other possible width and length of this plot of land that can exist with this given area? Explain why or why not.

5. The area of one section of peppers is $x^2 + bx + 24$ ft², where b is an unknown positive value.
   a. What are several different options for the value of b so that this area can be factored? Explain how you arrived at your answer(s).
   b. If the area of the section of peppers were $x^2 + bx - 24$ ft² instead, how would this change your answers to part a.?

For Problems 6 through 8, assume you are looking at a plot of land next to yours.

6. The total area of a square plot of land next to yours is $4x^2 + 20x + 25$. What is the length of one side of the square? Explain how you arrived at your answer.

7. The area of one section of tomatoes on the right side of the plot of land is $3x^2 - 2x - 5$ ft². What are the dimensions of this section? Explain how you arrived at your answer.

8. Assume you wanted to plant the cucumbers in the shape of a square. Is it possible for the area of a cucumber section to be $x^2 + 9$ ft² and have one side length of $x + 3$? Explain why or why not.

Chapter 13 Project: Planting Seeds

1. $3(2x - 1) = 6x - 3$ ft
2. $(4x + 1)(x - 4) = 4x^2 - 15x - 4$ ft^2, times 3 for each section: $12x^2 - 45x - 12$ ft^2
3. $x + 3$ ft
4. **a.** $(2x - 1)(4x + 2) = 8x^2 - 2$ ft^2

   b. $(4x - 2)(2x + 1)$ is the only other possibility. By fully factoring $8x^2 - 2$ and multiplying the constant factor into the first set of parentheses instead of the second set, as in the problem, a new possibility is created.
5. **a.** By factoring this trinomial, we can generate several possibilities: 10, 11, 14, and 25. 24 can be split into the factor pairs of 1 and 24, 2 and 12, 3 and 8, and 4 and 6. By adding these pairs, we arrive at the values for **b.**

   b. Due to the constant term becoming negative, this would change our answers to incorporate negative factors.
6. This is a perfect square trinomial. Factoring this results in $(2x + 5)^2$ Thus, the square should have a length of $(2x + 5)$ ft
7. This is a factorable trinomial. By factoring, we arrive at $(3x - 5)(x + 1)$ So, the dimensions are $3x - 5$ ft by $x + 1$ ft.
8. No, this is not possible because $x^2 + 9$ cannot be factored. Therefore, it is impossible for any value to be able to be multiplied by $x + 3$ to create $x^2 + 9$.

Chapter 13 Project: Building a Dog Pen

1. $x + 5$; $x + 10$ (the hypotenuse)
2. $x^2 - 10x - 75 = 0$
3. **a.** $(x - 15)(x + 5) = 0$

   b. $x = 15$ or $x = -5$

   c. No, the negative solution doesn't make sense. Length cannot be negative.

   d. Side lengths are 15 feet, 20 feet, and 25 feet.
4. **a.** 60 feet

   b. $154.80
6. 142.5 ft^2

CHAPTER 14

Rational Expressions

Connections

You may be familiar with the saying "many hands make light work," which is a quote attributed to John Heywood, an English playwright who lived shortly before Shakespeare. The truth behind this statement has proven itself time and time again. The effects of collaboration are everywhere: collaboration between team members helps companies meet deadlines, the collaboration of multiple organizations after a natural disaster allows more people to receive help, and thousands of collaborating solar cells produce more than enough energy to run the International Space Station.

As company managers look at how best to distribute their resources, they can use rational equations to determine exactly how much time each task will take a team of people to complete. For instance, a landscaping company can determine how many crew members to send to a location if they know the amount of work to be done and how long it took different crew members to complete different task on previous jobs. If a task can be worked on by multiple people at once, this can reduce the overall time for that task.

Suppose a landscaping company sends two crew members to mulch flower beds at a business complex. Past experience has shown that the more experienced crew member can complete the task alone in 3.5 hours, while the less experienced crew member can complete the task in 5 hours. How long would it take the two crew members to mulch the flower beds if they work together?

Objectives

A. Determine any restrictions on the variable in a rational expression.

B. Evaluate rational expressions for given values of the variable.

C. Reduce rational expressions to lowest terms.

14.1 Introduction to Rational Expressions

Rational expressions are fractions. So if you know how to work with fractions in arithmetic, you will find the same techniques used in Chapter 14. You will need to find a common denominator to add (or subtract) rational expressions, and division is accomplished by multiplying by the reciprocal of the divisor. The basic difference is that the numerators and denominators are polynomials and the skills of multiplying and factoring that you learned in Chapters 12 and 13 will be needed. If you keep in mind throughout Chapter 14 that you are following the same rules you learned for fractions in arithmetic, some of the difficult-looking expressions may seem considerably easier.

A Rational Expressions

The term **rational number** is the technical name for a fraction in which both the numerator and denominator are integers. Similarly, the term **rational expression** is the technical name for a fraction in which both the numerator and denominator are polynomials.

Rational Expressions

A **rational expression** is an algebraic expression that can be written in the form

$\frac{P}{Q}$, where P and Q are polynomials and $Q \neq 0$.

DEFINITION

Math Tip

Remember, the denominator of a rational expression can never be 0. Division by 0 is undefined.

Try dividing by numbers that are close to 0 (0.001, 0.00001, 0.000001, ect.) on your calculator and see what happens!

Examples of rational expressions are

$$\frac{4x^2}{9}, \quad \frac{y^2-25}{y^2+25}, \quad \text{and} \quad \frac{x^2+7x-6}{x^2-5x-14}.$$

As with rational numbers, the denominators of rational expressions cannot be 0. If a numerical value is substituted for a variable in a rational expression and the denominator assumes a value of 0, we say that the expression is **undefined** for that value of the variable. (These values are called **restrictions** on the variable.)

1. Determine which values of the variable, if any, will make the expression undefined.

a. $\frac{9}{5x-1}$

b. $\frac{x^2-25}{x^2-7x+12}$

c. $\frac{x+8}{x^2+16}$

Example 1 Finding Restrictions on the Variable

Determine which values of the variable, if any, will make the rational expression undefined.

a. $\frac{5}{3x-1}$ **b.** $\frac{x^2-4}{x^2-5x-6}$ **c.** $\frac{x+3}{x^2+36}$

Solution

a. $3x-1=0$ Set the denominator equal to 0.

$3x=1$ Solve the equation.

$x=\frac{1}{3}$

Thus, the expression $\frac{5}{3x-1}$ is undefined for $x=\frac{1}{3}$. Any other real number may be substituted for x in the expression. We write $x \neq \frac{1}{3}$ to indicate the restriction on the variable.

b. $x^2-5x-6=0$ Set the denominator equal to 0.

$(x-6)(x+1)=0$ Solve the equation by factoring.

$x-6=0$ or $x+1=0$

$x=6$ $\quad x=-1$

Thus, there are two restrictions on the variable: 6 and −1. We write $x \neq -1, 6$.

c. $x^2+36=0$ Set the denominator equal to 0.

$x^2=-36$ Solve the equation.

However, there is no real number whose square is −36. Thus, there are no restrictions on the variable.

Now work margin exercise 1.

Attention!

When the Numerator is 0

If the numerator of a rational expression has a value of 0 and the denominator is not 0 for that value of the variable, then the expression is defined and has a value of 0. If both numerator and denominator are 0, then the expression is **undefined**, just as in the case where only the denominator is 0.

B Evaluating Rational Expressions

While we cannot substitute values for the variable that will make a denominator 0, other values may be substituted. The following examples illustrate how to find the numerical value of a rational expression by substituting a value for the variable.

Example 2 Evaluating Rational Expressions

Find the value of each rational expression for the given value of the variable.

a. $\frac{2x}{x^2+1}; x=2$ **b.** $\frac{a-3}{a^2-5}; a=3$

Solution

a. $\frac{2x}{x^2+1}=\frac{2\cdot(2)}{(2)^2+1}=\frac{4}{4+1}=\frac{4}{5}$

Note that any real number may be substituted for x since the denominator x^2+1 is never 0 for any real number.

b. $\frac{a-3}{a^2-5}=\frac{(3)-3}{(3)^2-5}=\frac{0}{9-5}=\frac{0}{4}=0$

Note that a numerator may be 0.

Now work margin exercise 2.

2. Find the value of each rational expression for the given value of the variable.

a. $\frac{-x}{x^2-5}; x=4$

b. $\frac{z+1}{z^2+4}; z=-2$

C Reducing (or Simplifying) Rational Expressions

The rules for operating with rational expressions are essentially the same as those for operating with fractions in arithmetic. That is, simplifying, multiplying, and dividing rational expressions involve factoring and reducing. Addition and subtraction of rational expressions require common denominators. The basic rules for fractions were discussed in earlier sections and are summarized here for easy reference.

Summary of Arithmetic Rules for Rational Numbers (or Fractions)

A **fraction** (or **rational number**) is a number that can be written in the form $\frac{a}{b}$, where a and b are integers and $b \neq 0$. (Remember, no denominator can be 0.)

The Fundamental Principle: $\frac{a}{b} = \frac{a \cdot k}{b \cdot k}$, where $b, k \neq 0$

The **reciprocal** of $\frac{a}{b}$ is $\frac{b}{a}$ and $\frac{a}{b} \cdot \frac{b}{a} = 1$, where $a, b \neq 0$.

Multiplication: $\frac{a}{b} \cdot \frac{c}{d} = \frac{a \cdot c}{b \cdot d}$, where $b, d \neq 0$

Division: $\frac{a}{b} \div \frac{c}{d} = \frac{a}{b} \cdot \frac{d}{c}$, where $b, c, d \neq 0$

Addition: $\frac{a}{b} + \frac{c}{b} = \frac{a+c}{b}$, where $b \neq 0$

Subtraction: $\frac{a}{b} - \frac{c}{b} = \frac{a-c}{b}$, where $b \neq 0$

PROPERTIES

For rational expressions, each rule can be restated by replacing a and b with P and Q, where P and Q represent polynomials. In particular, the fundamental principle can be restated as follows.

The Fundamental Principle of Rational Expressions

If $\frac{P}{Q}$ is a rational expression and P, Q, and K are polynomials where $Q, K \neq 0$, then

$$\frac{P}{Q} = \frac{P \cdot K}{Q \cdot K}.$$

DEFINITION

The fundamental principle can be used to **reduce** (or **simplify**) a rational expression to **lowest terms** (for multiplication or division) and to **find equivalent fractions** (for addition or subtraction). Just as with rational numbers, a rational expression is said to be **reduced to lowest terms** if the numerator and denominator have no common factors other than 1 and −1.

Example 3 Reducing Rational Expressions

Use the fundamental principle of rational expressions to reduce each expression to lowest terms. State any restrictions on the variable by using the fact that no denominator can be 0. This restriction applies to denominators **before and after** a rational expression is reduced.

a. $\dfrac{2x-10}{3x-15}$ **b.** $\dfrac{x^2-3x-4}{x^2-16}$ **c.** $\dfrac{y-10}{10-y}$

Solution

a. $\dfrac{2x-10}{3x-15}=\dfrac{2\cancel{(x-5)}}{3\cancel{(x-5)}}=\dfrac{2}{3}\quad (x\neq 5)$

Note that $(x-5)$ is a common factor. The key word here is factor. We reduce using factors only.

b. $\dfrac{x^2-3x-4}{x^2-16}=\dfrac{\cancel{(x-4)}(x+1)}{(x+4)\cancel{(x-4)}}$

Reduce. The common factor is $(x-4)$.

$=\dfrac{x+1}{x+4}\quad (x\neq -4, 4)$

c. $\dfrac{y-10}{10-y}=\dfrac{y-10}{-y+10}$

$=\dfrac{1\cancel{(y-10)}}{-1\cancel{(y-10)}}$

Note that the expression $10-y$ is the opposite of $y-10$. When nonzero opposites are divided, the quotient is always -1.

$=\dfrac{1}{-1}=-1\quad (y\neq 10)$

Now work margin exercise 3.

3. Reduce each expression to lowest terms. State any restrictions on the variable.

a. $\dfrac{2x-6}{5x-15}$

b. $\dfrac{x^2-x-20}{x^2-25}$

c. $\dfrac{x-5}{5-x}$

Note

Note that the restrictions on the variable are determined before reducing.

In Example 3c., the result was -1. Notice that the expression $10-y$ is the opposite of $y-10$ for any value of y. That is,

$$10-y=-y+10=-1(y-10)=-(y-10).$$

When nonzero opposites are divided, the quotient is always -1.

$$\frac{-9}{9}=-1,\qquad \frac{23}{-23}=-1,\qquad \frac{x-3}{3-x}=\frac{\cancel{(x-3)}}{-1\cancel{(x-3)}}=\frac{1}{-1}=-1\ (x\neq 3)$$

Teaching Note

To help convince your students that -1 is always the result, you might want to have them pick several pairs of unequal values for x and a and evaluate $\dfrac{a-x}{x-a}$.

Opposites in Rational Expressions

For a polynomial P, $\dfrac{-P}{P}=-1$, where $P\neq 0$.

In particular, $\dfrac{a-x}{x-a}=\dfrac{-(x-a)}{x-a}=-1$, where $x\neq a$.

DEFINITION

Remember that the key word when reducing is **factor**. Many students make mistakes similar to the following when working with rational expressions.

Common Error

"Divide out" only common factors.

Wrong Solution

$\frac{4x+\cancel{8}}{\cancel{8}}$

8 is not a common factor.

$\frac{\overset{x}{\cancel{x^2}}-\overset{3}{\cancel{9}}}{\cancel{x}-\cancel{3}}$

3 and x are not common factors.

Correct Solution

$\frac{4x+8}{8}=\frac{\cancel{4}(x+2)}{\underset{2}{\cancel{8}}}$

4 is a common factor.

$\frac{x^2-9}{x-3}=\frac{(x+3)\cancel{(x-3)}}{\cancel{(x-3)}}$

$x-3$ is a common factor.

CAUTION

Margin Exercise Answers

1. a. $x\neq\frac{1}{5}$ **b.** $x\neq 3,4$ **c.** no restrictions **2. a.** $-\frac{4}{11}$ **b.** $-\frac{1}{8}$

3. a. $\frac{2}{5}; x\neq 3$ **b.** $\frac{x+4}{x+5}; x\neq -5,5$ **c.** $-1; x\neq 5$

14.1 Exercises

Concept Check

Fill-in-the-Blank. Complete each sentence using information found in this section.

1. The technical name for a fraction with integers in the numerator and denominator is __________ number.
2. To simplify a rational expression, divide out any common _______ from the numerators and denominators.
3. When a numerator and denominator are multiplied by the same number, this is an example of the ____________ principle of rational expressions.
4. Values that make an expression undefined cannot be used and are called ____________ on the variable.
5. The rules for rational expressions are the same as those for __________ in arithmetic.
6. The denominator of a rational expression can never equal _______.

14.1 Exercises

Concept Check

1. rational
2. factors
3. fundamental
4. restrictions
5. fractions
6. 0
7. True
8. True
9. False; A rational expression cannot have zero denominators.
10. False; All values except -5

Practice

1. $\frac{3x}{4y}; x\neq 0, y\neq 0$
2. $\frac{2y^3}{3x}; x\neq 0, y\neq 0$
3. $\frac{2x^3}{3y^3}; x\neq 0, y\neq 0$

True/False. Determine whether each statement is true or false. If a statement is false, explain how it can be changed so the statement will be true. (**Note:** There may be more than one acceptable change.)

7. A simplified rational expression cannot have any common factors other than 1 and −1 in both the numerator and denominator.

8. The difference between a rational number and a rational expression is that a rational expression generally has polynomials in the numerator and/or denominator.

9. While a rational number cannot have a zero denominator, a rational expression can have a zero denominator.

10. If a denominator is $x + 5$, it is defined for all values except 5.

Practice

Reduce each expression to lowest terms. State any restrictions on the variable(s). See Examples 1 and 3.

1. $\dfrac{9x^2y^3}{12xy^4}$
2. $\dfrac{18xy^4}{27x^2y}$
3. $\dfrac{20x^5}{30x^2y^3}$
4. $\dfrac{15y^4}{20x^3y^2}$
5. $\dfrac{x}{x^2-3x}$
6. $\dfrac{3x}{x^2+5x}$
7. $\dfrac{7x-14}{x-2}$
8. $\dfrac{4-2x}{2x-4}$
9. $\dfrac{9-3x}{4x-12}$
10. $\dfrac{2x-8}{16-4x}$
11. $\dfrac{6x^2+4x}{3xy+2y}$
12. $\dfrac{1+3y}{4x+12xy}$
13. $\dfrac{x^2+6x}{x^2+5x-6}$
14. $\dfrac{x^2-y^2}{3x^2+3xy}$
15. $\dfrac{x^2-4x-21}{x^2-9}$
16. $\dfrac{x^2-11x+18}{x^2-4}$
17. $\dfrac{xy-3y+2x-6}{y^2-4}$
18. $\dfrac{3x^2+14x-24}{18-9x-2x^2}$
19. $\dfrac{x^2+5x-14}{5x-2y+xy-10}$
20. $\dfrac{x^2+10x+24}{2x^2+x-28}$

Evaluate each rational expression for the given value of the variable. See Example 2.

21. $\dfrac{x-3}{3x^2};\ x=5$
22. $\dfrac{2x+1}{3x-2};\ x=1$
23. $\dfrac{5x^2}{x^2-4};\ x=-3$
24. $\dfrac{3y-4}{y^2+25};\ y=3$
25. $\dfrac{2x^2+5x}{x^2-1};\ x=0$
26. $\dfrac{n^3}{n^2-5n+6};\ n=-1$
27. $\dfrac{2m-7}{m^2+8m+12};\ m=2$
28. $\dfrac{-x+3}{x-3};\ x=-10$
29. $\dfrac{15-x}{x-15};\ x=1000$
30. $-\dfrac{16+x}{x^2-16};\ x=20$

4. $\dfrac{3y^2}{4x^3};\ x\neq 0, y\neq 0$
5. $\dfrac{1}{x-3};\ x\neq 0,3$
6. $\dfrac{3}{x+5};\ x\neq -5,0$
7. $7;\ x\neq 2$
8. $-1;\ x\neq 2$
9. $-\dfrac{3}{4};\ x\neq 3$
10. $-\dfrac{1}{2};\ x\neq 4$
11. $\dfrac{2x}{y};\ x\neq -\dfrac{2}{3}, y\neq 0$
12. $\dfrac{1}{4x};\ x\neq 0, y\neq -\dfrac{1}{3}$
13. $\dfrac{x}{x-1};\ x\neq -6,1$
14. $\dfrac{x-y}{3x};\ x\neq 0, x\neq -y$
15. $\dfrac{x-7}{x-3};\ x\neq -3,3$
16. $\dfrac{x-9}{x+2};\ x\neq -2,2$
17. $\dfrac{x-3}{y-2};\ y\neq -2,2$
18. $-\dfrac{3x-4}{2x-3};\ x\neq -6,\dfrac{3}{2}$
19. $\dfrac{x+7}{y+5};\ x\neq 2, y\neq -5$
20. $\dfrac{x+6}{2x-7};\ x\neq -4,\dfrac{7}{2}$
21. $\dfrac{2}{75}$
22. 3
23. 9
24. $\dfrac{5}{34}$
25. 0
26. $\dfrac{-1}{12}$
27. $\dfrac{-3}{32}$
28. −1
29. 1
30. $\dfrac{-3}{32}$

Applications

Solve.

31. The cost of renting a party room with tables, chairs, and simple decorations is \$200 plus \$15 per person attending.

a. Write a rational expression that represents the total price per person for renting the party room, where x is the number of people attending.

b. What is the price per person to rent the party room if 10 people are attending?

c. Determine which values of the variable will make the rational expression from part a. undefined.

d. Considering the context of the given problem, are there any additional restrictions on the variable? If so, explain why these restrictions are in place.

32. Amelia wants to join the Fit4Life gym, a local gym that offers a variety of different fitness classes. At Fit4Life gym, it costs \$85 for a lifetime membership, and then you pay \$8 per class. They are currently running a special where you get the first 10 classes for free.

a. Write a rational expression that represents the average price per class where x is the number of classes Amelia takes.

b. What is the price per class after taking 100 classes?

c. Determine which values of the variable will make the rational expression from part a. undefined.

d. Considering the context of the given problem, are there any additional restrictions on the variable? If so, explain why these restrictions are in place.

33. Columbus High School plans to buy scientific calculators for use in their Earth Science classes. If they buy the calculators in bulk from Math Supplies Plus, the cost per calculator depends on how many calculators are purchased. The cost per calculator c is determined by the equation $c = \frac{4(x+60)}{x+20}$, where x is the number of calculators purchased.

a. Find the cost per calculator if only 1 calculator is purchased.

b. Find the cost per calculator if 50 calculators are purchased.

c. Find the cost per calculator if 100 calculators are purchased.

d. What trend do you notice about the cost per calculator as the number of calculators purchased increases?

e. For what values of the variable is the price per calculator function undefined?

f. Considering the context of the equation, are there any additional restrictions on the variable? If so, explain why these restrictions are in place.

Applications

31. a. $p(x) = \frac{15x+200}{x}$

b. \$35 **c.** $x \neq 0$

d. The variable cannot be negative because you cannot have a negative quantity of people. There would also be a maximum number depending on the size of the room.

32. a. $p(x) = \frac{85+8(x-10)}{x} = \frac{5+8x}{x}$

b. \$8.05 **c.** $x \neq 0$

d. The variable should be greater than 10 since the first 10 classes are free.

33. a. \$11.62 **b.** \$6.29

c. \$5.33

d. The cost per calculator decreases as the number of calculators purchased increases.

e. $x \neq -20$

f. The value of x cannot be negative because you cannot buy a negative quantity of calculators

34. An annuity is a type of savings account that you put money into after equal periods of time to reach a goal amount. Annuities are a type of investment that is generally used to meet long-term savings goals such as college funds or retirement funds. The future value of an annuity is determined by the equation

$$FV = P\left[\frac{(1+r)^n - 1}{r}\right],$$

where FV is the future value of the annuity, P is the size of the periodic payment, r is the interest rate, and n is the number of payments or times the interest is compounded.

a. Determine the future value of an annuity if the monthly payment is \$100, the interest rate is 3%, and the payments are made for 60 months. Round your answer to the nearest cent.

b. Determine the future value of an annuity if the monthly payment is \$200, the interest rate is 3%, and the payments are made for 60 months. Round your answer to the nearest cent.

c. What was the total amount of money paid into the annuity from part a.?

d. What was the total amount of money paid into the annuity from part b.?

e. The regular payment in part b. is double the regular payment in part a. Is the future value from part b. double the future value from part a.? Why do you think this is?

35. The area of a rectangle (in square feet) is represented by the polynomial function $A(x) = 4x^2 - 4x - 15$. If the length of the rectangle is $(2x+3)$ feet, find a representation for the width.

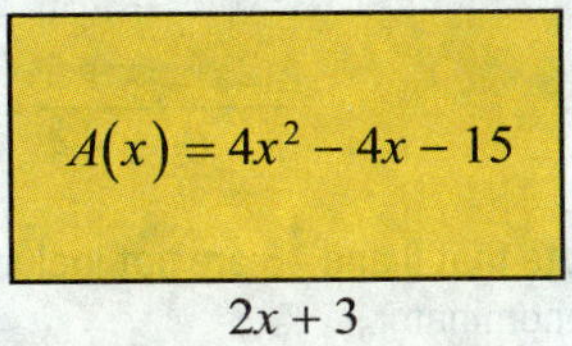

36. The area of a rectangle (in square feet) is represented by the polynomial function $A(x) = 3x^2 - x - 10$. If the length of the rectangle is $(3x+5)$ feet, find a representation for the width.

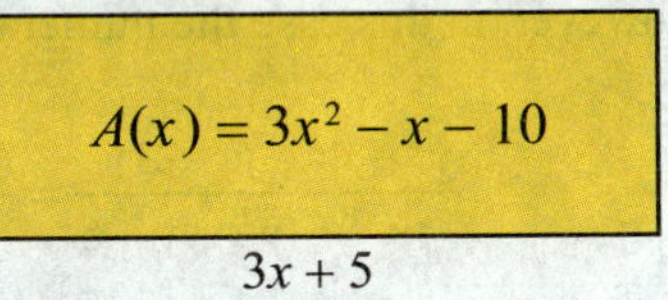

Writing & Thinking

37. **a.** Define the term rational expression.

b. Give an example of a rational expression that is undefined for $x = -2$ and $x = 3$ and has a value of 0 for $x = 1$. Explain how you determined this expression.

c. Give an example of a rational expression that is undefined for $x = -5$ and never has a value of 0. Explain how you determined this expression.

38. Write the opposite of each of the following expressions.

a. $3 - x$

b. $2x - 7$

c. $x + 5$

d. $-3x - 2$

---

34. **a.** \$16,305.34 **b.** \$32,610.69 **c.** \$6000 **d.** \$12,000 **e.** Yes. You put in double the amount. Answers will vary.

35. $(2x-5)$ feet

36. $(x-2)$ feet

Writing & Thinking

37. **a.** A rational expression is an algebraic expression that can be written in the form $\frac{P}{Q}$ where P and Q are polynomials and $Q \neq 0$. **b.** $\frac{x-1}{(x+2)(x-3)}$ Answers will vary. **c.** $\frac{1}{x+5}$ Answers will vary.

38. **a.** $x - 3$ **b.** $7 - 2x$ **c.** $-x - 5$ **d.** $3x + 2$

Objectives

A. Multiply rational expressions.

B. Divide rational expressions.

Teaching Note

It may benefit some students to introduce this section by using common fractions from arithmetic to demonstrate the principle: $\frac{2}{3}\cdot\frac{4}{5}=\frac{8}{15}$. Reinforce the extension from arithmetic whenever possible.

14.2 Multiplication and Division with Rational Expressions

A Multiplication with Rational Expressions

To multiply any two rational expressions, simply multiply the numerators and multiply the denominators.

Multiplying Rational Expressions

To multiply any two (or more) rational expressions,

1. completely factor each numerator and denominator,
2. multiply the numerators and multiply the denominators, keeping the expressions in factored form, and
3. "divide out" any common factors from the numerators and denominators. Remember that no denominator can have a value of 0.

PROCEDURE

No factoring is necessary in this example.

$$\frac{2x}{x-6}\cdot\frac{x+5}{x-4}=\frac{2x(x+5)}{(x-6)(x-4)} \text{ or } \frac{2x^2+10x}{(x-6)(x-4)} \quad (x\neq 4, 6)$$

Note that 0 and –5 are not included in the restriction list because neither 2*x* nor *x* – 5 are factors in the denominator.

However, in this case the numerators and denominators must be factored.

$$\frac{y^2-4}{y^3}\cdot\frac{y^2-3y}{y^2-y-6}=\frac{(y+2)(y-2)(y)(y-3)}{y^3(y-3)(y+2)}=\frac{y-2}{y^2} \quad (y\neq -2, 0, 3)$$

Note that 2 is not included in the restriction list because *y* – 2 is not a factor in the denominator. In this example, all the other restrictions relate to factors in both the numerator and denominator.

Multiplying Rational Expressions

If P, Q, R, and S are polynomials and $Q, S\neq 0$, then

$$\frac{P}{Q}\cdot\frac{R}{S}=\frac{P\cdot R}{Q\cdot S}.$$

DEFINITION

Example 1 Multiplying with Rational Expressions

Multiply and reduce, if possible. State any restrictions on the variables.

$$\frac{5x^2y}{9xy^3}\cdot\frac{6x^3y^2}{15xy^4}$$

Solution

$$\begin{aligned}\frac{5x^2y}{9xy^3}\cdot\frac{6x^3y^2}{15xy^4}&=\frac{\cancel{5}\cdot 2\cdot\cancel{3}\cdot x^5\cdot y^3}{3\cdot 3\cdot\cancel{3}\cdot\cancel{5}\cdot x^2\cdot y^7}\\&=\frac{2x^{5-2}y^{3-7}}{9}\\&=\frac{2x^3y^{-4}}{9}\\&=\frac{2x^3}{9y^4}\qquad (x\neq 0,\ y\neq 0)\end{aligned}$$

Now work margin exercise 1.

Example 2 Multiplying with Rational Expressions

Multiply and reduce, if possible. State any restrictions on the variables.

$$\frac{x}{x-2}\cdot\frac{x^2-4}{x^2}$$

Solution

$$\begin{aligned}\frac{x}{x-2}\cdot\frac{x^2-4}{x^2}&=\frac{\cancel{x}(x+2)\cancel{(x-2)}}{\cancel{(x-2)}\,\cancel{x^2}_{\,x}}\\&=\frac{x+2}{x}\qquad (x\neq 0,2)\end{aligned}$$

Now work margin exercise 2.

Example 3 Multiplying with Rational Expressions

Multiply and reduce, if possible. State any restrictions on the variables.

$$\frac{3x-3}{x^2+x}\cdot\frac{x^2+2x+1}{3x^2-6x+3}$$

Solution

$$\begin{aligned}\frac{3x-3}{x^2+x}\cdot\frac{x^2+2x+1}{3x^2-6x+3}&=\frac{\cancel{3}\cancel{(x-1)}\overset{(x+1)}{\cancel{(x+1)^2}}}{x\cancel{(x+1)}\cdot\cancel{3}\underset{(x-1)}{\cancel{(x-1)^2}}}\\&=\frac{x+1}{x(x-1)}\text{ or }\frac{x+1}{x^2-x}\qquad (x\neq -1,0,1)\end{aligned}$$

Now work margin exercise 3.

1. Multiply and reduce, if possible. State any restrictions on the variables.

$$\frac{7x^4y}{9x^2y^6}\cdot\frac{3x^3y^3}{14xy^4}$$

Math Tip

It is possible to use different strategies of simplifying when multiplying rational expressions. In Example 1, the quotient rule for exponents is used to subtract the exponents. In Example 2, cancellation is used instead. It is up to the solver how to proceed when simplifying.

2. Multiply and reduce, if possible. State any restrictions on the variables.

$$\frac{x}{x-3}\cdot\frac{x^2-9}{x^3}$$

3. Multiply and reduce, if possible. State any restrictions on the variables.

$$\frac{2x+2}{x^2-x}\cdot\frac{x^2-2x+1}{2x^2+4x+2}$$

4. Multiply and reduce, if possible. State any restrictions on the variables.

$$\frac{x^2-8x+12}{3x+9}\cdot\frac{x^2-25}{x^2-7x+10}$$

Example 4 Multiplying with Rational Expressions

Multiply and reduce, if possible. State any restrictions on the variables.

$$\frac{x^2-7x+12}{2x+6}\cdot\frac{x^2-4}{x^2-2x-8}$$

Solution

$$\frac{x^2-7x+12}{2x+6}\cdot\frac{x^2-4}{x^2-2x-8}=\frac{\cancel{(x-4)}(x-3)\cancel{(x+2)}(x-2)}{2(x+3)\cancel{(x-4)}\cancel{(x+2)}}$$

$$=\frac{(x-3)(x-2)}{2(x+3)} \text{ or } \frac{x^2-5x+6}{2(x+3)} \text{ or } \frac{x^2-5x+6}{2x+6}$$

$(x \neq -3, -2, 4)$

Now work margin exercise 4.

Attention!

As shown in Examples 3 and 4, there may be more than one correct form of an answer. After a rational expression has been reduced, the numerator and denominator may be multiplied out or left in factored form. **Generally, the denominator will be left in factored form and the numerator multiplied out**. As we will see in Section 14.4, this form makes the results easier to add and subtract. However, be aware that this form is just an option, and multiplying out the denominator is not an error.

5. Multiply and reduce, if possible. State any restrictions on the variables.

$$\frac{x^2+2x-3}{x^2}\cdot\frac{x}{x+3}$$

Completion Example 5 Multiplication with Rational Expressions

Multiply and reduce, if possible. State any restrictions on the variables.

$$\frac{x^2-36}{x^4}\cdot\frac{x}{x-6}$$

Solution

$(x \neq \underline{\ 0\ }, \underline{\ 6\ })$

$$\frac{x^2-36}{x^4}\cdot\frac{x}{x-6}=\frac{(\underline{x+6})(\underline{x-6})}{x^4}\cdot\frac{x}{x-6}$$

$$=\frac{\underline{x+6}}{\underline{x^3}}$$

Now work margin exercise 5.

B Division with Rational Expressions

To divide any two rational expressions, multiply the first fraction by the **reciprocal** of the second fraction.

Dividing Rational Expressions

If P, Q, R, and S are polynomials with $Q, R, S \neq 0$, then

$$\frac{P}{Q} \div \frac{R}{S} = \frac{P}{Q} \cdot \frac{S}{R}.$$

Note that $\frac{S}{R}$ is the reciprocal of $\frac{R}{S}$.

DEFINITION

Example 6 Dividing with Rational Expressions

Divide and reduce, if possible. Assume that no denominator has a value of 0.

$$\frac{12x^2y}{10xy^2} \div \frac{3x^4y}{xy^3}$$

Solution

$$\frac{12x^2y}{10xy^2} \div \frac{3x^4y}{xy^3} = \frac{12x^2y}{10xy^2} \cdot \frac{xy^3}{3x^4y}$$ The first step is to muliply by the reciprocal of the divisor.

$$= \frac{\cancel{2} \cdot 2 \cdot \cancel{3} \cdot x^3 \cdot y^4}{\cancel{2} \cdot 5 \cdot \cancel{3} \cdot x^5 \cdot y^3}$$ Now factor and multiply.

$$= \frac{2x^{3-5}y^{4-3}}{5}$$ Use the quotient rule for exponents to simplify.

$$= \frac{2x^{-2}y}{5} = \frac{2y}{5x^2}$$

Now work margin exercise 6.

6. Divide and reduce, if possible. Assume that no denominator has a value of 0.

$$\frac{15x^3y}{12xy^3} \div \frac{5x^5y}{xy^2}$$

Example 7 Dividing with Rational Expressions

Divide and reduce, if possible. Assume that no denominator has a value of 0.

$$\frac{x^2-y^2}{x^3} \div \frac{y-x}{xy}$$

Solution

$$\frac{x^2-y^2}{x^3} \div \frac{y-x}{xy} = \frac{x^2-y^2}{x^3} \cdot \frac{xy}{y-x}$$

$$= \frac{\overset{-1}{\cancel{(x-y)}}(x+y)\cancel{x}y}{\underset{x^2}{\cancel{x^3}}\cancel{(y-x)}}$$ Note that $\frac{x-y}{y-x} = -1$.

$$= \frac{-y(x+y)}{x^2} = \frac{-xy-y^2}{x^2}$$

Now work margin exercise 7.

7. Divide and reduce, if possible. Assume that no denominator has a value of 0.

$$\frac{x+y}{y^3} \div \frac{-x-y}{xy}$$

8. Divide and reduce, if possible. Assume that no denominator has a value of 0.

$$\frac{x^2-9x+18}{3x^2+19x+6} \div \frac{3x^2-17x-6}{x^2+x-30}$$

Example 8 Dividing with Rational Expressions

Divide and reduce, if possible. Assume that no denominator has a value of 0.

$$\frac{x^2-8x+15}{2x^2+11x+5} \div \frac{2x^2-5x-3}{4x^2-1}$$

Solution

$$\frac{x^2-8x+15}{2x^2+11x+5} \div \frac{2x^2-5x-3}{4x^2-1} = \frac{x^2-8x+15}{2x^2+11x+5} \cdot \frac{4x^2-1}{2x^2-5x-3}$$

$$= \frac{\cancel{(x-3)}(x-5)(2x-1)\cancel{(2x+1)}}{(2x+1)(x+5)\cancel{(x-3)}\cancel{(2x+1)}}$$

$$= \frac{(x-5)(2x-1)}{(2x+1)(x+5)} = \frac{2x^2-11x+5}{2x^2+11x+5}$$

Remember that you have the option of leaving the numerator and/or denominator in factored form.

Now work margin exercise 8.

9. Divide and reduce, if possible. Assume that no denominator has a value of 0.

$$\frac{3x+1}{x-4} \div \frac{3x^2-11x-4}{x^2-16}$$

Completion Example 9 Division with Rational Expressions

Divide and reduce, if possible. Assume that no denominator has a value of 0.

$$\frac{x^2-4}{2x-1} \div \frac{x^2-2x}{2x^2+x-1}$$

Solution

$$\frac{x^2-4}{2x-1} \div \frac{x^2-2x}{2x^2+x-1} = \frac{x^2-4}{2x-1} \cdot \frac{2x^2+x-1}{x^2-2x}$$

$$= \frac{(x+2)(x-2)}{2x-1} \cdot \frac{(2x-1)(x+1)}{x(x-2)}$$

$$= \frac{(x+2)(x+1)}{x}$$

$$= \frac{x^2+3x+2}{x}$$

Now work margin exercise 9.

14.2 Exercises

Concept Check

1. factor
2. reciprocal
3. numerator, denominator
4. 0
5. numerators, denominators
6. False; The reciprocal is $\frac{x+3}{x}$.

Completion Example Answers

5. $(x \neq 0, 6)$; $\frac{(x+6)(x-6)}{x^4} \cdot \frac{x}{x-6} = \frac{x+6}{x^3}$

9. $\frac{(x+2)(x-2)}{2x-1} \cdot \frac{(2x-1)(x+1)}{x(x-2)} = \frac{(x+2)(x+1)}{x}$

$$= \frac{x^2+3x+2}{x}$$

Margin Exercise Answers

1. $\frac{x^4}{6y^6}$; $x \neq 0, y \neq 0$ **2.** $\frac{x+3}{x^2}$; $x \neq 0, 3$ **3.** $\frac{x-1}{x(x+1)}$; $x \neq -1, 0, 1$

4. $\frac{x^2-x-30}{3(x+3)}; x \neq -3, 2, 5$ **5.** $\frac{x-1}{x}; x \neq 0, -3$ **6.** $\frac{1}{4x^2y}$ **7.** $-\frac{x}{y^2}$ **8.** $\frac{x^2-8x+15}{(3x+1)^2}$

9. $\frac{x+4}{x-4}$

14.2 Exercises

Concept Check

Fill-in-the-Blank. Complete each sentence using information found in this section.

1. To multiply two or more rational expressions, first completely ______ each numerator and denominator.
2. To divide any two rational expressions, multiply the first fraction by the _________ of the second fraction (the divisor).
3. After a rational expression has been reduced, it is typical to multiply out the _________ and leave the ___________ in factored form.
4. Remember that no rational expression can have a denominator with a value of_______.
5. When multiplying rational expressions, multiply the ______ and multiply the _______, keeping the expressions in factored form.

True/False. Determine whether each statement is true or false. If a statement is false, explain how it can be changed so the statement will be true. (**Note:** There may be more than one acceptable change.)

6. The reciprocal of $\frac{x}{x+3}$ is $\frac{-x-3}{x}$.
7. Dividing rational expressions is similar to dividing fractions.
8. There are no restrictions on the denominator $12x^2$.
9. Because $\frac{4x^2}{16x}$ reduces to $\frac{x}{4}$, there are no restrictions on the denominator.

Practice

Perform the indicated operations and reduce to lowest terms. Assume that no denominator has a value of 0.

1. $\frac{3ax^2}{4b} \cdot \frac{6b^2}{27x^2y}$
2. $\frac{18x^3}{5y^2} \cdot \frac{30y^3}{9x^4}$
3. $\frac{24x^3}{25y^2} \cdot \frac{10y^5}{18x}$
4. $\frac{16x^8}{3y^{11}} \cdot \frac{-21y^9}{10x^7}$
5. $\frac{x^2-9}{x^2+2x} \cdot \frac{x+2}{x-3}$
6. $\frac{16x^2-9}{3x^2-15x} \cdot \frac{6}{4x+3}$
7. $\frac{x^2+2x-3}{x^2+3x} \cdot \frac{x}{x+1}$
8. $\frac{4x+16}{x^2-16} \cdot \frac{x-4}{x}$

7. True
8. False; the restriction is 0.
9. False; the restriction is 0.

Practice

1. $\frac{ab}{6y}$
2. $\frac{12y}{x}$
3. $\frac{8x^2y^3}{15}$
4. $-\frac{56x}{5y^2}$
5. $\frac{x+3}{x}$
6. $\frac{8x-6}{x(x-5)}$
7. $\frac{x-1}{x+1}$
8. $\frac{4}{x}$
9. $-\frac{1}{x-8}$
10. $-\frac{3}{x-2}$
11. $\frac{x-2}{x}$
12. $\frac{4x+6}{x}$
13. $\frac{4x+20}{x(x+1)}$
14. $\frac{x+3}{(x+4)(x-1)}$
15. $\frac{x}{(x+3)(x-1)}$
16. $\frac{x^2-6x-7}{x+5}$
17. $-\frac{x+4}{x(x+1)}$
18. $\frac{3x^2-12x}{(x-7)(x+4)}$
19. $\frac{x+2y}{(x-3y)(x-2y)}$
20. $\frac{4x^2+6x}{(x+5)(x-1)}$

21. $\dfrac{x-1}{x(2x-1)}$

22. $\dfrac{3x^2-x}{(2x+3)(x+5)}$

23. $\dfrac{1}{x+1}$

24. $\dfrac{x+3}{2}$

25. $\dfrac{x+2}{x-2}$

26. $\dfrac{2x-1}{2x+1}$

27. $\dfrac{1}{3xy^6}$

28. $\dfrac{49y^3}{6x^5}$

29. $\dfrac{6y^7}{x^4}$

30. $\dfrac{1}{12x}$

31. $\dfrac{x}{12}$

32. $\dfrac{7x}{x+2}$

33. $\dfrac{6x+18}{x^2}$

34. $\dfrac{x}{6}$

35. $\dfrac{6}{5x}$

36. $\dfrac{x+3}{x+4}$

37. $\dfrac{3x+1}{x+1}$

38. $\dfrac{x+5}{x}$

39. $\dfrac{x-2}{2x-1}$

40. $\dfrac{x+2}{4(x+4)}$

41. $-\dfrac{x+4}{x(2x-1)}$

42. $\dfrac{x-2}{x+2}$

43. $\dfrac{x+1}{x-1}$

9. $\dfrac{x^2+6x-16}{x^2-64}\cdot\dfrac{1}{2-x}$

10. $\dfrac{4-x^2}{x^2-4x+4}\cdot\dfrac{3}{x+2}$

11. $\dfrac{x^2-5x+6}{x^2-4x}\cdot\dfrac{x-4}{x-3}$

12. $\dfrac{2x^2+x-3}{x^2+4x}\cdot\dfrac{2x+8}{x-1}$

13. $\dfrac{2x^2+10x}{3x^2+5x+2}\cdot\dfrac{6x+4}{x^2}$

14. $\dfrac{x+3}{x^2-16}\cdot\dfrac{x^2-3x-4}{x^2-1}$

15. $\dfrac{x}{x^2+7x+12}\cdot\dfrac{x^2-2x-24}{x^2-7x+6}$

16. $\dfrac{x^2-2x-3}{x+5}\cdot\dfrac{x^2-5x-14}{x^2-x-6}$

17. $\dfrac{8-2x-x^2}{x^2-2x}\cdot\dfrac{x-4}{x^2-3x-4}$

18. $\dfrac{3x^2+21x}{x^2-49}\cdot\dfrac{x^2-5x+4}{x^2+3x-4}$

19. $\dfrac{(x-2y)^2}{x^2-5xy+6y^2}\cdot\dfrac{x+2y}{x^2-4xy+4y^2}$

20. $\dfrac{4x^2+6x}{x^2+3x-10}\cdot\dfrac{x^2+4x-12}{x^2+5x-6}$

21. $\dfrac{2x^2+5x+2}{3x^2+8x+4}\cdot\dfrac{3x^2-x-2}{4x^3-x}$

22. $\dfrac{x^2+5x}{4x^2+12x+9}\cdot\dfrac{6x^2+7x-3}{x^2+10x+25}$

23. $\dfrac{x+2}{x^2-1}\cdot\dfrac{x^2-2x+1}{x^2+x-2}$

24. $\dfrac{x^2-9}{2x+16}\cdot\dfrac{x^2+6x-16}{x^2-5x+6}$

25. $\dfrac{x-2}{x+5}\cdot\dfrac{x^2+7x+10}{x^2-4x+4}$

26. $\dfrac{2x^2-7x+3}{x^2-9}\cdot\dfrac{3x^2+8x-3}{6x^2+x-1}$

27. $\dfrac{12x^2y}{9xy^9}\div\dfrac{4x^4y}{x^2y^3}$

28. $\dfrac{35xy^3}{24x^3y}\div\dfrac{15x^4y^3}{84xy^4}$

29. $\dfrac{45xy^4}{21x^2y^2}\div\dfrac{40x^4}{112xy^5}$

30. $\dfrac{x-3}{15x}\div\dfrac{4x-12}{5}$

31. $\dfrac{x-1}{6x+6}\div\dfrac{2x-2}{x^2+x}$

32. $\dfrac{7x-14}{x^2}\div\dfrac{x^2-4}{x^3}$

33. $\dfrac{6x^2-54}{x^4}\div\dfrac{x-3}{x^2}$

34. $\dfrac{x^2-25}{6x+30}\div\dfrac{x-5}{x}$

35. $\dfrac{2x-1}{x^2+2x}\div\dfrac{10x^2-5x}{6x^2+12x}$

36. $\dfrac{x+3}{x^2+3x-4}\div\dfrac{x+2}{x^2+x-2}$

37. $\dfrac{6x^2-7x-3}{x^2-1}\div\dfrac{2x-3}{x-1}$

38. $\dfrac{x^2-9}{2x^2+7x+3}\div\dfrac{x^2-3x}{2x^2+11x+5}$

39. $\dfrac{x^2-6x+9}{x^2-4x+3}\div\dfrac{2x^2-7x+3}{x^2-3x+2}$

40. $\dfrac{x^3+2x^2}{x^2+11x+28}\div\dfrac{4x^2}{x+7}$

41. $\dfrac{2x+1}{4x-x^2}\div\dfrac{4x^2-1}{x^2-16}$

42. $\dfrac{x^2-4x+4}{x^2+5x+6}\div\dfrac{x^2+2x-8}{x^2+7x+12}$

43. $\dfrac{x^2-x-6}{x^2+6x+8}\div\dfrac{x^2-4x+3}{x^2+5x+4}$

44. $\dfrac{x^2-x-12}{6x^2-25x-9}\div\dfrac{x^2-6x+8}{3x^2-17x-6}$

45. $\dfrac{6x^2+5x+1}{4x^3-3x^2} \div \dfrac{3x^2-2x-1}{3x^2-2x+1}$

46. $\dfrac{8x^2+2x-15}{3x^2+13x+4} \div \dfrac{2x^2+5x+3}{6x^2-x-1}$

47. $\dfrac{3x^2+13x+14}{4x^3-3x^2} \div \dfrac{6x^2-x-35}{4x^2+5x-6}$

48. $\dfrac{3x^2+2x}{9x^2-4} \div \dfrac{9x^2+6x-8}{9x^2-16}$

49. $\dfrac{x^2-8x+15}{x^2-9x+14} \div \dfrac{x^2+4x-21}{x-1}$

50. $\dfrac{6-11x-10x^2}{2x^2+x-3} \div \dfrac{5x^3-2x^2}{3x^2-5x+2}$

51. $\dfrac{x-6}{x^2-7x+6} \cdot \dfrac{x^2-3x}{x+3} \cdot \dfrac{x^2-9}{x^2-4x+3}$

52. $\dfrac{3x^2+11x+10}{2x^2+x-6} \cdot \dfrac{x^2+2x-3}{2x-1} \cdot \dfrac{2x-3}{3x^2+2x-5}$

53. $\dfrac{x^3+3x^2}{x^2+7x+12} \cdot \dfrac{2x^2+7x-4}{2x^2-x} \cdot \dfrac{x^2+4x-5}{2x^2-x-1}$

54. $\dfrac{x^2+2x-3}{x^2+10x+21} \cdot \dfrac{x^2+6x+5}{x^2-7x-8} \cdot \dfrac{x^2-x-56}{x^2-3x-40}$

55. $\dfrac{2x^2-5x+2}{4xy-2y+6x-3} \div \dfrac{xy-2y+3x-6}{2y^2+9y+9}$

56. $\dfrac{2xy-12x+y-6}{y^2-2y-24} \div \dfrac{2x^2+11x+5}{xy+5y+4x+20}$

Applications

Solve

57. Erik is building a cubby bookshelf; that is, a bookshelf divided into storage holes (cubbies) instead of shelves. He wants the height of the bookshelf to be $x^2-3x-10$ and the width to be x^2+5x+6. Each cubby hole in the bookshelf will have a height of $x+3$ and a width of $x-5$.

   a. Write a rational expression to determine how many cubbies high the bookshelf will be.

   b. Write a rational expression to determine how many cubbies wide the bookshelf will be.

   c. Multiply the rational expressions from parts a. and b. (and reduce to lowest terms) to obtain a rational expression that gives the total number of cubbies in the entire bookshelf.

58. The station manager at WSTB The AlterNation is planning a giveaway for the month.
The monthly budget for the station is decided by the expression $60x^2+330x+360$ and the budget is split evenly between $x+3$ things, including the giveaway. During the giveaway, the prizes will be given to every $x+3$ caller. The station usually receives $5x^2+65x+180$ calls during giveaways.

   a. Write a rational expression to determine how much of the budget will go to the giveaway.

   b. Write a rational expression to determine how many callers will win prizes during the giveaway.

   c. Find the rational expression used to determine the average amount the radio station can spend per prize by dividing the expression from part a. by the expression from part b. and reducing to lowest terms.

44. $\dfrac{x^2-3x-18}{(2x-9)(x-2)}$

45. $\dfrac{6x^3-x^2+1}{x^2(4x-3)(x-1)}$

46. $\dfrac{8x^2-14x+5}{(x+4)(x+1)}$

47. $\dfrac{x^2+4x+4}{x^2(2x-5)}$

48. $\dfrac{3x^2-4x}{(3x-2)^2}$

49. $\dfrac{x^2-6x+5}{(x-7)(x-2)(x+7)}$

50. $-\dfrac{3x-2}{x^2}$

51. $\dfrac{x^2-3x}{(x-1)^2}$

52. $\dfrac{x+3}{2x-1}$

53. $\dfrac{x^2+5x}{2x+1}$

54. $\dfrac{x-1}{x-8}$

55. 1

56. 1

Applications

57. a. $\dfrac{x^2-3x-10}{x+3}$

b. $\dfrac{x^2+5x+6}{x-5}$

c. $(x+2)^2=x^2+4x+4$

58. a. $\dfrac{60x^2+330x+360}{x+3}$

b. $\dfrac{5x^2+65x+180}{x+3}$

c. $\dfrac{6(2x+3)}{x+9}=\dfrac{12x+18}{x+9}$

Objectives

A. Find the least common multiple and add fractions.

B. Find the least common multiple of a list of polynomials.

C. Find equivalent rational expressions.

14.3 Least Common Multiple of Polynomials

A Review of Least Common Multiple and Adding Fractions

As stated in Chapter 2, the **least common multiple** (**LCM**) of two (or more) whole numbers is the smallest number that is a multiple of each of these numbers. For easy reference, the method for finding the LCM by using prime factorizations is restated here.

Finding the LCM of a Set of Counting Numbers

1. Find the prime factorization of each number.
2. List the prime factors that appear in any one of the prime factorizations.
3. Find the product of these primes using each prime the most number of times it appears in any one of the prime factorizations.

PROCEDURE

1. Find the LCM of 10, 25, and 40.

Example 1 Finding the LCM of a Set of Counting Numbers

Find the LCM of 15, 18, and 45.

Solution

Step 1: Prime factorizations:

$15 = 3 \cdot 5$ One 3, one 5

$18 = 2 \cdot 3 \cdot 3$ One 2, two 3s

$45 = 3 \cdot 3 \cdot 5$ Two 3s, one 5

Step 2: 2, 3, and 5 are the only prime factors.

Step 3: Using the most of each factor in any one prime factorization we have LCM $= 2 \cdot 3 \cdot 3 \cdot 5 = 90$.

Now work margin exercise 1.

From Chapter 2, recall that if two or more fractions have the same denominator, add the numerators and keep the denominator. Then, reduce if possible.

2. Add: $\frac{7}{18}+\frac{1}{18}+\frac{7}{18}$

Example 2 Adding Fractions with the Same Denominator

Add: $\frac{5}{12}+\frac{7}{12}+\frac{3}{12}$

Solution

$$\frac{5}{12}+\frac{7}{12}+\frac{3}{12}=\frac{5+7+3}{12}=\frac{15}{12}=\frac{\cancel{3}\cdot 5}{\cancel{3}\cdot 4}=\frac{5}{4}$$

Now work margin exercise 2.

If the fractions have different denominators, find the least common denominator (LCD), which is the least common multiple of the denominators. Change each fraction into an equivalent fraction with that denominator. Add the new fractions. Then reduce, if possible.

Example 3 Adding Fractions with Different Denominators

Add: $\frac{3}{4}+\frac{5}{8}+\frac{1}{10}$

Solution

Find the LCD.

$$\left.\begin{aligned}4&=2\cdot 2\\ 8&=2\cdot 2\cdot 2\\ 10&=2\cdot 5\end{aligned}\right\}\text{LCD}=2\cdot 2\cdot 2\cdot 5=40$$

Add by using the LCD. $\frac{3}{4}\cdot\frac{10}{10}+\frac{5}{8}\cdot\frac{5}{5}+\frac{1}{10}\cdot\frac{4}{4}=\frac{30}{40}+\frac{25}{40}+\frac{4}{40}=\frac{30+25+4}{40}=\frac{59}{40}$

Now work margin exercise 3.

3. Add: $\frac{1}{6}+\frac{3}{10}+\frac{7}{12}$

B Least Common Multiple of Polynomials

In Section 14.4, we will see how to add rational expressions. If the denominators are different, then the least common multiple (LCM) of the denominators (LCD) will be needed. The procedure for finding the LCM of a set of polynomials is similar to that for finding the LCM of a set of counting numbers.

Finding the LCM of a Set of Polynomials

1. Completely factor each polynomial (including prime factors for numerical factors).
2. Form the product of all factors that appear, using each factor the most number of times it appears in any one polynomial.

PROCEDURE

Math Tip

The LCM and the GCF of polynomials are different concepts—be careful not to confuse them. LCM stands for least common multiple and GCF stands for greatest common factor.

Consider the monomials $25x^2y^4$ and $100x^4y^2$.

$$\begin{aligned}25x^2y^4&=5\cdot 5\cdot x\cdot x\cdot y\cdot y\cdot y\cdot y\\&=5^2x^2y^4\end{aligned}$$

$$\begin{aligned}&100x^4y^2\\&=2\cdot 2\cdot 5\cdot 5\cdot x\cdot x\cdot x\cdot x\cdot y\cdot y\\&=2^2 5^2x^4y^2\end{aligned}$$

From this we know that the greatest common factor is $5^2x^2y^2$ and the least common multiple is $100x^4y^4$.

Example 4 Finding the LCM of Polynomials

Find the LCM of the polynomials $2x+8$ and x^2-16.

Solution

$$\left.\begin{aligned}2x+8&=2(x+4)\\ x^2-16&=(x+4)(x-4)\end{aligned}\right\}\text{LCM}=2(x+4)(x-4)=2(x^2-16)$$

Now work margin exercise 4.

4. Find the LCM of the polynomials $5x+15$ and x^2-9.

5. Find the LCM of the polynomials $4y^2 + 4y + 1$, $4y^2 + 2y$, and $8y - 4$.

Example 5 Finding the LCM of Polynomials

Find the LCM of the polynomials $y^2 + 10y + 25$, $3y^2 + 15y$, and $5y - 25$.

Solution

$$\left.\begin{aligned} y^2+10y+25&=(y+5)^2 \\ 3y^2+15y&=3y(y+5) \\ 5y-25&=5(y-5)\end{aligned}\right\} \text{LCM} = 3\cdot 5y(y+5)^2(y-5) = 15y(y+5)^2(y-5)$$

Now work margin exercise 5.

C Finding Equivalent Rational Expressions

To find a rational expression equivalent to a given rational expression, say $\frac{P}{Q}$, we choose R so that $Q \cdot R$ is the desired denominator and multiply by 1 as follows.

$$\frac{P}{Q} = \frac{P}{Q} \cdot 1 = \frac{P}{Q} \cdot \frac{R}{R} = \frac{P \cdot R}{Q \cdot R}$$

Examples 6 and 7 illustrate this technique.

6. Find the missing numerator that will make the rational expressions equivalent.

$$\frac{-2x}{10x-5} = \frac{?}{30x-15}$$

Example 6 Writing Equivalent Rational Expressions

Find the missing numerator that will make the rational expressions equivalent.

$$\frac{3x}{x+7} = \frac{?}{5x+35}$$

Solution

Factor the denominator on the right-hand side: $5x + 35 = 5(x+7)$

So, $\frac{3x}{x+7} = \frac{?}{5(x+7)}$.

Now we see that we need to multiply by $1 = \frac{5}{5}$ to get the equivalent expression with the desired denominator.

$$\frac{3x}{x+7} = \frac{3x}{x+7} \cdot 1 = \frac{3x}{x+7} \cdot \frac{5}{5} = \frac{3x \cdot 5}{(x+7) \cdot 5} = \frac{15x}{5x+35}$$

Now work margin exercise 6.

7. Find the missing numerator that will make the rational expressions equivalent.

$$\frac{9x}{x^2-1} = \frac{?}{x(x+1)(x-1)(x+3)}$$

Example 7 Writing Equivalent Rational Expressions

Find the missing numerator that will make the rational expressions equivalent.

$$\frac{2x}{x^2-9} = \frac{?}{x(x+3)(x-3)(x-1)}$$

Solution

The denominator on the right is already factored and we see that we need to multiply by $1 = \frac{x(x-1)}{x(x-1)}$ to get the equivalent expression with the desired denominator.

$$\frac{2x}{x^2-9} = \frac{2x}{(x+3)(x-3)} \cdot \frac{x(x-1)}{x(x-1)} = \frac{2x^2(x-1)}{x(x+3)(x-3)(x-1)}$$

Now work margin exercise 7.

Margin Exercise Answers

1. 200 **2.** $\frac{5}{6}$ **3.** $\frac{21}{20}$ **4.** $5(x^2-9)$ **5.** $4y(2y+1)^2(2y-1)$ **6.** $-6x$ **7.** $9x^2(x+3)$

14.3 Exercises

Concept Check

Fill-in-the-Blank. Complete each sentence using information found in this section.

1. The least common multiple (LCM) of two or more whole numbers is the ______ number that is a multiple of each of these numbers.

2. If two or more fractions have the same denominator, add the numerators and ______ the denominator.

3. When finding the LCM, the first step is to find the ______ ______ of each number.

4. To find a rational expression equivalent to a given rational expression $\frac{P}{Q}$, choose R so that $Q \cdot R$ is the desired ______.

5. When adding fractions with different denominators, you need to change each fraction into a/an ______ fraction with the denominator equal to the LCD of the fractions.

True/False. Determine whether each statement is true or false. If a statement is false, explain how it can be changed so the statement will be true. (**Note:** There may be more than one acceptable change.)

6. When adding fractions with different denominators, add the denominators.

7. The fraction $\frac{R}{R}$ is equivalent to 1.

8. The least common denominator (LCD) is the least common multiple of the denominators.

9. When finding the LCM of a set of polynomials, you only find the factors of any numerical terms.

14.3 Exercises

Concept Check

1. smallest

2. keep

3. prime factorization

4. denominator

5. equivalent

6. False; The least common denominator needs to be found.

7. True

8. True

9. False; You completely factor each polynomial, including prime factors for numerical terms.

Practice

Find the least common multiple (LCM) of each set of numbers. See Example 1.

1. 15, 25, 30

2. 18, 21, 63

3. 16, 24, 27

4. 35, 45, 63

5. 20, 30, 40, 50

6. 44, 55, 121

7. 5, 10, 15, 20, 30

8. 24, 36, 48, 54

Find the indicated sums and reduce, if possible. See Examples 2 and 3.

9. $\frac{5}{17}+\frac{6}{17}$

10. $\frac{3}{25}+\frac{7}{25}$

11. $\frac{5}{8}+\frac{7}{8}$

12. $\frac{5}{6}+\frac{5}{6}+\frac{1}{6}$

13. $\frac{1}{2}+\frac{7}{10}$

14. $\frac{7}{10}+\frac{1}{5}+\frac{3}{10}$

15. $\frac{1}{2}+\frac{1}{10}+\frac{1}{6}$

16. $\frac{2}{15}+\frac{7}{15}+\frac{2}{45}+\frac{1}{30}$

17. $\frac{11}{18}+\frac{13}{54}+\frac{5}{27}$

18. $\frac{3}{4}+\frac{9}{10}+\frac{7}{20}+\frac{1}{2}$

Find the least common multiple (LCM) of each set of polynomials. See Examples 4 and 5.

19. $x^2-25,\quad 7x+35$

20. $x^2-14x+49,\quad 9x-63$

21. $6y-24,\quad 3y-12,\quad 5y-20$

22. $20y+32,\quad 15y+24,\quad 45y+72$

23. $x^2-9,\quad x^2-6x+9$

24. $2x^2-50,\quad x^2-10x+25$

25. $y-3,\quad 3-y$

26. $22-x,\quad x-22$

27. $x^2-144,\quad 24-2x$

28. $30-3y,\quad y^2-20y+100$

29. $x^2+x-12,\quad x^2+9x+20$

30. $x^2-3x+2,\quad x^2-7x+6$

31. $x^2+5x-14,\quad xy-2y+3x-6$

32. $y^2+4y+3,\quad xy+3x-5y-15$

33. $2x^2-72,\quad x^2+9x+18$

34. $5x^2+5x-30,\quad 3x^2-9x+6$

35. $2xy-10y+12x-60,$
$3y^2+21y+18$

36. $8x^2-8y^2,\quad x^2-xy+3x-3y$

37. $x^2-4,\quad x^3-2x^2+4x-8$

38. $x^2-25,\quad x^3-5x^2+x-5$

Practice

1. 150
2. 126
3. 432
4. 315
5. 600
6. 2420
7. 60
8. 432
9. $\frac{11}{17}$
10. $\frac{2}{5}$
11. $\frac{3}{2}$
12. $\frac{11}{6}$
13. $\frac{6}{5}$
14. $\frac{6}{5}$
15. $\frac{23}{30}$
16. $\frac{61}{90}$
17. $\frac{28}{27}$
18. $\frac{5}{2}$
19. $7(x+5)(x-5)$
20. $9(x-7)(x-7)$
21. $30(y-4)$
22. $36(5y+8)$
23. $(x+3)(x-3)(x-3)$
24. $2(x+5)(x-5)(x-5)$
25. $-(y-3)$
26. $-(x-22)$
27. $-2(x+12)(x-12)$
28. $-3(y-10)(y-10)$
29. $(x+4)(x-3)(x+5)$
30. $(x-1)(x-2)(x-6)$
31. $(x-2)(y+3)(x+7)$
32. $(y+1)(y+3)(x-5)$

Write a rational expression on the right equivalent to the given rational expression on the left. See Examples 6 and 7.

39. $\frac{7}{2x+3}=\frac{?}{4(2x+3)}$

40. $\frac{2x}{x^2-4x}=\frac{?}{2x^2(x-4)}$

41. $\frac{11}{2x+6}=\frac{?}{6(x+3)(x-3)}$

42. $\frac{5}{7(x-10)}=\frac{?}{35(x-10)(x+10)}$

43. $\frac{3x}{4-x}=\frac{?}{x(x-4)}$

44. $\frac{4}{5x-x^2}=\frac{?}{x(x-5)(x+5)}$

45. $\frac{y-1}{y^2+5y}=\frac{?}{2y(y+3)(y+5)}$

46. $\frac{x+3}{2x^2-x-1}=\frac{?}{(2x+1)(x-1)(3x-2)}$

47. $\frac{x+1}{x^2+1}=\frac{?}{(x^2+1)(x+3)}$

48. $\frac{x+5}{x^2+6}=\frac{?}{(x^2+6)(x-5)}$

33. $2(x+6)(x-6)(x+3)$

34. $15(x-2)(x+3)(x-1)$

35. $6(x-5)(y+6)(y+1)$

36. $8(x+y)(x-y)(x+3)$

37. $(x+2)(x-2)(x^2+4)$

38. $(x+5)(x-5)(x^2+1)$

39. 28

40. $4x^2$

41. $33(x-3)$

42. $25(x+10)=25x+250$

43. $-3x^2$

44. $-4(x+5)=-4x-20$

45. $2(y+3)(y-1)$
$=2y^2+4y-6$

46. $(3x-2)(x+3)$
$=3x^2+7x-6$

47. $(x+1)(x+3)$
$=x^2+4x+3$

48. $(x-5)(x+5)=x^2-25$

Objectives

A. Add rational expressions.

B. Subtract rational expressions.

14.4 Addition and Subtraction with Rational Expressions

A Addition with Rational Expressions

To add rational expressions with the same denominator, proceed just as with fractions: add the numerators and keep the common denominator. Then reduce, if possible.

$$\frac{x^2+6}{x+2}+\frac{5x}{x+2}=\frac{x^2+5x+6}{x+2}=\frac{(x+2)(x+3)}{x+2}=x+3 \quad (x\neq -2)$$

Addition with Rational Expressions

For polynomials P, Q, and R, with $Q \neq 0$,

$$\frac{P}{Q}+\frac{R}{Q}=\frac{P+R}{Q}.$$

DEFINITION

1. Find each sum and reduce, if possible. State any restrictions on the variable.

   a. $\frac{x}{x^2-25}+\frac{5}{x^2-25}$

   b. $\frac{2}{x^2+8x+15}+\frac{3x+7}{x^2+8x+15}$

Example 1 Adding Rational Expressions with a Common Denominator

Find each sum and reduce, if possible. (**Note:** the importance of the factoring techniques we studied earlier will be important here.) State any restrictions on the variable.

a. $\frac{x}{x^2-1}+\frac{1}{x^2-1}$ **b.** $\frac{1}{x^2+7x+10}+\frac{2x+3}{x^2+7x+10}$

Solution

a.
$$\frac{x}{x^2-1}+\frac{1}{x^2-1}=\frac{x+1}{x^2-1}$$
$$=\frac{\overset{1}{x+1}}{(x+1)(x-1)}$$
$$=\frac{1}{x-1} \quad (x\neq -1, 1)$$

Remember, if we use the entire expression in the numerator (or denominator) to reduce, we are left with a factor of 1.

b.
$$\frac{1}{x^2+7x+10}+\frac{2x+3}{x^2+7x+10}=\frac{2x+4}{x^2+7x+10}$$
$$=\frac{2(x+2)}{(x+5)(x+2)}$$
$$=\frac{2}{x+5} \quad (x\neq -5, -2)$$

Now work margin exercise 1.

The LCM of a set of denominators is called the **least common denominator (LCD)**. To add rational expressions with different denominators, begin by changing each expression to an equivalent fraction with the LCD as the denominator. This is also called **building the fraction to higher terms**.

Use the following procedure when adding rational expressions with different denominators.

Adding Rational Expressions with Different Denominators

1. Find the LCD (the LCM of the denominators).
2. Rewrite each fraction in an equivalent form with the LCD as the denominator.
3. Add the numerators and keep the common denominator.
4. Reduce, if possible.

PROCEDURE

Example 2 Adding Rational Expressions with Different Denominators

Find each sum and reduce, if possible. Assume that no denominator has a value of 0.

a. $\dfrac{y}{y-3}+\dfrac{6}{y+4}$ **b.** $\dfrac{1}{x^2+6x+9}+\dfrac{1}{x^2-9}+\dfrac{1}{x+3}$

Solution

a. In this case, neither denominator can be factored, so the LCD is the product of the two given denominators. That is, $\text{LCD}=(y-3)(y+4)$.

Now, using the fundamental principle of rational expressions to write equivalent fractions, we have the following.

$$\begin{aligned}\frac{y}{y-3}+\frac{6}{y+4}&=\frac{y(y+4)}{(y-3)(y+4)}+\frac{6(y-3)}{(y+4)(y-3)}\\&=\frac{(y^2+4y)+(6y-18)}{(y-3)(y+4)}\\&=\frac{y^2+10y-18}{(y-3)(y+4)}\end{aligned}$$

The numerator is not factorable and the expression is reduced.

b. First, find the LCD.

$$\left.\begin{aligned}x^2+6x+9&=(x+3)^2\\x^2-9&=(x+3)(x-3)\\x+3&=x+3\end{aligned}\right\}\text{LCD}=(x+3)^2(x-3)$$

2. Find each sum and reduce, if possible. Assume that no denominator has a value of 0.

a. $\dfrac{x}{x+2}+\dfrac{3}{x+3}$

b. $\dfrac{1}{x^2+10x+25}+\dfrac{1}{x^2-25}+\dfrac{1}{x+5}$

Now, use the LCD and add as follows.

$$\frac{1}{x^2+6x+9}+\frac{1}{x^2-9}+\frac{1}{x+3}$$

$$=\frac{1}{(x+3)^2}+\frac{1}{(x+3)(x-3)}+\frac{1}{(x+3)}$$

$$=\frac{1\cdot(x-3)}{(x+3)^2\cdot(x-3)}+\frac{1\cdot(x+3)}{(x+3)(x-3)\cdot(x+3)}+\frac{1\cdot(x+3)(x-3)}{(x+3)\cdot(x+3)(x-3)}$$

$$=\frac{(x-3)+(x+3)+(x^2-9)}{(x+3)^2(x-3)}=\frac{x^2+2x-9}{(x+3)^2(x-3)}$$

Now work margin exercise 2.

3. Find the sum and reduce, if possible. Assume that no denominator has a value of 0.

$$\frac{s}{s+3}+\frac{4}{s+1}$$

Completion Example 3 Adding Rational Expressions

Find the sum and reduce, if possible. Assume that no denominator has a value of 0.

$$\frac{y}{y-5}+\frac{3}{y+6}$$

Solution

In this case, neither denominator can be factored, so the LCD is the product of the two given denominators. That is, LCD = ($y-5$)($y+6$).

$$\frac{y}{y-5}+\frac{3}{y+6}=\frac{y(y+6)}{(y-5)(y+6)}+\frac{3(y-5)}{(y+6)(y-5)}$$

$$=\frac{(y^2+6y)+(3y-15)}{(y+6)(y-5)}$$

$$=\frac{y^2+9y-15}{(y+6)(y-5)}$$

Now work margin exercise 3.

Attention!

Important note about the form of answers

In Examples 2 and 3, each denominator is left in factored form as a convenience for possibly reducing or adding to some other expression later. You may choose to multiply out these factors. Either form is correct. For consistency, denominators are left in factored form in the answers in the back of the text.

B Subtraction with Rational Expressions

When subtracting fractions, the placement of negative signs can be critical. For example, note how –2 can be indicated in three different forms.

$$-\frac{6}{3}=-2,\quad \frac{-6}{3}=-2,\ \text{and}\ \frac{6}{-3}=-2$$

Thus, we have $-\frac{6}{3}=\frac{-6}{3}=\frac{6}{-3}$.

With polynomials, we have the following statement about the placement of negative signs, which can be **very useful in subtraction**. We seldom leave the negative sign in the denominator.

Placement of Negative Signs

If P and Q are polynomials and $Q \neq 0$, then

$$-\frac{P}{Q}=\frac{P}{-Q}=\frac{-P}{Q}.$$

DEFINITION

To subtract rational expressions with a common denominator, proceed just as with fractions: subtract the numerators and keep the common denominator.

$$\frac{17}{x+7}-\frac{23}{x+7}=\frac{17-23}{x+7}=\frac{-6}{x+7}\quad\left(\text{or } -\frac{6}{x+7}\right)$$

Subtraction with Rational Expressions

For polynomials P, Q, and R, with $Q \neq 0$,

$$\frac{P}{Q}-\frac{R}{Q}=\frac{P-R}{Q}.$$

DEFINITION

Example 4 Subtracting Rational Expressions with a Common Denominator

Find each difference and reduce, if possible. Assume that no denominator has a value of 0.

a. $\frac{2x-5y}{x+y}-\frac{3x-7y}{x+y}$

b. $\frac{x^2}{x^2+4x+4}-\frac{2x+8}{x^2+4x+4}$

c. $\frac{x}{x-5}-\frac{3}{5-x}$

4. Find each difference and reduce, if possible. Assume that no denominator has a value of 0.

a. $\frac{4x-7y}{3x-y}-\frac{3x-9y}{3x-y}$

b. $\frac{x^2}{x^2+6x+9}-\frac{2x+15}{x^2+6x+9}$

c. $\frac{x}{x-4}-\frac{5}{4-x}$

Solution

a. $$\frac{2x-5y}{x+y}-\frac{3x-7y}{x+y}=\frac{2x-5y-(3x-7y)}{x+y}$$ Subtract the entire numerator.

$$=\frac{2x-5y-3x+7y}{x+y}$$

$$=\frac{-x+2y}{x+y}$$

b. $$\frac{x^2}{x^2+4x+4}-\frac{2x+8}{x^2+4x+4}=\frac{x^2-(2x+8)}{x^2+4x+4}$$ Subtract the entire numerator.

$$=\frac{x^2-2x-8}{x^2+4x+4}$$

$$=\frac{(x-4)\cancel{(x+2)}}{(x+2)\cancel{(x+2)}}$$ Factor and reduce.

$$=\frac{x-4}{x+2}$$

c. Each denominator is the **opposite** of the other. Multiply both the numerator and denominator of the second fraction by -1 so that both denominators will be the same, in this case $x-5$.

$$\frac{x}{x-5}-\frac{3}{5-x}=\frac{x}{x-5}-\frac{3}{(5-x)}\cdot\frac{(-1)}{(-1)}$$

$$=\frac{x}{x-5}-\frac{-3}{x-5}$$

$$=\frac{x-(-3)}{x-5}$$

$$=\frac{x+3}{x-5}$$

Now work margin exercise 4.

Common Error

Many beginning students make a mistake when subtracting rational expressions by not subtracting the entire numerator. They make a mistake similar to the following.

Wrong Solution

$$\frac{10}{x+5}-\frac{3-x}{x+5}=\frac{10-3-x}{x+5}=\frac{7-x}{x+5}$$

By using parentheses, you can avoid such mistakes.

Correct Solution

$$\frac{10}{x+5}-\frac{3-x}{x+5}=\frac{10-(3-x)}{x+5}=\frac{10-3+x}{x+5}=\frac{7+x}{x+5}=\frac{x+7}{x+5}$$

CAUTION

As with addition, if the rational expressions do not have the same denominator, find the LCM of the denominators (the LCD). Then, use the fundamental principle of rational expressions to change each expression to an equivalent fraction with the LCD as the denominator.

Example 5 Subtracting Rational Expressions with Different Denominators

Find each difference and reduce, if possible. Assume that no denominator has a value of 0.

a. $\frac{x+5}{x-5}-\frac{100}{x^2-25}$

b. $\frac{3x-12}{x^2+x-20}-\frac{x^2+5x}{x^2+9x+20}$

c. $\frac{x+1}{xy-3y+4x-12}-\frac{x-3}{xy+6y+4x+24}$

Solution

a. $\left.\begin{aligned} x-5&=x-5 \\ x^2-25&=(x+5)(x-5)\end{aligned}\right\}\text{LCD}=(x+5)(x-5)$

$$\begin{aligned}\frac{x+5}{x-5}-\frac{100}{x^2-25}&=\frac{(x+5)(x+5)}{(x-5)(x+5)}-\frac{100}{(x+5)(x-5)}\\&=\frac{(x^2+10x+25)-100}{(x+5)(x-5)}\\&=\frac{x^2+10x+25-100}{(x+5)(x-5)}\\&=\frac{x^2+10x-75}{(x+5)(x-5)}\\&=\frac{(x+15)\cancel{(x-5)}}{(x+5)\cancel{(x-5)}}\\&=\frac{x+15}{x+5}\end{aligned}$$

b. In this problem, both expressions can be reduced before looking for the LCD.

$$\begin{aligned}\frac{3x-12}{x^2+x-20}-\frac{x^2+5x}{x^2+9x+20}&=\frac{3\cancel{(x-4)}}{(x+5)\cancel{(x-4)}}-\frac{x\cancel{(x+5)}}{\cancel{(x+5)}(x+4)}\\&=\frac{3}{x+5}-\frac{x}{x+4}\end{aligned}$$

Now subtract these two expressions with $\text{LCD}=(x+5)(x+4)$.

$$\begin{aligned}\frac{3}{x+5}-\frac{x}{x+4}&=\frac{3(x+4)}{(x+5)(x+4)}-\frac{x(x+5)}{(x+4)(x+5)}\\&=\frac{(3x+12)-(x^2+5x)}{(x+5)(x+4)}\\&=\frac{3x+12-x^2-5x}{(x+5)(x+4)}\\&=\frac{-x^2-2x+12}{(x+5)(x+4)}\end{aligned}$$

---

5. Find each difference and reduce, if possible. Assume that no denominator has a value of 0.

a. $\frac{x+6}{x-6}-\frac{144}{x^2-36}$

b. $\frac{5x+y}{(x+y)^2}-\frac{y}{3x^2-3y^2}$

c. $\frac{4x-8}{x^2+4x-12}-\frac{x^2+6x}{x^2+9x+18}$

d. $\frac{y+4}{xy-2y+x-2}-\frac{y-5}{xy-2y+3x-6}$

c. $$\left.\begin{aligned} xy-3y+4x-12 &= y(x-3)+4(x-3) \\ &= (x-3)(y+4) \\ xy+6y+4x+24 &= y(x+6)+4(x+6) \\ &= (x+6)(y+4) \end{aligned}\right\}\text{LCD} = (x-3)(y+4)(x+6)$$

$$\begin{aligned} \frac{x+1}{xy-3y+4x-12}-\frac{x-3}{xy+6y+4x+24} &= \frac{(x+1)(x+6)}{(y+4)(x-3)(x+6)}-\frac{(x-3)(x-3)}{(x+6)(y+4)(x-3)} \\ &= \frac{x^2+7x+6-\left(x^2-6x+9\right)}{(y+4)(x-3)(x+6)} \\ &= \frac{x^2+7x+6-x^2+6x-9}{(y+4)(x-3)(x+6)} \\ &= \frac{13x-3}{(y+4)(x-3)(x+6)} \end{aligned}$$

Now work margin exercise 5.

6. Find the difference and reduce, if possible. Assume that no denominator has a value of 0.

$$\frac{x-4}{x^2-2x+1}-\frac{5}{x-1}$$

Completion Example 6 Subtracting Rational Expressions

Find the difference and reduce, if possible. Assume that no denominator has a value of 0.

$$\frac{x}{x^2-x-2}-\frac{1}{x-2}$$

Solution

$$\begin{aligned} \frac{x}{x^2-x-2}-\frac{1}{x-2} &= \frac{x}{(x-2)(x+1)}-\frac{1(x+1)}{(x-2)(x+1)} \\ &= \frac{x-(x+1)}{(x-2)(x+1)} \\ &= \frac{-1}{(x-2)(x+1)} \end{aligned}$$

Now work margin exercise 6.

7. Perform the indicated operations and reduce, if possible. Assume that no denominator has a value of 0.

$$\frac{9}{25-y^2}-\frac{2-y}{5-y}+\frac{y}{5+y}$$

Example 7 Adding and Subtracting Rational Expressions

Perform the indicated operations and reduce, if possible. Assume that no denominator has a value of 0.

$$\frac{15}{x^2-16}+\frac{x}{x+4}-\frac{x+3}{x-4}$$

Solution

$$\left.\begin{aligned} x^2-16 &= (x+4)(x-4) \\ x+4 &= x+4 \\ x-4 &= x-4 \end{aligned}\right\}\text{LCD} = (x+4)(x-4)$$

$$\frac{15}{x^2-16}+\frac{x}{x+4}-\frac{x+3}{x-4}=\frac{15}{(x+4)(x-4)}+\frac{x}{x+4}\cdot\frac{(x-4)}{(x-4)}-\frac{x+3}{x-4}\cdot\frac{(x+4)}{(x+4)}$$

$$=\frac{15+x(x-4)-(x+3)(x+4)}{(x+4)(x-4)}$$

$$=\frac{15+x^2-4x-(x^2+7x+12)}{(x+4)(x-4)}$$

$$=\frac{15+x^2-4x-x^2-7x-12}{(x+4)(x-4)}$$

$$=\frac{3-11x}{(x+4)(x-4)}$$

Now work margin exercise 7.

Completion Example Answers

3. LCD $=(y-5)(y+6)$;

$$\frac{y(y+6)}{(y-5)(y+6)}+\frac{3(y-5)}{(y+6)(y-5)}=\frac{(y^2+6y)+(3y-15)}{(y+6)(y-5)}=\frac{y^2+9y-15}{(y+6)(y-5)}$$

6. $\frac{x}{(x-2)(x+1)}-\frac{1(x+1)}{(x-2)(x+1)}=\frac{x-(x+1)}{(x-2)(x+1)}=\frac{-1}{(x-2)(x+1)}$

Margin Exercise Answers

1. a. $\frac{1}{x-5}$; $x\neq-5,5$ **b.** $\frac{3}{x+5}$; $x\neq-5,-3$ **2. a.** $\frac{x^2+6x+6}{(x+3)(x+2)}$ **b.** $\frac{x^2+2x-25}{(x+5)^2(x-5)}$

3. $\frac{s^2+5s+12}{(s+3)(s+1)}$ **4. a.** $\frac{x+2y}{3x-y}$ **b.** $\frac{x-5}{x+3}$ **c.** $\frac{x+5}{x-4}$ **5. a.** $\frac{x+18}{x+6}$ **b.** $\frac{15x^2-13xy-4y^2}{3(x+y)^2(x-y)}$

c. $\frac{-x^2-2x+12}{(x+6)(x+3)}$ **d.** $\frac{11y+17}{(x-2)(y+1)(y+3)}$ **6.** $\frac{-4x+1}{(x-1)^2}$ **7.** $\frac{8y-1}{(5-y)(5+y)}$

14.4 Exercises

Concept Check

Fill-in-the-Blank. Complete each sentence using information found in this section.

1. To add rational expressions with a common denominator, proceed just as with fractions: add the __________ and keep the common __________.

2. When finding the LCM for a set of polynomials, the first step is to ________ each polynomial.

3. Next, form the product of all factors that appear, using each factor the ________ number of times it appears in any one polynomial.

4. To add rational expressions with different denominators, first find the ________.

14.4 Exercises

Concept Check

1. numerator, denominator
2. factor
3. most
4. LCD

5. equivalent
6. reduce
7. True
8. False; The negative sign is seldom left in the denominator.
9. True
10. False; Using parentheses can help avoid mistakes and ensure that you subtract the entire numerator.

Practice

1. 3
2. 7
3. 2
4. 2
5. 1
6. $x+3$
7. $\frac{2}{x-1}$
8. $\frac{2x^2+3x+4}{(x-2)(x+2)}$
9. $\frac{14}{7-x}$
10. 6
11. 4
12. 4
13. $\frac{x^2-x+1}{(x+4)(x-3)}$
14. $\frac{5}{x-3}$
15. $\frac{x-2}{x+2}$
16. $\frac{-2x+3}{(2x+1)(x-1)}$
17. $\frac{4x+5}{2(7x-2)}$
18. $\frac{x+9}{2(2x+5)}$
19. $\frac{6x+15}{(x+3)(x-3)}$
20. $\frac{x^2+3x+1}{(x-5)(x+2)}$

5. Then, rewrite each fraction in a/an ____________ form with the LCD as the denominator.

6. The final step when adding or subtracting rational expressions is to ______, if possible.

True/False. Determine whether each statement is true or false. If a statement is false, explain how it can be changed so the statement will be true. (**Note:** There may be more than one acceptable change.)

7. The LCM of a set of denominators is called the least common denominator.

8. With polynomials, it is most common to place negative signs in the denominator.

9. As with addition, when subtracting rational expressions with different denominators, the first step is to find the LCM of the denominators.

10. You should not use parentheses when subtracting rational expressions.

Practice

Perform the indicated operations and reduce, if possible. Assume that no denominator has a value of 0.

1. $\frac{3x}{x+4}+\frac{12}{x+4}$

2. $\frac{7x}{x+5}+\frac{35}{x+5}$

3. $\frac{x-1}{x+6}+\frac{x+13}{x+6}$

4. $\frac{3x-1}{2x-6}+\frac{x-11}{2x-6}$

5. $\frac{3x+1}{5x+2}+\frac{2x+1}{5x+2}$

6. $\frac{x^2+3}{x+1}+\frac{4x}{x+1}$

7. $\frac{x-5}{x^2-2x+1}+\frac{x+3}{x^2-2x+1}$

8. $\frac{2x^2+5}{x^2-4}+\frac{3x-1}{x^2-4}$

9. $\frac{13}{7-x}-\frac{1}{x-7}$

10. $\frac{6x}{x-6}+\frac{36}{6-x}$

11. $\frac{3x}{x-4}+\frac{16-x}{4-x}$

12. $\frac{20}{x-10}-\frac{3}{10-x}$

13. $\frac{x^2+2}{x^2+x-12}+\frac{x+1}{12-x-x^2}$

14. $\frac{10}{x^2-x-6}-\frac{5x}{6+x-x^2}$

15. $\frac{x^2+2}{x^2-4}-\frac{4x-2}{x^2-4}$

16. $\frac{2x+5}{2x^2-x-1}-\frac{4x+2}{2x^2-x-1}$

17. $\frac{x+3}{7x-2}+\frac{2x-1}{14x-4}$

18. $\frac{3x+1}{4x+10}+\frac{4-x}{2x+5}$

19. $\frac{5}{x-3}+\frac{x}{x^2-9}$

20. $\frac{x+1}{x^2-3x-10}+\frac{x}{x-5}$

21. $\frac{x}{x-1}-\frac{4}{x+2}$

22. $\frac{x-1}{3x-1}-\frac{8+4x}{x+2}$

23. $\frac{x+2}{x+3}-\frac{4}{3-x}$

24. $\frac{x-1}{4-x}+\frac{3x}{x+5}$

25. $\frac{x+2}{3x+9}+\frac{2x-1}{2x-6}$

26. $\frac{x}{4x-8}-\frac{3x+2}{3x+6}$

27. $\dfrac{3x}{6+x}-\dfrac{2x}{x^2-36}$

28. $\dfrac{4}{x+5}-\dfrac{2x+3}{x^2+4x-5}$

29. $\dfrac{4x+1}{7-x}+\dfrac{x-1}{x^2-8x+7}$

30. $\dfrac{3x-4}{x^2-x-20}-\dfrac{2}{5-x}$

31. $\dfrac{4x}{x^2+3x-28}+\dfrac{3}{x^2+6x-7}$

32. $\dfrac{3x}{x^2+2x+1}-\dfrac{x}{x^2+9x+8}$

33. $\dfrac{3x+4}{2x^2-23x+30}-\dfrac{x+5}{2x^2-19x+24}$

34. $\dfrac{x+1}{x^2-3x+2}+\dfrac{6}{x^2-6x+8}$

35. $\dfrac{4x-1}{x^2-5x+4}+\dfrac{2x+7}{x^2-11x+28}$

36. $\dfrac{7x+3}{5x^2+27x+36}+\dfrac{3x-2}{5x^2+22x+24}$

37. $\dfrac{x-6}{7x^2-3x-4}+\dfrac{7-x}{7x^2+18x+8}$

38. $\dfrac{x+10}{x^2+5x+4}-\dfrac{4}{x^2+6x+8}$

39. $\dfrac{x-3}{4x^2-5x-6}-\dfrac{4x+10}{2x^2+x-10}$

40. $\dfrac{2x+1}{8x^2-37x-15}+\dfrac{2-x}{8x^2+11x+3}$

41. $\dfrac{3x}{4-x}+\dfrac{7x}{x+4}-\dfrac{x-3}{x^2-16}$

42. $\dfrac{x}{x+3}+\dfrac{x+1}{3-x}+\dfrac{x^2+4}{x^2-9}$

43. $-\dfrac{1}{2}+\dfrac{x-5}{x-3}+\dfrac{x-1}{x^2-5x+6}$

44. $-4+\dfrac{1-2x}{x+6}+\dfrac{x^2+1}{x^2+4x-12}$

45. $\dfrac{2}{x^2-4}-\dfrac{3}{x^2-3x+2}+\dfrac{x-1}{x^2+x-2}$

46. $\dfrac{5}{x^2+3x+2}+\dfrac{4}{x^2+6x+8}-\dfrac{6}{x^2+5x+4}$

47. $\dfrac{x}{x^2+4x-21}+\dfrac{1-x}{x^2+8x+7}+\dfrac{3x}{x^2-2x-3}$

48. $\dfrac{3x}{x^2+4x-5}-\dfrac{2}{3x^2+17x+10}-\dfrac{3}{3x^2-x-2}$

49. $\dfrac{3x+9}{x^2-5x+4}+\dfrac{49}{12+x-x^2}+\dfrac{3x+21}{x^2+2x-3}$

50. $\dfrac{5x+22}{x^2+8x+15}+\dfrac{4}{x^2+4x+3}+\dfrac{6}{x^2+6x+5}$

51. $\dfrac{x}{xy+x-2y-2}+\dfrac{x+2}{xy+x+y+1}$

52. $\dfrac{4x}{xy-3x+y-3}+\dfrac{x+2}{xy+2y-3x-6}$

53. $\dfrac{3y}{xy+2x+3y+6}+\dfrac{x}{x^2-2x-15}$

54. $\dfrac{2}{xy-4x-2y+8}+\dfrac{5y}{y^2-3y-4}$

55. $\dfrac{x+6}{2x-1}-\dfrac{3x^2+x-4}{2x^2-3x+1}$

56. $\dfrac{2x-5}{2x^2+2}+\dfrac{x^2-2x+5}{x^3+x^2+x+1}$

57. $\dfrac{x+1}{x^3-3x^2+x-3}+\dfrac{x^2-5x-8}{x^4-8x^2-9}$

58. $\dfrac{x+4}{x^3-5x^2+6x-30}-\dfrac{x-7}{x^3-2x^2+6x-12}$

59. $\dfrac{x-6}{3x^2+10x+3}-\dfrac{2x}{x^2+2x-3}+\dfrac{6x}{3x^2-2x-1}$

60. $\dfrac{x+1}{2x^2-x-1}+\dfrac{2x}{2x^2+5x+2}-\dfrac{2x}{x^2+x-2}$

21. $\dfrac{x^2-2x+4}{(x+2)(x-1)}$

22. $\dfrac{-11x+3}{3x-1}$

23. $\dfrac{-x^2-3x-6}{(x+3)(3-x)}$

24. $\dfrac{-2x^2+16x-5}{(4-x)(x+5)}$

25. $\dfrac{8x^2+13x-21}{6(x+3)(x-3)}$

26. $\dfrac{-9x^2+22x+16}{12(x-2)(x+2)}$

27. $\dfrac{3x^2-20x}{(x+6)(x-6)}$

28. $\dfrac{2x-7}{(x+5)(x-1)}$

29. $\dfrac{-4x}{x-7}$

30. $\dfrac{5x+4}{(x-5)(x+4)}$

31. $\dfrac{4x^2-x-12}{(x+7)(x-4)(x-1)}$

32. $\dfrac{2x^2+23x}{(x+1)^2(x+8)}$

33. $\dfrac{x-6}{(x-10)(x-8)}$

34. $\dfrac{x+5}{(x-1)(x-4)}$

35. $\dfrac{6x}{(x-1)(x-7)}$

36. $\dfrac{2x}{(x+2)(x+3)}$

37. $\dfrac{4x-19}{(7x+4)(x-1)(x+2)}$

38. $\dfrac{x+4}{(x+1)(x+2)}$

39. $\dfrac{-7x-9}{(4x+3)(x-2)}$

40. $\dfrac{x^2+10x-9}{(8x+3)(x-5)(x+1)}$

41. $\dfrac{4x^2-41x+3}{(x+4)(x-4)}$

42. $\dfrac{x^2-7x+1}{(x-3)(x+3)}$

43. $\dfrac{x-4}{2(x-2)}$

44. $\dfrac{-3(x^2+5x-1)}{(x+6)(x-2)}$

45. $\dfrac{x^2-4x-6}{(x+2)(x-2)(x-1)}$

46. $\dfrac{3}{(x+1)(x+2)}$

47. $\dfrac{3x^2+26x-3}{(x+7)(x-3)(x+1)}$

48. $\dfrac{9x^2+x-13}{(3x+2)(x+5)(x-1)}$

49. $\dfrac{6x+2}{(x-1)(x+3)}$

50. $\dfrac{5x+12}{(x+3)(x+1)}$

51. $\dfrac{2x^2+x-4}{(x-2)(y+1)(x+1)}$

52. $\dfrac{5x+1}{(y-3)(x+1)}$

53. $\dfrac{2x+4xy-15y}{(x+3)(y+2)(x-5)}$

54. $\dfrac{5xy-8y+2}{(y-4)(y+1)(x-2)}$

55. $\dfrac{-2x+2}{2x-1}$

56. $\dfrac{4x^2-7x+5}{2(x+1)(x^2+1)}$

57. $\dfrac{2x^2-x-5}{(x-3)(x+3)(x^2+1)}$

58. $\dfrac{14x-43}{(x^2+6)(x-5)(x-2)}$

59. $\dfrac{x^2+9x+6}{(3x+1)(x+3)(x-1)}$

60. $\dfrac{-1}{2x+1}$

Applications

61. a. $\dfrac{7x^2+3}{x-2}$ b. $\dfrac{4x^2+5}{x+2}$

c. $\dfrac{22x^3+12x^2+16x-8}{(x+2)(x-2)}$

62. $\dfrac{3x-3}{x(2x-3)}$

63. $\dfrac{5x+6}{2x(x+3)}$

64. $\dfrac{20}{x}$

65. a. $\dfrac{100}{x}$ b. $\dfrac{100}{2x+15}$

c. $\dfrac{100x+1500}{x(2x+15)}$

Applications

Solve.

61. A landscaper is hired to place large flowering bushes along the borders of a botanical garden. The property is in the shape of a rectangle that measures $7x^2 + 3$ feet long by $4x^2 + 5$ feet wide. The bushes are to be placed every $x + 2$ feet across the width of the property and every $x - 2$ feet along the length of the property.

a. Write a rational expression to determine how many bushes will go along one length of the property.

b. Write a rational expression to determine how many bushes will go along one width of the property.

c. Use the rational expressions from parts a. and b. to create a rational expression to determine how many bushes will be needed to line the entire property.

62. Two teams of set designers are jointly creating a set for a scene in a movie. In one hour, the first team can create $\frac{1}{x}$ of the set and the second team can create $\frac{1}{2x-3}$ of the set. If the two teams work together, how much of the set will be completed in one hour?

63. Three janitors work the night shift at the local hospital. Working alone, it takes Marla three more hours than it takes Tom, and it takes Bob twice as long as it takes Marla. So in one hour, Tom can clean $\frac{1}{x}$ of the building, Marla can clean $\frac{1}{x+3}$ of the building, and Bob can clean $\frac{1}{2x+6}$ of the building. If all three janitors work together, how much of the building can they clean in one hour?

64. Anna's average running speed is three times faster than her walking speed. Since $\text{time} = \dfrac{\text{distance}}{\text{rate}}$, the time it takes Anna to run 30 km is $\frac{30}{3x}$ and the time it takes Anna to walk 30 km is $\frac{30}{x}$. Find the difference between Anna's walking time and running time for the 30 km.

65. A car and a truck are both traveling to the same destination. The car is traveling 15 mph more than twice the speed of the truck. (Use the formula $\text{time} = \dfrac{\text{distance}}{\text{rate}}$.)

a. Write a rational expression to describe the time it will take the truck to travel 100 miles.

b. Write a rational expression to describe the time it will take the car to travel 100 miles.

c. Find a rational expression to describe the difference in travel time between the truck and the car for the 100 miles.

66. During Expedition 34 to the International Space Station, three crew members are tasked with unloading supplies from the SpaceX Dragon spacecraft. In one hour, Chris Hadfield can unload $\frac{1}{x}$ of the supplies, Thomas Marshburn can unload $\frac{1}{x+4}$ of the supplies, and Oleg Novitskiy can unload $\frac{1}{x-2}$ of the supplies. If they work together, what portion of the supplies will they unload in one hour?

a. Find the sum of the fractions of the supplies each crew member can unload in one hour.

b. If $x = 10$, what fraction of the supplies will be unloaded after one hour?

c. When $x = 10$, will more than half of the supplies be unloaded in one hour? Explain your answer.

67. Barbara's Bombtastic Bakery was a cupcake shop when it first opened up. The bakery space that Barbara rented came with most of the equipment that was needed, such as a commercial oven and a display case. This meant that she only had to buy a mixer for \$5000, cupcake pans for \$250, and various utensils for \$500. Barbara estimated that the ingredients for each cupcake would cost \$0.35.

a. What was the total amount that Barbara spent on equipment to bake the cupcakes?

b. The total cost to bake the cupcakes is equal to the sum of the total amount spent on equipment plus the total amount spent on ingredients. Create a function $C(x)$ to describe the total cost to bake the cupcakes and use the variable x to represent the number of cupcakes baked.

c. The average cost to bake each cupcake is calculated by dividing the total cost by the number of cupcakes baked. Create a formula $A(x)$ to describe the average cost to create each cupcake.

d. After Barbara's Bombtastic Bakery's first week of business, Barbara baked a total of 950 cupcakes. What was the average cost to bake each cupcake after the first week? Round your answer to the nearest cent.

e. After one month of business, Barbara baked a total of 3500 cupcakes. What was the average cost to bake each cupcake after the first month? Round your answer to the nearest cent.

Writing & Thinking

68. Discuss the steps in the process you go through when adding two rational expressions with different denominators. That is, discuss how you find the least common denominator when adding rational expressions and how you use this LCD to find equivalent rational expressions that you can add.

66. a. $\frac{3x^2+4x-8}{x(x+4)(x-2)}$

b. $\frac{83}{280}$

c. No, $\frac{83}{280} < \frac{140}{280} = \frac{1}{2}$

67. a. \$5750

b. $C(x) = 5750 + 0.35x$

c. $A(x) = \frac{5750+0.35x}{x}$

d. \$6.40 per cupcake

e. \$1.99 per cupcake

Writing & Thinking

68. Answers will vary.

Objectives

A. Simplify complex fractions by simplifying the numerator and denominator and then dividing.

B. Simplify complex fractions by multiplying by the LCM of all denominators and then simplifying.

C. Simplify complex algebraic expressions.

14.5 Simplifying Complex Fractions

A Simplifying Complex Fractions (First Method)

A **complex fraction** is a fraction in which the numerator and/or denominator are themselves fractions or the sum or difference of fractions. The following are examples of complex fractions.

$$\frac{\frac{1}{x+3}-\frac{1}{x}}{1+\frac{3}{x}} \quad \text{and} \quad \frac{x+y}{x^{-1}+y^{-1}}=\frac{x+y}{\frac{1}{x}+\frac{1}{y}}$$

The objective here is to develop techniques for simplifying complex fractions so that they are written in the form of a single reduced rational expression.

In a complex fraction such as $\frac{\frac{1}{x+3}-\frac{1}{x}}{1+\frac{3}{x}}$, the large fraction bar is a symbol of inclusion.

The expression could also be written as follows.

$$\frac{\frac{1}{x+3}-\frac{1}{x}}{1+\frac{3}{x}}=\left(\frac{1}{x+3}-\frac{1}{x}\right)\div\left(1+\frac{3}{x}\right)$$

Thus, a complex fraction indicates that the numerator is to be divided by the denominator.

Teaching Note

You will probably need to continually reinforce the concept that the fraction bar serves two important and distinct roles in the notation: on the one hand, it can translate as an operation (division), and on the other hand, the bar is a grouping symbol similar to the use of parentheses.

Simplifying Complex Fractions (First Method)

1. Simplify the numerator so that it is a single rational expression.
2. Simplify the denominator so that it is a single rational expression.
3. Divide the numerator by the denominator and reduce to lowest terms.

PROCEDURE

This method is used to simplify the complex fractions in Examples 1 and 2. Study the examples closely so that you understand what happens at each step.

Example 1 First Method for Simplifying Complex Fractions

Simplify the complex fraction. $\dfrac{\frac{3x}{y^2}}{\frac{12x}{7y}}$

Solution

Notice that the numerator and denominator are already simplified into a single rational expression. So we can jump to Step 3.

$$\frac{\frac{3x}{y^2}}{\frac{12x}{7y}} = \frac{\cancel{3}\,\cancel{x}}{\cancel{y^2}_{\,y}} \cdot \frac{7\cancel{y}}{\cancel{12}_{\,4}\,\cancel{x}}$$

To divide, multiply by the reciprocal of the denominator.

$$= \frac{7}{4y}$$

Now work margin exercise 1.

1. Simplify the complex fraction.

$$\frac{\frac{5x}{y^3}}{\frac{15x}{y^2}}$$

Example 2 First Method for Simplifying Complex Fractions

Simplify the complex fraction. $\dfrac{\frac{1}{x+3}-\frac{1}{x}}{1+\frac{3}{x}}$

Solution

$$\frac{\frac{1}{x+3}-\frac{1}{x}}{1+\frac{3}{x}} = \frac{\frac{1\cdot x}{(x+3)\cdot x}-\frac{1(x+3)}{x(x+3)}}{\frac{x}{x}+\frac{3}{x}}$$

Combine the fractions in the numerator and in the denominator separately.

Note that $1=\frac{x}{x}$.

$$= \frac{\frac{x-(x+3)}{x(x+3)}}{\frac{x+3}{x}}$$

$$= \frac{\frac{x-x-3}{x(x+3)}}{\frac{x+3}{x}}$$

$$= \frac{\frac{-3}{x(x+3)}}{\frac{x+3}{x}}$$

$$= \frac{-3}{\cancel{x}(x+3)} \cdot \frac{\cancel{x}}{x+3}$$

$$= \frac{-3}{(x+3)^2}$$

To divide, multiply by the reciprocal of the denominator.

Now work margin exercise 2.

2. Simplify the complex fraction.

$$\frac{\frac{1}{x+6}-\frac{1}{x}}{1+\frac{6}{x}}$$

3. Simplify the complex fraction.
$$\frac{3x-3y}{(3x)^{-1}-(3y)^{-1}}$$

Example 3 First Method for Simplifying Complex Fractions

Simplify the complex fraction. $\dfrac{x+y}{x^{-1}+y^{-1}}$

Solution

$$\frac{x+y}{x^{-1}+y^{-1}}=\frac{x+y}{\frac{1}{x}+\frac{1}{y}}$$

Recall that $y^{-1}=\frac{1}{y}$ and $x^{-1}=\frac{1}{x}$.

$$=\frac{\frac{x+y}{1}}{\frac{1}{x}\cdot\frac{y}{y}+\frac{1}{y}\cdot\frac{x}{x}}$$

Add the two fractions in the denominator.

$$=\frac{\frac{x+y}{1}}{\frac{y}{xy}+\frac{x}{xy}}$$

$$=\frac{\frac{x+y}{1}}{\frac{y+x}{xy}}$$

$$=\frac{\cancel{x+y}}{1}\cdot\frac{xy}{\cancel{y+x}}$$

Multiply by the reciprocal of the denominator.
Note: $x+y$ and $y+x$ are equal by the commutative property of addition.

$$=\frac{xy}{1}=xy$$

Now work margin exercise 3.

B Simplifying Complex Fractions (Second Method)

A second method is to find the LCM of the denominators in the fractions in both the original numerator and the original denominator, and then multiply **both** the numerator and denominator by this LCM.

Simplifying Complex Fractions (Second Method)

1. Find the LCM of all the denominators in the numerator and denominator of the complex fraction.
2. Multiply both the numerator and denominator of the complex fraction by this LCM.
3. Simplify **both** the numerator and denominator and reduce to lowest terms.

PROCEDURE

Example 4 Second Method for Simplifying Complex Fractions

Simplify the complex fraction. $\dfrac{\frac{1}{x+3}-\frac{1}{x}}{1+\frac{3}{x}}$

Solution

$$\frac{\frac{1}{x+3}-\frac{1}{x}}{1+\frac{3}{x}}=\frac{\left(\frac{1}{x+3}-\frac{1}{x}\right)\cdot x(x+3)}{\left(1+\frac{3}{x}\right)\cdot x(x+3)}$$

Multiply by $x(x+3)$, the LCM of $\{x, x+3\}$. This multiplication can be done because the net effect is that the fraction is multiplied by 1.

$$=\frac{\frac{1}{\cancel{x+3}}\cdot x\cancel{(x+3)}-\frac{1}{\cancel{x}}\cdot \cancel{x}(x+3)}{1\cdot x(x+3)+\frac{3}{\cancel{x}}\cdot \cancel{x}(x+3)}$$

$$=\frac{x-(x+3)}{x(x+3)+3(x+3)}$$

Factor the denominator by grouping.

$$=\frac{x-x-3}{(x+3)(x+3)}$$

$$=\frac{-3}{(x+3)^2}$$

Note that this matches the result found in Example 2 using Method 1.

Now work margin exercise 4.

4. Simplify the complex fraction.

$$\frac{\frac{1}{x+6}-\frac{1}{x}}{1+\frac{6}{x}}$$

Example 5 Second Method for Simplifying Complex Fractions

Simplify the complex fraction. $\dfrac{x+y}{x^{-1}+y^{-1}}$

Solution

$$\frac{x+y}{x^{-1}+y^{-1}}=\frac{\frac{x+y}{1}}{\frac{1}{x}+\frac{1}{y}}$$

$$=\frac{\left(\frac{x+y}{1}\right)xy}{\left(\frac{1}{x}+\frac{1}{y}\right)xy}$$

Multiply by xy, the LCM of $\{1, x, y\}$.

$$=\frac{(x+y)xy}{\frac{1}{\cancel{x}}\cdot \cancel{x}y+\frac{1}{\cancel{y}}\cdot x\cancel{y}}$$

$$=\frac{\cancel{(x+y)}xy}{\cancel{y+x}}=xy$$

Now work margin exercise 5.

5. Simplify the complex fraction.

$$\frac{3x-3y}{(3x)^{-1}-(3y)^{-1}}$$

C Simplifying Complex Algebraic Expressions

A **complex algebraic expression** is an expression that involves rational expressions and more than one operation. In simplifying such expressions, the rules for order of operations apply. As with complex fractions, the objective is to simplify the expression so that it is written in the form of a single reduced rational expression.

Example 6 Simplifying Complex Algebraic Expressions

Simplify the following expression.

$$\frac{4-x}{x+3}+\frac{x}{x+3}\div\frac{x}{x-3}$$

Solution

The rules for order of operations indicate that the division is to be done first, followed by the addition.

$$\begin{aligned}\frac{4-x}{x+3}+\frac{x}{x+3}\div\frac{x}{x-3}&=\frac{4-x}{x+3}+\frac{\cancel{x}}{x+3}\cdot\frac{x-3}{\cancel{x}}\\&=\frac{4-x}{x+3}+\frac{x-3}{x+3}\\&=\frac{4-x+x-3}{x+3}\\&=\frac{1}{x+3}\end{aligned}$$

Now work margin exercise 6.

6. Simplify the following expression.

$$\frac{6}{x+4}+\frac{x}{x+4}\div\frac{x}{x-4}$$

Math Tip

Pay close attention to the order of operations when simplifying complex algebraic expressions. It may be tempting to add the numerators of the first two terms in Example 6 since they share a common denominator. However, this would give us the wrong solution since division between the second and third terms needs to happen first.

Margin Exercise Answers

1. $\frac{1}{3y}$ **2.** $\frac{-6}{(x+6)^2}$ **3.** $-9xy$ **4.** $\frac{-6}{(x+6)^2}$ **5.** $-9xy$ **6.** $\frac{x+2}{x+4}$

14.5 Exercises

Concept Check

Fill-in-the-Blank. Complete each sentence using information found in this section.

1. The goal in simplifying complex fractions is to create a reduced ________ expression.

2. There are two methods of simplifying complex fractions. One method begins by simplifying the __________ and the _________ into single rational expressions.

3. A second method of simplifying complex fractions requires that the ___ of all denominators be found.

4. In a complex fraction, the large fraction bar is a symbol of ___________.

14.5 Exercises

Concept Check

1. rational
2. numerator, denominator
3. LCM
4. inclusion
5. complex algebraic
6. complex
7. True
8. False; Complex fractions can have fractions in either the numerator and/or the denominator.
9. True
10. False, The LCM is $6(x-6)$.

Practice

1. $\frac{4}{5xy}$
2. $12xy$
3. $\frac{8}{7x^2y}$
4. $\frac{6x}{y^4}$

5. An expression that involves rational expressions and more than one operation is called a/an ________ ________ expression.

6. A fraction in which the numerator and/or denominator are fractions or are the sums and/or differences of fractions is considered a/an _______ fraction.

True/False. Determine whether each statement is true or false. If a statement is false, explain how it can be changed so the statement will be true. (**Note:** There may be more than one acceptable change.)

7. When simplifying complex fractions, the answer should always be reduced to lowest terms.

8. Complex fractions are those fractions in which only the denominator consists of one or more fractions itself.

9. Sometimes finding the LCM of all denominators is an important first step for simplifying complex fractions.

10. The LCM of the denominators of $\frac{2}{x-6}$ and $\frac{x}{6}$ is 6.

Practice

Simplify the following complex fractions. See Examples 1 through 5.

1. $\dfrac{\frac{2x}{3y^2}}{\frac{5x^2}{6y}}$

2. $\dfrac{\frac{6x^2}{5y}}{\frac{x}{10y^2}}$

3. $\dfrac{\frac{12x^3}{7y^2}}{\frac{3x^5}{2y}}$

4. $\dfrac{\frac{9x^2}{7y^3}}{\frac{3xy}{14}}$

5. $\dfrac{\frac{x+3}{2x}}{\frac{2x-1}{4x^2}}$

6. $\dfrac{\frac{x-2}{6x}}{\frac{x+3}{3x^2}}$

7. $\dfrac{\frac{2x-1}{x}}{\frac{2}{x}+3}$

8. $\dfrac{2-\frac{3}{x}}{\frac{x^2-4}{x}}$

9. $\dfrac{\frac{3}{x}+\frac{1}{2x}}{1+\frac{2}{x}}$

10. $\dfrac{\frac{3}{x}+\frac{5}{2x}}{\frac{1}{x}+4}$

11. $\dfrac{1+\frac{1}{x}}{1-\frac{1}{x^2}}$

12. $\dfrac{\frac{2}{y}+1}{\frac{4}{y^2}-1}$

13. $\dfrac{\frac{1}{x}+\frac{1}{3x}}{\frac{x+6}{x^2}}$

14. $\dfrac{\frac{3}{x}-\frac{6}{x^2}}{\frac{x-2}{x^2}}$

15. $\dfrac{\frac{7}{x}-\frac{14}{x^2}}{\frac{1}{x}-\frac{4}{x^3}}$

16. $\dfrac{\frac{x}{3}+\frac{1}{9x^2}}{\frac{1}{27x^2}+\frac{x}{9}}$

17. $\dfrac{\frac{1}{3}+\frac{1}{x}}{\frac{1}{2}-\frac{1}{x}}$

18. $\dfrac{\frac{x}{y}-\frac{1}{3}}{\frac{6}{y}-\frac{2}{x}}$

19. $\dfrac{\frac{2}{x}+\frac{3}{4y}}{\frac{3}{2x}-\frac{5}{3y}}$

20. $\dfrac{\frac{4}{3x}-\frac{5}{y}}{\frac{1}{3}+\frac{3}{y}}$

21. $\dfrac{1+x^{-1}}{1-x^{-2}}$

22. $\dfrac{x^{-3}+1}{1-x^{-1}}$

23. $\dfrac{1}{x^{-1}+y^{-1}}$

24. $\dfrac{x-y}{x^{-2}-y^{-2}}$

25. $\dfrac{x^{-1}+y^{-1}}{x+y}$

26. $\dfrac{y^{-2}-x^{-2}}{x+y}$

27. $\dfrac{x^{-1}+y^{-1}}{x^{-1}-y^{-1}}$

5. $\dfrac{2x^2+6x}{2x-1}$
6. $\dfrac{x(x-2)}{2(x+3)}$
7. $\dfrac{2x-1}{2+3x}$
8. $\dfrac{2x-3}{(x-2)(x+2)}$
9. $\dfrac{7}{2(x+2)}$
10. $\dfrac{11}{2(1+4x)}$
11. $\dfrac{x}{x-1}$
12. $\dfrac{y}{2-y}$
13. $\dfrac{4x}{3(x+6)}$
14. 3
15. $\dfrac{7x}{x+2}$
16. 3
17. $\dfrac{2x+6}{3(x-2)}$
18. $\dfrac{x}{6}$
19. $\dfrac{24y+9x}{2(9y-10x)}$
20. $\dfrac{4y-15x}{x(y+9)}$
21. $\dfrac{x}{x-1}$
22. $\dfrac{x^3+1}{x^2(x-1)}$
23. $\dfrac{xy}{x+y}$
24. $\dfrac{-x^2y^2}{x+y}$
25. $\dfrac{1}{xy}$
26. $\dfrac{x-y}{x^2y^2}$
27. $\dfrac{y+x}{y-x}$

28. $\dfrac{xy}{y-x}$

29. $\dfrac{3-x}{x}$

30. $\dfrac{x^2-3x}{x-4}$

31. $\dfrac{x+1}{x+3}$

32. $\dfrac{x+1}{2x-3}$

33. $\dfrac{-1}{x(x+h)}$

34. $-\dfrac{2x+h}{x^2(x+h)^2}$

35. $\dfrac{-1}{x(x+h)}$

36. $-\dfrac{2x+h}{x^2(x+h)^2}$

37. $-(x-2y)(x-y)$

38. $\dfrac{2(2x-y)(2x+y)(x-y)}{2x+2y-1}$

39. $\dfrac{2x}{x^2+1}$

40. $\dfrac{2-x}{x+2}$

41. $\dfrac{(x-3)(x^2-2x+4)}{(x-4)(x-2)(x+1)}$

42. $\dfrac{(2-x^2)(x-1)}{(x-4)(x+1)^2}$

43. $\dfrac{-5}{x+1}$

44. $\dfrac{18}{5x}$

45. $\dfrac{29}{4(4x+5)}$

46. $\dfrac{14x-12}{x^2}$

47. $\dfrac{x^2-3x-6}{x(x-1)}$

48. $\dfrac{x^2+4x-3}{x(x+2)}$

49. $\dfrac{x^2-4x-2}{(x-4)(x+4)}$

28. $\dfrac{x^{-1}+y^{-1}}{x^{-2}-y^{-2}}$

29. $\dfrac{\frac{4}{x}-1}{1-\frac{1}{x-3}}$

30. $\dfrac{x+\frac{3}{x-4}}{1-\frac{1}{x}}$

31. $\dfrac{1-\frac{4}{x+3}}{1-\frac{2}{x+1}}$

32. $\dfrac{1+\frac{4}{2x-3}}{1+\frac{x}{x+1}}$

33. $\dfrac{\frac{1}{x+h}-\frac{1}{x}}{h}$

34. $\dfrac{\frac{1}{(x+h)^2}-\frac{1}{x^2}}{h}$

35. $\dfrac{\left(2+\frac{1}{x+h}\right)-\left(2+\frac{1}{x}\right)}{h}$

36. $\dfrac{\left(\frac{1}{(x+h)^2}-3\right)-\left(\frac{1}{x^2}-3\right)}{h}$

37. $\dfrac{x^2-4y^2}{1-\frac{2x+y}{x-y}}$

38. $\dfrac{8x^2-2y^2}{\frac{4x-1}{x-y}-2}$

39. $\dfrac{\frac{x+1}{x-1}-\frac{x-1}{x+1}}{\frac{x+1}{x-1}+\frac{x-1}{x+1}}$

40. $\dfrac{\frac{1}{x^2-1}-\frac{1}{x+1}}{\frac{1}{x-1}+\frac{1}{x^2-1}}$

41. $\dfrac{\frac{x}{x-4}-\frac{1}{x-1}}{\frac{x}{x-1}+\frac{2}{x-3}}$

42. $\dfrac{\frac{1}{x+1}-\frac{x}{x+2}}{\frac{x}{x+2}-\frac{2}{x-1}}$

Simplify the following complex algebraic expressions. See Example 6.

43. $\dfrac{1}{x+1}-\dfrac{3}{2x}\cdot\dfrac{4x}{x+1}$

44. $\dfrac{4}{x}-\dfrac{2}{x^2-2x}\cdot\dfrac{x-2}{5}$

45. $\left(\dfrac{8}{x}-\dfrac{3}{4x}\right)\div\dfrac{4x+5}{x}$

46. $\left(\dfrac{2}{x}+\dfrac{5}{x-3}\right)\div\dfrac{x}{2x-6}$

47. $\dfrac{x}{x-1}-\dfrac{3}{x-1}\cdot\dfrac{x+2}{x}$

48. $\dfrac{x+3}{x+2}+\dfrac{x}{x+2}\cdot\dfrac{x-3}{x^2}$

49. $\dfrac{x-1}{x+4}+\dfrac{x-6}{x^2+3x-4}\div\dfrac{x-4}{x-1}$

50. $\dfrac{x}{x+3}-\dfrac{3}{x+5}\div\dfrac{x-2}{x^2+3x-10}$

Applications

Solve.

51. To calculate the average rate of a two-part commute, where each part is the same distance, the following formula is used.

$$\frac{2d}{\frac{d}{r_1}+\frac{d}{r_2}}$$

In the formula, d is the commute distance traveled one way, r_1 is the rate, or speed, during the first part of the trip, and r_2 is the rate during the second part of the trip.

a. Simplify the expression.

b. Calculate the average rate of the trip if you can travel 35 miles per hour during the first part of the trip and 60 miles per hour during the second part of the trip. Round your answer to the nearest tenth.

c. Use the answer from part b. and the formula $d = rt$ to calculate how long the commute took if the total distance of the trip was 80 miles. Round your answer to the nearest tenth.

52. The average percent yield (APY) of an annuity is the annual interest rate earned in a given year that accounts for the effects of compounding. The APY acts as the interest rate for a simple interest account and is larger than the stated interest rate on the compound interest account. The formula to calculate the APY on an annuity after 2 years is

$$\text{APY} = \left(1+\frac{r}{2}\right)^2 - 1,$$

where r is the stated interest rate.

a. Simplify the expression for APY and write as a single rational expression.

b. Using the original formula, calculate the APY for an annuity whose interest rate is 6%. Do not round.

c. Using the expression in part a., calculate the APY for an annuity whose interest rate is 6%. Do not round.

d. Does the result from part c. match the result from part b.? Explain why or why not.

e. How much larger is the APY than the interest rate?

f. Why do you think the APY is larger than the interest rate? Write a complete sentence.

Writing & Thinking

53. Some complex fractions involve the sum (or difference) of complex fractions. Beginning with the outermost denominator, simplify each of the following expressions.

a. $1+\cfrac{1}{1+\cfrac{1}{1+\cfrac{1}{1+1}}}$ **b.** $2-\cfrac{1}{2-\cfrac{1}{2-\cfrac{1}{2-1}}}$ **c.** $x+\cfrac{1}{x+\cfrac{1}{x+\cfrac{1}{x+1}}}$

50. $\frac{-2x-9}{x+3}$

Applications

51. a. $\frac{2r_1r_2}{r_1+r_2}$

b. 44.2 miles per hour

c. 1.8 hours

52. a. $\frac{4r+r^2}{4}$

b. 0.0609 **c.** 0.0609

d. Yes; They are calculated from different forms of the same expression.

e. 0.009;

f. Answers will vary. It is larger because the interest is compounded along with the principal.

Writing & Thinking

53. a. $\frac{8}{5}$

b. 1

c. $\frac{x^4+x^3+3x^2+2x+1}{x^3+x^2+2x+1}$

Objectives

A. Solve proportions.

B. Solve equations that involve rational expressions.

C. Use the process of solving equations to manipulate formulas.

D. Use proportions to relate similar triangles.

14.6 Solving Rational Equations

A Solving Proportions

A **ratio** is a comparison of two numbers by division. Ratios are written in the form

$$a:b \quad \text{or} \quad \frac{a}{b} \quad \text{or} \quad a \text{ to } b.$$

For example, suppose the ratio of adjunct to tenure-track instructors at the local community college is 3 to 2. We can also write this ratio as $\frac{3 \text{ adjunct}}{2 \text{ tenure-track}}$ or as 3 adjunct : 2 tenure-track. This ratio does not mean that there are only 5 instructors at the college. There are many ratios that reduce to $\frac{3}{2}$. There might be 35 instructors at the college, 21 adjunct instructors and 14 tenure-track instructors, because $21+14=35$ and $\frac{21}{14}=\frac{3}{2}$. Until you know the total number of instructors at the college, you cannot know the exact numbers of adjunct and tenure-track instructors. But you do know that there are 3 adjunct instructors for every 2 tenure-track instructors.

A **proportion** is a special type of equation stating that two ratios are equal. For example,

$$\frac{21}{14}=\frac{3}{2} \quad \text{and} \quad \frac{3.5}{7}=\frac{3.75}{7.5} \quad \text{are proportions.}$$

Teaching Note

Another method of solving proportions is to "cross multiply". However, students often overlook the fact that cross multiplication works only with proportions and not with rational equations in general. In this text, we have tried to avoid this potential confusion by presenting only the method of clearing fractions.

Proportion

A **proportion** is an equation stating that two ratios are equal.

In symbols, $\frac{a}{b}=\frac{c}{d}$ is a proportion.

DEFINITION

Proportions may involve only numbers as in $\frac{2}{3}=\frac{10}{15}$. However, proportions can also be used to find unknown quantities in applications, and in such cases, will involve variables.

One method of solving proportions with variables is to "clear" the equation of fractions by first multiplying both sides of the equation by the LCM of the denominators, the LCD. This method is illustrated in Example 1.

Example 1 Solving Proportions

Solve the following proportions.

a. $\frac{x-5}{2x}=\frac{6}{3x}$

b. $\frac{3}{x-6}=\frac{5}{x}$

1. Solve the following proportions.

a. $\frac{x-7}{2x}=\frac{8}{4x}$

b. $\frac{4}{x-8}=\frac{8}{x}$

Solution

a. LCD = $6x$

Note: $x \neq 0$.

$$\frac{x-5}{2x} = \frac{6}{3x}$$

Multiply both sides by the LCD, $6x$.

$$\overset{3}{\cancel{6x}} \cdot \frac{x-5}{\cancel{2x}} = \overset{2}{\cancel{6x}} \cdot \frac{6}{\cancel{3x}}$$

$$3(x-5) = 2(6)$$

$$3x - 15 = 12$$

$$3x = 27$$

$$x = 9$$

Check

$$\frac{9-5}{2 \cdot 9} \stackrel{?}{=} \frac{6}{3 \cdot 9}$$

$$\frac{4}{18} \stackrel{?}{=} \frac{6}{27}$$

$$\frac{2}{9} = \frac{2}{9}$$

Thus, the solution is $x = 9$.

b. LCD = $x(x-6)$

Note: $x \neq 0, 6$.

$$\frac{3}{x-6} = \frac{5}{x}$$

Multiply both sides by the LCD, $x(x-6)$.

$$x\cancel{(x-6)} \cdot \frac{3}{\cancel{x-6}} = \cancel{x}(x-6) \cdot \frac{5}{\cancel{x}}$$

$$3x = 5x - 30$$

$$-2x = -30$$

$$x = 15$$

Check

$$\frac{3}{15-6} \stackrel{?}{=} \frac{5}{15}$$

$$\frac{3}{9} \stackrel{?}{=} \frac{5}{15}$$

$$\frac{1}{3} = \frac{1}{3}$$

Thus, the solution is $x = 15$.

Now work margin exercise 1.

Proportions can be used to solve many everyday types of word problems. Using the correct ratios of units on both sides of the proportion is critical. To set up the correct ratios, one of the following conditions must be true:

1. The numerators agree in type and the denominators agree in type.
2. The numerators correspond and the denominators correspond.

Example 2 illustrates one situation.

Example 2 Application: Solving Proportions

On an architect's scale drawing of a home, $\frac{1}{2}$ inch represents 10 feet. What length does a measure of $2\frac{1}{2}$ inches represent?

Solution

Set up a proportion representing the information. In this example the numerators are the same type (inches) and the denominators are the same type (feet).

2. On a scale of a map of a highway, 1 inch represents 60 miles. What length does a measure of $3\frac{3}{4}$ inches represent?

> **Note**
>
> Any of the following four equations could have been used to solve the problem in Example 2.
>
> **Agree in Type:**
>
> $$\frac{\frac{1}{2}\text{ inch}}{10\text{ feet}}=\frac{2\frac{1}{2}\text{ inches}}{x\text{ feet}}$$
>
> $$\frac{10\text{ feet}}{\frac{1}{2}\text{ inch}}=\frac{x\text{ feet}}{2\frac{1}{2}\text{ inches}}$$
>
> **Correspond:**
>
> $$\frac{\frac{1}{2}\text{ inch}}{2\frac{1}{2}\text{ inches}}=\frac{10\text{ feet}}{x\text{ feet}}$$
>
> $$\frac{2\frac{1}{2}\text{ inches}}{\frac{1}{2}\text{ inch}}=\frac{x\text{ feet}}{10\text{ feet}}$$

$\dfrac{\frac{1}{2}\text{ inch}}{10\text{ feet}}=\dfrac{2\frac{1}{2}\text{ inches}}{x\text{ feet}}$	LCD = $10x$
$\cancel{10}x\cdot\dfrac{\frac{1}{2}}{\cancel{10}}=10\cancel{x}\cdot\dfrac{\frac{5}{2}}{\cancel{x}}$	Multiply both sides by $10x$.
$\frac{1}{2}x=25$	Simplify.
$2\cdot\frac{1}{2}x=2\cdot25$	Multiply both sides by 2.
$x=50$	Simplify.

On this drawing, $2\frac{1}{2}$ inches represent 50 feet.

Now work margin exercise 2.

B Solving Rational Equations

An equation such as $\frac{3}{x}+\frac{1}{8}=\frac{13}{4x}$ that involves the sum of rational expressions is **not a proportion**. However, the method of finding the solution to the equation is similar in that we "clear" the fractions by multiplying both sides of the equation by the LCD of the fractions. In this example, the LCD is $8x$ and we can proceed as follows.

$\dfrac{3}{x}+\dfrac{1}{8}=\dfrac{13}{4x}$	Note that $x\neq0$.
$8x\left(\dfrac{3}{x}+\dfrac{1}{8}\right)=8x\cdot\dfrac{13}{4x}$	Multiply both sides by $8x$, the LCD.
$8\cancel{x}\cdot\dfrac{3}{\cancel{x}}+\cancel{8}x\cdot\dfrac{1}{\cancel{8}}=\overset{2}{\cancel{8x}}\cdot\dfrac{13}{\cancel{4x}}$	Use the distributive property and reduce.
$24+x=26$	Simplify.
$x=2.$	

As we have seen, rational expressions may contain variables in either the numerator or denominator or both. In any case, a general approach to solving equations that contain rational expressions is as follows.

> ### Solving an Equation Containing Rational Expressions
>
> 1. Find the LCD of the fractions.
> 2. List any restrictions on the variables.
> 3. Multiply both sides of the equation by this LCD and simplify.
> 4. Solve the resulting equation. (This equation will have only polynomials on both sides.)
> 5. Check each solution in the **original equation**. (Remember that no denominator can be 0. If one of the potential solutions matches a restriction on the variables, then that potential solution must be rejected.)
>
> PROCEDURE

Checking is particularly important when equations have rational expressions. Multiplying by the LCD may introduce solutions that are not solutions to the original equation. Such solutions are called **extraneous solutions** or **extraneous roots** and occur because multiplication by a variable expression may, in effect, be multiplying the original equation by 0.

Example 3 Solving Rational Equations

State any restrictions on the variable, then solve the equation.

$$\frac{1}{x-4}=\frac{3}{x^2-5x}$$

Solution

First, find the LCD of the fractions, and then multiply both sides of the equation by the LCD.

$$\left.\begin{aligned} x-4 &= x-4 \\ x^2-5x &= x(x-5) \end{aligned}\right\} \text{LCD} = x(x-5)(x-4)$$

$$\frac{1}{x-4}=\frac{3}{x(x-5)} \qquad \text{Restrictions: } x \neq 0, 4, 5$$

$$x(x-5)\cancel{(x-4)}\cdot\frac{1}{\cancel{x-4}}=\cancel{x}\cancel{(x-5)}(x-4)\cdot\frac{3}{\cancel{x}\cancel{(x-5)}}$$

$$x(x-5)=3(x-4)$$

$$x^2-5x=3x-12$$

$$x^2-8x+12=0$$

$$(x-6)(x-2)=0$$

$$x-6=0 \quad \text{or} \quad x-2=0$$

$$x=6 \qquad\qquad x=2$$

Since 6 and 2 are not restrictions, there are two solutions, $x = 6$ and $x = 2$.

Now work margin exercise 3.

3. State any restrictions on the variable, then solve the equation.

$$\frac{1}{x+2}=\frac{6}{x^2+7x}$$

Example 4 Solving Rational Equations

State any restrictions on the variable, then solve the equation.

$$\frac{x}{x-2}+\frac{x-6}{x(x-2)}=\frac{5x}{x-2}-\frac{10}{x-2}$$

Solution

First, find the LCD of the fractions and then multiply each term on both sides of the equation by the LCD.

$$\left.\begin{aligned} &x-2 \\ &x(x-2) \end{aligned}\right\} \text{LCD} = x(x-2)$$

4. State any restrictions on the variable, then solve the equation.

$$\frac{x}{x-2}+\frac{4x-4}{x(x-2)}=\frac{6x}{x-2}-\frac{8}{x-2}$$

$$\frac{x}{x-2}+\frac{x-6}{x(x-2)}=\frac{5x}{x-2}-\frac{10}{x-2}$$ Restrictions: $x \neq 0, 2$

$$x\cancel{(x-2)}\cdot\frac{x}{\cancel{x-2}}+\cancel{x}\cancel{(x-2)}\cdot\frac{x-6}{\cancel{x}\cancel{(x-2)}}=x\cancel{(x-2)}\cdot\frac{5x}{\cancel{x-2}}-x\cancel{(x-2)}\cdot\frac{10}{\cancel{x-2}}$$

$$\begin{aligned}x^2+x-6&=5x^2-10x\\0&=4x^2-11x+6\\0&=(4x-3)(x-2)\\4x-3=0 \quad &\text{or} \quad x-2=0\\4x=3 \quad & \quad \xcancel{x=2}\\x&=\frac{3}{4}\end{aligned}$$

The only solution is $x=\frac{3}{4}$, since 2 is a restricted value $(x \neq 0, 2)$ and thus not a solution. No denominator can be 0.

Math Tip

Notice that you could simplify the right-hand side of the equation first since the two fractions on the right-hand side share the same denominator. Doing this would leave you with the following rational equation to solve.

$$\frac{x}{x-2}+\frac{x-6}{x(x-2)}=\frac{5x-10}{x-2}$$

Now work margin exercise 4.

5. State any restrictions on the variable, then solve the equation.

$$\frac{4}{x^2-25}=\frac{2}{x^2}+\frac{2}{x^2-5x}$$

Example 5 Solving Rational Equations

State any restrictions on the variable, then solve the equation.

$$\frac{2}{x^2-9}=\frac{1}{x^2}+\frac{1}{x^2-3x}$$

Solution

First, find the LCD of the fractions and then multiply each term on both sides of the equation by the LCD.

$$\left.\begin{aligned}x^2-9&=(x+3)(x-3)\\x^2&=x^2\\x^2-3x&=x(x-3)\end{aligned}\right\}\ \text{LCD}=x^2(x+3)(x-3)$$

$$\frac{2}{(x+3)(x-3)}=\frac{1}{x^2}+\frac{1}{x(x-3)}$$ Restrictions: $x \neq 0, -3, 3$

$$x^2\cancel{(x+3)}\cancel{(x-3)}\cdot\frac{2}{\cancel{(x+3)}\cancel{(x-3)}}=\cancel{x^2}(x+3)(x-3)\cdot\frac{1}{\cancel{x^2}}$$

$$+\cancel{x^2}^{\,x}(x+3)\cancel{(x-3)}\cdot\frac{1}{\cancel{x}\cancel{(x-3)}}$$

$$\begin{aligned}2x^2&=(x+3)(x-3)+x(x+3)\\2x^2&=x^2-9+x^2+3x\\2x^2&=2x^2+3x-9\\9&=3x\\\xcancel{3}&\xcancel{=x}\end{aligned}$$

Note that 3 is one of the restrictions.

There is no solution. The solution set is the empty set, $\varnothing$. The original equation is a contradiction.

Math Tip

When solving rational equations, multiplying both sides of an equation by a variable term (the LCD in this case) may not strictly result in an equivalent equation. This is because it increases the degree of the polynomial, which can increase the number of potential solutions. These additional solutions may be extraneous. For example,

$$\begin{aligned}x&=3\\x\cdot x&=3\cdot x\\x^2&=3x\end{aligned}$$

Notice that $x^2=3x$ has two solutions, whereas $x=3$ has only one.

Now work margin exercise 5.

Completion Example 6 Solving Rational Equations

State any restrictions on the variable, then solve the equation.

$$\frac{x}{x-1}-\frac{3x+1}{x^2+4x-5}=\frac{x+2}{x+5}$$

Solution

Restrictions: $x \neq$ <u>−5</u>, <u>1</u>

$$\frac{x}{x-1}-\frac{3x+1}{x^2+4x-5}=\frac{x+2}{x+5}$$

$$(x+5)(\underline{x-1})\cdot\frac{x}{x-1}-(x+5)(\underline{x-1})\cdot\frac{3x+1}{(x+5)(x-1)}=(x+5)(\underline{x-1})\cdot\frac{x+2}{x+5}$$

$$(x+5)\cdot x-(\underline{3x+1})=(\underline{x-1})(x+2)$$

$$x^2+5x-\underline{3x}-\underline{1}=\underline{x^2+x-2}$$

$$\underline{2x-1}=\underline{x-2}$$

$$x=\underline{-1}$$

Now work margin exercise 6.

6. State any restrictions on the variable, then solve the equation.

$$\frac{4y^2}{y^2+3y-10}=\frac{y+1}{y+5}+\frac{3y}{y-2}$$

C Solving Formulas for Specified Variables

As discussed in Section 9.4, formulas are equations relating more than one variable. The equation is solved for one variable in terms of another variable (maybe more than one). The process of solving equations can be used to manipulate the formula so that it is solved for one of the other variables. As illustrated in Example 7, the resulting equation may involve a rational expression.

Example 7 Solving a Formula for a Specified Variable

The formula $S=2\pi r^2+2\pi rh$ is used to find the surface area (S) of a right circular cylinder, where r is the radius of the cylinder and h is the height of the cylinder. Solve the formula for h.

Solution

$S=2\pi r^2+2\pi rh$ Write the formula.

$S-2\pi r^2=2\pi rh$ Add $-2\pi r^2$ to both sides of the equation.

$\frac{S-2\pi r^2}{2\pi r}=h$ Divide both sides by $2\pi r$.

Thus, the formula solved for h is $h=\frac{S-2\pi r^2}{2\pi r}$ $\left(\text{or } h=\frac{S}{2\pi r}-r\right)$.

Now work margin exercise 7.

7. The surface area of a rectangular box is given by the formula $SA=2(lh)+2(wh)+2(lw)$. Solve the formula for l.

D Similar Triangles

Similar triangles, previously discussed in Chapter 6, are triangles that meet the following two conditions:

1. The measures of the corresponding angles are equal.
2. The lengths of the corresponding sides are proportional.

In similar triangles, corresponding sides are those sides opposite the equal angles. (See Figure 1.)

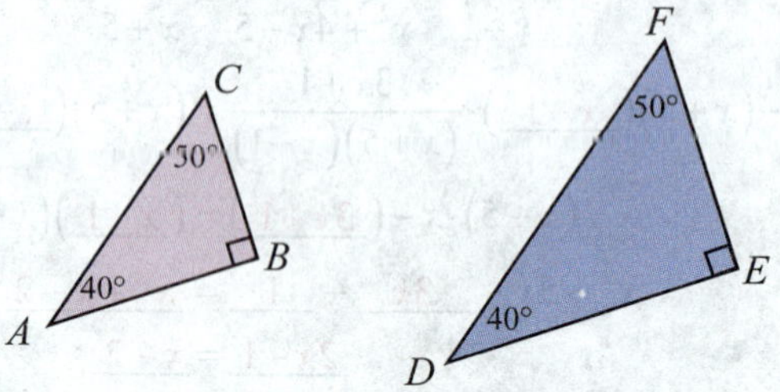

Figure 1

We write $\Delta ABC \sim \Delta DEF$. (~ is read "**is similar to**".) The lengths of the corresponding sides are proportional. Thus,

$$\frac{AB}{DE} = \frac{BC}{EF} \text{ and } \frac{AB}{DE} = \frac{AC}{DF} \text{ and } \frac{BC}{EF} = \frac{AC}{DF}.$$

In a pair of similar triangles, we can often find the length of an unknown side by setting up a proportion and solving. Example 8 illustrates such a situation.

8. Assume $\Delta ABC \sim \Delta PQR$. In ΔABC, $\overline{AB}$ has length of $x - 4$ and $\overline{BC}$ has length of 5. In ΔPQR, $\overline{PQ}$ has a length of 9 and $\overline{QR}$ has a length of x. Solve for x.

Example 8 Similar Triangles

In the figure shown, $\Delta ABC \sim \Delta PQR$. Find the lengths of $\overline{AB}$ and $\overline{QR}$.

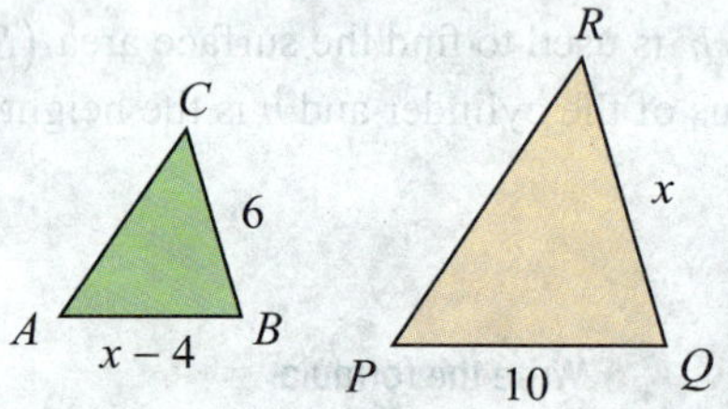

Solution

Set up a proportion involving corresponding sides and solve for x.

$$\frac{x-4}{10} = \frac{6}{x}$$

$$\cancel{10}x \cdot \frac{x-4}{\cancel{10}} = 10\cancel{x} \cdot \frac{6}{\cancel{x}}$$

$$x(x-4) = 10 \cdot 6$$

$$x^2 - 4x = 60$$

$$x^2-4x-60=0$$
$$(x-10)(x+6)=0$$
$$x-10=0 \quad \text{or} \quad x+6=0$$
$$x=10 \qquad x=-6$$

Because the length of a side cannot be negative, the only acceptable solution is $x=10$. Thus, $\overline{QR}=10$. Substituting 10 for x gives $\overline{AB}=10-4=6$.

Now work margin exercise 8.

Completion Example Answers

6. Restrictions: $x \neq -5, 1$

$$(x+5)(x-1)\cdot\frac{x}{x-1}-(x+5)(x-1)\cdot\frac{3x+1}{(x+5)(x-1)}=(x+5)(x-1)\cdot\frac{x+2}{x+5}$$
$$(x+5)\cdot x-(3x+1)=(x-1)(x+2)$$
$$x^2+5x-3x-1=x^2+x-2$$
$$2x-1=x-2$$
$$x=-1$$

Margin Exercise Answers

1. a. $x=11$ **b.** $x=16$ **2.** 225 miles **3.** $x\neq-7,-2,0;\ x=-4,3$ **4.** $x\neq0,2;\ x=-\frac{2}{5}$

5. $x\neq-5,0,5$; no solution **6.** $y\neq-5,2;\ y=-\frac{1}{7}$ **7.** $l=\frac{SA-2wh}{2h+2w}$ **8.** $x=9$

14.6 Exercises

Concept Check

Fill-in-the-Blank. Complete each sentence using information found in this section.

1. A comparison of two numbers by division is called a/an ________.

2. An equation stating that two ratios are equal is called a/an ________.

3. Solutions that are not actually solutions of the original equation are called __________ solutions.

4. One method of solving proportions is to clear the equation of ______ by first multiplying both sides of the equation by the LCD.

5. Rational expressions may contain ______ in either the numerator, the denominator, or both.

6. When solving an equation containing rational expressions, multiply both sides of the equation by the LCD and ______.

14.6 Exercises

Concept Check

1. ratio
2. proportion
3. extraneous
4. fractions
5. variables
6. simplify

7. False; It is not a proportion.

8. True

9. True

10. False; The solution should be discarded.

Practice

1. $x = 7$
2. $x = 1$
3. $x \neq 0, 2; x = 4$
4. $x \neq \frac{3}{2}, 3;\ x = -3$
5. $x \neq -3, 4; x = -10$
6. $x \neq -5, 2; x = -12$
7. $x \neq 0; x = 18$
8. $x \neq 0; x = 7$
9. $x \neq 6; x = -\frac{74}{9}$
10. $x \neq -\frac{2}{3}; x = \frac{62}{7}$
11. $x = \frac{1}{4}$
12. $x = 1$
13. $x = 6$
14. $x = 7$
15. $x = 4$
16. $x = 16$
17. $x \neq 0; x = \frac{10}{3}$
18. $x \neq 0; x = \frac{3}{2}$
19. $x \neq 0; x = -\frac{3}{4}$
20. $x \neq 0; x = \frac{1}{4}$
21. $x \neq 0; x = -\frac{3}{16}$
22. $x \neq 0; x = -\frac{2}{3}$
23. $x \neq -9, -\frac{1}{4}, 0; x = -2, 1$
24. $x \neq -1, 0, \frac{1}{4}; x = \frac{1}{3}, 4$
25. $x \neq \frac{3}{2}, 0, 6; x = \frac{3}{5}, 9$

True/False. Determine whether each statement is true or false. If a statement is false, explain how it can be changed so the statement will be true. (**Note:** There may be more than one acceptable change.)

7. An equation that involves the sum of rational expressions is also a proportion.

8. Multiplying by an LCD can cause extraneous roots.

9. A proportion is properly written if the numerators agree in type and the denominators agree in type.

10. When checking the solutions in the original equation, any solution that gives a 0 denominator cannot be checked.

Practice

State any restrictions on *x*, and then solve the proportions. See Example 1.

1. $\frac{4x}{7} = \frac{x+5}{3}$

2. $\frac{3x+1}{4} = \frac{2x+1}{3}$

3. $\frac{10}{x} = \frac{5}{x-2}$

4. $\frac{8}{x-3} = \frac{12}{2x-3}$

5. $\frac{4}{x-4} = \frac{2}{x+3}$

6. $\frac{3}{x+5} = \frac{6}{x-2}$

7. $\frac{x+2}{5x} = \frac{x-6}{3x}$

8. $\frac{x-4}{3x} = \frac{x-2}{5x}$

9. $\frac{5x+2}{x-6} = \frac{11}{4}$

10. $\frac{x+9}{3x+2} = \frac{5}{8}$

State any restrictions on *x*, and then solve the equations. See Examples 3 through 6.

11. $\frac{5x}{4} - \frac{1}{2} = -\frac{3}{16}$

12. $\frac{x}{6} - \frac{1}{42} = \frac{1}{7}$

13. $\frac{3x-1}{6} - \frac{x+3}{4} = \frac{7}{12}$

14. $\frac{x-2}{3} - \frac{x-3}{5} = \frac{13}{15}$

15. $\frac{2+x}{4} - \frac{5x-2}{12} = \frac{8-2x}{5}$

16. $\frac{4x+1}{5} = \frac{2x+3}{2} - \frac{x+2}{4}$

17. $\frac{2}{3x} = \frac{1}{4} - \frac{1}{6x}$

18. $\frac{1}{x} - \frac{8}{21} = \frac{3}{7x}$

19. $\frac{3}{5x} - \frac{1}{5} = \frac{3}{4x}$

20. $\frac{3}{8x} - \frac{7}{10} = \frac{1}{5x}$

21. $\frac{3}{4x} - \frac{1}{2} = \frac{7}{8x} + \frac{1}{6}$

22. $\frac{5}{3x} + \frac{1}{2} = \frac{7}{9x} - \frac{5}{6}$

23. $\frac{2}{4x+1} = \frac{4}{x^2+9x}$

24. $\frac{3}{4x-1} = \frac{4}{x^2+x}$

25. $\frac{9}{x^2-6x} = \frac{5}{2x-3}$

26. $\frac{-9}{x^2+5x} = \frac{8}{4-9x}$

27. $\frac{x}{x-4} - \frac{4}{2x-1} = 1$

28. $\frac{x}{x+3} + \frac{1}{x+2} = 1$

29. $\frac{x+2}{x+1}+\frac{x+2}{x+4}=2$

30. $\frac{3x-2}{x+4}+\frac{2x+5}{x-1}=5$

31. $\frac{2}{4x-1}+\frac{1}{x+1}=\frac{3}{x+1}$

32. $\frac{x-2}{x+4}-\frac{3}{2x+1}=\frac{x-7}{x+4}$

33. $\frac{x-2}{x-3}+\frac{x-3}{x-2}=\frac{2x^2}{x^2-5x+6}$

34. $\frac{x}{x-4}-\frac{12x}{x^2+x-20}=\frac{x-1}{x+5}$

35. $\frac{3x+5}{3x+2}+\frac{8x+16}{3x^2-4x-4}=\frac{x+2}{x-2}$

36. $\frac{3x+5}{3x+2}-\frac{4-2x}{3x^2+8x+4}=\frac{x+4}{x+2}$

37. $\frac{3}{3x-1}+\frac{1}{x+1}=\frac{4}{2x-1}$

38. $\frac{2}{x+1}+\frac{4}{2x-3}=\frac{4}{x-5}$

Solve each of the formulas for the specified variable. Assume no denominator has a value of 0. See Example 7.

39. $S=\frac{a}{1-r}$; solve for r (formula for the sum of an infinite geometric sequence)

40. $z=\frac{x-\bar{x}}{s}$; solve for x (formula used in statistics)

41. $z=\frac{x-\bar{x}}{s}$; solve for s (formula used in statistics)

42. $a_n=a_1+(n-1)d$; solve for d (formula for the nth term in an arithmetic sequence)

43. $m=\frac{y-y_1}{x-x_1}$; solve for y (formula for the slope of a line)

44. $v_{\text{avg}}=\frac{d_2-d_1}{t_2-t_1}$; solve for d_2 (formula for mean velocity)

45. $\frac{1}{R_{\text{total}}}=\frac{1}{R_1}+\frac{1}{R_2}$; solve for R_{total} (formula used in electronics)

46. $\frac{1}{x}=\frac{1}{t_1}+\frac{1}{t_2}$; solve for x (formula used in mathematics)

47. $A=P+Pr$; solve for P (formula used for compound interest)

48. $y=\frac{ax+b}{cx+d}$; solve for x (formula used in mathematics)

The following exercises show pairs of similar triangles. Find the lengths of the sides labeled with variables. See Example 8.

49. $\Delta JKL \sim \Delta JTB$

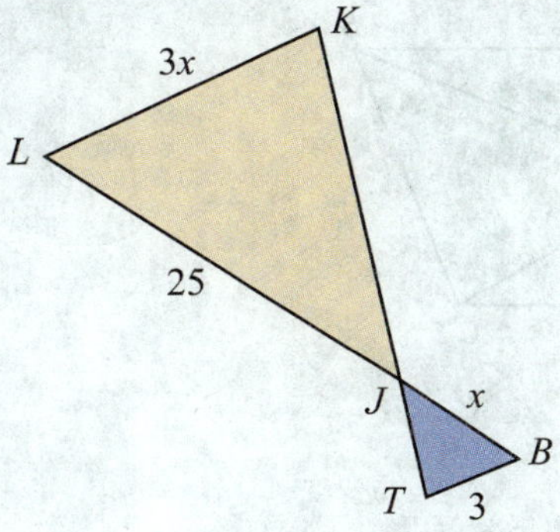

50. $\Delta QRP \sim \Delta TUS$

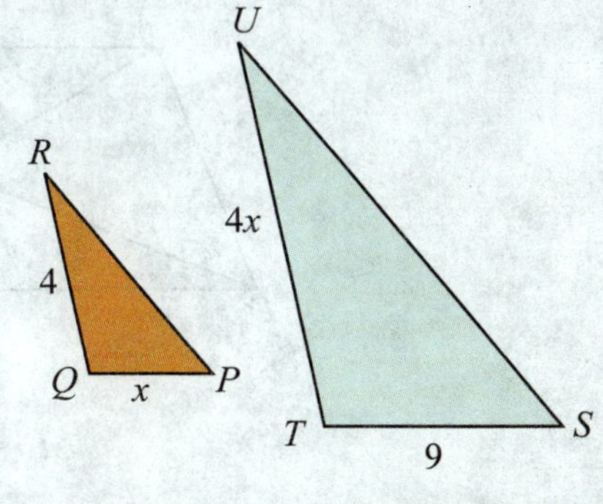

26. $x\neq-5,0,\frac{4}{9};\ x=\frac{9}{8},4$

27. $x\neq\frac{1}{2},4;\ x=-3$

28. $x\neq-3,-2;\ x=-\frac{3}{2}$

29. $x\neq-4,-1;\ x=2$

30. $x\neq-4,1;\ x=6$

31. $x\neq-1,\frac{1}{4};\ x=\frac{2}{3}$

32. $x\neq-4,-\frac{1}{2};\ x=1$

33. $x\neq2,3;\ x=\frac{13}{10}$

34. $x\neq-5,4;\ x=-2$

35. $x\neq-\frac{2}{3},2;$ no solution

36. $x\neq-2,-\frac{2}{3};$ no solution

37. $x\neq-1,\frac{1}{3},\frac{1}{2};\ x=\frac{1}{5}$

38. $x\neq-1,\frac{3}{2},5;\ x=\frac{11}{19}$

39. $r=\frac{S-a}{S}$

40. $x=zs+\bar{x}$

41. $s=\frac{x-\bar{x}}{z}$

42. $d=\frac{a_n-a_1}{n-1}$

43. $y=m(x-x_1)+y_1$

44. $d_2=v_{\text{avg}}(t_2-t_1)+d_1$

45. $R_{\text{total}}=\frac{R_1R_2}{R_1+R_2}$

46. $x=\frac{t_1t_2}{t_1+t_2}$

47. $P=\frac{A}{1+r}$

48. $x=\frac{b-yd}{yc-a}$

49. $LK=15,\ JB=5$

50. $QP=3,\ TU=12$

51. $AC = 2$, $ST = 12$

52. $DF = 10$, $GH = 4$

53. $ST = 8$, $TU = 12$, $QR = 24$

54. $AC = 6$, $ED = 12$; $CD = 18$

55. $AP = \frac{9}{2}$ in.; $PC = \frac{15}{2}$ in.

56. $AP = PC = 6$ in.

51. $\Delta ABC \sim \Delta RST$

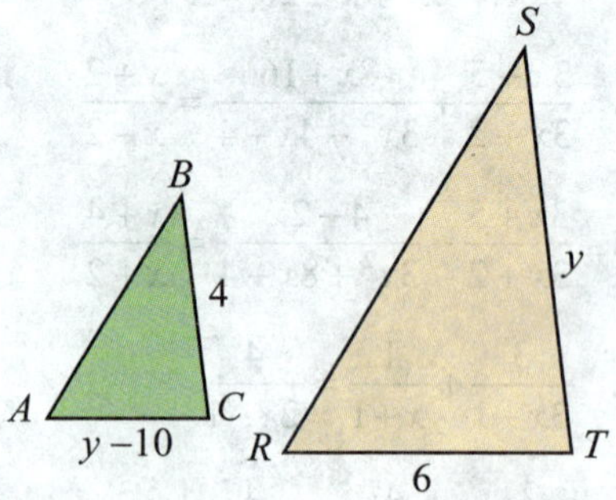

53. $\Delta SUT \sim \Delta PRQ$

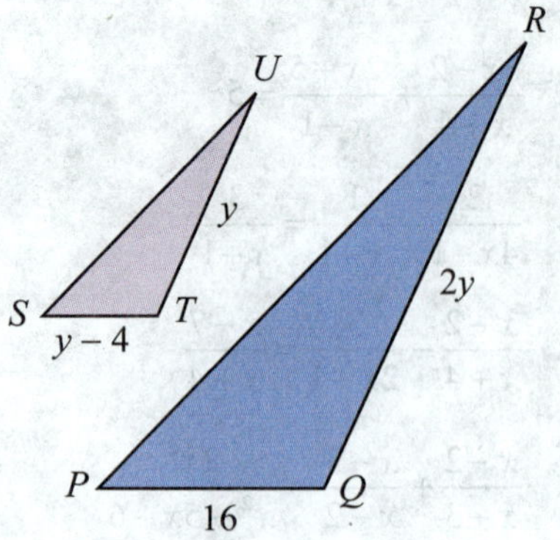

52. $\Delta FED \sim \Delta FGH$

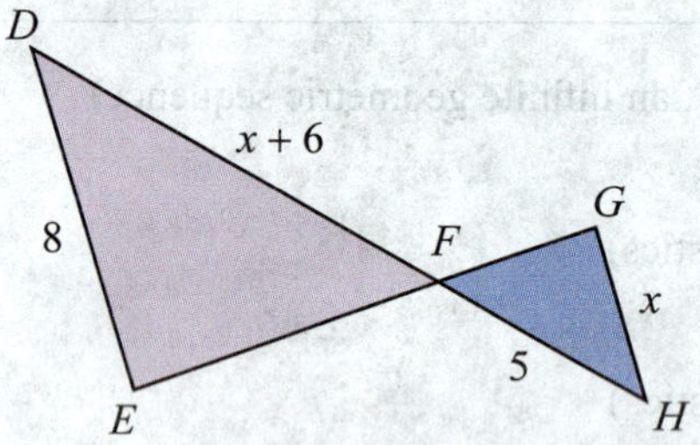

54. $\Delta ABC \sim \Delta DEC$

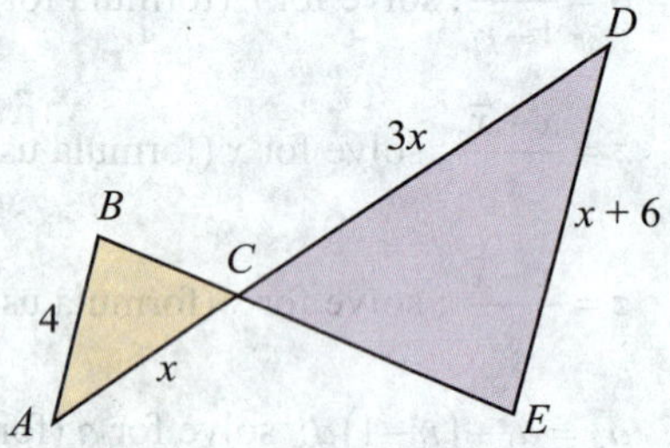

55. In the parallelogram $ABCD$, $AB = CD = 10$ in. Diagonal $AC = 12$ in. The point M on $\overline{AB}$ is 6 in. from A. Point P is the intersection of $\overline{DM}$ with $\overline{AC}$. The triangles APM and CPD are similar. (Symbolically, $\Delta APM \sim \Delta CPD$.) What are the lengths of $\overline{AP}$ and $\overline{PC}$?

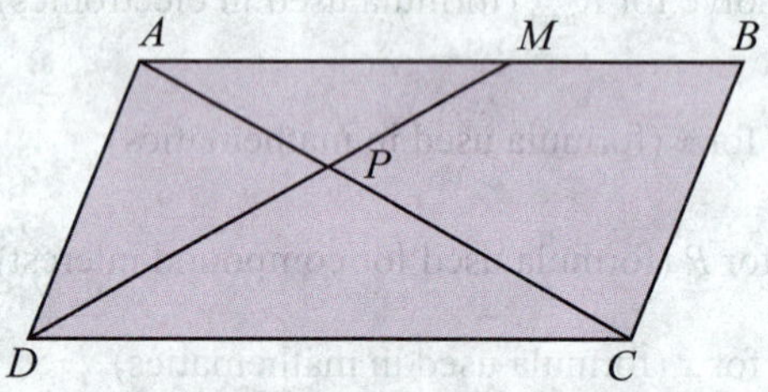

56. If, in the same parallelogram discussed in Exercise 55, the point P is the point of intersection of the two diagonals, $\overline{AC}$ and $\overline{DB}$, what are the lengths of $\overline{AP}$ and $\overline{PC}$?

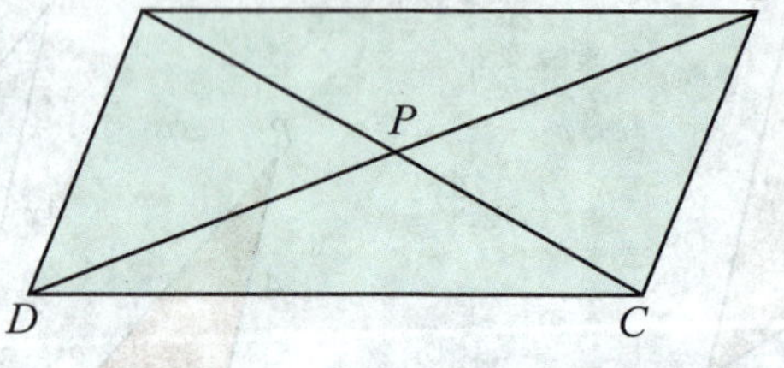

Applications

Solve.

57. Making a statistical analysis, Ana found 3 defective computers in a sample of 20 computers. If this ratio is consistent, how many defective computers does she expect to find in a batch of 2400 computers?

58. At the Bright-As-Day light bulb plant, 3 out of each 100 bulbs produced are defective. If the daily production is 4800 bulbs, how many are defective?

59. A university has a ratio of 1 professor for every 23 students. If there are 1600 faculty members at the university, how many students are enrolled there?

60. New York Yankees player Didi Gregorius has a recorded batting average of 15 hits for every 50 times at bat. If he maintains this average, how many at-bats will he need to achieve 111 hits? (Round to the nearest whole number.)

61. On a map of Maryland, one inch represents 4 miles. If there are 8.5 inches between Baltimore, MD, and Washington, DC, how far are the two cities from each other?

62. A floor plan is drawn to scale in which 1 inch represents 4 feet. What size will the drawing be for a room that is 30 feet by 40 feet? (**Hint:** Set up two proportions.)

63. The recipe for Nestle Tollhouse Chocolate Chip Cookies calls for 2 cups of chocolate chips to make 5 dozen cookies. If you want to bake 17 dozen cookies, how many cups of chocolate chips do you need?

64. The instructions for Never-Ice Antifreeze states that 4 quarts of antifreeze are needed for every 10 quarts of radiator capacity. If Sal's car has a 22-quart radiator, how many quarts of antifreeze will it need?

65. An architect is to draw plans for a city park. He intends to use a scale of $\frac{1}{2}$ inch to represent 25 feet. How many inches will be needed to use for the length and width of a rectangular playing field that is 50 yards by 125 yards? (**Note:** 1 yard = 3 feet.)

66. A test driver wants to increase the speed of the car he is driving by 3 miles per hour every 2 seconds. However, he can only check his speed every 5 seconds because he is busy with other tasks during the test drive.

a. By how much should he increase his speed in 5 seconds?

b. If he starts checking his speed at 40 miles per hour, how fast should he be going after 10 seconds?

67. Jack and Diane are decorating a nursery room for their baby, which will be born in a few months. In one hour, Jack can get $\frac{1}{6}$ of the nursery done and Diane can get $\frac{1}{12}$ of the nursery done. If they work together, they can get $\frac{1}{x}$ of the nursery done in one hour. Determine how many hours it will take Jack and Diane to decorate the nursery if they work together by solving the equation $\frac{1}{6}+\frac{1}{12}=\frac{1}{x}$ for x.

Applications

57. 360 defective computers

58. 144 defective bulbs

59. 36,800 students

60. 370 at bats

61. 34 miles

62. 7.5 inches by 10 inches

63. 6.8 cups

64. 8.8 quarts

65. Width = 3 inches; Length = 7.5 inches

66. **a.** 7.5 miles per hour **b.** 55 miles per hour

67. 4 hours

68. 5 hours

69. First group: 10 hours; Second group: 6 hours.

70. a. $x = 10$

b. 8 kilometers per mile

c. 12 kilometers per mile

d. 0.5 hour

71. a. $\frac{3x}{3} = \frac{16}{x}$

b. $x = -4, +4$

c. $x = -4$ does not make sense because length cannot be negative

d. The side of the small triangle is 4 feet long and the base of the large triangle is 12 feet long.

68. A local print shop has a big order of pamphlets to print, so they decide to use two of their printers for the one job. The newer printer can print the pamphlets four times as fast as the older printer. That means in one hour, the newer printer can complete $\frac{1}{x}$ of the print job and the older printer can complete $\frac{1}{4x}$ of the print job. Working together, the printers can complete the job in 4 hours. Determine how many hours it would take the newer printer to print all of the pamphlets by itself by solving the equation $\frac{1}{x} + \frac{1}{4x} = \frac{1}{4}$ for x.

69. Two groups of civil engineers are surveying an area to prepare for the construction of a shopping center. The first group is full of new college graduates, and it will take them four more hours than it takes the second group, which is full of seasoned professionals. The second group can complete the job in x hours. This means that in one hour, the first group can complete $\frac{1}{x+4}$ of the job and the second group can complete $\frac{1}{x}$ of the job. Working together, they can complete the surveying job in $\frac{15}{4}$ hours. Determine how many hours it would take each team to complete the job individually by solving the equation $\frac{1}{x+4} + \frac{1}{x} = \frac{4}{15}$ for x.

70. Terrence and Alicia are competing in a marathon where the average running speed is x kilometers per hour. Terrence is running 2 kilometers per hour slower than the average running speed. Alicia is running 2 kilometers per hour faster than the average running speed. After a certain amount of time, Terrence ran 4 kilometers and Alicia ran 6 kilometers.

a. Determine the speed of the average runner by solving the equation $\frac{4}{x-2} = \frac{6}{x+2}$ for x.

b. What was Terrence's average running speed?

c. What was Alicia's average running speed?

d. How long did it take Terrence to run 4 kilometers and Alicia to run 6 kilometers?

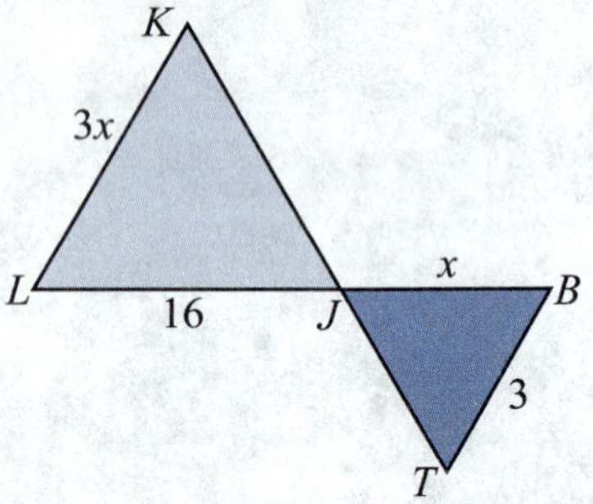

71. A team of gardeners is making two flower beds that are in the shape of similar triangles outside of an art museum. The apprentice gardener wasn't completely paying attention to the instructions given by the master gardener. All that he can remember is that the flower beds are isosceles triangles, the base of the small triangle is 3 feet wide, one side of the larger triangle is 16 feet long, and the base of the large triangle is three times the side length of the small triangle. The apprentice gardener needs to determine the unknown dimensions of the triangles.

a. Use the figure to write an equation to show that the side lengths are proportional.

b. Solve the equation from part a. for x.

c. Do any of the solutions from part b. not make sense in the context of the problem? If yes, explain why.

d. What are the lengths of the unknown sides of the triangles?

Writing & Thinking

In simplifying rational expressions, the result is a rational or polynomial expression. However, in solving equations with rational expressions, the goal is to find a value (or values) for the variable that will make the equation a true statement. Many students confuse these two ideas. To avoid confusing the techniques for adding and subtracting rational expressions with the techniques for solving equations, simplify the expression in part a. and solve the equation in part b. Explain, in your own words, the differences in your procedures. Assume no denominator has a value of 0.

72. **a.** $\frac{10}{x}+\frac{31}{x-1}+\frac{4x}{x-1}$

b. $\frac{10}{x}+\frac{31}{x-1}=\frac{4x}{x-1}$

73. **a.** $\frac{-4}{x^2-16}+\frac{x}{2x+8}-\frac{1}{4}$

b. $\frac{-4}{x^2-16}+\frac{x}{2x+8}=\frac{1}{4}$

74. **a.** $\frac{3x}{x^2-4}+\frac{5}{x+2}+\frac{2}{x-2}$

b. $\frac{3x}{x^2-4}+\frac{5}{x+2}=\frac{2}{x-2}$

75. **a.** $\frac{7}{5x}+\frac{2}{x-4}-\frac{3}{5x}$

b. $\frac{7}{5x}+\frac{2}{x-4}=\frac{3}{5x}$

76. **a.** $\frac{2}{x+9}-\frac{2}{x-9}+\frac{1}{2}$

b. $\frac{2}{x+9}-\frac{2}{x-9}=\frac{1}{2}$

Writing & Thinking

72. **a.** $\frac{4x^2+41x-10}{x(x-1)}$

b. $x=\frac{1}{4}, 10$

73. **a.** $\frac{x^2-8x}{4(x-4)(x+4)}$

b. $x=0, 8$

74. **a.** $\frac{10x-6}{(x-2)(x+2)}$

b. $x=\frac{7}{3}$

75. **a.** $\frac{14x-16}{5x(x-4)}$ **b.** $x=\frac{8}{7}$

76. **a.** $\frac{x^2-153}{2(x-9)(x+9)}$

b. $x=3, -3$

Objectives

A. Solve applications related to fractions.

B. Solve applications involving work.

C. Solve applications involving distance, rate, and time.

14.7 Applications: Rational Expressions

The following strategy for solving word problems is valid for all word problems that involve algebraic equations.

Solving a Word Problem Containing Rational Expressions

1. Read the problem carefully. Read it several times, if necessary.
2. Decide what is asked for and assign a variable to the unknown quantity. Draw a diagram or set up a chart whenever possible as a visual aid.
3. Form and solve an equation that relates the information provided.
4. Check your solution with the wording of the problem to be sure it makes sense.

PROCEDURE

A Problems Related to Fractions

We now introduce word problems involving rational expressions with problems relating the numerator and denominator of a fraction.

1. The denominator of a fraction is 3 more than the numerator. If both the numerator and denominator are increased by 4, the new fraction is equal to $\frac{3}{4}$. Find the original fraction.

Example 1 Solving Fractions

The denominator of a fraction is 8 more than the numerator. If both the numerator and denominator are increased by 3, the new fraction is equal to $\frac{1}{2}$. Find the original fraction.

Solution

Reread the problem to be sure that you understand all terminology used. Assign variables to the unknown quantities.

Let $n =$ original numerator,
$n+8 =$ original denominator, and
$\frac{n}{n+8} =$ original fraction.

$$\frac{n+3}{(n+8)+3} = \frac{1}{2}$$

The numerator and the denominator are each increased by 3, making a new fraction that is equal to $\frac{1}{2}$.

$$\frac{n+3}{n+11} = \frac{1}{2}$$

Multiply each term on both sides by $2(n+11)$, the LCD.

$$2(n+11)\cdot\frac{n+3}{n+11} = 2(n+11)\cdot\frac{1}{2}$$

$$2n+6 = n+11$$

$$n = 5 \quad \leftarrow \text{Original numerator}$$

$$n+8 = 13 \quad \leftarrow \text{Original denominator}$$

Check

$$\frac{(5)+3}{(13)+3}=\frac{8}{16}=\frac{1}{2} \quad \text{The original fraction is } \frac{5}{13}.$$

Now work margin exercise 1.

B Problems Related to Work

Problems involving work usually translate into equations involving rational expressions. The basic idea is to **represent what part of the work is done in one unit of time**. For example, if a man can dig a ditch in 3 hours, what part (of the ditch-digging job) can he do in one hour? The answer is that he can do $\frac{1}{3}$ of the work in one hour. If a fence was painted in 2 days, then $\frac{1}{2}$ of the work of painting the fence was done in 1 day. (These ideas assume a steady working pace.) In general, if the total work took x hours, then $\frac{1}{x}$ of the total work would be done in one hour.

Example 2 Application: Solving Work Problems

A carpenter can build a certain type of patio cover in 6 hours. His partner takes 8 hours to build the same cover. How long would it take them working together to build this type of patio cover?

Solution

Let x = number of hours to build the cover working together.

Person(s)	Time of Work (in Hours)	Part of Work Done in 1 Hour
Carpenter	6	$\frac{1}{6}$
Partner	8	$\frac{1}{8}$
Together	x	$\frac{1}{x}$

Part done in 1 hour by carpenter + Part done in 1 hour by partner = Part done in 1 hour together

$$\frac{1}{6} + \frac{1}{8} = \frac{1}{x}$$

$$\frac{1}{\cancel{6}}(\overset{4}{\cancel{24}}x)+\frac{1}{\cancel{8}}(\overset{3}{\cancel{24}}x)=\frac{1}{\cancel{x}}(24\cancel{x})$$

Multiply each term on both sides by $24x$, the LCD.

$$4x+3x=24$$

$$7x=24$$

$$x=\frac{24}{7}$$

2. An electrician can wire a house in 12 hours. Her assistant takes 18 hours to wire the same house. How long would it take them to wire the house together?

Together, they can build the patio cover in $\frac{24}{7}$ hours, or $3\frac{3}{7}$ hours.

(**Note:** This answer is reasonable because the time is less than either person would take working alone.)

Now work margin exercise 2.

3. A mother can carve a pumpkin twice as fast as her eight-year-old son. Together they can carve the pumpkin in 3 hours. How long would it take each of them working alone?

Example 3 Application: Solving Work Problems

A man can wax his car three times faster than his daughter can. Together they can do the job in 4 hours. How long would it take each of them working alone?

Solution

Let t = number of hours for the man alone to wax the car and
$3t$ = number of hours for the daughter alone to wax the car.

(**Note:** If the man can do the job 3 times faster, this means that the daughter would take 3 times longer, or $3t$ hours.)

Person(s)	Time of Work (in Hours)	Part of Work Done in 1 Hour
Man	t	$\frac{1}{t}$
Daughter	$3t$	$\frac{1}{3t}$
Together	4	$\frac{1}{4}$

Part done by man alone in 1 hour + Part done by daughter alone in 1 hour = Part done working together in 1 hour

$$\frac{1}{t} + \frac{1}{3t} = \frac{1}{4}$$

$$\frac{1}{\cancel{t}}(12\cancel{t}) + \frac{1}{\cancel{3t}}(\overset{4}{\cancel{12t}}) = \frac{1}{\cancel{4}}(\overset{3}{\cancel{12}}t)$$ Multiply each term on both sides by $12t$, the LCD.

$$12 + 4 = 3t$$

$$16 = 3t$$

$$t = \frac{16}{3}$$ ← Man's time

$$3t = (3)\left(\frac{16}{3}\right) = 16$$ ← Daughter's time

Check

Man's part in 1 hour: $\frac{1}{t}=\frac{1}{\left(\frac{16}{3}\right)}=\frac{3}{16}$

Daughter's part in 1 hour: $\frac{1}{3t}=\frac{1}{3\left(\frac{16}{3}\right)}=\frac{1}{16}$

Man's part in 4 hours: $\frac{3}{16}\cdot 4=\frac{3}{4}$

Daughter's part in 4 hours: $\frac{1}{16}\cdot 4=\frac{1}{4}$

$\frac{3}{4}+\frac{1}{4}=1$ car waxed in 4 hours

Working alone, the man takes $\frac{16}{3}$ hours, or $5\frac{1}{3}$ hours, and his daughter takes 16 hours.

Now work margin exercise 3.

Example 4 Application: Solving Work Problems

A man was told that his new pool would fill through an inlet valve in 3 hours. He knew something was wrong when the pool took 8 hours to fill. He found he had left the outlet valve open. How long would it take to drain the pool once it is completely filled and only the outlet valve is open?

Solution

Let t = time to drain pool with only the outlet valve open.

Note: We use the information gained when the pool was filled with both valves open. In that situation, the inlet and outlet valves worked against each other.

Valves	Hours to Fill or Drain	Part Filled or Drained in 1 Hour
Inlet	3	$\frac{1}{3}$
Outlet	t	$\frac{1}{t}$
Together	8	$\frac{1}{8}$

4. Janet's pool normally takes 4 hours to fill. Last time, it took 10 hours to fill and she found that she had left the drain open. How long would it take to drain the pool once it is completely filled and only the drain valve is open?

$$\underbrace{\text{Part filled by inlet in 1 hour}}_{\frac{1}{3}} - \underbrace{\text{Part emptied by outlet in 1 hour}}_{\frac{1}{t}} = \underbrace{\text{Part filled together in 1 hour}}_{\frac{1}{8}}$$

$$\frac{1}{3}(24t) - \frac{1}{t}(24t) = \frac{1}{8}(24t)$$

$$8t - 24 = 3t$$

$$5t = 24$$

$$t = \frac{24}{5}$$

The pool would drain in $\frac{24}{5}$ hours, or $4\frac{4}{5}$ hours. (**Note:** this is more time than the inlet valve would take to fill the pool. If the outlet valve worked faster than the inlet valve, then the pool would never have filled in the first place.)

Now work margin exercise 4.

C Problems Related to Distance-Rate-Time: $d = rt$

You may recall that the basic formula involving distance, rate, and time is $d = rt$. This relationship can also be stated in the forms $t = \frac{d}{r}$ and $r = \frac{d}{t}$.

If distance and rate are known or can be represented, then $t = \frac{d}{r}$ is the way to represent time. Similarly, if the distance and time are known or can be represented, then $r = \frac{d}{t}$ is the way to represent rate.

5. On Lake Wespanee, a man can paddle his kayak at a rate of 6 miles per hour. On the nearby Millitonka River, it takes him the same time to paddle 6 miles downstream as it does to paddle 3 miles upstream. What is the speed of the river's current?

Example 5 Application: Solving Distance-Rate-Time Problems

On Lake Itasca, a man can row his boat 5 miles per hour. On the nearby Mississippi River, it takes him the same time to row 5 miles downstream as it does to row 3 miles upstream. What is the speed of the river's current in miles per hour?

Solution

Let c = speed of the current.

Distance and rate are represented first in the following table. Then, the time going downstream and coming back upstream is represented in terms of distance and rate. Since the rate is in miles per hour, the distance is in miles and the time is in hours.

	Distance d	Rate r	Time $t = \frac{d}{r}$
Downstream	5	$5 + c$	$\frac{5}{5+c}$
Upstream	3	$5 - c$	$\frac{3}{5-c}$

Note

The current helps increase the rate while going downstream and decreases the rate while going back upstream.

$$\frac{5}{5+c}=\frac{3}{5-c} \qquad \text{The times are equal.}$$

$$\cancel{(5+c)}(5-c)\cdot\frac{5}{\cancel{5+c}}=(5+c)\cancel{(5-c)}\cdot\frac{3}{\cancel{5-c}}$$

$$25-5c=15+3c$$

$$-8c=-10$$

$$c=\frac{5}{4}$$

Check

$$\text{Time downstream}=\frac{5}{5+\frac{5}{4}}=\frac{5}{\frac{20}{4}+\frac{5}{4}}=\frac{5}{\frac{25}{4}}=5\cdot\frac{4}{25}=\frac{4}{5}\text{ hours}$$

$$\text{Time upstream}=\frac{3}{5-\frac{5}{4}}=\frac{3}{\frac{20}{4}-\frac{5}{4}}=\frac{3}{\frac{15}{4}}=3\cdot\frac{4}{15}=\frac{4}{5}\text{ hours}$$

The times are equal. The rate of the river current is $\frac{5}{4}$ mph, or $1\frac{1}{4}$ mph.

Now work margin exercise 5.

Example 6 Application: Solving Distance-Rate-Time Problems

If a passenger train travels three times as fast as a freight train, and the freight train takes 4 hours longer to travel 210 miles, what is the speed of each train?

Solution

Let r = rate of freight train in miles per hour and
$3r$ = rate of passenger train in miles per hour.

	Distance d	Rate r	Time $t=\frac{d}{r}$
Freight	210	r	$\frac{210}{r}$
Passenger	210	$3r$	$\frac{210}{3r}$

Note: If the rate is faster, then the time is shorter. Thus the fraction $\frac{210}{3r}$ is smaller than the fraction $\frac{210}{r}$.

6. If a commercial airplane travels twice as fast as a private aircraft, and the private aircraft takes $2\frac{1}{2}$ hours longer to travel 900 miles, what is the speed of each airplane?

$$\frac{210}{r}-\frac{210}{3r}=4$$ The difference between their times is 4 hours.

$$\frac{210}{r}-\frac{70}{r}=4$$

$$\frac{210}{\not{r}}\cdot\not{r}-\frac{70}{\not{r}}\cdot\not{r}=4\cdot r$$

$$210-70=4r$$

$$140=4r$$

$$35=r$$ Speed of the freight train.

$$105=3r$$ Speed of the passenger train.

Check

Time for freight train $=\frac{210}{35}=6$ hours

Time for passenger train $=\frac{210}{105}=2$ hours

$6-2=4$ hours difference in time

The freight train travels 35 mph, and the passenger train travels 105 mph.

Now work margin exercise 6.

Margin Exercise Answers

1. $\frac{5}{8}$ **2.** $\frac{36}{5}$ hours or $7\frac{1}{5}$ hours **3.** It takes the mom $\frac{9}{2}$ hours, or $4\frac{1}{2}$ hours, and her son takes 9 hours. **4.** The pool will drain in $\frac{20}{3}$ hours, or $6\frac{2}{3}$ hours. **5.** 2 mph
6. Commercial airplane: 360 mph; Private airplane: 180 mph

14.7 Exercises

Concept Check

Fill-in-the-Blank. Complete each sentence using information found in this section.

1. For problems involving work, you should represent what part of the work is done in one _____ of _____.

2. To solve any word problem, begin by _______ the problem carefully, possibly even several times.

3. Once a potential solution for a word problem has been found, ______ the solution with the original problem to make sure it makes sense.

4. The formula that relates distance, rate, and time is $d=rt$. This formula can be manipulated to represent the formula for rate, which is ______.

5. When solving word problems, it may be helpful to draw a/an ______ or set up a/an ______ as a visual aid.

14.7 Exercises

Concept Check

1. unit, time
2. reading
3. check
4. $r=\frac{d}{t}$
5. diagram, chart

True/False. Determine whether each statement is true or false. If a statement is false, explain how it can be changed so the statement will be true. (**Note:** There may be more than one acceptable change.)

6. The first step in solving application problems is to assign a variable to the unknown value.

7. If the total amount of work took 5 hours to do, then $\frac{1}{5}$ of the work can be done in one hour.

8. If you know the distance and rate, you can use the formula $t = d - r$ to represent time.

Applications

Solve.

1. The sum of two numbers is 117, and they are in the ratio of 8 to 5. Find the two numbers.

2. If 4 is subtracted from a certain number and the difference is divided by 2, the result is 1 more than $\frac{1}{5}$ of the original number. Find the original number.

3. What number must be added to both the numerator and denominator of $\frac{16}{21}$ to make the resulting fraction equal to $\frac{5}{6}$?

4. Find the number that can be subtracted from both the numerator and denominator of the fraction $\frac{69}{102}$ so that the result is $\frac{5}{8}$.

5. The denominator of a fraction exceeds the numerator by 7. If the numerator is increased by 3 and the denominator is increased by 5, the resulting fraction is equal to $\frac{1}{2}$. Find the original fraction.

6. The numerator of a fraction exceeds the denominator by 5. If the numerator is decreased by 4 and the denominator is increased by 3, the resulting fraction is equal to $\frac{4}{5}$. Find the original fraction.

7. One number is $\frac{3}{4}$ of another number. Their sum is 63. Find the numbers.

8. The sum of two numbers is 24. If $\frac{2}{5}$ of the larger number is equal to $\frac{2}{3}$ of the smaller number, find the numbers.

9. One number exceeds another by 5. The sum of their reciprocals is equal to 19 divided by the product of the two numbers. Find the numbers.

10. One number is 3 less than another. The sum of their reciprocals is equal to 7 divided by the product of the two numbers. Find the numbers.

Solve the following word problems. Remember to check each solution with the wording of the original problem to make sure it is reasonable.

11. It takes Rosa, traveling at 30 mph, 30 minutes longer to go a certain distance than it takes Melody traveling at 50 mph. Find the distance traveled.

12. ▦ It takes a plane, flying at 450 mph, 25 minutes longer to travel a certain distance than it takes a second plane to fly the same distance at 500 mph. Find the distance.

6. False; The first step is to read carefully.
7. True
8. False; Use the formula $t = \frac{d}{r}$.

Applications

1. 72, 45
2. 10
3. 9
4. 14
5. $\frac{6}{13}$
6. $\frac{12}{7}$
7. 36, 27
8. 15, 9
9. 7, 12
10. 2, 5
11. 37.5 miles
12. 1875 miles

13. $\frac{12}{5}$ or $2\frac{2}{5}$ hours
14. person: 505 minutes; machine: 5.05 minutes
15. $\frac{9}{5}$ or $1\frac{4}{5}$ hours
16. 45 minutes
17. 6 hours
18. Beth: 52 mph; Anna: 48 mph
19. Charles: 8 mph; Chase: 6 mph
20. commercial airliner: 300 mph; private plane: 120 mph
21. 63 mph
22. 6 hours
23. sailboat: 5 mph; cruise ship: 25 mph
24. 50 mph
25. 20 mph
26. $\frac{9}{4}$ or $2\frac{1}{4}$ hours

13. Kira needs 4 hours to complete the yard work. Her husband, Zackary, needs 6 hours to do the work. How long will the job take if they work together?

14. ▦ In 1921, automated wrapping machines were used to aid in the wrapping of Hershey Kisses® in the Hershey chocolate factory. The machine could wrap the candies 100 times faster than a person could. Together the machine and the person could wrap a crate full of Hershey Kisses® in 5 minutes. How long would it take each of them working alone?

15. Ben's secretary can address the weekly newsletters in $4\frac{1}{2}$ hours. Charlie's secretary needs only 3 hours. How long will it take if they both work on the job?

16. Working together, Greg and Cindy can clean the snow from the driveway in 20 minutes. It would have taken Cindy 36 minutes working alone. How long would it have taken Greg alone?

17. A carpenter and his partner can put up a patio cover in $3\frac{3}{7}$ hours. If the partner needs 8 hours to complete the patio alone, how long would it take the carpenter working alone?

18. ▦ Beth can travel 208 miles in the same length of time it takes Anna to travel 192 miles. If Beth's speed is 4 mph greater than Anna's, find both rates.

19. Charles can bike 32 miles in the same amount of time that his twin brother Chase can bike 24 miles. If Charles bikes 2 mph faster than Chase, how fast does each man bike?

20. ▦ A commercial airliner can travel 750 miles in the same amount of time that it takes a private plane to travel 300 miles. The speed of the airliner is 60 mph more than twice the speed of the private plane. Find the speed of each aircraft.

21. Gabriela drives her car 350 miles and has an average of a certain speed. If the average speed had been 9 mph less, she could have traveled only 300 miles in the same length of time. What was her average speed?

22. A family travels 18 miles down river and returns. It takes 8 hours to make the round trip. Their rate in still water is twice the rate of the river's current. How long will the return trip take?

23. Cruise ships travel 5 times faster than sailboats (in optimal wind conditions). If it takes 16 hours longer for a sailboat (with optimal wind conditions) to travel 100 miles from Charleston, SC, to Savannah, GA, what is the speed of each boat?

24. An airplane can fly 650 mph in still air. If it can travel 2800 miles with the wind in the same time it can travel 2400 miles against the wind, find the wind speed.

25. A one-engine plane can fly 120 mph in still air. If it can fly 490 miles with a tailwind in the same time that it can fly 350 miles against a headwind, what is the speed of the wind? (**Note:** A tailwind increases the speed of the plane and a headwind decreases the speed of the plane.)

26. Using a small inlet pipe, it takes 9 hours to fill a pool. Using a large inlet pipe, it only takes 3 hours. If both are used simultaneously, how long will it take to fill the pool?

27. An inlet pipe on a swimming pool can be used to fill a pool in 36 hours. The drain pipe can be used to empty the pool in 40 hours. If the pool is $\frac{2}{3}$ filled using the inlet pipe and then the drain pipe is accidentally opened, how long from that time will it take to fill the pool?

28. A contractor hires two bulldozers to clear the trees from a 20-acre tract of land. One works twice as fast as the other. It takes them 3 days to clear the tract working together. How long would it take each of them alone?

29. John, Raul, and Denny, working together, can clean their bait and tackle store in 6 hours. Working alone, Raul takes twice as long to clean the store as John does. Denny needs three times as long as John does. How long would it take each man working alone?

30. Francois rode his jet ski 36 miles downstream and then 36 miles back. The round trip took $5\frac{1}{4}$ hours. Find the speed of the jet ski in still water and the speed of the current if the speed of the current is $\frac{1}{7}$ the speed of the jet ski.

31. Momence, IL, is 12 miles upstream on the same side of the river from Kankakee, IL, on the Kankakee River. A motorboat that can travel 8 mph in still water leaves Momence and travels downstream toward Kankakee. At the same time, another boat that can travel 10 mph leaves Kankakee and travels upstream toward Momence. Each boat completes the trip in the same amount of time. Find the rate of the current.

32. Samantha rides the ski lift to the top of Blue Mountain, a distance of $1\frac{3}{4}$ kilometers (a little more than 1 mile). She then skis directly down the slope. If she skis five times as fast as the lift travels and the total trip takes 45 minutes, find the rate at which she skis.

33. A local print shop has a big order of pamphlets to print, so they decide to use two of their printers for the one job. The newest printer can print the pamphlets four times as fast as the old printer. Working together the printers can complete the job in 4 hours. How many hours would it take each printer to print all of the pamphlets by itself?

a. Use the table to set up a rational equation to describe the situation. Use the variable x to represent the time it takes the newest printer to complete the job.

Printer	Time of Work (in Hours)	Part of Work Done in 1 Hour
Newest		
Old		
Together		

b. Solve the equation from part a. for x.

c. Use the solution from part b. to answer the question in the problem statement.

27. 120 hours

28. $\frac{9}{2}$ or $4\frac{1}{2}$ days and 9 days

29. John: 11 hours; Raul: 22 hours; Denny: 33 hours

30. jet ski: 14 mph; current: 2 mph

31. 1 mph

32. 14 kilometers per hour

33. a. $\frac{1}{x}+\frac{1}{4x}=\frac{1}{4}$

Printer	Time of Work (in Hours)	Part of Work Done in 1 Hour
Newest	x	$\frac{1}{x}$
Old	$4x$	$\frac{1}{4x}$
Together	4	$\frac{1}{4}$

b. $x = 5$

c. It would take the newest printer 5 hours to complete the job and the old printer 20 hours to complete the job.

34. a. $\dfrac{350}{x+9} = \dfrac{300}{x}$

	Distance (d)	$\div$	Rate (r)	=	Time $\left(t = \dfrac{d}{r}\right)$
Car	350		$x+9$		$\dfrac{350}{x+9}$
Truck	300		x		$\dfrac{300}{x}$

b. $x = 54$

c. The truck is traveling at an average speed of 54 miles per hour and the car is traveling at an average speed of 63 miles per hour.

d. Car: 6 hours; Truck: 7 hours

Writing & Thinking

35. a. 5 and 7 b. 2 and 4

34. The Winston family is moving to another state. The family is driving to their new house in a car and all of their belongings are in a moving truck. The car is traveling at speed that is 9 miles per hour faster than the speed of the truck. After a certain amount of time, the family's car traveled 350 miles and the moving truck traveled 300 miles. What are the speeds of the car and the truck? Use the table to set up a rational equation to describe the situation. Use the variable x to represent the rate of the truck.

	Distance (d)	$\div$	Rate (r)	=	Time $\left(t = \dfrac{d}{r}\right)$
Car					
Truck					

a. Solve the equation from part a. for x.

b. Use the solution from part b. to answer the question in the problem statement.

c. If the Winston family's new home is 378 miles away, how long will it take the car and the truck to make the trip?

Writing & Thinking

35. If n is any integer, then $2n$ is an even integer and $2n + 1$ is an odd integer. Use these ideas to solve the following problems.

a. Find two consecutive odd integers such that the sum of their reciprocals is $\frac{12}{35}$.

b. Find two consecutive even integers such that the sum of the first and the reciprocal of the second is $\frac{9}{4}$.

14.8 Applications: Variation

Objectives

A. Solve problems involving direct variation.

B. Solve problems involving inverse variation.

C. Solve problems involving combined variation.

A Direct Variation

Suppose that you ride your bicycle at a steady rate of 15 miles per hour. If you ride for 1 hour, the distance you travel would be 15 miles. If you ride for 2 hours, the distance you travel would be 30 miles. This relationship can be written in the form of the formula $d = 15t$ (or $\frac{d}{t} = 15$), where d is the distance traveled and t is the time in hours. We say that distance and time **vary directly** (or are in **direct variation** or are **directly proportional**). The term proportional implies that the ratio is constant. In this example, 15 is the constant and is called the **constant of variation**. When two variables vary directly, an **increase in the value of one variable indicates an increase in the other**, and the ratio of the two quantities is constant.

Direct Variation

A variable quantity y **varies directly as** (or is **directly proportional to**) a variable x if there is a constant k such that

$$\frac{y}{x} = k \quad \text{or} \quad y = kx.$$

The constant k is called the **constant of variation**.

DEFINITION

Teaching Note

You might want to make the case for the study of variation (and proportion) as being extremely important to the sciences and technology. Many major relationships in chemistry, physics, and astronomy began by recognizing a variation relationship.

Example 1 Solving Using Direct Variation

If y varies directly as x, and $y = 6$ when $x = 2$, find y if $x = 6$.

Solution

$y = kx$	The general formula for direct variation
$6 = 2k$	Substitute the known values and solve for k.
$3 = k$	Use this value for k in the general formula.

So $y = 3x$. Thus, if $x = 6$, then $y = 3 \cdot 6 = 18$.

Now work margin exercise 1.

1. If y varies directly as x, and $y = 10$ when $x = 5$, find y if $x = 2$.

2. If a 25-lb weight is placed on the same spring mentioned in Example 2, calculate how far the spring will stretch.

Example 2 Application: Solving Using Direct Variation

A spring will stretch a greater distance as more weight is placed on the end of it. The distance (d) the spring stretches varies directly as the weight (w) placed at the end of the spring. This is a property of springs studied in physics and is known as Hooke's Law. If a weight of 10 lb stretches a certain spring 6 in., how far will the spring stretch with a weight of 15 lb? (**Note:** We assume that the weight is not so great as to break the spring.)

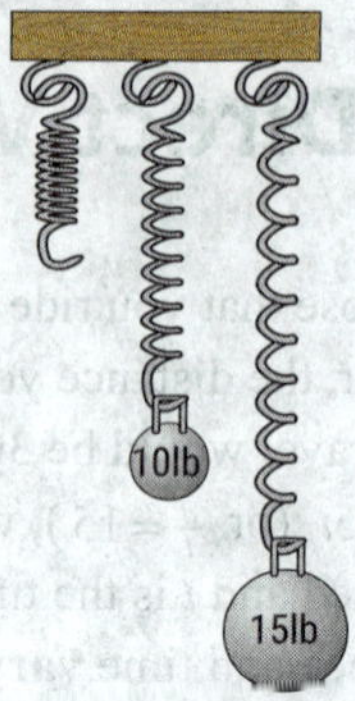

Solution

Because the two variables are directly proportional, the relationship can be indicated with the formula

$$d = k \cdot w,$$

where d = distance the spring stretches in inches,

w = weight in lb, and

k = constant of variation.

First, substitute the given information to find the value for k. (The value of k will depend on the particular spring. Springs made of different material or which are wound more tightly will have different values for k.)

$d = k \cdot w$

$6 = k \cdot 10$ Substitute the known values into the formula.

$\frac{3}{5} = k$ The constant of variation is $\frac{3}{5}$ (or 0.6).

$d = \frac{3}{5}w$ Use this value for k in the general formula.

If $w = 15$, we have $d = \frac{3}{5} \cdot 15 = 9$.

The spring will stretch 9 in. if a weight of 15 lb is placed at its end.

Now work margin exercise 2.

Listed here are several formulas involving direct variation.

$d = \frac{3}{5}w$ Hooke's Law for a certain spring where $k = \frac{3}{5}$.

$C = 2\pi r$ The circumference of a circle is directly proportional to the radius.

$A = \pi r^2$ The area of a circle varies directly as the radius squared.

$P = 62.5d$ Water pressure is proportional to the depth of the water.

B Inverse Variation

When two variables vary in such a way that their product is constant, we say that the two variables **vary inversely** (or are **inversely proportional**). For example, when a gas is placed in a container and pressure is increased on the gas, the product of the pressure and the volume of gas remains constant. The relationship between pressure and volume is modeled by the formula $V \cdot P = k$ (or, $V = \frac{k}{P}$).

Note that if a product of two variables is to remain constant, then **an increase in the value of one variable must be accompanied by a decrease in the other**. Or, in the case of a fraction with a constant numerator, if the denominator increases in value, then the fraction decreases in value. For the gas in the container, an increase in pressure indicates a decrease in the volume of gas.

Inverse Variation

A variable quantity y **varies inversely as** (or is **inversely proportional to**) a variable x if there is a constant k such that

$$\boldsymbol{x \cdot y = k} \quad \text{or} \quad \boldsymbol{y = \frac{k}{x}}.$$

The constant k is called the **constant of variation**.

DEFINITION

Example 3 Solving Using Inverse Variation

If y varies inversely as the cube of x, and $y = -1$ when $x = 3$, find y if $x = -3$.

Solution

$$y = \frac{k}{x^3}$$ The general formula for inverse variation adapted for x^3.

$$-1 = \frac{k}{3^3}$$ Substitute the known values and solve for k.

$$-1 = \frac{k}{27}$$

$$-27 = k$$ Use this value for k in the general formula.

So $y = \frac{-27}{x^3}$.

Thus, if $x = -3$, then $y = \frac{-27}{(-3)^3} = \frac{-27}{-27} = 1$.

Now work margin exercise 3.

3. If y varies inversely as x raised to the fifth power, and $y = -2$ when $x = 2$, find y when $x = -2$.

4. Assuming the same astronaut in Example 4 is on his way to the moon and is 300 miles from the surface of Earth, calculate his weight rounded to the nearest pound.

Example 4 Application: Solving Using Inverse Variation

The gravitational force (F) between an object and Earth is inversely proportional to the square of the distance (d) from the object to the center of Earth. Hence, we have the formula

$F \cdot d^2 = k$ or $F = \frac{k}{d^2}$, where F = force,

d = distance, and

k = constant of variation.

(As the distance of an object from Earth becomes larger, the gravitational force exerted by Earth on the object becomes smaller.)

If an astronaut weighs 200 pounds on the surface of Earth, what will he weigh (to the nearest pound) 100 miles above Earth? Assume that the radius of Earth is 4000 miles.

Solution

We know $F = 200 = 2 \times 10^2$ pounds

when $d = 4000 = 4 \times 10^3$ miles.

Use scientific notation to make values simpler to work with in the calculations.

$$2 \times 10^2 = \frac{k}{\left(4 \times 10^3\right)^2}$$

Substitute and solve for k.

$$k = \left(2 \times 10^2\right) \cdot \left(16 \times 10^6\right) = 32 \times 10^8 = 3.2 \times 10^9$$

So, $F = \frac{3.2 \times 10^9}{d^2}$.

When the astronaut is 100 miles above Earth,

$d = 4100 = 4.1 \times 10^3$ miles. Then

$$F = \frac{3.2 \times 10^9}{\left(4.1 \times 10^3\right)^2} = \frac{3.2 \times 10^9}{16.81 \times 10^6} \approx 0.190 \times 10^3 = 190 \text{ pounds.}$$

That is, 100 miles above Earth, the astronaut will weigh about 190 pounds.

Now work margin exercise 4.

C Combined Variation

If a variable varies either directly or inversely as more than one other variable, the variation is said to be **combined variation**. If the combined variation is all direct variation (the variables are multiplied), then it is called **joint variation**. For example, the volume of a cylinder varies jointly as its height and the square of its radius.

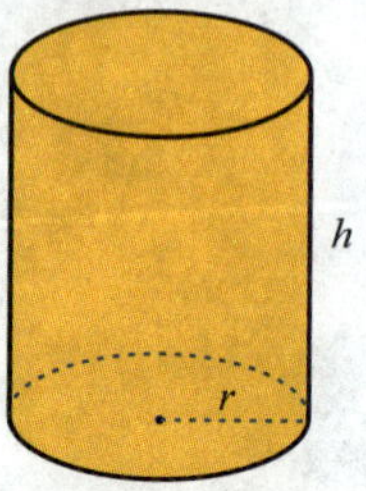

$V = kr^2h$,

where r = radius, h = height, and k = constant of variation

Using this information, what is the value of k, the constant of variation, if a cylinder has the approximate measurements $V = 198$ cubic feet, $r = 3$ feet, and $h = 7$ feet?

$$V = k \cdot r^2 \cdot h \quad \text{V varies jointly as } r^2 \text{ and } h.$$
$$198 = k \cdot 3^2 \cdot 7 \quad \text{Substitute the known values.}$$
$$\frac{198}{9 \cdot 7} = k$$
$$k = \frac{22}{7} \approx 3.14$$

Substituting the constant of variation, the formula is $V = \pi r^2 h$.

Example 5 Solving Using Joint Variation

If z varies jointly as x^2 and y, and $z = 18$ when $x = 2$ and $y = 4$, what is z when $x = 4$ and $y = 3$?

Solution

$$z = k \cdot x^2 \cdot y$$
$$18 = k \cdot 2^2 \cdot 4 \quad \text{Substitute the known values and solve for } k.$$
$$18 = k \cdot 16$$
$$\frac{9}{8} = k$$

So, $z = \frac{9}{8} x^2 y$. Substitute $\frac{9}{8}$ for k in the general formula

If $x = 4$ and $y = 3$, then

$$z = \frac{9}{8} \cdot 4^2 \cdot 3$$
$$z = 54.$$

Now work margin exercise 5.

5. If z varies jointly as x^3 and y^2, and $z = 108$ when $x = 2$ and $y = 3$, what is z when $x = 3$ and $y = 4$?

Example 6 Application: Solving Using Variation

The distance an object falls varies directly as the square of the time it falls (until it hits the ground and assuming little or no air resistance). If an object fell 64 feet in 2 seconds, how far would it have fallen by the end of 3 seconds?

Solution

$$d = k \cdot t^2 \quad \text{Where } d = \text{distance}, t = \text{time (in seconds), and } k = \text{constant of variation}$$
$$64 = k \cdot 2^2 \quad \text{Substitute the known values and solve for } k.$$
$$16 = k$$

So, $d = 16t^2$. Substitute 16 for k in the general formula.

If $t = 3$, then we have the following.

$$d = 16 \cdot 3^2$$
$$d = 144$$

The object would have fallen 144 feet in 3 seconds.

Now work margin exercise 6.

6. The distance an object falls varies directly as the square of the time it falls. If an object fell 64 feet in 2 seconds, how many feet will it fall in 5 seconds?

7. A volume of a gas varies inversely as the pressure on the gas. If a gas has a volume of 300 cubic inches under pressure of 8 pounds per square inch, what will be its volume if the pressure increases to 12 pounds per square inch?

Example 7 Application: Solving Using Variation

The volume of a gas in a container varies inversely as the pressure on the gas. If a gas has a volume of 200 cubic inches under a pressure of 5 pounds per square inch, what will be its volume if the pressure is increased to 8 pounds per square inch?

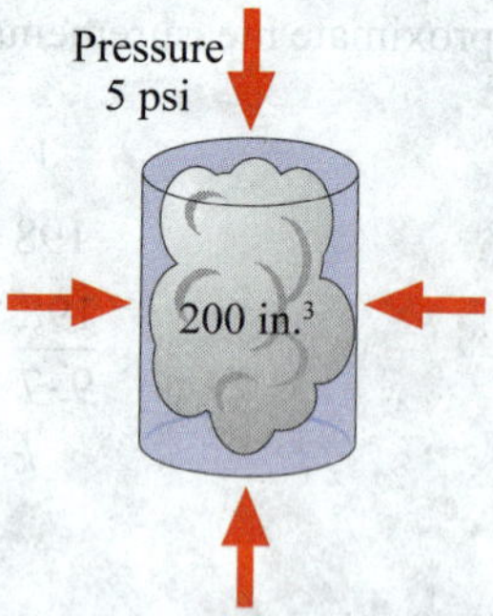

Solution

$$V = \frac{k}{P}$$ Where V = volume, P = pressure, and k = constant of variation

$$200 = \frac{k}{5}$$ Substitute the known values and solve for k.

$$k = 1000$$

So, $$V = \frac{1000}{P}$$ Substitute 1000 for k in the general formula.

$$V = \frac{1000}{8} = 125.$$

The volume will be 125 cubic inches.

Now work margin exercise 7.

8. The safe load a bridge support beam can carry varies jointly as the width w and the square of the depth d and inversely as the length l. A 5-inch-wide steel beam that is 12 inches deep and 15 feet long can support a load of 18,000 pounds. What is the safe load of a beam of the same material that is 4 inches wide, 8 inches deep, and 12 feet long?

Example 8 Application: Solving Using Variation

The safe load (L) of a wooden beam supported at both ends varies jointly as the width (w) and the square of the depth (d) and inversely as the length (l). A 3 in.-wide by 10 in.-deep beam that is 8 ft long supports a load of 9600 lb safely. What is the safe load of a beam of the same material that is 4 in. wide, 9 in. deep, and 12 ft long?

Solution

$$L = \frac{k \cdot w \cdot d^2}{l}$$ Where L = safe load, w = width, d = depth, and l = length

$$9600 = \frac{k \cdot 3 \cdot 10^2}{8}$$ Substitute the known values and solve for k.

$$9600 = \frac{k \cdot 300}{8}$$

$$k = \frac{9600 \cdot 8}{300}$$

$$k = 256$$

So, $$L = \frac{256 \cdot w \cdot d^2}{l}$$ Substitute 256 for k in the general formula.

$$L = \frac{256 \cdot 4 \cdot 9^2}{12} = \frac{256 \cdot 4 \cdot 81}{12} = 6912.$$

The safe load will be 6912 lb.

Now work margin exercise 8.

Margin Exercise Answers
1. $y = 4$ **2.** 15 cm **3.** $y = 2$ **4.** 173 pounds **5.** $z = 648$ **6.** 400 feet **7.** 200 cubic inches
8. 8000 pounds

14.8 Exercises

Concept Check

Fill-in-the-Blank. Complete each sentence using information found in this section.

1. If two variables are inversely proportional, an increase in the value of one variable must be accompanied by a/an ________ in the other.
2. If a variable varies (directly or inversely) with more than one other variable, this variation is said to be a/an __________ variation.
3. When two variables vary directly, an increase in one variable indicates a/an ________ in the other.
4. When two variables vary so that their product is constant, the two variables vary ________.
5. If there is a combined variation that is all direct variation, it is a/an ______ variation.
6. The letter k often represents the constant of ________.

True/False. Determine whether each statement is true or false. If a statement is false, explain how it can be changed so the statement will be true. (**Note:** There may be more than one acceptable change.)

7. The number of hamburgers eaten varies inversely with calories consumed.
8. The equation $y = \frac{k}{x}$ represents direct variation.
9. Distance and time varies directly, which means they are directly proportional.
10. The circumference of a circle varies directly with its radius.

Practice

Use the information given to find the unknown value. See Examples 1, 3, and 5.

1. If y varies directly as x, and $y = 3$ when $x = 9$, find y if $x = 7$.
2. If y is directly proportional to x^2, and $y = 3$ when $x = 2$, what is y when $x = 8$?
3. If y varies inversely as x, and $y = 5$ when $x = 8$, find y if $x = 20$.
4. If y is inversely proportional to x, and $y = 5$ when $x = 4$, what is y when $x = 2$?
5. If y varies inversely as x^2, and $y = -8$ when $x = 2$, find y if $x = 3$.
6. If y is inversely proportional to x^3, and $y = 40$ when $x = \frac{1}{2}$, what is y when $x = \frac{1}{3}$?
7. If y is directly proportional to the square root of x, and $y = 6$ when $x = \frac{1}{4}$, what is y when $x = 9$?

14.8 Exercises

Concept Check

1. decrease
2. combined
3. increase
4. inversely
5. joint
6. variation
7. False; Varies directly
8. False; Inverse variation
9. True
10. True

Practice

1. $\frac{7}{3}$
2. 48
3. 2
4. 10
5. $-\frac{32}{9}$
6. 135
7. 36

8. 180
9. 120
10. −27
11. $\frac{56}{3}$
12. $\frac{56}{3}$
13. 40
14. 81
15. 54
16. 32
17. $\frac{48}{5}$
18. 54
19. 27
20. 16

Applications

21. 400 feet
22. 6 in.
23. $35.10

8. If y is directly proportional to the square of x, and $y = 80$ when $x = 4$, what is y when $x = 6$?

9. z varies jointly as x and y, and $z = 60$ when $x = 2$ and $y = 3$. Find z if $x = 3$ and $y = 4$.

10. z varies jointly as x and y, and $z = -6$ when $x = 5$ and $y = 8$. Find z if $x = 12$ and $y = 15$.

11. z varies jointly as x and y^2, and $z = 63$ when $x = 5$ and $y = 3$. Find z if $x = \frac{10}{3}$ and $y = 2$.

12. z varies jointly as x^2 and y, and $z = 20$ when $x = 2$ and $y = 3$. Find z if $x = 4$ and $y = \frac{7}{10}$.

13. z varies directly as x and inversely as y^2. If $z = 5$ when $x = 1$ and $y = 2$, find z if $x = 2$ and $y = 1$.

14. z varies directly as x^3 and inversely as y^2. If $z = 24$ when $x = 2$ and $y = 2$, find z if $x = 3$ and $y = 2$.

15. z varies directly as $\sqrt{x}$ and inversely as y. If $z = 24$ when $x = 4$ and $y = 3$, find z if $x = 9$ and $y = 2$.

16. z varies directly as x^2 and inversely as $\sqrt{y}$. If $z = 108$ when $x = 6$ and $y = 4$, find z if $x = 4$ and $y = 9$.

17. s varies directly as the sum of r and t and inversely as w. If $s = 24$ when $r = 7$ and $t = 8$ and $w = 9$, find s if $r = 9$ and $t = 3$ and $w = 18$.

18. s varies directly as r and inversely as the difference of t and u. If $s = 36$ when $r = 12$ and $t = 9$ and $u = 6$, find s if $r = 18$ and $t = 11$ and $u = 8$.

19. L varies jointly as m and n and inversely as p. If $L = 6$ when $m = 7$ and $n = 8$ and $p = 12$, find L if $m = 15$ and $n = 14$ and $p = 10$.

20. W varies jointly as x and y and inversely as z. If $W = 10$ when $x = 6$ and $y = 5$ and $z = 2$, find W if $x = 12$ and $y = 6$ and $z = 3$.

Applications

Solve.

21. The distance a free-falling object falls is directly proportional to the square of the time it falls (before it hits the ground). If an object fell 256 feet in 4 seconds, how far would it have fallen by the end of 5 seconds?

22. The length a hanging spring stretches varies directly with the weight placed on the end. If a spring stretches 5 in. with a weight of 10 lb, how far will the spring stretch if the weight is increased to 12 lb?

23. The total price (P) of gasoline purchased varies directly as the number of gallons purchased. If 10 gallons are purchased for $23.40, what will be the price of 15 gallons?

24. Research shows that the value of gold and the value of the dollar are inversely proportional. In 2016, gold cost \$1200 per ounce and the dollar had a rating of 93 on the US dollar index. In 2017, the cost of gold was \$1300 per ounce. What was the 2017 rating of the dollar? (Round your answer to the nearest hundredth.)

25. The circumference of a circle varies directly as the diameter. A circular pizza pie with a diameter of 1 foot has a circumference of 3.14 feet. What will be the circumference of a pizza pie with a diameter of 1.5 feet?

26. The area of a circle varies directly as the square of its radius. A circular pizza pie with a radius of 6 in. has an area of 113.04 in.^{2} What will be the area of a pizza pie with a radius of 9 in.?

27. Several triangles have the same area. In this set of triangles, the height and base are inversely proportional. In one such triangle, the height is 5 m and the base is 12 m. Find the height of the triangle in this set with a base of 10 m.

28. If an astronaut weighs 250 pounds on the surface of the earth, what will the astronaut weigh 150 miles above the earth? Assume that the radius of the earth is 4000 miles, and round to the nearest tenth. (See Example 4.)

29. The elongation (E) in a wire when a mass (m) is hung at its free end varies jointly as the attached mass and the length (l) of the wire and inversely as the cross-sectional area (A) of the wire. The elongation is 0.0055 cm when a mass of 120 g is attached to a wire 330 cm long with a cross-sectional area of 0.4 cm^2. Find the elongation if a mass of 160 g is attached to the same wire.

30. When a mass of 240 oz is suspended by a wire 49 in. long whose cross-sectional area is 0.035 in.^{2}, the elongation of the wire is 0.016 in. Find the elongation if the same mass is suspended by a 28 in. wire of the same material with a cross-sectional area of 0.04 in.^{2} (See Exercise 29.)

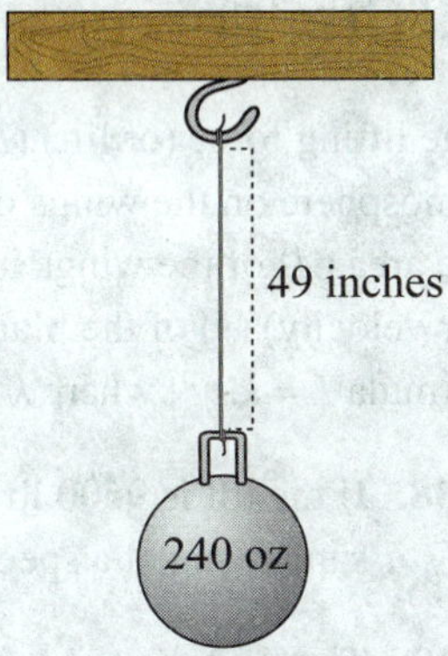

31. The safe load (L) of a wooden beam supported at both ends varies jointly as the width (w) and the square of the depth (d) and inversely as the length (l). A beam 4 in. wide, 6 in. deep, and 12 ft long supports a load of 4800 lb safely. What is the safe load of a beam of the same material that is 6 in. wide, 10 in. deep, and 15 ft long?

32. A wooden beam 2 in. wide, 8 in. deep, and 14 ft long holds up to 2400 lb. What load would a beam 3 in. wide, 6 in. deep, and 15 ft long, of the same material, support? (See Exercise 31.)

24. rating of 85.85
25. 4.71 feet
26. 254.34 in.^{2}
27. 6 m
28. 232.3 pounds
29. 0.0073 cm
30. 0.008 in.
31. 16,000 lb
32. 1890 lb

33. 9×10^{-11} N
34. 3.6×10^{-10} N
35. \$59.70
36. 2.5 seconds
37. a. $SL = \frac{kwd^3}{l^2}$ b. 200 c. 4320 pounds
38. 15,000 lb
39. 25,200 lb
40. 6400 lb
41. 18,000 lb

33. The gravitational force of attraction (F) between two bodies varies directly as the product of their masses (m_1 and m_2) and inversely as the square of the distance (d) between them. The gravitational force between a 5-kg mass and a 2-kg mass 1 m apart is 1.5×10^{-10} N. Find the force between a 24-kg mass and a 9-kg mass that are 6 m apart. (N represents a unit of force called a Newton.)

34. In Exercise 33, what is the force if the distance between the 24 kg mass and the 9 kg mass is cut in half?

35. The total price (P) of gasoline purchased varies directly as the number of gallons purchased. If 10 gallons are purchased for \$39.80, what will be the price of 15 gallons?

36. The distance that an object falls is directly proportional to the square of the time that has passed since the object started to fall. A rock falls a distance of 64 feet in 2 seconds. How long will it take the rock to fall a distance of 100 feet?

37. For a certain type of wooden beam that carries a load at its center, the safe load (SL) varies jointly as the width w and the cube of the depth (d) and inversely as the square of the length (l). A wooden beam that is 4 inches wide, 6 inches deep, and 12 feet long can safely support a load of 2400 pounds.

a. Set up the variation equation.

b. Determine the constant of variation.

c. How much weight can a wooden beam that is 5 inches wide, 6 inches deep, and 10 feet long safely support?

Solve the following lifting force problems.

Lifting Force

The lifting force (or lift) (L) in pounds exerted by the atmosphere on the wings of an airplane is related to the area (A) of the wings in square feet and the speed (or velocity) (v) of the plane in miles per hour by the formula $L = kAv^2$, where k is the constant of variation.

38. If the lift is 9600 lb for a wing area of 120 ft^2 at a speed of 80 mph, find the lift of the same wing at a speed of 100 mph.

39. The lift for a wing of area 280 ft^2 is 34,300 lb when the plane is traveling at 210 mph. What is the lift if the speed is decreased to 180 mph?

40. The lift for a wing with an area of 144 ft^2 is 10,000 lb when the plane is traveling at 150 mph. What is the lift if the speed is decreased to 120 mph?

41. A plane traveling 140 mph with wing area 195 ft^2 has 12,500 lb of lift exerted on the wings. Find the lift for the same plane traveling at 168 mph.

42. 1.8 ft^3
43. 200 cm^3
44. 1700 g per in.^2
45. 330 bar
46. 15 ohms
47. 5.2 ohms
48. 2.56 ohms
49. 10 ohms

Solve the following pressure problems.

Pressure

Boyle's Law states that if the temperature of a gas sample remains the same, the pressure (P) of the gas is related to the volume (V) by the formula $P = \frac{k}{V}$, where k is the constant of variation.

42. A pressure of 1600 lb per ft^2 is exerted by 2 ft^3 of air in a cylinder. If a piston is pushed into the cylinder until the pressure is 1800 lb per ft^2, what will be the volume of the air? Round to the nearest tenth.

43. The volume of gas in a container is 300 cm^3 when the pressure on the gas is 20 g per cm^2. What will be the volume if the pressure is increased to 30 g per cm^2?

44. The pressure in a canister of gas is 1360 g per in.^2 when the volume of gas is 5 in.^3 If the volume is reduced to 4 in.^3, what is the pressure?

45. A scuba diver is using a diving tank that can hold 6 liters of air. If the tank has a pressure rating of 220 bar when full, what is the pressure rating when the volume of gas is 4 liters?

Solve the following electricity problems.

Electricity

The resistance (R) (in ohms), in a wire is given by the formula $R = \frac{kL}{d^2}$, where k is the constant of variation, L is the length of the wire and d is the diameter.

46. The resistance of a wire 500 ft long with a diameter of 0.01 in. is 20 ohms. What is the resistance of a wire 1500 ft long with a diameter of 0.02 in.?

47. The resistance is 2.6 ohms when the diameter of a wire is 0.02 in. and the wire is 10 ft long. Find the resistance of the same type of wire with a diameter of 0.01 in. and a length of 5 ft.

48. Tristan's car stereo uses a 5-ft audio wire with diameter 0.025 in. and resistance of 1.6 ohms. What is the resistance of 8 ft of the same type of audio wire?

49. Nicole purchased a spool of wire with diameter 0.01 in. for the speakers in her home audio system. If the resistance of 15 ft of this wire is 6 ohms, what is the resistance of 25 ft of the wire?

50. 900 lb

51. $10\frac{2}{3}$ ft from the 120 lb weight or, $1\frac{1}{3}$ ft from the 960 lb weight.

52. 5 ft from the 300 lb weight (or 20 feet from the 75 lb weight)

53. 216 kilograms

Writing & Thinking

54. a. When two variables vary directly, an increase in the value of one variable indicates an increase in the other, and the ratio of the two quantities is constant.

b. Joint variation is when a variable varies directly with more than one other variable.

c. When two variables vary inversely, an increase in the value of one variable indicates a decrease in the other, and the product of the two quantities is constant.

d. Combined variation is when a variable varies directly or inversely with more than one variable.

Solve the following lever problems.

Levers

If a lever is balanced with weight on opposite sides of its balance point, then the following proportion exists:

$$\frac{W_1}{W_2}=\frac{L_2}{L_1} \text{ or } W_1L_1=W_2L_2$$

where $L_1+L_2=L$, the total length of the lever.

50. How much weight can be raised at one end of a bar 8 ft long by the downward force of 60 lb when the balance point is $\frac{1}{2}$ ft from the unknown weight?

51. Where should the balance point of a bar 12 ft long be located if a 120-lb force is to raise a load weighing 960 lb?

52. Find the location of the balance point of a 25-ft board that can raise a 300-lb package with a downward force of 75 lb.

53. How much weight can be raised on one end of a 17-meter board by 90 kilograms, if the balance point is 5 meters from the unknown weight?

Writing & Thinking

54. Explain, in your own words, the meaning of the following terms.

a. Direct variation

b. Joint variation

c. Inverse variation

d. Combined variation

Discuss an example of each type of variation that you have observed in your daily life.

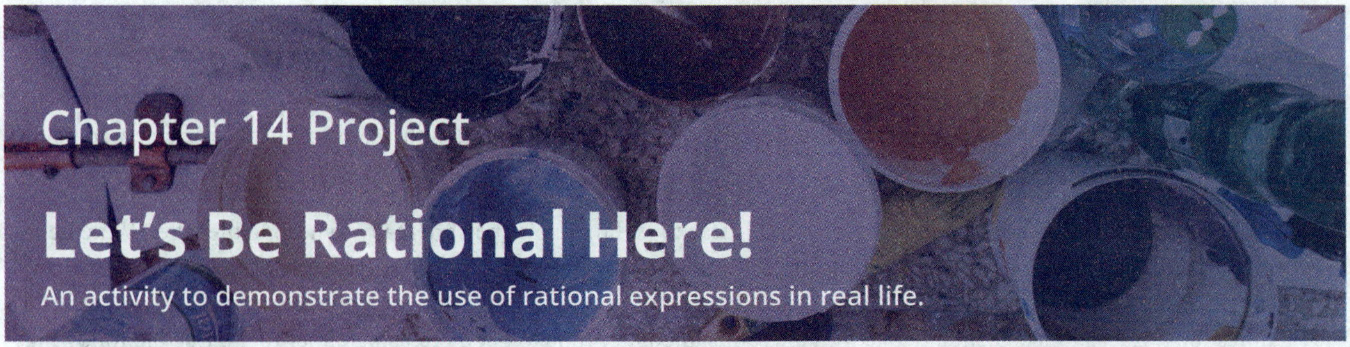

Chapter 14 Project

Let's Be Rational Here!

An activity to demonstrate the use of rational expressions in real life.

You may be surprised to learn how many different situations in real life involve working with rational expressions and rational equations. Hopefully after spending the day with Meghan and her friends you'll be convinced of their importance.

It's Saturday and Meghan has a list of things to accomplish today: revise her budget, paint the walls in the spare bedroom, travel to the lake with her friends and then water ski on the lake for the rest of the day, provided the weather stays nice.

1. Meghan decides to tackle the budget first. After reviewing her budget, she decides that she needs to get a part-time job to earn some extra money. She is remodeling the living room and would like to buy some new furniture. Letting x represent her new monthly combined salary, she estimates that $\frac{1}{4}$ of her new salary will be used for bills and approximately $\frac{1}{5}$ for her car payment. She would like to have \$1100 left over each month, of which \$100 will be saved for the new furniture. What must her new monthly salary be?

   a. Let the variable x represent Meghan's new monthly salary. Write an expression to represent the amount of her new salary used for bills. (Remember that the word **of** implies multiplication.)

   b. Write an expression to represent the amount of Meghan's new monthly salary used for her car payment.

   c. Write an equation that sums the expenditures and leftover balance, and set this sum equal to the new monthly salary of x.

   d. Find the LCD of the rational expressions from part c. and multiply it by each term in the equation to remove the fractions.

   e. Solve the equation from part d. to determine what Meghan's new monthly salary needs to be.

   f. Approximately how much of Meghan's new salary will be used to pay her car payment?

2. With the budget done, Meghan prepares to paint the spare bedroom. She painted her bedroom last week and it took her about 4 hours. Her housemate Ashley painted her bedroom a couple of weeks ago and it took her 6 hours. All the bedrooms are similar in size. Meghan realizes that if she gets Ashley to help her with the painting, it will take them less time and they can get to the lake sooner. How long will it take Meghan and Ashley working together to paint the spare bedroom? Use the table below to help you set up the problem.

Person	Time (in hours)	Part of Work Done in 1 Hour
Meghan	4	$\frac{1}{4}$
Ashley	6	$\frac{1}{6}$
Together	x	$\frac{1}{x}$

a. Fill in the missing information in column three of the table.Use the entries in the last column of the table to set up an equation to represent the sum of the amount of work done by both Meghan and Ashley in an hour.

b. Solve the equation to determine how long it will take the two girls to paint the spare bedroom when working together. Express the result in hours as a decimal rounded to the nearest tenth. Convert this measurement to hours and minutes.

3. After the spare bedroom is painted, the girls call Lucas to let him know they are ready to head to the lake. While he is preparing the boat for the lake, Lucas tries to decide which route they should travel. If he travels the highway, he can travel 20 mph faster than the scenic route. However, the highway is 30 miles longer than the scenic route, which is 60 miles long. Lucas thinks it should take him the same amount of time to get there using either route. Use the following table to help you organize the information for this problem.

	Distance (miles)	÷	Rate (mph)	=	Time (hours)
Highway	90		$x + 20$		$\frac{90}{x+20}$
Scenic Route	60		x		$\frac{60}{x}$

a. Fill in the missing information in column four of the table. (**Hint:** Recall that the formula that involves distance, rate, and time is $d = r \cdot t$.)

b. Since Lucas expects the time to be the same for each route, create an equation from the last column and solve for x.

c. How fast must Lucas travel on the highway to get to the lake in the same amount of time as traveling the scenic route?

Chapter 14 Project: Let's Be Rational Here!

1. a. $\frac{1}{4}x$

b. $\frac{1}{5}x$

c. $\frac{1}{4}x + \frac{1}{5}x + 1100 = x$

d. LCD = 20; $5x+4x+22000=20x$

e. \$2000

f. \$400

2. b. $\frac{1}{4} + \frac{1}{6} = \frac{1}{x}$

c. 2.4 hours; 2 hours 24 minutes

3. b. $\frac{90}{x+20} = \frac{60}{x}$

c. 60 mph

Chapter 14 Project

You Are What You Eat

An activity to demonstrate rational expressions in real life.

When you eat something, your body immediately reacts and responds to the food in a variety of different ways. For example, signals are sent to your brain from your stomach to help indicate when you're full and don't need any more food. Similarly, the central nervous system will calm your body down to indicate a state of relaxation and safety while eating. Amongst thousands of tiny operations your body does while eating, one such action your body takes is regulating the state of acidity inside the mouth based on the food that enters it. We are going to look at two common states our body experiences: eating food and taking an aspirin. Acidity is measured using a scale called pH.

This pH in a human can be determined by the formula $\text{pH}=\frac{20.4x}{x^2+36}+6.5,$ where x is the number of minutes that have passed since the food has been eaten. Normally, the body has a pH of about 7.35 to 7.45 on a scale of 0 to 14.

1. Simplify the equation so that the pH equation is expressed as a single rational expression.
2. Using your simplified equation, determine the acid level after 25 minutes. Round to the nearest hundredth.
3. Assume the pH in a human can be determined by the formula $\text{pH}=\frac{20.4x}{x^2+36}+6.5,$ but the formula for the pH in a rat can be found by the formula $\text{pH}=\frac{0.3(x^2-2x)}{x^3+x^2-6x}+6.5.$
   a. Simplify the equation for rat pH so that the equation is expressed as a single rational expression.
   b. Divide the pH formula for a rat by the pH formula for a human. Write the resulting expression as a product of two polynomials in the numerator and the denominator.
4. Does it take longer for the pH level of a rat or a human to level out and normalize? (**Hint:** Try graphing the equations using a graphing calculator or desmos.com/calculator.)
   a. Explain how you arrived at your answer.
   b. Hypothesize what this conclusion means for the pH level of a human and the pH level of a rat.

When an aspirin is taken, there is a concentration of the medication that can be found in the blood. The blood concentration of aspirin brand A is represented by $f(x)=\frac{2x}{3x^2-4x+5}$ where x is in hours. The concentration of aspirin brand B is represented by $g(x)=\frac{x}{2x^2-4x+10}$ where x is in hours.

5. Add the two functions together to find the new function that represents both aspirins in the bloodstream simultaneously. What is the new function? Make sure to distribute and simplify your answer as much as possible.
6. Which of the following three values is greatest? Explain your answer and show your work for each calculation. Round each value to the nearest hundredth.
   a. The concentration of aspirin brand A in the blood after 3 hours
   b. The concentration of aspirin brand B in the blood after 3 hours
   c. The concentration of both aspirin in the blood after 6 hours

Chapter 14 Project: You Are What You Eat

1. $\text{pH} = \dfrac{6.5x^2 + 20.4x + 234}{x^2 + 36}$

2. $\text{pH} = \dfrac{6.5(25)^2 + 20.4(25) + 234}{(25)^2 + 36} = 7.27$

3. a. $\text{pH} = \dfrac{0.3(x^2 - 2x) + 6.5\ (x^3 + x^2 - 6x)}{x^3 + x^2 - 6x} = \dfrac{6.5x^3 + 6.8x^2 - 39.6x}{x^3 + x^2 - 6x}$

   b. $\dfrac{(6.5x^3 + 6.8x^2 - 39.6x)(x^2 + 36)}{(x^3 + x^2 - 6x)(6.5x^2 + 20.4x + 234)}$

4. a. The rat pH takes a longer time to normalize. By plugging in larger and larger numbers for x, we can see what happens to the function over longer periods of time. The function for the human appears to hit a pH level of 7 around 40 minutes after eating. For a rat, the pH level appears to flatten off at 6.5.

   b. Answers will vary. Example: For a human, the pH can fluctuate between 6.5 and 7.5, but for a rat, it appears as though the pH level is lower, tends to stay around 6.5, and will not get much higher. Perhaps a rat has more acidity in its saliva and stomach acid due to the contents of food it may consume.

5. $f(x) + g(x) = \dfrac{7x^3 - 12x^2 + 25x}{6x^4 - 20x^3 + 56x^2 - 60x + 50}$

6. a. $f(3) = 0.30$

   b. $g(3) \approx 0.19$

   c. $f(3) + g(3) \approx 0.24$; The concentration of aspirin brand A is the highest. Aspirin brand B dissolves more rapidly than aspirin brand A.

CHAPTER 15

15

Roots, Radicals, and Complex Numbers

Connections

While roots and radicals may not be used in everyday situations such as grocery shopping, they are used in a variety of applications that relate to our day to day lives. One such application is the measurement of distances. In 1923, N. Korzenewsky, the leader of an expedition in the deserts of Turkestan, recorded seeing snow-capped mountains 750 km (466 miles) away. A similar but more reliable recorded observation was made in 1933 by Commander C. L. Garner of the Coast and Geodetic Survey. He wrote that instrumental measurements were made in both directions "between Mt. Shasta and Mt. St. Helena in California," which is a distance of 192 miles.

When you watch a sunset over the ocean, do you know how far you are from the horizon? On a clear day, a person looking over the ocean can see an island on the horizon up to 80 miles away (which is approximately the distance from San Diego, California, to San Clemente Island). However, the visible distance varies due to the clarity and temperature of the air. A general rule of thumb for calculating the distance in miles between you and the horizon is the formula $d = 1.32\sqrt{h}$, where h is the height in feet that your eyes are above the ground.

Suppose you and a group of friends are watching the sunset from the roof of a two-story building. One friend wonders how far away the horizon is. You know that your eyes are approximately 28 feet above the ground. How can you use this information to estimate the distance between you and the horizon?

Objectives

A. Evaluate square roots.

B. Evaluate cube roots.

C. Use a calculator to evaluate square and cube roots.

15.1 Evaluating Radicals

You should be familiar with the concept of **square roots** and the square root symbol (or **radical sign** $\sqrt{\ }$) from the discussions of square roots and the Pythagorean Theorem in Chapter 6. For example, $\sqrt{3}$ represents the square root of 3 and is the number whose square is 3.

A Square Roots

A number is **squared** when it is multiplied by itself. For example,

$$6^2 = 6 \cdot 6 = 36 \quad \text{and} \quad (-1.5)^2 = (-1.5)(-1.5) = 2.25.$$

If an integer is squared, the result is called a **perfect square.** The squares for the integers from 1 to 20 are shown in Table 1 for easy reference.

Squares of Integers from 1 to 20 (Perfect Squares)

Integers (n)	1	2	3	4	5	6	7	8	9	10
Squares (n^2)	1	4	9	16	25	36	49	64	81	100

Integers (n)	11	12	13	14	15	16	17	18	19	20
Squares (n^2)	121	144	169	196	225	256	289	324	361	400

Table 1

Now, we want to reverse the process of squaring. That is, given a number, we want to find a number that, when squared, will result in the given number. This is called **finding a square root** of the given number. In general,

if $b^2 = a$, then b is a square root of a.

For example,

- because $5^2 = 25$, 5 is a **square root** of 25, and we write $\sqrt{25} = 5$.
- because $9^2 = 81$, 9 is a **square root** of 81, and we write $\sqrt{81} = 9$.

Radical Terminology

The symbol $\sqrt{\ }$ is called a **radical sign**.

The number under the radical sign is called the **radicand**.

The complete expression, such as $\sqrt{64}$, is called a **radical** or **radical expression**.

DEFINITION

History of the Square Root

The idea of the square root was known by the Babylonians around 1700 BC. They created a square crossed by two diagonals to calculate the value √2 using base 60 numbers. Instead of the value 1.414213562373095 that appears on a calculator, the Babylonians obtained a value of 1;24,51,10 in base 60. Their result (1.41421296) is correct to five decimal places. In addition, the Greeks knew that the square root of a number that is not a perfect square is an irrational number. This knowledge dates back to around 380 BC.

Every positive real number has two square roots, one positive and one negative. The positive square root is called the **principal square root**. For example,

- because $(8)^2 = 64, \sqrt{64} = 8.$ ← the **principal square root**
- because $(-8)^2 = 64, -\sqrt{64} = -8.$ ← the **negative square root**

The $\sqrt{}$ sign is understood to represent the **positive square root** (or the **principal square root**) and $-\sqrt{}$ represents the **negative square root**. The number 0 has only one square root, namely 0.

Square Root

If a is a nonnegative real number, then

$\sqrt{a}$ is the **principal square root** of a,

and

$-\sqrt{a}$ is the **negative square root** of a.

DEFINITION

Note

Square roots of negative numbers are not real numbers. For example, $\sqrt{-4}$ is not a real number because there is no number times itself that results in the negative number −4. Consider how $2 \cdot 2 = 4$ and $(-2) \cdot (-2) = (4)$, these both result in a positive number. Later, we will discuss a special type of number called a complex number that can result in a negative number when multiplied by itself. These numbers are discussed in Sections 15.8 and 15.9.

Example 1 Evaluating Square Roots

a. Because $6^2 = 36$,

we have $\sqrt{36} = 6$ The **principal square root** of 36

and $-\sqrt{36} = -6.$ The **negative square root** of 36

b. Because $11^2 = 121$, we have $\sqrt{121} = 11$ and $-\sqrt{121} = -11$.

c. Because $0^2 = 0$, $\sqrt{0^2} = \sqrt{0} = 0$. (**Note:** This logic also applies to −0 since −0 = 0.)

d. $\sqrt{-25}$ is not a real number.

Now work margin exercise 1.

1. If real square roots exist for the following numbers, state them.

a. 81

b. 196

c. 49

d. −36

Example 2 Evaluating Square Roots

a. Because $\left(\frac{4}{5}\right)^2 = \frac{16}{25}$, we know that $\sqrt{\frac{16}{25}} = \frac{4}{5}$.

b. $-\sqrt{0.0009} = -0.03$ because $(-0.03)^2 = 0.0009$.

Now work margin exercise 2.

2. Evaluate the square roots.

a. $\sqrt{\frac{9}{64}}$

b. $-\sqrt{0.0016}$

Recall that **rational numbers** are numbers that can be expressed as the quotient of integers and are of the form $\frac{a}{b}$ where a and b are integers and $b \neq 0$. In decimal form, rational numbers are of the form of terminating decimals or repeating infinite decimals. As illustrated in Example 2, square roots of squared rational numbers are themselves rational numbers.

No rational number will square to give 2. So, $\sqrt{2}$ is an **irrational number** and a decimal representation is an infinite nonrepeating decimal. Your calculator will show the following.

$\sqrt{2} = 1.414213562...$ Accurate to 9 decimal places

$\sqrt{2} \approx 1.4142$ Rounded to the nearest ten-thousandth

To get a better idea of $\sqrt{2}$, we can compare as follows.

$1 < 2 < 4$ 1 and 4 are perfect squares.

and $\sqrt{1} < \sqrt{2} < \sqrt{4}$

which gives $1 < \sqrt{2} < 2$

We see that $\sqrt{2}$ is indeed between 1 and 2, and the value 1.4142 is reasonable.

3. The square root of 67 is approximately 8.1854. State why this is or is not a reasonable estimate.

Example 3 Estimating Square Roots

A calculator will give $\sqrt{30} \approx 5.4772$ rounded to the nearest ten-thousandth. Check that this is a reasonable estimate.

Solution

Because $25 < 30 < 36$, we have $\sqrt{25} < \sqrt{30} < \sqrt{36}$ and $5 < \sqrt{30} < 6$.

The approximation 5.4772 is between 5 and 6 and is reasonable.

Another approach is to square as follows.

$(5.4772)^2 = 29.99971984$, which is close to 30.

Now work margin exercise 3.

B Cube Roots

A number is **cubed** when it is used as a factor 3 times. For example,

$5^3 = 5 \cdot 5 \cdot 5 = 125$ and $(-30)^3 = (-30) \cdot (-30) \cdot (-30) = -27{,}000$.

If an integer is cubed, the result is called a **perfect cube**. The cubes for the integers from 1 to 10 are shown here for easy reference.

Perfect Cubes from 1 to 1000

Integer (n)	1	2	3	4	5	6	7	8	9	10
Cube (n^3)	1	8	27	64	125	216	343	512	729	1000

Table 2

The reverse of cubing is finding the **cube root**, symbolized $\sqrt[3]{\ }$. For example, because $5^3 = 125$, we know that the cube root of 125 is 5 and we write $\sqrt[3]{125} = 5$.

Cube Root

If a is a real number, then $\sqrt[3]{a}$ is the **cube root** of a.

DEFINITION

Example 4 Evaluating Cube Roots

a. Because $2^3 = 8$, $\sqrt[3]{8} = 2$.

b. Because $(-6)^3 = -216$, $\sqrt[3]{-216} = -6$.

The cube root of a negative number is a real number and is negative.

c. Because $\left(\frac{1}{3}\right)^3 = \frac{1}{27}$, $\sqrt[3]{\frac{1}{27}} = \frac{1}{3}$.

Now work margin exercise 4.

4. Evaluate the following radical expressions.

a. $\sqrt[3]{64}$

b. $\sqrt[3]{-125}$

c. $\sqrt[3]{\frac{1}{1000}}$

Attention!

In the cube root expression $\sqrt[3]{a}$, the number 3 is called the **index**. In a square root expression such as $\sqrt{a}$, the index is understood to be 2 and **is not written**. Expressions with square roots and cube roots (as well as other roots) are called **radical expressions.**

C Using a Graphing Calculator to Evaluate Expressions with Radicals

A TI-84 Plus graphing calculator can be used to find decimal approximations for radicals and expressions containing radicals. The displays in Example 5 illustrate the advantage of being able to see the entire expression being evaluated.

Example 5 Evaluating Radical Expressions with a Calculator

The following radical expressions are evaluated by using a TI-84 Plus graphing calculator. In each example, the steps (or keys to press) are shown. The TI-84 Plus gives answers rounded to at most ten decimal places. You may choose (through the [MODE] key) to have answers rounded to fewer places.

a. $\sqrt{17}$ **b.** $\sqrt[3]{100}$ **c.** $3\sqrt{20}$

5. Evaluate the following radical expressions with a calculator.

a. $\sqrt{45}$

b. $\sqrt[3]{50}$

c. $5\sqrt{30}$

Solution

a. $\sqrt{17}$

Step 1: Press [2nd] [x^2] to get the square root symbol $\sqrt{\ }$.

Step 2: Enter [1] [7] and the right-hand parenthesis [ )]. (**Note:** When the $\sqrt{\ }$ symbol appears, it will appear with a left-hand parenthesis. You should press the right-hand parenthesis to close the square root operation.)

Step 3: Press [ENTER].

The display will appear as follows.

```
√(17)
         4.123105626
```

b. $\sqrt[3]{100}$

Step 1: Press [3] [MATH] [5]. (**Note:** The [3] followed by [MATH] [5] indicates the cube root to the calculator. That is, the 3 takes the place of x in the expression $\sqrt[x]{\ }$.)

Step 2: Enter the left-hand parenthesis [(], then [1] [0] [0], and then finally the right hand parenthesis [)].

Step 3: Press [ENTER].

The display will appear as follows.

```
3*√(100)
         4.641588834
```

c. $3\sqrt{20}$ **Note:** This expression represents 3 times $\sqrt{20}$.

Step 1: Enter [3].

Step 2: Press [2nd] [x^2]. (This gives the $\sqrt{\ }$ symbol.)

Step 3: Enter [2] [0] and the right-hand parenthesis [)].

Step 4: Press [ENTER].

The display will appear as follows.

```
3√(20)
         13.41640786
```

Now work margin exercise 5.

Margin Exercise Answers

1. a. −9, 9 **b.** −14, 14 **c.** −7, 7 **d.** $\sqrt{-36}$ is not a real number **2. a.** $\frac{3}{8}$ **b.** −0.04 **3.** Because 64 < 67 < 81, we have $\sqrt{64} < \sqrt{67} < \sqrt{81}$ and $8 < \sqrt{67} < 9$. The approximation 8.1854 is between 8 and 9 and is reasonable. **4. a.** 4 **b.** −5 **c.** $\frac{1}{10}$ **5. a.** 6.708203932 **b.** 3.684031499 **c.** 27.38612788

15.1 Exercises

Concept Check

Fill-in-the-Blank. Complete each sentence using information found in this section.

1. In a radical expression that appears to have no index, the index is understood to be ___.
2. When a number is multiplied by itself, the product is said to be that number's ______.
3. To reverse the process of squaring, find the ______ ______ of the number.
4. If a is a nonnegative real number, then $\sqrt{a}$ is the ________ square root of a.
5. In $\sqrt{102}$, the symbol $\sqrt{\ }$ is called the _______ sign and 102 is the ________.
6. Square roots of negative numbers are not ____ numbers.

True/False. Determine whether each statement is true or false. If a statement is false, explain how it can be changed so the statement will be true. (**Note:** There may be more than one acceptable change.)

7. If a number is squared and the principal square root of the result is found, that square root is always equal to the original number.
8. There is no real number that can be a square root of a negative number.
9. The index is the number underneath the radical sign.
10. The cube root of –27 is a real number.

Practice

Simplify the following square roots and cube roots. See Examples 1, 2, and 4.

1. $\sqrt{9}$
2. $\sqrt{49}$
3. $\sqrt{81}$
4. $\sqrt{36}$
5. $\sqrt{289}$
6. $\sqrt{121}$
7. $\sqrt{169}$
8. $\sqrt{361}$
9. $\sqrt[3]{1}$
10. $\sqrt[3]{1000}$
11. $\sqrt[3]{125}$
12. $\sqrt[3]{343}$
13. $\sqrt[3]{216}$
14. $\sqrt[3]{512}$
15. $\sqrt{\frac{1}{4}}$
16. $\sqrt{\frac{9}{16}}$
17. $\sqrt[3]{\frac{27}{64}}$
18. $\sqrt[3]{\frac{1}{8}}$
19. $\sqrt{0.04}$
20. $\sqrt{0.0081}$
21. $-\sqrt{100}$
22. $-\sqrt{144}$
23. $-\sqrt{0.0016}$
24. $-\sqrt{0.000004}$
25. $\sqrt[3]{-27}$
26. $\sqrt[3]{-64}$
27. $\sqrt[3]{-125}$
28. $\sqrt[3]{729}$

15.1 Exercises

Concept Check

1. 2
2. (perfect) square
3. square root
4. principal
5. radical, radicand
6. real
7. False; If the original number is negative, the principal square root will not be the same as the original number.
8. True
9. False; The radicand is underneath the radical symbol.
10. True

Practice

1. 3
2. 7
3. 9
4. 6
5. 17
6. 11
7. 13
8. 19
9. 1
10. 10
11. 5
12. 7
13. 6
14. 8
15. $\frac{1}{2}$
16. $\frac{3}{4}$
17. $\frac{3}{4}$
18. $\frac{1}{2}$
19. 0.2
20. 0.09

21. -10
22. -12
23. -0.04
24. -0.002
25. -3
26. -4
27. -5
28. 9
29. $\frac{3}{5}$
30. $\frac{5}{9}$
31. $\sqrt{64} < \sqrt{74} < \sqrt{81}$ and $8 < \sqrt{74} < 9$ because $64 < 74 < 81$ or $(8.6023)^2 = 73.99956529$
32. $\sqrt{16} < \sqrt{18} < \sqrt{25}$ and $4 < \sqrt{18} < 5$ because $16 < 18 < 25$ or $(4.2426)^2 = 17.99965476$
33. $\sqrt{25} < \sqrt{32} < \sqrt{36}$ and $5 < \sqrt{32} < 6$ because $25 < 32 < 36$ or $(5.6569)^2 = 32.00051761$
34. $\sqrt{100} < \sqrt{110} < \sqrt{121}$ and $10 < \sqrt{110} < 11$ because $100 < 110 < 121$ or $(10.4881)^2 = 110.00024161$
35. >
36. <
37. <
38. <
39. =
40. =
41. =
42. >
43. Rational
44. Irrational
45. Rational
46. Rational
47. Irrational
48. Rational

29. $\sqrt{\frac{9}{25}}$

30. $\sqrt{\frac{25}{81}}$

Estimates of radicals are given (rounded to the nearest ten-thousandth). Show that these are reasonable estimates. See Example 3.

31. $\sqrt{74} \approx 8.6023$

32. $\sqrt{18} \approx 4.2426$

33. $\sqrt{32} \approx 5.6569$

34. $\sqrt{110} \approx 10.4881$

In each of the following problems, determine the symbol, <, >, or =, that makes the statement true.

35. $\sqrt{16}$ ___ $\sqrt[3]{27}$

36. $\sqrt[3]{64}$ ___ $\sqrt{125}$

37. $\sqrt{36}$ ___ $\sqrt[3]{343}$

38. $\sqrt[3]{125}$ ___ $\sqrt{64}$

39. $\sqrt{4}$ ___ $\sqrt{4}$

40. $\sqrt{4}$ ___ $\sqrt[3]{8}$

41. $\sqrt[3]{343}$ ___ $\sqrt{49}$

42. $\sqrt{25}$ ___ $\sqrt[3]{27}$

Use your knowledge of square roots and cube roots to determine whether each number is rational, irrational, or not a real number.

43. $\sqrt{4}$

44. $\sqrt{17}$

45. $\sqrt{169}$

46. $\sqrt[3]{8}$

47. $\sqrt{\frac{2}{9}}$

48. $-\sqrt{\frac{1}{4}}$

49. $\sqrt{-36}$

50. $\sqrt[3]{-27}$

51. $-\sqrt[3]{125}$

52. $\sqrt{-10}$

53. $\sqrt{1.68}$

54. $\sqrt{5.29}$

▦ Use a calculator to find the value of each radical expression rounded to the nearest ten-thousandth. See Example 5.

55. $\sqrt{39}$

56. $\sqrt{150}$

57. $\sqrt{6.23}$

58. $\sqrt{9.6}$

59. $\sqrt{\frac{1}{5}}$

60. $\sqrt{\frac{3}{8}}$

61. $4\sqrt{5}$

62. $6\sqrt{3}$

63. $-2\sqrt{17}$

64. $-3\sqrt{6}$

65. $\sqrt[3]{18}$

66. $\sqrt[3]{26}$

67. $2\sqrt[3]{15}$

68. $3\sqrt[3]{11}$

Applications

Use the formula $s = \sqrt[3]{V}$, which relates the length of the sides of a cube and the volume V, to answer the following questions.

69. The volume of a puzzle cube is 250 cubic inches. What is the length of one side? Round your answer to the nearest hundredth.

70. Three cubic blocks of different volumes were stacked on top of each other. The top block was 216 cubic centimeters. The middle block was 343 cubic centimeters, and the bottom block was 512 cubic centimeters. How tall was the stack of blocks?

Solve.

71. The area of a square tile is 16 square inches.

a. How long are the sides of the square tile?

b. How many tiles would be needed for a four-foot-long and four-inch-high backsplash in a newly designed bathroom?

72. Margaret needs to put a new gutter on one side of her roof. The shape of her roof is made up of two right triangles that are on opposite sides of a square. If the area of the square is 3600 ft² and the base of each of the triangles is 80 ft, what is the total length of the gutter she'll need to replace?

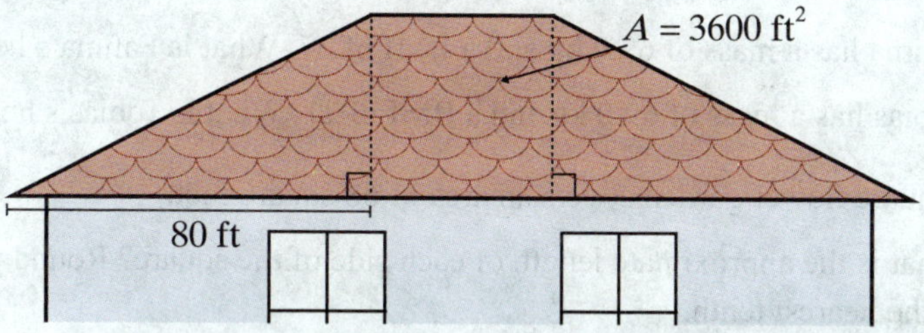

73. The volume of a volleyball is 523.33 cubic inches. Find the diameter of the ball using $\pi = 3.14$. Round your answer to the nearest inch. (**Note:** For a sphere, $V = \frac{4}{3}\pi r^3$.)

74. The volume of a child's building block is 64 cubic centimeters.

a. Assuming the building block is a perfect cube, find the length of each side of the block.

b. If a child stacks 5 blocks directly on top of each other, find the height of the structure that is created.

75. A pastry chef is making a batch of mini *petit fours*, which are little cakes, in the shape of cubes. To keep the nutritional value of each *petit four* consistent, the bakery manager wants each one to have a volume of 100 cm³. What should the side length be, to the nearest hundredth, for each *petit four*? (**Note:** For volume of a cube, $V = s^3$ where s = side length.)

49. Not a real number
50. Rational
51. Rational
52. Not a real number
53. Irrational
54. Rational
55. 6.2450
56. 12.2474
57. 2.4960
58. 3.0984
59. 0.4472
60. 0.6124
61. 8.9443
62. 10.3923
63. −8.2462
64. −7.3485
65. 2.6207
66. 2.9625
67. 4.9324
68. 6.6719

Applications

69. $5\sqrt{2}$ or 6.30 inches
70. 21 centimeters
71. **a.** 4 inches
b. 12 tiles
72. 220 ft
73. 10 inches
74. **a.** 4 cm
b. 20 cm
75. 4.64 cm

76. a. 0.64 seconds
b. 1.01 seconds
c. 1.09 seconds

77. a. 1.7 meters
b. 1.6 meters
c. 1.8 meters

78. a. 8.2 feet
b. 32.8 feet
c. $45.59

76. Isaac Newton fell asleep under an apple tree thinking about math. While he was sleeping, a squirrel knocked an apple off of a branch of the tree. The equation $t = \sqrt{\frac{2d}{9.8}}$ can be used to find the time t in seconds it takes for the apple to drop a certain distance d, where d is in meters. Round all answers to the nearest hundredth.

a. If the apple was connected to a branch 2 m above Newton's head, how long would it take before the apple hit Newton's head?

b. If the squirrel knocked a second apple off a branch that was 5 m above Newton's head, how long would it take before the apple hit Newton's head?

c. Suppose the second apple missed Newton's head and landed on the ground instead. If Newton's head was 0.8 m above the ground, how long would it take for the apple to hit the ground?

77. A person's Body Mass Index (BMI) is determined by the formula $B = \frac{m}{h^2}$ where B is the BMI, m is the person's mass in kilograms, and h is the person's height in meters. Having a BMI between 18.5 and 25 is considered optimal. To find a person's height based on their BMI and mass, the formula can be rearranged to $h = \sqrt{\frac{m}{B}}$. Round all answers to the nearest tenth.

a. Elias has a mass of 60.7 kg and a BMI of 21. What is Elias's height?

b. Fatima has a mass of 69.0 kg and a BMI of 27. What is Fatima's height?

c. Tobias has a mass of 69.0 kg and a BMI of 21. What is Tobias's height?

78. A square flower garden covers an area of 68 square feet.

a. What is the approximate length of each side of the square? Round your answer to the nearest tenth.

b. Use the answer from part a. to determine the amount of edging material needed to create a border around the flower garden.

c. The edging material costs $1.39 per foot. How much will the amount of edging material from part b. cost? Round your answer to the nearest cent.

79. Barbara's Bombtastic Bakery is installing a corner display stand for custom decorated cakes. The top of the display stand is designed in the shape of a right triangle, as shown.

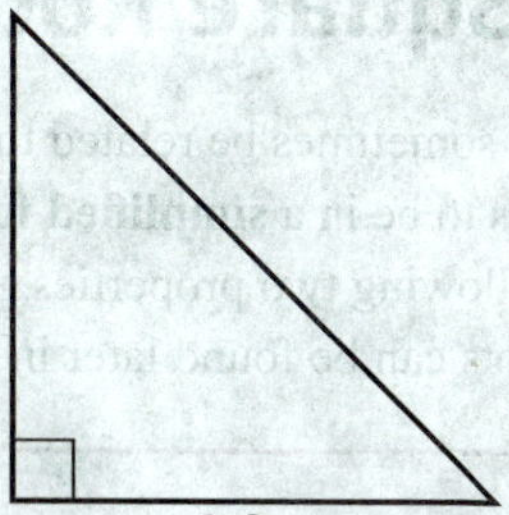

a. What is the length of the longest side of the display stand? Round your answer to the nearest tenth.

b. The top of the display has a decorative edge around all three sides to prevent the cakes from falling off. How much edging is required for the display?

c. If the top of the display stand is to be covered by square tiles that have a side length of 6 inches, how many tiles will be needed?

80. A glass company makes a paperweight in the shape of a cube that has a volume of 91.125 cubic inches.

a. What is the length of each side of the cube?

b. What is the area of the base of the cube?

c. What is the surface area of the cube?

Writing & Thinking

81. Discuss, in your own words, why the square root of a negative number is not a real number.

82. Discuss, in your own words, why the cube root of a negative number is a negative number.

79. **a.** 4.2 ft
b. 10.2 ft
c. 18 tiles

80. **a.** 4.5 inches
b. 20.25 square inches
c. 121.5 square inches

Writing & Thinking

81. There is no real number that results in a negative number when squared.

82. Cubing a negative real number is equivalent to multiplying a negative number by itself 3 times. The product of three negative numbers is negative.

Objectives

A. Simplify radical expressions that contain square roots.

B. Simplify square roots that contain variables.

C. Simplify radical expressions that contain cube roots.

15.2 Simplifying Radicals

A Simplifying Square Roots

Square roots and cube roots can sometimes be related to solutions of equations. If this is the case, we want the numbers to be in a **simplified form** for easier calculations and algebraic manipulations. The following two properties of square roots will help. Comments concerning simplifying cube roots can be found later in this section.

Properties of Square Roots

If a and b are **positive** real numbers, then

1. $\sqrt{ab} = \sqrt{a}\sqrt{b}$

2. $\sqrt{\frac{a}{b}} = \frac{\sqrt{a}}{\sqrt{b}}$

PROPERTIES

As an example, we know that $\sqrt{144} = 12$. However, in a situation where you may have forgotten this, you can proceed as follows using Property 1 of square roots.

$$\sqrt{144} = \sqrt{36 \cdot 4} = \sqrt{36} \cdot \sqrt{4} = 6 \cdot 2 = 12$$

Similarly, using Property 2, we can write the following.

$$\sqrt{\frac{49}{36}} = \frac{\sqrt{49}}{\sqrt{36}} = \frac{7}{6}$$

Simplest Form for Square Roots

A square root is considered to be in **simplest form** when the radicand has no perfect square as a factor.

DEFINITION

The number 200 is not a perfect square, and using a calculator we will find $\sqrt{200} \approx 14.1421$. Now to simplify $\sqrt{200}$, we can use Property 1 of square roots and any of the following three approaches.

Teaching Note

Illustrating three different approaches to get the same result should raise some interesting questions such as:

- Which one is better (or best)?
- Which one do you like?
- Which one is faster?
- Does it really make any difference?

Approach 1: Factor 200 as $4 \cdot 50$ because 4 is a perfect square.

$$\sqrt{200} = \sqrt{4 \cdot 50} = \sqrt{4} \cdot \sqrt{50} = 2\sqrt{50}$$

However, $2\sqrt{50}$ is **not in simplest form** because 50 has a perfect square factor, 25. Thus, to complete the process, we have the following

$$\sqrt{200} = 2\sqrt{50} = 2\sqrt{25 \cdot 2} = 2\sqrt{25} \cdot \sqrt{2} = 2 \cdot 5 \cdot \sqrt{2} = 10\sqrt{2}$$

Approach 2: Note that 100 is a perfect square factor of 200 and $200 = 100 \cdot 2$.

$$\sqrt{200} = \sqrt{100 \cdot 2} = \sqrt{100} \cdot \sqrt{2} = 10\sqrt{2}$$

Approach 3: Use prime factors.

$$\begin{aligned}\sqrt{200} &= \sqrt{2\cdot2\cdot2\cdot5\cdot5}\\ &= \sqrt{2\cdot2\cdot5\cdot5}\cdot\sqrt{2}\\ &= \sqrt{2\cdot2}\cdot\sqrt{5\cdot5}\cdot\sqrt{2}\\ &= 2\cdot5\cdot\sqrt{2}\\ &= 10\sqrt{2}\end{aligned}$$

(**Note:** Here you have a chance to see mathematics at work. You get to choose the approach that fits your way of thinking.)

> **Note**
>
> Of these three approaches, the second appears to be the easiest because it has the fewest steps. However, "seeing" the largest perfect square factor may be difficult. If you do not immediately see a perfect square factor, proceed by finding other factors or prime factors as illustrated.

Example 1 Simplifying Radical Expressions with Square Roots

Simplify each expression so that there are no perfect square factors in the radicand.

a. $\sqrt{48}$ **b.** $\sqrt{63}$ **c.** $\sqrt{\dfrac{75}{16}}$

Solution

a. $\sqrt{48} = \sqrt{16\cdot3} = \sqrt{16}\cdot\sqrt{3} = 4\sqrt{3}$ 16 is the largest perfect square factor.

b. $\sqrt{63} = \sqrt{9\cdot7} = \sqrt{9}\cdot\sqrt{7} = 3\sqrt{7}$ 9 is the largest perfect square factor.

c. $\sqrt{\dfrac{75}{16}} = \dfrac{\sqrt{75}}{\sqrt{16}} = \dfrac{\sqrt{25\cdot3}}{\sqrt{16}} = \dfrac{\sqrt{25}\cdot\sqrt{3}}{\sqrt{16}} = \dfrac{5\sqrt{3}}{4}$

Now work margin exercise 1.

1. Simplify each expression so that there are no perfect square factors in the radicand.

a. $\sqrt{98}$

b. $\sqrt{45}$

c. $\sqrt{\dfrac{12}{25}}$

B Square Roots with Variables

To simplify square root expressions that contain variables, such as $\sqrt{x^2}$, we must be aware of whether the variable represents a positive real number $(x > 0)$, zero $(x = 0)$, or a negative number $(x < 0)$.

$$\text{If } x = 0, \text{ then } \sqrt{x^2} = \sqrt{0^2} = \sqrt{0} = 0 = x.$$

$$\text{If } x = 5, \text{ then } \sqrt{x^2} = \sqrt{5^2} = \sqrt{25} = 5 = x.$$

$$\text{But, if } x = -5, \text{ then } \sqrt{x^2} = \sqrt{(-5)^2} = \sqrt{25} = 5 \neq x.$$

$$\text{In fact, if } x = -5, \text{ then } \sqrt{x^2} = \sqrt{(-5)^2} = \sqrt{25} = 5 = |-5| = |x|.$$

Thus, simplifying square roots with variables involves more detailed analysis than simplifying square roots with only constants. The following definition indicates the correct way to simplify $\sqrt{x^2}$.

> **Square Root of x^2**
>
> If x is a real number, then $\sqrt{x^2} = |x|$.
>
> **Note:** If $x \geq 0$ is given, then we can write $\sqrt{x^2} = x$.
>
> DEFINITION

Although using the absolute value when simplifying square roots is correct mathematically, we can avoid some confusion by assuming that the variable under the radical sign represents only positive real numbers or 0. This eliminates the need for absolute value signs.

Therefore, for the remainder of this text, we will assume that $x > 0$ and write $\sqrt{x^2} = x$ in all square root expressions, unless specifically stated otherwise.

2. Simplify each of the following square roots. Assume that all variables represent positive real numbers.

a. $\sqrt{36z^2}$

b. $\sqrt{75b^2}$

c. $\sqrt{45c^2d^2}$

Example 2 Simplifying Square Roots with Variables

Simplify each of the following square roots. Assume that all variables represent positive real numbers. (Note that, by making this assumption, we do not need to be concerned about the absolute value sign.)

a. $\sqrt{16y^2}$ **b.** $\sqrt{72a^2}$ **c.** $\sqrt{12x^2y^2}$

Solution

a. $\sqrt{16y^2} = 4y$

b. $\sqrt{72a^2} = \sqrt{36a^2} \cdot \sqrt{2} = 6a\sqrt{2}$

c. $\sqrt{12x^2y^2} = \sqrt{4x^2y^2} \cdot \sqrt{3} = 2xy\sqrt{3}$

Now work margin exercise 2.

To find the square root of an expression with even exponents, divide the exponents by 2.

$$x^2 \cdot x^2 = x^4 \qquad a^3 \cdot a^3 = a^6 \qquad y^5 \cdot y^5 = y^{10}$$

$$\downarrow \qquad\qquad \downarrow \qquad\qquad \downarrow$$

$$\sqrt{x^4} = x^2 \qquad \sqrt{a^6} = a^3 \qquad \sqrt{y^{10}} = y^5$$

To find the square root of an expression with odd exponents, factor the expression into two terms, one with exponent 1 and the other with an even exponent. For example,

$$x^3 = x^2 \cdot x \quad \text{and} \quad y^9 = y^8 \cdot y,$$

which means that

$$\sqrt{x^3} = \sqrt{x^2 \cdot x} = \sqrt{x^2} \cdot \sqrt{x} = x\sqrt{x} \quad \text{and} \quad \sqrt{y^9} = \sqrt{y^8 \cdot y} = \sqrt{y^8} \cdot \sqrt{y} = y^4\sqrt{y}.$$

3. Simplify each of the following square roots. Assume that all variables represent positive real numbers.

a. $\sqrt{16x^8}$

b. $\sqrt{100x^3y^3}$

c. $\sqrt{12x^8y^{12}}$

d. $\sqrt{\dfrac{25z^{18}}{y^8}}$

Example 3 Simplifying Square Roots with Variables

Simplify each of the following square roots. Look for perfect square factors and even powers of the variables. Assume that all variables represent positive real numbers.

a. $\sqrt{81x^4}$ **c.** $\sqrt{18a^4b^6}$

b. $\sqrt{64x^5y}$ **d.** $\sqrt{\dfrac{9a^{13}}{b^4}}$

Solution

a. $\sqrt{81x^4} = 9x^2$ — The exponent 4 is divided by 2.

b. $\sqrt{64x^5y} = \sqrt{64x^4} \cdot \sqrt{xy} = 8x^2\sqrt{xy}$

c. $\sqrt{18a^4b^6} = \sqrt{9a^4b^6} \cdot \sqrt{2} = 3a^2b^3\sqrt{2}$ — Each exponent is divided by 2.

d. $\sqrt{\dfrac{9a^{13}}{b^4}} = \dfrac{\sqrt{9a^{13}}}{\sqrt{b^4}} = \dfrac{\sqrt{9a^{12}} \cdot \sqrt{a}}{\sqrt{b^4}} = \dfrac{3a^6\sqrt{a}}{b^2}$ — Recall that *a* and *b* are greater than 0, so the denominator cannot be undefined.

Now work margin exercise 3.

C Simplifying Cube Roots

When simplifying expressions with cube roots, we need to be aware of perfect cube numbers and variables with exponents that are multiples of 3. (Multiples of 3 are 3, 6, 9, 12, 15, and so on.) **Thus, exponents are divided by 3 in simplifying cube root expressions.** For example,

$$x^2 \cdot x^2 \cdot x^2 = x^6 \qquad a^3 \cdot a^3 \cdot a^3 = a^9 \qquad y^5 \cdot y^5 \cdot y^5 = y^{15}$$

$$\downarrow \qquad\qquad \downarrow \qquad\qquad \downarrow$$

and $\sqrt[3]{x^6} = x^2$ $\quad \sqrt[3]{a^9} = a^3$ $\quad \sqrt[3]{y^{15}} = y^5$.

Simplest Form for Cube Roots

A cube root is considered to be in **simplest form** when the radicand has no perfect cube as a factor.

DEFINITION

When finding cube roots, we need not be concerned about positive and negative values for variables because cube roots of negative numbers are defined to be negative. For example,

we have $(-2)^3 = -8$, and therefore $\sqrt[3]{-8} = -2$.

Similarly, $(-5)^3 = -125$, and therefore $\sqrt[3]{-125} = -5$.

Thus, $\sqrt[3]{x^3} = x$ whether $x \geq 0$ or $x < 0$.

Example 4 Simplifying Radical Expressions with Cube Roots

Simplify each of the following radical expressions. Look for perfect cube factors and powers of the variables that are multiples of 3.

a. $\sqrt[3]{54x^6}$ **b.** $\sqrt[3]{-40x^4y^{13}}$ **c.** $\sqrt[3]{250a^8b^{11}}$

Solution

a. $\sqrt[3]{54x^6} = \sqrt[3]{27x^6} \cdot \sqrt[3]{2} = 3x^2\sqrt[3]{2}$

27 is a perfect cube and the exponent 6 is divisible by 3.

4. Simplify each of the following radical expressions.

a. $\sqrt[3]{48z^3}$

b. $\sqrt[3]{-81a^8b^{12}}$

c. $\sqrt[3]{686x^6y^{11}}$

b. $\sqrt[3]{-40x^4y^{13}} = \sqrt[3]{-8x^3y^{12}} \cdot \sqrt[3]{5xy} = -2xy^4\sqrt[3]{5xy}$

8 is a perfect cube and the exponents on the variables are separated so that one exponent on each variable is divisible by 3.

c. $\sqrt[3]{250a^8b^{11}} = \sqrt[3]{125a^6b^9} \cdot \sqrt[3]{2a^2b^2} = 5a^2b^3\sqrt[3]{2a^2b^2}$

125 is a perfect cube and the exponents on the variables are separated so that one exponent on each variable is divisible by 3.

Now work margin exercise 4.

Margin Exercise Answers

1. a. $7\sqrt{2}$ **b.** $3\sqrt{5}$ **c.** $\frac{2\sqrt{3}}{5}$ **2. a.** $6z$ **b.** $5b\sqrt{3}$ **c.** $3cd\sqrt{5}$ **3. a.** $4x^4$ **b.** $10xy\sqrt{xy}$ **c.** $2x^4y^6\sqrt{3}$ **d.** $\frac{5z^9}{y^4}$ **4. a.** $2z\sqrt[3]{6}$ **b.** $-3a^2b^4\sqrt[3]{3a^2}$ **c.** $7x^2y^3\sqrt[3]{2y^2}$

15.2 Exercises

Concept Check

Fill-in-the-Blank. Complete each sentence using information found in this section.

1. A cube root is considered in simplest form when the radicand has no perfect cube as a/an ________.
2. When simplifying with cube roots, look for variables with exponents that are multiples of ___.
3. To find the square root of an expression with even exponents, divide the exponents by ___.
4. A square root is in simplest form when the radicand has no ________ ________ as a factor.
5. If a and b are positive real numbers, then $\sqrt{ab} =$ ____.
6. If a and b are positive real numbers, then $\sqrt{\frac{a}{b}} =$ ____.

True/False. Determine whether each statement is true or false. If a statement is false, explain how it can be changed so the statement will be true. (**Note:** There may be more than one acceptable change.)

7. Any variable term with an exponent of 5 has a perfect cube factor within that variable term.
8. The simplest form of a radical expression can be found by using prime factorization.
9. If x is a real number, then $\sqrt{x^2} = x$.
10. The term $7b\sqrt[3]{6c^2}$ is in simplified form.

15.2 Exercises

Concept Check

1. factor
2. 3
3. 2
4. perfect square
5. $\sqrt{a}\sqrt{b}$
6. $\frac{\sqrt{a}}{\sqrt{b}}$
7. True
8. True
9. False; If x is a real number, then $\sqrt{x^2} = |x|$.
10. True

Practice

1. $2\sqrt{3}$
2. $-3\sqrt{5}$
3. $12\sqrt{2}$
4. $-3\sqrt{7}$
5. $-6\sqrt{2}$
6. $7\sqrt{2}$
7. $-2\sqrt{14}$
8. $9\sqrt{2}$
9. $-5\sqrt{5}$
10. -11
11. $\frac{1}{2}$
12. $\frac{4\sqrt{2}}{7}$
13. $-\frac{\sqrt{11}}{8}$
14. $-\frac{\sqrt{5}}{2}$
15. $\frac{2\sqrt{7}}{5}$
16. $\frac{7\sqrt{3}}{10}$
17. $6x$
18. $7y$
19. $2x\sqrt{2x}$
20. $3a^2\sqrt{2a}$

Practice

Simplify each of the following radical expressions. Assume that all variables represent positive real numbers.

1. $\sqrt{12}$
2. $-\sqrt{45}$
3. $\sqrt{288}$
4. $-\sqrt{63}$
5. $-\sqrt{72}$
6. $\sqrt{98}$
7. $-\sqrt{56}$
8. $\sqrt{162}$
9. $-\sqrt{125}$
10. $-\sqrt{121}$
11. $\sqrt{\frac{1}{4}}$
12. $\sqrt{\frac{32}{49}}$
13. $-\sqrt{\frac{11}{64}}$
14. $-\sqrt{\frac{125}{100}}$
15. $\sqrt{\frac{28}{25}}$
16. $\sqrt{\frac{147}{100}}$
17. $\sqrt{36x^2}$
18. $\sqrt{49y^2}$
19. $\sqrt{8x^3}$
20. $\sqrt{18a^5}$
21. $\sqrt{24x^{11}y^2}$
22. $\sqrt{20x^{15}y^3}$
23. $\sqrt{125x^3y^6}$
24. $\sqrt{8x^5y^4}$
25. $-\sqrt{18x^2y^2}$
26. $-\sqrt{32x^4y^8}$
27. $\sqrt{12ab^2c^3}$
28. $\sqrt{45a^2b^3c^4}$
29. $\sqrt{75x^4y^6z^8}$
30. $\sqrt{200x^2y^2z^2}$
31. $\sqrt{\frac{5x^4}{9}}$
32. $-\sqrt{\frac{7y^6}{16x^4}}$
33. $\sqrt{\frac{32a^5}{81b^{16}}}$
34. $\sqrt{\frac{75x^8}{121y^{12}}}$
35. $\sqrt{\frac{200x^8}{289}}$
36. $\sqrt{\frac{32x^{15}y^{10}}{169}}$
37. $\sqrt[3]{216}$
38. $\sqrt[3]{1}$
39. $\sqrt[3]{56}$
40. $\sqrt[3]{72}$
41. $\sqrt[3]{-1}$
42. $\sqrt[3]{-125}$
43. $\sqrt[3]{-128}$
44. $\sqrt[3]{-250}$
45. $\sqrt[3]{125x^4}$
46. $\sqrt[3]{64a^{12}}$
47. $\sqrt[3]{-8x^8}$
48. $\sqrt[3]{-512a^5}$
49. $\sqrt[3]{72a^6b^4}$
50. $\sqrt[3]{108ab^9}$
51. $\sqrt[3]{216x^6y^5}$
52. $\sqrt[3]{64x^9y^2}$
53. $\sqrt[3]{24x^5y^7z^9}$
54. $\sqrt[3]{250x^6y^9z^{15}}$
55. $\frac{\sqrt[3]{81}}{6}$
56. $\frac{\sqrt[3]{192}}{10}$
57. $\sqrt[3]{\frac{375}{8}}$
58. $\sqrt[3]{\frac{-48}{125}}$
59. $\sqrt[3]{\frac{125y^{12}}{27x^6}}$
60. $\sqrt[3]{\frac{x^6z^3}{64y^9}}$

21. $2x^5y\sqrt{6x}$
22. $2x^7y\sqrt{5xy}$
23. $5xy^3\sqrt{5x}$
24. $2x^2y^2\sqrt{2x}$
25. $-3xy\sqrt{2}$
26. $-4x^2y^4\sqrt{2}$
27. $2bc\sqrt{3ac}$
28. $3abc^2\sqrt{5b}$
29. $5x^2y^3z^4\sqrt{3}$
30. $10xyz\sqrt{2}$
31. $\frac{x^2\sqrt{5}}{3}$
32. $-\frac{y^3\sqrt{7}}{4x^2}$
33. $\frac{4a^2\sqrt{2a}}{9b^8}$
34. $\frac{5x^4\sqrt{3}}{11y^6}$
35. $\frac{10x^4\sqrt{2}}{17}$
36. $\frac{4x^7y^5\sqrt{2x}}{13}$
37. 6
38. 1
39. $2\sqrt[3]{7}$
40. $2\sqrt[3]{9}$
41. -1
42. -5
43. $-4\sqrt[3]{2}$
44. $-5\sqrt[3]{2}$
45. $5x\sqrt[3]{x}$
46. $4a^4$
47. $-2x^2\sqrt[3]{x^2}$
48. $-8a\sqrt[3]{a^2}$
49. $2a^2b\sqrt[3]{9b}$
50. $3b^3\sqrt[3]{4a}$
51. $6x^2y\sqrt[3]{y^2}$
52. $4x^3\sqrt[3]{y^2}$

53. $2xy^2z^3\sqrt[3]{3x^2y}$

54. $5x^2y^3z^5\sqrt[3]{2}$

55. $\frac{\sqrt[3]{3}}{2}$

56. $\frac{2\sqrt[3]{3}}{5}$

57. $\frac{5\sqrt[3]{3}}{2}$

58. $-\frac{2\sqrt[3]{6}}{5}$

59. $\frac{5y^4}{3x^2}$

60. $\frac{x^2z}{4y^3}$

Applications

61. $\sqrt{6} \approx 2.45$ amperes

62. $\sqrt{3} \approx 1.73$ amperes

63. 120 volts

64. $50\sqrt{5} \approx 111.80$ volts

65. **a.** 10 cm

b. $2\sqrt{30}$ cm

c. $5\sqrt{6}$ cm

Writing & Thinking

66. When $a < 0$

67. A cube root has no restrictions as the cube root of a negative number is negative.

Applications

Use the following two formulas associated with electricity to answer Exercises 61–64.

$I = \sqrt{\frac{P}{R}}$	P = power (in watts)
	I = current (in amperes)
	E = voltage (in volts)
$E = \sqrt{PR}$	R = resistance (in ohms, Ω)

61. What is the current in amperes of a light bulb that produces 150 watts of power and has a 25 Ω resistance?

62. If a light bulb has a resistance of 30 Ω and produces 90 watts of power, what is its current in amperes?

63. How many volts of electricity would Meghan need to produce 48 Ω of resistance from a 300 watt lamp?

64. A 5000 Ω resistor is rated at 2.5 watts. What is the maximum voltage of electricity that should be connected across it?

65. A nut company is determining how to package their new type of party mix. The marketing department is experimenting with different-sized cans for the party mix packaging. The designers use the equation $r = \sqrt{\frac{V}{h\pi}}$ to determine the radius of the can for a certain height h and volume V. The company decides they want the can to have a volume of 1200π cm^3. Keep your answers in simplified radical form.

a. Find the radius of the can if the height is 12 cm.

b. Find the radius of the can if the height is 10 cm.

c. Find the radius of the can if the height is 8 cm.

Writing & Thinking

66. Under what conditions is the expression $\sqrt{a}$ not a real number?

67. Explain why the expression $\sqrt[3]{y}$ is a real number regardless of whether $y > 0$, $y < 0$, or $y = 0$.

15.3 Rational Exponents

A nth Roots

In Sections 15.1 and 15.2, we restricted our discussions to radicals involving square roots and cube roots. In this section, we will expand on those ideas by discussing radicals indicating n^{th} roots in general and how to relate radical expressions to expressions with rational (fractional) exponents. For example, the fifth root of x can be written in radical form as $\sqrt[5]{x}$ or with a fractional exponent as $x^{\frac{1}{5}}$.

To understand roots in general, consider the following notation. (**Note:** In this discussion, we assume that n is an integer greater than 1.)

Type of Root	Radical Notation and Exponential Notation	Example
Square roots	If $b = \sqrt{a}$, then $b = a^{\frac{1}{2}}$.	If $3 = \sqrt{9}$, then $3 = 9^{\frac{1}{2}}$.
Cube roots	If $b = \sqrt[3]{a}$, then $b = a^{\frac{1}{3}}$.	If $5 = \sqrt[3]{125}$, then $5 = 125^{\frac{1}{3}}$.
Fourth roots	If $b = \sqrt[4]{a}$, then $b = a^{\frac{1}{4}}$.	If $2 = \sqrt[4]{16}$, then $2 = 16^{\frac{1}{4}}$.
n^{th} roots	If $b = \sqrt[n]{a}$, then $b = a^{\frac{1}{n}}$.	

Table 1

For example,

- because $\sqrt[4]{16} = 2$, we can say that $2 = 16^{\frac{1}{4}}$, and
- because $\sqrt[5]{243} = 3$, we can say that $3 = 243^{\frac{1}{5}}$.

The following notation is used for all radical expressions.

Radical Notation

If n is an integer greater than 1, then $\sqrt[n]{a} = a^{\frac{1}{n}}$ (assuming $\sqrt[n]{a}$ is a real number).

The expression $\sqrt[n]{a}$ is called a **radical**.

The symbol $\sqrt[n]{}$ is called a **radical sign**.

n is called the **index**.

a is called the **radicand**.

Note: If no index is given, it is understood to be 2. For example, $\sqrt{3} = \sqrt[2]{3} = 3^{\frac{1}{2}}$.

DEFINITION

Objectives

A. Evaluate n^{th} roots.

B. Translate between radical expressions and expressions with rational exponents.

C. Use the properties of rational exponents to simplify expressions.

D. Use a graphing calculator to evaluate expressions with rational exponents.

Teaching Note

You might want to use the power rule (a little early here) to help justify how fractional exponents relate to roots as in $\left(\sqrt{a}\right)^2 = a$ and $\left(a^{\frac{1}{2}}\right)^2 = a^{\frac{1}{2}(2)} = a^1 = a$. Similarly, $\left(\sqrt[3]{a}\right)^3 = a$ and $\left(a^{\frac{1}{3}}\right)^3 = a^{\frac{1}{3}(3)} = a^1 = a$.

Note

Special Notes about the Index n

For the expression $\sqrt[n]{a}$ (or $a^{\frac{1}{n}}$) to be a real number (assuming n is an integer greater than 1):

1. when a is nonnegative, n can be any index, and
2. when a is negative, n must be odd.
(If a is negative and n is even, then $\sqrt[n]{a}$ is not a real number.)

1. Evaluate each nth root.

a. $36^{\frac{1}{2}}$

b. $16^{\frac{1}{4}}$

c. $(-27)^{\frac{1}{3}}$

d. $(0.0016)^{\frac{1}{4}}$

e. $(-25)^{\frac{1}{2}}$

Example 1 Evaluating n^{th} Roots

a. $49^{\frac{1}{2}} = \sqrt{49} = 7$, because $7^2 = 49$.

b. $81^{\frac{1}{4}} = \sqrt[4]{81} = 3$, because $3^4 = 81$.

c. $(-8)^{\frac{1}{3}} = \sqrt[3]{-8} = -2$, because $(-2)^3 = -8$.

d. $(0.00001)^{\frac{1}{5}} = \sqrt[5]{0.00001} = 0.1$, because $(0.1)^5 = 0.00001$.

e. $(-16)^{\frac{1}{2}} = \sqrt{-16}$ is not a real number. Any even root of a negative number is not a real number.

Now work margin exercise 1.

Math Tip

An alternate way to solve Example 1 part d. is by considering the place value of a decimal number. Since 0.00001 is one *hundred-thousandth*, we can rewrite it as a fraction and solve as follows.

$$(0.00001)^{\frac{1}{5}} = \left(\frac{1}{100,000}\right)^{\frac{1}{5}} = \left(\frac{1^5}{10^5}\right)^{\frac{1}{5}} = \frac{1}{10} = 0.1$$

B Rational Exponents of the Form $a^{\frac{m}{n}}$

Earlier, we discussed the rules of exponents using only integer exponents. These same rules of exponents apply to rational exponents (fractional exponents) as well, and are repeated here for easy reference.

Summary of the Rules for Exponents

For nonzero real numbers a and b and rational numbers m and n,

1. The exponent 1: $a^1 = a$ (a is any real number.)
2. The exponent 0: $a^0 = 1$ $(a \neq 0)$
3. The product rule: $a^m \cdot a^n = a^{m+n}$
4. The quotient rule: $\dfrac{a^m}{a^n} = a^{m-n}$
5. Negative exponents: $a^{-n} = \dfrac{1}{a^n}$, $\dfrac{1}{a^{-n}} = a^n$
6. Power rule: $(a^m)^n = a^{mn}$
7. Power of a product: $(ab)^n = a^n b^n$
8. Power of a quotient: $\left(\dfrac{a}{b}\right)^n = \dfrac{a^n}{b^n}$

PROPERTIES

Teaching Note

Highlight the fact that m and n are not restricted to being positive integers—they may be negative and they may be rational numbers. Another very handy property of exponents (due to the commutative property of multiplication) is $(a^m)^n = (a^n)^m$. For example, $(81^3)^{\frac{1}{4}}$ is much easier to simplify if written as $\left(81^{\frac{1}{4}}\right)^3$.

Now, consider the problem of evaluating the expression $8^{\frac{2}{3}}$ where the exponent, $\frac{2}{3}$, is of the form $\frac{m}{n}$. Using the power rule for exponents, we can write

$$8^{\frac{2}{3}} = \left(8^{\frac{1}{3}}\right)^2 = (2)^2 = 4 \quad \text{or} \quad 8^{\frac{2}{3}} = \left(8^2\right)^{\frac{1}{3}} = (64)^{\frac{1}{3}} = 4.$$

The result is the same with either approach. That is, we can take the cube root first and then square the answer. Or we can square first and then take the cube root. In general, for an exponent of the form $\frac{m}{n}$, taking the nth root first and then raising this root to the exponent m is easier because the numbers are smaller.

For example,

$$81^{\frac{3}{4}} = \left(81^{\frac{1}{4}}\right)^3 = (3)^3 = 27$$

is easier to calculate and work with than

$$81^{\frac{3}{4}} = \left(81^3\right)^{\frac{1}{4}} = (531,441)^{\frac{1}{4}} = 27.$$

The fourth root of 81 is more commonly known than the fourth root of 531,441.

The General Form $a^{\frac{m}{n}}$

If n is an integer greater than 1, m is any integer, and $a^{\frac{1}{n}}$ is a real number, then

$$a^{\frac{m}{n}} = \left(a^{\frac{1}{n}}\right)^m = \left(a^m\right)^{\frac{1}{n}}.$$

In radical notation: $$a^{\frac{m}{n}} = \left(\sqrt[n]{a}\right)^m = \sqrt[n]{a^m}$$

DEFINITION

Example 2 Converting from Exponential Notation to Radical Notation

Assume that each variable represents a positive real number. Each expression is changed to an equivalent expression in either radical or exponential notation.

a. $x^{\frac{2}{3}} = \sqrt[3]{x^2}$ The index, 3, is the denominator in the rational exponent.

b. $3x^{\frac{4}{5}} = 3\sqrt[5]{x^4}$ The coefficient, 3, is not affected by the exponent.

c. $-a^{\frac{3}{2}} = -\sqrt{a^3}$ −1 is the understood coefficient.

d. $\sqrt[6]{a^5} = a^{\frac{5}{6}}$ The index, 6, is the denominator of the rational exponent.

e. $5\sqrt{x} = 5x^{\frac{1}{2}}$ In a square root, the index is understood to be 2.

f. $-\sqrt[3]{4} = -4^{\frac{1}{3}}$ The coefficient, −1, is not affected by the exponent.
We could also write $-4^{\frac{1}{3}} = -1 \cdot 4^{\frac{1}{3}}$.

Now work margin exercise 2.

2. Convert each expression to an equivalent radical or exponential form. Assume that all variables represent positive real numbers.

a. $x^{\frac{2}{5}}$

b. $8z^{\frac{6}{7}}$

c. $-b^{\frac{5}{4}}$

d. $\sqrt[7]{x^5}$

e. $3\sqrt{s}$

f. $-\sqrt[4]{5}$

Attention!

For the remainder of this chapter, we will assume that all variables represent positive real numbers unless otherwise stated.

C Simplifying Expressions with Rational Exponents

Expressions with rational exponents, such as

$$x^{\frac{2}{3}} \cdot x^{\frac{1}{6}}, \quad \frac{x^{\frac{3}{4}}}{x^{\frac{1}{3}}}, \quad \text{and} \quad \left(2a^{\frac{1}{4}}\right)^3,$$

can be simplified using the rules for exponents.

3. Simplify each expression using one or more of the rules for exponents.

a. $x^{\frac{1}{3}} \cdot x^{\frac{1}{4}}$

b. $\dfrac{a^{\frac{4}{9}}}{a^{\frac{2}{3}}}$

c. $\left(3b^{\frac{1}{5}}\right)^4$

d. $\left(64z^{-\frac{4}{9}}\right)^{-\frac{1}{2}}$

e. $(-49)^{-\frac{2}{4}}$

f. $16^{\frac{1}{2}}$

Example 3 Simplifying Expressions with Rational Exponents

Simplify each expression using one or more of the rules for exponents.

a. $x^{\frac{2}{3}} \cdot x^{\frac{1}{6}}$

b. $\dfrac{x^{\frac{3}{4}}}{x^{\frac{1}{3}}}$

c. $\left(2a^{\frac{1}{4}}\right)^3$

d. $\left(27y^{-\frac{9}{10}}\right)^{-\frac{1}{3}}$

e. $(-36)^{-\frac{1}{2}}$

f. $9^{\frac{2}{4}}$

g. $\left(\dfrac{49x^6y^{-2}}{z^{-4}}\right)^{\frac{1}{2}}$

Solution

a. $x^{\frac{2}{3}} \cdot x^{\frac{1}{6}} = x^{\frac{2}{3}+\frac{1}{6}}$ Add the exponents.

$= x^{\frac{4}{6}+\frac{1}{6}} = x^{\frac{5}{6}}$

b. $\dfrac{x^{\frac{3}{4}}}{x^{\frac{1}{3}}} = x^{\frac{3}{4}-\frac{1}{3}}$ Subtract the exponents.

$= x^{\frac{9}{12}-\frac{4}{12}} = x^{\frac{5}{12}}$

c. $\left(2a^{\frac{1}{4}}\right)^3 = 2^3 \cdot a^{\frac{1}{4}(3)} = 8a^{\frac{3}{4}}$

d. $\left(27y^{-\frac{9}{10}}\right)^{-\frac{1}{3}} = 27^{-\frac{1}{3}} \cdot y^{-\frac{9}{10}\left(-\frac{1}{3}\right)}$ Multiply the exponents of y and reduce.

$= \dfrac{y^{\frac{3}{10}}}{27^{\frac{1}{3}}} = \dfrac{y^{\frac{3}{10}}}{3}$

e. $(-36)^{-\frac{1}{2}} = \dfrac{1}{(-36)^{\frac{1}{2}}}$ This is not a real number. $(-36)^{\frac{1}{2}} = \sqrt{-36}$ is not real.

f. $9^{\frac{2}{4}} = 9^{\frac{1}{2}} = 3$ The exponent can be reduced as long as the expression is real.

g. $$\left(\frac{49x^6y^{-2}}{z^{-4}}\right)^{\frac{1}{2}} = \frac{\left(49x^6y^{-2}\right)^{\frac{1}{2}}}{\left(z^{-4}\right)^{\frac{1}{2}}}$$

$$= \frac{49^{\frac{1}{2}}x^{6\left(\frac{1}{2}\right)}y^{-2\left(\frac{1}{2}\right)}}{z^{-4\left(\frac{1}{2}\right)}}$$ Use the power rule four times.

$$= \frac{7x^3y^{-1}}{z^{-2}}$$ Simplify the exponents.

$$= \frac{7x^3z^2}{y}$$ Use the properties of negative exponents.

Now work margin exercise 3.

Example 4 shows how to use fractional exponents to simplify complicated-looking radical expressions. The results may seem surprising at first.

Example 4 Simplifying Radical Notation by Changing to Exponential Notation

Simplify each expression by first changing it into an equivalent expression with rational exponents. Then rewrite the answer in simplified radical form. Assume that all variables represent positive real numbers.

a. $\sqrt[4]{\sqrt[3]{x}}$ **b.** $\sqrt[3]{a}\sqrt{a}$ **c.** $\dfrac{\sqrt{x^3}\sqrt[3]{x^2}}{\sqrt[5]{x^2}}$

Solution

a. $$\sqrt[4]{\sqrt[3]{x}} = \left(\sqrt[3]{x}\right)^{\frac{1}{4}} = \left(x^{\frac{1}{3}}\right)^{\frac{1}{4}}$$ Note that $\frac{1}{3}\cdot\frac{1}{4}=\frac{1}{12}$.

$$= x^{\frac{1}{12}} = \sqrt[12]{x}$$

b. $$\sqrt[3]{a}\sqrt{a} = a^{\frac{1}{3}}\cdot a^{\frac{1}{2}}$$

$$= a^{\frac{1}{3}+\frac{1}{2}} = a^{\frac{2}{6}+\frac{3}{6}}$$

$$= a^{\frac{5}{6}} = \sqrt[6]{a^5}$$

c. $$\frac{\sqrt{x^3}\sqrt[3]{x^2}}{\sqrt[5]{x^2}} = \frac{x^{\frac{3}{2}}\cdot x^{\frac{2}{3}}}{x^{\frac{2}{5}}} = \frac{x^{\frac{3}{2}+\frac{2}{3}}}{x^{\frac{2}{5}}} = \frac{x^{\frac{9}{6}+\frac{4}{6}}}{x^{\frac{2}{5}}} = \frac{x^{\frac{13}{6}}}{x^{\frac{2}{5}}}$$

$$= x^{\frac{13}{6}-\frac{2}{5}} = x^{\frac{65}{30}-\frac{12}{30}} = x^{\frac{53}{30}}$$

$$= x^{\frac{30}{30}}\cdot x^{\frac{23}{30}} = x\cdot x^{\frac{23}{30}} = x\sqrt[30]{x^{23}}$$

Note that $\frac{53}{30}=\frac{30}{30}+\frac{23}{30}=1+\frac{23}{30}$.

Now work margin exercise 4.

4. Rewrite the expression in simplified radical form. Assume that all variables represent positive real numbers.

a. $\sqrt[5]{\sqrt[2]{x}}$

b. $\sqrt[4]{x}\cdot\sqrt{x}$

c. $\dfrac{\sqrt[3]{x^2}\cdot\sqrt{x^3}}{\sqrt[6]{x^5}}$

D Using a Graphing Calculator to Evaluate Expressions with Rational Exponents

The caret key [^] on the TI-84 Plus graphing calculator (and most graphing calculators) is used to indicate exponents. Using this key, roots of real numbers can be calculated with up to ten-digit accuracy. To set the number of decimal places you wish in any calculations, press the [MODE] key and highlight the digit opposite the word FLOAT that indicates the desired accuracy. If no digit is highlighted, then the accuracy will be to at most ten decimal places (the accuracy will vary; only ten digits can display in the answer).

CALCULATORS

To Find the Value of $a^{\frac{m}{n}}$ with a TI-84 Plus Graphing Calculator

1. Enter the value of the base, *a*.
2. Press the caret key [^].
3. Enter the fractional exponent enclosed in parentheses. (This exponent may be positive or negative.)
4. Press [ENTER].

5. Evaluate the following expressions using a TI-84 Plus graphing calculator.

a. $64^{\frac{4}{3}}$

b. $42^{\frac{5}{6}}$

Example 5 Evaluating Rational Exponents using a Calculator

Evaluate the following expressions using a TI-84 Plus graphing calculator.

a. $125^{\frac{4}{3}}$ **b.** $36^{\frac{3}{5}}$

Solution

a. To find $125^{\frac{4}{3}}$ proceed as follows.

Step 1: Enter the base, 125.

Step 2: Press the caret key [^].

Step 3: Enter the exponent in parentheses, $\frac{4}{3}$.

Step 4: Press [ENTER].

The display should appear as follows.

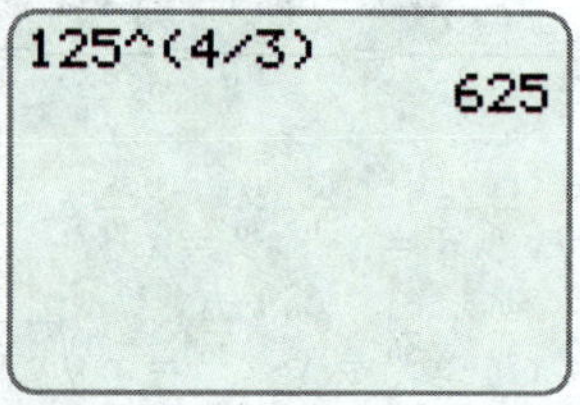

b. To find $36^{\frac{3}{5}}$ proceed as follows.

Step 1: Enter the base, 36.

Step 2: Press the caret key [^].

Step 3: Enter the exponent in parentheses, $\frac{3}{5}$.

Step 4: Press [ENTER].

The display should appear as follows.

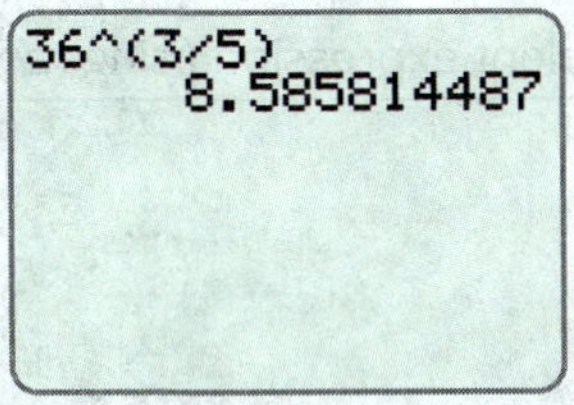

Now work margin exercise 5.

Margin Exercise Answers

1. a. 6 **b.** 2 **c.** −3 **d.** 0.2 **e.** not a real number **2. a.** $\sqrt[5]{x^2}$ **b.** $8\sqrt[7]{z^6}$ **c.** $-\sqrt[4]{b^5}$ **d.** $x^{\frac{5}{7}}$ **e.** $3s^{\frac{1}{2}}$ **f.** $-5^{\frac{1}{4}}$ **3. a.** $x^{\frac{7}{12}}$ **b.** $\frac{1}{a^{\frac{2}{9}}}$ **c.** $81b^{\frac{4}{5}}$ **d.** $\frac{z^{\frac{2}{9}}}{8}$ **e.** not a real number **f.** 4 **4. a.** $\sqrt[10]{x}$ **b.** $\sqrt[4]{x^3}$ **c.** $x\sqrt[3]{x}$ **5. a.** 256 **b.** 22.52722735

15.3 Exercises

Concept Check

Fill-in-the-Blank. Complete each sentence using information found in this section.

1. The expression $\sqrt[n]{a}$ is called a/an __________ expression, and n is called the __________.
2. In the expression $-\sqrt[5]{3}$, the coefficient is ___ and the index is ___.
3. The expression $\sqrt{a}$ would be rewritten as _____ in exponential notation.
4. An equivalent way to write $a^{\frac{m}{n}}$ is $\left(a^{\frac{1}{n}}\right)^m$ or ______.
5. The expression $a^{\frac{1}{3}}$ can be written as ____ in radical notation.
6. For n^{th} roots, if $b = \sqrt[n]{a}$, then $b =$ ____.

True/False. Determine whether each statement is true or false. If a statement is false, explain how it can be changed so the statement will be true. (**Note:** There may be more than one acceptable change.)

7. The same rules for exponents apply to both integer exponents and rational exponents.
8. If the cube root of 7 were to be converted into exponential notation it would be $\sqrt[3]{7}$.
9. Any expression to the power 0, such as $\left(\sqrt[4]{x}\right)^0$, is equal to 1.
10. The expression $y^{\frac{1}{2}}$ can be rewritten in radical notation as $\sqrt{y^2}$.

15.3 Exercises

Concept Check

1. radical, index
2. −1, 5
3. $a^{\frac{1}{2}}$
4. $\left(a^m\right)^{\frac{1}{n}}$
5. $\sqrt[3]{a}$
6. a^n
7. True
8. False; Exponential notation would be $7^{\frac{1}{3}}$.
9. True
10. False; $y^{\frac{1}{2}} = \sqrt{y}$.

Practice

1. $\sqrt[3]{8}$
2. $\sqrt{5}$
3. $-\sqrt[6]{x}$
4. $4\sqrt[4]{y^3}$
5. $\sqrt[5]{4z^2}$
6. $3^{\frac{1}{2}}$
7. $13^{\frac{1}{7}}$
8. $4x^{\frac{2}{3}}$
9. $(-9)^{\frac{1}{3}}$
10. $(4x)^{\frac{2}{5}}$
11. 3
12. 11
13. $\frac{1}{10}$
14. $\frac{1}{5}$
15. −512
16. Not a real number
17. −4
18. −4
19. Not a real number

20. $-\frac{2}{5}$
21. $\frac{3}{7}$
22. $\frac{5}{4}$
23. 16
24. $\frac{1}{4}$
25. $-\frac{1}{6}$
26. -5
27. $\frac{5}{2}$
28. $-\frac{27}{8}$
29. $\frac{1}{4}$
30. $\frac{9}{16}$
31. $\frac{3}{8}$
32. $\frac{2}{5}$
33. $-\frac{1}{1000}$
34. $-\frac{1}{16,807}$
35. 64
36. $\frac{1}{81}$
37. 8.5499
38. 2187
39. 10,000,000
40. 4.6416
41. 99.6055
42. 158.7401
43. 0.0922
44. 0.0131
45. 1.6083
46. 1.6785
47. 0.2236
48. 0.2
49. 7.7460

Practice

Write an equivalent expression using radical notation. See Example 2.

1. $8^{\frac{1}{3}}$

2. $5^{\frac{1}{2}}$

3. $-x^{\frac{1}{6}}$

4. $4y^{\frac{3}{4}}$

5. $(2z)^{\frac{2}{5}}$

Write an equivalent expression using exponential notation. See Example 2.

6. $\sqrt{3}$

7. $\sqrt[7]{13}$

8. $4\sqrt[3]{x^2}$

9. $\sqrt[3]{-9}$

10. $\sqrt[5]{16x^2}$

Simplify each numerical expression. See Example 3.

11. $9^{\frac{1}{2}}$

12. $121^{\frac{1}{2}}$

13. $100^{-\frac{1}{2}}$

14. $25^{-\frac{1}{2}}$

15. $-64^{\frac{3}{2}}$

16. $(-64)^{\frac{3}{2}}$

17. $(-64)^{\frac{1}{3}}$

18. $-(64)^{\frac{1}{3}}$

19. $\left(-\frac{4}{25}\right)^{\frac{1}{2}}$

20. $-\left(\frac{4}{25}\right)^{\frac{1}{2}}$

21. $\left(\frac{9}{49}\right)^{\frac{1}{2}}$

22. $\left(\frac{225}{144}\right)^{\frac{1}{2}}$

23. $64^{\frac{2}{3}}$

24. $8^{-\frac{2}{3}}$

25. $(-216)^{-\frac{1}{3}}$

26. $(-125)^{\frac{1}{3}}$

27. $\left(\frac{8}{125}\right)^{-\frac{1}{3}}$

28. $-\left(\frac{16}{81}\right)^{-\frac{3}{4}}$

29. $\left(-\frac{1}{32}\right)^{\frac{2}{5}}$

30. $\left(\frac{27}{64}\right)^{\frac{2}{3}}$

31. $3 \cdot 16^{-\frac{3}{4}}$

32. $2 \cdot 25^{-\frac{1}{2}}$

33. $-100^{-\frac{3}{2}}$

34. $-49^{-\frac{5}{2}}$

35. $\left[\left(\frac{1}{32}\right)^{\frac{2}{5}}\right]^{-3}$

36. $\left[(-27)^{\frac{2}{3}}\right]^{-2}$

Use a graphing calculator to find the value of each numerical expression accurate to the nearest ten-thousandth, if necessary. See Example 5.

37. $25^{\frac{2}{3}}$

38. $81^{\frac{7}{4}}$

39. $100^{\frac{7}{2}}$

40. $100^{\frac{1}{3}}$

41. $250^{\frac{5}{6}}$

42. $2000^{\frac{2}{3}}$

43. $24^{-\frac{3}{4}}$

44. $18^{-\frac{3}{2}}$

45. $\sqrt[9]{72}$

46. $\sqrt[8]{63}$

47. $\sqrt[4]{0.0025}$

48. $\sqrt[5]{0.00032}$

49. $\sqrt[4]{3600}$

50. $\sqrt[6]{4500}$

51. $\sqrt[5]{35.4}$

52. $\sqrt[10]{1.8}$

Simplify each algebraic expression. Assume that all variables represent positive real numbers. Leave the answers in exponential notation. See Example 3.

53. $\left(2x^{\frac{1}{3}}\right)^3$

54. $\left(3x^{\frac{1}{2}}\right)^4$

55. $\left(9a^4\right)^{-\frac{1}{2}}$

56. $\left(16a^3\right)^{-\frac{1}{4}}$

57. $8x^2 \cdot x^{\frac{1}{2}}$

58. $3x^3 \cdot x^{\frac{2}{3}}$

59. $5a^2 \cdot a^{-\frac{1}{3}} \cdot a^{\frac{1}{2}}$

60. $a^{\frac{2}{3}} \cdot a^{-\frac{3}{5}} \cdot a^0$

61. $\dfrac{x^{\frac{3}{4}}}{x^{\frac{1}{6}}}$

62. $\dfrac{a^{\frac{2}{3}}}{a^{\frac{1}{9}}}$

63. $\dfrac{x^{\frac{2}{5}}}{x^{-\frac{1}{10}}}$

64. $\dfrac{a^{\frac{1}{2}}}{a^{-\frac{2}{3}}}$

65. $\dfrac{a^{\frac{3}{4}} \cdot a^{\frac{1}{8}}}{a^2}$

66. $\dfrac{x^{\frac{2}{3}} \cdot x^{\frac{4}{3}}}{x^2}$

67. $\dfrac{a^{\frac{1}{2}} \cdot a^{-\frac{3}{4}}}{a^{-\frac{1}{2}}}$

68. $\dfrac{x^{\frac{2}{3}}x^{-1}}{x^{-\frac{3}{2}}}$

69. $\dfrac{a^{\frac{3}{2}}b^{\frac{4}{5}}}{a^{-\frac{1}{2}}b^2}$

70. $\dfrac{a^{\frac{3}{4}}b^{-\frac{1}{3}}}{a^{\frac{3}{2}}b^{\frac{1}{6}}}$

71. $\left(2x^{\frac{1}{2}}y^{\frac{1}{3}}\right)^3$

72. $\left(a^{\frac{1}{2}}a^{\frac{1}{3}}\right)^6$

73. $\left(4x^{-\frac{3}{4}}y^{\frac{1}{5}}\right)^{-2}$

74. $\left(81a^{-8}b^2\right)^{-\frac{1}{4}}$

75. $\left(-x^3y^6z^{-6}\right)^{\frac{2}{3}}$

76. $\left(9x^2y^{-4}z^{-3}\right)^{\frac{3}{2}}$

77. $\left(\dfrac{x^2y^{-3}}{z^4}\right)^{-\frac{1}{2}}$

78. $\left(\dfrac{27a^3b^6}{c^9}\right)^{-\frac{1}{3}}$

79. $\left(\dfrac{16a^{-4}b^3}{c^4}\right)^{\frac{3}{4}}$

80. $\left(\dfrac{-27a^2b^3}{c^{-3}}\right)^{\frac{1}{3}}$

81. $\dfrac{\left(x^{\frac{1}{4}}y^{\frac{1}{2}}\right)^3}{x^{\frac{1}{2}}y^{\frac{1}{4}}}$

82. $\dfrac{\left(x^{\frac{1}{2}}y\right)^{-\frac{1}{3}}}{x^{\frac{2}{3}}y^{-1}}$

83. $\dfrac{\left(8x^2y\right)^{-\frac{1}{3}}}{\left(5x^{\frac{1}{3}}y^{-\frac{1}{2}}\right)^2}$

84. $\dfrac{\left(25a^4b^{-1}\right)^{\frac{1}{2}}}{\left(2a^{\frac{1}{5}}b^{\frac{3}{5}}\right)^3}$

50. 4.0632

51. 2.0408

52. 1.0605

53. $8x$

54. $81x^2$

55. $\dfrac{1}{3a^2}$

56. $\dfrac{1}{2a^{\frac{3}{4}}}$

57. $8x^{\frac{5}{2}}$

58. $3x^{\frac{11}{3}}$

59. $5a^{\frac{13}{6}}$

60. $a^{\frac{1}{15}}$

61. $x^{\frac{7}{12}}$

62. $a^{\frac{5}{9}}$

63. $x^{\frac{1}{2}}$

64. $a^{\frac{7}{6}}$

65. $\dfrac{1}{a^{\frac{9}{8}}}$

66. 1

67. $a^{\frac{1}{4}}$

68. $x^{\frac{7}{6}}$

69. $\dfrac{a^2}{b^{\frac{6}{5}}}$

70. $\dfrac{1}{a^{\frac{3}{4}}b^{\frac{1}{2}}}$

71. $8x^{\frac{3}{2}}y$

72. a^5

73. $\dfrac{x^{\frac{3}{2}}}{16y^{\frac{2}{5}}}$

74. $\dfrac{a^2}{3b^{\frac{1}{2}}}$

75. $\dfrac{x^2y^4}{z^4}$

76. $\dfrac{27x^3}{y^6z^{\frac{9}{2}}}$

77. $\frac{y^{\frac{3}{2}}z^2}{x}$

78. $\frac{c^3}{3ab^2}$

79. $\frac{8b^{\frac{9}{4}}}{a^3c^3}$

80. $-3a^{\frac{2}{3}}bc$

81. $x^{\frac{1}{4}}y^{\frac{5}{4}}$

82. $\frac{y^{\frac{2}{3}}}{x^{\frac{5}{6}}}$

83. $\frac{y^{\frac{2}{3}}}{50x^{\frac{4}{3}}}$

84. $\frac{5a^{\frac{7}{5}}}{8b^{\frac{23}{10}}}$

85. $\frac{b^{\frac{5}{12}}}{a^{\frac{11}{12}}}$

86. $\frac{x^{\frac{7}{4}}}{y^{\frac{3}{4}}}$

87. $\frac{3x^{\frac{1}{2}}}{20y^{\frac{2}{3}}}$

88. $\frac{1}{8ab^2}$

89. $\sqrt[6]{x^5}$

90. $x\sqrt[15]{x^4}$

91. $\sqrt[12]{y^7}$

92. $x\sqrt[12]{x}$

93. $\sqrt[30]{x^{11}}$

94. $\sqrt[12]{a^5}$

95. $\sqrt[6]{y}$

96. $\sqrt[10]{x}$

97. $\sqrt[9]{x}$

98. $\sqrt[4]{a}$

99. $\sqrt[3]{7a}$

100. $\sqrt[3]{3x}$

85. $\left(\frac{a^{-3}b^{\frac{1}{3}}}{a^{\frac{1}{2}}b}\right)^{\frac{1}{2}} \cdot \left(\frac{ab^{\frac{1}{2}}}{a^{-\frac{2}{3}}b^{-1}}\right)^{\frac{1}{2}}$

86. $\left(\frac{x^2y^{\frac{1}{3}}}{x^{\frac{1}{2}}y^{\frac{3}{2}}}\right)^{\frac{1}{2}} \cdot \left(\frac{x^{-\frac{1}{2}}y^{\frac{2}{3}}}{x^{-1}y^{\frac{3}{4}}}\right)^2$

87. $\frac{\left(27xy^{\frac{1}{2}}\right)^{\frac{1}{3}}}{\left(25x^{-\frac{1}{2}}y\right)^{\frac{1}{2}}} \cdot \frac{\left(x^{\frac{1}{2}}y\right)^{\frac{1}{6}}}{\left(16x^{\frac{1}{3}}y\right)^{\frac{1}{2}}}$

88. $\frac{\left(4a^{-6}b\right)^{\frac{1}{2}}}{\left(7a^2b^3\right)^{-1}} \cdot \frac{\left(49a^4b^3\right)^{-\frac{1}{2}}}{\left(64a^{-3}b^6\right)^{\frac{2}{3}}}$

Simplify each expression by first changing it into an equivalent expression with rational exponents. Rewrite the answer in simplified radical form. Assume that all variables represent positive real numbers. See Example 4.

89. $\sqrt{x} \cdot \sqrt[3]{x}$

90. $\sqrt[3]{x^2} \cdot \sqrt[5]{x^3}$

91. $\frac{\sqrt[4]{y^3}}{\sqrt[6]{y}}$

92. $\frac{\sqrt[3]{x^4}}{\sqrt[4]{x}}$

93. $\frac{\sqrt[3]{x^2}\sqrt[5]{x^6}}{\sqrt{x^3}}$

94. $\frac{a\sqrt[4]{a}}{\sqrt[3]{a}\sqrt{a}}$

95. $\sqrt{\sqrt[3]{y}}$

96. $\sqrt[5]{\sqrt{x}}$

97. $\sqrt[3]{\sqrt[3]{x}}$

98. $\sqrt{\sqrt{a}}$

99. $\sqrt[15]{(7a)^5}$

100. $\sqrt[21]{(3x)^7}$

101. $\sqrt[4]{\sqrt[3]{\sqrt{x}}}$

102. $\sqrt[5]{\sqrt[4]{\sqrt[3]{x}}}$

103. $\left(\sqrt[3]{a^4bc^2}\right)^{15}$

104. $\left(\sqrt[4]{a^3b^6c}\right)^{12}$

Applications

Solve.

105. According to Kepler's Law of Periods, the square of the orbital period of a planet is directly proportional to the cube of the semi-major axis of its orbit, or the cube of the planet's maximum distance from the sun. Writing this relationship as an equation solved for the period of a planet gives us $p = d^{\frac{3}{2}}$ where p is the orbital period of a planet represented in Earth years and d is the planet's semi-major axis represented in astronomical units (AU). Using the given equation, if the planet Mercury has a semi-major axis of 0.39 AU, what is the orbital period of Mercury in Earth years? Round your answer to the nearest hundredth.

106. The orbital period of Saturn is about 29.5 Earth years.

a. Solving Kepler's Law of Periods (see Exercise 105) for the planet's maximum distance from the sun, gives the equation $d = p^{\frac{2}{3}}$. Use this equation to calculate the maximum distance away from the sun that Saturn reaches in astronomical units. Round your answer to the nearest hundredth.

b. Given that Earth's semi-major axis is 1 AU, about how many times further from the sun does Saturn travel than Earth?

107. The width of a rectangle is $\sqrt[3]{64^2}$ ft and the length is $216^{\frac{2}{3}}$ ft. What is the area of the rectangle?

108. The motion of a simple pendulum is represented by the following equation, where T = the pendulum period, L = length, and g = acceleration of gravity.

$$T = 2\pi\sqrt{\frac{L}{g}}$$

If the length of the pendulum is $320^{\frac{2}{3}}$ meters and the acceleration of gravity is equal to 9.8 meters per seconds squared, what is the period of the pendulum? Use $\pi = 3.14$ and round your answer to the nearest hundredth.

109. A crew of construction workers are disassembling the inside of a building and dropping things into dumpsters at the base of the building. The equation $v_a = \frac{(64d)^{\frac{1}{2}}}{2}$ is used to find the average velocity, or speed, in feet per second of an object that has fallen a distance of d feet.

a. What is the average velocity, to the nearest hundredth, of a lighting fixture that fell 25 feet?

b. What is the average velocity, to the nearest hundredth, of a ceiling tile that fell 80 feet?

110. Isaac Newton fell asleep under an apple tree thinking about math. While he was sleeping, a squirrel knocked an apple off of a branch of the tree. The equation $v = (19.8d)^{\frac{1}{2}}$ can be used to find the velocity v, in meters per second, of the apple after dropping a distance d, where d is in meters.

a. If the apple was connected to a branch 2 m above Newton's head, what was the velocity of the apple, to the nearest hundredth, when it hit Newton's head?

b. If the squirrel knocked a second apple off a branch that was 5 m above Newton's head, what was the velocity of the apple, to the nearest hundredth, when it hit Newton's head?

c. Suppose the second apple missed Newton's head and landed on the ground instead. If Newton's head was 0.8 m above the ground, what was the velocity of the apple, to the nearest hundredth, when it hit the ground?

101. $\sqrt[24]{x}$

102. $\sqrt[60]{x}$

103. $a^{20}b^{5}c^{10}$

104. $a^{9}b^{18}c^{3}$

Applications

105. 0.24 Earth years

106. a. 9.55 AU

b. Almost 10 times further

107. 576 ft^2

108. 13.72 seconds

109. a. 20 feet per second

b. 35.78 feet per second

110. a. 6.29 m/s

b. 9.95 m/s

c. 10.72 m/s

111. a. 50 kmph
b. 20 kmph

Writing & Thinking

112. No:

$\sqrt[5]{a}\cdot\sqrt{a}=a^{\frac{7}{10}}, \sqrt[5]{a^2}=a^{\frac{2}{5}},$
$a^{\frac{7}{10}}\neq a^{\frac{2}{5}}$

113. a. n must be an integer greater than 1.
b. n must be a number less than 0.
c. n must equal 0.
d. n must be a rational number that is not an integer.

111. An amusement park is creating signs to indicate the velocity of the roller coaster car on certain hills of the most popular rides. A roller coaster car gains kinetic energy as it goes down a hill. The velocity, or speed, of an object in kilometers per hour (kph) can be determined by $V=\left(\frac{2k}{m}\right)^{\frac{1}{2}}$, where k is the kinetic energy of the object in joules (J) and m is the mass of the object in kilograms (kg).

a. For the most popular roller coaster, the car has a mass of 300 kg and the car has a kinetic energy of 375,000 J on the first hill. What velocity does the car obtain on the first hill?

b. For the second most popular roller coaster, the car has a mass of 350 kg and the car has a kinetic energy of 70,000 on the first hill. What velocity does the car obtain on the first hill?

Writing & Thinking

112. Is $\sqrt[5]{a}\cdot\sqrt{a}$ the same as $\sqrt[5]{a^2}$? Explain why or why not.

113. Assume that x represents a positive real number. Describe what kind of number the exponent n must be for x^n to mean

a. a product.

b. a quotient.

c. 1.

d. a radical state.

15.4 Addition, Subtraction, and Multiplication with Radicals

Objectives

A. Add and subtract radical expressions.

B. Multiply radical expressions.

C. Use a graphing calculator to evaluate radical expressions.

A Addition and Subtraction with Radicals

Recall that to find the sum $2x^2+3x^2-8x^2$, you can use the distributive property.

$$\begin{aligned}2x^2+3x^2-8x^2&=(2+3-8)x^2\\&=-3x^2\end{aligned}$$

Recall that the terms $2x^2$, $3x^2$, and $-8x^2$ are called **like terms** because each term contains the same variable expression, x^2. Similarly, we have

$$\begin{aligned}2\sqrt{5}+3\sqrt{5}-8\sqrt{5}&=(2+3-8)\sqrt{5}\\&=-3\sqrt{5},\end{aligned}$$

where $2\sqrt{5}$, $3\sqrt{5}$, and $-8\sqrt{5}$ are called **like radicals** because each term contains the same radical expression, $\sqrt{5}$. **Like radicals** have the same index and radicand, or they can be simplified so that they have the same index and radicand.

The terms $2\sqrt{3}$ and $2\sqrt{7}$ are not like radicals because the radicands are not the same, and neither expression can be simplified. Therefore, a sum such as

$$2\sqrt{3}+2\sqrt{7}$$

cannot be simplified. That is, the terms cannot be combined.

Teaching Note

The fact that we cannot combine two exact real numbers into a simple, singular numerical form may be difficult for some students to follow. You may want to emphasize that, for example, $2\sqrt{3}+2\sqrt{7}$ is a single real number.

In some cases, radicals that do not appear to be like radicals can be simplified to have the same index and radicand. For example, $4\sqrt{12}$, $\sqrt{75}$, and $-\sqrt{108}$ do not appear to be like radicals at first. However, simplification of each radical allows the sum of these radicals to be found as follows.

$$\begin{aligned}4\sqrt{12}+\sqrt{75}-\sqrt{108}&=4\sqrt{4\cdot3}+\sqrt{25\cdot3}-\sqrt{36\cdot3}\\&=4\cdot2\sqrt{3}+5\sqrt{3}-6\sqrt{3}\\&=8\sqrt{3}+5\sqrt{3}-6\sqrt{3}\\&=(8+5-6)\sqrt{3}\\&=7\sqrt{3}\end{aligned}$$

Example 1 Adding and Subtracting with Radicals

Perform the indicated operation and simplify, if possible. Assume that all variables represent positive real numbers.

a. $\sqrt{32x}+\sqrt{18x}$

b. $\sqrt{12}+\sqrt{18}+\sqrt{27}$

c. $\sqrt[3]{5x}-\sqrt[3]{40x}$

d. $x\sqrt{4y^3}-5\sqrt{x^2y^3}$

1. Perform the indicated operation and simplify, if possible. Assume that all variables are positive.

a. $\sqrt{75a}+\sqrt{48a}$

b. $\sqrt{45}+\sqrt{20}+\sqrt{27}$

c. $\sqrt[3]{9x}-\sqrt[3]{72x}$

Solution

a. $\sqrt{32x}+\sqrt{18x}=\sqrt{16\cdot 2x}+\sqrt{9\cdot 2x}$
$=4\sqrt{2x}+3\sqrt{2x}$
$=(4+3)\sqrt{2x}$
$=7\sqrt{2x}$

b. $\sqrt{12}+\sqrt{18}+\sqrt{27}$
$=\sqrt{4\cdot 3}+\sqrt{9\cdot 2}+\sqrt{9\cdot 3}$
$=2\sqrt{3}+3\sqrt{2}+3\sqrt{3}$
$=(2+3)\sqrt{3}+3\sqrt{2}$
$=5\sqrt{3}+3\sqrt{2}$

Note that $\sqrt{3}$ and $\sqrt{2}$ are not like radicals. Therefore, the last expression is already fully simplified.

c. $\sqrt[3]{5x}-\sqrt[3]{40x}=\sqrt[3]{5x}-\sqrt[3]{8\cdot 5x}$
$=\sqrt[3]{5x}-2\sqrt[3]{5x}$
$=(1-2)\sqrt[3]{5x}$
$=-\sqrt[3]{5x}$

d. $x\sqrt{4y^3}-5\sqrt{x^2y^3}=x\sqrt{4y^2}\sqrt{y}-5\sqrt{x^2y^2}\sqrt{y}$
$=2xy\sqrt{y}-5xy\sqrt{y}$
$=-3xy\sqrt{y}$

Now work margin exercise 1.

2. Perform the indicated operation and simplify, if possible. Assume that all variables are positive.

a. $4\sqrt{7}-6\sqrt{7}+8\sqrt{7}$

b. $8\sqrt{x}-5\sqrt{3}-2\sqrt{3}-6\sqrt{x}$

Completion Example 2 Adding and Subtracting with Radicals

Perform the indicated operations and simplify, if possible. Assume that all variables represent positive real numbers.

a. $3\sqrt{6}+5\sqrt{6}-\sqrt{6}$ **b.** $\sqrt{a}+4\sqrt{b}+8\sqrt{a}+3\sqrt{b}$

Solution

a. $3\sqrt{6}+5\sqrt{6}-\sqrt{6}=(\underline{\ 3+5-1\ })\sqrt{6}=\underline{\ 7\ }\sqrt{6}$

b. $\sqrt{a}+4\sqrt{b}+8\sqrt{a}+3\sqrt{b}=(\underline{\ 1+8\ })\sqrt{a}+(\underline{\ 4+3\ })\sqrt{b}$
$=\underline{\ 9\ }\sqrt{a}+\underline{\ 7\ }\sqrt{b}$

Now work margin exercise 2.

Teaching Note
You might want to discuss with your students that an expression such as $2(\sqrt{5}+\sqrt{3})=2\sqrt{5}+2\sqrt{3}$, it does not equal $\sqrt{10}+\sqrt{6}$. Then show that $\sqrt{2}(\sqrt{5}+\sqrt{3}=)\sqrt{10}+\sqrt{6}$.

B Multiplication with Radicals

When multiplying expressions, we use the distributive property. To find the product of radicals, we proceed just as in multiplying polynomials. To illustrate the similarities, we show multiplication of a polynomial followed by multiplication of a similar radical expression.

$$5(x+y)=5x+5y$$

$$\sqrt{2}\left(\sqrt{7}+\sqrt{3}\right)=\sqrt{2}\sqrt{7}+\sqrt{2}\sqrt{3}=\sqrt{14}+\sqrt{6}$$ Recall that $\sqrt{a}\sqrt{b}=\sqrt{ab}$.

With binomials, the FOIL method gives the following.

$$(x+5)(x-7)=x^2-7x+5x-35=x^2-2x-35$$

$$\begin{aligned}\left(\sqrt{3}+5\right)\left(\sqrt{3}-7\right)&=\left(\sqrt{3}\right)^2-7\sqrt{3}+5\sqrt{3}+5(-7)\\&=3-7\sqrt{3}+5\sqrt{3}-35\\&=3-35+(-7+5)\sqrt{3}\\&=-32-2\sqrt{3}\end{aligned}$$

Example 3 Multiplying with Radicals

Multiply and simplify the following expressions. Assume that all variables represent positive real numbers.

a. $\sqrt{5}\cdot\sqrt{15}$

b. $\sqrt{7}\left(\sqrt{7}-\sqrt{14}\right)$

c. $\left(3\sqrt{7}-2\right)\left(\sqrt{7}+3\right)$

d. $\left(\sqrt{2x}+5\right)\left(\sqrt{2x}-5\right)$

e. $\left(1-\sqrt{3x+4}\right)^2$

Solution

a. $\sqrt{5}\cdot\sqrt{15}=\sqrt{5}\cdot\sqrt{5}\cdot\sqrt{3}=\left(\sqrt{5}\right)^2\cdot\sqrt{3}=5\sqrt{3}$

b. $$\begin{aligned}\sqrt{7}\left(\sqrt{7}-\sqrt{14}\right)&=\sqrt{7}\cdot\sqrt{7}-\sqrt{7}\cdot\sqrt{14}\\&=\left(\sqrt{7}\right)^2-\sqrt{7}\left(\sqrt{7}\cdot\sqrt{2}\right)\\&=\left(\sqrt{7}\right)^2-\left(\sqrt{7}\right)^2\sqrt{2}\\&=7-7\sqrt{2}\end{aligned}$$

c. $$\begin{aligned}\left(3\sqrt{7}-2\right)\left(\sqrt{7}+3\right)&=3\left(\sqrt{7}\right)^2+3\cdot3\sqrt{7}-2\sqrt{7}-2\cdot3\\&=3\cdot7+9\sqrt{7}-2\sqrt{7}-6\\&=21-6+(9-2)\sqrt{7}\\&=15+7\sqrt{7}\end{aligned}$$

d. $$\begin{aligned}\left(\sqrt{2x}+5\right)\left(\sqrt{2x}-5\right)&=\left(\sqrt{2x}\right)^2-(5)^2\\&=2x-25\end{aligned}$$ Difference of two squares: $(a+b)(a-b)=a^2-b^2$

e. $$\begin{aligned}\left(1-\sqrt{3x+4}\right)^2&=(1)^2-2(1)\left(\sqrt{3x+4}\right)+\left(\sqrt{3x+4}\right)^2\\&=1-2\sqrt{3x+4}+3x+4\\&=3x+5-2\sqrt{3x+4}\end{aligned}$$ Perfect square trinomial: $(a-b)^2=a^2-2ab+b^2$

Now work margin exercise 3.

3. Multiply and simplify the following expressions. Assume that all variables represent positive real numbers.

a. $\sqrt{3}\cdot\sqrt{75}$

b. $\sqrt{13}\left(\sqrt{26}-\sqrt{13}\right)$

c. $\left(2\sqrt{5}-1\right)\left(\sqrt{5}+2\right)$

d. $\left(\sqrt{3z}-\sqrt{5}\right)\left(\sqrt{3z}+\sqrt{5}\right)$

e. $\left(2-\sqrt{5x+1}\right)^2$

Math Tip

You can save time when simplifying by recalling the following special products.

Difference of two squares:
$(x+a)(x-a)=x^2-a^2$

Square of a binomial sum:
$(x+a)^2=x^2+2ax+a^2$

Square of a binomial difference:
$(x-a)^2=x^2-2ax+a^2$

4. Multiply and simplify the following expressions. Assume that all variables represent positive real numbers.

a. $4\sqrt{5}\left(\sqrt{6}-\sqrt{5}\right)$

b. $\left(\sqrt{s}-7\right)\left(9+\sqrt{s}\right)$

Completion Example 4 Multiplying with Radicals

Multiply and simplify the following expressions. Assume that all variables represent positive numbers.

a. $6\sqrt{3}\left(\sqrt{3}+2\sqrt{7}\right)$ **b.** $\left(\sqrt{x}+8\right)\left(\sqrt{x}-4\right)$

Solution

a. $6\sqrt{3}\left(\sqrt{3}+2\sqrt{7}\right)=\underline{6\sqrt{3}}\cdot\sqrt{3}+\underline{6\sqrt{3}}\cdot 2\sqrt{7}=\underline{18}+\underline{12\sqrt{21}}$

b. $\left(\sqrt{x}+8\right)\left(\sqrt{x}-4\right)=\sqrt{x}\cdot\underline{\sqrt{x}}+\sqrt{x}\cdot\left(\underline{-4}\right)+8\cdot\underline{\sqrt{x}}+8\cdot\left(\underline{-4}\right)$

$=\underline{x}+\underline{4\sqrt{x}}-\underline{32}$

Now work margin exercise 4.

C Evaluating Radical Expressions with a Graphing Calculator

Techniques for using a TI-84 Plus graphing calculator to evaluate radical expressions were illustrated in Section 15.1. These same basic techniques are used to evaluate numerical expressions that contain sums, differences, products, and quotients of radicals. Be careful to use parentheses to ensure that the rules for order of operations are maintained. In particular, sums and differences in numerators and denominators of fractions must be enclosed in parentheses. Study the following example carefully.

5. Use a TI-84 Plus graphing calculator to evaluate each expression. Round answers to the nearest ten–thousandth.

a. $4+5\sqrt{2}$

b. $\left(\sqrt{8}-1\right)\left(\sqrt{8}+1\right)$

Example 5 Using a Calculator to Evaluate Radical Expressions

Use a TI-84 Plus graphing calculator to evaluate each expression. Round answers to the nearest ten-thousandth.

a. $3+2\sqrt{5}$ **b.** $\left(\sqrt{2}+5\right)\left(\sqrt{2}-5\right)$

Solution

a. The display should appear as follows.

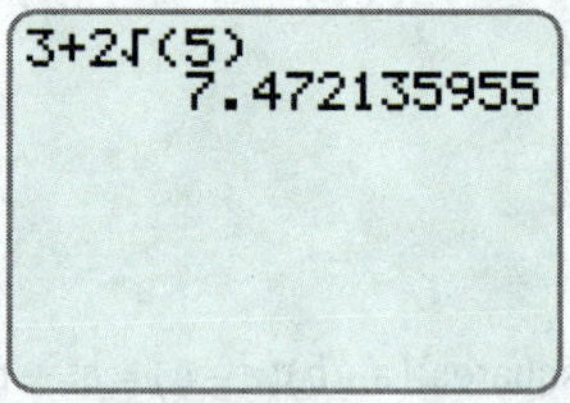

Thus, $3+2\sqrt{5}=7.4721359...\approx 7.4721.$

Rounded to the nearest ten-thousandth

b. The display should appear as follows.

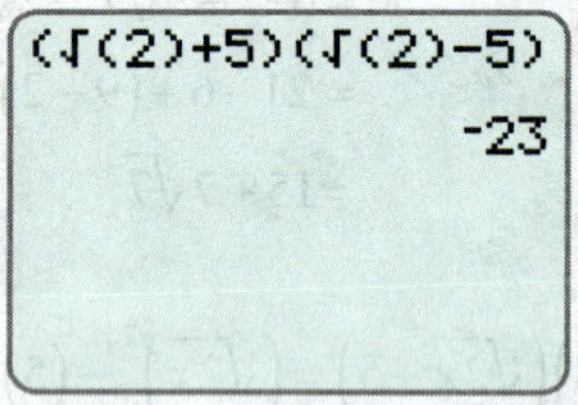

Note: The right parenthesis on 2 must be included. Otherwise, the calculator will interpret the expression as $\sqrt{(2+5)}$ (or $\sqrt{7}$) which is not intended.

Thus, $\left(\sqrt{2}+5\right)\left(\sqrt{2}-5\right)=-23.$

Now work margin exercise 5.

Completion Example Answers

2. a. $(3+5-1)\sqrt{6}=7\sqrt{6}$ **b.** $(1+8)\sqrt{a}+(4+3)\sqrt{b}=9\sqrt{a}+7\sqrt{b}$

4. a. $6\sqrt{3}\cdot\sqrt{3}+6\sqrt{3}\cdot 2\sqrt{7}=18+12\sqrt{21}$ **b.** $\sqrt{x}\cdot\sqrt{x}+\sqrt{x}\cdot(-4)+8\cdot\sqrt{x}+8\cdot(-4)=x+4\sqrt{x}-32$

Margin Exercise Answers

1. a. $9\sqrt{3a}$ **b.** $5\sqrt{5}+3\sqrt{3}$ **c.** $-\sqrt[3]{9x}$ **2. a.** $6\sqrt{7}$ **b.** $2\sqrt{x}-7\sqrt{3}$ **3. a.** 15 **b.** $13\sqrt{2}-13$ **c.** $8+3\sqrt{5}$ **d.** $3z-5$ **e.** $5x+5-4\sqrt{5x+1}$ **4. a.** $4\sqrt{30}-20$ **b.** $s+2\sqrt{s}-63$ **5. a.** 11.0711 **b.** 7

15.4 Exercises

Concept Check

Fill-in-the-Blank. Complete each sentence using information found in this section.

1. Like radicals have the same _____ and radicand or they can be simplified so that they do.

2. Sometimes two or more radicals that do not appear to be like radicals can be _________ so that they are like radicals.

3. To find the product of two binomials that contain radical terms, you can use the ______ method.

True/False. Determine whether each statement is true or false. If a statement is false, explain how it can be changed so the statement will be true. (**Note:** There may be more than one acceptable change.)

4. The radicals $\sqrt{a}$ and $\sqrt[3]{a}$ are like radicals.

5. The radicals $3\sqrt{a}$ and $\sqrt{a}$ are like radicals.

6. The sum $4\sqrt{3}+8\sqrt{5}$ cannot be simplified.

Practice

Simplify the following radical expressions. Assume that all variables represent positive real numbers. See Examples 1 and 2.

1. $3\sqrt{2}+5\sqrt{2}$
2. $7\sqrt{3}-2\sqrt{3}$
3. $4\sqrt{11}-3\sqrt{11}$
4. $6\sqrt{5}+\sqrt{5}$
5. $8\sqrt{10}-11\sqrt{10}$
6. $6\sqrt{17}-9\sqrt{17}$
7. $4\sqrt[3]{3}+9\sqrt[3]{3}$
8. $11\sqrt[3]{14}-6\sqrt[3]{14}$
9. $6\sqrt{11}-5\sqrt{11}-2\sqrt{11}$
10. $\sqrt{7}+6\sqrt{7}-2\sqrt{7}$
11. $\sqrt{a}+4\sqrt{a}-2\sqrt{a}$
12. $2\sqrt{x}-3\sqrt{x}+7\sqrt{x}$
13. $5\sqrt{x}+3\sqrt{x}-\sqrt{x}$
14. $6\sqrt{xy}-10\sqrt{xy}+\sqrt{xy}$

15.4 Exercises

Concept Check

1. index
2. simplified
3. FOIL
4. False; $\sqrt{a}$ and $\sqrt[3]{a}$ do not have the same index, so they are not like radicals.
5. True
6. True

Practice

1. $8\sqrt{2}$
2. $5\sqrt{3}$
3. $\sqrt{11}$
4. $7\sqrt{5}$
5. $-3\sqrt{10}$
6. $-3\sqrt{17}$
7. $13\sqrt[3]{3}$
8. $5\sqrt[3]{14}$
9. $-\sqrt{11}$
10. $5\sqrt{7}$
11. $3\sqrt{a}$
12. $6\sqrt{x}$
13. $7\sqrt{x}$
14. $-3\sqrt{xy}$

15. $4\sqrt{2}+3\sqrt{3}$
16. $8-\sqrt{5}$
17. $8\sqrt{b}-4\sqrt{a}$
18. $3\sqrt{x}+2\sqrt{y}$
19. $13\sqrt[3]{x}-2\sqrt[3]{y}$
20. $9\sqrt[3]{x}-\sqrt[3]{y}$
21. $5\sqrt{3}$
22. $\sqrt{2}$
23. 0
24. $12\sqrt{7}$
25. $17\sqrt[3]{2}$
26. $-7\sqrt[3]{2}$
27. $2\sqrt{2}-6\sqrt{3}$
28. $11\sqrt{3}-3\sqrt{6}$
29. $6+\sqrt{5}$
30. $8\sqrt{2}-4\sqrt{3}$
31. $\sqrt{3}-4\sqrt{2}$
32. $\sqrt{5}+7\sqrt{2}$
33. $5\sqrt[3]{2}-8\sqrt[3]{3}$
34. $\sqrt[3]{3}-8\sqrt[3]{2}$
35. $4\sqrt{2x}$
36. $9\sqrt{3x}$
37. $2y\sqrt{2y}$
38. $5x\sqrt{xy}$
39. $-4x\sqrt{3xy}$
40. $5x\sqrt{2x}$
41. $15x\sqrt{x}$
42. $5a\sqrt{b}$
43. $-xy^2\sqrt{x}$
44. $10x^6y^7\sqrt{2y}$
45. $4x^5y^{10}\sqrt{3}$
46. $12a^4$
47. $-8x^8y^2$
48. $9a^5\sqrt[3]{b}$
49. $xy^2\sqrt[3]{2}\left(-2x^2y^2-2x^3y+3\right)$

15. $3\sqrt{2}+5\sqrt{3}-2\sqrt{3}+\sqrt{2}$

16. $\sqrt{5}+\sqrt{4}-2\sqrt{5}+6$

17. $2\sqrt{a}+7\sqrt{b}-6\sqrt{a}+\sqrt{b}$

18. $4\sqrt{x}-3\sqrt{x}+2\sqrt{y}+2\sqrt{x}$

19. $6\sqrt[3]{x}-4\sqrt[3]{y}+7\sqrt[3]{x}+2\sqrt[3]{y}$

20. $5\sqrt[3]{x}+9\sqrt[3]{y}-10\sqrt[3]{y}+4\sqrt[3]{x}$

21. $\sqrt{12}+\sqrt{27}$

22. $\sqrt{32}-\sqrt{18}$

23. $3\sqrt{5}-\sqrt{45}$

24. $2\sqrt{7}+5\sqrt{28}$

25. $3\sqrt[3]{54}+8\sqrt[3]{2}$

26. $2\sqrt[3]{128}+5\sqrt[3]{-54}$

27. $\sqrt{50}-\sqrt{18}-3\sqrt{12}$

28. $2\sqrt{48}-\sqrt{54}+\sqrt{27}$

29. $2\sqrt{20}-\sqrt{45}+\sqrt{36}$

30. $\sqrt{18}-2\sqrt{12}+5\sqrt{2}$

31. $\sqrt{8}-2\sqrt{3}+\sqrt{27}-\sqrt{72}$

32. $\sqrt{80}+\sqrt{8}-\sqrt{45}+\sqrt{50}$

33. $5\sqrt[3]{16}-4\sqrt[3]{24}+\sqrt[3]{-250}$

34. $\sqrt[3]{192}-2\sqrt[3]{128}+\sqrt[3]{-81}$

35. $6\sqrt{2x}-\sqrt{8x}$

36. $5\sqrt{3x}+2\sqrt{12x}$

37. $5y\sqrt{2y}-y\sqrt{18y}$

38. $9x\sqrt{xy}-x\sqrt{16xy}$

39. $4x\sqrt{3xy}-x\sqrt{12xy}-2x\sqrt{27xy}$

40. $x\sqrt{32x}-x\sqrt{50x}+2x\sqrt{18x}$

41. $\sqrt{36x^3}+\sqrt{81x^3}$

42. $\sqrt{4a^2b}+\sqrt{9a^2b}$

43. $\sqrt{16x^3y^4}-\sqrt{25x^3y^4}$

44. $\sqrt{72x^{12}y^{15}}+\sqrt{18x^{12}y^{15}}+\sqrt{2x^{12}y^{15}}$

45. $\sqrt{12x^{10}y^{20}}+\sqrt{27x^{10}y^{20}}-\sqrt{3x^{10}y^{20}}$

46. $\sqrt[3]{8a^{12}}+\sqrt[3]{1000a^{12}}$

47. $\sqrt[3]{-27x^{24}y^6}+\sqrt[3]{-125x^{24}y^6}$

48. $\sqrt[3]{27a^{15}b}+\sqrt[3]{8a^{15}b}+\sqrt[3]{64a^{15}b}$

49. $\sqrt[3]{-16x^9y^{12}}-\sqrt[3]{16x^{12}y^9}+\sqrt[3]{54x^3y^6}$

50. $\sqrt[3]{54x^{13}y^3}+\sqrt[3]{8x^{23}y^6}+\sqrt[3]{3x^{13}y^3}$

Multiply the following radical expressions and then simplify the results. Assume that all variables represent positive real numbers. See Examples 3 and 4.

51. $\sqrt{2}\left(3-4\sqrt{2}\right)$

52. $2\sqrt{7}\left(\sqrt{7}+3\sqrt{2}\right)$

53. $3\sqrt{18}\cdot\sqrt{2}$

54. $2\sqrt{10}\cdot\sqrt{5}$

55. $-2\sqrt{6}\cdot\sqrt{8}$

56. $2\sqrt{15}\cdot 5\sqrt{6}$

57. $\sqrt{3}\left(\sqrt{2}+2\sqrt{12}\right)$

58. $\sqrt{2}\left(\sqrt{3}-\sqrt{6}\right)$

59. $\sqrt{y}\left(\sqrt{x}+2\sqrt{y}\right)$

60. $\sqrt{x}\left(\sqrt{x}-3\sqrt{y}\right)$

61. $\left(3+\sqrt{2}\right)\left(5-\sqrt{2}\right)$

62. $\left(\sqrt{6}+2\right)\left(\sqrt{6}-2\right)$

63. $\left(\sqrt{3x}-8\right)\left(\sqrt{3x}-1\right)$

64. $\left(6+\sqrt{2x}\right)\left(4+\sqrt{2x}\right)$

65. $\left(2\sqrt{7}+4\right)\left(\sqrt{7}-3\right)$

66. $\left(5\sqrt{3}-2\right)\left(2\sqrt{3}-7\right)$

67. $\left(\sqrt{5}+2\sqrt{2}\right)^2$

68. $\left(2\sqrt{5}+3\sqrt{2}\right)^2$

69. $\left(\sqrt{2}+\sqrt{3}\right)\left(\sqrt{5}-\sqrt{3}\right)$

70. $\left(\sqrt{6}+\sqrt{5}\right)\left(\sqrt{6}-\sqrt{2}\right)$

71. $\left(\sqrt{x}+\sqrt{6}\right)\left(\sqrt{x}-3\sqrt{6}\right)$

72. $\left(\sqrt{11}+\sqrt{3}\right)\left(\sqrt{11}-2\sqrt{3}\right)$

73. $\left(3\sqrt{7}+\sqrt{5}\right)\left(3\sqrt{7}-\sqrt{5}\right)$

74. $\left(7\sqrt{x}+\sqrt{2}\right)\left(7\sqrt{x}-\sqrt{2}\right)$

75. $\left(\sqrt{x}+5\sqrt{y}\right)^2$

76. $\left(3\sqrt{x}+\sqrt{y}\right)^2$

77. $\left(\sqrt{x+3}-5\right)^2$

78. $\left(\sqrt{x+2}+3\right)^2$

79. $\left(4-\sqrt{2x+3}\right)^2$

80. $\left(6-\sqrt{4x+1}\right)^2$

Use a graphing calculator to evaluate each expression. Round your answers to the nearest ten-thousandth, if necessary. See Example 5.

81. $13-\sqrt{75}$

82. $5-\sqrt{67}$

83. $\sqrt{900}+\sqrt{2.56}$

84. $\sqrt{1600}-\sqrt{1.69}$

85. $\left(\sqrt{7}+8\right)\left(\sqrt{7}-8\right)$

86. $\left(\sqrt{8}-\sqrt{5}\right)\left(\sqrt{8}+\sqrt{5}\right)$

87. $\left(2\sqrt{3}+5\sqrt{2}\right)\left(\sqrt{10}-3\sqrt{5}\right)$

88. $\left(6\sqrt{5}+5\sqrt{7}\right)\left(3\sqrt{2}-\sqrt{6}\right)$

Explain the error(s) made in each solution below.

89.
$$\begin{aligned}\sqrt{16}+\sqrt{48}&=\sqrt{16+48}\\&=\sqrt{64}\\&=8\end{aligned}$$

90.
$$\begin{aligned}\sqrt[3]{-125}+\sqrt[3]{98}&=\sqrt[3]{-125+98}\\&=\sqrt[3]{-27}\\&=3\end{aligned}$$

Applications

Solve.

91. For a complete radio circuit, $d=\sqrt{2g}+\sqrt{2h}$, where d equals the visual horizon distance and g and h are the heights of the radio antennas at the respective stations. What is d when $g = 75$ ft and $h = 85$ ft?

92. Mary is making a tile decoration for her wall. Using square tiles of different sizes, Mary created one decoration that is five tiles across, with sides touching. The first tile is 10 in.², the second is 20 in.², the third is 30 in.², the fourth is 20 in.², and the fifth is 10 in.² What is the length of the decoration?

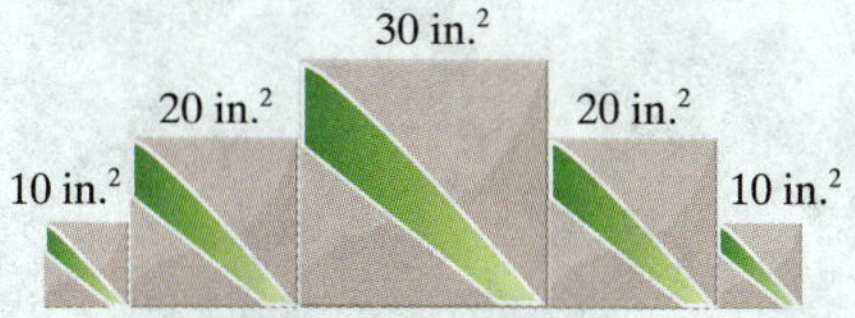

50. $3x^4y\sqrt[3]{2x}+2x^7y^2\sqrt[3]{x^2}$ $+x^4y\sqrt[3]{3x}$

51. $3\sqrt{2}-8$

52. $14+6\sqrt{14}$

53. 18

54. $10\sqrt{2}$

55. $-8\sqrt{3}$

56. $30\sqrt{10}$

57. $12+\sqrt{6}$

58. $\sqrt{6}-2\sqrt{3}$

59. $2y+\sqrt{xy}$

60. $x-3\sqrt{xy}$

61. $13+2\sqrt{2}$

62. 2

63. $3x-9\sqrt{3x}+8$

64. $24+10\sqrt{2x}+2x$

65. $2-2\sqrt{7}$

66. $44-39\sqrt{3}$

67. $13+4\sqrt{10}$

68. $38+12\sqrt{10}$

69. $\sqrt{10}+\sqrt{15}-\sqrt{6}-3$

70. $6+\sqrt{30}-2\sqrt{3}-\sqrt{10}$

71. $x-2\sqrt{6x}-18$

72. $5-\sqrt{33}$

73. 58

74. $49x-2$

75. $x+10\sqrt{xy}+25y$

76. $9x+6\sqrt{xy}+y$

77. $x+28-10\sqrt{x+3}$

78. $x+11+6\sqrt{x+2}$

79. $2x+19-8\sqrt{2x+3}$

80. $4x+37-12\sqrt{4x+1}$

81. 4.3397

82. −3.1854

83. 31.6

84. 38.7

85. −57

86. 3

87. -37.3569

88. 47.7788

89. $\sqrt{16}+\sqrt{48} \neq \sqrt{16+48}$, as radicals cannot be added by simply adding their radicands.

90. $\sqrt[3]{-125}+\sqrt[3]{98} \neq \sqrt[3]{-125+98}$ as radicals cannot be added by simply adding their radicands. Also, a negative sign is missing when evaluating the cube root, since $\sqrt[3]{-27}=-3$.

Applications

91. $5\sqrt{6}+\sqrt{170} \approx 25.29$ ft

92. $2\sqrt{10}+4\sqrt{5}+\sqrt{30} \approx 20.75$ in.

93. **a.** $7\sqrt{2t}$ dollars

b. $\sqrt{2t}$ dollars

94. $\dfrac{118+52\sqrt{7}}{9}$ square yards

95. **a.** $60\sqrt{6}\%$

b. $6x^4\sqrt{2}\%$

96. **a.** $4\sqrt{34}$ inches

b. 13 inches

c. $8\sqrt{34}+26$

93. Josue earns $\sqrt{32t}$ dollars when he works t hours. His roommate, Eric, earns $\sqrt{18t}$ dollars when he works t hours.

a. Find an expression that represents the total income that the two roommates earn after they each work t hours.

b. Find an expression that represents how much more money Josue will earn than Eric after they each work t hours.

94. The city planning committee is looking for places to build a community garden. One lot up for consideration has a length of $\dfrac{8+2\sqrt{b}}{a}$ and a width of $\dfrac{6+5\sqrt{b}}{a}$. If $a=3$ yards and $b=7$ yards, what is the area of this lot? Leave your answer in simplified radical form.

95. The owner of an apple orchard has a field of new trees that are growing and producing more apples each year. He wants to determine the average growth rate, or percentage increase, in the amount of apples produced by these trees over time. The amount of growth varies each year, so he decides to find the geometric mean, or average. The formula $g=\sqrt{a}\cdot\sqrt{b}$ is used to find the geometric mean of two numbers.

a. Over two years, the growth rate was 180% and 120%. Find the average growth rate for these two years. Leave your answer in simplified radical form.

b. Over two years, the growth rate was $8x^3\%$ and $9x^5\%$. Find the average growth rate for these two years. Leave your answer in simplified radical form.

96. A simple diamond-shaped kite is created from wooden dowel rods and nylon cloth. A diagram for the kite is shown.

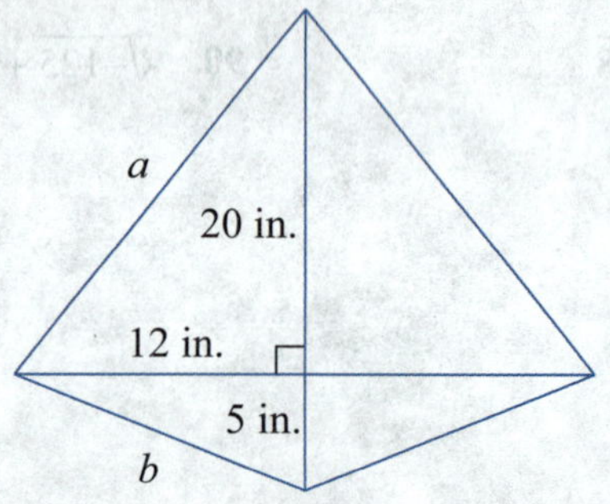

a. Determine the length of side a. Write your answer in simplified radical form.

b. Determine the length of side b. Write your answer in simplified radical form.

c. Calculate the perimeter of the kite. Write your answer in simplified radical form.

15.5 Rationalizing Denominators

Objectives

A. Rationalize denominators with one term in the denominator.

B. Rationalize denominators with a sum or difference in the denominator.

A Rationalizing Denominators with One Term

Each of the expressions

$$\frac{5}{\sqrt{3}},\ \frac{\sqrt{7}}{\sqrt{8x}},\ \text{and}\ \frac{2}{3-\sqrt{2}}$$

contains a radical in the denominator. Such expressions are not considered in simplest form because they are difficult to perform algebraic operations with. Calculations of sums and differences are much easier if the denominators are rational expressions. So, in simplifying, the objective is to find an equivalent fraction that has a rational number or an expression with no radicals for a denominator. That is, we want to simplify the expression by **rationalizing the denominator**. (**Note:** In this section we will deal only with radicals that involve square roots or cube roots; however, the ideas discussed here also apply to n^{th} roots in general.)

Rationalizing a Denominator Containing a Square Root or a Cube Root

1. If the denominator contains a square root, multiply both the numerator and denominator by an expression that will give a denominator with no square roots.
2. If the denominator contains a cube root, multiply both the numerator and denominator by an expression that will give a denominator with no cube roots.

PROCEDURE

The following examples illustrate this method. Each fraction is multiplied by 1 in the form $\frac{k}{k}$, where multiplication by k rationalizes the denominator.

Example 1 Rationalizing Denominators Containing Square Roots

Rationalize each denominator. Assume that all variables represent positive real numbers.

a. $\frac{5}{\sqrt{3}}$ **b.** $\frac{4}{\sqrt{x}}$ **c.** $\frac{3}{7\sqrt{2}}$ **d.** $\frac{\sqrt{7}}{\sqrt{8}}$

Solution

a. Multiply the numerator and denominator by $\sqrt{3}$ because $\sqrt{3}\cdot\sqrt{3}=3$ is a rational number.

$$\frac{5}{\sqrt{3}}=\frac{5\cdot\sqrt{3}}{\sqrt{3}\cdot\sqrt{3}}=\frac{5\sqrt{3}}{3}$$

1. Rationalize each denominator. Assume that all variables represent positive real numbers.

a. $\frac{8}{\sqrt{5}}$

b. $\frac{17}{\sqrt{y}}$

c. $\frac{4}{9\sqrt{3}}$

d. $\frac{\sqrt{11}}{\sqrt{18}}$

b. Multiply the numerator and denominator by $\sqrt{x}$ because $\sqrt{x} \cdot \sqrt{x} = x$. There is no guarantee that x is rational, but the radical sign does not appear in the denominator of the result.

$$\frac{4}{\sqrt{x}} = \frac{4 \cdot \sqrt{x}}{\sqrt{x} \cdot \sqrt{x}} = \frac{4\sqrt{x}}{x}$$

c. Multiply the numerator and denominator by $\sqrt{2}$ because $\sqrt{2} \cdot \sqrt{2} = 2$ is a rational number.

$$\frac{3}{7\sqrt{2}} = \frac{3 \cdot \sqrt{2}}{7\sqrt{2} \cdot \sqrt{2}} = \frac{3\sqrt{2}}{7 \cdot 2} = \frac{3\sqrt{2}}{14}$$

d. Multiply the numerator and denominator by $\sqrt{2}$ because $\sqrt{8} \cdot \sqrt{2} = \sqrt{16} = 4$ is a rational number. (**Note:** $8 \cdot 2 = 16$ and 16 is a perfect square.)

$$\frac{\sqrt{7}}{\sqrt{8}} = \frac{\sqrt{7} \cdot \sqrt{2}}{\sqrt{8} \cdot \sqrt{2}} = \frac{\sqrt{14}}{\sqrt{16}} = \frac{\sqrt{14}}{4}$$

Now work margin exercise 1.

In Example 1 part d. if the numerator and denominator are multiplied by $\sqrt{8}$, the results will be the same, but one more step will be added because the resulting fraction can be reduced.

$$\frac{\sqrt{7}}{\sqrt{8}} = \frac{\sqrt{7} \cdot \sqrt{8}}{\sqrt{8} \cdot \sqrt{8}} = \frac{\sqrt{56}}{8} = \frac{\sqrt{4} \cdot \sqrt{14}}{8} = \frac{\cancel{2}\sqrt{14}}{\underset{4}{\cancel{8}}} = \frac{\sqrt{14}}{4}$$

2. Rationalize each denominator. Assume that all variables represent positive real numbers.

a. $\dfrac{4}{\sqrt[3]{7}}$

b. $\dfrac{5}{\sqrt[3]{9x^5}}$

Example 2 Rationalizing Denominators Containing Cube Roots

Rationalize each denominator. Assume that all variables represent positive real numbers.

a. $\dfrac{3}{\sqrt[3]{5}}$ **b.** $\dfrac{7}{\sqrt[3]{32y}}$

Solution

a. Multiply the numerator and denominator by $\sqrt[3]{25}$ because $\sqrt[3]{5} \cdot \sqrt[3]{25} = \sqrt[3]{125} = 5$ is a rational number. (**Note:** $5 \cdot 25 = 125$ and 125 is a perfect cube.)

$$\frac{3}{\sqrt[3]{5}} = \frac{3 \cdot \sqrt[3]{25}}{\sqrt[3]{5} \cdot \sqrt[3]{25}} = \frac{3\sqrt[3]{25}}{\sqrt[3]{125}} = \frac{3\sqrt[3]{25}}{5}$$

b. Multiply the numerator and denominator by $\sqrt[3]{2y^2}$ because $\sqrt[3]{32y} \cdot \sqrt[3]{2y^2} = \sqrt[3]{64y^3} = 4y$. (**Note:** $32y \cdot 2y^2 = 64y^3$ and $64y^3$ is a perfect cube.)

$$\frac{7}{\sqrt[3]{32y}} = \frac{7 \cdot \sqrt[3]{2y^2}}{\sqrt[3]{32y} \cdot \sqrt[3]{2y^2}} = \frac{7\sqrt[3]{2y^2}}{\sqrt[3]{64y^3}} = \frac{7\sqrt[3]{2y^2}}{4y}$$

Now work margin exercise 2.

B Rationalizing Denominators with a Sum or Difference

If the denominator has a radical expression with a **sum or difference involving square roots,** such as

$$\frac{2}{4-\sqrt{2}} \text{ or } \frac{12}{3+\sqrt{5}},$$

then a different method is used for rationalizing the denominator. In this method, we think of the denominator in the form of $a-b$ or $a+b$. Thus,

$$\text{if } a-b=4-\sqrt{2}, \text{ then } a+b=4+\sqrt{2},$$

and

$$\text{if } a+b=3+\sqrt{5}, \text{ then } a-b=3-\sqrt{5}.$$

The two expressions $(a-b)$ and $(a+b)$ are called **conjugates** of each other, and their product $(a-b)(a+b)$ results in the **difference of two squares**.

$$(a-b)(a+b)=a^2-b^2$$

Rationalizing a Denominator Containing a Sum or Difference Involving Square Roots

If the denominator of a fraction contains a sum or difference involving a square root, rationalize the denominator by multiplying both the numerator and denominator by the **conjugate of the denominator**.

1. If the denominator is of the form $a-b$, multiply both the numerator and denominator by $a+b$.
2. If the denominator is of the form $a+b$, multiply both the numerator and denominator by $a-b$.

The new denominator will be the difference of two squares, and therefore will not contain a radical term.

PROCEDURE

Math Tip

When multiplying conjugates to remove a square root, notice that the middle term will cancel when the distributive property is applied. This will always happen and is a direct result of the b terms having opposite signs.

$$\left(\sqrt{a}+\sqrt{b}\right)\left(\sqrt{a}-\sqrt{b}\right)$$
$$=\sqrt{a}\sqrt{a}-\sqrt{a}\sqrt{b}+\sqrt{b}\sqrt{a}-\sqrt{b}\sqrt{b}$$
$$=\sqrt{a^2}-\cancel{\sqrt{ab}}+\cancel{\sqrt{ab}}-\sqrt{b^2}$$
$$=a-b$$

Furthermore, since both $\sqrt{a}$ and $\sqrt{b}$ are multiplied by themselves, we ensure that a perfect square is created under the radical signs to "undo" the radical when simplified.

Example 3 Rationalizing Denominators

Rationalize each denominator and simplify, if possible. Assume that all variables represent positive real numbers.

a. $\frac{2}{4-\sqrt{2}}$

b. $\frac{13}{5+2\sqrt{3}}$

c. $\frac{1}{\sqrt{7}-\sqrt{2}}$

d. $\frac{6}{1+\sqrt{x}}$ (Assume $x \neq 1$.)

e. $\frac{x-y}{\sqrt{x}-\sqrt{y}}$ (Assume $x \neq y$.)

3. Rationalize each denominator and simplify, if possible. Assume that all variables represent positive real numbers.

a. $\frac{5}{\sqrt{3}-5}$

b. $\frac{15}{3+2\sqrt{6}}$

c. $\frac{1}{\sqrt{8}+\sqrt{3}}$

d. $\frac{8}{1-\sqrt{z}}$ (Assume $x \neq 1$.)

e. $\frac{x+y}{\sqrt{x}+\sqrt{y}}$ (Assume $x \neq y$.)

Solution

a. Multiply the numerator and denominator by $4+\sqrt{2}$.

$$\frac{2}{4-\sqrt{2}}=\frac{2\left(4+\sqrt{2}\right)}{\left(4-\sqrt{2}\right)\left(4+\sqrt{2}\right)}$$

If $a-b=4-\sqrt{2}$, then $a+b=4+\sqrt{2}$.

$$=\frac{2\left(4+\sqrt{2}\right)}{4^2-\left(\sqrt{2}\right)^2}$$

The denominator is the difference of two squares.

$$=\frac{2\left(4+\sqrt{2}\right)}{16-2}$$

$$=\frac{\cancel{2}\left(4+\sqrt{2}\right)}{\cancel{14}_7}$$

The denominator is a rational number.

$$=\frac{4+\sqrt{2}}{7}$$

Note that the numerator is now irrational. However, this is generally preferred over having an irrational denominator.

b. Multiply the numerator and denominator by $5-2\sqrt{3}$.

$$\frac{13}{5+2\sqrt{3}}=\frac{13\left(5-2\sqrt{3}\right)}{\left(5+2\sqrt{3}\right)\left(5-2\sqrt{3}\right)}$$

$$=\frac{13\left(5-2\sqrt{3}\right)}{5^2-4\cdot 3}$$

$$=\frac{\cancel{13}\left(5-2\sqrt{3}\right)}{\cancel{13}}$$

$$=5-2\sqrt{3}$$

c. Multiply the numerator and denominator by $\sqrt{7}+\sqrt{2}$.

$$\frac{1}{\sqrt{7}-\sqrt{2}}=\frac{1\left(\sqrt{7}+\sqrt{2}\right)}{\left(\sqrt{7}-\sqrt{2}\right)\left(\sqrt{7}+\sqrt{2}\right)}$$

$$=\frac{\sqrt{7}+\sqrt{2}}{7-2}=\frac{\sqrt{7}+\sqrt{2}}{5}$$

d. Multiply the numerator and denominator by $1-\sqrt{x}$.

$$\frac{6}{1+\sqrt{x}}=\frac{6\left(1-\sqrt{x}\right)}{\left(1+\sqrt{x}\right)\left(1-\sqrt{x}\right)}$$

$$=\frac{6\left(1-\sqrt{x}\right)}{1-x}\text{ or }\frac{6-6\sqrt{x}}{1-x}$$

13. $\frac{\sqrt{4}}{\sqrt{3}}$

14. $\sqrt{\frac{3}{8}}$

15. $\sqrt{\frac{9}{2}}$

16. $\sqrt{\frac{3}{5}}$

17. $\sqrt{\frac{1}{x}}$

18. $\sqrt{\frac{x}{y}}$

19. $\sqrt{\frac{2x}{y}}$

20. $\sqrt{\frac{x}{4y}}$

21. $\frac{2}{\sqrt{2y}}$

22. $\frac{-10}{3\sqrt{5}}$

23. $\frac{21}{5\sqrt{7}}$

24. $\frac{x}{5\sqrt{x}}$

25. $\frac{-2y}{5\sqrt{2y}}$

26. $\frac{\sqrt[3]{35}}{\sqrt[3]{4}}$

27. $\frac{\sqrt[3]{10}}{\sqrt[3]{9}}$

28. $-\sqrt{\frac{2}{3y}}$

29. $-\sqrt{\frac{25}{x^3}}$

30. $\frac{\sqrt{8x}}{\sqrt{5y^2}}$

31. $\frac{\sqrt{4x}}{\sqrt{3y^2}}$

32. $\frac{\sqrt{16y^2}}{\sqrt{2y^3}}$

33. $\frac{\sqrt{24b}}{\sqrt{6b^2}}$

34. $\sqrt[3]{\frac{2y^3}{27x^2}}$

35. $\sqrt[3]{\frac{7x}{2y^4}}$

36. $\frac{\sqrt[3]{6a^4}}{\sqrt[3]{25a^2b^4}}$

37. $\frac{\sqrt[3]{x^5}}{\sqrt[3]{9xy}}$

38. $\frac{\sqrt[3]{24x}}{\sqrt[3]{9}}$

39. $\sqrt[3]{\frac{11}{4}}$

40. $\sqrt[3]{\frac{3x^3}{8y^2}}$

41. $\frac{\sqrt[3]{5a}}{\sqrt[3]{3b^4}}$

42. $\sqrt[3]{\frac{7x^4}{16x^2y^4}}$

43. $\frac{\sqrt[3]{a^5}}{\sqrt[3]{4ab}}$

44. $\frac{3}{1+\sqrt{2}}$

45. $\frac{2}{\sqrt{6}-2}$

46. $\frac{-11}{\sqrt{3}-4}$

47. $\frac{1}{\sqrt{5}-3}$

48. $\frac{7}{3-2\sqrt{2}}$

49. $\frac{-6}{5-3\sqrt{2}}$

50. $\frac{11}{2\sqrt{3}+1}$

51. $\frac{-\sqrt{3}}{\sqrt{2}+5}$

52. $\frac{\sqrt{2}}{\sqrt{7}+4}$

53. $\frac{7}{1-3\sqrt{5}}$

54. $\frac{-3\sqrt{3}}{6+\sqrt{3}}$

55. $\frac{1}{\sqrt{3}-\sqrt{5}}$

56. $\frac{-4}{\sqrt{7}-\sqrt{3}}$

57. $\frac{-5}{\sqrt{2}+\sqrt{3}}$

58. $\frac{7}{\sqrt{2}+\sqrt{5}}$

59. $\frac{4}{\sqrt{x}+1}$

60. $\frac{-7}{\sqrt{x}-3}$

61. $\frac{5}{6+\sqrt{y}}$

62. $\frac{x}{\sqrt{x}+2}$

63. $\frac{8}{2\sqrt{x}+3}$

64. $\frac{3\sqrt{x}}{\sqrt{2x}-5}$

65. $\frac{\sqrt{4y}}{\sqrt{5y}-\sqrt{3}}$

66. $\frac{\sqrt{3x}}{\sqrt{2}+\sqrt{3x}}$

67. $\frac{3}{\sqrt{x}-\sqrt{y}}$

68. $\frac{4}{2\sqrt{x}+\sqrt{y}}$

69. $\frac{x}{\sqrt{x}+2\sqrt{y}}$

70. $\frac{y}{\sqrt{x}-\sqrt{3y}}$

71. $\frac{\sqrt{3}+1}{\sqrt{3}-2}$

72. $\frac{\sqrt{2}+4}{5-\sqrt{2}}$

73. $\frac{\sqrt{5}-2}{\sqrt{5}+3}$

74. $\frac{1+\sqrt{3}}{3-\sqrt{3}}$

75. $\frac{\sqrt{x}+1}{\sqrt{x}-1}$

76. $\frac{\sqrt{x}-4}{\sqrt{x}+3}$

77. $\frac{\sqrt{x}+2}{\sqrt{3x}+y}$

78. $\frac{3-\sqrt{x}}{2\sqrt{x}+y}$

17. $\frac{\sqrt{x}}{x}$

18. $\frac{\sqrt{xy}}{y}$

19. $\frac{\sqrt{2xy}}{y}$

20. $\frac{\sqrt{xy}}{2y}$

21. $\frac{\sqrt{2y}}{y}$

22. $\frac{-2\sqrt{5}}{3}$

23. $\frac{3\sqrt{7}}{5}$

24. $\frac{\sqrt{x}}{5}$

25. $\frac{-\sqrt{2y}}{5}$

26. $\frac{\sqrt[3]{70}}{2}$

27. $\frac{\sqrt[3]{30}}{3}$

28. $-\frac{\sqrt{6y}}{3y}$

29. $-\frac{5\sqrt{x}}{x^2}$

30. $\frac{2\sqrt{10x}}{5y}$

31. $\frac{2\sqrt{3x}}{3y}$

32. $\frac{2\sqrt{2y}}{y}$

33. $\frac{2\sqrt{b}}{b}$

34. $\frac{y\sqrt[3]{2x}}{3x}$

35. $\frac{\sqrt[3]{28xy^2}}{2y^2}$

36. $\frac{\sqrt[3]{30a^2b^2}}{5b^2}$

37. $\frac{x\sqrt[3]{3xy^2}}{3y}$

38. $\frac{\sqrt[3]{72x}}{3}$

39. $\frac{\sqrt[3]{22}}{2}$

40. $\frac{x\sqrt[3]{3y}}{2y}$

41. $\frac{\sqrt[3]{45ab^2}}{3b^2}$

42. $\frac{\sqrt[3]{28x^2y^2}}{4y^2}$

43. $\frac{a\sqrt[3]{2ab^2}}{2b}$

44. $-3\left(1-\sqrt{2}\right)$

45. $\sqrt{6}+2$

46. $\frac{11\left(\sqrt{3}+4\right)}{13}$

47. $\frac{-\left(\sqrt{5}+3\right)}{4}$

48. $7\left(2\sqrt{2}+3\right)$

49. $\frac{-6\left(5+3\sqrt{2}\right)}{7}$

50. $2\sqrt{3}-1$

51. $\frac{\sqrt{3}\left(\sqrt{2}-5\right)}{23}$

52. $\frac{-\sqrt{2}\left(\sqrt{7}-4\right)}{9}$

53. $\frac{-7\left(1+3\sqrt{5}\right)}{44}$

54. $\frac{-6\sqrt{3}+3}{11}$

55. $\frac{-\left(\sqrt{3}+\sqrt{5}\right)}{2}$

56. $-\left(\sqrt{7}+\sqrt{3}\right)$

57. $5\left(\sqrt{2}-\sqrt{3}\right)$

58. $\frac{-7\left(\sqrt{2}-\sqrt{5}\right)}{3}$

59. $\frac{4\left(\sqrt{x}-1\right)}{x-1}$

60. $\frac{-7\left(\sqrt{x}+3\right)}{x-9}$

Identify the error(s) made in the following attempt to rationalize a denominator.

79.
$$\begin{aligned}\frac{y}{\sqrt{3}+y} &= \frac{y}{\sqrt{3}+y}\cdot\frac{\sqrt{3}+y}{\sqrt{3}+y} \\ &= \frac{y\left(\sqrt{3}+y\right)}{3+y^2} \\ &= \frac{y\sqrt{3}+y^2}{3+y^2}\end{aligned}$$

Applications

Solve.

80. Officers often need to recreate events that happen during accidents while they investigate, especially determining the initial speed of a car at the time of an accident. One way to do this is to use the formula $\frac{s}{\sqrt{l}}=k,$ where s is the initial speed of the vehicle in mph, l is the length of the skid marks left in feet, and k is a constant that depends on the driving conditions at the time of the accident.

a. Rationalize the denominator of the formula.

b. A driver claims that he was driving the speed limit, 55 mph, at the time of an accident. The skid marks on the road measured 176 feet. Officers estimate that the driving condition constant k based on the conditions at the time of the accident is $\sqrt{24}$. Based on the formula, is the driver's claim correct? If not, what was the driver's initial speed before the accident?

81. The radius of a cylinder can be expressed in terms of its volume and its height by $r=\sqrt{\frac{V}{\pi h}}$. Rationalize the denominator of this formula.

82. A company that sells computers learns that their income can be represented by the equation $I=\frac{6500p}{\sqrt{p-10}}$ dollars when they sell their computers for p dollars. Rationalize the denominator of this equation.

83. An intern at NASA needs to construct a cylinder to be used as a fuel cell for a scale model of a rocket. Her instructions are to make a fuel cell with a volume of 200π cm^3. To find the radius of the fuel cell for a certain height h and volume V, she uses the equation $r=\sqrt{\frac{V}{h\pi}}$. Keep all answers in simplified radical form.

a. Find the radius of the fuel cell if the height is 12 cm.

b. Find the radius of the fuel cell if the height is 15 cm.

c. Find the radius of the fuel cell if the height is 24 cm.

84. ▦ The formula $r=\sqrt{\frac{A}{P}}-1$ is used to determine the interest rate r of on an investment of initial value P that has a value of A after two years.

a. Rationalize the denominator of the fraction in the formula.

b. Find the interest rate on an investment with initial value \$1000 that has a value of \$1102.50 after 2 years.

85. ▦ A client tells his financial consultant that he has \$6000 to invest and would like to earn \$615 on his investment after 2 years. The client needs to know what the average interest rate of the investment will need to be to meet his expectations. The financial consultant can use the formula $A=P(r+1)^2$ to find the future amount A of an investment with a starting principle P and interest rate r after 2 years.

a. Solve the equation for r. Be sure to rationalize any denominators. (**Hint:** Divide both sides by P first and then take the square root of each side.)

b. Determine the interest rate that the \$6000 would need to be invested at to meet the client's expectations. Be sure to express the rate as a percent.

Writing & Thinking

86. In your own words, explain how to rationalize the denominator of a fraction containing the sum or difference of square roots in the denominator. Why does this work?

61. $\frac{5(6-\sqrt{y})}{36-y}$

62. $\frac{x(\sqrt{x}-2)}{x-4}$

63. $\frac{8(2\sqrt{x}-3)}{4x-9}$

64. $\frac{3\sqrt{x}(\sqrt{2x}+5)}{2x-25}$

65. $\frac{2\sqrt{y}(\sqrt{5y}+\sqrt{3})}{5y-3}$

66. $\frac{\sqrt{3x}(\sqrt{2}-\sqrt{3x})}{2-3x}$

67. $\frac{3(\sqrt{x}+\sqrt{y})}{x-y}$

68. $\frac{4(2\sqrt{x}-\sqrt{y})}{4x-y}$

69. $\frac{x(\sqrt{x}-2\sqrt{y})}{x-4y}$

70. $\frac{y(\sqrt{x}+\sqrt{3y})}{x-3y}$

71. $-(\sqrt{3}+1)(\sqrt{3}+2)$

72. $\frac{(\sqrt{2}+4)(5+\sqrt{2})}{23}$

73. $\frac{-(\sqrt{5}-2)(\sqrt{5}-3)}{4}$

74. $\frac{(1+\sqrt{3})(3+\sqrt{3})}{6}$

75. $\frac{(\sqrt{x}+1)^2}{x-1}$

76. $\frac{(\sqrt{x}-4)(\sqrt{x}-3)}{x-9}$

77. $\frac{(\sqrt{x}+2)(\sqrt{3x}-y)}{3x-y^2}$

78. $\frac{(3-\sqrt{x})(2\sqrt{x}-y)}{4x-y^2}$

79. The numerator and denominator were not multiplied by the conjugate of the denominator. The conjugate is $\sqrt{3}-y$. Also, the multiplication is incorrect: $(\sqrt{3}+y)(\sqrt{3}+y)\neq 3+y^2$.

Applications

80. a. $\frac{s\sqrt{l}}{l}=k$

b. No, the driver's initial speed was about 65 mph.

81. $r=\frac{\sqrt{V\pi h}}{\pi h}$

82. $I=\frac{6500p(\sqrt{p}+10)}{p-100}$

$=\frac{6500p^{\frac{3}{2}}+65,000p}{p-100}$

83. a. $\frac{5\sqrt{6}}{3}$ cm

b. $\frac{2\sqrt{30}}{3}$ cm

c. $\frac{5\sqrt{3}}{3}$ cm

84. a. $r=\frac{\sqrt{AP}}{P}-1$

b. 0.05 or 5%

85. a. $r=\frac{\sqrt{AP}}{P}-1$

b. 5%

Writing & Thinking

86. Multiply both the numerator and the denominator by the conjugate of the denominator. This works because multiplying the denominator by its conjugate results in an expression with no square roots. Answers will vary.

Objectives

A. Solve equations that contain one or more radical expressions.

15.6 Solving Radical Equations

A Solving Radical Equations

Each of the following equations involves at least one radical expression.

$$x+3=\sqrt{x+5} \qquad \sqrt{x}-\sqrt{2x-14}=1 \qquad \sqrt[3]{x+1}=5$$

If the radicals are square roots, we solve by isolating one of the square roots on one side of the equation and squaring both sides of the equation. If one of the radicals is some other root that can be isolated on one side of the equation, we solve by raising both sides of the equation to the integer power corresponding to the index of the radical. For example, for a cube root, both sides are raised to the third power.

Squaring both sides of an equation may introduce new solutions. For example, the first-degree equation $x=-3$ has only one solution: -3. However, squaring both sides gives the quadratic equation

$$x^2=(-3)^2 \qquad \text{or} \qquad x^2=9.$$

Teaching Note

Up to this point in algebra, students have solved equations by creating a series of equivalent equations. You might want to include a discussion about the procedures of this section that introduce additional variable factors into the series of equations, thereby creating equations that may not be entirely equivalent.

The quadratic equation $x^2=9$ has two solutions: 3 and -3. Thus, a new solution that is not a solution to the original equation has been introduced. Such a solution is called an extraneous solution.

When both sides of an equation are raised to a power, an extraneous solution may be introduced. Be sure to check all solutions in the original equation.

The following examples illustrate a variety of situations involving radicals. The steps used are related to the following general procedure.

Solving Equations with Radicals

1. Isolate one of the radicals on one side of the equation. (An equation may have more than one radical.)
2. Raise both sides of the equation to the power corresponding to the index of the radical.
3. If the equation still contains a radical, repeat Steps 1 and 2.
4. Solve the equation after all the radicals have been eliminated.
5. Be sure to check all possible solutions in the original equation and eliminate any extraneous solutions.

PROCEDURE

Example 1 Solving Equations with One Radical

Solve the equation: $\sqrt{x^2+13}=7$

Solution

The radical is by itself on one side of the equation, so square both sides.

$$\sqrt{x^2+13}=7$$

$$\left(\sqrt{x^2+13}\right)^2=7^2 \quad \text{Square both sides.}$$

$$x^2+13=49 \quad \text{This new equation contains no radical.}$$

$$x^2-36=0$$

$$(x+6)(x-6)=0 \quad \text{Solve by factoring.}$$

$$x=-6 \quad \text{or} \quad x=6$$

Check

Check **both answers** in the original equation.

$$\sqrt{(-6)^2+13}\stackrel{?}{=}7 \qquad \sqrt{(6)^2+13}\stackrel{?}{=}7$$

$$\sqrt{36+13}\stackrel{?}{=}7 \qquad \sqrt{36+13}\stackrel{?}{=}7$$

$$\sqrt{49}\stackrel{?}{=}7 \qquad \sqrt{49}\stackrel{?}{=}7$$

$$7=7 \qquad 7=7$$

Both –6 and 6 are solutions.

Now work margin exercise 1.

1. Solve the equation: $\sqrt{x^2-48}=4$

Example 2 Solving Equations with One Radical

Solve the equation: $\sqrt{y^2-10y-11}=1+y$

Solution

Since there is only one radical and it is by itself on one side of the equation, square both sides.

$$\sqrt{y^2-10y-11}=1+y$$

$$\left(\sqrt{y^2-10y-11}\right)^2=(1+y)^2 \quad \text{Square both sides.}$$

$$y^2-10y-11=1+2y+y^2$$

$$-12y-12=0 \quad \text{Simplifying gives a first-degree equation.}$$

$$-12y=12$$

$$y=-1$$

2. Solve the equation: $\sqrt{y^2+y-1}=y+3$

Check

$$\sqrt{(-1)^2-10(-1)-11} \stackrel{?}{=} 1+(-1)$$
$$\sqrt{1+10-11} \stackrel{?}{=} 0$$
$$\sqrt{0} \stackrel{?}{=} 0$$
$$0 = 0$$

There is one solution, -1.

Now work margin exercise 2.

3. Solve the equation:
$\sqrt{-x^2+x+12}-3=x$

Example 3 Solving Equations with One Radical

Solve the equation: $\sqrt{3x+13}+3=2x$

Solution

$$\sqrt{3x+13}+3=2x$$
$$\sqrt{3x+13}=2x-3 \quad \text{Isolate the radical.}$$
$$\left(\sqrt{3x+13}\right)^2=(2x-3)^2 \quad \text{Square both sides.}$$
$$3x+13=4x^2-12x+9$$
$$0=4x^2-15x-4$$
$$0=(4x+1)(x-4) \quad \text{Solve by factoring.}$$
$$x=-\frac{1}{4} \quad \text{or} \quad x=4$$

Check

Check **both answers** in the original equation.

$$\sqrt{3\left(-\frac{1}{4}\right)+13}+3 \stackrel{?}{=} 2\left(-\frac{1}{4}\right)$$
$$\sqrt{\frac{49}{4}}+3 \stackrel{?}{=} -\frac{1}{2}$$
$$\frac{7}{2}+3 \stackrel{?}{=} -\frac{1}{2}$$
$$\frac{13}{2} \neq -\frac{1}{2}$$

$$\sqrt{3(4)+13}+3 \stackrel{?}{=} 2(4)$$
$$\sqrt{25}+3 \stackrel{?}{=} 8$$
$$5+3 \stackrel{?}{=} 8$$
$$8 = 8$$

$-\frac{1}{4}$ is **not** a solution. The only solution is 4.

Now work margin exercise 3.

Math Tip

It is vital that you always check all the solutions using the original equation when solving equations that contain radicals. Extraneous solutions or no solution can be very common when solving these types of equations.

Example 4 Solving Equations with One Radical

Solve the equation: $\sqrt{x+1}=-3$

Solution

We could stop right here. There is **no real solution** to this equation because the radical on the left is nonnegative and cannot possibly equal –3, a negative number. Suppose we did not notice this relationship. Then, proceeding as usual, we will find an answer that is not a solution to the original equation.

$$\sqrt{x+1}=-3$$

$$\left(\sqrt{x+1}\right)^2=(-3)^2 \quad \text{Square both sides.}$$

$$x+1=9$$

$$x=8$$

Check

$$\sqrt{(8)+1}\stackrel{?}{=}-3$$

$$\sqrt{9}\stackrel{?}{=}-3$$

$$3\neq-3$$

So, 8 does **not** check, therefore there is **no solution**.

Now work margin exercise 4.

4. Solve the equation:
$\sqrt{x+5}=-9$

Note

It is possible that after checking the answers, you may find that none of the answers are solutions. In this case the answer is no solution.

Completion Example 5 Solving Equations with One Radical

Solve the equation: $\sqrt{5y+6}=3y-2$

Solution

$$\sqrt{5y+6}=3y-2$$

$$\left(\underline{\sqrt{5y+6}}\right)^2=\left(\underline{3y-2}\right)^2$$

$$\underline{5y+6}=\underline{9}\,y^2-\underline{12}\,y+\underline{4}$$

$$0=\underline{9}\,y^2-\underline{17}\,y-\underline{2}$$

$$0=\left(\underline{9y+1}\right)\left(\underline{y-2}\right)$$

$$\underline{9y+1}=0 \quad \text{or} \quad \underline{y-2}=0$$

$$y=\underline{-\frac{1}{9}} \qquad y=\underline{2}$$

The number $\underline{-\frac{1}{9}}$ does not check. The number $\underline{2}$ does check.

So, the solution is $\underline{2}$.

Now work margin exercise 5.

5. Solve the equation:
$\sqrt{3x+4}=2x+1$

6. Solve the equation:
$\sqrt{2x+5}=\sqrt{x+7}$

Example 6 Solving Equations with Two Radicals

Solve the equation: $\sqrt{x+4}=\sqrt{3x-2}$

Solution

There are two radicals on opposite sides of the equation. Squaring both sides will give a new equation with no radicals.

$$\sqrt{x+4}=\sqrt{3x-2}$$
$$\left(\sqrt{x+4}\right)^2=\left(\sqrt{3x-2}\right)^2$$
$$x+4=3x-2$$
$$6=2x \quad \text{Simplifying results in a first-degree equation.}$$
$$3=x$$

Check

$$\sqrt{(3)+4}\stackrel{?}{=}\sqrt{3(3)-2}$$
$$\sqrt{7}=\sqrt{7}$$

There is one solution, 3.

Now work margin exercise 6.

7. Solve the equation:
$2\sqrt{x}-\sqrt{3x-11}=2$

Example 7 Solving Equations with Two Radicals

Solve the equation: $\sqrt{x}-\sqrt{2x-14}=1$

Solution

Where there is a sum or difference of radicals, squaring is easier if the radicals are on different sides of the equation. Also, squaring both sides of the equation is easier if one of the radicals is by itself on one side of the equation.

$$\sqrt{x}-\sqrt{2x-14}=1$$
$$\sqrt{x}=1+\sqrt{2x-14} \quad \text{Isolate one of the radicals.}$$
$$\left(\sqrt{x}\right)^2=\left(1+\sqrt{2x-14}\right)^2 \quad \text{Square both sides.}$$
$$x=1+2\sqrt{2x-14}+(2x-14) \quad \text{Treat the right-hand side as the square of a binomial.}$$
$$x=2\sqrt{2x-14}+2x-13$$
$$-x+13=2\sqrt{2x-14} \quad \text{Simplify so that the radical is on one side by itself.}$$
$$(-x+13)^2=\left(2\sqrt{2x-14}\right)^2 \quad \text{Square both sides again.}$$
$$x^2-26x+169=4(2x-14)$$
$$x^2-26x+169=8x-56$$
$$x^2-34x+225=0$$
$$(x-9)(x-25)=0 \quad \text{Solve by factoring.}$$
$$x=9 \quad \text{or} \quad x=25$$

Check

Check **both answers** in the original equation.

$$\sqrt{(9)}-\sqrt{2(9)-14} \stackrel{?}{=} 1 \qquad \sqrt{(25)}-\sqrt{2(25)-14} \stackrel{?}{=} 1$$
$$3-\sqrt{4} \stackrel{?}{=} 1 \qquad 5-\sqrt{36} \stackrel{?}{=} 1$$
$$3-2 \stackrel{?}{=} 1 \qquad 5-6 \stackrel{?}{=} 1$$
$$1 = 1 \qquad -1 \neq 1$$

25 is **not** a solution. The only solution is 9.

Now work margin exercise 7.

Example 8 Solving Equations Containing a Cube Root

Solve the following equation containing a cube root. $\sqrt[3]{2x+1}+1=3$

8. Solve the following equation containing a cube root. $\sqrt[3]{3x+3}+2=5$

Solution

First, get the radical by itself on one side of the equation. Then, since this radical is a cube root, cube both sides of the equation.

$$\sqrt[3]{2x+1}+1=3$$
$$\sqrt[3]{2x+1}=2 \quad \text{Add } -1 \text{ to both sides.}$$
$$\left(\sqrt[3]{2x+1}\right)^3=2^3 \quad \text{Cube both sides.}$$
$$2x+1=8 \quad \text{Solve the equation.}$$
$$x=\frac{7}{2}$$

Check

Check in the original equation.

$$\sqrt[3]{2\left(\frac{7}{2}\right)+1}+1 \stackrel{?}{=} 3$$
$$\sqrt[3]{7+1}+1 \stackrel{?}{=} 3$$
$$\sqrt[3]{8}+1 \stackrel{?}{=} 3$$
$$2+1 \stackrel{?}{=} 3$$
$$3=3$$

There is one solution, $\frac{7}{2}$.

Now work margin exercise 8.

Completion Example Answers

5. $\left(\sqrt{5y+6}\right)^2=(3y-2)^2$

$5y+6=9y^2-12y+4$

$0=9y^2-17y-2$

$0=(9y+1)(y-2)$

$9y+1=0$ or $y-2=0$; $y=-\frac{1}{9}$; $y=2$

The number $-\frac{1}{9}$ does not check. The number 2 does check.

So, the solution is 2.

15.6 Exercises

Concept Check

1. isolate
2. index
3. radicals
4. extraneous
5. extraneous
6. False; Radical equations may also have one or no solution.
7. True
8. False; The process needs to be repeated until all radicals have been eliminated.
9. True

Practice

1. $x=3$
2. $x=5$
3. No solution
4. $x=13$
5. $x=-3$
6. $x=-14$
7. $x=14$
8. No solution
9. $x=9$
10. $x=6$
11. No solution
12. $x=10$
13. $x=6$
14. No solution
15. $x=-4, 1$
16. $x=-4, 9$
17. $x=-5, \frac{5}{2}$
18. $x=-\frac{7}{3}, 7$
19. $x=-2$
20. $x=2$
21. $x=2, 3$

Margin Exercise Answers

1. $x=-8, 8$ **2.** $y=-2$ **3.** $x=\frac{1}{2}, -3$ **4.** No solution **5.** $x=\frac{3}{4}$ **6.** $x=2$ **7.** $x=9, 25$ **8.** $x=8$

15.6 Exercises

Concept Check

Fill-in-the-Blank. Complete each sentence using information found in this section.

1. To solve equations with radicals, first ________ one of the radicals on one side of the equation.
2. Next, raise both sides of the equation to the power corresponding to the ________ of the radical.
3. Solve the equation after all ________ have been eliminated.
4. When both sides of an equation are raised to a power, a/an ________ solution may be introduced.
5. Once all possible solutions have been found for an equation, those solutions should be checked to eliminate any ________ solutions.

True/False. Determine whether each statement is true or false. If a statement is false, explain how it can be changed so the statement will be true. (**Note:** There may be more than one acceptable change.)

6. All radical equations will have two solutions.
7. If no true statements result when all possible solutions are checked, then there is no solution.
8. When solving equations with radicals, you should only have to raise both sides of the equation to a power one time.
9. A radical expression set equal to a negative value, such as $\sqrt{x+2}=-4$, has no real solution.

Practice

Solve the following equations. Be sure to check your answers in the original equation.

1. $\sqrt{8x+1}=5$
2. $\sqrt{7x+1}=6$
3. $\sqrt{3x+4}=-5$
4. $\sqrt{4x-3}=7$
5. $\sqrt{6-x}=3$
6. $\sqrt{11-x}=5$
7. $\sqrt{5x-6}=8$
8. $\sqrt{2x-5}=-1$
9. $\sqrt{5x+4}=7$
10. $\sqrt{3x-2}=4$
11. $\sqrt{x-4}+6=2$
12. $\sqrt{6x+4}+2=10$

13. $\sqrt{4x+1}+4=9$

14. $\sqrt{2x-7}+5=3$

15. $\sqrt{x(x+3)}=2$

16. $\sqrt{x(x-5)}=6$

17. $\sqrt{x(2x+5)}=5$

18. $\sqrt{x(3x-14)}=7$

19. $\sqrt{x+6}=x+4$

20. $\sqrt{x+7}=2x-1$

21. $\sqrt{x-2}=x-2$

22. $\sqrt{x+3}=x+3$

23. $x-2=\sqrt{3x-6}$

24. $x+6=\sqrt{2x+12}$

25. $\sqrt{x^2-16}=3$

26. $\sqrt{x^2-25}=12$

27. $5+\sqrt{x+5}-2x=0$

28. $x-2-\sqrt{x+4}=0$

29. $2x=\sqrt{7x-3}+3$

30. $x-\sqrt{3x-8}=4$

31. $\sqrt{2x+5}=\sqrt{4x-1}$

32. $\sqrt{5x-1}=\sqrt{x+7}$

33. $\sqrt{3x+2}=\sqrt{9x-10}$

34. $\sqrt{2+x}=\sqrt{2x-7}$

35. $\sqrt{2x-1}=\sqrt{x+1}$

36. $\sqrt{3x+2}=\sqrt{x+4}$

37. $\sqrt{x+2}=\sqrt{2x-5}$

38. $\sqrt{2x-5}=\sqrt{3x-9}$

39. $\sqrt{4x-3}=\sqrt{2x+5}$

40. $\sqrt{4x-6}=\sqrt{3x-1}$

41. $\sqrt{3x+1}=1-\sqrt{x}$

42. $\sqrt{x}=\sqrt{x+16}-2$

43. $\sqrt{x+4}=\sqrt{x+11}-1$

44. $\sqrt{1-x}+2=\sqrt{13-x}$

45. $\sqrt{x+1}=\sqrt{x+6}+1$

46. $\sqrt{x+4}=\sqrt{x+20}-2$

47. $\sqrt{x+5}+\sqrt{x}=5$

48. $\sqrt{x}+\sqrt{x-3}=3$

49. $\sqrt{2x+3}=1+\sqrt{x+1}$

50. $\sqrt{5x-18}-4=\sqrt{5x+6}$

51. $\sqrt{3x+1}-\sqrt{x+4}=1$

52. $\sqrt{3x+4}-\sqrt{x+5}=1$

53. $\sqrt{5x-1}=4-\sqrt{x-1}$

54. $\sqrt{2x-5}-2=\sqrt{x-2}$

55. $\sqrt{2x-1}+\sqrt{x+3}=3$

56. $\sqrt{2x+3}-\sqrt{x+5}=1$

57. $\sqrt[3]{4+3x}=-2$

58. $\sqrt[3]{2+9x}=9$

59. $\sqrt[3]{5x+4}=4$

60. $\sqrt[3]{7x+1}=-5$

22. $x=-3, -2$
23. $x=2, 5$
24. $x=-6, -4$
25. $x=-5, 5$
26. $x=-13, 13$
27. $x=4$
28. $x=5$
29. $x=4$
30. $x=8$
31. $x=3$
32. $x=2$
33. $x=2$
34. $x=9$
35. $x=2$
36. $x=1$
37. $x=7$
38. $x=4$
39. $x=4$
40. $x=5$
41. $x=0$
42. $x=9$
43. $x=5$
44. $x=-3$
45. No solution
46. $x=5$
47. $x=4$
48. $x=4$
49. $x=-1, 3$
50. No solution
51. $x=5$
52. $x=4$
53. $x=2$
54. $x=27$
55. $x=1$
56. $x=11$
57. $x=-4$
58. $x=\frac{727}{9}$
59. $x=12$
60. $x=-18$

Applications

61. a. $1814.06

b. $r = \sqrt{\frac{A}{P}} - 1$

62. 15 hours

63. a. 1 sec

b. 3 ft

64. a. 5.05 m

b. 31.57 m

c. 126.26 m

Applications

Solve.

61. When money is invested in an account earning an annual interest rate of r %, and the money is left in the account for two years, the interest rate, the principal (the initial amount invested), and the value accumulated after two years are related by the following formula.

$$(r+1)\sqrt{P} = \sqrt{A}$$

In this formula, r is the annual interest rate written as a decimal, A is the accumulated value, and P is the principal invested.

a. Suppose you originally invested your money at an annual interest rate of 5%, and at the end of 2 years, your account contained $2000. How much did you initially invest? Round your answer to the nearest cent.

b. Solve the formula for the annual interest rate so that you will have a formula for the interest rate in terms of the principal invested and the accumulated value.

62. You want to buy a used car for $8000. You already have $7840 saved up for the car. However, the person you are buying the car from has received several offers already. The current owner of the car says that he will hold the car for you for one week before taking another offer. At your after-school job, you make $\sqrt{1500x} + 10$ dollars when you work x hours. How many hours will you need to work in the next week in order to be able to buy the car? (Assume that all money from your after-school job can go straight to your car fund.) Round your answer to the nearest whole hour.

63. The hang time of an athlete can be represented by

$$t = 2\sqrt{\frac{2h}{g}},$$

where t is the hang time of the athlete in seconds, h is the height of the jump in feet, and g is the acceleration due to gravity. (The gravity constant g can be estimated by using 32 ft/sec^2.)

a. If Michael Jordan had a vertical jump of 48 inches, how long would he be in the air?

b. A volleyball player has a hang time of 0.866 seconds. How high is this player's vertical leap? Round your answer to the nearest foot.

64. A company claims their new line of dishware is indestructible, so product testers are dropping dinner plates from varying distances to determine the damage that happens on impact with the ground. The equation $v = \sqrt{19.8h}$ is used to calculate the velocity v of the dinner plate in meters per second when it is dropped from a certain height h in meters. Round all answers to the nearest hundredth.

a. From what height was the dinner plate dropped if its velocity on impact was 10 m/s?

b. From what height was the dinner plate dropped if its velocity on impact was 25 m/s?

c. From what height was the dinner plate dropped its velocity on impact was 50 m/s?

65. Giovanna drops a stone from the highest point of a bridge into the river below. When Giovanna dropped the stone, her arm was stretched out 2 m above the bridge. The equation $t = \frac{\sqrt{10(d+2)}}{7}$ is used to find the time t it takes for a stone to drop a distance of $d + 2$ meters. If it takes the stone 2.5 seconds to hit the water, what is the height of the bridge to the nearest hundredth?

66. A landscaper is designing a pond in the shape of a right triangle that has a square flower patch along each edge. She knows two of the flower patches will have side lengths of 5 ft and 13 ft and that the remaining flower patch must have a side length a that satisfies the equation $13 = \sqrt{a^2 + 5^2}$. What is the side length of the remaining flower patch?

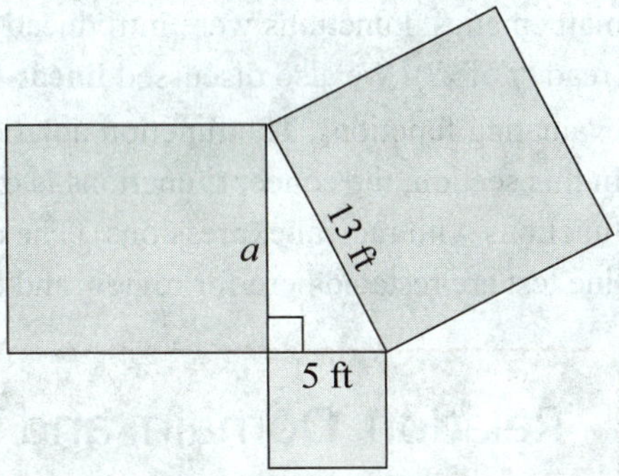

67. A farmer fell asleep under a tree in his apple orchard while thinking about pie. While he was sleeping, a squirrel knocked an apple off of a branch of the tree. The function $f(d) = \sqrt{\frac{2d}{9.8}}$ can be used to find the amount of time in seconds that it takes for the apple to drop a certain distance d, where d is in meters. Round your answers to the nearest hundredth.

a. If the apple was connected to a branch that was 2 meters above the farmer's head, how long would it take before the apple hit the top of the farmer's head?

b. If the squirrel knocked a second apple off of a branch that was 5 meters above the farmer's head, how long would it take before the apple hit the top of the farmer's head?

c. Suppose the second apple missed the farmer's head and landed on the ground instead. If the farmer's head was 0.8 meters above the ground, how long did it take for the apple to hit the ground?

68. A person's Body Mass Index (BMI) is determined by the formula $B = \frac{703w}{h^2}$, where B is the BMI, w is the person's weight in pounds, and h is the person's height in inches. Having a BMI between 18.5 and 24.9 is considered optimal. A BMI between 25 and 29.9 is considered overweight and a BMI over 30 is considered obese. A BMI below 18.5 is considered underweight.

a. Solve the BMI formula for the variable h.

b. How tall is a person who has a BMI of 20 and a weight of 120 pounds? Round to the nearest inch.

c. To be in the optimal BMI range with a weight of 200 pounds, what range in height should a person be? Round to the nearest inch. (**Hint:** Calculate the heights for the endpoints of the BMI range given.)

d. How tall is a person whose BMI is 30 and who weighs 150 pounds? Round to the nearest inch.

65. 28.63 m

66. 12 ft

67. **a.** 0.64 seconds
b. 1.01 seconds
c. 1.09 seconds

68. **a.** $h = \sqrt{\frac{703w}{B}}$
b. 65 inches
c. 75 inches to 87 inches
d. 59 inches

Objectives

A. Review the concepts of functions and function notation.

B. Find the domain and range of radical functions.

C. Evaluate radical functions.

D. Graph radical functions.

E. Use a graphing calculator to graph radical functions.

15.7 Functions with Radicals

A Review of Functions and Function Notation

The concept of functions is among the most important and useful ideas in all of mathematics. Functions were introduced earlier along with function notation, such as $f(x)$ (read "f of x"). We also discussed **linear functions** and the use of function notation in evaluating functions. The function notation $p(x)$ was used again to represent polynomials. In this section, the concept functions is expanded to include **radical functions** (that is functions with radical expressions). The definitions of relations, functions, and the vertical line test are restated here for review and easy reference.

Relation, Domain, and Range

A **relation** is a set of ordered pairs of real numbers.

The **domain** ***D*** of a relation is the set of all first coordinates in the relation.

The **range** ***R*** of a relation is the set of all second coordinates in the relation.

DEFINITION

Function

A **function** is a relation in which each domain element has exactly one corresponding range element.

DEFINITION

Teaching Note

You may want to discuss with your students that both of the characteristics listed (1 and 2) are equivalent to the definition of a function.

The definition of a function can also be stated in the following ways.

1. A function is a relation in which each first coordinate appears only once.

2. A function is a relation in which no two ordered pairs have the same first coordinate.

Vertical Line Test

If **any** vertical line intersects the graph of a relation at more than one point, then the relation is **not** a function.

DEFINITION

We have used the ordered pair notation, (x, y), to represent points on the graphs of relations and functions. For example, $y = 2x - 5$ represents a linear function and its graph is a line (as shown in Figure 1).

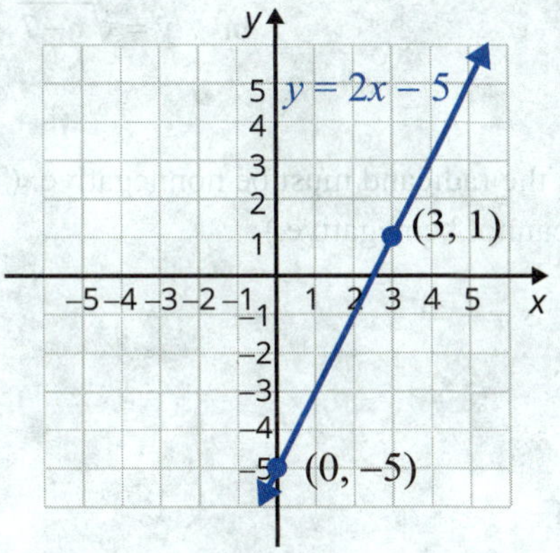

Figure 1

B Finding the Domain and Range of Radical Functions

We define **radical functions** as follows.

> ### Radical Function
>
> A **radical function** is a function of the form $y = \sqrt[n]{g(x)}$ in which the radicand contains a variable expression.
>
> The **domain** of such a function depends on the index n.
>
> **1.** If n is an even number, the domain is the set of all x such that $g(x) \geq 0$.
>
> **2.** If n is an odd number, the domain is the set of all real numbers, $(-\infty, \infty)$.
>
> The **range** of the function also depends on the index n.
>
> **1.** If n is an even number, the range is the set of all real numbers greater than or equal to 0, $[0, \infty)$.
>
> **2.** If n is an odd number, the range is the set of all real numbers, $(-\infty, \infty)$.
>
> DEFINITION

The following are examples of radical functions.

$$y = 3\sqrt{x}, \quad f(x) = \sqrt{2x+3}, \quad \text{and} \quad y = \sqrt[3]{x-7}$$

1. Determine the domain of each radical function.

a. $f(x)=\sqrt{5x+10}$

b. $y=\sqrt[3]{-x-9}$

Example 1 Finding the Domain of a Radical Function

Determine the domain of each radical function.

a. $f(x)=\sqrt{2x+3}$ **b.** $y=\sqrt[3]{x-7}$

Solution

a. Because the index is 2, the radicand must be nonnegative. (That is, the expression under the radical sign cannot be negative.)

Thus, we must have

$$2x+3\geq 0$$
$$2x\geq -3$$
$$x\geq -\frac{3}{2}$$

and the domain is the interval of real numbers $\left[-\frac{3}{2},\infty\right)$.

b. Because the index is 3, an odd number, the radicand may be any real number. Thus, the domain is the set of all real numbers $(-\infty,\infty)$.

Now work margin exercise 1.

C Evaluating Radical Functions

As illustrated in Chapter 10, function notation is particularly useful when evaluating functions for specific values of the variable. For example,

$$\text{if } f(x)=\sqrt{2x+3}, \text{ then } f\left(\frac{1}{2}\right)=\sqrt{2\left(\frac{1}{2}\right)+3}=\sqrt{1+3}=\sqrt{4}=2$$
$$\text{and } f(0)=\sqrt{2(0)+3}=\sqrt{0+3}=\sqrt{3}.$$

A calculator can be used to find decimal approximations. Such approximations are helpful when estimating the locations of points on a graph. For example,

$$\text{if } f(x)=\sqrt{x-5},$$
$$\text{then } f(8)=\sqrt{(8)-5}=\sqrt{3}\approx 1.7321 \quad \text{Accurate to the nearest ten-thousandth}$$
$$\text{and } f(25)=\sqrt{(25)-5}=\sqrt{20}\approx 4.4721. \quad \text{Accurate to the nearest ten-thousandth}$$

2. Evaluate each function for the given values of x.

a. $f(x)=5\sqrt{2x}$; find $f(x)$ values for $x=0, 2, 8$

b. $f(x)=\sqrt[3]{2x-3}$; find $f(x)$ values for $x=1, 2, 15$

Example 2 Evaluating Radical Functions

Complete each table by finding the corresponding $f(x)$ values for the given values of x.

a. $f(x)=\sqrt[3]{x-7}$

x	f(x)
7	?
6	?
−1	?

b. $f(x)=3\sqrt{x}$

x	f(x)
0	?
4	?
6	?

Solution

a.

x	$f(x)$
7	$\sqrt[3]{7-7}=\sqrt[3]{0}=0$
6	$\sqrt[3]{6-7}=\sqrt[3]{-1}=-1$
-1	$\sqrt[3]{-1-7}=\sqrt[3]{-8}=-2$

b.

x	$f(x)$
0	$3\sqrt{0}=0$
4	$3\sqrt{4}=3\cdot2=6$
6	$3\sqrt{6}\approx7.3485$

Now work margin exercise 2.

D Graphing Radical Functions

To graph a radical function, we need to be aware of its domain and plot at least a few points to see the nature of the resulting curve. Example 3 shows how to graph the radical function $y=\sqrt{x+5}$.

Example 3 Graphing a Radical Function

Graph the function $y=\sqrt{x+5}$.

Solution

For the domain, we have $x+5\geq0$, so $x\geq-5$.

To see the nature of the graph, we select a few values for x in the domain and find the corresponding values of y. Then we plot the points on a graph.

x	y
-5	0
-4	1
-3	$\sqrt{2}\approx1.41$
0	$\sqrt{5}\approx2.24$
4	3

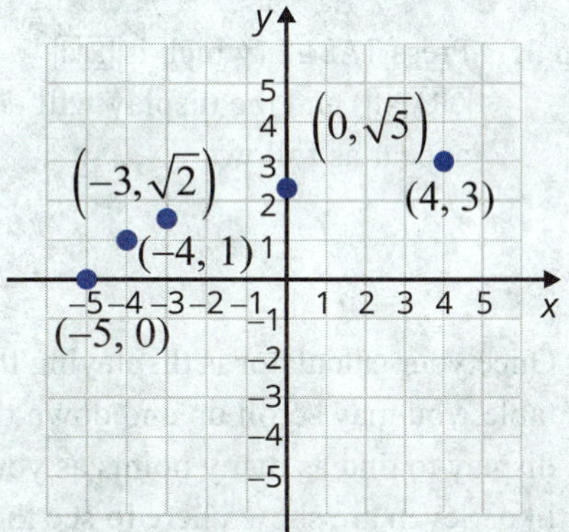

Note that $\sqrt{x+5}$ is the principal square root. This means that $y\geq0$. Thus, the point $(-5,0)$ is on the x-axis and is a zero of the function. The remaining points on the graph are above the x-axis. So, we can complete the graph by drawing a smooth curve that passes through the selected points. The graph of the function is shown here.

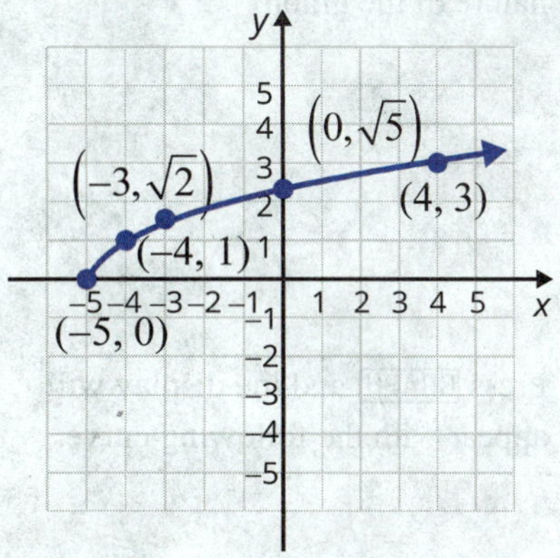

We see that the domain is $[-5,\infty)$ and the range is $[0,\infty)$.

Now work margin exercise 3.

3. Graph the function $y=-\sqrt{2x+1}$.

Math Tip

When asked to graph a radical function, with a bit of foresight you can choose x-values that are guaranteed to be easy to graph. Ask yourself, "What does the value of x need to be to cause the radicand to be a perfect root?" This way you can ensure that you can evaluate the radical without resorting to a decimal approximation.

E Using a Graphing Calculator to Graph Radical Functions

Example 4 shows how to use a TI-84 Plus graphing calculator to find many points on the graph of a radical function and then how to graph the function.

4. a. Use the TABLE feature of a TI-84 Plus graphing calculator to locate five points on the graph of the function $y=\sqrt[3]{2-4x}$.

b. Use a TI-84 Plus graphing calculator to graph the function.

Example 4 Using a TI-84 Plus to Graph a Radical Function

a. Use the TABLE feature of a TI-84 Plus graphing calculator to locate seven points on the graph of the function $y=\sqrt[3]{2x-3}$.

b. Plot seven points (approximately) on a graph and then connect them with a smooth curve.

c. Use a TI-84 Plus graphing calculator to graph the function.

Solution

a. Use the TABLE feature of a TI-84 Plus graphing calculator.

Step 1: Press [Y=] and enter the function as follows:

- **a.** Press [MATH].
- **b.** Choose 4: $\sqrt[3]{\ }$(.
- **c.** Enter 2X – 3) and press [ENTER].

Step 2: Press TBLSET (which is [2nd] [WINDOW]) and set the display as shown here.

Step 3: Press TABLE (which is [2nd] [GRAPH]) and the display will appear as follows.

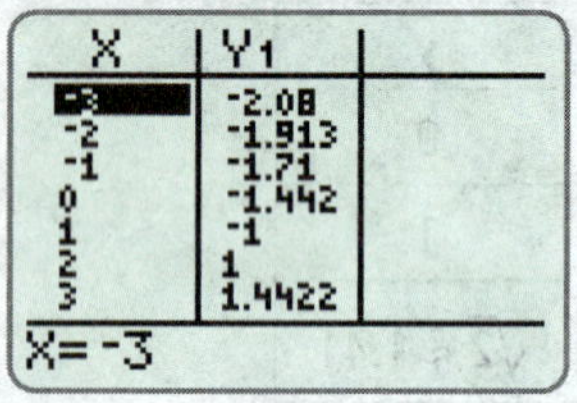

b. Once your calculator is displaying the table, you may scroll up and down the display to find as many points as you like. Seven are shown here to see the nature of the graph.

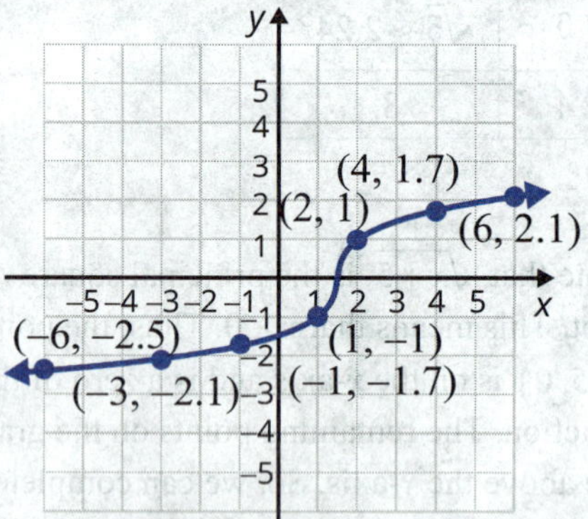

c. Press [GRAPH] and the display will appear with the following curve.

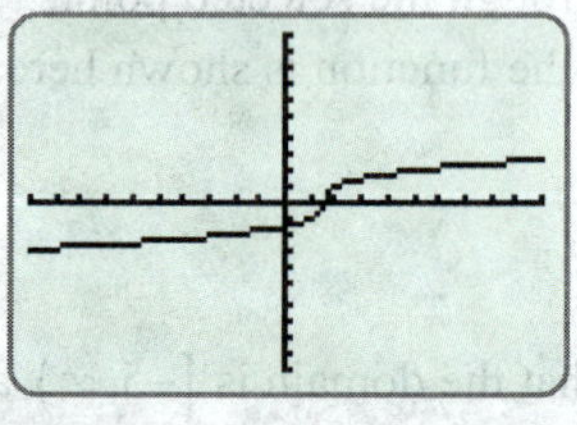

Now work margin exercise 4.

Note that you may need to adjust the window on your calculator. To adjust the window, use the following steps.

Press the WINDOW key and the standard window will be displayed. By default, the standard window displays a graph with x-values and y-values ranging from -10 to 10 with tic marks on the axes every 1 unit. This window can be changed at any time by changing the individual numbers or pressing the ZOOM key and selecting an option from the menu displayed. To return the screen back to the standard dimensions with x-values and y-values ranging from -10 to 10, press ZOOM and `ZStandard`. A square screen can be attained by pressing zoom and `ZSquare` or by pressing the window key and setting `Xmin` = −15 and `Xmax` = 15 to give the x-axis a length of 30 and the y-axis a length of 20 (a ratio of 3:2).

Margin Exercise Answers

1. a. $[-2,\infty)$ **b.** $(-\infty,\infty)$ **2. a.** $f(0)=0, f(2)=10, f(8)=20$ **b.** $f(1)=-1, f(2)=1, f(15)=3$

3.

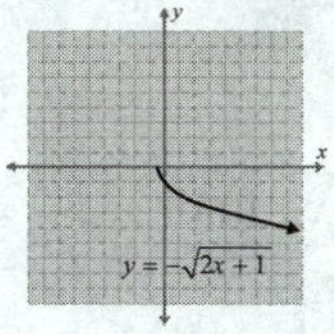

4. a.

x	y_1
−7	3.1072
−1	1.817
0	1.2599
2	−1.817
7	−2.962

b.

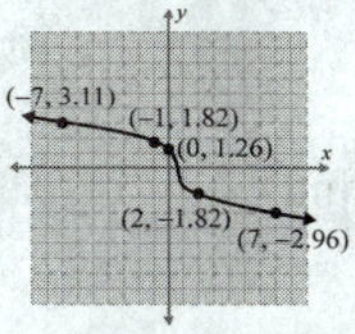

15.7 Exercises

Concept Check

Fill-in-the-Blank. Complete each sentence using information found in this section.

1. A relation is a set of __________ pairs of real numbers.

2. The set of all second coordinates of a relation is the relation's _______.

3. The domain D of a relation is the set of all _____ __________ in the relation.

4. A/An ___________ is a relation in which each domain element has exactly one corresponding range element.

5. If any ____________ line intersects the graph of a relation at more than one point, then the relation is not a/an ____________.

6. A radical function is a function of the form $y=\sqrt[n]{g(x)}$ in which the radicand contains a/an ________ _________.

True/False. Determine whether each statement is true or false. If a statement is false, explain how it can be changed so the statement will be true. (**Note:** There may be more than one acceptable change.)

7. If a radical function has an index that is an even number, then the domain is the set of all x such that $g(x)\geq 0$.

8. If a radical function has an odd numbered index, the domain is the set of all positive numbers.

15.7 Exercises

Concept Check

1. ordered
2. range
3. first coordinates
4. function
5. vertical, function
6. variable expression
7. True
8. False; The domain is the set of all real numbers.
9. True
10. True

Practice

1. a. $\sqrt{5}\approx 2.2361$
b. 3
c. $5\sqrt{2}\approx 7.0711$
d. 2

2. a. $\sqrt{5}\approx 2.2361$
b. $\sqrt{11}\approx 3.3166$
c. 5
d. $\sqrt{12.2}\approx 3.4928$

3. a. 3
b. −1
c. −2
d. $2\sqrt[3]{3}\approx 2.8845$

4. a. 0
b. 2
c. 0.1
d. $\sqrt[3]{6.5}\approx 1.8663$

5. $[-8,\infty)$

6. $\left[\frac{1}{2},\infty\right)$

7. $\left(-\infty,\frac{1}{2}\right]$

8. $\left(-\infty,\frac{1}{3}\right]$

9. $(-\infty,\infty)$

10. $(-\infty,\infty)$

11. $(-\infty,2]$

12. $[-4, \infty)$

13. $(-\infty, \infty)$

14. $(-\infty, \infty)$

15. $D = [-2, \infty); R = [0, \infty);$ zero $= (-2, 0)$

16. $D = (-\infty, 3]; R = (-\infty, 0];$ zero $= (3, 0)$

17. $D = [-1, \infty); R = (-\infty, 1];$ zero $= (0, 0)$

18. $D = [0, \infty); R = [-1, \infty);$ zero $= (1, 0)$

19. $D = [0, \infty); R = [3, \infty);$ zero = none

20. $D = [-2, \infty); R = (-\infty, -2];$ zero = none

21. E

22. D

23. B

24. F

25. A

26. C

27.

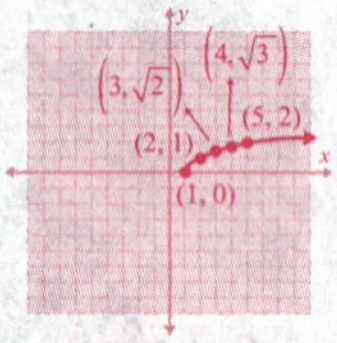

28.

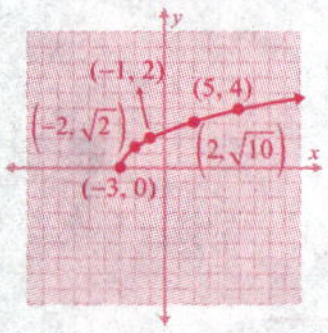

29.

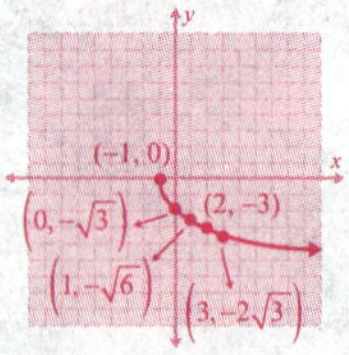

30.

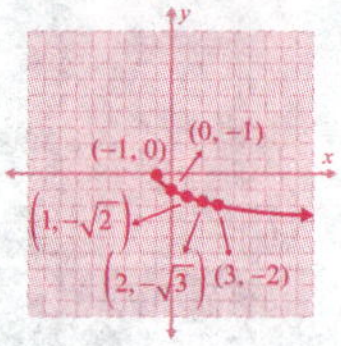

9. Both the domain and the range of a radical function depend on the index.

10. To graph a radical function, you must be aware of its domain and you should plot at least a few points to see the nature of the resulting curve.

Practice

Find each function value as indicated and round decimal values to the nearest ten-thousandth, if necessary. See Example 2.

1. Given $f(x) = \sqrt{2x+1}$, find

a. $f(2)$

b. $f(4)$

c. $f(24.5)$

d. $f(1.5)$

2. Given $f(x) = \sqrt{5-3x}$, find

a. $f(0)$

b. $f(-2)$

c. $f\left(-\frac{20}{3}\right)$

d. $f(-2.4)$

3. Given $g(x) = \sqrt[3]{x+6}$, find

a. $g(21)$

b. $g(-7)$

c. $g(-14)$

d. $g(18)$

4. Given $h(x) = \sqrt[3]{4-x}$, find

a. $h(4)$

b. $h(-4)$

c. $h(3.999)$

d. $h(-2.5)$

Use interval notation to indicate the domain of each radical function. See Example 1.

5. $y = \sqrt{x+8}$

6. $y = \sqrt{2x-1}$

7. $y = \sqrt{2.5-5x}$

8. $y = \sqrt{1-3x}$

9. $f(x) = \sqrt[3]{x+4}$

10. $f(x) = \sqrt[3]{6x}$

11. $g(x) = \sqrt{6-3x}$

12. $g(x) = \sqrt{x+4}$

13. $y = \sqrt[3]{2-5x}$

14. $y = \sqrt[3]{5x+9}$

Identify the domain, range, and any zeros from the graphs of the following radical functions.

15.

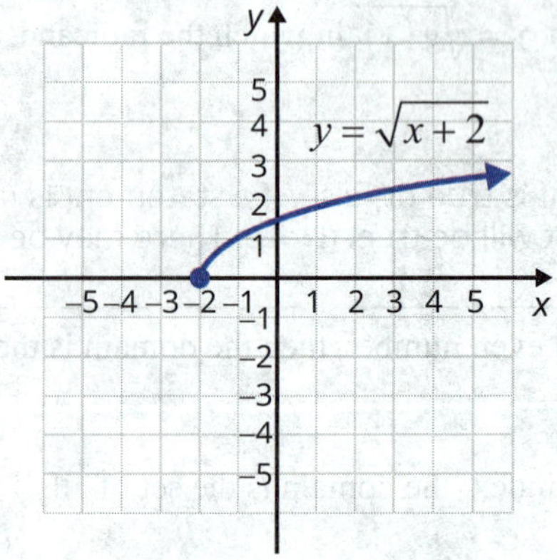

16.

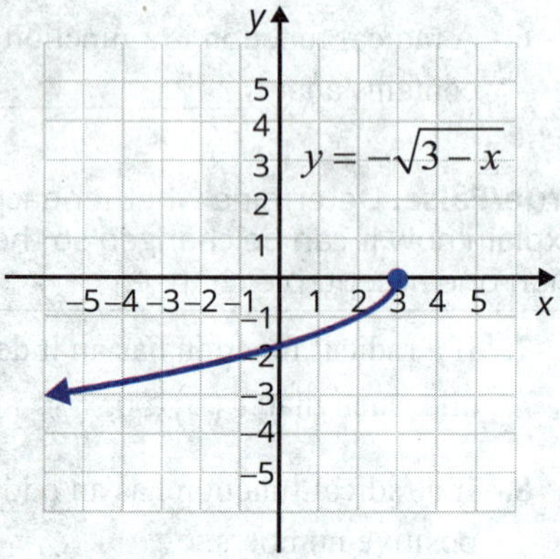

17.

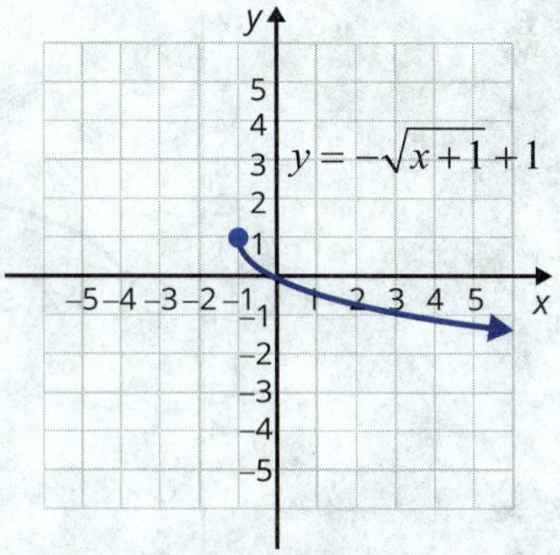

19.

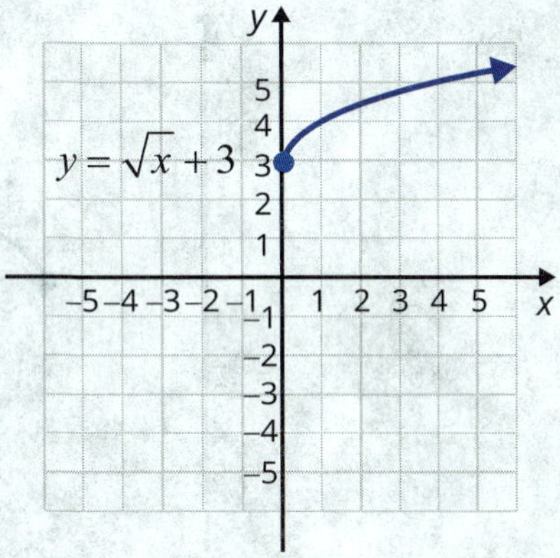

18.

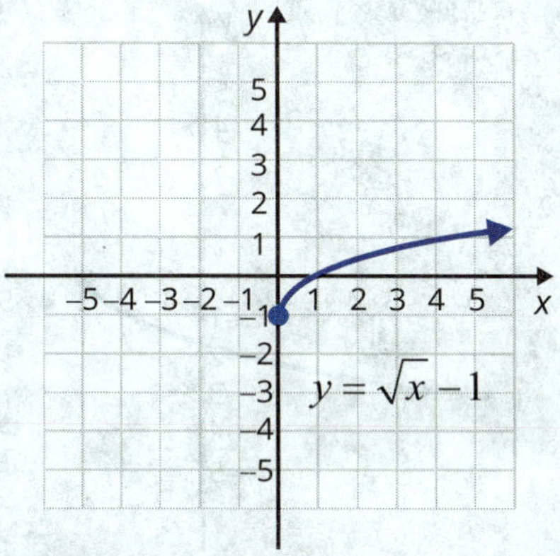

20.

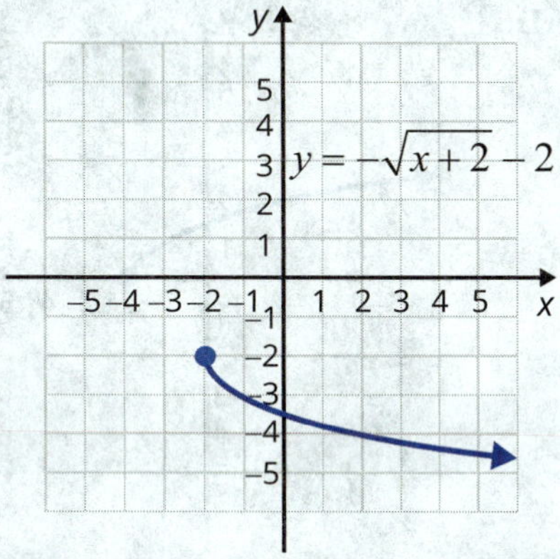

Match the functions given with the graphs of the functions (A) through (F).

21. $y=\sqrt{x-2}$

22. $y=\sqrt{2-x}$

23. $y=-\sqrt{x-3}$

24. $y=-\sqrt{3-x}$

25. $y=\sqrt{x+4}$

26. $y=\sqrt{x-4}$

A.

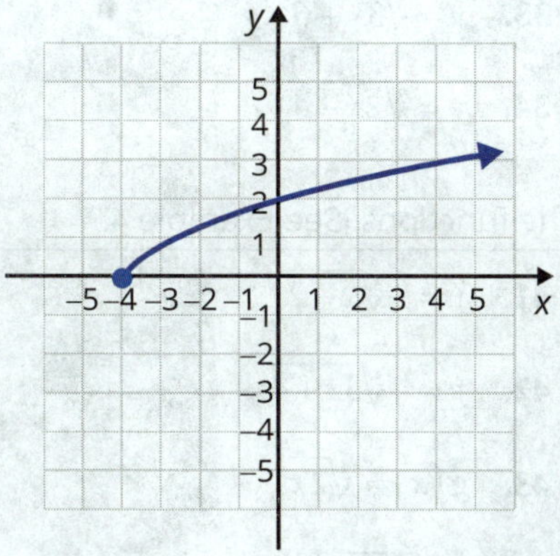

B.

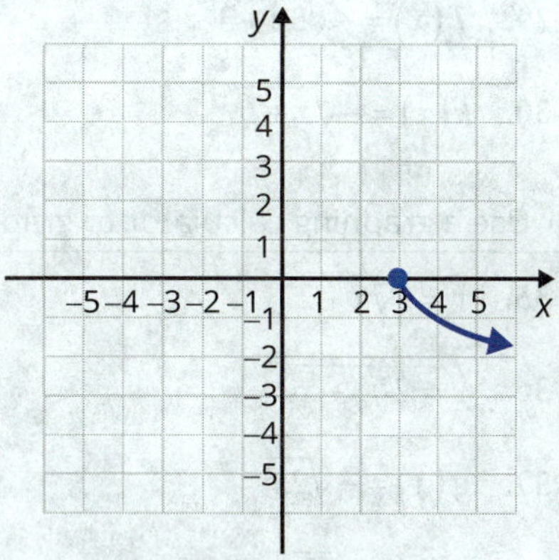

31.

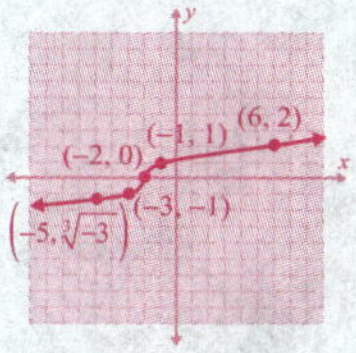

32.

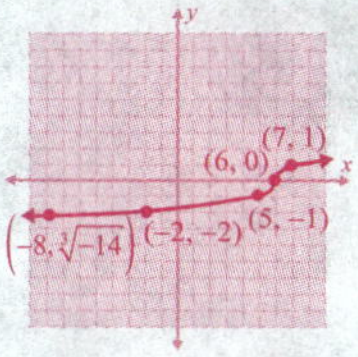

33.

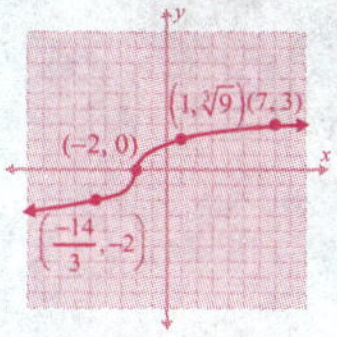

34.

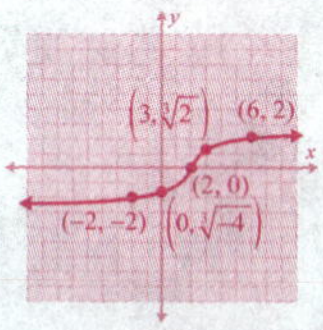

35.

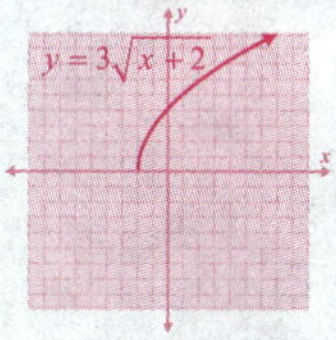

36.

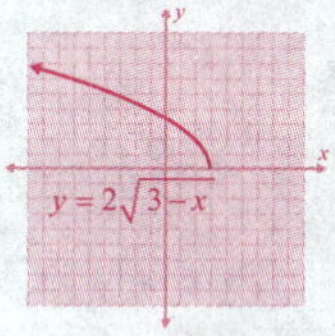

37.

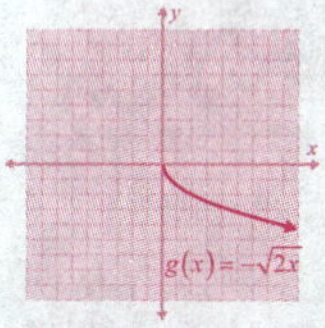

38.

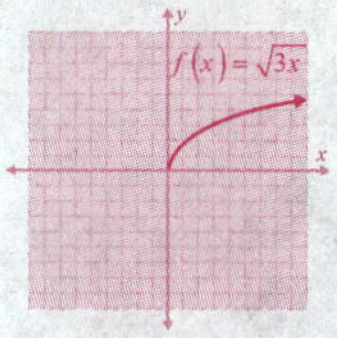

39.

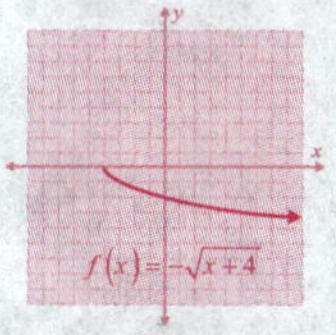

40.

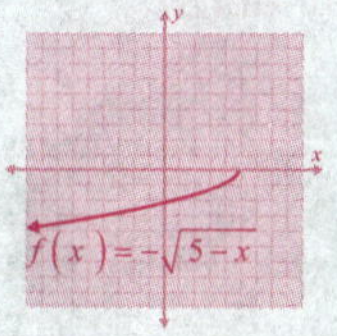

41.

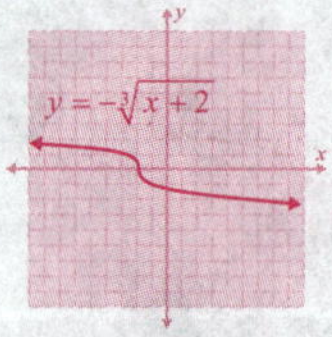

42.

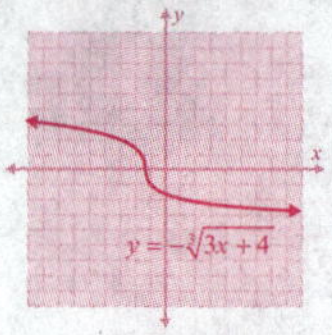

43.

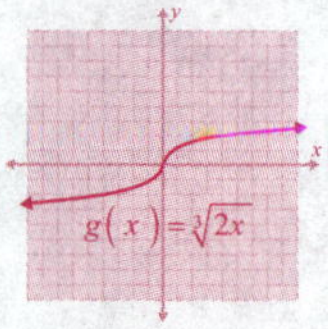

44.

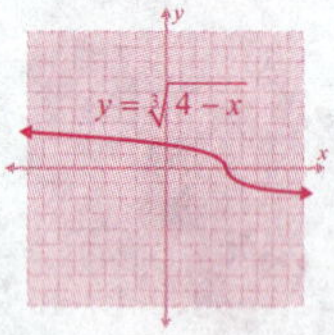

45.

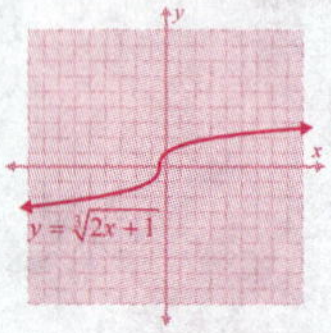

46.

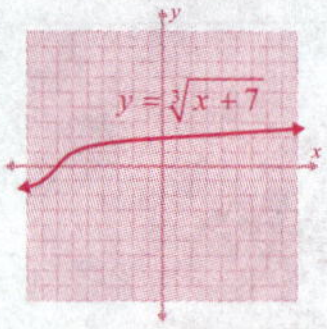

47. a.

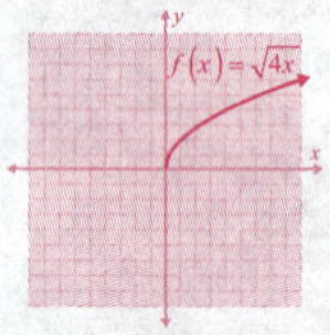

$D=[0,\infty)$; The graph of $\sqrt{x}$ is stretched vertically.

b.

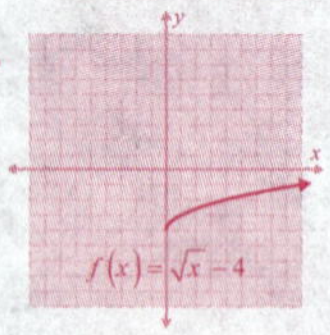

$D=[0,\infty)$; The graph of $\sqrt{x}$ is shifted down four units.

C.

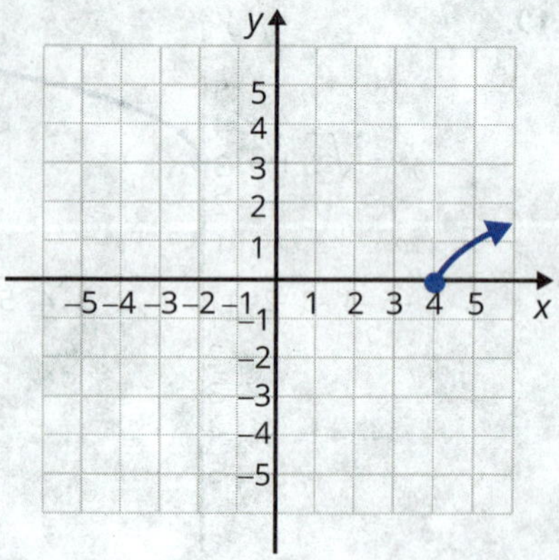

E.

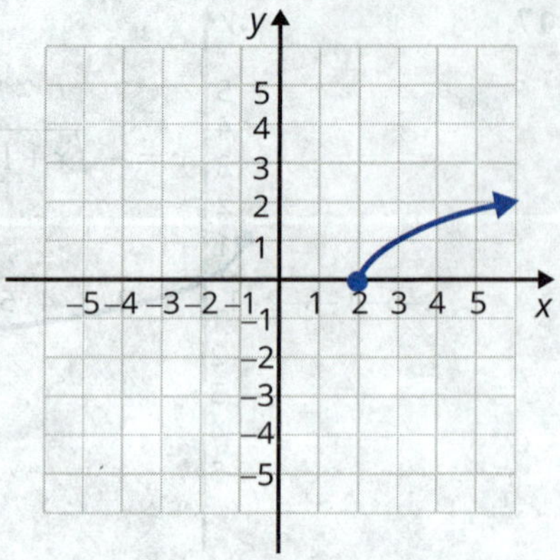

D.

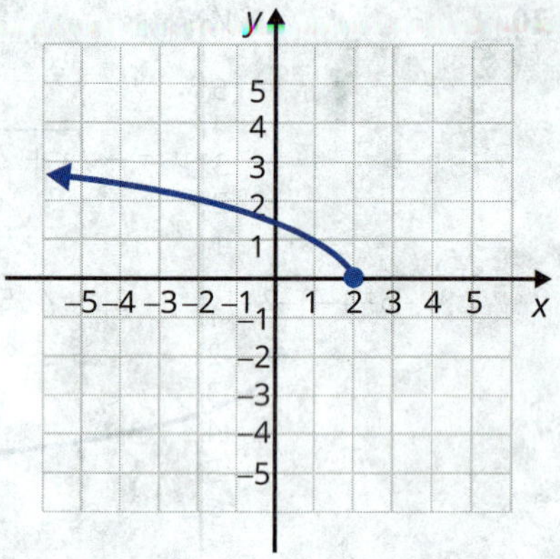

F.

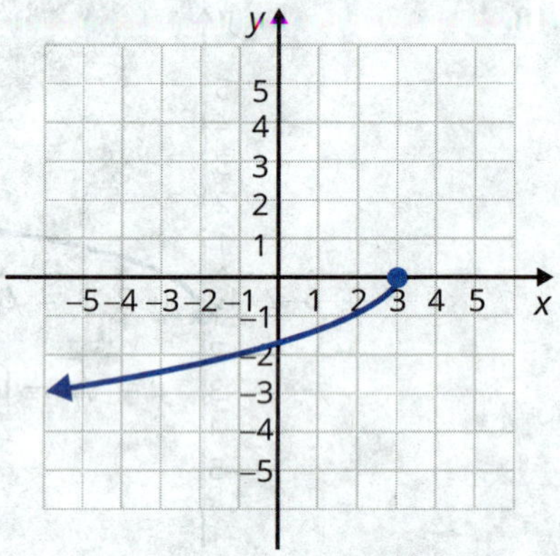

Determine at least 5 points for the given function and then sketch the graph of the function. See Example 3.

27. $y=\sqrt{x-1}$

28. $y=\sqrt{2x+6}$

29. $f(x)=-\sqrt{3x+3}$

30. $h(x)=-\sqrt{x+1}$

31. $f(x)=\sqrt[3]{x+2}$

32. $g(x)=\sqrt[3]{x-6}$

33. $y=\sqrt[3]{3x+6}$

34. $y=\sqrt[3]{2x-4}$

Use a graphing calculator to graph each of the functions. See Example 4.

35. $y=3\sqrt{x+2}$

36. $y=2\sqrt{3-x}$

37. $g(x)=-\sqrt{2x}$

38. $f(x)=\sqrt{3x}$

39. $f(x)=-\sqrt{x+4}$

40. $f(x)=-\sqrt{5-x}$

41. $y=-\sqrt[3]{x+2}$

42. $y=-\sqrt[3]{3x+4}$

43. $g(x)=\sqrt[3]{2x}$

44. $y=\sqrt[3]{4-x}$

45. $y=\sqrt[3]{2x+1}$

46. $y=\sqrt[3]{x+7}$

Complete the following problems.

47. Graph the following three radical functions. For each function, state the domain of the function using interval notation. Then, describe how the graph of the function differs from the graph of $f(x)=\sqrt{x}$.

a. $f(x)=\sqrt{4x}$ **b.** $f(x)=\sqrt{x}-4$ **c.** $f(x)=-\sqrt{x-4}$

48. Graph the following three radical functions. For each function, state the domain of the function using interval notation. Then, describe how the graph of the function differs from the graph of $f(x)=\sqrt{x}$.

a. $f(x)=-\sqrt{x}$ **b.** $f(x)=\sqrt{x+1}$ **c.** $f(x)=\sqrt{x}+1$

49. Using your results from Exercises 47 and 48, discuss any general conclusions you can draw from the differences between $f(x)=\sqrt{x}$ and $f(x)=-\sqrt{(ax+b)}+c$.

Applications

Solve.

50. The hang time of an athlete can be represented by

$$t=2\sqrt{\frac{2h}{g}},$$

where t is the hang time of the athlete in seconds, h is the height of the jump in feet, and g is the acceleration due to gravity. (The gravity constant g can be estimated by using 32 ft/sec^2.)

a. Using a graphing calculator, graph this equation.

b. Identify the domain and range of this function. Does this make sense in the context of the function?

c. Find the hang time of an athlete with a 24-inch vertical leap. Round your answer to the nearest hundredth.

51. The motion of a simple pendulum is represented by the following equation, where T = the pendulum period in seconds, L = length in meters, and g = acceleration of gravity.

$$T=2\pi\sqrt{\frac{L}{g}}$$

Use $g=9.8$ m/s^2 and $\pi=3.14$.

a. Using a graphing calculator, graph this equation.

b. Identify the domain and range of this function. Does this make sense in the context of the function?

c. Find the pendulum period of a simple pendulum with length 15 meters. Round your answer to the nearest hundredth.

c.

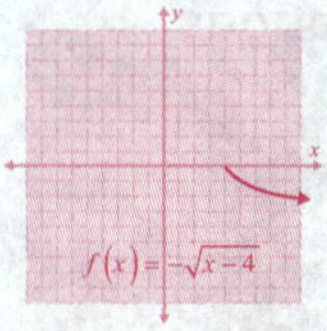

$D=[4,\infty)$; The graph of $\sqrt{x}$ is shifted to the right four units. and then reflected across the x–axis

48. a.

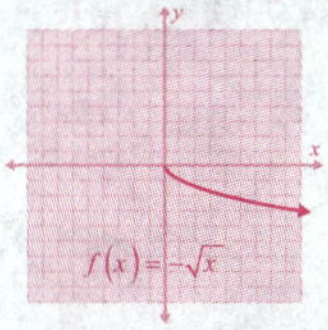

$D=[0,\infty)$; The graph of $\sqrt{x}$ is reflected across the x–axis.

b.

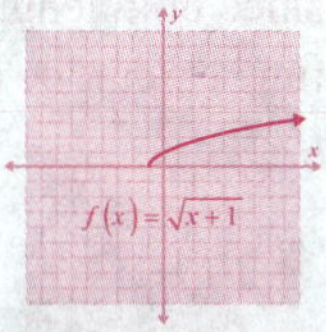

$D=[-1,\infty)$; The graph of $\sqrt{x}$ is shifted one unit to the left.

c.

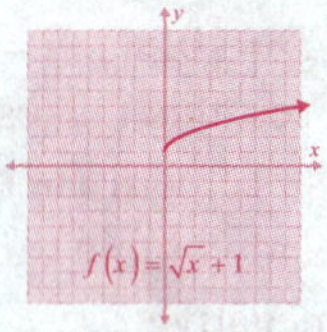

$D=[0,\infty)$; The graph of $\sqrt{x}$ is shifted one unit up.

49. A negative sign reflects the graph across the x–axis.

a. Affects how quickly the graph goes to infinity.

b. Causes the graph to move in a horizontal direction, and

c. Causes the graph to move in a vertical direction.

Applications

50. a.

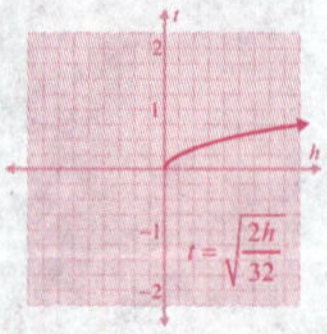

b. $D=[0,\infty); R=[0,\infty)$
Yes, this makes sense since time and height both cannot be negative

c. 0.71 seconds

51. a.

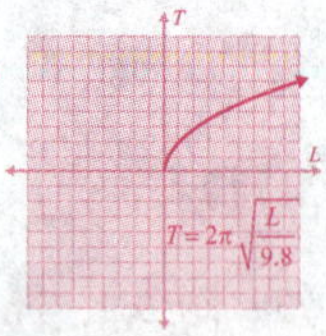

b. $D=[0,\infty); R=[0,\infty)$
Yes, this makes sense since both length and time cannot be negative

c. 7.77 seconds

52. a.

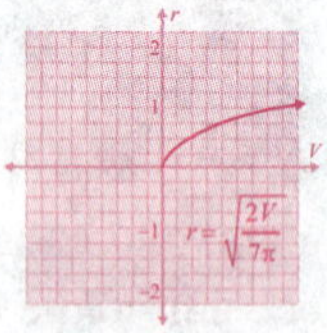

b. $D=[0,\infty); R=[0,\infty)$
Yes, this makes sense since both the radius and volume cannot be negative

c. 3.05 inches

Writing & Thinking

53. a. $\dfrac{1}{\sqrt{3+h}+\sqrt{3}}$

b. Slope of the line connecting $(3+h, f(3+h))$ and $(3, f(3))$

c. A line just touching the curve at one point

d. $\dfrac{1}{2\sqrt{3}}$; represents the slope of the line tangent to $f(x)$ at $x=3$.

54. The domain of g is $[0,\infty)$, while the domain of f is $(-\infty,\infty)$.

52. The relationship between the radius and volume of a cone of height 7 inches is as follows.

$$r=\sqrt{\frac{3V}{7\pi}}$$

In this equation, r is the radius of the cone in inches and V is the volume of the cone in cubic inches. Use $\pi = 3.14$.

a. Using a graphing calculator, graph this equation.

b. Identify the domain and range of this function. Does this make sense in the context of the function?

c. Find the radius of a cone whose height is 7 inches and whose volume is 68 cubic inches. Round your answer to the nearest hundredth.

Writing & Thinking

53. The graph of the radical function $f(x)=\sqrt{x}$ is shown with two values of x on the x–axis, 3 and 3 + h.

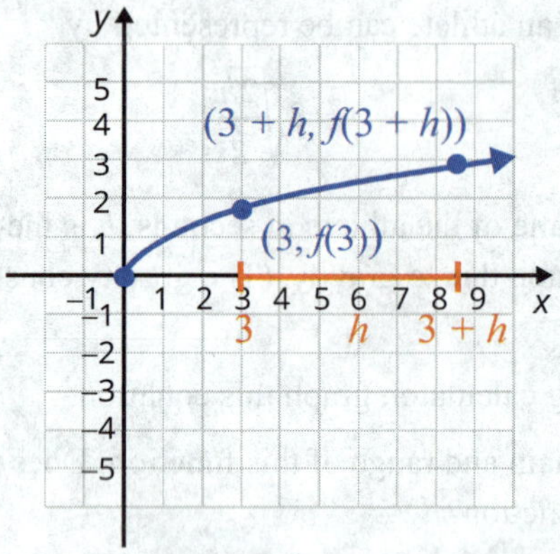

a. Rationalize the numerator of the expression $\dfrac{f(3+h)-f(3)}{h}=\dfrac{\sqrt{3+h}-\sqrt{3}}{h}$ by multiplying both the numerator and denominator by the conjugate of the numerator. Then simplify the resulting expression.

b. What do you think this expression represents graphically? (**Hint:** Two points determine a line.)

c. Using your results from parts a. and b., what do you see happening on the graph if the value of h shrinks slowly to 0?

d. Using your analysis from part c., what happens to the value of your simplified expression in part a. and what do you think this value represents?

54. Use your graphing calculator to graph the function $g(x)=\sqrt{x}\cdot\sqrt{x}$. Explain why the graph of this function differs from the graph of $f(x)=x$.

15.8 Introduction to Complex Numbers

A Square Roots of Negative Numbers

One of the properties of real numbers is that the square of any real number is nonnegative. That is, for any real number x, $x^2 \geq 0$. The square roots of negative numbers, such as $\sqrt{-4}$ and $\sqrt{-5}$, are not real numbers. However, they can be defined as part of the system of **complex numbers**.

Complex numbers include all the real numbers and the even roots of negative numbers. Earlier, we saw how these numbers occur as solutions to quadratic equations. At first, such numbers seem to be somewhat impractical because they are difficult to picture in any type of geometric setting and they are not solutions to the types of word problems that are familiar. However, complex numbers do occur quite naturally in trigonometry and higher-level mathematics and have practical applications in fields such as electrical engineering.

The first step in the development of complex numbers is to define $\sqrt{-1}$.

i and *i*²

$$i = \sqrt{-1} \quad \text{and} \quad i^2 = \left(\sqrt{-1}\right)^2 = -1$$

DEFINITION

Using the definition of $\sqrt{-1}$, the following definition for the square root of a negative number can be made.

$\sqrt{-a} = \sqrt{a}\,i$

If a is a positive real number, then

$$\sqrt{-a} = \sqrt{a} \cdot \sqrt{-1} = \sqrt{a}\,i.$$

Note: The number i is not under the radical sign. To avoid confusion, we sometimes write $i\sqrt{a}$.

DEFINITION

Objectives

A. Simplify square roots of negative numbers.

B. Identify the real and imaginary parts of complex numbers.

C. Solve equations containing complex numbers

D. Add and subtract complex numbers.

History of Imaginary Numbers

A complex number that can be written as the product of a real number and the imaginary unit (i) is referred to as an imaginary number. Although the Greek mathematician Hero of Alexandria is noted as the first to have conceived these numbers, the term was originally coined in the 17th century by René Descartes. Imaginary numbers weren't widely accepted until the 18th century when Leonhard Euler and Carl Friedrich Gauss worked with them.

Example 1 Finding the Square Roots of Negative Numbers

Simplify each radical.

a. $\sqrt{-25}$

b. $\sqrt{-36}$

c. $\sqrt{-24}$

d. $\sqrt{-45}$

Solution

a. $\sqrt{-25} = \sqrt{-1}\sqrt{25} = i \cdot 5 = 5i$ $\qquad (5i)^2 = 5^2 i^2 = 25(-1) = -25$

1. Simplify each radical.

a. $\sqrt{-100}$

b. $\sqrt{-49}$

c. $\sqrt{-18}$

d. $\sqrt{-72}$

b. $\sqrt{-36} = \sqrt{-1}\sqrt{36} = i \cdot 6 = 6i$

c. $\sqrt{-24} = \sqrt{-1}\sqrt{4 \cdot 6} = i \cdot 2 \cdot \sqrt{6} = 2\sqrt{6}\,i$ (or $2i\sqrt{6}$)

d. $\sqrt{-45} = \sqrt{-1}\sqrt{9 \cdot 5} = i \cdot 3 \cdot \sqrt{5} = 3\sqrt{5}\,i$ (or $3i\sqrt{5}$)

We can write $2\sqrt{6}\,i$ and $3\sqrt{5}\,i$ as long as we take care not to include the *i* under the radical sign.

Now work margin exercise 1.

B Real and Imaginary Parts of Complex Numbers

Note

The term "imaginary" is somewhat misleading. Complex numbers and imaginary numbers are no more "imaginary" than any other type of number. In fact, all the types of numbers that we have studied (whole numbers, integers, rational numbers, irrational numbers, and real numbers) are products of human imagination.

Complex Numbers

The **standard form** of a **complex number** is $\boldsymbol{a + bi}$, where a and b are real numbers. a is called the **real part**, and b is called the **imaginary part**.

If $b = 0$, then $a + bi = a + 0i = a$ is a real number.

If $a = 0$, then $a + bi = 0 + bi = bi$ is called a pure imaginary number (or an imaginary number).

Complex Number: $a + bi$

real part (a); imaginary part (b)

DEFINITION

2. Identify the real and imaginary parts of each complex number.

a. $5i$

b. $14 + \sqrt{7}i$

c. $\dfrac{6 - 11i}{5}$

d. -13

Example 2 Identifying Real and Imaginary Parts

Identify the real and imaginary parts of each complex number.

a. $4 - 2i$ **b.** $\dfrac{5 + 2i}{3}$ **c.** 7 **d.** $-\sqrt{3}\,i$

Solution

a. 4 is the real part; –2 is the imaginary part.

b. $\dfrac{5 + 2i}{3} = \dfrac{5}{3} + \dfrac{2}{3}i$ In standard form

Thus, $\dfrac{5}{3}$ is the real part; $\dfrac{2}{3}$ is the imaginary part.

c. $7 = 7 + 0i$ In standard form

Thus, 7 is the real part; 0 is the imaginary part. (Remember, if $b = 0$, the complex number is a real number.)

d. $-\sqrt{3}\,i = 0 - \sqrt{3}\,i$ In standard form

Thus, 0 is the real part; $-\sqrt{3}$ is the imaginary part. (If $a = 0$ and $b \neq 0$, then the complex number is a pure imaginary number.)

Now work margin exercise 2.

In general, if a is a real number, then we can write $a = a + 0i$. This means that a is a complex number. **Thus, every real number is a complex number.** Figure 1 illustrates the relationships among the various types of numbers we have studied.

Teaching Note

There are alternative models that illustrate the subset relationship with intersecting and nonintersecting sets of numbers. Students may not pick up on the subset relationships shown in Figure 1, and you may know of another diagram that clarifies the subset relationships. Seeing more than one approach helps many students.

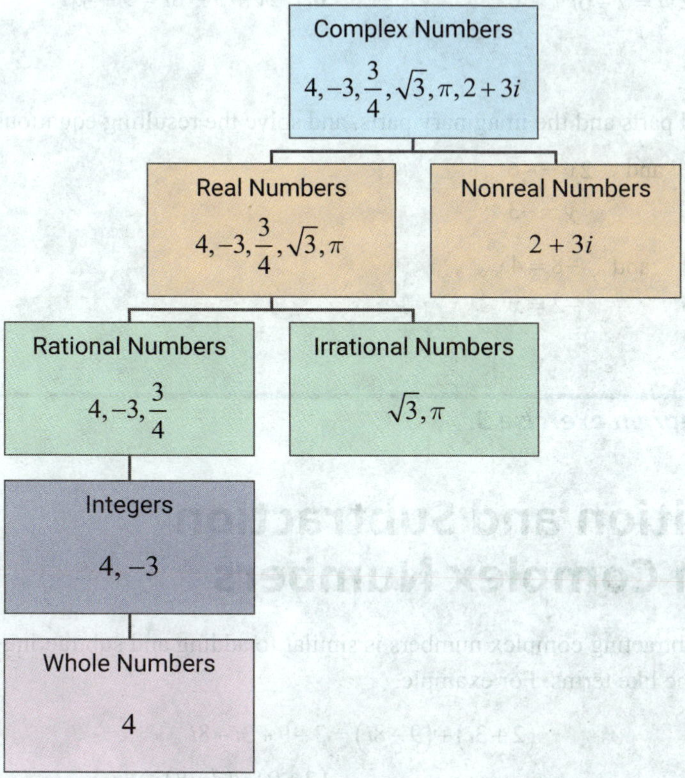

Figure 1

C Solving Equations Containing Complex Numbers

If two complex numbers are equal, then the real parts are equal and the imaginary parts are equal. For example, if

$$x + yi = 7 + 2i,$$

then $$x = 7 \quad \text{and} \quad y = 2.$$

This relationship can be used to solve equations involving complex numbers.

Equality of Complex Numbers

For complex numbers $a + bi$ and $c + di$,

$$\text{if } a + bi = c + di, \text{ then } a = c \text{ and } b = d.$$

DEFINITION

3. Solve each equation for x and y.

a. $(x-7)+4yi=3-8i$

b. $4y+7-5i=11+xi$

Example 3 Solving Equations

Solve each equation for x and y.

a. $(x+3)+2yi=7-6i$

b. $2y+3-8i=9+4xi$

Solution

Equate the real parts and the imaginary parts, and solve the resulting equations.

a. $x+3=7$ and $2y=-6$

$x=4$ $\quad y=-3$

b. $2y+3=9$ and $-8=4x$

$2y=6$ $\quad -2=x$

$y=3$

Now work margin exercise 3.

D Addition and Subtraction with Complex Numbers

Adding and subtracting complex numbers is similar to adding and subtracting polynomials. Simply combine like terms. For example,

$$\begin{aligned}(2+3i)+(9-8i)&=2+9+3i-8i\\&=(2+9)+(3-8)i\\&=11-5i.\end{aligned}$$

Similarly,

$$\begin{aligned}(5-2i)-(6+7i)&=5-2i-6-7i\\&=(5-6)+(-2-7)i\\&=-1-9i.\end{aligned}$$

Addition and Subtraction with Complex Numbers

For complex numbers $a+bi$ and $c+di$,

$$(a+bi)+(c+di)=(a+c)+(b+d)i$$

and

$$(a+bi)-(c+di)=(a-c)+(b-d)i.$$

DEFINITION

4. Find each sum or difference as indicated.

a. $(5-3i)+(2+7i)$

b. $\left(-7-\sqrt{2}i\right)-\left(-7+3\sqrt{2}i\right)$

c. $\left(4-\sqrt{5}i\right)+\left(\sqrt{2}+8i\right)$

Example 4 Adding and Subtracting with Complex Numbers

Find each sum or difference as indicated.

a. $(6-2i)+(1-2i)$

b. $\left(-8-\sqrt{2}\,i\right)-\left(-8+\sqrt{2}\,i\right)$

c. $\left(\sqrt{3}-2i\right)+\left(1+\sqrt{5}\,i\right)$

Solution

a. $(6-2i)+(1-2i)=(6+1)+(-2-2)i=7-4i$

b. $\left(-8-\sqrt{2}\,i\right)-\left(-8+\sqrt{2}\,i\right)=\left(-8-(-8)\right)+\left(-\sqrt{2}-\sqrt{2}\right)i$

$$=(-8+8)+\left(-2\sqrt{2}\right)i$$
$$=0-2\sqrt{2}i$$
$$=-2\sqrt{2}\,i \text{ (or } -2i\sqrt{2})$$

c. $\left(\sqrt{3}-2i\right)+\left(1+\sqrt{5}\,i\right)=\left(\sqrt{3}+1\right)+\left(-2+\sqrt{5}\right)i=\left(\sqrt{3}+1\right)+\left(\sqrt{5}-2\right)i$

Note: Here, the coefficients do not simplify. This means that the real part is $\sqrt{3}+1$ and the imaginary part is $\sqrt{5}-2$.

Now work margin exercise 4.

Margin Exercise Answers

1. a. $10i$ **b.** $7i$ **c.** $3i\sqrt{2}$ **d.** $6i\sqrt{2}$ **2. a.** real: 0; imaginary: 5 **b.** real: 14; imaginary: $\sqrt{7}$ **c.** real: $\frac{6}{5}$; imaginary: $-\frac{11}{5}$ **d.** real: −13; imaginary: 0 **3. a.** $x=10$ and $y=-2$ **b.** $y=1$ and $x=-5$ **4. a.** $7+4i$ **b.** $-4\sqrt{2}i$ **c.** $\left(4+\sqrt{2}\right)+\left(8-\sqrt{5}\right)i$

15.8 Exercises

Concept Check

Fill-in-the-Blank. Complete each sentence using information found in this section.

1. Adding and subtracting complex numbers is similar to adding and subtracting ________.
2. The square roots of negative numbers are not real numbers but are ________ numbers.
3. Complex numbers consist of two parts: a/an ________ part and a/an ________ part.
4. The standard form of a complex number is ________.
5. The expression $\sqrt{-a}$ can be rewritten as $\sqrt{a}\cdot\sqrt{-1}$ and ________.
6. The value i is defined to be ________.

True/False. Determine whether each statement is true or false. If a statement is false, explain how it can be changed so the statement will be true. (**Note:** There may be more than one acceptable change.)

7. An irrational number is a real number.
8. Every real number is a complex number.
9. If $a+bi=c+di$, then $a=d$ and $b=c$.
10. The square root of negative one is one.

15.8 Exercises

Concept Check

1. polynomials
2. complex
3. real, imaginary
4. $a+bi$
5. $i\sqrt{a}$
6. $\sqrt{-1}$
7. True
8. True
9. False; $a=c$, $b=d$
10. False; It is i.

Practice

1. Real part: 4, imaginary part: −3
2. Real part: $\frac{3}{4}$, imaginary part: 1
3. Real part: −11, imaginary part: $\sqrt{2}$
4. Real part: 6, imaginary part: $\sqrt{3}$
5. Real part: $\frac{3}{8}$, imaginary part: 0
6. Real part: 0, imaginary part: $\frac{4}{7}$
7. Real part: $\frac{4}{5}$, imaginary part: $\frac{7}{5}$
8. Real part: $\frac{1}{2}$, imaginary part: $-\frac{1}{4}$
9. Real part: $\frac{2}{3}$, imaginary part: $\sqrt{17}$
10. Real part: $-\sqrt{5}$, imaginary part: $\frac{\sqrt{2}}{2}$

11. $7i$
12. $11i$
13. $-8i$
14. $-13i$
15. $7\sqrt{3}$
16. $8\sqrt{2}$
17. $10i\sqrt{6}$
18. $12i\sqrt{11}$
19. $-12i\sqrt{3}$
20. $10\sqrt{7}$
21. $11\sqrt{2}$
22. $8i\sqrt{3}$
23. $10i\sqrt{10}$
24. $9i\sqrt{3}$
25. $x=6, y=-3$
26. $x=-1, y=-2$
27. $x=-2, y=\sqrt{5}$
28. $x=\frac{\sqrt{2}}{3}, y=-3$
29. $x=\sqrt{2}-3, y=1$
30. $x=-3, y=4+\sqrt{5}$
31. $x=1, y=4$
32. $x=-4, y=\frac{1}{2}$
33. $x=2, y=-6$
34. $x=2, y=-8$
35. $x=3, y=10$
36. $x=-3, y=7$
37. $x=-\frac{4}{3}, y=-3$
38. $x=-2, y=1$
39. $6+2i$
40. $10+5i$
41. $1+7i$
42. -9
43. $6-6i$
44. $13+3i$
45. $14i$
46. $-8-2i$
47. $\left(3+\sqrt{5}\right)-6i$
48. $4-\sqrt{2}$

Practice

Find the real part and the imaginary part of each of the complex numbers. See Example 2.

1. $4-3i$

2. $\frac{3}{4}+i$

3. $-11+\sqrt{2}\,i$

4. $6+\sqrt{3}\,i$

5. $\frac{3}{8}$

6. $\frac{4}{7}i$

7. $\frac{4+7i}{5}$

8. $\frac{2-i}{4}$

9. $\frac{2}{3}+\sqrt{17}\,i$

10. $-\sqrt{5}+\frac{\sqrt{2}}{2}i$

Simplify each radical. See Example 1.

11. $\sqrt{-49}$

12. $\sqrt{-121}$

13. $-\sqrt{-64}$

14. $-\sqrt{-169}$

15. $\sqrt{147}$

16. $\sqrt{128}$

17. $2\sqrt{-150}$

18. $4\sqrt{-99}$

19. $-2\sqrt{-108}$

20. $2\sqrt{175}$

21. $\sqrt{242}$

22. $\sqrt{-192}$

23. $\sqrt{-1000}$

24. $\sqrt{-243}$

Solve the equations for *x* and *y*. See Example 3.

25. $x+3i=6-yi$

26. $2x-8i=-2+4yi$

27. $\sqrt{5}-2i=y+xi$

28. $\sqrt{2}-2yi=3x+6i$

29. $\sqrt{2}+i-3=x+yi$

30. $\sqrt{5}\,i-3+4i=x+yi$

31. $3x+2-7i=i-2yi+5$

32. $x+yi+8=2i+4-3yi$

33. $x+2i=5-yi-3-4i$

34. $2x+3+6i=7-yi-2i$

35. $2+3i+x=5-7i+yi$

36. $11i-2x+4=10-3i+2yi$

37. $2x-2yi+6=6i-x+2$

38. $x+4-3x+i=8+yi$

Find each sum or difference as indicated. See Example 4.

39. $(2+3i)+(4-i)$

40. $(7-i)+(3+6i)$

41. $(4+5i)-(3-2i)$

42. $(-3+2i)-(6+2i)$

43. $(4-3i)+(2-3i)$

44. $(7+5i)+(6-2i)$

45. $(8+9i)-(8-5i)$

46. $(-6+i)-(2+3i)$

47. $(\sqrt{5}-2i)+(3-4i)$

48. $(4+3i)-(\sqrt{2}+3i)$

49. $(7+\sqrt{6}\,i)+(-2+i)$

50. $(\sqrt{11}+2i)+(5-7i)$

51. $(\sqrt{3}+\sqrt{2}\,i)-(5+\sqrt{2}\,i)$

52. $(\sqrt{5}+\sqrt{3}\,i)+(1-i)$

53. $(5+\sqrt{-25})-(7+\sqrt{-100})$

54. $(1+\sqrt{-36})-(-4-\sqrt{-49})$

55. $(13-3\sqrt{-16})+(-2-4\sqrt{-1})$

56. $(7+\sqrt{-9})-(3-2\sqrt{-25})$

57. $(4+i)+(-3-2i)-(-1-i)$

58. $(-2-3i)+(6+i)-(2+5i)$

59. $(7+3i)+(2-4i)-(6-5i)$

60. $(-5+7i)+(4-2i)-(3-5i)$

Writing & Thinking

61. Answer the following questions and give a brief explanation of your answer.
    a. Is every real number a complex number?
    b. Is every complex number a real number?

62. List 5 numbers that do and 5 numbers that do not fit each of the following categories (if possible).
    a. rational number
    b. integer
    c. real number
    d. pure imaginary number
    e. complex number
    f. irrational number

49. $5+(\sqrt{6}+1)i$

50. $(\sqrt{11}+5)-5i$

51. $\sqrt{3}-5$

52. $(\sqrt{5}+1)+(\sqrt{3}-1)i$

53. $-2-5i$

54. $5+13i$

55. $11-16i$

56. $4+13i$

57. 2

58. $2-7i$

59. $3+4i$

60. $-4+10i$

Writing & Thinking

61. a. Yes
    b. No

62. Answers will vary.
    a. Do: $\left\{0, -2, 3, \frac{4}{7}, \sqrt{16}\right\}$; Do not: $\left\{\sqrt{5}, \pi, \sqrt{99}, \sqrt{-4}, i\right\}$
    b. Do: $\{-8, -2, 0, 1, 58\}$; Do not: $\left\{2.3, \sqrt{6}, \frac{1}{2}, 3+i, 2\pi\right\}$
    c. Do: $\left\{-\sqrt{3}, 0, \frac{5}{6}, 4, \pi^2\right\}$; Do not: $\left\{-i, 2-3i, 4+i, \sqrt{-2}, 9i\right\}$
    d. Do: $\left\{-5i, -\sqrt{-4}, i, 3i, \sqrt{-36}\right\}$; Do not: $\left\{7, 2+i, -3, 4+4i, \pi\right\}$
    e. Do: $\left\{-2, 3+i, 4-\pi i, 55i, \frac{61}{100}\right\}$; Do not: None
    f. Do: $\left\{\pi, \sqrt{3}, \sqrt{5}+\sqrt{6}, -\sqrt{11}, 3\pi+4\right\}$; Do not: $\left\{4, 1.1, -\frac{3}{5}, \sqrt{9}, 0\right\}$

Objectives

A. Multiply with complex numbers.

B. Divide with complex numbers.

C. Simplify powers of i.

15.9 Multiplication and Division with Complex Numbers

A Multiplication with Complex Numbers

The product of two complex numbers can be found by using the FOIL method for multiplying two binomials. (See Section 12.6.) **Remember that $i^2 = -1$.**

$$\begin{aligned}(3+5i)(2+i) &= 6+3i+10i+5i^2 \\ &= 6+13i-5 \qquad 5i^2 = 5(-1) = -5 \\ &= 1+13i\end{aligned}$$

1. Find the following products.

   a. $(4i)(1-7i)$

   b. $(4+2i)(3+5i)$

   c. $(\sqrt{2}-i)(\sqrt{2}+i)$

Example 1 Multiplying with Complex Numbers

Find the following products.

a. $(3i)(2-7i)$ **b.** $(5+i)(2+6i)$ **c.** $(\sqrt{2}-i)(\sqrt{2}-i)$ **d.** $(-1+i)(2-i)$

Solution

a. $$\begin{aligned}(3i)(2-7i) &= 6i-21i^2 \\ &= 6i-21(-1) \qquad \text{Remember that } i^2 = -1. \\ &= 21+6i\end{aligned}$$

b. $$\begin{aligned}(5+i)(2+6i) &= 5(2+6i)+i(2+6i) \\ &= 10+30i+2i+6i^2 \\ &= 10+32i-6 \\ &= 4+32i\end{aligned}$$

c. $$\begin{aligned}(\sqrt{2}-i)(\sqrt{2}-i) &= (\sqrt{2})^2-\sqrt{2}\cdot i-\sqrt{2}\cdot i+i^2 \\ &= 2-2\sqrt{2}\,i-1 \\ &= 1-2i\sqrt{2}\end{aligned}$$

d. $$\begin{aligned}(-1+i)(2-i) &= -2+i+2i-i^2 \\ &= -2+3i+1 \\ &= -1+3i\end{aligned}$$

Now work margin exercise 1.

Common Error

Remember that $\sqrt{a} \cdot \sqrt{b} = \sqrt{ab}$ **only** if a and b are nonnegative real numbers.

Applying this rule to negative real numbers can lead to an error. The error can be avoided by first changing the radicals to imaginary form.

Wrong Solution

$$\begin{aligned}\sqrt{-6} \cdot \sqrt{-2} &= \sqrt{12} \\ &= \sqrt{4} \cdot \sqrt{3} \\ &= 2\sqrt{3}\end{aligned}$$

Correct Solution

$$\begin{aligned}\sqrt{-6} \cdot \sqrt{-2} &= \sqrt{6}\,i \cdot \sqrt{2}\,i \\ &= \sqrt{12}\,i^2 \\ &= 2\sqrt{3}(-1) \\ &= -2\sqrt{3}\end{aligned}$$

CAUTION

B Division with Complex Numbers

The two complex numbers $a + bi$ and $a - bi$ are called **complex conjugates**, or simply **conjugates,** of each other. As the following steps show, **the product of two complex conjugates will always be a nonnegative real number**.

$$\begin{aligned}(a+bi)(a-bi) &= a^2 - abi + abi - b^2i^2 \\ &= a^2 - b^2i^2 \\ &= a^2 + b^2\end{aligned}$$

The resulting product, $a^2 + b^2$, is a real number, and it is nonnegative since it is the sum of the squares of real numbers.

Remember that the form $a + bi$ is called the **standard form** of a complex number. The standard form allows for easy identification of the real and imaginary parts. For example,

$$\frac{1+3i}{5} \text{ is written as } \frac{1}{5} + \frac{3}{5}i \text{ in standard form,}$$

where the real part is $\frac{1}{5}$ and the imaginary part is $\frac{3}{5}$.

To write the fraction $\frac{1+i}{2-3i}$ in standard form, multiply both the numerator and denominator by $2 + 3i$ (the complex conjugate of the denominator) and simplify. This will give a positive real number in the denominator.

$$\begin{aligned}\frac{1+i}{2-3i} &= \frac{(1+i)(2+3i)}{(2-3i)(2+3i)} && \text{2 + 3i is the conjugate of the denominator.} \\ &= \frac{2+3i+2i+3i^2}{2^2+6i-6i-3^2i^2} \\ &= \frac{2+5i-3}{4-(9)(-1)} && \text{Remember that } i^2 = -1. \\ &= \frac{-1+5i}{13} = -\frac{1}{13} + \frac{5}{13}i\end{aligned}$$

Writing Fractions with Complex Numbers in Standard Form

1. Multiply both the numerator and denominator by the complex conjugate of the denominator.
2. Simplify the resulting products in both the numerator and denominator.
3. Write the simplified result in standard form.

PROCEDURE

Remember the following special product. We restate it here to emphasize its importance.

$$(a+bi)(a-bi)=a^2+b^2$$

2. Write the following fractions in standard form.

a. $\dfrac{8}{1-2i}$

b. $\dfrac{\sqrt{5}-2i}{\sqrt{5}+2i}$

c. $\dfrac{4-i}{4i}$

d. $\dfrac{\sqrt{3}+4i}{-\sqrt{3}+4i}$

Example 2 Dividing with Complex Numbers

Write the following fractions in standard form.

a. $\dfrac{4}{-1-5i}$

b. $\dfrac{\sqrt{3}+i}{\sqrt{3}-i}$

c. $\dfrac{6+i}{i}$

d. $\dfrac{\sqrt{2}+i}{-\sqrt{2}+i}$

Solution

a. $$\frac{4}{-1-5i}=\frac{4(-1+5i)}{(-1-5i)(-1+5i)}$$

Multiply the numerator and the denominator by the conjugate of the denominator.

$$=\frac{-4+20i}{(-1)^2-(5i)^2}=\frac{-4+20i}{1+25}$$

$$=\frac{-4+20i}{26}=-\frac{4}{26}+\frac{20}{26}i$$

$$=-\frac{2}{13}+\frac{10}{13}i$$

b. $$\frac{\sqrt{3}+i}{\sqrt{3}-i}=\frac{(\sqrt{3}+i)(\sqrt{3}+i)}{(\sqrt{3}-i)(\sqrt{3}+i)}$$

$$=\frac{3+2\sqrt{3}\,i+i^2}{(\sqrt{3})^2-i^2}=\frac{2+2\sqrt{3}\,i}{3+1}$$

$$=\frac{2+2\sqrt{3}\,i}{4}=\frac{2}{4}+\frac{2\sqrt{3}}{4}i$$

$$=\frac{1}{2}+\frac{\sqrt{3}}{2}i$$

c. $$\frac{6+i}{i}=\frac{(6+i)(-i)}{i(-i)}$$

Since $i=0+i$ and $-i=0-i$, the number $-i$ is the conjugate of i.

$$=\frac{-6i-i^2}{-i^2}=\frac{-6i+1}{1}$$

$$=1-6i$$

d. $$\frac{\sqrt{2}+i}{-\sqrt{2}+i}=\frac{\left(\sqrt{2}+i\right)\left(-\sqrt{2}-i\right)}{\left(-\sqrt{2}+i\right)\left(-\sqrt{2}-i\right)}$$

$$=\frac{-\left(\sqrt{2}\right)^2-\sqrt{2}\,i-\sqrt{2}\,i-i^2}{\left(-\sqrt{2}\right)^2-i^2}$$

$$=\frac{-2-2\sqrt{2}\,i+1}{2+1}=\frac{-1-2\sqrt{2}\,i}{3}$$

$$=-\frac{1}{3}-\frac{2\sqrt{2}}{3}i$$

Now work margin exercise 2.

C Powers of *i*

The powers of i form an interesting pattern. Regardless of the particular integer exponent, there are only four possible values for any power of i:

$$i,\quad -1,\quad -i,\quad \text{and}\quad 1.$$

The fact that these are the only four possibilities for powers of i becomes apparent from studying the provided list of powers.

$$
\begin{aligned}
i^1 &= i\\
i^2 &= -1\\
i^3 &= i^2\cdot i = -1\cdot i = -i\\
i^4 &= i^2\cdot i^2 = (-1)(-1) = 1\\
i^5 &= i^4\cdot i = 1\cdot i = i\\
i^6 &= i^4\cdot i^2 = (1)(-1) = -1\\
i^7 &= i^4\cdot i^3 = (1)(-i) = -i\\
i^8 &= i^4\cdot i^4 = 1\cdot 1 = 1
\end{aligned}
$$

Higher powers of i can be simplified using the fact that when i is raised to a power that is a multiple of 4, the result is 1. Thus, if n is a positive integer, then the generalized powers of i are true.

$$
\begin{aligned}
i^{4n} &= \left(i^4\right)^n = 1^n = 1\\
i^{4n+1} &= i^{4n}\cdot i = 1\cdot i = i\\
i^{4n+2} &= i^{4n}\cdot i^2 = (1)\cdot(-1) = -1\\
i^{4n+3} &= i^{4n}\cdot i^3 = (1)\cdot(-i) = -i
\end{aligned}
$$

Math Tip

Another approach to evaluating i^n is to divide n by 4 and look at the remainder.

If the remainder is 0, $i^n = i^0 = 1.$

If the remainder is 1, $i^n = i^1 = i.$

If the remainder is 2, $i^n = i^2 = -1.$

If the remainder is 3, $i^n = i^3 = -i.$

3. Simplify each power of i.

a. i^{37}

b. i^{14}

c. i^{-8}

Example 3 Simplifying Powers of i

Simplify each power of i.

a. $i^{45} = i^{44} \cdot i = \left(i^4\right)^{11} \cdot i = 1^{11} \cdot i = i$ $\quad i = 0 + i$ in standard form.

b. $i^{58} = i^{56} \cdot i^2 = \left(i^4\right)^{14} \cdot i^2 = 1^{14} \cdot (-1) = -1$ $\quad 1 = -1 + 0i$ in standard form.

c. $i^{-7} = \frac{1}{i^7} = \frac{1}{i^7} \cdot \frac{i}{i} = \frac{i}{i^8} = \frac{i}{1} = i$ $\quad i = 0 + i$ in standard form.

Now work margin exercise 3.

Margin Exercise Answers

1. a. $28+4i$ **b.** $2+26i$ **c.** 3 **2. a.** $\frac{8}{5}+\frac{16}{5}i$ **b.** $\frac{1}{9}-\frac{4\sqrt{5}}{9}i$ **c.** $-\frac{1}{4}-i$ **d.** $\frac{13}{19}-\frac{8\sqrt{3}}{19}i$ **3. a.** i **b.** -1 **c.** 1

15.9 Exercises

Concept Check

Fill-in-the-Blank. Complete each sentence using information found in this section.

1. When two complex numbers are multiplied, the _____ method can be used.
2. When two complex conjugates are multiplied, the answer will always be a nonnegative _______ _______.
3. There are 4 possible values for any power of i: ___, ___, ___, and ___.
4. The expressions $a + bi$ and $a - bi$ are __________ of each other.
5. The product of $(a+bi)(a-bi)$ is ________.
6. The expression $a + bi$ is called the _______ form of a complex number.

True/False. Determine whether each statement is true or false. If a statement is false, explain how it can be changed so the statement will be true. (**Note:** There may be more than one acceptable change.)

7. Regardless of the value of the exponent, the only possible values for any power of i are i and $-i$.
8. The product $\sqrt{a} \cdot \sqrt{b}$ can be rewritten as $\sqrt{ab}$ as long as a and b are real numbers.
9. When i is squared, the product is 1.
10. The conjugate of $4 - 5i$ is $4 + 5i$.

15.9 Exercises

Concept Check

1. FOIL
2. real number
3. $i, -1, -i, 1$
4. conjugates
5. $a^2 + b^2$
6. standard
7. False; $i, -i$, 1 and -1
8. False; The values of a and b must be nonnegative.
9. False; The product is -1.
10. True

Practice

1. $16 + 24i$
2. $-21 + 12i$
3. $-7\sqrt{2} + 7i$
4. $3 + 2i\sqrt{3}$
5. $3 + 12i$
6. $-28 - 24i$
7. $1 - i\sqrt{3}$
8. $-4 + 2i\sqrt{5}$
9. $3 + 2i\sqrt{3}$
10. $5\sqrt{2} + 10i$
11. $2 + 8i$
12. $5 + 44i$
13. $-7 - 11i$
14. $20 + 0i$
15. $13 + 0i$
16. $41 + 0i$
17. $34 + 13i$
18. $-3 - 7i$
19. $-24 + 70i$
20. $5 + 12i$
21. $5 - i\sqrt{3}$
22. $13 + i\sqrt{5}$
23. $23 - 10i\sqrt{2}$
24. $4 + 4i\sqrt{7}$
25. $21 + 0i$
26. $61 + 0i$

Practice

Perform the indicated operations and write each result in standard form. See Examples 1 and 2.

1. $8(2+3i)$
2. $-3(7-4i)$
3. $-7(\sqrt{2}-i)$
4. $\sqrt{3}(\sqrt{3}+2i)$
5. $3i(4-i)$
6. $-4i(6-7i)$
7. $-i(\sqrt{3}+i)$
8. $2i(\sqrt{5}+2i)$
9. $\sqrt{3}\,i(2-\sqrt{3}\,i)$
10. $5i(2-\sqrt{2}\,i)$
11. $(5+3i)(1+i)$
12. $(2+7i)(6+i)$
13. $(-3+5i)(-1+2i)$
14. $(6+2i)(3-i)$
15. $(2-3i)(2+3i)$
16. $(4+5i)(4-5i)$
17. $(4+3i)(7-2i)$
18. $(-2+5i)(i-1)$
19. $(5+7i)^2$
20. $(3+2i)^2$
21. $(\sqrt{3}+i)(\sqrt{3}-2i)$
22. $(2\sqrt{5}+3i)(\sqrt{5}-i)$
23. $(5-\sqrt{2}\,i)(5-\sqrt{2}\,i)$
24. $(\sqrt{7}+3i)(\sqrt{7}+i)$
25. $(4+\sqrt{5}\,i)(4-\sqrt{5}\,i)$
26. $(7+2\sqrt{3}\,i)(7-2\sqrt{3}\,i)$
27. $(\sqrt{5}+2i)(\sqrt{2}-i)$
28. $(2\sqrt{3}+i)(4+3i)$
29. $(3+\sqrt{5}\,i)(3+\sqrt{6}\,i)$
30. $(2-\sqrt{3}\,i)(3-\sqrt{2}\,i)$
31. $\dfrac{-3}{i}$
32. $\dfrac{7}{i}$
33. $\dfrac{5}{4i}$
34. $\dfrac{-3}{2i}$
35. $\dfrac{2+i}{-4i}$
36. $\dfrac{3-4i}{3i}$
37. $\dfrac{-4}{1+2i}$
38. $\dfrac{7}{5-2i}$
39. $\dfrac{6}{4-3i}$
40. $\dfrac{-8}{6+i}$
41. $\dfrac{2i}{5-i}$
42. $\dfrac{-4i}{1+3i}$
43. $\dfrac{2-i}{2+5i}$
44. $\dfrac{6+i}{3-4i}$
45. $\dfrac{2-3i}{-1+5i}$
46. $\dfrac{-3+i}{7-2i}$
47. $\dfrac{1+4i}{\sqrt{3}+i}$
48. $\dfrac{9-2i}{\sqrt{5}+i}$

27. $(2+\sqrt{10})+(2\sqrt{2}-\sqrt{5})i$
28. $(8\sqrt{3}-3)+(4+6\sqrt{3})i$
29. $(9-\sqrt{30})+(3\sqrt{5}+3\sqrt{6})i$
30. $(6-\sqrt{6})-(3\sqrt{3}+2\sqrt{2})i$
31. $0+3i$
32. $0-7i$
33. $0-\frac{5}{4}i$
34. $0+\frac{3}{2}i$
35. $-\frac{1}{4}+\frac{1}{2}i$
36. $-\frac{4}{3}-i$
37. $-\frac{4}{5}+\frac{8}{5}i$
38. $\frac{35}{29}+\frac{14}{29}i$
39. $\frac{24}{25}+\frac{18}{25}i$
40. $-\frac{48}{37}+\frac{8}{37}i$
41. $-\frac{1}{13}+\frac{5}{13}i$
42. $-\frac{6}{5}-\frac{2}{5}i$
43. $-\frac{1}{29}-\frac{12}{29}i$
44. $\frac{14}{25}+\frac{27}{25}i$
45. $-\frac{17}{26}-\frac{7}{26}i$
46. $-\frac{23}{53}+\frac{1}{53}i$
47. $\frac{4+\sqrt{3}}{4}+\left(\frac{4\sqrt{3}-1}{4}\right)i$
48. $\frac{9\sqrt{5}-2}{6}-\left(\frac{9+2\sqrt{5}}{6}\right)i$
49. $-\frac{1}{7}+\frac{4\sqrt{3}}{7}i$
50. $-\frac{1}{5}-\frac{2\sqrt{6}}{5}i$
51. $0+i$
52. $1+0i$
53. $-1+0i$
54. $0-i$
55. $0+i$
56. $0-i$
57. $1+0i$
58. $-1+0i$

59. $0 - i$

60. $0 + i$

61. $x^2 + 9$

62. $y^2 + 25$

63. $x^2 + 2$

64. $4x^2 + 7$

65. $5y^2 + 4$

66. $y^2 + 3$

67. $x^2 + 4x + 40$

68. $x^2 + 2x + 9$

69. $y^2 - 6y + 13$

70. $x^2 - 2x + 26$

Writing & Thinking

71. Given a complex number $(a+bi)$: $(a+bi)(a-bi)$ $= a^2 - abi + abi - b^2i^2$ $= a^2 + b^2$ which is the sum of squares of real numbers. Thus, the product must be a positive real number.

72. The property $\sqrt{a}\sqrt{b} = \sqrt{ab}$ is only true if a and b are positive. The actual value is -4, because $\sqrt{-4} \cdot \sqrt{-4} = 2i \cdot 2i$ $= 4i^2 = -4$.

73. $a^2 + b^2 = 1$

---

49. $\dfrac{\sqrt{3}+2i}{\sqrt{3}-2i}$

50. $\dfrac{\sqrt{6}-3i}{\sqrt{6}+3i}$

Simplify the following powers of i and write each result in standard form. Assume k is a positive integer. See Example 3.

51. i^{13}

52. i^{20}

53. i^{30}

54. i^{15}

55. i^{-3}

56. i^{-5}

57. i^{4k}

58. i^{4k+2}

59. i^{4k+3}

60. i^{4k+1}

Find the indicated products and simplify.

61. $(x+3i)(x-3i)$

62. $(y+5i)(y-5i)$

63. $\left(x+\sqrt{2}\,i\right)\left(x-\sqrt{2}\,i\right)$

64. $\left(2x+\sqrt{7}\,i\right)\left(2x-\sqrt{7}\,i\right)$

65. $\left(\sqrt{5}y+2i\right)\left(\sqrt{5}y-2i\right)$

66. $\left(y-\sqrt{3}\,i\right)\left(y+\sqrt{3}\,i\right)$

67. $[(x+2)+6i][(x+2)-6i]$

68. $\left[(x+1)-\sqrt{8}\,i\right]\left[(x+1)+\sqrt{8}\,i\right]$

69. $[(y-3)+2i][(y-3)-2i]$

70. $[(x-1)+5i][(x-1)-5i]$

Writing & Thinking

71. Explain why the product of every complex number and its conjugate is a nonnegative real number.

72. Explain why $\sqrt{-4} \cdot \sqrt{-4} \neq 4$. What is the correct value of $\sqrt{-4} \cdot \sqrt{-4}$?

73. What condition is necessary for the conjugate of a complex number, $a + bi$, to be equal to the reciprocal of this number?

Chapter 15 Project: Rationally Increasing Precision in Population Problems

1. **a.** $2^3 = 8$
b. $2^{3.1} \approx 8.5741877$
c. $2^{3.14} \approx 8.815240927$
d. $2^{3.141} \approx 8.821353305$
e. $2^{3.1415} \approx 8.824411082$
f. $2^{3.14159} \approx 8.824961595$
g. $2^{3.141592} \approx 8.824973829$
h. $2^{3.1415926} \approx 8.824977499$
i. $2^{3.14159265} \approx 8.824977805$

2. π

3. The results are increasing but slowing down as more digits are appended to the exponent. Thus, it seems to increase to a particular value.

4. **a.** $2^{\pi} \approx 8.824977827$. The results from Problem 1 are getting closer to this value as more digits are appended to the exponent.
b. This is the particular value that the calculated powers of 2 were increasing to.

5. **a.** 3.1
b. $300{,}000 \cdot 2^{3.1} \approx$ $300{,}000 \cdot 8.5741877 \approx$ 2,572,256.31.
c. x stands for how many times the population doubled (equivalently, how many groups of 25 years). Since the answer in Problem 5b is an estimate of the number of people, we should round to the nearest whole number, 2,572,256 people.
d. $3.14 \cdot 25 = 78.5$ years

Chapter 15 Project

Let's Get Radical!

An activity to demonstrate the use of radical expressions in real life.

There are many different situations in real life that require working with radicals, such as solving right-triangle problems, working with the laws of physics, calculating volumes, and solving investment problems. Let's take a look at a simple investment problem to see how radicals are involved.

The formula for computing compound interest for a principal P that is invested at an annual rate r and compounded annually is given by $A = P(1+r)^n$, where A is the accumulated amount in the account after n years.

1. Let's suppose that you have \$5000 to invest for a term of 2 years. If you want to make \$600 in interest, then at what interest rate should you invest the money?

   a. One way to approach this problem would be through trial and error, substituting various rates for r in the formula. This approach might take a while. Using the table below to organize your work, try substituting 3 values for r. Remember that rates are percentages and need to be converted to decimals before using the formula. Did you get close to \$5600 for the accumulated amount in the account after 2 years?

Annual Rate (r)	Principal (P)	Number of Years (n)	Amount, $A = P(1+r)^n$
	\$5000	2	
	\$5000	2	
	\$5000	2	

   b. Let's try a different approach. Substitute the value of 2 for n and solve this formula for r. Verify that you get the following result: $r = \sqrt{\frac{A}{P}} - 1$ (**Hint:** First solve for $(1+r)^2$ and then take the square root of both sides of the equation.) Notice that you now have a radical expression to work with. Substitute \$5000 for P and \$5600 for A (which is the principal plus \$600 in interest) to see what your rate must be. Round your answer to the nearest percent. 6%

2. Now, let's suppose that you won't need the money for 3 years.

   a. Use $n = 3$ years and solve the compound interest formula for r.

   b. What interest rate will you need to invest the principal of \$5000 at in order to have at least \$5600 at the end of 3 years? (To evaluate a cube root you may have to use the rational exponent of $\frac{1}{3}$ on your calculator.) Round to the nearest percent. 4%

   c. Compare the rates needed to earn at least \$600 when $n = 2$ years and $n = 3$ years. What did you learn from this comparison? Write a complete sentence.

3. Using the above formulas for compound interest when $n = 2$ years and $n = 3$ years, write the general formula for r for any value of n.

4. Using the formula from Problem 3, compute the interest rate needed to earn at least \$3000 in interest on a \$5000 investment in 7 years. Round to the nearest percent. 7%

5. Do an internet search on a local bank or financial institution to determine if the interest rate from Problem 4 is reasonable in the current economy. Using three to five sentences, briefly explain why or why not. Answers will vary.

Chapter 15 Project: Let's Get Radical!

1. Answers will vary. For example if $r = 5\%$, then $A = \$5,512.50$

2. a. $r = \sqrt[3]{\frac{A}{P}} - 1$

   c. The longer you can let an investment compound, the smaller the interest rate needed to reach the same goal.

3. $r = \sqrt[n]{\frac{A}{P}} - 1$

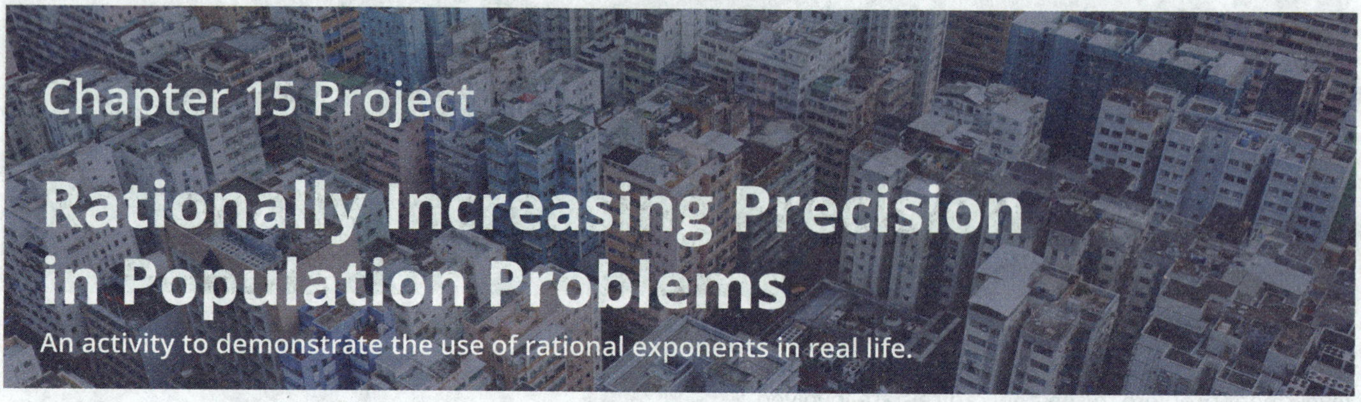

Chapter 15 Project

Rationally Increasing Precision in Population Problems

An activity to demonstrate the use of rational exponents in real life.

Sometimes we need to find the value of an exponential expression where the exponent is not an integer. This often happens when dealing with exponential growth. In the essay "Observations Concerning the Increase of Mankind, Peopling of Countries, etc.," written in 1751, Benjamin Franklin projected that the human population in the thirteen US colonies was doubling in size every twenty–five years. For example, if one year the population was 300,000 people, then to estimate the number of people 75 years later, calculate $300{,}000 \cdot 2^3 = 2{,}400{,}000$. Note that this works because $\frac{75}{25} = 3$, meaning the population would double 3 times in 75 years. Also notice that since $2 \cdot 2 \cdot 2 = 2^3 = 8$, we multiply 300,000 by 8 to get the final population.

If the timespan we want to estimate the future population for is not a multiple of 25, we can still calculate this value using rational exponents (with as much precision as we like). This investigation will suggest which types of numbers can be exponents, a topic which will be expanded upon in Chapter 17.

In the following investigation, do not round the exponents. Round the answers to have 10 digits, if necessary.

1. Calculate 2 to each of the following powers.

- **a.** 3
- **b.** 3.1
- **c.** 3.14
- **d.** 3.141
- **e.** 3.1415
- **f.** 3.14159
- **g.** 3.141592
- **h.** 3.1415926
- **i.** 3.14159265

2. The sequence of exponents in Problem 1 is approaching which special number? (**Hint:** See Section 6.4.)

3. In Problem 1, are the calculated powers of 2 increasing, decreasing, or is there no discernible pattern? If increasing (or decreasing), are they increasing (or decreasing) toward a particular value?

4. Consider raising the value 2 to the power of the special number found in Problem 2.

- **a.** If it is possible, state the result and compare it to the results of Problem 1. If this is not possible, explain why.
- **b.** How does this relate to your answer in Problem 3? If there is no relation, explain why.

5. Returning to Franklin's population prediction, consider if someone wanted to know the population 77.5 years later, instead of 75 years later.

- **a.** Determine the decimal form of the exponent $x = \frac{77.5}{25}$.
- **b.** Substitute the value of x found in part a. into the population equation and simplify: $300{,}000 \cdot 2^x$.
- **c.** Explain what x stands for. Interpret the answer to part b. In this case, does it make sense to round or not? If rounding does make sense, what place would you round to and what is the result?
- **d.** If the population starts at 300,000, how many years have passed if $300{,}000 \cdot 2^{3.14}$ provides an estimate of the population size?

CHAPTER 16

16 Quadratic Equations

Connections

The website Variety.com reported in 2020 that a whopping 102 million people worldwide watched Superbowl LIV, where the Kansas City Chiefs faced the San Francisco 49ers. Viewers might not have thought much about mathematics when watching the Chief's quarterback Pat Mahomes throwing a football over 50 yards, but physics and algebra can be used to explain part of his success. There is an entire field of study, named sports biomechanics, dedicated to understanding the performance of athletes.

According to a simplified model, the approximate horizontal distance d in yards that a football travels after it leaves a player's hands at a speed of v miles per hour is given by the following formula.

$$d = 0.022v^2$$

If we knew that Mahomes was able to complete a 50-yard pass, could we use the formula above to determine how fast the football was moving as it left his hand?

Objectives

A. Solve quadratic equations by factoring.

B. Solve quadratic equations using the square root method.

C. Solve problems related to right triangles and the Pythagorean Theorem.

16.1 Quadratic Equations: The Square Root Method

A Solving by Factoring

Not every polynomial can be factored so that the factors have integer coefficients, and not every polynomial equation can be solved by factoring. However, when the solutions of a polynomial equation can be found by factoring, the method depends on the **zero-factor law**, which is restated here for easy reference.

Zero-Factor Law

If the product of two (or more) factors is 0, then at least one of the factors must be 0. That is, if a and b are real numbers and $a \cdot b = 0$, then

$$a = 0 \text{ or } b = 0 \text{ or both.}$$

PROPERTIES

Also, as discussed in Section 13.6, polynomial equations of second degree are called **quadratic equations** and, because these are the equations of interest in this chapter, the definition is restated here.

Quadratic Equations

Quadratic equations are equations that can be written in the form

$$ax^2 + bx + c = 0,$$

where a, b, and c are real numbers and $a \neq 0$.

DEFINITION

The procedure for solving quadratic equations by factoring involves making sure that one side of the equation is 0 and then applying the zero-factor law. The following list of steps outlines the procedure.

Solving Quadratic Equations by Factoring

1. Add or subtract terms as necessary so that 0 is on one side of the equation and the equation is in the standard form $ax^2 + bx + c = 0$, where a, b, and c are real numbers and $a \neq 0$.
2. Factor completely. (If there are any fractional coefficients, multiply each term by the least common denominator so that all coefficients will be integers.)
3. Set each nonconstant factor equal to 0 and solve each linear equation for the unknown.
4. Check each solution, one at a time, in the original equation.

PROCEDURE

Example 1 Solving Quadratic Equations by Factoring

Solve the quadratic equation by factoring.

$x^2 - 15x = -50$

Solution

$$x^2 - 15x = -50$$

$$x^2 - 15x + 50 = 0 \quad \text{Add 50 to both sides. One side must be 0.}$$

$$(x-5)(x-10) = 0 \quad \text{Factor the left hand side.}$$

$$x - 5 = 0 \quad \text{or} \quad x - 10 = 0 \quad \text{Set each factor equal to 0.}$$

$$x = 5 \qquad x = 10 \quad \text{Solve each linear equation.}$$

Check

$$(5)^2 - 15\cdot(5) \stackrel{?}{=} -50 \qquad (10)^2 - 15\cdot(10) \stackrel{?}{=} -50$$

$$25 - 75 \stackrel{?}{=} -50 \qquad 100 - 150 \stackrel{?}{=} -50$$

$$-50 = -50 \qquad -50 = -50$$

Now work margin exercise 1.

1. Solve the quadratic equation by factoring.

$x^2 - 10x = -16$

Example 2 Solving Quadratic Equations by Factoring

Solve the quadratic equation by factoring.

$3x^2 - 24x = -48$

Solution

$$3x^2 - 24x = -48$$

$$3x^2 - 24x + 48 = 0 \quad \text{Add 48 to both sides. One side must be 0.}$$

$$3(x^2 - 8x + 16) = 0 \quad \text{Factor out the GCF, 3.}$$

$$3(x-4)^2 = 0 \quad \text{The trinomial is a perfect square.}$$

$$x - 4 = 0 \quad \text{Two factors are the same.}$$

$$x = 4 \quad \text{The solution is a double root.}$$

Check

$$3\cdot(4)^2 - 24\cdot(4) \stackrel{?}{=} -48$$

$$48 - 96 \stackrel{?}{=} -48$$

$$-48 = -48$$

Now work margin exercise 2.

2. Solve the quadratic equation by factoring.

$4x^2 - 24x = -36$

Now that complex numbers have been introduced and discussed, we will see that quadratic equations may have nonreal complex solutions. In particular, the sum of two squares can be factored as the product of complex conjugates. (Recall that this format was previously declared "not factorable" because real factors were implied.)

For example,

$$x^2+9=(x+3i)(x-3i).$$

As shown in Example 3, such factors can lead to nonreal solutions to a quadratic equation.

3. Solve the quadratic equation by factoring.

$x^2+25=0$

Example 3 Quadratic Equations Involving the Sum of Two Squares

Solve the quadratic equation by factoring.

$x^2+4=0$

Solution

$$\begin{aligned} x^2+4&=0 \\ (x+2i)(x-2i)&=0 \\ x+2i=0 \quad &\text{or} \quad x-2i=0 \\ x=-2i \quad & \quad\quad x=2i \end{aligned}$$

Note that x^2+4 is the sum of two squares and can be factored into the product of conjugates of complex numbers.

Check

$$\begin{aligned} (-2i)^2+4&\stackrel{?}{=}0 & (2i)^2+4&\stackrel{?}{=}0 \\ 4i^2+4&\stackrel{?}{=}0 & 4i^2+4&\stackrel{?}{=}0 \\ -4+4&\stackrel{?}{=}0 & -4+4&\stackrel{?}{=}0 \\ 0&=0 & 0&=0 \end{aligned}$$

Now work margin exercise 3.

B The Square Root Method

Teaching Note

You might want to explain how to use fractional exponents in solving equations. This may help your students understand solving higher degree equations later. For example:

$$x^2=13$$
$$\left(x^2\right)^{\frac{1}{2}}=13^{\frac{1}{2}}$$
$$|x|=\sqrt{13}$$
$$x=\pm\sqrt{13}$$

Consider the equation $x^2 = 13$. Allowing that the variable x might be positive or negative, we use the definition of a square root that $\sqrt{x^2}=|x|$. (See Section 15.1.) Taking the square root of both sides of the equation gives $|x|=\sqrt{13}$. So we have two solutions, $x=\sqrt{13}$ or $x=-\sqrt{13}$. To indicate both solutions, we write $x=\pm\sqrt{13}$. Similarly, for the equation $(x-3)^2=5$, the same process gives $x-3=\pm\sqrt{5}$.

This leads to the two equations and the two solutions, as follows.

$$\begin{aligned} x-3&=\sqrt{5} \quad \text{or} \quad & x-3&=-\sqrt{5} \\ x&=3+\sqrt{5} & x&=3-\sqrt{5} \end{aligned}$$

We can write the two solutions in the form $x=3\pm\sqrt{5}$.

This shows how the definition of a square root can be used to solve quadratic equations in certain forms. In particular, if one side of the equation is a squared expression and the other side is a constant, we can simply take the square root of both sides. If the constant is negative, then the solutions will involve nonreal numbers. This method uses the **square root property**.

Square Root Property

If $x^2 = c$, then $x = \pm\sqrt{c}$.

If $(x-a)^2 = c$, then $x - a = \pm\sqrt{c}$ (or $x = a \pm \sqrt{c}$).

Note: If c is negative $(c < 0)$, then the solutions will be nonreal.

PROPERTIES

Example 4 Using the Square Root Property

Solve using the square root property.

$(y+4)^2 = 8$

Solution

$(y+4)^2 = 8$ Note that $8 > 0$.

$y+4 = \pm\sqrt{8}$

$y = -4 \pm 2\sqrt{2}$

Now work margin exercise 4.

4. Solve using the square root property.

$(x+6)^2 = 12$

Example 5 Using the Square Root Property

Solve using the square root property.

$x^2 = -25$

Solution

$x^2 = -25$ Note that $-25 < 0$.

$x = \pm\sqrt{-25}$

$x = \pm 5i$

Now work margin exercise 5.

5. Solve using the square root property.

$x^2 = -16$

Teaching Note

Early on, remind students that the "±" symbol is an economical way of representing two solutions simultaneously.

Completion Example 6 Using the Square Root Property

Solve using the square root property.

$4(x+2)^2 = 80$

Solution

$4(x+2)^2 = 80$

$(x+2)^2 = \underline{20}$

$x+2 = \pm\underline{2\sqrt{5}}$

$x = \underline{-2} \pm \underline{2\sqrt{5}}$

Now work margin exercise 6.

6. Solve using the square root property.

$5(x-4)^2 = 60$

C The Pythagorean Theorem

The Pythagorean Theorem was first discussed in Section 6.7 and again in Section 13.7. In this section, we show how the Pythagorean Theorem can be used to solve a variety of problems that involve solving quadratic equations. The theorem is stated again for easy reference and to emphasize its importance.

The Pythagorean Theorem

In a right triangle, the square of the length of the hypotenuse is equal to the sum of the squares of the lengths of the two legs.

$$c^2 = a^2 + b^2$$

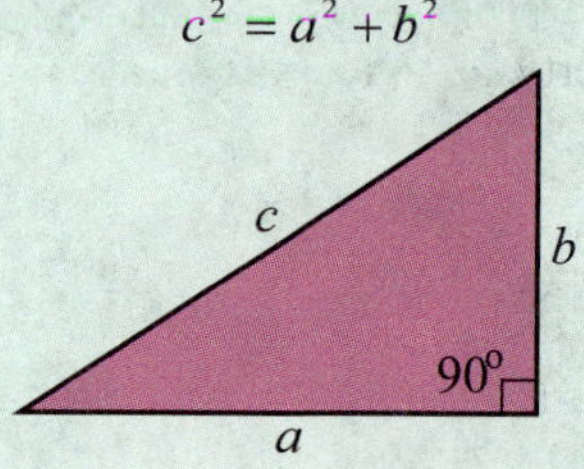

THEOREM

7. If the hypotenuse of a right triangle is 26 feet long and one leg is 24 feet long, what is the length of the other leg?

Example 7 The Pythagorean Theorem

If the hypotenuse of a right triangle is 15 cm long and one leg is 10 cm long, what is the length of the other leg?

Solution

Use the Pythagorean Theorem.

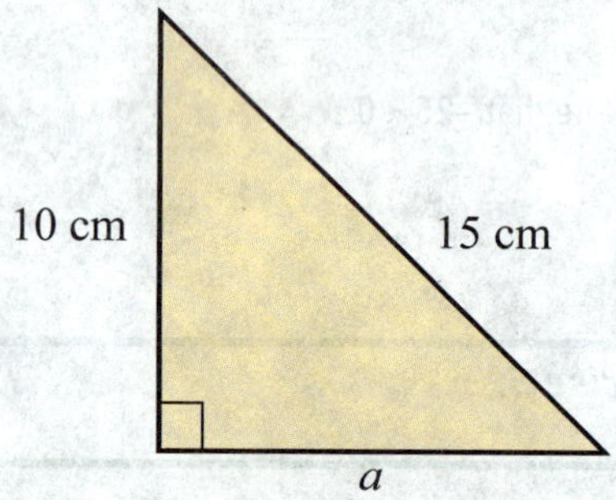

$$\begin{aligned} a^2 + 10^2 &= 15^2 \\ a^2 + 100 &= 225 \\ a^2 &= 125 \\ a &= \sqrt{125} \\ &= \sqrt{25} \cdot \sqrt{5} = 5\sqrt{5} \ (\approx 11.18) \end{aligned}$$

The other leg is $5\sqrt{5}$ cm long (or approximately 11.18 cm long).

The negative solution to the quadratic equation is not considered because length is not negative.

Now work margin exercise 7.

Example 8 The Pythagorean Theorem

The diagonal of a rectangle is twice the width. The length of the rectangle is 6 feet. Find the width of the rectangle.

Solution

Let x = width of the rectangle
and $2x$ = length of the diagonal.

Then, using the Pythagorean Theorem, we have

$$(2x)^2 = x^2 + 6^2$$
$$4x^2 = x^2 + 36$$
$$3x^2 = 36$$
$$x^2 = 12$$
$$x = \sqrt{12} = \sqrt{4} \cdot \sqrt{3} = 2\sqrt{3} \; (\approx 3.46).$$

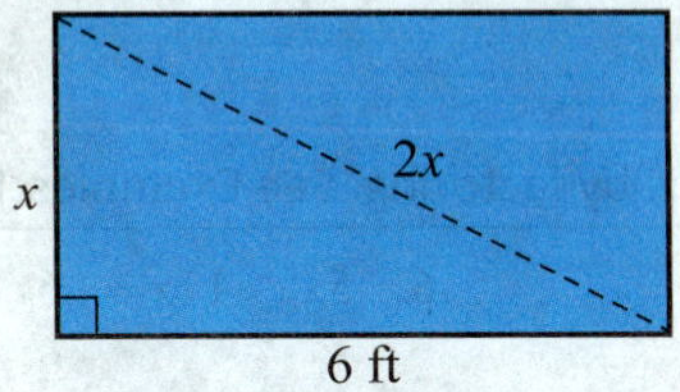

The width of the rectangle is $2\sqrt{3}$ feet or approximately 3.46 feet. The negative solution to the quadratic equation is not considered because length is not negative.

Now work margin exercise 8.

8. The diagonal of a rectangle is three times the width. The length of the rectangle is 10 inches. Find the width of the rectangle.

Completion Example Answers

6. 20; $\pm 2\sqrt{5}$; $-2 \pm 2\sqrt{5}$

Margin Exercise Answers

1. $x = 2, 8$ **2.** $x = 3$ **3.** $x = -5i, 5i$ **4.** $x = -6 \pm 2\sqrt{3}$ **5.** $\pm 4i$ **6.** $x = 4 \pm 2\sqrt{3}$

7. 10 feet **8.** $\frac{5\sqrt{2}}{2}$ inches ($\approx$ 3.54 inches)

16.1 Exercises

Concept Check

Fill-in-the-Blank. Complete each sentence using information found in this section.

1. In a right triangle, the square of the length of the __________ is equal to the sum of the _______ of the lengths of the two legs.
2. The two equations $x = \sqrt{c}$ or $x = -\sqrt{c}$ can be written as ___________.
3. When using the square root method, the polynomial is not set to zero. Instead, the squared expression is set equal to a/an ______ real number.
4. When using the square root method, you take the square root of _____ side(s).

16.1 Exercises

Concept Check

1. hypotenuse, squares
2. $x = \pm\sqrt{c}$
3. nonnegative
4. both

5. True
6. True
7. False; The square of a real number cannot be negative.

Practice

1. $x = 0, 11$
2. $x = 2, 8$
3. $x = -12, -3$
4. $x = -17, -1$
5. $x = 1, -\frac{5}{3}$
6. $x = -\frac{2}{5}, -3$
7. $x = -1, 3$
8. $x = 1, 8$
9. $x = \frac{3}{4}, 1$
10. $x = -\frac{1}{2}, -4$
11. $x = \pm 11$
12. $x = \pm 9$
13. $x = \pm 6$
14. $x = \pm 7$
15. $x = \pm\sqrt{35}$
16. $x = \pm\sqrt{42}$
17. $x = \pm 5i$
18. $x = \pm 9i$
19. $x = \pm\sqrt{62}$
20. $x = \pm 5\sqrt{3}$
21. $x = \pm 3\sqrt{5}$
22. $x = \pm 7\sqrt{2}$
23. $x = \pm 3\sqrt{2}$
24. $x = \pm 2\sqrt{3}$
25. $x = \pm\frac{2}{3}$
26. $x = \pm\frac{5}{2}$
27. $x = -1, 3$
28. $x = -6, 0$
29. $x = -2 \pm 5i$
30. $x = -1, 11$

True/False. Determine whether each statement is true or false. If a statement is false, explain how it can be changed so the statement will be true. (**Note:** There may be more than one acceptable change.)

5. Using the square root method on an equation of the form $x^2 = c$, where c is a nonnegative number, will always result in two distinct solutions.

6. Quadratic equations that are not easily solved using factoring might be solved by the square root method.

7. The square of a real number can be negative.

Practice

Solve the following quadratic equations by factoring. See Examples 1 through 3.

1. $x^2 = 11x$
2. $x^2 - 10x + 16 = 0$
3. $x^2 = -15x - 36$
4. $2x^2 + 36x + 34 = 0$
5. $9x^2 + 6x - 15 = 0$
6. $5x^2 + 17x = -6$
7. $(x+3)(x-1) = 4x$
8. $(x-7)(x-2) = 6$
9. $(2x-3)(2x+1) = 3x - 6$
10. $(x-2)(5x+4) = 3x^2 - 15x - 12$

Solve the following quadratic equations by using the square root method. Write each radical in simplest form. See Examples 4 through 6.

11. $x^2 = 121$
12. $x^2 = 81$
13. $3x^2 = 108$
14. $5x^2 = 245$
15. $x^2 = 35$
16. $x^2 = 42$
17. $x^2 + 25 = 0$
18. $x^2 + 81 = 0$
19. $x^2 - 62 = 0$
20. $x^2 - 75 = 0$
21. $x^2 - 45 = 0$
22. $x^2 - 98 = 0$
23. $3x^2 = 54$
24. $5x^2 = 60$
25. $9x^2 = 4$
26. $4x^2 = 25$
27. $(x-1)^2 = 4$
28. $(x+3)^2 = 9$
29. $(x+2)^2 = -25$
30. $(x-5)^2 = 36$
31. $(x-3)^2 = -4$
32. $(x+8)^2 = -9$
33. $(x+1)^2 = \frac{1}{4}$
34. $(x-9)^2 = -\frac{9}{25}$
35. $(x-2)^2 = \frac{1}{16}$
36. $(x-3)^2 = \frac{4}{9}$
37. $(x+2)^2 = -7$
38. $(x+8)^2 = 75$

39. $(5x-2)^2 = 63$

40. $(4x-3)^2 = 125$

41. $(3x+4)^2 + 3 = 30$

42. $(2x+1)^2 + 12 = 60$

43. $2(x-7)^2 = 24$

44. $3(x+11)^2 = 60$

45. $3(x-5)^2 + 5 = -25$

46. $2(x-6)^2 - 11 = 25$

47. $-3(x^2+15) - 7 = 20$

48. $-4(x^2+12) + 1 = 25$

The lengths of two sides are given for each of the following right triangles. Determine the length of the missing side. See Examples 7 and 8.

49. $a = 9$, $b = 12$, $c = ?$

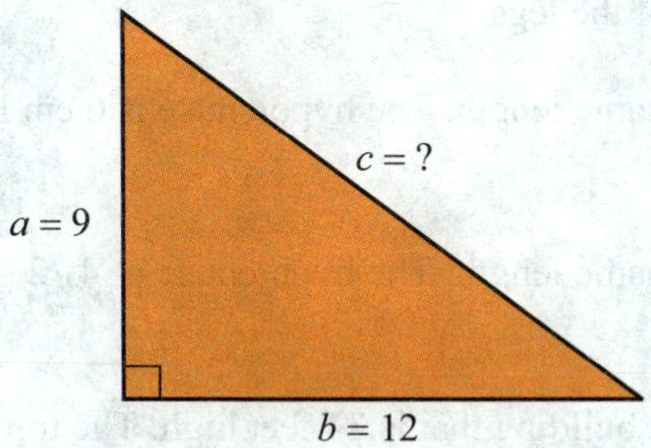

50. $a = 10$, $b = 24$, $c = ?$

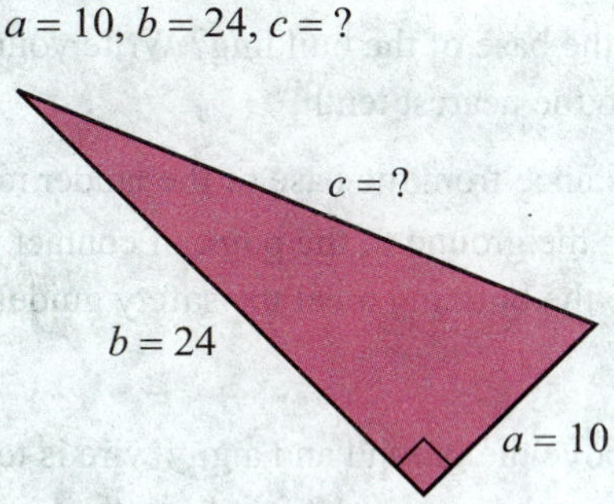

51. $a = 6$, $c = 12$, $b = ?$

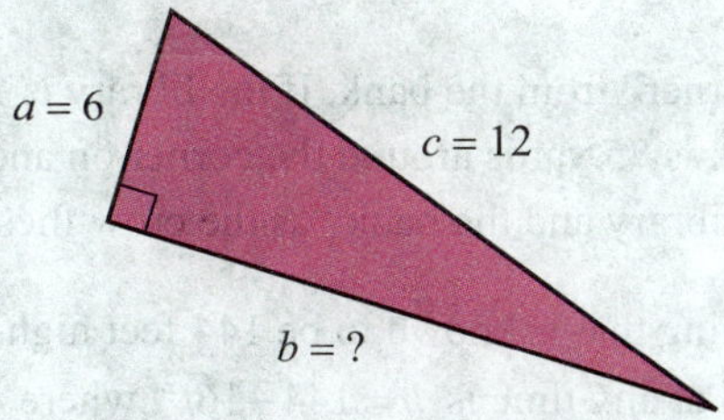

52. $b = 10$, $c = 30$, $a = ?$

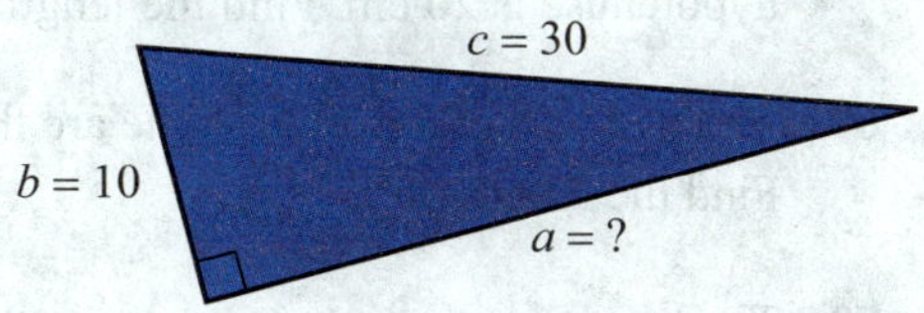

53. $a = 1$, $c = \sqrt{2}$, $b = ?$

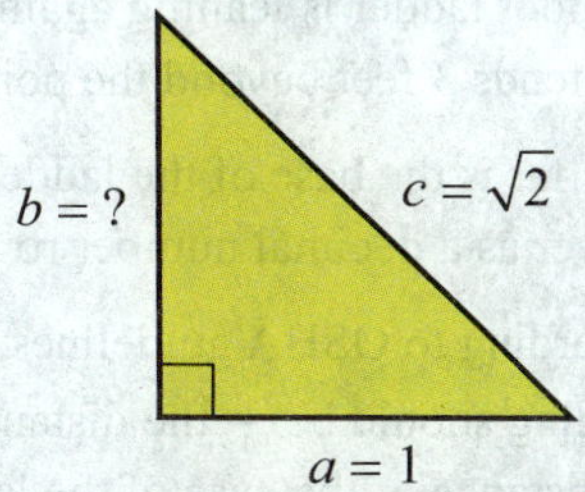

54. $b = 4$, $c = 4\sqrt{2}$, $a = ?$

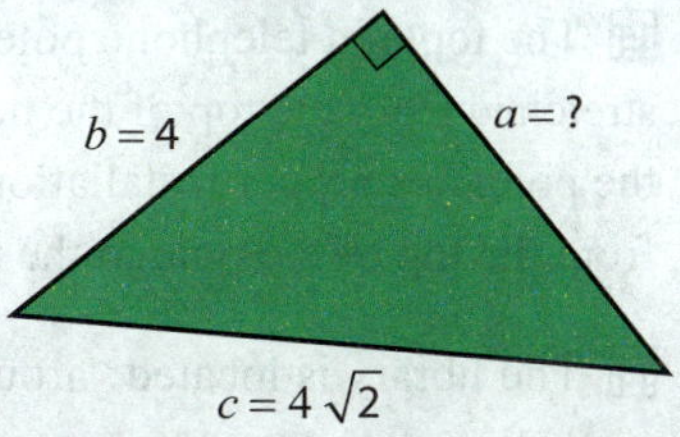

Use a calculator to solve the following quadratic equations. Round your answers to the nearest hundredth.

55. $x^2 = 647$

56. $x^2 = 378$

57. $19x^2 = 523$

58. $14x^2 = 795$

59. $6x^2 = 17.32$

60. $15x^2 = 229.63$

61. $2.1x^2 = 35.82$

62. $4.7x^2 = 118.34$

31. $x = 3 \pm 2i$

32. $x = -8 \pm 3i$

33. $x = -\frac{3}{2}, -\frac{1}{2}$

34. $x = 9 \pm \frac{3i}{5}$

35. $x = \frac{7}{4}, \frac{9}{4}$

36. $x = \frac{7}{3}, \frac{11}{3}$

37. $x = -2 \pm i\sqrt{7}$

38. $x = -8 \pm 5\sqrt{3}$

39. $x = \frac{2 \pm 3\sqrt{7}}{5}$

40. $x = \frac{3 \pm 5\sqrt{5}}{4}$

41. $x = \frac{-4 \pm 3\sqrt{3}}{3}$

42. $x = \frac{-1 \pm 4\sqrt{3}}{2}$

43. $x = 7 \pm 2\sqrt{3}$

44. $x = -11 \pm 2\sqrt{5}$

45. $x = 5 \pm i\sqrt{10}$

46. $x = 6 \pm 3\sqrt{2}$

47. $x = \pm 2i\sqrt{6}$

48. $x = \pm 3i\sqrt{2}$

49. $c = 15$

50. $c = 26$

51. $b = 6\sqrt{3}$

52. $a = 20\sqrt{2}$

53. $b = 1$

54. $a = 4$

55. $x = \pm 25.44$

56. $x = \pm 19.44$

57. $x = \pm 5.25$

58. $x = \pm 7.54$

59. $x = \pm 1.70$

60. $x = \pm 3.91$

61. $x = \pm 4.13$

62. $x = \pm 5.02$

63. 4.2361, −0.2361

64. 5.4340, −13.4340

65. 2.3229, −0.3229

66. 0.8606, 0.1394

Applications

67. The length of the leg is 4 feet and the hypotenuse is 8 feet.

68. $2\sqrt{10}$ cm and $6\sqrt{10}$ cm

69. $3\sqrt{2}$ cm

70. 4 m

71. **a.** $35^2 = x^2 + 34^2 \rightarrow 8.3$ ft

b. No, $\frac{1}{4}$ of 34 is 8.5, so the distance is too short by 0.2 feet.

72. 47.2 feet

73. 125 yards

74. 3 seconds

75. **a.** 640 ft; 384 ft

b. 144 ft; 400 ft

c. 7 seconds; $0 = -16(t+7)(t-7)$

Use a calculator to find the two values of each expression accurate to the nearest ten-thousandth.

63. $2 \pm \sqrt{5}$

64. $-4 \pm \sqrt{89}$

65. $\frac{2 \pm \sqrt{7}}{2}$

66. $\frac{5 \pm \sqrt{13}}{10}$

Applications

Solve.

67. The hypotenuse of a right triangle is twice the length of one of the legs. The length of the other leg is $4\sqrt{3}$ feet. Find the length of the leg and the hypotenuse.

68. One leg of a right triangle is three times the length of the other. The length of the hypotenuse is 20 cm. Find the lengths of the legs.

69. The two legs of a right triangle are the same length. The hypotenuse is 6 cm long. Find the length of the legs.

70. The two legs of a right triangle are the same length. The hypotenuse is $4\sqrt{2}$ m long. Find the length of the legs.

71. A 38-foot ladder is leaning against a building that is 34 feet high. The top of the ladder extends 3 feet beyond the point where it touches the building.

a. How far is the base of the ladder from the base of the building? Write your answer as a decimal number rounded to the nearest tenth.

b. According to OSHA guidelines, the distance from the base of the ladder to the building should be $\frac{1}{4}$ the distance from the ground to the point of contact. Does the distance of the base of the ladder to the building meet the safety guidelines? Explain your answer.

72. The top of a telephone pole is 40 feet above the ground and a guy wire is to be stretched from the top of the pole to a point on the ground 10 feet from the base of the pole. For proper installation, the wire needs to be 7 feet longer than the distance from the top of the pole to the point on the ground. How long should the wire be?

73. The library is located "around the corner" from the bank. If the library is 100 yards from the street corner and the bank is 75 yards around the corner on another street, what is the distance between the library and the bank "as the crow flies?"

74. A ball is dropped from the top of a building that is known to be 144 feet high. The formula for finding the height of the ball at any time is $h = 144 - 16t^2$ where t is measured in seconds. How many seconds will it take for the ball to hit the ground?

75. A ball is dropped from the top of a building that is 784 feet high. The height of the ball above ground level is given by the polynomial function $h(t) = -16t^2 + 784$ where t is measured in seconds.

a. How high is the ball after 3 seconds? 5 seconds?

b. How far has the ball traveled in 3 seconds? 5 seconds?

c. When will the ball hit the ground? Explain your reasoning in terms of factors.

76. A tennis ball is dropped from a building. The position of the ball after t seconds is given by the polynomial function $s(t) = -4.9t^2 + 490$, where s is the height in meters of the ball.

a. Find $s(0)$ What does this value represent in the context of this problem?

b. How high is the tennis ball 2 seconds after it has been dropped?

c. How long before the tennis ball hits the ground?

77. An oil spill from a ruptured pipeline is circular in shape and covers an area of about 8 square miles. About how many miles long (to the nearest tenth of a mile) is the diameter of the oil spill? (Use $\pi = 3.14$.)

78. A financial consultant is asked for advice about finances and savings plans. When a client invests money, they need to know which interest rate will meet their financial goals based on the amount invested. The financial consultant can use the formula $A = P(r+1)^n$ to find the future amount A, after n years, of an investment with a starting principal P invested at an interest rate of r.

a. A client has \$3000 to invest and would like to earn \$300 on his investment after 2 years. At what interest rate will the client need to invest his money? Round to the nearest hundredth of a percent.

b. Another client has \$5000 to invest and would like to earn \$750 on her investment after 2 years. At what interest rate will the client need to invest her money? Round to the nearest hundredth of a percent.

76. **a.** $s(0) = 490$; This means that the ball is dropped from a height of 490 meters. **b.** 470.4 meters **c.** 10 seconds

77. 3.2 miles

78. **a.** 4.88% **b.** 7.24%

Objectives

A. Solve quadratic equations by completing the square.

B. Write quadratic equations with given roots.

16.2 Quadratic Equations: Completing the Square

A Completing the Square

Recall that a perfect square trinomial is the result of squaring a binomial. Our objective here is to find the constant term of a perfect square trinomial when the first two terms are given. This is called **completing the square**. We will find this procedure useful in solving quadratic equations and in developing the quadratic formula.

The following examples illustrate the concept of a perfect square trinomial.

Perfect Square Trinomials		Equal Factors		Square of a Binomial
$x^2-8x+16$	$=$	$(x-4)(x-4)$	$=$	$(x-4)^2$
$x^2+20x+100$	$=$	$(x+10)(x+10)$	$=$	$(x+10)^2$
$x^2-9x+\frac{81}{4}$	$=$	$\left(x-\frac{9}{2}\right)\left(x-\frac{9}{2}\right)$	$=$	$\left(x-\frac{9}{2}\right)^2$
$x^2-2ax+a^2$	$=$	$(x-a)(x-a)$	$=$	$(x-a)^2$
$x^2+2ax+a^2$	$=$	$(x+a)(x+a)$	$=$	$(x+a)^2$

Table 1

The last two examples in Table 1 are in the form of formulas. We see two things in each case:

1. The leading coefficient (the coefficient of x^2) is 1.
2. The constant term is the square of $\frac{1}{2}$ of the coefficient of x.

For example, $\frac{1}{2}(2a)=a$ and the square of this result is the constant a^2.

What constant should be added to x^2-16x to get a perfect square trinomial? By following the ideas just discussed, we find that $\frac{1}{2}(-16)=-8$ and $(-8)^2=64$. Therefore, to complete the square, we add 64.

$$x^2-16x+64=(x-8)^2$$

By adding 64, we have completed the square for x^2-16x.

Teaching Note

Some of the power of algebra can be illustrated by pointing out how the general representation of the perfect square trinomial, $x^2+2ax+a^2$, leads to the procedure for completing the square. This is another example of the advantage in learning to "read" the algebraic form of expressions and equations.

Example 1 Completing the Square

Add the constant that will complete the square for each expression, and write the new expression as the square of a binomial.

a. $x^2 + 10x$ **b.** $x^2 - 7x$

Solution

a. $x^2 + 10x + \underline{\ ?\ } = (\underline{\ ?\ })^2$

$\frac{1}{2}(10) = 5$ and $(5)^2 = 25$

Find $\frac{1}{2}$ of the coefficient of x and square the result. Add this square constant to complete the square. The resulting trinomial will equal the square of a binomial.

So, add 25: $x^2 + 10x + 25 = (x+5)^2$

b. $x^2 - 7x + \underline{\ ?\ } = (\underline{\ ?\ })^2$

$\frac{1}{2}(-7) = -\frac{7}{2}$ and $\left(-\frac{7}{2}\right)^2 = \frac{49}{4}$

Find $\frac{1}{2}$ of the coefficient of x and square the result.

So, add $\frac{49}{4}$: $x^2 - 7x + \frac{49}{4} = \left(x - \frac{7}{2}\right)^2$

Now work margin exercise 1.

1. Add the constant that will complete the square for each expression, and write the new expression as the square of a binomial.

   a. $y^2 - 14y + \_\_ = (\_\_)^2$

   b. $x^2 + 9x + \_\_ = (\_\_)^2$

Now we want to use the process of completing the square to help in solving quadratic equations. This technique involves the following steps.

Solving a Quadratic Equation by Completing the Square

1. If necessary, divide or multiply on both sides of the equation so that the leading coefficient (the coefficient of x^2) is 1.
2. If necessary, isolate the constant term on one side of the equation.
3. Find the constant that completes the square of the polynomial and **add this constant to both sides**. Remember that the constant term is the square of $\frac{1}{2}$ of the coefficient of x. Rewrite the polynomial as the square of a binomial.
4. Use the square root property to find the solutions of the equation.

PROCEDURE

2. Solve by completing the square.

$x^2 - 10x = 31$

Example 2 Solving Quadratic Equations by Completing the Square

Solve by completing the square.

$x^2 - 8x = 25$

Solution

$x^2 - 8x = 25$ — The coefficient of x^2 is already 1 and the constant is isolated on one side of the equation.

$x^2 - 8x + 16 = 25 + 16$ — Complete the square: $\frac{1}{2}(-8) = -4$ and $(-4)^2 = 16$. Therefore, add 16 to both sides.

$(x-4)^2 = 41$ — Factor the polynomial.

$x - 4 = \sqrt{41}$ or $x - 4 = -\sqrt{41}$ — Use the square root property.

$x = 4 + \sqrt{41}$ $\quad$ $x = 4 - \sqrt{41}$

There are two real solutions: $4 + \sqrt{41}$ and $4 - \sqrt{41}$.

We write $x = 4 \pm \sqrt{41}$.

Now work margin exercise 2.

3. Solve by completing the square.

$2x^2 + 4x - 12 = 0$

Example 3 Solving Quadratic Equations by Completing the Square

Solve by completing the square.

$3x^2 + 6x - 15 = 0$

Solution

$3x^2 + 6x - 15 = 0$

$\frac{3x^2}{3} + \frac{6x}{3} - \frac{15}{3} = \frac{0}{3}$ — Divide each term by 3. **The leading coefficient must be 1.**

$x^2 + 2x - 5 = 0$

$x^2 + 2x = 5$ — Isolate the constant term.

$x^2 + 2x + 1 = 5 + 1$ — Complete the square: $\frac{1}{2}(2) = 1$ and $1^2 = 1$. Therefore, add 1 to both sides.

$(x+1)^2 = 6$ — Factor the polynomial.

$x + 1 = \pm\sqrt{6}$ — Use the square root property.

$x = -1 \pm \sqrt{6}$

Now work margin exercise 3.

Math Tip

Keep in mind that if an equation begins with a term of $-x^2$, it has a leading coefficient of -1. If you run into this scenario, you will need to divide (or multiply) the entire equation by -1 to clear the negative before you can solve by completing the square.

Example 4 Solving Quadratic Equations by Completing the Square

Solve by completing the square.

$2x^2+2x-7=0$

Solution

$2x^2+2x-7=0$ Divide each term by 2 so that the leading coefficient will be 1.

$x^2+x-\frac{7}{2}=0$

$x^2+x=\frac{7}{2}$ Isolate the constant term.

$x^2+x+\frac{1}{4}=\frac{7}{2}+\frac{1}{4}$ Complete the square: the coefficient of x is 1 and $\frac{1}{2}(1)=\frac{1}{2}$ and $\left(\frac{1}{2}\right)^2=\frac{1}{4}$. Therefore, add $\frac{1}{4}$ to both sides.

$\left(x+\frac{1}{2}\right)^2=\frac{15}{4}$ Factor the polynomial.

$x+\frac{1}{2}=\pm\sqrt{\frac{15}{4}}$ Use the square root property.

$x=-\frac{1}{2}\pm\frac{\sqrt{15}}{2}$

$x=\frac{-1\pm\sqrt{15}}{2}$

Now work margin exercise 4.

4. Solve by completing the square.

$2x^2+2x-9=0$

Example 5 Solving Quadratic Equations by Completing the Square

Solve by completing the square.

$x^2-2x+13=0$

Solution

$x^2-2x+13=0$

$x^2-2x=-13$ Isolate the constant term.

$x^2-2x+1=-13+1$ Complete the square: $\frac{1}{2}(-2)=-1$ and $(-1)^2=1$. Therefore, add 1 to both sides.

$(x-1)^2=-12$ Factor the polynomial.

$x-1=\pm\sqrt{-12}$ Use the square root property.

$x-1=\pm i\sqrt{12}=\pm 2i\sqrt{3}$

$x=1\pm 2i\sqrt{3}$ The solutions are nonreal complex conjugates.

Now work margin exercise 5.

5. Solve by completing the square.

$x^2+4x+11=0$

6. Solve the quadratic equation by completing the square.

$0 = 3x^2 - 6x - 15$

Completion Example 6 Completing the Square

Solve the quadratic equation by completing the square.

$2x^2 - 12x + 2 = 0$

Solution

$$
\begin{aligned}
2x^2 - 12x + 2 &= 0 \\
2x^2 - 12x &= \underline{-2} \\
x^2 - 6x &= \underline{-1} \\
x^2 - 6x + \underline{9} &= \underline{-1} + \underline{9} \\
(x - \underline{3})^2 &= \underline{8} \\
x - \underline{3} &= \pm\underline{\sqrt{8}} \\
x &= \underline{3} \pm \underline{2\sqrt{2}}
\end{aligned}
$$

Now work margin exercise 6.

B Writing Quadratic Equations with Known Roots

In Section 13.6, we found equations with known roots by setting the product of factors equal to 0 and simplifying. The same method is applied here with roots that are nonreal and roots that involve radicals.

7. Find a quadratic equation with the given roots.

$y = 4 + 3i$ and $y = 4 - 3i$

Example 7 Quadratic Equations with Known Roots

Find a quadratic equation with the given roots.

$y = 3 + 2i$ and $y = 3 - 2i$

Solution

$$
\begin{aligned}
y &= 3 + 2i & y &= 3 - 2i \\
y - 3 - 2i &= 0 & y - 3 + 2i &= 0
\end{aligned}
$$

Get 0 on one side of each equation.

Set the product of the two factors equal to 0 and simplify.

$(y - 3 - 2i)(y - 3 + 2i) = 0$

$[(y - 3) - 2i][(y - 3) + 2i] = 0$

Regroup the terms to represent the product of complex conjugates. This makes the multiplication easier.

$(y - 3)^2 - 4i^2 = 0$

Remember, $i^2 = -1$.

$y^2 - 6y + 9 + 4 = 0$

$y^2 - 6y + 13 = 0$

This equation has two solutions: $y = 3 + 2i$ and $y = 3 - 2i$.

Now work margin exercise 7.

Example 8 Quadratic Equations with Known Roots

Find a quadratic equation with the given roots.

$x = 5-\sqrt{2}$ and $x = 5+\sqrt{2}$

Solution

$$x = 5-\sqrt{2} \qquad x = 5+\sqrt{2}$$

$$x-5+\sqrt{2} = 0 \qquad x-5-\sqrt{2} = 0$$ Get 0 on one side of each equation.

Set the product of the two factors equal to 0 and simplify.

$$\left(x-5+\sqrt{2}\right)\left(x-5-\sqrt{2}\right) = 0$$ Regroup the terms to make the multiplication easier.

$$\left[(x-5)+\sqrt{2}\right]\left[(x-5)-\sqrt{2}\right] = 0$$

$$(x-5)^2-\left(\sqrt{2}\right)^2 = 0$$

$$x^2-10x+25-2 = 0$$

$$x^2-10x+23 = 0$$

This equation has two solutions: $x = 5-\sqrt{2}$ and $x = 5+\sqrt{2}$.

Now work margin exercise 8.

8. Find a quadratic equation with the given roots.
$x = 2-3\sqrt{2}$ and $x = 2+3\sqrt{2}$

Example 9 Quadratic Equations with Known Roots

Find a quadratic equation with the given roots.

$x = 3+i\sqrt{5}$ and $x = 3-i\sqrt{5}$

Solution

$$x = 3+i\sqrt{5} \qquad x = 3-i\sqrt{5}$$

$$x-3-i\sqrt{5} = 0 \qquad x-3+i\sqrt{5} = 0$$ Get 0 on one side of each equation.

Set the product of the two factors equal to 0 and simplify.

$$\left(x-3-i\sqrt{5}\right)\left(x-3+i\sqrt{5}\right) = 0$$ Regroup the terms to make the multiplication easier.

$$\left[(x-3)-i\sqrt{5}\right]\left[(x-3)+i\sqrt{5}\right] = 0$$

$$(x-3)^2-\left(i\sqrt{5}\right)^2 = 0$$

$$x^2-6x+9-i^2\left(\sqrt{5}\right)^2 = 0$$

$$x^2-6x+9-(-1)(5) = 0$$ Remember, $i^2 = -1$.

$$x^2-6x+14 = 0$$

This equation has two solutions: $x = 3+i\sqrt{5}$ and $x = 3-i\sqrt{5}$

Now work margin exercise 9.

9. Find a quadratic equation with the given roots.
$x = 1+i\sqrt{5}$ and $x = 1-i\sqrt{5}$

16.2 Exercises

Concept Check

1. trinomial
2. variables, constants
3. 1
4. add
5. square
6. True
7. False; It's also possible for there to be one solution.
8. True
9. True

Practice

1. $x^2-12x+\underline{36}=\left(\underline{x-6}\right)^2$
2. $y^2+14y+\underline{49}=\left(\underline{y+7}\right)^2$
3. $x^2+6x+\underline{9}=\left(\underline{x+3}\right)^2$
4. $x^2+8x+\underline{16}=\left(\underline{x+4}\right)^2$
5. $x^2-5x+\underline{\frac{25}{4}}=\left(\underline{x-\frac{5}{2}}\right)^2$
6. $x^2+7x+\underline{\frac{49}{4}}=\left(\underline{x+\frac{7}{2}}\right)^2$
7. $y^2+y+\underline{\frac{1}{4}}=\left(\underline{y+\frac{1}{2}}\right)^2$
8. $x^2+\frac{1}{2}x+\underline{\frac{1}{16}}=\left(\underline{x+\frac{1}{4}}\right)^2$
9. $x^2+\frac{1}{3}x+\underline{\frac{1}{36}}=\left(\underline{x+\frac{1}{6}}\right)^2$
10. $y^2+\frac{3}{4}y+\underline{\frac{9}{64}}=\left(\underline{y+\frac{3}{8}}\right)^2$
11. $2x^2+4x+2=2(x+1)^2$
12. $3x^2+18x+27=3(x+3)^2$
13. $x=-5, 1$
14. $x=-7, 1$
15. $y=-1\pm\sqrt{6}$
16. $x=4\pm\sqrt{13}$
17. $x=5\pm\sqrt{22}$
18. $z=-2\pm\sqrt{6}$

Completion Example Answers

6. $$2x^2-12x=\underline{-2}$$
$$x^2-6=\underline{-1}$$
$$x^2-6x+\underline{9}=\underline{-1}+\underline{9}$$
$$\left(x-\underline{3}\right)^2=\underline{8}$$
$$x-\underline{3}=\pm\underline{\sqrt{8}}$$
$$x=\underline{3}\pm\underline{2\sqrt{2}}$$

Margin Exercise Answers

1. a. $y^2-14y+\underline{49}=\left(\underline{y-7}\right)^2$ **b.** $x^2+9x+\underline{\frac{81}{4}}=\left(\underline{x+\frac{9}{2}}\right)^2$ **2.** $x=5\pm2\sqrt{14}$ **3.** $x=-1\pm\sqrt{7}$

4. $x=\frac{1\pm\sqrt{19}}{2}$ **5.** $x=-2\pm i\sqrt{7}$ **6.** $x=1\pm\sqrt{6}$ **7.** $y^2-8y+25=0$ **8.** $x^2-4x-14=0$

9. $x^2-2x+6=0$

16.2 Exercises

Concept Check

Fill-in-the-Blank. Complete each sentence using information found in this section.

1. Completing the square is the process of adding terms to binomials so that the result will be a perfect square ______.
2. To solve a quadratic equation by completing the square, arrange terms with ________ on one side of the equation and ________ on the other.
3. When solving by completing the square, the quadratic equation should have a leading coefficient of ______.
4. After finding the coefficient that completes the square of the polynomial, ______ this constant to both sides of the equation.
5. When completing the square, the constant term is the ______ of $\frac{1}{2}$ of the coefficient of x.

True/False. Determine whether each statement is true or false. If a statement is false, explain how it can be changed so the statement will be true. (**Note:** There may be more than one acceptable change.)

6. To get a leading coefficient of 1, multiply both sides of the equation by the reciprocal of the leading coefficient.
7. When solving a quadratic equation, there is either no solution or two solutions.
8. It's possible for the roots of a quadratic equation to be nonreal numbers.
9. The last step of solving a quadratic equation by completing the square is to use the square root property.

Practice

Add the correct constant to complete the square; then factor the trinomial as indicated. See Example 1.

1. $x^2-12x+\underline{\hspace{2em}}=(\underline{\hspace{3em}})^2$
2. $y^2+14y+\underline{\hspace{2em}}=(\underline{\hspace{3em}})^2$
3. $x^2+6x+\underline{\hspace{2em}}=(\underline{\hspace{3em}})^2$
4. $x^2+8x+\underline{\hspace{2em}}=(\underline{\hspace{3em}})^2$
5. $x^2-5x+\underline{\hspace{2em}}=(\underline{\hspace{3em}})^2$
6. $x^2+7x+\underline{\hspace{2em}}=(\underline{\hspace{3em}})^2$
7. $y^2+y+\underline{\hspace{2em}}=(\underline{\hspace{3em}})^2$
8. $x^2+\frac{1}{2}x+\underline{\hspace{2em}}=(\underline{\hspace{4em}})^2$
9. $x^2+\frac{1}{3}x+\underline{\hspace{2em}}=(\underline{\hspace{4em}})^2$
10. $y^2+\frac{3}{4}y+\underline{\hspace{2em}}=(\underline{\hspace{4em}})^2$
11. $2x^2+4x+\underline{\hspace{3em}}=2(\underline{\hspace{3em}})^2$
12. $3x^2+18x+\underline{\hspace{3em}}=3(\underline{\hspace{3em}})^2$

Solve the quadratic equations by completing the square. See Examples 2 through 6.

13. $x^2+4x-5=0$
14. $x^2+6x-7=0$
15. $y^2+2y=5$
16. $x^2+3=8x$
17. $x^2+3=10x$
18. $z^2+4z=2$
19. $x^2-4x-45=0$
20. $x^2-10x+21=0$
21. $x^2-3x-40=0$
22. $x^2+x-42=0$
23. $3x^2+x-4=0$
24. $2x^2+x-6=0$
25. $x^2-6x+10=0$
26. $x^2-2x+5=0$
27. $x^2+11=12x$
28. $x^2=6-x$
29. $y^2=10y-4$
30. $x^2=3-4x$
31. $z^2+3z-5=0$
32. $x^2-5x+5=0$
33. $x^2+x+2=0$
34. $y^2+3y+3=0$
35. $x^2+5x+2=0$
36. $4x^2+7x+2=0$
37. $3x^2-10x+5=0$
38. $3y^2+5y-3=0$
39. $3x^2+6x+18=0$
40. $4x^2+8x+16=0$
41. $2x-3=4x^2$
42. $2x+2=-6x^2$
43. $5y^2+15y+25=0$
44. $4x^2+20x+32=0$
45. $3y^2=4-y$
46. $2x^2+4=-9x$
47. $2x^2-8x+4=0$
48. $3x^2-18x+12=0$

19. $x=-5,9$
20. $x=3,7$
21. $x=-5,8$
22. $x=-7,6$
23. $x=-\frac{4}{3},1$
24. $x=\frac{3}{2},-2$
25. $x=3\pm i$
26. $x=1\pm 2i$
27. $x=1,11$
28. $x=-3,2$
29. $y=5\pm\sqrt{21}$
30. $x=-2\pm\sqrt{7}$
31. $z=\frac{-3\pm\sqrt{29}}{2}$
32. $x=\frac{5\pm\sqrt{5}}{2}$
33. $x=\frac{-1\pm i\sqrt{7}}{2}$ or $-\frac{1}{2}\pm\frac{\sqrt{7}}{2}i$
34. $y=\frac{-3\pm i\sqrt{3}}{2}$ or $-\frac{3}{2}\pm\frac{\sqrt{3}}{2}i$
35. $x=\frac{-5\pm\sqrt{17}}{2}$
36. $x=\frac{-7\pm\sqrt{17}}{8}$
37. $x=\frac{5\pm\sqrt{10}}{3}$
38. $y=\frac{-5\pm\sqrt{61}}{6}$
39. $x=-1\pm i\sqrt{5}$
40. $x=-1\pm i\sqrt{3}$
41. $x=\frac{1\pm i\sqrt{11}}{4}$ or $\frac{1}{4}\pm\frac{\sqrt{11}}{4}i$
42. $x=\frac{-1\pm i\sqrt{11}}{6}$ or $-\frac{1}{6}\pm\frac{\sqrt{11}}{6}i$

43. $y = \frac{-3 \pm i\sqrt{11}}{2}$ or $-\frac{3}{2} \pm \frac{\sqrt{11}}{2}i$

44. $x = \frac{-5 \pm i\sqrt{7}}{2}$ or $-\frac{5}{2} \pm \frac{\sqrt{7}}{2}i$

45. $y = -\frac{4}{3}, 1$

46. $x = -4, -\frac{1}{2}$

47. $x = 2 \pm \sqrt{2}$

48. $x = 3 \pm \sqrt{5}$

49. $x^2 - 7 = 0$

50. $x^2 - 6 = 0$

51. $x^2 - 2x - 2 = 0$

52. $z^2 - 6z + 7 = 0$

53. $y^2 + 4y - 16 = 0$

54. $x^2 - 2x - 11 = 0$

55. $x^2 + 16 = 0$

56. $x^2 + 49 = 0$

57. $y^2 + 6 = 0$

58. $y^2 + 5 = 0$

59. $x^2 - 4x + 5 = 0$

60. $x^2 + 6x + 13 = 0$

61. $x^2 - 2x + 3 = 0$

62. $x^2 - 4x + 7 = 0$

63. $x^2 + 10x + 49 = 0$

64. $y^2 - 8y + 34 = 0$

Applications

65. **a.** $p = 0, 90$

b. No income revenue is made if the price is set to \$0 or \$90 per frame.

66. **a.** $t = 0, 8$

b. That the golf ball will be on the ground at the beginning of the swing $(t = 0)$ and after 8 seconds.

Writing & Thinking

67. See Example 2.

Write a quadratic equation with integer coefficients that has the given roots. See Examples 7 through 9.

49. $x = \sqrt{7}, x = -\sqrt{7}$

50. $x = \sqrt{6}, x = -\sqrt{6}$

51. $x = 1 + \sqrt{3}, x = 1 - \sqrt{3}$

52. $z = 3 + \sqrt{2}, z = 3 - \sqrt{2}$

53. $y = -2 + 2\sqrt{5}, y = -2 - 2\sqrt{5}$

54. $x = 1 + 2\sqrt{3}, x = 1 - 2\sqrt{3}$

55. $x = 4i, x = -4i$

56. $x = 7i, x = -7i$

57. $y = i\sqrt{6}, y = -i\sqrt{6}$

58. $y = i\sqrt{5}, y = -i\sqrt{5}$

59. $x = 2 + i, x = 2 - i$

60. $x = -3 + 2i, x = -3 - 2i$

61. $x = 1 + i\sqrt{2}, x = 1 - i\sqrt{2}$

62. $x = 2 + i\sqrt{3}, x = 2 - i\sqrt{3}$

63. $x = -5 + 2i\sqrt{6}, x = -5 - 2i\sqrt{6}$

64. $y = 4 + 3i\sqrt{2}, y = 4 - 3i\sqrt{2}$

Applications

Solve.

65. A local frame shop determines that the revenue function for their custom framing service is $R(p) = 360p - 4p^2$, where p is the base price in dollars for each custom framing job.

a. Set the function equal to 0 and solve for p using the method of completing the square.

b. What do the solutions from part a. mean?

66. The height of a golf ball that is hit from the ground at a speed of 128 feet per second can be modeled with the expression $h(t) = -16t^2 + 128t$, where t is the time in seconds after the ball is hit.

a. Set the function equal to 0 and solve for t using the method of completing the square.

b. What do the solutions from part a. mean?

Writing & Thinking

67. Explain, in your own words, the steps involved in the process of solving a quadratic equation by completing the square.

16.3 Quadratic Equations: The Quadratic Formula

Objectives

A. Use the quadratic formula to solve quadratic equations.

B. Use the discriminant to determine the nature of a quadratic equation's solutions.

A The Quadratic Formula

The **quadratic formula** gives the roots of any quadratic equation in terms of the coefficients a, b, and c of the **general quadratic equation**,

$$ax^2 + bx + c = 0.$$

Therefore, if you memorize the quadratic formula, you can solve any quadratic equation by simply substituting the coefficients into the formula. To develop the quadratic formula, we solve the general quadratic equation by **completing the square**.

$ax^2 + bx + c = 0$	
$x^2 + \frac{b}{a}x + \frac{c}{a} = \frac{0}{a}$	Divide each term of the equation by a. The leading coefficient must be 1.
$x^2 + \frac{b}{a}x = -\frac{c}{a}$	Add $-\frac{c}{a}$ to both sides of the equation.
$x^2 + \frac{b}{a}x + \frac{b^2}{4a^2} = \frac{b^2}{4a^2} - \frac{c}{a}$	Add $\frac{b^2}{4a^2}$ to both sides to complete the square.
$\left(x + \frac{b}{2a}\right)^2 = \frac{b^2}{4a^2} - \frac{4ac}{4a^2}$	Factor the left side. $4a^2$ is the common denominator on the right-hand side.
$\left(x + \frac{b}{2a}\right)^2 = \frac{b^2 - 4ac}{4a^2}$	Simplify.
$x + \frac{b}{2a} = \pm\sqrt{\frac{b^2 - 4ac}{4a^2}}$	Use the square root property.
$x + \frac{b}{2a} = \pm\frac{\sqrt{b^2 - 4ac}}{2a}$	Simplify.
$x = -\frac{b}{2a} \pm \frac{\sqrt{b^2 - 4ac}}{2a}$	Solve for x.
$x = \frac{-b \pm \sqrt{b^2 - 4ac}}{2a}$	This equation is called the **quadratic formula**.

The Quadratic Formula

For the general quadratic equation $\boldsymbol{ax^2 + bx + c = 0}$ where $a \neq 0$ the solutions are

$$\boldsymbol{x = \frac{-b \pm \sqrt{b^2 - 4ac}}{2a}}.$$

FORMULA

Note

The Coefficient *a*

For convenience and without loss of generality, in the development of the quadratic formula (and in the examples and exercises) the leading coefficient a is positive. If a is a negative number, we can multiply both sides of the equation by –1. This will make the leading coefficient positive without changing any solutions of the original equation.

The quadratic formula should be memorized.

Applications of quadratic equations are found in such fields as economics, business, computer science, and chemistry and in almost all branches of mathematics. Most instructors assume that their students know the quadratic formula and how to apply it. You should recognize that the importance of the quadratic formula lies in the fact that **any** quadratic equation can be solved.

1. Solve using the quadratic formula.

$x^2 - 7x + 4 = 0$

Example 1 Using the Quadratic Formula

Solve using the quadratic formula.

$x^2 - 5x + 3 = 0$

Solution

Substitute $a = 1$, $b = -5$, and $c = 3$ into the formula.

$$x = \frac{-b \pm \sqrt{b^2 - 4ac}}{2a} = \frac{-(-5) \pm \sqrt{(-5)^2 - 4(1)(3)}}{2(1)}$$

$$= \frac{5 \pm \sqrt{25 - 12}}{2}$$

$$= \frac{5 \pm \sqrt{13}}{2}$$

Now work margin exercise 1.

2. Solve using the quadratic formula.

$5x^2 - 3x + 2 = 0$

Example 2 Using the Quadratic Formula

Solve using the quadratic formula.

$7x^2 - 2x + 1 = 0$

Solution

Substitute $a = 7$, $b = -2$, and $c = 1$ into the formula.

$$x = \frac{-b \pm \sqrt{b^2 - 4ac}}{2a} = \frac{-(-2) \pm \sqrt{(-2)^2 - 4(7)(1)}}{2(7)}$$

$$= \frac{2 \pm \sqrt{4 - 28}}{14}$$

$$= \frac{2 \pm \sqrt{-24}}{14}$$

$$= \frac{2 \pm 2i\sqrt{6}}{14}$$

$$= \frac{\cancel{2}\left(1 \pm i\sqrt{6}\right)}{\cancel{2} \cdot 7}$$

$$= \frac{1 \pm i\sqrt{6}}{7} \text{ or, in standard form, } \frac{1}{7} \pm \frac{\sqrt{6}}{7}i$$

The solutions are nonreal complex conjugates.

Now work margin exercise 2.

3. Solve using the quadratic formula.

$x^2 + 9 = 0$

Example 3 Using the Quadratic Formula

Solve using the quadratic formula.

$\frac{3}{4}x^2 - \frac{1}{2}x = \frac{1}{3}$

Solution

The quadratic formula is easier to use with integer coefficients. Multiply each term by the LCD, 12, so that the coefficients will be integers.

$$12 \cdot \frac{3}{4}x^2 - 12 \cdot \frac{1}{2}x = 12 \cdot \frac{1}{3}$$

$$9x^2 - 6x = 4$$

$$9x^2 - 6x - 4 = 0$$ To apply the formula, one side must be 0.

$$x = \frac{-(-6) \pm \sqrt{(-6)^2 - 4(9)(-4)}}{2(9)}$$ Substitute $a = 9$, $b = -6$, and $c = -4$ into the quadratic formula.

$$= \frac{6 \pm \sqrt{36 + 144}}{18}$$

$$= \frac{6 \pm \sqrt{180}}{18}$$

$$= \frac{6 \pm 6\sqrt{5}}{18}$$

$$= \frac{\cancel{6}\left(1 \pm \sqrt{5}\right)}{\cancel{6} \cdot 3} = \frac{1 \pm \sqrt{5}}{3}$$ Factor and reduce.

Now work margin exercise 3.

Example 4 Using the Quadratic Formula

Solve using the quadratic formula.

$$(3x - 1)(x + 2) = 4x$$

Solution

$$(3x - 1)(x + 2) = 4x$$

$$3x^2 + 5x - 2 = 4x$$ Multiply on the left-hand side.

$$3x^2 + x - 2 = 0$$ Subtract $4x$ from both sides so that one side is 0.

Now we see that $a = 3$, $b = 1$, and $c = -2$.

Substituting in the quadratic formula gives the following.

$$x = \frac{-(1) \pm \sqrt{(1)^2 - 4(3)(-2)}}{2(3)} = \frac{-1 \pm \sqrt{25}}{6} = \frac{-1 \pm 5}{6}$$

$$x = \frac{-1 + 5}{6} = \frac{4}{6} = \frac{2}{3} \quad \text{or} \quad x = \frac{-1 - 5}{6} = \frac{-6}{6} = -1$$

Note: Whenever the solutions are rational numbers, the equation can be solved by factoring. In this example, we could have solved as follows.

$$3x^2 + x - 2 = 0$$

$$(3x - 2)(x + 1) = 0$$

$$3x - 2 = 0 \quad \text{or} \quad x + 1 = 0$$

$$x = \frac{2}{3} \qquad x = -1$$

Now work margin exercise 4.

4. Solve using the quadratic formula.

$$(x + 1)(x - 1) = 3x$$

Common Error

A common mistake when simplifying fractions is to divide the denominator into only one of the terms in the numerator.

Wrong Solution

$$\frac{4+\cancel{2}\sqrt{3}}{\cancel{2}}=4+\sqrt{3}$$

Correct Solution

$$\frac{4+2\sqrt{3}}{2}=\frac{4}{2}+\frac{2\sqrt{3}}{2}=2+\sqrt{3}$$

$$\frac{4+2\sqrt{3}}{2}=\frac{\cancel{2}\left(2+\sqrt{3}\right)}{\cancel{2}}=2+\sqrt{3}$$

CAUTION

In Example 5, the equation is third-degree (a cubic equation) and one of the factors is quadratic. The quadratic formula can be applied to this factor.

5. Solve the following cubic equation by factoring and using the quadratic equation.

$3x^3-15x^2+6x=0$

Example 5 Solving with Cubic Equations

Solve the following cubic equation by factoring and using the quadratic formula.

$2x^3-10x^2+6x=0$

Solution

$$2x^3-10x^2+6x=0$$

$$2x\left(x^2-5x+3\right)=0 \qquad \text{Factor out } 2x.$$

$$2x=0 \quad \text{or} \quad x^2-5x+3=0 \qquad \text{Set each factor equal to 0.}$$

$$x=0 \qquad x=\frac{5\pm\sqrt{13}}{2}$$

Solve each equation. (The quadratic equation was solved in Example 1 using the quadratic formula.)

There are 3 solutions: $0,\ \frac{5+\sqrt{13}}{2}$, and $\frac{5-\sqrt{13}}{2}$.

Now work margin exercise 5.

B The Discriminant

The expression b^2-4ac, which is the part of the quadratic formula that lies under the radical sign, is called the **discriminant**. The discriminant identifies the number and type of solutions to a quadratic equation. Assuming a, b, and c are all real numbers, there are three possibilities: the discriminant is either positive, negative, or zero.

In Example 1, the discriminant was positive, $b^2-4ac=(-5)^2-4(1)(3)=13,$ and there were two real solutions: $x=\dfrac{5\pm\sqrt{13}}{2}$. In Example 2, the discriminant was negative, $b^2-4ac=(-2)^2-4(7)(1)=-24,$ and there were two nonreal solutions: $x=\dfrac{1\pm i\sqrt{6}}{7}$.

The discriminant gives the following information.

Discriminant	Nature of Solutions
$b^2-4ac>0$	Two real solutions
$b^2-4ac=0$	One real solution, $x=\dfrac{-b\pm 0}{2a}=-\dfrac{b}{2a}$
$b^2-4ac<0$	Two nonreal solutions

Table 1

In the case where $b^2-4ac=0$, we say $x=-\frac{b}{2a}$ is a **double root**. Additionally, if the discriminant is a perfect square, the equation is factorable.

Example 6 Finding the Discriminant

Find the discriminant and determine the nature of the solutions to each of the following quadratic equations.

a. $3x^2+11x-7=0$ **b.** $x^2+6x+9=0$ **c.** $x^2+1=0$

Solution

a. Substitute $a=3$, $b=11$, and $c=-7$ into the discriminant.

$$\begin{aligned} b^2-4ac &= (11)^2-4(3)(-7)\\ &= 121+84\\ &= 205>0 \end{aligned}$$

There are two real solutions.

b. Substitute $a=1$, $b=6$, and $c=9$ into the discriminant.

$$\begin{aligned} b^2-4ac &= (6)^2-4(1)(9)\\ &= 36-36\\ &= 0 \end{aligned}$$

There is one real solution, a double root.

c. Here $b=0$. We could write $x^2+0x+1=0$.

$$\begin{aligned} b^2-4ac &= (0)^2-4(1)(1)\\ &= 0-4\\ &= -4 \end{aligned}$$

There are two nonreal solutions.

Now work margin exercise 6.

6. Find the discriminant and determine the nature of the solutions to each of the following quadratic equations.

a. $x^2+15x-7=0$

b. $x^2+8x+16=0$

c. $x^2+9=0$

Teaching Note

You might want to discuss the fact that the discriminant will be 0 when the quadratic expression is a perfect square trinomial, which leads to a double root. (See Example 6b.)

Example 7 Understanding the Discriminant

Determine the value(s) for c such that $x^2+8x+c=0$ will have one real solution. (**Hint:** Set the discriminant equal to 0 and solve the equation for c.)

7. Determine the value(s) for c such that $x^2-6x+c=0$ will have one real solution.

Solution

$$b^2 - 4ac = 0$$
$$(8)^2 - 4(1)(c) = 0$$
$$64 - 4c = 0$$
$$-4c = -64$$
$$c = 16$$

Check

$$x^2 + 8x + (16) = 0$$
$$x^2 + 8x + 16 = 0$$
$$(x+4)^2 = 0$$
$$x = -4$$

There is only one real solution. Thus, -4 is a double root.

Now work margin exercise 7.

Example 8 Understanding the Discriminant

Determine the value(s) for a such that $ax^2 - 8x + 4 = 0$ will have two nonreal solutions. (**Hint:** Set the discriminant less than 0 and solve for a.)

Solution

$$b^2 - 4ac < 0$$
$$(-8)^2 - 4(a)(4) < 0$$
$$64 - 16a < 0$$
$$-16a < -64$$
$$a > 4$$

Thus, if a is any real number greater than 4, the discriminant will be negative and the equation will have two nonreal solutions.

Now work margin exercise 8.

8. Determine the value(s) for a such that $ax^2 - 8x + 1 = 0$ will have two nonreal solutions.

Margin Exercise Answers

1. $\frac{7 \pm \sqrt{33}}{2}$ **2.** $\frac{3 \pm i\sqrt{31}}{10}$ **3.** $\pm 3i$ **4.** $x = \frac{3 \pm \sqrt{13}}{2}$ **5.** $x = 0, \frac{5 \pm \sqrt{17}}{2}$ **6. a.** 253, there are two real solutions. **b.** 0, there is one real solution, a double root. **c.** -36, there are two nonreal solutions. **7.** $c = 9$ **8.** $a > 16$

16.3 Exercises

Concept Check

Fill-in-the-Blank. Complete each sentence using information found in this section.

1. The general form of a quadratic equation is ________.
2. In the quadratic formula, the expression $b^2 - 4ac$ is called the ________.
3. The quadratic formula is $x =$ ________.
4. When using the quadratic formula, the value of a cannot be ______.
5. To develop the quadratic formula, the general quadratic equation is solved by ______ the square.
6. In the case where the discriminant is zero, there is one real solution, also called a/an ______ root.

16.3 Exercises

Concept Check

1. $ax^2 + bx + c = 0$
2. discriminant
3. $\frac{-b \pm \sqrt{b^2 - 4ac}}{2a}$
4. 0
5. completing
6. double
7. True
8. True
9. False; Two real solutions
10. True

Practice

1. 68; Two real solutions
2. 5; Two real solutions
3. 0; One real solution
4. -11; Two nonreal solutions
5. -44; Two nonreal solutions
6. -23; Two nonreal solutions
7. 4; Two real solutions
8. 0; One real solution
9. 19,600; Two real solutions
10. -4356; Two nonreal solutions
11. -11; Two nonreal solutions
12. 49; Two real solutions
13. $x = -2 \pm 2\sqrt{2}$
14. $x = 3 \pm \sqrt{10}$
15. $x = -\frac{2}{3}$

True/False. Determine whether each statement is true or false. If a statement is false, explain how it can be changed so the statement will be true. (**Note:** There may be more than one acceptable change.)

7. The quadratic formula will always work when solving quadratic equations.

8. If the discriminant is a perfect square, the quadratic equation is factorable.

9. When using the quadratic formula, if the discriminant is greater than zero, there are infinite solutions.

10. If the discriminant is less than zero, there is no real solution.

Practice

Find the discriminant and determine the nature of the solutions of each quadratic equation. See Examples 6 through 8.

1. $x^2 + 6x - 8 = 0$
2. $x^2 + 3x + 1 = 0$
3. $x^2 - 8x + 16 = 0$
4. $x^2 + 3x + 5 = 0$
5. $4x^2 + 2x + 3 = 0$
6. $3x^2 - x + 2 = 0$
7. $5x^2 + 8x + 3 = 0$
8. $4x^2 + 12x + 9 = 0$
9. $100x^2 - 49 = 0$
10. $9x^2 + 121 = 0$
11. $3x^2 + x + 1 = 0$
12. $5x^2 - 3x - 2 = 0$

Solve each of the quadratic equations using the quadratic formula See Examples 1 through 4.

13. $x^2 + 4x - 4 = 0$
14. $x^2 - 6x - 1 = 0$
15. $9x^2 + 12x + 4 = 0$
16. $4x^2 - 20x + 25 = 0$
17. $x^2 - 2x + 7 = 0$
18. $x^2 - 2x + 3 = 0$
19. $2x^2 + 5x - 3 = 0$
20. $3x^2 - 7x + 4 = 0$
21. $4x^2 + 6x + 1 = 0$
22. $2x^2 - 3x - 1 = 0$
23. $4x^2 + 6x + 3 = 0$
24. $x^2 - 5x + 7 = 0$

Solve the given equations using any of the techniques discussed for solving quadratic equations: factoring, completing the square, or using the quadratic formula.

25. $x^2 + 3x - 5 = 0$
26. $x^2 - 7x - 3 = 0$
27. $x^2 + 4x + 3 = 0$
28. $x^2 + 14x + 49 = 0$
29. $x^2 + 8 = 0$
30. $x^2 - 7 = 0$
31. $x^2 - 5x + 2 = 0$
32. $3x^2 + 2x - 2 = 0$
33. $16x^2 + 8x = -1$
34. $6x^2 = 5x + 1$
35. $3x^2 - 4 = 0$
36. $4x^2 + 9 = 0$
37. $9x^2 - 12x + 4 = 0$
38. $9x^2 - 6x + 1 = 0$

16. $x = \frac{5}{2}$
17. $x = 1 \pm i\sqrt{6}$
18. $x = 1 \pm i\sqrt{2}$
19. $x = -3, \frac{1}{2}$
20. $x = 1, \frac{4}{3}$
21. $x = \frac{-3 \pm \sqrt{5}}{4}$
22. $x = \frac{3 \pm \sqrt{17}}{4}$
23. $x = \frac{-3 \pm i\sqrt{3}}{4}$ or $-\frac{3}{4} \pm \frac{\sqrt{3}}{4}i$
24. $x = \frac{5 \pm i\sqrt{3}}{2}$ or $\frac{5}{2} \pm \frac{\sqrt{3}}{2}i$
25. $x = \frac{-3 \pm \sqrt{29}}{2}$
26. $x = \frac{7 \pm \sqrt{61}}{2}$
27. $x = -3, -1$
28. $x = -7$
29. $x = \pm 2i\sqrt{2}$
30. $x = \pm\sqrt{7}$
31. $x = \frac{5 \pm \sqrt{17}}{2}$
32. $x = \frac{-1 \pm \sqrt{7}}{3}$
33. $x = -\frac{1}{4}$
34. $x = -\frac{1}{6}, 1$
35. $x = \pm\frac{2\sqrt{3}}{3}$
36. $x = \pm\frac{3}{2}i$
37. $x = \frac{2}{3}$
38. $x = \frac{1}{3}$
39. $x = \frac{-4 \pm i\sqrt{2}}{2}$ or $-2 \pm \frac{\sqrt{2}}{2}i$

40. $x = \dfrac{3 \pm i\sqrt{3}}{3}$ or $1 \pm \dfrac{\sqrt{3}}{3}i$

41. $x = \dfrac{7 \pm i\sqrt{51}}{10}$ or $\dfrac{7}{10} \pm \dfrac{\sqrt{51}}{10}i$

42. $x = \dfrac{5 \pm i\sqrt{23}}{8}$ or $\dfrac{5}{8} \pm \dfrac{\sqrt{23}}{8}i$

43. $x = -2, \dfrac{5}{3}$

44. $x = -2, -\dfrac{3}{2}$

45. $x = 3 \pm i\sqrt{2}$

46. $x = 4 \pm i\sqrt{2}$

47. $x = -1, 0$

48. $x = -\dfrac{4}{3}, 0$

49. $x = \dfrac{9 \pm \sqrt{65}}{2}, 0$

50. $x = \dfrac{11 \pm \sqrt{133}}{2}, 0$

51. $x = \dfrac{-3 \pm \sqrt{5}}{2}, 0$

52. $x = \dfrac{-5 \pm \sqrt{37}}{4}, 0$

53. $x = \dfrac{1}{2}, -3$

54. $x = -2 \pm \sqrt{6}$

55. $x = -\dfrac{1}{3}, 3$

56. $x = -1 \pm \sqrt{5}$

57. $x = \dfrac{2 \pm \sqrt{3}}{3}$

58. $x = \dfrac{8 \pm \sqrt{58}}{6}$

59. $x = \dfrac{1 \pm i\sqrt{3}}{6}$

60. $x = \dfrac{-3 \pm i}{4}$

61. $x = \dfrac{2 \pm i\sqrt{2}}{2}$

62. $x = \dfrac{1 \pm i\sqrt{11}}{4}$

63. $x = \dfrac{-7 \pm \sqrt{17}}{4}$

39. $2x^2 = -8x - 9$

40. $3x^2 = 6x - 4$

41. $5x^2 + 5 = 7x$

42. $4x^2 - 5x = -3$

43. $6x^2 + 2x = 20$

44. $10x^2 + 30 = -35x$

45. $3x^2 = 18x - 33$

46. $2x^2 = 16x - 36$

47. $x^2 + 4x = x - 2x^2$

48. $3x^2 + 4x = 0$

49. $x^3 - 9x^2 + 4x = 0$

50. $x^3 - 8x^2 = 3x^2 + 3x$

51. $x^3 + 3x^2 + x = 0$

52. $4x^3 + 10x^2 - 3x = 0$

53. $(2x+1)(x+3) = 2x+6$

54. $(x+5)(x-1) = -3$

55. $(3x-1)(x-2) = x+5$

56. $(x+4)(x-2) = -4$

First multiply each side of the equation by the LCD to get integer coefficients and then solve the resulting equation. See Example 3.

57. $3x^2 - 4x + \dfrac{1}{3} = 0$

58. $\dfrac{3}{4}x^2 - 2x + \dfrac{1}{8} = 0$

59. $2x^2 - \dfrac{2}{3}x + \dfrac{2}{9} = 0$

60. $2x^2 + 3x + \dfrac{5}{4} = 0$

61. $\dfrac{1}{2}x^2 - x + \dfrac{3}{4} = 0$

62. $\dfrac{2}{3}x^2 - \dfrac{1}{3}x + \dfrac{1}{2} = 0$

63. $\dfrac{1}{4}x^2 + \dfrac{7}{8}x + \dfrac{1}{2} = 0$

64. $\dfrac{5}{12}x^2 - \dfrac{1}{2}x - \dfrac{1}{4} = 0$

65. Determine the value(s) for c such that $x^2 - 8x + c = 0$ will have two real solutions.

66. Determine the value(s) for c such that $x^2 + 5x + c = 0$ will have two real solutions.

67. Determine the value(s) for c such that $x^2 + 9x + c = 0$ will have one real solution.

68. Determine the value(s) for c such that $x^2 - 7x + c = 0$ will have one real solution.

69. Determine the value(s) for a such that $ax^2 - 6x + 3 = 0$ will have two nonreal solutions.

70. Determine the value(s) for a such that $ax^2 + 4x - 2 = 0$ will have two nonreal solutions.

71. Determine the value(s) for a such that $ax^2 + x - 9 = 0$ will have two real solutions.

72. Determine the value(s) for a such that $ax^2 + 6x + 3 = 0$ will have two real solutions.

73. Determine the value(s) for a such that $ax^2 + 7x + 12 = 0$ will have one real solution.

74. Determine the value(s) for a such that $ax^2 - 2x + 8 = 0$ will have one real solution.

75. Determine the value(s) for c such that $3x^2 + 4x + c = 0$ will have two nonreal solutions.

76. Determine the value(s) for c such that $2x^2 + 3x + c = 0$ will have two nonreal solutions.

Solve the quadratic equations using the quadratic formula and your calculator. Write the solutions accurate to the ten-thousandth.

77. $0.02x^2 - 1.26x + 3.14 = 0$

78. $0.5x^2 + 0.07x - 5.6 = 0$

79. $\sqrt{2}x^2 - \sqrt{3}x - \sqrt{5} = 0$

80. $x^2 - 2\sqrt{10}x + 10 = 0$

81. $0.3x^2 + \sqrt{2}x + 0.72 = 0$

82. $\sqrt[3]{4}x^2 - \sqrt[4]{2}x - \sqrt{11} = 0$

83. $x^2 + 2\sqrt{15} - 15 = 0$

84. $0.05x^2 - \sqrt{30} = 0$

Applications

Solve.

85. An orange is thrown down from the top of a building that is 300 feet tall with an initial velocity of 6 feet per second. The distance of the object from the ground can be calculated using the equation $d = 300 - 6t - 16t^2$, where t is the time in seconds after the orange is thrown.

- **a.** On a balcony, a cup is sitting on a table located 100 feet from the ground. If the orange is thrown with the right aim to fall into the cup, how long will the orange fall? Round to the nearest hundredth. (**Hint:** The distance is 100 feet.)
- **b.** If the orange misses the cup and falls to the ground, how long will it take for the orange to splatter on the sidewalk? (**Hint:** What is the height of the orange when it hits the ground?)
- **c.** Approximately how much longer would it take for the orange to fall to the sidewalk than it would for the orange to fall into the cup?

86. Merida is practicing archery with her recurve bow. Her target is the top of a 3-foot tall bale of hay that is 400 feet away. She aims at a 45° angle and shoots the arrow with an initial velocity of 140 feet per second. The height of the arrow can be described by $h = 99t - 16t^2 + 5$, where 99 is the vertical velocity of the arrow, h is the height of the arrow, and t is the time in seconds that passes after the arrow leaves the bow.

- **a.** Solve the equation $3 = 99t - 16t^2 + 5$ to determine the time in seconds when the height of the arrow will be 3 feet. Round your answer to the nearest hundredth.
- **b.** When shot at a 45° angle, the horizontal velocity of the arrow is also 99 feet per second. Use this velocity to determine how long will it take the arrow to reach the bale of hay? Round your answer to the nearest hundredth. (**Hint:** Use the $d = rt$ formula.)
- **c.** Did Merida hit the target, undershoot the target, or overshoot the target? (**Hint:** Compare the answers from part a. and part b.)

Writing & Thinking

87. Find an equation of the form $Ax^4 + Bx^2 + C = 0$ that has the four roots ±2 and ±3. Explain how you arrived at this equation.

88. The surface area of a right circular cylinder can be found using the following formula: $S = 2\pi r^2 + 2\pi rh$, where r is the radius of the cylinder and h is the height. Estimate the radius of a circular cylinder of height 30 cm and surface area 300 cm². Explain how you used your knowledge of quadratic equations.

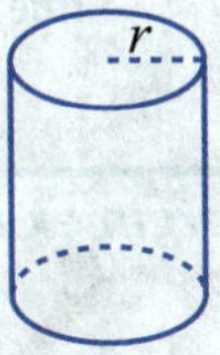

$h = 30$ cm

64. $x = \frac{3 \pm 2\sqrt{6}}{5}$

65. $c < 16$

66. $c < \frac{25}{4}$

67. $c = \frac{81}{4}$

68. $c = \frac{49}{4}$

69. $a > 3$

70. $a < -2$

71. $a > -\frac{1}{36}$

72. $a < 3$

73. $a = \frac{49}{48}$

74. $a = \frac{1}{8}$

75. $c > \frac{4}{3}$

76. $c > \frac{9}{8}$

77. $x \approx 2.5993, 60.4007$

78. $x \approx -3.4174, 3.2774$

79. $x \approx -0.7862, 2.0110$

80. $x \approx 3.1623$

81. $x \approx -4.1334, -0.5806$

82. $x \approx -1.1186, 1.8678$

83. $x \approx -2.6933, 2.6933$

84. $x \approx -10.4664, 10.4664$

Applications

85. **a.** 3.35 seconds
b. 4.15 seconds
c. 0.80 seconds

86. **a.** 6.21 seconds
b. 4.04 seconds
c. She overshot the target.

Writing & Thinking

87. $x^4 - 13x^2 + 36 = 0$; multiplied $(x-2)(x+2)(x-3)(x+3)$

88. $r \approx 1.5150$ cm. Answers will vary.

Objectives

Use quadratic equations to solve applications involving

A. The Pythagorean Theorem,

B. Work,

C. Distance-Rate-Time,

D. Projectiles,

E. Geometry, and

F. Cost per Person.

16.4 More Applications of Quadratic Equations

Application problems are designed to teach you to read carefully, to think clearly, and to translate from English to algebraic expressions and equations. The problems do not tell you directly to add, subtract, multiply, divide, or square. You must decide on a method of attack based on the wording of the problem and your previous experience and knowledge. Study the following examples and read the explanations carefully. They are similar to some, but not all, of the problems in the exercises.

A Pythagorean Theorem

Example 1 makes use of the Pythagorean Theorem that was discussed in Sections 6.7, 13.7, and 16.1.

1. In a right triangle, the hypotenuse is 10 ft long. One of the legs is 2 ft longer than the other leg. What are the lengths of the legs?

Teaching Note

It is important for students to develop the habit of drawing figures or constructing tables. Visualization can help students build the "connective tissue" for understanding and representing the relationships of variable expressions in a problem. You might consider requiring your students to sketch a figure or table for all the exercises.

Example 1 Using the Pythagorean Theorem

The length of a rectangle is 7 feet longer than the width. If one diagonal measures 13 feet, what are the dimensions of the rectangle?

Solution

Draw a diagram for problems involving geometric figures whenever possible.

Let x = width of the rectangle
and $x + 7$ = length of the rectangle.

Then, by the Pythagorean Theorem:

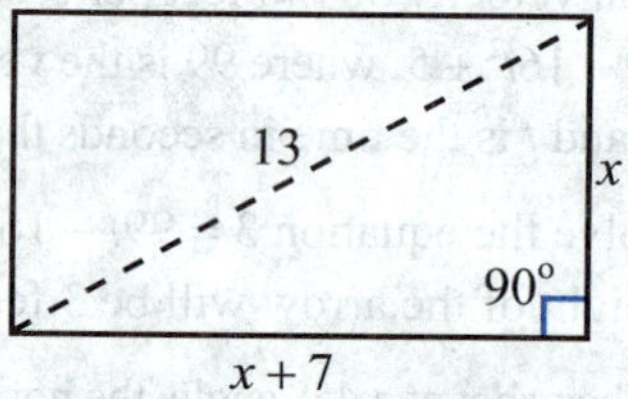

$$(x+7)^2 + x^2 = 13^2$$
$$x^2 + 14x + 49 + x^2 = 169$$
$$2x^2 + 14x + 49 - 169 = 0$$
$$2x^2 + 14x - 120 = 0$$
$$2(x^2 + 7x - 60) = 0$$
$$2(x-5)(x+12) = 0$$
$$x - 5 = 0 \quad \text{or} \quad x + 12 = 0$$
$$x = 5 \qquad \cancel{x = -12}$$

A negative number does not fit the conditions of the problem.

Thus, $x = 5$ and $x + 7 = 12$. The width is 5 feet and the length is 12 feet.

Check

$$5^2 + 12^2 \stackrel{?}{=} 13^2$$
$$25 + 144 \stackrel{?}{=} 169$$
$$169 = 169 \quad \text{True statement}$$

Now work margin exercise 1.

B Work Problems

Example 2 Application: Solving Work Problems

Working for a janitorial service, a woman and her daughter can clean a building in 5 hours. If the daughter were to do the job by herself, she would take 24 hours longer than her mother would take. How long would it take her mother to clean the building without the daughter's help?

Solution

Let x = time for mother to do the job alone
and $x + 24$ = time for daughter to do the job alone.

	Hours to Complete	Part Completed in 1 Hour
Mother	x	$\frac{1}{x}$
Daughter	$x+24$	$\frac{1}{x+24}$
Together	5	$\frac{1}{5}$

Part done by mother in 1 hour + Part done by daughter in 1 hour = Part done working together in 1 hour

$$\frac{1}{x} + \frac{1}{x+24} = \frac{1}{5}$$

$$(5x)(x+24)\frac{1}{x} + (5x)(x+24)\frac{1}{x+24} = (5x)(x+24)\frac{1}{5}$$

Multiply each term by the LCM of the denominators.

$$\begin{aligned} 5(x+24)+5x &= x(x+24) \\ 5x+120+5x &= x^2+24x \\ 0 &= x^2+24x-10x-120 \\ 0 &= x^2+14x-120 \\ 0 &= (x-6)(x+20) \end{aligned}$$

$$x-6=0 \quad \text{or} \quad x+20=0$$

$$x=6 \qquad \cancel{x=-20}$$

The mother could do the job alone in 6 hours.

Now work margin exercise 2.

2. Working together, Henry and his son, Bill, can paint one room in 2 hours. When Bill paints a room of the same size by himself, he takes 3 hours longer than his father. How long does it take his father to paint a room of this size?

Note

For reference, this problem is similar to those discussed in Section 14.7 on work. The difference is that in this section the resulting equations are quadratic.

C Distance-Rate-Time

3. A motorboat travels 10 mph in still water. The boat takes 4 hours longer to travel 48 miles going upstream than it does to travel 24 miles going downstream. Find the rate of the current.

Example 3 Application: Solving Distance-Rate-Time Problems

A small plane travels at a speed of 200 mph in still air. Flying with a tailwind, the plane is clocked over a distance of 960 miles. Flying against a headwind, it takes 2 hours more time to complete the return trip. What was the wind velocity?

Solution

The basic formula is $d = rt$ (distance = rate · time).

Also, $t = \frac{d}{r}$ and $r = \frac{d}{t}$.

Let x = wind velocity.

Then $200 + x$ = speed of airplane going with the wind (tailwind),

$200 - x$ = speed of airplane returning against the wind (headwind),

and 960 = distance each way.

Math Tip

Flying with a tailwind increases the speed of the airplane. Flying against a headwind decreases the speed of the airplane.

We know distance and can represent rate (or speed), so the formula $t = \frac{d}{r}$ is used for representing time.

	Distance	Rate	Time
Going	960	$200 + x$	$\frac{960}{200+x}$
Returning	960	$200 - x$	$\frac{960}{200-x}$

$$\underbrace{\frac{960}{200-x}}_{\text{Time returning}} - \underbrace{\frac{960}{200+x}}_{\text{Time going}} = \underbrace{2}_{\text{Difference in time}}$$

$$(200-x)(200+x)\frac{960}{200-x} - (200-x)(200+x)\frac{960}{200+x} = (200-x)(200+x)2$$

$$960(200+x) - 960(200-x) = 2(40{,}000 - x^2)$$

$$192{,}000 + 960x - 192{,}000 + 960x = 80{,}000 - 2x^2$$

$$2x^2 + 1920x - 80{,}000 = 0$$

$$x^2 + 960x - 40{,}000 = 0$$

$$(x-40)(x+1000) = 0$$

$$x - 40 = 0 \quad \text{or} \quad x + 1000 = 0$$

$$x = 40 \qquad \cancel{x = -1000}$$

The wind velocity was 40 mph.

Check

$$\frac{960}{200-40}-\frac{960}{200+40}\stackrel{?}{=}2$$

$$\frac{960}{160}-\frac{960}{240}\stackrel{?}{=}2$$

$$6-4\stackrel{?}{=}2$$

$$2=2 \quad \text{True statement}$$

Now work margin exercise 3.

D Projectiles

The formula $h=-16t^2+v_0t+h_0$ is used in physics and describes the height of a projectile such as a thrown ball, a bullet, or a rocket at a specific time. The variables in this formula are defined as follows.

h = height of object, in feet

t = time the object is in the air, in seconds

v_0 = beginning velocity, in feet per second

h_0 = beginning height ($h_0=0$ if the object is initially at ground level)

Example 4 Application: Projectiles

A bullet is fired straight up from ground level with a muzzle velocity of 320 feet per second.

a. When will the bullet hit the ground?

b. When will the bullet be 1200 feet above the ground?

Solution

a. In this problem, $v_0=320$ feet per second and

$h_0=0$ feet.

The bullet hits the ground when its height h equals 0.

$$h=-16t^2+v_0t+h_0$$

$$0=-16t^2+320t+0$$

$$0=t^2-20t \quad \text{Divide both sides by } -16.$$

$$0=t(t-20) \quad \text{Factor.}$$

$$t=0 \text{ or } t=20$$

The bullet hits the ground in 20 seconds. The solution $t=0$ confirms the fact that the bullet was fired from the ground.

4. A bullet is fired straight up from the ground level with a muzzle velocity of 480 feet per second.

a. When will the bullet hit the ground?

b. When will the bullet be 2000 feet above the ground?

b. Let $h = 1200$.

$$\begin{aligned} 1200 &= -16t^2 + 320t \\ 0 &= -16t^2 + 320t - 1200 \\ 0 &= t^2 - 20t + 75 && \text{Divide both sides by } -16. \\ 0 &= (t-5)(t-15) && \text{Factor.} \\ t &= 5 \text{ or } t = 15 \end{aligned}$$

Both solutions are meaningful. The bullet is at a height of 1200 feet twice; once at 5 seconds while going up and once at 15 seconds while coming down.

Now work margin exercise 4.

E Geometry

5. A square piece of cardboard has a small square, 1 in. by 1 in., cut from each corner. The edges are then folded up to form a box with a volume of 64 in.³ What are the dimensions of the box?

Example 5 Finding the Dimensions of a Box

A square piece of cardboard has a small square, 2 in. by 2 in., cut from each corner. The edges are then folded up to form a box with a volume of 512 in.³ What are the dimensions of the box? (**Hint:** The volume is the product of the length, width, and height: $V = lwh$.)

Solution

Draw a diagram illustrating the information.

Let x = length of one side of the square piece of cardboard.

Then $l = x - 4$, $w = x - 4$, and $h = 2$. (See diagram.)

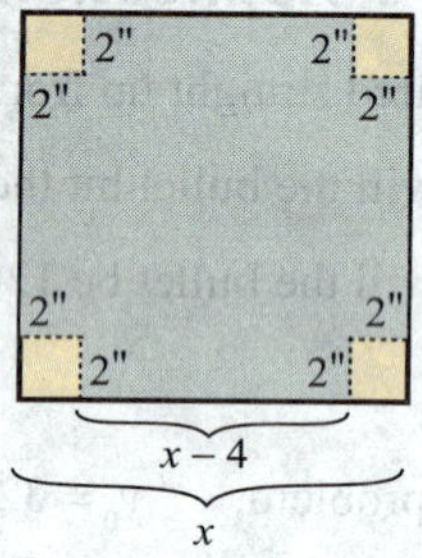

Length · Width · Height = Volume

$$\begin{aligned} (x-4)(x-4) \cdot 2 &= 512 \\ (x-4)(x-4) &= 256 \\ x^2 - 8x + 16 &= 256 \\ x^2 - 8x - 240 &= 0 \\ (x-20)(x+12) &= 0 \\ x - 20 = 0 \quad &\text{or} \quad x + 12 = 0 \\ x = 20 \quad & \quad \cancel{x = -12} \\ x - 4 &= 16 \end{aligned}$$

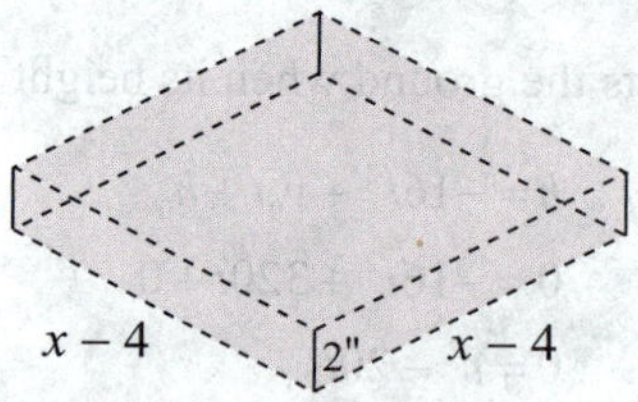

The dimensions of the box are 16 in. by 16 in. by 2 in.

Now work margin exercise 5.

Teaching Note

To help emphasize that many times there is more than one way to solve a problem, consider showing your students the following method of solving Example 5.

Let x = the length of the "inside square."

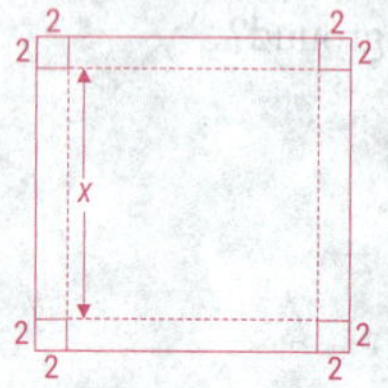

Then $V = 2(x)(x)$. So,

$2x^2 = 512$

$x^2 = 256$

$x = 16$

Discuss with your students whether they feel this approach is easier or harder.

Example 6 Finding the Dimensions of a Baseball Field

A little league baseball field is in the shape of a square with sides of 60 feet. What is the distance (to the nearest tenth of a foot) the catcher must throw from home plate to second base?

Solution

Since the distance from home plate to second base is the hypotenuse of a right triangle with sides of length 60 feet, we can use the Pythagorean Theorem as follows.

$$x^2 = 60^2 + 60^2$$

$$x^2 = 3600 + 3600$$

$$x^2 = 7200$$

$$x = \sqrt{7200} \qquad \text{We want only the positive square root.}$$

$$x = 84.9 \text{ ft} \qquad \text{Rounded to the nearest tenth of a foot}$$

The catcher must throw about 84.9 feet.

Now work margin exercise 6.

6. A major league baseball field is in the shape of a square with sides of 90 feet. What is the distance (to the nearest tenth of a foot) the catcher must throw from home plate to second base?

F Cost per Person

In the following example, note that the cost per person is found by dividing the total cost by the number of people going to the tournament. The cost per person changes because the total cost remains fixed while the number of people changes.

Teaching Note

You might need to refer your students back to the earlier section's discussion of equations with rational expressions.

Example 7 Application: Finding the Cost per Person

The members of a bowling club were going to rent a bus to drive them to a tournament at a total cost of \$2420, which was to be divided equally among the members. At the last minute, two of the members decided to drive their own cars. The cost to the remaining members increased by \$11 each. How many members rode the bus?

Solution

Let x = number of club members and
$x - 2$ = number of club members who rode the bus.

Final cost per member		Initial cost per member		Difference in cost per member
$\frac{2420}{x-2}$	$-$	$\frac{2420}{x}$	$=$	11

7. The members and family of a curling team were going to the national championships and were going to rent a bus to drive them there at a total cost of \$800, which was to be divided equally among the members. At the last minute, five members had an emergency and could not attend. The cost to the remaining members increased by \$8 each. How many members attended the national championship?

$$x(x-2)\frac{2420}{x-2} - x(x-2)\frac{2420}{x} = x(x-2)\cdot 11$$ LCD = $x(x-2)$

$$2420x - 2420(x-2) = 11x(x-2)$$

$$2420x - 2420x + 4840 = 11x^2 - 22x$$

$$0 = 11x^2 - 22x - 4840$$

$$0 = x^2 - 2x - 440$$ Divide both sides by 11.

$$0 = (x-22)(x+20)$$

$$x = 22 \text{ or } \cancel{x = -20}$$

$$x - 2 = 20$$

−20 does not fit the conditions. That is, the number of people in a club is a positive number.

Check

Final cost per member = $\frac{2420}{20} = \$121$

Initial cost per member = $\frac{2420}{22} = \$110$

$\$121 - \$110 = \$11$ Difference in cost per member

Twenty members rode the bus.

Now work margin exercise 7.

Margin Exercise Answers

1. 6 ft and 8 ft **2.** 3 hours **3.** 2 mph **4. a.** In 30 seconds **b.** At 5 seconds and at 25 seconds **5.** 8 in. by 8 in. by 1 in. **6.** 127.3 ft **7.** 20 members attended the championship.

16.4 Exercises

Concept Check

Fill-in-the-Blank. Complete each sentence using information found in this section.

1. Application problems are designed to teach you to _____ carefully, to _____ clearly, and to _____ from English to algebraic expressions and equations.
2. You must decide on a method of _____ based on the wording of the problem and your previous experience and knowledge.
3. Draw a/an _____ for problems involving geometric figures whenever possible.

True/False. Determine whether each statement is true or false. If a statement is false, explain how it can be changed so the statement will be true. (**Note:** There may be more than one acceptable change.)

4. Application problems will always directly tell you which operations to perform.
5. The basic formula for distance-rate-time problems is $d = rt$.

16.4 Exercises

Concept Check

1. read, think, translate
2. attack
3. diagram
4. False; Application problems do not tell you direction which operations to perform.
5. True

Applications

Solve.

1. A rectangle has a length 5 m less than twice its width. If the area is 63 m^2, find the dimensions of the rectangle.

2. The length of a rectangle is 2 cm less than 3 times its width. If the area of the rectangle is 225 cm^2, find the dimensions of the original rectangle.

3. The difference between two positive numbers is 9. If the smaller number is added to the square of the larger number, the result is 147. Find the numbers.

4. The difference between a positive number and 3 is four times the reciprocal of the number. Find the number.
(**Hint:** The reciprocal of x is $\frac{1}{x}$.)

5. The Wilsons have a rectangular swimming pool that is 10 ft longer than it is wide. The pool is completely surrounded by a concrete deck that is 6 ft wide. The total area of the pool and the deck is 1344 ft^2. Find the dimensions of the pool.

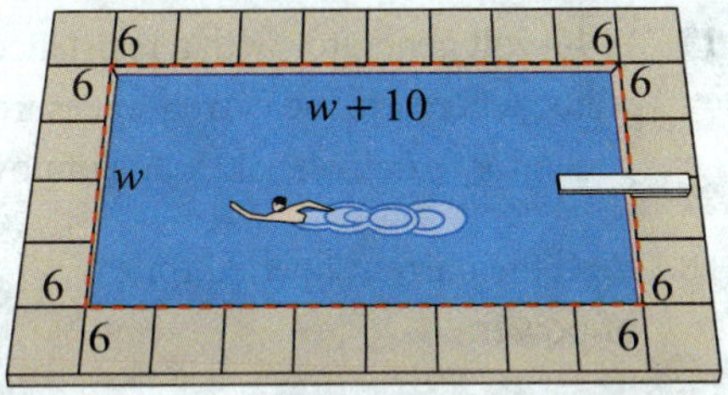

6. Each side of a square is increased by 10 cm. The area of the resulting square is 9 times the area of the original square. Find the length of the sides of the original square.

7. If 5 meters are added to each side of a square, the area of the resulting square is four times the area of the original square. Find the length of the sides of the original square.

8. The diagonal of a rectangle is 13 m. The length is 2 m more than twice the width. Find the dimensions of the rectangle.

9. The length of a rectangle is 4 m more than its width. If the diagonal is 20 m, what are the dimensions of the rectangle?

10. An orchard has 2030 trees. The number of trees in each row is 12 more than twice the number of rows. How many trees are in each row?

11. A rectangular auditorium seats 960 people. The number of seats in each row is 16 more than the number of rows. Find the number of seats in each row.

12. An apartment building has the same number of units on each floor. The building has five times as many units per floor as number of floors, and there are 405 units total. How many floors does the building have?

13. A farmer fenced in a 198-square-meter portion of his field with 58 meters of fencing. What are the length and width of the field? (**Hint:** The length plus the width is equal to half of the perimeter.)

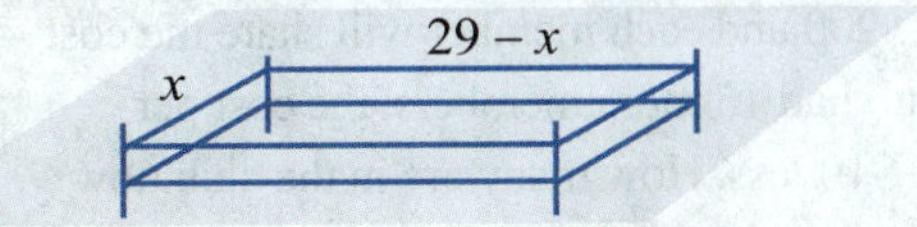

a. Write an equation to express the area of the fenced-in field.

b. Solve the equation from part a. for the variable.

c. Use the answer from part b. to determine the length and width of the fenced-in field.

Applications

1. $w(2w-5)=63$; $w=7$, the rectangle is 7 m by 9 m

2. $(3x-2)x=225$; $x=9$, the original rectangle is 9 cm by 25 cm

3. $x+(x+9)^2=147$; $x=3$, the numbers are 3 and 12

4. $x-3=\frac{4}{x}$; $x=4$

5. $(w+12)(w+22)=1344$; $w=20$, the pool is 20 ft by 30 ft

6. $(x+10)^2=9x^2$; $x=5$, the side of the original square is 5 cm

7. $(x+5)^2=4x^2$; $x=5$, the side of the original square is 5 m

8. $w^2+(2w+2)^2=169$; $w=5$, the rectangle is 5 m by 12 m

9. $w^2+(w+4)^2=400$; $w=12$, the rectangle is 12 m by 16 m

10. $x(2x+12)=2030$; $x=29$, there are 70 trees

11. $x(x+16)=960$; $x=24$, there are 40 seats

12. $x(5x)=405$; $x=9$, there are 9 floors

13. **a.** $x(29-x)=198$
b. $x=11$ or $x=18$
c. Either length = 11 meters and width = 18 meters, or length = 18 meters and width = 11 meters.

14. $8(w)(3w+4)=1590$; $w=7.5$, the dimensions are length = 26.5 ft; width = 7.5 ft; height = 8 ft

15. $(9+2x)(12+2x)-(9)(12)=162$; $x=3$, the frame is 3 in. thick

16. $(30+2x)(21+2x)-(30)(21)=756$; $x=6$, the frame is 6 in. thick

17. $40I-4I^2=100$; $I=5$, It needs a current of 5 amperes

18. $40I-4I^2=64$; $I=2$, It needs a current of 2 amperes or 8 amperes

19. $x(40-x)=336$; $x=12, 28$; He must sell 12 signs

20. $x(60-x)-3x=432$; $x=9, 48$; She must charge $12

21. **a.** $307.20

b. $(600+20x)(0.5-0.01x)=315$; $x=5, 15$; He must charge 45 cents or 35 cents

22. $x(125-x)=3750$; $x=50, 75$, the price of the shoes is $50 or $75

23. $(15-x)(700+70x)=10{,}920$; $x=2$ or 3, the rental rate is $840 or $910.

24. $\dfrac{900}{x}-10=\dfrac{900}{x+15}$; $x=30$, there are 30 people in the club

14. A large U-Haul truck is 8 ft tall. The length of the truck is 4 ft longer than three times the width. What are the dimensions of the truck if the volume is 1590 ft³?

15. A photograph 9 in. wide and 12 in. long is surrounded by a frame of uniform thickness. The area of the frame itself, not including the center, is 162 in.² Find the thickness of the frame.

16. The Mona Lisa is a famous painting by Leonardo da Vinci. The painting is 30 in. by 21 in. It is surrounded by a frame of uniform thickness whose area (not including the center) is 756 in.² Find the thickness of the frame.

17. A 40-volt generator with a resistance of 4 ohms delivers power externally of $40I-4I^2$ watts, where I is the current measured in amperes. Find the current needed for the generator to deliver 100 watts of power.

18. Find the current needed for the 40-volt generator in Exercise 17 to deliver 64 watts of power.

19. Raymond operates a small sign-making business. He finds that if he charges x dollars for each sign, he sells $40-x$ signs per week. What is the least number of signs he can sell to have an income of $336 in one week?

20. It costs Mrs. Snow $3 to build a picture frame. She estimates that if she charges x dollars each, she can sell $60-x$ frames per week. What is the lowest price necessary to make a profit of $432 each week?

21. Samuel operates a small peanut stand. He estimates that he can sell 600 bags of peanuts per day if he charges 50¢ for each bag. He determines that he can sell 20 more bags for each 1¢ reduction in price.

a. What would be his revenue for one day if he charged 48¢ per bag?

b. What should he charge in order to have receipts of $315?

22. A sporting goods store owner estimates that if he sells a certain model of basketball shoes for x dollars a pair, he will be able to sell $125-x$ pairs. Find the price if his sales are $3750. Is there more than one possible answer?

23. Mr. Prince owns a 15-unit apartment complex. If all units are rented, the rent for each apartment is $700 per month. Each time the rent is increased by $70, he will lose 1 tenant. What is the rental rate if he receives $10,920 monthly in rent? (**Hint:** Let x represent the number of empty units.)

24. The Ski Club is planning to charter a bus to a ski resort. The cost will be $900 and each member will share the cost equally. If the club had 15 more members, the cost per person would be $10 less. How many are in the club now? (**Hint:** If x = number in club now, $\frac{900}{x}$ = cost per person.)

25. A motorboat takes a total of 2 hours to travel 8 miles downstream and 4 miles back on a river that is flowing at a rate of 2 mph. Find the rate of the boat in still water.

	Rate	Time	Distance
Going	x	?	200
Returning	$x - 10$	?	200

26. A small motorboat travels 12 mph in still water. It takes 2 hours longer to travel 45 miles going upstream than it does going downstream. Find the rate of the current. (**Hint:** $12 + c$ = rate going downstream and $12 - c$ = rate going upstream.)

27. Recently Mr. and Mrs. Roberts spent their vacation in San Francisco, which is 540 miles from their home. Being a little reluctant to return home, the Roberts took 2 hours longer on their return trip and their average speed was 9 mph slower than when they were going. What was their average rate of speed as they traveled from home to San Francisco?

28. Lisa traveled to a college that is located 200 miles from the city where she works to train customers how to use the software that her company sells. Due to a traffic jam, her average speed returning was 10 miles per hour less than her average speed going to the college. The total travel time to and from the college was 9 hours. What was Lisa's average speed going to the college?

a. Use the table to set up a rational equation to describe the situation. Use the variable x to represent the average speed going to the college. (**Hint:** The sum of the times that it took Lisa to travel to and from the college is 9 hours.)

	Distance (d)	÷	Rate (r)	=	Time $\left(t = \frac{d}{r}\right)$
Going					
Returning					

b. Solve the equation from part a. Round your answer to the nearest tenth.

c. Which solution from part b. makes sense in the context of the situation? Explain your reasoning.

d. Use the answer from part c. to answer the question from the problem statement.

29. The Blumin Garden Club planned to give their president a gift of appreciation costing $120 and to divide the cost evenly. In the meantime, 5 members dropped out of the club. If it now costs each of the remaining members $2 more than originally planned, how many members initially participated in the gift buying? (**Hint:** If x = number in club initially, $\frac{120}{x}$ = cost per member.)

30. A manufacturing crew needs to assemble 1000 boxes per day, divided equally among the workers. One day, three workers call in sick, and the remaining members each need to assemble 75 more boxes than usual. How many workers are on the manufacturing crew?

31. A rectangular sheet of metal is 6 in. longer than it is wide. A box is to be made by cutting out 3 in. squares at each corner and folding up the sides. If the box has a volume of 336 in.^{3}, what were the original dimensions of the sheet metal? (See Example 5.)

32. A box is to be made out of a square piece of cardboard by cutting out 2 in. squares at each corner and folding up the sides. If the box has a volume of 162 in.^{3}, how big was the piece of cardboard? (See Example 5.)

25. $\frac{8}{x+2} + \frac{4}{x-2} = 2;$ $x = 6$, the speed of the boat in still water is 6 mph

26. $\frac{45}{12-c} - \frac{45}{12+c} = 2;$ $c = 3$, the rate of the current is 3 mph

27. $\frac{540}{x-9} - \frac{540}{x} = 2;$ their average speed to San Francisco was 54 mph

28. a. $\frac{200}{x} + \frac{200}{x-10} = 9$

	Distance (d)	Rate (r)	Time $t = \frac{d}{r}$
Going	200	x	$\frac{200}{x}$
Returning	200	$x - 10$	$\frac{200}{x-10}$

b. $x = \frac{40}{9}, 50$

c. 50; $\frac{40}{9} - 10$ is a negative value and velocity shouldn't be negative in this context.

d. Lisa had an average speed of 50 miles per hour going to the college and 40 miles per hour returning from the college.

29. $\frac{120}{x} + 2 = \frac{120}{x-5};$ $x = 20$, there were initially 20 members

30. $\frac{1000}{x-3} - \frac{1000}{x} = 75;$ $x = 8$, there are 8 workers

31. $3w(w+6) = 336;$ $w = 8$, the sheet metal was 14 in. by 20 in.

32. $2(x-4)^2 = 162;$ $x = 13$, the cardboard was 13 in. by 13 in.

33. $\frac{1}{x}+\frac{1}{x+8}=\frac{1}{3}$; $x=4$, the mother would take 4 hours

34. $\frac{1}{x}+\frac{1}{x+3}=\frac{1}{2}$; $x=3$, One employee takes 3 hours and the other 6 hours

35. $\frac{1}{x}+\frac{1}{x+30}=\frac{1}{8}$; $x=10$; the smaller pipe would take 40 min

36. $\frac{1}{x}+\frac{1}{x+6}=\frac{1}{4}$; $x=6$, the farmer would take 6 hours and the son would take 12 hours

37. a. $\frac{1}{x}+\frac{1}{x+6}=\frac{1}{4}$

Persons	Time of work (in Hours)	Part of work Done in 1 Hour
Jack	x	$\frac{1}{x}$
Diane	$x+6$	$\frac{1}{x+6}$
Together	4	$\frac{1}{4}$

b. $x=-4, 6$

c. 6 hours; Time cannot be negative in this context.

d. It would take Jack 6 hours to decorate the nursery and it would take Diane 12 hours to decorate the nursery.

38. a. 1 sec **b.** 2.5 sec

c. 3.0 sec

39. a. 4 sec

b. No, the projectile's maximum height is 256 ft.

c. 3 seconds, 5 seconds

d. 8 sec

33. A woman and her daughter can paint their cabin in 3 hours. Working alone it would take the daughter 8 hours longer than it would the mother. How long would it take the mother to paint the cabin alone?

34. Two employees together can prepare a large order in 2 hrs. Working alone, one employee takes three hours longer than the other. How long does it take each person working alone?

35. Two pipes can fill a tank in 8 minutes if both are turned on. If only one is used it would take 30 minutes longer for the smaller pipe to fill the tank than the larger pipe. How long will it take the smaller pipe to fill the tank?

36. A farmer and his son can plow a field with two tractors in 4 hours. If it would take the son 6 hours longer than the father to plow the field alone, how long would it take each if they worked alone?

37. Jack and Diane are decorating a nursery room for their baby, who will be born in a few months. Working together, they can completely decorate the nursery in 4 hours. Working alone, it would take Diane 6 hours longer to decorate the nursery than it would take Jack. How long would it take Jack and Diane to decorate the nursery by themselves?

a. Use the table to set up a rational equation to describe the situation. Use the variable x to represent the time it takes Jack to decorate the nursery by himself.

Person(s)	Time of Work (in Hours)	Part of Work Done in 1 Hour
Jack		
Diane		
Together		

b. Solve the equation from part a.

c. Which solution from part b. makes sense in the context of the situation? Explain your reasoning.

d. Use the answer from part c. to answer the question from the problem statement.

38. A ball is thrown upward with an initial velocity of 32 ft/sec from the edge of a cliff near the beach. The cliff is 50 ft above the beach. The height of the ball can be found by using the equation $h=-16t^2+32t+50$, where t is measured in seconds.

a. When will the ball be 66 feet above the beach?

b. When will the ball be 30 feet above the beach?

c. In about how many seconds will the ball hit the beach?

39. The height of a projectile fired upward from the ground with a velocity of 128 ft/sec is given by the formula $h=-16t^2+128t$.

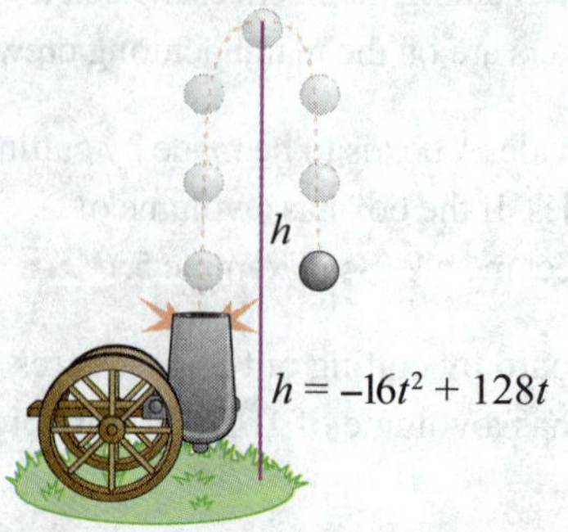

a. When will the projectile be 256 feet above the ground?

b. Will the projectile ever be 300 feet above the ground? Explain.

c. When will the projectile be 240 feet above the ground?

d. In how many seconds will the projectile hit the ground?

40. A ladder is 30 ft long and you want to place the base of the ladder 10 ft from the base of a building. About how far up the building (to the nearest tenth of a foot) will the ladder reach?

41. A flag pole is on top of a building and is held in place by steel cables attached to the top of the pole. If one such cable is 40 ft long and is attached at a point on the roof of the building 20 ft from the base of the flag pole, what is the length of the flag pole (to the nearest tenth of a foot)?

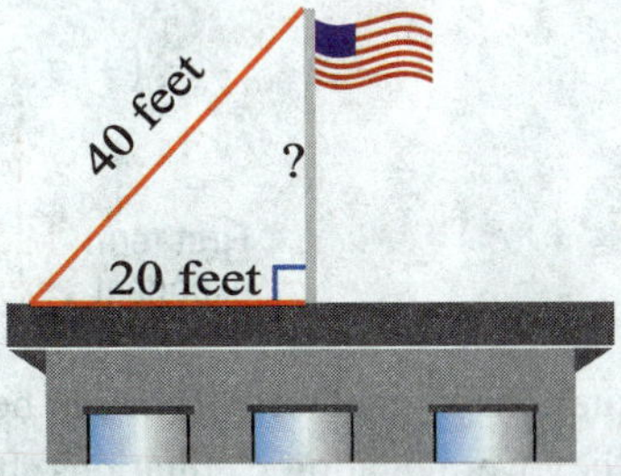

42. A landscaper was given the task to create a triangular flower garden in the corner of an office building. The landscaper has 12 feet of low fencing to use as a border along one side of the garden. The final garden will have the shape shown in the figure. The landscaper needs to know the remaining side lengths of the triangle to determine the area he will need to cover with fresh topsoil.

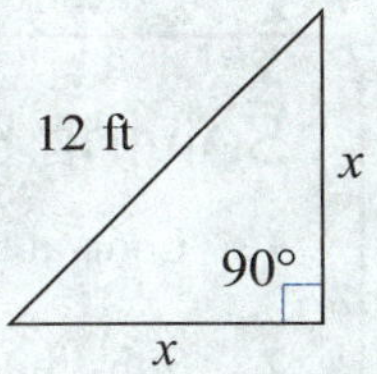

a. Use the Pythagorean Theorem to set up an equation which describes the relationship between the side lengths of the flower garden.

b. Solve the equation from part a. for the variable. Round your answer(s) to the nearest tenth.

c. Which solution from part b. makes sense in the context of the situation? Explain your reasoning.

d. Use the answer from part c. to determine the area that the landscaper will need to cover with topsoil.

Writing & Thinking

43. Suppose that you are to solve an applied problem and the solution leads to a quadratic equation. You decide to use the quadratic formula to solve the equation. Explain what restrictions you must be aware of when you use the formula.

40. 28.3 ft

41. 34.6 ft

42. **a.** $12^2 = x^2 + x^2$

b. $x = -8.5, 8.5$

c. 8.5; The side length cannot be negative.

d. 51 square feet

Writing & Thinking

43. a cannot be equal to zero and b^2 must be greater than or equal to $4ac$ to produce a real solution. Also, all the solutions found may not apply to the problem at hand. You must check that each answer makes sense in the context of the problem.

Objectives

A. Solve equations that can be written in quadratic form.

B. Solve rational equations that can be simplified to quadratic equations.

C. Solve higher-degree equations.

16.5 Equations in Quadratic Form

A Solving Equations in Quadratic Form

The general quadratic equation is $ax^2 + bx + c = 0$, where $a \neq 0$. The following equations are **not** quadratic equations,

$$x^4 - 7x^2 + 12 = 0 \quad \text{and} \quad x^{\frac{2}{3}} - 4x^{\frac{1}{3}} - 21 = 0$$

but they are in **quadratic form** because the degree of the middle term is one-half the degree of the first term. Specifically,

$$\frac{1}{2}(4) = 2 \quad \text{and} \quad \frac{1}{2}\left(\frac{2}{3}\right) = \frac{1}{3}.$$

First term exponent → 4; Middle term exponent → 2   First term exponent → $\frac{2}{3}$; Middle term exponent → $\frac{1}{3}$

Teaching Note

This technique might be introduced to students as a next step in the effort to solve a special family of higher order equations. Point out that solving equations of higher order requires various techniques and that many of these techniques have been around for hundreds of years.

Equations in quadratic form can be solved using the quadratic formula or by factoring as if they were quadratic equations. In each case, a substitution can be made to clarify the problem. The following list of steps outlines the procedure.

Solving Equations in Quadratic Form by Substitution

1. Look at the middle term.
2. Substitute a first-degree variable, such as u, for the variable expression in the middle term.
3. Substitute the square of this variable, u^2, for the variable expression in the first term.
4. Solve the resulting quadratic equation for u.
5. Substitute the results "back" for u in the beginning substitution and solve for the original variable.

PROCEDURE

The following examples illustrate how such a substitution will reveal the quadratic expression. Study these examples carefully and note the variety of algebraic manipulations used.

1. Solve the equation:

$x^4 - 9x^2 + 8 = 0$

Teaching Note

You may want to discuss solving Example 1 by factoring directly without substitution. This will probably depend on the ability level and mathematical sophistication of your students.

Example 1 Using Substitution to Solve Equations in Quadratic Form

Solve the equation: $x^4 - 7x^2 + 12 = 0$

Solution

$x^4 - 7x^2 + 12 = 0$

$u^2 - 7u + 12 = 0$ Substitute $u = x^2$ and $u^2 = \left(x^2\right)^2 = x^4$.

$(u - 3)(u - 4) = 0$ Solve for u by factoring.

$u = 3$ or $u = 4$

$x^2 = 3$ $\quad x^2 = 4$ Now substitute back x^2 for u.

$x = \pm\sqrt{3}$ $\quad x = \pm 2$ Solve quadratic equations for x.

There are four solutions: $x = -\sqrt{3}, \sqrt{3}, -2$, and 2.

Now work margin exercise 1.

Example 2 Using Substitution to Solve Equations in Quadratic Form

Solve the equation: $x^{\frac{2}{3}} - 4x^{\frac{1}{3}} - 21 = 0$

Solution

$$x^{\frac{2}{3}} - 4x^{\frac{1}{3}} - 21 = 0$$

$$u^2 - 4u - 21 = 0 \quad \text{Let } u = x^{\frac{1}{3}} \text{ and } u^2 = \left(x^{\frac{1}{3}}\right)^2 = x^{\frac{2}{3}}.$$

$$(u-7)(u+3) = 0 \quad \text{Solve for } u \text{ by factoring.}$$

$u = 7$ or $u = -3$

$x^{\frac{1}{3}} = 7$ $\quad x^{\frac{1}{3}} = -3$ Substitute back $x^{\frac{1}{3}}$ for u.

$\left(x^{\frac{1}{3}}\right)^3 = 7^3$ $\quad \left(x^{\frac{1}{3}}\right)^3 = (-3)^3$

$x = 343$ $\quad x = -27$ Cube both sides.

There are two solutions: $x = -27$ and 343.

Now work margin exercise 2.

2. Solve the equation:

$$x^{\frac{2}{3}} - 5x^{\frac{1}{3}} - 24 = 0$$

Example 3 Using Substitution to Solve Equations in Quadratic Form

Solve the equation: $x^{-4} - 7x^{-2} + 10 = 0$

Solution

$$x^{-4} - 7x^{-2} + 10 = 0 \quad \text{Let } u = x^{-2} \text{ and } u^2 = \left(x^{-2}\right)^2 = x^{-4}.$$

$$u^2 - 7u + 10 = 0 \quad \text{Solve for } u \text{ by factoring.}$$

$$(u-2)(u-5) = 0$$

$u = 2$ or $u = 5$

$x^{-2} = 2$ $\quad x^{-2} = 5$ Substitute back x^{-2} for u.

$\frac{1}{x^2} = 2$ $\quad \frac{1}{x^2} = 5$ Remember, $x^{-2} = \frac{1}{x^2}$.

$x^2 = \frac{1}{2}$ $\quad x^2 = \frac{1}{5}$ Reciprocals

$x = \pm\sqrt{\frac{1}{2}}$ $\quad x = \pm\sqrt{\frac{1}{5}}$

$x = \pm\frac{1}{\sqrt{2}}$ $\quad x = \pm\frac{1}{\sqrt{5}}$

3. Solve the equation:

$$x^{-4} - 10x^{-2} + 16 = 0$$

Rationalizing the denominators, we have $x = \pm\frac{\sqrt{2}}{2}, x = \pm\frac{\sqrt{5}}{5}$.

There are four solutions: $x = -\frac{\sqrt{2}}{2}, \frac{\sqrt{2}}{2}, -\frac{\sqrt{5}}{5}$, and $\frac{\sqrt{5}}{5}$.

Now work margin exercise 3.

4. Solve the equation:

$$(x-3)^2 + 5(x-3) - 14 = 0$$

Example 4 Using Substitution to Solve Equations in Quadratic Form

Solve the equation: $(x+2)^2 - (x+2) - 12 = 0$

Solution

$$(x+2)^2 - (x+2) - 12 = 0$$

$$u^2 - u - 12 = 0 \quad \text{Let } u = x+2.$$

$$(u-4)(u+3) = 0 \quad \text{Solve for } u \text{ by factoring.}$$

$$u = 4 \quad \text{or} \quad u = -3$$

$$x+2 = 4 \qquad x+2 = -3 \quad \text{Substitute back } x+2 \text{ for } u.$$

$$x = 2 \qquad x = -5$$

There are two solutions: $x = -5$ and 2.

Now work margin exercise 4.

B Rational Equations that Simplify to Quadratic Equations

Equations with rational expressions were first discussed in Section 14.6. In solving equations with rational expressions, first multiply every term on both sides of the equation by the least common multiple (LCM) of the denominators. This will "clear" the equation of fractions. In this section, the resulting equation will be a quadratic equation. Remember to check the restrictions on the variables. That is, no denominator can have a value of 0.

5. Solve the equation:

$$\frac{1}{4x-1} + \frac{1}{3x+2} = \frac{x}{3x+2}$$

Example 5 Solving Rational Equations that Simplify to Quadratic Equations

Solve the following equation containing rational expressions: $\frac{2}{3x-1} + \frac{1}{x+1} = \frac{x}{x+1}$

Solution

This equation is not in quadratic form. However, multiplying both sides of the equation by the LCM of the denominators, $(x+1)(3x-1)$, does give a quadratic equation. The restrictions on x are $x \neq -1, \frac{1}{3}$.

In this case, use the quadratic formula to solve the resulting quadratic equation.

$$\frac{2}{3x-1}+\frac{1}{x+1}=\frac{x}{x+1}$$

$$(x+1)\cancel{(3x-1)}\cdot\frac{2}{\cancel{3x-1}}+\cancel{(x+1)}(3x-1)\cdot\frac{1}{\cancel{x+1}}=\cancel{(x+1)}(3x-1)\cdot\frac{x}{\cancel{x+1}}$$

$$\begin{aligned}
2(x+1)+3x-1&=(3x-1)x\\
2x+2+3x-1&=3x^2-x\\
0&=3x^2-6x-1\\
x&=\frac{-(-6)\pm\sqrt{(-6)^2-4\cdot3(-1)}}{6}\\
&=\frac{6\pm\sqrt{48}}{6}\\
&=\frac{6\pm4\sqrt{3}}{6}\\
&=\frac{3\pm2\sqrt{3}}{3}
\end{aligned}$$

Since the values we found are not the restricted values, there are two solutions:

$\frac{3+2\sqrt{3}}{3}$ and $\frac{3+2\sqrt{3}}{3}$.

Now work margin exercise 5.

C Solving Higher-Degree Equations

Example 6 Solving Higher-Degree Equations

Solve the equation: $x^5-16x=0$

Solution

This equation can be solved by factoring and using the square root property.

$$x^5-16x=0$$

$$x(x^4-16)=0 \qquad \text{Factor out the common monomial, } x.$$

$$x(x^2+4)(x^2-4)=0 \qquad \text{Factor the difference of two squares.}$$

$$x=0 \quad \text{or} \quad x^2=-4 \quad \text{or} \quad x^2=4$$

$$x=\pm2i \qquad\qquad x=\pm2$$

There are five solutions: $x=0, -2i, 2i, -2, 2$.

Now work margin exercise 6.

6. Solve each equation:

$x^5-25x=0$

Example 7 Solving Higher-Degree Equations

Solve the equation: $x^3 - 27 = 0$

Solution

The polynomial is the difference of two cubes and can be factored. In this case, complex solutions can be found using the quadratic formula.

$$x^3 - 27 = 0$$
$$(x-3)(x^2+3x+9) = 0$$

$x - 3 = 0$

$x = 3$

$x^2 + 3x + 9 = 0$

Using the quadratic formula:

$$x = \frac{-3 \pm \sqrt{3^2 - 4 \cdot 1 \cdot 9}}{2}$$
$$= \frac{-3 \pm \sqrt{-27}}{2}$$
$$= \frac{-3 \pm 3i\sqrt{3}}{2} \text{ or } -\frac{3}{2} \pm \frac{3\sqrt{3}}{2}i$$

There are three solutions: $x = 3, \frac{-3+3i\sqrt{3}}{2}, \frac{-3-3i\sqrt{3}}{2}$.

Now work margin exercise 7.

7. Solve each equation:

$x^3 - 216 = 0$

Margin Exercise Answers

1. $\pm 1, \pm 2\sqrt{2}$ **2.** 512, −27 **3.** $\pm\frac{\sqrt{2}}{2}, \pm\frac{\sqrt{2}}{4}$ **4.** −4, 5 **5.** $\frac{2 \pm \sqrt{5}}{2}$ **6.** 0, $\pm\sqrt{5}$, $\pm i\sqrt{5}$
7. 6, $-3 \pm 3i\sqrt{3}$

16.5 Exercises

Concept Check

Fill-in-the-Blank. Complete each sentence using information found in this section.

1. An equation is in quadratic form when the degree of the ______ term is ______ the degree of the first term.
2. To solve an equation in quadratic form, a/an ______ can be made to clarify the problem.
3. When solving equations in quadratic form by substitution, substitute a/an ______-degree variable for the variable expression in the middle term.
4. When solving rational expressions, remember to check the ______ on the variables.

True/False. Determine whether each statement is true or false. If a statement is false, explain how it can be changed so the statement will be true. (**Note:** There may be more than one acceptable change.)

5. Equations in quadratic form can be solved using the quadratic equation or by factoring.
6. When solving higher-degree equations, you shouldn't factor out any common monomials.

16.5 Exercises

Concept Check

1. middle, one-half
2. substitution
3. first
4. restrictions
5. True
6. False; You should factor out any common monomial before using any solving techniques.
7. True
8. False; It can be anything as long as it is one-half of the degree of the first term.

Practice

1. $x = \pm 2, \pm 3$
2. $x = \pm 2, \pm 5$
3. $x = \pm 2, \pm\sqrt{5}$
4. $y = \pm 3, \pm\sqrt{2}$
5. $y = \pm\sqrt{7}, \pm 2i$
6. $y = \pm\sqrt{3}, \pm 2i$
7. $y = \pm\sqrt{5}, \pm i\sqrt{5}$
8. $x = \pm\sqrt{5}, \pm i\sqrt{5}$
9. $x = 4, \frac{25}{4}$
10. $x = \frac{1}{4}, 1$
11. $x = 1, 4$
12. $y = 1, 9$
13. $x = -8, \frac{1}{8}$
14. $x = -8, \frac{27}{8}$
15. $x = -27, -8, 0$
16. $x = 0, 49, 64$
17. $x = -17, -2$
18. $x = -2, 3$
19. $x = -\frac{7}{2}, -3$

7. The LCM of the denominators is used to clear rational expressions of any fractions.

8. The degree of the first term of an equation in quadratic form must be 2.

Practice

Solve the equations.

1. $x^4 - 13x^2 + 36 = 0$
2. $x^4 - 29x^2 + 100 = 0$
3. $x^4 - 9x^2 + 20 = 0$
4. $y^4 - 11y^2 + 18 = 0$
5. $y^4 - 3y^2 - 28 = 0$
6. $y^4 + y^2 - 12 = 0$
7. $y^4 - 25 = 0$
8. $4x^4 - 100 = 0$
9. $2x - 9x^{\frac{1}{2}} + 10 = 0$
10. $2x - 3x^{\frac{1}{2}} + 1 = 0$
11. $x^3 - 9x^{\frac{3}{2}} + 8 = 0$
12. $y^3 - 28y^{\frac{3}{2}} + 27 = 0$
13. $2x^{\frac{2}{3}} + 3x^{\frac{1}{3}} - 2 = 0$
14. $2x^{\frac{2}{3}} + x^{\frac{1}{3}} - 6 = 0$
15. $3x^{\frac{5}{3}} + 15x^{\frac{4}{3}} + 18x = 0$
16. $2x^2 - 30x^{\frac{3}{2}} + 112x = 0$
17. $(x+7)^2 + 5(x+7) = 50$
18. $(x-1)^2 + (x-1) - 6 = 0$
19. $(2x+3)^2 + 7(2x+3) + 12 = 0$
20. $(5x-4)^2 + 2(5x-4) - 8 = 0$
21. $(x-3)^2 - 2(x-3) - 15 = 0$
22. $(x+4)^2 - 2(x+4) = 3$
23. $(2x+1)^2 + (2x+1) = 0$
24. $(3x-5)^2 + (3x-5) - 2 = 0$
25. $x^4 - 2x^2 + 2 = 0$
26. $x^4 - 4x^2 + 5 = 0$
27. $x^4 - 2x^2 + 10 = 0$
28. $x^4 - 6x^2 + 13 = 0$
29. $x^4 - 4x^2 + 7 = 0$
30. $x^4 - 6x^2 + 11 = 0$
31. $x^{-2} - 12x^{-1} + 35 = 0$
32. $z^{-2} - 2z^{-1} - 24 = 0$
33. $3x^{-2} + x^{-1} - 24 = 0$
34. $2x^{-2} - 7x^{-1} + 6 = 0$
35. $x^{-1} + 5x^{-\frac{1}{2}} - 50 = 0$
36. $3y^{-1} - 7y^{-\frac{1}{2}} + 2 = 0$
37. $x^{-4} - 6x^{-2} + 5 = 0$
38. $3x^{-4} - 5x^{-2} + 2 = 0$
39. $3x^{-4} + 25x^{-2} - 18 = 0$
40. $2x^{-4} + 3x^{-2} - 20 = 0$
41. $\frac{2}{4x-1} + \frac{1}{x+1} = \frac{-x}{x+1}$
42. $\frac{3x-2}{15} - \frac{16-3x}{x+6} = \frac{x+3}{5}$
43. $\frac{2x}{x-4} - \frac{12x}{x^2+x-20} = \frac{x-1}{x+5}$
44. $\frac{x+1}{x+3} + \frac{2x-1}{x-2} = \frac{12x-2}{x^2+x-6}$
45. $\frac{x+5}{3x+2} - \frac{4-2x}{3x^2+8x+4} = \frac{x+4}{x+2}$
46. $\frac{x+5}{3x+4} + \frac{16x^2+5x+6}{3x^2-2x-8} = \frac{4x}{x-2}$

20. $x = 0, \frac{6}{5}$
21. $x = 0, 8$
22. $x = -5, -1$
23. $x = -1, -\frac{1}{2}$
24. $x = 1, 2$
25. $x = \pm\sqrt{1+i}, \pm\sqrt{1-i}$
26. $x = \pm\sqrt{2+i}, \pm\sqrt{2-i}$
27. $x = \pm\sqrt{1+3i}, \pm\sqrt{1-3i}$
28. $x = \pm\sqrt{3+2i}, \pm\sqrt{3-2i}$
29. $x = \pm\sqrt{2+i\sqrt{3}}, \pm\sqrt{2-i\sqrt{3}}$
30. $x = \pm\sqrt{3+i\sqrt{2}}, \pm\sqrt{3-i\sqrt{2}}$
31. $x = \frac{1}{7}, \frac{1}{5}$
32. $z = -\frac{1}{4}, \frac{1}{6}$
33. $x = -\frac{1}{3}, \frac{3}{8}$
34. $x = \frac{1}{2}, \frac{2}{3}$
35. $x = \frac{1}{25}$
36. $y = \frac{1}{4}, 9$
37. $x = \pm\frac{\sqrt{5}}{5}, \pm 1$
38. $x = \pm\frac{\sqrt{6}}{2}, \pm 1$
39. $x = \pm\frac{\sqrt{6}}{2}, \pm\frac{1}{3}i$
40. $x = \pm\frac{1}{2}i, \pm\frac{\sqrt{10}}{5}$
41. $x = -\frac{1}{4}$
42. $x = 9$
43. $x = -4, 1$
44. $x = -\frac{1}{3}, 3$
45. $x = -\frac{1}{2}$
46. $x = -\frac{2}{5}$

47. $x = -7, -\frac{3}{2}$

48. $x = -\frac{3}{4}$

49. $x = \frac{26}{5}$

50. $x = -3$

51. $x = 0, \pm 2\sqrt{2}, \pm 2i\sqrt{2}$

52. $x = 0, \pm\sqrt{6}, \pm i\sqrt{6}$

53. $x = 2, -1 \pm i\sqrt{3}$

54. $x = 5, \frac{-5 \pm 5i\sqrt{3}}{2}$ or $-\frac{5}{2} \pm \frac{5\sqrt{3}}{2}i$

55. $x = -10, 0, 5 \pm 5i\sqrt{3}$

56. $x = -3, 0, \frac{3 \pm 3i\sqrt{3}}{2}$ or $\frac{3}{2} \pm \frac{3\sqrt{3}}{2}i$

Writing & Thinking

57. **a.** $l^2 - l - 1 = 0$, $l = \frac{1+\sqrt{5}}{2}$ which is the golden ratio

b. 97.08 feet

c. Rectangle B is "golden."

58. Using u-substitution where $u = \sqrt{x}$, we get u values of -2, and 3. However, $\sqrt{x}$ never equals -2, so we are left with only one solution, $x = 9$.

47. $\frac{4x+1}{x-6} - \frac{3x^2-8x+20}{2x^2-13x+6} = \frac{3x+7}{2x-1}$

48. $\frac{3x+2}{x+3} + \frac{22x-31}{x^2-x-12} = \frac{3(x+4)}{x+3}$

49. $\frac{5(x-10)}{x-7} = \frac{2(x+1)}{x-4} + 3$

50. $2 + \frac{2-x}{x+2} = \frac{x-3}{x+5}$

51. $x^5 - 64x = 0$

52. $x^5 = 36x$

53. $8x^3 = 64$

54. $x^3 - 125 = 0$

55. $x^5 + 1000x^2 = 0$

56. $2x^5 + 54x^2 = 0$

Writing & Thinking

57. One of the most studied and interesting visual and numerical concepts in algebra is the **Golden Ratio**. Ancient Greeks thought (and many people still do) that a rectangle was most aesthetically pleasing to the eye if the ratio of its length to its width is the Golden Ratio (about 1.618). In fact, the Parthenon, built by Greeks in the fifth century BC, utilizes the Golden Ratio. A rectangle is "golden" if its length l and width w satisfy the equation $\frac{l}{w} = \frac{w}{l-w}$.

a. By letting $w = 1$ unit in the equation above, we get the equation $\frac{l}{1} = \frac{1}{l-1}$. Solve this equation for the positive value of l (which is the algebraic expression for the golden ratio).

b. Suppose that an architect is constructing a building with a rectangular front that is to be 60 feet high. About how long should the front be if he wants the appearance of a golden rectangle? (Assume $w = 60$ feet and that you are looking for l.) Round to the hundredths place.

c. Consider rectangle A and rectangle B. Which seems most pleasing to your eye? Measure the length and width of each rectangle and see if you chose the golden rectangle.

Rectangle A

Rectangle B

58. Consider the following equation: $x - x^{\frac{1}{2}} - 6 = 0$
In your own words, explain why, even though it is in quadratic form, this equation has only one solution.

16.6 Graphing Quadratic Functions

Objectives

A. Graph quadratic functions of the form $y = ax^2$.

B. Graph quadratic functions of the form $y = ax^2 + k$.

C. Graph quadratic functions of the form $y = a(x-h)^2$.

D. Graph quadratic functions of the form $y = a(x-h)^2 + k$.

A Graphing Quadratic Functions of the Form $y = ax^2$

In this section, we expand our interest in functions to include **quadratic functions**.

Quadratic Functions

A **quadratic function** is any function that can be written in the form

$$y = ax^2 + bx + c,$$

where a, b, and c are real numbers and $a \neq 0$.

DEFINITION

Notice that in function notation we can write $y = f(x)$. This means that throughout this discussion, y can be replaced by $f(x)$ or a function notation such as $g(x)$ or $h(x)$.

We begin with a table of values and graph the quadratic function $y = x^2$ (or $f(x) = x^2$).

x	$x^2 = y$
−3	$(-3)^2 = 9$
−2	$(-2)^2 = 4$
−1	$(-1)^2 = 1$
0	$(0)^2 = 0$
1	$(1)^2 = 1$
2	$(2)^2 = 4$
3	$(3)^2 = 9$

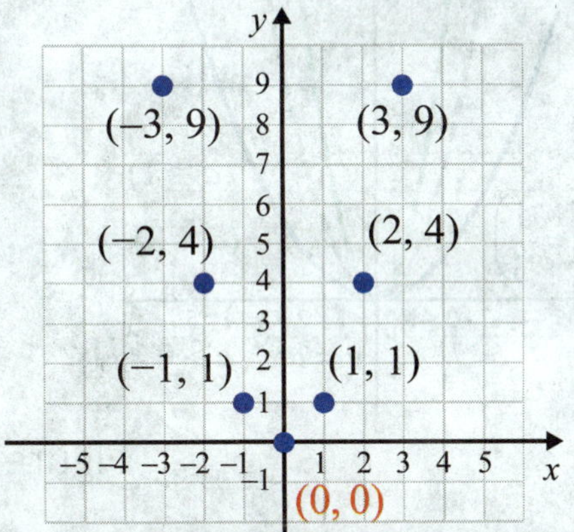

Figure 1

The complete graph of $y = x^2$ is shown in Figure 2. The curve is called a **parabola**. The point $(0, 0)$ is the "turning point" of the parabola and is called the **vertex**. The line $x = 0$ (the y-axis) is the **line of symmetry** or **axis of symmetry**. That is, the curve is a "mirror image" of itself with respect to the line $x = 0$.

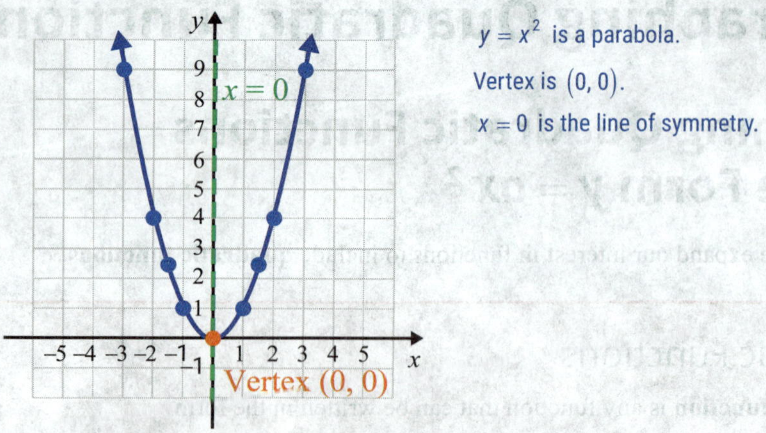

Figure 2

For any real number x, $x^2 \geq 0$. So, $ax^2 \geq 0$ if $a > 0$ and $ax^2 \leq 0$ if $a < 0$. This means that the graph of $y = ax^2$ is above the x-axis if $a > 0$ and below the x-axis if $a < 0$. The **vertex** is at the origin $(0, 0)$ in either of these cases and is the one point where each graph touches (or is tangent to) the x-axis. Figure 3 illustrates graphs in relation to the graphs of $y = x^2$ and $y = -x^2$.

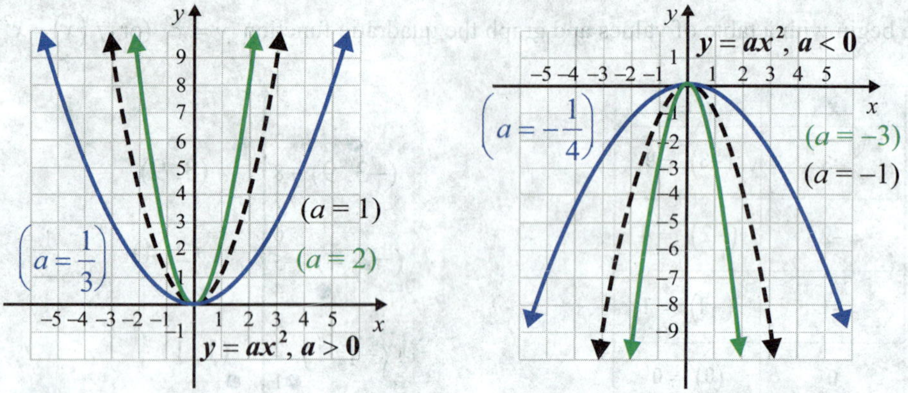

Figure 3

Figure 3 illustrates the following characteristics of quadratic functions of the form $y = ax^2$.

a. The vertex is at $(0, 0)$.

b. If $a > 0$, the parabola "opens upward."

c. If $a < 0$, the parabola "opens downward."

d. The bigger $|a|$ is, the narrower the opening.

e. The smaller $|a|$ is, the wider the opening.

f. The line $x = 0$ (the y-axis) is the line of symmetry.

Example 1 Graphing Quadratic Functions

Graph each quadratic function, setting up a table of values for x and y as an aid. Choose values of x on each side of the line of symmetry.

a. $y = 3x^2$ **b.** $y = -\frac{1}{3}x^2$

Solution

a.

$y = 3x^2$

x	y
$-\frac{4}{3}$	$\frac{16}{3}$
-1	3
$-\frac{1}{2}$	$\frac{3}{4}$
0	0
$\frac{1}{2}$	$\frac{3}{4}$
1	3
$\frac{4}{3}$	$\frac{16}{3}$

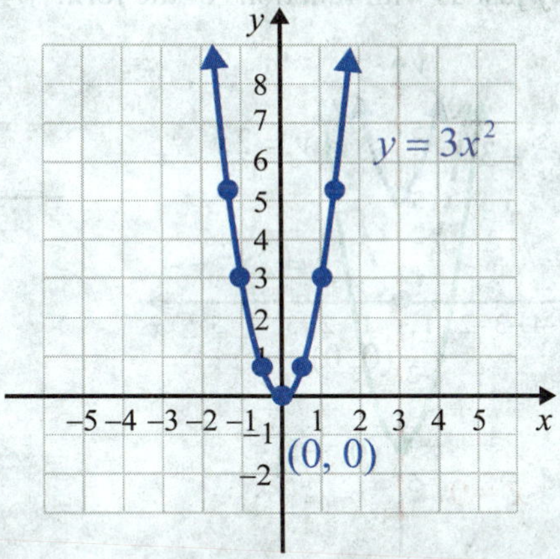

b.

$y = -\frac{1}{3}x^2$

x	y
-4	$-\frac{16}{3}$
-3	-3
-2	$-\frac{4}{3}$
0	0
2	$-\frac{4}{3}$
3	-3
4	$-\frac{16}{3}$

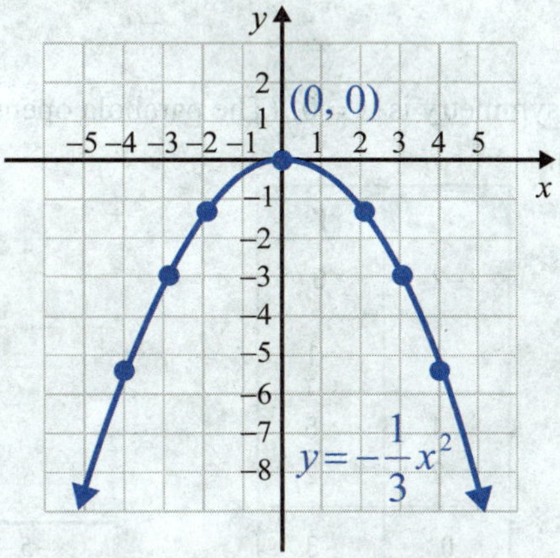

Now work margin exercise 1.

1. Graph each quadratic function, setting up a table of values for x and y as an aid. Choose values of x on each side of the line of symmetry.

a. $y = -4x^2$

b. $y = \frac{1}{4}x^2$

16.6 Exercises

Concept Check

1. parabola
2. vertex
3. $\geq$
4. domain
5. range
6. True
7. False; It is a horizontal shift (or horizontal translation).
8. False; The bigger $|a|$ is, the narrower the opening of the parabola is.

Practice

1. a.

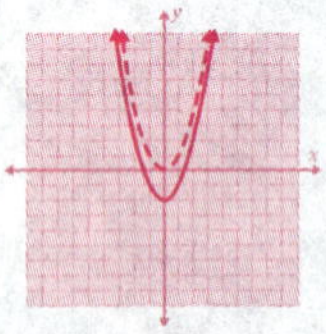

B Graphing Quadratic Functions of the Form $y = ax^2 + k$

The graph of a function of the form

$$y = ax^2 + k$$

is a **vertical shift** (or **vertical translation**) of the graph of $y = ax^2$. The shift is up k units if $k > 0$ or down $|k|$ units if $k < 0$. The **vertex** is at the point $(0, k)$. The line $x = 0$ is the **line of symmetry** just as with functions of the form $y = ax^2$. (See Figure 4.)

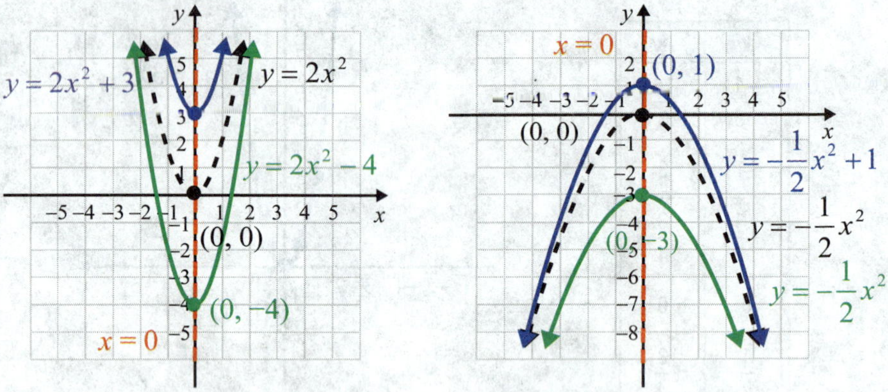

Figure 4

2. Find the line of symmetry and vertex for the quadratic function $y = 3x^2 + 5$. Then graph the function.

b.

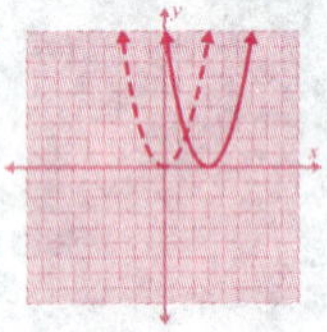

c.

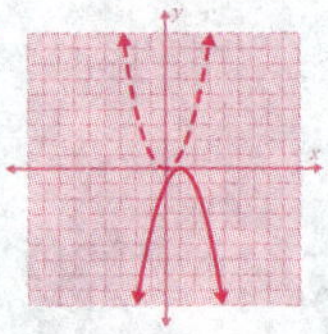

d.

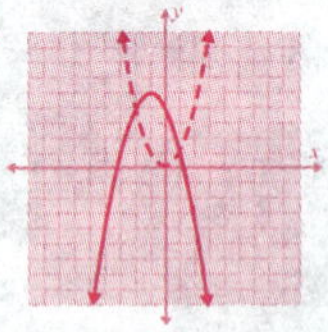

Example 2 Graphing Quadratic Functions

Find the line of symmetry and vertex for the quadratic function $y = 2x^2 - 3$. Then graph the function, setting up a table of values for x and y as an aid. Choose values of x on each side of the line of symmetry.

Solution

The line of symmetry is $x = 0$. (The parabola opens upward since a is positive.) The vertex is at $(0, -3)$.

x	y
-2	5
-1	-1
$-\frac{1}{2}$	$-\frac{5}{2}$
0	-3
$\frac{1}{2}$	$-\frac{5}{2}$
1	-1
2	5

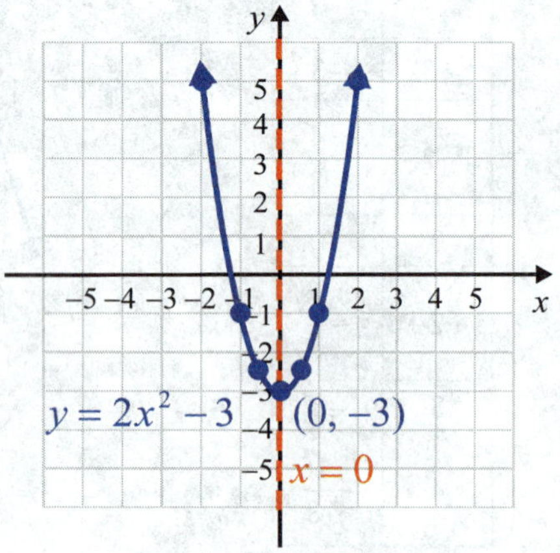

Now work margin exercise 2.

C Graphing Quadratic Functions of the Form $y = a(x-h)^2$

The graph of a function of the form

$$y = a(x-h)^2$$

is a **horizontal shift** (or **horizontal translation**) of the graph of $y = ax^2$. The shift is to the right if $h > 0$ and to the left if $h < 0$. Notice that if $h = -3$, then $y = a(x-h)^2$ gives $y = a(x-(-3))^2$ or $y = a(x+3)^2$, which is a shift of 3 units to the left. Thus, if h is negative, the expression $(x-h)^2$ appears with a plus sign. If h is positive, the expression $(x-h)^2$ appears with a minus sign. In either case, the line $x = h$ is the **line of symmetry** and the **vertex** occurs at $(h, 0)$. (See Figure 5.)

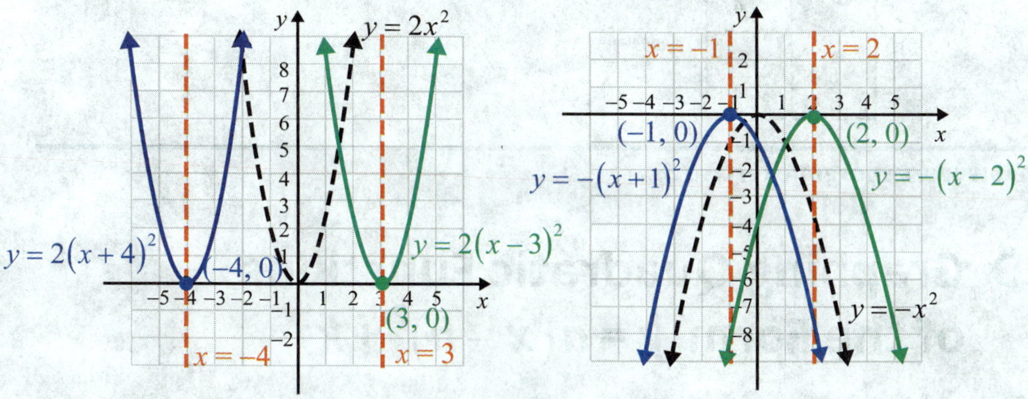

Figure 5

Example 3 Graphing Quadratic Functions

Find the line of symmetry and vertex for the quadratic function $y = -\left(x - \frac{5}{2}\right)^2$. Then graph the function, setting up a table of values for x and y as an aid. Choose values of x on each side of the line of symmetry.

Solution

The line of symmetry is $x = \frac{5}{2}$ and the vertex is at $\left(\frac{5}{2}, 0\right)$. Note that the parabola opens downward since a is negative.

2. a.

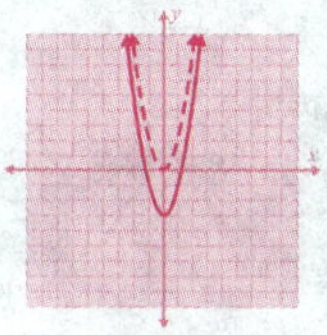

b.

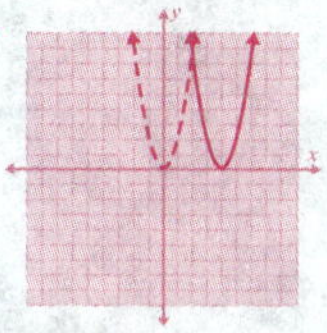

c.

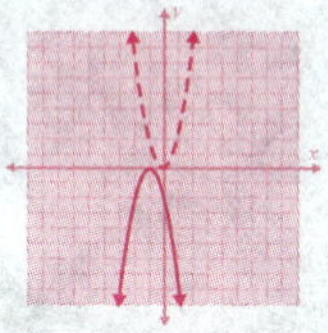

d.

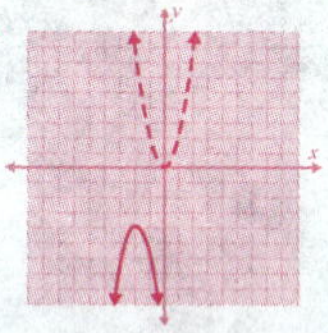

3. a.

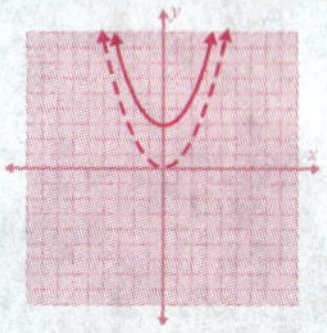

3. Find the line of symmetry and vertex for the quadratic function $y = (x-2)^2$. Then graph the function.

b.

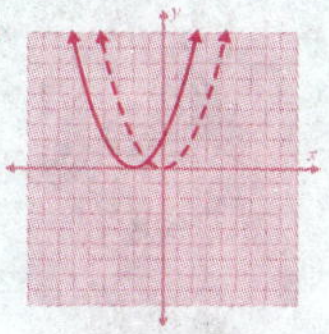

c.

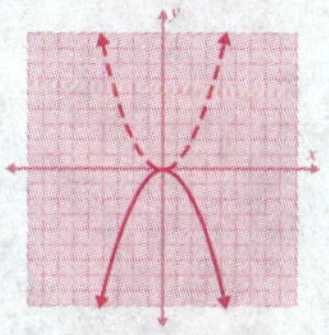

d.

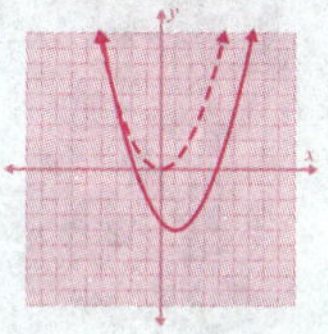

x	y
0	$-\frac{25}{4}$
1	$-\frac{9}{4}$
2	$-\frac{1}{4}$
$\frac{5}{2}$	0
3	$-\frac{1}{4}$
4	$-\frac{9}{4}$
5	$-\frac{25}{4}$

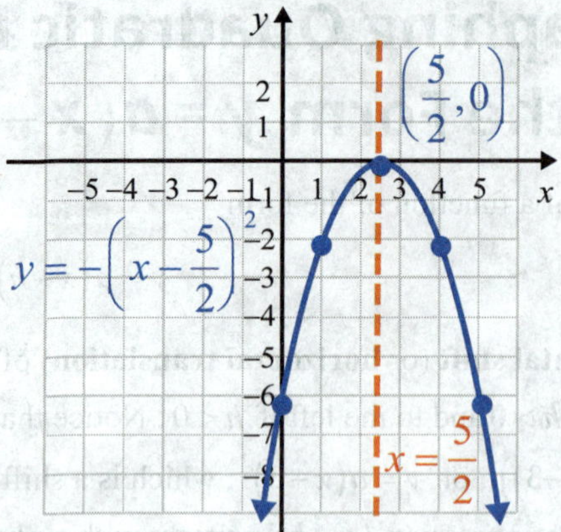

Now work margin exercise 3.

D Graphing Quadratic Functions of the Form $y = a(x-h)^2 + k$

The graph of a function of the form

$$y = a(x-h)^2 + k$$

is a combination of both a vertical shift of k units and a horizontal shift of h units of the graph of $y = ax^2$. The **vertex** is at (h, k). The **line of symmetry** is $x = h$.

For example, the graph of the function $y = -2(x-3)^2 + 5$ is a vertical shift of 5 units (5 units up) and a horizontal shift of 3 units (3 units to the right) of the graph of $y = -2x^2$. The line of symmetry is $x = 3$ and the vertex is at $(3, 5)$.

The graph of the function $y = \left(x + \frac{1}{2}\right)^2 - 2$ is a vertical shift of –2 units (2 units down) and a horizontal shift of $-\frac{1}{2}$ units ($\frac{1}{2}$ units to the left) of the graph of $y = x^2$. The line of symmetry is $x = -\frac{1}{2}$ and the vertex is at $\left(-\frac{1}{2}, -2\right)$. Both of these graphs are illustrated in Figure 6.

4. a.

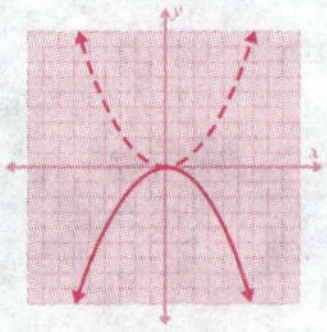

b.

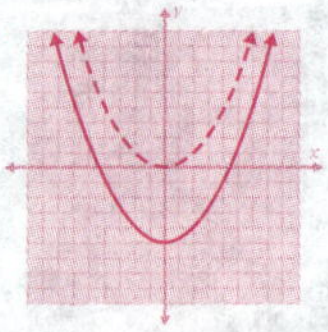

c.

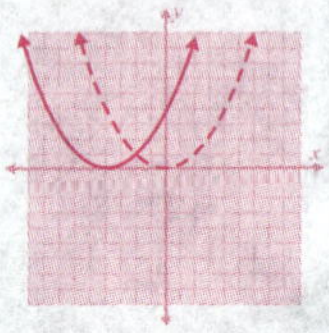

d.

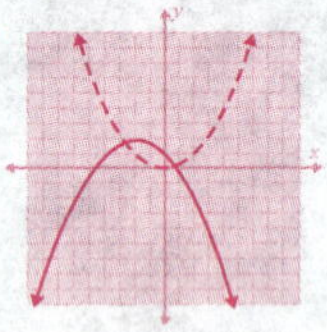

5. $x = 0; (0, -4)$

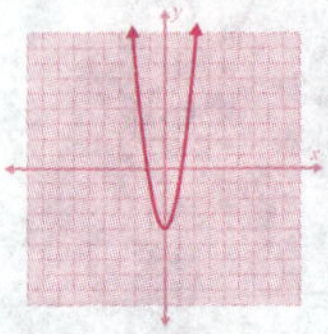

6. $x = 0; (0, 6)$

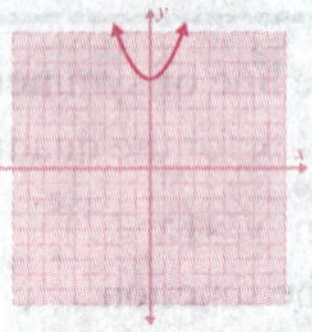

7. $x = 0; (0, -9)$

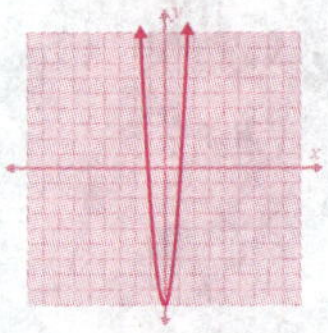

8. $x = 0; (0, -1)$

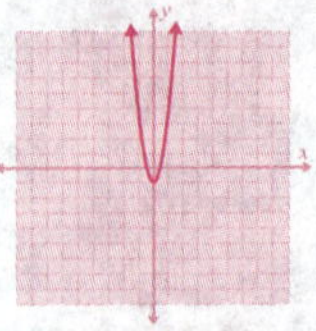

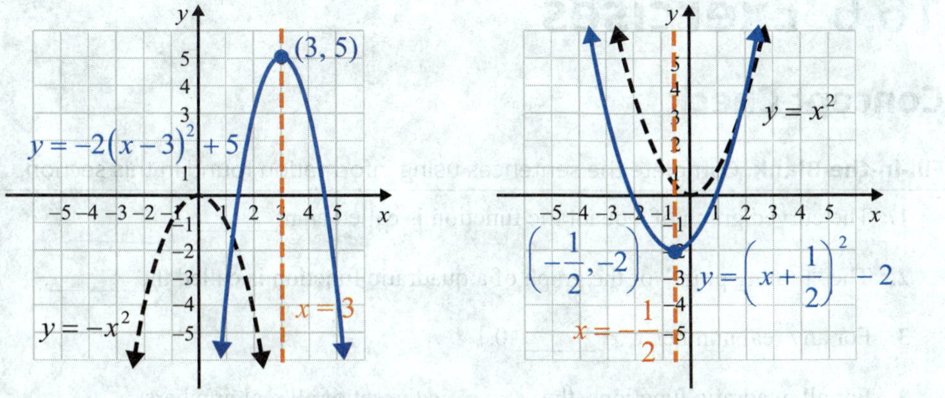

Figure 6

Example 4 Graphing Quadratic Functions

Find the line of symmetry and vertex for the quadratic function $y=-(x+2)^2+1$. Then graph the function, setting up a table of values for x and y as an aid. Choose values of x on each side of the line of symmetry.

Solution

The line of symmetry is $x=-2$. (The parabola opens downward since a is negative.) The vertex is at $(-2,1)$.

x	y
−5	−8
−4	−3
−3	0
−2	1
−1	0
0	−3
1	−8

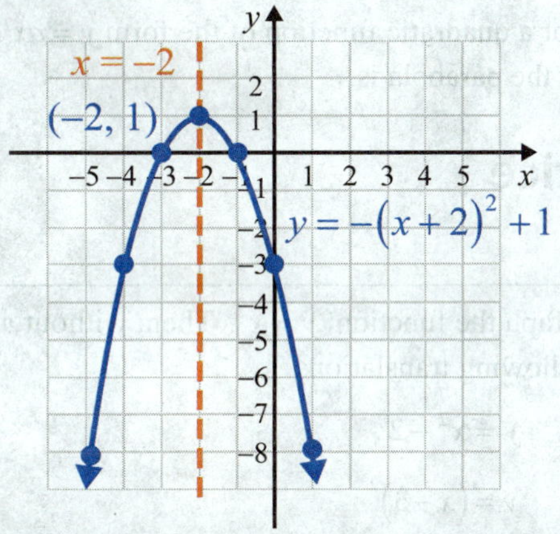

Now work margin exercise 4.

Margin Exercise Answers

1. a.

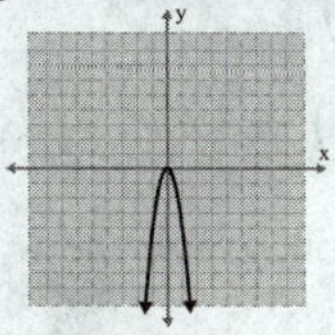

b.

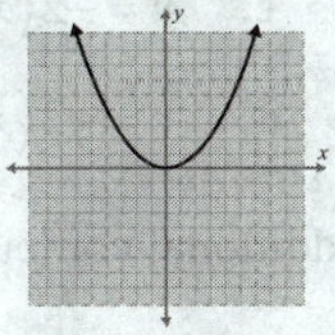

2. $x=0;(0,5)$

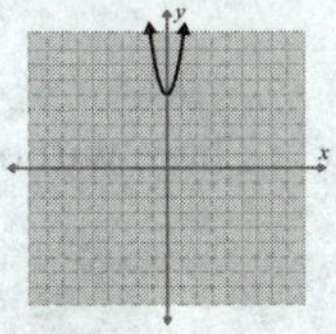

3. $x=2;(2,0)$

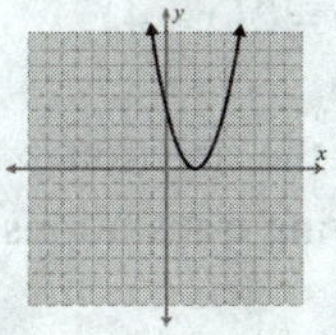

4. $x=1;(1,-3)$

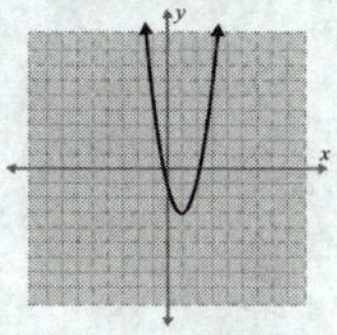

9. $x=0;(0,1)$

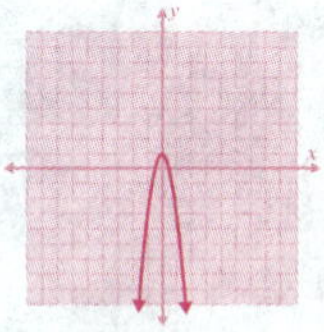

10. $x=0;(0,-2)$

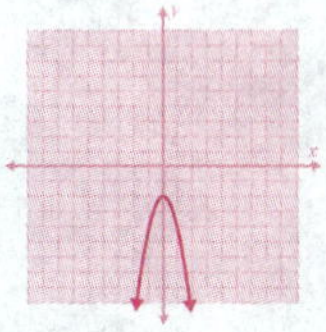

4. Find the line of symmetry and vertex for the quadratic function $y=2(x-1)^2-3$. Then graph the function.

11. $x=0;(0,5)$

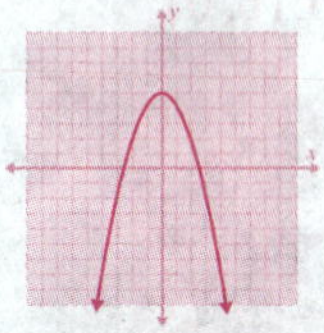

12. $x=0;(0,-3)$

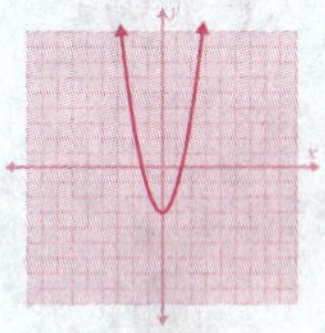

13. $x=-1;(-1,0)$

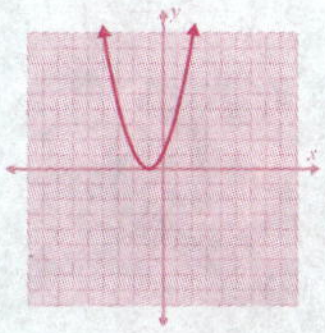

14. $x=1;(1,0)$

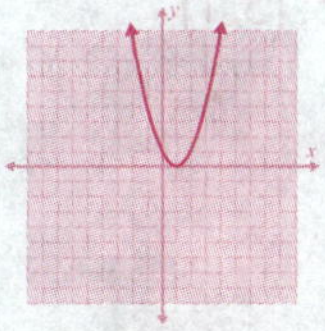

15. $x=4;(4,0)$

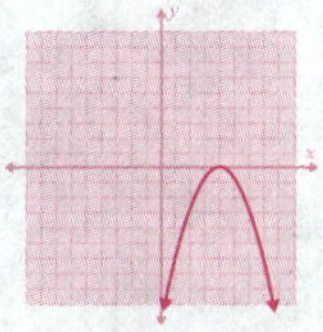

16. $x=-2; (-2, 0)$

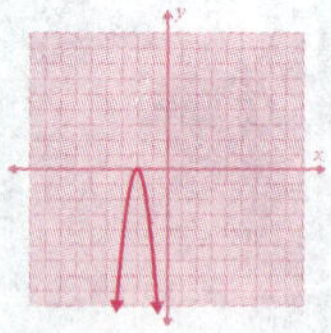

17. $x=5; (5, 0)$

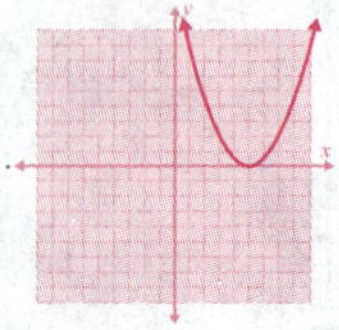

18. $x=-3; (-3, 0)$

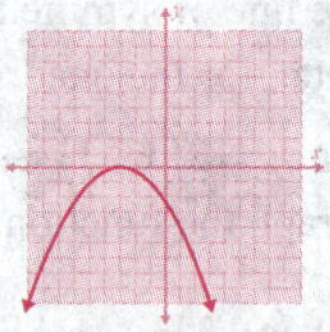

19. $x=6; (6, 0)$

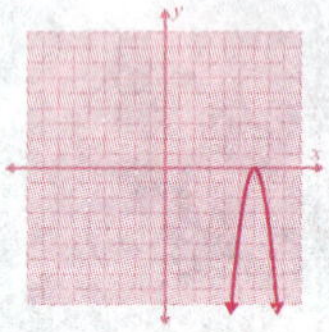

20. $x=-7; (-7, 0)$

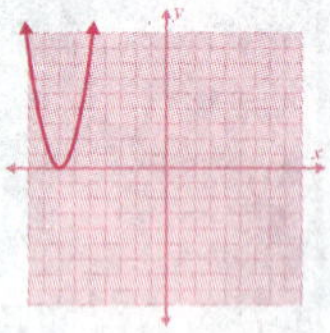

21. $x=-3; (-3, -2)$

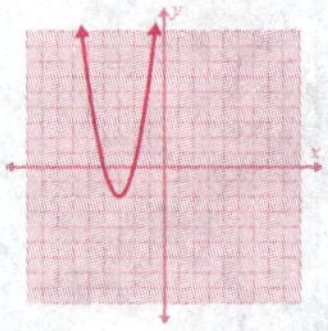

22. $x=5; (5, 1)$

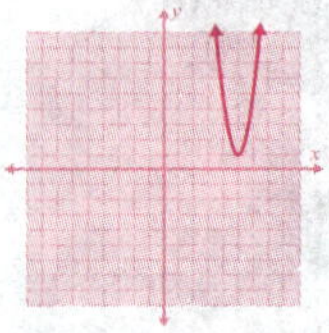

16.6 Exercises

Concept Check

Fill-in-the-Blank. Complete the sentences using information found in this section.

1. The curved graph of a quadratic function is called a/an ______.
2. The "turning point" of the graph of a quadratic function is called the ______.
3. For any real number x, x^2 ______ 0.
4. For all quadratic functions, the ______ is the set of all real numbers.
5. The ______ of the function $y = ax^2$ depends on the value of a.

True/False. Determine whether each statement is true or false. If a statement is false, explain how it can be changed so the statement will be true. (**Note:** There may be more than one acceptable change.)

6. The graph of a quadratic function is a mirror of itself across the line, or axis, of symmetry.
7. The graph of $y = a(x-h)^2$ is a vertical shift (or vertical translation) of the graph of $y = ax^2$.
8. For a quadratic function of the form $y = ax^2$, the bigger $|a|$ is, the wider the opening of the parabola is.

Practice

Solve.

1. Graph the function $y = x^2$. Then, without additional computation, graph the following translations.
   - **a.** $y = x^2 - 2$
   - **b.** $y = (x-3)^2$
   - **c.** $y = -(x-1)^2$
   - **d.** $y = 5-(x+1)^2$
2. Graph the function $y = 2x^2$. Then, without additional computation, graph the following translations.
   - **a.** $y = 2x^2 - 3$
   - **b.** $y = 2(x-4)^2$
   - **c.** $y = -2(x+1)^2$
   - **d.** $y = -2(x+2)^2 - 4$
3. Graph the function $y = \frac{1}{2}x^2$. Then, without additional computation, graph the following translations.
   - **a.** $y = \frac{1}{2}x^2 + 3$
   - **b.** $y = \frac{1}{2}(x+2)^2$
   - **c.** $y = -\frac{1}{2}x^2$
   - **d.** $y = \frac{1}{2}(x-1)^2 - 4$

4. Graph the function $y = \frac{1}{4}x^2$. Then, without additional computation, graph the following translations.

a. $y = -\frac{1}{4}x^2$

b. $y = \frac{1}{4}x^2 - 5$

c. $y = \frac{1}{4}(x+4)^2$

d. $y = 2 - \frac{1}{4}(x+2)^2$

For each of the quadratic functions, determine the line of symmetry and the vertex. Then, graph the function.

5. $y = 3x^2 - 4$

6. $y = \frac{2}{3}x^2 + 6$

7. $y = 7x^2 - 9$

8. $y = 5x^2 - 1$

9. $y = -4x^2 + 1$

10. $y = -2x^2 - 2$

11. $y = -\frac{3}{4}x^2 + 5$

12. $y = \frac{5}{3}x^2 - 3$

13. $y = (x+1)^2$

14. $y = (x-1)^2$

15. $y = -\frac{2}{3}(x-4)^2$

16. $y = -5(x+2)^2$

17. $y = \frac{1}{2}(x-5)^2$

18. $y = -\frac{1}{4}(x+3)^2$

19. $y = -4(x-6)^2$

20. $y = 2(x+7)^2$

21. $y = 2(x+3)^2 - 2$

22. $y = 4(x-5)^2 + 1$

23. $y = \frac{3}{4}(x+2)^2 - 6$

24. $y = -2(x+1)^2 - 4$

25. $y = \frac{1}{3}(x+1)^2 - 2$

26. $y = -\frac{3}{2}(x-4)^2 - 1$

27. $y = -3(x-3)^2 + 3$

28. $y = 5(x+3)^2 - 6$

Writing & Thinking

29. Explain why the shape of the parabola of a quadratic of the form $y = ax^2$ gets narrower as the value of $|a|$ increases. (**Hint:** Pick two values of a and compare the value of y for different values of x.)

23. $x = -2$; $(-2, -6$

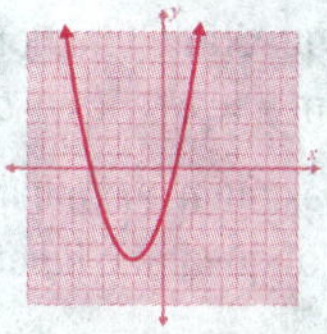

24. $x = -1$; $(-1, -4)$

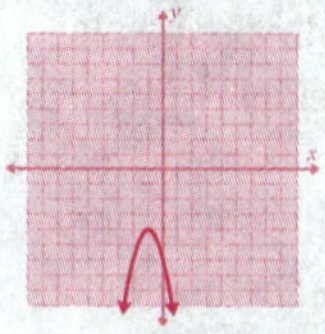

25. $x = -1$; $(-1, -2)$

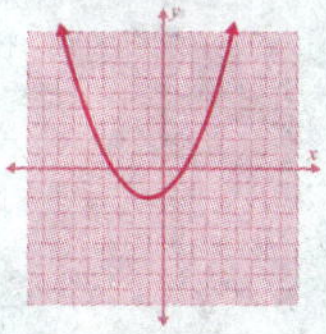

26. $x = 4$; $(4, -1)$

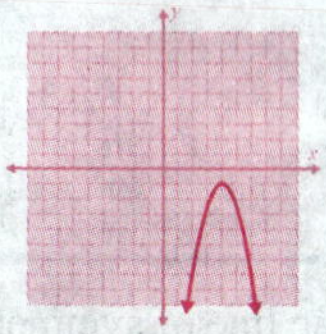

27. $x = 3$; $(3, 3)$

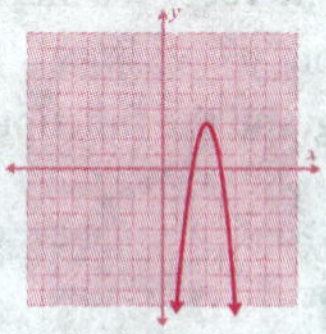

28. $x = -3$; $(-3, -6$

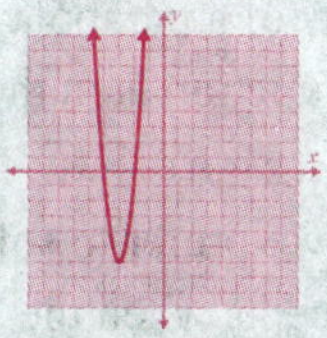

Writing & Thinking

29. The parabola is narrower for larger values of a because the value of y increases faster for increasing values of x.

Objectives

A. Graph quadratic functions of the form $y = ax^2 + bx + c$.

B. Find maximum and minimum values of quadratic functions.

Teaching Note

This is another good place in this development to make the case for learning to "read" information from an equation so that key characteristics can be known about these functions before (or without) graphing.

16.7 More on Graphing Quadratic Functions and Applications

A Graphing Quadratic Functions of the Form $y = ax^2 + bx + c$

The general form of a quadratic function is $y = ax^2 + bx + c$. This form does not give as much information about the graph of the corresponding parabola compared to the form $y = a(x-h)^2 + k$. With the general form, we need to find how the coefficients a, b, and c relate to the vertex and the line of symmetry. We can accomplish this by algebraically changing the general form from $y = ax^2 + bx + c$ to the more useful form $y = a(x-h)^2 + k$.

Note that the square of $(x-h)$ is a perfect square trinomial (see Section 13.4).

$$(x-h)^2 = x^2 - 2hx + h^2$$

So, given a function in the form $y = x^2 - 2hx + h^2$, we could write it in the familiar form $y = (x-h)^2$. However, given a function such as $y = x^2 - 10x + 21$, we see that $x^2 - 10x + 21$ is not a perfect square. So, we need to complete the square (see Section 16.2), as illustrated in the following examples.

1. Find the line of symmetry, the vertex, the x-intercepts, and y-intercept for the function $y = x^2 - 4x + 3$. Then graph the function.

Math Tip

It is often quicker to find x-intercepts of a parabola when the function is in the form $y = a(x-h)^2 + k$. Similarly, we prefer to solve for y-intercepts when the parabola is in general form $y = ax^2 + bx + c$. While you will see this preference in the text, it is still possible to solve for intercepts using either form.

Example 1 Graphing Quadratic Functions of the Form $y = ax^2 + bx + c$

Find the line of symmetry, the vertex, the x-intercepts (zeros), and y-intercept for the function $y = x^2 - 10x + 21$. Then, graph the function.

Solution

$y = x^2 - 10x + 21$ Write the function.

$y = (x^2 - 10x + 25 - 25) + 21$ Complete the square: $\frac{1}{2}(-10) = -5$ and $(-5)^2 = 25$

$y = (x^2 - 10x + 25) - 25 + 21$ $x^2 - 10x + 25$ is a perfect square trinomial.

$y = (x-5)^2 - 4$ Simplify.

In this form, we see that the line of symmetry is $x = 5$ and the vertex is at $(5, -4)$.

By setting $y = 0$, we can find the ***x*-intercepts (zeros)** of the function by solving for x.

$$(x-5)^2 - 4 = 0$$
$$(x-5)^2 = 4$$
$$x - 5 = \pm\sqrt{4}$$
$$x = 5 \pm 2$$

This gives $x = 7$ or $x = 3$, and the zeros (x-intercepts) are at $(7, 0)$ and $(3, 0)$.

To find the ***y*-intercept**, substitute $x = 0$ in the function.

$$y = (0)^2 - 10(0) + 21 = 21$$

The y-intercept is at $(0, 21)$.

The figure shows the graph with vertex, line of symmetry, and intercepts labeled. Note that the y-intercept lies outside of the given grid.

Summary:

line of symmetry is $x = 5$.

vertex: $(5, -4)$

x-intercepts: $(3, 0)$ and $(7, 0)$

y-intercept: $(0, 21)$

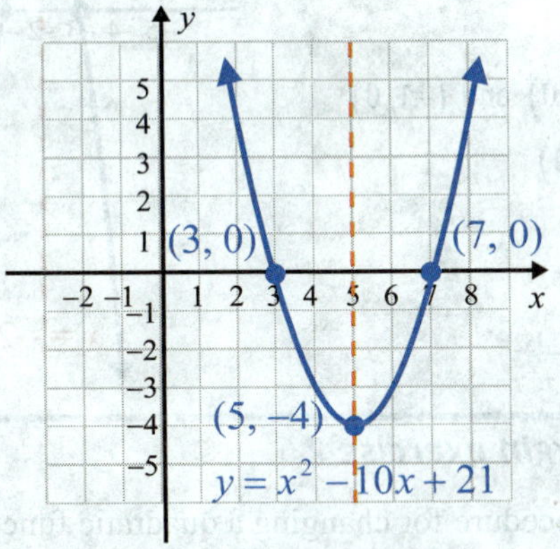

Now work margin exercise 1.

Example 2 Graphing Quadratic Functions of the Form $y = ax^2 + bx + c$

Find the vertex, the line of symmetry, the x-intercepts, and y-intercept for the function $y = -3x^2 - 12x - 9$. Then, graph the function.

Solution

$y = -3x^2 - 12x - 9$	Write the function.
$y = -3(x^2 + 4x) - 9$	Factor -3 from the first two terms.
$y = -3(x^2 + 4x + 4 - 4) - 9$	Complete the square: $\frac{1}{2}(4) = 2$ and $(2)^2 = 4$
$y = -3(x^2 + 4x + 4) + 12 - 9$	Multiply: $(-3)(-4) = +12$
$y = -3(x + 2)^2 + 3$	Simplify.

In this form, we see that $x = -2$ is the line of symmetry and the vertex is at $(-2, 3)$.

By setting $y = 0$, we can find the **x-intercepts** by solving for x.

$$-3(x+2)^2 + 3 = 0$$
$$-3(x+2)^2 = -3$$
$$(x+2)^2 = 1$$
$$x + 2 = \pm\sqrt{1}$$
$$x = -2 \pm 1$$

This gives $x = -1$ or $x = -3$ and the zeros (x-intercepts) are at $(-3, 0)$ and $(-1, 0)$.

To find the y-intercept, substitute $x = 0$ in the function.

$$y = -3(0)^2 - 12(0) - 9 = -9$$

The y-intercept is at $(0, -9)$.

16.7 Exercises

Concept Check

1. vertex
2. zeros
3. x-axis
4. upward, lowest, minimum
5. downward, highest, maximum
6. zeros, 0
7. False; The line of symmetry is at $x = -\frac{b}{2a}$
8. False; The vertex is either the highest or lowest point on the parabola.

2. Find the line of symmetry, the vertex, the x-intercepts, and y-intercept for the function $y = -2x^2 + 4x + 6$. Then graph the function.

9. True
10. True

Practice

1. $y = 2(x-1)^2; x = 1;$ Vertex: $(1, 0)$; x-int: $(1, 0)$; y-int: $(0, 2)$

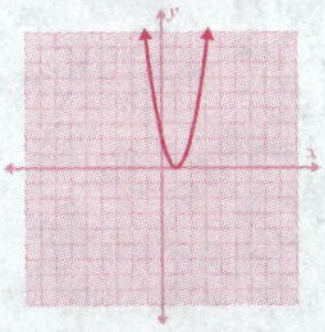

2. $y = -3(x-2)^2; x = 2;$ Vertex: $(2, 0)$; x-int: $(2, 0)$; y-int: $(0, -12)$

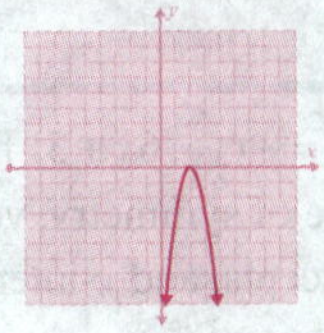

3. $y=(x-1)^2-4;\ x=1;$

Vertex: $(1,-4)$;

x-int: $(-1,0),(3,0)$;

y-int: $(0,-3)$

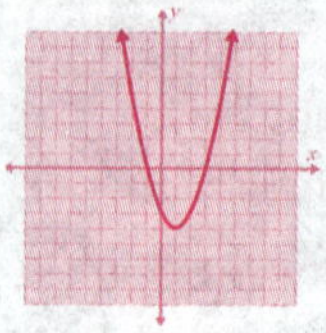

4. $y=(x-2)^2+1;\ x=2;$

Vertex: $(2,1)$; x-int: None;

y-int: $(0,5)$

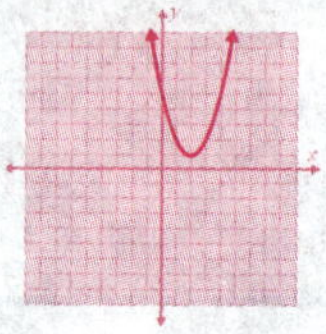

5. $y=(x+3)^2-4;\ x=-3;$

Vertex: $(-3,-4)$;

x-int: $(-5,0),(-1,0)$;

y-int: $(0,5)$

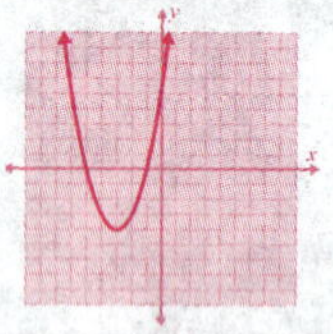

The figure shows the graph with vertex, line of symmetry, and intercepts labeled.

Summary:

line of symmetry is $x=-2$.

vertex: $(-2,3)$

x-intercepts: $(-3,0)$ and $(-1,0)$

y-intercept: $(0,-9)$

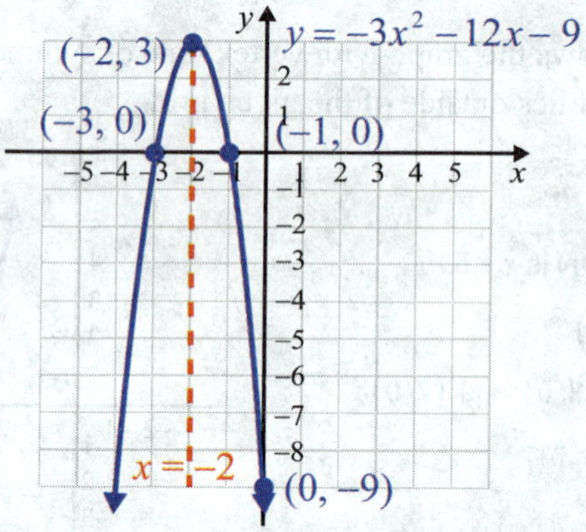

Now work margin exercise 2.

The algebraic procedure for changing a quadratic function from the general form $y=ax^2+bx+c$ to the form $y=a(x-h)^2+k$ is as follows.

$y=ax^2+bx+c$ Write the function.

$=a\left(x^2+\frac{b}{a}x\right)+c$ Factor a from just the first two terms.

$=a\left(x^2+\frac{b}{a}x+\frac{b^2}{4a^2}-\frac{b^2}{4a^2}\right)+c$ Complete the square of $x^2+\frac{b}{a}x$.

$\frac{1}{2}\left(\frac{b}{a}\right)=\frac{b}{2a}$ and $\left(\frac{b}{2a}\right)^2=\frac{b^2}{4a^2}$

Add and subtract $\frac{b^2}{4a^2}$ inside the parentheses.

$=a\left(x^2+\frac{b}{a}x+\frac{b^2}{4a^2}\right)-\frac{b^2}{4a}+c$ Multiply $a\left(\frac{-b^2}{4a^2}\right)$ and write this term outside the parentheses.

$=a\left(x+\frac{b}{2a}\right)^2+\frac{4ac-b^2}{4a}$ Write the square of the binomial and add the last two terms to get the form $y=a(x-h)^2+k$.

In terms of the coefficients a, b, and c,

$$x=-\frac{b}{2a}\quad \text{is the } \textbf{line of symmetry}$$

and

$$(h,k)=\left(-\frac{b}{2a},\ \frac{4ac-b^2}{4a}\right)\ \text{is the } \textbf{vertex}. \text{ (In function notation, } \left(-\frac{b}{2a},\ f\left(-\frac{b}{2a}\right)\right).\text{)}$$

3. For $y=-x^2-6x+3$, find the line of symmetry, vertex, x-intercepts, and y-intercept.

Example 3 Graphing Quadratic Functions of the Form $y = ax^2 + bx + c$

For $y=-x^2-4x+2$, find the line of symmetry, vertex, x-intercepts, and y-intercept.

Solution

For $y=-x^2-4x+2,$ we have $a=-1$, $b=-4$, and $c=2$. If $x=-\dfrac{b}{2a}=-\dfrac{-4}{2(-1)}=-2,$ then

$$\begin{aligned} y &= -(-2)^2-4(-2)+2 \\ &= -4+8+2 \\ &= 6. \end{aligned}$$

So the vertex is at $(-2, 6)$ and the line of symmetry is $x=-2$.

If $x=0$, then $y=-(0)^2-4\cdot 0+2=2$ and the y-intercept is at $(0, 2)$.

If $y=0$, then

$$-x^2-4x+2=0$$

$$\begin{aligned} x &= \frac{4\pm\sqrt{(-4)^2-4(-1)(2)}}{2(-1)} \quad \text{Quadratic formula} \\ &= \frac{4\pm\sqrt{24}}{-2} \\ &= \frac{4\pm 2\sqrt{6}}{-2} \\ &= -2\pm\sqrt{6}. \end{aligned}$$

The x-intercepts are at $(-2+\sqrt{6}, 0)$ and $(-2-\sqrt{6}, 0)$.

Summary:

line of symmetry is $x=-2$.

vertex: $(-2, 6)$

x-intercepts: $(-2+\sqrt{6}, 0)$ and $(-2-\sqrt{6}, 0)$

y-intercept: $(0, 2)$

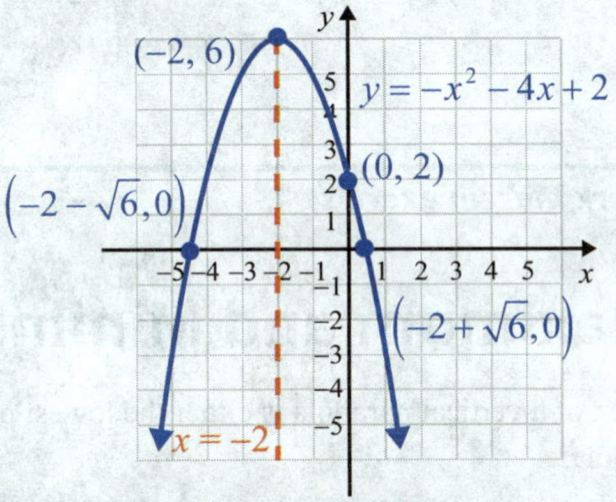

Now work margin exercise 3.

Example 4 Graphing Quadratic Functions of the Form $y = ax^2 + bx + c$

For $y=2x^2-6x+5,$ find the line of symmetry, vertex, x-intercepts, and y-intercept.

Solution

For $y=2x^2-6x+5,$ we have $a=2$, $b=-6$, and $c=5$. If $x=-\dfrac{b}{2a}=-\dfrac{-6}{2(2)}=\dfrac{3}{2},$ then

$$\begin{aligned} y &= 2\left(\frac{3}{2}\right)^2-6\left(\frac{3}{2}\right)+5 \\ &= \frac{9}{2}-9+5=\frac{1}{2}. \end{aligned}$$

Note

Rather than memorize the formula for the coordinates of the vertex, you should just remember that the x-coordinate of the vertex is $x=-\dfrac{b}{2a}$. Substituting this value for x in the function will give the y-value for the vertex.

6. $y=(x-4)^2-4;\ x=4;$
Vertex: $(4, -4);$
x-int: $(2, 0), (6, 0);$
y-int: $(0, 12)$

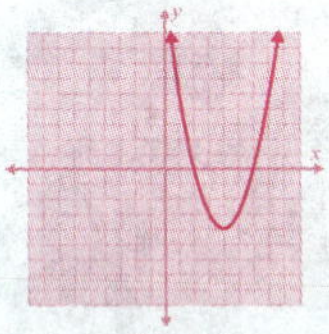

7. $y=2(x-2)^2-3;\ x=2;$
Vertex: $(2, -3);$
x-int: $\left(\dfrac{4+\sqrt{6}}{2}, 0\right),$ $\left(\dfrac{4-\sqrt{6}}{2}, 0\right);$
y-int: $(0, 5)$

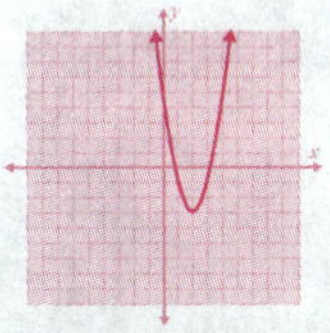

4. For $y=x^2-2x+4$ find the line of symmetry, vertex, x-intercepts, and y-intercept.

8. $y = 2(x-3)^2 - 2;\ x = 3;$
Vertex: $(3, -2)$;
x-int: $(2,0), (4,0)$;
y-int: $(0, 16)$

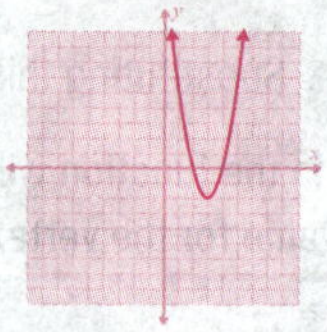

9. $y = -3(x+2)^2 + 3;$
$x = -2$; Vertex: $(-2, 3)$;
x-int: $(-1,0), (-3,0)$;
y-int: $(0, -9)$

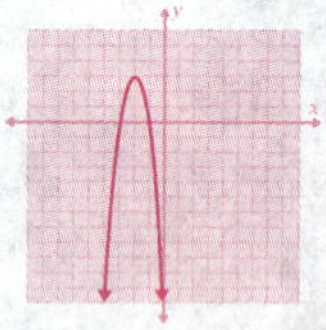

10. $y = 3(x-1)^2 - 4;$
$x = 1$; Vertex: $(1, -4)$;
x-int: $\left(\frac{3+2\sqrt{3}}{3}, 0\right)$, $\left(\frac{3-2\sqrt{3}}{3}, 0\right)$;
y-int: $(0, -1)$

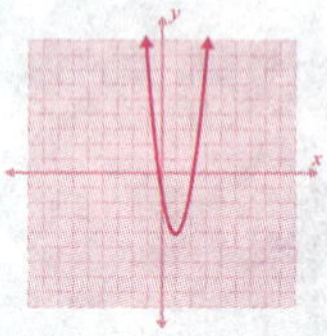

11. $y = 5(x-1)^2 + 3;$
$x = 1$; Vertex: $(1, 3)$;
x-int: None; y-int: $(0, 8)$

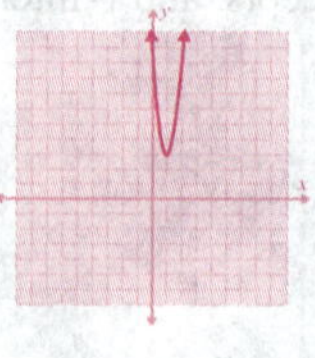

So we have the vertex at $\left(\frac{3}{2}, \frac{1}{2}\right)$ and the line of symmetry is $x = \frac{3}{2}$.

If $x = 0$, then $y = 2(0)^2 - 6(0) + 5 = 5$ and the y-intercept is at $(0,5)$.

If $y = 0$, then

$$2x^2 - 6x + 5 = 0$$

$$x = \frac{6 \pm \sqrt{(-6)^2 - 4(2)(5)}}{2(2)} \qquad \text{Quadratic formula}$$

$$= \frac{6 \pm \sqrt{-4}}{4} \quad \text{or} \quad x = \frac{3}{2} \pm \frac{1}{2}i. \qquad \text{Nonreal complex numbers}$$

There are no real zeros because the discriminant is negative. The graph will not cross the x-axis.

Summary:

vertex: $\left(\frac{3}{2}, \frac{1}{2}\right)$

line of symmetry is $x = \frac{3}{2}$.

x-intercepts: none

y-intercept: $(0,5)$

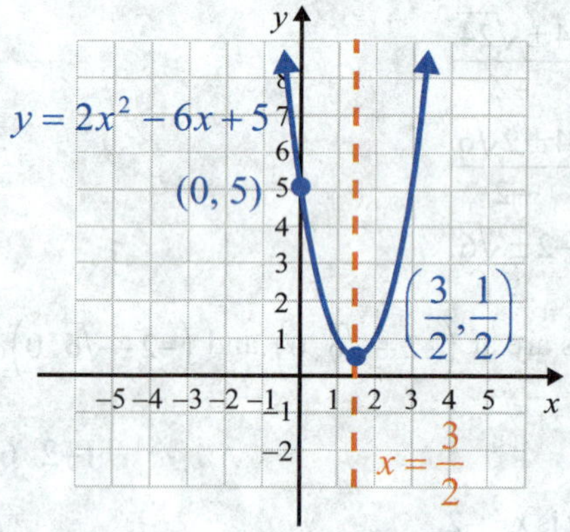

Now work margin exercise 4.

B Maximum and Minimum Values

The vertex of a vertical parabola is either the lowest point or the highest point on the parabola.

Minimum and Maximum Values

For $y = a(x-h)^2 + k$,

1. If $a > 0$, then the parabola opens upward and (h, k) is the lowest point and the y-value k is called the **minimum value** of the function.

2. If $a < 0$, then the parabola opens downward and (h, k) is the highest point and y-value k is called the **maximum value** of the function.

DEFINITION

If the function is in the general quadratic form $y = ax^2 + bx + c$, then the maximum or minimum value can be found by letting $x = -\frac{b}{2a}$ and solving for y.

The concepts of maximum and minimum values of a function help not only in graphing but also in solving many types of applications. Two applications involving quadratic functions are shown in the following examples. Other types of applications are covered in more advanced courses in mathematics.

Example 5 Application: Finding Minimum and Maximum Values

A ball is thrown vertically upward from the ground with an initial velocity of 64 ft per second. Using the function $h = -16t^2 + v_0 t + h_0$, where h is the height of the object after time t, v_0 is the initial velocity, and h_0 is the initial height, determine how long it will take for the ball to reach its maximum height and what that height will be.

Solution

For this problem, we are told that $v_0 = 64$ ft per second and $h_0 = 0$ ft (ground level), so the quadratic function modeling the path of the ball is $h = -16t^2 + 64t$.

For this function $a = -16$, $b = 64$, and the maximum height occurs at the point where

$t = -\frac{b}{2a} = -\frac{64}{-32} = 2$ sec.

For $t = 2$,

$$\begin{aligned} h &= -16t^2 + 64t \\ &= -16(2)^2 + 64(2) \\ &= 64 \text{ ft} \end{aligned}$$

Thus, it will take the ball 2 seconds to reach its maximum height of 64 feet.

Now work margin exercise 5.

5. A ball is thrown vertically upward from the ground with an initial velocity of 48 ft per second. Using the function $h = -16t^2 + v_0 t + h_0$, where h is the height of the object after time t, v_0 is the initial velocity, and h_0 is the initial height, determine how long it will take for the ball to reach its maximum height and what that height will be.

Math Tip

Notice that we use $t = -\frac{b}{2a}$ since t is the independent variable of the function $h = -16t^2 + v_0 t + h_0$.

Example 6 Application: Finding Minimum and Maximum Values

A rancher is going to build three sides of a rectangular corral next to a river. He has 240 feet of fencing and wants to enclose the maximum area possible. What are the dimensions of the fencing that gives the corral its maximum area?

Solution

Let x represent the length of the two equal sides of the rectangular corral and $240 - 2x$ represent the length of the third side of the rectangular corral.

Since area equals length times width, the area A of the corral is represented by the quadratic function $A = x(240 - 2x) = 240x - 2x^2$.

6. A towing company is going to build a rectangular fenced-in area against the side of a building to store impounded cars. The manager has 360 yards of fencing and wants to enclose the maximum area possible inside the lot. What are the dimensions of the lot with the maximum area and what is this area?

12. $y=-4(x-2)^2+5$; $x=2$; Vertex: $(2,5)$; x-int: $\left(\frac{4+\sqrt{5}}{2},0\right)$, $\left(\frac{4-\sqrt{5}}{2},0\right)$; y-int: $(0,-11)$

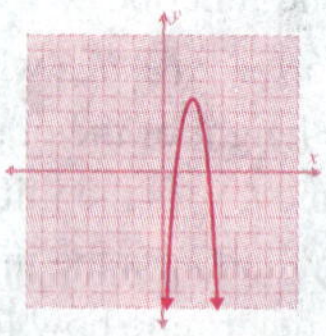

13. $y=-\left(x+\frac{5}{2}\right)^2+\frac{17}{4}$; $x=-\frac{5}{2}$; Vertex: $\left(-\frac{5}{2},\frac{17}{4}\right)$; x-int: $\left(\frac{-5+\sqrt{17}}{2},0\right)$, $\left(\frac{-5-\sqrt{17}}{2},0\right)$; y-int: $(0,-2)$

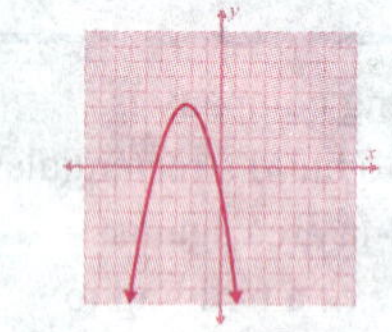

14. $y=\left(x+\frac{3}{2}\right)^2-\frac{13}{4}$; $x=-\frac{3}{2}$; Vertex: $\left(-\frac{3}{2},-\frac{13}{4}\right)$; x-int: $\left(\frac{-3+\sqrt{13}}{2},0\right)$, $\left(\frac{-3-\sqrt{13}}{2},0\right)$; y-int: $(0,-1)$

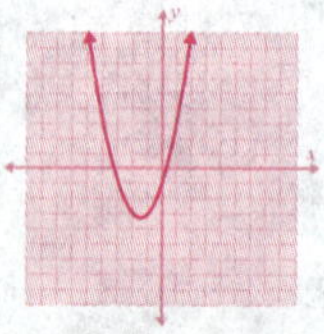

The maximum area occurs at the point where $x=-\frac{b}{2a}=-\frac{240}{-4}=60.$

Two sides of the rectangle are 60 feet and the third side is $240-2(60)=120$ feet.

The maximum area possible is $60(120)=7200$ square feet.

Now work margin exercise 6.

Margin Exercise Answers

1. $x=2$; vertex: $(2,-1)$; x-int: $(1,0),(3,0)$; y-int: $(0,3)$;

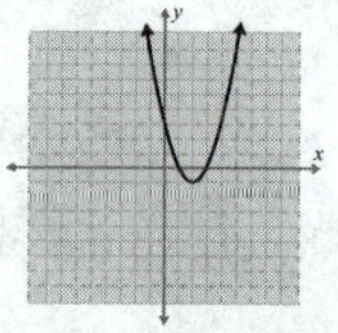

2. $x=1$; vertex: $(1,8)$; x-int: $(3,0),(-1,0)$; y-int: $(0,6)$;

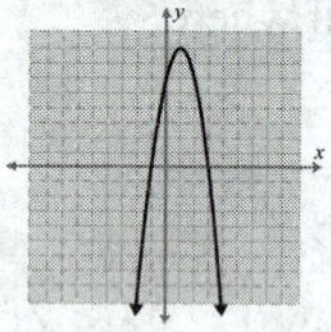

3. $x=-3$; vertex: $(-3,12)$; x-int: $\left(-3+2\sqrt{3},0\right),\left(-3-2\sqrt{3},0\right)$; y-int: $(0,3)$
4. $x=1$; vertex: $(1,3)$; x-int: none; y-int: $(0,4)$; **5.** It will take the ball 1.5 sec to reach its maximum height of 36 ft. **6.** Two sides of the lot are 90 yards and the third side is 180 yards for a maximum area of 16,200 square yards.

16.7 Exercises

Concept Check

Fill-in-the-Blank. Complete each sentence using information found in this section.

1. Quadratic functions of the form $y=a(x-h)^2+k$ have a/an ______ at (h,k).

2. The points where a parabola crosses the x-axis, if any, are the x-intercepts. These points are also called the ______ of the function.

3. If the solutions of a quadratic function are nonreal complex numbers, then the graph does not cross the ______.

4. If $a>0$, then the parabola opens ______ and (h,k) is the ______ point and the y-value k is called the ______ value of the function.

5. If $a<0$, then the parabola opens ______ and (h,k) is the ______ point and the y-value k is called the ______ value of the function.

6. The x-intercepts, or ______, of a function can be found by substituting ______ for y and solving the resulting quadratic equation.

True/False. Determine whether each statement is true or false. If a statement is false, explain how it can be changed so the statement will be true. (**Note:** There may be more than one acceptable change.)

7. Quadratic functions of the form $y = a(x - h)^2 + k$ have a line of symmetry at $x = \frac{b}{2a}$.

8. The vertex of a vertical parabola is the lowest point on the parabola.

9. The maximum or minimum value of a quadratic function written in general form can be found by letting $x = -\frac{b}{2a}$ and solving for y.

10. When the solutions to a quadratic function are nonreal, the entire graph lies either completely above or below the x-axis.

Practice

Rewrite each quadratic function in the form $y = a(x-h)^2 + k$. Find the line of symmetry, the vertex, the x-intercepts, and the y-intercept. Graph the function. See Examples 1 and 2.

1. $y = 2x^2 - 4x + 2$

2. $y = -3x^2 + 12x - 12$

3. $y = x^2 - 2x - 3$

4. $y = x^2 - 4x + 5$

5. $y = x^2 + 6x + 5$

6. $y = x^2 - 8x + 12$

7. $y = 2x^2 - 8x + 5$

8. $y = 2x^2 - 12x + 16$

9. $y = -3x^2 - 12x - 9$

10. $y = 3x^2 - 6x - 1$

11. $y = 5x^2 - 10x + 8$

12. $y = -4x^2 + 16x - 11$

13. $y = -x^2 - 5x - 2$

14. $y = x^2 + 3x - 1$

15. $y = 2x^2 + 7x + 5$

16. $y = 2x^2 + x - 3$

For each quadratic function use the formula $x = -\frac{b}{2a}$ to find the line of symmetry and the vertex. Then find the x-intercepts and the y-intercept. Graph the function. See Examples 3 and 4.

17. $y = -3x^2 + 6x - 3$

18. $y = x^2 - 2x - 8$

19. $y = x^2 - 4x + 3$

20. $y = x^2 + 6x + 5$

21. $y = -2x^2 + 8x - 9$

22. $y = 5x^2 + 10x + 7$

23. $y = x^2 + 2x + 1$

24. $y = x^2 + 8x + 7$

25. $y = -x^2 - 2x - 2$

26. $y = 2x^2 + 4x - 6$

15. $y = 2\left(x + \frac{7}{4}\right)^2 - \frac{9}{8}$;
$x = -\frac{7}{4}$;
Vertex: $\left(-\frac{7}{4}, -\frac{9}{8}\right)$;
x-int: $\left(-\frac{5}{2}, 0\right), (-1, 0)$;
y-int: $(0, 5)$

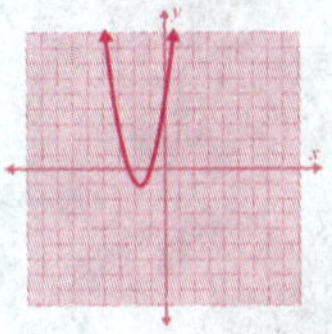

16. $y = 2\left(x + \frac{1}{4}\right)^2 - \frac{25}{8}$;
$x = -\frac{1}{4}$;
Vertex: $\left(-\frac{1}{4}, -\frac{25}{8}\right)$;
x-int: $\left(-\frac{3}{2}, 0\right), (1, 0)$;
y-int: $(0, -3)$

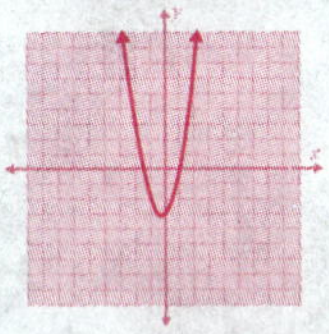

17. $x = 1$; Vertex: $(1, 0)$;
x-int: $(1, 0)$; y-int: $(0, -3)$

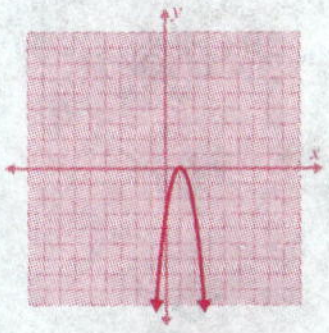

18. $x = 1$; Vertex: $(1, -9)$;
x-int: $(-2, 0), (4, 0)$;
y-int: $(0, -8)$

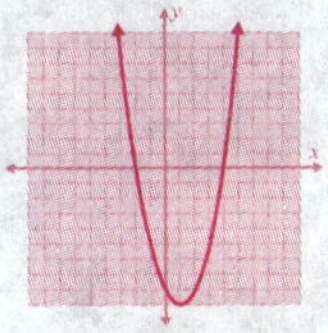

19. $x = 2$; Vertex: $(2, -1)$; x-int: $(3, 0), (1, 0)$; y-int: $(0, 3)$

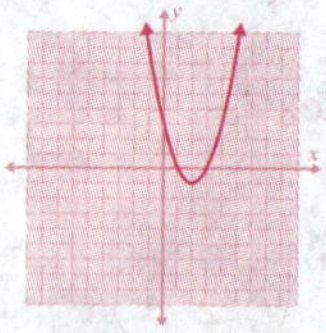

20. $x = -3$; Vertex: $(-3, -4)$; x-int: $(-1, 0), (-5, 0)$; y-int: $(0, 5)$

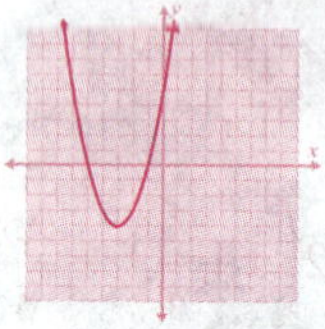

21. $x = 2$; Vertex: $(2, -1)$; x-int: None; y-int: $(0, -9)$

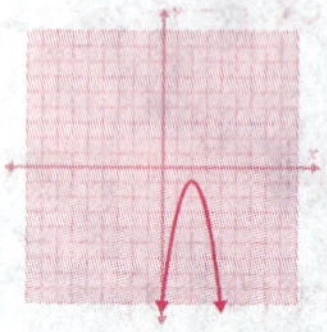

22. $x = -1$; Vertex: $(-1, 2)$; x-int: None; y-int: $(0, 7)$

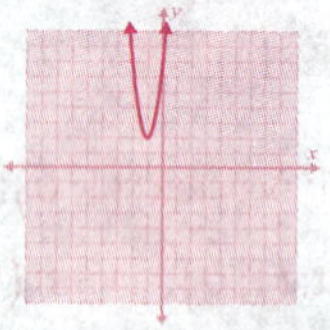

23. $x = -1$; Vertex: $(-1, 0)$; x-int: $(-1, 0)$; y-int: $(0, 1)$

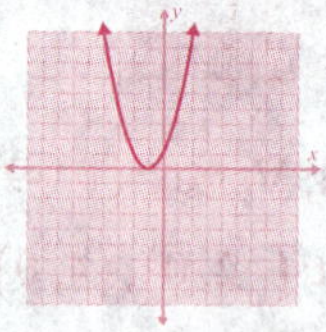

24. $x = -4$; Vertex: $(-4, -9)$; x-int: $(-7, 0), (-1, 0)$; y-int: $(0, 7)$

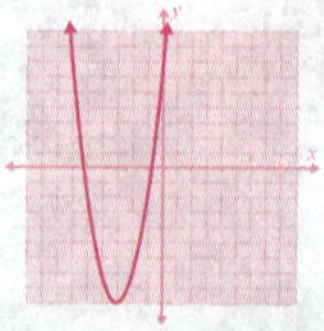

Graph the two given functions and answer the following questions:

a. Are the graphs the same?
b. Do the functions have the same zeros?
c. Briefly, discuss your interpretation of the results in parts a. and b.

27. $\begin{cases} y = x^2 - 3x - 10 \\ y = -x^2 + 3x + 10 \end{cases}$

28. $\begin{cases} y = x^2 - 5x + 6 \\ y = -x^2 + 5x - 6 \end{cases}$

29. $\begin{cases} y = 2x^2 - 5x - 3 \\ y = -2x^2 + 5x + 3 \end{cases}$

30. $\begin{cases} y = -4x^2 - 15x + 4 \\ y = 4x^2 + 15x - 4 \end{cases}$

Use the CALC features of the calculator to find the zeros of the function. (**Hint:** The `zero` item on the CALC menu will locate the zeros of the function.) Round answers to nearest ten-thousandth.

31. $y = x^2 - 2x - 2$

32. $y = 3x^2 + x - 1$

33. $y = -2x^2 + 2x + 5$

34. $y = -x^2 - 2x + 7$

35. $y = x^2 + 3x + 3$

36. $y = -4x^2 - x - 6$

Use a graphing calculator to graph each function by pressing [Y=] and entering the function. Find the coordinates of the maximum as follows. Round answers to nearest ten-thousandth.

Step 1: Press CALC ([2nd] [TRACE]).

Step 2: Press or choose `maximum`.

Step 3: Follow the directions for moving the cursor to `Left Bound?`, `Right Bound?`, and `Guess?`. (Press [ENTER] each time.)

37. $y = 4x - x^2$

38. $y = 1 - 2x - x^2$

39. $y = -8 + 4x - x^2$

40. $y = 3 - 2x - x^2$

Use a graphing calculator to graph each function by pressing [Y=] and entering the function. Find the coordinates of the minimum as follows. Round answers to nearest ten-thousandth.

Step 1: Press CALC ([2nd] [TRACE]).

Step 2: Press or choose `minimum`.

Step 3: Follow the directions for moving the cursor to `Left Bound?`, `Right Bound?`, and `Guess?`. (Press [ENTER] each time.)

41. $y = x^2 - 8x + 15$

42. $y = x^2 + 10x + 22$

43. $y = 2x^2 + 4x + 3$

44. $y = 3x^2 - 6x + 5$

Applications

Use the function $h = -16t^2 + v_0 t + h_0$, where h is the height of the object after time t, v_0 is the initial velocity, and h_0 is the initial height. See Example 5.

45. A ball is thrown vertically upward from the ground with an initial velocity of 112 ft/s.

a. When will the ball reach its maximum height?

b. What will be the maximum height?

46. A water rocket is launched from the ground and has an initial velocity of 104 ft/s.

a. When will the rocket reach its maximum height?

b. What will be the maximum height?

47. A stone is projected vertically upward from a platform that is 20 feet high at a rate of 160 feet per second.

a. When will the stone reach its maximum height?

b. What will be the maximum height?

48. A cannonball is projected vertically upward from a platform that is 32 feet high at a rate of 128 feet per second.

a. When will the cannonball reach its maximum height?

b. What will be the maximum height?

Solve.

49. A retailer sells fitness trackers. He estimates that by selling them for x dollars each, he will be able to sell $100 - x$ fitness trackers each month.

a. What price will yield maximum revenue?

b. What will be the maximum revenue?

50. Mrs. Richey can sell 72 picture frames each month if she charges $24 each. She estimates that for each $1 increase in price, she will sell 2 fewer frames.

a. Find the price that will yield maximum revenue.

b. What will be the maximum revenue?

51. A store owner estimates that by charging x dollars each for a certain lamp, he can sell $40 - x$ lamps each week. What price will give him maximum sales revenue?

52. The band Pumpkin Riot estimates that by selling T-shirts for x dollars each, they can sell $250 - 10x$ T-shirts per show they play. Determine the price per T-shirt that will give the band maximum sales revenue.

53. A nature reserve plans to fence off an area of land to restore balance to the native plant and wildlife populations. The fence will be on three sides and form a rectangle, with a river along the fourth side. The planning committee determines they have enough funds to install 600 yards of fencing.

a. What dimensions should the fence have to enclose the maximum area?

b. What is the maximum area that can be enclosed?

25. $x = -1$; Vertex: $(-1, -1)$; x-int: None; y-int: $(0, -2)$

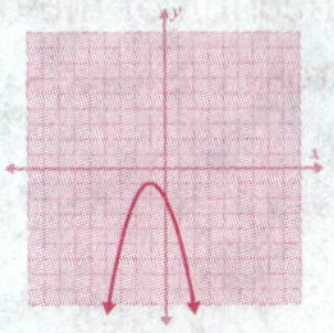

26. $x = -1$; Vertex: $(-1, -8)$; x-int: $(-3, 0)$, $(1, 0)$; y-int: $(0, -6)$

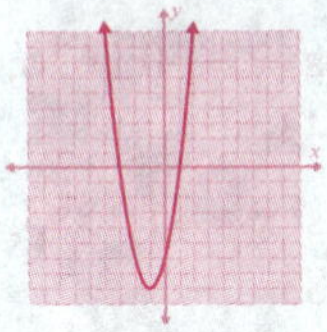

27. a. No **b.** Yes

c. One function is a reflection of the other across the x-axis.

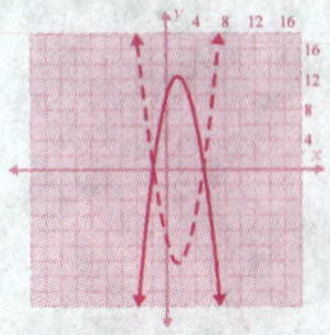

28. a. No **b.** Yes

c. One function is a reflection of the other across the x-axis.

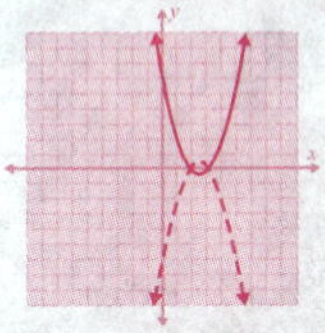

29. a. No **b.** Yes

c. One function is a reflection of the other across the x-axis.

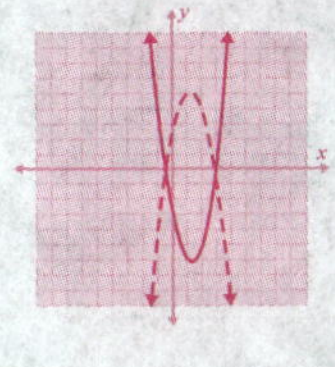

54. A contractor is to build a six-foot-high brick wall to enclose a rectangular garden. The wall will be on three sides of the rectangle while the fourth side is a building. The owner wants to enclose the maximum area but only wants to pay for 150 feet of wall. What dimensions should the contractor make the garden?

Writing & Thinking

55. Discuss the following features of the general quadratic function $y = ax^2 + bx + c$.

a. What type of curve is its graph?

b. What is the value of x at its vertex?

c. What is the equation of the line of symmetry?

d. Does the graph always cross the x-axis? Explain.

56. Discuss the discriminant of the general quadratic equation $ax^2 + bx + c = 0$ and how the value of the discriminant is related to the graph of the corresponding quadratic function $y = ax^2 + bx + c$.

30. **a.** No **b.** Yes

c. One function is a reflection of the other across the x-axis.

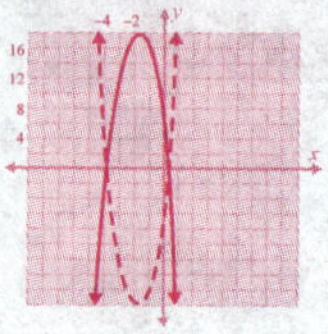

31. Zeros: $x \approx -0.7321, 2.7321$

32. Zeros: $x \approx -0.7676, 0.4343$

33. Zeros: $x \approx -1.1583, 2.1583$

34. Zeros: $x \approx -3.8284, 1.8284$

35. No real zeros

36. No real zeros

37.

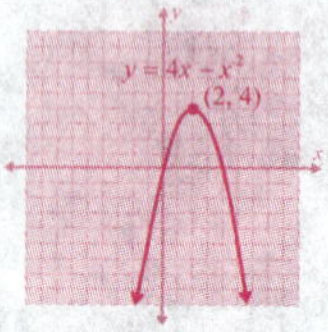

38.

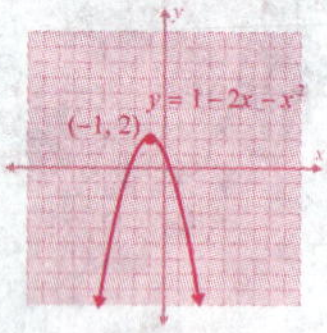

39.

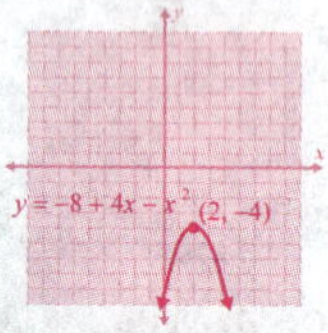

40.

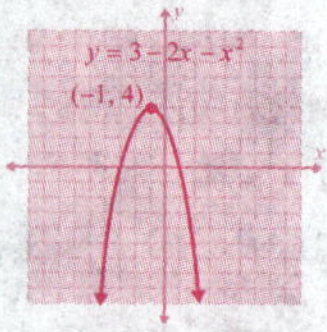

41.

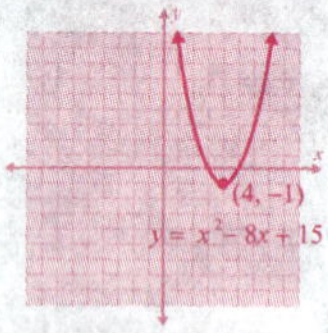

42.

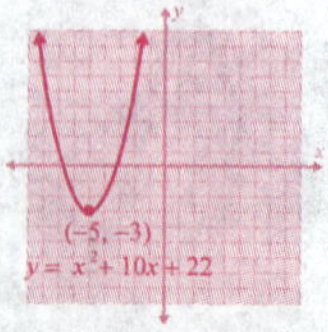

43.

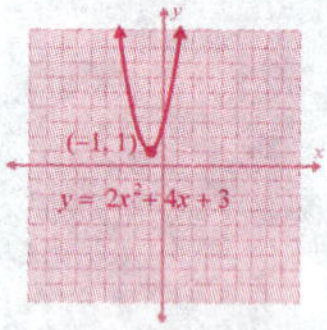

44.

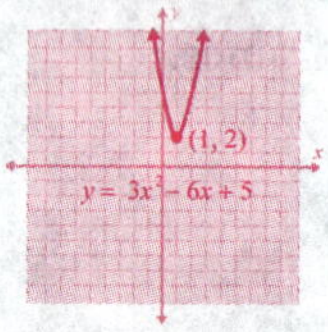

Applications

45. **a.** 3.5 s **b.** 196 ft

46. **a.** 3.25 s **b.** 169 ft

47. **a.** 5 s **b.** 420 ft

48. **a.** 4 s **b.** 288 ft

49. **a.** \$50 **b.** \$2500

50. **a.** \$30 **b.** \$1800

51. \$20

52. \$12.50

53. **a.** Two sides are 150 yards and one side is 300 yards

b. 45,000 yd²

54. Two sides are 37.5 feet and one side is 75 feet

Writing & Thinking

55. **a.** Parabola

b. $x = -\dfrac{b}{2a}$

c. $x = -\dfrac{b}{2a}$

d. No. A graph can be entirely above or below the x-axis.

56. If discriminant > 0, then graph crosses x-axis twice; if discriminant < 0, then graph is entirely above or entirely below the x-axis; if discriminant $= 0$, then graph touches the x-axis at one point (is tangent to).

16.8 Solving Polynomial and Rational Inequalities

Objectives

A. Solve quadratic inequalities.

B. Solve rational inequalities.

C. Use a graphing calculator as an aid in solving quadratic inequalities.

In this section, we will develop algebraic and graphical techniques for solving inequalities with an emphasis on **quadratic inequalities** and **rational inequalities**. For example, we will see that inequalities such as

$$x^2+3x+2\geq 0,\quad x^2-2x>8,\quad x^3+4x^2-5x<0,\quad \text{and}\quad \frac{x+3}{x-2}>0$$

can be solved by factoring or using the quadratic formula and then analyzing the graphs of the corresponding functions. The solution sets to these inequalities consist of intervals on the real number line.

Teaching Note

Many common situations routinely and naturally use the inequality concept. You might want to remind students that common problems dealing with budgets, capacities, sales quotas, profit, fluid pressure, memberships, etc., are more likely to be expressed in terms of "at least," "at most," "not greater than," "not less than," "minimum," and "maximum" than as discrete numbers.

A Quadratic Inequalities

The technique of factoring to solve inequalities is based on the simple idea that for values of x on either side of a number a, the sign for an expression of the form $(x-a)$ changes.

If $x>a$, then $(x-a)$ is positive.

If $x<a$, then $(x-a)$ is negative.

Graphically, we have the following.

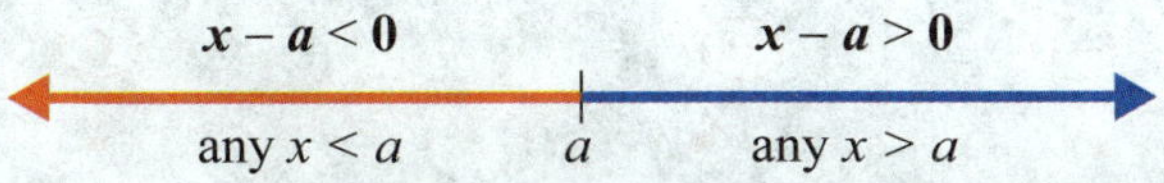

To solve a polynomial inequality algebraically, rearrange terms so that one side of the inequality is equal to 0 and then factor the polynomial on the other side. Locate the points where each factor is 0, and then analyze the sign of the polynomial in the intervals on each side of these points by testing a point in each interval.

Solving a Polynomial Inequality Algebraically

1. Arrange the terms so that one side of the inequality is 0.
2. Factor the polynomial expression, if possible, and find the points where each factor is 0. (Use the quadratic formula, if necessary.)
3. Mark each of these points on a number line. These are the interval **endpoints**.
4. Test one point from each interval to determine the sign of the polynomial expression for all points in that interval.
5. The solution consists of those intervals where the test points satisfy the original inequality.
6. Use brackets for endpoints that are included in the solution and parentheses for endpoints that are not.

PROCEDURE

The following examples illustrate this procedure.

1. Solve the inequality.

$x^2 + 3x < 10$

Example 1 Solving Polynomial Inequalities by Factoring

Solve the inequality by factoring and using a number line. Then, graph the solution set on a number line.

$$x^2 - 2x > 8$$

Solution

$$x^2 - 2x > 8$$

$$x^2 - 2x - 8 > 0 \quad \text{Add } -8 \text{ to both sides so that one side is 0.}$$

$$(x+2)(x-4) > 0 \quad \text{Factor.}$$

Set each factor equal to 0 to locate the interval endpoints.

$$x + 2 = 0 \qquad x - 4 = 0$$

$$x = -2 \qquad x = 4$$

We want to determine where the product is positive. Test one point from each of the intervals to determine if the product is positive or negative in that interval.

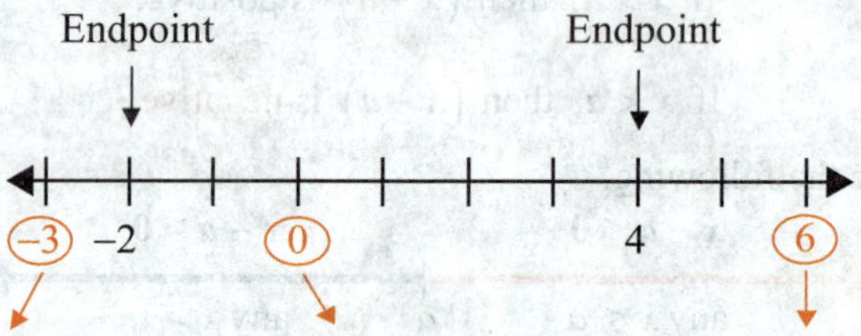

Test $x = -3$:

$(-3+2)(-3-4)$

$= (-1)(-7) = 7 > 0$

This means that $(x+2)(x-4) > 0$ if $x < -2$.

Test $x = 0$:

$(0+2)(0-4)$

$= (2)(-4) = -8 < 0$

This means that $(x+2)(x-4) < 0$ if $-2 < x < 4$.

Test $x = 6$:

$(6+2)(6-4)$

$= (8)(2) = 16 > 0$

This means that $(x+2)(x-4) > 0$ if $x > 4$.

Graphically, the solution set is shown here.

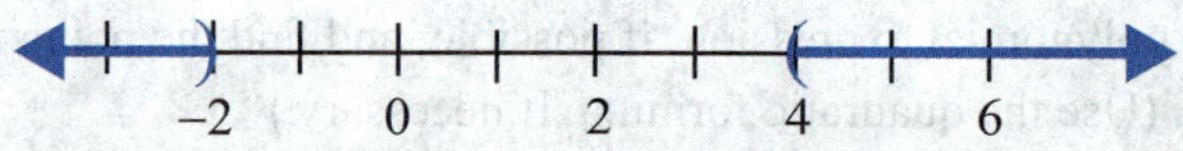

The solution set can be written in two ways.

algebraic notation	or	interval notation
$x < -2$ or $x > 4$		$(-\infty, -2) \cup (4, \infty)$

Now work margin exercise 1.

> **Math Tip**
>
> Consider using the following strategies when choosing test points to simplify the calculations:
>
> - If possible, always use 0 as a point.
> - If possible, choose integers values.
> - Don't use the values of the interval endpoints. (If you accidentally use one of these values, it will not tell you any information about the value of the interval.)

2. Solve the inequality.

$2x^2 + 7x \geq 4$

Example 2 Solving Polynomial Inequalities by Factoring

Solve the inequality by factoring and using a number line. Then, graph the solution set on a number line.

$$2x^2 + 15 \leq 13x$$

Solution

$$2x^2 + 15 \le 13x$$
$$2x^2 - 13x + 15 \le 0$$
$$(2x-3)(x-5) \le 0$$

Set each factor equal to 0 to locate the interval endpoints.

$$2x - 3 = 0 \qquad x - 5 = 0$$
$$x = \frac{3}{2} \qquad x = 5$$

We want to determine where the product is negative or zero. Test one point from each of the intervals to determine if the product is positive or negative in that interval.

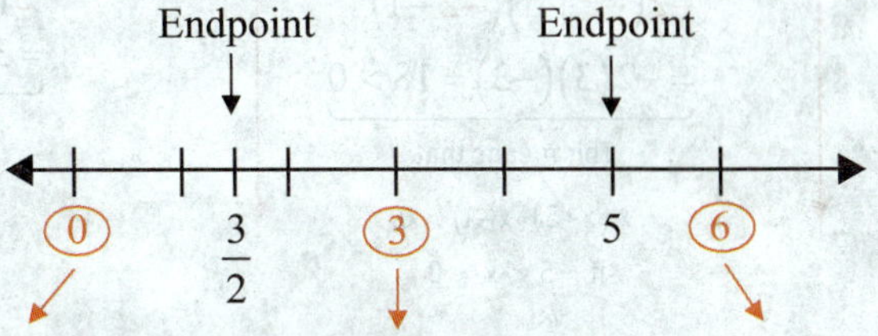

Test $x = 0$:
$$(2 \cdot 0 - 3)(0 - 5) = (-3)(-5) = 15 > 0$$
This means that $(2x-3)(x-5) > 0$ if $x < \frac{3}{2}$.

Test $x = 3$:
$$(2 \cdot 3 - 3)(3 - 5) = (3)(-2) = -6 < 0$$
This means that $(2x-3)(x-5) < 0$ if $\frac{3}{2} < x < 5$.

Test $x = 6$:
$$(2 \cdot 6 - 3)(6 - 5) = (9)(1) = 9 > 0$$
This means that $(2x-3)(x-5) > 0$ if $x > 5$.

The solution set includes both endpoints since the inequality ($\le$) includes 0.

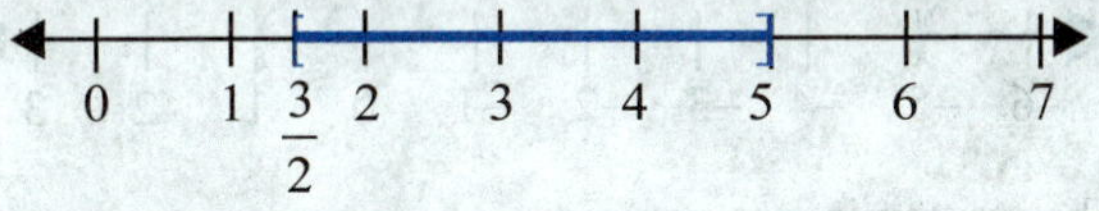

The solution set can be written in two ways.

algebraic notation	or	interval notation
$\frac{3}{2} \le x \le 5$		$\left[\frac{3}{2}, 5\right]$

Now work margin exercise 2.

Example 3 Solving Polynomial Inequalities by Factoring

Solve the inequality by factoring and using a number line. Then, graph the solution set on a number line.

$$x^3 + 4x^2 - 5x < 0$$

Solution

$$x^3 + 4x^2 - 5x < 0$$
$$x(x^2 + 4x - 5) < 0$$
$$x(x+5)(x-1) < 0$$

3. Solve the inequality.

$$x^3 - 6x^2 - 7x > 0$$

Set each factor equal to 0 to locate the interval endpoints.

$$x = 0 \qquad x + 5 = 0 \qquad x - 1 = 0$$
$$x = -5 \qquad x = 1$$

We want to determine where the product is negative. Test one point from each of the intervals to determine if the product is positive or negative in that interval.

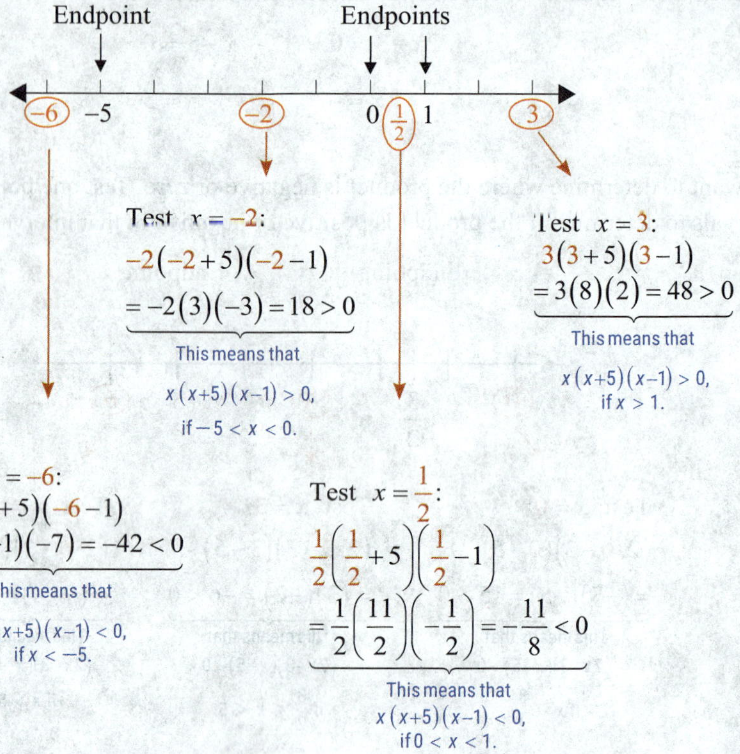

Graphically, the solution set is shown here.

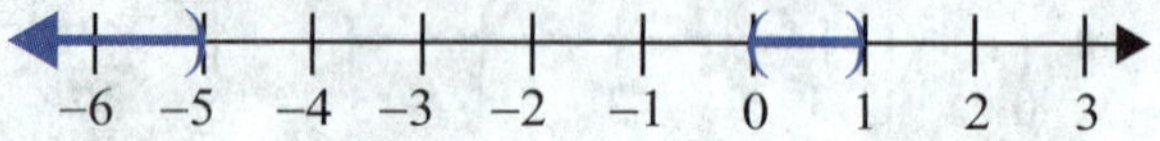

The solution set can be written in two ways.

algebraic notation	or	interval notation
$x < -5$ or $0 < x < 1$		$(-\infty, -5) \cup (0, 1)$

Now work margin exercise 3.

4. Solve the inequality.

$x^2 - x - 5 < 0$

Example 4 Solving Polynomial Inequalities Using the Quadratic Formula

Solve the inequality by using the quadratic formula and a number line. Then, graph the solution set on a number line.

$$x^2 - 2x - 1 > 0$$

Solution

The quadratic expression $x^2 - 2x - 1$ cannot be factored with integer coefficients. Use the quadratic formula and find the roots of the equation $x^2 - 2x - 1 = 0$. Then, use these roots as endpoints for the intervals. The test points can themselves be integers.

$$x^2 - 2x - 1 = 0$$

$$x = \frac{2 \pm \sqrt{(-2)^2 - 4(1)(-1)}}{2(1)} \quad \text{Use the quadratic formula.}$$

$$x = \frac{2 \pm \sqrt{4+4}}{2}$$

$$x = \frac{2 \pm 2\sqrt{2}}{2} = 1 \pm \sqrt{2}$$

The endpoints are $x = 1 - \sqrt{2}$ and $x = 1 + \sqrt{2}$.

We want to determine where the product is positive. Test one point from each interval.

Note: With a calculator, you can determine that $1 - \sqrt{2} \approx -0.414$ and $1 + \sqrt{2} \approx 2.414$.

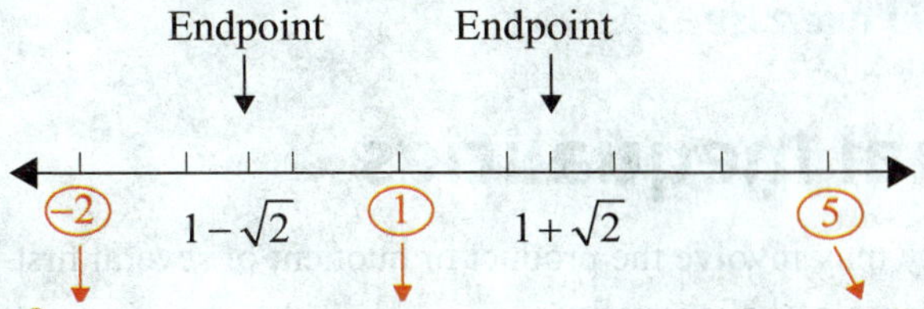

Test $x = -2$:
$(-2)^2 - 2(-2) - 1$
$= 4 + 4 - 1 = 7 > 0$
This means that if $x < 1 - \sqrt{2}$, then $x^2 - 2x - 1 > 0$.

Test $x = 1$:
$1^2 - 2(1) - 1$
$= 1 - 2 - 1 = -2 < 0$
This means that if $1 - \sqrt{2} < x < 1 + \sqrt{2}$, then $x^2 - 2x - 1 < 0$.

Test $x = 5$:
$5^2 - 2(5) - 1$
$= 25 - 10 - 1 = 14 > 0$
This means that if $x > 1 + \sqrt{2}$, then $x^2 - 2x - 1 > 0$.

Graphically, the solution set is shown here.

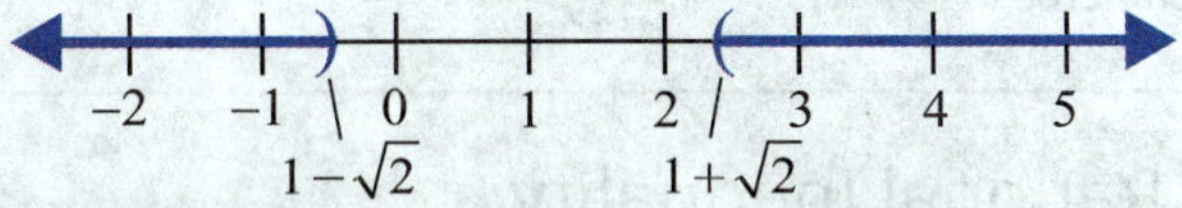

The solution set can be written in two ways.

algebraic notation	or	interval notation
$x < 1 - \sqrt{2}$ or $x > 1 + \sqrt{2}$		$(-\infty, 1 - \sqrt{2}) \cup (1 + \sqrt{2}, \infty)$

Now work margin exercise 4.

Example 5 Solving Polynomial Inequalities Using the Quadratic Formula

Solve the inequality by using the quadratic formula and a number line. Then graph the solution set on a number line.

$$x^2 - 2x + 13 > 0$$

Solution

To find where $x^2 - 2x + 13 > 0$, use the quadratic formula.

$$x = \frac{2 \pm \sqrt{(-2)^2 - 4(1)(13)}}{2(1)} = \frac{2 \pm \sqrt{-48}}{2}$$

$$= \frac{2 \pm 4i\sqrt{3}}{2} = 1 \pm 2i\sqrt{3}$$

5. Solve the inequality.

$$x^2 + 3x + 6 < 0$$

Since these values are nonreal, the polynomial is either always positive or always negative for all real values of x. Therefore, we only need to test one point. If that point satisfies the inequality, then the solution set is all real numbers. If it does not, then there is no solution. In this example, we test $x = 0$ because the polynomial is easy to evaluate for $x = 0$.

$$(0)^2 - 2(0) + 13 = 13 > 0$$

Since the real number 0 satisfies the inequality, the solution set is all real numbers. In interval notation we write $(-\infty, \infty)$, and in algebraic notation we write $\mathbb{R}$. Graphically, the solution set is shown here.

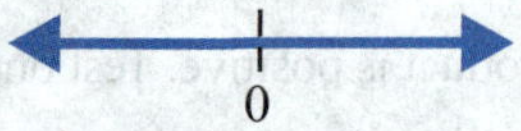

Now work margin exercise 5.

B Rational Inequalities

A rational inequality may involve the product or quotient of several first-degree expressions. For example, the inequality

$$\frac{x+3}{x-2} > 0$$

involves the two first-degree expressions $x + 3$ and $x - 2$.

Inequalities of this form can be solved using a similar procedure to that used for quadratic inequalities outlined here.

Solving a Rational Inequality

1. Simplify the inequality so that one side is 0 and the other side has a single fraction with both the numerator and denominator in factored form.
2. Find the points that cause the factors in the numerator or in the denominator to be 0.
3. Mark each of these points on a number line. These are the interval endpoints.
4. Test one point from each interval to determine the sign of the rational expression for all points in that interval.
5. The solution consists of those intervals where the test points satisfy the original inequality.
6. Use brackets for endpoints that are included in the solution and parentheses for endpoints that are not.

PROCEDURE

6. Solve the inequality.

$$\frac{x-5}{x+1} > 0$$

Example 6 Solving Rational Inequalities

Solve the rational inequality and graph the solution set on a number line.

$$\frac{x+3}{x-2} > 0$$

Solution

Set each factor in the numerator and the denominator equal to 0 to locate the interval endpoints.

$$x+3=0 \qquad x-2=0$$
$$x=-3 \qquad x=2$$

We want to determine where the quotient is positive. Test one point from each of the intervals to determine if the quotient is positive or negative in that interval.

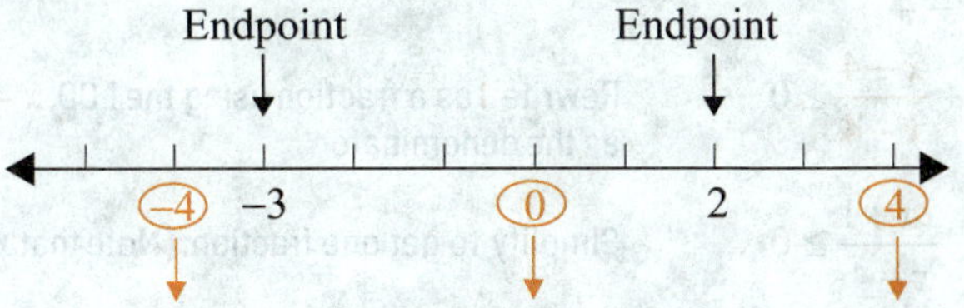

Test $x=-4$: $\frac{-4+3}{-4-2}=\frac{-1}{-6}=\frac{1}{6}>0$ This means that $\frac{x+3}{x-2}>0$ if $x<-3$.

Test $x=4$: $\frac{4+3}{4-2}=\frac{7}{2}>0$ This means that $\frac{x+3}{x-2}>0$ if $x>2$.

Test $x=0$: $\frac{0+3}{0-2}=-\frac{3}{2}<0$ This means that $\frac{x+3}{x-2}<0$ if $-3<x<2$.

Graphically, the solution set is shown here.

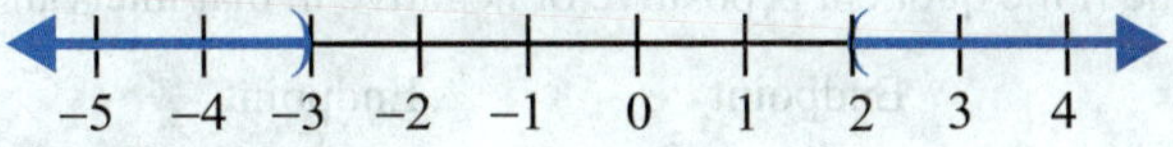

The solution set can be written in two ways.

algebraic notation	or	interval notation
$x<-3$ or $x>2$		$(-\infty,-3)\cup(2,\infty)$

Now work margin exercise 6.

Example 7 **Solving Rational Inequalities**

Solve the rational inequality and graph the solution set on a number line.

$$\frac{x+3}{x-2}<0$$

Solution

For this inequality, we want to know where the quotient is negative. Using the graph and test points from Example 5, we know that

$$\frac{x+3}{x-2}<0 \text{ if } -3<x<2.$$

Graphically, the solution set is shown here.

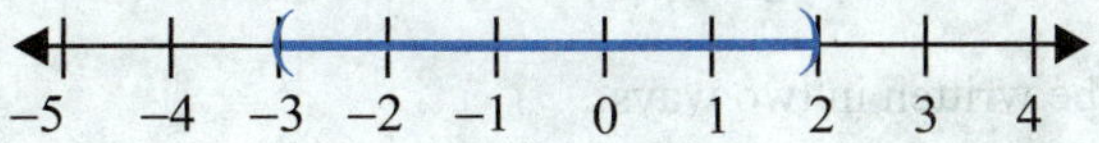

The solution set can be written in two ways.

algebraic notation	or	interval notation
$-3<x<2$		$(-3, 2)$

Now work margin exercise 7.

Math Tip

When solving a rational inequality, you need to be aware of when the denominator may equal zero and be sure not to include this endpoint in the interval.

When solving an inequality that uses the strict inequalities < and >, we know it will not be possible to include a value that would lead to a 0 denominator. However, if the inequality uses ≤ or ≥, it is possible that one of our endpoints should not be included. We will need to use a parenthesis instead of a bracket at the particular endpoint in interval notation to reflect that it is not included.

7. Solve the inequality.

$$\frac{x+4}{x+6}\le 0$$

16.8 Exercises

Concept Check

1. graphs
2. sign, changes
3. endpoints
4. sign
5. bracket, parenthesis
6. denominator
7. False; The goal is to get 0 on one side of the inequality and factor the other side.
8. True

8. Solve the inequality.

$$\frac{x+3}{x-5} \le -2$$

> **Note**
>
> Notice that in the first step, we do **not** multiply by the denominator $x - 4$. The reason is that the variable expression is positive for some values of x and negative for other values of x. Therefore, if we did multiply by $x - 4$, we would not be able to determine whether the inequality should stay as $\ge$ or be reversed to $\le$.

9. False; The solution consists of all intervals where the test points satisfy the original inequality.

10. True

Practice

1. $(-2, 6)$

2. $(-\infty, -4) \cup (2, \infty)$

3. $\left(-\infty, \frac{2}{3}\right) \cup (5, \infty)$

4. $\left[-1, -\frac{1}{4}\right]$

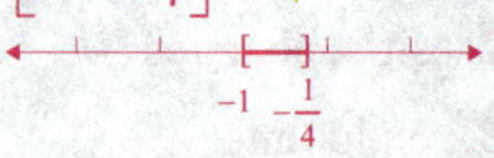

> **Note**
>
> See Section 10.5 to review graphing functions and finding zeros (x-intercepts) using a TI-84 Plus graphing calculator.

Example 8 Solving Rational Inequalities

Solve the following rational inequality.

$$\frac{x+5}{x-4} \ge -1$$

Solution

$$\frac{x+5}{x-4} + 1 \ge 0 \qquad \text{One side must be 0.}$$

$$\frac{x+5}{x-4} + \frac{x-4}{x-4} \ge 0 \qquad \text{Rewrite 1 as a fraction using the LCD, } x-4\text{, as the denominator.}$$

$$\frac{2x+1}{x-4} \ge 0 \qquad \text{Simplify to get one fraction. (Note that } x \ne 4.)$$

Set each linear expression equal to 0 to find the interval endpoints.

$$2x+1=0 \qquad x-4=0$$
$$x=-\frac{1}{2} \qquad x=4$$

We want to determine where the quotient is positive or zero. Test one point from each of the intervals to determine if the quotient is positive or negative in that interval.

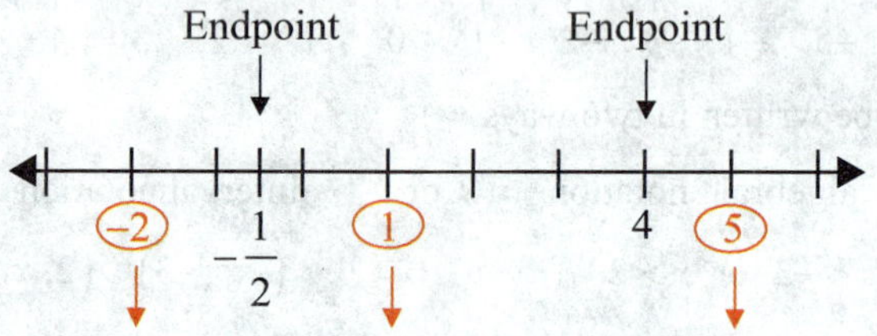

Test $x = -2$:

$$\frac{2(-2)+1}{-2-4} = \frac{-3}{-6} = \frac{1}{2} > 0$$

This means that $\frac{2x+1}{x-4} > 0$ if $x < -\frac{1}{2}$.

Test $x = 1$:

$$\frac{2(1)+1}{1-4} = \frac{3}{-3} = -1 < 0$$

This means that $\frac{2x+1}{x-4} < 0$ if $-\frac{1}{2} < x < 4$.

Test $x = 5$:

$$\frac{2(5)+1}{5-4} = \frac{11}{1} = 11 > 0$$

This means that $\frac{2x+1}{x-4} > 0$ if $x > 4$.

The solution set includes the endpoint $-\frac{1}{2}$ since the inequality ($\ge$) includes 0. However, the endpoint 4 is not included in the solution set because the rational inequality is undefined (the denominator equals 0) when $x = 4$.

Graphically, the solution set is shown here.

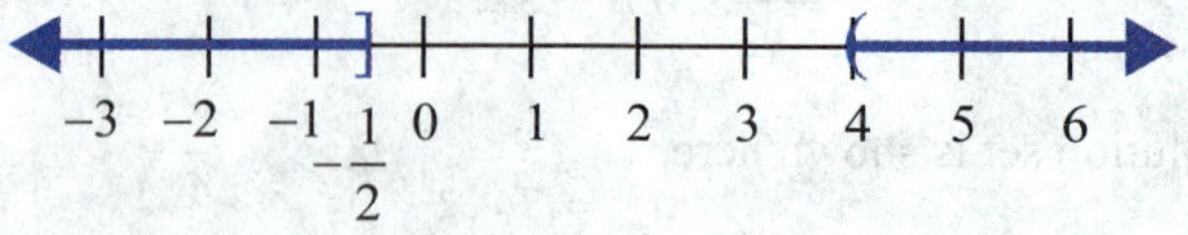

The solution set can be written in two ways.

algebraic notation	or	interval notation
$x \le -\frac{1}{2}$ or $x > 4$		$\left(-\infty, -\frac{1}{2}\right] \cup (4, \infty)$

Now work margin exercise 8.

C Solving Quadratic Inequalities Using a Graphing Calculator

To solve quadratic (or other polynomial) inequalities with a TI-84 Plus graphing calculator by graphing the related polynomial function, look for the intervals of x for which the graph of the function is above the x-axis (the y-values will be positive) and where it is below the x-axis (the y-values will be negative). The y-values represent the values of the related polynomial function.

Solving a Polynomial Inequality Using a Graphing Calculator

1. Arrange the terms so that one side of the inequality is 0.
2. Form a function by letting $y =$ [*the polynomial*], and graph the function. Be sure to set the window so that all of the zeros are easily seen. This may be difficult if large numbers are involved.
3. Use the CALC key ([2nd] [TRACE]) and select `zero` to find (or approximate) the real zeros of the function, if there are any.
   - **a.** The values of x for which the y-values are above the x-axis satisfy $y > 0$.
   - **b.** The values of x for which the y-values are below the x-axis satisfy $y < 0$.
4. Endpoints of intervals are included if the inequality includes 0, as in $y \geq 0$ or $y \leq 0$.

PROCEDURE

Example 9 Solving a Quadratic Inequality Using a Graphing Calculator

Use a graphing calculator to solve the quadratic inequality. Graph the solution set on a real number line.

$$x^2 - 2x > 8$$

Solution

This inequality was solved algebraically in Example 1. We repeat the solution here using a graphing calculator to show how the two methods are related.

Manipulate the inequality so that one side is 0.

$$x^2 - 2x > 8$$

$$x^2 - 2x - 8 > 0$$

Press [Y=] and enter the function.

$$y = x^2 - 2x - 8$$

5. $(-\infty, -7] \cup \left[\frac{5}{2}, \infty\right)$

6. $\left[\frac{3}{5}, 3\right]$

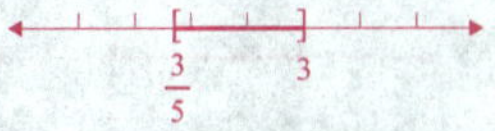

7. $\left[-2, -\frac{1}{3}\right]$

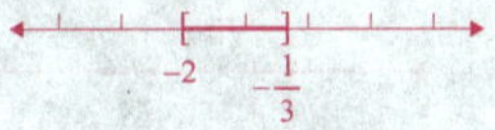

8. $\left(-\infty, \frac{8}{3}\right) \cup (4, \infty)$

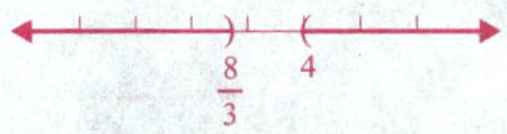

9. $\left(-\infty, -\frac{4}{3}\right) \cup (0, 5)$

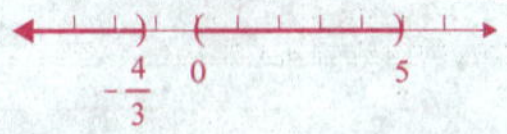

10. $(-\infty, -4) \cup \left(-\frac{5}{2}, 1\right)$

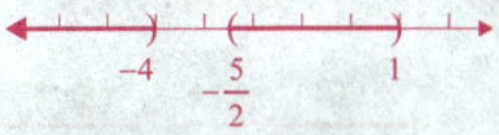

9. Use a graphing calculator to solve the inequality.

$x^2 > 3x + 4$

11. $\{-2\}$

12. $(-\infty, -2) \cup \left(\frac{6}{5}, \infty\right)$

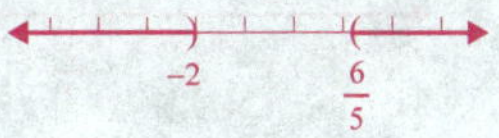

13. $\left(-\infty, -\frac{5}{2}\right) \cup (3, \infty)$

14. $\left(-\infty, -\frac{2}{3}\right) \cup \left(\frac{1}{2}, \infty\right)$

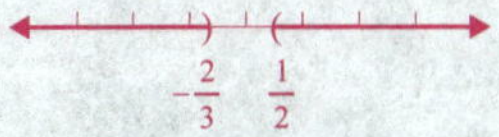

Press GRAPH to graph the function.

The graph will appear as illustrated here. (In this case, the graph is a parabola.)

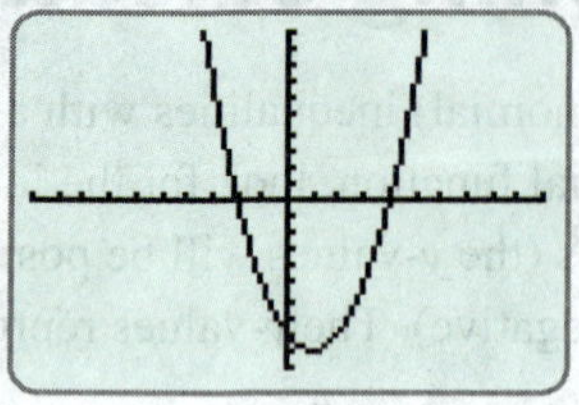

Find the zeros (or estimate the zeros) as follows:

Step 1: Press CALC (2nd TRACE).

Step 2: Press or choose `zero`.

Step 3: Follow the directions for moving the cursor to `Left Bound?`, `Right Bound?`, and `Guess?` for each point. (Press ENTER each time.)

Note: If there are no real zeros, then the graph will be entirely above or entirely below the x-axis. The solution is then dependent on the nature of the inequality. It will be either the empty set (no solution) or the entire set of real numbers $(-\infty, \infty)$.

In this case, the zeros are $x=-2$ and $x=4$ (just as we found in Example 1). Because we want to know where $y>0$, we look at the graph and choose the intervals for x where the curve is above the x-axis. (**Note:** Endpoints are not included because the inequality does not include 0.)

Thus, the solution set consists of the union of two intervals: $(-\infty,-2)\cup(4,\infty)$.

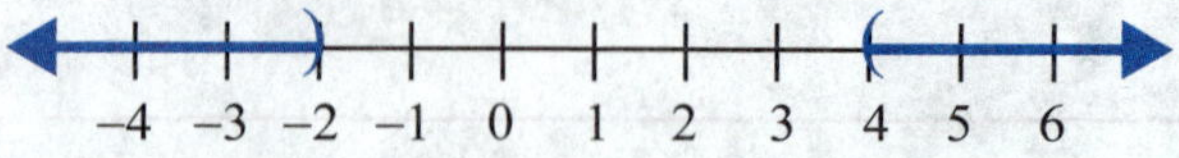

Now work margin exercise 9.

Example 10 Solving a Quadratic Inequality Using a Graphing Calculator

Use a graphing calculator to solve the quadratic inequality. Graph the solution set on a real number line.

$$2x^2+3x-10\le 0$$

Solution

Press Y= and enter the function.

$$y=2x^2+3x-10$$

Press GRAPH to graph the function.

The graph will appear as illustrated here. (In this case, the graph is a parabola.)

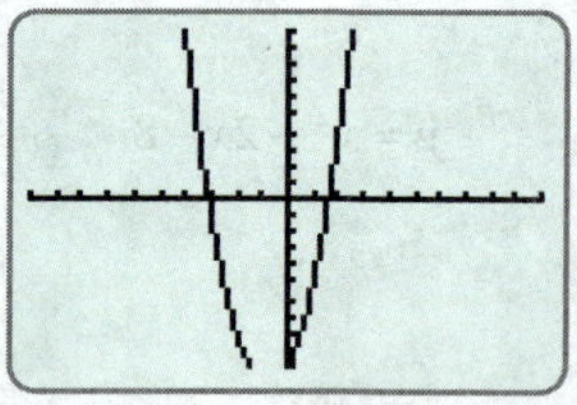

10. Use a graphing calculator to solve the inequality.

$2x^2+x-5<0$

15. $\left(-\frac{1}{4},\frac{3}{2}\right)$

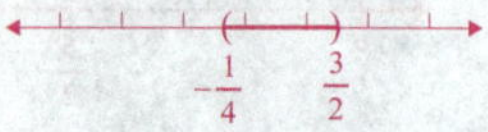

16. $\left(-2,\frac{5}{2}\right)$

17. $\left(-\infty,\frac{1}{2}\right]\cup[2,\infty)$

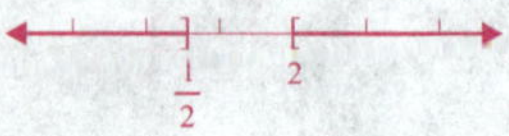

18. $\left(-\frac{3}{5},2\right)$

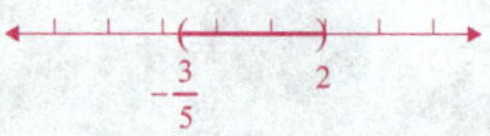

19. $\left(-\frac{2}{3},-\frac{1}{2}\right)$

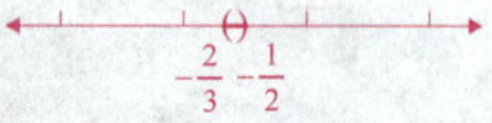

20. $\left(-\infty,\frac{1}{3}\right]\cup[3,\infty)$

21. $\left(-\infty,\frac{5}{2}\right)\cup\left(\frac{5}{2},\infty\right)$

22. $\left[-\frac{2}{3},\frac{7}{5}\right]$

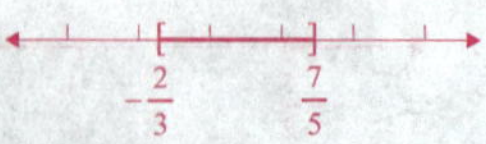

23. $\left[-\frac{5}{2},\frac{7}{4}\right]$

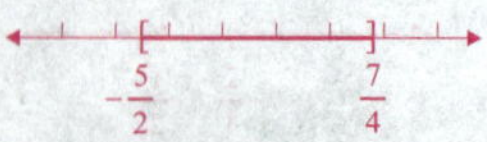

Find the zeros (or estimate the zeros) as follows:

Step 1: Press CALC ([2nd] [TRACE]).

Step 2: Press or choose `zero`.

Step 3: Follow the directions for moving the cursor to `Left Bound?`, `Right Bound?`, and `Guess?` for each point. (Press [ENTER] each time.)

In this case, the zeros are estimates: $x \approx -3.1085$ and $x \approx 1.6085$.

Because we want to know where $y \le 0$, we look at the graph and find the intervals for x where the curve is below the x-axis and include the endpoints because 0 is included in the inequality.

Thus, the solution set is the closed interval $[-3.1085, 1.6085]$.

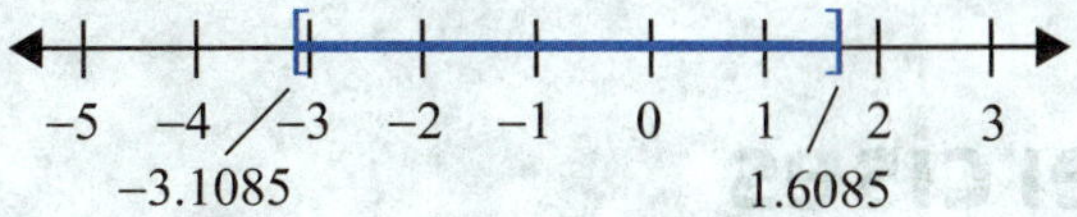

Now work margin exercise 10.

Example 11 Solving a Higher-Degree Polynomial Inequality Using a Graphing Calculator

Solve the third-degree polynomial inequality using a graphing calculator.

$$x^3 + 2x^2 - 11x - 12 > 0$$

Solution

Press [Y=] and enter the function.

$$y = x^3 + 2x^2 - 11x - 12$$

Press [GRAPH] to graph the function.

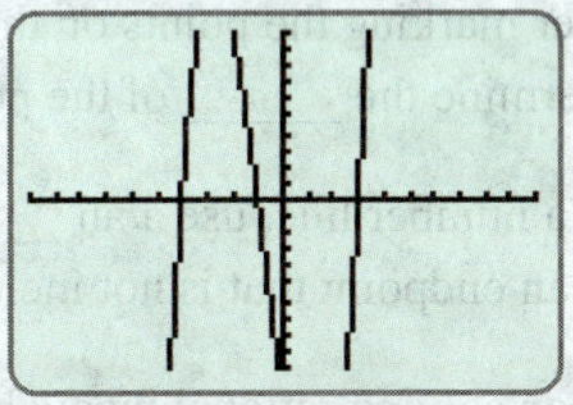

The graph will appear as illustrated here. (In this case, the graph is not a parabola.)

Find the zeros (or estimate the zeros) as follows:

Step 1: Press CALC ([2nd] [TRACE]).

Step 2: Press or choose `zero`.

Step 3: Follow the directions for moving the cursor to `Left Bound?`, `Right Bound?`, and `Guess?` for each point. (Press [ENTER] each time.)

In this case, there are three zeros: $x = -4$, $x = -1$, and $x = 3$.

Because we want to know where $y > 0$, we look at the graph and find the intervals for x where the curve is above the x-axis. Do not include the endpoints because 0 is not included in the inequality.

---

24. $(-7, 0) \cup (1, \infty)$

25. $(-1, 0) \cup (3, \infty)$

26. $(-\infty, 0)$

27. $(0, 1) \cup (4, \infty)$

28. $[0, 1] \cup [3, \infty)$

11. Solve the third-degree polynomial inequality using a graphing calculator.

$$x^3 + x^2 - 10x + 8 < 0$$

29. $(-\infty, -1) \cup (4, \infty)$

30. $(-3, 2)$

31. $(-\infty, -2) \cup (-1, 1) \cup (2, \infty)$

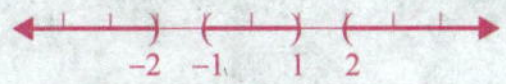

32. $(-4, -3) \cup (3, 4)$

33. $[-3, -2] \cup [2, 3]$

34. $(-\infty, -4] \cup [4, \infty)$

35. $(-\infty, -4] \cup [2, \infty)$

36. $\left(-1, \dfrac{5}{3}\right)$

37. $\left(-\dfrac{2}{3}, 2\right)$

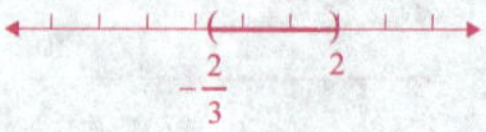

38. $(-\infty, 0) \cup (1, \infty)$

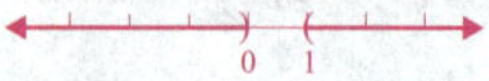

39. $(-\infty, -1-\sqrt{5}) \cup (-1+\sqrt{5}, \infty)$

40. $(4-\sqrt{2}, 4+\sqrt{2})$

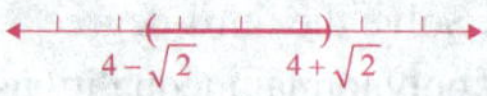

41. $(-\infty, -3-\sqrt{2}] \cup [-3+\sqrt{2}, \infty)$

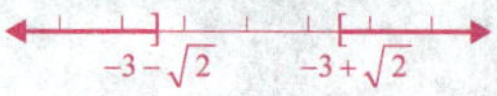

42. $\left(\dfrac{-2-\sqrt{10}}{2}, \dfrac{-2+\sqrt{10}}{2}\right)$

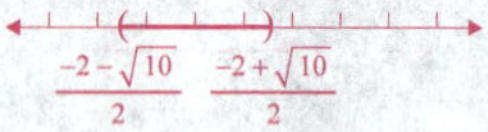

43. $\left(\dfrac{-5-\sqrt{13}}{6}, \dfrac{-5+\sqrt{13}}{6}\right)$

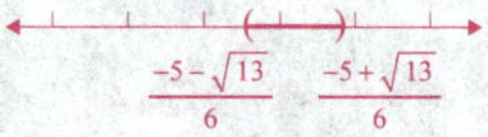

44. $\left(-\infty, -\dfrac{5}{3}\right] \cup [-1, \infty)$

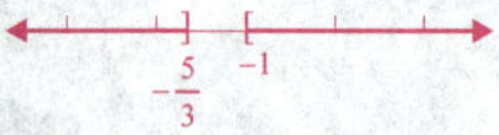

45. $\left(-\infty, -\dfrac{1}{2}\right] \cup [0, 4]$

Thus, the solution set is the union of two intervals: $(-4, -1) \cup (3, \infty)$.

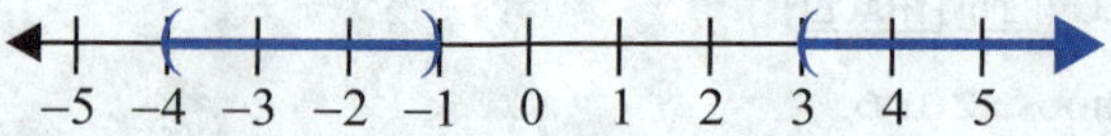

Now work margin exercise 11.

Margin Exercise Answers

1. $(-5, 2)$ **2.** $(-\infty, -4] \cup \left[\dfrac{1}{2}, \infty\right)$ **3.** $(-1, 0) \cup (7, \infty)$ **4.** $\left(\dfrac{1-\sqrt{21}}{2}, \dfrac{1+\sqrt{21}}{2}\right)$

5. $\varnothing$ (No Solution) **6.** $(-\infty, -1) \cup (5, \infty)$ **7.** $(-6, -4]$ **8.** $\left[\dfrac{7}{3}, 5\right)$

9. $(-\infty, -1) \cup (4, \infty)$ **10.** $(-1.8508, 1.3508)$ **11.** $(-\infty, -4) \cup (1, 2)$

16.8 Exercises

Concept Check

Fill-in-the-Blank. Complete each sentence using information found in this section.

1. Quadratic and rational inequalities can be solved by factoring or using the quadratic formula and then analyzing the ______ of the corresponding functions.
2. The technique of factoring to solve inequalities is based on the idea that for values of x on either side of a number a, the ______ for an expression of the form $(x-a)$ ______.
3. When solving a polynomial inequality algebraically, mark the points where each factor is 0 on the number line. These are the interval ______.
4. After marking the points on the number line, test one point from each interval to determine the ______ of the polynomial expression for all points in that interval.
5. On a number line, use a/an ______ for an endpoint that is included and a/an ______ for an endpoint that is not included.
6. When solving rational inequalities, mark the points on a number line where each factor is 0 or causes the ______ to be 0.

True/False. Determine whether each statement is true or false. If a statement is false, explain how it can be changed so the statement will be true. (**Note:** There may be more than one acceptable change.)

7. When solving a polynomial inequality algebraically, the goal is to get the constants on one side of the inequality and to factor the polynomial on the other side.
8. Test points are used to determine which intervals on the number line satisfy the original inequality.
9. The solution of a polynomial inequality is a single interval.
10. If an endpoint causes the denominator of a rational inequality to be 0, it should be marked with a parenthesis.

Practice

Solve the quadratic (and higher degree) inequalities algebraically. Write the answers in interval notation, and then graph each solution set on a number line. (**Note:** You may need to use the quadratic formula to find endpoints of intervals.) See Examples 1 through 5.

1. $(x-6)(x+2)<0$
2. $(x+4)(x-2)>0$
3. $(3x-2)(x-5)>0$
4. $(4x+1)(x+1)\le 0$
5. $(x+7)(2x-5)\ge 0$
6. $(x-3)(5x-3)\le 0$
7. $(3x+1)(x+2)\le 0$
8. $(x-4)(3x-8)>0$
9. $x(3x+4)(x-5)<0$
10. $(x-1)(x+4)(2x+5)<0$
11. $x^2+4x+4\le 0$
12. $5x^2+4x-12>0$
13. $2x^2>x+15$
14. $6x^2+x>2$
15. $8x^2<10x+3$
16. $2x^2<x+10$
17. $2x^2-5x+2\ge 0$
18. $15y^2-21y-18<0$
19. $6y^2+7y<-2$
20. $3x^2+3\ge 10x$
21. $4z^2-20z+25>0$
22. $15x^2-11x-14\le 0$
23. $8x^2+6x\le 35$
24. $7x<6x^2+x^3$
25. $x^3>2x^2+3x$
26. $x^3<6x^2-9x$
27. $x^3>5x^2-4x$
28. $4x^2\le x^3+3x$
29. $(x+2)(x-2)>3x$
30. $(x+4)(x-1)<2x+2$
31. $x^4-5x^2+4>0$
32. $x^4-25x^2+144<0$
33. $y^4-13y^2+36\le 0$
34. $y^4-13y^2-48\ge 0$
35. $(x+1)^2-9\ge 0$
36. $(3x-1)^2-16<0$
37. $(2x-3)(3x+2)-(3x+2)<0$
38. $2(x-1)(x-3)>(x-1)(x-6)$
39. $x^2+2x-4>0$
40. $x^2-8x+14<0$
41. $x^2+6x+7\ge 0$
42. $2x^2+4x-3<0$
43. $3x^2+5x+1<0$
44. $3x^2+8x+5\ge 0$
45. $2x^3\le 7x^2+4x$
46. $2x^2>9x-8$
47. $x^2-2x+2>0$
48. $x^2+3x+3<0$
49. $2x-1>3x^2$
50. $6x-10<x^2$

46. $\left(-\infty, \frac{9-\sqrt{17}}{4}\right) \cup \left(\frac{9+\sqrt{17}}{4}, \infty\right)$

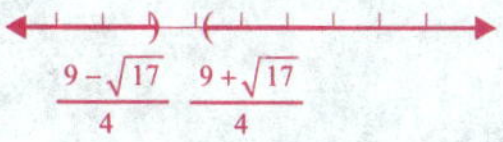

47. $(-\infty, \infty)$

48. $\varnothing$

49. $\varnothing$

50. $(-\infty, \infty)$

51. **a.** $x=\frac{7\pm\sqrt{89}}{2}$

b. $\left(-\infty, \frac{7-\sqrt{89}}{2}\right) \cup \left(\frac{7+\sqrt{89}}{2}, \infty\right)$

c. $\left(\frac{7-\sqrt{89}}{2}, \frac{7+\sqrt{89}}{2}\right)$

52. **a.** $x=-6, 1$

b. $(-\infty, -6)\cup(1, \infty)$

c. $(-6, 1)$

53. **a.** $x=-5, 1$

b. $(-5, 1)$

c. $(-\infty, -5)\cup(1, \infty)$

54. **a.** $x=-1\pm\sqrt{6}$

b. $\left(-1-\sqrt{6}, -1+\sqrt{6}\right)$

c. $\left(-\infty, -1-\sqrt{6}\right) \cup \left(-1+\sqrt{6}, \infty\right)$

55. $(-\infty, -4]\cup[0, \infty)$

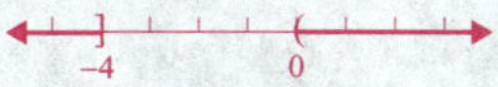

56. $(-\infty, 0]\cup(4, \infty)$

57. $(-\infty, -6)$

58. $(-\infty, -1)$

59. $(-\infty,-9)\cup(-3,\infty)$

60. $\left(-\frac{3}{2},4\right)$

61. $\left(2,\frac{5}{2}\right)$

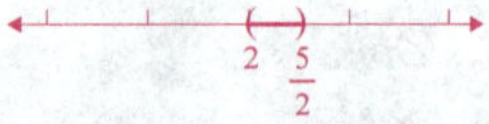

62. $(-\infty,-2)\cup\left[\frac{4}{3},\infty\right)$

63. $(7,\infty)$

64. $(1,\infty)$

65. $\left[\frac{7}{5},4\right)$

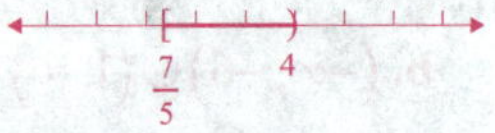

66. $\left(-\infty,\frac{1}{4}\right)\cup\left(\frac{5}{9},\infty\right)$

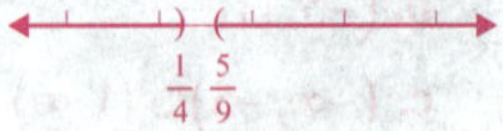

67. $\left(-9,-\frac{4}{3}\right)$

68. $\left(-\frac{28}{3},-5\right)$

69. $(-\infty,-4]\cup[0,3)$

70. $(-3,-1)\cup(2,\infty)$

71. $(-2,0)\cup[5,\infty)$

The graph of a quadratic function is given. Use the information in the graph to solve the related equations and inequalities in parts a. through c.

51. $y=x^2-7x-10$

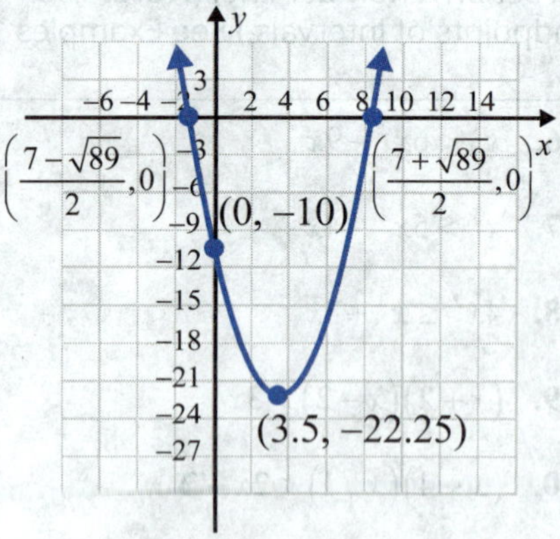

a. $x^2-7x-10=0$

b. $x^2-7x-10>0$

c. $x^2-7x-10<0$

53. $y=-x^2-4x+5$

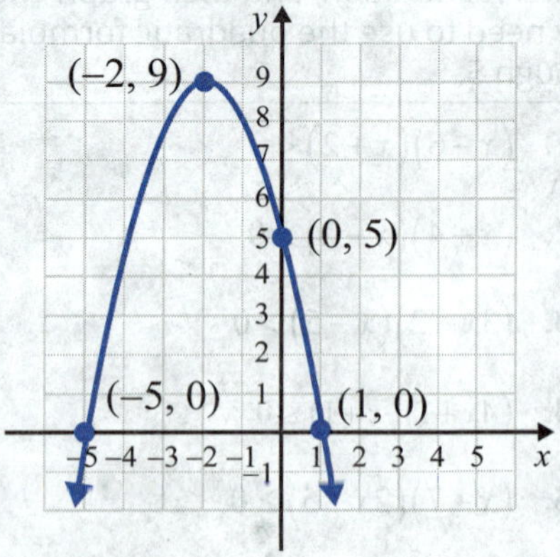

a. $-x^2-4x+5=0$

b. $-x^2-4x+5>0$

c. $-x^2-4x+5<0$

52. $y=x^2+5x-6$

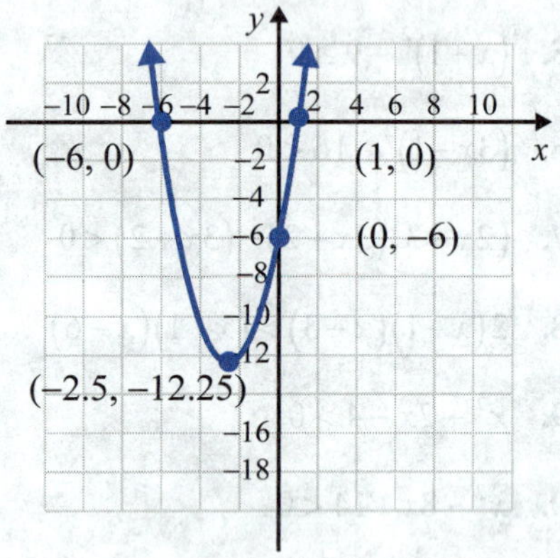

a. $x^2+5x-6=0$

b. $x^2+5x-6>0$

c. $x^2+5x-6<0$

54. $y=-3x^2-6x+15$

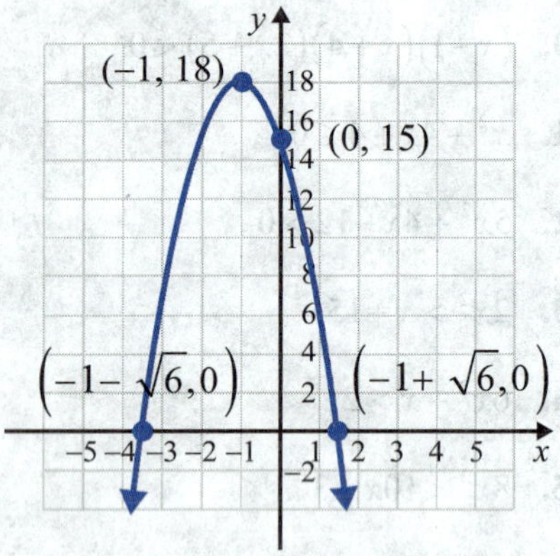

a. $-3x^2-6x+15=0$

b. $-3x^2-6x+15>0$

c. $-3x^2-6x+15<0$

Solve the rational inequalities algebraically. Write the answers in interval notation, and then graph each solution set on a number line. See Examples 6 through 8.

55. $\frac{x+4}{2x}\geq 0$

56. $\frac{x}{x-4}\geq 0$

57. $\frac{x+6}{x^2}<0$

58. $\frac{3x^2}{x+1}<0$

59. $\frac{x+3}{x+9}>0$

60. $\frac{2x+3}{x-4}<0$

61. $\frac{3x-6}{2x-5}<0$

62. $\frac{4-3x}{2x+4}\leq 0$

63. $\frac{x+5}{x-7}\geq 1$

64. $\frac{2x+3}{x-1}>2$

65. $\dfrac{2x+5}{x-4} \le -3$

66. $\dfrac{3x+2}{4x-1} < 3$

67. $\dfrac{5-2x}{3x+4} < -1$

68. $\dfrac{8-x}{x+5} < -4$

69. $\dfrac{x(x+4)}{x-3} \le 0$

70. $\dfrac{(x+3)(x-2)}{x+1} > 0$

71. $\dfrac{x-5}{x(x+2)} \ge 0$

72. $\dfrac{-(x-3)^2}{(x-1)(x-4)} < 0$

Use a graphing calculator to solve the inequalities. Write the answers in interval notation, and then graph each solution set on a number line. (Estimate endpoints, when necessary, to 4 decimal places.)

73. $x^2 > 10$

74. $20 \ge x^2$

75. $x^2 - 2.5x + 6.25 < 0$

76. $x^2 + 2x \ge -1$

77. $x^3 - 9x < 0$

78. $x^3 - 4x^2 + 4x \le 0$

79. $2x^3 - 5x + 4 \ge 0$

80. $x^3 - 4x^2 + 3 < 0$

81. $-x^4 + 6x^2 - 3 > 0$

82. $x^4 - 2x^3 - x^2 - 1 < 0$

Applications

Solve.

83. A high school student is selling T-shirts to raise money for the band. She realizes that the number of shirts she sells each week can be modeled by $f(x) = -x^2 + 12x - 17$, where x is the amount she charges per shirt. Solve the inequality $-x^2 + 12x - 17 \ge 10$ to find the range she can charge per shirt and sell at least 10 shirts in a week.

84. Maria tracked the nighttime temperatures for a week and noticed that the temperature, in Celsius, could be modeled by $f(x) = \frac{1}{2}x^2 - 4x + 6$, where x is the number of hours after midnight. Maria has plants that might die if they are left out when the temperature drops below freezing (0 degrees Celsius). Solve the inequality $\frac{1}{2}x^2 - 4x + 6 < 0$ to find timeframe in which her plants will be in danger.

Writing & Thinking

85. Use a graphing calculator to graph the rational function $y = \dfrac{x^2 + 3x - 4}{x}$.

a. Use the graph to find the solution set for $y > 0$.

b. Use the graph to find the solution set for $y < 0$.

c. Explain the effect of $x = 0$ on the graph and why $x = 0$ is not included in either parts a. or b.

86. In your own words, explain why (as in Example 5), when the quadratic formula gives nonreal values, the quadratic polynomial is either always positive or always negative.

86. Because the polynomial cannot be 0 for any real number, it cannot be negative for some number and positive for another. This corresponds to the graph not crossing the x-axis.

72. $(-\infty, 1) \cup (4, \infty)$

73. $(-\infty, -3.1623) \cup (3.1623, \infty)$

74. $[-4.4721, 4.4721]$

75. $\varnothing$

76. $(-\infty, \infty)$

77. $(-\infty, -3) \cup (0, 3)$

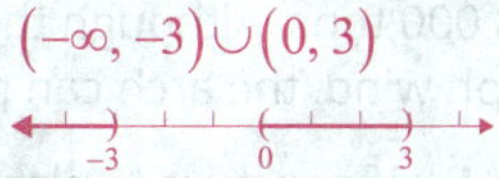

78. $(-\infty, 0] \cup \{2\}$

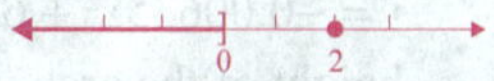

79. $[-1.8868, \infty)$

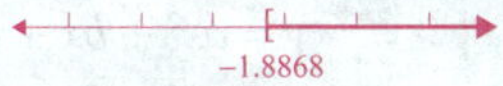

80. $(-\infty, -0.7913) \cup (1, 3.7913)$

81. $(-2.3344, -0.7420) \cup (0.7420, 2.3344)$

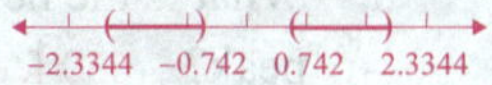

82. $(-0.8445, 2.4710)$

Applications

83. $[3, 9]$

84. From 2:00 a.m. to 6:00 a.m., or $(2, 6)$

Writing & Thinking

85. a. $(-4, 0) \cup (1, \infty)$

b. $(-\infty, -4) \cup (0, 1)$

c. The function is undefined at $x = 0$.

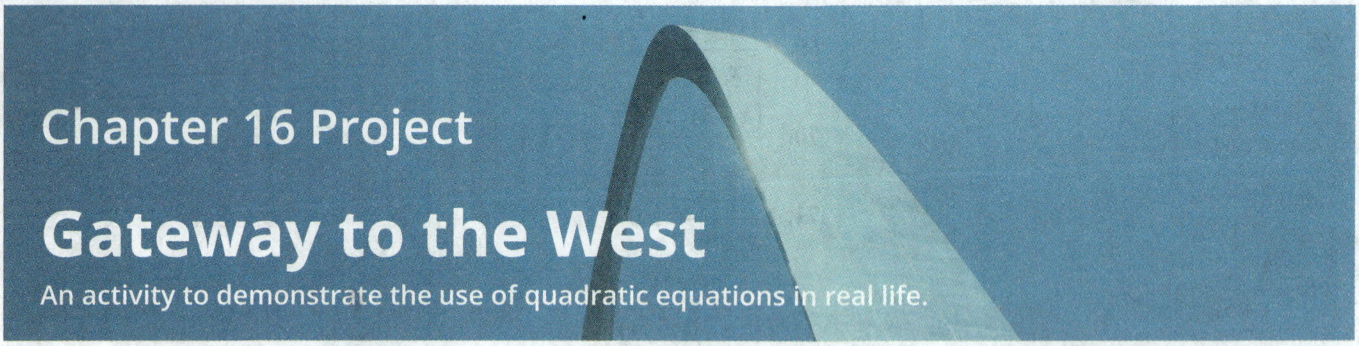

Chapter 16 Project

Gateway to the West

An activity to demonstrate the use of quadratic equations in real life.

The Gateway Arch on the St. Louis riverfront in Missouri serves as an iconic monument symbolizing the westward expansion of American pioneers, such as Lewis and Clark. A nationwide competition was held to choose an architect to design the monument and the winner was Eero Saarinen, a Finnish American who immigrated to the United States with his parents when he was 13 years old. Construction began in 1962, and the monument was completed in 1965. The Gateway Arch is the tallest monument in the United States. It is constructed of stainless steel and weighs more than 43,000 tons. Although the arch is heavy, it was built to sway with the wind to prevent it from being damaged. In a 20 mph wind, the arch can move up to 1 inch. In a 150 mph wind, the arch can move up to 18 inches.

1. If you were to place the Gateway Arch on a coordinate plane centered around the y-axis, then the equation $y = -0.00635x^2 + 630$ could be used to model the height of the arch in feet.

a. The general form for a quadratic function is $y = ax^2 + bx + c$. Identify the values for a, b, and c from the Gateway Arch equation.

b. Find the vertex of the Gateway Arch equation. (0, 630)

c. Does the vertex represent a maximum or a minimum? Explain your answer based on the coefficients of the Gateway Arch equation.

d. What is the height of the Gateway Arch at its peak? 630 feet

e. Write the equation for the axis of symmetry of the Gateway Arch equation.

f. Find the x-intercepts of the Gateway Arch equation. Round to the nearest integer.

2. Using the coordinate plane below and the information from Problem 1, graph the Gateway Arch equation.

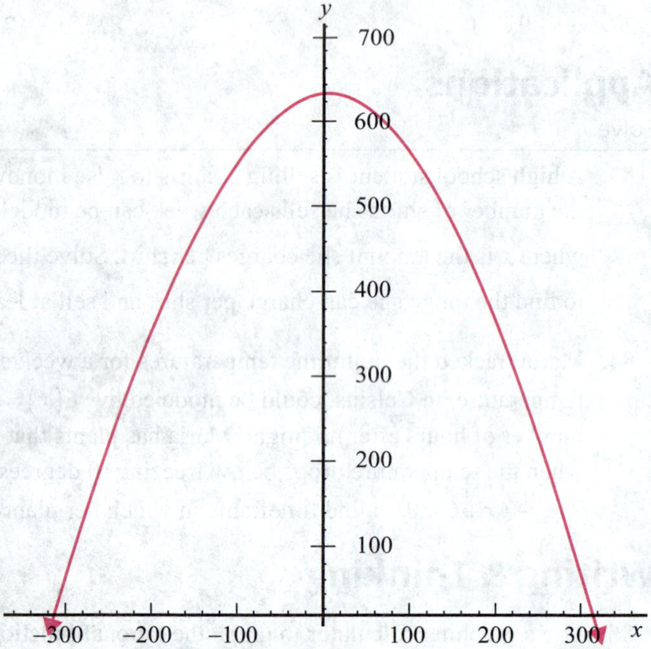

3. How far apart are the legs of the Gateway Arch at its base? 630 feet

4. The Gateway Arch equation is a mathematical model. Look up the actual values for the height of the Gateway Arch and the distance between the legs of the arch at its base on the internet and describe how they compare to the values calculated using the equation. The answers from the internet should be the same as Problems 1d and 3.

Chapter 16 Project: Gateway to the West

1. a. $a = -0.00635$, $b = 0$, $c = 630$

c. The vertex represents a maximum since $a < 0$.

e. $x = 0$, or the y-axis

f. (315, 0) and (−315, 0)

Chapter 16 Project

Determining Product Pricing to Maximize Revenue

An activity to demonstrate the use of quadratic equations in real life.

A company determines that if p is the price charged for an electric bicycle they manufacture, then the number of bikes that will sell n is a function of p: $n(p) = 3750 - 3p$. This is because for each \$1 increase in the price, three fewer bikes are sold. The revenue R earned is also a function of p because it is the number of bikes sold times the price per bike: $R(p) = p(3750 - 3p) = -3p^2 + 3750p$. Using methods learned in this chapter, we will investigate just how much revenue the bike manufacturer can earn. Is it boundless or is there a maximum?

For simplicity, in order to compare variable names used within the chapter, let's replace p with x and R with y, so $y = x(3750 - 3x)$. Notice that the graph of this equation is a parabola.

1. Solve the equation $x(3750 - 3x) = 0$ for x. Notice that by setting $y = 0$, you are solving for the x-coordinates of the two x-intercepts. State the coordinates of the two x-intercepts.

2. Use the equation $y = x(3750 - 3x)$ for the following.

a. Choose an x-value smaller than the smallest x-value of the x-intercepts found in Problem 1 and calculate y.

b. Choose an x-value larger than the largest x-value of the x-intercepts found in Problem 1 and calculate y.

c. What do you notice about y-values found in part a. and b? Recall that y is revenue and x is price. Can you conclude an initial interval for price that the company should stay within?

3. Calculate the mean of the two x-values found in Problem 1.

4. Pick two new values of x between the smallest x-value from Problem 1 and the answer to Problem 3. Next, pick two values of x strictly between the answer to Problem 3 and the largest x-value from Problem 1.

5. Arrange these four values from least to greatest and include your answer from Problem 3, for a total of five unique values.

6. a. Evaluate $y = x(3750 - 3x)$ for these five values of x from Problem 5. This will give you five points on the parabola.

b. For the smallest of the x-values from part a, explain in words what the (x, y) coordinates represent. Include the values.

7. Carefully choose a horizontal and vertical scale and plot the five points found in Problem 6 part a. Also plot the x-intercepts found in Problem 1. Use these points to sketch the parabola.

8. Based on your values and the graph, state the (x, y) coordinates of the vertex (or your best approximation).

9. Is the y-coordinate of the vertex a minimum or maximum value of y? Why do you think this is, based both on the context and on the function?

10. Algebraically change the equation $y = x(3750 - 3x) = -3x^2 + 3750x$ to the form $y = a(x - h)^2 + k$. (**Hint:** You will need to complete the square. See Section 16.7.)

11. Does the equation found in Problem 10 support your approximation of the vertex found in Problem 8?

12. Now return to the question about revenue from selling electric bicycles.

a. What is the lower bound on how much revenue the company can earn?

b. What is the upper bound on how much revenue the company can earn? What price is charged for that to be the revenue earned?

c. How many bikes need to be sold to reach the maximum revenue found in part b? (**Hint:** Recall the number of bikes sold is given by $n(p) = 3750 - 3p$ where p is the price per bike.)

Chapter 16 Project: Determining Product Pricing to Maximize Revenue

1. $x = 0$ and $x = 1250$; x-intercepts: (0, 0) and (1250, 0)
2. a. Answers will vary.

   b. Answers will vary.

   c. The values are negative. For revenue to be nonnegative, we must have $0 \le x \le 1250$.
3. 625
4. Answers will vary. For example, we can choose $x = 200$ and $x = 400$ for the first pair. For the second pair, we can choose $x = 800$ and $x = 1000$.
5. Answers will vary. For example: $x = 200$, $x = 400$, $x = 625$, $x = 800$, and $x = 1000$.
6. a. Answers will vary. For example, (200, 630000), (400, 1020000), (625, 1171875), (800, 1080000), and (1000, 750000)

   b. Answers will vary. For example, producing and selling 200 bikes will result in a revenue of $630,000.
7. Points will vary.

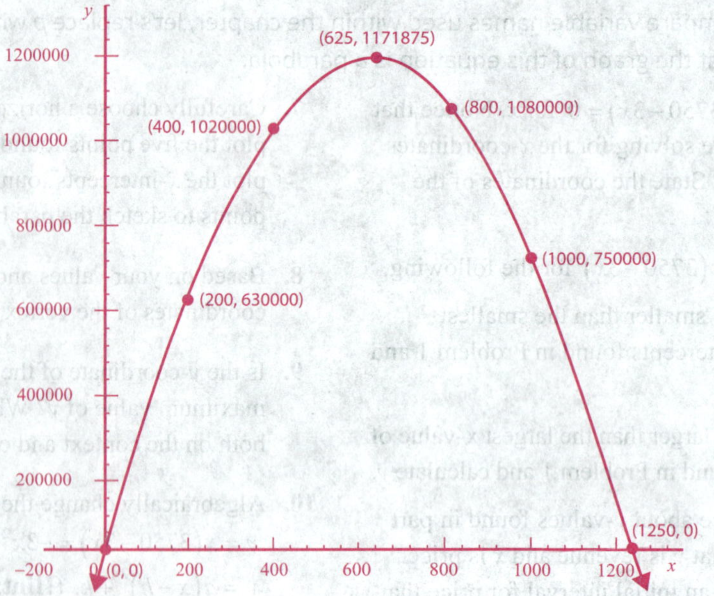

8. (625, 1171875)
9. A maximum. Based on the function, this is because the coefficient of x^2, $a = -3$ is negative and so the parabola opens downward
10. $y = -3(x - 625)^2 + 1{,}171{,}875$
11. Answers may vary. The equation in Problem 10 shows the vertex is (625, 1171875).
12. a. Revenue cannot be negative, so it must be at a minimum zero. That is due to the context.

    b. The upper bound on revenue is $1,171,875 and that is obtained by charging a price of $625.

    c. The number of bikes to reach this maximum revenue is $n(625) = 3750 - 3(625) = 1875$. This could also be determined by dividing the maximum revenue by the price: $\frac{1171875}{625} = 1875$.

CHAPTER 17

17 Exponential and Logarithmic Functions

Connections

While they may not seem common, exponential and logarithmic functions are used in a variety of fields. Exponential functions can be used to describe the relationship between musical notes in music theory and to model population growth in ecology. Logarithmic functions can be used to describe the intensity of an earthquake in geology and to measure the acidity or alkalinity of solutions in chemistry.

A common situation where you might find yourself working with exponential functions in your everyday life is in financial situations, such as calculating the amount of interest owed on your student loans or credit cards. When making large purchases, such as buying a car or a house, you'll often have the option to choose between different financing options. Being comfortable with exponential functions can help you make the decision that is best for you and your circumstances.

Suppose you're buying a car that costs $15,000 and the dealership has two financing options available. If you put $5000 down, you can finance the rest at 2.5% interest for 6 years. Alternatively, you could put $0 down and finance the entire purchase at 5.5% for 4 years. What are some factors that might cause you to go with the first option? Under what circumstances might you go with the second option? Which option would result in you paying less overall for the car?

Objectives

A. Find the sum, difference, product, and quotient of two functions.

B. Graph the sum of two functions.

C. Use a graphing calculator to graph the sum of two functions.

17.1 Algebra of Functions

As discussed throughout this text, functions are an important topic in mathematics. Function notation $f(x)$ is particularly helpful in evaluating functions and describing graphical relationships. Operating algebraically with functions as well as understanding and finding the **composition** and **inverses** of functions rely heavily on function notation. The concepts of composite and inverse functions form the basis of the relationship between logarithmic and exponential functions.

Logarithms are exponents. Traditionally, logarithmic and exponential values were calculated with the extensive use of printed tables and techniques for estimating values not found in the tables. As some of your teachers will tell you, this was a long and detailed process. These tables are no longer printed in textbooks. Now, handheld calculators have programs stored in their electronic memories that calculate the values in these tables with even greater accuracy, and complicated expressions can be evaluated by pressing a few keys.

Of all the topics discussed in algebra, logarithmic functions and exponential functions probably have the most value in terms of applied problems. Learning curves, important in business and education, can be described with logarithmic and exponential functions. Exponential growth and decay are basic concepts in biology and medicine. (Cancer cells grow exponentially and radium decays exponentially.) Computers use logarithmic and exponential concepts in their design and implementation. These concepts are likely to be encountered in almost any field of study.

A Algebraic Operations with Functions

Teaching Note

As you begin this topic, you might want to remind students that most functions they have seen up to this point have been polynomial or rational functions. This chapter introduces them to two related functions that are not based on polynomials.

If two (or more) functions have the same domain, then we can perform the operations of addition, subtraction, multiplication, and division with these functions. For example, consider the following quadratic functions.

$$f(x)=2x^2-1 \quad \text{and} \quad g(x)=x^2+2x-5$$

Both have the same domain: $\mathbb{R}$ = all real numbers = $(-\infty,\infty)$. This means that we can

a. choose any value for x from the common domain,

b. evaluate each function for that value of x, and

c. perform operations with those functional values.

The functional values are the *y*-values. For example, if we choose $x=3$, then

$$f(3)=2(3)^2-1=17 \quad \text{and} \quad g(3)=(3)^2+2(3)-5=10.$$

Now, we can easily find the sum and difference of the given functions.

$$f(3)+g(3)=17+10=27 \quad \text{and} \quad f(3)-g(3)=17-10=7$$

However, if we want to find, say $f(5)+g(5)$ and $f(5)-g(5)$, we would need to again evaluate both functions, this time for $x=5$. Another way to find sums and differences of functions is to find the algebraic sum (or difference) of the two expressions. Then these new expressions will allow us to find the sum (or difference) directly for any value of x that is in the original domain of both functions. For example, we find the sum $f+g$ as follows.

$$\begin{aligned}(f+g)(x) &= f(x)+g(x)\\ &= \left(2x^2-1\right)+\left(x^2+2x-5\right)\\ &= 3x^2+2x-6\end{aligned}$$

With this new function, we find $(f+g)(3)$ directly.

$$(f+g)(3)=3(3)^2+2(3)-6=27$$

Similarly,

$$\begin{aligned}(f-g)(x) &= f(x)-g(x)\\ &= \left(2x^2-1\right)-\left(x^2+2x-5\right)\\ &= x^2-2x+4,\end{aligned}$$

and we have

$$(f-g)(3)=(3)^2-2(3)+4=7.$$

Similar notation is used for the product and quotient of two functions. One important condition is that both functions **must have the same domain**. If not, then the algebraic sums, differences, products, and quotients are **restricted to portions of the domains that are in common. Also, in the case of quotients, no denominator can be 0.**

Algebraic Operations with Functions

If $f(x)$ and $g(x)$ represent two functions and x is a value in the **domain of both functions**, then we define the following operations.

1. **Sum of two functions:** $(f+g)(x)=f(x)+g(x)$
2. **Difference of two functions:** $(f-g)(x)=f(x)-g(x)$
3. **Product of two functions:** $(f\cdot g)(x)=f(x)\cdot g(x)$
4. **Quotient of two functions:** $\left(\dfrac{f}{g}\right)(x)=\dfrac{f(x)}{g(x)}$, where $g(x)\neq 0$

DEFINITION

Example 1 Using Algebraic Operations with Functions

Let $f(x)=3x^2+x-4$ and $g(x)=x-6$. Perform the following operations.

a. $(f+g)(x)$

b. $(f-g)(x)$

c. $(f\cdot g)(x)$

d. Evaluate each of the functions found in parts a. through c. for $x=2$.

Solution

a. $(f+g)(x)=\left(3x^2+x-4\right)+(x-6)=3x^2+2x-10$

1. Let $f(x)=2x^2+3x-9$ and $g(x)=x+3$. Perform the following operations.

a. $(f+g)(x)$

b. $(f-g)(x)$

c. $(f\cdot g)(x)$

b. $(f-g)(x)=(3x^2+x-4)-(x-6)=3x^2+x-4-x+6=3x^2+2$

c. $$\begin{aligned}(f\cdot g)(x)&=(3x^2+x-4)(x-6)\\&=3x^3-18x^2+x^2-6x-4x+24\\&=3x^3-17x^2-10x+24\end{aligned}$$

Teaching Note
You may want to give your students several examples involving radicals.

d. Evaluating each of these functions for $x=2$ gives the following results.

$$(f+g)(2)=3(2)^2+2(2)-10=12+4-10=6$$

$$(f-g)(2)=3(2)^2+2=12+2=14$$

$$\begin{aligned}(f\cdot g)(2)&=3(2)^3-17(2)^2-10(2)+24\\&=24-68-20+24\\&=-40\end{aligned}$$

Now work margin exercise 1.

2. Let $f(x)=2x^2-x$ and $g(x)=x-1$. Perform the following operations.

a. $(g-f)(x)$

b. $\left(\frac{g}{f}\right)(x)$

Example 2 Using Algebraic Operations with Functions

Let $f(x)=x^2-x$ and $g(x)=2x+1$. Find the following functions.

a. $(f+g)(x)$ **b.** $(g-f)(x)$ **c.** $\left(\frac{f}{g}\right)(x)$

d. Evaluate each of the functions found in parts a. through c. for $x = 3$.

Solution

a. $(f+g)(x)=(x^2-x)+(2x+1)=x^2+x+1$

b. $(g-f)(x)=(2x+1)-(x^2-x)=2x+1-x^2+x=-x^2+3x+1$

c. $\left(\frac{f}{g}\right)(x)=\frac{x^2-x}{2x+1}$ where $2x+1\neq 0$ $\left(\text{or } x\neq -\frac{1}{2}\right)$

d. Evaluating each of these functions for $x=3$ gives the following results.

$$(f+g)(3)=(3)^2+(3)+1=9+3+1=13$$

$$(g-f)(3)=-(3)^2+3(3)+1=-9+9+1=1$$

$$\left(\frac{f}{g}\right)(3)=\frac{(3)^2-(3)}{2(3)+1}=\frac{9-3}{6+1}=\frac{6}{7}$$

Now work margin exercise 2.

Note that, except for Example 2c, the domain for all of the functions discussed in Examples 1 and 2 is the set of all real numbers, $(-\infty,\infty)$. In Example 2c, we noted that the denominator cannot equal 0. In Example 3, we show how the domain may need to be limited before performing algebra with functions that contain radical expressions.

Example 3 Using Algebraic Operations with Functions with Limited Domains

Let $f(x)=x+5$ and $g(x)=\sqrt{x-2}$. Find the following functions and state the domain of each function.

a. $(f+g)(x)$ **b.** $\left(\frac{f}{g}\right)(x)$

Solution

a. $(f+g)(x)=x+5+\sqrt{x-2}$

The domain of f is the set of all real numbers. However, the domain of the sum is restricted to the domain of g, the radical function. In this case we must have $x-2\geq 0$. Thus, in interval notation, the domain is $[2,\infty)$.

b. $\left(\frac{f}{g}\right)(x)=\frac{x+5}{\sqrt{x-2}}$

For this function, the denominator cannot be 0, so $x\neq 2$. Therefore, we must have $x-2>0$ and the domain, in interval notation, is $(2,\infty)$.

Note: The domain can become smaller than, but never larger than, the domain of the two original functions.

Now work margin exercise 3.

3. Let $f(x)=x-1$ and $g(x)=\sqrt{x+3}$. Find the following functions.

a. $(f+g)(x)$

b. $\left(\frac{g}{f}\right)(x)$

Example 4 Using Algebraic Operations with Functions

Sally is analyzing the finances for her bakery. She finds that the bakery's revenue can be represented by the function $23x + 18$ when they sell x cakes. The cost function can be represented by $8x - 4$ when they make x cakes. Using the formula profit = revenue − cost, find an expression that represents the bakery's profits when they sell x cakes, assuming they make and sell at least one cake.

Solution

Plugging the algebraic expressions into the formula profit = revenue − cost and simplifying, we obtain the following.

$$\begin{aligned}\text{profit}&=(23x+18)-(8x-4)\\&=23x+18-8x+4\\&=15x+22\end{aligned}$$

Thus, the profit function for the bakery can be represented by $15x + 22$ when they sell x cakes.

Now work margin exercise 4.

4. The dimensions of a rectangle are constantly changing. The length of the rectangle can be represented by the formula $2x^2 - 4x + 5$ after x seconds have gone by. The width of the rectangle can be represented by $x^2 - 6x + 10$ after x seconds have gone by. Find an expression that represents the perimeter of the rectangle after x seconds have gone by.

B Graphing the Sum of Two Functions

In this section, we discuss how to graph the sum of two functions. (Graphing the difference, product, and quotient can be accomplished in a similar manner.) Remember that, in any case, algebraic operations with functions can be performed only over a common domain.

We begin with two functions f and g that have the same domain and only a finite number of ordered pairs.

$$f=\{(-2,1),(0,4),(3,5)\}$$

$$g=\{(-2,3),(0,1),(3,2)\}$$

Note that the domain of both functions is $\{-2,0,3\}$.

Both functions have only three points and are graphed in Figure 1.

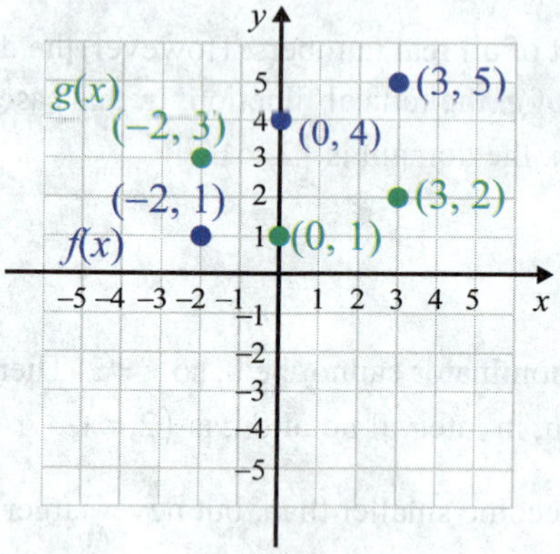

Figure 1

Teaching Note

To test their understanding, you might have your students graph the difference of the two functions in Figure 1. Also, you might want to draw several pairs of functions on the board (or on hand-out sheets) and have the students graph the sums, differences, products, and quotients of the functions. Start with a finite number of points. I have found that the students are fascinated with this process and treat it as a fun and rewarding hands-on exercise. It can definitely help in their general understanding about functions and graphs.

To add the two functions graphically, look at each of the x-values on the x-axis and the points directly above (or below) them and add the corresponding y-values. This process will give new points, and these new points represent a function that is the sum of the two original functions.

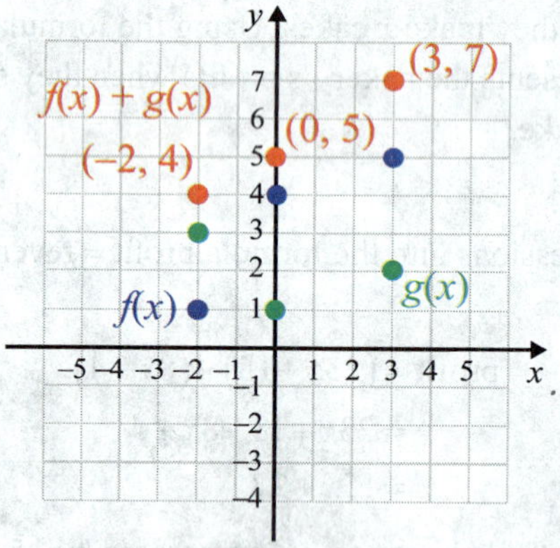

Figure 2

From the points in the Figure 2, we get the following function.

$$\begin{aligned}(f+g)(x)&=f(x)+g(x)\\&=\{(-2,1+3),(0,4+1),(3,5+2)\}\\&=\{(-2,4),(0,5),(3,7)\}\end{aligned}$$

In general, graphing the sum, difference, product, or quotient of two functions will involve an infinite number of points. Certainly not all of these points can be plotted one at a time.

However, by making a table of a few key points and joining these points with smooth curves or line segments, the general nature of the result can be found. In fact, if the two functions consist of line segments, then the sum will also consist of line segments. Table 1 and Figures 3 and 4 illustrate such a case.

When operating with functions, the operations are performed with the y-values for each value of x in the common domain. (Don't mess with the x-values!)

Points Illustrated in Figure 3 and Figure 4

x	$f(x)$	$g(x)$	$f(x)+g(x)$
−3	1	−3	$1+(-3)=-2$
−2	1	0	$1+0=1$
−1	1	3	$1+3=4$
0	−2	3	$(-2)+3=1$
1	1	3	$1+3=4$
2	1	4	$1+4=5$
3	0	5	$0+5=5$

Table 1

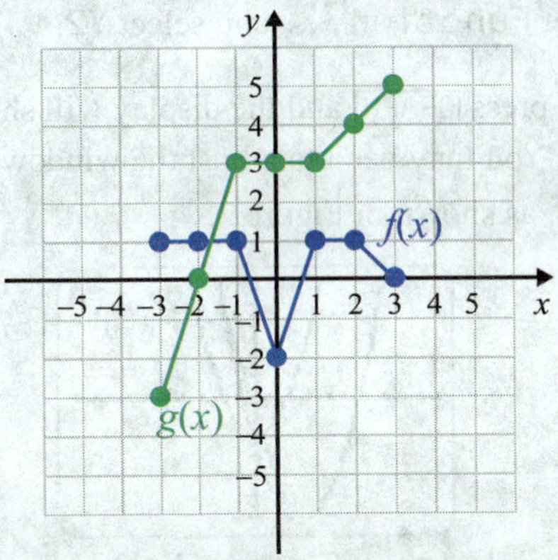

Figure 3

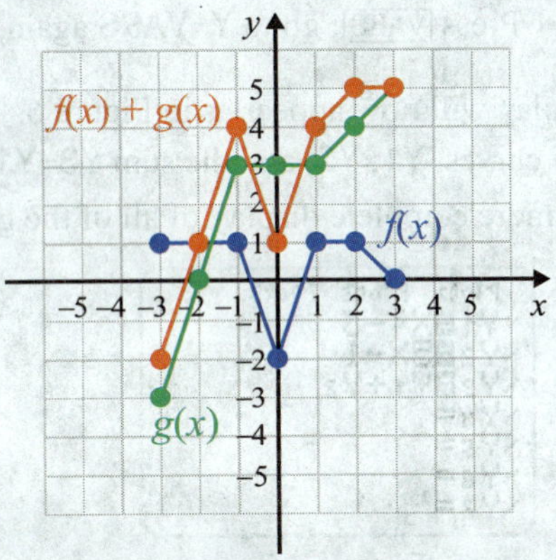

Figure 4

C Using a Graphing Calculator to Graph the Sum of Two Functions

A graphing calculator can be used to graph the sum (difference, product, or quotient) of two functions. First, press [Y=] and enter the functions as Y1 and Y2, then assign Y3 to be the sum of the first two. Figures 5a and 5b show the displays for entering two functions and their graphs.

Using a TI-84 Plus graphing calculator, we graph

$$f(x)=x^2-3 \text{ (a parabola) and } g(x)=5x-1 \text{ (a straight line).}$$

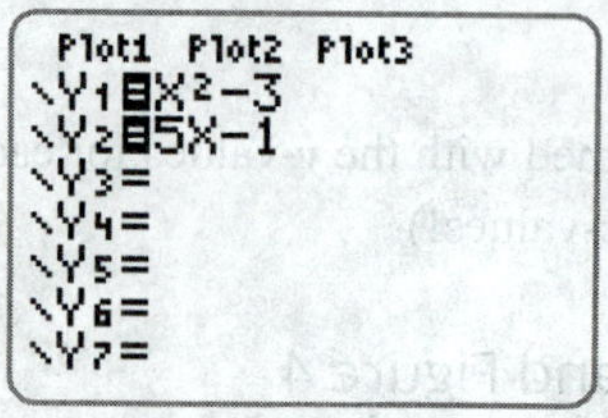

The Two Functions Entered

Figure 5a

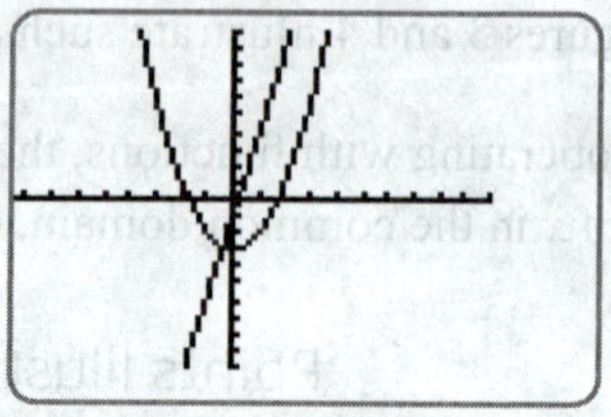

The Two Functions Graphed

Figure 5b

Now, we can graph the sum of the two functions, $f(x)+g(x)$, (or Y1+Y2 on the calculator) by assigning Y3=Y1+Y2 as follows. (This saves performing the actual algebraic operations, as the calculator does it for us.)

Step 1: Press [Y=] and select Y3.

Step 2: Press the [VARS] key.

Step 3: Move the cursor to the Y-VARS at the top of the display.

Step 4: Select `Function...` and press [ENTER].

Step 5: Select Y1 and press [ENTER].

Step 6: Press the [+] key.

Step 7: Press [VARS], go to Y-VARS again, select `Function...` and select Y2.

The display will now appear as in Figure 6. Now, press [GRAPH] and the display will show all three curves, Y1, Y2, and the sum Y3=Y1+Y2. You may need to adjust the window to allow a more complete display of all of the graphs, as shown in Figure 7.

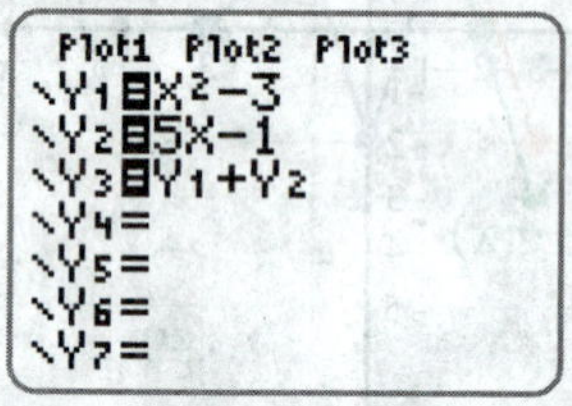

Figure 6

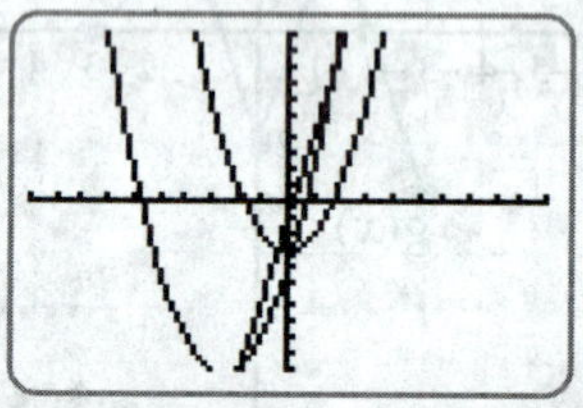

Figure 7

If this graph is somewhat confusing (because three graphs are shown), you can turn off the first two graphs and show just the sum. To turn off a graph, go to the highlighted equal sign, =, next to the equation and hit [ENTER]. You will see the following screens.

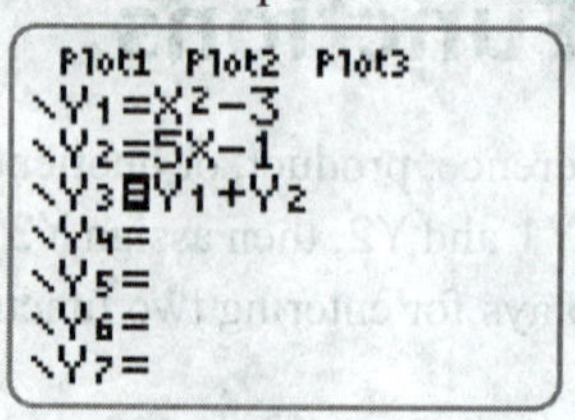

Figure 8a

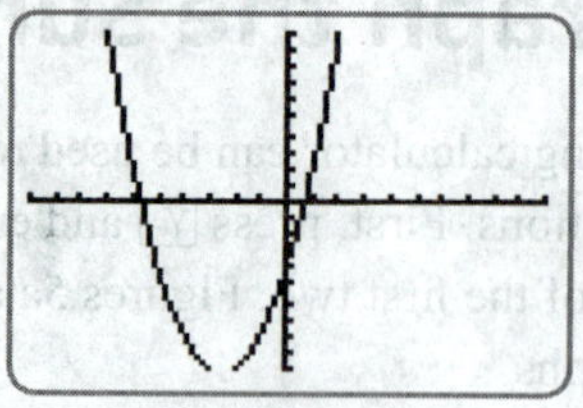

Figure 8b

17.1 Exercises

Concept Check

1. composition, inverses
2. exponents
3. domain
4. 0
5. y-values, x
6. infinite
7. True
8. True
9. False; The operations are restricted to the portions of the domain that are in common.
10. False; The sum of the two functions will also be line segments.

Practice

1. **a.** $2x-3$ **b.** 7 **c.** $x^2-3x-10$ **d.** $\frac{x+2}{x-5},\ x\neq 5$
2. **a.** $3x+4$ **b.** $x-4$ **c.** $2x^2+8x$ **d.** $\frac{2x}{x+4},\ x\neq -4$
3. **a.** x^2+3x-4 **b.** x^2-3x+4 **c.** $3x^3-4x^2$ **d.** $\frac{x^2}{3x-4},\ x\neq\frac{4}{3}$
4. **a.** x^2+x-2 **b.** $-x^2+x-4$ **c.** x^3-3x^2+x-3 **d.** $\frac{x-3}{x^2+1}$
5. **a.** x^2+x-12 **b.** x^2-x-6 **c.** $x^3-3x^2-9x+27$ **d.** $x+3,\ x\neq 3$

Margin Exercise Answers

1. a. $2x^2+4x-6$ **b.** $2x^2+2x-12$ **c.** $2x^3+9x^2-27$ **2. a.** $-2x^2+2x-1$ **b.** $\frac{x-1}{2x^2-x},x\neq\frac{1}{2},0$ **3. a.** $x-1+\sqrt{x+3}$, $[-3,\infty)$ **b.** $\frac{\sqrt{x+3}}{x-1}$, $[-3,1)\cup(1,\infty)$ **4.** $6x^2-20x+30$

17.1 Exercises

Concept Check

Fill-in-the-Blank. Complete each sentence using information found in this section.

1. Operating algebraically with functions, as well as understanding and finding the ______ and ______ of functions, relies heavily on function notation.

2. Logarithms are ______.

3. If two or more functions have the same ______, then we can perform the operations of addition, subtraction, multiplication, and division with these functions.

4. In the case of finding the quotient of functions, no denominator can be ______.

5. When operating with functions, the operations are performed with the ______ for each value of ______ in the common domain.

6. In general, graphing the sum of two functions will involve a/an ______ number of points.

True/False. Determine whether each statement is true or false. If a statement is false, explain how it can be changed so the statement will be true. (**Note:** There may be more than one acceptable change.)

7. One way to find the sum of two functions is to find the algebraic sum of the two expressions.

8. The function $(f+g)(x)$ means the same as $f(x)+g(x)$.

9. If functions do not have the same domain, any algebraic sums, differences, products, and quotients are restricted to portions of the range that are in common.

10. If two functions have graphs that consist of line segments, the sum of the two functions will produce a graph that is a continuous line.

Practice

For the following pairs of functions find, **a.** $(f+g)(x)$, **b.** $(f-g)(x)$, **c.** $(f\cdot g)(x)$, and **d.** $\left(\frac{f}{g}\right)(x)$. See Examples 1 and 2.

1. $f(x)=x+2,\ g(x)=x-5$
2. $f(x)=2x,\ g(x)=x+4$
3. $f(x)=x^2,\ g(x)=3x-4$
4. $f(x)=x-3,\ g(x)=x^2+1$
5. $f(x)=x^2-9,\ g(x)=x-3$
6. $f(x)=x^2-25,\ g(x)=x+5$
7. $f(x)=2x^2+x,\ g(x)=x^2+2$
8. $f(x)=x^3+6x,\ g(x)=x^2+6$
9. $f(x)=x^2+4x+1, g(x)=x^2-4x+1$
10. $f(x)=x^3-x^2,\ g(x)=6-x^2$

6. a. x^2+x-20
 b. x^2-x-30
 c. $x^3+5x^2-25x-125$
 d. $x-5,\ x\neq -5$

7. a. $3x^2+x+2$
 b. x^2+x-2
 c. $2x^4+x^3+4x^2+2x$
 d. $\dfrac{2x^2+x}{x^2+2}$

8. a. x^3+x^2+6x+6
 b. x^3-x^2+6x-6
 c. x^5+12x^3+36x
 d. x

9. a. $2x^2+2$
 b. $8x$
 c. x^4-14x^2+1
 d. $\dfrac{x^2+4x+1}{x^2-4x+1}, x\neq 2\pm\sqrt{3}$

10. a. x^3-2x^2+6
 b. x^3-6
 c. $-x^5+x^4+6x^3-6x^2$
 d. $\dfrac{x^3-x^2}{6-x^2},\ x\neq\pm\sqrt{6}$

11. 9
12. 8
13. $-a^2-a-1$
14. $\dfrac{-a+3}{a^2+4}$
15. 27
16. 1.75
17. $\dfrac{8}{5}$
18. 78
19. −31
20. $\dfrac{4}{5}$
21. $\sqrt{2x-6}+x+4$, $D=[3,\infty)$
22. $x-1$; $D=(-\infty,1)\cup(1,\infty)$
23. $3x^2-19x-14$; $D=(-\infty,\infty)$

24. $2;\ D=(-\infty,\infty)$

25. $\dfrac{x-5}{\sqrt{x+3}};\ D=(-3,\infty)$

26. $2x\sqrt{2-x}-8\sqrt{2-x};$ $D=(-\infty,2]$

27. $-3x\sqrt{x-3};D=[3,\infty)$

28. $5-x+\sqrt{4-x};$ $D=(-\infty,4]$

29. $\sqrt[3]{x+3}+\sqrt{5+x};$ $D=[-5,\infty)$

30. $\sqrt{x-1}-\sqrt[3]{2x+1};$ $D=[1,\infty)$

31. a.

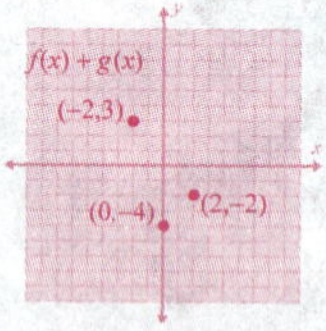

b.

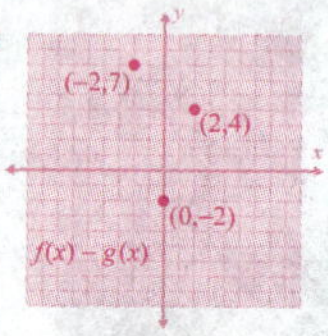

32. a.

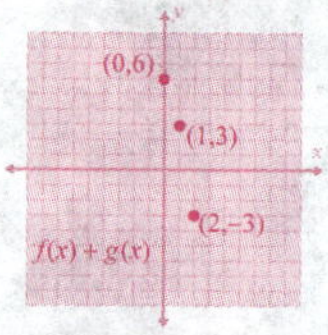

b.

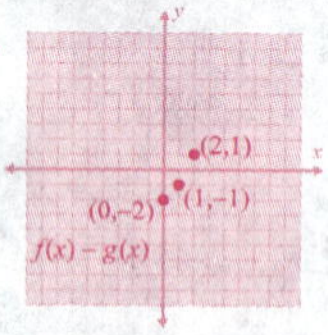

33. a.

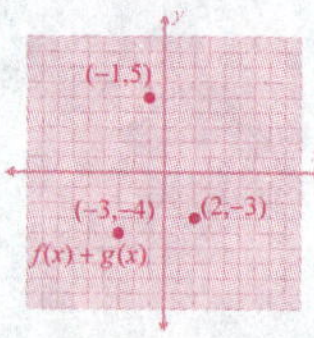

b.

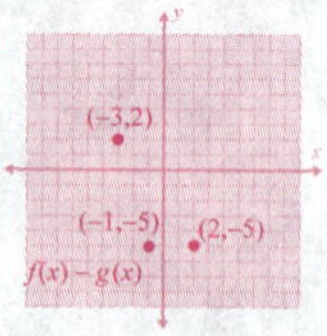

Let $f(x)=x^2+4$ and $g(x)=-x+3$. Find the values of the indicated expressions. See Examples 1 and 2.

11. $f(2)+g(2)$

12. $f(2)\cdot g(2)$

13. $g(a)-f(a)$

14. $\dfrac{g(a)}{f(a)}$

15. $(f+g)(-4)$

16. $(f-g)(0.5)$

17. $\left(\dfrac{f}{g}\right)(-2)$

18. $(f\cdot g)(-3)$

19. $(g-f)(-6)$

20. $\left(\dfrac{g}{f}\right)(-1)$

Find the indicated functions and state their domains in interval notation. See Example 3.

21. If $f(x)=\sqrt{2x-6}$ and $g(x)=x+4$, find $(f+g)(x)$.

22. If $f(x)=x^2-2x+1$ and $g(x)=x-1$, find $\left(\dfrac{f}{g}\right)(x)$.

23. Find $f(x)\cdot g(x)$ given that $f(x)=3x+2$ and $g(x)=x-7$.

24. Find $f(x)-g(x)$ given that $f(x)=x^2$ and $g(x)=x^2-2$.

25. For $f(x)=x-5$ and $g(x)=\sqrt{x+3}$, find $\dfrac{f(x)}{g(x)}$.

26. For $f(x)=2x-8$ and $g(x)=\sqrt{2-x}$, find $f(x)\cdot g(x)$.

27. If $f(x)=-\sqrt{x-3}$ and $g(x)=3x$, find $(f\cdot g)(x)$.

28. If $f(x)=-\sqrt{4-x}$ and $g(x)=5-x$, find $(g-f)(x)$.

29. If $f(x)=\sqrt[3]{x+3}$ and $g(x)=\sqrt{5+x}$, find $f(x)+g(x)$.

30. If $f(x)=\sqrt{x-1}$ and $g(x)=\sqrt[3]{2x+1}$, find $f(x)-g(x)$.

For the following pairs of functions, graph **a.** the sum $(f+g)$ and **b.** the difference $(f-g)$ on two different graphs.

31. $f=\{(-2,5),(0,-3),(2,1)\}$
$g=\{(-2,-2),(0,-1),(2,-3)\}$

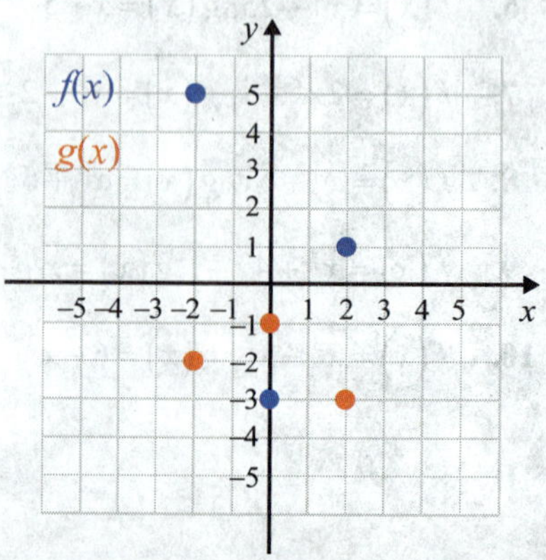

32. $f=\{(0,2),(1,1),(2,-1)\}$
$g=\{(0,4),(1,2),(2,-2)\}$

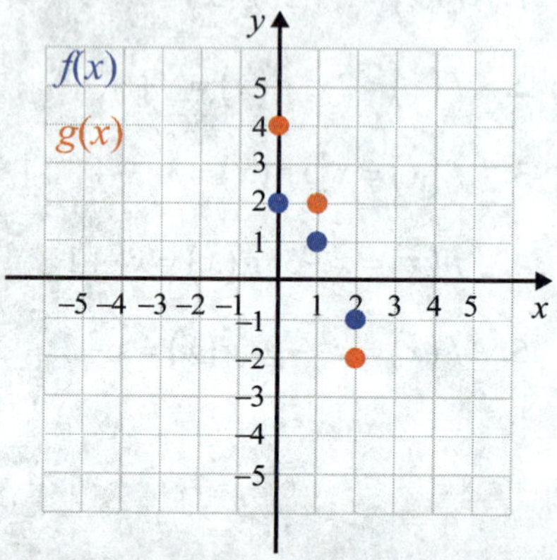

33. $f=\{(-3,-1),(-1,0),(2,-4)\}$

$g=\{(-3,-3),(-1,5),(2,1)\}$

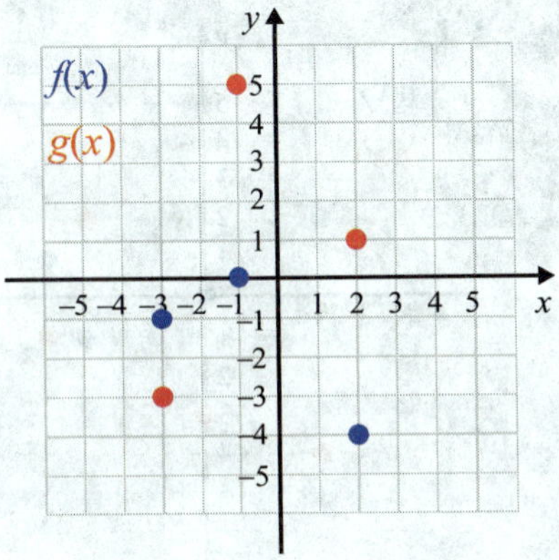

34. $f=\{(-4,4),(-2,-5),(1,3)\}$

$g=\{(-4,1),(-2,0),(1,-2)\}$

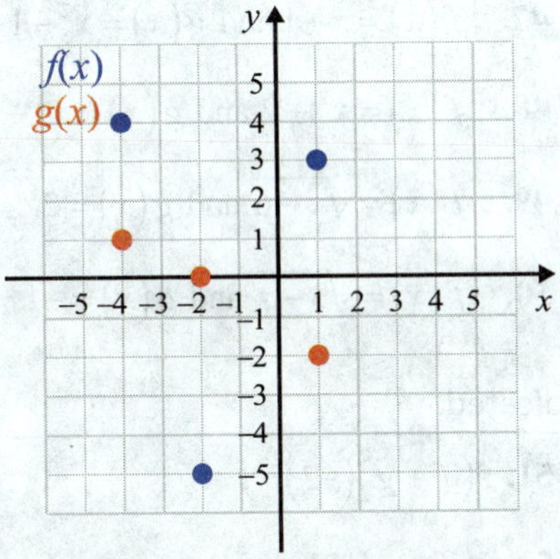

35. $f=\{(-2,-5),(-1,4),(0,-1),(1,5)\}$

$g=\{(-2,-1),(-1,0),(0,-4),(1,1)\}$

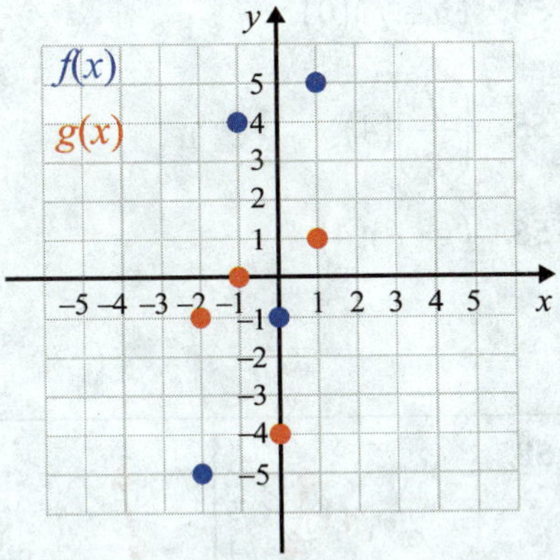

36. $f=\{(-2,3),(1,2),(2,1),(3,0)\}$

$g=\{(-2,-4),(1,-3),(2,-2),(3,-1)\}$

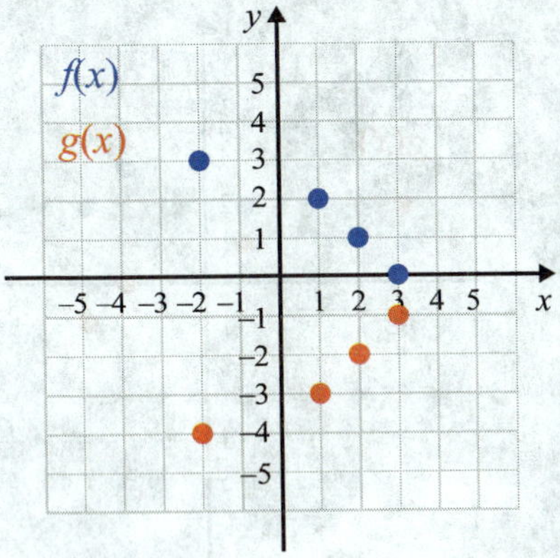

37. $f=\{(-3,1),(-1,2),(1,1),(3,2)\}$

$g=\{(-3,-3),(-1,-4),(1,3),(3,4)\}$

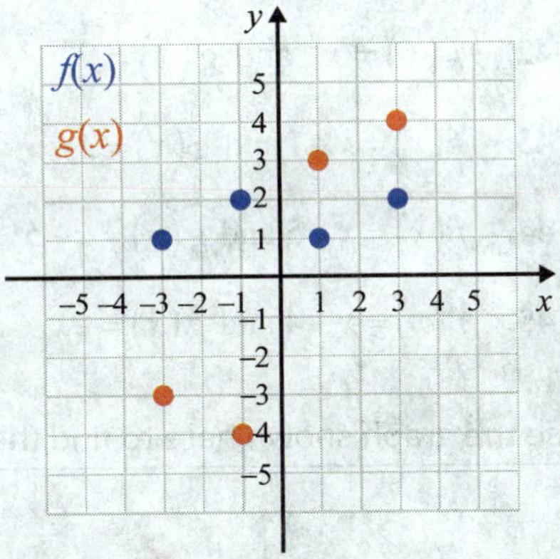

38. $f=\{(-4,6),(-2,0),(0,-2),(3,-3)\}$

$g=\{(-4,0),(-2,-4),(0,1),(3,3)\}$

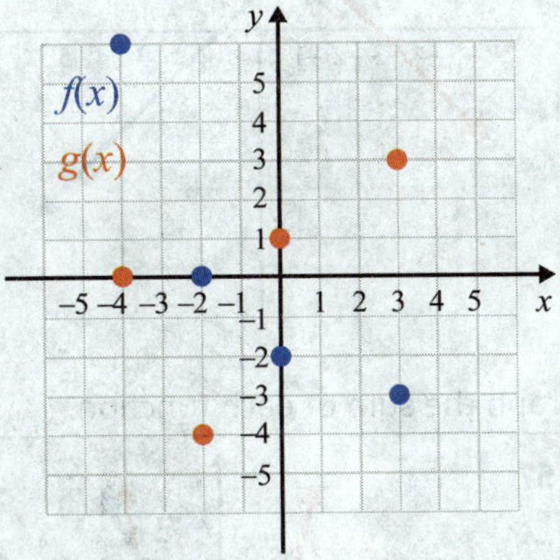

34. a.

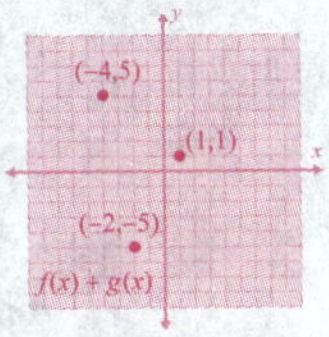

b.

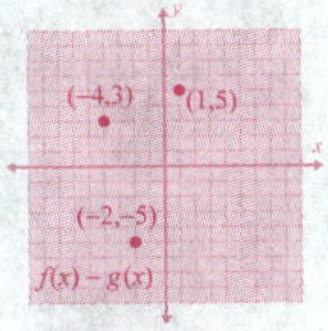

35. a.

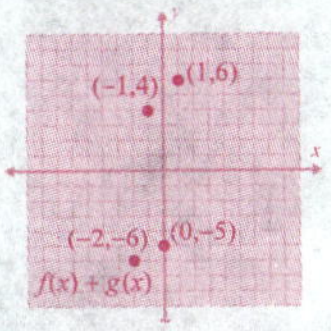

b.

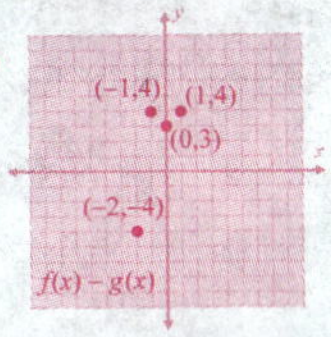

36. a.

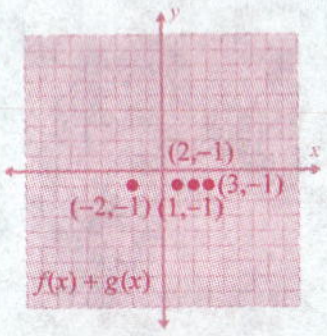

b.

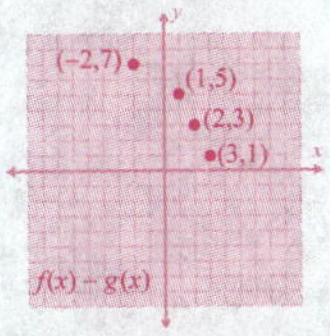

37. a.

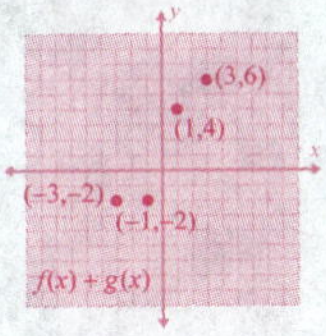

b.

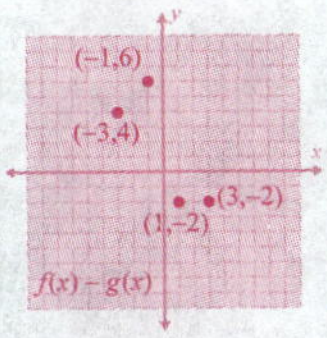

38. a.

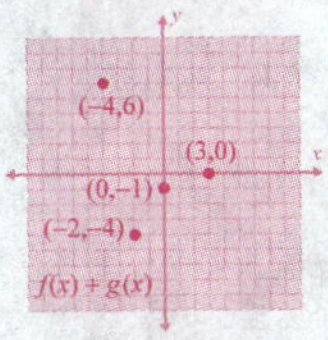

b.

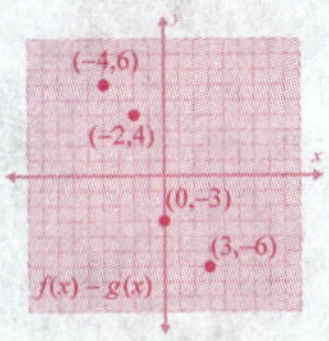

39. a.

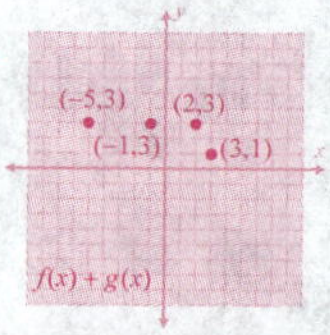

b.

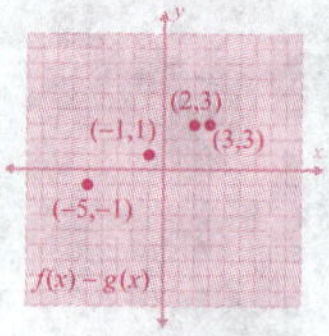

40. a.

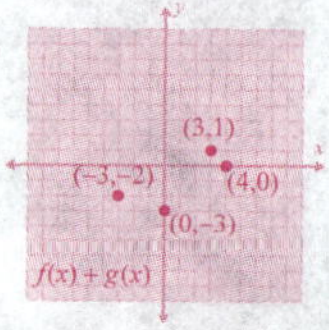

b.

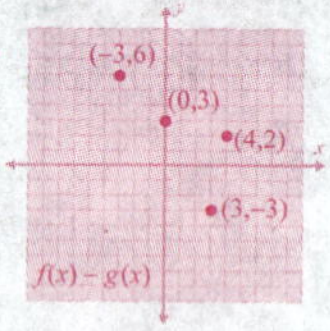

41.

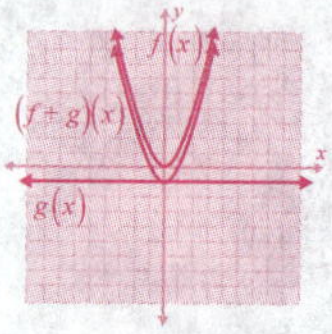

42.

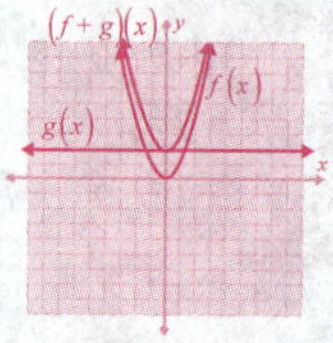

43.

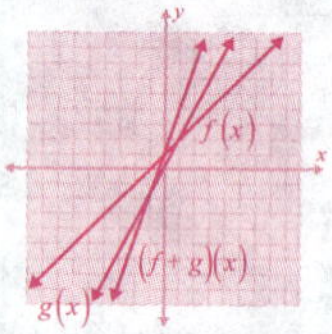

44.

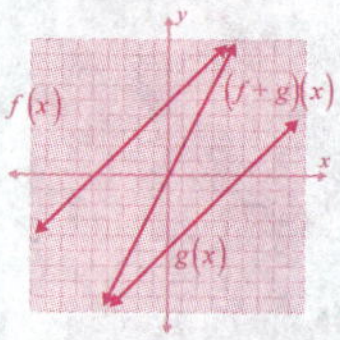

45.

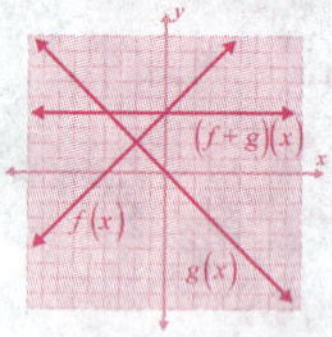

46.

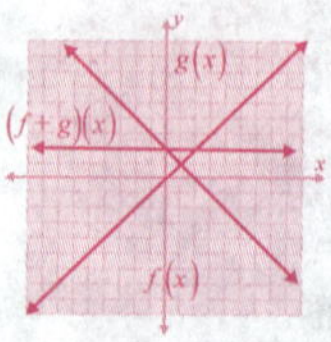

39. $f=\{(-5,1),(-1,2),(2,3),(3,2)\}$
$g=\{(-5,2),(-1,1),(2,0),(3,-1)\}$

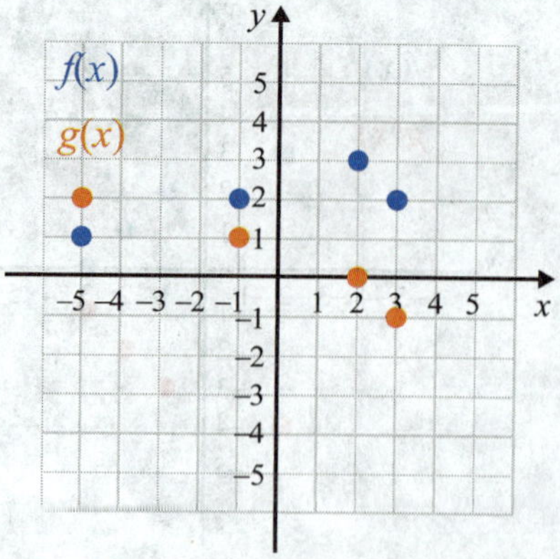

40. $f=\{(-3,2),(0,0),(3,-1),(4,1)\}$
$g=\{(-3,-4),(0,-3),(3,2),(4,-1)\}$

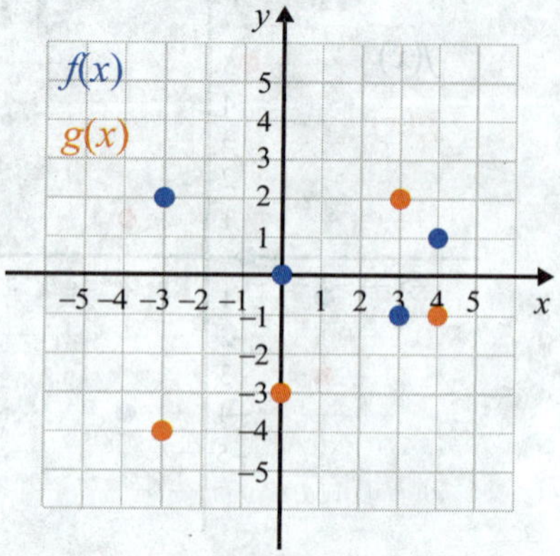

Graph each pair of functions and the sum of these functions on the same set of axes.

41. $f(x)=x^2$ and $g(x)=-1$

42. $f(x)=x^2$ and $g(x)=2$

43. $f(x)=x+1$ and $g(x)=2x$

44. $f(x)=x+5$ and $g(x)=x-5$

45. $f(x)=x+4$ and $g(x)=-x$

46. $f(x)=2-x$ and $g(x)=x$

47. $f(x)=x+1$ and $g(x)=x^2-1$

48. $f(x)=x^2+2$ and $g(x)=x^2-2$

49. $f(x)=\sqrt{x-6}$ and $g(x)=2$

50. $f(x)=\sqrt{3-x}$ and $g(x)=-1$

Use the graph shown here to find the values indicated.

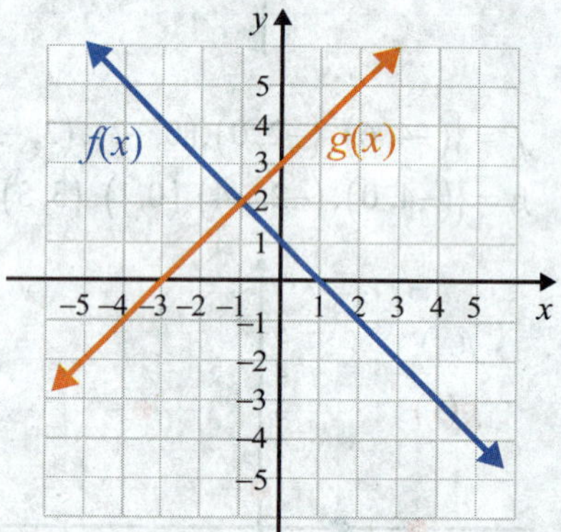

51. $(f+g)(-2)$

52. $(f-g)(2)$

53. $(f\cdot g)(3)$

54. $(g-f)(0)$

55. $\left(\frac{f}{g}\right)(4)$

56. $(g\cdot f)(4)$

Graph the sum of each function.

57.

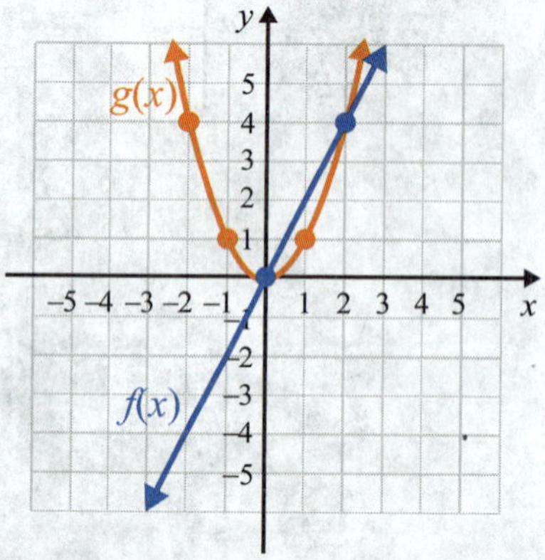

58.

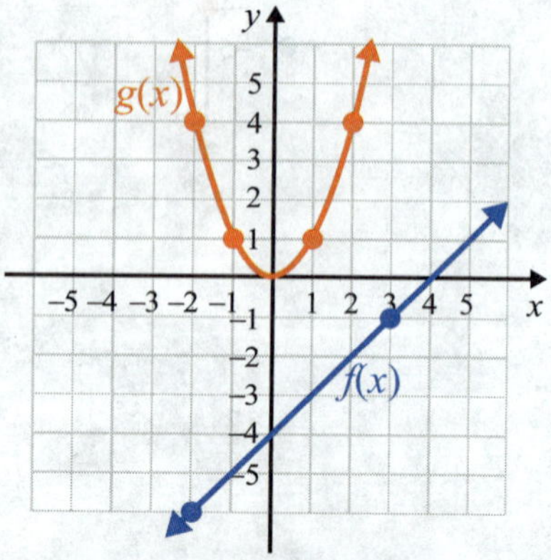

59.

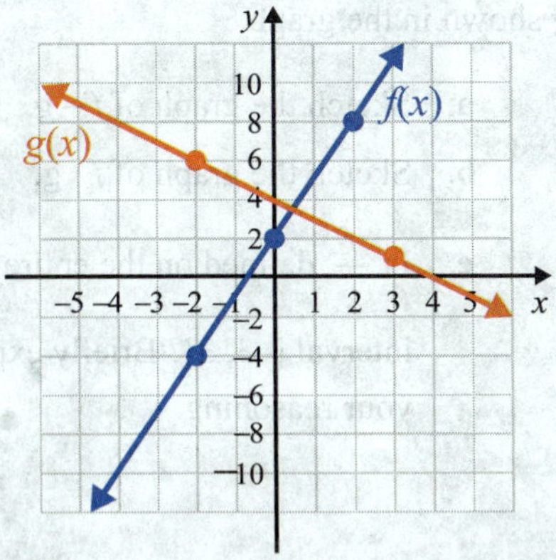

61.

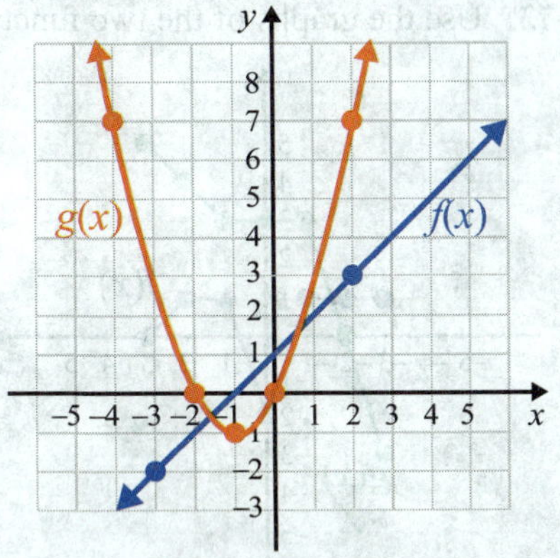

60.

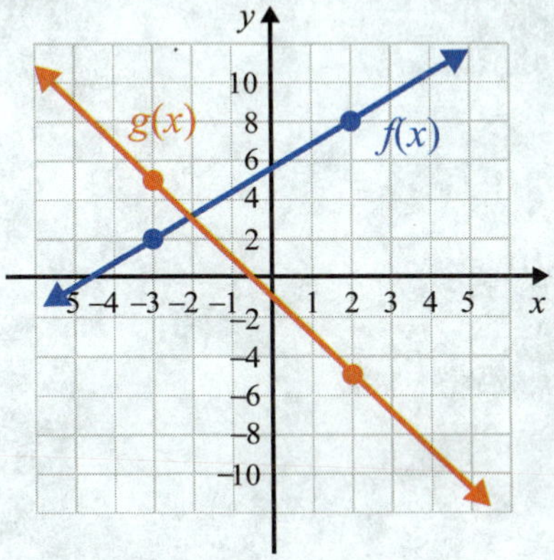

62.

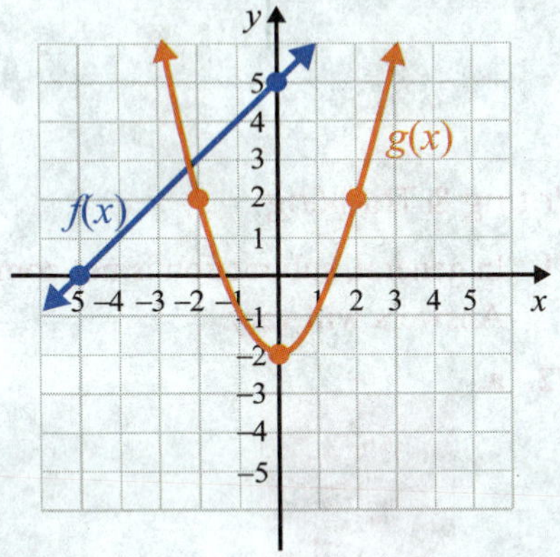

Use a graphing calculator to graph each pair of functions and the sum of these functions on the same set of axes.

63. $f(x)=x^2$ and $h(x)=2x+1$

64. $g(x)=x^2+x$ and $h(x)=3x+4$

65. $f(x)=\sqrt{x+4}$ and $g(x)=-2$

66. $f(x)=-\sqrt{x-1}$ and $g(x)=3$

67. $f(x)=\sqrt[3]{x+5}$ and $h(x)=2x$

68. $h(x)=\sqrt[3]{x-1}$ and $g(x)=x-1$

69. $g(x)=7-x^2$ and $h(x)=x^2-3$

70. $f(x)=x^2+5$ and $g(x)=4-x^2$

Writing & Thinking

71. Explain why, in general, $(f-g)(x)\neq(g-f)(x)$ if $f(x)\neq g(x)$.

72. Given the two functions f and g,

$$f=\{(-2,0),(-1,1),(0,4),(2,4),(3,5),(4,1)\}$$
$$g=\{(-2,3),(-1,4),(0,1),(2,-1),(3,2),(4,6)\},$$

find and graph the following.

a. $f-g$ **b.** $f\cdot g$ **c.** $\dfrac{f}{g}$

47.

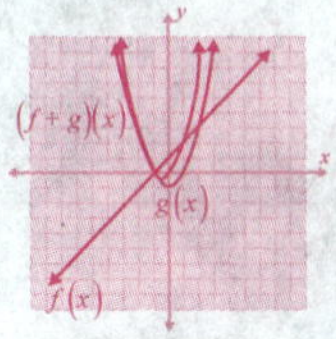

48.

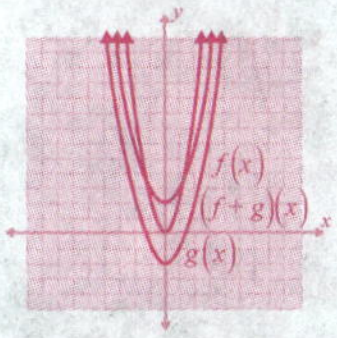

49.

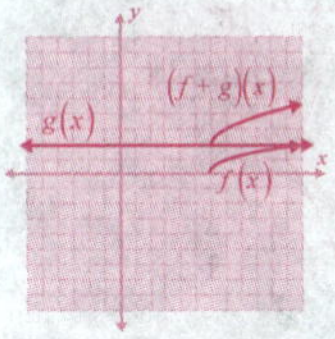

50.

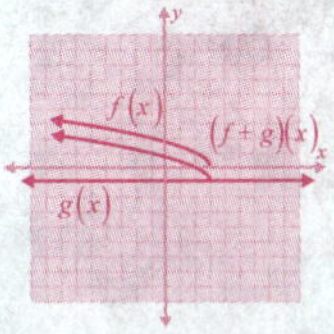

51. 4

52. −6

53. −12

54. 2

55. $-\dfrac{3}{7}$

56. −21

57.

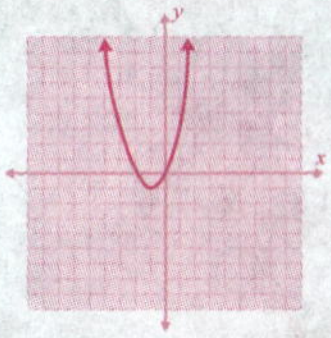

58.

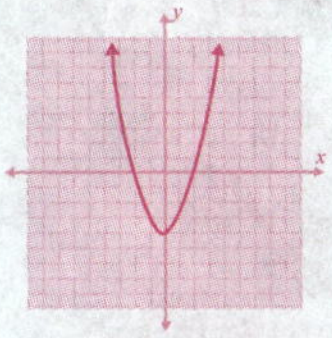

59.

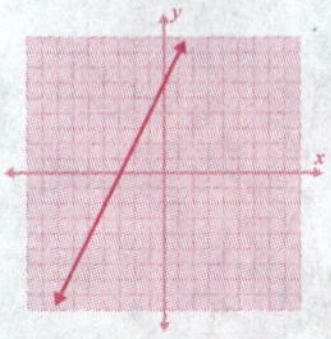

60.

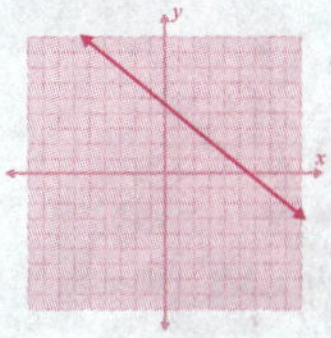

73. Use the graphs of the two functions f and g shown in the graph.

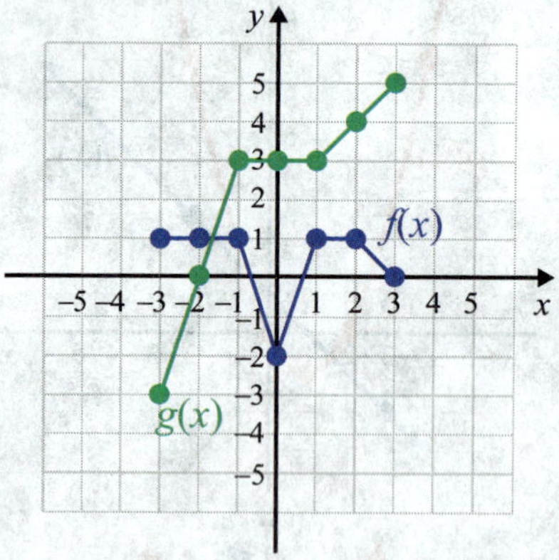

a. Sketch the graph of $f - g$.

b. Sketch the graph of $f \cdot g$.

c. Is $\frac{f}{g}$ defined on the entire interval [–3, 3]? Briefly explain your reasoning.

61.

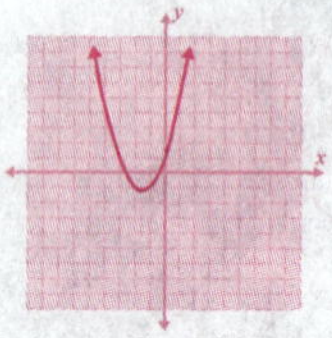

62.

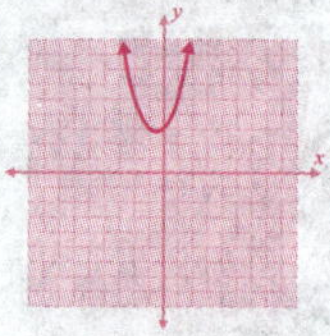

63.

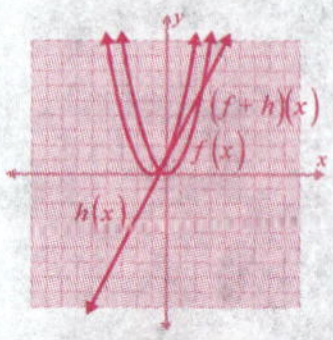

64.

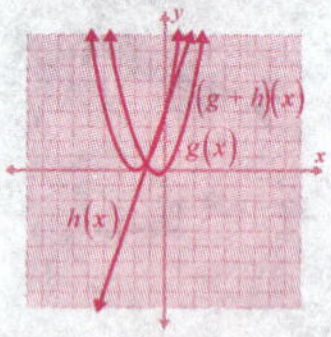

65.

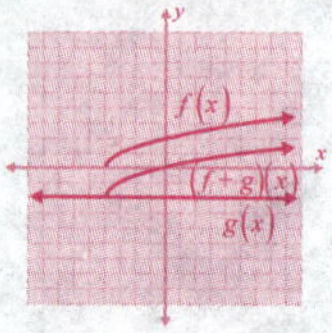

66.

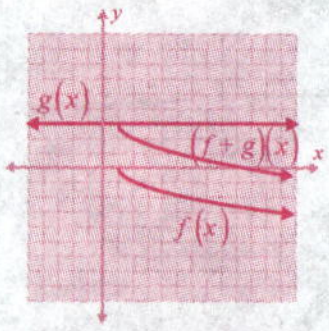

67.

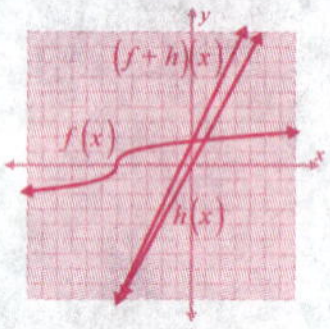

68.

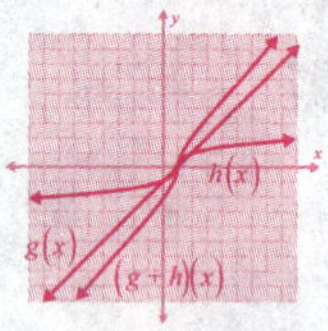

69.

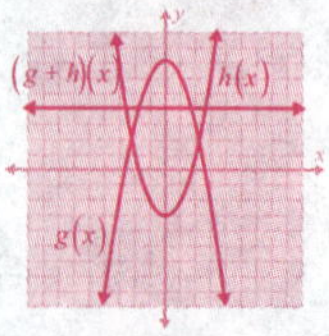

70.

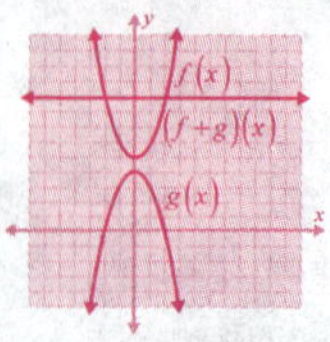

Writing & Thinking

71. In general, subtraction is not commutative. Answers will vary.

72. a.

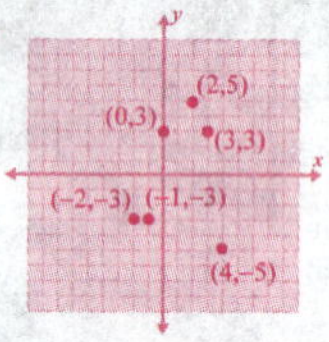

b.

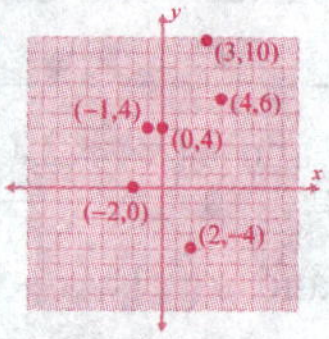

c.

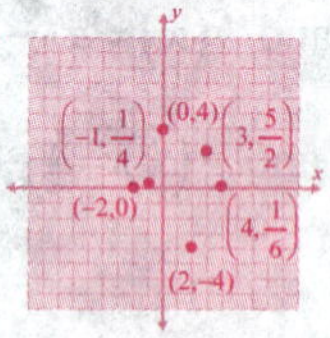

73. a.

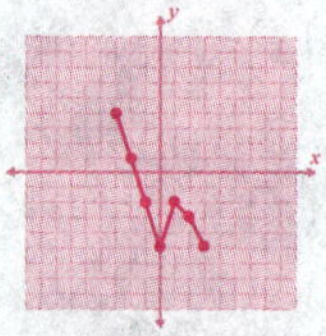

b.

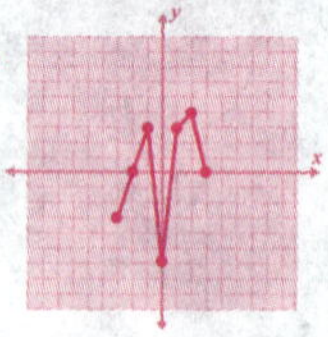

c. $\frac{f}{g}$ is undefined for $x = -2$ as $g(-2) = 0$.

17.2 Composition of Functions and Inverse Functions

Objectives

A. Evaluate a function at a given algebraic expression.

B. Form the composition of two functions.

C. Use the horizontal line test to determine if a function is one-to-one.

D. Determine whether or not two functions are inverses of each other.

E. Find and graph the inverse of a one-to-one function.

A Evaluating a Function at an Algebraic Expression

In Section 10.5, we introduced function notation and how to evaluate a function at a given value of x. Later, in Section 12.4, we expanded this idea and evaluated a function at a given algebraic expression. This is accomplished by replacing the variable with the given expression everywhere the variable appears. The following example reviews this important idea.

Example 1 Evaluating a Function at an Algebraic Expression

Find the following function values when $f(x)=x^2-5x+8$.

a. $f(2a)$ **b.** $f(g+3)$

Solution

a. Substitute $2a$ in for x everywhere x appears and simplify.

$$\begin{aligned} f(2a) &= (2a)^2-5(2a)+8 \\ &= 4a^2-10a+8 \end{aligned}$$

b. Substitute $g+3$ in for x everywhere x appears and simplify.

$$\begin{aligned} f(g+3) &= (g+3)^2-5(g+3)+8 \\ &= g^2+6g+9-5g-15+8 \\ &= g^2+g+2 \end{aligned}$$

Now work margin exercise 1.

1. Find the following function values when $g(x)=x-8x^2$.

a. $g(r+1)$

b. $g(b^2)$

B Composition of Two Functions

Previously, function notation, $f(x)$, proved useful in evaluating functions and in indicating arithmetic operations with functions. This same notation is needed in developing the concept of **a function of a function**, called the **composition** of two functions. Suppose that,

$$g(x)=2x-4 \quad \text{and} \quad f(x)=x^2-4x+1.$$

Now for $x=3$, we have

$$g(3)=2\cdot(3)-4=2 \quad \text{and} \quad f(2)=(2)^2-4\cdot(2)+1=-3.$$

This illustrates the **composition** $f(g(3))$ (read "f of g of 3"), which can be written as

$$f(g(3))=f(2)=-3.$$

We see that $g(3)$, which equals 2, replaces x in the function $f(x)$.

Math Tip

A diagram can be used to show that composition can be thought of as a shortcut when used to evaluate functions.

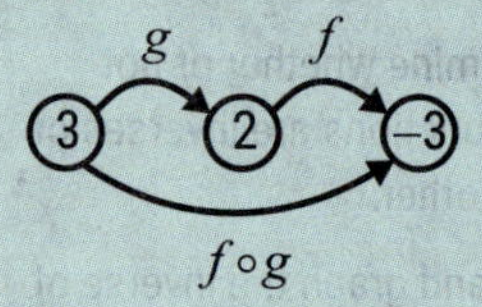

$f \circ g$ takes 3 directly to −3.

When given two functions $f(x)$ and $g(x)$, a new function $f(g(x))$, called the **composition** (or **composite**) of *f* and *g*, is found by substituting the expression for $g(x)$ in place of *x* in the function *f*. In this case, the value of $g(x)$ must be in the domain of *f*. Thus, for

$g(x)=2x-4$ and $f(x)=x^2-4x+1$, the composition

$f(g(x))$ (read "*f* of *g* of *x*") is found as follows.

$$\begin{aligned} f(x) &= x^2-4x+1 \\ f(g(x)) &= (g(x))^2-4(g(x))+1 && \text{Replace the } x \text{ in } f(x) \text{ with } g(x). \\ &= (2x-4)^2-4(2x-4)+1 && \text{Replace } g(x) \text{ with } 2x-4. \\ &= 4x^2-16x+16-8x+16+1 && \text{Simplify.} \\ &= 4x^2-24x+33 \end{aligned}$$

Teaching Note

An overdue question to put to students is whether the sum, difference, product, and quotient of two functions is always a function. You also might want to raise the same question about composition. Is the composition of two functions always a function?

The composition of *g* and *f* (reversing the order of *f* and *g*) is indicated by

$g(f(x))$ (read "*g* of *f* of *x*")

and is found by substituting the expression for $f(x)$ in place of *x* in the function *g*.

$$\begin{aligned} g(x) &= 2x-4 \\ g(f(x)) &= 2(f(x))-4 && \text{Replace the } x \text{ in } g(x) \text{ with } f(x). \\ &= 2(x^2-4x+1)-4 && \text{Replace } f(x) \text{ with } x^2-4x+1. \\ &= 2x^2-8x+2-4 && \text{Simplify.} \\ &= 2x^2-8x-2 \end{aligned}$$

As we can see with these examples, in general, $f(g(x)) \neq g(f(x))$ and substitutions must be done carefully and accurately. The following definition shows another notation (a small raised circle) often used to indicate the composition of functions.

Composite Function

For two functions *f* and *g*, the **composite function** $f \circ g$ is defined as follows.

$$(f \circ g)(x) = f(g(x))$$

The domain of $f \circ g$ consists of those values of *x* in the domain of *g* for which $g(x)$ is in the domain of *f*.

DEFINITION

2. If $f(x)=x^2$ and $g(x)=x^2+2$ form the compositions $(f \circ g)(x)$ and $(g \circ f)(x)$.

Example 2 Forming Compositions

Form the compositions **a.** $(f \circ g)(x)$ and **b.** $(g \circ f)(x)$ if $f(x)=5x+2$ and $g(x)=3x-7$.

Solution

a. $(f \circ g)(x) = f(g(x)) = 5 \cdot g(x)+2 = 5(3x-7)+2 = 15x-33$

b. $(g \circ f)(x) = g(f(x)) = 3 \cdot f(x)-7 = 3(5x+2)-7 = 15x-1$

Note: Both $(f \circ g)(x)$ and $(g \circ f)(x)$ are defined for all real numbers.

Now work margin exercise 2.

Example 3 Forming Compositions

Form the composite functions **a.** $(f \circ g)(x)$ and **b.** $(g \circ f)(x)$ if $f(x)=\sqrt{x-3}$ and $g(x)=x^2+4$.

Solution

a. $(f \circ g)(x)=\sqrt{g(x)-3}$

$=\sqrt{(x^2+4)-3}=\sqrt{x^2+1}$

Note: For the expression under the radical to be defined, we must have $g(x)\geq 3$. Because $x^2+4\geq 3$ for all real numbers, the domain of $(f \circ g)(x)$ is all real numbers.

b. Now, for $(g \circ f)(x)$ the domain is restricted by $f(x)=\sqrt{x-3}$. That is, the domain is restricted to $x\geq 3$ regardless of the simplified result.

$(g \circ f)(x)=(f(x))^2+4$

$=(\sqrt{x-3})^2+4=x-3+4=x+1$

Now work margin exercise 3.

3. If $f(x)=\sqrt{5-x}$ and $g(x)=x^2-2$ form the compositions $(f \circ g)(x)$ and $(g \circ f)(x)$.

Example 4 Finding Compositions

Find **a.** $f(g(x))$ and **b.** $g(f(x))$ if $f(x)=\sqrt{x+3}$ and $g(x)=2x-5$.

Solution

a. $f(g(x))=\sqrt{g(x)+3}$

$=\sqrt{(2x-5)+3}=\sqrt{2x-2}$

Note: $f(g(x))$ is defined only for $2x-2\geq 0$ or $x\geq 1$.

b. $g(f(x))=2(f(x))-5$

$=2(\sqrt{x+3})-5=2\sqrt{x+3}-5$

Note: $g(f(x))$ is defined only for $x\geq -3$.

Now work margin exercise 4.

4. If $f(x)=3x+3$ and $g(x)=\sqrt{x+3}$ form the compositions $f(g(x))$ and $g(f(x))$.

C One-to-One Functions

By the definition of a function, there is only one corresponding y-value for each x-value in a function's domain. Graphically, the **vertical line test** can be used to help determine whether a graph represents a function. Now in order to develop the concept of **inverse functions**, we need to study functions that have only one x-value for each y-value in the range. Such functions are said to be **one-to-one functions** (or **1-1 functions**).

Consider the following **functions**.

$$f=\{(1,2),(2,4),(3,6),(4,8),(5,10)\} \quad \text{and} \quad g=\{(-2,6),(0,6),(1,5),(2,4),(4,1)\}$$

Both sets of ordered pairs are functions because each value of x appears only once. In the function f, each y-value appears only once. But, in function g, the y-value 6 appears twice: in $(-2, 6)$ and $(0, 6)$. Figure 1 illustrates both functions.

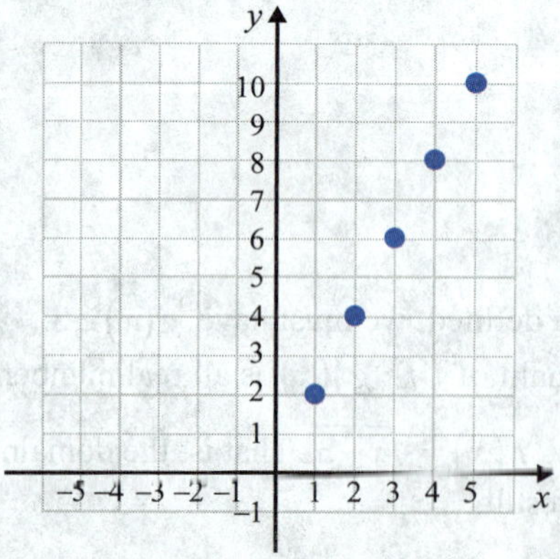

Function f is a one-to-one function.

Figure 1a

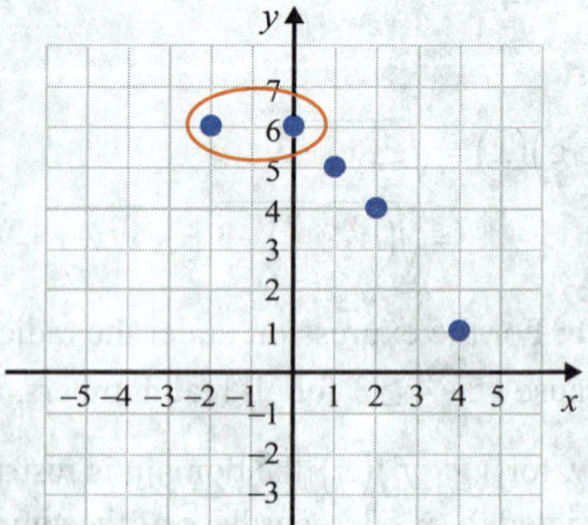

Function g is not a one-to-one function.

Figure 1b

One-to-One Functions

A function is a **one-to-one function** (or **1-1 function**) if for each value of y in the range there is only one corresponding value of x in the domain.

DEFINITION

Graphically, as illustrated in Figure **1b**, if a horizontal line intersects the graph of a function in more than one point, then it is **not** one-to-one. This is, in effect, the **horizontal line test.**

Horizontal Line Test

A function is one-to-one if no **horizontal line** intersects the graph of the function at more than one point.

DEFINITION

The graphs in Figure 2 illustrate the concept of one-to-one functions. (Note that each graph passes the vertical line test and is indeed a function.)

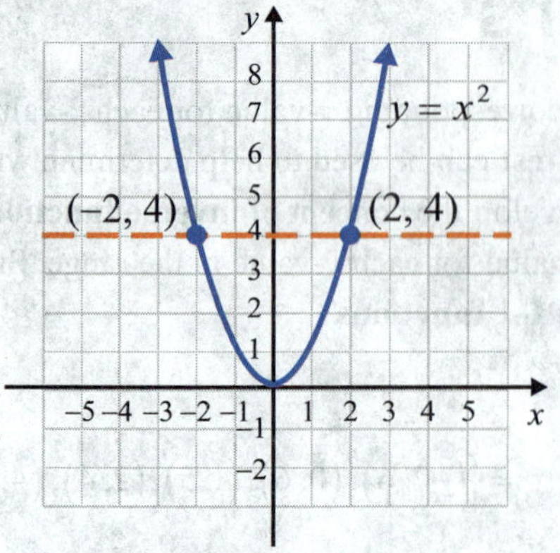

Not one-to-one

Figure 2a

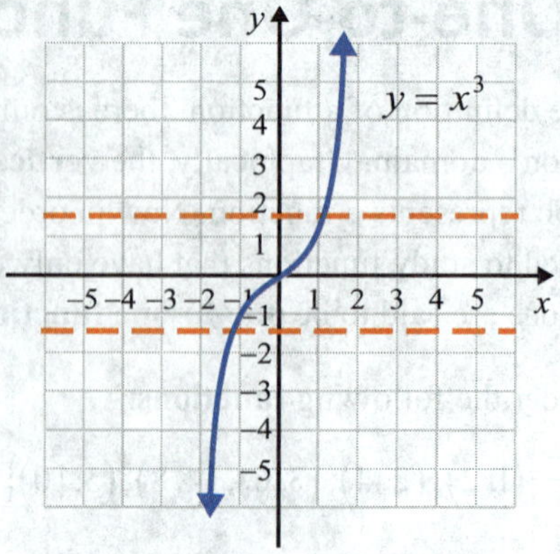

One-to-one

Figure 2b

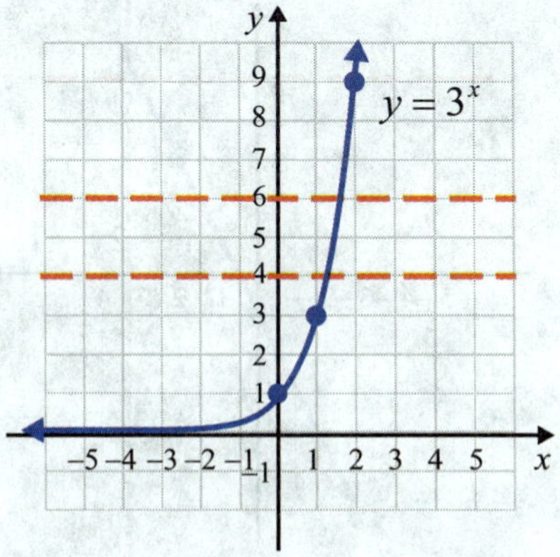

One-to-one

Figure 2c

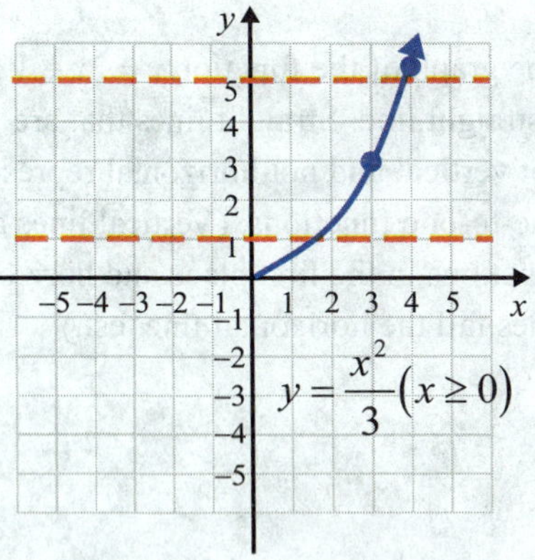

One-to-one

Figure 2d

Example 5 Determining One-to-One Functions

Determine whether each function is one-to-one.

a. $y = \sqrt{x+5}$

b. $f = \{(-3,4),(-2,1),(0,4),(3,1)\}$

c. $y = 2x - 1$

d. $y = -x^2 + 1$

Solution

a. The horizontal line test shows that this function is one-to-one.

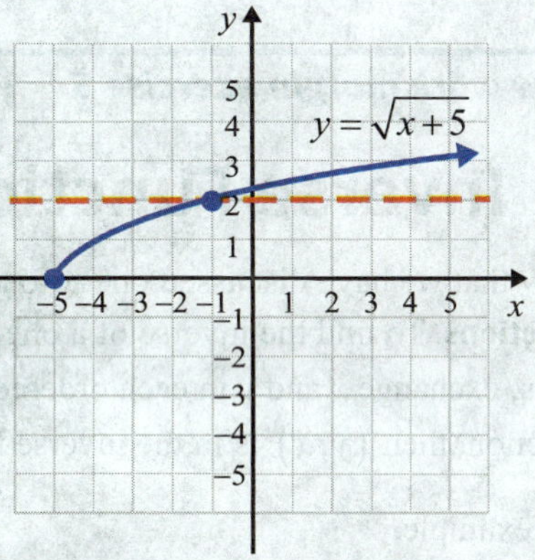

b. This function is not one-to-one. Both y-values, 4 and 1, have more than one corresponding x-value.

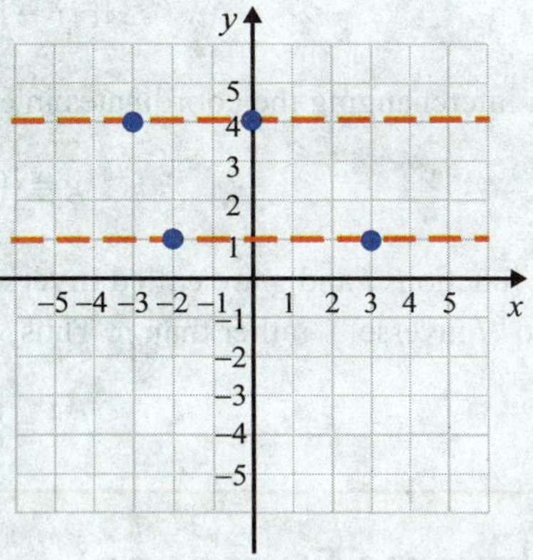

5. Determine whether the following functions are one-to-one.

a. $y = \sqrt{3x+3}$

b. $y = x^2 + x + 17$

c. The graph of the function $y = 2x - 1$ is a straight line. Straight lines that are not vertical and not horizontal represent one-to-one functions. (Vertical lines are not functions in the first place and horizontal lines fail the horizontal line test.)

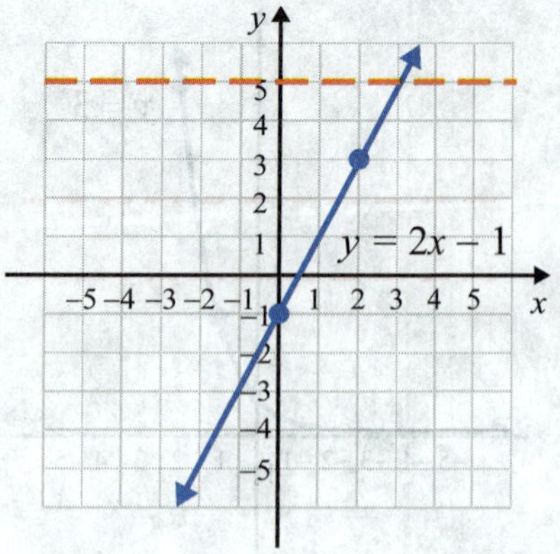

d. The graph of the function $y = -x^2 + 1$ is a parabola and the horizontal line test shows that the function is not one-to-one.

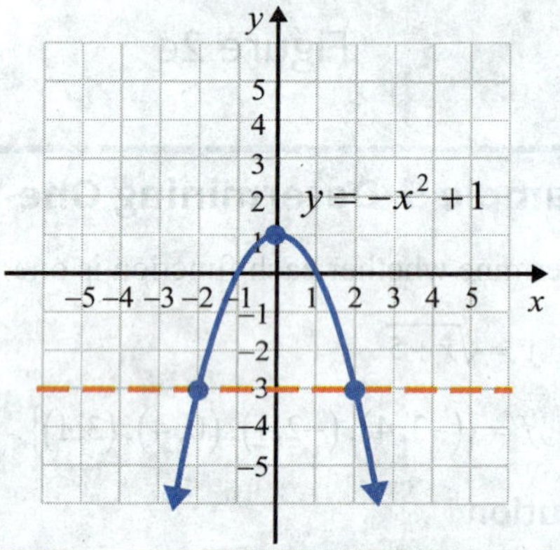

Now work margin exercise 5.

D Inverse Functions

Now that we have discussed one-to-one functions, we can develop the concept of **inverse functions**. To find the inverse of a one-to-one function represented by a set of ordered pairs, exchange x and y in each ordered pair. That is, if (x, y) is in the original one-to-one function, then (y, x) is in the inverse function.

For example,

$$\text{if } f = \{(-1, 1), (0, 2), (1, 4)\},$$

then interchanging the coordinates in each ordered pair gives

$$g = \{(1, -1), (2, 0), (4, 1)\}.$$

The functions f and g are called **inverses** of each other. If g is the inverse of f, we write f^{-1} (read "f inverse") rather than g. Thus, in this example we can write

$$f^{-1} = \{(1, -1), (2, 0), (4, 1)\}.$$

Inverse Functions

If f is a one-to-one function with ordered pairs of the form (x, y), then its **inverse function**, denoted as f^{-1}, is also a one-to-one function with ordered pairs of the form (y, x).

DEFINITION

Only one-to-one functions have inverse functions. If a function is not one-to-one, then interchanging the x- and y-values would yield a relation that is not a function. Graphically, the function must satisfy the horizontal line test. If it does not, then the inverse will not pass the vertical line test and will not be a function.

The graph of any point (b, a) is the reflection of the point (a, b) across the line $y = x$. Thus, the points of the inverse function f^{-1} are reflections of the points of the function f across the line $y = x$. We say that the graphs are **symmetric about the line $y = x$**. Figure 3 illustrates these reflections and the symmetry.

We see that the domain, denoted D_f, and range, denoted R_f, of the two functions are interchanged.

$$D_f = R_{f^{-1}} = \{-1, 0, 1\}$$
$$R_f = D_{f^{-1}} = \{1, 2, 4\}$$

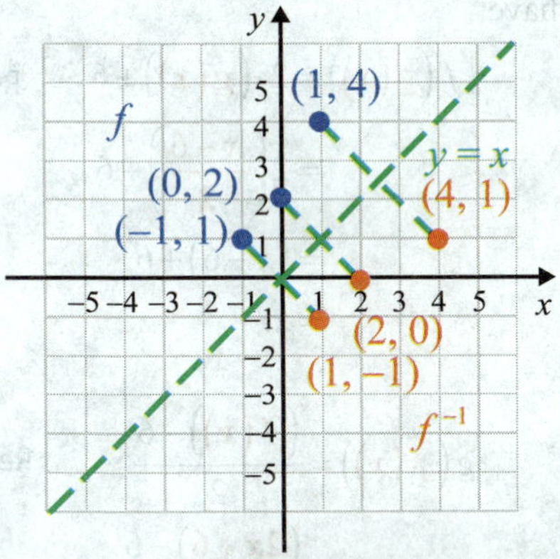

Figure 3

Note

The notation $f^{-1}(x)$ represents the inverse of a one-to-one function. This inverse is a new function in which the x- and y-values have been interchanged. $f^{-1}(x)$ does NOT mean $\dfrac{1}{f(x)}$ because the -1 is NOT an exponent.

In general, every one-to-one function has an inverse function (or inverse), and the graph of the inverse function of any one-to-one function f can be found by reflecting the graph of f across the line $y = x$. Figure 4 shows two more illustrations of this concept.

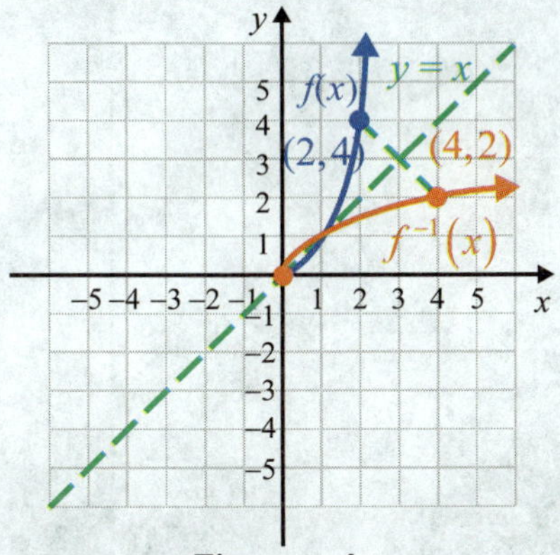

Figure 4a

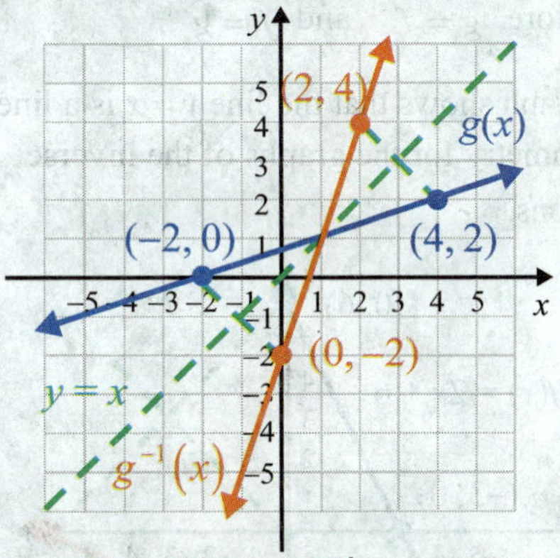

Figure 4b

Teaching Note

Have students do paper folds along $y = x$ to manually reflect the inverse graph. Onion-skin paper is good for this. Transparent acetates are especially ideal for showing the reflection by tracing the original function and the axes and line $y = x$, and then lifting the image, rotating, and repositioning the axes and line $y = x$. The x-axis becomes the y-axis, etc. The inverse graph is now in position.

The following definition of inverse functions helps to determine whether two functions are inverses of each other.

Determining Whether Two Functions Are Inverses

If f and g are one-to-one functions and

$$f(g(x)) = x \quad \text{for all } x \text{ in } D_g, \text{ and}$$
$$g(f(x)) = x \quad \text{for all } x \text{ in } D_f,$$

then f and g are **inverse functions**.

That is, $g = f^{-1}$ and $f = g^{-1}$.

PROCEDURE

17.2 Exercises

Concept Check

1. variable, variable
2. expression, function
3. $\neq$
4. domain, domain
5. one-to-one
6. one-to-one, inverse
7. False; The vertical line test determines whether a graph represents a function.
8. True

6. Show that f and g are inverse functions.

$f(x)=5x-2$ and

$g(x)=\dfrac{x+2}{5}$

9. True

10. False; The notation represents the inverse of a one-to-one function.

Practice

1. **a.** $8r-5$

 b. $24a-13$

2. **a.** $4g-26$

 b. $4h^2+26$

3. **a.** $5x^2-20x+24$

 b. $45n^4+4$

4. **a.** $81p^4+8$

 b. $16s^8+8$

5. **a.** $3n^2-6n-9$

 b. $48y^6+24y^3-9$

6. **a.** $25k^2-10k+7$

 b. x^2+6

7. $f(g(2))=14$

 $g(f(2))=\dfrac{15}{2}$

8. $f(g(4))=\dfrac{21}{4}$

 $g(f(4))=5$

9. $(f\circ g)(-5)=49$

 $(g\circ f)(-1)=5$

10. $(f\circ g)(3)=10$

 $(g\circ f)(-2)=-1$

11. $f(g(x))=\sqrt{x^2}=|x|$

 $g(f(x))=(\sqrt{x})^2=x$

Example 6 Verifying Inverse Functions

Use the definition of inverse functions to show that f and g are inverse functions.

$$f(x)=2x+6 \text{ and } g(x)=\frac{x-6}{2}.$$

Solution

$f(x)=2x+6$ and $g(x)=\dfrac{x-6}{2}$. The domain of both functions is the set of all real numbers.

We have

$$\begin{aligned} f(g(x)) &= 2\cdot(g(x))+6 && \text{Replace the } x \text{ in } f(x) \text{ with } g(x) \text{ and simplify.}\\ &= 2\left(\frac{x-6}{2}\right)+6\\ &= (x-6)+6\\ &= x. \end{aligned}$$

Also,

$$\begin{aligned} g(f(x)) &= \frac{(f(x))-6}{2} && \text{Replace the } x \text{ in } g(x) \text{ with } f(x) \text{ and simplify.}\\ &= \frac{(2x+6)-6}{2}\\ &= \frac{2x}{2}\\ &= x. \end{aligned}$$

Therefore, $g=f^{-1}$ and $f=g^{-1}$.

The graph shows that the line $y=x$ is a line of symmetry for the graphs of the inverse functions.

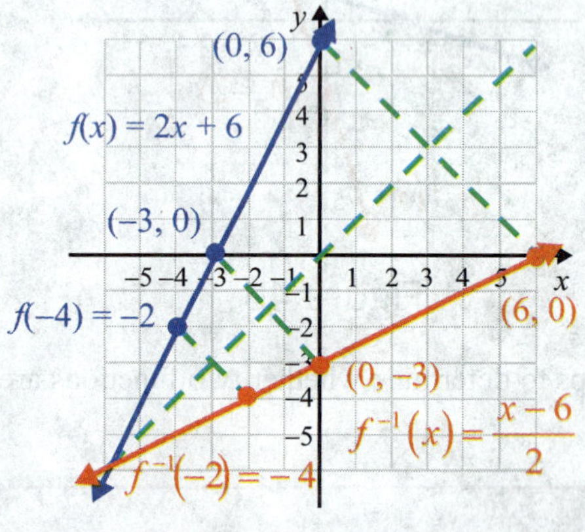

Now work margin exercise 6.

7. Show that f and g are inverse functions.

$f(x)=x^2-3$ for $x\ge 0$, and

$g(x)=\sqrt{x+3}$

Example 7 Verifying Inverse Functions

Use the definition of inverse functions to show that f and g are inverse functions.

$$f(x)=\sqrt{x-3} \text{ and } g(x)=x^2+3 \text{ for } x\ge 0.$$

Solution

$f(x)=\sqrt{x-3}$ and $g(x)=x^2+3$ for $x\geq 0$. The domain of f is the interval $[3,\infty)$ and the domain of g is the interval $[0,\infty)$.

We have,

$$\begin{aligned} f(g(x)) &= \sqrt{g(x)-3} && \text{Replace the } x \text{ in } f(x) \text{ with } g(x) \text{ and simplify.} \\ &= \sqrt{(x^2+3)-3} \\ &= \sqrt{x^2} \\ &= x \quad \text{for } x\geq 0. \end{aligned}$$

Also,

$$\begin{aligned} g(f(x)) &= (f(x))^2+3 && \text{Replace the } x \text{ in } g(x) \text{ with } f(x) \text{ and simplify.} \\ &= \left(\sqrt{x-3}\right)^2+3 \\ &= x-3+3 \\ &= x \quad \text{for } x\geq 3. \end{aligned}$$

Therefore, $g=f^{-1}$ and $f=g^{-1}$.

The graph shows that the line $y=x$ is a line of symmetry for the graphs of the inverse functions. Note that the figure only shows a portion of each graph and the domains have been restricted accordingly.

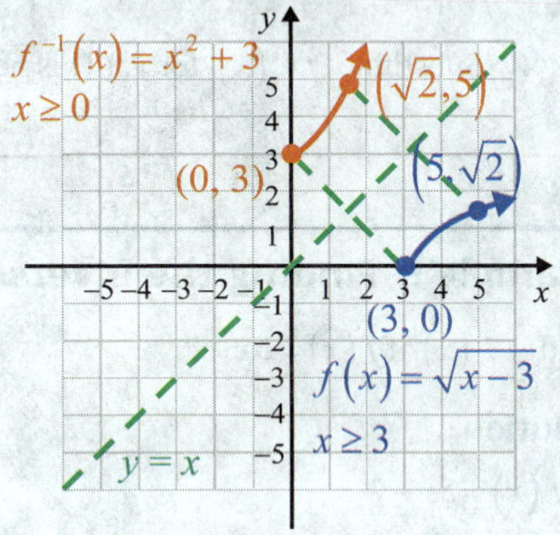

Now work margin exercise 7.

Example 8 Evaluating Compositions of Inverses

Given $f(x)=\sqrt{x+4}$ for $x\geq -4$ and $f^{-1}(x)=x^2-4$ for $x\geq 0$, evaluate the indicated compositions.

a. $f\left(f^{-1}(2)\right)$

b. $f\left(f^{-1}(5)\right)$

c. $f^{-1}\left(f(-2)\right)$

d. $f^{-1}\left(f(-5)\right)$

Solution

a. $f^{-1}(2)=(2)^2-4=0$, so $f\left(f^{-1}(2)\right)=f(0)=\sqrt{(0)+4}=\sqrt{4}=2$

b. $f^{-1}(5)=(5)^2-4=21$, so $f\left(f^{-1}(5)\right)=f(21)=\sqrt{(21)+4}=\sqrt{25}=5$

c. $f(-2)=\sqrt{(-2)+4}=\sqrt{2}$, so $f^{-1}\left(f(-2)\right)=f^{-1}\left(\sqrt{2}\right)=\left(\sqrt{2}\right)^2-4=2-4=-2$

Math Tip

A parabola is not usually considered one-to-one as it fails the horizontal line test. However, if you restrict the domain starting at the vertex, the resulting curve is a one-to-one function.

12. $f(g(x))=x$
$g(f(x))=x$

13. $f(g(x))=\sqrt{x-2}$
$g(f(x))=\sqrt{x}-2$

14. $f(g(x))=\sqrt{x^2-9}$
$g(f(x))=x-9$

15. $f(g(x))=\frac{1}{x^2}-1$
$g(f(x))=\frac{1}{(x-1)^2}$

16. $f(g(x))=\frac{1}{(x^2+1)^2}$
$g(f(x))=\frac{1}{x^4}+1$

17. $f(g(x))=x^3+3x^2+4x+3$
$g(f(x))=x^3+x+2$

18. $f(g(x))=(2x-1)^3$
$g(f(x))=2x^3-1$

8. Given
$f(x)=\sqrt{x+2}$ for $x\geq -2$
and
$f^{-1}(x)=x^2-2$ for $x\geq 0$
evaluate the indicated compositions.

a. $f\left(f^{-1}(4)\right)$

b. $f^{-1}\left(f(-1)\right)$

19. $f(g(x)) = \frac{1}{|x|}$

$g(f(x)) = \frac{1}{x}$

20. $f(g(x)) = \frac{1}{\sqrt{x^2 - 4}}$

$g(f(x)) = \frac{1}{x} - 4$

21.

$f(g(x)) = \frac{1}{x^2 + 7x - 8}$

$g(f(x)) = \left(\frac{1}{x}\right)^2 + 7\left(\frac{1}{x}\right) - 8$

22.

$f(g(x)) = \frac{1}{x^2 + x - 2}$

$g(f(x)) = \frac{1}{(x+1)^2} + \frac{1}{x+1} - 3$

23. $f(g(x)) = (2x - 6)^{3n}$

$g(f(x)) = 2x^{3n} - 6$

d. $f(-5)$ does not exist because -5 is not in the domain of f. Therefore, $f^{-1}(f(-5))$ does not exist.

Now work margin exercise 8.

E Finding the Inverse of a One-to-One Function

In Examples 6 and 7 the given functions were inverses of each other. It is important to be able to find the inverse of a given one-to-one function. The following procedure shows one method for finding the inverse using the fact that if an ordered pair (x, y) belongs to the function f, then (y, x) belongs to f^{-1}.

Finding the Inverse of a One-to-One Function

1. Let $y = f(x)$. (In effect, substitute y for $f(x)$.)
2. Interchange x and y.
3. In the new equation, solve for y in terms of x.
4. Substitute $f^{-1}(x)$ for y. (This new function is the inverse of f.)

PROCEDURE

9. Find $f^{-1}(x)$ if $f(x) = 7x + 2$.

Example 9 Finding the Inverse

Find $f^{-1}(x)$ if $f(x) = 5x - 7$.

Solution

$f(x) = 5x - 7$

$y = 5x - 7$ — Substitute y for $f(x)$.

$x = 5y - 7$ — Interchange x and y.

$x + 7 = 5y$ — Solve for y in terms of x.

$\frac{x+7}{5} = y$

$f^{-1}(x) = \frac{x+7}{5}$ — Substitute $f^{-1}(x)$ for y.

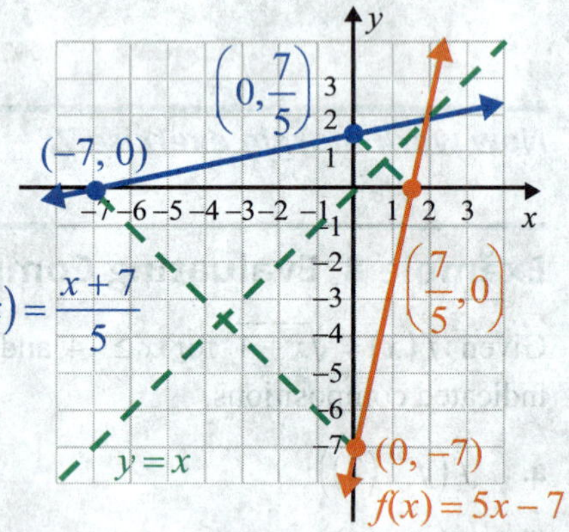

Now work margin exercise 9.

24. $f(g(x)) = (4x + 7)^{\frac{1}{3}}$

$g(f(x)) = 4x^{\frac{1}{3}} + 7$

25. $f(g(x)) = (x - 8)^{\frac{3}{2}}$

$g(f(x)) = \sqrt{x^3 - 8}$

26. $f(g(x)) = \frac{1}{x^3} + 1$

$g(f(x)) = \frac{1}{x^3 + 1}$

27. **a.** 21

b. 2

c. No; $f(g(x)) \neq x$ and $g(f(x)) \neq x$

28. **a.** 6

b. −4

c. yes; $f(g(x)) = g(f(x)) = x$

Example 10 Finding the Inverse

Find $g^{-1}(x)$ if $g(x)=x^2-2$ for $x \geq 0$.

Solution

$g(x)=x^2-2$	For $x \geq 0$ (Note that g is one-to-one for $x \geq 0$.)
$y=x^2-2$	Substitute y for $g(x)$.
$x=y^2-2$	Interchange y and x.
$\pm\sqrt{x+2}=y$	Solve for y in terms of x.
$g^{-1}(x)=\sqrt{x+2}$	Take the positive square root because we must have $y \geq 0$. (The domain of g is $x \geq 0$, so the range of g^{-1} is $y \geq 0$.)

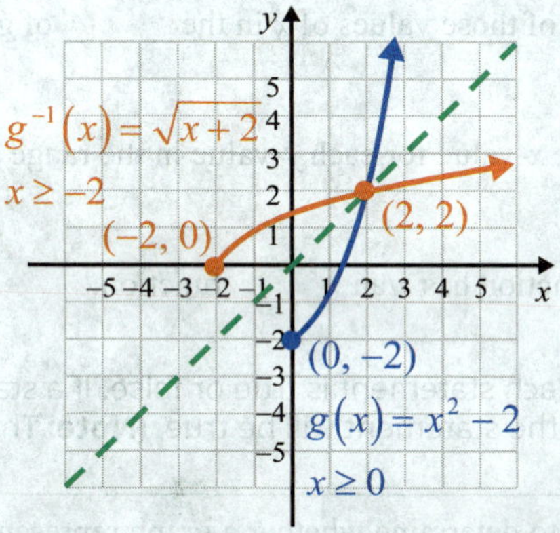

Now work margin exercise 10.

10. Find $f(x)$ if $f^{-1}(x)=\sqrt{2x+2}$ for $x \geq -1$.

Margin Exercise Answers

1. a. $-8r^2-15r-7$ **b.** b^2-8b^4

2. $(f \circ g)(x)=(x^2+2)^2$ or x^4+4x^2+4 and $(g \circ f)(x)=x^4+2$

3. $(f \circ g)(x)=\sqrt{7-x^2}$ and $(g \circ f)(x)=3-x$ **4.** $f(g(x))=3\sqrt{x+3}+3$ and $g(f(x))=\sqrt{3x+6}$

5. a. one-to-one **b.** not one-to-one **6.** $f(g(x))=5\left(\frac{x+2}{5}\right)-2$
$=x+2-2$
$=x$
$g(f(x))=\frac{(5x-2)+2}{5}$
$=\frac{5x}{5}$
$=x$

7. $f(g(x))=\left(\sqrt{x+3}\right)^2-3$
$=x+3-3$
$=x$ for $x \geq -3$
$g(f(x))=\sqrt{(x^2-3)+3}$
$=\sqrt{x^2}$
$=x$ for $x \geq 0$

8. a. 4 **b.** −1 **9.** $f^{-1}(x)=\frac{x-2}{7}$

10. $f(x)=\frac{x^2-2}{2}$ for $x \geq 0$

29. a. −9

b. does not exist

c. $f(4)=\frac{1}{9}$ and $\frac{1}{9}$ is in the domain of g, so $g(f(4))$ is defined. However, $g(2)=-\frac{1}{2}$ and $-\frac{1}{2}$ is not in the domain of f, so $f(g(2))$ is not defined.

30. a. 1

b. does not exist

c. $f(109)=10$ and 10 is in the domain of g, so $g(f(109))$ is defined. However, $g(9)=0$ and 0 is not in the domain of f, so $f(g(9))$ is not defined.

31.

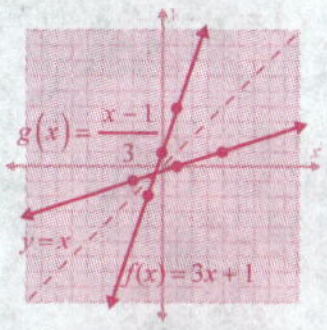

32.

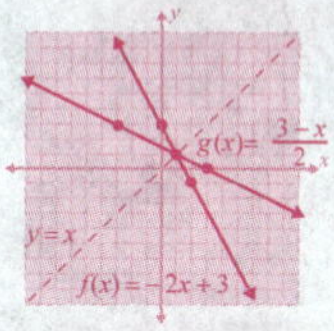

33.

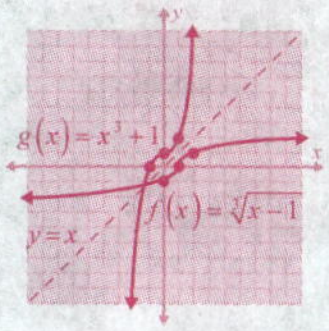

34.

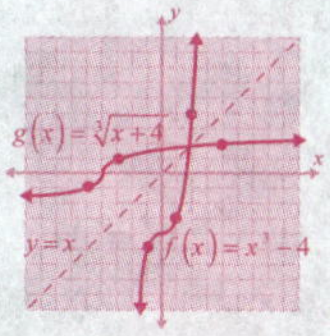

35.

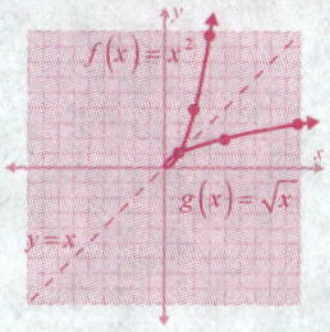

36. 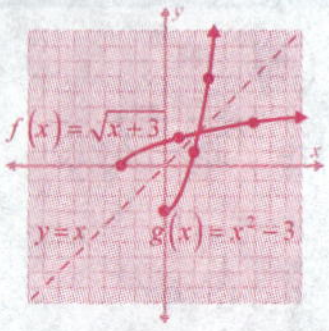

37. 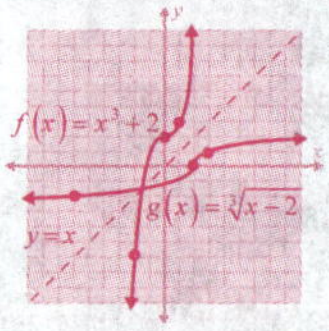

38. 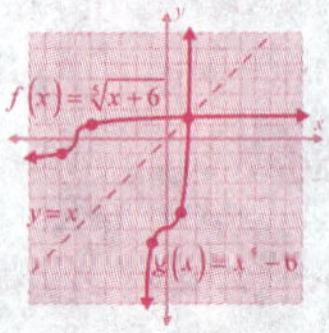

39. 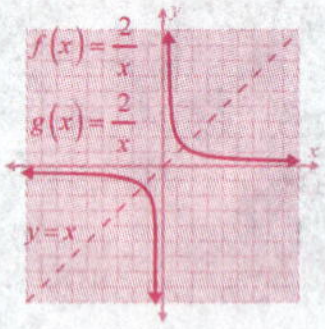

40. 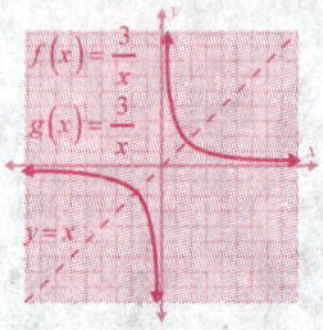

41. $f^{-1}(x)=\frac{x+3}{2}$

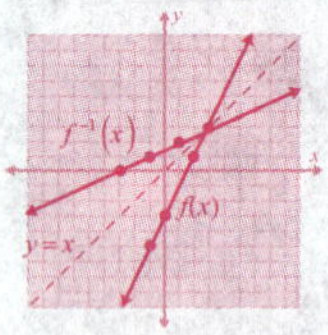

42. $f^{-1}(x)=\frac{x+5}{2}$

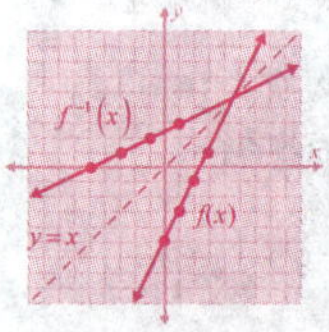

43. $g^{-1}(x)=x$

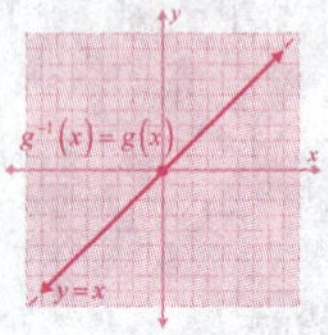

17.2 Exercises

Concept Check

Fill-in-the-Blank. Complete each sentence using information found in this section.

1. To evaluate a function at an algebraic expression, replace the ______ with the expression everywhere the ______ appears.

2. Given two functions $f(x)$ and $g(x)$ a new function $f(g(x))$, called the composition of f and g, is found by substituting the ______ for $g(x)$ into the place of x in the ______ f.

3. In general, $f(g(x))$ ______ $g(f(x))$.

4. The domain of $f \circ g$ consists of those values of x in the ______ of g for which $g(x)$ is in the ______ of f.

5. Functions that have only one x-value for each y-value in the range are said to be ________ functions.

6. In general, every ______ function has a/an ______ function.

True/False. Determine whether each statement is true or false. If a statement is false, explain how it can be changed so the statement will be true. (**Note:** There may be more than one acceptable change.)

7. The vertical line test is used to determine whether a graph represents a vertical line.

8. In a one-to-one function, each x-value corresponds to exactly one y-value.

9. The horizontal line test is used to determine whether a graph of a function is one-to-one.

10. The notation $f^{-1}(x)$ means $\frac{1}{f(x)}$.

Practice

Find the indicated function values for each function given. See Example 1.

1. $f(x)=8x-5$
   a. $f(r)$
   b. $f(3a-1)$

2. $r(x)=4x-6$
   a. $r(g-5)$
   b. $r(h^2+8)$

3. $g(y)=5y^2+4$
   a. $g(x-2)$
   b. $g(3n^2)$

4. $h(y)=y^4+8$
   a. $h(3p)$
   b. $h(2s^2)$

5. $f(c)=3c^2+6c-9$
   a. $f(n-2)$
   b. $f(4y^3)$

6. $b(t)=t^2-2t+7$
   a. $b(5k)$
   b. $b(x+1)$

Find the following function compositions.

7. $f(x)=3x+5,\ g(x)=\dfrac{x+4}{2}$ Find **a.** $f(g(2))$ and **b.** $g(f(2))$.

8. $f(x)=\dfrac{1}{4}x+1,\ g(x)=6x-7$ Find **a.** $f(g(4))$ and **b.** $g(f(4))$.

9. $f(x)=x^2,\ g(x)=2x+3$ Find **a.** $(f\circ g)(-5)$ and **b.** $(g\circ f)(-1)$.

10. $f(x)=x^2+1,\ g(x)=x-6$ Find **a.** $(f\circ g)(3)$ and **b.** $(g\circ f)(-2)$.

Form the compositions $f(g(x))$ and $g(f(x))$ for each pair of functions. See Examples 2 through 4.

11. $f(x)=\sqrt{x},\quad g(x)=x^2$

12. $f(x)=\dfrac{1}{x},\quad g(x)=\dfrac{1}{x}$

13. $f(x)=\sqrt{x},\quad g(x)=x-2$

14. $f(x)=\sqrt{x},\quad g(x)=x^2-9$

15. $f(x)=x-1,\quad g(x)=\dfrac{1}{x^2}$

16. $f(x)=\dfrac{1}{x^2},\quad g(x)=x^2+1$

17. $f(x)=x^3+x+1,\quad g(x)=x+1$

18. $f(x)=x^3,\quad g(x)=2x-1$

19. $f(x)=\dfrac{1}{\sqrt{x}},\quad g(x)=x^2$

20. $f(x)=\dfrac{1}{\sqrt{x}},\quad g(x)=x^2-4$

21. $f(x)=\dfrac{1}{x},\quad g(x)=x^2+7x-8$

22. $f(x)=\dfrac{1}{x+1},\quad g(x)=x^2+x-3$

23. $f(x)=x^{3n},\quad g(x)=2x-6$

24. $f(x)=x^{\frac{1}{3}},\quad g(x)=4x+7$

25. $f(x)=x^3,\quad g(x)=\sqrt{x-8}$

26. $f(x)=x^3+1,\quad g(x)=\dfrac{1}{x}$

Solve.

27. For the functions $f(x)=6x-3$ and $g(x)=\dfrac{1}{3}x+3$, find:

a. $f(g(3))$ **b.** $g(f(0))$

c. Does it appear that *f* and *g* are inverses of each other? Explain.

28. For the functions $h(x)=-2x+4$ and $g(x)=\dfrac{4-x}{2}$, find:

a. $h(g(6))$ **b.** $g(h(-4))$

c. Does it appear that *h* and *g* are inverses of each other? Explain.

29. Given $f(x)=\dfrac{1}{2x+1}$ and $g(x)=-\dfrac{1}{x}$, find:

a. $g(f(4))$ **b.** $f(g(2))$

c. Explain the different results from parts a. and b.

30. Given $f(x)=\sqrt{x-9}$ and $g(x)=x-9$, find:

a. $g(f(109))$ **b.** $f(g(9))$

c. Explain the different results from parts a. and b.

44. $g^{-1}(x)=\dfrac{1-x}{4}$

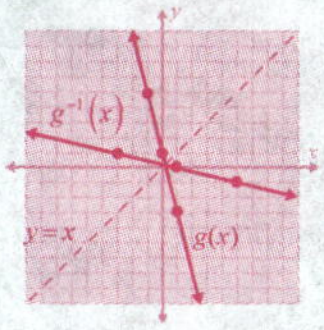

45. $f^{-1}(x)=\dfrac{x-1}{5}$

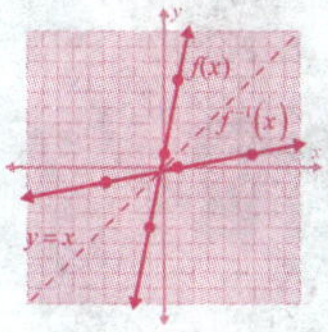

46. $g^{-1}(x)=\dfrac{1-x}{3}$

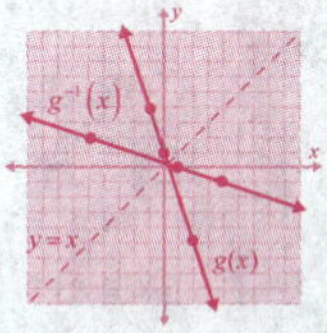

47. $g^{-1}(x)=\dfrac{3(x-2)}{2}$

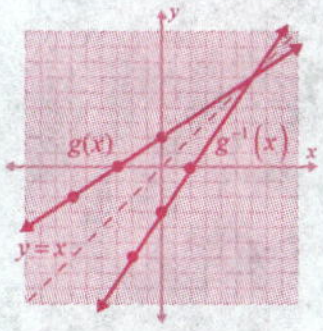

48. $f^{-1}(x)=-2(x+3)$

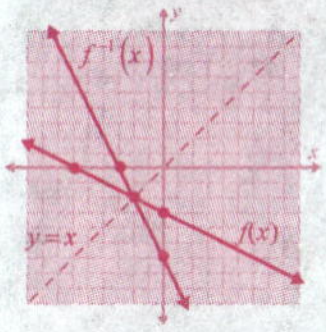

49. $f^{-1}(x)=-x-2$

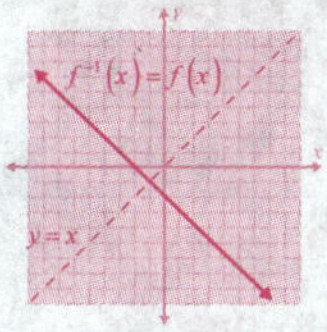

50. $f^{-1}(x)=\dfrac{4-x}{2}$

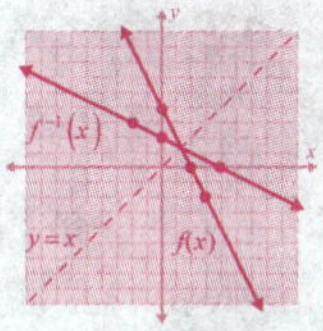

51. $f^{-1}(x) = \sqrt{x-1}$

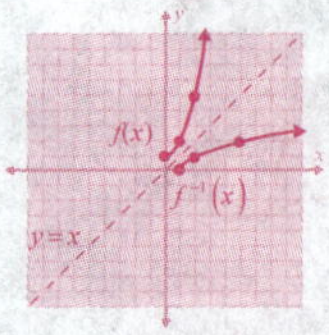

52. $f^{-1}(x) = \sqrt{x+1}$

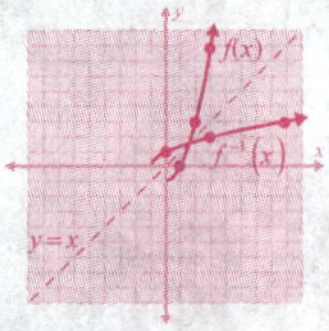

53. $f^{-1}(x) = x^2, x \le 0$

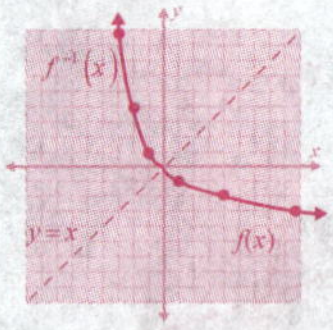

54. $f^{-1}(x) = x^2 + 2, x \le 0$

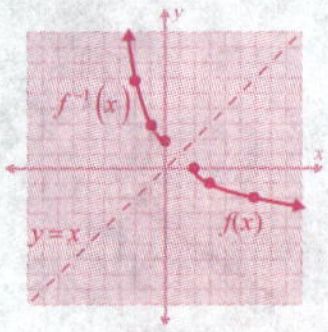

55. One-to-one

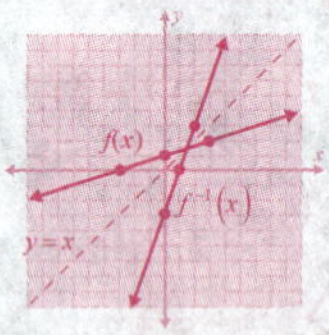

56. One-to-one

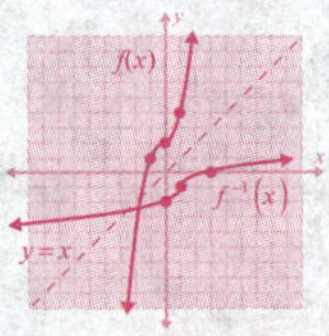

57. Not one-to-one

58. One-to-one

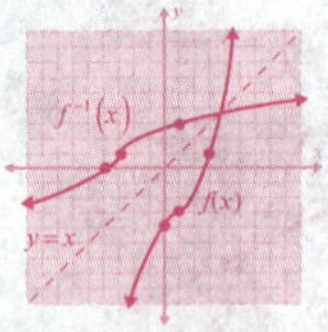

59. Not one-to-one

60. Not one-to-one

Show that the given one-to-one functions are inverses of each other. Then graph both functions on the same set of axes and show the line $y = x$ as a dotted line on each graph. (You may use a calculator as an aid in finding the graphs.) See Examples 6 and 7.

31. $f(x) = 3x + 1$ and $g(x) = \dfrac{x-1}{3}$

32. $f(x) = -2x + 3$ and $g(x) = \dfrac{3-x}{2}$

33. $f(x) = \sqrt[3]{x-1}$ and $g(x) = x^3 + 1$

34. $f(x) = x^3 - 4$ and $g(x) = \sqrt[3]{x+4}$

35. $f(x) = x^2$ for $x \ge 0$ and $g(x) = \sqrt{x}$

36. $f(x) = \sqrt{x+3}$ and $g(x) = x^2 - 3$ for $x \ge 0$

37. $f(x) = x^3 + 2$ and $g(x) = \sqrt[3]{x-2}$

38. $f(x) = \sqrt[5]{x+6}$ and $g(x) = x^5 - 6$

39. $f(x) = \dfrac{2}{x}$ and $g(x) = \dfrac{2}{x}$

40. $f(x) = \dfrac{3}{x}$ and $g(x) = \dfrac{3}{x}$

Find the inverse of the given function. Then graph both functions on the same set of axes and show the line $y = x$ as a dotted line on the graph. See Examples 9 and 10.

41. $f(x) = 2x - 3$

42. $f(x) = 2x - 5$

43. $g(x) = x$

44. $g(x) = 1 - 4x$

45. $f(x) = 5x + 1$

46. $g(x) = -3x + 1$

47. $g(x) = \dfrac{2}{3}x + 2$

48. $f(x) = -\dfrac{1}{2}x - 3$

49. $f(x) = -x - 2$

50. $f(x) = -2x + 4$

51. $f(x) = x^2 + 1, x \ge 0$

52. $f(x) = x^2 - 1, x \ge 0$

53. $f(x) = -\sqrt{x}, x \ge 0$

54. $f(x) = -\sqrt{x-2}, x \ge 2$

Using the horizontal line test, determine which of the graphs are graphs of one-to-one functions. If the graph represents a one-to-one function, graph its inverse by reflecting the graph of the function across the line $y = x$. (**Hint:** If a function is one-to-one, label a few points on the graph and use the fact that the x- and y-coordinates are interchanged on the graph of the inverse.) See Example 5.

55.

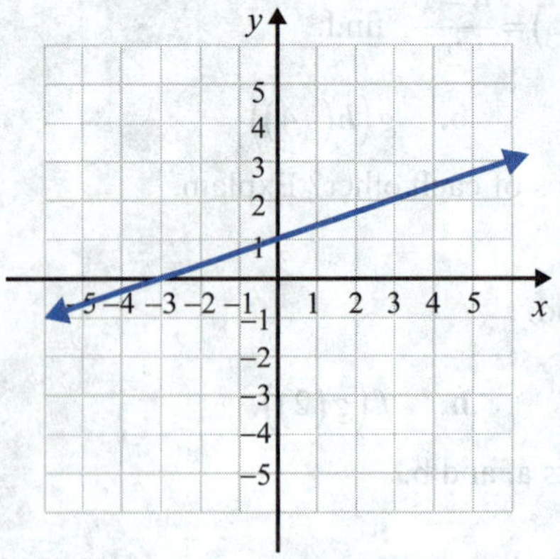

56.

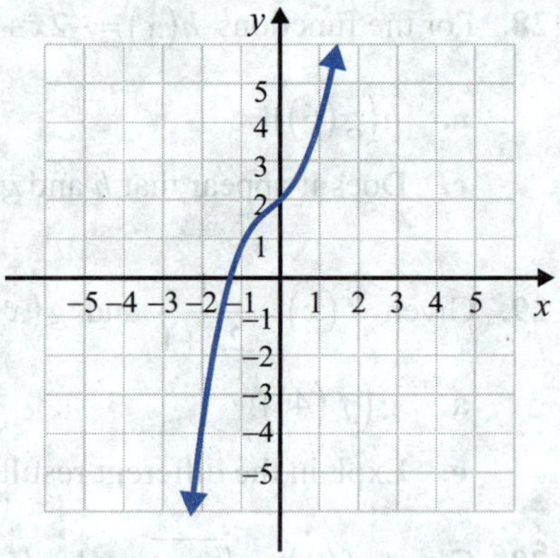

57.

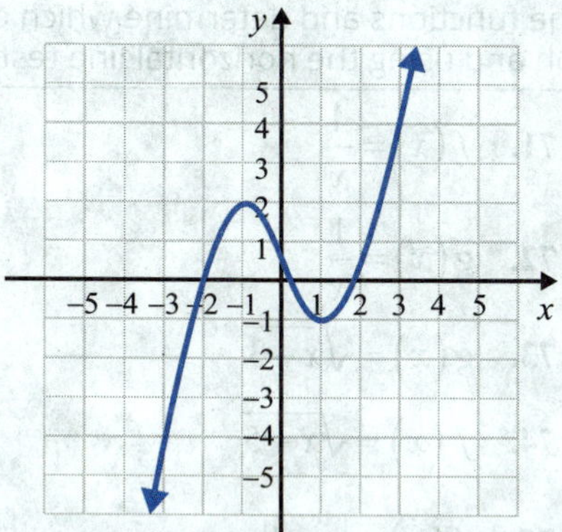

58.

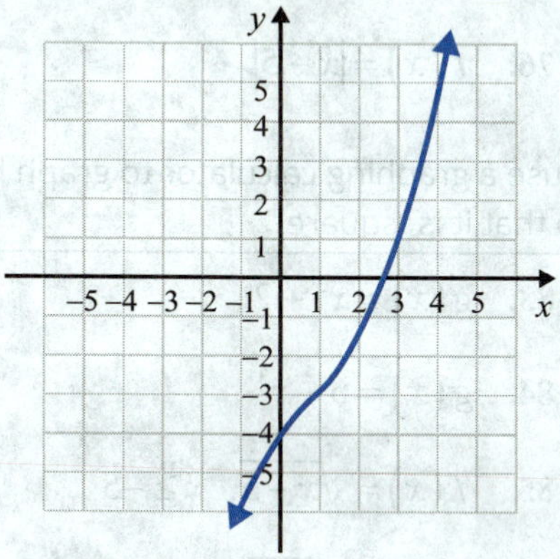

59.

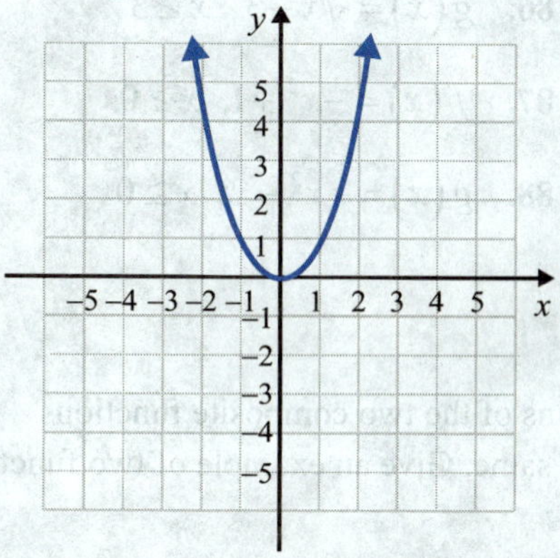

60.

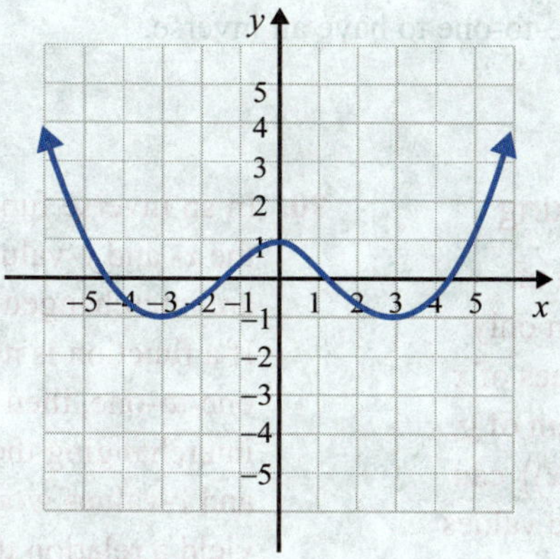

61.

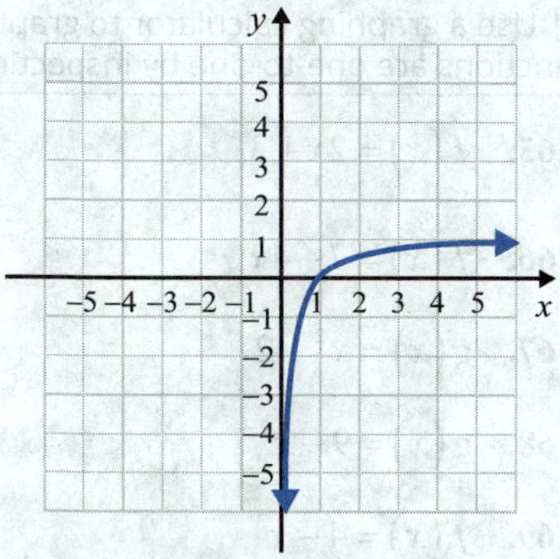

62.

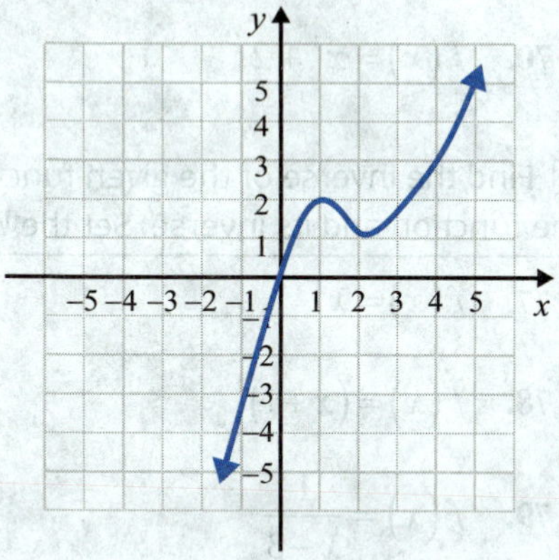

63.

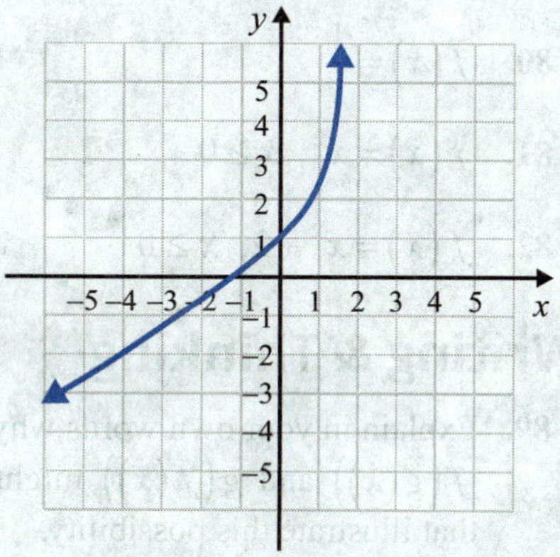

64.

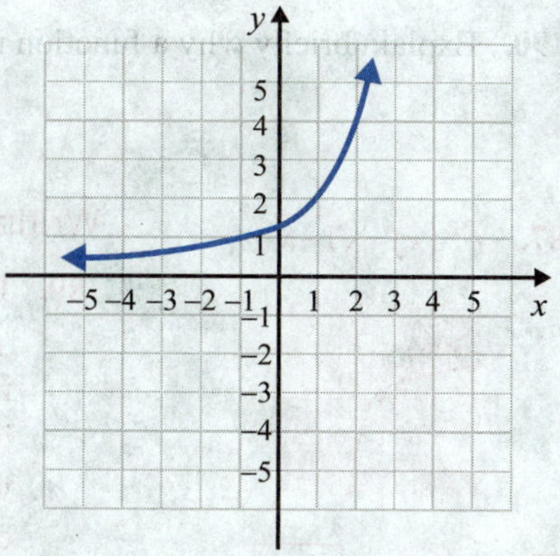

61. One-to-one

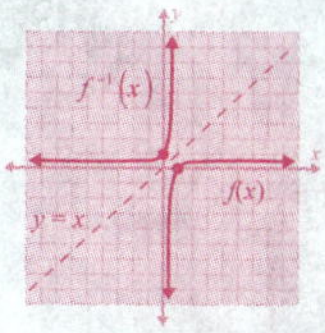

62. Not one-to-one

63. One-to-one

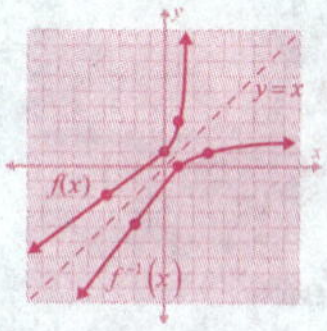

64. One-to-one

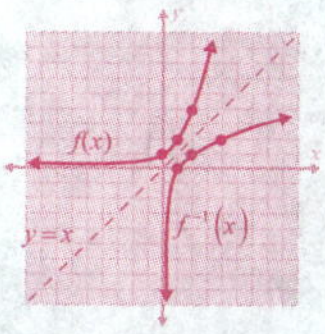

65. One-to-one

66. One-to-one

67. Not one-to-one

68. Not one-to-one

69. One-to-one

70. One-to-one

71. One-to-one

72. One-to-one

73. One-to-one

74. One-to-one

75. Not one-to-one

76. Not one-to-one

77. $f^{-1}(x)=\sqrt[3]{x}$

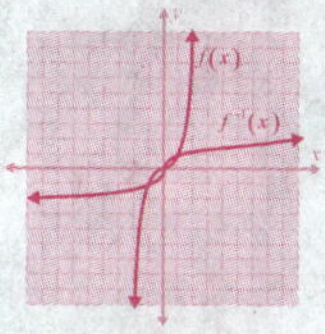

78. $f^{-1}(x)=\sqrt[3]{x}-1$

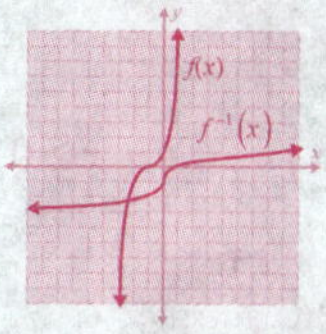

Use a graphing calculator to graph each of the functions and determine which of the functions are one-to-one by inspecting the graph and using the horizontal line test.

65. $f(x)=2x+3$

66. $f(x)=7-4x$

67. $g(x)=x^2-2$

68. $g(x)=9-x^2$

69. $f(x)=4-x^3$

70. $f(x)=x^3+2$

71. $f(x)=\dfrac{4}{x}$

72. $g(x)=\dfrac{1}{x}$

73. $g(x)=\sqrt{x-3}$

74. $f(x)=\sqrt{x+5}$

75. $f(x)=|x+1|$

76. $f(x)=|x-5|$

Find the inverse of the given function. Then use a graphing calculator to graph both the function and its inverse. Set the `WINDOW` so that it is "square."

77. $f(x)=x^3$

78. $f(x)=(x+1)^3$

79. $f(x)=\dfrac{1}{x-3}$

80. $f(x)=\dfrac{1}{x}$

81. $f(x)=x^2,\ x\geq 0$

82. $f(x)=x^2+2,\ x\geq 0$

83. $g(x)=x^3+2$

84. $g(x)=6-x^3$

85. $f(x)=\sqrt{x+5},\ x\geq -5$

86. $g(x)=\sqrt{x-3},\ x\geq 3$

87. $f(x)=-x^2+1,\ x\geq 0$

88. $g(x)=-x^2-2,\ x\geq 0$

Writing & Thinking

89. Explain in your own words why the domains of the two composite functions $f(g(x))$ and $g(f(x))$ might not be the same. Give an example of two functions that illustrate this possibility.

90. Explain briefly why a function must be one-to-one to have an inverse.

79. $f^{-1}(x)=\dfrac{1}{x}+3$

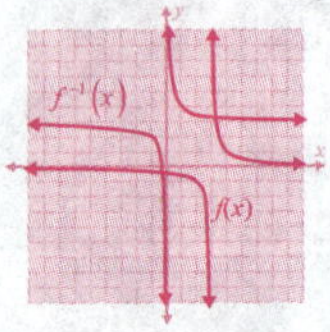

80. $f^{-1}(x)=\dfrac{1}{x}$

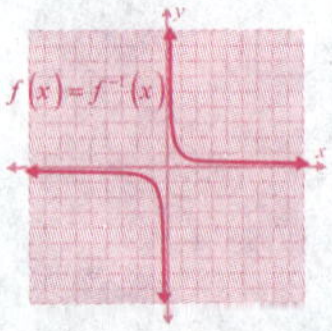

81. $f^{-1}(x)=\sqrt{x}$

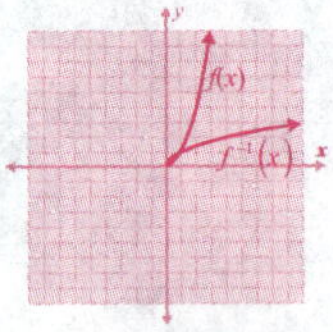

82. $f^{-1}(x)=\sqrt{x-2}$

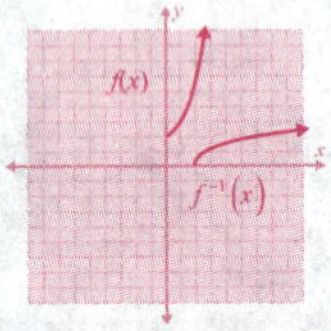

83. $g^{-1}(x)=\sqrt[3]{x-2}$

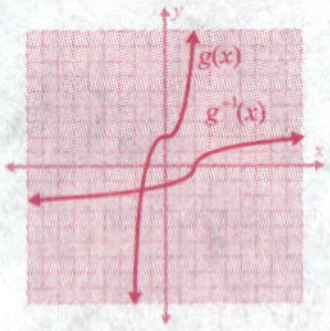

84. $g^{-1}(x)=\sqrt[3]{6-x}$

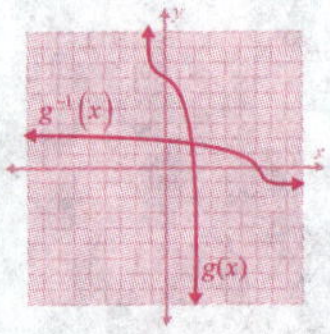

85. $f^{-1}(x)=x^2-5,\ x\geq 0$

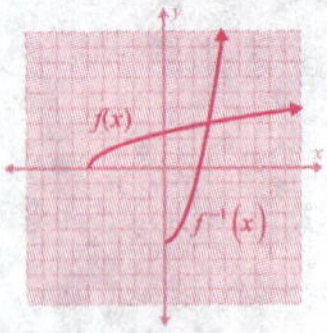

86. $g^{-1}(x)=x^2+3,\ x\geq 0$

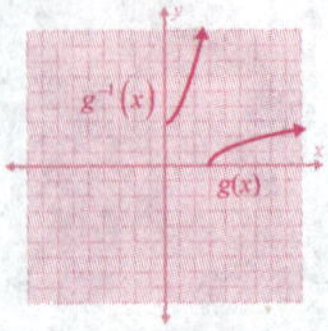

87. $f^{-1}(x)=\sqrt{1-x}$

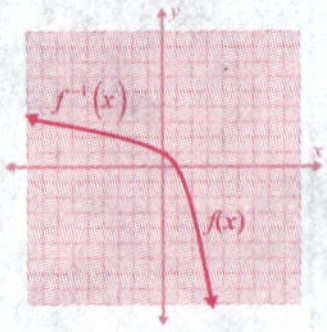

88. $g^{-1}(x)=\sqrt{-2-x}$

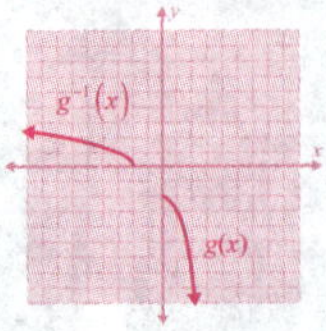

Writing & Thinking

89. The domain of $f(g(x))$ can only include values of x in the domain of g, while $g(f(x))$ can only include values of x in the domain of f. For example, $f(x)=\sqrt{x}, g(x)=-x$. Answers will vary.

90. In an inverse function, the x- and y-values are interchanged. If a function is not one-to-one, then interchanging the x- and y-values would yield a relation that is not a function.

17.3 Exponential Functions

A Introduction to Exponential Functions

You may have read that the population of the world is growing exponentially or studied the exponential growth of bacteria in a biology class. Radioactive materials decay exponentially and never actually disappear. The graph in Figure 1 illustrates that exponential growth has a relatively slow beginning and then builds at an exceedingly rapid rate. This can be extremely important to a doctor trying to curb the growth of "bad" bacteria in a patient.

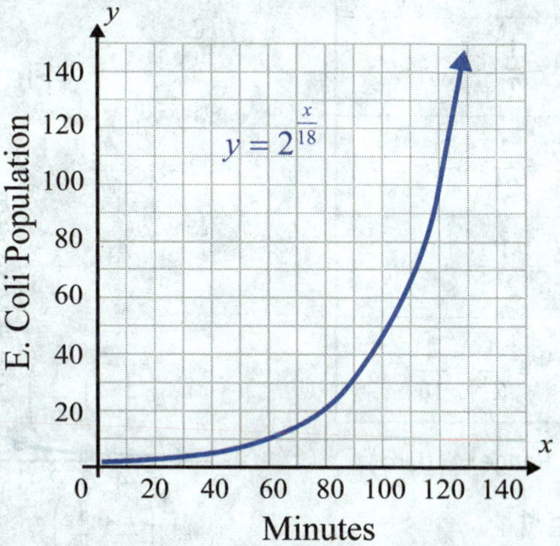

The population of certain strains of E. coli doubles every 18 minutes under optimal conditions.

Figure 1

Quadratic functions have a variable base and a constant exponent, as in $f(x) = x^2$. However, in **exponential functions**, the base is constant and the variable is in the exponent, as in $f(x) = 2^x$. As we will see, these two types of functions have major differences in their characteristics. Exponential functions are defined as follows.

> **Exponential Functions**
>
> An **exponential function** is a function of the form
>
> $$f(x) = b^x,$$
>
> where $b > 0$, $b \neq 1$, and x is any real number.
>
> DEFINITION

Examples of exponential functions are

$$f(x) = 2^x,\ f(x) = 3^x, \text{ and } y = \left(\frac{1}{3}\right)^x.$$

B Exponential Growth

The following table of values and the graphs of the corresponding points give a very good idea of what the graph of the **exponential growth** function $y = 2^x$ looks like (see Figure 2a). Because we know that 2^x is defined for all real exponents, points such as $(\sqrt{2}, 2^{\sqrt{2}})$, $(\pi, 2^{\pi})$, and $(\sqrt{5}, 2^{\sqrt{5}})$ are on the graph. The graph for $f(x) = 2^x$ is a smooth curve, as shown in Figure 2b.

Objectives

A. Know the characteristics of an exponential function.

B. Recognize exponential growth functions.

C. Recognize exponential decay functions.

D. Use exponential growth and decay functions to solve applications.

E. Use exponential functions and the number *e* to solve compound interest applications.

Note

The two conditions $b > 0$ and $b \neq 1$ found in the definition are important. We must have $b > 0$ so that b^x is defined for all real numbers x. For example, we do not consider $y = (-2)^x$ to be an exponential because $(-2)^{\frac{1}{2}}$ is not a real number. Also, $b \neq 1$ because the function $y = 1^x = 1$ is not considered an exponential function for all real numbers x.

Teaching Note

At any time in this discussion you may want your students to graph these functions with the TI-84 Plus graphing calculator. Make sure that they understand the limitations on the display in terms of the horizontal asymptote and the appearance of the graph touching the *x*-axis.

x	$y = 2^x$
3	$2^3 = 8$
2	$2^2 = 4$
1	$2^1 = 2$
$\frac{1}{2}$	$2^{\frac{1}{2}} = \sqrt{2} \approx 1.4142$
0	$2^0 = 1$
$-\frac{1}{2}$	$2^{-\frac{1}{2}} = \frac{1}{\sqrt{2}} \approx 0.7071$
−1	$2^{-1} = \frac{1}{2}$
−2	$2^{-2} = \frac{1}{2^2} = \frac{1}{4}$
−3	$2^{-3} = \frac{1}{2^3} = \frac{1}{8}$
−4	$2^{(-4)} = \frac{1}{2^4} = \frac{1}{16}$

Domain $= (-\infty, \infty)$ Range $= (0, \infty)$

Table 1

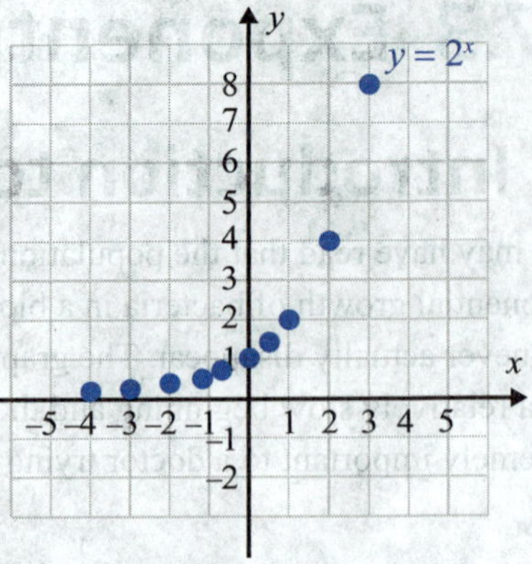

Figure 2a

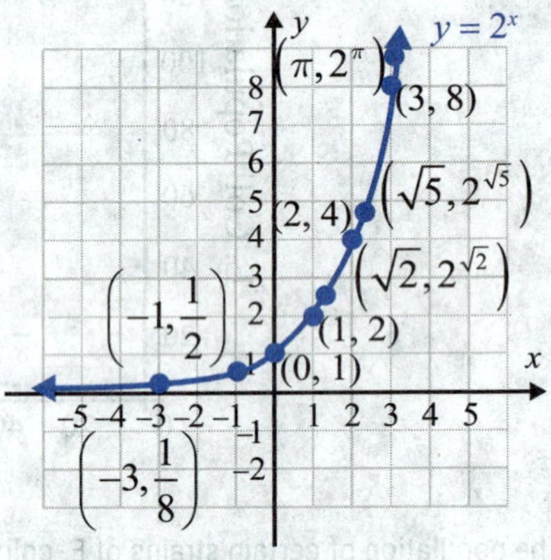

Figure 2b

Table 2 shows the values of $y = 3^x$, and Figure 3 shows the graph of the function $y = 3^x$. Notice that the graphs of $y = 2^x$ and $y = 3^x$ are quite similar, but that the graph of $y = 3^x$ rises faster. That is, the **exponential growth is faster if the base is larger**.

x	$y = 3^x$
2	$3^2 = 9$
1	$3^1 = 3$
0	$3^0 = 1$
−1	$3^{-1} = \frac{1}{3} \approx 0.3333$
−3	$3^{-3} = \frac{1}{27} \approx 0.0370$

Domain $= (-\infty, \infty)$ Range $= (0, \infty)$

Table 2

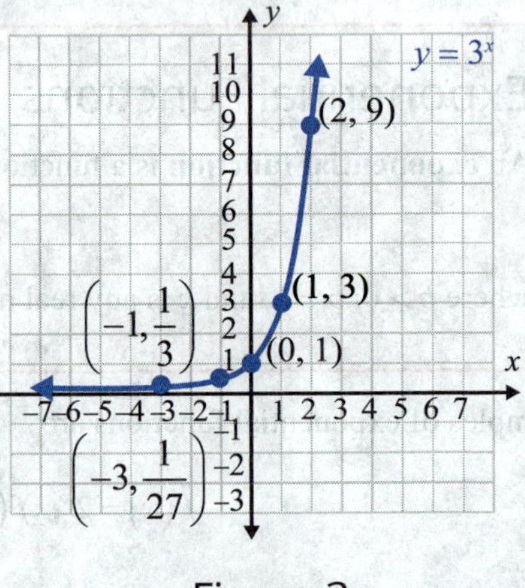

Figure 3

In both graphs, the curves tend to get *very close* to the line $y = 0$ (the x-axis) without ever touching the x-axis. When this happens, the line is called an **asymptote**. If the line is horizontal, as in the cases of exponential growth and (as we will see) exponential decay, the line is called a **horizontal asymptote**. We say that the curve (or function) approaches the line **asymptotically**. In mathematics, this phenomenon happens frequently.

C Exponential Decay

Now consider the **exponential decay** function $f(x)=\left(\frac{1}{2}\right)^x$. The values in Table 3 and the graph of the corresponding points shown in Figure 4 indicate the nature of the graph of this function.

x	$y=\left(\frac{1}{2}\right)^x=2^{-x}$
-3	$2^{-(-3)}=2^3=8$
-2	$2^{-(-2)}=2^2=4$
-1	$2^{-(-1)}=2^1=2$
$-\frac{1}{2}$	$2^{-\left(-\frac{1}{2}\right)}=2^{\frac{1}{2}}=\sqrt{2}\approx 1.4142$
0	$2^{-0}=2^0=1$
$\frac{1}{2}$	$2^{-\frac{1}{2}}=\frac{1}{\sqrt{2}}\approx 0.7071$
1	$2^{-1}=\frac{1}{2}$
2	$2^{-2}=\frac{1}{2^2}=\frac{1}{4}$
3	$2^{-3}=\frac{1}{2^3}=\frac{1}{8}$

Domain $=(-\infty,\infty)$ Range $=(0,\infty)$

Table 3

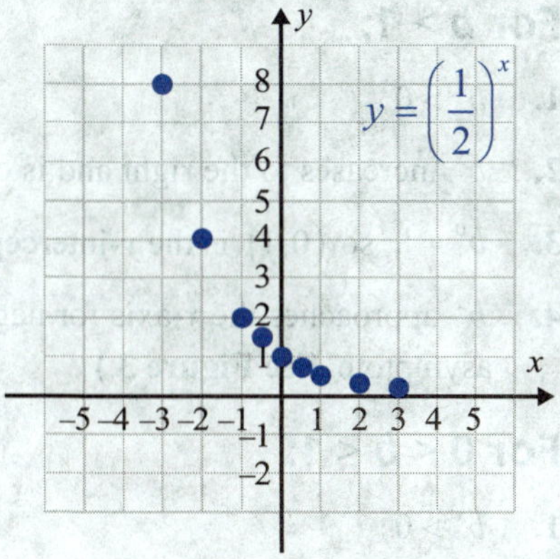

Figure 4

Figures 5a and 5b show the complete graphs of the two exponential decay functions

$$y=\left(\frac{1}{2}\right)^x=2^{-x} \text{ and } y=\left(\frac{1}{3}\right)^x=3^{-x}.$$

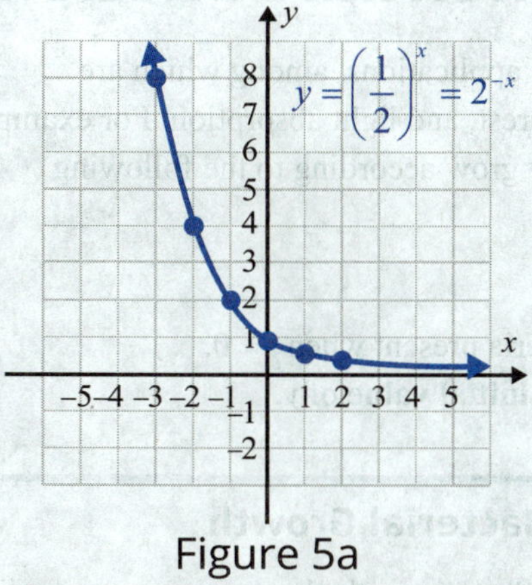

Figure 5a

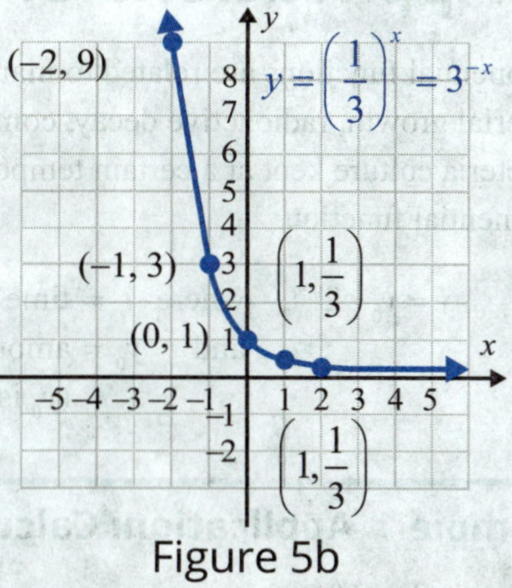

Figure 5b

Note again, in Figures 5a and 5b, that the line $y=0$ (the x-axis) is a **horizontal asymptote** for both curves. For exponential growth, the curves approach the asymptote as x moves further in the negative direction, as in Figures 2b and 3. For exponential decay, the graph approaches the asymptote as x moves further in the positive direction, as in Figures 5a and 5b.

Note

Because $\frac{1}{2}=2^{-1}$, $\frac{1}{3}=3^{-1}$, $\frac{1}{4}=4^{-1}$, and so on, for fractions between 0 and 1, we can write an exponential function with a fractional base between 0 and 1 (these are exponential decay functions) in the form of an exponential function with a base greater than 1 and a negative exponent. Thus, we write

$$y=\left(\frac{1}{2}\right)^x=\left(2^{-1}\right)^x=2^{-x}$$

and

$$y=\left(\frac{1}{3}\right)^x=\left(3^{-1}\right)^x=3^{-x}.$$

The following general concepts are helpful in understanding the graphs and the nature of exponential functions, both exponential growth and exponential decay.

General Concepts of Exponential Functions

For $b > 1$:

1. $b^x > 0$
2. b^x increases to the right and is called an **exponential growth function**.
3. $b^0 = 1$, so $(0, 1)$ is the y-intercept.
4. b^x approaches the x-axis for negative values of x. (The x-axis is a horizontal asymptote. **See Figure 3.**)

For $0 < b < 1$:

1. $b^x > 0$
2. b^x decreases to the right and is called an **exponential decay function**.
3. $b^0 = 1$, so $(0, 1)$ is the y-intercept.
4. b^x approaches the x-axis for positive values of x. (The x-axis is a horizontal asymptote. **See Figures 5a and 5b.**)

DEFINITION

As with all functions, exponential functions can be multiplied by constants, shifted horizontally, and shifted vertically. (**Note:** We did this with parabolas in Chapter 16 and more on this topic will be discussed in Chapter 18.) Thus,

$$y = a \cdot b^{r \cdot x}, \quad y = b^{x-h}, \quad \text{and} \quad y = b^x + k,$$

and various combinations of these expressions are all exponential functions.

D Applications of Exponential Functions

Exponential functions are related to many practical applications, among which are bacterial growth, radioactive decay, compound interest, and light absorption. For example, a bacteria culture kept at a certain temperature may grow according to the following exponential function.

$y = y_0 \cdot 2^{0.5t}$, where t = time in hours
and y_0 = amount of bacteria present when $t = 0$.
(y_0 is called the **initial value** of y.)

1. Using the same function and initial value given in Example 1, how many bacteria will be present at the end of 36 hours?
$t = 36$ hours and $y_0 = 10{,}000$

Example 1 Application: Calculating Bacterial Growth

A scientist has 10,000 bacteria present when $t = 0$. She knows the bacteria grow according to the function $y = y_0 \cdot 2^{0.5t}$, where t is measured in hours. How many bacteria will be present at the end of one day?

Solution

Substitute $t = 24$ hours and $y_0 = 10{,}000$ into the function.

$$
\begin{aligned}
y &= 10{,}000 \cdot 2^{0.5(24)} = 10{,}000 \cdot 2^{12} \\
&= 10{,}000(4096) \\
&= 40{,}960{,}000 \\
&= 4.096 \times 10^{7}
\end{aligned}
$$

At the end of one day, there will be 40,960,000 bacteria.

CALCULATORS

Calculating Bacterial Growth

You could also use your calculator to calculate the result by entering the numbers as shown in the following display. Press ENTER to get the result.

```
10000*2^(0.5*24)
        40960000
```

Now work margin exercise 1.

Example 2 Application: Calculating Bacterial Growth

Use the formula for exponential growth, $y = y_0 b^t$, to determine the exponential function that fits the following information: $y_0 = 5000$ bacteria with 135,000 bacteria present after 3 days.

Solution

Use $y = y_0 b^t$ where t is measured in days. Substitute 135,000 for y, 3 for t, and 5000 for y_0, then solve for b.

$$
\begin{aligned}
135{,}000 &= 5000b^3 \\
27 &= b^3 \\
\sqrt[3]{(27)} &= \sqrt[3]{(b^3)} \\
3 &= b
\end{aligned}
$$

The function is $y = 5000 \cdot 3^t$.

Now work margin exercise 2.

2. Determine the exponential growth function that fits the following information: $y_0 = 7000$ bacteria with 112,000 bacteria present after 4 days.

E Compound Interest and the Number *e*

The topic of compound interest (which is interest paid on interest) leads to a particularly interesting (and useful) exponential function. The formula $A = P(1+r)^t$ can be used for finding the value (or amount) accumulated when a principal P is invested and interest is compounded once a year. If compounding is performed more than once a year, we use the following formula to find A.

Teaching Note

You might want to discuss interest rates with your students. Point out that interest rates fluctuate with the economy.

Compound Interest

Compound interest on a principal P invested at an annual interest rate r (in decimal form) for t years that is compounded n times per year can be calculated using the following formula, where A is the amount accumulated.

$$A = P\left(1+\frac{r}{n}\right)^{nt}$$

FORMULA

3. Find the value of $2000 invested at 5% for 2 years if the interest is compounded annually.

Example 3 Application: Calculating Compound Interest

If P dollars are invested at a rate of interest r (in decimal form) compounded annually (once a year, $n = 1$) for t years, the formula for the amount A becomes $A = P\left(1+\frac{r}{1}\right)^{1t} = P(1+r)^t$. Find the value of $1000 invested at $r = 6\% = 0.06$ for 3 years.

Solution

We have $P = 1000$, $r = 0.06$, and $t = 3$.

$$\begin{aligned} A &= 1000(1+0.06)^3 \\ &= 1000(1.06)^3 \\ &= 1000(1.191016) \approx 1191.02 \end{aligned}$$

The account will have $1191.02 invested in it after 3 years.

Now work margin exercise 3.

CALCULATORS

Caclulating Compound Interest

To use your calculator to evaluate $(1.06)^3$, enter 1.06, press the [^] key, enter 3, then press the [ENTER] key.

4. Find the value of $2000 invested at $r = 5\%$ for 2 years if the interest is compounded monthly.

Example 4 Application: Calculating Compound Interest

What will be the value of a principal investment of $1000 invested at 6% for 3 years if interest is compounded monthly (12 times per year)?

Solution

Use the formula for compound interest.

We have $P = 1000$, $r = 0.06$, $n = 12$, and $t = 3$.

$$\begin{aligned} A &= 1000\left(1+\frac{0.06}{12}\right)^{12(3)} \\ &= 1000(1+0.005)^{36} \\ &= 1000(1.005)^{36} \\ &\approx 1000(1.196680525) \quad \text{Simplify using a calculator.} \\ &= 1196.68 \end{aligned}$$

The value of the account after 3 years will be $1196.68.

Now work margin exercise 4.

Math Tip

Remember that applications involving money should have answers rounded to the nearest cent.

Example 5 Application: Calculating Compound Interest

Find the value of A if $1000 is invested at 6% for 3 years and interest is compounded daily (365 times per year).

Solution

Use the formula for compound interest.

We have $P = 1000$, $r = 0.06$, $n = 365$, and $t = 3$.

$$A = 1000\left(1+\frac{0.06}{365}\right)^{365(3)}$$

$$\approx 1000(1.000164384)^{1095} \quad \text{Simplify using a calculator.}$$

$$\approx 1000(1.197200227)$$

$$\approx 1197.20$$

After three years, there will be $1197.20 in the account.

Now work margin exercise 5.

5. Find the value of $2000 invested at $r = 5\%$ for 2 years if the interest is compounded daily.

Examples 3, 4, and 5 illustrate the effects of compounding interest more frequently over 3 years. The formula gives the following results.

$A = \$1191.02$ for $n = 1$ (once a year)

$A = \$1196.68$ for $n = 12$ (monthly)

$A = \$1197.20$ for $n = 365$ (daily)

These numbers might not seem very dramatic, only a difference of $6.18 for 3 years; but, if you use your calculator, in 20 years you will see a difference of $112.65 for a $1000 investment. An investment of $10,000 for 20 years at 9% will show a difference of $4438.94. The results show that more frequent compounding will result in higher income.

If interest is **compounded continuously** (which is even faster than every second), then the irrational number e $(e \approx 2.718)$ can be shown to be the base of the corresponding exponential function for calculating interest. Table 4 shows how the expression $\left(1+\frac{1}{n}\right)^n$ changes as n takes on larger and larger values. The number e is the **limit** (or **limiting value**) of the expression "as n approaches infinity" $(n \to \infty)$ and we write $e = \lim_{n\to\infty}\left(1+\frac{1}{n}\right)^n$. Study the following table to help understand the ideas.

Teaching Note

Have students notice how little change is occuring in this sequence of values (3rd column) and that this suggests that there might be a boundary ("limit") to the sequence—and there is—e! Stress that the concept of limit presented in the definition of e is precise, but beyond the scope of this text.

Values of the expression $(1+\frac{1}{n})^n$ as n approaches infinity ($n \to \infty$)

n	$\left(1+\frac{1}{n}\right)$	$\left(1+\frac{1}{n}\right)^n$
1	$\left(1+\frac{1}{1}\right)=2$	$(2)^1=2$
2	$\left(1+\frac{1}{2}\right)=1.5$	$(1.5)^2=2.25$
5	$\left(1+\frac{1}{5}\right)=1.2$	$(1.2)^5=2.48832$
10	$\left(1+\frac{1}{10}\right)=1.1$	$(1.1)^{10}\approx 2.59374246$
100	$\left(1+\frac{1}{100}\right)=1.01$	$(1.01)^{100}\approx 2.704813829$
1000	$\left(1+\frac{1}{1000}\right)=1.001$	$(1.001)^{1000}\approx 2.716923932$
10,000	$\left(1+\frac{1}{10{,}000}\right)=1.0001$	$(1.0001)^{10{,}000}\approx 2.718145927$
100,000	$\left(1+\frac{1}{100{,}000}\right)=1.00001$	$(1.00001)^{100{,}000}\approx 2.718268237$
↓		↓
∞		$e = 2.718281828459\ldots$

Table 4

This gives us the following definition of e. (Be aware that these are very sophisticated mathematical concepts and it will take some careful reading to understand how to arrive at e and the formula $A = Pe^{rt}$.)

Tech Tip

The number e is on the TI-84 Plus calculator above the divide key.

The Number e

The number e is defined to be

$$e = 2.718281828459\ldots$$

DEFINITION

As we know, the formula for compound interest is

$$A = P\left(1+\frac{r}{n}\right)^{nt}.$$

This formula for compound interest takes a different form when interest is **compounded continuously**. The new form involves the number e and is stated as follows.

Continuously Compounded Interest

Continuously compounded interest on a principal P invested at an annual interest rate r for t years can be calculated using the following formula, where A is the amount accumulated.

$$A = Pe^{rt}$$

FORMULA

As illustrated in Example 6, a calculator is needed to use the formula for continuously compounded interest.

Example 6 Using a Graphing Calculator to Calculate Continuously Compounded Interest

Find the value of \$1000 invested at 6% for 3 years if interest is compounded continuously. (In this case, $P = \$1000$, $r = 6\% = 0.06$, and $t = 3$.)

Solution

To find the value of $A = Pe^{rt} = 1000e^{0.06 \cdot 3}$ enter the numbers as shown and press [ENTER] to get the result.

```
1000*e^(0.06*3)
            1197.217363
```

The entire exponent must be in parentheses.

Note: Press [2nd] and [LN] and e^(will appear on the display.)

Thus, the value of \$1000 compounded continuously at 6% for 3 years will be \$1197.22. (Notice that from Example 4 there is only a 54 cent gain in A when \$1000 is compounded continuously instead of monthly at 6% for 3 years.)

Now work margin exercise 6.

6. Find the value of \$1500 invested at 5.5% for 4 years if interest is compounded continuously.

Margin Exercise Answers

1. $y = 2,621,440,000$ or 2.62144×10^9 **2.** $y = 7000 \cdot 2^t$ **3.** $A = \$2205$ **4.** $A = \$2209.88$ **5.** $A = \$2210.33$ **6.** \$1869.12

17.3 Exercises

Concept Check

Fill-in-the-Blank. Complete each sentence using information found in this section.

1. In an exponential function, the base is a/an ______ and the exponent is a/an ______.

2. Exponential growth is faster if the base is ______.

3. Exponential decay functions have a base between ____ and ____.

4. The y-intercept of any exponential function is _____.

17.3 Exercises

Concept Check

1. constant, variable
2. larger
3. 0, 1
4. (0, 1)
5. compound
6. continuously
7. False; $b > 0$ and $b \neq 1$
8. True
9. True

10. False; e is defined to be 2.71828…

Practice

1.
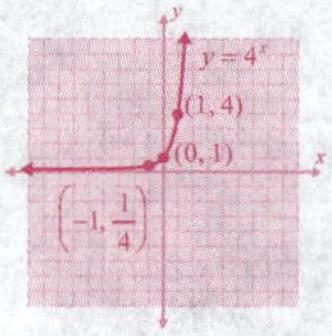

2.
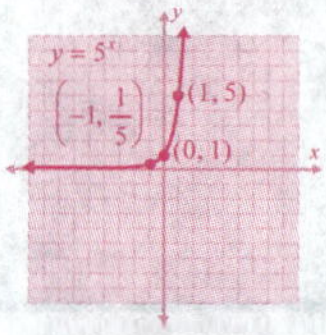

3.
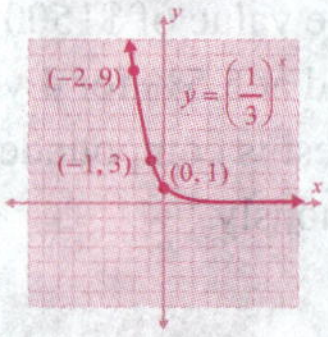

4.
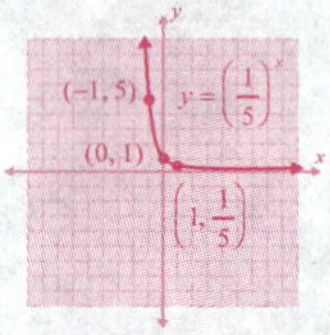

5.
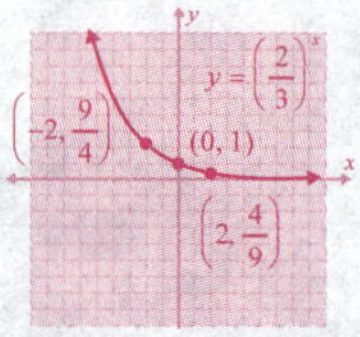

6.
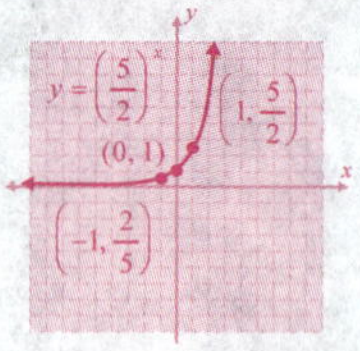

7.
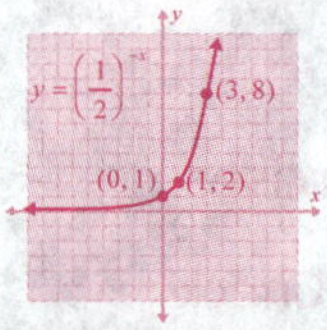

8.
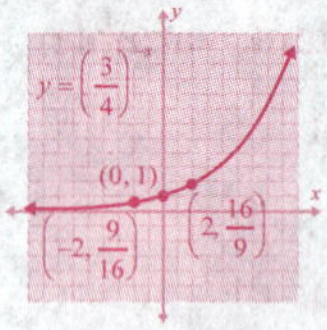

9.
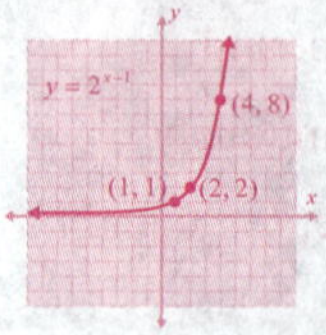

5. The formula $A = P\left(1+\dfrac{r}{n}\right)^{nt}$ is used to calculate ________ interest.

6. The formula $A = Pe^{rt}$ is used to calculate ________ compounded interest.

True/False. Determine whether each statement is true or false. If a statement is false, explain how it can be changed so that the statement will be true. (**Note:** There may be more than one acceptable change.)

7. For all exponential functions $f(x) = x^b$, $b < 0$.

8. The function $f(x) = 5^x$ is an example of an exponential growth model.

9. In an exponential decay function, b^x approaches the x-axis for positive values of x.

10. The number e is defined to be approximately 3.14159.

Practice

Sketch the graph of each exponential function and label three points on each graph. (Note that some of the graphs are shifts, horizontal or vertical, of the basic exponential functions. These are similar to the shifts performed on parabolas in Chapter 16.)

1. $y = 4^x$

2. $y = 5^x$

3. $y = \left(\dfrac{1}{3}\right)^x$

4. $y = \left(\dfrac{1}{5}\right)^x$

5. $y = \left(\dfrac{2}{3}\right)^x$

6. $y = \left(\dfrac{5}{2}\right)^x$

7. $y = \left(\dfrac{1}{2}\right)^{-x}$

8. $y = \left(\dfrac{3}{4}\right)^{-x}$

9. $y = 2^{x-1}$

10. $y = 3^{x+1}$

11. $f(x) = 2^x + 1$

12. $f(x) = 3^x - 1$

13. $f(x) = -4^{-x}$

14. $g(x) = -2^{-x}$

15. $f(x) = 2^{0.5x}$

16. $g(x) = 10^{0.5x}$

17. $f(x) = 4^{-x} - 1$

18. $g(x) = 10^{-x} - 3$

19. $f(x) = 3 \cdot \left(\dfrac{1}{2}\right)^{x+1}$

20. $y = -4 \cdot \left(\dfrac{1}{3}\right)^{x-1}$

Find the following function values.

21. If $f(t) = 3 \cdot 4^t$, what is the value of $f(2)$?

22. For $f(x) = 3 \cdot 10^{2x}$, find the value of $f(0.5)$.

Use your calculator to find each value as indicated. Round your answer to the nearest hundredth.

23. Find $f(2)$ if $f(x) = 27.3 \cdot e^{-0.4x}$.

24. Find $f(3)$ if $f(x) = 41.2 \cdot e^{-0.3x}$.

25. Find $f(9)$ if $f(t) = 2000 \cdot e^{0.08t}$.

26. Find $f(22)$ if $f(t) = 2000 \cdot e^{0.05t}$.

Solve.

27. Use a graphing calculator to graph each of the following functions. In each case the x-axis is a horizontal asymptote.

a. $y=e^x$ **b.** $y=e^{-x}$ **c.** $y=e^{-x^2}$

Applications

Solve.

28. A biologist knows that, in the laboratory, bacteria in a culture grow according to the function $y=y_0\cdot 5^{0.2t}$, where y_0 is the initial number of bacteria present and t is time measured in hours. How many bacteria will be present in a culture at the end of 5 hours if there were 5000 present initially?

29. Referring to Exercise 28, how many bacteria were present initially if, at the end of 15 hours, there were 2,500,000 bacteria present?

30. Four thousand dollars is deposited into a savings account with a rate of 8% per year. Find the total amount A on deposit at the end of 5 years if the interest is compounded

a. annually. **c.** quarterly. **e.** continuously.

b. semiannually. **d.** daily.

31. Find the amount A in a savings account if \$2000 is invested at 7% for 4 years and the interest is compounded

a. annually. **c.** quarterly. **e.** continuously.

b. semiannually. **d.** daily.

32. Find the value of \$1800 invested at 6% for 3 years if the interest is compounded continuously.

33. Find the value of \$2500 invested at 5% for 5 years if the interest is compounded continuously.

34. The revenue function is given by $R(x)=x\cdot p(x)$ dollars, where x is the number of units sold and $p(x)$ is the unit price. If $p(x)=25(2)^{\frac{-x}{5}}$, find the revenue if 15 units are sold.

35. Referring to Exercise 34, if $p(x)=40(3)^{\frac{-x}{6}}$, find the revenue if 12 units are sold.

36. A radio station knows that during an intense advertising campaign, the number of people N who will hear a commercial is given by $N=A\left(1-2^{-0.05t}\right)$, where A is the number of people in the broadcasting area and t is the number of hours the commercial has been run. If there are 500,000 people in the area, how many will hear a commercial during the first 20 hours?

37. Bethany invested \$45,000 in a retirement fund that earns 8% interest and is compounded continuously. How much money will the account be worth after:

a. 10 years

b. 20 years

c. 40 years

10.
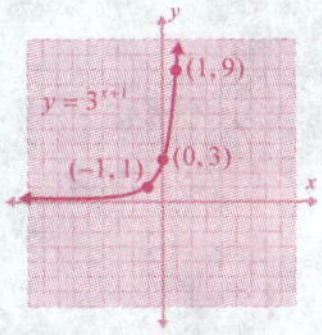

11.
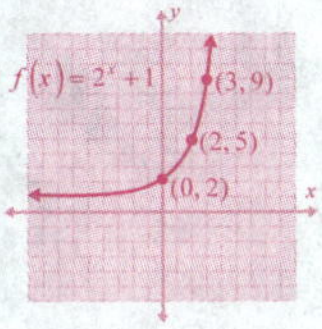

12.
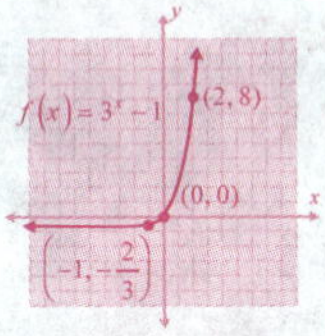

13.
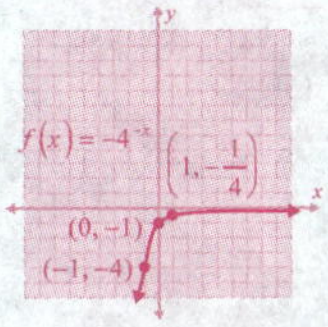

14.
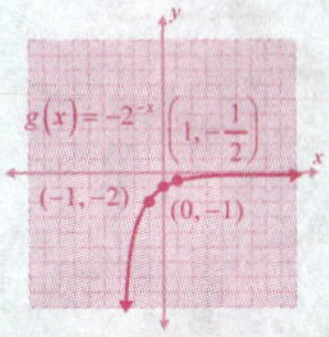

15.
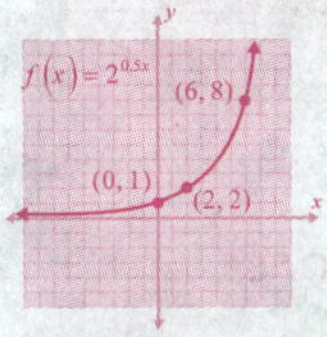

16.
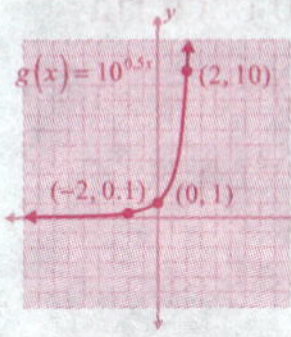

17.
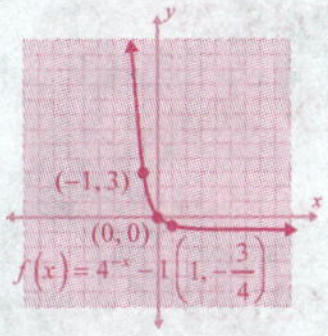

18.
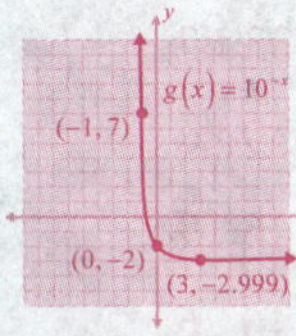

19.
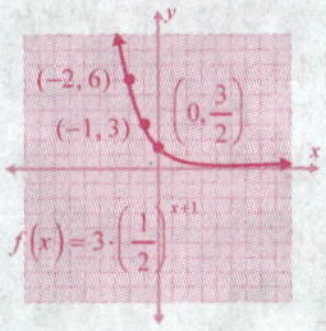

38. Statistics show that the fractional part of flashlight batteries f that are still good after t hours of use is given by $f = 4^{-0.02t}$. What fractional part of the batteries are still operating after 150 hours of use?

$t = 0$

$t = 150$

39. If a principal P is invested at a rate r compounded continuously, the interest earned is given by $I = A - P$.

a. Find the interest earned in 20 years on $10,000 invested at 10% and compounded continuously.

b. Find the interest earned in 20 years on $10,000 invested at 5% and compounded continuously.

c. Explain why the interest earned at 5% is not just one-half of the interest earned at 10% in parts a. and b.

40. The value V of a machine at the end of t years is given by $V = C(1-r)^t$, where C is the original cost and r is the rate of depreciation. Find the value of a machine at the end of 4 years if the original cost was $1200 and $r = 0.20$.

41. Referring to Exercise 40, find the value of a machine at the end of 3 years if the original cost was $2000 and $r = 0.15$.

42. A cancer patient is given a dose of 50 mg of a particular drug. In five days, the amount of the drug in her system is reduced to 1.5625 mg. If the drug decays (or is absorbed) at an exponential rate, find the function that represents the amount of the drug at a given time. (**Hint:** Use the formula $y = y_0 b^{-t}$ and solve for b.)

43. Determine the exponential function that fits the following information concerning exponential growth of cancer cells: $y_0 = 10{,}000$ cancer cells, and there are 160,000 cancer cells present after 4 days. (**Hint:** Use the formula $y = y_0 b^t$ and solve for b.)

Writing & Thinking

44. Discuss, in your own words, the symmetrical relationship of the graphs of the two exponential functions $y = 10^x$ and $y = 10^{-x}$.

45. Discuss, in your own words, the symmetrical relationship of the graphs of the two exponential functions $y = 10^x$ and $y = -10^x$.

20.

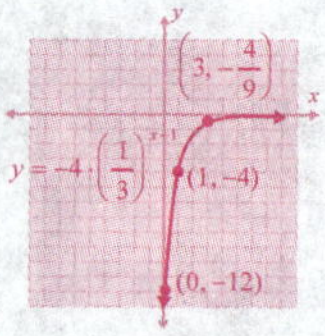

21. 48

22. 30

23. 12.27

24. 16.75

25. 4108.87

26. 6008.33

27. a.

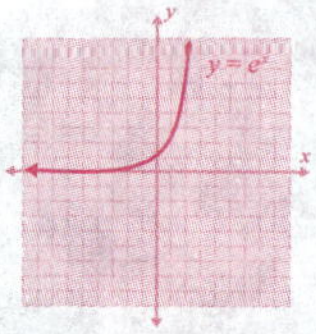

b.

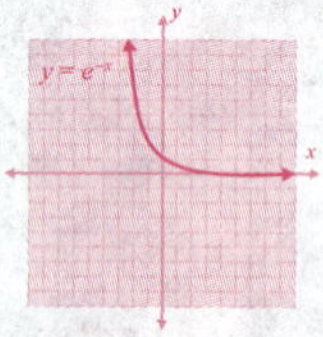

c.

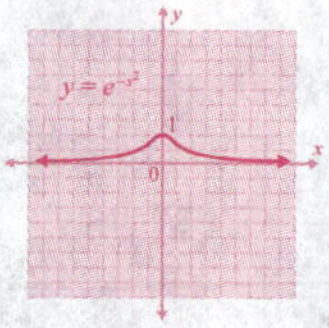

Applications

28. 25,000 bacteria

29. 20,000 bacteria

30. a. $5877.31

b. $5920.98

c. $5943.79

d. $5967.03

e. $5967.30

31. a. $2621.59

b. $2633.62

c. $2639.86

d. $2646.19

e. $2646.26

32. $2154.99

33. $3210.06

34. $46.88

35. $53.33

36. 250,000 people

Collaborative Learning

46. The following formula can be used to calculate monthly mortgage payments:

$$A = \frac{P\left(1+\frac{r}{12}\right)^n \cdot \frac{r}{12}}{\left(1+\frac{r}{12}\right)^n - 1}$$

where

A = the monthly payment,

P = amount initially borrowed (the mortgage),

r = the annual interest rate (in decimal form), and

n = the total number of monthly payments (12 times the number of years).

With the class divided into teams of 3 or 4 students, each team should complete one table (using different values for r and for P). Discuss the results as a class. Explain what this might mean for you personally.

For annual rate r = _____ and initial mortgage P = ________

Length of Mortgage (in years)	Monthly Payment A	Total Cost of Mortgage n times A
15		
20		
25		
30		

37. a. \$100,149.34 **b.** \$222,886.46 **c.** \$1,103,963.86

38. $\frac{1}{64}$

39. a. \$63,890.56 **b.** \$17,182.82 **c.** Answers will vary.

40. \$491.52

41. \$1228.25

42. $y = 50(2)^{-t}$

43. $y = 10{,}000 \cdot 2^t$

Writing & Thinking

44. The graphs are reflections of each other across the y-axis.

45. The graphs are reflections of each other across the x-axis.

Collaborative Learning

46. Answers will vary.

Objectives

A. Translate expressions between exponential and logarithmic form.

B. Evaluate logarithms by using their basic properties.

C. Use the definitions of exponential and logarithmic functions to solve equations.

D. Graph exponential and logarithmic functions on the same set of axes.

History

Scottish mathematician John Napier (1550–1617) is best known for his discovery of logarithms. He named logarithms from the Greek words *logos* "proportion, word" and *arithmos* "number". The main focus in his hobby of mathematics was to devise methods to aid in calculations in a time before computers. As a result, his discovery of logarithms contributed to the advance of science, particularly in astronomy where very large numbers are common. Another notable French mathematician, Pierre-Simon Laplace, called logarithms "...[a]n admirable artifice which, by reducing to a few days the labour of many months, doubles the life of the astronomer, and spares him the errors and disgust inseparable from long calculations."

17.4 Logarithmic Functions

A Introduction to Logarithms

Exponential functions of the form $y = b^x$ are one-to-one functions and, therefore, have inverses. To find the inverse of a function, we interchange x and y in the equation and solve for y. Thus, for the function

$$y = b^x,$$

interchanging x and y gives the inverse function

$$x = b^y.$$

Figure 1 shows the graphs of these two functions with $b > 1$. Each function is a reflection of the other across the line $y = x$. Note that the exponential function $y = b^x$ has the line $y = 0$ (the x-axis) as a horizontal asymptote, and the inverse function $x = b^y$ has the line $x = 0$ (the y-axis) as a vertical asymptote.

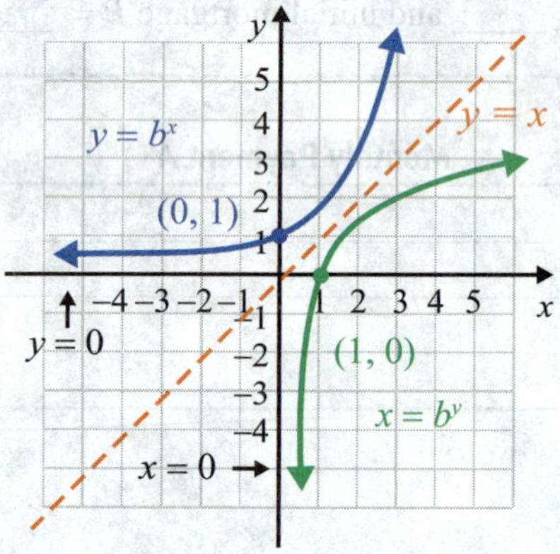

Figure 1

To solve the inverse equation $x = b^y$ for y, mathematicians created a name for y: **logarithm** (abbreviated as log). This means that the inverse of an exponential function is a logarithmic function.

Definition of Logarithm (base *b*)

For $x > 0$, $b > 0$, and $b \neq 1$,

$$x = b^y \text{ is equivalent to } y = \log_b x.$$

$y = \log_b x$ is read "y is the logarithm (base b) of x."

DEFINITION

A logarithm is an exponent. Example 1 shows the relationship between exponential forms and logarithmic forms of equations.

Example 1 Translating between Exponential and Logarithmic Form

	Exponential Form		Logarithmic Form	
a.	$2^3 = 8$	$\Leftrightarrow$	$\log_2 8 = 3$	The base is 2. The logarithm is 3.
b.	$2^4 = 16$	$\Leftrightarrow$	$\log_2 16 = 4$	The base is 2. The logarithm is 4.
c.	$10^3 = 1000$	$\Leftrightarrow$	$\log_{10} 1000 = 3$	The base is 10. The logarithm is 3.
d.	$3^0 = 1$	$\Leftrightarrow$	$\log_3 1 = 0$	The base is 3. The logarithm is 0.
e.	$5^{-1} = \frac{1}{5}$	$\Leftrightarrow$	$\log_5 \frac{1}{5} = -1$	The base is 5. The logarithm is –1.

Notice that in each case the base of the exponent is the base of the logarithm.

Now work margin exercise 1.

It is important to remember that a logarithm is an exponent. For example,

logarithm ↓ $10^2 = 100$ and logarithm ↓ $\log_{10} 100 = 2$ and logarithm ↓ $100 = 10^{\log_{10} 100}$

are all equivalent. In words,

2 is the **exponent** of base 10 needed to get 100 $(10^2 = 100)$, and

2 is equal to the **logarithm** base 10 of 100 $(\log_{10} 100 = 2)$.

B Basic Properties of Logarithms

We know that exponents are logarithms, and from our previous knowledge of exponents, we can make the following equivalent statements for logarithms with base b.

$$b^0 = 1 \quad \Leftrightarrow \quad \log_b 1 = 0$$

$$b^1 = b \quad \Leftrightarrow \quad \log_b b = 1$$

Also, directly from the definition of $y = \log_b x$, we can make two more general statements.

$$x = b^{\log_b x} \quad \text{and} \quad \log_b b^x = x$$

In summary, we have the following four basic properties of logarithms.

1. Express each equation in logarithmic form.

a. $4^1 = 4$

b. $4^3 = 64$

c. $4^{-1} = \frac{1}{4}$

Express each equation in exponential form.

d. $\log_3 9 = 2$

e. $\log_3 81 = 4$

f. $\log_3 \left(\frac{1}{3}\right) = -1$

Math Tip

Notice that the third logarithm $100 = 10^{\log_{10} 100}$ is a result of substituting the second logarithm $\log_{10} 100$ in for the exponent 2 in the first logarithm $10^2 = 100$ (since $\log_{10} 100 = 2$).

Basic Properties of Logarithms

For $b > 0$ and $b \neq 1$,

1. $\log_b 1 = 0$ — Regardless of the base, the logarithm of 1 is 0.
2. $\log_b b = 1$ — The logarithm of the base is always 1.
3. $x = b^{\log_b x}$ — For $x > 0$
4. $\log_b b^x = x$

PROPERTIES

2. Use the four basic properties of logarithms to evaluate each expression.

a. $\log_8 1$

b. $\log_6 6$

c. $10^{\log_{10} 30}$

d. $\log_2 64$

e. $\log_{10} 0.001$

Example 2 Evaluating Logarithms

Use the four basic properties of logarithms to evaluate each expression.

a. $\log_3 1$ c. $10^{\log_{10} 20}$ e. $\log_{10} 0.01$

b. $\log_8 8$ d. $\log_2 32$

Solution

a. $\log_3 1 = 0$ By Property 1

b. $\log_8 8 = 1$ By Property 2

c. $10^{\log_{10} 20} = 20$ By Property 3

d. $\log_2 32 = \log_2 2^5$ Write 32 as 2^5, so that the base is 2.

$= 5$ By Property 4

e. $\log_{10} 0.01 = \log_{10} \frac{1}{100}$

$= \log_{10} \frac{1}{10^2}$

$= \log_{10} 10^{-2}$ Write $\frac{1}{10^2}$ as 10^{-2}, so that the base is 10.

$= -2$ By Property 4

Now work margin exercise 2.

C Solving Logarithmic Equations

3. Solve by first changing the equation to exponential form:

$\log_8 x = \frac{2}{3}$

Example 3 Solving Logarithmic Equations

Solve by first changing the equation to exponential form: $\log_{16} x = \frac{3}{4}$

Solution

$\log_{16} x = \frac{3}{4}$

$x = 16^{\frac{3}{4}}$ Write the equation in exponential form and solve for x.

$x = \left(16^{\frac{1}{4}}\right)^3 = 2^3 = 8$

Thus, $\log_{16} 8 = \frac{3}{4}$.

Now work margin exercise 3.

Example 4 Solving Logarithmic Equations

Solve by first changing the equation to exponential form: $\log_4 8 = x$

Solution

$\log_4 8 = x$

$4^x = 8$ Write the equation in exponential form and solve for x.

$\left(2^2\right)^x = 2^3$ Use the common base, 2.

$2^{2x} = 2^3$

$2x = 3$ The exponents are equal because the bases are the same.

$x = \dfrac{3}{2}$

Thus, $\log_4 8 = \dfrac{3}{2}$.

Now work margin exercise 4.

4. Solve by first changing the equation to exponential form: $\log_4 32 = x$

D Graphs of Logarithmic Functions

As illustrated in Figure 2, because logarithmic functions are the inverses of exponential functions, the graphs of logarithmic functions can be found by reflecting the corresponding exponential functions across the line $y = x$. Figure 2a shows how the graphs of $y = 2^x$ and $y = \log_2 x$ are related. Figure 2b shows how the graphs of $y = 10^x$ and $y = \log_{10} x$ are related. Note that in the graphs of both logarithmic functions the values of y are negative when x is between 0 and 1 $(0 < x < 1)$.

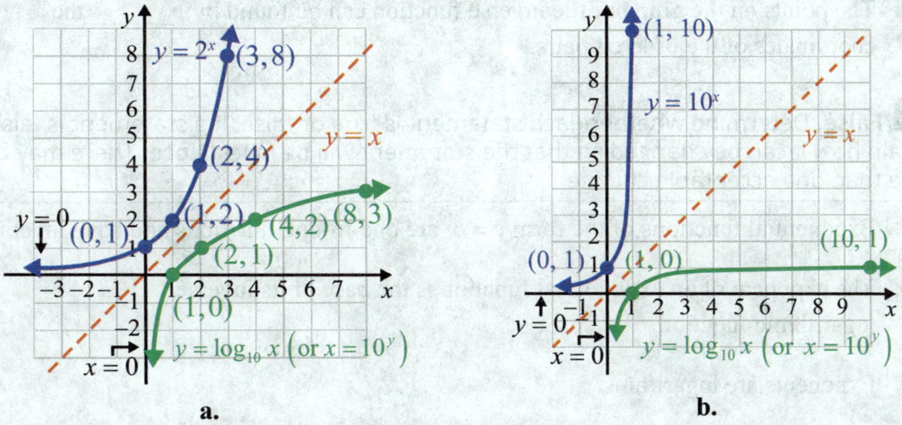

Figure 2

Recall that points on the graphs of inverse functions can be found by reversing the coordinates of ordered pairs. This means that the domain and range of a function and its inverse are interchanged. Thus, for exponential functions and logarithmic functions, we have the following.

- For the exponential function $y = b^x$,

  the domain is all real x, and

  the range is all $y > 0$. (The graph is above the x-axis.)

  There is a horizontal asymptote at $y = 0$.

- For the logarithmic function $y = \log_b x \left(\text{or } x = b^y\right)$,
  the domain is all $x > 0$, and (The graph is to the right of the y-axis.)
  the range is all real y.
  There is a vertical asymptote at $x = 0$.

Margin Exercise Answers

1. a. $\log_4 4 = 1$ **b.** $\log_4 64 = 3$ **c.** $\log_4\left(\frac{1}{4}\right) = -1$ **d.** $3^2 = 9$ **e.** $3^4 = 81$ **f.** $3^{-1} = \frac{1}{3}$ **2. a.** 0 **b.** 1 **c.** 30 **d.** 6 **e.** -3 **3.** 4 **4.** $\frac{5}{2}$

17.4 Exercises

Concept Check

Fill-in-the-Blank. Complete each sentence using information found in this section.

1. The function $x = b^y$ is equivalent to $y =$ ______.
2. The line $y = 0$ is the ______ asymptote of $y = b^x$.
3. The inverse of an exponential function is a/an ______ function.
4. Regardless of the base, the logarithm of 1 is ______.
5. The graph of a logarithmic function can be found by ______ the corresponding exponential function across the line $y = x$.
6. The points on the graph of the inverse function can be found by ______ the coordinates of the ordered pairs.

True/False. Determine whether each statement is true or false. If a statement is false, explain how it can be changed so that the statement will be true. (**Note:** There may be more than one acceptable change.)

7. Exponential functions of the form $y = b^x$ are one-to-one functions and have inverses.
8. The exponent of an exponential function is the base of its inverse logarithmic function.
9. Exponents are logarithms.
10. The logarithm of the base is always 1.

Practice

Express each equation in logarithmic form. See Example 1.

1. $7^2 = 49$
2. $3^3 = 27$
3. $5^{-2} = \frac{1}{25}$
4. $2^{-5} = \frac{1}{32}$
5. $1 = \pi^0$
6. $6^0 = 1$
7. $10^2 = 100$
8. $10^1 = 10$
9. $10^k = 23$

17.4 Exercises

Concept Check

1. $\log_b x$
2. horizontal
3. logarithmic
4. 0
5. reflecting
6. reversing
7. True
8. False; The base of the exponential function is the base of the logarithmic function.
9. True
10. True

Practice

1. $\log_7 49 = 2$
2. $\log_3 27 = 3$
3. $\log_5 \frac{1}{25} = -2$
4. $\log_2 \frac{1}{32} = -5$
5. $\log_\pi 1 = 0$
6. $\log_6 1 = 0$
7. $\log_{10} 100 = 2$
8. $\log_{10} 10 = 1$
9. $\log_{10} 23 = k$
10. $\log_4 11.6 = k$
11. $\log_{\frac{2}{3}} \frac{4}{9} = 2$
12. $\log_{\frac{3}{4}} \frac{9}{16} = 2$
13. $3^2 = 9$
14. $5^3 = 125$
15. $9^{\frac{1}{2}} = 3$
16. $b^{\frac{2}{3}} = 4$
17. $7^{-1} = \frac{1}{7}$
18. $\left(\frac{1}{2}\right)^{-3} = 8$
19. $10^{1.74} = N$
20. $2^x = 42.3$

10. $4^k = 11.6$ **11.** $\left(\frac{2}{3}\right)^2 = \frac{4}{9}$ **12.** $\left(\frac{3}{4}\right)^2 = \frac{9}{16}$

Express each equation in exponential form. See Example 1.

13. $\log_3 9 = 2$ **17.** $\log_7 \frac{1}{7} = -1$ **21.** $\log_b 18 = 4$

14. $\log_5 125 = 3$ **18.** $\log_{1/2} 8 = -3$ **22.** $\log_b 39 = 10$

15. $\log_9 3 = \frac{1}{2}$ **19.** $\log_{10} N = 1.74$ **23.** $\log_n y^2 = x$

16. $\log_b 4 = \frac{2}{3}$ **20.** $\log_2 42.3 = x$ **24.** $\log_b a = x^2$

Use the four basic properties of logarithms to evaluate each expression. See Example 2.

25. $\log_3 81$ **27.** $\log_7 1$ **29.** $\log_4 \frac{1}{64}$

26. $7^{\log_7 15}$ **28.** $5^{\log_5 25}$ **30.** $\log_{12} 12$

Solve by first changing each equation to exponential form. See Examples 3 and 4.

31. $\log_4 x = 2$ **37.** $\log_{36} x = -\frac{1}{2}$ **43.** $\log_8 8^{3.7} = x$

32. $\log_3 x = 4$ **38.** $\log_{81} x = -\frac{3}{4}$ **44.** $\log_{10} 10^{1.52} = x$

33. $\log_{14} 196 = x$ **39.** $\log_x 32 = 5$ **45.** $\log_5 5^{\log_5 25} = x$

34. $\log_{25} 125 = x$ **40.** $\log_x 121 = 2$ **46.** $\log_4 4^{\log_2 8} = x$

35. $\log_5 \frac{1}{125} = x$ **41.** $\log_8 x = \frac{5}{3}$

36. $\log_3 \frac{1}{9} = x$ **42.** $\log_{16} x = \frac{3}{4}$

Graph each function and its inverse on the same set of axes. Label two points on each graph.

47. $f(x) = 6^x$ **50.** $y = \left(\frac{1}{4}\right)^x$ **53.** $y = \log_{1/2} x$

48. $f(x) = 2^x$ **51.** $f(x) = \log_4 x$ **54.** $y = \log_{1/3} x$

49. $y = \left(\frac{2}{3}\right)^x$ **52.** $f(x) = \log_5 x$ **55.** $y = \log_8 x$

56. $y = \log_7 x$

57. Consider the function $y = c(3^x)$ where c is a constant greater than zero. List the following:

a. The domain of the function.

b. The range of the function.

c. Any asymptotes of the graph of the function.

d. Give c two different values and sketch the graphs of both functions.

21. $b^4 = 18$

22. $b^{10} = 39$

23. $n^x = y^2$

24. $b^{x^2} = a$

25. 4

26. 15

27. 0

28. 25

29. −3

30. 1

31. $x = 16$

32. $x = 81$

33. $x = 2$

34. $x = \frac{3}{2}$

35. $x = -3$

36. $x = -2$

37. $x = \frac{1}{6}$

38. $x = \frac{1}{27}$

39. $x = 2$

40. $x = 11$

41. $x = 32$

42. $x = 8$

43. $x = 3.7$

44. $x = 1.52$

45. $x = 2$

46. $x = 3$

47.

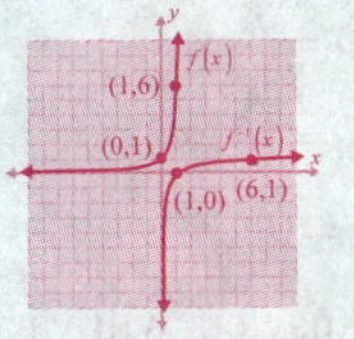

48.

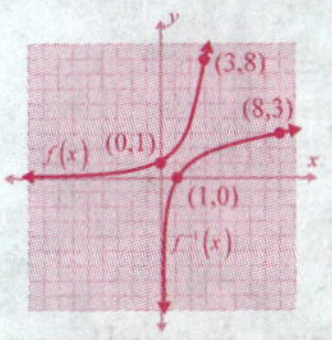

49.

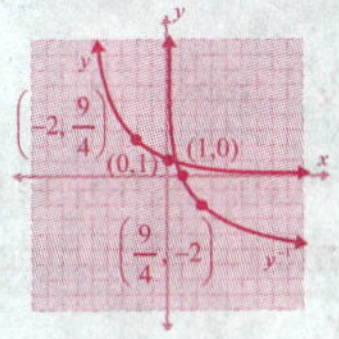

58. Consider the function $y = c\left(3^{-x}\right)$ where c is a constant greater than zero. List the following:

a. The domain of the function.

b. The range of the function.

c. Any asymptotes of the graph of the function.

d. Give c two different values and sketch the graphs of both functions.

Writing & Thinking

59. Discuss, in your own words, the symmetrical relationship of the graphs of the two functions $y = 10^x$ and $y = \log_{10} x$.

60. Discuss, in your own words, the symmetrical relationship of the graphs of the two logarithmic functions $y = \log_{10} x$ and $y = -\log_{10} x$.

60. The two functions are symmetric about the x-axis.

50. 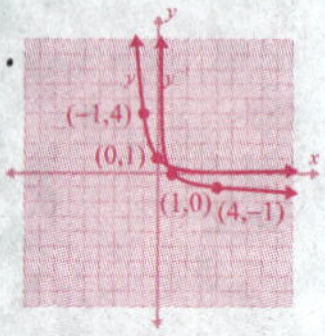

51. 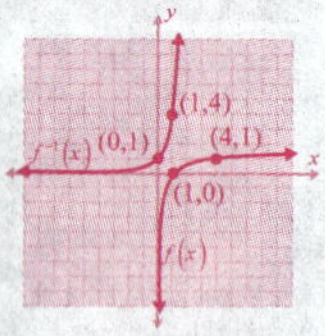

52. 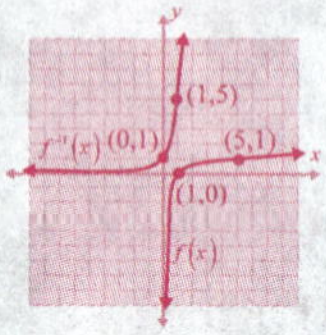

53. 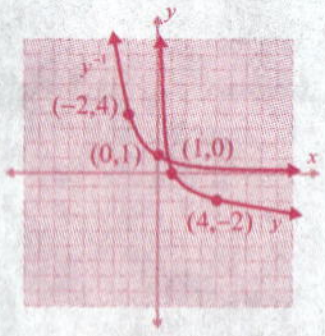

54. 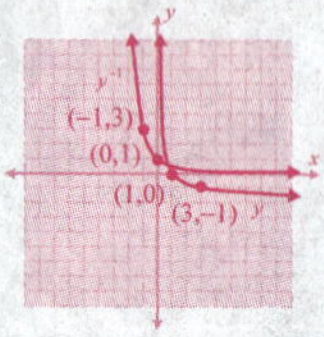

55. 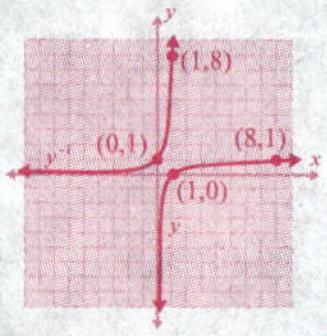

56. 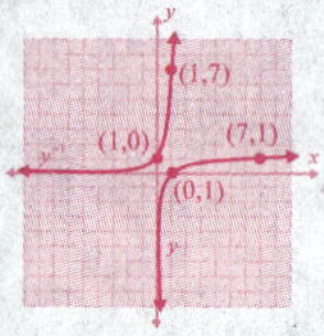

57. **a.** $D = (-\infty, \infty)$

b. $R = (0, \infty)$

c. $y = 0$

d. Answers will vary.

58. **a.** $D = (-\infty, \infty)$

b. $R = (0, \infty)$

c. $y = 0$

d. Answers will vary.

Writing & Thinking

59. The two functions are symmetric about the line $y = x$.

17.5 Properties of Logarithms

While calculators are effective in giving numerical evaluations, many do not simplify or solve equations. In this section, we will discuss several properties (or rules) of logarithms that are helpful in solving equations and simplifying expressions that involve logarithms or exponential functions.

The following four basic properties have been discussed.

For $b > 0$ and $b \neq 1$,

1. $\log_b 1 = 0$

2. $\log_b b = 1$

3. $x = b^{\log_b x}$ for $x > 0$

4. $\log_b b^x = x$

With these four properties as a basis, we now develop three more properties (or rules) for logarithms: the product rule, the quotient rule, and the power rule.

Objectives

A. Use the product rule for logarithms to simplify logarithmic expressions.

B. Use the quotient rule for logarithms to simplify logarithmic expressions.

C. Use the power rule for logarithms to simplify logarithmic expressions.

D. Use the rules for logarithms to expand logarithmic expressions and to write expanded forms as single logarithms.

Teaching Note

It doesn't hurt to review the product rule for exponents before getting into this section, as the product rule is the same in this context. Similarly, the power rule and quotient rule are the same.

A The Product Rule for Logarithms

Because logarithms are exponents, their properties are similar to those of exponents. In fact, the properties of exponents are used to prove the rules for logarithms. Study the lesson carefully and you will see that logarithms are handled the same as exponents.

Consider the following analysis.

We know that $6 = 2 \cdot 3$.

Using property 3,

$10^{\log_{10} 6} = 6,\qquad 10^{\log_{10} 2} = 2,\qquad \text{and}\qquad 10^{\log_{10} 3} = 3.$

Next, use the properties of exponents.

$$6 = 2 \cdot 3$$

$$\downarrow \quad\quad \downarrow \quad\quad \downarrow$$

$$10^{\log_{10} 6} = 10^{\log_{10} 2} \cdot 10^{\log_{10} 3} = 10^{\log_{10} 2 + \log_{10} 3}$$

Equating the exponents (with the same base) gives the following result.

$$\log_{10} 6 = \log_{10} 2 + \log_{10} 3$$

$$\downarrow \quad\quad \downarrow \quad\quad \downarrow$$

$$0.7782 \approx 0.3010 + 0.4771$$ Note that the fourth digit is off due to rounding.

This technique can be used to prove the following **product rule for logarithms**.

Note

Using a calculator to find the values of logarithms is discussed in the next section. To illustrate some of the properties of logarithms in this section, the following base 10 logarithmic values, accurate to 4 decimal places are used.

$\log_{10} 2 \approx 0.3010$

$\log_{10} 3 \approx 0.4771$

$\log_{10} 5 \approx 0.6990$

$\log_{10} 6 \approx 0.7782$

Also, because the logarithms are rounded to 4 decimal places, a slight difference may occur in the answers if a calculator is used to find the logarithms in some of the illustrations.

Product Rule for Logarithms

For $b > 0$, $b \neq 1$, and $x, y > 0$,

$$\log_b xy = \log_b x + \log_b y.$$

In words, the logarithm of a product is equal to the sum of the logarithms of the factors.

PROPERTIES

Proof of the Product Rule

$$\overbrace{b^{\log_b(xy)} = xy}^{\text{Property 3}} = \overbrace{x \cdot y = b^{\log_b x} \cdot b^{\log_b y}}^{\text{Property 3 again}} \overbrace{= b^{\log_b x + \log_b y}}^{\text{Add exponents}}$$

Thus,

$$b^{\log_b(xy)} = b^{\log_b x + \log_b y}.$$

Equating the exponents gives the product rule for logarithms.

$$\log_b xy = \log_b x + \log_b y$$

1. Simplify each expression using the product rule.
   a. $\log_{10} 10{,}000$
   b. $\log_{10} 40$
   c. $\log_{10} 60$

Example 1 Using the Product Rule of Logarithms

Simplify each expression using the product rule.

a. $\log_{10} 1000$ **b.** $\log_{10} 30$ **c.** $\log_{10} 30$

Solution

a.
$$\begin{aligned}\log_{10} 1000 &= \log_{10}(10 \cdot 10 \cdot 10)\\ &= \log_{10} 10 + \log_{10} 10 + \log_{10} 10\\ &= 1 + 1 + 1\\ &= 3\end{aligned}$$

(We knew this result from another approach: $\log_{10} 1000 = \log_{10} 10^3 = 3$. But the approach shown helps to confirm that the product rule works.)

b.
$$\begin{aligned}\log_{10} 30 &= \log_{10} 10 \cdot 3\\ &= \log_{10} 10 + \log_{10} 3\\ &\approx 1 + 0.4771 = 1.4771\end{aligned}$$

c.
$$\begin{aligned}\log_{10} 30 &= \log_{10}(2 \cdot 3 \cdot 5)\\ &= \log_{10} 2 + \log_{10} 3 + \log_{10} 5\\ &\approx 0.3010 + 0.4771 + 0.6990 = 1.4771\end{aligned}$$

Note that in b. and c., using different factors of 30 gives the same result.

Now work margin exercise 1.

B The Quotient Rule for Logarithms

Now consider the problem of finding the logarithm of a quotient. For example, to find $\log_{10}\frac{3}{2}$, we can proceed as follows.

$$10^{\log_{10}\frac{3}{2}} = \frac{3}{2}$$
$$= \frac{10^{\log_{10}3}}{10^{\log_{10}2}}$$
$$= 10^{\log_{10}3 - \log_{10}2}$$

Equating the exponents gives the following.

$$\log_{10}\frac{3}{2} = \log_{10}3 - \log_{10}2$$
$$\approx 0.4771 - 0.3010$$
$$= 0.1761$$

These ideas lead to the following **quotient rule for logarithms**. The proof is left as an exercise for the student.

Quotient Rule for Logarithms

For $b > 0$, $b \neq 1$, and $x, y > 0$,

$$\log_b \frac{x}{y} = \log_b x - \log_b y.$$

In words, the logarithm of a quotient is equal to the difference between the logarithm of the numerator and the logarithm of the denominator.

PROPERTIES

Example 2 Using the Quotient Rule of Logarithms

Simplify each expression using the quotient rule.

a. $\log_{10}\frac{1}{2}$ **b.** $\log_2\frac{1}{8}$ **c.** $\log_{10}\frac{15}{2}$

Solution

a. $\log_{10}\frac{1}{2} = \log_{10}1 - \log_{10}2$

$= 0 - \log_{10}2 \approx -0.3010$

b. $\log_2\frac{1}{8} = \log_2 1 - \log_2 8$

$= \log_2 1 - \log_2 2^3$

$= 0 - 3 = -3$

Note: Writing $\frac{1}{8}$ as 2^{-3} produces the same result.

2. Simplify each expression using the quotient rule.

a. $\log_{10}\frac{1}{3}$

b. $\log_3\frac{1}{9}$

c. $\log_{10}\frac{12}{5}$

c. $$\begin{aligned}\log_{10}\frac{15}{2} &= \log_{10}15-\log_{10}2\\ &= \log_{10}(3\cdot 5)-\log_{10}2\\ &= \log_{10}3+\log_{10}5-\log_{10}2\\ &\approx 0.4771+0.6990-0.3010=0.8751\end{aligned}$$

Now work margin exercise 2.

C The Power Rule for Logarithms

The next property of logarithms involves a number raised to a power and the multiplication property of exponents. For example,

$$\begin{aligned}10^{\log_{10}2^3} &= 2^3\\ &= \left(10^{\log_{10}2}\right)^3\\ &= 10^{3(\log_{10}2)}. \quad \text{By the power rule for exponents}\end{aligned}$$

Now equating the exponents in the first and last expressions gives

$$\log_{10}2^3 = 3\cdot\log_{10}2.$$

These ideas lead to the following **power rule for logarithms**.

Power Rule for Logarithms

For $b>0$, $b\neq 1$, $x>0$, and any real number r,

$$\log_b x^r = r\cdot\log_b x.$$

In words, the logarithm of a number raised to a power is equal to the product of the exponent and the logarithm of the number.

PROPERTIES

Math Tip

Be careful not to confuse the power rule for logarithms with the power rule for exponents. The power rule for exponents states that if a is a nonzero real number and m and n are integers, then we have the following.

$$\left(a^m\right)^n = a^{mn}$$

Proof of the Power Rule for Logarithms

$$b^{\log_b x^r} = \underbrace{x^r = \left(b^{\log_b x}\right)^r}_{\text{Property 3 twice}} = \underbrace{b^{r\cdot\log_b x}}_{\text{Multiply exponents}}$$

Equating the exponents gives the result called the power rule for logarithms.

$$\log_b x^r = r\cdot\log_b x$$

3. Simplify each expression using the power rule.

a. $\log_{10}\sqrt{3}$

b. $\log_{10}27$

c. $\log_{10}36$

Example 3 Using the Power Rule

Simplify each expression using the power rule.

a. $\log_{10}\sqrt{2}$ **b.** $\log_{10}8$ **c.** $\log_{10}25$

Solution

a. $\log_{10}\sqrt{2} = \log_{10} 2^{\frac{1}{2}} = \frac{1}{2}\cdot\log_{10} 2 \approx \frac{1}{2}(0.3010) = 0.1505$

b. $\log_{10} 8 = \log_{10} 2^3 = 3\cdot\log_{10} 2 \approx 3(0.3010) = 0.9030$

c. $\log_{10} 25 = \log_{10} 5^2 = 2\cdot\log_{10} 5 \approx 2(0.6990) = 1.3980$

Now work margin exercise 3.

Summary of Properties of Logarithms

For $b > 0$, $b \neq 1$, $x, y > 0$, and any real number r,

1. $\log_b 1 = 0$
2. $\log_b b = 1$
3. $x = b^{\log_b x}$
4. $\log_b b^x = x$
5. $\log_b xy = \log_b x + \log_b y$ The product rule
6. $\log_b \frac{x}{y} = \log_b x - \log_b y$ The quotient rule
7. $\log_b x^r = r\cdot\log_b x$ The power rule

PROPERTIES

D Using the Properties of Logarithms

Example 4 Using the Properties of Logarithms to Expand Expressions

Use the properties of logarithms to expand each expression as much as possible.

a. $\log_b 2x^3$ **b.** $\log_b \frac{xy^2}{z}$ **c.** $\log_b (xy)^{-3}$ **d.** $\log_b \sqrt{3x}$

Solution

a. $\log_b 2x^3 = \log_b 2 + \log_b x^3$ Product rule
$= \log_b 2 + 3\log_b x$ Power rule

b. $\log_b \frac{xy^2}{z} = \log_b xy^2 - \log_b z$ Quotient rule
$= \log_b x + \log_b y^2 - \log_b z$ Product rule
$= \log_b x + 2\log_b y - \log_b z$ Power rule

4. Use the properties of logarithms to expand each expression as much as possible.

a. $\log_b 3x^2$

b. $\log_b \frac{x^3 y}{z}$

c. $\log_b (mn)^{-2}$

d. $\log_b \sqrt{2a}$

c. $\log_b (xy)^{-3} = -3\log_b xy$ Power rule

$= -3(\log_b x + \log_b y)$ Product rule

$= -3\log_b x - 3\log_b y$

d. $\log_b \sqrt{3x} = \log_b (3x)^{\frac{1}{2}} = \frac{1}{2}\log_b 3x$ Power rule

$= \frac{1}{2}(\log_b 3 + \log_b x)$ Product rule

$= \frac{1}{2}\log_b 3 + \frac{1}{2}\log_b x$

Now work margin exercise 4.

5. Use the properties of logarithms to write each expression as a single logarithm.

a. $3\log_b x - 4\log_b y$

b. $\frac{1}{2}\log_d 9 - \log_d 4 - \log_d x$

c. $\log_a (y+2) + \log_a (y-2)$

d. $\log_b \sqrt[4]{y} + \log_b \sqrt{y}$

Example 5 Using the Properties of Logarithms to Write Expanded Forms as Single Logarithms

Use the properties of logarithms to write each expression as a single logarithm. Note that the bases must be the same when simplifying sums and differences of logarithms.

a. $2\log_b x - 3\log_b y$

b. $\frac{1}{2}\log_a 4 - \log_a 5 - \log_a y$

c. $\log_a (x+1) + \log_a (x-1)$

d. $\log_b \sqrt{x} + \log_b \sqrt[3]{x}$

Solution

a. $2\log_b x - 3\log_b y = \log_b x^2 - \log_b y^3$ Power rule

$= \log_b \left(\frac{x^2}{y^3}\right)$ Quotient rule

b. $\frac{1}{2}\log_a 4 - \log_a 5 - \log_a y$

$= \log_a 4^{\frac{1}{2}} - \log_a 5 - \log_a y$ Power rule

$= \log_a 2 - (\log_a 5 + \log_a y)$ $(4^{\frac{1}{2}} = 2)$ Factor out –1.

$= \log_a 2 - \log_a 5y$ Product rule

$= \log_a \left(\frac{2}{5y}\right)$ Quotient rule

c. $\log_a (x+1) + \log_a (x-1) = \log_a [(x+1)(x-1)]$ Product rule

$= \log_a (x^2 - 1)$ Multiply

d. $\log_b \sqrt{x} + \log_b \sqrt[3]{x} = \log_b x^{\frac{1}{2}} + \log_b x^{\frac{1}{3}}$

$= \frac{1}{2}\log_b x + \frac{1}{3}\log_b x$ Power rule

$= \left(\frac{1}{2} + \frac{1}{3}\right)\log_b x$ Distributive property

$= \frac{5}{6}\log_b x$ $\frac{1}{2} + \frac{1}{3} = \frac{3}{6} + \frac{2}{6} = \frac{5}{6}$

$= \log_b x^{\frac{5}{6}}$ Power rule

OR

$$\begin{aligned}\log_b \sqrt{x} + \log_b \sqrt[3]{x} &= \log_b x^{\frac{1}{2}} + \log_b x^{\frac{1}{3}} \\ &= \log_b \left(x^{\frac{1}{2}} \cdot x^{\frac{1}{3}}\right) \quad \text{Product rule} \\ &= \log_b x^{\left(\frac{1}{2}+\frac{1}{3}\right)} \\ &= \log_b x^{\frac{5}{6}}\end{aligned}$$

Now work margin exercise 5.

Common Misunderstandings about Logarithms

There is no logarithmic property for the logarithm of a sum or a difference.

$\log_b(x+y)$ Cannot be simplified

$\log_b(x-y)$ Cannot be simplified

Also,

$\log_b(xy) \neq \log_b x \cdot \log_b y$ The log of a product does not equal the product of the logs.

$\log_b \frac{x}{y} \neq \frac{\log_b x}{\log_b y}$ The log of a quotient does not equal the quotient of the logs.

CAUTION

Margin Exercise Answers

1. a. 4 **b.** 1.6021 **c.** 1.7782 **2. a.** −0.4771 **b.** −2 **c.** 0.3802 **3. a.** 0.2386 **b.** 1.4314 **c.** 1.5563

4. a. $\log_b 3 + 2\log_b x$ **b.** $3\log_b x + \log_b y - \log_b z$ **c.** $-2\log_b m - 2\log_b n$ **d.** $\frac{1}{2}\log_b 2 + \frac{1}{2}\log_b a$

5. a. $\log_b\left(\frac{x^3}{y^4}\right)$ **b.** $\log_d\left(\frac{3}{4x}\right)$ **c.** $\log_a\left(y^2-4\right)$ **d.** $\log_b y^{\frac{3}{4}}$

17.5 Exercises

Concept Check

Fill-in-the-Blank. Complete each sentence using information found in this section.

1. Because logarithms are ______, their properties are similar to those of ______.

2. The logarithm of a product is equal to the ______ of the logarithms of the factors.

3. The logarithm of a quotient is equal to the ______ ______ the logarithm of the numerator and the logarithm of the denominator.

4. The logarithm of a number raised to a power is equal to the ______ of the exponent and the logarithm of the number.

17.5 Exercises

Concept Check

1. exponents, exponents
2. sum
3. difference between
4. product

5. sum, difference
6. 0, 1
7. True
8. False; The exponent r can be any real number.
9. True
10. False; The expression $\log_5 \frac{4}{3} = \log_5 4 - \log_5 3$

Practice

1. a. 2
   b. 3
   c. 5
   d. 6
2. a. 12
   b. 18
   c. 15
   d. 2.5
3. 5
4. 2
5. −2
6. −3
7. $\frac{1}{2}$
8. $\frac{3}{2}$
9. 10
10. 17
11. $\sqrt{3}$
12. 5
13. $\log_b 5 + 4\log_b x$
14. $\log_b 3 + 2\log_b x + \log_b y$
15. $\log_b 2 - 3\log_b x + \log_b y$
16. $\log_5 x + 2\log_5 y - \log_5 z$
17. $\log_6 2 + \log_6 x - 3\log_6 y$
18. $\log_3 x + \log_3 y - \log_3 4 - \log_3 z$
19. $2\log_b x - \log_b y - \log_b z$
20. $\log_3 x + 2\log_3 y - 2\log_3 z$
21. $-2\log_5 x - 2\log_5 y$
22. $8\log_b x + 4\log_b y$
23. $\frac{1}{3}\log_6 x + \frac{2}{3}\log_6 y$

5. There is no logarithmic property for the logarithm of a/an ______ or a/an ______.

6. When dealing with logarithms of the form $\log_b x$, we assume that $b >$ ______ and $b \neq$ ______.

True/False. Determine whether each statement is true or false. If a statement is false, explain how it can be changed so the statement will be true. (**Note:** There may be more than one acceptable change.)

7. The properties of exponents are used to prove the properties of logarithms.

8. The power rule for logarithms states that the exponent r must be a positive integer.

9. The log of a product does not equal the product of the logs.

10. The expression $\log_5 \frac{4}{3}$ is equivalent to $\frac{\log_5 4}{\log_5 3}$.

Practice

Use the following logarithms (accurate to 4 decimal places) in Exercises 1 and 2. See Example 1.

$$\log_{10} 2 \approx 0.3010 \qquad \log_{10} 3 \approx 0.4771$$
$$\log_{10} 5 \approx 0.6990 \qquad \log_{10} 6 \approx 0.7782$$

1. Find the values of the following expressions.
   a. $10^{0.3010}$
   b. $10^{0.4771}$
   c. $10^{0.6990}$
   d. $10^{0.7782}$

2. Find the values of the following expressions.
   a. $10^{0.3010+0.7782}$
   b. $10^{0.4771+0.7782}$
   c. $10^{0.4771+0.6990}$
   d. $10^{0.6990-0.3010}$

Use your knowledge of logarithms and exponents to find the value of each expression.

3. $\log_2 32$
4. $\log_3 9$
5. $\log_4 \frac{1}{16}$
6. $\log_5 \frac{1}{125}$
7. $\log_3 \sqrt{3}$
8. $\log_2 \sqrt{8}$
9. $5^{\log_5 10}$
10. $3^{\log_3 17}$
11. $6^{\log_6 \sqrt{3}}$
12. $5^{\log_5 5}$

Use the properties of logarithms to expand each expression as much as possible. See Example 4.

13. $\log_b 5x^4$
14. $\log_b 3x^2 y$
15. $\log_b 2x^{-3} y$
16. $\log_5 xy^2 z^{-1}$

17. $\log_6 \frac{2x}{y^3}$

18. $\log_3 \frac{xy}{4z}$

19. $\log_b \frac{x^2}{yz}$

20. $\log_3 \frac{xy^2}{z^2}$

21. $\log_5 (xy)^{-2}$

22. $\log_b (x^2 y)^4$

23. $\log_6 \sqrt[3]{xy^2}$

24. $\log_5 \sqrt{2x^3 y}$

25. $\log_3 \sqrt{\frac{xy}{z}}$

26. $\log_6 \sqrt[3]{\frac{x^2}{y}}$

27. $\log_5 21x^2 y^{\frac{2}{3}}$

28. $\log_b 15x^{-\frac{1}{2}} y^{\frac{1}{3}}$

29. $\log_6 \frac{x}{\sqrt{x^3 y^5}}$

30. $\log_3 \frac{1}{\sqrt{x^4 y}}$

31. $\log_b \left(\frac{x^3 y^2}{z}\right)^{-3}$

32. $\log_4 \left(\frac{x^{-\frac{1}{2}} y}{z^2}\right)^{-2}$

Use the properties of logarithms to write each expression as a single logarithm of a single expression. See Example 5.

33. $2\log_b 3 + \log_b x - \log_b 5$

34. $\frac{1}{2}\log_b 25 + \log_b 3 - \log_b x$

35. $\log_2 7 + \log_2 9 + 2\log_2 x$

36. $\log_5 4 + \log_5 6 + \log_5 y$

37. $2 \log_b x + \log_b y$

38. $\log_2 x + 3 \log_2 y$

39. $3\log_5 y - \frac{1}{2}\log_5 x$

40. $3 \log_{10} x - 2 \log_{10} y$

41. $\frac{1}{2}(\log_5 x - \log_5 y)$

42. $\frac{1}{3}(\log_{10} x - 2\log_{10} y)$

43. $\log_2 x - \log_2 y + \log_2 z$

44. $\log_b x - 2 \log_b y - 2 \log_b z$

45. $\log_b x + 2\log_b y - \frac{1}{2}\log_b z$

46. $-\frac{2}{3}\log_2 x - \frac{1}{3}\log_2 y + \frac{2}{3}\log_2 z$

47. $2\log_5 x + \log_5 (2x+1)$

48. $\log_b (3x+1) + 2\log_b x$

49. $\log_2 (x-1) + \log_2 (x+3)$

50. $\log_{10} (x+3) + \log_{10} (x-3)$

51. $\log_b (x^2 - 2x - 3) - \log_b (x-3)$

52. $\log_2 (x-4) - \log_2 (x^2 - 2x - 8)$

53. $\log_{10} (x+6) - \log_{10} (2x^2 + 9x - 18)$

54. $\log_5 (3x^2 + 5x - 2) - \log_5 (3x-1)$

24. $\frac{1}{2}\log_5 2 + \frac{3}{2}\log_5 x + \frac{1}{2}\log_5 y$

25. $\frac{1}{2}\log_3 x + \frac{1}{2}\log_3 y - \frac{1}{2}\log_3 z$

26. $\frac{2}{3}\log_6 x - \frac{1}{3}\log_6 y$

27. $\log_5 21 + 2\log_5 x + \frac{2}{3}\log_5 y$

28. $\log_b 15 - \frac{1}{2}\log_b x + \frac{1}{3}\log_b y$

29. $-\frac{1}{2}\log_6 x - \frac{5}{2}\log_6 y$

30. $-2\log_3 x - \frac{1}{2}\log_3 y$

31. $-9 \log_b x - 6 \log_b y + 3 \log_b z$

32. $\log_4 x - 2 \log_4 y + 4 \log_4 z$

33. $\log_b \left(\frac{9x}{5}\right)$

34. $\log_b \left(\frac{15}{x}\right)$

35. $\log_2 63x^2$

36. $\log_5 24y$

37. $\log_b x^2 y$

38. $\log_2 xy^3$

39. $\log_5 \left(\frac{y^3}{\sqrt{x}}\right)$ or $\log_5 \left(\frac{y^3\sqrt{x}}{x}\right)$

40. $\log_{10} \left(\frac{x^3}{y^2}\right)$

41. $\log_5 \sqrt{\frac{x}{y}}$

42. $\log_{10} \sqrt[3]{\frac{x}{y^2}}$

43. $\log_2 \left(\frac{xz}{y}\right)$

44. $\log_b \left(\frac{x}{y^2 z^2}\right)$

45. $\log_b\left(\dfrac{xy^2}{\sqrt{z}}\right)$ or $\log_b\left(\dfrac{xy^2\sqrt{z}}{z}\right)$

46. $\log_2\sqrt[3]{\dfrac{z^2}{x^2y}}$

47. $\log_5\left(2x^3+x^2\right)$

48. $\log_b\left(3x^3+x^2\right)$

49. $\log_2\left(x^2+2x-3\right)$

50. $\log_{10}\left(x^2-9\right)$

51. $\log_b\left(x+1\right)$

52. $\log_2\left(\dfrac{1}{x+2}\right)$

53. $\log_{10}\left(\dfrac{1}{2x-3}\right)$

54. $\log_5\left(x+2\right)$

Writing & Thinking

55. Answers will vary.

56. Answers will vary.

Writing & Thinking

55. Prove the quotient rule for logarithms: For $b>0$, $b\neq 1$, and $x, y>0$,

$$\log_b\frac{x}{y}=\log_b x-\log_b y.$$

56. Prove the following property of logarithms: For $b>0$, $b\neq 1$, and $x>0$, $\log_b b^x=x$.

17.6 Common Logarithms and Natural Logarithms

Objectives

A. Identify common logarithms (base 10) and use a calculator to find their approximate values.

B. Use a calculator to find the inverse of a common logarithm.

C. Identify natural logarithms (base *e*) and use a calculator to find their approximate values.

D. Use a calculator to find the inverse of a natural logarithm.

Base 10 logarithms (called **common logarithms**) and base *e* logarithms (called **natural logarithms**) are the two most commonly used logarithms. Common logarithms occur frequently because of their close relationship with the decimal system and are used in chemistry, astronomy, and physical science. Natural logarithms are an important tool in calculus and appear "naturally" in growth and decay problems. Both common and natural logarithms have designated keys on calculators.

A Common Logarithms (Base 10 Logarithms)

The abbreviated notation $\log x$ is used to indicate common logarithms (base 10 logarithms). That is, whenever the base notation is omitted, the base is understood to be 10.

$$\log_{10} x = \log x$$

Example 1 Translating between Exponential Form and Common Logarithms

	Exponential Form		Logarithmic Form	
a.	$10^4 = 10{,}000$	$\Leftrightarrow$	$\log 10{,}000 = 4$	This is a common logarithm that equals 4.
b.	$10^{-2} = 0.01$	$\Leftrightarrow$	$\log 0.01 = -2$	This is a common logarithm that equals −2.
c.	$10^x = 6.3$	$\Leftrightarrow$	$\log 6.3 = x$	This is a common logarithm that equals *x*.

Now work margin exercise 1.

1. Translate $10^2 = 100$ into logarithmic form.

The TI-84 Plus graphing calculator has the designated key LOG that can be used in the following 3-step process to find the values of common logarithms.

CALCULATORS

Using a Calculator to Evaluate Common Logarithms (Base 10)

To evaluate a **common logarithm** on a TI-84 Plus graphing calculator:

1. Press the LOG key. `log(` will appear on the display.
2. Enter the number and a right-hand parenthesis,).
3. Press ENTER.

2. Use a TI-84 Plus graphing calculator to find the values of the following common logarithms, accurate to the nearest ten-thousandth.

a. log 400

b. log 5000

Example 2 Evaluating Common Logarithms Using a Calculator

Use a TI-84 Plus graphing calculator to find the approximate values of the following common logarithms.

a. $\log 200$

b. $\log 50{,}000$

c. $\log 0.0006$

Solution

From the display, we see the results (accurate to 9 decimal places).

a. $\log 200 \approx 2.301029996$

b. $\log 50{,}000 \approx 4.698970004$

c. $\log 0.0006 \approx -3.22184875$

Now work margin exercise 2.

Example 2c shows a negative value for the logarithm. This is because 0.0006 is between 0 and 1. In fact, the logarithm (with base greater than 1) of any number between 0 and 1 will always be negative. See Figure 1.

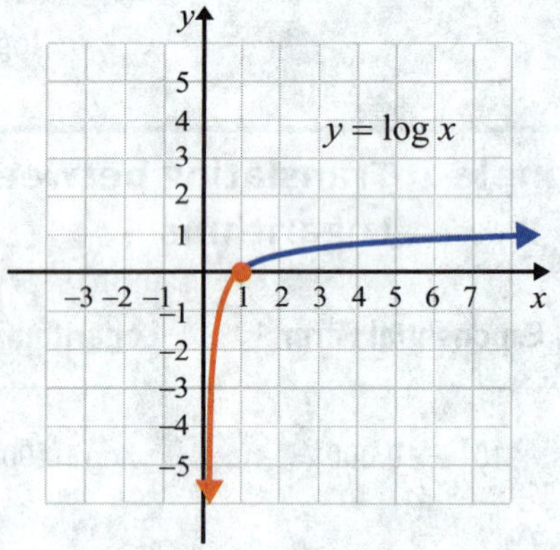

Figure 1

Logarithms are exponents and they may be positive, negative, or 0. For example,

$$10^{-2} = \frac{1}{100} = 0.01 \quad \text{and} \quad \log 0.01 = -2.$$

However, there are no logarithms of negative numbers or 0. The domains of logarithmic functions are positive real numbers. The logarithm of a negative number is undefined. For example, $\log(-2)$ is **undefined.** That is, there is no real number x such that $10^x = -2$. If $\log(-2)$ is entered in a calculator, an error message will appear as shown here.

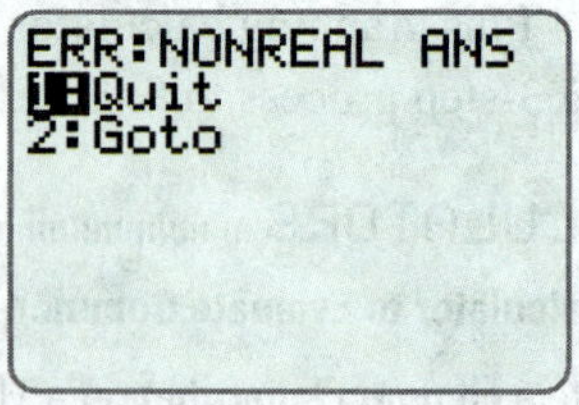

Figure 2

B Finding the Inverse of a Common Logarithm (Base 10)

Because logarithms are exponents, if the value of a common logarithm is known, an exponent is known. Finding the value of the related exponential expression is called finding the **inverse of the logarithm** (or finding the **antilog**). For example,

if $\log x = N$, then $x = 10^N$ and the number x is called the **inverse log of** N.

Finding the inverse log of N is a three-step process.

CALCULATORS

Using a Calculator to Find the Inverse Log of *N*

To evaluate the **inverse log of *N*** on a TI-84 Plus graphing calculator:

1. Press [2nd] and [LOG]. The expression 10^(will appear on the display.
2. Enter the value of *N* and a right-hand parenthesis [)].
3. Press [ENTER].

Example 3 Using a Calculator to Find the Inverse Log of *N*

Use a TI-84 Plus graphing calculator to find the inverse log of N for each expression. (That is, find the value of x.)

a. $\log x = 5$

b. $\log x = -3$

c. $\log x = 2.4142$

d. $\log x = 16.5$

Solution

The calculator gives the following results.

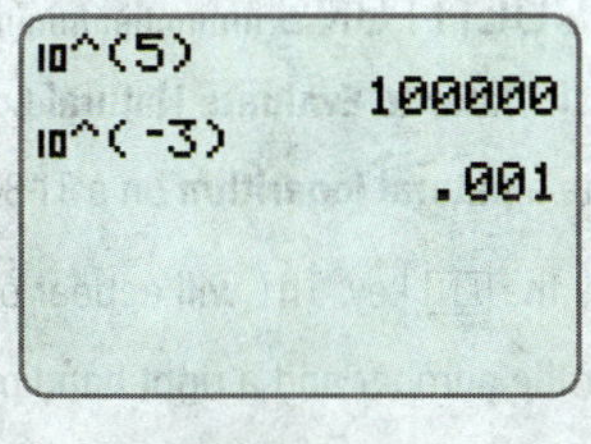

a. $x = 10^5 = 100{,}000$

b. $x = 10^{-3} = 0.001$

c. $x = 10^{2.4142} \approx 259.5374301$

d. $x = 10^{16.5} \approx 3.16227766\text{E}16$

The letter E in the solution is the calculator version of scientific notation. Thus,
$3.16227766\text{E}16 = 3.16227766 \times 10^{16}$.

Now work margin exercise 3.

3. Use a TI-84 Plus graphing calculator to find the inverse log of N for each expression, accurate to the nearest ten-thousandth.

a. $\log x = 2.5$

b. $\log x = 3.333$

c. $\log x = \frac{1}{2}$

C Natural Logarithms (Base *e* Logarithms)

The number $e \approx 2.718281828$ is an irrational number that appears quite naturally in continuously compounded interest. Base *e* logarithms are called **natural logarithms**. The notation for natural logarithms is shortened to $\ln x$ (read "L N of x" or "natural log of x"). That is,

$$\log_e x = \ln x.$$

Figure 3 shows the graphs of the two functions $y = e^x$ and its inverse $y = \ln x$.

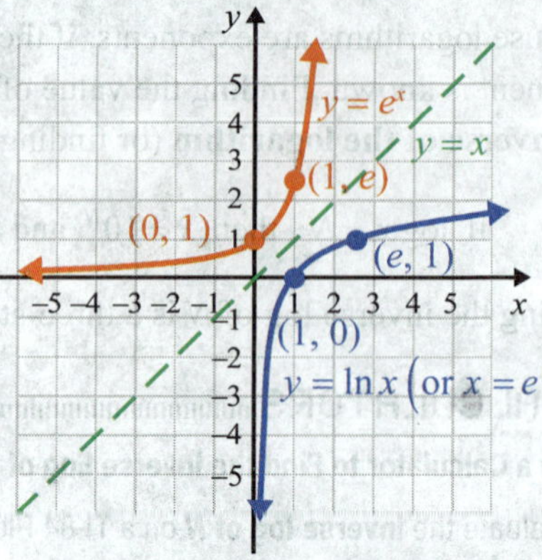

Figure 3

4. Translate each of the following expressions into logarithmic form using natural logarithms.

a. $e^x = 2$

b. $e^4 = c$

Example 4 Translating between Exponential Form and Natural Logarithms

	Exponential Form		Logarithmic Form	
a.	$e^t = 3.21$	$\Leftrightarrow$	$\ln 3.21 = t$	This is a natural logarithm that equals t.
b.	$e^0 = 1$	$\Leftrightarrow$	$\ln 1 = 0$	This is a natural logarithm that equals 0.
c.	$e^2 = n$	$\Leftrightarrow$	$\ln n = 2$	This is a natural logarithm that equals 2.

Now work margin exercise 4.

CALCULATORS

Using a Calculator to Evaluate Natural Logarithms (Base *e*)

To evaluate a **natural logarithm** on a TI-84 Plus graphing calculator:

1. Press the [LN] key. ln(will appear on the display.
2. Enter the number and a right-hand parenthesis, [)].
3. Press [ENTER].

5. Use a TI-84 Plus graphing calculator to find the values of the following natural logarithms accurate to the nearest ten-thousandth.

a. $\ln 2$

b. $\ln 0.64$

Example 5 Evaluating Natural Logarithms Using a Calculator

Use a TI-84 Plus graphing calculator to find the approximate values of the following natural logarithms.

a. $\ln 1$

b. $\ln 3$

c. $\ln 0.02$

Solution

From the display, we see the results (accurate to 9 decimal places).

a. $\ln 1 = 0$

b. $\ln 3 \approx 1.098612289$

c. $\ln 0.02 \approx -3.912023005$

Now work margin exercise 5.

D Finding the Inverse of a Natural Logarithm (Base *e*)

Again, because logarithms are exponents, if we know the value of a natural logarithm, we know an exponent. Finding the value of the related exponential expression is called finding the **inverse of the logarithm**. For example,

if $\ln x = N$, then $x = e^N$ and the number x is called the **inverse ln of *N***.

Example 6 Using a Calculator to Find the Inverse ln of *N*

Use a TI-84 Plus graphing calculator to find the inverse ln of N for each expression. (That is, find the value of x.)

a. $\ln x = 3$

b. $\ln x = -1$

c. $\ln x = -0.1$

d. $\ln x = 50$

Solution

The calculator gives the following results.

a. $x = e^3 \approx 20.08553692$

b. $x = e^{-1} \approx 0.3678794412$

c. $x = e^{-0.1} \approx 0.904837418$

d. $x = e^{50} \approx 5.184705529\text{E}21$
$= 5.184705529 \times 10^{21}$

Now work margin exercise 6.

6. Use a TI-84 Plus graphing calculator to find the inverse ln of N for each expression, accurate to the nearest ten-thousandth.

a. $\ln x = 4$

b. $\ln x = -\dfrac{1}{100}$

CALCULATORS

Using a Calculator to Find the Inverse ln of *N*

To evaluate the **inverse ln of *N*** on a TI-84 Plus graphing calculator:

1. Press [2nd] and [LN]. The expression e^(will appear on the display.
2. Enter the value of N and a right-hand parenthesis [)].
3. Press [ENTER].

Margin Exercise Answers

1. $\log 100 = 2$ **2. a.** 2.6021 **b.** 3.6990 **3. a.** 316.2278 **b.** 2152.7817 **c.** 3.1623 **4. a.** $\ln 2 = x$ **b.** $\ln c = 4$ **5. a.** 0.6931 **b.** -0.4463 **6. a.** $x = 54.5982$ **b.** 0.9900

17.6 Exercises

Concept Check

1. 10
2. e
3. negative
4. inverse
5. $\ln x$
6. undefined
7. False; When the base is omitted, it is understood to be 10.
8. True
9. False; Both common logarithms and natural logarithms have inverses.
10. True

Practice

1. $\log x = 1.5$
2. $\log 23 = k$
3. $\log \frac{1}{1000} = -3$
4. $\log 0.0001 = -4$
5. $\ln 27 = x$
6. $\ln 12.4 = k$
7. $\ln 1 = 0$
8. $\ln x = 4$
9. $\log 3.2 = x$
10. $\log x = y$
11. $10^0 = 1$
12. $10^2 = 100$
13. $10^y = 5.4$
14. $e^x = 40.1$
15. $e^{1.54} = x$
16. $10^{25.3} = x$
17. $e^1 = e$
18. $10^1 = 10$
19. $e^a = x$
20. $10^x = a$

17.6 Exercises

Concept Check

Fill-in-the-Blank. Complete each sentence using information found in this section.

1. Base ______ logarithms are called common logarithms.
2. Base ______ logarithms are called natural logarithms.
3. The logarithm with base greater than 1 of any number between 0 and 1 will always be ______.
4. Finding the value of the related exponential expression is called finding the ______ of the logarithm.
5. The notation for natural logarithms is shortened to ______.
6. The logarithm of a negative number is ______.

True/False. Determine whether each statement is true or false. If a statement is false, explain how it can be changed so the statement will be true. (**Note:** There may be more than one acceptable change.)

7. Whenever the base of a logarithm is omitted, it is understood to be 1.
8. Logarithms of negative numbers or 0 do not exist.
9. Common logarithms have an inverse while natural logarithms do not.
10. Given $\log x = 4$, the inverse log of 4 is $x = 10^4 = 10{,}000$.

Practice

Express each equation in logarithmic form. See Examples 1 and 4.

1. $10^{1.5} = x$
2. $10^k = 23$
3. $10^{-3} = \frac{1}{1000}$
4. $10^{-4} = 0.0001$
5. $e^x = 27$
6. $e^k = 12.4$
7. $e^0 = 1$
8. $e^4 = x$
9. $10^x = 3.2$
10. $10^y = x$

Express each equation in exponential form. See Examples 1 and 4.

11. $\log 1 = 0$
12. $\log 100 = 2$
13. $\log 5.4 = y$
14. $\ln 40.1 = x$
15. $\ln x = 1.54$
16. $\log x = 25.3$
17. $\ln e = 1$
18. $\log 10 = 1$
19. $\ln x = a$
20. $\log a = x$

Use a calculator to evaluate the logarithms accurate to the nearest ten-thousandths place. See Examples 2 and 5.

21. $\log 173$

22. $\log 396$

23. $\log 88.4$

24. $\log 0.0061$

25. $\log 0.0573$

26. $\log(-8.47)$

27. $\ln 37.5$

28. $\ln 96$

29. $\ln(-14.9)$

30. $\ln 157.6$

31. $\ln 0.00461$

32. $\ln 0.0139$

Use a calculator to find the value of x in each equation accurate to the nearest ten-thousandths place. See Examples 3 and 6.

33. $\log x = 2.31$

34. $\log x = -3$

35. $\log x = -1.7$

36. $\log x = 4.1$

37. $2 \log x = -0.038$

38. $5 \log x = 9.4$

39. $\ln x = 5.17$

40. $\ln x = 4.9$

41. $\ln x = -8.3$

42. $\ln x = 6.74$

43. $0.2 \ln x = 0.0079$

44. $3 \ln x = -0.066$

Writing & Thinking

45. Explain the difference in the meaning of the expressions $\log x$ and $\ln x$.

46. The function $y = \log x$ is defined only for $x > 0$. Discuss the function $y = \log(-x)$. That is, does this function even exist? If it does exist, what is its domain? Sketch its graph and the graph of the function $y = \log x$.

47. What is the domain of the function $y = \ln|x|$? Graph the function.

21. 2.2380

22. 2.5977

23. 1.9465

24. −2.2147

25. −1.2418

26. Error (undefined)

27. 3.6243

28. 4.5643

29. Error (undefined)

30. 5.0601

31. −5.3795

32. −4.2759

33. 204.1738

34. 0.0010

35. 0.0120

36. 12,589.2541

37. 0.9572

38. 75.8578

39. 175.9148

40. 134.2898

41. 0.0002

42. 845.5607

43. 1.0403

44. 0.9782

Writing & Thinking

45. $\log x$ is a base 10 logarithm. $\ln x$ is a base e logarithm.

46. The function $y = \log(-x)$, does exist. Its domain is $(-\infty, 0)$.

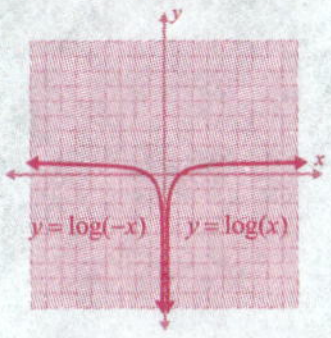

47. $D = (-\infty, 0) \cup (0, \infty)$

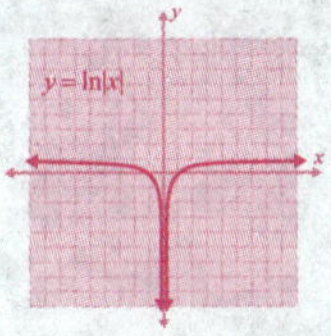

Objectives

A. Solve exponential equations in which the bases are the same.

B. Solve exponential equations in which the bases are not the same.

C. Solve logarithmic equations.

D. Use the change-of-base formula and a calculator to evaluate logarithmic expressions.

17.7 Logarithmic and Exponential Equations and Change-of-Base

A Solving Exponential Equations with the Same Base

All the properties of exponents that have been discussed are valid for real exponents, not just integers or fractions. For example, $2^{\sqrt{2}} \cdot 2^3 = 2^{\sqrt{2}+3}$ is a valid operation and gives a real number. To be complete in our development of the properties of exponents, the properties for real exponents and positive real bases are stated here.

Properties of Real Exponents

If a and b are positive real numbers and x and y are any real numbers, then

1. $b^0 = 1$

2. $b^{-x} = \dfrac{1}{b^x}$

3. $b^x \cdot b^y = b^{x+y}$

4. $\dfrac{b^x}{b^y} = b^{x-y}$

5. $\left(b^x\right)^y = b^{xy}$

6. $(ab)^x = a^x b^x$

7. $\left(\dfrac{a}{b}\right)^x = \dfrac{a^x}{b^x}$

PROPERTIES

The following properties are used in solving equations containing exponents and logarithms.

Properties of Equations with Exponents and Logarithms

For $b > 0$ and $b \neq 1$,

1. If $b^x = b^y$, then $\boldsymbol{x = y}$.

2. If $\boldsymbol{x = y}$, then $b^x = b^y$.

3. If $\log_b x = \log_b y$, then $\boldsymbol{x = y}$ ($x > 0$ and $y > 0$).

4. If $\boldsymbol{x = y}$, then $\log_b x = \log_b y$ ($x > 0$ and $y > 0$).

PROPERTIES

Example 1 Solving Exponential Equations with the Same Base

Solve each equation for x.

a. $2^{x^2-7} = 2^{6x}$ **b.** $8^{4-2x} = 4^{x+2}$

Solution

a.

$2^{x^2-7} = 2^{6x}$ Both bases are 2.

$x^2 - 7 = 6x$ The exponents are equal because the bases are the same.

$x^2 - 6x - 7 = 0$ Solve for x.

$(x-7)(x+1) = 0$

$x = 7$ or $x = -1$

b.

$8^{4-2x} = 4^{x+2}$ Here the bases are different.

$\left(2^3\right)^{4-2x} = \left(2^2\right)^{x+2}$ Rewrite both sides so that the bases are the same.

$2^{12-6x} = 2^{2x+4}$ Use the property $\left(b^x\right)^y = b^{xy}$.

$12 - 6x = 2x + 4$ The exponents are equal because the bases are the same.

$8 = 8x$ Solve for x.

$1 = x$

Now work margin exercise 1.

1. Solve each equation for x.

a. $4^{x^2+3} = 4^{4x}$

b. $4^{\frac{x}{2}} = 2^{x^2-2}$

B Solving Exponential Equations with Different Bases

As Example 1 demonstrated, if the bases are the same in an exponential equation, there is no need to involve logarithms. However, when the bases are not the same, equations are solved by taking the logarithm of both sides. (See Property 4.)

Example 2 Solving Exponential Equations with Different Bases

Solve each of the following exponential equations by taking the log (or ln) of both sides or using the definition of logarithm as an exponent.

a. $10^{3x} = 2.1$ **b.** $6^x = 18$ **c.** $5^{2x-1} = 10^x$

Solution

a. First, take the log of both sides Then use the definition of logarithm as an exponent.

$10^{3x} = 2.1$

$\log\left(10^{3x}\right) = \log 2.1$ Take the log of both sides.

$3x \log 10 = \log 2.1$ Power rule

$3x = \log 2.1$ By the definition of a common logarithm

$x = \dfrac{\log 2.1}{3}$

Use a calculator to find a decimal approximation:

$$x = \frac{\log 2.1}{3} \approx \frac{0.3222}{3} = 0.1074.$$

2. Solve each of the following exponential equations.

a. $4^x = 3$

b. $5^x = 20$

b. The base is 6, not 10 or e, but we can solve by taking the log of both sides or by taking the ln of both sides. The result is the same.

Taking the log of both sides:

$$6^x = 18$$
$$\log 6^x = \log 18$$
$$x \cdot \log 6 = \log 18$$
$$x = \frac{\log 18}{\log 6}$$

Using a calculator:

$$x = \frac{\log 18}{\log 6} \approx \frac{1.2553}{0.7782}$$
$$\approx 1.6131$$

Taking the ln of both sides:

$$6^x = 18$$
$$\ln 6^x = \ln 18$$
$$x \cdot \ln 6 = \ln 18$$
$$x = \frac{\ln 18}{\ln 6}$$

Using a calculator:

$$x = \frac{\ln 18}{\ln 6} \approx \frac{2.8904}{1.7918}$$
$$\approx 1.6131$$

c.

$$5^{2x-1} = 10^x$$
$$\log 5^{2x-1} = \log 10^x \quad \text{Take the log of both sides.}$$
$$(2x-1)\log 5 = x \log 10 \quad \text{Power rule}$$
$$2x \cdot \log 5 - 1 \cdot \log 5 = x \cdot 1 \quad \text{Distributive property; } \log 10 = 1$$
$$2x \log 5 - x = \log 5 \quad \text{Arrange } x\text{-terms on one side.}$$
$$x(2\log 5 - 1) = \log 5 \quad \text{Factor out the } x.$$
$$x = \frac{\log 5}{2\log 5 - 1}$$

As a decimal approximation,

$$x = \frac{\log 5}{2\log 5 - 1} \approx \frac{0.6990}{2(0.6990) - 1} \approx 1.7563. \quad \text{Using rounded values}$$

Math Tip

Logarithms may look complicated, but it is important to remember that they represent real numbers. Once you recognize this, you can apply the properties you know about real numbers, such as the distribution property.

Now work margin exercise 2.

C Solving Logarithmic Equations

All the various properties of logarithms can be used to solve equations that involve logarithms. Remember that logarithms are defined only for positive real numbers, so each answer should be checked in the original equation.

3. Use the properties of logarithms to solve the following equations.

a. $\log 2x = 2$

b. $\log(x+2) + \log(x+2) = 0$

Example 3 Solving Logarithmic Equations

Use the properties of logarithms to solve the following equations.

a. $\log 5x = 3$

b. $\log(x-1) + \log(x-4) = 1$

c. $\log x - \log(x-1) = \log 3$

d. $\ln(x^2 - x - 6) - \ln(x+2) = 2$

Solution

a. $\log 5x = 3$

$5x = 10^3$ — Definition of a common logarithm

$5x = 1000$

$x = 200$ — Checking will show that the equation is defined at $x = 200$.

b. $\log(x-1) + \log(x-4) = 1$

$\log[(x-1)(x-4)] = 1$ — Product rule

$(x-1)(x-4) = 10^1$ — Definition of a common logarithm

$x^2 - 5x + 4 = 10$

$x^2 - 5x - 6 = 0$

$(x-6)(x+1) = 0$ — Solve by factoring.

$x = 6$ or ~~$x = -1$~~ — Checking $x = -1$ yields $\log(-1-1) = \log(-2)$, which is undefined.

c. $\log x - \log(x-1) = \log 3$

$\log\left(\dfrac{x}{x-1}\right) = \log 3$ — Quotient rule

$\dfrac{x}{x-1} = 3$ — If $\log_b x = \log_b y$, then $x = y$.

$x = 3(x-1)$ — Solve for x.

$x = 3x - 3$

$3 = 2x$

$\dfrac{3}{2} = x$

d. $\ln(x^2 - x - 6) - \ln(x+2) = 2$

$\ln\left(\dfrac{x^2 - x - 6}{x+2}\right) = 2$ — Quotient rule

$\ln\left(\dfrac{(x+2)(x-3)}{(x+2)}\right) = 2$ — Factor the numerator.

$\ln(x-3) = 2$ — Simplify.

$x - 3 = e^2$ — Change to exponential form with base e.

$x = 3 + e^2$

Or, using a calculator, $x = 3 + e^2 \approx 3 + 7.3891 = 10.3891$.

Now work margin exercise 3.

D Change-of-Base

Because a calculator can be used to evaluate common logarithms and natural logarithms, most of the examples have been restricted to base 10 or base e expressions. If an equation involves logarithms of other bases, use the following formula to rewrite each logarithm using any base.

Change-of-Base Formula

For $a, b, x > 0$ and $a, b \neq 1$,

$$\log_b x = \frac{\log_a x}{\log_a b}.$$

FORMULA

The change-of-base formula can be derived using properties of logarithms as follows.

$$b^{\log_b x} = x \quad \text{Property 3 in Section 17.5}$$

$$\log_a\left(b^{\log_b x}\right) = \log_a x \quad \text{Take the } \log_a \text{ of both sides.}$$

$$\log_b x\left(\log_a b\right) = \log_a x \quad \text{By the power rule using } \log_b x \text{ as the exponent}$$

$$\log_b x = \frac{\log_a x}{\log_a b} \quad \text{Divide both sides by } \log_a b \text{ to arrive at the change-of-base formula.}$$

4. Use the change-of-base formula to evaluate $\log_2 4$.

Example 4 Using the Change-of-Base Formula

Use the change-of-base formula to evaluate the following expressions.

a. $\log_2 3.42$ **b.** $\log_3 0.3333$

Solution

a. This expression can be evaluated using either base 10 or base e since both are easily available on a calculator.

$$\log_2 3.42 = \frac{\ln 3.42}{\ln 2} \quad \text{Change-of-base formula}$$

$$\approx \frac{1.2296}{0.6931} \approx 1.7741 \quad \text{Using rounded values}$$

(Use a calculator to show that $\frac{\log 3.42}{\log 2}$ gives the same result.)

In exponential form: $2^{1.7741} \approx 3.42$.

b. $$\log_3 0.3333 = \frac{\log 0.3333}{\log 3} \quad \text{Change-of-base formula}$$

$$\approx \frac{-0.4772}{0.4771} \approx -1.0002 \quad \text{Using rounded values}$$

Now work margin exercise 4.

Example 5 Using the Change-of-Base Formula

Use the change-of-base formula to find the value of x (accurate to the nearest ten-thousandth) in the equation $5^x = 16$.

Solution

Because the base is 5, we can take $\log_5$ of both sides.

(This method is not necessary, but it does show how the change-of-base formula can be used.)

$$5^x = 16$$
$$\log_5 5^x = \log_5 16$$
$$x = \log_5 16$$
$$x = \frac{\ln 16}{\ln 5} \quad \text{Change-of-base formula}$$
$$x \approx \frac{2.7726}{1.6094} \approx 1.7228$$

Now work margin exercise 5.

5. Use the change-of-base formula to find the value of x.

$8^x = 25$

Margin Exercise Answers

1. a. $x = 1$ or $x = 3$ **b.** $x = -1$ or $x = 2$ **2. a.** $x = \frac{\log 3}{\log 4} \approx 0.7925$ **b.** $x = \frac{\log 20}{\log 5} \approx 1.8614$

3. a. $x = 50$ **b.** $x = -1$ **4.** $\frac{\ln 4}{\ln 2} = 2$ **5.** $\frac{\ln 25}{\ln 8} \approx 1.5480$

17.7 Exercises

Concept Check

Fill-in-the-Blank. Complete each sentence using information found in this section.

1. When solving exponential equations, if the bases are not the same, the equations are solved by taking the ______ of both sides.

2. Logarithms are defined only for ______ ______ numbers, so each answer should be checked in the original equation.

3. For any positive real number b, b^0 is equal to ______.

4. For any positive real number b and any real numbers x and y, $\left(b^x\right)^y$ is equivalent to ______.

5. For any positive real number b and any value of x, b^{-x} is equivalent to ______.

6. For $b > 0$ and $b \neq 1$, if $\log_b x = \log_b y$, then $x =$ ______.

17.7 Exercises

Concept Check

1. logarithm
2. positive real
3. 1
4. b^{xy}
5. $\frac{1}{b^x}$
6. y
7. False; The change of base formula is $\log_b x = \frac{\log_a x}{\log_a b}$.
8. True
9. True
10. True

Practice

1. $x = 11$
2. $x = 5$
3. $x = 9$
4. $x = 2$
5. $x = \frac{7}{2}$
6. $x = \frac{7}{12}$

7. $x = \frac{6}{5}$
8. $x = \frac{8}{3}$
9. $x = \frac{7}{2}$
10. $x = -2$
11. $x = -5$
12. $x = -8$
13. $x = -3$
14. $x = -5$
15. $x = -1, 4$
16. $x = -3, 3$
17. $x = -1, \frac{3}{2}$
18. $x = -2, -1$
19. $x = -2, 3$
20. $x = -4, -1$
21. $x = -2$
22. $x = 2$
23. $x = -\frac{3}{2}, 0$
24. $x = -\frac{5}{2}, 0$
25. $x = -1, 3$
26. $x = -1, 3$
27. $x \approx 0.7154$
28. $x \approx 0.9934$
29. $x \approx 7.5098$
30. $x \approx -4.0572$
31. $x \approx -1.5947$
32. $x \approx -0.6879$
33. $x \approx 0.2473$
34. $t \approx 1.3037$
35. $x \approx -7.7003$
36. $t \approx -653.0008$
37. $t \approx -1.5193$
38. $t \approx 0.9061$
39. $x \approx 3.3219$
40. $x \approx 6.1918$
41. $x \approx -1.4307$
42. $x \approx -0.3944$
43. $x = 1$

True/False. Determine whether each statement is true or false. If a statement is false, explain how it can be changed so the statement will be true. (**Note:** There may be more than one acceptable change.)

7. The change of base formula is $\log_b x = \frac{\log_a b}{\log_a x}$, for $a, b, x > 0$ and $b \neq 1$.
8. If the terms in an exponential equation all have the same base, there is no need to use logarithms to solve the equation.
9. Exponential equations with different bases can be solved by taking either the log of both sides or the ln of both sides.
10. If $5^x = 5^y$, then x is equal to y.

Practice

Use the properties of exponents and logarithms to solve each of the equations. If necessary, use a calculator and round answers to the nearest ten-thousandth. See Examples 1 through 3.

1. $2^4 \cdot 2^7 = 2^x$
2. $3^7 \cdot 3^{-2} = 3^x$
3. $\left(3^5\right)^2 = 3^{x+1}$
4. $\left(2^x\right)^3 = 2^{x+4}$
5. $\frac{10^4 \cdot 10^{\frac{1}{2}}}{10^x} = 10$
6. $\left(10^2\right)^x = \frac{10 \cdot 10^{\frac{2}{3}}}{10^{\frac{1}{2}}}$
7. $2^{5x} = 4^3$
8. $7^{3x} = 49^4$
9. $(25)^x = 5^3 \cdot 5^4$
10. $10^x \cdot 10^8 = 100^3$
11. $8^{x+3} = 2^{x-1}$
12. $100^{2x+1} = 1000^{x-2}$
13. $27^x = 3 \cdot 9^{x-2}$
14. $16^x = 2 \cdot 8^{2x+3}$
15. $2^{3x+5} = 2^{x^2+1}$
16. $10^{x^2+x} = 10^{x+9}$
17. $10^{2x^2+3} = 10^{x+6}$
18. $3^{x^2+5x} = 3^{2x-2}$
19. $\left(3^{x+1}\right)^x = \left(3^{x+3}\right)^2$
20. $\left(10^x\right)^{x+3} = \left(10^{x+2}\right)^{-2}$
21. $3^{x+4} = 9$
22. $2^{5x-8} = 4$
23. $4^{x^2-x} = \left(\frac{1}{2}\right)^{5x}$
24. $25^{x^2+x} = \left(\frac{1}{5}\right)^{3x}$
25. $5^{2x-x^2} = \frac{1}{125}$
26. $10^{x^2-2x} = 1000$
27. $10^{3x} = 140$
28. $10^{2x} = 97$
29. $10^{0.32x} = 253$
30. $10^{-0.48x} = 88.6$
31. $4 \cdot 10^{-0.94x} = 126.2$
32. $3 \cdot 10^{-2.1x} = 83.5$
33. $e^{3x} = 2.1$
34. $e^{4t} = 184$

35. $e^{-0.5x} = 47$

36. $e^{-0.006t} = 50.3$

37. $3e^{-0.12t} = 3.6$

38. $5e^{2.4t} = 44$

39. $2^x = 10$

40. $3^{x-2} = 100$

41. $5^{2x} = \frac{1}{100}$

42. $7^{3x} = \frac{1}{10}$

43. $5^{1-x} = 1$

44. $12^{5x+2} = 1$

45. $4^{2x+5} = 0.01$

46. $4^{2-3x} = 0.1$

47. $14^{3x-1} = 10^3$

48. $12^{2x+7} = 10^4$

49. $7^x = 9$

50. $2^x = 20$

51. $3^{3x} = 23$

52. $5^{2x} = 23$

53. $6^{2x-1} = 14.8$

54. $4^{7-3x} = 26.3$

55. $5\log x = 7$

56. $3\log x = 13.2$

57. $4\log x - 6 = 0$

58. $2\log x - 15 = 0$

59. $4\log x^{\frac{1}{2}} + 8 = 0$

60. $\frac{2}{3}\log x^{\frac{2}{3}} + 9 = 0$

61. $5\ln x - 8 = 0$

62. $2\ln x + 3 = 0$

63. $\ln x^2 + 2.2 = 0$

64. $\ln x^2 - 41.6 = 0$

65. $\log x + \log 2x = \log 18$

66. $\log(x+4) + \log(x-4) = \log 9$

67. $\log x^2 - \log x = 2$

68. $\log x + \log x^2 = 3$

69. $\ln(x-3) + \ln x = \ln 18$

70. $\ln(x+5) + \ln(x-1) = \ln 16$

71. $\log(x-15) = 2 - \log x$

72. $\log(2x-17) = 2 - \log x$

73. $\log(3x-5) + \log(x-1) = 1$

74. $\log(2x-3) + \log(x+3) = 3$

75. $\log(x-3) - \log(x+1) = 1$

76. $\ln(x+1) + \ln(x-1) = 0$

77. $\log(x^2-9) - \log(x-3) = -2$

78. $\ln(x^2+4x-5) - \ln(x+5) = -2$

79. $\log(x^2-4x-5) - \log(x+1) = 2$

80. $\log(x^2-x-12) - \log(x-4) = -2$

81. $\ln(x^2-4) = 3 + \ln(x+2)$

82. $\ln(x^2+2x-3) = 1 + \ln(x-1)$

83. $\log\sqrt[3]{x^2+2x+20} = \frac{2}{3}$

84. $\log\sqrt{x^2-24} = \frac{3}{2}$

44. $x = -\frac{2}{5}$

45. $x \approx -4.1610$

46. $x \approx 1.2203$

47. $x \approx 1.2058$

48. $x \approx -1.6467$

49. $x \approx 1.1292$

50. $x \approx 4.3219$

51. $x \approx 0.9513$

52. $x \approx 0.9741$

53. $x \approx 1.2520$

54. $x \approx 1.5472$

55. $x \approx 25.1189$

56. $x \approx 25{,}118.8643$

57. $x \approx 31.6228$

58. $x \approx 3.1623 \times 10^7$

59. $x = 0.0001$

60. $x \approx \pm 5.6234 \times 10^{-21}$

61. $x \approx 4.9530$

62. $x \approx 0.2231$

63. $x \approx \pm 0.3329$

64. $x \approx \pm 1.0798 \times 10^9$

65. $x = 3$

66. $x = 5$

67. $x = 100$

68. $x = 10$

69. $x = 6$

70. $x = 3$

71. $x = 20$

72. $x = \frac{25}{2}$

73. $x \approx 3.1893$

74. $x \approx 21.7236$

75. No solution

76. $x = \sqrt{2}$

77. No solution

78. $x \approx 1.1353$

79. $x = 105$

80. No solution

81. $x \approx 22.0855$

82. No solution

83. $x = -10, 8$
84. $x = \pm 32$
85. 2.2619
86. 2.5850
87. 0.3223
88. 1.5342
89. −0.6279
90. −0.0841
91. 2.4391
92. 3.8278
93. −1.2222
94. 0.3954
95. $x \approx 2.3219$
96. $x \approx 1.0480$
97. $x \approx 1.5480$
98. $x \approx 3.1133$
99. $x \approx 0.8390$
100. $x \approx 0.7345$

Writing & Thinking

101. Answers will vary. $x = \frac{1}{2}$

102. a. $5^x \cdot 5^2$

b. $3^x \cdot 3^{-2}$

103. $7 \cdot 7^x = 7^{1+x}$ and $49^x = 7^{2x}$. Since $1 + x \neq 2x$, in general, $7 \cdot 7^x \neq 49^x$. Answers will vary.

Use the change-of-base formula to evaluate each of the expressions or solve the equations. Round answers to the nearest ten-thousandth. See Examples 4 and 5.

85. $\log_3 12$

86. $\log_4 36$

87. $\log_5 1.68$

88. $\log_{11} 39.6$

89. $\log_8 0.271$

90. $\log_7 0.849$

91. $\log_{15} 739$

92. $\log_2 14.2$

93. $\log_{20} 0.0257$

94. $\log_9 2.384$

95. $2^x = 5$

96. $3^{2x} = 10$

97. $9^{2x-1} = 100$

98. $5^{x-1} = 30$

99. $4^{3-x} = 20$

100. $6^{4-3x} = 25$

Writing & Thinking

101. Solve the following equation for x two different ways: $a^{2x-1} = 1$.

102. Rewrite each of the following expressions as products.

a. 5^{x+2}

b. 3^{x-2}

103. Explain, in your own words, why $7 \cdot 7^x \neq 49^x$ when $x \neq 1$. Show each of the expressions $7 \cdot 7^x$ and 49^x as a single exponential expression with base 7.

17.8 Applications: Exponential and Logarithmic Functions

Objectives

A. Solve applied problems using logarithmic and exponential equations.

A Applications

In Section 17.3, we found that the number *e* appears in a surprisingly natural way in the formula for continuously compounding interest,

$$A = Pe^{rt},$$

which was developed from the formula for compounding *n* times per year:

$$A = P\left(1+\frac{r}{n}\right)^{nt}.$$

There are many formulas that involve exponential functions. A few are shown here and in the exercises.

$A = A_0 e^{-0.04t}$ This is a formula for decomposition of radium where *t* is in centuries.

$A = A_0 e^{-0.1t}$ This is one formula for skin healing where *t* is measured in days.

$A = A_0 2^{-\frac{t}{5600}}$ This formula is used for carbon-14 dating to determine the age of fossils where *t* is measured in years.

$T = Ae^{-kt} + C$ This is Newton's law of cooling where *C* is the constant temperature of the surrounding medium. The values of *A* and *k* depend on the particular object that is cooling.

Example 1 Application: Calculating Exponential Growth

Suppose that the formula $y = y_0 e^{0.4t}$ represents the number of bacteria present after *t* days, where y_0 is the initial number of bacteria. In how many days will the bacteria double in number?

Solution

$$y = y_0 e^{0.4t}$$

$$2y_0 = y_0 e^{0.4t} \quad 2y_0 \text{ is double the initial number present.}$$

$$2 = e^{0.4t}$$

$$\ln 2 = \ln e^{0.4t} \quad \text{Take the natural log of both sides.}$$

$$\ln 2 = 0.4t \cdot (1) \quad \text{Power rule of logarithms; } \ln e = 1$$

$$t = \frac{\ln 2}{0.4} \approx \frac{0.6931}{0.4}$$

$$t \approx 1.73 \text{ days}$$

The number of bacteria will double in approximately 1.73 days. Note that this number is completely independent of the number of bacteria initially present. That is, if $y_0 = 10$ or $y_0 = 1000$, the doubling time is the same, namely 1.73 days.

Now work margin exercise 1.

1. Suppose that the formula $y = y_0 e^{0.3t}$ represents the number of bacteria present after *t* days, where y_0 is the initial number of bacteria. In how many days will the bacteria double in number?

2. If \$600 is invested at a rate of 5% compounded continuously, in how many years will it grow to \$2400?

Example 2 Application: Calculating Continuously Compounded Interest

If \$1000 is invested at a rate of 6% compounded continuously, in how many years will it grow to \$5000?

Solution

$$A = Pe^{rt}$$ Formula for continuously compounded interest

$$5000 = 1000e^{0.06t}$$ Substitute $A = 5000$, $P = 1000$, and $r = 0.06$.

$$5 = e^{0.06t}$$

$$\ln 5 = \ln e^{0.06t}$$ Take the natural log of both sides.

$$\ln 5 = 0.06t \cdot (1)$$ Power rule of logarithms; $\ln e = 1$

$$t = \frac{\ln 5}{0.06} \approx \frac{1.6094}{0.06}$$

$$t \approx 26.82 \text{ years}$$

\$1000 will grow to \$5000 in approximately 26.82 years.

Now work margin exercise 2.

3. The 2017 Chiapas earthquake in Mexico measured 8.2 on the Richter scale. What was the intensity of this earthquake?

Example 3 Application: Using the Richter Scale

The magnitude of an earthquake is measured on the Richter scale as a logarithm of the intensity of the shock wave. For magnitude R and intensity I, the formula is $R = \log I$. The 1994 earthquake in Northridge, California, measured 6.7 on the Richter scale. What was the intensity of this earthquake?

Solution

Substitute 6.7 for R in the formula and solve for I.

$$6.7 = \log I$$

$$I = 10^{6.7}$$

$$I \approx 5.01 \times 10^{6}$$

The intensity of the 1994 earthquake in Northridge, California, was approximately 5,010,000.

Now work margin exercise 3.

4. The strongest recorded earthquake in Mexico's history occurred in 1787 and measured 8.6 on the Richter scale. How much stronger was this earthquake than the 2017 Chiapas earthquake (see margin exercise 3)?

Example 4 Application: Using the Richter Scale

The Long Beach earthquake in 1933 measured 6.2 on the Richter scale. How much stronger was the Northridge earthquake than the 1933 Long Beach earthquake?

Solution

The comparative sizes of the earthquakes can be found by finding the ratio of the intensities. For the Long Beach earthquake, $6.2 = \log I$ and $I = 10^{6.2}$.

Therefore, the ratio of the two intensities is

$$\frac{I \text{ for Northridge}}{I \text{ for Long Beach}} = \frac{10^{6.7}}{10^{6.2}} = 10^{0.5} \approx 3.16.$$

Thus, the Northridge earthquake had an intensity about 3.16 times the intensity of the Long Beach earthquake.

Now work margin exercise 4.

Example 5 Application: Using the Half-Life of Radium

The half-life of a substance is the time needed for the substance to decay to one-half of its original amount. The half-life of radium-226, a common isotope of radium, is 1600 years. If 10 grams are present today, how many grams will remain in 500 years?

Solution

The model for radioactive decay is $y = y_0e^{-kt}$. Since the half-life is 1600 years, if we assume $y_0 = 10$ g, then y would be 5g after 1600 years. We solve for k as follows.

$$5 = 10e^{-k(1600)} \quad \text{Substitute } y = 5, y_0 = 10, \text{ and } t = 1600.$$

$$\frac{5}{10} = e^{-1600k}$$

$$\ln 0.5 = \ln e^{-1600k} \quad \text{Take the natural log of both sides.}$$

$$\ln 0.5 = -1600k \cdot (1) \quad \text{Power rule for logarithms; } \ln e = 1$$

$$k = \frac{\ln 0.5}{-1600} \approx \frac{-0.6931}{-1600} \approx 0.0004332$$

The model is $y = 10e^{-0.0004332t}$.

Substituting $t = 500$ gives

$$y = 10e^{(-0.0004332)(500)} = 10e^{-0.2166} \approx 10(0.8053) \approx 8.05.$$

Thus, there will still be about 8.05g of the radium-226 remaining after 500 years.

Now work margin exercise 5.

5. The half-life of plutonium is 24,100 years. If 12 grams are present today, how many grams will remain in 10,000 years?

Example 6 Application: Using Newton's Law of Cooling

Suppose that the room temperature is 70 °F, and the temperature of a cup of tea is 150 °F when it is placed on the table. In 5 minutes, the tea cools to 120 °F. How long will it take for the tea to cool to 100 °F?

Solution

Using the formula $T = Ae^{-kt} + C$ (Newton's law of cooling), first find A and then k. We know that $C = 70$ and that $T = 150$ when $t = 0$.

Find A by substituting these values.

$$150 = Ae^{-k(0)} + 70$$

$$150 = A \cdot (1) + 70 \quad e^{-k(0)} = e^0 = 1$$

$$80 = A$$

Therefore, the formula can be written as $T = 80e^{-kt} + 70$.

6. Suppose that the temperature outside is 65 °F, and the temperature of a cup of coffee is 185 °F when it is placed on the picnic table. In 5 minutes, the coffee cools to 145 °F. How long will it take for the coffee to cool to 125 °F?

Since $T = 120$ when $t = 5$, substituting these values allows us to find k.

$$120 = 80e^{-k(5)} + 70$$
$$50 = 80e^{-5k}$$
$$\frac{50}{80} = e^{-5k}$$
$$\ln\frac{5}{8} = \ln e^{-5k} \qquad \text{Take the natural log of both sides.}$$
$$\ln 0.625 = -5k$$
$$k = \frac{\ln 0.625}{-5} \approx \frac{-0.4700}{-5} = 0.094$$

The formula can now be written as $T = 80e^{-0.094t} + 70$.

With all the constants in the formula known, now we can find t when $T = 100$.

$$100 = 80e^{-0.094t} + 70$$
$$30 = 80e^{-0.094t}$$
$$\frac{30}{80} = e^{-0.094t}$$
$$\ln\frac{3}{8} = \ln e^{-0.094t} \qquad \text{Take the natural log of both sides.}$$
$$\ln 0.375 = -0.094t$$
$$t = \frac{\ln 0.375}{-0.094}$$
$$\approx \frac{-0.9808}{-0.094} \approx 10.43 \text{ minutes}$$

The tea will cool to 100 °F in about 10.43 minutes.

Now work margin exercise 6.

Margin Exercise Answers

1. 2.31 days **2.** 27.73 years **3.** $I = 10^{8.2} \approx 1.58 \times 10^{8}$ **4.** 2.51 **5.** $y = 12e^{-0.00002876t}$; 9.00 grams **6.** $T = 120e^{-0.0811t} + 65$; 8.55 minutes

17.8 Exercises

Concept Check

Fill-in-the-Blank. Complete each sentence using information found in this section.

1. The formula $A = A_0e^{-0.04t}$ is for the ______ of radium where t is in ______.

2. The formula $A = A_0 2^{-\frac{t}{5600}}$ is used for carbon-14 dating to determine the age of ______, where t is measured in ______.

3. The magnitude of an earthquake is measured on the ______ scale.

17.8 Exercises

Concept Check

1. decomposition, centuries

2. fossils, years

3. Richter

True/False. Determine whether each statement is true or false. If a statement is false, explain how it can be changed so the statement will be true. (**Note:** There may be more than one acceptable change.)

4. In Newton's law of cooling, the variable C is the constant temperature of the medium surrounding the cooling object.

5. The formula $A = A_0e^{-0.1t}$ is used for skin healing, where t is measured in hours.

Applications

Solve. If necessary, round answers to the nearest hundredth (unless otherwise specified).

1. If Kim invests \$2000 at a rate of 7% compounded continuously, what will be her balance after 10 years?

2. Find the amount of money that will be accumulated in a savings account if \$3200 is invested at 6.5% for 6 years and the interest is compounded continuously.

3. Four thousand dollars is invested at 6% compounded continuously. How long will it take for the balance to be \$8000?

4. How long does it take \$1000 to double if it is invested at 5% compounded continuously?

5. The reliability of a certain type of flashlight battery is given by $f = e^{-0.03x}$, where f is the fractional part of the batteries produced that last x hours. What fraction of the batteries produced are good after 40 hours of use?

6. From Exercise 5, how long will at least one-half of the batteries last?

7. The concentration of a drug in the blood stream is given by $C = C_0e^{-0.8t}$, where C_0 is the initial dosage and t is the time in hours elapsed after administering the dose. If 20 mg of the drug is given, how much time elapses until 5 mg of the drug remains?

8. Using the formula in Exercise 7, determine the amount of the drug present after 3 hours if 0.60 mg is given.

9. One law for skin healing is $A = A_0e^{-0.1t}$, where A is the number of cm² of unhealed area after t days and A_0 is the number of cm² of the original wound. Find the number of days needed to reduce the wound to one-third the original size.

10. A swarm of bees grows according to the formula $P = P_0e^{0.35t}$, where P_0 is the number present initially and t is the time in days. How many bees will be present in 6 days if there were 1000 present initially? (Round to the nearest integer.)

11. If inversion of raw sugar is given by $A = A_0e^{-0.03t}$, where A_0 is the initial amount and t is the time in hours, how long will it take for 1000 lb of raw sugar to be reduced to 800 lb?

12. Atmospheric pressure P is related to the altitude, h by the formula $P = P_0e^{-0.00004h}$, where P_0 the pressure at sea level, is approximately 15 lb per in.² Determine the pressure at 5000 in.

4. True
5. False; The variable t is measured in days.

Applications

1. \$4027.51
2. \$4726.34
3. 11.55 years
4. 13.86 years
5. $f \approx 0.30$
6. 23.10 hours
7. 1.73 hours
8. 0.05 mg
9. 10.99 days
10. 8166 bees
11. 7.44 hours
12. 12.28 lb per in.²

13. 2350.02 years
14. 15.40 hours
15. 2.31 days
16. 4.87 hours
17. 39.65 minutes
18. 8.66 years
19. 7.00 years
20. 9.00 years
21. 5600 years
22. 294.41 days
23. **a.** 13.86 years
b. 6.93 years
24. **a.** 27.47 years
b. 13.73 years
25. 100

13. A radioactive substance decays according to $A = A_0 e^{-0.0002t}$, where A_0 is the initial amount and t is the time in years. If $A_0 = 640$ grams, find the time for A to decay to 400 grams.

14. A substance decays according to $A = A_0 e^{-0.045t}$, where t is in hours and A_0 is the initial amount. Determine the half-life of the substance.

15. An employee is learning to assemble remote-control units. The number of units per day he can assemble after t days of intensive training is given by $N = 80\left(1 - e^{-0.3t}\right)$. How many days of training will be needed before the employee is able to assemble 40 units per day?

16. A scientist collects a lava sample and measures that its temperature is 1650°. To safely analyze the sample, it must be no warmer than 500°. The scientist stores the sample in a cooling chamber with a temperature of 50° and finds that in 2 hours, the lava has cooled to 1000°. When will the lava sample be safe to analyze?

17. The temperature of a carrot cake is 350° when it is removed from the oven. The temperature in the room is 72°. In 10 minutes, the cake cools to 280°. How long will it take for the cake to cool to 160°?

18. How long does it take $10,000 to double if it is invested at 8% compounded quarterly?

19. If $1000 is deposited at 6% compounded monthly, how long before the balance is $1520?

20. The value V of a machine at the end of t years is given by $V = C(1-r)^t$, where C is the original cost of the machine and r is the rate of depreciation. A machine that originally cost $12,000 is now valued at $3800. How old is the machine if $r = 0.12$?

21. The formula $A = A_0 2^{-\frac{t}{5600}}$ is used for carbon–14 dating to determine the age of fossils where t is measured in years. Determine the half-life of carbon–14.

22. Radioactive iodine has a half-life of 60 days. If an accident occurs at a nuclear plant and 30 grams of radioactive iodine are present, in how many days will 1 gram be present? (Round k to at least 7 decimal places.)

23. If a principal P is doubled, then $A = 2P$. Use the formula for continuously compounded interest to find the time it takes the principal to double in value if the rate of interest is **a.** 5% **b.** 10% (Note that the time for doubling the principal is completely independent of the principal itself.)

24. If a principal P is tripled, then $A = 3P$. Use the formula for continuously compounded interest to find the time it takes the principal to triple in value if the interest rate is **a.** 4% **b.** 8% (Note that the time for tripling the principal is completely independent of the principal itself.)

25. The 1906 earthquake in San Francisco measured 8.6 on the Richter scale. In 1971, an earthquake in the San Fernando Valley measured 6.6 on the Richter scale. How many times greater was the 1906 earthquake than the 1971 earthquake? (See Example 4.)

26. In 1985, an earthquake in Mexico measured 8.1 on the Richter scale. How many times greater was this earthquake than the one in Landers, California, in 1992 that measured 7.3 on the Richter scale? (See Example 4.)

27. Population does not generally grow in a linear fashion. In fact, the population of many species grows exponentially, at least for a limited time. Using the exponential model $y = y_0 e^{kt}$ for population growth, estimate the population of a state in 2040 if the population was 5 million in 2010 and 6 million in 2020. (Assume that t is measured in years and $t = 0$ corresponds to 2010.)

28. Suppose that a lake is stocked with 500 fish, and biologists predict that the population of these fish will be approximated by the function $P(t) = 500\ln(2t + e)$ where t is measured in years. What will the fish population be in 3 years? in 5 years? in 10 years? (Round answers to the nearest integer.)

29. Sales representatives of a new type of computer predict that sales can be approximated by the function $S(t) = 1000 + 500\ln(3t + e)$ where t is measured in years. What are the predicted sales in 2 years? in 5 years? in 10 years? Round to the nearest integer.

30. In chemistry, the pH of a solution is a measure of the acidity or alkalinity of a solution. Water has a pH of 7 and, in general, acids have a pH less than 7 and alkaline solutions have a pH greater than 7. The model for pH is $\text{pH} = -\log[\text{H}^+]$ where $[\text{H}^+]$ is the hydrogen ion concentration in moles per liter of a solution.

a. Find the pH of a solution with a hydrogen ion concentration of 8.6×10^{-7}.

b. Find the hydrogen ion concentration $[\text{H}^+]$ of a solution if the pH of the solution is 4.5. Write the answer in scientific notation.

31. A decibel (abbreviated dB) is a unit used to measure the loudness of sound. The decibel level D of a sound of intensity I is measured by comparing it to a barely audible sound of intensity I_0 with the following formula: $D = 10\log\left(\frac{I}{I_0}\right)$. Sounds measuring over 85 dB are not considered safe.

a. Find the decibel level of a rock concert with an intensity of $6.24 \times 10^{11} I_0$.

b. What is the intensity level of 85 dB?

c. What is the intensity level of 60 dB (normal conversation)?

26. 6.31

27. 8.64 million

28. 1083, 1272, 1562

29. 2083, 2437, 2744

30. **a.** 6.07

b. 3.16×10^{-5}

31. **a.** 117.95 dB

b. $3.16 \times 10^{8} I_0$

c. $10^{6} I_0$

Chapter 17 Project: Twice as Loud but Way More than Twice as Intense

1. **a.** Since $\text{pH} = -\log[\text{H}^+]$, we get $\text{pH} = -\log(10^{-7}) = 7$. Pure water has a pH of 7.

b. Since $\text{pH} = -\log[\text{H}^+]$, we get $\text{pH} = -\log(10^{-11}) = 11$. Bleach has a pH of 11.

c. $\frac{11}{7} \approx 1.6$. Bleach has pH 1.6 times that of water.

d. $\frac{10^{-7}}{10^{-11}} = 10^4$ times larger.

e. Answers will vary. Citric acid has a pH between 3 and 6 which corresponds to an $[\text{H}^+]$ ion concentration between 10^{-3} and 10^{-6}.

f. Because $-\log(10^{-k}) = -(-k) = k$.

g. Larger values of k make for smaller values of 10^{-k}. Since $10^{-k} = \frac{1}{10^k}$, we know that the larger the value of k, the larger the value we divide by. We know 1 divided by large values will result in overall small values.

Note: Answers continued after the project

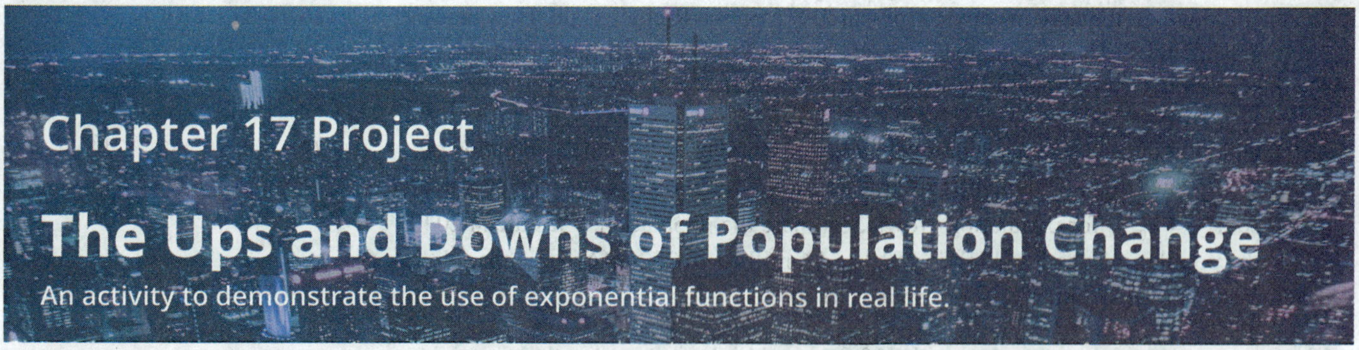

Chapter 17 Project

The Ups and Downs of Population Change

An activity to demonstrate the use of exponential functions in real life.

A midsized city with a population of 125,000 people experienced a population boom after several tech startups produced very successful products. The population of the city grew by 7% per year for several years. The population of the city can be modeled by the equation $P = P_0e^{rt}$, where P is the future population, P_0 is the initial population of the city, r is the rate of growth (or decay), and t is the time passed, in years.

1. Write the equation of the model for the city's population growth in terms of t. $P = 125{,}000e^{0.07t}$

2. Find the population of the city and then determine the number of new citizens after 5 years of growth. Round your answer to the nearest person.

3. Find the population of the city and then determine the number of new citizens after 10 years of growth. Round your answer to the nearest person.

4. Compare the number of new citizens after 5 years and after 10 years. Is the number of new citizens after 10 years twice the number of new citizens after 5 years? Explain your answer.

After 10 years, the population became too large for the city infrastructure (such as roads and freeways) to handle. The city raised taxes to improve the infrastructure, but the updates took a few years. Due to the traffic congestion in the city and the increased taxes, several of the large companies moved to neighboring cities. This loss of jobs in the city lead to a decline in population at a rate of 5% per year.

5. Write the equation of the model for the city's population decay in terms of t. $P = 251{,}719e^{-0.05t}$

6. It took 4 years to finish the roadway project in the city and people continued to move away at a steady rate. Find the population of the city when the project was complete.

7. What percent of the population moved away during those 4 years? Round your answer to the nearest percent. 18%

After the city infrastructure was updated, a few companies moved back to the city. As a result, the population started to grow by 2% per year.

8. Write the equation of the model for the city's population growth in terms of t. $P = 206{,}090e^{0.02t}$

9. At this rate of growth, how long would it takefor the city population to reach 250,000 citizens? Round your answer to the nearest year. 10 years

Chapter 17 Project: The Ups and Downs of Population Change

2. Population after 5 years is 177,383 people. There are 52,383 new citizens.
3. Population after 10 years is 251,719 people. There are 126,719 new citizens.
4. The population growth after 10 years is not twice the growth after 5 years. This is because the population grows exponentially, not linearly.
6. The population after the roadway project is finished is 206,090 people.

Chapter 17 Project

Twice as Loud but Way More than Twice as Intense

An activity to demonstrate the use of logarithms in real life.

In this chapter you have learned about logarithms and their properties. In this activity, you will explore applications of logarithms within physics and chemistry.

1. In chemistry, the acidity of a substance is measured using its pH. More acidic solutions have a lower pH and more alkaline solutions have a higher pH. The formula for the pH is given by $\text{pH} = -\log\left[\text{H}^+\right]$ where $\left[\text{H}^+\right]$ is the hydrogen ion concentration in moles per liter. The smaller the value of $\left[\text{H}^+\right]$, the larger the pH.

- **a.** Pure water is considered a neutral substance. Compute the pH of water knowing that $\left[\text{H}^+{}_{\text{water}}\right] = 10^{-7}$.
- **b.** Bleach is a very alkaline solution. Compute the pH of bleach knowing that $\left[\text{H}^+{}_{\text{bleach}}\right] = 10^{-11}$.
- **c.** How many times is the pH of bleach larger than the pH of water? (**Hint:** divide the pH of bleach by that of water.) Round to the nearest tenth.
- **d.** How many times is the concentration of hydrogen ions in water larger than the concentration of hydrogen ions in bleach?
- **e.** Find an example of an acidic substance; that is, find a substance with pH less than 7. Use its pH to compute its hydrogen ion concentration $\left[\text{H}^+\right]$.
- **f.** For any natural number k suppose there is a substance for which $\left[\text{H}^+\right] = 10^{-k}$. Explain why this substance would have a pH equal to k.
- **g.** Explain why smaller values of $\left[\text{H}^+\right]$ yield larger pH values.

2. The more energy a sound wave carries, the louder the wave sounds to a listener. For a wave with intensity I_{wave}, measured in watts per square meter (W/m²), its loudness L_{wave} in decibels (dB) is given by the equation

$$L_{\text{wave}} = 10\log\left(\frac{I_{\text{wave}}}{I_0}\right),$$

where I_0 is a constant equal to 10^{-12} W/m².

- **a.** What is the loudness in decibels of normal conversation, knowing that it has intensity $I_{\text{conversation}} = 10^{-6}$?
- **b.** Busy traffic is about 72 dB. How many times is busy traffic louder than normal conversation? In other words, how many times is L_{traffic} larger than $L_{\text{conversation}}$?
- **c.** How many times is busy traffic more intense than normal conversation? In other words, how many times is I_{traffic} larger than $I_{\text{conversation}}$? Round to the nearest tenth.
- **d.** A 60 dB sound is twice as loud as a 30 dB sound. Is it correct to say that a 60 dB sound is twice as intense as a 30 dB sound? Explain.

2. a. $L_{\text{conversation}} = 10\log\left(\frac{10^{-6}}{10^{-12}}\right) = 10\log\left(10^6\right) = 60 \text{ dB}$

b. $\frac{L_{\text{traffic}}}{L_{\text{conversation}}} = \frac{72 \text{ dB}}{60 \text{ dB}} = 1.2$. So, traffic is 1.2 times louder than normal conversation.

c. We know that $L_{\text{traffic}} = 10\log\left(\frac{I_{\text{traffic}}}{I_0}\right)$.

So, $72 = 10\log\left(\frac{I_{\text{traffic}}}{10^{-12}}\right)$. Solving the equation, we get $I_{\text{traffic}} = 10^{-4.8}$. So, we have that $\frac{I_{\text{traffic}}}{I_{\text{conversation}}} = \frac{10^{-4.8}}{10^{-6}} \approx 15.8$ So, busy traffic is 15.8 times more intense than normal conversation.

d. No, a 60-decibel sound will be about 1000 times more intense than a 30-decibel sound.

CHAPTER 18

18 Conic Sections

18.1 Translations and Reflections

18.2 Parabolas as Conic Sections

18.3 Distance Formula, Midpoint Formula, and Circles

18.4 Ellipses and Hyperbolas

18.5 Nonlinear Systems of Equations

CHAPTER 18 PROJECTS

What's in a Logo?

Conic Sections in Medicine

Connections

Halley's Comet is visible from Earth every 74 to 79 years. The comet has an elliptical orbit around the Sun, just like the orbits of the planets in our solar system. Ellipses, along with parabolas, hyperbolas, and circles, are conic sections. Conic sections occur naturally all around us. If you've seen a shadow on a wall cast by a circular lamp shade, you've seen a hyperbolic curve. If you've seen a fountain with a jet of water going upward and then falling back down, you've seen a parabola.

Not only do conic sections occur naturally, they are also useful in a variety of applications. Early camera makers discovered that circular lenses did the most efficient job of focusing light onto film. Car headlights use a parabolic shape to concentrate the light beam to improve safety. Medical specialists have used the ellipse to create a device that treats kidney stones with shock waves instead of surgery. The hyperboloid is the design standard for all nuclear cooling towers and some coal-fired power plants because its structure can withstand high winds, it requires less material than other shapes, and it maximizes cooling efficiency.

Suppose the town you live in is holding an architectural competition to design a public monument that will be part of the downtown renovation project. You submit a proposal to build a circular fountain at the center of town. You've mapped the town on a coordinate grid and found the origin, where you plan to place the center of the fountain. If your proposal states that one point on the edge of the fountain will be at the point $(0, 5)$, is this enough information to find the equation of the circle? What is another point on the edge of the fountain?

Objectives

A. Evaluate functions at a given expression.

B. Find the difference quotient for a given function.

C. Graph horizontal and vertical translations of given graphs of functions.

D. Graph reflections and translations of given graphs of functions.

18.1 Translations and Reflections

A Evaluating Functions at a Given Expression

We have already used function notation $f(x)$ (read "f of x") in Chapter 12 to represent and to evaluate polynomials, in Chapter 15 to represent and evaluate radical functions, and in Chapter 17 in dealing with inverse functions. In this section, function notation is used to indicate a new value or a new function when the single variable x is replaced by an expression such as $(a+1)$ or $(x+h)$. For example,

if $f(x)=3x+5$, then

$f(2)=3(2)+5=11$ — x is replaced by 2.

$f(a)=3(a)+5$ — x is replaced by a.

$f(a+1)=3(a+1)+5$ — x is replaced by $a+1$.

$f(x+h)=3(x+h)+5$ — x is replaced by $x+h$.

These substitutions for x relate to formulas and to graphs of functions.

1. For the function $f(x)=3x^2+5$, find:

a. $f(-2)$

b. $f(a)$

c. $f(a+2)$

Example 1 Using $f(x)$ Notation

For the function $f(x)=2x^2-4$, find:

a. $f(3)$

b. $f(a)$

c. $f(a+1)$

Solution

a.
$$\begin{aligned} f(3)&=2(3)^2-4\\ &=2\cdot 9-4\\ &=18-4=14\end{aligned}$$

b.
$$\begin{aligned} f(a)&=2(a)^2-4\\ &=2a^2-4\end{aligned}$$

c. Replace x with $a+1$ and simplify.

$$\begin{aligned} f(a+1)&=2(a+1)^2-4\\ &=2(a^2+2a+1)-4\\ &=2a^2+4a+2-4\\ &=2a^2+4a-2\end{aligned}$$

Now work margin exercise 1.

B The Difference Quotient: $\dfrac{f(x+h)-f(x)}{h}$

The formula

$$\frac{f(x+h)-f(x)}{h}$$

is called the **difference quotient**. As shown in Figure 1, a geometric interpretation of the difference quotient is the **slope** of a line through two points on the graph of a function.

Teaching Note

You might want to ask your students if they can picture the point Q sliding down the curve, getting closer and closer to P. Then ask what is happening to the secant line and what is happening to h? Give two or three other curves on the board and you can have a lively class discussion.

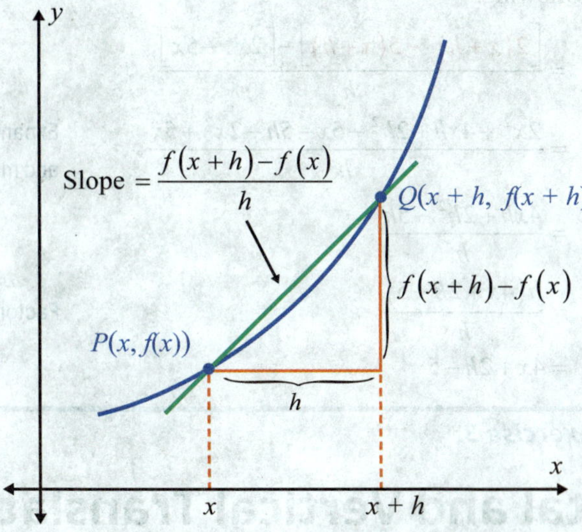

Figure 1

> **Note**
>
> The line illustrated in Figure 1 (through two points on the graph) is called a **secant line.**

Note carefully that $f(x+h)$ is not the same as $f(x)+h$. For example,

if $f(x)=x^2+2x+3,$

then $f(x+h)=(x+h)^2+2(x+h)+3=x^2+2xh+h^2+2x+2h+3$

and $f(x)+h=x^2+2x+3+h.$

Example 2 Finding the Difference Quotient $\frac{f(x+h)-f(x)}{h}$

Find the difference quotient for the function $f(x)=2-6x.$

Solution

$f(x+h)=2-6(x+h)$ and $f(x)=2-6x$

Substituting gives the following.

$$\frac{f(x+h)-f(x)}{h}=\frac{[2-6(x+h)]-[2-6x]}{h}=\frac{2-6x-6h-2+6x}{h}=\frac{-6h}{h}=-6.$$

Now work margin exercise 2.

2. Find the difference quotient for the function $f(x)=8x-12.$

Example 3 Finding the Difference Quotient $\frac{f(x+h)-f(x)}{h}$

Find the difference quotient for the function $f(x)=2x^2-5x.$

Solution

$f(x+h)=2(x+h)^2-5(x+h)$ and $f(x)=2x^2-5x$

3. Find the difference quotient for the function $f(x)=x^2+3x-10.$

Substituting gives the following.

$$\frac{f(x+h)-f(x)}{h} = \frac{\left[2(x+h)^2-5(x+h)\right]-\left[2x^2-5x\right]}{h}$$

$$= \frac{2x^2+4xh+2h^2-5x-5h-2x^2+5x}{h}$$ Expand $(x+h)^2$ and multiply by 2.

$$= \frac{4xh+2h^2-5h}{h}$$

$$= \frac{h(4x+2h-5)}{h}$$ Factor out h.

$$= 4x+2h-5$$

Now work margin exercise 3.

C Horizontal and Vertical Translations

Math Tip

Remember that a positive value of h shifts the graph to the right while a negative value of h shifts the graph to the left. Similarly, a positive value of k shifts the graph up while a negative value of k shifts the graph down.

Teaching Note

You might ask your students what they think will happen if the leading coefficient 2 is changed.

In Section 16.6, the graphs of parabolas of the form $y = ax^2$ were discussed. We found that the graph of $y = a(x-h)^2$ is a **horizontal translation (horizontal shift)** of h units and the graph of $y = ax^2 + k$ is a **vertical translation (vertical shift)** of k units of the graph of $y = ax^2$.

In function notation,

if $f(x) = 2x^2$,

then $f(x-3) = 2(x-3)^2$ is a horizontal translation of 3 units to the right,

and $f(x)-4 = 2x^2-4$ is a vertical translation of 4 units down.

Figure 2 shows the graphs of the three functions and their relationships. Note that the curves themselves are identical, only their positions are changed.

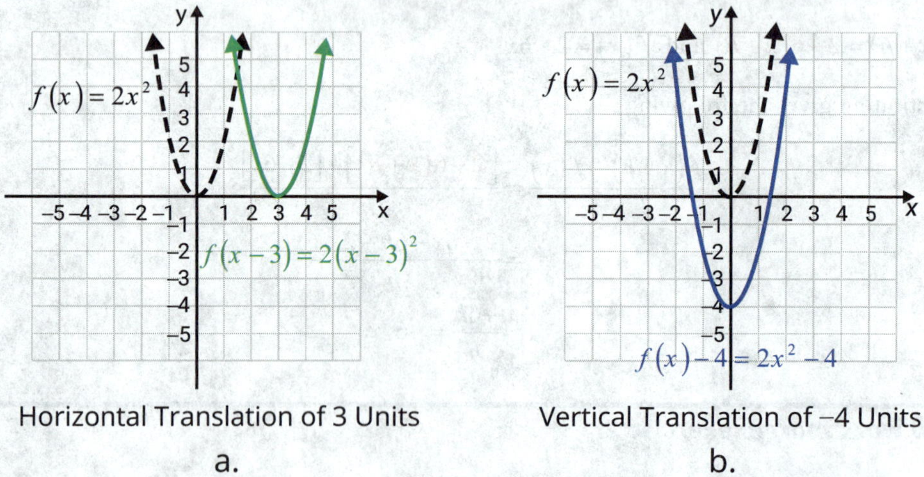

Figure 2

The following general approach can be used to help graph a horizontal and/or vertical translation of any function.

Horizontal and Vertical Translations

Given the graph of $y = f(x)$, the graph of $y = f(x-h)+k$ is

1. a horizontal translation of h units and
2. a vertical translation of k units

of the graph of $y = f(x)$.

Draw the graph of $y = f(x)$ in relation to (h, k) as if (h, k) were the origin, $(0, 0)$. This new graph will be the graph of $y = f(x-h)+k$.

PROCEDURE

Another function to use in illustrating translations is the function $f(x)=|x|$. First, we need to know what the graph of $f(x)=|x|$ or $y=|x|$, looks like. The definition of $|x|$ can be written as follows.

$$f(x)=|x|=\begin{cases} x & \text{if } x \geq 0 \\ -x & \text{if } x < 0 \end{cases}$$

This means the graph can be analyzed in two pieces.

First Piece

The graph of $f(x)=x$ is a line, as shown in Figure 3a, but we want only the part where $x \geq 0$, as shown in Figure 3b.

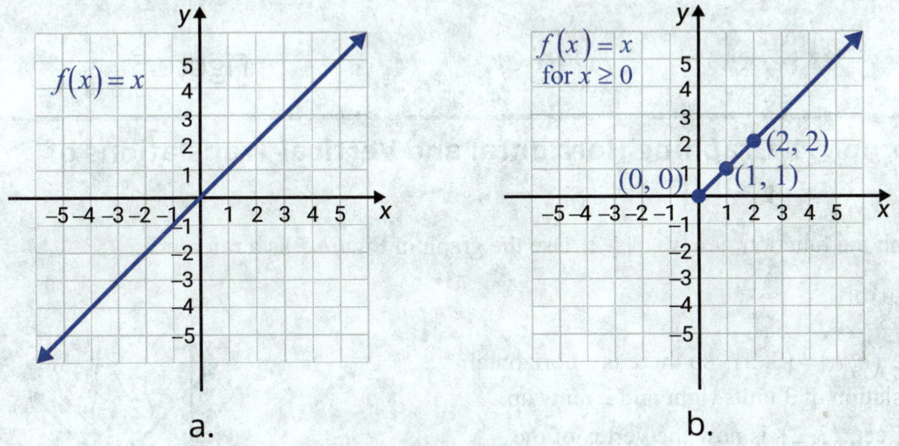

Figure 3

Second Piece

The graph of $f(x) = -x$ is also a line, as shown in Figure 4a, but we want only the part where $x < 0$, as shown in Figure 4b.

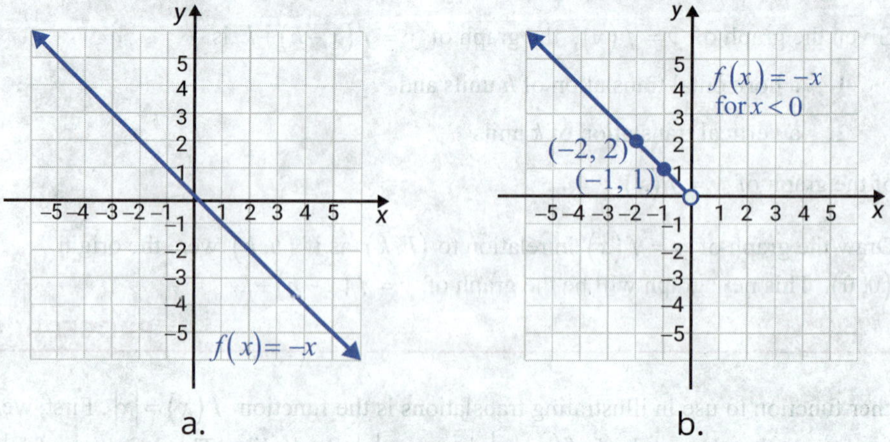

Figure 4

Combining the two graphs in Figures 3b and 4b gives the graph of $f(x) = |x|$ as shown in Figure 5.

Now, using the graph of the function $f(x) = |x|$ or $y = |x|$, we can examine related horizontal and vertical shifts as in Example 4.

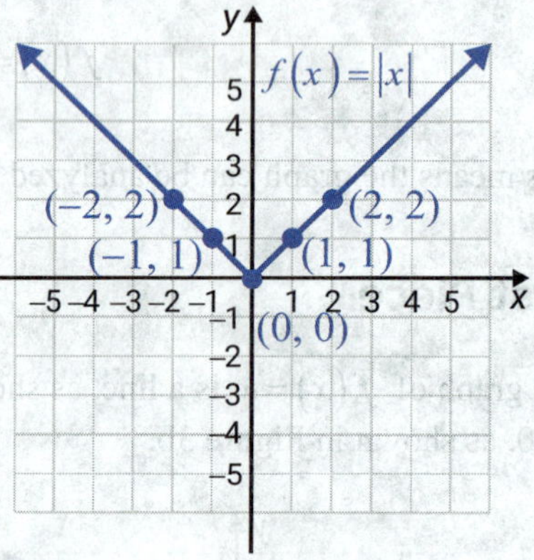

Figure 5

Teaching Note

You can explain how points other than $(0, 0)$ can be translated by moving 3 units right and 2 units up. For example, referencing Figure 5, we see $(1, 1)$ goes to $(4, 3)$ and $(-1, 1)$ goes to $(2, 3)$. This can be done for any point on the graph of the function.

4. Graph the function $y = |x - 1| + 3$.

Example 4 Graphing Horizontal and Vertical Translations of $y = |x|$

Graph the function $y = |x - 3| + 2$. Use the graph in Figure 5 as a reference.

Solution

Here $(h, k) = (3, 2)$, so there is a horizontal translation of 3 units right and 2 units up. In effect, $(3, 2)$ is now the vertex of the new graph just as $(0, 0)$ is the vertex of the original graph. You should check that the points shown on the graph here do indeed satisfy the function.

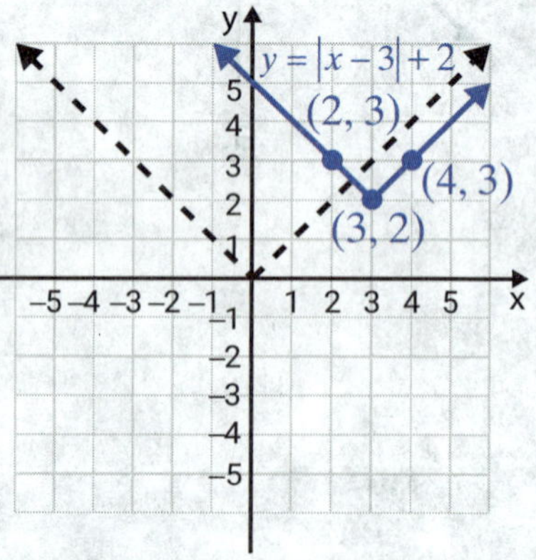

Now work margin exercise 4.

Example 5 Graphing Horizontal and Vertical Translations of $y = |x|$

Graph the function $y = |x+4| - 1$.

Solution

Here $(h, k) = (-4, -1)$, so the horizontal translation is –4 (4 units left) and the vertical translation is –1 (1 unit down). The effect is that the vertex is now at the point $(-4, -1)$.

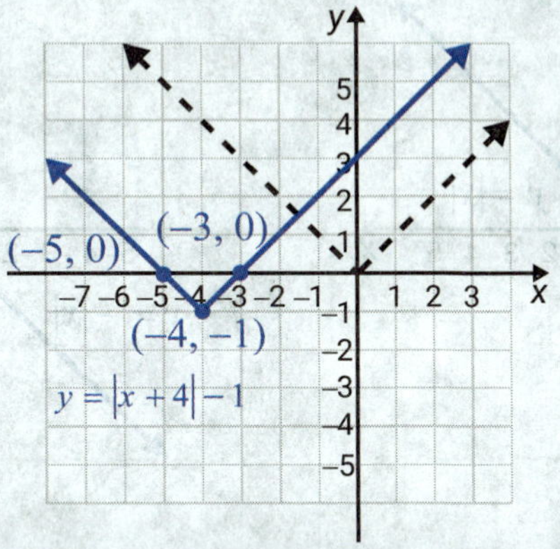

Now work margin exercise 5.

5. Graph the function $y = |x+2| - 5$.

Example 6 Graphing Horizontal and Vertical Translations of $y = |x|$

Graph the function $y = |x+2| + 7$.

Solution

Here $(h, k) = (-2, 7)$, so the horizontal translation is –2 (2 units left) and the vertical translation is 7 (7 units up). The effect is that the vertex is now at the point $(-2, 7)$.

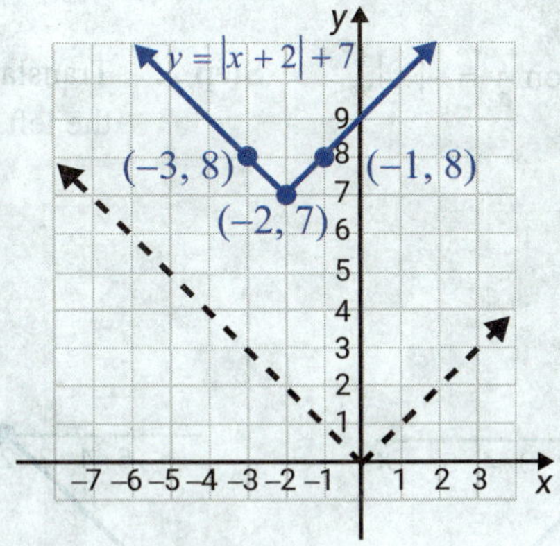

Now work margin exercise 6.

6. Graph the function $y = |x+1| + 6$.

D Reflections and Translations

In general, the graph of $y=-f(x)$ is a **reflection across the *x*-axis** of the graph of $y=f(x)$. If we examine the function $y=|x|$ and the graph of the function $y=-|x|$, we see that the graph of each function is the mirror image of the other across the *x*-axis. The first "opens" upward and the second "opens" downward, as illustrated in Figure 6. Both have the same vertex of $(0, 0)$.

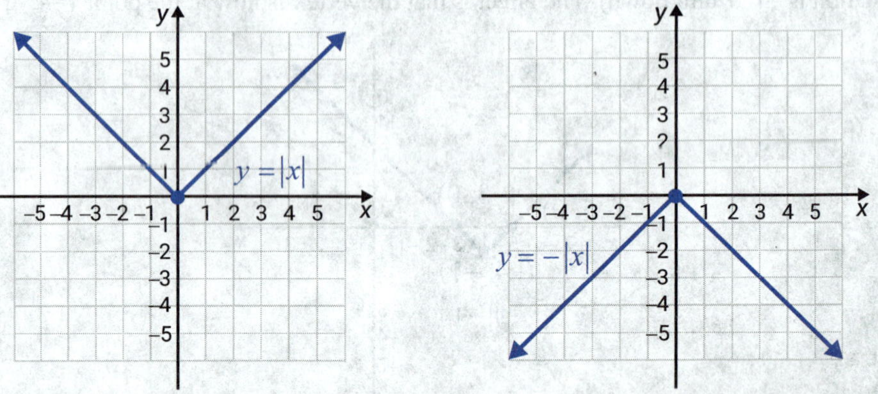

a. b.

Figure 6

Teaching Note

To reinforce this idea, you might consider reviewing the discussions concerning the two functions $y=x^2$ and $y=-x^2$ in Section 16.6. Students should see the similarities.

Example 7 Graphing Reflections and Translations of $y = |x|$

Graph the function $y=-|x+2|+5$.

Solution

The reflection is performed first, followed by the translations.

Here $(h, k)=(-2, 5)$, and the graph is reflected across the *x*-axis. We show step-by-step how to "arrive" at the graph. (You may do these steps mentally and graph only the last step.)

Step 1: Graph the reflection $y=-|x|$.

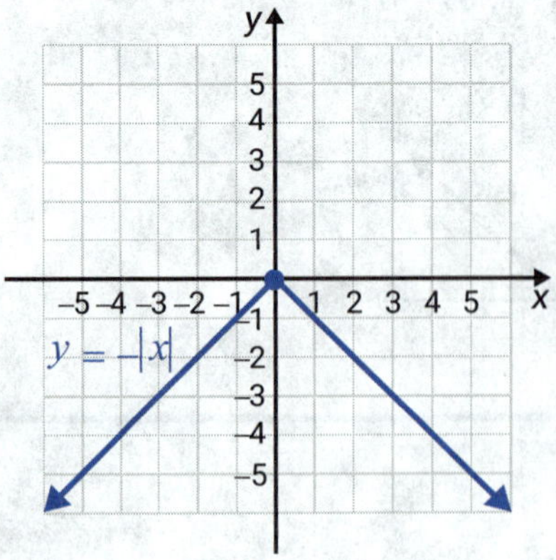

Step 2: Translate the graph 2 units to the left.

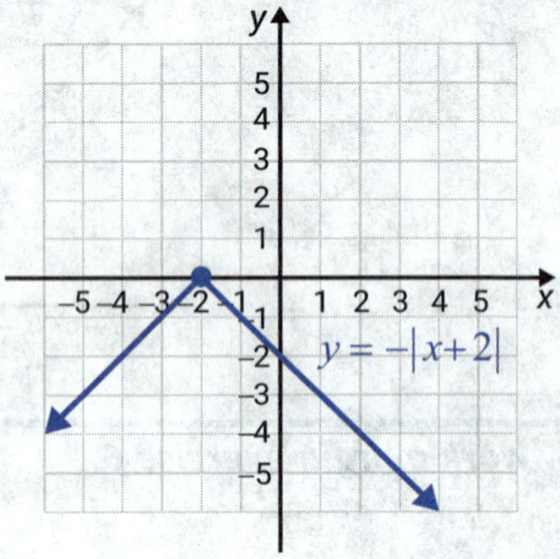

18.1 Exercises

Concept Check

1. slope, two
2. vertical
3. horizontal
4. reflection, *x*-axis
5. shape, direction
6. False; The two functions are different.
7. False; Translation does not change the shape of the graph of a function.

7. Graph the function $y=-|x+1|-2$.

8. False; Translations can be combined to form new functions.

Practice

1. **a.** 0 **b.** a^2-6a+5
   c. $x^2+2xh+h^2-4$
   d. $2x+h$
2. **a.** 0 **b.** $1-a^2+2a$
   c. $2-x^2-2xh-h^2$
   d. $-2x-h$
3. **a.** 0 **b.** $2a^2-11a+14$
   c. $2x^2+4xh-3x+2h^2-3h$
   d. $4x+2h-3$
4. **a.** 44 **b.** $3a^2+11a+10$
   c. $3x^2+6xh-x+3h^2-h$
   d. $6x+3h-1$
5. 1
6. 2
7. −2

Step 3: Translate the graph 5 units up.

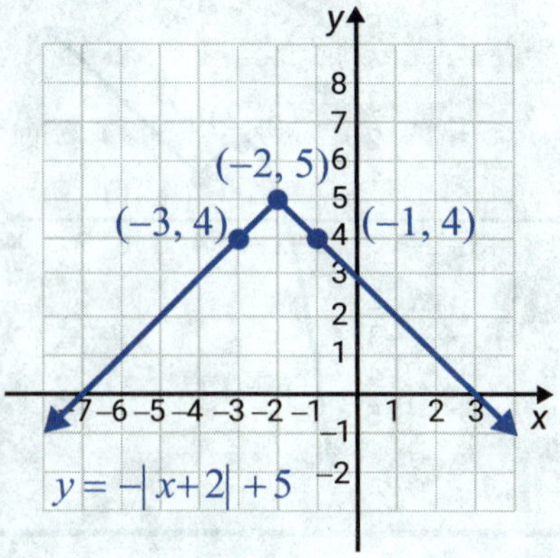

Now work margin exercise 7.

In a translation (horizontal or vertical), the shape of the graph of a function is unchanged. However, translations do change the position of the graph in the coordinate system. Examples 8 and 9 illustrate two such cases.

Teaching Note

In graphing reflections and translations, the reflections are performed first. You might have your students do some translations first, followed by the reflections and show them how points on such a graph do not satisfy the new function. One example would be $y = -\sqrt{x-2} + 1$ to compare with Example 8.

Example 8 Graphing Translations of a Function Given Its Graph

The graph of $y = \sqrt{x}$ is given. Graph the function $y = \sqrt{x-2} + 1$.

Solution

If $y = \sqrt{x}$ is written $y = f(x)$, then $y = \sqrt{x-2} + 1$ is the same as $y = f(x-2) + 1$. So $(h, k) = (2, 1)$, and there is a horizontal translation of 2 units to the right and a vertical translation of 1 unit up.

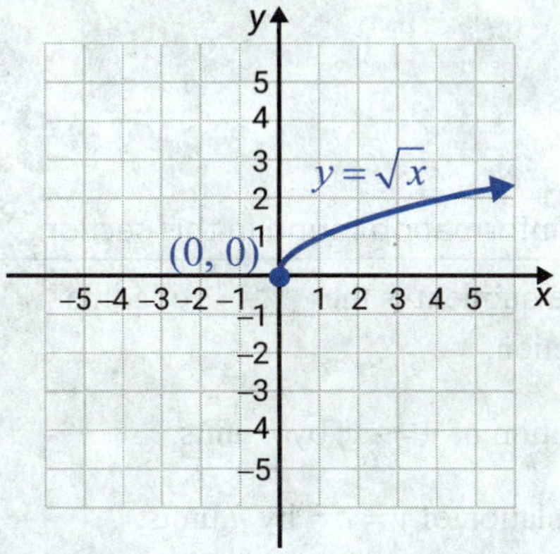

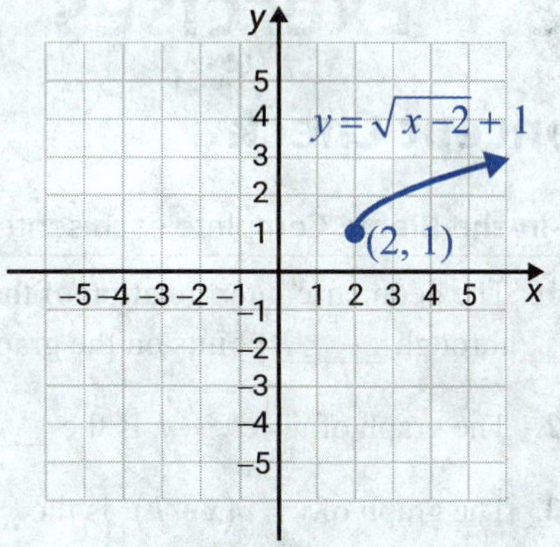

Now work margin exercise 8.

8. Given the function $y = \sqrt{x}$, graph $y = \sqrt{x+2} + 3$.

8. 4

9. The results are the coefficients of x.

10. The results equal the slopes of the lines.

11.

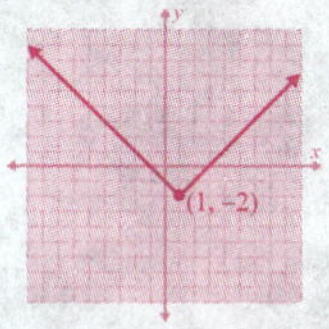

12.

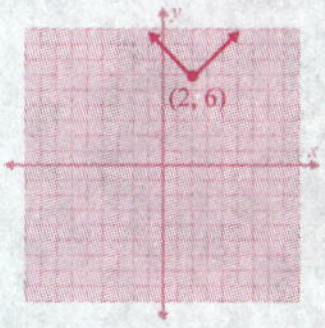

Example 9 Graphing Translations of a Function Given Its Graph

The graph of $y = f(x)$ is given. Graph the function $y = f(x-3) - 2$.

Solution

Here $(h, k) = (3, -2)$, so translate the graph horizontally 3 units to the right and vertically −2 units down. (Add 3 to each x-value and −2 to each y-value.)

9. Given $f(x)$ from Example 9, graph the function $y = f(x-3) + 1$.

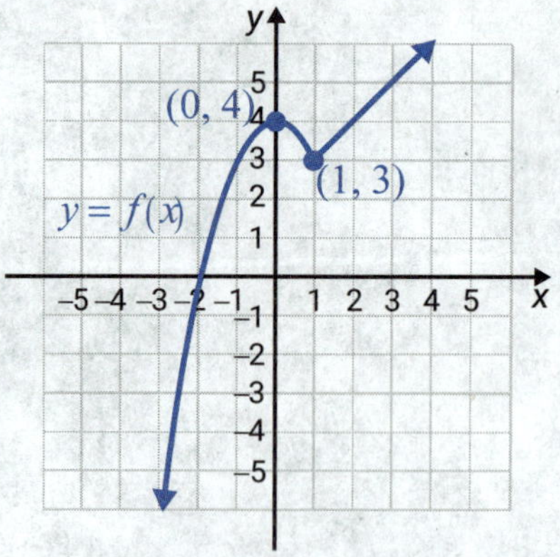

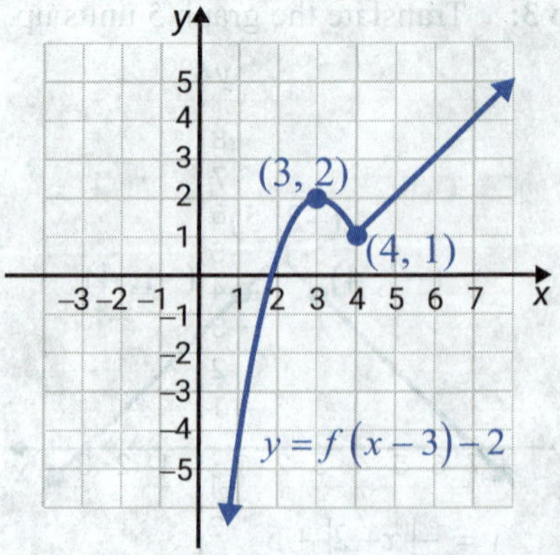

Now work margin exercise 9.

Margin Exercise Answers

1. a. 17 **b.** $3a^2+5$ **c.** $3a^2+12a+17$ **2.** 8 **3.** $2x+h+3$ **4.**

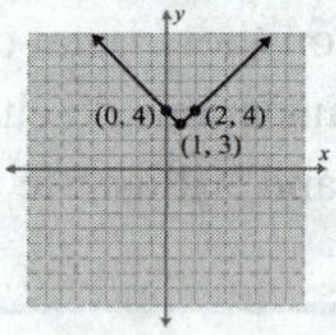

5.

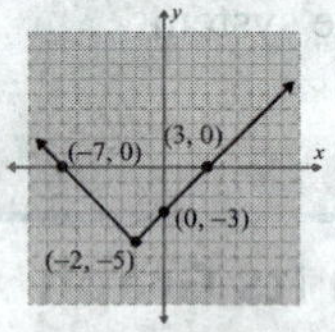

6.

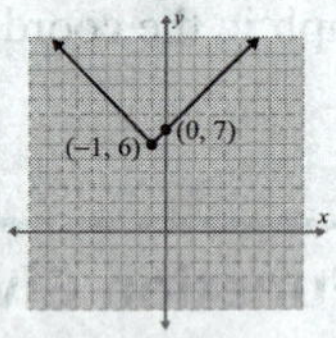

7.

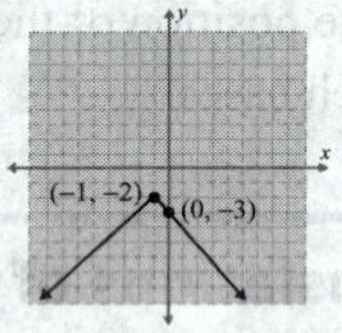

8.

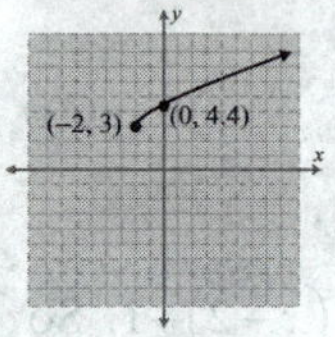

9.

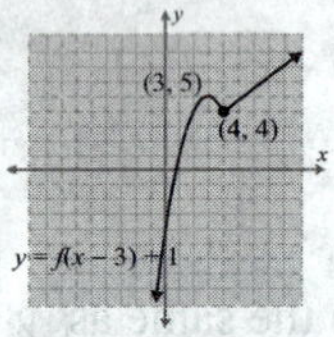

18.1 Exercises

Concept Check

Fill-in-the-Blank. Complete each sentence using information found in this section.

1. The geometric interpretation of the difference quotient is the ______ of a line through ______ points on the graph of a function.

2. The graph of $y = ax^2 + k$ is the ______ translation of $y = ax^2$ by k units.

3. The graph of $y = a(x-h)^2$ is the ______ translation of $y = ax^2$ by h units.

4. In general, the graph of $y = -f(x)$ is the ______ across the ______ of the graph $y = f(x)$.

5. For the function $y = ax^2$, changing the coefficient a can have an effect on the ______ and ______ of the graph of the base function.

13.

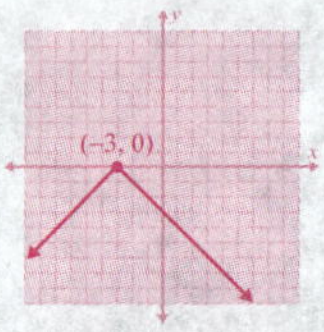

14.

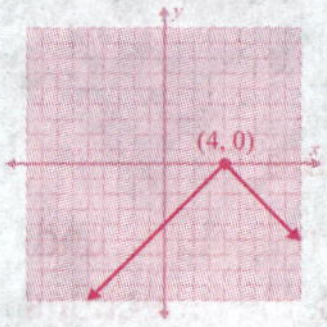

15.

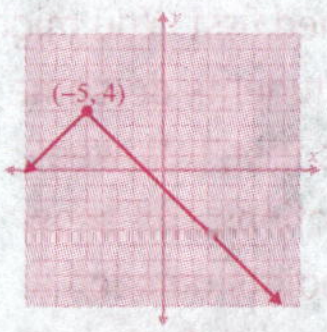

16.

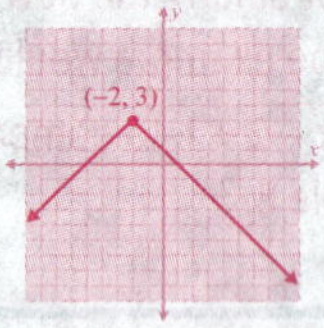

17.

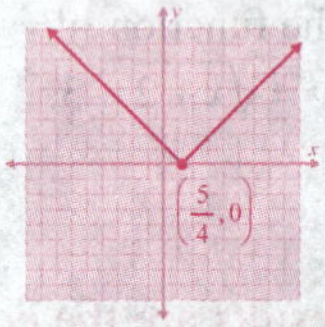

18.

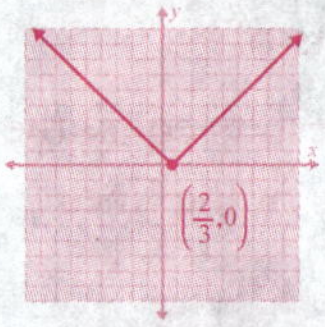

19.

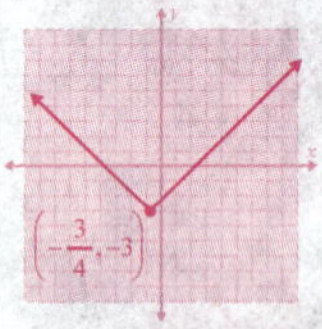

20.

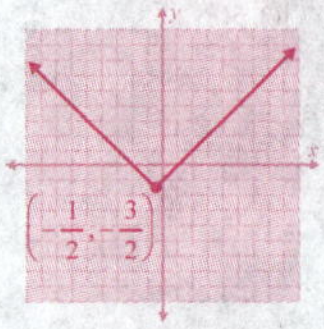

21.

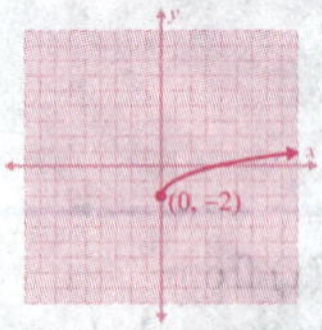

22.

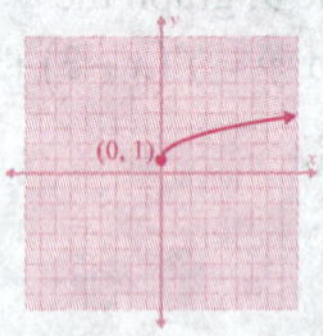

True/False. Determine whether each statement is true or false. If a statement is false, explain how it can be changed so the statement will be true. (**Note:** There may be more than one acceptable change.)

6. Algebraically, the two functions $f(x+h)$ and $f(x)+h$ represent the same thing.

7. Translation changes the shape of the graph of a function.

8. More than one translation cannot be applied to a function at the same time.

Practice

Evaluate the function at the given expressions. See Example 1.

1. For $f(x)=x^2-4$, find:

a. $f(-2)$

b. $f(a-3)$

c. $f(x+h)$

d. $\dfrac{f(x+h)-f(x)}{h}$

2. For $g(x)=2-x^2$, find:

a. $g(\sqrt{2})$

b. $g(a-1)$

c. $g(x+h)$

d. $\dfrac{g(x+h)-g(x)}{h}$

3. For $f(x)=2x^2-3x$, find:

a. $f(0)$

b. $f(a-2)$

c. $f(x+h)$

d. $\dfrac{f(x+h)-f(x)}{h}$

4. For $f(x)=3x^2-x$, find:

a. $f(4)$

b. $f(a+2)$

c. $f(x+h)$

d. $\dfrac{f(x+h)-f(x)}{h}$

Find and simplify the difference quotient, $\dfrac{f(x+h)-f(x)}{h}$, for each function. See Examples 2 and 3.

5. $f(x)=x+7$

6. $f(x)=2x-3$

7. $f(x)=5-2x$

8. $f(x)=4x-3$

9. What particular trend, if there is one, do you notice about the results in Exercises 5 through 8?

10. Analyze, in your own words, how the results in Exercise 9 relate to the graphs of the functions in relation to the secant line discussion in the text.

23.

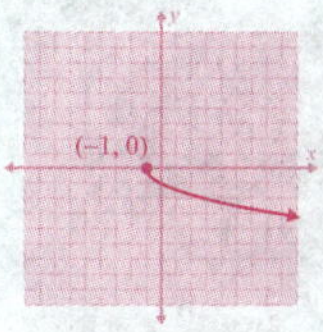

24.

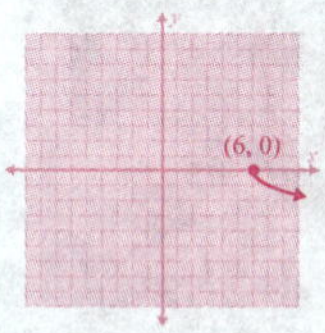

25.

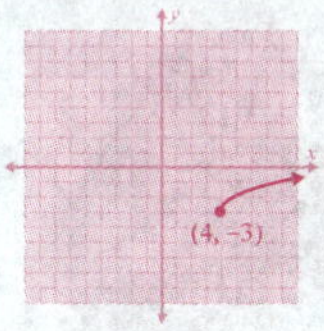

26.

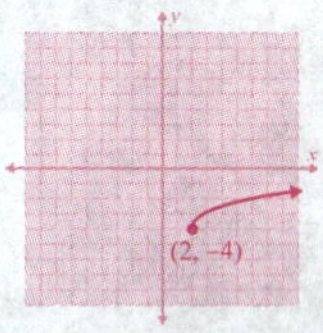

27.

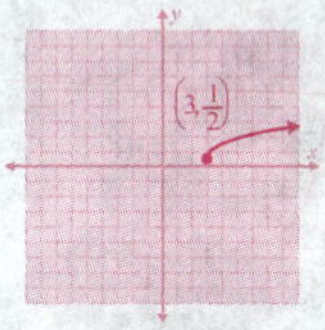

28.

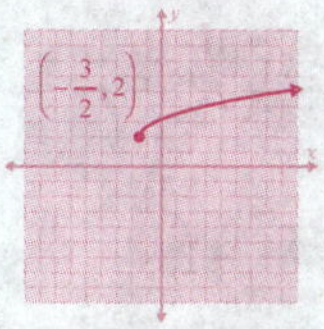

29.

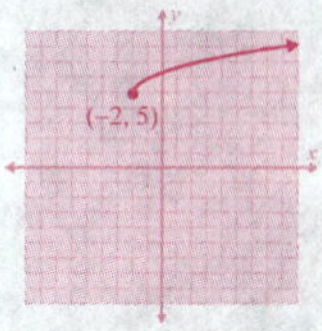

30.

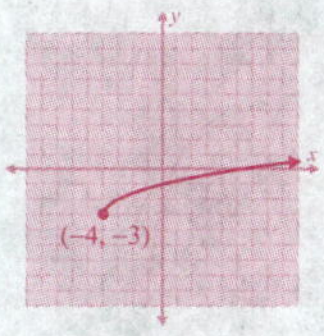

31.

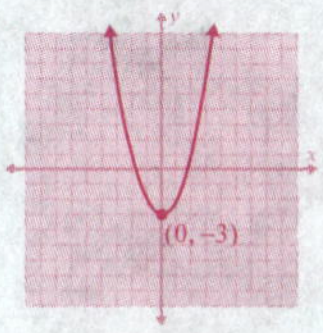

32.

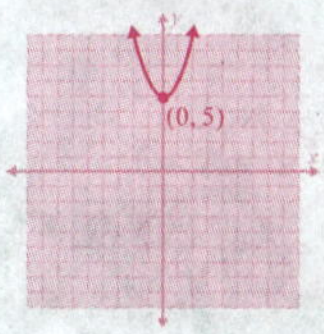

33.

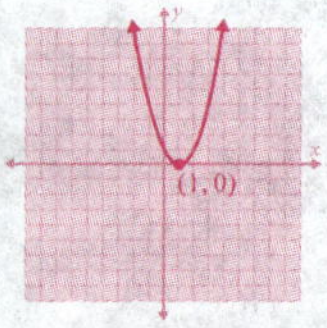

34.

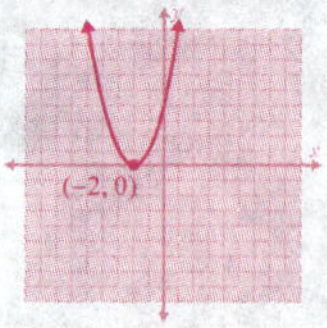

35.

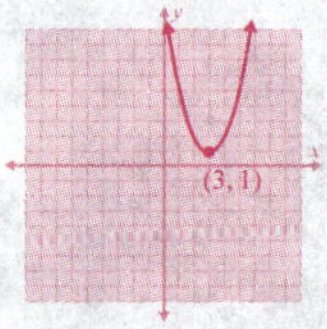

36.

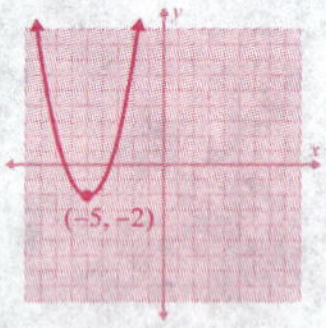

37.

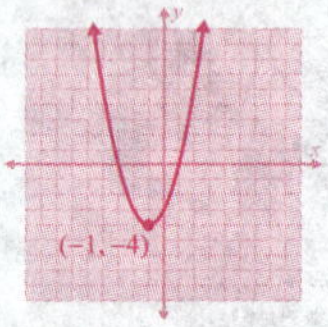

38.

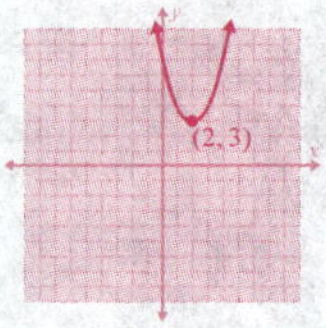

39.

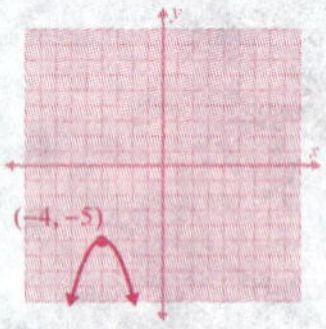

40.

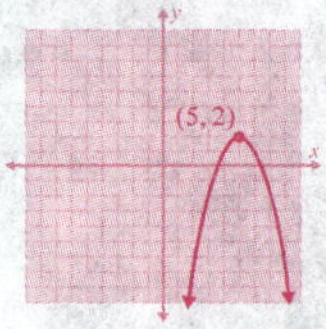

41.

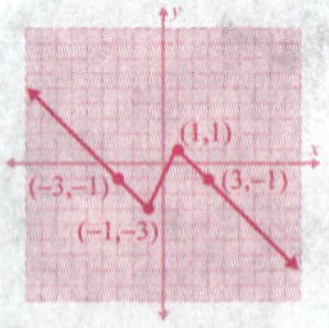

42.

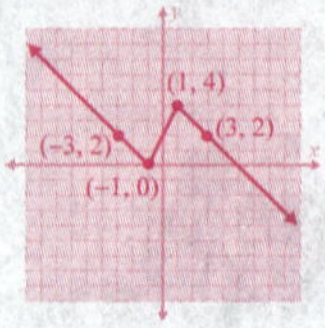

The graph of $y = |x|$ is given along with a few points as aids. Graph the functions using your understanding of reflections and translations with no additional computations. See Examples 4 through 7.

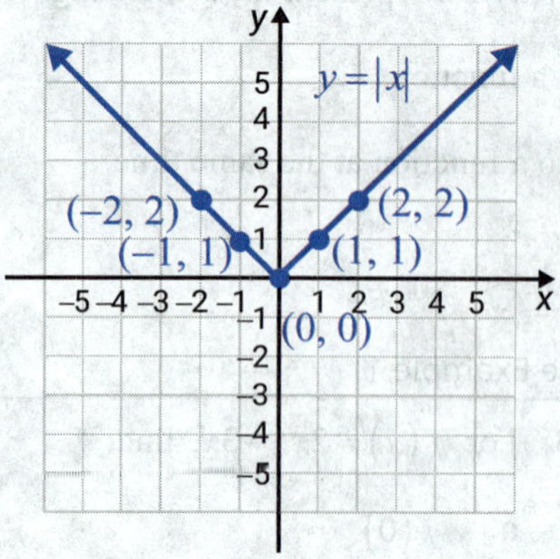

11. $y = |x-1| - 2$

12. $y = |x-2| + 6$

13. $y = -|x+3|$

14. $y = -|x-4|$

15. $y = -|x+5| + 4$

16. $y = -|x+2| + 3$

17. $y = \left|x - \frac{5}{4}\right|$

18. $y = \left|x - \frac{2}{3}\right|$

19. $y = \left|x + \frac{3}{4}\right| - 3$

20. $y = \left|x + \frac{1}{2}\right| - \frac{3}{2}$

The graph of $y = \sqrt{x}$ is given along with a few points as aids. Graph the functions using your understanding of reflections and translations with no additional computations. See Example 8.

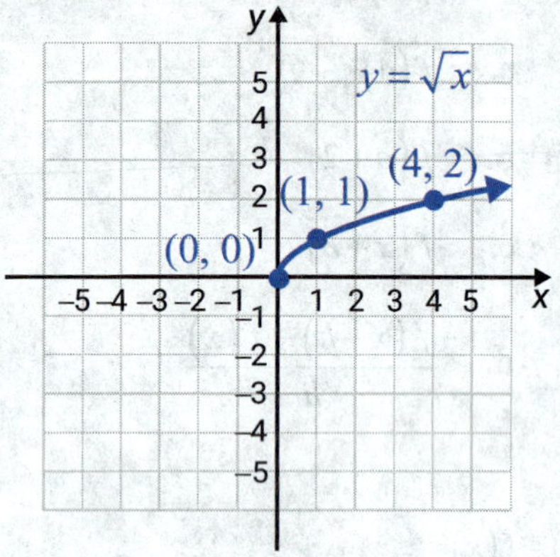

21. $y = \sqrt{x} - 2$

22. $y = \sqrt{x} + 1$

23. $y = -\sqrt{x+1}$

24. $y = -\sqrt{x-6}$

25. $y = \sqrt{x-4} - 3$

26. $y = \sqrt{x-2} - 4$

27. $y = \sqrt{x-3} + \frac{1}{2}$

28. $y = \sqrt{x + \frac{3}{2}} + 2$

29. $y = 5 + \sqrt{x+2}$

30. $y = \sqrt{x+4} - 3$

Using the graph of $y = x^2$, graph the functions using your understanding of reflections and translations with no additional computations.

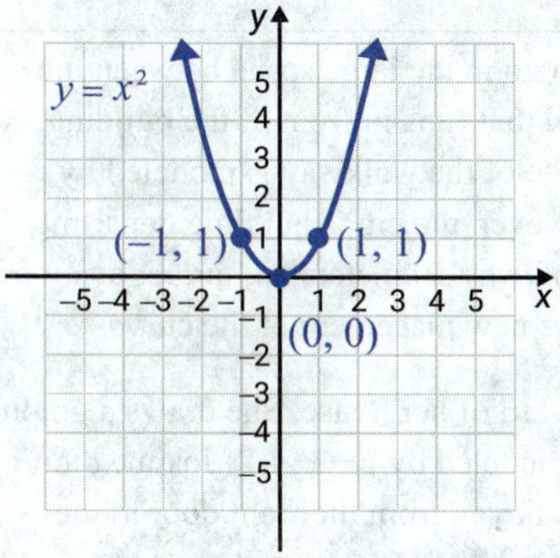

31. $y = x^2 - 3$

32. $y = x^2 + 5$

33. $y = (x-1)^2$

34. $y = (x+2)^2$

35. $y = (x-3)^2 + 1$

36. $y = (x+5)^2 - 2$

37. $y = (x+1)^2 - 4$

38. $y = (x-2)^2 + 3$

39. $y = -(x+4)^2 - 5$

40. $y = -(x-5)^2 + 2$

The graph of a function $y = f(x)$ is given with the coordinates of four points. Graph the functions using your understanding of reflections and translations ith no additional computations. Label the new points that correspond to the four labeled points. See Example 9.

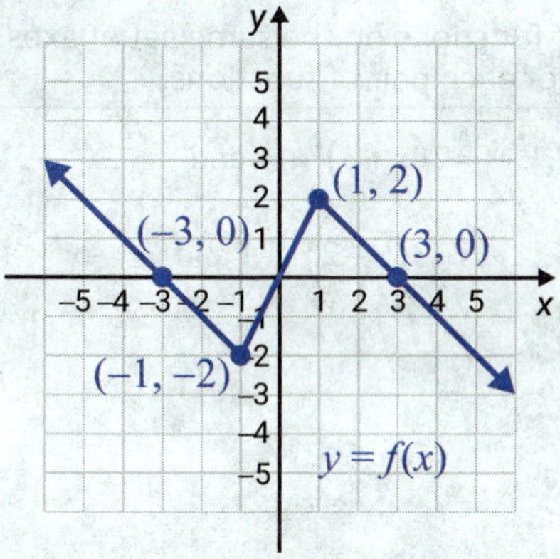

41. $y = f(x) - 1$

42. $y = f(x) + 2$

43. $y = f(x-3)$

44. $y = f(x+1)$

45. $y = -f(x-3)$

46. $y = -f(x+1)$

47. $y = f(x+5) + 3$

48. $y = f(x-1) + 5$

49. $y = f(x+2) - 4$

50. $y = f(x+3) + 2$

The graph of $y = \log(x)$ is given. Graph the functions and state the domain and range of each function.

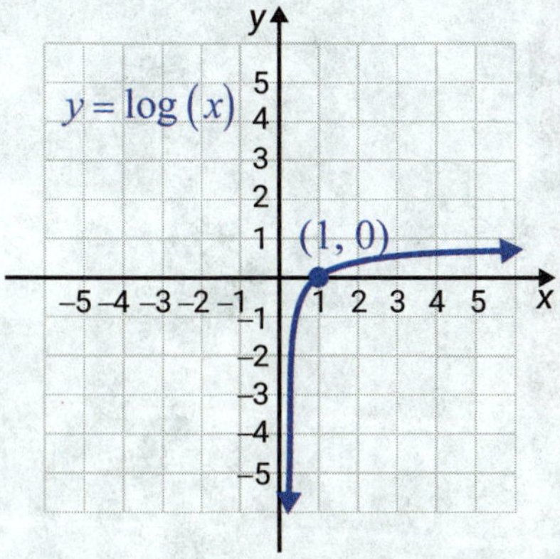

51. $y = \log(x+1)$

52. $y = \log(2-x)$

53. $y = -\log x$

54. $y = \log(-x)$

55. $y = 1 + \log x$

56. $y = -3 - \log x$

43.

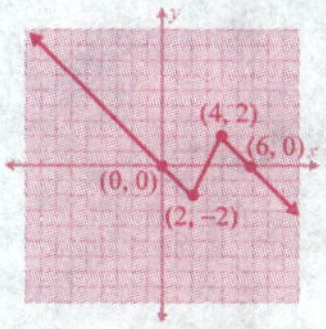

44.

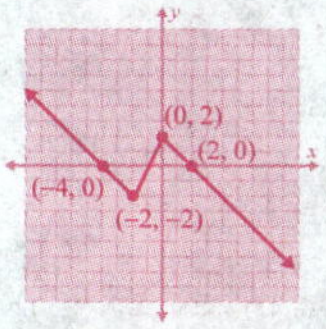

45.

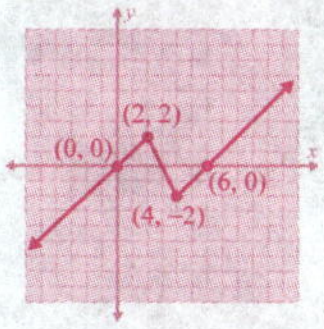

46.

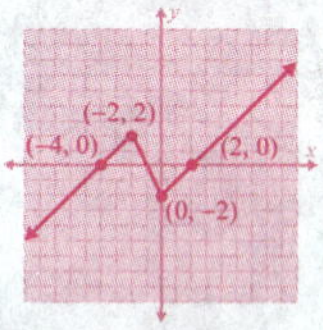

47.

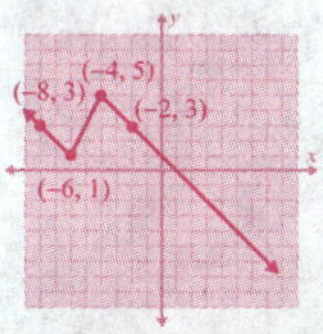

48.

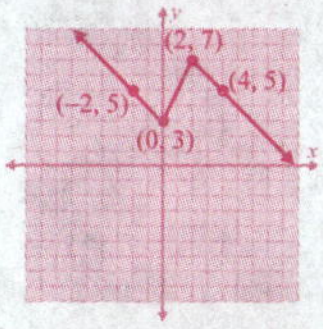

49.

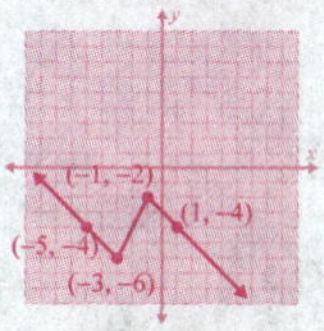

50.

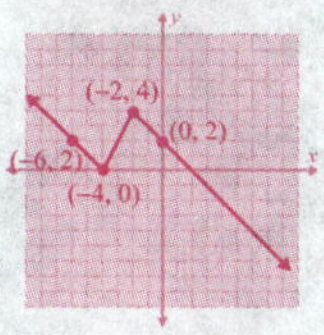

51.

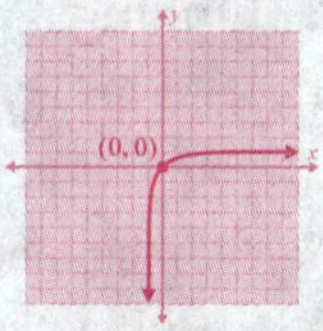

$D = (-1, \infty); R = (-\infty, \infty)$

52.

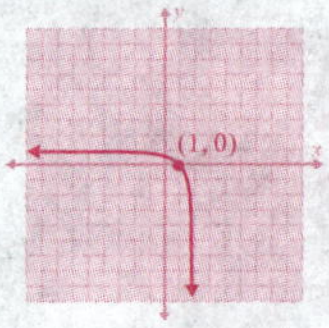

$D=(-\infty, 2); R=(-\infty, \infty)$

53.

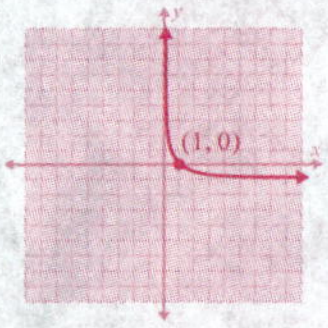

$D=(0, \infty); R=(-\infty, \infty)$

54.

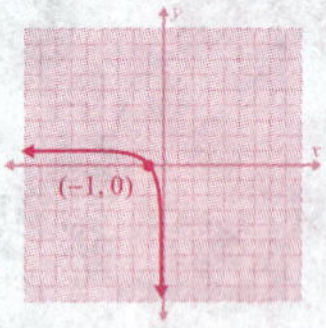

$D=(-\infty, 0); R=(-\infty, \infty)$

55.

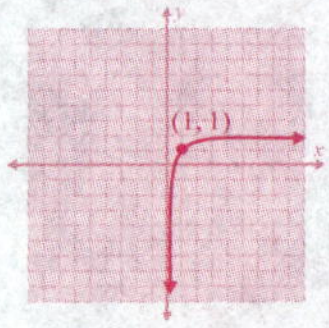

$D=(0, \infty); R=(-\infty, \infty)$

56.

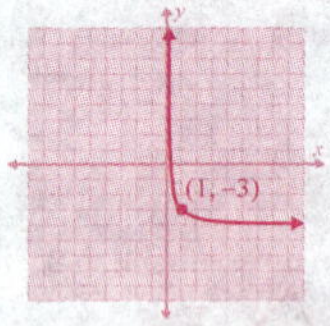

$D=(0, \infty); R=(-\infty, \infty)$

Applications

57. $y = x^4 - 5x^2 + 2x + 10$

58. $y = (x-10)^2 + 2$

Writing & Thinking

59. The graph of the function $y = f(x-h)+k$ is the graph of the function $y = f(x)$ shifted h units to the right and k units up.

Applications

Solve.

57. A landscaper is planning a scenic walkway to go in the large space between the front entrance of a building and the main road that runs in front of the building. He sketches a graph of the area. On it, the edge of the walkway is modeled by $y = x^4 - 5x^2 + 2x$. In order to get a permit, however, the city's building department tells the landscaper that he must move the walkway 10 meters up, farther away from the road. Write the function that represents the new placement of the curve.

58. Joanne wants to put a flowerbed in the front yard of her house. She draws a graph of the land, and the edge of the flowerbed is modeled by $y = x^2 + 2$. Joanne didn't realize that her husband planned to build a walkway from the front door to the driveway. To accommodate the plans for the walkway, Joanne will need to move her flowerbed ten feet to the right. Write the function that represents the new placement of the curve.

Writing & Thinking

59. Explain, in your own words, how the graph of the function $y = f(x-h)+k$ represents a horizontal and a vertical shift of the graph of the function $y = f(x)$.

Use a graphing calculator to graph each pair of functions on the same set of axes. Then discuss the differences between the graphs of each pair of functions.

60. $y = 2x^2$ and $y = -3x^2$

61. $y = 4x^2$ and $y = -x^2$

62. $y = x^2 + 5$ and $y = (x-1)^2$

63. $y = (x+1)^2$ and $y = x^2 - 4$

64. $y = 2(x+3)^2 - 4$ and $y = 2x^2 + 3$

65. $y = -3(x-2)^2 + 1$ and $y = -x^2 + 1$

60.

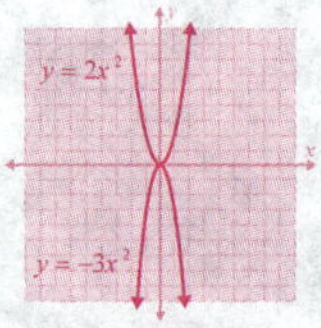

61.

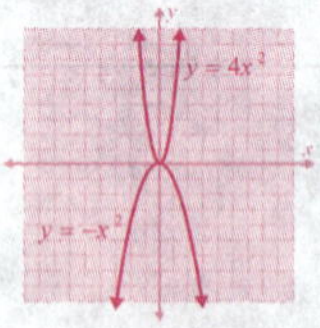

62.

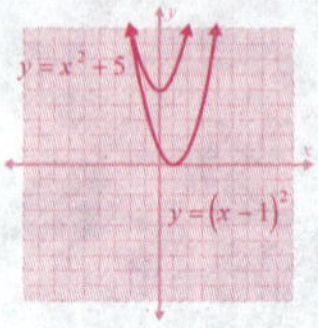

63.

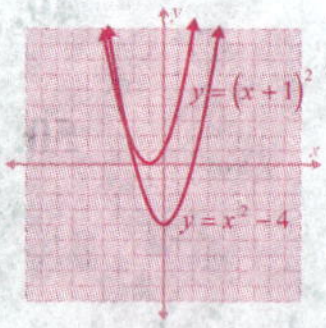

64.

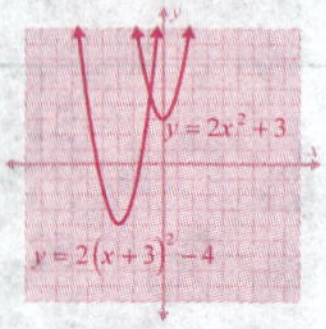

65.

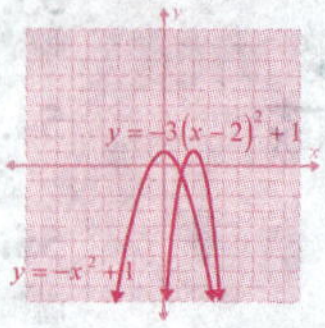

18.2 Parabolas as Conic Sections

A double circular cone is a three-dimensional figure that can be generated by choosing one point on a line and rotating the line in a circular fashion about this point. (You can visualize a double cone by holding a meter stick (or a yard stick) with two fingers and rotating the stick so that each end moves in a circle. Note that the double cone has a top portion and a bottom portion. If a plane intersects a double cone, the intersections will be curves (or a single point) on the plane. (See Figure 1.) These curves are called **conic sections**. The four conic sections we will discuss in this chapter are circles, ellipses, parabolas, and hyperbolas.

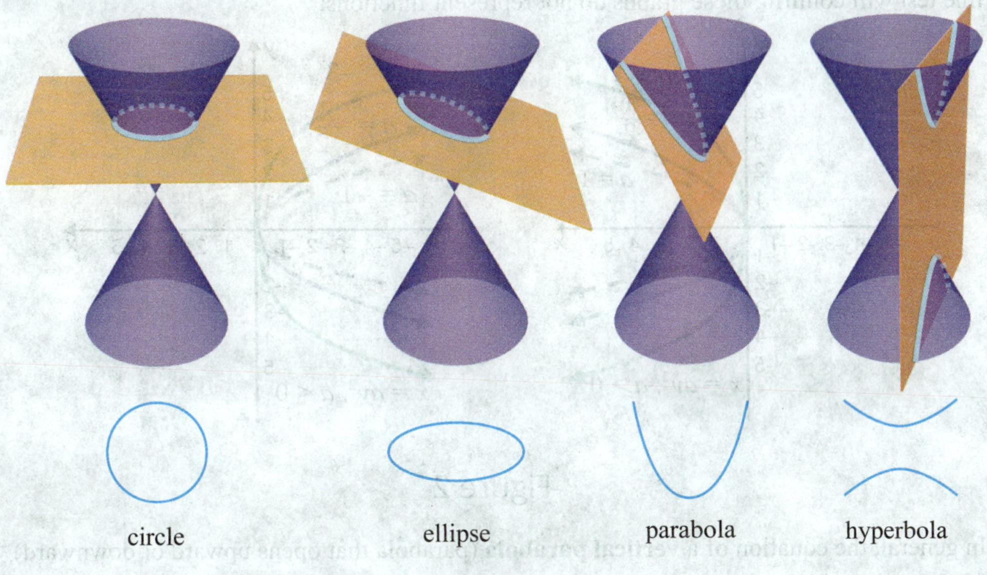

Figure 1

The corresponding algebraic equations for these conic sections are called quadratic equations because they are second-degree in x and/or y. Careful examination of these equations can show exactly what type of curve it represents and where the curve is located with respect to a Cartesian coordinate system. The technique is similar to that used in Chapter 10 in discussing lines.

By looking at a linear equation, you can identify

a. the **slope** of the line,

b. the ***y*-intercept** of the line, and

c. **points** on the line.

By looking at a quadratic equation, you will be able to determine if the graph is

a. a **circle** and identify the center and radius,

b. an **ellipse** and identify the center and intercepts,

c. a **parabola** and identify the vertex and line of symmetry, or

d. a **hyperbola** and identify the vertices and asymptotes.

Objectives

A. Graph horizontal parabolas and find their vertices, y-intercepts, and lines of symmetry.

B. Use a graphing calculator to graph horizontal parabolas.

Note

The study of geometry with algebraic equations is called analytic geometry. If you continue your studies in mathematics, you will find that analytic geometry can be applied in three dimensions as well as in two dimensions. Related figures would be spheres, ellipsoids (football shapes), and paraboloids (the reflective surfaces of telescopes).

A Horizontal Parabolas

As discussed in Section 16.6, the equations of **quadratic functions** are of the basic form $y = ax^2$, and the corresponding graphs are parabolas with vertex at the origin. Not all parabolas are functions. Parabolas that open upward or downward are functions, but those that open to the left or to the right are not functions.

The basic form for equations of parabolas that open left or right with vertex at the origin is $x = ay^2$, and several graphs of equations of this type are shown in Figure 2. As the vertical line test will confirm, these graphs do not represent functions.

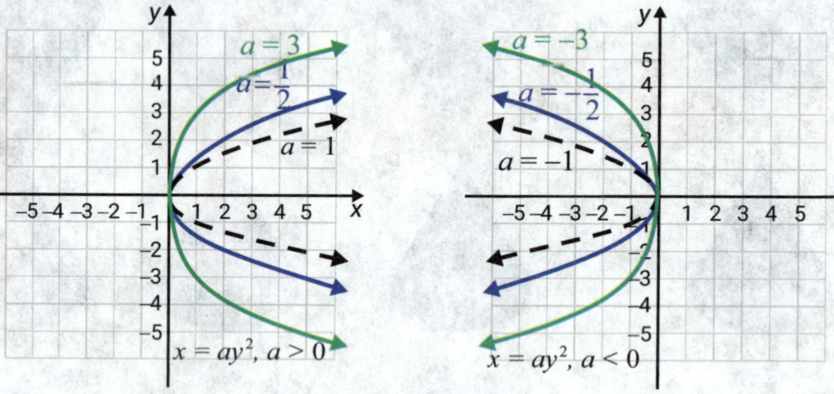

Figure 2

In general, the equation of a **vertical parabola** (parabola that opens upward or downward) can be written in the form

$$y = ax^2 + bx + c \quad \text{or} \quad y = a(x-h)^2 + k, \quad \text{where } a \neq 0.$$

A parabola opens downward if $a < 0$ or upward if $a > 0$ and has its vertex at (h, k). The line $x = h$ is the line of symmetry.

By exchanging the roles of x and y, the equations of **horizontal parabolas** (parabolas that open to the left or right) can be written in the following form.

Equations of Horizontal Parabolas

Equations of horizontal parabolas (parabolas that open to the left or right) can be written in the form

$$\boldsymbol{x = ay^2 + by + c} \quad \text{or} \quad \boldsymbol{x = a(y-k)^2 + h}, \quad \text{where } a \neq 0.$$

The parabola opens to the left if $a < 0$ and to the right if $a > 0$.

The vertex is at (h, k).

The line $y = k$ is the line of symmetry.

DEFINITION

18.2 Exercises

Concept Check

1. circle, ellipse, parabola, hyperbola
2. left, right
3. vertical
4. horizontal
5. 0, y-intercepts
6. (h,k)
7. True
8. False; Parabolas open down if $a < 0$ and open up if $a > 0$.
9. False; $y = k$ is the line of symmetry for a horizontal parabola.

Practice

1. **a.** $(4, 0)$ **b.** None **c.** $y = 0$

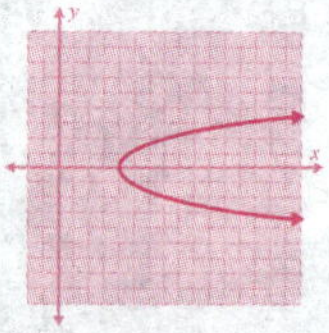

2. **a.** $(-5, 0)$ **b.** $(0, \sqrt{5}), (0, -\sqrt{5})$ **c.** $y = 0$

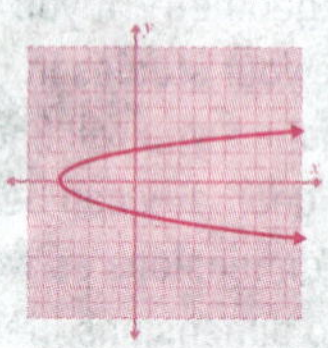

3. **a.** $(0, -3)$ **b.** $(0, -3)$ **c.** $x = 0$

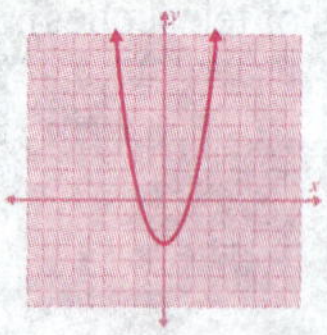

4. **a.** $(0, 2)$ **b.** $(0, 2)$ **c.** $x = 0$

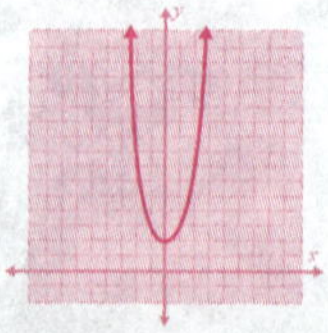

In a manner similar to the discussion in Section 18.1, adding h to the right hand side and replacing y with $(y-k)$ in the equation $x = ay^2$ creates the equation $x = a(y-k)^2 + h$ whose graph is a horizontal translation of h units and a vertical translation of k units of the graph of $x = ay^2$.

For example, the graph of $x = 2(y-3)^2 - 1$ is shown in Figure 3 along with a table of y- and x-values in Table 1. The vertex is at $(h, k) = (-1, 3)$, and the line of symmetry is $y = 3$. The y-values in the table are chosen on each side of the line of symmetry.

y	$2(y-3)^2 - 1 = x$
3	$2(3-3)^2 - 1 = -1$
4	$2(4-3)^2 - 1 = 1$
2	$2(2-3)^2 - 1 = 1$
5	$2(5-3)^2 - 1 = 7$
1	$2(1-3)^2 - 1 = 7$

Table 1

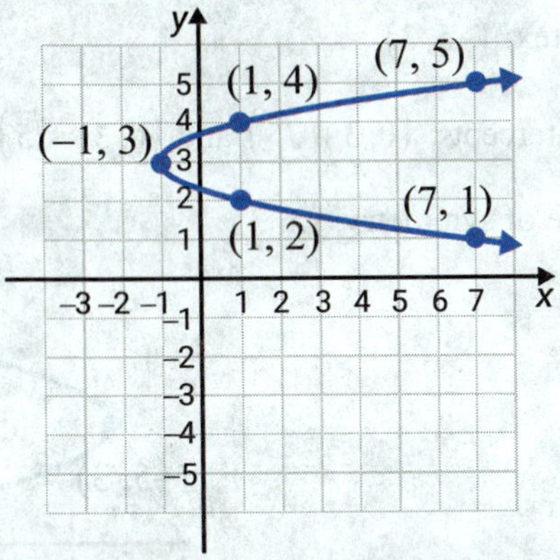

Figure 3

5. **a.** $(3, 0)$ **b.** None
c. $y = 0$

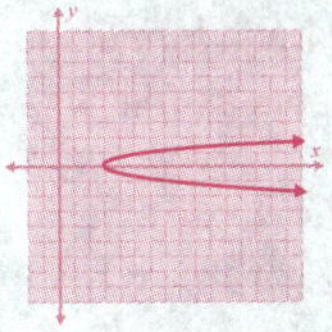

6. **a.** $(1, 0)$ **b.** None
c. $y = 0$

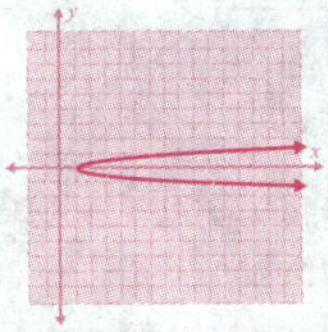

7. **a.** $(0, 3)$ **b.** $(0, 3)$
c. $y = 3$

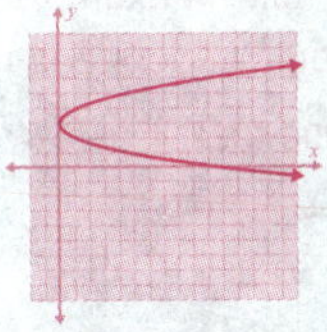

8. **a.** $(0, 2)$ **b.** $(0, 2)$
c. $y = 2$

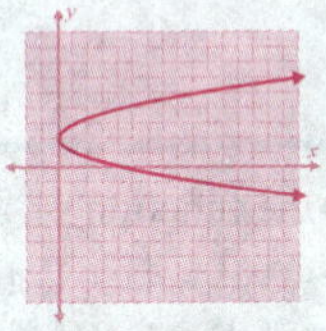

The graph of an equation of the form $x = ay^2 + by + c$ can be found by completing the square (as in Section 16.7) and writing the equation in the form

$$x = a(y-k)^2 + h.$$

Also, by setting $x = 0$ and solving the following quadratic equation

$$0 = ay^2 + by + c,$$

we can determine the **y-intercepts** (the points, if any, where the graph intersects the y-axis).

Example 1 Sketching Horizontal Parabolas

For $x = y^2 - 6y + 4$, find the vertex, the y-intercepts, and the line of symmetry. Then sketch the graph.

Solution

To find the vertex, complete the square.

$$x = y^2 - 6y + 4$$

$$x = (y^2 - 6y + 9) - 9 + 4$$

Complete the square: $\frac{1}{2}(-6) = -3$ and $(-3)^2 = 9$.
Therefore, add $0 = 9 - 9$.

$$x = (y-3)^2 - 5$$

The vertex is at $(-5, 3)$.

1. For $x = y^2 - 4y + 1$, find the vertex, the y-intercepts, and the line of symmetry. Then sketch the graph.

9. **a.** $(4, -2)$ **b.** None
c. $y = -2$

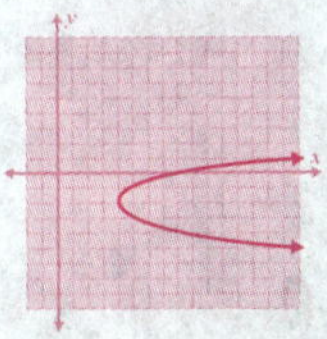

Teaching Note

In Example 1, you could also use the quadratic formula to find the y-intercepts as follows.

$$y^2-6y+4=x$$
$$y^2-6y+4=0$$
$$y=\frac{6\pm\sqrt{(-6)^2-4(1)(4)}}{2}$$
$$y=\frac{6\pm\sqrt{20}}{2}$$
$$y=3\pm\sqrt{5}$$

You will probably want to remind your students that they can use their calculators to estimate the values of the intercepts when radicals are involved.

For example, in Example 1 they should understand how to determine that $3+\sqrt{5}\approx 5.2$ and $3-\sqrt{5}\approx 0.8$.

To find the y-intercepts, let $x=0$ and use the square root method as follows.

$$(y-3)^2-5=0$$
$$(y-3)^2=5$$
$$y-3=\pm\sqrt{5}$$
$$y=3\pm\sqrt{5}$$

Since $a=1$, the parabola has the same shape as $x=y^2$.

Vertex: $(-5,3)$

y-intercepts: $(0,3+\sqrt{5})$ and $(0,3-\sqrt{5})$

Line of symmetry: $y=3$

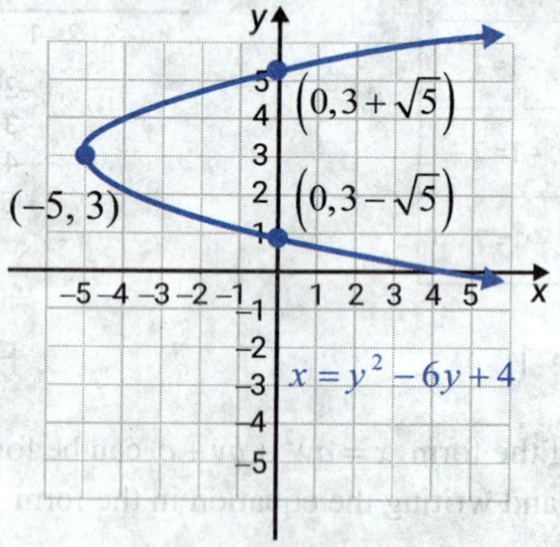

Now work margin exercise 1.

2. For the function $x=-3y^2+4y-1$, find the vertex, the y-intercepts, and the line of symmetry. Then sketch the graph.

Example 2 Sketching Horizontal Parabolas

For $x=-2y^2-4y+6$, find the vertex, the y-intercepts, and the line of symmetry. Then sketch the graph.

Solution

To find the vertex, complete the square.

$$x=-2y^2-4y+6$$
$$x=-2(y^2+2y)+6$$
$$x=-2(y^2+2y+1-1)+6$$

Complete the square: $\frac{1}{2}(2)=1$ and $1^2=1$.
Therefore, add $0=1-1$ inside the parentheses.

$$x=-2(y^2+2y+1)+2+6$$

Multiply $-2(-1)=2$ and write this term outside the parentheses

$$x=-2(y^2+2y+1)+8$$
$$x=-2(y+1)^2+8$$

The vertex is at $(8,-1)$.

10. a. $(-3,5)$ **b.** $(0,5+\sqrt{3}),(0,5-\sqrt{3})$

c. $y=5$

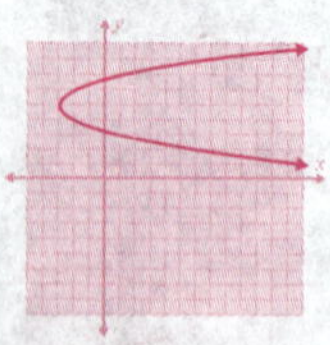

11. a. $(1,-1)$ **b.** $(0,0)$

c. $x=1$

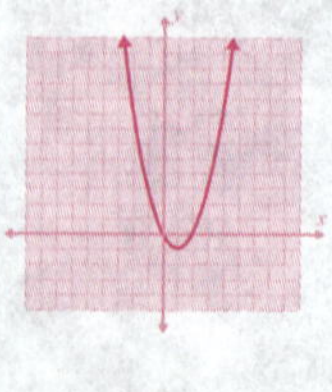

To find the y-intercepts, let $x = 0$ and factor.

$$-2y^2 - 4y + 6 = 0$$
$$-2(y^2 + 2y - 3) = 0$$
$$-2(y+3)(y-1) = 0$$
$$y = -3 \quad \text{or} \quad y = 1$$

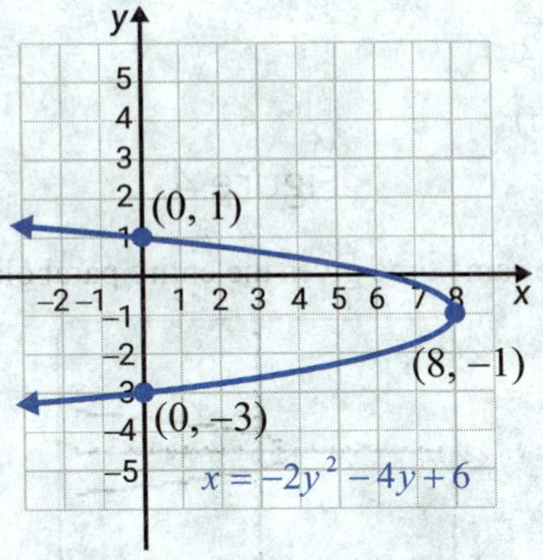

Since $a = -2$, the graph opens to the left and is slightly narrower than $x = y^2$.

Vertex: $(8, -1)$

y-intercepts: $(0, -3)$ and $(0, 1)$

Line of symmetry: $y = -1$

Now work margin exercise 2.

B Using a Graphing Calculator to Graph Horizontal Parabolas

Horizontal parabolas are not functions and the graphing calculator is designed to graph only functions. Therefore, to graph a horizontal parabola, solve the given equation for y. For example, the graph of the equation $x = y^2 - 2$ is a horizontal parabola opening right with its vertex at $(-2, 0)$. To graph this equation using a graphing calculator, we must first solve for y since equations must be entered with y (to the first power) on the left-hand side. By using the square root property, two functions designated as y_1 and y_2 can be found as follows.

$$x = y^2 - 2 \qquad \text{First solve for } y^2.$$
$$y^2 = x + 2$$
$$\begin{cases} y_1 = \sqrt{x+2} \\ y_2 = -\sqrt{x+2} \end{cases} \qquad \text{Solving for } y \text{ gives two equations that represent two functions.}$$

12. **a.** $(3, 5)$ **b.** $(0, 14)$
c. $x = 3$

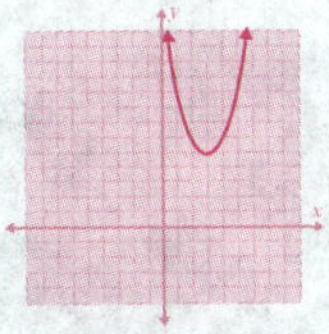

13. **a.** $(0, -2)$ **b.** $(0, -2)$
c. $y = -2$

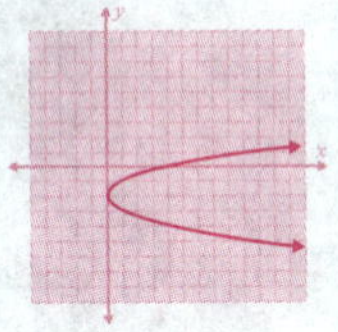

14. **a.** $(0, 4)$ **b.** $(0, 4)$
c. $y = 4$

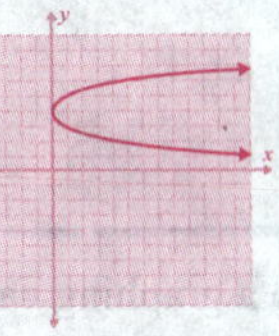

15. **a.** $(0, 5)$ **b.** $(0, 5)$
c. $y = 5$

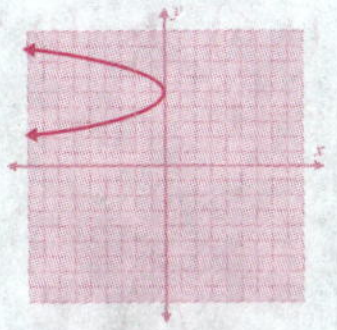

16. **a.** $(0, -3)$ **b.** $(0, -3)$
c. $y = -3$

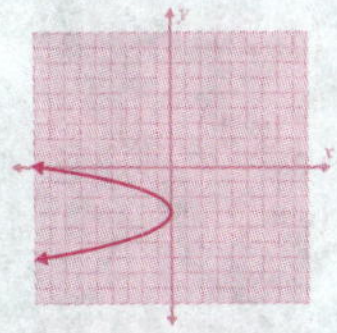

17. **a.** $(-3, -4)$ **b.** $(0, 5)$
c. $x = -3$

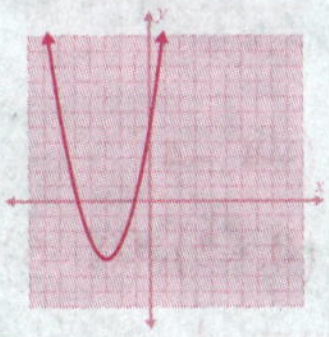

18. **a.** $(-2, 2)$ **b.** $(0, 6)$

c. $x = -2$

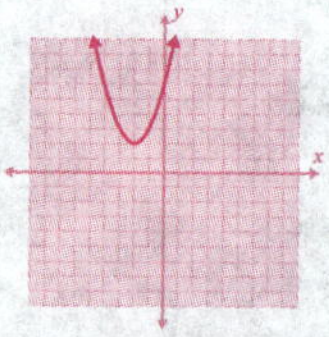

19. **a.** $(-2, 9)$ **b.** $(0, 5)$

c. $x = -2$

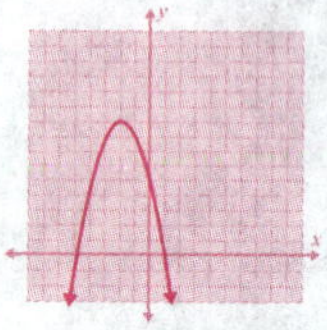

Graphing these equations individually gives the upper and lower halves of the parabola.

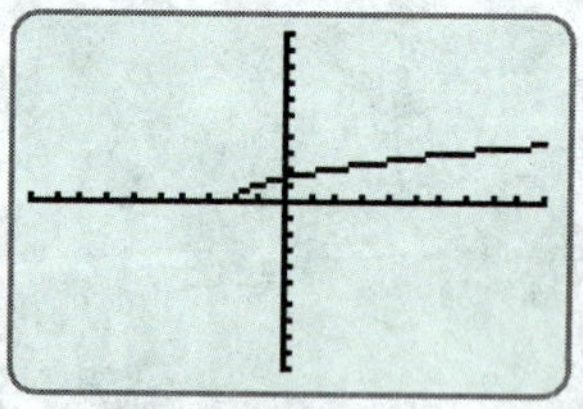

$y_1 = \sqrt{x+2}$

(upper half)

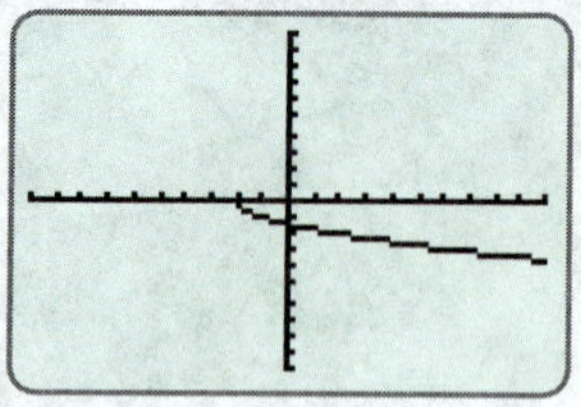

$y_2 = -\sqrt{x+2}$

(lower half)

Figure 4

Graphing both halves at the same time gives the entire parabola $x = y^2 - 2$.

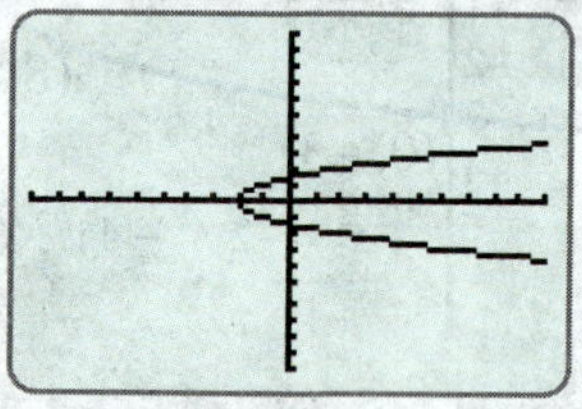

Figure 5

3. Use a graphing calculator to graph the function $x = y^2 + 6y + 10$ and estimate the y-intercepts.

20. **a.** $(1, 6)$ **b.** $(0, 5)$

c. $x = 1$

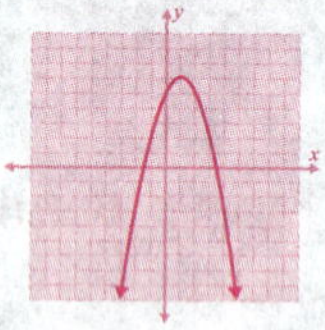

21. **a.** $(1, 2)$

b. $(0, 1), (0, 3)$

c. $y = 2$

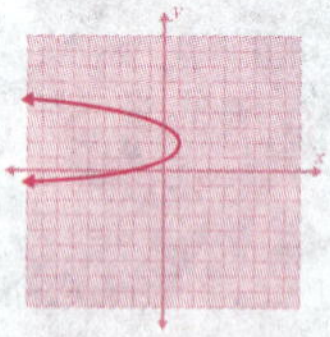

22. **a.** $(-4, -4)$

b. $(0, -2), (0, -6)$

c. $y = -4$

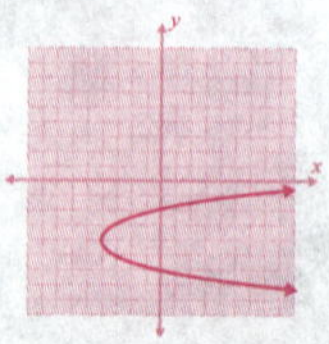

Example 3 Using a Graphing Calculator to Graph Horizontal Parabolas

Use a graphing calculator to graph the horizontal parabola $x = y^2 - 4y + 5$. Estimate the y-intercepts using the zoom and trace features of the calculator.

Solution

To solve for y, complete the square and use the square root property as follows.

$$x = y^2 - 4y + 5$$

$$y^2 - 4y = x - 5$$

$$y^2 - 4y + 4 = x - 5 + 4$$ Complete the square: $\frac{1}{2}(-4) = -2$ and $(-2)^2 = 4$. Therefore, add 4 to both sides.

$$(y-2)^2 = x - 1$$

$$y - 2 = \pm\sqrt{x-1}$$ Use the square root property.

$$\begin{cases} y_1 = 2 + \sqrt{x-1} \\ y_2 = 2 - \sqrt{x-1} \end{cases}$$ Graph both of these equations.

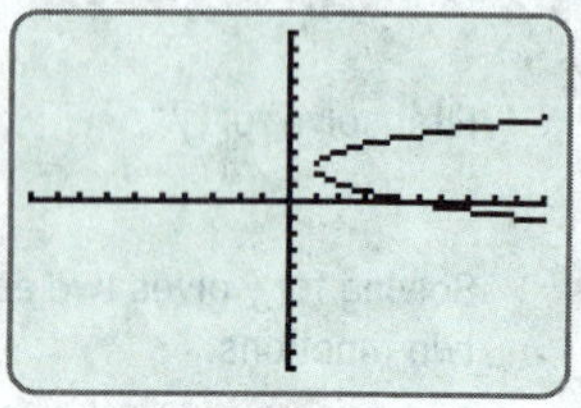

From this graph, we can determine that there are no y-intercepts.

Now work margin exercise 3.

Margin Exercise Answers

1. vertex: $(-3, 2)$ y-intercepts: $(0, 2-\sqrt{3})$, $(0, 2+\sqrt{3})$ line of symmetry: $y = 2$;

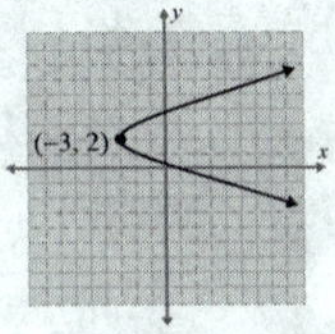

2. vertex: $\left(\frac{1}{3}, \frac{2}{3}\right)$; y-intercepts: $\left(0, \frac{1}{3}\right)$ and $(0, 1)$ line of symmetry: $y = \frac{2}{3}$

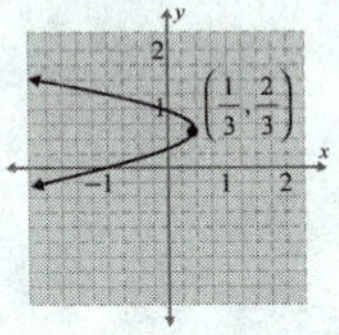

3.

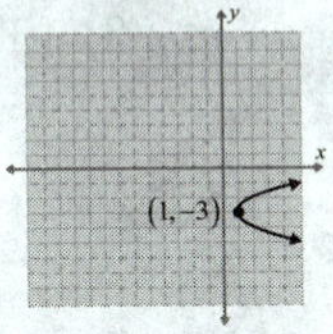

y-intercepts: none

18.2 Exercises

Concept Check

Fill-in-the-Blank. Complete each sentence using information found in this section.

1. Four conic sections are the ______, ______, ______, and ______.
2. The basic form of a parabola that opens ______ or ______ is $x = ay^2$.
3. The equations of ______ parabolas can be written in the form $y = a(x-h)^2 + k$, where $a \neq 0$.
4. The equations of ______ parabolas can be written in the form $x = a(y-h)^2 + k$, where $a \neq 0$.
5. By setting $x =$ ____ and solving $0 = ay^2 + by + c$, we can determine the ______.
6. The vertex of a parabola is at the point ______.

True/False. Determine whether each statement is true or false. If a statement is false, explain how it can be changed so the statement will be true. (**Note:** There may be more than one acceptable change.)

7. Not all parabolas are functions.
8. Parabolas open down if $a > 0$ and open up if $a < 0$.
9. The line $x = h$ is the line of symmetry for a horizontal parabola.

Practice

For the given equations, **a.** find the vertex, **b.** find the y-intercept, **c.** find the line of symmetry, and **d.** sketch the graph. See Examples 1 and 2.

1. $x = y^2 + 4$

2. $x = y^2 - 5$

3. $y + 3 = x^2$

4. $y - 2 = x^2$

23. a. $\left(-\frac{1}{4}, -\frac{9}{8}\right)$

b. $(0, -1)$

c. $x = -\frac{1}{4}$

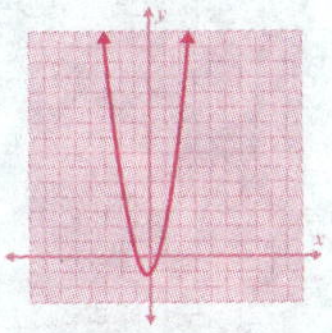

24. a. $\left(\frac{1}{4}, \frac{25}{8}\right)$ **b.** $(0, 3)$

c. $x = \frac{1}{4}$

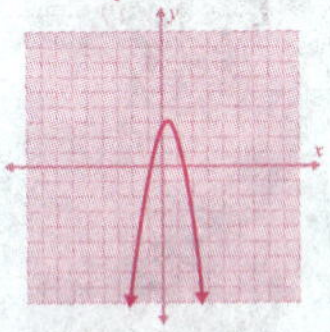

25. a. $\left(\frac{9}{8}, \frac{5}{4}\right)$

b. $\left(0, \frac{1}{2}\right), (0, 2)$

c. $y = \frac{5}{4}$

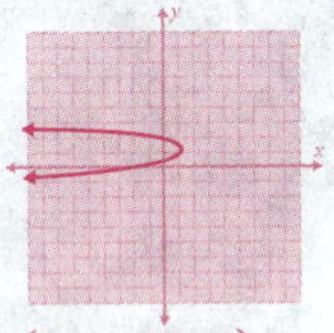

26. a. $\left(-\frac{1}{12}, -\frac{5}{6}\right)$

b. $\left(0, -\frac{2}{3}\right), (0, -1)$

c. $y = -\frac{5}{6}$

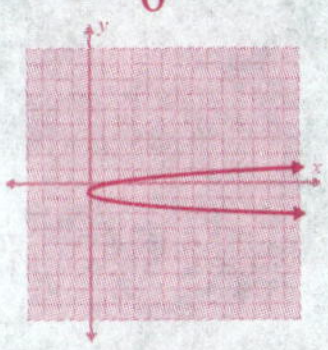

27. **a.** $(-8,-1)$

b. $\left(0,-1+\dfrac{2\sqrt{6}}{3}\right)$, $\left(0,-1-\dfrac{2\sqrt{6}}{3}\right)$

c. $y=-1$

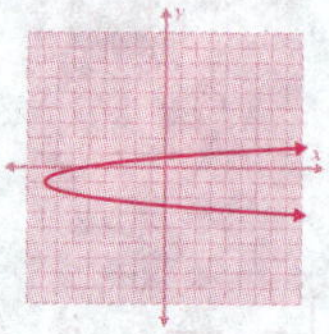

28. **a.** $\left(-16,\dfrac{1}{2}\right)$

b. $\left(0,\dfrac{5}{2}\right),\left(0,-\dfrac{3}{2}\right)$

c. $y=\dfrac{1}{2}$

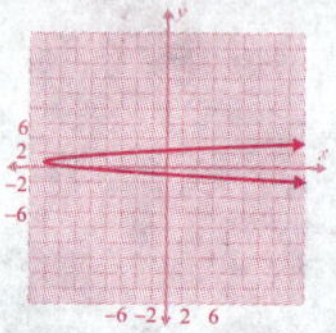

29. **a.** $\left(\dfrac{3}{2},0\right)$ **b.** $(0,9)$

c. $x=\dfrac{3}{2}$

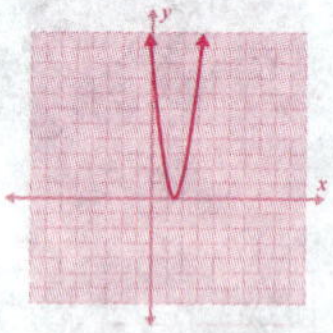

30. **a.** $(1,7)$ **b.** $(0,2)$

c. $x=1$

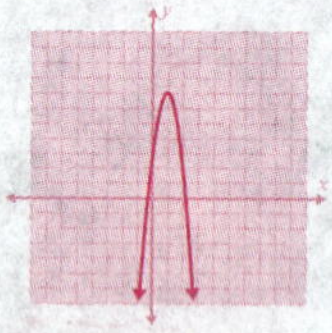

31. $(0,1.225)$, $(0,-1.225)$

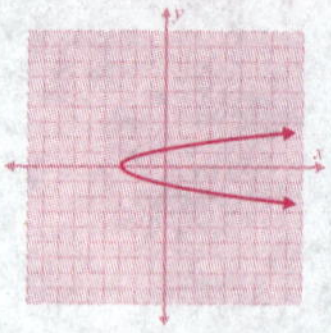

5. $x=2y^2+3$

6. $x=3y^2+1$

7. $x=(y-3)^2$

8. $x=(y-2)^2$

9. $x-4=(y+2)^2$

10. $x+3=(y-5)^2$

11. $y+1=(x-1)^2$

12. $y-5=(x-3)^2$

13. $x=y^2+4y+4$

14. $x=y^2-8y+16$

15. $x=-y^2+10y-25$

16. $x=-y^2-6y-9$

17. $y=x^2+6x+5$

18. $y=x^2+4x+6$

19. $y=-x^2-4x+5$

20. $y=-x^2+2x+5$

21. $x=-y^2+4y-3$

22. $x=y^2+8y+12$

23. $y=2x^2+x-1$

24. $y=-2x^2+x+3$

25. $x=-2y^2+5y-2$

26. $x=3y^2+5y+2$

27. $x=3y^2+6y-5$

28. $x=4y^2-4y-15$

29. $y=4x^2-12x+9$

30. $y=-5x^2+10x+2$

Use a graphing calculator to graph each of the parabolas. Use the trace and zoom features of the calculator to estimate the *y*-intercepts of the parabola. See Example 3. (See Section 10.5 to review the trace and zoom features on a TI-84 Plus graphing calculator.)

31. $x=2y^2-3$

32. $x=-3y^2+1$

33. $x=-y^2+2y$

34. $x=y^2-5y$

35. $x=2y^2+y+1$

36. $x=-y^2-4y+1$

37. $x=4y^2+8y-7$

38. $x=3y^2+3y+2$

39. $x=-2y^2+4y+3$

40. $x=-5y^2-10y-4$

Use your knowledge of parabolas and equations to match the equation with the graph.

41. $x = 2(y-3)^2 + 3$

42. $x = -(y+1)^2 + 5$

43. $x = -y^2 - 1$

44. $x = y^2 - 6$

a.

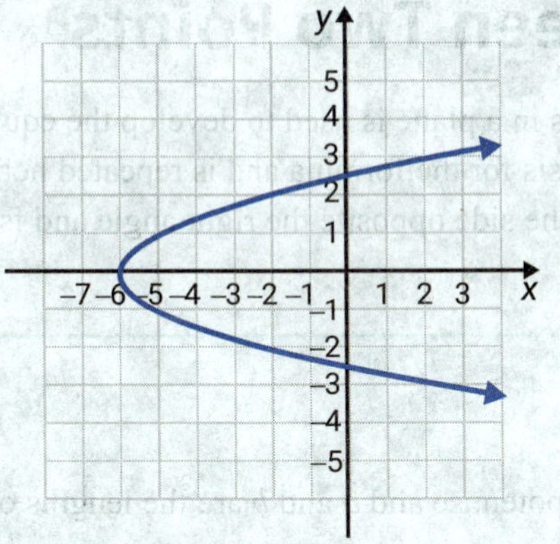

c.

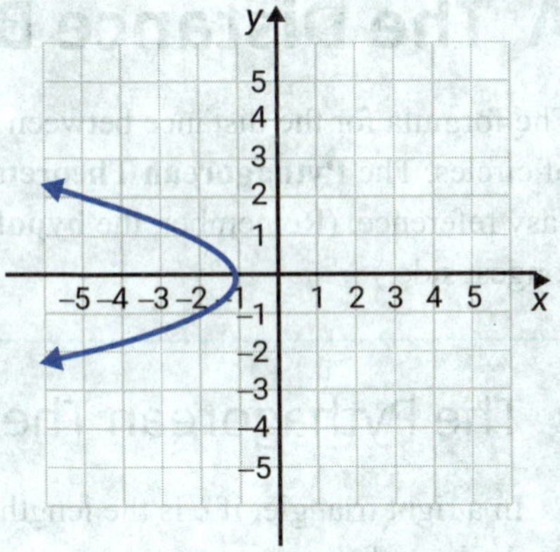

b.

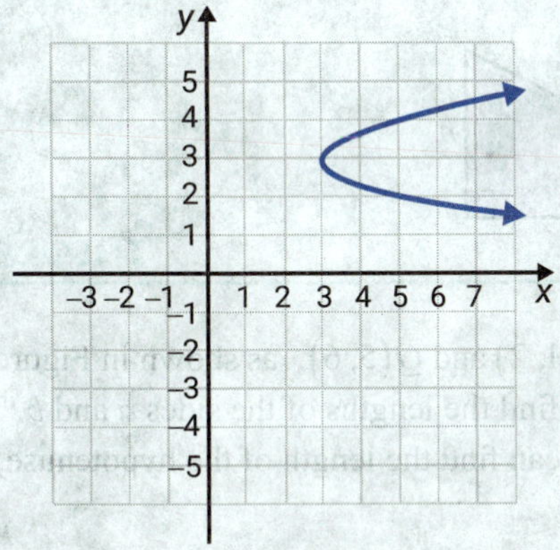

d.

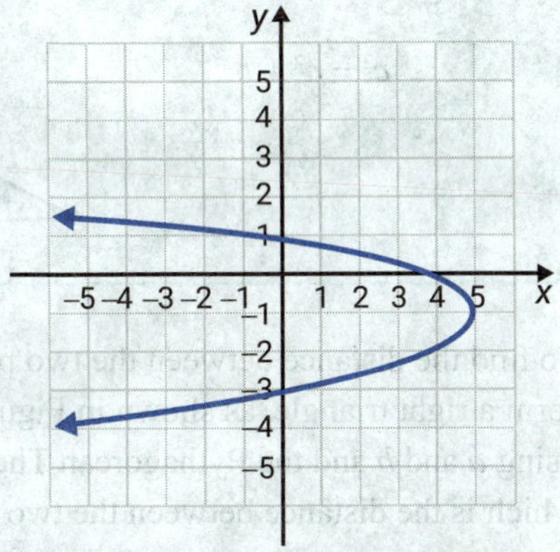

Writing & Thinking

45. For $x = ay^2 + by + c$ we know that the graph of the parabola opens to the right if $a > 0$ and to the left if $a < 0$. Discuss which values of a will cause the parabola to be wider and which will cause it to be narrower than the graph of $x = y^2$.

32. $(0, 0.577), (0, -0.577)$

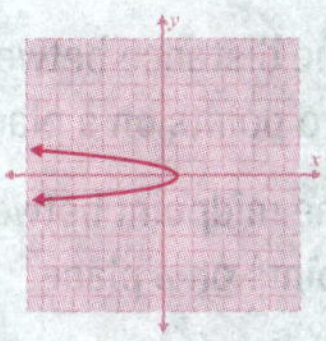

33. $(0, 2), (0, 0)$

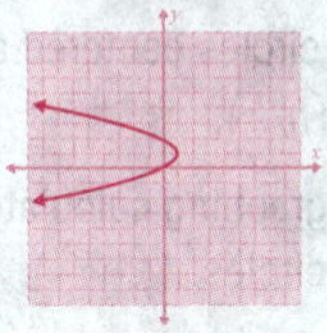

34. $(0, 5), (0, 0)$

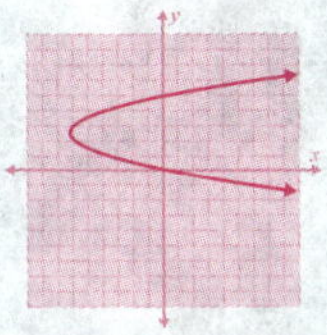

35. No y-intercept

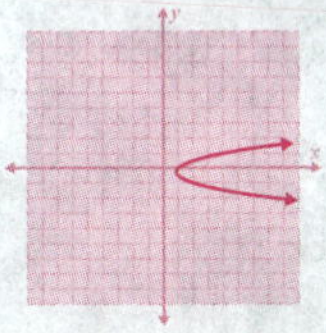

36. $(0, 0.236), (0, -4.236)$

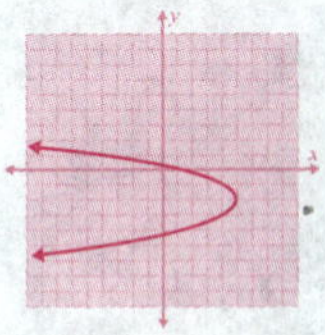

37. $(0, 0.658), (0, -2.658)$

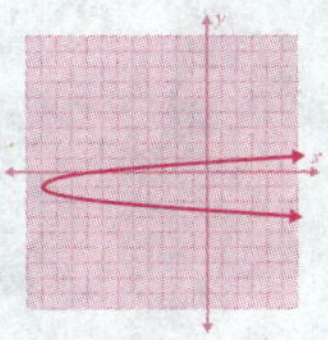

38. No y-intercept

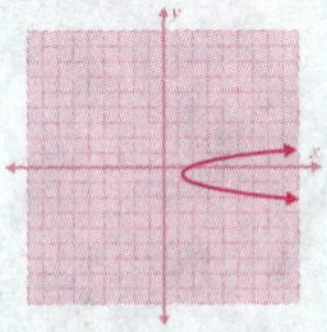

39. $(0, 2.581), (0, -0.581)$

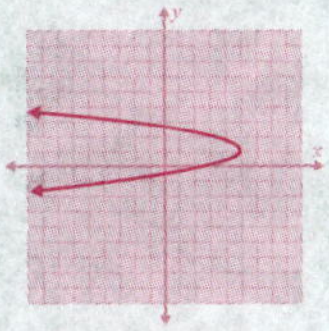

40. $(0, -0.553), (0, -1.447)$

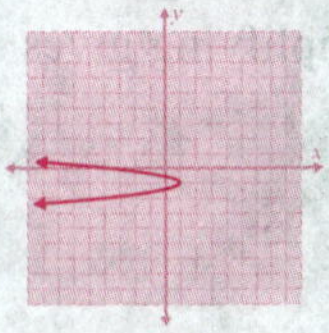

41. b

42. d

43. c

44. a

Writing & Thinking

45. If $|a| < 1$, then the parabola will be wider than the graph of $x = y^2$. If $|a| > 1$, then the parabola will be narrower than the graph of $x = y^2$.

Objectives

A. Find the distance between any two points on a plane.

B. Find the midpoint between two points on a plane.

C. Write the equation of a circle given its center and radius.

D. Graph circles centered at the point (h, k).

E. Use a graphing calculator to graph circles.

18.3 Distance Formula, Midpoint Formula, and Circles

A The Distance Between Two Points

The formula for the distance between two points in a plane is used to develop the equations of circles. The **Pythagorean Theorem** is the basis for the formula and is repeated here for easy reference. (Remember, the **hypotenuse** is the side opposite the right angle and is the longest side.)

The Pythagorean Theorem

In a right triangle, if c is the length of the hypotenuse and a and b are the lengths of the legs, then

$$c^2 = a^2 + b^2.$$

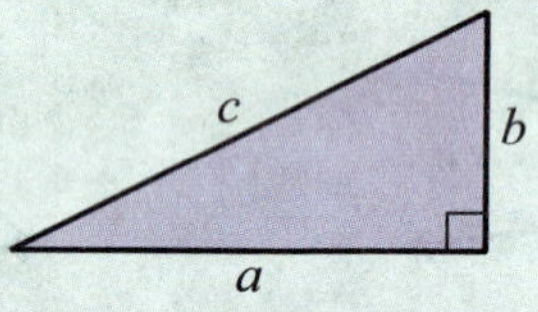

THEOREM

To find the distance between the two points $P(-1, 2)$ and $Q(5, 6)$, as shown in Figure 1a, form a right triangle, as shown in Figure 1b, and find the lengths of the sides a and b. Then using a and b and the Pythagorean Theorem, we can find the length of the hypotenuse, which is the distance between the two points.

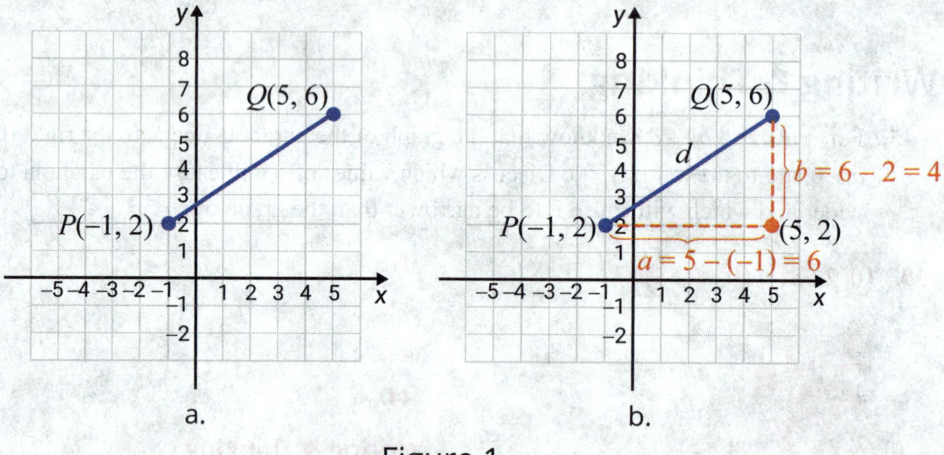

Figure 1

From Figure 1b, we know $a = 6$ and $b = 4$. Substituting these values into the equation of the Pythagorean Theorem (using d for distance in place of c) and simplifying gives the following.

$$d^2 = a^2 + b^2 = 6^2 + 4^2 = 36 + 16 = 52$$

Taking the positive square root of both sides gives the distance *d*.

$$d = \sqrt{52} = 2\sqrt{13} \quad (2\sqrt{13} \approx 7.2111 \text{ estimated with a calculator})$$

In general, for points $P(x_1, y_1)$ and $Q(x_2, y_2)$ in a plane, with $a = |x_2 - x_1|$ and $b = |y_2 - y_1|$, the Pythagorean Theorem produces the following distance formula. (See Figure 2.)

The Distance Formula

For two points $P(x_1, y_1)$ and $Q(x_2, y_2)$ in a plane, the distance between the points is

$$d = \sqrt{(x_2 - x_1)^2 + (y_2 - y_1)^2}.$$

FORMULA

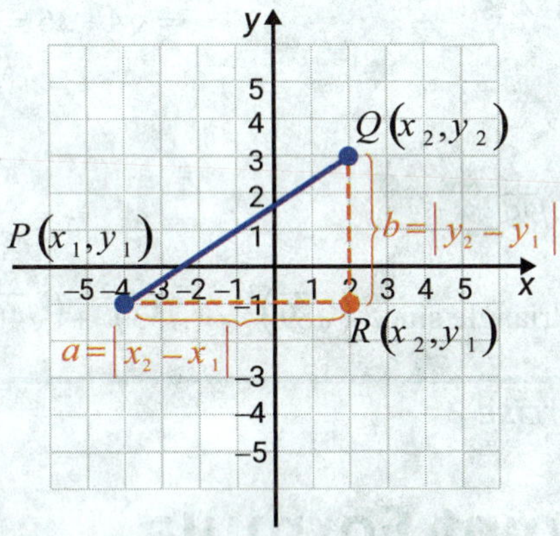

$$d = \sqrt{(x_2 - x_1)^2 + (y_2 - y_1)^2}$$

Figure 2

Notice in Figure 2 that the calculations for *a* and *b* involve absolute values. These absolute values guarantee nonnegative values for *a* and *b* to represent the lengths of the legs. In the distance formula, the absolute values are disregarded because $x_2 - x_1$ and $y_2 - y_1$ are squared. In the calculation of *d*, be sure to add the squares before taking the square root.

Math Tip

When using the distance formula, the order of the points (and the order of the subtraction) does not affect the final answer due to the squaring step. The only thing that matters is that you find the difference of the *x*-values and the difference of the *y*-values separately. For example, consider the two points (1,2) and (–5,3).

Order of subtraction:

$$a^2 = (x_1 - x_2)^2 = (2-3)^2$$
$$= (3-2)^2 = (x_2 - x_1)^2 = 1$$
$$b^2 = (y_1 - y_2)^2 = (1-(-5))^2$$
$$= (-5-1)^2 = (y_2 - y_1)^2$$
$$= 36$$

Once you have these values, the commutative property of addition verifies that the order of the points does not matter.

Order of points:

$$\sqrt{a^2 + b^2} = \sqrt{1+36}$$
$$= \sqrt{36+1}$$
$$= \sqrt{b^2 + a^2} = \sqrt{37}$$

So no matter which order you choose, the distance will be $\sqrt{37}$.

Example 1 Using the Distance Formula

Find the distance between the two points $(3, 4)$ and $(-2, 7)$.

Solution

$$d = \sqrt{[3-(-2)]^2 + (4-7)^2}$$
$$= \sqrt{5^2 + (-3)^2} = \sqrt{25+9} = \sqrt{34}$$

Now work margin exercise 1.

1. Find the distance between the two points $(-1, -6)$ and $(3, -3)$.

2. Use the distance formula and the Pythagorean Theorem to determine whether the triangle with vertices at $D(-1, 4)$, $E(-4, -2)$, and $F(4, -6)$ is a right triangle.

Example 2 Determining If a Triangle Is a Right Triangle

Use the distance formula (3 times) and the Pythagorean Theorem to determine whether the triangle with vertices at $A(-5, -1)$, $B(2, 1)$, and $C(0, 7)$ is a right triangle.

Solution

Find the lengths of the three line segments $\overline{AB}$, $\overline{AC}$, and $\overline{BC}$, and decide whether the Pythagorean Theorem is satisfied.

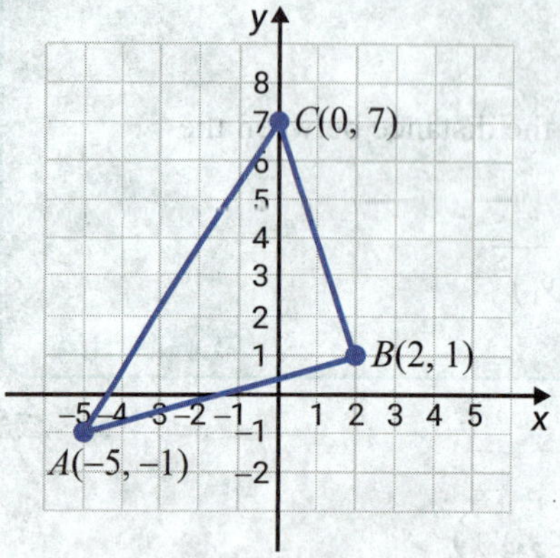

$$AB = \sqrt{(-5-2)^2 + (-1-1)^2} = \sqrt{(-7)^2 + (-2)^2}$$
$$= \sqrt{49+4} = \sqrt{53}$$
$$AC = \sqrt{(-5-0)^2 + (-1-7)^2} = \sqrt{(-5)^2 + (-8)^2}$$
$$= \sqrt{25+64} = \sqrt{89}$$
$$BC = \sqrt{(2-0)^2 + (1-7)^2} = \sqrt{(2)^2 + (-6)^2}$$
$$= \sqrt{4+36} = \sqrt{40}$$

The longest side is $AC = \sqrt{89}$.

The triangle is not a right triangle since $\left(\sqrt{89}\right)^2 \neq \left(\sqrt{53}\right)^2 + \left(\sqrt{40}\right)^2$ as $89 \neq 53 + 40$.

Now work margin exercise 2.

B The Midpoint Formula

Another useful formula that involves the coordinates of two points is the formula for finding the **midpoint** of the segment joining two points. These two points are called **endpoints** of the segment. The midpoint is found by **averaging** the corresponding coordinates of the endpoints.

Midpoint Formula

The formula for the **midpoint** between two points $P(x_1, y_1)$ and $Q(x_2, y_2)$ is

$$\left(\frac{x_1 + x_2}{2}, \frac{y_1 + y_2}{2}\right).$$

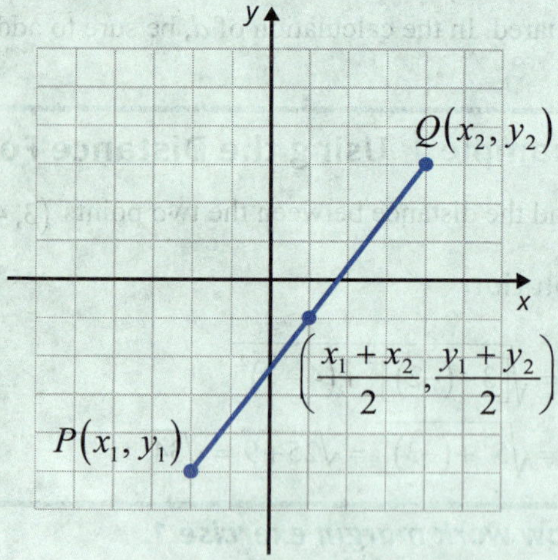

FORMULA

Example 3 Finding the Midpoint

Find the coordinates of the midpoint of the line segment joining the two points $P(-4, 6)$ and $Q(1, 2)$.

Solution

The midpoint is

$$\left(\frac{-4+1}{2}, \frac{6+2}{2}\right) = \left(-\frac{3}{2}, 4\right).$$

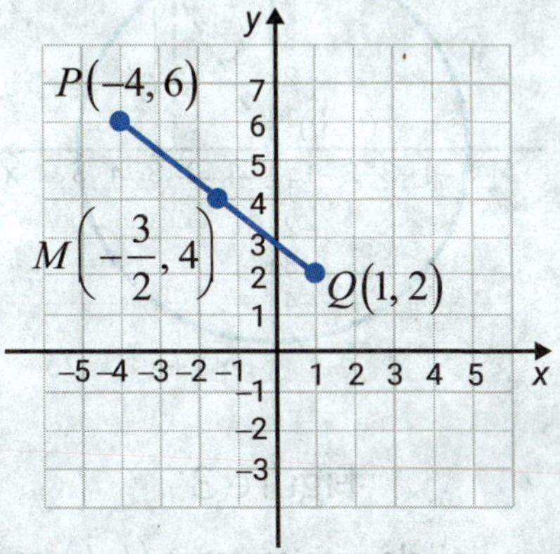

Now work margin exercise 3.

3. Find the coordinates of the midpoint of the line segment joining the two points $A(3,1)$ and $B(2,-7)$.

C Equations of Circles

Circles and the terms related to circles (center, radius, and diameter) are defined as follows.

Circle, Center, Radius, and Diameter

A **circle** is the set of all points in a plane that are a fixed distance from a fixed point.

The fixed point is called the **center** of the circle.

The distance from the center to any point on the circle is called the **radius** of the circle.

The distance from one point on the circle to another point on the circle measured through the center is called the **diameter** of the circle.

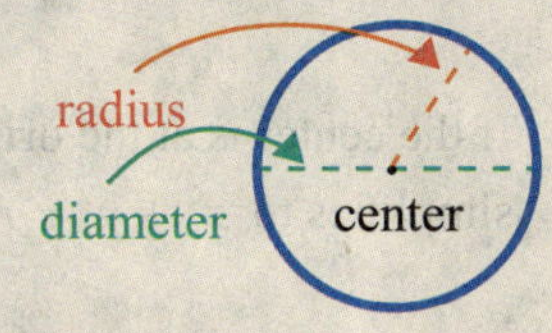

Note: The diameter is twice the length of the radius.

DEFINITION

The distance formula is used to find the equation of a circle. For example, to find the equation of the circle with its center at the origin $(0, 0)$ and radius 5, for any point on the circle (x, y) the distance from (x, y) to $(0, 0)$ must be 5. Therefore, using the distance formula, we get the following.

Teaching Note

While discussing terms related to circles, you might want to point out the dual use of some terms. "The" radius most often refers to a length, while "a" radius most often refers to a particular line segment. (Similarly with diameter.)

$$\sqrt{(x_2 - x_1)^2 + (y_2 - y_1)^2} = d$$
$$\sqrt{(x - 0)^2 + (y - 0)^2} = 5$$
$$\sqrt{x^2 + y^2} = 5$$
$$x^2 + y^2 = 25 \quad \text{Square both sides.}$$

Thus, as shown in Figure 3, all points on the circle satisfy the equation $x^2 + y^2 = 25$.

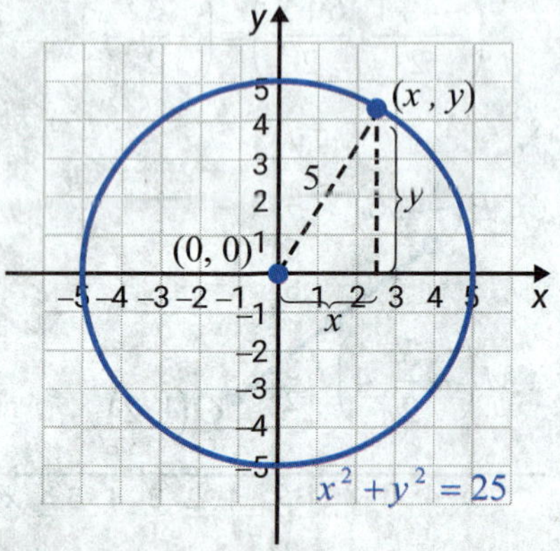

Figure 3

In general, any point (x, y) on a circle with center at (h, k) must satisfy the equation

$$\sqrt{(x - h)^2 + (y - k)^2} = r.$$

Squaring both sides of this equation gives the **standard form** for the equation of a circle.

$$(x - h)^2 + (y - k)^2 = r^2$$

Equation of a Circle

The equation of a circle with radius r and center at (h, k) is

$$(x - h)^2 + (y - k)^2 = r^2.$$

If the center is at the origin, $(0, 0)$, the equation simplifies to $x^2 + y^2 = r^2$.

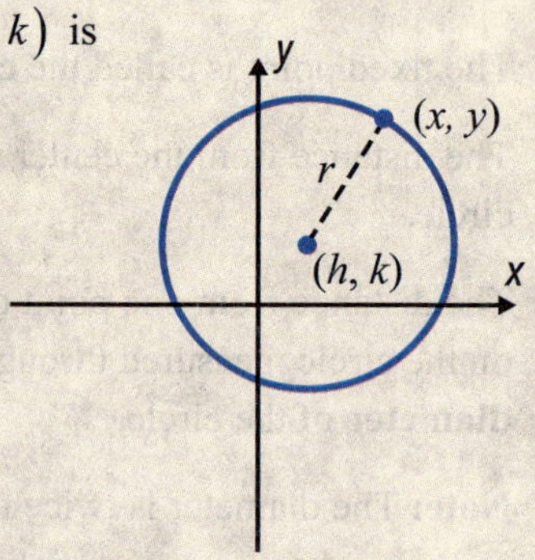

DEFINITION

Example 4 Finding the Equation of a Circle

Find the equation of the circle with its center at the origin and radius $\sqrt{3}$. Are the points $\left(\sqrt{2},1\right)$ and $(1,2)$ on the circle?

Solution

The equation of the circle is $x^2+y^2=3$.

To determine whether or not the points $\left(\sqrt{2},1\right)$ and $(1,2)$ are on the circle, substitute each of these points into the equation.

Substituting $\left(\sqrt{2},1\right)$ gives $\left(\sqrt{2}\right)^2+(1)^2=2+1=3$.

Substituting $(1,2)$ gives $(1)^2+(2)^2=1+4=5\neq 3$.

Therefore, $\left(\sqrt{2},1\right)$ is on the circle, but $(1,2)$ is not on the circle.

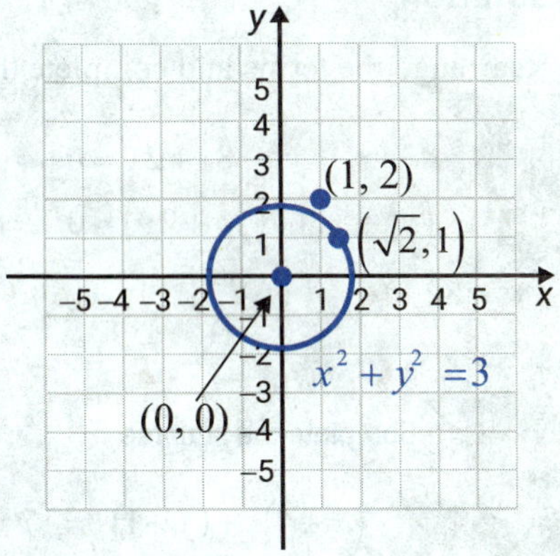

Now work margin exercise 4.

4. Find the equation of the circle with center at the origin and radius $\sqrt{5}$ Are the points $\left(\sqrt{2},\sqrt{3}\right)$ and $(1,2)$ on the circle?

Example 5 Finding the Equation of a Circle

Find the equation of the circle with center at $(5,2)$ and radius 3. Is the point $(5,5)$ on the circle?

Solution

The equation of the circle is $(x-5)^2+(y-2)^2=9$.

Substituting $(5,5)$ gives $(5-5)^2+(5-2)^2=0^2+3^2=9$.

Therefore, $(5,5)$ is on the circle.

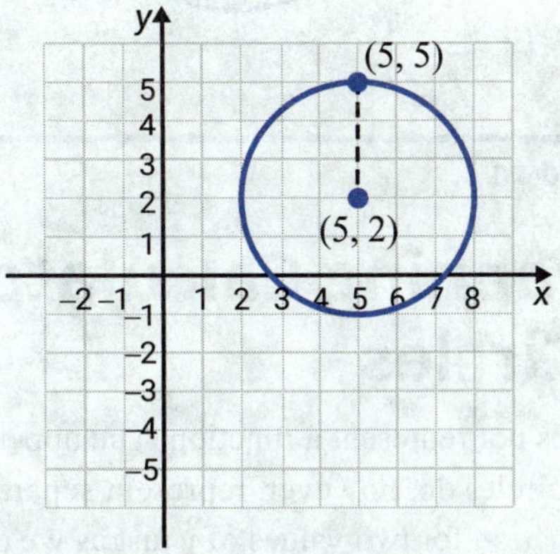

Now work margin exercise 5.

5. Find the equation of the circle with center at $(3,-2)$ and radius 4. Is the point $(7,-2)$ on the circle?

18.3 Exercises

Concept Check

1. $\sqrt{(x_2-x_1)^2+(y_2-y_1)^2}$
2. squares
3. averaging
4. distance, point
5. twice
6. $(x-h)^2+(y-k)^2$
7. True
8. False; This distance is called the radius.
9. True
10. True

D Graphing Circles

Example 6 Graphing a Circle

Show that $x^2 + y^2 - 8x + 2y = 0$ represents a circle. Find its center and radius. Then graph the circle.

Solution

Rearrange the terms and complete the square for $x^2 - 8x$ and $y^2 + 2y$.

$$x^2 + y^2 - 8x + 2y = 0$$

$$\left(x^2 - 8x\right) + \left(y^2 + 2y\right) = 0$$

$$\left(x^2 - 8x + 16\right) + \left(y^2 + 2y + 1\right) = 16 + 1$$ Add 16 and 1 to both sides.

Complete the squares

$$(x-4)^2 + (y+1)^2 = 17$$ Standard form for the equation of a circle

$h = 4 \quad k = -1 \quad r^2 = 17$

Center is at $(4, -1)$ and radius $= \sqrt{17}$.

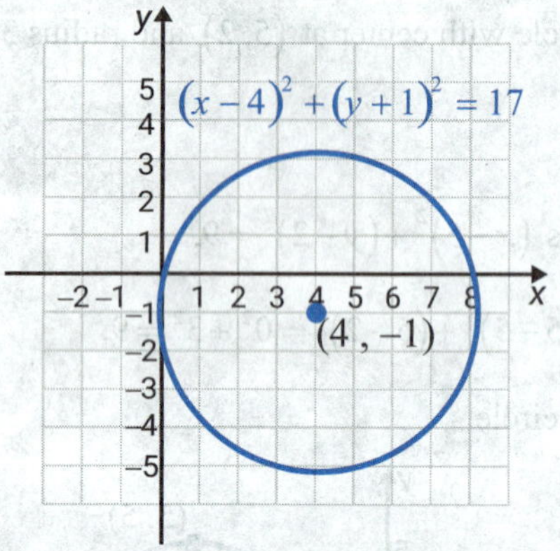

Now work margin exercise 6.

E Using a Graphing Calculator to Graph Circles

The equation of a circle does not represent a function. The upper half (upper semicircle) and the lower half (lower semicircle) do, however, represent separate functions. Therefore, to graph a circle, solve the equation for two values of y just as we did for horizontal parabolas in Section 18.2. For example, consider the circle with equation $x^2 + y^2 = 4$.

$$x^2 + y^2 = 4$$

$$y^2 = 4 - x^2$$ First solve for y^2.

$$\begin{cases} y_1 = \sqrt{4 - x^2} \\ y_2 = -\sqrt{4 - x^2} \end{cases}$$ Solving for y gives two equations that represent two functions.

---

6. Show that $x^2 + y^2 + 6x - 2y = 0$ represents a circle. Find its center and radius. Then graph the circle.

Math Tip

When stating the radius of the circle, we discard the negative solution of $r^2 = 17$ since a negative radius does not fit the context.

Practice

1. $5; (4, 5.5)$
2. $13; (3.5, 6)$
3. $13; (3, 4.5)$
4. $5; (-4, 1.5)$
5. $\sqrt{29}; (2, 4.5)$
6. $5\sqrt{2}; (0.5, -1.5)$
7. $3; (5.5, -3)$
8. $7; (1.5, 6)$
9. $\sqrt{13}; (6, -3.5)$
10. $\sqrt{85}; (7, -0.5)$
11. $17; (-3, -4.5)$
12. $13; (0.5, 2)$
13. $x^2 + y^2 = 16$
14. $x^2 + y^2 = 36$
15. $x^2 + y^2 = 3$
16. $x^2 + y^2 = 7$
17. $x^2 + y^2 = 11$
18. $x^2 + y^2 = 13$
19. $x^2 + y^2 = \frac{4}{9}$
20. $x^2 + y^2 = \frac{49}{16}$
21. $x^2 + (y-2)^2 = 4$
22. $x^2 + (y-5)^2 = 25$
23. $(x-4)^2 + y^2 = 1$
24. $(x+3)^2 + y^2 = 16$

Graphing both of these functions gives the circle pictured in Figure 4.

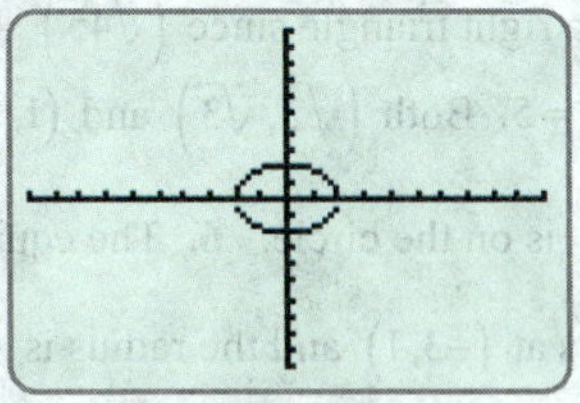

Figure 4

The screen on a TI-84 Plus graphing calculator is rectangular and not a square. The ranges for the standard viewing window are from −10 for both `Xmin` and `Ymin` to 10 for both `Xmax` and `Ymax`. That is, the horizontal scale (x-values) and the vertical scale (y-values) are the same. Since the rectangular screen is in the approximate ratio of 3 to 2, the graph of a circle using the **standard window** will appear flattened as the circle does in Figure 4.

To get a more realistic picture of a circle press the WINDOW key and set the `Xmin` and `Xmax` values to –6 and 6, respectively. Set the `Ymin` and `Ymax` values to –4 and 4, respectively. Since the numbers 6 and 4 are in the ratio of 3 to 2, the screen is said to show a "square window," and the graphs of y_1 and y_2 will now give a more realistic picture of a circle, as shown in Figure 5. Alternatively, pressing the ZOOM key and choosing option `ZSquare` automatically "squares" the window.

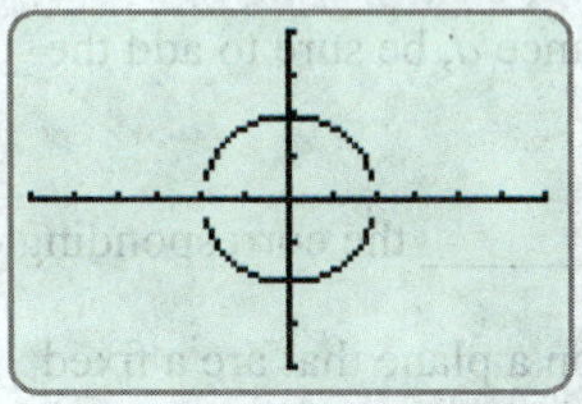

Figure 5

Example 7 Using a Graphing Calculator to Graph Circles

Use a TI-84 Plus graphing calculator to graph the circle $x^2 + y^2 = 9$.

Solution

Press the WINDOW key and set the values to –6 and 6 for `Xmin` and `Xmax` and –4 and 4 for `Ymin` and `Ymax`, respectively.

Solving for y^2 gives: $y^2 = 9 - x^2$

Solving for y_1 and y_2 gives: $\begin{cases} y_1 = \sqrt{9 - x^2} \\ y_2 = -\sqrt{9 - x^2} \end{cases}$

Graphing both y_1 and y_2 gives the following graph of the circle.

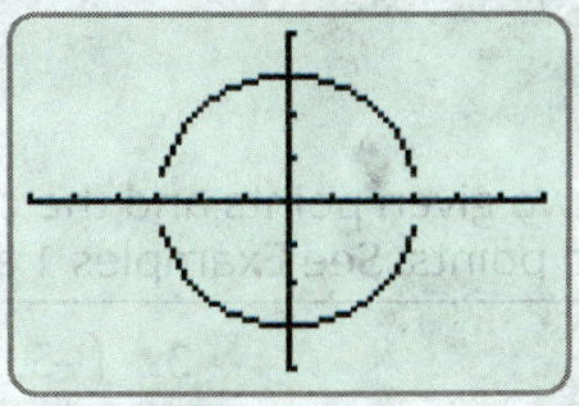

Now work margin exercise 7.

25. $(x+2)^2 + y^2 = 8$

26. $(x-5)^2 + y^2 = 2$

27. $(x-3)^2 + (y-1)^2 = 36$

28. $(x+1)^2 + (y-2)^2 = 25$

29. $(x-3)^2 + (y-5)^2 = 12$

30. $(x-4)^2 + (y+2)^2 = 14$

31. $(x-7)^2 + (y-4)^2 = 10$

32. $(x+3)^2 + (y-2)^2 = 7$

33. $x^2 + y^2 = 9$

Center: $(0,0)$; $r = 3$

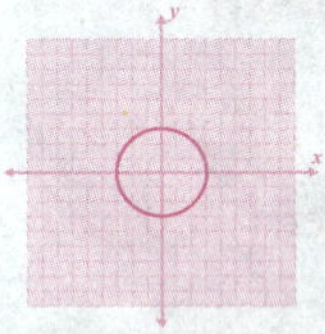

34. $x^2 + y^2 = 16$

Center: $(0,0)$; $r = 4$

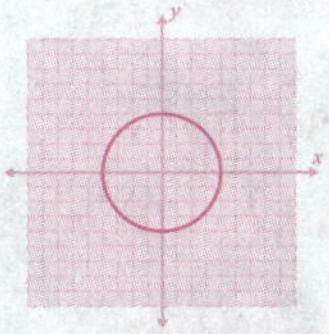

7. Use a graphing calculator to graph the circle $x^2 + y^2 = 36$.

35. $x^2 + y^2 = 49$

Center: $(0,0)$; $r = 7$

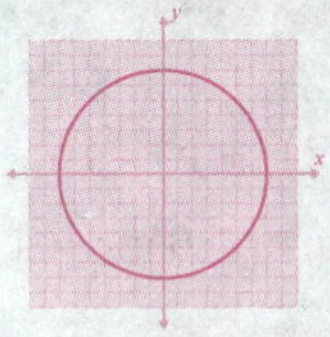

36. $x^2 + y^2 = 25$

Center: $(0,0)$; $r = 5$

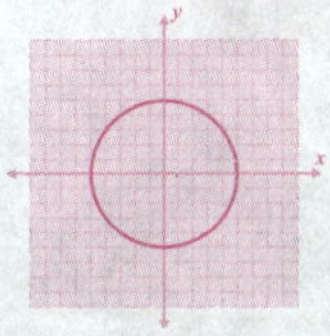

37. $x^2 + y^2 = 18$

Center: $(0,0); r = 3\sqrt{2}$

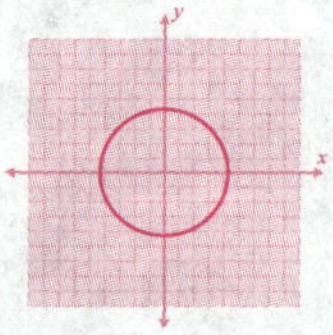

38. $x^2 + y^2 = 12$

Center: $(0,0); r = 2\sqrt{3}$

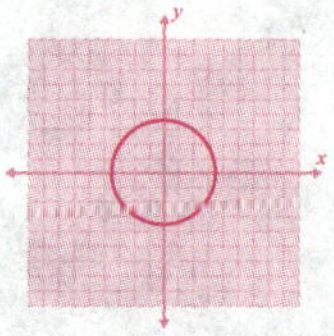

39. $(x+1)^2 + y^2 = 9$

Center: $(-1,0); r = 3$

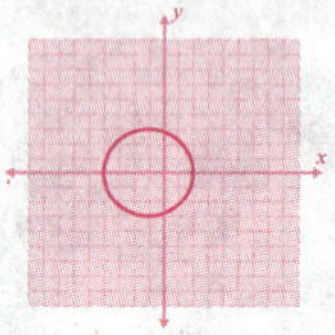

40. $(x+3)^2 + y^2 = 9$

Center: $(-3,0); r = 3$

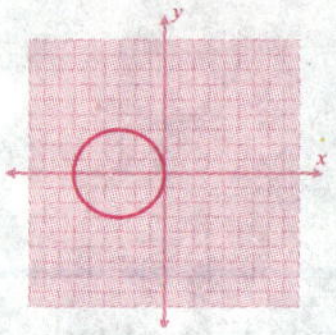

41. $x^2 + (y-2)^2 = 4$

Center: $(0,2); r = 2$

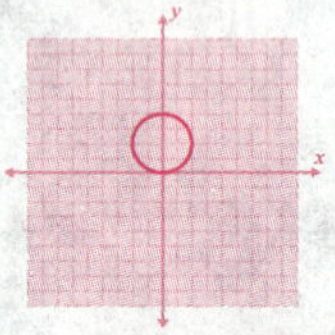

42. $(x-2)^2 + y^2 = 16$

Center: $(2,0); r = 4$

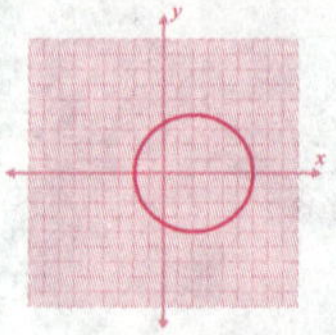

Margin Exercise Answers

1. Distance: 5 **2.** Triangle DEF is a right triangle since $\left(\sqrt{45}\right)^2 + \left(\sqrt{80}\right)^2 = \left(\sqrt{125}\right)^2$.

3. Midpoint: $\left(\frac{5}{2}, -3\right)$ **4.** $x^2 + y^2 = 5$. Both $\left(\sqrt{2}, \sqrt{3}\right)$ and $(1,2)$ are on the circle.

5. $(x-3)^2 + (y+2)^2 = 16$, $(7,-2)$ is on the circle. **6.** The equation can be written in the form $(x+3)^2 + (y-1)^2 = 10$. The center is at $(-3,1)$ and the radius is $\sqrt{10}$.

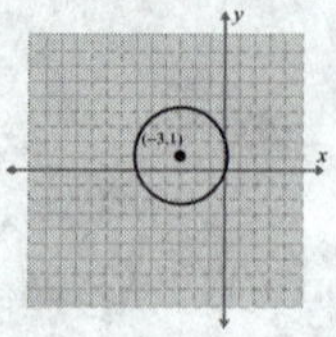

7.

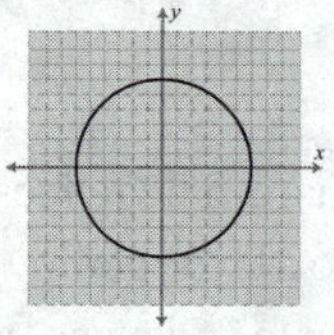

18.3 Exercises

Concept Check

Fill-in-the-Blank. Complete the sentences using information found in this section.

1. The distance between two points $P(x_1, y_1)$ and $Q(x_2, y_2)$ in a plane can be determined with the formula $d =$ ____________.
2. When calculating the distance d, be sure to add the ______ before taking the square root.
3. The midpoint is found by ______ the corresponding coordinates of the endpoints.
4. A circle is a set of points on a plane that are a fixed ______ from a fixed ______.
5. The diameter is ______ the length of the radius.
6. The equation of a circle with radius r and center at (h, k) is $r^2 =$ ______.

True/False. Determine whether each statement is true or false. If a statement is false, explain how it can be changed so the statement will be true. (**Note:** There may be more than one acceptable change.)

7. The Pythagorean Theorem is used to derive the distance formula.
8. The distance from the center of a circle to any point on the circle is called the diameter of the circle.
9. To find the equation of a circle with center at $(2, 5)$ and radius 6, the distance formula can be used.
10. The hypotenuse of a right triangle is the longest side of the triangle.

Practice

Find the distance between the two given points and the coordinates of the midpoint of the line segment joining the two points. See Examples 1 and 3.

1. $(2,4), (6,7)$

2. $(1,0), (6,12)$

3. $(-3,2), (9,7)$

4. $(-6,3), (-2,0)$

5. $(1,7),(3,2)$
6. $(-2,1),(3,-4)$
7. $(4,-3),(7,-3)$
8. $(-2,6),(5,6)$
9. $(5,-2),(7,-5)$
10. $(6,4),(8,-5)$
11. $(-7,3),(1,-12)$
12. $(3,8),(-2,-4)$

Find equations for each of the circles. See Examples 4 and 5.

13. Center $(0,0)$; $r=4$
14. Center $(0,0)$; $r=6$
15. Center $(0,0)$; $r=\sqrt{3}$
16. Center $(0,0)$; $r=\sqrt{7}$
17. Center $(0,0)$; $r=\sqrt{11}$
18. Center $(0,0)$; $r=\sqrt{13}$
19. Center $(0,0)$; $r=\dfrac{2}{3}$
20. Center $(0,0)$; $r=\dfrac{7}{4}$
21. Center $(0,2)$; $r=2$
22. Center $(0,5)$; $r=5$
23. Center $(4,0)$; $r=1$
24. Center $(-3,0)$; $r=4$
25. Center $(-2,0)$; $r=\sqrt{8}$
26. Center $(5,0)$; $r=\sqrt{2}$
27. Center $(3,1)$; $r=6$
28. Center $(-1,2)$; $r=5$
29. Center $(3,5)$; $r=\sqrt{12}$
30. Center $(4,-2)$; $r=\sqrt{14}$
31. Center $(7,4)$; $r=\sqrt{10}$
32. Center $(-3,2)$; $r=\sqrt{7}$

Write each of the equations in standard form. Find the center and radius of the circle and then sketch the graph. See Example 6.

33. $x^2+y^2=9$
34. $x^2+y^2=16$
35. $x^2=49-y^2$
36. $y^2=25-x^2$
37. $x^2+y^2=18$
38. $x^2+y^2=12$
39. $x^2+y^2+2x=8$
40. $x^2+y^2+6x=0$
41. $x^2+y^2-4y=0$
42. $x^2+y^2-4x=12$
43. $x^2+y^2+2x+4y=11$
44. $x^2+y^2+4x+4y=8$
45. $x^2+y^2-4x+10y+20=0$
46. $x^2+y^2-6x-8y+9=0$
47. $x^2+y^2-4x-6y+5=0$
48. $x^2+y^2+10x-2y+14=0$

Use the Pythagorean Theorem to decide if the triangle determined by the given points is a right triangle. See Example 2.

49. $A(1,-2), B(7,1), C(5,5)$
50. $A(-5,-1), B(2,1), C(-1,6)$

43. $(x+1)^2+(y+2)^2=16$
Center: $(-1,-2)$; $r=4$

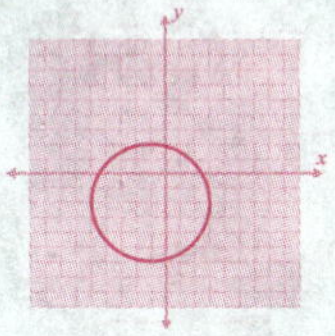

44. $(x+2)^2+(y+2)^2=16$
Center: $(-2,-2)$; $r=4$

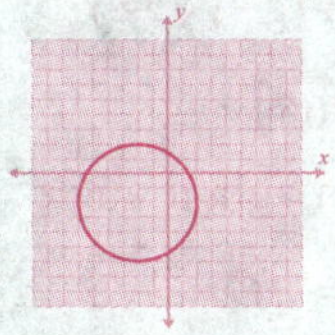

45. $(x-2)^2+(y+5)^2=9$
Center: $(2,-5)$; $r=3$

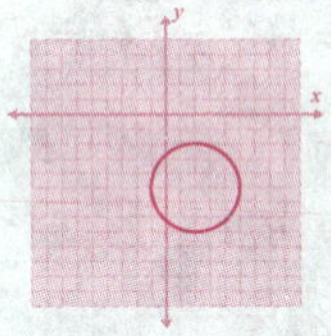

46. $(x-3)^2+(y-4)^2=16$
Center: $(3,4)$; $r=4$

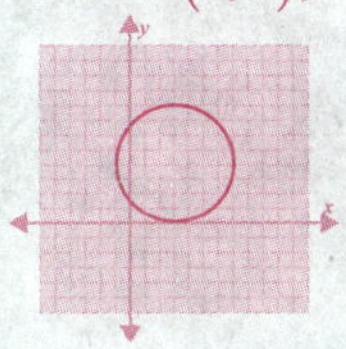

47. $(x-2)^2+(y-3)^2=8$
Center: $(2,3)$; $r=2\sqrt{2}$

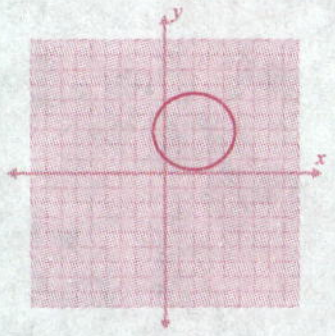

48. $(x+5)^2+(y-1)^2=12$
Center: $(-5,1)$; $r=2\sqrt{3}$

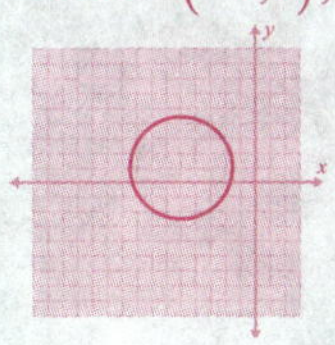

49. $\overline{AB} = 3\sqrt{5}, \overline{AC} = \sqrt{65}, \overline{BC} = 2\sqrt{5}$, $(3\sqrt{5})^2 + (2\sqrt{5})^2 = (\sqrt{65})^2$; Right triangle

50. $\overline{AB} = \sqrt{53}, \overline{AC} = \sqrt{65}, \overline{BC} = \sqrt{34}$, $(\sqrt{53})^2 + (\sqrt{34})^2 \neq (\sqrt{65})^2$; Not a right triangle

51. $\overline{AB} = \overline{AC} = 4\sqrt{5}$

52. $\overline{AB} = \overline{BC} = 2\sqrt{10}$

53. $\overline{AB} = \overline{AC} = \overline{BC} = 4$

54. $\overline{AB} = \overline{AC} = \overline{BC} = 8$

55. $\overline{AC} = \overline{BD} = \sqrt{61}$

56. $\overline{AC} = \overline{BD} = \sqrt{34}$

57. $10 + 4\sqrt{5}$

58. $18 + \sqrt{82}$

59.

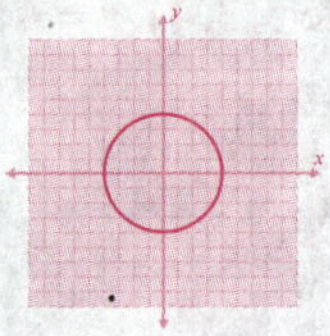

60.

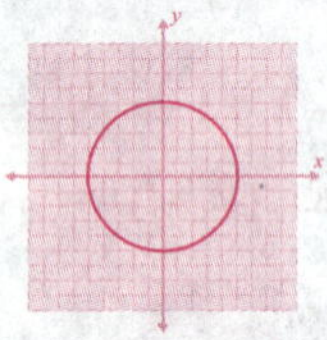

61.

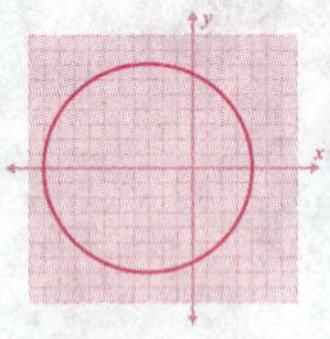

62.

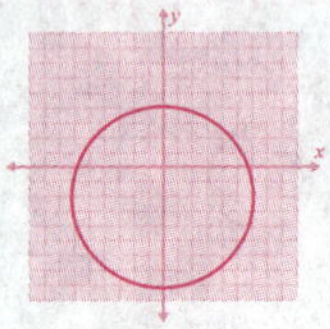

63.

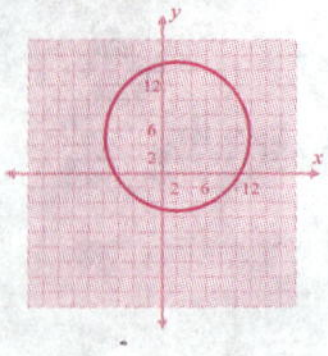

Show that the triangle determined by the given points is an isosceles triangle (has two equal sides).

51. $A(1,1), B(5,9), C(9,5)$

52. $A(1,-4), B(3,2), C(9,4)$

Show that the triangle determined by the given points is an equilateral triangle (all sides equal).

53. $A(1,0), B(3,\sqrt{12}), C(5,0)$

54. $A(0,5), B(0,-3), C(\sqrt{48},1)$

Show that the lengths of the diagonals (*AC* and *BD*) of the rectangle *ABCD* are equal.

55. $A(2,-2), B(2,3), C(8,3), D(8,-2)$

56. $A(-1,1), B(-1,4), C(4,4), D(4,1)$

Find the perimeter of the triangle determined by the given points.

57. $A(-5,0), B(3,4), C(0,0)$

58. $A(-6,-1), B(-3,3), C(6,4)$

Use a graphing calculator to graph the circles. Be sure to set a square window.

59. $x^2 + y^2 = 16$

60. $x^2 + y^2 = 25$

61. $(x+3)^2 + y^2 = 49$

62. $x^2 + (y+2)^2 = 36$

63. $(x-2)^2 + (y-5)^2 = 100$

64. $(x-1)^2 + (y+3)^2 = 64$

Applications

Solve.

65. The roadways in Descartesville are laid out such that streets run east to west and avenues run north to south. The north to south avenues and the east to west streets are numbered sequentially, beginning with 1. For example, a person standing on the corner of 1st Street and 1st Avenue may be considered to be at standing at (1, 1). Suppose a person begins at the corner of 1st Street and 2nd Avenue and walks to the corner of 9th Street and 8th Avenue. If the person were able to walk directly from the beginning corner to the ending corner, the distance traveled would be the same as the distance of how many blocks?

66. A bored greens keeper at the Descartesville Golf Course decides to sketch the entire golf course on a coordinate grid, where each unit on the grid corresponds to 100 yards. He notices that the first hole's tee box is located at the point $(-8, -10)$ and the hole's cup is located at the point $(-11, -6)$. What is the straight-line distance in yards between the first hole's tee box and the hole's cup?

67. A group of math majors at Homestate University decide to make a map of the campus on a coordinate grid. Each unit on the grid corresponds to 100 yards. If the calculus class is held in the Math building located at (5, 12) on the graph and the cafeteria is located at (9, 14) on the graph, how many hundreds of yards do students have to walk if they follow a straight line from the calculus class to the cafeteria?

68. A conservation society used a grant to purchase a 1000-acre tract of land in Descartesville and plans to turn the land into a camping ground. The society uses a coordinate grid to plan the design of the camp where each unit on the grid corresponds to 1000 yards. The plans include a dock on the lake to be located at $(6, 15)$ on the grid and a picnic pavilion to be located at $(10, 8)$. How many thousands of yards will campers have to walk if they follow a straight line from the dock to the picnic pavilion?

Writing & Thinking

69. For a given line and a point not on the line, a parabola is defined as the set of all points that are the same distance from the point and the line. The point is called the focus and the line is called the directrix. See the figure provided.

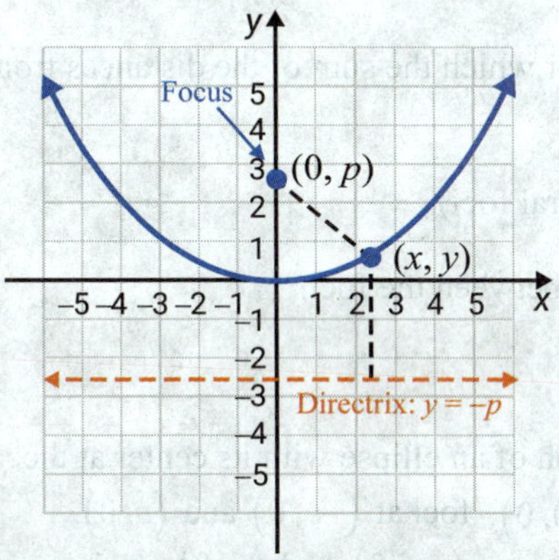

a. Suppose that (x, y) is any point on a parabola and $(0, p)$ is the focus. Find the distance from (x, y) to the focus.

b. Suppose that (x, y) is any point on the same parabola in part a. and the line $y = -p$ is the directrix. Find the distance from (x, y) to the directrix.

c. Show that the equation of the parabola is $x^2 = 4py$.

70. Using the equation developed in Exercise 69, find the equation of the parabola with focus at $(0, 2)$ and line $y = -2$ as directrix. Draw the graph.

71. For a given line and a point not on the line, a parabola is defined as the set of all points that are the same distance from the point and the line. The point is called the focus and the line is called the directrix. See the figure provided.

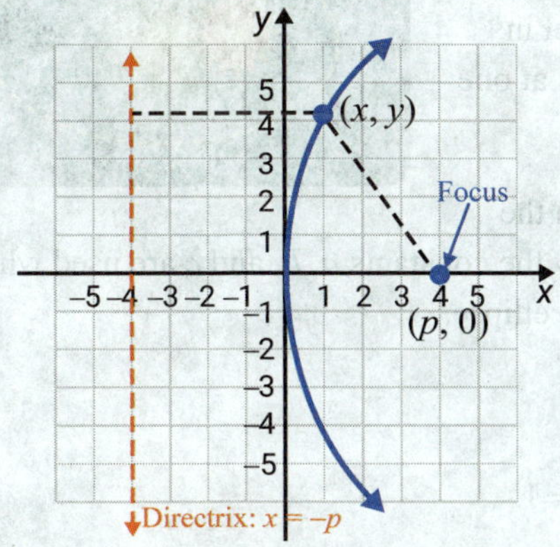

a. Suppose that (x, y) is any point on a parabola and $(p, 0)$ is the focus. Find the distance from (x, y) to the focus.

b. Suppose that (x, y) is any point on the same parabola in part a. and the line $x = -p$ is the directrix. Find the distance from (x, y) to the directrix.

c. Show that the equation of the parabola is $y^2 = 4px$.

72. Using the equation developed in Exercise 71, find the equation of the parabola with focus at $(-3, 0)$ and line $x = 3$ as directrix. Draw the graph.

64.

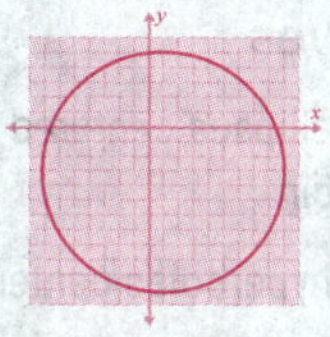

Applications

65. 10 blocks

66. 500 yards

67. $\sqrt{20} = 2\sqrt{5}$ hundred yards

68. $\sqrt{65}$ thousand yards

Writing & Thinking

69. a. $d = \sqrt{x^2 + (y-p)^2}$

b. $d = |y+p|$

c.
$$y + p = \sqrt{x^2 + (y-p)^2}$$
$$(y+p)^2 = x^2 + (y-p)^2$$
$$y^2 + 2py + p^2 = x^2 + y^2 - 2py + p^2$$
$$x^2 = 4py$$

70. $x^2 = 8y$ or $y = \frac{1}{8}x^2$

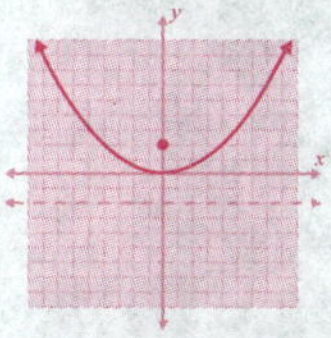

71. a. $d = \sqrt{(x-p)^2 + y^2}$

b. $d = |x+p|$

c.
$$x + p = \sqrt{(x-p)^2 + y^2}$$
$$(x+p)^2 = (x-p)^2 + y^2$$
$$x^2 + 2px + p^2 = x^2 - 2px + p^2 + y^2$$
$$y^2 = 4px$$

72. $y^2 = -12x$ or $x = -\frac{1}{12}y^2$

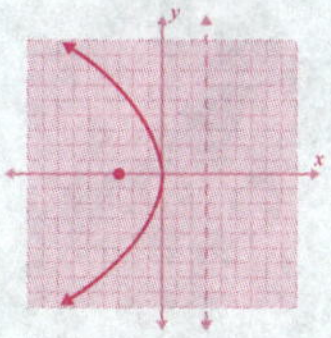

Objectives

A. Graph ellipses centered at the origin.

B. Graph hyperbolas centered at the origin.

C. Find the equations for the asymptotes of hyperbolas.

D. Graph ellipses and hyperbolas centered at the point (h,k).

18.4 Ellipses and Hyperbolas

A Graphing Ellipses

Ellipses are conic sections that are oval in shape. To draw an ellipse, begin with two fixed points and a distance greater than the distance between the two points. Exercise 45 in the set of exercises is designed to help you understand how points in an elliptical path can be traced by using the following formal definition. You might want to go directly to that exercise now.

Ellipse

An **ellipse** is the set of all points in a plane for which the sum of the distances from two fixed points is constant.

Each of the fixed points is called a **focus** (plural foci).

The **center** of an ellipse is the point midway between the foci.

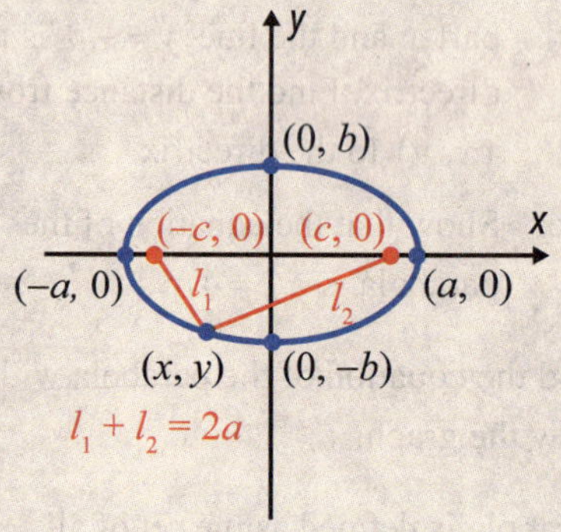

The graph of an ellipse with its center at the origin $(0, 0)$, foci at $(-c, 0)$ and $(c, 0)$, x-intercepts at $(-a, 0)$ and $(a, 0)$, and y-intercepts at $(0, -b)$ and $(0, b)$ (where $a^2 > b^2$).

DEFINITION

Ellipses have many practical applications in the sciences, particularly in astronomy. For example, the planets in our solar system (including earth) travel in elliptical orbits (not circular orbits) and the sun is at one of the foci of each ellipse.

Initially, we will study ellipses with foci either on the x-axis or on the y-axis. Pay close attention to how the constants a, b, and c are used when defining ellipses. Note that the foci are not on the ellipse.

As an example, consider the equation

$$\frac{x^2}{25} + \frac{y^2}{9} = 1.$$

Several points that satisfy this equation are given in Table 1 and then are graphed in Figure 1.

x	y
5	0
–5	0
0	3
0	–3
3	$\frac{12}{5}$
3	$-\frac{12}{5}$
–3	$\frac{12}{5}$
–3	$-\frac{12}{5}$

Table 1

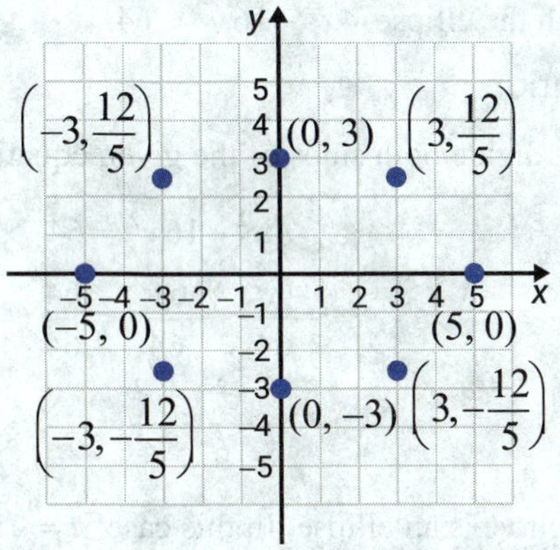

Figure 1

Joining the points in Figure 1 with a smooth curve produces the ellipse shown in Figure 2. The points $(5, 0)$ and $(-5, 0)$ are the x-intercepts, and the points $(0, 3)$ and $(0, -3)$ are the y-intercepts.

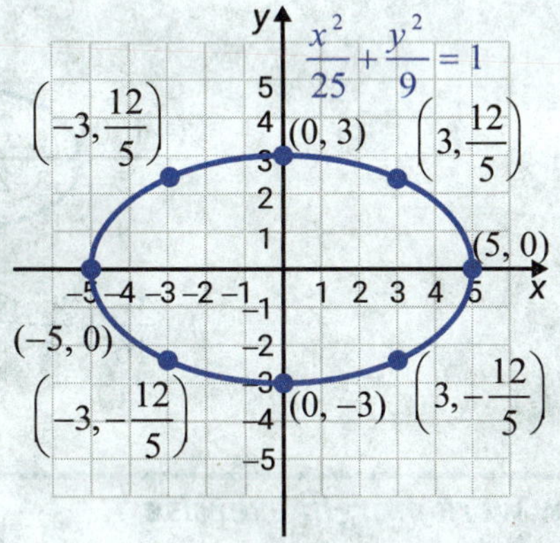

Figure 2

Equation of an Ellipse

The standard form for the equation of an ellipse with its center at the origin is

$$\frac{x^2}{a^2}+\frac{y^2}{b^2}=1.$$

The points $(a, 0)$ and $(-a, 0)$ are the ***x*-intercepts** (called **vertices**).

The points $(0, b)$ and $(0, -b)$ are the ***y*-intercepts** (called **vertices**).

When $a^2 > b^2$:

1. The segment of length $2a$ joining the x-intercepts is called the **major axis**.
2. The segment of length $2b$ joining the y-intercepts is called the **minor axis**.

When $b^2 > a^2$:

1. The segment of length $2b$ joining the y-intercepts is called the **major axis**.
2. The segment of length $2a$ joining the x-intercepts is called the **minor axis**.

Note: In either case, the foci lie on the major axis.

DEFINITION

Teaching Note

You might want to discuss the situation when $a^2 = b^2$ and call a circle a degenerate ellipse. As the two foci move closer together, the ellipse degenerates into a circle with radius r.

1. Graph the ellipse $4x^2 + 9y^2 = 36$.

Example 1 Graphing an Ellipse with a Horizontal Major Axis

Graph the ellipse $4x^2 + 16y^2 = 64$.

Solution

First, divide both sides of the given equation by 64 to find the standard form.

$$4x^2 + 16y^2 = 64$$

$$\frac{4x^2}{64} + \frac{16y^2}{64} = \frac{64}{64}$$

$$\frac{x^2}{16} + \frac{y^2}{4} = 1 \qquad \text{Standard form: } \frac{x^2}{a^2} + \frac{y^2}{b^2} = 1$$

Math Tip

When determining the length of the major or minor axis, we discard the negative solution since a length cannot be negative. In Example 1, this is why we state $a = \sqrt{16} = 4$ is the length of the major axis from the quadratic $a^2 = 16$.

The curve is an ellipse. In this case $a = \sqrt{16} = 4$ and the major axis has length $2a = 8$. Also, $b = \sqrt{4} = 2$ and the minor axis has length $2b = 4$. The endpoints of the major axis are $(-4, 0)$ and $(4, 0)$. The endpoints of the minor axis are $(0, -2)$ and $(0, 2)$. The major and minor axes intersect at the center of the ellipse, $(0, 0)$.

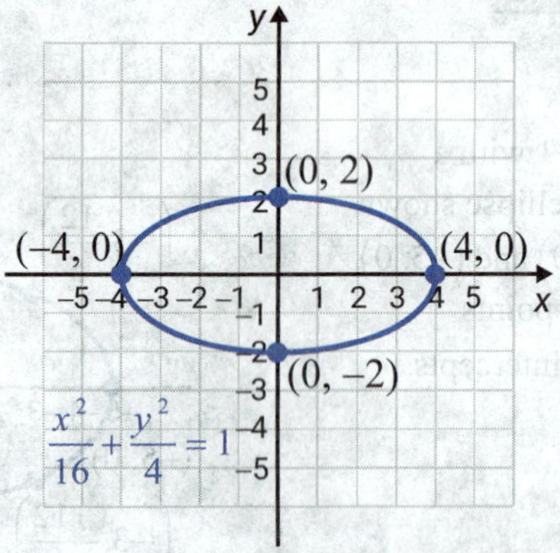

Now work margin exercise 1.

2. Graph the ellipse $\frac{x^2}{4} + \frac{y^2}{16} = 1$.

Example 2 Graphing an Ellipse with a Vertical Major Axis

Graph the ellipse $\frac{x^2}{1} + \frac{y^2}{9} = 1$.

Solution

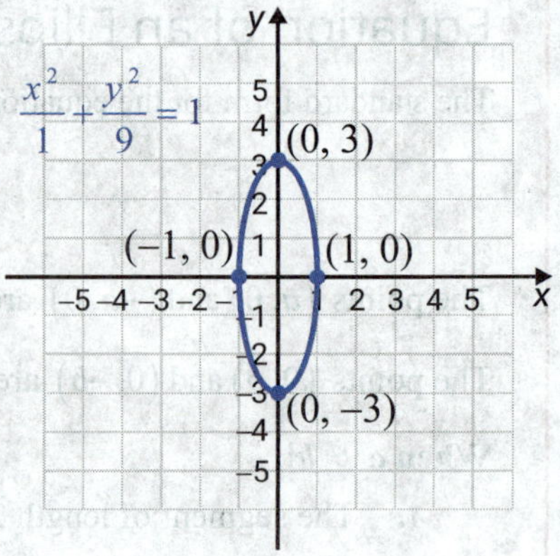

The equation is in standard form with $a^2 = 1$ and $b^2 = 9$.

Because $b^2 > a^2$, we know the major axis is vertical. That is, the ellipse is elongated along the y-axis.

The points $(0, -3)$ and $(0, 3)$ are the endpoints of the major axis.

The points $(-1, 0)$ and $(1, 0)$ are the endpoints of the minor axis.

Now work margin exercise 2.

In the equation of an ellipse,

$$\frac{x^2}{a^2}+\frac{y^2}{b^2}=1,$$

the coefficients for x^2 and y^2 are both positive. If one of these coefficients is negative, then the equation represents a **hyperbola**.

B Graphing Hyperbolas

Notice how the following definition of a hyperbola differs from the definition of an ellipse. Instead of the sum of the distances from two fixed points, we find the difference of the distances from two fixed points.

Hyperbola

A **hyperbola** is the set of all points in a plane such that the absolute value of the difference of the distances from two fixed points is constant.

Each of the fixed points is called a **focus** (plural foci).

The **center** of a hyperbola is the point midway between the foci.

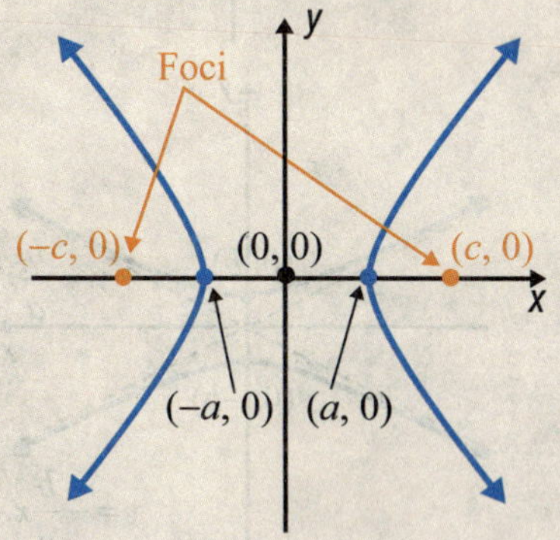

The graph of a hyperbola with its center at the origin $(0, 0)$, foci along the x-axis at $(-c, 0)$ and $(c, 0)$, and x-intercepts at $(-a, 0)$ and $(a, 0)$. There are no y-intercepts.

DEFINITION

Several points that satisfy the equation

$$\frac{x^2}{16}-\frac{y^2}{9}=1$$

are shown in Table 2, and the curves joining these points (a hyperbola) are shown in Figure 3.

x	y
4	0
−4	0
5	$\frac{9}{4}$
5	$-\frac{9}{4}$
−5	$\frac{9}{4}$
−5	$-\frac{9}{4}$

Table 2

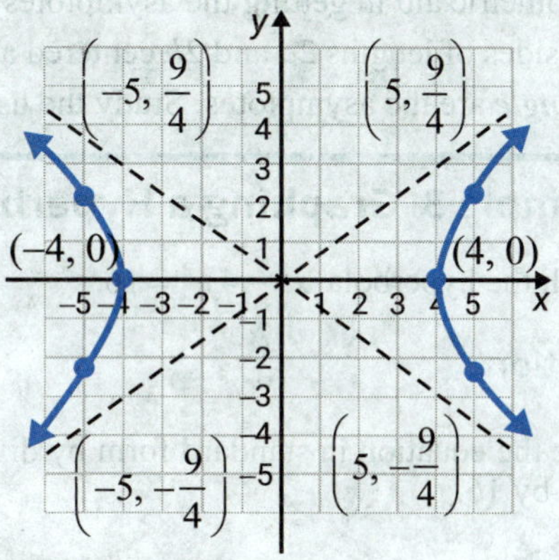

Figure 3

18.4 Exercises

Concept Check

1. oval
2. sum, constant
3. focus
4. difference, constant
5. minor
6. major
7. True
8. False; The foci lie on the major axis.
9. False; The midway point between the foci is the center.

Practice

1. $\frac{x^2}{36}+\frac{y^2}{4}=1$

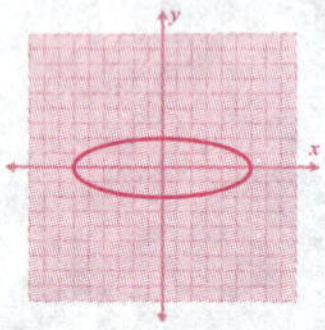

2. $\frac{x^2}{16}+\frac{y^2}{4}=1$

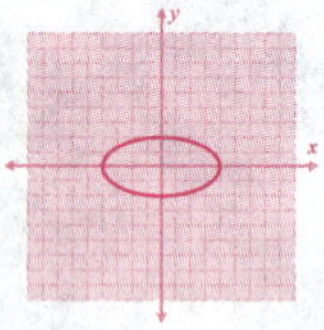

3. Graph the hyperbola $x^2-4y^2=4$.

3. $\frac{x^2}{25}+\frac{y^2}{4}=1$

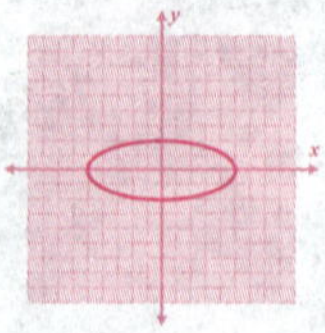

The two dotted lines shown in Figure 3 are called asymptotes. (Asymptotes were discussed in Chapter 17 with exponential and logarithmic functions.) These lines are not part of the hyperbola, but they serve as guidelines for sketching the graph. The curves get closer and closer to these lines without ever touching them. In this figure the equations of the asymptotes are

$$y=\frac{3}{4}x \quad \text{and} \quad y=-\frac{3}{4}x.$$

Equations of Hyperbolas

In general, there are two standard forms for equations of hyperbolas with their centers at the origin.

1. $\frac{x^2}{a^2}-\frac{y^2}{b^2}=1$

   x-intercepts (vertices) at $(a, 0)$ and $(-a, 0)$

   No y-intercepts

   Asymptotes: $y=\frac{b}{a}x$ and $y=-\frac{b}{a}x$

   The curves "open" left and right.

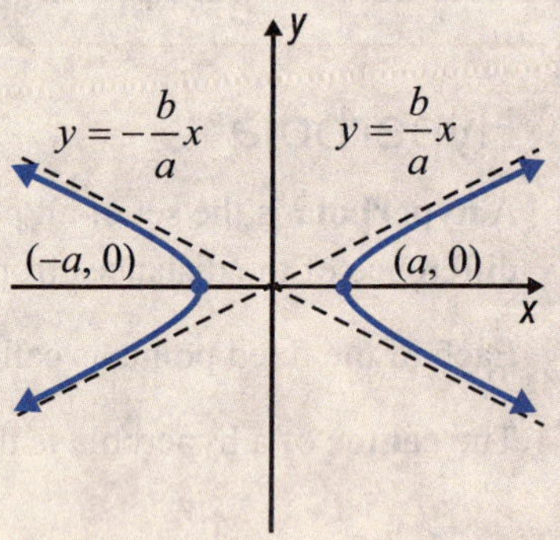

2. $\frac{y^2}{b^2}-\frac{x^2}{a^2}=1$

   y-intercepts (vertices) at $(0, b)$ and $(0, -b)$

   No x-intercepts

   Asymptotes: $y=\frac{b}{a}x$ and $y=-\frac{b}{a}x$

   The curves "open" up and down.

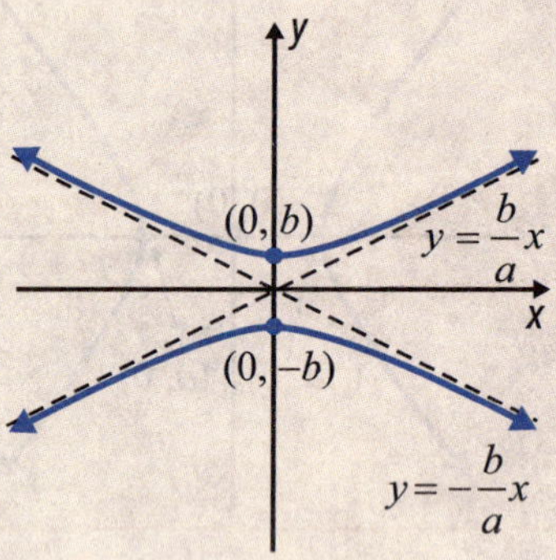

DEFINITION

C Asymptotes of Hyperbolas

A geometric aid in getting the asymptotes in the correct positions is to draw a rectangle with sides of lengths $2a$ and $2b$ centered at the origin. The diagonals of this *fundamental rectangle* are the asymptotes. Study the use of these rectangles in Examples 3 and 4.

Example 3 Graphing a Hyperbola Opening Left and Right

Graph the hyperbola $x^2-4y^2=16$.

Solution

Write the equation in standard form by dividing both sides by 16.

$$\frac{x^2}{16}-\frac{y^2}{4}=1$$

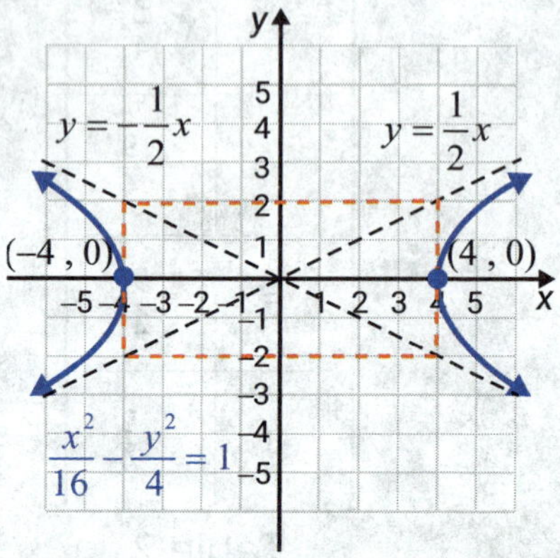

Here $a^2 = 16$ and $b^2 = 4$. So, using $a = 4$ and $b = 2$, the asymptotes are $y = \frac{2}{4}x = \frac{1}{2}x$ and $y = -\frac{2}{4}x = -\frac{1}{2}x$.

The vertices are $(-4, 0)$ and $(4, 0)$ and the curve opens left and right.

(Note that the fundamental rectangle has sides of lengths $2a = 8$ and $2b = 4$.)

Now work margin exercise 3.

Example 4 Graphing a Hyperbola Opening Up and Down

Graph the hyperbola $\frac{y^2}{1} - \frac{x^2}{9} = 1$.

Solution

Here $a^2 = 9$ and $b^2 = 1$. So $a = 3$ and $b = 1$ and the asymptotes are $y = \frac{1}{3}x$ and $y = -\frac{1}{3}x$. The vertices are $(0, -1)$ and $(0, 1)$ and the curve opens up and down.

(Note that the fundamental rectangle has sides of lengths $2a = 6$ and $2b = 2$.)

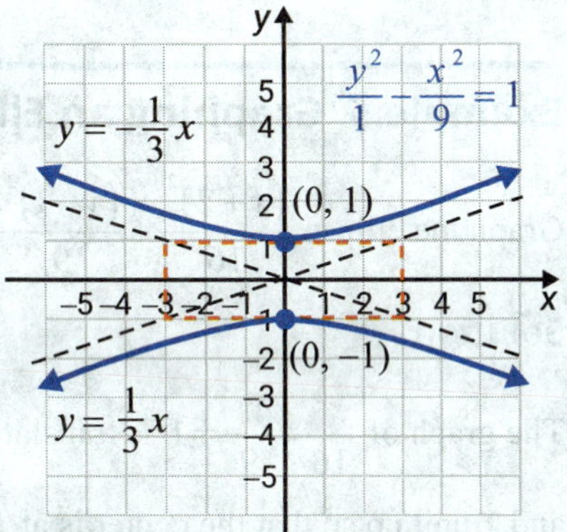

Now work margin exercise 4.

D Ellipses and Hyperbolas with Centers at (h, k)

The discussion of translations in Section 18.1 indicates that replacing x with $x - h$ and y with $y - k$ in an equation gives the corresponding graph a horizontal shift of h units and a vertical shift of k units. These same ideas were used in the discussion of circles in Section 18.3 for graphs of circles with center at (h, k).

$$\left[(x-h)^2 + (y-k)^2 = r^2\right]$$

This procedure is also used to write the equations of ellipses and hyperbolas with centers at (h, k). That is, the graphs of ellipses and hyperbolas can be translated to center at (h, k) by replacing x with $x - h$ and y with $y - k$ in the standard forms. The roles of a^2 and b^2 remain the same.

Ellipse with Center at (h, k)

The equation of an ellipse with its center at (h, k) is

$$\frac{(x-h)^2}{a^2} + \frac{(y-k)^2}{b^2} = 1,$$

where a and b are distances from (h, k) to the vertices.

DEFINITION

---

4. $\frac{x^2}{9} + \frac{y^2}{4} = 1$

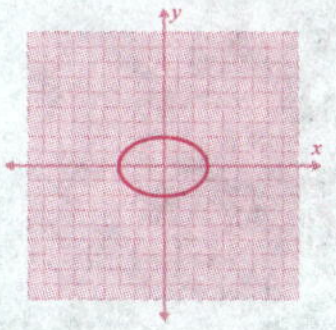

4. Graph the hyperbola $\frac{y^2}{9} - \frac{x^2}{4} = 1$.

5. $\frac{x^2}{1} + \frac{y^2}{16} = 1$

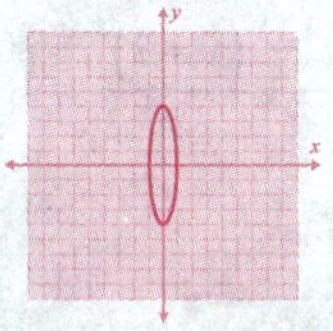

6. $\frac{x^2}{1} + \frac{y^2}{4} = 1$

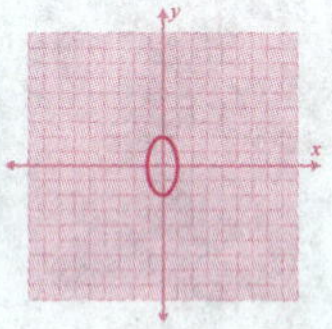

7. $\frac{x^2}{1} - \frac{y^2}{1} = 1$

8. $\frac{x^2}{4} - \frac{y^2}{4} = 1$

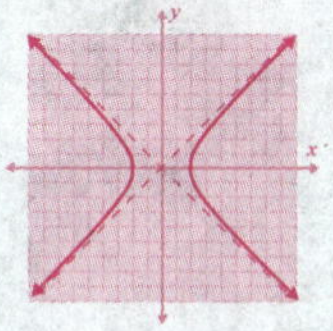

9. $\frac{x^2}{1} - \frac{y^2}{9} = 1$

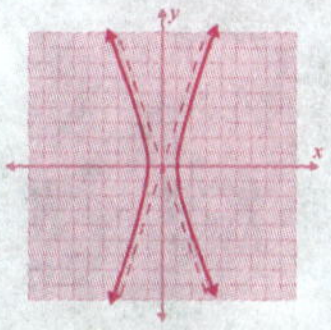

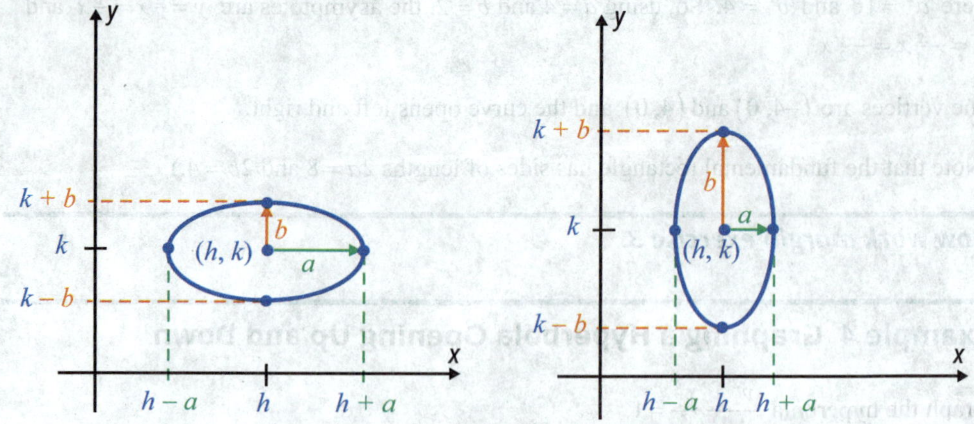

Figure 4

Example 5 Graphing an Ellipse with Center at (*h*, *k*)

Graph the ellipse $\frac{(x+2)^2}{16}+\frac{(y-1)^2}{9}=1$.

Solution

The graph of $\frac{x^2}{16}+\frac{y^2}{9}=1$ is translated 2 units left and 1 unit up so that the center is at $(-2,1)$ with $a=4$ and $b=3$. The graph is shown here with the center and vertices labeled.

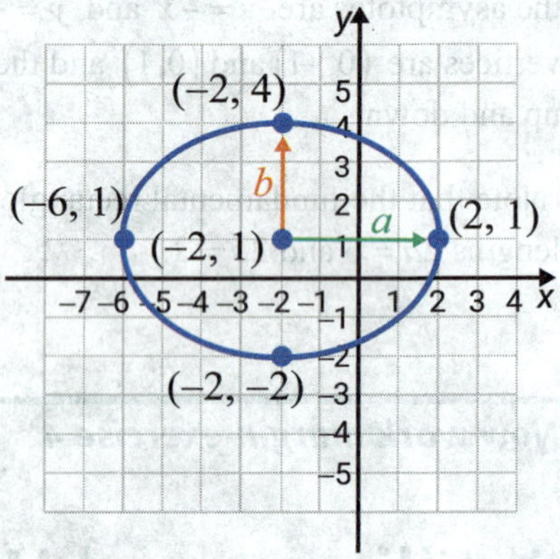

Now work margin exercise 5.

> ## Hyperbola with Center at (*h*, *k*)
>
> The equation of a hyperbola with its center at (h,k) is
>
> $$\frac{(x-h)^2}{a^2}-\frac{(y-k)^2}{b^2}=1 \quad \text{or} \quad \frac{(y-k)^2}{b^2}-\frac{(x-h)^2}{a^2}=1,$$
>
> where a and b are distances from (h,k). to the vertices.
>
> **DEFINITION**

10. $\frac{x^2}{1}-\frac{y^2}{4}=1$

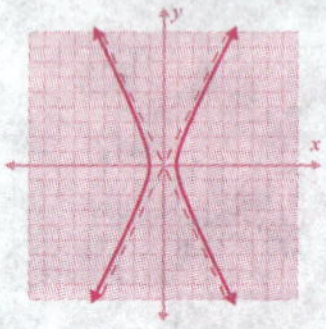

11. $\frac{x^2}{9}-\frac{y^2}{4}=1$

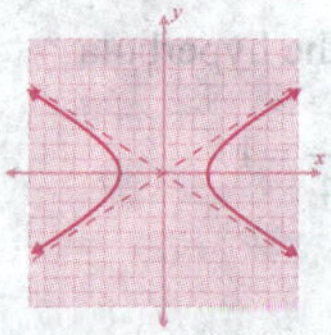

5. Graph the ellipse $\frac{(x-1)^2}{9}+\frac{(y+3)^2}{4}=1$.

12. $\frac{x^2}{16}-\frac{y^2}{9}=1$

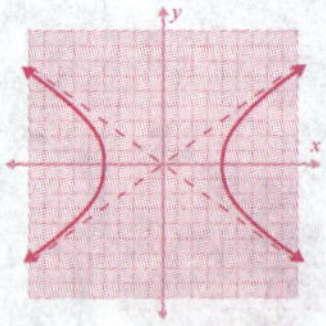

13. $\frac{x^2}{4}+\frac{y^2}{8}=1$

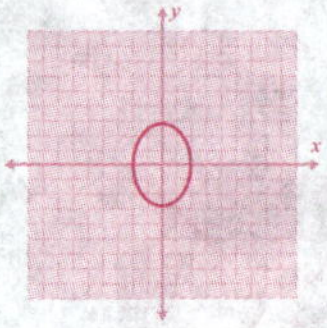

14. $\frac{x^2}{4}+\frac{y^2}{12}=1$

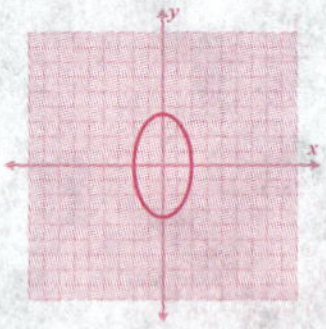

15. $\frac{x^2}{20}+\frac{y^2}{4}=1$

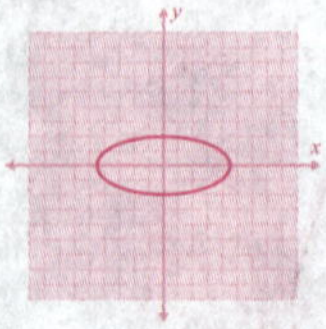

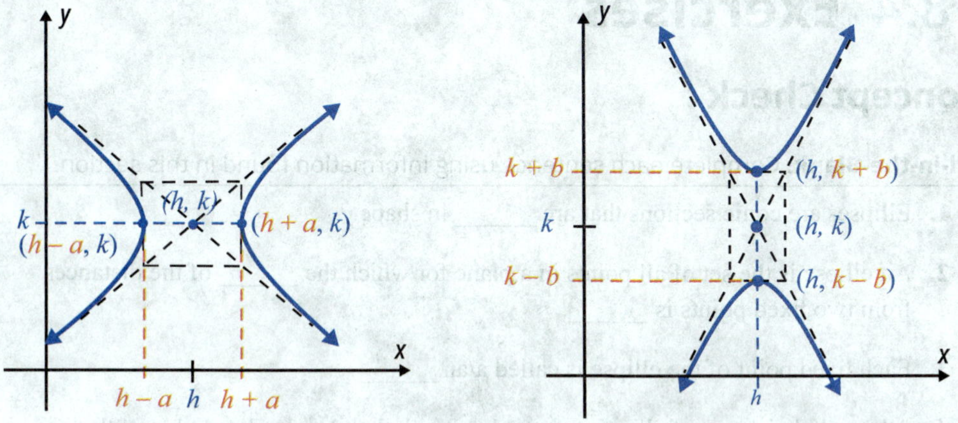

Figure 5

Example 6 Graphing a Hyperbola with Center at (*h*, *k*)

Graph the hyperbola $\frac{(x-3)^2}{25}-\frac{(y+4)^2}{36}=1.$

Solution

The graph of $\frac{x^2}{25}-\frac{y^2}{36}=1$ is translated 3 units right and 4 units down so that the center is at $(3,-4)$ with $a=5$ and $b=6$. The graph is shown here with its asymptotes. The center and the vertices are labeled.

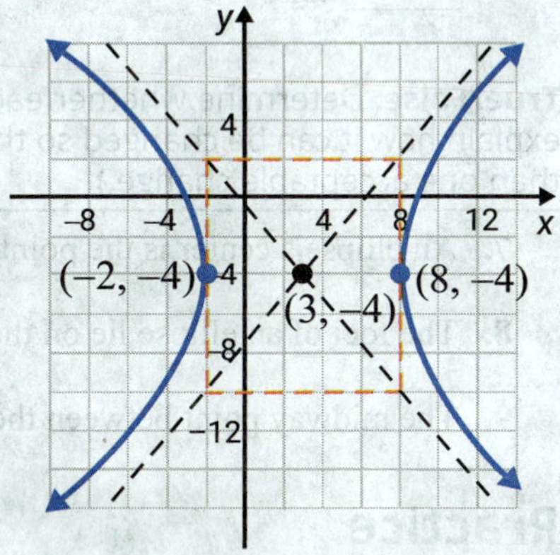

Now work margin exercise 6.

Margin Exercise Answers

1. $\frac{x^2}{9}+\frac{y^2}{4}=1$

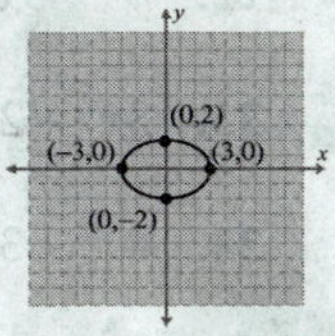

2.

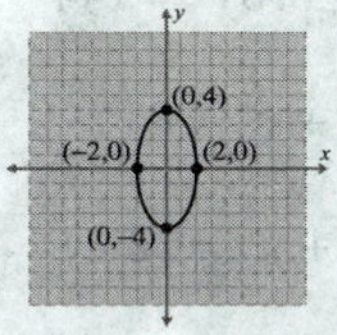

3.

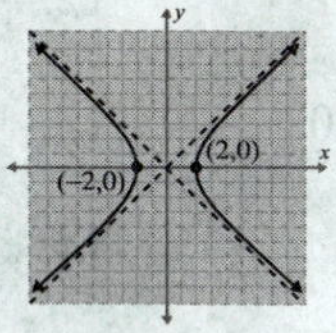

4.

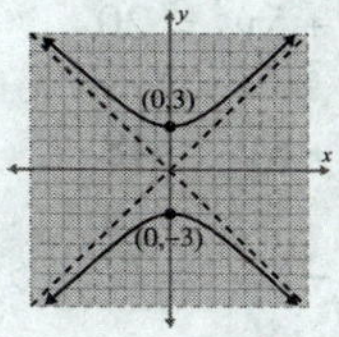

5.

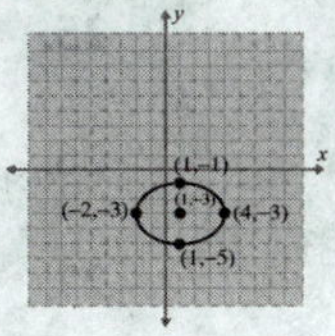

6.

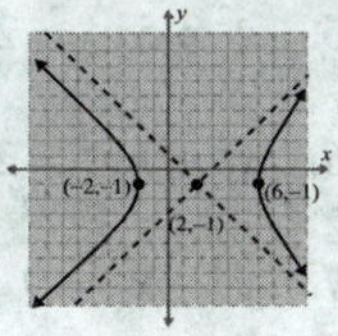

16. $\frac{x^2}{28}+\frac{y^2}{4}=1$

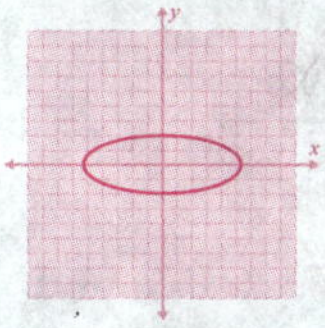

17. $\frac{y^2}{9}-\frac{x^2}{9}=1$

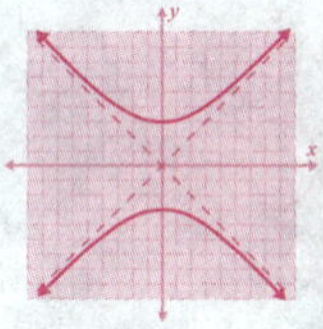

6. Graph the hyperbola $\frac{(x-2)^2}{16}-\frac{(y+1)^2}{25}=1.$

18. $\frac{y^2}{16}-\frac{x^2}{16}=1$

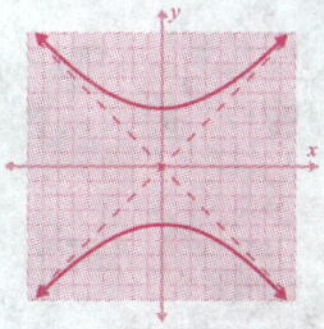

19. $\frac{y^2}{8}-\frac{x^2}{4}=1$

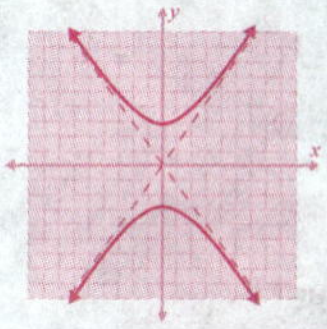

20. $\frac{y^2}{12}-\frac{x^2}{4}=1$

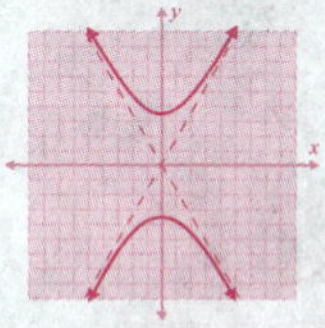

21. $\frac{y^2}{18}-\frac{x^2}{9}=1$

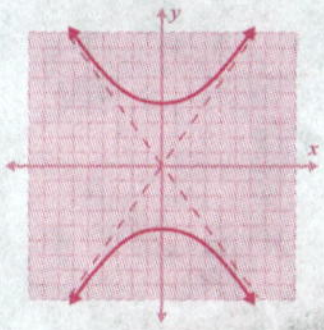

22. $\dfrac{y^2}{20}-\dfrac{x^2}{4}=1$

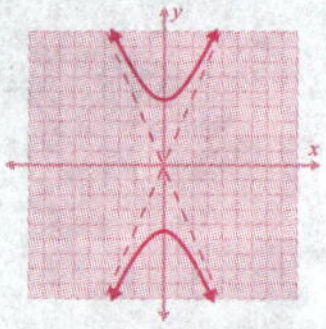

23. $\dfrac{x^2}{6}+\dfrac{y^2}{9}=1$

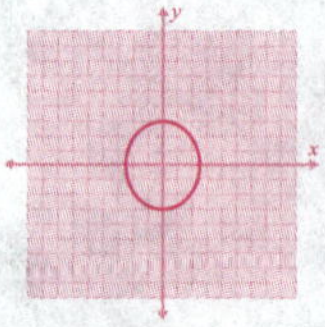

24. $\dfrac{x^2}{3}+\dfrac{y^2}{4}=1$

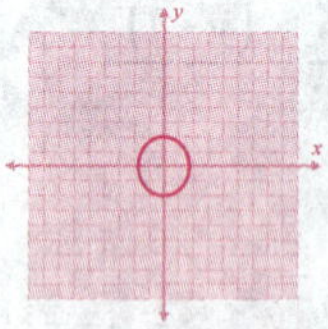

25. $\dfrac{x^2}{5}+\dfrac{y^2}{4}=1$

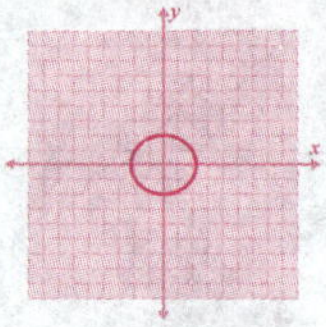

26. $\dfrac{x^2}{16}+\dfrac{y^2}{6}=1$

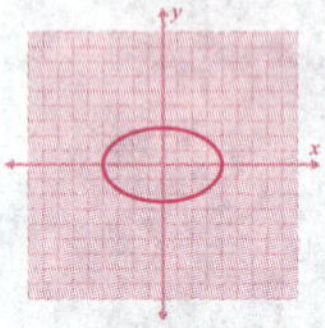

27. $\dfrac{y^2}{8}-\dfrac{x^2}{9}=1$

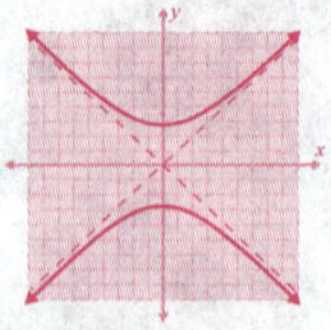

28. $\dfrac{x^2}{7}-\dfrac{y^2}{4}=1$

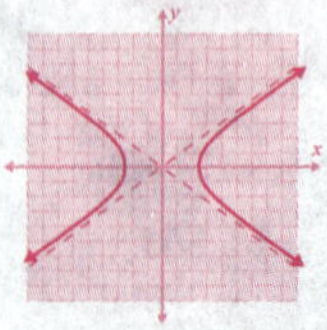

18.4 Exercises

Concept Check

Fill-in-the-Blank. Complete each sentence using information found in this section.

1. Ellipses are conic sections that are ______ in shape.
2. An ellipse is the set of all points in a plane for which the ______ of the distances from two fixed points is ______.
3. Each fixed point of the ellipse is called a/an ______.
4. A hyperbola is a set of all points in a plane such that the absolute value of the ______ of the distances from two fixed points is ______.
5. When $a^2 > b^2$, the segment of length $2b$ joining the y-intercepts is called the ______ axis.
6. When $b^2 > a^2$, the segment of length $2b$ joining the y-intercepts is called the ______ axis.

True/False. Determine whether each statement is true or false. If a statement is false, explain how it can be changed so the statement will be true. (**Note:** There may be more than one acceptable change.)

7. An ellipse's center is the point midway between the two foci.
8. The foci of an ellipse lie on the ellipse.
9. The midway point between the foci of a hyperbola is called the origin.

Practice

Write each of the equations in standard form. Then sketch the graph. For hyperbolas, graph the asymptotes as well. See Examples 1 through 4.

1. $x^2+9y^2=36$
2. $x^2+4y^2=16$
3. $4x^2+25y^2=100$
4. $4x^2+9y^2=36$
5. $16x^2+y^2=16$
6. $36x^2+9y^2=36$
7. $x^2-y^2=1$
8. $x^2-y^2=4$
9. $9x^2-y^2=9$
10. $4x^2-y^2=4$
11. $4x^2-9y^2=36$
12. $9x^2-16y^2=144$
13. $2x^2+y^2=8$
14. $3x^2+y^2=12$
15. $x^2+5y^2=20$
16. $x^2+7y^2=28$
17. $y^2-x^2=9$
18. $y^2-x^2=16$
19. $y^2-2x^2=8$
20. $y^2-3x^2=12$
21. $y^2-2x^2=18$
22. $y^2-5x^2=20$
23. $3x^2+2y^2=18$
24. $4x^2+3y^2=12$
25. $4x^2+5y^2=20$
26. $3x^2+8y^2=48$
27. $9y^2-8x^2=72$
28. $4x^2-7y^2=28$
29. $3y^2-4x^2=36$
30. $3x^2-5y^2=75$

Match the equations with the given graphs. See Examples 5 and 6.

31. $\dfrac{(x-1)^2}{4}+\dfrac{(y-3)^2}{25}=1$

32. $\dfrac{(x+1)^2}{4}+\dfrac{(y+3)^2}{25}=1$

33. $\dfrac{(x-1)^2}{25}+\dfrac{(y-3)^2}{4}=1$

34. $\dfrac{(x+1)^2}{25}+\dfrac{(y+3)^2}{4}=1$

35. $\dfrac{(x+1)^2}{25}-\dfrac{(y+3)^2}{4}=1$

36. $\dfrac{(y+3)^2}{4}-\dfrac{(x+1)^2}{25}=1$

a.

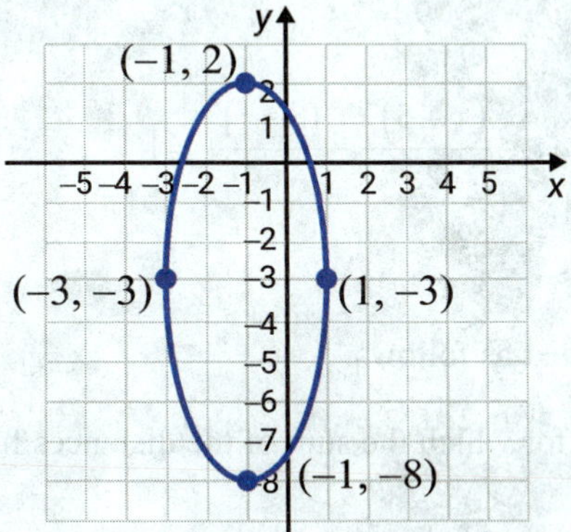

d.

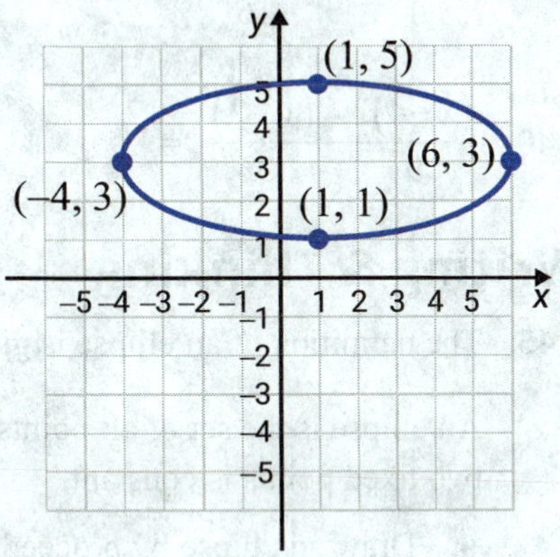

b.

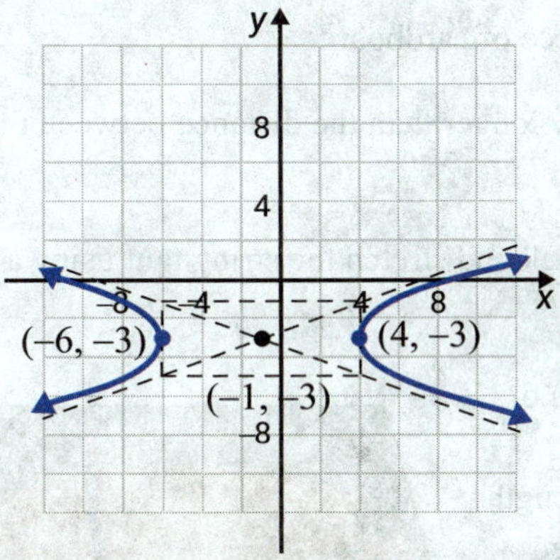

e.

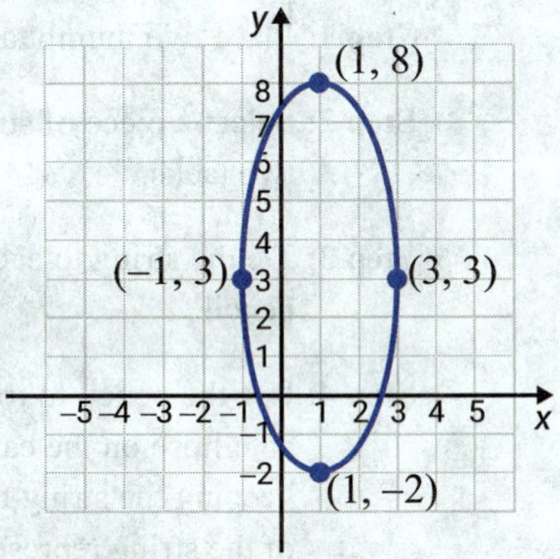

c.

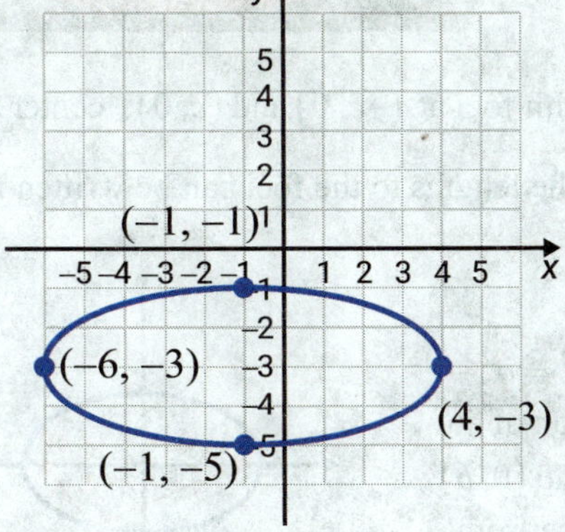

f.

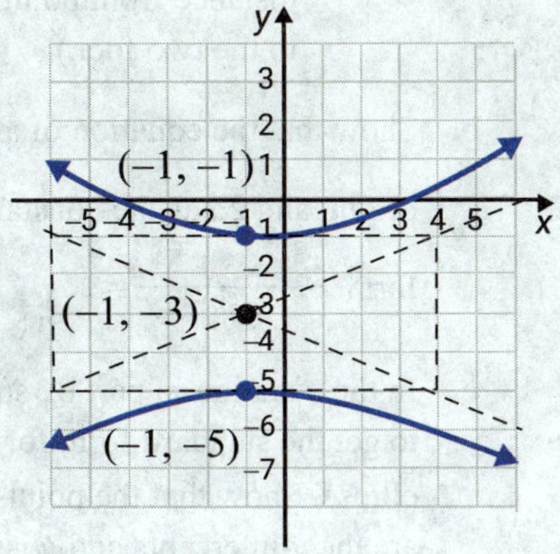

29. $\dfrac{y^2}{12}-\dfrac{x^2}{9}=1$

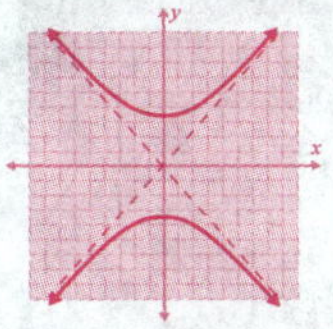

30. $\dfrac{x^2}{25}-\dfrac{y^2}{15}=1$

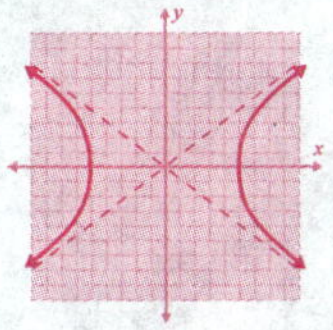

31. e

32. a

33. d

34. c

35. b

36. f

37.

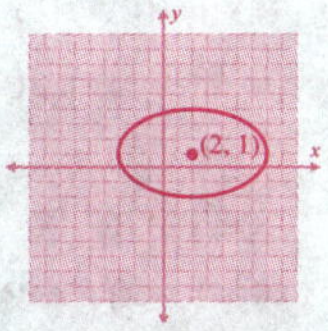

38.

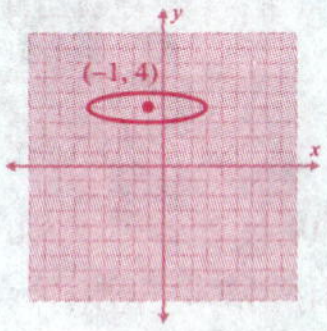

39.

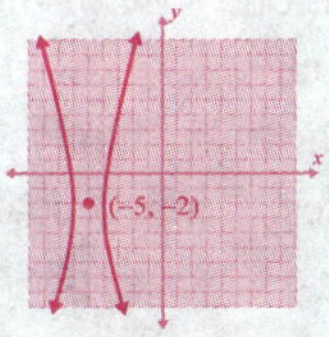

40.

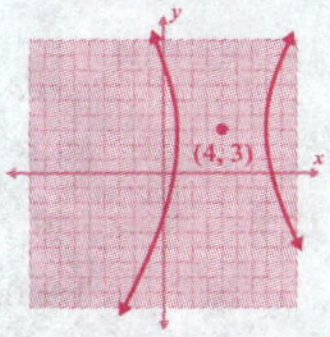

41.

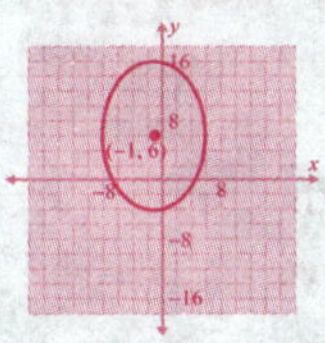

42.

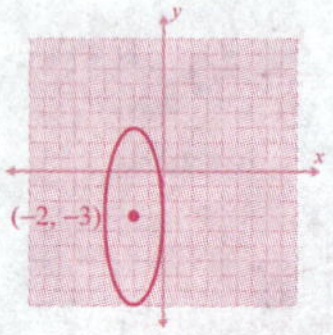

43.

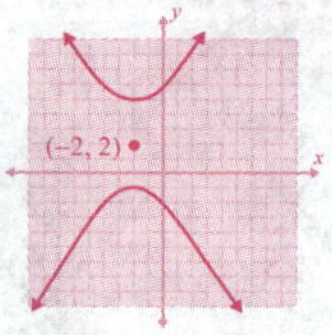

44.

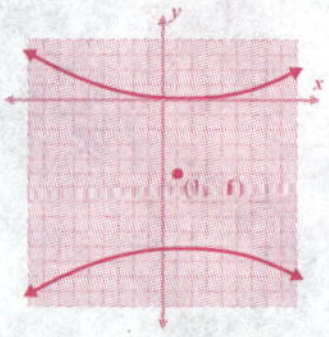

Writing & Thinking

45. b. Pick an arbitrary point (x, y) located on the ellipse. The sum of the distances from (x, y) to the points $(-c, 0)$ and $(c, 0)$ (the foci) will equal $2a$. Using the distance formula we get the equation $\sqrt{(x+c)^2+(y-0)^2}+\sqrt{(x-c)^2+(y-0)^2}=2a.$ Now to remove the radicals, simplify so that one side of the equation is a single radical and square both sides. Then repeat the previous step and simplify. This gives $\frac{x^2}{a^2}+\frac{y^2}{a^2-c^2}=1.$

c. At the y-intercept $(0, b)$ a right triangle is formed with hypotenuse a and sides b and c. The Pythagorean Theorem gives $a^2=b^2+c^2$.

Use your knowledge of translations to graph each of the following equations. These graphs are ellipses and hyperbolas with centers at points other than the origin. See Examples 5 and 6.

37. $\frac{(x-2)^2}{25}+\frac{(y-1)^2}{9}=1$

38. $\frac{(x+1)^2}{16}+\frac{(y-4)^2}{1}=1$

39. $\frac{(x+5)^2}{1}-\frac{(y+2)^2}{16}=1$

40. $\frac{(x-4)^2}{9}-\frac{(y-3)^2}{36}=1$

41. $\frac{(x+1)^2}{49}+\frac{(y-6)^2}{100}=1$

42. $\frac{(x+2)^2}{4}+\frac{(y+3)^2}{36}=1$

43. $\frac{(y-2)^2}{9}-\frac{(x+2)^2}{4}=1$

44. $\frac{(y+5)^2}{25}-\frac{(x-1)^2}{64}=1$

Writing & Thinking

45. The definition of an ellipse is given in the text as follows.

An ellipse is the set of all points in a plane for which the sum of the distances from two fixed points is constant.

a. Draw an ellipse by proceeding as follows.

Step 1: Place two thumb tacks in a piece of cardboard.

Step 2: Select a piece of string slightly longer than the distance between the two tacks.

Step 3: Tie the string to each thumb tack and stretch the string taut using a pencil.

Step 4: Use the pencil to trace the path of an ellipse on the cardboard by keeping the string taut. (The length of the string represents the fixed distance from points on the ellipse to the two foci.)

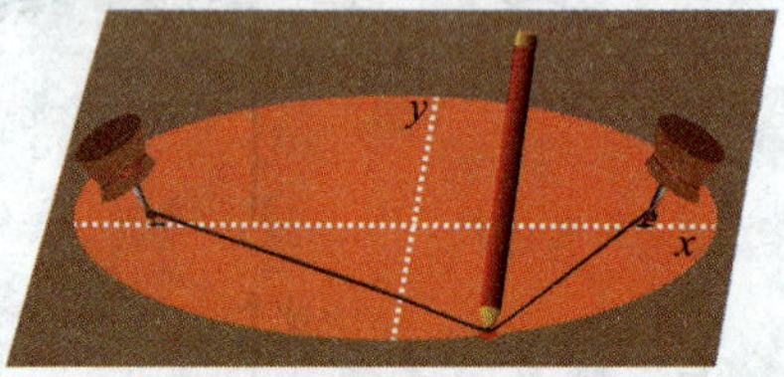

b. Show that the equation of an ellipse with foci at $(-c, 0)$ and $(c, 0)$, center at the origin, and $2a$ as the constant sum of the lengths to the foci can be written in the form $\frac{x^2}{a^2}+\frac{y^2}{a^2-c^2}=1.$

c. In the equation in part b., substitute $b^2=a^2-c^2$ to get the standard form for the equation of an ellipse. Show that the points $(0, -b)$ and $(0, b)$ are the y-intercepts and a is the distance from each y-intercept to a focus.

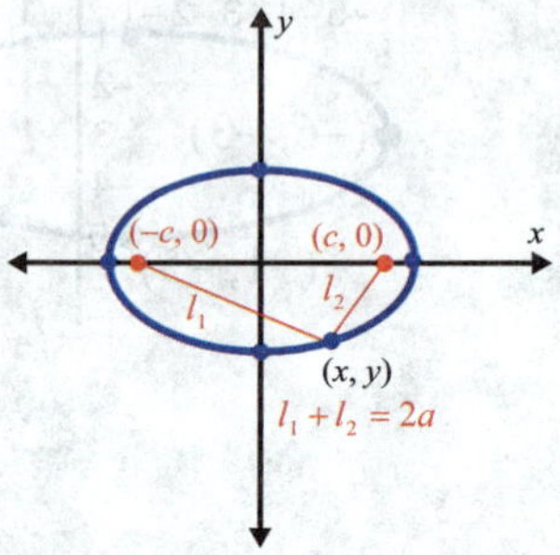

18.5 Nonlinear Systems of Equations

Objectives

A. Solve systems of equations where one or both equations are nonlinear.

The equations for the conic sections that we have discussed each have at least one term that is second-degree. These equations are called **nonlinear equations**. (**Note:** Only the equations for parabolas that can be written in the form $y = ax^2 + bx + c$ are **quadratic equations**.) A summary of the equations with their related graphs is shown in Figure 1.

Summary of Equations and Related Graphs

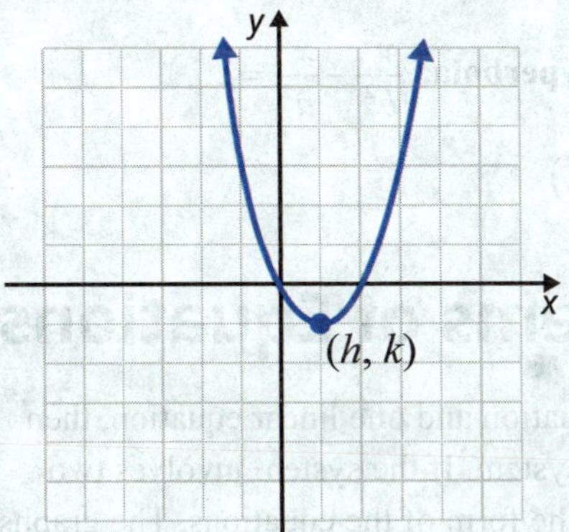

Parabola: $y = a(x-h)^2 + k$

$a > 0$, opens upward

$a < 0$, opens downward

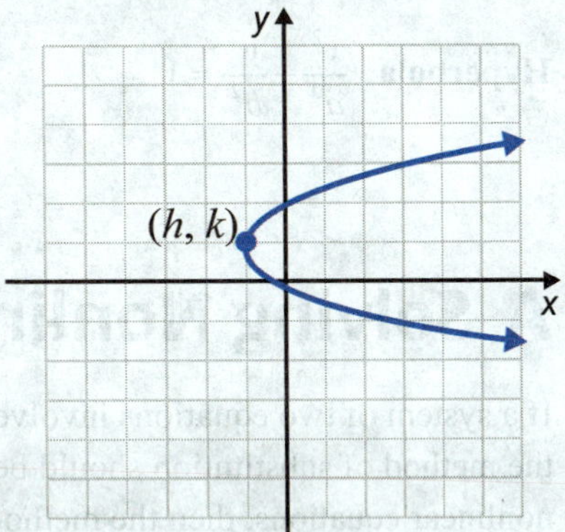

Parabola: $x = a(y-k)^2 + h$

$a > 0$, opens right

$a < 0$, opens left

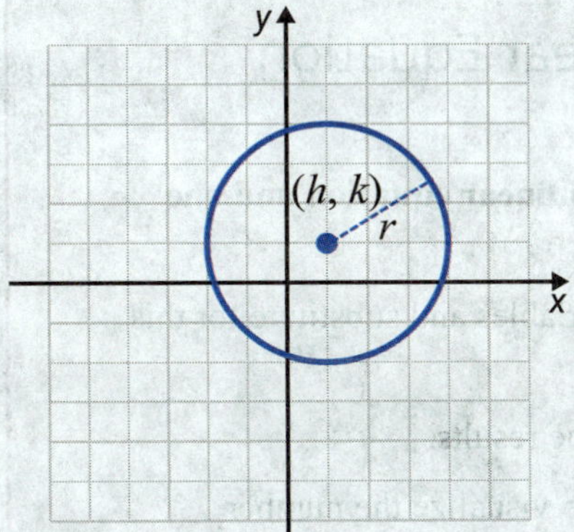

Circle: $(x-h)^2 + (y-k)^2 = r^2$

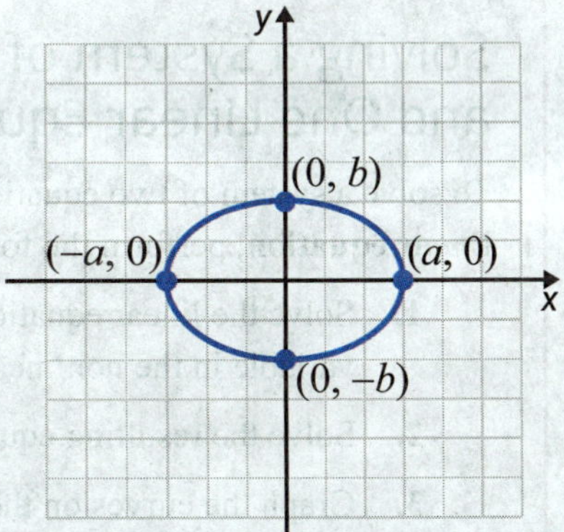

Ellipse: $\dfrac{x^2}{a^2} + \dfrac{y^2}{b^2} = 1$

Figure 1

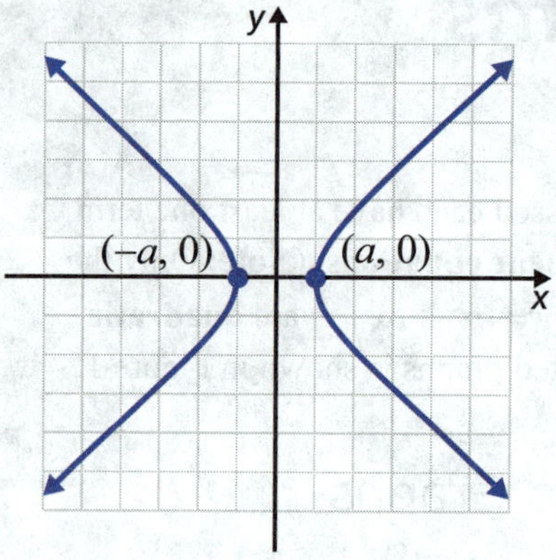

Hyperbola: $\frac{x^2}{a^2} - \frac{y^2}{b^2} = 1$

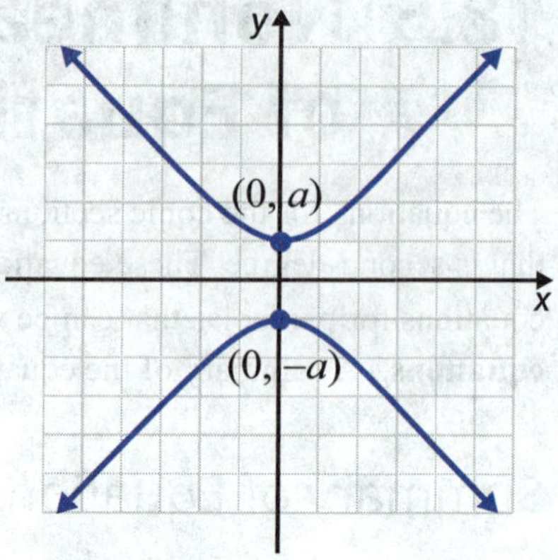

Hyperbola: $\frac{y^2}{b^2} - \frac{x^2}{a^2} = 1$

Figure 1 (cont.)

A Solving Nonlinear Systems of Equations

If a system of two equations involves one nonlinear equation and one linear equation, then the method of substitution should be used to solve the system. If the system involves two nonlinear equations, then the method used depends on the form of the equations. The graphs of the curves are particularly useful for approximating solutions and determining the exact number of solutions.

Examples 1 through 3 illustrate three possibilities and emphasize the value of sketching the graphs to visualize the number of solutions and approximating these solutions.

Solving a System of One Nonlinear Equation and One Linear Equation

To solve a system of two equations that has **one nonlinear equation and one linear equation**, perform the following steps.

1. Solve the linear equation for one of the variables and substitute for this variable in the nonlinear equation.
2. Solve the resulting equation and analyze the results.
3. Graph the curves on the same set of axes to visualize the number of solutions and check that the solutions are reasonable and satisfy both equations.

PROCEDURE

Example 1 Solving a System of One Nonlinear and One Linear Equation

Solve the following system of equations and graph both curves on the same set of axes.

A circle and a line: $\begin{cases} x^2 + y^2 = 25 & \text{Circle} \\ x + y = 5 & \text{Line} \end{cases}$

Solution

Solve $x + y = 5$ for y (or x). Then substitute into the other equation.

$$y = 5 - x$$
$$x^2 + (5 - x)^2 = 25$$
$$x^2 + 25 - 10x + x^2 = 25$$
$$2x^2 - 10x = 0$$
$$2x(x - 5) = 0 \qquad \text{Now, solve for } x.$$

$x = 0$, $y = 5 - 0 = 5$ or $x = 5$, $y = 5 - 5 = 0$

The solutions (points of intersection) are $(0, 5)$ and $(5, 0)$.

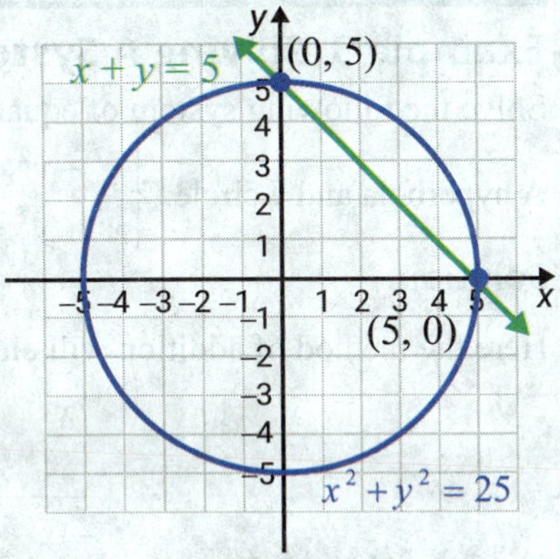

Now work margin exercise 1.

1. Solve the system and graph both curves.

$\begin{cases} x - y = 6 \\ x^2 + y^2 = 36 \end{cases}$

Math Tip

You might recognize that a system involving a linear and a nonlinear equation has the potential to intersect at multiple points. However, it is also possible for the graphs to intersect at only one point or to not intersect at all.

Example 2 Solving a System of One Nonlinear and One Linear Equation

Solve the following system of equations and graph both curves on the same set of axes.

A line and a parabola: $\begin{cases} x + y = -7 & \text{Line} \\ y = x^2 - 4x - 5 & \text{Parabola} \end{cases}$

Solution

Solve the linear equation for y (or x), then substitute into the quadratic equation. (In this case, the quadratic equation is already solved for y, so we could have chosen to make the substitution into the linear equation instead.)

$$y = -x - 7$$
$$(-x - 7) = x^2 - 4x - 5$$
$$0 = x^2 - 3x + 2$$
$$0 = (x - 2)(x - 1)$$

$x = 2$, $y = -2 - 7 = -9$ or $x = 1$, $y = -1 - 7 = -8$

The solutions are $(2, -9)$ and $(1, -8)$.

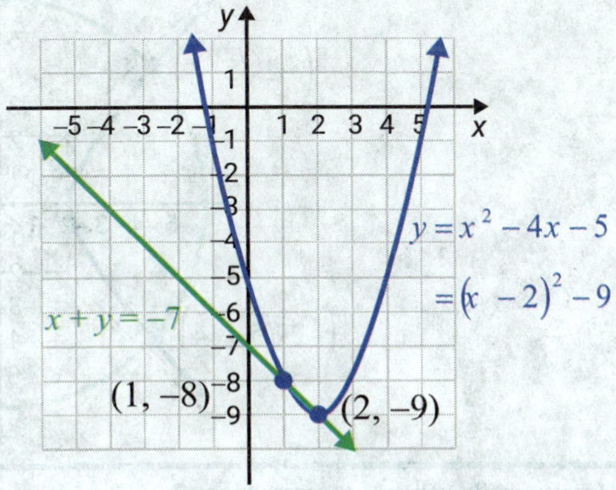

Now work margin exercise 2.

2. Solve the system and graph both curves.

$\begin{cases} y = x + 3 \\ y = -x^2 + 9 \end{cases}$

Solving a System of Two Nonlinear Equations

To solve a system of two equations that has **two nonlinear equations**:

1. The method used depends on the form of the equations.
2. Substitution may work or addition may work.
3. Graph the curves on the same set of axes to visualize the number of solutions and check that the solutions are reasonable and satisfy both equations.

PROCEDURE

3. Solve the system and graph both curves.

$$\begin{cases} x^2 + y^2 = 25 \\ y = x^2 - 5 \end{cases}$$

Example 3 Solving A System of Two Nonlinear Equations

Solve the following system of equations and graph both curves on the same set of axes.

A hyperbola and a circle: $\begin{cases} x^2 - y^2 = 4 & \text{Hyperbola} \\ x^2 + y^2 = 36 & \text{Circle} \end{cases}$

Solution

Here the method of addition will eliminate y^2.

$$\begin{aligned} x^2 - y^2 &= 4 \\ x^2 + y^2 &= 36 \\ \hline 2x^2 &= 40 \\ x^2 &= 20 \\ x &= \pm\sqrt{20} = \pm 2\sqrt{5} \end{aligned}$$

If $x = 2\sqrt{5}$: $\left(2\sqrt{5}\right)^2 + y^2 = 36$ If $x = -2\sqrt{5}$: $\left(-2\sqrt{5}\right)^2 + y^2 = 36$

$20 + y^2 = 36$ $20 + y^2 = 36$

$y^2 = 16$ $y^2 = 16$

$y = \pm 4$ $y = \pm 4$

There are four points of intersection:

$$\left(2\sqrt{5}, 4\right), \left(2\sqrt{5}, -4\right), \left(-2\sqrt{5}, 4\right), \text{ and } \left(-2\sqrt{5}, -4\right)$$

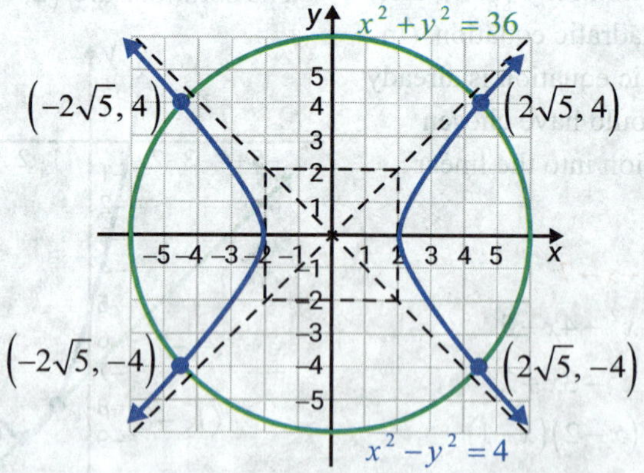

Now work margin exercise 3.

Attention!

In the examples and the exercises, the curves intersect. However, there are many situations where the curves do **not** intersect. This can be confirmed both algebraically and graphically.

Margin Exercise Answers

1. $(0,-6)$ and $(6,0)$. **2.** $(-3,0)$ and $(2,5)$. **3.** $(0,-5)$, $(-3,4)$, and $(3,4)$.

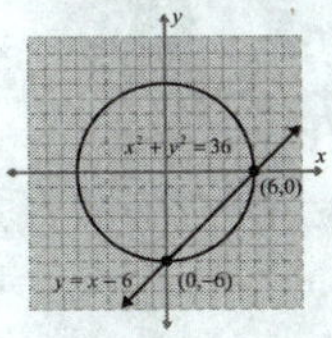

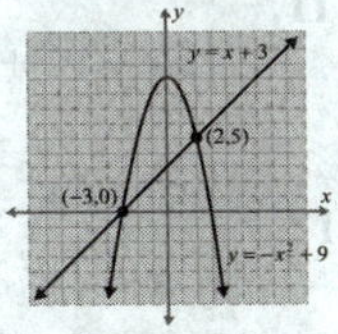

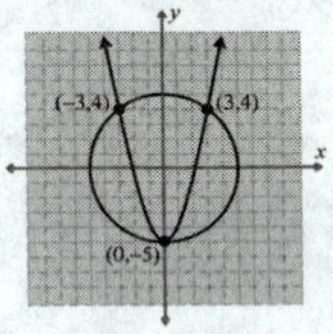

18.5 Exercises

Concept Check

Fill-in-the-Blank. Complete each sentence using information found in this section.

1. The equations for the conic sections presented in this lesson all have at least one term that is ______-degree.
2. If a system of two equations has one nonlinear equation and one linear equation, then the method of ______ should be used to solve the system.
3. If a system of two equations has two nonlinear equations, then either the ______ method or the ______ method will work.
4. When solving a nonlinear system of equations, the final step is to ______ the curves on the same set of axes.
5. Solutions to a system of two equations must satisfy ______ equations.

True/False. Determine whether each statement is true or false. If a statement is false, explain how it can be changed so the statement will be true. (**Note:** There may be more than one acceptable change.)

6. Equations that have at least one second-degree term are called linear equations.
7. Graphing a system of equations is useful for approximating solutions.
8. Graphing a system of equations will determine the exact number of solutions.
9. If a system of equations has two nonlinear equations, the curves must intersect.

18.5 Exercises

Concept Check

1. second
2. substitution
3. substitution, addition
4. graph
5. both
6. False; These equations are called nonlinear equations.
7. True
8. True
9. False; There are many situations where the curves do not intersect.

Practice

1. $(-3,10),(1,2)$

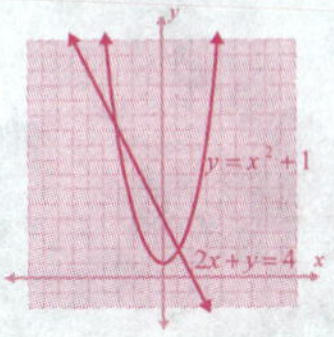

2. $(-2,-1),(3,-6)$

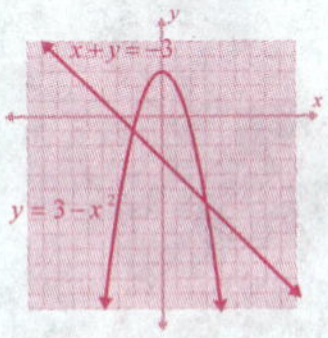

3. $(1,1),(2,0)$

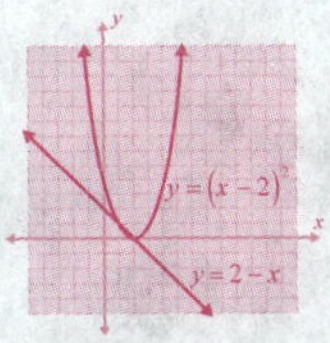

4. $(0,-5),(-5,0)$

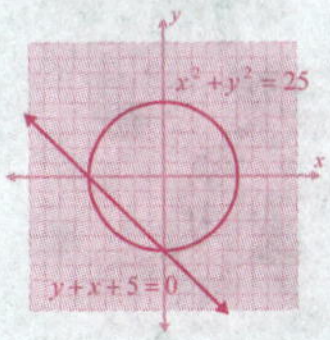

5. $(-2,-4),(4,2)$

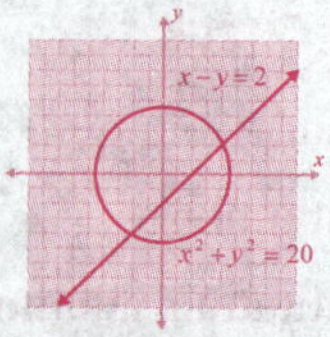

6. $(-5,3),(5,-3)$

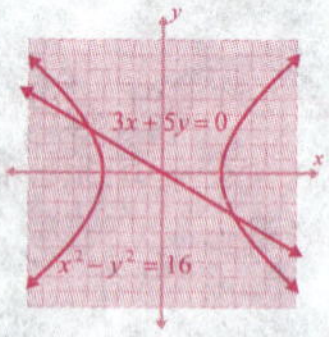

7. $(5,3)$

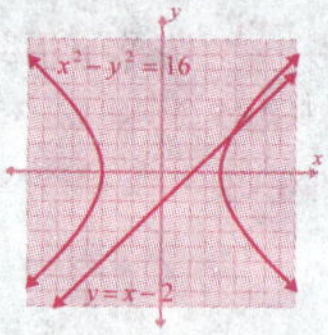

8. $(3,1),(-3,-1)$

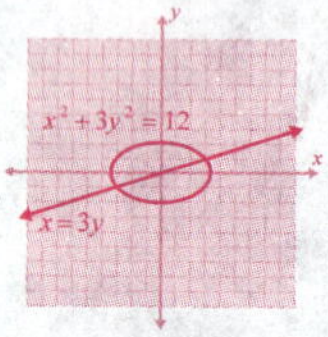

9. $(-3,0),(3,0)$

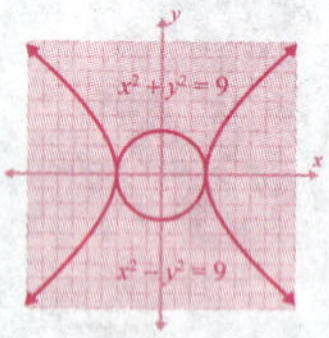

10. $(0,3)$

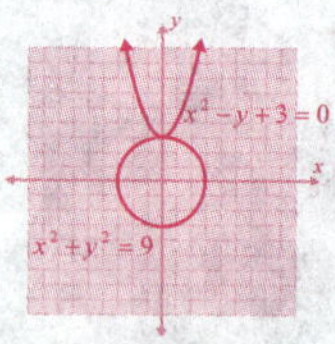

11. $(2,3),(2,-3)$

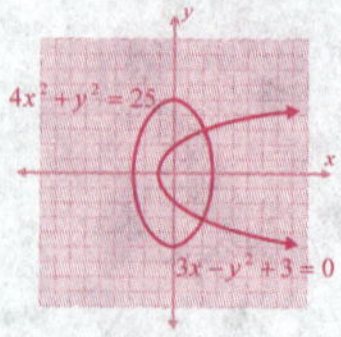

Practice

Solve each system of equations. Sketch the graphs.

1. $\begin{cases} y = x^2 + 1 \\ 2x + y = 4 \end{cases}$

2. $\begin{cases} y = 3 - x^2 \\ x + y = -3 \end{cases}$

3. $\begin{cases} y = 2 - x \\ y = (x-2)^2 \end{cases}$

4. $\begin{cases} x^2 + y^2 = 25 \\ y + x + 5 = 0 \end{cases}$

5. $\begin{cases} x^2 + y^2 = 20 \\ x - y = 2 \end{cases}$

6. $\begin{cases} x^2 - y^2 = 16 \\ 3x + 5y = 0 \end{cases}$

7. $\begin{cases} y = x - 2 \\ x^2 = y^2 + 16 \end{cases}$

8. $\begin{cases} x^2 + 3y^2 = 12 \\ x = 3y \end{cases}$

9. $\begin{cases} x^2 + y^2 = 9 \\ x^2 - y^2 = 9 \end{cases}$

10. $\begin{cases} x^2 + y^2 = 9 \\ x^2 - y + 3 = 0 \end{cases}$

11. $\begin{cases} 4x^2 + y^2 = 25 \\ 3x - y^2 + 3 - 0 \end{cases}$

12. $\begin{cases} x^2 - 4y^2 = 9 \\ x + 2y^2 = 3 \end{cases}$

13. $\begin{cases} x^2 + y^2 + 4x - 2y = 4 \\ x + y = 2 \end{cases}$

14. $\begin{cases} x^2 - y^2 = 9 \\ x^2 + y^2 - 2x - 3 = 0 \end{cases}$

15. $\begin{cases} x^2 - y^2 = 5 \\ x^2 + 4y^2 = 25 \end{cases}$

16. $\begin{cases} 2x^2 - 3y^2 = 6 \\ 2x^2 + y^2 = 22 \end{cases}$

Solve each system of equations.

17. $\begin{cases} x^2 - y^2 = 20 \\ x^2 - 9y = 0 \end{cases}$

18. $\begin{cases} x^2 + 5y^2 = 16 \\ x^2 + y^2 = 4x \end{cases}$

19. $\begin{cases} x^2 + y^2 = 10 \\ x^2 + y^2 - 4y + 2 = 0 \end{cases}$

20. $\begin{cases} x^2 + y^2 = 20 \\ 4x + 8 = y^2 \end{cases}$

21. $\begin{cases} x^2 + y^2 - 4y = 16 \\ x - y = 0 \end{cases}$

22. $\begin{cases} y = x^2 + 2x + 2 \\ 2x + y = 2 \end{cases}$

23. $\begin{cases} 4y + 10x^2 + 7x - 8 = 0 \\ 6x - 8y + 1 = 0 \end{cases}$

24. $\begin{cases} x^2 + y^2 - 4x + 6y = -3 \\ 2x - y - 2 = 0 \end{cases}$

25. $\begin{cases} 2x^2 - y^2 = 7 \\ 2x^2 + y^2 = 29 \end{cases}$

26. $\begin{cases} 4x^2 + y^2 = 11 \\ y = 4x^2 - 9 \end{cases}$

27. $\begin{cases} x^2 - y^2 - 2y = 22 \\ 2x + 5y + 5 = 0 \end{cases}$

28. $\begin{cases} x^2 + y^2 - 6y = 0 \\ 2x^2 - y^2 + 15 = 0 \end{cases}$

29. $\begin{cases} y = x^2 - 2x + 3 \\ y = -x^2 + 2x + 3 \end{cases}$

30. $\begin{cases} y^2 = x^2 - 5 \\ 4x^2 - y^2 = 32 \end{cases}$

Use a graphing calculator to graph and estimate the solution(s) to each system of equations. If necessary, round values to two decimal places.

31. $\begin{cases} y = x^2 + 3 \\ x + y = 4 \end{cases}$

32. $\begin{cases} y = 1 - x^2 \\ x + y = -4 \end{cases}$

33. $\begin{cases} y = 3 - 2x \\ y = (x-1)^2 \end{cases}$

34. $\begin{cases} x^2 + y^2 = 36 \\ y = x + 5 \end{cases}$

35. $\begin{cases} x^2 + y^2 = 10 \\ x - y = 1 \end{cases}$

36. $\begin{cases} x^2 + y^2 = 4 \\ x^2 - y^2 = 3 \end{cases}$

12. $(-5,-2),(-5,2),(3,0)$

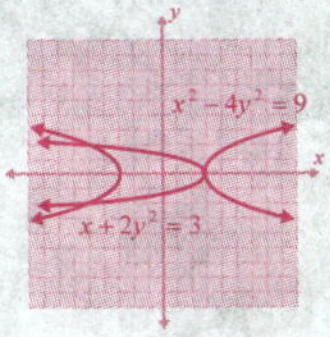

13. $(-2,4),(1,1)$

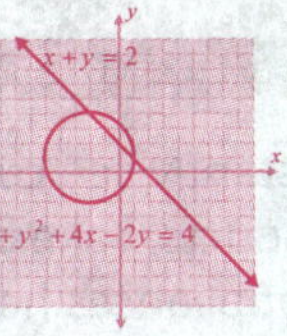

14. $(3,0)$

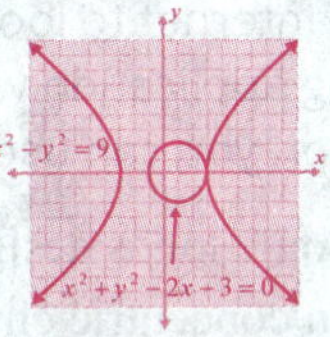

15. $(-3,-2),(3,2),$
$(-3,2),(3,-2)$

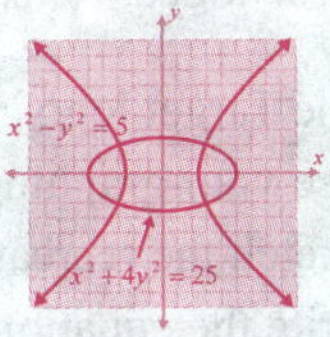

16. $(-3,-2),(3,2),$
$(-3,2),(3,-2)$

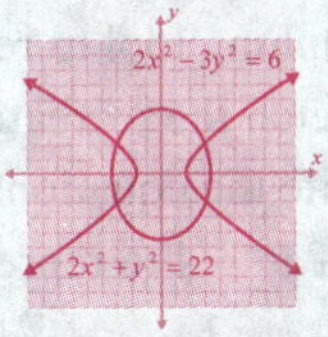

17. $(-3\sqrt{5},5),(-6,4),$
$(3\sqrt{5},5),(6,4)$

18. $(1,-\sqrt{3}),(1,\sqrt{3}),(4,0)$

19. $(1,3),(-1,3)$

20. $(2,4),(2,-4)$

21. $(4,4),(-2,-2)$

22. $(-4,10),(0,2)$

23. $\left(-\frac{3}{2},-1\right),\left(\frac{1}{2},\frac{1}{2}\right)$

24. $(1,0),(-1,-4)$

25. $(-3,\sqrt{11}),(-3,-\sqrt{11}),$
$(3,\sqrt{11}),(3,-\sqrt{11})$

26. $\left(\frac{\sqrt{7}}{2},-2\right),\left(-\frac{\sqrt{7}}{2},-2\right),$
$\left(\frac{\sqrt{10}}{2},1\right),\left(-\frac{\sqrt{10}}{2},1\right)$

27. $(5,-3),(-5,1)$

28. $(\sqrt{5},5),(-\sqrt{5},5)$

29. $(0,3),(2,3)$

30. $(3,2),(3,-2),$
$(-3,2),(-3,-2)$

31. $(-1.62,5.62),(0.62,3.38)$

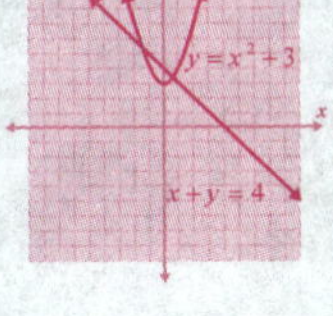

32. $(2.79,-6.79),$
$(-1.79,-2.21)$

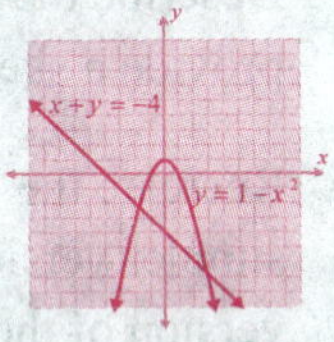

33. $(1.41,0.17),(-1.41,5.83)$

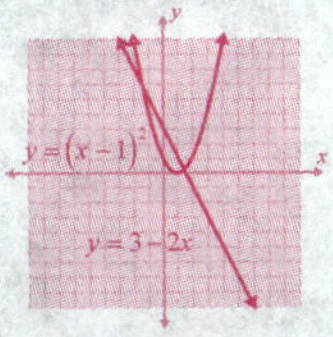

34. $(0.93,5.93),$
$(-5.93,-0.93)$

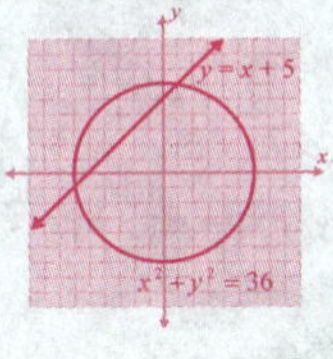

35. $(2.68,1.68),$
$(-1.68,-2.68)$

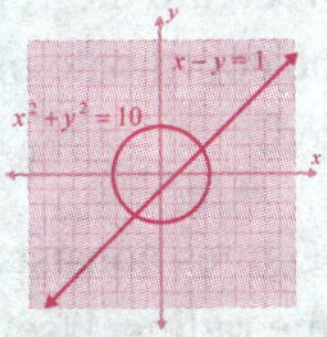

36. $(-1.87,0.71),$
$(1.87,0.71),$
$(-1.87,-0.71),$
$(1.87,-0.71)$

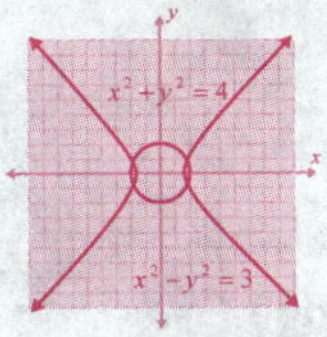

Chapter 18 Project

What's in a Logo?

An activity to demonstrate the use of conic sections in real life.

You constantly see company logos in your day-to-day life. Companies use logos for brand recognition and to advertise their products. Take a look around you as you go throughout your day and you'll see company logos everywhere. Your kitchen appliances have logos on them, your car has a logo on it, your smartphone has a logo, and logos appear in all forms of advertisements. To make a logo clean and easy to remember, a company may use conic sections in the design.

Use the internet to look up or research the logos mentioned in the problems that follow. A graphing calculator or graphing application will be needed to recreate the logos indicated in the problems. If you don't have a graphing calculator, you can use the free graphing application located at www.desmos.com.

1. Some logos designs, such as the Target logo, are based on concentric circles. Circles are concentric when they share the same center. Perform an internet search for other logos that use concentric circles. Draw three of these logos. Answers will vary.

2. The Target logo consists of three concentric circles. Use your graphing tool to recreate the Target logo. Sketch the recreation on the coordinate plane and write the equations you used.

3. The logo for the Olympic Games consists of five circles that all have the same radius. Use your graphing tool to recreate the Olympic Games logo. Sketch the recreation on the coordinate plane and write the equations you used. Use the equation $x^2 + y^2 = 4$ to represent the center ring on the top row.

4. The Toyota logo consists of three ellipses. Use your graphing tool to recreate the Toyota logo.Sketch the recreation on the coordinate plane and write the equations you used. Answers will vary.

5. Find an existing logo or create your own logo that consists of more than one type of conic section. Describe the conic sections used in the logo and provide a sketch of the logo. Answers will vary.

6. Recreate the logo by graphing conic sections. Create a simpler version of the logo, if necessary. Sketch your version of the logo on the coordinate plane and write the equations you used. Answers will vary.

Chapter 18 Project: What's in a Logo?

2. Answers will vary. Possible answer: $x^2 + y^2 = 2$, $x^2 + y^2 = 4$, $x^2 + y^2 = 8$

3. Answers will vary. Possible answer: $x^2 + y^2 = 4$, $(x + 4)^2 + y^2 = 4$, $(x - 4)^2 + y^2 = 4$, $(x + 2)^2 + (y + 2)^2 = 4$, $(x - 2)^2 + (y + 2)^2 = 4$

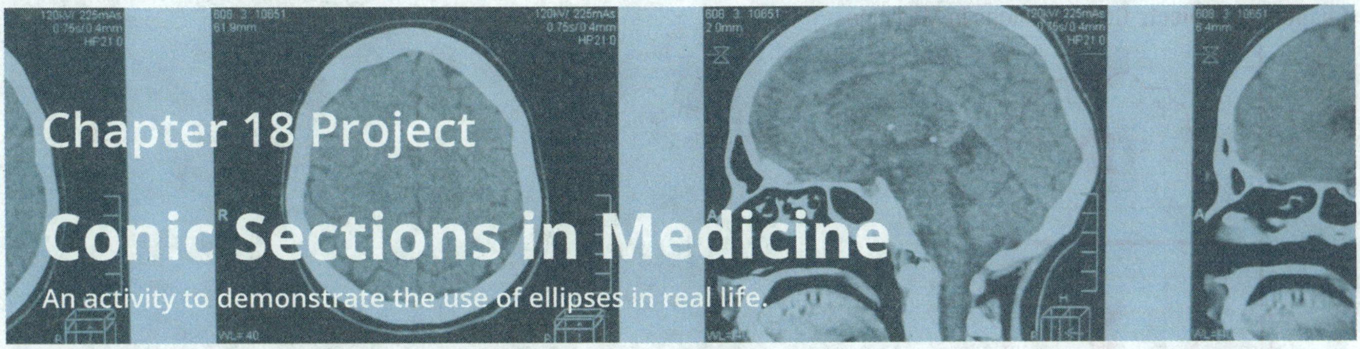

Chapter 18 Project

Conic Sections in Medicine

An activity to demonstrate the use of ellipses in real life.

The following image shows a rough sketch of the cross-section of a human brain with a grid overlaid. The shape is roughly elliptical, with the major axis being vertical.

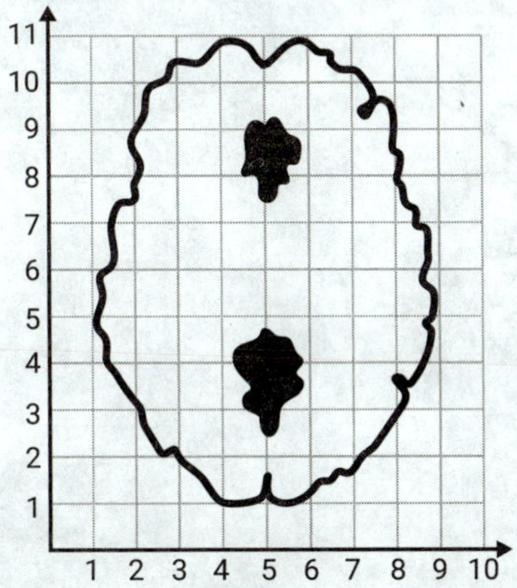

1. a. Sketch a line segment on the vertical grid lines so that the segment going through the cross-section is as long as it can be and the portions to the left and right of the line segment are roughly symmetric.

   b. Sketch a horizontal line segment on the horizontal grid lines so that the segment going through the cross-section is as long as it can be and the portions above and below the line segment are roughly symmetric.

2. Let the intersection of the two segments be the center of the ellipse. State the coordinates of the following as (x, y) ordered pairs.

   a. The center

   b. The vertices on the major axis

   c. The vertices on the minor axis

3. State the lengths of the major and minor axes.

4. Write the equation of the ellipse.

5. Sketch the ellipse over the grid with the brain cross-section. (As a check, the point (8, 9) should be very close to the ellipse, but not on the ellipse.)

6. Given $x = 8$, use your equation of the ellipse to solve for the larger of the two values of y. Write this point in (x, y) form with the y-value as a simplified square root and rounded to the nearest tenth.

7. Suppose there is a tumor indicated in the scan and the closest part of the tumor to the point determined in Problem 6 is at (6, 8).

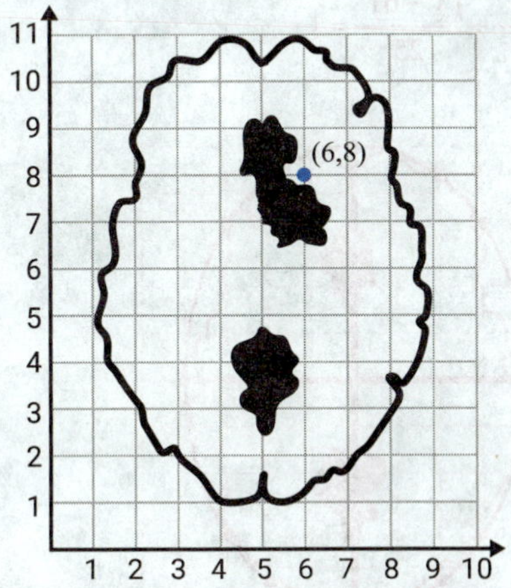

   a. Calculate the distance from the point found in Problem 6 to the point (6, 8) and round the answer to the nearest tenth.

   b. Suppose after chemotherapy that the closest part of the tumor to the point determined in Problem 6 is (5, 7.5) Calculate the distance from this point to the point found in Problem 6.

   c. Compare the distances from parts a. and b., and provide a probable reason for the differing values.

Chapter 18 Project: Conic Sections in Medicine

1. 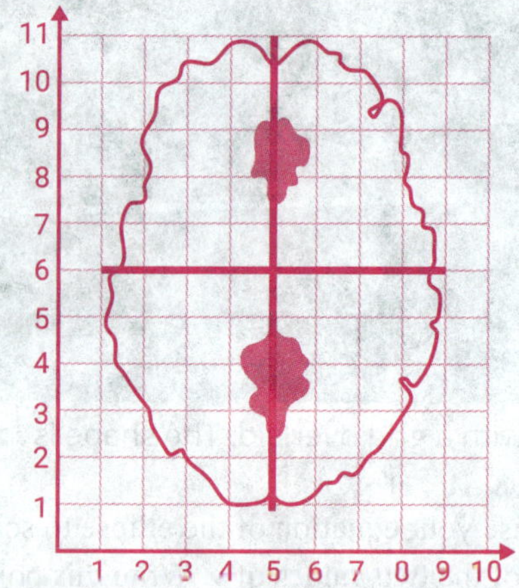

2. **a.** The center is at (5,6).

b. The vertices on the major axis are at (5,1) and (5,11).

c. The vertices on the minor axis are at (1,6) and (9,6).

3. The major axis is 5 units long, while the minor axis is 4 units long.

4. $\frac{(x-5)^2}{16}+\frac{(y-6)^2}{25}=1$

5. 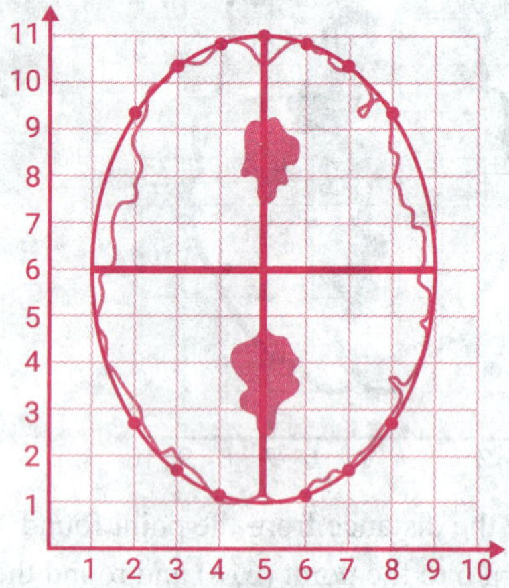

6. $\left(8, 6+\frac{5\sqrt{7}}{4}\right), (8, 9.3)$

7. **a.** $d=\sqrt{(x_2-x_1)^2+(y_2-y_1)^2}=\sqrt{(8-6)^2+(9.3-8)^2}$
≈ 2.4 units.

b. $d=\sqrt{(x_2-x_1)^2+(y_2-y_1)^2}=\sqrt{(8-5)^2+(9.3-7.5)^2}$
≈ 3.5 units.

c. The distance in part b. is greater. Since it is now further from a point on the edge of the brain, this may indicate the chemotherapy has caused the tumor to shrink.

Answer Key

Chapter 1: Whole Numbers

1.1 Exercises

Concept Check

1. 1

3. Counting

5. 5000 + 0 + 10 + 2

7. True

9. True

Practice

1. **a.** 6 **b.** 8 **c.** 0

3. **a.** Ten thousands
b. Hundreds
c. Ones

5. 2: ten thousands, 4: thousands, 6: hundreds, 8: ones

7. 2: billions, 4: hundred millions, 3: millions, 1: hundred thousands, 8: ten thousands, 9: thousands, 5: hundreds

9. 30 + 7

11. 500 + 60 + 3

13. 1000 + 800 + 90 + 2

15. 20,000 + 5000 + 600 + 50 + 8

17. Eight hundred thirty-five

19. Thirty thousand, two hundred one

21. Six hundred eighty-three thousand, one hundred

23. Sixteen million, three hundred two thousand, five hundred ninety

25. 580

27. 2005

29. 3834

31. 78,902

33. 63,065

35. 400,736

37. 82,700,000

39. 35,960,013

Applications

41. Eighty-two thousand, one hundred three

43. One hundred ten; two thousand six hundred fifty

45. Three hundred fifty-two thousand, one hundred forty-three; nine hundred twelve thousand, fifty

47. Ten thousand, six hundred sixty-seven; sixty-six thousand, four hundred ninety-six

49. 8520; 2000

51. 44; 218,792

53. 193,046; 184,327

55. 10,533,871; 3,154,000

57. No; 480,105

59. Sam Burns

61. Will Zalatoris

63. Seven million, twelve thousand, six hundred seventy-two

65. *Wicked*

67. *Plaza Suite,* seven thousand, seven hundred ninety-one

69. *Plaza Suite, Moulin Rouge! The Musical, MJ The Musical,* and *Six*

71. *Wicked*

73. 40,000,000 + 7,000,000 + 500,000 + 40,000 + 7000 + 200 + 30 + 1

75. $183,651,655; One hundred eighty-three million, six hundred fifty-one thousand, six hundred fifty-five dollars

Writing & Thinking

77. 10^{100}

79. Hyphens are used for two-digit numbers larger than 20 that do not end in a zero.

1.2 Exercises

Concept Check

1. sum

3. associative

5. perimeter

7. False; A polygon has three or more sides.

9. False; Borrowing must occur.

Practice

1. 16

3. 10

5. 25

7. Commutative property of addition

9. Associative property of addition

11. Additive identity

13. 78 **15.** 74

17. 361 **19.** 689

21. 16,072

23. 10,500

25. 53,904

27. 329,134

29. 835

31. 529

33. 2,140,589

35. 11 **37.** 144

39. 13 **41.** 25

43. 94 **45.** 376

47. 593 **49.** 1531

51. 1424 **53.** 694

55. 2,806,644

57. 1,006,958

59. 45 cm

61. 108 ft

63. 31 m

Applications

65. 250 yards

67. $1053

69. 2762 miles

71. **a.** 125
b. Yes, 125 ≤ 125

73. 222

75. 18 gallons

77. $480,000

79. $67

81. $39,100

83. Company A; $7000

85. **a.** Utilities $262; Car costs $585
b. $2482 **c.** $1018

Writing & Thinking

87. Commutative property of addition, associative property of addition, and additive identity property; Examples will vary.

89. Borrowing is required when a digit is smaller than the digit being subtracted; Examples will vary.

1.3 Exercises

Concept Check

1. product

3. 0 (or zero)

5. grouping

7. False; The numbers being multiplied are called factors.

9. True

Practice

1. 168 **3.** 162

5. 854 **7.** 2808

9. 1760 **11.** 2352

13. 4233 **15.** 1054

17. 7632 **19.** 2850

21. 21,500 **23.** 43,680

25. 3,305,565

27. 6,036,009

29. 0

31. 8300

33. Zero-factor law

35. Associative property of multiplication

37. Commutative property of multiplication

39. Multiplicative identity

41. 900

43. 2500

45. 4000

47. 160,000

49. 1,500,000

51. 320,000,000

53. $3 \cdot 9 + 3 \cdot 7 = 27 + 21 = 48$

55. $6 \cdot 3 + 6 \cdot 11$
$= 18 + 66 = 84$

57. 25 square yards

59. 63 square meters

61. 216 square yards

Applications

63. $2975

65. 360 minutes; 2520 minutes

67. 9250 calories

69. $284,400

71. 234

73. $23

75. 648 square feet

77. 10,080 square feet

Writing & Thinking

79. Zero times any number is 0.

81. The distributive property distributes multiplication to two (or more) numbers that are being added; Examples will vary.

1.4 Exercises

Concept Check

1. Divisor

3. 1

5. Undefined

7. True

9. False; $\frac{0}{7} = 0$.

Practice

1. 7 3. 12

5. 0 7. 5 R2

9. Undefined

11. 40 13. 41

15. 44 R2 17. 9

19. 24 21. 50

23. 15 R12

25. 15 R5

27. 28 R7

29. 2 R2

31. 20 R3

33. 3 R6

35. 400 R6

37. 61 R15

39. 196 R370

41. 407

43. $704 \div 22 = 32$; $704 \div 32 = 22$

45. $1575 \div 45 = 35$; $1575 \div 35 = 45$

Applications

47. 3 cookies

49. 19 students

51. 6 hours

53. 22 teams

55. 427 square feet

57. 5423 dozen

59. $4837 per piano

61. 720 pounds

63. 9 games with $6 left over

Writing & Thinking

65. The main terms used in division are: dividend, divisor, quotient, and remainder. Examples will vary.

67. Division can be thought of as the reverse of multiplication and as a result, can be used to check the result of a multiplication problem. Answers will vary.

1.5 Exercises

Concept Check

1. Right 3. Place

5. 0s 7. True

9. True

Practice

1. 80;

3. 660;

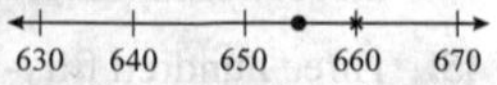

5. 500;

7. **a.** 2 **b.** 6 **c.** 6, 2, 3, 6 **d.** 1430, ten

9. 30 11. 30

13. 500 15. 1000

17. 100 19. 600

21. 4200 23. 10,000

25. 7000 27. 10,000

29. 62,000 31. 14,000

33. 10,000 35. 60,000

37. 130,000

39. 1,600,000

41. 220; 223

43. 800; 789

45. 21,000; 21,141

47. 500; 487

49. 20,000; 20,804

51. 210,000; 181,300

53. **a.** C **b.** D **c.** B **d.** E **e.** A

55. **a.** C **b.** A **c.** E **d.** D **e.** B

57. 240; 224

59. 600; 544

61. 40,000; 43,680

63. 10; 12

65. 20; 20 R13

67. 60; 61 R15

Applications

69. 4560 widgets

71. $8600

73. $26,000

75. 100,000 miles

77. $40,000; $35,316

79. $1100; $1076

81. 330 sets; 316 sets

83. $6000; $7634

85. 40 vans; 36 vans

87. $3000; $3160

Writing & Thinking

89. Estimation uses rounded values to find an approximate sum, difference, product, etc. Answers will vary.

91. Rounding is used to find another number close to the given number. Estimation uses rounded values to find an approximate sum, difference, product, etc. Answers will vary.

1.6 Exercises

Concept Check

1. multiplication

3. read

5. check, reasonable

7. True

9. False; Quotient indicates division.

Applications

1. 1103 calories

3. 420 syringes

5. 648 miles

7. 64 people

9. 300 pens

11. 768 beads

13. 26 cans 15. $1175

17. $691 19. $384

21. $160 23. $255

25. $85 27. $150

29. Blue car for $9112; $33 cheaper

31. 174 cm

33. 82 in.

35. 380 sq in.

37. 348 sq ft

39. 54 41. 485

43. 6 **45.** \$321

47. 7 hours **49.** \$125

51. 21 points

53. \$76.60/share; \$100 profit

55. **a.** 12 pairs of shoes

b. 254 pairs of shoes

57. 3873 miles

Writing & Thinking

59. **1.** Read the problem carefully. **2.** Draw any type of figure or diagram that might be helpful and decide what operations are needed. **3.** Perform the operations to solve the problem. **4.** Check your work.

61. Answers will vary. Averages may be used when looking at attendance, money made per month, commute time, etc.

1.7 Exercises

Concept Check

1. Cubed

3. Base, exponent

5. 1

7. False; Equals 81

9. False; 7^0 is 1.

Practice

1. 11^3

3. 7^4

5. $2^3 \cdot 3^2$

7. $5^3 \cdot 7^2$

9. $2 \cdot 3^2 \cdot 11^2$

11. **a.** 4 **b.** 2 **c.** 16

13. **a.** 2 **b.** 3 **c.** 8

15. **a.** 1 **b.** 6 **c.** 1

17. **a.** 5 **b.** 3 **c.** 125

19. **a.** 2 **b.** 4 **c.** 16

21. **a.** 9 **b.** 2 **c.** 81

23. **a.** 7 **b.** 2 **c.** 49

25. **a.** 3 **b.** 5 **c.** 243

27. **a.** 30 **b.** 2 **c.** 900

29. **a.** 20 **b.** 3 **c.** 8000

31. **a.** 1 **b.** 57 **c.** 1

33. **a.** 4 **b.** 0 **c.** 1

35. **a.** 13 **b.** 1 **c.** 13

37. **a.** 22 **b.** 0 **c.** 1

39. 21 **41.** 19

43. 28 **45.** 21

47. 2 **49.** 12

51. 12 **53.** 9

55. 0 **57.** 5

59. 46 **61.** 13

63. 74 **65.** 9

67. 132 **69.** 261

71. 61 **73.** 994

Applications

75. **a.** No. Here it shows that we are only dividing the old trading cards by 6 friends versus both the old and new trading cards by 6 friends.

b. 522; $\dfrac{15 \cdot 10 \cdot 20 + 132}{6}$

77. **a.** No. The formula shows that we are using $\dfrac{126}{4}$ yards of fabric to make only skirts, not full dresses. Adding 2 yards of silk after will only be enough for 1 bodice.

b. 20; $\dfrac{126}{4+2}$

Writing & Thinking

79. If addition is within parentheses (or other grouping symbols), addition would be performed first.

Collaborative Learning

81. 1.024E13, 1E20; Answers will vary.

1.8 Exercises

Concept Check

1. 0, 5 **3.** 4

5. 2 **7.** True

9. False; 7605 is divisible by 5.

Practice

1. 3, 5

3. 2, 3, 4, 6, 9

5. 3

7. 3, 9

9. 3, 5

11. None

13. None

15. 2, 3, 4, 5, 6, 10

17. 2, 5, 10

19. 3

21. 2

23. 3, 5

25. 2, 3, 4, 6

27. 3, 9

29. 3, 5

31. 2, 3, 4, 6, 9

33. 2

35. None

37. 2, 3, 4, 5, 6, 9, 10

39. 3, 5

41. 2, 3, 6

43. 2, 3, 4, 5, 6, 9, 10

45. 2, 3, 4, 6

47. 102, 108, 114, 120; Answers will vary.

49. 120, 150, 180, 210; Answers will vary.

51. 120, 150, 180, 210; Answers will vary.

Applications

53. 3 students would get \$3375 each; 9 students would be \$1125 each.

55. 4 team members would work 110 hours each; 8 team members would work 55 hours each.

Writing & Thinking

57. **a.** 30, 45; Answers will vary.

b. 9, 12; Answers will vary.

c. 10, 25; Answers will vary.

59. For example: 15,030. The number is divisible by 3 and 9 because digits add up to 9 and the number is divisible by 5 and 10 because the number ends in 0.

61. Answers will vary. Any number that ends in 5 is divisible by 5 and not divisible by 10. Two examples are 15 and 25.

1.9 Exercises

Concept Check

1. two

3. 1

5. itself

7. False; A prime number has exactly 2 factors.

9. False; 231 is a composite number.

Practice

1. A number whose only factors are 1 and itself. Answers will vary.

3. Prime

5. Composite; 1, 3, 7, 9, 21, 63

7. Prime

9. Composite; 1, 5, 41, 205

11. Composite; 1, 11, 13, 143

13. 4, 6 **15.** 1, 12

17. 3, 8 **19.** 4, 4

21. $2^3 \cdot 3$ **23.** 3^3

25. 2^4 **27.** $2^2 \cdot 3^2$

29. $2^2 \cdot 5$ **31.** $2 \cdot 5 \cdot 7$

33. 37 is prime

35. $2 \cdot 3 \cdot 5^2$

37. $2 \cdot 3 \cdot 5 \cdot 7$

39. $2^3 \cdot 3^2 \cdot 5$

41. $2^3 \cdot 5^2 \cdot 11$

43. $2^3 \cdot 5^3$

45. **a.** $2 \cdot 3^2$

b. 1, 2, 3, 6, 9, 18

47. **a.** $2 \cdot 3 \cdot 5$

b. 1, 2, 3, 5, 6, 10, 15, 30

49. **a.** $2^2 \cdot 3 \cdot 7$

b. 1, 2, 3, 4, 6, 7, 12, 14, 21, 28, 42, 84

51. **a.** $5^2 \cdot 7$

b. 1, 5, 7, 25, 35, 175

53. **a.** $2^2 \cdot 3 \cdot 5^2$

b. 1, 2, 3, 4, 5, 6, 10, 12, 15, 20, 25, 30, 50, 60, 75, 100, 150, 300

Applications

55. 1, 2, 3, 4, 6, 8, 12, 24

57. 1, 2, 3, 4, 6, 8, 9, 12, 18, 24, 36, 72

Writing & Thinking

59. No, some odd numbers are the product of two or more odd prime factors, for example, $3 \cdot 3 = 9$, $3 \cdot 5 = 15$, $3 \cdot 7 = 21$, etc.

61. The tests for divisibility can be used to find relatively small factors quickly. Knowing these factors makes finding the prime factorization easier.

Collaborative Learning

63. **a.** 700 has 18 factors. They are: 1, 2, 4, 5, 7, 10, 14, 20, 25, 28, 35, 50, 70, 100, 140, 175, 350, and 700.

b. 660 has 24 factors. They are: 1, 2, 3, 4, 5, 6, 10, 11, 12, 15, 20, 22, 30, 33, 44, 55, 60, 66, 110, 132, 165, 220, 330, and 660.

c. 450 has 18 factors. They are: 1, 2, 3, 5, 6, 9, 10, 15, 18, 25, 30, 45, 50, 75, 90, 150, 225, and 450.

Chapter 2: Fractions and Mixed Numbers

2.1 Exercises

Concept Check

1. improper

3. mixed

5. False; The numerator is 11.

7. True

Practice

1. **a.** $\frac{2}{3}$ **b.** $\frac{1}{3}$

3. **a.** $\frac{4}{7}$ **b.** $\frac{3}{7}$

5. **a.** $\frac{1}{6}$ **b.** $\frac{5}{6}$

7. **a.** $\frac{7}{12}$ **b.** $\frac{5}{12}$

9. **a.** $\frac{7}{30}$ **b.** $\frac{23}{30}$

11.

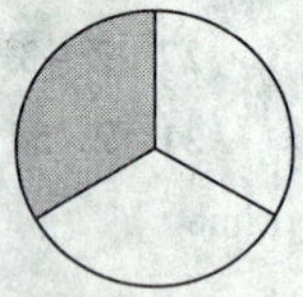

13.

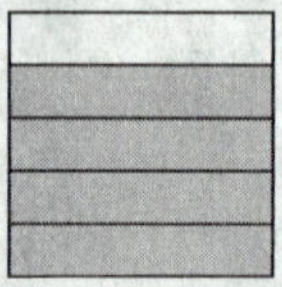

15. $\frac{5}{4}$

17. $\frac{10}{4}$

19. 0

21. Undefined

23.

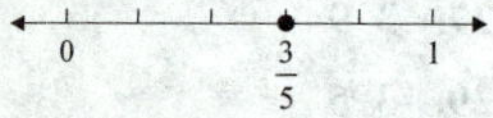

25.

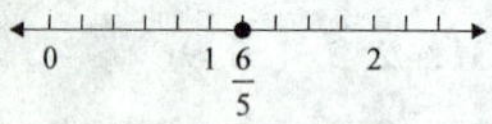

27. Mixed number

29. Proper fraction

31. **a.** $1\frac{5}{12}$ **b.** $\frac{17}{12}$

33. **a.** $1\frac{5}{12}$ **b.** $\frac{17}{12}$

35. $2\frac{3}{8}$ in.

37. $1\frac{7}{8}$ in.

39.

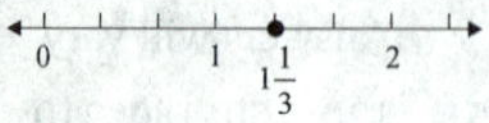

41.

43. $\frac{8}{5}$ **45.** $\frac{9}{4}$

47. $\frac{68}{10}$ **49.** $\frac{37}{8}$

51. $\frac{34}{7}$ **53.** $\frac{17}{15}$

55. $\frac{128}{12}$ **57.** $\frac{701}{100}$

59. $1\frac{1}{3}$ **61.** $6\frac{1}{2}$

63. $2\frac{1}{2}$ **65.** $2\frac{7}{10}$

67. $4\frac{5}{8}$ **69.** 3

71. $1\frac{15}{20}$ **73.** $1\frac{85}{100}$

Applications

75. $\frac{9}{20}$; $\frac{11}{20}$

77. $\frac{23}{45}$

79. $\frac{43}{60}$

81. **a.** $\frac{1}{8}$ **b.** $\frac{1}{512}$ **c.** $\frac{159}{512}$

83. **a.** $\frac{35}{12}$ **b.** $2\frac{11}{12}$

Writing & Thinking

85. The two parts are the numerator and the denominator. The denominator represents the number of pieces in a whole and the numerator represents the number of these pieces being considered.

87. Multiply the denominator by the whole number and add the numerator. This number is the new numerator. The denominator stays the same.

2.2 Exercises

Concept Check

1. 1

3. factors

5. common

7. True

9. False; The statement $\frac{1}{3} \cdot \frac{2}{5} = \frac{2}{5} \cdot \frac{1}{3}$ is an example of the commutative property of multiplication.

Practice

1. $\frac{1}{16}$ **3.** $\frac{4}{25}$

5. 0 **7.** $\frac{12}{1}$ or 12

9. $\frac{12}{35}$ **11.** $\frac{15}{32}$

13. $\frac{45}{2}$ **15.** $\frac{27}{28}$

17. $\frac{343}{216}$ **19.** $\frac{1}{6}$

21. $\frac{4}{45}$ **23.** $\frac{2}{9}$

25. $\frac{1}{3}$ **27.** $\frac{3}{4}$

29. $\frac{5}{11}$ **31.** 0

33. $\frac{2}{5}$ **35.** $\frac{12}{35}$

37. $\frac{1}{2}$ **39.** $\frac{2}{3}$

41. $\frac{2}{3}$ **43.** $\frac{25}{76}$

45. 4 **47.** $\frac{6}{25}$

49. $\frac{3}{7}$ **51.** $\frac{12}{13}$

53. $\frac{10}{9}$ **55.** $\frac{3}{2}$

57. $\frac{1}{4}$ **59.** $\frac{8}{9}$

61. $\frac{4}{35}$ **63.** $\frac{1}{3}$

65. $\frac{9}{16}$ **67.** 1

69. $\frac{5}{77}$ **71.** $\frac{1}{6}$

73. $\frac{10}{3}$ or $3\frac{1}{3}$

75. $\frac{28}{45}$

77. $\frac{21}{16}$ or $1\frac{5}{16}$

79. $\frac{9}{4}$

Applications

81. $\frac{1}{12}$ **83.** $\frac{3}{8}$

85. $\frac{9}{20}$ **87.** 6 in.

89. 120 square feet

91. 5 ft

93. **a.** 35 women
b. 21 women
c. Yes
d. Pass by 3 votes

95. **a.** 1500 voters
b. 2500 voters
c. 1000 Democrats
d. 2500 voters

Writing & Thinking

97. No. If a fraction is less than 1 then its product with another number will be less than that other number. So, if the other number is less than 1, the product will be less than 1. Answers will vary.

99. To multiply two fractions, multiply the numerators, multiply the denominators, and then reduce the product to lowest terms. Examples will vary.

2.3 Exercises

Concept Check

1. reciprocal

3. $\frac{1}{5}$

5. reciprocal

7. False; The reciprocal of 1 is 1.

9. False; The reciprocal of 12 is $\frac{1}{12}$.

Practice

1. $\frac{4}{3}$

3. 3

5. $\frac{1}{2}$

7. No reciprocal

9. $\frac{7}{12}$

11. $\frac{8}{9}$

13. $\frac{5}{7}$

15. 0

17. Undefined

19. $\frac{7}{5}$ **21.** $\frac{1}{3}$

23. $\frac{7}{6}$ **25.** $\frac{4}{9}$

27. $\frac{7}{16}$ **29.** $\frac{10}{9}$

31. $\frac{4}{5}$ **33.** $\frac{8}{3}$

35. 1 **37.** $\frac{32}{21}$

39. $\frac{10}{3}$ **41.** $\frac{12}{35}$

43. $\frac{7}{60}$ **45.** $\frac{7}{5}$

47. $\frac{41}{12}$ **49.** $\frac{5}{7}$

51. 98

Applications

53. $\frac{12}{25}$

55. 200 years

57. **a.** More **b.** Less
c. 200 passengers

59. **a.** More **b.** Less
c. 8000 steel rods per week

Writing & Thinking

61. $0=\frac{0}{1}$ and the reciprocal would be $\frac{1}{0}$ but division by 0 is undefined. So 0 has no reciprocal.

63. $12 \div 3 = 4$ and $12 \cdot \frac{1}{3} = 4$. We see that dividing by 3 is the same as multiplying by $\frac{1}{3}$, the reciprocal of 3.

65. No. For example, $\frac{4}{5} \neq \frac{5}{4}$. Answers will vary.

2.4 Exercises

Concept Check

1. Mixed

3. $\frac{16}{3}$

5. $\frac{2}{3}$

7. True

9. False; The mixed number $4\frac{1}{5}$ is equal to $\frac{21}{5}$.

Practice

1. $2\frac{1}{6}$ **3.** $\frac{2}{3}$

5. $3\frac{1}{5}$ **7.** $13\frac{1}{3}$

9. 2 **11.** 35

13. $34\frac{2}{7}$ **15.** $12\frac{1}{4}$

17. 6 **19.** $11\frac{33}{35}$

21. $1\frac{3}{10}$ **23.** $7\frac{13}{21}$

25. $8\frac{2}{3}$ **27.** 126

29. $2\frac{1}{4}$ **31.** 18

33. $11\frac{2}{3}$ **35.** $3\frac{1}{9}$

37. 4 **39.** $5\frac{1}{5}$

41. $1\frac{1}{2}$ **43.** $3\frac{1}{7}$

45. $1\frac{6}{25}$ **47.** $2\frac{2}{5}$

49. $1\frac{5}{6}$ **51.** $\frac{5}{7}$

53. $18\frac{1}{2}$ **55.** $29\frac{1}{3}$

57. $\frac{3}{8}$ **59.** $\frac{25}{32}$

61. $\frac{3}{8}$ **63.** $\frac{40}{63}$

65. $\frac{9}{32}$

Applications

67. **a.** $22\frac{1}{2}$ gallons **b.** \$45

69. 63 miles

71. **a.** $\frac{1}{4}$ **b.** $\frac{5}{8}$ of a cup

73. $7\frac{1}{2}$ ft^2

75. **a.** $34\frac{2}{3}$ in. **b.** $75\frac{1}{9}$ in.^{2}

77. $6\frac{1}{2}$ feet

79. $2\frac{12}{13}$ inches per hour

81. **a.** 1 sq ft **b.** 23 sq ft

Writing & Thinking

83. Change each mixed number to an improper fraction and then multiply the two fractions. Then, change the result to a mixed number. For example,
$$2\frac{1}{11} \cdot 5\frac{1}{2} = \frac{23}{11} \cdot \frac{11}{2} = \frac{23}{2} = 11\frac{1}{2}.$$

85. $$2\frac{1}{5} \div 1\frac{4}{5} = \frac{11}{5} \div \frac{9}{5} = \frac{11}{5} \cdot \frac{5}{9} = \frac{11 \cdot \cancel{5}}{\cancel{5} \cdot 9} = \frac{11}{9}$$
or $1\frac{2}{9}$; **1.** Convert the mixed numbers into improper fractions, **2.** Find the reciprocal of the divisor, **3.** Multiply the numerators and the denominators and reduce

if possible, and **4.** Change the answer to mixed number form. (**Note:** The mixed number form is preferred, but the fraction form is acceptable.)

2.5 Exercises

Concept Check

1. multiples

3. prime factorization

5. 100

7. True

9. True

Practice

1. 5, 10, 15, 20, 25, 30, 35, 40, 45, 50, 55, 60

3. 12, 24, 36, 48, 60, 72, 84, 96, 108, 120, 132, 144

5. 30 **7.** 15

9. 30 **11.** 30

13. 200 **15.** 150

17. 40 **19.** 112

21. 196 **23.** 66

25. 10 **27.** 36

29. 60 **31.** 90

33. 120

35. a. LCM = 150

b. $150 = 10 \cdot 15 = 15 \cdot 10 = 25 \cdot 6$

37. a. LCM = 120

b. $120 = 6 \cdot 20 = 24 \cdot 5 = 30 \cdot 4$

39. a. LCM = 108

b. $108 = 12 \cdot 9 = 18 \cdot 6 = 27 \cdot 4$

41. a. LCM = 2520

b. $1260 = 20 \cdot 63 = 28 \cdot 45 = 45 \cdot 28$

43. 15 **45.** 4

47. 42 **49.** 15

51. 40 **53.** 60

55. 44 **57.** 32

59. 81 **61.** 90

63. 42 **65.** 110

Applications

67. a. 60 minutes

b. 4, 3, and 2 trips, respectively

69. a. 90 hours

b. 6 and 5 orbits, respectively

71. a. 840 days

b. 105, 70, 60, and 56 trips, respectively

73. a. 840 seconds

b. 24 laps, 21 laps, and 20 laps, respectively

75. a. 120 hours

b. 15 orbits and 8 orbits, respectively

77. Every 180 days

79. 60 years

81. $12

Writing & Thinking

83. Since the LCM is constructed using the prime factors of each number in the set, by definition, each number will divide the LCM.

85. Multiplying the two numbers together will give the LCM if those two numbers have no common factors. If they have any factors in common, then you would only use that common factor once. Examples will vary.

2.6 Exercises

Concept Check

1. denominators

3. denominators

5. equivalent

7. True

9. False; When subtracting fractions, subtract the numerators and keep the common denominator.

Practice

1. $\frac{5}{14}$ **3.** 2

5. $\frac{1}{5}$ **7.** $\frac{3}{4}$

9. $\frac{7}{9}$ **11.** $\frac{11}{8}$

13. $\frac{62}{45}$ **15.** $\frac{13}{12}$

17. $\frac{3}{4}$ **19.** $\frac{7}{10}$

21. 1 **23.** $\frac{17}{20}$

25. $\frac{5}{6}$ **27.** $\frac{11}{20}$

29. $\frac{17}{20}$ **31.** $\frac{317}{1000}$

33. $\frac{631}{100}$ **35.** $\frac{3}{5}$

37. $\frac{1}{3}$ **39.** $\frac{8}{13}$

41. $\frac{1}{2}$ **43.** $\frac{13}{30}$

45. $\frac{3}{5}$ **47.** $\frac{1}{12}$

49. $\frac{5}{16}$ **51.** $\frac{13}{20}$

53. $\frac{11}{18}$ **55.** $\frac{1}{10}$

57. $\frac{23}{16}$ **59.** 0

61. $\frac{3}{50}$ **63.** $\frac{1}{50}$

Applications

65. 1 ounce

67. $\frac{23}{20}$ inches

69. $\frac{5}{8}$ inch

71. a. $\frac{5}{18}$ **b.** $6000

73. $\frac{7}{24}$

75. $\frac{11}{36}$

77. a. $\frac{3}{4}$

b. 0 (the pie is totally consumed)

79. $\frac{23}{45}$ of the assignment

81. $\frac{1}{10}$

Writing & Thinking

83. The LCM finds the least common multiple of a set of numbers. The LCD does the same thing for the set of numbers determined by the denominators.

85. Answers will vary. Adding with fractions may be used when cooking or when measuring sewing or construction materials. Subtracting with fractions may be used when cooking or when measuring sewing or construction materials.

2.7 Exercises

Concept Check

1. Mixed number, whole numbers

3. Reduced (or simplified)

5. Improper, 1

7. True

9. False; LCDs are required when adding or subtracting mixed numbers.

Practice

1. 10 **3.** $28\frac{1}{2}$

5. $14\frac{5}{7}$ **7.** $7\frac{2}{3}$

9. $10\frac{3}{8}$ **11.** $8\frac{10}{21}$

13. $5\frac{17}{26}$ **15.** $9\frac{3}{4}$

17. $8\frac{2}{3}$ **19.** $9\frac{29}{40}$

21. $12\frac{7}{27}$ **23.** $12\frac{17}{24}$

25. $12\frac{47}{60}$ **27.** $4\frac{2}{3}$

29. $9\frac{7}{20}$ **31.** $4\frac{1}{2}$

33. 4 **35.** $4\frac{3}{5}$

37. $3\frac{5}{8}$ **39.** $6\frac{1}{6}$

41. $3\frac{11}{20}$ **43.** $6\frac{7}{12}$

45. $16\frac{5}{24}$ **47.** $\frac{1}{4}$

49. $1\frac{1}{8}$ **51.** $\frac{1}{3}$

53. $12\frac{9}{70}$ **55.** $3\frac{29}{32}$

57. $12\frac{25}{28}$ **59.** $2\frac{16}{21}$

Applications

61. $8\frac{7}{12}$ hours

63. $12\frac{1}{8}$ inches

65. $900 million

67. $4\frac{13}{20}$ parts

69. $\frac{3}{5}$ hours

71. $\frac{5}{6}$ hour

73. $\frac{3}{4}$ hour

75. $3\frac{1}{4}$ pounds

77. **a.** $\$378\frac{9}{10}$ million

b. $\$27\frac{1}{2}$ million

c. $\$48\frac{1}{5}$ million

Writing & Thinking

79. Fractions should be first in case the fraction being subtracted is larger than the other fraction and 1 needs to be borrowed from the whole number.

81. **a.** 6 **b.** 600 **c.** 60,000

83. **a.** 4 **b.** 11 **c.** 100

2.8 Exercises

Concept Check

1. Numerators

3. Multiplication, division

5. Divide

7. False; According to the rules for order of operations, multiplication and division occur from left to right.

Practice

1. $\frac{3}{4}$ by $\frac{1}{12}$

3. $\frac{7}{10}$ by $\frac{1}{6}$

5. $\frac{17}{20}$ by $\frac{1}{20}$

7. $\frac{13}{20}$ by $\frac{1}{40}$

9. Equal

11. $\frac{21}{25}$ by $\frac{11}{400}$

13. $\frac{11}{48}$ by $\frac{1}{60}$

15. $\frac{3}{8}, \frac{2}{5}, \frac{1}{2}; \frac{1}{8}$

17. $\frac{1}{4}, \frac{1}{3}, \frac{1}{2}; \frac{1}{4}$

19. $\frac{7}{9}, \frac{31}{36}, \frac{17}{18}; \frac{1}{6}$

21. $\frac{8}{9}, \frac{9}{10}, \frac{11}{12}; \frac{1}{36}$

23. $\frac{5}{42}, \frac{1}{3}, \frac{3}{7}; \frac{13}{42}$

25. $\frac{20}{10{,}000}, \frac{3}{1000}, \frac{1}{100}; \frac{1}{125}$

27. $\frac{2}{3}$

29. $\frac{1}{5}$

31. $\frac{19}{4} = 4\frac{3}{4}$

33. $\frac{1}{4}$

35. $\frac{89}{9} = 9\frac{8}{9}$

37. $\frac{8}{39}$

39. $\frac{5}{7}$

41. $\frac{15}{64}$

43. $\frac{14}{45}$

45. $\frac{5}{3} = 1\frac{2}{3}$

47. $\frac{3}{16}$

49. $\frac{29}{36}$

51. 1

53. $\frac{1}{5}$

55. $\frac{138}{65} = 2\frac{8}{65}$

57. $\frac{172}{9} = 19\frac{1}{9}$

59. $\frac{5}{6}$

61. $\frac{791}{120} = 6\frac{71}{120}$

63. $\frac{63}{19} = 3\frac{6}{19}$

65. $\frac{3}{2} = 1\frac{1}{2}$

67. $\frac{22}{45}$

69. $\frac{47}{40} = 1\frac{7}{40}$

71. $\frac{223}{32} = 6\frac{31}{32}$

73. $\frac{2123}{360} = 5\frac{323}{260}$

Applications

75. $\frac{543}{40} = 13\frac{23}{40}$

77. 43 inches

79. 375

81. 17 pounds

83. **a.** $\frac{1}{3}$ kg **b.** $6

85. $\frac{91}{48} = 1\frac{43}{48}$ feet

Writing & Thinking

87. **a.** Yes; If both fractions are greater than one half, the sum will be greater than one.

b. No; Multiplying a number by a fraction between 0 and 1 results in a product that is less than the original number.

89. **a.** Less than **b.** Less than

Chapter 3: Decimal Numbers

3.1 Exercises

Concept Check

1. and

3. two, eight hundredths

5. right

7. True

9. False; On a number line, any number to the right of another number is larger than that other number.

Practice

1. 6.5

3. 18.76

5. 56.03

7. 37.498

9. 87.003

11. Nine tenths

13. Twenty and seven tenths

15. One and fifty-three hundredths

17. Nineteen and one hundred two thousandths

19. Eight hundred and nine thousandths

21. 0.3 **23.** 17.9

25. 0.23 27. 6.028

29. 0.4502 31. 0.27

33. 0.163 35. 24.295

37. 0.01

39.

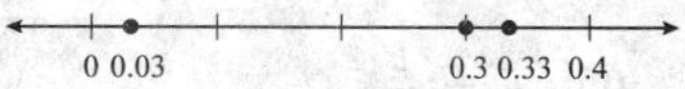

41.

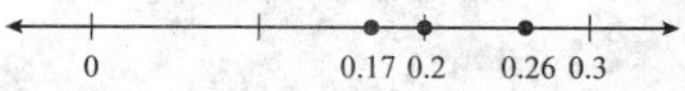

43.

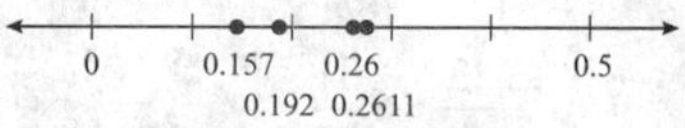

45. 7, 8, 8, 7, 8, 8, 34.8

47. 5, 2, 2, 5, 2, 3.0065

49. 8.6 51. 10.0

53. 1.68 55. 0.08

57. 0.057 59. 3.003

61. 4 63. 30

65. 5200 67. 400

69. 103,000 71. 51,000

Applications

73. One hundred fourteen and eight tenths

75. Nine-hundred fourteen thousandths; One and nine hundredths; Thirty-nine and thirty-seven hundredths; Three and thirty-seven hundredths

77. Two and eight-hundred twenty-five ten-thousandths

79. Two and seventy-one thousand eight hundred twenty-eight hundred-thousandths

81. Nine and fifty-eight hundredths; Nineteen and nineteen hundredths; Forty-three and three hundredths

83. Thirty-five and eight tenths; twenty-six and nine tenths; Eighteen and nine tenths; Twelve and three tenths; Seven and two tenths

Writing & Thinking

85. Answers will vary. Example: If the bill is $129, then that would be 129 and one hundred twenty-nine.

87. Moving left to right, compare digits with the same place value. When one compared digit is larger, the corresponding number is larger.

3.2 Exercises

Concept Check

1. vertically

3. whole

5. True

7. False; In subtracting decimal numbers, line up the decimal points and corresponding digits vertically.

Practice

1. 50.085 3. 7.8798

5. 599.07 7. 156.305

9. 237.74 11. 9.83

13. 72.31 15. 11.5

17. 19.541 19. 40.313

21. 11.131 23. 17.8

25. 2.13 27. 32.7

29. 45.01 31. 4.7974

33. 8.93 35. 1.44

37. 0.9757 39. 4.895

41. 28.475 43. 2.9434

45. 17.5616 47. 10 cm

49. 61.2 cm 51. 18.84 m

Applications

53. **a.** $94.85 **b.** $5.15

55. 82.83 feet

57. $671.75

59. **a.** 39.9 million

b. 31.9 million

61. **a.** 6.97 feet **b.** 2.73 feet

63. **a.** 12.15 gallons

b. 3.25 gallons

65. $6089.90

67. 12.1778 tons

69. No, he is short $0.19.

71. **a.** 758.99909 sec

b. 761.00073 sec

73. 0.02458 liter

75. **a.** $67.22

b. $32.78

Writing & Thinking

77. Decimal numbers need to be aligned vertically so that numbers with the same place value are being added together. If not, then a 60 may be added to a 7 as if it were a 6 being added to a 7, giving 13, not the value of 67 that it should be.

79. **a.** Answers will vary.

b. Answers will vary.

81. Sophia did not align the decimal points in the addends vertically, so she was not adding corresponding place values.

3.3 Exercises

Concept Check

1. whole numbers

3. 3, right

5. 10, 100, 1000

7. True

9. False; Multiplying by 100 requires that the decimal point be moved 2 places to the right.

Practice

1. 0.42 3. 0.04

5. 21.6 7. 0.42

9. 0.004 11. 0.108

13. 0.073 15. 0.0276

17. 0.0486 19. 0.0034

21. 0.3375 23. 1.500

25. 1.4000 27. 3.975

29. 1.200 31. 5.063

33. 0.0648 35. 0.18

37. 65.3980 39. 0.04336

41. 4.35 43. 346

45. 1610 47. 4178.2

49. 380 51. 7.9

53. 61.5 55. 50

Applications

57. 12.8 ft; 10.24 ft^2

59. 26.6 cm; 29.40 cm^2

61. $240.90

63. $936,945

65. $621.40

67. 2.8926 in.

69. $932.60

71. Discount: $45.30; Total: $288.67

73. $1320

Writing & Thinking

75. In multiplication with decimal numbers, placement of the decimal point must be considered. Otherwise, multiplication with whole numbers and decimal numbers are the same.

3.4 Exercises

Concept Check

1. whole

3. whole numbers

5. left

7. True

9. False; Dividing by a power of 10 involves dividing by 10, 100, 1000, etc.

Practice

1. 2.34 3. 0.57

5. 0.18 7. 0.08

9. 2.36 11. 2056

13. 6.54 15. 20

17. 56.9 19. 0.7

21. 0.9 23. 728.7

25. 21.0 27. 0.01

29. 5.78 31. 12.12

33. 0.56 35. 3.8

37. 13.81 39. 0.785

41. 0.794

43. 0.5036

45. 3.164963

47. 0.0186

49. 0.045621

51. 1.17385

53. 0.000154

Applications

55. $239.56

57. $295

59. 5630 at bats

61. 4.4 yards per carry

63. 67.92

Writing & Thinking

65. In division with decimal numbers, placement of the decimal point must be considered. Also, with whole numbers a remainder is considered while, with decimal numbers, rounding the quotient is considered. Otherwise, division with whole numbers and decimal numbers are the same.

67. Answers will vary.

3.5 Exercises

Concept Check

1. leftmost

3. decimal point

5. exponential expressions

7. False; When estimating the product of decimal numbers, round the numbers to the leftmost nonzero digit.

9. True

Practice

1. 34; 32.82

3. 19; 21.79

5. 106; 102.46

7. 3; 3.31

9. 188; 207.45

11. 20; 26.08

13. 1; 1.55

15. 10; 9.09

17. 2.4; 2.621

19. 7000; 6548.06

21. 6; 4.86

23. 0.72; 0.7452

25. 6; 6.2144

27. 5; 5.17

29. 2; 2.032

31. 0.06; 0.0486

33. 5.4; 4.9077

35. 2; 2.05

37. 10; 9.45

39. 10,000; 9446.67

41. 2000; 2115

43. 75; 78.44

45. 3.5; 3.09

47. 0.2; 0.24

49. 4.75 **51.** 3.08

53. 15.26 **55.** 77.08

57. 62.12 **59.** 6.38

61. 13.7 **63.** 14.31

65. 26.88 **67.** 16.11

Applications

69. **a.** 39 pounds

b. 35.43 pounds

71. **a.** 15 feet **b.** 15.03 feet

73. **a.** $5000 **b.** $5959.80

75. **a.** $900 **b.** $891.50

c. $10,698.00

77. **a.** 600 miles **b.** 480.7

79. **a.** $4 per pound

b. $5.11 per pound

81. **a.** 15 mpg **b.** 22 mpg

83. **a.** $740 **b.** $772

85. **a.** $99 **b.** $98.32

c. $49.16

Writing & Thinking

87. Answers will vary. The estimate says you are paying $4000 more with the payment plan than with cash. This is a significant overestimate since the payment and the number of months were both rounded up.

3.6 Exercises

Concept Check

1. numerator, denominator

3. repeating, nonrepeating

5. terminating

7. True

9. False; In some cases, fractions can be converted to decimal form without losing accuracy.

Practice

1. $\frac{9}{10}$ **3.** $\frac{526}{1000}$

5. $\frac{51}{10}$ **7.** $\frac{815}{100}$

9. $\frac{1}{5}$ **11.** $\frac{9}{50}$

13. $\frac{9}{40}$

15. $\frac{25}{4}$ or $6\frac{1}{4}$

17. 0.05 **19.** 0.625

21. $0.\overline{6}$ **23.** $0.\overline{27}$

25. 6.67 **27.** 0.48

29. 0.03 **31.** 1.43

33. 0.7 **35.** 11.59

37. 1.13 **39.** 1

41. 6.50 **43.** 23.41

45. 120.31 **47.** 2.64

49. 1.16 **51.** 1.80

53. 0.28 is larger; $0.00\overline{72}$

55. 3.3 is larger; $0.\overline{1571428}$

57. $3\frac{2}{3}$ is larger; $0.0\overline{16}$

59. $\frac{7}{10}, \frac{3}{4}, 0.76$

61. $\frac{5}{16}$,0.3126,0.314

Applications

63. $93\frac{4}{5}$

65. $30\frac{3}{10}$; $28\frac{2}{5}$

67. $21\frac{1}{2}$

69. 0.58

71. 2.9

73. 17.92 inches

75. 20.2 miles

77. 60 students that drive to school and have a job are taking AP English

79. 0.9 ounces

81. 25 slides

Writing & Thinking

83. For the numerator, write the whole number formed by all the digits of the decimal number, and for the denominator, write the power of 10 that corresponds to the rightmost digit. Reduce the fraction, if possible.

85. This problem could be solved by converting all numbers to fractions or all numbers to decimal numbers. Answers will vary.

Chapter 4: Ratios, Proportions, and Percents

4.1 Exercises

Concept Check

1. ratio
3. reduced
5. 1
7. True
9. False; The ratio 8:2 can be reduced to the ratio 4:1.

Practice

1. $\frac{9}{14}$
3. $\frac{4}{3}$
5. $\frac{4}{5}$
7. $\frac{11}{16}$
9. $\frac{7}{24}$
11. $\frac{1}{2}$
13. $\frac{4}{1}$ or 4
15. $\frac{1 \text{ minute}}{40 \text{ seconds}}$ or $\frac{3}{2}$
17. $\frac{3}{4}$
19. $\frac{4}{7}$
21. $\frac{\$2 \text{ profit}}{\$5 \text{ invested}}$
23. $\frac{1 \text{ teacher}}{12 \text{ students}}$
25. $\frac{1 \text{ hit}}{5 \text{ times at bat}}$
27. $\frac{10 \text{ children}}{3 \text{ families}}$
29. $\frac{5 \text{ scholarships}}{42 \text{ students}}$
31. 60 miles per hour
33. 25 miles per gallon
35. 42 words per minute
37. 90 gallons per person
39. \$12.75 per hour
41. 113.7¢/oz; 66.6¢/oz; 12 oz at \$7.99
43. 250¢/pair; 275¢/pair; 5 pairs at \$12.50
45. 33.3¢/oz; 36.1¢/oz; 12 oz at \$3.99
47. 19.9¢/oz; 16.8¢/oz; 12.2¢/oz; 10.6¢/oz; 64 oz at \$6.79

Applications

49. $\frac{3 \text{ fat grams}}{10 \text{ grams}}$
51. $\frac{12 \text{ clear days}}{61 \text{ cloudy days}}$
53. $\frac{4}{5}$
55. $\frac{10}{7}$
57. $\frac{36}{5}$
59. $\frac{3}{2}$
61. $\frac{1}{12}$
63. $\frac{4}{5}$
65. $\frac{4}{15}$
67. a. $\frac{7}{3}$ b. $\frac{3}{7}$
69. 25 miles per gallon
71. 74 applicants per opening
73. 5 iPads per classroom
75. \$62.50 per person
77. We can expect the player to get 3 hits for every 10 times at bat.
79. 50 trash bags for \$25.98
81. \$0.53 per pound
83. \$0.25 per can
85. \$0.36 per ounce
87. a. 31.5 miles per gallon
    b. No; 31.5 miles per gallon is lower than the advertised amount.
89. 63 miles per hour is lower than the posted speed limit.

Writing & Thinking

91. The ratio can be written as $\frac{5}{3}$, 5:3, or 5 to 3. The preferred method is as a fraction because it is easier to simplify and manipulate in mathematical operations.
93. a. A ratio with different units in the numerator and denominator (the units do not cancel out); Answers will vary.
    b. A rate with a denominator value of 1 unit; Answers will vary.
    c. The average price for each unit of an item containing multiple units; Answers will vary.

4.2 Exercises

Concept Check

1. cross products or two ratios
3. rf and kt
5. units
7. False; Cross products are used to determine if a proportion is true.
9. False; When using proportions to solve a word problem, there are many correct ways to set up the proportion.

Practice

1. True
3. True
5. True
7. True
9. False
11. True
13. True
15. False
17. True
19. False
21. $x = 10$
23. $x = 50$
25. $D = 100$
27. $A = \frac{21}{2}$ or 10.5
29. $x = 1$
31. $y = 6$
33. $x = \frac{1}{4}$ or 0.25
35. $R = 50$
37. $A = 27.3$
39. $B = 6.75$
41. $B = 7.8$
43. $x = 1.5$
45. $B = 90$
47. $x = 25$ hr
49. $x = 20$ ounces
51. $x = 135$ min

Applications

53. They are the same.
55. They are the same.
57. They are different.
59. They are different.
61. \$25
63. 180 minutes or 3 hours
65. \$4.17
67. $1\frac{1}{2}$ cups or 1.5 cups
69. 33 hits
71. 437.5 grams
73. 20 yards
75. \$400
77. 9 hours
79. 118.8 pounds
81. \$160,000
83. 109,500 times

Writing & Thinking

85. A proportion has been set up correctly if the same units are in the same location in both ratios.
87. Traci's proportion has inches in the numerator in one ratio and in the denominator in the other. The units should be the same in the numerators and the same in the denominators. This means she should have used the proportion $\frac{4 \text{ inches}}{300 \text{ miles}} = \frac{x}{750 \text{ miles}}$ or $\frac{300 \text{ miles}}{4 \text{ inches}} = \frac{750 \text{ miles}}{x}$

4.3 Exercises

Concept Check

1. 83
3. 2; right
5. True

7. False; A decimal number that is between 0.01 and 0.10 is between 1% and 10%.

Practice

1. 60% 3. 65%
5. 12.5% 7. 100%
9. 20% 11. 9%
13. 15% 15. 125%
17. 13.4% 19. 0.48%
21. 0.5% 23. 2.14%
25. 2% 27. 10%
29. 36% 31. 40%
33. 2.5% 35. 0.5%
37. 12.8% 39. 112%
41. 230% 43. 200%
45. 0.02 47. 0.18
49. 0.25 51. 0.2
53. 1.25 55. 1.2
57. 0.065 59. 0.173
61. 0.0026 63. 0.0137

Applications

65. 12% 67. 2%
69. 6.5% 71. 0.085
73. 0.45 75. 5.23

Writing & Thinking

77. Percent means per centum or per 100. For example, fifty-eight percent means 58 out of 100. Percent can be written as a fraction with 100 in the denominator as in 58/100. The decimal equivalent, 0.58 is read as "fifty-eight hundredths," indicating percent can be written using the hundredths place, another connection..
79. Both conversions involve moving the decimal point two places. The difference is whether the decimal point moves left or right. When changing a percent to decimal, the decimal point moves two places to the left and when changing a decimal to a percent, the decimal point moves two places to the right.

4.4 Exercises

Concept Check

1. 100
3. 2; right
5. False; Fractions that have denominators other than 100 can be changed to a percent.
7. False; When changing from a percent to a mixed number, the fraction should be reduced.

Practice

1. 3% 3. 16%
5. 50% 7. 75%
9. 5% 11. 40%
13. 26% 15. 87.5%
17. 4% 19. 12.5%
21. 58.3% 23. 116.7%
25. 125% 27. 105%
29. 210% 31. 206.7%
33. $\frac{1}{20}$ 35. $\frac{3}{20}$
37. $\frac{17}{100}$ 39. $\frac{1}{4}$
41. $\frac{33}{100}$ 43. 1
45. $1\frac{1}{5}$ 47. $1\frac{1}{4}$
49. $\frac{1}{8}$ 51. $\frac{3}{400}$
53. $\frac{1}{200}$ 55. $\frac{1}{8}$
57. $\frac{1}{6}$ 59. $\frac{1}{3}$
61. **a.** 0.625 **b.** 62.5%
63. **a.** $\frac{9}{100}$ **b.** 9%
65. **a.** $\frac{9}{25}$ **b.** 0.36

Applications

67. $\frac{7}{20}$
69. $\frac{6}{25}$
71. $\frac{16}{125}$
73. $\frac{7}{20{,}000}$
75. 4%
77. 0.000075%
79. **a.** 0.0011 **b.** 0.11% **c.** Yes
81. 9.8%

Writing & Thinking

83. 100% = 1 so anytime there is a mixed number, which has a value greater than 1, the percentage will be greater than 100%. Proper fractions (numerator is smaller than denominator) have a value less than 1 and therefore the percentage will be less than 100%.

4.5 Exercises

Concept Check

1. percent, 100
3. 172.3
5. $\frac{95}{100} = \frac{A}{60}$
7. False; In the proportion $\frac{P}{100} = \frac{63}{180}$, the base is 200.
9. False; The base is not always larger than the amount.

Practice

1. 7.5 3. 15
5. 64 7. 47
9. 900 11. 150
13. 62 15. 20%
17. 150% 19. 150
21. 30 23. 24
25. 33.3% 27. 40
29. 70 31. 230
33. 3.6 35. 17.5
37. 512 39. 65
41. 81 43. 66.5%
45. 12.82 47. 125%
49. 33.6 51. 105
53. 115.61 55. 400
57. 200 59. 28
61. 120

Applications

63. 56,000 fans
65. 47.31%
67. $97,600
69. **a.** 93.75% **b.** 6.25%; Keep this player
71. 27 aces
73. Army: 13.44%; Marine Corps: 6.66%; Navy: 15.90%; Air Force: 19.21%; Coast Guard: 16.43%

Writing & Thinking

75. Proportions would work for mixed numbers because a mixed number can be rewritten as a fraction. The only additional step required would be to change the mixed number to an improper fraction and then solve the proportion as normal.

4.6 Exercises

Concept Check

1. base
3. multiplication or times
5. amount, *A*
7. True
9. True

Practice

1. 7 3. 3.1
5. 9 7. 42
9. 150 11. 20%
13. 150 15. 50
17. 70 19. 36
21. 33.3% 23. 35%
25. 12.5 27. 18
29. 110 31. 180%
33. 614 35. 38
37. 80 39. 200%
41. 75 43. 16.32
45. 58.5 47. 11.4

49. 24 **51.** 72

53. 95% **55.** 29%

57. 175% **59.** 72

61. 16 **63.** 10

65. 25 **67.** 28

69. 165.6

Applications

71. 6800 **73.** 4.8%

75. 34.29 g **77.** 40

79. **a.** 61.728% **b.** 56.790%
c. 57.407% **d.** 49.383%
e. 58.642% **f.** 55.556%
g. 54.658% **h.** 50.617%
i. 58.642% **j.** 55.556%
k. 66.049% **l.** 65.432%

Writing & Thinking

81. The three parts are the rate, base, and amount. The rate is the percent but should be written in decimal number form. The amount is the number that is part of the whole and the base is the whole.

83. The amount is the number that is often near the word "is." The base is the number that often follows the word "of." The rate is the number written either as a fraction or as a decimal number that has not been identified as the amount or the base, and usually appears before the word "of."

4.7 Exercises

Concept Check

1. look back or check

3. (total) sale or selling

5. appreciation

7. True

9. True

Applications

1. **a.** \$30 **b.** \$270

3. **a.** \$465 **b.** \$15,035

5. \$16.88; \$5.63

7. \$61.25

9. **a.** \$161 **b.** \$5474

11. \$52

13. **a.** \$6.93 **b.** \$122.38

15. 1.5%

17. **a.** \$250 **b.** 20%
c. \$216

19. \$11,700

21. \$875

23. \$28,000

25. \$770

27. \$2520

29. \$112,700

31. \$10,000

33. 5.9%

35. 2.484%

37. 215,966 people

39. **a.** \$200.00 **b.** 11.11%

41. 38%

43. **a.** \$7 **b.** $33\frac{1}{3}\%$
c. 25%

45. **a.** \$180 **b.** 40%
c. $28\frac{4}{7}\%$ or 28.6%

47. **a.** \$500 **b.** 25% **c.** 20%

49. **a.** \$1075 **b.** \$258
c. 24% **d.** 30%

51. \$1.89

53. \$1.39; \$1.75

55. \$8.64

57. \$31.80

59. **a.** \$3.60 **b.** \$22.50

61. \$111.60

Writing & Thinking

63. Sales tax and tips are percentages of some item or service. The percent is the rate, while the cost of the item being purchased is the base. The amount is then the sales tax itself, which is being compared to the base. A sales tax might be 8%, as in "what is 8% of the cost of the item purchased?"

65. For both types of percent of profit, the profit (the difference between selling price and cost) must be determined first. The profit is then used as a numerator. In profit based on cost, the denominator is the cost. In profit based on selling price, the selling price is the denominator. The fraction is then divided and the result is a percent of profit, either based on cost or selling price.

4.8 Exercises

Concept Check

1. interest

3. simple

5. quarterly

7. True

9. False; Compound interest is earned on the principal and interest earned.

Applications

1. \$30 **3.** \$32

5. \$200 **7.** \$100

9. \$37.50 **11.** \$2500

13. \$3000 **15.** 1 year

17. 72 days **19.** \$9

21. \$3.33 **23.** 10%

25. \$1030 **27.** \$730

29. \$337,500

31. **a.** \$1000
b. 9 months or $\frac{3}{4}$ year

33. \$1030

35. 240 days or $\frac{2}{3}$ year

37. **a.** \$16 **b.** \$100
c. 30 days or $\frac{1}{12}$ year
d. 8.5%

39. **a.** \$100 **b.** \$2100, \$105
c. \$2205, \$110.25
d. \$315.25

41. **a.** \$30
b. \$9030, \$30.10
c. \$9060.10, \$30.20
d. \$9090.30, \$30.30
e. \$120.60
f. \$9120.60

43. \$509.58

45. \$9090.30

47. \$189.24

49. **a.** \$1051.56 **b.** \$51.56

51. **a.** \$28,051.03
b. \$329.40

53. **a.** \$634.13 **b.** No
c. Monthly compounding allows interest earned the previous month to gain interest the following month, but semiannual compounding waits six months for this process to begin.

55. **a.** \$1493.42 **b.** \$93.42

57. **a.** \$67,952.39 **b.** More
c. \$184,675.81

59. **a.** \$1645.31 **b.** 645.31

61. **a.** \$10,560.33
b. \$5560.33

63. **a.** \$275,470.73
b. \$250,470.73

65. Milk: \$4.01; Bread: \$3.35

67. \$12,665.57

69. 11.4%

71. \$8541.94

Writing & Thinking

73. The simple interest formula is $I = P \cdot r \cdot t$ where I is interest, P is principal, r is rate, and t is time. Interest is the amount of money paid for the use of money. The principal is the starting amount invested. Rate is the interest rate and should be written as a decimal or fraction. Time is the amount of time, in years, that interest is being earned on the principal. Time can be written as a decimal or fraction. When a decimal is used, it should only be when it is a terminating decimal so that no rounding is required, which could change the value calculated.

75. Answers will vary.

a. $13,498.03

b. About 7 years.

Collaborative Learning

77. a.

Monthly Income	4%	6%	8%
$2000	$2433.31	$2676.45	$2938.66
$2500	$3041.63	$3345.56	$3673.32
$4000	$4866.61	$5352.90	$5877.31

b. Answers will vary. The total increase in income after 5 years is similar for the lower income with an 8% yearly pay raise and the higher income with the 4% yearly pay raise.

c. Answers will vary.

Chapter 5: Measurement

5.1 Exercises

Concept Check

1. pint, 16

3. yard

5. converted

7. True

9. True

Practice

1. 12 **3.** 1

5. 1 **7.** 5280

9. 1 **11.** 8

13. 36 **15.** 300

17. 2 **19.** 21

21. 2 **23.** 1.5

25. $\frac{3\text{ ft}}{1\text{ yd}}$; 21

27. $\frac{1\text{ qt}}{2\text{ pt}}$; 3

29. $\frac{1\text{ gal}}{4\text{ qt}}$; 3.25

31. $\frac{1\text{ ft}}{12\text{ in.}}$; 1.5

33. $\frac{5280\text{ ft}}{1\text{ mi}}$; 15,840

35. $\frac{1\text{ mi}}{5280\text{ ft}}$; 1.5

37. 8 **39.** 32,000

41. 4 **43.** 88

45. 1.5 **47.** 150

Applications

49. $\frac{7}{8}$ or 0.875 square feet

51. $93.22

53. $\frac{1}{2}$ or 0.5 miles

55. 15 miles per hour

57. The small bag; $0.005¢/oz

59. 126,720 inches

61. $65

63. 5

Writing & Thinking

65. Colby would need to know that there are 3 feet in a yard and 5280 feet in a mile.

67. A unit fraction is a fraction with different units that is equivalent to 1. It can be used to convert between units. When used, the numerator should contain the same units as the desired outcome and the denominator should have the same unit as the unit being converted.

5.2 Exercises

Concept Check

1. threes

3. meter

5. multiply

7. True

Practice

1.–9. Answers will vary.

11. Meters

13. Meters

15. Millimeters

17. 300 **19.** 80

21. 1500 **23.** 3.6

25. 0.82 **27.** 0.0525

29. 0.75 **31.** 0.245 m

33. 0.23 m **35.** 10 km

37. 200 m

39. 0.00679 km

41. 150.3

43. 30 000 000

45. 960

47. 5

49. 500 000

51. 1300; 130 000

53. 115 000; 11 500 000

55. 400; 40 000

57. 670

59. 20 000

61. 575 ha

63. 956; 95 600

65. 0.0625; 0.000625

Applications

67. 3.2 cm^2

69. 50 mm

71. 44 000 square meters

73. 6 gigahertz

75. 500 gigabytes

77. 75 hectares

79. 70 000 square meters

Writing & Thinking

81. Converting among the U.S. customary systems requires knowledge of the equivalencies and there is no consistency among them so they must be memorized. Within the metric system, typically all that is required for conversions is to simply move the decimal point the desired number of places according to the prefixes.

83. Each category of metric units has a base unit. The prefixes determine how many or what fraction of the base unit is being used. For example, the basic unit of length is meter and a millimeter is 1/1000 of a meter, a centimeter is 1/100 of a meter, and a kilometer is 1000 meters.

5.3 Exercises

Concept Check

1. cubic

3. weight

5. kilogram

7. False; Volume is measured in cubic units.

9. False; A metric ton and a US customary ton are not equal (a metric ton weighs about 2200 US pounds).

Practice

1. Milliliters **3.** Liters

5. Milliliters **7.** 2000

9. 0.019 **11.** 13 000

13. 0.5 **15.** 6300

17. 0.0764 **19.** 0.95

21. 1250 **23.** 5300 mL

25. Kilograms

27. Grams

29. Milligrams

31. 2000 **33.** 7580

35. 540 **37.** 2

39. 0.0345 **41.** 0.091

43. 4 600 000

45. 2.963

47. 5000 kg

49. 96 000 mg

51. 75 kg

53. 0.0016 g

55. 0.000 34 kg

57. 7 000 000 g

59. m **61.** L

63. km **65.** L

67. g **69.** 0.006

71. 0.4

73. 9 000 000 cm^3

Applications

75. 60 doses

77. Yes. The total amount of solution to be disposed is 3.7 L.

79. 60 000 000 000 grains

81. 57.8 L

Writing & Thinking

83. You could change to milliliters by multiplying by 1000 or by using a unit fraction where the numerator is 1000 mL and the denominator is the given measure in liters.

85. You would probably use kilograms because a gram is approximately the weight of a paperclip and a kilogram is about 2.2 pounds.

5.4 Exercises

Concept Check

1. Fahrenheit

3. 2.54

5. 0.946

7. False; Water freezes at 32 degrees Fahrenheit.

9. False; A 5K (km) run is shorter than a 5 mile run.

Practice

1. 77 **3.** 50

5. 45 **7.** 122

9. 0 °C **11.** 59 °F

13. 2.74 **15.** 11.99

17. 83.82 **19.** 27.88

21. 32.19 **23.** 2.74 m

25. 96.6 km **27.** 124.22 mi

29. 19.69 in. **31.** 35.56 cm

33. 19.35 **35.** 55.74

37. 83.61 **39.** 405

41. 741.32 acres

43. 53.82 ft^2

45. 4.65 $in.^2$

47. 3.79 **49.** 1.06

51. 9.46 **53.** 78

55. 11.10 **57.** 10.57 qt

59. 189.25 L **61.** 72.75

63. 992.23 **65.** 4.56

67. 3.53 **69.** 453.59 g

71. 264.55 lb

Applications

73. 177 °C

75. 9.7 miles per hour

77. 226.3 km

79. 3035.14 m^2

81. **a.** 11 145.6 cm^2

b. 1.116 m^2

83. 7 cans

85. Answers will vary.
Small: 5.4 ounces;
Medium: 10.8 ounces or 10.9 ounces;
Large: 1 pound 5 ounces or (21 ounces)

87. 2.7 mL

Writing & Thinking

89. One meter is equivalent to 1.09 yards so they are close in length.

91. Because conversions between the metric system and the US system require rounding, it is possible that two answers could be right although they might be slightly different. If Kai used the US to Metric equivalent and Kristen used the Metric to US equivalent, they would both have a correct answer but their answers will be slightly different answers because they used different approximations.

Chapter 6: Geometry

6.1 Exercises

Concept Check

1. line

3. vertex

5. straight

7. perpendicular

9. equilateral

11. True

13. True

15. True

Practice

1. 35°

3. 80°

5. Acute

7. Obtuse

9. Acute

11. **a.** Obtuse **b.** Acute **c.** Right

13. **a.** 180° **b.** 90° **c.** 30° **d.** 150°

15. **a.** 135° **b.** 90° **c.** 70° **d.** 45°

17. **a.** 150°

b. Yes; ∠2 and ∠3 are supplementary.

c. ∠1 and ∠3; ∠2 and ∠4

d. ∠1 and ∠2; ∠2 and ∠3; ∠3 and ∠4; ∠1 and ∠4

19. $m\angle 2 = 138°$; $m\angle 3 = 42°$; $m\angle 4 = 138°$

21. **a.** $m\angle 2 = 70°$; $m\angle 3 = 90°$; $m\angle 4 = 20°$; $m\angle 5 = 70°$

b. ∠3

c. ∠2 and ∠5

23. **a.** 125°; ∠1 and ∠3 are vertical angles.

b. 55°; ∠8 and ∠6 are vertical angles.

c. $m\angle 7 = 125°$; ∠6 and ∠7 are supplementary angles.

d. Yes; ∠2 and ∠6 are corresponding angles.

25. Scalene

27. Right

29. Isosceles

31. Isosceles and right

33. Acute

35. Acute

Applications

37. Yes, since 25 < 12 + 15.

39. **a.** $m\angle Z = 80°$

b. Acute

c. $\overline{YZ}$

d. $\overline{XZ}$ and $\overline{XY}$

e. No, no angle is 90°

Writing & Thinking

41. A ray is similar to a line in that it has at least one end that continues infinitely. An angle is formed by two rays that have a common endpoint, called a vertex. A line has no endpoint and continues indefinitely in opposite directions in the same plane.

43. a. A right angle
b. An acute angle
c. An obtuse angle

6.2 Exercises

Concept Check

1. polygon
3. rectangle
5. triangle
7. a. True
b. False; Not all rectangles have four equal sides.
9. True
11. a. B b. C c. A
d. E e. D

Practice

1. 44 cm 3. 116 cm
5. 18 km 7. 143 in.
9. 48.1 yd 11. 40 cm
13. 45 cm 15. 34 cm
17. 200 yd 19. 36 cm
21. 35 ft 23. 43 cm
25. 36 m 27. 40 ft
29. 50 in. 31. 108 ft

Applications

33. 114 cm
35. 38.4 in.
37. a. 4605 ft
b. 15.7 minutes
39. a. 548 feet b. $8220
41. $13\frac{4}{5}$ inches
43. $17\frac{1}{2}$ km

Writing & Thinking

45. Some of the polygons are: triangle (3 sides), square (4 sides), rectangle (4 sides), parallelogram (4 sides), and trapezoid (4 sides).
47. Perimeter is the distance around a figure. Formulas for the perimeter of: triangle $(P = a + b + c)$, square $(P = 4s)$, rectangle $(P = 2l + 2w)$, trapezoid $(P = a + b + c + d)$, and parallelogram $(P = 2a + 2b)$.

6.3 Exercises

Concept Check

1. square
3. parallelogram
5. square
7. False; The $(b + c)$ in the trapezoid area formula represents the sum of the lengths of the two parallel bases.
9. False; The area formula for a triangle is $A = \frac{1}{2}bh$.

Practice

1. 81 ft^2
3. 525 km^2
5. 27.37 ft^2
7. $\frac{5}{27}$ $in.^2$
9. 165 cm^2
11. 48 $in.^2$
13. 99 ft^2
15. 162 yd^2
17. 1925 cm^2
19. 196 $in.^2$
21. 48 cm^2
23. 160 $in.^2$
25. 60 $in.^2$
27. 107 m^2
29. 99 $in.^2$
31. 220 mm^2
33. 36 cm^2
35. 192 m^2
37. 99 $in.^2$
39. a. 12 ft; 6 ft^2
b. 30 cm; 30 cm^2
c. 48 in.; 96 $in.^2$
41. a. 70 cm b. 220 cm^2
43. a. 32 m b. 48 m^2

Applications

45. 204 square feet
47. 3136 square inches.
49. a. 30 ft^2 b. 30 ft
51. a. 2350 square feet
b. 11.75 pounds
53. 123 square feet
55. 426 square meters

Writing & Thinking

57. The expression $(b + c)$ represents the sum of the two bases in a trapezoid. The bases are the two sides that are parallel in a trapezoid.
59. The area of the interior triangle is equal to half of the area of the rectangle no matter where the point is chosen.

6.4 Exercises

Concept Check

1. diameter
3. radius
5. π
7. True
9. False; The length of the radius of a circle is half of the length of the diameter.

Practice

1. a. 31.4 ft b. 78.5 ft^2
3. a. 19.468 yd
b. 30.1754 yd^2
5. a. 3.14 m b. 0.785 m^2
7. a. 188.4 cm b. 2826 cm^2
9. a. 4.71 ft b. 1.76625 ft^2
11. a. 37.68 cm
b. 113.04 cm^2
13. a. 94.2 in. b. 706.5 $in.^2$
15. a. 35.98 in.
b. 76.93 $in.^2$
17. a. 10.71 cm
b. 7.065 cm^2
19. a. 13.71 in.
b. 12.5325 $in.^2$
21. a. 21.42 m b. 32.13 m^2
23. a. 14.28 m b. 11.14 m^2
25. a. 19.42 cm
b. 26.13 cm^2
27. a. 36.56 m b. 38.88 m^2
29. 7536 m^2
31. 0.86 m^2

Applications

33. a. 63.6 square inches
b. $0.13 per square inch
35. a. 61.68 in.
b. 226.08 $in.^2$
37. 15,307.5 ft^2
39. 314.96 $in.^2$
41. a. 4.57 square feet
b. 10.34 feet
43. 62,800 in. (a little less than one mile)
45. 90.7 $in.^2$
47. 4710 ft

Writing & Thinking

49. One diameter is equal to two radii. Thus, $d = 2r$ and $C = 2\pi r = \pi d$.
51. Pi is a nonrepeating nonterminating decimal and therefore, when using π, a rounded number must be used for calculations. Anytime a rounded number is used in calculations, the result will be an approximation.

Collaborative Learning

53. a. 4

b. Answers will vary somewhat, but they should be about 3.3.

c. Answers will vary. The numbers will be come closer to π as the number of sides increases.

6.5 Exercises

Concept Check

1. volume
3. surface area
5. right circular cylinder
7. True
9. True
11. a. C b. E c. D d. B e. A

Practice

1. 70 in.^{3}
3. 381.51 cm^3
5. 12.56 mm^3
7. 60 in.^{3}
9. 14.13 ft^3
11. 401.92 m^3
13. 376.8 ft^3
15. 2289.06 cm^3
17. 224 cm^3
19. 113.04 in.^{3}
21. 9106 dm^3
23. 56.52 ft^3
25. 1017.36 mm^2
27. 122 in.^{2}
29. 226.08 m^2
31. $V = 70$ in.^{3}; $V = 1147.09$ cm^3
33. $V = 12.56$ dm^3; $V = 0.012\ 56$ m^3

Applications

35. 2,596,902 m^3
37. 10.39 in.^{3}
39. 800 ft^3
41. 13 cm
43. a. 3.375 ft^3 b. 13.5 ft^2
45. 1536 square inches

Writing & Thinking

47. Volume is measured in cubic units. Volume takes up a three-dimensional space and the units can be thought of as small cubes which leads to the concept of cubic units.
49. Volume is more important because it determines how many packages or what sized packages the driver can fit into the truck. Three-dimensional space (volume) is more important for this job.

6.6 Exercises

Concept Check

1. shape
3. proportional
5. False; Similar triangles have corresponding sides that are proportional.
7. False; If $\Delta ABC \cong \Delta DEF$, then $AC = DF$.

Practice

1. The triangles are not similar. The corresponding sides are not proportional.
3. $\Delta PQR \sim \Delta SUT$. All pairs of corresponding sides are proportional to the ratio 1:2.
5. $\Delta ABC \sim \Delta EDC$. The corresponding angles have the same measure.
7. $x = 50°$; $y = 70°$
9. $x = 50°$; $y = 50°$
11. $x = 20°$; $y = 100°$
13. $x = 4.8$; $y = 7.2$
15. $x = 7.5$; $y = 15$
17. $x = 6$; $y = 4$
19. $x = 95°$; $y = 20$
21. $x = 25°$; $y = 20°$
23. $x = 25°$; $y = 40°$
25. $x = 12$; $y = 10$
27. $x = 4.8$; $y = 2.5$
29. Congruent by SAS
31. Congruent by ASA
33. Not congruent
35. Congruent by SSS

Applications

37. 7.5 feet 39. 125 yd
41. 480 ft 43. 48 ft
45. 6.9 ft

Writing & Thinking

47. a. The triangles are similar since the three pairs of corresponding sides are proportional.
b. $m\angle A = m\angle D$, $m\angle B = m\angle E$, $m\angle C = m\angle F$
49. She needs to calculate the ratios of each pair of corresponding sides. Answers will vary.

6.7 Exercises

Concept Check

1. radical, radical
3. radicand
5. right
7. True
9. True

Practice

1. Yes, $16 = 4^2$
3. Not a perfect square
5. Yes, $400 = 20^2$
7. Yes, $121 = 11^2$
9. 144 11. 400
13. 6 15. 15
17. 14 19. 30
21. 2.8284 23. 4.8990
25. 5.8310 27. 28.2843
29. 15.5563 31. 0.5
33. 1.23 35. 1.4
37. 0.05 39. 0.0548
41. a. Answers will vary. $\sqrt{39} > \sqrt{36} = 6$, so $\sqrt{39} > 6$
b. Answers will vary. $\sqrt{39} < \sqrt{49} = 7$, so $\sqrt{39} < 7$
43. a. 7 and 8 b. 7.0711
45. a. 3 and 4 b. 3.3166
47. a. 7 and 8 b. 7.7460
49. Yes, $6^2 + 8^2 = 10^2$
51. No, $3^2 + 4^2 \neq 6^2$
53. Yes, $5^2 + 12^2 = 13^2$
55. $c = 2.23$
57. $c = 5$
59. $c = 20.62$
61. $c = 14.14$ cm

Applications

63. 2.24 miles
65. 58.31 ft
67. 522.0 feet
69. a. 20 in. b. 25 in.^{2} c. 7.0711 in.
71. 64.9 inches
73. a. 17 feet b. 51 feet

Writing & Thinking

75. A perfect square is the result when a number is squared. For example 5 squared is 25, therefore 25 is the perfect square of 5. Then if one finds the square root of 25, the answer is 5 because the square root is the original number that was squared to get the result of 25.
77. 28.28 m
79.

m	n	$a = 2nm$	$b = m^2 - n^2$	$c = m^2 + n^2$	Pythagorean Triple?
5	1	10	24	26	Yes: $10^2 + 24^2 = 26^2$
7	1	14	48	50	Yes: $14^2 + 48^2 = 50^2$
3	2	12	5	13	Yes: $12^2 + 5^2 = 13^2$
7	2	28	45	53	Yes: $28^2 + 45^2 = 53^2$
5	3	30	16	34	Yes: $30^2 + 16^2 = 34^2$
11	3	66	112	130	Yes: $66^2 + 112^2 = 130^2$
13	7	182	120	218	Yes: $182^2 + 120^2 = 218^2$

Chapter 7: Statistics, Graphs, and Probability

7.1 Exercises

Concept Check

1. statistics

3. data

5. odd

7. True

9. False; The number that appears the greatest number of times in a set of data is the mode.

Practice

1. a. 58 **b.** 57
c. 57 **d.** 4

3. a. \$48,625 **b.** \$46,500
c. \$63,000 **d.** \$43,000

5. a. 83.8 **b.** 83.5
c. 82 **d.** 12

7. a. 17.6 in. **b.** 14.9 in.
c. None **d.** 19.7 in.

9. a. 96,360 m^3
b. 84,500 m^3
c. None **d.** 82,700 m^3

11. a. \$4892.60 **b.** \$5100
c. None **d.** \$2890

13. a. 22.3 **b.** 21
c. 23 **d.** 19

15. a. 15.425 **b.** 14.5
c. 17 and 3 **d.** 44

17. a. 1135 miles
b. 980 miles
c. None **d.** 2020 miles

19. a. 12,371,400
b. 11,872,500
c. None **d.** 6,955,000

Applications

21. 79

23. a. 3.48 million farms
b. 2.23 million farms

25. a. 734,616.7 **b.** 320,725

Writing & Thinking

27. The first step to finding the median is always to arrange the data in order. The median is the middle number. If there is an even number of item, average the two middle numbers to find the median.

29. Answers will vary. Examples include: median income or home value for a particular area, mean temperature at a certain level in the ocean, mode of days having a particular heat index in a particular geographic location, or the range of SAT or ACT scores for a graduating class.

Collaborative Learning

31. Answers will vary.

7.2 Exercises

Concept Check

1. graphs **3.** circle
5. upper **7.** True
9. True

Applications

1. a. Social Science
b. Chemistry & Physics, Humanities
c. About 3300
d. About 21.2%

3. a. Sue **b.** Bob and Sue
c. 85.7%
d. Bob and Sue, Bob and Sue, Yes, in most cases
e. No, the vertical scales represent two different types of quantities

5. a. News: 300 min; Movies: 120 min; Sitcoms: 156 min; Soaps: 180 min; Drama: 144 min; Children's shows: 120 min; Commercials: 180 min
b. News **c.** 480 min

7. a. Taxes **b.** 25% **c.** 1:2

9. a. February and May
b. 6 inches **c.** March
d. 3.58 inches

11. a. August **b.** 4
c. April, May, July, August
d. 1 **e.** 14% **f.** 13%

13. a. West: 5%; Northeast: 28%; Midwest: 35%; South: 32%
b. West: 22%; Northeast: 19%; Midwest: 23%; South: 36%
c. South **d.** 5%
e. West **f.** 5%, 1900
g. 36%, 2000
h. Midwest

15. a. 8 **b.** 3 **c.** Eight class
d. 2 **e.** 27, 29 **f.** 50
g. 10 **h.** 16%

Writing & Thinking

17. The four types of graphs are **1.** bar graphs, used for comparative amounts; **2.** circle graphs, also known as pie charts, to help understand percents or parts of a whole; **3.** line graphs, to indicate tendencies or trends over time; and **4.** histograms to indicate data in a range or interval of numbers (called classes).

19. Both bar graphs and histograms use bars to indicate information. A bar graph uses categories and has spaces between the bars to separate the categories. A histogram uses boundaries of intervals instead of categories and the bars have no space between the classes.

7.3 Exercises

Concept Check

1. horizontal

3. width

5. 360

7. False; In creating a vertical bar graph, all bar widths should be the same.

9. True

Applications

1.

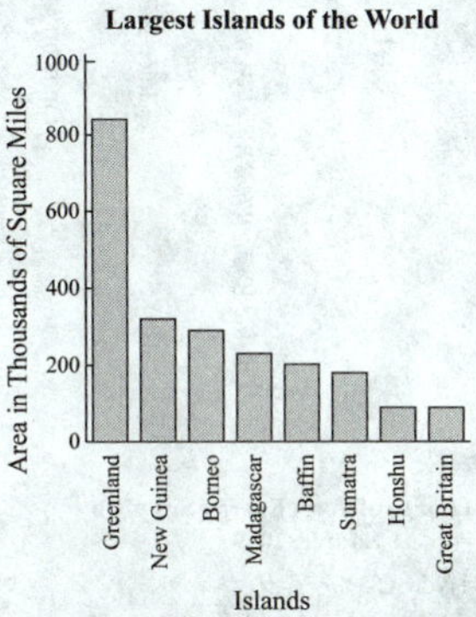

3.

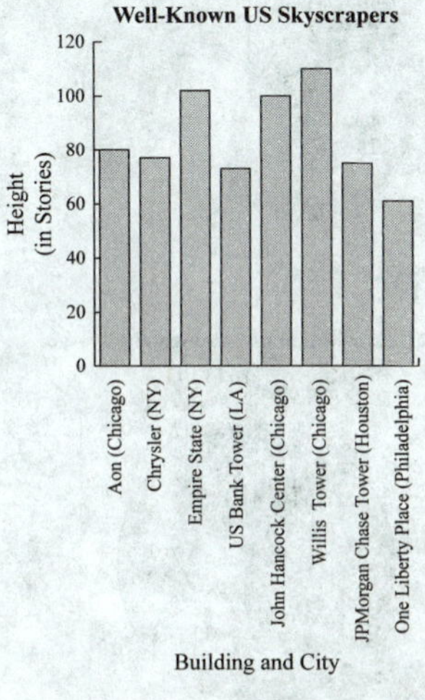

5.

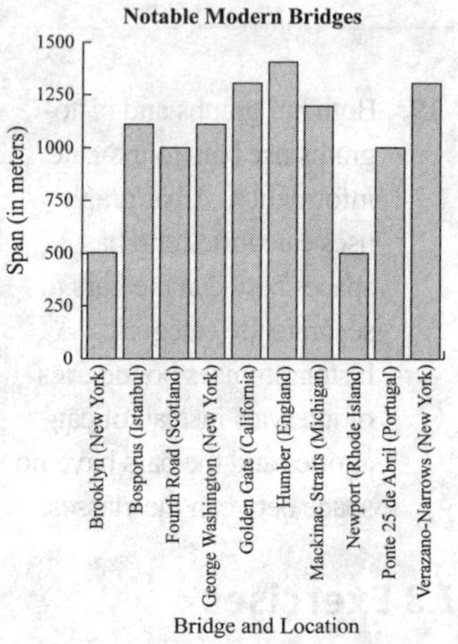

7.

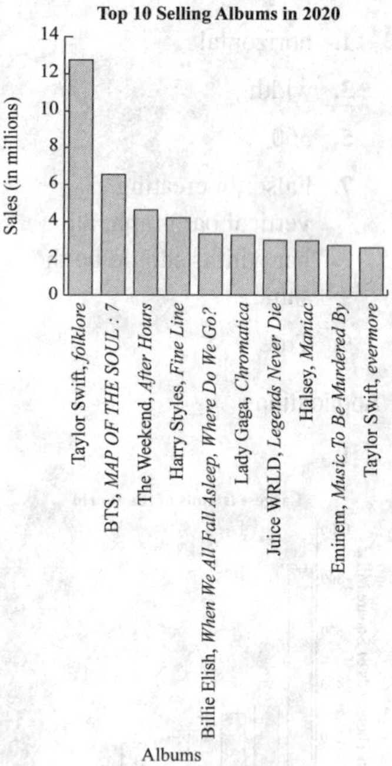

9.

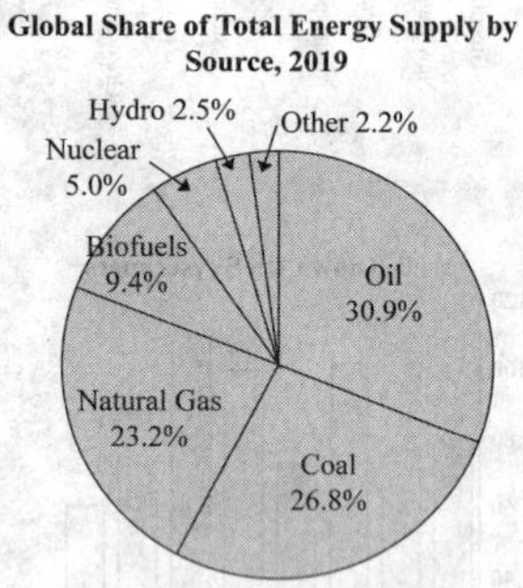

11.

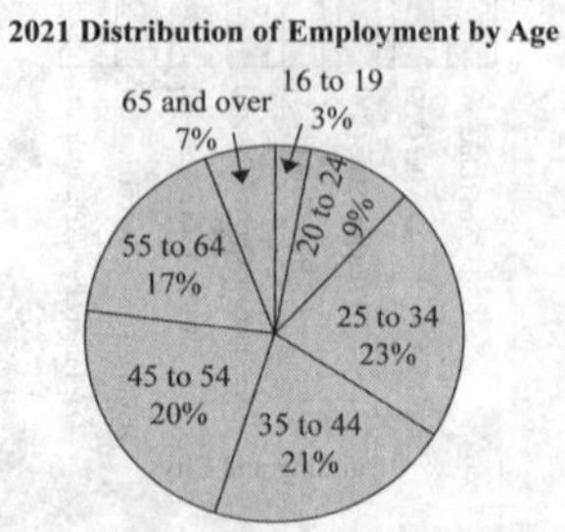

13.

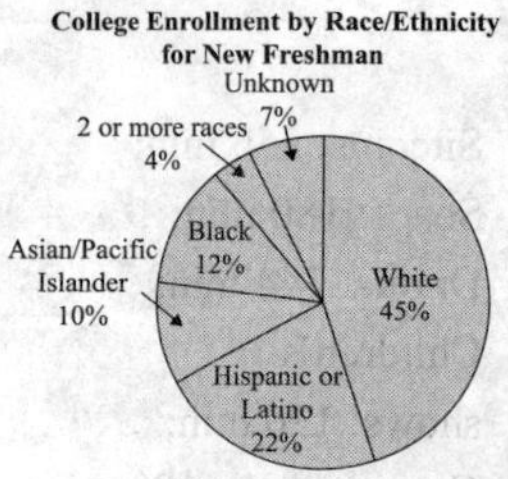

Writing & Thinking

15. Answers will vary. A person could mislabel or not label the graph at all. This mistake will make it difficult to properly interpret or understand the information presented. If the person does not pay attention to neatness and organization, the graph may be difficult to read. A person might forget to use appropriate titles which can decrease the understanding or purpose of the graph.

7.4 Exercises

Concept Check

1. tree

3. probability

5. sample space

7. False; The individual result of an experiment is an outcome.

9. True

Applications

1.

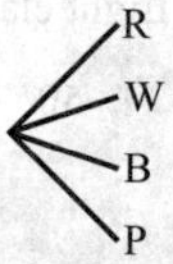

$S = \{R, W, B, P\}$
R = red, W = white,
B = blue, P = purple

3.

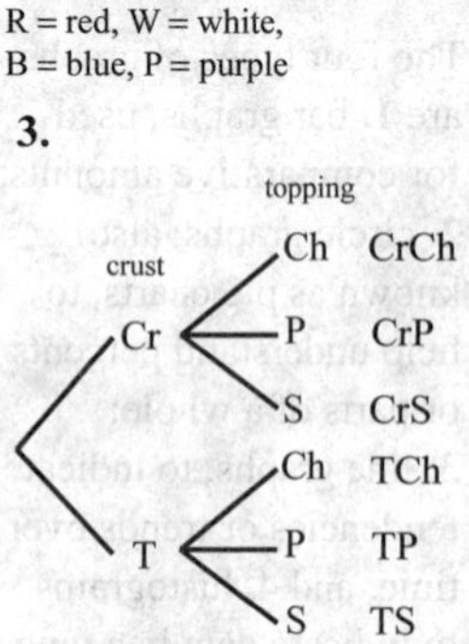

$S = \{CrCh, CrP, CrS, TCh, TP, TS\}$
Cr = crispy, T = thick,
Ch = cheese, P = pepperoni,
S = sausage

5.

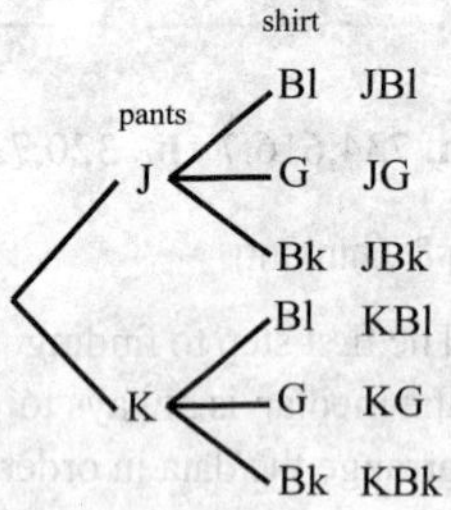

$S = \{JBl, JG, JBk, KBl, KG, KBk\}$
J = jeans, K = khaki pants, Bl = blue,
G = green, Bk = black

7.

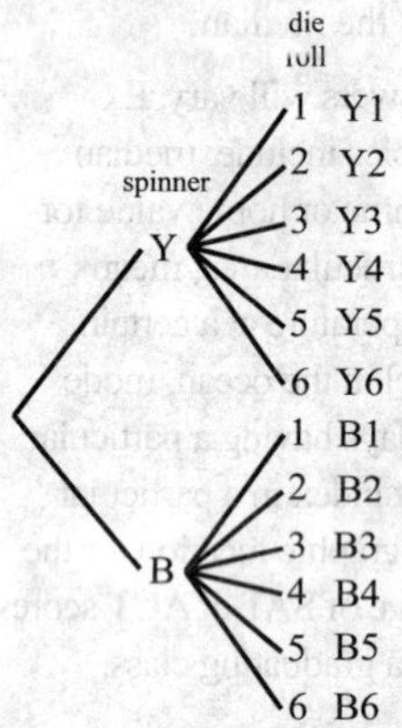

$S = \{Y1, Y2, Y3, Y4, Y5, Y6$
$B1, B2, B3, B4, B5, B6\}$
Y = yellow, B = blue

9.

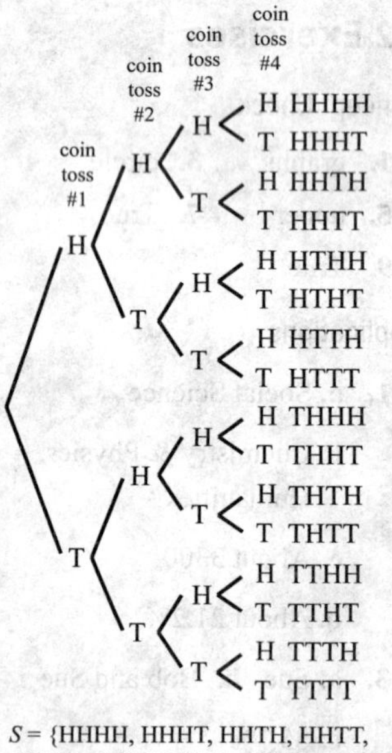

$S = \{HHHH, HHHT, HHTH, HHTT,$
$HTHH, HTHT, HTTH, HTTT, THHH$
$THHT, THTH, THTT, TTHH, TTHT,$
$TTTH, TTTT\}$

11. $\frac{2}{5}$ **13.** $\frac{1}{20}$

15. $\frac{3}{4}$ **17.** $\frac{1}{6}$

19. 0 **21.** $\frac{1}{36}$

23. $\frac{1}{4}$ **25.** $\frac{1}{4}$

27. $\frac{1}{2}$ **29.** $\frac{1}{2}$

31. 1 **33.** $\frac{1}{13}$

35. $\frac{1}{4}$ **37.** $\frac{1}{52}$

39. $\frac{2}{13}$

Writing & Thinking

41. Chance experiments include, but are not limited to, tossing a coin, spinning a bottle, drawing a card from a standard deck of cards, picking numbers in the lottery, choosing straws, and picking colored marbles.

43. Probabilities are between 0 and 1, inclusive. The sum of the probabilities of the outcomes in a sample space is 1. An event has probability 0 if it can never occur, such as rolling a 7 on a six-sided die. An event has probability 1 if it will always occur, such as rolling one of the numbers 1, 2, 3, 4, 5, or 6 on a six-sided die.

Chapter 8: Introduction to Algebra

8.1 Exercises

Concept Check

1. integers

3. irrational

5. graph

7. True

9. True

Practice

1. 3

3. 0

5. −2

7. 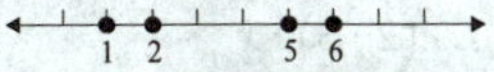

9. 

11. 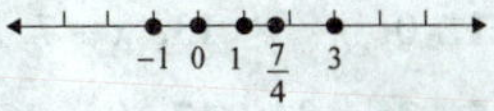

13. 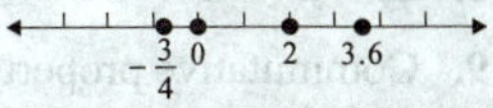

15. 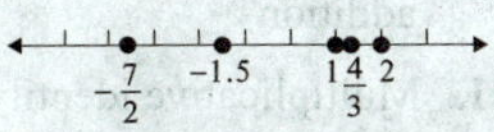

17. 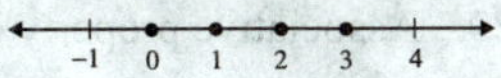

19. 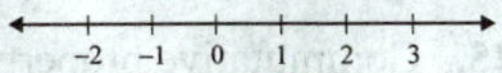

21. 4, 8

23. −7, −2, 0, 4, 8

25. $-7, -2, -\frac{5}{3}, -1.4, 0, \frac{3}{5}, 4, 5.9, 8$

27. < **29.** >

31. < **33.** =

35. > **37.** <

39. = **41.** >

43. True

45. False; $-9 < -8.5$

47. True

49. True

51. False; $-|-3| > -|4|$

53. $x = -5, 5$

55. $x = -2, 2$

57. No solution

59. $x = -23, 23$

61. Sometimes. Examples will vary.

63. Always. Examples will vary.

Applications

65. −4500 meters

67. +8844 meters

69. **a.** $|x| \le 3$

b. 

71. **a.** 

b. $-\frac{5}{2}$ has the furthest distance from zero.

73. **a.** $x = -5$ millimeters, 5 millimeters

b. The length of the board can be 5 mm longer or 5 mm shorter than 1 meter.

c. 1.05 m or 0.95 m

75. **a.** Fresn

b. El Centro

Writing & Thinking

77. If y is a negative number then $-y$ represents a positive number. For example, if $y = -2$, then $-y = -(-2) = 2$.

8.2 Exercises

Concept Check

1. positive

3. negative

5. equation

7. True

Practice

1. 13 **3.** −4

5. 0 **7.** 5

9. −13 **11.** −8

13. −10 **15.** 0

17. 17 **19.** −4.6

21. −29 **23.** −9

25. −15.4 **27.** −3

29. −54 **31.** −2.5

33. $\frac{3}{7}$ **35.** $\frac{5}{8}$

37. $-\frac{7}{4}$ **39.** −7

41. −0.5 **43.** 76

45. −310 **47.** 27

49. 2.3 **51.** 18

53. 14 **55.** 35

57. 84 **59.** 3807

61. −97,714

Applications

63. Profit: \$6000

65. **a.** −0.3 cm, −0.1 cm

b. 0.02 cm

c. Increase

67. 275 feet

69. 8 feet

71. 8

73. The solution is −\$97.86.

Writing & Thinking

75. $|0| + |0| = 0$

77. Sometimes. Examples will vary.

79. Sometimes. Examples will vary.

81. Never. Examples will vary.

83. Never. Examples will vary.

8.3 Exercises

Concept Check

1. additive inverse

3. left

5. subtract

7. False; The sum of a number and its additive inverse is zero.

9. True

Practice

1. −11 **3.** 6

5. −4.7 **7.** 0

9. $\frac{5}{7}$ **11.** 5

13. −10 **15.** 7

17. 3 **19.** 12

21. 24 **23.** 5.9

25. −4 **27.** 10

29. $\frac{7}{20}$ **31.** $-\frac{1}{4}$

33. $\frac{3}{40}$ **35.** −15

37. −1 **39.** −3

41. 1 **43.** 0

45. $-8 < -5$

47. $-3.1 < 3.1$

49. $\frac{19}{4} = \frac{19}{4}$

51. $0 > -27$

53. $-237 > -248$

55. $1044 > -39$

57. $-15{,}254 > -35{,}090$

59. 1 **61.** −7

63. −26 **65.** 11

Applications

67. −18 °F (a decrease of 18 degrees Fahrenheit)

69. \$7 (an increase of 7 dollars)

71. −14,787 ft (a decrease of 14,787 feet)

73. **a.** Negative **b.** Positive

c. May and June

d. Answer will vary. March and April or April and May

75. \$250,404

77. \$847

79. \$165

81. −6 °C (or 6 °C below 0 °C)

83. 2316 points

85. 1220 gallons

87. a. 1454 people

b. 198,580 people

89. The solution is 90.85 g

91. 3 °F

93. 4238 points

Writing & Thinking

95. Add the opposite of the second number to the first number.

97. Answers will vary. When the absolute value of the number being subtracted is greater than the absolute value of the other number, the difference will be positive.

8.4 Exercises

Concept Check

1. average, mean

3. positive

5. 0

7. True

9. False; The product and quotient will be positive.

Practice

1. −12 3. 48

5. 56 7. −21

9. 56 11. 26

13. −70 15. −24

17. −288 19. 0

21. −9 23. −7.31

25. $-\frac{1}{6}$ 27. $\frac{18}{5}$

29. $-\frac{12}{5}$ 31. $-\frac{9}{2}$

33. 4 35. −6

37. 2 39. 0

41. −3 43. −17

45. Undefined

47. 5 49. −0.6

51. 20 53. −2.3

55. $-\frac{1}{12}$ 57. $\frac{9}{22}$

59. $-\frac{5}{12}$

61. False; $(-4)(6) < 3 \cdot 8$

63. True

65. False; $6(-3) = (-14) + (-4)$

67. True 69. True

71. 116,640 73. 22,032

75. 1682 77. −45.5

79. −9.5 81. −54.605

Applications

83. −12

85. $2975

87. 8928 slices of bread

89. $442,800

91. 3 cookies

93. 22 teams

95. 17 grams

97. $10,740 per year

99. 1152 bags

101. 427 sq ft

103. 5423 dozen

105. 2 slices

107. 33,087.5 accidents

109. a. −8 psi

b. −10.5 psi

c. 5.25 hours (or 5 hours 15 minutes)

111. 26.6 calls

113. 80

115. 78.6 inches

117. 31.5 hours

119. 18,631.8 square miles

Writing & Thinking

121. Positive; The product of two negative numbers will be positive so an even number of negatives will give a positive product.

123. Answers will vary.

8.5 Exercises

Concept Check

1. grouping symbols

3. left, right

5. −1

7. True

9. True

Practice

1. a. 36 b. 16

3. −25 5. −10

7. −45 9. −137

11. 152 13. −189

15. 143 17. −36

19. −10 21. $\frac{11}{30}$

23. $\frac{1}{24}$ 25. $\frac{31}{24}$

27. $\frac{10}{3}$ 29. $\frac{41}{32}$

31. $\frac{3}{5}$ 33. −5.7

35. −3.5 37. 24

39. −5 41. 1

43. $\frac{13}{64}$ 45. 42.45

47. 67.77 49. 15.41

Applications

51. a. $\$11 \cdot \left(1\frac{1}{4}\right) + \$8.50$

b. $22.25

c. $31

53. a. $-\$42 - \$35 - (3 \cdot \$5)$

b. −$92

55. a. $\left(12\frac{1}{2} + 15\frac{3}{4} + 11\frac{3}{4} - 1\frac{1}{4}\right) \div 6$

b. $6\frac{11}{24}$ pounds of potatoes

57. a. $2 \cdot \$225 + 4 \cdot \$20 + 4 \cdot \$15$

b. $147.50

Writing &Thinking

59. Smaller; When any positive number is multiplied by a fraction (or decimal) between 0 and 1, the result will be smaller. This is what is happening when a number between 0 and 1 is squared. Answers will vary.

8.6 Exercises

Concept Check

1. reciprocal

3. zero factor

5. opposite

7. False; The commutative property of addition allows the order to change.

9. False; The additive identity of all numbers is 0.

Practice

1. $3 + 7$

3. $4 \cdot 19$

5. $30 + 48$

7. $(2 \cdot 3) \cdot x$

9. $(3 + x) + 7$

11. 0 13. $x + 7$

15. $2x - 24$ 17. 0

19. Commutative property of addition

21. Multiplicative identity

23. Associative property of addition

25. Commutative property of multiplication

27. Commutative property of multiplication

29. Multiplicative inverse

31. Additive inverse

33. Multiplicative identity

35. Zero-factor law

37. Associative property of addition

39. $6(11) = 66$ and $6 \cdot 3 + 6 \cdot 8 = 66$

41. $10(-7) = -70$ and $10 \cdot 2 - 10 \cdot 9 = -70$

43. Commutative property of multiplication; $6 \cdot 4 = 4 \cdot 6 = 24$

45. Associative property of addition; $8 + (5 + (-2)) = (8 + 5) + (-2) = 11$

47. Distributive property; $5(4 + 18) = 5(4) + 90 = 110$

49. Associative property of multiplication;
$(6\cdot(-2))\cdot 9 = 6\cdot(-2\cdot 9)$
$= -108$

51. Commutative property of addition;
$3+(-34)=(-34)+3=-31$

53. Commutative property of addition;
$2(3+4)=2(4+3)=14$

55. Commutative property of addition;
$5+(4-15)=(4-15)+5$
$=-6$

57. Associative property of multiplication;
$(3\cdot 4)\cdot 5=3\cdot(4\cdot 5)=60$

Applications

59. **a.** \$118.25

b. $\$11\cdot\left(6\frac{1}{2}\right)+\$11\cdot\left(4\frac{1}{4}\right)$

c. Distributive property

61. **a.** \$85.04 – \$28.79 – \$50.00 – \$12.16 or \$85.04 – (\$28.79 + \$50.0 + \$12.16)

b. –\$5.91

c. –\$5.91 + \$5.91 = 0

d. The additive inverse property

8.7 Exercises

Concept Check

1. 1

3. constant

5. coefficients

7. True

9. False; In the term $12a$, 12 is the coefficient.

Practice

1. –5, 3, and 8 are like terms; $7x$ and $9x$ are like terms.

3. $-x^2$ and $2x^2$ are like terms; $-6xy$ and $5xy$ are like terms; $3x^2y$ and $5x^2y$ are like terms.

5. 24, –6, and 8.3 are like terms; $1.5xyz$, $-1.4xyz$, and xyz are like terms.

7. $15x$ **9.** $3x$

11. $5y^2$ **13.** $10x$

15. $-2c$ **17.** $7x+2$

19. $2x^2+2x$

21. $3x^2+20$

23. $13x^2-2y$

25. $7x-1$

27. $-2n^2+2n-3$

29. $-x^2+3x+45$

31. $4n+3$ **33.** $7a-8b$

35. $8x+y$ **37.** $-2y$

39. $\frac{8x}{3}$ **41.** –4

43. 4 **45.** –2

47. $3x+4$; 13

49. $5y+1$; 16

51. $-3.1a^2+4a$; –20.4

53. $3.7x+1.1$; 15.9

55. $8a$; –16

57. $3b$; –3

59. $-x^2+5x-7$; –21

61. $3y^2-y$; 4

63. $4y^3-4y^2+5y+1$; –12

65. $5x^2+9x-22$; –20

67. $2a^2-2a$; 4

69. $9ab-7a+2b$; 21

71. $14a-15b+2c+2$; 24

73. $10a+10b+10c$; 0

Applications

75. \$50,000

77. 375 pounds

79. **a.** $47s+143m$

b. 6170 boxes

Writing & Thinking

81. A number is a constant and its value does not change regardless of where it appears in an expression or equation. A variable, usually represented by a letter, represents an unknown number or numbers.

83. **a.** For $x=3$,
$4x^2-5(x+2)+3x+10+2x$
$=4(3)^2-5(3+2)+3(3)+10+2(3)$
$=4(9)-5(5)+9+10+6$
$=36-25+9+10+6$
$=36$

b.
$4x^2-5(x+2)+3x+10+2x$
$=4x^2-5x-10+3x+10+2x$
$=4x^2$
For $x=3$,
$4x^2=4(3)^2$
$=4(9)=36$

The second method, as it requires fewer calculations.

8.8 Exercises

Concept Check

1. ambiguous

3. subtraction

5. multiplication

7. True

9. False; Subtraction is indicated by the phrase "five less than a number."

Practice

1. $x+6$ **3.** $y-4$

5. $\frac{2m}{10}$ **7.** $6y-4$

9. $4x+2x$ **11.** $15-2y$

13. $3x-5x$ **15.** $9(x+2)$

17. $4(x+1)-13$

19. $3(x+6)+8$

21. $3(7-x)-4$

23. $\frac{x}{2}-18$

25. **a.** $x-6$ **b.** $6-x$

27. **a.** $3x-5$ **b.** $5-3x$

29. $24d$ **31.** $\$3.15x$

33. $365y$ **35.** $7t+3$

37. $7t+3$

39. $\$20+\$0.15m$

41. $2w+2(2w-3)=6w-6$

43. Four times a number

45. A number increased by five

47. Four times a number decreased by seven

49. Seven times the sum of a number and one

51. Negative two times the difference between a number and eight

53. Five times the sum of twice a number and three

55. The quotient of six and the difference between a number and one

57. Six times a number plus the number minus one

59. Eight increased by twice the difference between a number and one

61. The sum of three times a number and seven; three times the sum of a number and seven

63. The product of seven and a number minus three; seven times the difference between a number and three

Writing & Thinking

65. The Commutative Property of Addition and Multiplication permits the order of items being added or multiplied to change and still have the same result. This property does not hold true for subtraction or division. Therefore, order is important for subtraction and division problems or the answer will change or be incorrect.

67. Answers will vary.

Chapter 9: Solving Linear Equations and Inequalities

9.1 Exercises

Concept Check

1. equation
3. addition, equality
5. dividing
7. True
9. True

Practice

1. $x=-2$ is a solution
3. $x=4$ is not a solution
5. $x=-4$ is a solution
7. $x=-18$ is a solution
9. $x=-28$ is a solution
11. $x=7$
13. $y=-4$
15. $x=-19$
17. $n=37$
19. $z=-6$
21. $x=5$
23. $y=-5.9$
25. $x=-1.2$
27. $x=\frac{11}{20}$
29. $x=9$
31. $y=8$
33. $x=20$
35. $y=10$
37. $x=-8$
39. $x=12$
41. $n=8$
43. $y=2.1$
45. $x=\frac{20}{9}$
47. $x=-13.3$
49. $y=-12$
51. $x=-\frac{2}{5}$
53. $x=-4$
55. $x=\frac{5}{8}$
57. $n=9.7$
59. $x=-4$
61. $y=-50.753$
63. $x=-17.214$
65. $x=246$
67. $x=-153.17$

Applications

69. 1945 *kanji* characters
71. **a.** Answers will vary.
    b. $x=12.5$
    c. The garden should be 12.5 feet wide.
73. 4.24 light years
75. 11,500 words
77. 2250 students
79. **a.** Answers will vary.
    b. $x=11{,}550$
    c. Clara will need a loan for $11,550.

Writing & Thinking

81. **a.** Yes. It is stating that $6+3$ is equal to 9.
    b. No. If we substitute 4 for x, we get the statement $9=10$, which is not true.

9.2 Exercises

Concept Check

1. like
3. multiplication
5. substituting
7. False; Subtract 3 from both sides.
9. True

Practice

1. $x=-3$
3. $x=2$
5. $x=2$
7. $x=2$
9. $y=-1$
11. $t=-1$
13. $x=-0.12$
15. $x=4$
17. $x=0$
19. $y=0$
21. $x=-2$
23. $y=-6$
25. $n=6$
27. $n=8$
29. $x=0$
31. $x=-7$
33. $x=-\frac{1}{8}$
35. $x=\frac{13}{2}$
37. $x=-\frac{21}{5}$
39. $x=-\frac{8}{15}$
41. $y=\frac{28}{5}$
43. $x=2$
45. $y=\frac{7}{5}$
47. $x=-4.5$
49. $x=-44$
51. $x=2$
53. $x=-4$
55. $y=0.5$
57. $x=1.5$
59. $x=0.2$
61. $x=6.1$
63. $x=1.12$

Applications

65. 14,000 tickets per hour
67. 52 pages
69. 240 cookies of each remaining variety
71. 135 yards
73. 379.7 feet
75. **a.** Lowest temperature; Change in temperature per hour; Current temperature
    b. $x=4$
    c. The low temperature occurred 4 hours ago.

Writing & Thinking

77. **a.** The 4 should have been multiplied by 3 so that the 3 was distributed over the entire left-hand side of the equation; Correct answer is $x=15$.
    b. 3 should be subtracted from each side, not from each term, and $5x-3$ doesn't simplify to $2x$; Correct answer is $x=\frac{8}{5}$.

9.3 Exercises

Concept Check

1. identity
3. conditional
5. real, $\mathbb{R}$
7. True
9. True

Practice

1. $x=-5$
3. $n=3$
5. $y=6$
7. $x=3$
9. $n=0$
11. $y=0$
13. $z=-1$
15. $y=\frac{1}{5}$
17. $x=-3$
19. $x=-4$
21. $x=-21$
23. $y=0$
25. $y=1$
27. $x=-\frac{3}{2}$
29. $x=\frac{1}{4}$
31. $x=\frac{3}{17}$
33. $x=-\frac{1}{4}$
35. $x=\frac{8}{5}$

37. $x = \frac{2}{3}$

39. $x = 6$

41. $x = -11$

43. $x = \frac{1}{2}$

45. $x = -5$

47. $n = -1.5$

49. $x = 0$

51. Conditional

53. Conditional

55. Contradiction

57. Identity

59. Conditional

61. $x = -50.21$

63. $x = 1.067$

Applications

65. 20 guests

67. 0.34 hours

69. **a.** The area of the first flyer
b. The width of the second flyer
c. $x = 5$
d. Solution should be correct if part c. is correct.
e. The first flyer has a width of 5 inches.

71. 1800 square feet

73. 240 sundaes

Writing & Thinking

75. **a.** $5x + 1$ **b.** $x = 6$
c. Answers will vary.

9.4 Exercises

Concept Check

1. mathematically

3. time

5. substitute

7. False; Case matters in formulas.

9. True

Applications

1. $192

3. $10,000

5. **a.** $183.75 **b.** $3683.75

7. 2 seconds

9. 4 milliliters

11. $1030

13. 14 rafters

15. 336 in. or 28 ft

17. $1030

19. 230 calculators

21. $5 million

23. $2400

25. 7 hours

27. 1.625 mph

29. $b = P - a - c$

31. $m = \frac{F}{a}$

33. $w = \frac{A}{l}$

35. $n = \frac{R}{p}$

37. $P = A - I$

39. $m = 2A - n$

41. $t = \frac{I}{Pr}$

43. $b = \frac{P - a}{2}$

45. $\beta = 180° - \alpha - \gamma$

47. $h = \frac{V}{lw}$

49. $b = \frac{2A}{h}$

51. $h = \frac{V}{\pi r^2}$

53. $g = \frac{mv^2}{2K}$

55. $y = \frac{6 - 2x}{3}$

57. $x = \frac{11 - 2y}{5}$

59. $b = \frac{2A - hc}{h}$ or $b = \frac{2A}{h} - c$

61. $x = \frac{8R + 36}{3}$ or $x = \frac{8R}{3} + 12$

63. $y = -x - 12$

65. $C = nt + 9$

67. $C = 325n + 5400$

69. **a.** $I = Prt$
b. $P = \$8000$; $I = \$600$; $t = 0.5$
c. r **d.** $r = 15\%$

71. **a.** $A = \frac{1}{2}bh$ **b.** $b = 3h$
c. $A = \frac{1}{2}(3h)h = \frac{3}{2}h^2$
d. $h^2 = \frac{2}{3}A$
e. Take the square root of both sides
f. 10 **g.** 30 feet

Writing & Thinking

73. **a.** 1 **b.** −1 **c.** 1.5
d. −12

9.5 Exercises

Concept Check

1. consecutive

3. $n + 2, n + 4$

5. 2

7. True

9. False; Integers that are not even

Practice

1. x = savings account balance in dollars (or other unit of money); $\frac{3}{5}x$

3. s = the length of a side of the square; $4s = 20$

5. $x - 5 = 13 - x$; 9

7. $36 = 2x + 4$; 16

9. $7x = 2x + 35$; 7

11. $3x + 14 = 6 - x$; −2

13. $\frac{2x}{5} = x + 6$; −10

15. $4(x - 5) = x + 4$; 8

17. $\frac{2x + 5}{11} = 4 - x$; 3

19. $2x + 3x = 4(x + 3)$; 12

21. $n + (n + 2) = 60$; 29, 31

23. $n + (n + 1) + (n + 2) = 69$; 22, 23, 24

25. $n + (n + 1) + (n + 2) + (n + 3) = 74$; 17, 18, 19, 20

27. $171 - n = (n + 1) + (n + 2)$; 56, 57, 58

29. $208 - 3n = (n + 1) + (n + 2) + (n + 3) - 50$; 42, 43, 44, 45

31. $n + 2(n + 2) = 4(n + 4) - 54$; 42, 44, 46

33. $(n + 2) + (n + 4) - n = 66$; 60, 62, 64

35. $2n + 3(n + 2) = 2(n + 4) + 7$; 3, 5, 7

Applications

37. **a.** The unknown value is the number of postcards purchased.
b. $p = 9$
c. Brooke purchased 9 postcards.

39. **a.** The number of slices each pizza was cut into
b. $\frac{1}{4}p + \frac{1}{2}p + \frac{3}{8}p = 9$
c. $p = 8$
d. Each pizza was cut into 8 slices.

41. $(c - 58.96) + c = 96.94$; Flash drive: $18.99; Printer: $77.95

43. $7x = 91{,}399$; 13,057

45. $x + x + (x + 3) + (x + 3 + 5) = 19$; 2 magazines

47. $50 - 2x = 10.50$; $19.75

49. $800 + 50(x - 2) = 1450$; 15 hours

51. $19.99 + 0.65x = 127.24$; 165 miles

53. $n + 4n - 4 + 2n + 7 = 59$; 8 cm, 28 cm, 23 cm

55. $x + 3x - 1 + 2x + 5 = 64$; 10 inches, 29 inches, 25 inches

Note: Answers for 57 through 61 will vary.

57. Twice a number increased by 3 equals 9; $x = 3$

59. Find 3 consecutive even integers such that the sum of the first and the third is 3 times the second; $n = -2, 0, 2$

61. Find a number such that twice the sum of the number and 2, decreased by 6 is equal to the sum of the number and 4 minus the number; $n = 3$

9.6 Exercises

Concept Check

1. rate, time

3. 12%

5. one

7. False; The value of r should be written as a decimal number.

9. True

Applications

1. **a.** Original price
b. x; $0.20x$ **c.** \$119.95

3. **a.** 2 liters **b.** $0.25x$
c. $x + 20$
d. $0.20(x + 20) = 2 + 0.25x$
e. 40 liters

5. **a.** x is the number of quarters in Brian's pocket; $0.10(2q) + 0.25q = 2.70$; $q = 6$, so Brian has 6 quarters and 12 dimes.
b. Check against the language of the problem: 12 is twice as much as 6, and 6 quarters (\$1.50) and 12 dimes (\$1.20) add up to \$2.70.

7. 1.68 mph

9. 8.75 hours

11. 3 hours

13. 50 mph; 300 miles

15. 7 hours

17. 36 mph; 60 mph

19. Day: 56 mph; Night: 69 mph

21. 4.5 miles

23. **a.**

	Rate (mph)	·	Time (min)	=	Distance (miles)
Tortoise	10		t		$10t$
Achilles	25		$t-2$		$25(t-2)$

b. $10t = 25(t - 2)$
c. $3\frac{1}{3}$ hours
d. $1\frac{1}{3}$ hours
e. Yes

25. \$7400 at 5.5%; \$2600 at 6%

27. \$7000 at 6%; \$9000 at 8%

29. \$5500 at 8%; \$6500 at 10%

31. **a.**

	Principal (\$)	·	Rate	=	Interest (\$)
High-Risk Fund	P		0.08		$0.08P$
Low-Risk Fund	$3600 - P$		0.04		$0.04(3600 - P)$

b. $0.08P + 0.04(3600 - P) = 198$
c. $P = \$1350$
d. High-risk fund: \$1350; Low-risk fund: \$2250
e. Answer will vary.

33. \$6720 at 5.5%; \$5280 at 7%

35. \$600 at 2.5%; \$800 at 4%

37. \$11,000 at 4%; \$9500 at 5%

39. \$40.50

41. \$113.75

43. **a.** \$1140
b. $x - 0.05x$ or $0.95x$
c. $x - 0.05x - 2850 = 1140$ or $0.95x - 2850 = 1140$
d. \$4200
e. Robin's Refurbished Wrecks should sell the used car for \$4200.

45. 94

47. 4 phone calls

49. \$20

51. **a.** Not possible
b. 192

53. **a.** 3.2 inches
b. 7.1 inches
c. 6.9 inches

55. **a.** 80.4 million
b. 26 million
c. 75 million

Writing & Thinking

57. If each equal side is 9 cm long, that would make the perimeter more than 18 cm; Correct answer: 6 cm

59. He would not be able to paddle faster than the current, so he would not be able to return upriver at all; Correct answer: $\frac{2}{3}$ mph

9.7 Exercises

Concept Check

1. interval

3. closed

5. addition

7. True

9. False; Only one value in the solution set needs to be checked.

Practice

1. $(-1, \infty)$
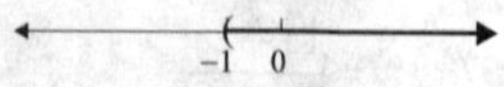

3. $(-\infty, 5]$

5. $[-5, -1]$
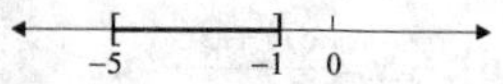

7. $[-7, -4)$
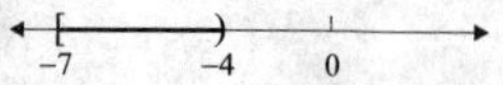

9. 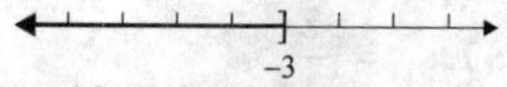

Half-open interval

11. 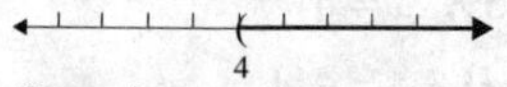

Open interval

13. 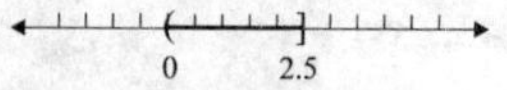

Half-open interval

15. 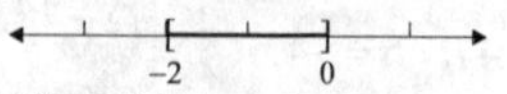

Closed interval

17. 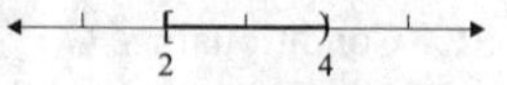

Half-open interval

19. $(4, \infty)$
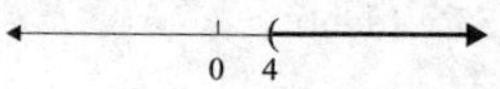

21. $(-\infty, 4]$
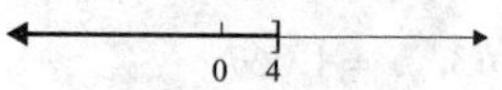

23. $(-1, \infty)$
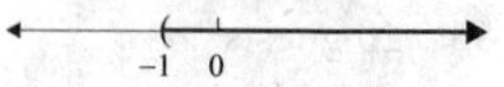

25. $(-\infty, 7]$
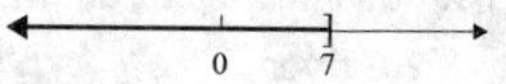

27. $(4, \infty)$
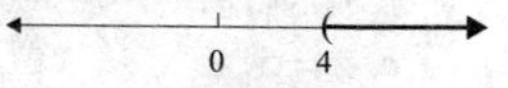

29. $(-\infty, 3]$
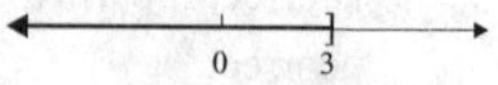

31. $(-2, \infty)$
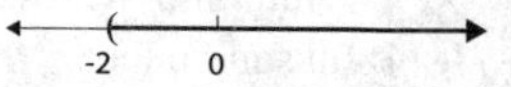

33. $(-\infty, 7]$
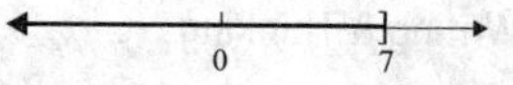

35. $(-\infty, 1)$
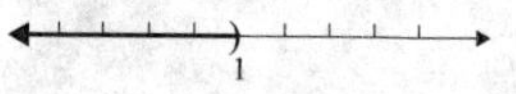

37. $(2, \infty)$
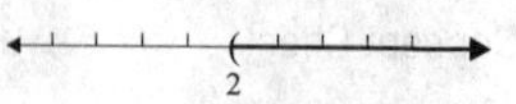

39. $\left(\frac{8}{3}, \infty\right)$
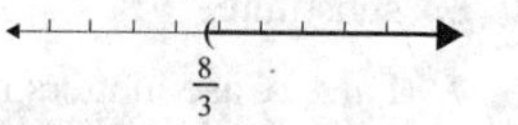

41. $[-0.4, \infty)$
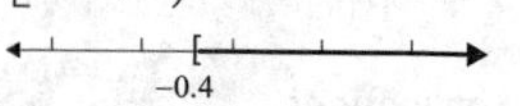

43. $(-\infty, 2)$

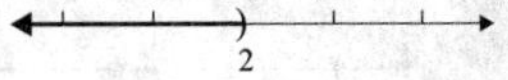

45. $[3, \infty)$

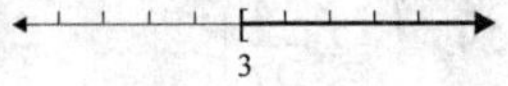

47. $(-\infty, 2]$

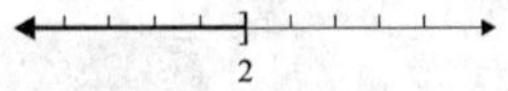

49. $\left(-\frac{1}{3}, \infty\right)$

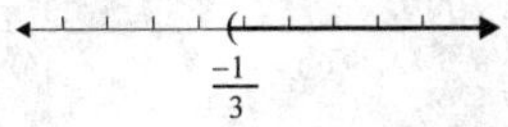

51. $\left(-\infty, -\frac{11}{2}\right)$

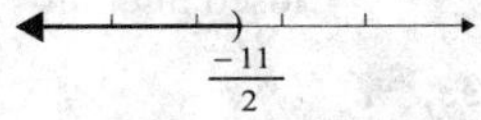

53. $[1, \infty)$

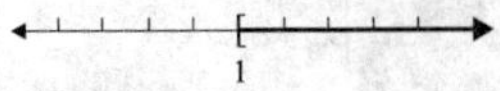

55. $\left(\frac{9}{2}, \infty\right)$

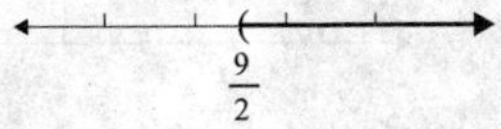

57. $(-\infty, -0.6)$

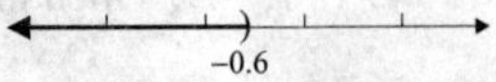

59. $(6.5, \infty)$

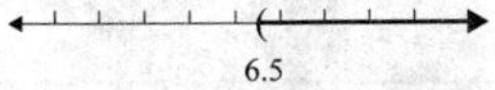

61. $(-\infty, -13)$

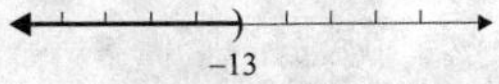

63. $\left(-\infty, -\frac{10}{9}\right]$

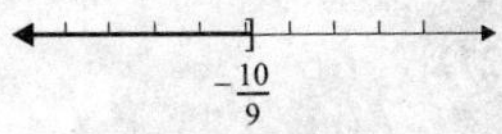

65. $[-4, \infty)$

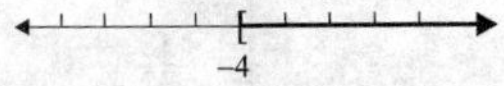

67. $(-\infty, 1]$

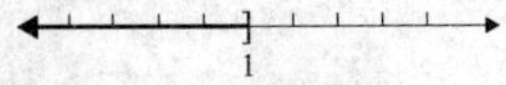

69. $(-17, \infty)$

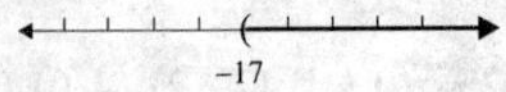

71. $(-\infty, -30)$

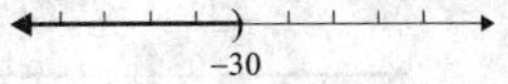

73. $\left(-\infty, -\frac{29}{2}\right]$

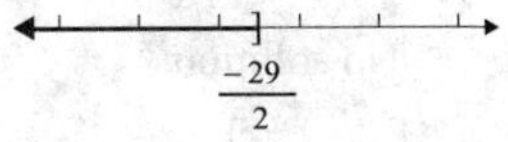

75. $x \geq 58$

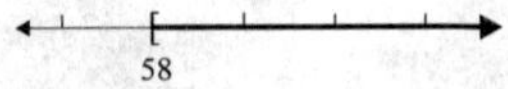

77. $x \leq 5$

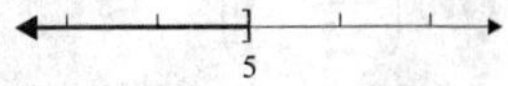

Applications

79. **a.** The student would need a score higher than 102 points, which is not possible. Thus he cannot earn an A in the course.
    b. The student must score at least 192 points to earn an A in the course.

81. He must sell at least 10 cars.

83. **a.** $15 - x$
    b. $8x + 5(15 - x)$
    c. 8 large arrangements

85. $100 - 5(x - 3) \geq 70$; 9 unexcused absences

87. 22,500 tickets

89. 8 times

91. **a.** $150 + 60 + 12.50c \leq 400$
    b. If you solve the inequality you get $c \leq 15.2$, but you can't buy 0.2 ink cartridges, so $c \leq 15$.
    c. Jeph can buy at most 15 ink cartridges.

93. **a.** $\frac{92 + 74 + 80 + 72 + E}{5} \geq 80$
    b. $E \geq 82$
    c. Andrew needs to earn at least an 82 on the fifth exam.

9.8 Exercises

Concept Check

1. elements
3. such that
5. roster
7. False; The union of two sets contains elements that belong to either one set, the other set, or both sets.
9. True

Practice

1. Union: $\{1, 2, 3, 4, 6, 8\}$; Intersection: $\{2, 4\}$

3. Union: $\{-4, -3, -2, -1, 0, 1, 2, 4, 6\}$; Intersection: $\{0\}$

5. Union: $\{1, 2, 4, 8, 16, 32, 64\}$; Intersection: $\{1, 4, 16\}$

7. $(3, \infty) \cup (-\infty, 2]$

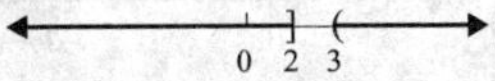

9. $(3, 8)$

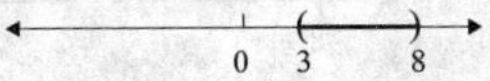

11. $(-\infty, -6]$

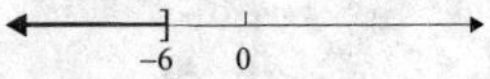

13. $(-\infty, 5)$

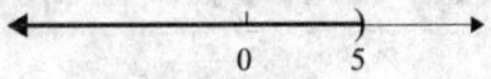

15.

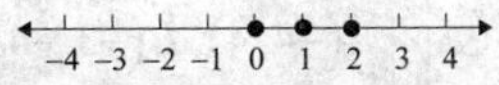

17.

19. $\{x \mid 3 \leq x < 5\}$

21. $\{x \mid x \geq -2.5\}$

23. **a.** $\{x \mid -3 \leq x \leq 1\}$
    b. $[-3, 1]$

25. **a.** $\{x \mid -8 < x \leq -2\}$
    b. $(-8, -2]$

27. **a.** $\{x \mid x \leq 1\}$ **b.** $(-\infty, 1]$

29. **a.** $\{x \mid -4 < x < 4\}$
    b. $(-4, 4)$

31. $(-1, 6)$

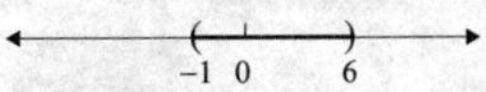

33. $(-\infty, 2] \cup [6, \infty)$

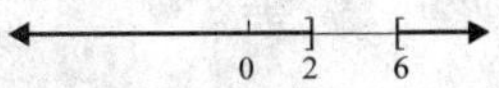

35. $(-1, 0)$

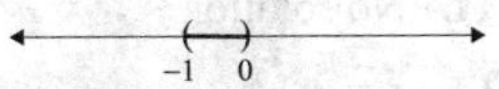

37. $\left[-\frac{16}{3}, \frac{1}{2}\right]$

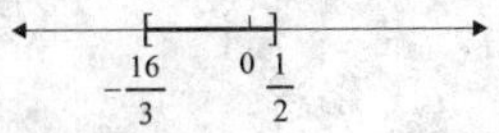

39. $(-\infty, 0.4] \cup [5, \infty)$

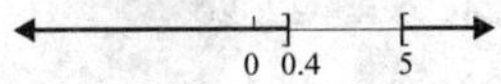

41. $(-9, 1)$

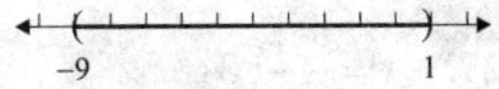

43. $\left[\frac{1}{2}, \frac{3}{2}\right]$

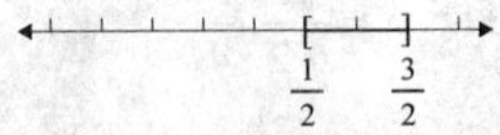

45. $[3, 15]$

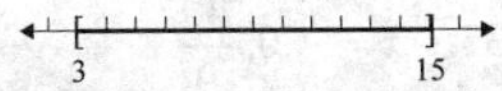

47. $(-10, -5)$

49. $(-2.8, -0.3)$

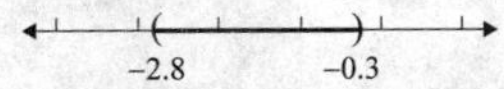

Writing & Thinking

51. Answers will vary. Answers should mention that *and* indicates the intersection and *or* indicates the union.

9.9 Exercises

Concept Check

1. standard
3. distance
5. False; Equations involving absolute value can have more than one solution.
7. True

Practice

1. $x = -8, 8$
3. No solution
5. $x = -5, -1$

7. $x=-\frac{4}{3},\frac{5}{3}$

9. $n=-2,\frac{2}{3}$

11. No solution

13. $x=2$

15. $x=-\frac{3}{2},2$

17. $x=-\frac{1}{5},1$

19. $x=-\frac{1}{3},3$

21. $x=-22, 26$

23. $x=-8, 4$

25. $x=-6, 0$

27. $x=-\frac{1}{3},3$

29. $x=1$

31. $x=-\frac{5}{2},\frac{3}{4}$

33. $x=-\frac{20}{7},\frac{4}{5}$

35. $x=-20,\frac{40}{9}$

9.10 Exercises

Concept Check

1. standard

3. or

5. False; Only one statement/inequality must be true.

7. False; Must be greater than 2

Practice

1. $(-\infty,\infty)$

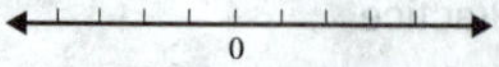

3. $\left[-\frac{4}{5},\frac{4}{5}\right]$

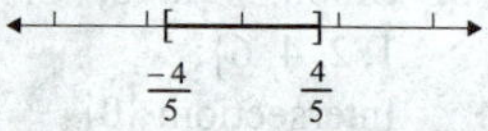

5. $(-\infty,1)\cup(5,\infty)$

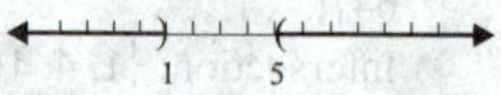

7. $[-10,-2]$

9. $(-\infty,-8]\cup[-2,\infty)$

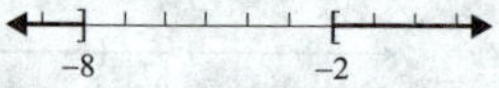

11. $\left(-\infty,-\frac{1}{2}\right]\cup\left[\frac{3}{2},\infty\right)$

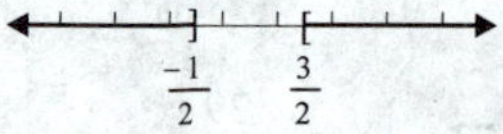

13. 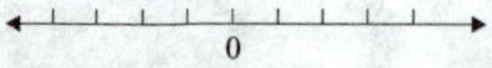

No solution

15. $\left[-2,-\frac{1}{2}\right]$

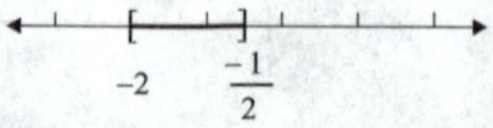

17. $\left(-\frac{5}{3},-1\right)$

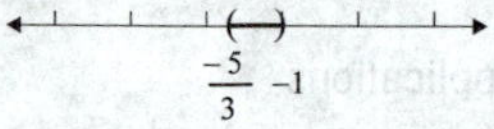

19. $\left(-\infty,-\frac{2}{3}\right]\cup[6,\infty)$

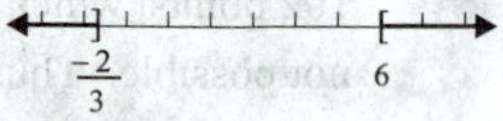

21. $[-1,10]$

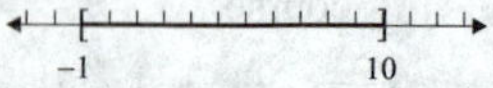

23. $(-\infty,\infty)$

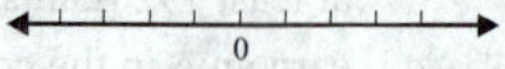

25. $\left(\frac{1}{2},\frac{7}{2}\right)$

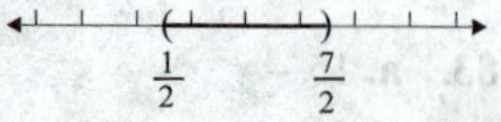

27. $\left(-\infty,-\frac{5}{2}\right)\cup(0,\infty)$

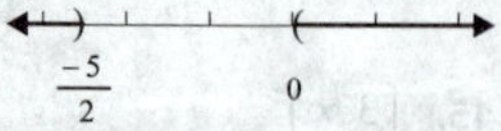

29. $\left(-2,-\frac{4}{7}\right)$

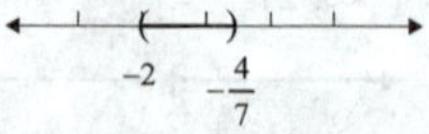

Writing & Thinking

31. a.

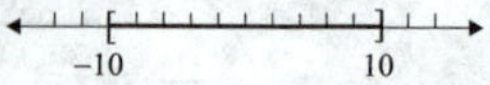

b. $|x|\le 10$

c. $[-10,10]$, Closed interval

33. a.

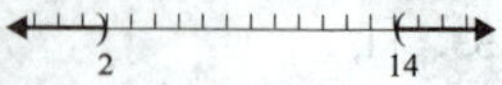

b. $|x-8|>6$

c. $(-\infty,2)\cup(14,\infty)$

35. a.

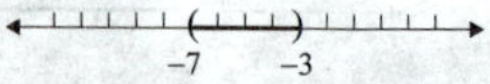

b. $|x+5|<2$

c. $(-7,-3)$, Open interval

Chapter 10: Graphing Linear Equations and Inequalities

10.1 Exercises

Concept Check

1. quadrants

3. III

5. origin

7. True

9. True

Practice

1. $\left\{\begin{array}{l}A(-5,1), B(-3,3),\\ C(-1,1), D(1,2),\\ E(2,-2)\end{array}\right\}$

3. $\left\{\begin{array}{l}A(-3,-2),\\ B(-1,-3), C(-1,3),\\ D(0,0), E(2,1)\end{array}\right\}$

5. $\left\{\begin{array}{l}A(-4,4), B(-3,-4),\\ C(0,-4), D(0,3),\\ E(4,1)\end{array}\right\}$

7. $\left\{\begin{array}{l}A(-3,-5), B(-1,4),\\ C(0,-1), D(3,1),\\ E(6,0)\end{array}\right\}$

9. $\left\{\begin{array}{l}A(-5,0), B(-2,2),\\ C(-1,-4), D(0,6),\\ E(2,0)\end{array}\right\}$

11. 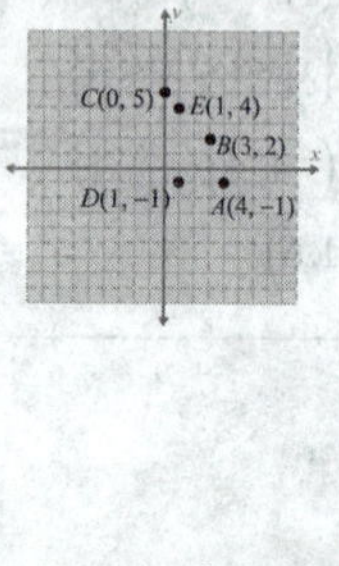

13. 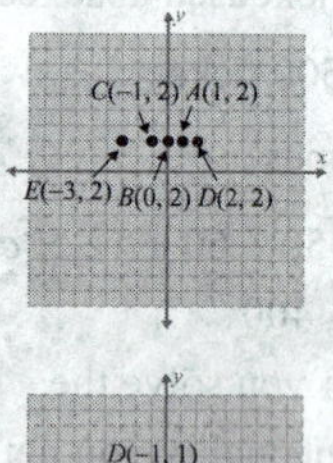

15. 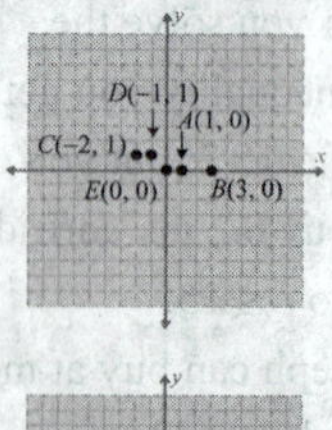

17. 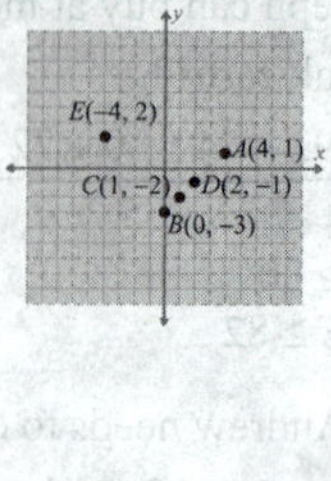

19. 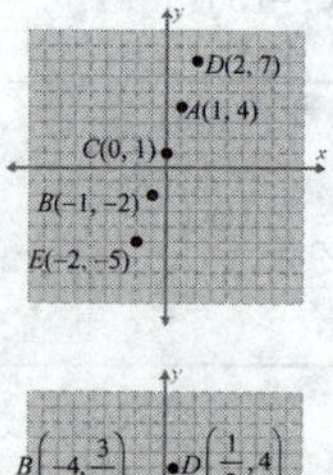

21. 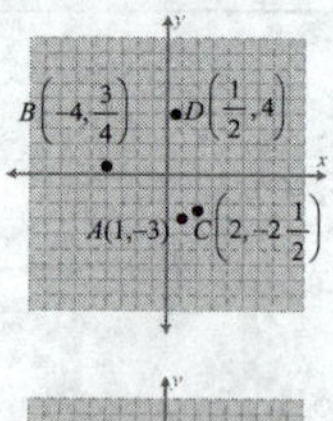

23. 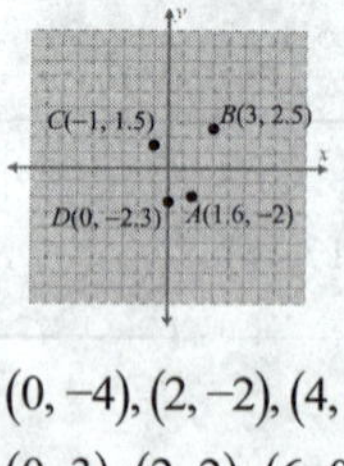

25. $(0,-4), (2,-2), (4,0), (1,-3)$

27. $(0,3), (2,2), (6,0), (-2,4)$

29. $(0,-8), (1,-4), (2,0), (3,4)$

31. $(0, 2)$, $\left(-1, \frac{8}{3}\right)$, $(3, 0)$, $(6, -2)$

33. $\left(0, -\frac{7}{4}\right)$, $(1, -1)$, $\left(\frac{7}{3}, 0\right)$, $\left(3, \frac{1}{2}\right)$

35.

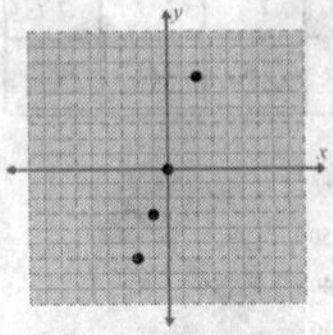

$(0, 0)$, $(-1, -3)$, $(-2, -6)$, $(2, 6)$

37.

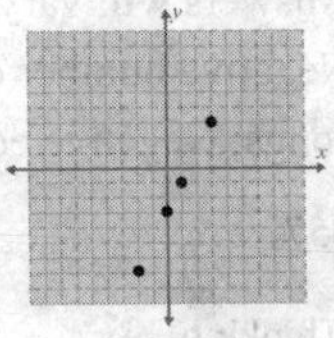

$(0, -3)$, $(1, -1)$, $(-2, -7)$, $(3, 3)$

39.

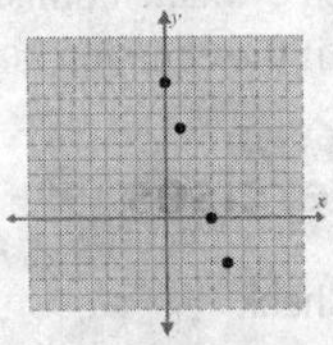

$(0, 9)$, $(3, 0)$, $(1, 6)$, $(4, -3)$

41.

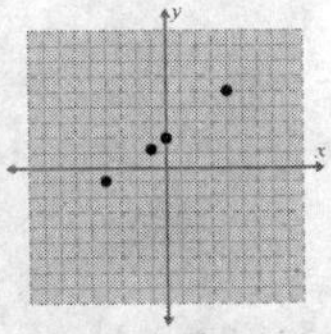

$(0, 2)$, $(4, 5)$, $(-4, -1)$, $\left(-1, \frac{5}{4}\right)$

43.

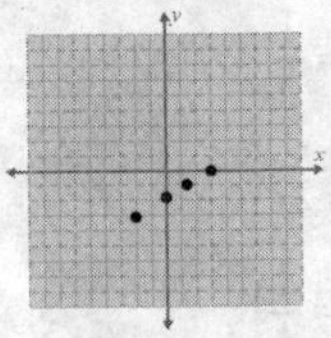

$\left(0, -\frac{9}{5}\right)$, $(3, 0)$, $(-2, -3)$, $\left(\frac{4}{3}, -1\right)$

45.

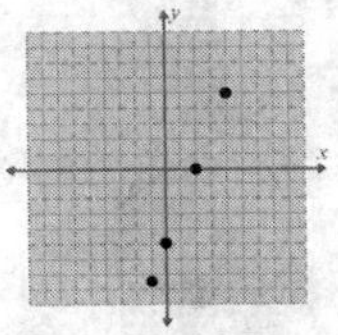

$(0, -5)$, $(2, 0)$, $\left(-1, -\frac{15}{2}\right)$, $(4, 5)$

47.

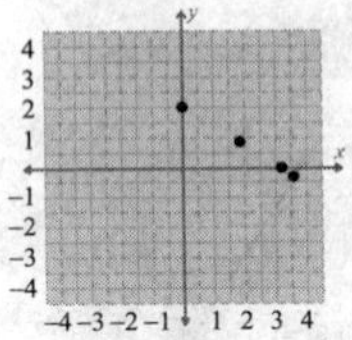

$(0, 2)$, $(3.2, 0)$, $(1.92, 0.8)$, $(3.52, -0.2)$

49. b, c, d

51. a, d

53. a, c, d

55. For example, $(-3, -3)$, $(0, 1)$, and $(3, 5)$.

57. For example, $(-3, -1)$, $(0, 0)$, and $(3, 1)$.

59. For example, $(-2, 3)$, $(0, 3)$, and $(4, 3)$.

61. For example, $(-4, -4)$, $(0, -3)$, and $(4, -2)$.

63. For example, $(-3, -3)$, $(-2, 0)$, and $(-1, 3)$.

Applications

65. a.

D	*E*
100	85
200	170
300	255
400	340
500	425

b.

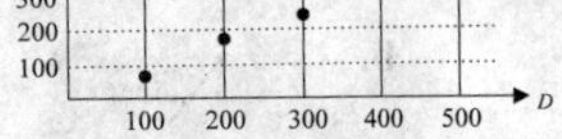

67. a.

t	*d*
1	16
2	64
3.5	196
4	256
4.5	324
5	400

b.

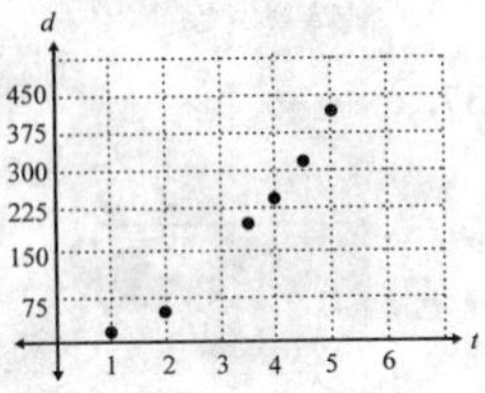

c. The time *t* is squared in the equation.

69.

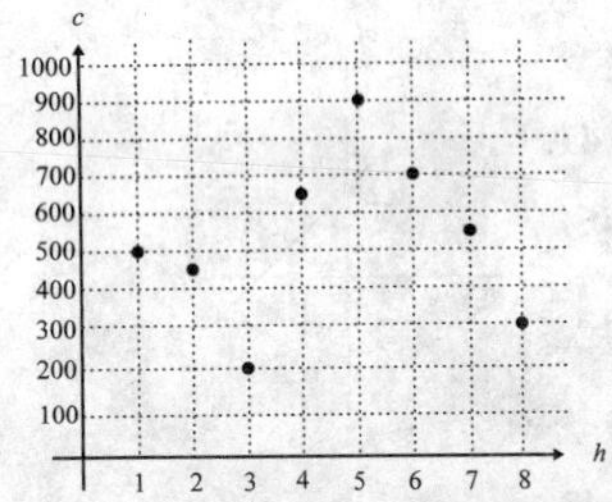

71. a.

r (mph)	*t* (hours)
25	6
50	3
60	2.5

b.

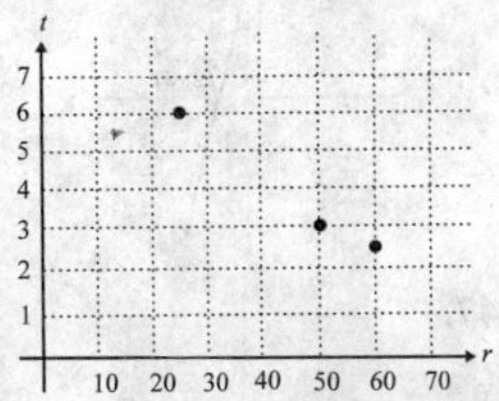

c. The time decreases as the rate increases.

d.

r (mph)	*t* (hours)
30	5
60	2.5
150	1
300	0.5
1500	0.1

e. Answers will vary. Rate increases as the time decreases.

Writing & Thinking

73. a.

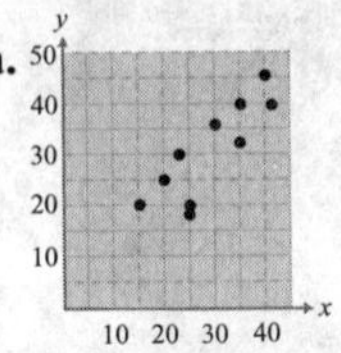

b. Yes, the more push-ups a person can do it appears the more sit-ups he/she can do.

c. 24 sit-ups, 33 sit-ups, 36 sit-ups, 45 sit-ups. Answers will vary.

75. Answers will vary. Not all scatter plots can be used to predict information related to the two variables graphed because not all variables are related.

10.2 Exercises

Concept Check

1. *y*
3. infinite
5. line
7. True
9. False; Horizontal lines have *y*-intercepts.

Practice

1. **a.**
3. **d.**
5. **f.**

7.

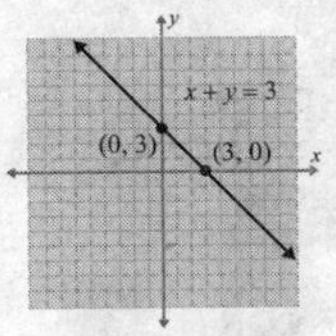

9.

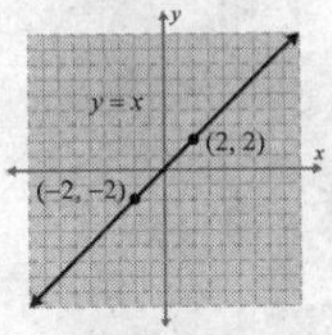

11.

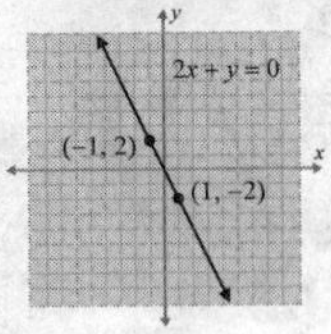

13.

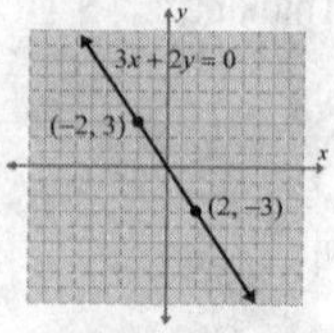

15.

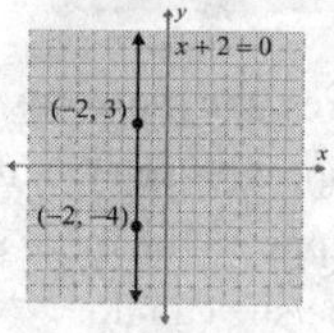

17.

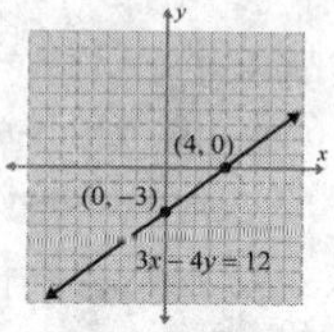

19.

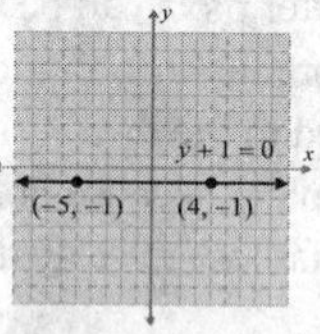

21.

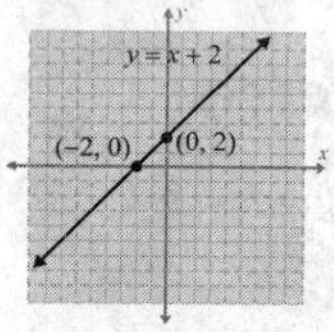

23.

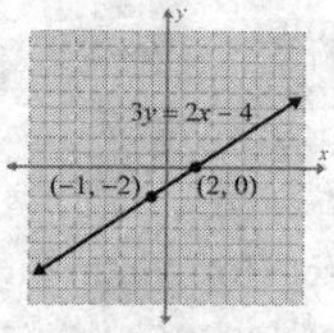

25.

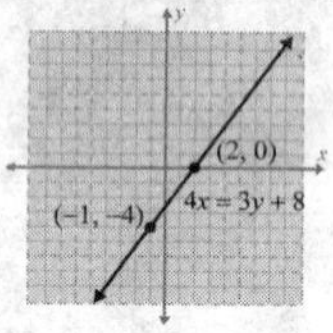

27.

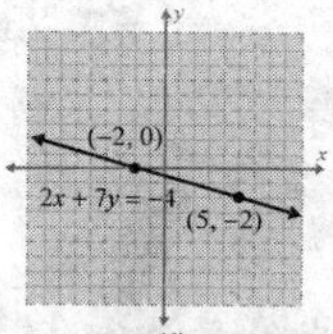

29.

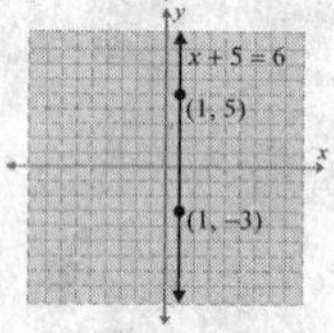

31.

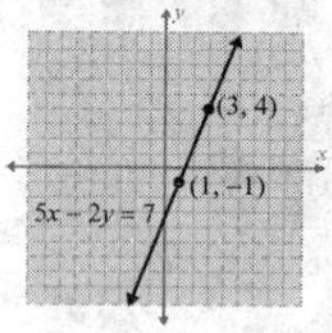

33.

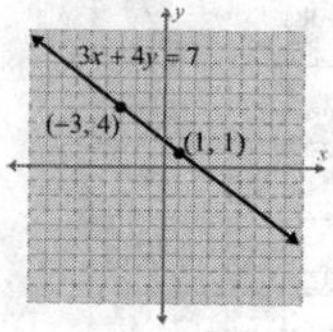

35.

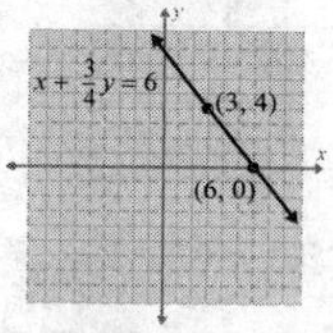

37.

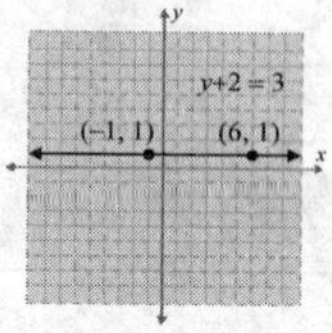

39.

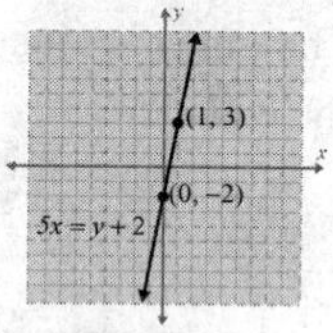

41.

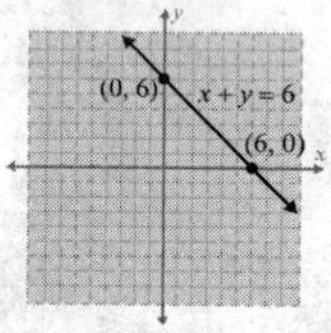

43.

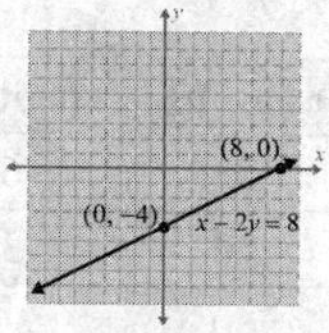

45.

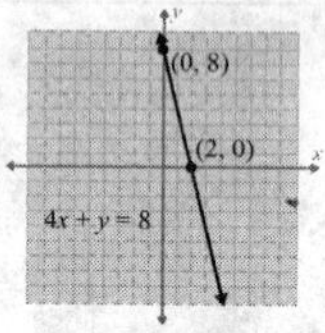

47.

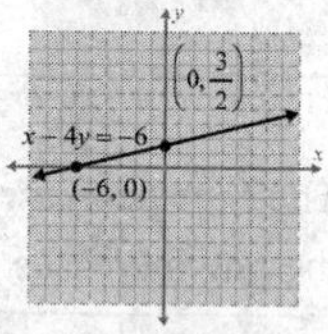

49.

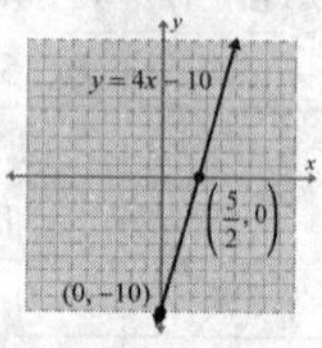

51.

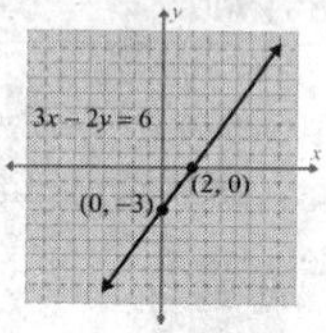

53.

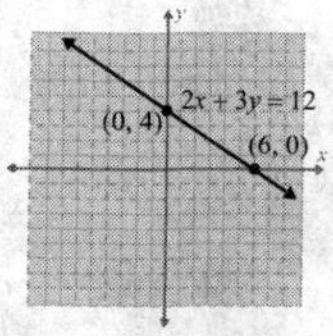

55.

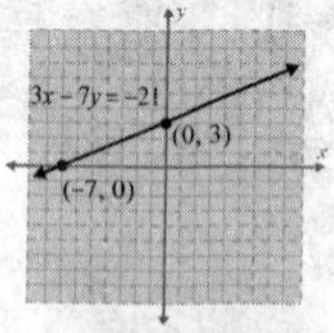

57.

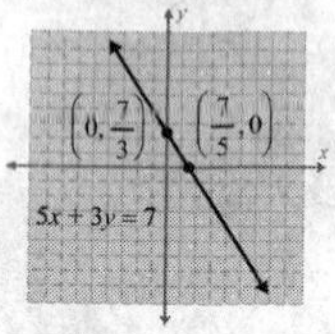

59.

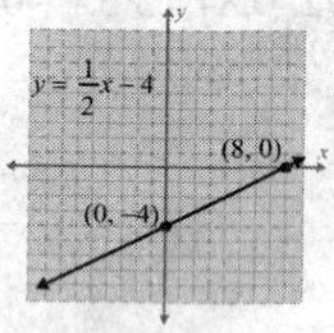

61.

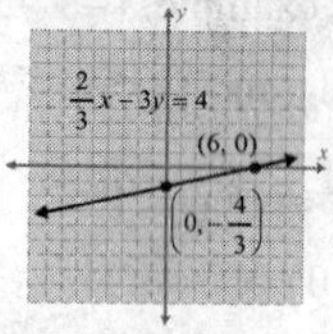

63.

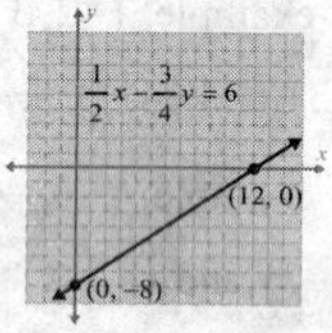

Applications

65. The x-intercept is $(12, 0)$ meaning that after 12 days, there will be 0 mg of potassium remaining in the bottle of sports drink. The y-intercept is $(0, 360)$ meaning that the original amount of potassium in the sports drink is 360 mg.

67. **a.** C-intercept: $(30, 0)$
P-intercept: $(0, 30)$

b.

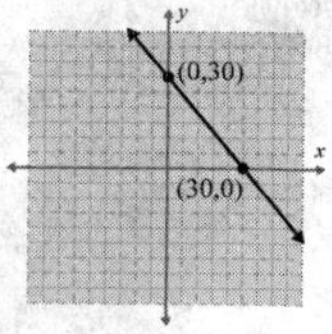

c. Answers will vary. Negative number of cookies; Total amount of cookies is over 30 dozen.

d. 14 dozen chocolate chip cookies

69. **a.**

n	5	10	37
C	25	30	57

b.

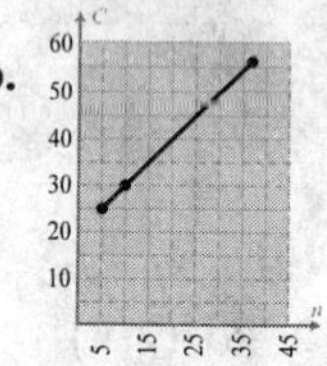

c. Answers will vary. Negative number of songs; Negative cost.

d. $57

Writing & Thinking

71. Substitute the x and y values into the equation. Then evaluate both sides to see if the equation is true.

10.3 Exercises

Concept Check

1. run
3. positive
5. 0
7. True
9. True

Practice

1. $m = 5$
3. $m = -\frac{1}{7}$
5. $m = 0$
7. $m = \frac{1}{2}$
9. $m = 2$
11. $m = \frac{1}{5}$
13.

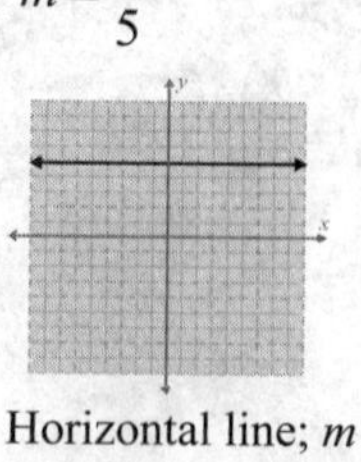

Horizontal line; $m = 0$

15.

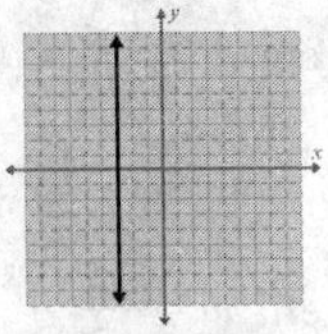

Vertical line; m is undefined

17.

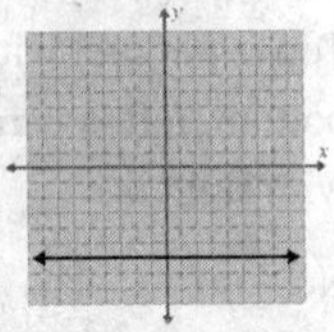

Horizontal line; $m = 0$

19.

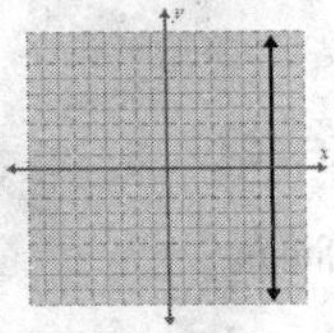

Vertical line; m is undefined

21.

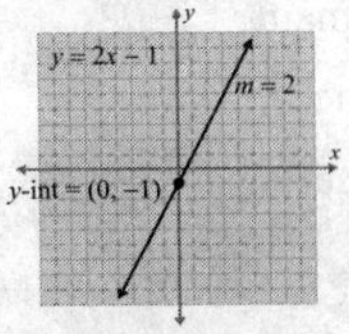

23.

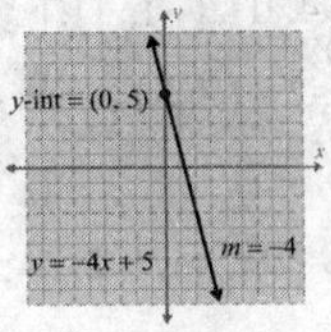

25.

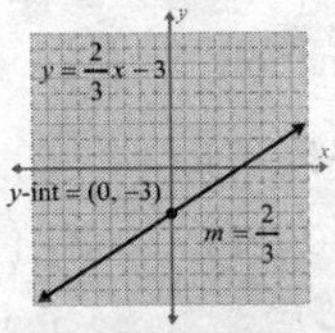

27.

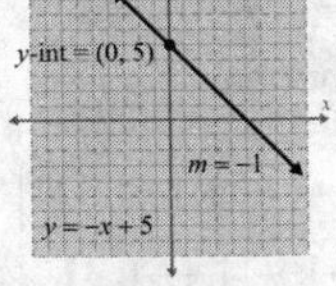

29.

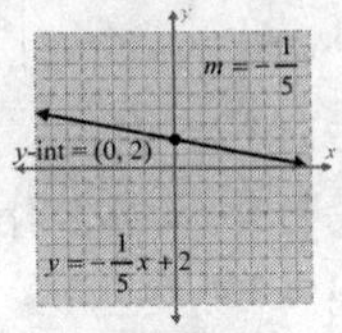

31.

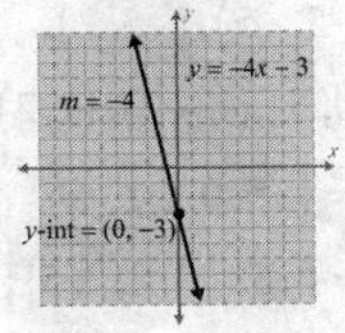

33.

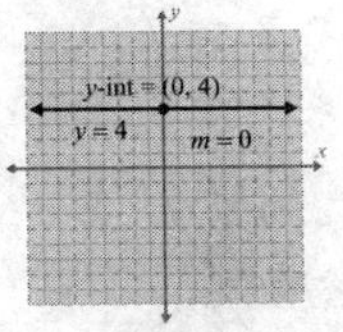

35.

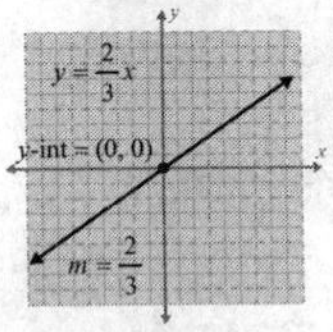

37.

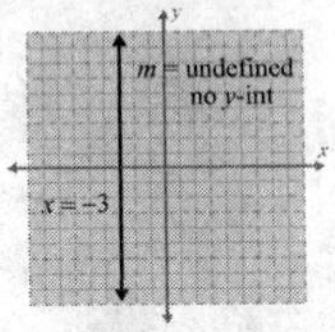

39.

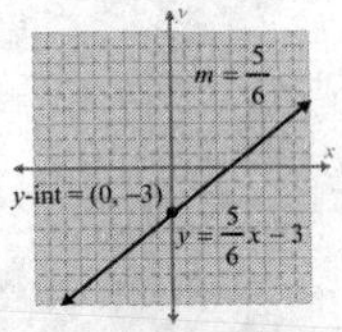

41.

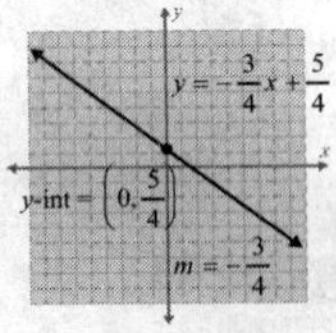

43.

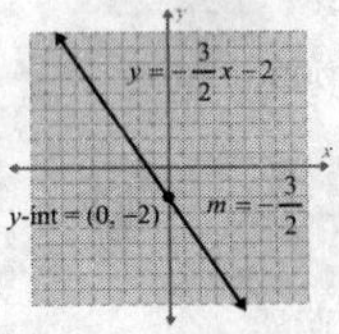

45.

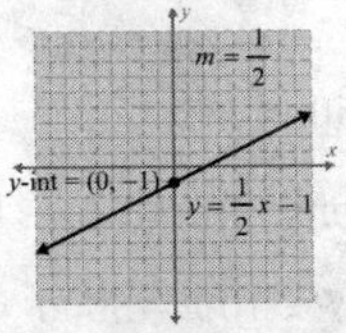

47.

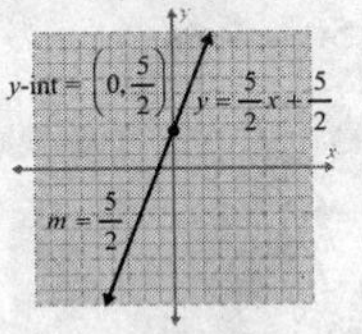

49. Answers will vary.

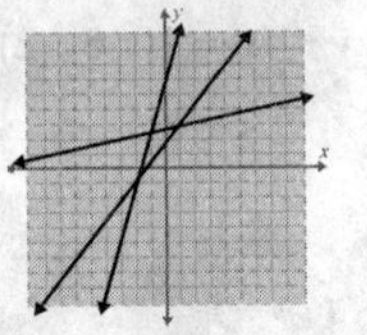

51. Answers will vary.

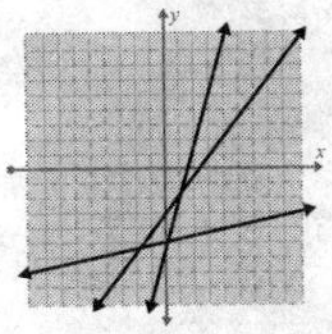

53. $y = -\dfrac{1}{2}x + 3$

55. $y = \dfrac{2}{5}x - 3$

57. $y = 4x - 5$

59. $y = x - 4$

61. $y = -\dfrac{5}{6}x - 3$

63. **a.** $m = -\dfrac{1}{6}$ **b.** $\left(0, \dfrac{5}{2}\right)$

c. $y = -\dfrac{1}{6}x + \dfrac{5}{2}$

65. **a.** $m = 0$ **b.** $(0, -6)$

c. $y = -6$

67. **a.** $m = \dfrac{1}{2}$ **b.** $(0, -3)$

c. $y = \dfrac{1}{2}x - 3$

69. **a.** $m = -\dfrac{1}{3}$ **b.** $(0, 2)$

c. $y = -\dfrac{1}{3}x + 2$

71. Yes

73. No

75. Yes

Applications

77. $\dfrac{3}{4}$

79. $\dfrac{11}{10}$

81. 54 mph

83. \$4000/year

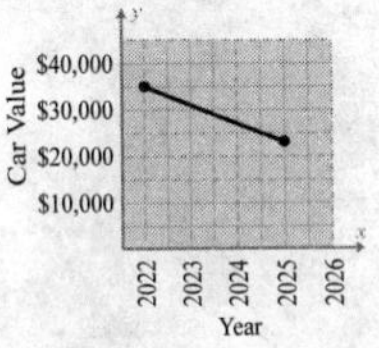

85. −5.66

87. **a.** Slope is \$15; Units are dollars and paintings

b. $(0, 30)$

c. $y = 15x + 30$

d.

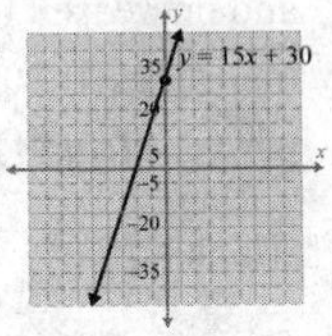

e. Answers will vary. x cannot be negative since you can't sell a negative number of paintings.

f. \$90

89. **a.** and **b.**

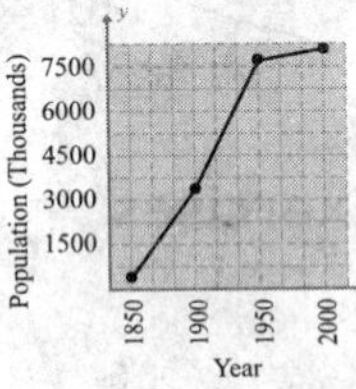

c. 58,433.1; 89,095.1; 2326.42

d. The population of New York increased by 58,433 people/year from 1850-1900; 89,095 ppy from 1900-1950; and 2326 ppy from 1950-2000.

91. **a.** and **b.**

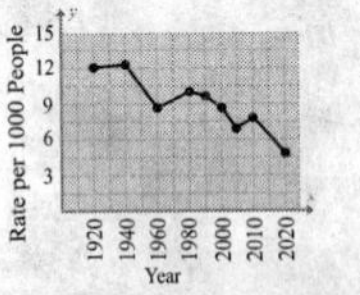

c. −0.18, 0.075, −0.02, −0.13, −0.14, −0.17

d. # of marriages decreased 0.18 marriages/1000 people from '40–'60, increased 0.075 marriages/1000 people from '60–'80, decreased 0.02 marriages/1000 people from '80–'90, decreased 0.13 marriages/1000 people from '90–'00, decreased 0.14 marriages/1000 people from '00–'10, and decreased 0.17 marriages/1000 people per year from '10–'20.

Writing & Thinking

93. **a.** The x-axis

b. The y-axis

95. A grade of 12% means that the slope of the road is 0.12. For every 100 feet of horizontal distance (run) there is a vertical distance (rise) of 12 feet.

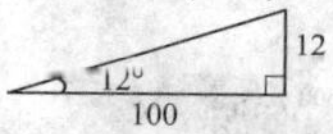

10.4 Exercises

Concept Check

1. negative reciprocals

3. same **5.** slope

7. True **9.** True

Practice

1. **a.** $m = 2$ **b.** $(3,1)$

c.

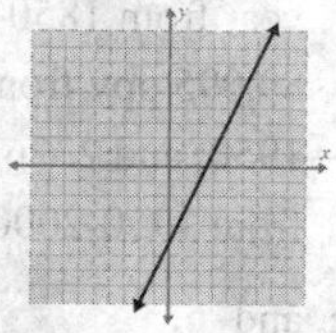

3. **a.** $m = -5$ **b.** $(0,-2)$

c.

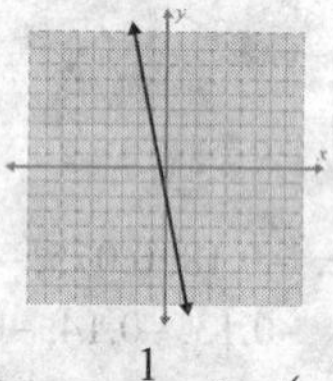

5. **a.** $m = -\frac{1}{4}$ **b.** $(-2,3)$

c.

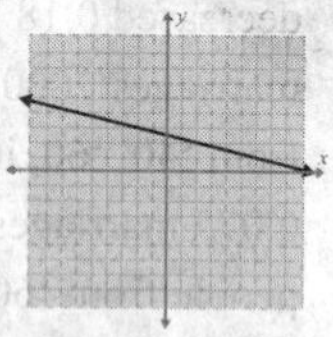

7. $2x + y = -3$

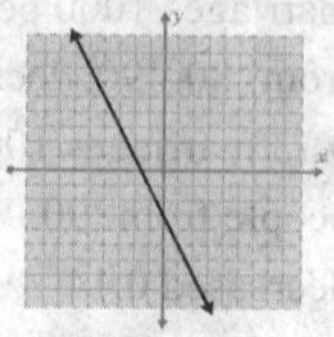

9. $y = -2$

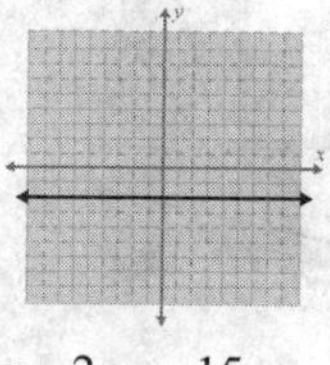

11. $x - 2y = -15$

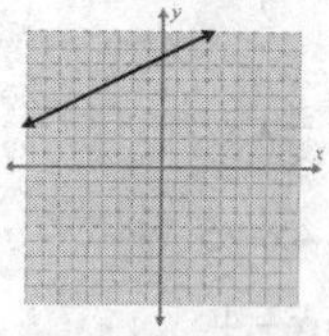

13. $3x - 5y = -29$

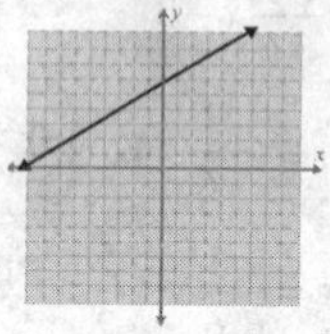

15. $2x - 3y = -5$

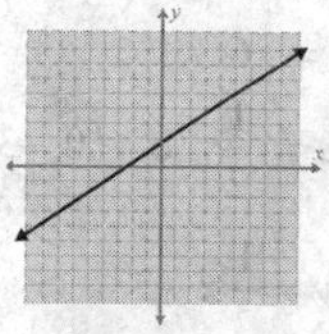

17. $y = \frac{1}{2}x + \frac{9}{2}$

19. $y = -\frac{1}{7}x + \frac{2}{7}$

21. $y = -\frac{5}{4}x + 2$

23. $y = -5$

25. $y = -x + 4$

27. $x - y = 1$

29. $3x + y = 2$

31. $3x - 4y = -14$

33. $7x + 3y = -5$

35. $y = 6$ **37.** $y = 7$

39. $y = 7$ **41.** $y = 2x$

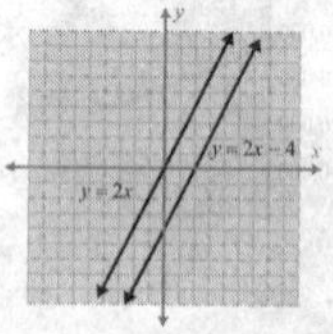

43. $y = 5x + 2$

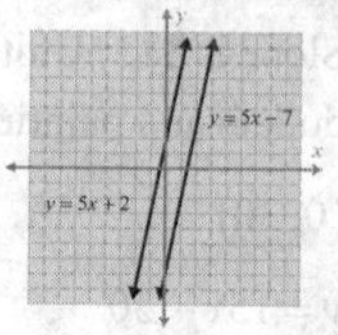

45. $y = -\frac{3}{5}x + \frac{7}{5}$

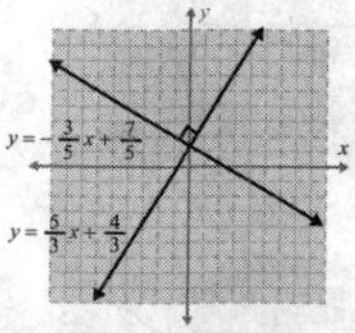

47. $y = -\frac{1}{3}x$

49. $y = -\frac{1}{2}x - 2$

51.

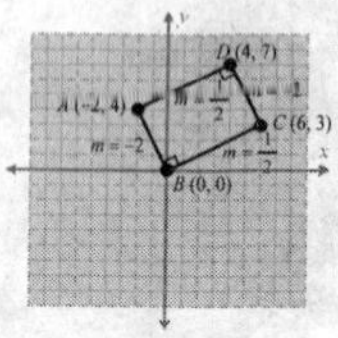

53.

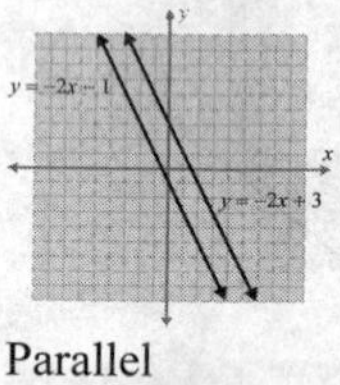

Parallel

55.

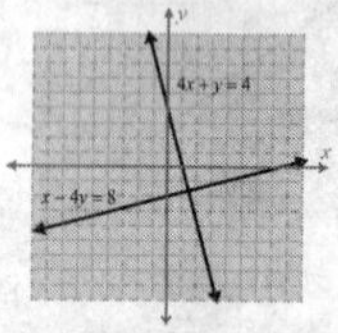

Perpendicular

57.

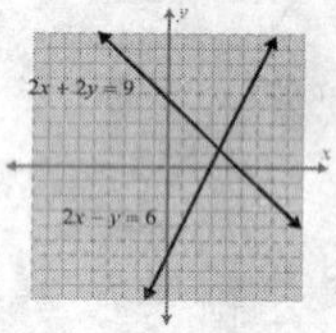

Neither

Applications

59. **a.** $P = 100t - 5000$

b. 50 tickets

61. **a.**

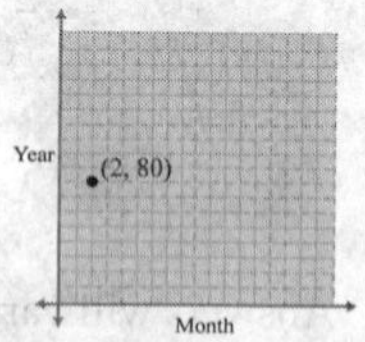

b. $y - 80 = 10(x - 2)$

63. **a.** $f = 5 + 2m$ **b.** \$35

65. **a.**

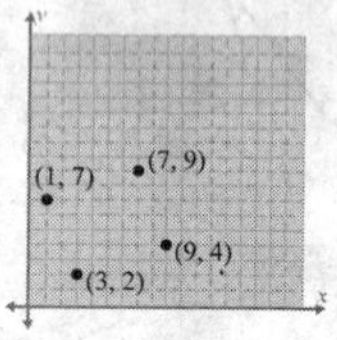

b. Top: $m = \frac{1}{3}$;

Bottom: $m = \frac{1}{3}$;

Left: $m = -\frac{5}{2}$

Right: $m = -\frac{5}{2}$

c. Top is parallel with bottom, left is parallel with right.

d. No

e. Parallelogram

10.5 Exercises

Concept Check

1. function

3. domain

5. domain

7. True

9. True

Practice

1. $\left\{(-4,0),(-1,4),(1,2),(2,5),(6,-3)\right\}$;

$D = \{-4, -1, 1, 2, 6\}$;

$R = \{-3, 0, 2, 4, 5\}$;

Function

3. $\left\{(-5,-4),(-4,-2),(-2,-2),(1,-2),(2,1)\right\}$;

$D = \{-5, -4, -2, 1, 2\}$;

$R = \{-4, -2, 1\}$;

Function

5. $\left\{(-4,-3),(-4,1),(-1,-1),(-1,3),(3,-4)\right\}$;

$D = \{-4, -1, 3\}$;

$R = \{-4, -3, -1, 1, 3\}$;

Not a function

7. $\left\{(-5,-5),(-5,3),(0,5),(1,-2),(1,2)\right\}$;

$D = \{-5, 0, 1\}$;

$R = \{-5, -2, 2, 3, 5\}$;

Not a function

9. 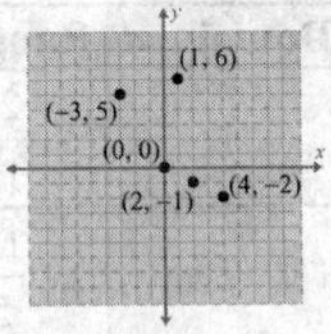

$D=\{-3, 0, 1, 2, 4\}$;
$R=\{-2, -1, 0, 5, 6\}$;
Function

11. 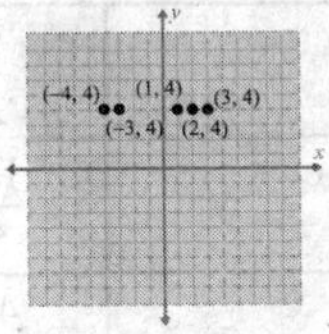

$D=\{-4, -3, 1, 2, 3\}$;
$R=\{4\}$;
Function

13. 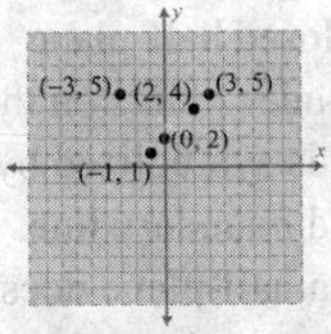

$D=\{-3, -1, 0, 2, 3\}$;
$R=\{1, 2, 4, 5\}$;
Function

15. 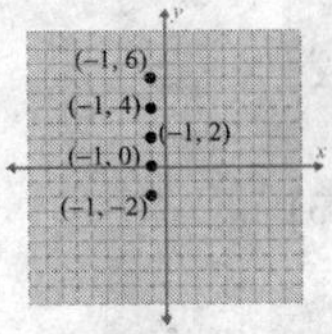

$D=\{-1\}$;
$R=\{-2, 0, 2, 4, 6\}$;
Not a function

17. Function;
$D=(-\infty,\infty)$;
$R=[0,\infty)$

19. Function;
$D=(-\infty,\infty)$;
$R=(-\infty,\infty)$

21. Not a function;
$D=(-\infty,\infty)$;
$R=(-\infty,\infty)$

23. Not a function;
$D=(-\infty,\infty)$;
$R=(-\infty,\infty)$

25. Function;
$D=[-5,5]$;
$R=[-2,2]$

27. Not a function;
$D=\{-3\}$;
$R=(-\infty,\infty)$

29. $\left\{(-9,-26), \left(-\frac{1}{3},0\right),(0,1), \left(\frac{4}{3},5\right),(2,7)\right\}$

31. $\left\{(-2,-11), (-1,-2),(0,1), (1,-2),(2,-11)\right\}$

33. $D=(-\infty,\infty)$

35. $D=(-\infty,0)\cup(0,\infty)$
or $x\neq 0$

37. $D=(-\infty,3)\cup(3,\infty)$
or $x\neq 3$

39. **a.** −4 **b.** −16 **c.** −10

41. **a.** 0 **b.** 12 **c.** 56

43. **a.** −3 **b.** 0 **c.** 3

45. $f(1)=3$

47. $f(4)=0$

49. 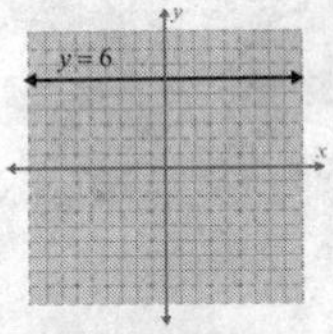

51. 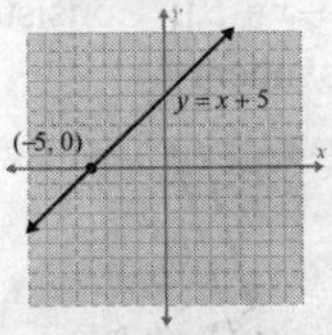

53. 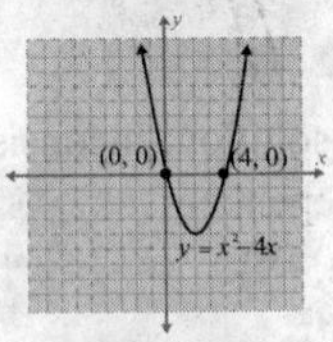

55. 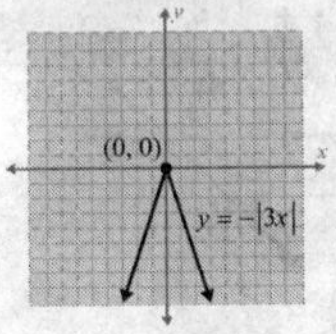

57. 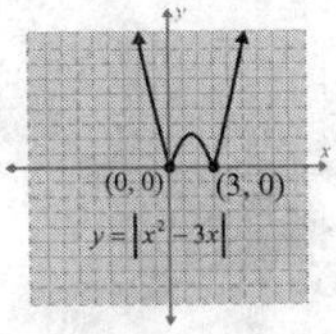

59. 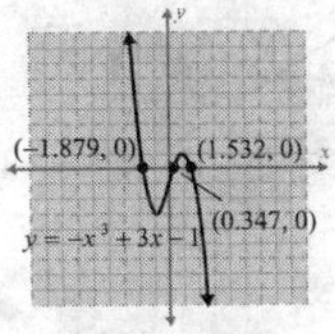

61. 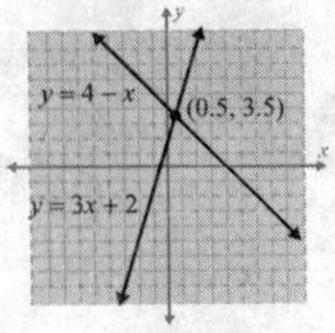

63. 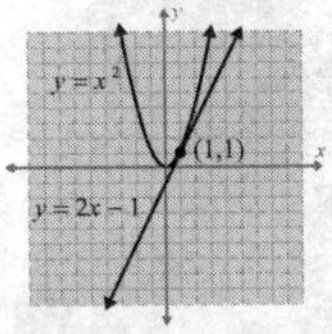

65. y-intercept = $(0,-5)$
(should be $(0,5)$)

67. y-intercept = $(0,-8)$
(should be $(0,-2)$)

69. Slope = −3 (should be the reciprocal, $-\frac{1}{3}$) and y-intercept = $(0,2)$ (should be $(0,0)$)

Applications

71. **a.** (0, 250), represents Ariella's base weekly salary
b. 0.15; Represents the commission rate
c. $y = 0.15x + 250$
d. $f(x) = 0.15x + 250$
e. Domain is all real numbers; Range is all real numbers
f. Answers will vary. Some students may say sales cannot be negative, some will allow negative sales due to returns.
g. $1000

10.6 Exercises

Concept Check

1. test **3.** open
5. closed **7.** True
9. False; The boundary line is solid if the inequality uses ≤ or ≥.

Practice

1. 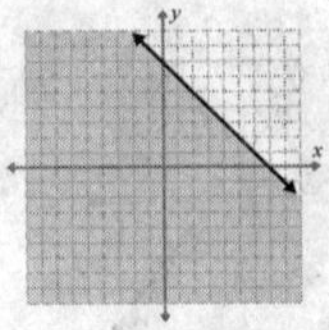

3. 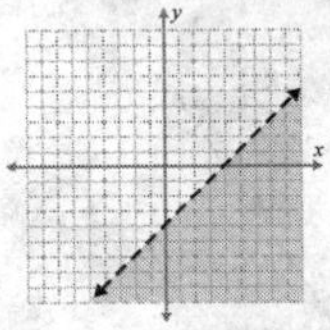

5. 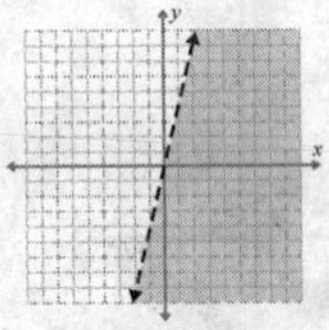

7. 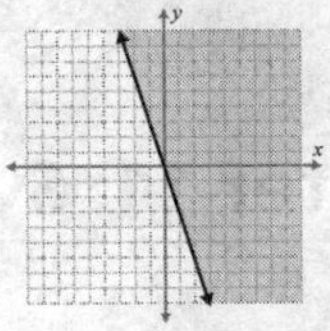

9. 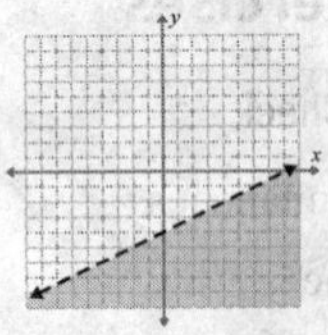

11. 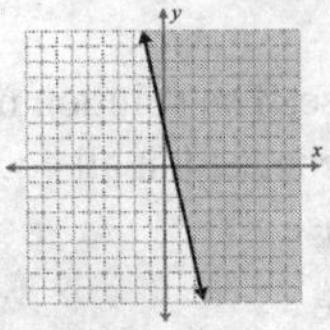

13. 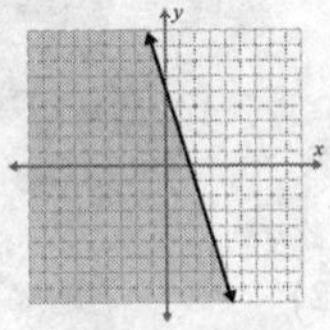

15. 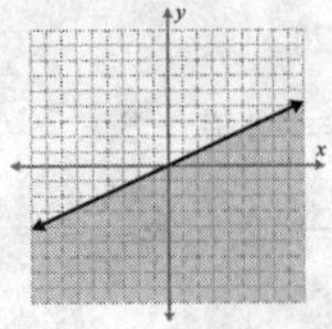

17.

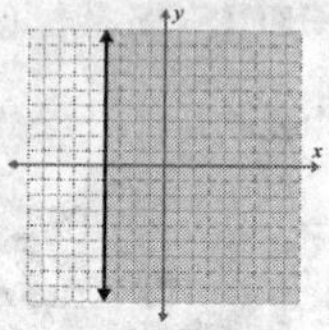

19.

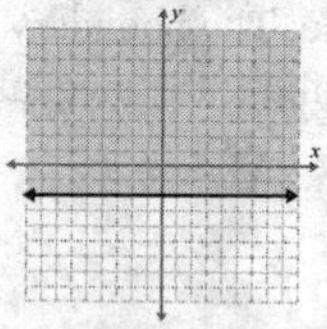

21.

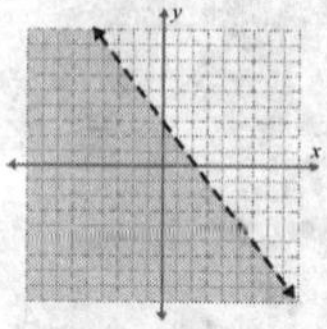

23.

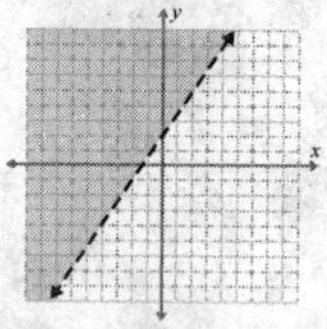

25.

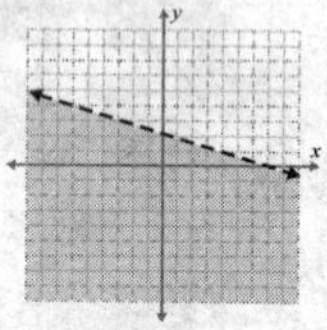

27.

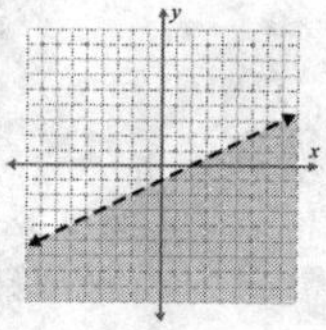

29.

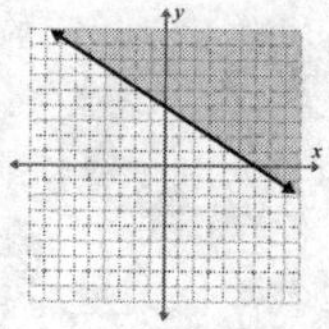

31.

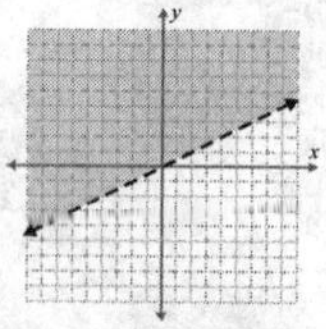

33.

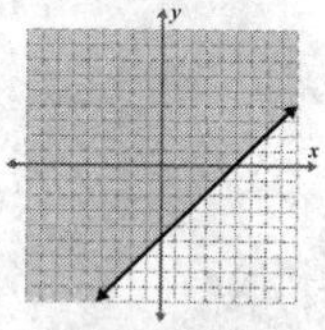

35.

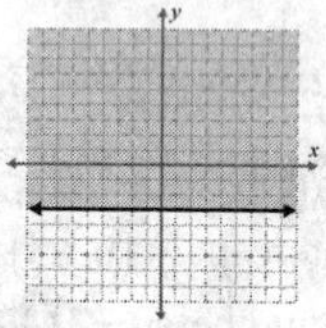

37.

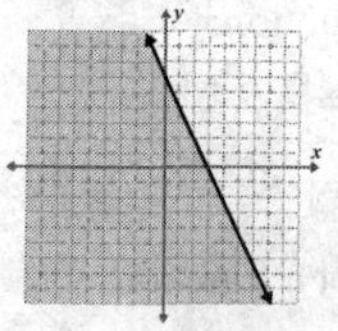

39.

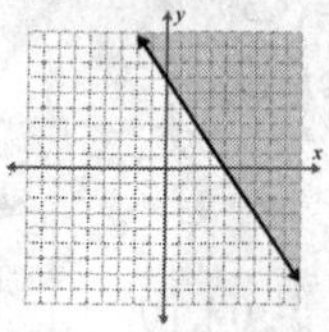

Applications

41. **a.** $x + y \geq 75$

b.

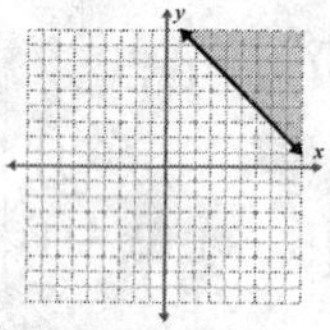

c. No; $67 < 75$

d. Answers will vary; Any points with x or $y < 0$ or > 50.

43. **a.**

Domain, d	Range, m
1	$\frac{23}{10}$
2	$\frac{13}{5}$
3	$\frac{29}{10}$
4	$\frac{16}{5}$
5	$\frac{7}{2}$
6	$\frac{19}{5}$
7	$\frac{41}{10}$

b. 6.2

Writing & Thinking

45. A closed half-plane includes points on the line and is symbolized by a solid boundary line. An open half-plane does not include points on the line and is symbolized by a dashed boundary line.

Chapter 11: Systems of Linear Equations

11.1 Exercises

Concept Check

1. simultaneous

3. same

5. one

7. False; The solution must be checked in all equations.

9. True

Practice

1. c

3. a, c

5. $(0, 2)$

7. $(4, 2)$

9. $m_1 = -2, b_1 = 3,$

$m_2 = -2, b_2 = \frac{5}{2}$

11. $m_1 = \frac{1}{2}, b_1 = 3,$

$m_2 = \frac{1}{2}, b_2 = -\frac{1}{2}$

13.

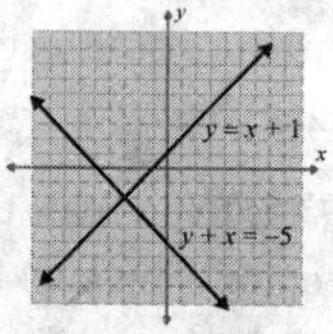

$(-3, -2)$; Consistent; Independent

15.

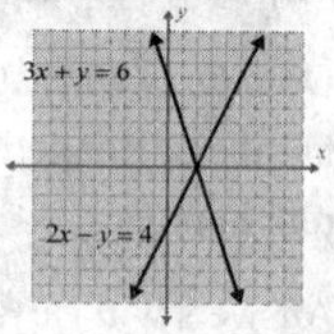

$(2, 0)$; Consistent; Independent

17.

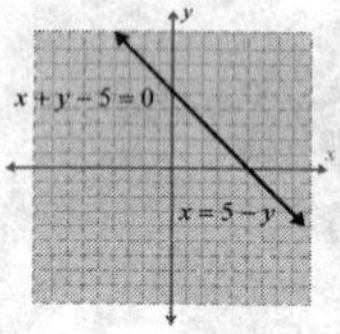

$(x, -x + 5)$; Consistent; Dependent

19.

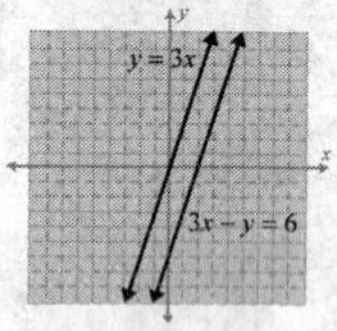

No solution; Inconsistent; Independent

21. $(-3, -8)$

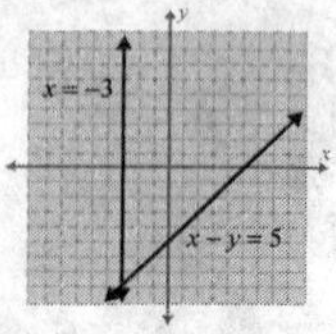

23. $(2, 3)$

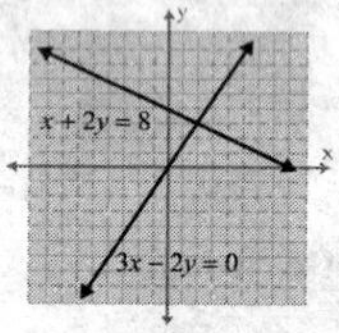

25. $\left(x, \frac{5}{4}x - \frac{5}{4}\right)$

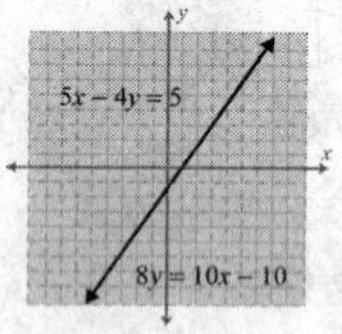

27. $(-1, -1)$

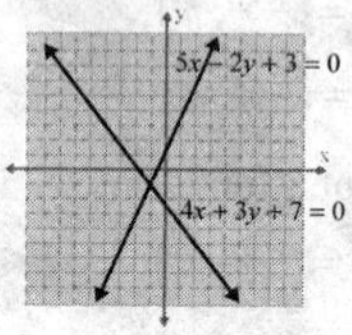

29. $(5,-1)$

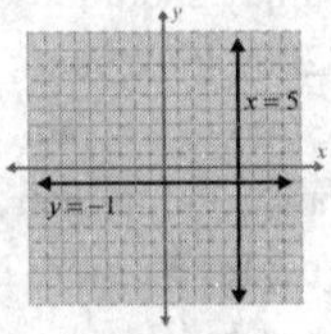

31. No solution

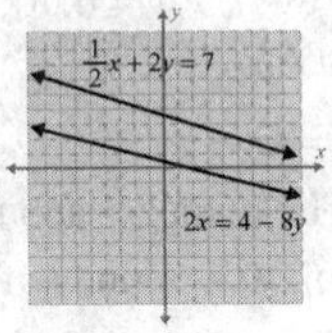

33. $(1,3)$

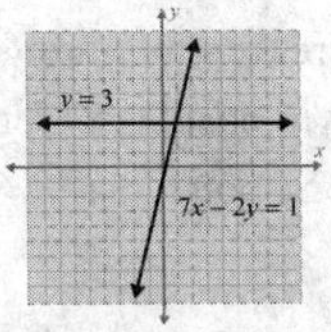

35. $\left(x, \frac{1}{2}x+2\right)$

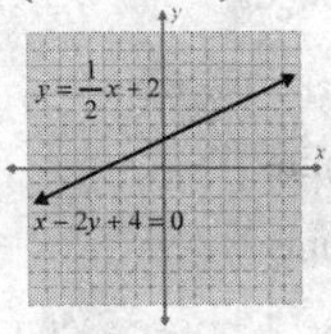

37. $(1,1)$

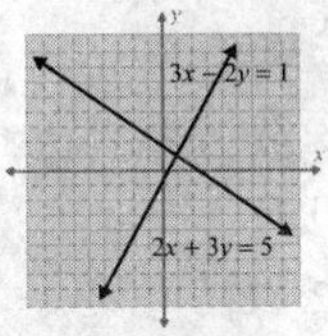

39. No solution

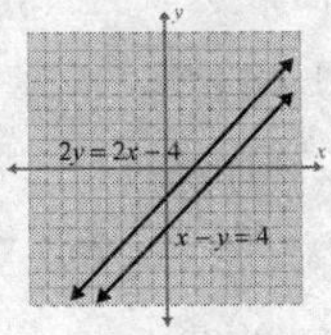

41. $(1,-1)$

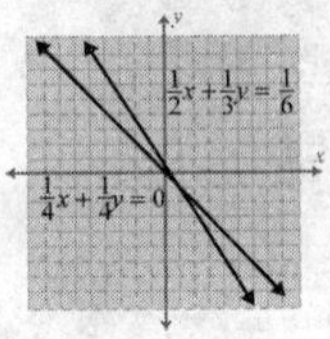

43. $(1,4)$

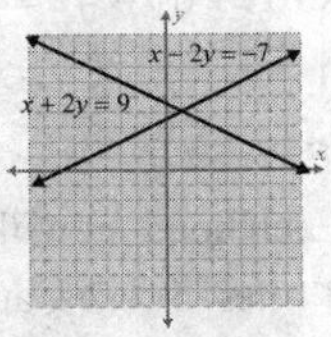

45. $(1.5, 2)$

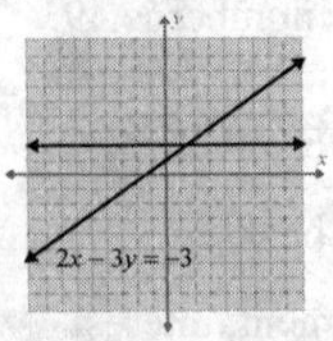

47. $(1.5, -3)$

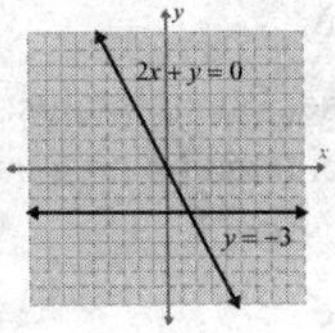

49. $(1.4, 2.1)$

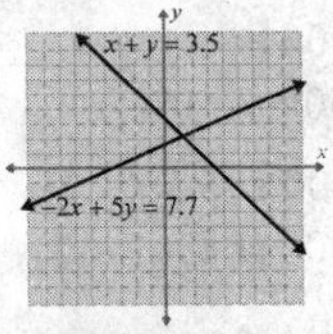

Applications

51.

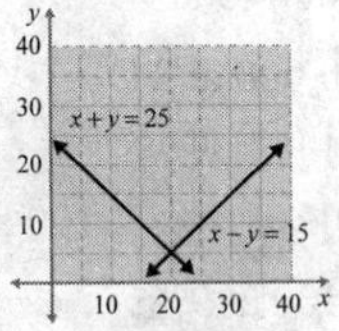

$(20, 5)$

53.

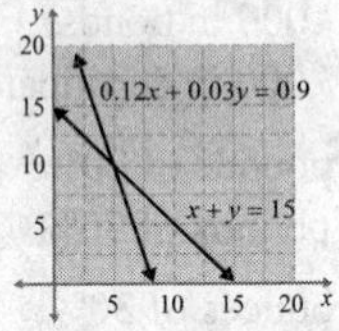

5 gallons of 12%; 10 gallons of 3%

55. 134 hours

57. **a.** Fit4life: $y = 20 + 10x$; Workout Nation: $y = 45 + 5x$

b.

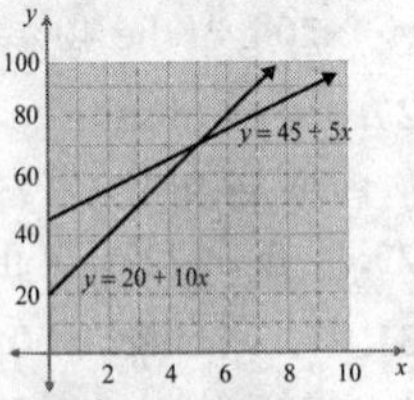

c. $(5, 70)$

d. The gym memberships will cost the same amount per month if 5 classes are taken.

e. Workout Nation

Writing & Thinking

59. The solution to a consistent system of linear equations is a single point, which is easily written as an ordered pair.

11.2 Exercises

Concept Check

1. variables

3. no

5. back substitution

7. True

9. True

Practice

1. $(2,4)$

3. $(x, 3x-7)$

5. $(-6,-2)$

7. $(4,1)$

9. No solution

11. $(3,2)$

13. $(4,-5)$

15. $(3,-2)$

17. $\left(2, \frac{5}{2}\right)$

19. $\left(\frac{1}{2}, -4\right)$

21. $\left(\frac{7}{2}, -\frac{1}{2}\right)$

23. $(x, 2-3x)$

25. $(-2,1)$

27. $\left(-\frac{4}{5}, -\frac{7}{5}\right)$

29. $\left(\frac{11}{7}, \frac{8}{7}\right)$

31. $\left(x, \frac{1}{6}x+\frac{10}{3}\right)$

33. $(3,3)$

35. $(10,20)$

37. No solution

39. $\left(x, -\frac{3}{2}x+12\right)$

41. $\left(2, \frac{5}{3}\right)$

Applications

43. $(20,5)$

45. 5 gallons of 12%; 10 gallons of 3%

47. 3 months

49. **a.** $\begin{cases} c = 45 + 0.15t \\ c = 55 + 0.10t \end{cases}$

b. $t = 200$, $c = 75$

c. The plans will both cost $75 when 200 text messages are sent.

d. More than 200 text messages

Writing & Thinking

51. Answers will vary.

11.3 Exercises

Concept Check

1. eliminate

3. standard

5. combining

7. False; The solution always needs to be checked in both original equations.

9. False; Both methods give exact solutions.

Practice

1. $(4,3)$

3. $\left(1, -\frac{3}{2}\right)$

5. No solution

7. $(x, 3-x)$

9. $(-2,-3)$

11. $(7,5)$

13. $(1,-5)$

15. $\left(x, \frac{1}{2}x-2\right)$

17. $\left(\frac{22}{7}, -\frac{2}{7}\right)$

19. $(2,-2)$

21. $(x, 2x-4)$

23. $(2,-1)$

25. $(5,-6)$

27. $(3,-1)$

29. $(2,4)$

31. $(-6,2)$

33. No solution

35. $(20,10)$

37. $\left(2,\frac{10}{9}\right)$

39. No solution

41. $\left(-\frac{45}{7},\frac{92}{7}\right)$

43. $y=5x-7;\ m=5,\ b=-7$

45. $y=-3;\ m=0,\ b=-3$

47. $y=\frac{1}{2}x+\frac{3}{2};$

$m=\frac{1}{2},\ b=\frac{3}{2}$

Applications

49. $y=25x+50$

51. \$4000 at 10%; \$6000 at 6%

53. 40 liters of 30% solution (x), 60 liters of 40% solution (y)

55. **a.** $\begin{cases} y=3500+1.5x \\ y=1000+2x \end{cases}$

b. $x=5000,\ y=11{,}000$

c. The cards will earn the same amount of rewards points if you spend \$5000.

d. City

Writing & Thinking

57. Answers will vary.

11.4 Exercises

Practice

1. 23, 33

3. 15, 21

5. 37, 50

Applications

7. 80°, 100°

9. 55°, 55°, 70°

11. The rate of the boat is 10 mph and the rate of the current is 2 mph.

13. Bolt's speed was 10.32 meters per second and the wind speed was 0.12 meters per second.

15. He traveled $1\frac{1}{2}$ hours at 52 mph and 2 hours at 56 mph.

17. Marcos traveled at 40 mph and Cana traveled at 51 mph.

19. Steve traveled at 28 mph and Tim traveled at 7 mph.

21. The westbound train was traveling 45 mph and the eastbound train was traveling 40 mph.

23. He jogged 12 miles.

25. 20 nickels and 10 dimes

27. 52 nickels and 130 pennies

29. 800 adults and 2700 children attended.

31. $l=14$ meters; $w=8$ meters

33. 100 yards × 45 yards

35. $l=33$ meters; $w=17$ meters

37. Priscilla was 21 years old and Elvis was 32 years old.

39. She bought 10 paperbacks and 5 hardbacks.

41. 5000 general admission and 7500 reserved tickets were sold.

43. 22 at \$625 and 25 at \$550

45. 30 dozen (360 balls)

47. The store sold 22 of the \$95 jackets and 18 of the \$120 jackets.

49. One Big Mac costs \$4.11 and one order of medium French fries costs \$1.84.

51. They will produce 7 of Model X and 10 of Model Y.

53. **a.** $\begin{cases} 5c+10a=1400 \\ c+a=200 \end{cases}$

b. $c=120,\ a=80$

c. 120 children and 80 adults visited the petting zoo on Tuesday.

Writing & Thinking

55. The number is 49.

11.5 Exercises

Concept Check

1. payment, interest

3. paid or earned

5. time

7. False; When interest is calculated on an annual basis, $t=1$.

9. True

Applications

1. \$5500 at 6%; \$3500 at 10%

3. \$7400 at 5.5%; \$2600 at 6%

5. \$450

7. \$3500 in each or \$7000 total

9. \$20,000 at 24%; \$11,000 at 18%

11. \$800 at 5%; \$2100 at 7%

13. \$8500 at 9%; \$3500 at 11%

15. \$87,000 in bonds; \$37,000 in certificates

17. 20 pounds of 20%; 30 pounds of 70%

19. 20 ounces of 30%; 30 ounces of 20%

21. 450 pounds of 35%; 1350 pounds of 15%

23. 20 pounds of 40%; 30 pounds of 15%

25. 10 g of pure acid; 20 g of the 40% solution

27. 10 oz of salt; 50 oz of the 4% solution

29. 2 lb of 72%; 4 lb of 42%

31. 7.11 ounces of 0.5% solution; 0.89 ounces of the 5% solution

33. **a.** $\begin{cases} x+y=5000 \\ 0.04x+0.09y=350 \end{cases}$

Account	Principal	·	Interest Rate	=	Interest
Account A	x		0.04		$0.04x$
Account B	y		0.09		$0.09y$
Totals	\$5000		N/A		\$350

b. $x=2000,\ y=3000$

c. Sanjay should invest \$2000 in Account A and \$3000 in Account B to earn \$350 in interest.

d. Yes; Invest more in Account B.

35. **a.** $5x+7.50y=6.50;$ $x+y=1$

b. $x=0.4,\ y=0.6$

c. The manager should mix 0.4 pounds of cashews with 0.6 pounds of almonds to create a mixture which will cost \$6.50 for a one-pound bag.

11.6 Exercises

Concept Check

1. $Ax+By+Cz=D$

3. infinite

5. octants

7. False; Choose 2 equations, eliminate 1 variable.

9. True

Practice

1. $(1,0,1)$

3. $(1,2,-1)$

5. Infinite number of solutions

7. $(4,1,1)$

9. $(1,2,-1)$

11. $(-2,3,1)$

13. No solution

15. $(3,-1,2)$

17. $(2,1,-3)$

19. $\left(\frac{1}{2},\frac{1}{3},-1\right)$

Applications

21. **a.** $\begin{cases} x+y+z=16 \\ 10x+6y+4z=92 \\ x+y=z \end{cases}$

b. $(3,5,8)$

c. Each bouquet will have 3 lilies, 5 roses, and 8 daisies.

23. 34, 6, 27

25. 18 ones, 16 fives, 12 tens

27. 19 cm, 24 cm, 30 cm

29. Bananas: \$0.60/lb; Apples: \$1.60/lb; Grapes: \$3.60/lb

31. Savings: \$30,000; Bonds: \$55,000; Stocks: \$15,000

33. 3 liters of 10%, 4.5 liters of 30%, 1.5 liters of 40%

Writing & Thinking

35. No. Graphically, the three planes intersect at one point (one solution), or in a line (infinitely many solutions) or, they do not have a common intersection (no solution).

37. $A = 4, B = 2, C = -1$

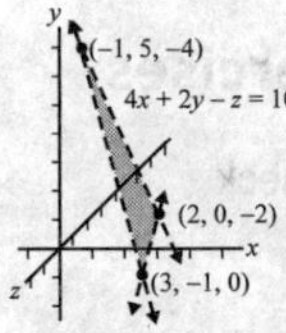

11.7 Exercises

Concept Check

1. matrix, matrices
3. dimension
5. augmented
7. False; A matrix that has 3 rows and 5 columns is a 3 x 5 matrix.
9. False; Interchanging two equations in a system of linear equations will not change the solution.

Practice

1. $\begin{bmatrix} 2 & 2 \\ 5 & -1 \end{bmatrix}, \left[\begin{array}{cc|c} 2 & 2 & 13 \\ 5 & -1 & 10 \end{array}\right]$

3. $\begin{bmatrix} 7 & -2 & 7 \\ -5 & 3 & 0 \\ 0 & 4 & 11 \end{bmatrix}, \left[\begin{array}{ccc|c} 7 & -2 & 7 & 2 \\ -5 & 3 & 0 & 2 \\ 0 & 4 & 11 & 8 \end{array}\right]$

5. $\begin{bmatrix} 3 & 1 & -1 & 2 \\ 1 & -1 & 2 & -1 \\ 0 & 2 & 5 & 1 \\ 1 & 3 & 0 & 3 \end{bmatrix}, \left[\begin{array}{cccc|c} 3 & 1 & -1 & 2 & 6 \\ 1 & -1 & 2 & -1 & -8 \\ 0 & 2 & 5 & 1 & 2 \\ 1 & 3 & 0 & 3 & 14 \end{array}\right]$

7. $\begin{cases} -3x + 5y = 1 \\ -x + 3y = 2 \end{cases}$

9. $\begin{cases} x + 3y + 4z = 1 \\ 2x - 3y - 2z = 0 \\ x + y = -4 \end{cases}$

11. a. $\begin{bmatrix} -1 & 7 \\ 1 & 4 \end{bmatrix}$

b. $\begin{bmatrix} 1 & 4 \\ 2 & -14 \end{bmatrix}$

13. a. $\begin{bmatrix} 1 & 3 & 7 \\ 4 & -1 & 6 \\ -8 & -2 & 5 \end{bmatrix}$

b. $\begin{bmatrix} 1 & 3 & 7 \\ -16 & 0 & -7 \\ 4 & -1 & 6 \end{bmatrix}$

15. $(-1, 2)$

17. $(-1, -1)$

19. $(-1, -2, 3)$

21. $(1, 0, 1)$

23. $(2, 1, -1)$

25. $(-2, 9, 1)$

27. No solution

29. Infinite number of solutions

31. $(1, -3, 2)$

33. $(0, -4)$

35. $(2, 1, 7)$

37. $(2, -3, 4)$

39. $\left(\frac{13}{12}, \frac{5}{4}, \frac{8}{3}\right)$

Applications

41. 52, 40, 77

43. Bacon: \$3.09/lb; Eggs: \$4.03/doz; Bread: \$1.40/loaf

45. a. $\begin{cases} u + c = 12 \\ 2500u + 625c = 13{,}125 \end{cases}$

b. $\left[\begin{array}{cc|c} 1 & 1 & 12 \\ 2500 & 625 & 13{,}125 \end{array}\right]$

c. University: 3, Community College: 9

Writing & Thinking

47. Solving the second equation for z, we can back substitute into the first equation, eliminating z. The result is the equation $x + 5y = 6$, which means the system has an infinite number of solutions.

11.8 Exercises

Concept Check

1. test-point
3. closed
5. three
7. False; When boundary lines are parallel, the solution is either the strip between the boundary lines, a half-plane, or there is no solution.
9. True

Practice

1.

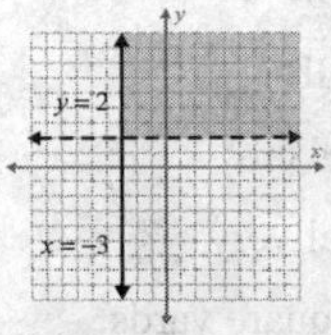

3.

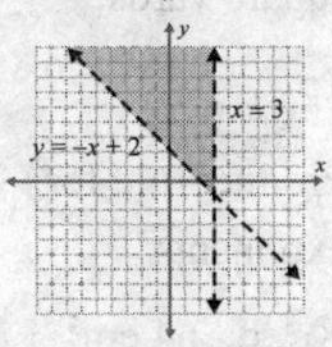

5.

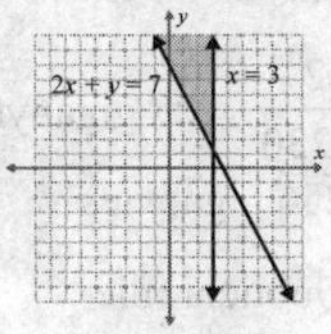

7.

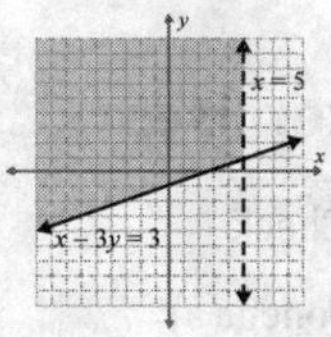

9.

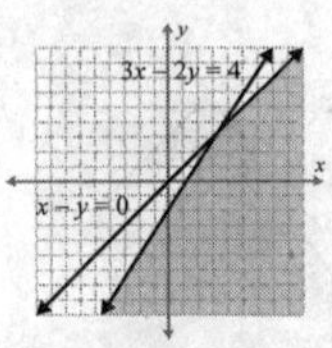

11.

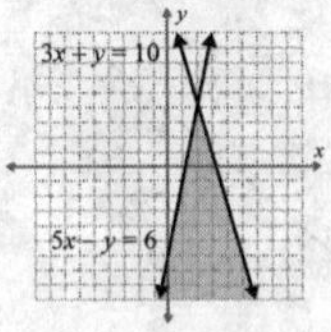

13.

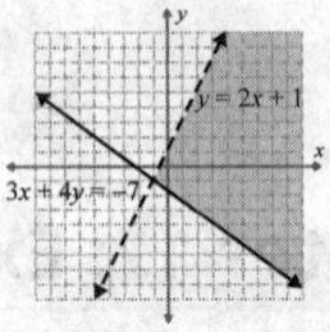

15.

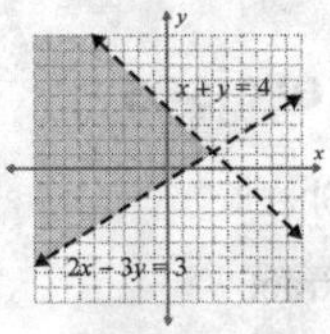

17.

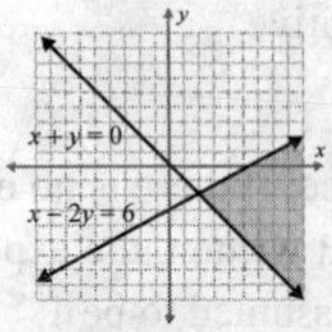

19.

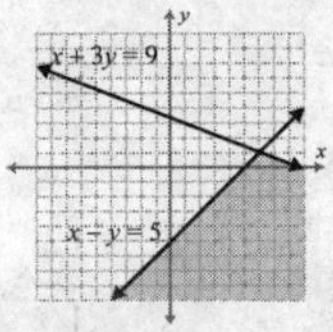

21. No solution

23.

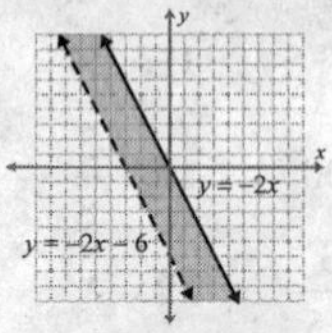

25.

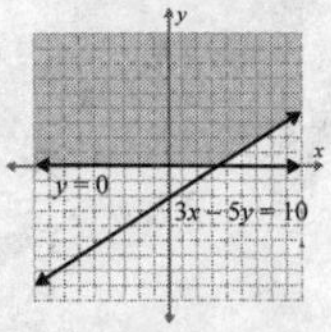

27.

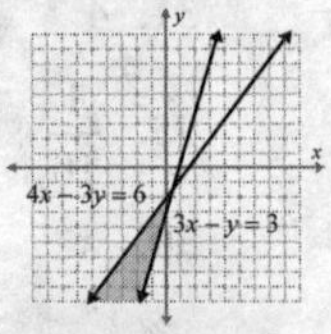

29.

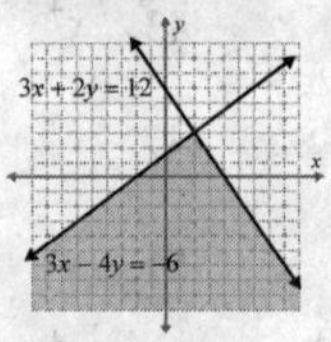

31.

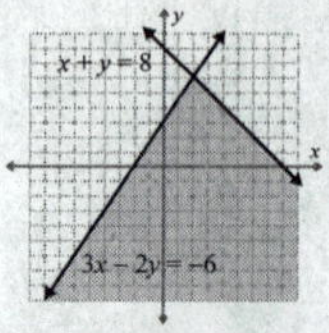

33.

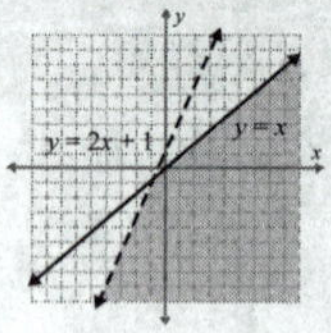

Applications

35. **a.** $\begin{cases} 0.1x + 0.3y + 1.2 < 5 \\ x + 2y < 30 \end{cases}$

b.

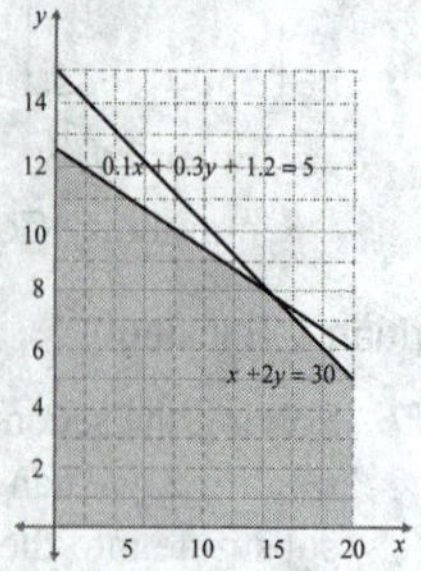

c. Answers will vary.

d. Answers will vary For example, any negative values of x and y since you cannot have a negative number of cookies.

Writing & Thinking

37. The solutions of the two inequalities do not overlap.

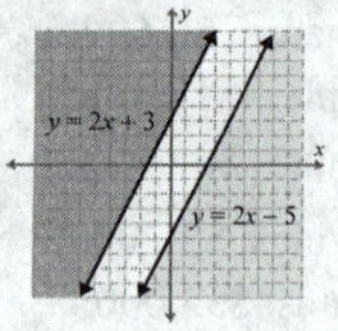

Chapter 12: Exponents and Polynomials

12.1 Exercises

Concept Check

1. subtract

3. positive

5. a

7. False; If there is no exponent written, the exponent is assumed to be 1.

9. True

Practice

1. 27 **3.** 512

5. $\frac{1}{3}$ **7.** $\frac{1}{25}$

9. 16 **11.** 24

13. −500 **15.** $\frac{3}{8}$

17. $-\frac{3}{25}$ **19.** x^5

21. y^2 **23.** $\frac{1}{x^3}$

25. $\frac{2}{x}$ **27.** $-\frac{8}{y^2}$

29. $\frac{5x^6}{y^4}$ **31.** 4

33. 49 **35.** $\frac{1}{10}$

37. $\frac{1}{8}$ **39.** x^2

41. x^2 **43.** x^4

45. $\frac{1}{x^4}$ **47.** x^6

49. x^2 **51.** y^2

53. $3x^3$ **55.** x^4

57. $36x^3$ **59.** $-14x^5$

61. $-12x^6$ **63.** $4y$

65. $3y^2$ **67.** $-2y^2$

69. $\frac{1}{x^2}$ **71.** 1000

73. 1 **75.** $-18x^5y^7$

77. $-\frac{2y^2}{x}$

79. $12a^3b^9c$

81. $-4a^{10}b^3c$

83. $\frac{5}{x}$

85. 1

87. 0.3906

89. 8875.147264

Applications

91. 2^8 GB

93. 10^{-24} cm^3

95. 1 delivery

97. 3^9 square yards

99. **a.** $(2x)^3$

b. Product rule for exponents

c. $8x^3$ **d.** 64 cm^3

101. **a.** $\frac{2^{38}}{230} = 28$ gigabytes

b. 256 gigabytes

c. Quotient rule

12.2 Exercises

Concept Check

1. exponent(s)

3. multiply, keep

5. coefficient

7. True

9. True

Practice

1. −81 **3.** −16

5. 1,000,000

7. a^6 **9.** $\frac{1}{x^{10}}$

11. $\frac{1}{256}$ **13.** $9y^2$

15. $16x^2y^2$ **17.** $\frac{1}{x^6y^6}$

19. $36x^6$ **21.** $-108x^6$

23. $\frac{5x^2}{y}$ **25.** $-\frac{2y^6}{27x^{15}}$

27. $\frac{a^4}{b^4}$ **29.** $\frac{4}{9}$

31. $\frac{x^6}{y^6}$ **33.** $\frac{27x^3}{y^3}$

35. 1 **37.** $4x^4y^4$

39. $\frac{y^2}{x^2}$ **41.** $\frac{1}{3xy^2}$

43. $-\frac{x^3y^6}{27}$ **45.** m^2n^4

47. $-\frac{y^2}{49x^2}$ **49.** $\frac{25x^6}{y^2}$

51. $\frac{x^6}{y^6}$ **53.** $\frac{36y^{14}}{x^4}$

55. $\frac{49y^4}{x^6}$ **57.** $\frac{24y^5}{x^7}$

59. $\frac{4x^3}{3}$ **61.** $\frac{16x^{13}}{243y^6}$

63. $96x^2y^4z^4$

65. 19.4481

67. $208.5136x^4$

69. 22.7024

12.3 Exercises

Concept Check

1. 1, 10

3. decimal point

5. False; The decimal point should be moved 3 places to the left.

7. False; 4000 written in scientific notation is 4.0×10^3.

Practice

1. 8.6×10^4

3. 3.62×10^{-2}

5. 1.83×10^7

7. 2.368×10^{-10}

9. 9×10^{-7}

11. 3.28×10^{-11}

13. 0.042

15. 7,560,000

17. 0.00006132

19. 30,670,000,000

21. 7,205,000,000

23. 0.00000691

25. $(3 \times 10^2)(1.5 \times 10^{-4})$; 4.5×10^{-2}

27. $\left(3\times10^{-4}\right)\left(2.5\times10^{-6}\right)$; 7.5×10^{-10}

29. $\left(2.34\times10^{10}\right)\left(5.5\times10^{9}\right)$; 1.287×10^{20}

31. $\dfrac{3.9\times10^{3}}{3\times10^{-3}}$; 1.3×10^{6}

33. $\dfrac{1.25\times10^{2}}{5\times10^{4}}$; 2.5×10^{-3}

35. $\dfrac{1.3\times10^{-12}}{2.6\times10^{-8}}$; 5×10^{-5}

37. $\dfrac{\left(8.4\times10^{-3}\right)\left(3\times10^{-3}\right)}{\left(2.1\times10^{-1}\right)\left(6\times10\right)}$; 2×10^{-6}

39. $\dfrac{\left(5.4\times10^{0}\right)\left(3\times10^{-3}\right)\left(5\times10\right)}{\left(1.5\times10\right)\left(2.7\times10^{-3}\right)\left(2\times10^{2}\right)}$; 1×10^{-1}

41. $\dfrac{\left(1.4\times10^{-2}\right)\left(9.22\times10^{2}\right)}{\left(3.5\times10^{3}\right)\left(2.0\times10^{6}\right)}$; 1.844×10^{-9}

43. $\dfrac{(2.5\times10)(3.75\times10^{-5})}{(4\times10^{10})(7.5\times10^{-6})}$; 3.125×10^{-9}

45. $\dfrac{\left(1.4\times10^{-7}\right)\left(7\times10^{13}\right)}{4\times10}$; 2.45×10^{5}

47. $\dfrac{\left(1.95\times10^{-5}\right)\left(2.65\times10^{3}\right)\left(7.56\times10^{10}\right)}{\left(1.5\times10^{7}\right)\left(1.3\times10^{-13}\right)}$; 2.0034×10^{15}

49. 3E8

51. 8.5E7

53. 3E22

55. 1.6051E15

57. 1E2

59. 1.02E22

Applications

61. 1.67×10^{-24} grams

63. 60,000,000,000,000 cells

65. 5.98×10^{27} grams

67. 4.0678×10^{16} m

69. 6.5×10^{-19} grams

71. 2.592×10^{6} seconds

73. No. Should be 5.2×10^{5}

12.4 Exercises

Concept Check

1. descending

3. leading

5. binomial

7. False; Degree of 0

9. True

Practice

1. Monomial

3. Binomial

5. Trinomial

7. Not a polynomial

9. Binomial

11. $4y$; First-degree monomial; Leading coefficient 4

13. x^3+3x^2-2x; Third-degree trinomial; Leading coefficient 1

15. $-2x^2$; Second-degree monomial; Leading coefficient −2

17. 0; Monomial of no degree; Leading coefficient 0

19. $6a^5-7a^3-a^2$; Fifth-degree trinomial; Leading coefficient 6

21. $2y^3+4y$; Third-degree binomial; Leading coefficient 2

23. 4; Monomial of degree 0; Leading coefficient 4

25. $2x^3+3x^2-x+1$; Third-degree polynomial; Leading coefficient 2

27. $4x^4-x^2+4x-10$; Fourth-degree polynomial; Leading coefficient 4

29. $9x^3+4x^2+8x$; Third-degree trinomial; Leading coefficient 9

31. **a.** −4 **b.** −16 **c.** −10

33. −16

35. −41

37. 4379

39. 13

41. −28

43. $3a^4+5a^3-8a^2-9a$

45. $3a+11$

47. $10a+25$

Applications

49. **a.** First degree polynomial
b. 5 feet per second
c. 50 feet per second
d. 95 feet per second

51. **a.** First degree polynomial
b. 9 games **c.** 11 games
d. 29 games

53. **a.** \$20,250 **b.** \$16,000

55. **a.** 12 feet **b.** 28 feet
c. 72 feet
d. No. Answers will vary.

57. \$45,950 in 2020; \$24,950 in 2026

59. 61 after 4 hours; 250,122 after 2 days (48 hours)

Writing & Thinking

61. He was correct. The expression is a sum/difference of monomials.

63. **a.**

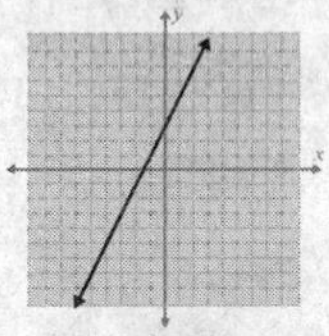

b.

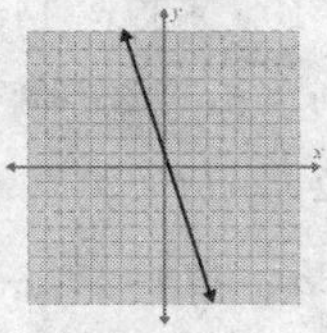

c.

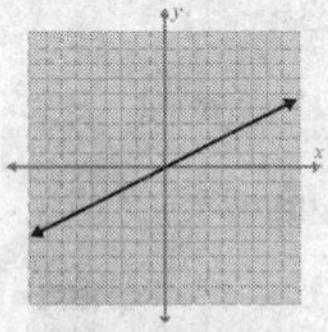

65. **a.**

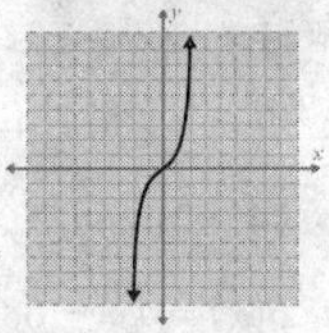

b.

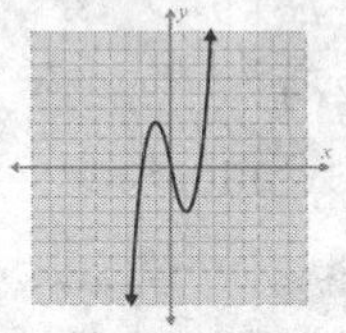

c.

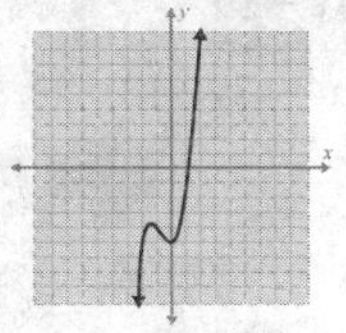

12.5 Exercises

Concept Check

1. sign

3. like

5. variables

7. False; All terms are subtracted.

9. False; They are like terms because they have the same variable raised to the same exponent.

Practice

1. $3x^2+7x+2$

3. $2x^2+11x-7$

5. $3x^2$

7. $x^2-5x+17$

9. $-x^2+2x-7$

11. $2x^3+4x^2-3x-7$

13. $-x^2+7x-3$

15. $-x^3+2x^2+3x-4$

17. $4x^3+7$

19. $2x^3+11x^2-4x-3$

21. x^2+x+6

23. $-3x^2-6x-2$

25. $3x^2-4x-8$

27. $-x^4-2x^3-16$

29. $-4x^4-7x^3-11x^2+5x+13$

31. $-4x^2+6x+1$

33. $3x^4-7x^3+3x^2-5x+9$

35. $6x^2-7x+18$

37. $8x^4+6x^2+15$

39. $2x^3+4x^2+3x-10$

41. $4x-13$

43. $-7x+9$

45. $2x+17$

47. $3x^3+13x^2+9$

49. $8x^2-x-2$

51. $3x-19$

53. $x^2-2x+13$

55. $7x-3$

57. $4x^2+3x+5$

59. $7x-18$

61. $x^2+11x-8$

63. $7x^3-x^2-7$

Applications

65. **a.** $3.50x^3+4x^2-30$ dollars

b. $6.50x^3+4x^2-20$ dollars

c. $460

67. **a.** $27x-40$ dollars

b. $284

Writing & Thinking

69. Add the opposite of each term in the polynomial being subtracted to the original polynomial and combine like terms.

12.6 Exercises

Concept Check

1. distributive

3. FOIL

5. like

7. False; The product of $(a+b)$ and $(c+d)$ is $ac+ad+bc+bd$.

Practice

1. $-6x^5-15x^3$

3. $4x^7-12x^6+4x^5$

5. $-y^5+8y-2$

7. $-4x^8+8x^7-12x^4$

9. $25x^5-5x^4+10x^3$

11. $a^7+2a^6-5a^3+a^2$

13. x^2+x-12

15. $a^2-2a-48$

17. x^2-3x+2

19. $3t^2-3t-60$

21. x^3+11x^2+24x

23. $2x^2-7x-4$

25. $6x^2+17x-3$

27. $4x^2-9$

29. $16x^2+8x+1$

31. y^3+2y^2+y+12

33. $3x^2-8x-35$

35. $5x^3+6x^2-22x-9$

37. $2x^4+7x^3+5x^2+x-15$

39. $3x^2+2x-8$

41. $2x^2+3x-5$

43. $7x^2-13x-2$

45. $6x^2-13x-8$

47. $4x^2+12x+9$

49. $x^3+3x^2-4x-12$

51. $4x^2-49$

53. x^3+1

55. $49a^2-28a+4$

57. x^3+8x^2+15x

59. $6x^3-2x^2-4x$

61. $2x^3+x^2-5x-3$

63. $x^3+6x^2+11x+6$

65. a^4-a^2+2a-1

67. $t^4+6t^3+13t^2+12t+4$

69. $6x^2-x-2$

71. $9a^2-25$

73. $-10x^2+39x-14$

75. x^3+5x^2+8x+4

77. $2y^2-61$

79. $3a^2-17a+11$

81. $-3x-21$

83. a^2+9a-1

Applications

85. $6x^2+15x$ square inches

87. $R(x)=-8x^2+70x+1500$

89. **a.** $A(x)=280-4x^2$

b. $14-2x$ and $20-2x$

c. $B(x)=4x^2-68x+280$

d. $V(x)=4x^3-68x^2+280x$

Writing & Thinking

91. **a.** Answers will vary.

b. Answers will vary.

12.7 Exercises

Concept Check

1. difference

3. trinomial

5. False; The product will be a binomial.

7. True

Practice

1. $x^2-14x+49$; Perfect square trinomial

3. $x^2+8x+16$; Perfect square trinomial

5. x^2-9; Difference of two squares

7. x^2-81; Difference of two squares

9. $2x^2+x-3$

11. $9x^2-24x+16$; Perfect square trinomial

13. $25x^2-4$; Difference of two squares

15. $9x^2-12x+4$; Perfect square trinomial

17. $x^2-16x+64$; Perfect square trinomial

19. $16x^2-25$; Difference of two squares

21. $25x^2-81$; Difference of two squares

23. $x^2-8x+16$; Perfect square trinomial

25. $4x^2-49$; Difference of two squares

27. $10x^4-11x^2-6$

29. $49x^2+14x+1$; Perfect square trinomial

31. $x^2+10x+25$

33. x^4-1

35. x^4+6x^2+9

37. x^6-4x^3+4

39. $9x^2-4$

41. $16x^2-9$

43. $25x^2+30x+9$

45. $36x^2-60x+25$

47. $x^2-1.96$

49. $x^2-5x+6.25$

51. $x^2-4.6225$

53. $x^2+2.48x+1.5376$

55. $2.0164x^2+27.264x+92.16$

57. $129.96x^2-12.25$

59. $93.24x^2+142.46x-104.04$

Applications

61. **a.** $A(x)=900-4x^2$

b. $P(x)=4(30-2x)+8x=120$

63. **a.** $f(x)=15x^6-30x^5+15x^4$

b. 0.234375

65. **a.** $A(x)=150-4x^2$

b. $P(x)=2(15-2x)+2(10-2x)+8x=50$

c. $V(x)=(10-2x)(15-2x)x=4x^3-50x^2+150x$

67. **a.** $A(x)=400-4x^2$

b. $20-2x$ and $20-2x$

c. $B(x)=400-80x+4x^2$

d. $V(\mathrm{I})=400x-80x^2+4x^3$

12.8 Exercises

Concept Check

1. rational

3. numerator, simplify

5. 0

7. False; Descending order

9. True

Practice

1. y^2-2y+3

3. $2x^2-3x+1$

5. $10x^3-11x^2+x$

7. $-4x^2+7x-\frac{5}{2}$

9. $y^3-\frac{7}{2}y^2-\frac{15}{2}y+4$

11. $x-6+\frac{4}{x+4}$

13. $3x-4-\frac{7}{2x-1}$

15. $3x+4+\frac{1}{7x-1}$

17. $x-9$

19. $x^2-x-\frac{3}{x-8}$

21. $4x^2-6x+9-\frac{17}{x+2}$

23. $x^2+7x+55+\frac{388}{x-7}$

25. $2x^2-9x+18-\frac{30}{x+2}$

27. $7x^2+2x+1$

29. $x^2+2x+2-\frac{12}{2x+3}$

31. $x^2+3x+2-\frac{2}{x-4}$

33. $2x^2+x-3+\frac{18}{5x+3}$

35. $2x^2-8x+25-\frac{98}{x+4}$

37. $3x^2+2x-5-\frac{1}{3x-2}$

39. $3x^2-2x+5$

41. $x^3+2x+5+\frac{17}{x-3}$

43. $x^3+2x^2+6x+9+\frac{23}{x-2}$

45. $x^3+\frac{1}{2}x^2-\frac{3}{4}x-\frac{3}{8}+\frac{45}{16\left(x-\frac{1}{2}\right)}$

47. $3x+5+\frac{x-1}{x^2+2}$

49. $x^2+x-4+\frac{-8x+17}{x^2+4}$

51. $2x+3+\frac{6}{3x^2-2x-1}$

53. $3x^2-10x+12+\frac{-x-14}{x^2+x+1}$

55. $x^2-2x+7+\frac{-17x+14}{x^2+2x-3}$

57. x^2+3x+9

59. $x^5-x^4+x^3-x^2+x-1$

61. $x^4+x^3+x^2+x+1+\frac{2}{x-1}$

63. $x^4-\frac{1}{2}x^3-\frac{3}{4}x^2+\frac{3}{8}x+\frac{13}{16}-\frac{13}{32\left(x+\frac{1}{2}\right)}$

Applications

65. **a.** x^2-4x-5 square inches
b. $x^2-3x-10$ square inches

Writing & Thinking

67. $3x^3+4x^2+8x+12$; Multiply the divisor $(3x-2)$ by the quotient, $x^2+2x+4+\frac{20}{3x-2}$, to get the original polynomial.

69. **a.** 19; $2x^2-4x+2+\frac{19}{x-2}$
b. −5; $2x^2-10x+20-\frac{5}{x+1}$
c. 55; $2x^2+10+\frac{55}{x-4}$
Yes; $R=P(a)$ when $P(x)$ is divided by x.

12.9 Exercises

Concept Check

1. binomial

3. remainder

5. one

7. True

9. True

Practice

1. **a.** $x-9$
b. $c=3$; $P(3)=0$

3. **a.** $x^2-4x+33-\frac{265}{x+8}$
b. $c=-8$; $P(-8)=-265$

5. **a.** $4x^2-6x+9-\frac{17}{x+2}$
b. $c=-2$; $P(-2)=-17$

7. **a.** $x^2+7x+55+\frac{388}{x-7}$
b. $c=7$; $P(7)=388$

9. **a.** $2x^2-2x+6-\frac{27}{x+3}$
b. $c=-3$; $P(-3)=-27$

11. **a.** $x^3+2x+5+\frac{17}{x-3}$
b. $c=3$; $P(3)=17$

13. **a.** $x^3+2x^2+6x+9+\frac{23}{x-2}$
b. $c=2$; $P(2)=23$

15. **a.** $x^3+\frac{1}{2}x^2-\frac{3}{4}x-\frac{3}{8}+\frac{45}{16\left(x-\frac{1}{2}\right)}$
b. $c=\frac{1}{2}$; $P\left(\frac{1}{2}\right)=\frac{45}{16}$

17. **a.** $x^4+x^3+x^2+x+1$
b. $c=1$; $P(1)=0$

19. **a.** $x^3-\frac{14}{5}x^2+\frac{56}{25}x-\frac{224}{125}+\frac{3396}{625\left(x+\frac{4}{5}\right)}$
b. $c=-\frac{4}{5}$; $P\left(-\frac{4}{5}\right)=\frac{3396}{625}$

Applications

21. 41 °F

Collaborative Learning

23. **a.** Answers will vary. $x^2-\frac{7}{2}x+\frac{13}{4}+\frac{73}{4(2x-1)}$; $2x^2-7x+\frac{13}{2}+\frac{73}{4\left(x-\frac{1}{2}\right)}$
b. $\frac{P(x)}{x-\frac{b}{a}}=\frac{aP(x)}{ax-b}$
c. If a polynomial $P(x)$ is divided by $(ax-b)$, then the remainder will be $P\left(\frac{b}{a}\right)$. Answers will vary.

Chapter 13: Factoring Polynomials

13.1 Exercises

Concept Check

1. product, factors

3. greatest common factor, largest

5. negative / opposite

7. True

9. False; Binomials can be common factors.

Practice

1. 5

3. 8

5. 1

7. $10x^3$

9. $4a^2$

11. $13ab$

13. $15xy^2z^2$

15. x^4

17. $-4y$

19. $3x^3$

21. $2x^2y$

23. $m+9$

25. $x-6$

27. $b+1$

29. $3y+4x+1$

31. $11(x-11)$

33. $4y(4y^2+3)$

35. $-3a(2x-3y)$

37. $5xy(2x-5)$

39. $-2yz(9yz-1)$

41. $8(y^2-4y+1)$

43. $x(2y^2-3y-1)$

45. $4m^2(2x^3-3y+z)$

47. $-7x^2z^3(8x^2+14xz+5z^2)$

49. $x^4y^2(15+24x^2y^4-32x^3y)$

51. $(y+3)(7y^2+2)$

53. $(x-4)(3x+1)$

55. $(x-2)(4x^3-1)$

57. $(2y+3)(10y-7)$

59. $(x-2)(a-b)$

61. $(b+c)(x+1)$

63. $(x^2+6)(x+3)$

65. Not factorable

67. $(3-b)(x+y)$

69. $(y-4)(5x+z)$

71. $(z^2+3)(a+1)$

73. $(6x+1)(a+2)$

75. $(x+1)(y+1)$

77. $(2y-7z)(5x-y)$

79. $(3x-4u)(y-2v)$

81. Not factorable

83. $(2c-3d)(3a+b)$

Applications

85. She can make 30 identical treat bags, each containing 5 pieces of candy A, 6 pieces of candy B, and 11 pieces of candy C.

87. **a.** 32 feet **b.** $16x(3-x)$

c. 32 feet

d. Yes. They are equivalent expressions.

13.2 Exercises

Concept Check

1. constant, middle

3. positive

5. different

7. False; One factor is negative, one is positive.

9. True

Practice

1. $\{1,15\},\{-1,-15\},\{3,5\},\{-3,-5\}$

3. $\{1,20\},\{-1,-20\},\{4,5\},\{-4,-5\},\{2,10\},\{-2,-10\}$

5. $\{1,-6\},\{6,-1\},\{2,-3\},\{3,-2\}$

7. $\{1,16\},\{-1,-16\},\{4,4\},\{-4,-4\},\{8,2\},\{-8,-2\}$

9. $\{1,-10\},\{10,-1\},\{5,-2\},\{2,-5\}$

11. 4, 3

13. −7, 2

15. 8, −1

17. −6, −6

19. −5, −4

21. $x+1$

23. $p-10$

25. $a+6$

27. $(x-4)(x+3)$

29. $(y+6)(y-5)$

31. Not factorable

33. $(x-4)(x-4)$

35. $(x+4)(x+3)$

37. $(y-1)(y-2)$

39. Not factorable

41. $(x+8)(x-9)$

43. $(z-6)(z-9)$

45. $x(x+7)(x+3)$

47. $5(x-4)(x+3)$

49. $10y(y-3)(y+2)$

51. $4p^2(p+1)(p+8)$

53. $2x^2(x-9)(x+2)$

55. $2(x^2-x-36)$

57. $2a^2(a-10)(a+6)$

59. $3y^3(y-8)(y+1)$

61. $x(x-2)(x-8)$

63. $5(a^2+2a-6)$

65. $20a^2(a+1)(a+1)$

Applications

67. Base $= x+48$; Height $= x$

69. $x+5$

71. **a.** 0 feet

b. $-16(t-3)(t+2)$

c. 0 feet

d. Yes. They are equivalent expressions.

Writing & Thinking

73. If the sign of the constant term is positive, the signs in the factors will both be positive or both be negative. If the sign of the constant term is negative, the sign in one factor will be positive and the sign in the other factor will be negative.

13.3 Exercises

Concept Check

1. factors

3. factorable

5. product, sum

7. False; The middle term should be the sum of the inner and outer products.

9. False; The first step is to multiply a and c.

Practice

1. $(x+2)(x+3)$

3. $(2x-5)(x+1)$

5. $(6x+5)(x+1)$

7. $-(x-2)(x-1)$

9. $(x-5)(x+2)$

11. $-(x-14)(x+1)$

13. Not factorable

15. $-x(2x+1)(x-1)$

17. $(t-1)(4t+1)$

19. $(5a-6)(a+1)$

21. $(7x-2)(x+1)$

23. $(2x-3)(4x+1)$

25. $(3x+4)(3x-5)$

27. $2(2x-5)(3x-2)$

29. $(3x-1)(x-2)$

31. $(3x-1)(3x-1)$

33. $(3y+2)(2y+1)$

35. $(x-1)(x-45)$

37. Not factorable

39. $2b(4a-3)(a-2)$

41. Not factorable

43. $(4x-1)(4x-1)$

45. $(8x-3)(8x-3)$

47. $2(3x-5)(x+2)$

49. $5(2x+3)(x+2)$

51. $-2(9x^2-36x+4)$

53. $-15(3y+4)(y-2)$

55. $3(2x-5)(2x-5)$

57. $3x(2x-1)(x+2)$

59. $3x(2x-9)(2x-9)$

61. $9xy^3(x^2+x+1)$

63. $4xy(3y-4)(4y-3)$

65. $7y^2(y-4)(3y-2)$

Writing & Thinking

67. This is not an error, but the trinomial is not completely factored. The completely factored form of this trinomial is $2(x+2)(x+3)$.

69. $x(5-2x)(25-2x)$; The height is x, the width is $(5-2x)$, and the length is $(25-2x)$.

13.4 Exercises

Concept Check

1. binomial

3. $2ax$, $-2ax$

5. $(x+a)(x^2-ax+a^2)$
7. True
9. False; The sum of two squares is not factorable.

Practice

1. $(x-5)(x+5)$
3. $(9-y)(9+y)$
5. $2(x-8)(x+8)$
7. $4(x-2)(x+2)(x^2+4)$
9. Not factorable
11. $(y-8)^2$
13. $-4(x-5)(x+5)$
15. $(3x-5)(3x+5)$
17. $(y-5)^2$
19. $(2x-1)^2$
21. $(5x+3)^2$
23. $(4x-5)^2$
25. $4x(x-4)(x+4)$
27. $2xy(x+8)^2$
29. $(y+3)^2$
31. $(x-10)^2$
33. $(x^2+5y)^2$
35. $(x-5)(x^2+5x+25)$
37. $(y+6)(y^2-6y+36)$
39. $(x+3y)(x^2-3xy+9y^2)$
41. Not factorable
43. $4(x-2)(x^2+2x+4)$
45. $2(3x-y)$ $(9x^2+3xy+y^2)$
47. $y(x+y)(x^2-xy+y^2)$
49. $x^2y^2(1-y)(1+y+y^2)$
51. $3xy(2x+3y)$ $(4x^2-6xy+9y^2)$
53. (x^2-y^3) $(x^4+x^2y^3+y^6)$
55. $(3x+y^2)$ $(9x^2-3xy^2+y^4)$
57. $(2x+y)(4x^2-2xy+y^2)$
59. $8(y-1)(y^2+y+1)$
61. $(3x-y)(3x+y)$
63. $(x-2y)(x+2y)$ (x^2+4y^2)
65. $(x-y-9)(x-y+9)$
67. $(x-y-6)(x-y+6)$
69. $(4x+1-y)(4x+1+y)$
71. **a.** x^2-16

b.

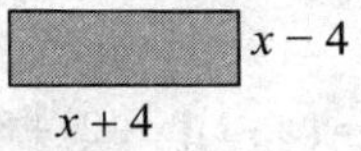

Writing & Thinking

73. **a.** $xy+xy+x^2+y^2$
$=x^2+2xy+y^2$
$=(x+y)^2$

b. $(x+y)(x+y)$
$=(x+y)^2$

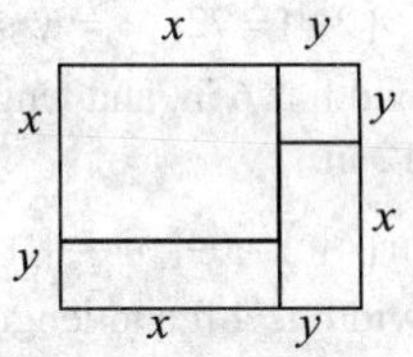

75. For a 3-digit integer: abc
$=100a+10b+c$
$=(99+1)a+(9+1)b+c$
$=9(11a+b)+a+b+c$
So, if the sum $(a+b+c)$ is divisible by 3 (or 9), then the number abc will be divisible by 3 (or 9).
For a 4-digit integer: $abcd$
$=1000a+100b+10c+d$
$=(999+1)a+(99+1)b$
$+(9+1)c+d$
$=9(111a+11b+c)+a$
$+b+c+d$
So, if the sum $(a+b+c+d)$ is divisible by 3 (or 9), then the number $abcd$ will be divisible by 3 (or 9).

13.5 Exercises

Concept Check

1. trial-and-error, ac
3. original
5. monomial
7. True

Practice

1. $(m+6)(m+1)$
3. $(x+9)(x+2)$
5. $(x-10)(x+10)$
7. $(m-3)(m+2)$
9. Not factorable
11. $(8a-1)(8a+1)$
13. $(x+5)^2$
15. $(x+12)(x-3)$
17. $3(a+6)(a-2)$
19. $-5(x-6)(x-8)$
21. Not factorable
23. $x(x-6)(x+2)$
25. $-2a(a+8)(a-7)$
27. $4x(2x-5)(2x+5)$
29. $-(x-5)(3x-2)$
31. $(2x-1)(3x-4)$
33. $(4m+3)(3m-2)$
35. $2(2x-1)(x-3)$
37. $(4x-7)(2x+5)$
39. $(5x+6)(4x-9)$
41. $-(5x-7)(3x+2)$
43. $-(2a-3)(4a-5)$
45. $(4y+5)(5y-4)$
47. $(6x-1)(3x-2)$
49. $-6(5x-4)(5x+4)$
51. $3(4n^2-20n-25)$
53. $a(21a^2-13a-2)$
55. $3x(3x-2)(4x+5)$
57. $2x(2x-1)(4x-11)$
59. $5(24m^2+2m+15)$
61. $(y-4)(x+3)$
63. $(x+2y)(x-6)$
65. $-(x^2-5)(x-8)$
67. $(x+5)(x^2-5x+25)$
69. $x^4(y-1)(y^2+y+1)$
71. $(2a^2+3b^2)$ $(4a^4-6a^2b^2+9b^4)$
73. (x^2y-5) $(x^4y^2+5x^2y+25)$
75. $(x-3)(x+3)(x+7)$
77. $(3x+y+6)(3x-y-6)$
79. $(y+10+7x)(y+10-7x)$

13.6 Exercises

Concept Check

1. zero
3. factor
5. substituting
7. True
9. True

Practice

1. $x=2,3$
3. $x=-2,\frac{9}{2}$
5. $x=-3$
7. $x=-5$
9. $x=0,2$
11. $x=-6$
13. $x=-1,4$
15. $x=-3,4$
17. $x=-3,0$
19. $x=2,4$
21. $x=-4,3$
23. $x=-\frac{1}{2},3$
25. $x=-\frac{2}{3},2$
27. $x=-\frac{1}{2},4$
29. $x=-2,\frac{4}{3}$
31. $x=\frac{3}{2}$
33. $x=0,\frac{8}{5}$
35. $x=-2,2$
37. $x=1$
39. $x=2$
41. $x=-3,3$

43. $x=-5, 10$

45. $x=-6, -2$

47. $x=\frac{1}{2}$

49. $x=0, 2, 4$

51. $x=-\frac{2}{3}, -\frac{1}{2}, 0$

53. $x=-10, 10$

55. $x=-5, 5$

57. $x=-4$

59. $x=3$

61. $x=-1, 3$

63. $x=-8, -2$

65. $x=-5, 2$

67. $x=-5, 7$

69. $x=-6, 2$

71. $x=-1, \frac{2}{3}$

73. $x=-\frac{3}{2}, 4$

75. $y^2-y-6=0$

77. $2x^2+11x+5=0$

79. $8x^2-10x+3=0$

81. $x^3-x^2-6x=0$

83. $y^3-4y^2-3y+18=0$

Applications

85. **a.** 640 ft; 384 ft

b. 144 ft; 400 ft

c. 7 seconds; $0=-16(t+7)(t-7)$

87. **a.** The ball will hit the ground.

b. $0=-16t^2+16t+96$

c. $t=-2, 3$

d. The ball will hit the ground after −2 seconds and after 3 seconds.

e. No, time cannot be negative.

89. **a.** π in.^{2}

b. 4π in.^{2}

c. 16π in.^{2}

d. 64π in.^{2}

e. The area gets 4 times larger.

Writing & Thinking

91. This allows for use of the zero-factor law, which says that for the product to equal zero one of the factors must equal zero. Answers will vary.

13.7 Exercises

Concept Check

1. variable, unknown

3. consecutive

5. 1

7. False; The sum of the squares of the lengths of the legs is equal to the hypotenuse squared.

9. False: Only with right triangles

Applications

1. $x(x+8)=-16$; $x=-4$, so the numbers are −4 and 4.

3. $x^2=7x$; $x=0, 7$

5. $x^2+3x=28$; $x=4$

7. $x(x+3)=40$; $x=-8, 5$, so the numbers are −8 and −5 or 5 and 8.

9. $(x+5)^2+x^2=53$; $x=2$, so the numbers are 2 and 7.

11. $x+(x+4)^2=38$; $x=2$, so the integers are 2 and 6.

13. $x(2x-5)=x+56$; $x=-4$

15. $x(x+1)=72$; $x=8$, so the integers are 8 and 9.

17. $x^2+(x+1)^2=85$; $x=6$, so the integers are 6 and 7.

19. $x(x+2)=63$; $x=-9, 7$, so the integers are −9 and −7 or 7 and 9.

21. $4x+(x+1)^2=41$; $x=4$, so the integers are 4 and 5.

23. $2x(x+1)=(x+1)(x+2)+88$; $x=10$, so the integers are 10, 11, and 12.

25. $6x(x+2)=(x+1)+(x+3)^2$; $x=-2, 1$, so the integers are −2, −1, 0, and 1 or 1, 2, 3, and 4.

27. $w(2w)=72$; $w=6$, so width is 6 in. and length is 12 in.

29. $w(4w)=64$; $w=4$, so width is 4 ft and length is 16 ft.

31. $l(l-4)=45$; $l=9$, so width is 5 ft and length is 9 ft.

33. $\frac{1}{2}b(b-4)=16$; $b=8$, so base is 8 ft and height is 4 ft.

35. $\frac{1}{2}(h+6)h=20$; $h=4$, so the base is 10 in.

37. $w(16-w)=48$; $w=4$, 12, so the rectangle is 4 in. by 12 in.

39. $r(r+13)=140$; $r=7$, so there are 7 trees in each row.

41. $r(r+7)=144$; $r=9$, so there are 9 rows.

43. $n(n+1675)=8400$; $n=5$, so there are 5 floors.

45. $(w+11)(w+4)=98$; $w=3$, so the rectangle was 3 cm by 10 cm.

47. $w=10$, 15, so width is 10 ft and length is 30 ft or width is 15 ft and length is 20 ft.

49. $h^2+(h-34)^2=(h+2)^2$; $h=48$, so the height of the pole is 48 ft.

51. $x^2+(x-49)^2=(x+1)^2$; $x=60$, so height is 60 ft.

53. $l^2+(l-28)^2=(l+8)^2$; $l=60$, so the length of the mat is 60 inches.

55. **a.** Figures will vary.

b. $(17-x)^2+x^2=13^2$

c. $x=12$, $x=5$

d. Yes

e. The wire is attached to the pole 5 feet from the ground (12 feet from the top) and attached to the stake which is 12 feet from the pole. Or the wire is attached to the pole 12 feet from the ground (5 feet from the top) and attached to the stake which is 5 feet from the pole.

f. Answers will vary.

57. \$1.50 per pound

59. \$16 or \$20 per reel

Writing & Thinking

61. 128 in.^{2}; We can use the Pythagorean Theorem to find the length and the width (by finding the diagonals of the interior square) instead.

Chapter 14: Rational Expressions

14.1 Exercises

Concept Check

1. rational
3. fundamental
5. fractions
7. True
9. False; A rational expression cannot have a zero denominator.

Practice

1. $\frac{3x}{4y}; x \neq 0, y \neq 0$
3. $\frac{2x^3}{3y^3}; x \neq 0, y \neq 0$
5. $\frac{1}{x-3}; x \neq 0, 3$
7. $7; x \neq 2$
9. $-\frac{3}{4}; x \neq 3$
11. $\frac{2x}{y}; x \neq -\frac{2}{3}, y \neq 0$
13. $\frac{x}{x-1}; x \neq -6, 1$
15. $\frac{x-7}{x-3}; x \neq -3, 3$
17. $\frac{x-3}{y-2}; y \neq -2, 2$
19. $\frac{x+7}{y+5}; x \neq 2, y \neq -5$
21. $\frac{2}{75}$
23. 9
25. 0
27. $\frac{-3}{32}$
29. 1

Applications

31. **a.** $p(x) = \frac{15x+200}{x}$
b. \$35 **c.** $x \neq 0$
d. The variable cannot be negative because you cannot have a negative quantity of people. There would also be a maximum number depending on the size of the room.

33. **a.** \$11.62 **b.** \$6.29
c. \$5.33
d. The cost per calculator decreases as the number of calculators purchased increases.
e. $x \neq -20$
f. The value of x cannot be negative because you cannot buy a negative quantity of calculators

35. $(2x-5)$ feet

Writing & Thinking

37. **a.** A rational expression is an algebraic expression that can be written in the form $\frac{P}{Q}$ where P and Q are polynomials and $Q \neq 0$.
b. $\frac{x-1}{(x+2)(x-3)}$
Answers will vary.
c. $\frac{1}{x+5}$
Answers will vary.

14.2 Exercises

Concept Check

1. factor
3. numerator, denominator
5. numerators, denominators
7. True
9. False; the restriction is 0.

Practice

1. $\frac{ab}{6y}$
3. $\frac{8x^2y^3}{15}$
5. $\frac{x+3}{x}$
7. $\frac{x-1}{x+1}$
9. $-\frac{1}{x-8}$
11. $\frac{x-2}{x}$
13. $\frac{4x+20}{x(x+1)}$
15. $\frac{x}{(x+3)(x-1)}$
17. $-\frac{x+4}{x(x+1)}$
19. $\frac{x+2y}{(x-3y)(x-2y)}$
21. $\frac{x-1}{x(2x-1)}$
23. $\frac{1}{x+1}$
25. $\frac{x+2}{x-2}$
27. $\frac{1}{3xy^6}$
29. $\frac{6y^7}{x^4}$
31. $\frac{x}{12}$
33. $\frac{6x+18}{x^2}$
35. $\frac{6}{5x}$
37. $\frac{3x+1}{x+1}$
39. $\frac{x-2}{2x-1}$
41. $-\frac{x+4}{x(2x-1)}$
43. $\frac{x+1}{x-1}$
45. $\frac{6x^3-x^2+1}{x^2(4x-3)(x-1)}$
47. $\frac{x^2+4x+4}{x^2(2x-5)}$
49. $\frac{x^2-6x+5}{(x-7)(x-2)(x+7)}$
51. $\frac{x^2-3x}{(x-1)^2}$
53. $\frac{x^2+5x}{2x+1}$
55. 1

Applications

57. **a.** $\frac{x^2-3x-10}{x+3}$
b. $\frac{x^2+5x+6}{x-5}$
c. $(x+2)^2 = x^2+4x+4$

14.3 Exercises

Concept Check

1. smallest
3. prime factorization
5. equivalent
7. True
9. False; You completely factor each polynomial, including prime factors for numerical terms.

Practice

1. 150
3. 432
5. 600
7. 60
9. $\frac{11}{17}$
11. $\frac{3}{2}$
13. $\frac{6}{5}$
15. $\frac{23}{30}$
17. $\frac{28}{27}$
19. $7(x+5)(x-5)$
21. $30(y-4)$
23. $(x+3)(x-3)(x-3)$
25. $-(y-3)$
27. $-2(x+12)(x-12)$
29. $(x+4)(x-3)(x+5)$
31. $(x-2)(y+3)(x+7)$
33. $2(x+6)(x-6)(x+3)$
35. $6(x-5)(y+6)(y+1)$
37. $(x+2)(x-2)(x^2+4)$
39. 28

41. $33(x-3)$

43. $-3x^2$

45. $2(y+3)(y-1)$
$= 2y^2 + 4y - 6$

47. $(x+1)(x+3)$
$= x^2 + 4x + 3$

14.4 Exercises

Concept Check

1. numerator, denominator
3. most
5. equivalent
7. True
9. True

Practice

1. 3
3. 2
5. 1
7. $\frac{2}{x-1}$
9. $\frac{14}{7-x}$
11. 4
13. $\frac{x^2-x+1}{(x+4)(x-3)}$
15. $\frac{5}{x-3}$
17. $\frac{4x+5}{2(7x-2)}$
19. $\frac{6x+15}{(x+3)(x-3)}$
21. $\frac{x^2-2x+4}{(x+2)(x-1)}$
23. $\frac{-x^2-3x-6}{(x+3)(3-x)}$
25. $\frac{8x^2+13x-21}{6(x+3)(x-3)}$
27. $\frac{3x^2-20x}{(x+6)(x-6)}$
29. $\frac{-4x}{x-7}$
31. $\frac{4x^2-x-12}{(x+7)(x-4)(x-1)}$
33. $\frac{x-6}{(x-10)(x-8)}$
35. $\frac{6x}{(x-1)(x-7)}$
37. $\frac{4x-19}{(7x+4)(x-1)(x+2)}$
39. $\frac{-7x-9}{(4x+3)(x-2)}$
41. $\frac{4x^2-41x+3}{(x+4)(x-4)}$
43. $\frac{x-4}{2(x-2)}$
45. $\frac{x^2-4x-6}{(x+2)(x-2)(x-1)}$
47. $\frac{3x^2+26x-3}{(x+7)(x-3)(x+1)}$
49. $\frac{6x+2}{(x-1)(x+3)}$
51. $\frac{2x^2+x-4}{(x-2)(y+1)(x+1)}$
53. $\frac{2x+4xy-15y}{(x+3)(y+2)(x-5)}$
55. $\frac{-2x+2}{2x-1}$
57. $\frac{2x^2-x-5}{(x-3)(x+3)(x^2+1)}$
59. $\frac{x^2+9x+6}{(3x+1)(x+3)(x-1)}$

Applications

61. **a.** $\frac{7x^2+3}{x-2}$ **b.** $\frac{4x^2+5}{x+2}$
c. $\frac{22x^3+12x^2+16x-8}{(x+2)(x-2)}$

63. $\frac{5x+6}{2x(x+3)}$

65. **a.** $\frac{100}{x}$ **b.** $\frac{100}{2x+15}$
c. $\frac{100x+1500}{x(2x+15)}$

67. **a.** \$5750
b. $C(x) = 5750 + 0.35x$
c. $A(x) = \frac{5750+0.35x}{x}$
d. \$6.40 per cupcake
e. \$1.99 per cupcake

14.5 Exercises

Concept Check

1. rational
3. LCM
5. complex algebraic
7. True
9. True

Practice

1. $\frac{4}{5xy}$
3. $\frac{8}{7x^2y}$
5. $\frac{2x^2+6x}{2x-1}$
7. $\frac{2x-1}{2+3x}$
9. $\frac{7}{2(x+2)}$
11. $\frac{x}{x-1}$
13. $\frac{4x}{3(x+6)}$
15. $\frac{7x}{x+2}$
17. $\frac{2x+6}{3(x-2)}$
19. $\frac{24y+9x}{2(9y-10x)}$
21. $\frac{x}{x-1}$
23. $\frac{xy}{x+y}$
25. $\frac{1}{xy}$
27. $\frac{y+x}{y-x}$
29. $\frac{3-x}{x}$
31. $\frac{x+1}{x+3}$
33. $\frac{-1}{x(x+h)}$
35. $\frac{-1}{x(x+h)}$
37. $-(x-2y)(x-y)$
39. $\frac{2x}{x^2+1}$
41. $\frac{(x-3)(x^2-2x+4)}{(x-4)(x-2)(x+1)}$
43. $\frac{-5}{x+1}$
45. $\frac{29}{4(4x+5)}$
47. $\frac{x^2-3x-6}{x(x-1)}$
49. $\frac{x^2-4x-2}{(x-4)(x+4)}$

Applications

51. **a.** $\frac{2r_1r_2}{r_1+r_2}$
b. 44.2 miles per hour
c. 1.8 hours

Writing & Thinking

53. **a.** $\frac{8}{5}$
b. 1
c. $\frac{x^4+x^3+3x^2+2x+1}{x^3+x^2+2x+1}$

14.6 Exercises

Concept Check

1. ratio
3. extraneous
5. variables
7. False; It is not a proportion.
9. True

Practice

1. $x = 7$
3. $x \neq 0, 2; x = 4$
5. $x \neq -3, 4; x = -10$
7. $x \neq 0; x = 18$
9. $x \neq 6; x = -\frac{74}{9}$
11. $x = \frac{1}{4}$
13. $x = 6$
15. $x = 4$
17. $x \neq 0; x = \frac{10}{3}$
19. $x \neq 0; x = -\frac{3}{4}$
21. $x \neq 0; x = -\frac{3}{16}$
23. $x \neq -9, -\frac{1}{4}, 0; x = -2, 1$
25. $x \neq \frac{3}{2}, 0, 6; x = \frac{3}{5}, 9$

27. $x \neq \frac{1}{2}, 4; x = -3$

29. $x \neq -4, -1; x = 2$

31. $x \neq -1, \frac{1}{4}; x = \frac{2}{3}$

33. $x \neq 2, 3; x = \frac{13}{10}$

35. $x \neq -\frac{2}{3}, 2$; no solution

37. $x \neq -1, \frac{1}{3}, \frac{1}{2}; x = \frac{1}{5}$

39. $r = \frac{S-a}{S}$

41. $s = \frac{x-\bar{x}}{z}$

43. $y = m(x - x_1) + y_1$

45. $R_{\text{total}} = \frac{R_1 R_2}{R_1 + R_2}$

47. $P = \frac{A}{1+r}$

49. $LK = 15, JB = 5$

51. $AC = 2, ST = 12$

53. $ST = 8, TU = 12, QR = 24$

55. $AP = \frac{9}{2}$ in.; $PC = \frac{15}{2}$ in.

Applications

57. 360 defective computers

59. 36,800 students

61. 34 miles

63. 6.8 cups

65. Width = 3 inches; Length = 7.5 inches

67. 4 hours

69. First group: 10 hours; Second group: 6 hours.

71. a. $\frac{3x}{3} = \frac{16}{x}$
b. $x = -4, +4$
c. $x = -4$ does not make sense because length cannot be negative
d. The side of the small triangle is 4 feet long and the base of the large triangle is 12 feet long.

Writing & Thinking

73. a. $\frac{x^2 - 8x}{4(x-4)(x+4)}$
b. $x = 0, 8$

75. a. $\frac{14x-16}{5x(x-4)}$ b. $x = \frac{8}{7}$

14.7 Exercises

Concept Check

1. unit, time
3. check
5. diagram, chart
7. True

Applications

1. 72, 45
3. 9
5. $\frac{6}{13}$
7. 36, 27
9. 7, 12
11. 37.5 miles
13. $\frac{12}{5}$ or $2\frac{2}{5}$ hours
15. $\frac{9}{5}$ or $1\frac{4}{5}$ hours
17. 6 hours
19. Charles: 8 mph; Chase: 6 mph
21. 63 mph
23. sailboat: 5 mph; cruise ship: 25 mph
25. 20 mph
27. 120 hours
29. John: 11 hours; Raul: 22 hours; Denny: 33 hours
31. 1 mph
33. a. $\frac{1}{x} + \frac{1}{4x} = \frac{1}{4}$

Printer	Time of Work (in Hours)	Part of Work Done in 1 Hour
Newest	x	$\frac{1}{x}$
Old	$4x$	$\frac{1}{4x}$
Together	4	$\frac{1}{4}$

b. $x = 5$
c. It would take the newest printer 5 hours to complete the job and the old printer 20 hours to complete the job.

Writing & Thinking

35. a. 5 and 7 b. 2 and 4

14.8 Exercises

Concept Check

1. decrease
3. increase
5. joint
7. False; Varies directly
9. True

Practice

1. $\frac{7}{3}$
3. 2
5. $-\frac{32}{9}$
7. 36
9. 120
11. $\frac{56}{3}$
13. 40
15. 54
17. $\frac{48}{5}$
19. 27

Applications

21. 400 feet
23. \$35.10
25. 4.71 feet
27. 6 m
29. 0.0073 cm
31. 16,000 lb
33. 9×10^{-11} N
35. \$59.70
37. a. $SL = \frac{kwd}{l^2}$
b. 200 c. 4320 pounds
39. 25,200 lb
41. 18,000 lb
43. 200 cm^3
45. 330 bar
47. 5.2 ohms
49. 10 ohms
51. $10\frac{2}{3}$ ft from the 120 lb weight or, $1\frac{1}{3}$ ft from the 960 lb weight.
53. 216 kilograms

Chapter 15: Roots, Radicals, and Complex Numbers

15.1 Exercises

Concept Check

1. 2
3. square root
5. radical, radicand
7. False; If the original number is negative, the principal square root will not be the same as the original number.
9. False; The radicand is underneath the radical symbol.

Practice

1. 3
3. 9
5. 17
7. 13
9. 1
11. 5
13. 6
15. $\frac{1}{2}$
17. $\frac{3}{4}$
19. 0.2
21. −10
23. −0.04
25. −3
27. −5
29. $\frac{3}{5}$
31. $\sqrt{64} < \sqrt{74} < \sqrt{81}$ and $8 < \sqrt{74} < 9$ because $64 < 74 < 81$ or $(8.6023)^2 = 73.99956529$

33. $\sqrt{25} < \sqrt{32} < \sqrt{36}$ and $5 < \sqrt{32} < 6$ because $25 < 32 < 36$ or $(5.6569)^2 = 32.00051761$

35. $>$ **37.** $<$

39. $=$ **41.** $=$

43. Rational

45. Rational

47. Irrational

49. Not a real number

51. Rational

53. Irrational

55. 6.2450

57. 2.4960

59. 0.4472

61. 8.9443

63. −8.2462

65. 2.6207

67. 4.9324

Applications

69. $5\sqrt{2}$ or 6.30 inches

71. **a.** 4 inches

b. 12 tiles

73. 10 inches

75. 4.64 cm

77. **a.** 1.7 meters

b. 1.6 meters

c. 1.8 meters

79. **a.** 4.2 ft

b. 10.2 ft

c. 18 tiles

Writing & Thinking

81. There is no real number that results in a negative number when squared.

15.2 Exercises

Concept Check

1. factor

3. 2

5. $\sqrt{a}\sqrt{b}$

7. True

9. False; If x is a real number, then $\sqrt{x^2} = |x|$.

Practice

1. $2\sqrt{3}$

3. $12\sqrt{2}$

5. $-6\sqrt{2}$

7. $-2\sqrt{14}$

9. $-5\sqrt{5}$

11. $\frac{1}{2}$

13. $-\frac{\sqrt{11}}{8}$

15. $\frac{2\sqrt{7}}{5}$

17. $6x$

19. $2x\sqrt{2x}$

21. $2x^5y\sqrt{6x}$

23. $5xy^3\sqrt{5x}$

25. $-3xy\sqrt{2}$

27. $2bc\sqrt{3ac}$

29. $5x^2y^3z^4\sqrt{3}$

31. $\frac{x^2\sqrt{5}}{3}$

33. $\frac{4a^2\sqrt{2a}}{9b^8}$

35. $\frac{10x^4\sqrt{2}}{17}$

37. 6

39. $2\sqrt[3]{7}$

41. −1

43. $-4\sqrt[3]{2}$

45. $5x\sqrt[3]{x}$

47. $-2x^2\sqrt[3]{x^2}$

49. $2a^2b\sqrt[3]{9b}$

51. $6x^2y\sqrt[3]{y^2}$

53. $2xy^2z^3\sqrt[3]{3x^2y}$

55. $\frac{\sqrt[3]{3}}{2}$

57. $\frac{5\sqrt[3]{3}}{2}$

59. $\frac{5y^4}{3x^2}$

Applications

61. $\sqrt{6} \approx 2.45$ amperes

63. 120 volts

65. **a.** 10 cm

b. $2\sqrt{30}$ cm

c. $5\sqrt{6}$ cm

Writing & Thinking

67. A cube root has no restrictions as the cube root of a negative number is negative.

15.3 Exercises

Concept Check

1. radical, index

3. $a^{\frac{1}{2}}$

5. $\sqrt[3]{a}$

7. True

9. True

Practice

1. $\sqrt[3]{8}$ **3.** $-\sqrt[6]{x}$

5. $\sqrt[5]{4z^2}$ **7.** $13^{\frac{1}{7}}$

9. $(-9)^{\frac{1}{3}}$ **11.** 3

13. $\frac{1}{10}$ **15.** −512

17. −4

19. Not a real number

21. $\frac{3}{7}$ **23.** 16

25. $-\frac{1}{6}$ **27.** $\frac{5}{2}$

29. $\frac{1}{4}$ **31.** $\frac{3}{8}$

33. $-\frac{1}{1000}$ **35.** 64

37. 8.5499

39. 10,000,000

41. 99.6055

43. 0.0922

45. 1.6083

47. 0.2236

49. 7.7460

51. 2.0408

53. $8x$ **55.** $\frac{1}{3a^2}$

57. $8x^{\frac{5}{2}}$ **59.** $5a^{\frac{13}{6}}$

61. $x^{\frac{7}{12}}$ **63.** $x^{\frac{1}{2}}$

65. $\frac{1}{a^{\frac{9}{8}}}$ **67.** $a^{\frac{1}{4}}$

69. $\frac{a^2}{b^{\frac{6}{5}}}$ **71.** $8x^{\frac{3}{2}}y$

73. $\frac{x^{\frac{3}{2}}}{16y^{\frac{2}{5}}}$ **75.** $\frac{x^2y^4}{z^4}$

77. $\frac{y^{\frac{3}{2}}z^2}{x}$ **79.** $\frac{8b^{\frac{9}{4}}}{a^3c^3}$

81. $x^{\frac{1}{4}}y^{\frac{5}{4}}$ **83.** $\frac{y^{\frac{2}{3}}}{50x^{\frac{4}{3}}}$

85. $\frac{b^{\frac{5}{12}}}{a^{\frac{11}{12}}}$ **87.** $\frac{3x^{\frac{1}{2}}}{20y^{\frac{2}{3}}}$

89. $\sqrt[6]{x^5}$ **91.** $\sqrt[12]{y^7}$

93. $\sqrt[30]{x^{11}}$ **95.** $\sqrt[6]{y}$

97. $\sqrt[9]{x}$ **99.** $\sqrt[3]{7a}$

101. $\sqrt[24]{x}$

103. $a^{20}b^5c^{10}$

Applications

105. 0.24 Earth years

107. 576 ft²

109. **a.** 20 feet per second

b. 35.78 feet per second

111. **a.** 50 kmph

b. 20 kmph

Writing & Thinking

113. **a.** n must be an integer greater than 1.

b. n must be a number less than 0.

c. n must equal 0.

d. n must be a rational number that is not an integer.

15.4 Exercises

Concept Check

1. index
3. FOIL
5. True

Practice

1. $8\sqrt{2}$
3. $\sqrt{11}$
5. $-3\sqrt{10}$
7. $13\sqrt[3]{3}$
9. $-\sqrt{11}$
11. $3\sqrt{a}$
13. $7\sqrt{x}$
15. $4\sqrt{2}+3\sqrt{3}$
17. $8\sqrt{b}-4\sqrt{a}$
19. $13\sqrt[3]{x}-2\sqrt[3]{y}$
21. $5\sqrt{3}$
23. 0
25. $17\sqrt[3]{2}$
27. $2\sqrt{2}-6\sqrt{3}$
29. $6+\sqrt{5}$
31. $\sqrt{3}-4\sqrt{2}$
33. $5\sqrt[3]{2}-8\sqrt[3]{3}$
35. $4\sqrt{2x}$
37. $2y\sqrt{2y}$
39. $-4x\sqrt{3xy}$
41. $15x\sqrt{x}$
43. $-xy^2\sqrt{x}$
45. $4x^5y^{10}\sqrt{3}$
47. $-8x^8y^2$
49. $xy^2\sqrt[3]{2}\left(-2x^2y^2-2x^3y+3\right)$
51. $3\sqrt{2}-8$
53. 18
55. $-8\sqrt{3}$
57. $12+\sqrt{6}$
59. $2y+\sqrt{xy}$
61. $13+2\sqrt{2}$
63. $3x-9\sqrt{3x}+8$
65. $2-2\sqrt{7}$
67. $13+4\sqrt{10}$
69. $\sqrt{10}+\sqrt{15}-\sqrt{6}-3$
71. $x-2\sqrt{6x}-18$
73. 58
75. $x+10\sqrt{xy}+25y$
77. $x+28-10\sqrt{x+3}$
79. $2x+19-8\sqrt{2x+3}$
81. 4.3397
83. 31.6
85. −57
87. −37.3569
89. $\sqrt{16}+\sqrt{48}\neq\sqrt{16+48}$, as radicals cannot be added by simply adding their radicands.

Applications

91. $5\sqrt{6}+\sqrt{170}\approx 25.29$ ft
93. **a.** $7\sqrt{2t}$ dollars
    b. $\sqrt{2t}$ dollars
95. **a.** $60\sqrt{6}\%$
    b. $6x^4\sqrt{2}\%$

15.5 Exercises

Concept Check

1. denominator
3. rational
5. conjugate
7. True
9. False; You would need to multiply by $\sqrt[3]{a^2}$.

Practice

1. $\dfrac{5\sqrt{2}}{2}$
3. $\dfrac{-3\sqrt{7}}{7}$
5. $2\sqrt{3}$
7. 3
9. 3
11. $\dfrac{1}{3}$
13. $\dfrac{2\sqrt{3}}{3}$
15. $\dfrac{3\sqrt{2}}{2}$
17. $\dfrac{\sqrt{x}}{x}$
19. $\dfrac{\sqrt{2xy}}{y}$
21. $\dfrac{\sqrt{2y}}{y}$
23. $\dfrac{3\sqrt{7}}{5}$
25. $\dfrac{-\sqrt{2y}}{5}$
27. $\dfrac{\sqrt[3]{30}}{3}$
29. $-\dfrac{5\sqrt{x}}{x^2}$
31. $\dfrac{2\sqrt{3x}}{3y}$
33. $\dfrac{2\sqrt{b}}{b}$
35. $\dfrac{\sqrt[3]{28xy^2}}{2y^2}$
37. $\dfrac{x\sqrt[3]{3xy^2}}{3y}$
39. $\dfrac{\sqrt[3]{22}}{2}$
41. $\dfrac{\sqrt[3]{45ab^2}}{3b^2}$
43. $\dfrac{a\sqrt[3]{2ab^2}}{2b}$
45. $\sqrt{6}+2$
47. $\dfrac{-\left(\sqrt{5}+3\right)}{4}$
49. $\dfrac{-6\left(5+3\sqrt{2}\right)}{7}$
51. $\dfrac{\sqrt{3}\left(\sqrt{2}-5\right)}{23}$
53. $\dfrac{-7\left(1+3\sqrt{5}\right)}{44}$
55. $\dfrac{-\left(\sqrt{3}+\sqrt{5}\right)}{2}$
57. $5\left(\sqrt{2}-\sqrt{3}\right)$
59. $\dfrac{4\left(\sqrt{x}-1\right)}{x-1}$
61. $\dfrac{5\left(6-\sqrt{y}\right)}{36-y}$
63. $\dfrac{8\left(2\sqrt{x}-3\right)}{4x-9}$
65. $\dfrac{2\sqrt{y}\left(\sqrt{5y}+\sqrt{3}\right)}{5y-3}$
67. $\dfrac{3\left(\sqrt{x}+\sqrt{y}\right)}{x-y}$
69. $\dfrac{x\left(\sqrt{x}-2\sqrt{y}\right)}{x-4y}$
71. $-\left(\sqrt{3}+1\right)\left(\sqrt{3}+2\right)$
73. $\dfrac{-\left(\sqrt{5}-2\right)\left(\sqrt{5}-3\right)}{4}$
75. $\dfrac{\left(\sqrt{x}+1\right)^2}{x-1}$
77. $\dfrac{\left(\sqrt{x}+2\right)\left(\sqrt{3x}-y\right)}{3x-y^2}$
79. The numerator and denominator were not multiplied by the conjugate of the denominator. The conjugate is $\sqrt{3}-y$. Also, the multiplication is incorrect: $\left(\sqrt{3}+y\right)\left(\sqrt{3}+y\right)\neq 3+y^2$.

Applications

81. $r=\dfrac{\sqrt{V\pi h}}{\pi h}$
83. **a.** $\dfrac{5\sqrt{6}}{3}$ cm
    b. $\dfrac{2\sqrt{30}}{3}$ cm
    c. $\dfrac{5\sqrt{3}}{3}$ cm
85. **a.** $r=\dfrac{\sqrt{AP}}{P}-1$
    b. 5%

15.6 Exercises

Concept Check

1. isolate
3. radicals
5. extraneous
7. True
9. True

Practice

1. $x=3$
3. No solution
5. $x=-3$

7. $x = 14$

9. $x = 9$

11. No solution

13. $x = 6$

15. $x = -4, 1$

17. $x = -5, \dfrac{5}{2}$

19. $x = -2$ 21. $x = 2, 3$

23. $x = 2, 5$ 25. $x = -5, 5$

27. $x = 4$ 29. $x = 4$

31. $x = 3$ 33. $x = 2$

35. $x = 2$ 37. $x = 7$

39. $x = 4$ 41. $x = 0$

43. $x = 5$ 45. No solution

47. $x = 4$ 49. $x = -1, 3$

51. $x = 5$ 53. $x = 2$

55. $x = 1$ 57. $x = -4$

59. $x = 12$

Applications

61. **a.** $1814.06

b. $r = \sqrt{\dfrac{A}{P}} - 1$

63. **a.** 1 sec

b. 3 ft

65. 28.63 m

67. **a.** 0.64 seconds

b. 1.01 seconds

c. 1.09 seconds

15.7 Exercises

Concept Check

1. ordered
3. first coordinates
5. vertical, function
7. True
9. True

Practice

1. **a.** $\sqrt{5} \approx 2.2361$

b. 3

c. $5\sqrt{2} \approx 7.0711$

d. 2

3. **a.** 3

b. -1

c. -2

d. $2\sqrt[3]{3} \approx 2.8845$

5. $[-8, \infty)$

7. $\left(-\infty, \dfrac{1}{2}\right]$

9. $(-\infty, \infty)$

11. $(-\infty, 2]$

13. $(-\infty, \infty)$

15. $D = [-2, \infty); R = [0, \infty);$ zero $= (-2, 0)$

17. $D = [-1, \infty); R = (-\infty, 1];$ zero $= (0, 0)$

19. $D = [0, \infty); R = [3, \infty);$ zero = none

21. E

23. B

25. A

27.
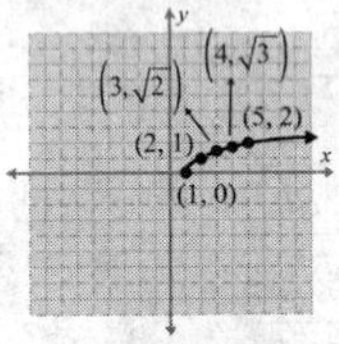

29.
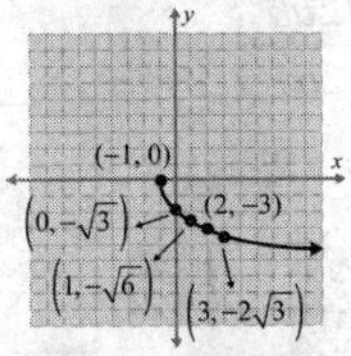

31.
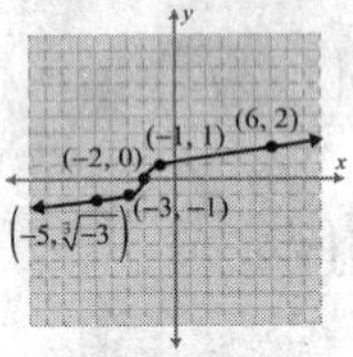

33.
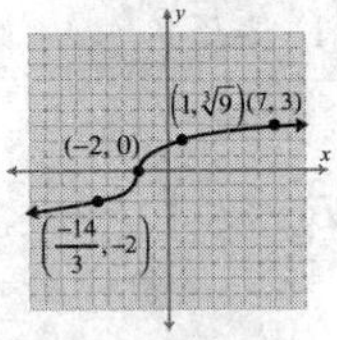

35.
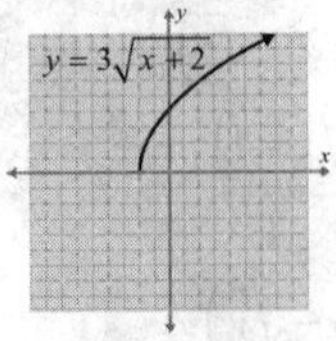

37.
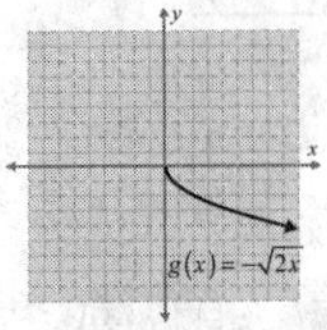

39.
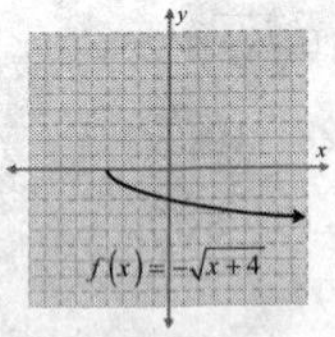

41.
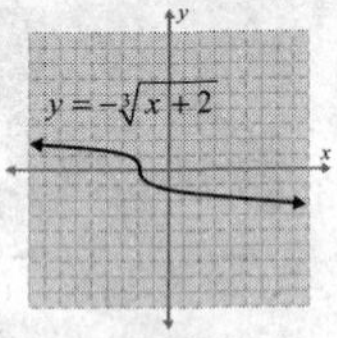

43.
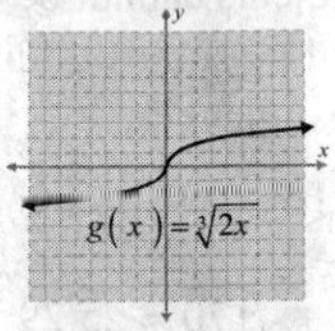

45.
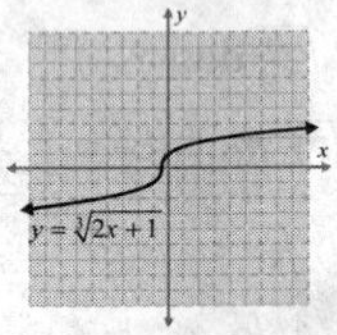

47. **a.**
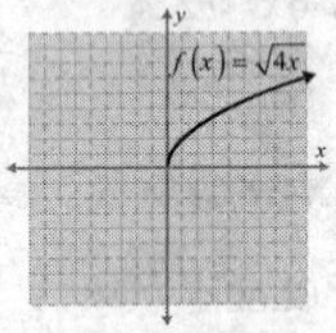

$D = [0, \infty)$; The graph of $\sqrt{x}$ is stretched vertically.

b.
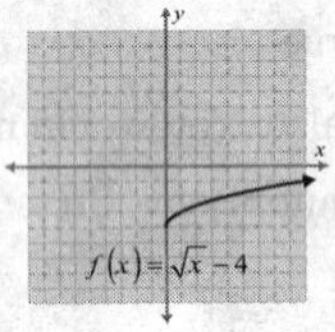

$D = [0, \infty)$; The graph of $\sqrt{x}$ is shifted down four units.

c.
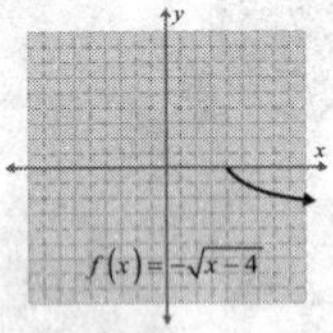

$D = [4, \infty)$; The graph of $\sqrt{x}$ is shifted to the right four units. and then reflected across the x-axis

49. A negative sign reflects the graph across the x-axis.

a. Affects how quickly the graph goes to infinity.

b. Causes the graph to move in a horizontal direction, and

c. Causes the graph to move in a vertical direction.

Applications

51. **a.**
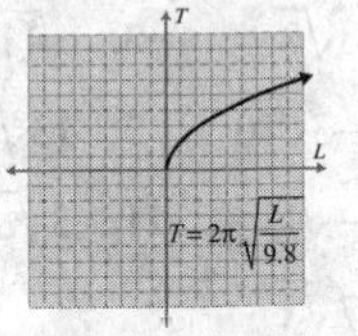

b. $D = [0, \infty); R = [0, \infty)$ Yes, this makes sense since both length and time cannot be negative

c. 7.77 seconds

Writing & Thinking

53. **a.** $\dfrac{1}{\sqrt{3+h} + \sqrt{3}}$

b. Slope of the line connecting $(3+h, f(3+h))$ and $(3, f(3))$

c. A line just touching the curve at one point

d. $\dfrac{1}{2\sqrt{3}}$; represents the slope of the line tangent to $f(x)$ at $x = 3$.

15.8 Exercises

Concept Check

1. polynomials
3. real, imaginary
5. $i\sqrt{a}$
7. True
9. False; $a = c, b = d$

Practice

1. Real part: 4, imaginary part: -3

3. Real part: -11, imaginary part: $\sqrt{2}$

5. Real part: $\dfrac{3}{8}$, imaginary part: 0

7. Real part: $\frac{4}{5}$, imaginary part: $\frac{7}{5}$

9. Real part: $\frac{2}{3}$, imaginary part: $\sqrt{17}$

11. $7i$

13. $-8i$

15. $7\sqrt{3}$

17. $10i\sqrt{6}$

19. $-12i\sqrt{3}$

21. $11\sqrt{2}$

23. $10i\sqrt{10}$

25. $x=6, y=-3$

27. $x=-2, y=\sqrt{5}$

29. $x=\sqrt{2}-3, y=1$

31. $x=1, y=4$

33. $x=2, y=-6$

35. $x=3, y=10$

37. $x=-\frac{4}{3}, y=-3$

39. $6+2i$

41. $1+7i$

43. $6-6i$

45. $14i$

47. $(3+\sqrt{5})-6i$

49. $5+(\sqrt{6}+1)i$

51. $\sqrt{3}-5$

53. $-2-5i$

55. $11-16i$

57. 2

59. $3+4i$

Writing & Thinking

61. **a.** Yes

b. No

15.9 Exercises

Concept Check

1. FOIL

3. $i, -1, -i, 1$

5. a^2+b^2

7. False; $i, -i$, 1 and -1

9. False; The product is -1.

Practice

1. $16+24i$

3. $-7\sqrt{2}+7i$

5. $3+12i$

7. $1-i\sqrt{3}$

9. $3+2i\sqrt{3}$

11. $2+8i$

13. $-7-11i$

15. $13+0i$

17. $34+13i$

19. $-24+70i$

21. $5-i\sqrt{3}$

23. $23-10i\sqrt{2}$

25. $21+0i$

27. $(2+\sqrt{10})+(2\sqrt{2}-\sqrt{5})i$

29. $(9-\sqrt{30})+(3\sqrt{5}+3\sqrt{6})i$

31. $0+3i$

33. $0-\frac{5}{4}i$

35. $-\frac{1}{4}+\frac{1}{2}i$

37. $-\frac{4}{5}+\frac{8}{5}i$

39. $\frac{24}{25}+\frac{18}{25}i$

41. $-\frac{1}{13}+\frac{5}{13}i$

43. $-\frac{1}{29}-\frac{12}{29}i$

45. $-\frac{17}{26}-\frac{7}{26}i$

47. $\frac{4+\sqrt{3}}{4}+\left(\frac{4\sqrt{3}-1}{4}\right)i$

49. $-\frac{1}{7}+\frac{4\sqrt{3}}{7}i$

51. $0+i$

53. $-1+0i$

55. $0+i$

57. $1+0i$

59. $0-i$

61. x^2+9

63. x^2+2

65. $5y^2+4$

67. $x^2+4x+40$

69. $y^2-6y+13$

Writing & Thinking

71. Given a complex number $(a+bi)$: $(a+bi)(a-bi)$
$=a^2-abi+abi-b^2i^2$
$=a^2+b^2$
which is the sum of squares of real numbers. Thus, the product must be a positive real number.

73. $a^2+b^2=1$

Chapter 16: Quadratic Equations

16.1 Exercises

Concept Check

1. hypotenuse, squares

3. nonnegative

5. True

7. False; The square of a real number cannot be negative.

Practice

1. $x=0, 11$

3. $x=-12, -3$

5. $x=1, -\frac{5}{3}$

7. $x=-1, 3$

9. $x=\frac{3}{4}, 1$

11. $x=\pm 11$

13. $x=\pm 6$

15. $x=\pm\sqrt{35}$

17. $x=\pm 5i$

19. $x=\pm\sqrt{62}$

21. $x=\pm 3\sqrt{5}$

23. $x=\pm 3\sqrt{2}$

25. $x=\pm\frac{2}{3}$

27. $x=-1, 3$

29. $x=-2\pm 5i$

31. $x=3\pm 2i$

33. $x=-\frac{3}{2}, -\frac{1}{2}$

35. $x=\frac{7}{4}, \frac{9}{4}$

37. $x=-2\pm i\sqrt{7}$

39. $x=\frac{2\pm 3\sqrt{7}}{5}$

41. $x=\frac{-4\pm 3\sqrt{3}}{3}$

43. $x=7\pm 2\sqrt{3}$

45. $x=5\pm i\sqrt{10}$

47. $x=\pm 2i\sqrt{6}$

49. $c=15$

51. $b=6\sqrt{3}$

53. $b=1$

55. $x=\pm 25.44$

57. $x=\pm 5.25$

59. $x=\pm 1.70$

61. $x=\pm 4.13$

63. 4.2361, −0.2361

65. 2.3229, −0.3229

Applications

67. The length of the leg is 4 feet and the hypotenuse is 8 feet.

69. $3\sqrt{2}$ cm

71. **a.** $35^2=x^2+34^2 \rightarrow 8.3$ ft

b. No, $\frac{1}{4}$ of 34 is 8.5, so the distance is too short by 0.2 feet.

73. 125 yards

75. **a.** 640 ft; 384 ft

b. 144 ft; 400 ft

c. 7 seconds; $0=-16(t+7)(t-7)$

77. 3.2 miles

16.2 Exercises

Concept Check

1. trinomial
3. 1
5. square
7. False; It's also possible for there to be one solution.
9. True

Practice

1. $x^2-12x+\underline{36}=\left(\underline{x-6}\right)^2$
3. $x^2+6x+\underline{9}=\left(\underline{x+3}\right)^2$
5. $x^2-5x+\underline{\frac{25}{4}}=\left(\underline{x-\frac{5}{2}}\right)^2$
7. $y^2+y+\underline{\frac{1}{4}}=\left(\underline{y+\frac{1}{2}}\right)^2$
9. $x^2+\frac{1}{3}x+\underline{\frac{1}{36}}=\left(\underline{x+\frac{1}{6}}\right)^2$
11. $2x^2+4x+2=2(x+1)^2$
13. $x=-5,1$
15. $y=-1\pm\sqrt{6}$
17. $x=5\pm\sqrt{22}$
19. $x=-5,9$
21. $x=-5,8$
23. $x=-\frac{4}{3},1$
25. $x=3\pm i$
27. $x=1,11$
29. $y=5\pm\sqrt{21}$
31. $z=\frac{-3\pm\sqrt{29}}{2}$
33. $x=\frac{-1\pm i\sqrt{7}}{2}$ or $-\frac{1}{2}\pm\frac{\sqrt{7}}{2}i$
35. $x=\frac{-5\pm\sqrt{17}}{2}$
37. $x=\frac{5\pm\sqrt{10}}{3}$
39. $x=-1\pm i\sqrt{5}$
41. $x=\frac{1\pm i\sqrt{11}}{4}$ or $\frac{1}{4}\pm\frac{\sqrt{11}}{4}i$
43. $y=\frac{-3\pm i\sqrt{11}}{2}$ or $-\frac{3}{2}\pm\frac{\sqrt{11}}{2}i$
45. $y=-\frac{4}{3},1$
47. $x=2\pm\sqrt{2}$
49. $x^2-7=0$
51. $x^2-2x-2=0$
53. $y^2+4y-16=0$
55. $x^2+16=0$
57. $y^2+6=0$
59. $x^2-4x+5=0$
61. $x^2-2x+3=0$
63. $x^2+10x+49=0$

Applications

65. **a.** $p=0$, 90

    b. No income revenue is made if the price is set to \$0 or \$90 per frame.

Writing & Thinking

67. See Example 2.

16.3 Exercises

Concept Check

1. $ax^2+bx+c=0$
3. $\frac{-b\pm\sqrt{b^2-4ac}}{2a}$
5. completing
7. True
9. False; Two real solutions

Practice

1. 68; Two real solutions
3. 0; One real solution
5. −44; Two nonreal solutions
7. 4; Two real solutions
9. 19,600; Two real solutions
11. −11; Two nonreal solutions
13. $x=-2\pm2\sqrt{2}$
15. $x=-\frac{2}{3}$
17. $x=1\pm i\sqrt{6}$
19. $x=-3,\frac{1}{2}$
21. $x=\frac{-3\pm\sqrt{5}}{4}$
23. $x=\frac{-3\pm i\sqrt{3}}{4}$ or $-\frac{3}{4}\pm\frac{\sqrt{3}}{4}i$
25. $x=\frac{-3\pm\sqrt{29}}{2}$
27. $x=-3$, 1
29. $x=\pm2i\sqrt{2}$
31. $x=\frac{5\pm\sqrt{17}}{2}$
33. $x=-\frac{1}{4}$
35. $x=\pm\frac{2\sqrt{3}}{3}$
37. $x=\frac{2}{3}$
39. $x=\frac{-4\pm i\sqrt{2}}{2}$ or $-2\pm\frac{\sqrt{2}}{2}i$
41. $x=\frac{7\pm i\sqrt{51}}{10}$ or $\frac{7}{10}\pm\frac{\sqrt{51}}{10}i$
43. $x=-2,\frac{5}{3}$
45. $x=3\pm i\sqrt{2}$
47. $x=-1,0$
49. $x=\frac{9\pm\sqrt{65}}{2},0$
51. $x=\frac{-3\pm\sqrt{5}}{2},0$
53. $x=\frac{1}{2},-3$
55. $x=-\frac{1}{3},3$
57. $x=\frac{2\pm\sqrt{3}}{3}$
59. $x=\frac{1\pm i\sqrt{3}}{6}$
61. $x=\frac{2\pm i\sqrt{2}}{2}$
63. $x=\frac{-7\pm\sqrt{17}}{4}$
65. $c<16$
67. $c=\frac{81}{4}$
69. $a>3$
71. $a>-\frac{1}{36}$
73. $a=\frac{49}{48}$
75. $c>\frac{4}{3}$
77. $x\approx2.5993, 60.4007$
79. $x\approx-0.7862, 2.0110$
81. $x\approx-4.1334, -0.5806$
83. $x\approx-2.6933, 2.6933$

Applications

85. **a.** 3.35 seconds

    b. 4.15 seconds

    c. 0.80 seconds

Writing & Thinking

87. $x^4-13x^2+36=0$; multiplied $(x-2)(x+2)(x-3)(x+3)$

16.4 Exercises

Concept Check

1. read, think, translate
3. diagram
5. True

Applications

1. $w(2w-5)=63$; $w=7$, the rectangle is 7 m by 9 m
3. $x+(x+9)^2=147$; $x=3$, the numbers are 3 and 12
5. $(w+12)(w+22)=1344$; $w=20$, the pool is 20 ft by 30 ft
7. $(x+5)^2=4x^2$; $x=5$, the side of the original square is 5 m
9. $w^2+(w+4)^2=400$; $w=12$, the rectangle is 12 m by 16 m
11. $x(x+16)=960$; $x=24$, there are 40 seats

13. **a.** $x(29-x)=198$

b. $x=11$ or $x=18$

c. Either length = 11 meters and width = 18 meters, or length = 18 meters and width = 11 meters.

15. $(9+2x)(12+2x)-(9)(12)=162$; $x=3$, the frame is 3 in. thick

17. $40I-4I^2=100$; $I=5$, It needs a current of 5 amperes

19. $x(40-x)=336$; $x=12$, 28; He must sell 12 signs

21. **a.** \$307.20

b. $(600+20x)(0.5-0.01x)=315$; $x=5$, 15; He must charge 45 cents or 35 cents

23. $(15-x)(700+70x)=10{,}920$; $x=2$ or 3, the rental rate is \$840 or \$910.

25. $\frac{8}{x+2}+\frac{4}{x-2}=2$; $x=6$, the speed of the boat in still water is 6 mph

27. $\frac{540}{x-9}-\frac{540}{x}=2$; their average speed to San Francisco was 54 mph

29. $\frac{120}{x}+2=\frac{120}{x-5}$; $x=20$, there were initially 20 members

31. $3w(w+6)=336$; $w=8$, the sheet metal was 14 in. by 20 in.

33. $\frac{1}{x}+\frac{1}{x+8}=\frac{1}{3}$; $x=4$, the mother would take 4 hours

35. $\frac{1}{x}+\frac{1}{x+30}=\frac{1}{8}$; $x=10$; the smaller pipe would take 40 min

37. **a.** $\frac{1}{x}+\frac{1}{x+6}=\frac{1}{4}$

Persons	Time of work (in Hours)	Part of work Done in 1 Hour
Jack	x	$\frac{1}{x}$
Diane	$x+6$	$\frac{1}{x+6}$
Together	4	$\frac{1}{4}$

b. $x=-4, 6$

c. 6 hours; Time cannot be negative in this context.

d. It would take Jack 6 hours to decorate the nursery and it would take Diane 12 hours to decorate the nursery.

39. **a.** 4 sec

b. No, the projectile's maximum height is 256 ft.

c. 3 seconds, 5 seconds

d. 8 sec

41. 34.6 ft

Writing & Thinking

43. a cannot be equal to zero and b^2 must be greater than or equal to $4ac$ to produce a real solution. Also, all the solutions found may not apply to the problem at hand. You must check that each answer makes sense in the context of the problem.

16.5 Exercises

Concept Check

1. middle, one-half

3. first

5. True

7. True

Practice

1. $x=\pm2, \pm3$

3. $x=\pm2, \pm\sqrt{5}$

5. $y=\pm\sqrt{7}, \pm2i$

7. $y=\pm\sqrt{5}, \pm i\sqrt{5}$

9. $x=4, \frac{25}{4}$

11. $x=1, 4$

13. $x=-8, \frac{1}{8}$

15. $x=-27, -8, 0$

17. $x=-17, -2$

19. $x=-\frac{7}{2}, -3$

21. $x=0, 8$

23. $x=-1, -\frac{1}{2}$

25. $x=\pm\sqrt{1+i}, \pm\sqrt{1-i}$

27. $x=\pm\sqrt{1+3i}, \pm\sqrt{1-3i}$

29. $x=\pm\sqrt{2+i\sqrt{3}}, \pm\sqrt{2-i\sqrt{3}}$

31. $x=\frac{1}{7}, \frac{1}{5}$

33. $x=-\frac{1}{3}, \frac{3}{8}$

35. $x=\frac{1}{25}$

37. $x=\pm\frac{\sqrt{5}}{5}, \pm1$

39. $x=\pm\frac{\sqrt{6}}{2}, \pm\frac{1}{3}i$

41. $x=-\frac{1}{4}$

43. $x=-4, 1$

45. $x=-\frac{1}{2}$

47. $x=-7, -\frac{3}{2}$

49. $x=\frac{26}{5}$

51. $x=0, \pm2\sqrt{2}, \pm2i\sqrt{2}$

53. $x=2, -1\pm i\sqrt{3}$

55. $x=-10, 0, 5\pm5i\sqrt{3}$

Writing & Thinking

57. **a.** $l^2-l-1=0$, $l=\frac{1+\sqrt{5}}{2}$ which is the golden ratio

b. 97.08 feet

c. Rectangle B is "golden."

16.6 Exercises

Concept Check

1. parabola

3. $\geq$

5. range

7. False; It is a horizontal shift (or horizontal translation).

Practice

1. **a.**

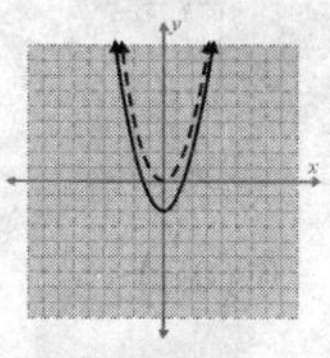

b.

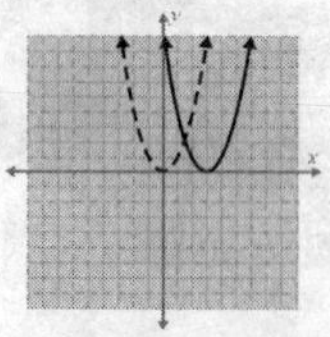

c.

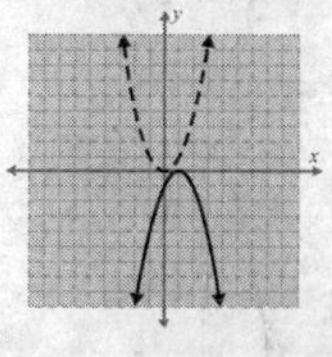

d.

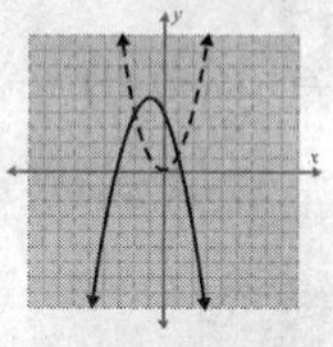

3. **a.**

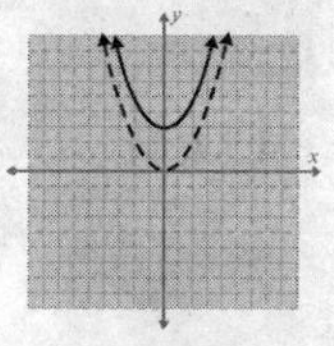

b.

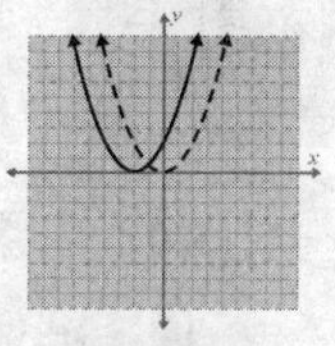

c.

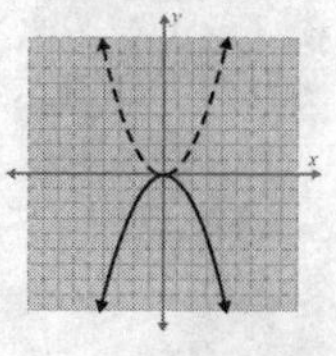

d.

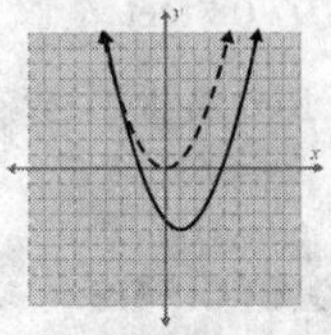

5. $x=0; (0,-4)$

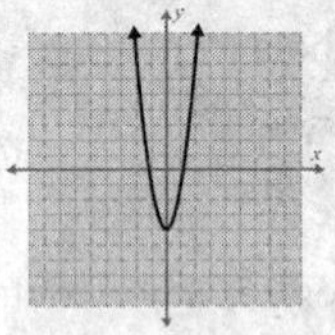

7. $x=0; (0,-9)$

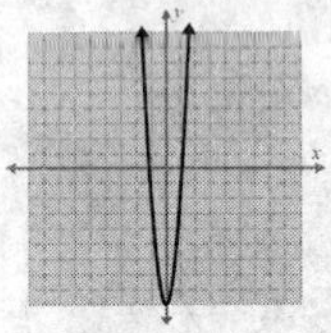

9. $x=0; (0,1)$

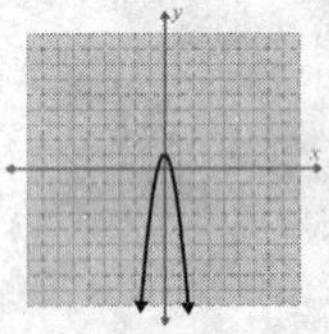

11. $x=0; (0,5)$

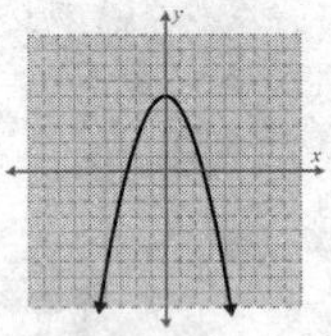

13. $x=-1; (-1,0)$

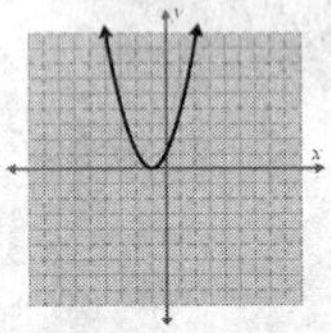

15. $x=4; (4,0)$

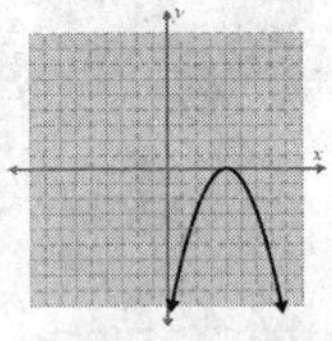

17. $x=5; (5,0)$

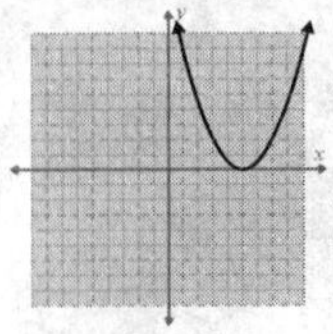

19. $x=6; (6,0)$

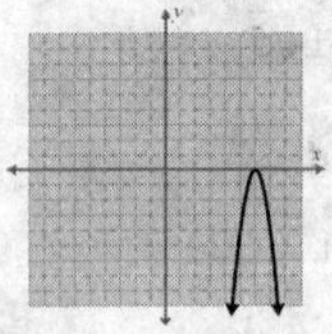

21. $x=-3; (-3,-2)$

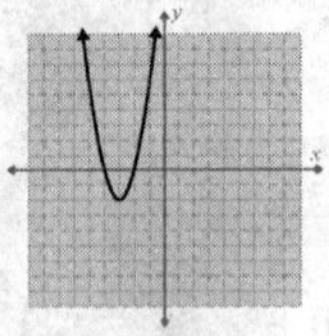

23. $x=-2; (-2,-6)$

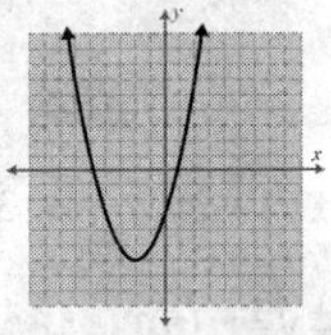

25. $x=-1; (-1,-2)$

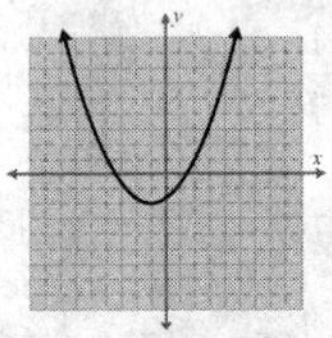

27. $x=3; (3,3)$

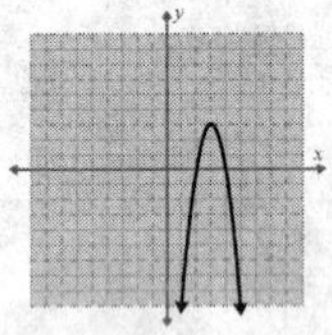

Writing & Thinking

29. The parabola is narrower for larger values of a because the value of y increases faster for increasing values of x.

16.7 Exercises

Concept Check

1. vertex

3. x-axis

5. downward, highest, maximum

7. False; The line of symmetry is at $x=-\frac{b}{2a}$.

9. True

Practice

1. $y=2(x-1)^2; x=1;$
Vertex: $(1,0)$; x-int: $(1,0)$;
y-int: $(0,2)$

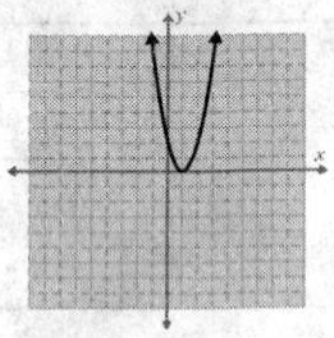

3. $y=(x-1)^2-4; x=1;$
Vertex: $(1,-4)$;
x-int: $(-1,0),(3,0)$;
y-int: $(0,-3)$

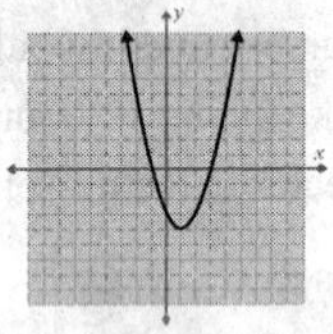

5. $y=(x+3)^2-4; x=-3;$
Vertex: $(-3,-4)$;
x-int: $(-5,0),(-1,0)$;
y-int: $(0,5)$

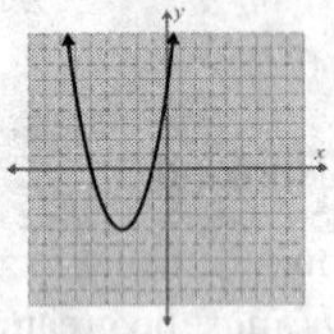

7. $y=2(x-2)^2-3; x=2;$
Vertex: $(2,-3)$;
x-int: $\left(\frac{4+\sqrt{6}}{2},0\right)$, $\left(\frac{4-\sqrt{6}}{2},0\right)$;
y-int: $(0,5)$

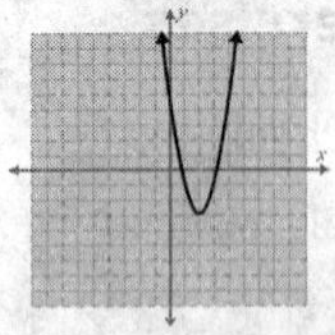

9. $y=-3(x+2)^2+3;$
$x=-2$; Vertex: $(-2,3)$;
x-int: $(-1,0),(-3,0)$;
y-int: $(0,-9)$

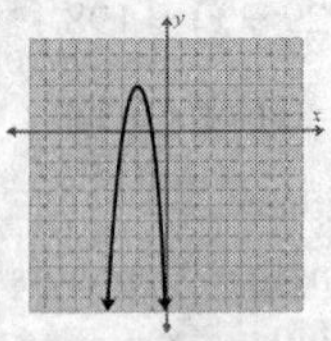

11. $y=5(x-1)^2+3;$
$x=1$; Vertex: $(1,3)$;
x-int: None; y-int: $(0,8)$

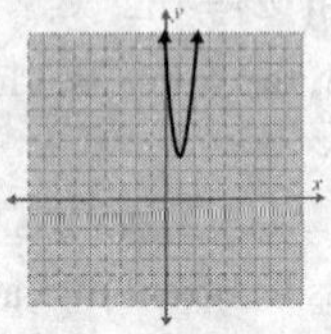

13. $y=-\left(x+\frac{5}{2}\right)^2+\frac{17}{4};$
$x=-\frac{5}{2};$
Vertex: $\left(-\frac{5}{2},\frac{17}{4}\right)$;
x-int: $\left(\frac{-5+\sqrt{17}}{2},0\right)$, $\left(\frac{-5-\sqrt{17}}{2},0\right)$;
y-int: $(0,-2)$

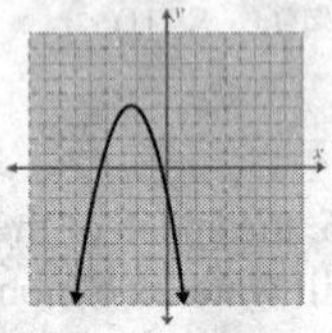

15. $y=2\left(x+\frac{7}{4}\right)^2-\frac{9}{8};$
$x=-\frac{7}{4};$
Vertex: $\left(-\frac{7}{4},-\frac{9}{8}\right)$;
x-int: $\left(-\frac{5}{2},0\right),(-1,0)$;
y-int: $(0,5)$

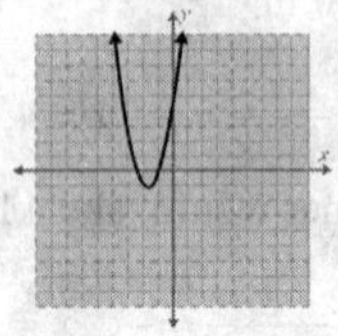

17. $x = 1$; Vertex: $(1, 0)$; x-int: $(1, 0)$; y-int: $(0, -3)$

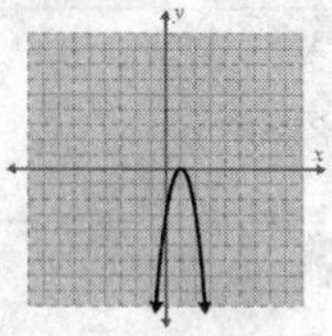

19. $x = 2$; Vertex: $(2, -1)$; x-int: $(3, 0), (1, 0)$; y-int: $(0, 3)$

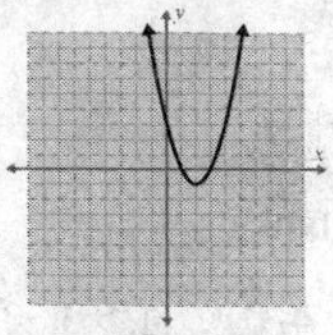

21. $x = 2$; Vertex: $(2, -1)$; x-int: None; y-int: $(0, -9)$

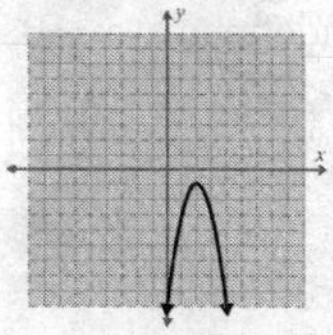

23. $x = -1$; Vertex: $(-1, 0)$; x-int: $(-1, 0)$; y-int: $(0, 1)$

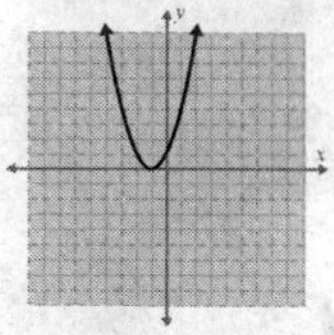

25. $x = -1$; Vertex: $(-1, -1)$; x-int: None; y-int: $(0, -2)$

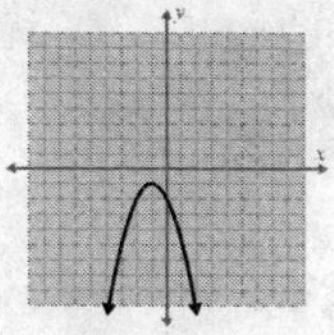

27. **a.** No **b.** Yes

c. One function is a reflection of the other across the x-axis.

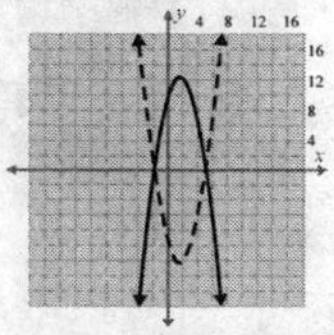

29. **a.** No **b.** Yes

c. One function is a reflection of the other across the x-axis.

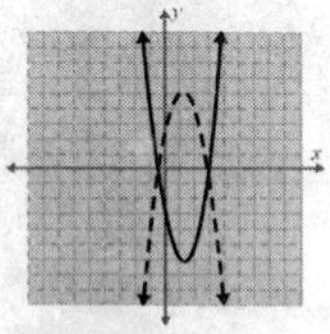

31. Zeros: $x \approx -0.7321$, 2.7321

33. Zeros: $x \approx -1.1583$, 2.1583

35. No real zeros

37.

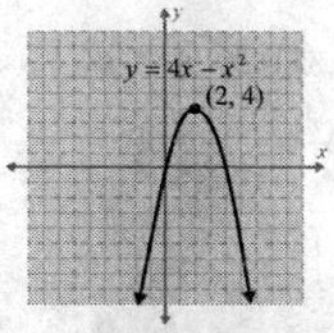

39.

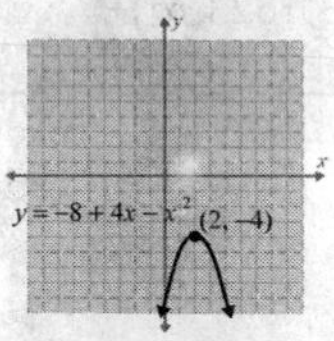

41.

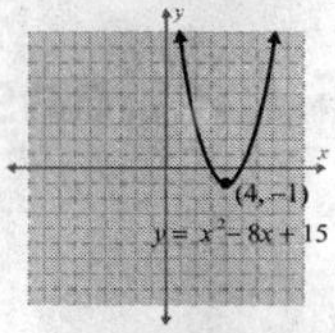

43.

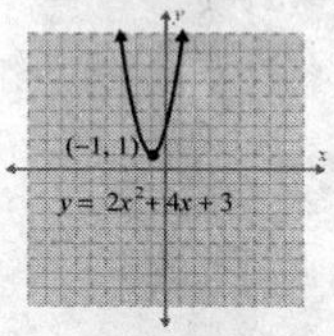

Applications

45. **a.** 3.5 s **b.** 196 ft

47. **a.** 5 s **b.** 420 ft

49. **a.** \$50 **b.** \$2500

51. \$20

53. **a.** Two sides are 150 yards and one side is 300 yards;

b. 45,000 yd^2

Writing & Thinking

55. **a.** Parabola **b.** $x = -\dfrac{b}{2a}$

c. $x = -\dfrac{b}{2a}$

d. No. A graph can be entirely above or below the x-axis.

16.8 Exercises

Concept Check

1. graphs

3. endpoints

5. bracket, parenthesis

7. False; The goal is to get 0 on one side of the inequality and factor the other side.

9. False; The solution consists of all intervals where the test points satisfy the original inequality.

Practice

1. $(-2, 6)$

3. $\left(-\infty, \dfrac{2}{3}\right) \cup (5, \infty)$

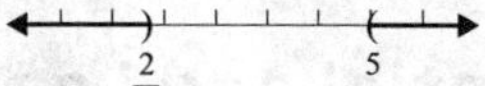

5. $(-\infty, -7] \cup \left[\dfrac{5}{2}, \infty\right)$

7. $\left[-2, -\dfrac{1}{3}\right]$

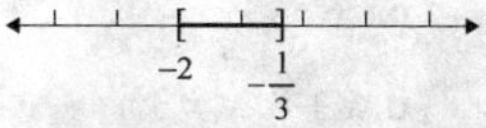

9. $\left(-\infty, -\dfrac{4}{3}\right) \cup (0, 5)$

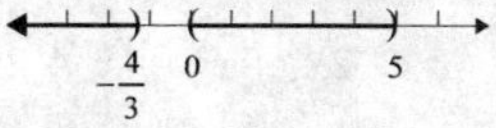

11. $\{-2\}$

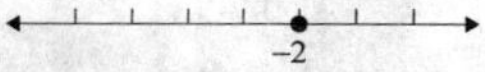

13. $\left(-\infty, -\dfrac{5}{2}\right) \cup (3, \infty)$

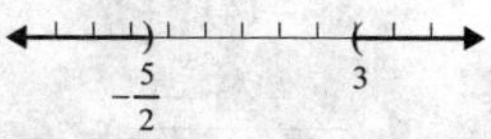

15. $\left(-\dfrac{1}{4}, \dfrac{3}{2}\right)$

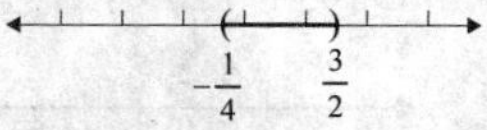

17. $\left(-\infty, \dfrac{1}{2}\right] \cup [2, \infty)$

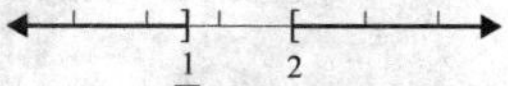

19. $\left(-\dfrac{2}{3}, -\dfrac{1}{2}\right)$

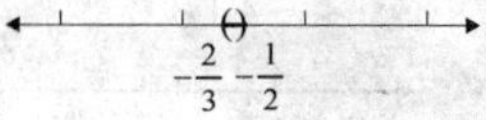

21. $\left(-\infty, \dfrac{5}{2}\right) \cup \left(\dfrac{5}{2}, \infty\right)$

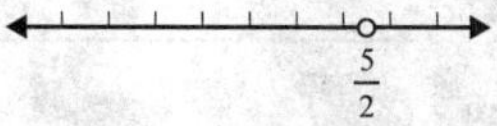

23. $\left[-\dfrac{5}{2}, \dfrac{7}{4}\right]$

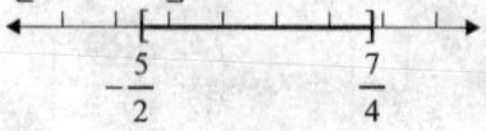

25. $(-1, 0) \cup (3, \infty)$

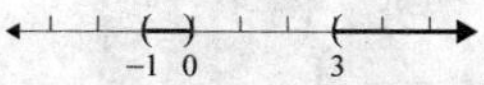

27. $(0, 1) \cup (4, \infty)$

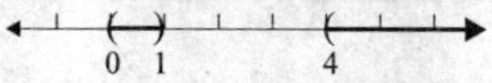

29. $(-\infty, -1) \cup (4, \infty)$

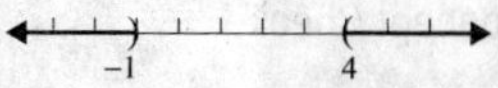

31. $(-\infty, -2) \cup (-1, 1) \cup (2, \infty)$

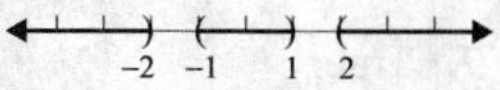

33. $[-3, -2] \cup [2, 3]$

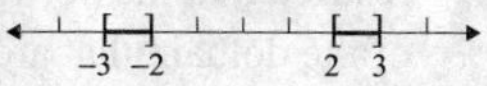

35. $(-\infty, -4] \cup [2, \infty)$

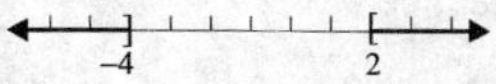

37. $\left(-\dfrac{2}{3}, 2\right)$

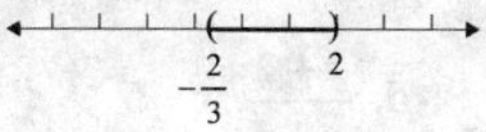

39. $\left(-\infty, -1 - \sqrt{5}\right) \cup \left(-1 + \sqrt{5}, \infty\right)$

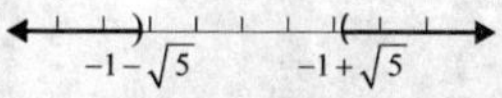

41. $(-\infty, -3-\sqrt{2}] \cup [-3+\sqrt{2}, \infty)$

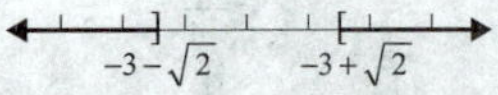

43. $\left(\dfrac{-5-\sqrt{13}}{6}, \dfrac{-5+\sqrt{13}}{6}\right)$

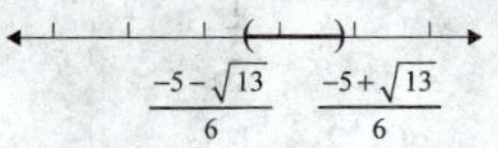

45. $\left(-\infty, -\dfrac{1}{2}\right] \cup [0, 4]$

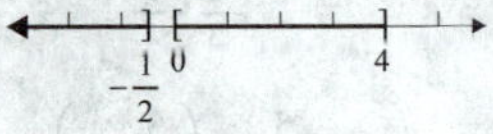

47. $(-\infty, \infty)$

49. $\varnothing$

51. **a.** $x = \dfrac{7 \pm \sqrt{89}}{2}$

b. $\left(-\infty, \dfrac{7-\sqrt{89}}{2}\right) \cup \left(\dfrac{7+\sqrt{89}}{2}, \infty\right)$

c. $\left(\dfrac{7-\sqrt{89}}{2}, \dfrac{7+\sqrt{89}}{2}\right)$

53. **a.** $x = -5, 1$

b. $(-5, 1)$

c. $(-\infty, -5) \cup (1, \infty)$

55. $(-\infty, -4] \cup [0, \infty)$

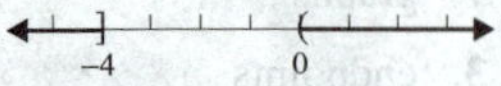

57. $(-\infty, -6)$

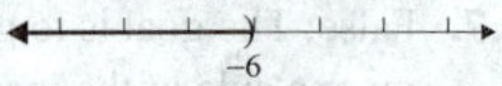

59. $(-\infty, -9) \cup (-3, \infty)$

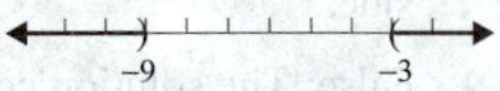

61. $\left(2, \dfrac{5}{2}\right)$

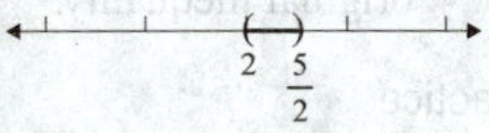

63. $(7, \infty)$

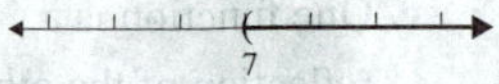

65. $\left[\dfrac{7}{5}, 4\right)$

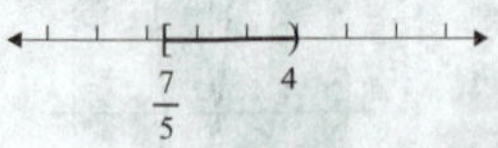

67. $\left(-9, -\dfrac{4}{3}\right)$

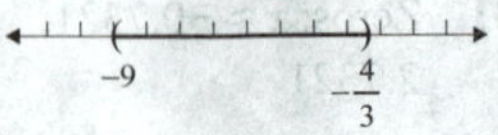

69. $(-\infty, -4] \cup [0, 3)$

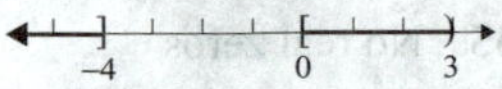

71. $(-2, 0) \cup [5, \infty)$

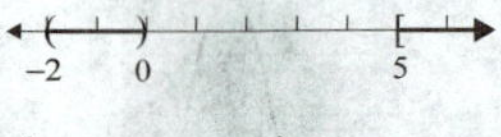

73. $(-\infty, -3.1623) \cup (3.1623, \infty)$

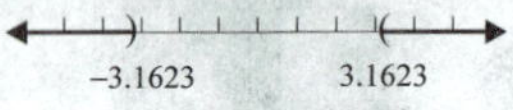

75. $\varnothing$

77. $(-\infty, -3) \cup (0, 3)$

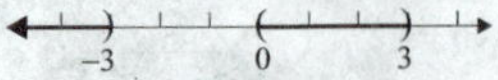

79. $[-1.8868, \infty)$

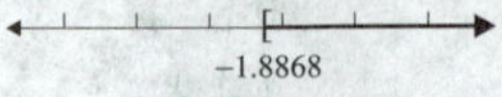

81. $(-2.3344, -0.7420) \cup (0.7420, 2.3344)$

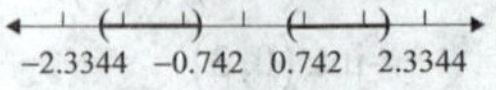

Applications

83. $[3, 9]$

Writing & Thinking

85. **a.** $(-4, 0) \cup (1, \infty)$

b. $(-\infty, -4) \cup (0, 1)$

c. The function is undefined at $x = 0$.

Chapter 17: Exponential and Logarithmic Functions

17.1 Exercises

Concept Check

1. composition, inverses
3. domain
5. y-values, x
7. True
9. False; The operations are restricted to the portions of the domain that are in common.

Practice

1. **a.** $2x - 3$

b. 7

c. $x^2 - 3x - 10$

d. $\dfrac{x+2}{x-5}, x \neq 5$

3. **a.** $x^2 + 3x - 4$

b. $x^2 - 3x + 4$

c. $3x^3 - 4x^2$

d. $\dfrac{x^2}{3x-4}, x \neq \dfrac{4}{3}$

5. **a.** $x^2 + x - 12$

b. $x^2 - x - 6$

c. $x^3 - 3x^2 - 9x + 27$

d. $x + 3, x \neq 3$

7. **a.** $3x^2 + x + 2$

b. $x^2 + x - 2$

c. $2x^4 + x^3 + 4x^2 + 2x$

d. $\dfrac{2x^2 + x}{x^2 + 2}$

9. **a.** $2x^2 + 2$

b. $8x$

c. $x^4 - 14x^2 + 1$

d. $\dfrac{x^2 + 4x + 1}{x^2 - 4x + 1}, x \neq 2 \pm \sqrt{3}$

11. 9

13. $-a^2 - a - 1$

15. 27

17. $\dfrac{8}{5}$

19. -31

21. $\sqrt{2x-6} + x + 4$, $D = [3, \infty)$

23. $3x^2 - 19x - 14$; $D = (-\infty, \infty)$

25. $\dfrac{x-5}{\sqrt{x+3}}$; $D = (-3, \infty)$

27. $-3x\sqrt{x-3}$; $D = [3, \infty)$

29. $\sqrt[3]{x+3} + \sqrt{5+x}$; $D = [-5, \infty)$

31. **a.**

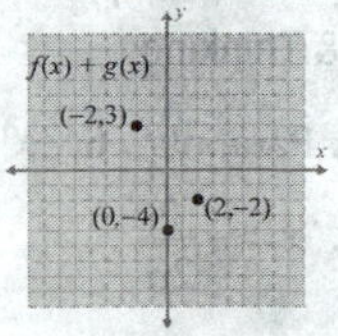

b.

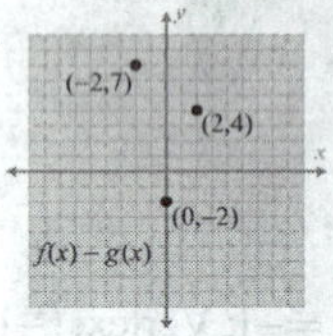

33. **a.**

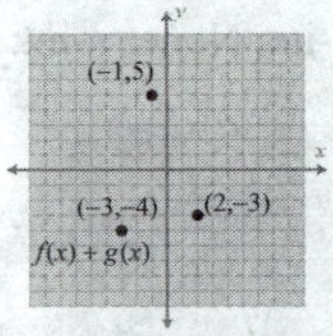

b.

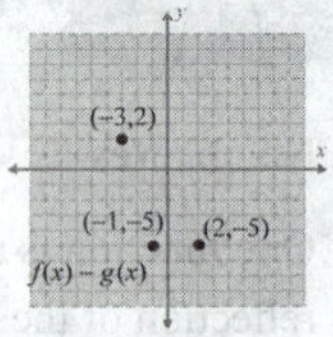

35. **a.**

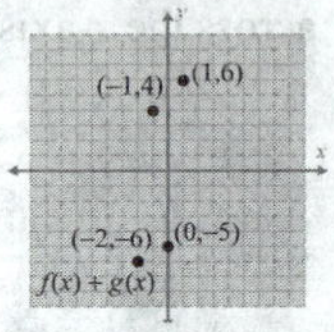

b.

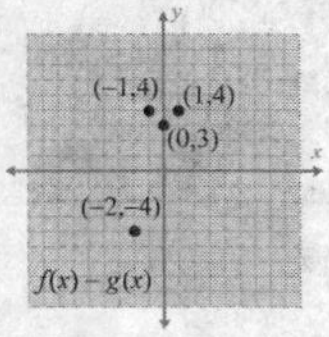

37. a.

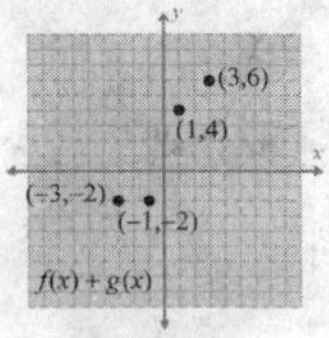

b.

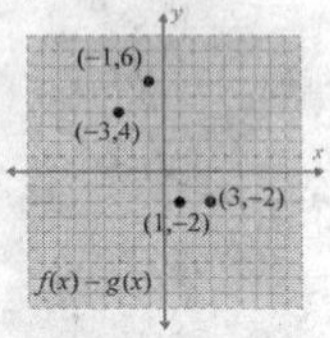

39. a.

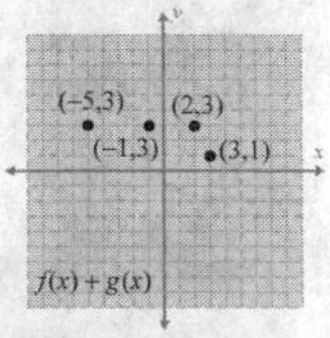

b.

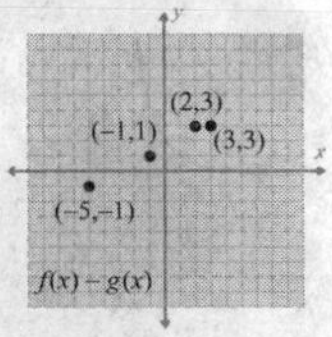

41.

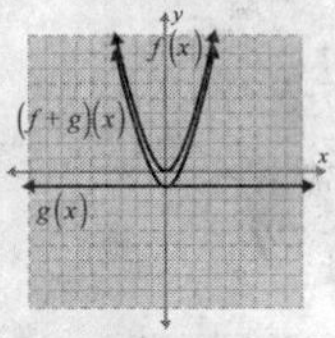

43.

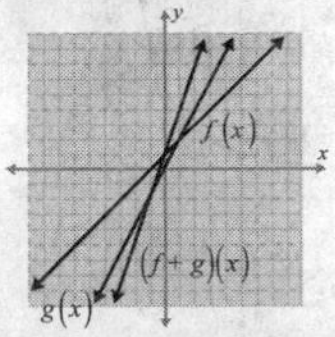

45.

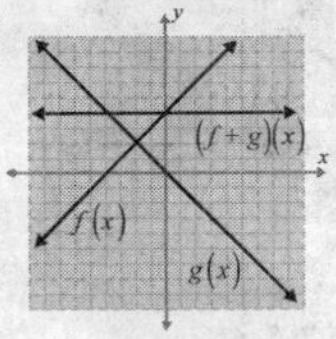

47.

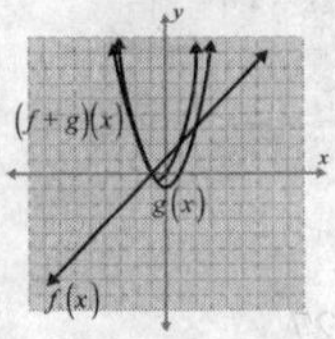

49.

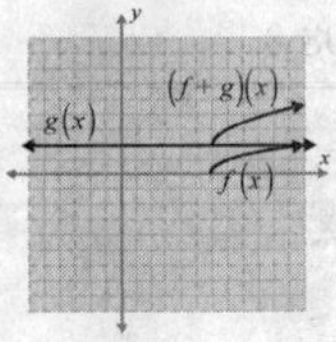

51. 4

53. −12

55. $-\frac{3}{7}$

57.

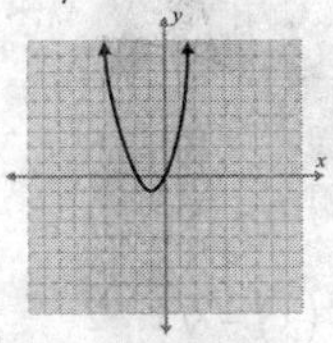

59.

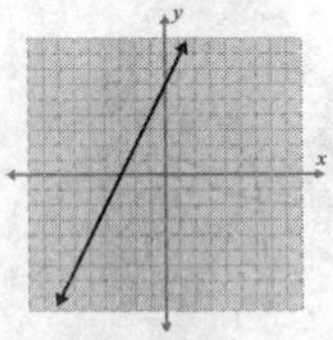

61.

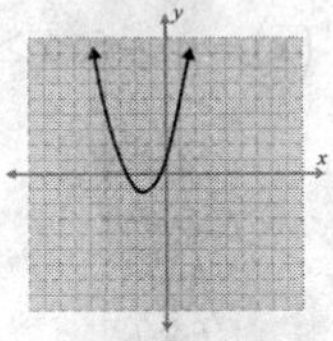

63.

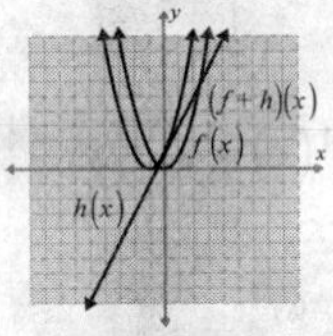

65.

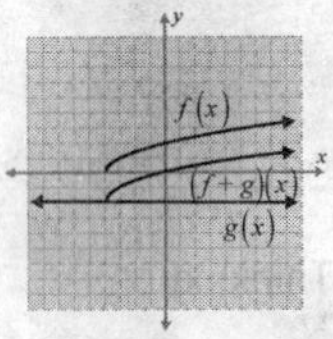

67.

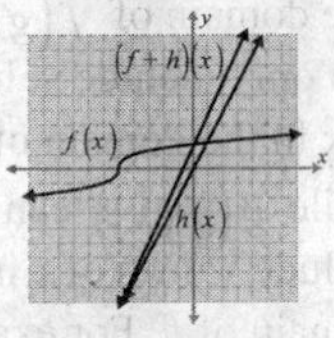

69.

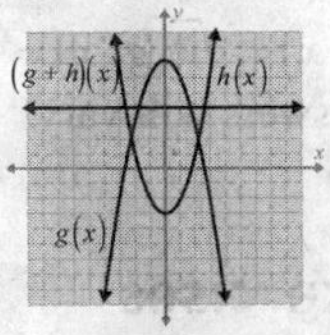

Writing & Thinking

71. In general, subtraction is not commutative. Answers will vary.

73. a.

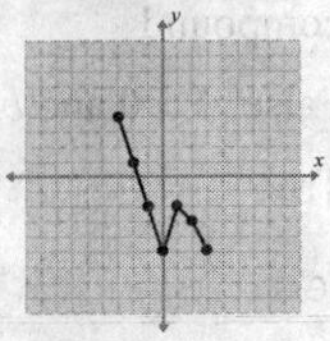

b.

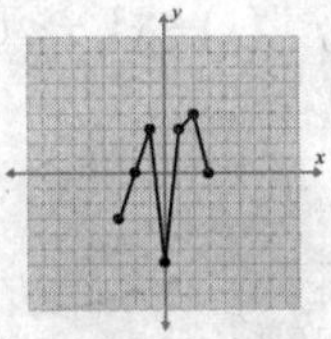

c. $\frac{f}{g}$ is undefined for $x = -2$ as $g(-2) = 0$.

17.2 Exercises

Concept Check

1. variable, variable
3. $\neq$
5. one-to-one
7. False; The vertical line test determines whether a graph represents a function.
9. True

Practice

1. a. $8r - 5$

b. $24a - 13$

3. a. $5x^2 - 20x + 24$

b. $45n^4 + 4$

5. a. $3n^2 - 6n - 9$

b. $48y^6 + 24y^3 - 9$

7. $f(g(2)) = 14$

$g(f(2)) = \frac{15}{2}$

9. $(f \circ g)(-5) = 49$

$(g \circ f)(-1) = 5$

11. $f(g(x)) = \sqrt{x^2} = |x|$

$g(f(x)) = (\sqrt{x})^2 = x$

13. $f(g(x)) = \sqrt{x - 2}$

$g(f(x)) = \sqrt{x} - 2$

15. $f(g(x)) = \frac{1}{x^2} - 1$

$g(f(x)) = \frac{1}{(x-1)^2}$

17. $f(g(x)) = x^3 + 3x^2 + 4x + 3$

$g(f(x)) = x^3 + x + 2$

19. $f(g(x)) = \frac{1}{|x|}$

$g(f(x)) = \frac{1}{x}$

21.

$f(g(x)) = \frac{1}{x^2 + 7x - 8}$

$g(f(x)) = \left(\frac{1}{x}\right)^2 + 7\left(\frac{1}{x}\right) - 8$

23. $f(g(x)) = (2x - 6)^{3n}$

$g(f(x)) = 2x^{3n} - 6$

25. $f(g(x)) = (x - 8)^{\frac{3}{2}}$

$g(f(x)) = \sqrt{x^3 - 8}$

27. a. 21

b. 2

c. No; $f(g(x)) \neq x$ and $g(f(x)) \neq x$

29. a. −9

b. does not exist

c. $f(4) = \frac{1}{9}$ and $\frac{1}{9}$ is in the domain of *g*, so $g(f(4))$ is defined. However, $g(2) = -\frac{1}{2}$ and $-\frac{1}{2}$ is not in the domain of *f*, so $f(g(2))$ is not defined.

31.

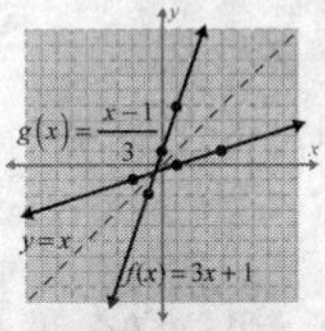

33.

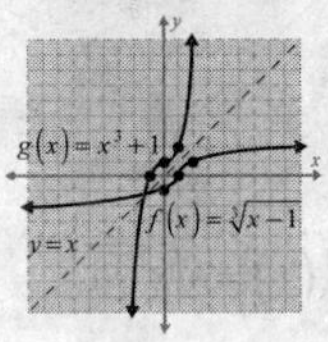

35.

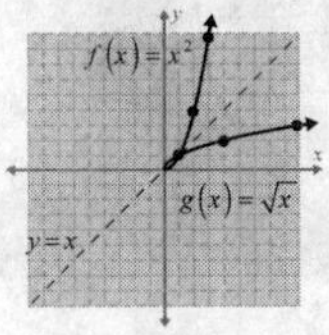

37.

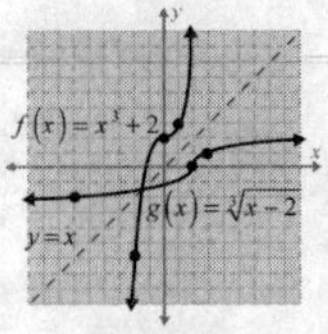

39.

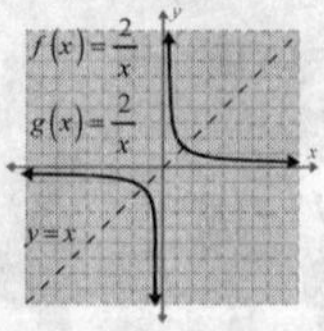

41. $f^{-1}(x)=\dfrac{x+3}{2}$

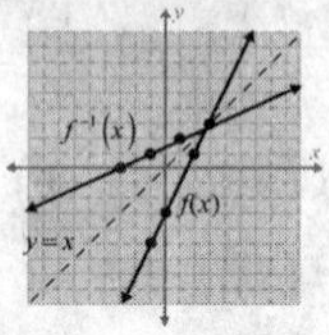

43. $g^{-1}(x)=x$

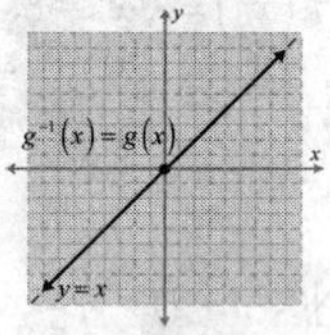

45. $f^{-1}(x)=\dfrac{x-1}{5}$

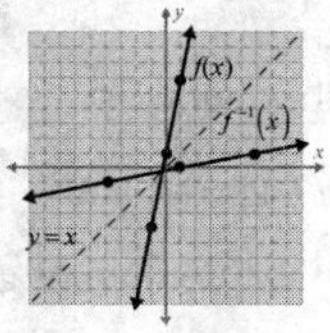

47. $g^{-1}(x)=\dfrac{3(x-2)}{2}$

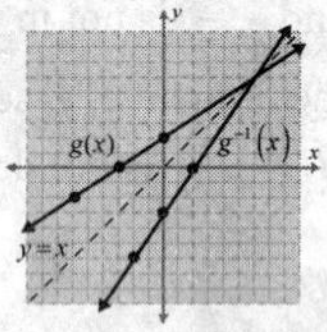

49. $f^{-1}(x)=-x-2$

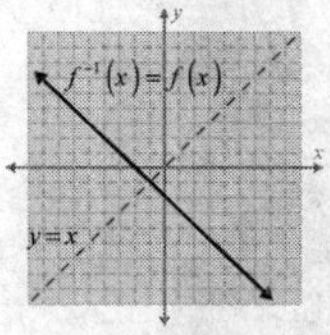

51. $f^{-1}(x)=\sqrt{x-1}$

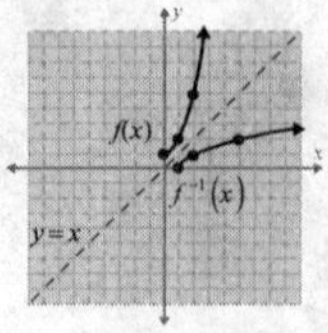

53. $f^{-1}(x)=x^2, x\le 0$

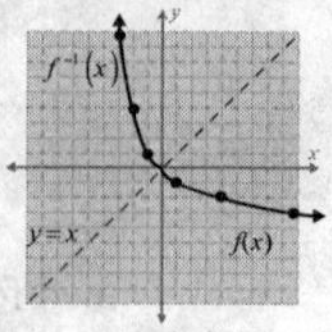

55. One-to-one

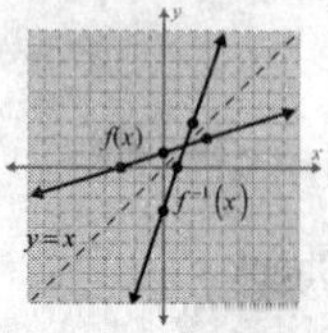

57. Not one-to-one

59. Not one-to-one

61. One-to-one

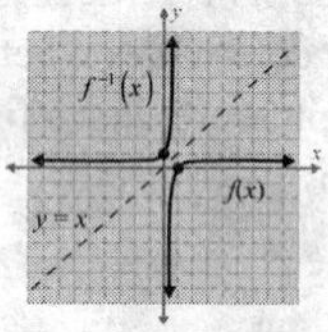

63. One-to-one

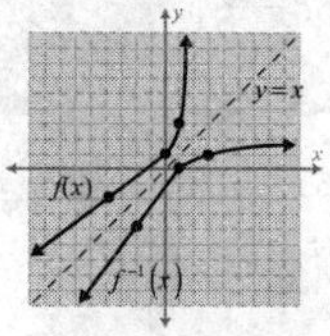

65. One-to-one

67. Not one-to-one

69. One-to-one

71. One-to-one

73. One-to-one

75. Not one-to-one

77. $f^{-1}(x)=\sqrt[3]{x}$

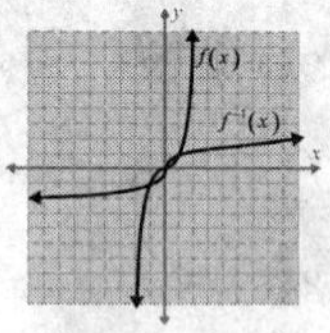

79. $f^{-1}(x)=\dfrac{1}{x}+3$

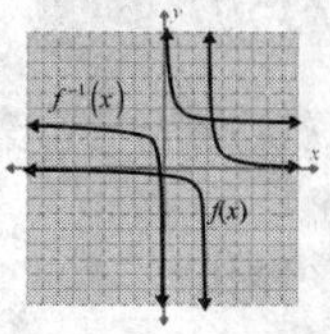

81. $f^{-1}(x)=\sqrt{x}$

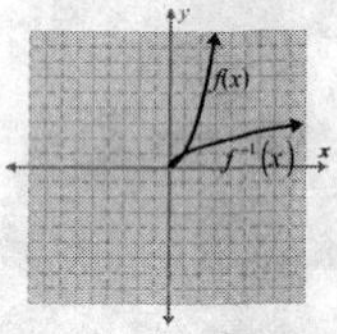

83. $g^{-1}(x)=\sqrt[3]{x-2}$

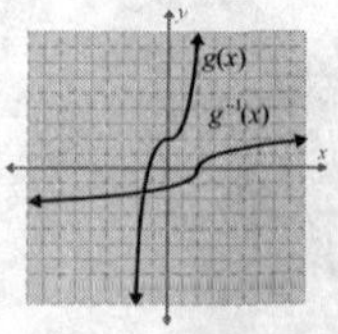

85. $f^{-1}(x)=x^2-5,\ x\ge 0$

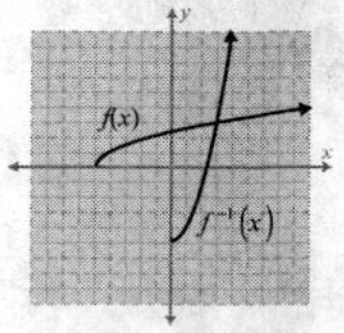

87. $f^{-1}(x)=\sqrt{1-x}$

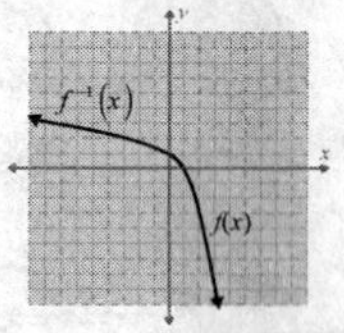

Writing & Thinking

89. The domain of $f(g(x))$ can only include values of x in the domain of g, while $g(f(x))$ can only include values of x in the domain of f. For example, $f(x)=\sqrt{x}, g(x)=-x$. Answers will vary.

17.3 Exercises

Concept Check

1. constant, variable

3. 0, 1

5. compound

7. False; $b>0$ and $b\ne 1$

9. True

Practice

1.

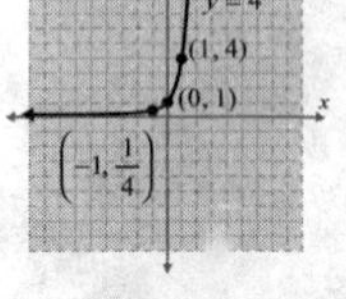

3.

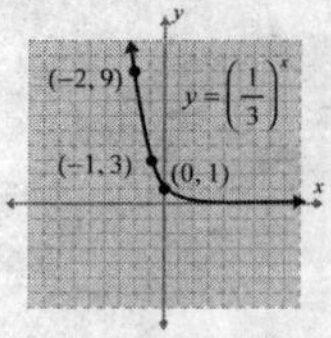

5.

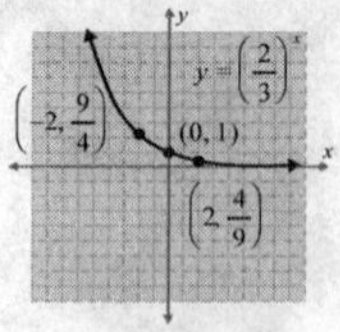

7.

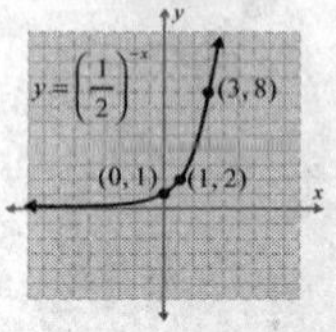

9.

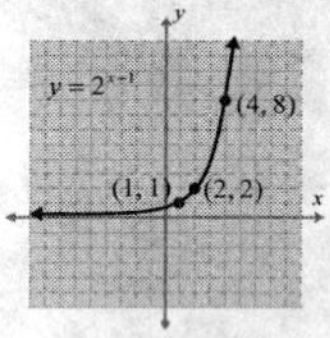

11.

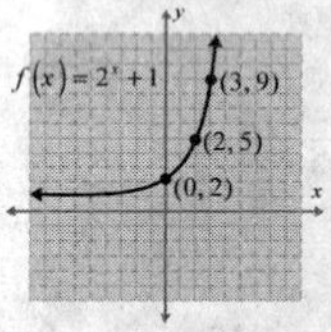

13.

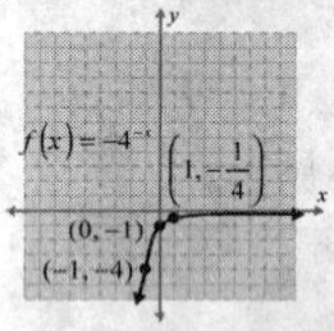

15.

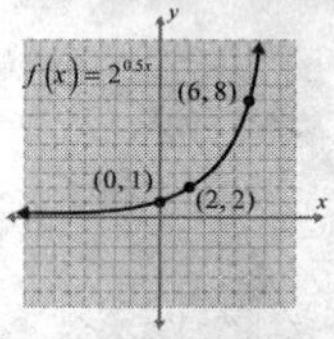

17.

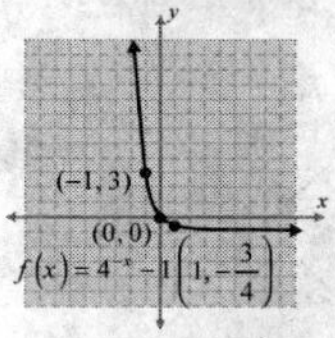

19.

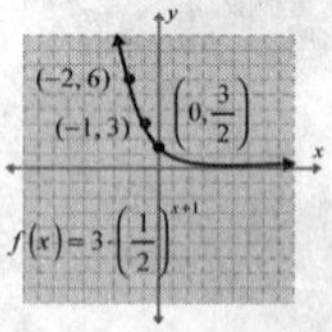

21. 48

23. 12.27

25. 4108.87

27. a.

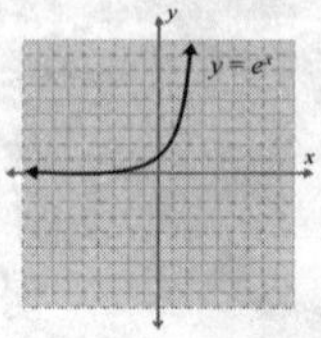

b.

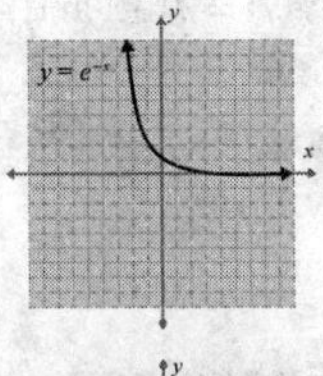

c.

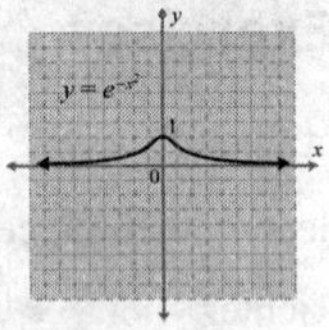

Applications

29. 20,000 bacteria

31. **a.** \$2621.59

b. \$2633.62

c. \$2639.86

d. \$2646.19

e. \$2646.26

33. \$3210.06

35. \$53.33

37. **a.** \$100,149.34

b. \$222,886.46

c. \$1,103,963.86

39. **a.** \$63,890.56

b. \$17,182.82

c. Answers will vary.

41. \$1228.25

43. $y = 10{,}000 \cdot 2^t$

Writing & Thinking

45. The graphs are reflections of each other across the x-axis.

17.4 Exercises

Concept Check

1. $\log_b x$

3. logarithmic

5. reflecting

7. True

9. True

Practice

1. $\log_7 49 = 2$

3. $\log_5 \frac{1}{25} = -2$

5. $\log_\pi 1 = 0$

7. $\log_{10} 100 = 2$

9. $\log_{10} 23 = k$

11. $\log_{\frac{2}{3}} \frac{4}{9} = 2$

13. $3^2 = 9$

15. $9^{\frac{1}{2}} = 3$

17. $7^{-1} = \frac{1}{7}$

19. $10^{1.74} = N$

21. $b^4 = 18$

23. $n^x = y^2$

25. 4

27. 0

29. -3

31. $x = 16$

33. $x = 2$

35. $x = -3$

37. $x = \frac{1}{6}$

39. $x = 2$

41. $x = 32$

43. $x = 3.7$

45. $x = 2$

47.

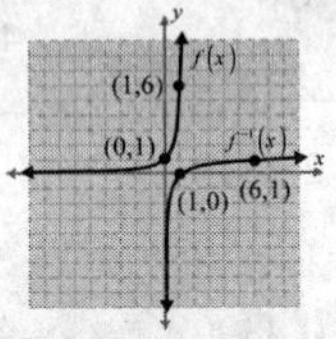

49.

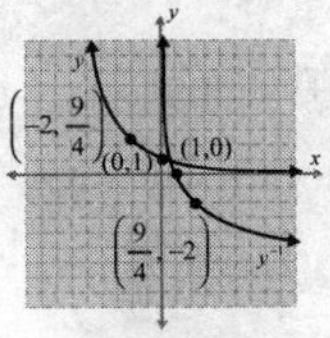

51.

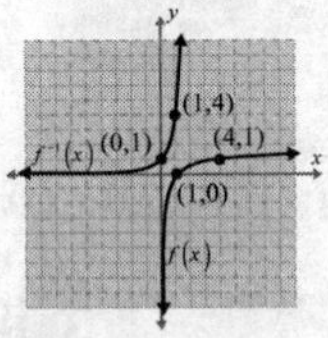

53.

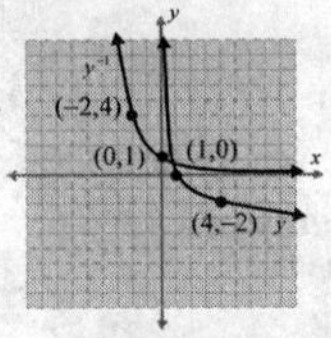

55.

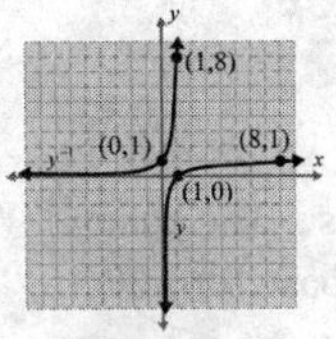

57. **a.** $D = (-\infty, \infty)$

b. $R = (0, \infty)$

c. $y = 0$

d. Answers will vary.

Writing & Thinking

59. The two functions are symmetric about the line $y = x$.

17.5 Exercises

Concept Check

1. exponents, exponents

3. difference between

5. sum, difference

7. True

9. True

Practice

1. **a.** 2 **b.** 3

c. 5 **d.** 6

3. 5

5. -2

7. $\frac{1}{2}$

9. 10

11. $\sqrt{3}$

13. $\log_b 5 + 4\log_b x$

15. $\log_b 2 - 3\log_b x + \log_b y$

17. $\log_6 2 + \log_6 x - 3\log_6 y$

19. $2\log_b x - \log_b y - \log_b z$

21. $-2\log_5 x - 2\log_5 y$

23. $\frac{1}{3}\log_6 x + \frac{2}{3}\log_6 y$

25. $\frac{1}{2}\log_3 x + \frac{1}{2}\log_3 y - \frac{1}{2}\log_3 z$

27. $\log_5 21 + 2\log_5 x + \frac{2}{3}\log_5 y$

29. $-\frac{1}{2}\log_6 x - \frac{5}{2}\log_6 y$

31. $-9\log_b x - 6\log_b y + 3\log_b z$

33. $\log_b\left(\frac{9x}{5}\right)$

35. $\log_2 63x^2$

37. $\log_b x^2 y$

39. $\log_5\left(\frac{y^3}{\sqrt{x}}\right)$ or $\log_5\left(\frac{y^3\sqrt{x}}{x}\right)$

41. $\log_5 \sqrt{\frac{x}{y}}$

43. $\log_2\left(\frac{xz}{y}\right)$

45. $\log_b\left(\frac{xy^2}{\sqrt{z}}\right)$ or $\log_b\left(\frac{xy^2\sqrt{z}}{z}\right)$

47. $\log_5\left(2x^3 + x^2\right)$

49. $\log_2\left(x^2 + 2x - 3\right)$

51. $\log_b(x+1)$

53. $\log_{10}\left(\frac{1}{2x-3}\right)$

Writing & Thinking

55. Answers will vary.

17.6 Exercises

Concept Check

1. 10

3. negative

5. $\ln x$

7. False; When the base is omitted, it is understood to be 10.

9. False; Both common logarithms and natural logarithms have inverses.

Practice

1. $\log x = 1.5$

3. $\log \frac{1}{1000} = -3$

5. $\ln 27 = x$

7. $\ln 1 = 0$

9. $\log 3.2 = x$

11. $10^0 = 1$

13. $10^y = 5.4$

15. $e^{1.54} = x$

17. $e^1 = e$

19. $e^a = x$

21. 2.2380

23. 1.9465

25. −1.2418

27. 3.6243

29. Error (undefined)

31. −5.3795

33. 204.1738

35. 0.0120

37. 0.9572

39. 175.9148

41. 0.0002

43. 1.0403

Writing & Thinking

45. $\log x$ is a base 10 logarithm. $\ln x$ is a base e logarithm.

47. $D = (-\infty, 0) \cup (0, \infty)$

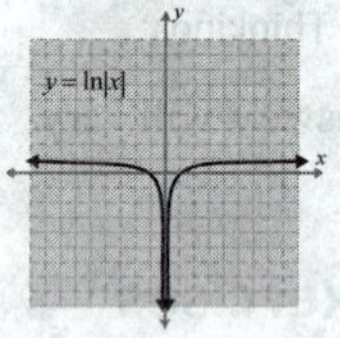

17.7 Exercises

Concept Check

1. logarithm

3. 1

5. $\dfrac{1}{b^x}$

7. False; The change of base formula is $\log_b x = \dfrac{\log_a x}{\log_a b}$.

9. True

Practice

1. $x = 11$

3. $x = 9$

5. $x = \dfrac{7}{2}$

7. $x = \dfrac{6}{5}$

9. $x = \dfrac{7}{2}$

11. $x = -5$

13. $x = -3$

15. $x = -1, 4$

17. $x = -1, \dfrac{3}{2}$

19. $x = -2, 3$

21. $x = -2$

23. $x = -\dfrac{3}{2}, 0$

25. $x = -1, 3$

27. $x \approx 0.7154$

29. $x \approx 7.5098$

31. $x \approx -1.5947$

33. $x \approx 0.2473$

35. $x \approx -7.7003$

37. $t \approx -1.5193$

39. $x \approx 3.3219$

41. $x \approx -1.4307$

43. $x = 1$

45. $x \approx -4.1610$

47. $x \approx 1.2058$

49. $x \approx 1.1292$

51. $x \approx 0.9513$

53. $x \approx 1.2520$

55. $x \approx 25.1189$

57. $x \approx 31.6228$

59. $x = 0.0001$

61. $x \approx 4.9530$

63. $x \approx \pm 0.3329$

65. $x = 3$

67. $x = 100$

69. $x = 6$

71. $x = 20$

73. $x \approx 3.1893$

75. No solution

77. No solution

79. $x = 105$

81. $x \approx 22.0855$

83. $x = -10, 8$

85. 2.2619

87. 0.3223

89. −0.6279

91. 2.4391

93. −1.2222

95. $x \approx 2.3219$

97. $x \approx 1.5480$

99. $x \approx 0.8390$

Writing & Thinking

101. Answers will vary. $x = \dfrac{1}{2}$

103. $7 \cdot 7^x = 7^{1+x}$ and $49^x = 7^{2x}$. Since $1 + x \neq 2x$, in general, $7 \cdot 7^x \neq 49^x$. Answers will vary.

17.8 Exercises

Concept Check

1. decomposition, centuries

3. Richter

5. False; The variable t is measured in days

Applications

1. $4027.51

3. 11.55 years

5. $f \approx 0.30$

7. 1.73 hours

9. 10.99 days

11. 7.44 hours

13. 2350.02 years

15. 2.31 days

17. 39.65 minutes

19. 7.00 years

21. 5600 years

23. **a.** 13.86 years
b. 6.93 years

25. 100

27. 8.64 million

29. 2083, 2437, 2744

31. **a.** 117.95 dB
b. $3.16 \times 10^8 I_0$
c. $10^6 I_0$

Chapter 18: Conic Sections

18.1 Exercises

Concept Check

1. slope, two

3. horizontal

5. shape, direction

7. False; Translation does not change the shape of the graph of a function.

Practice

1. **a.** 0 **b.** $a^2 - 6a + 5$
c. $x^2 + 2xh + h^2 - 4$
d. $2x + h$

3. **a.** 0 **b.** $2a^2 - 11a + 14$
c. $2x^2 + 4xh - 3x + 2h^2 - 3h$
d. $4x + 2h - 3$

5. 1

7. −2

9. The results are the coefficients of x.

11.

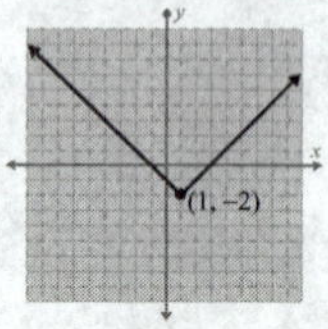

13.

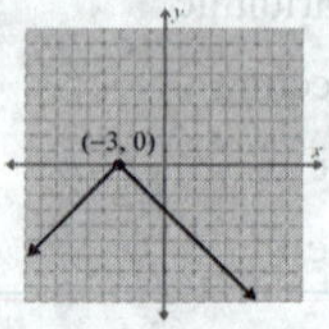

15.

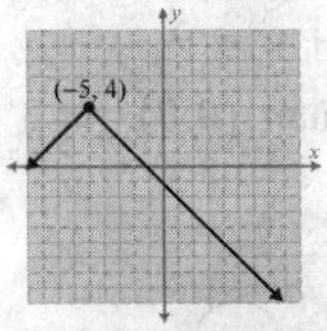

17.

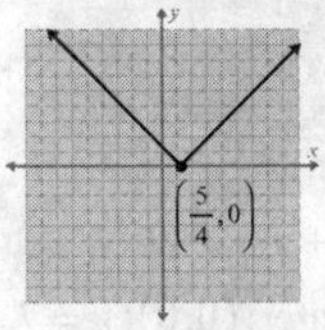

19.

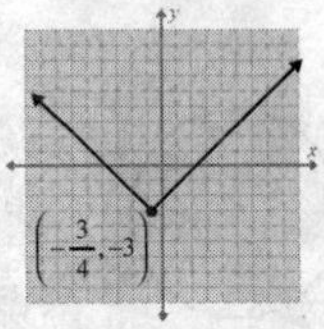

21.

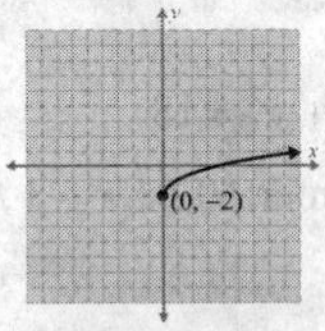

23.

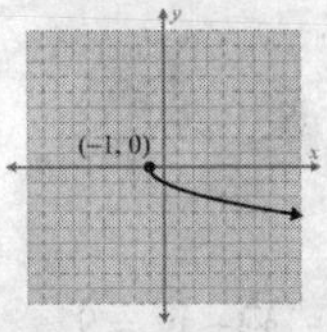

25.

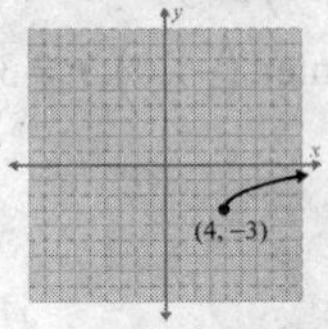

27.

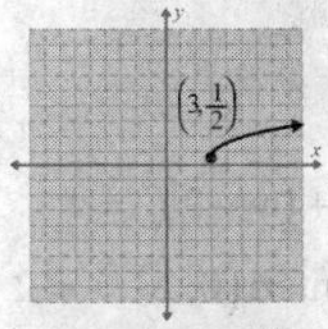

29.

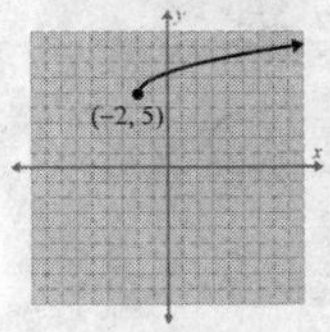

31.

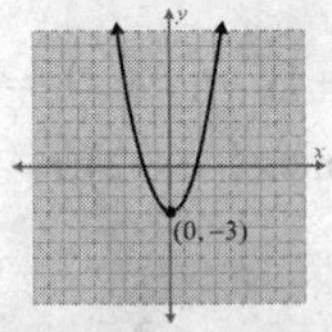

33.

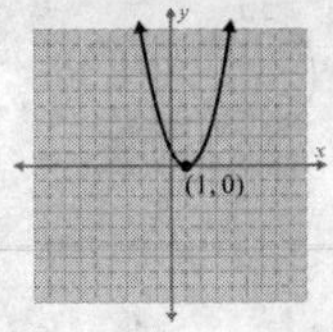

35.

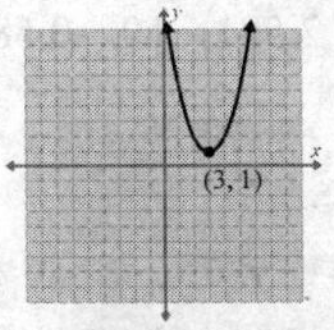

37.

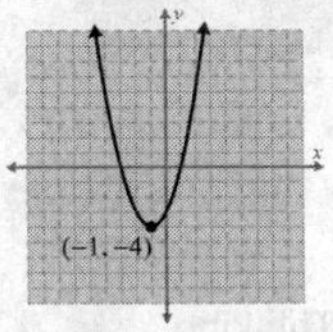

39.

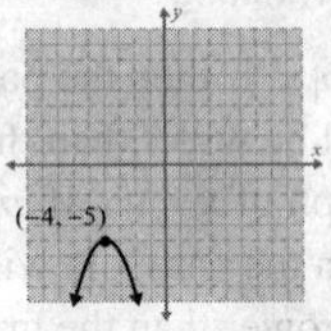

41.

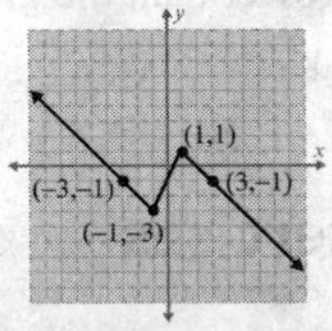

43.

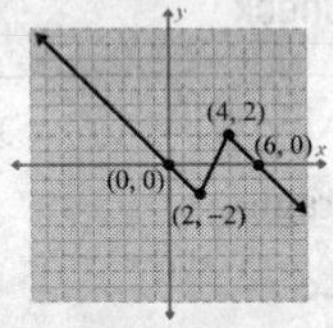

45.

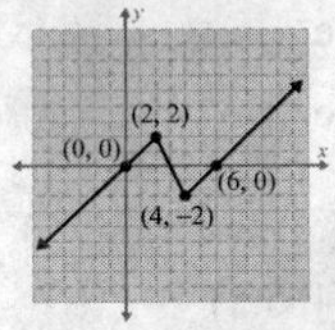

47.

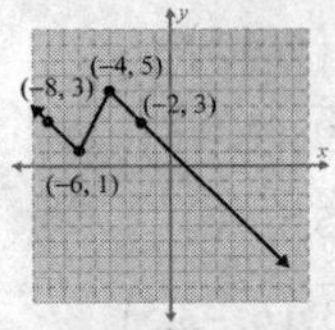

49.

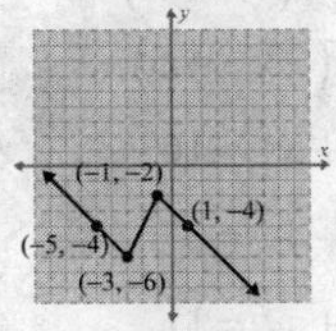

51.

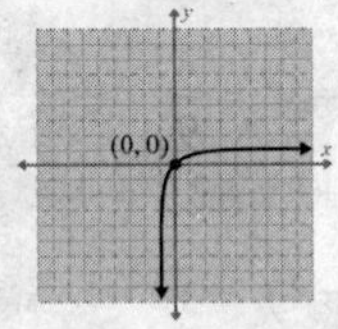

$D=(-1,\infty);R=(-\infty,\infty)$

53.

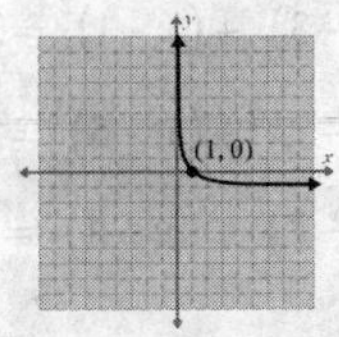

$D=(0,\infty);R=(-\infty,\infty)$

55.

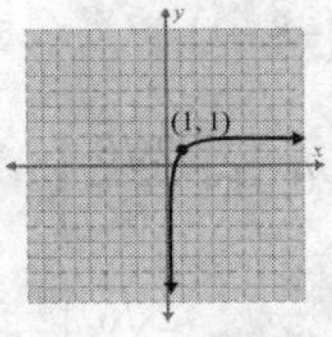

$D=(0,\infty);R=(-\infty,\infty)$

Applications

57. $y=x^4-5x^2+2x+10$

Writing & Thinking

59. The graph of the function $y=f(x-h)+k$ is the graph of the function $y=f(x)$ shifted h units to the right and k units up.

61.

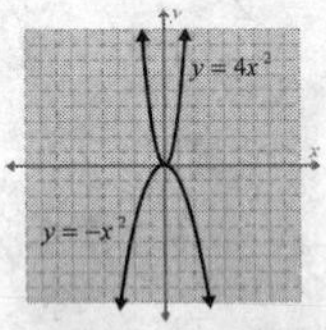

63.

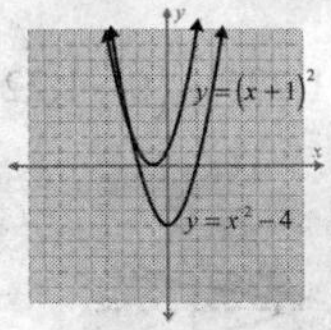

65.

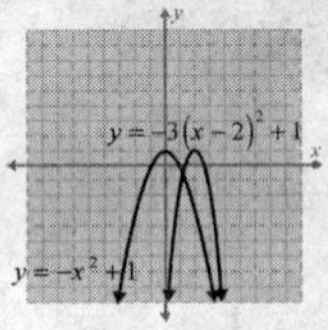

18.2 Exercises

Concept Check

1. circle, ellipse, parabola, hyperbola
3. vertical
5. 0, y-intercepts
7. True
9. False; $y=k$ is the line of symmetry for a horizontal parabola.

Practice

1. a. $(4,0)$ **b.** None
c. $y=0$

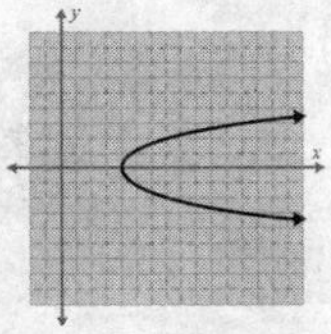

3. a. $(0,-3)$ **b.** $(0,-3)$
c. $x=0$

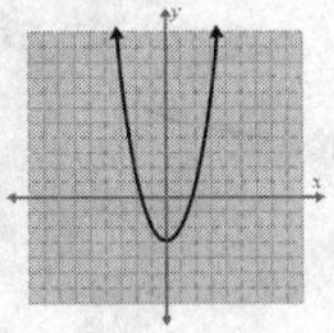

5. a. $(3,0)$ **b.** None
c. $y=0$

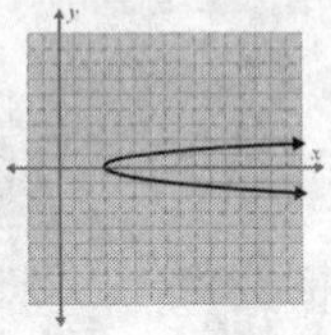

7. a. $(0,3)$ **b.** $(0,3)$
c. $y=3$

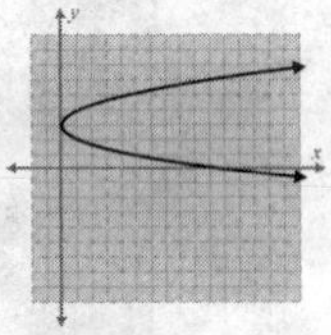

9. a. $(4,-2)$ **b.** None
c. $y=-2$

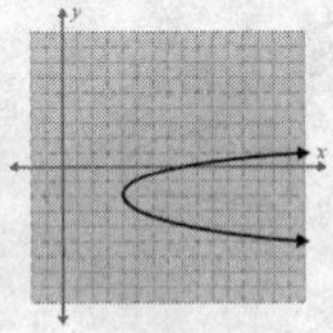

11. a. $(1,-1)$ **b.** $(0,0)$
c. $x=1$

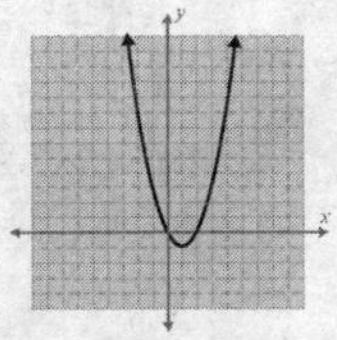

13. a. $(0,-2)$ **b.** $(0,-2)$
c. $y=-2$

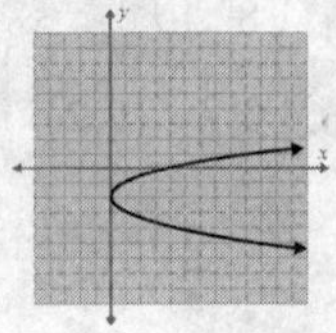

15. **a.** $(0, 5)$ **b.** $(0, 5)$

c. $y = 5$

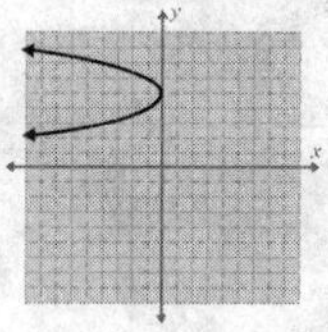

17. **a.** $(-3, -4)$ **b.** $(0, 5)$

c. $x = -3$

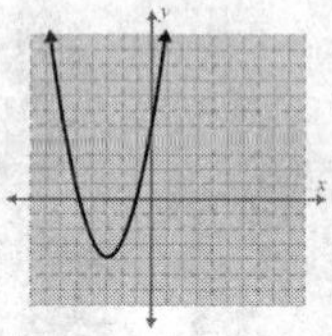

19. **a.** $(-2, 9)$ **b.** $(0, 5)$

c. $x = -2$

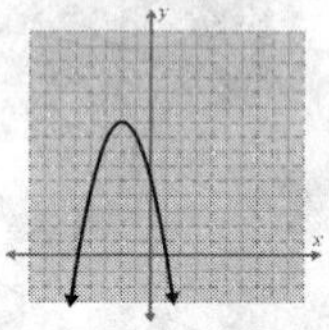

21. **a.** $(1, 2)$ **b.** $(0, 1), (0, 3)$

c. $y = 2$

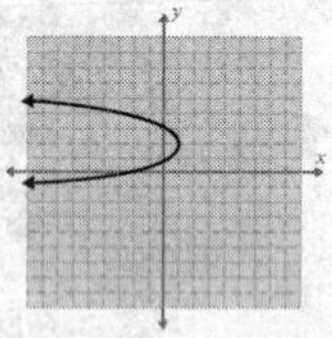

23. **a.** $\left(-\frac{1}{4}, -\frac{9}{8}\right)$ **b.** $(0, -1)$

c. $x = -\frac{1}{4}$

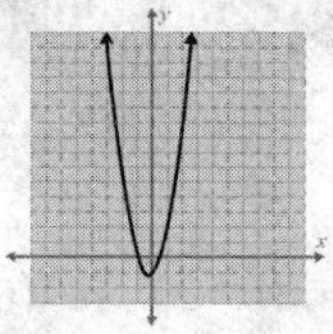

25. **a.** $\left(\frac{9}{8}, \frac{5}{4}\right)$

b. $\left(0, \frac{1}{2}\right), (0, 2)$

c. $y = \frac{5}{4}$

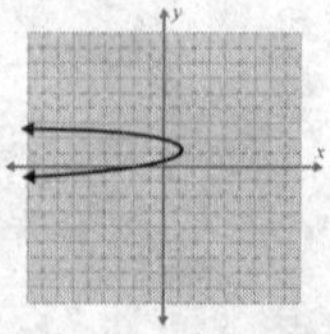

27. **a.** $(-8, -1)$

b. $\left(0, -1 + \frac{2\sqrt{6}}{3}\right)$, $\left(0, -1 - \frac{2\sqrt{6}}{3}\right)$

c. $y = -1$

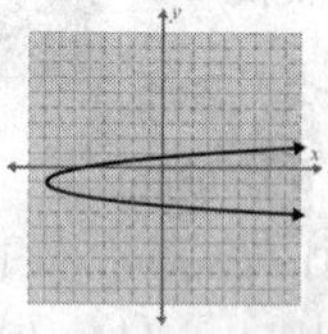

29. **a.** $\left(\frac{3}{2}, 0\right)$ **b.** $(0, 9)$

c. $x = \frac{3}{2}$

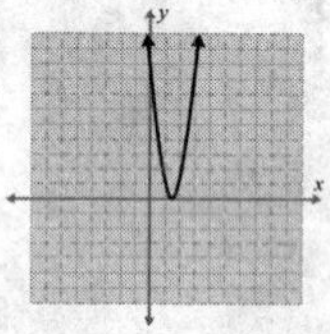

31. $(0, 1.225)$, $(0, -1.225)$

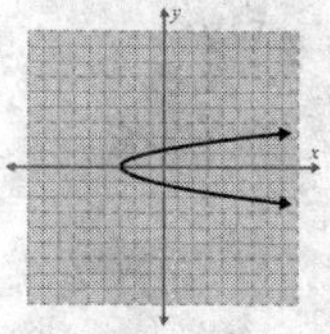

33. $(0, 2), (0, 0)$

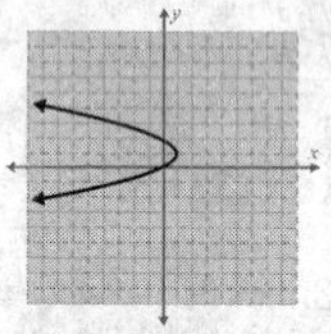

35. No y-intercept

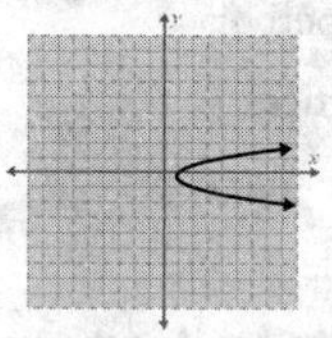

37. $(0, 0.658)$, $(0, -2.658)$

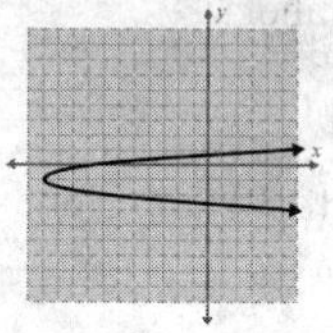

39. $(0, 2.581)$, $(0, -0.581)$

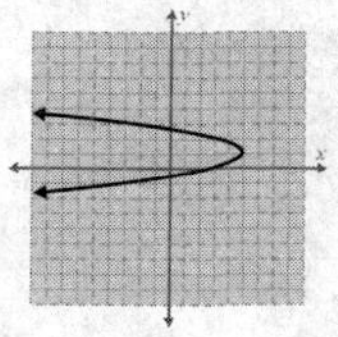

41. b

43. c

Writing & Thinking

45. If $|a| < 1$, then the parabola will be wider than the graph of $x = y^2$. If $|a| > 1$, then the parabola will be narrower than the graph of $x = y^2$.

18.3 Exercises

Concept Check

1. $\sqrt{(x_2 - x_1)^2 + (y_2 - y_1)^2}$

3. averaging

5. twice

7. True

9. True

Practice

1. $5; (4, 5.5)$

3. $13; (3, 4.5)$

5. $\sqrt{29}; (2, 4.5)$

7. $3; (5.5, -3)$

9. $\sqrt{13}; (6, -3.5)$

11. $17; (-3, -4.5)$

13. $x^2 + y^2 = 16$

15. $x^2 + y^2 = 3$

17. $x^2 + y^2 = 11$

19. $x^2 + y^2 = \frac{4}{9}$

21. $x^2 + (y - 2)^2 = 4$

23. $(x - 4)^2 + y^2 = 1$

25. $(x + 2)^2 + y^2 = 8$

27. $(x - 3)^2 + (y - 1)^2 = 36$

29. $(x - 3)^2 + (y - 5)^2 = 12$

31. $(x - 7)^2 + (y - 4)^2 = 10$

33. $x^2 + y^2 = 9$

Center: $(0, 0)$; $r = 3$

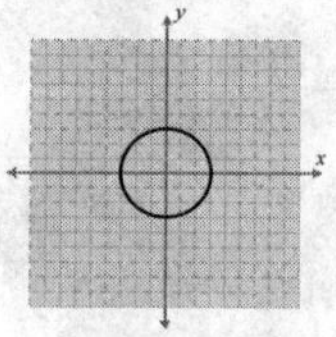

35. $x^2 + y^2 = 49$

Center: $(0, 0)$; $r = 7$

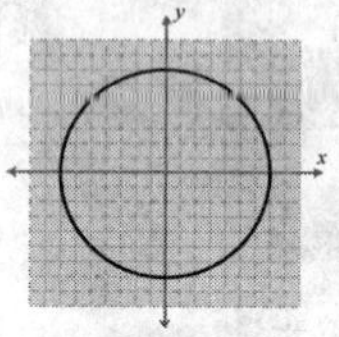

37. $x^2 + y^2 = 18$

Center: $(0, 0)$; $r = 3\sqrt{2}$

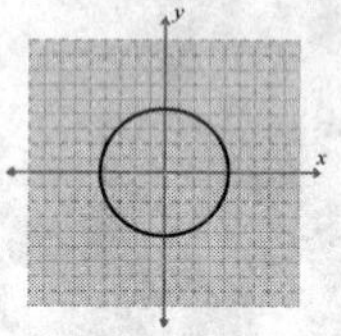

39. $(x + 1)^2 + y^2 = 9$

Center: $(-1, 0)$; $r = 3$

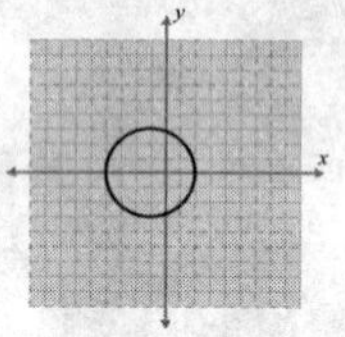

41. $x^2 + (y - 2)^2 = 4$

Center: $(0, 2)$; $r = 2$

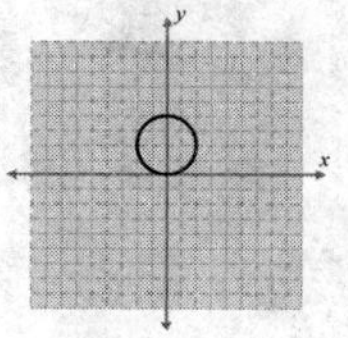

43. $(x + 1)^2 + (y + 2)^2 = 16$

Center: $(-1, -2)$; $r = 4$

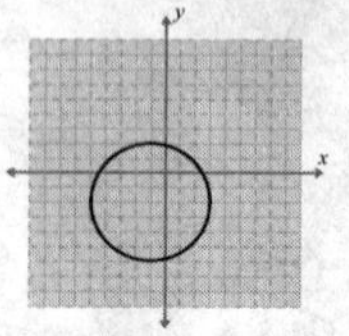

41.

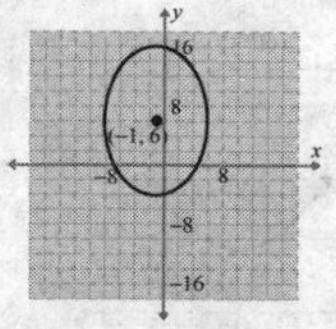

43.

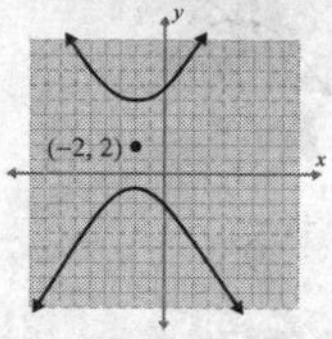

Writing & Thinking

45. **b.** Pick an arbitrary point (x, y) located on the ellipse. The sum of the distances from (x, y) to the points $(-c, 0)$ and $(c, 0)$ (the foci) will equal $2a$. Using the distance formula we get the equation

$$\sqrt{(x+c)^2+(y-0)^2}+\sqrt{(x-c)^2+(y-0)^2}=2a.$$

Now to remove the radicals, simplify so that one side of the equation is a single radical and square both sides. Then repeat the previous step and simplify. This gives $\frac{x^2}{a^2}+\frac{y^2}{a^2-c^2}=1.$

c. At the y-intercept $(0, b)$ a right triangle is formed with hypotenuse a and sides b and c. The Pythagorean Theorem gives $a^2=b^2+c^2$.

18.5 Exercises

Concept Check

1. second
3. substitution, addition
5. both
7. True
9. False; There are many situations where the curves do not intersect.

Practice

1. $(-3, 10), (1, 2)$

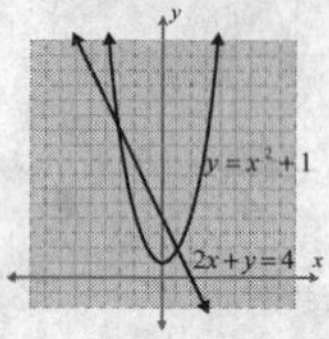

3. $(1, 1), (2, 0)$

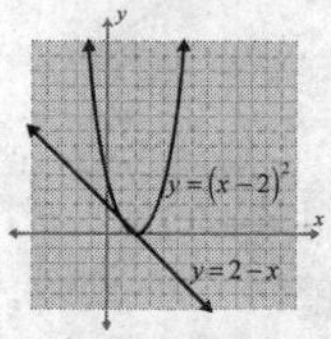

5. $(-2, -4), (4, 2)$

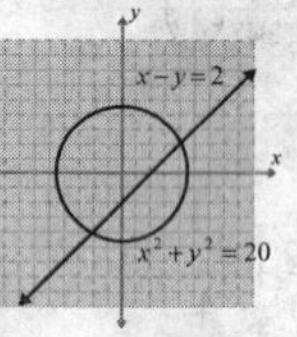

7. $(5, 3)$

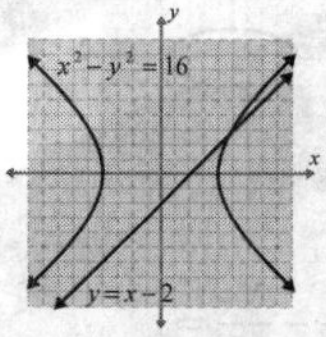

9. $(-3, 0), (3, 0)$

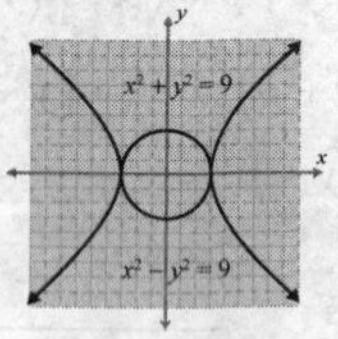

11. $(2, 3), (2, -3)$

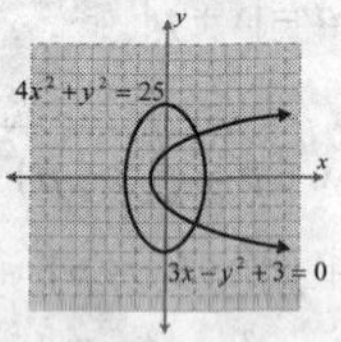

13. $(-2, 4), (1, 1)$

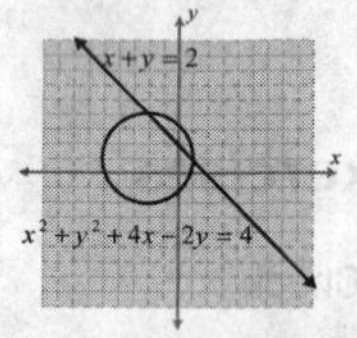

15. $(-3, -2), (3, 2), (-3, 2), (3, -2)$

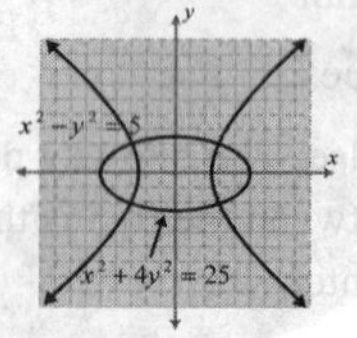

17. $(-3\sqrt{5}, 5), (-6, 4), (3\sqrt{5}, 5), (6, 4)$

19. $(1, 3), (-1, 3)$

21. $(4, 4), (-2, -2)$

23. $\left(-\frac{3}{2}, -1\right), \left(\frac{1}{2}, \frac{1}{2}\right)$

25. $(-3, \sqrt{11}), (-3, -\sqrt{11}), (3, \sqrt{11}), (3, -\sqrt{11})$

27. $(5, -3), (-5, 1)$

29. $(0, 3), (2, 3)$

31. $(-1.62, 5.62), (0.62, 3.38)$

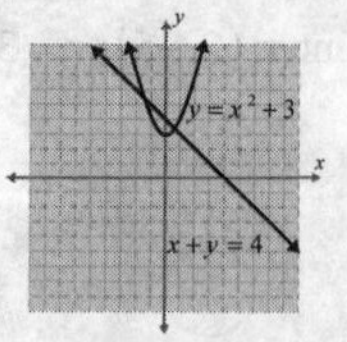

33. $(1.41, 0.17), (-1.41, 5.83)$

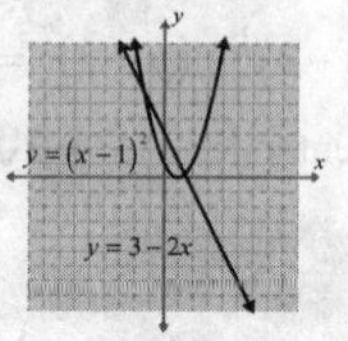

35. $(2.68, 1.68), (-1.68, -2.68)$

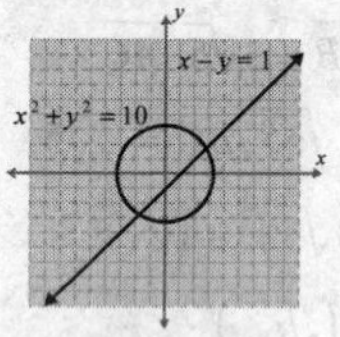

Index

A

B

C

D

E

Q

R

S

Geometry

P = Perimeter, A = Area, C = Circumference, V = Volume, SA = Surface Area

Perimeter and Area

Rectangle

$P = 2l + 2w$

$A = lw$

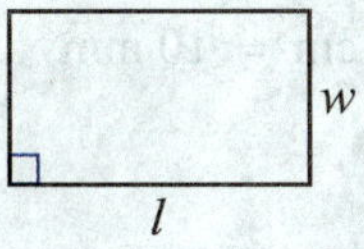

Square

$P = 4s$

$A = s^2$

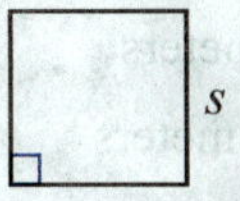

Triangle

$P = a + b + c$

$A = \frac{1}{2}bh$

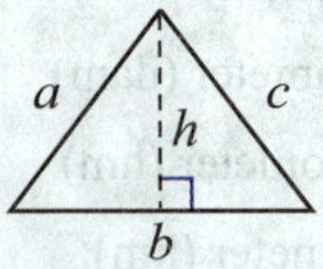

Parallelogram

$P = 2a + 2b$

$A = bh$

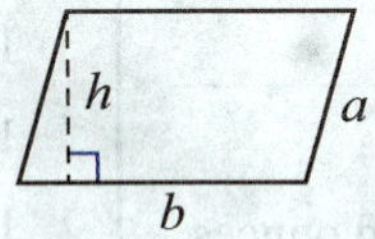

Trapezoid

$P = a + b + c + d$

$A = \frac{1}{2}h(b + c)$

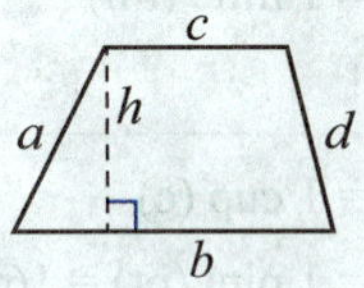

Circle

$C = 2\pi r = \pi d$

$A = \pi r^2$

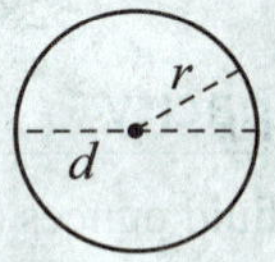

Volume and Surface Area

Rectangular Solid

$V = lwh$

$SA = 2lw + 2wh + 2lh$

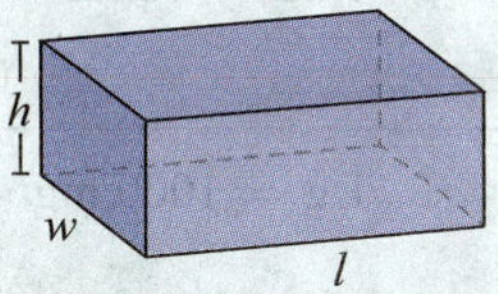

Rectangular Pyramid

$V = \frac{1}{3}lwh$

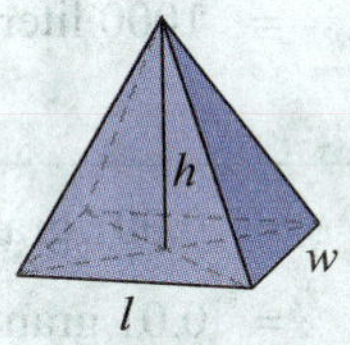

Right Circular Cone

$V = \frac{1}{3}\pi r^2 h$

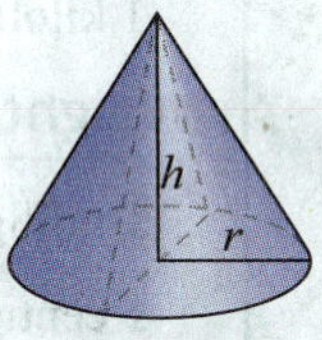

Right Circular Cylinder

$V = \pi r^2 h$

$SA = 2\pi r^2 + 2\pi rh$

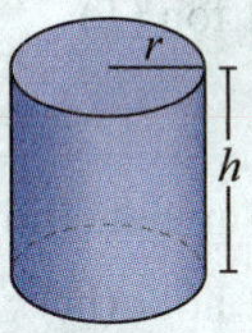

Sphere

$V = \frac{4}{3}\pi r^3$

$SA = 4\pi r^2$

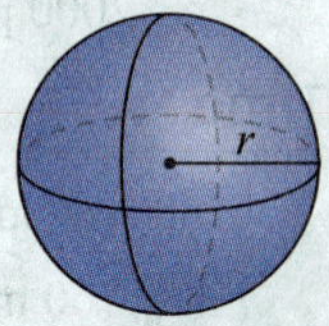

Angles Classified by Measure

Acute

$0° < m\angle A < 90°$

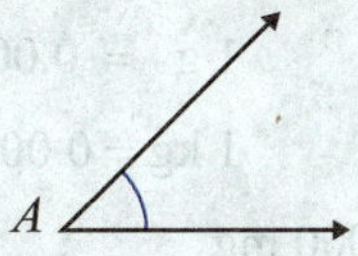

Right

$m\angle A = 90°$

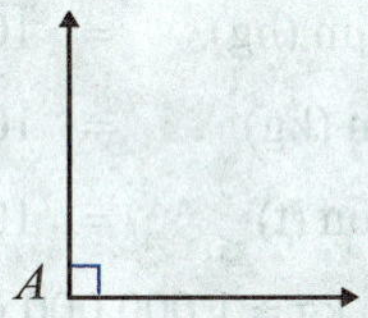

Obtuse

$90° < m\angle A < 180°$

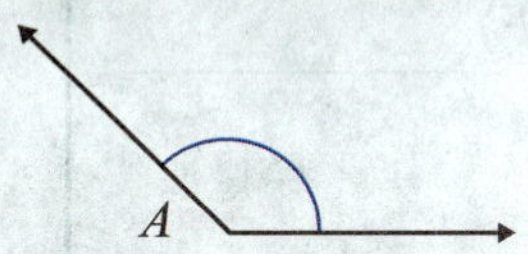

Straight

$m\angle A = 180°$

Triangles Classified by Sides

Scalene

No two sides have equal lengths.

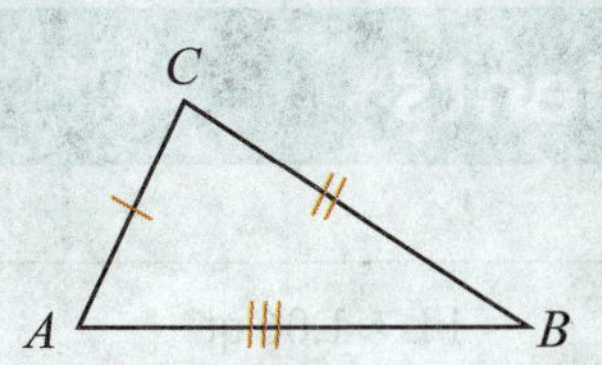

Isosceles

At least two sides have equal lengths.

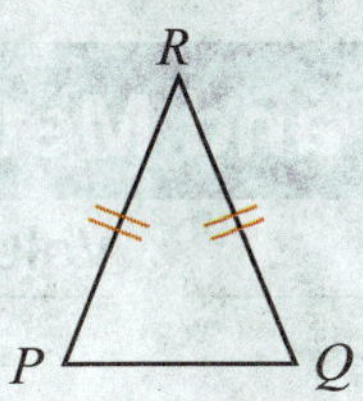

Equilateral

All three sides have equal lengths.

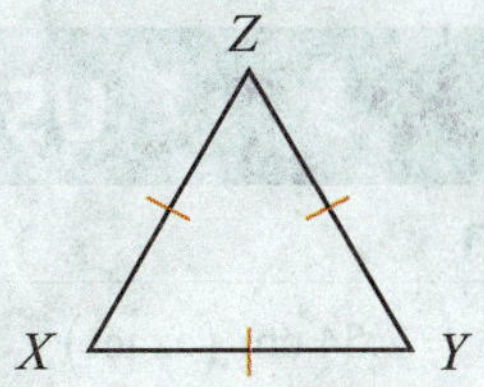

Triangles Classified by Angles

Acute

All three angles are acute.

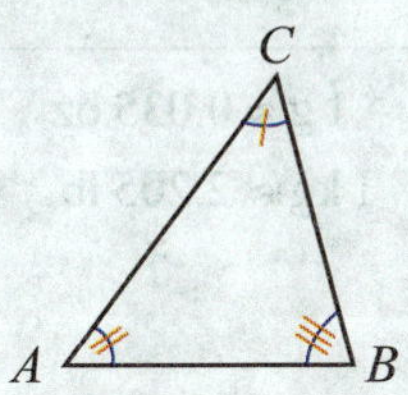

Right

One angle is a right angle.

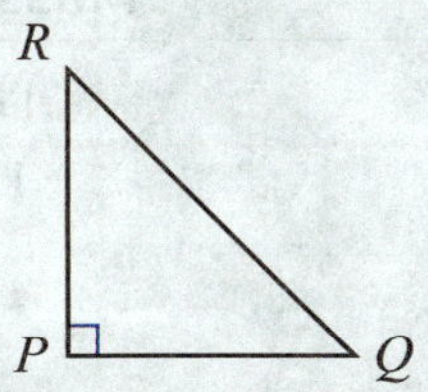

Obtuse

One angle is obtuse.

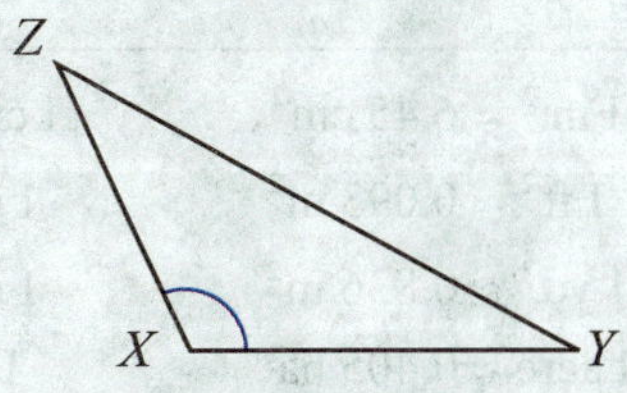

US Customary System of Measurement

Length

12 inches (in.) = 1 foot (ft)
36 inches = 1 yard (yd)
3 feet = 1 yard
5280 feet = 1 mile (mi)

Capacity

8 fluid ounces (fl oz) = 1 cup (c)
2 cups = 1 pint (pt) = 16 fluid ounces
2 pints = 1 quart (qt)
4 quarts = 1 gallon (gal)

Weight

16 ounces (oz) = 1 pound (lb)
2000 pounds = 1 ton (T)

Time

60 seconds (sec) = 1 minute (min)
60 minutes = 1 hour (hr)
24 hours = 1 day
7 days = 1 week

Temperature

Celsius (C) to Fahrenheit (F)

$$F = \frac{9C}{5} + 32$$

Fahrenheit (F) to Celsius (C)

$$C = \frac{5(F - 32)}{9}$$

Metric System of Measurement

Length

1 millimeter (mm)	= 0.001 meter		1 m = 1000 mm
1 centimeter (cm)	= 0.01 meter		1 m = 100 cm
1 decimeter (dm)	= 0.1 meter		1 m = 10 dm
1 meter (m)	= 1.0 meter		
1 dekameter (dam)	= 10 meters		1 cm = 10 mm
1 hectometer (hm)	= 100 meters		
1 kilometer (km)	= 1000 meters		

Capacity (Liquid Volume)

1 milliliter (mL)	= 0.001 liter	1 L = 1000 mL
1 liter (L)	= 1.0 liter	
1 hectoliter (hL)	= 100 liters	
1 kiloliter (kL)	= 1000 liters	1 kL = 10 hL

Weight

1 milligram (mg)	= 0.001 gram	1 g = 1000 mg
1 centigram (cg)	= 0.01 gram	
1 decigram (dg)	= 0.1 gram	
1 gram (g)	= 1.0 gram	
1 dekagram (dag)	= 10 grams	
1 hectogram (hg)	= 100 grams	
1 kilogram (kg)	= 1000 grams	1 g = 0.001 kg
1 metric ton (t)	= 1000 kilograms	1 kg = 0.001 t

1 t = 1000 kg = 1,000,000 g = 1,000,000,000 mg

Land Area

1 are (a) = 100 m^2
1 hectare (ha) = 100 a = 10 000 m^2

US Customary and Metric Equivalents

Length

1 in. = 2.54 cm (exact)	1 cm ≈ 0.394 in.
1 ft ≈ 0.305 m	1 m ≈ 3.28 ft
1 yd ≈ 0.914 m	1 m ≈ 1.09 yd
1 mi ≈ 1.61 km	1 km ≈ 0.62 mi

Volume

1 qt ≈ 0.946 L	1 L ≈ 1.06 qt
1 gal ≈ 3.785 L	1 L ≈ 0.264 gal

Area

1 $in.^2$ ≈ 6.45 cm^2	1 cm^2 ≈ 0.155 $in.^2$
1 ft^2 ≈ 0.093 m^2	1 m^2 ≈ 10.764 ft^2
1 yd^2 ≈ 0.836 m^2	1 m^2 ≈ 1.196 yd^2
1 acre ≈ 0.405 ha	1 ha ≈ 2.47 acres

Mass

1 oz ≈ 28.35 g	1 g ≈ 0.035 oz
1 lb ≈ 0.454 kg	1 kg ≈ 2.205 lb

Notation and Terminology

Exponents

$\underbrace{a \cdot a \cdot a \cdot a \cdot \ldots \cdot a}_{n \text{ factors}} = a^n$ (a = base, n = exponent)

Fractions

$\frac{a}{b}$ ← numerator (a), ← denominator (b)

Least Common Multiple (LCM)

Given a set of counting numbers, the smallest number that is a multiple of each of these numbers.

Ratios

$\frac{a}{b}$ or $a:b$ or a to b

A comparison of two quantities by division.

Proportions

$\frac{a}{b} = \frac{c}{d}$ A statement that two ratios are equal.

Greatest Common Factor (GCF)

Given a set of integers, the largest integer that is a factor (or divisor) of all of the integers.

Types of Numbers

Natural Numbers (Counting Numbers):

$\mathbb{N} = \{1, 2, 3, 4, 5, 6, \ldots\}$

Whole Numbers: $\mathbb{W} = \{0, 1, 2, 3, 4, 5, 6, \ldots\}$

Integers: $\mathbb{Z} = \{\ldots, -4, -3, -2, -1, 0, 1, 2, 3, 4, \ldots\}$

Rational Numbers: A number that can be written in the form $\frac{a}{b}$ where a and b are integers and $b \neq 0$.

Irrational Numbers: A number that can be written as an infinite nonrepeating decimal.

Real Numbers: All rational and irrational numbers.

Complex Numbers: All real numbers and the even roots of negative numbers. The standard form of a complex number is $a + bi$, where a and b are real numbers, a is called the real part and b is called the imaginary part.

Absolute Value

$|a|$ The distance a real number a is from 0.

Equality and Inequality Symbols

$=$ "is equal to"

$\neq$ "is not equal to"

$<$ "is less than"

$>$ "is greater than"

$\leq$ "is less than or equal to"

$\geq$ "is greater than or equal to"

Sets

The **empty set** or **null set** (symbolized $\varnothing$ or $\{\ \}$): A set with no elements.

The **union** of two (or more) sets (symbolized $\cup$): The set of all elements that belong to either one set or the other set or to both sets.

The **intersection** of two (or more) sets (symbolized $\cap$): The set of all elements that belong to both sets.

The word **or** is used to indicate union and the word **and** is used to indicate intersection.

Algebraic and Interval Notation for Intervals

Type of Interval	Algebraic Notation	Interval Notation	Graph
Open Interval	$a < x < b$	(a, b)	
Closed Interval	$a \leq x \leq b$	$[a, b]$	
Half-Open Interval	$a \leq x < b$ $a < x \leq b$	$[a, b)$ $(a, b]$	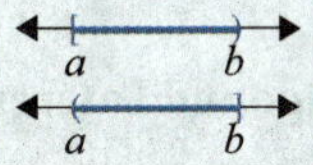
Open Interval	$x > a$ $x < b$	(a, ∞) $(-\infty, b)$	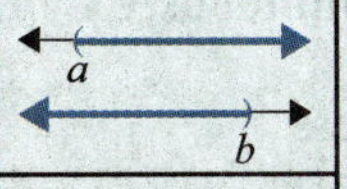
Half-Open Interval	$x \geq a$ $x \leq b$	$[a, \infty)$ $(-\infty, b]$	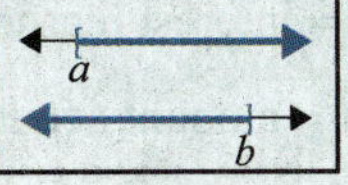

Radicals

The symbol $\sqrt{\ }$ is called a **radical sign**.

The number under the radical sign is called the **radicand**.

The complete expression, such as $\sqrt{64}$, is called a **radical** or **radical expression**.

In a cube root expression $\sqrt[3]{a}$, the number 3 is called the index. In a square root expression such as $\sqrt{a}$, the index is understood to be 2 and is not written.

The Imaginary Number *i*

$i = \sqrt{-1}$ and $i^2 = \left(\sqrt{-1}\right)^2 = -1$

Formulas and Theorems

Percent

$\frac{P}{100} = \frac{A}{B}$ (the percent proportion),

where

$P\% =$ percent (written as the ratio $\frac{P}{100}$)
$B =$ base (number we are finding the percent of)
$A =$ amount (a part of the base)

$R \cdot B = A$ (the basic percent equation),
where

$R =$ rate or percent (as a decimal or fraction)
$B =$ base (number we are finding the percent of)
$A =$ amount (a part of the base)

Profit

Profit: The difference between selling price and cost.

$$\text{Profit} = \text{Selling Price} - \text{Cost}$$

Percent of Profit:

1. Percent of profit **based on cost:** $\frac{\text{Profit}}{\text{Cost}}$
2. Percent of profit **based on selling price:** $\frac{\text{Profit}}{\text{Selling Price}}$

Interest

Simple Interest: $I = P \cdot r \cdot t$

Compound Interest: $A = P\left(1 + \frac{r}{n}\right)^{nt}$

Continuously Compounded Interest: $A = Pe^{rt}$

where

$I =$ interest (earned or paid)

$A =$ amount accumulated

$P =$ principal (the amount invested or borrowed)

$r =$ annual interest rate in decimal or fraction form

$t =$ time (years or fraction of a year)

$n =$ number of compounding periods in 1 year

$e = 2.718281828459\ldots$

The Pythagorean Theorem

In a right triangle, the square of the length of the hypotenuse is equal to the sum of the squares of the lengths of the two legs. $c^2 = a^2 + b^2$

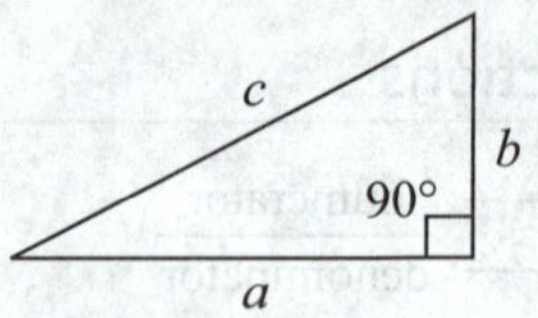

Probability of an Event

$$\text{probability of an event} = \frac{\text{number of outcomes in event}}{\text{number of outcomes in sample space}}$$

Distance-Rate-Time

$d = rt$ The **distance traveled** d equals the product of the rate of speed r and the time t.

Special Products

1. $x^2 - a^2 = (x+a)(x-a)$ Difference of two squares
2. $x^2 + 2ax + a^2 = (x+a)^2$ Square of a binomial sum
3. $x^2 - 2ax + a^2 = (x-a)^2$ Square of a binomial difference
4. $x^3 + a^3 = (x+a)(x^2 - ax + a^2)$ Sum of two cubes
5. $x^3 - a^3 = (x-a)(x^2 + ax + a^2)$ Difference of two cubes

Change-of-Base Formula for Logarithms

For $a, b, x > 0$ and $a, b \neq 1$, $\log_b x = \frac{\log_a x}{\log_a b}$.

Distance Between Two Points

The distance d between points $P(x_1, y_1)$ and $Q(x_2, y_2)$ is $d = \sqrt{(x_2 - x_1)^2 + (y_2 - y_1)^2}$.

Midpoint Formula

The midpoint between points $P(x_1, y_1)$ and $Q(x_2, y_2)$ is $\left(\frac{x_1 + x_2}{2}, \frac{y_1 + y_2}{2}\right)$.

Principles and Properties

Properties of Addition and Multiplication

Property	Addition	Multiplication
Commutative Property	$a+b=b+a$	$ab=ba$
Associative Property	$(a+b)+c=a+(b+c)$	$a(bc)=(ab)c$
Identity	$a+0=0+a=a$	$a\cdot 1=1\cdot a=a$
Inverse	$a+(-a)=0$	$a\cdot\frac{1}{a}=1\ (a\neq 0)$

Zero-Factor Law: $a\cdot 0=0\cdot a=0$

Distributive Property: $a(b+c)=a\cdot b+a\cdot c$

Addition (or Subtraction) Principle of Equality

$A=B$, $A+C=B+C$, and $A-C=B-C$ have the same solutions (where A, B, and C are algebraic expressions).

Multiplication (or Division) Principle of Equality

$A=B$, $AC=BC$, and $\frac{A}{C}=\frac{B}{C}$ have the same solutions (where A and B are algebraic expressions and C is any nonzero constant, $C\neq 0$).

Properties of Exponents

For nonzero real numbers a and b and integers m and n:

The exponent 1	$a=a^1$
The exponent 0	$a^0=1$
The product rule	$a^m\cdot a^n=a^{m+n}$
The quotient rule	$\frac{a^m}{a^n}=a^{m-n}$
Negative exponents	$a^{-n}=\frac{1}{a^n}$
Power rule	$(a^m)^n=a^{mn}$
Power of a product	$(ab)^n=a^nb^n$
Power of a quotient	$\left(\frac{a}{b}\right)^n=\frac{a^n}{b^n}$

Zero-Factor Law

If a and b are real numbers, and $a\cdot b=0$, then $a=0$ or $b=0$ or both.

Properties of Rational Numbers (or Fractions)

If $\frac{P}{Q}$ is a rational expression and P, Q, R, and K are polynomials where $Q, R, S, K\neq 0$, then

The Fundamental Principle	$\frac{P}{Q}=\frac{P\cdot K}{Q\cdot K}$
Multiplication	$\frac{P}{Q}\cdot\frac{R}{S}=\frac{P\cdot R}{Q\cdot S}$
Division	$\frac{P}{Q}\div\frac{R}{S}=\frac{P}{Q}\cdot\frac{S}{R}$
Addition	$\frac{P}{Q}+\frac{R}{Q}=\frac{P+R}{Q}$
Subtraction	$\frac{P}{Q}-\frac{R}{Q}=\frac{P-R}{Q}$

Properties of Radicals

If a and b are positive real numbers, n is a positive integer, m is any integer, and $\sqrt[n]{a}$ is a real number then

1. $\sqrt[n]{ab}=\sqrt[n]{a}\cdot\sqrt[n]{b}$
2. $\sqrt[n]{\frac{a}{b}}=\frac{\sqrt[n]{a}}{\sqrt[n]{b}}$
3. $\sqrt[n]{a}=a^{\frac{1}{n}}$
4. $a^{\frac{m}{n}}=\left(a^{\frac{1}{n}}\right)^m=\left(a^m\right)^{\frac{1}{n}}$

   or, in radical notation,

   $a^{\frac{m}{n}}=\left(\sqrt[n]{a}\right)^m=\sqrt[n]{a^m}$

Properties of Logarithms

For $b>0$, $b\neq 1$, $x, y>0$, and any real number r,

1. $\log_b 1=0$
2. $\log_b b=1$
3. $x=b^{\log_b x}$
4. $\log_b b^x=x$
5. $\log_b xy=\log_b x+\log_b y$ The product rule
6. $\log_b\frac{x}{y}=\log_b x-\log_b y$ The quotient rule
7. $\log_b x^r=r\cdot\log_b x$ The power rule

Properties of Equations with Exponents and Logarithms

For $b>0$, $b\neq 1$,

1. If $b^x=b^y$, then $x=y$.
2. If $x=y$, then $b^x=b^y$.
3. If $\log_b x=\log_b y$, then $x=y$ ($x>0$ and $y>0$).
4. If $x=y$, then $\log_b x=\log_b y$ ($x>0$ and $y>0$).

Equations and Inequalities

Linear Equation in *x* (First-Degree Equation in *x*)

$ax + b = c$, where a, b, and c are real numbers and $a \neq 0$.

Types of Equations and their Solutions

Conditional: Finite Number of Solutions

Identity: Infinite Number of Solutions

Contradiction: No Solution

Linear Inequalities

Linear inequalities have the following forms where a, b, and c are real numbers and $a \neq 0$:

$ax + b < c$ and $ax + b \leq c$

$ax + b > c$ and $ax + b \geq c$

Compound Inequalities

The inequalities $c < ax + b < d$ and $c \leq ax + b \leq d$ are called **compound linear inequalities**.

(This includes $c < ax + b \leq d$ and $c \leq ax + b < d$ as well.)

Absolute Value Equations

For statements 1 and 2, $c > 0$:

1. If $|x| = c$, then $x = c$ or $x = -c$.

2. If $|ax + b| = c$, then $ax + b = c$ or $ax + b = -c$.

3. If $|a| = |b|$, then either $a = b$ or $a = -b$.

4. If $|ax + b| = |cx + d|$, then either $ax + b = cx + d$ or $ax + b = -(cx + d)$.

Absolute Value Inequalities

For $c > 0$:

1. If $|x| < c$, then $-c < x < c$.

2. If $|ax + b| < c$, then $-c < ax + b < c$.

3. If $|x| > c$, then $x < -c$ **or** $x > c$.

4. If $|ax + b| > c$, then $ax + b < -c$ **or** $ax + b > c$.

(These statements hold true for $\leq$ and $\geq$ as well.)

Quadratic Equation

An equation that can be written in the form $ax^2 + bx + c = 0$, where a, b, and c are real numbers and $a \neq 0$.

Quadratic Formula

The solutions of the general quadratic equation

$$ax^2 + bx + c = 0, \text{ where } a \neq 0, \text{ are } x = \frac{-b \pm \sqrt{b^2 - 4ac}}{2a}.$$

The Discriminant

The expression $b^2 - 4ac$, the part of the quadratic formula that lies under the radical sign, is called the **discriminant**.

If $b^2 - 4ac > 0$, there are two real solutions.

If $b^2 - 4ac = 0$, there is one real solution, $x = -\frac{b}{2a}$.

If $b^2 - 4ac < 0$, there are two nonreal solutions.

Systems of Linear Equations

Systems of Linear Equations (Two Variables)

The system is...

consistent, and the equations are **independent**.
(One solution)

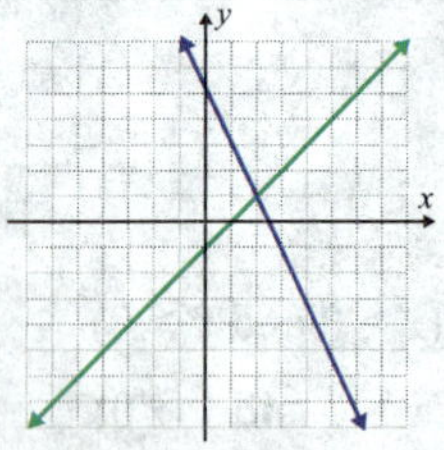

inconsistent, and the equations are **independent**.
(No solution)

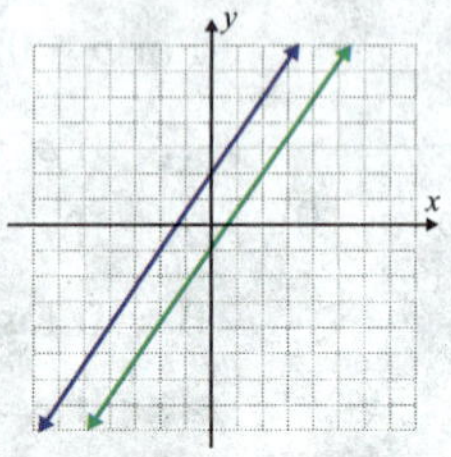

consistent, and the equations are **dependent**.
(Infinite number of solutions)

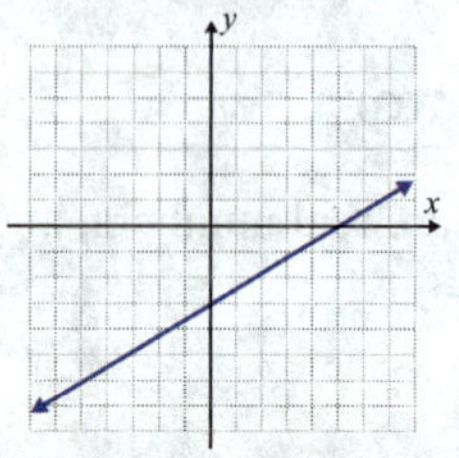

Functions

Function, Relation, Domain, and Range

A relation is a set of ordered pairs of real numbers.

The domain ***D*** of a relation is the set of all first coordinates in the relation.

The range ***R*** of a relation is the set of all second coordinates in the relation.

A function is a relation in which each domain element has exactly one corresponding range element.

One-to-One Functions

A function is a one-to-one function if for each value of *y* in the range there is only one corresponding value of *x* in the domain.

Algebraic Operations with Functions

1. $(f+g)(x)=f(x)+g(x)$

2. $(f-g)(x)=f(x)-g(x)$

3. $(f\cdot g)(x)=f(x)\cdot g(x)$

4. $\left(\frac{f}{g}\right)(x)=\frac{f(x)}{g(x)}$

5. $(f\circ g)(x)=f(g(x))$

Inverse Functions

If *f* is a one-to-one function with ordered pairs of the form (x, y), then its inverse function, denoted as f^{-1}, is also a one-to-one function with ordered pairs of the form (y, x).

If *f* and *g* are one-to-one functions and $f(g(x))=x$ for all x in D_g and $g(f(x))=x$ for all x in D_f, then *f* and *g* are inverse functions.

Graphs of Functions

The Cartesian Coordinate System

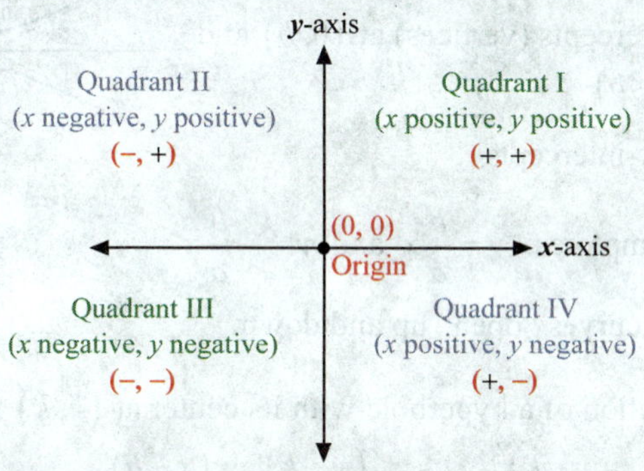

Linear Functions (Lines)

Standard form:

$Ax + By = C$ Where *A* and *B* do not both equal 0

Slope of a line:

$m=\dfrac{y_2-y_1}{x_2-x_1}$ Where $x_1 \neq x_2$

Slope-intercept form:

$y = mx + b$ With slope *m* and *y*-intercept (0, *b*)

Point-slope form:

$y-y_1=m(x-x_1)$ With slope *m* and point (x_1, y_1) on the line

Horizontal line, slope 0: $y=b$

Vertical line, undefined slope: $x=a$

Parallel lines have the same slope.

Perpendicular lines have slopes that are negative reciprocals of each other.

Quadratic Functions (Parabolas)

Parabolas of the form $y=ax^2+bx+c$:

1. Vertex: $\left(-\dfrac{b}{2a}, f\left(-\dfrac{b}{2a}\right)\right)$.
2. Line of Symmetry: $x=-\dfrac{b}{2a}$

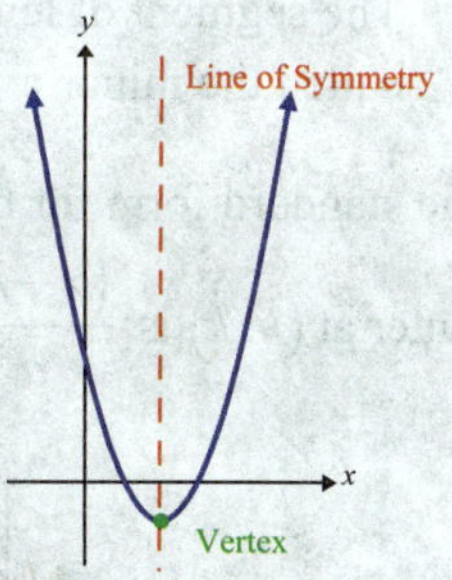

Parabolas of the form $y=a(x-h)^2+k$:

1. Vertex: (h, k)
2. Line of Symmetry: $x=h$
3. The graph is a horizontal shift of *h* units and a vertical shift of *k* units of the graph of $y=ax^2$.

In both cases:

1. If $a>0$, the parabola "opens upward."
2. If $a<0$, the parabola "opens downward."

Conic Sections

Equations of a Horizontal Parabola

$x = ay^2 + by + c$ or $x = a(y-k)^2 + h$ where $a \neq 0$.

The parabola opens left if $a < 0$ and right if $a > 0$.

The vertex is at (h, k).

The line $y = k$ is the line of symmetry.

Equation of an Ellipse

The standard form for the equation of an ellipse with its center at the origin is $\dfrac{x^2}{a^2} + \dfrac{y^2}{b^2} = 1$.

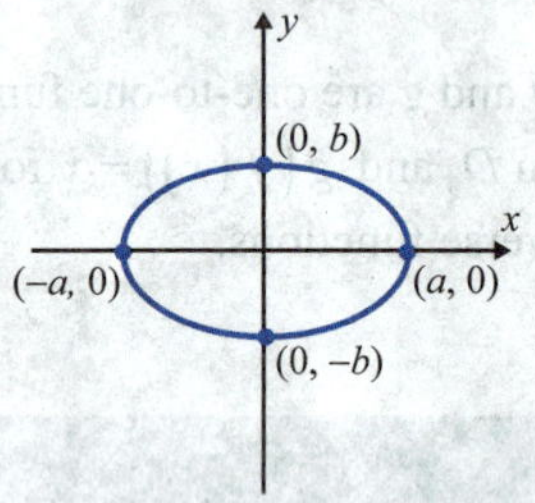

The points $(a, 0)$ and $(-a, 0)$ are the x-intercepts (called vertices).

The points $(0, b)$ and $(0, -b)$ are the y-intercepts (called vertices).

When $a^2 > b^2$:

- The segment of length $2a$ joining the x-intercepts is called the major axis.
- The segment of length $2b$ joining the y-intercepts is called the minor axis.

When $b^2 > a^2$:

- The segment of length $2b$ joining the y-intercepts is called the major axis.
- The segment of length $2a$ joining the x-intercepts is called the minor axis.

The standard form for the equation of an ellipse with its center at (h, k) is $\dfrac{(x-h)^2}{a^2} + \dfrac{(y-k)^2}{b^2} = 1$.

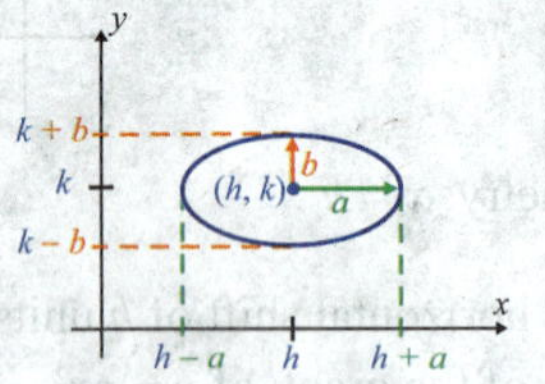

Equation of a Circle

The equation of a circle with radius r and center (h, k) is $(x-h)^2 + (y-k)^2 = r^2$.

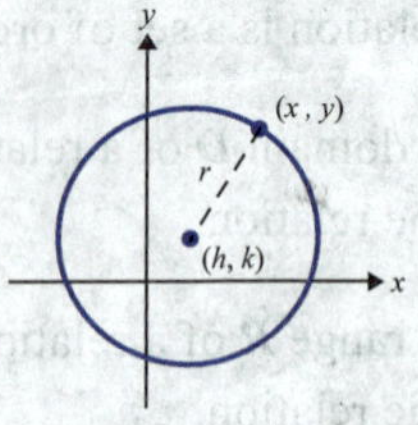

Equation of a Hyperbola

In general, there are two standard forms for equations of hyperbolas with their centers at the origin.

1. $\dfrac{x^2}{a^2} - \dfrac{y^2}{b^2} = 1$

x-intercepts (vertices) at $(a, 0)$ and $(-a, 0)$

No y-intercepts

Asymptotes: $y = \dfrac{b}{a}x$ and $y = -\dfrac{b}{a}x$

The curves "open" left and right.

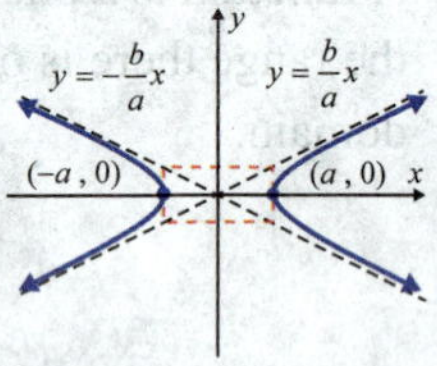

2. $\dfrac{y^2}{b^2} - \dfrac{x^2}{a^2} = 1$

y-intercepts (vertices) at $(0, b)$ and $(0, -b)$

No x-intercepts

Asymptotes: $y = \dfrac{b}{a}x$ and $y = -\dfrac{b}{a}x$

The curves "open" up and down.

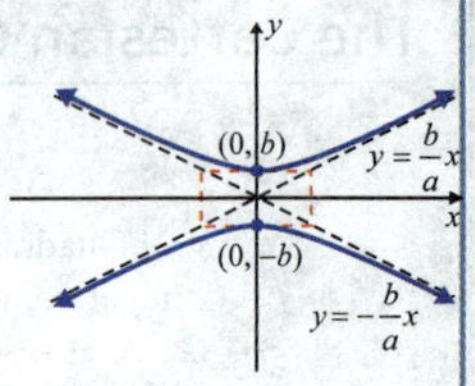

The equation of a hyperbola with its center at (h, k) is

$$\frac{(x-h)^2}{a^2} - \frac{(y-k)^2}{b^2} = 1 \text{ or } \frac{(y-k)^2}{b^2} - \frac{(x-h)^2}{a^2} = 1.$$

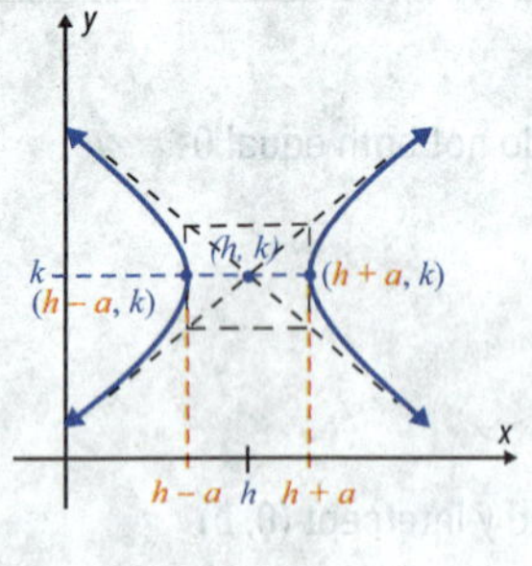

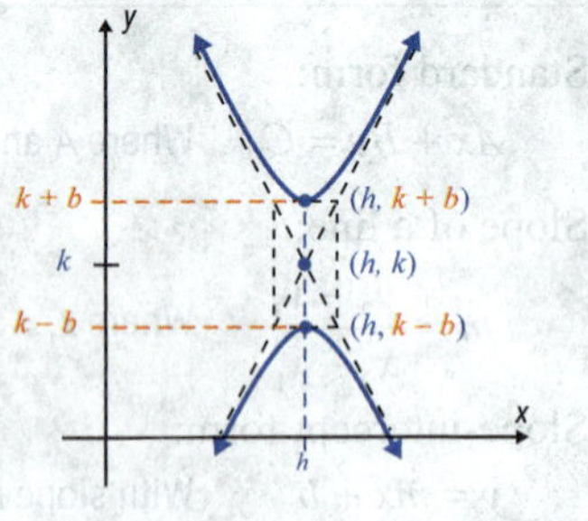